Experience the Excitement of Scientific Discovery

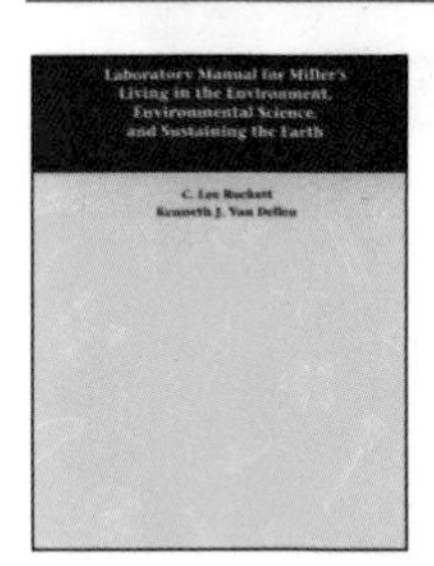

Laboratory Manual
ISBN 0-534-17809-X
by C. Lee Rockett and Kenneth J. Van Dellen
The real way to get a feel for science is to *do* what scientists do. With this manual's interesting activities, laboratory exercises and projects you'll experience the fun of discovery. Designed for use with a minimum of sophisticated equipment, the exercises involve various areas of science, such as biology, chemistry, geology, physics, and meteorology to help you understand the interaction among the different environmental elements. 116 pages.

Green Lives, Green Campuses

...vestigate ...npus and ...here's no ...epts understandable and real. Packed with self-assessment, brainstorming, and evaluation exercises to strengthen your critical thinking and understanding of environmental concepts and issues. 96 pages.

Eye-Opening, Fascinating Reading About Environmental Science

Watersheds 2: Ten Cases in Environmental Ethics
ISBN 0-534-51181-3
by Lisa H. Newton and Catherine K. Dillingham
Get the full stories behind envionmental issues you've been reading about in the newspapers— toxic waste from nuclear weapons facilities, pesticides, global climate change, our disappearing tropical rainforests, and more. 240 pages.

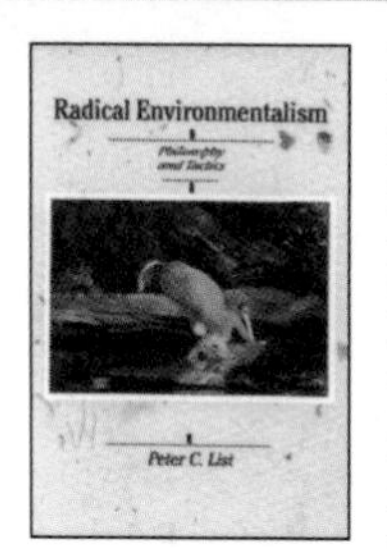

Radical Environmentalism: Philosophy and Tactics
ISBN 0-534-17790-5
by Peter List
Find out about the unsettling ecotactics that are more like guerilla warfare than plans for environmental change! Includes the most important radical ecophilosophies and examples of ecotactics that radical groups espouse. Balanced between "radical" selections and critiques of radical tactics. 288 pages.

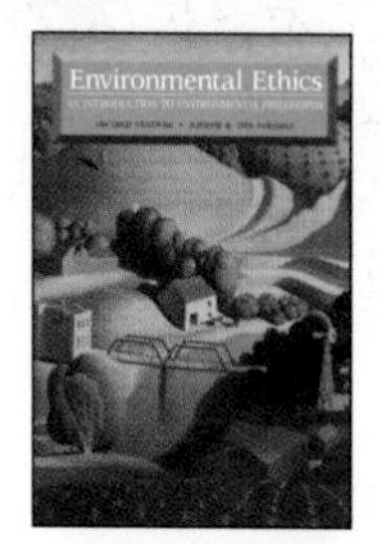

Environmental Ethics: An Introduction to Environmental Philosophy, 2nd Edition
ISBN 0-534-50508-2
by Joseph Desjardins
Learn about the ethical issues that come up when human beings interact with their natural environment . . . when we all try to share a planet that seems to be growing smaller by the day. Includes real-life cases. 272 pages.

And Even More Tips for Your Success

Your College Experience: Strategies for Success, 3rd Edition
by John N. Gardner and A. Jerome Jewler
Thousands of students swear by the valuable advice in this best-selling book. Tips for managing your time, taking tests, writing and speaking, note taking, reading and memory, critical thinking, and campus resources—including the library. Ask your instructor about special custom options so you can get just the chapters you need.

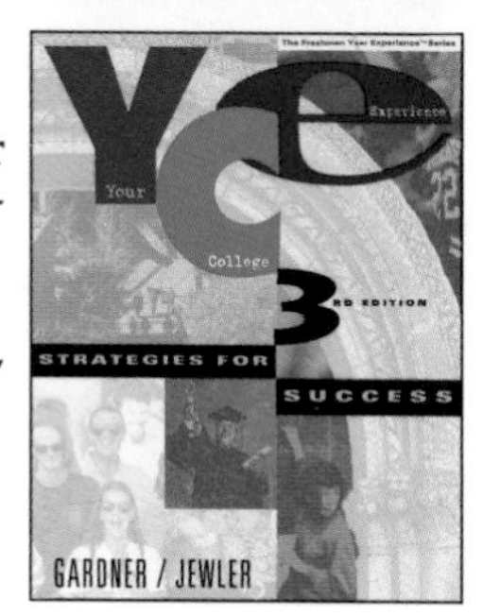

Study Skills for Science Students
ISBN 0-314-03983-X
by Daniel D. Chiras
Written specifically for science students, this 86-page book discusses how to develop good study habits, sharpen memory, learn more quickly, get the most out of lectures, prepare for tests, produce excellent term papers, and improve critical thinking skills.

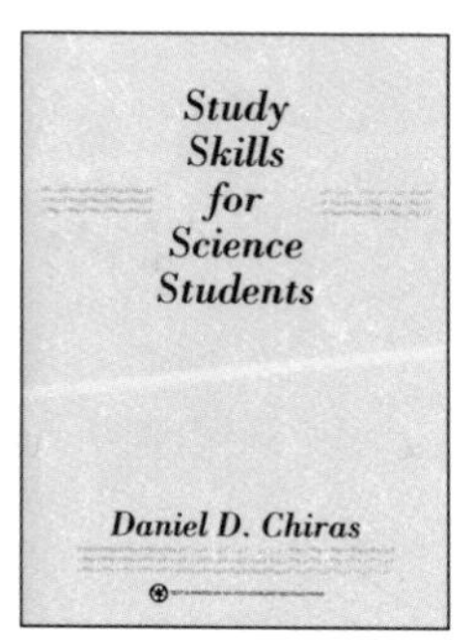

PLACE STAMP HERE

Tami Cueny
Marketing Manager
Brooks/Cole Publishing Company
10 Davis Drive
Belmont,CA 94002-3098

Name____

School____

Phone____ **E-mail**____

Did you use:

InfoTrac College Edition?____

The Brooks/Cole Biology Resource Center?____

Neither?____

How useful were these services? Please answer in the space below and continue your comments on the back of this postcard if necessary.

May we quote your remarks? Yes____ No____

Living in the Environment

AG 11

26.1
26.-

4.5

27.

30.

11/00
4645

Living in the Environment

Principles, Connections, and Solutions

ELEVENTH EDITION

G. Tyler Miller, Jr.

President, Earth Education and Research

Adjunct Professor of Human Ecology
St. Andrews Presbyterian College

Brooks/Cole Publishing Company

I(T)P® *An International Thomson Publishing Company*

Pacific Grove • Albany • Belmont • Boston • Cincinnati • Johannesburg • London • Madrid •
Melbourne • Mexico City • New York • Scottsdale • Singapore • Tokyo • Toronto

Publisher: *Jack Carey*
Assistant Editor: *Kristin Milotich*
Editorial Assistant: *Susan Lussier*
Production Coordinator: *Tessa McGlasson Avila*
Production Management: *Electronic Publishing Services Inc., NYC*
Marketing: *Tami Cueny*
Cover Design: *Vernon T. Boes*
Cover Photo: *Kennan Ward*
Interior Illustration: *Electronic Publishing Services Inc., NYC; Precision Graphics; Sarah Woodward; Darwin and Vally Hennings; Tasa Graphic Arts, Inc.; Alexander Teshin Associates; John and Judith Waller; Raychel Ciemma; and Victor Royer*
Typesetting: *Electronic Publishing Services Inc., NYC*
Printing and Binding: *Von Hoffmann Printing Company*

Title Page Photographs: *Left: Invertebrate sea star (E. R. Degginger). Right: Red nakari monkey (Roy P. Fontaine/Photo Researchers, Inc.). Bottom: Crater Lake, Oregon (Jack Carey).*
Part Opening Photographs:
Part I: *Composite satellite view of Earth. © Tom Van Sant/The GeoSphere Project*
Part II: *Endangered green sea turtle. © David B. Fleetham*
Part III: *View from space showing major biomes and surface features of most of North America.* NOAA
Part IV: *Burning of a tropical forest area in Brazil's Amazon Basin. A. Giardet/The Environmental Picture Library*
Part V: *Highly endangered Florida panther. John Cancalosi/Peter Arnold, Inc.*
Part VI: *Anti-pollution demonstration by Louisana residents against Monsanto Chemical Co. Sam Kitner/SIPA Press*

COPYRIGHT © 2000 by Brooks/Cole Publishing Company
A Division of International Thomson Publishing Inc.
The ITP logo is a registered trademark used herein under license.

Two trees have been planted in a tropical rain forest for every tree used to make this book, courtesy of G. Tyler Miller, Jr., and Brooks/Cole Publishing Company. The author also sees that 50 trees are planted to compensate for the paper he uses and that several hectares of tropical rain forest are protected.

For more information, contact:
BROOKS/COLE PUBLISHING COMPANY
511 Forest Lodge Road
Pacific Grove, CA 93950
USA

International Thomson Publishing Europe
Berkshire House 168-173
High Holborn
London WC1V 7AA
England

Thomas Nelson Australia
102 Dodds Street
South Melbourne, 3205
Victoria, Australia

Nelson Canada
1120 Birchmount Road
Scarborough, Ontario
Canada M1K 5G4

International Thomson Editores
Seneca 53
Col. Polanco
11560 México, D. F., México

International Thomson Publishing GmbH
Königswinterer Strasse 418
53227 Bonn
Germany

International Thomson Publishing Asia
60 Albert Street
#15-01 Albert Complex
Singapore 189969

International Thomson Publishing Japan
Hirakawacho Kyowa Building, 3F
2-2-1 Hirakawacho
Chiyoda-ku, Tokyo 102
Japan

All rights reserved. No part of this work may be reproduced, stored in a retrieval system, or transcribed, in any form or by any means—electronic, mechanical, photocopying, recording, or otherwise—without the prior written permission of the publisher, Brooks/Cole Publishing Company, Pacific Grove, California 93950. You can request permission to use material from this text through the following phone and fax numbers:
Phone: 1-800-730-2214 Fax: 1-800-730-2215

Printed in the United States of America.

10 9 8 7 6 5 4 3 2

11/00 8380
4645
LLYFRGELL
LLYSFASI
LIBRARY

Library of Congress Cataloging-in-Publication Data

Miller, G. Tyler (George Tyler).
Living in the environment : principles, connections, and solutions / G. Tyler Miller.
p. cm.
Includes bibliographical references and index.
ISBN 0-534-56268-X (hardcover). — ISBN 0-534-56269-8 (pbk.)
1. Environmental sciences 2. Human ecology. 3. Environmental policy. I. Title
GE105.M547 1999
363.7—DC21 99-17666
CIP

PREFACE

For Instructors and Students

How Did I Become Involved with Environmental Problems? In 1966 I heard a scientist give a lecture on the problems of overpopulation and environmental abuse. Afterward I went to him and said, "If even a fraction of what you have said is true, I will feel ethically obligated to give up my research on the corrosion of metals and devote the rest of my life to research and education on environmental problems and solutions. Frankly, I don't want to believe a word you have said, and I'm going into the literature to try to prove that your statements are either untrue or grossly distorted."

After 6 months of study I was convinced of the seriousness of these problems. Since then, I have been studying, teaching, and writing about them. This book summarizes what I have learned in over three decades of trying to understand environmental principles, problems, connections, and solutions.

What Is My Philosophy of Education? In our lifelong pursuit of knowledge, I believe we should do three things. The first is to question everything and everybody, as any good scientist does.

Second, each of us should develop a list of principles, concepts, and rules to serve as guidelines in making decisions, and continually evaluate and modify this list on the basis of experience. This is based on my belief that the purpose of our lifelong pursuit of education is to learn how to sift through mountains of facts and ideas to find the few that are most useful and worth knowing. We need to be *wisdom seekers*, not information vessels. This requires a firm commitment to learning how to think logically and critically. This book is full of facts and numbers, but they are useful only to the extent that they lead to an understanding of key ideas, scientific laws, concepts, principles, and connections.

Third, I believe in interacting with what I read as a way to use and continually sharpen my critical thinking skills. I do this by marking key sentences and paragraphs with a highlighter or pen. I put an asterisk in the margin next to something I think is important and double asterisks next to something that I think is especially important. I write comments in the margins, such as *Beautiful*, *Confusing*, *Misleading*, or *Wrong*. I fold down the top corner of pages with highlighted passages and the top and bottom corners of especially important pages. This way, I can flip through a book and quickly review the key passages. I urge you to interact in such ways with this book.

Why Is There a Need for More Scientific Content? This book is designed for introductory courses on environmental science. It treats environmental science as an *interdisciplinary* study, combining ideas and information from natural sciences (such as biology, chemistry, and geology) and social sciences (such as economics, politics, and ethics) to present a general idea of how nature works and how things are interconnected. It is a study of *connections in nature*.

Since its first edition in 1975 this book has led the way in using scientific laws, principles, models, and concepts to help us understand environmental and resource problems and possible solutions, and how these concepts, problems, and solutions are connected. I have introduced only the concepts and principles necessary for understanding the material in the book, and I have tried to present them simply but accurately.

A number of people have criticized environmental science teachers and textbooks for **(1)** not being grounded in basic science, **(2)** treating key environmental issues (such as global warming, ozone depletion, and biodiversity loss) superficially, **(3)** giving an unbalanced, biased view of environmental issues, and **(4)** focusing mostly on bad environmental news without giving the good environmental news.

I generally agree with these criticisms and believe that *there is a need to expand the fundamental scientific content of environmental science courses*. In each new edition of this book I have done this. This latest edition takes a major step forward in further expanding scientific content (see changes in this edition listed on p. vii).

Since the first edition of this book I have also *emphasized in-depth treatment of key environmental issues*. Many instructors have written me thanking me for being the only author in this field to give students an in-depth, balanced, and up-to-date view of major environmental problems such as global warming and ozone depletion (Chapter 19) and loss of biodiversity (a major integrating theme of this book; see especially Chapters 6, 7, 9, 21, 23, 24, and 25) without doing so in a complex and hard-to-understand manner.

To help ensure that the material is accurate and up to date, I have consulted more than 10,000 research sources in the professional literature. In writing this book, I have also benefited from the more than 200 experts and teachers (see list on pp. xi) who have provided detailed reviews of this and my other xi, xiii three books in this field.

There are at least two sides to all problems and controversies. The challenge for authors of environmental science textbooks and teachers of environmental science courses is to give a fair and balanced view of opposing viewpoints without injecting personal bias. The reason for doing this is to allow students to make up their own minds and to end up with their own opinions and biases. Studying a subject as important as environmental science and ending up with no conclusions, opinions, and beliefs means that both the teacher and student have failed. But such conclusions should be based on using critical thinking to evaluate opposing ideas.

In this edition I continue to scour every sentence to identify and reduce bias as much as possible. Other examples of my efforts to give a balanced presentation of opposing viewpoints are the pros and cons of reducing birth rates (pp. 265–268), the Pro/Con box on genetic engineering (pp. 462–463), Section 19-2 on global warming, Section 19-5 on ozone depletion, discussion of acid deposition (pp. 478–483), the asbestos problem (p. 484), problems with the Superfund law (pp. 603–604), food aid (p. 304), pros and cons of pesticides (Sections 21-2 and 21-3), use of public lands (pp. 613 and 621), and evaluating claims of environmentalists and anti-environmentalists (Section 28-5 and Guest Essay on p. 792). Even more important are the thousands of sentences and phrases that have been omitted or reworded to eliminate any hint of bias.

Bias is subtle, however, and there is still room for improvement. I invite instructors and students to write me and point out any remaining bias.

How Do I Deal with the Bad News/Good News Dilemma? Critics have charged environmentalists with wallowing in doom and gloom and not reporting or rejoicing in the many improvements in environmental quality that have taken place over the past several decades. I generally agree with this criticism and in this edition I have emphasized the good environmental news along with the bad environmental news in Chapter 1 (Table 1-2, p. 30) and throughout the book.

Environmental good news should inspire us and make us feel good about what has been accomplished by the hard and dedicated work of millions of people. At the same time, environmental bad news should challenge us to maintain and strengthen past environmental improvements and to deal with the real and more complex environmental problems we face.

Rosy optimism and gloom-and-doom pessimism are traps; both usually lead to denial, indifference, and inaction. I have tried to avoid these two extremes and give a realistic yet hopeful view of the future. This book is filled with technological and scientific advances that have led to environmental improvements (for example, see the Individuals Matter boxes on pp. 363, 423, and 528, the openings to Chapters 12, 16, and 20, and the Guest Essay on p. 407) and stories of people who have acted to help sustain the earth's life-support systems for us and for all life, and whose actions inspire us to do better (for example, see the Individuals Matter boxes on pp. 249, 591, and 627, the openings to Chapters 3, 24, and 28, and the Guest Essay on p. 308). It's an exciting and challenging time to be alive as we struggle to enter into a new, more cooperative relationship with the planet that is our only home.

What Are Some Key Features of This Book? The book is divided into six major parts (see Brief Contents, p. xiv). After Parts I and II have been covered, the rest of the book can be used in almost any order. In addition, most chapters and many sections within these chapters can be moved around or omitted to accommodate courses with different lengths and emphases.

After major scientific concepts and environmental problems are discussed, various solutions proposed by a variety of scientists, environmental activists, and analysts are given. A range of possible solutions, some of them controversial, are provided to encourage students to think critically and make up their own minds.

Since its first edition this book has consistently used simple systems analysis to categorize proposed solutions to environmental problems as either **(1)** *input (prevention)* solutions such as pollution prevention and waste reduction or **(2)** *output (cleanup)* solutions such as pollution control and waste management. Both approaches are needed, but so far most emphasis has been on output or management solutions. There is a growing awareness of the need to put more emphasis on input or prevention approaches—another major theme of this book.

Each chapter begins with a brief *Earth Story*, a case study designed to capture interest and set the stage for the material that follows. In addition to these 29 case studies, 74 other case studies are found throughout the book (some in special boxes and others within the text); they provide a more in-depth look at specific environmental problems and their possible solutions. Twenty *Guest Essays* (three of them new) present an individual researcher's or activist's point of view, which is then evaluated through critical thinking questions.

Other special boxes found in the text include *Pro/Con boxes* that present both sides of controversial environmental issues; *Connections boxes* that show connections in nature and among environmental concepts, problems, and solutions; *Solutions boxes* that summarize a variety of solutions (some of them controversial) to environmental problems proposed by various analysts; *Spotlight boxes* that highlight and give insights into key environmental problems and concepts; and *Individuals Matter boxes* that describe what people have done to help solve environmental problems. To encourage critical thinking and integrate it throughout the book, all boxes (except Individuals Matter) end with critical thinking questions.

This book is an integrated study of environmental problems, connections, and solutions. The 8 integrating themes in this book are *biodiversity and earth capital; sustainability; pollution prevention and waste reduction; population and exponential growth; energy and energy efficiency; connections in nature; solutions to environmental problems; and individual action.*

I hope you will start by looking at the brief table of contents (p. xiv) to get an overview of this book. Then I suggest that you look at the concepts and connections diagram inside the back cover, which shows the major components and relationships found in environmental science. In effect, it is a map of the book. Then, to get a summary of the book's key principles, I urge you to read the list of key principles found in the Epilogue (pp. 814–815)

The book's 540 illustrations are designed to present complex ideas in understandable ways and to relate learning to the real world. They include 395 full-color diagrams (60 of them maps) and 145 carefully selected color photographs.

I have not cited specific sources of information; this is rarely done for an introductory-level text in any field, and it would interrupt the flow of the material. Instead, the Internet sites and readings listed at the end of the book for each chapter provide backup for almost all the information in this book and serve as springboards to further information and ideas. This edition also has a greatly expanded and improved interactive World Wide Web site that can be used as a source of further information and ideas (see description on p. ix), a new *BioLink* CD-ROM, and the *InfoTrac College Edition* online university library of articles (p. ix).

Instructors wanting shorter books covering this material with a different emphasis and organization can use one of my three other books written for various types of environmental science courses: *Environmental Science*, 7th edition (566 pages, Brooks/Cole, 1999, a shorter version of this book); *Sustaining the Earth: An Integrated Approach*, 4th edition (356 pages, Brooks/Cole, 2000, a shorter book with a different integrated approach); and *Environment: Problems and Solutions* (150 pages, Brooks/Cole, 1994, a very short introduction that assumes a background in scientific concepts).

What Are the Major Changes in the Eleventh Edition? Major changes include

- Updated and revised material throughout the book.
- A continuing effort to reduce bias (see p. vi).
- Significantly increased coverage of basic ecological concepts. The number of chapters on ecological principles has been increased from 4 to 7 and the order of topics has been rearranged to improve scientific content and flow. **Paul M. Rich**, associate professor of Ecology and Evolutionary Biology and Environmental Studies at the University of Kansas, was brought in as a coauthor of Chapters 4 through 10 and 23 through 26.
- Greatly increased *scientific content*, with the addition or expansion of material on geographic information systems (GIS), bioinformatics, principles of ecological sustainability, overloading of the nitrogen cycle, selective pressure, convergence, geographic ecology, sea otters, coral reefs, flying foxes, community processes, population ecology, logistic population growth, controversy over roles of predators in controlling population size, conservation biology, and ecological decline in Yellowstone National Park.
- Addition or expansion of many topics, including the information revolution, controversy over the definition of organic food, food supply and demand, growing more food in urban areas, decline of the bluefin tuna population, sustainable fishery management, importance of agricultural research, water conservation in Los Angeles, improving water efficiency in China, wasting water in Las Vegas, effects of dams on marine life in the Black Sea, metal resources from deep ocean deposits, a solar village in Columbia (Gaviotas), advances in solar cell technology, the spread of genetic resistance of bacteria to antibiotics, drug-resistant bacteria in meat, Lyme disease, hospital-acquired infections, controversy over reducing air pollution emissions of ultrafine particles, effects of global warming on Kirkland's warblers and penguins, controversy over the role of

developing countries in reducing carbon dioxide emissions, possible signs of global warming, the 1997 meeting on global warming in Kyoto, Japan, a rise in blooms of toxic tides, oyster production and cleaning up the Chesapeake Bay, using methane gas from landfills to heat schools, source separation of solid waste in Georgia schools, intergenerational equity, GIS and urban planning, solutions to the excessive snow goose population, sustainable agriculture in tropical forests, invasions by brown tree snakes, using gap analysis to preserve biodiversity, using simple stoves to save forests in India, and microloans to the poor.

- Three new Guest Essays.
- *Online Biology Resource Center* (see description on p. ix) at http://www.brookscole.com/biology.
- Revision and updating of *BioLink*, an instructor presentation tool (see description on p. ix).
- *InfoTrac College Edition*, a fully searchable online university library (see description on p. ix).
- *Online Regional Articles for Environmental Science* (see description on p. ix).
- *Thomson World Class Course software* (see description on p. ix).
- *CNN Today Videos for Environmental Science* (see description on p. ix).
- Booklet on *Study Skills for Science Students* (see description on p. ix).
- Revision and updating of the *Instructor's Manual and Test Items* booklet.

Welcome to Controversy and Challenge There are no easy solutions to the environmental problems and challenges we face. We will never have complete agreement about what we should do about environmental problems because science advances through continuous controversy and careful scrutiny or its results until there is a general consensus about their validity. What is important is not what the experts disagree on (the frontiers of scientific knowledge that are still being developed, tested, and argued about), but what they generally agree on—the *scientific consensus*—about concepts, problems, and possible solutions.

Despite considerable research, we still know little about how nature works at a time when we are altering nature at an accelerating pace. This uncertainty, as well as the complexity and importance of these issues to current and future generations of humans and other species, makes many of these issues highly controversial. Intense controversy also arises because environmental science is an interdisciplinary blend of natural and social sciences that sometimes questions the ways we view and act in the world around us.

Study Aids Each chapter begins with a few general questions to reveal how it is organized and what students will be learning. When a new term is introduced and defined, it is printed in boldface type. A glossary of all key terms is located at the end of the book.

Factual recall questions (with answers) are listed at the bottom of most pages. You might cover the answer (on the right-hand page) with a piece of paper and then try to answer the question on the left-hand page. (These questions are not necessarily related to the chapter in which they are found.) There are also one or two exercises at the bottom of pages in each chapter to help students learn how to use the new *InfoTrac* system available on the World Wide Web (see p. ix).

Each chapter and box (except Individuals Matter boxes) ends with a set of questions to encourage students to think critically and apply what they have learned to their lives. Some ask students to take sides on controversial issues and to back up their conclusions and beliefs. The end-of-chapter questions are followed by several projects that individuals or groups can carry out. Many additional projects are given in the *Instructor's Manual* and in the *Green Lives, Green Campuses* and *Critical Thinking and the Environment* supplements available with this book.

Readers who become especially interested in a particular topic can consult the list of further readings for each chapter, given in the back of the book. Appendix 1 contains a list of important publications and some key environmental organizations and government and international agencies.

Students can also access World Wide Web for material in the book marked with the icon . Access to this system is at http://www.brookscole.com/biology. With this new feature, students can click on a chapter in the Hypercontents listed and find resources that couldn't be listed in the book. If you can't find a topic you are looking for, try our search page. These resources are updated constantly. Because the World Wide Web is like an image gallery, library, and information booth, each link reference is accompanied by a description to help guide you in your selections. In the Web site for this text, you will also find Cool Events, Critical Thinking Questions, Tips on Surfing, Interactive Quizzes for each chapter, and much more. Happy surfing.

Help Me Improve This Book Let me know how you think this book can be improved; if you find any errors, bias, or confusing explanations please send them to Jack Carey, Biology Publisher, Brooks/Cole Publishing Company, 10 Davis Drive, Belmont, CA 94002. He will forward them to me. Most errors can be corrected in subsequent printings of this edition, rather than waiting for a new edition.

Supplements The following supplements are available:

- *BioLink*, an instructor presentation tool that allows for the quick, easy assembly of media files into a multimedia presentation. Through an easy-to-use interface, instructors can assemble, edit, publish, and present custom lectures built from an extensive multimedia database. It includes all illustrations from the book and art from other Brooks/Cole biology books and CD-ROMs. BioLink also has a browser with an easy drag-and-drop feature that allows file export into such presentation tools as PowerPoint. Upon its completion, a file or lecture created with BioLink can be posted to the Web, where students can access it for reference or study.
- *InfoTrac College Edition* is a fully searchable online university library that is available free with each copy of *Living in the Environment.* It gives students access to full-length articles from over 700 scholarly and popular journals, updated daily, and dating back as much as 4 years. An online *Student Guide to InfoTrac College Edition* is located on the Brooks/Cole Biology Resource Center Web Site. It has Critical Thinking questions and a set of electronic readings for each chapter to provide deeper examination of the material.
- The *Online Regional Articles for Environmental Science* is a collection of online articles from *InfoTrac College Edition* that are correlated to each chapter in the book and organized by region: west, southwest, rocky mountains, midwest, east, and south. These regional articles allow students to learn about local environmental issues in the region in which they live. Students can access the *Online Regional Articles* on the Biology Resource Center site or directly at http://www.upcloser.com
- *The Brooks/Cole Biology Resource Center* is arranged by chapter. Every month it has new BioUpdates on relevant applications and hyperlinks. It also has an average of 40 practice quiz questions per chapter, descriptions of degrees and careers in biology and environmental science, a student feedback site, clip art, ideas for teaching on the Web, and a forum in which instructors can share ideas on teaching courses. It also includes flashcards for all glossary terms, critical thinking exercises, newsgroups, a variety of search engines, and Internet exercises for each chapter. An event of the quarter will include an ongoing experiment in which students and instructors can participate. The address for the Brooks/Cole Biology Resource Center is http://www.brookscole.com/biology.
- **Thomson World Class Course* software enables instructors to create their own Web sites. Instructors can post course information, office hours, related Internet links, downloaded materials, lesson information, assignments, and sample tests or quizzes. More information is available at http://www.worldclasslearning.com.
- **CNN Today Videos for Environmental Science* are short clips of current news footage that make great lecture launchers. A new tape is offered every year.
- **Study Skills for Science Students*, by Dan Chiras, is an 86-page booklet that explains how to develop good study habits, sharpen memory and learning, prepare for tests, and produce term papers.
- *Internet Booklet*, by Daniel J. Kurland and Jane Heinze-Fry. An introduction to the Internet and World Wide Web, plus selected sites to visit and learning exercises.
- *Instructor's Manual and Test Items*, written and updated by Jane Heinze-Fry (Ph.D. in science and environmental education). For each chapter, it has goals and objectives; one or more concept maps; key terms; teaching suggestions; multiple-choice test questions with answers; projects, field trips, and experiments; term paper and report topics; and a list of audiovisual materials and computer software. The test items are available electronically for Mac and Windows.
- *Critical Thinking and the Environment: A Beginner's Guide*, by Jane Heinze-Fry and G. Tyler Miller, Jr. An introduction to different critical thinking approaches, with questions by chapter using these approaches.
- *Green Lives, Green Campuses*, written by Jane Heinze-Fry. This hands-on workbook contains projects to help students evaluate the environmental impact of their own lives and to guide them in making an environmental audit of their campus.
- *Laboratory Manual* by C. Lee Rockett (Bowling Green State University) and Kenneth J. Van Dellen (Macomb Community College).
- A set of 100 color acetates and more than 600 black-and-white transparency masters for making overhead transparencies or slides of line art (including concept maps for each chapter), available to adopters.
- *Watersheds: Classic Cases in Environmental Ethics*, 2d ed., by Lisa H. Newton and Catherine K. Dillingham (Wadsworth, 1997). Nine balanced case studies that amplify material in this book.

- *Environmental Ethics*, by Joseph R. Des Jardins (Wadsworth, 1993). A very useful survey of environmental ethics. Brief case studies and many specific examples are included.
- *Radical Environmentalism*, by Peter C. List (Wadsworth, 1993). A series of readings on environmental politics and philosophy.
- *A Beginner's Guide to Scientific Method*, 2d ed., by Stephen S. Carey (Wadsworth, 1998). A concise, hands-on introduction that helps students develop critical thinking skills essential to understanding the scientific process.
- *The Game of Science*, 5th edition, by Garvin McCain and Erwin M. Segal (Brooks/Cole, 1988). An accurate, lively, and up-to-date view of what science is, who scientists are, and how they approach science.

Annenberg/CPB Television Course This textbook is being offered as part of the Annenberg/CPB Project television series *Race to Save the Planet*, a 10-part public broadcasting series and a college-level telecourse examining the major environmental questions facing the world today. The series takes into account the wide spectrum of opinion about what constitutes an environmental problem and discusses the controversies about appropriate remedial measures. It analyzes problems and emphasizes the successful search for solutions. The course develops a number of key themes that cut across a broad range of environmental issues, including sustainability, the interconnections of the economy and the ecosystem, short-term versus long-term gains, and the trade-offs involved in balancing problems and solutions. A study guide and a faculty guide, both available from Brooks/Cole Publishing Company, integrate the telecourse and this text.

For further information about available television course licenses and duplication licenses, contact PBS Adult Learning Service, 1320 Braddock Place, Alexandria, VA 22314-1698 (1-800-ALS-AL5-8).

For information about purchasing videocassettes and print material, contact the Annenberg/CPB Collection, P.O. Box 2284, South Burlington, VT 05407-2284 (1-800-LEARNER).

Acknowledgments I wish to thank the many students and teachers who responded so favorably to the ten previous editions of *Living in the Environment*, the seven editions of *Environmental Science*, the three editions of *Sustaining the Earth: An Integrated Approach*, and the first editions of *Resource Conservation and Management* and *Environment: Problems and Solutions*—and who corrected errors and offered many helpful suggestions for improvement. I am also deeply indebted to the reviewers, who pointed out errors and suggested many important improvements in this book. Any errors and deficiencies left are mine.

The members of the talented production team, listed on the copyright page, have made vital contributions as well. Their labors of love are also gifts to helping sustain the earth. I especially appreciate the competence and cheerfulness of production editor Rob Anglin and the helpful inputs by copyeditor Carol Anne Peschke. My thanks also go to Brooks/Cole's hard-working sales staff, and to Kristin Milotich for her help and efficiency.

My deep gratitude also goes to Paul M. Rich, who served as coauthor for Chapters 4 through 10 and 23 through 26. In addition to being delightful to work with, he brings great depth of knowledge about ecological concepts and biodiversity. My thanks also go to C. Lee Rockett and Kenneth J. Van Dellen for developing the *Laboratory Manual* to accompany this book. I also wish to thank the people who have translated this book into five different languages for use throughout much of the world.

My special thanks go Jane Heinze-Fry for her insightful analysis, for being such a delight to work with, and for her outstanding work on the *Instructor's Manual*, concept mapping, *Green Lives, Green Campuses, Environmental Articles*, and the new *Critical Thinking and the Environment: A Beginner's Guide* and *Internet Booklet*.

My deepest thanks go to Jack Carey, biology publisher at Brooks/Cole, for his encouragement, help, 33 years of friendship, and superb reviewing system. It helps immensely to work with the best and most experienced editor in college textbook publishing.

G. Tyler Miller, Jr.

Coauthor, Guest Essayists, and Reviewers

Coauthor **Paul M. Rich,** associate professor of Ecology and Evolutionary Biology and Environmental Studies at the University of Kansas, is coauthor of chapters 4 through 10 and 23 through 26.

Guest Essayists The following are authors of Guest Essays: **Lester R. Brown**, president, Worldwatch Institute; **Alberto Ruz Buenfil**, environmental activist, writer, and performer; **Robert D. Bullard**, professor of sociology, University of California, Riverside; **Herman E. Daly**, senior research scholar, School of Public Affairs, University of Maryland; **Lois Marie Gibbs**, director, Citizens' Clearinghouse for Hazardous Wastes; **Garrett Hardin**, professor emeritus of human ecology, University of California, Santa Barbara; **Paul Hawken**, environmental author and business leader; **Jane-Heinze Fry**, author and consultant in environmental education; **Allison Gannett**, consultant and environmental designer; **John Harte**, professor of energy and resources, University of California, Berkeley; **Paul F. Kamitsuka**, infectious disease expert and physician; **Amory B. Lovins**, energy policy consultant and director of research, Rocky Mountain Institute; **Bobbi S. Low**, professor of resource ecology, University of Michigan; **Lester W. Milbrath**, director of the research program in environment and society, State University of New York, Buffalo; **Peter Montague**, director, Environmental Research Foundation; **Norman Myers**, consultant in environment and development; **David W. Orr**, professor of environmental studies, Oberlin College; **David Pimentel**, professor of entomology, Cornell University; **Andrew (Andy) C. Revkin**, environmental journalist, *New York Times*, and environmental author; and **Vandana Shiva**, director, Research Foundation for Science, Technology, and Natural Resource Policy, Dehra Dun, India.

Cumulative Reviewers Barbara J. Abraham, Hampton College; Donald D. Adams, State University of New York at Plattsburgh; Larry G. Allen, California State University, Northridge; James R. Anderson, U.S. Geological Survey; Kenneth B. Armitage, University of Kansas; Gary J. Atchison, Iowa State University; Marvin W. Baker, Jr., University of Oklahoma; Virgil R. Baker, Arizona State University; Ian G. Barbour, Carleton College; Albert J. Beck, California State University, Chico; W. Behan, Northern Arizona University; Keith L. Bildstein, Winthrop College; Jeff Bland, University of Puget Sound; Roger G. Bland, Central Michigan University; Georg Borgstrom, Michigan State University; Arthur C. Borror, University of New Hampshire; John H. Bounds, Sam Houston State University; Leon F. Bouvier, Population Reference Bureau; Daniel J. Bovin, Université Laval; Michael F. Brewer, Resources for the Future, Inc.; Mark M. Brinson, East Carolina University; Patrick E. Brunelle, Contra Costa College; Terrence J. Burgess, Saddleback College North; David Byman, Pennsylvania State University, Worthington Scranton; Lynton K. Caldwell, Indiana University; Faith Thompson Campbell, Natural Resources Defense Council, Inc.; Ray Canterbery, Florida State University; Ted J. Case, University of San Diego; Ann Causey, Auburn University; Richard A. Cellarius, Evergreen State University; William U. Chandler, Worldwatch Institute; F. Christman, University of North Carolina, Chapel Hill; Preston Cloud, University of California, Santa Barbara; Bernard C. Cohen, University of Pittsburgh; Richard A. Cooley, University of California, Santa Cruz; Dennis J. Corrigan; George Cox, San Diego State University; John D. Cunningham, Keene State College; Herman E. Daly, The World Bank; Raymond F. Dasmann, University of California, Santa Cruz; Kingsley Davis, Hoover Institution; Edward E. DeMartini, University of California, Santa Barbara; Charles E. DePoe, Northeast Louisiana University; Thomas R. Detwyler, University of Wisconsin; Peter H. Diage, University of California, Riverside; Lon D. Drake, University of Iowa; T. Edmonson, University of Washington; Thomas Eisner, Cornell University; Michael Esler, Southern Illinois University; David E. Fairbrothers, Rutgers University; Paul P. Feeny, Cornell University; Nancy Field, Bellevue Community College; Allan Fitzsimmons, University of Kentucky; Andrew J. Friedland, Dartmouth College; Kenneth O. Fulgham, Humboldt State University; Lowell L. Getz, University of Illinois at Urbana–Champaign; Frederick F. Gilbert, Washington State University; Jay Glassman, Los Angeles Valley College; Harold Goetz, North Dakota State University; Jeffery J. Gordon, Bowling Green State University; Eville Gorham, University of Minnesota; Michael Gough, Resources for the Future; Ernest M. Gould, Jr., Harvard University; Katharine B. Gregg, West Virginia Wesleyan College; Peter Green, Golden West College; Paul K. Grogger, University of Colorado at Colorado Springs; L. Guernsey, Indiana State University; Ralph Guzman, University of California, Santa Cruz; Raymond Hames, University of Nebraska, Lincoln; Raymond E. Hampton, Central Michigan University; Ted L. Hanes, California State University, Fullerton; William S. Hardenbergh, Southern Illinois University at Carbondale; John P. Harley, Eastern Kentucky University; Neil A. Harriman, University of Wisconsin, Oshkosh; Grant A. Harris, Washington State University; Harry S. Hass, San Jose City College; Arthur N. Haupt, Population Reference Bureau; Denis A. Hayes, environmental consultant; Gene Heinze-Fry, Department of Utilities, State of Massachusetts; Jane Heinze-Fry, environmental educator; John G. Hewston, Humboldt State University; David L. Hicks, Whitworth College; Eric Hirst, Oak Ridge National Laboratory; S. Holling, University of British Columbia; Donald Holtgrieve, California State University, Hayward; Michael H. Horn, California State University, Fullerton; Mark A. Hornberger, Bloomsberg University; Marilyn Houck, Pennsylvania State University; Richard D. Houk, Winthrop College; Robert J. Huggett, College of William and Mary; Donald Huisingh, North Carolina State University; Marlene K. Hutt, IBM; David R. Inglis, University of Massachusetts; Robert Janiskee, University of South Carolina; Hugo H. John, University of Connecticut; Brian A. Johnson, University of Pennsylvania, Bloomsburg; David I. Johnson, Michigan State University; Agnes Kadar, Nassau Community College; Thomas L. Keefe, Eastern Kentucky University; Nathan Keyfitz, Harvard University; David Kidd, University of New Mexico; Edward J. Kormondy, University of Hawaii–Hilo/West Oahu College; John V. Krutilla, Resources for the Future, Inc.; Judith Kunofsky, Sierra Club; E. Kurtz; Theodore Kury, State University of New York, Buffalo; Steve Ladochy, University of Winnipeg; Mark B. Lapping, Kansas State University; Tom Leege, Idaho Department of Fish and Game; William S. Lindsay, Monterey Peninsula College; E. S. Lindstrom, Pennsylvania State University; M. Lippiman, New York University Medical Center; Valerie A. Liston, University of Minnesota; Dennis Livingston, Rensselaer Polytechnic Institute; James P. Lodge, air pollution consultant; Raymond C. Loehr, University of Texas at Austin; Ruth Logan, Santa Monica City College; Robert D. Loring, DePauw University; Paul F. Love, Angelo State University; Thomas Lovering, University of California, Santa Barbara; Amory B. Lovins, Rocky Mountain Institute; Hunter Lovins, Rocky Mountain Institute; Gene A. Lucas, Drake University; Claudia Luke; David Lynn; Timothy F. Lyon, Ball State University; Melvin G. Marcus, Arizona State University; Gordon E. Matzke, Oregon State University; Parker Mauldin, Rockefeller Foundation; Theodore R. McDowell, California State University; Vincent E. McKelvey, U.S. Geological Survey; John G. Merriam, Bowling Green State University; A. Steven Messenger, Northern Illinois University; John Meyers, Middlesex Community College; Raymond W. Miller, Utah State University; Arthur B. Millman, University of Massachusetts, Boston; Rolf Monteen, California Polytechnic State University; Ralph Morris, Brock University, St. Catherines, Ontario, Canada; William W. Murdoch, University of California, Santa Barbara; Norman Myers, environmental consultant; Brian C. Myres, Cypress College; A. Neale, Illinois State University; Duane Nellis, Kansas State University; Jan Newhouse, University of Hawaii, Manoa; John E. Oliver, Indiana State University; Eric Pallant, Allegheny College; Charles F. Park, Stanford University; Richard J. Pedersen, U.S. Department of Agriculture, Forest Service; David Pelliam, Bureau of Land Management, U.S. Department of Interior; Rodney Peterson, Colorado State University; William S. Pierce, Case Western Reserve University; David Pimentel, Cornell University; Peter Pizor, Northwest Community College; Mark D. Plunkett, Bellevue Community College; Grace L. Powell, University of Akron; James H. Price, Oklahoma College; Marian E. Reeve, Merritt College; Carl H. Reidel, University of Vermont; Roger Revelle, California State University, San Diego; L. Reynolds, University of Central Arkansas; Ronald R. Rhein, Kutztown University of Pennsylvania; Charles Rhyne, Jackson State University; Robert A. Richardson, University of Wisconsin; Benjamin F. Richason III, St. Cloud State University; Ronald Robberecht, University of Idaho; William Van B. Robertson, School of Medicine, Stanford University; C. Lee Rockett, Bowling Green State University; Terry D. Roelofs, Humboldt State University; Richard G. Rose, West Valley College; Stephen T. Ross, University of Southern Mississippi; Robert E. Roth, The Ohio State University; Floyd Sanford, Coe College; David Satterthwaite, I.E.E.D., London; Stephen W. Sawyer, University of Maryland; Arnold Schecter, State University of New York, Syracuse; Frank Schiavo, San Jose State University; William H. Schlesinger, Ecological Society of America; Stephen H. Schneider, National Center for Atmospheric Research; Clarence A. Schoenfeld, University of Wisconsin, Madison; Henry A. Schroeder, Dartmouth Medical School; Lauren A. Schroeder, Youngstown State University; Norman B. Schwartz, University of Delaware; George Sessions, Sierra College; David J. Severn, Clement Associates; Paul Shepard, Pitzer College and Claremont Graduate School; Michael P. Shields, Southern Illinois University at Carbondale; Kenneth Shiovitz; F. Siewert, Ball State University; E. K. Silbergold, Environmental Defense Fund; Joseph L. Simon, University of South Florida; William E. Sloey, University of Wisconsin, Oshkosh; Robert L. Smith, West Virginia University; Howard M. Smolkin, U.S. Environmental Protection Agency; Patricia M. Sparks, Glassboro State College; John E. Stanley, University of Virginia; Mel Stanley, California State Polytechnic University, Pomona; Norman R. Stewart, University of Wisconsin, Milwaukee; Frank E. Studnicka, University of Wisconsin, Platteville; Chris Tarp, Contra Costa College; William L. Thomas, California State University, Hayward; John D. Usis, Youngstown State University;

Tinco E. A. van Hylckama, Texas Tech University; Robert R. Van Kirk, Humboldt State University; Donald E. Van Meter, Ball State University; Gary Varner, Texas A&M University; John D. Vitek, Oklahoma State University; Lee B. Waian, Saddleback College; Warren C. Walker, Stephen F. Austin State University; Thomas D. Warner, South Dakota State University; Kenneth E. F. Watt, University of California, Davis; Alvin M. Weinberg, Institute of Energy Analysis, Oak Ridge Associated Universities; Brian Weiss; Anthony Weston, SUNY at Stony Brook; Raymond White, San Francisco City College; Douglas Wickum, University of Wisconsin, Stout; Charles G. Wilber, Colorado State University; Nancy Lee Wilkinson, San Francisco State University; John C. Williams, College of San Mateo; Ray Williams, Rio Hondo College; Roberta Williams, University of Nevada Las Vegas; Samuel J. Williamson, New York University; Ted L. Willrich, Oregon State University; James Winsor, Pennsylvania State University; Fred Witzig, University of Minnesota at Duluth; George M. Woodwell, Woods Hole Research Center; Robert Yoerg, Belmont Hills Hospital; Hideo Yonenaka, San Francisco State University; Malcolm J. Zwolinski, University of Arizona.

Brief Contents

Detailed Contents

Temperate deciduous forest, fall, Rhode Island

Temperate deciduous forest, winter, Rhode Island

Heather Angel/Biofotos

Water hyacinth, Florida

Alan Watson

Mangrove swamp, Colombia

Endangered Indian tiger

Michael Greco/Picture Group

Sulfur dioxide emissions from coal-burning power plant

Gene Alexander/USDA

Tree farm, North Carolina

Saguaro cacti, Arizona

Burning tropical forest in Brazil's Amazon Basin

U.S. Windpower

Wind farm in California

Silvestris Fotoservice/SIPA Press

Czechoslavakian forest damaged by acid deposition

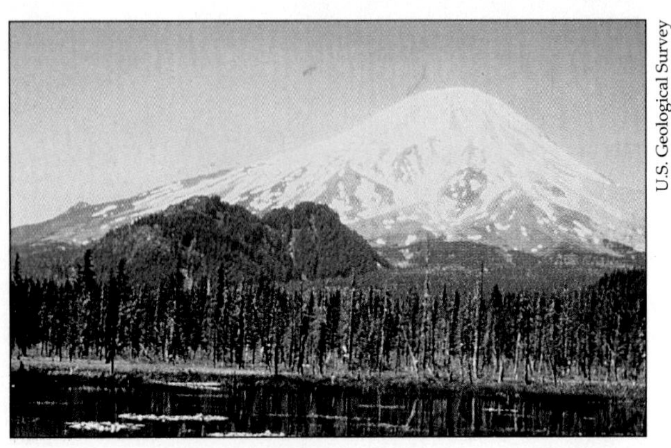

Mt. St. Helens volcano before the eruption

Mt. St. Helens volcano after 1980 eruption

Clown fish and sea anemones, coral sea, Australia

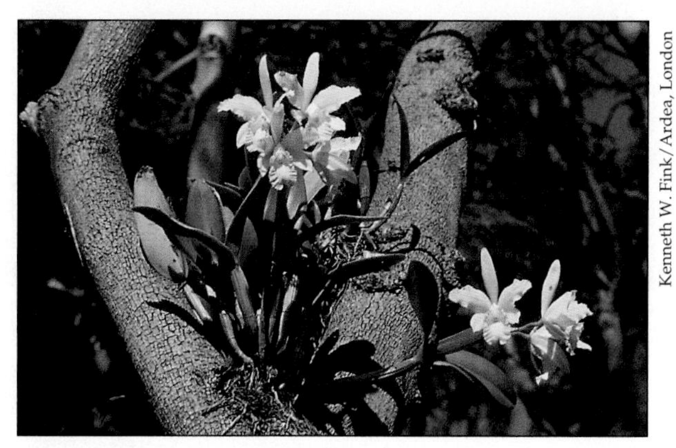

White orchid epiphyte, tropical forest, Latin America

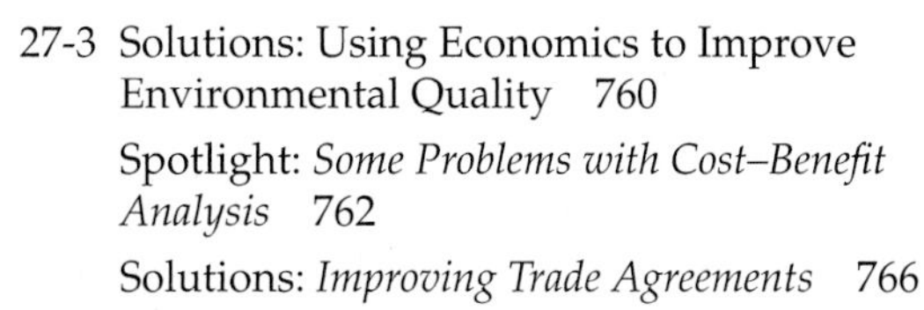

PART I

ENVIRONMENTAL PROBLEMS AND SCIENTIFIC PRINCIPLES

The environmental crisis is an outward manifestation of a crisis of mind and spirit. There could be no greater misconception of its meaning than to believe it is concerned only with endangered wildlife, human-made ugliness,

and pollution. These are part of it, but more importantly, the crisis is concerned with the kind of creatures we are and what we must become in order to survive.

LYNTON K. CALDWELL

1 ENVIRONMENTAL PROBLEMS, THEIR CAUSES, AND SUSTAINABILITY

Living in an Exponential Age

Once there were two kings from Babylon who enjoyed playing chess, with the winner claiming a prize from the loser. After one match, the winning king asked the loser to pay him by placing one grain of wheat on the first square of the chessboard, two on the second, four on the third, and so on. The number of grains was to double each time until all 64 squares were filled.

The losing king, thinking he was getting off easy, agreed with delight. It was the biggest mistake he ever made. He bankrupted his kingdom and still could not produce the 2^{63} grains of wheat he had promised. In fact, it's probably more than all the wheat that has ever been harvested!

This is an example of **exponential growth**, in which a quantity increases by a fixed percentage of the whole in a given time. As the losing king learned, exponential growth is deceptive. It starts off slowly, but after only a few doublings it grows to enormous numbers because each doubling is more than the total of all earlier growth.

Here is another example. Fold a piece of paper in half to double its thickness. If you could do this 42 times, the stack would reach from the earth to the moon, 386,400 kilometers (240,000 miles) away. If you could double it 50 times, the folded paper would almost reach the sun, 149 million kilometers (93 million miles) away!

The environmental problems we face—population growth, wasteful use of resources, destruction and degradation of wildlife habitats, extinction of plants and animals, poverty, and pollution—are interconnected and are growing exponentially. For example, world population has more than doubled in only 48 years, from 2.5 billion in 1950 to 5.9 billion in 1998. Unless death rates rise sharply, it may reach 8 billion by 2025, 10–11 billion by 2050, and 14 billion by 2100 (Figure 1-1). Global economic output, much of it environmentally damaging, has increased almost sixfold since 1950.

Each year more forests, grasslands, coral reefs, and wetlands disappear or are seriously degraded, and some deserts grow larger. Water tables are falling, some rivers are running dry because of excessive withdrawal of water, and a growing number of countries are

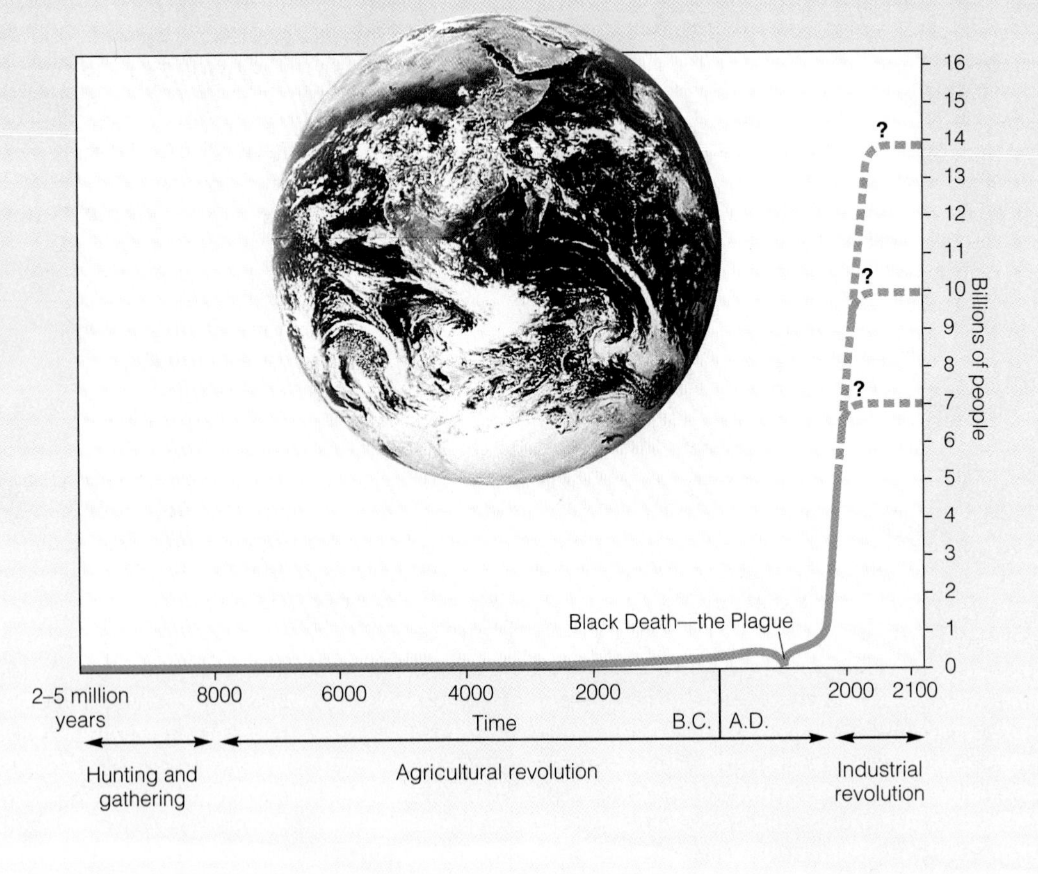

Figure 1-1 The *J*-shaped curve of past exponential world population growth, with projections beyond 2100. Notice that exponential growth starts off slowly, but as time passes the curve becomes increasingly steep. World population has more than doubled in only 47 years, from 2.5 billion in 1950 to 5.9 billion in 1998. Unless death rates rise sharply, it may reach 8 billion by 2025, 10–11 billion by 2050, and 14 billion by 2100. (This figure is not to scale.) (Data from World Bank and United Nations; photo courtesy of NASA)

squabbling over access to shared but limited water supplies. Vital topsoil is washed or blown away from farmland, cleared forests, and construction sites, clogging streams, lakes, and reservoirs with sediment. Many grasslands have been overgrazed and fisheries overharvested to the point of collapse. Oceans, streams, and the atmosphere are used as trash cans for a variety of wastes, many of them toxic. We drive an estimated two to eight wildlife species to extinction every hour, mostly because of loss of their habitats.

Within the next 40–50 years, the earth's climate may become warm enough to disrupt agricultural productivity, alter water distribution, drive countless species to extinction, and cause economic chaos because of the release of heat-trapping gases into the lower atmosphere from burning fossil fuels and the destruction of forests. Extracting and burning fossil fuels (oil, coal, and natural gas) also pollute the air and water and disrupt the land. Other chemicals we add to the air drift into the upper atmosphere and deplete a gas (ozone) that filters out much of the sun's harmful ultraviolet radiation. Toxic wastes from factories and mines poison the air, water, and soil. Agricultural pesticides contaminate some of our drinking water and food.

There is also some exciting good news. Mainly because of improved sanitation and medical advances, average human life expectancy has doubled and global infant mortality has dropped by almost two-thirds during this century. Since the 1960s global food production has outpaced population growth, thanks mostly to new high-yield forms of agriculture.

Because of improved mining technology, there have been significant increases in proven deposits of virtually all of the earth's fossil fuel and mineral resources since 1950. Since 1970 air and water pollution levels in most industrialized countries have dropped because of new pollution control laws and technologies. Recently, industrial nations have developed international treaties to phase out production of chemicals that deplete ozone in the upper atmosphere.

It is encouraging that a growing number of analysts, including several prominent economists, are beginning to recognize that the current economic model used by most nations will not lead to long-term sustainability for economies, human cultures, and many of the world's wild species. The challenge is to develop new models for dealing with the environmental problems we face and for living on the earth more sustainably, as discussed throughout the rest of this book.

Alone in space, alone in its life-supporting systems, powered by inconceivable energies, mediating them to us through the most delicate adjustments, wayward, unlikely, unpredictable, but nourishing, enlivening, and enriching in the largest degree—is this not a precious home for all of us? Is it not worth our love?

BARBARA WARD AND RENÉ DUBOS

This chapter is an overview of environmental problems, their root causes, and the controversy over their seriousness. It discusses these questions:

- What is earth capital? What is a sustainable society?
- How fast is the human population increasing?
- What are the earth's main types of resources? How can they be depleted or degraded?
- What are the principal types of pollution? How can pollution be reduced and prevented?
- What are the root causes of the environmental problems we face?
- How serious are environmental problems, and is our current course sustainable?
- What major effects have hunter–gatherer societies, nonindustrialized agricultural societies, and industrialized societies had on the environment?
- What major human-centered environmental worldviews guide most industrial societies?
- What are some life-centered and earth-centered worldviews?
- How can we live more sustainably?

1-1 LIVING SUSTAINABLY

What Are Environment, Ecology, and Environmental Science? Many people confuse the terms *environment* and *ecology*, and even use them interchangeably. They represent different but related concepts.

The term **environment** refers to all external conditions and factors that affect living organisms. By contrast, **ecology** is the study of the relationships between living organisms and their environment.

Environmental science is interdisciplinary study that examines the role of humans on the earth. It uses concepts and information from ecology, chemistry, geology, engineering, economics, politics, ethics, philosophy—just about every discipline you can imagine—to help us understand how the earth works and how we are affecting the earth's life-support systems for us and other species.

What Are Solar Capital and Earth Capital? Our existence, lifestyles, and economies depend completely on the sun and the earth, a blue and white island in the black void of space. We can think of energy from the sun as **solar capital,** and we can think of the planet's air, water, soil, wildlife, minerals, and natural purification, recycling, and pest control processes as **earth capital** (Figure 1-2). The term *environment* is often used to describe these life-support systems; in effect it's another term for the external conditions and factors making up solar capital and earth capital that affect us and other organisms.

The concept of earth capital means we and all other organisms are interdependent and interconnected parts of nature and are completely dependent on nature. Our survival and health, our economies, and the survival and health of all living things depend on the earth and its natural systems (Figure 1-2). The air you breathe, the water you drink, the food you eat, and all of your possessions are derived from solar energy and the earth's air, water, soil, plants and animals,

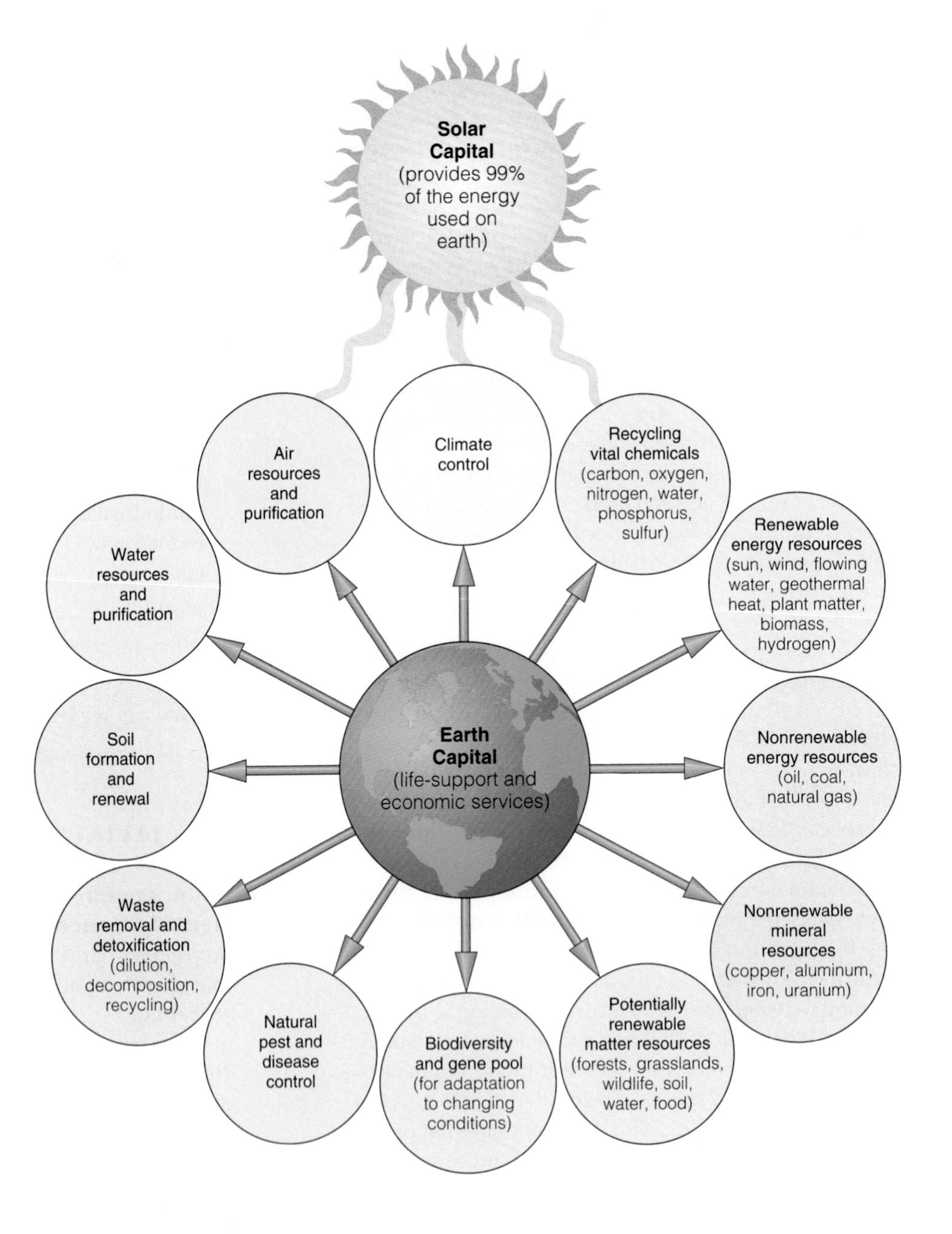

Figure 1-2 Solar and earth capital consist of the life-support resources and processes provided by the sun and the planet for use by us and other species. These two forms of capital support and sustain all life and all economies on the earth.

Q: How many people are there in the world?

minerals, energy resources (such as oil and coal), and life-sustaining processes.

Environmentalists, many leading scientists, and a growing number of prominent economists believe that we are depleting and degrading the earth's natural capital at an accelerating rate as our population (Figure 1-1) and demands on the earth's resources and natural processes increase exponentially (Figure 1-3). For example, according to Lester R. Brown (Guest Essay, p. 34), from 1950 to 1997 "the use of lumber tripled, that of paper increased sixfold, the fish catch increased nearly fivefold, grain consumption nearly tripled, fossil fuel burning nearly quadrupled, and air and water pollutants multiplied severalfold." During the last 50 years, the United States alone has used more resources than the rest of humanity in all previous years.

Others, mostly some economists and business leaders, disagree. They contend that there are no limits to human population growth and economic growth that can't be overcome by human ingenuity and technology.

What are Sustainability and Carrying Capacity? **Sustainability** is the ability of a specified system to survive and function over a specified time. There are several types of sustainability. A **sustainable resource harvest,** such as a sustainable supply of fish or timber, means that a certain quantity of that resource can be harvested each year (or other time interval) over a specified period. A **sustainable earth** means that the earth's supplies of resources and the processes that make up earth capital (Figure 1-2) are used and maintained over a specified period. A **sustainable society** manages its economy and population size without exceeding all or part of the planet's ability to absorb environmental insults, replenish its resources, and sustain human and other forms of life over a specified period, usually hundreds to thousands of years.

A sustainable society learns how to live within the **carrying capacity**: the maximum number of organisms a local, regional, or global environment can support over a specified period. This capacity depends on the available resource supplies and the ability of the environment to absorb, detoxify, or recycle wastes produced by resource use.

Carrying capacity is rarely a fixed quantity. It varies with **(1)** location, **(2)** time (including short-term seasonal changes and long-term global changes in factors such as climate), and **(3)** the types of technology used to extract and process resources and to deal with the environmental problems caused by population growth and resource use.

Living sustainably means living off of income and not depleting the capital that supplies the income. Imagine that you inherit $1 million. If you invest this capital at 10% interest, you will have a sustainable annual income of $100,000; that is, you can spend up to $100,000 a year without touching your capital.

Suppose you develop a taste for diamonds or a yacht, or all your relatives move in with you. If you spend $200,000 a year, your $1 million will be gone early in the 7th year; even if you spend just $110,000 a year, you will be bankrupt early in the 18th year. The lesson here is a very old one: *Don't eat the goose that lays the golden egg.* Deplete your capital, and you move from a sustainable to an unsustainable lifestyle.

The same lesson applies to earth capital (Figure 1-2). With the help of solar energy, natural processes developed over billions of years can indefinitely renew the topsoil, water, air, forests, grasslands, and wildlife on which we and other forms of life depend, as long as we don't use these potentially renewable resources faster than they are replenished.

Some of earth's natural processes also provide flood prevention, build and renew soil, slow soil erosion, and keep the populations of at least 95% of the species we consider pests under control. Living sustainably involves not disrupting or diminishing these and other natural processes and services provided by nature (Figure 1-2).

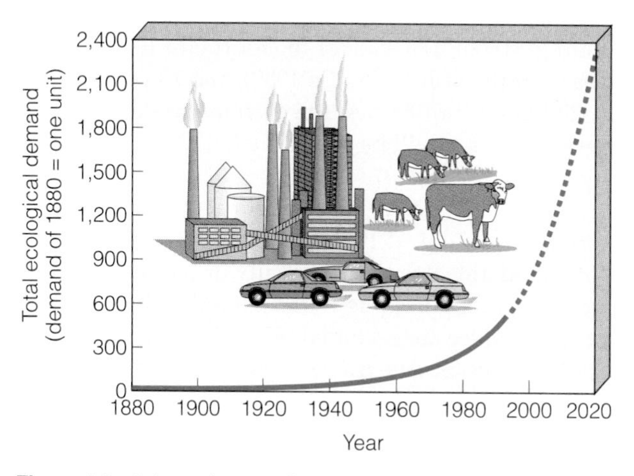

Figure 1-3 *J*-shaped curve of exponential growth in the total ecological demand on the earth's resources from agriculture, mining, and industry between 1880 and 1996. Projections to 2020 assume that resource use will continue to increase at the current rate of 5.5% per year. At that rate, our total ecological demand on the earth's resources doubles every 13 years. If global economic output grew by only 3% a year, resource consumption would still double every 23 years. (Data from United Nations, World Resources Institute, and Carrol Wilson, *Man's Impact on the Global Environment*, Cambridge, Mass.: MIT Press, 1970)

A: Approximately 5.9 billion in mid-1998 (approximately 84 million added per year)

1-2 GROWTH AND THE WEALTH GAP

What Is the Difference Between Linear Growth and Exponential Growth? Suppose you hop on a train that accelerates by 1 kilometer (0.6 mile) an hour every second. After 5 seconds, you would be traveling at 5 kilometers (3 miles) per hour; after 30 seconds, your speed would be 30 kilometers (19 miles) per hour. This is an example of **linear growth,** in which a quantity increases by a constant amount per unit of time, as in 1, 2, 3, 4, 5—or 1, 3, 5, 7, 9—and so on. If plotted on a graph, such growth in speed or growth of money in a savings account yields a straight line sloping upward (Figure 1-4).

But suppose the train has a magic motor strong enough to double its speed every second. After 6 seconds, you would be moving at 64 kilometers (40 miles) per hour; after only 30 seconds, you would be traveling a billion kilometers (620 million miles) per hour!

This is an example of the astounding power of **exponential growth,** in which a quantity increases by a fixed percentage of the whole in a given time as each increase is applied to the base for further growth. Any quantity growing by a fixed percentage, even as small as 0.001% or 0.1%, is undergoing exponential growth and will experience extraordinary growth as its base of growth doubles again and again. If plotted on a graph, continuing exponential growth eventually yields a graph shaped somewhat like the letter J (Figure 1-4).

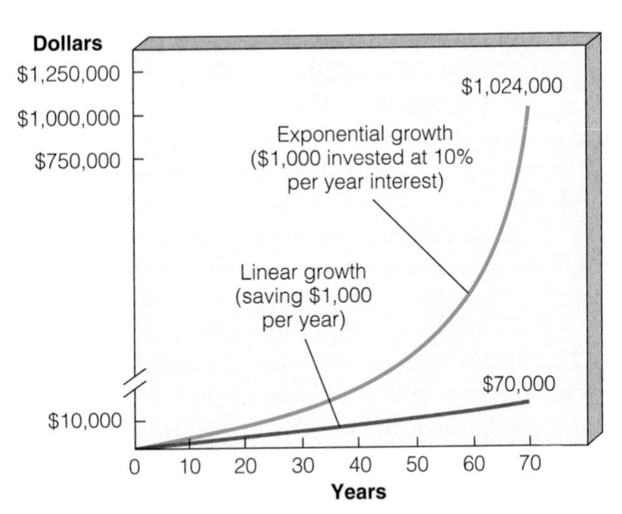

Figure 1-4 Linear and exponential growth. If you save $1,000 a year for a lifetime of 70 years, the resulting linear growth will allow you to save $70,000 (lower curve). If you invest $1,000 each year at 10% interest for 70 years and reinvest the interest, your money will grow exponentially to $1,024,000 (upper curve). If resource use, economic growth, or money in a savings account grows exponentially for 70 years (a typical human lifetime) at a rate of 10% a year, it will experience 1,023-fold increase.

How long does it take to double resource use, population size, or money in a savings account that is growing exponentially? A quick way to calculate this **doubling time** in years is to use the **rule of 70**: 70/percentage growth rate = doubling time in years (a formula derived from the basic mathematics of exponential growth). For example, in 1998 the world's population grew by 1.43%. If that rate continues, the earth's population will double in about 49 years (70/1.43 = 49 years)—more growth than has occurred in *all* of human history. In the example of exponential growth of savings by 10% a year (Figure 1-4), your money would double roughly every 7 years (70/10 = 7).

Even though supplies of many resources seem large, exponential growth in their use can deplete them in a short time. For example, suppose we have an exhaustible resource so plentiful that its estimated supplies would last 1 billion years at its current rate of consumption. If the rate of use of this resource increases exponentially by 5% per year, however, the billion-year supply would last only 500 years.

How Rapidly Is the Human Population Growing? The increasing size of the human population is one example of exponential growth. If such exponential growth continues (even at a low percentage of annual growth), eventually the population growth curve rounds a bend and heads almost straight up, creating a *J*-shaped curve (Figure 1-1 and Spotlight, right).

It took at least 60,000 years to reach 1 billion people, 130 years to add the second billion, 30 years for the third (in 1960), 17 years for the fourth (in 1977), and 12 years for the fifth billion (in 1989), and 10 years for the sixth billion (in 1999). At the current growth rates, the seventh billion will be added by 2012, and the eighth by 2025.

Recent studies by researchers at Conservation International suggest that roughly 48% of the earth's total land area has been partially or totally modified by human activities (Figure 1-5). If uninhabitable areas of rock and ice are excluded, *73% of the habitable area of the planet has been altered by human activities*. What will happen to the earth's remaining diversity of wildlife habitats and wildlife species if the human population increases from 5.9 billion to 8 billion between 1998 and 2025 and perhaps to 11 billion by 2050?

What Is Economic Growth? Virtually all countries seek **economic growth**: an increase in their capacity to provide goods and services for people's final use. Such growth is normally achieved by increasing the flow or **throughput** of matter and energy resources used to produce goods and services through

Q: How many people are added to the world's population each day?

an economy. This is accomplished by means of population growth (more consumers and producers), more consumption per person, or both.

Economic growth is usually measured by an increase in a country's **gross national product (GNP)**: the market value in current dollars of all goods and services produced within and outside of a country by the country's businesses for final use during a year. **Gross domestic product (GDP)** is the market value in current dollars of all goods and services produced *within* a country for final use during a year. To show one person's slice of the economic pie, economists often calculate the **per capita GNP**: the GNP divided by the total population.

The United Nations broadly classifies the world's countries as economically developed or developing. The **developed countries** are highly industrialized. Most (except the countries of the former Soviet Union) have high average per capita GNPs (above $4,000). These countries, with 1.2 billion people (20% of the world's population in 1998), command about 85% of the world's wealth and income and use about 88% of its natural resources. They generate about 75% of its pollution and wastes (including about 90% of the world's estimated hazardous waste). Three developed countries—the United States, Japan, and Germany—together account for more than half of the world's economic output.

All other nations are classified as **developing countries**, with low to moderate industrialization and per capita GNPs. Most are in Africa, Asia, and Latin America. Their 4.7 billion people (80% of the world's population in 1998) have only about 15% of the wealth and income and use only about 12% of the world's natural resources. In this context, **development** is the change from a society that is largely rural, agricultural, illiterate, and poor, with a rapidly growing population, to one that is mostly urban, industrial, educated, and wealthy, with a slow-growing or stationary population.

Figure 1-6 compares various characteristics of developed countries and developing countries. More than 95% of the projected increase in world population is expected to take place in developing countries (Figure 1-7), *where 1 million people are added every 4 days.* By 2010, the combined population of Asia and Africa is projected to be 5.3 billion—almost as many as now live on the entire planet. The primary reason for such rapid population growth in developing countries (1.7% compared to 0.1% in developed countries) is the *large percentage of people who are under age 15* (35% compared to 19% in developed countries in 1998) and who will be moving into their prime reproductive years over the next several decades.

SPOTLIGHT

Current Exponential Growth of the Human Population

The current growth of the world's population is the result of roughly 2.6 times as many births (about 136 million) as deaths (some 51 million) per year. The relentless ticking of this population clock means that in 1998 the world's population of 5.9 billion grew by 84 million people (5.9 billion × 0.0143 = 84 million), an average increase of 230,000 people a day, 9,600 an hour.

At this 1.43% annual rate of exponential growth, it takes only about

- 5 days to add the number of Americans killed in all U.S. wars
- 4 months to add as many people as live in the Los Angeles basin
- 2 years to add the 167 million people killed in all wars fought in the past 200 years
- 3 years to add 270 million people (the population of the United States in 1998)
- 15 years to add 1.24 billion people (the population of China, the world's most populous country, in 1998).

Critical Thinking

Some argue that current population growth is good because it provides more workers, consumers, and problem solvers to keep the global economy growing. Others argue that current population growth threatens economies and the earth's life-support systems through increased pollution and environmental degradation. What is your position? Why?

Between 1950 and 1997 the global output of goods and services increased from $5 trillion to $29 trillion. Between 1990 and 1997 it grew by $5 trillion—equal to all of the economic growth from the beginning of civilization to 1950. According to Lester R. Brown (Guest Essay, p. 34),

> *While economic indicators such as investment, production, and trade are consistently positive, the key environmental indicators are increasingly negative. Forests are shrinking, water tables are falling, soils are eroding, wetlands are disappearing, fisheries are collapsing, rangelands are deteriorating, rivers are running dry, temperatures are rising and plant and*

A: An average of about 230,000

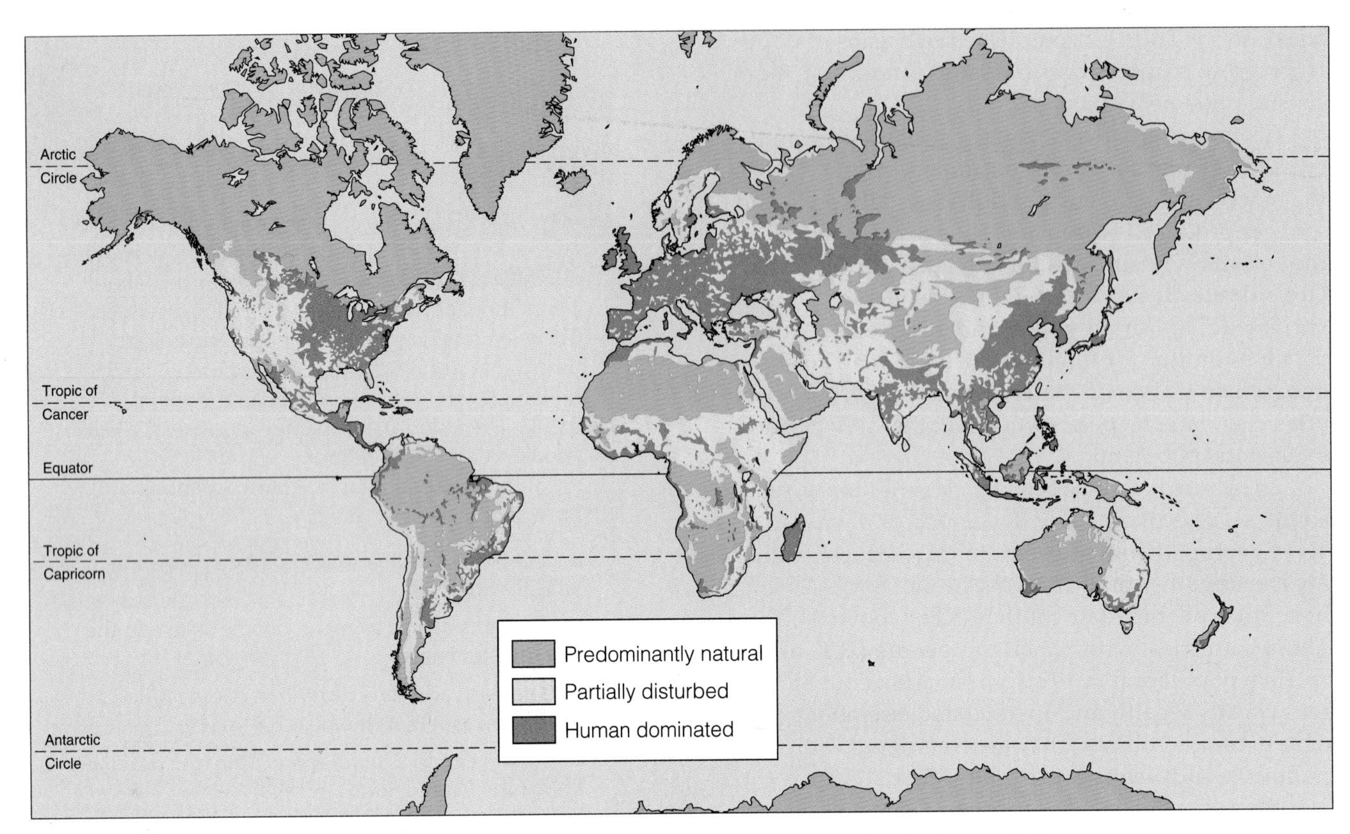

Figure 1-5 The degree of human disturbance of the earth's land area. Green indicates undisturbed areas. Yellow indicates partially disturbed areas, including secondary forest growth after cutting and shifting cultivation. Red indicates seriously disturbed natural ecosystems, including deforestation, farmland, overgrazed grasslands, and urban areas. Excluding uninhabitable areas of rock, ice, desert, and steep mountain terrain, *only about 27% of the planet's land area remains undisturbed by human activities.* (Data from Conservation International)

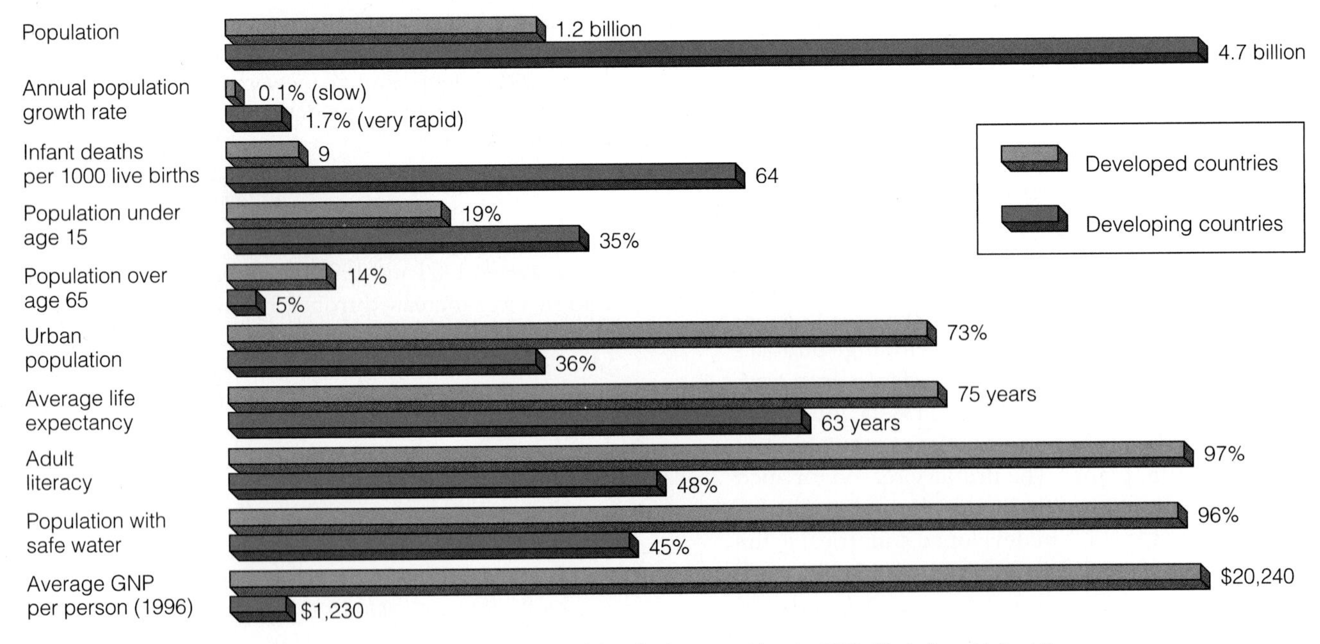

Figure 1-6 Some characteristics of developed countries and developing countries in 1998. (Data from United Nations and Population Reference Bureau)

Q: What is the projected population of the world in 2050?

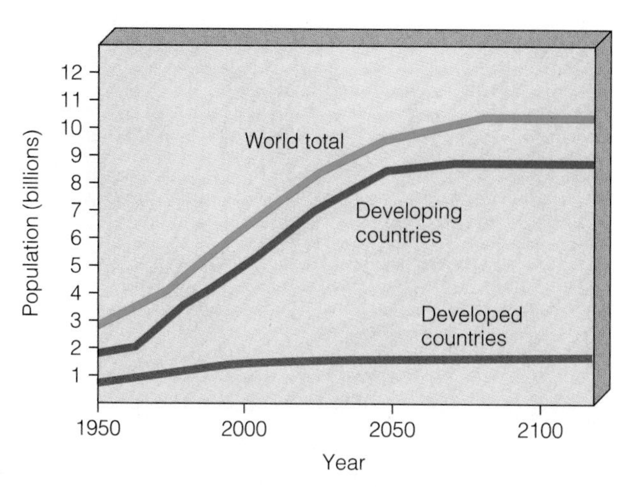

Figure 1-7 Past and projected population size for developed countries, developing countries, and the world, 1950–2120. Over 95% of the projected addition of 3.6 billion people between 1990 and 2030 is projected to occur in developing countries. (Data from United Nations)

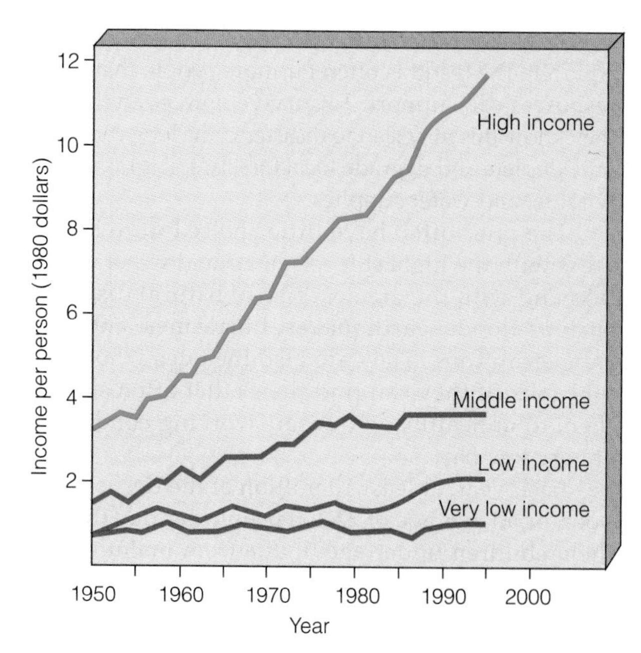

Figure 1-8 The wealth gap: changes in the distribution of global per capita GNP in high-income, middle-income, low-income, and very-low-income countries, 1950–96. Instead of trickling down, most of the income from economic growth has flowed up, with the situation worsening since 1980. More than 1 billion people survive on less than a dollar a day. (Data from United Nations)

animal species are disappearing. The global economy as now structured cannot continue to expand much longer if the natural systems on which it depends continues to deteriorate at the current rate. . . .

As a result humanity now faces a challenge that rivals any in its history: restoring balance with nature while expanding economic opportunities for the billions of people whose basic needs—for food and clean water, for example—are still not being met.

What Is the Wealth Gap? Since 1960, and especially since 1980, the gap between the per capita GNP of the rich, middle income, and poor has widened (Figure 1-8). Today, one person in five lives in luxury, the next three get by, and the fifth struggles to survive on less than $1 a day. One person in six is hungry or malnourished (Figure 1-9) or severely undernourished and lacks clean drinking water, decent housing (Figure 1-10), and adequate health care. One of every three people lacks enough fuel to keep warm and to cook food and more than half of humanity lacks sanitary toilets.

Daily life for an estimated 1.3 billion desperately poor people in developing countries is a harsh struggle for survival. Parents, some with nine or more children, are struggling to live on a cash income the equivalent of $1 a day or less. Having many children makes good sense to most poor parents because their children are a form of economic security, helping them grow food, gather fuel (mostly wood and dung), haul drinking water, tend livestock, work, or beg in the streets. The desperately poor tend to have many offspring because many of their children die at an early age. The two or three who live to adulthood will help their parents survive in old age (their 50s or 60s). Another 1.7 billion poor people struggle to survive on a cash income of about $3 per day.

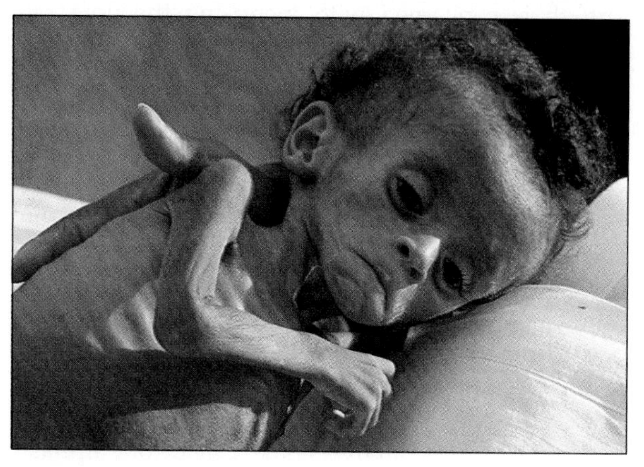

Figure 1-9 One in every three children under age 5, such as this Brazilian child, suffers from malnutrition. Each day, at least 13,700 children die prematurely from hunger or hunger-related diseases—an average of 10 preventable deaths each minute. Some analysts put the estimated death toll at twice this number. (John Bryson/Photo Researchers)

A: 10–11 billion (about twice the current population)

However, when many poor families have several children, the result is often far more people than local resources can support. To survive now, even though they know this may lead to disaster in the long run, they may deplete and degrade local forests, soil, grasslands, wildlife, and water supplies.

The poor often have little choice but to live in areas with the highest levels of air and water pollution and with the greatest risk of natural disasters such as floods, earthquakes, hurricanes, and volcanic eruptions. They are also the ones who must take jobs (if they can find them) that often subject them to unhealthy and unsafe working conditions at very low pay.

Each year, at least 10 million of the desperately poor, or an average of 27,400 people per day (half of them children under age 5), die from malnutrition (lack of protein and other nutrients needed for good health) or related diseases and from contaminated drinking water. *This premature dying of human beings is equivalent to 69 jumbo jet planes, each carrying 400 passengers, crashing every day with no survivors.* A 1997 study by Johns Hopkins University researchers put this annual death toll from poverty at about 18 million a year, or an average of 49,300 premature deaths per day.

What Is Sustainable Development? Economic growth is concerned with increasing the flow rate (throughput) and quantity of goods produced. **Economic development** involves using economic systems to improve the *quality* of people's lives and the environment.

Some economists and business leaders talk about sustainable economic growth. However, most ecologists, environmental scientists, and a growing number of economists contend that the current form of economic growth is not sustainable in the long run because of the limits imposed by finite supplies of resources and the capacity of the environment to absorb, detoxify, and recycle our wastes.

Some of these analysts have called for emphasis on **sustainable development,** which was defined in *Our Common Future*, a 1987 report of the World Commission on Environment and Development as meeting present needs without preventing future generations of humans and other species from meeting their needs. This idea assumes that we have a right to use the earth's resources and earth capital (Figure 1-2) to meet our needs but that we have an obligation to pass on the earth's resources and services to future generations in as good or better shape than these conditions were passed on to us. This ethical concept that future generations should receive undiminished earth capital and economic opportunity is called **intergenerational equity** or **fairness**.

Figure 1-10 One-sixth of the people in the world have inadequate housing, and at least 100 million have no housing at all. These homeless people in Calcutta, India, must sleep on the street. (United Nations)

To some people, however, sustainable development means continuing present patterns of earth-degrading economic growth with only minor modifications. For this reason, many environmental scientists prefer to use the terms *sustainable use of the planet* or *sustainability*, instead of *sustainable development*.

1-3 RESOURCES

What Is a Resource? Ecological Versus Economic Resources An **ecological resource** is anything required by an organism for normal maintenance, growth, and reproduction. Examples include habitat, food, water, and shelter.

An **economic resource** is anything obtained from the environment (the earth's life-support systems) to meet human needs and wants. Examples include food, water, shelter, manufactured goods, transportation, communication, and recreation. On our short human time scale, we classify the material resources we get from the environment as renewable, potentially renewable, or nonrenewable (Figure 1-11; also see the orange boxes in the "Concepts and Connections" diagram inside the back cover).

Some resources, such as solar energy, fresh air, wind, fresh surface water, fertile soil, and wild edible plants, are directly available for use by us

Q: Where does everything that supports your life come from?

and other organisms. Other resources, such as petroleum (oil), iron, groundwater (water found underground), and modern crops, aren't directly available. They become useful to us only with some effort and technological ingenuity. Petroleum, for example, was a mysterious fluid until we learned how to find, extract, and convert (refine) it into gasoline, heating oil, and other products that could be sold at affordable prices.

What Are Renewable Resources? Solar energy is called a **renewable** or **perpetual resource** because on a human time scale this solar capital (Figure 1-2) is essentially inexhaustible. It is expected to last at least 6 billion years as the sun completes its life cycle.

A **potentially renewable resource*** can be replenished fairly rapidly (hours to several decades) through natural processes. Examples of such resources are forest trees, grassland grasses, wild animals, fresh lake and stream water, groundwater, fresh air, and fertile soil.

One important potentially renewable resource for us and other species is **biological diversity,** or **biodiversity,** which consists of the different life forms (species) that can best survive the variety of conditions currently found on the earth. Kinds of biodiversity include **(1) genetic diversity** (variety in the genetic makeup among individuals within a single species), **(2) species diversity** (variety among the species or distinct types of living organisms found in different habitats of the planet, Figure 1-12), and **(3) ecological diversity** (variety of forests, deserts, grasslands, streams, lakes, oceans, wetlands, and other biological communities).

This rich variety of genes, species, and biological communities gives us food, wood, fibers, energy, raw materials, industrial chemicals, and medicines, all of which pour hundreds of billions of dollars into the world economy each year. The earth's vast inventory of life-forms and biological communities also provides free recycling, purification, and natural pest control services (Figure 1-2).

However, potentially renewable resources can be depleted. The highest rate at which a potentially renewable resource can be used *indefinitely* without reducing its available supply is called its **sustainable yield.** If a resource's natural replacement rate is exceeded, the available supply begins to shrink, a process known as

*Most sources use the term *renewable resource.* I have added the word *potentially* to emphasize that these resources can be depleted if we use them faster than natural processes renew them.

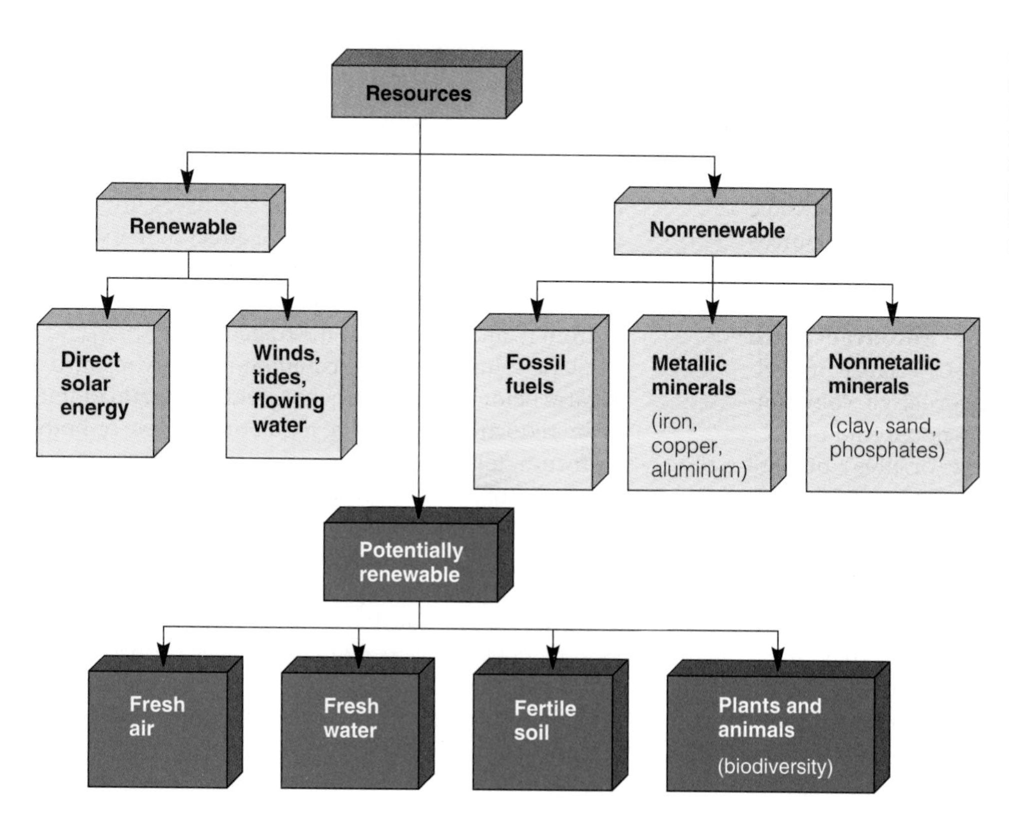

Figure 1-11 Major types of material resources. This scheme isn't fixed; potentially renewable resources can become nonrenewable resources if used for a prolonged period at a faster rate than they are renewed by natural processes.

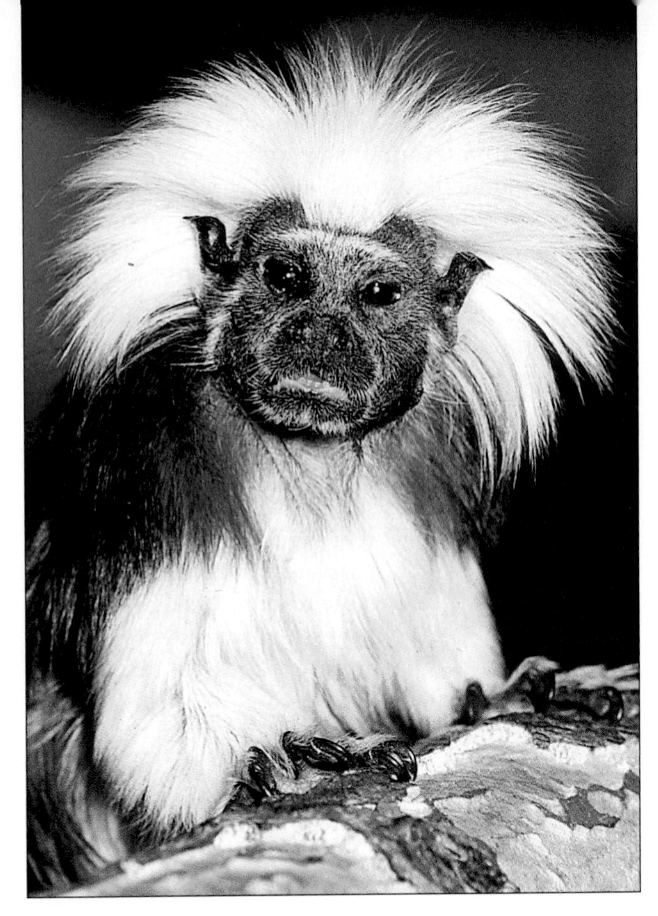

Figure 1-12 Two species found in tropical forests are part of the earth's precious biodiversity. On the left is the world's largest flower, the flesh flower (*Rafflesia arnoldi*), growing in a tropical rain forest in Sumatra. The flower of this leafless plant can be as large as 1 meter (3.3 feet) in diameter and weigh 7 kilograms (15 pounds). The plant gives off a smell like rotting meat, presumably to attract flies and beetles that pollinate its flower. After blossoming for a few weeks, the flower dissolves into a slimy black mass. On the right is a cotton top tamarin. (Left, Mitschuhiko Imanori/Nature Production; right, Gary Milburn/Tom Stack & Associates)

environmental degradation. Several types of environmental degradation can change potentially renewable resources into nonrenewable or unusable resources (Figures 1-13 and 1-14).

Connections: Renewable Resources and the Tragedy of the Commons One cause of environmental degradation is the overuse of **common-property resources,** which are owned by no one (or jointly by everyone in a country or area) but are available to all users free of charge. Most are potentially renewable. Examples include clean air, the open ocean and its fish, migratory birds, publicly owned lands (such as national forests, national parks, and wildlife refuges), gases of the lower atmosphere, and space.

In 1968, biologist Garrett Hardin (Guest Essay, p. 266) called the degradation of common-property resources the **tragedy of the commons**. It happens because each user reasons, "If I don't use this resource, someone else will. The little bit I use or pollute is not enough to matter." With only a few users, this logic works. However, the cumulative effect of many people trying to exploit a common-property resource eventually exhausts or ruins it. Then no one can benefit from it, and therein lies the tragedy.

One solution is to use common-property resources at rates below their sustainable yields or overload limits by reducing population, regulating access, or both. Unfortunately, it is difficult to determine the sustainable yield of a forest, grassland, or an animal population, partly because yields vary with weather, climate, and unpredictable biological factors, and because tracking such data is expensive.

These uncertainties mean that *it is best to use a potentially renewable resource at a rate well below its estimated sustainable yield*. This is a prevention or precautionary approach designed to reduce the risk of environmental degradation. This approach is rarely used because it requires hard-to-enforce regulations

Q: How many years does it take to add 1 billion people at current growth rates?

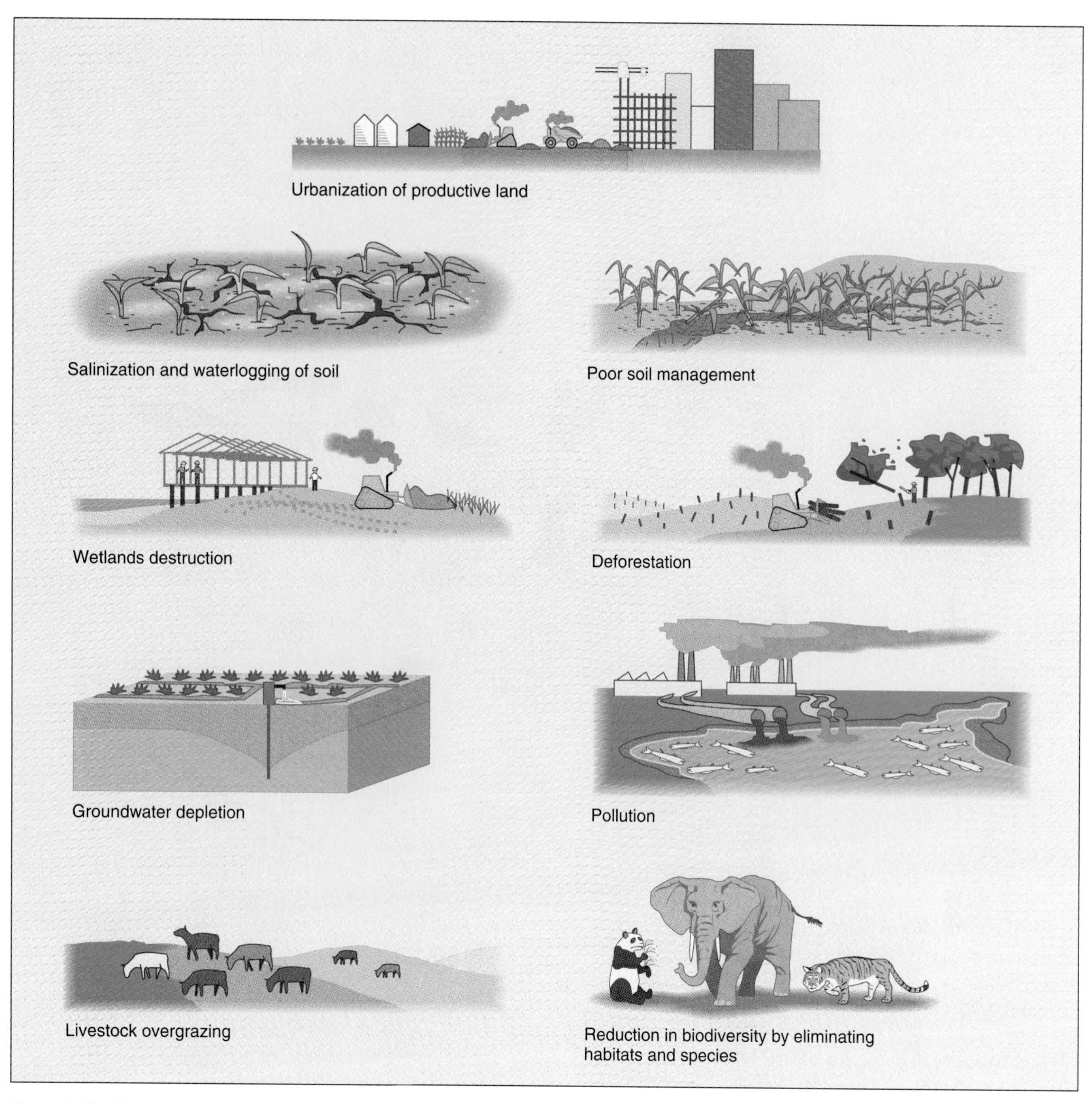

Figure 1-13 Major types of environmental degradation that can convert potentially renewable resources into nonrenewable resources.

that restrict resource use and thus conflict with the drive for short-term profit or pleasure.

Another approach is to convert common-property resources to private ownership. The reasoning behind this is that owners of land or some other resource have a strong incentive to see that their investment is protected. However, this approach is not practical for global common resources, such as the atmosphere, the open ocean, and migratory birds that cannot be divided up and converted to private property. Experience has also shown that private ownership can lead to short-term exploitation and environmental degradation instead of long-term sustainability.

Some believe that privatization is a better way to protect nonrenewable and potentially renewable resources found on publicly owned lands than relying on command-and-control government regulations and

A: About 12 years

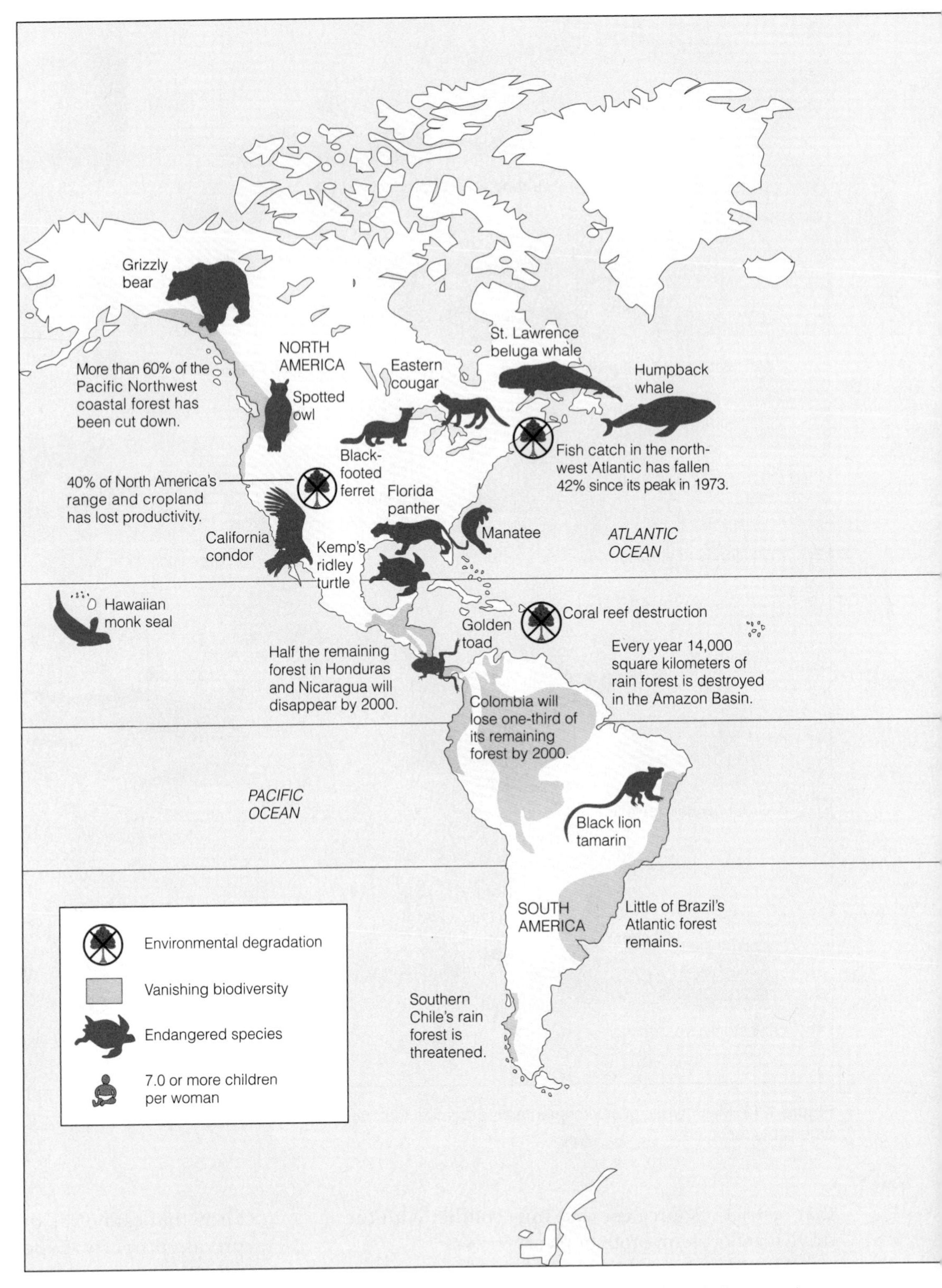

Figure 1-14 Examples of how some of the earth's natural systems that support all life and all economies are being assaulted at an accelerating rate as a result of the exponential growth of population and resource use and the resulting environmental degradation and loss of biodiversity. (Data from The World Conservation Union, World Wildlife Fund, Conservation International, United Nations, Population Reference Bureau, U.S. Fish and Wildlife Service, and Daniel Boivin)

Q: How long does it take to add the number of people killed in the wars fought during the past 200 years?

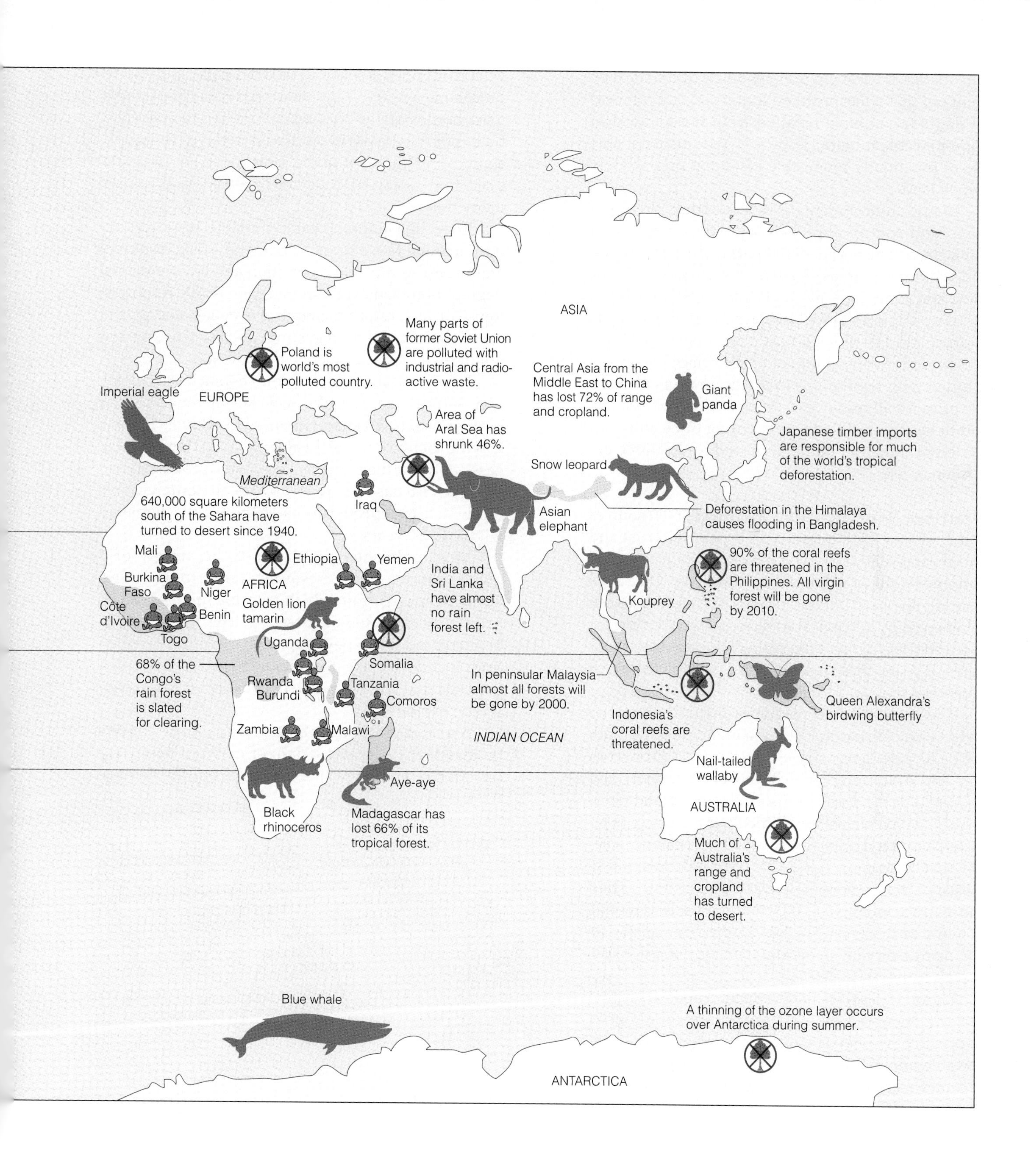

A: About 2 years

bureaucracies. Most environmentalists disagree. They point out that widespread pollution and environmental degradation have resulted from the removal of nonrenewable mineral resources and unsustainable use of potentially renewable resources on privately owned lands.

Many environmentalists agree that the command-and-control approach to use of resources on public lands (such as national parks, wildlife refuges, and wilderness areas) has some serious problems. They and some free-market economists are seeking *users-pay* solutions to replace the current *taxpayers-pay* approach to use of such publicly owned resources. This would involve a mix of marketplace incentives coupled with regulations that require users to pay a fair price for all resources extracted from public lands and to be responsible for preventing or cleaning up any environmental damage caused by resource extraction or use.

What Are Nonrenewable Resources? Resources that exist in a fixed quantity in the earth's crust and thus theoretically can be completely used up are called **nonrenewable,** or **exhaustible, resources**. On a time scale of millions to billions of years, such resources can be renewed by geological processes. However, on the much shorter human time scale of hundreds to thousands of years, these resources can be depleted much faster than they are formed.

These exhaustible resources include *energy resources* (coal, oil, natural gas, and uranium, which cannot be recycled), *metallic mineral resources* (iron, copper, and aluminum, which can be recycled), and *nonmetallic mineral resources* (salt, clay, sand, and phosphates), which are usually difficult or too costly to recycle). A **mineral** is any hard, usually crystalline material that is formed naturally. Soil and most rocks consist of two or more minerals. We know how to find and extract more than 100 nonrenewable minerals from the earth's crust. We convert these raw materials into many everyday items and then we discard, reuse, or recycle them.

Figure 1-15 shows the production and exhaustion cycle of a nonrenewable energy or mineral resource. In practice, we never completely exhaust a nonrenewable mineral resource. However, such a resource becomes *economically depleted* when the costs of exploiting what is left exceed its economic value. At that point, we have five choices: recycle or reuse existing supplies (except for nonrenewable energy resources, which cannot be recycled or reused), waste less, use less, try to develop a substitute, or do without and wait millions of years for more to be produced.

Some nonrenewable material resources, such as copper and aluminum, can be recycled or reused to extend supplies. **Recycling** involves collecting and reprocessing a resource into new products. For example, glass bottles can be crushed and melted to make new bottles or other glass items. **Reuse** involves using a resource over and over in the same form. For example, glass bottles can be collected, washed, and refilled many times.

Recycling nonrenewable metallic resources requires much less energy, water, and other resources and produces much less pollution and environmental degradation than exploiting virgin metallic resources. Reuse of such resources requires even less energy and other resources than recycling, and it results in less pollution and environmental degradation.

Nonrenewable *energy* resources, such as coal, oil, and natural gas, can't be recycled or reused. Once burned, the useful energy in these fossil fuels is gone, leaving behind waste heat and polluting exhaust gases. Most of the per capita economic growth shown in Figure 1-8 has been fueled by cheap nonrenewable oil, which is expected to be economically depleted within 40–80 years.

Most published estimates of the supply of a given nonrenewable resource refer to **reserves**: known deposits from which a usable mineral can be profitably extracted at current prices. Reserves can be increased when new deposits are found or when price increases make it profitable to extract identified deposits that were previously considered too expensive to exploit.

Some environmentalists and resource experts believe that the greatest danger may not be the exhaustion of nonrenewable resources but the damage

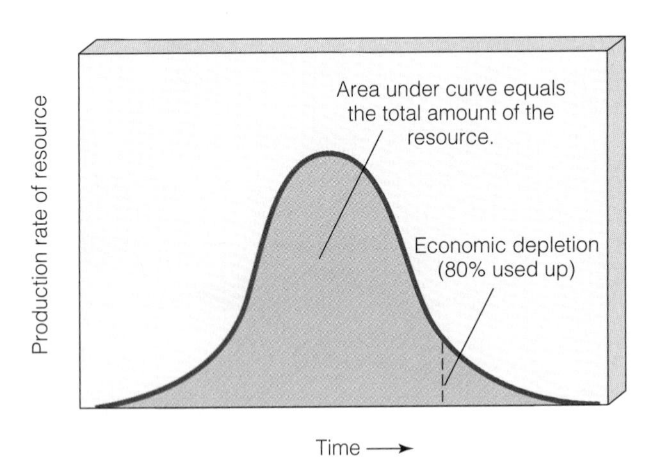

Figure 1-15 Full production and exhaustion cycle of a nonrenewable resource such as copper, iron, oil, or coal. Usually, a nonrenewable resource is considered economically depleted when 80% of its total supply has been extracted and used. Normally, it costs too much to extract and process the remaining 20%.

Q: How long will it take to double the population of a country with a population growth rate of 2% per year?

that their extraction, processing, and conversion to products do to the environment in the form of energy use, land disturbance, soil erosion, water pollution, and air pollution

1-4 POLLUTION

What Is Pollution, and Where Does It Come From? Any addition to air, water, soil, or food that threatens the health, survival, or activities of humans or other living organisms is called **pollution**. Most pollutants are solid, liquid, or gaseous by-products or wastes produced when a resource is extracted, processed, made into products, or used. Pollution can also take the form of unwanted energy emissions, such as excessive heat, noise, or radiation.

Pollutants can enter the environment naturally (for example, from volcanic eruptions) or through human (anthropogenic) activities (for example, from burning coal). Most pollution from human activities occurs in or near urban and industrial areas, where pollutants are concentrated. Industrialized agriculture is also a major source of pollution. Some pollutants contaminate the areas where they are produced; others are carried by winds or flowing water to other areas. Pollution does not respect local, state, or national boundaries.

Some pollutants come from single, identifiable sources, such as the smokestack of a power plant, the drainpipe of a meat-packing plant, or the exhaust pipe of an automobile. These are called **point sources**. Other pollutants come from dispersed (and often difficult to identify) **nonpoint sources**. Examples are the runoff of fertilizers and pesticides (from farmlands, golf courses, and suburban lawns and gardens) into streams and lakes and pesticides sprayed into the air or blown by the wind into the atmosphere. It is much easier and cheaper to identify and control pollution from point sources than from widely dispersed nonpoint sources.

What Types of Harm Are Caused by Pollutants? Unwanted effects of pollutants include disruption of life-support systems for humans and other species, damage to wildlife, damage to human health, damage to property, and nuisances such as noise and unpleasant smells, tastes, and sights.

Three factors determine how severe the harmful effects of a pollutant are. One is its *chemical nature*: how active and harmful it is to living organisms. Another is its **concentration**: the amount per unit of volume or weight of air, water, soil, or body weight.

A concentration of 1 **part per million (ppm)** corresponds to 1 part pollutant per 1 million parts of the gas, liquid, or solid mixture in which the pollutant is found; 1 **part per billion (ppb)** refers to 1 part of pollutant per 1 billion parts of the medium in which it is found; 1 **part per trillion (ppt)** means that 1 part of pollutant is found in 1 trillion parts of its medium. Table 1-1 shows some equivalents for these units of concentration. In a gas mixture the reference is usually to parts per million, billion, or trillion by *volume*; in liquids and solids the reference is generally to parts per million, billion, or trillion by *weight*.

Parts per million, billion, or trillion may seem like negligible amounts of pollution. Nevertheless, concentrations of *some* pollutants at such low levels can have serious effects on people, other animals, and plants.

One way to lower the concentration of a pollutant is to dilute it in a large volume of air or water. Until we started overwhelming the air and waterways with pollutants, dilution was *the* solution to pollution. Now it is only a partial solution.

The third factor is a pollutant's **persistence**: how long it stays in the air, water, soil, or body. **Degradable,** or **nonpersistent, pollutants** are broken down completely or reduced to acceptable levels by natural physical, chemical, and biological processes. Complex chemical pollutants broken down (metabolized) into simpler chemicals by living organisms (usually by specialized bacteria) are called **biodegradable pollutants**.

Table 1-1 Equivalents of Some Trace Concentration Units

Unit	1 part per million	1 part per billion	1 part per trillion
Time	1 minute in 2 years	1 second in 32 years	1 second in 320 centuries
Money	1¢ in $10,000	1¢ in $10,000,000	1¢ in $10,000,000,000
Weight	1 pinch of salt in 10 kilograms (22 lbs.) of potato chips	1 pinch of salt in 10 tons of potato chips	1 pinch of salt in 10,000 tons of potato chips
Volume	1 drop of in 1,000 liters (265 gallons) of water	1 drop in 1,000,000 liters (265,000 gallons) of water	1 drop in 1,000,000,000 liters (265,000,000 gallons) of water

A: 35 years (70/% growth rate = 70/2 = 35 years)

Human sewage in a river, for example, is biodegraded fairly quickly by bacteria if the sewage is not added faster than it can be broken down.

Many of the substances we introduce into the environment take decades or longer to degrade. Examples of these **slowly degradable**, or **persistent, pollutants** include the insecticide DDT and most plastics.

Nondegradable pollutants cannot be broken down by natural processes. Examples include the toxic elements lead and mercury. The best ways to deal with nondegradable pollutants (and slowly degradable pollutants) are to avoid releasing them into the environment or to recycle or reuse them. Removing them from contaminated air, water, or soil is expensive, and sometimes impossible.

We know little about the possible harmful effects of 90% of the 72,000 synthetic chemicals now in commercial use and the roughly 1,000 new ones added each year. Our knowledge about the effects of the other 10% of these chemicals is limited, mostly because it is quite difficult, time-consuming, and expensive to get this information. Even if we determine the main health and other environmental risks associated with a particular chemical, we know little about its possible interactions with other chemicals or about the effects of such interactions on human health, other organisms, and life-support processes.

Solutions: What Can We Do About Pollution?

There are two basic approaches to dealing with pollution: prevent it from reaching the environment or clean it up if it does (Figure 1-16). **Pollution prevention**, or **input pollution control**, is a throughput solution. It slows or eliminates the production of pollutants, often by switching to less harmful chemicals or processes. Pollution can be prevented (or at least reduced) by the four *R*s of resource use: *refuse (don't use)*, *reduce*, *reuse*, and *recycle*.

Pollution cleanup, or **output pollution control**, involves cleaning up pollutants after they have been produced. However, environmentalists have identified three major problems with relying primarily on pollution cleanup. First, *it is often only a temporary bandage as long as population and consumption levels continue to grow*

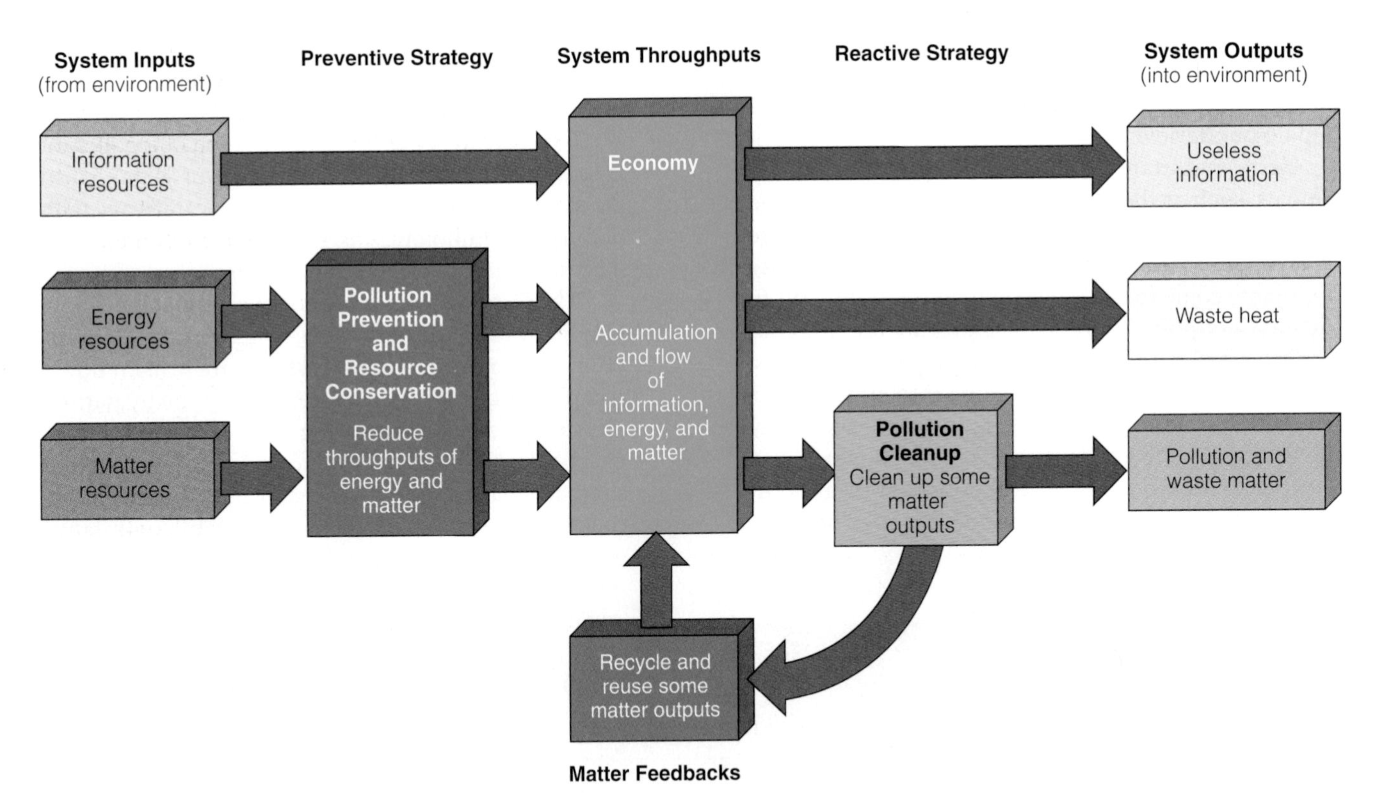

Figure 1-16 Inputs, throughputs, and outputs of a system (an economy) with two strategies for reducing pollution. Pollution prevention, or input pollution control, is based on reducing or eliminating pollution and conserving resources by reducing the throughputs (flows) of matter and energy resources through an economy. Throughput can also be reduced by recycling or reusing some or most of the output of pollution or waste matter. Pollution cleanup, or output pollution control, involves trying to reduce pollutants to acceptable levels after they have been produced. Both approaches are needed, but environmentalists and a growing number of economists believe that much greater emphasis should be placed on pollution prevention.

Vitousek, Peter M. 1997. "Human Domination of Earth's Ecosystems." *Science*, vol. 275, no. 5325, 494(6).

without corresponding improvements in pollution control technology. For example, adding catalytic converters to cars has reduced air pollution, but increases in the number of cars and in the total distance each travels (increased throughput) have reduced the effectiveness of this cleanup approach.

Second, *pollution cleanup often removes a pollutant from one part of the environment only to cause pollution in another.* We can collect garbage, but the garbage is then either burned (perhaps causing air pollution and leaving a toxic ash that must be put somewhere), dumped into streams, lakes, and oceans (perhaps causing water pollution), or buried (perhaps causing soil and groundwater pollution). Third, *once pollutants have entered and become dispersed in the air and water (and in some cases, the soil) at harmful levels, it usually costs too much to reduce them to acceptable concentrations.*

Both pollution prevention and pollution cleanup are needed, but environmentalists and some economists urge us to emphasize prevention because it works better and is cheaper than cleanup. For widely dispersed and difficult-to-identify nonpoint pollution, hazardous wastes, and slowly degradable and nondegradable pollutants, pollution prevention is the most effective (perhaps the only) approach. As Benjamin Franklin reminded us long ago, "An ounce of prevention is worth a pound of cure."

An increasing number of businesses have found that *pollution prevention pays*. So far, however, about 99% of environmental spending in the United States (and in most other countries) is devoted to pollution cleanup and only 1% to pollution prevention, a situation that environmental scientists and some economists believe must be reversed as soon as possible.

Both pollution prevention and pollution cleanup can be encouraged either by the *carrot approach* of using incentives such as various subsidies and tax write-offs or by the *stick approach* of regulations and taxes. Most analysts believe that a combination of both approaches is probably best because excessive regulation and too much taxation can incite resistance and cause political backlash. Achieving the right balance is difficult.

The countries of the former Soviet Union, many eastern European countries such as Poland (Case Study, p. 20), China, and most developing countries near the bottom of the economic ladder are far behind in both pollution control and prevention. Consider some of the environmental horrors found in various parts of the former Soviet Union: Of every 10 barrels of oil produced each day, about 1 is spilled—equivalent to one huge oil tanker spill every 6 hours. Indiscriminate use of pesticides has led to severe contamination of soil and water in many regions. The once-huge Aral Sea is disappearing because most of its water has been diverted for agriculture. Severe erosion has depleted soil on large tracts of farmland.

Deposits of long-lived radioactive materials as a result of the explosion of Chernobyl nuclear power plant accident in 1986 has left large areas uninhabitable. There are thousands of unregulated dump sites containing nuclear and toxic wastes. Some 14–16% of all territory in Russia has been designated as ecological disaster zones. Past abuses and the rapid and unpredictable political and economic changes are also taking an increasing toll on the incredible biodiversity found in these countries.

As these countries struggle to transform their economies, little money is available for repairing such ecological damage. Even with adequate resources it would take decades to clean up such widespread ecological devastation.

1-5 ENVIRONMENTAL AND RESOURCE PROBLEMS: CAUSES AND CONNECTIONS

What Are Key Environmental Problems and Their Root Causes? We face a number of interconnected environmental and resource problems (Figure 1-17). The first step in dealing with these problems is to identify their underlying causes. According to environmentalists, these include the following:

- Rapid population growth based in part on the natural tendency of organisms to reproduce and expand their populations in response to available supplies of resources (biological imperialism)
- Rapid and wasteful use of resources with too little emphasis on pollution prevention and waste reduction
- Simplification and degradation of parts of the earth's life-support systems
- Poverty, which can drive poor people to use potentially renewable resources unsustainably for short-term survival and often exposes the poor to health risks and other environmental risks
- Failure of economic and political systems to encourage earth-sustaining forms of economic development and discourage earth-degrading forms of economic growth
- Failure of economic and political systems to have market prices include the overall environmental cost of an economic good or service
- Our urge to dominate and manage nature for our use with far too little knowledge about how nature works

Hint: Enter the search terms *environment, effect of humans* using the Subject Guide.

Pollution, Environmental Degradation, and Environmental Progress in Poland

CASE STUDY

In the past 50 years, Poland has experienced a massive deterioration of environmental quality and loss of biodiversity. In 1985, the Polish Academy of Sciences described heavily industrialized Poland as the most polluted country in the world.

Most of the country's coal-burning industrial and power plants have few or no pollution control devices. Air pollution in nearly every major Polish city (especially in the heavily industrialized southwest) is reportedly 50 times legal limits; sometimes motorists must turn on their lights during the day to see. By 1990, an estimated 32% of the country's forests showed signs of moderate to severe damage from air pollution.

About half of Poland's cities have no wastewater treatment systems. More than 75% of the water in the Vistula River (Poland's largest river) is unsuitable for agricultural or industrial use. In 1991, 19% of the local drinking water in Poland was too polluted to drink, even after disinfection.

It is estimated that a quarter of Poland's soil is too contaminated to grow food safe for livestock or people. In 1988, the government declared five villages in the industrial region of Silesia unfit for habitation because of toxic metals in the soil and water, and it paid the villagers to relocate permanently. Environmental pollution, loss and fragmentation of wildlife habitat, and industrialization of crop lands and forests have led to considerable loss of biodiversity.

In the early 1900s the country enacted pioneering environmental protection and conservation laws that are stronger than those found in the United States and most developed countries. By the 1930s the Polish government had established 6 national parks and 180 nature preserves. Poland's 1949 Nature Conservation Act had provisions that in 1969 became the basis for the National Environmental Policy Act (NEPA) in the United States.

Poland has been an environmental paradox. Despite having a tradition of some of the world's best and most progressive environmental and conservation laws for many decades, it has experienced some of the worst pollution and environmental degradation in Europe. The primary reason is that the centrally planned economy during the communist era ignored the country's pioneering environmental laws and emphasized production based on heavy industry, mining, and relying on highly polluting and inefficient use of coal as the primary energy resource.

After decades of lax enforcement, Poland is attempting to reduce pollution significantly by enforcing its existing environmental laws. In 1990 Poland became the first country in eastern Europe to approve a comprehensive national environmental policy. A Ministry of Environmental Protection, Natural Resources, and Forestry is now responsible for planning, coordinating, and enforcing all environmental and resource conservation policies.

In 1992, the government shut down 18 of the most polluting industrial plants. If the money is available, the country hopes to build a network of new wastewater treatment plants. The government has raised energy prices and increased the price of coal relative to oil and natural gas. This policy should decrease coal use, increase the use of less-polluting natural gas (which supplies only 2% of the country's energy), and encourage improvements in energy efficiency. So far regulatory authorities and regulated industries have emphasized negotiated solutions rather than confrontation.

However, environmental improvements have been hampered by economic recession, lack of funds, staggering cleanup costs, inadequate outside aid, and the country's $43-billion debt to Western developed countries. Also, Polish labor unions are more concerned about fair wages and employment security than environmental protection.

Between 1989 and 1992 the volume of rubbish generated nationwide increased by 40% because of the importation of heavily packaged foreign goods. As a result, Poland's once highly effective recycling and reuse systems (driven by national shortages of raw materials) are being weakened by the introduction of throwaway packaging and other goods.

With adequate aid and debt relief from developed countries, Poland could rapidly implement modern pollution control, pollution prevention, and energy-efficient technologies while creating new jobs. Without such aid and without sustained action by Poland's citizens and officials, environmental quality may not improve. Other eastern European countries face similar problems.

Critical Thinking

1. What factors might result in Poland eventually having a more sustainable environmental society than the United States?

2. How can adopting the economic models of most of today's developed countries hinder environmental progress in Poland?

Q: What % of the world's population growth between now and 2050 is expected to take place in developing countries?

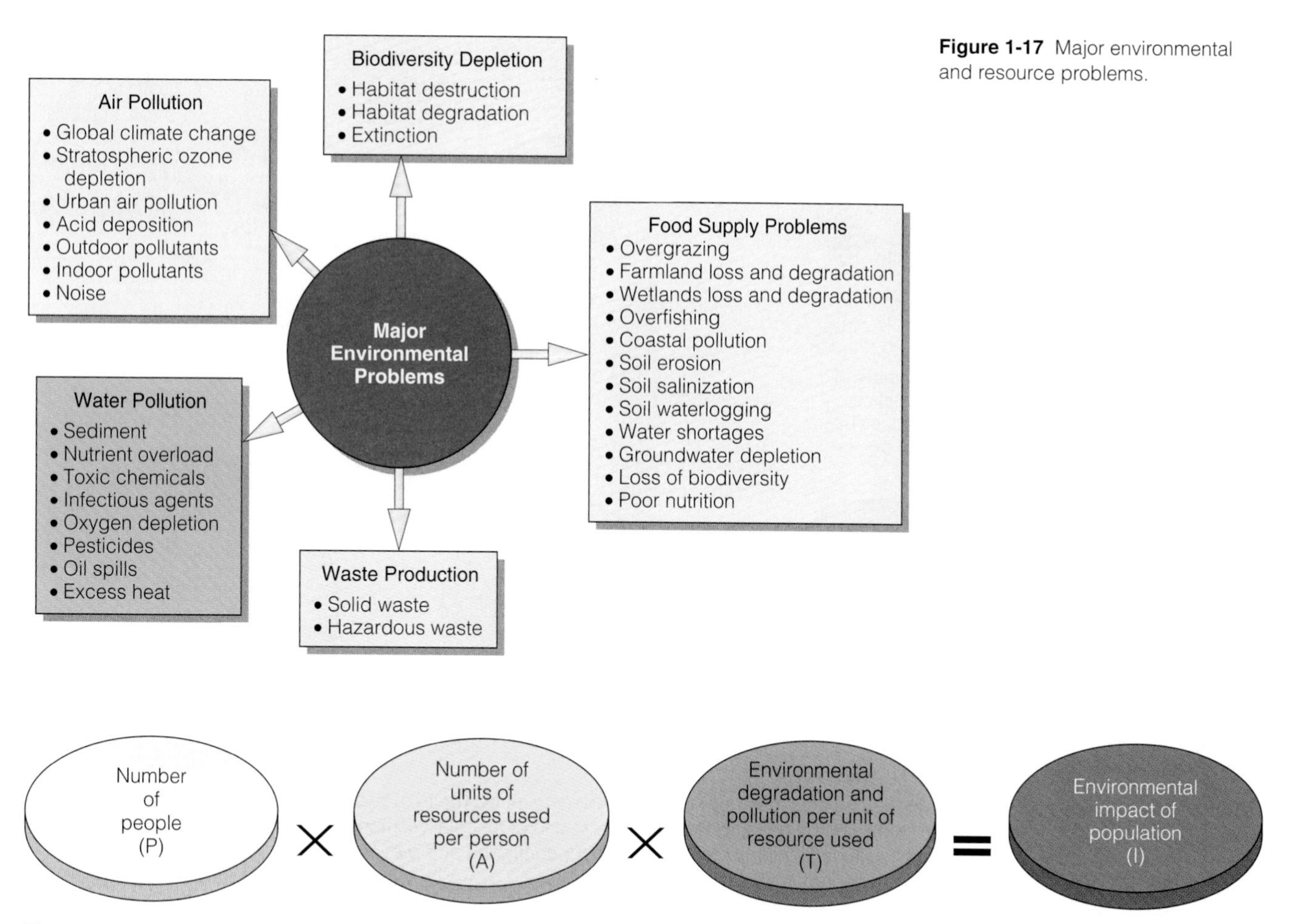

Figure 1-17 Major environmental and resource problems.

Figure 1-18 Simplified model of how three factors—population, affluence, and technology—affect the environmental impact of a population. According to this model, the damage we do to the earth is equal to the number of people there are, multiplied by the amount of resources each person uses, multiplied by the amount of pollution, resource waste, and environmental degradation involved in extracting, making, and using each unit of resource.

How Are Environmental Problems and Their Root Causes Connected? Once we have identified environmental problems and their root causes, the next step is to understand how they are connected to one another. The three-factor model in Figure 1-18 is a good starting point.

According to this simple model, a given area's total environmental degradation and pollution—that is, the environmental impact of population—depends on three factors: the number of people (population size, *P*), the average number of units of resources each person uses (per capita consumption or affluence, *A*), and the amount of environmental degradation and pollution produced for each unit of resource used (the environmental destructiveness of the technologies used to provide and consume resources, *T*). This model, developed in the early 1970s by biologist Paul Ehrlich and physicist John Holdren, can be summarized in simplified form as

Population × Affluence × Technology = Environmental impact

or

$$P \times A \times T = I$$

Figure 1-19 shows how the three factors depicted in Figure 1-18 can interact in developing countries and developed countries. In developing countries, population size and the resulting degradation of potentially renewable resources (as the poor struggle to stay alive) tend to be the key factors in total environmental impact.

In developed countries, high rates of per capita resource use (and the resulting high levels of pollution and environmental degradation per person) are believed to be the key factors determining overall environmental impact. For example, it is estimated that the

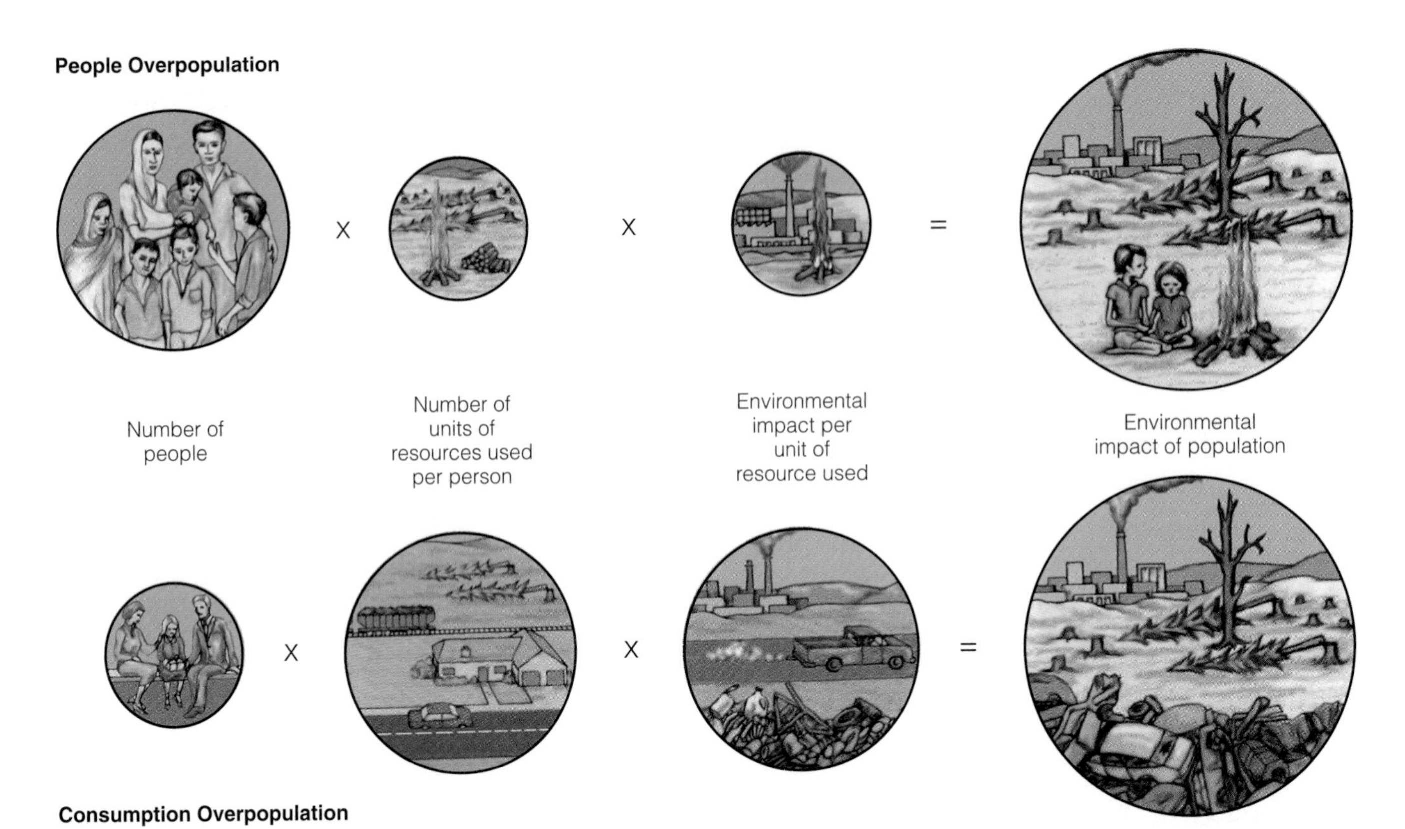

Figure 1-19 Environmental impact of developing countries (top) and developed countries (bottom) based on the relative importance of the factors in the model shown in Figure 1-18. Circle size shows the relative importance of each factor. The size of the *T* factor can be reduced by improved technology for controlling and preventing pollution, resource waste, and environmental degradation.

average U.S. citizen consumes 35 times as much as the average citizen of India and 100 times as much as the average person in the world's poorest countries. *Thus, poor parents in a developing country would need 70–200 children to have the same lifetime environmental impact as 2 children in a typical U.S. family.* Many of the world's leading scientists warn that if we keep adding 80–90 million people each year (all of whom want to become affluent), our life-support systems in many parts of the world will be overwhelmed.

Some forms of technology, such as polluting factories and motor vehicles and energy-wasting devices, increase environmental impact by raising the *T* factor in the equation. Other technologies, such as pollution control, solar cells, and more energy-efficient devices, can lower environmental impact by decreasing the *T* factor in the equation.

The three-factor model in Figure 1-18 can help us understand how key environmental problems and some of their causes are connected. However, the interconnected problems we face involve a number of poorly understood interactions among many more factors than those in the three-factor model (Figure 1-20). As a result, there is an urgent need for greatly increased interdisciplinary research designed to explore the connections between the physical, chemical, and biological environment, human health, the economy, political systems, social justice, national and global security, and the worldviews that guide our actions toward one another and the earth.

1-6 CULTURAL CHANGES AND SUSTAINABILITY

What Major Human Cultural Changes Have Taken Place? Fossil and anthropological evidence suggests that the current form of our species, *Homo sapiens*, has walked the earth for only about 60,000 years (some recent evidence suggests 90,000 years), an instant in the planet's estimated 4.6-billion-year existence. Until about 12,000 years ago, we were mostly **hunter–gatherers** who moved as needed to find enough food for survival. Since then, there have been two major cultural shifts: the agricultural revolution, which began 10,000–12,000 years ago, and the industrial revolution, which began about 275 years ago.

These cultural revolutions have given us much more energy (Figure 1-21) and new technologies with

Meisner, Mark. 1997. "Green on the Screen." *Alternatives Journal*, vol. 23, no. 1, 9(2).

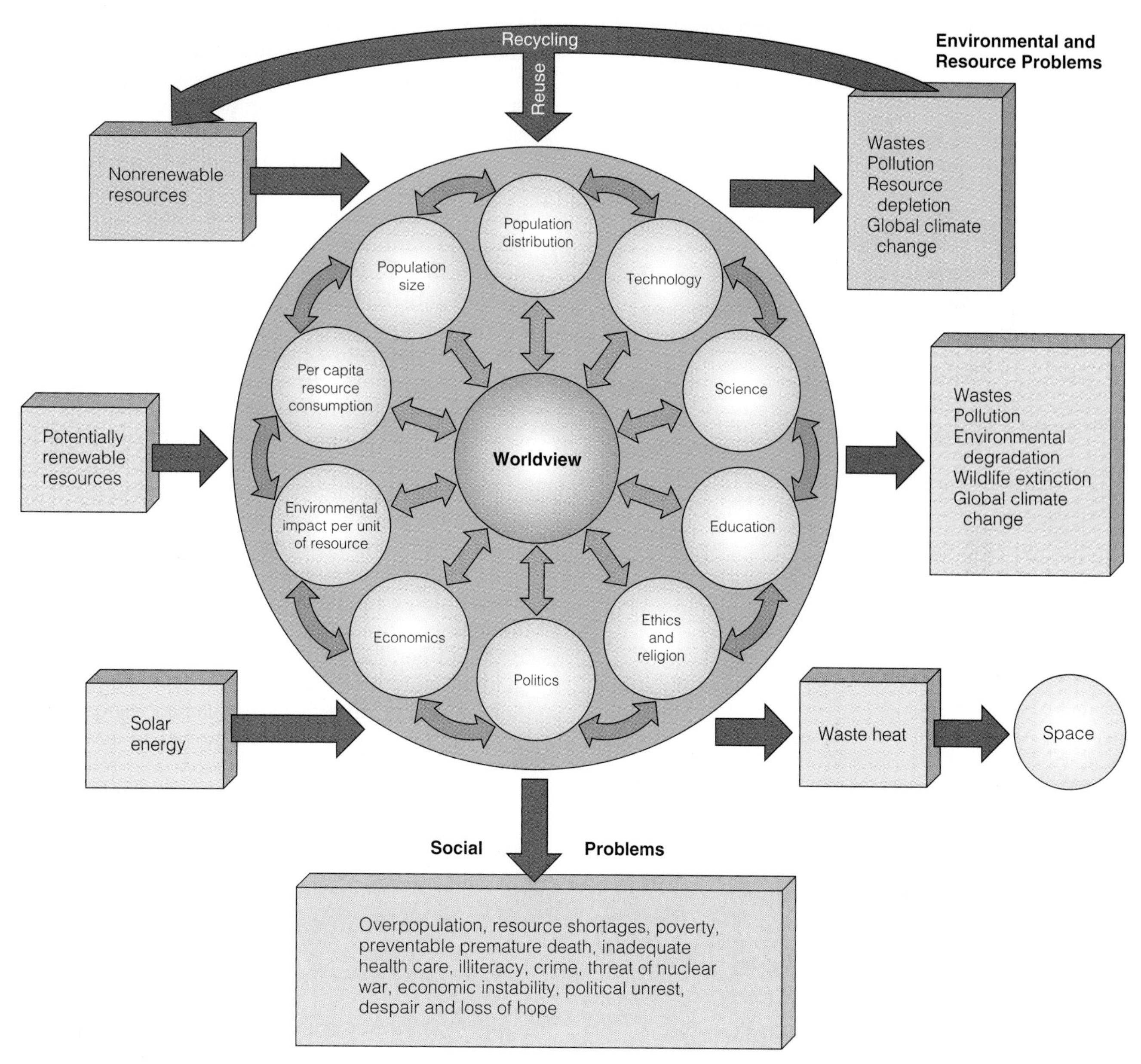

Figure 1-20 Environmental, resource, and social problems are caused by a complex, poorly understood mix of interacting factors, as illustrated by this simplified model.

which to alter and control more of the planet to meet our basic needs and increasing desires. By expanding food supplies, lengthening life spans, and raising living standards for many people, each cultural shift contributed to the expansion of the human population. More people fed, bred, and spread (Figure 1-22). However, the results include skyrocketing resource use, pollution, and accelerating environmental degradation.

How Did Ancient Hunting-and-Gathering Societies Affect the Environment? During most of our 60,000-year existence, we were hunter–gatherers who survived by collecting edible wild plant parts, hunting, fishing, and scavenging meat from animals killed by other predators. Archaeological and anthropological evidence indicates that our hunter–gatherer ancestors typically lived in small bands (of fewer than 50 people) who worked together to get enough food to survive. Most groups were nomadic, picking up their few possessions and moving from place to place as needed to find enough food.

The earliest hunter–gatherers (and those still living this way today) survived through *earth wisdom*: expert knowledge of their natural surroundings. They discovered that a variety of plants and animals could be eaten and used as medicines. They knew where to

Hint: Enter the search terms *environment, information* using Key Words.

find water, how to predict the weather with reasonable accuracy, how plant availability changed throughout the year, and how some game animals migrated to get enough food.

These dwellers in nature had only three energy sources: **(1)** sunlight captured by plants (which also served as food for the animals they hunted), **(2)** fire, and **(3)** their own muscle power. Because of high infant mortality and an estimated average life expectancy of 30–40 years, hunter–gatherer populations grew very slowly (Figure 1-22).

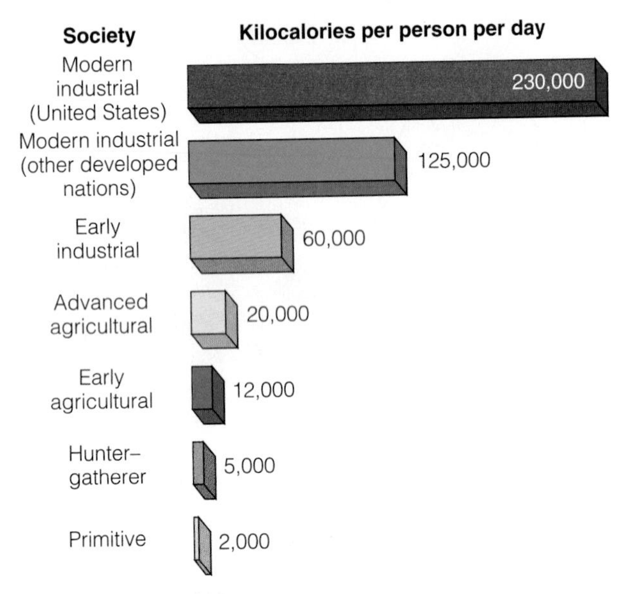

Figure 1-21 Average direct and indirect per capita daily energy use at various stages of human cultural development. A *calorie* is the amount of energy needed to raise the temperature of 1 gram of water 1°C (1.8°F); a *kilocalorie* is 1,000 calories. Food calories or calories expended during exercise are kilocalories, sometimes designated *Calories* (with a capital *C*).

As hunter–gatherers gradually improved their tools and hunting practices, their harmful effects on the environment increased. Some worked together to hunt herds of reindeer, mammoths, European bison, and other big game. Some advanced hunter–gatherers used fire to flush game from forest thickets and grasslands toward waiting hunters. They also stampeded herds into traps or over cliffs, killing many more animals than they needed. Some learned that burning vegetation promoted the growth of plants they could eat or that were favored by certain game animals.

Advanced hunter–gatherers had a greater impact on their environment than did early hunter–gatherers. Their use of fire, in particular, converted forests into grasslands. There is some evidence that they contributed to—perhaps even caused—the extinction of some large animals, including the mastodon, saber-toothed tiger, giant sloth, cave bear, mammoth, and giant bison. Other researchers attribute such extinctions mostly to rapid changes in climate, especially ice ages. As these early humans moved to new areas, many probably carried plant seeds and roots with them. This altered the distribution of plants and, in some cases, the types of animals feeding on the newly introduced plants.

Figure 1-22 Expansion of the earth's carrying capacity for humans. Technological innovation has led to major cultural changes, and we have displaced and depleted numerous species that compete with us for—and provide us with—resources. Dashed lines represent three alternative futures: **(1)** uninhibited human population growth, **(2)** population stabilization, and **(3)** growth followed by a crash and stabilization at a much lower level.

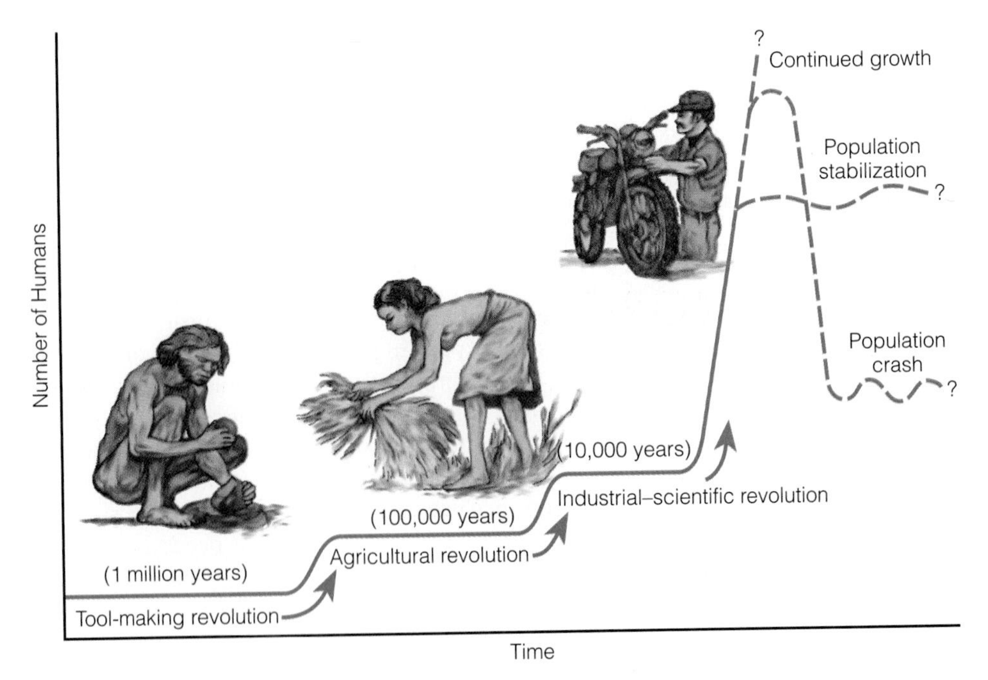

Bjorner, Susanne. 1998. "The Empty Glass and Too Much Water." *Online,* vol. 22, no.2, 8(1).

Hunter–gatherers exploited their environment to survive, but their environmental impact was usually limited and local. They relied on potentially renewable resources, and their use of resources was low. Most of the damage they caused was easily repaired by natural processes because of their small groups and frequent migrations.

Both early and advanced hunter–gatherers were *dwellers in nature* who survived and attempted to live sustainably through low resource use per person and learning to work with nature and with one another in small groups. Some people tend to make sweeping generalizations and to romanticize the wide variety of hunter–gatherer groups, but evidence reveals that some groups were peaceful and some were aggressive; some respected nature and some did not.

How Has the Agricultural Revolution Affected the Environment? Some 10,000–12,000 years ago, a cultural shift known as the **agricultural revolution** began in several regions of the world. It involved a gradual move from a lifestyle based on nomadic hunting and gathering to one centered on settled agricultural communities in which people domesticated wild animals and cultivated wild plants.

Plant cultivation probably developed in many areas, especially in the tropical forests of southeast Asia, northeast Africa, and Mexico. People discovered that they could grow various wild food plants from roots or tubers (fleshy underground stems). Often they planted a mixture of food crops and tree crops, an ancient and sustainable form of **agroforestry**. To prepare the land for planting, they cleared small patches of tropical forests by cutting down trees and other vegetation and then burning the underbrush (Figure 1-23). The ashes fertilized the nutrient-poor soils in this **slash-and-burn cultivation**.

These early growers also used various forms of **shifting cultivation** (Figure 1-23). After a plot had been used for several years, the soil would be depleted of nutrients or reinvaded by the forest. Then the growers moved and cleared a new plot. They learned that each abandoned patch had to be left fallow (unplanted) for 10–30 years before the soil became fertile enough to grow crops again. While patches were regenerating, growers used them for tree crops, medicines, fuelwood, and other purposes. In this manner, early growers practiced sustainable cultivation within tropical forests.

These early practices involved **subsistence farming,** in which a family grew only enough food to feed itself. Their dependence mostly on human muscle power and crude stone or stick tools meant that growers could cultivate only small plots; thus, they had little impact on their environment. However, the main reason that this form of agriculture was sustainable was that the population size and density of these early farmers were low. This meant that people could move on to other areas and leave abandoned plots unplanted for the several decades needed to restore soil fertility.

The invention of the metal plow pulled by domesticated animals allowed farmers to cultivate larger plots of land and to break up fertile grassland soils, which previously could not be cultivated because of their dense root systems. In some arid regions, early farmers increased crop output by diverting water from nearby streams into hand-dug ditches and canals to irrigate crops.

The gradual shift from hunting and gathering to farming had several significant effects:

- *Using domesticated animals to plow fields, haul loads, and perform other tasks increased the average energy use per person and thus the ability to expand agriculture* (Figures 1-21 and 1-22).
- *Birth rates rose faster than death rates, and population increased, mostly because the larger, more reliable food supply could support more people* (Figure 1-22).
- *People cleared increasingly larger fields and built irrigation systems to transfer water from one place to another.*
- *People began accumulating material goods.* Nomadic hunter–gatherers could not carry many possessions in their travels, but farmers living in one place could acquire as much as they could afford.
- *Farmers could grow more than enough food for their families.* They could store the excess for a "rainy day" or use it to barter with craftspeople who specialized in weaving, tool making, and pottery.
- *Urbanization—the formation of villages, towns, and cities—became practical.* With fewer people needed to provide food, many people left farms and moved to villages, where they took up crafts and other occupations. Some villages grew into towns and cities, which served as centers for trade, government, and religion.
- *Conflict between societies became more common as ownership of land and water rights became crucial economic issues.* Armies and their leaders rose to power and conquered large areas of land and water supplies. These rulers forced powerless people (slaves and landless peasants) to do the hard, disagreeable work of producing food and constructing

Hint: Enter the search terms *environment, information*" using the Subject Guide.

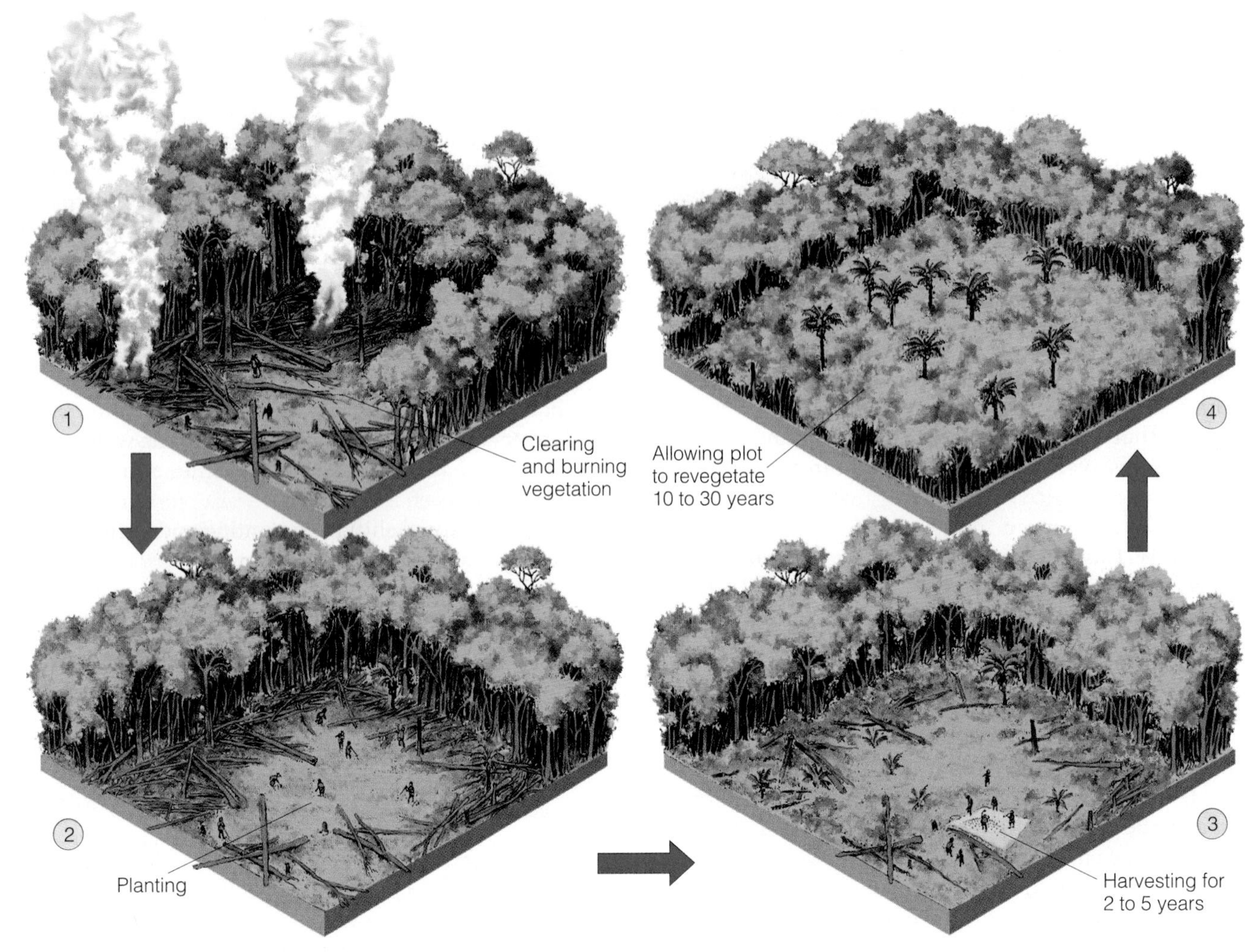

Figure 1-23 The first crop-growing technique may have been a combination of slash-and-burn and shifting cultivation in tropical forests. This method is sustainable only if small plots of the forest are cleared, cultivated for no more than 5 years, and then allowed to regenerate for 10 to 30 years to renew soil fertility. Indigenous cultures have developed many variations of this technique and have found ways to make some nondestructive uses of former plots while they are being regenerated.

things such as irrigation systems, temples, and walled fortresses.

- *The survival of wild plants and animals, once vital to humanity, became less important.* Wild animals, which competed with livestock for grass and fed on crops, became enemies to be killed or driven from their habitats. Wild plants invading cropfields became weeds to be eliminated.

The growing populations of these emerging civilizations needed more food and more wood for fuel and building materials, so people cut down vast forests and plowed up large expanses of grassland. Such extensive land clearing degraded or destroyed the habitats of many wild plants and animals, causing or hastening their extinction.

Many of these cleared lands were poorly managed. Soil erosion, salt buildup in irrigated soils, and overgrazing of grasslands by huge herds of livestock helped turn fertile land into desert; topsoil washed into streams, lakes, and irrigation canals. The gradual degradation of the vital resource base of soil, water, forests, grazing land, and wildlife converted many productive landscapes into barren regions. This degradation was a factor in the downfall of many great civilizations in the Middle East, North Africa, and the Mediterranean (Guest Essay, p. 34). Historian Henry Kissinger reminds us, "As a historian, you have to be conscious of the fact that every civilization that has ever existed has ultimately collapsed."

The shift of people to towns and cities, the emergence of specialized occupations and new technologies,

Q: How many people are added to the population of developing countries every 4 days?

and the expansion of commerce and trade greatly increased the demand for metals and other nonrenewable mineral resources. The expansion of mining degraded land and water. Increased production and use of material goods created growing volumes of wastes. Towns and cities concentrated sewage and other wastes, polluted the air and water, and greatly increased the spread of diseases.

The spread of agriculture meant that most of the world's human population gradually shifted from being hunter–gatherers, working with nature in order to survive, to becoming shepherds, farmers, and urban dwellers trying to tame and manage nature to survive and prosper.

How Has the Industrial Revolution Affected the Environment? The next great cultural shift, the **industrial revolution,** began in England in the mid-1700s and spread to the United States in the 1800s. It multiplied per capita energy consumption and thus the power of humans to shape the earth to their will and promote economic growth. Production, commerce, trade, and distribution of goods all expanded rapidly.

The industrial revolution represented a shift from dependence on potentially renewable wood (with supplies dwindling in some areas because of unsustainable cutting) and flowing water to dependence on nonrenewable fossil fuels. Coal-fired steam engines were invented to pump water and perform other tasks. Eventually an array of new machines were developed, powered by coal (and later by oil and natural gas). The new machines in turn led to a switch from small-scale, localized production of handmade goods to large-scale production of machine-made goods in centralized factories within rapidly growing industrial cities.

Factory towns grew into cities as rural people came to the factories for work. There they worked long hours under noisy, dirty, and hazardous conditions. Other workers toiled in dangerous coal mines. In these early industrial cities, coal smoke belching out of chimneys was so heavy that many people died of lung ailments. Ash and soot covered everything, and on some days the smoke was so thick that it blotted out the sun.

Fossil-fuel–powered farm machinery, commercial fertilizers, and new plant-breeding techniques increased per acre crop yields. This development helped protect biodiversity by reducing the need to expand the area of cropland to grow food. Because fewer farmers were needed, more people migrated to cities. With a larger and more reliable food supply, the size of the human population began the sharp increase that continues today.

After World War I (1914–18), more efficient machines and mass-production techniques were developed. This led to the basis of today's advanced industrial societies in places such as the United States, Canada, Japan, Australia, and western Europe.

Advanced industrial societies provide a variety of benefits to most people living in them, including mass production of many useful and economically affordable products; a sharp increase in agricultural productivity; lower infant mortality and higher life expectancy because of better sanitation, hygiene, nutrition, and medical care; a decrease in the rate of population growth; better health, birth control methods, and education; methods for controlling pollution; and greater average income and old-age security.

These important benefits of industrialized societies have been accompanied by the resource and environmental problems we face today. Industrialization also isolates more people from nature and reduces understanding of the important ecological and economic services nature provides (Figure 1-2).

How Might the Information Revolution Affect the Environment? We are in the midst of a new cultural shift, the **information revolution,** in which new technologies such as the telephone, radio, television, and computers are enabling people to deal with more information more rapidly.

There are four major areas in which information technology is advancing:

- *Information collection is becoming increasingly automated.* For example, up-to-date weather information is continuously gathered from thousands of weather stations, as well as ground-based radar systems and weather satellites.

- *Information storage involves building larger and larger databases.* For example, most municipalities in the United States are now developing geographical information system (GIS) databases that consist of digital maps of such features as property ownership, zoning, roads, and buried water, sewer, electrical, telephone, and gas lines.

- *Information communication involves nearly instantaneous transmission of large amounts of data.* Radio, television, and the telephone have become part of everyday life for most people in developed countries. Increasingly, overnight mail services, the Internet, and very rapid transmission of information through networks of computers are emerging as new powerful ways to communicate. Computers have become our main tool for processing and using massive amounts of information. They are used in businesses,

A: About 1 million

supermarket checkout systems, cars (as tiny microprocessors that regulate fuel–air mixtures during combustion and monitor and diagnose engine malfunctions), and personal computers that give people the ability to do sophisticated computing virtually anywhere.

- *Information processing and use involve making sense of information and doing useful things with it.*

What the information revolution means for the environment is not yet clear. On the positive side, we can understand and respond to environmental problems more effectively. The powerful information technologies summarized in Figure 1-24 are helping us to understand how the earth, economies, and other complex systems work and to evaluate how such systems might be affected by our actions. They also illustrate that we are globally interconnected with one another and with nature.

On the negative side, we are increasingly being faced with an overload of information (infoglut), much of it useless. This can lead to confusion, distraction, and a sense of hopelessness as we try to identify useful information and ideas in a rapidly growing ocean of information.

In addition, we have not armed people with effective guidelines and critical thinking skills for separating the small amount of useful information (wisdom) from an avalanche of incorrect, meaningless, and misleading information. Finding nuggets of wisdom or useful information in large amounts of data is not a new problem. However, the amount of information we are expected to evaluate today is enormous and is growing at a rapid rate.

A growing number of analysts believe that what we need is a *wisdom revolution*, not an information revolution. Such analysts see an urgent need to revise the education system at all levels to train people

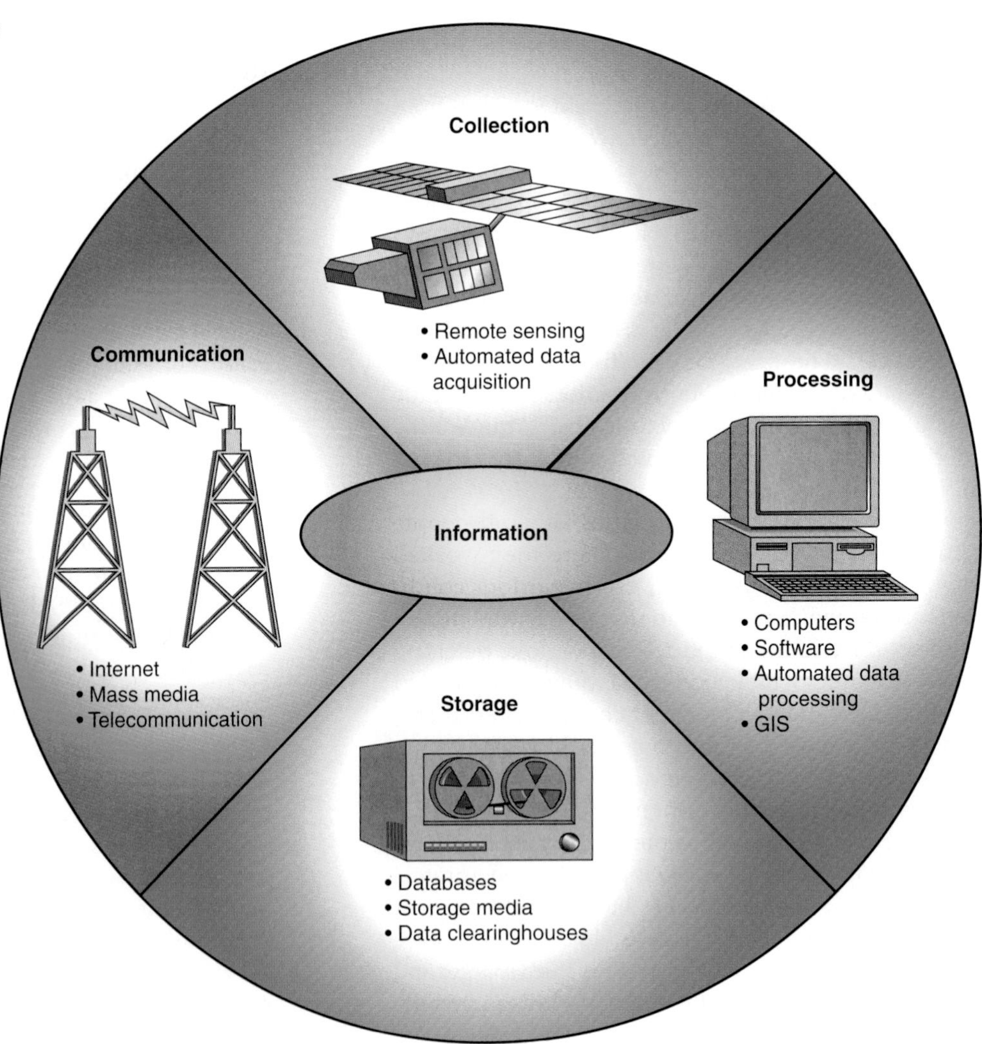

Figure 1-24 Major components of the information revolution.

Q: What percentage of the world's resources are used by people in developed countries?

how to evaluate and filter out incorrect, misleading, or useless information. This involves recognizing that a major purpose of education is to learn as much as possible in the form of wisdom or useful information instead of clogging our brains with trivial or useless information. Dealing with this urgent problem will be a major factor in **(1)** improving environmental quality and human health, psychological well-being, and environmental literacy; **(2)** guiding decision making in the face of uncertainty, and **(3)** learning how to minimize the ecological footprints of human activities.

1-7 IS OUR PRESENT COURSE SUSTAINABLE?

Are Things Getting Better or Worse? There are conflicting views about how serious our population and environmental problems are and what should be done about them (Table 1-2). Some analysts, mostly economists, contend that the world is not overpopulated and that people are our most important resource, as consumers and producers to fuel continued economic growth and as sources of technological innovation.

They also believe that human ingenuity and technological advances will allow us to clean up pollution to acceptable levels, find substitutes for any resources that become scarce, and keep expanding the earth's ability to support more humans, as we have done in the past. They don't acknowledge any insurmountable limits to population growth or economic growth. They accuse most scientists and environmentalists of exaggerating the seriousness of the problems we face and of failing to appreciate the progress (Table 1-2, left) that has been made in improving quality of life and protecting the environment.

On the other hand, environmentalists, many leading scientists, and a small number of economists contend that we are depleting and degrading the earth's natural capital at an accelerating rate and that this is leading to serious environmental and economic harm (Table 1-2, right). They warn that we are modifying the earth's physical, chemical, and biological systems in new ways, at faster rates, and over larger areas than at any time in human history. These analysts are encouraged by the progress that has been made but point out how much more must be done to help make the earth more sustainable for present and future human generations and for other species that support us and other forms of life.

What Is the Scientific Consensus About Environmental Problems? On November 18, 1992, some 1,680 of the world's senior scientists from 70 countries, including 102 of the 196 living scientists who are Nobel laureates, signed and sent an urgent warning to government leaders of all nations. According to this warning,

> *The environment is suffering critical stress. . . . Our massive tampering with the world's interdependent web of life—coupled with the environmental damage inflicted by deforestation, species loss, and climate change—could trigger widespread adverse effects, including unpredictable collapses of critical biological systems whose interactions and dynamics we only imperfectly understand. Uncertainty over the extent of these effects cannot excuse complacency or delay in facing the threats. . . . No more than one or a few decades remain before the chance to avert the threats we now confront will be lost and the prospects for humanity immeasurably diminished. . . . Whether industrialized or not, we all have but one lifeboat. No nation can escape injury when global biological systems are damaged. . . . We must recognize the earth's limited capacity to provide for us.*

Also in 1992, the prestigious U.S. National Academy of Sciences and the Royal Society of London issued a joint report, their first ever, which began,

> *If current predictions of population growth prove accurate and patterns of human activity on the planet remain unchanged, science and technology may not be able to prevent either irreversible degradation of the environment or continued poverty for much of the world.*

According to a 1991 assessment from the Ecological Society of America, "environmental problems resulting from human activities have begun to threaten the sustainability of the Earth's life support systems....Among the most critical challenges facing humanity are the conservation, restoration and wise management of the Earth's resources."

Some past short-term prophecies of environmental doom by a small number of scientists have not been borne out and critics have used this fact to urge people to ignore all warnings from scientists and environmentalists. However, the more recent warnings just cited are not the views of a small number of scientists but the consensus of the mainstream scientific community, consisting of most of the world's key researchers on environmental problems. Some analysts also point out that many of the warnings of environmental doom made in the 1960s and 1970s were averted because people realized that they might come

A: About 88% (even though only 20% of the world's people live in developed countries)

Table 1-2 Good and Bad News About Environmental Problems

Some Good News	Some Bad but Challenging News
Annual world population growth slowed from 2.2% to 1.43% between 1963 and 1998.	The world's population is still growing rapidly and is projected to increase from 5.9 billion to 8 billion between 1998 and 2025.
Global life expectancy rose from 33 to 66 years between 1900 and 1998.	Life expectancy in developing countries is 12 years less than in developed countries.
Annual global infant mortality dropped by 65% between 1900 and 1998.	Infant mortality in developing countries is eight times higher than in developed countries and could be improved in many developed countries.
Global food production has outpaced population growth since 1978.	Since 1980 population growth has exceeded grain production in 69 developing countries and the global per capita fish catch declined by 7.5% between 1988 and 1995. Future food production may be limited by its harmful environmental impacts.
The percentage of the world's population that are hungry has been reduced since 1960.	Because of population growth there are now more hungry people (800 million) than ever.
Biodiversity loss has been reduced by higher crop yields on less land.	The challenge is to increase high yields while sharply reducing the growing environmental impacts of such intensive production.
Biodiversity loss has been reduced by increased urbanization, with more people living on less land.	This effect is misleading because the higher resource use per person in cities must be supplied by croplands, forests, grasslands, fisheries, and mines elsewhere. This often diminishes biodiversity and increases environmental degradation in these areas that sustain otherwise unsustainable cities.
At least 5% of the world's land area has been set aside to protect wildlife and wildlife habitats.	This is half of the estimated minimum needed. Many wildlife sanctuaries exist on paper only and receive little protection and many important types of biological communities are not protected. Every hour an estimated two to eight wildlife species become extinct because of human activities; this rate is expected to increase over the next few decades.
Total forest area in temperate industrial countries increased during the 1980s.	Much of this increase came from replacing sustainable, biologically diverse old-growth forests with simplified and vulnerable tree farms. This has led to reduced forest biodiversity.
Tropical deforestation has been overblown and enough wood, paper, and food can be produced by converting much of such diverse forests to tree farms, cropland, and grazing land.	If current deforestation rates continue, within 30 to 50 years little of these forests will remain. This will lead to a massive loss of biodiversity (because such forests contain at least 50% of the world's terrestrial species).
Conservation tillage and no–till cultivation can cut soil erosion by about 75%.	These techniques are widely used in the United States but are rarely used in other countries. Topsoil is eroding faster than it forms on about one-third of the world's cropland. Deforestation and construction are increasing soil erosion in many countries.
Food industry scientists contend that the benefits of pesticides outweigh their harmful effects.	Most environmental and health scientists in this field contend that the harmful effects of pesticides outweigh their beneficial effects. Reliance mostly on chemical pesticides (all of which eventually fail because of genetic resistance by pests) should be replaced with an ecological approach called integrated pest management (IPM).

true and worked to pass legislation and carry out individual acts to help keep them from coming true (a result of prevention).

Whom Should We Believe? There is no easy answer to this question. It depends mostly on how each of us believes the world works and what role the human species can and should play on this planet and on our ability to use critical thinking to evaluate opposing claims. People with widely differing fundamental beliefs can take the same data, be logically consistent, and arrive at quite different con-

Q: What percentage of the global income goes to the richest 20% of the world's people?

Table 1-2 *(continued)*	
Some Good News	**Some Bad but Challenging News**
A very small minority of scientists contend that possible global warming is not a serious problem.	The vast majority of scientists in this field contend that the possible global warming is one of the world's most serious environmental and economic problems.
Some economists say that reducing the threat of global warming by shifting from nonrenewable fossil fuels to renewable solar-based energy sources and reducing deforestation (especially in the tropics) over the next few decades will be too costly.	In the long run it will cost much more (ecologically and economically) if such shifts are not made. Even if global warming were not a serious problem, these shifts are needed to reduce pollution and environmental degradation and to protect biodiversity.
A very small minority of scientists say that depletion of ozone in the stratosphere (the atmosphere's second layer) is not a serious problem.	The vast majority of scientists in this field contend that ozone depletion in the stratosphere is a very serious environmental problem.
Water is recycled by a natural cycle and the earth has much more water than it needs. Dams and water transfer projects can meet projected water shortages.	Earth's water is not distributed equally and much of it is inaccessible, unnecessarily wasted, or polluted. Currently, about 1.2 billion people lack access to clean drinking water. By 2025 at least 3 billion people in 90 countries are expected to face chronic water shortages. Large dams and water transfer projects lead to serious environmental degradation.
Since 1950 proven supplies (reserves) of virtually all nonrenewable fossil fuel and key mineral resources have increased significantly.	The exponential increase in the use of fossil fuel and mineral resources is causing massive land degradation, water pollution, and air pollution. These harmful environmental effects—not supplies or costs—may limit future resource use.
Adjusted for inflation, most nonrenewable fossil fuel and key mineral resources cost less today than in 1950.	These resource prices are low because most of the harmful environmental and health costs associated with their production are not included in their market prices. Such artificially low resource prices encourage waste, pollution, and environmental degradation and discourage waste reduction and pollution prevention.
In the United States, the recycling and composting of solid municipal waste (produced by households and businesses) increased from 7% to 27% between 1970 and 1998.	At least 60% of all municipal solid waste could be reused, recycled, or composted. Only about 1.5% of the estimated solid waste produced in the United States is municipal solid waste. The other 98.5% comes mostly from mining, fossil fuel production, agriculture, and industrial activities, with very little of it being recycled, reused, or composted.
Some scientists contend that the health risks from exposure to toxic and hazardous chemicals are overblown.	The harmful health effects of most chemicals are unknown and a growing body of evidence suggests that their harmful effects (especially to the nervous, immune, and endocrine systems) have been underestimated.
Since 1970, air and water pollution levels in most industrialized countries have dropped significantly for most pollutants.	Gains in pollution control in industrialized countries by laws enacted in the 1970s are being threatened by more people using more resources and a lack of emphasis on pollution prevention. Little progress has been made in reducing air and water pollution in most developing countries.

clusions (Table 1-2) because they start with different assumptions and are often seeking answers to different questions.

What we believe and do is also influenced by whether we have an optimistic ("the glass is half-full") or pessimistic ("the glass is half-empty") outlook about the future. People with a realistic but hopeful outlook recognize that the future is represented by the top half of the glass and that what we put into this part of the glass is the key issue. They rejoice in how much has been accomplished in dealing with environmental problems since 1970 but

A: About 85%

recognize that we should also roll up our sleeves and work together to see that such accomplishments are sustained and expanded.

1-8 ENVIRONMENTAL WORLDVIEWS AND SUSTAINABILITY

How Do Major Environmental Worldviews Differ? There are conflicting views about how serious our environmental problems are and what we should do about them. These conflicts arise mostly out of differing **environmental worldviews**: how people think the world works, what they think their role in the world should be, and what they see as right and wrong environmental behavior (**environmental ethics**).

Most people in today's industrial–consumer societies have a **planetary management worldview,** which has become increasingly common during the past 50 years. The basic environmental beliefs of this worldview include the following:

- *We are the planet's most important species, and we are in charge of the rest of nature.* This idea crops up when people talk about "our" planet or "our" earth and when people talk about "saving the earth."
- *There is always more.* The earth has an essentially unlimited supply of resources, to which we gain access via science and technology. If we deplete a resource, we will find substitutes. To deal with pollutants, we can invent technology to clean them up, dump them into space, or move into space ourselves. If we extinguish other species, we can use genetic engineering to create new and better ones.
- *All economic growth is good, more economic growth is better, and the potential for economic growth is essentially limitless.*
- *Our success depends on how well we can understand, control, and manage the earth's life-support systems for our benefit.*

People with this (and related) environmental worldviews seek answers to questions such as, "How can we keep economic growth or throughput of resources growing exponentially?" "How can we become better managers of the entire planet?" "How can we control and manage the pollutants and wastes we produce and the environmental degradation we cause?" This worldview is widely supported because it is said to be the primary driving force behind the major improvements in the human condition since the beginning of the industrial revolution (Table 1-2, left). Several variations of this environmental worldview are discussed in Section 29-1.

A small but growing number of people question the planetary management worldview and are searching for a better one. One environmental worldview is known as the **earth-wisdom worldview**. It is based on the following major beliefs, which are the opposite of those making up the planetary management worldview:

- *Nature exists for all of the earth's species, not just for us.* We need the earth, but the earth does not need us.
- *There is not always more.* The earth's limited resources should not be wasted, but instead used sustainably for us and all species.
- *Some forms of economic growth are environmentally beneficial and should be encouraged, but some are environmentally harmful and should be discouraged.*
- *Our success depends on learning to cooperate with one another and with the rest of nature by learning how to work with the earth.* Management of resources is essential to human survival. However, such management should involve learning as much as we can about how the earth works, sustains itself, and adapts to changing conditions and then using these lessons from nature to guide our actions.

People with such environmental worldviews seek answers to the following questions: "How can we design and use economic and political systems to encourage earth-sustaining forms of development and discourage earth-degrading ones?" "What is sustainability, and what do we need to do to live sustainably on the planet?" "How can we produce fewer pollutants and wastes and not cause so much environmental degradation?" Several variations of this worldview are discussed in Section 29-2.

Do We Need to Make a New Cultural Change? Most environmentalists believe that there is an urgent need to begin shifting from our present array of industrialized and partially industrialized societies to a variety of more sustainable earth-wisdom societies throughout the world, by launching an environmental or sustainability revolution to take place over the next 50 years (Guest Essay, p. 34). This new cultural change calls for us to change the ways we view and treat the earth (and thus ourselves) by using environmental science to learn more about how nature sustains itself and by mimicking such processes in human cultural systems. Environmental scientists call for us to move toward a society that is

Q: How many of the world's people attempt to survive on an annual income of about $1 per day?

ecologically and economically sustainable and socially just for current and future generations of humans and other species.

A growing number of environmental scientists call for us to shift our efforts from pollution cleanup to pollution prevention, from waste disposal (mostly burial and burning) to waste prevention and reduction, from species protection to habitat protection, and from increased resource use to increased resource conservation.

They urge us to use our political and economic systems to reward earth-sustaining economic activities and discourage those that harm the earth, as discussed in Sections 27-3 and 27-5. In addition, they call for much greater emphasis on social justice because economically and racially disadvantaged groups often suffer the most from the harmful consequences of pollution and environmental degradation. They also believe that we should allow many parts of the world we have damaged to heal, help restore severely damaged areas, and protect remaining wild areas from destructive forms of economic development.

Others, mostly economists and business leaders, contend that environmentalists have exaggerated the problems we face. They believe that the solution to our environmental problems involves technological innovation and greatly increased economic growth through participation in a global economy based on mostly unregulated free-market capitalism, as discussed in Section 27-1.

How Can We Live More Sustainably? Working with the Earth This chapter has presented an overview of the serious problems most environmentalists and many of the world's most prominent scientists believe we face (also summarized in the yellow boxes of the "Concepts and Connections" diagram inside the back cover) and their root causes. It has also summarized key arguments about how serious environmental problems are and the cultural changes that have influenced how we have treated the earth.

The rest of this book presents a more detailed analysis of these problems, the controversies they have created, and solutions proposed by various scientists, environmentalists, and other analysts. The Spotlight box gives some guidelines various analysts have suggested for living more sustainably by working with the earth.

The *bad news* is the environmental problems we face and their root causes, as outlined in this chapter. The *good news* is how much has been done and that it's not too late to replace our earth-degrading actions with earth-sustaining ones, as discussed throughout this book. The key is *earth wisdom*: learning as much as we can about how the earth sustains itself and adapts to ever-changing environmental conditions and integrating such lessons from nature into the ways we think and act.

Another key to dealing with our environmental problems and challenges lies in recognizing that individuals matter. Anthropologist Margaret Mead has summarized our potential for change: "Never doubt that a small group of thoughtful, committed citizens can change the world. Indeed, it is the only thing that ever has."

What's the use of a house if you don't have a decent planet to put it on?

HENRY DAVID THOREAU

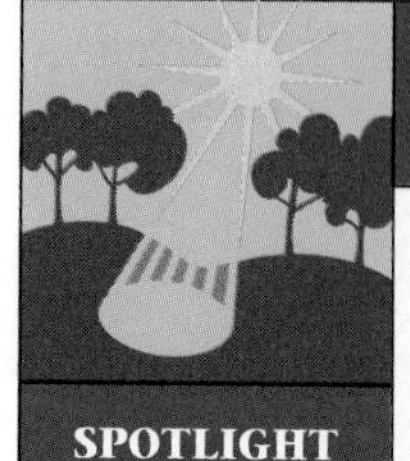

SPOTLIGHT

Some Guidelines for Working with the Earth

- Leave the earth as good as or better than we found it.
- Take no more than we need.
- Try not to harm life, air, water, or soil.
- Sustain biodiversity.
- Help maintain the earth's capacity for self-repair.
- Don't use potentially renewable resources (soil, water, forests, grasslands, and wildlife) faster than they are replenished.
- Don't waste resources.
- Don't release pollutants into the environment faster than the earth's natural processes can dilute or assimilate them.
- Emphasize pollution prevention and waste reduction.
- Slow the rate of population growth.
- Reduce poverty.

Specific things you can do to work with the earth by trying to implement such guidelines are listed in Appendix 5.

Critical Thinking

Which of these guidelines do you agree with and which do you disagree with? Why? Can you add any other guidelines?

A: About 1.3 billion, or 1 in 5 people

GUEST ESSAY

Launching the Environmental Revolution*

Lester R. Brown

Lester R. Brown is president of the Worldwatch Institute, a private nonprofit research institute he founded in 1974 that is devoted to analysis of global environmental issues. Under his leadership, the institute publishes the annual State of the World Report, *considered by environmentalists and world leaders to be the best way to become informed about key environmental issues. It also publishes monographs on specific topics,* World Watch *magazine, and a series of Environmental Alert books. He is author of a dozen books, recipient of the McArthur Foundation Genius Award, and winner of the United Nations' 1989 environment prize. He has been described by the* Washington Post *as "one of the world's most influential thinkers."*

Our world of the late-1990s faces potentially convulsive change. The question is, In what direction will it take us? Will the change come from strong worldwide initiatives that reverse the degradation of the planet and restore hope for the future, or will it come from continuing environmental deterioration that leads to economic decline and social instability?

Muddling through will not work. Either we will turn things around quickly or the self-reinforcing internal dynamic of the deterioration-and-decline scenario will take over. The policy decisions we make in the years immediately ahead will determine whether our children live in a world of development or decline.

There is no precedent for the rapid and substantial change we need to make. Building an environmentally sustainable future depends on restructuring the global economy, major shifts in human reproductive behavior, and dramatic changes in values and lifestyles. Doing all this quickly adds up to a revolution that is driven and defined by the need to restore and preserve the earth's environmental systems. If this *environmental revolution* succeeds, it will rank with the agricultural and industrial revolutions as one of the great economic and social transformations in human history.

Like the agricultural revolution, it will dramatically alter population trends. Although the former set the stage for enormous increases in human numbers, this revolution will succeed only if it stabilizes human population size, reestablishing a balance between people and natural systems on which they depend. In contrast to the industrial revolution, which was based on a shift to fossil fuels, this new transformation will be based on a shift away from fossil fuels.

The two earlier revolutions were driven by technological advances—the first by the discovery of farming and the second by the invention of the steam engine, which converted the energy in coal into mechanical power. The environmental revolution, though it will obviously need new technologies, will be driven primarily by the restructuring of the global economy so that it does not destroy or degrade its natural support systems.

The pace of the environmental revolution needs to be far faster than that of its predecessors. The agricultural revolution began some 10,000 years ago, and the industrial revolution has been under way for about two centuries. But if the environmental revolution is to succeed, it must be compressed into a few decades.

Progress in the agricultural revolution was measured almost exclusively in the growth in food output that eventually enabled farmers to produce a surplus that could feed city dwellers. Similarly, industrial progress was gained by success in expanding the output of raw materials and manufactured goods. The environmental revolution will be judged by whether it can shift the world economy into an environmentally sustainable development path, one that leads to greater economic security, healthier lifestyles, and a worldwide improvement in the human condition.

Many still do not see the need for such an economic and social transformation. They see the earth's deteriorating physical condition as a peripheral matter that can be dealt with by minor policy adjustments. But 30 years of effort have failed to stem the tide of environmental degradation. There is now too much evidence on too many fronts to take these issues lightly.

Already the planet's degradation is damaging human health, slowing the growth in world food production, and reversing economic progress in dozens of countries. By the age of 10, thousands of children living in southern California's Los Angeles basin have respiratory systems that are permanently impaired by polluted air. Some 300,000 people in the former Soviet Union are being treated for radiation sickness caused by the Chernobyl

*Updated excerpt from Brown's expanded version of these ideas in "Launching the Environmental Revolution," State of the World 1992 (New York: WW Norton, 1992).

Q: How many children under age 5 die each day in poor countries of causes that could be prevented?

nuclear power plant accident. The accelerated depletion of ozone in the stratosphere in the northern hemisphere will lead to an estimated additional 20,000 skin cancer fatalities over the next half century in the United States alone. Worldwide, millions of lives are at stake. These examples, and countless others, show that our health is closely linked to that of the planet.

A scarcity of new cropland and fresh water plus the negative effects of soil erosion, air pollution, and hotter summers on crop yields is slowing the growth of the world grain harvest. Combined with continuing rapid population growth, this has reversed the steady rise in grain output per person that the world had become accustomed to. Between 1950 and 1984, the historical peak year, world grain production per person climbed by nearly 40%. Since then, it has fallen roughly 1% a year, with the drop concentrated in poor countries. With food imports in these nations restricted by rising external debt, there are far more hungry people today than ever before.

On the economic front, the signs are equally ominous: Soil erosion, deforestation, and overgrazing are adversely affecting productivity in the farming, forestry, and livestock sectors, slowing overall economic growth in agriculturally based economies. The World Bank reports that after three decades of broad-based economic gains, incomes fell during the 1980s in 40 developing countries. Collectively, these nations contain more than 800 million people—almost three times the population of North America and nearly one-sixth that of the world. In Nigeria, the most populous country in the ill-fated group, the incomes of 123 million people fell a painful 29%, exceeding the fall in U.S. incomes during the depression decade of the 1930s.

Anyone who thinks these environmental, agricultural, and economic trends can easily be reversed need only look at population projections. Those of us born before the middle of this century have seen the world population more than double to 5.9 billion. We have witnessed the environmental effects of adding 3 billion people, especially in developing countries. We can see the loss of tree cover, the devastation of grasslands, the soil erosion, the crowding and poverty, the land hunger, and the air and water pollution associated with this addition of people. But what if 4.2 billion more people are added by 2050, over 90% of them in developing countries, as now projected by UN population experts?

The decline in living standards that was once predicted by some ecologists from the combination of continuing rapid population growth, spreading environmental degradation, and rising external debt has become a reality for one-sixth of humanity. Moreover, if a more comprehensive system of national economic accounting were used–one that incorporated losses of natural capital, such as topsoil and forests, the destruction of productive grasslands, the extinction of plant and animal species, and the health costs of air and water pollution, nuclear radiation, and increased ultraviolet radiation–it might well show that most of humanity suffered a decline in living conditions in the 1980s and 1990s.

Today, we study archaeological sites of civilizations that were undermined by environmental deterioration. The wheatlands that made North Africa the granary of the Roman Empire are now largely desert. The early civilizations of the Tigris-Euphrates Basin declined as the waterlogging and salting of irrigation systems slowly shrank their food supply. And the collapse of the Mayan civilization that flourished in the Guatemalan lowlands from the third century B.C. to the ninth century A.D. may have been triggered by deforestation and soil erosion.

No one knows for certain why centers of Mayan culture and art fell into neglect, or whether the population of 1 million to 3 million moved or died off, but recent progress in deciphering hieroglyphs in the area adds credence to the environmental decline hypothesis. One of those involved with the project, Linda Schele of the University of Texas, observes: "They were worried about war at the end. Ecological disasters, too. Deforestation. Starvation. I think the population rose to the limits their technology could bear. They were so close to the edge, if anything went wrong, it was all over."

Whether the Mayan economy had become environmentally unsustainable before it actually began to decline, we do not know. We do know that ours is.

Critical Thinking

1. Do you agree with the author that we need to bring about an environmental revolution within a few decades? Explain.

2. Do you believe that this can be done by making minor adjustments in the global economy or that the global economy must be restructured to put less strain on the earth's natural systems?

A: At least 14,000 (600 during your lunch hour)

CRITICAL THINKING*

1. Is the world overpopulated? Explain. Is the United States overpopulated? Explain.

2. **(a)** Do you believe that the society you live in is on an unsustainable path? Explain. What about the world as a whole? **(b)** Do you believe that it is possible for the society you live in to become a sustainable society within the next 50 years? Explain. **(c)** Would you classify yourself as a technological optimist, environmental pessimist, or hopeful environmental realist? Explain.

3. Do you favor instituting policies designed to reduce population growth and stabilize **(a)** the size of the world's population as soon as possible and **(b)** the size of the U.S. population (or the population of the country where you live) as soon as possible? Explain. If you agree that population stabilization is desirable, what three major policies do you believe should be implemented to accomplish this goal?

4. Explain why you agree or disagree with the following propositions:
   a. High levels of resource use by the United States and other developed countries are more beneficial than harmful.
   b. The economic growth from high levels of resource use in developed countries provides money for more financial aid to developing countries for reducing pollution, environmental degradation, and poverty.
   c. Stabilizing population is not desirable because without more consumers, economic growth would stop.

5. Explain why you agree or disagree with the following proposition: The world will never run out of potentially renewable resources and most currently used nonrenewable resources because technological innovations will produce substitutes, reduce resource waste, or allow use of lower grades of scarce nonrenewable resources.

6. Would you support a sharp increase in local, state, and federal taxes if you could be sure the money were used to help improve the environment? Explain.

7. Would you support greatly increasing the amount of land designated as wilderness to protect it from economic development, even if the land contained valuable minerals, oil, natural gas, timber, or other resources? Explain.

*A detailed discussion of critical thinking techniques is found in the *Critical Thinking and the Environment* supplement that can be used with this textbook. Also see the Guest Essay by Jane Heinze-Fry on p. 46.

8. Suppose enforcement of government pollution control regulations meant that you would lose your job. If you had a choice, would you choose unemployment and a cleaner environment, or employment and a dirtier environment? Can you think of ways to avoid this dilemma? Explain.

9. One bacterium is capable of leaving 16,777,216 offspring in only 24 hours. Imagine that you are a bacterium living in such a colony after 1 hour, 5 hours, and 22 hours. Describe what your life might be like at each of these times in the history of this bacterial colony. Suppose some bacterium warned members of the colony after 5 hours that they would run out of food (nutrients) if they didn't learn how to slow down their population growth or consume fewer nutrient resources. How do think this warning might be re- ceived? How is this situation like, and how does it differ from, that of the human population at this time in history?

10. Do you believe that your current lifestyle is sustainable? If the answer is yes, explain why and include the impact of the world's other 5.9 billion people on your ability to sustain your current lifestyle. If your answer is no, explain why and indicate ten things you could do now to make your lifestyle more sustainable. Which of these things do you actually plan to do?

11. Which is more important: the survival of people alive today, or the sustainability of the earth's ecosystem services on which future human life and economies depend? Explain.

12. Do you agree or disagree with the list on p. 19 of root causes of environmental problems proposed by various analysts? Explain. Can you add any other root causes?

13. Try to come up with analogies to complete the following statements: **(a)** Depleting earth capital is like _______. **(b)** Living sustainably is like _______. **(c)** Exceeding the earth's ability to support humans is like _______. **(d)** Exponential growth is like _______. **(e)** Living in a more developed country is like _______. **(f)** Living off of potentially renewable resources is like _______. **(g)** Burning or burying waste materials that are potential resources is like _______. **(h)** Not including the harmful environmental and health cost of any product in its market price is like _______. **(i)** Thinking that we are in charge of the earth is like _______. **(j)** Thinking that we will never run out of resources is like _______.

14. Would we be better off if agricultural practices had never been developed and we were still hunter–gatherers? Explain.

15. List the most important benefits and drawbacks of an advanced industrial society such as the United States. Do the benefits outweigh the drawbacks? Explain. What are the alternatives?

PROJECTS*

1. What are the major resource and environmental problems in: **(a)** the city, town, or rural area where you live and **(b)** the state where you live? Which of these problems affect you directly?

2. Roughly what percentages of key resources such as water, food, and energy used by your local community come from the following places: nearby, another state, another country? How do these inputs affect the long-term environmental and economic sustainability of your community?

*These are either laboratory exercises or individual or class projects. Additional exercises and projects are found in the laboratory manual and in the *Green Lives, Green Campuses* workbook that can be used with this textbook.

3. Make a list of the resources you truly need. Then make another list of the resources you use each day only because you want them. Finally, make a third list of resources you want and hope to use in the future. Compare your lists with those compiled by other members of your class, and relate the overall result to the tragedy of the commons.

4. Write two-page scenarios describing what your life and that of any children you choose to have might be like 50 years from now if **(a)** we continue on our present path, or **(b)** we shift to sustainable societies throughout most of the world.

5. Make a concept map of this chapter's major ideas using the section heads and subheads and the key terms (in boldface). Look at the inside back cover and in Appendix 4 for information about concept maps.

A: About 72,000 (with about 1,000 new ones added each year)

2 CRITICAL THINKING: SCIENCE, MODELS, AND SYSTEMS

Two Islands: Can We Treat This One Better?

Easter Island (Rapa Nui) is a small island isolated in the great expanse of the South Pacific more than 3,200 kilometers (2,000 miles) west of the nearest continent. Its rather mild climate and volcanic origins that covered it with a fertile soil should make the island a paradise.

It was first colonized by Polynesians about 2,500 years ago, as confirmed in 1994 by DNA extracted from 12 Easter Island skeletons. Somehow they sailed nonstop southeast in small canoes over at least 2,000 kilometers (1,200 miles) of open water, bringing with them animals (dogs, chickens, and rats) and typical Polynesian food plants such as bananas, taro, sweet potatoes, and sugarcane. Skeletons show that the islanders also used porpoises, seabirds, and land birds as sources of food.

Using pollen analysis, scientists have determined that when the Polynesians arrived the island was a subtropical forest containing a variety of large trees (especially an abundant species of palm) and woody bushes. The civilization they developed was based on the island's trees, which were used for shelter, tools, boats, fuel, food, rope, and clothing. Using these resources, they developed an impressive civilization and a technology capable of making large stone structures, including their famous statues (Figure 2-1).

The people flourished, the population peaking at about 10,000 (with estimates ranging from 7,000 to 20,000). But they were unable to stop themselves from using up the precious trees—an example of the tragedy of the commons (p. 12). Each person who cut a tree reaped immediate personal benefits while helping to doom the civilization as a whole. As they started to run out of the wood that supported them, the starving people turned to warfare and, possibly, cannibalism. Both the population and the civilization

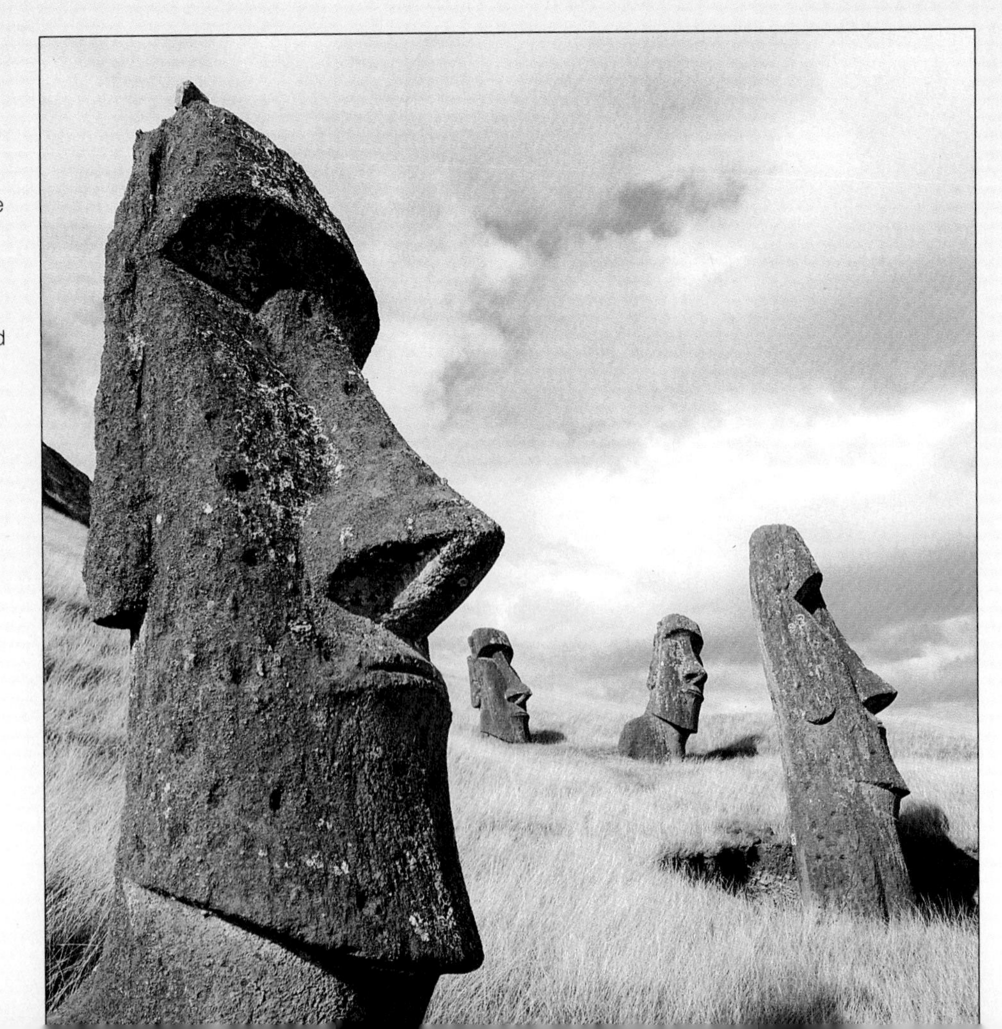

Figure 2-1 These massive stone figures on Easter Island are the remains of the technology created by an ancient civilization of Polynesians that collapsed because the people used up the trees (especially large palm trees) that were the basis of their livelihood. More than 200 of these stone statues once stood on huge stone platforms lining the coast. At least 700 additional statues were abandoned in rock quarries or on ancient roads between the quarries and the coast. No one knows how the early islanders (with no wheels, no draft animals, and no sources of energy except their own muscles) transported these gigantic structures for miles before erecting them, but it is presumed that they did this by felling large trees and using them to roll and erect the statues. (George Holton/Photo Researchers, Inc.)

collapsed. When Dutch explorers first reached the island on Easter Day, 1722, they found only about 2,000 inhabitants, struggling under primitive conditions on a mostly barren island.

Why didn't the islanders realize that they were consuming the resources that supported them and stop before it was too late? Because the trees vanished slowly over decades, only older people could have recognized what was happening. It is also likely that anyone who tried to warn about the dangers of excessive deforestation was overridden by tree cutters, carvers, statue movers, chiefs, and others whose jobs depended on continued deforestation—not unlike the cries today that protecting the earth's remaining old-growth forests will jeopardize jobs for loggers and mill workers and profits for logging companies.

Like Easter Island at its peak, the planet earth is, in its own way, an isolated island (in the vastness of space) with a thriving technological civilization and no other suitable planet to migrate to. As on Easter Island, our population is growing, and we are consuming exhaustible and potentially renewable resources at a rapid pace. Will the humans on Earth Island recreate the tragedy of Easter Island on a grander scale, or will we learn how to live sustainably on this planet that is our only home?

The two main goals of environmental science are to understand the extremely complex interactions of the earth's populations and resources and to learn how both can be sustained indefinitely. The example of Easter Island shows that to succeed, we must try to understand more about how nature works and sustains itself. Then we must use such information to *anticipate* how our actions may be altering natural processes (in ways that could make life unsustainable for us and other species) and to adopt actions to *prevent* such an environmental catastrophe.

Fortunately, our flair for inventing technologies that empower us to degrade the earth's life-support systems may also empower us to anticipate problems and prevent them. Environmental scientists successfully predicted the thinning of the earth's protective ozone layer over 20 years ago, and actions to stop this destruction have been undertaken on an international scale. Have we acted in time? What other key resources (like the trees of Easter Island) are we using unsustainably and placing in jeopardy? These are urgent questions.

Science is an adventure of the human spirit. It is essentially an artistic enterprise, stimulated largely by curiosity, served largely by disciplined imagination, and based largely on faith in the reasonableness, order, and beauty of the universe.

WARREN WEAVER

This chapter introduces the techniques and tools scientists use to understand earth's life-support systems and to discover actions that help protect the environmental processes supporting all life. This chapter will answer the following questions:

- What are science and technology? What is environmental science, and what are some of its limitations?
- What are models, and how can they be useful in understanding complex systems?
- What are inputs, throughputs, and outputs of systems? How do various types of feedback loops influence the behavior of systems?
- What are some of the ways complex systems behave, and what are the implications of such behavior for our future and that of the environment?

2-1 SCIENCE, TECHNOLOGY, AND ENVIRONMENTAL SCIENCE

What Is Science and What Do Scientists Do? **Science** is a pursuit of knowledge about how the world works. It is an attempt to discover order in nature and use that knowledge to make predictions about what should happen in nature. Science is based on the assumption that there is discoverable order in nature. As Albert Einstein once said, "The whole of science is nothing more than a refinement of everyday thinking." Figure 2-2 and the Guest Essay on p. 46 summarize the more systematic version of the everyday critical thinking process used by scientists.

The first thing scientists must do is ask a question or identify a problem to be investigated. Then scientists working on this problem collect **scientific data**, or facts, by making observations and taking measurements. Sometimes the observations are made with the unaided senses of sight, hearing, smell, and touch. Often scientists extend these senses by using devices such as microscopes, chemical analyzers, and radiation detectors.

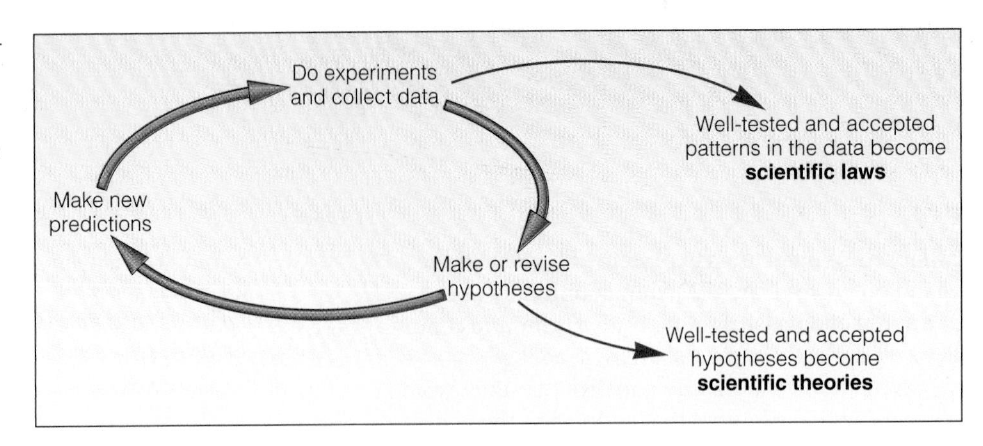

Figure 2-2 What scientists do: a summary of the scientific process, a form of critical thinking. Facts (data) are gathered and verified by repeated experiments; data are analyzed to see whether there is a consistent pattern of behavior that can be summarized as a scientific law; hypotheses are proposed to explain the data; deductions or predictions are made and tested to evaluate each hypothesis. A hypothesis that is supported by a great deal of evidence and is widely accepted by the scientific community becomes a scientific theory.

The procedure a scientist uses to study some phenomenon under known conditions is called an **experiment**. Some experiments are conducted in the laboratory, but others are carried out in nature. The resulting scientific data or facts must be verified or confirmed by repeated observations and measurements, ideally by several different investigators. This concept of *reproducibility* is important in science to detect errors in measurement, unconscious bias by investigators, and occasional cheating. This is done by repeating observations and measurements and by publishing the results of scientific research for others to examine, repeat, verify, and criticize.

The primary goal of science is not facts themselves, but a new idea, principle, or model that connects and explains certain facts and leads to useful predictions about what should happen in nature. Scientists working on a particular problem try to come up with a variety of possible explanations, or **scientific hypotheses**, of what they (or other scientists) observe in nature. A scientific hypothesis is a *tentative explanation*—often an educated guess or hunch based on past experience or intuition—about what has been observed and measured, connections in a chain of events, or cause and effect relationships among observed events or data.

To be accepted, a scientific hypothesis not only must explain scientific data and phenomena but also should make predictions that can be used to test the validity of the hypothesis. Once a scientific hypothesis is invented, experiments are conducted (and repeated to be sure they are reproducible) to test the deductions or predictions. Experiments can eliminate (disprove) various hypotheses, but they can never prove that any hypothesis is the best (most useful) or the only explanation. All scientists can say is that experiments strongly support the validity and predictions of a particular hypothesis.

One method scientists use to test a hypothesis is to develop a **model**, an approximate representation or simulation of a system being studied. There are many types of models: mental, conceptual, graphic, physical, and mathematical, as discussed in Section 2-2.

Scientists are skeptics; they want lots of evidence before they are willing to accept the accuracy of any data or the usefulness of a particular hypothesis or model. Usually a succession of scientists working on the same problem, often working around the world, subject each other's data, hypotheses, and models to careful scrutiny. The process takes this form:

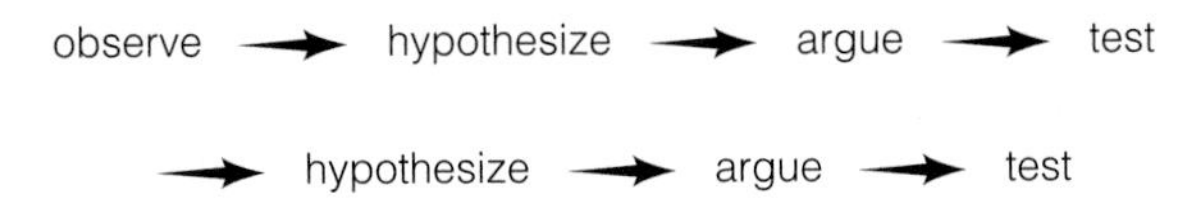

It continues until some general consensus or agreement is reached.

If many experiments by different scientists support a particular hypothesis, it becomes a **scientific theory**: an idea, principle, or model that usually ties together and explains many facts that previously appeared to be unrelated and is supported by a great deal of evidence.

Nonscientists often use the word *theory* incorrectly when they mean to refer to a *scientific hypothesis*, a tentative explanation that needs further evaluation. The statement, "Oh, that's just a theory," made in everyday conversation implies a lack of knowledge and careful testing—the opposite of the scientific meaning of the word. To scientists, theories are not to be taken lightly. They are ideas or principles stated with a high degree of certainty because they are supported by a great deal of evidence.

Another important result of science is a **scientific law**: a description of what we find happening in nature over and over in the same way, without known exception. For example, after making thousands of observations and measurements over many decades, scientists discovered what is called the *second law of energy* or *thermodynamics*. One simple way of stating this law is that heat always flows spontaneously from hot

Soule, Michael E., and Daniel Press. 1998. "What Is Environmental Studies?" *BioScience*, vol. 48, no. 5, 397(9).

to cold—something you learned the first time you touched a hot object.

Scientific laws describe what we find happening in nature in the same way, whereas theories are widely accepted explanations of data and laws. In 1783, after conducting numerous experiments on chemical reactions, French chemist Antoine Lavosier formulated the *law of conservation of matter*. This law, which has been confirmed by hundreds of thousands of experiments, applies to any physical change (such as melting ice to form water) or any chemical change (such as hydrogen and oxygen combining to form water). According to this law (based on experimental data) we can change matter from one physical or chemical form to another, but no matter is created or destroyed by such processes.

There was no explanation of this law until the early 1800s, when chemist John Dalton came up with the *atomic theory of matter*. According to Dalton's hypothesis (which later became a widely accepted scientific theory), all matter is made up of extremely small particles called *atoms* and these atomic building blocks cannot be created, destroyed, or subdivided by physical and chemical changes. He explained the law of conservation of matter by reasoning that no matter is destroyed or created in physical and chemical changes because the atoms involved cannot be created or destroyed.

The scientific process requires not only logical reasoning, but also imagination, creativity, and intuition. According to physicist Albert Einstein, "There is no completely logical way to a new scientific idea." Intuition, imagination, and creativity are as important in science as they are in poetry, art, music, and other great adventures of the human spirit that awaken us to the wonder, mystery, and beauty of life, the earth, and the universe.

What Is the Difference Between Accuracy and Precision in Scientific Measurements? If you make a scientific measurement, how do you know whether it is correct? All scientific observations and measurements have some degree of uncertainty because people and measuring devices are not perfect. Scientists take great pains to reduce the errors in observations and measurements as much as possible by **(1)** using standard procedures in making observations and measurements, **(2)** testing (calibrating) measuring devices using samples whose values are known, and **(3)** repeating their measurements several times and then finding the average value of these measurements (reproducibility).

In determining the uncertainty involved in a measurement it is important to distinguish between accuracy and precision. **Accuracy** is the extent to which a measurement agrees with the accepted or correct value for that quantity, based on careful measurements by many people over a long time. **Precision** is a measure of *reproducibility*, or how closely a series of measurements of the same quantity agree with *one another*.

The dartboard analogy shown in Figure 2-3 illustrates the difference between precision and accuracy. *Accuracy* depends on how close the darts are to the bull's-eye. *Precision* depends on how close the darts are to each other. Note that good precision is necessary for accuracy but does not guarantee it. Three closely spaced darts may be far from the bull's-eye.

How Do Scientists Learn About Nature? We often hear about *the* scientific method, but in reality there are many **scientific methods**: ways scientists gather data and formulate and test scientific hypotheses, models, theories, and laws (Figure 2-2). Instead of being a recipe, each scientific method involves trying to answer a set of questions, with no particular guidelines for answering them:

- What is the question about nature to be answered?
- What relevant facts and data are already known?
- What new data (observations and measurements) should be collected, and how should this be done?
- After the data are collected, can they be used to formulate a scientific law?

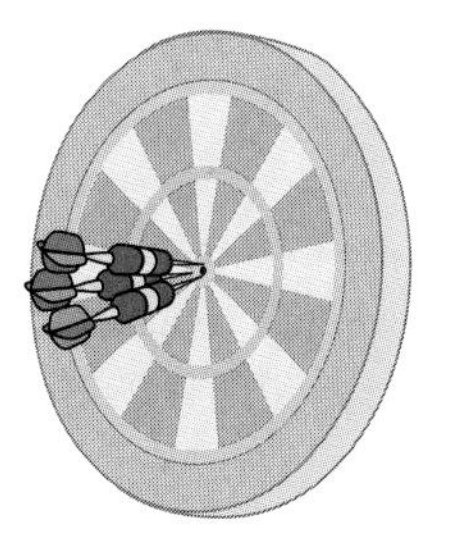

Good accuracy and good precision

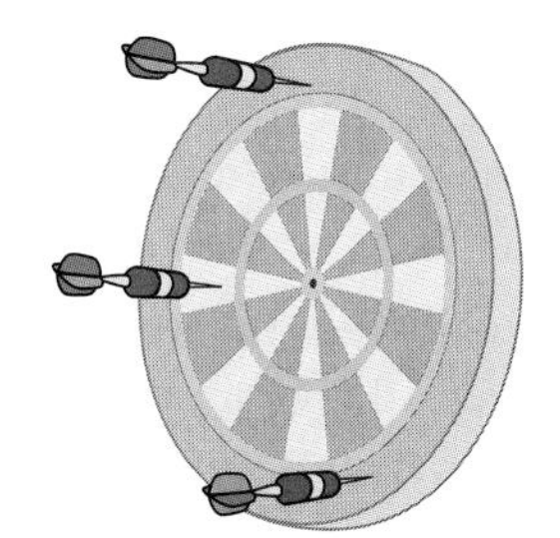

Poor accuracy and poor precision

Poor accuracy and good precision

Figure 2-3 The distinction between accuracy and precision. In scientific measurements, a measuring device that has not been calibrated to determine its accuracy may give precise or reproducible results that are inaccurate.

Hint: Enter the search term *environmental sciences* using the Subject Guide.

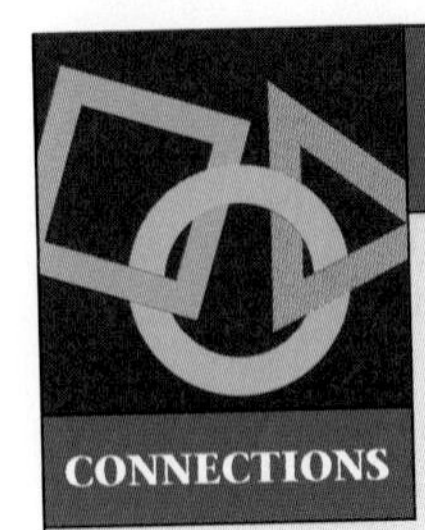

CONNECTIONS

What's Harming the Robins?

Suppose a scientist has observed an abnormality in the growth of robin embryos in a certain area. She knows that the area has been sprayed with a pesticide and suspects that the chemical may be causing the abnormalities she has observed.

To test this hypothesis, the scientist carries out a *controlled experiment*. She maintains two groups of robin embryos of the same age in the laboratory. Each group is exposed to exactly the same conditions of light, temperature, food supply, and so on, except that the embryos in the experimental group are exposed to a known amount of the pesticide in question.

The embryos in both groups are then examined over an identical period of time for the abnormality. If there is a statistically significantly larger number of the abnormalities in the experimental group than in the control group, then the results support the idea that the pesticide is the culprit.

To be sure there were no errors in the procedure, the experiment should be repeated several times by the original researcher, and ideally by one or more other scientists. Conducting such replications is a standard procedure in science.

Critical Thinking

Can you find flaws in this experiment that might lead you to question the scientist's conclusions? (*Hint:* What other factors in nature—not the laboratory—and in the embryos themselves could possibly explain the results?)

- How can a hypothesis be invented that explains the data and predicts new facts? Is this the simplest and only reasonable hypothesis?
- What new experiments can be done to test the hypothesis (and modify it if necessary) so it can become a scientific theory?

New discoveries happen in many ways. Some follow this sequence:

data → law → hypothesis → theory

At other times, scientists simply follow a hunch, a bias, or a belief and then do experiments to test their idea or hypothesis.

Other discoveries occur when an experiment gives totally unexpected results, and the scientist investigates to find out what happened. For example, in 1938 the plastic Teflon was discovered accidentally (by *serendipity*, or looking for one thing and finding a different thing) by DuPont chemists. The chemists were trying to produce compounds to use as refrigerants. In one case they stored a gaseous compound used to make a refrigerant in a cylinder that also happened to contain small amounts of oxygen gas as an impurity. The two gases reacted, and the chemists opened the cylinder to find out why it did not contain the gaseous compound they thought was there. To their surprise, they discovered a waxy, white, slippery solid that was heat-resistant and insoluble in almost all liquids—a plastic we now know as Teflon.

Most processes or parts of nature that scientists seek to understand are influenced by a number of *variables* or *factors*. One way scientists test a hypothesis about the effects of a particular variable is to conduct a *controlled experiment*. This is done by setting up two groups: an *experimental group*, in which the chosen variable is changed in a known way, and a *control group*, in which the chosen variable is not changed. The experiment is designed so that all components of each group are as identical as possible and experience the same conditions, except for the single factor being varied in the experimental group. If the experiment is designed properly, any difference between the two groups should result from a variable that was changed in the experimental group (Connections, left).

Another example of a controlled experiment, called a *double-blind* experiment, is used to test most new drugs. One group of patients is given the new drug (the experimental group). A second, similar set of patients (the control group) is given a *placebo*, a harmless starch pill similar in shape, size, and color to the pill being tested. To avoid bias, neither the patients nor the physicians involved know who is getting the experimental drug and who is getting the placebo, which is why this is called a double-blind approach. The secret code used to identify the experimental and control patients is revealed only after the experiment is over and the results are in.

A basic problem is that many of nature's components and processes, especially those investigated by environmental scientists, involve a huge number of variables interacting in often poorly understood ways. In such cases, it is very difficult or impossible to carry out meaningful controlled experiments.

How Valid Are the Results of Science? Scientists can disprove things, and they can establish that a particular model, theory, or law has a very high degree of validity and is extremely useful in explaining how nature works and in predicting what will happen in nature. However, like scholars in virtually any field, scientists cannot prove that their ideas are *absolutely* true.

Q: What percentage of environmental spending in the United States is devoted to preventing pollution?

When people say that something has or has not been "scientifically proven," either they don't understand and the nature of science, or they are trying to mislead us by falsely implying that science yields absolute proof or certainty. Although it may be extremely low, there is always some degree of uncertainty involved in any scientific model, theory, or law. The goal of the rigorous scientific process is to reduce the degree of uncertainty as much as possible. However, the more complex the system being studied, the greater the degree of uncertainty or unpredictability about its behavior.

The standard for evaluating a scientific model, theory, or law is not absolute truth or proof. Instead, the validity of such ideas is based on how *useful* they are in helping us understand how nature works and in making predictions about what will happen in nature.

What Types of Reasoning Do Scientists Use? Scientists arrive at certain conclusions with varying degrees of certainty by using inductive and deductive reasoning. **Inductive reasoning** involves using observations and facts to arrive at generalizations or hypotheses and is widely used in science. It goes from the specific to the general. For example, suppose we observe that a variety of different types of birds fly. We might then use inductive reasoning to conclude that *all birds fly*. However, there is no certainty that this generalization is true. All we are saying is that all of the types of birds that we (or other observers) have seen fly or that there is a high degree of certainty or probability that all birds fly. There may be types of birds that cannot fly.

Similarly, when we arrive at a conclusion based on a series of measurements all we are saying is that the conclusion is valid only within the degree of uncertainty inherent in any measurement. For example, we might measure the masses of chemicals before and after a chemical reaction and conclude that in such a change no matter is created or destroyed—a generalization or scientific law known as the *law of conservation of matter*. However, this conclusion is valid only within the detection limits of the instruments used to make the measurements of mass.

For example, suppose that we measured the masses of the chemicals found before and after a chemical change using an analytical balance that can measure mass within 0.001 gram. Creation or destruction of a smaller amount of mass than 0.001 gram could not be detected using such an instrument. Thus, all we could say is that matter cannot be created or destroyed within the measurement limits of the devices we used to measure mass and arrive at this deductive conclusion (in this case, 0.001 gram).

Deductive reasoning involves using logic to arrive at a specific conclusion based on a generalization or premise. It goes from the general to the specific. For example:

> *Generalization or premise:* All birds have feathers.
> *Example:* Eagles are birds.
> *Deductive conclusion:* All eagles have feathers.

This conclusion of this syllogism (a series of logically connected statements) is valid as long as the premise is correct and we don't use faulty logic to arrive at the conclusion.

Deductive reasoning can be used to arrive at a logically correct but untrue statement. For example,

> *Generalization or premise:* All animals with feathers are mammals.
> *Example:* Eagles have feathers.
> *Deductive conclusion:* Eagles are mammals.

In this case the conclusion must be valid if the first two statements are valid. However, although the second statement in this syllogism is true, the first statement is not, so the conclusion is false.

Are Scientists Always Objective? Scientists are human beings who have conscious and unconscious values, biases, and beliefs that can influence how they design and interpret data and the reliability they attach to various scientific hypotheses and theories. Part of the critical thinking process for scientists and those evaluating the results of scientific research is to understand that scientists do have biases and to try to identify them.

The idea that scientists cannot always reach the goal of complete objectivity and that the results of science are not entirely value-free should not be taken to mean that shoddy thinking is acceptable in science. The standards of evidence in science are very high. The heart of science is open communication. Scientific results are published so that other scientists can correct mistaken information, challenge fuzzy or biased thinking, modify faulty conclusions, and discover rare but occasional cheating (faking scientific results).

The ideas that science cannot establish absolute proof and that scientists as human beings are not always completely objective does not mean that the results of scientific research are not to be taken seriously. Despite its limitations, the high standards of evidence required for reaching scientific conclusions mean that science is the best way we have come up with to get reliable knowledge about how nature works.

Science has revealed many of nature's secrets and relationships, sparked our imagination and sense of wonder and awe. It has also challenged old ways of thinking and demonstrated the fuzzy thinking inherent in many of our worldviews and pseudoscientific or false conclusions based on unfounded ideas. Science has also shown that the world we live in is a holistic, interdependent, interconnected, interactive

A: About 1%; the other 99% is spent on pollution cleanup system in which the whole is determined by the *functions* of its parts, not merely by the parts themselves.

How Does Frontier Science Differ from Consensus Science? News reports often focus on new scientific "breakthroughs" and on disputes among scientists over the validity of preliminary (untested) data, hypotheses, and models (which are by definition tentative). This aspect of science, controversial because it has not been widely tested and accepted, is called **frontier science**.

The media focus on frontier science because its "breakthroughs" and scientific controversies make good news stories. Just because something is in the realm of frontier science does not mean that it isn't worthy of serious consideration. Instead, such matters need further study to determine their reliability.

By contrast, **consensus science** consists of data, theories, and laws that are widely accepted by scientists considered experts in the field involved. This aspect of science is very reliable but is rarely considered newsworthy. One way to find out what scientists generally agree on is to seek out reports by scientific bodies such as the U.S. National Academy of Sciences and the British Royal Society that attempt to summarize consensus among experts in key areas of science.

To give a sense of balance and fairness, the media often quote one or more of the very small number of scientists who criticize the consensus view of the vast majority of scientists in a particular field. Instead of balance, this usually leads to a biased presentation because a minority view is given about the same weight as the consensus view held by most scientists.

Often, fundamental advances in scientific knowledge are made when a creative scientist proposes and uses experiments to establish a new hypothesis that eventually alters the consensus view about a particular topic. However, merely criticizing the consensus view without coming up with a new view and using scientific experiments to establish that view is not very useful and often misleads the public. Science thrives on new ideas but demands that they be backed up by hard evidence before they can be widely accepted as part of consensus science.

What Is Technology? **Technology** is the creation of new products and processes intended to improve our efficiency, chances for survival, comfort level, and quality of life. Whereas science is a search for understanding of how the natural world works, technology involves finding ways to control or manage the natural world primarily for the benefit of humans.

In many cases, technology develops from known scientific laws and theories. Scientists invented the laser, for example, by applying knowledge about the internal structure of atoms. Applied scientific knowledge about chemistry has given us nylon, pesticides, laundry detergents, pollution control devices, and countless other products. Applications of theories in nuclear physics led to nuclear weapons and nuclear power plants.

However, some technologies arose long before anyone understood the underlying scientific principles. For example, aspirin, extracted from the bark of a willow tree, relieved pain and fever long before anyone found out how it did so. Similarly, photography was invented by people who had no inkling of its chemistry, and farmers crossbred new strains of crops and livestock long before biologists understood the principles of genetics. In fact, much of science is an attempt to understand and explain why various technologies work.

Science and technology usually differ in the way the information and ideas they produce are shared. Many of the results of scientific research are published and distributed freely to be tested, challenged, verified, or modified. In contrast, many technological discoveries are kept secret until the new process or product is patented. However, the basis of some technology is published in journals and enjoys the same kind of public distribution and peer review as science.

What Is Environmental Science and What Are Its Limitations? **Environmental science** is the study of how we and other species interact with one another and with the nonliving environment (matter and energy). It is a *physical and social science* that integrates knowledge from a wide range of disciplines including physics, chemistry, biology (especially ecology), geology, meteorology, geography, resource technology and engineering, resource conservation and management, demography (the study of population dynamics), economics, politics, sociology, psychology, and ethics. In other words, it is a study of how the parts of nature and human societies operate and interact—a study of *connections* and *interactions* (see inside back cover of this book).

There is controversy over some of the knowledge provided by environmental science, for much of it falls into the realm of frontier science. One problem involves *arguments over the validity of data*. There is no way to measure accurately how many metric tons of soil are eroded worldwide, how many hectares of tropical forest are cut, how many species become extinct, or how many metric tons of certain pollutants are emitted into the atmosphere or aquatic systems each year.

We may legitimately argue over the numbers, but the point environmental scientists want to make is that the trends in these phenomena are significant enough to be evaluated and addressed. Such environmental data should not be dismissed because they are "only estimates" (which are all we can ever have). However, this does not relieve investigators of the responsibility of getting the best estimates possible and pointing out that they *are* estimates.

Q: What percentage of the world's population lives in the United States?

Another limitation is that *most environmental problems involve so many variables and such complex interactions that we don't have enough information or sufficiently sophisticated models to aid in understanding them very well.* Much progress has been made during the past 50 years (and especially the past 25 years), but we still know much too little about how the earth works, its current state of environmental health, and the effects of our activities on its life-support systems.

Reputable scientists in a field may state contradictory opinions about the meaning of the data from experiments (especially frontier scientific results) and the validity of various hypotheses. This is especially true in environmental science, in which hypotheses and predictions are almost always based on inadequate data and understanding of the complex systems involved. However, the limited but growing knowledge and wisdom based on the findings of environmental science are key factors in helping us make more informed personal, national, and global decisions that can enable societies to become more sustainable.

Because environmental problems won't go away, at some point we must evaluate the available (but always inadequate) information and make political and economic decisions. Without sufficient scientific evidence, such decisions are often based primarily on individual and societal values, which is why differing environmental worldviews are at the heart of most environmental controversies (Section 1-8 and Chapter 29). People with different worldviews and values can take the same information, use inductive reasoning to come to completely different conclusions, and still be logically consistent.

Understanding and evaluating our worldviews and trying to uncover the biases we all have is a prerequisite to critical thinking (Guest Essay, p. 46). For as American psychologist William James said, "A great many people think they are thinking when they are merely rearranging their prejudices."

2-2 SYSTEMS AND SYSTEM MODELS

What Is a System? A **system** is set of components that function and interact in some regular and theoretically predictable manner. A system has a *structure*, consisting of its components or parts that fit together to make a whole, and a *function*, or what the system does. For example, the circulatory system in our bodies consists of various parts (heart, arteries, veins, capillaries, and blood) that function together to move blood through the body (carrying oxygen from the lungs, carbon dioxide to the lungs, nutrients from the digestive tract, and waste products to the kidneys and liver).

The earth consists of an incredible number of interacting systems based on a variety of climates, soil types, and communities of plants, animals, and decomposers (which break down plant and animal wastes and dead plant and animal matter for use again). These processes, which connect the earth's living and nonliving components, are discussed in more detail in Chapters 4 and 5.

How Do Scientists Use Models in the Creative Process? Understanding and living in the environment requires us to be creative (Figure 2-4)—to constantly try new things. Of course, trying everything out to see whether it works is time-consuming, expensive, and sometimes dangerous.

Over time, people have learned the value of using models as approximate representations or simulations of real systems to help find out which ideas or hypotheses work. Among the many kinds of models, people use the following:

- ***Mental models*** to perceive the world, control their bodies, and think about things.
- ***Conceptual models*** to describe the general relationships among components of a system. The diagram inside the back cover of this book is a conceptual model of environmental science. Most of the diagrams

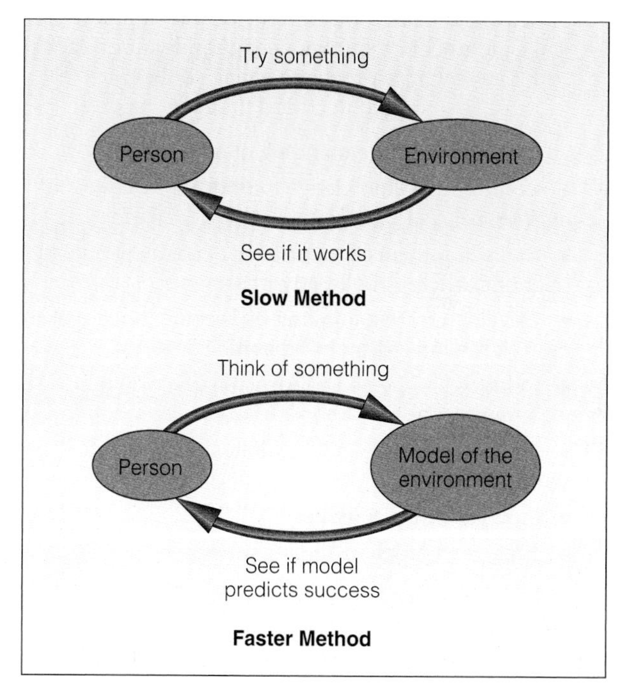

Figure 2-4 The essence of the creative process: To determine whether a creation is useful, you must try it out; if it works, you have succeeded; if it doesn't, you must try again. This process is usually accurate, but it can be slow. When creating solutions to environmental problems (such as industrialization and rapid population growth), it may take decades to find out whether a proposed solution is a success. Using models enables people to speed up evaluation of the creative process, but the models must be good enough to neither reject too many good ideas nor encourage too many bad ideas.

A: 4.6% (270 million people in mid-1998)

Critical Thinking and Environmental Studies

GUEST ESSAY

Jane Heinze-Fry

Jane Heinze-Fry has a Ph.D. in environmental education and is the author of the instructor's manual and several other supplements that accompany this textbook. She is author of Green Lives, Green Campuses *and* Critical Thinking and Environmental Studies: A Beginner's Guide *(with G. Tyler Miller as coauthor). Previously, she taught and directed environmental studies at Sweet Briar College. She also taught biology to students at the junior high, high school, and college levels, and has been teaching environmental science at Emerson College and Bunker Hill Community College in Boston. Her interdisciplinary orientation is reflected in her concept maps, including the ones inside the back cover and in Appendix 4 of this textbook.*

Defining *critical thinking* is not easy. Critical thinking has come to mean different things to different people. In my opinion, *critical thinking* is thinking that moves learners **(1)** to more meaningful learning instead of rote learning; **(2)** to higher levels of learning (application, analysis, synthesis, and evaluation as opposed to knowledge and comprehension); **(3)** to apply concepts and principles to real-world experience and situations; **(4)** to make judgments about knowledge and values claims; and **(5)** to enhance problem-solving skills.

Learners engaged in critical thinking make efforts to

- Clarify their understanding of new concepts
- Connect new knowledge to prior knowledge and experience
- Make judgments about claims made by people in scientific articles, books, advertisements
- Develop creative thinking by learning how to exercise both right- and left-brain capabilities
- Relate what they have learned to their own life experiences
- Understand and evaluate their environmental worldview
- Take positions on issues
- Engage in problem-solving processes

Whenever we are faced with new information, we need to evaluate it by using critical thinking. Do we believe the information or not, and why? Do the claims seem reasonable or exaggerated? Here are some rules for evaluating evidence and claims:

1. Gather all the information you can.
2. Be sure that all key terms and concepts are defined and that you understand these definitions.
3. Question how the information (data) was obtained.
   - Were the studies involved well designed and carried out?
   - Was there an experimental group and a control group? Were the control and experimental groups treated identically except for the variable changed in the experimental group?
   - Did the investigators repeat their experiments several times and get essentially the same results?
   - Were the results verified by one or more other investigators?
4. Question the conclusions derived from the data.
   - Do the data support the claims, conclusions, and predictions?
   - Are there other possible interpretations? Are there more reasonable interpretations?
   - Do the conclusions involve a correlation or apparent connection between two or more variables, or do they imply a strong cause and effect relationship between such variables?
   - Are the conclusions based on the results of original research by experts in the field involved, or are they conclusions drawn by reporters or scientists in other fields?
5. Try to determine the assumptions and biases of the investigators and then question them.
   - Do the investigators have a monetary or political advantage in the outcome of the investigation or issue involved?
   - What are the underlying basic assumptions or worldviews of the investigators? Would investi-

in this chapter and throughout much of this book are conceptual models.

- ***Graphic models*** to compile and display data in meaningful patterns. A map is an example of a graphic model.
- ***Physical models*** (miniature versions of large systems) to try out designs and ideas. Examples are scale models of airplanes, buildings, and landscapes.
- ***Mathematical models,*** which consist of one or more mathematical equations to describe the behavior of a system. People use such models to predict, on paper or in a computer program, the results of experiments or designs.

The bottom of Figure 2-4 illustrates the use of models as a faster means of manipulating the envi-

Q: What percentage of the world's mineral resources and nonrenewable energy is used by the United States?

gators with different basic assumptions or worldviews take the same data and come to different conclusions?

6. Expect and tolerate uncertainty. Recognize that science is a dynamic process that provides only a certain degree of probability or certainty and that the more complex the system or process being investigated, the greater the degree of uncertainty.

- Are the data, claims, and conclusions based on the tentative results of *frontier science* or the more reliable and widely accepted results of *consensus science*?

7. Look at the big picture (think holistically).

- How do the results and conclusions fit into the whole system (the earth, ecosystem, economy, political system) involved?
- What additional data and experiments are needed to relate the results to the whole system?

8. Based on these steps, take a position by either rejecting or conditionally accepting the claims.

- Reject claims based on no evidence, insufficient evidence, or evidence coming from questionable sources.
- If evidence does not support a claim, reject it and state the conclusion you would draw.
- If the evidence supports the claims, conditionally accept the claims with the understanding that your support may change if new evidence disproves the claim.

In addition to these rules to enhance making reasoned judgments, there are a number of other strategies to improve your critical thinking skills. Some strategies focus on improving thinking, others on attitudes and values, others on actions.

Thinking strategies focus on learning how to use your left brain (which is good at logic and analysis) and your right brain (which is good at visualizing and creating). They include constructing models, clarifying concepts, brainstorming, defining problems, creating alternative solutions, visualizing future possibilities, and exercising your creative brain paths.

Attitudes and values focus on establishing an ecological identity, reflecting on the interaction of lifestyle and the environment, understanding and evaluating your worldview, and using value analysis for evaluating proposed environmental policies and making personal environmental decisions. *Action strategies* focus on problem solving. They include techniques for visualizing problems, redefining problems, considering alternative solutions, creating plans of action, and developing strategies for implementing action plans.

In the environmental course you are taking, there are many opportunities to develop your critical thinking skills. Your textbook offers critical thinking questions at the end of chapters, Guest Essays, and most boxed material. If your course uses the supplement *Critical Thinking and Environmental Studies: A Beginner's Guide*, you will learn the critical thinking strategies mentioned here. Courses using the supplement *Green Lives, Green Campuses* can offer you additional opportunities to use critical thinking in evaluating your environmental lifestyle and making an environmental audit of your campus.

Learning how to think critically is essential in helping you evaluate the validity and usefulness of what you read in newspapers, magazines, and books (such as this textbook), what you hear in lectures and speeches, and what you see and hear on the news and in advertisements.

Critical Thinking

1. Can you come up with an example in which critical thinking has helped you make a major change in one or more of your beliefs or helped you make an important personal decision? Can you think of a decision that may have come out better if you had used critical thinking skills such as those discussed in this essay?

2. Rote learning often involves the "memorize and spit back" strategy. Meaningful learning (including critical thinking) goes far beyond memorization and requires us to evaluate the validity of what we learn. Currently, about what percentage of your learning involves rote learning and what percentage involves critical thinking as defined at the beginning of this essay?

ronment. Notice the similarity between the scientific process (Figure 2-2) and the creative process (Figure 2-4). Basically, science, music, invention, poetry, politics, and all creative endeavors share the same process. The heart of all human activity is to think up new things and try them out, using models where possible to reduce the number of costly or dangerous experiments and to understand complex processes.

What Are Mental Models? To guide our perceptions and help us make predictions, we use *mental models*. Most of our mental models are built into the structure of our nervous systems, and we are usually unaware of them.

We don't interpret the world according to direct knowledge of reality, but according to mental models, which people often mistake for reality. For example, we all share a built-in mental model that the world is continuous, unless our eyes tell us differently. This built-in

A: About 25%

Figure 2-5 How to see your built-in mental model that the world is continuous unless your eyes tell you it isn't. Cover your left eye and look at the X. Slowly move the page toward or away from your eye until you can't see the triangle. The triangle is now in your blind spot. Notice that you don't see a blank space or a black hole or a white area in your blind spot; you see a continuation of the pattern surrounding the blind spot. Your mental model (which fills in the missing information) says that the unseen area (blind spot) is just like its surroundings.

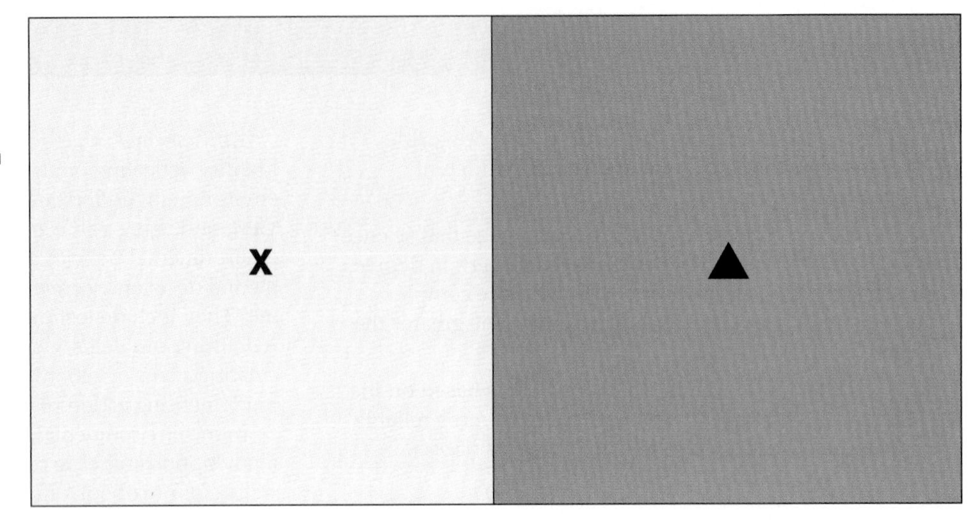

mental model tells us what to "see" in a part of the eye that doesn't actually see anything—the blind spot. This blind spot model causes us to *perceive* made-up information (a continuation of surrounding patterns), and it *predicts* that if we look at a previously hidden spot it will look like its surroundings. To see how it works, see Figure 2-5 and follow the instructions.

Not all mental models are built-in; some we learn or make up. For example, most people believe that the automobiles driving down the street will not veer onto the sidewalk and hit pedestrians. If we believed otherwise, we would act like frightened squirrels, always stopping to look around and proceeding only if no automobiles were operating nearby.

We use mental models of our surroundings to perceive what we believe to be true and to predict what may happen. Unfortunately, all of our mental models are approximations, contained only in the lump of nervous tissue we call the brain. Sophisticated as the brain is, it is very small and simple compared with the complexities of the outside world.

What Are Mathematical Models, and Why Are They Important in Environmental Science?

Some of the most powerful technologies invented by humans are mathematical models, which are used to supplement our mental models. *Mathematical models* are equations that help us perceive and predict various things. For example, equations that describe the movement of air, moisture, and heat in the atmosphere (weather models) are used to predict the weather for the next few days. Sometimes, television weather forecasters refer to such models when they tell us what to expect from tomorrow's weather; usually they translate the results of these mathematical models into weather maps, a type of graphic model.

The big difference between mental models and mathematical models is the way we get information from them. To get a perception or prediction from a mental model, we need only think; we get estimates or predictions from such models without even being conscious of them. In contrast, to get a prediction from a mathematical model, we must do some calculating—perhaps in our heads, perhaps with pencil and paper, perhaps with a computer.

Mathematical models vary in size and sophistication. Weather models are fairly complicated; to calculate predictions from them takes the use of powerful computers. Other mathematical models such as the *rule of 70* (70/percentage growth rate = doubling time in years) are simpler. Using this relationship (derived from the mathematical equation for exponential growth), we can make a simple calculation to estimate how many days until we run out of money, or how long it might take for the world's population to double.

Like mental models, mathematical models are imperfect approximations of reality. They tend to make predictions ranging from fairly accurate to very accurate, depending on the model and the accuracy of the data used to formulate the model.

The basic process for developing mathematical models is essentially trial and error, similar to the processes depicted in Figures 2-2 and 2-4. Making a mathematical model usually requires going many times through three familiar steps: **(1)** Make a guess and write down some equations, **(2)** compute the predictions implied by the equations, and **(3)** compare the predictions with observations, the predictions of mental models, existing experimental data, and scientific hypotheses, laws, and theories.

Mathematical models are important because they can give us improved perceptions and predictions, especially concerning matters for which our mental models are weak. Research has shown that people's mental models tend to be especially unreliable **(1)** when

Q: What percentage of the world's pollution is produced by the United States?

there are many interacting variables, **(2)** when we attempt to extrapolate from too few experiences to a general case, **(3)** when consequences follow actions only after long delays, **(4)** when the consequences of actions lead to other consequences, **(5)** when responses are especially variable from one time to the next, and **(6)** when controlled experiments (Connections, p. 42) are impossible, too slow, or too expensive to conduct. Under such conditions, a good mathematical model can do better than most mental models.

Most of our effects on the environment have the following characteristics: **(1)** We have only a few experiences on which to base the generalizations, **(2)** the responses occur only after long delays (such as possible ozone depletion in the stratosphere or second layer of the earth's atmosphere), **(3)** the responses cause other, delayed responses (such as the development of skin cancers 10–20 years after exposure to increased ultraviolet radiation), and **(4)** there is a great deal of variability in what happens because the environment is full of diversity and unpredictability.

These are exactly the kinds of situations in which our mental models tend to be unreliable and in which mathematical models, *if they are sufficiently good approximations of reality*, can help us make better predictions. Better predictions can lead to better decisions.

How Can We Use Models to Predict New Behavior? After building and testing a model, scientists apply it to a useful purpose by predicting what will happen under a variety of alternative conditions. In effect, they use mathematical models to answer *if–then* questions: "*If* we do such and such, *then* what is likely to happen now and in the future?"

A good model can be used to test and predict the implications of many different courses of action, looking for those that will ensure our survival and life quality with minimum risk. Using mathematical models of some aspect of the environment, we can gain the equivalent of hundreds or thousands of years of experience in a few weeks or months, just as pilots can use an aircraft simulator to gain experience without risking their lives.

2-3 SOME BASIC COMPONENTS AND BEHAVIORS OF SYSTEM MODELS

What Are Inputs, Throughputs, and Outputs of System Models? Any system being studied or modeled has one or more **inputs** of things such as matter, energy, or information. Once such inputs enter a system they can accumulate. An example of such an **accumulation** in the environment is a *population*: a group of individuals of the same species occupying a given area at a given time. The size of a particular population represents the accumulation that is the net difference between births and deaths of its members (assuming that there is no net loss or gain from migration).

Inputs flow through a system at a certain rate. Such **flows** or **throughputs** of matter, energy, or information through a system are represented by arrows. Forms of matter, energy, or information flowing out of a system are called **outputs** and end up in *sinks* in the environment. Examples of such sinks are the atmosphere, bodies of water, underground water, soil, and land surfaces. In system diagrams, inputs, accumulations, and outputs are often represented by boxes or circles. Figure 2-6 is a generalized diagram of the

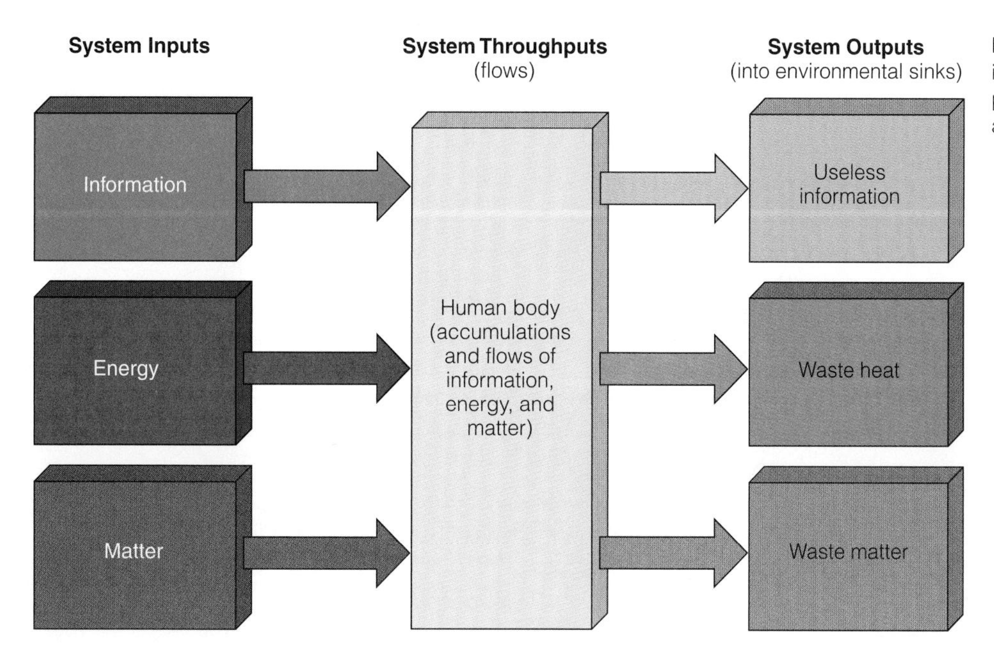

Figure 2-6 Generalized diagram of inputs, throughputs (flows), and outputs of information, matter, and energy in a system.

A: About 25%

basic components of a system such as your body. Figure 1-16 is a generalized diagram of inputs, throughputs, and outputs for an economic system.

What Are Feedback Loops? A **feedback loop** occurs when one change leads to some other change, which eventually either reinforces or slows the original change. Feedback loops determine how things happen over time.

Feedback loops occur when an output of matter, energy, or information is fed back into the system as an input (Figure 1-16). For example, recycling aluminum cans involves melting aluminum and feeding it back into an economic system to make new aluminum products. This feedback loop of matter reduces the need to find, extract, and process virgin aluminum ore. It also reduces the flow of waste matter (discarded aluminum cans) into the environment.

All change is driven or controlled by feedback loops, and every feedback loop has at least one accumulation in it. Feedback loops can be either positive or negative.

What Are Positive Feedback Loops? A **positive feedback loop** is a runaway cycle in which a change in a certain direction provides information that causes a system to change further in the same direction. The environment is full of positive feedback loops, some desirable and some undesirable from a human standpoint.

For example, as long as there are more human births than deaths on this planet, the human population will grow exponentially (Figures 1-1 and 2-7). More births and fewer deaths mean an ever-growing accumulation of people. This form of positive feedback can be viewed as desirable or undesirable depending on whether one sees population growth as an economic benefit (more workers and consumers) or an environmental drawback (more crowding, pollution, and resource depletion and degradation).

Similarly, if you deposit money in a bank at compound interest, the interest increases the balance, which leads to more interest, and so it goes (Figure 1-4). Notice from Figure 2-7 that in exponential growth, not only does the curve rise, but the slope of the curve (rate of change) rises proportionally.

Another result of one or more positive feedback loops is accelerating destruction leading to *catastrophic collapse*, an undesirable effect of positive feedback. For example, a positive population feedback loop also can drive the faster and faster depletion of natural resources. This is what happened on Easter Island (p. 38). The first people to inhabit the island developed a thriving culture that inventively used the local palm trees to meet virtually all of their material needs. The positive feedback loop of population growth, sustained by the trees, grew so fast that more and more people consumed all the trees. Eventually the accelerating destruction of trees was mirrored in the population of people: When the trees were gone, most of the people died.

Here are some other examples of positive feedback loops:

- *Industrialization*: Machines and technology lead to surplus wealth, which is invested in more machines and more technology to help create more wealth.
- *Greenhouse effect*: Water vapor in the atmosphere traps heat there, which raises the temperature of the oceans, which evaporates more water into the atmosphere, which raises the temperature of the atmosphere, and so on—unless other processes in the earth's complex climate control system counteract this form of runaway feedback.
- *Resource consumption*: Consumption of resources fuels industrialization and technology, which increase and consume still more resources.
- *Ponzi "investment" scheme*: Promises of high-return investments attract investors. But the investors' money is not invested; rather, it is used to pay high rates of

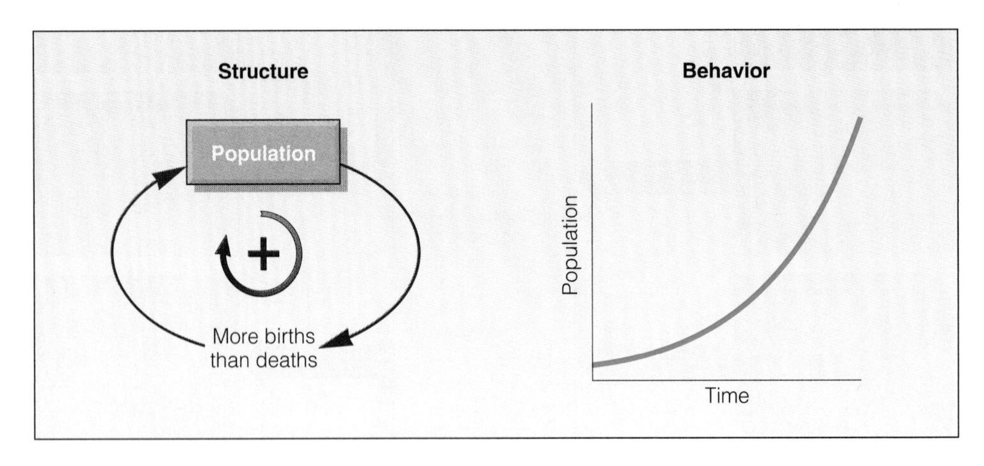

Figure 2-7 Structure and behavior of a simple positive feedback loop. As long as the number of human births exceeds human deaths, the human population will grow exponentially in a self-reinforcing positive feedback loop. Such growth will continue until the lack of one or more critical resources such as food, clean water, or clean air limits the number of people. Then population growth will level off or drop to a lower level through greatly reduced birth rates or a sharp rise in death rates (a dieback).

Q: How much of the world's land area has been set aside to protect wildlife?

interest, attracting more "investment." Later investors typically lose most or all of their investment, whereas the initiators of this scheme to fleece investors come away with huge profits.

Positive loops always have finite lives; neither exponential growth nor catastrophic collapse can continue forever in a finite world with finite resources.

What Are Negative Feedback Loops? In a **negative feedback loop**, one change leads to a lessening of that change. To survive, each of us must maintain our body temperature within a certain range, regardless of whether the temperature outside is steamy or freezing (Figure 2-8). This phenomenon is called **homeostasis**: the maintenance of favorable internal conditions despite fluctuations in external conditions. Homeostatic systems consist of one or more negative feedback loops that help maintain constant internal conditions when changes occur.

Here are other examples of negative feedback loops:

- *Engine of an automobile*: When the driver presses the accelerator to a new position and holds it there, more gas is fed to the engine, increasing its power. The acceleration from the engine accumulates and increases the speed of the car. The higher speed creates higher wind resistance, which reduces the net acceleration of the car. The car gently approaches the equilibrium speed dictated by the position of the gas pedal.
- *Driver of an automobile*: You decide you want to go at a specific speed (say, 55 mph), so you press on the accelerator until the car increases speed. As your speed approaches your goal, you let off the gas for less acceleration.
- *Pollution control*: The more emissions of a pollutant are eliminated, the less it bothers people, so its control can become less aggressive.

How Can Coupled Positive and Negative Loops Limit Growth and Maintain Stability? Most systems contain one or a series of *coupled positive and negative feedback loops*. Remember that a positive feedback loop leads to something that continues to increase. A negative (or corrective) feedback loop coupled to a positive feedback loop can dampen or even halt a positive feedback loop of runaway growth. Generally, at any given time, one of the two loops dominates the other. Which loop controls depends on the state of the system.

The temperature-regulating system of your body involves coupled negative and positive feedback loops. Normally a negative feedback regulates your body temperature (Figure 2-8). However, if your body temperature exceeds 42°C (108°F), your built-in negative feedback temperature control system breaks down as your body's metabolism (the heat-generating chemical reactions necessary for life) produces more heat than your sweat-dampened skin can get rid of. Then a

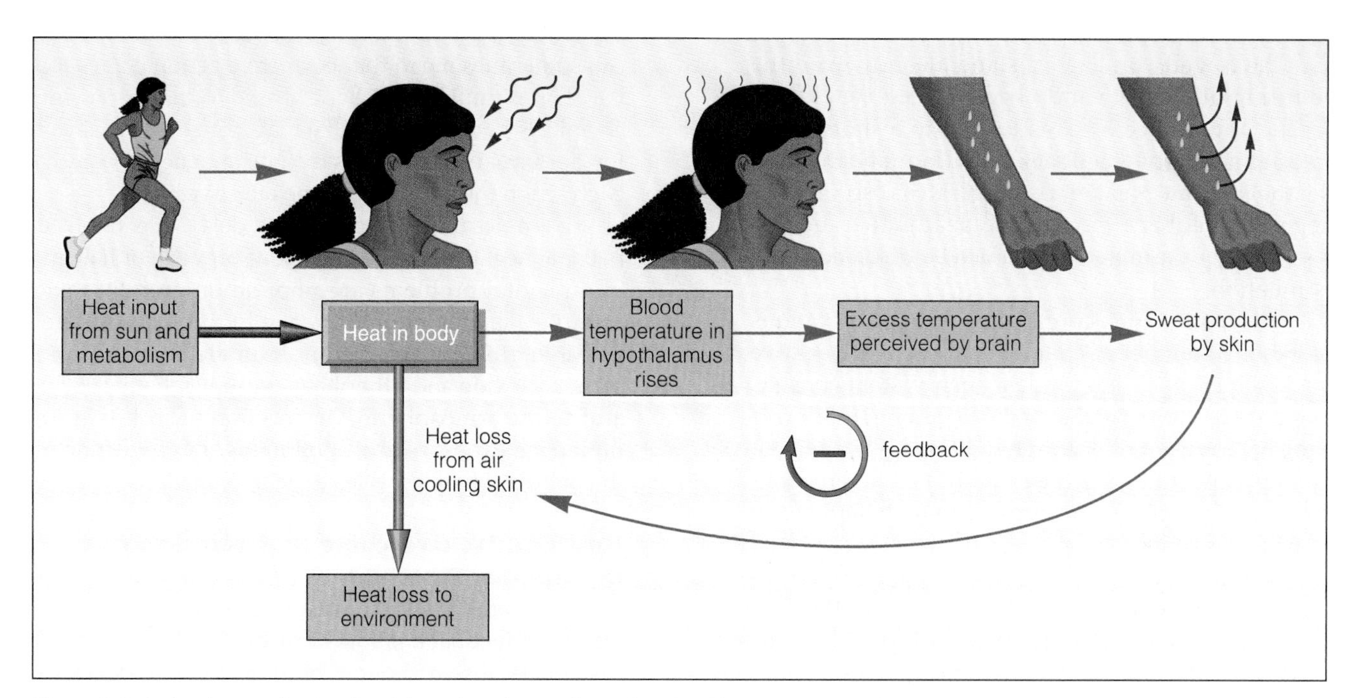

Figure 2-8 A simple negative feedback loop: keeping cool on a hot day. The temperature of the human body is controlled by a system that uses a negative feedback loop to counteract external and internal heat sources and maintain homeostasis. As the brain senses a rise in body temperature, sweating increases. As the sweat evaporates, it removes heat from the body. Skin sensors detect the cooling and feed this information back to the brain. This negative feedback loop then stops sweat production.

A: About 5%

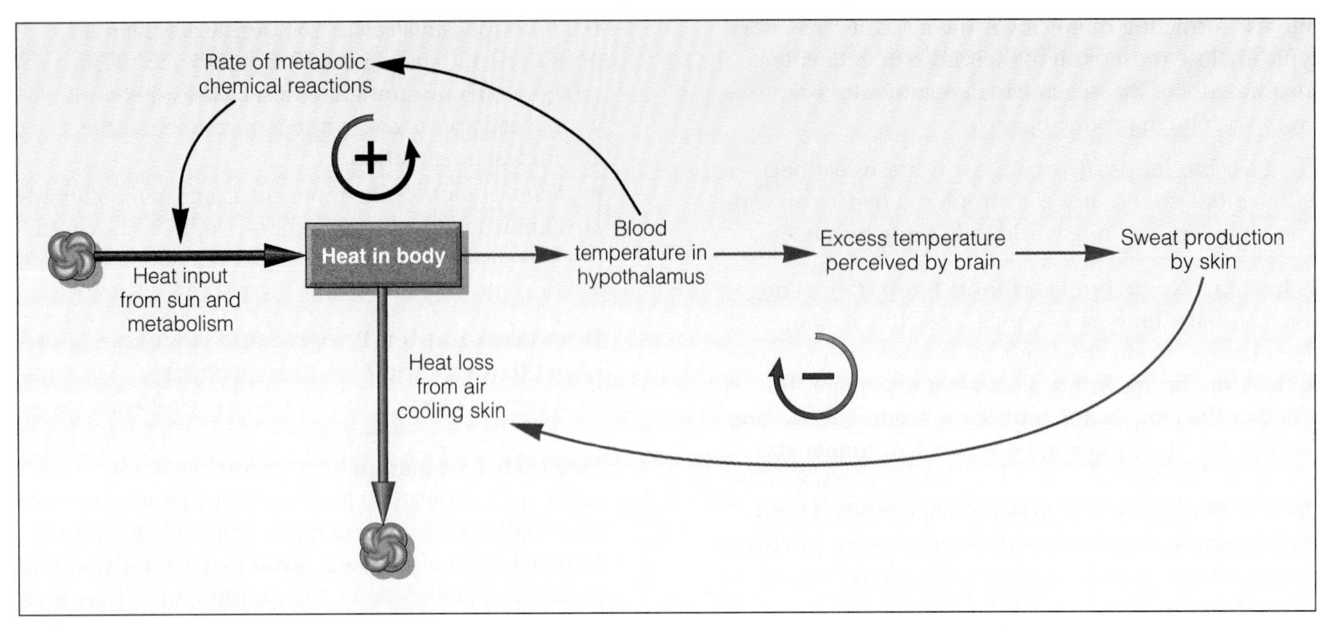

Figure 2-9 Coupled negative and positive feedback loops involved in temperature control of the human body. Homeostasis works in a limited range only; above a certain body temperature body metabolic rates get out of control and generate large amounts of heat. This positive (runaway) feedback loop generates more heat than the negative feedback loop can get rid of and body temperature increases out of control, resulting in death.

positive feedback loop caused by overloading the system (Figure 2-9) overwhelms the negative feedback loop. In practical terms, these conditions produce a net gain in body heat, which speeds up your metabolism even further, producing even more body heat, and so on, until death from heatstroke.

The tragedy on Easter Island also involved the coupling of positive and negative feedback loops. As the abundance of trees turned to a shortage of trees, the positive feedback loop (more births than deaths) became weaker as death rates rose, and the negative feedback loop (more deaths than births) eventually dominated and caused a dieback of the population.

All ecosystems contain homeostatic systems based on a number of coupled positive and negative feedback loops. The grandest such system consists of all life on the earth. Since the beginning of life, organisms have not only adapted to changing environmental conditions but also modified the environment in ways that have proved to be beneficial to some forms of life.

The idea that life on the earth helps sustain its own environment is a modified version of the *Gaia* (pronounced GUY-uh) *hypothesis*, proposed in the early 1970s by British chemist and inventor James Lovelock and American biologist Lynn Margulis. However, the Gaia hypothesis does not mean that the entire earth is alive or is somehow purposefully directing its own activities.

2-4 SOME BEHAVIORS OF COMPLEX SYSTEMS

How Do Time Delays Affect Complex Systems? Complex systems often show **time delays** between the input of a stimulus and the response to it. A long delay can sometimes mean that corrective action comes too late. For example, a smoker exposed to cancer-causing chemicals in cigarette smoke may not get lung cancer for 20–30 years; by then it's too late for a negative feedback action (not smoking) to be effective.

Examples in which prolonged delays are involved in using negative feedback to slow, prevent, or halt environmental problems are population growth, toxic dump leaks, depletion of ozone in the stratosphere, possible global warming and climate change (from carbon dioxide and other heat-trapping chemicals we add to the atmosphere), destruction of forests from prolonged exposure to air pollutants, and reduction of the population of a species to near extinction.

How Can We Overcome Resistance to Change in Complex Systems? Because of negative feedback, systems have a built-in resistance to change. Negative feedback is found throughout any natural system, because without it the system would be unstable and would have long ago disappeared. Therefore, when we attempt to change a complex system, even for its own benefit, we often encounter a frustrating resistance to change. Because of negative feedback or

 Amundson, Ronald, and Hans Jenny. 1997. "On a State Factor Model of Ecosystems." *BioScience*, vol. 47, no. 8, 536(8).

corrective feedback, biological, chemical, and physical components throughout the system can shift to absorb our efforts and cancel much of the attempted change.

Complex systems, such as economic and political systems, are also resistant to change. However, a careful analysis of such systems usually reveals a few components for which small changes (or incentives) can yield a large improvement. Such *leverage* often comes from making it easy and rewarding for people to do the right thing. For example, levying taxes on pollution emissions (Figure 2-10) gives business leaders incentives to find ways to reduce emissions (and thus taxes), either by inventing new technologies for removing pollutants (*pollution control*) or by redesigning the process to reduce or eliminate production of pollutants (*pollution prevention*). The cleaner environment then leads to healthier people, animals, and plants. It eventually reduces the cost of things such as health care, insurance, and cleaning. Thus, pollution taxes would eventually pay for themselves.

Taxing emissions of pollutants is a *polluter-pays* approach. Understandably, it is resisted by polluters because it means that they would have to bear the costs of pollution. Then they would have to raise the price of their products, unless they can come up with innovations for cleaning up or preventing pollution. The current *taxpayer- or consumer-pays* approach passes the costs of pollution on to taxpayers and people who buy products and to future generations. This means that they eventually pay for the costs of pollution in the form of poorer health, higher medical bills, and higher

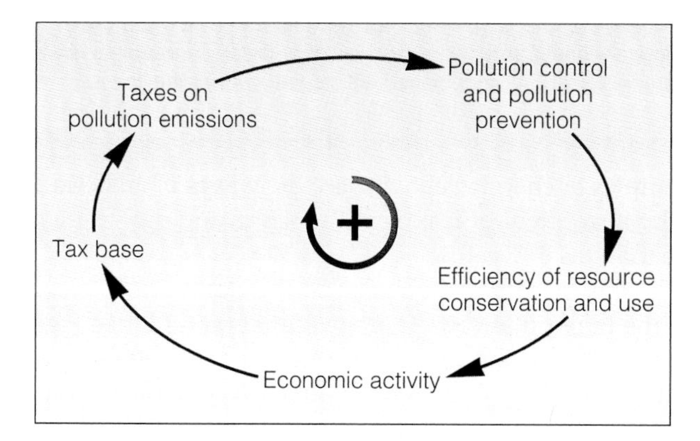

Figure 2-10 Leverage comes from activating positive feedback loops at crucial points. For example, taxes on polluting emissions can encourage environmental improvement in ways that produce a net decrease in pollution. This can lead to more efficient use of resources (less waste) and an increase in economic activity. Then the tax base grows, providing more tax revenue and allowing still more improvement. Taxing pollution can cause the prices of goods and services to increase, but at the same time decrease costs to consumers because of poor health and high insurance costs.

health insurance costs. However, these costs are hidden because they are not included in the market prices of things people buy.

Taxing pollutant emissions changes this system by incorporating most or all of the environmental costs of products into their market prices—something called *full-cost pricing*. Consumers would pay more for most things, but their overall costs in the form of higher health and insurance costs should decrease.

A major goal of environmental science is to discover the positive feedback loops and leverage points in complex natural, economic, and political systems that can amplify constructive action that helps sustain the earth's life support systems and thus our own species and economies.

What Is Synergy, and How Can It Be Used to Bring About Change? In arithmetic, 1 plus 1 always equals 2. But in nature, 1 plus 1 may add up to more than 2 because of synergistic interactions. A **synergistic interaction** occurs when two or more processes interact so that the combined effect is greater than the sum of their separate effects. **Synergy** can result when two people work together in accomplishing a task. For example, suppose you and I need to move a 140-kilogram (300-pound) tree that has fallen across the road. By ourselves, each of us can lift only, say, 45 kilograms (100 pounds). However, if we cooperate and use our muscles properly, together we can move the tree out of the way. That's using synergy to solve a problem.

Most political change is brought about by a small group of people working together (synergizing) and expanding their efforts. Research in the social sciences suggests that most political changes or changes in cultural beliefs are brought about by only about 5% (and rarely more than 10%) of a population working together in an organized way.

Synergy amplifies the action of positive feedback loops and thus can be an amplifier of change we believe is favorable. But synergy can also amplify harmful changes (Connections, p. 54).

By identifying potentially harmful synergistic interactions and the leverage points that can activate them, we can anticipate, counteract, and even prevent some environmental problems. Thus, we may be able to counter harmful synergisms, promote beneficial ones, and improve the quality of life on the earth.

How Does Chaos Limit Predictability? Systems are defined as having orderly and theoretically predictable behavior. However, increasing evidence indicates that the behavior of some systems appears to be random, chaotic, and unpredictable.

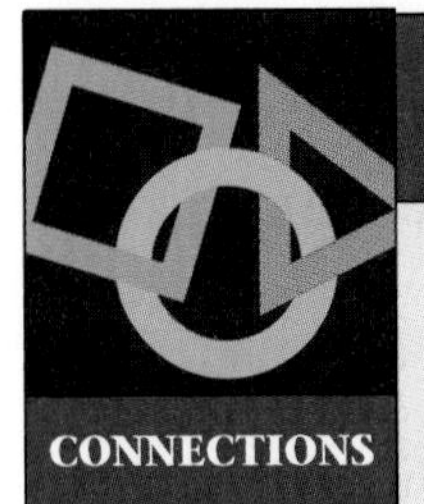

Some Harmful Synergistic Interactions

The consensus of atmospheric scientists is that human inputs of various chemicals into the atmosphere are causing depletion of ozone in the second layer of the atmosphere (the stratosphere), especially above Antarctica, and are projected to cause warming in the troposphere (the lowest layer of the earth's atmosphere). There is evidence that these two human-caused problems can interact synergistically to make matters worse.

As the troposphere warms, the stratosphere cools, increasing ice-cloud formation above the south pole. These ice clouds can increase stratospheric ozone depletion, providing solid surfaces on which ozone-depleting chemical reactions occur, leading to a loss of ozone or *ozone thinning* above Antarctica several months a year. Thus, projected global warming in the troposphere could worsen ozone depletion in the stratosphere.

To make matters worse, warming of the troposphere from burning fossil fuels and deforestation speeds up chemical reactions that lead to smog formation. A reduction in beneficial ozone in the stratosphere also lets more harmful ultraviolet radiation reach the earth's surface and produce more smog in the troposphere.

Thus, global warming in the troposphere and depletion of beneficial ozone in the stratosphere can interact synergistically to produce more harmful ozone in the troposphere. Furthermore, plants and animals weakened by any one of these threats are more vulnerable to disease, heat, and other stresses as a result of further synergistic interactions.

Critical Thinking

Pollution that forms tiny particles and droplets of certain chemicals in the troposphere can reflect some incoming solar radiation. This can lead to some cooling of the troposphere and thus help counteract the warming of the troposphere from other chemicals we inject into the atmosphere–an example of a corrective negative feedback loop. Should we increase the levels of harmful pollutants into the troposphere to help counteract projected global warming? Explain.

Sometimes the noisy or unpredictable behavior of systems comes from within the system itself; such a system is said to be generating **chaos**. A chaotic system follows a pattern that never repeats itself; its behavior generally maintains the same overall appearance, but the details seem to vary endlessly. Examples of chaotic behavior are the waves of an ocean, the movement of leaves in the wind, and day-to-day variations in weather.

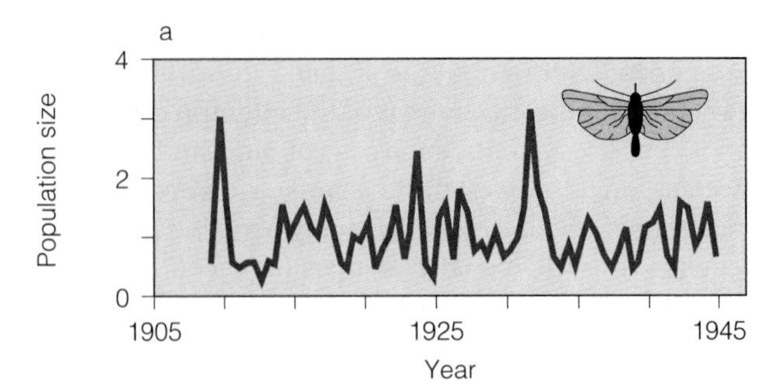

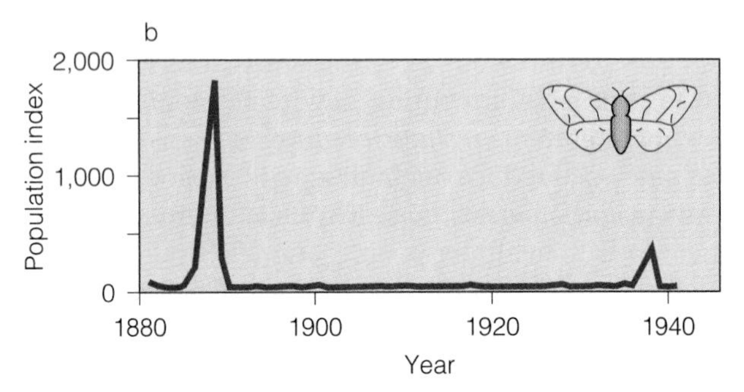

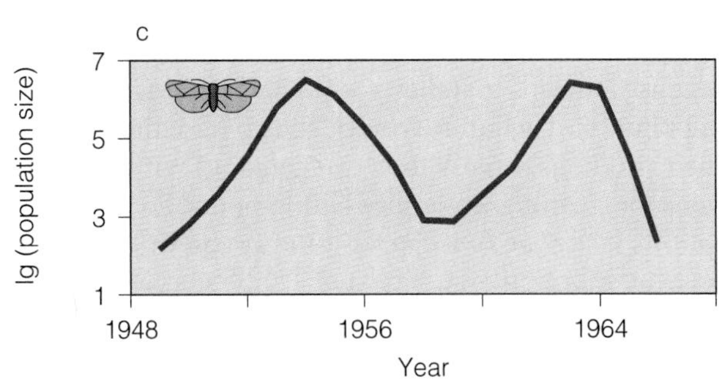

Figure 2-11 Population data for three moth species: **(a)** *Chilo suppressalis*, Japan, **(b)** large pine moth, *Dendrolimus pini*, Germany, and **(c)** larch budmoth, *Zeiraphera diniana*, Switzerland. These patterns are typical of dynamic systems influenced by noise or chaos. It is not known whether the effects shown here are caused by chaotic behavior or orderly and potentially predictable behaviors that scientists have not yet discovered.

Chaos can also be seen in the fluctuations over time in populations of many animals and plants. Figure 2-11 shows year-to-year fluctuations of the populations of three moth species. In each case, the overall shape of the curve is recognizably consistent, but the details vary from one cycle to the next. Some scientists attribute such behavior to chaos in such systems. Other scientists contend that the behavior may be due to orderly, nonchaotic behavior whose details and interactions are still poorly understood. In either case, we are unable to predict how such systems behave.

Chaotic systems can be extremely sensitive to even small disturbances. This means that if you could start the system over again and make a small change in some

Q: Over the last 50 years what has happened to the total area of forest and woodlands in the United States?

variable, the small change would quickly grow into a large unpredictable change in overall behavior. This is sometimes called the *butterfly effect*, alluding to the possibility that a single butterfly flapping its wings can cause minute changes in the movement of air that can eventually initiate a cascade of highly unpredictable changes in an entire forest, or even the entire planet. The practical consequence of this chaotic effect is that even if you have a nearly perfect model of a system, the tiniest error in measurement or the smallest error in approximation may cause the predicted future behavior of the system to be quite different from its actual behavior.

Scientific knowledge is a body of statements of varying degrees of certainty—some most unsure, some nearly sure, but none absolutely certain.

RICHARD FEYNMAN

CRITICAL THINKING

1. Do you think cheating, in the form of faking or doctoring the results of experiments, is more or less likely to occur in science than in other areas of knowledge? Explain.

2. To what extent are scientists responsible for the applications of knowledge they discover? Should scientists abandon research because of its possible harmful uses? Explain.

3. Respond to the following statements:

a. It has never been scientifically proven that anyone has ever died from smoking cigarettes.

b. The greenhouse theory—that certain gases (such as water vapor and carbon dioxide) trap heat in the atmosphere and thus influence the average temperature and climate of the earth—is not a reliable idea because it is only a scientific theory.

4. In a courtroom, a lawyer asks an expert witness who is a scientist, "Can you state that emissions of sulfur dioxide produced by burning sulfur-containing coal and oil contribute without any doubt to human respiratory ailments?" What conclusion might you draw if the witness answers "Yes"? Explain. What conclusion might you draw if the witness answers "No"? Explain.

5. Design a controlled experiment to test each of the following hypotheses:

a. Exposure to ultraviolet radiation from the sun can cause skin cancer in humans.

b. Carbon dioxide gas traps heat in the atmosphere.

c. Cutting down forests without replacing them increases the amount of carbon dioxide entering the atmosphere because trees remove carbon dioxide from the atmosphere.

d. Chlorofluorocarbons (CFCs) released into the atmosphere are destroying some of the ozone in the stratosphere.

In each case, explain the limitations or hidden assumptions built into each proposed experiment that might limit the validity of your findings. Do these limitations disprove the hypotheses? Explain. How could you improve the experiments?

6. See whether you can find an advertisement or an article describing or using some aspect of science in which **(a)** the concept of scientific proof is misused by implying that some scientific model or theory is not valid because it has not been proven absolutely, **(b)** the term *theory* is used when it should have been *hypothesis*, and **(c)** a consensus scientific finding is dismissed or downplayed because it is "only a theory."

7. **(a)** Give four examples of negative feedback loops not discussed in this chapter. Give two that are beneficial and two that are detrimental. Compare your examples with those of other members of your class. **(b)** Give four examples of positive feedback loops not discussed in this chapter. Include two that are beneficial and two that are detrimental. Compare your examples with those of other members of your class.

8. Think of four examples of actions that help at first but have undesirable consequences later on. Think of four actions that hurt at first but end up as net beneficial changes.

9. Try to come up with analogies to complete the following statements: **(a)** A scientific theory is like _______. **(b)** Frontier science is like _______. **(c)** Negative feedback is like _______. **(d)** Being in a state of homeostasis is like _______. **(e)** Using synergy to solve a problem is like _______.

10. How does a scientific law (such as the law of conservation of matter) differ from a societal law (such as one imposing maximum speed limits for vehicles)? Can each be broken?

PROJECTS

1. As a class, identify a change that would improve the quality of the environment on your campus. Work together to come up with a plan for bringing about this change. Cooperate (synergize) to bring about the proposed change, and use feedback to modify your plan as needed to accomplish your goal.

2. Develop the words and images for a public service advertisement urging what we should do about a specific environmental problem. If you have access to a video camera, develop a 30-second public service commercial about what we should do about a particular environmental problem. In developing such a commercial, narrow your focus to one problem and one message. Compare your print ads or video commercials with ones developed by other members of your class and critique one another's work.

3. Make a concept map of this chapter's major ideas using the section heads and subheads and the key terms (in boldface). Look at the inside of the back cover and in Appendix 4 for information about concept maps.

A: It has remained about the same

3 MATTER AND ENERGY RESOURCES: TYPES AND CONCEPTS

Saving Energy, Money, and Jobs in Osage, Iowa

Osage, Iowa (population about 4,000), has developed into the energy-efficiency capital of the United States. Its transformation began in 1974 when Wes Birdsall, general manager of Osage Municipal Gas and Electric Company, started urging the townspeople to save energy and reduce their natural gas and electric bills. The utility would also save money by not having to add new electrical generating facilities.

Birdsall started his crusade by telling homeowners about the importance of insulating walls and ceilings and plugging leaky windows and doors. These repairs provided jobs and income for people selling and installing insulation, caulking, and energy-efficient windows. He also advised people to replace their incandescent light bulbs with more efficient fluorescent bulbs and to turn down the temperature on water heaters and wrap them with insulation—economic boons to the local hardware and lighting stores. The utility company even gave away free water heater blankets. Birdsall also suggested saving water and fuel by installing low-flow shower heads.

Birdsall stepped up his efforts by offering to give every building in town a free thermogram—an infrared scan that shows where heat escapes (Figure 3-1). When people could see the energy (and their money) hemorrhaging out of their buildings, they took action

Figure 3-1 An infrared photo showing heat loss (red, white, and yellow colors) around the windows, doors, roofs, and foundations of houses and stores in Plymouth, Michigan. Wes Birdsall provided similar thermograms of houses in Osage, Iowa. The average house in the United States has heat leaks and air infiltration equivalent to leaving a window wide open during the heating season. Because of poor design, most office buildings and houses in this country waste about half the energy used to heat and cool them. Americans pay about $300 billion a year for this wasted heat—more than the entire annual military budget. (VANSCANr Continuous Mobile Thermogram by Daedalus Enterprises, Inc.)

to plug the leaks, again helping the local economy. Birdsall then stepped up his campaign even more, announcing that no new houses could be hooked up to the company's natural gas line unless they met minimum energy-efficiency standards.

Since 1974 the town has reduced its natural gas consumption by 45%, no mean feat in a locale with frigid winter temperatures. In addition, the utility company saved enough money to prepay all its debt, accumulate a cash surplus, and cut inflation-adjusted electricity rates by a third (which attracted two new factories to the area). Furthermore, each household saves more than $1,000 per year; this money supports jobs, and most of it circulates in the local economy. Before the town's energy-efficiency revolution, about $1.2 million a year left town to buy energy. The town's reduced fossil-fuel use also eases local and regional air pollution and the threat of global warming.

Osage's success in making energy efficiency a way of life earned the town a National Environmental Achievement Award in 1991. What are your local utility companies and community doing to save energy and stimulate the local economy?

The laws of thermodynamics control the rise and fall of political systems, the freedom or bondage of nations, the movements of commerce and industry, the origins of wealth and poverty, and the general physical welfare of the human race.

FREDERICK SODDY (NOBEL LAUREATE, CHEMISTRY)

This chapter examines what is going on in the world from a physical and chemical standpoint. Chapters 4 through 10 examine how key physical and chemical processes are integrated into the biological systems we call life. This chapter will answer the following questions:

- What are the basic forms of matter? What is matter made of? What makes matter useful to us as a resource?
- What are the major forms of energy? What makes energy useful to us as a resource?
- What are physical and chemical changes? What scientific law governs changes of matter from one physical or chemical form to another?
- What three main types of nuclear changes can matter undergo?
- What two scientific laws govern changes of energy from one form to another?
- How are the scientific laws governing changes of matter and energy from one form to another related to resource use and environmental disruption?

3-1 MATTER: FORMS, STRUCTURE, AND QUALITY

What Are Nature's Building Blocks? **Matter** is anything that has mass (the amount of material in an object) and takes up space. For example, this book has a certain amount of material—its mass—and occupies a certain volume.

Matter includes the solids, liquids, and gases around us and within us. Matter is found in two *chemical forms*: **elements** (the distinctive building blocks of matter that make up every material substance) and **compounds** (two or more different elements held together in fixed proportions by attractive forces called *chemical bonds*). Various elements, compounds, or both can be found together in **mixtures**.

All matter is built from the 112 known chemical elements. (Ninety-two of them occur naturally and the other 20 have been synthesized in laboratories.) Each has properties that make it unique, just as each of the 26 letters in the alphabet is different from the others. To simplify things, chemists represent each element by a one- or two-letter symbol; hydrogen (H), carbon (C), oxygen (O), nitrogen (N), phosphorus (P), sulfur (S), chlorine (Cl), fluorine (F), bromine (Br), sodium (Na), calcium (Ca), lead (Pb), mercury (Hg), and uranium (U) are but a few.

If we had a supermicroscope capable of looking at individual elements and compounds, we could see that they are made up of three types of building blocks: **atoms** (the smallest unit of matter that is unique to a particular element), **ions** (electrically charged atoms or combinations of atoms), and **molecules** (combinations of two or more atoms of the same or different elements held together by chemical bonds). Because ions and molecules are formed from atoms, *atoms are the ultimate building blocks for all matter*.

Some elements are found in nature as molecules. Examples are nitrogen and oxygen, which together make up about 99% of the volume of the air we breathe. Two atoms of nitrogen (N) combine to form a nitrogen gas molecule, with the shorthand formula N_2 (read as "N-two"). The subscript after the element's symbol indicates the number of atoms of that element in a molecule. Similarly, most of the oxygen gas in the atmosphere exists as O_2 (read as "O-two") molecules. A small amount of oxygen, found mostly in the second layer of the atmosphere (stratosphere), exists as O_3 (read as "O-three") molecules; this gaseous form of oxygen is called *ozone*.

Elements can combine to form an almost limitless number of compounds, just as the letters of the alphabet can be combined to form almost a million English words. So far, chemists have identified more than 10 million compounds.

Matter is also found in three *physical states*: solid, liquid, and gas. For example, water exists as ice, liquid water, and water vapor, depending on its temperature and pressure. The three physical states of matter differ in the spacing and orderliness of its atoms, ions, or molecules, with solids having the most compact and orderly arrangement and gases the least compact and orderly arrangement (Figure 3-2).

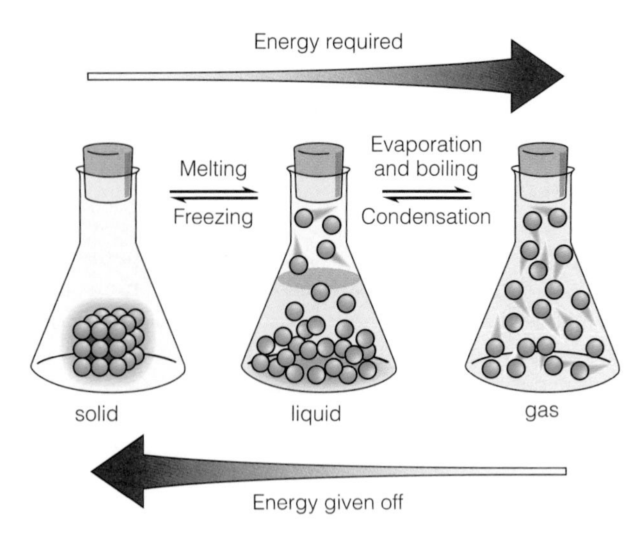

Figure 3-2 Comparison of the solid, liquid, and gaseous physical states of matter.

What Are Atoms and Ions? If we increased the magnification of our supermicroscope, we would find that each different type of atom contains a certain number of *subatomic particles*. The main building blocks of an atom are positively charged **protons** (p), uncharged **neutrons** (n), and negatively charged **electrons** (e). Protons and neutrons are the heaviest particles, with electrons having an almost negligible mass compared to them. Other subatomic particles have been identified in recent years, but they need not concern us here.

Each atom has an extremely small center, or **nucleus**, containing protons and neutrons. Almost all of an atom's mass comes from its protons and neutrons, which are tightly packed inside its tiny nucleus. Because protons are positively charged and neutrons have no charge, the nucleus carries a positive charge equal to its number of protons.

Surrounding the positively charged nucleus is a region called an *electron cloud*. It consists of one or more negatively charged electrons moving rapidly through a large volume of space (compared to the volume of the nucleus). We can describe the locations of electrons only in terms of their probability of being at any given place outside the nucleus. This is analogous to saying that a certain number of tiny gnats are found somewhere in a cloud, without being able to identify their exact positions.

Atoms are mostly free space. You can get a crude, greatly magnified model of a hydrogen atom (with one proton in its nucleus and one electron outside the nucleus) by imagining a spherical cloud with a diameter of about 0.5 kilometer (0.3 mile). Hidden from view at the center of this gigantic cloud is a grape (representing the nucleus) with a single gnat (representing the atom's single electron) moving around at extremely high speeds somewhere inside the cloud. Because of the atoms' incredibly small size, we can touch or see only pieces of matter containing unimaginably large numbers of atoms, so we experience them as if they

Q: How much of the municipal solid waste produced in the United States is recycled?

were not mostly free space. For example, more than 3 million hydrogen atoms could sit side by side on the period at the end of this sentence.

Each atom has an equal number of positively charged protons (inside its nucleus) and negatively charged electrons (outside its nucleus). Because these electrical charges cancel one another, *the atom as a whole has no net electrical charge.*

Each element has its own specific **atomic number**, equal to the number of protons in the nucleus of each of its atoms. The simplest element, hydrogen (H), has only 1 proton in its nucleus, so its atomic number is 1. Carbon (C), with 6 protons, has an atomic number of 6, whereas uranium (U), a much larger atom, has 92 protons and an atomic number of 92.

Because atoms are electrically neutral, the atomic number of an atom tells us the number of positively charged protons in its nucleus and the equal number of negatively charged electrons outside its nucleus. For example, an uncharged hydrogen atom with an atomic number of 1 has 1 positively charged proton in its nucleus and 1 negatively charged electron outside its nucleus and thus no overall electrical charge. Similarly, each atom of uranium with an atomic number of 92 has 92 protons in its nucleus and 92 electrons outside, and thus no net electrical charge. How many protons and electrons are there in an atom of sodium (Na) with an atomic number of 11? In an atom of chlorine (Cl) with an atomic number of 17?

Because electrons have so little mass compared with the mass of a proton or a neutron, *most of an atom's mass is concentrated in its nucleus.* We describe the mass of an atom in terms of its **mass number**: the total number of neutrons and protons in its nucleus. A hydrogen atom with 1 proton and no neutrons in its nucleus has a mass number of 1, and an atom of uranium with 92 protons and 143 neutrons in its nucleus has a mass number of 235.

Although all atoms of an element have the same number of protons in their nuclei, they may have different numbers of uncharged neutrons in their nuclei, and thus may have different mass numbers. Various forms of an element having the same atomic number but a different mass number are called **isotopes** of that element. Isotopes are identified by attaching their mass numbers to the name or symbol of the element. For example, hydrogen has three isotopes: hydrogen-1 (H-1), hydrogen-2 (H-2, common name *deuterium*), and hydrogen-3 (H-3, common name *tritium*). A natural sample of an element contains a mixture of its isotopes in a fixed proportion or percentage abundance by weight (Figure 3-3).

Atoms of some elements can lose or gain one or more electrons to form *ions*: atoms or groups of atoms with one or more net positive (+) or negative (–) electrical charges. For example, an atom of sodium (Na) (atomic number 11) with 11 positively charged protons and 11 negatively charged electrons can lose one of its electrons. It then becomes a sodium ion with a positive charge of 1 (Na^+) because it now has 11 positive charges (protons) but only 10 negative charges (electrons). An atom of chlorine (Cl) (with an atomic number of 17) can gain an electron and become a chlorine ion with a negative charge of 1 (Cl^-) because it has 17 positively charged protons and 18 negatively charged electrons.

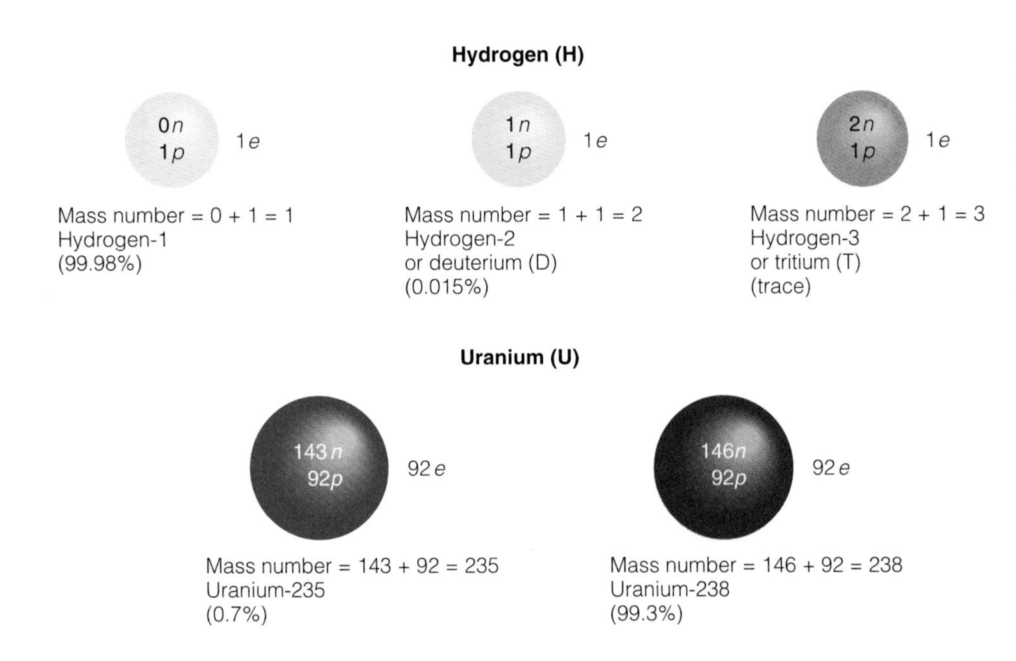

Figure 3-3 Isotopes of hydrogen and uranium. All isotopes of hydrogen have an atomic number of 1 because each has one proton in its nucleus; similarly, all uranium isotopes have an atomic number of 92. However, each isotope of these elements has a different mass number because its nucleus contains a different number of neutrons. Figures in parentheses indicate the percentage abundance by weight of each isotope in a natural sample of the element.

A: About 27% (up from 7% in 1970)

The number of positive or negative charges on an ion is shown as a superscript after the symbol for an atom or a group of atoms. Examples of other positive ions are calcium ions (Ca^{2+}) and ammonium ions (NH_4^+); other common negative ions are nitrate ions (NO_3^-), sulfate ions (SO_4^{2-}), and phosphate ions (PO_4^{3-}).

How Can Elements Be Arranged in the Periodic Table According to Their Chemical Properties? Chemists have developed a way to classify the elements according to their chemical behavior, in what is called the *periodic table of elements* (Figure 3-4). Each of the horizontal rows in the table is called a *period*. Each vertical column lists elements with similar chemical properties and is called a *group*.

The partial periodic table in Figure 3-4 shows how the elements can be classified as *metals, nonmetals*, and *metalloids*. Most of the elements found to the left and at the bottom of the table are *metals*, which usually conduct electricity and heat and are shiny. Examples are sodium (Na), calcium (Ca), aluminum (Al), iron (Fe), lead (Pb), and mercury (Hg). Atoms of such metals achieve a more stable state by losing one or more of their electrons to form positively charged ions such as Na^+, Ca^{2+}, and Al^{3+}.

Nonmetals, found in the upper right of the table, do not conduct electricity very well and are usually not shiny. Examples are hydrogen (H)*, carbon (C), nitrogen (N), oxygen (O), phosphorus (P), sulfur (S), chlorine (Cl), and fluorine (F). Atoms of some nonmetals such as chlorine, oxygen, and sulfur tend to gain one or more electrons lost by metallic atoms to form negatively charged ions such as O^{2-}, S^{2-}, and Cl^-. Atoms of nonmetals can also combine with one another to form molecules in which they share one or more pairs of their electrons.

The elements arranged in a diagonal staircase pattern between the metals and nonmetals have a mixture of metallic and nonmetallic properties and are called *metalloids*. Figure 3-4 also identifies the elements required as *nutrients* for all or some forms of life and elements that are moderately or highly toxic to all or most forms of life. Six nonmetallic elements—carbon (C) oxygen (O), hydrogen (H), nitrogen (N), sulfur (S), and phosphorus (P)—make up about 99% of the atoms of all living things.

What Holds the Atoms and Ions in Compounds Together? Most matter exists as compounds. Chemists use a shorthand **chemical formula** to

*Hydrogen, a nonmetal, is placed by itself above the center of the table because it does not fit very well into any of the groups.

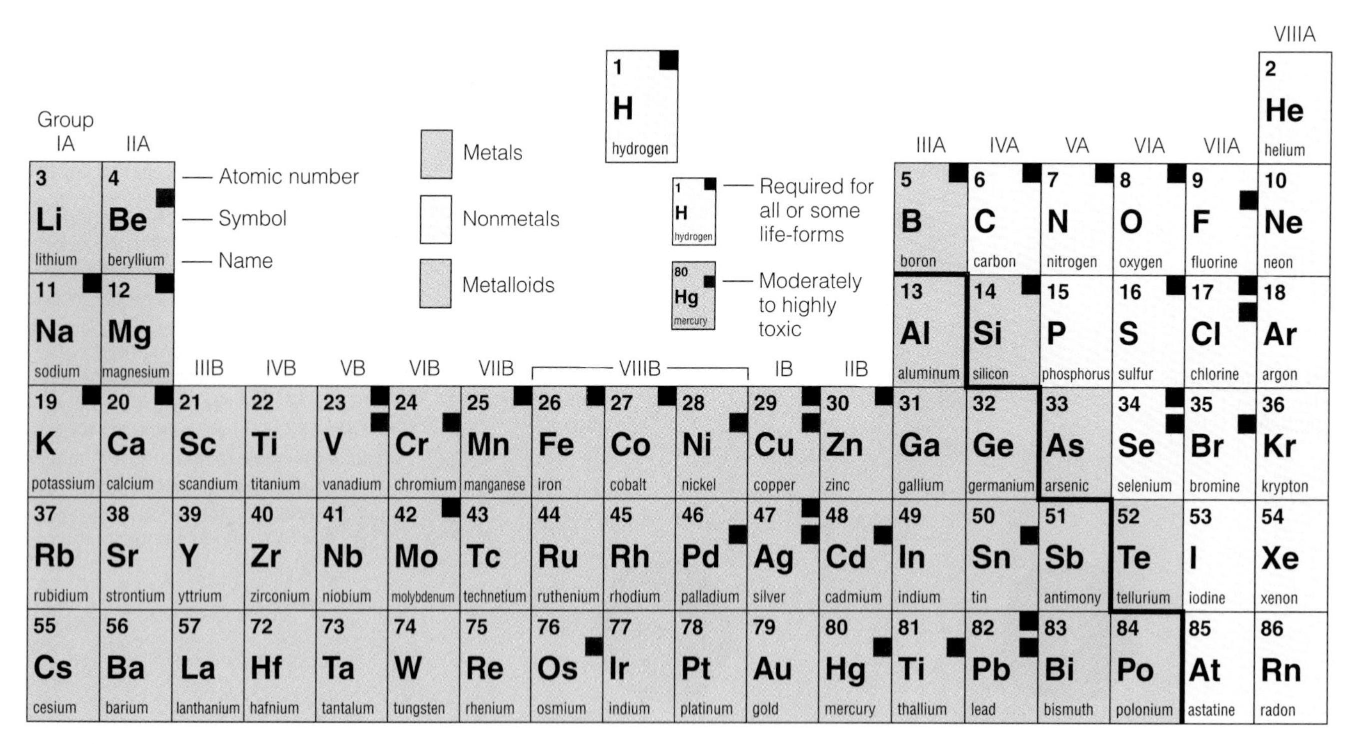

Figure 3-4 Periodic table of elements. Elements in the same vertical column, called a group, have similar chemical properties. To simplify matters at this introductory level, only 72 of the 112 known elements are shown.

Q: What percentage of the municipal solid waste in the United States could be recycled?

show the number of atoms (or ions) of each type in a compound. The formula contains the symbols for each of the elements present and uses subscripts to represent the number of atoms or ions of each element in the compound's basic structural unit. Compounds made up of oppositely charged ions are called *ionic compounds*, and those made up of molecules of uncharged atoms are called *covalent* or *molecular compounds*.

Sodium chloride (table salt), an ionic compound, is represented as NaCl. It consists of a network of oppositely charged *ions* (Na^+ and Cl^-) held together by the forces of attraction between opposite charges. The strong forces of attraction between such oppositely charged ions are called *ionic bonds*. Ionic compounds, such as sodium chloride, tend to exist at room temperature and pressure as solids consisting of a highly ordered, three-dimensional array of opposite-charge ions (Figure 3-5). Because ionic compounds consist of ions formed from atoms of metallic (positive ions) and nonmetallic (negative ions) elements, they can be described as *metal–nonmetal compounds*.

Water, a *covalent* or *molecular compound*, consists of molecules made up of uncharged atoms of hydrogen (H) and oxygen (O). Each water molecule consists of two hydrogen atoms chemically bonded to an oxygen atom, yielding H_2O (read as "H-two-O") molecules. Figure 3-6 shows the chemical formulas and shapes of the molecules for several common covalent compounds, formed when atoms of one or more nonmetallic elements (Figure 3-4) combine with one another. The bonds between the atoms in such molecules are called *covalent bonds* and are formed when the atoms in the molecule share one or more pairs of their electrons. Because they are formed from atoms of nonmetallic elements, molecular or covalent compounds can be described as *nonmetal–nonmetal compounds*. Most molecular or covalent compounds are gases or liquids at typical room temperatures and normal atmospheric pressure.

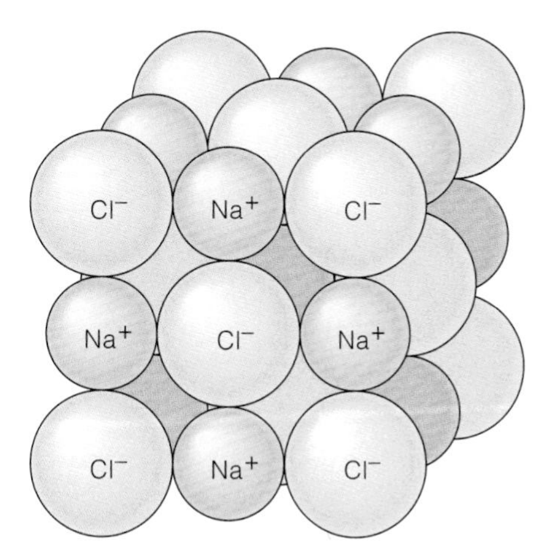

Figure 3-5 A solid crystal of an ionic compound such as sodium chloride consists of a three-dimensional array of opposite-charged ions held together by ionic bonds resulting from the strong forces of attraction between opposite electrical charges. They are formed when an electron is transferred from a metallic atom such as sodium (Na) to a nonmetallic element such as chlorine (Cl). Such compounds tend to exist as solids at normal room temperature and atmospheric pressure.

Ionic and covalent bonds form between the ions or atoms *within* a compound. There are also weaker forces of attraction *between* the molecules of covalent compounds (such as water) when the electrons shared by two atoms are shared unequally. For example, an oxygen atom has a much greater attraction for electrons than does a hydrogen atom. Thus, in a water molecule the electrons shared between the oxygen atom and its two hydrogen atoms are pulled closer to the oxygen atom, but not actually transferred to the oxygen atom. As a result, the oxygen atom in a water molecule has a slightly negative partial charge and its two hydrogen atoms have a slightly positive partial charge. The slightly positive hydrogen atoms in one water molecule are then attracted to the slightly negative oxygen atoms in other water molecules. These forces of attraction *between* water molecules are called *hydrogen bonds* (Figure 3-7). Hydrogen bonds also form between other covalent molecules or portions of such molecules containing hydrogen and nonmetallic atoms with a strong ability to attract electrons.

What Are Organic Compounds? Table sugar, vitamins, plastics, aspirin, penicillin, and many other important materials have one thing in common: They are *organic compounds*, containing carbon atoms combined with each other and with atoms of one or more other elements such as hydrogen, oxygen, nitrogen, sulfur, phosphorus, chlorine, and fluorine. Virtually all organic compounds are molecular compounds held together by covalent bonds. Organic compounds can be either natural or synthetic (such as plastics and many drugs made by humans).

Among the millions of known organic (carbon-based) compounds are

- *Hydrocarbons*: compounds of carbon and hydrogen atoms. An example is methane (CH_4, Figure 3-6), the main component of natural gas.
- *Chlorinated hydrocarbons*: compounds of carbon, hydrogen, and chlorine atoms. Examples are DDT ($C_{14}H_9Cl_5$, an insecticide) and toxic PCBs (such as

A: At least 60%

Figure 3-6 Chemical formulas and shapes for some molecular compounds formed when atoms of one or more nonmetallic elements combine with one another. The bonds between the atoms in such molecules are called covalent bonds. Molecular compounds tend to exist as gases or liquids at normal room temperature and atmospheric pressure.

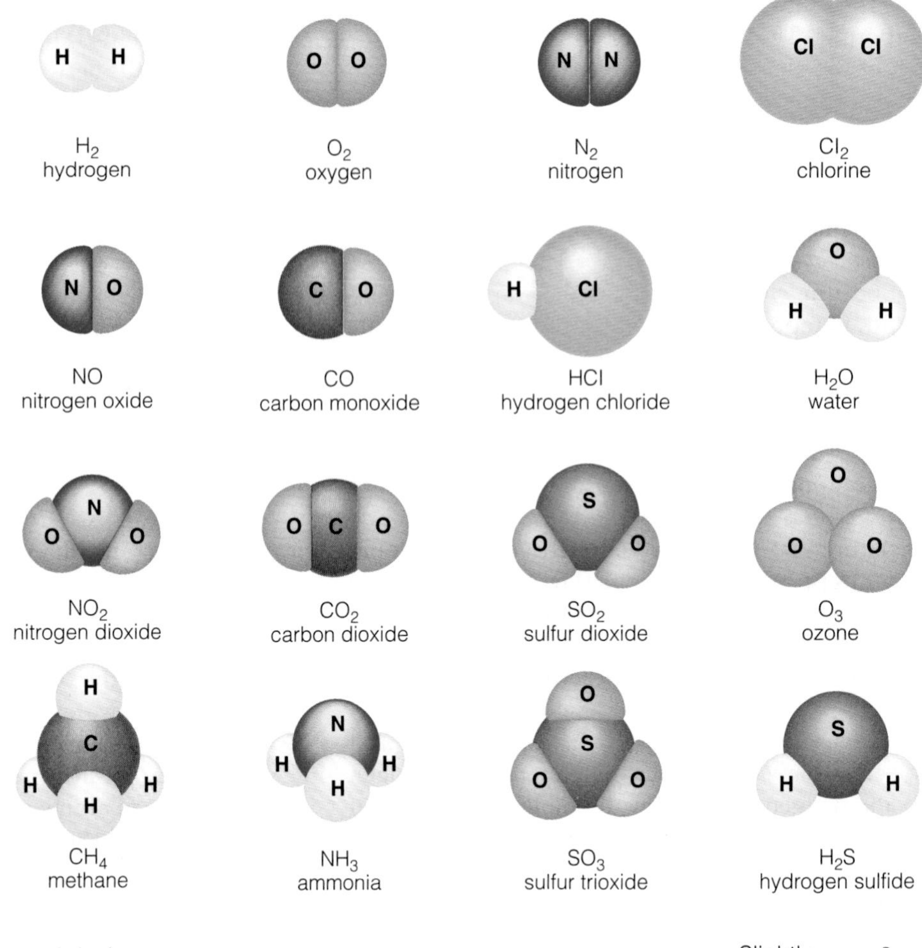

Figure 3-7 Hydrogen bonds. Slightly unequal sharing of electrons in the water molecule creates a molecule with a slightly negatively charged end and a slightly positively charged end. Because of this polarity, hydrogen atoms of one water molecule are attracted to oxygen atoms of another water molecule. These forces of attraction between water molecules are called *hydrogen bonds*.

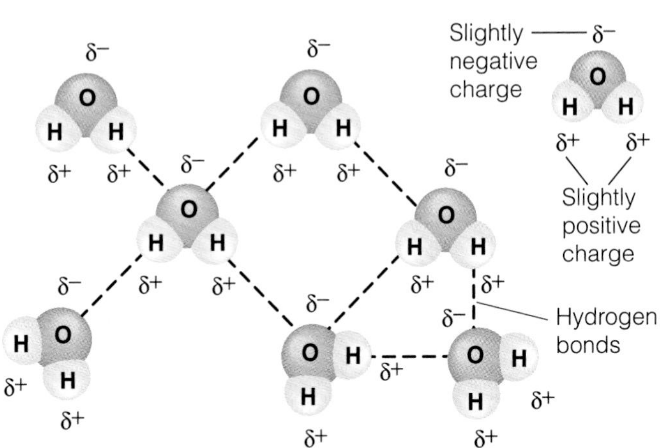

$C_{12}H_5Cl_5$), oily compounds used as insulating materials in electric transformers.

- *Chlorofluorocarbons* (CFCs): compounds of carbon, chlorine, and fluorine atoms. An example is Freon-12 (CCl_2F_2), until recently widely used as a coolant in refrigerators and air conditioners, as an aerosol propellant, and as a foaming agent for making some plastics.
- *Simple carbohydrates* (simple sugars): certain types of compounds of carbon, hydrogen, and oxygen atoms. An example is glucose ($C_6H_{12}O_6$), which most plants and animals break down in their cells to obtain energy.

Larger and more complex organic compounds, called *polymers*, consist of a number of basic structural or molecular units (*monomers*) linked by chemical bonds, somewhat like cars linked in a freight train. The three major types of organic polymers are complex carbohydrates, proteins, and nucleic acids.

Complex carbohydrates are made by linking a number of simple carbohydrate molecules such as glucose

Q: What percentage of the tested rivers and lakes in the United States are fishable and swimmable?

($C_6H_{12}O_6$). Examples are the complex starches in rice and potatoes and cellulose found in the walls around plant cells. Simple and complex carbohydrates are broken down in cells to supply energy.

Proteins are produced in cells by linking different sequences of about 20 different monomers known as *alpha-amino acids,** whose number and sequence in each protein are specified by the genetic code found in DNA molecules in an organism's cells. Most animals, including humans, can make about 10 of these alpha-amino acids in their cells. Sufficient quantities of the other 10, known as *essential alpha-amino acids,* must be obtained from food to prevent protein deficiency diseases. Various protein molecules important for cell structure and energy storage act as *enzymes* to control the rate at which all chemical reactions in a cell (cellular metabolism) take place. From only about 20 alpha-amino acid molecules, earth's life-forms can make tens of millions of different protein molecules.

Nucleic acids are made by linking hundreds to thousands of five different types of monomers, called *nucleotides.* Each nucleotide consists of a phosphate group, a sugar molecule containing five carbon atoms (deoxyribose in DNA molecules and ribose in RNA molecules), and one of four different nucleotide bases (represented by *A, G, C,* and *T,* the first letters in each of their names) (Figure 3-8). In the cells of living organisms, these nucleotide units combine in different numbers and sequences to form nucleic acids such as various types of DNA and RNA.

In each DNA molecule, two strands of nucleotides are held together like a spiral staircase, forming a double helix (Figure 3-9). The two strands are held together by hydrogen bonds between the nucleotide bases (A, G, C, and T) in one strand and those in the other strand. DNA molecules can unwind and replicate themselves. They contain the hereditary instructions for assembling new cells and for assembling the proteins each cell needs to survive and reproduce. Each sequence of three nucleotides in a DNA molecule carries the code or instruction for making a specific alpha-amino acid; various alpha-amino acids are then linked in cells to form specific protein molecules. RNA molecules carry instructions (provided by DNA molecules) for producing proteins within cells.

Genes consist of specific sequences of nucleotides in a DNA molecule. Each gene carries the codes (each consisting of three nucleotides) required to make various proteins. These coded units of genetic information about specific traits are passed on from parents to offspring during reproduction. There are approximately 75,000 genes in each human cell. Occasionally one or more of the nucleotide bases in a gene sequence are deleted, added, or replaced. Such changes, called **gene mutations**, can be helpful or harmful to an organism and its offspring, or not affect them at all.

Chromosomes are combinations of genes that make up a single DNA molecule, together with a number of proteins. A human cell contains 46 chromosomes (23 pairs), which together make up an individual's entire genetic endowment. Genetic information coded in your chromosomal DNA is what makes you different from an oak leaf, an alligator, or a flea, and from your parents as well. The relationships of genetic material to cells are depicted in Figure 3-10.

Another class of large biologically important molecules that are not polymers are called *lipids.* They include molecules of *fats* (such as those found in butter, lard, and bacon fat), *oils* (such as olive, corn, and coconut oils), *waxes* (such as earwax and beeswax), *phospholipids* (important components of cell membranes), and various *steroids* (including certain hormones, vitamins, and cholesterol). Lipids can serve as energy storage molecules, regulators of certain cellular functions (hormones), nutrients (certain vitamins), and waterproof coverings around cells.

*Most textbooks incorrectly refer to the 20 amino acids. In fact, there are virtually an unlimited number of amino acids. What is meant is the 20 particular alpha-amino acids specified by the genetic code and contained in DNA molecules in an organism's cells.

What Are Inorganic Compounds? All other compounds are called *inorganic compounds.* Some of the inorganic compounds discussed in this book are

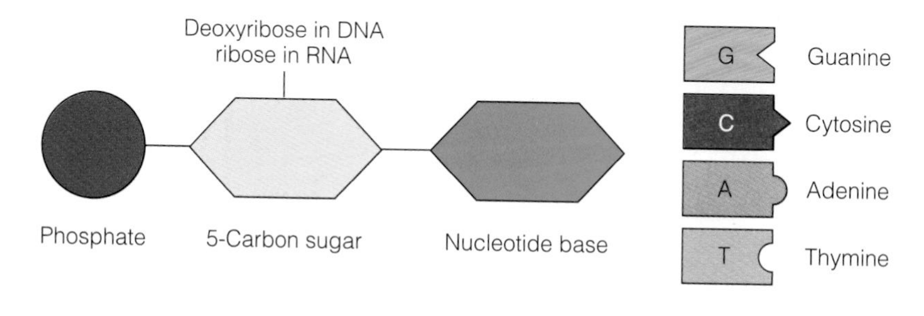

Figure 3-8 Generalized structure of nucleotide molecules linked in various numbers and sequences to form large nucleic acid molecules such as various types of DNA (deoxyribose nucleic acid) and RNA (ribose nucleic acid). In DNA the 5-carbon sugar in each nucleotide is deoxyribose; in RNA it is ribose. The four basic nucleotides used to make various forms of DNA molecules differ in the types of nucleotide bases they contain: guanine (G), cytosine (C), adenine (A), and thymine (T).

A: About 62% (compared to about 36% in 1972)

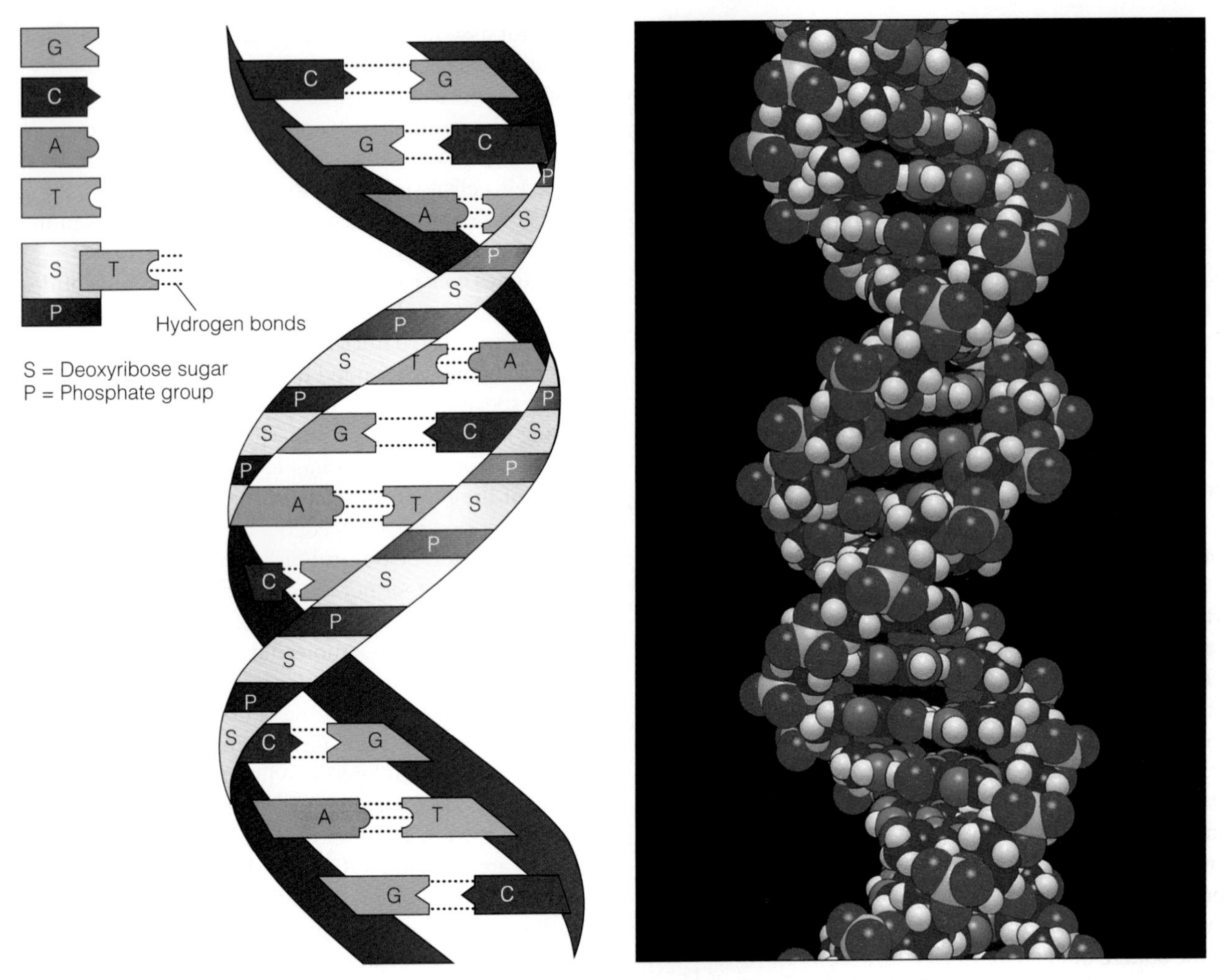

Figure 3-9 Portion of the double helix of a DNA molecule. The helix is composed of two spiral (helical) strands of nucleotides, each containing a unit of phosphate (P), deoxyribose (S), and one of four nucleotide bases: guanine (G), cytosine (C), adenine (A), and thymine (T). The two strands are held together by hydrogen bonds formed between various pairs of the nucleotide bases. Guanine (G) bonds with cytosine (C), and adenine (A) with thymine (T).

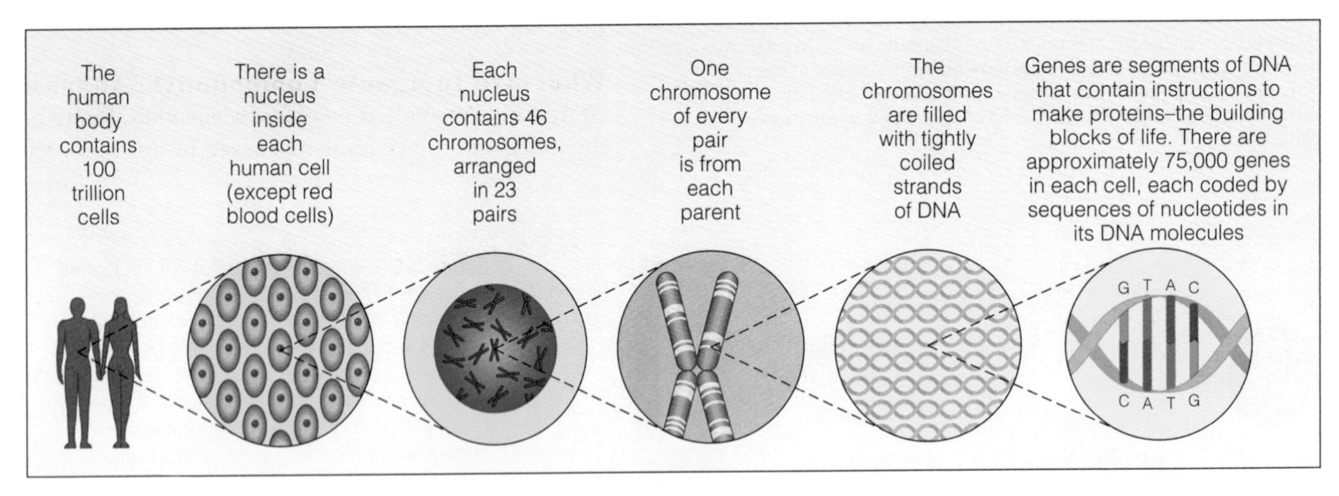

Figure 3-10 Relationships among cells, nuclei, chromosomes, DNA, and genes.

Q: What percentage of the wetlands in the United States have been lost or seriously degraded?

ionic compounds (Figure 3-5) such as sodium chloride (table salt, NaCl), sodium bicarbonate (baking soda, $NaHCO_3$), and sodium hydroxide (caustic soda, NaOH). Others are covalent compounds (Figure 3-6) such as hydrogen (H_2), nitrogen (N_2), oxygen (O_2), ozone (O_3), hydrogen chloride (HCl), water (H_2O), nitrous oxide (N_2O), nitric oxide (NO), carbon monoxide (CO), carbon dioxide (CO_2),* nitrogen dioxide (NO_2), sulfur dioxide (SO_2), ammonia (NH_3), hydrogen sulfide (H_2S), sulfuric acid (H_2SO_4), and nitric acid (HNO_3).

The earth's **crust** or outermost layer is composed mostly of inorganic minerals and rocks. A **mineral** is an element or an inorganic compound that occurs naturally and is solid. It usually has a crystalline internal structure made up of an orderly, three-dimensional arrangement of atoms or ions (Figure 3-5). Some minerals consist of a single element, such as gold, silver, diamond (carbon), or sulfur. However, most of the over 2,000 identified minerals occur as inorganic compounds formed by various combinations of the eight elements that make up 98.5% by weight of the earth's crust (Figure 3-11). Examples are salt, mica, and quartz.

Rock is any material that makes up a large, natural, continuous part of the earth's crust. Some kinds of rock, such as limestone (calcium carbonate, or $CaCO_3$) and quartzite (silicon dioxide, or SiO_2), contain only one mineral, but most rocks consist of two or more minerals. The crust is the source of virtually all the nonrenewable resources we use: fossil fuels, metallic minerals, and nonmetallic minerals (Figure 1-11). It is also the source of soil and of the elements that make up our bodies and those of other living organisms.

*Classifying compounds as organic or inorganic is somewhat arbitrary. All organic compounds contain one or more carbon atoms, but CO and CO_2 are classified as inorganic compounds.

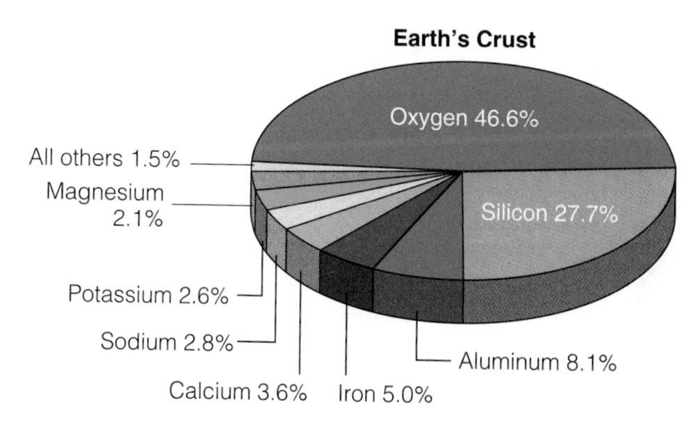

Figure 3-11 Composition by weight of the earth's crust. Various combinations of only eight elements make up the bulk of most minerals.

What Is Matter Quality? From a human standpoint, we can classify matter according to its quality or usefulness to us. **Matter quality** is a measure of how useful a matter resource is, based on its availability and concentration. **High-quality matter** is organized, concentrated, and usually found near the earth's surface, and has great potential for use as a matter resource; **low-quality matter** is disorganized, dilute, and often deep underground or dispersed in the ocean or the atmosphere, and usually has little potential for use as a matter resource (Figure 3-12).

An aluminum can is a more concentrated, higher-quality form of aluminum than aluminum ore containing the same amount of aluminum. That's why it takes less energy, water, and money to recycle

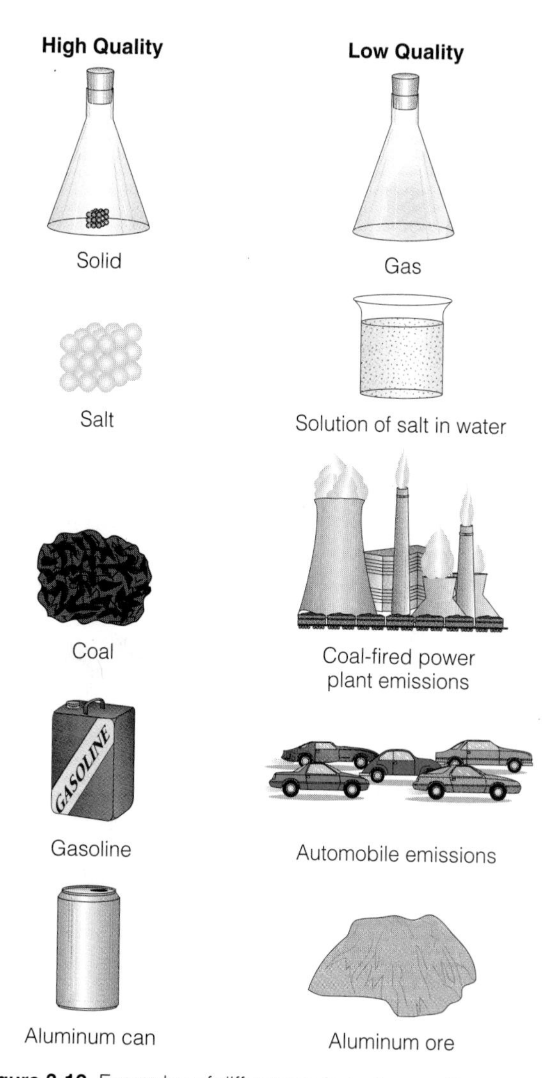

Figure 3-12 Examples of differences in matter quality. High-quality matter (left-hand column) is fairly easy to extract and is concentrated; low-quality matter (right-hand column) is more difficult to extract and is more dispersed than high-quality matter.

A: About 55%

an aluminum can than to make a new can from aluminum ore.

Entropy is a measure of the disorder or randomness of a system. The greater the disorder of a sample of matter, the higher its entropy; the greater its order, the lower its entropy. Thus, an aluminum can has a lower entropy (more order) than aluminum ore with the same amount of aluminum mixed with other materials. Similarly, a piece of ice in which the water molecules are held in an ordered solid structure has a lower entropy (more order) than the highly dispersed and rapidly moving water molecules in water vapor (Figure 3-2).

3-2 ENERGY: FORMS AND QUALITY

What Different Forms of Energy Do We Encounter? **Energy** is the capacity to do work and transfer heat. Work is performed when an object—be it a grain of sand, this book, or a giant boulder—is moved over some distance. Work, or matter movement, also is needed to boil water (to change it into the more dispersed and faster-moving water molecules in steam) or to burn natural gas to heat a house or cook food. Energy is also the heat that flows automatically from a hot object to a cold object when they come in contact. Touch a hot stove, and you experience this energy flow in a painful way.

Energy comes in many forms: light; heat; electricity; chemical energy stored in the chemical bonds in coal, sugar, and other materials; the mechanical energy of moving matter such as flowing water, wind (air masses), and joggers; and nuclear energy emitted from the nuclei of certain isotopes.

Scientists classify energy as either kinetic or potential. **Kinetic energy** is the energy that matter has because of its mass and its speed or velocity. It is energy in action or motion. Wind (a moving mass of air), flowing streams, falling rocks, heat flowing from a body at a high temperature to one at a lower temperature, electricity (flowing electrons), moving cars—all have kinetic energy.

Electromagnetic radiation is a form of kinetic energy consisting of a wide band or spectrum of electromagnetic waves that differ in wavelength (distance between successive peaks or troughs) and energy content (Figure 3-13). Examples are radio waves, TV waves, microwaves, infrared radiation, visible light, ultraviolet radiation, X rays, gamma rays, and cosmic rays. Infrared radiation is now being used to show us where energy is leaking out of our houses and buildings (Figure 3-1). Satellite scans use infrared radiation, ultraviolet radiation, and microwaves to show where plant life is found on land and at sea and to identify temperature changes in the upper layers of the oceans.

Cosmic rays, gamma rays, X rays, and ultraviolet radiation have enough energy to knock electrons from atoms and change them to positively charged ions. The resulting highly reactive electrons and ions can disrupt living cells, interfere with body processes, and cause many types of sickness, including various cancers. These potentially harmful forms of electro-

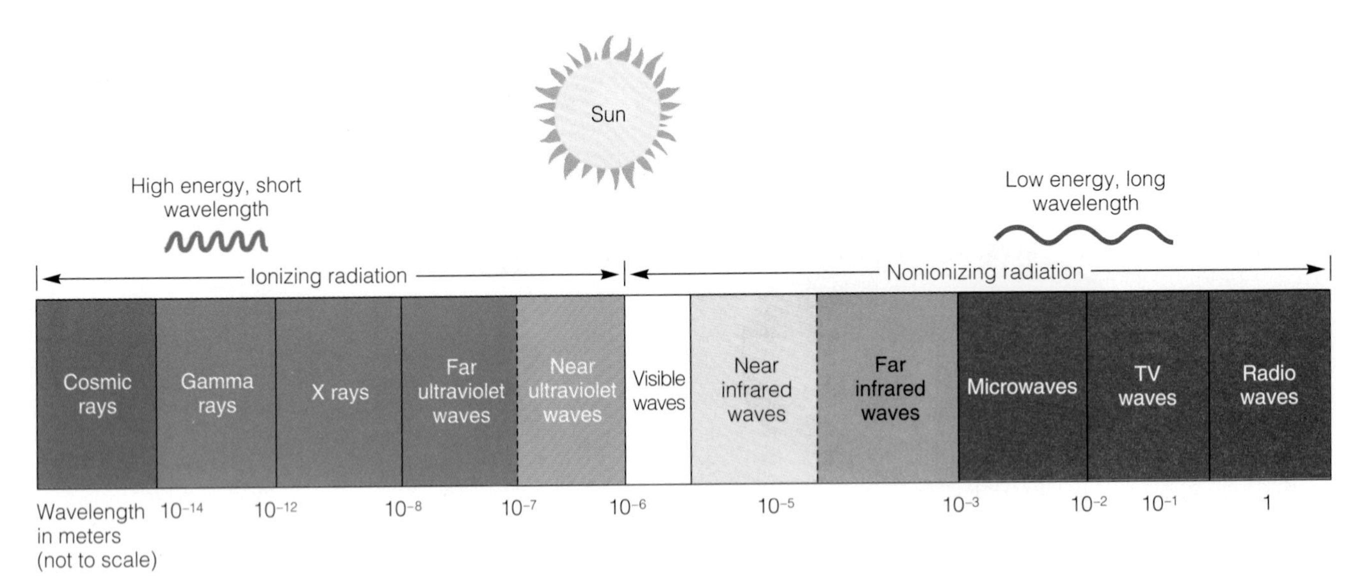

Figure 3-13 The electromagnetic spectrum: the range of electromagnetic waves, which differ in wavelength (distance between successive peaks or troughs) and energy content.

Q: What percentage of the world's species are found in tropical forests?

magnetic radiation are called **ionizing radiation**. The other forms of electromagnetic radiation do not contain enough energy to form ions and are called **nonionizing radiation**.

Heat is the total kinetic energy of all the moving atoms, ions, or molecules within a given substance, excluding the overall motion of the whole object. **Temperature** is a measure of the average speed of motion of the atoms, ions, or molecules in a sample of matter at a given moment. A substance can have a high heat content (much mass and many moving atoms, ions, or molecules) but a low temperature (low average molecular speed). For example, the total heat content of a lake is enormous, but its average temperature is low. Another substance can have a low heat content and a high temperature; a cup of hot coffee, for example, has a much lower heat content than a lake, but its temperature is much higher.

Potential energy is stored energy that is potentially available for use. A rock held in your hand, an unlit stick of dynamite, still water behind a dam, the gasoline in a car tank, and the nuclear energy stored in the nuclei of atoms all have potential energy because of their position or the position of their parts. Potential energy can be changed to kinetic energy. When you drop a rock, its potential energy changes into kinetic energy. When you burn gasoline in a car engine, the potential energy stored in the chemical bonds of its molecules changes into heat, light, and mechanical (kinetic) energy that propels the car.

What Is Energy Quality? From a human standpoint, the measure of an energy source's ability to do useful work is called its **energy quality** (Figure 3-14). **High-quality energy** is organized or concentrated and can perform much useful work. Examples are electricity, coal, gasoline, concentrated sunlight, nuclei of uranium-235 used as fuel in nuclear power plants, and heat concentrated in small amounts of matter so that its temperature is high.

By contrast, **low-quality energy** is disorganized or dispersed and has little ability to do useful work. An example is heat dispersed in the moving molecules of a large amount of matter (such as the atmosphere

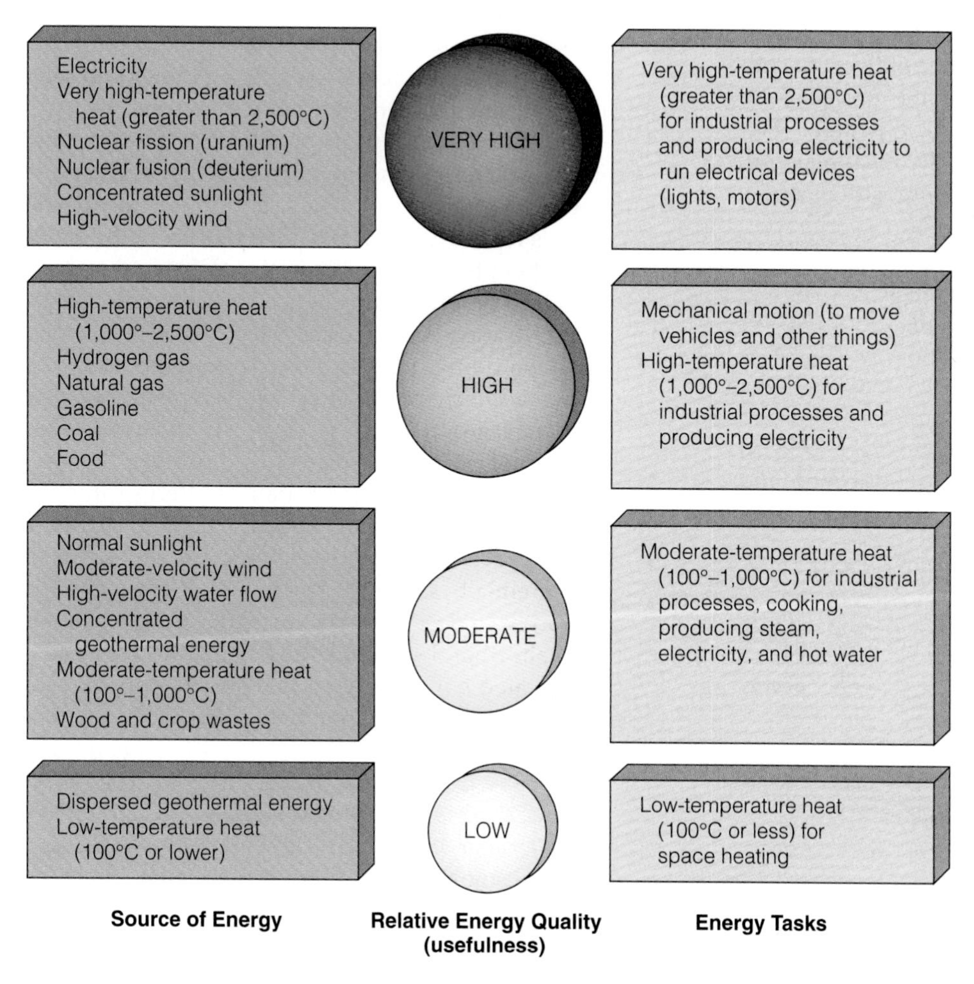

Figure 3-14 Categories of the quality (usefulness for performing various energy tasks) of different sources of energy. *High-quality energy* is concentrated and has great ability to perform useful work; *low-quality energy* is dispersed and has little ability to do useful work. To avoid unnecessary energy waste, it is best to match the quality of an energy source with the quality of energy needed to perform a task.

A: At least 50% (some biologists say 90%)

or a large body of water) so that its temperature is low. Thus, even though the total amount of heat stored in the Atlantic Ocean is greater than the amount of high-quality chemical energy stored in all the oil deposits of Saudi Arabia, the ocean's heat is so widely dispersed that it can't be used to move things or to heat things to high temperatures.

We use energy to accomplish certain tasks, each requiring a certain minimum energy quality. It makes sense to match the quality of an energy source with the quality of energy needed to perform a particular task (Figure 3-14) because doing so saves energy and usually money.

3-3 PHYSICAL AND CHEMICAL CHANGES AND THE LAW OF CONSERVATION OF MATTER

What Is the Difference Between a Physical and a Chemical Change? A **physical change** involves no change in chemical composition. Cutting a piece of aluminum foil into small pieces is one example. Changing a substance from one physical state to another is a second example. When solid water (ice) is melted or liquid water is boiled, none of the H_2O molecules involved are altered; instead, the molecules are organized in different spatial (physical) patterns (Figure 3-2). In a **chemical change** or **chemical reaction**, on the other hand, the chemical compositions of the elements or compounds are altered.

Chemists use shorthand chemical equations to represent what happens in a chemical reaction. A chemical equation shows the chemical formulas for the *reactants* (initial chemicals) and the *products* (chemicals produced), with an arrow placed between them. For example, when coal burns completely, the solid carbon (C) it contains combines with oxygen gas (O_2) from the atmosphere to form the gaseous compound carbon dioxide (CO_2):

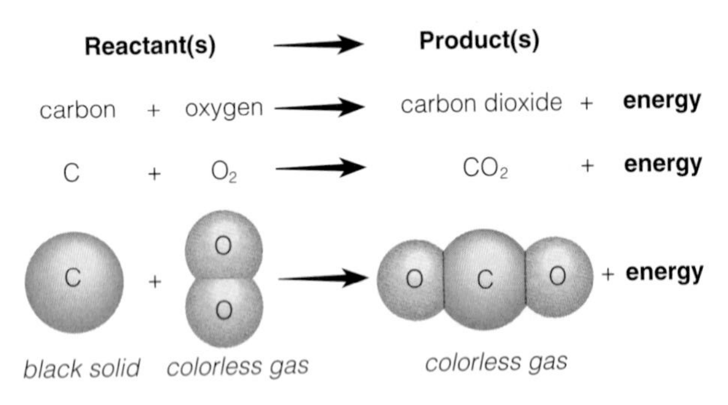

Energy is given off in this reaction, making coal a useful fuel. The reaction also shows how the complete burning of coal (or any of the carbon-containing compounds in wood, natural gas, oil, and gasoline) gives off carbon dioxide gas, which is a key gas that can lead to projected global warming of the lower atmosphere (troposphere).

Physical and chemical changes illustrate the connections between matter and energy: *All physical or chemical changes either require or give off energy*. For example, it takes energy to convert liquid water to steam. However, when the steam contacts a cooler object (such as your skin), it releases energy as it is converted back to liquid water, which is why steam can cause severe burns.

What Is the Law of Conservation of Matter? Why There Is No "Away" The earth loses some gaseous molecules to space, and it gains small amounts of matter from space, mostly in the form of occasional meteorites and cosmic dust. These losses and gains of matter are minute compared with the earth's total mass, which means that *the earth has essentially all the matter it will ever have*. In terms of matter, the earth is essentially a closed system (although in terms of energy it is an open system, receiving energy from the sun and emitting heat back into space). Over billions of years natural processes have evolved for continuously cycling key chemicals between living things and their nonliving environment (soil, air, and water).

People commonly talk about consuming or using up material resources, but the truth is that *we don't consume matter*—we only use some of the earth's resources for a while. We take materials from the earth, carry them to another part of the globe, and process them into products that are used and then discarded, burned, buried, reused, or recycled.

In so doing, *we may change various elements and compounds from one physical or chemical form to another, but in no physical and chemical change can we create or destroy any of the atoms involved*. All we can do is rearrange them into different spatial patterns (physical changes) or different combinations (chemical changes). The italicized statement, based on many thousands of measurements, is known as the **law of conservation of matter**. In describing chemical reactions, chemists use a shorthand system to make sure atoms are neither created nor destroyed, as required by the law of conservation of matter (Spotlight, right).

The law of conservation of matter means that there really is no "away" in "to throw away." *Everything we think we have thrown away is still here with us in one form or another*. We can collect dust and soot from the smokestacks of industrial plants, but these solid wastes must then be put somewhere. We can remove substances from polluted water at a sewage

Winters, Jeffrey. 1997. "Let There Be Matter." *Discover*, vol. 18, no. 12, 40(1).

Keeping Track of Atoms

SPOTLIGHT

In keeping with the law of conservation of matter, each side of a chemical equation must have the same number of *atoms* of each element involved. When this is the case, the equation is said to be *balanced.* The equation for the burning of carbon ($C + O_2 \rightarrow CO_2$) is balanced because there is one atom of carbon and two atoms of oxygen on both sides of the equation.

Now consider the following chemical reaction: When electricity is passed through water (H_2O), the latter can be broken down into hydrogen (H_2) and oxygen (O_2), as represented by the following equation:

H_2O	$\rightarrow$	H_2	+	O_2
2 H atoms 1 O atom		2 H atoms		2 O atoms

This equation is unbalanced because there is one atom of oxygen on the left but two atoms on the right.

We can't change the subscripts of any of the formulas to balance this equation because then we would be changing the arrangements of the atoms involved. Instead, we could use different numbers of the *molecules* involved to balance the equation. For example, we could use two water molecules:

$2\,H_2O$	$\rightarrow$	H_2	+	O_2
4 H atoms 2 O atom		2 H atoms		2 O atoms

This equation is still unbalanced because even though the numbers of oxygen atoms on both sides are now equal, the numbers of hydrogen atoms are not.

We can correct this by having the reaction produce two hydrogen molecules.

$2\,H_2O$	$\rightarrow$	$2\,H_2$	+	O_2
4 H atoms 2 O atoms		4 H atoms		2 O atoms

Now the equation is balanced, and the law of conservation of matter has not been violated. We see that for every two molecules of water through which we pass electricity, two hydrogen molecules and one oxygen molecule are produced.

See whether you can balance the chemical equation for the reaction of nitrogen gas (N_2) with hydrogen gas (H_2) to form ammonia gas (NH_3).

Critical Thinking

1. Balancing equations is based on the law of conservation of matter. Do you believe that this is an ironclad law of nature or one that through new scientific discoveries could be overthrown? Explain.
2. Imagine that you have the power to revoke the law of conservation of matter. List the three major ways this would affect your life.

treatment plant, but the gooey sludge must be burned (producing some air pollution), buried (possibly contaminating underground water supplies), or cleaned up and applied to the land as fertilizer (dangerous if the sludge contains nondegradable toxic metals such as lead and mercury). Banning use of the pesticide DDT in the United States but still selling it abroad means that it can return to the United States as DDT residues in imported coffee, fruit, and other foods, or as fallout from air masses moved long distances by winds—something environmentalists call the *circle of poison*.

Even though we can make the environment cleaner and convert some potentially harmful chemicals into less harmful physical or chemical forms, the law of conservation of matter means that we will always be faced with the problem of what to do with some quantity of wastes. By placing much greater emphasis on pollution prevention, waste reduction, and more efficient use of resources, we can greatly reduce the amount of wastes we add to the environment (Guest Essay, p. 70).

3-4 NUCLEAR CHANGES

What Is Natural Radioactivity? In addition to physical and chemical changes, matter can undergo a third type of change known as a **nuclear change**. This occurs when nuclei of certain isotopes spontaneously change or are made to change into one or more different isotopes. Three types of nuclear change are natural radioactive decay, nuclear fission, and nuclear fusion.

The law of conservation of matter does not apply to nuclear changes because they convert a small but measurable amount of the mass in a nucleus into energy. This type of change is governed by the **law of conservation of matter and energy**: In any nuclear change, the total amount of matter and energy involved remains the same.

Hint: Enter the search term *matter* using the Subject Guide and then select *Periodical References*.

A New Environmentalism for the 1990s and Beyond

GUEST ESSAY

Peter Montague

Peter Montague is director of the Environmental Research Foundation in Washington, D.C., which studies and informs the public about environmental problems and the technologies and policies that might help solve them. He has served as project administrator of a hazardous waste research program at Princeton University and has taught courses in environmental impact analysis at the University of New Mexico. He is the coauthor of two books on toxic heavy metals in the natural environment and is editor of Rachel's Environment and Health Weekly, *an informative and readable newsletter on environmental problems and solutions.*

Environmentalism as we have known it for almost three decades is dead. The environmentalism of the 1970s advocated strict numerical controls on releases into the environment of *dangerous wastes* (any unwanted or uncontrolled materials that can harm living things or disrupt ecosystems). But industry's ability to create new hazards quickly outstripped government's ability to establish adequate controls and enforcement programs.

After so many years of effort by government and by concerned citizens (the environmental movement), the overwhelming majority of dangerous chemicals is still not regulated in any way. Even those few that are covered by regulations have not been adequately controlled.

In short, the *pollution management* approach to environmental protection has failed and stands discredited; *pollution prevention* is our only hope. An ounce of prevention really is worth a pound of cure.

Here, in list form, is the situation facing environmentalists today:

- *All waste disposal—landfilling, incineration, deep-well injection—is polluting because "disposal" means dispersal into the environment.* Once wastes are created, they cannot be contained or controlled because of the scientific laws of matter and energy. The old environmentalism failed to recognize this important truth and thus squandered enormous resources trying to achieve the impossible. While we in the United States currently spend about $90 billion per year on pollution control, the global environment is increasingly threatened by a buildup of heat because of heat-trapping gases we emit into the atmosphere. At least half the surface of the planet is being subjected to damaging ultraviolet radiation from the sun as a result of ozone-depleting chemicals we have discharged into the atmosphere. Vast regions of the United States, Canada, and Europe are suffering from loss of forests, crop productivity, and fish as a result of a mixture of air pollutants, produced mostly by burning fossil fuels in power plants and automobiles. Soil and water are dangerously polluted at thousands of locales where municipal garbage and industrial wastes have been (and continue to be) dumped or incinerated; thousands of such sites remain to be discovered, according to U.S. government estimates.
- *The inevitable result of our reliance upon waste treatment and disposal systems has been an unrelenting buildup of toxic synthetic materials in humans and other forms of life worldwide.* For example, breast milk of women in industrialized countries is so contaminated with pesticides and industrial hydrocarbons that, if human milk were bottled and sold commercially, it could be banned by the Food

Natural radioactive decay is a nuclear change in which unstable isotopes spontaneously emit fast-moving particles, high-energy radiation, or both at a fixed rate. The unstable isotopes are called **radioactive isotopes** or **radioisotopes**. Radioactive decay into various isotopes continues until the original isotope is changed into a stable isotope that is not radioactive.

Radiation emitted by radioisotopes is damaging ionizing radiation. The most common form of ionizing energy released from radioisotopes is **gamma rays**, a form of high-energy electromagnetic radiation (Figure 3-13). High-speed ionizing particles emitted from the nuclei of radioactive isotopes are most commonly of two types: **alpha particles** (fast-moving, positively charged chunks of matter that consist of two protons and two neutrons) and **beta particles** (high-speed electrons). Figure 3-15 depicts the relative penetrating power of alpha, beta, and gamma ionizing radiation. All of us are exposed to small amounts of harmful ionizing radiation from both natural and human sources.

Each type of radioisotope spontaneously decays at a characteristic rate into a different isotope. This rate of decay can be expressed in terms of **half-life**: the time needed for *one-half* of the nuclei in a radioisotope to decay and emit their radiation to form a different isotope (Figure 3-16). The decay continues, often producing a series of different radioisotopes, until a nonradioactive isotope is formed. Each radioisotope has a characteristic half-life, which may range from a few millionths of a second to several billion years (Table 3-1).

Q: How many wild species are driven to premature extinction *each hour* by human activities?

and Drug Administration as unsafe for human consumption. If a whale today beaches itself on U.S. shores and dies, its body must be treated as a "hazardous waste" because whales contain concentrations of PCBs (polychlorinated biphenyl, a class of industrial toxins) legally defined as hazardous.

- *The ability of humans and other life forms to adapt to changes in their environment is strictly limited by the genetic code each form of life inherits.* Continued contamination at a rate hundreds of times faster than we can adapt will subject humans to increasingly widespread sickness and degradation of the species, and could ultimately lead to extinction.
- *Damage to humans (and to other life forms) is abundantly documented.* Birds, fish, and humans in industrialized countries are enduring steadily rising levels of cancer, genetic mutations, and damage to their nervous, immune, and hormonal systems as a result of pollution.

If we will but look, the handwriting is on the wall everywhere.

To deal with these problems, industrial societies must abandon their reliance upon waste treatment and disposal and upon the regulatory system of numerical standards created to manage the damage that results from relying on waste disposal instead of waste prevention. We must—relatively quickly—move the industrialized and industrializing countries to new technical approaches accompanied by new industrial goals—namely, "clean production" or zero discharge systems.

The concept of "clean production" involves industrial systems that avoid or eliminate dangerous wastes and dangerous products and minimize the use and waste of raw materials, water, and energy. Goods manufactured in a clean production process must not damage natural ecosystems throughout their entire life cycle, including **(1)** raw materials selection, extraction, and processing; **(2)** product conceptualization, design, manufacture, and assembly; **(3)** materials transport during all phases; **(4)** industrial and household usage; and **(5)** reintroduction of the product into industrial systems or into the environment when it no longer serves a useful function.

Clean production does not rely on "end-of-pipe" pollution controls such as filters or scrubbers or chemical, physical, or biological treatment. Measures that pretend to reduce the volume of waste by incineration or concentration, that mask the hazard by dilution, or that transfer pollutants from one environmental medium to another are also excluded from the concept of "clean production."

A new industrial pattern, and a new environmentalism, is thus emerging. It insists that the long-term well-being of humans and other species must be factored into our production and consumption plans. These new requirements are not optional; human survival and life quality depend upon our willingness to make, and pay for, the necessary changes.

Critical Thinking

1. Do you agree with the author that the *pollution management* approach to environmental protection practiced during the past 29 years has failed and must be replaced with a *pollution prevention* approach? Explain.

2. List key economic, health, consumption, and lifestyle changes you might experience as a consequence of putting much greater emphasis on pollution prevention. What changes might the next generation face?

An isotope's half-life cannot be changed by temperature, pressure, chemical reactions, or any other known factor. Half-life can be used to estimate how long a sample of a radioisotope must be stored in a safe container before it decays to what is considered a safe level. A general rule is that such decay takes about 10 half-lives. Thus, people must be protected from radioactive waste containing iodine-131 (which concentrates in the thyroid gland) for 80 days (10×8 days). In contrast, plutonium-239 (which is produced in nuclear reactors and used as the explosive in some nuclear weapons) can cause lung cancer when its particles are inhaled in minute amounts; it must be stored safely for 240,000 years ($10 \times 24{,}000$ years)—about four times longer than the latest version of our species has existed.

What Are Some Useful Applications of Radioisotopes? Scientists can use radioisotopes to estimate the age of ancient rocks, bones, and fossils. One common method, called *radiocarbon dating*, uses radioactive carbon-14 to estimate the age of plants, wood, teeth, bone, fossils, and other carbon-containing substances from dead plants and animals.

Radioisotopes are also used as *tracers* in pollution detection, agriculture, and industry. Suppose that a leak occurs somewhere in an underground pipeline carrying water or some other chemical. One way to locate the leak (without digging up the pipeline until the leak is found) is to mix a harmless and fairly rapidly decaying radioisotope with the material being transported in the pipeline. Then a radiation detector can be used to scan the ground above the pipeline until a

A: 2 to 8 and the rate is expected to increase over the next few decades

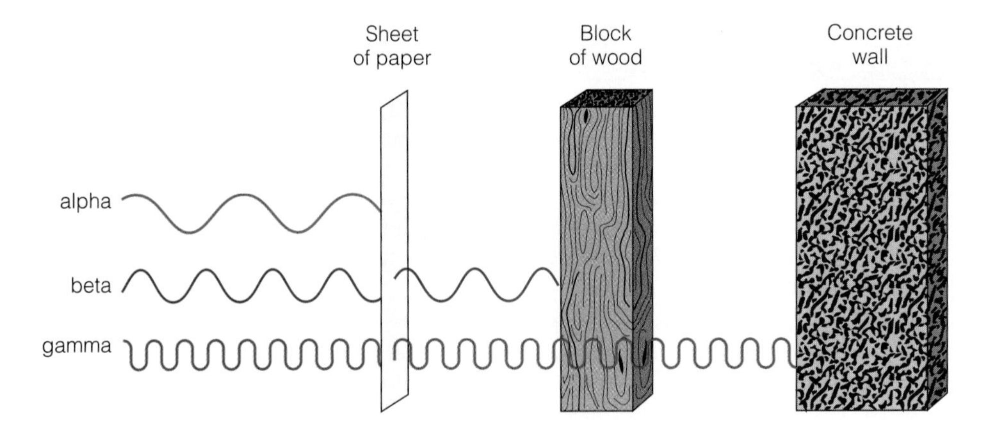

Figure 3-15 The three principal types of ionizing radiation emitted by radioactive isotopes differ considerably in their penetrating power.

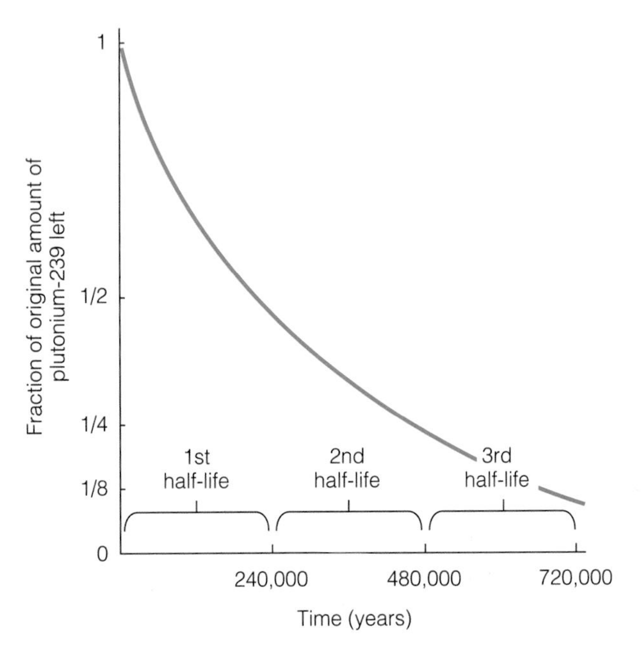

Figure 3-16 The radioactive decay of plutonium-239, which is produced in nuclear reactors and used as the explosive in some nuclear weapons, has a half-life of 240,000 years. The amount of radioactivity emitted by a radioactive isotope decreases by one-half for each half-life that passes. Thus, after three half-lives, amounting to 720,000 years, one-eighth of a sample of plutonium-239 would still be radioactive.

high level of radioactivity is measured (where the pipe is leaking).

The U.S. Department of Agriculture has used radioisotopes to expose millions of screwworm flies (an insect pest that can kill cattle) in laboratories to produce sterile male flies. Once released in large numbers into the natural population, these flies can outnumber the fertile males. Matings then produce no offspring and help reduce the size of the pest population.

A special branch of medicine, called *nuclear medicine*, uses radioisotopes for diagnosis and treatment of various diseases. For example, radioactive sodium-24 can be injected in a salt solution into the bloodstream. Then its radioactivity can be measured to detect constrictions or blockages in the blood vessels. Excessive exposure to radioactivity can cause cancer, but it is also useful in treating cancer by using high doses of radioactivity from radioisotopes (such as cobalt-60 or cesium-137) to kill rapidly growing cancer cells. Because the radioactivity also kills noncancerous cells, people exposed to radioactivity to kill cancer cells often have nausea, diarrhea, hair loss, and a low white blood cell count (which reduces their ability to fight infections).

What Is Nuclear Fission? Splitting Nuclei

Nuclear fission is a nuclear change in which nuclei of certain isotopes with large mass numbers (such as uranium-235) are split apart into lighter nuclei when struck by neutrons; each fission releases two or three more neutrons and energy (Figure 3-17). Each of these neutrons, in turn, can cause an additional fission. For these multiple fissions to take place, enough fissionable nuclei must be present to provide the **critical mass** needed for efficient capture of these neutrons.

Multiple fissions within a critical mass form a **chain reaction**, which releases an enormous amount of energy (Figure 3-18). Living cells can be damaged by the ionizing radiation released by the radioactive lighter nuclei and by high-speed neutrons produced by nuclear fission.

In an atomic bomb, an enormous amount of energy is released in a fraction of a second in an uncontrolled nuclear fission chain reaction. This reaction is initiated by an explosive charge, which suddenly

Q: What percentage of the world's land has been partially or totally modified by human activities?

Table 3-1 Half-Lives of Selected Radioisotopes

Isotope	Half-Life	Radiation Emitted
Potassium-42	12.4 hours	Alpha, beta
Iodine-131	8 days	Beta, gamma
Cobalt-60	5.27 years	Beta, gamma
Hydrogen-3 (tritium)	12.5 years	Beta
Strontium-90	28 years	Beta
Carbon-14	5,370 years	Beta
Plutonium-239	24,000 years	Alpha, gamma
Uranium-235	710 million years	Alpha, gamma
Uranium-238	4.5 billion years	Alpha, gamma

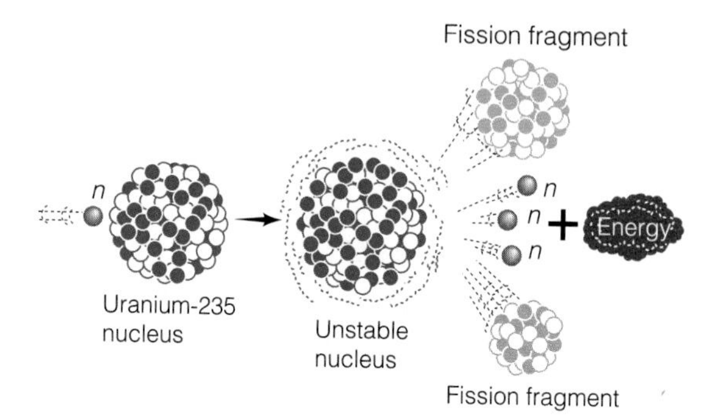

Figure 3-17 Fission of a uranium-235 nucleus by a neutron (*n*).

pushes two masses of fissionable fuel together, causing the fuel to reach the critical mass needed for a chain reaction.

In the reactor of a nuclear power plant, the rate at which the nuclear fission chain reaction takes place is controlled so that under normal operation only one of every two or three neutrons released is used to split another nucleus. In conventional nuclear fission reactors, the splitting of uranium-235 nuclei releases heat, which produces high-pressure steam to spin turbines and thus generates electricity.

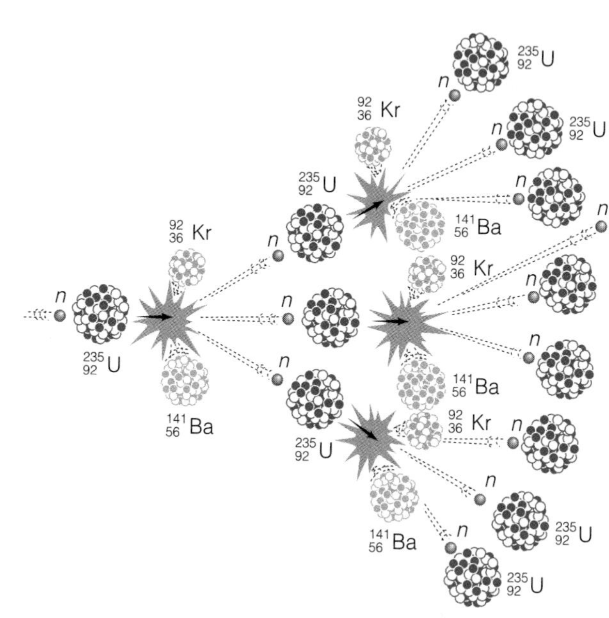

Figure 3-18 A nuclear chain reaction initiated by one neutron triggering fission in a single uranium-235 nucleus. This figure illustrates only a few of the trillions of fissions caused when a single uranium-235 nucleus is split within a critical mass of uranium-235 nuclei. The elements krypton (Kr) and barium (Ba), shown here as fission fragments, are only two of many possibilities.

What Is Nuclear Fusion? Forcing Nuclei to Combine **Nuclear fusion** is a nuclear change in which two isotopes of light elements, such as hydrogen, are forced together at extremely high temperatures until they fuse to form a heavier nucleus, releasing energy in the process. Temperatures of at least 100 million °C are needed to force the positively charged nuclei (which strongly repel one another) to fuse.

Nuclear fusion is much more difficult to initiate than nuclear fission, but once started it releases far more energy per unit of fuel than does fission. Fusion of hydrogen nuclei to form helium nuclei is the source of energy in the sun and other stars.

After World War II, the principle of *uncontrolled nuclear fusion* was used to develop extremely powerful hydrogen, or thermonuclear, weapons. These weapons use the D–T fusion reaction in which a hydrogen-2, or deuterium (D), nucleus and a hydrogen-3 (tritium, T) nucleus are fused to form a larger, helium-4 nucleus, a neutron, and energy (Figure 3-19).

Scientists have also tried to develop *controlled nuclear fusion*, in which the D–T reaction is used to produce heat that can be converted into electricity. Despite more than 40 years of research, this process is still in the laboratory stage. Even if it becomes technologically and economically feasible, it probably won't be a practical source of energy until 2050 or later.

A: About 48% and 73% if areas of rock and ice are excluded

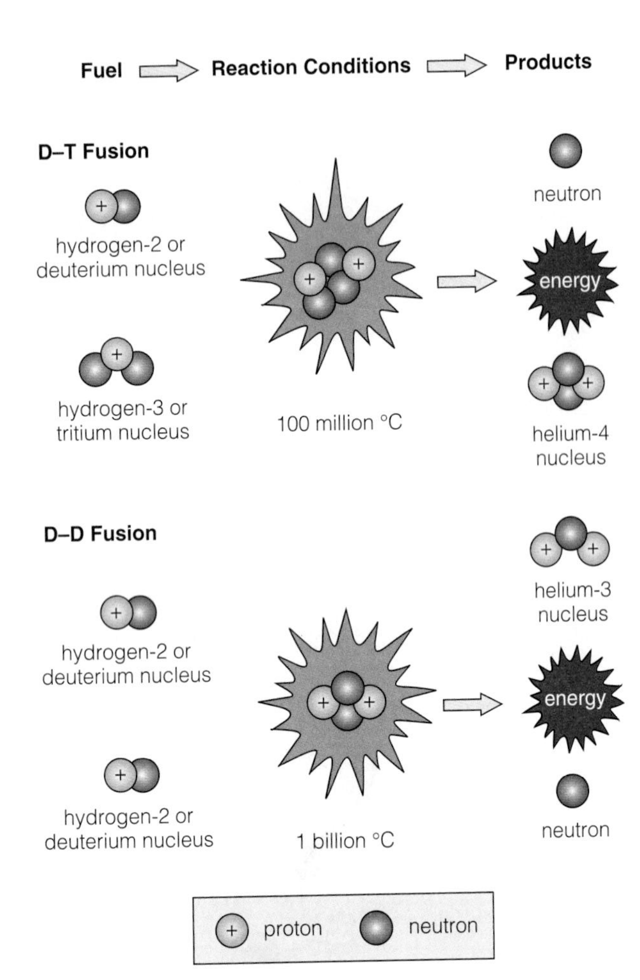

Figure 3-19 The deuterium–tritium (D–T) and deuterium–deuterium (D–D) nuclear fusion reactions, which take place at extremely high temperatures.

3-5 THE TWO IRONCLAD LAWS OF ENERGY

What Is the First Law of Energy? You Can't Get Something for Nothing Scientists have observed energy being changed from one form to another in millions of physical and chemical changes, but they have never been able to detect either the creation or destruction of any energy (except in nuclear changes). The results of their experiments have been summarized in the **law of conservation of energy**, also known as the **first law of energy** or the **first law of thermodynamics**: *In all physical and chemical changes, energy is neither created nor destroyed, but it may be converted from one form to another.*

This scientific law tells us that when one form of energy is converted to another form in any physical or chemical change *energy input always equals energy output*. No matter how hard we try or how clever we are, we can't get more energy out of a system than we put in; in other words, *we can't get something for nothing in terms of energy quantity.*

What Is the Second Law of Energy? You Can't Even Break Even Because the first law of energy states that energy can be neither created nor destroyed, it's tempting to think that there will always be enough energy, yet if we fill a car's tank with gasoline and drive around, or use a flashlight battery until it is dead, something has been lost. If it isn't energy, what is it? The answer is *energy quality* (Figure 3-14), the amount of energy available that can perform useful work.

Countless experiments have shown that when energy is changed from one form to another, a decrease in energy quality always occurs. The results of these experiments have been summarized in what is called the **second law of energy** or the **second law of thermodynamics**: *When energy is changed from one form to another, some of the useful energy is always degraded to lower-quality, more dispersed (higher-entropy), less useful energy*. This degraded energy usually takes the form of heat given off at a low temperature to the surroundings (environment). There it is dispersed by the random motion of air or water molecules and becomes even more disorderly and less useful. Another way to state the second law of energy is that *heat always flows spontaneously from hot (high-quality energy) to cold (low-quality energy).*

Basically, this law says that in any energy conversion, we always end up with less usable energy than we started with. So not only can we not get something for nothing in terms of energy quantity, *we can't even break even in terms of energy quality because energy always goes from a more useful to a less useful form.* The more energy we use, the more low-grade energy (heat)—or entropy (disorder)—we add to the environment. No one has ever found a violation of this fundamental scientific law.

Consider three examples of the second energy law in action. First, when a car is driven, only about 10% of the high-quality chemical energy available in its gasoline fuel is converted into mechanical energy (to propel the vehicle) and electrical energy (to run its electrical systems); the remaining 90% is degraded to low-quality heat that is released into the environment and eventually lost into space. Second, when electrical energy flows through filament wires in an incandescent light bulb, it is changed into about 5% useful light and 95% low-quality heat that flows into the environment; this so-called *light bulb* is really a *heat bulb*. Third, in living systems, solar energy is converted into chemical energy (photosynthesis and food) and then into

Kestenbaum, David. 1998. "Gentle Force of Entropy Bridges Disciplines." *Science*, vol. 279, no. 5358, 1849(1).

mechanical energy (moving, thinking, and living); high-quality energy is degraded during this change of forms (Figure 3-20).

The second law of energy also means that *we can never recycle or reuse high-quality energy to perform useful work.* Once the concentrated energy in a serving of food, a liter of gasoline, a lump of coal, or a chunk of uranium is released, it is degraded to low-quality heat that is dispersed into the environment. We can heat air or water at a low temperature and upgrade it to high-quality energy, but the second law of energy tells us that it will take more high-quality energy to do this than we get in return.

Connections: How Does the Second Energy Law Affect Life? To form and maintain the highly ordered arrangement of molecules and the organized biochemical processes in your body, you must continually get and use high-quality matter and energy resources from your surroundings. As you use these resources, you add low-quality (high-entropy) heat and low-quality (high-entropy) waste matter to your surroundings. Your body continuously gives off heat equal to that of a 100-watt incandescent light bulb; this is why a closed room full of people gets warm. You also continuously break down solid, large molecules (such as glucose) into smaller molecules of carbon dioxide gas and water vapor, which are dispersed in the atmosphere.

Planting, growing, processing, and cooking food all require high-quality energy and matter resources that add low-quality (high-entropy) heat and waste materials to the environment. In addition, enormous amounts of low-quality heat and waste matter are added to the environment when concentrated deposits of minerals and fuels are extracted from the earth's crust, processed, and used. Thus, *all forms of life are tiny pockets of order (low entropy) maintained by creating a sea of disorder (high entropy) in their environment.*

3-6 CONNECTIONS: MATTER AND ENERGY LAWS AND ENVIRONMENTAL PROBLEMS

What Are High-Throughput or High-Waste Societies? As a result of the law of conservation of matter and the second law of energy, individual

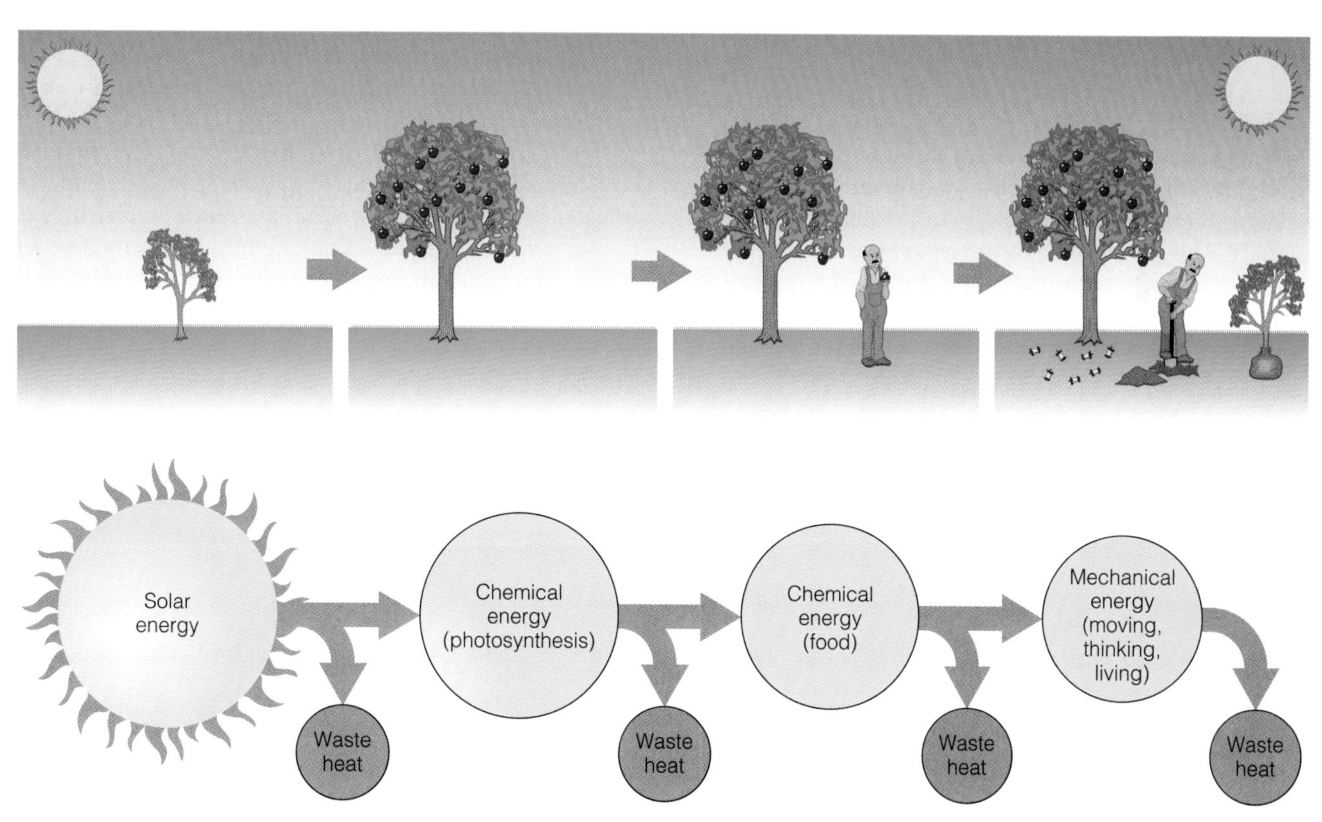

Figure 3-20 The second energy law in action in living systems. Each time energy is changed from one form to another, some of the initial input of high-quality energy is degraded, usually to low-quality heat that disperses into the environment.

Hint: Enter the search term *energy* using the Subject Guide and then select the hotkey to *force and energy.*

resource use automatically adds some waste heat and waste matter to the environment. Your individual use of matter and energy resources and your additions of waste heat and matter to the environment may seem small and insignificant. But there are 1.2 billion people in developed countries rapidly using large quantities of matter and energy resources. Meanwhile, the 4.7 billion people in developing countries hope to be able to use more of these resources, and each year there are about 84 million new consumers of the earth's energy and matter resources.

Most of today's advanced industrialized countries are **high-waste** or **high-throughput societies** that attempt to sustain ever-increasing economic growth by increasing the *throughput* of matter and energy resources in their economic systems (Figure 3-21). These resources flow through the economies of such societies to planetary *sinks* (air, water, soil, organisms), where pollutants and wastes end up and can accumulate to harmful levels.

However, the scientific laws of matter and energy discussed in this chapter tell us that if more and more people continue to use and waste more and more energy and matter resources at an increasing rate, eventually the capacity of the environment to dilute and degrade waste matter and absorb waste heat will be exceeded. Thus, *at some point high-waste or high-throughput societies become unsustainable.*

What Are Matter-Recycling Societies? A stopgap solution to this problem is to convert an unsustainable high-throughput society to a **matter-recycling society**. The goal of such a conversion is to allow economic growth to continue without depleting matter resources or producing excessive pollution and environmental degradation. As we have learned, however, there is no free lunch when it comes to energy and energy quality.

Even though recycling matter saves energy, the two laws of energy tell us that *recycling matter resources always requires expenditure of high-quality energy (which cannot be recycled) and adds waste heat to the environment.* For the long run, a matter-recycling society based on continuing population growth and per capita resource consumption must have an inexhaustible supply of affordable high-quality energy, and its environment must have an infinite capacity to absorb and disperse waste heat and to dilute and degrade waste matter.

There is also a limit to the number of times some materials, such as paper fiber, can be recycled before they become unusable. Changing to a matter-recycling society is an important way to buy some time, but it does not allow more and more people to use more and more resources indefinitely, even if all of them were somehow perfectly recycled.

What Are Low-Waste Societies? Learning from Nature The three scientific laws governing matter and energy changes indicate that the best long-term solution to our environmental and resource problems is to shift from a society based on maximizing matter and energy flow (throughput) to a sustainable **low-waste society** or **low-throughput society**. The major features of such a society are summarized in Figure 3-22.

According to many scientists, because of the three basic scientific laws of matter and energy, we all depend on one another and on the rest of nature for our survival; we are all in it together. In the next seven chapters, we apply these laws to living systems and look at some biological principles that can teach us how to work with nature.

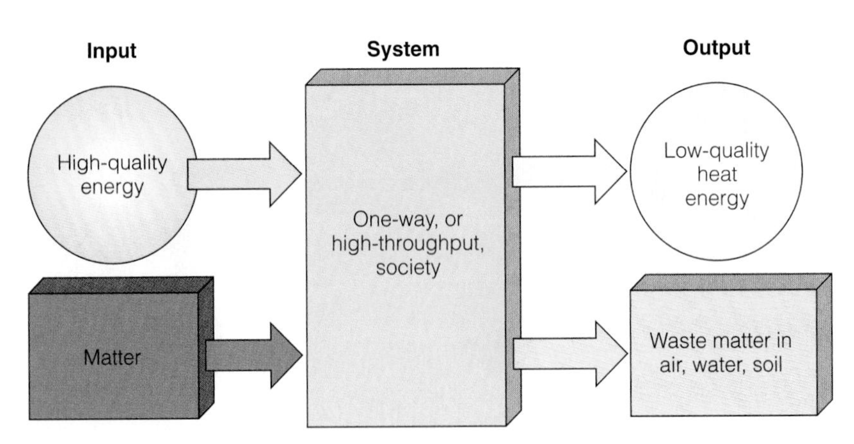

Figure 3-21 Conceptual model of the eventually unsustainable high-waste or high-throughput societies of most developed countries, which are based on maximizing the rates of energy and matter flow. This process is rapidly converting the world's high-quality matter and energy resources into waste, pollution, and low-quality heat.

Q: What percentage of the world's cropland is suffering from soil erosion?

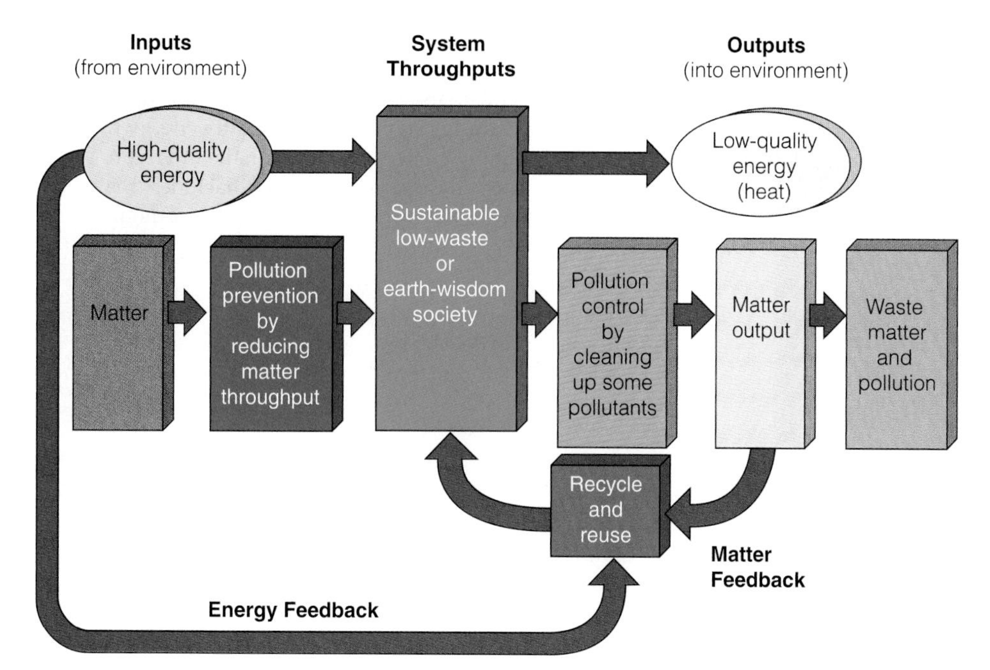

Figure 3-22 Conceptual model of a sustainable low-waste or low-throughput society, based on energy flow and matter recycling, that works with nature to reduce throughput. This is done by **(1)** reusing and recycling most nonrenewable matter resources, **(2)** using potentially renewable resources no faster than they are replenished, **(3)** using matter and energy resources efficiently, **(4)** reducing unnecessary consumption, **(5)** emphasizing pollution prevention and waste reduction, and **(6)** controlling population growth.

The second law of thermodynamics holds, I think, the supreme position among laws of nature. . . . If your theory is found to be against the second law of thermodynamics, I can give you no hope.

ARTHUR S. EDDINGTON

CRITICAL THINKING

1. Explain why we don't really consume anything and why we can never really throw matter away.

2. A tree grows and increases its mass. Explain why this isn't a violation of the law of conservation of matter.

3. If there is no "away," why isn't the world filled with waste matter?

4. Respond to the following statement: The limitations on what we can and cannot do, imposed by the law of conservation of matter and the first and second laws of energy, can be overcome because these scientific laws may be overturned sometime in the future.

5. Someone wants you to invest money in an automobile engine that will produce more energy than the energy in the fuel (such as gasoline or electricity) you use to run the motor. What is your response? Explain.

6. Use the second energy law to explain why a barrel of oil can be used as a fuel only once.

7. Explain how a community that wastes fewer energy and matter resources helps sustain its local economy and jobs.

8. Use the law of conservation of matter to explain why a matter-recycling society will sooner or later be necessary.

9. Use the first and second laws of energy to explain why, in the long run, we will need a low-waste or low-throughput society, not just a matter-recycling society.

10. Peter Montague (Guest Essay, p. 70), a number of other environmentalists, and some anti-environmentalists believe that we have been sidetracked economically and politically for 25 years, placing the primary emphasis on pollution control instead of pollution prevention. Do you agree or disagree with this assessment? Explain. If we emphasized pollution prevention, would this mean that we could have fewer and less prescriptive environmental regulations? Explain. What five things can you do to help ensure that we move into a pollution prevention era?

11. Try to come up with analogies to complete the following statements: **(a)** Being a molecule of oxygen gas is like _______. **(b)** Thinking we could violate the law of conservation of matter is like _______. **(c)** A nuclear fission chain reaction is like _______. **(d)** Living in a low-waste society is like _______.

A: About 33%

12. **(a)** Imagine that you have the power to violate the law of conservation of energy (the first energy law) for 1 day. What are the three most important things you would do with this power? **(b)** Repeat this process, imagining that you have the power to violate the second law of energy for 1 day.

PROJECTS

1. If you have the use of a sensitive balance, try to demonstrate the law of conservation of mass in a physical change. Weigh a container with a lid (a glass jar will do), add an ice cube and weigh it again, and then allow the ice to melt and weigh it again.

2. Use the library or Internet to find examples of various perpetual motion machines and inventions that allegedly violate the two laws of energy (thermodynamics) by producing more high-quality energy than the high-quality energy needed to make them run. What has happened to these schemes and machines (many of them developed by scam artists to attract money from investors)?

3. Make a concept map of this chapter's major ideas using the section heads and subheads and the key terms (in boldface). Look at the inside of the back cover and in Appendix 4 for information about concept maps.

ECOLOGICAL PRINCIPLES

how wide a circle of disturbance we produce in the harmonies of nature when we throw the smallest pebble into the ocean of organic life.

George Perkins Marsh

4 ECOLOGY, ECOSYSTEMS, AND FOOD WEBS*

Have You Thanked the Insects Today?

Insects have a bad reputation. We classify many insect species as pests because they compete with us for food, spread human diseases (such as malaria), and invade our lawns, gardens, and houses. Some people have "bugitis," fear all insects, and think that the only good bug is a dead bug. However, this view fails to recognize the vital roles insects play in helping sustain life on earth.

*Paul M. Rich, associate professor of Ecology and Evolutionary Biology and Environmental Studies at the University of Kansas, is coauthor of this chapter.

A large proportion of the earth's plant species (including many trees) depend on insects to pollinate their flowers (Figure 4-1, bottom). Plants also benefit when insects help loosen the soil around their roots and decompose dead tissue into nutrients the plants need. In turn, we and other land-dwelling animals depend on plants for food, either by eating them or by consuming animals that eat them. If there were no pollinating insects, there would be very few fruits and vegetables for us and plant-eating animals to eat.

Some insects, such as the praying mantis (Figure 4-1, top), feed on some of the insect species we classify as pests. Indeed, insects that eat other insects help control the populations of at least half the species of insects we call pests. This free pest control service is an important part of the earth capital that helps sustain us.

Suppose all insects disappeared today. Within a year most of the earth's amphibians, reptiles, birds, and mammals would become extinct because of the disappearance of so much plant life. The earth would be covered with rotting vegetation and animal carcasses being decomposed by unimaginably huge hordes of bacteria and fungi. The land, largely devoid of animal life, would be covered by mats of wind-pollinated vegetation and intermittent clumps of small trees and bushes.

Fortunately, this is not a realistic scenario because insects, which have been around for at least 400 million years, are phenomenally successful forms of life. They were the first animals to invade the land and, later, the air. Today they are by far the planet's most diverse, abundant, and successful group of animals. Scientists have identified about 950,000 insect species (compared to only 4,100 known mammal species),

Figure 4-1 Insects play important roles in helping sustain life on earth. The bright green caterpillar moth feeding on pollen in a crocus (bottom) and other insects pollinate flowering plants that serve as food for many plant eaters. The praying mantis eating a grasshopper (top) and many other insect species help control the populations of at least half of the insect species we classify as pests. (Top, M. H. Sharp/Photo Researchers, Inc.; bottom: Stephen Hopkins/Planet Earth Pictures)

but some researchers believe that up to 100 million unidentified insect species exist.

Insects can rapidly evolve new genetic traits, such as resistance to pesticides. They also have an exceptional ability to evolve into new species when faced with new environmental conditions, and they are extremely resistant to extinction (see cartoon). Because of their ability to quickly develop genetic resistance to pesticides, our efforts to eradicate the insects we don't like using chemical warfare will always fail eventually. Moreover, many of the pesticides we use reduce the populations of insect species that help keep pest insect species under control.

Some people have wondered whether insects will take over the world if the human race extinguishes itself. This is the wrong question. Insects are already in charge. The approximately billion billion (10^{18}) insects alive at any given time have already dominated much of the earth's land surface for millions of years, and they're likely to be here for millions of years more. Insects can thrive without newcomers such as us, but we and most other land organisms would quickly perish without them.

Learning about the roles insects play in nature requires us to understand how insects and other organisms living in a biological *community* (such as a forest or pond) interact with one another and with the nonliving environment. *Ecology* is the science that studies such relationships and interactions in nature, as discussed in this chapter and the six chapters that follow.

Students who can begin early in their lives to think of things as connected, even if they revise their views every year, have begun the life of learning.

Mark Van Doren

This chapter focuses on answering the following questions:

- What is ecology?
- What basic processes keep us and other organisms alive?
- What are the major nonliving and living parts of an ecosystem?
- What happens to energy in an ecosystem?
- How do scientists study ecosystems?

 4-1 ECOLOGY AND LIFE

What Is Ecology? The term *ecology* was coined in 1869 by German biologist Ernst Haeckel, and came into general use in the late 1800s, when scientists such as M.I.T. chemist Ellen Swallow began calling themselves ecologists. **Ecology** (from the Greek words *oikos*, "house" or "place to live," and *logos*, "study of") is the study of relationships between organisms and their environment. Ecology examines how organisms interact with their nonliving environment (including such factors as sunlight, temperature, moisture, and vital nutrients) and with each other. A key word is *interact*. Ecologists focus on trying to understand the interactions among organisms, populations, communities, ecosystems, and the ecosphere (Figure 4-2).

An **organism** is any form of life. Organisms can be classified into **species**, groups of organisms that resemble one another in appearance, behavior, chemistry, and genetic endowment. Organisms that reproduce sexually are classified in the same species if, under natural conditions, they can actually or potentially breed with one another and produce live, fertile offspring.

We don't know how many species exist on the earth. Estimates range from 5 million to 100 million, most of which are insects (left) and microorganisms. So far biologists have identified and named only about 1.8 million species. Biologists know a fair amount about

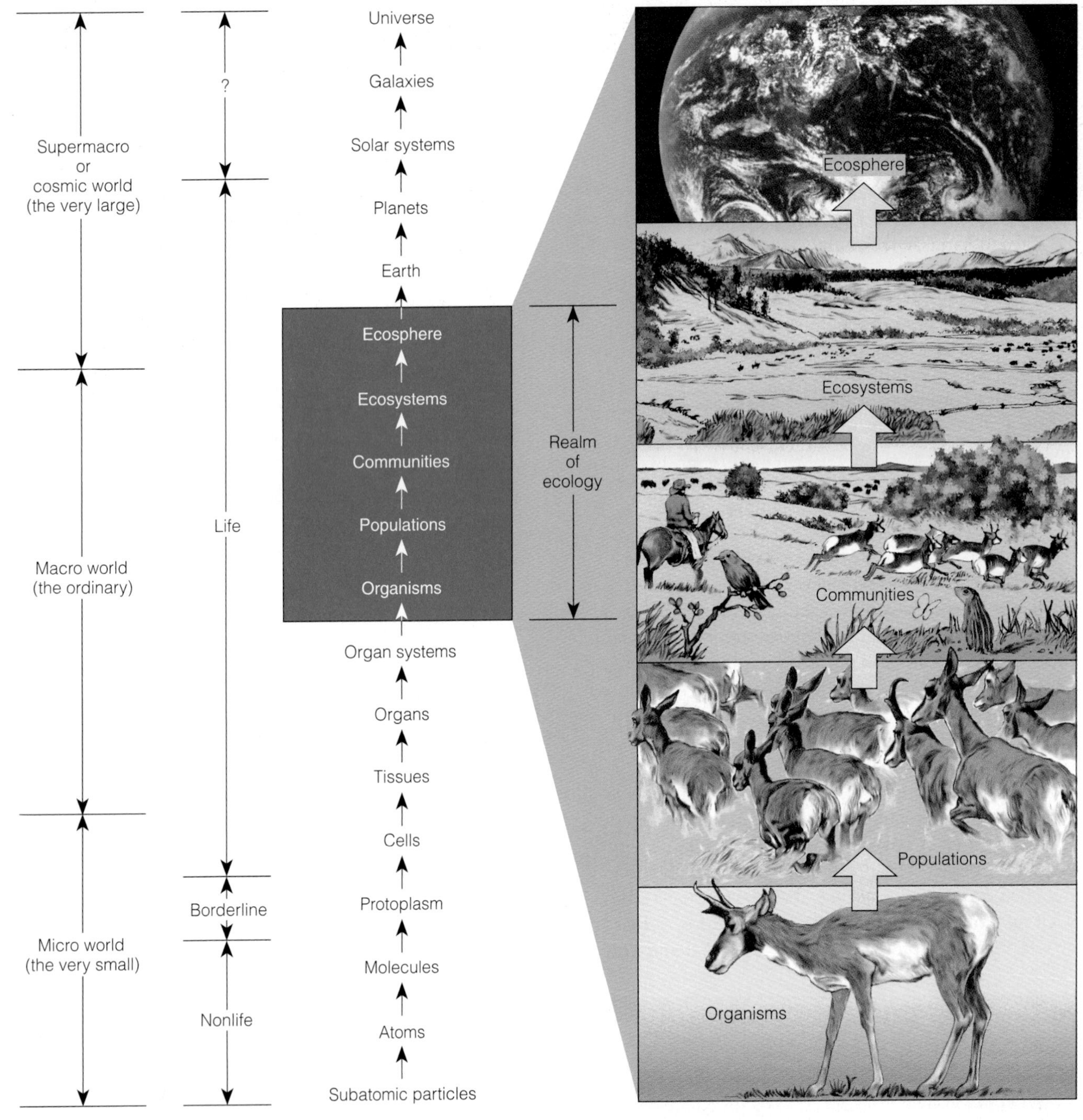

Figure 4-2 Conceptual model of levels of organization of matter in nature. Note that ecology focuses on five levels of this hierarchical model.

roughly one-third of the known species, but understand the detailed roles and interactions of only a few.

Each species is the result of a long evolutionary history, involving the storage of an immense amount of unique and irreplaceable genetic information about how to survive under specific environmental conditions. According to biologist Edward O. Wilson, if the genetic information encoded in the DNA found in the approximately 100,000 genes of a house mouse were translated into a printed text, it would fill all of the books in all 15 editions of the *Encyclopedia Britannica* published since 1768.

Ecologists make a distinction between wild species and domesticated species. A **wild species** is one that exists as a population of individuals in a natural habitat, ideally similar to the one in which its ancestors evolved. A **domesticated species**—such as cows, sheep, food crops, animals in zoos, plants in

Rykiel, Ed. 1997. "Ecosystem Science for the Twenty-First Century." *BioScience*, vol. 47, no. 10, 705(3).

Figure 4-3 A population of monarch butterflies wintering in Michoacán, Mexico. The geographic distribution of this butterfly coincides with that of the milkweed plant, on which monarch larvae and caterpillars feed. (Frans Lanting/Bruce Coleman Ltd.)

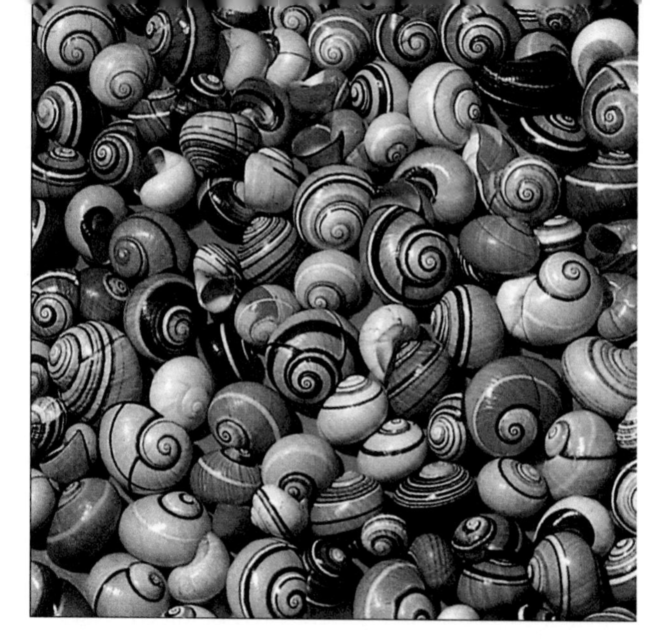

Figure 4-4 The genetic diversity among individuals of one species of Caribbean snail is reflected in the variations in shell color and banding patterns. (Alan Solem)

arboretums, and lawn grasses, flowers, and trees—is one that has been plucked from its normal ecological environment to support the needs and wants of humans and thus plays a much weaker ecological and evolutionary role than wild species. Domesticated plants and animals need to be externally fed and propagated. They are usually subjected to artificial selection, in which humans choose traits that correspond to only a small subset of the genetic diversity present in the wild ancestors of the domesticated species.

A **population** consists of a group of interacting individuals of the same species that occupy a specific area at the same time (Figure 4-3). Examples are all sunfish in a pond, all white oak trees in a forest, and all people in a country. In most natural populations, individuals vary slightly in their genetic makeup, which is why they don't all look or behave exactly alike—a phenomenon called **genetic diversity** (Figure 4-4). Populations are dynamic groups that change in size, age distribution, density, and genetic composition as a result of changes in environmental conditions.

The place where a population (or an individual organism) normally lives is known as its **habitat**. It may be as large as an ocean or prairie or as small as the underside of a rotting log or the intestine of a termite. Populations of all the different species occupying in a particular place make up a **community**, or **biological community**: a complex interacting network of plants, animals, and microorganisms.

An **ecosystem** is a community of different species interacting with one another *and* with their nonliving environment of matter and energy. The size of an ecosystem is somewhat arbitrary; it is defined by the particular system we wish to study. The unit of study may be small, such as a particular stream or field or a patch of woods, desert, or marsh. Or the units may be large, generalized types of terrestrial (land) ecosystems such as a particular type of grassland, forest, or desert. Ecosystems can be natural or artificial (human-created). Examples of human-created ecosystems are cropfields, farm ponds, and reservoirs or artificial lakes created behind dams.

All of the earth's ecosystems together make up what we call the **ecosphere**, or **biosphere**.

What Is Life? All life shares a set of basic characteristics that enable growth, survival, and reproduction:

- *Living organisms are made of cells that have highly organized internal structures and functions.* The **cell** is the basic unit of life. Each cell is bounded by an outer membrane or wall and contains genetic material (DNA) and other structures needed to perform its life functions. Organisms may consist of a single cell (bacteria, for instance), or they may contain many cells.
- *Living organisms have characteristic types of deoxyribonucleic acid (DNA) molecules in each cell.* DNA (Figures 3-9 and 3-10) is the stuff of which genes, the basic units of heredity, are made. These self-replicating molecules contain the instructions for making new cells and for assembling proteins and other molecules each cell needs for survival and reproduction (Figure 3-10).
- *Living organisms capture and transform matter and energy from their environment to supply their needs for survival, growth, and reproduction.* The complete set of chemical reactions that carries out this role in cells and organisms is called **metabolism**.

Hint: Enter the search terms *ecosystem, science* using Key Words.

- *Living organisms maintain favorable internal conditions, despite changes in their external environment, through homeostasis* (Section 2-3), *if not overstressed.*

- *Living organisms perpetuate themselves through* **reproduction**: *the production of offspring by one or more parents.* **Asexual reproduction** generally occurs by simple cell division and is common in single-cell organisms. In this case, the mother cell divides to produce two identical daughter cells that are clones or replicas of the mother cell. **Sexual reproduction** occurs in organisms that produce offspring by combining sex cells or gametes (such as ovum and sperm) from both parents. This produces offspring that have combinations of traits from each of their parents. Sexual reproduction usually gives the offspring a greater chance of survival under changing environmental conditions than the genetic clones produced by asexual reproduction.

- *Living organisms adapt to changes in environmental conditions through the process of* **evolution**. Evolution involves changes in the genetic composition of organisms through time. When evolution continues long enough, it can result in new species. Ongoing processes of extinction and evolution, in response to environmental changes over millions of years, have led to the diversity of life-forms found on the earth today. This *biodiversity,* a vital part of earth capital, sustains life and provides the genetic raw material for adaptation to future environmental changes.

4-2 EARTH'S LIFE-SUPPORT SYSTEMS

What Are the Major Parts of Earth's Life-Support Systems? We can think of the earth as being made up of several layers or concentric spheres (Figure 4-5).

As the primitive earth cooled over eons, its interior separated into three major concentric zones, which geologists identify as the core, the mantle, and the crust (Figure 4-5). The earth's innermost zone, the **core**, is made mostly of iron (with perhaps some nickel). The core has a solid inner part, surrounded by a liquid core of molten material.

The earth's core is surrounded by a thick, solid zone called the **mantle**. This largest zone of the earth's interior is rich in the elements iron (its major constituent), silicon, oxygen, and magnesium. Most of the mantle is solid rock, but under its rigid outermost part there is a zone of very hot, partly melted rock that flows like soft plastic. This plastic region of the mantle is called the *asthenosphere*.

The outermost and thinnest zone of the earth is called the **crust**. Only eight elements make up 98.5% of the weight of the earth's crust (Figure 3-11). The crust contains nonrenewable fossil fuels and minerals, as well as potentially renewable soil chemicals (nutrients) required for plant life. The **lithosphere** is the earth's crust and upper mantle.

The **atmosphere** is a thin envelope of air around the planet. Its inner layer, the **troposphere**, extends only about 17 kilometers (11 miles) above sea level, but contains most of the planet's air, mostly nitrogen (78%) and oxygen (21%), and is where weather occurs. The next layer, stretching 17–48 kilometers (11–30 miles) above the earth's surface, is the **stratosphere**. Its lower portion contains enough ozone (O_3) to filter out most of the sun's harmful ultraviolet radiation, thus allowing life to exist on land and in the surface layers of bodies of water.

The **hydrosphere** consists of the earth's liquid water (both surface and underground), ice (polar ice, icebergs, ice in permafrost), and water vapor in the atmosphere.

The **ecosphere**, or **biosphere**, is the portion of the earth in which living (biotic) organisms exist and interact with one another and with their nonliving (abiotic) environment. The ecosphere includes most of the hydrosphere and parts of the lower atmosphere and upper lithosphere, reaching from the deepest ocean floor, 20 kilometers (12 miles) below sea level, to

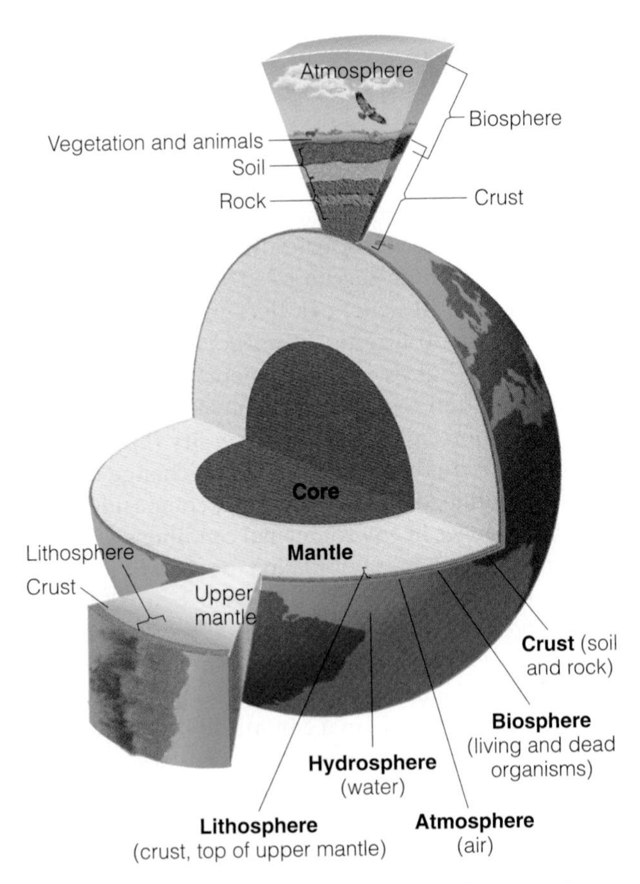

Figure 4-5 The general structure of the earth. The atmosphere consists of several layers, including the troposphere (innermost layer) and the stratosphere (second layer).

Q: How many people lack access to clean drinking water?

the tops of the highest mountains. If the earth were an apple, the ecosphere would be no thicker than the apple's skin. *The goal of ecology is to understand the interactions in this thin, life-supporting global skin of air, water, soil, and organisms.*

What Sustains Life on Earth? Life on the earth depends on three interconnected factors (Figure 4-6):

- The *one-way flow of high-quality (usable) energy* from the sun, first through materials and living things in their feeding interactions, then into the environment as low-quality energy (mostly heat dispersed into air or water molecules at a low temperature), and eventually back into space as infrared radiation.
- The *cycling of matter or nutrients* (all atoms, ions, or molecules needed for survival by living organisms) through parts of the ecosphere. Each chemical cycle can be viewed as a system with *inputs* of matter and solar energy (used to drive the cycle) and their flow or *throughput* through parts of the ecosphere. In each chemical cycle, the *output* of degraded energy flows as heat into the atmosphere and eventually back into space, and the *output* of matter is returned to the system (*feedback*) to be cycled again and again.
- *Gravity*, which allows the planet to hold onto its atmosphere and causes the downward movement of chemicals in the matter cycles.

In real life there are two types of systems: closed and open. In a **closed system**, energy—but not matter—is exchanged between the system and its environment. The earth is a closed system that can receive energy from the sun but loses essentially no matter into space and receives almost negligible amounts of matter from space (mostly cosmic dust and occasional meteorites that strike the earth). Because the earth is closed to significant inputs of matter, essentially all of the nutrients used by organisms are already present on earth, and must be recycled again and again for life to continue.

In an **open system**, such as a living organism, both matter and energy are exchanged between the system and the environment. You are an open system. You take matter (air, water, food) and energy into your body and transform and use these inputs to stay alive and healthy. At the same time, you put waste matter (urine, feces, dead skin cells, exhaled air, evaporated perspiration) and low-quality heat into the environment. You and other organisms remain alive and healthy only if the

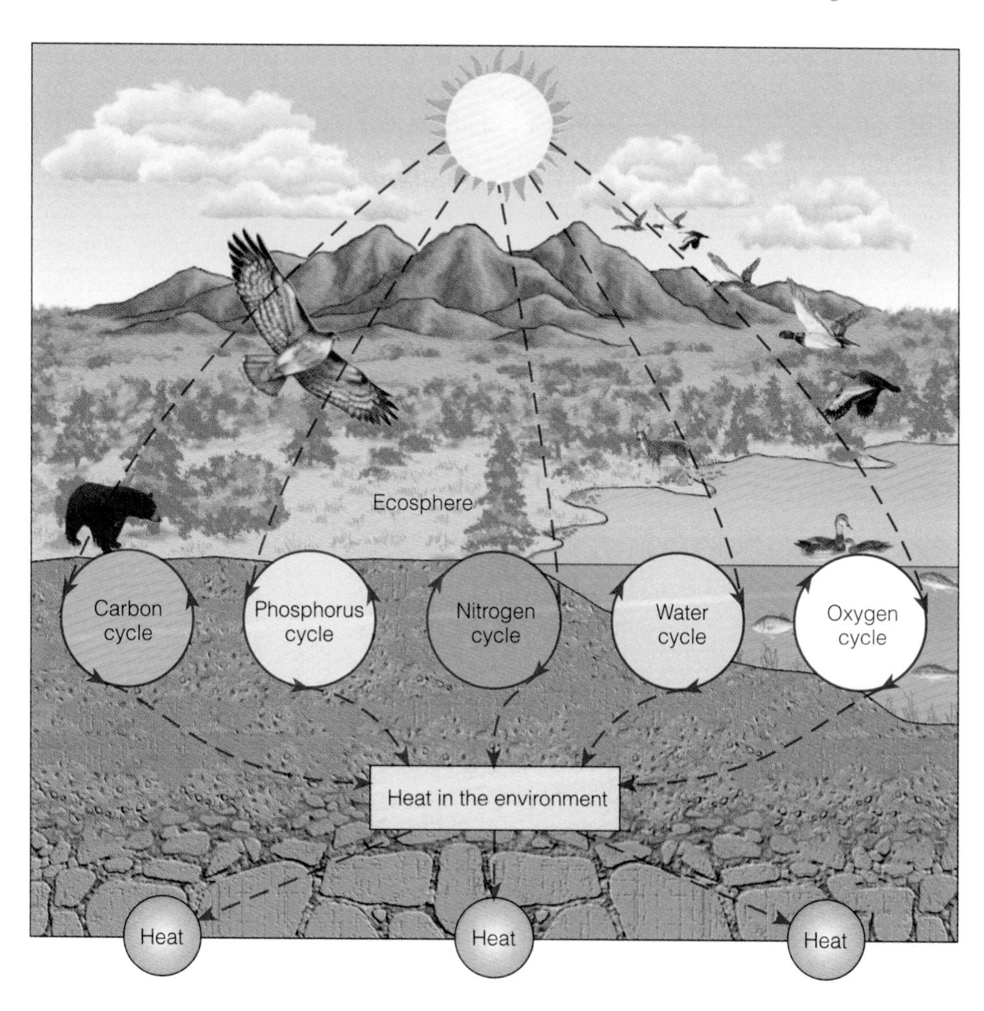

Figure 4-6 Life on the earth depends on the *one-way flow of energy* (dashed lines) from the sun through the ecosphere, the *cycling of crucial elements* (solid lines around circles), and *gravity*, which keeps atmospheric gases from escaping into space and draws chemicals downward in the matter cycles. This simplified conceptual model depicts only a few of the many cycling elements.

A: About 1 billion, or one in six

input of certain types of matter and energy is balanced by an output of certain types of matter and energy.

Thus, whereas life on earth as a whole is maintained by *energy flow and matter cycling*, the life of an individual organism is maintained by the processes of *energy flow and matter flow*.

How Does the Sun Help Sustain Life on Earth? The sun is a middle-aged star that lights and warms the planet. It also supplies the energy for *photosynthesis*, the process used by green plants and some bacteria to synthesize the compounds that keep them alive and feed most other organisms. Solar energy also powers the cycling of matter and drives the climate and weather systems that distribute heat and fresh water over the earth's surface.

The sun is a gigantic fireball of hydrogen (72%) and helium (28%) gases. Temperatures and pressures in its inner core are so high that hydrogen nuclei fuse to form helium nuclei (Figure 3-19), releasing enormous amounts of energy. In this fusion process, about 4.1 billion kilograms (4.2 million tons) of the mass in the sun's hydrogen nuclei are converted into energy every second. However, we need not worry about the sun running out of hydrogen fuel. In the normal life cycle of stars, the sun is entering middle age. It has probably been in existence for at least 6 billion years and has enough fuel to provide the earth with energy for at least another 6.5 billion years.

The sun, then, is really a gigantic nuclear fusion (thermonuclear) reactor running on hydrogen fuel. This enormous reactor radiates energy in all directions as electromagnetic radiation (Figure 3-13). Moving at the speed of light, this radiation makes the 150-million-kilometer (93-million-mile) trip between the sun and the earth in slightly more than 8 minutes.

Because the earth is a tiny sphere in the vastness of space, it receives only about one-billionth of this output of energy. Much of this energy is either reflected away or absorbed by chemicals in its atmosphere. Most of what reaches the troposphere is visible light, infrared radiation (heat), and the small amount of ultraviolet radiation that is not absorbed by ozone in the stratosphere. About 28% of the solar energy reaching the troposphere is reflected back into space by clouds, chemicals, dust, and the earth's surface land and water (Figure 4-7).

Most of the remaining 72% of solar energy warms the troposphere and land, evaporates water and cycles it through the ecosphere, and generates winds. A tiny fraction (about 0.023%) is captured by green plants and bacteria, fueling photosynthesis to make the organic compounds that most life-forms need to survive.

Most of this unreflected solar radiation is degraded into infrared radiation (which we experience as heat) as it interacts with the earth. How fast this heat flows through the atmosphere and back into space is affected by tropospheric heat-trapping (greenhouse) gases, such as water vapor, carbon dioxide, methane, nitrous oxide, and ozone. Without this atmospheric thermal blanket, known as the **natural greenhouse effect**, the earth would be nearly as cold as Mars, and life as we know it could not exist.

How Do Nutrient Cycles Sustain Life? Any atom, ion, or molecule an organism needs to live, grow, or reproduce is called a **nutrient**. Some elements

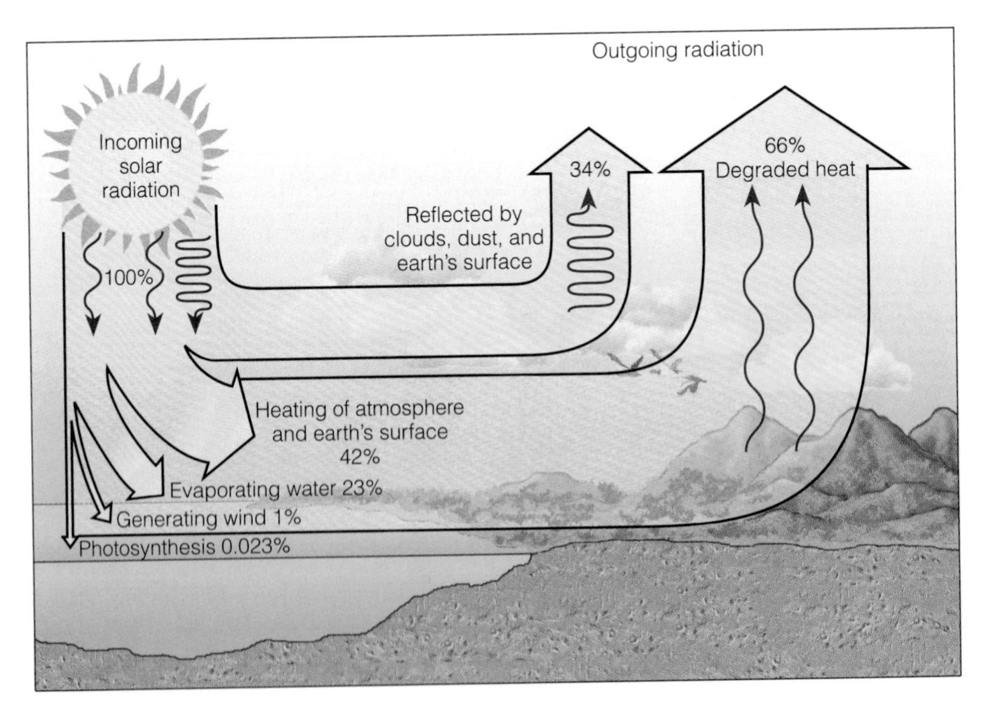

Figure 4-7 Conceptual model of the flow of energy to and from the earth. The ultimate source of energy in most ecosystems is sunlight.

Q: What percentage of the world's households don't have running water?

(such as carbon, oxygen, hydrogen, nitrogen, phosphorus, sulfur, potassium, calcium, magnesium, and iron), called **macronutrients**, are needed in fairly large amounts, whereas others (such as sodium, zinc, copper, chlorine, and iodine), called **micronutrients**, are needed in small or even trace amounts (Figure 3-4).

These nutrient atoms, ions, and molecules are continuously cycled from the nonliving environment (air, water, soil, rock) to living organisms (biota) and then back again in what are called **nutrient cycles**, or **biogeochemical cycles** (literally, "life–earth–chemical" cycles). These cycles, driven directly or indirectly by incoming solar energy and gravity, include the hydrologic (water), carbon, oxygen, nitrogen, phosphorus, and sulfur cycles (Figure 4-6). These cycles are discussed in detail in Chapter 5.

Because the earth is essentially a closed system, the planet's chemical cycles are vital for all life, and they explain why without death there could be no life. The cycle of reproduction, growth, death, and decay of organisms keeps renewing the chemicals that support all life.

The earth's chemical cycles also connect past, present, and future forms of life. Some of the carbon atoms in your skin may once have been part of a leaf, a dinosaur's skin, or a layer of limestone rock. Some of the oxygen molecules you just inhaled may have been inhaled by your grandmother, by Plato, or by a hunter–gatherer who lived 25,000 years ago.

4-3 ECOSYSTEM CONCEPTS

What Are the Major Types of Ecosystems?

Viewed from outer space, the earth resembles an enormous jigsaw puzzle consisting of large masses of land and vast expanses of ocean (p. 1 and Figure 1-1). Biologists have classified the terrestrial (land) portion of the ecosphere into **biomes**, large regions (such as forests, deserts, and grasslands) characterized by a distinct climate and specific life-forms, especially vegetation adapted to it (Figure 4-8). **Climate**—long-term weather—is the main factor determining what type of life, especially what plants, will thrive in a given land area. Each biome consists of many ecosystems whose communities have adapted to differences in climate, soil, and other factors throughout the biome.

Marine and freshwater portions of the ecosphere can be divided into **aquatic life zones**, each contain-

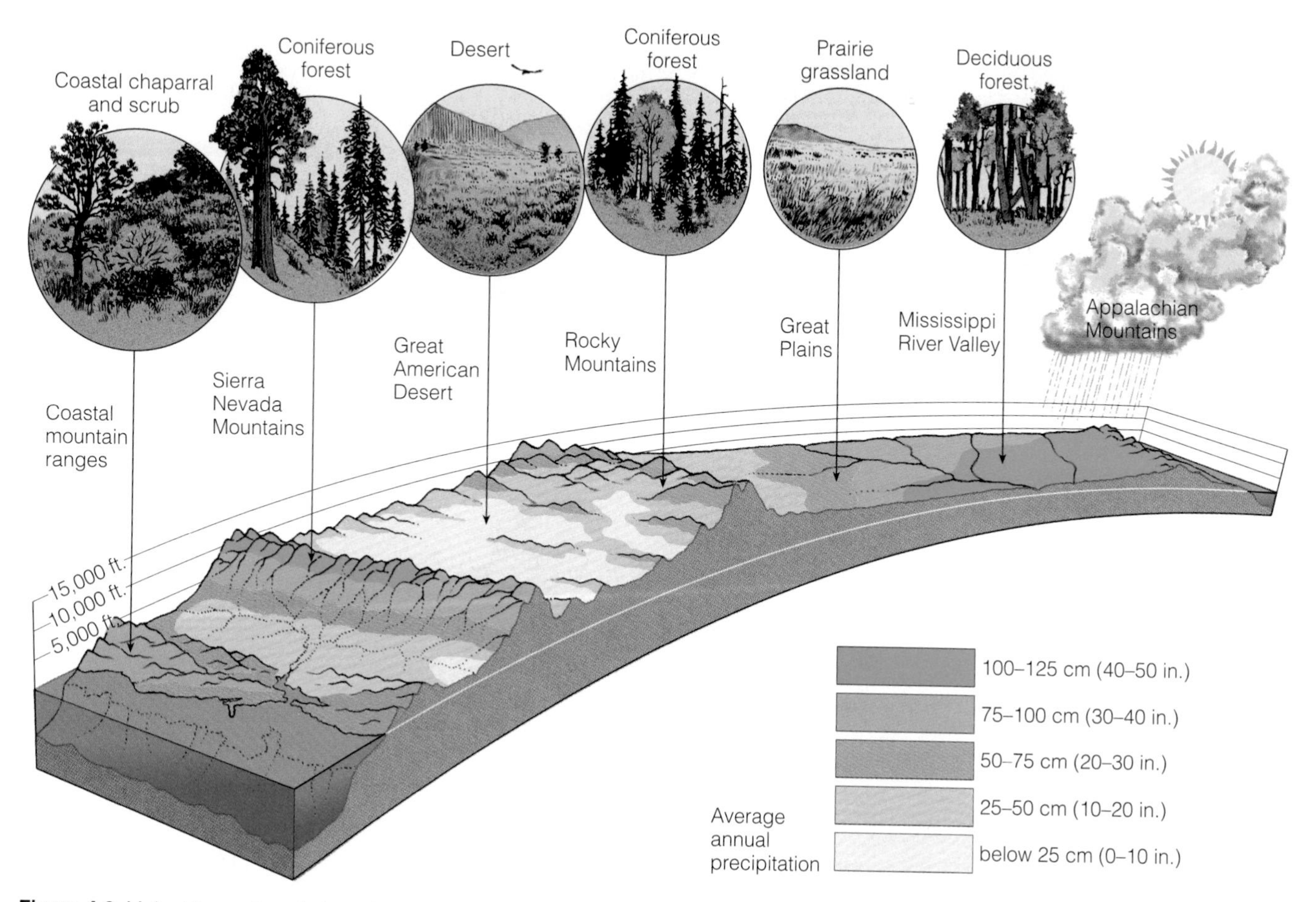

Figure 4-8 Major biomes found along the 39th parallel across the United States. The differences reflect changes in climate, mainly differences in average annual precipitation and temperature (not shown).

A: About 66%

ing numerous ecosystems. Aquatic life zones are the aquatic equivalent of biomes. Examples include freshwater life zones (such as lakes and streams) and ocean or marine life zones (such as estuaries, coastlines, coral reefs, and the deep ocean). The earth's major land biomes and aquatic life zones are discussed in more detail in Chapter 7.

How Do We Recognize the Boundaries of an Ecosystem? For convenience, scientists often consider an ecosystem under study as an isolated unit. However, natural ecosystems rarely have distinct boundaries and are not truly self-contained, self-sustaining systems. Instead, one ecosystem tends to merge with the next in a transitional zone called an **ecotone**, a region containing a mixture of species from adjacent regions and often species not found in either of the bordering ecosystems. For example, a marsh or wetland found between dry land and the open water of a lake or ocean is an ecotone (Figure 4-9). Another example is the zone of grasses,

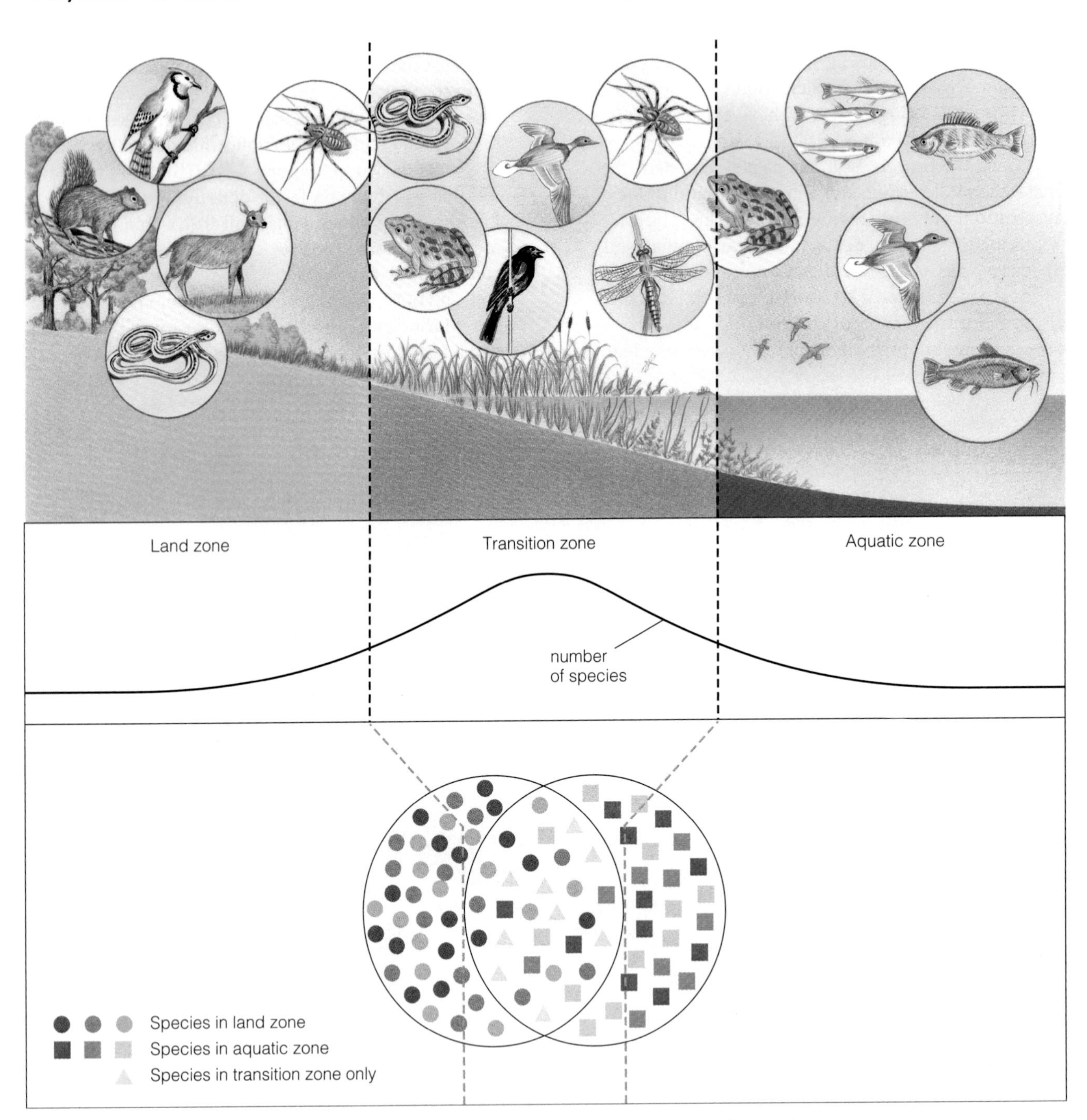

Figure 4-9 Ecosystems rarely have sharp boundaries. Two adjacent ecosystems such as dry land and an open lake often contain a marsh—an ecotone or transitional zone—between them. This zone contains a mixture of species found in each ecosystem and contains some species not found in either ecosystem.

Q: How long has the latest version of the human species been on the earth?

small shrubs, and scattered small trees found between a grassland and a forest. An ecotone often contains both a greater number of species and a higher population density than either adjacent ecosystem.

What Are the Major Components of Ecosystems? The ecosphere and its ecosystems can be separated into two parts: **(1) abiotic**, or nonliving, components (water, air, nutrients, and solar energy), and **(2) biotic**, or living, components (plants, animals, and microorganisms, sometimes called biota). Figures 4-10 and 4-11 are greatly simplified diagrams of some of the biotic and abiotic components in a freshwater aquatic ecosystem and a terrestrial ecosystem.

What Are the Major Nonliving Components of Ecosystems? The nonliving, or abiotic, components of an ecosystem are the physical and chemical factors that influence living organisms. Some important physical factors affecting land ecosystems are sunlight, temperature, precipitation, wind, latitude (distance from the equator), altitude (distance above sea level), frequency of fire, and nature of the soil. For aquatic ecosystems, major physical factors include water currents, concentrations of dissolved nutrients (such as nitrogen and phosphorus), and the amount of suspended solid material.

Important chemical factors affecting ecosystems are the supply of water and air in the soil and the supply of plant nutrients or toxic substances dissolved in soil moisture (or in water in aquatic habitats). In aquatic ecosystems, salinity and the level of dissolved oxygen are also major chemical factors.

Different species thrive under different physical conditions. Some need bright sunlight; others thrive better in shade. Some require a hot environment, others a cool or cold one. Some do best under wet conditions; others under dry conditions.

Each population in an ecosystem has a **range of tolerance** to variations in its physical and chemical environment (Figure 4-12). For example, trout thrive in colder water than do bass or perch, which in turn need colder water than catfish. Individuals within a population may also have slightly different tolerance ranges for temperature or other factors because of small differences in genetic makeup, health, and age. Thus, although a trout population may do best within a narrow band of temperatures (*optimum level or range*), a few individuals can survive both above and below that band. As Figure 4-12 shows, tolerance has its limits, beyond which none of the trout can survive.

These observations are summarized in the **law of tolerance**: *The existence, abundance, and distribution of a species in an ecosystem are determined by whether*

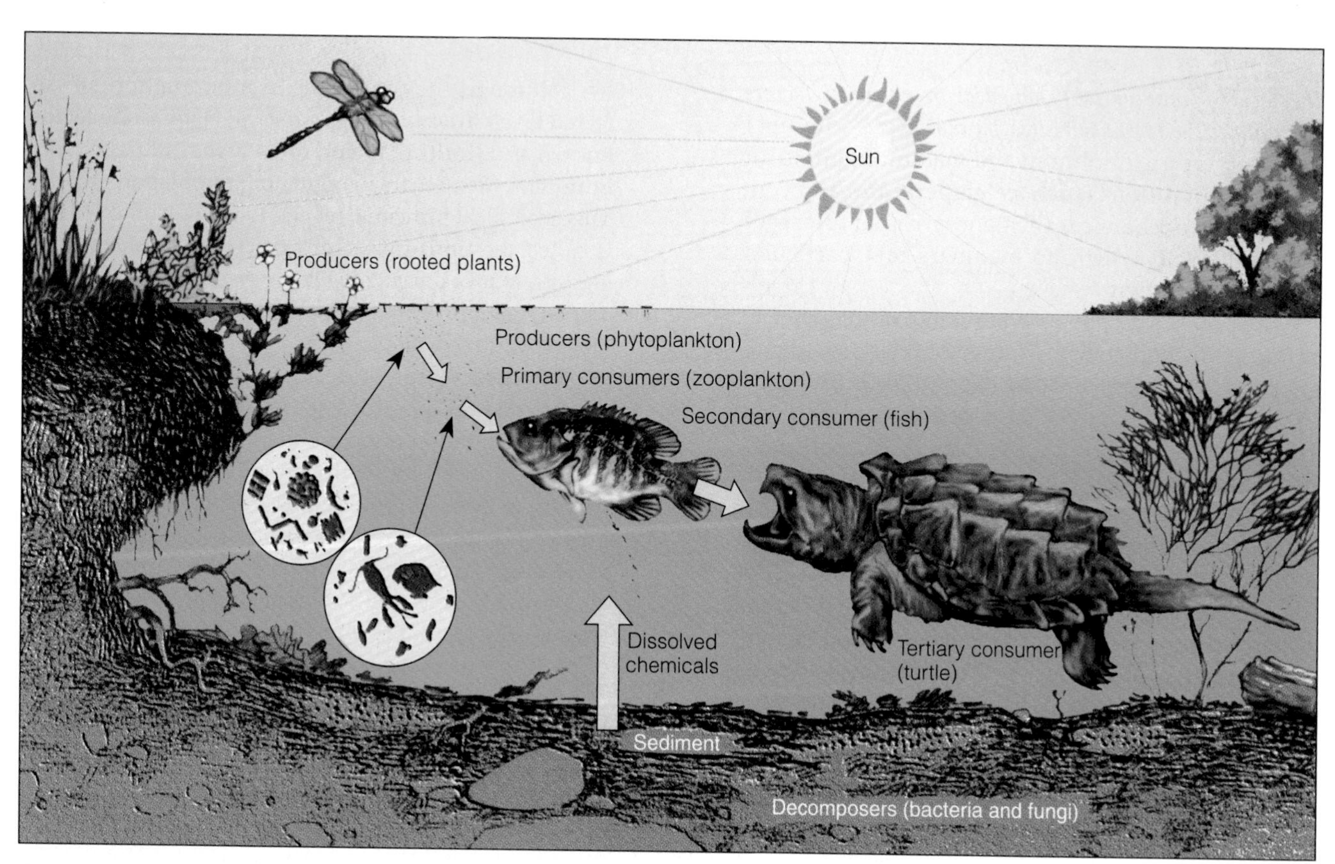

Figure 4-10 Major components of a freshwater aquatic ecosystem.

A: About 60,000 years, and possibly 90,000 years

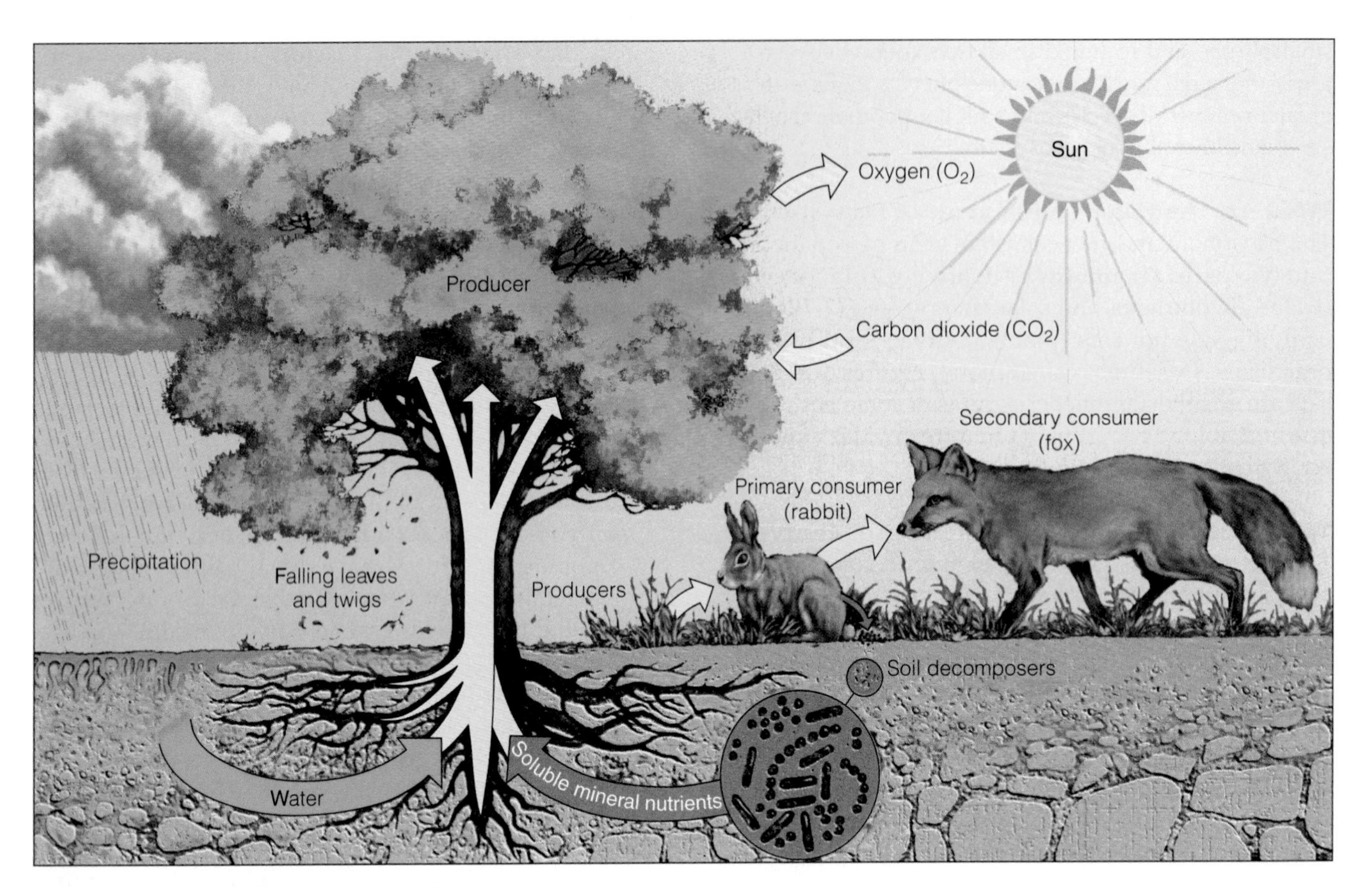

Figure 4-11 Major components of a terrestrial ecosystem.

the levels of one or more physical or chemical factors fall within the range tolerated by that species. In other words, there are minimum and maximum limits for physical conditions (such as temperature) and concentrations of chemical substances, called **tolerance limits**, beyond which no members of a particular species can survive.

A species may have a wide range of tolerance to some factors and a narrow range of tolerance to others. Most organisms are least tolerant during juvenile or reproductive stages of their life cycles. Highly tolerant species can live in a variety of habitats with widely different conditions.

Humans often alter the physical and chemical components of parts of the earth. A large dam, for example, converts the part of a stream above the dam to a deep reservoir or artificial lake with quite different water temperatures, water flow, penetration of sunlight, and other abiotic factors. The flow rate and temperature of the stream below the dam are also changed because the dam regulates the flow of water and releases extremely cold water from the bottom of its huge reservoir. As a result, many of the fish species that lived in the stream above and below the dam perish because their tolerance limits are exceeded.

Although the organisms in a population are affected by a variety of environmental factors, one factor, known as a **limiting factor**, often turns out to be more important than others in regulating population growth. This ecological principle, related to the law of tolerance, is called the **limiting factor principle**: *Too much or too little of any abiotic factor can limit or prevent growth of a population, even if all other factors are at or near the optimum range of tolerance.*

Limiting factors in terrestrial (land) ecosystems can include temperature, water, light, and soil nutrients. Precipitation often is the limiting factor. Lack of water in a desert limits the growth of plants. Soil nutrients can also be limiting factors on land. Suppose a farmer plants corn in phosphorus-poor soil. Even if water, nitrogen, potassium, and other nutrients are at optimum levels, the corn will stop growing when it uses up the available phosphorus. Here, the amount of phosphorus in the soil limits corn growth.

Too much of an abiotic factor can also be limiting. For example, plants can be killed by too much water or too much fertilizer, a common mistake of many beginning gardeners.

The limiting factor for a particular population can sometimes change. At the beginning of a plant's

Q: When did humans start shifting from hunting and gathering to agriculture?

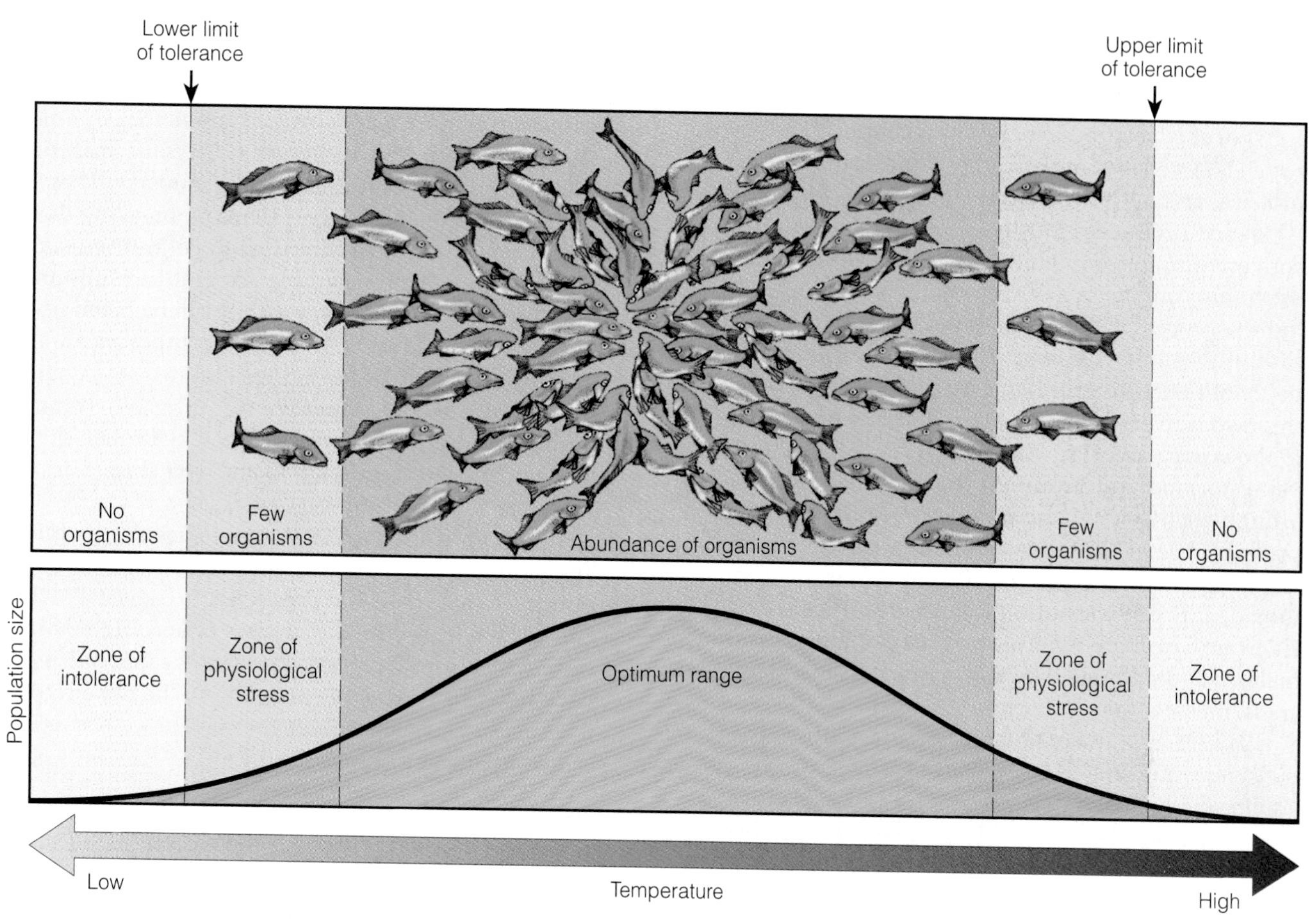

Figure 4-12 Range of tolerance for a population of organisms to an abiotic environmental factor—in this case, temperature.

growing season, for example, temperature may be the limiting factor; later on, the supply of a particular nutrient may limit growth; if a drought occurs, water may be the limiting factor.

Important limiting factors for aquatic ecosystems include temperature, sunlight, **dissolved oxygen content** (the amount of oxygen gas dissolved in a given volume of water at a particular temperature and pressure), and availability of nutrients. Another limiting factor in aquatic ecosystems is **salinity** (the amounts of various salts dissolved in a given volume of water). Different layers or depths in aquatic ecosystems can have different limiting factors. For example, sunlight is typically a limiting factor for growth of algae populations in the lower layers of lake and ocean communities, but not at the surface layer; nutrients are often limiting factors for algae growing in the surface layer.

What Are the Major Living Components of Ecosystems? Living organisms in ecosystems are usually classified as either *producers* or *consumers*, based on how they get food. **Producers**, sometimes called **autotrophs** (self-feeders), make their own food from compounds obtained from their environment. On land, most producers are green plants. In freshwater and marine ecosystems, algae and plants are the major producers near shorelines; in open water the dominant producers floating and drifting are *phytoplankton*, most of them microscopic. Only producers make their own food; all other organisms are consumers, which depend directly or indirectly on food provided by producers.

Most producers capture sunlight to make sugars (such as glucose, $C_6H_{12}O_6$) and other complex organic compounds from inorganic (abiotic) nutrients in the environment. This process is called **photosynthesis**. In most green plants, *chlorophyll* (a pigment molecule that gives plants their green color) traps solar energy for use in photosynthesis and converts it into chemical energy. Although a sequence of hundreds of chemical changes takes place during photosynthesis, the overall reaction can be summarized as follows:

carbon dioxide + water + **solar energy** → glucose + oxygen

$6\,CO_2 + 6\,H_2O + \textbf{solar energy} \rightarrow C_6H_{12}O_6 + 6\,O_2$

A: 10,000–12,000 years ago

Green primary producers absorb only about half of the light energy falling on them and reflect or transmit the rest. Plants then manage to convert only about 1–5% of this absorbed energy into chemical energy, which is stored in complex carbohydrates, lipids (fats), proteins, and nucleic acids in plant tissues.

A few producers, mostly specialized bacteria, can convert simple compounds from their environment into more complex nutrient compounds without sunlight, a process called **chemosynthesis**. In one such case, the source of energy is heat generated by the decay of radioactive elements deep in the earth's core; this heat is released at hot-water (hydrothermal) vents in the ocean's depths, where new crust is constantly being formed and reformed. In the pitch darkness around such vents, large populations of specialized producer bacteria use this geothermal energy to convert dissolved hydrogen sulfide (H_2S) and carbon dioxide into organic nutrient molecules. These bacteria in turn become food for a variety of aquatic animals, including huge tube worms and various clams, crabs, mussels, and barnacles.

In 1995, researchers found bacteria subsisting on rock and water about 1,000 meters (3,200 feet) down in aquifers within volcanic rocks near the Columbia River in Washington. They survive by getting dissolved CO_2 from the groundwater and appear to get energy by using hydrogen (H_2) generated in a reaction between iron-rich minerals in the rock and groundwater.

All other organisms in an ecosystem are **consumers** or **heterotrophs** ("other-feeders"), which get their energy and nutrients by feeding on other organisms or their remains. There are several classes of consumers, depending on their primary source of food. **Herbivores** (plant eaters) are called **primary consumers** because they feed directly on producers. **Carnivores** (meat eaters) feed on other consumers; those called **secondary consumers** feed only on primary consumers (herbivores). Most secondary consumers are animals, but a few (such as the Venus's-flytrap plant) trap and digest insects. **Tertiary (higher-level) consumers** feed only on other carnivores. **Omnivores** are consumers that eat both plants and animals; examples are pigs, rats, foxes, bears, cockroaches, and humans.

Other consumers, called **scavengers**, feed on dead organisms that were killed by other organisms or died naturally. Vultures, flies, crows, hyenas, and some species of sharks, beetles, and ants are examples of scavengers. **Detritivores** (detritus feeders and decomposers) live off **detritus** (pronounced di-TRI-tus), or parts of dead organisms and cast-off fragments and wastes of living organisms (Figure 4-13). **Detritus feeders**, such as crabs, carpenter ants, termites, earthworms, and wood beetles, extract nutrients from partly decomposed organic matter in leaf litter, plant debris, and animal dung.

Decomposers, mostly certain types of bacteria and fungi (Figure 4-14), are consumers that complete the breakdown and recycling of organic materials from the remains or wastes of all organisms. They recycle organic matter in ecosystems by breaking down dead organic material (detritus) to get nutrients and releasing the resulting simpler inorganic compounds into the soil and water, where they can be taken up as nutrients by producers. In turn, decomposers are important food sources for worms and insects living in the soil and water. When we say that something is **biodegradable**, we mean that it can be broken down by decomposers. Figures 4-10 and 4-11 show various types of producers and consumers.

Both producers and consumers use the chemical energy stored in glucose and other organic compounds to fuel their life processes. In most cells, this energy is released by the process of **aerobic respiration**, which uses oxygen to convert organic nutrients back into carbon dioxide and water. The net effect of the hundreds of steps in this complex process is represented by the following reaction:

$$\text{glucose} + \text{oxygen} \longrightarrow \text{carbon dioxide} + \text{water} + \textbf{energy}$$

$$C_6H_{12}O_6 + 6\,O_2 \longrightarrow 6\,CO_2 + 6\,H_2O + \textbf{energy}$$

Although the detailed steps differ, the net chemical change for aerobic respiration is the opposite of that for photosynthesis, which takes place during the day, when sunlight is available; aerobic respiration can happen day or night.

Some decomposers get the energy they need through the breakdown of glucose (or other organic compounds) in the absence of oxygen. This form of cellular respiration is called **anaerobic respiration** or **fermentation**. Instead of carbon dioxide and water, the end products of this process are compounds such as methane gas (CH_4), ethyl alcohol (C_2H_6O), acetic acid (the main component of vinegar, $C_2H_4O_2$), and hydrogen sulfide (H_2S, when sulfur compounds are broken down).

The survival of any individual organism depends on the *flow of matter and energy* through its body. However, an ecosystem as a whole survives primarily through a combination of *matter recycling* (rather than one-way flow) and *one-way energy flow* (Figure 4-15). Decomposers complete the cycle of matter by breaking down detritus into inorganic nutrients that are usable by producers. Without decomposers, the entire world would soon be knee-deep in plant litter, dead animal bodies, animal wastes, and garbage. Most life as we know it would no longer exist.

Q: When did the industrial revolution begin?

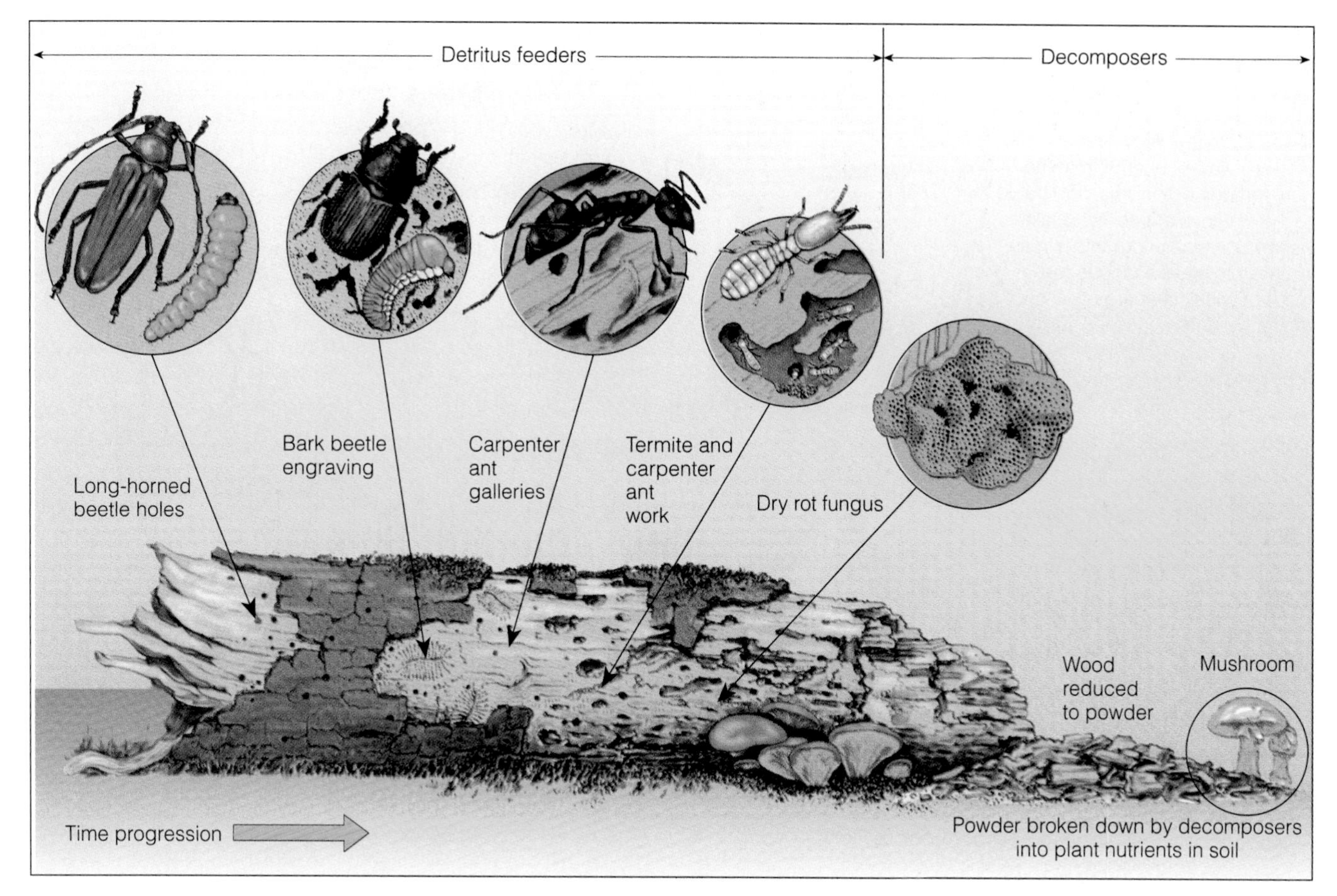

Figure 4-13 Some detritivores, called *detritus feeders*, directly consume tiny fragments of this log. Other detritivores, called *decomposers* (mostly fungi and bacteria), digest complex organic chemicals in fragments of the log into simpler inorganic nutrients. If these nutrients are not washed away or otherwise removed from the system, they can be used again by producers near this location.

Figure 4-14 Two examples of decomposers are shelf fungi (right) and *Boletus luridus* mushrooms (left). The visible part of these decomposing fungi, called the fruiting body or reproductive body, is only a small part of the entire organism. Beneath the fruiting body is an extensive network of microscopic filaments called *mycelia*. The mycelia penetrate into dead leaves, wood, or other detritus and secrete chemicals called *enzymes* that speed up the breakdown of the detritus material into simpler organic nutrients, which are absorbed into fungal cells. (Right S. Flegler/Visuals Unlimited; left, Hans Reinhard/Bruce Coleman Ltd.)

A: About 275 years ago

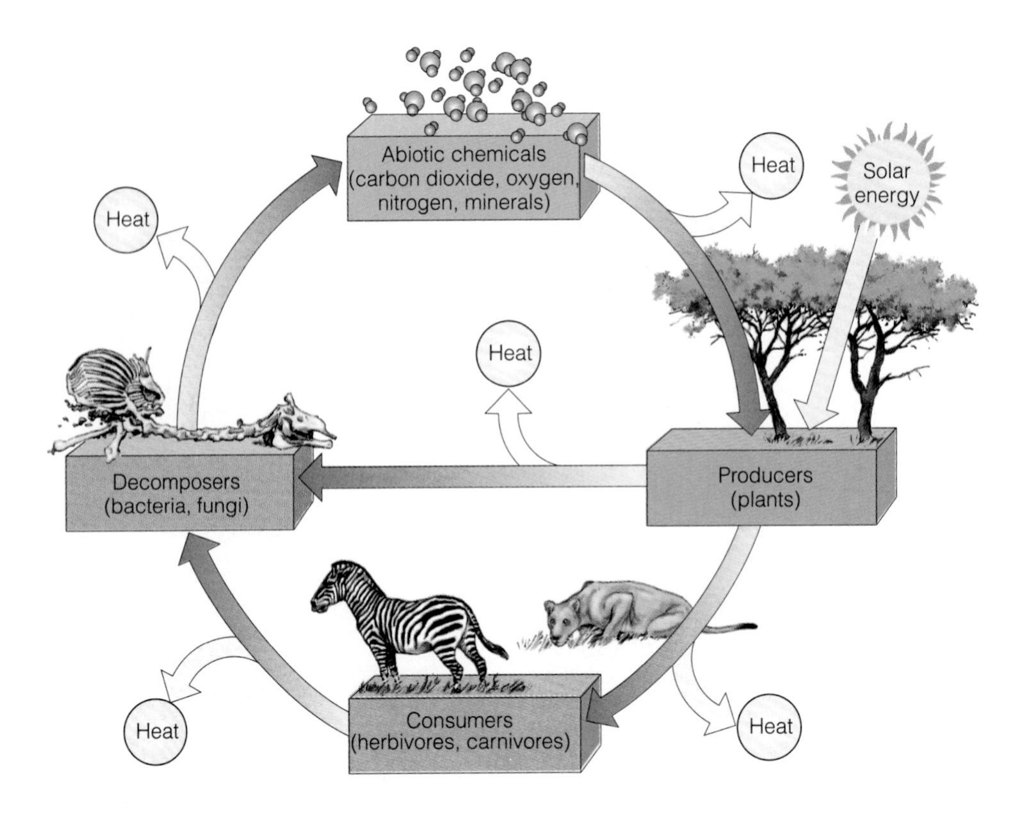

Figure 4-15 Conceptual model showing how an ecosystem's main structural components (energy, chemicals, and organisms) are linked by matter recycling and the one-way flow of energy from the sun, through organisms, and back to the environment as low-quality heat. Each type of organism in an ecosystem plays a unique role in the processes of energy flow and matter cycling.

4-4 FOOD WEBS AND ENERGY FLOW IN ECOSYSTEMS

What Are Food Chains and Food Webs? All organisms, whether dead or alive, are potential sources of food for other organisms. A caterpillar eats a leaf, a robin eats the caterpillar, and a hawk eats the robin. When leaf, caterpillar, robin, and hawk have all died, they in turn are consumed by decomposers. As a result, *there is little waste in natural ecosystems.*

This sequence of organisms, each of which is a source of food for the next, is called a **food chain**. It determines how energy and nutrients move from one organism to another through the ecosystem (Figure 4-16). Energy enters most ecosystems as high-quality sunlight, which is converted to nutrients by photosynthesizing producers (mostly plants). The energy is then passed on to consumers and eventually to decomposers. As each organism uses the high-quality chemical energy in its food to move, grow, and reproduce, this energy is converted into low-quality heat that flows into the environment in accordance with the second energy law.

Ecologists assign each of the organisms in an ecosystem to a *feeding level*, or **trophic level** (from the Greek word *trophos*, "nourishment"), depending on whether it is a producer or a consumer and on what it eats or decomposes. Producers belong to the first trophic level, primary consumers to the second trophic level, secondary consumers to the third, and so on. Detritivores, or decomposers, process detritus from all trophic levels.

Real ecosystems are more complex than this. Most consumers feed on more than one type of organism, and most organisms are eaten by more than one type of consumer. Because most species participate in several different food chains, the organisms in most ecosystems form a complex network of interconnected food chains called a **food web** (Figure 4-17).

Food chains and webs provide avenues for the one-way flow of energy and the cycling of nutrients through ecosystems (Figure 4-15). In most ecosystems, energy and nutrients typically move through two interconnected types of food webs (Figure 4-18). In **grazing food webs**, energy and nutrients move from plants to herbivores (grazers), then through an array of carnivores, and eventually to decomposers. In **detrital food webs**, organic waste material or detritus is the major food source, and energy flows mainly from producers (plants) to decomposers and detritivores. In many terrestrial ecosystems (such as forests) and in aquatic ecosystems (such as streams

Q: What is a scientific theory?

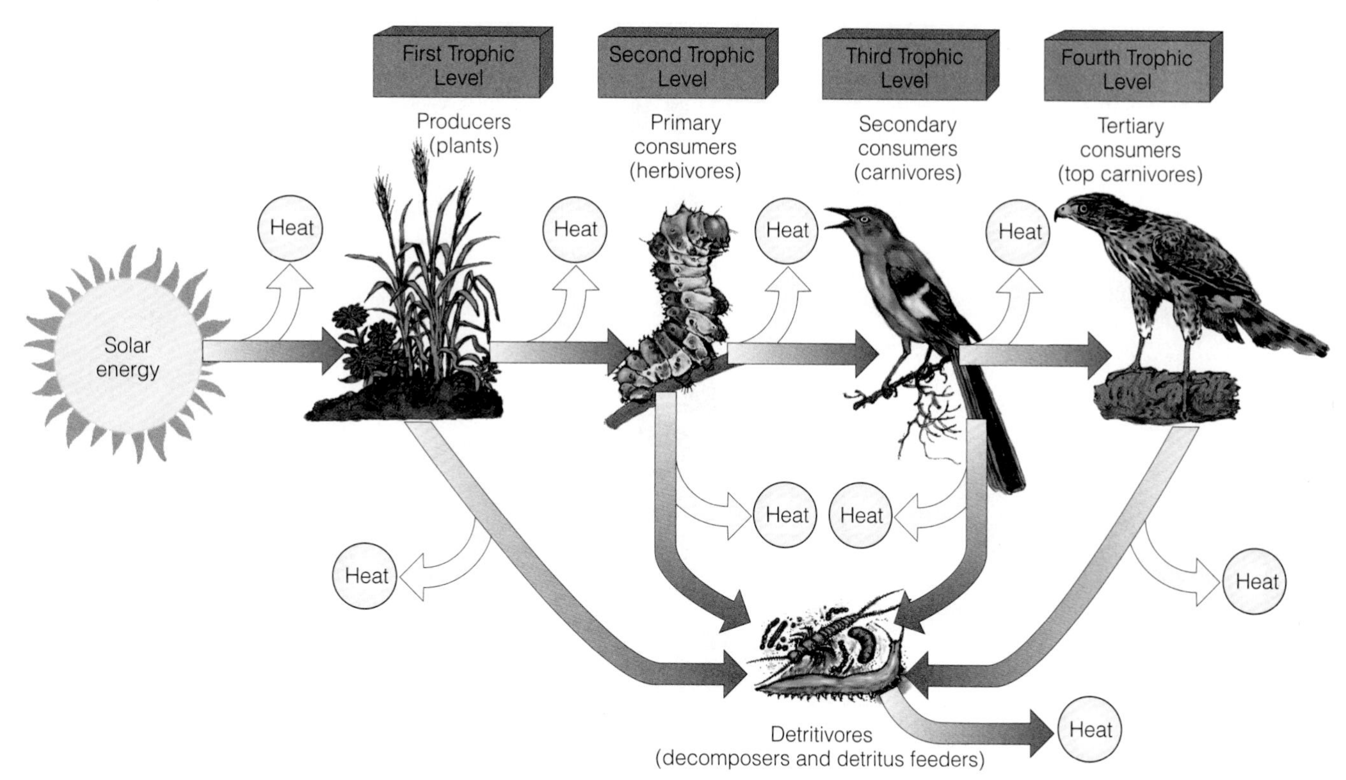

Figure 4-16 Conceptual model of a food chain. The arrows show how chemical energy in food flows through various *trophic levels* or energy transfers; most of the energy is degraded to heat, in accordance with the second law of energy. Food chains rarely have more than four trophic levels.

and marshes), detrital pathways predominate. In the deep ocean, most of the energy flows through grazing food webs.

How Can We Represent the Energy Flow and Storage of Biomass in an Ecosystem? Each trophic level in a food chain or web contains a certain amount of **biomass**, the dry weight of all organic matter contained in its organisms. Biomass is measured as the *dry weight* of organisms because the water they contain is not a source of energy or nutritional value. Biomass represents the chemical energy stored in the organic matter of a trophic level. For example, the biomass of the first trophic level is the total dry weight of all producers. The biomass of the second trophic level is the total dry weight of all herbivores, and so on.

In a food chain or web, energy stored in biomass is transferred from one trophic level to another, with some usable energy degraded and lost to the environment as low-quality heat in each transfer. At each successive trophic level, some of the available biomass is neither eaten, digested, nor absorbed; it simply goes through the intestinal tract of the consumer and is expelled as fecal waste.

Thus, only a small portion of what is eaten and digested is actually converted into an organism's bodily material or biomass, and the amount of usable energy available to each successive trophic level declines. The percentage of usable energy transferred as biomass from one trophic level to the next varies from 5% to 20% (that is, a loss of 80–95%), depending on the types of species and the ecosystem involved. The percentage of energy transferred from one trophic level to another is called **ecological efficiency**. Assuming 10% ecological efficiency (90% loss) at each trophic transfer, if green plants in an area manage to capture 10,000 units of energy from the sun, then only about 1,000 units of energy will be available to support herbivores, and only about 100 units to support carnivores.

The more trophic levels or steps in a food chain or web, the greater the cumulative loss of usable energy as energy flows through the various trophic levels. The **pyramid of energy flow*** in Figure 4-19 illustrates this energy loss for a simple food chain, assuming a 90% energy loss with each transfer. Figure 4-20 shows the pyramid of energy flow during 1 year for an aquatic

*Because such pyramids represent energy flow, not energy storage, they should not be called pyramids of energy (a common error in many biology and environmental science textbooks).

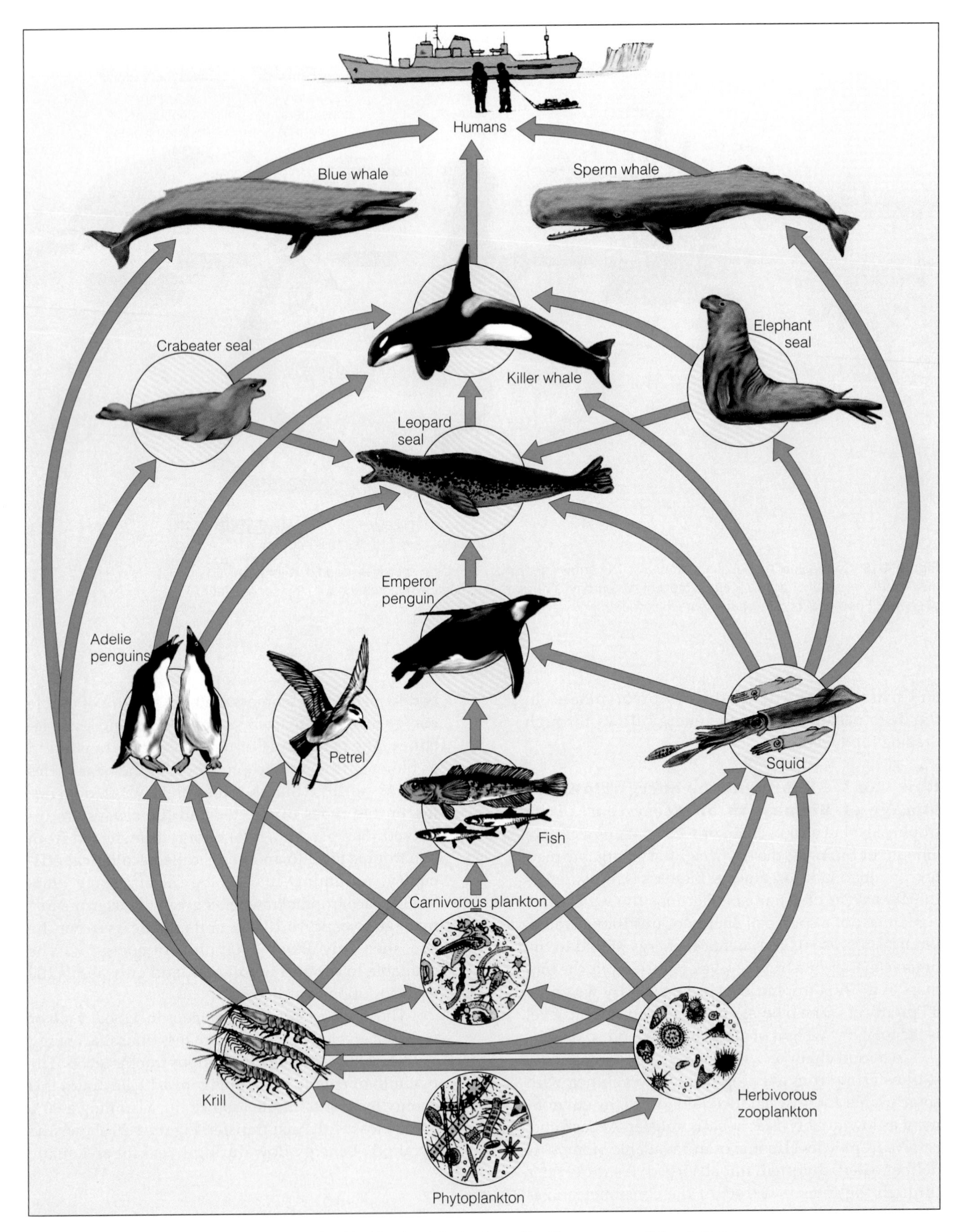

Figure 4-17 Conceptual model of a greatly simplified food web in the Antarctic. Many more participants in the web, including an array of decomposer organisms, are not depicted here.

Q: Are scientific laws and theories *absolutely* true?

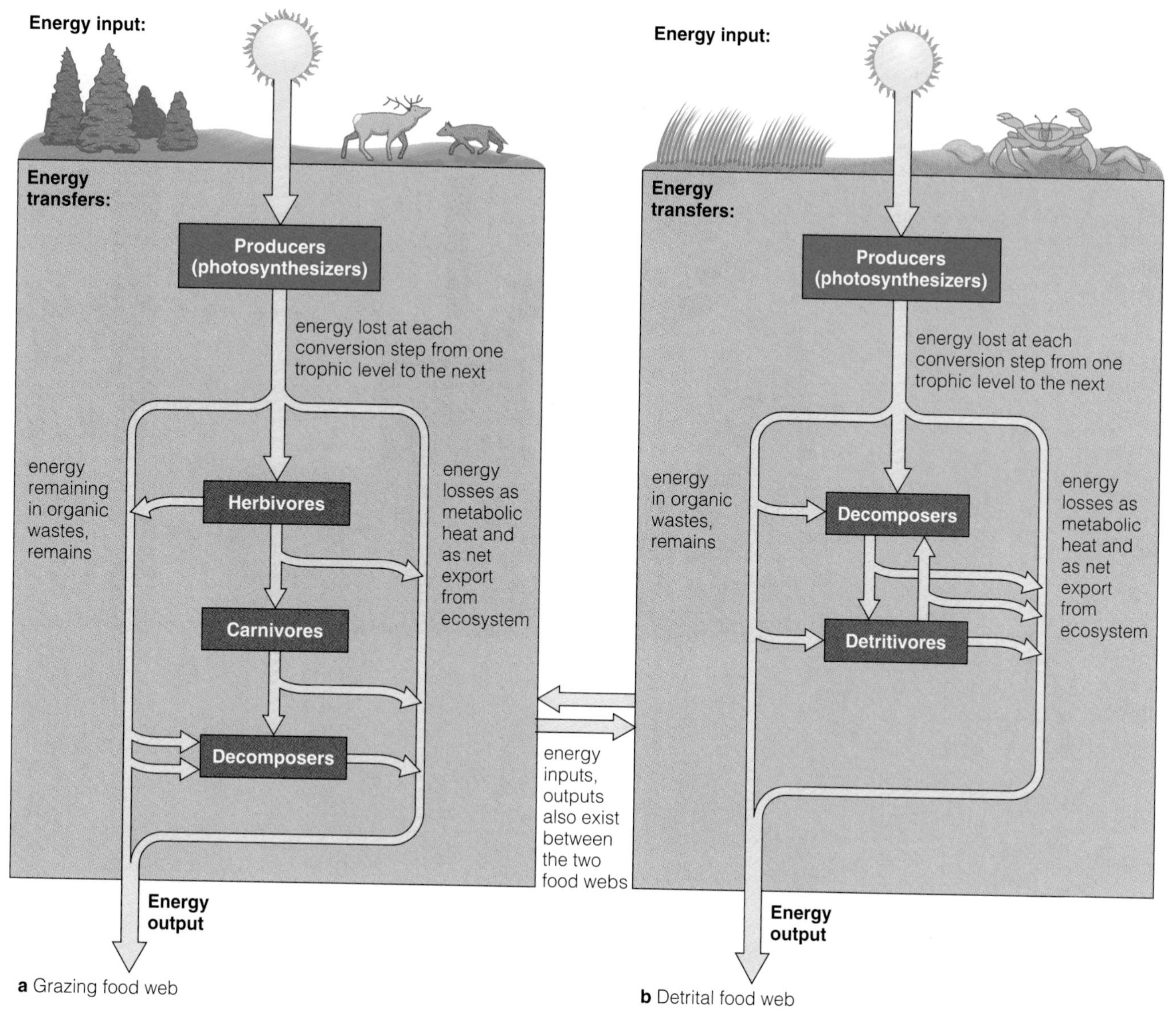

Figure 4-18 Conceptual models of energy inputs, outputs, and transfers for two interconnected types of food webs in land ecosystems. (Used by permission from Cecie Starr, *Biology: Concepts and Applications*, 2d ed., Belmont, Calif.: Wadsworth, 1994)

ecosystem in Silver Springs, Florida. Pyramids of energy flow always have an upright pyramidal shape because of the automatic degradation of energy quality required by the second law of energy.

Energy flow pyramids explain why the earth can support more people if they eat at lower trophic levels by consuming grains, vegetables, and fruits directly (for example, grain → human), rather than passing such crops through another trophic level and eating grain-eaters (grain → steer → human).*

*To avoid malnutrition, anyone eating a strictly or mostly vegetarian diet must consume enough beans or other sources of vegetable protein to provide the essential amino acids not made by the human body.

The large loss in energy between successive trophic levels also explains why food chains and webs rarely have more than four or five trophic levels. In most cases, too little energy is left after four or five transfers to support organisms feeding at these high trophic levels. This explains why top carnivores such as eagles, hawks, tigers, and white sharks are few in number and are usually the first to suffer when the ecosystems that support them are disrupted. This makes such species especially vulnerable to extinction.

The storage of biomass at various trophic levels in an ecosystem can be represented by a **pyramid of biomass** (Figure 4-21). Ecologists estimate biomass by harvesting organisms from random patches

A: No, but they have a very high degree of validity based on numerous experiments and tests.

Figure 4-19 Generalized pyramid of energy flow, showing the decrease in usable energy available at each succeeding trophic level in a food chain or web. This conceptual model assumes a 10% ecological efficiency (90% loss in usable energy to the environment, in the form of low-quality heat) with each transfer from one trophic level to another. In nature, ecological efficiency varies from 5% to 20%. Because of the degradation of energy quality required by the second law of energy, these models always have a pyramidal shape.

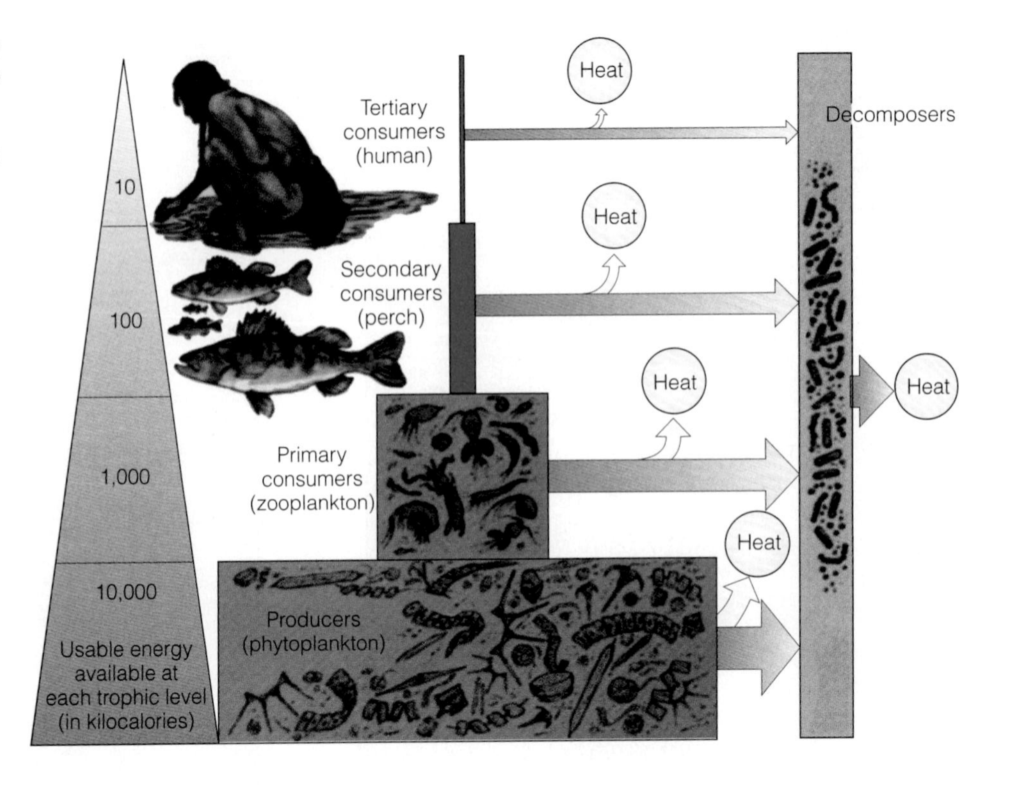

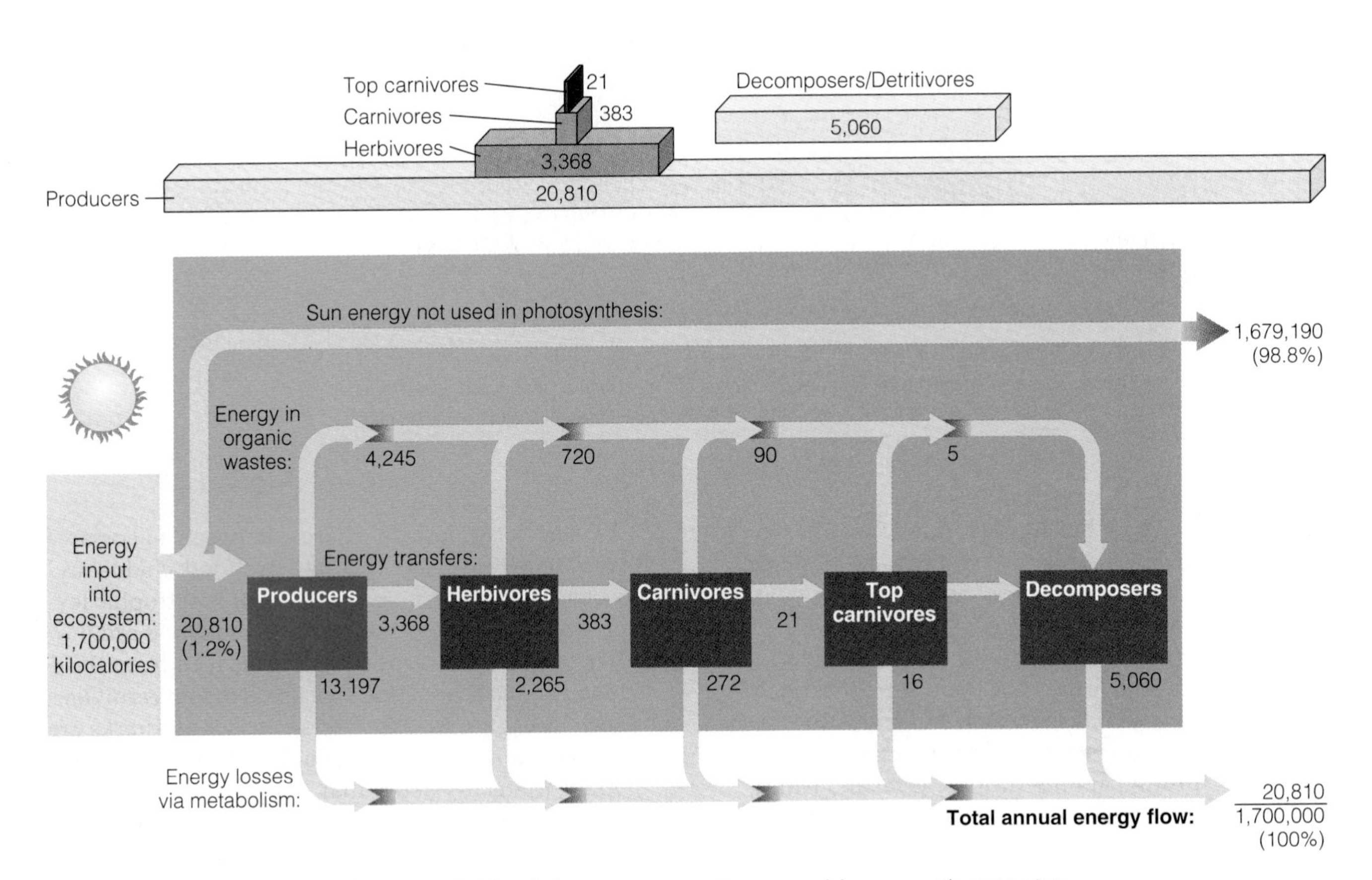

Figure 4-20 Annual pyramid of energy flow (in kilocalories per square meter per year) for an aquatic ecosystem in Silver Springs, Florida. The pyramid is constructed by using the data on energy flow through this ecosystem shown in the bottom drawing. (Used by permission from Cecie Starr, *Biology: Concepts and Applications*, 2nd ed., Belmont, Calif.: Wadsworth, 1994)

or narrow strips in an ecosystem. The sample organisms are then sorted according to trophic levels, dried, and weighed. These data are used to plot a pyramid of biomass. Because the typical relationship involves many producers but not so many primary consumers and just a few secondary consumers, the graph usually looks like an upright pyramid (Figure 4-21, left). However, such graphs for some ecosystems do not have the typical upright pyramid shape (Figure 4-21, right).

By estimating the number of organisms at each trophic level, ecologists can also create a graphic display called a **pyramid of numbers** for an ecosystem (Figure 4-22, left). For example, 1,000 metric tons of grass might support 27,000,000 grasshoppers, which might support 90,000 frogs, which might support 300 trout, which in turn could feed 1 person for about 30 days. By eliminating trout from their diet, 30 people could survive for 30 days by consuming 100 frogs a day. If the food situation were more desperate, frogs could be eliminated from the diet and 900 people could survive for 30 days by eating about 1,000 grasshoppers a day. (Fried grasshoppers are a delicacy in some countries and contain more protein per unit of weight than most conventional forms of meat and meat products.) These graphs for some ecosystems do not have the typical upright pyramid shape (Figure 4-22, right).

How Rapidly Do Producers in Different Ecosystems Produce Biomass? The *rate* at which an ecosystem's producers convert solar energy into chemical energy as biomass is the ecosystem's **gross primary productivity (GPP)**. In effect, it is the rate at which plants or other producers can use photosynthesis to make more plant material (biomass). Figure 4-23 shows how this productivity varies in different parts of the earth.

Figure 4-24 shows that gross primary productivity is generally greatest in the shallow waters near continents; along coral reefs where abundant light, heat, and nutrients stimulate the growth of algae; and where upwelling currents bring nitrogen and phosphorus from the ocean bottom to the

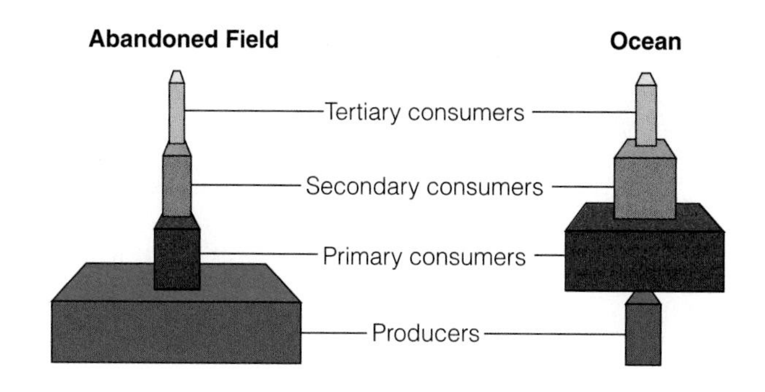

Figure 4-21 Generalized graphs of biomass of organisms in the various trophic levels for two ecosystems. The size of each tier in this conceptual model represents the dry weight per square meter of all organisms at that trophic level. For most land ecosystems, the total biomass at each successive trophic level decreases, yielding a pyramid of biomass with a large base of producers, topped by a series of increasingly smaller biomasses at higher trophic levels (left). In the open waters of aquatic ecosystems (right), the biomass of primary consumers (zooplankton) can actually exceed that of producers (which are microscopic phytoplankton that grow and reproduce rapidly, not large plants that grow and reproduce slowly). The zooplankton eat the phytoplankton almost as fast as they are produced. As a result, the producer population is never very large, and the graph is not an upright pyramid.

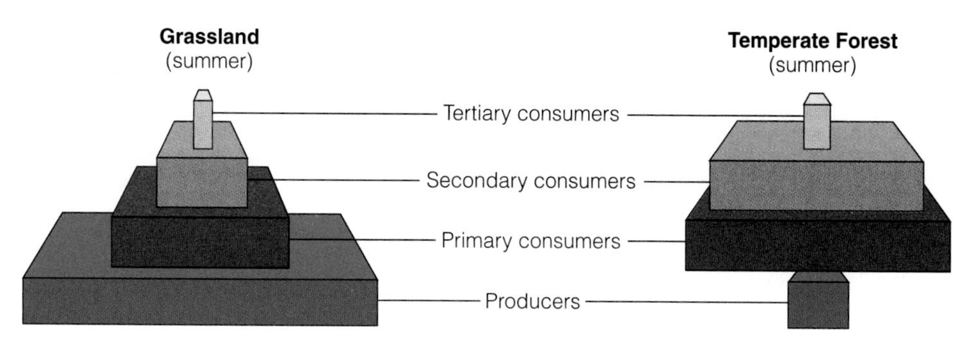

Figure 4-22 Generalized graphs of numbers of organisms in the various trophic levels for two ecosystems. Numbers of organisms for grasslands and many other ecosystems taper off from the producer level to the higher trophic levels, forming an upright pyramid (left). For other ecosystems, however, the graph can take a different shape. For example, a temperate forest (right) has a few large producers (the trees) that support a much larger number of small primary consumers (insects) that feed on the trees.

surface. High-nutrient, upwelling regions of the ocean make up only one-thousandth of the total ocean area, but they have high enough primary productivities to produce nearly one-half of the world's marine fish. Most of the open ocean has a low gross primary productivity.

To stay alive, grow, and reproduce, an ecosystem's producers must use some of the total biomass they produce for their own respiration. Only what is left, called **net primary productivity (NPP)**, is available for use as food by other organisms (consumers) in an ecosystem:

Net primary productivity	=	Rate at which producers store **chemical energy** stored as biomass (produced by photosynthesis)	−	Rate at which producers use **chemical energy** stored as biomass (through aerobic respiration)

Net primary productivity is the *rate* at which energy for use by consumers is stored in new biomass (cells, leaves, roots, and stems). It is measured in units of the energy or biomass available to consumers in a specified area over a given time. It is typically measured in kilocalories per square meter per year ($kcal/m^2/yr$) or in grams of biomass created per square meter per year ($g/m^2/yr$). Ultimately, the planet's total net primary productivity limits the number of consumers, including humans, that can survive on the earth. In other words, *the earth's total net primary productivity is the upper limit determining the planet's carrying capacity for all species.*

Various ecosystems and life zones differ in their net primary productivity (Figure 4-24). Estuaries, swamps and marshes, and tropical rain forests are highly productive; open ocean, tundra (arctic and alpine grasslands), and desert are the least productive. It is tempting to conclude that to feed the world's hungry millions we should harvest plants in estuaries, swamps, and marshes, or clear tropical forests and plant crops. Most plants in estuaries, swamps, and marshes cannot be eaten by people, however, and they are vital food sources (and spawning areas) for fish, shrimp, and other aquatic life that provide us and other consumers with protein.

In tropical forests, most nutrients are stored in the vegetation rather than in the soil. When the trees are removed, the nutrient-poor soils are rapidly depleted

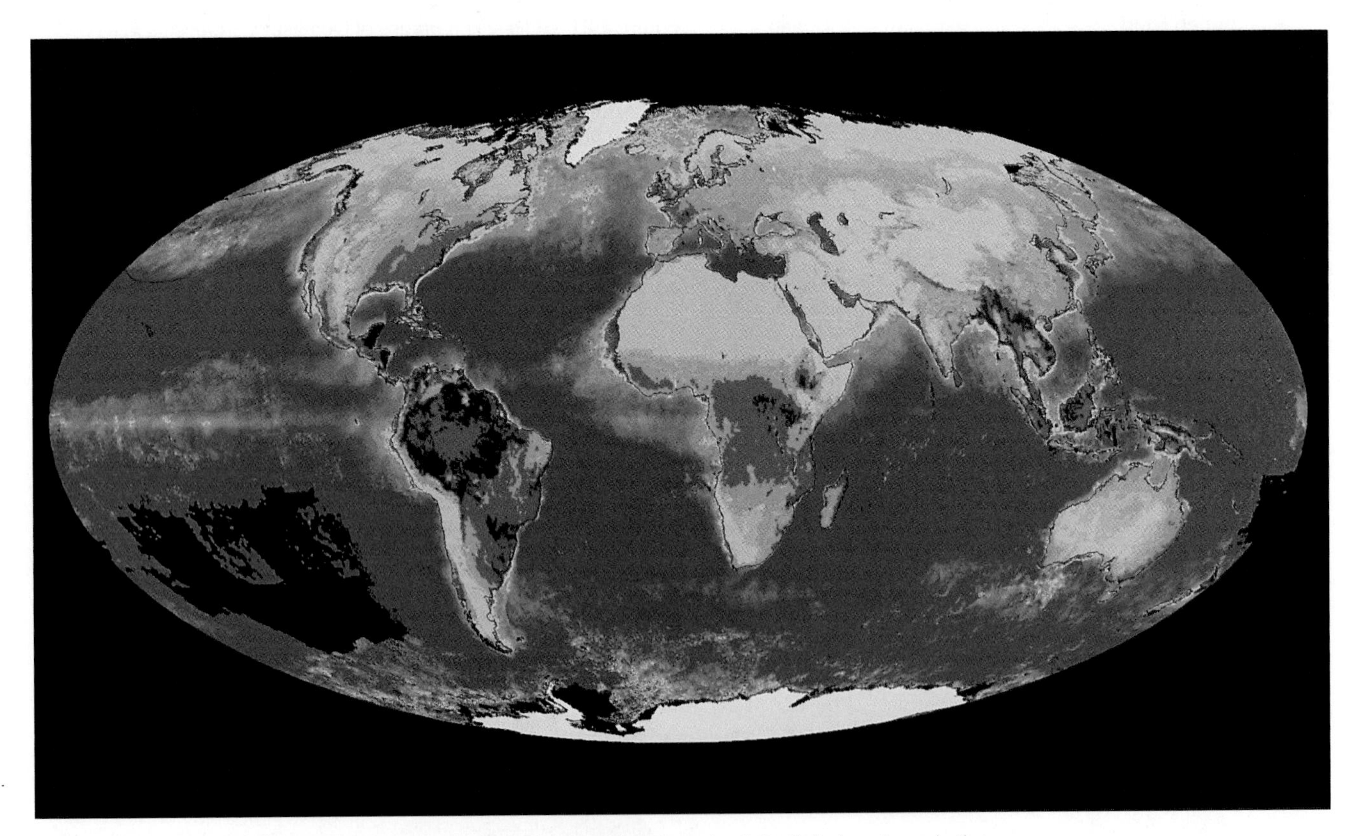

Figure 4-23 Three years of satellite data on the earth's gross primary productivity. Rain forests and other highly productive areas appear as dark green, deserts as yellow. The concentration of phytoplankton, a primary indicator of ocean productivity, ranges from red (highest) to orange, yellow, green, and blue (lowest). (Gene Carl Feldman, Compton J. Tucker-NASA/Goddard Space Flight Center)

Q: What is throughput?

of their nutrients by frequent rains and growing crops. Crops can be grown only for a short time without massive and expensive applications of commercial fertilizers. This explains why many ecologists urge us to protect, not clear, large areas of tropical forest to supply food.

Agricultural land is a highly modified ecosystem in which we try to increase the net primary productivity and biomass of selected crop plants by adding water (irrigation) and nutrients (fertilizers). Nitrogen as nitrate (NO_3^-) and phosphorus as phosphate (PO_4^{3-}) are the most common nutrients in fertilizers because they are most often the nutrients limiting crop growth. Despite such inputs, the net primary productivity of agricultural land is not particularly high compared to that of other ecosystems (Figure 4-24).

Figure 4-25 shows the total global net primary productivity (in billions of kilocalories per year) for various types of ecosystems and life zones. An estimated 59% of the earth's annual net primary productivity occurs on land; the remaining 41% is produced in oceans and other aquatic systems.

Even though the open ocean has a low net primary productivity (Figure 4-24), its overall contribution is high (Figure 4-25) because it covers about 71% of the planet's surface. Tropical rain forests cover only about 2% of the earth's surface but have such a high net primary productivity (Figure 4-24) that they make a major contribution to the planet's overall net primary productivity (Figure 4-25).

Because the open ocean contributes so much of the earth's overall net primary productivity, why not harvest its primary producers (floating and drifting phytoplankton) to help feed the rapidly growing human population? The second law of energy explains why this won't work. Harvesting the widely dispersed tiny floating producers in the open ocean would require enormous amounts of energy for pumping, filtration, and processing. This would take much more fossil fuel and other types of energy than the food energy we would get.

Excluding uninhabitable areas of rock, ice, desert, and steep mountain terrain, humans have taken over, disturbed, or degraded about 73% of the earth's land surface (Figure 1-5). Ecologists have estimated that humans now use, waste, or destroy about 27% of the earth's total potential net primary productivity and

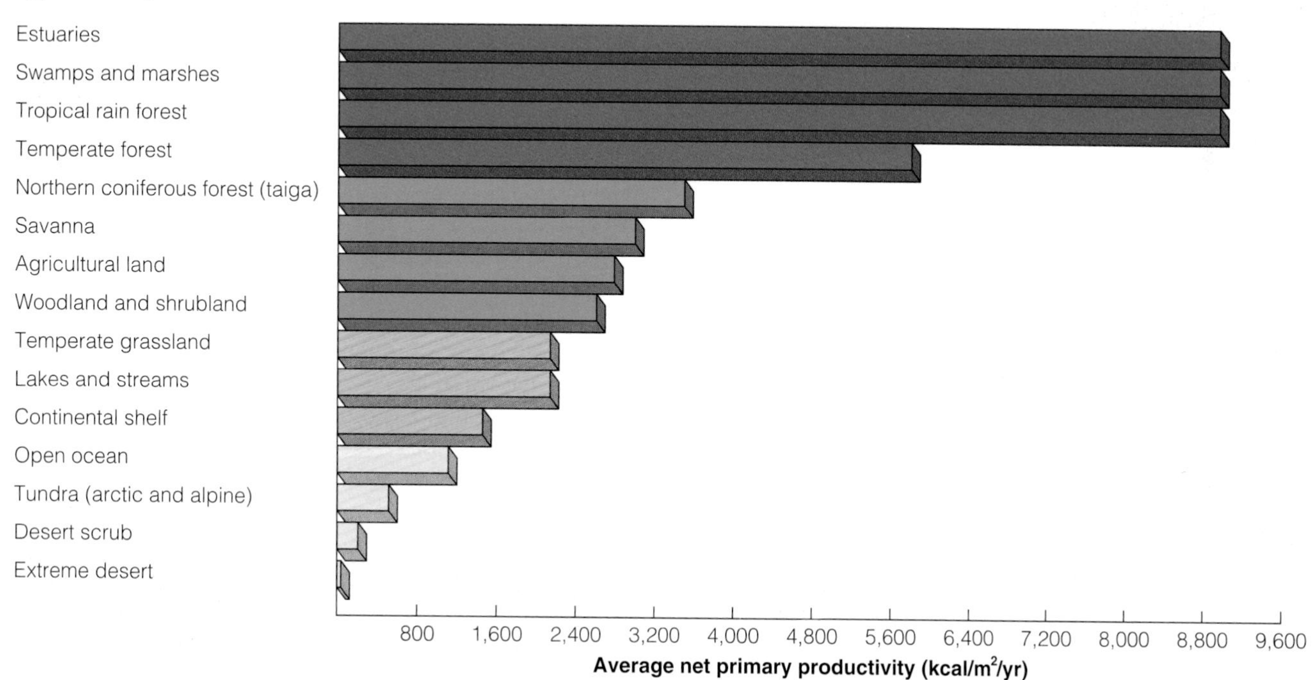

Figure 4-24 Estimated annual average net primary productivity per unit of area in major life zones and ecosystems, expressed as kilocalories of energy produced per square meter per year (kcal/m²/yr). (Data from R. H. Whittaker, *Communities and Ecosystems*, 2d ed., New York: Macmillan, 1975)

A: The flow of matter, energy, or information through a system (Figure 2-6)

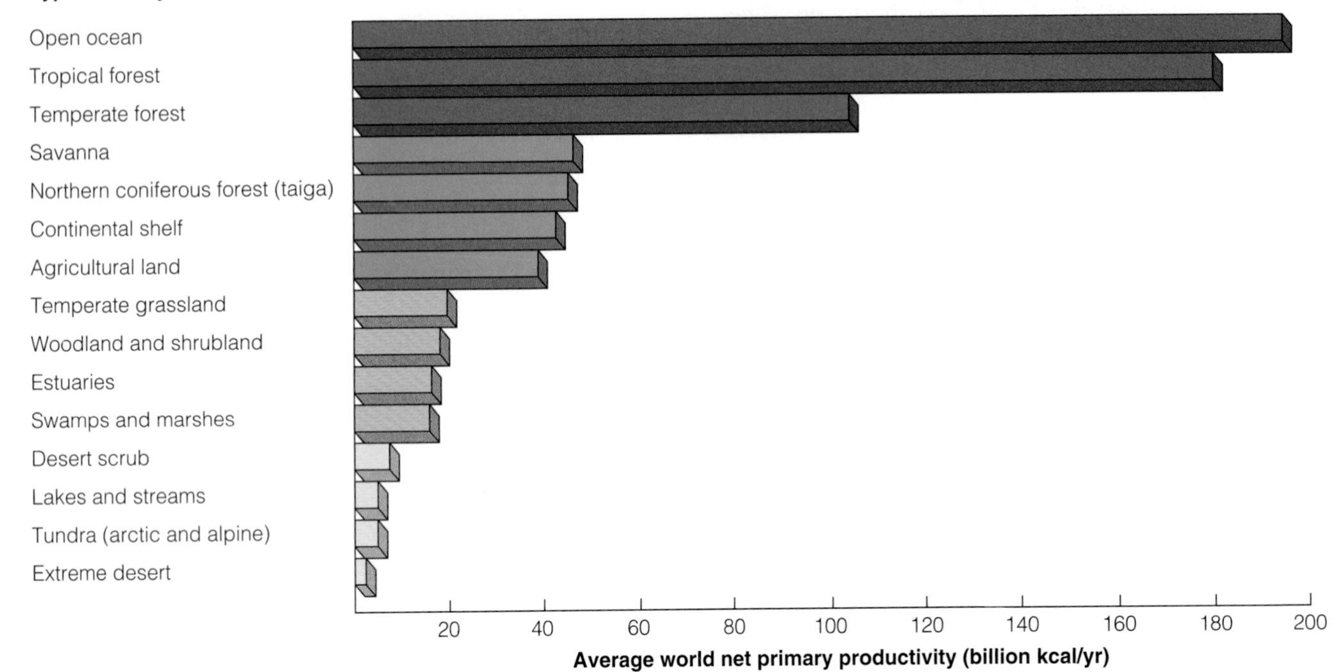

Figure 4-25 Estimated annual contributions of various types of ecosystems and life zones to the world's overall net primary productivity, in billions of kilocalories per year. This figure shows the *total* net primary productivity per year throughout the world for each type of ecosystem or life zone. Figure 4-24 gives the average net primary productivity per unit of area (square meter) per year of each of these ecosystems or life zones. (Data from R. H. Whittaker, *Communities and Ecosystems*, 2d ed., New York: Macmillan, 1975)

40% of the net primary productivity of the planet's terrestrial ecosystems.

This is the main reason why we are crowding out or eliminating the habitats and food supplies of a growing number of other species. If current estimates of our use of the earth's annual net primary productivity are reasonably correct, what will happen to us and to other species if the human population doubles over the next 40–50 years and per capita consumption of resources rises sharply?

4-5 HOW DO ECOLOGISTS LEARN ABOUT ECOSYSTEMS?

What Is Field Research? Ecologists try to unravel some of the workings of ecosystems in three general ways: field research, laboratory research, and system analysis. Each approach has its advantages and disadvantages.

Field research, sometimes called muddy-boots biology, involves going into nature and observing and measuring the structure of ecosystems and what happens in them. Most of what we know about the structure and functioning of ecosystems and how ecosystems change with changing environmental conditions has come from such research.

However, because the parts and interactions among the components of even simple ecosystems are so complex, it is expensive, time-consuming, and difficult to get such information by carrying out field experiments. Because of the large number of interacting variables in nature, it is often difficult or impossible to set up meaningful controlled experiments in which only one variable is varied.

Increasingly, ecologists are using new technologies to collect field data. *Remote sensing* from aircraft and satellites has allowed ecologists to collect data at scales that can include many interacting ecosystems and even whole biomes (Figure 4-26). *Geographic information systems* (*GIS*) have enabled ecologists to organize information gathered across broad geographic regions in a spatial database and perform analyses at a scale that was previously difficult to achieve (Figure 4-26). Even on the "muddy-boots" scale of individual organisms, ecologists are using sensors (such as solar radiation sensors and

Friedrich, R.L., and R.V. Blystone, "Internet Teaching Resources for Remote Sensing and GIS." *BioScience*, v. 48, n. 3, 187(6).

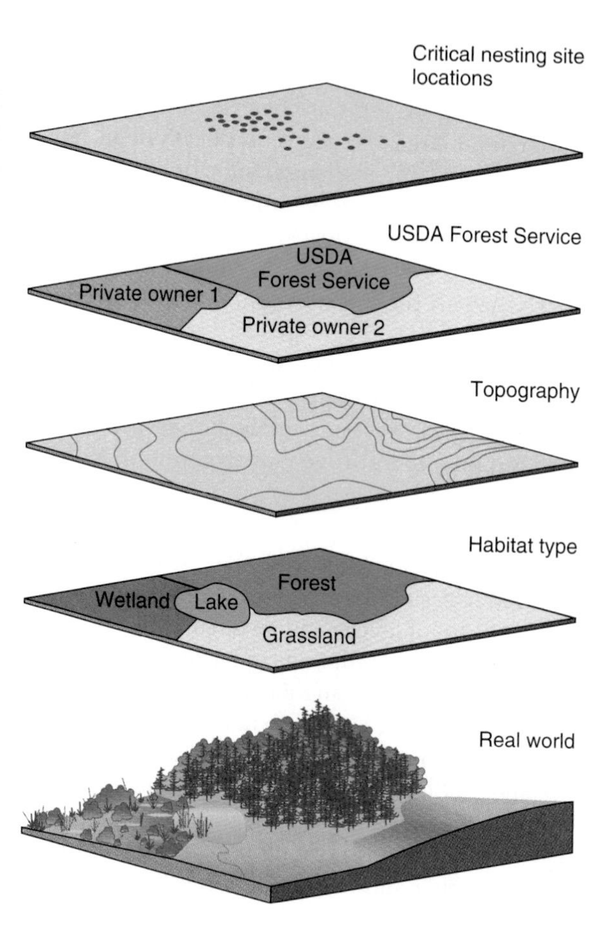

Figure 4-26 New technologies are enabling scientists to collect field information more effectively across broad geographic scales. **(a)** Remote sensing involves use of sensors to collect information about a system from a distance. Remote sensing is providing such information about the distribution of natural ecosystems, properties of those ecosystems, and changes through time. **(b)** Geographic information systems (GIS) provide the computer technology for organizing, storing, and analyzing complex data collected over broad geographical areas. GIS enables scientists to overlay many layers of data (for example, soils, topography, distribution of endangered populations, and land protection status). (Landsat 7 illustration Courtesy of NASA)

infrared gas analyzers), along with miniature computers, to gather field data more efficiently.

What Is Laboratory Research? In the past 50 years, ecologists have increasingly supplemented field research by using *laboratory research* to set up, observe, and make measurements of model ecosystems and populations under laboratory conditions. Such simplified systems have been set up in containers such as culture tubes, bottles, aquarium tanks, and greenhouses, and in chambers where temperature, light, CO_2, humidity, and other variables can be carefully controlled. In such systems, it is easier for scientists to carry out controlled experiments. Often such laboratory experiments are quicker and cheaper than similar experiments in the field.

Such laboratory research has led to enormous amounts of information and many fruitful hypotheses. Often laboratory research can isolate cause and effect relationships that can be tested further in the field or by systems analysis using computer simulations. Indoor and outdoor chambers can also be used to study the effects of increasing levels of atmospheric CO_2 and ultraviolet light on populations, diversity, and the productivity of producers (including important agricultural crops). This can help us project the consequences of ozone depletion in the stratosphere and possible global warming. Such experiments can be carried out in the field only by waiting for such changes to occur, far too late to prevent possible harmful consequences.

Currently, more than half of ecological research involves laboratory experiments and a much smaller proportion is carried out in the field. However, there is the important question of whether what scientists observe and measure in a simplified, controlled system under laboratory conditions takes place in the same way in the more complex and dynamic conditions found in nature. Thus, the results of laboratory research must be coupled with and supported by field research.

What Is Systems Analysis? Since the late 1960s ecologists have made increasing use of *systems analysis* to simulate ecosystems and study their structure and function. Systems analysis is a useful tool for ecosystem research and management because ecosystems are complex and dynamic and involve inputs,

Hint: Enter the search terms *geographic information systems, remote sensing* using the Subject Guide.

accumulations, flows (throughputs), outputs, multiple coupled feedback loops, and time delays. The advantage of systems analysis is that it can help us understand large and very complex systems (such as rivers, oceans, forests, grasslands, or cities) that cannot be adequately studied and modeled in field and laboratory research. Most such analysis is carried out by a team of investigators because a single person rarely has all the knowledge and skills needed for this approach.

Figure 4-27 outlines the major stages of systems analysis. Ecologist Charles Southwick has described how systems analysis might be used to find out how to improve water quality in a severely polluted stream:

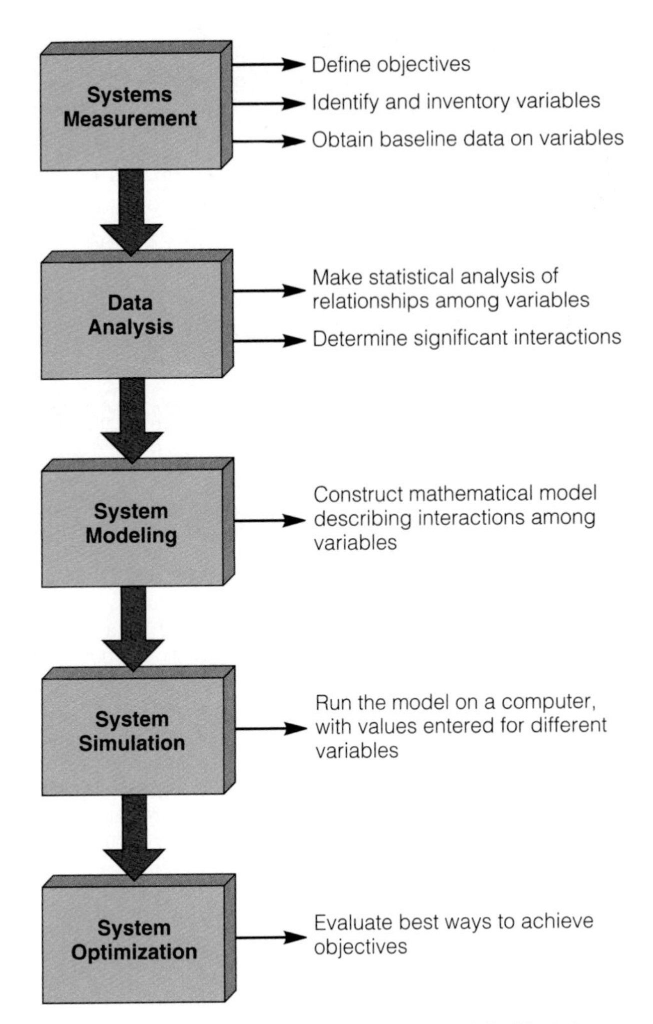

Figure 4-27 Major stages of system analysis. (Modified data from Charles Southwick)

- *Systems measurement.* State the physical and chemical characteristics (temperature, dissolved oxygen, clarity, nutrient levels, and so on) and biological characteristics (populations of bacteria, algae, crustaceans, insects, fish, and so on) desired in the stream. Use field research to make a detailed inventory of the polluted river's existing physical, chemical, and biotic conditions. Identify the major influences on the stream through field studies of its watershed, geologic formations, changes in stream flows, weather and climate patterns, land-use patterns, and the types and amounts of plant nutrients and toxic pollutants flowing into the stream. For a small stream, it would take a team of 5–10 scientists several years to get this baseline data based mostly on field research.
- *Data analysis.* Use computer and statistical analysis of the data to discover relationships among variables and determine how the physical, chemical, and biotic conditions of the stream are affected by land-use patterns, topography, climate and weather, variations in stream flow, and inputs of water, plant nutrients, and toxic chemicals.
- *Systems modeling.* Create a set of mathematical equations representing the major interacting factors influencing the current conditions in the stream. Analysis of existing field and laboratory research on stream quality may help identify key variables and interactions. Test this model by running it on a computer to see whether it reproduces the stream's conditions. Modify the model until it does reproduce the conditions.
- *Systems simulation.* Evaluate ways to improve the quality of the stream by making computer runs of the model in which one or more variables are modified. This involves using the model to answer questions such as "What might happen if we do so and so?" For example, the simulation would estimate and predict changes that might take place if agricultural practices were changed to reduce erosion and runoff of pesticides, if sewage treatment plants were installed, or if discharges from industrial plants were halted or sharply reduced.
- *Systems optimization.* Add economic data for various solutions to the problem and use computer runs to determine the alternatives that improve stream quality at minimum cost. Use this information to develop an economically and politically acceptable plan for improving the quality of the stream.

The simulations and predictions made using ecosystem models are no better than the data and assumptions used to develop the models. Thus, careful field and laboratory ecological research must be used to provide the baseline data and causal relationships

Q: What is a positive feedback loop?

among key variables needed to develop and test ecosystem models. Often systems analysis also reveals surprising results and new questions that must be verified by a combination of field and laboratory research. Thus, meaningful ecological research involves a combination of field research, laboratory research, and systems analysis.

4-6 ECOSYSTEM SERVICES AND SUSTAINABILITY

What Are Ecosystem Services? Most of us spend the bulk of our lives in human-made houses, office buildings, factories, cars, and other artificial environments that insulate us from raw nature. This insulated life can give us the illusion that we live our lives apart from nature and that we can control or manipulate nature at will to meet our needs and wants.

The truth is that *none of us live apart from nature.* We depend on nature for food, air, water, and virtually everything we use. Ecosystems provide us (and other species) with a number of *natural services.* These **ecosystem services**, which constitute earth capital (Figure 1-2), support life on the earth and are essential to the quality of human life and to the functioning of the world's economies. In the chapters that follow we will explore ways in which ecosystems do the following:

- Control and moderate climate
- Provide us with and renew air, water, and soil
- Recycle vital nutrients through chemical cycling
- Provide us with renewable and nonrenewable energy sources and nonrenewable minerals
- Furnish us with food, fiber, medicines, timber, and paper
- Pollinate crops and other plant species
- Absorb, dilute, or detoxify many pollutants and toxic chemicals
- Help control populations of pests and disease organisms
- Slow soil erosion and help prevent flooding
- Provide the biodiversity of genes and species needed to adapt to ever-changing environmental conditions through evolution and genetic engineering

All natural and human-altered ecosystems that have their functional integrity intact provide some or all of these services free of charge. Most people tend to take these life-saving and money-saving services for granted or are unaware of them until the ecosystems that provide such services are degraded or destroyed.

Why Is Biodiversity Such an Important Ecosystem Service? As environmental conditions have changed over billions of years, many species have become extinct, and new ones have formed. The result of these changes is **biological diversity**, or **biodiversity**: the many forms of life the conditions currently found on the earth. As you learned in Chapter 1, biodiversity includes **genetic diversity** (variability in the genetic makeup among individuals in a single species, Figure 4-4), **species diversity** (the variety of species in different habitats on the earth), and **ecological diversity** (the variety of biological communities that interact with one another and with their nonliving environments).

Another term for biodiversity is **wildness**: the existence of wild gene pools, species, and ecosystems that are completely or mostly undisturbed by human activities. In the words of Henry David Thoreau, "In wildness is the preservation of the world."

We are utterly dependent on this mostly unknown biocapital. This rich variety of genes, species, and ecosystems gives us food, wood, fibers, energy, raw materials, industrial chemicals, and medicines, and it pours hundreds of billions of dollars yearly into the global economy.

The earth's life-forms and ecosystems also provide recycling, purification, and natural pest control. Every species here today contains genetic information that represents thousands to millions of years of adaptation to the earth's changing environmental conditions and is the raw material for future adaptations. Biodiversity is nature's insurance policy against disasters.

Some people also include *human cultural diversity* as part of the earth's biodiversity. The variety of human cultures represents numerous social and technological solutions that have enabled us to survive and adapt to and work with the earth.

What Are the Two Basic Principles of Ecosystem Sustainability? In this chapter we have seen that almost all natural ecosystems and the ecosphere itself achieve *sustainability* through two basic processes: **(1)** by using renewable solar energy as their energy source and **(2)** by recycling reasonably efficiently the nutrients its organisms need for survival, growth, and reproduction. What is common to all ecosystems is not their physical structure but the existence of these two life-sustaining processes (Figure 4-15).

A: An input of information into a system in which a change leads to change in the same direction

These two principles for sustainability arise from the structure and function of natural ecosystems and from the law of conservation of mass and the two laws of energy discussed in Chapter 3 (Figure 3-22). Thus, the results of basic research in the physical and biological sciences provide us with the same guidelines or lessons from nature on how we as humans might live sustainably on the earth.

Why Is an Understanding of Ecology Essential for Environmental Studies? We have seen that the essential features of the living and nonliving parts of individual ecosystems, and of the ecosphere as a whole, are *interdependence* and *connectedness*. Without the services performed by diverse communities of species, we would be starving, gasping for breath, and drowning in our own wastes. Understanding interdependence and connectedness is essential for solving environmental problems and ensuring the sustainability of a high-quality life for human beings and for the survival and good health of other organisms in the ecosphere.

The problems of the human future range far beyond ecology, yet ecology is an essential part of them.

ROBERT H. WHITTAKER

CRITICAL THINKING

1. **(a)** A bumper sticker asks, "Have you thanked a green plant today?" Give two reasons for appreciating a green plant. **(b)** Trace the sources of the materials that make up the bumper sticker and decide whether the sticker itself is a sound application of the slogan. **(c)** Explain how decomposers help keep you alive.

2. Using the second law of energy, explain why there is such a sharp decrease in usable energy as energy flows through a food chain or web. Doesn't an energy loss at each step violate the first law of energy? Explain.

3. Using the second law of energy, explain why many poor people in developing countries live mostly on a vegetarian diet.

4. Using the second law of energy, explain why, on a per weight basis, steak costs more than corn.

5. Why can the total amount of animal flesh on the earth never exceed the total amount of plant flesh, even if all animals were vegetarians?

6. Which causes a larger loss of energy from an ecosystem: a herbivore eating a plant or a carnivore eating an animal? Explain.

7. Why are there fewer lions than mice in an African ecosystem supporting both types of animals?

8. Imagine that you are having a discussion with a maker of pesticides who believes that the only good insect is a dead insect, or with a child who has been culturally conditioned to have "bugitis" (a fear of all insects). **(a)** How would you explain to these people the value of insects? **(b)** Should we never kill insects? Under what circumstances should they be killed? Which kinds? **(c)** Should a species of insects we have classified as a pest be wiped from the face of the earth? Is it likely that we could even do this? Explain.

9. Try to come up with analogies to complete the following statements: **(a)** The atmosphere is like _______. **(b)** Genetic diversity is like _______. **(c)** Being a producer is like _______. **(d)** A food chain is like _______. **(e)** Net primary productivity is like _______.

10. Pick a particular producer, primary consumer, secondary consumer, and decomposer in an ecosystem. Come up with two important questions you would like to ask each organism.

11. Which biome or aquatic zone has the highest net primary productivity (NPP), and which has the lowest primary productivity on a per area basis? Which biome has the highest total NPP on a global scale, and which has the lowest? Are the answers to these two questions the same? Why or why not?

PROJECTS

1. Visit several types of nearby aquatic and terrestrial ecosystems. For each ecosystem, try to determine **(a)** the major producers, consumers, detritivores, and decomposers; **(b)** whether grazing or detrital food webs predominate; and **(c)** the shapes of the pyramids of energy flow, biomass, and numbers.

2. Write a brief scenario describing the sequence of consequences to us and to other forms of life (identify some of these organisms) if **(a)** all decomposers and detritus feeders were somehow eliminated and **(b)** all producers on land and in the upper zone of aquatic ecosystems were eliminated by drastic increases in ultraviolet radiation because of loss of the protective stratospheric ozone layer.

3. Keep track of the number of kilocalories (Calories) of energy you consume in each type of food you eat for a day. Classify each type of food as a plant (or producer such as corn, beans, rice, or potatoes) or as an animal or animal product (such as milk, eggs, or cheese). Assume that about 1% of the solar energy striking vegetation is converted to chemical energy and that there is a 10% ecological efficiency (90% loss) of energy for each trophic level transfer. Calculate the number of kilocalories of

Q: What is a negative feedback loop?

solar energy used to provide you with the kilocalories consumed for each item you ate. Add up the inputs of solar energy for your entire daily diet to calculate how many solar kilocalories it took to feed you for a day. This is your daily solar connection.

4. Write a brief scenario describing the consequences to us and to other forms of life (identify some of these organisms) if **(a)** all decomposers and detritus feeders were somehow eliminated or **(b)** all producers on land and in the upper zone of aquatic ecosystems were eliminated by drastic increases in ultraviolet radiation because of severe depletion of ozone in the stratosphere.

5. Make a concept map of this chapter's major ideas using the section heads and subheads and the key terms (in boldface). Look at the inside back cover and in Appendix 4 for information about concept maps.

A: An input of information into a system in which one change leads to a lessening of that change

5 NUTRIENT CYCLES AND SOILS*

Why Should We Care About Nutrient Cycling in Ecosystems?

In a mature, undisturbed ecosystem, measurements have shown that nutrients recycle over and over with only small losses into the air and in water runoff. This leads to an important question: *How do human activities such as deforestation affect the efficiency of nutrient cycles in ecosystems?*

To help answer this question, since the 1960s F. H. Bormann of Yale University, Gene Likens of Cornell University, and their colleagues have been carrying out a classic field experiment at the Hubbard Brook Experimental Forest in the White Mountains of central New Hampshire. This mature forest is an ideal place to set up experimental ecosystems because it consists of several valleys or watersheds, each drained only by a small creek running down its middle. Bedrock impenetrable to water is close to the surface and prevents seepage of water from one forested hillside, valley, and creek ecosystem to another.

The researchers designed a *controlled experiment* to compare the loss of water and nutrients from an uncut forest ecosystem (the control system) to one that was stripped of its trees (the experimental system). To do this, *V*-shaped concrete catchment dams were built across the creeks at the bottom of several valleys (Figure 5-1). The dams were anchored on impenetrable bedrock so that all of the surface water leaving each forested valley ecosystem had to flow across the dams, where its volume and dissolved nutrient content could be measured.

The first project was to measure the amounts of water that entered and left an undisturbed (control) forest and the amount of dissolved nutrients in this inflow and outflow. For several years precipitation gauges were used to measure the amount of water

*__Paul M. Rich__, associate professor of Ecology and Evolutionary Studies and Environmental Studies at the University of Kansas, is coauthor of this chapter.

Figure 5-1 Classic controlled field experiment on the effects of deforestation on the loss of water and nutrients from a forest ecosystem. The experiment has been carried out since the 1960s at the Hubbard Brook Experimental Forest in New Hampshire by F. H. Bormann and Gene Likens. *V*-notched dams were built into the impenetrable bedrock at the bottom of several forested valleys (left) so that all water and minerals flowing from each valley could be collected and measured for volume and mineral content. Baseline data were collected on several control forested valleys. Then all of the trees in one valley were cut (right) and the flow of water and minerals from this experimental valley were measured for 3 years.

that fell as rain or snow, and the collected water was analyzed to determine the input of nutrients in this precipitation into each valley. Then the amount of water leaving a valley by flowing through its concrete catchment dam was measured and the water was analyzed for its nutrient content (Figure 5-1, left).

These baseline data showed that an undisturbed mature forest ecosystem is very efficient at retaining chemical nutrients. The amount of nutrients leaving such forest ecosystems by way of the creeks in several valleys was found to be roughly equal to the dissolved nutrients entering the ecosystem in rain and snow. Moreover, this nutrient inflow and outflow was small compared to the quantity of nutrients being recycled *within* the forest ecosystem.

The next experiment was to disturb the system and observe any changes that occurred. One winter the investigators cut down all of the trees and shrubs in one valley. They left them where they fell and were careful not to disturb the soil. The cut area was then sprayed with herbicides to prevent regrowth. The inflow and outflow of water and nutrients in this modified experimental valley (Figure 5-1, right) were then compared with those in the control valley for three years.

With no plants to absorb and transpire water from the soil, water runoff in the deforested valley increased by 30–40%. As this excess water ran rapidly over the surface of the ground, it eroded soil and carried nutrients out of the ecosystem. Overall, the loss of minerals from the cut forest was six to eight times the loss in a nearby undisturbed forest. For example, chemical analysis of the water flowing through the dams showed a 60-fold rise in the concentration of nitrate ions (NO_3^-) (Figure 5-2).

So much nitrogen as nitrate (NO_3^-) was lost from the experimental valley that the water flowing out of it was unsafe to drink and the overfertilized stream below this valley became covered with populations of cyanobacteria and algae. After a few years, however, vegetation grew back and nitrate levels returned to normal (Figure 5-2).

Similar studies in other experimental forests have verified the evidence from the Hubbard Brook studies that deforestation severely alters water rainage and nutrient cycling in forest ecosystems in ways that increase the likelihood of nutrient loss from a forest, of downstream flooding, and of overfertilization of downstream waters. Such studies have shown that understanding *nutrient cycles* is fundamental to our understanding of human influences on the environment.

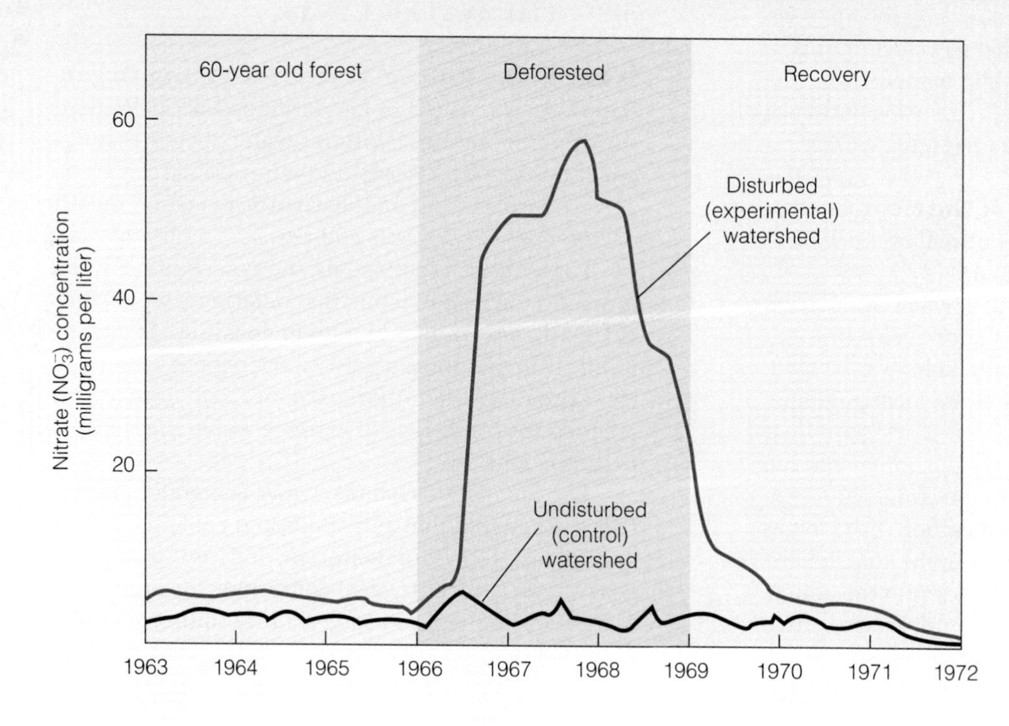

Figure 5-2 Loss of nitrate ions (NO_3^-) from a deforested watershed in the Hubbard Brook Experimental Forest in New Hampshire. The concentration of nitrate ions in runoff from the deforested experimental watershed was 60 times greater than in a control unlogged watershed. (Data from F. H. Bormann and Gene Likens)

The earth's thin film of living matter is sustained by grand-scale cycles of energy and chemical elements.

G. EVELYN HUTCHINSON

This chapter focuses on answering the following questions:

- What happens to matter in an ecosystem?
- How do the water cycle and water's unique properties relate to life and climate?
- How does the carbon cycle relate to life, energy flow, and climate?
- How does the nitrogen cycle relate to life and growth?
- How does the phosphorus cycle relate to life and growth?
- How does the sulfur cycle relate to life and growth?
- What are the three major types of rocks and how are they cycled?
- What are soils?
- What are the relationships between nutrient cycling and ecosystem sustainability?

5-1 MATTER CYCLING IN ECOSYSTEMS

What Are Nutrient Cycles? An important implication of the law of conservation of matter is that it is possible to do bookkeeping of the flow of important atoms, ions, and molecules as they continuously cycle between the nonliving environment and living organisms. These *nutrient cycles*, or *biogeochemical cycles*, involve natural processes that recycle nutrients in various chemical forms in a cyclic manner from the nonliving environment to living organisms and back to the nonliving environment again (Figure 4-6).

What Are the Major Types of Nutrient Cycles? There are three general types of nutrient cycles: hydrologic, atmospheric, and sedimentary.

In the *hydrologic*, or *water, cycle*, water in the form of ice, liquid water, and water vapor cycles through the ecosphere. In this case, the hydrosphere is the main storehouse. This cycle operates at the local, regional, and global levels.

In an *atmospheric cycle*, a large portion of a given element exists in gaseous form in the atmosphere. Examples are nitrogen gas (N_2) and carbon dioxide gas (CO_2), which cycle fairly rapidly from the atmosphere, through soil and organisms, and back into the atmosphere. Because they involve the atmosphere, such cycles operate at local, regional, and global levels.

In a *sedimentary cycle*, an element does not have a gaseous phase or its gaseous compounds don't make up a significant portion of its supply. In this case, the earth's crust is the main storehouse. Such elements cycle quite slowly, moving mostly from the land to sediments in the seas, and then back to the land through long-term geological uplifting of the earth's crust over millions to hundreds of million of years. Phosphorus and most nonrenewable solid minerals are circulated in such cycles. The slow rate of cycling of such nutrients explains, for example, why availability of phosphorus in soil often limits plant growth. Because they have no (or little) circulation in the atmosphere, such cycles tend to operate only on a local and regional basis.

Figure 5-3 shows the major nonliving and living storehouses for the carbon, nitrogen, phosphorus, and sulfur cycles.

Scientists have developed various methods for determining how (and how quickly) chemicals are cycled in various ecosystems and in the ecosphere. Often, they use radioactive tracers (Section 3-4). A radioactive isotope of a particular element (such as carbon, nitrogen, phosphorus, or sulfur) is used to follow the element through the various living and nonliving components of an ecosystem. Samples of the element or its compounds are taken from each phase of the cycle and an instrument is used to measure the radioactivity of each sample. This reveals how much of the element exists in each phase of the cycle and how rapidly it is transferred through the cycle's various phases.

5-2 THE WATER CYCLE

What Is the Role of Water for Living Organisms? In the words of Leonardo da Vinci, "Water is the driver of nature." Without water, the other nutrient cycles would not exist in their present forms, and current forms of life on the earth—consisting mostly of water-containing cells and tissues—could not exist.

For terrestrial ecosystems, the availability of water is among the major factors that determine what kinds of organisms can live in a given location. At one extreme, lush rain forests grow in the regions with highest rainfall. At another extreme, only organisms adapted for conditions of extreme water stress live in desert regions.

For aquatic ecosystems, water is literally the matrix, the material that surrounds and contains aquatic organisms. Flows of water in and out of aquatic ecosystems affect physical conditions for organisms by influencing temperature, salinity, and availability of nutrients.

Q: What is homeostasis?

Element	Main nonliving storehouse	Main forms in living organisms	Other nonliving storehouse
Carbon (C)	***Atmospheric:*** carbon dioxide (CO_2)	Carbohydrates $(CH_2O)_n$ and all other organic molecules	***Hydrologic:*** dissolved carbonate (CO_3^{2-}) and bicarbonate (HCO_3^-) ***Sedimentary:*** carbon containing minerals in rocks
Nitrogen (N)	***Atmospheric:*** nitrogen gas (N_2)	Proteins and other nitrogen-containing organic molecules	***Hydrologic:*** dissolved ammonium (NH_4^+), nitrate (NO_3^-), and nitrate (NO_2^-) in water and soils
Phosphorus (P)	***Sedimentary:*** phosphate (PO_4^{3-}) containing minerals in rocks	DNA, other nucleic acids (e.g, ATP), and phospholipids	***Hydrologic:*** dissolved phosphate (PO_4^{3-})
Sulfur (S)	***Sedimentary:*** rocks (e.g., iron disulfide and pyrite) and minerals (e.g., sulfate [SO_4^{2-}])	Sulfur-containing amino acids in most proteins, some vitamins	***Atmospheric:*** hydrogen sulfide (HgS), sulfur dioxide (SO_2), sulfur trioxide (SO_3), and sulfuric acid (H_2SO_4) ***Hydrologic:*** sulfate (SO_4^{2-}) and sulfuric acid (H_2SO_4)

Figure 5-3 Major nonliving and living storehouses of elemental nutrients.

How Is Water Cycled in the Ecosphere?

The **hydrologic cycle**, or **water cycle**, which collects, purifies, and distributes the earth's fixed supply of water, is shown in simplified form in Figure 5-4. The main processes in this water recycling and purifying cycle are *evaporation* (conversion of water into water vapor, Figure 3-2), *transpiration* (evaporation from leaves of water extracted from soil by roots and transported throughout the plant), *condensation* (conversion of water vapor into droplets of liquid water, Figure 3-2), *precipitation* (rain, sleet, hail, and snow), *infiltration* (movement of water into soil), *percolation* (downward flow of water through soil and permeable rock formations to groundwater storage areas called aquifers), and *runoff* (downslope surface movement back to the sea to resume the cycle).

The water cycle is powered by energy from the sun and by gravity. Incoming solar energy evaporates water from oceans, streams, lakes, soil, and vegetation. About 84% of water vapor in the atmosphere comes from the oceans, which cover about 71% of the earth's surface; the rest comes from land. On a global scale, the amount of water vapor entering the atmosphere is equal to the amount returning to the earth's surface as precipitation.

Water's unique properties (Spotlight, p. 113) play an important role in the water cycle.

The amount of water vapor air can hold depends on temperature; warm air is capable of holding more water vapor than cold air. **Absolute humidity** is the amount of water vapor found in a certain mass of air; it is usually expressed as grams of water per kilogram of air. **Relative humidity** is the amount of water vapor in a certain mass of air, expressed as a percentage of the maximum amount it could hold at that temperature. For example, a relative humidity of 60% at 27°C (80°F) means that each kilogram (or other unit of mass) of air contains 60% of the maximum amount of water vapor it could hold at that temperature.

Winds and air masses transport water vapor over various parts of the earth's surface, often over long distances. Falling temperatures cause the water vapor to condense into tiny droplets that form clouds or fog. For precipitation to occur, air must contain **condensation nuclei**: tiny particles on which droplets of water vapor can collect. Volcanic ash, soil dust, smoke, sea

A: The maintenance of favorable internal conditions in a system despite changes in external conditions

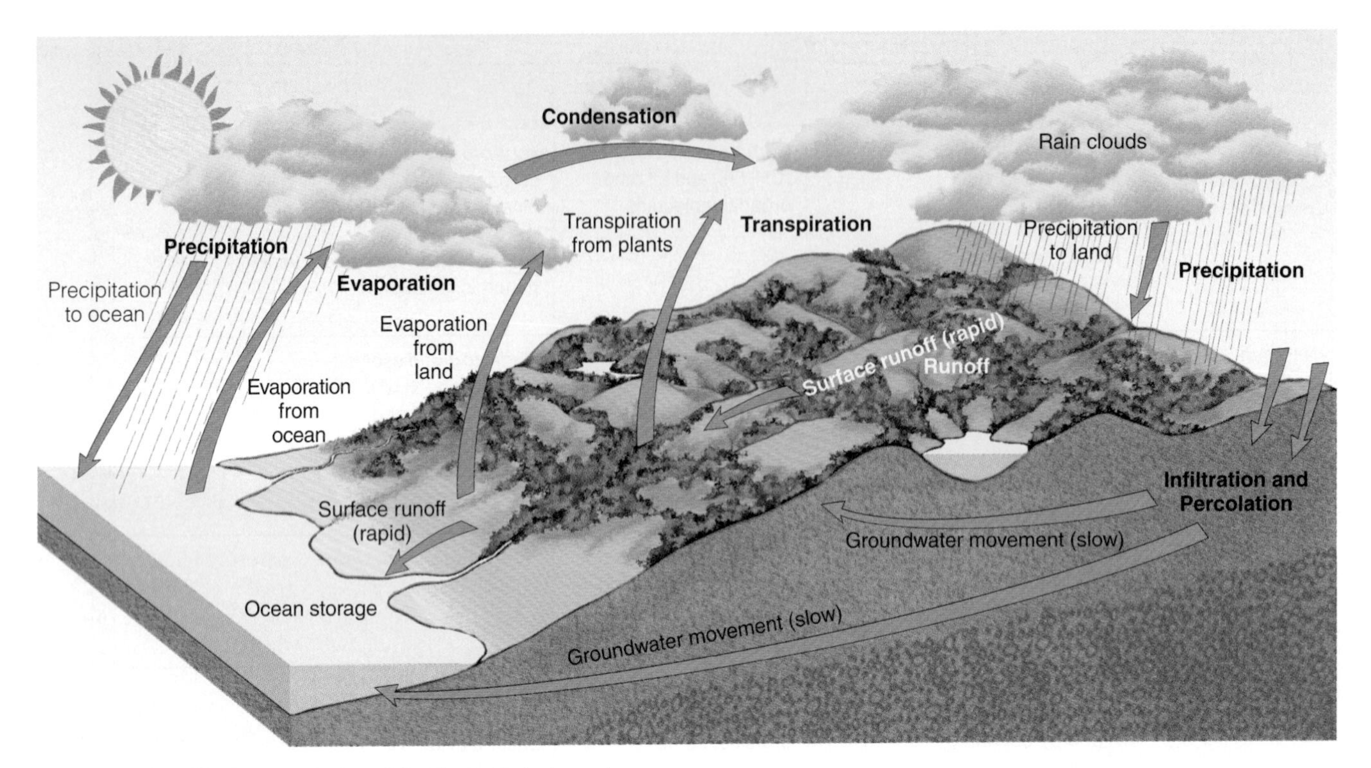

Figure 5-4 Simplified conceptual model of the hydrologic cycle.

salts, and particulate matter emitted by factories, coal-burning power plants, and vehicles are sources of such particles. The temperature at which condensation occurs is called the **dew point**.

It takes millions of tiny water droplets adhering to a condensation nucleus to produce a drop of rain or a snowflake that will fall to the earth's surface. About 77% of this precipitation falls back into the sea, and the rest over land. Some of the fresh water returning to the earth's surface as precipitation becomes locked in glaciers. Most of the precipitation falling on terrestrial ecosystems becomes *surface runoff* flowing into streams and lakes, which eventually carry water back to the oceans, where it can be evaporated to cycle again.

Besides replenishing streams and lakes, surface runoff also causes soil erosion, which moves soil and weathered rock fragments from one place to another. Water is thus the primary sculptor of the earth's landscape. Also, rainwater that is acidic, either naturally or from pollution, reacts chemically with certain atoms and ions on the earth's surface, releasing water-soluble atoms and ions that are carried by streams to the sea. This is one of the reasons the ocean is so salty. Because water dissolves many nutrient compounds, it is a major medium for transporting nutrients within and among ecosystems.

Some of the water returning to the land soaks into (infiltrates) the soil and porous rock and then percolates downward, dissolving minerals from porous rocks on the way. This water is stored as groundwater in the pores and cracks of rocks.

Where the pores are joined, a network of water channels allows water to flow through the porous rock. Such water-laden rock is called an *aquifer*, and the level of the earth's land crust to which it is filled is called the *water table*. This underground water flows slowly downhill through rock pores and seeps out into streams and lakes or comes out in springs. Eventually, this water evaporates or reaches the sea to continue the cycle. The average circulation rate of underground water in the hydrologic cycle is extremely slow (300–4,600 years) compared with that of water in lakes (13 years), streams (13 days), and the atmosphere (9 days). The turnover time for water in the ocean is about 37,000 years, and for ice in glaciers it is about 16,000 years.

Throughout the hydrologic cycle many natural processes act to purify water. Evaporation and subsequent precipitation act as a natural distillation process, which removes impurities dissolved in water. As water flows above ground through streams and lakes, and below ground in aquifers, it is naturally filtered and purified by chemical and biological processes. Thus, the hydrologic cycle can also be viewed as a cycle of natural renewal of water quality.

How Are Humans Influencing the Water Cycle? Humans intervene in the water cycle in three main ways. *First*, we withdraw large quantities of fresh water from streams, lakes, and underground sources.

Field, C. B., et al. 1998. "Primary Production of the Biosphere." *Science*, vol. 281, no. 5374, 237(4).

Water's Importance and Unique Properties

SPOTLIGHT

We live on the water planet, with a precious film of water—most of it salt water—covering about 71% of the earth's surface. The earth's organisms are made up mostly of water; a tree is about 60% water by weight, and most animals (including humans) are about 50–65% water.

Each of us needs only a dozen or so cupfuls of water per day to survive, but huge amounts of water are needed to supply us with food, shelter, and our other needs and wants. Water also plays a key role in sculpting the earth's surface, moderating climate, and diluting pollutants. In fact, without water the earth would have no oceans, no life as we know it, and no people.

Water is a remarkable substance with a unique combination of properties:

- *There are strong forces of attraction called hydrogen bonds between molecules of water* (Figure 3-7). These attractive forces between water molecules are the major factor determining water's unique properties.
- *Water exists as a liquid over a wide temperature range because of the strong forces of attraction between water molecules.* Its high boiling point of 100°C (212°F) and low freezing point of 0°C (32°F) mean that water remains a liquid in most climates on the earth.
- *Liquid water changes temperature very slowly because it can store a large amount of heat without a large change in temperature.* This high heat capacity helps protect living organisms from the shock of abrupt temperature changes; it also moderates the earth's climate and makes water an excellent coolant for car engines, power plants, and heat-producing industrial processes.
- *It takes a lot of heat to evaporate liquid water because of the strong forces of attraction between its molecules.* Water's ability to absorb large amounts of heat as it changes into water vapor—and to release this heat as the vapor condenses back to liquid water—is a primary factor in distributing heat throughout the world and thus plays an important role in determining the climates of various areas. This property makes evaporation of water an effective cooling process, which is why you feel cooler when perspiration or bathwater evaporates from your skin.
- *Liquid water can dissolve a wide variety of compounds.* This enables it to carry dissolved nutrients into the tissues of living organisms, to flush waste products out of those tissues, to serve as an all-purpose cleanser, and to help remove and dilute the water-soluble wastes of civilization. Water's superiority as a solvent also means that it is easily polluted by water-soluble wastes.
- *The strong attractive forces between the molecules of liquid water cause its surface to contract (high surface tension) and also to adhere to and coat a solid (high wetting ability).* Together these properties allow water to rise through a plant from the roots to the leaves (capillary action).
- *Unlike most liquids, water expands when it freezes and becomes ice.* This means that ice has a lower density (mass per unit of volume) than liquid water. Thus ice floats on water, and as air temperatures fall below freezing, bodies of water freeze from the top down instead of from the bottom up. Without this property, lakes and streams in cold climates would freeze solid, and most of their current forms of aquatic life could not exist. Because water expands upon freezing, it can also break pipes, crack engine blocks (which is why we use antifreeze), and break up streets and fracture rocks (thus forming soil).

Critical Thinking

1. What would happen to your body if suddenly your water molecules no longer formed hydrogen bonds (Figure 3-7) with one another?

2. What would happen to plant life if the planet's water molecules did not form hydrogen bonds with one another?

In heavily populated or heavily irrigated areas, withdrawals have led to groundwater depletion or intrusion of ocean salt water into underground water supplies. The availability of water resources in various parts of the world is discussed in Chapter 13.

Second, we clear vegetation from land for agriculture, mining, road and building construction, and other activities. This increases runoff and reduces infiltration that recharges groundwater supplies; it also increases the risk of flooding and accelerates soil erosion and landslides.

Third, we modify water quality, particularly by adding nutrients (such as phosphates) and other pollutants and by changing ecological processes that naturally purify water. Chapter 20 examines issues of water pollution and water quality.

5-3 THE CARBON CYCLE

What Is the Role of Carbon for Living Organisms? Carbon is essential to life as we know it. It is the basic building block of the carbohydrates, fats, proteins, nucleic acids such as DNA and RNA, and all of the other organic compounds necessary for life.

Hint: Enter the search term *carbon cycle* using the Subject Guide.

Carbon is sometimes said to be the currency for energy exchange in living systems because most of the chemical energy needed for life is stored in organic compounds as bonds between carbon atoms and as bonds between carbon atoms and other atoms. Thus, following the movement of carbon through the food web also allows us to follow the flow of energy.

As a heat-trapping (greenhouse) gas, carbon dioxide is a key component of nature's thermostat. If the carbon cycle removes too much CO_2 from the atmosphere, the earth will cool; if the cycle generates too much, the earth will get warmer. Thus, even slight changes in the carbon cycle can affect climate and ultimately the types of life that can exist on various parts of the planet.

How Is Carbon Cycled in the Ecosphere? The **carbon cycle**, a global gaseous cycle, is based on carbon dioxide gas, which makes up only 0.036% of the volume of the troposphere and is also dissolved in water.

Terrestrial producers remove CO_2 from the atmosphere and aquatic producers remove it from the water. They then use photosynthesis to convert CO_2 into complex carbohydrates such as glucose ($C_6H_{12}O_6$). Although the concentration of CO_2 in the atmosphere is low, carbon cycles at a rapid rate because photosynthetic producers have a high demand for CO_2.

The cells in oxygen-consuming producers, consumers, and decomposers carry out aerobic respiration, which breaks down glucose and other complex organic compounds and converts the carbon back to CO_2 in the atmosphere or water, for reuse by producers. This linkage between photosynthesis in producers and aerobic respiration in producers, consumers, and decomposers circulates carbon in the ecosphere and is a major part of the global carbon cycle (Figure 5-5). Oxygen and hydrogen, the other elements in carbohydrates, cycle almost in step with carbon.

Most carbon cycles fairly rapidly through the air, water, and biota, but carbon stored for many decades as biomass in the wood of trees and in other organic materials cycles much more slowly. Decomposition eventually recycles this carbon as CO_2 in the atmosphere, although fires recycle it much faster.

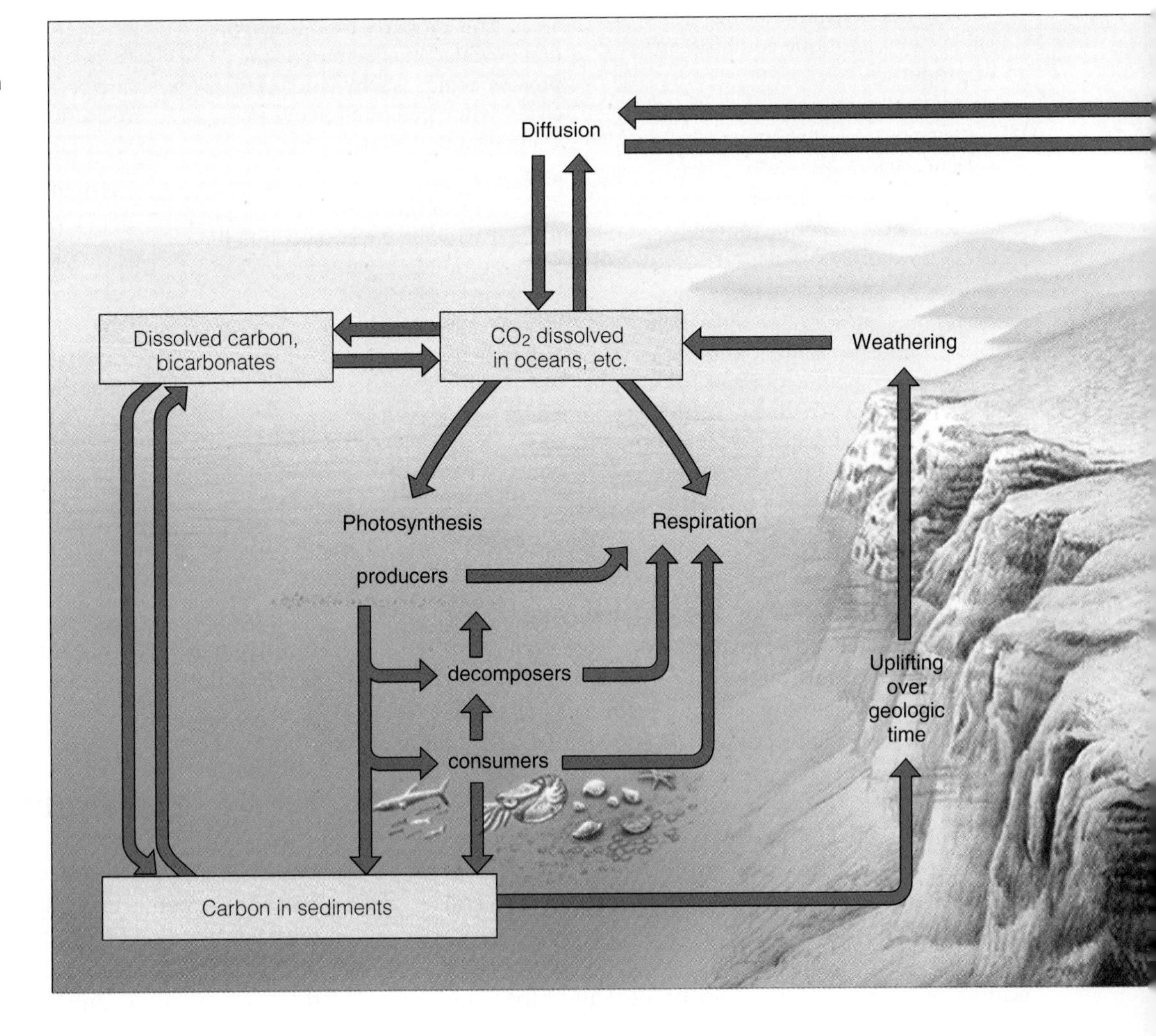

Figure 5-5 Simplified conceptual model of the global carbon cycle. The left portion shows the movement of carbon through marine ecosystems, and the right portion shows its movement through terrestrial ecosystems. Carbon reservoirs are shown as boxes. (Modified by permission from Cecie Starr and Ralph Taggart, *Biology: The Unity and Diversity of Life*, 8th ed., Belmont, Calif.: Wadsworth, 1998)

Q: What are the three basic building blocks of matter?

Highly productive wetlands such as bogs and swamps store large amounts of carbon. Under certain conditions, some deposits of dead plant matter and bacteria accumulate faster than they are decomposed in wetlands and other ecosystems. These deposits are locked away in underground deposits. Over millions of years such buried organic matter is compressed between layers of sediment, where it forms carbon-containing fossil fuels such as coal and oil (Figure 5-5). This carbon is not released to the atmosphere as CO_2 for recycling until these fuels are extracted and burned or until long-term geological processes expose these deposits to air.

In the short time period of a few hundred years, we have been extracting and burning fossil fuels that took millions of year to form from dead plant matter. This is why fossil fuels are nonrenewable resources on a human time scale. Some CO_2 also enters the atmosphere from aerobic respiration and from volcanic eruptions, which free carbon from rocks deep in the earth's crust.

The largest storage reservoir for the earth's carbon is sedimentary rocks such as limestone ($CaCO_3$) on ocean floor sediments and on continents. This carbon reenters the cycle very slowly, when some of the sediments dissolve and form dissolved CO_2 gas that can enter the atmosphere. Geologic processes can also bring bottom sediments to the surface, exposing the carbonate rock to chemical attack by oxygen and converting them to carbon dioxide gas. Acidic rain falling on exposed limestone rock also releases carbon dioxide back into the atmosphere.

The oceans are the second largest storage reservoir in the carbon cycle. Oceans also play a major role in regulating the level of carbon dioxide in the atmosphere. Some carbon dioxide gas, which is readily soluble in water, stays dissolved in the sea, some is removed by photosynthesizing producers, and some reacts with seawater to form carbonate ions (CO_3^{2-}) and bicarbonate ions (HCO_3^-). As water warms, more dissolved CO_2 returns to the atmosphere, just as more carbon dioxide fizzes out of a carbonated beverage when it warms.

In marine ecosystems, some organisms take up dissolved CO_2 molecules, carbonate ions, or bicarbonate

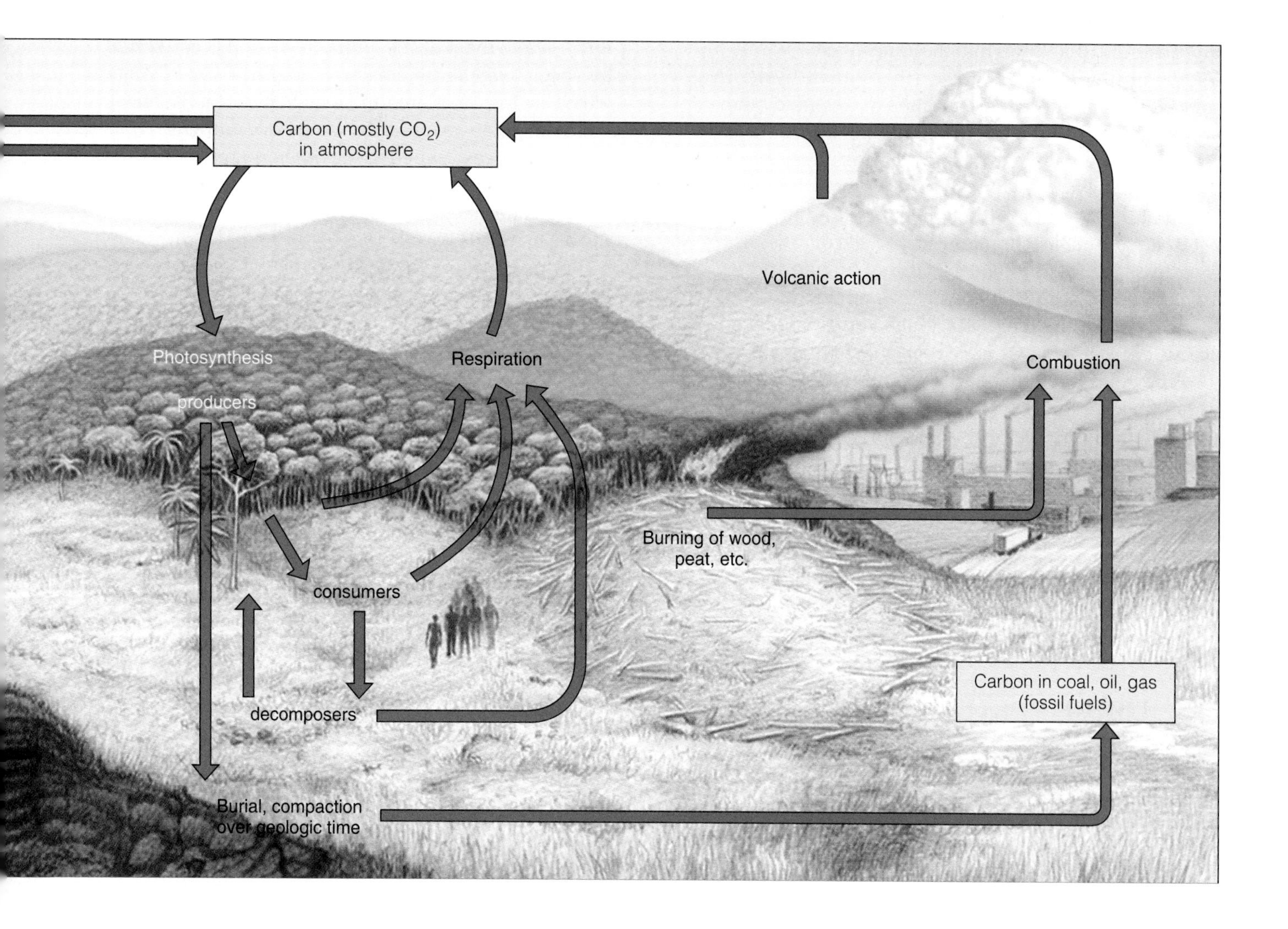

ions from ocean water. These ions can then react with calcium ions (Ca^{2+}) in seawater to form slightly soluble carbonate compounds such as calcium carbonate ($CaCO_3$) to build the shells and skeletons of marine organisms. When these organisms die, tiny particles of their shells and bone drift slowly to the ocean depths and are buried for eons (as long as 400 million years) in deep bottom sediments (Figure 5-5), where under immense pressure they are converted into limestone rock. About 55 times more carbon is stored in dissolved carbon compounds and in insoluble marine sediments than is stored in the atmosphere. The average residence time of CO_2 is about 3 years in the atmosphere, 25–30 years in soils, and about 1,500 years in the oceans.

How Are Humans Influencing the Carbon Cycle? Since 1800 and especially since 1950, as world population and resource use have soared, we have disturbed the carbon cycle in two ways that add more carbon dioxide to the atmosphere than oceans and plants so far have been able to remove: **(1)** Forest and brush removal has left less vegetation to absorb CO_2 through photosynthesis and **(2)** burning fossil fuels and wood produces CO_2 that flows into the atmosphere.

Computer models of the earth's climate systems suggest that increased concentration of CO_2 (and other heat-trapping gases we're adding to the atmosphere) could enhance the planet's natural greenhouse effect. This natural warming effect, caused mostly by water vapor and CO_2 in the atmosphere, is a key factor in producing livable climates on the earth over the roughly 60,000 years since the current human species arrived on the scene. However, measurements show that we are adding more CO_2 to the atmosphere than oceans and plants so far have been able to remove.

The release of this carbon-containing material (fossil fuels and vegetation) over only a few hundred years is upsetting the heat-absorbing balance developed over millions of years. This could alter climate patterns for hundreds to thousands of years as the carbon cycle adjusts to these rapid inputs. It could also disrupt global food production and wildlife habitats and possibly raise the average sea level, as discussed in more detail in Chapter 19.

5-4 THE NITROGEN CYCLE

What Is the Role of Nitrogen for Living Organisms? Organisms use nitrogen to make vital organic compounds such as amino acids, proteins, DNA, and RNA. In both terrestrial and aquatic ecosystems, nitrogen is typically in short supply and limits the rate of primary production.

How Is Nitrogen Cycled in the Ecosphere? Bacteria in Action Although chemically unreactive nitrogen gas (N_2) makes up 78% of the volume of the troposphere, it cannot be absorbed and used directly as a nutrient by multicellular plants or animals. Fortunately, lightning and certain bacteria convert nitrogen gas into compounds that can enter food webs as part of the **nitrogen cycle** (Figure 5-6), a global gaseous cycle that is the most complex of the earth's chemical cycles.

In the first step in the nitrogen cycle, called *nitrogen fixation*, specialized bacteria convert gaseous nitrogen (N_2) to ammonia (NH_3) by the reaction $N_2 + 3H_2 \rightarrow 2NH_3$. This is done mostly by cyanobacteria in soil and water and by *Rhizobium* bacteria living in small nodules (swellings) on the root systems of legumes, a huge family of plants with about 18,000 species, including soybeans, alfalfa, and clover (Figure 5-7).

Plants can use ammonia or ammonium ions (NH_4^+) formed when ammonia reacts with water as a source of nitrogen. However, in a two-step process called *nitrification*, most of the ammonia in soil is converted by specialized aerobic bacteria to nitrite ions (NO_2^-), which are toxic to plants and then to nitrate ions (NO_3^-), which are easily taken up by plants as a nutrient.

In a process called *assimilation*, plant roots then absorb inorganic ammonia, ammonium ions, and nitrate ions formed by nitrogen fixation and nitrification in soil water. They use these ions to make nitrogen-containing organic molecules such as DNA, amino acids, and proteins. Animals in turn get their nitrogen by eating plants or plant-eating animals.

After nitrogen has served its purpose in living organisms, vast armies of specialized decomposer bacteria convert the nitrogen-rich organic compounds, wastes, cast-off particles, and dead bodies of organisms into simpler nitrogen-containing inorganic compounds such as ammonia (NH_3) and water-soluble salts containing ammonium ions (NH_4^+). This process is known as *ammonification*.

In a process called *denitrification*, other specialized bacteria (mostly anaerobic bacteria in waterlogged soil or in the bottom sediments of lakes, oceans, swamps, and bogs) then convert NH_3 and NH_4^+ back into nitrite (NO_2^-) and nitrate (NO_3^-) ions and then into nitrogen gas (N_2) and nitrous oxide gas (N_2O). These are then released to the atmosphere to begin the cycle again.

Although the nitrogen cycle provides large amounts of ammonium and nitrate ions for use by terrestrial plants, compounds containing these ions are

Q: What are the basic building blocks of atoms?

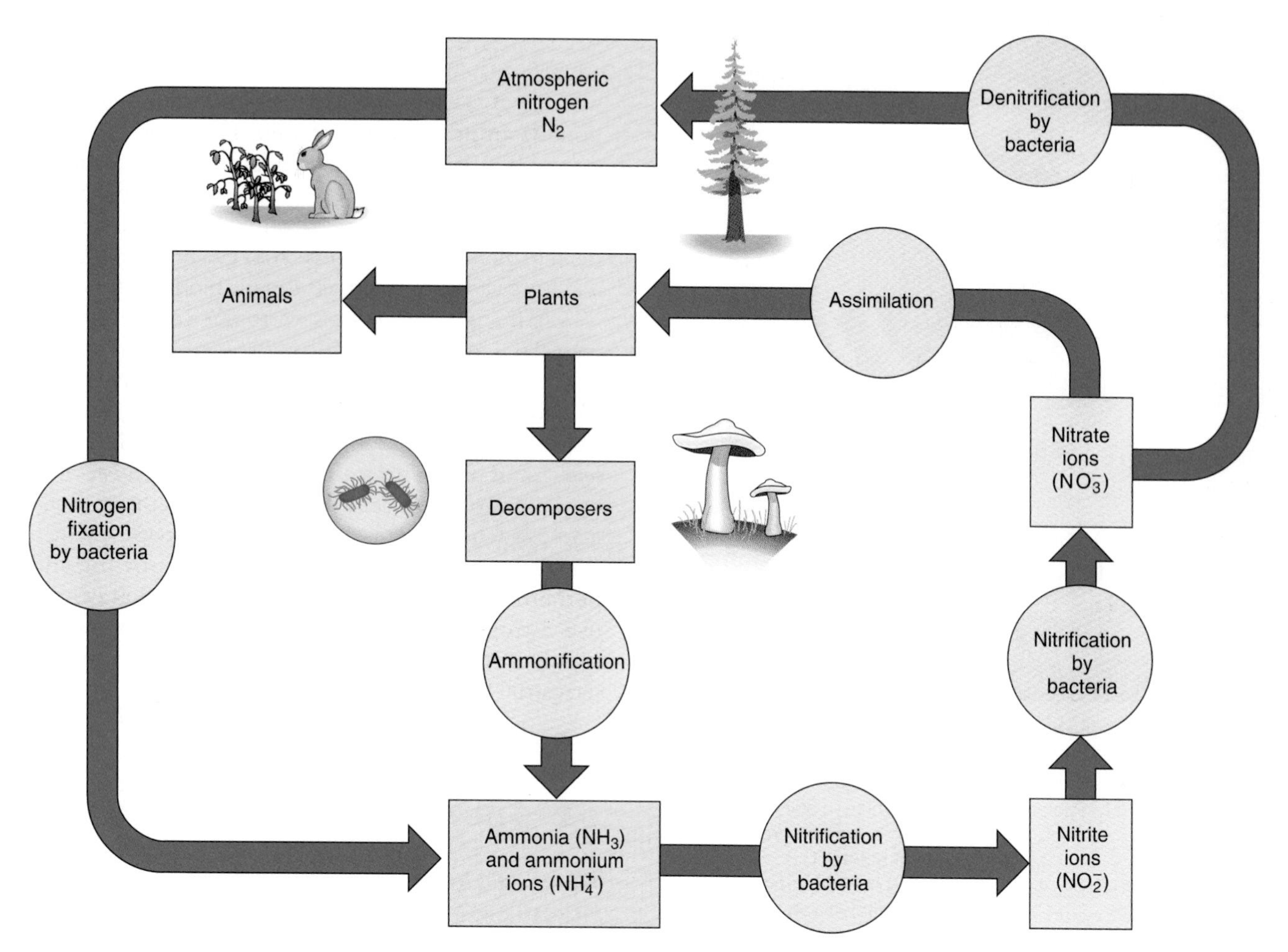

Figure 5-6 Simplified conceptual model of the nitrogen cycle in a terrestrial ecosystem. Nitrogen reservoirs are shown as boxes and processes changing one form of nitrogen to another are shown in circles.

soluble in water and thus can easily be removed (leached) by water flowing across or percolating down through soil. Thus, nitrogen is often a factor limiting plant growth.

How Are Humans Influencing the Nitrogen Cycle? Until about 50 years ago, nitrogen fixation by microorganisms was the only source of nitrogen usable by plants. However, during World War II German chemist Fritz Haber developed a chemical process in which nitrogen and hydrogen gas are combined at high temperatures and pressures to form gaseous ammonia ($N_2 + 3H_2 \rightarrow 2NH_3$). The ammonia made by this Haber process can be converted into salts such as ammonium nitrate that can be sold as commercial inorganic fertilizer. Coupled with irrigation, this additional and cheap input of nitrogen into the soil has revolutionized agriculture by increasing crop yields.

Humans intervene in the nitrogen cycle in several ways. *First*, we emit large quantities of nitric oxide (NO) into the atmosphere when we burn any fuel. Most of this NO is produced when nitrogen and oxygen molecules in the air combine at high temperatures: $N_2 + O_2 \rightarrow 2NO$. According to a 1997 report from the Ecological Society of America, human sources now account for 80% of all atmospheric NO. In the atmosphere, this nitric oxide combines with oxygen to form nitrogen dioxide gas (NO_2), which can react with water vapor to form nitric acid (HNO_3). Droplets of HNO_3 dissolved in rain or snow are components of *acid deposition* (commonly called *acid rain*), which along with other air pollutants can damage and weaken trees and upset aquatic systems (Section 18-3). Nitric oxide also plays a role in creating ozone near the ground from the formation of photochemical smog (Section 18-2).

A: Protons, neutrons, and electrons

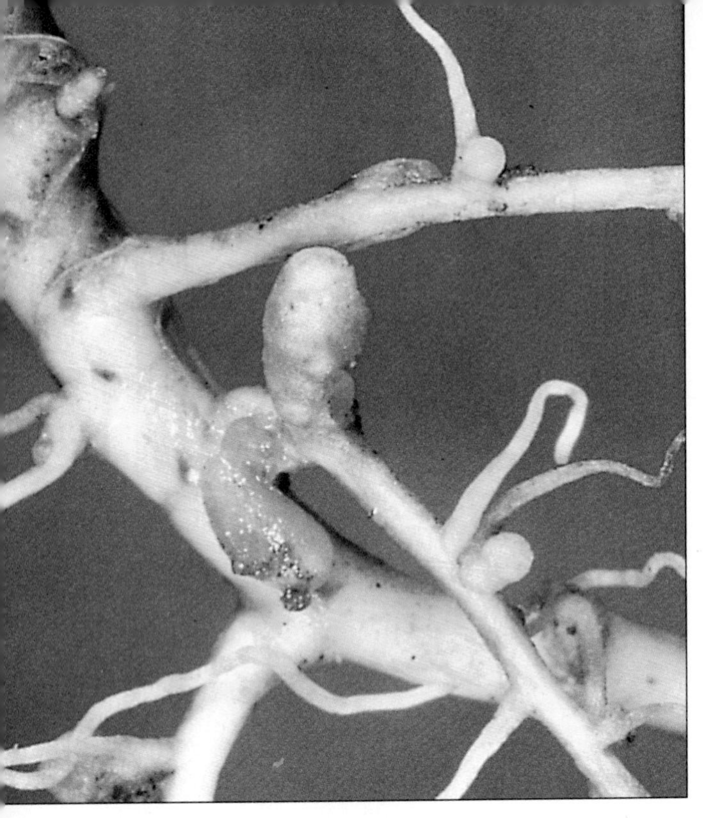

Figure 5-7 Plants in the legume family (which includes alfalfa, clover, peas, beans, mesquite, and acacias) have root nodules where *Rhizobium* bacteria fix nitrogen; that is, they convert nitrogen gas (N_2) into ammonia (NH_3), which in soil water forms ammonium ions (NH_4^+), which are converted to nitrate ions (NO_3^-) that are taken up by the roots of plants. This process benefits both species: The bacteria convert atmospheric nitrogen into a form usable by the plants, and the plants provide the bacteria with some simple carbohydrates. (E. R. Degginger)

Second, human activities emit heat-trapping nitrous oxide gas (N_2O) into the atmosphere through the action of anaerobic bacteria on livestock wastes and commercial inorganic fertilizers applied to the soil. Emissions of this gas are rising and account for a few percent of the greenhouse gases that can cause global warming (Section 19-2). When N_2O reaches the stratosphere it contributes to depletion of the earth's ozone shield that filters out harmful ultraviolet radiation from the sun.

Third, we remove nitrogen from the earth's crust when we mine nitrogen-containing mineral deposits (such as ammonium nitrate or NH_4NO_3) for fertilizers, deplete nitrogen from topsoil by harvesting nitrogen-rich crops, and leach water-soluble nitrate ions from soil through irrigation.

Fourth, we remove nitrogen from topsoil when we burn grasslands and clear forests before planting crops. At the same time, we emit nitrogen oxides into the atmosphere.

Fifth, we add excess nitrogen compounds to aquatic systems in agricultural runoff, discharge of municipal sewage, and deposition of nitrogen compounds from the atmosphere. This excess of plant nutrients stimulates rapid growth of photosynthesizing algae and other aquatic plants. The subsequent breakdown of dead algae by aerobic decomposers can deplete the water of dissolved oxygen and can disrupt aquatic systems and reduce aquatic biodiversity.

Sixth, we add excess nitrogen compounds to terrestrial ecosystems through *atmospheric deposition,* which involves the movement of reactive nitrogen compounds, such as nitric acid (HNO_3) and nitrogen dioxide (NO_2), from the atmosphere onto plant leaves and other surfaces. This nitrogen then becomes available for plant and microbial growth, and can lead to such problems as weedy plant species, which can better use nitrogen for growth, outgrowing and perhaps eliminating other plant species that cannot use nitrogen as well. Thus, our current excessive inputs of nitrogen into the atmosphere can reduce terrestrial biodiversity.

5-5 THE PHOSPHORUS CYCLE

What Is the Role of Phosphorus for Living Organisms? Phosphorus, mainly in the form of the ions (PO_4^{3-} and HPO_4^{2-}), is an essential nutrient of both plants and animals. It is a part of DNA molecules, which carry genetic information; nucleic acids (ATP and ADP) that store chemical energy for use by organisms in cellular respiration; certain fats (phospholipids) in cell membranes of plant and animal cells; and the bones, teeth, and shells of animals (calcium phosphate compounds).

How Is Phosphorus Cycled in the Ecosphere? Phosphorus circulates through water, the earth's crust, and living organisms in the **phosphorus cycle** (Figure 5-8). In this sedimentary cycle, phosphorus moves slowly from phosphate deposits on land and in shallow ocean sediments to living organisms, and then much more slowly back to the land and ocean.

Bacteria are less important here than in the nitrogen cycle. Unlike carbon and nitrogen, very little phosphorus circulates in the atmosphere because at the earth's normal temperatures and pressures, phosphorus and its compounds are not gases. Phosphorus is found in the atmosphere only as small particles of dust. In contrast to the carbon cycle, the phosphorus cycle is slow, and on a short human time scale much phosphorus flows one way from the land to the oceans.

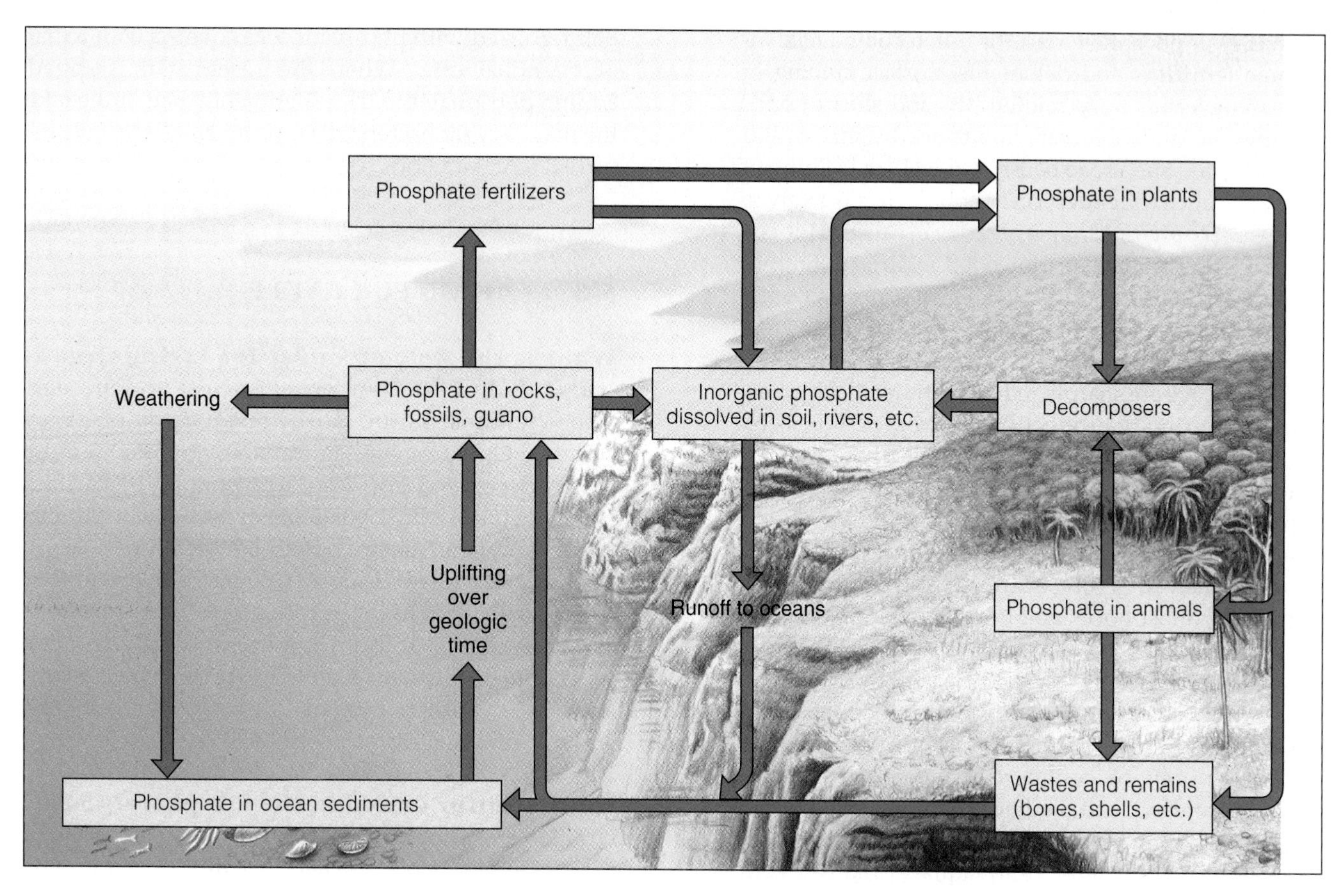

Figure 5-8 Simplified conceptual model of the phosphorus cycle. Phosphorus reservoirs are shown as boxes; the processes that change one form of phosphorus to another are shown in unboxed print. (Modified by permission from Cecie Starr and Ralph Taggart, *Biology: The Unity and Diversity of Life*, 8th ed., Belmont, Calif.: Wadsworth, 1998)

Phosphorus is typically found as phosphate salts containing phosphate ions (PO_4^{3-}) in terrestrial rock formations and ocean bottom sediments. Phosphorus released by the slow breakdown, or *weathering*, of terrestrial phosphate rock deposits is dissolved in soil water and then taken up by plant roots. Wind can also carry phosphate particles long distances.

Because most soils contain little phosphate, it is often the limiting factor for plant growth on land unless phosphorus (as phosphate salts mined from the earth) is applied to the soil as a fertilizer. Phosphorus also limits the growth of producer populations in many freshwater streams and lakes because phosphate salts are only slightly soluble in water.

Phosphorus cycles much more rapidly through the living components of ecosystems than it does through geological formations. Animals get phosphorus by eating producers or animals that have eaten producers. Animal wastes and the decay of dead animals and producers return much of this phosphorus as phosphate to the soil, to streams, and eventually to the ocean bottom as insoluble deposits of phosphate rock. This phosphorus remains there for millions to hundreds of millions of years before being returned to the land by geologic processes that over millions of years may push up and expose the seafloor. Weathering then slowly releases phosphorus from the exposed rocks and continues the cycle.

Some phosphate returns to the land as guano, phosphate-rich manure typically of fish-eating birds such as pelicans and cormorants. This return is small, though, compared with the phosphate transferred from the land to the oceans each year by natural processes and human activities. Severe erosion, especially by human activities, accelerates this transfer of phosphate from terrestrial ecosystems to the sea.

How Are Humans Influencing the Phosphorus Cycle? Humans intervene in the phosphorus cycle in three main ways. *First*, we mine large

A: The usefulness of a matter resource based on its availability and concentration quantities of phosphate rock for use in commercial inorganic fertilizers and detergents. Surface mining of phosphate leaves huge mining pits and slurry ponds that mar the landscape, can pollute nearby surface and groundwater, and are expensive to reclaim. Mining of land and underwater deposits of phosphate is likely to increase sharply in coming decades because **(1)** it is a limited but crucial element needed in large quantities by living organisms and as a source of commercial fertilizer and phosphate detergents and **(2)** it is returned very slowly from the ocean to the land.

Second, we are sharply reducing the available phosphate and primary productivity of tropical forests by cutting them. In such ecosystems, hardly any phosphorus is found in the soil. Rather, it's in the ecosystem's plant and animal life, which is rapidly recycled from dead plants and animals by hordes of decomposers. When such forests are cut and burned, most remaining phosphorus and other nutrients are readily washed away by heavy rains, and the land becomes unproductive.

Third, we add excess phosphate to aquatic ecosystems in runoff of animal wastes from livestock feedlots, runoff of commercial phosphate fertilizers from cropland, and discharge of municipal sewage. Too much of this nutrient causes explosive growth of cyanobacteria, algae, and aquatic plants, with surface mats of such species blocking sunlight. As a result, plants rooted on the bottom may die because sunlight becomes the limiting factor. Huge numbers of cyanobacteria (formerly known as blue-green algae), algae, and plants age and die in an overfertilized lake. Then decomposing aerobic bacteria feeding on their dead cells use up so much dissolved oxygen that some species of fish and other aquatic animals may suffocate.

In 1974, P. H. Schindler and his colleagues carried out a controlled field experiment on a lake in Ontario, Canada, that dramatically demonstrated this overfertilization of lakes. They hypothesized that phosphorus in the form of phosphates (PO_4^{3-}) was the limiting factor in the lake's primary productivity. To test this hypothesis, they found a lake with an hourglass shape and stretched a vinyl curtain across the lake's narrowest portion in the center of the hourglass. This divided the lake into two sections, one serving as the *control* side and the other as the *experimental* side.

The researchers then fertilized both halves of the lake with compounds containing carbon and nitrogen nutrients, but they also added phosphate compounds to the experimental half. Without excess phosphorus, the control half of the lake showed no significant changes in primary productivity or populations of organisms. However, within 2 months, the experimental half fertilized with phosphate was covered with a mat of cyanobacteria. When the researchers stopped adding phosphates to the experimental half of the lake the mat of cyanobacteria disappeared, and the lake returned to its previous condition.

5-6 THE SULFUR CYCLE

What Is the Role of Sulfur for Living Organisms? Sulfur is a component of most proteins and some vitamins. Sulfur, mostly taken up by plants as sulfate (SO_4^{2-}) dissolved in water, is common in most plant tissues and becomes incorporated in various organic compounds, particularly sulfur-containing amino acids that are the building blocks for proteins. Bonds between sulfur molecules of sulfur-containing amino acids help to give proteins their three-dimensional structure and are important for the proper functioning of structural proteins and enzymes. Many animals, including humans, depend on plants for sulfur-containing amino acids.

How Is Sulfur Cycled in the Ecosphere? Sulfur circulates through the ecosphere in the **sulfur cycle** (Figure 5-9). Much of the earth's sulfur is tied up underground in rocks (such as iron disulfide or pyrite) and minerals, including sulfate (SO_4^{2-}) salts (such as hydrous calcium sulfate or gypsum) buried deep under ocean sediments.

Sulfur also enters the atmosphere from several natural sources. Hydrogen sulfide (H_2S), a colorless, highly poisonous gas with a rotten-egg smell, is released from active volcanoes and by the breakdown of organic matter in swamps, bogs, and tidal flats caused by decomposers that don't use oxygen (anaerobic decomposers). Sulfur dioxide (SO_2), a colorless, suffocating gas, also comes from volcanoes. Particles of sulfate (SO_4^{2-}) salts, such as ammonium sulfate, enter the atmosphere from sea spray.

In the atmosphere, sulfur dioxide reacts with oxygen to produce sulfur trioxide gas (SO_3). Some of the sulfur dioxide then reacts with water droplets in the atmosphere to produce tiny droplets of sulfuric acid (H_2SO_4). Sulfur dioxide also reacts with other chemicals in the atmosphere such as ammonia to produce tiny particles of sulfate salts. These droplets and particles fall to the earth as components of acid deposition, which along with other air pollutants can harm trees and aquatic life. Droplets of sulfuric acid are also produced from dimethylsulfide (DMS) emitted into the atmosphere by many species of plankton.

Q: What is entropy?

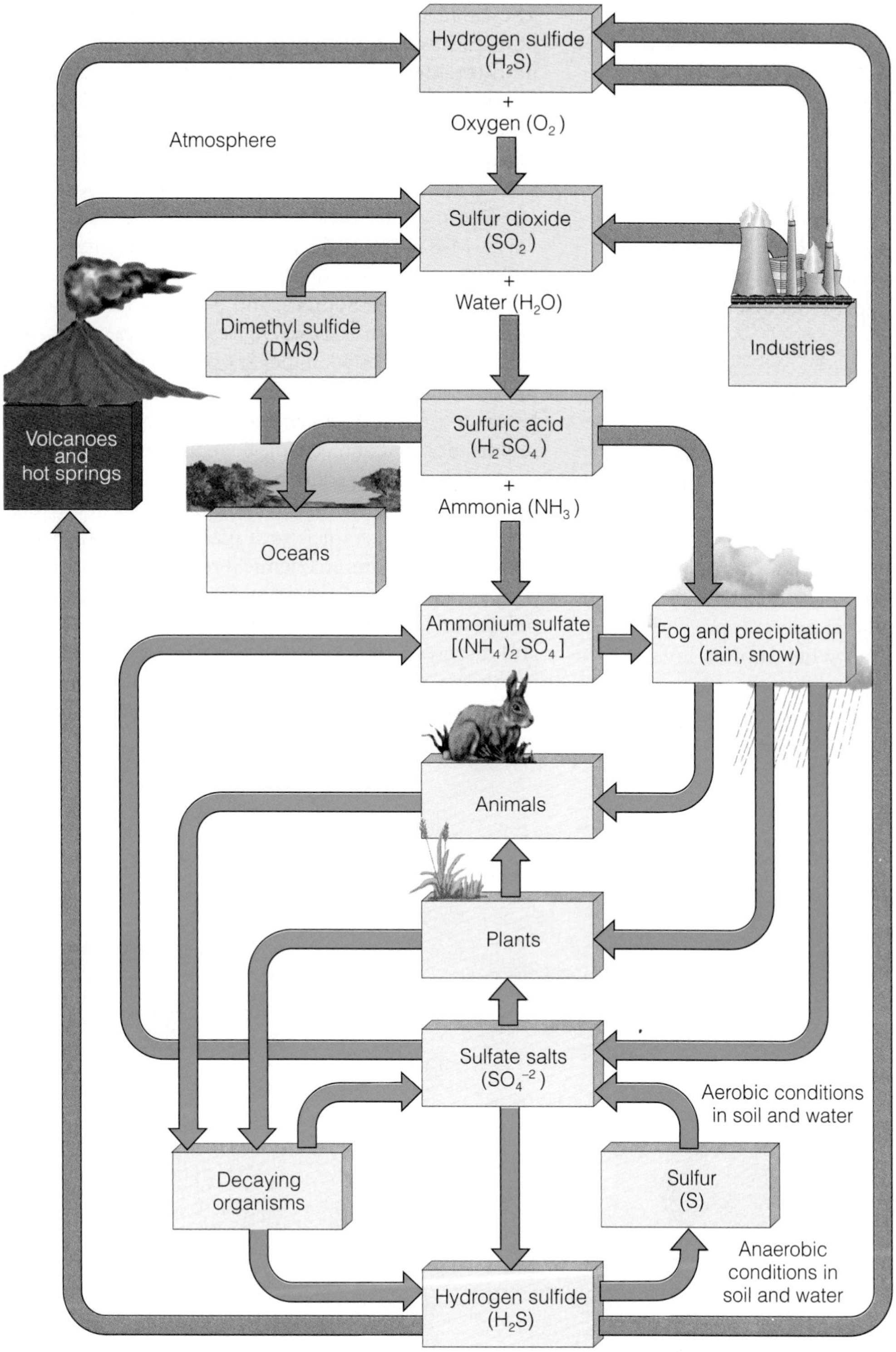

Figure 5-9 Simplified conceptual model of the sulfur cycle.

How Are Humans Influencing the Sulfur Cycle? About a third of all sulfur (including 99% of the sulfur dioxide) that reaches the atmosphere comes from human activities. We modify the sulfur cycle through various activities that release sulfur compounds into the atmosphere: **(1)** by burning sulfur-containing coal and oil to produce electric power, producing about two-thirds of the human inputs of sulfur dioxide; **(2)** by refining petroleum; **(3)** by using smelting to convert sulfur compounds of metallic minerals into free metals such as copper, lead, and zinc; and **(4)** by using other industrial processes.

A: A measure of the disorder or randomness of a system

5-7 THE ROCK CYCLE

How Are the Earth's Rocks Cycled? Rocks are constantly exposed to various physical and chemical conditions that can change them over time. The interaction of processes that change rocks from one type to another is called the **rock cycle** (Figure 5-10).

Recycling material over millions of years, this slowest of the earth's cyclic processes is responsible for concentrating the planet's nonrenewable mineral resources on which humans depend, as discussed in greater detail in Chapter 14.

What Are the Three Major Types of Rock? Geologic processes constantly redistribute the chemical elements within the earth and at the earth's surface. Based on the way it forms, rock is placed in three broad classes: igneous, sedimentary, and metamorphic.

Igneous rock can form below the earth's surface, or on it, when magma (molten rock) wells up from the upper mantle or deep crust, cools, and hardens into rock. Examples are granite (formed underground) and lava rock (formed above ground when molten lava cools and hardens).

Although often covered by sedimentary rocks or soil, igneous rocks form the bulk of the earth's crust. They also are the main source of many nonfuel mineral resources. Granite and its relatives are used for monuments and as decorative stone in buildings, basalt is used as crushed stone where gravel is scarce, and volcanic rocks are used in landscaping. Many popular gemstones, such as diamond, tourmaline, garnet, ruby, and sapphire, are part of igneous rocks.

Sedimentary rock forms from sediment in several ways. Most such rocks are formed when preexisting rocks are weathered and eroded into small pieces, transported from their sources, and deposited in a body of surface water. In *mechanical weathering,* a large rock mass is broken into smaller fragments of the original material, similar to the results you would get by using a hammer to break a rock into small fragments. The most important agent of mechanical weathering is *frost wedging,* in which water collects in pores and cracks of rock, expands upon freezing, and splits off pieces of the rock.

In *chemical weathering,* a mass of rock is decomposed by one or more chemical reactions, resulting in products that are chemically different from the original material. The products usually include both solid and dissolved components. Most chemical weathering involves a reaction of rock material with oxygen, carbon dioxide, and moisture in the atmosphere and the ground (Figure 5-11).

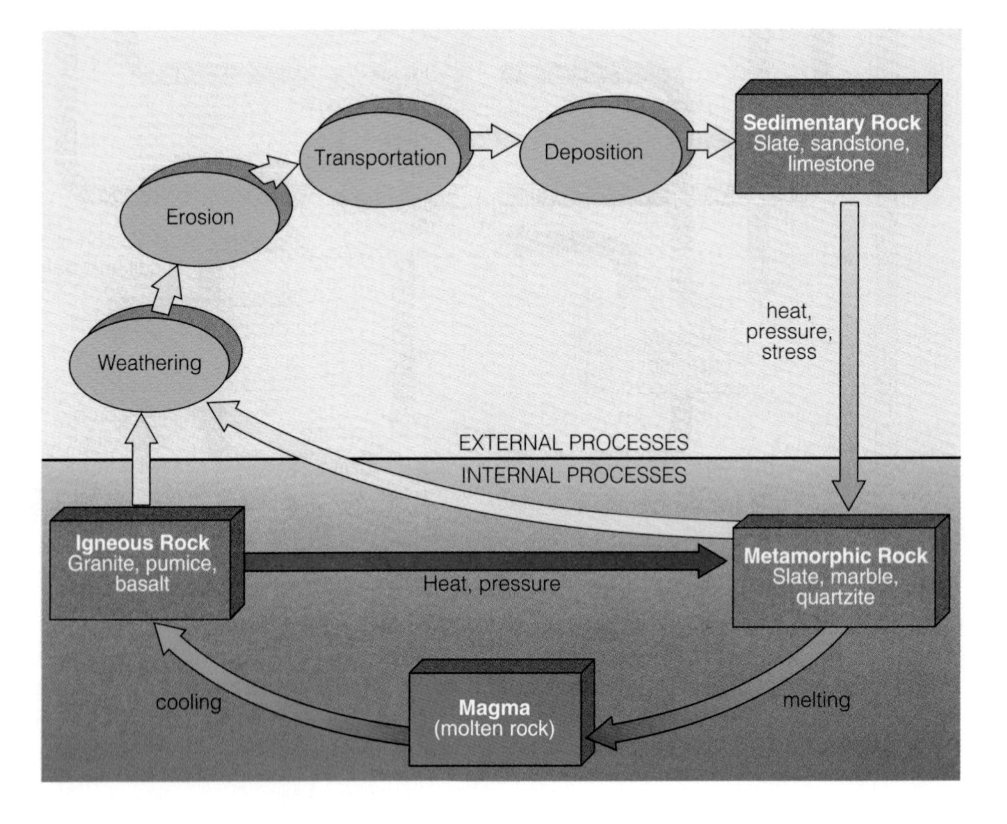

Figure 5-10 The rock cycle, the slowest of the earth's cyclic processes. The earth's materials are recycled over millions of years by three processes: melting, erosion, and metamorphism, which produce igneous, sedimentary, and metamorphic rocks. Rock of any of the three classes can be converted to rock of either of the other two classes (or can even be recycled within its own class).

Zimer, Carl. 1997. "The Web Below." *Discover,* vol. 18, no. 11, 44(1).

Figure 5-11 Chemical weathering of rock deposited by a glacier. Air and water have weathered the surface of the granitelike rock, producing a shell of weathering products that surrounds it like the rind of an orange. Weathering has also occurred in a fracture cut through the rock. The quartz grains of the parent rock are still present in the weathering rind, but some of the silicates have reacted with oxygen and carbonic acid to form clay, brown iron oxide, and soluble products. This is how soil is formed and how some trace elements are added to soil and groundwater. (The lens cap in the photo shows the relative size of the rock.) (Kenneth J. Van Dellen)

Figure 5-12 These rocks near Negaunee, Michigan, which were originally layers of mud that included some layers of sand, later became shale with sandstone beds. As a result of a mountain-building event about 1.75 billion years ago, they have been tilted to near vertical and have metamorphosed to slate (dark) with quartzite (light). (Kenneth J. Van Dellen)

Disintegration of rock by mechanical weathering accelerates chemical weathering by increasing the surface area that can be attacked by chemical weathering agents. This is similar to the way granulated sugar dissolves much faster than a large chunk of sugar. Because chemical weathering is also aided by higher temperatures and precipitation, it occurs most rapidly in the tropics and next most rapidly in temperate climates.

As deposited layers from weathering and erosion become buried and compacted, the resulting pressure causes their particles to bond together to form sedimentary rocks such as sandstone and shale. Some sedimentary rocks, such as dolomite and limestone, are formed from the compacted shells, skeletons, and other remains of dead organisms. Two types of coal—lignite and bituminous coal—are sedimentary rocks derived from plant remains.

In most places, sedimentary rocks are not more than 100 meters (330 feet) thick, but they cover nearly three-fourths of the earth's land surface. Besides making up much of the planet's scenic landscape, some sedimentary rocks are important resources. Limestone, for example, is used as crushed stone, as building stone, as flux in blast furnaces for smelting iron ore, and (with shale) for making Portland cement.

Metamorphic rock is produced when a preexisting rock is subjected to high temperatures (which may cause it to melt partially), high pressures, chemically active fluids, or a combination of those agents. Examples are anthracite (a form of coal), slate (Figure 5-12), and marble. Talc, asbestos, graphite, titanium, and some gems are also found in metamorphic rocks.

5-8 SOIL: THE BASE OF LIFE

What Are the Major Layers Found in Mature Soils? The material we call **soil** is a complex mixture of eroded rock, mineral nutrients, decaying organic matter, water, air, and billions of living organisms, most of them microscopic decomposers (Figure 5-13). Although soil is a potentially renewable resource, it is produced very slowly by the weathering of rocks (Figure 5-11), deposit of sediments by erosion, and decomposition of organic matter in dead organisms.

Mature soils are arranged in a series of zones called **soil horizons**, each with a distinct texture and composition that varies with different types of soils. A cross-sectional view of the horizons in a soil is called a **soil profile**. Most mature soils have at least three of the possible horizons (Figure 5-13).

The top layer, the *surface-litter layer*, or *O horizon*, consists mostly of freshly fallen and partially decomposed leaves, twigs, animal waste, fungi, and other organic materials. Normally, it is brown or black. The *topsoil layer*, or *A horizon*, is a porous mixture of partially decomposed organic matter, called **humus**, and

Hint: Enter the search term *soil fungi* using the Subject Guide.

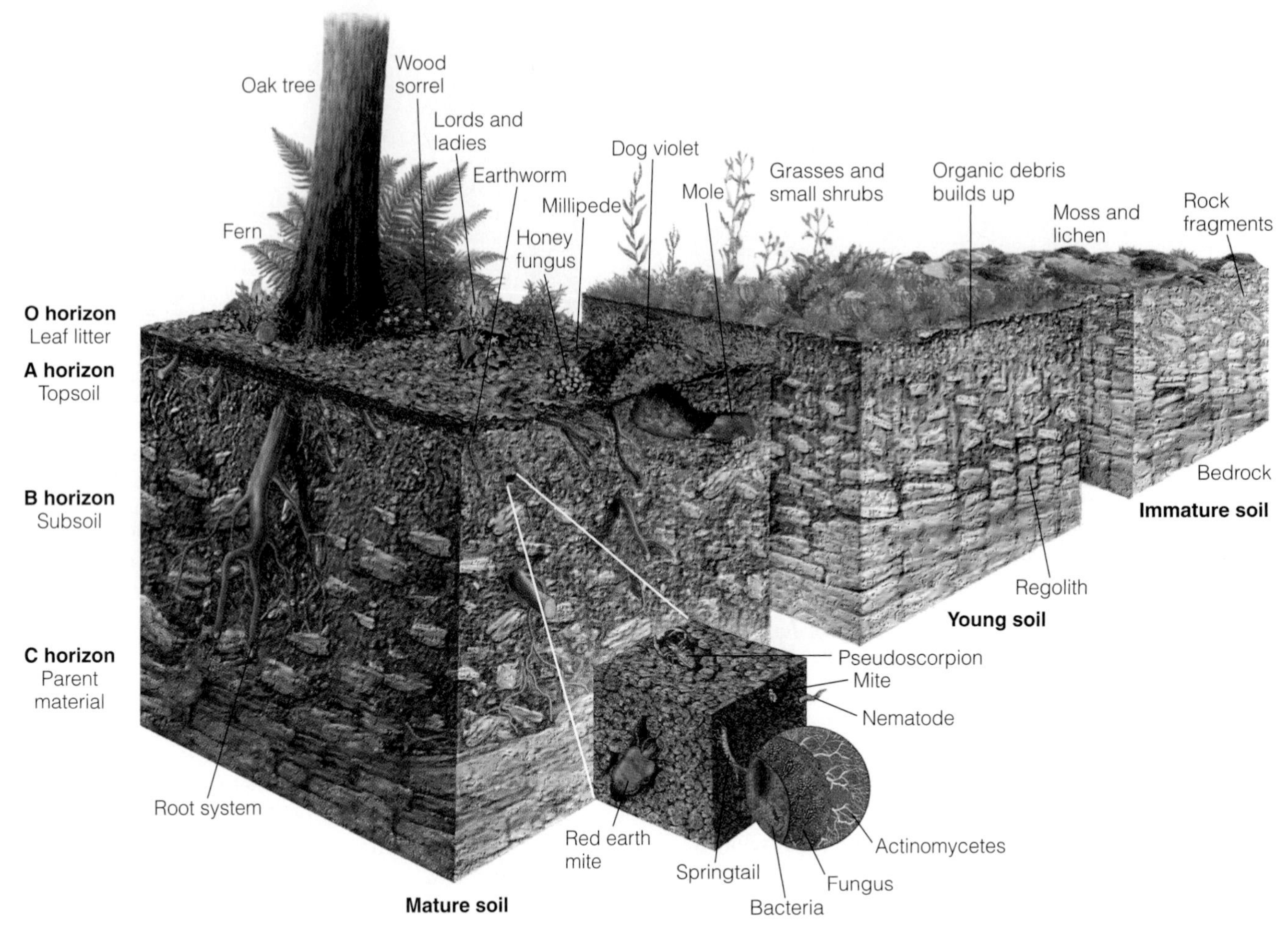

Figure 5-13 Formation and generalized profile of soils. Horizons, or layers, vary in number, composition, and thickness, depending on the type of soil. (Used by permission of Macmillan Publishing Company from Derek Elsom, *Earth*, New York: Macmillan, 1992. Copyright ©1992 by Marshall Editions Developments Limited)

some inorganic mineral particles. It is usually darker and looser than deeper layers. The roots of most plants and most of a soil's organic matter are concentrated in these two upper layers. As long as these layers are anchored by vegetation, soil stores water and releases it in a nourishing trickle instead of a devastating flood.

The two top layers of most well-developed soils teem with bacteria, fungi, earthworms, and small insects that interact in complex food webs (Figure 5-14). Bacteria and other decomposer microorganisms found by the billions in every handful of topsoil recycle the nutrients we and other land organisms need (Figure 5-15). They break down some complex organic compounds into simpler inorganic compounds soluble in water. Soil moisture carrying these dissolved nutrients is drawn up by the roots of plants and transported through stems and into leaves.

Some organic litter in the two top layers is broken down into a sticky, brown residue of partially decomposed organic material (humus). Because humus is only slightly soluble in water, most of it stays in the topsoil layer. A fertile soil that produces high crop yields has a thick topsoil layer with lots of humus, which helps topsoil hold water and nutrients taken up by plant roots.

The color of its topsoil often tells us a lot about how useful a soil is for growing crops. For example, dark brown or black topsoil is nitrogen rich and high in organic matter. Gray, bright yellow, or red topsoils are low in organic matter and need nitrogen enrichment to support most crops.

The *B horizon (subsoil)* and the *C horizon (parent material)* contain most of a soil's inorganic matter, mostly

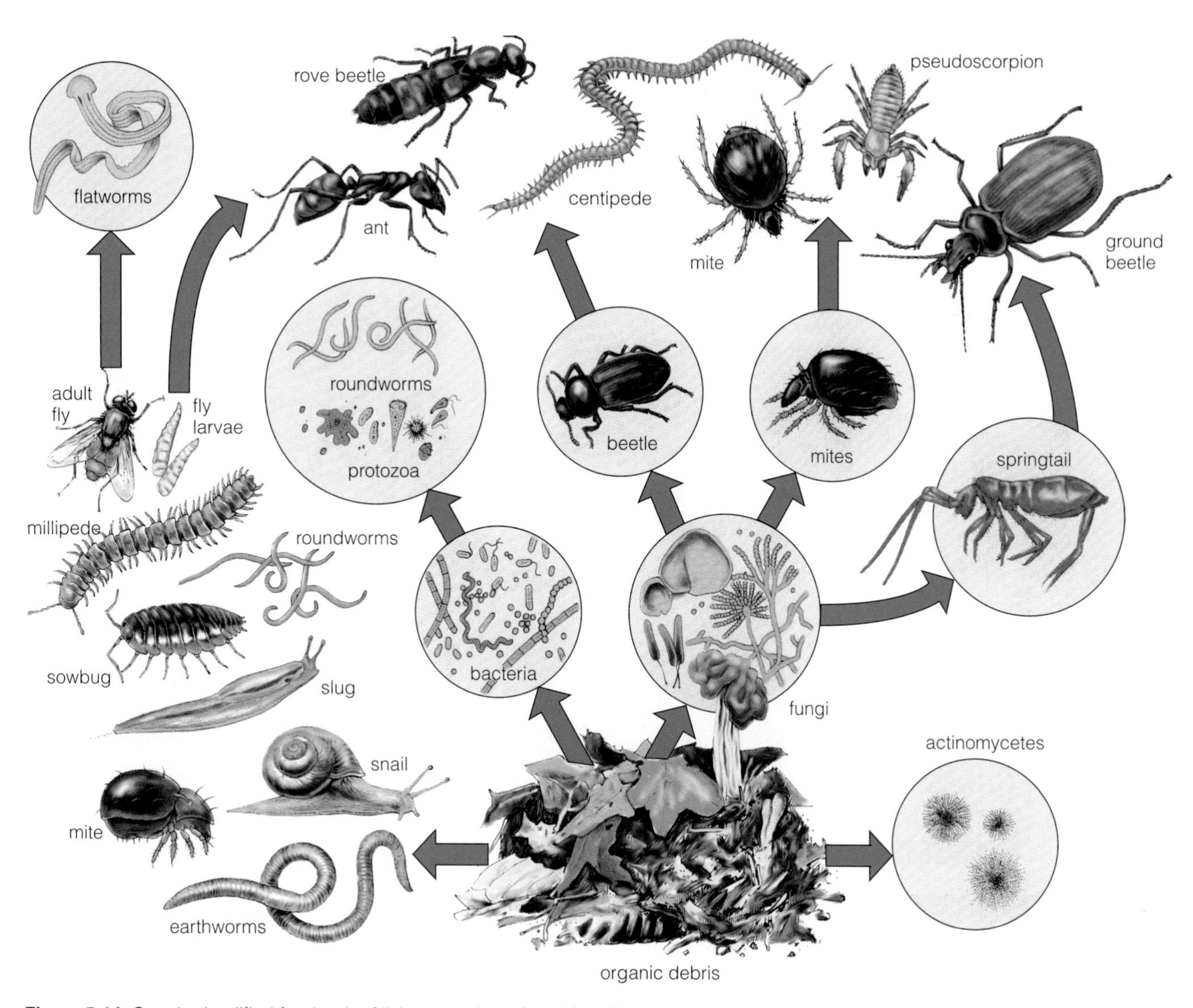

Figure 5-14 Greatly simplified food web of living organisms found in soil.

broken-down rock consisting of varying mixtures of sand, silt, clay, and gravel. The C horizon lies on a base of unweathered parent rock called *bedrock*.

The spaces, or pores, between the solid organic and inorganic particles in the upper and lower soil layers contain varying amounts of air (mostly nitrogen and oxygen gas) and water. Plant roots need oxygen for cellular respiration.

Some of the precipitation that reaches the soil percolates through the soil layers and occupies many of the soil's open spaces or pores. This downward movement of water through soil is called **infiltration**. As the water seeps down, it dissolves various soil components in upper layers and carries them to lower layers in a process called **leaching**.

Soils develop and mature slowly. It can take 200 to 1,000 years to develop an inch (2.5 cm) of topsoil (A horizon). Well-developed soil profiles, in which distinct O, A, B, and C horizons are visible, are characteristic of older, well-established terrestrial ecosystems. Five important soil types, each with a distinct profile, are shown in Figure 5-16. Most of the world's crops are grown on soils exposed when grasslands and deciduous forests are cleared.

How Do Soils Differ in Texture and Porosity? Soils vary in their content of *clay* (very fine particles), *silt* (fine particles), *sand* (medium-size particles), and *gravel* (coarse to very coarse particles). The relative amounts of the different sizes and types of mineral particles determine **soil texture**, as depicted in Figure 5-17. Soils with roughly equal mixtures of clay, sand, silt, and humus are called **loams**.

To get an idea of a soil's texture, take a small amount of topsoil, moisten it, and rub it between

A: The capacity to do work and transfer heat

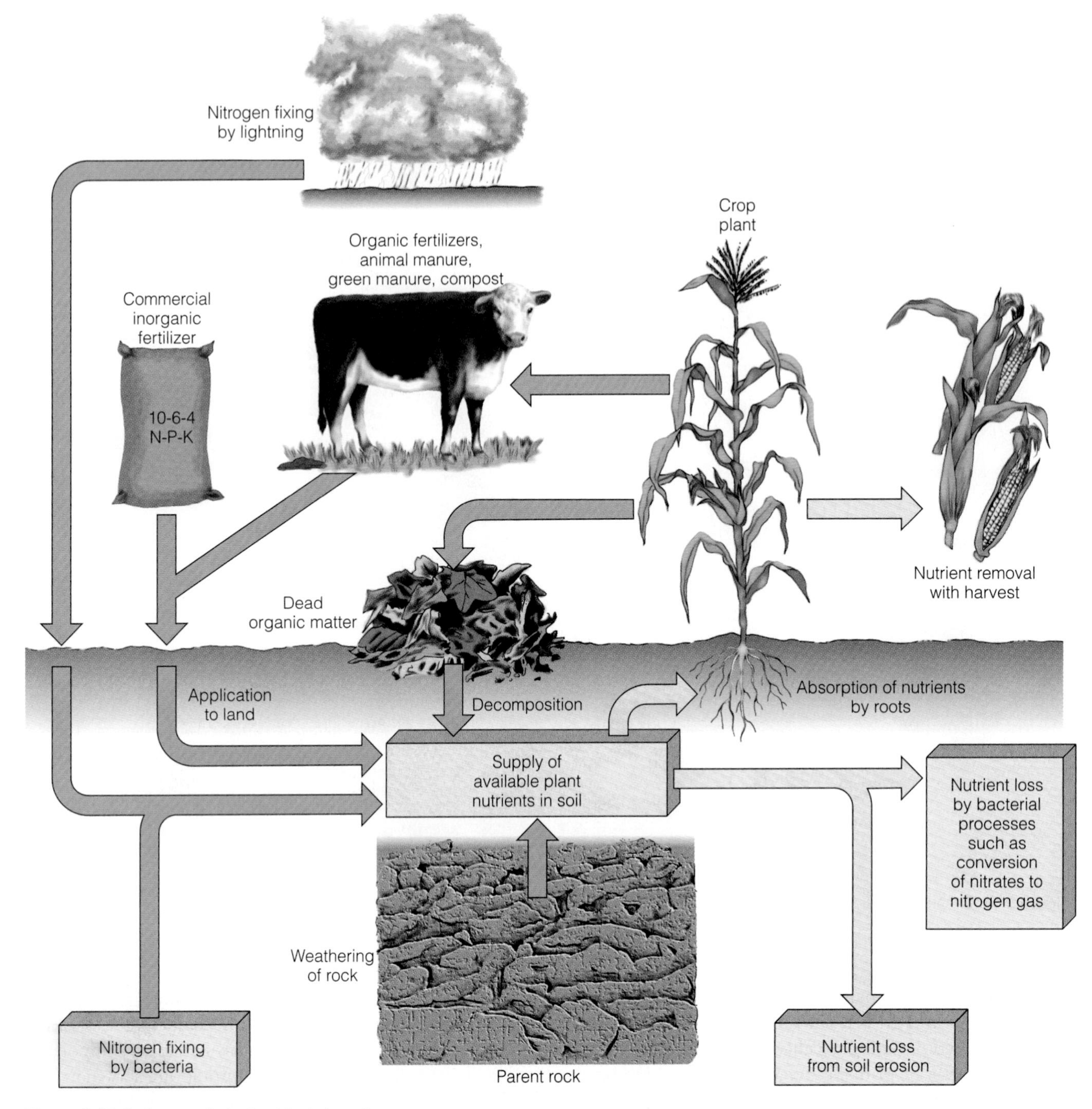

Figure 5-15 Pathways of plant nutrients in soils.

your fingers and thumb. A gritty feel means that it contains a lot of sand. A sticky feel means a high clay content, and you should be able to roll it into a clump. Silt-laden soil feels smooth, like flour. A loam topsoil, which is best suited for plant growth, has a texture between these extremes—a crumbly, spongy feeling—with many of its particles clumped loosely together.

Soil texture helps determine **soil porosity**, a measure of the volume of pores or spaces per volume of soil and of the average distances between those spaces. A porous soil has many pores and can hold more water and air than a less porous soil. The average size of the spaces or pores in a soil determines **soil permeability**: the rate at which water and air move from upper to lower soil layers. Soil porosity is also

Q: What is kinetic energy?

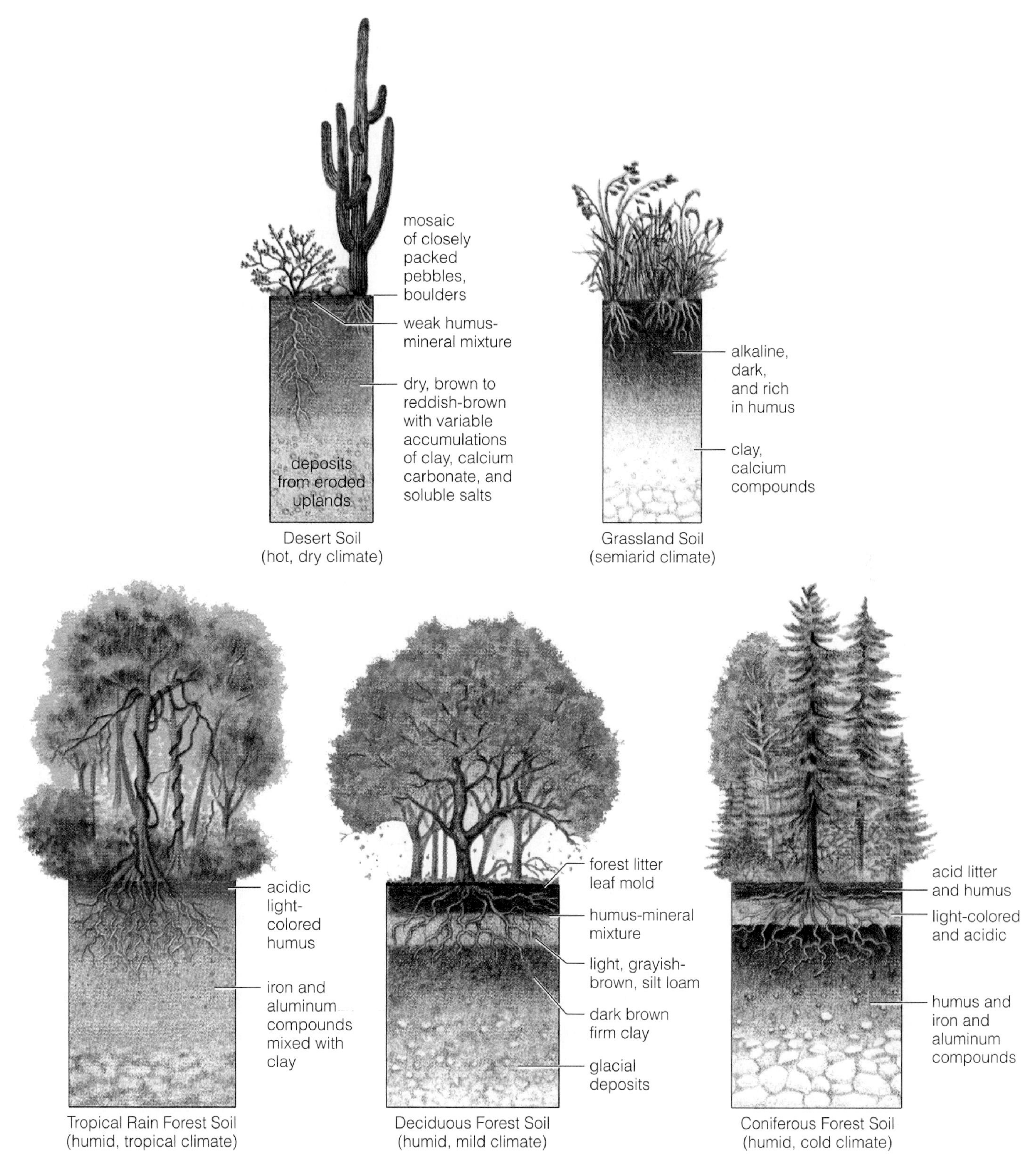

Figure 5-16 Soil profiles of the principal soil types typically found in five different biomes.

A: The energy that matter has because of its mass and speed

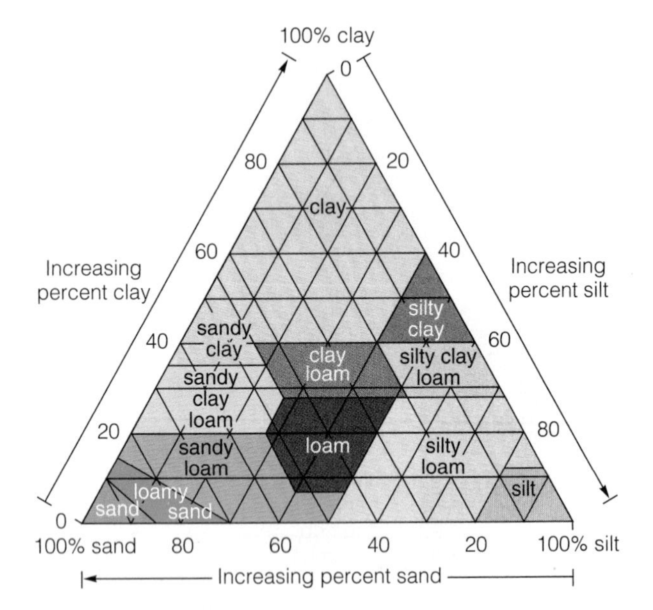

Figure 5-17 Soil texture depends on the proportions of clay, silt, and sand particles in the soil. Soil texture affects soil porosity, the average number and spacing of pores in a given volume of soil. Loams—roughly equal mixtures of clay, sand, silt, and humus—are the best soils for growing most crops. (Data from Soil Conservation Service)

influenced by **soil structure**: how soil particles are organized and clumped together.

Soil texture, porosity, and permeability determine a soil's *water-holding capacity*, *aeration* or *oxygen content* (the ability of air to move through the soil), and *workability* (how easily it can be cultivated). Table 5-1 compares the main physical and chemical properties of sand, clay, silt, and loam soils.

Loams are the best soils for growing most crops because they hold lots of water, but not too tightly for plant roots to absorb. Sandy soils are easy to work, but water flows rapidly through them. They are useful for growing irrigated crops or those with low water requirements, such as peanuts and strawberries.

The particles in clay soils are very small and easily compacted. When these soils get wet, they form large, dense clumps, which is why wet clay can be molded into bricks and pottery. Clay soils are more porous and have a greater water-holding capacity than sandy soils, but the pore spaces are so small that these soils have a low permeability. Because little water can infiltrate to lower levels, the upper layers can easily become too waterlogged for most crops.

How Does Soil Acidity (pH) Affect Plant Growth? A numerical scale of pH values is used to compare the acidity and alkalinity in water solutions (Figure 5-18). The pH of a soil influences the uptake of soil nutrients by plants, which vary in the pH ranges they can tolerate. For example, the uptake of nitrogen and phosphorus by plants is reduced in acidic soils with a pH below 5.5 and severely reduced in soils with a pH of 4 or lower.

When soils are too acidic, the acids can be partially neutralized by an alkaline substance such as lime. Because lime speeds up the decomposition of organic matter in the soil, however, manure or another organic fertilizer should also be added to maintain soil fertility.

In dry regions such as much of the western and southwestern United States, rain does not leach away calcium and other alkaline compounds, so soils in such areas may be too alkaline (pH above 7.5) for some crops. For example, uptake of phosphorus and zinc by plants is reduced in soils with a pH of 7.5 to 8.5. Adding sulfur, which is gradually converted into sulfuric acid by soil bacteria, reduces soil alkalinity.

Burning fossil fuels, especially coal, releases sulfur dioxide and nitrogen oxides into the atmosphere. These gases form acidic compounds that return to the earth's surface as *acid deposition* (Section 18-3). As

Table 5-1 Useful Properties of Soils with Different Textures

Soil Texture	Nutrient-Holding Capacity	Water-Infiltration Capacity	Water-Holding Capacity	Aeration	Workability
Clay	Good	Poor	Good	Poor	Poor
Silt	Medium	Medium	Medium	Medium	Medium
Sand	Poor	Good	Poor	Good	Good
Loam	Medium	Medium	Medium	Medium	Medium

Q: What is potential energy?

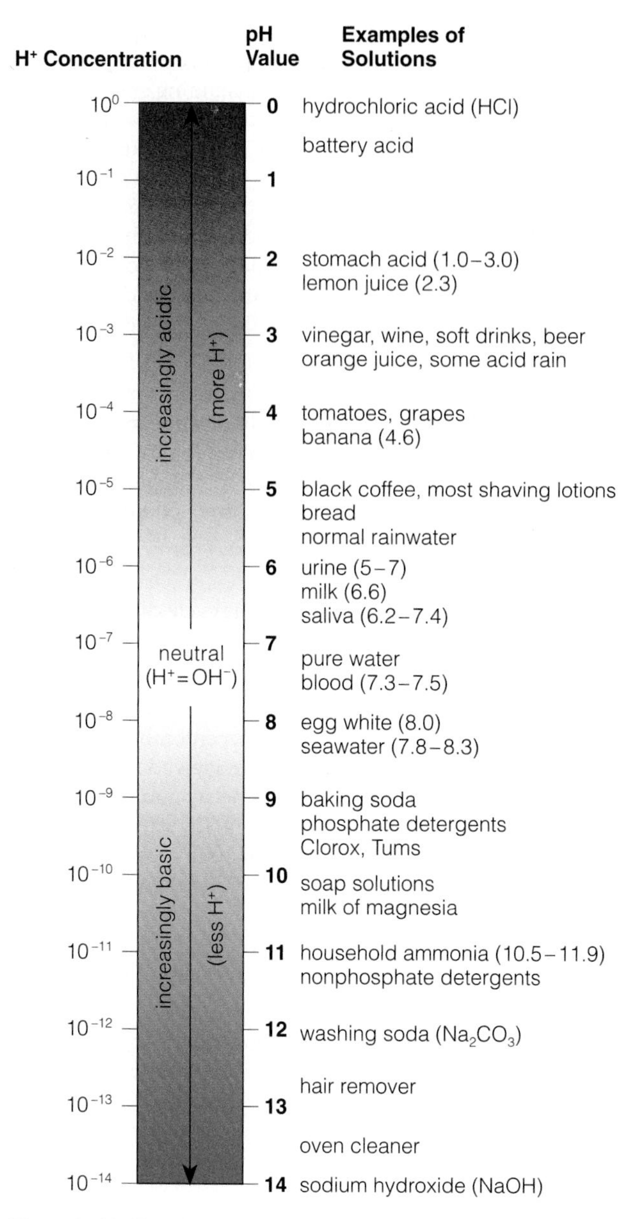

Figure 5-18 The pH scale, used to measure acidity and alkalinity of water solutions. Values shown are approximate. A solution with a pH less than 7 is acidic, a neutral solution has a pH of 7, and one with a pH greater than 7 is basic. The lower the pH below 7, the more acidic the solution. Each whole-number drop in pH represents a 10-fold increase in acidity.

acidic rain or melted acidic snow infiltrates the soil, the positively charged hydrogen ions (H^+) in the acids are attracted to negatively charged particles of minerals and humus in the topsoil layer, displacing some of the potassium (K^+), calcium (Ca^{2+}), magnesium (Mg^{2+}), and ammonium (NH_4^+) ions that were attached to those particles. The resulting loss of soil fertility can reduce crops and tree growth and make them more vulnerable to drought, disease, and pests.

The serious environmental problem of soil erosion and ways to reduce such erosion through soil conservation are discussed in Chapter 14.

5-9 NUTRIENT CYCLING AND ECOSYSTEM SUSTAINABILITY

Are Ecosystems Completely Self-Contained? The general concepts of energy flow and nutrient cycling seem to imply that ecosystems are virtually self-sustaining, closed systems. At the ecosphere level this is true, but most ecosystems exchange water and nutrients with nearby ecosystems.

As long as they are not disturbed by human activities such as clearing, mature ecosystems such as old-growth forests can approach being closed systems because they gain and lose fairly small amounts of matter and efficiently cycle most of the matter they contain. In contrast, immature communities and ecosystems such as young forests and cleared grasslands or forests are fairly leaky systems that cycle matter inefficiently and lose it from processes such as erosion.

Rainfall can erode soil and dissolve some soil nutrients, which are carried away as sediment and in surface runoff of water into another land ecosystem or into a stream, lake, or ocean. Wind can also transfer nutrients to other ecosystems by blowing away topsoil and depositing the nutrients elsewhere and by blowing gaseous molecules released by one ecosystem over other ecosystems, where they can be absorbed. However, because of the law of conservation of matter (Section 3-3), the nutrients lost from one ecosystem must enter one or more other ecosystems.

How Does Nutrient Cycling Relate to Ecosystem Sustainability? In this chapter we applied the law of the conservation of matter to understand the major nutrient cycles. We observed that we can follow the transformations and flows of matter, and that given time natural ecosystems tend to come into a balance, wherein nutrients are recycled with reasonable efficiency. We also observed that humans are changing nutrient cycles. In general, humans are accelerating rates of the flow of matter (causing nutrient loss from soils, for example).

One of the results is redistribution of matter in the ecosphere; for example, human activity is moving more carbon into the atmosphere in the form of heat-trapping carbon dioxide. According to a 1997 report by the Ecological Society of America, human activities now fix at least as much nitrogen as natural

A: Stored energy that is potentially available for use

processes. These scientists warn that this doubling of the normal flow of nitrogen in the nitrogen cycle is a serious global problem that contributes to global warming, ozone depletion, air pollution, and loss of biodiversity.

What effects might such disturbances of the major nutrient cycles have on ecosystem sustainability? This is not an easy question to answer. Part of the answer must come from careful study of natural ecosystems, in which human impacts on nutrient cycles are minimal, and part from determination of historical changes in more heavily affected ecosystems. Because of the global nature of nutrient cycles and the interconnectedness of ecosystems, even isolated ecosystems are being influenced by human activities. Developing a better understanding of these connections and effects is vital for learning how to work with the earth.

All things come from earth, and to earth they all return.
MENANDER (342–290 B.C.)

CRITICAL THINKING

1. **(a)** How would you set up a self-sustaining aquarium for tropical fish? **(b)** Suppose you have a balanced aquarium sealed with a clear glass top. Can life continue in the aquarium indefinitely as long as the sun shines regularly on it? **(c)** A friend cleans out your aquarium and removes all the soil and plants, leaving only the fish and water. What will happen?

2. In an essay titled "Odyssey," published in the book *A Sand County Almanac*, Aldo Leopold describes the voyage of a single atom "X" as it travels through the ecosphere. Trace the hypothetical voyage of a single water molecule as it evaporates from the ocean and goes through several cycles of change. How would this odyssey differ from that of a carbon atom in the carbon dioxide you just exhaled? a phosphorus atom in a phosphate ion dissolved in a river? a nitrogen atom that is part of a molecule of nitrogen gas in the air?

3. Carbon, nitrogen, and phosphorus are cycled in the ecosphere. Why do farmers not need to apply carbon to grow their crops, but often need to supply fertilizer containing nitrogen and phosphorus?

4. Carbon dioxide (CO_2) in the atmosphere fluctuates significantly on a daily and a seasonal basis. Why are CO_2 levels higher during the day than at night? In the northern hemisphere, why are CO_2 levels higher during the summer than during the winter? How do you expect CO_2 levels to fluctuate in the southern hemisphere?

5. In the 1970s James Lovelock, a British atmospheric scientist, put forth the *Gaia hypothesis*, a controversial hypothesis that asserts that the earth itself is not simply an environment for life, but is essentially a living organism, a self-organizing, self-sustaining system. Lovelock was careful to emphasize that the earth's self-regulation is automatic, requiring no conscious guidance. Part of this self-sustaining nature of the earth involves the cyclic nature of the movement of nutrients. Given what you know about the two energy laws, the law of the conservation of matter, and the behavior of natural systems, why do you think that nutrients tend to move in cycles? Do you agree with the Gaia hypothesis? Why or why not?

6. Imagine that all nitrogen-fixing organisms (*Rhizobium*, cyanobacteria, etc.) suddenly disappeared. What would be the long- and short-term ecological effects? What would you expect to happen to nitrogen levels in soils? in streams and lakes? in the atmosphere? Would it be practical for humans to intervene by adding chemical fertilizers containing nitrogen?

7. In what ways is the phosphorus cycle different from other major nutrient cycles? Is the phosphorus cycle truly cyclic, from the perspective of a human lifetime?

8. Explain in ecological terms how the human practice of embalming a corpse and then entombing it in an airtight metal container could be viewed as contrary to the earth's natural processes of decay and recycling. Even though such practices are done for important cultural, religious, and health reasons, a growing number of people prefer to be cremated and then have their ashes spread on the land or at sea; they want their atoms back in action in the earth's vital chemical cycles as soon as possible after their death. What is your preference? Why?

9. Imagine that you are an igneous rock. Act as a reporter and send in a written report on what you experience as you move through various parts of the rock cycle (Figure 5-10). Repeat this experience, assuming in turn that you are a sedimentary rock and then a metamorphic rock.

10. How do the soil profiles in plowed farmland differ from those of unplowed natural areas? What properties of soils change by the act of plowing, both in the short term and in the long term?

11. Why might ecologists define *sustainability* in a different way than economists or politicians? Why is nutrient cycling an important consideration for ecosystem sustainability? What are other ecological considerations for sustainability, and how might they relate to nutrient cycling?

PROJECTS

1. Examine the labels of various fertilizers available in a garden supply store. Find out where each of the major nutrients and other nutrients come from. Why do different fertilizers have different amounts of nitrogen, phosphorus, potassium, and other nutrients? Do all of the fertilizers contain chemical salts, or do some have nutrients in the form of organic compounds? How might the form of the nutrient affect its movement and availability for plant growth?

2. Do a Web search to find out how commercial water purification systems work. Various processes are used, including *prefiltration* or *settling chambers*, *millipore filtration*, *reverse osmosis*, *ozone treatment*, and *distillation*. How does each of the processes correspond to natural processes in the water cycle?

3. On a map of your state, mark where igneous, sedimentary, and metamorphic rocks occur. (You may have to combine two types in certain areas.)

4. Get permission to dig a series of *soil pits* or to view roadcuts and construction sites where you can see soil profiles. Draw the profiles for each site. How easily can you see boundaries between the soil horizons? Why might soil profiles differ in the depth and distinctness of each of the horizons?

5. Make a concept map of this chapter's major ideas using the section heads and subheads and the key terms (in boldface). Look at the inside back cover and in Appendix 4 for information about concept maps.

A: A measure of an energy source's ability to do useful work

6 EVOLUTION AND BIODIVERSITY: ORIGINS, NICHES, AND ADAPTATION*

Why Should We Care About Alligators?

The American alligator (Figure 6-1), North America's largest reptile, has no natural predators except humans. The lineage of this species goes back 200 million years. Hunters once killed large numbers of these animals for their exotic meat and their supple belly skin, used to make shoes, belts, and pocketbooks.

Other people considered alligators to be useless, dangerous vermin and hunted them for sport or out of hatred. Between 1950 and 1960, hunters wiped out 90% of the alligators in Louisiana, and by the 1960s the alligator population in the Florida Everglades was also near extinction.

People who say "So what?" are overlooking the alligator's important ecological roles in subtropical wetland ecosystems. Alligators dig deep depressions, or gator holes, that collect fresh water during dry spells, serve as refuges for aquatic life, and supply fresh water and food for many animals. Large alligator nesting mounds provide nesting and feeding sites for species of herons and egrets. Alligators also eat large numbers of predatory gar fish and thus help maintain populations of game fish such as bass and bream.

As alligators move from gator holes to nesting mounds, they help keep areas of open water free of invading vegetation. Without these ecosystem services,

*__Paul M. Rich__, associate professor of Ecology and Evolutionary Biology and Environmental Studies at the University of Kansas, is coauthor of this chapter.

Figure 6-1 The American alligator plays important ecological roles in its marsh and swamp habitats in the southeastern United States. After being classified as an endangered species in 1967, it has recovered enough to be changed from an endangered to a threatened species—an outstanding success story in wildlife conservation. (M. H. Sharp/Photo Researchers, Inc.)

freshwater ponds and coastal wetlands found in the alligator's habitat would be filled in by shrubs and trees, and dozens of species would disappear. Some ecologists classify the North American alligator as a *keystone species* because of its important ecological roles in helping maintain the structure, function, and sustainability of its natural ecosystems.

In 1967 the U.S. government placed the American alligator on the endangered species list. Protected from hunters, the alligator population made a strong comeback in many areas by 1975—too strong, according both to those who find alligators in their backyards and swimming pools and to duck hunters, whose retriever dogs are sometimes eaten by alligators. Large alligators have also been known to eat pigs, deer, and even cattle, dragging them under water to drown before dismembering them.

In 1977 the U.S. Fish and Wildlife Service reclassified the American alligator from an *endangered* to a *threatened* species in Florida, Louisiana, and Texas, where 90% of the animals live. In 1987 this reclassification was extended to seven other states. Alligators now number perhaps 3 million, most in Florida and Louisiana. It is generally illegal to kill members of a threatened species, but limited kills by licensed hunters are allowed in some areas of Florida, Louisiana, and South Carolina to control the population. The comeback of the American alligator is an important success story in wildlife conservation.

The increased demand for alligator meat and hides has created a booming business in alligator farms, especially in Florida. By controlling diet and other conditions, alligator farm operators have quadrupled the species's reproductive rate, doubled its growth rate, and reduced mortality from 35% to 1%. Such success reduces the need for illegal hunting of wild alligators.

Each species represents a long chain of evolution, and each species plays a unique ecological role in communities and ecosystems. Understanding such roles is important for protecting species from premature extinction, for returning captive populations of endangered species to the wild, and in helping us predict the effects of human actions on wild species. Understanding evolution, the evolutionary origins of biodiversity, and the ecological roles of species is the subject of this chapter.

There is a grandeur to this view of life . . . that, whilst this planet has gone cycling on . . . endless forms most beautiful and most wonderful have been, and are being, evolved.

CHARLES DARWIN

This chapter focuses on answering the following questions:

- What major types of life are found on the earth?
- How do scientists account for the emergence of life on the earth?
- What is evolution and how has it led to the current diversity of organisms on the earth?
- How does evolution affect the way organisms fit into their environment?
- What is the ecological niche and how does it relate to adaptation?
- How do extinction of species and formation of new species affect biodiversity?
- Are humans changing the course of evolution and the future of biodiversity on the earth?

6-1 MAJOR TYPES OF LIFE ON THE EARTH

What Types of Organisms Are Found on the Earth? Before we learn about how life evolved on the earth, it is useful to have a general knowledge of the earth's *biodiversity*, the many types of life we find on this planet. Recall that all forms of life consist of one or more cells. Each cell contains two types of components: **(1)** genes (containing specific DNA molecules, Figure 3-9), which determine what form the cell will take (plant or animal, for example) and its functions (for example, what proteins it produces), and **(2)** other parts, which protect the cell and carry out the instructions encoded in the cell's DNA molecules (by making proteins and converting energy into forms usable by the cell, Figure 3-10).

On the basis of their cell structure, biologists classify all organisms as either eukaryotic or prokaryotic. All organisms except bacteria are **eukaryotic**: Their cells are surrounded by a membrane and have a *nucleus* (a membrane-bounded structure containing genetic material in the form of DNA) and several other internal parts (Figure 6-2a).

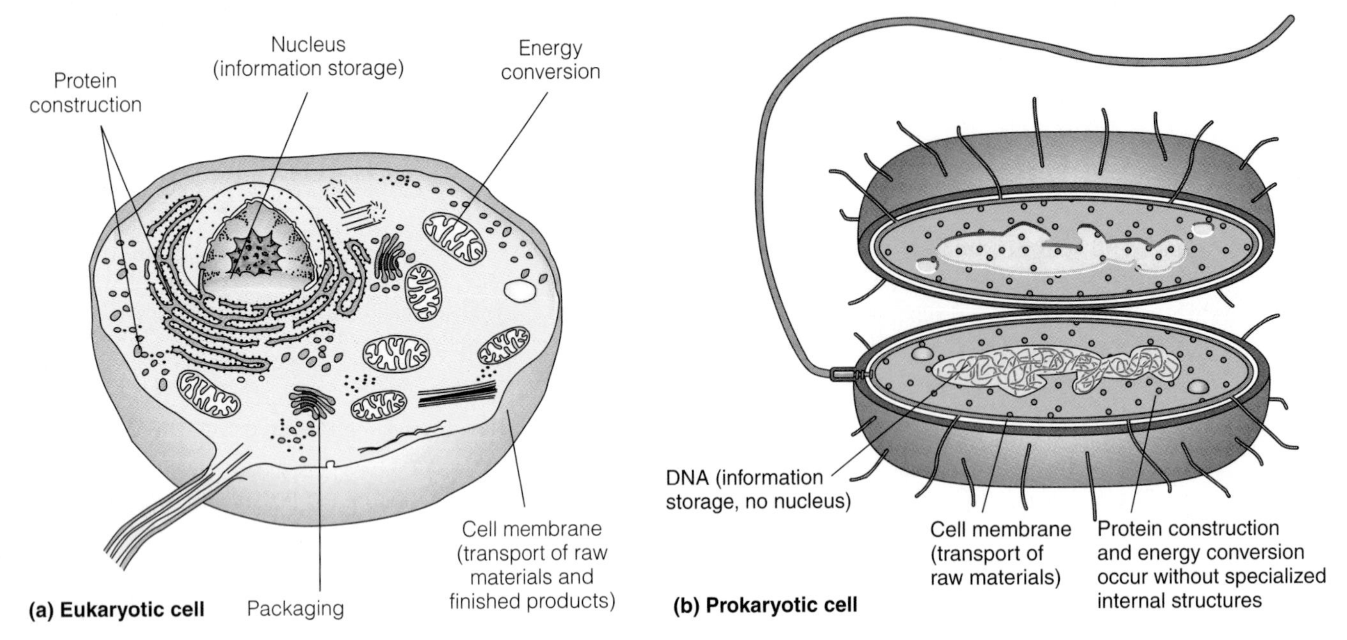

Figure 6-2 **(a)** Generalized structure of a eukaryotic cell. The parts and internal structure of cells in various types of organisms such as plants and animals differ somewhat from this generalized model. **(b)** Generalized structure of a prokaryotic cell. Note that prokaryotic cells lack a nucleus. (Adapted from Cecie Starr and Ralph Taggart, *Biology: The Unity and Diversity of Life*, 8th ed., Belmont, Calif.: Wadsworth, 1998).

Bacterial cells are **prokaryotic** (which means "before nucleus"): They are surrounded by a membrane but have no distinct nucleus or other internal parts enclosed by membranes (Figure 6-2b). Although most familiar organisms are eukaryotic, they could not exist without hordes of microscopic prokaryotic organisms (bacteria).

Scientists group organisms into various categories based on their common characteristics, a process called *taxonomic classification*. The largest category is the *kingdom*, which includes all organisms that have one or perhaps several common features. The earth's organisms are commonly classified into five kingdoms: monera, protists, fungi, plants, and animals (Figure 6-3). Most monera, fungi, and protists are **microorganisms**: organisms that are so small that they can be seen only by using a microscope.

Monera (**bacteria** and **cyanobacteria**) are single-celled, microscopic prokaryotic organisms. Most bacteria play a vital role as decomposers, which break down the tissue of dead organisms into simpler compounds that serve as nutrients for the bacteria and that are eventually reused as nutrients by plants. A few bacteria—*Streptococcus* (which causes strep throat) and *Salmonella* (which causes food poisoning), for example—can cause diseases in humans. However, your body is inhabited by billions of beneficial bacteria that help keep you healthy by aiding in food digestion and by crowding out disease-causing bacteria.

Protists (protista) are mostly single-celled eukaryotic organisms such as diatoms, dinoflagellates, amoebas, golden brown and yellow-green algae, and protozoans. Some protists cause human diseases such as malaria, sleeping sickness, and Chagas's disease.

Fungi are mostly many-celled (sometimes microscopic) eukaryotic organisms such as mushrooms (Figure 4-13), molds, mildews, and yeasts. Many fungi are decomposers. Other fungi kill various plants and cause huge losses of crops and valuable trees.

Plants (plantae) are mostly many-celled eukaryotic organisms such as red, brown, and green algae and mosses, ferns, conifers, and flowering plants (whose flowers produce seeds that perpetuate the species). Some plants such as corn and marigolds are **annuals**, which complete their life cycles in one growing season; others are **perennials**, which can live for more than two years, such as roses, grapes, elms, and magnolias.

Animals (animalia) are also many-celled, eukaryotic organisms. Most, called **invertebrates**, have no backbones. They include sponges, jellyfish, worms, arthropods (insects, shrimp, and spiders), mollusks (snails, clams, and octopuses), and echinoderms (sea urchins and sea stars). Insects play roles that are vital

Figure 6-3 Five kingdoms of the living world, a scheme for classifying the earth's diverse species into major groups. Most biologists believe that protocells gave rise to single-celled prokaryotes, and that these in turn evolved into the more complex protists, fungi, plants, and animals that make up the earth's stunning biodiversity. This is also a greatly simplified overview of *macroevolution*: changes in the genetic makeup of populations through successive generations over hundreds of millions of years in which one species leads to the appearance of many other species and groups of species.

Q: What is the law of conservation of matter?

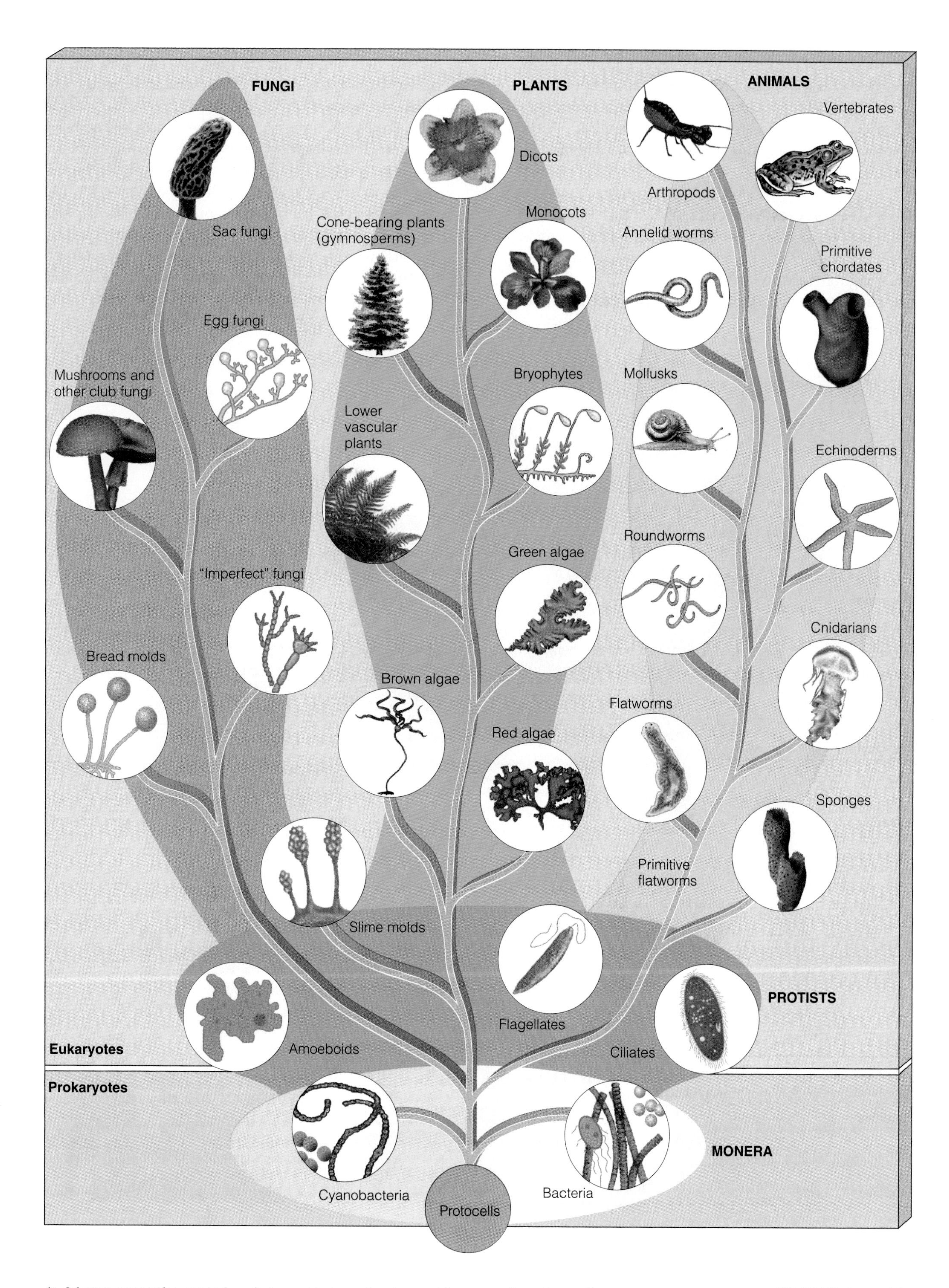

A: Matter cannot be created or destroyed but can be changed from one form to another to our existence (p. 80). **Vertebrates** (animals with backbones and a brain protected by skull bones) include fishes (sharks and tuna), amphibians (frogs and salamanders), reptiles (crocodiles and snakes), birds (eagles and robins), and mammals (bats, elephants, whales, and humans).

How Are Species Named? Within each kingdom, biologists have created subcategories based on anatomical, physiological, and behavioral characteristics. Kingdoms are divided into *phyla*, which are divided into subgroups called *classes*. Classes are subdivided into *orders*, which are further divided into *families*. Families consist of *genera* (singular, *genus*), and each genus contains one or more *species*. Note that the word *species* is both singular and plural. Figure 6-4 shows this detailed taxonomic classification for the current human species.

Most people call a species by its common name, such as *robin* or *grizzly bear*. Biologists use scientific names (derived from Latin) consisting of two parts (printed in italics, or underlined) to describe a species. The first word is the capitalized name (or abbreviation) for the genus to which the organism belongs. This is followed by a lowercase name that distinguishes the

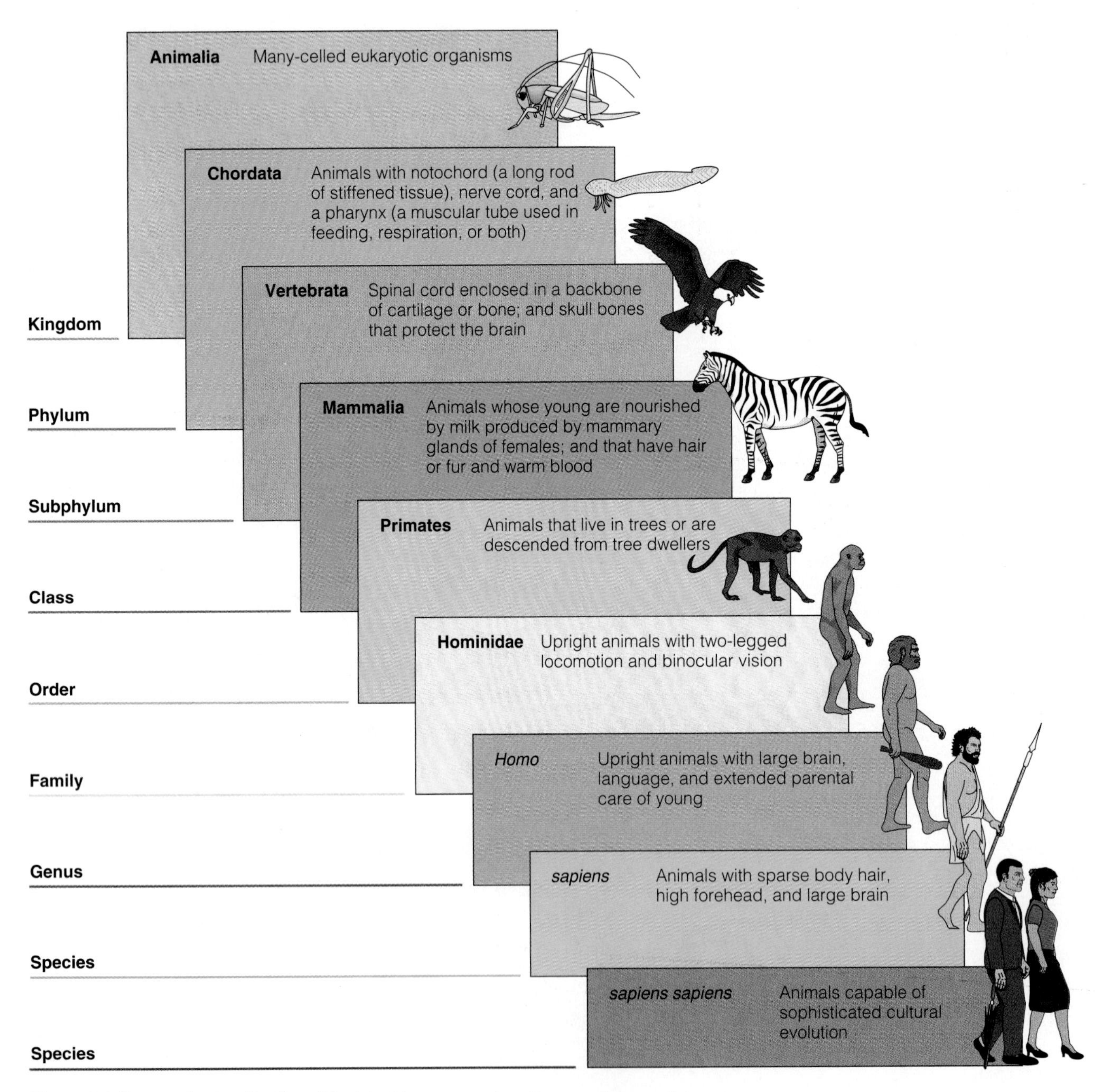

Figure 6-4 Taxonomic classification of the latest human species, *Homo sapiens sapiens*.

Q: What is the first law of energy or thermodynamics?

species from other members of the same genus. For example, the scientific name of the robin is *Turdus migratorius* (Latin for "migratory thrush") and the grizzly bear goes by the scientific name *Ursus horribilis* (Latin for "horrible bear").

What Is a Species? In general terms, a *species* (the Latin word for "appearance" or "kind") is defined as a distinct type of organism. Most of us can identify many different species based on their distinctive external appearance. We can easily distinguish frogs from fish, robins from hawks, and oak trees from pine trees. However, we may have a harder time distinguishing between two types of frogs or two types of pine trees. Also, we might not recognize that all of the shells of different colors and sizes of Caribbean snails shown in Figure 4-4 belong to a single species.

To improve their ability to distinguish species, biologists have refined the concept of species to include reproductive behavior. If members of two populations that reproduce sexually have the potential to interbreed with one another and produce live, fertile offspring, they belong to the same species. In other words, each sexually reproducing species is *reproductively isolated* from every other species in nature.

This definition of species, according to the potential ability to interbreed, has been useful, but also has some limitations. It is often impractical or impossible to observe or test whether interbreeding occurs in the wild. Because populations of similar organisms living in different places don't have an *opportunity* to interbreed, we don't know whether they have the *potential* to interbreed in nature. Another problem is that this definition does not apply to the many forms of life that reproduce asexually (including various animals, plants, fungi, and microorganisms). These species must be identified by their appearance, biochemical traits, and DNA sequences. Despite these difficulties, the species concept is the most useful and widely accepted way to identify most of the earth's forms of life.

6-2 ORIGINS OF LIFE

How Did Life Emerge on the Earth? How did life on the earth evolve to its present system of diverse kingdoms and species (Figure 6-3), living in an interlocking network of matter cycles, energy flow, and species interactions? We don't know the full answer to this question, but a growing body of evidence suggests what might have happened.

Evidence about the earth's early history comes from chemical analysis and measurements of radioactive elements in primitive rocks and fossils. Chemists have also conducted laboratory experiments showing how simple inorganic compounds in the earth's early atmosphere might have reacted to produce organic molecules such as amino acids, simple sugars, and other building-block molecules for large biopolymer molecules (such as proteins, complex carbohydrates, RNA, and DNA, Figures 3-9 and 3-10) needed for life.

When this diverse and continually accumulating evidence is pieced together, an important idea emerges: *The evolution of life is linked to the physical and chemical evolution of the earth.* It becomes apparent that the earth has the right physical and chemical conditions for life as we know it to exist (Connections, p. 139).

A second conclusion from this body of evidence is that life on the earth evolved in two phases over the past 4.7–4.8 billion years (Figures 6-5 and 6-6): *chemical evolution* of the organic molecules, biopolymers, and systems of chemical reactions needed to form the first protocells (about 1 billion years) and *biological evolution* of single-celled organisms (first prokaryotes and then eukaryotes) and then multicellular organisms (about 3.7–3.8 billion years).

Trying to piece together what happened on earth billions of years ago is an incredibly difficult form of scientific detective work. It should not be surprising that this exciting and ongoing investigation is filled with new and often unexpected discoveries and many controversial hypotheses.

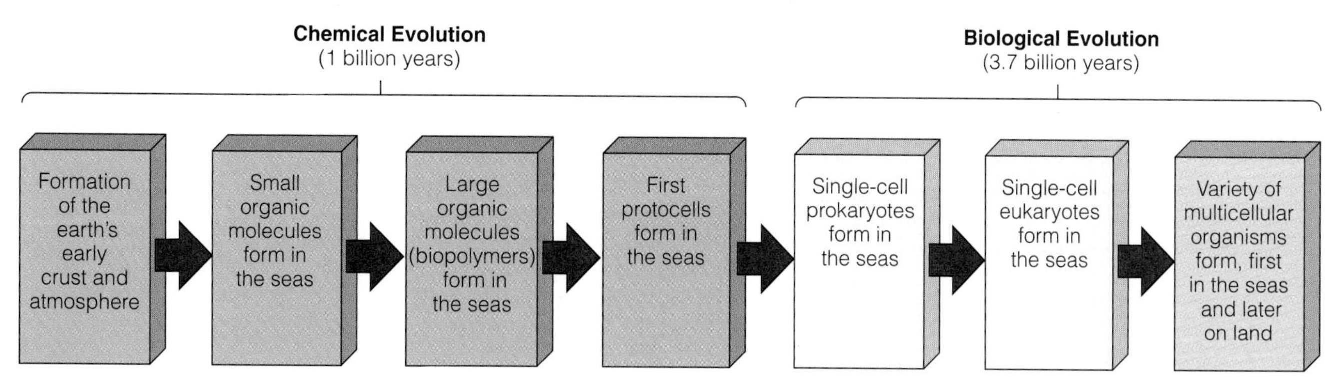

Figure 6-5 Summary of the evolution of the earth and its life.

A: Energy cannot be created or destroyed but can be changed from one form to another

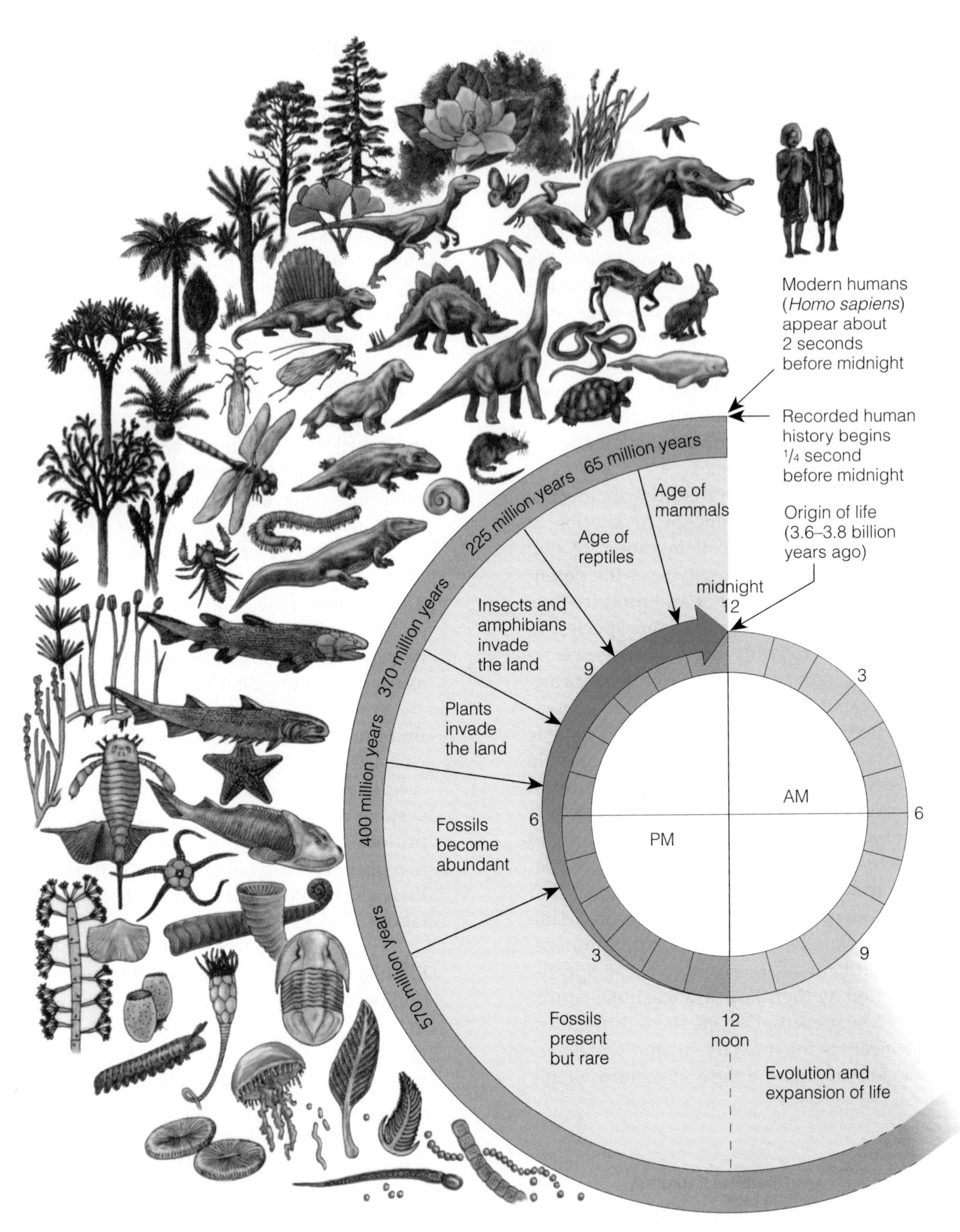

Figure 6-6 Greatly simplified overview of the biological evolution of life on the earth, which was preceded by about 0.5–1 billion years of chemical evolution. The early span of biological evolution on the earth, between about 3.7 billion and about 570 million years ago, was dominated by microorganisms (mostly bacteria and, later, protists) that lived in water. Plants and animals evolved first in the seas and moved onto land about 400 million years ago. Humans arrived on the scene only a very short time ago. If we compress the earth's roughly 3.7-billion-year history of biological evolution to a 24-hour time scale, the first human species (*Homo habilis*) appeared about 47–94 seconds before midnight and our species (*Homo sapiens sapiens*) appeared about 1.4 seconds before midnight. Agriculture began only 0.25 second before midnight, and the industrial revolution has been around for only 0.007 second. (Adapted from George Gaylord Simpson and William S. Beck, *Life: An Introduction to Biology*, 2d ed., New York: Harcourt Brace Jovanovich, 1965)

Q: What is the second law of energy or thermodynamics?

How Did Chemical Evolution Take Place? Here is an overview of how scientists believe life *may* have formed and evolved on the earth, based on the current evidence. Some 4.6–4.7 billion years ago a cloud of cosmic dust condensed into planet earth, which soon turned molten from meteorite impacts and from the heat of the radioactive decay of elements in its interior. As cooling took place, the outermost portion of the molten sphere solidified to form a thin, hardened crust of rocks, devoid of atmosphere and oceans.

Volcanic eruptions and comets hitting the lifeless earth pierced its thin crust, releasing water vapor and other gases from the molten interior. Eventually the crust cooled enough for the water vapor to condense and fall to the surface as rain. This rain eroded minerals from rocks, and solutions of these minerals collected in depressions to form the early oceans that covered most of the globe.

Research using radioisotopes suggests that most of the planet's first atmosphere had formed by 4.4 billion years ago. The exact composition of this primitive atmosphere is unknown, but atmospheric scientists believe it was dominated by carbon dioxide (CO_2), nitrogen (N_2), and water vapor (H_2O). Trace amounts of methane (CH_4), ammonia (NH_3), hydrogen sulfide (H_2S), and hydrogen chloride (HCl) were also probably present (although scientists disagree over the relative amounts of these gases).

Whatever the composition of this primitive atmosphere, scientists agree that it had no oxygen gas (O_2) because this element is so chemically reactive that it would have combined into compounds. (The only reason today's atmosphere has so much O_2 is that plants and some aerobic bacteria produce it in vast quantities through photosynthesis. But this is getting ahead of the story.)

Energy from electrical discharges (lightning), heat from volcanoes, and intense ultraviolet (UV) light and other forms of solar radiation was readily available for the synthesis of biologically important organic molecules from the inorganic chemicals found in the earth's primitive atmosphere—an idea first proposed in 1923 by Russian biochemist Alexander Oparin.

In a number of experiments conducted since 1953, various mixtures of gases believed to have been in the earth's early atmosphere have been put in closed, sterilized glass containers. Then they were subjected to spark discharges to simulate lightning and heat. In these experiments, compounds necessary for life—various amino acids (the building-block molecules of proteins), simple carbohydrates, nucleic acids (the building-block molecules of DNA and RNA), and other small organic compounds—formed from the inorganic gaseous molecules. These experiments supported Oparin's hypothesis, although many details are missing or hotly debated.

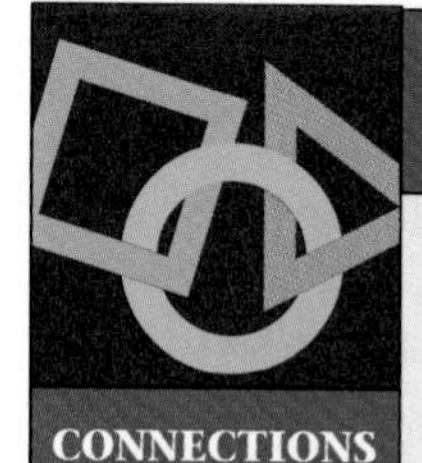
CONNECTIONS

Earth: The Just-Right, Resilient Planet

Like Goldilocks tasting porridge at the Three Bears' house, life on the earth as we know it requires a certain temperature range: Venus is much too hot and Mars is much too cold, but the earth is *just right.* (Otherwise, you wouldn't be reading these words.)

Life as we know it depends on liquid water. Again, temperature is crucial; life on the earth requires average temperatures between the freezing and boiling points of water, between 0°C and 100°C (32°F and 212°F) at the earth's range of atmospheric pressures.

The earth's orbit around the sun is the right distance from the sun to provide these conditions. If the earth were much closer, it would be too hot—like Venus—for water vapor to condense to form rain. If it were much farther away, its surface would be so cold—like Mars—that its water would exist only as ice. The earth also spins on a tilted axis (Figure 7-3, p. 159); if it didn't, the side facing the sun would be too hot and the other side too cold for water-based life to exist. So far, the temperature has been, like Baby Bear's porridge, just right.

The earth is also the right size; that is, it has enough gravitational mass to keep its iron–nickel core molten and to keep the gaseous molecules in its atmosphere from flying off into space. (A much smaller earth would be unable to hold onto an atmosphere consisting of such light molecules as N_2, O_2, CO_2, and H_2O.) The slow transfer of its internal heat (geothermal energy) to the surface also helps keep the planet at the right temperature for life. And thanks to the development of photosynthesizing bacteria over 2 billion years ago, an ozone sunscreen protects us and many other forms of life from an overdose of ultraviolet radiation.

On a time scale of millions of years, the earth is also enormously resilient and adaptive. Its average temperatures have remained between the freezing and boiling points of water even though the sun's energy output has increased by about 30% over the 3.7 billion years since life arose. In short, the earth is just right for life as we know it.

Critical Thinking

1. Which do you believe is in greater danger: the earth or the human species? Explain.

2. Do you believe that humans can learn enough about the earth to manage it (mostly for the human species) on a global scale? Explain. If not, what are the alternatives?

A: When energy is changed from one form to another, there is always a loss in energy quality

Another possibility is that simple organic molecules necessary for life formed on dust particles in space and reached the earth on meteorites or comets (or on countless interplanetary dust particles floating around in space when the earth was formed). Another hypothesis is that life could have arisen around the mineral-rich and very hot *hydrothermal vents*, which sit atop cracks in the ocean floor leading to subterranean chambers of molten rock. We don't know which of these processes might have produced the organic molecules necessary for life, but all of these hypotheses are reasonable explanations of how this could have happened.

Once these building-block organic molecules formed in the early atmosphere, they were removed by rain. Then they accumulated and underwent countless chemical reactions in the earth's warm, shallow waters. After several hundred million years of different chemical combinations in this hot organic soup, conglomerates of proteins, RNA, and other biopolymers may have combined to form membrane-bound *protocells*: small globules that could take up materials from their environment and grow and divide (much like living cells). Again, there are several hypotheses explaining how this could have happened. With these forerunners of living cells, the stage was set for the drama of biological evolution (Figure 6-6).

How Did Life First Evolve? Over time, it is believed that the protocells evolved into single-celled, bacterialike prokaryotes having the properties we describe as life (although the details of how this might have happened are hotly debated by scientists). These anaerobic cells probably evolved either in the muddy sediments of tidal flats or at least 10 meters (30 feet) below the ocean's surface, protected from the intense UV radiation that bathed the earth at that time.

Scientists believe that these single-celled anaerobic bacteria multiplied and underwent genetic changes (mutated) for about a billion years in the earth's warm, shallow seas. The result was a variety of new types of prokaryotic cells.

Scientists contend that during this long early period, life could not have survived on land, a hypothesis supported by a lack of land fossils. There was no ozone layer to shield the DNA and other molecules of early life from bombardment by intense ultraviolet radiation.

About 2.3–2.5 billion years ago, something happened in the ocean that drastically changed the earth: the evolution of photosynthetic prokaryotes called *cyanobacteria* (Connections, right). These cells could remove carbon dioxide from the water and (powered by sunlight) combine it with water to make the carbohydrates they needed. In the process, they released oxygen (O_2) into both the ocean and the atmosphere.

Recent research indicates that levels of O_2 in the atmosphere began increasing 2.0–2.1 billion years ago and reached their current concentration about 1.5 billion years ago. If this hypothesis is correct, some of the oxygen molecules you just breathed into your lungs were probably released into the atmosphere by cyanobacteria living about 2 billion years ago.

The resulting *oxygen revolution* took place over about half a billion years and opened the way for the evolution of a great variety of oxygen-using (aerobic) bacteria. Later came more complex organisms, first in the seas and then on land (after a protective ozone layer formed in the stratosphere) (Figure 6-6). Fossil evidence indicates that at least 1.2 billion years ago the first *eukaryotic cells* (with nuclei) emerged in earth's shallow seas. Because eukaryotes could reproduce sexually, they produced a variety of offspring with different genetic characteristics. Genetic changes in these eukaryotic cells eventually spawned an amazing variety of protists, fungi, and eventually plants and animals (Figure 6-3).

As oxygen accumulated in the atmosphere, some was converted by incoming solar energy into ozone (O_3), which began forming in the lower stratosphere. This shield protected life-forms from the sun's deadly UV radiation, allowing green plants to live closer to the ocean's surface. Fossil evidence suggests that about 400–500 million years ago UV levels were low enough for the first plants to exist on the land. Over the next several hundred million years a variety of land plants and animals arose, followed by mammals (and eventually) the first humans (Figure 6-6).

How Do We Know What Organisms Lived in the Past? Most of what we know of the earth's life history comes from **fossils**: mineralized or petrified replicas of skeletons, bones, teeth, shells, leaves, and seeds, or impressions of such items. Such fossils give us physical evidence of organisms that lived long ago and show us what their internal structures looked like. Artists with a background in paleobiology (the science specializing in extracting information from the fossil record) can add flesh or plant tissue to the fossil outlines and thus draw pictures or make 3-D models of what such species might have looked like.

Despite its importance, the fossil record is uneven and incomplete. Some life forms left no fossils, some fossils have decomposed, and others are yet to be found. So far we have found fossils representing only about 1% of the species believed to have ever lived. Examining the few fossils we have in order to hypothesize about the other species believed to have lived on the earth has limitations—somewhat like a blind person trying to describe what an elephant looks like by being able to feel only the tip of its tail.

Gould, S. J. 1998. "Stretching to Fit: How Life Explores and Colonizes the Landscape." *The Sciences*, vol. 38, no. 4, 12(3).

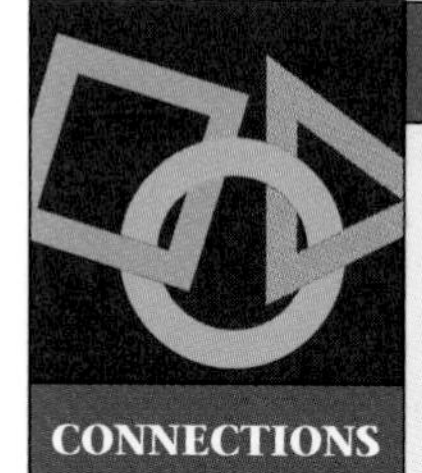

CONNECTIONS

Life, the Atmosphere, and Banded Iron Formation

The earth's rocks and the fossils in them help us decipher some of the planet's history. The oldest known indication of life is found in 3.5-billion-year-old rocks that contain peculiar limestone structures called stromatolites (Figure 6-7), produced by and covered with a mat of fossilized marine cyanobacteria.

Chemical analysis of these ancient rocks shows that the cyanobacteria were photosynthetic. Because photosynthesis releases oxygen, dissolved oxygen gradually accumulated in the seawater. As the oxygen level increased, the gas eventually began to escape into the atmosphere, leading to the modern oxygen-rich troposphere and ultraviolet-filtering stratospheric ozone. Together, these made possible the existence of life on land. One strand of evidence that the atmospheric oxygen content was changing is that red sediments, colored by oxidized iron, appear in rocks approximately 2.5–2.8 billion years old.

The most economically important deposits of iron ore owe their existence to ancient cyanobacteria. Oxygen released into the ancient oceans by cyanobacteria combined with dissolved iron. This combination changed the iron to a less soluble form, which precipitated out of solution, settled to the bottom of the ocean, and became part of sediments. Over millions of years, these sediments formed the thick deposits of the Lake Superior Banded Iron Formation that are mined today in Minnesota and Michigan. This is a striking example of how life, the atmosphere, and the earth's crust have interacted over billions of years—another example of connections in nature.

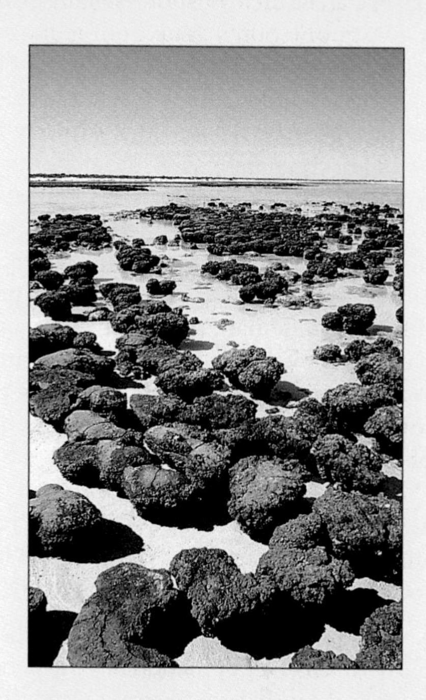

Figure 6-7 Stromatolites in Shark's Bay, Australia. These limestone structures are similar to those deposited as much as 3.5 billion years ago by photosynthetic cyanobacteria (formerly known as blue-green algae). A matlike community of cyanobacteria covers the outer surface of stromatolites. Tiny particles of calcium carbonate collect among the filamentous organisms when the tide is high and water covers the structures; this slowly builds up thin layers of rock over time. About 2 billion years ago, stromatolites were probably a common feature on the earth, but today they exist in only a few places. (C. A. Henley/Biofotos)

Critical Thinking

State two major differences that would have resulted from the evolution of life on the earth if photosynthetic cyanobacteria had not evolved.

Other sources of information include chemical and radioactive dating (Section 3-4) of fossils, nearby ancient rocks, and material in cores drilled out of buried ice. Another record of evolutionary history is written in the DNA of organisms alive today. Unfortunately, ancient mineralized fossil specimens rarely include any of the original tissues, cells, or molecules found in the organism's body. (This is why the recreation of ancient dinosaurs from their DNA in the movie *Jurassic Park* is essentially a scientific fairy tale).

6-3 EVOLUTION AND ADAPTATION

What Is Evolution? Within limits, populations can adapt to changes in environmental conditions. The major driving force of adaptation to environmental change is **biological evolution**, or **evolution**: the change in a population's genetic makeup through successive generations. Note that *populations, not individuals, evolve by becoming genetically different.*

According to the **theory of evolution**, all species descended from earlier, ancestral species. This theory explains the way life has changed over the past 3.7 billion years, and why life is so diverse today.

Biologists use the term ***microevolution*** to describe the small genetic changes that occur in a population. The term ***macroevolution*** is used to describe long-term, large-scale evolutionary changes among groups of species, wherein new species are formed from ancestral species and other species are lost through extinction.

How Does Microevolution Work? The raw material of microevolution is the *genetic variability* in a population. Recall that **(1)** the genetic information is contained in various sequences of chemical units (called *nucleotides*)

Hint: Enter the search terms *evolution, environmental* using the Subject Guide.

in DNA molecules (Figures 3-8 and 3-9) and that **(2)** the *genes* consist of segments of DNA and code for traits that can be passed on to offspring (Figure 3-10).

A population's **gene pool** is the sum total of all genes possessed by the individuals of the population of a species and microevolution is a change in a population's gene pool over time. Although members of a population generally have the same number and kinds of genes, a particular gene may have two or more different molecular forms, called **alleles**. Sexual reproduction leads to a random shuffling or recombination of alleles, such that each individual in a population has a different combination of alleles. Without such genetic variability, evolution as we know it could not occur.

Microevolution works through a combination of four processes that change the genetic composition of a population:

- **Mutation**, which involves random changes in the structure or number of DNA molecules in a cell and is the ultimate source of genetic variability in a population
- **Natural selection**, which occurs when some individuals of a population have genetically based traits that cause them to survive and produce more offspring than other individuals
- **Gene flow**, which involves movement of genes between populations and can lead to changes in the genetic composition of local populations
- **Genetic drift**, which involves change in the genetic composition of a population by chance and is especially important for small populations

What Is the Role of Mutation in Microevolution? Genetic variability in a population originates through *mutations*: random changes in the structure or number of DNA molecules in a cell. One way mutations occur is by exposure to external agents such as radioactivity, X rays, and natural and human-made chemicals (called *mutagens*). Another source of mutations is random mistakes that are sometimes made in coded genetic instructions when DNA molecules are copied (each time a cell divides and whenever an organism reproduces). Mutations can occur in any cells, but only those in reproductive cells are passed on to offspring.

Some mutations are harmless, but many are harmful, altering traits in such a way that an individual cannot survive (lethal mutations). Every so often a mutation is beneficial. The result is new genetic traits that give their bearer and its offspring better chances for survival and reproduction, either under existing environmental conditions or when such conditions change.

It is important to understand that mutations are **(1)** random and unpredictable, **(2)** the only source of totally new genetic raw material (alleles), and **(3)** very rare events. Once created by mutation, however, new alleles can be shuffled together or recombined randomly to create new combinations of genes in populations of sexually reproducing species.

What Role Does Natural Selection Play in Microevolution? The process of *natural selection* occurs when some individuals of a population have genetically based traits that cause them to better survive and produce offspring. This idea was developed in 1846 by Charles Darwin and published in 1859 in his now-famous book *On the Origin of Species by Means of Natural Selection*. Darwin recognized that three conditions are necessary for evolution by natural selection to occur:

- There must be natural *variability* for a trait in a population.
- The trait must be *heritable*, meaning that it must have a genetic basis such that it can be passed from one generation to another.
- The trait must somehow lead to **differential reproduction**, meaning that it must enable individuals with the trait to leave more offspring than other members of the population.

In natural populations we can observe variability of almost any characteristic, including size, color, shape, and behavior (Figure 4-4). Only part of this observed variability is heritable. The other type of variability is nonheritable, and results from the environment in which the organism is growing. For example, if we observe that an American robin is larger than others in the population, it may be because it inherited genes that make it larger or it may be because it grew up in a place where there were more worms to eat. Typically, observable variation in a population consists of a combination of heritable and nonheritable, environmentally influenced traits.

Natural selection causes any allele or set of alleles that result in a beneficial trait to become more common in succeeding generations, and other alleles to become less common. Depending on environmental conditions, different traits become more or less beneficial for survival and reproduction than other traits. Thus, certain traits increase in the population, depending on environmental conditions. A heritable trait that enables organisms to better survive and reproduce under a given set of environmental conditions is called an **adaptation**, or **adaptive trait**. A factor in a population's environment that causes natural selection to occur is known as a **selective pressure**.

It is important to understand that environmental conditions do not create favorable heritable characteristics. Instead, natural selection favors some individuals over others by acting on inherited genetic

Cartmill, Matt. 1998. "Oppressed by Evolution." *Discover*, vol. 19, no. 3, 78(6).

variations (alleles) already present in the gene pool of a population.

What Is an Example of Evolution by Natural Selection? One of the best documented examples of evolution by natural selection involves camouflage coloration in the peppered moth (*Biston betularia*), which is found in England (Figure 6-8).

This example illustrates the important points of evolution by natural selection. Natural selection occurred because **(1)** there were two color forms (*variability*), **(2)** color form was genetically based (*heritability*), and **(3)** there was greater survival and reproduction by one of the color forms (*differential reproduction*). In this case, first an environmental change occurred: Soot caused a change in the background color of tree trunks. Then the environmental change led to a change in selective force: Predators were able to find and eat the moths with the coloration that no longer blended in with the background.

What Are the Three Types of Natural Selection? Biologists recognize three types of natural selection that take place under quite different environmental conditions. Each type leads to a different genetic outcome for the populations involved (Figure 6-9). In *directional natural selection* (Figure 6-9, left), changing environmental conditions cause allele frequencies to shift so that individuals with traits at one end of the normal range become more common than midrange forms. Examples of this "it pays to be different" type of natural selection include the changes in the varieties of peppered moths (Figure 6-8) and the evolution of genetic resistance to pesticides among insects and to antibiotics among disease-carrying bacteria. This type of natural selection is most common during periods of environmental change or when members of a population migrate to a new habitat with different environmental conditions.

Stabilizing natural selection tends to eliminate individuals on both ends of the genetic spectrum and favor individuals with an average genetic makeup (Figure 6-9, center). This "it pays to be average" type of natural selection occurs when an environment changes little and when most members of the population are well adapted to that environment.

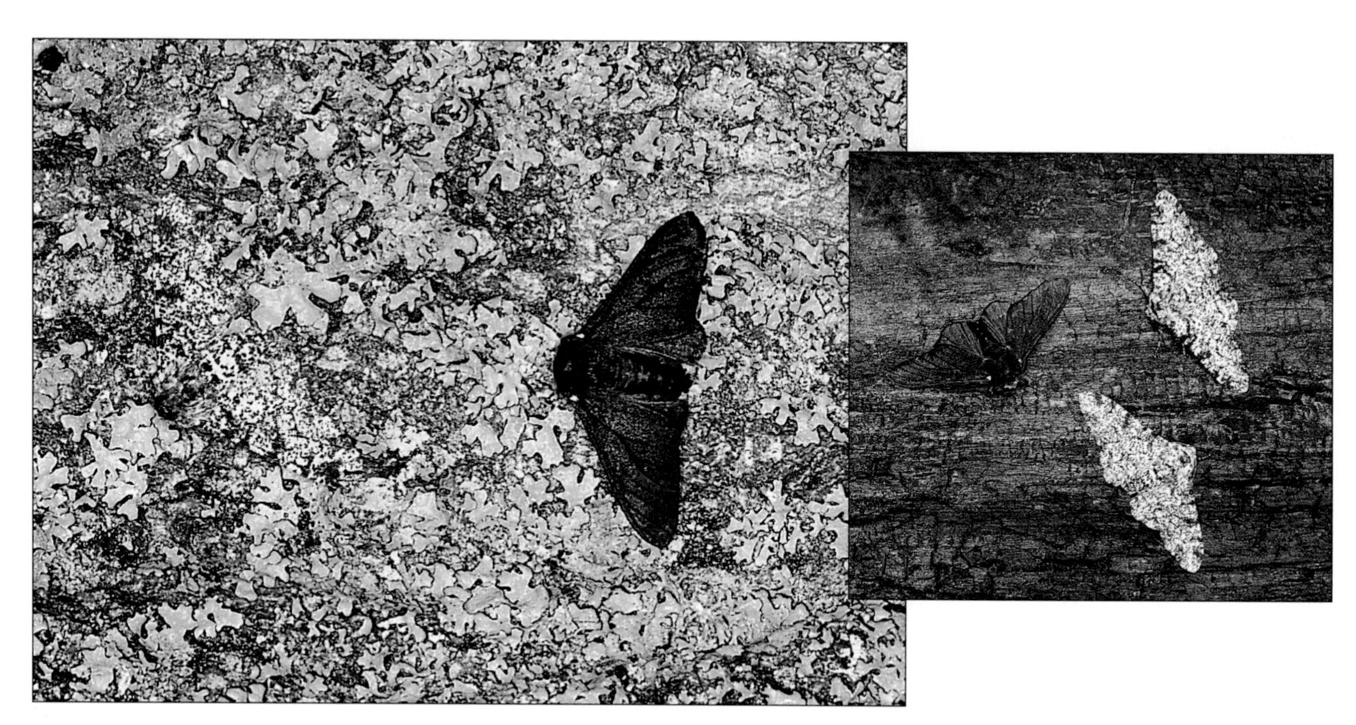

Figure 6-8 Two varieties of peppered moths found in England illustrate one kind of adaptation: camouflage. Before the industrial revolution in the mid-1800s, the speckled light-gray form of this moth was prevalent. When these night-flying moths rested on light-gray lichens on tree trunks during the day, their color camouflaged them from their predators (left). A dark-gray form also existed but was quite rare. However, during the industrial revolution, when soot and other pollutants from factory smokestacks began killing lichens and darkening tree trunks, the dark form became the common one, especially near industrial cities. In this new environment, the dark form blended in with the blackened trees, whereas the light form was highly visible to predators. Through natural selection, the dark form began to survive and reproduce at a greater rate than its light-colored kin. (Both varieties appear in each photo. Can you spot them?) This is an example of directional natural selection (Figure 6-9, left). (Left, Michael Tweedie/NHPA; right, Kim Taylor/Bruce Coleman Ltd.)

Hint: Enter the search terms *evolution, political aspects* using the Subject Guide.

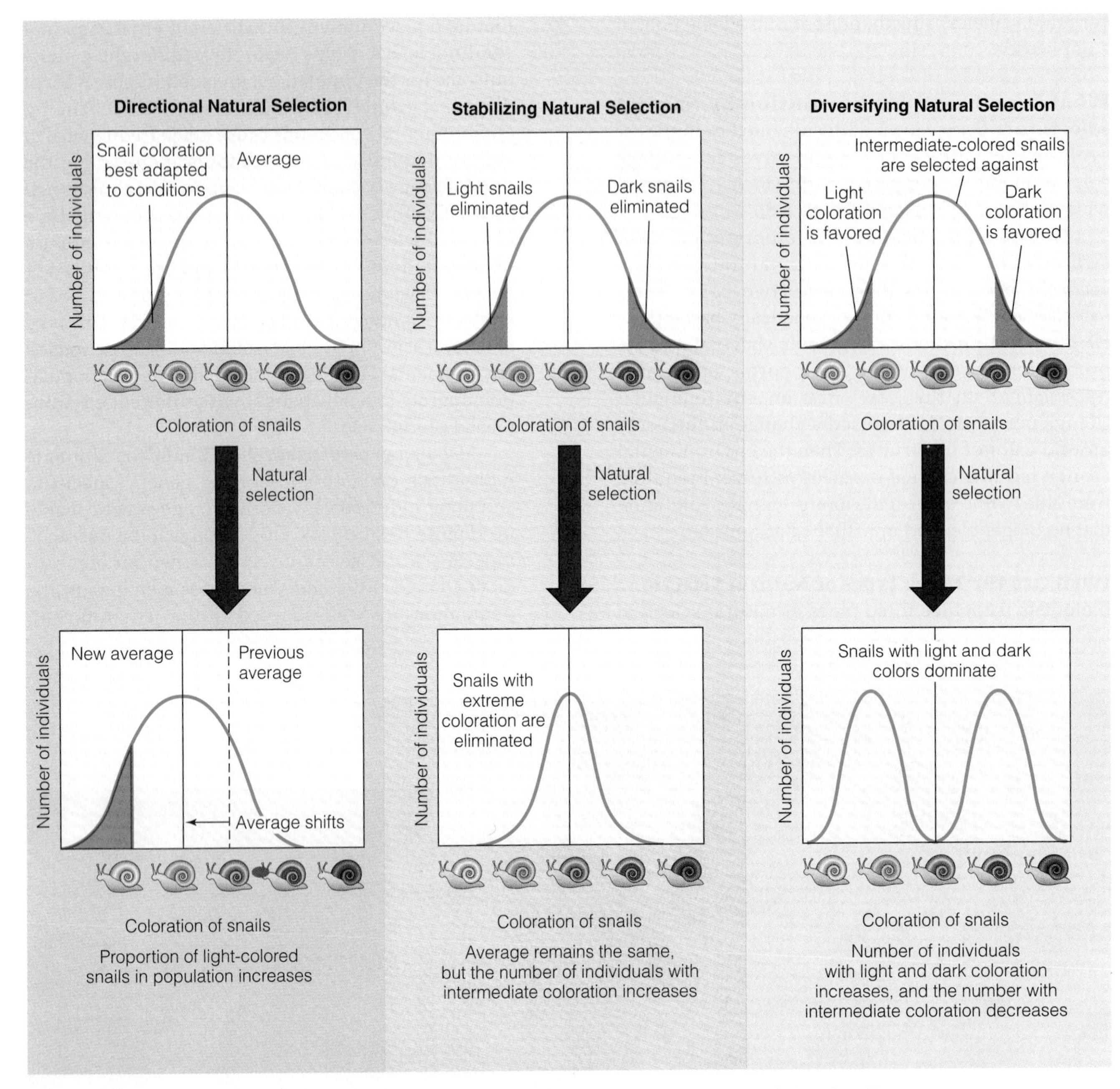

Figure 6-9 Three ways in which natural selection can occur, using the trait of coloration in a population of snails. In *directional natural selection*, changing environmental conditions select organisms with alleles that deviate from the norm so that their offspring (lighter-colored snails) make up a larger proportion of the population. In *stabilizing selection*, environmental factors eliminate fringe individuals (light- and dark-colored snails) and increase the number of individuals with average genetic makeup (intermediate-colored snails). In *diversifying natural selection*, environmental factors favor individuals with uncommon traits (light- and dark-colored snails) and greatly reduce those with average traits (intermediate-colored snails).

Individuals with unusual alleles have no advantage and tend to be eliminated.

Diversifying natural selection (also called disruptive natural selection) occurs when environmental conditions favor individuals at both extremes of the genetic spectrum and eliminate or sharply reduce individuals with normal or intermediate genetic traits (Figure 6-9, right). In this "it doesn't pay to be normal" type of natural selection, a population is split into two groups. This can occur when a shift in the food supply selects against the average individuals. For example, most members of a population of finches have a certain beak length and width that allows them to eat certain fruits, seeds, and insects. Suppose that an environmental dis-

Q: What is the basic unit of life?

turbance eliminates most foods except those that can be eaten only by the few birds with wider or stronger beaks. Then more birds with these extreme variations survive than those with intermediate beaks, and disruptive natural selection will have occurred.

What Is Coevolution? So far our focus has been on natural selection resulting from interactions between *individuals* in a population with abiotic environmental conditions (such as changes in climate or catastrophic events such as fires, floods, or volcanic eruptions). Some biologists have proposed that interactions between *species* can also result in microevolution in each of their populations. According to this hypothesis, when populations of two different species interact over a long time, changes in the gene pool of one species can lead to changes in the gene pool of the other species. This process is called **coevolution**.

Suppose, for example, that certain individuals in a population of carnivores (such as owls) become better at hunting prey (such as mice). Because of genetic variation, certain individuals of the prey have traits that allow them to escape or hide from their predators, and they pass these adaptive traits on to some of their offspring. But a few individuals in the predator population may also have traits (such as better eyesight or quicker reflexes) that allow them to successfully hunt even the better adapted prey, and they pass these traits on to some of their offspring.

Similarly, individual plants in a population may evolve defenses such as camouflage, thorns, or poisons, against efficient herbivores. In turn, some herbivores in the population may have genetic characteristics that enable them to overcome these defenses and produce more offspring than those lacking them.

Flowering plants have evolved many means for attracting pollinators and "tricking" them to carry pollen from one flower to another (Figure 1-11, left); likewise, pollinators have evolved forms and behaviors that enable them to gain food from flowers in the form of sugar-rich nectar, protein-rich pollen, or edible flower parts.

In coevolution, adaptation follows adaptation in something like an ongoing, long-term arms race among individuals in interacting populations of different species.

6-4 ADAPTATION AND THE ECOLOGICAL NICHE

What Is the Ecological Niche? If asked what role a certain species such as an alligator (Figure 6-1) plays in an ecosystem, an ecologist would describe its **ecological niche**, or simply **niche** (pronounced nitch), the species's way of life or functional role in an ecosystem. A species's niche involves everything that affects its survival and reproduction. This includes **(1)** the range of tolerance for various physical and chemical conditions, such as temperature or water availability (Figure 4-12); **(2)** the types of resources it uses, such as food or nutrient requirements; **(3)** how it interacts with other living and nonliving components of the ecosystems in which it is found, such as what it eats or is eaten by; and **(4)** the role it plays in the flow of energy and cycling of matter in an ecosystem.

The ecological niche of a species is different from the **habitat** of the species. Habitat is the actual location where a species lives. Ecologists often say that a niche is like a species's occupation, whereas habitat is like its address.

The concept of the ecological niche is examined in more detail in Section 9-1.

How Does the Ecological Niche Relate to Adaptation? Evolution by natural selection leads to a remarkable fit between organisms and their environment. In terms of the ecological niche of a particular species, this fit involves having a set of traits that enables individuals to survive and reproduce in a particular environment. Species that have similar niches tend to evolve similar sets of traits, even if they are unrelated species growing in different parts of the world. The resemblance among species belonging to different taxonomic groups as the result from adaptation to similar environments is known as **convergence**.

Convergent evolution has led to similar organisms in different parts of the world where the selective pressures have been similar. For example, in desert biomes, shrubs tend to converge on a set of traits that enable them to grow in hot, dry conditions. Examples of such traits include deep roots to access water, small leaves to reduce heat loads, hairs or waxy coatings that protect against intense sunlight, tolerance to high daytime temperatures and low night temperatures, and flowering during wet periods, when insect pollinators are most active. In another example, herbivores from different parts of the world tend to display convergence for a set of traits that enable them to forage and digest plant matter efficiently, escape predators, and migrate or become dormant when their food plants are scarce. Chapter 7 discusses the biomes of the world, where similar selective pressures have led to similar sets of ecological niches.

The ecological niche of a species is determined in part by selective pressures that have acted within the environment in which the species evolved. However, although evolution can lead to a strong fit between organisms and their environment, these adaptations are not perfect.

What Limits Adaptation? Shouldn't evolution lead to perfectly adapted organisms? Shouldn't adaptations to new environmental conditions allow our skin to become more resistant to the harmful effects of ultraviolet radiation, our lungs to cope with air pollutants, and our livers to become better at detoxifying pollutants? The answer to these questions is *no* because there are limits to adaptations in nature.

First, *a change in environmental conditions can lead to adaptation only for traits already present in the gene pool of a population.* Environmental change in the form of increased pollution did not produce the dark form of the peppered moth (Figure 6-8); the dark trait was already present in the population (probably as a result of a mutation). Natural selection favored individuals that already had this trait because it made them more suited (adapted) to the changed environment.

Second, *because each organism must do many things, its adaptations are usually compromises.* Seals spend part of their life on rocks and could probably walk better if they had legs instead of flippers, but then they could not swim nearly as well. This and the role of chance are two reasons why evolution doesn't lead to perfectly adapted organisms. Even if organisms were perfectly adapted to their current environment, this wouldn't last if environmental conditions changed (as they usually do).

Third, *even if a beneficial heritable trait is present in a population, that population's ability to adapt can be limited by its reproductive capacity.* If members of a population can't reproduce quickly enough to adapt to a particular environmental change, all of them can die. Populations of genetically diverse species that reproduce quickly—such as weeds, mosquitoes, rats, or bacteria—often can adapt to a change in environmental conditions in a short time. In contrast, populations of species such as elephants, tigers, sharks, and humans, which cannot produce large numbers of offspring rapidly, take a long time (typically thousands or even millions of years) to adapt through natural selection.

Finally, *even if a favorable genetic trait is present in a population, most of its members would have to die or become sterile so that individuals with the trait could predominate and pass the trait on*—hardly a desirable solution to the environmental problems humans face.

6-5 SPECIATION, EXTINCTION, AND BIODIVERSITY

What Is Macroevolution? Macroevolution is concerned with how evolution takes place above the level of species and over much longer periods than microevolution. Macroevolutionary patterns include **(1)** *genetic persistence*, or the inheritance of DNA molecules from the origin of the first cells through all subsequent lines of descent, which is the basis of the unity of life; **(2)** *genetic divergence*, or long-term changes in lineages of species, which is the basis of the diversity of life; and **(3)** *genetic losses*, or the steady loss (background extinction) or abrupt, catastrophic loss (mass extinction) of lineages.

Macroevolution is sometimes described as a tree in which the branches consist of neat linear sequences of species. However, it is better described as a dense shrub with many branches that lead to genetic dead ends or extinctions. The modern horse (genus *Equus*), for example, emerged through a 60-million-year evolution of species from a dog-sized woodland browser species (genus *Eohippus*) through a series of increasingly large grassland-dwelling grazers. As Figure 6-10 shows, this was not a linear progression. Along the way, many species evolved and many species became extinct.

How Do New Species Evolve? Under certain circumstances natural selection can lead to an entirely new species. In this process, called **speciation**, two species arise from one.

The most common mechanism of speciation (especially among animals) takes place in two phases: geographic isolation and reproductive isolation. **Geographic isolation** occurs when two populations of a species or two groups of the same population become physically separated for fairly long periods into areas with different environmental conditions. For example, part of a population may migrate in search of food and then begin living in another area with different environmental conditions (Figure 6-11). Populations may also become separated by a physical barrier (such as a mountain range, stream, lake, or road), by a change such as a volcanic eruption or earthquake, or when a few individuals are carried to a new area by wind or water.

The second phase of speciation is **reproductive isolation**. It occurs as mutation and natural selection operate independently in two geographically isolated populations and change the allele frequencies in different ways—a process called *divergence*. If divergence continues long enough, members of the geographically and reproductively isolated populations may become so different in genetic makeup that they can't interbreed—or if they do, they can't produce live, fertile offspring. Then one species has become two, and *speciation* has occurred through *divergent evolution*.

In a few rapidly reproducing organisms this type of speciation may occur within hundreds of years; with most species, however, it takes from tens of thousands to millions of years. Given this time scale, it is difficult to observe and document the appearance of a new species. As a result, there are many controversial hypotheses about the details of speciation.

Tuxill, J., and C. Bright. 1998. "Protecting Nature's Diversity: Mending Strands." *The Futurist*, vol. 32, no. 5, 46(6).

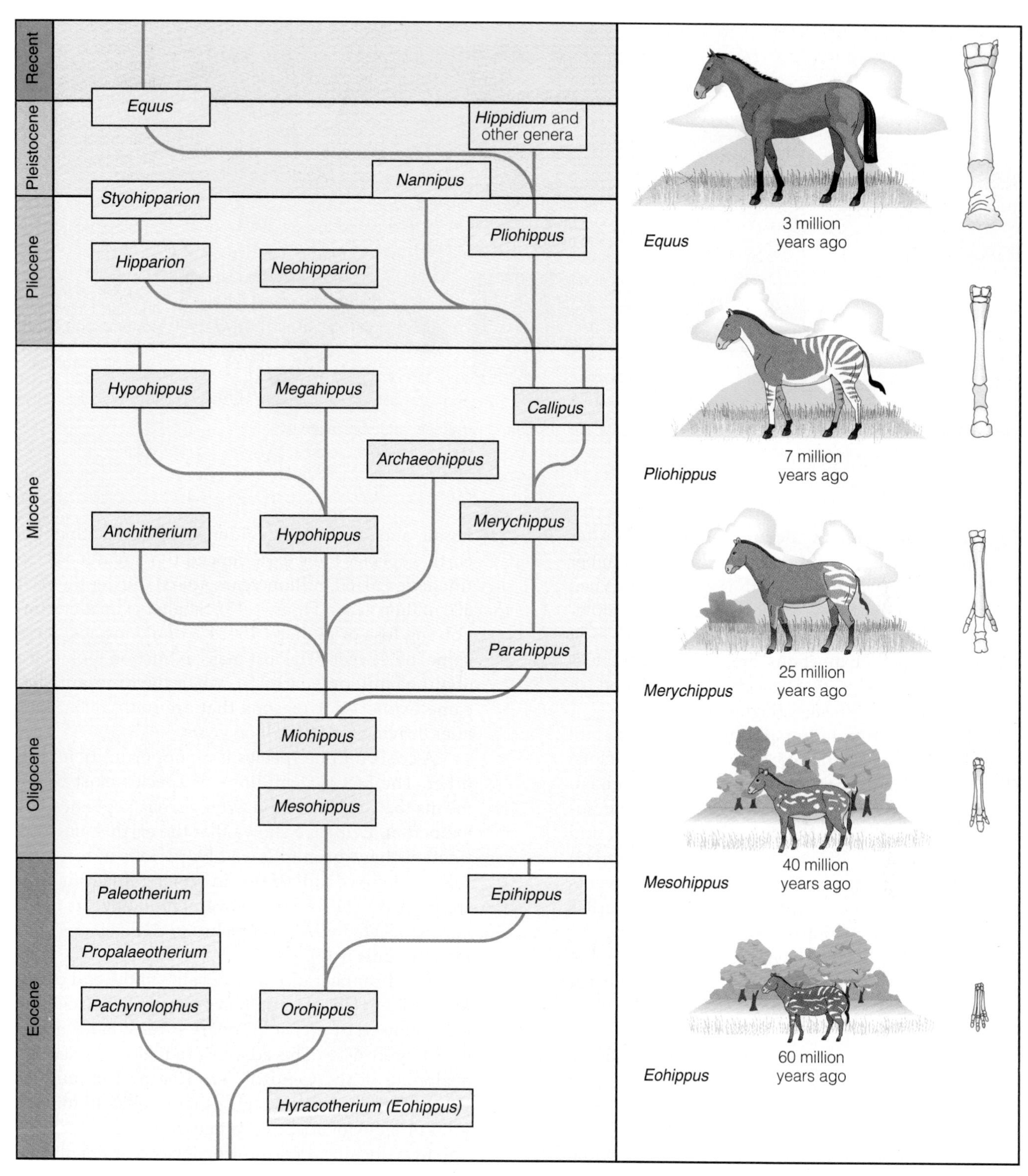

Figure 6-10 The 60-million-year branched evolution of the genus of the horse. An excellent fossil record shows that the *Equus* genus of large modern horse species (with one toe and large teeth that allow it to graze on tough grass) evolved from the ancient *Eohippus* genus of small, dog-sized species (with four toes and small teeth that allowed them to browse on the soft leaves of trees and shrubs in woodlands). Note that this is not a simple linear progression in which early species evolved smoothly into modern horse species. Instead, many of the species became extinct along the way.

Hint: Enter the search terms *extinction, prevention* using the Subject Guide.

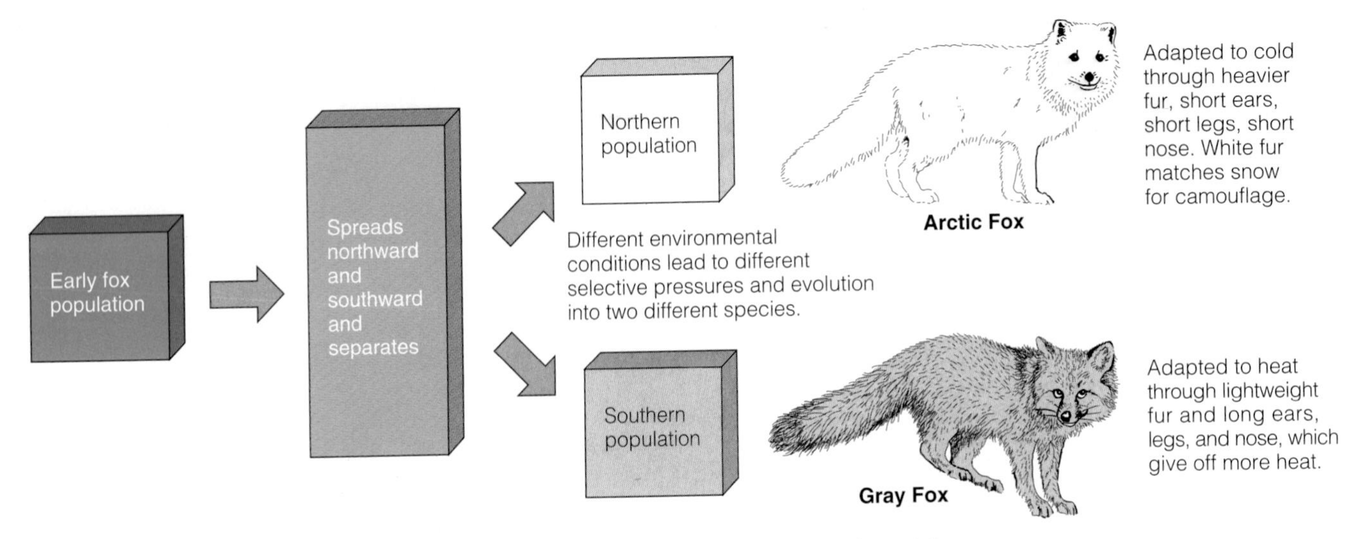

Figure 6-11 How geographic isolation can lead to reproductive isolation, divergence, and speciation.

How Do Species Become Extinct? After evolution, the second process affecting the number and types of species on the earth is **extinction**. When environmental conditions change, a species may either evolve (become better adapted) or cease to exist (become extinct). Extinctions are a permanent loss of genetic diversity that affects the evolution of macroevolutionary lineages (Figure 6-10).

The earth's long-term patterns of speciation and extinction have been affected by several major factors: **(1)** large-scale movements of the continents (continental drift) over millions of years (Figure 6-12), **(2)** gradual climate changes caused by continental drift and slight shifts in the earth's orbit around the sun, and **(3)** rapid climate change caused by catastrophic events (such as large volcanic eruptions and huge meteorites and asteroids crashing into the earth). Such events create dust clouds that shut down or sharply reduce photosynthesis long enough to eliminate huge numbers of producers—and the consumers feeding on them shortly thereafter.

Extinction is the ultimate fate of all species, just as death is for all individual organisms. Biologists estimate that 99.9% of all the species that have ever existed are now extinct.

Some species inevitably disappear at some low rate, called **background extinction**, as local conditions change. In contrast, **mass extinction** is an abrupt rise in extinction rates above the background level. It is a catastrophic, widespread (often global) event in which large groups of existing species (perhaps 25–70%) are wiped out.

Most mass extinctions are believed to result from global climate changes that kill many species and leave behind those able to adapt to the new conditions. Fossil and geological evidence indicates that the earth's species have experienced five great mass extinctions (20–60 million years apart) during the past 500 million years (Figure 6-13). Smaller extinctions (involving loss of perhaps 15–24% of all species) have come in between. The last mass extinction took place about 65 million years ago, when the dinosaurs became extinct—for reasons that are hotly debated—after thriving for 140 million years.

A crisis for one species is an opportunity for another. The fact that millions of species exist today means that speciation, on average, has kept ahead of extinction. Evidence shows that the earth's mass extinctions have been followed by periods of recovery called **adaptive radiations**, in which numerous new species have evolved over several million years to fill new or vacated ecological niches in changed environments (Figure 6-13).

The disappearance of dinosaurs at the end of the Mesozoic era (about 65 million years ago), for example, was followed by an evolutionary explosion for mammals (Figure 6-14). This adaptive radiation marked the beginning of the Cenozoic era (the past 65 million years). Fossil records suggest that it takes 10 million years or more for adaptive radiations to rebuild biological diversity after a mass extinction. According to biologists, our species owes its existence to the mass extinction 65 million years ago, when an incredible variety of dinosaur species died and a small, primitive mammal that eventually led to us happened to survive.

Does Macroevolution Take Place Gradually or in Bursts? Until recently it was widely accepted that macroevolutionary change occurred gradually over many millions of years as a result of steady and

Q: What are the major parts of the earth's life-support system?

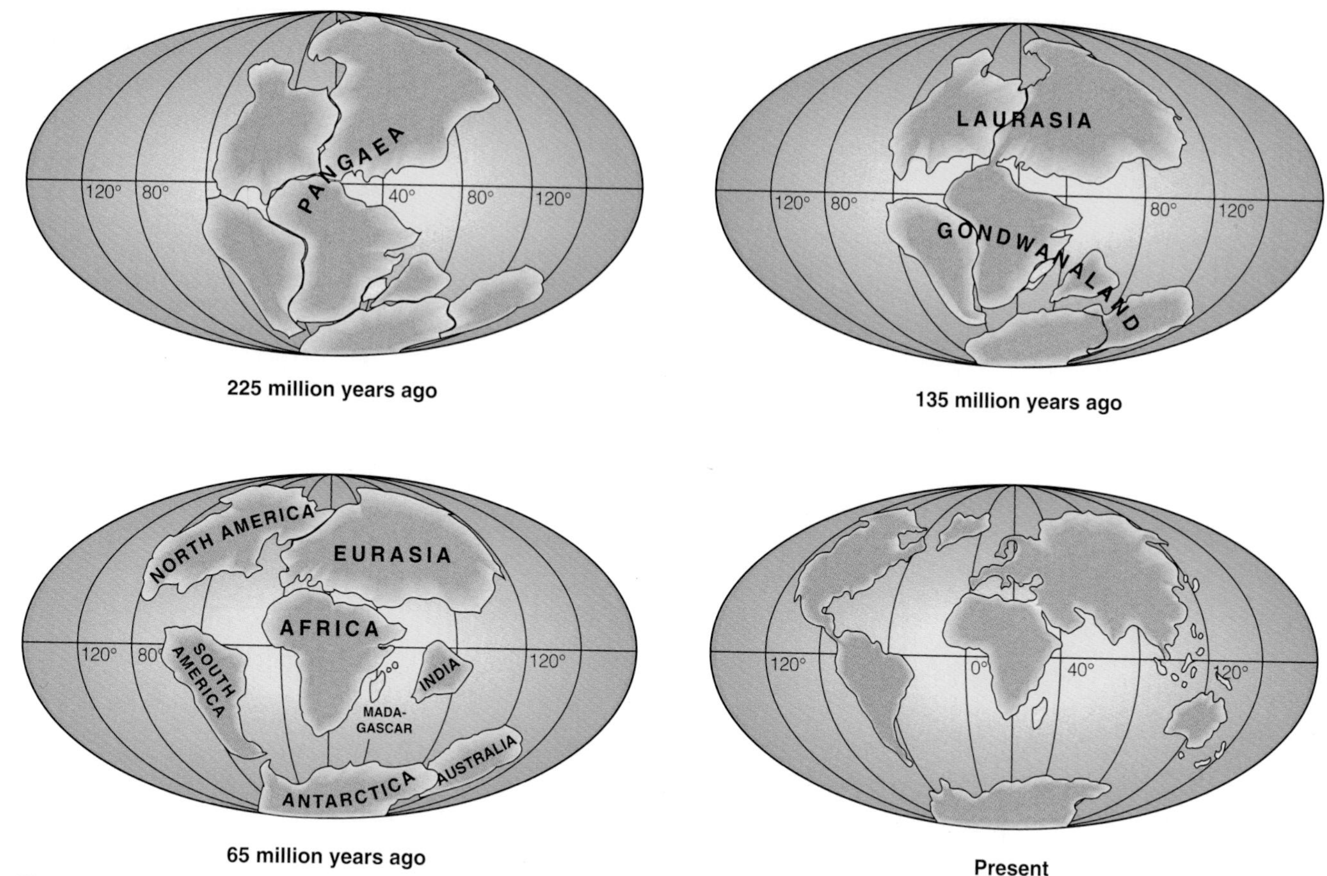

Figure 6-12 Continental drift, the extremely slow movement of continents over millions of years on several gigantic plates (discussed in more detail in Section 14-1). This plays a role in the extinction of species and the rise of new species. Populations are geographically and eventually reproductively isolated as land masses float apart and new coastal regions are created. Rock and fossil evidence indicates that about 200–250 million years ago all of the earth's present-day continents were locked together in a supercontinent called Pangaea. About 180 million years ago, Pangaea began splitting apart as the earth's huge plates separated and eventually resulted in today's locations of the continents.

small microevolutionary changes. This is called the *gradualist model of evolution*. However, fossils rarely document such gradual changes in lineage. Instead, most species appear suddenly, persist for a fairly long time with little change, and then become extinct.

In the early 1970s, Stephen Jay Gould of Harvard University and Niles Eldredge of the American Museum of Natural History offered an alternative explanation of how macroevolution takes place called the *punctuated equilibrium hypothesis*. According to this model, evolution consists of long periods of little change in species (equilibrium) punctuated with brief periods of rapid change (thousands to tens of thousands of years). During periods of rapid change, more species become extinct and more new species arise.

There is intense debate over which of these models of evolution best explains the data. However, most evolutionary biologists believe that the rate and mode of evolution vary and fall on a spectrum between the extremes of the gradualist and punctuated equilibrium models.

What Are Three Common Misconceptions About Evolution? The theory of evolution has been under attack ever since its key ideas were published by Charles Darwin in 1859. Some of the attackers have distorted the meaning of evolution, either from a lack of understanding how evolution works or in deliberate attempts to mislead.

One misconception arises from the interpretation of the expression "survival of the fittest" (which biologists almost never use), sometimes used to describe how natural selection works. This has often been misinterpreted as "survival of the strongest." To biologists, however, *fitness* is a measure of reproductive success, so that the fittest individuals are those that leave the most descendants. For example, members of a population that are better at hiding from their

A: Atmosphere, hydrosphere, lithosphere, and ecosphere (biosphere)

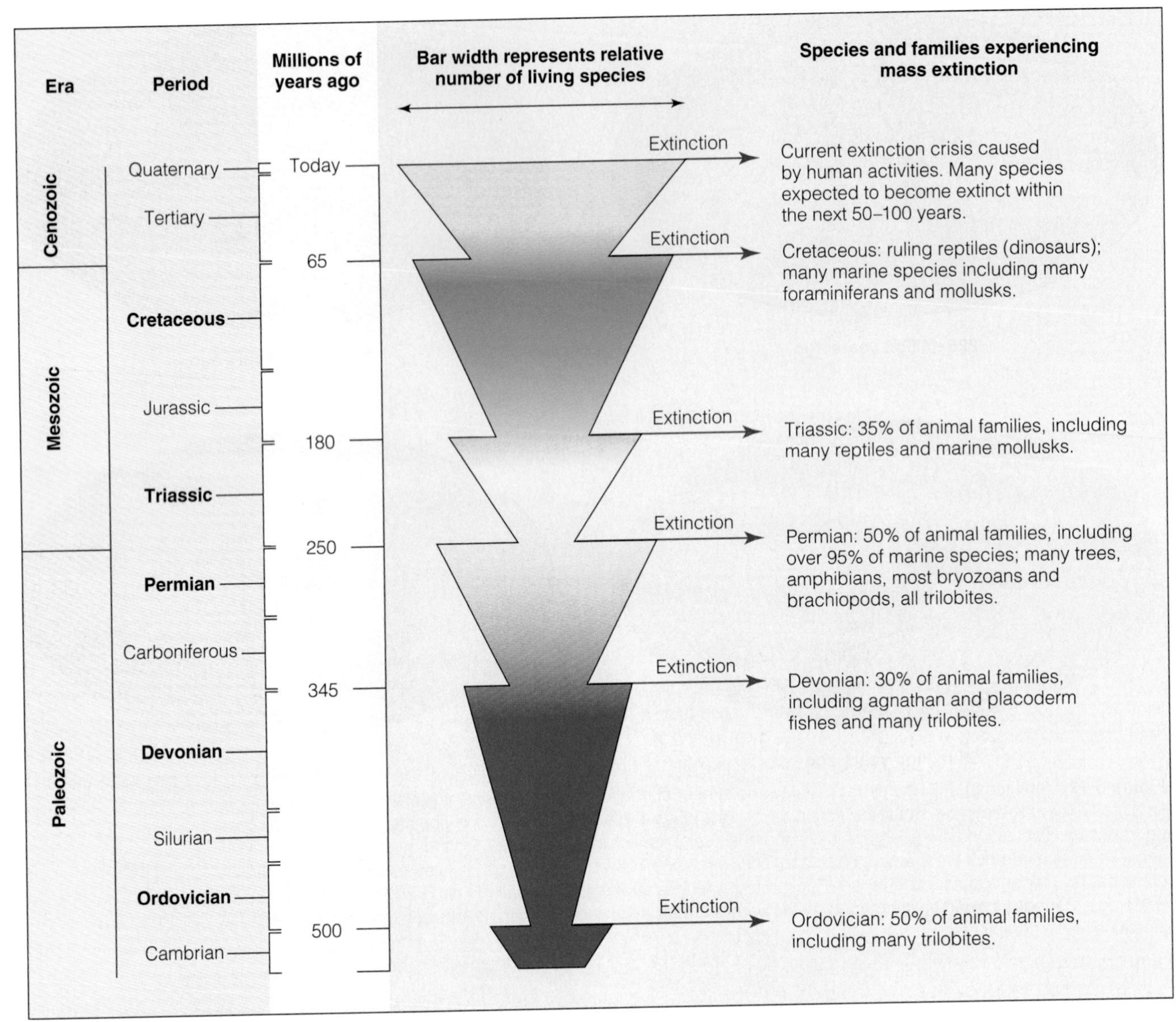

Figure 6-13 Over millions to hundreds of millions of years, macroevolution has consisted of dramatic exits (extinctions) and grand entrances (speciation and radiations) of large groups of species. Fossil and radioactive dating evidence indicate that five major mass extinctions (indicated by arrows) have taken place over the past 500 million years. Mass extinctions leave large numbers of niches unoccupied and create new ones. As a result, each mass extinction has been followed by periods of recovery (represented by the wedge shapes) called *adaptive radiations* in which (over 10 million years or more) new species evolve to fill new or vacated ecological niches. Many scientists say that we are now in the midst of a sixth mass extinction, caused primarily by overhunting and the increasing elimination, degradation, and fragmentation of wildlife habitats as a result of human activities.

predators are more fit (can live to produce more offspring) than those that aren't as good at hiding, regardless of how strong they are.

Instead of tooth-and-claw competition, natural selection favors populations of species that *avoid* direct competition by producing offspring that can occupy niches different from those of other species. In the Darwinian world, peaceful coexistence, resulting from the evolution of organisms that can occupy different niches, is the key to the survival of populations of different species.

Some people have misinterpreted the theory of evolution's assertion that all species share a common ancestry to mean that "humans evolved from apes." The theory of evolution makes no such claim. Instead, it states that apes and humans are descended from a common ancestor. In other words, at some time in the distant past a particular population of organisms had

Q: On what three factors does life on the earth depend?

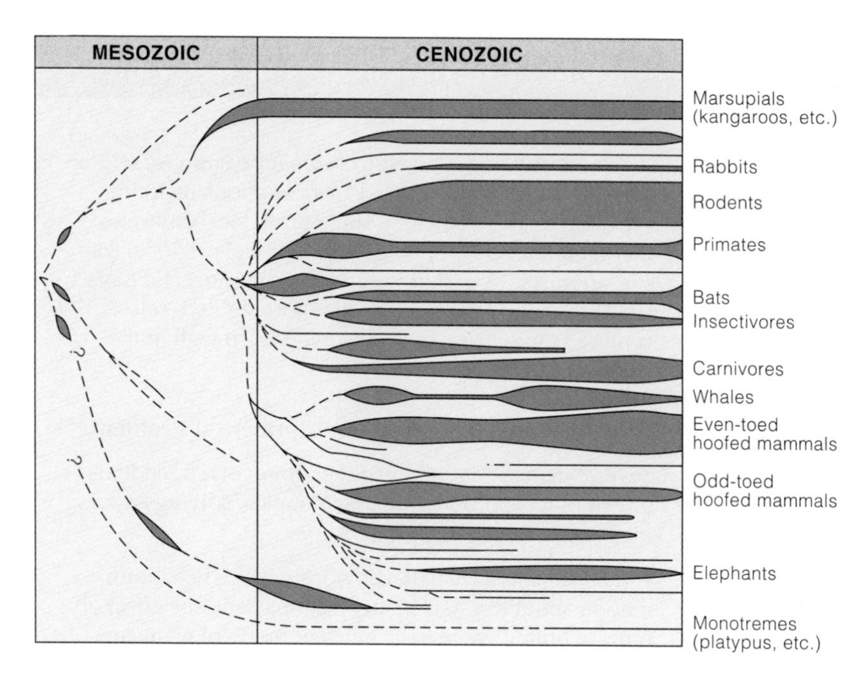

Figure 6-14 Adaptive radiation of mammals began in the first 10–12 million years of the Cenozoic era (which began about 65 million years ago) and continues today. This evolution of a large number of new species is thought to have resulted when huge numbers of new and vacated ecological niches became available after the mass extinction of dinosaurs near the end of the Mesozoic era. (Used by permission from Cecie Starr and Ralph Taggart, *Biology: The Unity and Diversity of Life*, 8th ed., Belmont, Calif.: Wadsworth, 1998)

descendants, with some of them evolving into apes and others evolving into the human species.

A third misconception is that evolution involves some grand plan of nature in which species become progressively more perfect. This ignores the fact that the mutations and other processes that drive microevolution occur as a result of random, unpredictable events. From a scientific standpoint, there is no plan or goal of perfection in the evolutionary process.

How Do Speciation and Extinction Affect Biodiversity? Speciation minus extinction equals *biodiversity*, the planet's genetic raw material for future evolution in response to changing environmental conditions. In this long-term give-and-take between extinction and speciation, mass extinctions temporarily reduce biodiversity. However, they also create evolutionary opportunities for surviving species to undergo adaptive radiations to fill unoccupied and new niches (Figure 6-14).

Although extinction is a natural process, humans have become a major force in the premature extinction of species. As population and resource consumption increase over the next 50 years and we take over more and more of the planet's surface (Figure 1-5), we may cause the extinction of up to a quarter of the earth's current species, each the product of millions to billions of years of evolution. If this happens it will constitute a sixth mass extinction, caused by us. Mostafa K. Tolba, director of the United Nations Environment Programme, has said, "If Charles Darwin were alive today, his work would most likely focus not on the origins but, rather, on the obituaries of species."

On our short time scale, such catastrophic losses cannot be recouped by formation of new species; it took tens of millions of years after each of the earth's five great mass extinctions for life to recover to the previous level of biodiversity. Genetic engineering cannot stop this loss of biodiversity because genetic engineers do not create new genes. Rather, they transfer existing genes or gene fragments from one organism to another and thus rely on natural biodiversity for their raw material.

Humans have been around for only a wink of the earth's long history, but we have become a highly adaptable species. Many scientists fear that our immense and rapidly growing abilities to exploit nature could backfire. By reducing and degrading the earth's life-support systems (earth or ecological capital), they contend that we could make our own species more vulnerable to extinction, or at least to a massive dieback.

6-6 SUSTAINABILITY AND EVOLUTION

How Does Our Time Frame Influence Our Thinking? Earth is constantly changing. Throughout the earth's history the atmosphere has changed, the climate has changed, the geography has changed, the types and numbers of organisms have changed, and continental drift has changed the positions of the earth's continents (Figure 6-12). Some of these changes are so slow that we are not aware of them; others are much faster.

A: One-way flow of energy from the sun, cycling of matter, and gravity (Figure 4-6)

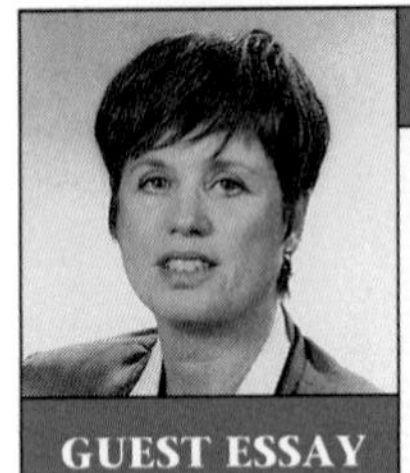

GUEST ESSAY

We Haven't Evolved to Be Environmental Altruists, but We Can Solve Environmental I

Bobbi S. Low

Bobbi S. Low, an evolutionary and behavioral ecologist, is a professor of resource ecology at the University of Michigan. She works on issues of resource control and fertility, and on sex-related differences in resource use.

Our Evolutionary Inheritance

We've created a series of environmental messes: acid deposition, possible global climate change, leaking toxic dumps, water pollution, and soil erosion. Often, we identify workable solutions, but we can't transcend our differences long enough to get the job done. We call these "social traps," but they are really evolutionary traps.

What strikes me, as an evolutionary biologist, is that the more we study the problem, the more it seems that precisely those behaviors we have evolved—the things that enhanced our survivorship and reproduction in past environments—are making solutions difficult! If this is true and if we understand it, we can figure out how to solve our problems.

At the heart of evolutionary history is the "selfish gene." Individuals live and die, but their genes (or their "replicates") can be immortal. This fact leads to interesting behavioral complexities. Obviously, competitive behavior can perpetuate genes in all species. But in social species like us, cooperative behaviors are often quite effective in passing our genes on—*if* those we help are our relatives, with whom we share genes, or friends who will help us in return (reciprocity).

In our evolutionary past, we mostly lived in small groups of related families. Even when societies became larger and more complex, extended families (often including a network of friends) remained central. These were the people who mattered most to us, not the rest of the world. Our main concerns were getting enough resources from the environment to meet our needs, maintaining satisfactory and stable friendships, finding mates, and raising families. Predicting the effect of our actions decades in the future was never a priority; in fact, uncertainties in the environment usually made such long-term planning futile. Most of the time, our populations and technology were sufficiently limited that we did only local damage to the environment.

As a result, we evolved to strive for resources; seldom (if ever) were we "rewarded" for conscious restraint. We evolved to be efficient, short-term, local environmental managers, not long-term regional or global conservationists. Now that we are so numerous, and have such effective technologies, our short-sighted, self-centered tendencies—which served us so well in the past—cause us difficulties.

Impediments to Solving Environmental Problems

Several factors—involving either our evolved tendencies or external conditions—interact to make solving environmental problems difficult:

- *Limited information.* It's obvious that we need information about the state of any resource, and the effect of our use of it, if we're to be efficient resource managers. The potentially usable supplies of a number of resources (coal, oil, natural gas, and water in aquifers) are still poorly known.
- *Discounting the future.* Perhaps because we had little control over changing environments in the past, we have evolved to "discount" future benefits if current costs are involved, and to discount future costs if we can get benefits now. Like Wimpy in the old Popeye cartoon, "We'll gladly pay you Tuesday for a hamburger today." The time frame over which people are willing to pay now for future benefits is very short: about 3–5 years. Benefits any farther in the future are not considered worth paying for.
- *Externalizing costs.* Even better than a hamburger today, paid for later, is a hamburger we get somebody else to pay for! So it's not surprising that much effort is expended in exporting ("externalizing") our harmful costs such as pollution, and having someone else pay the resulting environmental and health bills—another example of our "selfish genes" in action.
- *Common property versus private ownership.* Some resources that we call "commons" are open to everyone and can be degraded by overuse. The classic case of the resulting "tragedy of the commons," described by Garret Hardin, was the English grazing common: land on which everyone grazed their sheep and cattle. Now, if I put an extra sheep on the land, I may exceed the carrying capacity and hurt the land. But because we all share in that

Weather changes in a matter of minutes or hours, and climate over decades to centuries. Populations can change their size and age distribution in response to changes in environmental conditions within hours to decades. As environmental conditions have changed, some of the earth's species have become extinct and new species have arisen (Figures 6-6, 6-11, 6-13, and 6-14). Speciation usually takes thousands to millions of years, depending mainly on the reproductive ability of the species involved.

Some geological changes in the earth's interior and crust have taken place over millions or billions of cost (and my part of the total cost is small) while I alone profit from the extra sheep, exploiting the commons for my individual gain is tempting. Many current environmental problems involve overexploitation of the commons: possible global warming, ozone depletion, acid deposition, whaling, toxic disposal. A frequently proposed solution to the commons dilemma is privatization, because private owners have an incentive to keep the resource workable and healthy. But not everyone shares the same interest. If I owned all the whales in the world, I might decide to exterminate them and bank the money! If private Northwest logging interests make the decisions on logging old growth forests, we might have no habitat for spotted owls and other species dependent on the existence of large areas of old-growth forests.

Strategies for Working with Our Evolved Behaviors

One way to solve our problems is to design strategies that work with, rather than against, our evolved tendencies. Here are some known strategies that work:

- *Getting information.*
- *Using persuasion and telling success stories.* It can help to exhort ourselves (and others) to "do the right thing" partly because we have evolved in complex social groups, where the opinions of others matter. So far, relatively local problems in which we can see progress are the best candidates for this approach.
- *Accomplishing small wins.* When we "think globally and act locally" by working on local environmental issues, we are more likely to see results, feel reinforced, and continue our efforts.
- *Using economic incentives.* States that require returnable deposits on beverage containers have fewer bottles and cans lying about. Because businesses measure success by profit, using economic incentives to make it possible for them to "do well by doing good" is an effective strategy that works with our "selfish genes."
- *Establishing regulations.* Economists prefer strategies that work in the marketplace but keep people from externalizing costs. For example, we can place an economic value on each unit of pollution and allow companies to trade their pollution permits in the marketplace. But that's not always possible, and regulations are often difficult to monitor and enforce. Governmental regulations can also be outstripped by new technological advances, which can have such perverse effects as making it illegal to adopt newer, cheaper, more effective technology.
- *Communicating and forming coalitions.* Participating in local recycling is a common success story. The costs and benefits are local, and we see results quickly. Successful programs typically have both economic and social incentives. People are more likely to cooperate for the good of the group (even if it might mean a bit less for themselves) if they can establish communication and get to know the others in the group.

What Each of Us Can Do

All of these strategies can be useful and can be combined for greater effectiveness. Here are some things you and I can do with surprisingly little effort:

- *Continue to gather information.*
- *Look at your habits in light of this information.* What things would require little effort, now that you know they are important to change? Recycling, turning off the lights, conserving water—every little bit helps.
- *Cooperate with others to solve problems.* Join and support organizations (coalitions) whose goals you support.
- *Contact your elected representatives about things that matter to you.* You'll be appealing to their self-interest, because they rely on votes to get re-elected.
- *Vote, even (perhaps especially) in local elections.* Many decisions that affect your daily life are made locally, including the ecological issues in your own backyard.

Critical Thinking

1. Do you believe that we have "selfish" genes? If so, do you believe that we can avoid some of the environmentally harmful actions we take in the name of survival and self-interest by using the strategies listed in this essay? Which strategies do you think will work best? Which ones might work best for you?

2. Choose one small-scale and one large-scale environmental problem and propose solutions to each of them. Which evolved human traits contribute to the problems? Which proposed solutions are likely to be most and least successful, and why?

years; changes from erosion, volcanic eruption, and some subsidence (sinking of the earth's crust) can happen very rapidly. We, and the other species who inhabit the earth today, are the beneficiaries of the processes of evolution and geological change of the landscape, each acting over eons.

In the finger snap of geological time that humans have been on the earth, we have had a powerful impact on the environment, accelerating natural processes and introducing processes and changes that would not have occurred without us. One of the major problems we face is that exponential growth (p. 2) of

A: The study of how organisms interact with one another and with their environment the human population and human alteration of the planet is radically increasing the rate of species extinction. Biologists estimate that the current human-accelerated extinction rate of species is 1,000 to 10,000 times higher than natural extinction rates.

Can We Heal the Earth? We can help heal some of the geological and ecological wounds we have inflicted. However, such earth healing takes time and lots of money, which explains why prevention of earth-degrading processes is the best approach.

Large-scale air pollution and water pollution from factories, farms, and lawns could be prevented, and the damaged ecosystems renewed, in a matter of decades. Desertification could be reversed within a century, and most tropical forests can grow back within 1,000 years. Soil cover in severely eroded areas can be renewed in 10,000 years. However, it will take many millions of years before a new burst of speciation will be able to expand into the empty niches we have created in only a few decades, by our actions that produce extinctions of life on the earth.

To deny future generations even part of their natural heritage millions of years old, to destroy it for all time, is just morally wrong. . . . Evolution is not going to replace this heritage in any period that has meaning for the human mind.

EDWARD O. WILSON

CRITICAL THINKING

1. Why do biologists use scientific names for species? What would happen if biologists agreed on a standard set of easier-to-understand common names, instead of the binomial (two-word) Latin names they now use for naming species? Are scientific names actually important for understanding?

2. Why are most taxonomic classifications of organisms are designed as "natural" classification systems, meaning that they are designed to reflect our understanding of the evolutionary relationships between organisms? What kinds of characteristics of organisms are most useful to use in such classifications? Why?

3. From what you know about evolution by natural selection, are all of the necessary conditions present for evolution to occur by natural selection in the abiotic part of the ecosphere? Which conditions are or are not present?

4. British atmospheric scientist James Lovelock's controversial Gaia hypothesis postulates that the physical and chemical conditions of the earth are made livable by the presence of life itself and that this is in contrast to the conventional view that life has adapted to the planetary conditions as it and they evolved separately. Though largely discredited in its initial form, the Gaia hypothesis provides a system perspective to understanding larger-scale processes occurring on the earth. In what ways has life modified the abiotic part of the earth? What feedbacks do these modifications have in terms of the course of biotic evolution?

5. Someone tells you not to worry about air pollution because through natural selection the human species will evolve lungs that can detoxify pollutants. How would you reply?

6. How would you respond to someone who says that he or she doesn't believe in evolution because it is "just a theory"?

7. Try to come up with analogies to complete the following statements: **(a)** Evolution is like _______. **(b)** Genetic variability is like _______. **(c)** Mutations are like _______. **(d)** Alleles are like _______. **(e)** Differential reproduction is like _______. **(f)** An ecological niche is like _______. **(g)** A habitat is like _______. **(h)** Speciation is like _______. **(i)** Extinction is like _______. **(j)** Adaptive radiation is like _______.

8. The peppered moth (*Biston betularia*) responded to industrial pollution in England by short-term evolution (Figure 6-8). Consider species such as northern spotted owls, California condors, or grizzly bears. Would you expect such species that are being heavily affected by human activity to be able to respond by short-term evolution? Why or why not? Consider coyotes, deer, and cockroaches. Would you expect these species to respond to human impacts by short-term evolution?

9. Some scientists argue that the *gradualist model of evolution* and the *punctuated equilibrium hypothesis* are not truly alternative theories because they simply operate at different time scales. These scientists argue that even during periods of rapid evolutionary change, the changes would still appear gradual on a human time scale. What do you think?

10. What would happen if extinction had never occurred during evolutionary history? Do you think we would have more species than we do now? Why? What do you think would happen to biodiversity if extinction rates occurred throughout evolutionary history at the same rate they have been occurring during the past 50 years?

11. How would you respond to someone who says that extinction is a natural process and that we shouldn't worry about the loss of biodiversity?

12. Why is it important for somebody who is interested in environmental science to understand the basics of evolution?

PROJECTS

1. Go to the library and locate a replica of the first edition of Charles Darwin's *On the Origin of Species by Means of Natural Selection*. Why did Darwin feel compelled to publish his book more quickly than he had originally planned, and why does he consider the book only an "abstract"? Find a later edition of the

Q: What five levels of the organization of matter are the focus of ecology?

book. What did Charles Darwin change in later editions of his book?

2. An important adaptation of humans is a strong opposable thumb, which allows us to grip and manipulate things with our hands. As a demonstration of the importance of this trait, fold each of your thumbs into the palm of its hand and then tape them securely in that position for an entire day. After the demonstration, make a list of the things you could not do without the use of your thumbs.

3. The Texas horned lizard of North America (*Phrynosoma cornutum*) and the thorny devil lizard (*Moloch horridus*) of Australia are distantly related lizard species that both live in desert environments and both eat ants as their main source of food. Do a Web search for these species and compare photographs and information about their natural history. Describe major characteristics of the ecological niche and the appearance of each of these species.

4. Make a concept map of this chapter's major ideas, using the section heads and subheads and the key terms (in boldface). Look inside the back cover and in Appendix 4 for information about concept maps.

7 GEOGRAPHICAL ECOLOGY, CLIMATE, AND BIOMES*

Connections: Blowing in the Wind

Environmental science is a study of connections. One of the things that connects all life on the earth is wind, a vital part of the planet's circulatory system. Without wind, most of the earth would be uninhabitable: The tropics would be unbearably hot, and most of the rest of the planet would freeze.

Winds also transport nutrients from one place to another. Dust rich in phosphates blows across the Atlantic from the Sahara Desert in Africa (Figure 7-1), helping to replenish rain forest soils in Brazil. Iron-rich dust blowing from China's Gobi Desert falls into the Pacific Ocean between Hawaii and Alaska six or seven times a year and stimulates the growth of phytoplankton, the minute producers that support ocean food webs. That's the good news.

The bad news is that wind also transports harmful substances. Wind-blown particles of reddish-brown soil from Africa blanket Florida's sky, making it difficult for the state to meet federal air-pollution standards. Sulfur compounds and soot from oil-well fires in Kuwait have been detected over Wyoming, and deposits of DDT and PCBs have been turning up in isolated Antarctica for decades. Cesium-137 blow-

*Paul M. Rich, associate professor of Ecology and Evolutionary Biology and Environmental Studies at the University of Kansas, is coauthor of this chapter.

Figure 7-1 Some of the dust shown here blowing from Africa's Sahara Desert can end up as soil nutrients in Amazonian rain forests. (NOAA. USGS/NMD EROS Data Center)

ing from the 1986 Chernobyl nuclear power-plant disaster in Ukraine has made lichens (the food of Lapland's reindeer) radioactive. As a result, much of the reindeer meat, milk, and cheese has become unfit to eat for the herders who depend on it.

There's mixed news as well. On the one hand, clouds of particles from volcanic eruptions ride the winds, circle the globe, and change the earth's climate for a while. After Indonesia's Tambora volcano blew up in 1815, the ash in the atmosphere significantly reduced the amount of sunlight reaching the earth. As a result, distant Europe had a "year without a summer" in 1816. Emissions from the 1991 eruption of Mount Pinatubo in the Philippines—the largest eruption this century—cooled the earth slightly for 3 years, temporarily masking signs of possible global warming. On the other hand, volcanic ash, like the blowing desert dust, adds valuable trace minerals to the soil where it settles.

The lesson, once again, is that *there is no "away"* and wind—acting as part of the planet's circulatory system for heat, moisture, and plant nutrients—is one reason. Movement of soil particles from one place to another by wind and water is a natural phenomenon, but when we disturb the soil and leave it unprotected we hasten the process. As the Roman poet Virgil wrote over 2,000 years ago, "Before we plow an unfamiliar patch, it is well to be informed about the winds."

Until recently, wind has been mostly invisible, but now we can "see" it. Satellite pictures allow us to chart the courses of blowing dust clouds (Figure 7-1). Atmospheric chemists use laser probes from aircraft to locate and identify substances in swirling wind plumes.

Other sensors can detect trace gases caught in wind currents and record how their concentrations change over fractions of a second. These new technologies revealed the intimate connection between African desert and Amazonian rain forest. In coming years such technologies should help us learn more about the important roles of wind in the ecosphere.

Wind is also an important factor in climate through its influence on global air circulation patterns. Climate, in turn, is crucial for determining what kinds of terrestrial life are found in the major biomes of the different geographical regions of the ecosphere, as we shall see in this chapter.

To do science is to search for repeated patterns, not simply to accumulate facts, and to do the science of geographical ecology is to search for patterns of plant and animal life that can be put on a map.

ROBERT H. MACARTHUR

This chapter addresses several broad questions about geographic patterns of ecology:

- What key factors determine variations in the earth's weather and climate?
- How does climate determine the major geographical patterns of ecology on the earth?
- What are the major types of desert biomes, and how are they being affected by human activities?
- What are the major types of grassland biomes, and how are they being affected by human activities?
- What are the major types of forest biomes, and how are they being affected by human activities?
- Why are mountain and arctic biomes important, and how are they being affected by human activities?
- Why is it important to have a geographical perspective of ecology?

7-1 WEATHER AND CLIMATE: A BRIEF INTRODUCTION

How Does Weather Differ from Climate? At every moment at any spot on the earth, the *troposphere* (the inner layer of the atmosphere containing most of the earth's air) has a particular set of physical properties. Examples are temperature, pressure, humidity, precipitation, sunshine, cloud cover, and wind direction and speed. These short-term properties of the troposphere at a given place and time are what we call **weather**.

Climate is the average long-term weather of an area; it is a region's general pattern of atmospheric or weather conditions, including seasonal variations and weather extremes (such as hurricanes or prolonged drought or rain) averaged over a long period (at least 30 years). The two main factors determining an area's climate are *temperature*, with its seasonal variations, and the amount and distribution of *precipitation*. Figure 7-2 is a generalized map of the earth's major climate zones.

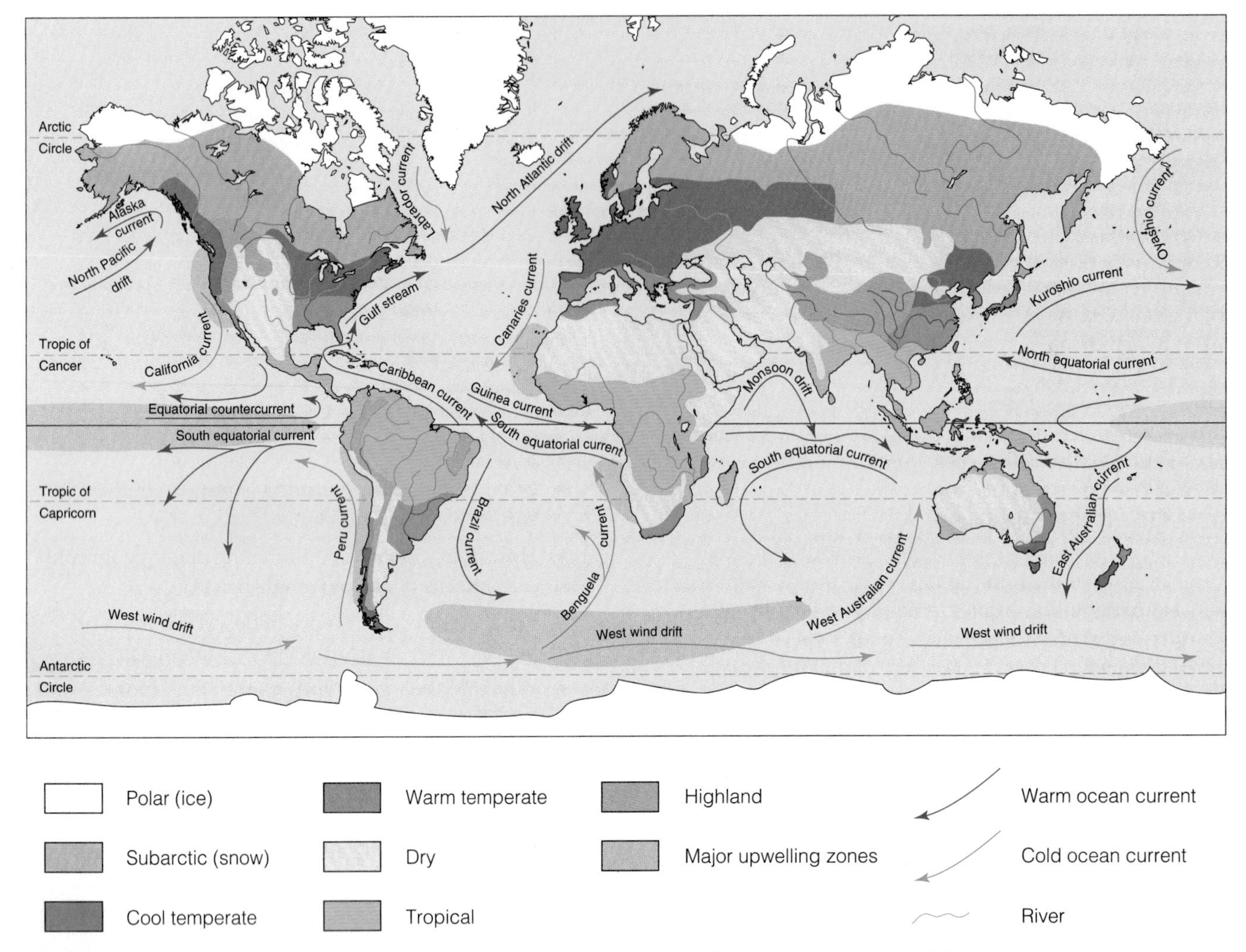

Figure 7-2 Generalized map of global climate zones, showing the major contributing ocean currents and drifts. Large variations in climate are dictated mainly by temperature (with its seasonal variations) and by the quantity and distribution of precipitation.

Why Is Weather So Changeable? Masses of air that are either warm or cold, wet or dry, and contain air at either high or low pressure are constantly moving across the land and the sea. Weather changes as one air mass replaces or meets another.

The most dramatic changes in weather occur along a **front**, the boundary between two air masses with different temperatures and densities. A **warm front** is the boundary between an advancing warm air mass and the cooler one it is replacing. Because warm air is less dense than cool air, an advancing warm front rises up over a mass of cool air.

As the warm front rises, its moisture starts to condense and it produces many layers of clouds at different altitudes. High, wispy clouds are the first signs of an advancing warm front. Gradually the clouds thicken, descend to a lower altitude, and often release their moisture as rainfall. A moist warm front can bring days of cloudy skies and drizzle.

A **cold front** is the leading edge of an advancing mass of cold air. Because cold air is more dense than warm air, an advancing cold front stays close to the ground and wedges underneath less dense warmer air. An approaching cold front produces rapidly moving towering clouds called *thunderheads*. As the overlying mass of warm air is pushed upward it cools, and its water vapor condenses and falls to the earth's surface as precipitation. As a cold front passes through, we often experience high surface winds and thunderstorms. After the front passes through we usually experience cooler temperatures and a clear sky.

Weather is also affected by changes in atmospheric pressure. An air mass with high pressure, called a *high*, contains cool, dense air that descends toward the earth's surface and becomes warmer. Fair weather follows as long as the high-pressure air mass remains over an area.

In contrast, a low-pressure air mass, called a *low*, produces cloudy, sometimes stormy weather. This

Q: What is the sun's source of energy?

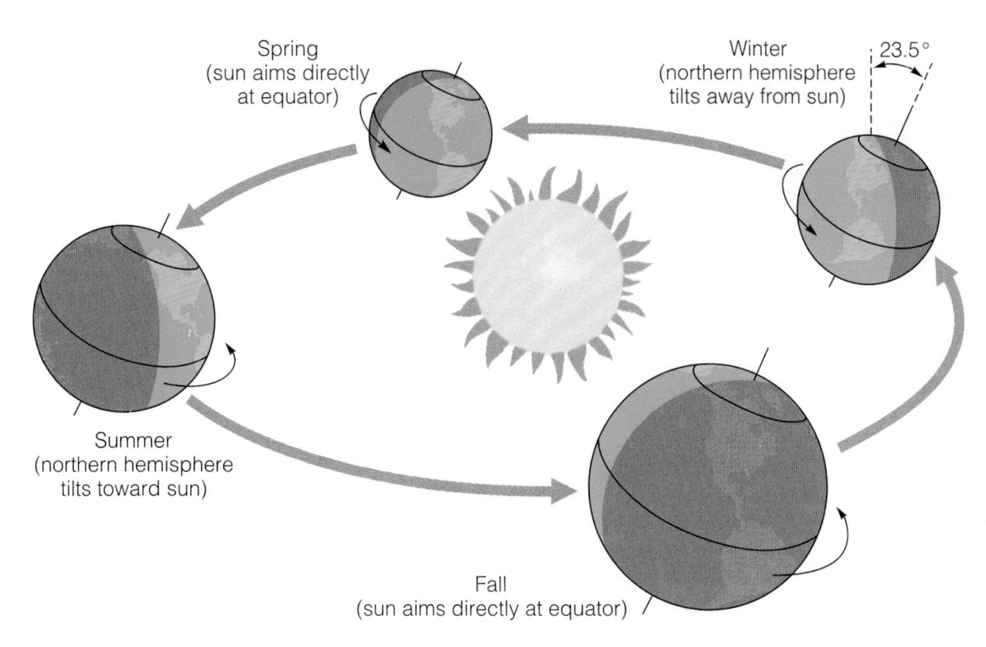

Figure 7-3 The effects of the earth's tilted axis. As the planet makes its annual revolution around the sun on an axis tilted about 23.5°, various regions are tipped toward or away from the sun. This produces the variations in the amount of solar energy reaching the earth and causes the seasons.

happens because less dense warm air spirals inward toward the center of a low-pressure air mass. Because of its low pressure and low density, the center of the low rises, and its warm air expands and cools. This causes moisture to condense (because cool air cannot hold as much moisture as warm air). Precipitation often follows.

In addition to normal weather, we sometimes experience *weather extremes*. Examples are violent storms called *tornadoes* (which form over land) and *tropical cyclones* (which form over warm ocean waters and sometimes pass over coastal land). Tropical cyclones are called *hurricanes* in the Atlantic and *typhoons* in the Pacific.

Meteorologists use devices such as weather balloons, aircraft, ships, radar, and satellites to obtain data on variables such as atmospheric pressures, precipitation, temperatures, wind speeds, and locations of air masses and fronts. These data are fed into computer models to draw weather maps for each of seven levels of the troposphere, ranging from the ground to 19 kilometers (12 miles) up. Computer models use the map data to forecast the weather in each box of the seven-layer grid for the next 12 hours. Other computer models project the weather for the next several days by calculating the probabilities that air masses, winds, and other factors will move and change in certain ways.

How Does the Global Circulation of Air Affect Regional Climates? The two most important factors determining a region's climate are its *average temperature* and *average precipitation*. The temperature and precipitation patterns that lead to different climates (Figure 7-2) are caused primarily by the way air circulates over the earth's surface.

Several factors determine global air circulation patterns. One is the *uneven heating of the earth's surface*. Air is heated much more at the equator (where the sun's rays strike directly throughout the year) than at the poles (where sunlight strikes at an angle and is thus spread out over a much greater area). These differences help explain why tropical regions near the equator are hot, polar regions are cold, and temperate regions in between generally have intermediate average temperatures (Figure 7-2).

Second, *seasonal changes occur because the earth's axis* (an imaginary line connecting the north and south poles) is tilted; as a result, various regions are tipped toward or away from the sun as the earth makes its annual revolution (Figure 7-3). This creates opposite seasons in the northern and southern hemispheres.

Third, *the earth rotates on its axis*, which prevents air currents from moving due north and south from the equator. Forces in the atmosphere created by this rotation deflect winds (moving air masses) to the right in the northern hemisphere and to the left in the southern hemisphere, in what is called the *Coriolis effect*. The result is six huge convection cells of swirling air masses—three north and three south of the equator—that convey or move heat and water from one area to another (Figure 7-4).

A fourth factor involves *long-term variations in the amount of solar energy striking the earth*. Such variation occurs because of occasional changes in solar output, slight planetary shifts in which the earth's axis wobbles (22,000-year cycle) and tilts (44,000-year cycle) as it revolves around the sun, and minute changes in the shape of its orbit around the sun (100,000-year cycle).

Finally, climate and global air circulation are affected by the *properties of air and water*. When heated by the sun, ocean water evaporates and removes heat

A: Nuclear fusion

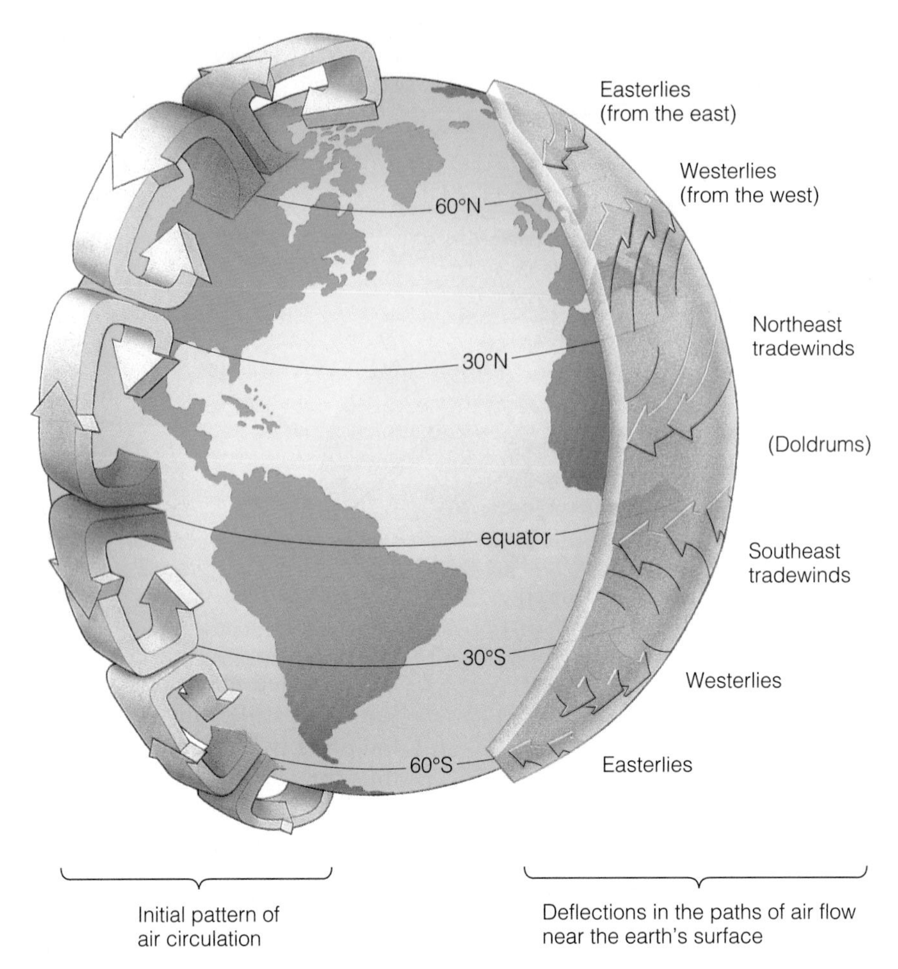

Figure 7-4 Formation of prevailing surface winds, which disrupt the general flow of air from the equator to the poles and back to the equator. As the earth rotates, its surface turns faster beneath air masses at the equator and slower beneath those at the poles. This deflects air masses moving north and south to the west or east, creating six huge convection cells in which air swirls upward and then descends toward the earth's surface at different latitudes. The direction of air movement in these cells sets up belts of prevailing winds that distribute air and moisture over the earth's surface. These winds affect the general types of climate found in different areas and drive the circulation of ocean currents. (Used by permission from Cecie Starr and Ralph Taggart, *Biology: The Unity and Diversity of Life*, 8th ed., Belmont, Calif.: Wadsworth, 1998)

from the oceans to the atmosphere, especially near the hot equator. This moist, hot air expands, becomes less dense (weighs less per unit of volume), and rises in fairly narrow vortices. These upward spirals create an area of low pressure at the earth's surface.

As this moisture-laden air rises, it cools and releases moisture as condensation (because cold air can hold less water vapor than warm air). When water vapor condenses it releases heat, which radiates into space. The resulting cooler, drier air becomes denser, sinks (subsides), and creates an area of high pressure. As this air mass flows across the earth's surface, it picks up heat and moisture and begins to rise again (Figure 7-5). The resulting small and giant convection cells circulate air, heat, and moisture both vertically and from place to place in the troposphere, leading to different climates and patterns of vegetation (Figure 7-6).

How Do Ocean Currents Affect Regional Climates? The factors just listed, plus differences in water density, cause warm and cold ocean currents (Figure 7-2). These currents are driven by the wind and the earth's rotation and, along with air masses, redistribute heat received from the sun. Generally, cold currents flow from the polar areas toward the equator and warm currents flow away from the equator. The deeper currents are driven partly by cooling (which makes water denser and causes it to sink) and partly by increased salinity (which has the same effect).

Ocean currents, like air currents, redistribute heat and thus influence climate and vegetation, especially near coastal areas. For example, without the warm Gulf Stream, which transports 25 times more water than all the world's rivers, the climate of northwestern Europe would be subarctic. Currents also help mix

Q: What are the three components of biodiversity?

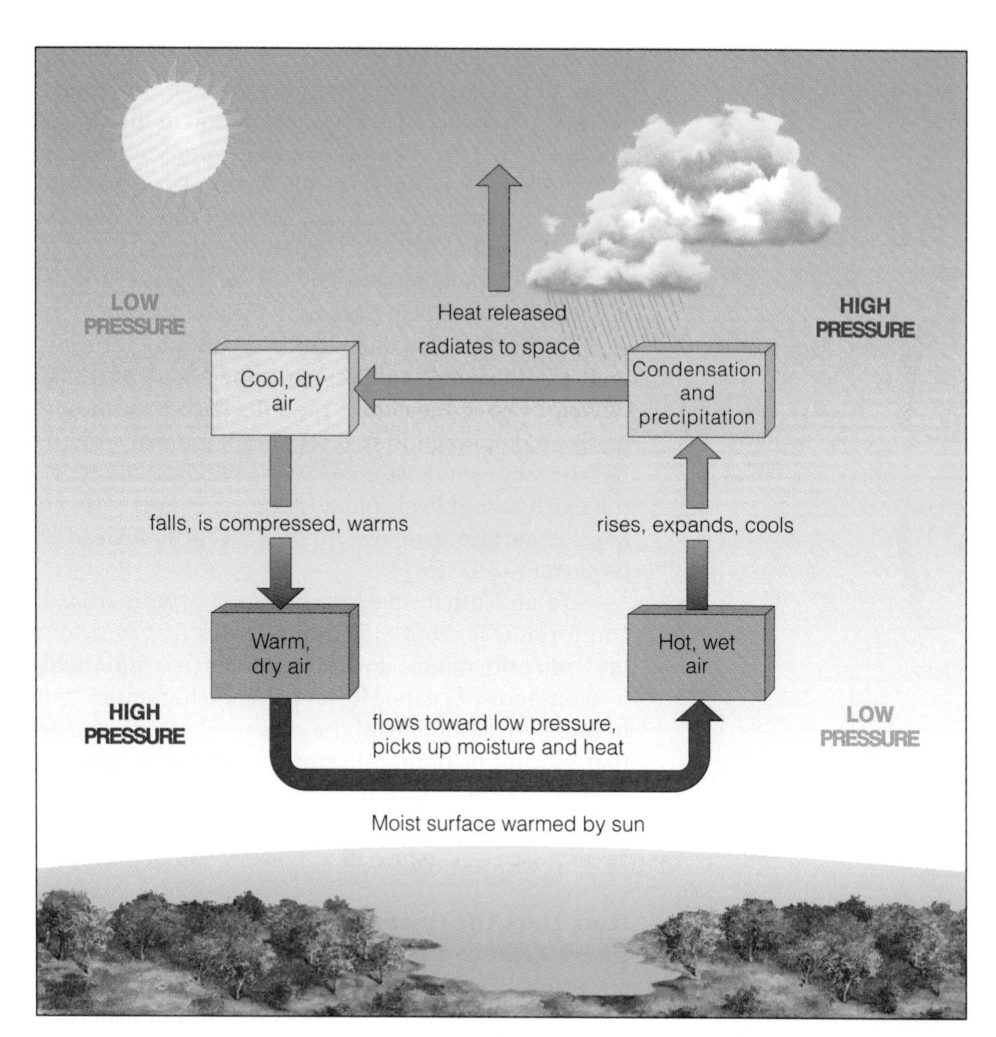

Figure 7-5 Distribution of heat and water occurs because vertical convection currents stir up air in the troposphere and transport heat and water from one area to another in circular convection cells. The relative humidity increases as the air rises (right side) and decreases as it falls (left side).

ocean waters and distribute nutrients and dissolved oxygen needed by aquatic organisms.

Along some steep western coasts of continents, almost constant trade winds blow offshore, pushing surface water away from the land. This outgoing surface water is replaced by an **upwelling** of cold, nutrient-rich bottom water (Figure 7-7). Upwellings, whether far from shore or near shore (Figure 7-2), bring plant nutrients from the deeper parts of the ocean to the surface and support large populations of phytoplankton, zooplankton, fish, and fish-eating seabirds.

What Is the El Niño–Southern Oscillation? Every few years, in the Pacific Ocean normal coastal upwelling (Figure 7-8, left) is affected by changes in climate patterns called the *El Niño–Southern Oscillation*, or *ENSO* (Figure 7-8, right). In an ENSO, the prevailing westerly winds weaken or cease (for reasons that are not clear) and surface water along the South and North American coasts becomes warmer. The normal upwelling of cold, nutrient-rich water is suppressed, which reduces primary productivity and causes a sharp decline in the populations of some fish species. A strong ENSO can trigger extreme weather changes over at least two-thirds of the globe, especially in lands along the Pacific and Indian Oceans.

How Does the Chemical Makeup of the Atmosphere Lead to the Greenhouse Effect? Small amounts of carbon dioxide and water vapor, as well as trace amounts of ozone, methane, nitrous oxide, chlorofluorocarbons, and other gases in the troposphere, play a key role in determining the earth's average temperatures and thus its climates.

Together, these gases, known as **greenhouse gases**, act somewhat like the glass panes of a greenhouse: They allow light, infrared radiation, and some ultraviolet radiation from the sun (Figure 3-13) to pass through the troposphere. The earth's surface then absorbs much of this solar energy and degrades it to longer-wave, infrared radiation (that is, heat), which then rises into the troposphere (Figure 4-7). Some of this heat escapes into space; some is absorbed by molecules of greenhouse gases, warming the air; and some radiates back toward the earth's surface. This natural trapping of heat in the troposphere is called the **greenhouse effect** (Figure 7-9).

The amount of heat trapped in the troposphere depends primarily on the concentrations of greenhouse

A: Genetic diversity, species diversity, and ecosystem diversity

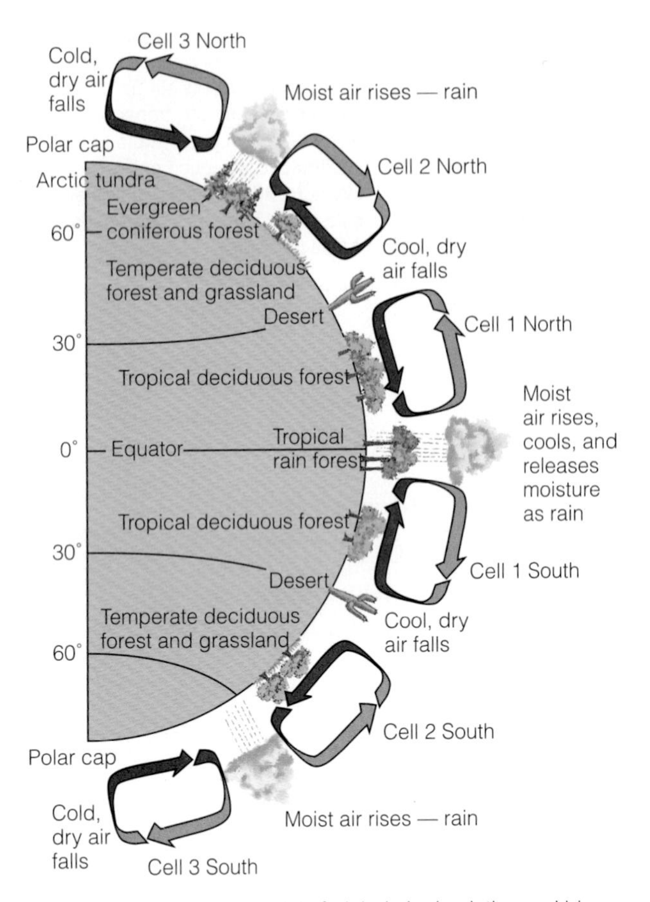

Figure 7-6 Conceptual model of global air circulation and biomes. Heat and moisture are distributed over the earth's surface by vertical convection currents that form into six large convection cells (called Hadley cells) at different latitudes. The direction of air flow and the ascent and descent of air masses in these convection cells determine the earth's general climatic zones. The uneven distribution of heat and moisture over the planet's surface leads to the forests, grasslands, and deserts that make up the earth's biomes.

gases and the length of time they stay in the atmosphere. The primary heat-trapping gas in the atmosphere is water vapor; because its concentration in the atmosphere is high (1–5%), inputs of water vapor from human activities have little effect on this chemical's greenhouse effects. By contrast, the concentration of carbon dioxide in the atmosphere is so small (0.036%) that fairly large input of CO_2 from human activities can significantly affect the amount of heat trapped in the atmosphere.

The basic theory behind the greenhouse effect is well established; satellites equipped with infrared detectors have measured the effects of greenhouse gases on outgoing infrared radiation. Indeed, without its current greenhouse gases (especially water vapor), the earth would be a cold and lifeless planet with an average surface temperature of –18°C (0°F) instead of its current 15°C (59°F).

We and other species currently benefit from a comfortable level of greenhouse gases that typically undergo only minor, slow fluctuations over hundreds to thousands of years. However, crude but improving mathematical models of the earth's climate indicate that natural or human-induced global warming (or cooling) taking place over a few decades could be disastrous for human societies and many forms of life, as discussed in Chapter 19.

How Does the Chemical Makeup of the Atmosphere Lead to the Ozone Layer? In a band of the stratosphere 17–26 kilometers (11–16 miles) above the earth's surface, oxygen (O_2) is continuously converted to ozone (O_3) and back to oxygen by a sequence of reactions initiated by ultraviolet radiation from the sun ($3O_2 + UV \rightleftharpoons 2O_3$). The result is a thin veil of protective ozone at very low concentrations (up to 10 parts

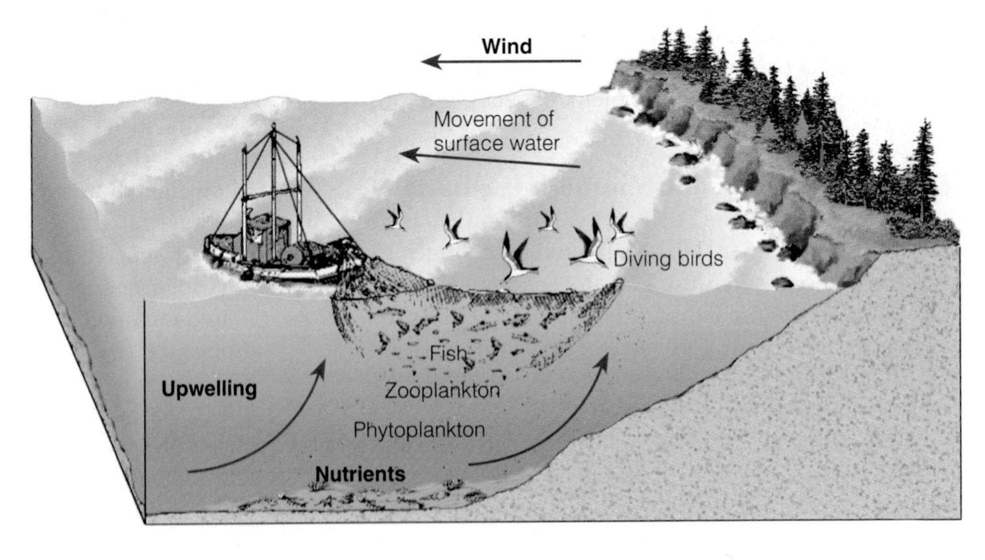

Figure 7-7 A *shore upwelling* (shown here) occurs when deep, cool, nutrient-rich waters are drawn up to replace surface water moved away from a steep coast by wind-driven currents. Such areas support large populations of phytoplankton, zooplankton, fish, and fish-eating birds. *Equatorial upwellings* occur in the open sea near the equator when northward and southward currents interact to push deep waters and their nutrients to the surface, thus greatly increasing primary productivity in such areas (Figure 7-2).

Q: What two major types of organisms are found in ecosystems?

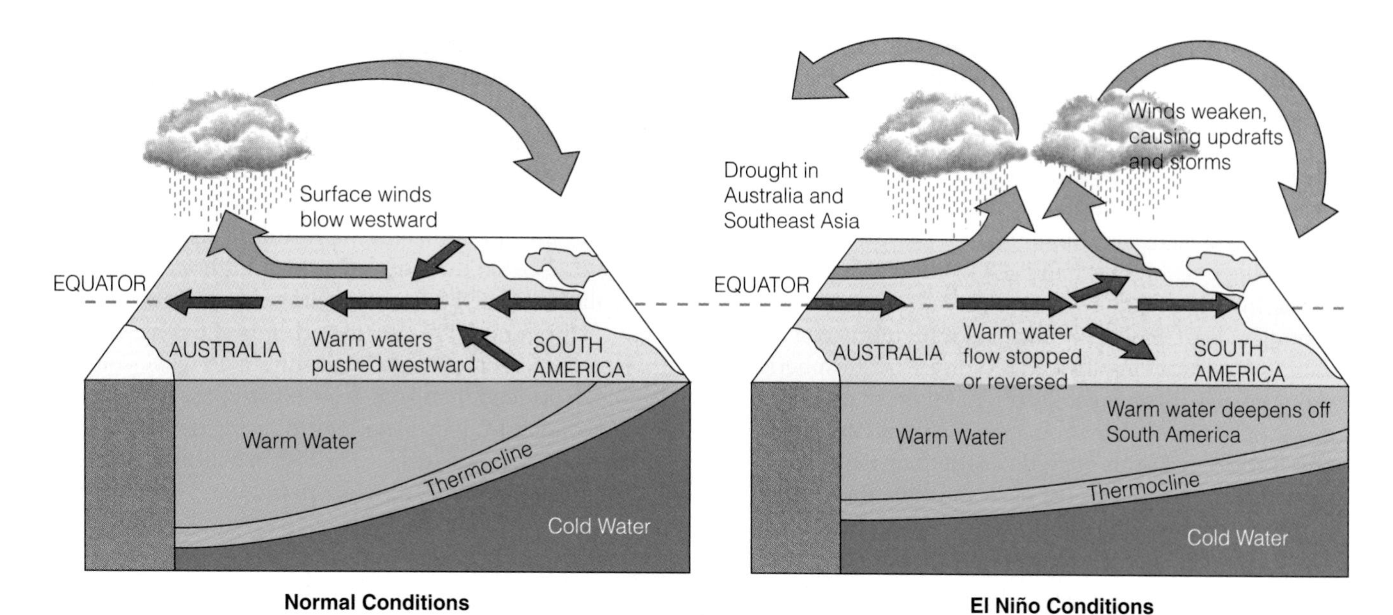

Figure 7-8 Normal surface winds blowing westward cause shore upwellings of cold, nutrient-rich bottom water in the tropical Pacific Ocean near the coast of Peru (left). The warm and cold water are separated by a zone of gradual temperature change called the thermocline. Every few years a climate shift known as the El Niño–Southern Oscillation (ENSO) disrupts this pattern. Westward surface winds weaken, which depresses the coastal upwellings and warms the surface waters off South America (right). ENSOs typically last for several months to over a year and occur every 3 or 4 years, although the interval has been as long as 7 years. When an ENSO lasts 12 months or longer, it severely disrupts populations of plankton, fish, and seabirds in upwelling areas and can trigger extreme weather changes over much of the globe.

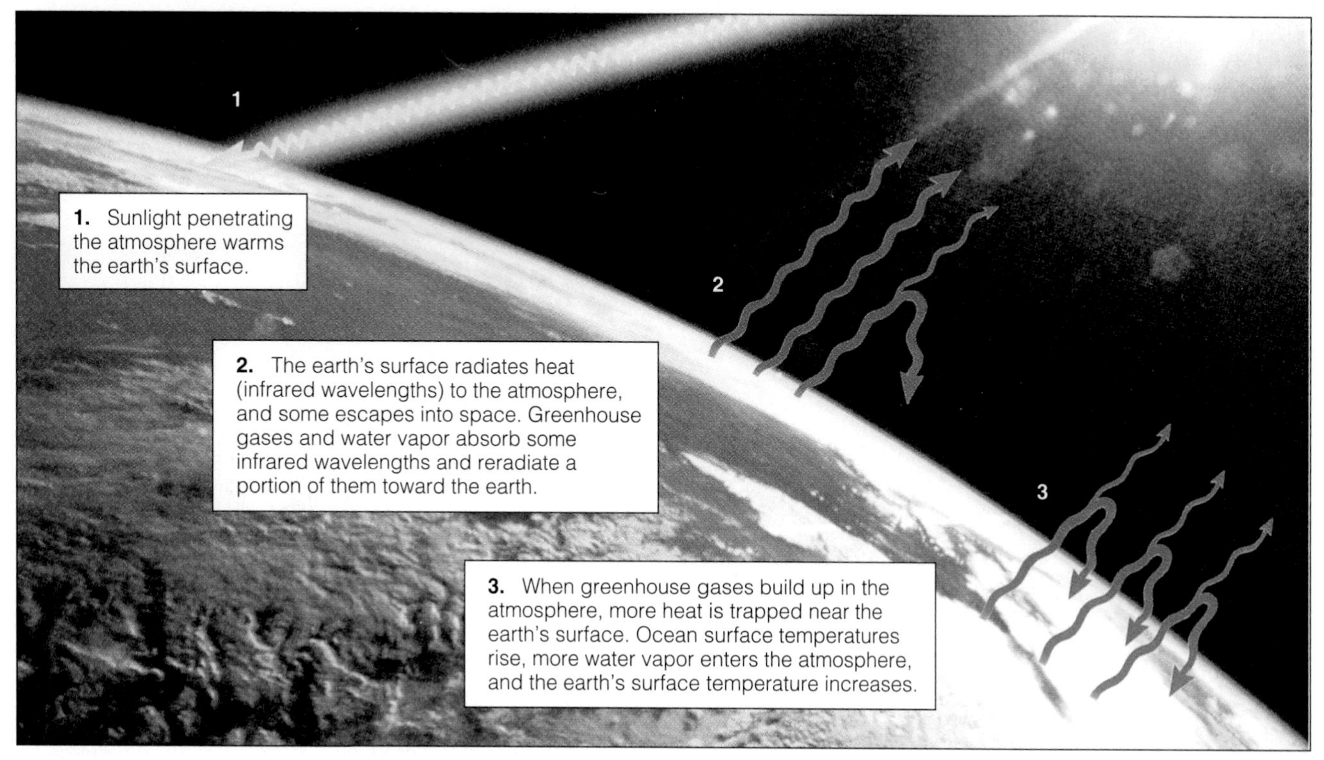

Figure 7-9 Conceptual model of the greenhouse effect. Without the atmospheric warming provided by this natural effect, the earth would be a cold and mostly lifeless planet. According to the widely accepted greenhouse theory, when concentrations of greenhouse gases in the atmosphere rise, the average temperature of the troposphere also rises. (Used by permission from Cecie Starr and Ralph Taggart, *Biology: The Unity and Diversity of Life*, 8th ed., Belmont, Calif.: Wadsworth, 1998)

per million). Normally, the average levels of ozone in this lifesaving layer don't change much because the rate of ozone destruction is equal to its rate of formation.

Ultraviolet (UV) radiation reaching the stratosphere is composed of three bands: A, B, and C. The ozone layer blocks out nearly all the highest-energy, shortest-wavelength radiation (UV-C), approximately half of the next highest band (UV-B), and a small part of the lowest-energy radiation (UV-A). Besides preventing at least 95% of the sun's harmful ultraviolet radiation from reaching the earth's surface, stratospheric ozone creates warm layers of air that prevent churning gases in the troposphere from entering the stratosphere. This *thermal cap* is important in determining the average temperature of the troposphere and thus the earth's current climates. There is considerable evidence that chemicals added to the atmosphere by human activities are decreasing levels of protective ozone in the stratosphere, as discussed in more detail in Chapter 19.

How Do Topography and Other Features of the Earth's Surface Modify Climate to Form Microclimates? Various topographic features of the earth's surface create local climatic conditions, or **microclimates**, that differ from the general climate of a region. For example, mountains interrupt the flow of prevailing surface winds and the movement of storms. When moist air blowing inland from an ocean reaches a mountain range, it cools as it is forced to rise and expand. This causes the air to lose most of its moisture, in the form of rain and snow, on the windward (wind-facing) slopes. As the drier air mass flows down the leeward (away from the wind) slopes, it is compressed, becomes warmer, and thus can hold more moisture. This air draws moisture out of the plants and soil over which it passes, rather than giving it up as precipitation. The lower precipitation and the resulting semiarid or arid conditions on the leeward side of high mountains together are called the **rain shadow effect** (Figure 7-10).

Topography also causes different locations to receive different amounts of solar radiation. For example, in the northern hemisphere, south-facing slopes receive more solar radiation than north facing slopes, leading to higher temperatures and drier conditions for the plants and animals that live there. West-facing slopes tend to receive more solar radiation in the afternoon, and east-facing slopes receive more in the morning.

Vegetation also creates microclimates because it takes up and releases water, changes the movement of wind near the ground, and casts shadows. For example, compared with nearby areas of open land, forests are warmer in winter and cooler in summer, and they have lower wind speeds and higher humidity.

Cities also create distinct microclimates. Bricks, concrete, asphalt, and other building materials absorb and hold heat, and buildings block wind flow. Motor vehicles and the climate control systems of buildings release large quantities of heat and pollutants; thus, cities tend to have more haze and smog, higher temperatures, and lower wind speeds than the surrounding countryside.

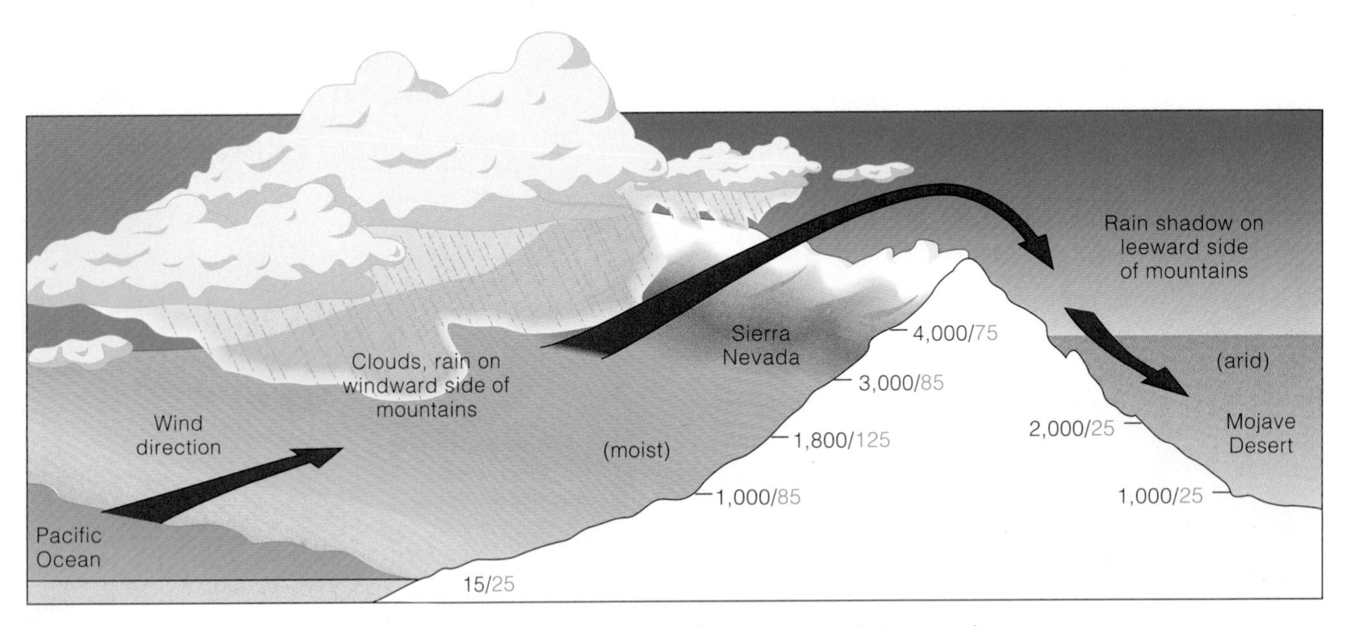

Figure 7-10 The rain shadow effect is a reduction of rainfall on the side of high mountains facing away from prevailing surface winds. It occurs when warm moist air in prevailing onshore winds loses most of its moisture as rain and snow on the windward slopes of a mountain range. This leads to semiarid and arid conditions on the leeward side of the mountain range and the land beyond. The Mojave Desert, east of the Sierra Nevada in California, is produced by this effect. Blue numbers represent average annual precipitation (in centimeters); black numbers are elevations (in meters). (Used by permission from Cecie Starr, *Biology: The Unity and Diversity of Life*, 8th ed., Belmont, Calif.: Wadsworth, 1998)

Q: How do most producer organisms get the nutrients they need?

7-2 BIOMES: CLIMATE AND LIFE ON LAND

Why Are There Different Organisms in Different Places? Why is one area of the earth's land surface a desert, another a grassland, and another a forest? Why are there different types of deserts, grasslands, and forests?

The general answer to these questions is differences in *climate* (Figure 7-2). Such differences result primarily from differences in average temperature and precipitation caused by global air circulation (Figure 7-6). Figures 7-11 and 7-12 and p. 1 show distributions of *biomes*: terrestrial regions with characteristic types of natural, undisturbed ecological communities adapted to the climate of the region. By comparing Figure 7-11 with Figure 7-2, you can see how the world's major biomes vary with climate. Figure 4-8 shows major biomes in the United States as one moves through different climates along the 39th parallel.

For plants, *precipitation is generally the limiting factor that determines whether a land area is desert, grassland, or forest.* Taken together, average annual precipitation and temperature (along with soil type) are the most important factors in producing tropical, temperate, or polar deserts, grasslands, and forests (Figure 7-12).

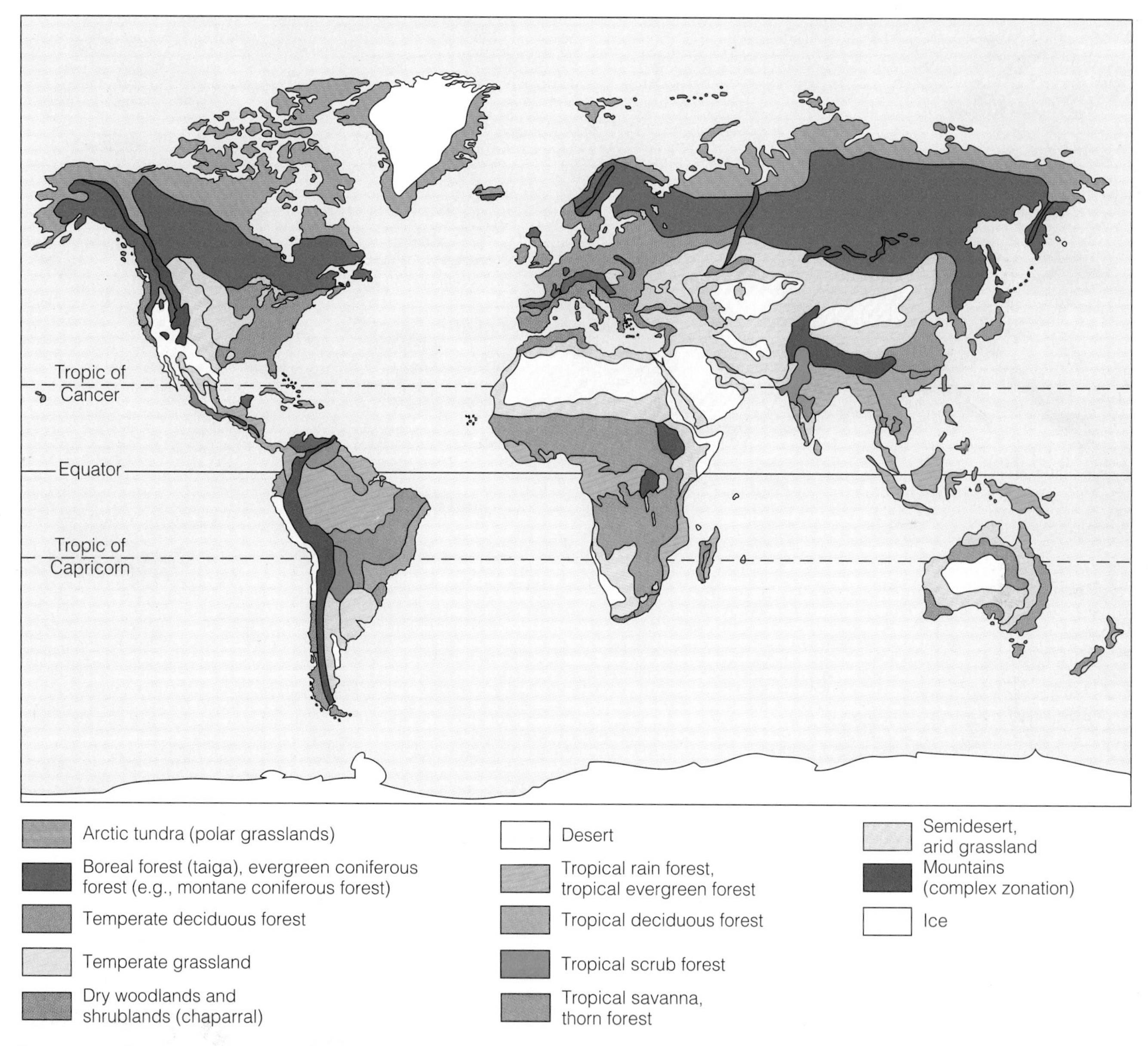

Figure 7-11 The earth's major biomes—the main types of natural vegetation in different undisturbed land areas—result primarily from differences in climate. Each biome contains many ecosystems whose communities have adapted to differences in climate, soil, and other environmental factors. In reality, people have removed or altered much of this natural vegetation for farming, livestock grazing, harvesting lumber and fuelwood, mining, and constructing villages and cities, thereby altering the biomes (Figure 1-5).

A: They produce them through photosynthesis

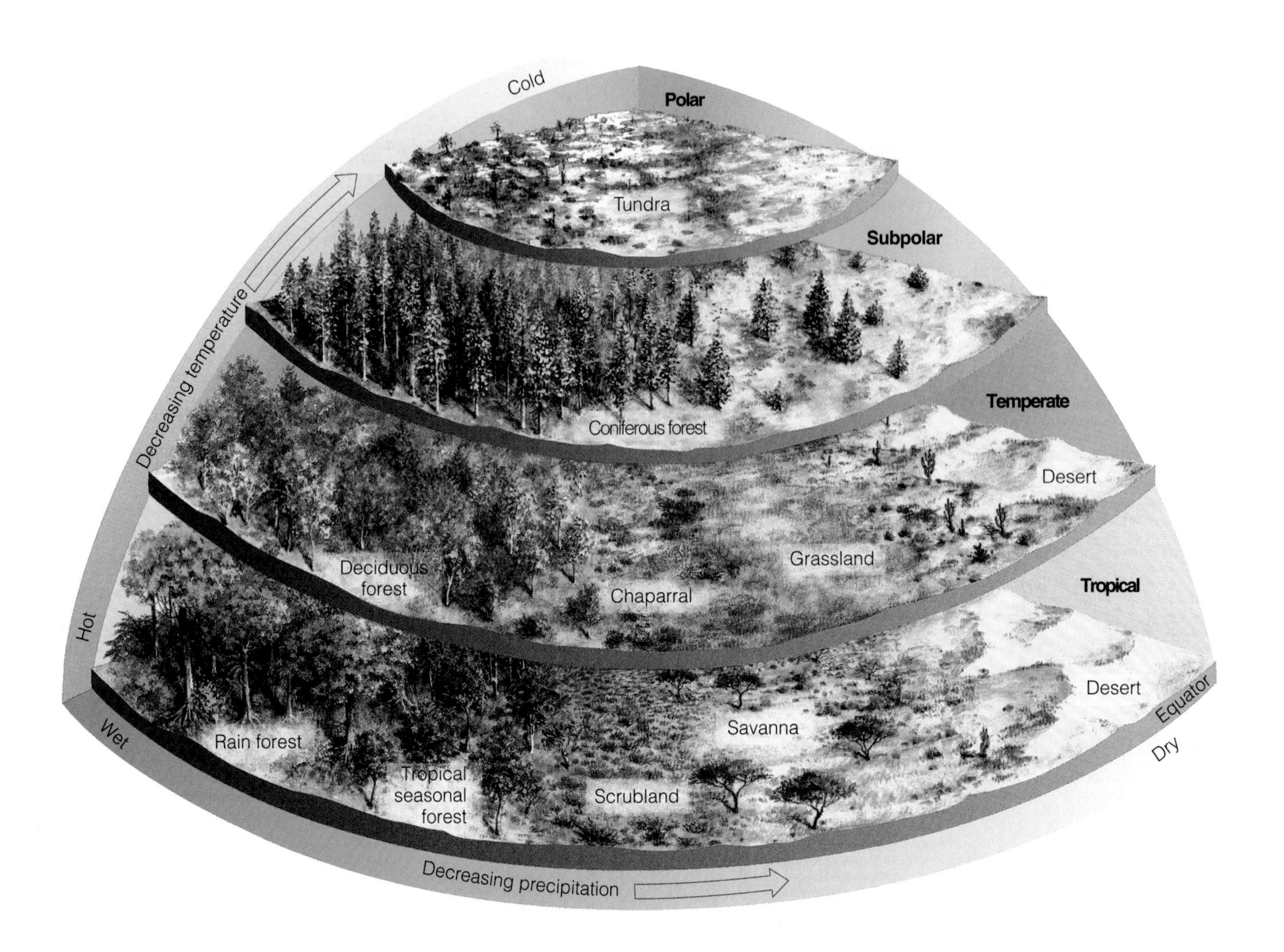

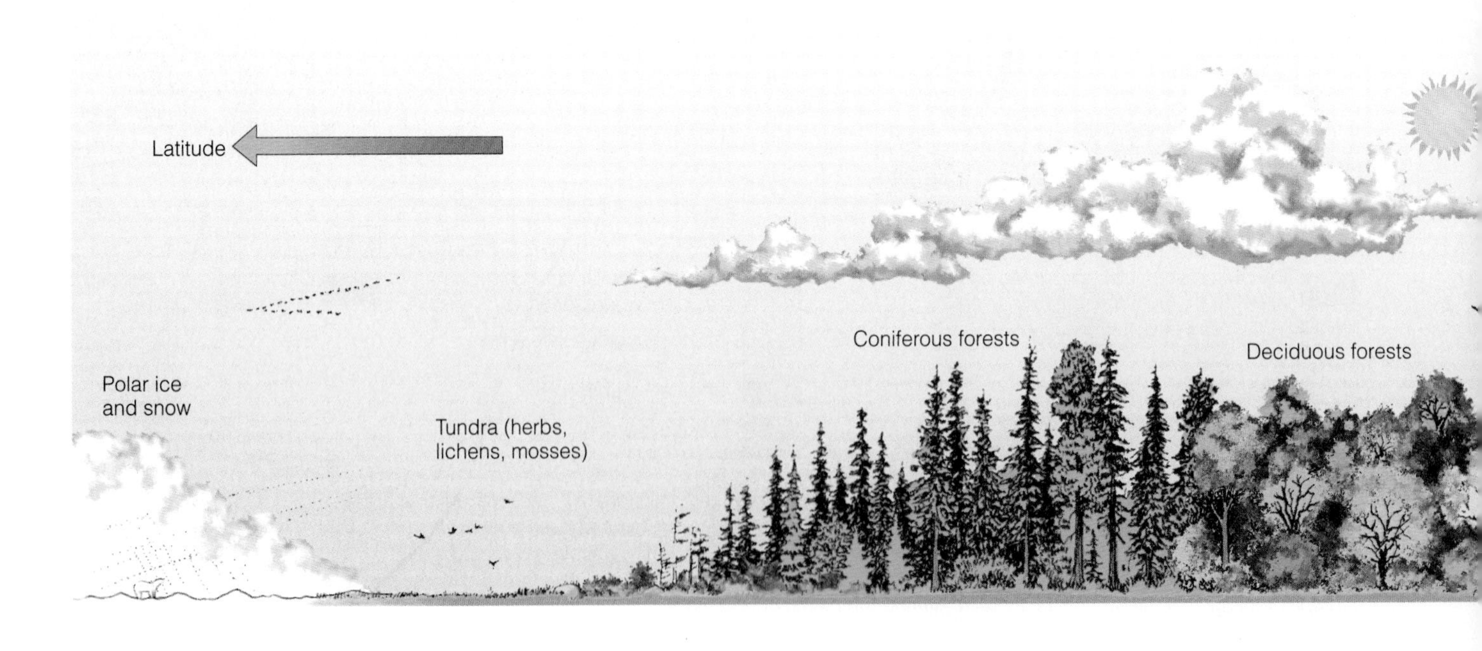

Q: What is earth's slowest cycle involving changes in matter?

On maps such as the one in Figure 7-11, biomes are presented as having sharp boundaries and as being covered with the same general type of vegetation. In reality, biomes don't have sharp boundaries, but instead blend into one another in transitional zones or ecotones (Figure 4-9). Also, the types and numbers of organisms in a biome vary from one location to another because of small variations in climate (microclimate), soil type, history, and natural and human-caused disturbances (Figure 1-5). As a result, biomes are not uniform, but rather consist of a mosaic of patches, all with somewhat different ecological communities, but all with similarities unique to the biome.

Climate and biomes both vary with **latitude** (distance from the equator) and **altitude** (elevation above sea level). If you travel from the equator toward either pole, you will generally encounter colder climates and zones of vegetation adapted to those climates. Similarly, as elevation above sea level increases, climate becomes colder. Thus, if you climb a tall mountain from its base to its summit, you can observe changes in plant life similar to those you would encounter in traveling from equator to poles (Figure 7-13).

Figure 7-12 Average precipitation and average temperature, acting together as limiting factors over a period of 30 or more years, determine the type of desert, grassland, or forest biome in a particular area. Although the actual situation is much more complex, this simplified diagram explains how climate determines the types and amounts of natural vegetation found in an area left undisturbed by human activities. (Used by permission of Macmillan Publishing Company, from Derek Elsom, *The Earth*, New York: Macmillan, 1992. Copyright © 1992 by Marshall Editions Developments Limited)

Figure 7-13 Generalized effects of latitude and altitude on climate and biomes. Parallel changes in vegetation type occur when we travel from the equator to the poles or from lowlands to mountaintops.

Why Do Plant Sizes, Forms, and Survival Strategies Differ? Arctic soils are wet and nutrient rich. So why are there no trees in the Arctic, and why are the plants there so close to the ground? Why don't desert plants such as cacti have leaves? Why do trees in most forests found in both the warm tropics and in cold areas such as Canada and Sweden keep their leaves year-round, whereas most trees in temperate forests lose their leaves in winter?

The plant communities found in various biomes have distinct physical appearances depending on the types, sizes, and forms of their plant species. Climate and soil type (Figure 5-16) play a role in determining these characteristics. The size and form of a plant species (and whether it keeps its leaves year-round) tend to represent adaptations for gathering sunlight for photosynthesis and for maintaining the optimum temperature in a particular environment.

Plants exposed to cold air year-round or during winter have traits that keep them from losing too much heat and water. For example, trees or tall plants in the cold, windy arctic grasslands (tundra) would lose too much of their heat for survival.

Desert plants exposed to the sun all day long must be able to lose enough heat so that they don't overheat and die. They must also conserve enough water for survival. **Succulent** (fleshy) **plants**, such as the saguaro (pronounced sah-WAH-row) cactus (see photo in table of contents), survive in dry climates by having a vertical orientation of most surfaces, no leaves, and the ability to store water and synthesize food in their expandable, fleshy tissue. The plant's shape and lack of leaves give it a small surface area exposed to sunlight and a large area away from the sun for radiating heat out. Succulent plants also reduce water loss by opening their pores (stomata) to take up carbon dioxide (CO_2) only at night. The carbon is stored as acids in the plant tissue and used for photosynthesis during the day.

Trees of wet tropical rain forests tend to be **broadleaf evergreen plants**, which keep most of their broad leaves year-round. The large surface area of the leaves allows them to collect ample sunlight for photosynthesis and also radiate out heat during the hot summer.

In a climate with a cold (and sometimes dry) winter, keeping such leaves would cause plants to lose too much heat and water for survival. In such climates, **broadleaf deciduous plants**, such as oak and maple trees, survive drought and cold by shedding their leaves and becoming dormant during such periods.

If we move further north to areas such as Canada and Sweden, where summers are cool and short, this strategy is less successful. Instead evolution has favored **coniferous** (cone-bearing) **evergreen plants** (such as spruces, pines, and firs). These plants keep some of their narrow pointed leaves (needles) all year. The waxy coating, shape, and clustering of conifer needles slow down heat loss and evaporation during the long, cold winter. Additionally, by keeping their leaves all winter, such trees are ready to take advantage of the brief summer without having to take time to grow new needles.

7-3 DESERT BIOMES

What Are the Major Types of Deserts? A **desert** is an area where evaporation exceeds precipitation. Precipitation is typically less than 25 centimeters (10 inches) a year and is often scattered unevenly throughout the year. Deserts have sparse, widely spaced, mostly low vegetation, with the density of plants determined primarily by the frequency and amount of precipitation.

Deserts cover about 30% of the earth's land, and are situated mainly between tropical and subtropical regions north and south of the equator, at about 30° north and 30° south latitude (Figure 7-11). In these areas, air that has lost its moisture over the tropics falls back toward the earth (Figure 7-6). The largest deserts are in the interiors of continents, far from moist sea air and moisture-bearing winds. Other, more local deserts form on the downwind sides of mountain ranges because of the rain shadow effect (Figure 7-10).

The baking sun warms the ground in the desert during the day. At night, however, most of this heat quickly escapes because desert soils (Figure 5-16) have little vegetation and moisture and the skies are usually clear. This explains why in a desert you may roast during the day but shiver at night.

Low rainfall combined with different average temperatures creates tropical, temperate, and cold deserts (Figures 7-12 and 7-14). In *tropical deserts*, such as the southern Sahara (Arabic for "the desert") in Africa, temperatures are usually high year-round. Average annual rainfall is less than 2 centimeters (0.8 inch), and rain typically falls during only one or two months of the year, if at all (Figure 7-14, left). Chile's Atacama tropical desert has had no measurable precipitation in over 28 years. These driest places on earth typically have few plants and a hard, windblown surface strewn with rocks and some sand.

Daytime temperatures in *temperate deserts* are hot in summer and cool in winter, and these deserts have more precipitation than tropical deserts (Figure 7-14, center). Examples are the Mojave, Sonoran, and Chihuahuan deserts, which occupy much of the American southwest and northern and western Mexico. The vegetation is sparse, consisting mostly of widely dispersed, drought-resistant shrubs and cacti or other succulents. Animals are adapted to the lack of water and temperature variations (Figure 7-15). In *cold deserts*, such as the Gobi Desert in China, winters are cold and summers are warm or hot; precipitation is low (Figure 7-14, right).

In the semiarid zones between deserts and grasslands, we find *semidesert*. This biome is dominated by thorn trees and shrubs adapted to long dry spells followed by brief, sometimes heavy rains.

How Do Desert Plants and Animals Survive? Adaptations for survival in the desert have two themes: "Beat the heat" and "Every drop of water counts." Desert stoneplants avoid predators by looking like stones, and their light coloration also reflects heat. Some

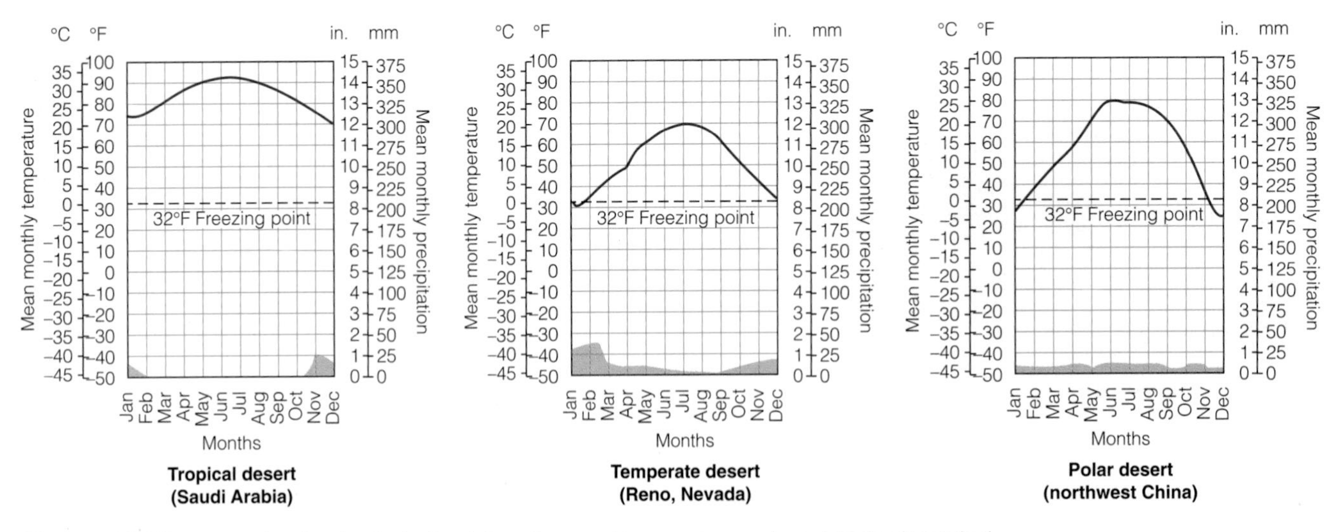

Figure 7-14 Climate graphs showing typical variations in annual temperature and precipitation in tropical, temperate, and polar (cold) deserts.

 Cohn, J. P. 1996. "The Sonoran Desert: Changing Face of the Desert Keeps Communities Dynamic." *BioScience*, vol. 46, no. 2, 84(4).

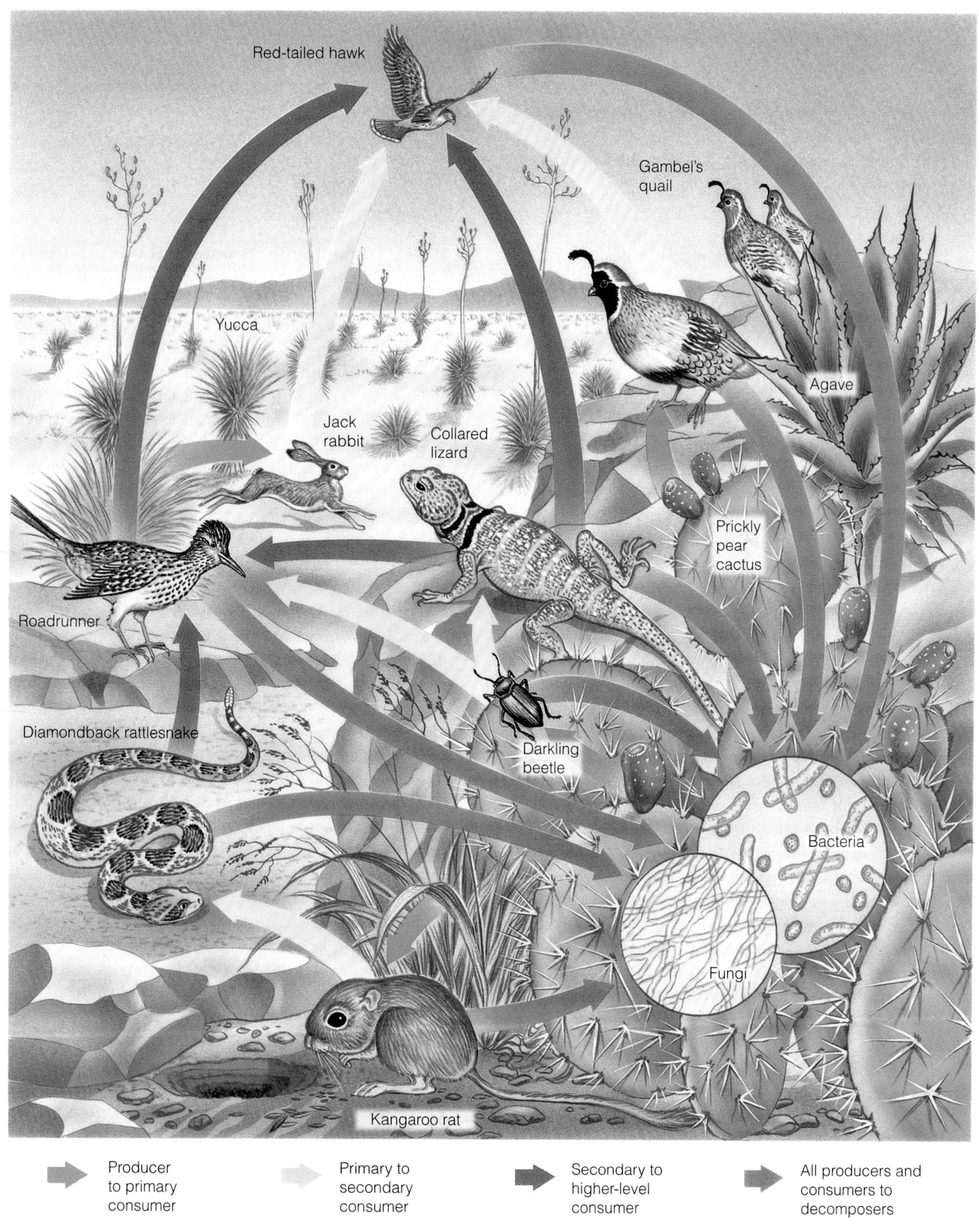

Figure 7-15 Some components and interactions in a temperate desert biome. When these organisms die, their organic matter is broken down by decomposers into minerals used by plants. Transfers of matter and energy among producers, primary consumers (herbivores), and secondary (or higher-level) consumers (carnivores) are indicated by colored arrows.

Hint: Enter the search term *desert plants* using the Subject Guide.

desert plants are evergreens with wax-coated leaves (creosote bush) that minimize transpiration. Most desert perennials tend to have small leaves (coachman's whip) or no leaves (cacti), which helps them conserve water. Perennial shrubs such as mesquite and creosote plants grow deep roots to tap into groundwater, and they drop their leaves to survive in a dormant state during long dry spells.

Other perennials such as short (prickly pear; Figure 7-15) and tall (saguaro) cacti spread their shallow roots wide to quickly collect water after brief showers; they then store it in their spongy tissues. Most of these succulents are armed with sharp spines to discourage herbivores from feeding on their water-storing fleshy tissue. The spines also reduce overheating by reflecting some sunlight and by providing shade and insulation. Some desert plants, such as the creosote bush and sagebrush, also secrete toxins in the soil. This reduces competition for water and soil nutrients from nearby plants of other species.

Many desert plants are annual wildflowers and grasses that store much of their biomass in seeds during dry periods and remain inactive (sometimes for years) until they receive enough water to germinate. Shortly after a rain, in a frenzy of biological activity, the seeds germinate, grow, carpet the desert with a dazzling array of colorful flowers, produce new seed, and die—all in only a few weeks.

Other, less visible desert plants are mosses and lichens, which can tolerate extremely high temperatures, dry out completely, and become dormant until the next rain falls. In museums, some desert moss specimens have been known to recover and grow after 250 years without water.

If you visit a desert during the daytime you may see only a few lizards, a bird or two, and some insects. However, deserts have a surprisingly large number of animal inhabitants that come out mostly in the cool of night. Most desert animals are small. They beat the heat and reduce water loss by evaporative cooling. They hide in cool burrows or rocky crevices by day and come out at night or in the early morning. Birds, ants, rodents, and other seed-eating herbivores are common, feeding on the multitudes of seeds produced by the desert's annual plants. Some deserts have a few large grazing animals such as gazelle and the endangered Arabian oryx.

Major carnivores in temperate North American deserts are coyotes, kit and gray foxes, and various species of snakes and owls that come out mainly at night to prey on the desert's many rodent species. The few daytime animals, such as fast-moving lizards and some snake species, are preyed upon mostly in the early morning and late afternoon by hawks and roadrunners.

Desert animals have physical adaptations for conserving water (Spotlight, right). Insects and reptiles have thick outer coverings to minimize water loss through evaporation. They also reduce water loss by having dry feces and by excreting a dried concentrate of urine. Some of the smallest desert animals, such as spiders and insects, get their water only from dew or from the food they eat. Some desert animals become dormant during periods of extreme heat or drought and are active only during the cooler months of the year. Arabian oryxes survive by licking the dew that accumulates at night on rocks and on one another's hair.

What Impacts Do Humans Have on Desert Ecosystems? Many people tend to view deserts as essentially useless. They don't realize that these ecosystems are important and can easily be disrupted. Deserts take a long time to recover because of their slow plant growth, low species diversity, slow nutrient cycling (because of little bacterial activity in their soils), and shortages of water. Vegetation destroyed by livestock overgrazing and off-road vehicles may take decades to grow back. For example, tracks left by tanks practicing in the California desert during World War II are still visible. Vehicles can also collapse underground burrows where many desert animals live.

Many rapidly growing, large desert cities have sprung up in countries such as Saudi Arabia and Egypt and in the southwestern United States (for example, Palm Springs, California; Las Vegas, Nevada; and Phoenix, Arizona). Increasingly, residents of such cities with four-wheel-drive vehicles and motorcycles are destroying fragile desert soil, plants, and animal burrows.

In some areas, such as parts of southern California, desert soil is rich enough in mineral nutrients to grow irrigated crops. However, when deserts are irrigated with large volumes of water, salts may accumulate in the soil as the water is evaporated. This salt buildup, called *salinization*, eventually limits crop production.

The expansion of desert cities and the growing demand for underground water (for drinking, washing, watering lawns and golf courses, and irrigating cropland) are depleting some aquifers and causing desert land over some aquifers to subside (sink).

Some desert areas are disrupted and polluted by the extraction of minerals and the buildup of nearby mining towns and cities. Most of the world's oil is extracted from desert areas in Saudi Arabia and in other Middle Eastern oil-producing countries. Africa's Sahara contains important iron ore and phosphate deposits. The deserts of Iran, North America, South America, and Australia are rich in a wide range of minerals including copper, gold, and silver. Diamonds are mined in some deserts of Namibia, Botswana, and Australia. Deserts are also sources of building materials such as road stone, lime, gypsum, and sand.

Traditionally, deserts and semideserts have been used by nomadic tribes to graze domesticated live-

Q: How much of the high-quality energy is transferred from one trophic level to another in a food chain or web?

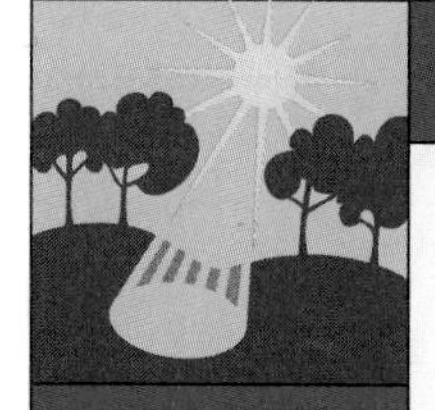

SPOTLIGHT

The Kangaroo Rat: Water Miser and Keystone Species

The kangaroo rat (Figure 7-16) is a remarkable mammal superbly adapted for conserving water in its desert environment. As the desert's chief seed eater, it is also a keystone species that helps support other desert species and helps keep desert shrubland from becoming grassland.

This rodent comes out of its burrow only at night, when the air is cool and water evaporation has slowed. It seeks dry seeds that it quickly stuffs into its cheek pouches.

After a night of foraging it empties its cache of seeds into its cool burrow, where they soak up water exhaled in the rodent's breath. When the rodent eats these seeds, it gets this water back.

The kangaroo rat does not drink water; its water comes from the recycled moisture in the seeds and from water produced when sugars in the seeds undergo aerobic respiration during digestion.

Some of the water vapor in the rat's breath also condenses on the cool inside surface of its nose. This condensed water then diffuses back to its body.

Kangaroo rats have no sweat glands, so they don't lose water by perspiration. In addition, they save water by excreting hard, dry feces and thick, nearly solid urine produced by their extremely efficient kidneys.

Figure 7-16 This nocturnal kangaroo rat of the California desert is an expert in water conservation; it is also a keystone species because it consumes such vast quantities of seeds that it slows conversion of desert shrubland to grassland. (B. & C. Calhoun/ Bruce Coleman Ltd.)

Critical Thinking

Water is scarce in much of the southwestern United States, where the kangaroo rat lives. However, this area has one of the highest rates of human population growth. As this happens, what ecological lesson can we learn from the kangaroo rat about how to survive in this area (and other water-short areas throughout the world)?

stock animals such as cattle, goats, and camels. These nomads get enough vegetation and water by moving livestock from place to place and by withdrawing water by hand from scattered deep wells. In recent years, this normally sustainable grazing system has been disrupted by using gasoline-powered pumps to withdraw large amounts of underground water for livestock. The result has been an increase in groundwater wells, an explosive increase inboth cattle and human populations, overgrazing, groundwater depletion, and disruption of the sustainable nomadic life.

The largest and most untapped resource of deserts is abundant sunlight. Over the next 40–50 years many analysts expect industrialized societies to make a transition from reliance on nonrenewable fossil and nuclear fuels to greatly increased use of renewable solar energy. If such a shift takes place, large areas of many deserts near urban areas will be covered with arrays of solar collectors and solar cells used to produce electricity and hydrogen gas (Section 16-7). Great care must be taken to do this in ways that don't seriously disrupt desert ecosystems.

Because large areas of desert are not populated and have little water, they are becoming attractive as sites for storage of toxic and radioactive wastes and for underground testing of nuclear weapons. They are also used for maneuvers by heavy tanks and other military vehicles. Learning how to use the resources of deserts sustainably with minimal harmful impact is an important challenge.

7-4 GRASSLAND, TUNDRA, AND CHAPARRAL BIOMES

What Are the Major Types of Grasslands?

A region with enough average annual precipitation to allow grass (and in some areas, a few shrubs or trees) to prosper, but with precipitation so erratic that drought and fire prevent large stands of trees from

A: 5–20% (that is, a loss of 80–95%)

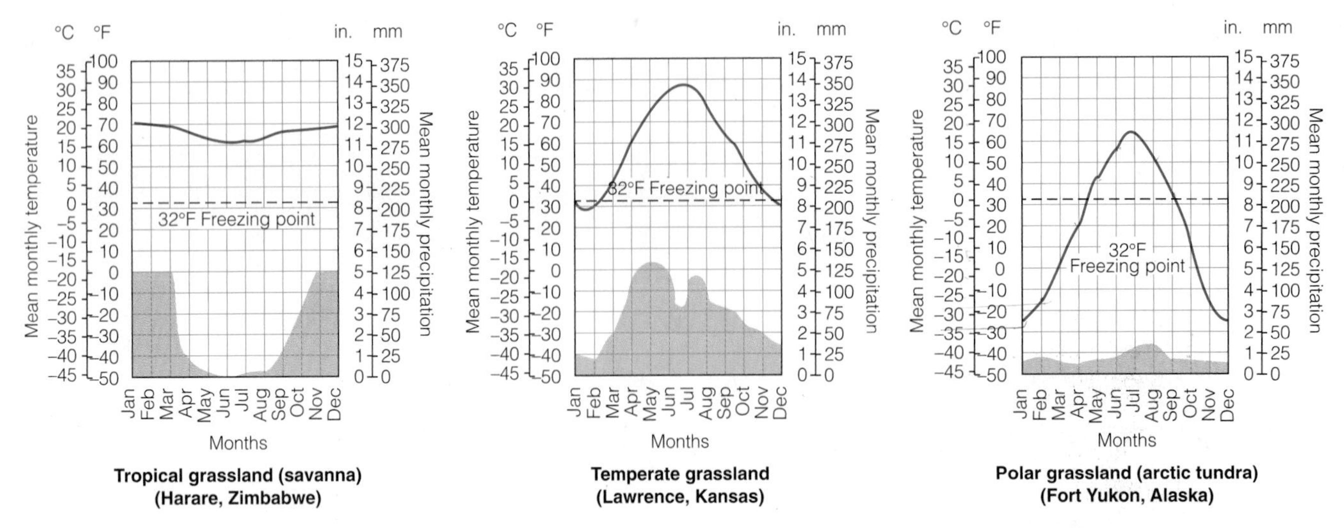

Figure 7-17 Climate graphs showing typical variations in annual temperature and precipitation in tropical, temperate, and polar (arctic tundra) grasslands.

growing, is called a **grassland**. Most grasslands are found in the interiors of continents (Figure 7-11).

Grasslands persist because of a combination of seasonal drought, grazing by large herbivores, and periodic fires, all of which keep large numbers of shrubs and trees from invading and becoming established. Grasses (many of them perennials) in these biomes are renewable resources, if not overgrazed, because grass plants grow out from the bottom; thus, their stems can grow again after being nibbled off by grazing animals. The three main types of grasslands—tropical, temperate, and polar (tundra)—result from combinations of low average precipitation and various average temperatures (Figures 7-12 and 7-17).

What Are Tropical Grasslands and Savannas? *Tropical grasslands* are found in areas with high average temperatures, low to moderate precipitation, and a prolonged dry season. They occur in a wide belt on either side of the equator beyond the borders of tropical rain forests (Figure 7-11).

Savannas are tropical and subtropical grasslands that are warm all year; they have alternating wet and dry seasons and abundant rain the rest of the year (Figure 7-17, left) but experience occasional prolonged droughts. The largest savannas occur in central and southern Africa, but they are also found in central South America, Australia, and Southeast Asia.

Savanna vegetation consists of wind-pollinated grasses, small broadleaf deciduous plants (forbs) pollinated by insects, and occasional deciduous trees, such as flat-top acacias in the African savanna (Figure 7-18). In Australian savanna (bush), the predominant tree is the eucalyptus (whose leaves are eaten by herbivores such as the koala), and the predominant grazers are native kangaroos and nonnative rabbits introduced by settlers in 1859.

African tropical savannas contain enormous herds of *grazing* (grass- and herb-eating) and *browsing* (twig- and leaf-nibbling) hoofed animals, including wildebeests, gazelles, zebras, giraffes, and antelopes. These and other large herbivores have evolved specialized eating habits that minimize interspecific competition for resources (resource partitioning): Giraffes eat leaves and shoots from the tops of trees, elephants eat leaves and branches further down, Thompson's gazelles and wildebeests prefer short grass, and zebras graze on longer grass and stems.

During the dry season many large grazing animals migrate to more productive areas. Smaller animals become dormant during the dry season or survive by eating plant seeds. The large grazing animals (and many small ones) are preyed upon by predators such as cheetahs, lions, hyenas, eagles, and hawks.

What Are Temperate Grasslands? *Temperate grasslands* cover vast expanses of plains and gently rolling hills in the interiors of North and South America, Europe, and Asia (Figure 7-11). Winters there are bitterly cold and summers are hot and dry (Figure 7-17, center). Annual precipitation averages only 25–100 centimeters (10–39 inches) and falls unevenly through the year.

Because the aboveground parts of most of the grasses die and decompose each year, organic matter accumulates to produce a deep, fertile soil (Figure 5-16). This soil is held in place by a thick network of intertwined roots of drought-tolerant perennial grasses—unless the topsoil is plowed up and allowed to blow away by prolonged exposure to high winds found in these biomes.

Types of temperate grasslands are the *tall-grass prairies* (Figure 7-19) and *short-grass prairies* of the Midwestern and Western United States and Canada, the

Frank, Douglas A., et al. 1998. "The Ecology of the Earth's Grazing Ecosystems." *BioScience*, vol. 48, no. 7, 513(10).

Figure 7-18 Serengeti tropical savanna in Tanzania, Africa, an example of one type of a tropical grassland biome. Most savannas consist of grasslands punctuated by stands of deciduous shrubs and trees, which shed their leaves during the dry season and thus avoid excessive water loss. More large, hoofed, plant-eating mammals (ungulates), such as the wildebeest shown here, live in this biome than anywhere else. The more than 1 million wildebeests (and many other large herbivores) living in the Serengeti biome undergo spectacular migrations to find enough water and high-quality grasses as the seasons change. (Jonathan Scott/Planet Earth Pictures)

South American *pampas*, the African *veldt*, and the *steppes* of central Europe and Asia. Here winds blow almost continuously and evaporation is rapid, often leading to fires in the summer and fall. Before the plow and the cow arrived, the tall-grass prairie predominated in the humid eastern portion of the Great North American Prairie. The short-grass prairie predominated in the arid west, with mixed-grass prairie between. In the past, period wildfires kept shrubs and trees expanding into the prairie, especially at the ecotone between forest and prairie. Indigenous people used fire as a management tool to maintain and expand the prairie.

Huge numbers and varieties of beetles, spiders, grasshoppers, and other insects and invertebrate animals live among the plants in North American temperate grasslands. Ants and earthworms are abundant in the soil. Primary consumers include a variety of small animals such as prairie dogs, prairie voles, deer mice, jackrabbits, squirrels, and meadowlarks. Most of these animals live in burrows or escape predators by being able to run or hop swiftly. These small animals are preyed upon by carnivores such as coyotes, bobcats, kit foxes, ferrets, snakes, and hawks. Wolves and pumas once preyed upon these larger animals, but most have now been killed or driven out by farmers, ranchers, and hunters.

What Is Arctic Tundra? *Polar grasslands*, or *arctic tundra*, covering about 10% of the earth's land area, occur just south of the Arctic polar ice cap (Figure 7-11). During most of the year these treeless plains are bitterly cold, swept by frigid winds, and covered with ice and snow (Figure 7-17, right). Winters are long and dark, and the scant precipitation falls mostly as snow.

This biome is carpeted with a thick, spongy mat of low-growing plants, primarily grasses, mosses, lichens, and dwarf woody shrubs (Figure 7-20). These hardy pioneer plants are adapted to the lack of sunlight and water, to freezing temperatures, and to constant high winds. Most of the annual growth of these plants occurs during the 6- to 8-week summer, when sunlight shines almost around the clock.

To retain water and survive the winter cold, most tundra plants grow close to the ground, and some have leathery evergreen leaves coated by waxes that reduce heat loss. Other plants survive the long, cold winter underground as roots, stems, bulbs, and tubers; some, such as lichens, dehydrate during winter to avoid frost damage.

One effect of the extreme cold is **permafrost**, a perennially frozen layer of the soil that forms when the water there freezes. In summer, water near the surface thaws, but the permafrost soil layer below stays frozen and prevents liquid water at the surface from seeping into the ground. Thus, during the brief summer the soil above the permafrost layer remains waterlogged, forming a large number of shallow lakes, marshes, bogs, ponds, and other seasonal wetlands. Hordes of mosquitoes, blackflies, and other insects thrive in these shallow surface pools. They feed large colonies of migratory birds, especially waterfowl, that return from the south to nest and breed in the bogs and ponds. In North American tundra, caribou herds also arrive to feed on the summer vegetation, bringing with them their wolf predators. In arctic tundra found in Europe and Asia, reindeer occupy the grazing niche of caribou.

The arctic tundra's permanent animal residents are mostly small herbivores such as lemmings, hares, voles, and ground squirrels, which burrow underground to escape the cold. They are eaten by predators such as the lynx, weasel, snowy owl, and arctic fox. Most tundra animals do not hibernate because the summer is too short for them to accumulate adequate fat reserves. Animals in this biome survive the intense winter cold through adaptations such as thick coats of fur (arctic wolf, arctic fox, and musk oxen), feathers (snowy owl), compact bodies to expose as little surface as possible to the air, and living underground (arctic lemming).

Because of the cold, decomposition is slow; partially decomposed organic matter forms soggy peat bogs (the source of gardener's peat moss), which contain about 95% of this biome's carbon. The soil is poor in organic matter and in nitrates, phosphates, and other minerals that decomposers liberate in richer soils.

Hint: Enter the search terms *biotic communities, research* using the Subject Guide.

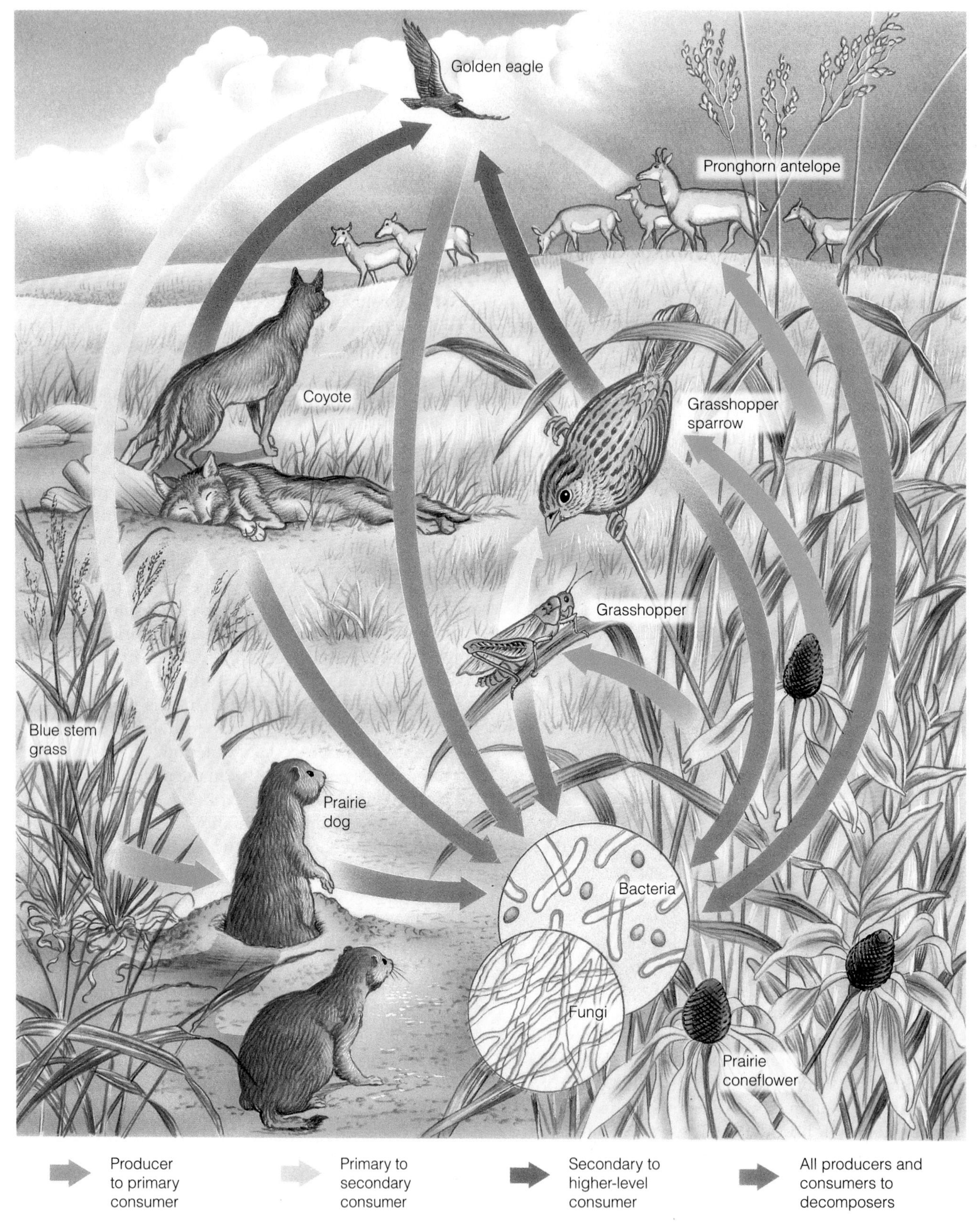

Figure 7-19 Some components and interactions in a temperate tall-grass prairie ecosystem in North America. When these organisms die, their organic matter is broken down by decomposers into minerals used by plants. Transfers of matter and energy among producers, primary consumers (herbivores), and secondary (or higher-level) consumers (carnivores) are indicated by colored arrows.

Q: What are the three most productive types of ecosystems?

Figure 7-20 Some components and interactions in an arctic tundra (polar grassland) ecosystem. When these organisms die, their organic matter is broken down by decomposers into minerals used by plants. Transfers of matter and energy among producers, primary consumers (herbivores), and secondary (or higher-level) consumers (carnivores) are indicated by colored arrows.

A: Estuaries, swamps and marshes, and tropical rain forests

What Is Alpine Tundra? Another type of tundra, called *alpine tundra*, occurs above the limit of tree growth but below the permanent snow line on high mountains (Figure 7-13). The vegetation there is similar to that found in arctic tundra, but it gets more sunlight than arctic vegetation and has no permafrost layer. For a few weeks each summer the land blazes with color as wildflowers burst into bloom. The small plants that survive in this biome are grazed by herbivores such as elk and mountain goats and sheep, while golden eagles soar above looking for marmots and ground squirrels.

What Impacts Do Humans Have on Grassland Ecosystems? Today about 50 million people, mostly in parts of Africa and Asia, survive by raising livestock on tropical and temperate grasslands. Traditionally, most of these nomadic pastoralists move their livestock each year as the seasons change, similar to the migrations of wild grazing animals living on grasslands. When governments and aid agencies encourage the drilling of wells, the usually sustainable balance between the pastoralists' livestock and grasses is upset. The number of livestock increases and the pastures around wells are overgrazed, trampled by thousands of hooves, and converted into less productive desert and semidesert.

Because of its fairly high net primary productivity (Figure 4-24), some areas of savanna are being converted into cropland. Much of the carbon removed from the atmosphere by savanna producers is locked up in the soil, dead plant matter, and roots and underground stems. Thus, deliberately burning savanna, plowing up its grasses and roots, and converting it into cropland releases large quantities of carbon dioxide into the atmosphere. This practice may contribute to the greenhouse effect as much as (if not more than) the more publicized clearing and burning of tropical rain forests.

Because of their fertile soils and ability to support cereal crops (which are domesticated grasses), many temperate grasslands in North America, western Europe, and Ukraine have been cleared of most of their primarily perennial native grasses. They are used instead for growing annual crops that must be replanted each year (Figure 7-21). These grasslands, sometimes called the world's breadbaskets, also directly or indirectly supply most of the world's meat. Large herds of bison and pronghorn antelope, which used to roam over the temperate grasslands of North America, have been replaced by cattle and sheep that feed on grasses. Cattle, hogs, and chickens fatten on corn grown on former grasslands.

As long as temperate grasslands keep their fertile soil and the climate does not change, they can continue producing much of the world's cereal grains. However, poor farming practices, overgrazing, mismanagement, and occasional prolonged droughts lead to severe wind erosion and loss of topsoil, which can convert temperate grasslands into desert or semidesert shrubland. For example, the mesquite- and cactus-covered desert and semidesert in Texas were productive grasslands before they were overgrazed by large herds of cattle.

The arctic tundra is a fragile environment, easily damaged by human actions such as mining and oil and gas development. Its low rate of decomposition, shallow soil, short growing season, and slow plant growth rate make this biome especially vulnerable to disruption; vegetation destroyed by human activities can take decades to grow back. Damage by spills of oil or toxic waste may take far longer.

Because of its inhospitable climate, few humans have invaded the arctic tundra. However, oil exploration and drilling in the tundra of Alaska and especially in Siberia have exposed these fragile ecosystems to air pollution, oil spills, toxic wastes, and disruption of soil and vegetation by vehicles.

What Is Chaparral? Some temperate areas have a biome known as *temperate shrubland* or *chaparral* (Figure 7-11). This biome occurs along coastal areas with what is called a *Mediterranean climate*: winters are mild and moderately rainy and summers are long, hot, and dry. It is found mainly along parts of the Pacific coast of North America and in the coastal hills of Chile, the Mediterranean, southwestern Africa, and southwestern Australia.

This biome is usually dominated by an almost impenetrably dense growth of drought-resistant, spiny evergreen shrubs (with leathery leaves that resist water loss) and that have large underground root systems. Many of these shrubs produce compounds that leach into the soil and inhibit the germination and growth of competing plant species. Some areas also have a sprinkling of small, drought-resistant trees such as pines and scrub oak.

During the long, hot dry season chaparral vegetation is dormant and becomes very dry and brittle. In the fall, fires started by lightning or by human activities spread with incredible swiftness through the dry brush and litter of leaves and fallen branches.

Research reveals that chaparral is adapted to and maintained by periodic fires. Many of the shrubs store food reserves in their fire-resistant root crown and have seeds that sprout only after a hot fire. With the first rain, annual grasses and wildflowers spring up and use nutrients released by the fire. New shrubs grow quickly and crowd out the grasses. Within a decade or two after a fire the natural chaparral community is restored.

People like living in this biome because of its favorable climate, but those living in chaparral assume the high risk of losing their homes (and possibly their lives) to the frequent fires associated with it. After fires often comes the hazard of flooding; when heavy rains come, great torrents of water pour off the unprotected burned hillsides to flood lowland areas.

Q: What percentage of the world's net primary productivity on land is used, wasted, or destroyed by the human population?

Figure 7-21 Replacement of a temperate grassland with a monoculture crop near Blythe, California. When the tangled root network of natural grasses is removed, the fertile topsoil is subject to severe wind erosion unless it is covered with some type of vegetation. If global warming accelerates over the next 50 years, many of these grasslands may become too hot and dry for farming, thus threatening the world's food supply. (National Archives/EPA Documerica)

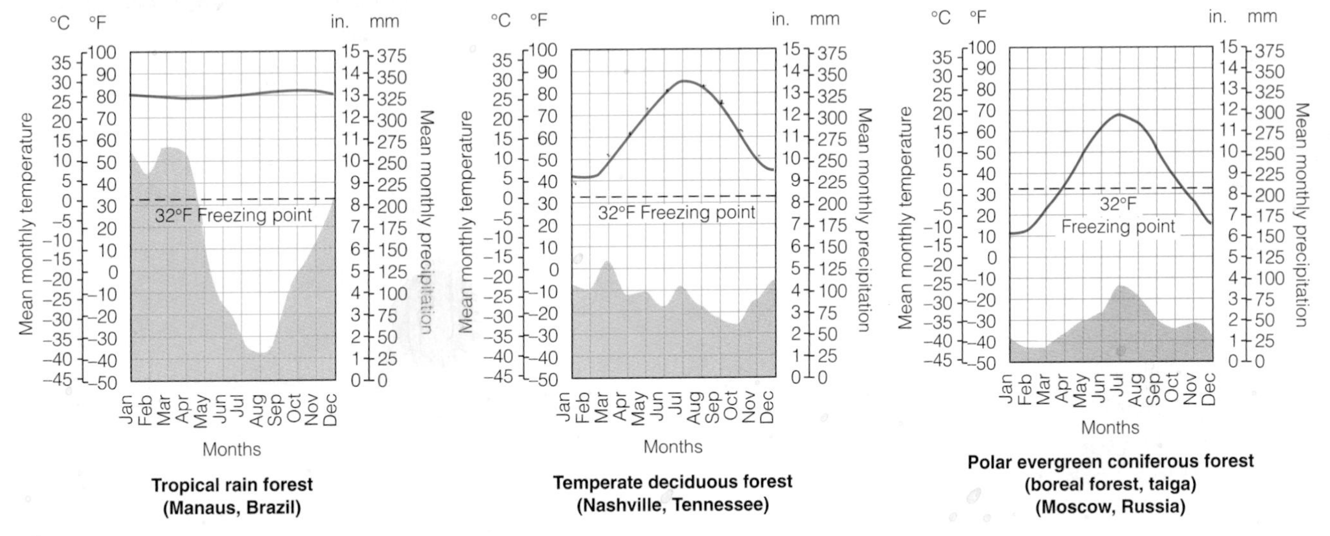

Figure 7-22 Climate graphs showing typical variations in annual temperature and precipitation in tropical, temperate, and polar forests.

7-5 FOREST BIOMES

What Are the Major Types of Forests? Undisturbed areas with moderate to high average annual precipitation tend to be covered with **forest**, which contains various species of trees and smaller forms of vegetation. The three main types of forest—*tropical*, *temperate*, and *boreal* (polar)—result from combinations of this precipitation level and various average temperatures (Figures 7-12 and 7-22).

What Are Tropical Rain Forests? *Tropical rain forests* are a type of broadleaf evergreen forest (Figure 7-23) found near the equator (Figure 7-11), where hot, moisture-laden air rises and dumps its moisture (Figure 7-6). The world's largest tropical rain forest is in the Amazon River basin in South America.

Tropical rain forests have a warm annual mean temperature (which varies little, daily or seasonally), high humidity, and heavy rainfall almost daily (Figure 7-22, left). Typically, this biome receives 250–450 centimeters (100–180 inches) of rainfall per year. The consistently wet and warm climate means that water and temperature are not the main limiting factors, as in other biomes; instead, soil nutrients are the main limiting factors.

There are several types of tropical rain forests, depending on variations in altitude, temperature, and average rainfall. What they all have in common is their incredible biological diversity. Unlike what we find in other forest biomes, in most tropical forests no one or two tree species dominate. Often, two individuals of the same tree species may be hundreds of meters apart.

The struggle for sunlight leads to several levels of plant height above the floor of a tropical rain forest (Figure 7-24). In the upper story, or *emergent layer*, trees up to 60 meters (200 feet) tall rise above the surrounding vegetation and capture direct sunlight. Beneath the emergent layer is a continuous *canopy* of trees whose dense leaves and branches overlap so that only dim light penetrates to the third layer: the *understory* of small trees adapted to low light. Beneath this is a *shrub layer* of shrubs and short plants adapted to even less light. The dark and wet *forest floor* is open and free of vegetation. The popular image of a rain forest floor as a tangled, almost impenetrable jungle is accurate only along river banks,

A: About 40% (27% including terrestrial and aquatic productivity)

Figure 7-23 Some components and interactions in a tropical rain forest ecosystem. When these organisms die, their organic matter is broken down by decomposers into minerals used by plants. Transfers of matter and energy among producers, primary consumers (herbivores), and secondary (or higher-level) consumers (carnivores) are indicated by colored arrows.

Q: What are the three major ways ecologists learn about ecosystems?

Figure 7-24 Stratification of specialized plant and animal niches in various layers of a tropical rain forest. The presence of these specialized niches enables species to avoid or minimize competition for resources and results in the coexistence of a great variety of species (biodiversity).

near clearings, or where a large tree has fallen and sunlight reaches the ground.

Many plants have evolved specialized ways to grow in tropical rain forests. Climbing vines, called *lianas*, most rooted in the soil (Figure 5-16), wind upward around the trunks of larger trees until their leaves reach the sunlit canopy. Orchids, bromeliads, and other *epiphytes* attach themselves to the trunks and branches of canopy trees. The highly perched epiphytes obtain nutrients from bits of organic matter falling from the canopy. Many plants dwelling in the pale light of the understory and shrub layer (such as philodendrons) survive by using huge, dark green leaves to capture enough sunlight. This ability to thrive under low indoor light levels makes them good house plants.

The roots of even the largest trees tend to be shallow and spread out in the nutrient-poor, moist, and thin layer of soil (Figure 5-16). Many of the large trees are supported by large bulges at their bases called *buttresses*. Virtually all plant roots have a mutualistic relationship with mycorrhizae (fungi) that quickly transfer nutrients from decomposed leaves and other organic debris back to the trees.

The stratification of specialized plant and animal niches in various layers of a tropical rain forest enables coexistence of a great variety of species (biodiversity). Although tropical rain forests cover only about 2% of the earth's land surface, they are habitats for 50–80% of the earth's terrestrial species.

The complex vegetation in tropical rain forests provides niches for a large variety of animals. Much of the animal life, particularly insects, bats, and birds, lives in the sunny canopy layer, with its abundant shelter and supplies of leaves, flowers, and fruits (Figure 7-23). Ecologists climb trees and build platforms and boardwalks to study life in the canopy.

Multitudes of tiny animals live on the ground, alongside vast populations of termites and decomposers. Dropped leaves, fallen trees, and dead animals are decomposed quickly because of the warm, moist conditions and hordes of decomposers. This rapid recycling of scarce soil nutrients is why there is usually little litter on the ground (Figure 5-16). Instead of being stored in the soil, most minerals released by decomposition are quickly taken up by plants. Thus, most of a tropical rain forest's nutrients are stored in the biomass of its living organisms.

Because of the dense vegetation, little wind blows in tropical rain forests, eliminating the possibility of wind pollination. Many of the plants have evolved elaborate flowers (Figure 1-12, left) that attract particular insects, birds, or bats as pollinators.

What Are Tropical Deciduous Forests? Moving a little further from the equator, we find *tropical deciduous forests* (sometimes called *tropical monsoon forests* or *tropical seasonal forests*). These are usually located between tropical rain forests and tropical savannas. These forests are warm year-round, and most of their plentiful rainfall occurs during a wet (monsoon) season that is followed by a long dry season.

Tropical deciduous forests have a lower canopy than tropical rain forests. They contain a mixture of deciduous trees (which lose their leaves to survive the dry season) and drought-tolerant evergreen trees (which retain most of their leaves year-round). Where the dry season is especially long, we find *tropical scrub forests* (Figure 7-11) containing mostly small deciduous trees and shrubs.

What Are Temperate Deciduous Forests? *Temperate deciduous forests* (Figure 7-25) grow in areas with moderate average temperatures that change significantly with the season (Figure 7-12). These areas have long and warm summers, cold but not too severe winters, and abundant precipitation, often spread fairly evenly throughout the year (Figure 7-22, center). This biome is dominated by a few species of broadleaf deciduous trees, such as oak, hickory, maple, beech, poplar, and sycamore. They survive cold winters by dropping their leaves in the fall and becoming dormant (see photo in table of contents). Each spring they grow new leaves that change in the fall into a blazing array of reds and golds (see photo in table of contents) before dropping.

Compared with tropical rain forests, temperate deciduous forests have a simpler structure and contain few tree species. However, the penetration of more sunlight supports a richer diversity of plant life at ground level. Because of the fairly low rate of decomposition, these forests accumulate a thick layer of slowly decaying leaf litter that is a storehouse of nutrients.

The temperate deciduous forests of the eastern United States were once home for such large predators as bears, wolves, foxes, wildcats, and mountain lions (pumas). Today most of the predators have been killed or displaced, and the dominant mammal species is often the white-tailed deer, along with smaller mammals such as squirrels, rabbits, opossums, raccoons, and mice.

Warblers, robins, and other bird species migrate to the forests during the summer to feed and breed. However, many of these species are declining in numbers because of loss or fragmentation of their summer and winter habitats, which also makes them more vulnerable to predators and parasitic cowbirds. Small mammals and birds are preyed upon by owls and hawks (and in remote areas) by bobcats and foxes.

What Are Evergreen Coniferous Forests? *Evergreen coniferous forests*, also called *boreal forests* (meaning "northern forests") and *taigas* (pronounced "TIE-guhs"), are found just south of the arctic tundra in northern regions across North America, Asia, and Europe (Figure 7-11). They cover approximately 11% of the earth's land. In this subarctic climate, winters are long, dry, and extremely cold; in the northernmost boreal forest, sunlight is available only 6–8 hours a day. Summers are short, with mild to warm temperatures (Figure 7-22, right), and the sun typically shines 19 hours a day.

Most boreal forests are dominated by a few species of evergreen conifer trees, such as spruce, fir, cedar, hemlock, and pine. The small, needle-shaped, waxy-coated leaves of these trees can withstand the intense cold and drought of winter when snow blankets the ground. Plant diversity is low in these forests because few species can survive the winters when soil moisture is frozen.

Beneath the stands of trees, there is a deep layer of partially decomposed conifer needles and leaf litter. Decomposition is slow because of the low temperatures, the waxy coating of conifer needles, and the high acidity. As the conifer needles decompose, they make the thin, nutrient-poor soil (Figure 5-16) acidic and prevent most other plants (except certain shrubs) from growing on the forest floor. Because there is often little understory, strolls on the soft carpet of needles in these forests can be quite pleasant.

These biomes contain a variety of wildlife (Figure 7-26). The animal populations of these forests consist mostly of seed eaters (such as squirrels and nutcrackers), insect herbivores, and larger browsers (such as elk and moose). Predators include wolves, wolverines, grizzly bears, and black bears. In North America, caribou migrate from the tundra to this biome in winter.

During the brief summer the soil becomes waterlogged, forming acidic bogs, or *muskegs*, in low-lying areas of these forests. Warblers and other insect-eating birds feed on hordes of flies, mosquitoes, and caterpillars.

Q: What is the ecological niche of a species?

Figure 7-25 Some components and interactions in a temperate deciduous forest ecosystem. When these organisms die, their organic matter is broken down by decomposers into minerals used by plants. Transfers of matter and energy among producers, primary consumers (herbivores), and secondary (or higher-level) consumers (carnivores) are indicated by colored arrows.

A: Its total way of life or its functional role in an ecosystem; everything it does to survive and reproduce

Figure 7-26 Some components and interactions in an evergreen coniferous (boreal or taiga) ecosystem. When these organisms die, their organic matter is broken down by decomposers into minerals used by plants. Transfers of matter and energy among producers, primary consumers (herbivores), and secondary (or higher-level) consumers (carnivores) are indicated by colored arrows.

What Are Temperate Rain Forests? In scattered coastal temperate areas with ample rainfall or moisture from dense ocean fogs, we find *temperate rain forests* or *coastal coniferous forests*. Along the coast of North America, from Canada to northern California, these biomes are dominated by dense stands of large conifers such as Sitka spruce, Douglas fir, and magnificent redwoods. Nearness to the ocean moderates the temperature; winters are mild and summers cool. The trees in these moist forests depend on frequent rains and moisture from summer fog that rolls in off the Pacific.

What Impacts Do Humans Have on Forest Ecosystems? *Tropical rain forests* can sustain themselves indefinitely, as long as the climate does not change and they are not cleared or seriously degraded. Unfortunately, as we will see in Chapter 24, these storehouses of biodiversity are being cleared or degraded at an alarming rate. More than half of the world's people live in the tropics and subtropics, and their numbers and demands for food, wood, and mineral resources from tropical rain forests are growing rapidly. Within 50 years only scattered fragments of these diverse forests might remain, causing a massive irreversible loss of the earth's vital biodiversity within your lifetime. Many *tropical deciduous forests* are also being cleared for timber, grazing land, and agriculture, subjecting them to erosion that, in turn, leads to desertification.

Because of their nutrient-rich soils (Figure 5-16) and favorable climate, temperate deciduous forests in Europe, Asia, and North America were among the first biomes to be cleared and converted to cropland. All but about 0.1% of the original stands of temperate deciduous forests in North America have been cleared for farms, orchards, timber, and urban development. Some are regrowing and others have been converted to managed *tree farms* or *tree plantations*, where a single species is grown for timber, pulpwood, or Christmas trees (see photo in table of contents).

In settled areas of temperate rain forests in North America, farmers, ranchers, and hunters have virtually eliminated large predators, such as timber wolves, which can prey on livestock and large game animals. Loggers have removed the trees from large areas of such forests in North America, and many of these remaining ancient forests may soon be cut.

Most of the boreal forests that once covered Finland and Sweden have been cut, and some have been replaced with even-aged tree plantations. Within a decade the vast boreal forests of Siberia and Russia may also disappear. Boreal forests in Canada are also being cleared rapidly, mostly for export to Japan to make paper. Because trees grow slowly in the cold northern climate, these forests can take a long time to recover from disturbance.

Large-scale mining for peat, iron ore, gold, diamonds, and other minerals in boreal forests (especially in Russia) has disrupted and polluted parts of these forests. Boreal forests, especially in Canada and Russia, have also suffered from air pollution and acid deposition (Section 18-3) from ore-smelting plants, pulp and paper mills, and other industrial plants. The acid soils of boreal forests cannot neutralize acid deposition. Huge areas of boreal forest have also been drowned in water by vast hydroelectric schemes such as the Angara River project in Siberia and the James Bay project in Canada (Section 13-4).

Because they contain stands of large, commercially valuable trees, many of the world's ancient temperate rain forests are being clear-cut for their lumber, further depleting biodiversity. There is an intense struggle in the United States between logging companies and environmentalists: Should the few remaining old-growth temperate rain forests and coniferous evergreen forests in the Pacific Northwest (many of them on public lands) be clear-cut for timber, or should they be preserved as important oases of biodiversity (as discussed in Section 24-5)?

7-6 MOUNTAIN BIOMES

Why Are Mountains Ecologically Important? Some of the world's most spectacular and important environments are mountains, which make up about 20% of the earth's land surface. What makes mountains such a varied and distinct biome is that dramatic changes in altitude, climate, soil, and vegetation take place over a very short distance (Figure 7-13). It is estimated that each 100-meter (300-foot) gain in elevation on a mountainside is roughly equivalent to a 100-kilometer (62-mile) change in latitude.

Above a certain altitude, known as the *snow line*, temperatures are so cold that the mountain is almost permanently covered by snow and ice (except in places too steep for snow to cling to). Because each mountain has its own set of microclimate and other conditions, changes in altitude don't always exactly reflect how vegetation changes with latitude.

The effect of gravity on the soil, snow, and rock found on the steep inclines of mountains also causes dramatic changes in mountain soils and vegetation. Mountain soils are especially prone to erosion when the vegetation holding them in place is removed by natural disturbances, such as landslides and avalanches, or human activities. Compared to less productive lowland ecosystems, mountain ecosystems also usually take much longer (often hundreds of years) to recuperate from heavy losses of vegetation and soil.

Many freestanding mountains are islands of biodiversity surrounded by a sea of lower-elevation landscapes transformed by human activities. As a result, many mountain areas contain species found nowhere

A: Interspecific competition, predation, parasitism, mutualism, and commensalism else on earth; they are also sanctuaries for animal species driven from lowland areas.

The patterns of life we find on mountainsides are also affected by variables such as sunlight and precipitation. Because it receives more sunlight and precipitation, one side of a mountain may have different plant and animal life from the other side (Figure 7-10).

Many people take mountain ranges for granted, not recognizing the important roles they play in the ecosphere. The ice and snow of mountaintops help regulate the earth's climate by reflecting solar radiation back into space. Sea levels depend on the melting of glacial ice, most of which is locked up in Antarctica, the most mountainous of all continents.

Mountain regions also contain the majority of the world's forests. Forests harbor much of the earth's biodiversity, and their vegetation helps moderate global climate by absorbing much of the greenhouse gas carbon dioxide found in the atmosphere. Mountains also play a critical role in the hydrologic cycle (Figure 5-4). They gradually release melting ice, snow, and water stored in the soils and vegetation of mountainsides to small streams. These streams flow into larger streams (rivers) that in turn empty into the ocean to begin the cycle again.

What Impacts Do Humans Have on Mountain Ecosystems? Mountains are home to at least 570 million people—10% of the world's population. An additional 2 billion people depend on mountains for much of their food, fresh water, hydroelectricity, timber, and mineral resources. For example, about 1 billion Bangladeshis, Indians, and Chinese depend directly on water flowing from the Himalayas. Streams originating in the Sierra Nevada supply water for most of California's 31 million people.

Many of the world's increasingly endangered tribal and indigenous people live in mountain ecosystems. Despite their ecological, economic, and cultural importance, the fate of mountain ecosystems has not been a high priority of governments or many environmental organizations.

Mountain ecosystems are coming under increasing environmental pressure from several major trends:

- *Rapidly increasing population, especially in developing countries.* This is forcing landless poor people, refugees, and minority populations to migrate uphill and try to survive on less stable soils. These newcomers often use mountain soils and forests unsustainably in a desperate struggle to survive or because of a lack of knowledge about how to grow food, raise livestock, and harvest wood in these habitats.
- *Increased commercial extraction of timber and mineral resources.* This is often done with little regard for the resulting environmental consequences such as excessive soil erosion, flooding in the valleys below, loss of wildlife habitat, and air and water pollution from mining activities.
- *A growing number of hydroelectric dams and reservoirs.* Rivers in mountains are attractive sites for dams and reservoirs because the elevation and slope of mountains increase the force of flowing water. These reservoirs flood mountain slopes, and the dams alter the types and abundance of species in rivers.
- *Environmental degradation from the global boom in skiing, trekking, and other forms of recreation and tourism in mountainous areas.* In 1993, U.S. ski resorts hosted over 50 million skiers, and populations of many small ski towns have doubled since 1980. About 100 million people per year visit the European Alps; by 1996 these mountainous areas contained 500 golf courses, which require fertilizers and pesticides that can run off and harm vegetation and fish downstream.
- *Increased air pollution from growing urban and industrial centers and increased use of automobiles.* Trees and other vegetation at high elevations (especially conifers such as spruce) are bathed year-round in air pollutants such as ozone and acidic compounds, carried there by prevailing winds from cities, factories, and coal-burning power plants. A growing number of high-elevation forests in parts of the United States and Europe have suffered severe damage from air pollutants.
- *Changes in climate and levels of UV radiation brought about by human activities.* If global warming occurs as projected during the next century, many species of mountain plants could be displaced by opportunist lowland species that have higher rates of growth and reproduction. Similarly, any increase in UV radiation brought about by depletion of the ozone layer may have a pronounced effect on mountain life, which is already exposed to high levels of ultraviolet radiation because of altitude.
- *Increased warfare in mountainous areas.* In 1993 some 22 of 28 major armed conflicts took place primarily in mountains.

7-7 NEW PERSPECTIVES ON GEOGRAPHICAL ECOLOGY

What Lessons Can Be Learned from a Geographical Perspective of Ecology? In this chapter we examined the relationship between weather, climate, and the distribution of the earth's biomes. Weather results from the uneven heating of the troposphere from beneath, which over time leads to different climates according to geographic location. The geographic distribution of climate, in turn, leads to the geographic distribution of biomes.

Q: What is ecological succession?

The most general lesson is that *everything is connected.* A general climate map for the earth (Figure 7-2) can be drawn based on broad patterns of temperature and precipitation. The distribution of temperature and precipitation results from the interplay between incoming solar energy and the geometry of the earth's spin and orbit (Figure 7-3), which in turn leads to large-scale patterns of air circulation (Figure 7-4) and ocean currents (Figure 7-2).

A general biome map for the earth (Figure 7-11) can be drawn that largely follows the general climate map. This is because temperature and precipitation tolerances are major factors in determining the ecological niches of biological organisms. Because of similar climate, species within biomes in very different parts of the world have similar ecological niches and display similar adaptations.

Understanding the general characteristics of each biome leads to a global view, in which one gains a general understanding of the range of biodiversity on the earth and how it is distributed. All of the biomes play a role in global cycles, such as the carbon cycle (Figure 5-5). Although the greatest concentration of biodiversity lies in the tropics, all of the biomes have their own unique sets of species that must be understood and protected. And all biomes are connected.

Why Does a Geographical Perspective Help Us Understand Ecological Processes? Ecology involves understanding connections in time and space. Maps are an excellent way of representing complex information and understanding complex relationships.

The challenges for geographical ecology are many. To solve environmental problems we need accurate maps of species distributions, changes in climate, and changes in species distribution. We need accurate maps to understand how human activities in one part of the earth relate to changes in other parts of the earth. A geographical perspective means a deeper ecological understanding of connections in time and space.

When we try to pick out anything by itself, we find it hitched to everything else in the universe.

JOHN MUIR

CRITICAL THINKING

1. List a limiting factor for each of the following ecosystems: **(a)** a desert; **(b)** the arctic tundra; **(c)** the alpine tundra; **(d)** the floor of a tropical rain forest; **(e)** a temperate deciduous forest

2. Why do deserts and arctic tundra support a much smaller biomass of animals than do tropical forests?

3. What type of biome do you live in or near? What are its major threats?

4. Suppose that global warming shifts the climate of the temperate grasslands (which now provide most of the world's food) northward to temperate evergreen forest biomes. If we were to clear these forests and plant wheat, explain why wheat might not grow despite the favorable climate.

5. Why do you think there are no amphibians and reptiles in arctic tundra?

6. Why might **(a)** the microclimate of a north-facing slope differ from that of a south-facing slope, **(b)** a ridgetop differ from that of a valley bottom, and **(c)** an isolated mountaintop differ from a mountaintop surrounded by other mountains?

7. How would you respond to someone who claims that it is not important to protect areas of temperate and polar biomes because most of the world's biodiversity is in the tropics?

8. Drilling of wells in desert areas has allowed many traditional nomadic tribes to raise more livestock by not having to migrate. Is this desirable or undesirable from **(a)** an ecological standpoint, **(b)** an economic standpoint, and **(c)** a cultural standpoint? Explain.

9. How might technologies such as remote sensing and geographical information systems (GIS) (Figure 4-26) help our understanding of ecology and our environment? Both remote sensing and GIS involve expensive technology. Do you think it is worth investing in such technologies to help solve environmental problems, or are there better ways we could spend our money? Explain.

PROJECTS

1. How has the climate changed in the area where you live during the past 50 years? Investigate the beneficial and harmful effects of these changes. How have these changes benefited or harmed you personally?

2. What effects have human activities over the past 50 years had on the characteristic vegetation and animal life normally found in the biome you live in? How is your own lifestyle affecting this biome?

3. Visit the ecotone between the biome in which you live and one of the adjacent biomes. What does the ecotone look like and how broad is it? What factors lead to the presence of one biome versus the other?

4. Go to the library and peruse the section devoted to geography and cartography. Locate maps of vegetation and climate. How were these maps produced? How do the ways in which they were produced differ?

5. Make a concept map of this chapter's major ideas, using the section heads and subheads and the key terms (in boldface). Look at the inside of the front cover and in Appendix 4 for information about concept maps.

A: Change in plant types (and thus other species) in response to changing environmental conditions

8 AQUATIC ECOLOGY*

Why Should We Care About Coral Reefs?

In the shallow coastal zones of warm tropical and subtropical oceans we often find **coral reefs** (Figure 8-1). These beautiful natural wonders are among the world's oldest and most diverse and productive ecosystems and are home for one-fourth of all marine species.

Coral reefs are formed by massive colonies of tiny animals called *polyps* that are close relatives of jellyfish. They slowly build reefs by secreting a protective crust of limestone (calcium carbonate) around their soft bodies. When they die, their empty crusts or outer skeletons remain as a platform for more reef growth. These coral builders work slowly, taking five years or more to create 2.5 centimeters (1 inch) of coral. The resulting tangled maze of cracks, crevices, and caves formed by different types of coral provides shelter and niches for a huge variety of marine plants and animals, including many species of colorful reef fish, sea anemones, starfish, and algae.

Coral reefs are actually a joint venture between the polyps and tiny single-celled algae called *zooxanthellae* that live in the tissues of the polyps. In this *mutualistic relationship*, the microscopic photosynthesizing algae provide the polyp with food and oxygen. The polyps in turn provide a well-protected home for the algae and make nutrients such as nitrogen and

***Paul M. Rich**, associate professor of Ecology and Evolutionary Biology and Environmental Studies at the University of Kansas, is coauthor of this chapter.

Figure 8-1 A healthy coral reef in the Philippines covered by colorful algae (left) and a bleached coral reef in the Bahamas that has lost most of its algae (right) because of changes in the environment (such as cloudy water or extreme temperatures). With the algae gone, the white limestone of the coral skeleton becomes visible. If the environmental stress is not removed, the corals die. These diverse and productive ecosystems are being damaged and destroyed at an alarming rate. (Left, Karl & Jill Wallin/FPG International; right, Robert Wicklund)

phosphorus available that are usually scarce in tropical waters. The algae and other producers give corals most of their bright colors and provide plentiful food for a variety of marine life.

By forming limestone shells, coral polyps take up carbon dioxide as part of the carbon cycle. The reefs also act as natural barriers that help protect 15% of the world's coastlines from battering waves and storms. In addition, the reefs build beaches, atolls, and islands, and they provide food, jobs, and building materials for some of the world's poorest countries. They also act as nurseries for many species of fish.

Despite their ecological importance, coral reefs are disappearing and being degraded at an alarming rate. They are vulnerable to damage because they grow slowly, are easily disrupted, and thrive only in clear, warm, and fairly shallow water of constant high salinity.

The biggest threats to many of the world's coral reefs come from human activities resulting from population growth, especially the deposition of eroded soil produced by deforestation, construction, agriculture, mining, dredging, and poor land management along increasingly populated coastlines. The suspended soil sediment that washes downriver to the sea or erodes from coastal areas smothers coral polyps or blocks their sunlight. Such silting is one cause of coral reef bleaching. It occurs when a reef loses its colorful algae and other producers, exposing the colorless coral animals and the underlying white skeleton of calcium carbonate (Figure 8-1, right).

Other threats to coral reefs include increased ultraviolet radiation resulting from depletion of stratospheric ozone; global warming; runoff of toxic pesticides, industrial chemicals, and fertilizers; commercial fishing boats' use of cyanide or dynamite to stun fish so they can be harvested; removing coral for building material, aquariums, and jewelry; oil spills; and damage from tourists and recreational divers.

Coral reefs are sometimes said to be the aquatic equivalent of tropical rain forests because they harbor such a high biodiversity, with myriad ecological interrelationships. This chapter examines limiting factors, characteristics, and biodiversity of the major saltwater and freshwater aquatic life zones.

If there is magic on this planet, it is contained in water.

LOREN EISELEY

This chapter addresses the following general questions about aquatic ecology:

- What are the basic types of aquatic life zones, and what factors influence the kinds of life they contain?
- What are the major types of saltwater life zones, and how are they being affected by human activities?
- What are the major types of freshwater life zones, and how are they being affected by human activities?
- How can we help sustain aquatic life zones?

8-1 LIMITING FACTORS IN AQUATIC ENVIRONMENTS

What Are the Two Major Types of Aquatic Life Zones? The aquatic equivalents of biomes are called *aquatic life zones*. The major types of organisms found in aquatic environments are determined by the water's *salinity* (the amounts of various salts such as sodium chloride [NaCl] dissolved in a given volume of water). As a result, aquatic life zones are divided into two major types: *saltwater* or *marine* (particularly estuaries, coastlines, coral reefs, coastal marshes, mangrove swamps, the ocean above the continental shelf, and the deep ocean) and *freshwater* (particularly lakes and ponds, streams and rivers, and inland wetlands).

What Are the Main Kinds of Organisms in Aquatic Life Zones? Saltwater and freshwater life zones contain several major types of organisms: weakly swimming, free-floating **plankton**; strongly swimming **nekton**; bottom-dwelling **benthos**; and *decomposers* (mostly bacteria). Plankton are divided into three categories: phytoplankton, nanoplankton, and zooplankton. *Phytoplankton* are free-floating, microscopic, photosynthetic cyanobacteria and many types of algae; they are the producers that support most aquatic food chains (Figure 4-10) and food webs (Figure 4-17). Also included are smaller, more recently discovered (and poorly understood) producers called *nanoplankton*. *Zooplankton* are nonphotosynthetic primary consumers (herbivores) that feed on phytoplankton or secondary consumers that feed on other zooplankton.

Nekton are larger, more strongly swimming consumers such as fish, turtles, and whales. *Benthos* are bottom-dwelling creatures. Some, such as barnacles and oysters, anchor themselves to one spot; some burrow into the sand (many worms); others walk about on the bottom (lobsters). *Decomposers* break down the organic compounds in the dead bodies and wastes of aquatic organisms into simple nutrient compounds for use by producers.

How Do Water's Unique Properties Affect Aquatic Life? Living in water has its advantages. Water, with its many unique properties (Spotlight, p. 113), presents an environment for life that is very different from terrestrial environments. Water's buoyancy provides physical support. Limited fluctuations in temperature greatly reduce the risks of drying out or becoming overheated or too cold. Required nutrients are dissolved and readily available. And potentially toxic metabolic wastes secreted by aquatic organisms are diluted and dispersed.

What Factors Limit Life at Different Depths in Aquatic Life Zones? Most aquatic life zones can be divided into surface, middle, and bottom layers. Important environmental factors determining the types and numbers of organisms found in these layers are *temperature, access to sunlight for photosynthesis, dissolved oxygen content*, and *availability of nutrients* such as carbon (as dissolved CO_2 gas), nitrogen (as NO_3^-), and phosphorus (mostly as PO_4^{3-}) for producers.

The water temperature of aquatic ecosystems usually falls with increasing depth, mainly because of decreasing exposure to sunlight. However, because the temperature of water is less subject to change than the temperature of air, most aquatic organisms have evolved the ability to withstand only a narrow range of temperatures. A limited range of tolerance to temperature (Figure 4-12) is typical of most aquatic organisms. This means that small changes in water temperature can have a significant effect on the performance and survival of aquatic organisms.

Sunlight is a limiting factor for aquatic producers because it can penetrate only to a depth of about 30 meters (100 feet) below the water surface. Thus, photosynthesis is confined mostly to this upper layer or **euphotic zone**. Suspended materials (mostly clay and silt) that make water cloudy decrease the penetration of sunlight and can significantly reduce the depth of the euphotic zone.

O_2 enters an aquatic system from the atmosphere and through photosynthesis by aquatic producers; it is removed by aerobic respiration of producers, consumers, and decomposers. CO_2 enters an aquatic system from the atmosphere and through aerobic respiration by producers, consumers, and decomposers; it is removed by photosynthesizing producers. This removal of heat-trapping CO_2 from the atmosphere by aquatic producers helps keep the earth's average atmospheric temperature from rising as a result of the natural greenhouse effect (Figure 7-9). Some dissolved CO_2 forms carbonate ions (CO_3^{2-}). They are stored for long periods in sediments, minerals, and the shells and skeletons of living aquatic animals as part of the carbon cycle (Figure 5-5).

The concentration of oxygen in the atmosphere varies little, but the amount of oxygen dissolved in water can vary widely, depending on such factors as temperature, number of producers (which add O_2), and number of consumers and aerobic decomposers (which remove O_2). Many aquatic organisms, especially fish, die when dissolved oxygen levels fall below 3–5 ppm. The concentrations of dissolved O_2 and CO_2 in water vary in different ways with depth (Figure 8-2) because of differences in the rates of

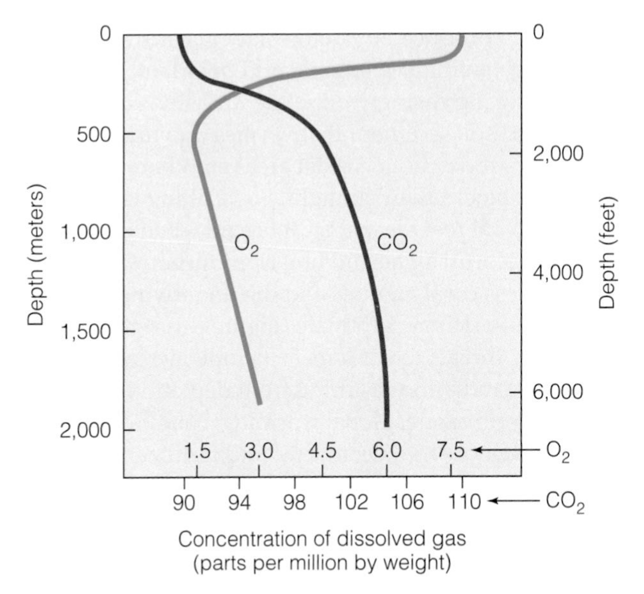

Figure 8-2 Variations in concentrations of dissolved oxygen (O_2) and carbon dioxide (CO_2) in parts per million (ppm) with water depth. Dissolved O_2 is high near the surface because oxygen-producing photosynthesis takes place there. Because photosynthesis cannot take place below the sunlit layer, O_2 levels fall because of aerobic respiration by aquatic animals and decomposers. In contrast, levels of dissolved CO_2 are low in surface layers because producers use CO_2 during photosynthesis and high in deeper, dark layers where aquatic animals and decomposers produce CO_2 through aerobic respiration.

Wagener, S. M., et al. 1998. "Rivers and soils: Parallels in Carbon and Nutrient Processing." *BioScience*, v. 48, n.2, 104(5).

photosynthesis (which produces O_2 and consumes CO_2) and aerobic respiration (which consumes oxygen and produces CO_2).

In shallow waters in streams, ponds, and near coasts, there are usually ample supplies of nitrates and phosphate nutrients for producers. In the open ocean nitrates, phosphates, iron, and other nutrients are often in short supply and limit net primary productivity (Figure 4-24). However, net primary productivity is much higher in parts of the ocean where upwellings (Figures 7-2 and 7-7) bring such nutrients from the ocean bottom to the surface for use by producers. As a general rule, phosphorus (in the form of phosphates, PO_4^{3-}) is the most limiting nutrient in freshwater ecosystems, and nitrogen (in the form nitrates, NO_3^-) is the most limiting nutrient in saltwater ecosystems.

8-2 SALTWATER LIFE ZONES

Why Are the Oceans Important? Saltwater oceans cover about 71% of the earth's surface (Figure 8-3). The oceans play key roles in the survival of virtually all life on the earth. Because solar heat is distributed through ocean currents (Figure 7-2) and because ocean water evaporates as part of the global hydrologic cycle, oceans play a major role in regulating the earth's climate. They also participate in other important nutrient cycles.

By serving as a gigantic reservoir for carbon dioxide (Figure 8-2), oceans help regulate the temperature of the troposphere. Oceans provide habitats for about 250,000 known species of marine plants and animals, which are food for many other organisms (including humans). In addition, many human-produced wastes that flow into or are dumped into the ocean are dispersed by currents (and often diluted to less harmful levels).

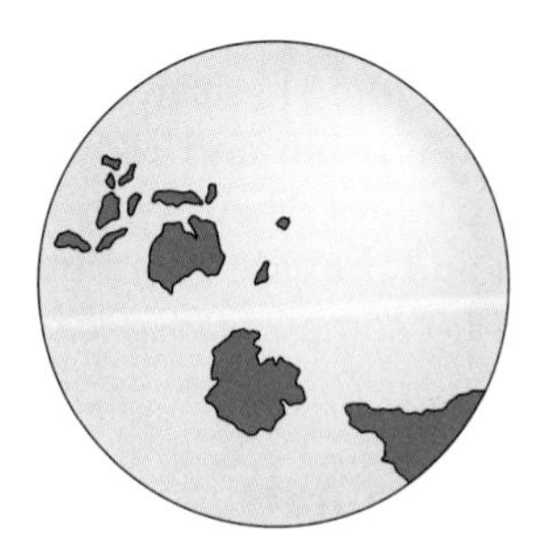

Figure 8-3 The ocean planet. The salty oceans cover about 71% of the earth's surface. About 97% of the earth's water is in the interconnected oceans, which cover 90% of the planet's mostly ocean southern hemisphere (left) and 50% of its land–ocean northern hemisphere (right). The average depth of the world's oceans is 3.8 kilometers (2.4 miles).

Oceans have two major life zones: the coastal zone and the open sea (Figure 8-4). The **coastal zone** is the warm, nutrient-rich, shallow water that extends from the high-tide mark on land to the gently sloping, shallow edge of the *continental shelf* (the submerged part of the continents). Although it makes up less than 10% of the ocean's area, the coastal zone contains 90% of all marine species and is the site of most large commercial marine fisheries. Most ecosystems found in the coastal zone have a very high primary productivity and net primary productivity per unit of area because of the zone's ample supplies of sunlight and plant nutrients (deposited from land and stirred up by wind and ocean currents).

What Are Estuaries and Tidal Ecosystems? One highly productive area in the coastal zone is an **estuary**, a partially enclosed area of coastal water where seawater mixes with fresh water and nutrients from rivers, streams, and runoff from land (Figure 8-5). It is an ecotone (Figure 4-9) where large volumes of fresh water from land and salty ocean water mix. The constant water movement stirs up the nutrient-rich silt, making it available to producers.

Temperature and salinity levels vary widely in estuaries because of the daily rhythms of the tides and seasonal variations in the flow of fresh water into the estuary. After a heavy rainfall, streams and rivers deposit huge amounts of fresh water into an estuary, raising the temperature and reducing the salinity of estuarine waters. During high tides, large volumes of cold, salty water flow into an estuary and then flow out during low tides.

According to one estimate, just 0.4 hectares (1 acre) of tidal estuary provides an estimated $75,000 worth of free waste treatment and has a value of about $83,000 when recreation and fish for food are included. By comparison, 0.4 hectare (1 acre) of prime farmland in Kansas has a top value of about $1,200 and an annual production value of about $600.

What Are Coastal Wetlands? Areas of coastal land that are covered all or part of the year with salt water are called **coastal wetlands**. They are breeding grounds and habitats for a variety of waterfowl and other wildlife; they also serve as popular areas for recreational activities such as boating, fishing, and hunting. They also help maintain the quality of coastal waters by diluting, filtering, and settling out sediments, excess nutrients, and pollutants. In addition, coastal wetlands protect lives and property during floods by absorbing and slowing the flow of

Hint: Enter the search term *stream ecology* using the Subject Guide.

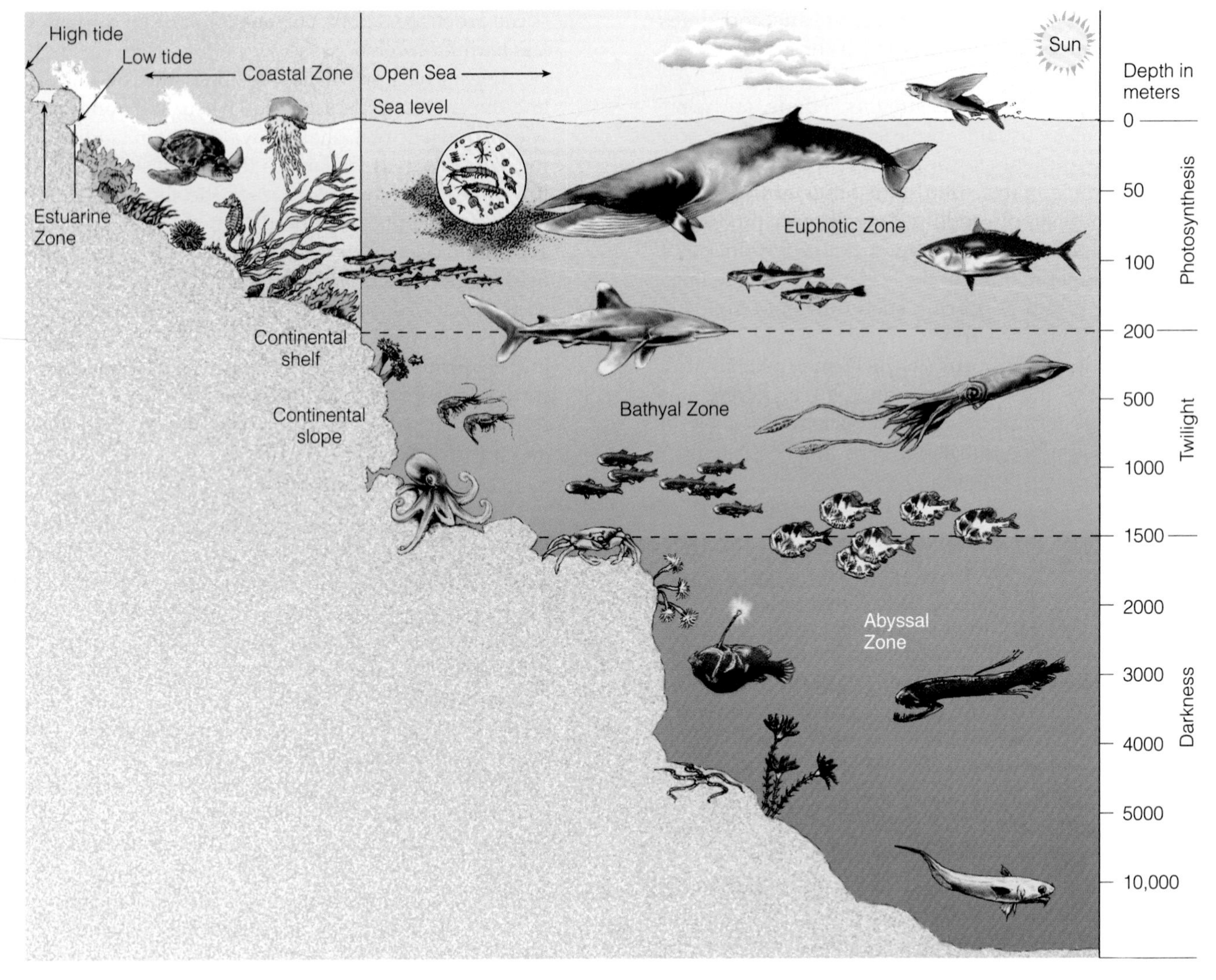

Figure 8-4 Major life zones in an ocean. (Not drawn to scale. Actual depths of zones may vary.)

water, and during storms they buffer shores against damage and erosion.

About 3% of the wetland area in the United States consists of coastal wetlands, which extend inland from estuaries. (The other 97% consists of inland wetlands.) In temperate areas, including the United States, coastal wetlands usually consist of a mixture of *bays, lagoons, salt flats, mud flats, and salt marshes* (Figure 8-6). Grasses are the dominant vegetation. These highly productive ecosystems serve as nurseries and habitats for shrimp and many other aquatic animals. About two-thirds of the U.S. major commercial fisheries depend on coastal estuaries and marshes for nursery or spawning grounds.

What Are Mangrove Swamps? Along warm tropical coasts where there is too much silt for coral reefs to grow we find highly productive **mangrove swamps** (see photo in table of contents), dominated by salt-tolerant trees or shrubs known as mangroves, of which there are about 55 species worldwide. These swamps help protect the coastline from erosion and reduce damage from typhoons and hurricanes. They also trap sediment washed off the land and provide breeding, nursery, and feeding grounds for some 2,000 species of fish, invertebrates, and plants.

What Are Rocky and Sandy Shores? The area of shoreline between low and high tides is called the **intertidal zone**. The organisms that live in this stressful zone must be able to avoid being swept away or crushed by waves, or being immersed during high tides and left high and dry (and much hotter) at low tides. They must also cope with changing

 Denny, M. W., and R. T. Paine. 1998. "Celestial Mechanics, Sea-Level Changes, and Ecology." *Biological Bulletin*, vol. 194, no. 2, 108(8).

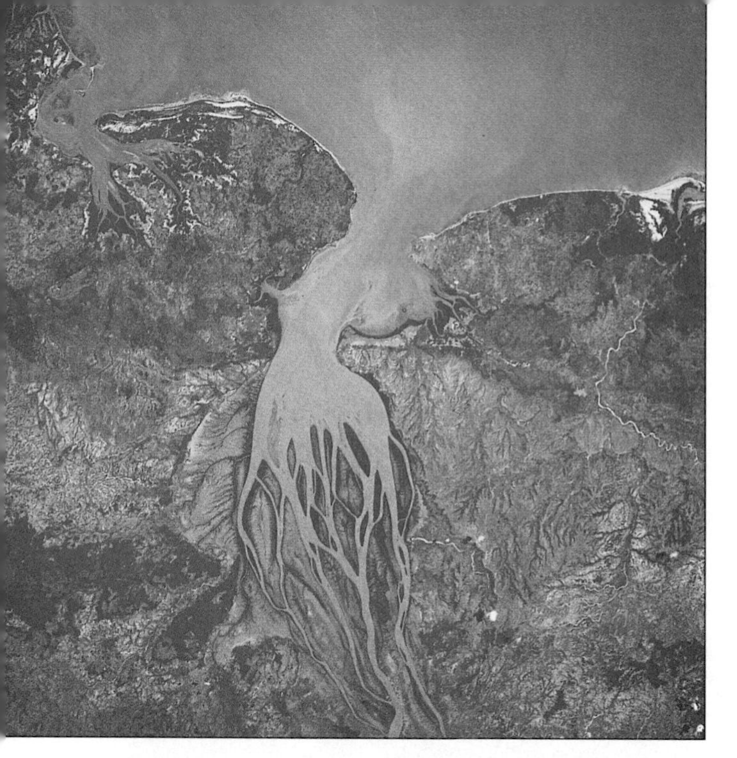

Figure 8-5 View from the space shuttle of the sediment plume at the mouth of Madagascar's Betsiboka River as it flows through the estuary and into the Mozambique Channel, which separates this huge island from the African coast. Topography, heavy rainfall, and the clearing of forests for agriculture make Madagascar the world's most eroded country. (NASA)

levels of salinity when heavy rains dilute salt water. To deal with such stresses, most intertidal organisms hold onto something, dig in, or hide in protective shells.

Some coasts have steep *rocky shores* pounded by waves. Visitors are often surprised at the variety of life found in the numerous pools and other niches in the rocks in the intertidal zone of rocky shores (Figure 8-7, top).

Other coasts have gently sloping *barrier beaches*, or *sandy shores*, with niches for different marine organisms, including crabs, lugworms, clams, ghost shrimp, sand dollars, and flounder (Figure 8-7, bottom). Most of them are hidden from view and survive by burrowing, digging, and tunneling in the sand. These sandy beaches and their adjoining coastal wetlands are also home to a variety of shorebirds that feed in specialized niches on crustaceans, insects, and other organisms (Figure 9-4).

If not destroyed by human activities, one or more rows of natural sand dunes on barrier beaches (with the sand held in place by the roots of grasses) serve as the first line of defense against the ravages of the sea (Figure 8-8, p. 196). However, such beaches are prime sites for development. When coastal developers remove the protective dunes or build behind the first set of dunes, storms can flood and even sweep away seaside buildings and severely erode the unprotected beaches.

What Are Barrier Islands? Along some coasts (such as most of North America's Atlantic and Gulf coasts) are **barrier islands**: long, thin, low offshore islands of sediment that generally run parallel to the shore. These islands help protect the mainland, estuaries, lagoons, and coastal wetlands by dispersing the energy of approaching storm waves. Their low-lying beaches are constantly shifting: Gentle waves build them up, and storms flatten and erode them. Longshore currents, which run parallel to the beaches, constantly take sand from one area and deposit it in another. Sooner or later many of the structures humans build on low-lying barrier islands, such as Atlantic City, Miami Beach, and Ocean City, Maryland (Figure 8-9, p. 196), are damaged or destroyed by flooding, severe beach erosion, or major storms (including hurricanes).

What Are Coral Reefs? Coral reefs form in clear, warm coastal waters of the tropics and subtropics. We saw in the chapter opening (Figure 8-1) that coral reefs are among the most biologically diverse life zones. Coral reefs are also ecologically complex in terms of the many interactions among the diverse organisms that live there (Figure 8-10, p. 197).

What Impacts Do Humans Have on Coastal Zones? In our desire to live near the coast, we are destroying or degrading the very resources that make coastal areas so valuable and enjoyable. Currently, nearly two-thirds of the world's population—some 3.9 billion people—live along coasts or within 160 kilometers (100 miles) of a coast. By 2025, it's estimated that 75%, or 6.2 billion people, will reside on or near coastal areas.

Since 1900, the world has lost approximately half of its coastal wetlands, primarily through coastal development. In the past 200 years nearly 55% of the area of estuaries and coastal wetlands in the United States has been destroyed or damaged, primarily because of dredging and filling and waste contamination. California alone has lost 91% of its original coastal wetlands, but Florida has lost the largest area of such wetlands in the United States. A 1994 study by researchers at the University of California at Berkeley estimated that the values of the long-term earth capital services provided by California's remaining 184,000 hectares (454,000 acres) of wetlands is $124191.5 billion.

Coastal ecosystems are particularly vulnerable to toxic contamination because they trap pesticides,

Hint: Enter the search term *intertidal zonation* using the Subject Guide.

Figure 8-6 Some components and interactions in a salt marsh ecosystem. When these organisms die, their organic matter is broken down by decomposers into minerals used by plants. Transfers of matter and energy among consumers (herbivores) and secondary (or higher-level) consumers (carnivores) are indicated by colored arrows.

Q: What is climate?

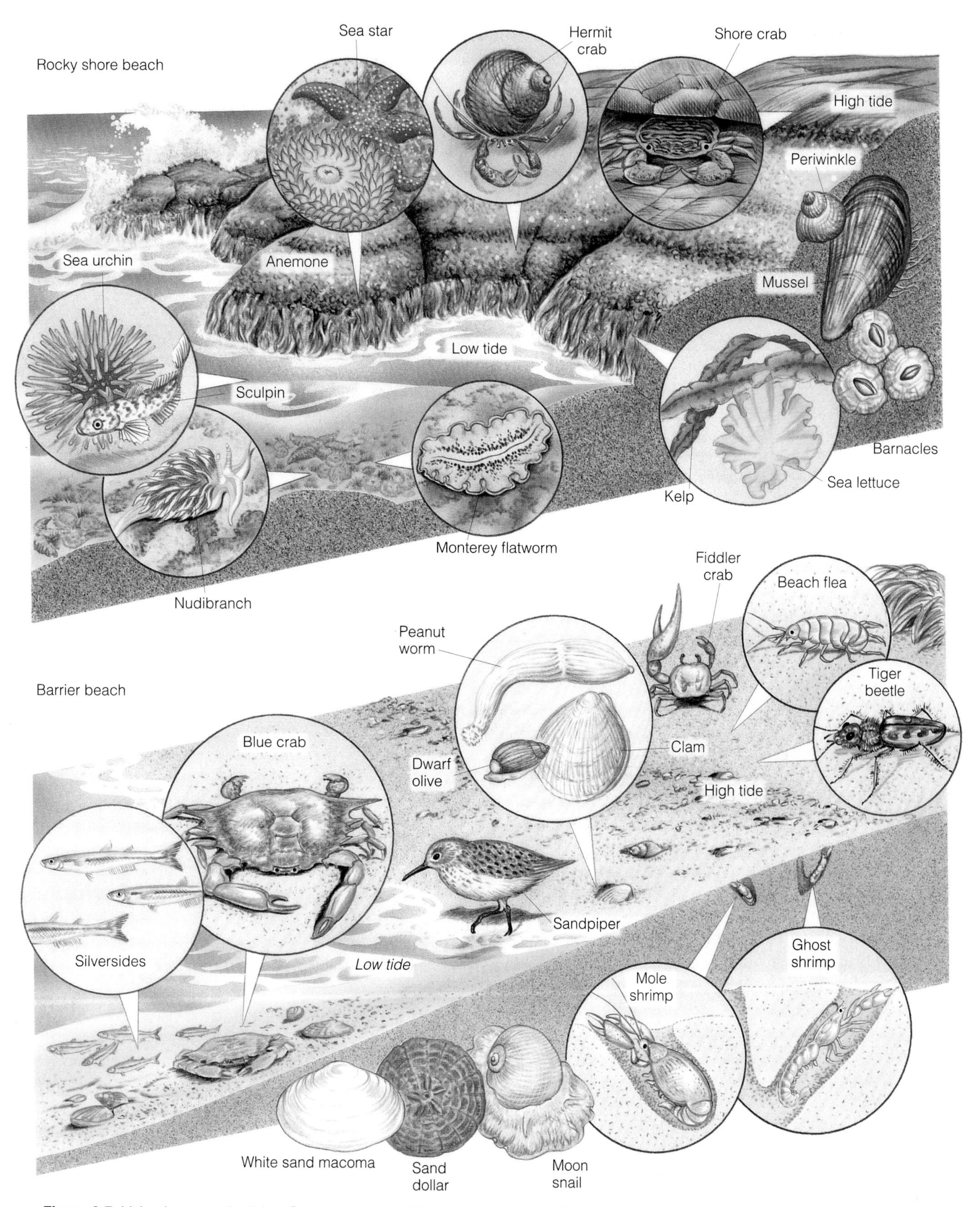

Figure 8-7 Living between the tides. Some organisms with specialized niches found in various zones on rocky shore beaches (top) and barrier or sandy beaches (bottom).

A: Average long-term weather of an area

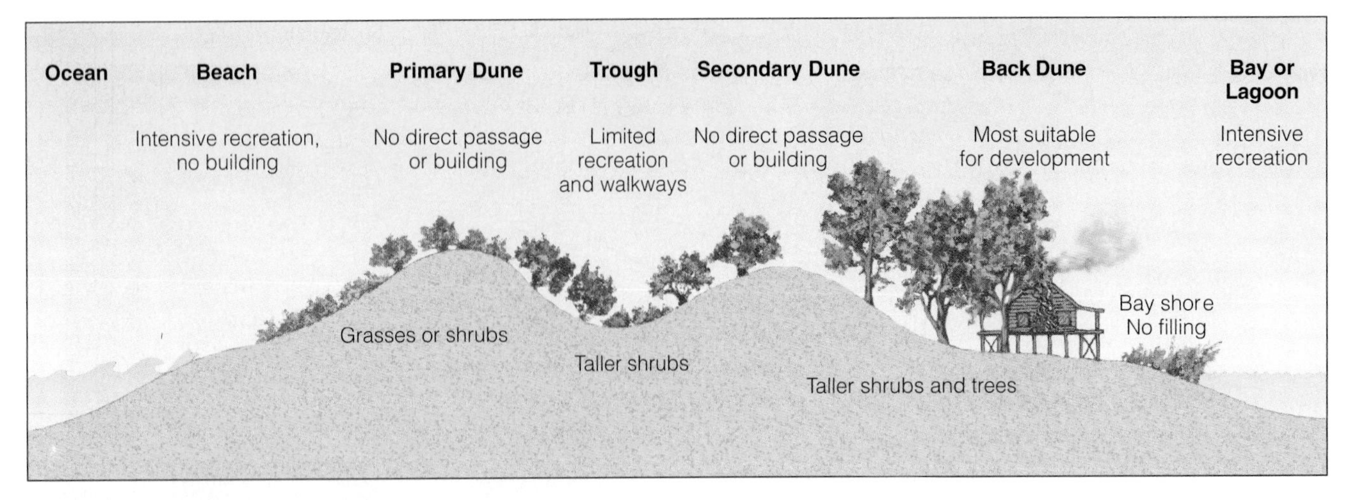

Figure 8-8 Primary and secondary dunes on gently sloping sandy beaches play an important role in protecting the land from erosion by the sea. The roots of various grasses that colonize the dunes help hold the sand in place. Ideally, construction and development should be allowed only behind the second strip of dunes; walkways to the beach should be built over the dunes to keep them intact. This not only helps preserve barrier beaches; it also protects structures from being damaged and washed away by wind, high tides, beach erosion, and flooding from storm surges. This type of protection is rare, however, because the short-term economic value of oceanfront land is considered to be much higher than its long-term ecological and economic values.

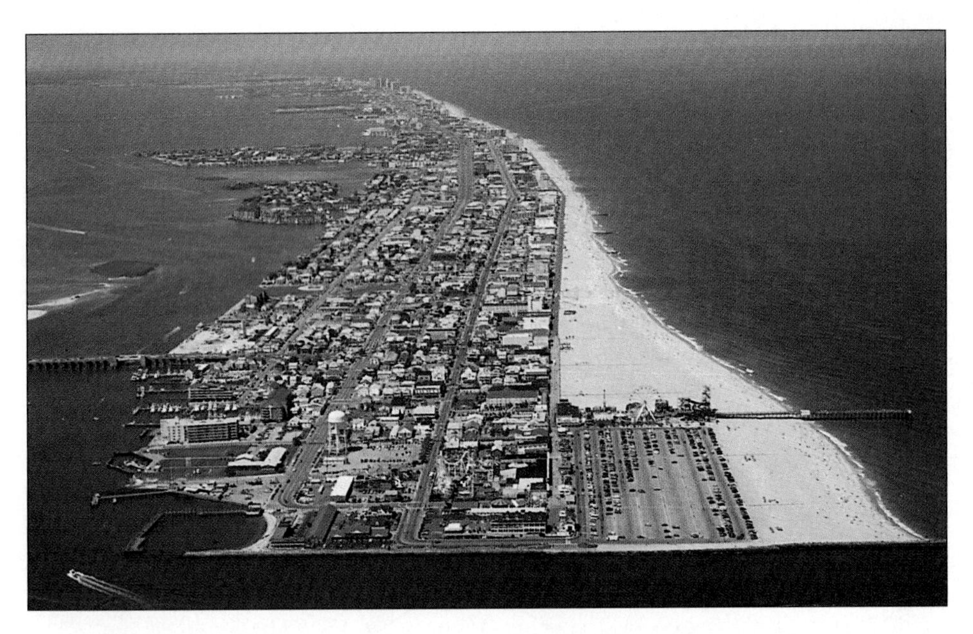

Figure 8-9 A developed barrier island: Ocean City, Maryland, host to 8 million visitors a year. To keep up with shifting sands, taxpayers spend millions of dollars to pump sand onto the beaches and to rebuild natural sand dunes; they may end up spending millions more to keep buildings from sinking. Barrier islands lack effective protection against flooding and damage from severe storms; within a few hours, barrier islands may be cut in two or destroyed by a hurricane. If global warming raises average sea levels, as projected, most of these valuable pieces of real estate will be under water. (G. H. Demetrakas/ O. C. Camera)

heavy metals, and other pollutants, concentrating them to high levels. On any given day 37% of U.S. coastal shellfish beds are closed to commercial or sport fishing, usually because of contamination from sewage treatment plants, septic tank systems, and urban runoff.

Marine biologists estimate that humans have directly or indirectly caused the death of 10% of the world's coral reefs (p. 188), especially those in Southeast Asia and the Caribbean. Another 30% of the remaining reefs are in critical condition and 30% more are threatened; only 30% are stable. If the current rates of destruction continue, another 60% of the reefs could be gone in the next 20–40 years. Some 300 coral reefs in 65 countries are protected as reserves or parks, and another 600 have been recommended for protection. The good news is that protected coral reefs can often recover (Connections, p. 198).

However, protecting reefs is difficult and expensive, and only half the countries with coral reefs have

Figure 8-10 Some components and interactions in a coral reef ecosystem. When these organisms die, their organic matter is broken down by decomposers into minerals used by plants. Transfers of matter and energy among producers, primary consumers (herbivores), and secondary (or higher-level) consumers (carnivores) are indicated by colored arrows.

A: A natural process in which certain gases trap heat in the lower atmosphere (troposphere)

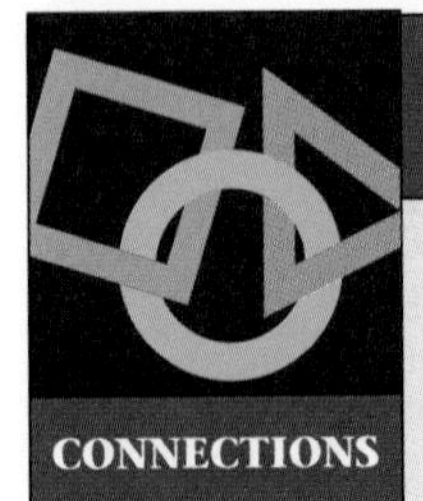

Coral Partners May Enhance Reef Survival

CONNECTIONS

All of the news about coral reefs is not bad. In 1998 researchers found that in some cases coral bleaching (Figure 8-1, left) may not be as fatal to reefs as once thought.

Researchers have been puzzled because corals in shallow water in some areas might be healthy, while at a slightly greater depth there is extensive bleaching. Marine biologist Rob Rowan studied DNA extracted from coral at different depths and locations and identified up to three different species of algae (which he labeled A, B, and C) in the same coral.

He also noticed a pattern in the distribution of these algae species. Two species, which he called A and B, preferred areas on coral with lots of light, and species C was adapted to lower light at greater depths. He hypothesized that coral with species A and B in shallow water may have some resistance to bleaching. In deeper and darker water species C can grow on top of the corals. However, any changes in water condition or depth that allow in more light can kill the C species of algae and lead to bleaching.

Recent research also suggests that bleaching may help a type of coral to switch to new algae species more adapted to the changed water conditions. Additional research indicates that some types of coral that appear dead from bleaching may have a secret reserve of algae that may eventually allow the coral to come back to life.

However, scientists caution that we still have much to learn about the complex interactions between various types of algae and coral.

More good news is the growing evidence that coral reefs can recover when given a chance. When localities or nations have imposed restrictions on reef fishing or reduced inputs of nutrients and other pollutants, reefs have rebounded.

Critical Thinking

Someone tells you that the research described in this presentation means that we don't need to worry so much about protecting coral reefs. How would you respond?

set aside reserves. Unless we act now to protect the world's diminishing coral reefs, biologists warn that another important part of the earth's vital biodiversity will be greatly depleted. Success in doing this will depend on improving public understanding of the ecological and economic (tourism) value of coral reefs and involving local people in their management and protection.

Since the mid-1960s, some tropical coastal countries have lost half or more of their mangrove forests because of industrial logging for timber and fuelwood, conversion to ponds for raising fish and shellfish (aquaculture), conversion to rice fields and other agricultural land, and urban development. Mangroves are disappearing fastest in Asia, especially in the Philippines, Thailand, Bangladesh, Indonesia, and Java.

In the opinion of biologist G. Carleton Ray, "The coastal zone may be the single most important portion of our planet. The loss of its biodiversity may have repercussions far beyond our worst fears."

Case Study: What Can Be Done About Beach Erosion? An estimated 70% of the world's beaches are eroding. Beach erosion is a serious problem along most of the gently sloping beaches of barrier islands and mainland shores, with 30% of U.S. shoreline experiencing significant erosion. The main cause of this problem is that the sea level has been rising gradually for the past 12,000 years or so, mainly because the warmer climate since the most recent ice age has melted much ice and expanded the volume of seawater. (Other causes of rising sea levels are extracting groundwater, redirecting rivers, draining wetlands, and other human activities that divert more water to the oceans.)

Engineers have tried several methods to halt or reduce beach erosion (Figure 8-11), but at best these attempts are only temporary solutions because of the dynamic nature of shorelines. Beach erosion in one place and beach buildup in another is a natural process that we can do little to control.

Many coastal zone ecologists call for banning or severely limiting the construction of seawalls, breakwaters, groins, and jetties (essentially long groins used to protect harbors and inlets) because in the long run they can cause more damage than they prevent. These analysts also favor prohibiting development on remaining beach areas or allowing such development only behind protective dunes (Figure 8-8).

Since 1965 governments, developers, and communities in the United States have spent an estimated $3.5 billion replenishing beaches that have been eroded. The U.S. government alone spends $150 million a year on beach replenishment. Under present policy the government (taxpayers) usually pays 65% of the cost of beach replenishment. Some critics have called for elimination of this federal subsidy or for reducing the government's share to no more than 35% of the costs. Many analysts also believe that federal flood insurance subsidies, which greatly reduce the financial risk of building structures at the sea's edge, should be eliminated.

Q: What is a greenhouse gas?

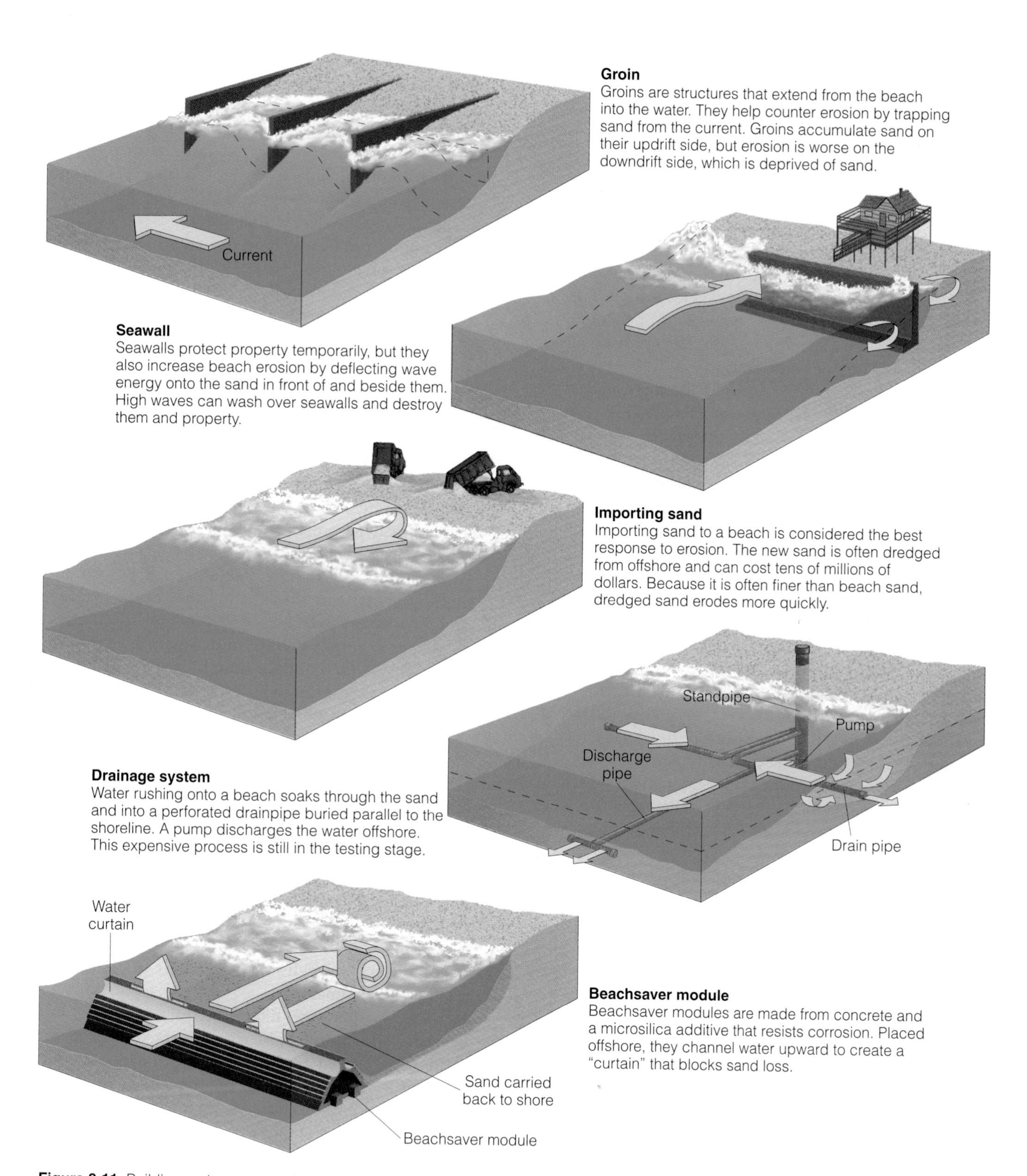

Figure 8-11 Building groins or seawalls and importing sand to reduce beach erosion make matters worse or provide only an expensive temporary fix. Two new techniques for reducing beach erosion shown here—a drainage system and beachsaver modules that act as an artificial reef—are being evaluated for effectiveness and cost.

A: A gas (such as CO_2 and water vapor) that traps heat in the lower atmosphere (troposphere)

What Biological Zones Are Found in the Open Sea? The sharp increase in water depth at the edge of the continental shelf separates the coastal zone from the **open sea**, which is divided into three vertical zones—euphotic, bathyal, and abyssal—based primarily on the penetration of sunlight (Figure 8-4). This vast volume of ocean contains only about 10% of all marine species. The lighted upper (*euphotic*) zone is where photosynthesis occurs in the open sea. Here nutrient levels are low (except around upwellings, Figure 7-7) and levels of dissolved oxygen are high (Figure 8-2). Although the euphotic zone of the open ocean makes up 90% of ocean surface, it produces only about 10% of the world's commercial fish.

The dimly lit *bathyal zone* and the dark *abyssal zone* are found only in the open sea and do not contain photosynthesizing producers because of a lack of sunlight. Water in the abyssal zone is very cold and has little dissolved oxygen (Figure 8-2). The high levels of nutrients on the ocean floor support about 98% of the 250,000 identified species living in the ocean.

Dead and decaying organisms fall to the ocean floor to feed microscopic decomposers and scavengers such as crabs and sea urchins. Some of these organisms (such as many worms) are *deposit feeders*, which take mud into their guts and extract nutrients from it. Others (such as oysters and mussels) are *filter feeders*, which pass water through or over their bodies and extract nutrients from it. On portions of the dark, deep ocean floor near hydrothermal vents, scientists have recently found communities of organisms in which specialized microscopic bacteria use chemosynthesis to produce food for themselves and other organisms feeding on them.

Except at an occasional equatorial upwelling, average primary productivity and net primary productivity per unit of area are quite low in the open sea (Figure 4-24). However, because the open sea covers so much of the earth's surface, it makes the largest contribution to the earth's overall net primary productivity (Figure 4-25). Without strong international agreements, the open ocean can be overexploited and degraded because of the tragedy of the commons (Section 1-3).

8-3 FRESHWATER LIFE ZONES

What Are Freshwater Life Zones? **Freshwater life zones** occur where water with a dissolved salt concentration of less than 1% by volume accumulates on or flows through the surfaces of terrestrial biomes. Examples are *standing* (lentic) bodies of fresh water such as lakes, ponds, and inland wetlands and *flowing* (lotic) systems such as streams and rivers. These bodies of water cover only a small part of the earth's surface, and their locations are largely unrelated to climate. Although only about 1% of the earth's surface is covered by fresh water, about 41% of the world's known fish species live in this water.

Runoff from nearby land provides freshwater life zones with an almost constant input of organic material, inorganic nutrients, and pollutants. Thus, these life zones are closely connected to their surrounding terrestrial biomes.

What Life Zones Are Found in Freshwater Lakes? Large natural bodies of standing fresh water—formed when precipitation, runoff, or groundwater seepage fills depressions in the earth's surface—are called **lakes**. Causes of such depressions include glaciation (the Great Lakes of North America), crustal displacement accompanied by or causing earthquakes (Lake Nyasa in East Africa), and volcanic activity (Crater Lake in Oregon). Lakes are fed by rainfall, melting snow, and the streams that drain the surrounding watershed.

Lakes normally consist of distinct zones (Figure 8-12), providing habitats and niches for different species, that are defined by their depth and distance from shore. The *littoral zone* is the shallow area near the shore, to the depth at which rooted plants stop growing. Because of abundant sunlight and the nutrients it gets from the surrounding land, the littoral zone is the most productive zone of a lake. It has a high biological diversity, containing a variety of phytoplankton, rooted plants that extend above the water's surface (such as cattails and water lilies), totally submerged rooted plants (such as muskgrass), and various species of floating plants (such as duckweed). The littoral zone also contains large numbers of decomposers, as well as frogs, snails, insects, fish, and other consumers.

The *limnetic zone*, like the euphotic zone of the ocean, is the open, sunlit water surface layer away from the shore that extends to the depth penetrated by sunlight. It contains varying amounts of phytoplankton, zooplankton, and fish, depending on the nutrients available. Small fish that feed on the plankton are consumed by larger fish such as bass and bluegill.

The *profundal zone* is the deep, open water where it is too dark for photosynthesis. It is inhabited by fish adapted to its cooler, darker water. The *benthic zone*, at the bottom of a lake, is inhabited mostly by decomposers (bacteria and fungi), detritus-feeding clams, wormlike insect larvae, and catfish.

 Wallace, J. B., et al. 1997. "Multiple Trophic Levels of a Forest Stream Linked to Terrestrial Litter Inputs." *Science*, v. 276, no 5322, 102(3).

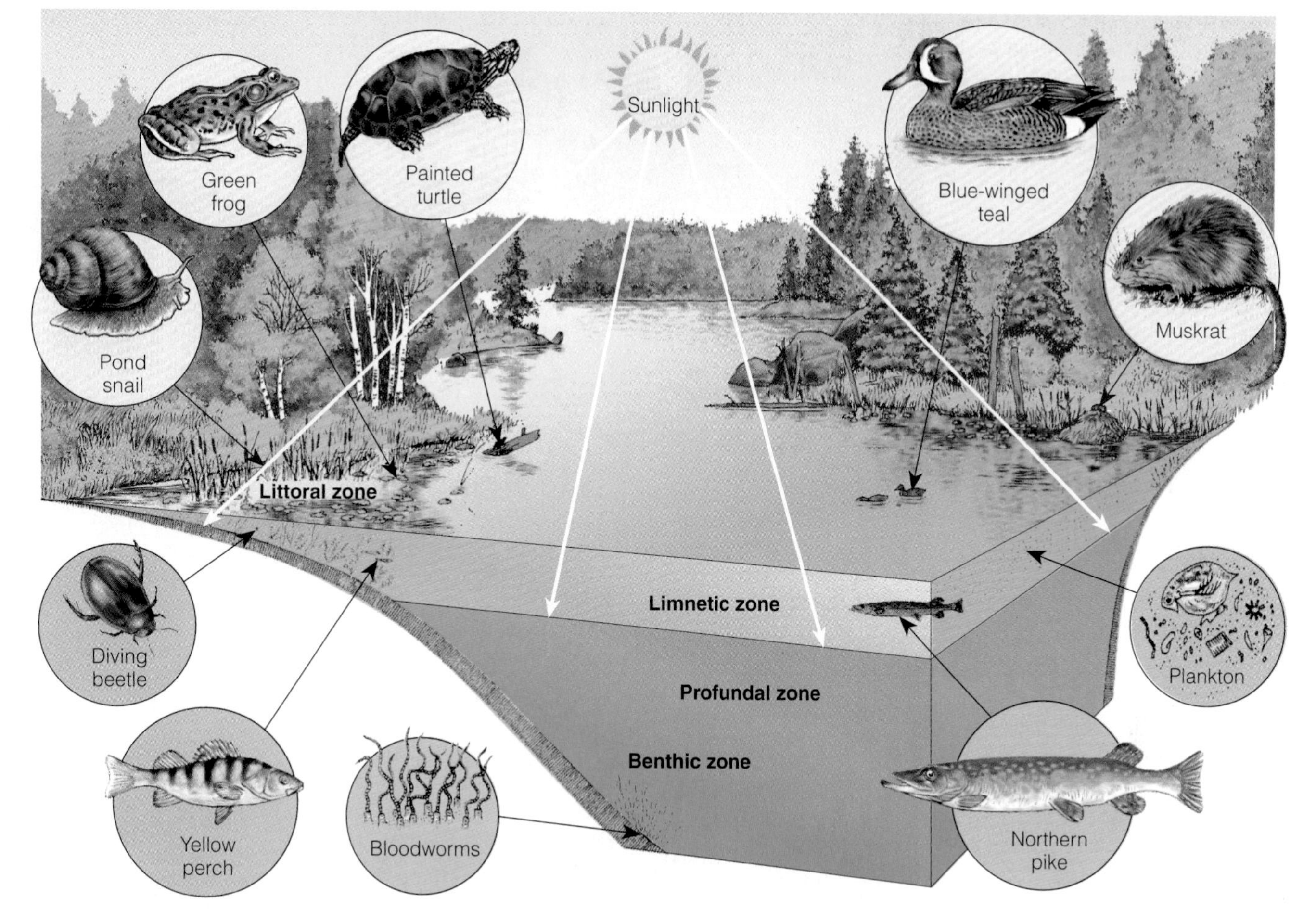

Figure 8-12 The distinct zones of life in a temperate-zone lake.

How Do Plant Nutrients Affect Lakes? Ecologists classify lakes according to their nutrient content and their primary productivity. A newly formed lake generally has a small supply of plant nutrients and is called an **oligotrophic** (poorly nourished) **lake** (Figure 8-13, bottom). This type of lake is often deep, with steep banks. Because of its low net primary productivity, such a lake usually has crystal-clear blue or green water, with small populations of phytoplankton and fish, such as smallmouth bass and trout.

Over time, sediment washes into an oligotrophic lake and plants grow and decompose to form bottom sediments. A lake with a large or excessive supply of nutrients (mostly nitrates and phosphates) needed by producers is called a **eutrophic** (well-nourished) **lake** (Figure 8-13, top). Such lakes are typically shallow, and their water is generally a murky brown or green with very poor visibility. Because of their high levels of nutrients, these lakes have a high net primary productivity. They contain large populations of phytoplankton (especially cyanobacteria), many zooplankton, and diverse populations of fish (such as bass, sunfish, and yellow perch). In warm summer months the bottom layer of a eutrophic lake is often depleted of dissolved oxygen. Many lakes fall somewhere between the two extremes of nutrient enrichment and are called **mesotrophic lakes**.

What Seasonal Changes Occur in Temperate Lakes? Most substances become denser as they go from gaseous to liquid to solid physical states. Water doesn't follow this typical behavior; it is densest as a liquid at 4°C (39°F). In other words, solid ice at 0°C (32°F) is less dense than liquid water at 4°C (39°F), which is why ice floats on water. This is fortunate for us and for most freshwater organisms; otherwise, lakes and other bodies of standing fresh water in cold climates would freeze from the bottom up instead of from the surface down, which would push fish and other organisms to the top, killing them.

Hint: Enter the search term *rivers* using the Subject Guide.

Figure 8-13 A eutrophic, or nutrient-rich, lake (top), and an oligotrophic, or nutrient-poor, lake (bottom). Mesotrophic lakes fall between these two extremes of nutrient enrichment.

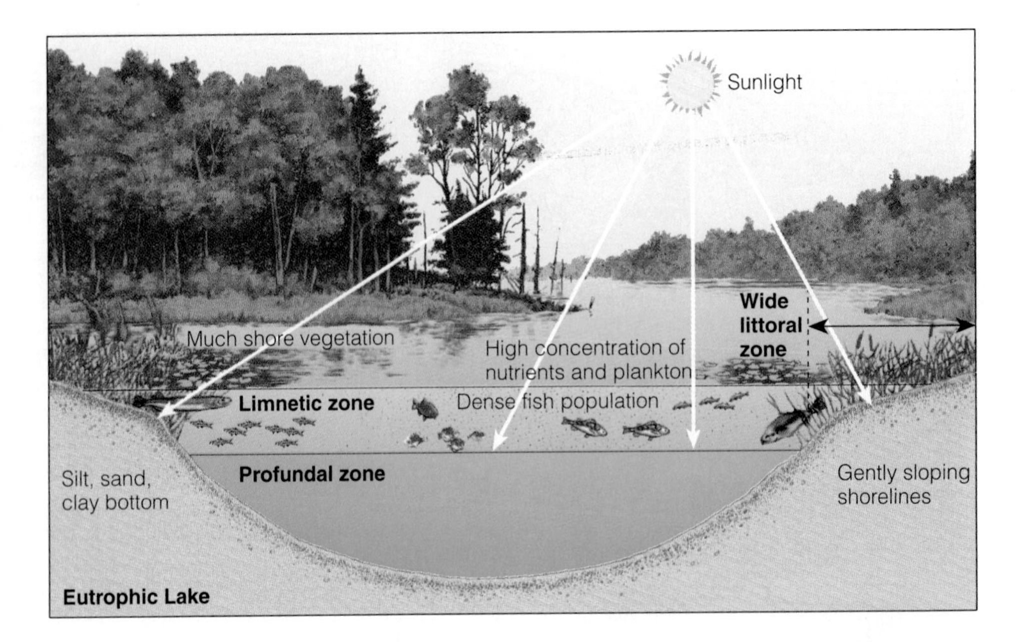

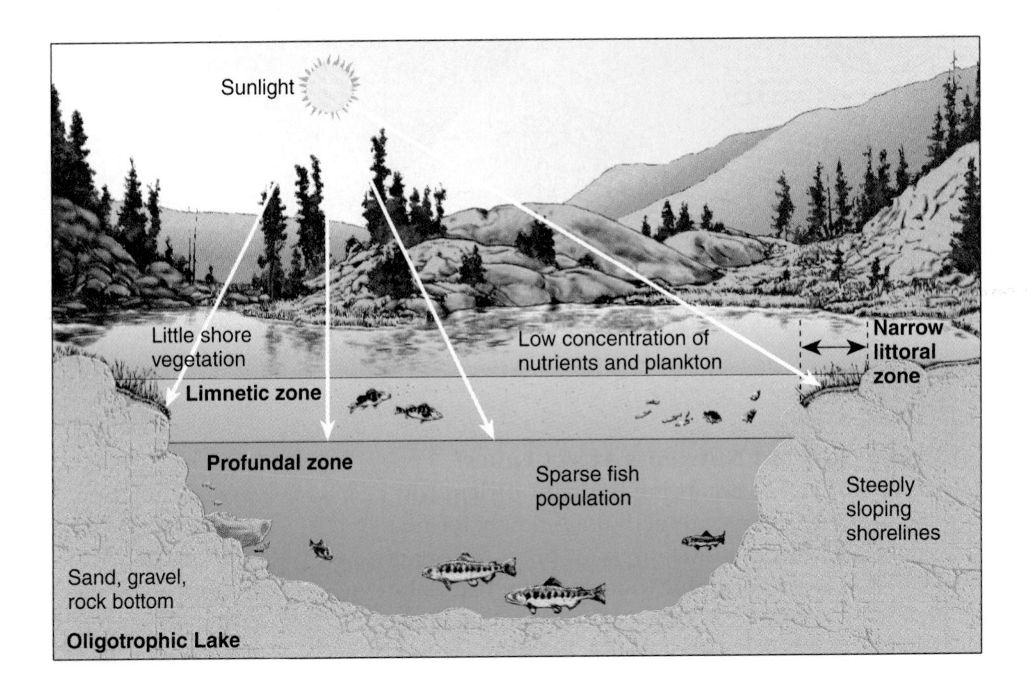

This unusual property of water causes *thermal stratification* of deep lakes in northern temperate areas with warm summers and cold winters (Figure 8-14). During the summer such lakes become stratified into different temperature layers that resist mixing because summer sunlight warms surface waters, making them less dense (Figure 8-14, top left). These lakes have an *epiliminon* (an upper layer of warm water with high levels of dissolved oxygen) and a *hypoliminon* (a lower layer of colder, denser water, usually with a lower concentration of dissolved oxygen because it is not exposed to the atmosphere). These layers are separated by a middle layer called a *thermocline,* where the water temperature changes rapidly with increased depth. The thermocline acts as a barrier to the transfer of nutrients and dissolved oxygen from the epiliminon to the hypoliminon.

Q: What two gases make up 99% of the volume of air in the atmosphere's innermost layer (troposphere)?

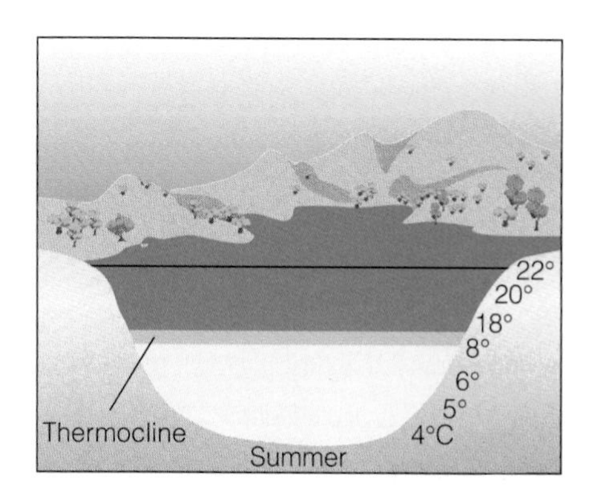

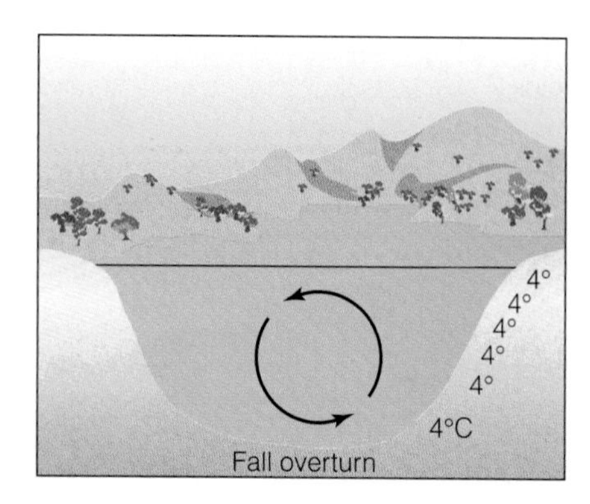

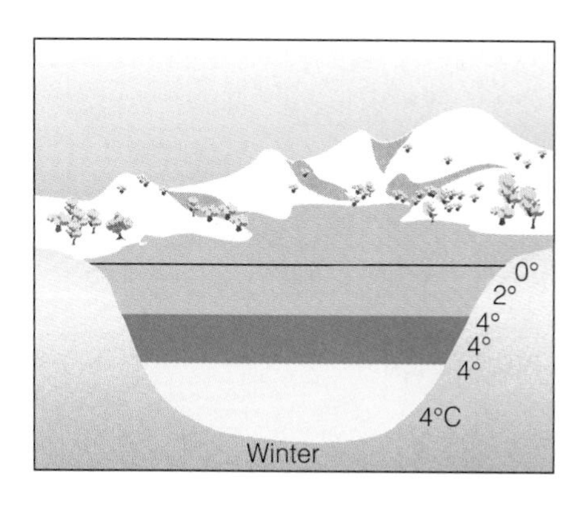

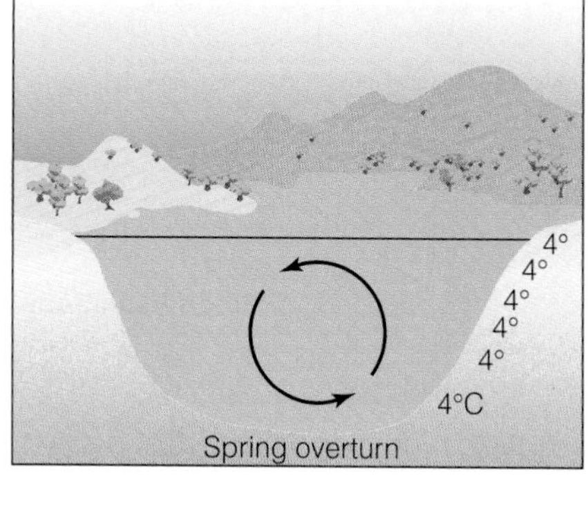

Dissolved O_2 concentration — High — Medium — Low

Figure 8-14 During the summer and winter the water in temperate-zone lakes becomes stratified into different temperature layers, which don't mix. Twice a year, in the fall and spring, the waters at all layers of these lakes mix in overturns that equalize the temperature at all depths. These overturns bring oxygen from the surface water to the lake bottom and also bring nutrients from the lake bottom to the surface waters.

In fall, when temperatures begin to drop, the surface layer becomes more dense. It sinks to the bottom when it cools to 4°C (39°F), and the thermocline disappears (Figure 8-14, top right). This mixing, or *fall overturn*, brings nutrients from bottom sediments to the surface and sends dissolved oxygen from the surface to the bottom.

In *spring*, the ice melts and the lake's surface water warms to 4°C (39°F), reaching maximum density. It then sinks through and below the cooler, less dense water. Winds blowing across the lake's surface cause strong vertical currents that mix the surface and bottom water, bringing dissolved oxygen from the surface to the bottom and nutrients from the bottom to the surface. During this brief *spring overturn*, both the temperature of the lake and dissolved oxygen levels are roughly the same at all depths (Figure 8-14, bottom right).

What Are the Zones in Freshwater Streams and Rivers? Precipitation that doesn't sink into the ground or evaporate is **surface water**. It becomes **runoff** when it flows into streams and eventually to the ocean as part of the hydrologic cycle (Figure 8-15). This entire land area, which delivers water, sediment, and dissolved substances via small streams to a larger stream or river and ultimately to the sea, is called a **watershed**, or a **drainage basin**.

The downward flow of surface water and groundwater from mountain highlands to the sea takes place in three zones (Figure 8-15). Because of different environmental conditions in each zone, a *river system* is actually a series of different ecosystems. In the first, narrow zone, headwater or mountain highland streams of cold, clear water rush over waterfalls and rapids. As this turbulent water flows and tumbles downward, it dissolves large amounts of oxygen from the air. Here plants such as algae and mosses are attached to rocks. The fish are coldwater fish such as trout, which need lots of dissolved oxygen.

There are few phytoplankton in these rapidly flowing waters. Many fish and other animals in headwater streams have compact and flattened bodies that allow them to live under stones, where the current is reduced and where they are less visible to predators.

A: Nitrogen (78%) and oxygen (21%)

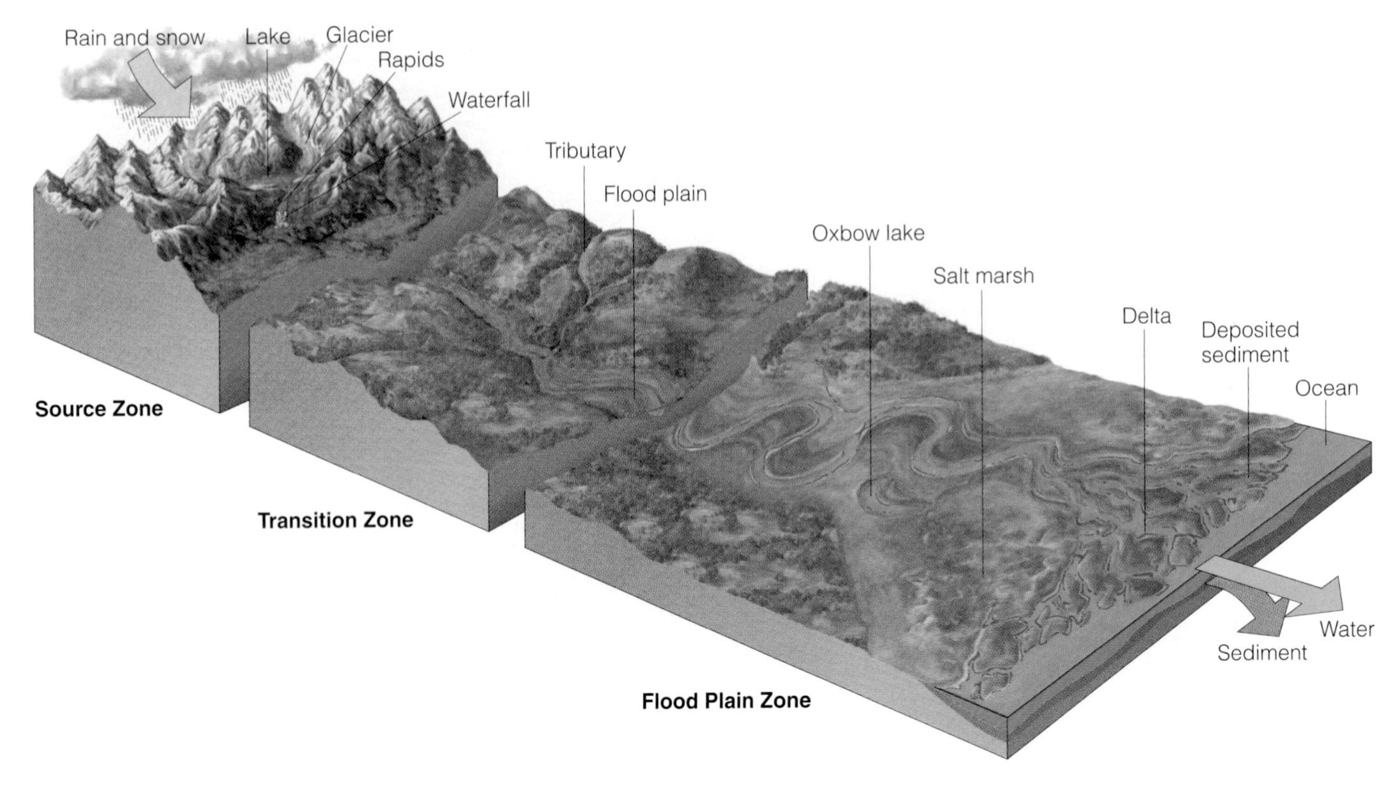

Figure 8-15 The three zones in the downhill flow of water: mountain (headwater) streams; wider, lower-elevation streams; and rivers, which empty into the ocean.

As anglers know, the best place to catch trout is in a deep bottom hole or along the downstream side of large rocks.

In the second zone, the headwater streams merge to form wider, deeper streams that flow down gentler slopes with fewer obstacles. The warmer water and other conditions in this zone support more producers (phytoplankton and a variety of cool-water and warm-water fish species such as black bass) with slightly lower oxygen requirements.

In the third zone, streams join into wider and deeper rivers that meander across broad, flat valleys. Water in this zone usually has higher temperatures and less dissolved oxygen than water in the first two zones. These slow-moving rivers sometimes support fairly large populations of producers such as algae and cyanobacteria. Rooted aquatic plants are found along the shores. Because of increased erosion runoff over a larger area, water in this zone is often muddy and contains high concentrations of suspended particulate matter (silt).

The main channels of these slow-moving, wide, and murky rivers support distinctive varieties of fish (carp and catfish), whereas their backwaters support species similar to those present in lakes. Humans sometimes straighten, deepen, and widen meandering streams to improve navigation and to help reduce flooding and bank erosion, but such *stream channelization* is controversial. At its mouth, a river may divide into many channels as it flows through coastal wetlands and estuaries, where the river water mixes with ocean water (Figure 8-5).

As streams flow downhill, they become powerful shapers of land. Over millions of years the friction of moving water levels mountains and cuts deep canyons; the rock and soil the water removes are deposited as sediment in low-lying areas.

Streams are fairly open ecosystems that receive many of their nutrients from bordering land ecosystems. Such nutrient inputs come from falling leaves, animal feces, insects, and other forms of biomass washed into streams during heavy rainstorms or by melting snow. To protect a stream or river system from excessive inputs of nutrients and pollutants, one must protect its watershed—the land around them.

Why Are Freshwater Inland Wetlands Important? Lands covered with fresh water all or part of the time (excluding lakes, reservoirs, and streams) and located away from coastal areas are called **inland**

Q: What gas in the stratosphere keeps 95% of the sun's harmful ultraviolet radiation from reaching the earth's surface?

wetlands. There is no typical inland wetland. They include marshes, prairie potholes (depressions carved out by glaciers), swamps (dominated by trees and shrubs), mud flats, floodplains, bogs (rain-fed, peat-rich areas), wet meadows, and the wet arctic tundra in summer. Some wetlands are huge; others are small.

Some wetlands are covered with water year-round; others, such as prairie potholes, floodplain wetlands, and bottomland hardwood swamps are *seasonal wetlands*, usually underwater or soggy for only a short time each year. Some stay dry for years before filling with water again. In such cases, only the composition of the soil or the presence of plants such as cattails, bulrushes, or red maples may indicate that a given area is really a wetland.

In the United States, developers and farmers have been pressuring members of Congress to revise the definition of *wetlands* protected under the Clean Water Act. According to the U.S. Army Corps of Engineers and the EPA, under this revised law 60–75% of the country's existing wetlands would lose federal protection and become eligible for development. A 1995 report by the National Academy of Sciences calls such political redefining of wetlands unscientific.

Inland wetlands play important ecological and economic roles. First, they provide food and habitats for fish, migratory waterfowl, and other wildlife. Second, they improve water quality by filtering, diluting, and degrading toxic wastes, excess nutrients, sediments, and other pollutants. This is why inland and coastal wetlands have been called "nature's kidneys." The Audubon Society conservatively estimates that inland wetlands in the United States provide water-quality protection worth at least $1.6 billion per year.

Third, floodplain wetlands near rivers reduce flooding and erosion by absorbing stormwater and releasing it slowly and by absorbing overflows from streams and lakes. According to Audubon Society estimates, if the remaining wetlands in the United States were destroyed, additional flood-control costs would be $7.7 billion to $31 billion per year.

Inland wetlands also help replenish groundwater supplies, a primary water source for over 50% of the U.S. population. They play significant roles in the global carbon, nitrogen, sulfur, and water cycles. Additionally, they provide recreation (especially waterfowl hunting) and are used to grow crops such as blueberries, cranberries, and rice.

What Impacts Do Humans Have on Inland Wetlands? Despite the ecological importance of year-round and seasonal inland wetlands, many are drained, dredged, filled in, or covered over. Each year some 475 square kilometers (183 square miles) of inland wetland in the United States are lost, about 80% to agriculture and the rest to mining, forestry, oil and gas extraction, highways, and urban development. Other countries have suffered similar losses. For example, Canada, with 24% of the world's inland wetlands, has lost 14% of its original wetlands, mostly to agriculture.

A federal permit is now required in the United States to fill wetlands or deposit dredged or fill material in them. According to the U.S. Fish and Wildlife Service, this law has helped cut the average annual loss of wetlands by 75% since 1969. However, there are continuing attempts to weaken it by using unscientific criteria to classify areas as wetlands. Only about 8% of remaining inland wetlands is under federal protection, and federal, state, and local protection of wetlands is weak.

Even though the stated goal of current U.S. federal policy is zero net loss in the function and value of wetlands, this policy allows destruction of existing wetlands as long as an equal area of the same type of wetland is created or restored, a policy known as *mitigation banking*. However, experience has shown that at least half of the attempts to create new wetlands fail to replace lost ones, and most of those that are created bear little resemblance to natural wetlands. Restoring and creating wetlands is also expensive.

Because the best and cheapest course of action is prevention, environmentalists call for several strategies to protect existing wetlands. One is comprehensive land-use planning in which local, state, and federal officials work together to steer developers away from wetlands. Another is to consider mitigation banking as a last resort and to require that a new wetland be created and evaluated *before* any existing wetland can be destroyed. However, many developers, farmers, and resource extractors vigorously oppose wetlands protection.

8-4 SUSTAINABILITY OF AQUATIC ECOSYSTEMS

Why Is an Understanding of Aquatic Life Zones Important? In this chapter we have seen that we live on a water planet, a planet whose surface is more than 71% covered by water. We also saw that the two main kinds of aquatic zones, *saltwater* and *freshwater*, have similar environmental factors that limit aquatic life, particularly *temperature, access to sunlight for photosynthesis, dissolved oxygen for respiration*, and *availability of nutrients for plant growth*. Our survey of the various saltwater environments (*continental shelf, estuaries and coastal wetlands, coral reefs, rocky and*

A: Ozone (O_3)

sandy shores, and *open ocean*) and the various freshwater environments (*lakes and ponds*, *streams and rivers*, and *inland wetlands*) gives us the opportunity to grapple with both the general principles of ecology and the unique characteristics and diversity of each life zone.

As an example for understanding ecological processes and limiting resources, the seasonal stratification of temperate freshwater lakes (Figure 8-14) is ideal because it demonstrates how different resources are limiting in different places (nutrients at the upper layer and light and oxygen in the lower layer) and at different times (for example, during spring turnover versus when the lake is stratified). As examples of extremes of diversity and productivity, coral reefs are the most biologically diverse of the aquatic zones (the aquatic counterpart to tropical rain forests), estuaries are the most productive, and the open ocean is the most expansive and least productive (the aquatic counterpart to deserts).

The lessons are many. As in terrestrial biomes (Chapter 7), again we see the grand lesson that everything is connected.

How Sustainable Are Aquatic Ecosystems? From an ecological perspective, addressing the question of sustainability of aquatic ecosystems requires an understanding of connections—of how human activities affect energy and nutrient flow, trophic relationships, and biodiversity. The bad news in terms of human impacts is that aquatic ecosystems are directly connected to everything upstream, and therefore accumulate not only direct abuses from human activities, but also indirect abuses from anything that occurs in their watersheds. Each stream, river, and lake reflects the sum of all that occurs in the watersheds above, and the ocean is the ultimate receptacle of human activity. Recent research also shows that many of the chemicals reaching aquatic systems come from the atmosphere.

The good news is that aquatic ecosystems are constantly renewed. Water is purified by natural hydrologic processes (Figure 5-4), nutrients cycle in and out, and populations of biological organisms can be replenished, given sufficient opportunity. However, these life-sustaining processes work only if they are not overloaded with pollutants and excessive nutrients and are not overfished.

All at last returns to the sea—to Oceanus, the ocean river, like the ever-flowing stream of time, the beginning and the end.

RACHEL CARSON

CRITICAL THINKING

1. List a limiting factor for each of the following:
   - a. the surface layer of a tropical lake
   - b. the surface layer of the open sea
   - c. an alpine stream
   - d. a large, muddy river
   - e. the bottom of a deep lake

2. Consider the differences in selective pressures in aquatic versus temperate environments. Why do terrestrial organisms evolve tolerances to broader temperature ranges than aquatic organisms? Why do aquatic plants tend to be very small (e.g., phytoplankton), whereas most terrestrial plants tend to be large (e.g., trees) and have more specialized structures for growth (e.g., stems and leaves)? Why are some aquatic animals (especially marine mammals) extremely large compared with terrestrial animals?

3. Come up with analogies to complete the following statements: **(a)** An estuary is like _______. **(b)** A mangrove swamp is like _______. **(c)** The ocean above the continental shelf is like _______. **(d)** An inland wetland is like _______. **(e)** A stream is like _______.

4. How do you expect productivity to change through the year for a temperate freshwater lake? How would you expect the biomass of primary producers versus consumers to change through the year for the same lake? Do you expect biomass to correlate with productivity?

5. Describe the changes in physical environment for organisms that live in the intertidal zone. In many respects the intertidal environment fluctuates between extremes. Do you expect such extreme fluctuations to lead to a higher or a lower diversity of species than in other zones?

6. The deep oceans are vast and are located far away from human habitats; why not use them as the depository for our radioactive and other hazardous wastes? Give reasons for your response.

7. What factors in your lifestyle contribute to the destruction and degradation of coastal and inland wetlands?

8. You are a defense attorney arguing in court for sparing an undeveloped old-growth tropical rain forest and a coral reef from severe degradation or destruction by development. Write your closing statement for the defense of each of these ecosystems. If the judge decides you can save only one of the ecosystems, which one would you choose, and why?

Q: What are the three major types of biomes?

PROJECTS

1. Search for information about mangrove trees, using the Internet and your library. Are the different species of mangrove trees closely related? What characteristics do they have in common? What characteristics do they have that make them a good place for fish to breed?

2. If possible, visit a nearby lake, pond, or reservoir. Would you classify it as oligotrophic, mesotrophic, or eutrophic? What are the primary factors contributing to its nutrient enrichment? Which of these factors are related to human activities?

3. Examine a topographic map for the area around a stream or lake near where you live. Define the watershed for the stream. Does the map include all of the watershed? What human activities occur in the watershed? What influence, if any, do you expect these activities to have on the ecology of the stream or lake?

4. Make a concept map of this chapter's major ideas, using the section heads and subheads and the key terms (in boldface). Look at the inside of the front cover and in Appendix 4 for information about concept maps.

9 COMMUNITY PROCESSES: SPECIES INTERACTIONS AND SUCCESSION*

Flying Foxes: Keystone Species of the Old World Tropics

The durian (Figure 9-1a) is one of the most prized fruits in southeast Asia. It is about the size of a football, is covered with sharp spikes, and has an unforgettable odor. The odor is so strong and disturbing that it is illegal to transport durians in the subways of Singapore and many hotels ban durians in their rooms. The custard-like flesh has been described as "exquisite," "sensual," "intoxicating," and "the world's finest fruit."

Durian fruits come from a wild tree that grows in the tropical rainforest. The tree depends on flying foxes for pollination of its flowers. Flying foxes (Figure 9-1b), also known as Old World fruit bats (family Pteropidae), feed on nectar, pollen, and edible parts of the large flowers that hang high in the durian trees (Figure 9-1c). In the process of feeding, the flying foxes pollinate the flowers with pollen carried unintentionally on their bodies from other durian trees. Pollination by flying foxes is an example of a *mutualism*—an interaction in which both species benefit—that has come into existence because of the *coevolution* of flowering plants and their pollinators.

In nature, durian fruits are eaten by tigers, gibbons, pigs, and other large mammals. The seeds pass unharmed through their guts and are defecated in a new location, where they germinate away from the parent tree. Dispersal of such seeds by large mammals is another example of *mutualism*.

*__Paul M. Rich__, associate professor of Ecology and Evolutionary Biology and Environmental Studies at the University of Kansas, is coauthor of this chapter.

(a)

(b)

(c)

Figure 9-1 Flying foxes play a key role in tropical rain forests of the Old World. **(a)** The durian, a highly prized tropical fruit. (Christer Fredriksson/Bruce Coleman Ltd.) **(b)** Flying foxes (Dr. Merlin Tuttle/Photo Researchers). **(c)** A flying fox pollinating a durian flower.

Attempts have been made to pollinate durian flowers artificially, but success has been low. Thus, flying foxes are essential for continued availability of this prized fruit.

There are nearly 200 species of flying foxes in Africa, southeast Asia, Australia, and islands of the South Pacific and Indian Ocean. Flying foxes are much larger than the bats of the New World. Most have wingspans of about 1 meter (39 inches) and some have wingspans up to 2 meters (78 inches).

Unlike New World bats, most flying foxes rely on vision to navigate during flight, rather than on echolocation (navigation by bouncing high-frequency sounds off of objects). A few species of flying foxes do rely primarily on echolocation. A few species of flying foxes feed exclusively on flowers, but most feed on both flowers and fruit.

Flying foxes congregate in large numbers when they feed and sleep. Accounts of spectacular flights containing hundreds of thousands or even a million flying foxes were once common.

Many species of flying foxes are now listed as *endangered*, and most populations are much smaller than historic numbers. Populations have declined for a variety of reasons, including deforestation and hunting. Flying foxes are easy to hunt because they tend to congregate. A large commercial market for bat meat exists in China and other parts of Asia. Flying fox populations have been intentionally reduced as a "pest control" measure, to keep them from eating fruits that are being grown commercially, though these fruits are usually picked green.

Flying foxes are recognized as *keystone species* in tropical forest ecosystems. They are important not only for the plant species they pollinate and disperse, but also for all of the species that depend on those species. Rain forest ecologists are concerned that the decline of flying fox populations could lead to a cascade of linked extinctions. Studies have also shown that flying foxes have an extraordinary economic importance to products including drinks, fruits, other foods, medicine, timber, fibers, tannins, dyes, medicines, animal fodder, and fuel.

The story of flying foxes and durians illustrates the unique role of each species in an ecosystem (*ecological niches*) and the intricacy of relationships between different species (*species interactions* and *ecosystem structure*). This chapter deals with these topics and the topic of *ecological succession*.

What is this balance of nature that ecologists talk about?
STUART L. PIMM

This chapter focuses on answering the following questions:

- What roles (ecological niches) do different types of organisms play in ecosystems?
- How can we classify species according to their roles?
- How do species interact with one another?
- How do communities and ecosystems change as environmental conditions change?
- Does species diversity increase the stability of ecosystems?
- What determines the number of species in an area?

9-1 THE ECOLOGICAL NICHE

What Role Does a Species Play in an Ecosystem? Each species has a particular *ecological niche* or role that it plays in an ecosystem. Sometimes the niche of a species is called its *lifestyle* or *way of life*. The niche of a species differs from its *habitat*, which refers to the actual physical location where organisms making up a species live.

Theoretically, the ecological niche of a particular organism can be defined by the ranges of *conditions* and *resources* within which the organism can live. **Conditions** are the many physical or chemical attributes of the environment that, though not being consumed, influence biological processes and population growth. Examples are temperature, salinity, and acidity. **Resources** are substances that can be consumed by an organism and, as a result, become unavailable to other organisms. Examples include food, water, and nesting sites for animals, and water, nutrients, and solar radiation for plants.

In the 1950's G. Evelyn Hutchinson developed the concept of the ecological niche as an "n-dimensional hypervolume". This is a fancy way of saying that theoretically many conditions and resources influence the maintenance, growth, and reproduction of an organism. If we were to try to represent each of these n conditions and resources as a different axis in a graph, we would have many more than three axes; because three axes would represent a three-dimensional volume, *n* axes would represent an "*n*-dimensional hypervolume".

Practically, it is not possible to describe all of the conditions and resources that influence a particular organism. However, it is possible to determine specific niche parameters that are most important for an organism in a particular circumstance. For example, temperature tolerance (Figure 4-12) may be the most important niche parameter when looking at the northern limit of where a plant species can grow.

What Is the Difference Between the Fundamental Niche and the Realized Niche? A species's **fundamental niche** is the full potential range of conditions and resources it could theoretically use if there were no direct competition from other species. However, the niche of a species in a particular ecosystem tends to overlap with niches of other species. Niche overlap leads to competition. As a result, a species usually occupies only part of its fundamental niche in a particular ecosystem—what ecologists call its **realized niche**. By analogy, you may be capable of being president of a particular company (your *fundamental professional niche*), but competition from others may mean that you may become only a vice president (your *realized professional niche*).

Why Is Understanding the Niches of Species So Important? One reason for wanting to understand the niches of the earth's 5–100 million species is scientific curiosity: the desire to understand as much as we can about how the earth works.

Understanding a species's niche is also vital so that we can work to prevent it from becoming prematurely extinct. Conserving a species in its native habitat means being sure that all the requirements of its niche are present. Returning an endangered species to the wild is not merely a matter of turning it loose in a certain area. Conservation biologists need to know enough about the niche of a species to be sure that the area has the conditions and resources needed for the species to survive and reproduce.

Understanding the niches of species is also useful in assessing the environmental impact of human changes in terrestrial and aquatic systems. For example, how will clearing a forest, plowing up a grassland, filling in a wetland, or dumping pollutants into a lake or stream change the niches of various species?

Unfortunately, we have little knowledge about the ecological roles of most of the earth's species because determining the niche of a species is a difficult, expensive, and time-consuming process. As we seek to preserve the earth's biodiversity, ecologists urge greatly expanded research in understanding the niches of species. Such basic research underpins efforts to protect ecosystems (Chapters 23 and 24), keep species from becoming prematurely extinct (Chapter 25), and evaluate the effects of our activities on ecosystems, communities, and species.

Is It Better to Be a Generalist or a Specialist Species? Species can be broadly classified as generalists or specialists, according to their niches. **Generalist species** have broad niches: They can live in many different places, eat a variety of foods, and tolerate a wide range of environmental conditions. Flies, cockroaches (Spotlight, right), mice, rats, white-tailed deer, black bears, raccoons, coyotes, bullfrogs, copperheads, robins, channel catfish, and humans are all generalist species.

Specialist species have narrow niches: They may be able to live in only one type of habitat, tolerate only a narrow range of climatic and other environmental conditions, or use only one or a few types of food. This makes them more prone to becoming endangered when environmental conditions change. Examples of specialists are *tiger salamanders*, which can breed only in fishless ponds so their larvae won't be eaten; *red-cockaded woodpeckers*, which carve nest-holes almost exclusively in old (at least 75 years) longleaf pines; *spotted owls*, which require old-growth forests in the Pacific Northwest for food and shelter; *koalas*, which live in a fairly limited habitat in Australia and feed exclusively on Eucalyptus leaves; and China's giant pandas (Case Study, p. 210).

In a tropical rain forest, an incredibly diverse array of species survives by occupying specialized ecological niches in various distinct layers of vegetation exposed to different levels of light (Figures 7-23 and 7-24). The widespread clearing and degradation of such forests is threatening large numbers of such specialized species to extinction.

Is it better then to be a generalist than a specialist? It depends. When environmental conditions are fairly constant, as in a tropical rain forest, specialists have an advantage because they have fewer competitors. When environments are changing rapidly, however, the generalist is usually better off than the specialist.

9-2 SOME GENERAL TYPES OF SPECIES

How Can We Classify the Roles Various Species Play in Ecosystems? When examining ecosystems, ecologists often apply particular labels—such as *native*, *nonnative*, *indicator*, or *keystone*—to various species to clarify some of the ecological roles they play.

Power, Mary E., et al. 1996. "Challenges in the Quest for Keystones." *BioScience*, vol. 46, no. 8, 609(12).

Any given species may function as more than one of these four types in a particular ecosystem.

Why Can Nonnative Species Cause Problems? Species that normally live and thrive in a particular ecosystem are known as **native species**. Others that migrate into an ecosystem or are deliberately or accidentally introduced into an ecosystem by humans are called **nonnative species**, **exotic species**, or **alien species**. Some of these introduced species (such as crop species and game for sport hunting) are beneficial to humans, but some thrive and crowd out many native species.

From a human standpoint, introduction of nonnative species can become a nightmare. In 1957, wild African bees were imported to Brazil to help increase honey production. Instead, these bees have displaced domestic honeybees and reduced the honey supply. Since then these nonnative bee species, popularly known as "killer bees," have moved northward into Central America (killing 150 people in Mexico since 1986); by the summer of 1994 they had become established in Texas (one death in 1994), Arizona (one death in 1993), New Mexico, and Puerto Rico. They are now heading north at 240 kilometers (150 miles) per year, although they will be stopped eventually by cold winters in the central United States.

Although they are not the killer bees portrayed in some horror movies, these bees are aggressive and unpredictable. They have killed thousands of domesticated animals and an estimated 1,000 people in the western hemisphere. Fortunately, most people not allergic to bee stings can run away. Most people killed by these honeybees have died because they fell down or became trapped and could not flee.

What Are Indicator Species? Species that serve as early warnings that a community or an ecosystem is being damaged are called **indicator species**. For example, research indicates that a major factor in the current decline of migratory, insect-eating songbirds in North America is loss or fragmentation of habitat. The tropical forests of Latin America and the Caribbean that are the birds' winter habitats are rapidly disappearing. Their summer habitats in North America are also disappearing or are being fragmented into patches that make the birds more vulnerable to attack by predators and parasites. Birds are excellent biological indicators because they are found almost everywhere and respond quickly to environmental change.

The presence or absence of trout species in water at temperatures within their range of tolerance (Figure 4-12) is an indicator of water quality because trout require clean water with high levels of dissolved oxygen. Some amphibians (frogs, toads, and salamanders), which live part of their lives in water and part on land, are also classified as indicator species (Connections, p. 211).

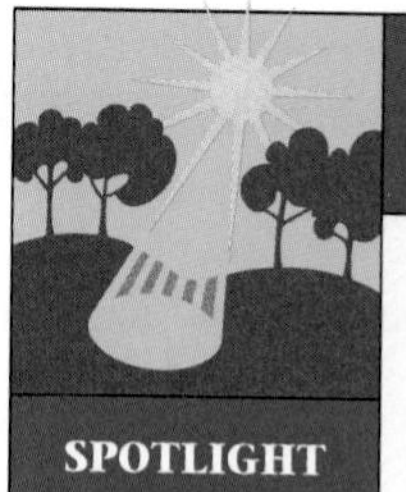

SPOTLIGHT

Cockroaches: Nature's Ultimate Survivors

Cockroaches, the bugs many people love to hate, have been around for about 350 million years and are one of the great success stories of evolution. The major reason they are so successful is that they are *generalists*, able to eat almost anything (including algae, dead insects, salts in tennis shoes, electrical cords, glue, paper, soap, and—when times are bad—other, weaker cockroaches). The 4,000 known cockroach species can live and breed almost anywhere except polar regions.

Some species can go for months without food, last a month without water, and withstand massive doses of radiation. One species can survive being frozen for 48 hours. The antennae of most cockroach species (which can detect minute movements of air), the vibration sensors in their knee joints, and their lightning-fast response times (faster than you can blink) allow them to evade predators and a human foot in hot pursuit. Some even have wings.

High reproductive rates also aid the survival of cockroaches. In only a year, a single Asian cockroach (especially prevalent in Florida) and its young can add about 10 million new cockroaches to the world. Their high reproductive rate also helps them quickly develop genetic resistance to almost any poison we throw at them. Most cockroaches also sample food before it enters their mouths and learn to shun foul-tasting poisons.

Only about 25 species of cockroach live in homes, but such species can carry viruses and bacteria that cause such diseases as hepatitis, polio, typhoid fever, plague, and salmonella. They can cause people to have allergic reactions ranging from watery eyes to severe wheezing. Indeed, about 60% of the 12 million Americans suffering from asthma are allergic to dead or live cockroaches.

Critical Thinking

How do you feel about cockroaches? If you could, would you exterminate them? What might be some ecological consequences of doing this?

What Are Keystone Species? Some ecologists call species whose roles in an ecosystem are much more important than their abundance or biomass

Hint: Enter the search term *keystone species* using the Subject Guide.

The Giant Panda: Specialized and Endangered

CASE STUDY

The highly endangered giant panda (Figure 9-2) is a specialist species that feeds almost exclusively on various types of bamboo. To survive, these animals must spend most of each day eating up to one-third of their body weight in bamboo.

Today only about 800 giant pandas survive in the wild in about 20 isolated "habitat islands" of bamboo forest in southwestern China (12 of them set aside as protected reserves). These isolated populations of 10–50 animals are vulnerable to extinction by illegal hunting (a panda pelt brings $40,000 or more in Hong Kong and Japan), habitat loss, inbreeding, and capture for sale to zoos.

Giant pandas are also vulnerable to extinction because only about one cub per female survives every other year, and young pandas must be cared for by their mothers for up to 22 months after birth.

Pandas are also quite finicky about picking mates, which becomes critical in light of their low numbers and isolated habitats. Another threat to the remaining pandas is that bamboo dies off in cycles of 15–120 years (depending on the species) and takes several years to grow to edible size.

When bamboo was abundant this was no problem. The pandas moved to another area. However, as China's human population has soared, people trying to survive have pushed the pandas into smaller and smaller areas in the country's western mountains.

Today, the few remaining pandas are confined to islands of forest dominated by a few bamboo species. When these plants undergo a dieback, the pandas have no food.

About 220 giant pandas are found in zoos and research centers in China and in other countries, but more captive pandas die than are born. Within your lifetime this specialized species may disappear from the wild and then slowly disappear from zoos and be gone forever.

Critical Thinking

Do you believe that zoos should be allowed to exhibit giant pandas (one of their biggest money-makers) captured from the wild? Explain.

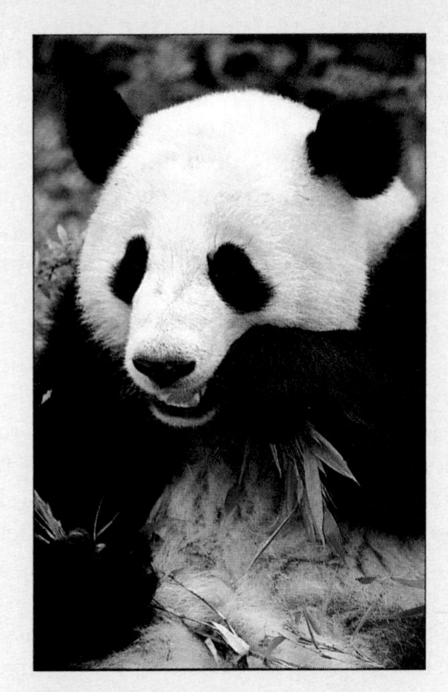

Figure 9-2 The giant panda, found in the wild in western China, is a highly endangered species because of a combination of habitat loss, low birth rate, and its highly specialized diet consisting mostly of bamboo stalks and leaves. (Richard Laird/FPG International)

would suggest **keystone species**, although this designation is controversial.* Such species play pivotal roles in the structure, function, and integrity of an ecosystem because they are critically linked to a large number of other species.

In tropical forests, various species of bees, bats (p. 206), ants, and hummingbirds play keystone roles by pollinating flowering plants, dispersing seed, or both. Have you thanked a *dung beetle* today? You should, because without these keystone species that rapidly remove, bury, and recycle animal wastes (dung) we would be up to our eyeballs in such waste, and many plants would be starved for nutrients. The dung these beetles bury also establishes new plants because it contains seeds that have passed through the digestive tracts of fruit-eating animals. In addition, they churn up and aerate the soil, making it more suitable for plant life. Dung beetle larvae feed on parasitic worms and maggots that live in the dung. This helps reduce populations of microorganisms that spread disease to wild and domesticated animals and to humans and other vertebrates (animals with backbones).

Sea otters are keystone species that keep sea urchins from depleting kelp beds in offshore waters from Alaska to southern California (p. 234). If the sea otter is removed, too little kelp is left to support the diverse community of crustaceans, mollusks, fish, and marine mammals such as fish-eating harbor seals. In addition to sea urchins, sea otters feed on abalone, crabs, and mollusks. This makes them unpopular with people who make a living harvesting crabs and abalone.

*All species play some role in their ecosystems and thus are important. Whereas some scientists consider all species equally important, others consider certain species to be more important than others, at least in helping maintain the ecosystems they are a part of.

Q: What factor is most important in determining whether a land area is a desert, grassland, or forest?

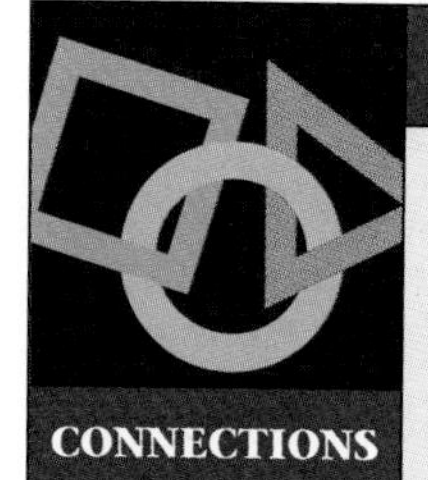

CONNECTIONS

Why Are Amphibians Vanishing?

Amphibians first appeared about 350 million years ago. These cold-blooded creatures range in size from a frog that can sit on your thumb to a Japanese salamander that is about 1.5 meters (5 feet) long.

Fossil records suggest that frogs and toads, the oldest of today's amphibians, existed as long as 150 million years ago; such longevity testifies to their adaptability. Within the last decade or two, however, hundreds of the world's estimated 5,100 amphibian species (including 2,700 species of frogs and toads) have been vanishing or declining in diverse locations, even in protected wildlife reserves and parks.

Scientists have identified a variety of possible causes for such declines. For some species, such as Costa Rican golden toads, die-backs may be caused by prolonged drought, which dries up breeding pools so that few tadpoles survive. Dehydration can also weaken amphibians, making them more susceptible to fatal viruses, bacteria, and fungi. Introduction of exotic predatory fish, such as stocked populations of bass, can quickly clear a stream of tadpoles and frog eggs.

In other cases, the culprit may be pollution. Because amphibians live part of their lives in water and part on land, they are exposed to water, soil, and air pollutants. Their insect diet guarantees them abundant food, but in farming areas it also exposes them to pesticides. The soft, permeable skin that allows them to absorb oxygen from water also makes them extremely sensitive to waterborne pollutants. In some regions the leading culprit in amphibian losses may be increased acidity of the water in lakes and ponds from acid deposition (acid rain).

Their eggs may also be sensitive to increases in ultraviolet radiation caused by reductions in stratospheric ozone. Especially vulnerable amphibian species may be those living at cooler and higher elevations, where the ozone layer is thinnest, and where individuals must bask in the sunlight to stay warm. Recent evidence suggests that environmental pollutants that mimic estrogens and disrupt the immune and endocrine systems of amphibians may play a role in their population declines in some locales. In 1998 scientists discovered that many frogs from a variety of areas were being killed by a previously unknown type of aquatic fungus.

In Asia and France, where frog legs are a delicacy, overhunting may play a part. In other areas, immigration or introduction of nonnative predators and competitors (such as fish) can threaten amphibian populations.

Loss of habitat—or its fragmentation into pieces too small to support populations of some amphibians—is a problem in some places. Once small, isolated populations decline to a certain level, they may not be able to recover or can be wiped out more easily by a chance event.

Scientists are concerned about amphibians' decline for three reasons:

- It suggests that the world's environmental health is deteriorating rapidly, because amphibians are generally tough survivors.
- Adult amphibians play important roles in the world's ecosystems. For example, amphibians eat more insects (including mosquitoes) than do birds. In some habitats, extinction of certain amphibian species could also result in extinction of other species, such as reptiles, birds, aquatic insects, fish, mammals, and other amphibians that feed on them or their larvae.
- From a human perspective amphibians represent a genetic storehouse of pharmaceutical products waiting to be discovered. Hundreds of secretions from amphibian skin have been isolated, and some of these compounds are being used as painkillers and in treating burns and heart attacks. Others are being evaluated for their use as antiviral and antibacterial medicines.

As indicator species, amphibians may be sending us an important message. They don't need us, but we and other species need them.

Critical Thinking

On an evolutionary time scale all species eventually become extinct. Some suggest that the widespread disappearance of amphibians is the result of natural responses to changing environmental conditions. Others contend that these losses are caused mostly by human actions and that such declines are a warning of possible harm for our own species and other species. What is your position? Why?

In the sandhills of Florida and other southern U.S. states, *gopher tortoises* are considered keystone species because the large burrows they dig become cool refuges for some 40 other species, including the gray fox, opossum, and an array of insects, frogs, snakes, and mice. Research has shown that in areas of Florida where gopher tortoises have become locally extinct (mostly because of hunting), some 37 species of invertebrates no longer exist, and Florida mice and gopher frogs are rare. The American alligator is also considered a keystone species in Florida and several other southern states (p. 132).

Some ecologists consider large grazing animals, such as endangered elephants and rhinoceroses, as

A: Climate, based mostly on average temperature and average precipitation keystone species in the savanna grasslands and woodlands of Africa. By pushing over, breaking, or uprooting trees, elephants create forest openings. This promotes the growth of grasses and other forage plants that benefit smaller grazing species such as antelopes. It also accelerates rates of nutrient cycling. Grazing pressure from rhinoceroses transforms medium-tall grasslands into patches of short and tall grassland, which improves food quality for smaller, more selective grazing animals.

Beavers are viewed as keystone species because by building dams they can change a fast-moving stream into a pond. This attracts fish (such as bluegill), muskrats, herons, and ducks that prefer deeper, slower-moving water, as well as woodpeckers that feed on dead trees emerging from the pond. If the hard-working beavers leave or are removed, the pond will silt up and the beaver pond community will be replaced by different plants and animals. Some keystone species—including the wolf, leopard, lion, giant anteater, great white shark, and giant armadillo—are top predators that exert a stabilizing effect on their ecosystems by feeding on and regulating the populations of certain species.

The loss of a keystone species can lead to population crashes and extinctions of other species that depend on it for certain services, a ripple or domino effect that spreads throughout an ecosystem. According to biologist Edward O. Wilson, "The loss of a keystone species is like a drill accidentally striking a power line. It causes lights to go out all over."

9-3 SPECIES INTERACTIONS: COMPETITION AND PREDATION

How Do Species Interact? An Overview

When different species in an ecosystem have activities or resource requirements in common, they may interact with one another. Members of these species may be harmed by, benefit from, or be unaffected by the interaction. There are three basic types of interactions among species: *interspecific competition*, *predation*, and *symbiosis*.

Interspecific competition (competition between species) occurs when two or more species compete for food, space, or any other limited resource. This competition harms the competing species to varying degrees, depending on which is the best competitor.

In **predation**, members of one species (the *predator*) feed directly on all or part of a living organism of another species (the *prey*). However, they do not live on or in the prey. The prey may or may not die from the interaction. In this interaction the predator benefits, and the individual prey is clearly harmed.

Symbiosis (from the Greek word for "living together") is a long-lasting relationship in which species live together in an intimate association. There are three general types of symbiosis: *parasitism*, *mutualism*, and *commensalism*.

Parasitism occurs when one species (the *parasite*) feeds on part of another organism (the *host*) by living on or in the host for a significant portion of the host's life. In this symbiotic relationship the parasite benefits and the host is harmed. In **mutualism** two species involved in a symbiotic relationship interact in ways that benefit both. **Commensalism** is a symbiotic interaction that benefits one species but neither harms nor helps the other species much, if at all.

These interactions tend to regulate the populations of species and can help them to survive changes in environmental conditions, as discussed in more detail in Section 10-2.

How Do Species Compete with One Another?

As long as commonly used resources are abundant, different species can share them. This allows each species to come closer to occupying the *fundamental niche* it could theoretically use if there were no competition from other species.

However, most species face competition from other species for one or more limited resources (such as food, sunlight, water, soil nutrients, space, nest sites, and good places to hide). Whenever parts of the fundamental niches of different species overlap significantly, *interspecific competition* results. The more the niches of the two species overlap, the more they compete with one another. With significant niche overlap, one of the competing species must **(1)** migrate to another area (if possible), **(2)** shift its feeding habits or behavior through natural selection and evolution, **(3)** suffer a sharp population decline, or **(4)** become extinct in that area. Competition among species can be an agent of natural selection that may lead to the evolution of new adaptations and even new species (Section 6-3).

Species compete in two ways. In **interference competition**, one species may limit another's access to some resource, regardless of its abundance. This often takes the form of behavior in which members of a species establish a *territory* they defend against other invading species. For example, one species of hummingbird may defend patches of spring wildflowers from which it gets nectar by chasing away members of other hummingbird species.

Some species carry out "aggressive" acts against competing species. In coral reefs (p. 186), tiny coral animals (close relatives of jellyfish) kill nearby coral species by poisoning them and by growing over them. In desert and grassland habitats, many plants release chemicals into the soil that either prevent

Yu, D. W., and D. W. Davidson. 1997. "Studies in Cecropia–Ant Relationships." *Ecological Monographs*, vol. 67, no. 3, 273(22).

the growth of competing species or reduce their seeds' germination rates.

In **exploitation competition**, competing species have roughly equal access to a specific resource but differ in how fast or efficiently they exploit it. The species that can use the resource more quickly gets more of the resource and hampers the growth, reproduction, or survival of the other species.

What Is the Principle of Competitive Exclusion? Sometimes one species eliminates another species in a particular area through competition for limited resources. In 1934, Russian ecologist G. P. Gause demonstrated this effect by carrying out a laboratory experiment in which two closely related species of single-celled, bacteria-eating *Paramecium* were grown, first separately and then together in culture tubes (Figure 9-3).

The graph on the left of Figure 9-3 shows that when both species were grown under identical conditions in separate containers with ample supplies of food (bacteria), both grew rapidly and established stable populations. However, the smaller *Paramecium aurelia* grew faster (red curve) than the larger *Paramecium caudatum* (green curve), indicating that the former used the available food supply more efficiently than the latter. The graph on the right shows that when both species were grown together in a culture tube with a limited amount of bacteria, the smaller *Paramecium aurelia* (red curve) outmultiplied and eliminated the smaller *Paramecium caudatum* (green curve).

This research, which has been supported by various laboratory and field experiments using other species of animals, showed that two species that require the same resource (i.e., with identical niches with respect to a particular resource) cannot coexist indefinitely in an ecosystem in which there is not enough of that resource to meet the needs of both species. This finding is called the **competitive exclusion** principle, sometimes described as the "one-niche, one-species, one-place" principle.

Some people are surprised to learn that competition among plant species is often more intense than that among animals. Some plants displace others by having leaf and root systems that allow them to absorb more sunlight and soil nutrients than their competitors (*exploitation competition*). Other plants, such as broom sedge grass, produce chemicals that inhibit the growth or germination of seeds of competing species (*interference competition*).

Case Study: The Fire Ant Strong interspecific competition can result from the accidental or deliberate introduction of nonnative species into an ecosystem. In the late 1930s, for example, extremely aggressive red fire ants were accidentally introduced into the United States in Mobile, Alabama, perhaps arriving on shiploads of lumber imported from South America or by hitching a ride in the soil-containing ballast of cargo ships. Without natural predators, these ants have spread rapidly by land and water (they can float) throughout the South from Texas to Florida and as far north as Tennessee and Virginia. They have also hitched rides on truckloads of produce going to Arizona, California, New Mexico, Oregon, and Washington. So far, the ants' aversion to frost has kept them out of the Midwest and Northwest, but this might not last long because of genetic changes in some of their populations.

Wherever the fire ant has gone, up to 90% of native ant populations have been sharply reduced or wiped out. Without natural predators to control

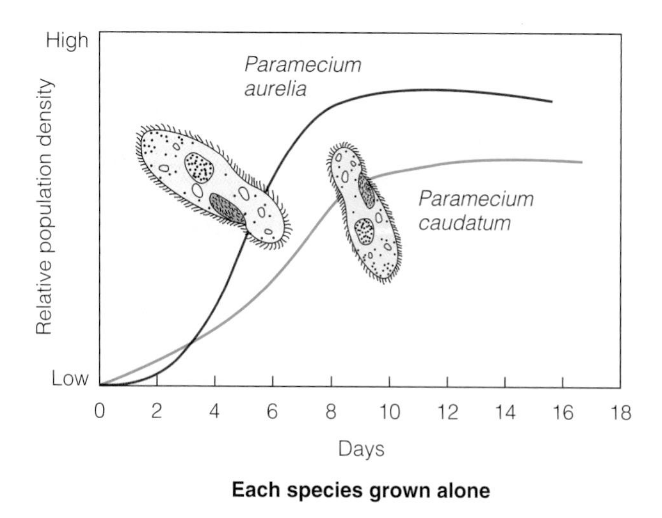

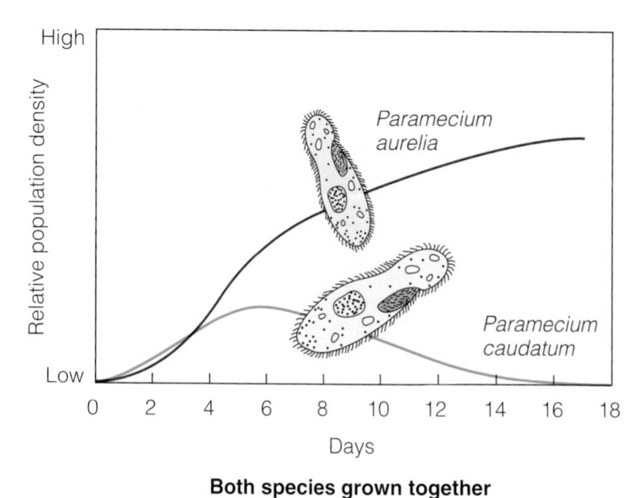

Figure 9-3 The results of G. F. Gause's classic laboratory experiment with two similar, single-celled, bacteria-eating species of *Paramecium* (that reproduce asexually) support the competitive exclusion principle that similar species cannot indefinitely occupy the same ecological niche.

Hint: Enter the search terms *cecropia, ants* using Key Words.

their populations, the invaders reduce the access of other ant species to resources by beating them in direct combat and by having 10 times more colonies (large, foot-high mounds) per area than native ant species. Interference competition by fire ants has also reduced populations of many other species, including ladybugs and many species of spiders, ticks, and cockroaches.

Their extremely painful stings have also killed deer fawn, lizards, birds, livestock, pets, and at least 100 people allergic to their venom. In the South, they have made parents afraid to have their small children play in backyards. They have invaded cars and caused accidents by attacking drivers. They have made cropfields unplowable, disrupted phone service and electrical power, and caused some fires by chewing through underground cables.

Mass spraying of pesticides in the 1950s and 1960s temporarily reduced populations of fire ants. In the end, however, this chemical warfare hastened the advance of the fire ant by reducing populations of many native ant species. The rapidly multiplying fire ants also developed genetic resistance to heavily used pesticides. Researchers at the U.S. Department of Agriculture are evaluating the use of a tiny parasitic fly that lays its eggs on the fire ant's body. After the eggs hatch, the larvae eat their way through the ant's head. Before releasing these biological control agents, however, researchers must be sure they will not cause problems for native ant species.

How Have Some Species Reduced or Avoided Competition? Over a time scale long enough for evolution to occur, species that compete for the same resources may evolve adaptations that reduce or avoid competition or overlap of their fundamental niches, a process that increases biological diversity instead of leading to local extinction. One way this happens is through **resource partitioning**, the dividing up of scarce resources so that species with similar requirements use them at different times, in different ways, or in different places (Figure 9-4). In effect, negative feedback loops (Section 2-3) evolve that allow them to share the wealth, with each competing species occupying a realized niche that makes up only part of its fundamental niche.

For example, where lions and leopards live in the same area, lions take mostly larger animals as prey; leopards take smaller ones. Hawks and owls feed on similar prey, but hawks hunt during the day; owls hunt at night. Some bird species feed on the ground, whereas others seek food in trees and shrubs. Ecologist Robert H. MacArthur studied the feeding habits of five species of warblers (small insect-eating birds) that coexist in the forests of the northeastern United States and in the adjacent area of Canada. Although they appear to be competing for the same food resources, MacArthur found that the bird species reduce competition by spending at least half their time hunting for insects in different parts of trees (Figure 9-5).

On an evolutionary time scale, closely related and anatomically similar competing species may also parti-

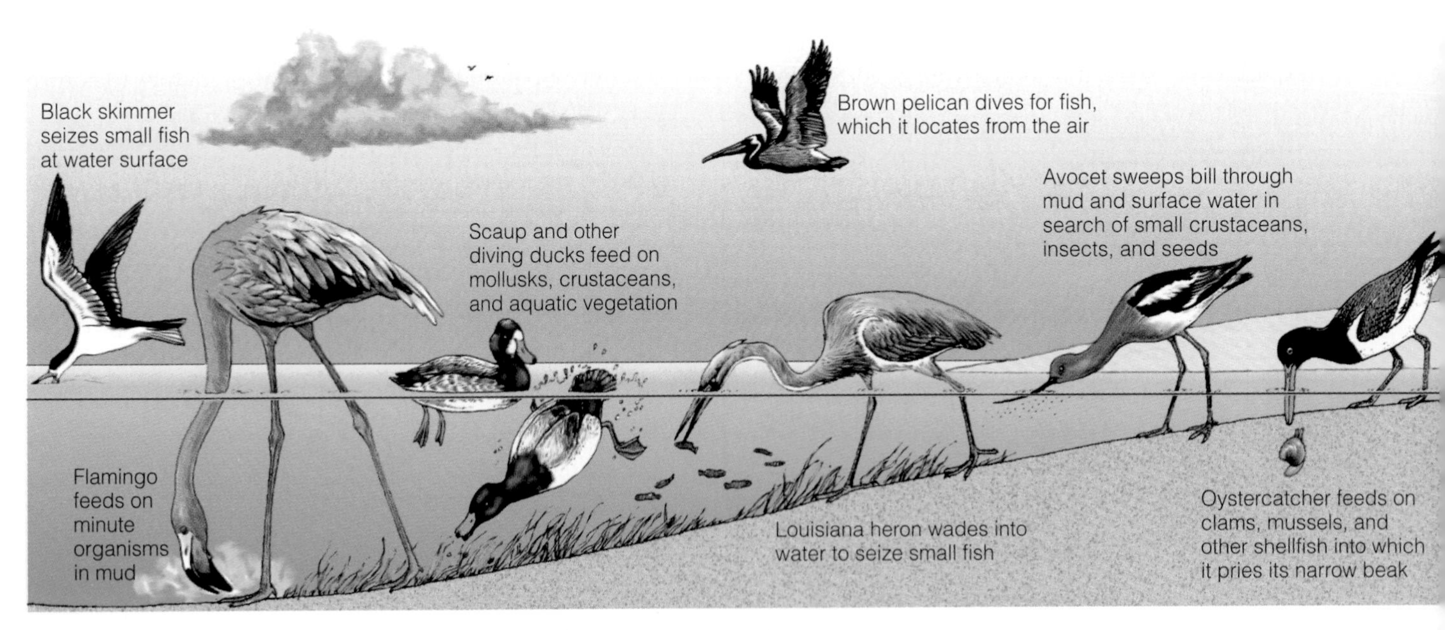

Figure 9-4 Specialized feeding niches of various species of birds in a coastal wetland. Such resource partitioning reduces competition and allows sharing of limited resources.

Q: What percentage of the earth's tropical forests have been cleared or damaged?

Figure 9-5 Resource partitioning of five species of common insect-eating warblers in spruce forests of Maine. Each species minimizes competition with the others for food by spending at least half its feeding time in a distinct portion (shaded areas) of the spruce trees; each also consumes somewhat different insect species from the others. (After R. H. MacArthur, "Population Ecology of Some Warblers in Northeastern Coniferous Forests," *Ecology*, vol. 36, 533–36, 1958.)

tion resources and lessen competition through **character displacement**. Such species develop physical or behavioral characteristics or adaptations that allow them to use different resources. In birds, for instance, bill sizes of similar species found in the same ecosystem often differ. Field research has shown that one species of ground finch with a longer, thinner bill feeds on large insects, whereas another with a shorter, thicker bill feeds on seeds or small insects (Figure 9-6). Where these same bird species occur alone, their beak size is often similar.

How Do Predator and Prey Species Interact?

Recall that in *predation*, members of a *predator* species feed on members of a *prey* species, but they do not live on or in the prey. Together, the two kinds of organisms, such as lions (the predator or hunter) and zebras (the prey or hunted), are said to have a **predator–prey relationship**, as depicted in Figures 4-10, 4-11, 4-16, and 4-17. When people hunt, fish, or pick and eat vegetation, they act as predators. When we buy meat or vegetables from a grocery store we are indirect predators who have someone else kill, skin, and cut up cattle,

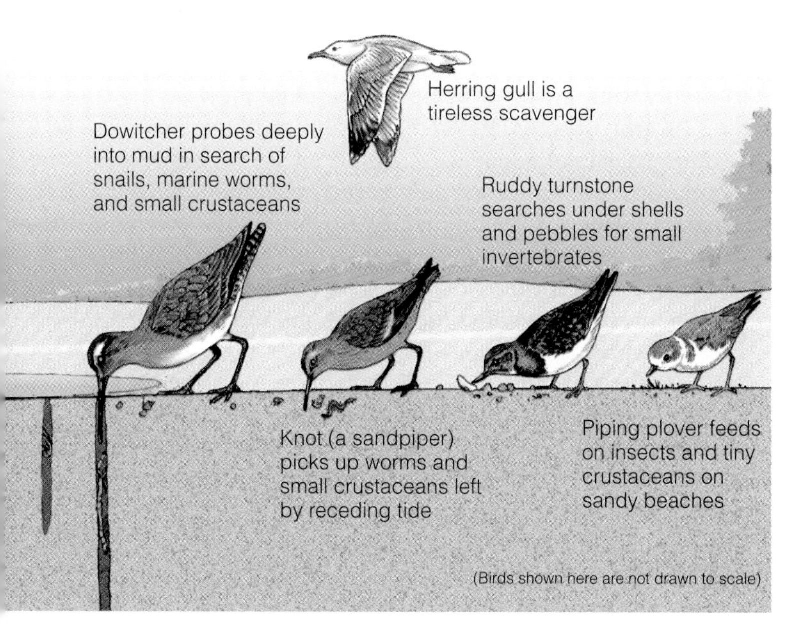

A: About 56%

Figure 9-6 Character displacement. Over many generations closely related species occupying the same fundamental niche in an ecosystem may reduce or avoid competition by undergoing physical adaptations. One example is changes in bill size that allow the species coexisting in the same ecosystem to eat different types of seeds or insects. When the same species occur alone, their beak sizes are often similar.

chickens, or other sources of meat for us and pick and process vegetables and other plants for us.

A predator–prey relationship is a positive feedback system (Figure 2-7), at least for the victor (predator), who gets more of scarce resources than does the loser (prey). This *win–lose* system is limited by negative feedback when the prey population falls below the level needed to support the predator population.

We normally use the term *predator* for animals (carnivores) that feed on other animals. However, the term can also be applied to animals (herbivores) that feed on plants and even to carnivorous plants such as the Venus flytrap and aquatic bladderworts, which capture living prey (insects) in a leaf that closes like a steel trap.

Prey organisms may or may not be killed by their predators. Most carnivores such as cheetahs and lions hunt and kill their prey. However, herbivores such as zebras or mule deer that eat plant parts often harm the plant but don't necessarily kill it.

At the individual level, members of the prey species are clearly harmed, but at the population level, predation can benefit the prey species because predators often kill the sick, weak, and aged members (Case Study, p. 217). Reducing the prey population gives remaining prey greater access to the available food supply. It can also improve the genetic stock of the prey population, which enhances its chances of reproductive success and long-term survival. The effects of predation on populations of predator and prey species are discussed in more detail in Section 10-2.

Some people tend to view predators with contempt, often not recognizing their own roles as indirect predators. This view is often based on cultural conditioning through stories about the Big Bad Wolf trying to prey on the Three Little Pigs or Little Red Riding Hood. When a hawk tries to capture and feed on a rabbit, many people tend to side with the rabbit. Yet the hawk (like all predators) is merely trying to get enough food to feed itself and its young; in the process, it is playing an important role in its ecosystem.

How Do Predators Increase Their Chances of Getting a Meal? Predators have a variety of methods that help them capture prey. Herbivores can simply walk, swim, or fly up to their plant prey.

Carnivores feeding on mobile prey have two main options: *pursuit* and *ambush*. Some, such as the cheetah, catch prey by being able to run fast; others, such as the American bald eagle, fly and have keen eyesight; still others, such as wolves and African lions, cooperate in capturing their prey by hunting in packs. Humans have invented tools (weapons and traps) to capture prey.

Other predators ambush their prey. A stationary frog ambushes flying insects by flicking out its long sticky tongue. Many spiders and some praying mantises (Figure 4-1, bottom) sit in flowers that match their color and ambush visiting insects. The color of a lion allows it to blend in with savanna grass and have a better chance of getting close enough to its prey to make a kill. The white coloration of ermines (a type of weasel) and snowy owls found in snow-covered areas enables them to hide and ambush their prey. The alligator snapping turtle ambushes fish prey by lying camouflaged

Q: What percentage of the earth's land surface is covered by tropical rain forests?

CASE STUDY

Why Are Sharks Important Species?

The world's 350 shark species range in size from the dwarf dog shark, about the size of a large goldfish, to the whale shark, the world's largest fish at 18 meters (60 feet) long. Various shark species, feeding at the top of food webs, cull injured and sick animals from the ocean and thus play an important ecological role. Without such shark species the oceans would be overcrowded with dead and dying fish.

Many people, influenced by movies and popular novels, think of sharks as people-eating monsters. But the two largest species—the whale shark and the basking shark—sustain their enormous bulk by filtering out and swallowing huge quantities of *plankton* (small free-floating sea creatures).

Every year, members of a few species of shark—mostly great white, bull, tiger, gray reef, lemon, and blue—injure about 50–100 people worldwide and kill between 5 and 12. Most attacks are by great white sharks, which feed on sea lions and other marine mammals and sometimes mistake surfers and divers in wet suits for their usual prey. A typical oceangoer is 150 times more likely to be killed by lightning than by a shark.

For every shark that injures a person, we kill 500,000 to 1 million sharks, for a total of 50 to 100 million sharks each year. Sharks are killed mostly for their fins, widely used in Asia as a soup ingredient and as a pharmaceutical cure-all.

Sharks are also killed for their livers, meat (especially mako and thresher), and jaws (especially great whites), or just because we fear them. Some sharks (especially blue, mako, and oceanic whitetip) die when they are trapped in nets deployed to catch swordfish, tuna, shrimp, and other commercially important species.

Sharks also help save human lives. In addition to providing people with food, they are helping us learn how to fight cancer (which sharks almost never get), bacteria, and viruses. Their highly effective immune system is being studied because it allows wounds to heal without becoming infected, and their blood is being studied in connection with AIDS research.

Chemicals extracted from shark cartilage (the elastic connective tissue attached to bones) have killed several types of cancerous tumors in laboratory animals. This research someday might help prolong your life or the life of a loved one.

Sharks have several natural traits that make them prone to population declines from overfishing. Unlike most other fish, they have only a few offspring (between 2 and 10) once every year or two. Depending on the species, sharks require 10 to 15 years (and in some cases, 24 years) to reach sexual maturity and begin reproducing. Sharks also have long gestation (pregnancy) periods—as much as 24 months for some species.

Several steps are being taken to reduce the slaughter of sharks. In January 1993, the National Marine Fisheries Service imposed limits on the commercial and sport fishing catch for 39 of the most vulnerable species in U.S. waters. All sharks killed must be brought ashore whole, a requirement designed to reduce finning. New laws in South Africa ban hunting of great whites within 320 kilometers (200 miles) of its coast.

With more than 400 million years of evolution behind them, sharks have had a long time to get things right. Preserving this evolutionary genetic wisdom begins with the recognition that sharks don't need us, but we and other species need them.

Critical Thinking

Do you fear sharks? Why? How do you feel about killing sharks in large numbers? Why?

on its stream-bottom habitat and dangling its worm-shaped tongue to entice fish into its powerful jaws.

Drops of sticky digestive fluid looking somewhat like pollen on the tentacles of a carnivorous sundew plant (found in many bogs in the United States) lure insect prey.

How Do Prey Defend Themselves Against Predators? Prey species have various protective mechanisms; otherwise, they would easily be captured and eaten. Some can run, swim, or fly fast; others have highly developed sight or a sense of smell that alerts them to the presence of a predator. Still others have protective shells (armadillos, which roll themselves up into an armor-plated ball, and turtles) or thick bark (giant sequoia); some have spines (porcupines) or thorns (cacti and rosebushes). Many lizards have brightly colored tails that break off when they are attacked, often giving them enough time to escape.

Other prey species camouflage themselves by having certain shapes or colors (Figure 9-7) or the ability to change color (chameleons and cuttlefish). A brilliant green tree frog tends to be invisible against a background of lush rain forest leaves. Similarly, an arctic hare in its white winter fur blends into the snow. Some insect species have evolved shapes that look like twigs (stick caterpillars), thorns (treehoppers), dead leaves, or bird droppings on leaves.

A: About 2% (about 6% is covered by all types of tropical forests combined)

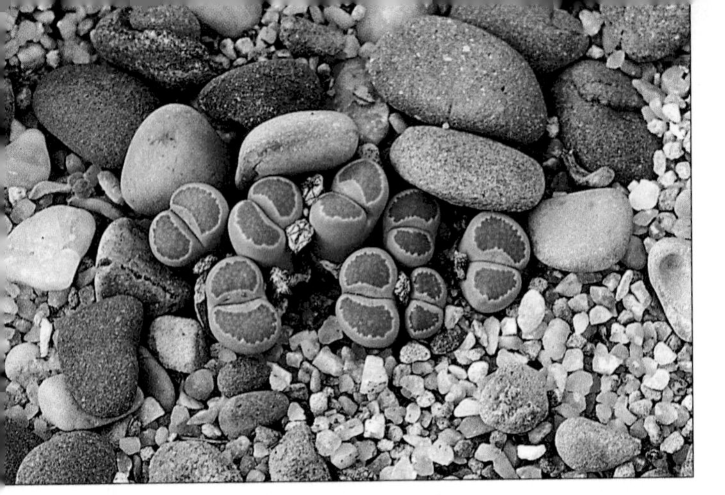

Figure 9-7 Predator avoidance by camouflage. The African stoneplant avoids predators by closely resembling nearby stones. Can you find it? (Heather Angel)

Figure 9-8 Predator avoidance by chemical warfare. When a bombardier beetle is assaulted by an ant, it defends itself by spewing out a boiling hot irritant from special glands. (Thomas Eisner)

Figure 9-9 Predators, Beware. Warning coloration on this poison dart frog found in a rain forest in Costa Rica can send a signal to experienced predators not to eat its poisonous flesh. The skin of these frogs oozes a toxin that is one of the strongest animal poisons known. Members of some hunter–gatherer tribes numb or kill prey by using arrow tips dipped in these secretions. (M. P. L. Fogden/Bruce Coleman, Ltd.)

Figure 9-10 The viceroy butterfly (right) gains some protection from predators by looking like the poisonous monarch butterfly (left). For years it was believed that, unlike the monarch butterfly, the viceroy butterfly was not foul-tasting to birds. However, recent taste-test research by Lincoln Brower with wild blackbirds indicates that they find viceroy butterflies as foul-tasting as the monarch butterflies the viceroys mimic. (© Breck P. Kent/Animals Animals)

Chemical warfare is another common strategy. Some prey species discourage predators with chemicals that are poisonous (oleander plants), irritating (bombardier beetles, Figure 9-8), foul smelling (skunks, skunk cabbages, and stinkbugs), or bad tasting (buttercups and monarch butterflies). To date, scientists have identified over 10,000 defensive chemicals made by plants, including cocaine, caffeine, nicotine, cyanide, opium, strychnine, peyote, and rotenone (used as an insecticide).

Many bad-tasting, bad-smelling, toxic, or stinging prey species have evolved *warning coloration*, brightly colored advertising that enables experienced predators to recognize and avoid them. Examples are brilliantly colored poisonous frogs (Figure 9-9) and red-yellow-and-black-striped coral snakes, foul-tasting monarch butterflies (Figure 9-10, left) and grasshoppers, and stinging sea slugs. Other butterfly species, such as the nonpoisonous viceroy, gain some protection by looking and acting like the poisonous monarch, a protective device known as *mimicry* (Figure 9-10, right).

Q: What percentage of North America's original temperate deciduous forestland has been cleared?

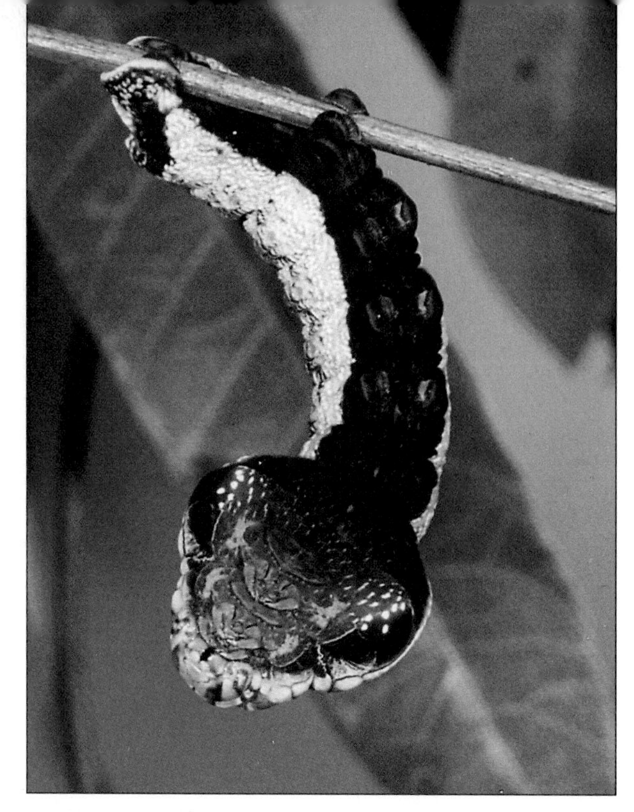

Figure 9-11 Predator avoidance by behavior. When touched, this snake caterpillar alters its body shape to look like the head of a snake. This puffed-up head "strikes" at whatever touches it. (Lincoln P. Brower)

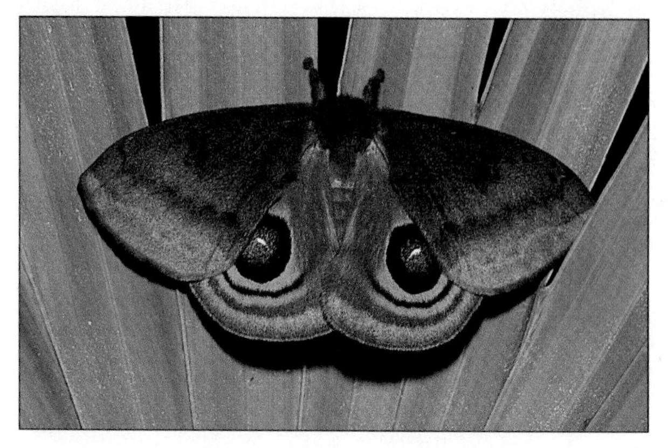

Figure 9-12 Deceptive coloration. The hindwing markings of this io moth in Florida's Osceola National Park resemble the eyes of a much larger animal. When the moth moves its forewings, this may startle potential predators long enough for the moth to escape. (NHPA/Rod Planck)

Some prey species use behavioral strategies to avoid predation. Some attempt to scare off predators by puffing up (blowfish), spreading their wings (peacocks and short-eared owls), or mimicking a predator (Figure 9-11). To help fool or frighten would-be predators, some moths have wings that look like the eyes of much larger animals (Figure 9-12). Other prey gain some protection by living in large groups (schools of fish, herds of antelope, flocks of birds).

9-4 SYMBIOTIC SPECIES INTERACTIONS: PARASITISM, MUTUALISM, AND COMMENSALISM

What Are Parasites and Why Are They Important? Recall that *parasitism* is an interaction in which a member of one species (the *parasite*) obtains its nourishment by living on, in, or near a member of another species (its *host*) over an extended time. Parasitism can be viewed as a special form of predation, but unlike a conventional predator, a parasite **(1)** is usually smaller than its host (prey), **(2)** remains closely associated with, draws nourishment from, and may gradually weaken its host over time, and **(3)** rarely kills its host.

The small number of parasites that routinely kill their hosts as part of their life cycle are called *parasitoids*. Examples are parasitic wasps, which lay their eggs in the larvae of another insect species. When the young hatch, they kill their living host by feeding on its tissues. Because parasitoids kill specific host species, farmers sometimes use them to help control populations of pest species, as discussed in more detail in Section 21-5.

Some parasites, such as tapeworms (often yards long) and plasmodium microorganisms that cause malaria, live *inside* their hosts and are called *endoparasites*. Some biologists classify disease-causing bacteria and viruses as endoparasites. Other parasites, such as lice, ticks, mosquitoes, mistletoe plants, fungi (that cause diseases such as athlete's foot), and lampreys (Figure 9-13), attach themselves to the *outside* of their hosts and are called *ectoparasites*. Some parasites, especially ectoparasites, move from one host to another, as fleas and ticks do; others, such as tapeworms, spend their adult lives with a single host.

All species have parasites, and you and most hosts support hundreds of parasites of many species. Many parasites also have parasites. Deer are hosts to parasitic protozoa (common single-celled animals) in their guts, fungi on their hooves, and ticks, mites, and flies on their skin. Oak trees typically harbor various parasitic mushrooms, lichens, mosses, and insects.

From the host's point of view parasites are harmful, but parasites play important ecological roles. Collectively, the incredibly complex matrix of parasitic relationships in an ecosystem acts somewhat like a glue that helps hold the species in an ecosystem together. Parasites living within their hosts also help dampen drastic swings in population sizes and

A: About 99.9%

Figure 9-13 Parasitism. Sea lampreys use their suckerlike mouths to attach themselves to their fish hosts. They then bore a hole in the fish with their teeth and feed on its blood. (Tom Stack)

parasites promote biodiversity by helping prevent some organisms from becoming too plentiful.

Donald Winsor speculates that "if not for parasites appearing very early in evolutionary history, the enormous biodiversity we witness today would not have evolved. Instead, just a few successful species would have taken over Earth a long time ago." It is becoming increasingly clear that we cannot understand how a community or ecosystem functions without understanding its complex and important web of parasitic interactions.

How Do Species Interact So That Both Species Benefit? *Mutualism* is a symbiotic relationship in which both interacting species benefit. These benefits could include having pollen dispersed for reproduction (p. 206), being supplied with food, or receiving protection. It is tempting to think of mutualism as an example of cooperation between species, but it actually involves each species benefiting by exploiting the other.

In Section 2-3 you learned that positive or runaway feedback loops can cause a system to collapse. But positive feedback isn't always harmful. Mutualism is a positive feedback loop in which different species are involved in a *win–win* relationship.

The pollination relationship between flowering plants and insects is one of the most common forms of mutualism. Butterflies and bees depend on flowers for food in the form of nectar and pollen. In turn, the flowering plants depend on bees, bats (Figure 9-1c), or other pollinators to carry their male reproductive cells (sperm in pollen grains) to the female flowering parts of other flowers of the same species.

There are many examples of nutritional mutualism. One involves *lichens*, hardy species that can grow on trees or barren rocks. Lichens actually consist of two organisms—colorful photosynthetic algae and chlorophyll-lacking fungi—living together in an intimate mutualistic relationship. The fungi provide a home for the algae, and their bodies collect and hold moisture and mineral nutrients used by both species. The algae, through photosynthesis, provide sugars as food for themselves and the fungi. The fungi and algae of lichens cannot live apart and are therefore an example of *obligatory mutualism*.

Plants in the legume family have root nodules (Figure 5-7), where *Rhizobium* bacteria convert atmospheric nitrogen into a form usable by the plants, and the plants provide the bacteria with some simple sugars. Other important nutritional mutualistic relationships exist between animals and the vast armies of bacteria in their digestive systems that break down (digest) their food. The bacteria gain a safe home with a steady food supply; the animal gains more efficient access to a large source of energy. In addition to helping digest your food, bacteria in your intestines synthesize vitamin K and the B-complex vitamins, which you can't make yourself.

Termites have a nutritional mutualistic relationship with tiny protozoans found in their gut. These microbes digest the wood the termites eat, and the insects use some of the resulting sugar as food. Neither termites nor protozoans could live without the other. Japanese scientists have recently discovered termite species that can eat steel and concrete.

Many mutualistic relationships provide a combination of nutrition and protection. One involves birds that ride on the backs of large animals such as African buffalo, elephants, and rhinoceroses (Figure 9-14). The birds remove and eat parasites from the animal's body and often make noises warning the animal when predators approach.

Another such example is the mutualistic relationship between stinging ants and certain species of acacia trees in dry areas of Central and South America and East Africa. These trees, called bull horn acacias, have hollow horns of swollen thorns that house colonies of certain species of stinging ants. The ants get nutrients from sugar-rich nectar secreted from glands near the base of the tree's leaves; they are protected within the horn of thorns. The ants sense vibrations caused when an insect or larger herbivore touches the tree's swollen thorns; they then swarm out, sting, and drive away the intruder, sometimes killing and devouring plant-eating insects. This helps protect the tree from herbivores and provides additional food for the ants.

The ants also help the acacias compete with other plants by cutting away branches of other plants that touch the trees. In addition to destroying nearby competing plants, this creates a tunnel of light through

Q: What percentage of the world's people live in mountain biomes?

Figure 9-14 Mutualism. Oxpeckers (or tickbirds) feed on the parasitic ticks that infest large, thick-skinned animals such as this endangered black rhinoceros in Kenya. The rhino benefits by having parasites removed from its body, and oxpeckers benefit by having a dependable source of food. The oxpecker also serves as a sentinel for the rhino by making a fierce hissing sound when alarmed. (Joe McDonald/Tom Stack & Associates)

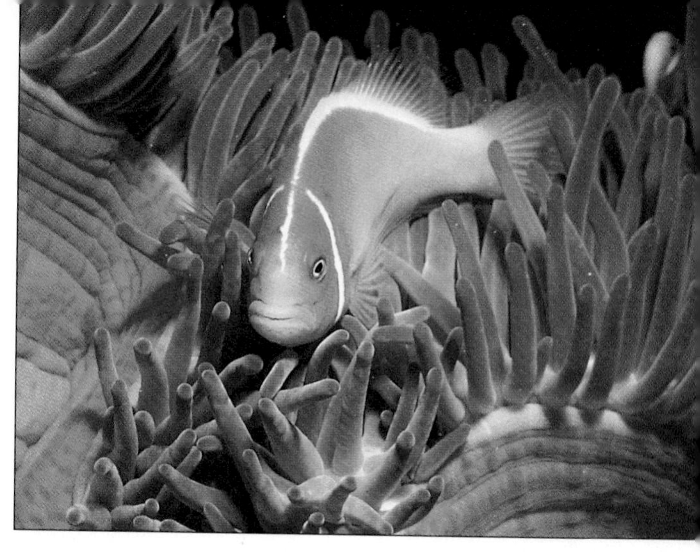

Figure 9-16 This clownfish in the Coral Sea, Australia, has a mutualistic relationship with deadly sea anemones, whose tentacles quickly paralyze most other fish that touch them. The clownfish gains protection and food by feeding on scraps left over from fish killed by the sea anemone, which, in turn, gains protection by the clownfish against various fish that feed on sea anemones. (Carl Roessler/FPG International)

Figure 9-15 An illustration of the importance of the mutualistic relationship between mycorrhizae (fungi) and the roots of a juniper tree. The juniper seedlings to the left were grown for 6 months in sterilized soil inoculated with a mycorrhizae fungus; the seedlings to the right were grown for the same length of time in sterilized soil without the fungus. (F. B. Reeves)

which the acacia can get sunlight and grow, even in lush forests. A series of experiments in the 1950s and 1960s by tropical biologist Daniel Janzen demonstrated the mutualistic nature of this relationship. When the ants on experimental trees were poisoned, the trees died.

Nutritional and protection mutualism also occurs between most plants and minute fungi called mycorrhizae (from the Greek for "root fungus") living on their roots (Figure 9-15). The fungi get nutrition from a plant's roots and in turn benefit the plant by sending out billions of hairlike extensions into the soil around the roots. These extensions help the plant absorb more water and nutrients than its roots could on their own. Some of these fungi also enhance plant growth by producing a growth hormone. Others protect plants by sopping up potential toxins, and they defend a plant's root system against small roundworms by using their filament nets to trap and digest them.

In another example of mutualism, various species of clownfish gain protection from sea anemones, marine animals with stinging tentacles that paralyze most fish that touch them (Figure 9-16). The clownfish are not harmed by sea anemones. They gain protection by living among the deadly tentacles, and they feed on the detritus left from the meals of the anemone. The sea anemones benefit because the clownfish protect them from some predators that eat anemones.

Mutualism is more common when resources are scarce. For example, in soils with low nitrogen content, legumes and other plants with nitrogen-fixing organisms in their roots are favored over plants that don't house nitrogen-fixing organisms. In other words, when the going gets tough, the tough often survive by evolving mutually beneficial relationships with other species.

How Do Species Interact So That One Benefits but the Other Is Not Harmed? *Commensalism* involves a symbiotic association in which one species benefits while the other is neither helped nor harmed to any significant degree. For example, redwood sorrel, a small herb, benefits from growing in the shade of

A: About 10%

Figure 9-17 Commensalism. This white orchid (an epiphyte or air plant from the tropical forests of Latin America) roots in the fork of a tree rather than the soil without penetrating or harming the tree. In this interaction, the epiphytes gain access to water, other nutrient debris, and sunlight; the tree apparently remains unharmed. (Kenneth W. Fink/Ardea London)

tall redwood trees, with no known effects on the redwood trees.

Another example is the commensalistic relationship between various trees and other plants called *epiphytes* (such as various types of orchids and bromeliads) that attach themselves to the trunks or branches of large trees (Figure 9-17) in tropical and subtropical forests. These so-called air plants benefit by having a solid base on which to grow and by living in an elevated spot that gives them better access to sunlight. Their position in the tree allows them to get most of their water from the humid air and from rain collecting in their usually cupped leaves. They also absorb nutrient salts falling from the tree's upper leaves and limbs and from the dust in rainwater. Commensalism exists in cases in which an epiphyte does not harm the tree in which it grows. However, in some cases epiphytes become very abundant and can block light, which becomes an example of interspecific competition.

9-5 ECOSYSTEM STRUCTURE AND ECOLOGICAL SUCCESSION

What Is Ecosystem Structure? When ecologists describe the structure of an ecosystem, they usually do this in terms of four characteristics:

- *Physical appearance*: relative sizes and stratification of its species (Figure 9-18)
- *Niche structure*: the number of ecological niches and how they resemble or differ from each other
- *Species diversity or richness*: the number of different species
- *Species abundance*: the number of individuals of each species, especially the numbers of rare species relative to common species

How Do Ecosystems Respond to Change? One characteristic of all ecosystems is that their structures are constantly changing in response to changing environmental conditions. The gradual and fairly predictable change in species composition of a given area is called **ecological succession**. Succession involves a complex set of species interactions over time. During succession some species colonize and their populations become more numerous, whereas populations of other species decline and even disappear.

Ecologists recognize two types of ecological succession: primary and secondary, depending on the conditions present at the beginning of the process. **Primary succession** involves the gradual establishment of biotic communities in an area that has not been occupied by life before. In contrast, **secondary succession**, the more common type of succession, involves the *reestablishment* of a biotic community in an area where a biotic community was previously present.

What Is Primary Succession? Primary succession begins with an essentially lifeless area, where there is no soil in a terrestrial ecosystem or no bottom sediment in an aquatic ecosystem (Figure 9-19). Examples of such barren areas include bare rock exposed by a retreating glacier or severe soil erosion, newly cooled lava, an abandoned highway or parking lot, or a newly created shallow pond or reservoir.

Establishment of a new biological community is generally slow. Before a community of plants (producers), consumers, and decomposers can become established on land, there must be *soil*: a complex mixture of rock particles, decaying organic matter, air, water, and living organisms (Figure 5-13). Depending mostly on the climate, it takes natural processes several hundred to several thousand years to produce fertile soil.

Soil formation begins when hardy **pioneer species** attach themselves to inhospitable patches of bare rock. Examples are wind-dispersed lichens and mosses, which can withstand lack of moisture and soil nutrients and hot and cold temperature extremes found in such habitats.

These species can extract nutrients from dust in rain or snow and from bare rock. They start the process of soil formation on patches of bare rock by trapping wind-blown soil particles and tiny pieces of detritus, by producing tiny bits of organic matter, and

Q: What percentage of the earth's surface is covered by oceans?

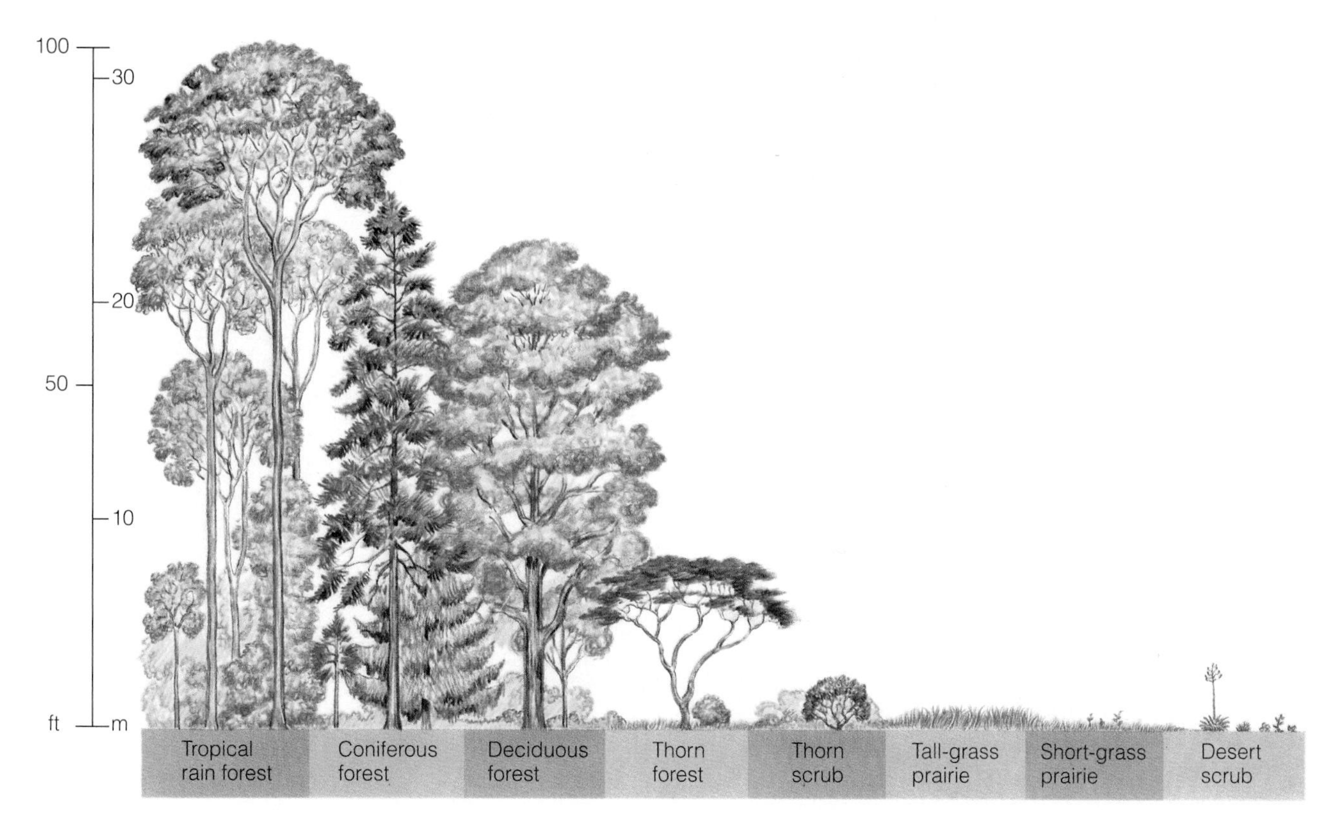

Figure 9-18 Generalized physical appearance showing the types, relative sizes, and stratification of plant species in various terrestrial communities or ecosystems.

by secreting mild acids that slowly fragment and break down the rock. This chemical breakdown is hastened by physical weathering such as the fragmentation of rock when water freezes in cracks and expands.

As very thin patches of soil accumulate, they contain enough nutrients and retain enough water to support some species of bacteria, fungi, insects, and small worms. As these organisms grow, reproduce, and die, they add organic material to the developing soil. Eventually the community of lichens and mosses is replaced by a community of small perennial grasses (plants that live for more than 2 years without having to reseed) and herbs (ferns in tropical areas), whose seeds germinate after being blown in by the wind (or carried there in the droppings of birds or on the coats of animals passing through).

These **early successional plant species** grow close to the ground, can establish large populations quickly under harsh conditions, and have short lives—species that biologist Edward O. Wilson calls "nature's sprinters." Some of their roots penetrate into the rock and help break it up into more soil particles, and the decay of their wastes and dead bodies adds more nutrients to the soil. Soil buildup is hastened by the arrival of soil-dwelling microorganisms (many of them blown in by the wind) and small animals.

After hundreds of years the soil may be deep and fertile enough to store enough moisture and nutrients to support the growth of less hardy **midsuccessional plant species** of herbs, grasses, and low shrubs. These, in turn, are replaced by trees that require lots of sunlight and are adapted to the area's climate and soil.

As these tree species grow and create shade, they are replaced by **late successional plant species** (mostly trees) that can tolerate shade. Unless fire, flooding, severe erosion, tree cutting, climate change, or other natural or human processes disturb the area, what was once bare rock becomes a fairly stable and complex forest community (Figure 9-19).

The specific composition of pioneer, early successional, midsuccessional, and late successional communities—and the rates of primary succession—varies from one site to another. Generally, primary succession occurs fastest in humid tropical areas and slowest in dry polar areas.

A: About 71%

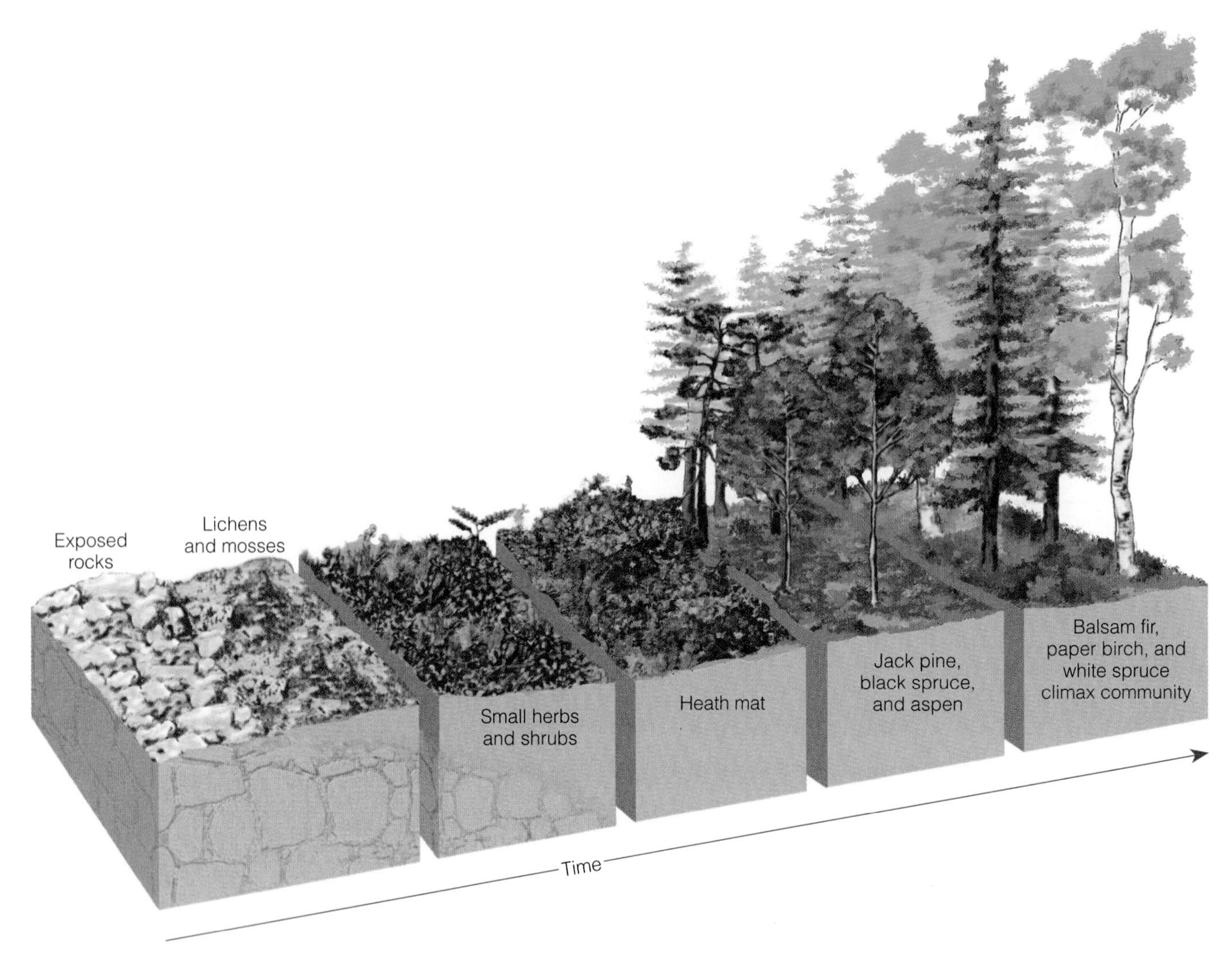

Figure 9-19 Primary succession over several hundred years of plant communities on bare rock exposed by a retreating glacier on Isle Royal in northern Lake Superior.

Primary succession also occurs in aquatic ecosystems such as a shallow pond or reservoir created by humans or by some natural change. This is illustrated by succession in a pond at the south end of Lake Michigan (Figure 9-20).

What Is Secondary Succession? Secondary succession begins in an area where the natural community of organisms has been disturbed, removed, or destroyed, but the soil or bottom sediment remains. Candidates for secondary succession include abandoned farmlands, burned or cut forests, heavily polluted streams, and land that has been dammed or flooded. Because some soil or sediment is present, new vegetation can usually begin to germinate within a few weeks.

In the central (Piedmont) region of North Carolina, European settlers cleared the mature native oak and hickory forests and replanted the land with crops. Some of the land was subsequently abandoned because of erosion and loss of soil nutrients. Figure 9-21 shows how such abandoned farmland has undergone secondary succession.

Descriptions of ecological succession usually focus on changes in vegetation. However, these changes in turn affect food and shelter for various types of animals. Thus, as succession proceeds the numbers and types of animals and decomposers also change. Figure 9-22 shows some of the wildlife species likely to be found at various stages of secondary ecological succession in areas with a temperate climate.

Because primary and secondary succession involve changes in community structure (Figure 9-18), it is not surprising that the various stages of succession have different patterns of species diversity, trophic structure, niches, nutrient cycling, and energy flow and efficiency (Table 9-1 and Figure 9-23).

Q: What percentage of the earth's coral reefs have been destroyed by human activities?

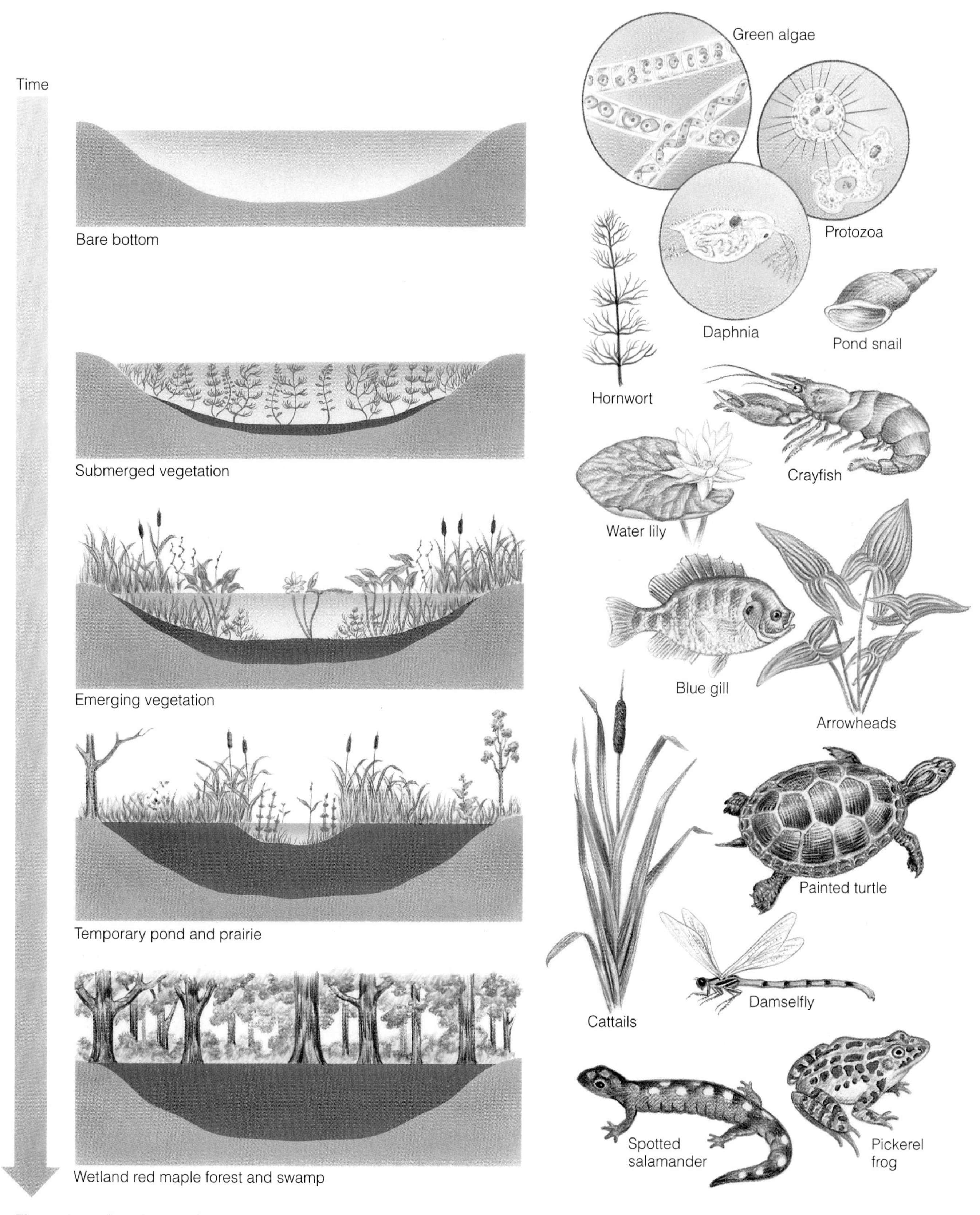

Figure 9-20 Greatly simplified view of primary succession in a newly created pond in a temperate area.. Nutrient-rich bottom sediment is shown in dark brown.

A: About 10% and another 30% are in critical condition

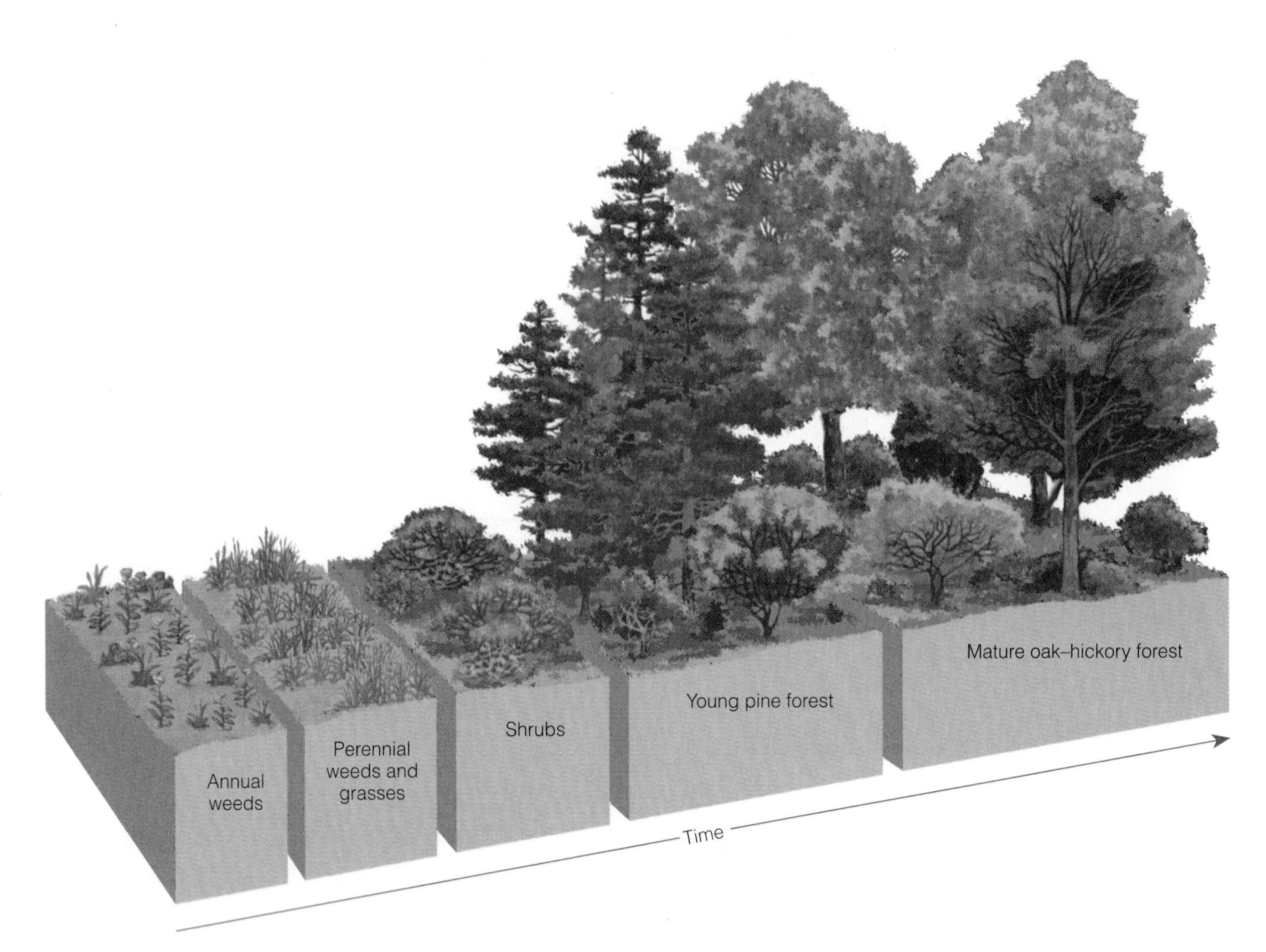

Figure 9-21 Secondary ecological succession of plant communities on an abandoned farm field in North Carolina. It took about 150–200 years after the farmland was abandoned for the area to be covered with a mature oak and hickory forest. A new disturbance such as deforestation or fire would create conditions favoring pioneer species, and in the absence of new disturbances, secondary succession would again occur over time, although not necessarily in the same sequence shown here.

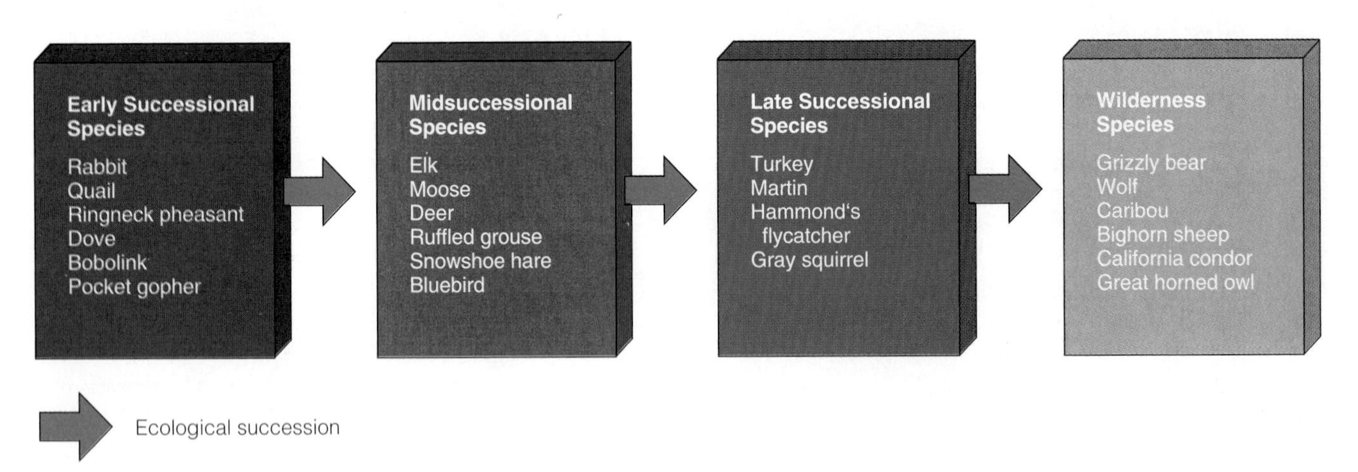

Figure 9-22 Examples of wildlife species typically found in areas at different stages of ecological succession in areas of the United States with a temperate climate.

Q: What percentage of the world's people live on or near coastal areas?

Table 9-1 Ecosystem Characteristics at Immature and Mature Stages of Ecological Succession		
Characteristic	**Immature Ecosystem (Early Successional Stage)**	**Mature Ecosystem (Late Successional Stage)**
Ecosystem Structure		
Plant size	Small	Large
Species diversity	Low	High
Trophic structure	Mostly producers, few decomposers	Mixture of producers, consumers, and decomposers
Ecological niches	Few, mostly generalized	Many, mostly specialized
Community organization (number of interconnecting links)	Low	High
Ecosystem Function		
Biomass	Low	High
Net primary productivity	High	Low
Food chains and webs	Simple, mostly plant → herbivore with few decomposers	Complex, dominated by decomposers
Efficiency of nutrient recycling	Low	High
Efficiency of energy use	Low	High

How Do Species Replace One Another in Ecological Succession? Ecologists have identified three factors that affect how and at what rate succession occurs: *facilitation*, *inhibition*, and *tolerance*.

Sometimes pioneer species make an area suitable for species with different niche requirements. As lichens and mosses gradually build up soil on a rock in primary succession, for example, herbs and grasses can colonize the site. Similarly, plants such as legumes (Figure 5-7) add nitrogen to the soil, making it more suitable for other plants found at a later stage of succession. Recent research indicates that this process, called *facilitation*, plays an important role in primary succession when pioneer species first begin building soil. In another example, facilitation in both primary and secondary succession results when early successional plants produce shaded conditions that are favorable for germination and growth of later successional species.

A common process governing secondary succession (and primary succession after soil has been built up) is *inhibition*. This occurs when early species hinder the establishment and growth of other species. Succession then can proceed only when a fire, bulldozer, or other disturbance removes most of the existing species.

For example, in the early stages of secondary succession on abandoned farm fields in central North Carolina (Figure 9-21), horseweed inhibited the growth of aster plants that followed them. The horseweed shaded the asters and released toxic chemicals when its roots decayed. Field experiments have confirmed this inhibitory effect by demonstrating that the growth of asters is enhanced when horseweed plants are removed.

In other cases, late successional plants are largely unaffected by plants at earlier stages of succession, a phenomenon known as *tolerance*. Tolerance may explain why late successional plants can thrive in mature communities without eliminating some early successional and midsuccessional plants. There is no consensus among ecologists about whether most of the stages of secondary succession occur because of inhibition, tolerance, or some combination of these two processes.

What Is the Role of Disturbance in Ecological Succession? All forms of life face environmental conditions that change, sometimes gradually, sometimes suddenly (Table 9-2). A **disturbance** is a discrete event in time that disrupts an ecosystem or community. Examples of *natural disturbance* include fires, hurricanes, tornadoes, droughts, and floods. Examples of *human-caused disturbances* include deforestation, overgrazing, and plowing. Disturbances are less

A: Almost 66%

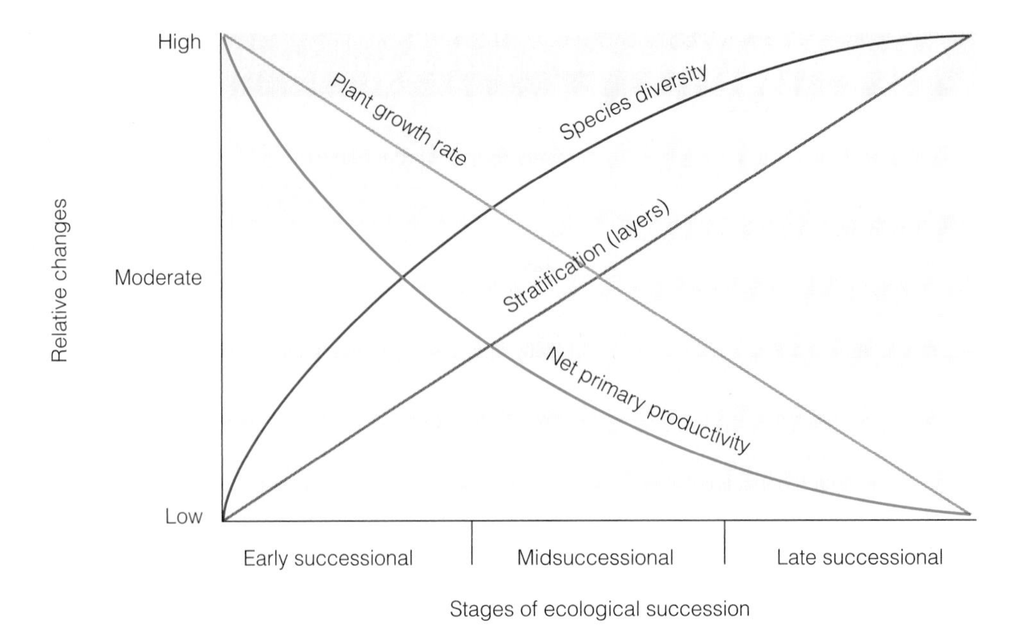

Figure 9-23 Some generalized changes taking place during ecological succession.

frequent and less predictable than natural cycles, such as the daily or seasonal changes in availability of solar radiation. Disturbances such as those shown in Table 9-2 play a major role in ecological succession by converting a particular stage of succession to an earlier stage.

Natural disturbance is an important part of all ecosystems. The immediate effect of a natural disturbance is to change conditions and release resources. When a large tree falls in a tropical forest, this local disturbance leads to increased availability of light and nutrients for growth of plants in the understory. When a log hits a rock in an intertidal ecosystem, this local disturbance dislodges or kills the organisms that are growing on the rock and provides space for colonization of new intertidal organisms. After a fire burns a coniferous forest community, nutrients are released and light becomes abundant near the ground, enabling fast-growing herbaceous plants and new conifer trees to colonize. In all of these cases, discrete events change the ecosystem or community structure in a way that makes resources available to colonizing species, which starts secondary succession.

We tend to think of large terrestrial communities undergoing succession as being covered with a predictable blanket of vegetation. However, a close look at almost any ecosystem reveals that it consists of an ever-changing, irregular mosaic of patches at different stages of succession, the result of a variety of small and medium-sized disturbances. Examples include open spaces created when trees or other plants die and the irregular effects of burning, overgrazing, and drought. This irregular "moth-eaten patchwork quilt" of vegetation increases the diversity of plant and animal life and continually provides sites where early successional species can gain a foothold.

According to the *intermediate disturbance hypothesis*, communities that experience fairly frequent, moderate disturbances have the greatest diversity of species. It is hypothesized that in such communities, disturbances are large enough to create openings for opportunistic species but mild and infrequent enough to allow the survival of some mature species. This mixture of early and late successional plant species, along with those in between, leads to a large number of different species. Some field experiments have supported this hypothesis.

What Is the Role of Fire in Succession? The vegetation characteristics of many ecosystems are maintained at a certain stage of succession by fires started by lightning. Examples of fire-maintained communities include savanna, temperate grasslands, chaparral, southern pine forests, western forests containing giant sequoia trees, and northern coniferous forests. In such ecosystems, occasional fires burn away much of the low-lying vegetation and small trees. A burst of new vegetation follows.

In forests maintained by occasional fires, most of the large trees are fire resistant and suffer little harm. Indeed, some species of conifers (such as the jack pine and giant sequoia) have seeds that are released or germinate only after being exposed to the intense heat created by a fire. This adaptation ensures regeneration

Q: What percentage of the earth's coastal and inland wetlands have been destroyed or polluted?

of such species after a fire has destroyed many of their competitors. Conifers (cone-bearing trees) in such an area are often about the same size because they all germinated after the most recent fire.

Suppressing fires in fire-maintained communities alters their structure and function. In some forests, lack of fire allows buildup of large quantities of highly flammable underbrush and undergrowth. This can convert what would have been fairly harmless ground fires into fires intense enough to destroy the larger fire-resistant species needed for regeneration. An entirely different ecosystem may then develop on such sites. Suppression of fires in grasslands can allow succession to shrublands and woodlands.

How Do Agriculture and Plantation Forestry Relate to Succession? The primary goal of agriculture is high productivity by selected plant species such as wheat, rice, and corn. In modern industrialized agriculture, this is achieved by replacing species-rich late successional communities (such as mature grasslands and forests) with an early successional community. Such a community often consists of a single crop (monoculture) of opportunistic species that puts most of its primary productivity into edible parts instead of tall grass, trunks, thick limbs, and roots. Herbicides keep out other species (such as weeds) that would compete with the food crops for soil nutrients.

Similarly, timber companies attempt to increase timber productivity by replacing diverse forests with farms or plantations of a single, fast-growing species; they also sometimes use herbicides to kill competing plants. In the southeastern United States, for example, row after row of fast-growing pine trees now stand where late successional oak and hickory forests (Figure 9-21) once stood. Controlled burning is often used to prevent oaks and hickories from invading these pine plantations. In tropical areas, species-rich rain forests have been cut down and replaced with single-species plantations of banana, rubber, and cocoa plants. Most homeowners spend lots of time and money establishing and keeping monoculture grass lawns at an early stage of secondary succession (and protecting them from invasions by opportunistic species such as crabgrass and other weeds).

How Predictable Is Succession? It is tempting to conclude that ecological succession is an orderly sequence in which each stage leads predictably to the next, more stable stage. According to this classic view, succession proceeds until an area is occupied by a predictable type of *climax community* dominated by a few long-lived plant species (which Edward O. Wilson calls "nature's long-distance runners"). Theoretically, these

Table 9-2 Unfavorable Changes Affecting Ecosystems

Catastrophic	
Natural	Drought
	Flood
	Fire
	Volcanic eruption
	Earthquake
	Hurricane/Tornado
	Landslide
	Change in stream course
	Disease
Human-Caused	Deforestation
	Overgrazing
	Plowing
	Erosion
	Pesticide application
	Fire
	Mining
	Toxic contamination
	Urbanization
	Water and air pollution
	Loss and degradation of wildlife habitat
Gradual	
Natural	Climatic changes
	Immigration
	Adaption/evolution
	Ecological succesion
	Disease
Human-Caused	Salinization and waterlogging of soils from irrigation
	Soil compaction
	Depletion of groundwater
	Water and air pollution
	Loss and degradation of wildlife habitat
	Elimination of "pests" and predators
	Introduction of exotic species
	Overhunting/overfishing
	Toxic contamination
	Urbanization
	Excessive tourism

*Many changes can be either catastrophic or gradual.

A: 25–50% (55% of those in the United States, 91% of those in California)

dominant climax species can replace and sustain themselves indefinitely—unless the dominant vegetation is eliminated by natural factors or human activities.

However, research has shown that the exact sequences of species and community types that appear during primary or secondary succession can be highly variable, chaotic, and unpredictable. When the trees in a forest suddenly die, whether from a natural disaster or human intervention, it's not a certainty that the same species or mixture of species will return, or sometimes even that a forest will eventually reappear—at least on our short-term (25- to 500-year) view of what happens in nature.

When all trees are removed, soil erosion increases. This soil loss can mean that the area can no longer support a mature forest until enough new soil builds up over hundreds or thousands of years. Tree removal, especially over large areas of tropical rain forests, can also make the local climate drier because of the sharp drop in water added to the atmosphere from transpiration by trees. The resulting drier climate can bake soil and make it unsuitable for tree growth for long periods of time.

According to chaos theory (Section 2-4), random, unpredictable events that magnify small changes in environmental conditions can play key roles in the nature and rate of ecological succession. It's somewhat like starting a long ocean voyage in which by chance you deviate slightly from the correct bearing. By the time you reach land, you will be a long way from your intended destination.

Research indicates that there is no ecological plan leading to ecological balance or equilibrium in an ideally adapted and stable climax community. Rather, succession reflects the ongoing struggle by each species for enough light, nutrients, food, and space to survive and to give it a reproductive advantage over other species. In other words, ecological succession consists of a mixture of species, each "doing its own thing" by attempting to occupy as much of its fundamental niche as possible.

We do know general patterns of how succession proceeds. In a temperate deciduous forest we know that succession tends toward an oak–hickory climax community (Figure 9-21). However, even if succession were entirely an orderly process, our knowledge of ecosystems is so limited that we could not predict the exact course of a given succession. We don't know the normal range for most of the variables in a given ecosystem, and we can't predict the effects of small, random, chaotic events on changes in ecosystem structure and function. This explains why a growing number of ecologists prefer terms such as *biotic change* instead of *succession* (which implies an ordered and predictable sequence of changes). Many ecologists have also replaced the term *climax community* with terms such as *mature community*.

9-6 ECOLOGICAL STABILITY AND SUSTAINABILITY

What Is Stability? All living systems, from single-celled organisms to the ecosphere, contain complex networks of negative and positive feedback loops (Section 2-3) that interact to provide some degree of stability or sustainability over each system's expected life span. This stability is maintained only by constant dynamic change in response to changing environmental conditions. For example, in a mature tropical rain forest, some trees will die and others will take their place. Unless the forest is cut or burned, however, you will still recognize it as a tropical rain forest 50 or 100 years from now.

It is useful to distinguish among three aspects of stability in living systems. **Inertia**, or **persistence**, is the ability of a living system to resist being disturbed or altered. **Constancy** is the ability of a living system such as a population to maintain a certain size or keep its numbers within the limits imposed by available resources. **Resilience** is the ability of a living system to bounce back after an external disturbance that is not too drastic.

Populations, communities, and ecosystems are so complex and variable that ecologists have little understanding of how they maintain inertia, constancy, and resilience while continually responding to changes in environmental conditions. Ecologists also find it difficult to predict which one or combination of environmental factors (Table 9-2) will stress ecosystems beyond their range of tolerance.

However, scientists have learned that the signs of ill health in stressed ecosystems include **(1)** a drop in primary productivity, **(2)** increased nutrient losses, **(3)** decline or extinction of indicator species, **(4)** increased populations of insect pests or disease organisms, **(5)** decline in species diversity, and **(6)** the presence of contaminants.

Does Species Diversity Increase Ecosystem Stability? In the 1960s, most ecologists believed that the greater the species diversity and the accompanying web of feeding and biotic interactions in an ecosystem, the greater its stability. According to this hypothesis, an ecosystem with a diversity of species and feeding paths has more ways to respond to most environmental stresses because it does not have "all its eggs in one basket." However, most recent research indicates that there are many exceptions to this intuitively appealing idea.

Of course, there is a minimum threshold of species diversity below which ecosystems cannot function; no ecosystem can function without some plants and decomposers. (Note that we and other animal

Q: What is carrying capacity?

consumers are not absolutely necessary for life to continue on the earth.) Beyond this it is difficult to know whether simple ecosystems are less stable than complex ones, or to identify the threshold below which complex ecosystems fail. In part because some species play redundant roles (niches) in ecosystems, we don't know how many or which species can be eliminated before the entire ecosystem begins to lose stability or collapse.

Research indicates that ecosystems with more species tend to have higher net primary productivities than simpler ecosystems and can also be more resilient. For example, a recent study of grasslands showed that species-rich fields had a smaller decline in net primary productivity during prolonged drought and rebounded more quickly than species-poor fields in the same area.

This supports the idea that some level of biodiversity provides insurance against catastrophe. But there is uncertainty over how much biodiversity is needed in various ecosystems. For example, some recent research suggests that average annual net primary productivity reaches a peak at 10–40 producer species. Many ecosystems contain more producer species than this, but we can't distinguish between those that are essential and those that aren't.

Part of the problem is that ecologists disagree on how to define *stability* and *diversity*. Does an ecosystem need both high inertia and high resilience to be considered stable? Evidence suggests that some ecosystems have one of these properties but not the other. For example, tropical rain forests have high species diversity and high inertia; that is, they are resistant to significant alteration or destruction. However, once a large tract of tropical forest is severely degraded, the ecosystem's resilience is so low that the forest may not be restored. Nutrients (which are stored primarily in the vegetation, not in the soil) and other factors needed for recovery may no longer be present. Such a large-scale loss of forest cover may so change the local or regional climate that forests can no longer be supported.

Grasslands, by contrast, are much less diverse than most forests, and because they burn easily they have low inertia. However, because most of their plant matter is stored in underground roots (Figure 5-16), these ecosystems have high resilience and recover quickly. A grassland can be destroyed only if its roots are plowed up and something else is planted in its place, or if it is severely overgrazed by livestock or other herbivores.

Another difficulty is that populations, communities, and ecosystems are rarely, if ever, at equilibrium. Instead, nature is in a continuing state of disturbance, fluctuation, and change; this means that an ecologist studying an ecosystem is trying to investigate a moving target. Indeed, ecologists recognize that disturbances of ecosystems are integral parts of the way nature works. Clearly, we have a long way to go in understanding how the factors in natural communities and ecosystems respond to changes in environmental conditions.

What Determines the Number of Species in an Ecosystem? There are no simple answers to this question. Two factors affecting the species diversity (or richness) of an ecosystem are its size and degree of isolation. In the 1960s Robert MacArthur and Edward O. Wilson began studying communities on islands to discover why large islands tend to have more species of a certain category (such as insects, birds, or ferns) than do small islands.

To explain these differences in species diversity with island size, MacArthur and Wilson proposed what is called the **species equilibrium model** or the **theory of island biogeography**. According to this model, the number of species found on an island is determined by a balance between two factors: the *immigration rate* of species to the island from other inhabited areas and the *extinction rate* of species established on the island. The model predicts that at some point the rates of immigration and extinction will reach an equilibrium point (Figure 9-24a) that determines the island's average number of different species (species diversity).

The model also predicts that immigration and extinction rates (and thus species diversity) are, in turn, affected by two important variables: the *size of the island* (Figure 9-24b) and its *distance from a mainland source of immigrant species* (Figure 9-24c). According to the model, a small island tends to have a lower species diversity than a large one for two reasons: First, a small island is a smaller "target" for potential colonizers, which results in lower immigration rates compared to a larger island. Second, because a small island normally has fewer resources and less diverse habitats, it should have a higher extinction rate (because of increased competitive exclusion) than a larger island.

The model also predicts that an island's distance from a mainland source of new species is important in determining its species diversity. Assume that we have two islands of about equal size and that all other factors are roughly the same. According to the model, the island closest to a mainland source of immigrant species will have a higher immigration rate and thus a higher species diversity (assuming that extinction rates on both islands are about the same).

MacArthur and Wilson's original model or scientific hypothesis has been tested and supported by a

A: The number of individuals of a species that can be sustained indefinitely in a given area

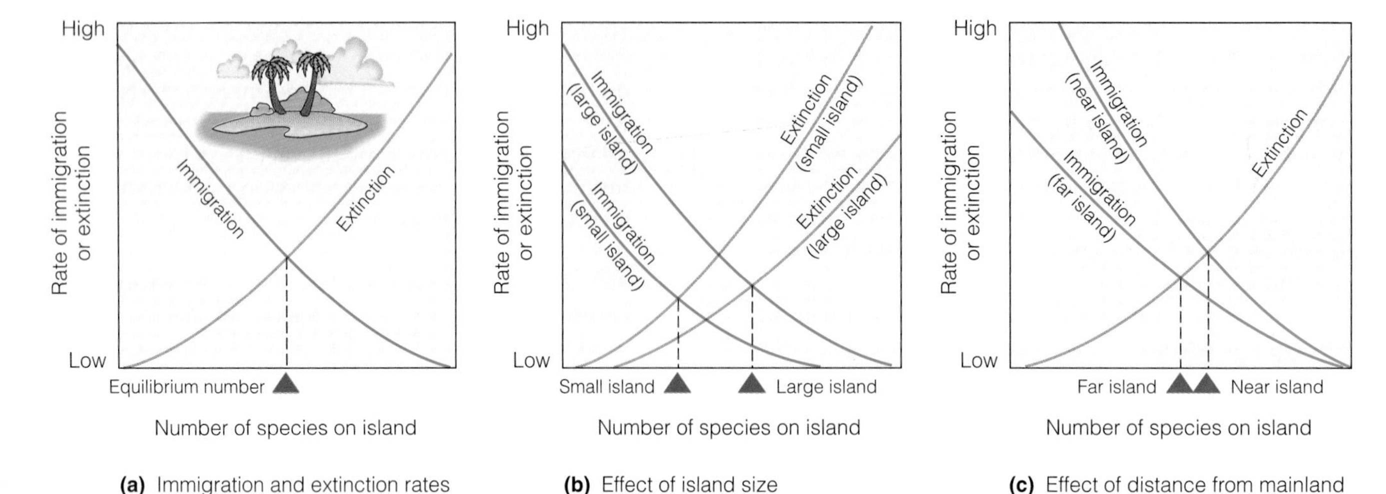

(a) Immigration and extinction rates **(b)** Effect of island size **(c)** Effect of distance from mainland

Figure 9-24 The species equilibrium model or theory of island biogeography developed by Robert MacArthur and Edward O. Wilson. **(a)** The equilibrium number of species (blue triangle) on an island is determined by a balance between the immigration rate of new species and the extinction rate of species already on the island. **(b)** With time large islands will have a larger equilibrium number of species than smaller islands because of higher immigration rates and lower extinction rates on large islands. **(c)** Assuming equal extinction rates, an island near a mainland will have a larger equilibrium number of species than a more distant island because the immigration rate to a near island is higher than that to a more distant one.

series of field experiments. As a result, biologists have accepted it well enough to elevate it to the status of an important and useful scientific theory. In recent years, it has even been applied to conservation efforts to protect wildlife on land (Connections, p. 245).

We sang the songs that carried in their melodies all the sounds of nature—the running waters, the sighing of winds, and the calls of the animals. Teach these to your children that they may come to love nature as we love it.

GRAND COUNCIL FIRE OF AMERICAN INDIANS

CRITICAL THINKING

1. In what ways are humans generalists? In what ways are we specialists? Do you think humans should become more dependent on a greater variety of foods and energy resources to promote long-term survival and sustainability? Explain and discuss the short- and long-term economic and political consequences of such a shift.

2. By analogy, use the concepts of generalist and specialist to evaluate the roles of humans in today's societies. In general, the role of college and graduate education is to create specialists in a particular field or a narrow portion of a field. What are the pros and cons of relying mainly on this approach? Is there a need for more generalists? Or are they people who may know a lot about many things (and about connections among things) but not enough about anything in particular? Are they able to serve a useful role and to make a satisfactory living in today's increasingly specialized societies? Explain your answers.

3. As well as you can, describe the major differences between the ecological niches of humans and cockroaches. Are these two species in competition? If so, how do they manage to coexist?

4. Imagine that after you die you are somehow given the choice to live again as any organism you choose except a human being. What organism would you choose to be, and why?

5. Suppose that somehow all species of parasites were eliminated from the earth. How would this affect you?

6. **(a)** What would you do if your yard and home were invaded by fire ants? **(b)** What would you do if your home were invaded by large numbers of cockroaches?

7. Try to come up with analogies to complete the following statements: **(a)** Being a specialized species is like _______. **(b)** Being a keystone species is like _______. **(c)** Being a predator is like _______. **(d)** Being a parasite is like _______. **(e)** Mutualism is like _______. **(f)** Resource partitioning is like _______. **(g)** Undergoing primary succession is like _______. **(h)** Undergoing secondary succession is like _______.

8. Some butterfly species mimic other butterfly species that taste bad to predators such as birds (Figure 9-10). Design an experiment to determine whether a particular mimic species of butterfly **(a)** tastes bad like its model (called *Müllerian mimicry* after Fritz Müller, who first recorded this phenomenon in 1878) or **(b)** doesn't taste bad but gains some protection by looking and acting like its bad-tasting model (called *Batesian mimicry* after Henry W. Bates, who first described this phenomenon in the 1860s).

9. Most domesticated animals and plants that we use for food benefit from a mutualistic relationship with us. In exchange for food, we care for them and ensure their continued reproduction. However, through crossbreeding and genetic engineering (to increase food production) we have eliminated many of the adaptive traits they need for survival in the wild. How do you feel about this form of mutualism we have imposed on other species? How would your life be changed if we decided to end this mutualistic relationship? Do you believe that domesticated animals we raise should be treated as humanely as possible during the time we are fattening them up for slaughter? Explain your answer. What effect might this have on the price and availability of meat produced from such animals?

10. Science is usually defined as a search for order and predictability in nature (Section 2-1). However, the more scientists look for order and predictability in the structure and functioning of communities and ecosystems, the more they are discovering the role of chaos and unpredictability. Does this mean that science cannot reveal very much about how nature works and that we are wasting large amounts of money, time, and talent chasing a mirage of order and predictability? Explain your answer. Closely examine the assumptions built into this view. Evaluate whether there is some useful middle ground between complete order and complete chaos.

11. If the world is mostly chaotic and unpredictable, does this mean that we should put more or less emphasis on preventing pollution, reducing waste, protecting large areas from human disturbance, and helping heal areas we have disturbed? Explain.

12. Mature temperate forests are usually dominated by one or a small number of tree species (Figures 9-19 and 9-21). However, tropical forests tend to be dominated by a large diversity of different tree species. Explain these differences. (Hint: Consider seasonal changes in climate and the roles of plant-eating insects in the two ecosystems.)

PROJECTS

1. Conduct field studies and try to find the research results of others to help you identify the major native, non-native, indicator, and keystone species in the natural areas around where you live.

2. Make field studies, consult research papers, and interview people to identify and evaluate **(a)** the effects of the deliberate introduction of a beneficial nonnative species into the area where you live and **(b)** the effects of the deliberate or accidental introduction of a harmful species into the area where you live.

3. Use the library or Internet to find and describe two species not discussed in this textbook that are engaged in **(a)** a commensalistic interaction, **(b)** a mutualistic interaction, and **(c)** a parasite–host relationship.

4. Use the library or Internet to list as many as possible of the parasites likely to be found in your body.

5. Visit a nearby land area or aquatic system (such as a wetland) and look for and record signs of ecological succession. Study the area carefully to see whether you can find patches that are at different stages of succession because of various disturbances.

6. Make a concept map of this chapter's major ideas, using the section heads and subheads and the key terms (in boldface). Look at the inside back cover and in Appendix 4 for information about concept maps.

10 POPULATION DYNAMICS, CARRYING CAPACITY, AND CONSERVATION BIOLOGY*

Sea Otters: Back from the Brink of Extinction

Sea otters (*Enhydra lutris*) live in kelp beds in shallow waters along the Pacific coast (Figure 10-1a). They are one of the few animals known to use tools. They use stones to pry shellfish off rocks underwater and to break open the shells while swimming on their backs and using their bellies as a table. Each day a sea otter consumes about 25% of its weight in sea urchins (Figure 10-1b), clams, mussels, crabs, abalone, and about 40 other species of benthic organisms.

Sea otter fur is the thickest fur of any mammal, with about 130,000 hairs per square centimeter (850,000 per square inch). Compare that to 20,000 hairs on the whole head of a typical human. Unlike all other marine mammals, sea otters lack a blubber layer to keep them warm. Instead, their thick fur traps an insulating layer of tiny air bubbles close to the body. This luxuriously warm and beautiful fur almost led to the extinction of the southern race of sea otters because it was so prized by fur hunters.

Before European settlers arrived, about 1 million sea otters lived along the 9,600 kilometers (6,000 miles) of Pacific coastline, from Baja Mexico, up the California coast through the Aleutian Islands, and across to northern Japan. By the early 1900s the southern sea otter (*Enhydra lutris nereis*), the southern race that ranged from Baja Mexico up the California coast, was believed to be extinct. Hunters had killed every southern sea otter they could find.

After fur hunters caused the local extinction of southern sea otters, kelp beds began to die. In the absence of sea otter predation, sea urchin population exploded and consumed much of the kelp. Sea otters are *keystone species* of kelp bed ecosystems.

In 1938 a population of about 300 southern sea otters was discovered to be living along the Big Sur coast of northern California. Since that time the population has made a slow but steady comeback. There are now about 2,300 southern sea otters. Although that is still a small population, the southern sea otter is no longer in as immediate danger of extinction. Southern sea otters are legally protected as a *threatened species* under the U.S. Endangered Species Act

***Paul M. Rich**, associate professor of Ecology and Evolutionary Biology and Environmental Studies at the University of Kansas, is coauthor of this chapter.

(a)

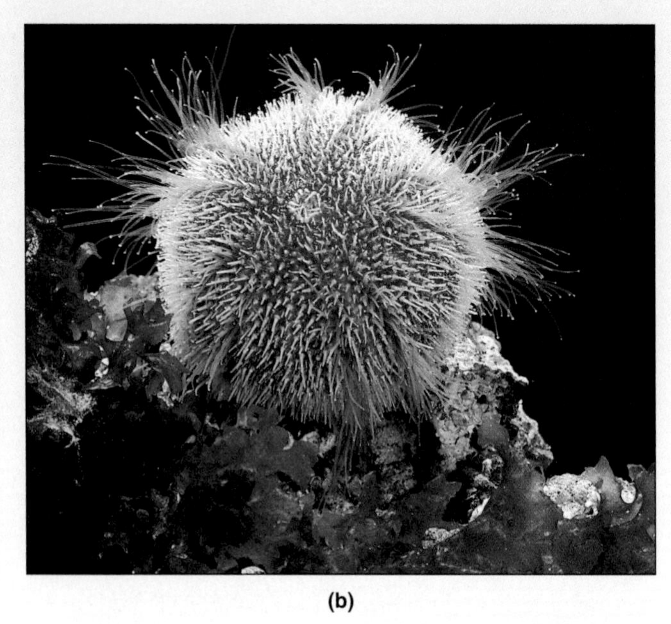

(b)

Figure 10-1 **(a)** A southern sea otter. (Jeff Foott Productions/Bruce Coleman Ltd.) **(b)** A sea urchin. (Jane Burton/Bruce Coleman Ltd.)

(Section 25-5) and as a *depleted species* under the Marine Mammal Protection Act.

In the 1980s the slow growth of the southern sea otter population was stopped for several years when sea otters started getting caught in fishing nets. Laws were passed to limit the use of fishing nets near shore, and the population started to increase again. During the past couple of years there has again been a decline in the population. The reason for the decline is not yet known, but federal biologists have observed a possible increase in infectious disease and shellfish-borne parasites. It is hoped that the decline is only temporary. One concern is that the population was once reduced to such a small size that it now lacks sufficient genetic diversity to respond to environmental changes.

Increasing pollution of coastal waters is a threat to sea otter populations. Shellfish tend to concentrate toxins that are released into coastal waters. High levels of toxins may directly kill sea otters, or lower levels over longer terms may reduce their resistance to disease and parasites.

Oil spills are another threat to sea otter populations. If the fur becomes coated with oil, it loses its insulating properties, and the otter dies from losing too much body heat (hypothermia). Sea otters can also become poisoned by fumes from the oil and by eating oil-contaminated shellfish.

The local extinction of most southern sea otter populations and the near extinction of the entire race helped us to better understand the ecological importance of this species. The slow comeback of the population is a lesson about how populations can recover. *Population dynamics*, *conservation biology*, and *restoration of damaged ecosystems* are the subject of this chapter.

Abalone fishers have been concerned that the increasing sea otter populations may threaten their livelihood, and offer evidence that the harvest of abalones has declined. Conservationists contend that excessive commercial harvest of red abalone and export of much of the catch is threatening abalone populations with extinction in central and southern California. They also note that for the most part the current range of the southern sea otter does not overlap with the range of the commercial abalone fishery. Biological studies have shown that by eating organisms such as sea urchins, sea otters can actually favor abalone populations. Nonetheless sea otters do like to feed on abalones, a case of *competition* with humans.

Through the animal and vegetable kingdoms, nature has scattered the seeds of life abroad with the most profuse and liberal hand. She has been comparatively sparing in the room and nourishment necessary to rear them.

Thomas R. Malthus

This chapter focuses on answers to the following questions:

- What are the major characteristics of a population?
- How do populations change in size, density, and makeup in response to environmental stress?
- What different reproductive strategies do species use to enhance their survival?
- What is conservation biology?
- What impacts do human activities have on populations, communities, and ecosystems?
- What efforts are being made to restore ecosystems damaged by human activities?

10-1 CHARACTERISTICS OF POPULATIONS

What Are the Major Characteristics of a Population? Populations are dynamic; they change in size, density, dispersion, and age distribution (proportion of individuals of each age in a population) in response to environmental stress or changes in environmental conditions (Table 10-1). These changes are called **population dynamics**. These characteristics can be used to describe a population and to project how its size is likely to change in the future.

Population size is the number of individuals in a population at a given time. Very small populations can become extinct because they are vulnerable to disease, predation, and natural catastrophes. By contrast, when populations become too large, many individuals may die of starvation, predation, or disease.

Population density is the number of individuals of a population in a certain space at a given time. For terrestrial ecosystems population density is usually expressed as the number of individuals per unit area, such as the number of white-footed mice per hectare. For aquatic ecosystems, population density is usually expressed as the number of individuals per unit volume, such as the number of fathead minnows per

Table 10-1 Some Effects of Environmental Stress
Organism Level
Physiological and biochemical changes
Psychological disorders
Behavioral changes
Fewer or no offspring
Genetic defects in offspring (mutagenic effects)
Birth Defects (teratogenic effects)
Cancers (Carcinogenic effects)
Death
Population Level
Population increase or decrease
Change in age structure (old, young, and weak may die)
Survival of strains genetically resistant to stress
Loss of genetic diversity and adaptability
Extinction
Community–Ecosystem Level
Disruption of energy flow
Decrease or increase in solar energy uptake and heat output
Changes in trophic structure in food chains and webs
Disruption of chemical cycles
Depletion of essential nutrients
Excessive addition of nutrients
Simplification
Reduction in species diversity
Reduction or elimination of habitats and filled ecological niches
Less complex food webs
Possibility of lowered stability
Possibility of ecosystem collapse

cubic meter of pond water. Population density may vary in time and space depending on availability of food or other resources, microclimate and other conditions, habitat availability and quality, and interactions with populations of other species.

Population dispersion is the spatial pattern in which the members of a population are found in their habitat (Figure 10-2). The most common pattern is *clumping*, in which members of a population exist in clumps throughout their habitat (Figures 4-3, 7-18, and 10-2, left). One reason for clumping is that resources are usually patchy, not uniform. Populations of some animal species (such as herds of grazing animals, schools of fish, flocks of birds, and troops of baboons) may also clump for protection against predators, before or during migration, during mating season, or because they live in social groups.

The members of a population may also be *uniformly dispersed* over their habitat, but such a pattern is rare in nature. It occurs mostly when individuals of the same species compete for resources that are scarce and spread fairly evenly or when individuals are antagonistic to one another and defend their access to resources by physical or chemical means. An example is the desert creosote bush, which competes for scarce water by excreting toxic chemicals that prevent seedlings of other creosote bushes from growing near it (Figure 10-2, middle).

Individuals in some populations may be *randomly dispersed* in an unpredictable way over their habitat (Figure 10-2, right). This can occur if resources and other environmental conditions are fairly uniform throughout a habitat and if the members of the population do not attract or repel one another most of the time. Like uniform dispersions, random dispersions are rare because resources are seldom uniformly available in nature.

All three types of population dispersion may vary in response to mating habits or seasonal changes that alter the distribution of resources. Population

Figure 10-2 Generalized dispersion patterns for individuals in a population throughout their habitat. The most common pattern is one in which members of a population exist in clumps (left).

Clumped
(elephants)

Uniform
(creosote bush)

Random
(dandelions)

Thorson, Bruce. 1998. "Boom and Bust." *Canadian Geographic*, vol. 118, no. 2, 68(7).

dispersion is also affected by the movement of individuals out of (emigration) or into (immigration) a population's normal geographical range.

Age structure is the proportion of individuals in each age group in a population. Common age groups are *prereproductive* (younger than the age of sexual maturity), *reproductive*, and *postreproductive* (older than the maximum age of reproduction). A population with a large percentage of its individuals in the prereproductive and reproductive categories has a high potential for growth.

10-2 POPULATION DYNAMICS AND CARRYING CAPACITY

What Limits Population Growth? Four variables—births, deaths, immigration, and emigration—govern changes in population size. A population gains individuals by birth and immigration and loses them by death and emigration:

$$\text{Population change} = \begin{pmatrix} \text{Births} \\ + \\ \text{Immigration} \end{pmatrix} - \begin{pmatrix} \text{Deaths} \\ + \\ \text{Emigration} \end{pmatrix}$$

These variables in turn depend on changes in resource availability or on other environmental changes (Figure 10-3). If the number of individuals added from births and immigration equals the number lost to deaths and immigration, then there is **zero population growth**.

Populations vary in their capacity for growth, also known as the **biotic potential** of the population. The **intrinsic rate of increase (r)** is the rate at which a population could grow if it had unlimited resources. The intrinsic rate of increase is expressed as the number of new individuals per existing individual per unit time,

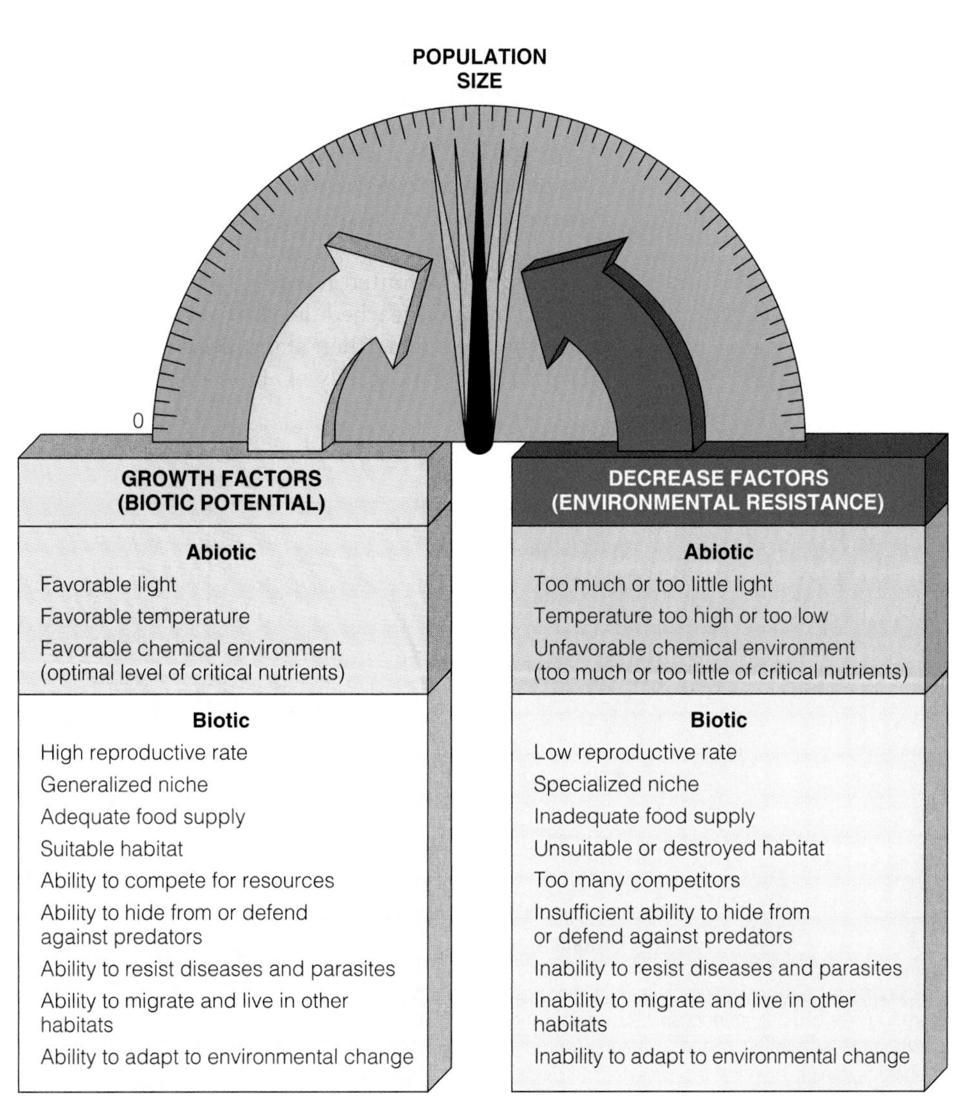

Figure 10-3 Factors that tend to increase or decrease populations. Population size at any given time is determined by the balance between growth factors (biotic potential) and decrease factors (environmental resistance).

Hint: Enter the search term *lemmings* using the Subject Guide.

such as the number of new pepper moths per existing peppered moth per year. Generally, individuals in populations with a high intrinsic rate of increase *reproduce early in life, have short generation times* (the time between successive generations), *can reproduce many times* (have a long reproductive life), and *produce many offspring each time they reproduce.*

The intrinsic rate of increase of many species depends on having a certain minimum population size. If a population declines below the *critical size* needed to support a breeding population, certain individuals may not be able to locate mates, genetically related individuals may interbreed and produce weak or malformed offspring, and the genetic diversity may be too low to enable adaptation to new environmental conditions. Then the intrinsic rate of increase falls and extinction is likely unless humans intervene to help the species recover.

No population can grow indefinitely. In the real world, a rapidly growing population reaches some size limit imposed by a shortage of one or more limiting factors, such as light, water, space, or nutrients. This is why the planet is not covered completely with cockroaches, crabgrass, oak trees, people, or any species. *There are always limits to population growth in nature.*

Environmental resistance consists of all the factors acting jointly to limit the growth of a population. The population size of a species in a given place and time is determined by the interplay between its biotic potential and environmental resistance (Figure 10-3).

Together biotic potential and environmental resistance determine the **carrying capacity (K)**, the number of individuals of a given species that can be sustained indefinitely in a given space (area or volume).

What Is the Difference Between Exponential and Logistic Growth? A population that does not have resource limitations grows exponentially. *Exponential growth* starts out slowly and then proceeds faster and faster as the population increases. If number of individuals is plotted against time, this sequence yields a *J*-shaped exponential growth curve (Figure 10-4a).

Because of environmental resistance, population growth tends to decrease as a population gets larger, and eventually to level off at the population's carrying capacity. In most cases, the size of a population fluctuates slightly above and below the carrying capacity. **Logistic growth** involves exponential population growth when the population is small and a steady decrease in population growth with time as the population approaches the carrying capacity. A plot of number of individuals against time yields a sigmoid or *S*-shaped logistic growth curve (Figure 10-4b).

A classic case of logistic growth involves increase of the sheep population on the island of Tasmania, south of Australia, in the early 19th century (Figure 10-5a). Resources were abundant when English immigrants introduced sheep to the island, and for several decades the sheep population expanded exponentially. As the number and density of sheep rose, competition for limited resources increased and the sheep population reached the carrying capacity of the land. Their numbers then stabilized and fluctuated around a carrying capacity of about 1.6 million sheep.

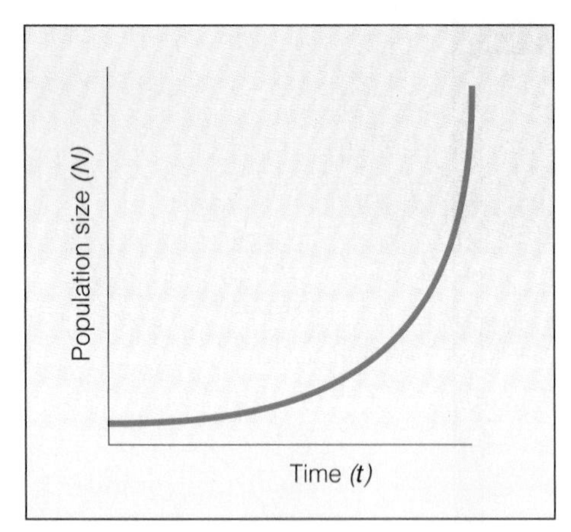

(a) Exponential growth

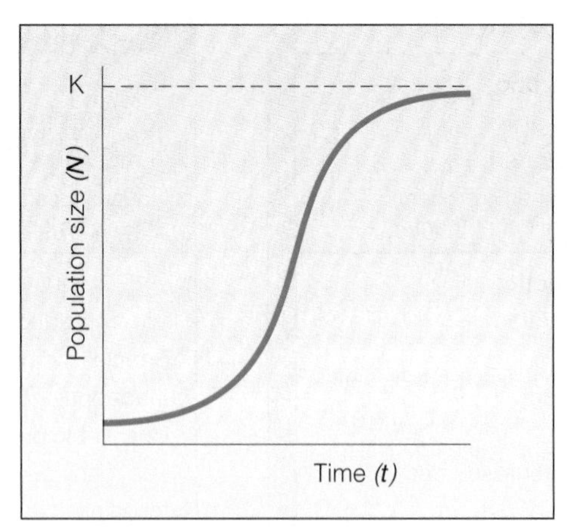

(b) Logistic growth

Figure 10-4 Theoretical population growth curves. **(a)** *Exponential growth*, in which the population grows faster with time. Exponential growth occurs when resources are not limiting and a population can grow at its *intrinsic rate of increase* (r). Exponential growth of a population cannot continue forever because eventually some factor limits population growth. **(b)** *Logistic growth*, in which the growth rate decreases as the population gets larger. With time, the population size stabilizes at the *carrying capacity* (K).

What Happens if the Population Size Exceeds the Carrying Capacity? The populations of some species don't make such a smooth transition from exponential growth to logistic growth. Instead, they temporarily use up their resource base (for example, by eating more plants or animals than can be replenished). Thus, the population temporarily *overshoots* or exceeds the carrying capacity of its habitat.

This overshoot occurs because of a *reproductive time lag*, the period required for the birth rate to fall and the death rate to rise in response to resource overconsumption; in other words, the corrective negative feedback does not take effect immediately.

Unless the excess individuals switch to new resources or move to an area with more favorable conditions, the population will suffer a *dieback* or *crash*. In some cases, a population exceeding its carrying capacity may begin an ecological roller-coaster ride, with its population alternately exceeding and falling below its carrying capacity size because of reproductive time lag. Sometimes a population degrades the environment by consuming potentially renewable resources faster than they are renewed, thus lowering the area's carrying capacity. Unless the species can migrate to find enough food, its population size will crash to a level where it can survive at the area's lower carrying capacity.

A classic case of such a population crash occurred when 26 reindeer (24 of them female) were introduced in 1910 onto an island of the Aleutian chain off the southwest coast of Alaska (Figure 10-5b). When the reindeer were first introduced, lichens, mosses, and other food sources were plentiful. By 1935 the herd's population had soared to 2,000, overshooting the island's carrying capacity. So many reindeer eating vegetation faster than it could be replenished caused overgrazing and resulted in a population crash, with the herd plummeting to only 8 reindeer by 1950.

Humans are not exempt from overshoot and dieback. Ireland experienced a population crash after a fungus destroyed the potato crop in 1845. About 1 million people died and 3 million people emigrated to other countries. The human population on Easter Island also crashed, apparently because of overconsumption of the island's trees (Figure 2-1).

Technological, social, and other cultural changes have extended the earth's carrying capacity for the human species (Figure 1-22). We have increased food production and used large amounts of energy and matter resources to make normally uninhabitable areas of the earth habitable. However, there is growing concern about how long we will be able to keep doing this on a planet with a finite size and resources but an exponentially growing population and per capita resource use.

Carrying capacity is not a simple, fixed quantity, but rather a variable determined by many factors. Examples include competition within and among species, immigration and emigration, natural and human-caused catastrophic events, and seasonal fluctuations in the supply of food, water, hiding places, and nesting sites.

How Does Population Density Affect Population Growth? *Density-independent population controls* affect a population's size regardless of its

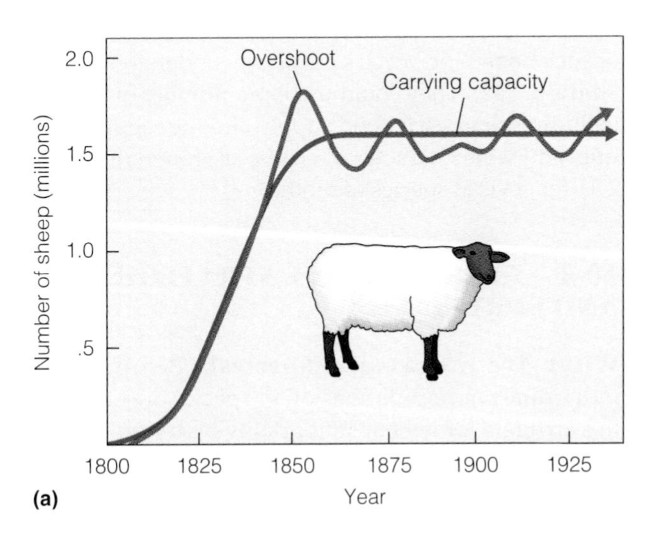

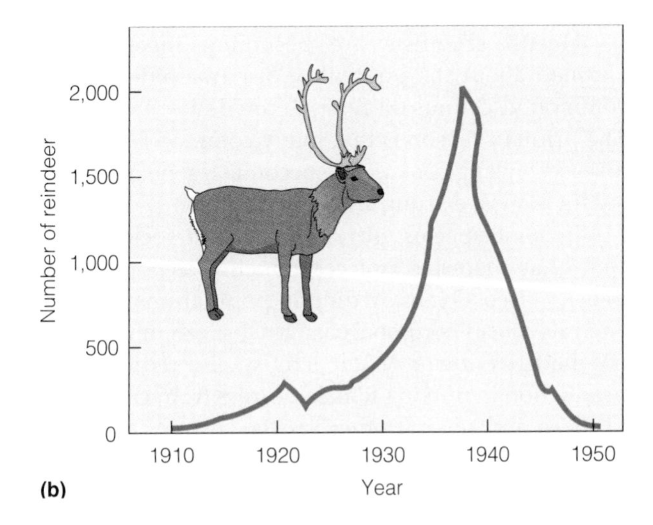

Figure 10-5 **(a)** *Logistic growth* of a sheep population on the island of Tasmania between 1800 and 1925. After sheep were introduced in 1800 their population grew exponentially because of ample food. By 1855 they overshot the land's carrying capacity. Their numbers then stabilized and oscillated around a carrying capacity of about 1.6 million sheep. Sheep are large and long-lived, and reproduce slowly. As a result, their population size does not vary much from year to year once they reach the carrying capacity of their habitat (assuming that other environmental factors do not lower the carrying capacity). **(b)** *Exponential growth*, *overshoot*, and *population crash* of reindeer introduced to a small island off the southwest coast of Alaska.

A: No

population density. Examples include floods, hurricanes, earthquakes, landslides, severe drought, unseasonable weather, fire, destruction of habitat (such as clearing a forest of its trees or filling in a wetland), and pesticide spraying. For example, a severe freeze in late spring can kill many individuals in a plant population, regardless of its density.

Some limiting factors have a greater effect as a population's density increases. Examples of such *density-dependent population controls* are competition for resources, predation, parasitism, and disease. Dense populations generally have lower birth rates and higher death rates, which lead to slower growth rates than in less dense populations.

As prey populations become more dense, their members compete more for limited resources. This can lead to a larger number of weakened individuals that are easy targets for predators. Individuals in a dense population are also more likely to be infected by parasites or contagious disease organisms. You are more likely to catch the flu in a crowded city than on a farm in a rural area. Organisms in a densely populated environment may even die of poisoning by their own waste products, as fish do in overcrowded ponds.

Infectious disease is a classic example of density-dependent population control. An example is the *bubonic plague*, which swept through Europe during the 14th century. The bacterium that causes this disease normally lives in rodents. It was transferred to humans by fleas that fed on infected rodents and then bit humans. The disease spread like wildfire through crowded cities, where sanitary conditions were poor and rats were abundant. At least 25 million people in European cities died from the disease.

Health scientists are becoming increasingly alarmed about the possibility of new epidemics of common infectious disease in crowded urban areas. The primary reason is that many common strains of disease-causing bacteria are becoming genetically resistant to most existing antibiotics (Section 17-5).

In some species, physiological and sociological control mechanisms limit reproduction as population density rises. Overcrowding in populations of mice and rats causes hormonal changes that can inhibit sexual maturity, lower sexual activity, and reduce milk production in nursing females. Stress from crowding in these and several other species also reduces the number of offspring produced per litter through such mechanisms as spontaneous abortion. In some species crowding may lead to population control through cannibalism and killing of the young.

What Kinds of Population Change Curves Do We Find in Nature? In nature we find that over time species have three general types of *population cycles*: stable, irruptive, and cyclic (Figure 10-6).

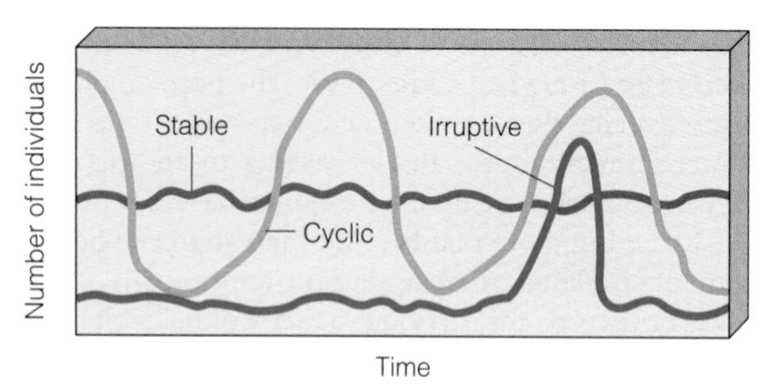

Figure 10-6 General types of idealized population change curves found in nature. A species with a *stable* population fluctuates slightly above and below its carrying capacity. A species with a *cyclic* population size undergoes boom–bust cycles and one with an *irruptive* pattern has an occasional sharp increase in size, followed by a sharp decline.

A species whose population size fluctuates slightly above and below its carrying capacity is said to have a fairly stable population size (Figure 10-5a). Such stability is characteristic of many species found in undisturbed tropical rain forests, where there is little variation in average temperature and rainfall.

Some species, such as the raccoon, normally have a fairly stable population that may occasionally explode, or *irrupt*, to a high peak, and then crash to a more stable lower level or in some cases to a very low level (Figure 10-5b). The population explosion is caused by some factor that temporarily increases carrying capacity for the population, such as more food or fewer predators.

Other species undergo sharp increases in their numbers followed by seemingly periodic crashes. Predators are sometimes blamed, but the actual causes of such boom–bust cycles are poorly understood (Case Study, p. 242). They could involve a number of factors, including changes in food supply and the random and unpredictable character of chaos, as shown in Figure 2-11 for several species of moth.

10-3 REPRODUCTIVE STRATEGIES AND SURVIVAL

What Are r-Strategist Species? Reproductive individuals in populations of all species are engaged in a struggle for genetic immortality by trying to have as many members of the next generation as possible carrying their genes. Each species has a characteristic mode of reproduction. At one extreme are species that reproduce early. They have many (usually small) offspring each time they reproduce, reach reproductive age rapidly, and have short generation times. They give their offspring little or no parental care or protection to help them survive. The result of this *many-small-and-unprotected-young strategy* is that most of the

Q: What are r-strategists?

tiny, helpless offspring die or are eaten before reaching reproductive age. Species with this reproductive strategy overcome the massive loss of their offspring by producing so many young that a few will survive to reproduce many offspring to begin the cycle again.

Species with such a capacity for a high intrinsic rate of increase (r) are called **r-strategists** (Figure 10-7). Algae, bacteria, rodents, annual plants (such as dandelions), many bony fish, and most insects (p. 80) are examples. Such species tend to be *opportunists*, reproducing and dispersing rapidly when conditions are favorable or when a new habitat or niche becomes available for invasion, as in the early stages of ecological succession (Figures 9-19 and 9-21). Changing or unfavorable environmental conditions can cause such populations to crash. Hence, most r-strategists go through irregular and unstable boom–bust cycles.

What Are K-Strategist Species? At the other extreme are **K-strategists**, species that tend to reproduce late and have few offspring with long generation times. Typically these offspring develop inside their mothers (where they are safe). They are fairly large and mature slowly. They are cared for and protected by one or both parents until they reach reproductive age. This *few-but-large-young reproductive strategy* results in a few big and strong individuals that can compete for resources and reproduce a few young to begin the cycle again (Figure 10-8).

K-strategists tend to be *competitors* who have big bodies, live for a long time, and spend little of their energy on reproduction. Such species are common in fairly stable late successional environments where competition for resources is intense. Such species are called K-strategists because they tend to do well in competitive conditions when their population size is near their habitat's carrying capacity (K). Their populations typically follow a logistic growth curve (Figure 10-4b). Examples are most large mammals (such as elephants, whales, and humans), birds of prey, and large and long-lived plants (such as the saguaro cactus, redwood trees, and most tropical rain forest trees).

Many K-strategist species, especially those with long generation times and low reproductive rates (such as elephants, rhinoceroses, and sharks), are prone to extinction. For example, some shark species are now threatened with extinction because of overfishing (Case Study, p. 217).

In agriculture we raise both r-strategists (crops) and K-strategists (livestock). Many organisms have reproductive strategies between the extremes of r-strategists and K-strategists, or they change from one extreme to the other under certain environmental conditions.

The reproductive strategy of a species may give it a temporary advantage, but *the availability of suitable habitat for individuals of a population in a particular area is what determines its ultimate population size.* Regardless of how fast a species can make babies, there can be no more dandelions than there is dandelion habitat and no more zebras than there is zebra habitat in a particular area.

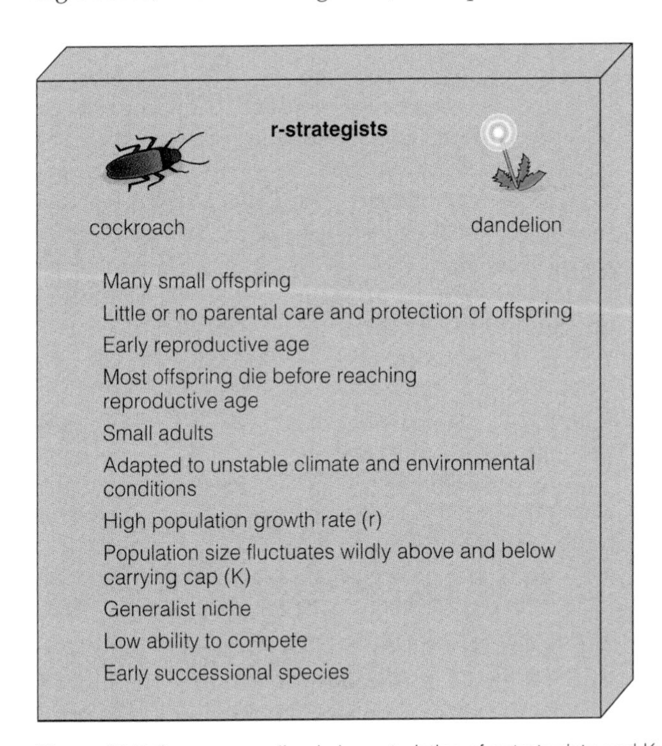

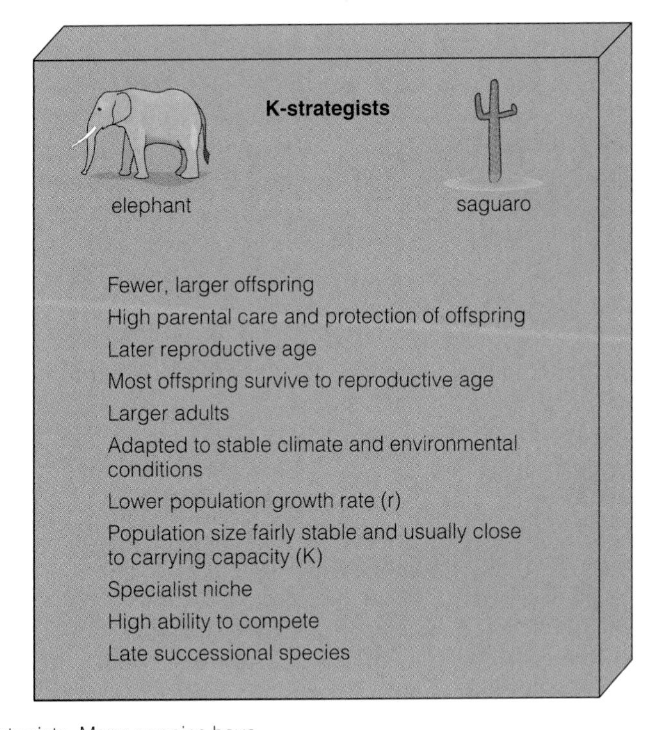

Figure 10-7 Some generalized characteristics of r-strategists and K-strategists. Many species have characteristics between these two extremes.

A: Species with a capacity for a high rate of population growth

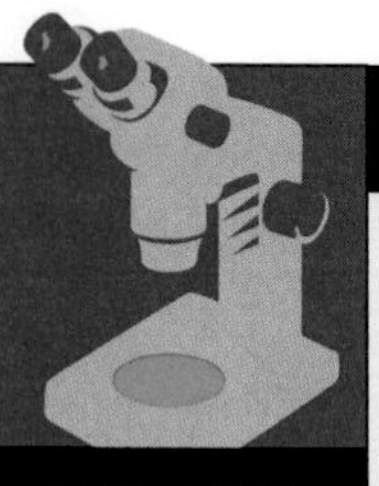

CASE STUDY

What Is the Role of Predation in Controlling Population Size?

There is considerable evidence that large predators tend to keep prey populations healthy and strong by killing off the weak, diseased, and old members. However, studies reveal that the role of predators in controlling prey populations is not always so clear-cut.

For decades, predation was the popular explanation for the correlation and time lag between the 10-year population cycles of the snowshoe hare and its predator, the Canadian lynx (Figure 10-8). According to this hypothesis, lynx preying upon hares periodically reduced their population. The shortage of hares then reduced the lynx population, which allowed the hare population to build up again. At some point the lynx population could increase to take advantage of the increased supply of hares—and start the cycle again.

Recent research has cast serious doubt on this appealing hypothesis because snowshoe hare populations have been found to have similar 10-year boom-and-bust cycles on islands where lynx are absent. Researchers now hypothesize that the periodic crashes in the hare population occur when large numbers of hares consume food plants faster than they can be replenished and have a decrease in the quantity and quality of their food. Once the hare population crashes, the plants recover, and the hare population begins rising again. Instead of lynx controlling hare populations, the changing hare population size apparently causes fluctuations in the lynx population.

Many wildlife managers believe that the way to control populations of large grazing animals such as mule deer and moose is by introducing predators such as wolves and mountain lions. Big game hunters believe that the best way to have plenty of large grazing game animals is to kill off such predators. Research suggests that both of these notions may be overly simplistic.

The idea that predators control populations of large herbivores gained popularity in 1906 when President Theodore Roosevelt declared the Kaibab Plateau in northern Arizona a federal wildlife game refuge. Sheep, cattle, and horses that had been competing with the mule deer for forage were banned. Hunting of the mule deer was also banned, and between 1906 and 1931 hunters were hired to kill off populations of predators such as mountain lions, wolves, coyotes, and bobcats.

Then came reports that the population of mule deer (estimated to be about 4,000 in 1904) had increased to as high as 100,000 by 1930 but then crashed to about 10,000, presumably because of overgrazing. The famous wildlife conservationist Aldo Leopold and other biologists blamed the large population increase and subsequent crash on elimination of the mule deer's natural predators.

However, research by Australian biologist Graham Caughey reported in 1970 has cast doubt on this notion. First, Caughey found that the population estimates of mule deer on the Kaibab were so unreliable

What Are Survivorship Curves? Individuals of species with different reproductive strategies tend to have different *life expectancies.* One way to represent the age structure of a population is with a **survivorship curve**, which shows the number of survivors of each age group for a particular species. There are three generalized types of survivorship curves: late loss, early loss, and constant loss (Figure 10-9).

Late loss curves are typical for K-strategists (such as elephants and humans) that produce few young and care for them until they reach reproductive age (thus reducing juvenile mortality). *Early loss* curves are typical for r-strategist species (such as most annual plants and most bony fish species) with many offspring, high juvenile mortality, and high survivorship once the surviving young reach a certain age and size.

Species with *constant loss* survivorship curves typically have intermediate reproductive strategies with a fairly constant rate of mortality in all age classes and thus a steadily declining survivorship curve. Examples include many types of songbirds, lizards, and small mammals that face a fairly constant threat from starvation, predation, and disease throughout their lives.

A table of the numbers of individuals at each age from a survivorship curve is called a *life table.* It shows the projected life expectancy and probability of death for individuals at each age. Insurance companies use life tables of human populations in various countries or regions to determine policy costs for their customers. Because life tables show that women in the United States survive an average of 7 years longer than men, a 65-year-old man will normally pay more for life insurance than a 65-year-old woman.

10-4 CONSERVATION BIOLOGY

What Is Conservation Biology? **Conservation** involves sensible and careful use of natural resources by humans. **Conservation biology** is an interdisciplinary science that originated in the 1970s to deal

Glover, James M., and Joseph Dadey. 1997. "Islands of Nature." *Parks & Recreation*, vol. 32, no. 6, 28(6).

that we cannot be sure whether their population rose to as high as 100,000 and then crashed. Second, even if such changes did take place, there is no convincing evidence that it was caused primarily by removal of the mule deer's predators. Caughey analyzed all known cases of the introduction of large ungulates (including mule deer) into new habitats and found that a population crash occurred every time, with or without predators.

Possible explanations for the rise in mule deer population on the Kaibab include **(1)** availability of more food because of reduced competition after sheep and cattle were removed from the plateau, **(2)** changes in weather patterns or the frequency of fires, which increased the supply of edible vegetation, and **(3)** decreased disease. Because data on these (or perhaps other) factors are not available, we don't really know what factor or combination of factors led to the changes in the population size of mule deer on the Kaibab Plateau.

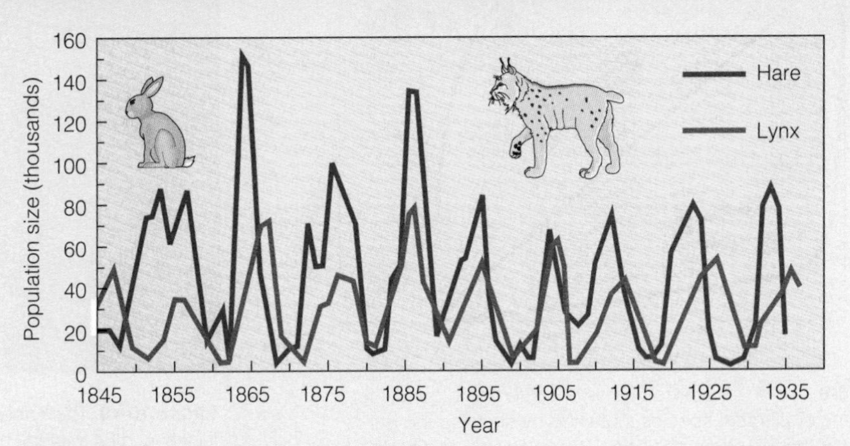

Figure 10-8 Population cycles for the snowshoe hare and Canadian lynx. At one time it was widely believed that these curves provided circumstantial evidence that these predator and prey populations regulated one another. More recent research suggests that the periodic swings in the hare population are caused by the hares themselves as they overconsume the plants they eat, die back, and then slowly recover when the plant supply is replenished. Instead of the lynx population size controlling hare population size, the rise and fall of the hare population apparently controls the size of the lynx population. (Data from D. A. MacLulich)

Much more research is needed to determine what factors control large herbivore populations. Nature is full of surprises.

Critical Thinking

Suppose that the hypothesis that predators such as the gray wolf don't necessarily control populations of their prey is valid. How might this affect current efforts to restore gray wolves to the Yellowstone area in the United States and to reintroduce natural predators to help control populations of species such as deer? Explain.

with problems in maintaining the earth's biodiversity, including the genetic, species, and ecosystem components of life on the earth. Its goals are to investigate human impacts on biodiversity and to develop practical approaches to maintaining biodiversity. As an interdisciplinary applied science, conservation biology uses a fundamental understanding of ecology, economics, politics, and other disciplines to solve practical problems related to how to ensure the continued existence of populations of wild species.

Conservation biology is dedicated to protecting ecosystems and to finding practical ways to prevent premature extinctions of species as a result of human activities. It differs from *wildlife management*, which is devoted primarily to the manipulation of the population sizes of various animal species, especially game species prized by hunters and fishers (Section 25-5).

Conservation biology rests on three underlying principles: **(1)** Biodiversity and ecological integrity are useful and necessary to all life on earth and should not be reduced by human actions, **(2)** humans should not cause or hasten the premature extinction of populations and species or disrupt vital ecological processes, and **(3)** the best way to preserve earth's biodiversity and ecological integrity is to protect intact ecosystems that provide sufficient habitat for sustaining natural populations of species.

Conservation biologists believe that we must preserve **ecological integrity**, the conditions and natural processes (such as energy flow and matter cycling in ecosystems, Figure 4-15) that generate and maintain biodiversity and allow evolutionary change as a key mechanism for adapting to changes in environmental conditions.

This *scientific approach* recognizes that protecting wildlife populations means protecting their habitats and not disrupting the complex interactions among species in an ecosystem. It is based on Aldo Leopold's ethical principle that something is right when it tends to maintain the earth's life-support systems for us and other species, and wrong when it doesn't.

Hint: Enter the search term *island biogeography* using the Subject Guide.

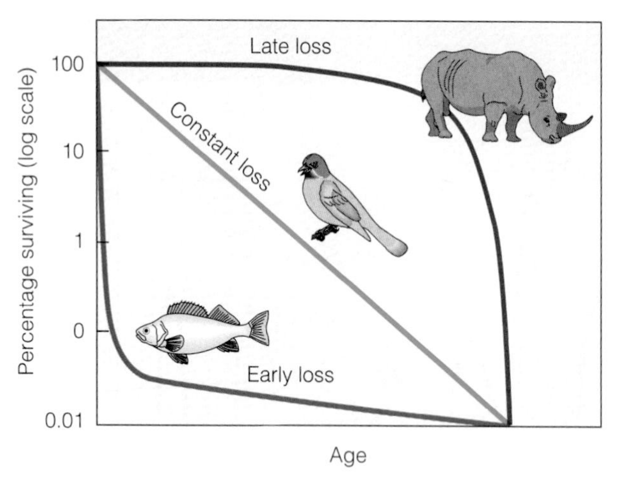

Figure 10-9 Three generalized types of survivorship curves for populations of different species, obtained by showing the percentages of the members of a population surviving at different ages. For a *late loss* population (such as elephants, rhinoceroses, and humans), there is typically high survivorship to a certain age, then high mortality. A *constant loss* population (such as many songbirds) shows a fairly constant death rate at all ages. For an *early loss* population (such as annual plants and many bony fish species), survivorship is low early in life. These generalized survivorship curves only approximate the behavior of species.

Figure 10-10 The Kansas Biotic Succession Facility was established in 1984 to study how habitat fragmentation influences secondary succession. An "archipelago" of three habitat patch sizes was arrayed in a successional field and the patches have been maintained by regular mowing to form a closely mown "sea" of grassy turf. Ongoing studies are examining the dynamics of the plant and small mammal communities. The rates of succession and species composition are being influenced by both patch size and the distance to a nearby extensive forest that serves as the source for new colonizers. (Photograph courtesy of Robert Holt)

Chapter 23 examines methods of sustaining ecosystems through land management practices, Chapter 24 examines loss of biodiversity caused by deforestation, and Chapter 25 examines issues of premature extinction, with an emphasis on rare and endangered species.

How Can Conservation Biology Help Prevent Premature Extinction and Ensure Ecosystem Integrity? Conservation biology seeks answers to three questions:

- *What is the status of natural populations, and which species are in danger of extinction?*
- *What is the status of the integrity of ecosystems, and what ecosystem services* (Section 4-6) *of value to humans and other species are we in danger of losing?*
- *What measures can we take to ensure that we maintain habitat of the quality and size needed to ensure viable populations of wild species, and that ecosystem integrity can be sustained?*

These are challenging questions, and the answers require extensive field research and a strong grounding in ecological theory. To understand the status of natural populations requires measurement of the current population size, determination of how population size is likely to change with time, and evaluation of whether existing populations are likely to be sustainable. *Ensuring viable populations of wild species ultimately requires protection of sufficient suitable habitat.*

What Is Habitat Fragmentation? **Habitat fragmentation** is the process by which human activity breaks natural ecosystems into smaller and smaller pieces of land called *habitat fragments.* The land between habitat fragments is typically not suitable to support populations of many wild species. One concern is that the remaining habitat often is not of a sufficient *size* or *quality* to maintain viable population of wild species. Another concern is that even if habitat is not lost, it often becomes degraded through human activities such as grazing, logging, mining, hunting, or recreation.

Habitat fragmentation has had big impacts on populations of species that require large areas of continuous habitat. Examples include large predators, such as jaguars or grizzly bears, and migratory species, such as the American bison. Populations of other species that require special habitats, such as the giant panda (Case Study, p. 210), also tend to be severely affected.

Habitat fragments are often compared to islands (Connections, right). Like actual islands, habitat islands are separated from each other and from more extensive areas of habitat by an inhospitable environment (Figure 10-10).

What Are Corridors? A solution to habitat fragmentation involves protection of **corridors,** long areas of land that connect habitat that would otherwise become fragmented. Corridors can permit movement of migratory animals and ensure a diverse gene pool by

Q: What are K-strategists?

Using Island Biogeography Theory to Protect Mainland Communities and Species

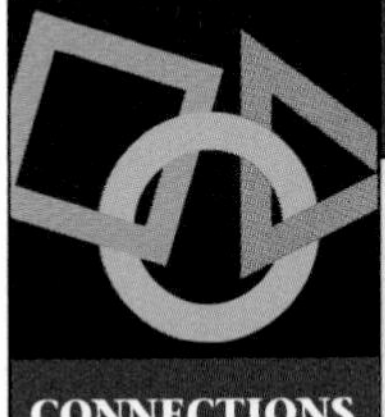

CONNECTIONS

Ecologists and conservation biologists are now applying the species equilibrium model or theory of island biogeography (Figure 9-24) to communities on the mainland. Because of widespread disruption and fragmentation by humans, most remaining wildlife sanctuaries on land are *habitat islands* surrounded by an inhospitable "sea" of disturbed and unsuitable habitat. According to the species equilibrium model, the species diversity in these terrestrial patches or islands should be determined by their size and by their distance from other patches that serve as sources of colonists.

Conservation biologists are using the model to help them locate (and try to protect) areas in the most danger of losing much of their species diversity. Researchers are also using the model to estimate how big a nature preserve should be in a particular area to prevent it from losing species.

Of course, the size of a preserve also depends on which species we hope to protect. Members of endangered species such as tigers and grizzly bears, for example, feed over a large home range. They need a much larger area of undisturbed habitat than many other species to preserve a viable population.

For example, the average grizzly bear has a home range of 250–1,000 square kilometers (100–400 square miles). Thus, a minimum viable population of 50–90 grizzly bears needs a reserve of up to 90,000 square kilometers (35,000 square miles) for survival. This is much larger than the 53,500-square-kilometer (20,600-square-mile) Greater Yellowstone ecosystem, one of the world's largest protected habitat islands (containing a national park and six national forests).

Conservation biologists are also using the species equilibrium model to estimate how closely a series of small wildlife reserves should be spaced to allow the possibility of immigration from one preserve to another if a species in one reserve becomes locally extinct.

In addition, they are using the model to estimate the size and number of protected corridors needed to connect various reserves and encourage the spread of protected species among them. Much more research must be done to answer such practical questions, but progress is being made through this application of ecological theory to wildlife conservation.

Critical Thinking

How would you respond to the statement that the theory of island biogeography should not be taken too seriously because it is only a theory?

permitting some interbreeding between different populations of the same species.

Conservation efforts to protect corridor areas between core natural reserves have been initiated in many locations throughout both temperate and tropical regions. For example, a major conservation effort in Costa Rica has focused on protecting land in a corridor between La Selva Biological Reserve in the tropical lowlands and Braulio Carrillo National Park in the nearby mountains. This successful effort has ensured that species that migrate up and down the mountains have sufficient suitable habitat.

Corridors have been criticized by some conservation biologists because they may be too narrow to be of value. They may also give a false sense of security because they don't necessarily substitute for more continuous habitat. Still, corridors are one of the best solutions to the problem of habitat fragmentation.

What Is Bioinformatics? Good conservation biology depends on good information. Increasingly, efforts of conservation biologists are focused on building computer databases that store useful information concerning biodiversity. Recently, a new discipline called *bioinformatics* has developed that concerns itself with providing tools for storage and access to key biological information, and with building the actual databases that contain the needed biological information.

10-5 HUMAN IMPACTS ON ECOSYSTEMS: LEARNING FROM NATURE

How Have Humans Modified Natural Ecosystems? To survive and support growing numbers of people, humans have greatly increased the number and area of the earth's natural systems that we have modified, cultivated, built on, or degraded (Figure 1-5). We have used technology to severely alter much of the rest of nature in several ways:

- *Fragmenting and degrading habitat.* The landscape is rapidly being transformed from one with vast expanses of continuous forest, grassland, or other natural ecosystems into a patchwork of remnant habitat fragments surrounded by land developed for farms, urban areas, and other human uses. The remaining habitat, even when it is continuous, is often degraded and unsuitable for many native species.

A: Species with a capacity for a low to moderate rate of population growth

- *Simplifying natural ecosystems.* Humans eliminate some wildlife habitats by plowing grasslands, clearing forests, and filling in wetlands, often replacing their thousands of interrelated plant and animal species with one crop or one kind of tree—called *monocultures*—or with buildings, highways, and parking lots. Then we spend a lot of time, energy, and money trying to protect such monocultures from invasion by opportunist species of plants (weeds), pests (mostly insects), and pathogens (fungi, viruses, or bacteria that harm the plants we want to grow).

- *Strengthening some populations of pest species and disease-causing bacteria by speeding up natural selection and causing genetic resistance through overuse of pesticides.*

- *Eliminating some predators.* Ranchers, who don't want bison or prairie dogs competing with their sheep or cattle for grass, want to eradicate those species. They also want to eliminate wolves, coyotes, eagles, and other predators that occasionally kill sheep. Big game hunters also push for elimination of predators that prey on game species.

- *Deliberately or accidentally introducing new species,* some beneficial and some harmful to us and other species. In the late 1800s several Chinese chestnut trees brought to the United States were infected with a fungus that spread to the American chestnut, once found throughout much of the eastern United States. Between 1910 and 1940, this accidentally introduced fungus virtually eliminated the American chestnut.

- *Overharvesting potentially renewable resources.* Ranchers and nomadic herders sometimes allow livestock to overgraze grasslands until erosion converts these ecosystems to less productive semideserts or deserts. Farmers sometimes deplete the soil of nutrients by excessive crop growing. Species of fish are overharvested. Wildlife species with economically valuable parts (such as elephant tusks, rhinoceros horns, and tiger skins) are endangered by illegal hunting (poaching).

- *Interfering with the normal chemical cycling and energy flows (throughputs) in ecosystems.* Soil nutrients can be easily eroded from monoculture crop fields, tree farms, construction sites, and other simplified ecosystems and can overload and disrupt other ecosystems such as lakes and coastal ecosystems. Chemicals such as chlorofluorocarbons (CFCs) released into the atmosphere can increase the flow of ultraviolet energy reaching the earth by reducing ozone levels in the stratosphere. Emissions of carbon dioxide and other greenhouse gases—from burning fossil fuels and from clearing and burning forests and grasslands—may trigger global climate change by disrupting energy flow through the atmosphere.

To survive we must exploit and modify parts of nature. However, we are beginning to understand that any human intrusion into nature has multiple effects, most of them unpredictable (Connections, right).

The challenge is to maintain a balance between simplified, human-altered ecosystems and the neighboring, more complex natural ecosystems on which we and other forms of life depend, and to slow down the rates at which we are altering nature for our purposes. If we simplify and degrade too much of the planet to meet our needs and wants, what's at risk is not the earth but our own species. According to biodiversity expert E. O. Wilson, "If this planet were under surveillance by biologists from another world, I think they would look at us and say, 'Here is a species in the mid-stages of self-destruction.'" The evolutionary lesson to be learned from nature is that no species can get "too big for its britches," at least not for long.

What Can We Learn from Nature About Living Sustainably? After billions of years of trial and error, the rest of nature has solved the problem of dynamic sustainability by being able to change and adapt to new conditions through the biological evolution of populations. But there is growing concern that humans, as newcomers on the scene, are beginning to threaten this sustainability for our own species and millions of others.

Scientific research indicates that living systems have six key features: *interdependence, diversity, resilience, adaptability, unpredictability,* and *limits*. Many biologists believe that the best way for us to live sustainably is to learn about and mimic the processes and adaptations by which nature sustains itself. Here are some basic ecological lessons from nature:

- Most ecosystems use sunlight as their primary source of energy. Plants (producers) capture nonpolluting and virtually inexhaustible solar energy and convert it to chemical energy that keeps them, consumers (herbivorous and carnivorous animals), and decomposers alive.

- Ecosystems replenish nutrients and dispose of wastes by recycling chemicals. There is virtually no waste in nature. The waste outputs and decayed flesh of one organism are resource inputs for other organisms.

- Soil, water, air, plants, and animals are renewed through natural processes.

- Energy is always required to produce or maintain an energy flow or to recycle chemicals.

- Biodiversity takes various forms in different parts of the world because species diversity, genetic diversity, and ecological diversity have evolved over billions of years under different environmental conditions.

Q: When did the first forms of life arise on the earth?

▪ Complex networks of positive and negative feedback loops give organisms and populations information and control mechanisms for adapting, within limits, to changing conditions.

▪ The population size and growth rate of all species are controlled by their interactions with other species and with their nonliving environment. In nature there are always limits to population growth.

▪ Organisms, except perhaps humans, generally use only what they *need* (not merely want) to survive, stay healthy, and reproduce.

Currently, humans are violating these principles of sustainability. No one knows how long we can continue doing this.

Biologists have formulated several important principles that can help guide us in our search for more sustainable lifestyles:

▪ *We are part of, not apart from, the earth's dynamic web of life.*

▪ *Our lives, lifestyles, and economies are totally dependent on the sun and the earth.*

▪ *We can never do merely one thing*—what biologist Garrett Hardin calls the **first law of human ecology**.

▪ *Everything is connected to everything else; we are all in it together.* We are connected to all living organisms through the long evolutionary history contained in our DNA. We are connected to the earth through our interactions with air, water, soil, and other living organisms making up the ever-changing web of life. The destiny of all species is a shared one. The primary goal of ecology is to discover which connections in nature are the strongest, most important, and most vulnerable to disruption.

We need not—and indeed cannot—stop growing food or building cities. Indeed, concentrating people in cities and increasing food supplies by raising the yields per area of cropland are both ways to help protect much of the earth's biodiversity from being destroyed or degraded, as long as the harmful environmental side effects of such activities are kept under control. The key lesson is that we need earth wisdom, care, restraint, humility, cooperation, and love as we alter the ecosphere to meet our needs and wants. Earth care is self-care, and we can change our ways.

According to environmentalist David Brower, we need to focus on "global CPR—that's conservation, preservation, and restoration." This means **(1)** building societies based on conservation, not waste, **(2)** preserving what we can't replace, and **(3)** working with nature to help restore what we have degraded or destroyed.

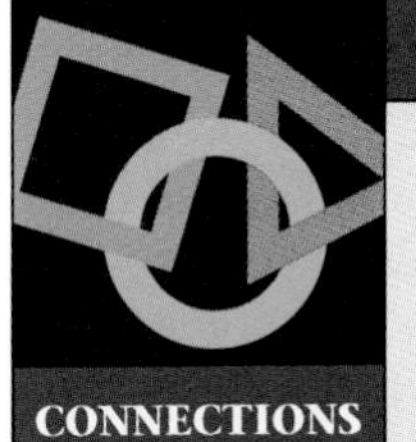

Ecological Surprises

Malaria once infected 9 out of 10 people in North Borneo, now known as Brunei. In 1955 the World Health Organization (WHO) began spraying the island with dieldrin (a DDT relative) to kill malaria-carrying mosquitoes. The program was so successful that the dreaded disease was virtually eliminated.

Other, unexpected things began to happen, however. The dieldrin also killed other insects, including flies and cockroaches living in houses. At first the islanders applauded this turn of events, but then small lizards that also lived in the houses died after gorging themselves on dieldrin-contaminated insects. Next, cats began dying after feeding on the lizards. Then, in the absence of cats, rats flourished and overran the villages. When the people became threatened by sylvatic plague carried by rat fleas, WHO parachuted healthy cats onto the island to help control the rats.

Then the villagers' roofs began to fall in. The dieldrin had killed wasps and other insects that fed on a type of caterpillar that either avoided or was not affected by the insecticide. With most of its predators eliminated, the caterpillar population exploded, munching its way through its favorite food: the leaves used in thatched roofs.

Ultimately, this episode ended happily: Both malaria and the unexpected effects of the spraying program were brought under control. Nevertheless, the chain of unforeseen events emphasizes the unpredictability of interfering with an ecosystem.

A more recent example of unanticipated interactions occurred in June 1995, when a pair of woodpeckers canceled a flight of the space shuttle Discovery at Cape Canaveral, Florida. A pair of yellow-shafted flickers punched holes in the space vehicle's insulation.

Investigation revealed that the problem may have been caused by an earlier flight of the space shuttle that killed four great horned owls, a natural predator of the flicker. It's possible that those four owls in the Cape Canaveral wildlife refuge were the only things keeping the population of local woodpeckers under control.

Attempts to scare the flickers away with horned owl decoys and taped horned owl screeches haven't worked. NASA wildlife officials may import more great horned owls in an attempt to solve this unanticipated problem.

Critical Thinking

Do you believe that the beneficial effects of spraying pesticides in North Borneo outweighed the resulting unexpected and harmful effects? Explain.

A: 3.7–3.8 billion years ago (probably bacterial cells)

10-6 SOLUTIONS: WORKING WITH NATURE TO HELP ECOSYSTEMS HEAL

How Can We Rehabilitate and Restore Damaged Ecosystems? The ideal goal is to reduce and minimize the damage we do to nature—a *prevention strategy*. Meanwhile, we can work with nature to help heal some of the ecological wounds we have inflicted.

Much of this environmental damage is at least partially reversible. Degraded ecosystems can be protected and allowed to replenish themselves through ecological succession. Forests can be replanted, streams can be cleaned up, and wetlands can be restored (Individuals Matter, right). Researchers are creating a new discipline of *restoration ecology* devoted to renewing, repairing, or reconstructing damaged ecosystems.

Farmer and philosopher Wendell Berry says we should try to answer three questions in deciding whether and how to modify or rehabilitate natural ecosystems: **(1)** What is here? **(2)** What will nature permit us to do here? and **(3)** What will nature help us do here?

When a degraded ecosystem is abandoned and not redisturbed, in most cases it will eventually at least partially recover through secondary ecological succession (Figure 9-21). But such *natural restoration* often takes a long time on a human time scale.

By studying how natural ecosystems recover, scientists are learning how to speed up repair operations. *Rehabilitation* involves making degraded land productive again by doing such things as stopping soil erosion and allowing the land once again to produce food or wood for fuel and timber.

Active restoration is more ambitious. Its goal is to take a degraded site and reestablish a diverse, dynamic community of organisms consistent with the climate and soil of an area. Instead of trying to return an ecosystem to some predisturbance state, it involves working with nature to restore the biodiversity and ecological processes that occur in wild ecosystems. Often, it's not necessary to plant anything; we need only to find and protect the strongest types of natural growth and allow natural ecological succession to proceed. Another approach is *replacement* of a degraded ecosystem with another type of ecosystem, such as replacing a degraded area of forest with a productive pasture.

Repairing and protecting ecosystems require that a pool of suitable existing species be available to occupy a wide variety of habitats and niches. This is why it is so important to preserve biodiversity throughout the world and to protect sustainable patches of ecosystems that harbor species that recolonize ecosystems after a disturbance.

Scientists and concerned citizens have successfully restored or rehabilitated some damaged ecosystems. Examples include constructing artificial reefs in Lake Erie and along the Atlantic and Pacific coasts and rehabilitating damaged wetlands and grasslands.

For example, the world's largest tall-grass prairie restoration project is being carried out at the Fermi National Accelerator Laboratory in Batavia, Illinois, by scientists Ray Schulenberg and Robert Betz. They found remnants of virgin Illinois prairie in old cemeteries, on embankments, and on other patches of land. In 1972 they transplanted these remnants by hand to a 4-hectare (10-acre) patch at the Fermi Laboratory site. Each year since then, volunteers have carefully prepared more land, sowed it with native prairie plants, weeded it by hand, and used controlled burning to maintain established communities. Today more than 180 hectares (445 acres) have been restored with prairie plants. New native species are introduced each year, with the goal of eventually establishing the 150–200 species that once flourished on the entire site.

What Are the Limits of Ecosystem Rehabilitation and Restoration? Working with nature to help ecosystems heal isn't easy. It takes lots of money and decades of hard work, but the long-term costs of doing nothing are much higher. Each piece of land or aquatic system poses unique challenges.

Some environmentalists worry that environmental restoration could encourage continuing environmental destruction by giving us the idea that we can undo any ecological harm we do. But ecological restoration is imperfect at best and falls far short of fully restoring damaged ecosystems. Preventing ecosystem damage in the first place is much cheaper and more effective.

Another concern is government policies that allow developers to destroy one ecosystem or wetland if they protect, restore, or "create" a similar one of roughly the same size. This tradeoff or *mitigation* approach is preferable to wanton destruction of ecosystems. However, by legitimizing further destruction, it partially defeats the main purpose of ecosystem restoration, *repairing previous damage*. The goal is to prevent ecosystem destruction and degradation so that expensive and only partially effective ecological restoration is not needed.

The dedicated scientists and volunteers who are carrying out ecological rehabilitation and restoration projects are inspiring examples of people who care for the earth. Their actions also remind us that it is easier and cheaper not to harm the earth in the first place.

If we love our children, we must love the earth with tender care and pass it on, diverse and beautiful, so that on a warm spring day 10,000 years hence they can feel peace in a sea of grass, can watch a bee visit a flower, can hear a sandpiper call in the sky, and can find joy in being alive.

HUGH H. ILTIS

Q: When did the first photosynthesizing cells arise on the earth?

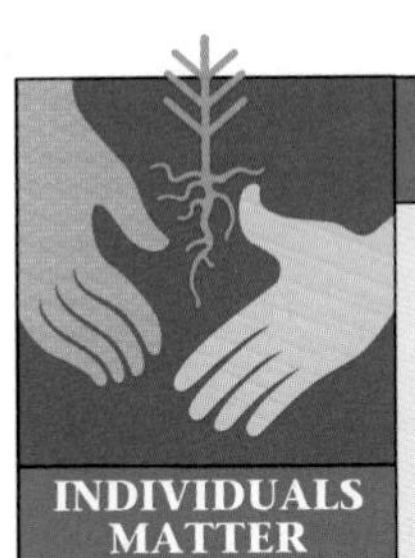

Earth Healing: Restoration of a Wetland

Some people's deepest fears are linked with swamps, marshes, quicksand, and the "things" that lurk there. Driven by such fears and ignorance about wetlands, as well as by a desire for land and profit, we have drained swamps and marshes relentlessly for centuries.

Belatedly, we have begun to question such campaigns against nature as we learn more about the ecological importance of natural wetlands for us and for other species. For example, by removing pollutants from water, wetlands are nature's free sewage treatment plants. Wetlands also provide free flood control and are key habitats for a large number of species, many of them important sources of seafood for us. Can we turn back the clock to restore or rehabilitate lost marshes?

California rancher Jim Callender decided to try. His motives were mixed: He wanted a place to hunt ducks, as well as a restored wetland habitat. In 1982 he bought 20 hectares (50 acres) of Sacramento Valley ricefield that had been a marsh until the early 1970s. The previous owner had destroyed it, bulldozing, draining, leveling, uprooting the native plants and spraying with chemicals to kill the snails and other food of the waterfowl.

Aided by a water bank program of the U.S. Fish and Wildlife Service (USFWS), by the sale of his agricultural rights in the form of a conservation easement (to USFWS), and by the guidance of biologists from the California Waterfowl Association, Callender and his friends set out to restore the marshland. They hollowed out low areas, built up islands, replanted tules and bulrushes, reintroduced smartweed and other plants needed by birds, and planted fast-growing Peking willows.

Figure 10-11 Snow geese migrating along the Pacific flyway in eastern Oregon. (Pat and Tom Leeson/Photo Researchers, Inc.)

After 6 years of care, handplanting, and annual seeding with a mixture of watergrass, smartweed, and rice, the marsh is once again a part of the Pacific flyway used by migratory waterfowl (Figure 10-11). Many birds pass through on their way south in autumn and north in the spring, and some mallards and wood ducks nest there.

A variety of hawks, shorebirds, and songbirds also live or forage in the marsh. The native insects and snails are back in force, and local mammals have returned as well, including deer, muskrat, and beaver.

The tens of millions of migrating birds that once darkened the December sky in the Sacramento Valley may never come again. But Jim Callender and a few others have shown that at least part of the continent's wetlands heritage can be reclaimed with planning and hard work. Such earth healing is vital, but the real challenge is to protect remaining wetlands (and other undisturbed ecosystems) from harm in the first place—a prevention approach.

CRITICAL THINKING

1. After an environmental change such as a change in climate, some species would be expected to have higher carrying capacities and others lower carrying capacities. Explain why. What other factors might cause the carrying capacity of a particular species to change?

2. Why are pest species likely to be extreme r-strategists? Why are many endangered species likely to be extreme K-strategists?

3. Predict the type of survivorship curve you would expect given descriptions of the following organisms:
   a. This organism is an annual plant. It lives only 1 year. During that time, it sprouts, reaches

A: 2.3–2.5 billion years ago

maturity, produces many wind-dispersed seeds, and dies.

b. This organism is a mammal. It reaches maturity after 10 years. It bears one young every 2 years. The young are protected by the parents and the rest of the herd.

4. If you could come back as a member of a particular type of species, what type of survivorship curve (Figure 10-9) would you like to have, and why?

5. Why might secondary succession occur at different rates and with different sets of species in a small habitat fragment as compared with a large habitat fragment? in a habitat fragment that is close to a source of colonizer as compared with a habitat fragment that is far away? Would you expect as many species on these "habitat islands" as in a larger, more continuous habitat? Explain.

6. Ecologists observe that in naturally functioning ecosystems there is essentially no waste and that *waste* is a human concept arising from ultimately unsustainable high-waste societies (Figure 3-21). Do you agree with this assessment? If so, what should we do about this situation? What can you do (or what are you doing) about this situation?

7. Explain why a simplified ecosystem such as a cornfield is usually much more vulnerable to harm from insects and plant diseases than a more complex, natural ecosystem such as a grassland.

8. A bumper sticker reads "Nature always bats last." Explain what this means in ecological terms. What is its lesson for the human species?

9. Try to come up with analogies to complete the following statements: **(a)** Biotic potential is like _______. **(b)** Being an r-strategist is like _______. **(c)** Habitat fragmentation is like _______. **(d)** Relying on monoculture crops and tree farms is like _______. **(e)** Restoring a degraded ecosystem is like _______.

10. Given current environmental conditions, if you had a choice would you rather be an r-strategist or a K-strategist? Explain your answer. What implications does your decision have for your current lifestyle?

11. Suppose a hurricane blows down most of the trees in a forest. Timber company officials offer to salvage the fallen trees and replant a tree plantation to reduce the chances of fire and to improve the area's appearance with an agreement that they can harvest the trees when they reach maturity and plant another tree farm. Others argue that the damaged forest should be left alone because hurricanes and other natural events are part of nature and the dead trees will serve as a source of nutrients for natural recovery through ecological succession. What do you think should be done, and why?

PROJECTS

1. Use the library and Internet to choose one wild plant species and one animal species and analyze the factors that are likely to limit the populations of each of the species. How are the factors that limit animal and plant populations similar? How are they different?

2. Use the principles of sustainability found in nature (p. 246–247) to evaluate the sustainability of the following parts of human systems: **(a)** transportation, **(b)** cities, **(c)** agriculture, **(d)** manufacturing, **(e)** waste disposal, and **(f)** your own lifestyle. Compare your analysis with those made by other members of your class.

3. Try to trace the known and potential short- and long-term environmental impacts of the following human activities on wildlife, energy flows through ecosystems and the ecosphere, and on the water, carbon, and nitrogen cycles: **(a)** driving a gasoline-burning car; **(b)** growing food as monocultural crops; **(c)** cutting down diverse forests and replacing them with tree plantations; **(d)** eating meat produced by raising domesticated livestock; **(e)** increasing crop yields by applying large amounts of commercial inorganic fertilizer; **(f)** building and using a shopping center; **(g)** using air conditioning; **(h)** producing electricity by burning a fossil fuel such as coal; **(i)** producing electricity using nuclear power; **(j)** using large quantities of chemical pesticides; **(k)** using large quantities of antibiotics; **(l)** burning trash in incinerators; **(m)** building sewage treatment plants and discharging the resulting effluent into nearby streams, lakes, or coastal zones; **(n)** encouraging people to live in cities; **(o)** encouraging people to spread out and live in rural areas. Compare your analyses with those of other members of your class to help reveal the many connections and effects we have on the ecosphere.

4. Use the library and Internet to find three examples of ecosystem restoration. For each example, explain how the restoration was achieved. What ecological principles were used in restoring each of the ecosystems? How successful was each of the restoration efforts? How could you improve each of the efforts?

5. Make a concept map of this chapter's major ideas, using the section heads and subheads and the key terms (in boldface). Look inside the back cover and in Appendix 4 for information about concept maps.

PART III

HUMAN POPULATION, RESOURCES, AND SUSTAINABILITY

The problems to be faced are vast and complex, but come down to this: 5.9 billion people are breeding exponentially. The process of fulfilling their wants and needs is stripping earth of its biotic capacity to produce life; a climactic burst of consumption by a single species is overwhelming the skies, earth, waters, and fauna.

PAUL HAWKEN

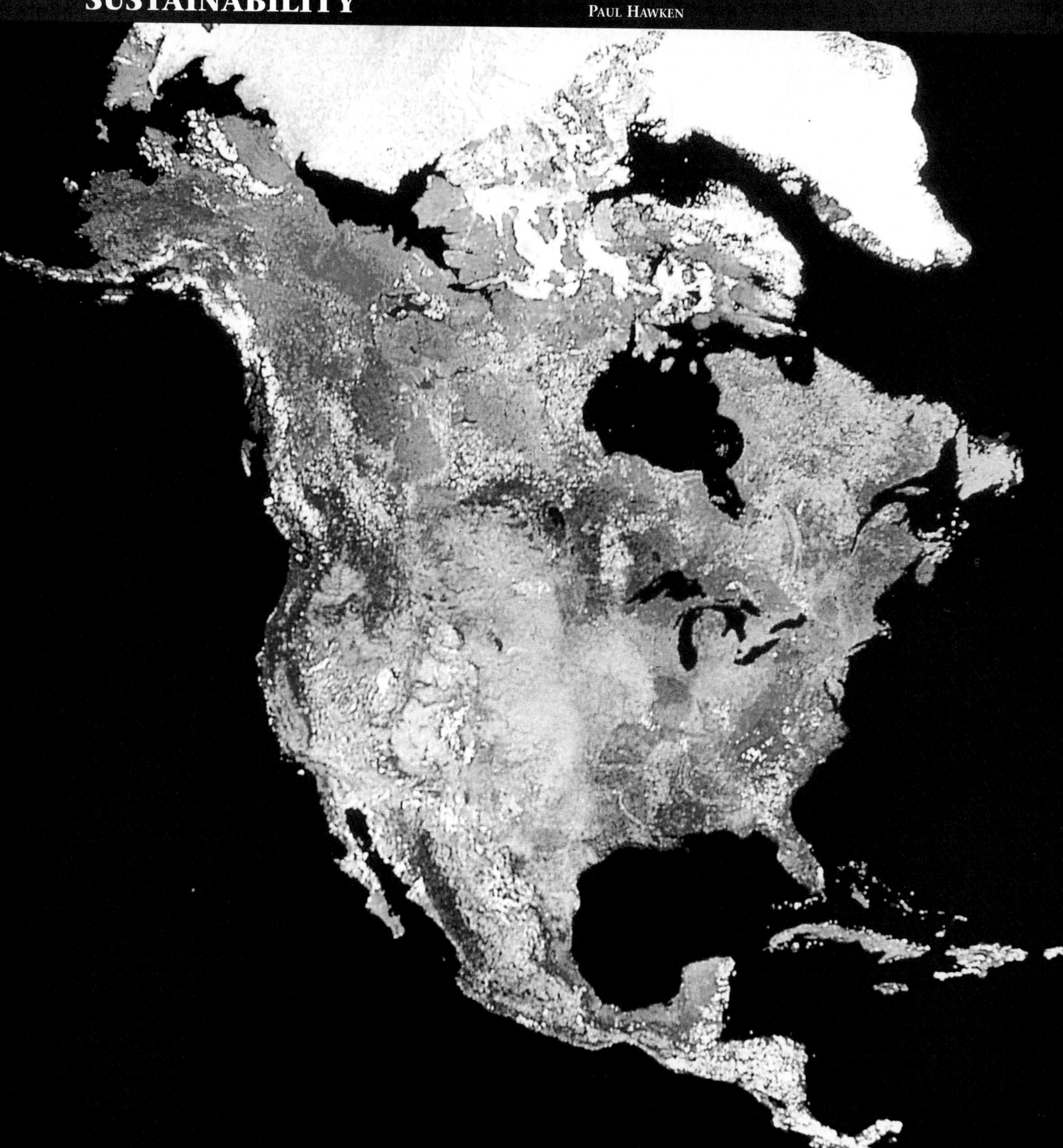

11 HUMAN POPULATION: GROWTH, DEMOGRAPHY, AND CARRYING CAPACITY

Slowing Population Growth in Thailand

Can a country sharply reduce its population growth in only 15 years? Thailand did.

In 1971, Thailand adopted a national policy to control and reduce its population growth. When the program began the country's population was growing at a rate of 3.2% per year and the average Thai family had 6.4 children. Fifteen years later, the country's population growth rate had been cut in half to 1.6%. By 1998 the rate had fallen to 1.1%, and the average number of children per family was 2.0. Thailand's population is projected to grow from 61 million in 1998 to 71 million by 2025.

There are several reasons for this impressive feat: the creativity of the government-supported family-planning program, high literacy among women (90%), an increasing economic role for women, advances in women's rights, better health care for mothers and children, the openness of the Thai people to new ideas, the willingness of the government to encourage and financially support family planning and to work with the private, nonprofit Population and Community Development Association (PCDA), and support of family planning by the country's religious leaders (95% of Thais are Buddhist). Buddhist scripture teaches that "many children make you poor."

This remarkable transition was catalyzed by the charismatic leadership of Mechai Viravidaiya, a public relations genius and former government economist who launched the PCDA in 1974 to help make family planning a national goal. PCDA workers handed out condoms anywhere there was a crowd—at festivals, movie theaters, even traffic jams. Humorous songs were written about condom use and the reasons to have no more than two children. Mechai also persuaded traffic police to hand out condoms on New Year's Eve, now known as "Cops and Rubbers Day."

The PCDA also had birth control carts dispensing birth control pills and spermicidal foam at bus stations and public events and opened vasectomy clinics throughout the country. It paid the insurance of taxi drivers willing to dispense condoms and birth control pills in their cabs. On the Thai king's birthday, the PCDA offers free vasectomies; sterilization is now the most widely used form of birth control in Thailand. Between 1971 and 1998 the percentage of married women using modern birth control rose from 15% to 70%—higher that the 61% usage in developed countries and the 49% usage in developing countries.

Mechai helped establish a German-financed revolving loan to enable people participating in family-planning programs to install toilets and drinking water

Figure 11-1 This policeman in Bangkok, Thailand, is wearing a mask to reduce his intake of air polluted mainly by automobiles. Bangkok is one of the world's most car-clogged cities, with car commutes averaging 3 hours per day. Roughly one of every nine of its residents has respiratory ailments of some sort. (NHPA/Martin Harvey)

systems. Low-rate loans were also offered to farmers practicing family planning. The government also offers loans to individuals from a fund that increases as their village's level of contraceptive use rises.

All is not completely rosy. Although Thailand has done well in slowing population growth and raising per capita income, it has been less successful in reducing pollution and improving public health, especially maternal health and control of AIDS and other sexually transmitted diseases.

Its capital, Bangkok, remains one of the world's most polluted and congested cities. It is plagued with notoriously high levels of traffic congestion and air pollution (Figure 11-1). The typical motorist in Bangkok spends 44 days per year sitting in traffic, costing $2.3 billion in lost work time.

Humans have undergone several cultural changes, developing various technologies that have enabled them to tap into the earth's ecosystem services or earth capital (Figure 1-2), including various nonrenewable and renewable resources (Figures 1-11 and 1-21). As a result of these cultural revolutions, the overall carrying capacity for humans on earth has expanded and allowed the human population to grow rapidly (Figures 1-1, 1-7, and 1-22).

However, the combination of population growth (discussed in this chapter), increased resource use per person (discussed in the next five chapters), and environmental degradation and pollution per unit of resource used (discussed in Parts IV and V) has led to an exponential increase in the environmental impacts of the human population on the earth's resources and ecosystem services (Figures 1-3, 1-5, 1-13, 1-14, 1-17–1-19, and 3-21). Thus, population growth has a multiplier effect on resource use and on the long-term sustainability of human economies and societies.

Birth rates and death rates are coming down worldwide, but death rates have fallen more sharply than birth rates. As a result, there are more births than deaths; every time your heart beats three more babies are added to the world's population. At this rate, each day we share the earth and its resources with about 236,000 more people than the day before. Every year there are about 84 million more mouths to feed, and at this rate 12 years from now there will be 1 billion more people.

If this trend continues or is not slowed down, the number of people on the earth will at least double, unless death rates rise sharply because of unsustainable use and distribution of the world's resources.

We shouldn't delude ourselves: the population explosion will come to an end before very long. The only remaining question is whether it will be halted through the humane method of birth control, or by nature wiping out our surplus.

PAUL EHRLICH

This chapter is devoted to answering the following questions:

- How is population size affected by birth, death, fertility, and migration rates?
- How is population size affected by the percentage of males and females at each age level?
- How can population growth be slowed?
- What success have India and China had in slowing population growth?

11-1 FACTORS AFFECTING HUMAN POPULATION SIZE

How Is Population Size Affected by Birth Rates and Death Rates? Populations grow or decline through the interplay of three factors: births, deaths, and migration. **Population change** is calculated by subtracting the number of people leaving a population (through death and emigration) from the number entering it (through birth and immigration) during a specified period of time (usually a year):

$$\text{Population change} = \left(\begin{matrix}\text{Births}\\+\\\text{Immigration}\end{matrix}\right) - \left(\begin{matrix}\text{Deaths}\\+\\\text{Emigration}\end{matrix}\right)$$

When births plus immigration exceed deaths plus emigration, population increases; when the reverse is true, population declines. When these factors balance out, population size remains stable, a condition known as **zero population growth (ZPG)**.

Instead of using the total numbers of births and deaths per year, demographers use two statistics: the **birth rate**, or **crude birth rate** (the number of live births per 1,000 people in a population in a given year), and the **death rate**, or **crude death rate** (the number of deaths per 1,000 people in a population in a given year). Figure 11-2 shows the crude birth and death rates for various groupings of countries in 1998.

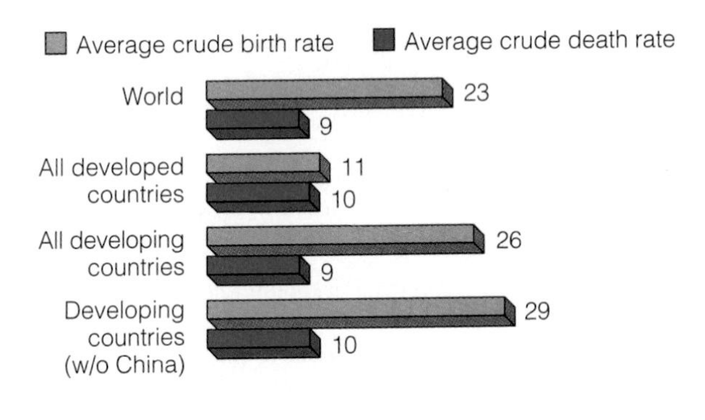

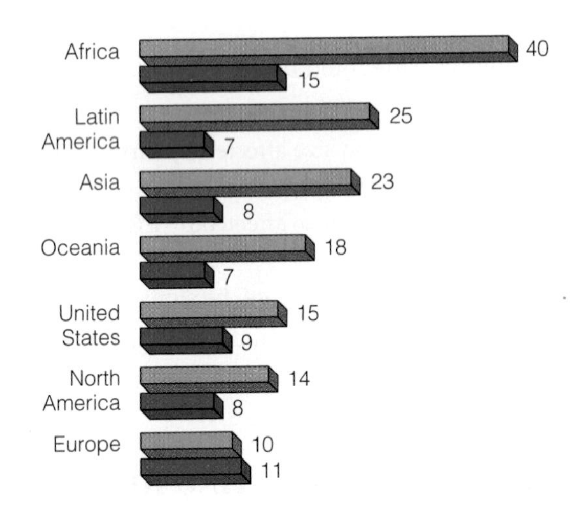

Figure 11-2 Average crude birth and death rates for various groupings of countries in 1998. (Data from Population Reference Bureau)

The rate of the world's annual population change (excluding migration) is usually expressed as a percentage*

$$\text{Annual rate of natural population change (\%)} = \frac{\text{Birth rate} - \text{Death rate}}{1{,}000 \text{ people}} \times 100 = \frac{\text{Birth rate} - \text{Death rate}}{10}$$

The rate of the world's annual population growth (natural increase) dropped 35% between 1963 and 1998, from 2.2% to 1.43%. This is good news, but during the same period the population base rose by about 82%, from 3.2 billion to 5.93 billion. This 35% drop in the rate of population increase is roughly analogous to learning that the truck heading straight at you has slowed from 100 kilometers per hour to 65 while its weight increased by 85%. In other words, the problem of exponential population growth has not disappeared; it's just occurring at a slower rate.

Figure 11-3 presents the annual rates of population change for major parts of the world in 1998. An annual natural increase rate of 1–3% may seem small, but such exponential growth rates lead to enormous increases in population size over a 100-year period. For example, a population growing by 3% a year will increase its population 19-fold in a century.

The current annual population increase rate of 1.43% adds about 84 million people per year (5.9 billion × 1.43% = 84 million), equal to adding another Los Angeles every 3 weeks, another New York City every month, a Germany every year, and a United States every 3 years. Despite the drop in the rate of population growth, the larger base of population means that 84 million people were added in 1998, whereas only 69 million were added in 1963 when the world's population growth rate reached its peak.

In numbers of people, China (with 1.24 billion in 1998, one of every five people in the world) and India (with 989 million) dwarf all other countries (Figure 11-4). Together they make up 38% of the world's population. The United States, with 270 million people in 1998, has the world's third largest population but only 4.6% of the world's people. Figure 11-5 gives projected population growth in various regions between 1998 and 2025; more than 95% of this growth is projected to take place in developing countries, where hunger and poverty have become a way of life for almost a billion people. In 1950 the world population was 2.5 billion and 78% of the annual increase of 45 million took place in developing countries. Now, about 84 million people are added each year, with 98% of this growth occurring in developing countries.

How Have Global Fertility Rates Changed?

Two types of fertility rates affect a country's population size and growth rate. The first type, **replacement-level fertility**, is the number of children a couple must bear to replace themselves. It is slightly higher than two children per couple (2.1 in developed countries and as high as 2.5 in some developing countries in 1997), mostly because some female children die before reaching their reproductive years.

Lowering fertility rates to replacement level does not mean an immediate halt in population growth (zero population growth); there are so many future parents already alive that if each had an average of 2.1

*Crude birth and death rates that have not been rounded off to the nearest whole number are often used to calculate natural change; the result is then rounded off to the nearest tenth of a percent. Consequently, use of the rounded-off crude birth and death rate figures shown in Figure 11-2 will not always produce the rounded-off percentage growth figures shown in Figure 11-3.

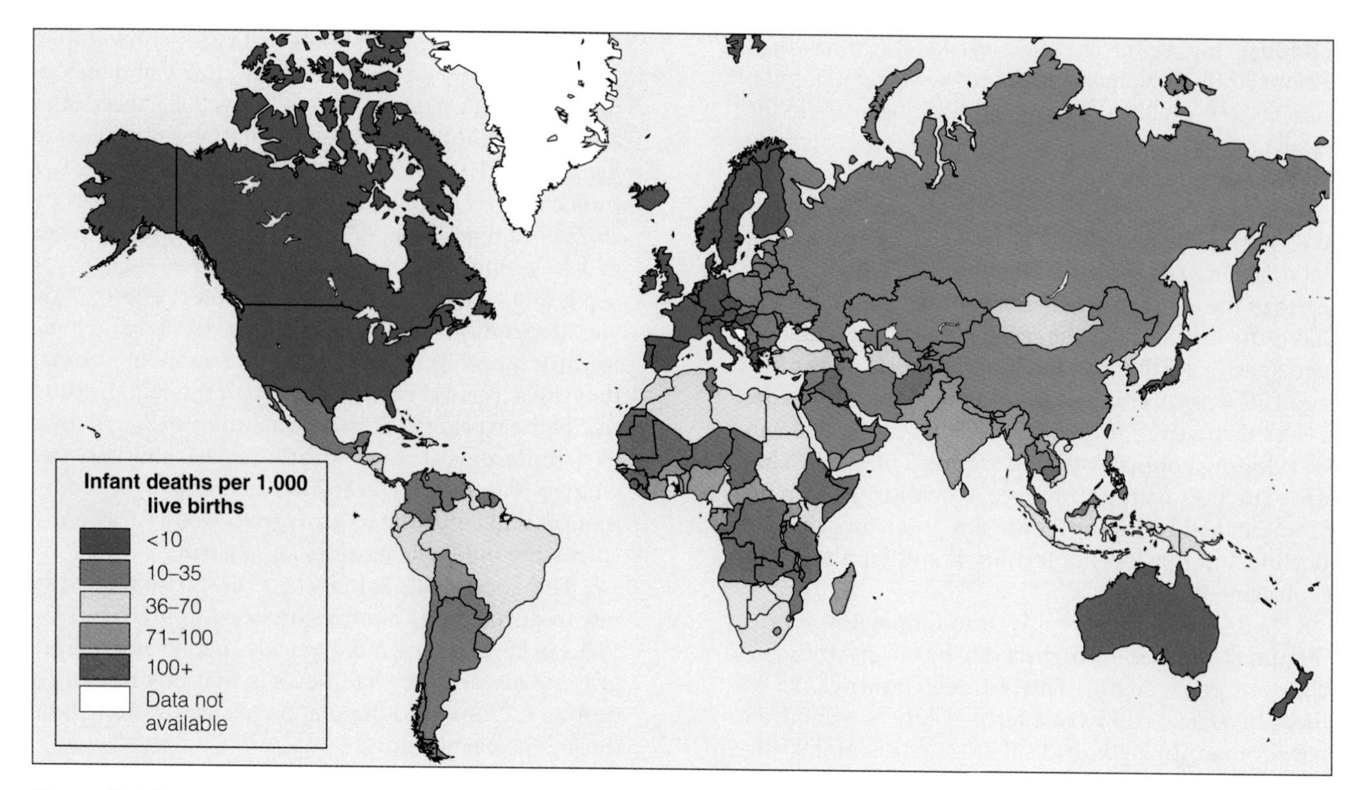

Figure 11-3 Average annual rate of population change (natural increase) in 1998. (Data from Population Reference Bureau)

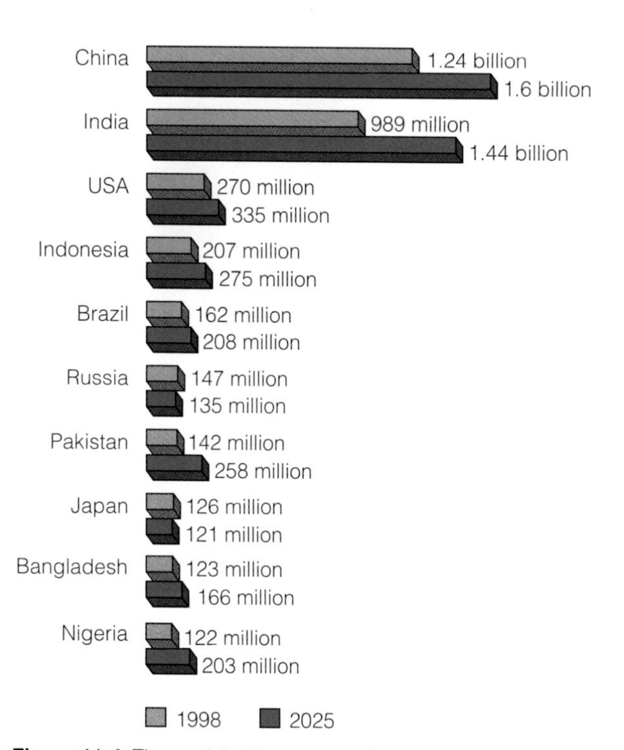

Figure 11-4 The world's 10 most populous countries in 1998, with projections of their population size in 2025. In 1998, more people lived in China than in all of Europe, Russia, North America, Japan, and Australia combined. (Data from World Bank)

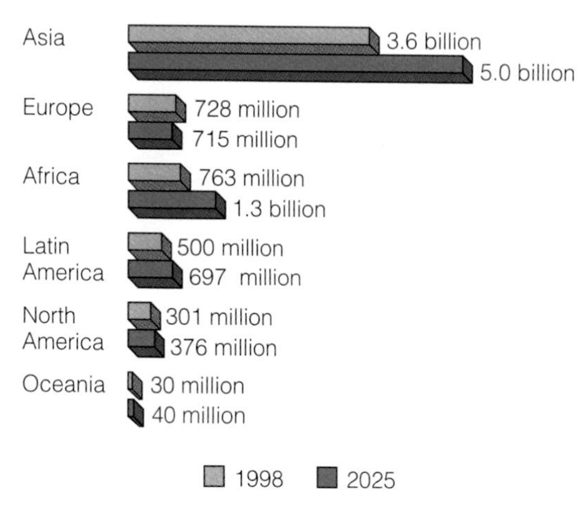

Figure 11-5 Population projections by region, 1998–2025. (Data from United Nations and Population Reference Bureau)

children and their children also had 2.1 children, the population would continue to grow for 50 years or more (assuming that death rates don't rise). Because of high population growth rates in the past, large numbers of young women and men are beginning to have children. The resulting population increases resulting from a large number of people entering their

A: We don't know; estimates range from 5 million to 100 million childbearing years is called *population momentum*. About 3 billion women—equal to the world's population in 1960—will enter their childbearing years in the next few years, thus causing population growth to continue for decades.

The second type of fertility rate, and the most useful measure of fertility for projecting future population change, is the **total fertility rate (TFR)**: an estimate of the average number of children a woman will have during her childbearing years under current age-specific birth rates. In 1998, the worldwide average TFR was 2.9 children per woman. It was 1.6 in developed countries (down from 2.5 in 1950) and 3.3 in developing countries (down from 6.5 in 1950). This drop in the average number of children born to women in developing countries is an impressive decline, but this level of fertility is still far above the replacement level.

TFRs vary considerably throughout the world (Figure 11-6), with the highest rate by far in Africa (5.6 children per woman). Thirty-three countries, 25 of them in Africa, still have a fertility rate of 6 children per woman. If the world's TFR remains at 2.9, the human population would reach 694 billion by the year 2150, some 120 times the current population and a sobering illustration of the enormous power of exponential population growth.

Population experts expect TFRs in developed countries to remain around 1.6 and those in developing countries to drop to around 2.3 by 2025; these rates are the basis of the medium population projections in Figure 11-5. That is good news, but it will still lead to a projected world population of around 8 billion by 2025, with more than 95% of this growth taking place in developing countries (Figure 1-6).

United Nations population projections to 2050 vary depending on the world's projected average total fertility rate (Figure 11-7). A key variable in population projections is the time at which the total fertility rate of the world (or of individual countries) will drop to a replacement level of around 2.1 children per woman (Figure 11-8); even after that time, world population will continue to grow for 50–60 years before stabilizing unless death rates rise sharply.

The good news is that the average total fertility rate in developing countries dropped from 6.5 in 1950 to 3.3 in 1998, falling more quickly than ever in demographic history. The bad news is that this rate must drop to 2.1 for world population to stabilize sometime during the next century.

How Have Fertility Rates Changed in the United States? The population of the United States has grown from 76 million in 1900 to 270 million in

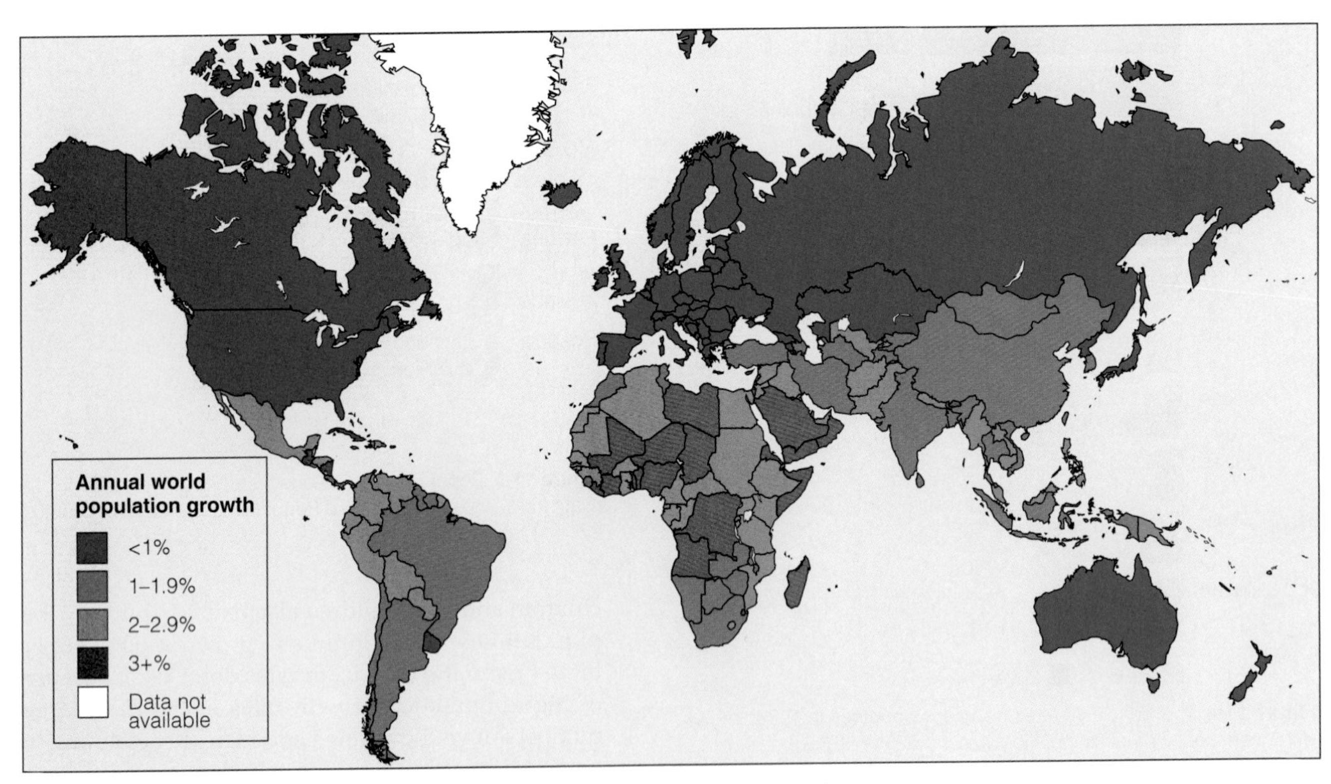

Figure 11-6 Total fertility rates (TFRs) in 1998. (Data from Population Reference Bureau)

Q: How many of earth's species have been identified?

1998—a 2.6-fold increase—even though the country's total fertility rate has oscillated wildly (Figure 11-9). In 1957, the peak of the post–World War II baby boom, the TFR reached 3.7 children per woman. Since then it has generally declined, remaining at or below replacement level since 1972. However, the United States has the highest fertility rate and the highest immigration rate of any industrialized country.

The drop in the total fertility rate has led to a decline in the rate of population growth in the United States. However, the country's population is still growing faster than that of most developed countries and is not even close to zero population growth. Including immigration, the U.S. population of 270 million grew by 1.17% in 1998—more than double the mean rate of the world's industrialized nations. This growth added about 3.1 million people: 1.8 million more births than deaths (accounting for about 60% of the growth), 935,000 legal immigrants and refugees, and an estimated 400,000 illegal immigrants. This is equivalent to adding another California every 10 years.

According to U.S. Bureau of Census projections, the U.S. population will increase from 270 million to 383 million by the year 2050—a 41% increase, with no stabilization on the horizon. This is a moderate projection. A less conservative estimate projects a population of 507 million by 2050, almost double the population in 1998. Because of a high per capita rate of resource use, each addition to the U.S. population has an enormous environmental impact. A disproportionate amount of the U.S. population increase ends up in the Pacific Northwest, where the population growth rate is higher than that of India.

The main reasons for this projected growth are **(1)** the large number of baby-boom women who are still in their childbearing years (even though the total fertility rate has remained at or below replacement level for 21 years, there has been a large increase in the number of potential mothers), **(2)** an increase in the number of unmarried mothers (including teenagers), **(3)** a continuation of higher fertility rates for women in some racial and ethnic groups than for Caucasian women, **(4)** high levels of legal and illegal immigration (which accounts for about 43% of current U.S. population growth), and **(5)** inadequate family-planning services (especially for the poor, who often can't afford such services). If immigration continues at current levels, new immigrants and their descendants would account for 80 million or 71% of the projected 112-million increase in the U.S. population between 1998 and 2050.

Case Study: Increasing Fertility Rates and Environmental Problems in California Consider population growth and environmental impact in

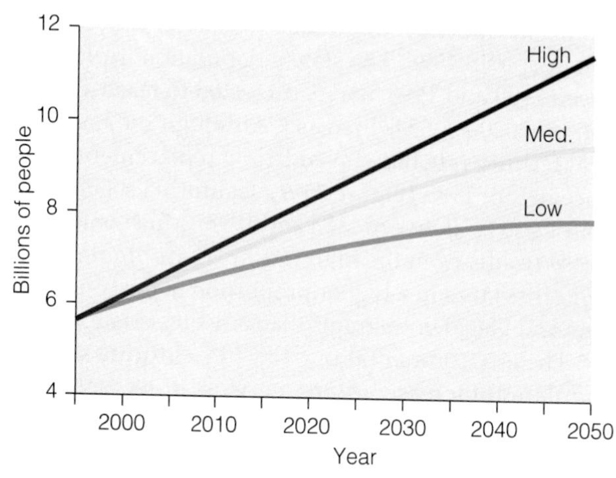

Figure 11-7 UN population projections to 2050 assuming that the world's total fertility rate will fall in a range between 2.5 (high), 2.1 (medium), and 1.7 (low) children per woman.

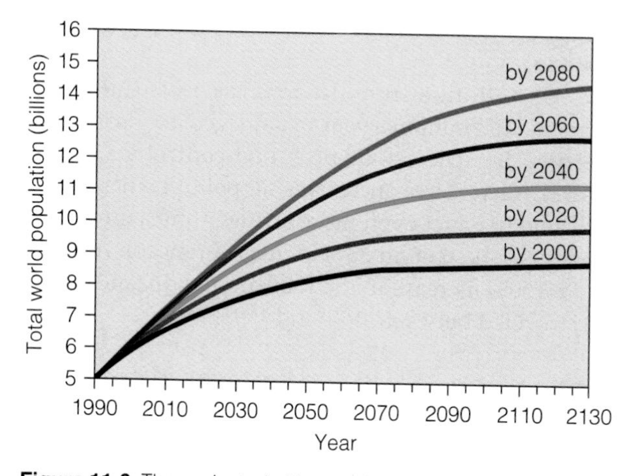

Figure 11-8 The projected ultimate size of the human population varies with the year by which a total fertility rate of 2.1 children per woman (replacement-level fertility) is reached. In 1998 the world's average total fertility rate was 2.9 children per woman. (Data from United Nations)

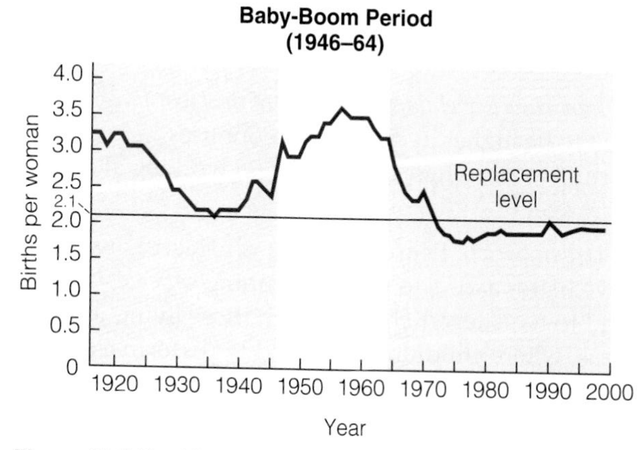

Figure 11-9 Total fertility rate for the United States between 1917 and 1998. (Data from Population Reference Bureau and U.S. Census Bureau)

A: About 1.8 million

California, the most populous U. S. state, with 33 million people in 1998. The state's population tripled between 1950 and 1998, and is projected to reach 49 million by 2020—a 63% increase. Although the national total fertility rate has stayed below replacement level since before 1985 (Figure 11-9), California's TFR rose from 1.9 to 2.5 between 1985 and 1998. (Much of this increase results from the high fertility of its immigrants; California takes in a higher proportion of the country's legal and illegal immigrants than any other state.)

The environmental impacts of California's large population increase are many. Most of its wetlands are gone. Its proportion of endangered and threatened species is higher than in any of the other lower 48 states. Its sewage systems are nearing capacity. Water shortages from occasional prolonged droughts remain a long-range threat, especially in southern California. A growing portion of its irrigated cropland is losing productivity because of salinization and waterlogging of soil.

Air pollution from urban areas costs California farmers $100 million a year in damaged and lost crops. Despite the strictest air-pollution control standards in the United States, increasing air pollution has forced the state to enact even more stringent measures, primarily because of an 83% increase in motor vehicles and a 150% increase in the total annual distance vehicles traveled between 1971 and 1995.

What Factors Affect Birth Rates and Fertility Rates? Among the most significant and interrelated factors affecting a country's average birth rate and total fertility rate are the following:

- *Average level of education and affluence.* Birth and fertility rates are usually lower in developed countries, where levels of education and affluence are higher than in developing countries. In developing countries, women with no education generally have two more children than women with a secondary education.
- *Importance of children as a part of the labor force.* Rates tend to be higher in developing countries (especially in rural areas, where children begin working at an early age).
- *Urbanization.* People living in urban areas usually have better access to family-planning services and tend to have fewer children than those living in rural areas, where children are needed to perform essential tasks.
- *Cost of raising and educating children.* Rates tend to be lower in developed countries, where raising children is much more costly because children don't enter the labor force until their late teens or early 20s. On average, a child born in the United States in 1997 will cost $2.7 million to raise to age 18 for a high-income family, $1.5 million for a middle-income family, and $762 thousand for a lower-income family.
- *Educational and employment opportunities for women.* Rates tend to be low when women have access to education and paid employment outside the home. Total fertility rates tend to decline as the female literacy rate increases. College graduates in the United States typically have 1.6 to 2.0 children. High school graduates average 2.7 children and high school dropouts average 3.2 children.
- *Infant mortality rate.* In areas with low infant mortality rates, people tend to have fewer children because fewer children die at an early age.
- *Average age at marriage* (or more precisely, the average age at which women have their first child). Women normally have fewer children when their average age at marriage is 25 or older.
- *Availability of private and public pension systems.* Pensions eliminate the need of parents to have many children to help support them in old age.
- *Availability of legal abortions.* There are an estimated 30 million legal abortions and 11–22 million illegal abortions worldwide each year.
- *Availability of reliable methods of birth control* (Figure 11-10).
- *Religious beliefs, traditions, and cultural norms.* In some countries, these factors favor large families and strongly oppose abortion and some forms of birth control.

What Factors Affect Death Rates? The rapid growth of the world's population over the past 100 years is not the result of a rise in the crude birth rate; rather, it has been caused largely by a decline in crude death rates (especially in developing countries; Figure 11-11). As United Nations consultant Peter Adamson puts it, "It's not that people suddenly started breeding like rabbits; it's just that they stopped dying like flies." More people started living longer (and fewer infants died) because of increased food supplies and distribution, higher living standards, better nutrition, improvements in medical and public health technology (such as immunizations and antibiotics), improvements in sanitation and personal hygiene, and safer water supplies (which have curtailed the spread of many infectious diseases).

Two useful indicators of overall health in a country or region are **life expectancy** (the average number of years a newborn infant can expect to live) and the

Q: What is biological evolution?

infant mortality rate (the number of babies out of every 1,000 born each year that die within a year of birth). In most cases, a low life expectancy in an area is the result of high infant mortality. Life expectancy has increased since 1965 to an average of 75 years in developed countries and 63 years in developing countries in 1998. Globally, life expectancy increased from 48 years in 1955 to 66 years in 1998, and is projected to reach 73 years by 2025. But in the world's 41 poorest countries, mainly in Asia and Africa, life expectancy is only about 50 years.

Because it reflects the general level of nutrition and health care, infant mortality is probably the single most important measure of a society's quality of life (Figure 11-12). A high infant mortality rate usually indicates insufficient food (undernutrition), poor nutrition (malnutrition), and a high incidence of infectious disease (usually from contaminated drinking water). Between 1965 and 1998, the world's infant mortality rate dropped from 20 per 1,000 live births to 8 in developed countries, and from 118 to 64 in developing countries. This is an impressive achievement, but it still means that at least 9 million infants die of preventable causes during their first year of life—an average of 24,700 unnecessary infant deaths per day.

The U.S. infant mortality rate of 7.0 per 1,000 in 1998 was low and down from 16.1 in 1975. Despite this improvement, 32 other countries had lower rates in 1998. Three factors that keep the U.S. infant mortality rate higher than it could be are inadequate health care (for poor women during pregnancy and for their babies after birth), drug addiction among pregnant women, and the high birth rate among teenage women.

Some good news is that the live birth rate for U.S. teens age 15–19 dropped 12% between 1991 and 1996. The bad news is that the United States still has the highest teenage pregnancy rate of any industrialized country. Each year nearly 1 million teenage girls become pregnant in the United States and more than 200,000 of them have abortions. Babies born to teenagers are more likely to have low birth weights, the most important factor in infant deaths.

How Is Migration Related to Environmental Degradation? Environmental Refugees The population of a given geographic area is also affected by movement of people into (immigration) and out of (emigration) that area. Population movement, both within and between countries, is usually desirable. Typically, people voluntarily move from less affluent areas of low opportunity to more affluent areas of higher opportunity. The economies of many receiving countries can benefit from migrant labor. However, legal and illegal immigration of large numbers of

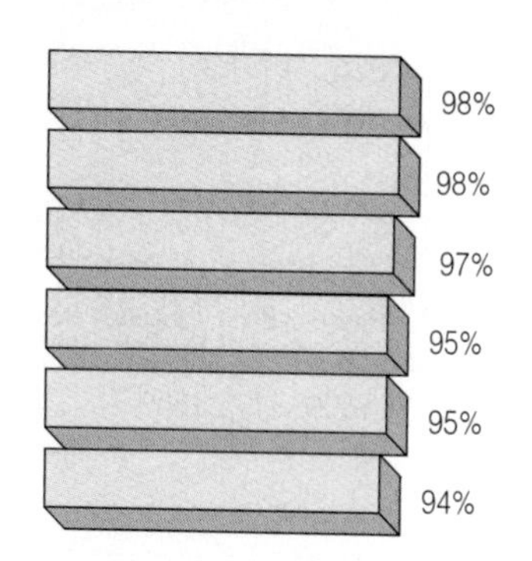

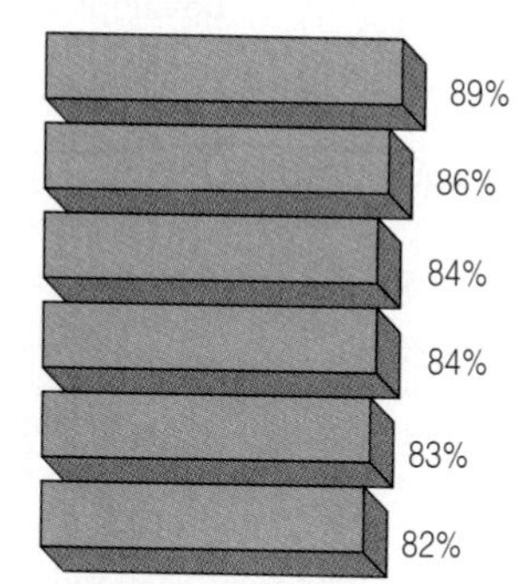

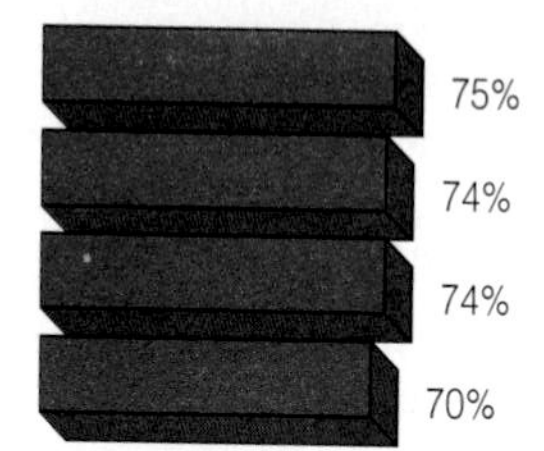

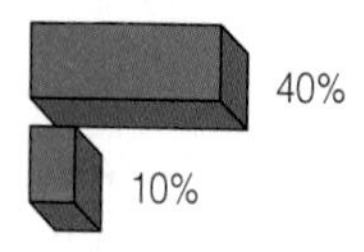

Figure 11-10 Typical effectiveness of birth control methods in the United States. Percentages are based on the number of undesired pregnancies per 100 couples using a specific method as their sole form of birth control for a year. For example, a 94% effectiveness rating for oral contraceptives means that for every 100 women using the pill regularly for 1 year, 6 will get pregnant. Effectiveness rates tend to be lower in developing countries, primarily because of lack of education. (Data from Alan Guttmacher Institute)

A: The change in the genetic makeup of a population through successive generations

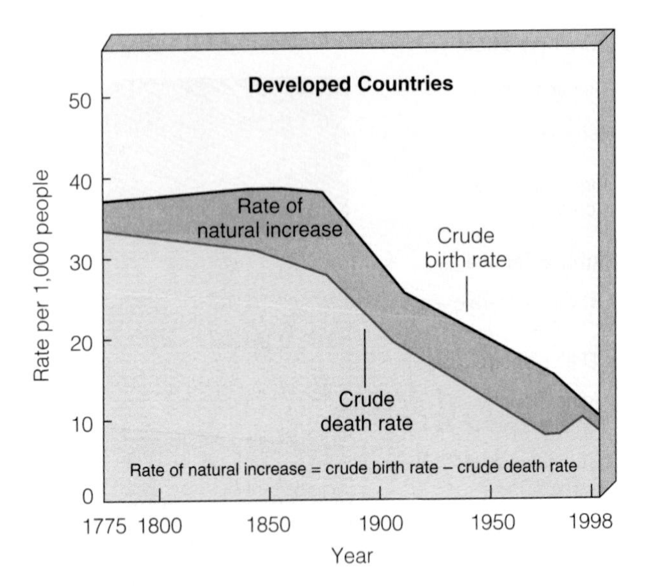

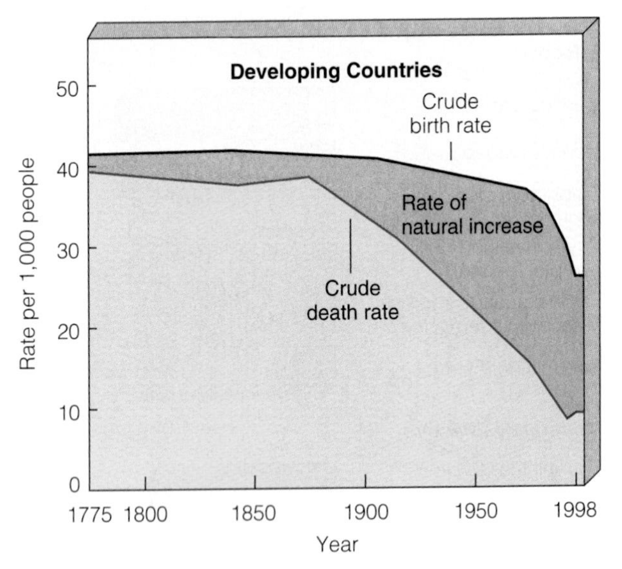

Figure 11-11 Changes in crude birth and death rates for developed and developing countries between 1775 and 1998. (Data from Population Reference Bureau and United Nations)

unskilled workers into many countries are increasingly being seen as an unwanted economic burden and a source of social disruption.

Some migration is involuntary and involves refugees displaced by armed conflict, environmental degradation, or natural disaster. According to Worldwatch Institute researcher Michael Renner, in 1995 an estimated 27 million international *environmental refugees*, moved from one country to another because of problems such as drought, desertification, deforestation, soil erosion, and resource shortages (compared with about 23 million *traditional refugees* fleeing from political oppression, religious persecution, ethnic strife, and war). Another 20 million refugees were displaced within the borders of their own countries in 1995. During the past decade some 50 million people have been left homeless by natural disasters including earthquakes, hurricanes, floods, and landslides.

This study and another by the Climate Institute project that the number of environmental refugees could reach 200 million in the next century if global warming projections are correct. Dealing with today's 27 million international environmental refugees and 20 million domestic refugees is a monumental problem that affects the environmental, economic, and military security of a growing number of countries (see the Guest Essay by Norman Myers, p. 530). To put this in perspective, World War II produced about 7 million homeless refugees in Europe.

Most countries influence their rates of population growth to some extent by restricting immigration; only a few countries accept large numbers of immigrants or refugees. Only about 1% of the annual population growth in developing countries is absorbed by developed countries through international migration. Thus, population change for most countries is determined mainly by the difference between their birth rates and death rates.

Migration within countries, especially from rural to urban areas, plays an important role in the population dynamics of cities, towns, and rural areas, as discussed in Section 26-1. In addition, during the past decade 80 to 90 million people were displaced by infrastructure projects such as road and dam construction.

11-2 POPULATION AGE STRUCTURE

What Are Age Structure Diagrams? As mentioned earlier, even if the replacement-level fertility rate of 2.1 were magically achieved globally tomorrow, the world's population would keep growing for at least another 50 years, stabilizing at about 8.6 billion (assuming no increase in death rates). The reason for this is the **age structure** of a population, or the proportion of the population (or of each sex) at each age level.

Demographers typically construct a population age structure diagram by plotting the percentages or numbers of males and females in the total population in each of three age categories: *prereproductive* (ages 0–14), *reproductive* (ages 15–44), and *postreproductive* (ages 45 and up). Figure 11-13 presents generalized age structure diagrams for countries with rapid, slow, zero, and negative population growth rates.

How Does Age Structure Affect Population Growth? Any country with many people below age 15 (represented by a wide base in Figure 11-13, left)

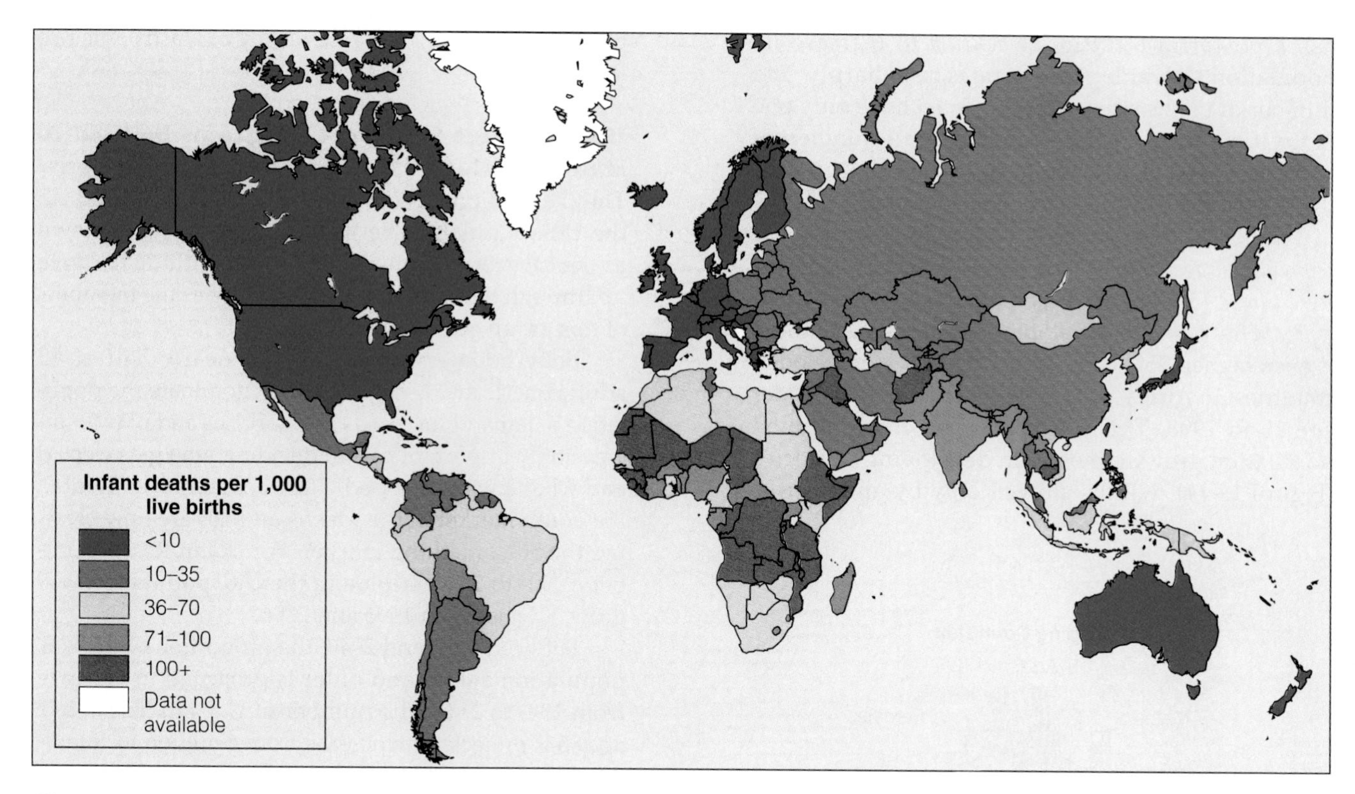

Figure 11-12 Infant mortality rates in 1998. (Data from Population Reference Bureau)

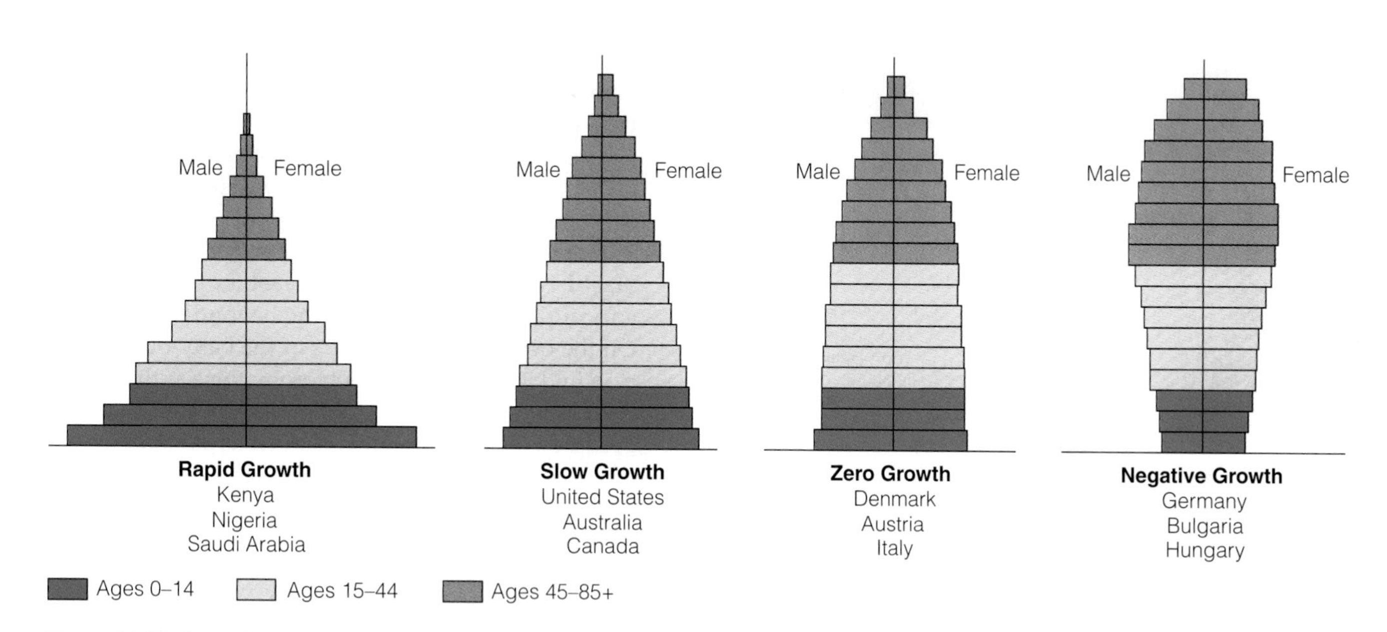

Figure 11-13 Generalized population age structure diagrams for countries with rapid, slow, zero, and negative population growth rates. (Data from Population Reference Bureau)

A: No; it involves survival of those that can leave the most descendants has a powerful built-in *momentum* to increase its population size unless death rates rise sharply. The number of births rises even if women have only one or two children because of the large number of women who will soon be moving into their reproductive years.

In 1998, half the world's 3.0 billion women were in the reproductive age group, and 32% of the people on the planet were under 15 years old, poised to move into their prime reproductive years. In developing countries the number is even higher: 35% compared with 19% in developed countries. In Africa, 44% of the population was under age 15 in 1998. This powerful force for continued population growth, mostly in developing countries (Figure 11-14), will be slowed only by an effective program to reduce birth rates, or by a catastrophic rise in death rates.

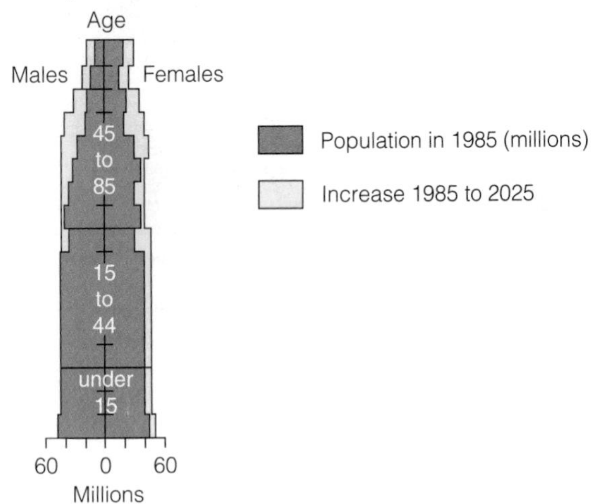

Figure 11-14 Age structure and projected population growth in developing countries and developed countries, 1985–2025. (Data from Population Reference Bureau)

How Can Age Structure Diagrams Be Used to Make Population and Economic Projections? The 78-million-person increase that occurred in the U.S. population between 1946 and 1964, known as the *baby boom* (Figure 11-9), will continue to move up through the country's age structure as the members of this group grow older (Figure 11-15).

Baby boomers now make up nearly half of all adult Americans. As a result, they dominate the population's demand for goods and services and play an increasingly important role in deciding who gets elected and what laws are passed. Baby boomers who created the youth market in their teens and 20s are now creating the 50-something market. For example, the number of 50- to 59-year-olds in the U.S. population will grow 50% between 1996 and 2006.

Between 1996 and 2040, the proportion of the U.S. population age 65 and older is projected to increase from 13% to 21%. The number of U.S. citizens above age 85 is projected to increase from 4 million to 18 million between 1995 and 2040. The economic burden of helping support so many retired baby boomers will fall on the *baby-bust generation*: people born since 1965 (when total fertility rates fell sharply and have remained below 2.1 since 1970; Figure 11-9). Retired baby boomers may use their political clout to force the smaller number of people in the baby-bust generation to pay higher income, health-care, and Social Security taxes. This could lead to much resentment and conflicts between the two generations.

In other respects the baby-bust generation should have an easier time than the baby-boom generation. Fewer people will be competing for educational opportunities, jobs, and services, and labor shortages may drive up their wages, at least for jobs requiring education or technical training beyond high school. On the other hand, members of the baby-bust group may find it difficult to get job promotions as they reach middle age because most upper-level positions will be occupied by members of the much larger baby-boom group. Many baby boomers may delay retirement because of improved health and the need to accumulate adequate retirement funds.

From these few projections we can see that any booms or busts in the age structure of a population create social and economic changes that ripple through a society for decades.

What Are Some Effects of Population Decline? Populations can decline for decades after replacement level is reached if they don't have a youth-dominated

Q: Does evolution assert that humans evolved from apes?

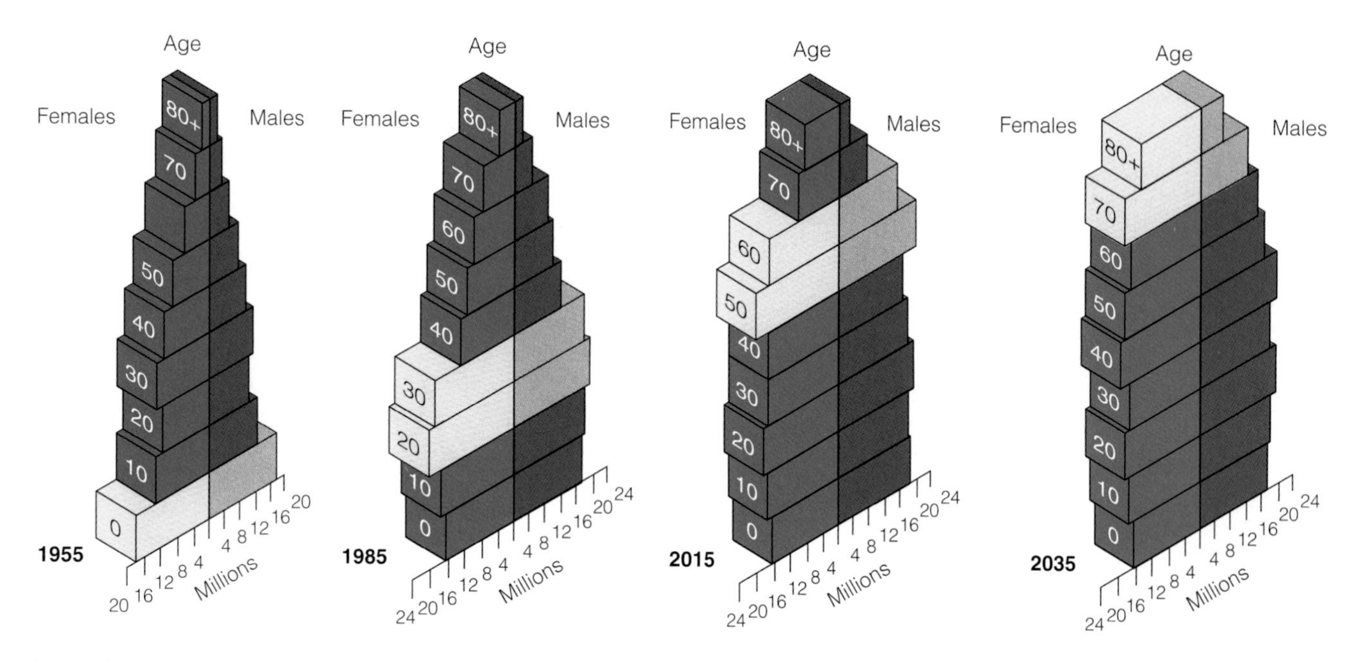

Figure 11-15 Tracking the baby-boom generation in the United States. (Data from Population Reference Bureau and U.S. Census Bureau)

age structure. When more people are in their postreproductive years than in their reproductive and prereproductive years, there are more deaths than births, even at replacement-level fertility.

The populations of most of the world's countries are projected to grow throughout most of the 21st century. By 1998, however, 37 countries with 684 people—12% of humanity—had roughly stable populations (annual growth rates below 0.3%) or declining populations. In other words, about one-eighth of humanity has achieved a stable population. As the projected age structure of the world's population changes between 1998 and 2150 and the percentage of people age 65 or older increases, more and more countries will begin experiencing population declines.

If population decline is gradual, its negative effects can usually be managed. But rapid population decline, like rapid population growth, can lead to severe economic and social problems. Countries undergoing rapid population decline have a sharp rise in the proportion of older people, who consume a large share of medical care, Social Security, and other costly public services. A country with a declining population can also face labor shortages unless it relies on greatly increased automation, immigration of foreign workers, or both.

Fearing that declining populations will threaten their economic well-being and national security, some European countries have offered economic incentives to encourage more births. In some cases such incentives have slowed the rate of decline, but not enough to prevent a population decrease. Massive immigration is also a solution to population decline and labor shortages, but so far most European countries have opposed this approach.

Case Study: The Graying of Japan In only 7 years, between 1949 and 1956, Japan (Figure 11-16) cut its birth, total fertility, and population growth rates in half. The main reason was widespread access to family planning implemented by the post–World War II U.S. occupation forces and the Japanese government. Since 1956 these rates have declined further, mostly because of access to family-planning services and other factors: cramped housing, high land prices, late marriage ages, and high costs of education.

In 1949 Japan's total fertility rate was 4.5; in 1998 it was 1.4, one of the world's lowest. If this trend continues and immigration doesn't rise, Japan's population should begin decreasing around 2006 and could shrink to 65–96 million by 2090, depending on its fertility rate and immigration rate (which is currently negligible).

As Japan approaches zero population growth, it is beginning to face some of the problems of an aging population. Japan's universal health insurance and pension systems used about 43% of the national

A: No; it states that humans and apes are descended from a common ancestor

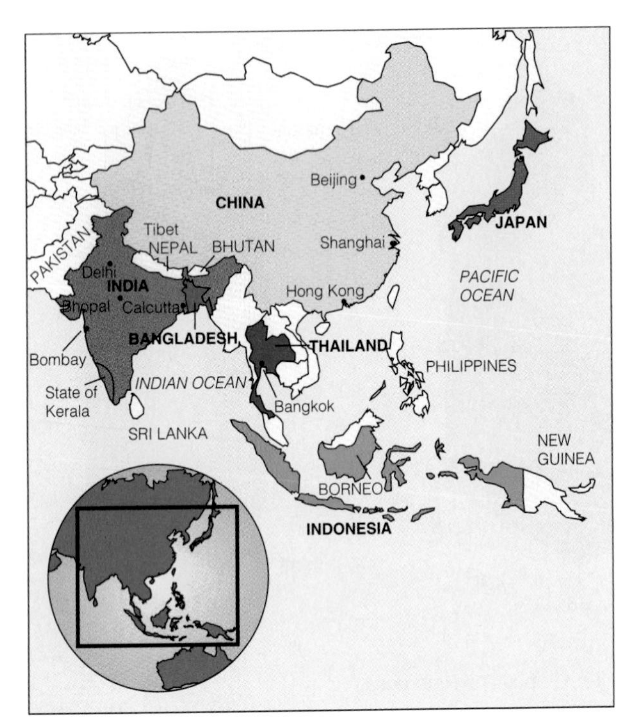

Figure 11-16 Where are Japan, Thailand, Indonesia, India, China, and Bangladesh? Some of the countries highlighted here are discussed in other chapters.

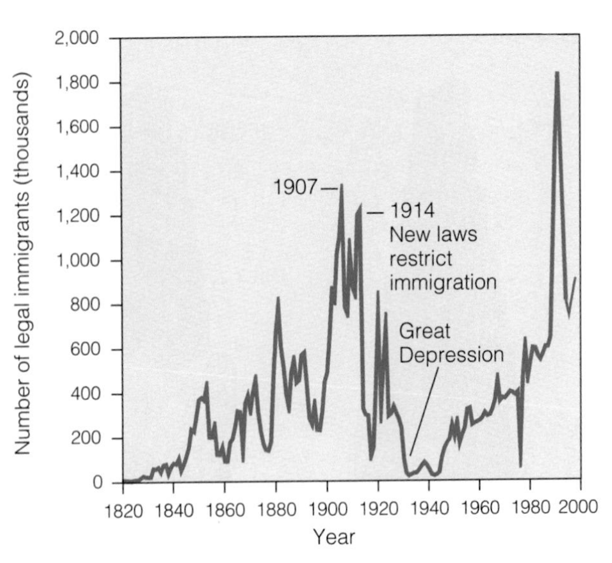

Figure 11-17 Legal immigration to the United States, 1820–1997. The large increase in immigration since 1989 resulted mostly from the Immigration Reform and Control Act of 1986, which granted legal status to illegal immigrants who could show that they had been living in the country for several years. (Data from U.S. Immigration and Naturalization Service)

income in 1997. This economic burden is projected to rise to 60% or higher in 2020. Japanese economists worry that the steep taxes needed to fund these services could discourage economic growth.

Since 1980 Japan has been feeling the effects of a declining workforce. This is one reason it has invested heavily in automation and encouraged women to work outside the home (about 43% were employed in 1998).

The population of Japan is 99% Japanese. Fearing a breakdown in its social cohesiveness, the government has been unwilling to increase immigration to provide more workers. Despite this official policy, the country is becoming increasingly dependent on illegal immigrants to keep its economic engines running. How Japan deals with these problems will be watched closely by other countries as they make the transition to zero population growth and, eventually, to population decline.

11-3 SOLUTIONS: INFLUENCING POPULATION SIZE

Case Study: Immigration in the United States A country can increase or maintain its population size by encouraging immigration, or it can decrease or maintain its population by encouraging emigration. Only a few countries—chiefly Canada, Australia, and the United States—allow large annual increases in population from immigration.

In 1998, the United States received about 935,000 legal immigrants and refugees (Figure 11-17) and 400,000 illegal immigrants, together accounting for 43% of the country's population growth. The Immigration and Naturalization Service estimates that there are about 1–3 million illegal immigrants in the United States. Currently, more than 75% of all legal immigrants live in six states: California, Florida, Illinois, New York, New Jersey, and Texas. If illegal immigrants are included, this figure rises to about 90%.

Does immigration help or hinder the U.S. economy? Economists have conflicting views, but many say the answer is both, depending on the time frame used. In times of high unemployment, legal and illegal immigrants willing to work for lower wages can either take jobs away from native-born workers or lower their wages. But many immigrants take low-paying, often menial jobs (such as crop picker, house cleaner, gardener, and janitor) that many native-born Americans refuse to do.

In the long run, working immigrants can boost the economy by paying taxes, increasing the supply of goods and services with their labor, and increasing

Ehrlich, P. R., and A. H. Ehrlich. 1997. "The Population Explosion." *Environmental Law*, vol. 27, no. 4, 1187(22).

the demand for goods and services by spending their wages. According to a 1997 study by the National Academy of Sciences, the work and taxes paid by immigrants add as much as $10 billion per year to the overall U.S. economy. The study estimated that during a lifetime each immigrant will pay an average of $80,000 more in taxes than he or she costs in services.

However, there is growing friction between the federal government and states with high levels of immigration because the federal government gets most of the taxes paid by immigrants and the states (and cities) bear most of the costs. In California, for example, the average household pays an extra $1,178 in taxes per year because of immigrants.

Some elected state and federal leaders propose that the federal government crack down much harder on illegal immigrants, screen refugees more carefully, and reimburse states and localities for all or most costs resulting from refugees and illegal immigration. Other observers call for a constitutional amendment that would deny citizenship to the children of illegal immigrants as a way to reduce the attractiveness of illegal immigration.

In 1986 Congress passed a law that prohibits the hiring of illegal immigrants and requires employers to examine the identity documents of all new employees. Employers who knowingly hire illegal immigrants can be fined, and repeat offenders can be imprisoned for up to 6 months. The bill also authorized funds to enforce the new law, to increase the border patrol staff by 50%, and to deport illegal immigrants.

Most analysts argue that the law has not worked. Critics charge that illegal immigrants easily get around the law with counterfeit documents and fake job histories because employers are not responsible for verifying the authenticity of documents. Some critics call for development of a fingerprint-based, tamper-resistant ID card for all Americans to facilitate the detection of illegal immigrants.

Despite more funds and increased border patrol agents, the Immigration and Naturalization Service doesn't have enough money or staff to enforce the new law effectively or to patrol more than a small fraction of the 3,140-kilometer (1,950-mile) U.S.–Mexico border. With nearly 60% of Mexico's labor force unemployed or underemployed, many Mexican immigrants think being caught and sent back (often repeatedly) is a minor risk compared with remaining in poverty.

Between 1820 and 1960, most legal immigrants to the United States came from Europe; since then, most have come from Asia and Latin America. If current trends continue, by 2050 about 25% of Americans will be of Hispanic origin (up from 10% in 1993) and the Caucasian population as a percentage will decline from 78% in 1988 to 56% in 2050.

In 1995 the U.S. Commission on Immigration Reform recommended reducing the number of legal immigrants and refugees to about 700,000 per year for a transition period and then 550,000 a year. Some demographers and environmentalists go further and call for lowering the annual ceiling for legal immigrants and refugees into the United States to 300,000–450,000, or for limiting legal immigration to about 20% of annual population growth.

Most of these analysts also support efforts to sharply reduce illegal immigration, although some are concerned that a crackdown on illegal immigrants can also lead to discrimination against legal immigrants. Proponents argue that such policies would allow the United States to stabilize its population sooner and help reduce the country's enormous environmental impact (Figure 1-19). Others oppose reducing current levels of legal immigration, arguing that it would diminish the historical role of the United States as a place of opportunity for the world's poor and oppressed. What is your opinion on this issue?

What Are the Pros and Cons of Reducing Births? Because raising the death rate is not desirable, lowering the birth rate is the focus of most efforts to slow population growth. Today about 93% of the world's population (and 91% of the peo-ple in developing countries) live in countries with fertility reduction programs. The funding for and effectiveness of these programs vary widely; few governments spend more than 1% of the national budget on them.

The unprecedented projected doubling of the human population from 5 to 10 billion or more between 1985 and 2050 raises some important questions. Can we provide enough food, energy, water, sanitation, education, health care, and housing for twice as many people? Can we reduce already serious poverty so that people can get enough food and other basic necessities without being forced to use potentially renewable resources unsustainably to survive? Can the world provide an adequate standard of living for twice as many people without causing massive environmental damage?

There is intense controversy concerning these questions, whether the earth is overpopulated, and what measures, if any, should be taken to slow population growth. To some the planet is already overpopulated, but others claim that if everyone existed at a minimum survival level, the earth could support 20–48 billion people. This would require that everyone exist on a diet of grain only, that all arable land be

Hint: Enter the search term *zero population growth* using the Subject Guide.

Moral Implications of Cultural Carrying Capacity

GUEST ESSAY

Garrett Hardin

As longtime professor of human ecology at the University of California at Santa Barbara, Garrett Hardin made important contributions to relating ethics to biology. He has raised hard ethical questions, sometimes taken unpopular stands, and forced people to think deeply about environmental problems and their possible solutions. He is best known for his 1968 essay "The Tragedy of the Commons," which has had a significant impact on the disciplines of economics and political science and on the management of potentially renewable resources. His many books include Promethean Ethics, Filters Against Folly: How to Survive Despite Economists, Ecologists, and the Merely Eloquent, *and* Living Within Limits *(see Further Readings).*

For many years, Angel Island in San Francisco Bay was plagued with too many deer. A few animals transplanted there in the early 1900s lacked predators and rapidly increased to nearly 300 deer—far beyond the carrying capacity of the island. Scrawny, underfed animals tugged at the heartstrings of Californians, who carried extra food for them from the mainland to the island.

Such well-meaning charity worsened the plight of the deer. Excess animals trampled the soil, stripped the bark from small trees, and destroyed seedlings of all kinds. The net effect was to lower the island's carrying capacity, year by year, as the deer continued to multiply in a deteriorating habitat.

State game managers proposed that the excess deer be shot by skilled hunters. "How cruel!" some people protested. Then the managers proposed that coyotes be introduced onto the island. Though not big enough to kill adult deer, coyotes can kill fawns, thereby reducing the size of the herd. However, the Society for the Prevention of Cruelty to Animals was adamantly opposed to this proposal.

In the end, it was agreed that some deer would be transported to other areas suitable for deer. A total of 203 animals were caught and trucked many miles away. From the fate of a sample of animals fitted with radio collars, it was estimated that 85% of the transported deer died within a year (most of them within 2 months) from various causes: predation by coyotes, bobcats, and domestic dogs; shooting by poachers and legal hunters; and being run over by automobiles.

The net cost (in 1982 dollars) for relocating each animal surviving for a year was $2,876. The state refused to continue financing the program, and no volunteers stepped forward to pay future bills.

Angel Island is a microcosm of the planet as a whole. Organisms reproduce exponentially, but the environment doesn't increase at all. The moral is a simple ecological commandment: *Thou shalt not transgress the carrying capacity.*

Now let's examine the situation for humans. A competent physicist has placed global human carrying capacity at 50 billion, about 10 times the current world population. Before you give in to the temptation to urge women to have more babies, consider what Robert Malthus said nearly 200 years ago: "There should be no more people in a country than could enjoy daily a glass of wine and piece of beef for dinner."

A diet of grain or bread and water is symbolic of minimum living standards; wine and beef are symbolic of higher living standards that make greater demands on the environment. When land that could produce plants for direct human consumption is used to grow grapes for wine or corn for cattle, more energy is expended to feed the human population. Because carrying capacity is defined as the *maximum* number of animals (humans) an area can support, using part of the area to support such cultural luxuries as wine and beef reduces the carrying capacity. This reduced capacity is called the *cultural carrying capacity,* and it is always smaller than simple carrying capacity.

Energy is the common "coin of the realm" for all competing demands on the environment. Energy saved by giving up a luxury can be used to produce more food staples and support more people. We could increase the simple carrying capacity of the earth by giving up any (or all) of the following "luxuries": street lighting, vacations, private cars, air conditioning, and

cultivated, and that much of the earth's crust be mined to a depth of 1.6 kilometers (1 mile).

Other analysts believe the planet could support 7–12 billion people at a decent standard of living by distributing land and food more equally. However, even if such optimistic estimates were technologically, politically, and environmentally feasible, many analysts doubt that our social and political structures could adapt to such a crowded and stressful world.

Others believe that asking how many people the world can support is the wrong question, equivalent to asking how many cigarettes one can smoke before getting lung cancer. Instead, they say, we should be asking what the *optimum sustainable population* of the

Q: According to the theory of evolution, do species become progressively more perfect?

artistic performances of all sorts. But what we consider "luxuries" depends on our values as individuals and societies, and values are largely matters of choice. At one extreme, we could maximize the number of human beings living at the lowest possible level of comfort. Or we could try to optimize the quality of life for a much smaller human population.

The carrying capacity of the earth is a scientific question. It may be possible to support 50 billion people at a bread-and-water level. Is that what we choose? The question, "What is the cultural carrying capacity?" requires that we debate questions of value, about which opinions differ.

An even greater difficulty must be faced. So far, we have been treating carrying capacity as a *global* issue, as if there were some global sovereignty capable of enforcing a solution on all people. But there is no global sovereignty ("one world"), nor is there any prospect of one in the foreseeable future. Thus, we must ask how some 200 nations are to coexist in a finite global environment if different sovereignties adopt different standards of living.

Consider a protected redwood forest that produces neither food for humans nor lumber for houses. Because people must travel many kilometers to visit it, the forest is a net loss in the national energy budget. However, for those fortunate enough to wander through the cathedral-like aisles beneath an evergreen vault, a redwood forest does something precious for the human spirit. But then intrudes an appeal from a distant land, where millions are starving because their population has overshot the carrying capacity; we are asked to save lives by sending food. So long as we have surpluses, we may safely indulge in the pleasures of philanthropy. But after we have run out of our surpluses, then what?

A spokesperson for the needy from that land makes a proposal: "If you would only cut down your redwood forests, you could use the lumber to build houses and then grow potatoes on the land, shipping the food to us. Since we are all passengers together on Spaceship Earth, are you not duty bound to do so? Which is more precious, trees or human beings?"

This last question may sound ethically compelling, but let's look at the consequences of assigning a preemptive and supreme value to human lives. At least 2 billion people in the world are poorer than the 39 million "legally poor" in America, and their numbers are increasing by about 40 million per year. Unless this increase is halted, sharing food and energy on the basis of need would require the sacrifice of one amenity after another in rich countries. The ultimate result of sharing would be complete poverty everywhere on the earth in order to maintain the earth's simple carrying capacity. Is that the best humanity can do?

To date, there has been overwhelmingly negative reaction to all proposals to make international philanthropy conditional on the cessation of population growth by poor, overpopulated recipient nations. Foreign aid is governed by two apparently inflexible assumptions:

- The right to produce children is a universal, irrevocable right of every nation, no matter how hard it presses against the carrying capacity of its territory.
- When lives are in danger, the moral obligation of rich countries to save human lives is absolute and undeniable.

Considered separately, each of these two well-meaning doctrines might be defensible; taken together, they constitute a fatal recipe. If humanity gives maximum carrying capacity precedence over problems of cultural carrying capacity, the result will be universal poverty and environmental ruin.

Or do you see an escape from this harsh dilemma?

Critical Thinking

1. What population size would allow the world's people to have good quality of life? What do you believe is the cultural carrying capacity of the United States? Should the United States have a national policy to establish this population size as soon as possible? Explain.

2. Do you support the two principles this essay lists as the basis of foreign aid to needy countries? If not, what changes would you make in the requirements for receiving such aid?

earth might be, based on the planet's *cultural carrying capacity* (Guest Essay, above). Such an optimum level would allow most people to live in reasonable comfort and freedom without impairing the ability of the planet to sustain future generations. No one knows what this optimum population might be. Some consider it a meaningless concept; some put it at 20 billion, others at 8 billion, and others at a level below today's population size.

Those who don't believe that the earth is overpopulated point out that the average life span of the world's 5.92 billion people is longer today than at any time in the past. Those holding this view say that talk of a population crash is alarmist, that the world can

A: No; from a scientific standpoint, there is no plan or goal of perfection in evolution support billions more people, and that people are the world's most valuable resource for solving the problems we face. To these analysts, the primary cause of poverty is not population growth but the lack of free and productive economic systems in developing countries.

They argue that without more babies, developed countries with declining populations will face a shortage of workers, taxpayers, scientists and engineers, consumers, and soldiers needed to maintain healthy economic growth, national security, and global power and influence. These analysts urge the governments of the United States and other developed countries to give tax breaks and other economic incentives to couples who have more than two children.

Some people view any form of population regulation as a violation of their religious beliefs, whereas others see it as an intrusion into their privacy and personal freedom. They believe that all people should be free to have as many children as they want. Some developing countries and some members of minorities in developed countries regard population control as a form of genocide to keep their numbers and power from rising.

Proponents of slowing and eventually stopping population growth point out that we fail to provide the basic necessities for one out of six people on the earth today. If we can't (or won't) do this now, how will we be able to do this for twice as many people within the next 45 years?

Proponents of slowing population growth also consider overpopulation as a threat to the earth's life-support systems. They contend that if we don't sharply lower birth rates, we are deciding by default to raise death rates for humans and greatly increase environmental harm (Figure 1-19). In 1992, for example, the highly respected U.S. National Academy of Sciences and the Royal Society of London issued the following joint statement: "If current predictions of population growth and patterns of human activity on the planet remain unchanged, science and technology may not be able to prevent either irreversible degradation of the environment or continued poverty for much of the world."

Proponents of this view recognize that population growth is not the only cause of our environmental and resource problems. However, they argue that adding several hundred million more people in developed countries and several billion more in developing countries can only intensify many environmental and social problems.

Rapid population growth in developing countries is also a major cause of unemployment and underemployment. To provide jobs for the projected 1 billion more people of working age in developing countries by 2025, these countries will have to create at least 33 million full-time jobs each year for the next three decades. The United States, with an economy 50% larger than the combined economies of all developing countries, often has difficulty in generating 2 million new jobs per year.

Those favoring slowing population growth point out that technological innovation, not sheer numbers of people, is the key to military and economic power. Otherwise, England, Germany, Japan, and Taiwan, with fairly small populations, would have little global economic and military power and China and India would rule the world.

Proponents of slowing and eventually halting population growth believe that the United States and other developed countries should establish an official goal of stabilizing their populations as soon as possible. Proponents also believe that developed countries have a better chance of influencing developing countries to reduce their population growth more rapidly if the more affluent countries officially recognize the need to stabilize their own populations and reduce their unnecessary waste of the earth's resources.

These analysts believe that people should have the freedom to produce as many children as they want. However, such freedom would apply only if it did not reduce the quality of other people's lives now and in the future, either by impairing the earth's ability to sustain life or by causing social disruption. They point out that limiting the freedom of individuals to do anything they want in order to protect the freedom of other individuals is the basis of most laws in modern societies. What is your opinion on this issue?

How Can Computer Models Be Used to Evaluate Limits to Growth and Make the Transition to More Sustainable Societies? Since 1970, systems analysts have used *system dynamics computer modeling* to mimic the behavior of complex systems and to make projections about how key variables interact. First, system analysts develop mathematical equations that represent the interactions of key variables resulting from feedback loops, time delays, synergistic interactions, and other properties of complex systems (Sections 2-3 and 2-4); then they feed the equations into a computer to project future dynamic behavior of the system and to test the potential effects of various policy decisions. One of the best things about such models is that we can ask, "What if we do such and such?" and then run the model to learn what the results might be.

Such models are no better than the assumptions built into them and the accuracy of the data used. Despite their many limitations, they are a useful

Q: What percentage of all species that have ever lived on earth have become extinct?

tool for evaluating possible implications of current trends and proposed changes in environmental and economic policies.

In the early 1970s Jay Forrester, Donella Meadows, Dennis Meadows, Jørgen Randers, and their associates developed dynamic computer models to evaluate the global limits to human population growth and industrialization. These admittedly crude models examined the dynamic interaction of five major variables: population, pollution, use of nonrenewable resources, industrial output per capita, and food output per capita.

In 1972 the projections of the model, published in *The Limits to Growth*, indicated that if economic, resource use, and population trends in the early 1970s continued unchanged, the limits to physical growth on this planet would be reached within 100 years and result in economic and ecological collapse. The model also demonstrated that certain policy changes could forestall such collapse.

The projections of this model challenged the basic assumption of today's industrial societies that new advances in technology place no physical limits on industrial and population growth. *Limits to Growth* (with sales of 9 million copies in 29 languages) generated intense debate and research.

Twenty years later, in 1992, the authors updated their work in *Beyond the Limits: Confronting Global Collapse, Envisioning a Sustainable Future.* Their updated model projects that the world has already overshot some limits, and that if current trends continue unchanged, we face global economic and environmental collapse sometime in the 21st century.

By including in the model various policy decisions on population, resource use, pollution control, and per capita industrial and food output, the researchers were able to evaluate a range of possible courses of action by asking *what if* questions, such as the following:

Question 1: *What if the world's population and industrial output continue to expand exponentially at 1990 rates with no major policy changes?* Figure 11-18 summarizes the results projected by the model in this business-as-usual scenario. This status-quo run projects that current practices would lead to overshoot and collapse of the system sometime within the next 100 years. In this case the collapse would be caused by depletion of nonrenewable resources and environmental overload.

Another run of the model shows that doubling the projected supply of nonrenewable resources would delay overshoot and collapse by about 20 years. If

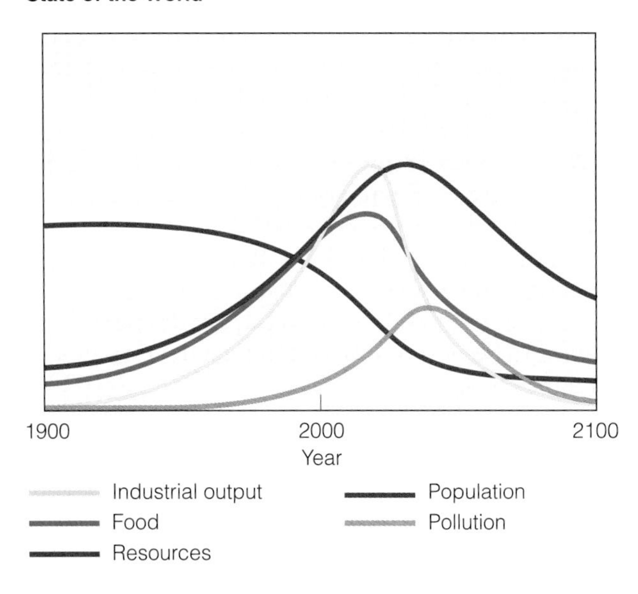

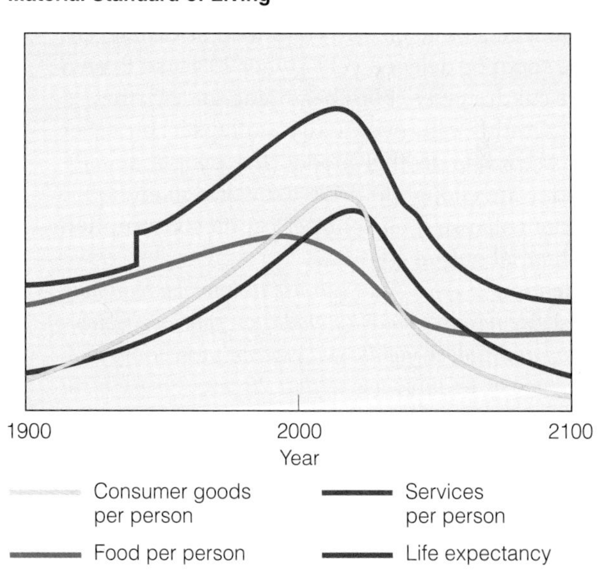

Figure 11-18 Plots of a computer model projecting what might happen if the world's population and economy continue growing exponentially at 1990 levels, assuming no major policy changes or technological innovations. This scenario projects that the world has already overshot some of its limits and that if current trends continue unchanged, we face global economic and environmental collapse sometime in the next century. (Used by permission from Donella Meadows et al., *Beyond the Limits: Confronting Global Collapse, Envisioning a Sustainable Future,* White River Junction, Vt.: Chelsea Green, 1992)

A: About 99.9%

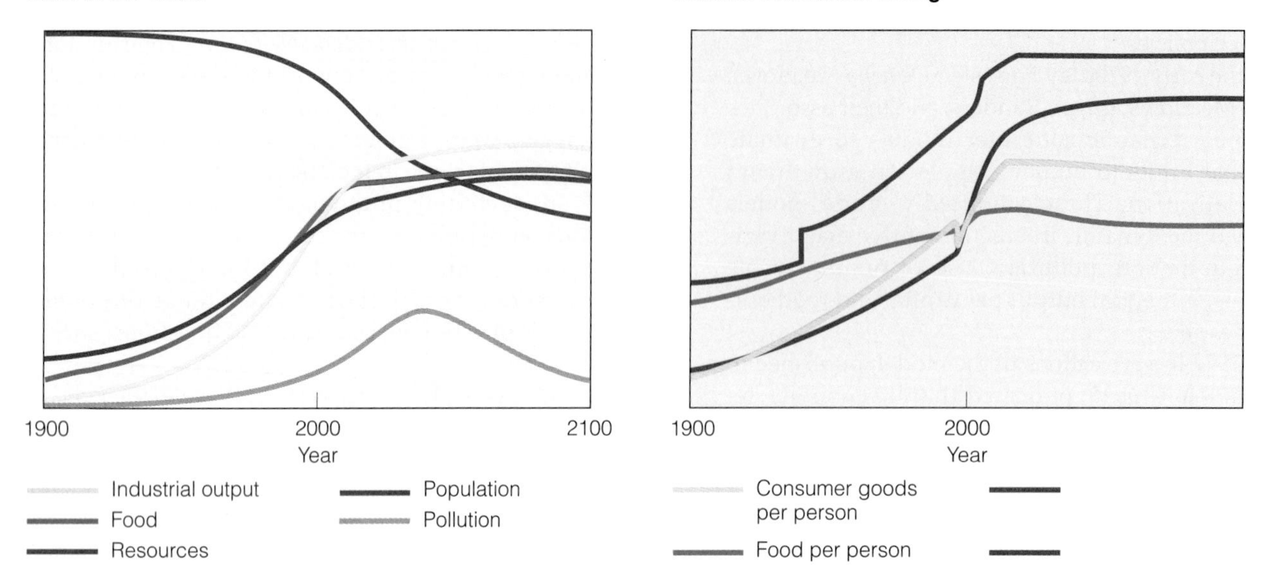

Figure 11-19 Computer-generated scenario projecting how we can avoid overshoot and collapse and make a fairly smooth transition to a sustainable future. It assumes that **(1)** technology allows us to double supplies of nonrenewable resources, double crop and timber yields, cut soil erosion in half, and double the efficiency of resour use within 20 years; **(2)** 100% effective birth control was made available to everyone by 1995; **(3)** no couple has more than two children, beginning in 1995; and **(4)** per capita industrial output is stabilized at 1990 levels. Another computer run projects that waiting until 2015 to implement these changes would lead to collapse and overshoot sometime around 2075, followed by a transition to sustainability by 2100. (Used by permission from Donella Meadows et al., *Beyond the Limits: Confronting Global Collapse, Envisioning a Sustainable Future,* White River Junction, Vt.: Chelsea Green, 1992)

everyone had no more than two children beginning in 1995 and if economic growth continued at 1990 levels, collapse could be delayed yet another 20 years. Even if no more children were born beginning in 1995, the system would still eventually collapse. In other words, according to this model, the people already alive today are enough to drive the world to economic and social collapse if current per capita consumption and industrial output continue.

Question 2: *What if we can use technology to double nonrenewable resource supplies, pollution control effectiveness, crop and timber yields, soil erosion protection, and the efficiency of resource use within 20 years, make birth control methods available to everyone by 1995, and stabilize current per capita industrial output at 1990 levels?* Figure 11-19 shows that doing all of these things avoids overshoot and collapse, and results in a fairly smooth transition to a sustainable future.

The developers of this model emphasize that the model gives only projections, not predictions. Instead of viewing their projections as models of doom, the models' developers see them as challenges, opportunities, and possible guidelines for achieving economically and environmentally sustainable societies. Critics of such models consider them too simplistic to be of much use, and they believe that advances in technology can prevent overshoot and environmental collapse.

How Can Economic Development Help Reduce Births? Demographers have examined the birth and death rates of western European countries that industrialized during the 19th century, and from these data they developed a hypothesis of population change known as the **demographic transition**: As countries become industrialized, first their death rates and then their birth rates decline.

According to this hypothesis, the transition takes place in four distinct stages (Figure 11-20). In the *preindustrial stage,* harsh living conditions lead to a high birth rate (to compensate for high infant mortality) and a high death rate. Thus there is little population growth.

In the *transitional stage,* industrialization begins, food production rises, and health care improves. Death rates drop and birth rates remain high, so the population grows rapidly (typically 2.5–3% a year).

In the *industrial stage,* industrialization is widespread. The birth rate drops and eventually

Q: What are the six key features of living systems?

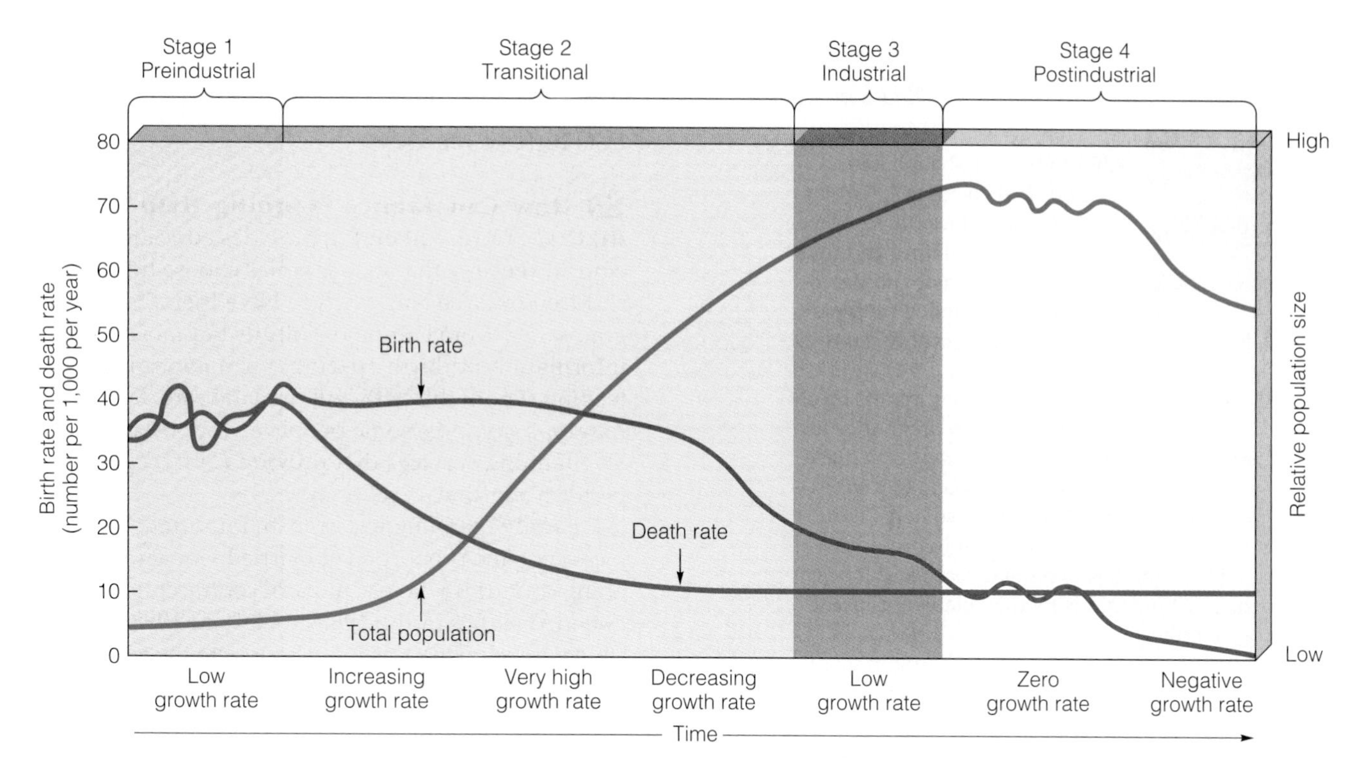

Figure 11-20 Generalized model of the demographic transition that may take place as countries become more industrialized. During the phases of this hypothesized transition, as countries industrialize their death rates decline and then their birth rates decline until their population stops growing. Population size can eventually decrease if birth rates become lower than death rates.

approaches the death rate. Reasons for this convergence of rates include better access to birth control, decline in the infant mortality rate, increased job opportunities for women, and the high costs of raising children who don't enter the workforce until after high school or college. Population growth continues, but at a slower and perhaps fluctuating rate, depending on economic conditions. Most developed countries are now in this third stage, and a few developing countries are entering this stage.

In the *postindustrial stage*, the birth rate declines even further, equaling the death rate and thus reaching zero population growth. Then the birth rate falls below the death rate and total population size slowly decreases. Emphasis shifts from unsustainable to sustainable forms of economic development. Some 37 countries (most of them in western Europe) containing about 12% of the world's population have entered this stage. To most population experts, the challenge is to help the remaining 88% of humanity reach this stage.

In most developing countries today, death rates have fallen much more than birth rates. In other words, these developing countries—mostly in Southeast Asia, Africa, and Latin America—are still in the transitional stage, halfway up the economic ladder, with high population growth rates. Some economists believe that developing countries will make the demographic transition over the next few decades without increased family-planning efforts.

However, despite encouraging declines in fertility, some population analysts fear that the still-rapid population growth in many developing countries will outstrip economic growth and overwhelm local life-support systems. This could cause many of these countries to be caught in a *demographic trap*, something that is currently happening in a number of developing countries, especially in Africa.

A poor country with a population growth rate of 2.5% per year needs an economic growth rate of 5% per year in order to achieve the 2.5% per capita economic growth often regarded as the minimum required to make the demographic transition. Many poor countries have had little total or per capita economic growth since 1980 (Figure 1-8). Countries with high rates of population growth that are caught in a demographic trap are likely to cause widespread environmental degradation by exceeding the carrying capacity of local ecosystems.

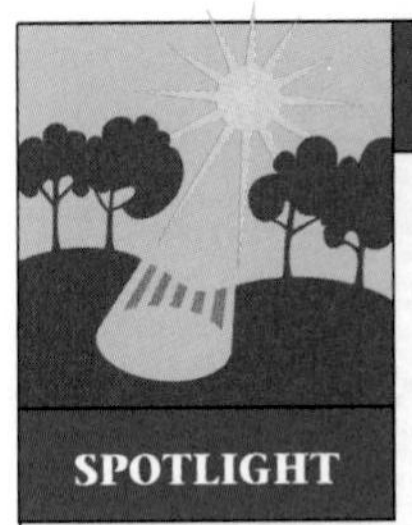

Breast-feeding

SPOTLIGHT

Although it is not a reliable method of birth control, regular breast-feeding does prevent some women from ovulating and thus delays conception of another child. Pregnancy is rare among nursing mothers who have not yet resumed their monthly menstrual cycles.

Generally, women who feed their infants breast milk are about 98% protected from pregnancy for 6 months after delivery. However, *any supplement to breast milk can interfere with this natural contraceptive protection.* Breast-feeding also helps reduce infant mortality rates; mother's milk provides a baby with antibodies to resist disease and it is usually the most nutritious food available for infants in poor families.

However, breast-feeding is declining in many developing countries. This has happened mostly because corporations have promoted the use of infant formulas instead of mother's milk by telling poor mothers that formula is more modern and better for their babies. Buying infant formula when free breast milk is available is an unnecessary expense for a poor family struggling to survive. Moreover, poor people who lack fuel often prepare formula with unboiled, contaminated water and use unsterilized bottles. In such cases, using formula can lead to more infant illnesses and deaths.

Critical Thinking

Should corporations continue their efforts to encourage poor (and other) mothers to use formula instead of breast milk? Explain. If you believe that this practice is undesirable, how would you restrict it?

Analysts also point out that some of the conditions that allowed developed countries to develop are not available to many of today's developing countries. Even with large and growing populations, many developing countries do not have enough skilled workers to produce the high-tech products needed to compete in the global economy. Most low- and middle-income developing countries also lack the capital and resources needed for rapid economic development. Furthermore, the amount of economic assistance for developing countries, which are struggling under tremendous debts, has been decreasing since 1980. Indeed, since the mid-1980s, the developing countries have paid developed countries $40–50 billion a year more (mostly in debt interest) than they have received from these countries.

How Can Family Planning Help Reduce Births? **Family planning** provides educational and clinical services that help couples choose how many children to have and when to have them. Such programs vary from culture to culture, but most provide information on birth spacing, birth control, breast-feeding (Spotlight, left), and prenatal care. Mostly on religious grounds, some people are opposed to family-planning services that provide contraceptives or perform abortions.

Family planning has been an important factor in increasing the proportion of married women in developing countries who use modern contraception. It has gone from 10% in the 1960s to 49% in 1998 (36%, if China is excluded); this is fairly close to the 61% usage by married women in developed countries. Family planning is also responsible for at least 40% of the drop in total fertility rates in developing countries, from 6 in 1960 to 3.3 in 1998 (3.9 if China is excluded).

Family-planning programs save a society money by reducing the need for children's social services. Proponents also argue that providing access to family planning throughout the world would bring about a sharp decline in the estimated 50 million legal and illegal abortions per year.

Family planning, coupled with improved prenatal care and health services, can also help reduce the risk of childbearing. Each year at least 600,000 women die from pregnancy-related causes, with 99% of such deaths occurring in developing countries.

The effectiveness of family planning varies with program design and funding. It has been a significant factor in reducing birth and fertility rates in populous countries such as China (Section 11-4), Indonesia, Brazil, and Bangladesh. Bangladesh still has a long way to go in slowing its population growth from 123 million in 1988 to a projected 166 million by 2025. However, its total fertility rate dropped from 6.3 children per woman in the mid-1970s to 3.3 in 1998.

Two major factors underlying this significant fertility decline are **(1)** a sixfold increase in contraceptive use by married women during the last two decades as a result of a government-supported family-planning program and **(2)** couples seeing fewer benefits of large families because of lack of available land, unstable farming conditions, and extensive flooding (Case Study, p. 334) and government-supported mass media messages promoting the economic and health advantages of limiting or spacing births. Important to this success has been the work of 35,000 women who provide contraceptive methods and counseling to women at their homes as part of the government program.

Family planning has also played a major role in reducing population growth in Japan, Thailand (p. 252), Mexico, South Korea, Taiwan, and several other countries with moderate to small populations. These successful programs, based on committed leadership, local implementation, and wide availability of contraceptives, demonstrate that population rates can decrease significantly within 15–30 years.

Despite such successes, family planning has had moderate to poor results in more populous developing countries such as India, Egypt, Pakistan, and Nigeria. Results have also been poor in 79 less populous developing countries, especially in Africa and Latin America, with high or very high population growth rates.

According to UN studies, an estimated 300 million women in developing countries want to limit the number and determine the spacing of their children, but they lack access to services. Extending family-planning services to these women and to those who will soon be entering their reproductive years could prevent an estimated 5.8 million births a year and over 5 million abortions a year. Other analysts call for expanding existing family-planning programs to include teenagers and sexually active unmarried women, who are often excluded. Currently, U.S. teenagers account for about one-fifth of the abortions, or about 242,000 per year.

Some analysts urge that programs be broadened to educate males about the importance of having fewer children and taking more responsibility for raising them. Proponents also argue that much more research is needed on developing new, more effective, and acceptable methods of birth control for men. What do you think about this proposal?

Family planning could be provided in developing countries to all couples who want it for about $17 billion a year, the equivalent of less than a week's worth of worldwide military expenditures. If developed countries provided one-third of this $17 billion, each person in the developed countries would spend only $4.80 a year. This would help reduce world population by about 2.7 billion people, shrinking average family size from 3.0 to 2.1 children. However, in 1996 the U.S. Congress cut international family-planning assistance funds by 87% from 1995 levels and by 60% for 1997. According to the Alan Guttmacher Institute, the 1996 funding cuts increased abortions and led to 134,000 additional infant deaths and 8,000 deaths among women during pregnancy and childbirth.

Many analysts call for stepped up family-planning efforts in developed countries, especially the United States because of its higher growth rate than other industrialized, high-consumption countries. Elimination of unplanned births in the United States would result in at least 1 million fewer births and about 900,000 fewer abortions per year, almost half of them among teenagers. Some analysts urge pro-choice and pro-life groups to join forces in greatly reducing unplanned births, especially among teenagers. What do you think?

How Can Economic Rewards and Penalties Be Used to Help Reduce Births? Some population experts argue that family planning, even coupled with economic development, cannot lower birth and fertility rates quickly enough to avoid a sharp rise in death rates in many developing countries. They point to studies showing that most couples in developing countries want three or four children, which is well above the replacement-level fertility required to bring about eventual population stabilization.

These analysts believe that we must go beyond family planning and offer economic rewards and penalties to help slow population growth. About 20 countries offer small payments to people who agree to use contraceptives or to be sterilized; however, such payments are most likely to attract people who already have all the children they want. Some of these countries also pay doctors and family-planning workers for each sterilization they perform and each IUD they insert.

Some countries, including China, penalize couples who have more than one or two children by raising their taxes, charging other fees, or eliminating income tax deductions for a couple's third child (as in Singapore, Hong Kong, and Ghana). Families who have more children than the prescribed limit may also lose health-care benefits, food allotments, and job options. However, programs that withhold food or increase the cost of raising children punish innocent children for the actions of their parents.

Economic rewards and penalties designed to lower birth rates work best if they encourage (rather than mandate) people to have fewer children, reinforce existing customs and trends toward smaller families, do not penalize people who produced large families before the programs were established, and increase a poor family's economic status. Once a country's population growth is out of control, however, it may be forced to use coercive methods to prevent mass starvation and hardship, as has been the case for China (Section 11-4).

How Can Empowering Women Help Reduce Births? Studies show that women tend to have fewer and healthier children and live longer when they have access to education and to paying jobs outside the home, and when they live in societies in which their individual rights are not suppressed.

Women, roughly half of the world's population, do almost all of the world's domestic work and child care and provide more health care with little or no pay than all the world's organized health services combined. They also do more than half the work associated with growing food, gathering fuelwood, and hauling water. As one Brazilian woman put it, "For poor women the only holiday is when you are asleep." Women's unpaid work, at an estimated value of $11 trillion annually (almost half of the annual total global output), is not included in a country's GDP.

Women work two-thirds of all hours worked, but receive only one-tenth of the world's income and own a mere 0.01% of the world's property. In most developing countries, women don't have the legal right to own land or borrow money to increase agricultural productivity. Women also make up 70% of the world's poor and almost two-thirds of the more than 960 million adults who can neither read nor write.

Women are almost universally excluded from economic and political decision making. They hold only 14% of the world's administrative and managerial positions and occupy only 10% of parliamentary seats.

Most analysts believe that women everywhere should have full legal rights and the opportunity to become educated and earn income outside the home. This would not only slow population growth but also promote human rights and freedom. However, empowering women by seeking gender equality will require some major social changes that will be difficult to achieve in male-dominated societies.

11-4 CASE STUDIES: SLOWING POPULATION GROWTH IN INDIA AND CHINA

What Success Has India Had in Controlling Its Population Growth? The world's first national family-planning program began in India (Figure 11-16) in 1952, when its population was nearly 400 million. In 1998, after 46 years of population control efforts, India was the world's second most populous country, with a population of 989 million—3.5 times as many people as the United States.

In 1952, India added 5 million people to its population; in 1998 it added 18 million—49,300 more mouths to feed each day. In 1998, more than 30% of the world's births took place in India. With 36% of its people under age 15, India's population is projected to reach 1.4 billion by 2025 and possibly 1.9 billion before leveling off early in the 22nd century. In 1998, women in India averaged 3.5 children, down from 5.3 in 1970. Despite this decline in fertility, India's population is growing exponentially at about 1.9% a year and India is expected to overtake China as the world's most populous country by the middle of the next century.

India's people are among the poorest in the world, with an average per capita income of about $340 a year. About 52% of the population live on incomes of less than $1 a day and 30% of the population have incomes of about $100 a year, or 27¢ a day. Nearly half of India's labor force is unemployed or can find only occasional work.

Although India is currently self-sufficient in food-grain production, about 40% of its population today suffers from malnutrition, mostly because of poverty. Nearly two-thirds of Indian children younger than 5 are malnourished. Life expectancy is only 59 years, and the infant mortality rate is 72 deaths per 1,000 live births. Children who reach school age stay in school only about 3.5 years if they are boys and 1.5 years if they are girls. As a result, about 48% of India's people older than 15 are illiterate. About 70% of Indians have no access to toilets and 30% have no supply of safe water.

Some analysts fear that India's already serious malnutrition and health problems will worsen as its population continues to grow rapidly. With 17% of the world's people, India has just 2.3% of the world's land resources and 1.7% of the world's forests. About half of India's cropland is degraded as a result of soil erosion, waterlogging, salinization, overgrazing, and deforestation. About 70% of India's water is seriously polluted, and sanitation services are often inadequate.

Without its long-standing family-planning program, India's population and environmental problems would be growing even faster. Still, to its supporters the results of the program have been disappointing because of poor planning, bureaucratic inefficiency, the low status of women (despite constitutional guarantees of equality), extreme poverty, and a lack of administrative and financial support.

Even though the government has provided information about the advantages of small families for years, Indian women still have an average of 3.5 children because most couples believe they need many children to do work and care for them in old age. Many social and cultural norms favor large families, including the strong preference for male children; some couples keep having children until they produce

1998. "Record Number of Young People Reaching Childbearing Years." *Economic Review*, vol. 29, no. 7, 24(3).

one or more boys. These factors in part explain why even though 90% of Indian couples know of at least one modern birth control method, only 36% actually use one.

What Success Has China Had in Controlling Its Population Growth? China has the world's largest population (Figure 11-4). Although China is roughly the same size as the United States, it has about 4.6 times as many people.

Since 1970, China (Figure 11-16) has made impressive efforts to feed its people and bring its population growth under control. Between 1972 and 1998 China achieved a remarkable drop in its crude birth rate, from 32 to 17 per 1,000 people, and its total fertility rate dropped from 5.7 to 1.8 children per woman. Since 1985 its infant mortality rate has been almost one-half the rate in India and its illiteracy rate of 18.5% is about almost one-third that of India. Life expectancy in China is 71 years, 12 years higher than in India. China's per capita income of $750 is almost twice that in India. Despite these achievements, with the world's largest population (1.24 billion) and a growth rate of 1.0%, China had about 12 million more mouths to feed in 1998. Its population is projected to reach 1.6 billion by 2025.

To achieve its sharp drop in fertility, China has established the most extensive, intrusive, and strict population control program in the world. Couples are strongly urged to postpone the age at which they marry and to have no more than one child. Married couples have ready access to free sterilization, contraceptives, and abortion. Paramedics and mobile units ensure access even in rural areas.

Couples who pledge to have no more than one child are given extra food, larger pensions, better housing, free medical care, and salary bonuses; their child will be given free school tuition and preferential treatment in employment when he or she enters the job market. Couples who break their pledge lose all the benefits. The result is that 81% of married women in China are using modern contraception, compared to 60% in developed countries and only 36% in other developing countries.

Government officials realized in the 1960s that the only alternative to strict population control was mass starvation. China is a dictatorship, and thus, unlike India, it has been able to impose a consistent population policy throughout society. Moreover, Chinese society is fairly homogeneous and has a widespread common written language, which aids in educating people about the need for family planning and in implementing policies for slowing population growth.

China's large and still growing population has an enormous environmental impact that could reduce its ability to produce enough food and threaten the health of many of its people. China has 21% of the world's population, but only 7% of its fresh water and cropland, 3% of its forests, and 2% of its oil. According to the Chinese Academy of Sciences, China's timber resources could be depleted by 2016.

It is encouraging that between 1986 and 1996 the Chinese government almost doubled its expenditures on environmental protection. However, most of the nation's rivers, especially in urban areas, are seriously polluted, and air pollution in many of its cities is causing widespread health problems.

Most countries prefer to avoid the coercive elements of China's program. Coercion is not only incompatible with democratic values and notions of basic human rights, but ineffective in the long run because sooner or later people resist. However, other parts of this program could be used in many developing countries. Especially useful is the practice of localizing the program rather than asking people to go to distant centers. Perhaps the best lesson for other countries is to act to curb population growth before the choice is between mass starvation and coercive measures that severely restrict human freedom.

11-5 CUTTING GLOBAL POPULATION GROWTH AND SUSTAINABILITY

How Could Population Growth Be Reduced to More Sustainable Levels? Many of the world's leading scientists and other analysts have concluded that we are exceeding the carrying capacity for humans in parts of the world and eventually for the entire world. If their analysis is correct, then reducing the current rate of population growth in both developed and developing countries and eventually stabilizing the world's population are essential goals for an ecologically and economically sustainable future.

The good news is that the experience of Japan, Thailand (p. 252), South Korea, Taiwan, and China indicates that a country can achieve replacement-level fertility within 15–30 years. Such experience also suggests that the best way to slow population growth is a combination of investing in family planning, reducing poverty, and elevating the status of women.

Because countries differ in population growth rates, use and availability of resources, and social structure, the combination of these factors must be tailored to each country's situation. Furthermore, most analysts believe

Hint: Enter the search term *demographic transition* using the Subject Guide.

that government policy makers should devise policies that minimize the environmental impact of population growth in their efforts to achieve sustainability.

How Are Governments Planning to Reduce Population Growth? In 1994 the United Nations held its third once-in-a-decade Conference on Population and Development in Cairo, Egypt. One of the conference's goals was to encourage action to stabilize the world's population at 7.8 billion by 2050, instead of the projected 11–12.5 billion. The major goals of the resulting 20-year population plan, endorsed by 180 governments, are by 2015 to

- Provide universal access to family-planning services and reproductive health care
- Improve the health care of infants, children, and pregnant women
- Encourage development and implementation of national population policies as part of social and economic development policies
- Bring about more equitable relationships between men and women, with emphasis on improving the status of women and expanding education and job opportunities for young women
- Increase access to education, especially for girls
- Increase the involvement of men in child-rearing responsibilities and family planning
- Take steps to eradicate poverty
- Reduce and eliminate unsustainable patterns of production and consumption

Many analysts applaud these goals, but some call them wishful thinking. Even if they wanted to, most governments could not afford to implement many of these goals. However, proponents of reduced population growth argue that the possible alternatives of greatly increased pollution, environmental degradation, and a massive dieback of people if we exceed the earth's carrying capacity for humans should spur us to implement the goals of the 1994 UN Population Conference as rapidly as possible. What is your opinion about this issue?

Short of thermonuclear war itself, rampant population growth is the gravest issue the world faces over the decades immediately ahead. If we do not act, the problem will be solved by famine, riots, insurrection, and war.

Robert S. McNamara

CRITICAL THINKING

1. Why are falling birth rates not necessarily a reliable indicator of future population growth trends?

2. Why is it rational for a poor couple in India to have five or six children? What changes might induce such a couple to consider their behavior irrational?

3. Are there physical limits to growth on earth? Are there social limits to growth? Explain your answers.

4. Do you believe that the population of your own country is too high? Explain. What about the population of the area in which you live?

5. Evaluate the claims made by those opposing the reduction of births and those promoting a reduction in births, as discussed on pp. 265–268. Which position do you support, and why?

6. Explain why you agree or disagree with each of the following proposals:

- **a.** The number of legal immigrants and refugees allowed into the United States each year should be sharply reduced.
- **b.** Illegal immigration into the United States should be sharply decreased. If you agree, how would you go about achieving this?
- **c.** Families in the United States should be given financial incentives to have more children to prevent population decline.
- **d.** The United States should adopt an official policy to stabilize its population and reduce unnecessary resource waste and consumption as rapidly as possible.
- **e.** Everyone should have the right to have as many children as they want.

7. Some people have proposed that the earth could solve its population problem by shipping people off to space colonies, each containing about 10,000 people. Assuming that such large-scale, self-sustaining space stations could be built (which can't be done with existing technology), how many people would have to be shipped off each day to provide living spaces for the approximately 84 million people being added to the population each year? Current space shuttles can handle about 6 to 8 passengers. Assuming that this capacity can be increased to 100 passengers per shuttle, how many shuttles would have to be launched per day to take care of the 84 million people being added each year? According to your calculations, determine whether this proposal is a logical solution to the earth's population problem.

8. Try to come up with analogies to complete the following statements: **(a)** Achieving zero population growth is like _______. **(b)** Living in a country whose population is growing rapidly is like _______. **(c)** Encouraging emigration to help reduce population growth is like _______. **(d)** Trying to slow population growth is like _______. **(e)** Using economic penalties to slow population growth

Q: What percentage of the world's people die each year?

is like _______. **(f)** Reducing births by empowering women is like _______.

9. Why has China been more successful than India in reducing its rate of population growth? Do you agree with China's current population control policies? Explain. What alternatives, if any, would you suggest?

10. Congratulations—you have just been put in charge of the world. List the five most important features of your population policy.

PROJECTS

1. Survey members of your class to determine how many children they plan to have. Tally the results and compare them for males and females.

2. Assume that your entire class (or manageable groups of your class) is charged with coming up with a plan for halving the world's rate of population growth within the next 20 years. Develop a detailed plan that would achieve this goal, including any differences between policies in developing countries and developed countries. Justify each part of your plan. Predict what problems you might face in implementing the plan, and devise strategies for dealing with these problems.

3. Prepare an age structure diagram for your community. Use the diagram to project future population growth and economic and social problems.

4. Make a concept map of this chapter's major ideas, using the section heads and subheads and the key terms (in boldface). Look at the inside back cover and in Appendix 4 for information about concept maps.

A: About 0.9%

12 FOOD RESOURCES

Perennial Crops on the Kansas Prairie

When you think about farms in Kansas, you probably picture seemingly endless fields of wheat or corn plowed up and planted each year. By 2040 the picture might change, thanks to pioneering work at the nonprofit Land Institute near Salina, Kansas (Figure 12-1).

The Institute, headed by plant geneticist Wes Jackson and founded in 1978, is experimenting with an ecological approach to agriculture on the Midwestern prairie. The goal is to raise food by mimicking many of the natural conditions of the prairie without losing fertile grassland soil (Figure 5-16)—an approach called *natural system agriculture*. Crops are to be grown by planting a mix of *perennial* grasses, legumes (a source of nitrogen fertilizer), sunflowers, grain crops, and plants that provide natural insecticides in the same field (polyculture). Because these plants are perennials, the soil doesn't have to be plowed up and prepared each year to replant them.

In the natural grasslands of the Midwest, soil is rarely left exposed to wind and rain; instead, it is protected and sustained by a community of perennial plants with deep roots that anchor the soil in place (Figure 5-16), retain water (thus reducing the need for irrigation), and enrich the soil when they die and decay. The Institute is attempting to duplicate this process by planting polycultures of edible perennial plants and thus creating a self-sustaining, grain-producing prairie that mimics the natural one.

Institute researchers believe that perennial polyculture can be blended with modern agriculture to reduce the massive and growing harmful environmental effects of industrialized agriculture. Perennial polyculture is especially suitable for marginal land, leaving prime, flat land available for raising annual (monoculture) crops.

By eliminating yearly plowing and planting, perennial polyculture requires much less labor than conventional monoculture or diversified organic farms that grow annual crops. This reduces depletion of nonrenewable fossil fuels and the resulting air

Figure 12-1 The Land Institute in Salina, Kansas, is a farm, a prairie laboratory, and a school dedicated to changing the way we grow food. It advocates growing a diverse mixture of edible perennial plants to supplement traditional annual monoculture crops. (Terry Evans)

and water pollution and potential global warming. It also lessens soil erosion because the unplowed soil is not exposed to wind and rain. Pollution caused by chemical fertilizers and pesticides is also decreased and the reduced need for irrigation helps protect rivers and aquifers from being depleted faster than they can be renewed.

If the Institute and similar groups doing such earth-sustaining research succeed, within a few decades many people may be eating food made from a mix of perennials such as **(1)** *eastern gamma grass* (a warm-season grass that is a relative of corn with three times as much protein as corn and twice as much as wheat), **(2)** *mammoth wildrye* (a cool-season grass that is distantly related to rye, wheat, and barley), **(3)** *Illinois bundleflower* (a wild nitrogen-producing legume that can enrich the soil and whose seeds may serve as livestock feed), and **(4)** *Maximillian sunflower* (which produces seeds with as much protein as soybeans).

In 1992, Jackson's work at the Institute earned him a "genius grant" from the MacArthur Foundation. Jackson estimates that his work will take 25 years and cost more than $100 million, or $4 million a year—a drop in the bucket compared to the estimated loss of topsoil worth $44 billion a year for the United States alone.

Jackson's perennial crops will come none too soon. Global food production has increased substantially over the past two decades, but producing food and other agricultural products by conventional means uses more soil, water, plant, animal, and energy resources—and causes more pollution and environmental damage—than any other human activity. To feed the 8 billion people projected by 2025, we must produce and distribute as much food during the next 26 years as was produced since agriculture began about 10,000 years ago.

Producing food at a faster rate than the population grows will not solve hunger problems unless the poor have enough land to grow their own food or enough income to buy the food they need. Currently, chronic hunger and catastrophic famine result mainly from a lack of *access* to food, not from a shortage of food.

The basic question is, "How can the growing human population produce and distribute enough food to meet *everyone's* basic nutritional needs without degrading the soil, water, air, and biodiversity that support all food production?"

There are two spiritual dangers in not owning a farm. One is the danger of supposing that breakfast comes from the grocery, and the other that heat comes from the furnace.

ALDO LEOPOLD

The discussion in this chapter answers several general questions:

- How is the world's food produced?
- What are the world's food problems?
- What can we do to help solve the world's food problems?
- How can we design and shift to more sustainable agricultural systems?

12-1 HOW IS FOOD PRODUCED?

What Plants and Animals Feed the World? The multitude of species of plants and animals (species diversity) and the varieties of plants and animals (genetic diversity) that provide us with food are an important part of the planet's biodiversity. Biologists estimate that although the earth has perhaps 30,000 plant species with parts that people can eat, only 15 plant and 8 animal species supply 90% of our food.

Just three grain crops—wheat, rice, and corn—provide almost half of the calories that people consume. These three, and most of our other food crops, are *annuals*, whose seeds must be replanted each year.

Two out of three of the world's people survive primarily on grains (mainly rice, wheat, and corn), mostly because they can't afford meat. As incomes rise, people consume even more grain, but indirectly in the form of meat (mostly beef, pork, and chicken), eggs, milk, cheese, and other products of grain-eating domesticated livestock.

What Are the Major Types of Food Production? There are two major types of agricultural systems: industrialized and traditional. **Industrialized agriculture**, or **high-input agriculture**, uses large amounts of fossil fuel energy, water, commercial fertilizers, and pesticides to produce huge quantities of single crops (monocultures) or livestock animals for sale. Practiced on about 25% of all cropland, mostly in developed countries (Figure 12-2), industrialized agriculture has spread since the mid-1960s to some

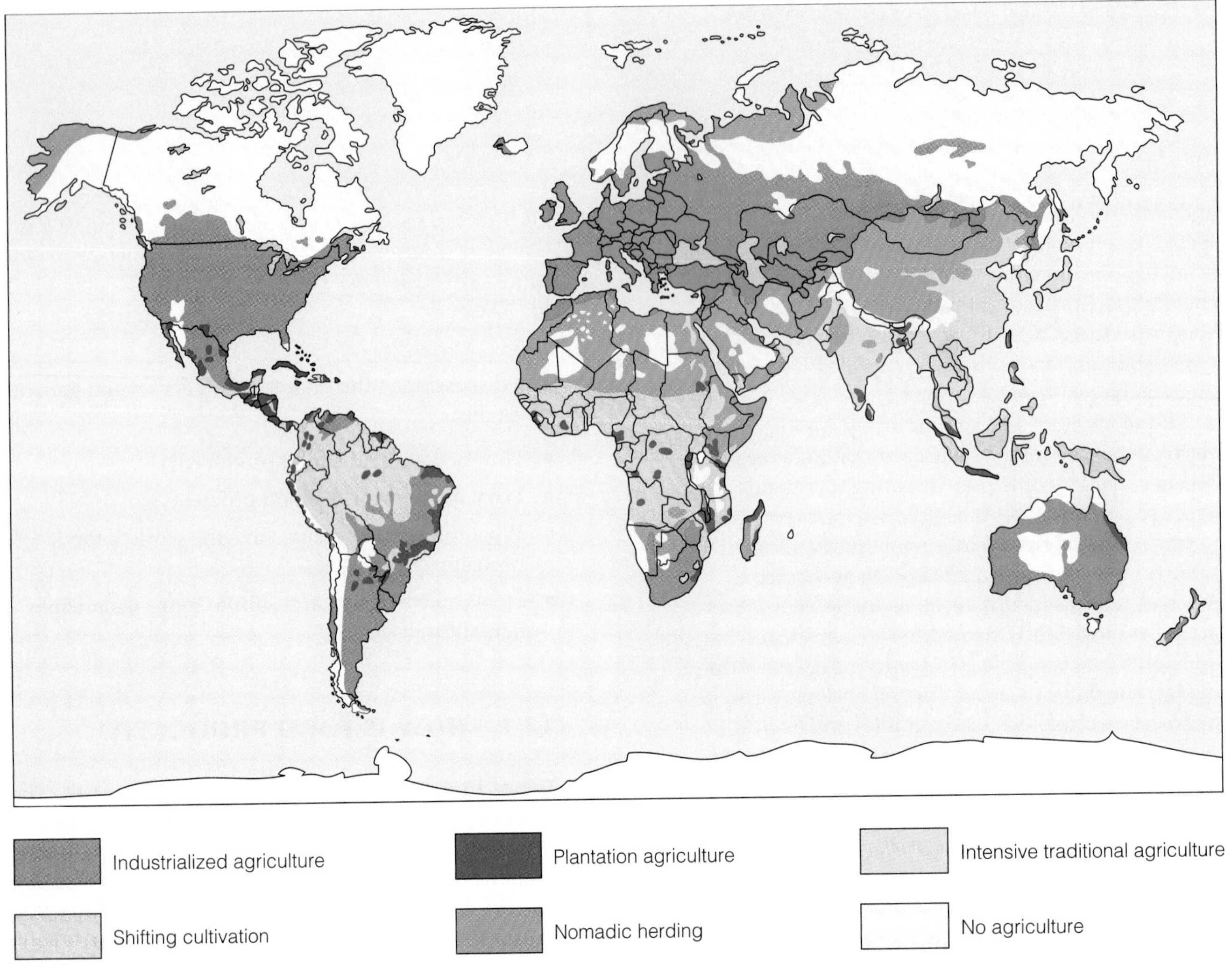

Figure 12-2 Generalized locations of the world's principal types of food production.

developing countries. **Plantation agriculture**, a form of industrialized agriculture practiced primarily in tropical developing countries, grows cash crops such as bananas, coffee, and cacao, mostly for sale in developed countries.

Traditional agriculture consists of two main types, which together are practiced by about 2.7 billion people in developing countries—almost half the people on the earth. **Traditional subsistence agriculture** typically produces only enough crops or livestock for a farm family's survival; in good years there may be a surplus to sell or to put aside for hard times. Subsistence farmers use primarily human labor and draft animals. Examples of this type of agriculture include numerous forms of shifting cultivation in tropical forests (Figure 1-23) and nomadic livestock herding.

In **traditional intensive agriculture**, farmers increase their inputs of human and draft labor, fertilizer, and water to get a higher yield per area of cultivated land to produce enough food to feed their families and to sell for income. Figure 12-3 shows the inputs of land, human and animal labor, fossil fuel energy, and capital needed to produce one unit of food energy in various types of food production.

How Have Green Revolutions Increased Food Production? Farmers can produce more food either by farming more land or by getting higher yields per unit of area from existing cropland. From the beginning of agriculture until the middle of this century, increases in food production came from expanding the cultivated area. Since 1950 most of the increase in global food production has resulted from increased yields per unit of area of cropland in an agricultural system called the **green revolution**.

This process involves three steps: **(1)** developing and planting monocultures (Figure 7-21) of selectively bred or genetically engineered high-yield varieties of key crops such as rice, wheat, and corn with emphasis on shifting more of plant growth to seeds (the part used for food); **(2)** lavishing fertilizer, pesticides, and water on crops to produce high yields; and **(3)** often

Q: What two countries have the world's largest populations?

increasing the intensity and frequency of cropping. This approach dramatically increased crop yields in most developed countries between 1950 and 1970 in what is considered the *first green revolution* (Figure 12-4).

A *second green revolution* has been taking place since 1967 (Figure 12-4), when fast-growing dwarf varieties of rice and wheat, specially bred for tropical and subtropical climates, were introduced into several developing countries. With sufficient fertile soil and enough fertilizer, water, and pesticides, yields of these new plants (Figure 12-5) can be two to five times those of traditional wheat and rice varieties. The fast growth also allows farmers to grow two or even three crops a year (multiple cropping) on the same land.

New corn varieties can be planted closer together to increase the number of ears harvested per area of cropland. In addition, new herbicides to control weeds have eliminated the traditional need to plant corn rows far enough apart to allow mechanical weed controllers to move through fields early in the growing season.

Between 1970 and 1992, India doubled its total food production (primarily by use of high-yield varieties of grain) and increased per capita food production by 18%—an impressive achievement considering its large population increase during this period. Without the green revolution, India would have faced massive famines in the 1970s and 1980s.

Producing more food on less land is also an important way to protect biodiversity by saving large areas of forests, grasslands, wetlands, and mountain terrain from being used to grow food. Since 1950 high-yield agriculture has saved an estimated 9–31 million square kilometers (3.5–12 million square miles) of such land from destruction or degradation by farming. According to Dennis Avery, without the two major green revolutions the world would have lost wild land equal to the combined land area of the United States, Europe, and Brazil.

These yield increases depend not only on having fertile soil and ample water, but also on extensive use of fossil fuels to run machinery, produce and apply inorganic fertilizers and pesticides, and pump water for irrigation. Between 1950 and 1990, agricultural use of fossil fuels increased 4-fold, irrigated area expanded 2.5-fold, use of commercial fertilizer rose 10-fold, and use of pesticides increased 30-fold. All told, green-revolution agriculture now uses about 8% of the world's oil output.

These high inputs of energy, water, fertilizer, and pesticides on high-yield crop varieties have yielded dramatic results by allowing crop plants to reach more of their full genetic potential. At some point, however, additional inputs become useless because no more output can be squeezed from the soil and the crop varieties—the principle of diminishing returns in action.

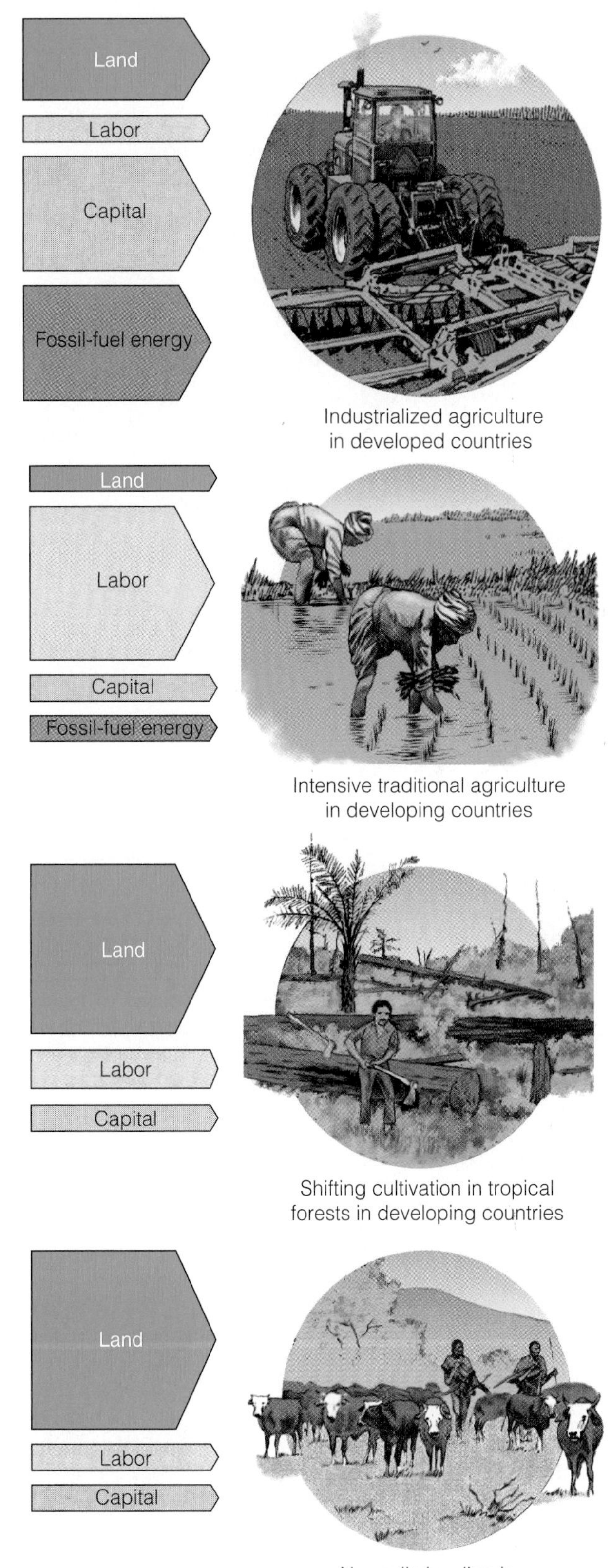

Figure 12-3 Relative inputs of land, labor, capital, and fossil-fuel energy in four agricultural systems. An average of 60% of the people in developing countries are directly involved in producing food, compared with only 8% in developed countries (2% in the United States).

A: China (1.24 billion) and India (989 million), together making up 38% of the world's population

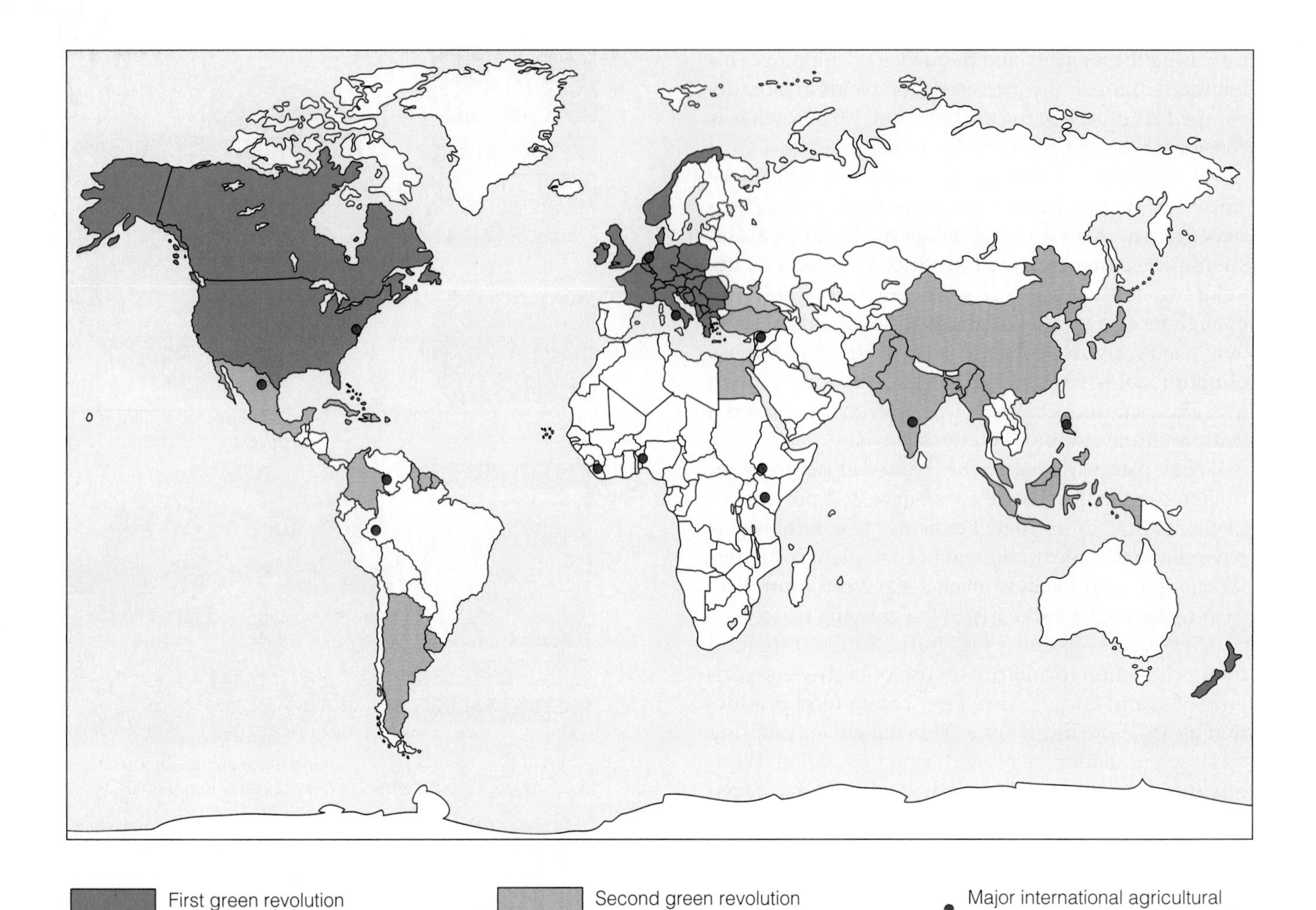

Figure 12-4 Countries whose crop yields per unit of land area increased during the two green revolutions. The first took place in developed countries between 1950 and 1970; the second has occurred since 1967 in developing countries with enough rainfall or irrigation capacity. Several agricultural research centers and gene or seed banks play a key role in developing high-yield crop varieties.

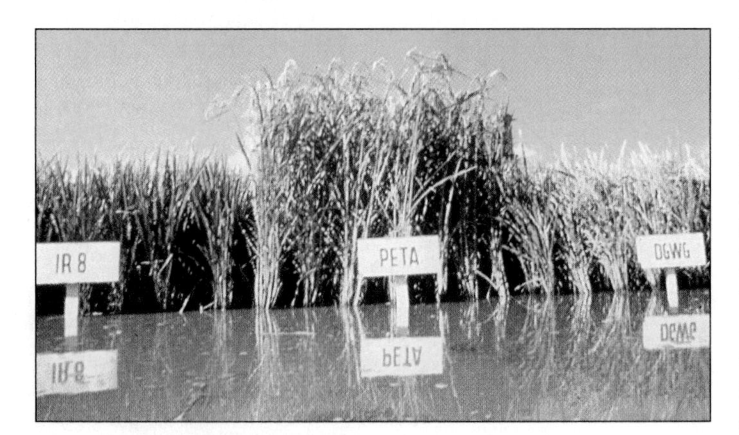

Figure 12-5 A high-yield, semidwarf variety of rice called IR-8 (left), a part of the second green revolution, was produced by crossbreeding two parent strains of rice: PETA from Indonesia (center) and DGWG from China (right). The shorter and stiffer stalks of the new variety allow the plants to support larger heads of grain without toppling over, increase the amount of plant growth going into food-producing seed, and increase the benefit of applying more fertilizer. (International Rice Research Institute, Manila)

Case Study: Food Production in the United States Since 1940 U.S. farmers have more than doubled crop production without cultivating more land, a result of industrialized agriculture using green-revolution techniques in a favorable climate on some of the world's most fertile and productive soils. This has also kept large areas of forests, grasslands, wetlands, and easily erodible land from being converted to farmland.

Farming has become *agribusiness* as big companies and larger family-owned farms have taken control of most U.S. food production. Only about 650,000 Americans are full-time farmers. However, about 9% of the population is involved in the U.S. agricultural system, from growing and processing food to distributing it and selling it at the supermarket.

In terms of total annual sales, agriculture is the biggest industry in the United States, bigger than the automotive, steel, and housing industries combined. It generates about 18% of the country's gross national product and 19% of all jobs in the private sector, employing more people than any other industry.

The U.S. agricultural system is highly productive. Currently each U.S. farmer feeds and clothes about 140 people (105 at home and 35 abroad), up from 58 people in 1976. Today U.S. farms, with only 0.3% of the world's farm labor force, produce about 25% of the world's food and nearly half of the world's grain exports. U.S. residents spend an average of only 10–12% of their income on food (down from 21% in 1940), compared to 18% in Japan and 40–70% in most developing countries.

This industrialization of agriculture was made possible by the availability of cheap energy, most of it from oil. Agriculture consumes about 17% of all commercial energy in the United States each year (Figure 12-6). On average, a piece of food eaten in the United States has traveled 2,100 kilometers (1,300 miles). Processing food also requires large amounts of energy.

Most plant crops in the United States provide more food energy than the energy used to grow them. However, if we include livestock as well as crops, the U.S. food production system currently uses about three units of fossil fuel energy to produce one unit of food energy.

Energy efficiency is much lower if we look at the whole U.S. food system. Considering the energy used to grow, store, process, package, transport, refrigerate, and cook all plant and animal food, *an average of about 10 units of nonrenewable fossil fuel energy are needed to put 1 unit of food energy on the table.* By comparison, every unit of energy from human labor in subsistence farming provides at least 1 unit of food energy; with traditional intensive farming, each unit of energy provides up to 10 units of food energy.

How Are Livestock Produced and What Are the Environmental Consequences? The world's rangelands make up the second land-based food system. For thousands of years domesticated animals such as cattle, horses, oxen, sheep, chickens, and pigs have played important roles in the human economy by providing food, fertilizer, fuel, clothing, and transport. Whereas only about 10% of the world's land is suitable for producing crops, about 20% is used for grazing cattle and sheep.

Meat and meat products are good sources of high-quality protein. Traditionally, when both crops and livestock are grown on diversified farms (such as those found in the United States before the recent shift to industrial monocultures), the livestock returned nutrients to the soil as manure, provided draft power, and grazed on fallow fields.

During the past 50 years the global livestock population has exploded as increased affluence has led to rising production and consumption of meat, mainly in developed countries and more recently in middle-income developing countries. Between 1950 and 1996, world meat production increased fourfold and per capita meat production rose by 29%.

Currently about 1.2 billion people in developed countries live high on the food chain by having a diet based on high consumption of meat and meat-based products. As a result, the world's developed countries, with one-fifth of the world's population, use their affluence to consume more than half of the world's grain.

Roughly another 1 billion people living in the poorest developing countries (especially in most of Africa), where incomes are not rising, consume mostly grain and live low on the food chain. The remaining 3.7 billion people live in low- or middle-income countries (especially in Asia) where rising incomes allow them to move further up the food chain by consuming more meat and meat products.

An increasing amount of livestock production in developed countries is industrialized; large numbers of cattle are typically brought to crowded feedlots, where they are fattened up for about 4 months before

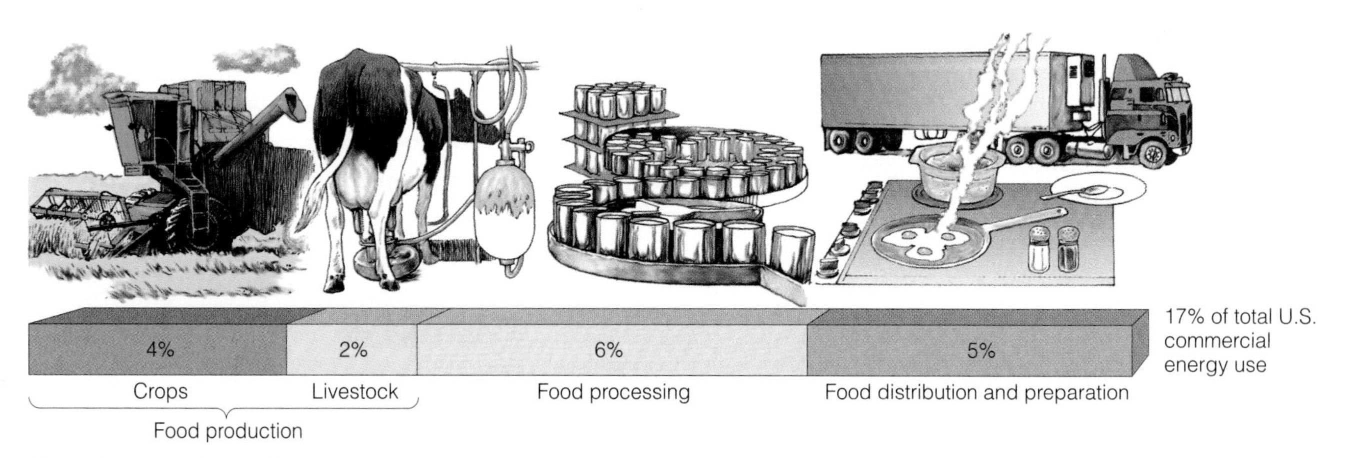

Figure 12-6 In the United States, industrialized agriculture uses about 17% of all commercial energy.

A: 1.43% (down from 2.2% in 1963)

slaughter. Most pigs and chickens in developed countries spend their entire lives in densely populated pens and cages and are fed mostly grain grown on cropland.

The meat-based diet of affluent people in developed countries and developing countries has enormous effects on resource use, environmental degradation, pollution, and disease. More than half of the world's cropland (19% in the United States) is used to produce livestock feed grain (mostly field corn, sorghum, and soybeans). In 1995 livestock and fish raised for food consumed about 37% of the world's grain production (70% in the United States). Livestock use more than half the water withdrawn each year in the United States; most of this water irrigates crops fed to livestock and washes manure from crowded livestock pens away. Livestock, poultry, and fish vary widely in the efficiency with which they convert grain into animal protein (Figure 12-7).

About 14% of U.S. topsoil loss is directly associated with livestock grazing. Globally, overgrazing of sparse vegetation and trampling of the soil by too many livestock is the major cause of desertification in arid and semiarid areas (Figure 14-19). The United Nations estimates that 70% of the world's dry rangeland is at least moderately desertified. If everyone in the world became true vegetarians (vegans) and all other factors stayed the same, the world's current oil reserves would last about 260 years instead of 40–80 years.

Cattle belch out 12–15% of all the methane released into the atmosphere (Figure 19-2c). Some of the nitrogen in commercial inorganic fertilizer used to grow livestock feed is converted to nitrous oxide, another greenhouse gas (Figure 19-2d). Nitrogen in manure escapes into the atmosphere as gaseous ammonia (NH_3), a pollutant that contributes to acid deposition (Figure 18-7). Livestock in the United States produce 21 times more waste (manure) than is produced by the country's human population. Only about half of this nutrient-rich livestock waste is recycled to the soil.

Some environmentalists have called for reducing livestock production (especially cattle) as a way to feed more humans. According to Lester Brown, if Americans cut their grain intake by just 16% (taking them back to their 1975 level) this would save enough grain to provide a subsistence diet for nearly 900 million people, or more than the total hungry today.

Other analysts say this won't work. They argue that reducing livestock production would decrease its environmental impact (an increasingly important concern), but it would not free up much land or grain to feed more people. Cattle and sheep that graze on rangeland use a resource (grass) that humans can't eat, and most of this land is not suitable for growing crops. Moreover, because of poverty, insufficient economic aid, and the nature of global economic and food distribution systems, very little if any additional grain grown on land used to raise livestock or livestock feed would reach the world's hungry people.

What Is Traditional Agriculture? Agrodiversity in Action Traditional farmers in developing countries today grow about 20% of the world's food on about 75% of its cultivated land. Many traditional farmers simultaneously grow several crops on the same plot, a practice known as **interplanting**. Such crop diversity reduces the chance of losing most or all of their year's food supply to pests and other misfortunes.

Common interplanting strategies found throughout the world, mostly in developing countries, include the following:

- **Polyvarietal cultivation**, in which a plot is planted with several varieties of the same crop.
- **Intercropping**, in which two or more different crops are grown at the same time on a plot (for example, a carbohydrate-rich grain that uses soil nitrogen alongside a protein-rich legume that puts it back).
- **Agroforestry**, or **alley cropping**, in which crops and trees are planted together (Figure 14-21c). For example, a grain or legume crop can be planted around fruit-bearing orchard trees or in rows between fast-growing trees or shrubs that can be used for fuelwood or for adding nitrogen to the soil.
- **Polyculture**, a more complex form of intercropping in which many different plants maturing at various times are planted together. If cultivated properly, these plots can provide food, medicines, fuel, and natural pesticides and fertilizers on a sustainable basis.

With polyculture, root systems at different depths in the soil capture nutrients and moisture efficiently

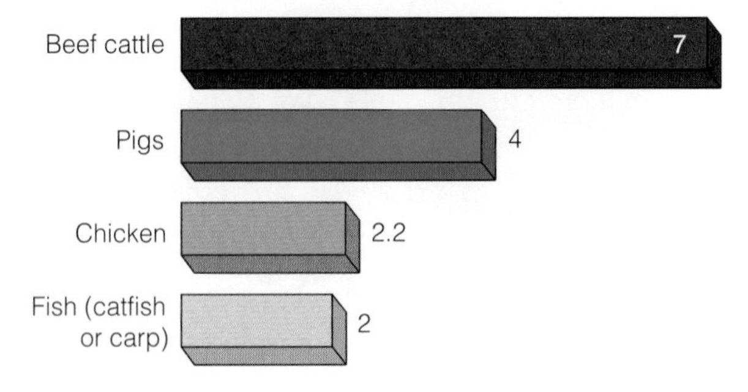

Figure 12-7 Efficiency of converting grain into animal protein. Data in kilograms of grain per kilogram of body weight added. (Data from U.S. Department of Agriculture)

Brown, Lester. 1998. "Food Scarcity: An Environmental Wakeup Call." *Futurist*, vol. 32, no. 1, 34(5).

and minimize the need for fertilizer or irrigation water. Year-round plant coverage also protects the soil from wind and water erosion. The presence of various habitats for insects' natural predators means that crops need not be sprayed with insecticides. In addition, weeds have trouble competing with the multitude of crop plants and thus can be removed fairly easily by hand, without herbicides. Crop diversity also provides insurance against bad weather: If one crop fails because of too much or too little rain, another crop may survive or even thrive.

Recent ecological research on crop yields of 14 artificial ecosystems found that on average polyculture (with four or five different crop species) produces higher yields per unit of area than high-input monoculture (Figure 7-21). This important finding has major implications for development of high-yield sustainable agriculture in developing countries by combining the techniques of traditional high-yield interplanting with modern inputs of fertilizer and irrigation.

12-2 WORLD FOOD PROBLEMS AND CHALLENGES

How Much Has Food Production Increased? Figure 12-8 shows the success story of global agriculture. Between 1950 and 1990, world grain production almost tripled (Figure 12-8, left) and per capita production rose by about 36% (Figure 12-8, right), helping reduce global hunger and malnutrition. During the same period, average food prices adjusted for inflation dropped by 25% and the amount of food traded in the world market quadrupled.

Despite these impressive achievements in food production, population growth is outstripping food production and distribution in areas that support about 2 billion people. Since 1978 grain production has lagged behind population growth in 88 developing countries. Food production in Africa has been rising steadily since 1961, but not fast enough to keep up with population growth; between 1974 and 1996 per capita production fell by 20%.

Since 1950 global grain production has been rising but its rate of growth has slowed (Figure 12-8, left) and the rate of growth in per capita grain production has declined (Figure 12-8, right). More than 100 countries regularly import grain from the United States, Canada, Australia, Argentina, western Europe, New Zealand, Thailand, and a few other countries.

Major factors leading to this slowdown in the growth of per capita grain production (from an average of 2.1% a year between 1950 and 1990 to about 1.1% a year between 1990 and 1997*) are **(1)** population growth (growing at about 1.5% a year), **(2)** increasing affluence, which increases the demand for food (especially meat produced by feeding livestock grain), **(3)** degradation and loss of cropland, mostly because of erosion and to a lesser extent because of industrialization and urbanization (especially in Asia), **(4)** little growth in irrigation since 1980 (shrinking the irrigated area per person by 5% between 1980 and 1995), and **(5)** a 10% decline in global fertilizer use between 1989 and 1997 (amounting to a 20% drop in per capita fertilizer use).

Unless death rates rise sharply, we seem destined to have a population of around 8 billion people by 2025. To provide this many people with even a meatless subsistence diet will require doubling food

*The average is 0.7% a year. However, it rises to 1.1% if sharply dropping yields in the former Soviet Union between 1990 and 1995 are excluded during the breakup of the country and the phasing in of new economic reforms.

Grain production (millions of tons)
Year
Total World Grain Production

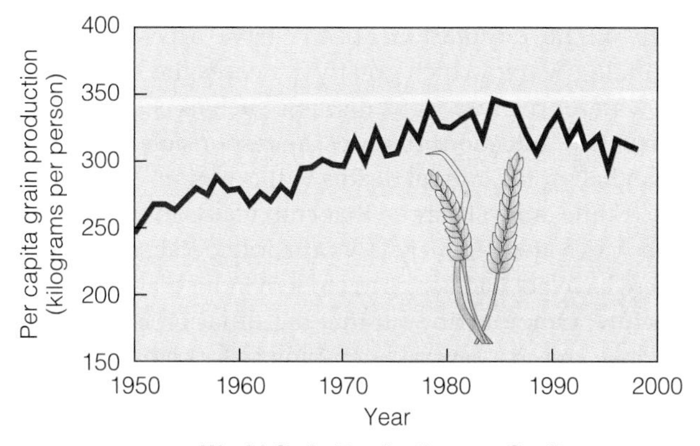

Figure 12-8 Total worldwide grain production of wheat, corn, and rice, and per capita grain production, 1950–96. (Data from U.S. Department of Agriculture and Worldwatch Institute)

production and distribution between 1990 and 2025. This will require that grain production grow by 2% per year, compared to the 1.1% growth per year between 1990 and 1997. Actually, food production will have to be increased even more because projected increasing affluence in many developing countries will raise meat consumption and require higher amounts of grain for animal feed.

How Many People Can the World Support? There is no definitive answer to this question because the carrying capacity of the earth for humans depends on **(1)** the quality of life or cultural carrying capacity per person (Guest Essay, p. 266), **(2)** whether future food production can be increased and is sustainable enough to meet projected food needs, and **(3)** the length of the food chain based on the percentages of the world's population that are grain eaters and meat eaters.

According to Lester Brown (Guest Essay, p. 34), if the slower growth rate of grain productivity of the 1990s continues, we may have to make the difficult transition from a half-century dominated by *food surpluses* to a period dominated by *food scarcity*. Any drop in annual grain production by major food-exporting countries (especially the United States) has severe economic consequences for food-importing nations (Connections, right).

Case Study: The Food Crisis in Sub-Saharan Africa Sub-Saharan Africa includes all of Africa's countries except South Africa and the six countries north of the Sahara desert. In this area, which contains 82% of Africa's people, per capita food production dropped 30% between 1960 and 1994 and is projected to fall another 30% during the next 25 years.

Some 30 million people in sub-Saharan Africa typically suffer from famine, and there are 1 million more mouths to feed every 3 weeks. Thousands die each day from malnutrition or hunger-related diseases. Almost half of the population doesn't have access to safe drinking water, which greatly increases the incidence of waterborne infectious diseases such as cholera and dysentery. Infectious diseases are responsible for more than half of the annual deaths in this region.

Some analysts argue that crop yields in sub-Saharan Africa are extremely low primarily because farmers there have not had access to modern industrialized agriculture. Others point out that the situation is not that simple, and that several interacting factors are to blame.

One such factor is the very rapid population growth rate of 2.9%, which is projected to double the region's population from 600 million in 1998 to 1.2 billion by 2025. The region has the world's youngest age structure (Figure 11-13). Furthermore, most of the high-quality land is already being used to grow crops and crop and livestock production is limited in a vast area by the tsetse fly (Figure 12-9). In addition, sub-Saharan Africa is afflicted by severe soil erosion (Figure 14-17), nutrient-poor soils, lack of water for irrigation, severe and increasing desertification (which affects more than one-fourth of the region, Figure 14-19), and deforestation in some areas, mostly because of growing fuelwood shortages (Figure 24-16).

To make matters worse, political turmoil and frequent wars within and between many countries reduce agricultural production and hinder the development of food storage and distribution systems. These conflicts also create large numbers of environmental refugees and divert to military uses substantial funds for basic food, health, and education needs.

If current trends are not reversed fairly quickly, most of the region is expected to undergo increasingly severe famine, disease, social chaos, and ecological deterioration—a prime example of overshoot and dieback when a human population and its resource consumption exceed an area's carrying capacity. Just staying even will require doubling food production and reducing average family size from 6.4 children to 3.1 by 2020—a formidable task even in areas with better climates and soils and less social turmoil and violence.

Despite sub-Saharan Africa's severe problems, there are some hopeful developments that, if nurtured, expanded, and transferred to other areas, could

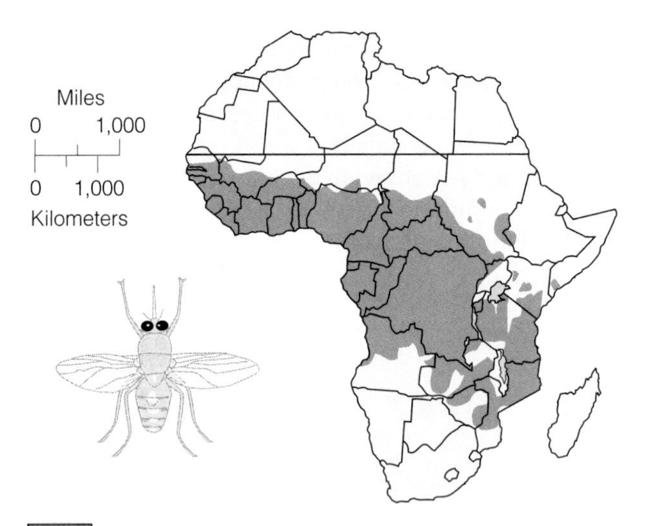

Figure 12-9 Portions of Africa in which the tsetse fly has blocked the development of agriculture. The fly transmits a protozoan parasite that causes incurable sleeping sickness in humans and a fatal disease in livestock. A $120-million eradication program has been proposed, but many scientists doubt it can succeed. Others point out that developing this land would cause extinction of many forms of wildlife, thereby contributing to the planet's growing biodiversity crisis.

Q: How many people are added to the world's population each year?

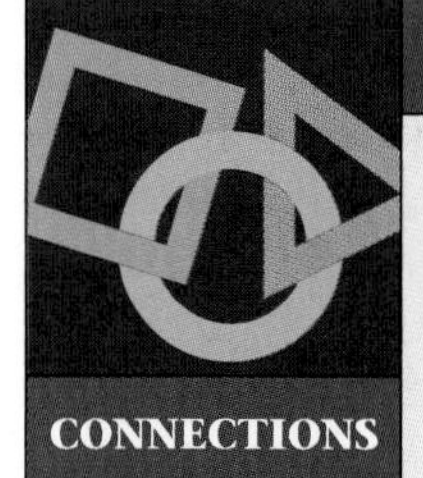

What Happens if Food Demand Exceeds the Supply?

CONNECTIONS

The demand for grain is rising. In most developed countries and rapidly industrializing middle-income countries this increasing demand is fueled by rising affluence. In the poorest developing countries (most in Africa), population growth is the major cause of the rising demand for grain.

So far the supply of food for export (primarily by the United States) has kept up with the demand. However, if bad weather or changes in climate lower food production or crop yields reach their biological limits, the demand for grain is likely to exceed the supply. Then real grain prices (after adjusting for inflation) will rise.

Just one poor harvest year can lead to a sharp rise in grain prices if global emergency grain stocks are not enough to meet at least 70 days of world consumption. Whenever such stocks fall below 60 days, grain prices become highly volatile. For example, when grain stocks dropped to 52 days of consumption in 1996, the world price of wheat and corn more than doubled.

If the demand for grain exceeds the supply, this will greatly increase the economic and political power of a small number of grain-exporting countries, especially the United States, which controls almost half of such exports—a larger share of the world's grain exports than the share of oil exports controlled by Saudi Arabia. Any drops in available food exports from the United States because of bad weather or long-term changes in climate can cause economic chaos in the more than 100 countries that import grain. This is a very risky situation for much of the world's population because the U.S. grain harvest, largely rain-fed, varies widely from year to year and is highly vulnerable to drought and heat waves.

If real global grain prices double, the number of poor and malnourished people in the world would rise sharply. Ways to prevent this are limited. There is little idle land that can be used to increase production. World grain stocks have dropped below the needed 60 days of consumption since 1996 and reached a low of only 57 days in 1998. Irrigation could be increased because of higher food prices, but the potential for increased irrigation is limited in many areas by the sustainable yield of aquifers. Levels of fertilizers could be increased, but yields are limited because existing varieties are reaching their physiological limits for absorbing more fertilizer.

Between the late 1940s and 1993, food aid provided a safety net for needy countries. However, between 1993 and 1996 the international budget for food aid was cut in half. According to U.S. Department of Agriculture estimates, global food aid needs in 2005 will be four times the amount available in 1996.

If grain production fails to keep up with demand, China, Japan, Taiwan, South Korea, and most other rapidly industrializing countries of Asia that depend on imports for a large share of their grain will have to divert an increasing amount of their income to pay for food imports. The biggest loser will be most of Africa, with low rates of economic growth and the fastest population growth rate of any continent (Figure 11-3).

If there are unfavorable climate changes in major U.S. food-growing areas as a result of projected global warming (Figure 19-5 and Sections 19-2 and 19-3), the entire global food production system could be thrown into severe economic and political chaos. This could also lead to social chaos as hordes of hungry people and environmental refugees illegally migrate to other countries in a desperate search for food and work.

Critical Thinking

The United States is by far the world's largest food exporter. What harmful economic and environmental effects might the United States experience if its ability to supply food for export exceeds the demand?

increase food production. In Kenya, soil erosion has been slowed and fuelwood increased by the Green Belt Movement, which since the 1970s has planted more than 10 million trees (Figure 24-23). In the Machakos district of southeastern Kenya, terracing (Figure 14-21a) is used to grow crops on almost 70% of the arable land, making the area one of Africa's leaders in soil and water conservation. Recently, food production there has outpaced population growth by 3% per year.

Even though Nigeria, Africa's most populous nation, has a population growth rate of 3.1%, food production there has kept up with population growth because of increases in land productivity. Nigeria has developed new varieties of cassava, yams, and sweet potatoes that give high yields without fertilizers or pesticides.

The late Ryoichi Sasakawa, a Japanese industrialist–philantropist, Norman Bourlag (Nobel prize–winning father of the second green revolution), and former U.S. president Jimmy Carter have created Sasakawa Global 2000 to develop and demonstrate new green-revolution techniques for Africa. By 1997 the organization had set up almost 400,000 demonstration plots in Africa—most in sub-Saharan Africa—where green-revolution techniques (that can double

A: About 84 million

or triple yields) are compared to current, traditional practices. Bourlag believes that productivity gains will result from using a variety of techniques including breeding strains that better resist crop diseases, tolerate metallic or acidic soils, and provide better nutrition.

As a result of government commitment to such green-revolution techniques, grain production in Ethiopia almost doubled between 1994 and 1996. However, the overall success of such efforts in sub-Saharan Africa will depend on strong government support, increased financial aid from international lending agencies, and efforts to reduce the area's economic and social problems.

How Serious Are Undernutrition, Malnutrition, and Overnutrition? People who cannot grow or buy enough food to meet their basic energy needs suffer from **undernutrition**. People who receive less than 90% of their minimum daily calorie intake of food on a long-term basis are said to be *chronically undernourished*. Such people are not starving to death, but they often don't have enough energy for an active, productive life and are more susceptible to infectious diseases.

People getting less than 80% of their minimum calorie intake are considered to be *seriously undernourished*. Children in this category are likely to suffer from mental retardation and stunted growth. They are also much more susceptible to infectious diseases such as measles and diarrhea, which kill one child in four in developing countries.

To maintain good health and resist disease, people need not only a certain number of calories, but also the proper amounts of protein (from animal or plant sources), carbohydrates, fats, vitamins, and minerals. People who are forced to live on a low-protein, high-carbohydrate diet consisting only of grains such as wheat, rice, or corn often suffer from **malnutrition**—deficiencies of protein and key *micronutrients* such as various vitamins (that cannot be made in the human body) and minerals that are needed in small quantities for good health. Many of the world's desperately poor people, especially children, suffer from both undernutrition and malnutrition.

The two most common nutritional deficiency diseases are marasmus and kwashiorkor. *Marasmus* (from the Greek word *marasmos*, "to waste away") occurs when a diet is low in both calories and protein (Figure 1-9). Most victims are either nursing infants of malnourished mothers or children who do not get enough food after being weaned from breast-feeding. If the child is treated in time with a balanced diet, most of these effects can be reversed. In practice, however, relief efforts are often too little or too late to save many children from death.

Kwashiorkor (meaning "displaced child" in a West African dialect) is a severe protein deficiency occurring in infants and children ages 1–3, usually after the arrival of a new baby deprives them of breast milk. The displaced child's diet changes to grain or sweet potatoes, which provide enough calories but not enough protein. If it is caught soon enough, most of the harmful effects can be cured with a balanced diet; otherwise, if the child survives, stunted growth and mental retardation result.

Here's some good news. *Between 1970 and 1995 the worldwide proportion of people suffering from chronic undernutrition fell from 36% to 14%.* Also, despite population growth, *the estimated number of chronically malnourished people fell from 940 million in 1970 to 840 million in 1995 (including 200 million children).*

Despite this progress, about one of every five people in developing countries (including one of every three children below age 5) is chronically undernourished or malnourished; 87% of them live in Asia and Africa. Such people are disease prone, and adults are too weak to work productively or think clearly. As a result, their children also tend to be underfed and malnourished. If these children survive to adulthood, many are locked in a tragic malnutrition–poverty cycle (Figure 12-10) that can be perpetuated for generations.

Critics point out that mass dieoffs from famine projected by some environmentalists have not occurred. However, during the past 25 years the number of people dying prematurely from undernutrition, malnutrition, or normally nonfatal diseases such as measles and diarrhea worsened by malnutrition has averaged at least 10 million per year.* This amounts to at least one-quarter of a billion people dying prematurely within a single generation, half of them children under age 5. However, children don't have to die prematurely because of undernutrition and malnutrition (Solutions, p. 290). Environmental expert Norman Myers (Guest Essay, p. 530) asks how many more deaths it would take for this phenomenon to qualify as a starvation disaster.

Although balanced diets, vitamin-fortified foods, and vitamin supplements have slashed the number of vitamin-deficiency diseases in developed countries, millions of cases occur each year in developing countries. An estimated 750 million to 1 billion people lack the energy to lead a healthy and productive life because of preventable vitamin and mineral deficiencies. For example, about 6 million people in developing countries suffer from a deficiency of vitamin A. Each year up to 500,000 children go blind because their diet lacks vitamin A. Half of these children die within 6 months of losing their sight.

*According to a 1997 Johns Hopkins University study, this death toll is at least 18 million people per year.

Q: What continent has the world's highest population growth rate?

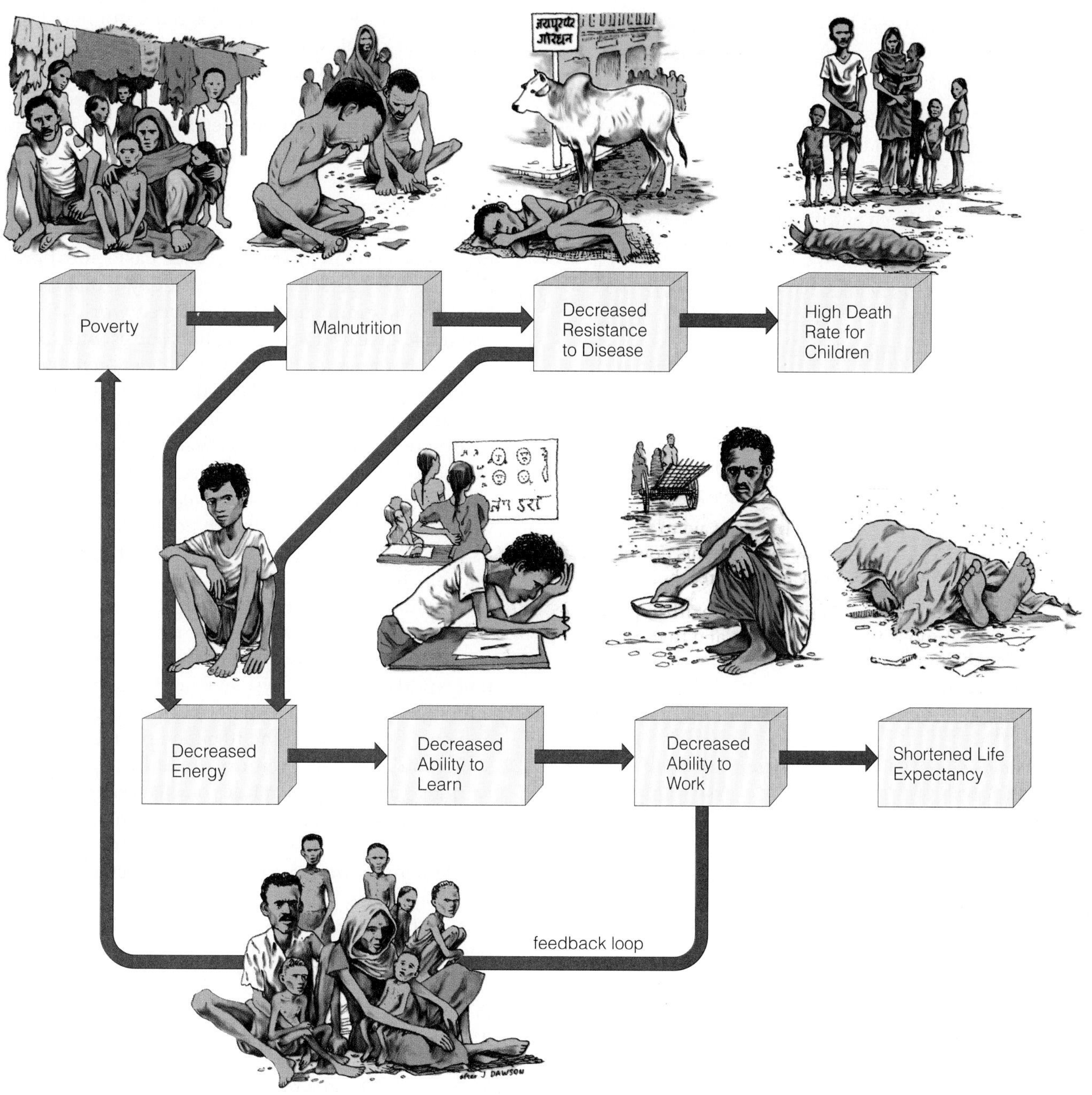

Figure 12-10 Interactions among poverty, malnutrition, and disease form a tragic cycle that tends to perpetuate such conditions in succeeding generations of families.

Other nutritional-deficiency diseases are caused by the lack of certain minerals. For example, too little iron (a component of hemoglobin that transports oxygen in the blood) causes anemia. This mineral deficiency causes fatigue, makes infection more likely, increases a woman's chances of dying in childbirth, increases an infant's chances of dying from infection during its first year of life, and cripples efforts to improve primary school education because developing brains need adequate iron to learn. In tropical regions of Asia, Africa, and Latin America, iron-deficiency anemia affects about 350 million people. Millions of poor people with diets low in zinc and iron cannot fight off diarrhea, malaria, and pneumonia, three of the world's leading killers.

Elemental iodine, found in seafood and crops grown in iodine-rich soils, is essential for the functioning of the thyroid gland, which produces a hormone that controls the body's rate of metabolism. According to the World Health Organization, some 1.6 billion

A: Africa (2.5% in 1998)

Saving Children

SOLUTIONS

Officials of the United Nations Children's Fund (UNICEF) estimate that between one-half and two-thirds of childhood deaths from nutrition-related causes could be prevented at an average annual cost of only $5–10 per child—only 10–19¢ per week. This life-saving program would involve the following simple measures:

- Immunizing children against childhood diseases such as measles
- Encouraging breast-feeding
- Preventing dehydration from diarrhea by giving infants a mixture of sugar and salt in a glass of water
- Preventing blindness by giving people a vitamin A capsule twice a year at a cost of about 75¢ per person
- Providing family-planning services to help mothers space births at least 2 years apart
- Increasing education for women, with emphasis on nutrition, sterilization of drinking water, and child care

Critical Thinking

How much money (if any) would you be willing to spend each year to help implement such a program for saving children? Why has little money been allocated for such a program?

people—almost one of every four—get too little iodine in their diet. Lack of iodine can cause goiter, an abnormal enlargement of the thyroid gland in the neck, which leads to deafness if untreated. Worldwide, an estimated 100 million children born to mothers with severe iodine deficiency suffer from stunted physical growth and severe mental retardation; another 20 million have some degree of brain damage from too little iodine.

Whereas an estimated 17% of the people in developing countries suffer from undernutrition and malnutrition, about 15% of the people in developed countries (32% in the United States) suffer from **overnutrition**—an excessive intake of food, especially fats, that can cause obesity (excess body fat).

Overnutrition is associated with at least two-thirds of the deaths in the United States each year, mostly from coronary heart disease, cancer, stroke, and diabetes. A study of thousands of Chinese villagers indicates that the healthiest diet for humans is largely vegetarian, with only 10–15% of calories coming from fat—in contrast to the typical meat-based diet, in which 40% of the calories come from fat. Polls report that 7% of Americans call themselves vegetarians. However, two-thirds of these eat meat regularly and most of them eat dairy products and eggs. Fewer than 500,000 Americans are vegans, who eat no meat and meat products.

Do We Produce Enough Food to Feed the World's People? The *good news* is that we produce more than enough food to meet the basic nutritional needs of every person on the earth today. If distributed equally, the grain currently produced worldwide is enough to give everyone a meatless subsistence diet.

The *bad news* for those not getting enough food is that food is not distributed equally among the world's people because of differences in soil, climate, political and economic power, and average per capita income throughout the world. Most agricultural experts agree that *the principal cause of hunger and malnutrition is and will continue to be poverty,* which prevents poor people from growing or buying enough food regardless of how much is available.

If everyone ate the diet typical of a person in a developed country, with 30–40% of the calories coming from animal products, estimates suggest that the current world agricultural system would support only 2.5 billion people—less than half the current population and only one-fourth of the 10 billion people projected sometime in the next century. In addition, scientists warn that the current emphasis on increasing food production by focusing on caloric output from rice, wheat, and corn can worsen micronutrient malnutrition.

Moreover, increases in global and per capita food production often hide large differences in food supply and quality among and within countries. For example, despite impressive gains in total and per capita food production since 1970, roughly 40% of India's population suffers from malnutrition because they are too poor to buy or grow enough food to meet their basic needs.

Food is also unevenly distributed within families. In poor families the most food goes to men working outside the home. Children ages 1–5 and women (especially pregnant women and nursing mothers) are the most likely to be underfed and malnourished.

Developed countries also have pockets of poverty, hunger, and malnutrition. According to a 1997 U.S. Department of Agriculture study, as many as 11 million Americans (excluding 600,000 homeless people) go hungry each year and 2 million of them suffer preventable severe hunger.

What Are the Environmental Effects of Producing Food? *Agriculture has a greater harmful impact on air, soil, water, and biodiversity resources than any other human activity* (Figure 12-11). A dramatic casualty of intensive, industrialized agriculture is the drastically

Q: What continent has the largest annual increase in population?

Biodiversity Loss

Loss and degradation of habitat from clearing grasslands and forests and draining wetlands

Fish kills from pesticide runoff

Killing of wild predators to protect livestock

Loss of genetic diversity from replacing thousands of wild crop strains with a few monoculture strains

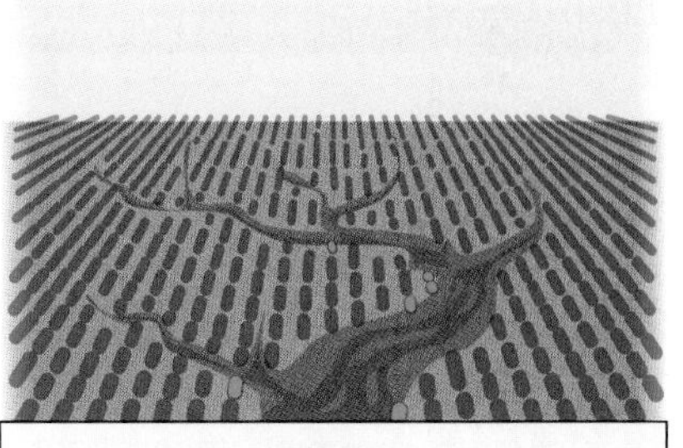

Soil

Erosion

Loss of fertility

Salinization

Waterlogging

Desertification

Air Pollution

Greenhouse gas emissions from fossil fuel use

Other air pollutants from fossil fuel use

Pollution from pesticide sprays

Water

Aquifer depletion

Increased runoff and flooding from land cleared to grow crops

Sediment pollution from erosion

Fish kills from pesticide runoff

Surface and groundwater pollution from pesticides and fertilizers

Overfertilization of lakes and slow-moving rivers from runoff of nitrates and phosphates from fertilizers, livestock wastes, and food processing wastes

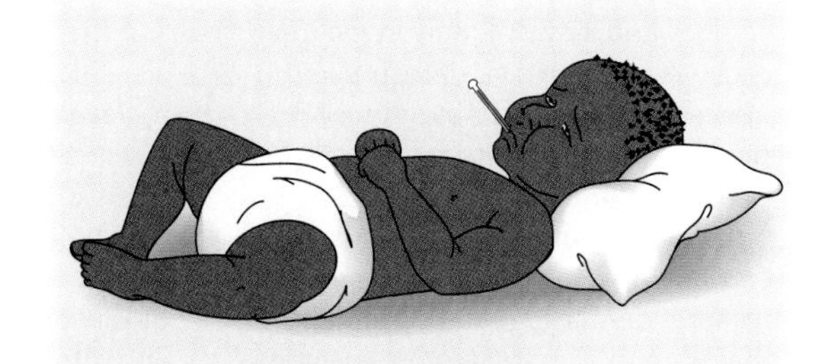

Human Health

Nitrates in drinking water

Pesticide residues in drinking water, food, and air

Contamination of drinking and swimming water with disease organisms from livestock wastes

Bacterial contamination of meat

Figure 12-11 Major environmental effects of food production.

shrinking Aral Sea in Uzbekistan (Figure 13-14). The results of this ecodisaster are a sharp drop in the agricultural and aquatic productivity of the region, increased poverty and health problems, and economic losses of an estimated $110 billion per year.

Another example of unsustainable farming has taken place in Kazakhstan, the largest wheat producer in Southeast Asia. Since 1980, when wheat production peaked, this semiarid area has lost one-third of its cropland to erosion (mostly from wind erosion). Soil scientists estimate that because of unsustainable use of the land, the area currently can support only about half as much grain production as it did in its peak production year of 1980.

Food prices in the United States are deceptively low because they do not include the harmful environmental and health costs associated with U.S. food production. David Pimentel (Guest Essay, p. 364) has estimated that the harmful costs not included in the prices of food in the United States are $150–200 billion per year. According to a 1990 UN study, degradation of irrigated cropland, rain-fed cropland, and rangeland now costs the world more than $42 billion a year in lost crop and livestock output; this loss is roughly equal to the annual value of the entire U.S. grain harvest.

Many analysts believe that it is possible to produce enough food to feed the 8 billion people projected by 2025 through new advances in agricultural

Can China's Population Be Fed?

CASE STUDY

Since 1970 China has made significant progress in feeding its people and slowing its rate of population growth (Section 11-4). In addition, since 1980 the Chinese economy has quadrupled. During the 1990s China has lifted many of its people out of poverty by becoming the world's second largest economy (after the United States). With its economy growing very rapidly (typically at 9–14% a year), China could become the world's largest economy by 2010.

But with such a rapidly increasing demand for food and other resources and 12 million more people each year, a shortage of resources may slow China's ability to feed its people and sustain its rapid economic growth.

There is growing concern that crop yields may not be able to keep up with demand. A basic problem is that with 21% of the world's people, China has only 7% of the world's cropland and fresh water, 3% of its forests, and 2% of its oil. This concern was highlighted when China shifted from being a net exporter of grain in 1994 to being the world's second largest importer of grain (after Japan) in 1995, but was able to grow enough food without such imports in 1996 and 1997.

Despite the country's huge area, much of western China is desert and unfit for agriculture. Thus, most of its cropland is concentrated in the eastern part of the country (Figure 12-12). Most of China's people—five times the population of the United States—live on the country's southern and eastern coasts in an area about the size of the United States east of the Mississippi River.

China irrigates 60% of its cropland, but water tables from aquifer depletion are dropping under about 10% of its cultivated area. Using more irrigation to raise wheat yields is becoming more difficult because of aquifer depletion, diminished response to fertilizer because of lower soil moisture, and the growing use of irrigation water to support the country's rapidly growing cities. Indeed, lack of irrigation water may be the factor that ultimately limits food production in China.

Since 1950 China has lost an area of cropland roughly equal to the area of Argentina, mostly because of population growth, industrialization, and urbanization, trends that are expected to get worse. Government officials talk of the need to build 600 *new* cities and 1 million *new* factories by 2010 to accommodate the country's booming urban population and industrialization. This could lead to another 5% loss of the country's cropland by 2010.

According to Worldwatch Institute projections, China's grain production is likely to fall by at least 20% between 1990 and 2030. Even if China's booming economy resulted in no increases in meat consumption, this 20% drop would mean that by 2030 China would need to import more than the world's grain exports in 1993 (roughly half from the United States). However, if the increased demand for meat led to a rise in per capita grain consumption equal to the current level in Taiwan (one-half the current U.S. level), by 2030 China would have to import more grain than the entire current grain output of the United States.

The Worldwatch Institute warns that if either of these scenarios is correct, no country or combination of countries has the potential to supply even a small fraction of China's potential food supply deficit. This is not even taking into account the huge food deficits that are projected in other parts of the world by 2030, especially in Africa and India.

However, according to a 1997 study by the International Food Policy Institute, China should be able to feed its population and begin exporting grain again by 2020 if the government invests in expanding irrigation and increasing agricultural research. As China's population growth rate declines, the institute projects that the resulting decrease in direct grain consumption will offset the rapid increase in production of cereal grains to feed meat-producing animals. Recent satellite surveys also show that China has far more potential cropland than previously thought.

Other analysts believe that serious and rapidly growing environmental problems may also limit China's economic growth and its ability to feed its people. China now consumes more grain, red meat, and fertilizer, and produces more steel than the United States, leading to significant and growing pollution and environmental degradation.

technology and by spreading the use of existing green-revolution or high-yield techniques. Other analysts disagree. They have serious doubts about the ability of new food production technologies and food distribution systems to keep up with current levels of population growth, mostly because the harmful environmental effects of agriculture will reduce yields (Case Study, above).

According to Norman Myers (Guest Essay, p. 530), the ability to produce more food will be limited by a combination of soil erosion, desertification, salinization and waterlogging of irrigated lands, water deficits and droughts, and loss of wild species that provide the genetic resources for new foods and improved forms of existing foods. According to Worldwatch Institute estimates, between 1945 and

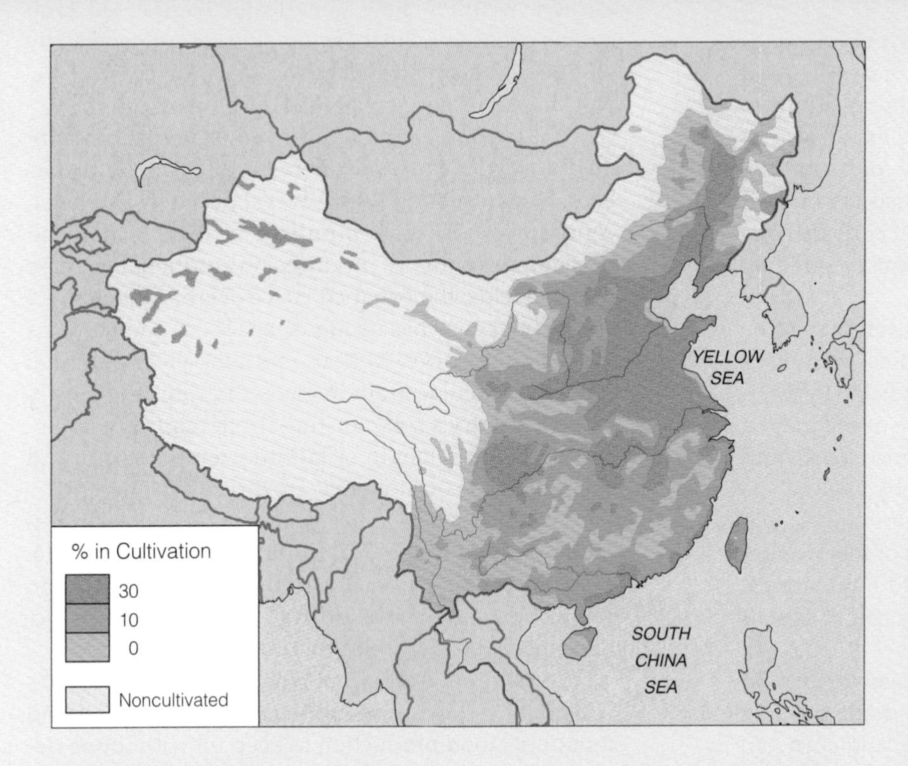

Figure 12-12 Most of China's arable land suitable for growing crops is found in the eastern part of the country. Because of population growth and environmental degradation, the amount of land available for agriculture has declined and is expected to decline more because of China's rapid economic growth, increasing affluence, growing urbanization, and increasing pollution and environmental degradation.

With its limited oil reserves and huge supplies of coal, China gets 75% of its energy from burning highly polluting coal (compared to 22% in the United States). According to one estimate, by 2035 China's SO_2 emissions (mostly from burning coal) will exceed those of all other industrialized countries.

China ranks second in the world (after the United States) in emissions of CO_2, the world's major greenhouse gas (Figure 19-2a). Within the next 25 years China is expected to become the world's largest emitter of CO_2, making it the major contributor to projected global warming (Figure 19-5).

An estimated 80% of the country's industrial and domestic waste is discharged untreated into rivers. As a result, 25% of its rivers are too polluted to use for irrigation. According to a top Chinese environmental official, all but 5 of China's more than 500 cities suffer from severe air pollution, which helps to explain why respiratory disease is the leading cause of death in China's urban areas. According to a 1997 World Bank study, an estimated 2.03 million people die prematurely in China annually from the effects of air and water pollution.

So far China has concentrated on rapid industrialization and devoted little attention to sustainable use of its resources and to reducing pollution and environmental degradation. However, some Chinese officials are beginning to realize that without such policies its economic growth cannot be sustained.

If China begins acting now to chart a new course, it has a unique opportunity to leapfrog over the traditional Western forms of economic development and show the world how to build an environmentally sustainable economy over the next few decades.

Critical Thinking

If the scenarios about China's growing dependence on food imports are valid, how might this affect **(a)** world food prices, **(b)** your life, and **(c)** the harmful environmental impacts of food production (Figure 12-11). What actions, if any, do you suggest for dealing with this potential problem?

1990, erosion, salinization, waterlogging, desertification, and other forms of environmental degradation eliminated an area of land from food production equal to the cropland of two Canadas.

This trend is expected to accelerate as modern industrialized farming and environmentally unsound subsistence farming increase in coming decades. According to David Pimentel (Guest Essay, p. 364), an area of cropland larger than the land area of India is expected to lose most of its agricultural productivity from environmental degradation between 1994 and 2013. Two other environmental constraints that are likely to limit food production in the future are increased ultraviolet radiation from ozone-layer depletion and projected global warming (Chapter 19).

A: 2.9 (1.6 in developed countries and 3.3 in developing countries) in 1998

12-3 INCREASING WORLD FOOD PRODUCTION

Is Increasing Crop Yields the Answer? Agricultural experts expect most future increases in food yields per hectare on existing cropland to result from improved strains of plants and from expansion of green-revolution technology to new parts of the world. For example, in 1994 crop scientists announced that they had developed new strains of corn that can increase crop yields up to 40% in regions plagued by droughts and acidic soils.

In addition, a new strain of rice developed in the Philippines by the International Rice Research Institute is expected to be commercially available by 2001. This new variety diverts a larger share of the plants' carbohydrates into grain and less into stems and leaves and could increase rice yields by as much as 20%. However, the future of such important research may be hampered by drops in funding of the Rice Research Institute from donor nations and foundations. In 1997 the institute had to lay off nearly half of its staff because of a 20% drop in funding.

Currently, plant breeders have raised the *harvest index*—the share of a plant's photosynthetic product going into seed in today's wheat, rice, and corn—to more than 50%. However, scientists believe that the physiological limit is around 60%. Thus, plant breeders using traditional techniques have exploited most of the genetic potential for increasing the harvest index. Increasing the efficiency of photosynthesis could increase crop yields, but despite decades of research scientists have had little success in doing this.

Scientists are working to create new green revolutions—actually *gene revolutions*—by using genetic engineering and other forms of biotechnology. Over the next 20–40 years they hope to breed high-yield plant strains that are more resistant to insects and disease, thrive on less fertilizer, make their own nitrogen fertilizer (as do legumes, Figure 5-7), do well in slightly salty soils, can withstand drought, and can use solar energy more efficiently during photosynthesis. But according to Donald Duvick, former director of research at Pioneer HiBred International (one of the world's largest seed producers), "No breakthroughs are in sight. Biotechnology, while essential to progress, will not produce sharp upward swings in yield potential except for isolated crops in certain situations."

Several factors have limited the success of the green and gene revolutions to date, and may continue to do so. Without huge amounts of fertilizer and water, most green-revolution crop varieties produce yields that are no higher (and are sometimes lower) than those from traditional strains; this is why the second green revolution has not spread to many arid and semiarid areas such as much of Africa and Australia (Figure 12-4). Without ample water, good soil, and favorable weather, new genetically engineered crop strains could fail. Furthermore, the cost of genetically engineered crop strains is too high for most of the world's subsistence farmers in developing countries.

It is encouraging that every country that has initiated a two- to four-fold increase in grain yields has been able to sustain it for several decades. However, continuing to increase inputs of fertilizer, water, and pesticides eventually produces no additional increase in crop yields; the *J*-shaped curve of crop productivity slows down, reaches its limits, levels off, and becomes an *S*-shaped curve. At that point, the yield potential for any particular grain in a country depends mostly on that country's soil moisture (from rainfall or ability to irrigate), day length or latitude, temperatures, and solar intensity—factors that are difficult to alter.

Grain yields per hectare are still increasing in many parts of the world, but at a much slower rate. This has been a key factor in the slowdown in grain production and the decline in per capita grain production since 1990 (Figure 12-8).

Some think new genetic technological advances will allow yields to rise to their former high rates and thus allow food production to keep up with future demand. Others believe that this is unlikely because of built-in biological limits to increases in plant productivity and the serious harmful environmental effects of high-yield agriculture (Figure 12-11).

There has also been a tendency to use the highest yields achieved for grains such as rice, wheat, and corn as the basis for estimating yield increases elsewhere. Lester Brown (Guest Essay, p. 34) argues that this is unrealistic because maximum attainable yields vary with variations in temperature, precipitation, day length (based on latitude), solar intensity, and inherent soil fertility.

For example, regardless of inputs, rice yields in Japan are higher than those in Indonesia because the days are much longer in Japan during the summer growing season and because Japan is able to irrigate 99% of its riceland. Similarly, wheat yields are much higher in Western Europe than in Russia and Canada, where harsh winters prevent farmers from growing winter wheat. And rice yields in California are much higher than those in Japan because of California's much greater solar intensity. Grain yields in arid and semiarid Africa have not increased as much as those in Asia because of the lack of rainfall or irrigation water. As a result, African farmers cannot use enough fertilizer to more fully exploit the genetic potential of their crops.

Moreover, Indian economist Vandana Shiva (Guest Essay, p. 674) contends that overall gains in grain yields from new green- and gene-revolution

Q: What is the size of the U.S. population?

varieties may be much lower than claimed. The reason is that the yields are based on comparisons between the output per hectare of old and new *monoculture* varieties, rather than between the even higher yields per hectare for *polyculture* cropping systems and the new monoculture varieties that often replace them.

In addition to leveling off, yields may even start dropping for a number of reasons: The soil erodes, loses fertility, and becomes salty and waterlogged (Figure 14-20); underground and surface water supplies become depleted and polluted with pesticides and nitrates from fertilizers; and populations of rapidly breeding pests develop genetic immunity to widely used pesticides.

Connections: Will Loss of Genetic Diversity Limit Crop Yields? Some agricultural scientists think that new genetically engineered or crossbred varieties will enable yields of key crops to continue rising. Other scientists question whether this is possible, primarily because of the environmental impacts of current forms of industrialized agriculture and the accelerating loss of biodiversity, which can limit the genetic raw material needed for future green and gene revolutions.

In India, which once had 30,000 varieties of rice, more than 75% of the rice production now comes from 10 varieties. In the United States, about 97% of the food plant varieties that were available to farmers in the 1940s no longer exist, except perhaps as a handful of seeds in a seed bank or in the backyards of a few gardeners. In other words, we are rapidly shrinking the world's genetic "library" just when we need it more than ever.

Scientists can crossbreed varieties of animal and plant life and genetic engineers can move genes from one organism to another, but they need the genetic materials in the earth's existing species to work with. We are losing much of this genetic diversity as a small number of specially bred monoculture varieties of key crops have replaced thousands of strains of various crops and natural areas have been cleared.

The UN Food and Agriculture Organization estimates that by the year 2000, two-thirds of all seed planted in developing countries will be of uniform strains. Such genetic uniformity increases the vulnerability of food crops to pests, diseases, and harsh weather. Many biologists argue that this decreased variability, plus growing species extinction, can severely limit the potential of future green and gene revolutions.

In the mid-1970s a valuable wild corn species, the only known perennial strain of corn, was barely saved from extinction. When this strain was discovered in south central Mexico, only a few thousand stalks survived in three tiny patches that were about to be cleared by squatter cultivators and commercial loggers.

Crossbreeding this perennial strain with commercial varieties could reduce the need for yearly plowing and sowing, which would reduce soil erosion and save water and energy. Even more important, this wild corn has a built-in genetic resistance to four of the eight major corn viruses, and it grows in cooler and damper habitats than established commercial strains. Overall, the economic benefits of cultivating this barely rescued wild plant could total several billion dollars per year.

Wild varieties of the world's most important plants can be collected and stored in gene or seed banks, agricultural research centers, and botanical gardens. However, space and money severely limit the number of species that can be preserved. Many plants (such as potatoes) cannot be stored successfully as seed in gene banks. Power failures, fires, or unintentional disposal of seeds can also cause irreversible losses.

In addition, because stored seeds don't remain alive indefinitely, periodically they must be planted (germinated) and new seeds collected for storage. Unless this is done, seed banks become seed morgues. Moreover, stored plant species stop evolving; thus, they may be difficult to reintroduce into their native habitats, which may have changed in the meantime.

Because of these limitations, ecologists and plant scientists warn that the only effective way to preserve the genetic diversity of most plant and animal species is to protect representative ecosystems throughout the world from agriculture and other forms of development (Chapters 23 and 24).

Will People Try New Foods? Some analysts recommend greatly increased cultivation of less widely known plants to supplement or replace such staples as wheat, rice, and corn. One of many possibilities is the winged bean, a protein-rich legume now common only in New Guinea and Southeast Asia (Figure 12-13). Insects—called *microlivestock*—are also important potential sources of protein, vitamins, and minerals in many parts of the world (Figure 12-14). There are about 450 edible species of insects, including black ant larvae (served in tacos in Mexico) and giant waterbugs (crushed into vegetable dip in Thailand). Two basic problems are getting farmers to take the financial risk of cultivating new types of food crops and convincing consumers to try new foods.

David Pimentel (Guest Essay, p. 364) and plant scientists at the Land Institute in Salina, Kansas (p. 278), believe that we could rely more on polycultures of perennial crops, which are better adapted to regional soil and climate conditions than most annual food crops. Using perennials would also eliminate the need to till soil and replant seeds each year; it would greatly

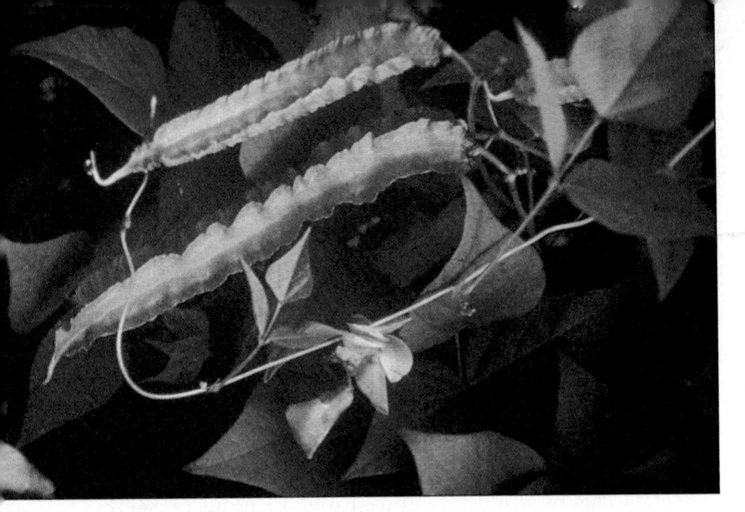

Figure 12-13 The winged bean, a protein-rich annual legume from the Philippines, is only one of many currently unfamiliar plants that could become important sources of food and fuel. Its edible winged pods, spinachlike leaves, tendrils, and seeds contain as much protein as soybeans, and its edible roots contain more than four times the protein of potatoes. Its seeds can be ground into flour or used to make a caffeine-free beverage that tastes like coffee. Indeed, this plant produces so many different edible parts that it has been called a supermarket on a stalk. Because of nitrogen-fixing nodules in its roots, this fast-growing plant needs little fertilizer. (Larry Mellichamp/Visuals Unlimited)

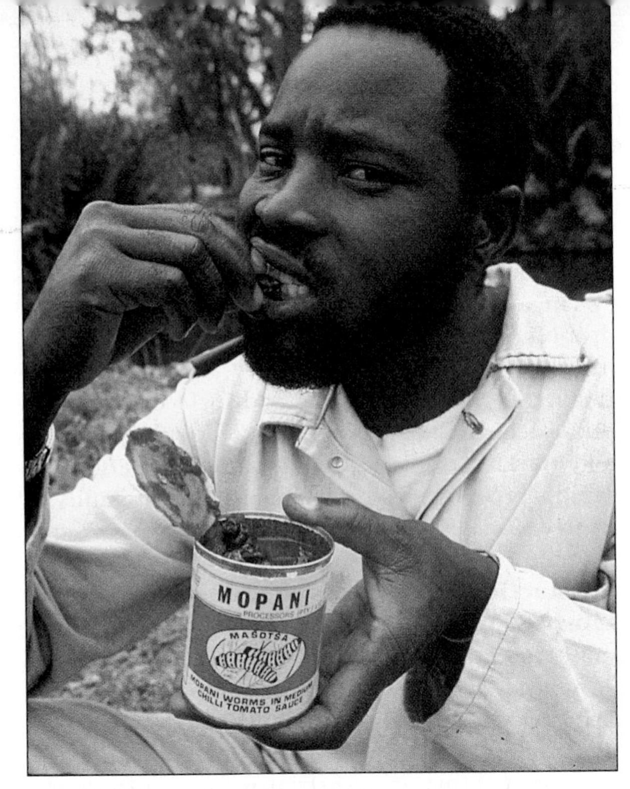

Figure 12-14 Insects are important food items in many parts of the world. *Mopani*—emperor moth caterpillars—are among several insects eaten in South Africa. However, this food is so popular that the caterpillars (known as mopane worms) are being overharvested. Kalahari Desert dwellers eat cockroaches, lightly toasted butterflies are a favorite food in Bali, and French-fried ants are sold on the streets of Bogota, Colombia. Most of these insects are 58–78% protein by weight—three to four times as protein-rich as beef, fish, or eggs. (Anthony Bannister/Natural History Photographic Agency)

reduce energy use, save water, and reduce soil erosion and sediment water pollution. Of course, widespread use of perennials would reduce the profits of agribusinesses selling annual seeds, fertilizers, and pesticides, which explains why they don't favor this approach.

Is Cultivating More Land the Answer? According to the World Bank, about 36% of the world's land is devoted to raising crops. Today, nearly all of the world's best agricultural land is in use. Between 1980 and 1990 the area of the world's cropland expanded by only 2%.

Theoretically, the world's cropland could be more than doubled by clearing tropical forests and irrigating arid land, mostly in Africa and Latin America (Figure 12-15). However, many analysts believe that this potential for agricultural expansion is often overestimated because much of the land is marginal land, where cultivation is unlikely to be sustainable.

Clearing rain forests to grow crops and graze livestock, for example, can have disastrous ecological consequences, as discussed in Section 24-3. Furthermore, most of this cleared land has nutrient-poor soils and cannot support crop growth for more than a couple of years. In addition, potential cropland in savanna and other semiarid land in Africa cannot be used for farming or livestock grazing because of the presence of 22 species of the tsetse fly (Figure 12-9).

Some researchers hope to develop new methods of intensive cultivation in tropical areas. But other scientists argue that it makes more ecological and economic sense to combine various ancient methods of shifting cultivation (Figure 1-23), followed by fallow periods long enough to restore soil fertility, with various forms of polyculture. Some scientists also recommend increased plantation cultivation of rubber trees, oil palms, and banana trees, which are adapted to tropical climates and soils.

Much of the world's potentially cultivable land lies in dry areas, especially in Australia and Africa. Large-scale irrigation in these areas would require expensive dam projects with a mixture of beneficial and harmful impacts (Figure 13-10) and large inputs of fossil fuel to pump water long distances. Large-scale irrigation could also deplete groundwater supplies by removing water faster than it is replenished. The land would need constant and expensive maintenance to prevent erosion, groundwater contamination, salinization (Figures 12-16 and 14-20), and waterlogging (Figure 14-20). Expanding wetland production of rice could also accelerate projected global warming by increasing atmospheric emissions of methane.

Q: What is the average number of children per woman in the United States?

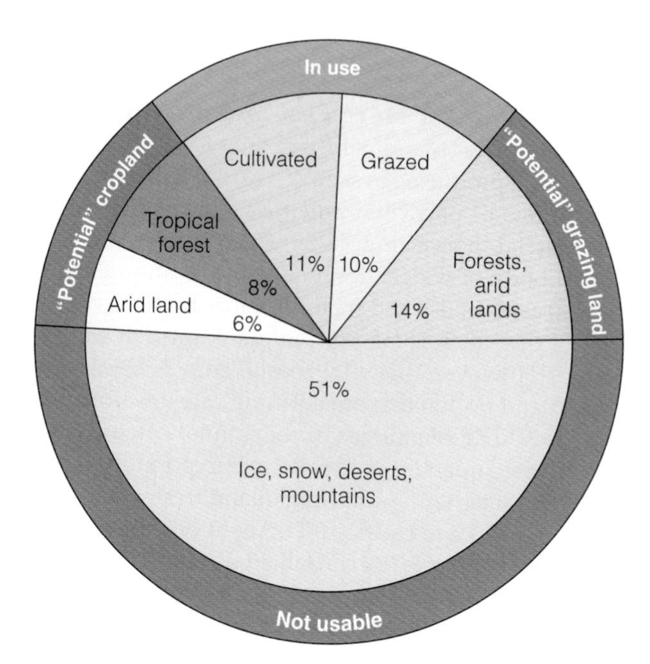

Figure 12-15 Classification of the earth's land. Theoretically, we could double the amount of cropland by clearing tropical forests and irrigating arid lands. However, converting these lands into cropland would destroy valuable forest resources, reduce earth's biodiversity, affect water quality and quantity, and cause other serious environmental problems, usually without being cost-effective.

Figure 12-16 Because of high evaporation, poor drainage, and severe salinization, white alkaline salts have displaced crops that once grew on this heavily irrigated land in Colorado. (Soil Conservation Service)

Thus, much of the new cropland that could be developed would be on land that is marginal for raising crops, requiring expensive inputs of fertilizer, water, and energy. Furthermore, these potential increases in cropland would not offset the projected loss of almost one-third of today's cultivated cropland due to erosion, overgrazing, waterlogging, salinization, mining, and urbanization. The United Nations estimated that between 1945 and 1992, approximately 20% of the earth's land used for growing crops and grazing livestock has suffered moderate to extreme degradation from erosion and desertification.

Even if it is financially feasible, such expansion would reduce wildlife habitats and thus the world's biodiversity and ecological integrity. According to the UN Food and Agriculture Organization (FAO), cultivating all potential cropland in developing countries would reduce forests, woodlands, and permanent pasture by 47%.

In addition to providing wildlife habitats and conserving water and soils, these forests store 20–50 times more carbon as biomass than crops and pasture do. Clearing these forests would release a huge amount of carbon dioxide into the atmosphere and accelerate possible global warming.

For the reasons discussed in this section, *many analysts believe that a major economically profitable and environmentally sustainable expansion of cropland is unlikely over the next few decades.* If this assessment is correct, the world's grainland area per person, which dropped by almost half between 1950 and 1997, is expected to decline further (Figure 12-17). Then the critical question is whether crop yields per area of cropland can be increased enough to offset the shrinkage of cropland per person and keep up with projected population growth.

Can We Grow More Food in Urban Areas? Food experts project that people in urban areas could live more sustainably and save money by growing food in empty lots, on rooftops, and in their own backyards. Currently, urban gardens provide about 15% of the world's food but this could be increased. A study by the UN Center for Human Settlements estimated that up to 50% of the total area in many cities in developing countries is vacant public land that could be used to produce food.

According to Worldwatch Institute estimates, at least 200 million people now grow some of their food and provide 800 million people with at least some of their food. Farmers in Accra, Ghana, provide the city with about 90% of its vegetables. In Singapore urban farmers supply 80% of the city's poultry and 25% of its vegetables. Farmers in or near 18 of China's largest

A: 2.0 in 1998

cities provide urban dwellers with 85% of their vegetables and more than half of their meat and poultry.

Recycling nutrient-rich animal and human wastes to grow food in urban areas can also greatly reduce water pollution from runoff of plant nutrients. In China human waste is treated and sold to farmers as fertilizer. For more than 50 years sewage-fed lagoons (ponds) in Calcutta, India, have provided the city's people with one-tenth of the fish they consume. This wastewater-fed aquaculture system also is a cost-effective way to treat sewage wastes that are usually discharged into nearby rivers and lakes.

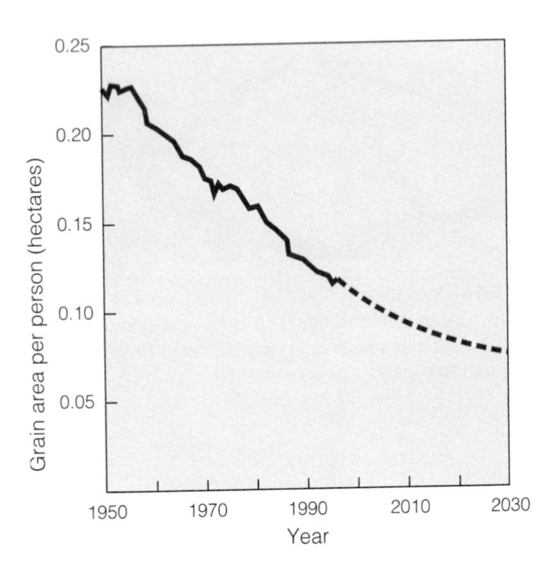

Figure 12-17 Average grain area per person worldwide, 1950–97, with projections to 2030. (Data from U.S. Department of Agriculture and Worldwatch Institute)

12-4 CATCHING OR RAISING MORE FISH

Can We Harvest More Fish and Shellfish? In addition to cropland and grazing land, the third major food producing system consists of **fisheries**: concentrations of particular aquatic species suitable for commercial harvesting in a given ocean area or inland body of water. About 70% of the annual commercial catch of fish and shellfish comes from the ocean; 99% of this catch is taken from plankton-rich coastal waters. However, this vital coastal zone is being disrupted and polluted at an alarming rate (Sections 8-2 and 20-3). The remainder of the annual catch comes from using aquaculture to raise fish in ponds and underwater cages (20%) and from inland freshwater fishing from lakes and rivers (10%). About one-third of the world fish harvest is not consumed directly by humans and is used primarily as animal feed, fish meal, and oils.

Here is some *good news*. Between 1950 and 1996 the annual world fish catch (marine plus freshwater harvest) increased 4.9-fold (Figure 12-18, left). And between 1950 and 1988 the per capita world fish catch more than doubled (Figure 12-18, right).

However, mostly because the rate of population growth exceeded the rate of growth of the world's fish catch, the per capita catch fell by 6% between 1989 and 1996 (Figure 12-18, right). Because of overfishing, pollution, and population growth, the world's fish catch is not expected to increase significantly; it may even decline as population growth exceeds the growth of the catch.

How Are Overfishing and Habitat Degradation Affecting Fish Harvests? Fish are potentially renewable resources as long as the annual har-

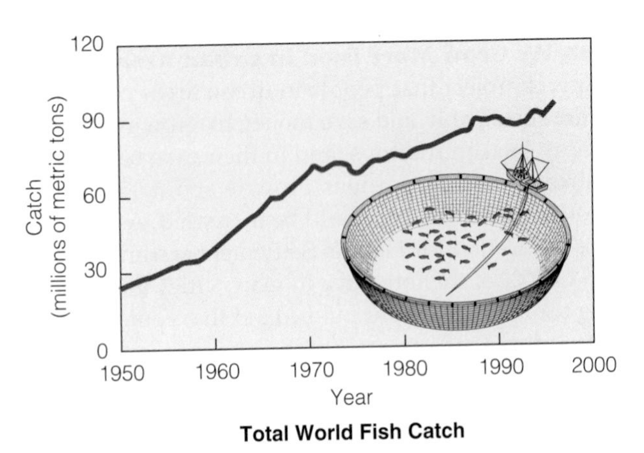

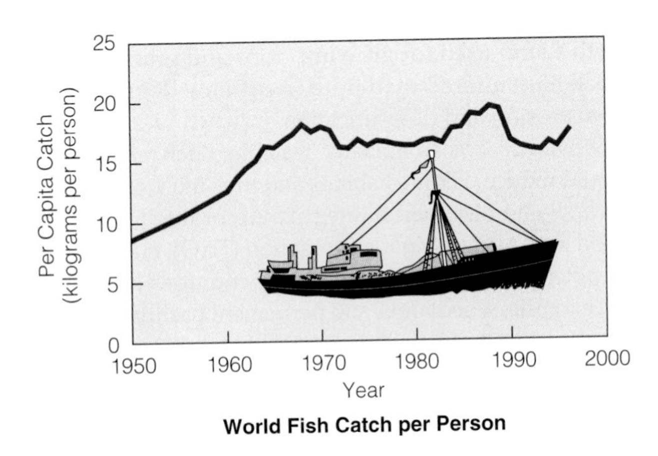

Figure 12-18 World fish catch (marine plus freshwater harvest) (left) and catch per person (right), 1950–96. Worldwide per capita fish catch did not rise much between 1968 and 1989 and has dropped and leveled off since then. Scientists estimate that the sustainable yield of the world's marine fishery is 100 million metric tons (110 million tons)—an amount that is already being exceeded if the unused bycatch is added to the annual fish catch. (Data from UN Food and Agriculture Organization and Worldwatch Institute)

Q: How many people were added to the U.S. population in 1998?

vest leaves enough breeding stock to renew the species for the next year. Ideally, an annual **sustainable yield**—the size of the annual catch that could be harvested indefinitely without a decline in the population of a species—should be established for each species to avoid depleting the stock.

However, determining sustainable yields is difficult. Estimating mobile aquatic populations isn't easy, and sustainable yields shift from year to year because of changes in climate, pollution, and other factors. Furthermore, sustainably harvesting the entire annual surplus of one species may severely reduce the populations of other species that rely on it for food. Indeed, declines in one species can lead to a series of effects throughout a marine ecosystem by altering predator–prey relationships, making the system vulnerable to invasion by alien species, and modifying community structure, function, and productivity—another example of connections in nature.

Overfishing is the taking of so many fish that too little breeding stock is left to maintain numbers; that is, overfishing is a harvest that exceeds the estimated sustainable yield. Then fish stocks begin to shrink. Prolonged overfishing leads to **commercial extinction**: reduction of a species to the point at which it's no longer profitable to hunt for them. Fishing fleets then move to a new species or a new region, hoping that the overfished species will eventually recover.

Experts warn that unprecedented forces—among them a burgeoning global seafood market and highly efficient large-scale industrial fishing fleets—are depleting stocks to the point at which they may not recover or may take decades to recover. In effect, the global fishing industry is depleting the renewable natural capital (fish stocks) instead of living off the renewable income naturally produced by stocks that are not overfished (Connections, p. 300).

According to the UN Food and Agriculture Organization, *11 of the world's 15 major oceanic fishing areas have been fished at or beyond their estimated maximum sustainable yield for commercially valuable species and are in a state of decline.* As a result, 69% of the world's commercial fish stocks are in decline and in need of urgent management.

Another biological indicator of overfishing is the average size of captured fish. When a fish stock is under stress the catch increasingly consists of younger and smaller fish because the largest ones have already been caught. Depleting future breeding populations leads to a declining population and harvest in coming years.

Populations of large fish such as sharks (Case Study, p. 217), swordfish, marlin, and tuna that are top predators feeding at the top of food webs are being decimated. In turn, this affects the populations of marine plant species that support such fish. For example, during the past 20 years the average size of swordfish caught by baited hooks on longlines has dropped from 120 kilograms (260 pounds) to 30 kilograms (66 pounds). As a result, the breeding populations of swordfish have been cut in half and the catch now consists of mostly small, immature fish.

In the western Atlantic, the breeding population of bluefin tuna, the largest tuna species, is thought to consist of perhaps 40,000 adults, down from an estimated 250,000 two decades ago. In Tokyo a single large bluefin tuna can fetch $50,000 or more at an auction—somewhat like winning the lottery for those who catch such fish. This provides a strong incentive to deplete the already decimated population of this species.

According to the U.S. National Fish and Wildlife Foundation, 14 major commercial fish species in U.S. waters (accounting for one-fifth of the world's annual catch and half of all U.S. stocks) are so depleted that even if all fishing stopped immediately it would take up to 20 years for stocks to recover.

Degradation and destruction of wetlands, estuaries, coral reefs, salt marshes, and mangroves (half-submerged ocean forests), and pollution of coastal areas from the land and the air are also serious and growing threats to populations of fish and shellfish. About 70% of the U.S. fish catch consists of species that depend on increasingly threatened and polluted estuaries for at least part of their life cycle.

For example, the survival of an estimated 90% of the commercial fish in the Bay of Bengal depends on mangroves, which are increasingly being removed or degraded. For anadromous species such as white perch, shad, and salmon (Figure 12-19), which live part of their lives in fresh water and part in salt water, habitat degradation of spawning streams and migration routes, not overfishing, is the leading cause of depletion.

Driving and enhancing these threats is a rapidly growing population that lives in or near coastal areas. Currently, 3.8 million people—or 64% of the world's population—live within 100 kilometers (62 miles) of a coastline and two-thirds of the world's largest cities are coastal. During the next 30 years 6.3 million are expected to live in or near densely populated coastal areas.

In the 1970s and 1980s extensive investment in fishing fleets, aided by government and development agency subsidies, significantly boosted the fish catch (Figure 12-18, left). In the late 1990s, however, food analysts expect such investments and subsidies to hasten the collapse of ocean fisheries—an example of harmful runaway or positive feedback.

Because there are now too many fishing boats competing for a declining supply of fish, the number of clashes over access to fisheries has increased in recent years. These include conflicts over turbot off Canada's east coast, cod wars between the fishing fleets of Iceland and Norway, squid wars in the southwest Atlantic, tuna wars in the Northeast Atlantic, and

A: About 3.1 million (44% from legal immigrants, refugees, and illegal immigrants)

Commercial Fishing and the Tragedy of the Commons

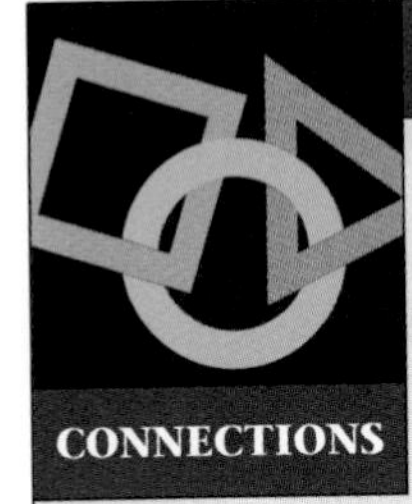

CONNECTIONS

Today the commercial fishing industry is dominated by industrial fishing fleets with factory trawlers the size of football fields. They are equipped with satellite positioning equipment (that allows them to return precisely to productive fishing areas), sonar (allowing them to fish in deeper waters), radar (permitting them to operate in the fog and the dark), massive nets (large enough to swallow a dozen Boeing 747 jets), and spotter planes. Most of these industrial fishing fleets can remain at sea for months because they have factory ships that can process and freeze their catches, thus giving them an advantage over smaller land-based boats.

Between 1975 and 1996, the size of the industrial fishing fleet expanded twice as fast as the rise in catches. As a result, there are now too many boats fishing for a declining number of fish. This leads to overfishing—an example of the tragedy of the commons (p. 12).

This global "sea grab" is the result of a number of interacting factors, including **(1)** technological improvements in fishing efficiency that allow ships to locate and harvest huge fish catches in deeper waters and to fish at lower levels of aquatic food webs, **(2)** difficulty in monitoring and restricting access to fisheries owned by no one, **(3)** growing demand for seafood, **(4)** subsidies by government and development agencies that lead to an excessive number of fishing vessels and encourage fishers to remain in a declining industry, and **(5)** lack of knowledge by consumers about the global fishing crisis and its harmful impacts on aquatic ecosystems and biodiversity.

Approximately 40% of the fish (most at low levels of food webs) caught by supertrawlers is considered unwanted *bycatch* and is ground up and thrown back into the ocean or dumped overboard dead or dying to save freezer space for more profitable species. For example, in 1993 shrimp trawlers in the Gulf of Mexico caught and threw away an estimated 34 million red snappers—11 times the annual commercial catch of this species. As a result of this waste, the productivity of the red snapper fishery has fallen.

A 1998 study by Daniel Pauly warned that the current harvesting of species at increasingly lower trophic levels in ocean food webs decreases chances for recovery of species at the top of ocean food webs. Eventually this can lead to abrupt declines in plankton-eating species, wholesale collapse of marine ecosystems, a drop in aquatic biodiversity, and a loss of high-quality protein for humans. According to Pauly, "If things go unchecked, we might end up with a marine junkyard dominated by plankton."

Because of the overcapacity of the fishing fleet and overfishing, it costs the global fishing industry about $125 billion a year to catch $70 billion worth of fish. Most of the $54 billion dollar annual deficit of the industry is made up by government subsidies such as fuel-tax exemptions, price controls, low-interest loans, and grants for fishing gear. Critics contend that such subsidies accelerate overfishing.

As international agreements allow foreign-owned industrialized fleets greater access to coastal waters in many developing countries in Africa and Asia, small- and middle-size fishers (who are rarely consulted about such agreements) are being deprived of their livelihoods and key sources of nutrition for their families and local communities. Nearly 1 billion people in Asia and coastal developing countries depend on fish to supply a majority of their protein needs. Some analysts consider this transfer of fish from developing countries to developed countries a serious international environmental justice issue.

Critics of this view point out that export of high-value fish brings in foreign exchange income for exporting countries. If distributed to the poor and lower middle class by governments, this income can contribute to better nutrition, education, family planning, and other social development programs.

Critical Thinking

1. Do you believe that government subsidies for the fishing industry should be eliminated? Explain. How would you feel about eliminating such subsidies if your livelihood depended on fishing?

2. Should governments in coastal developing countries encourage foreign-owned fishing fleets to harvest and export fish taken from their waters in exchange for foreign exchange income and thus put many local fishers out of business? Explain. What are the alternatives?

crab and salmon wars in the north Pacific—all warnings of serious declines in fish stocks from overfishing.

An even greater threat than overfishing is the possibility of global climate change over the next 50–100 years, which can warm ocean waters and enhance the effects of habitat degradation, pollution, and ultraviolet radiation (Sections 19-2 and 19-3). Because fish cannot regulate their internal temperature, they are particularly vulnerable to changes in the temperature of their water environments.

Is Aquaculture the Answer? Aquaculture, in which fish and shellfish are raised for food, supplies about 20% of the world's commercial fish harvest. Aquaculture production increased 3.3-fold between 1984 and 1996; by 2005 it may account for one-third of

Q: What factor contributed the most to the world's rapid population growth over the past 100 years?

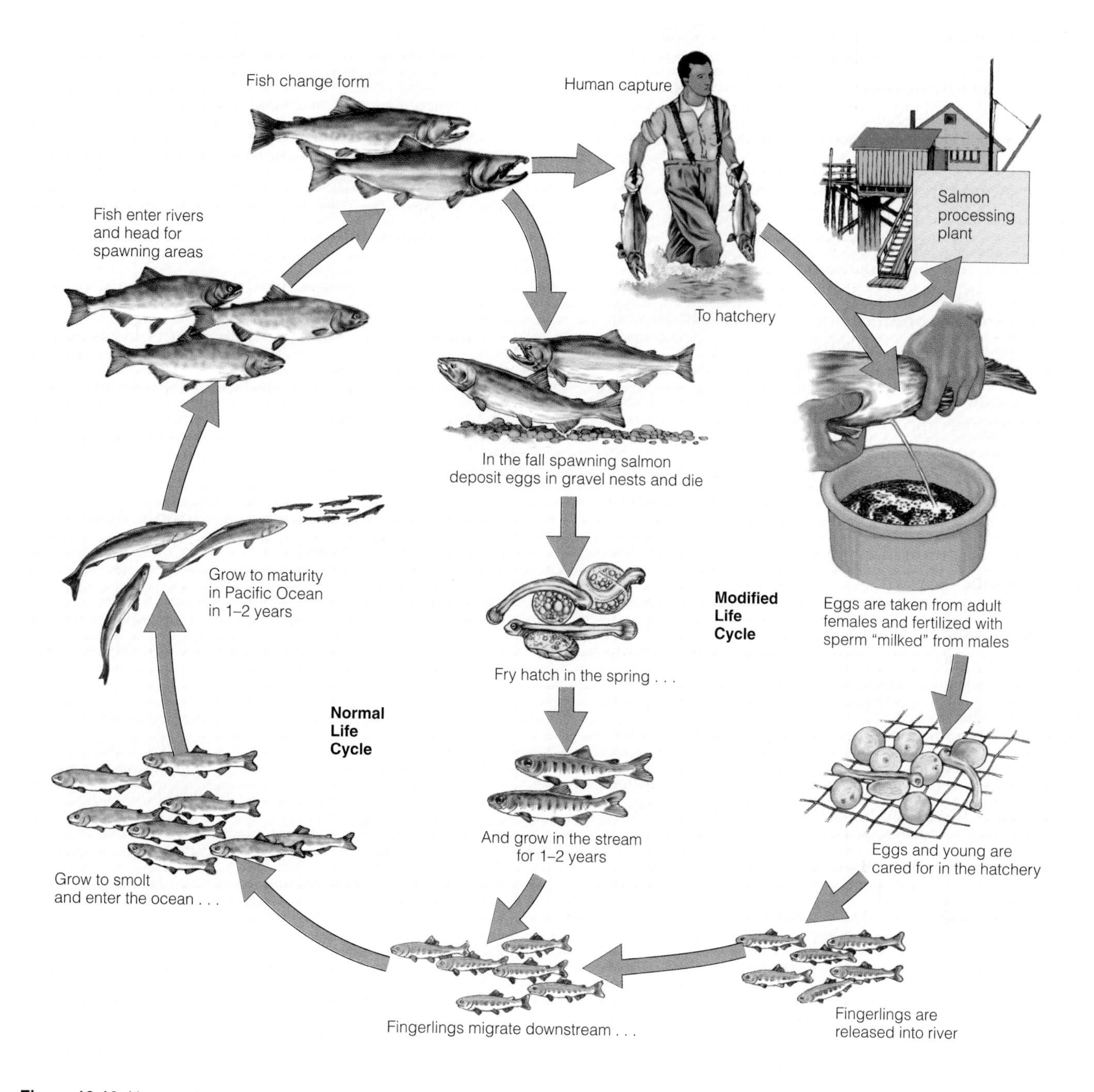

Figure 12-19 Normal life cycle of wild salmon (left) and human-modified life cycle of hatchery-raised salmon (right). Salmon spend part of their lives in fresh water and part in salt water.

the world's fish harvest. China is the world leader in aquaculture (producing almost half of the world's output), followed by India and Japan.

There are two basic types of aquaculture. **Fish farming** involves cultivating fish in a controlled environment, often a pond or tank, and harvesting them when they reach the desired size. **Fish ranching** involves holding anadromous species such as salmon (that live part of their lives in fresh water and part in salt water) in captivity for the first few years of their lives (usually in fenced-in areas or floating cages in coastal lagoons and estuaries), releasing them, and then harvesting the adults when they return to spawn (Figure 12-19).

Species cultivated in developing countries (mostly by inland aquaculture) include carp, tilapia, milkfish,

A: A decline in death rates

clams, and oysters, all of which feed on phytoplankton and other aquatic plants. In developed countries and some rapidly developing countries in Asia, aquaculture is used mostly to stock lakes and streams with game fish or to raise expensive fish and shellfish such as oysters, catfish, crayfish, rainbow trout, shrimp, and salmon. Aquaculture now produces 90% of all oysters, one-third of all salmon (75% in the United States), and one-quarter of the shrimp and prawns (50% in the U.S.) sold in the global marketplace.

Aquaculture has several advantages. It is highly efficient and can produce high yields in a small volume of water. Because little fuel is needed, yields and profits are not closely tied to the price of oil (as they are in commercial marine fishing). Crossbreeding and genetic engineering can help increase yields.

Some people in the aquaculture industry talk of a *blue revolution* that would do for fish farmers and fish ranchers what the green revolution did for grain growers. They project that freshwater and saltwater aquaculture production could double during the next 10 years.

There are problems, however. Fish farms are essentially *aquatic feedlots*, requiring large inputs of land, feed, water, and energy and producing large outputs of wastes. Large-scale aquaculture also requires considerable capital and scientific knowledge, which are in short supply in most developing countries. In addition, scooping out huge ponds for fish and shrimp farming in some developing countries has destroyed ecologically important mangrove forests (see photo in table of contents). Pesticide runoff from nearby croplands can kill fish in aquaculture ponds, and dense populations make fish more vulnerable to bacterial and viral infections. For example, all the major shrimp-farming countries (Thailand, China, Indonesia, India, and Ecuador) have experienced widespread epidemics in their fish ponds. Moreover, most shrimp farms in countries such as Thailand are not registered, lack wastewater treatment and sedimentation ponds, and rarely test waters where wastes are discharged for oxygen levels.

On average, an aquaculture fish pond lasts 5 years before it is too contaminated to use. As a result, shrimp aquaculture farmers buy and flood farms and villages, extract profits of 40–50% return on their investment, then move on, leaving barren, salty land that can't be used to raise shrimp, rice, or anything else.

Chemicals used to keep nets and cages free of unwanted marine life can be toxic to nearby free-ranging marine animals. Escaped farm-raised fish may also breed with wild fish and degrade the genetic stock of such species. Moreover, without adequate pollution control, waste outputs from shrimp farming and other large-scale aquaculture operations can contaminate nearby estuaries, surface water, and groundwater and eliminate some native aquatic species. A typical salmon farm with 75,000 fish produces as much organic waste as a city of 20,000 people.

Critics warn that with current feeding practices, fish farming is not a long-term way out of overfishing because most farmed fish are fed fish meal made from unpopular fish such as menhaden or herring. As the fishing industry scoops up more and more of the world's fish species at increasingly lower levels on food webs, species available for fish meal may also be depleted.

Fish farmers also see their profits being eaten up by herons, terns, cranes, and other predators attracted to this concentrated and easily eaten source of food. They respond with poisons and shotguns, which reduce populations of fish-eating birds and other predators.

Can We Develop a More Sustainable Approach to Fishery Management? There is widespread agreement that our current use of the world's potentially renewable fisheries is unsustainable in the long run (Connections, p. 300). General guidelines for more sustainable management of global fisheries include the following:

- *Shifting the burden of proof to the fishing industry by requiring those who profit from harvesting publicly owned resources in territorial waters and the open ocean to show that their harvests are sustainable before allowing them to operate.*
- *Setting and strictly monitoring and enforcing conservative quotas for fisheries that are set well below their estimated maximum sustainable yields.*
- *Establishing and dividing up fishing quotas based on fairness, local needs and conditions, the best available scientific data, and inputs from local communities and fishers.*
- *Strengthening commitment to protection of marine biodiversity and integrated coastal management programs that promote both sustainable fishing and the ecological health of marine ecosystems (Guest Essay, p. 308).*
- *Sharply reducing or eliminating fishing subsidies to shrink the size of global fishing fleets, encourage free-market competition, sustain small-scale local fishers, reduce overfishing, and allow economically and biologically depleted stocks to recover.*

Ways to manage fisheries to reduce overfishing are discussed in more detail in Section 25-6.

It is doubtful that such principles will be put into practice without a public outcry by citizens demanding that policymakers implement such policies. The most important tool consumers have for more

Q: Globally, how many environmental refugees are there?

sustainable management of fisheries is to vote with their buying power by asking food suppliers where fish came from and how it was harvested and by buying only fish products that have been certified to have been produced sustainably and whose stocks have not been depleted.

12-5 AGRICULTURAL POLICY, FOOD AID, AND LAND REFORM

How Do Government Agricultural Policies Affect Food Production? Agriculture is a financially risky business. Whether farmers have a good year or a bad year is determined by factors over which they have little control: weather, crop prices, crop pests and diseases, interest rates, and the global market. Because of the need for reliable food supplies despite fluctuations in these factors, most governments provide various forms of assistance to farmers and consumers.

One approach is to *keep food prices artificially low.* This makes consumers happy but means that farmers may not be able to make a living. Many governments in developing countries keep food prices in cities lower than in the countryside to prevent political unrest. With food prices lower in the cities, more rural people migrate to urban areas, aggravating urban problems and unemployment and increasing the chances of political unrest—another example of harmful positive feedback in action.

A second approach is to *give farmers subsidies to keep them in business and to encourage them to increase food production.* In developed countries government price supports and other subsidies for agriculture total more than $300 billion per year (including $100 billion per year in the United States). If government subsidies are too generous and the weather is good, farmers may produce more food than can be sold; food prices and profits then drop because of the surplus. Large amounts of food then become available for export or food aid to developing countries, depressing world food prices; the low prices reduce the financial incentive for farmers in developing countries to increase domestic food production. Moreover, the taxes citizens in developed countries pay to provide agricultural subsidies more than offset the lower food prices they enjoy.

Government agricultural subsidies in developing countries have also had perverse effects because of positive feedback. In India, for example, the government gave farmers free electricity, but this encouraged them to overpump irrigation water, which led to salinization and waterlogging of soils and in some cases to depletion of aquifers. Government subsidies by the Brazilian government to ranchers and the landless poor have encouraged destruction of rain forests and soil degradation, without significantly increasing food production (Section 24-3).

A third policy is to *eliminate most or all price controls and subsidies,* allowing market competition to be the primary factor determining food prices and thus the amount of food produced. Some analysts call for phasing out all government price controls and subsidies over, say, 5–10 years and letting farmers respond to market demand. However, these analysts urge that any phaseout of farm subsidies, which in effect subsidize manufacturers of farm chemicals and machinery, should be coupled with increased aid for the poor and the lower middle class, who would suffer the most from any increase in food prices.

Many environmentalists believe that instead of eliminating all subsidies, they should be used only to reward farmers and ranchers who protect the soil, conserve water, reforest degraded land, protect and restore wetlands, and conserve wildlife.

Another way that governments and private organizations deal with a lack of food production and hunger is through food aid (Pro/Con, p. 304).

How Do We Ensure That the Poor Benefit? Increasing per capita food production is a big task, but making sure the food reaches hungry people is an even bigger one. *Food security*—ensuring that all people have the physical and economic access to enough food to survive, function, and work—differs from food production.

A basic problem is that most poor farmers don't have enough land, money, or credit to buy the seed, fertilizer, irrigation water, pesticides, equipment, and fuel that the new plant varieties require. As a result, the second green revolution (Figure 12-4) has bypassed more than 1 billion poor people in developing countries. In addition, farmers switching to green- or gene-revolution farming greatly increase their debt load and financial risk and thus can lose what little land they have.

Current forms of the green revolution also displace many poor subsistence farmers from their land, and mechanization reduces the need for landless farm workers. This increases the migration of the rural poor to already overburdened cities—another example of connections and harmful positive feedback in action.

How Important Is Land Distribution Reform? Some analysts believe that land distribution reform is an important factor in reducing world hunger, malnutrition, poverty, and environmental degradation. Such reform usually involves giving the landless rural poor in developing countries either ownership or free use of enough land to produce their own food and, ideally, enough surplus to provide some income.

A: About 25 million (projected to reach 50 million by 2010)

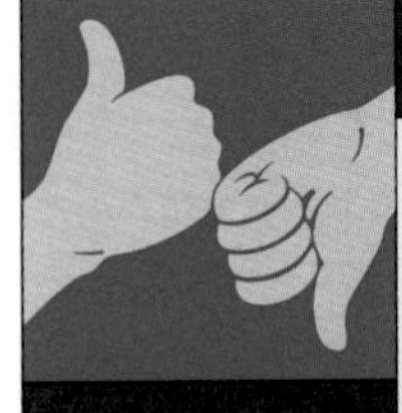

PRO/CON

How Useful Is International Food Aid?

Most people view international food aid as a humanitarian effort to prevent people from dying prematurely. However, some analysts contend that giving food to starving people in countries with high population growth rates does more harm than good in the long run. By not helping people grow their own food, they argue, food relief can condemn even greater numbers to premature death from starvation and disease in the future—another example of a harmful positive feedback process.

Biologist Garrett Hardin (Guest Essay, p. 266) has suggested that we use the concept of *lifeboat ethics* to decide which countries get food relief. His basic premise is that there are already too many people in the lifeboat we call earth. Thus, if food relief is given to countries that are not reducing their populations, the effect is to add more people to an already overcrowded lifeboat. Sooner or later the overloaded boat will sink and most of the passengers will drown.

Large amounts of food relief can also depress local food prices, decrease food production, and stimulate mass migration from farms to already overburdened cities. In addition, food relief discourages local and national governments from investing in the rural agricultural development needed to enable farmers to grow enough food for the population.

Another problem is that much food relief does not reach hunger victims. Transportation networks and storage facilities are often inadequate, so some of the food rots or is devoured by pests before it can reach the hungry. Moreover, officials often steal some of the food and sell it for personal profit; usually some must be given to officials as bribes for approving the unloading and transporting of the remaining food to the hungry.

Providing food and other forms of aid during famines can also be risky and costly. For example, supplying starving people in Somalia in 1992 required a UN peacekeeping force that probably cost at least 10 times more than the food that was distributed. This illustrates the strong linkages between environmental security and economic and military security (Guest Essay, p. 530).

Most critics are not against providing aid, but they believe that such aid should help countries control population growth and grow enough food to feed their populations by using sustainable agricultural methods (Section 12-6). Temporary food relief, they believe, should be given only when there is a complete breakdown of an area's food supply because of natural disaster.

Current food aid is enough to meet the minimum calories for only about one-fourth of the world's malnourished people and such needs are projected to double by 2010.

Critical Thinking

Is sending food to famine victims helpful or harmful in the short run? in the long run? Explain. Are there any conditions you would attach to providing such aid? Explain.

Proponents argue that such reform would increase agricultural productivity in developing countries and reduce the need to farm and degrade marginal land. It would also help reduce migration of poor people to overcrowded urban areas by creating employment in rural areas.

The UN projects that 1.24 billion people will be landless or near-landless by the year 2000. The world's most unequal land distribution is in Latin America, where 7% of the population owns 93% of the farmland. Most of this land is either used to grow crops for export (such as sugar, tea, coffee, bananas, or beef) or left idle on huge estates. In 1998, hundreds of landless poor people in Brazil occupied government buildings in 10 state capitals to press their demands for land reform.

Land reform is difficult in countries where government leaders are heavily influenced by wealthy and powerful landowners. They strongly oppose land reform because it violates the private-property rights of those whose land is redistributed. Another problem is that giving peasants title to land they clear and farm in tropical forests leads to destruction and degradation of these important reservoirs of biodiversity.

How Important Is Agricultural Research? According to Dennis Avery, past agricultural research has saved perhaps 1 billion lives from famine, increased food calories by one-third for 4 billion people in developing countries, and kept millions of square kilometers of wild land from being cleared and plowed under.

Currently the world's estimated total investment in agricultural research is only $15 billion a year—a tiny fraction of the value of the goods and services provided by agriculture. The average annual rate of increase of funds for agricultural research fell from 4.4% between 1971 and 1981 to 2.8% between 1981 and 1991. In addition, very little of this research is spent on

Loftus, M., and M. Marcus. 1998. "Chefs and Grocers Sell More Organic Food." *U.S. News & World Report,* vol. 124, no. 19, 74(3).

sustainable agriculture and improving nutrition—both areas in need of greatly increased funding for research.

Many agricultural scientists believe that it is possible to provide enough food to sustain the projected population growth over the next three decades. However, they warn that this will happen only if the world supports a greatly increased budget for agricultural research and development, which is not taking place. According to environmental scientists, a significantly larger portion of the world's agricultural research must be devoted to developing more sustainable methods for growing more food and finding ways to meet everyone's basic nutritional needs.

12-6 SOLUTIONS: SUSTAINABLE AGRICULTURE

What Is Sustainable Agriculture? Many agricultural scientists and experts believe that a key to reducing world hunger, poverty, and the harmful environmental effects of both industrialized and subsistence agriculture (Figure 12-11) is to develop systems of **sustainable agriculture**, or **low-input agriculture**, and phase them in over the next three decades.

Low-input farming reduces waste of irrigation water and uses less pesticides and inorganic fertilizer (Spotlight, p. 306). To maintain and restore soil fertility, farmers rely on good soil conservation practices (Section 14-7) and use manure, compost, and other forms of organic matter. They also emphasize biological and physical methods for controlling pests and use chemical pesticides only as a last resort (and in the smallest amounts possible). A growing number of farmers are discovering that low-input farming is often more profitable than high-input farming because they spend less money on inputs of irrigation water, fertilizer, and pesticides.

Most proponents of more sustainable agriculture are not opposed to high-yield agriculture; indeed, they see it as vital for protecting the earth's biodiversity and ecological integrity by reducing the need to cultivate new and often marginal land. Instead, they believe that current research and economic incentives should be redirected to encourage increases in yield per hectare without depleting or degrading soil, water, and biodiversity.

General guidelines for sustainable agricultural systems suggested by various analysts include the following:

- *Combine traditional high-yield polyculture and modern monoculture methods for growing crops.*
- *Breed locally available crops that have higher contents of essential micronutrients to help reduce malnutrition.* The goal would be to develop new strains of grain and other foods that take up and retain more of the essential minerals (especially iron and zinc) from the soil.
- *Measure the success of food policies not in terms of food production but in terms of human nutrition and health.* The goal should be not to produce more food but to produce more better nourished, healthier, and productive people.
- *Grow more perennial crops* (p. 278).
- *Minimize soil erosion, salinization, and waterlogging.*
- *Reduce destruction of forests, grasslands, and wetlands for producing foods by emphasizing increased yields per area of cropland using sustainable methods.*
- *Stabilize aquifers by reducing the rate of water removal to the rate of recharge.*
- *Reduce water waste in irrigation* (Section 13-6).
- *Reduce overfishing by implementing the principles listed on p. 302.*
- *Reduce use and waste of fossil fuels* (Section 16-2) *and shift to an energy-efficient solar–hydrogen economy* (Section 16-7).
- *Reduce loss of agricultural land to car-based transportation systems and pollution by encouraging a shift to transportation systems based on rail, bus, electric scooters, bicycles, and walking* (Section 26-3).
- *Increase use of organic fertilizers, solar, wind, and biomass energy to grow and process crops.*
- *Emphasize biological pest control and integrated pest management* (Section 21-5).
- *Protect existing prime cropland from environmental degradation and conversion to urban or industrial uses* (Section 26-4).
- *Subsidize sustainable farming and phase out subsidies for unsustainable farming.*
- *Shift to full-cost pricing* (Section 27-3) *that includes the harmful environmental effects of agriculture* (Figure 12-11) *in food prices.*
- *Educate the public about the hidden environmental and health costs they are paying for food and the need to gradually incorporate these costs into market prices.*
- *Reduce food waste.* According to a 1997 USDA study, an estimated 27% of the food produced in the United States (not including crop losses) is wasted by grocery stores, restaurants, and consumers. This amounts to each American throwing out an average of 166 kilograms (365 pounds) of food per year—roughly equal to each person throwing away the equivalent of 1,000 quarter-pound hamburgers a year.
- *Greatly increase research on sustainable agriculture and improving human nutrition.*

Hint: Enter the search term *organic food* using the Subject Guide.

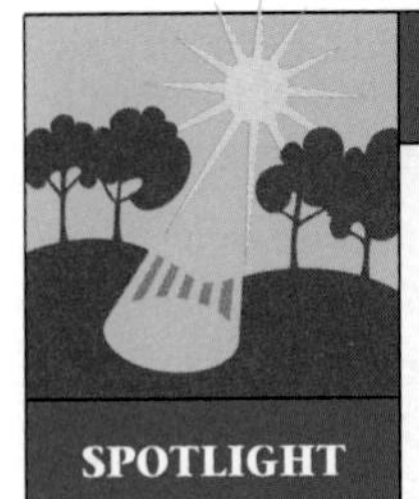

SPOTLIGHT

What Is Organic Food?

Organic food is normally defined as plant or animal food grown or produced without the use of synthetic fertilizers, synthetic pesticides, and antibiotics. For years the U.S. Department of Agriculture considered organic growers a fringe group and ignored them. So organic food growers developed their own standards for certifying food as being organic, set up inspection programs, and developed markets.

The results have been astonishing. In 1980 sales of certified organic food were $78 million. By 1998 annual sales had reached almost $4 billion and could rise to $10 billion over the next few years.

Today there are 11 state certification systems and 33 private ones. Such a diverse system creates problems and conflicts. In 1990 organic growers celebrated when the U.S. Congress passed the Organic Food Production Act, designed to come up with uniform national standards for defining what *organic* means and regulating sales.

A National Organic Standards Board made up of organic farmers and sellers, ecologists, and representatives from consumer groups proposed national certification rules to the U.S. Department of Agriculture. This agency then passed the recommendations on to the Office of Management and Budget (OMB), as is required for all regulations to ensure that they are cost-effective and fair.

In 1998, the board's recommendations came back with major changes. Organic farmers were especially alarmed about modified regulations that allow **(1)** genetically engineered plants and animals, **(2)** food fertilized with municipal sewage sludge (which can be contaminated with pesticides, other harmful organic chemicals, and toxic metals such as lead and mercury), **(3)** food raised organically and then preserved by being zapped with radiation, and **(4)** food grown hydroponically in nutrient-rich water without soil to be certified as organic food.

When these proposed changes were announced, the U.S. Department of Agriculture was bombarded by over 200,000 angry letters, faxes, and e-mails of protests from organic farmers, consumers, and environmentalists. As a result, USDA officials announced that food that has been genetically engineered, fertilized with municipal sludge, or zapped with radiation cannot be labeled as organic food.

Critical Thinking

Do you agree or disagree with the four revised standards for classifying food as organic? Explain.

- *Set up demonstration projects throughout each country so that farmers can see how sustainable agricultural systems work.*
- *Establish training programs in sustainable agriculture for farmers and government agricultural officials and encourage the creation of college curricula in sustainable agriculture and human nutrition.*
- *Reduce poverty so that the poor can have enough money or land to supply their basic food needs.*
- *Provide poor women with access to credit, markets, and technical food-growing advice, as well as education and health care.*
- *Educate people about the right combination of foods to eat for good nutrition and how to prepare foods to reduce losses of essential vitamins and minerals.*
- *Slow population growth to help all of the world's countries reach the more sustainable postindustrial stage of the demographic transition* (Figure 11-20). According to the World Bank and the United Nations, countries that have raised their crop yields the fastest have generally brought their birth rates per woman down the fastest.
- *Integrate agriculture, population, urban and rural, energy, health, nutrition, climate, water resource, soil resource, land use, pollution, and biodiversity protection policies.*

Can We Make the Transition to Sustainable Agriculture? A growing number agricultural analysts believe that over the next 30 years we must make a transition from unsustainable and environmentally harmful (Figure 12-11) agriculture to more sustainable forms of agriculture.

In developed countries, including the United States, even a partial shift to more environmentally sustainable food production will not be easy. It will be opposed by agribusiness, by successful farmers with large investments in unsustainable forms of industrialized agriculture, and by specialized farmers unwilling to learn the demanding art of farming sustainably. It might also be resisted by many consumers unwilling or unable to pay higher prices for food, because full-cost accounting would include agriculture's harmful environmental and health costs in the market prices of food.

Despite such difficulties, many environmentalists believe that a new *eco-agricultural revolution* could take place throughout most of the world over the next 30 years. Whether it does occur is primarily a political and ethical issue. Some actions you can take to help promote sustainable agriculture are listed in Appendix 5.

Q: What developed country has the highest rate of teenage pregnancy?

The need to bring birthrates well below death rates, increase food production while protecting the environment, and distribute food to all who need it is the greatest challenge our species has ever faced.

PAUL AND ANNE EHRLICH

CRITICAL THINKING

1. What are the biggest advantages and disadvantages of **(a)** labor-intensive subsistence agriculture, **(b)** energy-intensive industrialized agriculture, and **(c)** sustainable agriculture?

2. Summarize the advantages and limitations of each of the following proposals for increasing world food supplies and reducing hunger over the next 30 years: **(a)** cultivating more land by clearing tropical forests and irrigating arid lands, **(b)** catching more fish in the open sea, **(c)** producing more fish and shellfish with aquaculture, and **(d)** increasing the yield per area of cropland.

3. What are the three most important things that you believe should be done to reduce hunger in the country where you live? In the world?

4. Some people argue that starving people could get enough food by eating nonconventional plants and insects; others point out that most starving people don't know what plants and insects are safe to eat and can't take a chance on experimenting when even the slightest illness could kill them. If you had no money to grow or buy food, would you collect and eat protein-rich grasshoppers, moths, beetles, or other insects?

5. If the demand for imported food increases as projected, few African countries will be able to compete with more affluent Asian countries in a bidding contest for grain. Africa uses grain more efficiently by using it mostly for direct human consumption. Asians feed much of their food imports to livestock to supply a growing demand for meat and meat products as their affluence increases.

- **a.** Does Africa have a greater moral claim on surplus food than Asia because of its more efficient use of grain? Explain.
- **b.** Does the United States, with one of the world's highest levels of meat consumption, have a moral responsibility to cut its grain consumption to make more grain available for export to countries whose people get most of their food by direct consumption of grain? Explain.

6. Should all price supports and other government subsidies paid to farmers be eliminated? Explain. Try to consult one or more farmers in answering this question.

7. Should governments phase in agricultural tax breaks and subsidies to encourage farmers to switch to more sustainable farming? Explain. At the same time, should governments phase in higher taxes and reduce subsidies to discourage farmers from using unsustainable, earth-degrading forms of farming, and then use the resulting revenue to encourage earth-sustaining farming? Explain.

8. Try to come up with analogies to complete the following statements: **(a)** Using polyculture instead of monoculture to produce food is like _______. **(b)** People who don't eat meat are like _______. **(c)** Growing more food by clearing tropical forests is like _______. **(d)** Overfishing is like _______. **(e)** Making a transition to sustainable agriculture is like _______. **(f)** Aquaculture is like _______.

PROJECTS

1. If possible, visit both a conventional industrialized farm and an organic or low-input farm. Compare soil erosion and other forms of land degradation, use and costs of energy, use and costs of pesticides and inorganic fertilizer, use and costs of natural pest control and organic fertilizer, yields per hectare for the same crops, and overall profit per hectare for the same crops.

2. Use health and other local government records to determine how many people in your community suffer from undernutrition or malnutrition. Has this problem increased or decreased since 1980? What are the basic causes of this hunger problem, and what is being done to alleviate it? Share the results of your study with local officials and then present your own plan for improving efforts to reduce hunger in your community.

3. Gather information from your local planning office to determine how much cropland in your area has been lost to urbanization since 1980. What policies, if any, do your state and local community have for promoting preservation of cropland?

4. Try to gather data showing the harmful environmental effects of nearby agriculture on your local community. What things are being done to reduce these effects?

5. Make a survey in the nearest urban area to estimate what percentage of the food is grown by urban dwellers. Survey unused land and use it to estimate how much it could contribute to urban food production. Use these data to draw up a plan for increasing urban food production and present it to city officials.

6. Make a concept map of this chapter's major ideas, using the section heads and subheads and the key terms (in boldface). Look at the inside back cover and in Appendix 4 for information about concept maps.

A: United States

Mazunte: A Farming and Fishing Ecological Reserve in Mexico

GUEST ESSAY

Alberto Ruz Buenfil

Alberto Ruz Buenfil, an international environmental activist, writer, and performer, is the founder of Huehuecoytl, a land-based ecovillage in the mountains of Mexico. His articles on ecology and alternative living have appeared in publications in the United States, Canada, Mexico, Japan, and Europe. His book, Rainbow Nation Without Borders: Toward an Ecotopian Millennium *(New Mexico: Bear, 1991), has been published in English, Italian, and Spanish.*

The world's eight known species of sea turtles are all officially listed as endangered or threatened. Seven of these species nest on Mexico's Pacific and Atlantic coastlines, making Mexico the world's most important turtle nesting country. The Pacific coasts of southern Mexico, especially the shores of the state of Oaxaca, are the main sites for turtle nesting, reproduction, and conservation; they also contain some of Mexico's last reserves of wetlands.

Only recently have we begun recognizing the ecological values of wetlands, deltas, and coastal ecosystems [Section 8-2]. They provide habitats for a rich diversity of wildlife, maintain water supplies, protect shorelines from erosion, and play a role in regulating global climate.

Swamps, marshes, and bogs were once considered wastelands to be drained, filled, and turned into "productive land," especially for developing and constructing urban and tourist areas. During the past few decades, the coasts of Oaxaca have not escaped such exploitation, especially after business interests discovered the beaches of Mazunte, Zipolite, San Agustinilo, Puerto Angel, and Puerto Escondido, and the magnificent bays of Hautulco.

The villages of Mazunte and San Agustinilo were founded in the late 1960s, basically to bring in cheap labor from neighboring indigenous villages to provide workers for a slaughterhouse making products from various species of turtles nesting in the area. Members of nearly 200 indigenous farming families became fishermen and employees of the new turtle meat factory, which was fully operational in the 1970s and 1980s. According to some of these workers, nearly 2,000 turtles were killed and quartered every day during those years. At night, dozens of poachers came to collect turtle eggs from their nests.

In the 1980s, this situation came to the attention of two of the first environmental organizations to speak up in defense of species, forests, and natural resources in Mexico. They began denouncing the massacre of the area's turtles, and they pointed out the danger that some of the species might be exterminated. With support from other international organizations, they campaigned for almost 10 years, until a 1990 decree by the president of Mexico made it illegal to exploit the turtles and led to the closing of San Agustinilo's turtle slaughterhouse.

The Mexican government provided some funds, boats, and freezers to compensate for the loss of jobs. However, only about 5% of the indigenous population benefited from this compensation. Since then, most of the people have been living on the verge of starvation, and some have illegally killed protected turtles to survive. What had seemed to be an important environmental victory turned into a nightmare for a large population of indigenous people. Understandably, these people had no use for ecologists or environmentalists.

In 1990, a group called ECOSOLAR A.C. began efforts to change this situation by implementing a plan for sustainable development of the coast of Oaxaca. They were successful in obtaining funding for this project from different national and international institutions. By 1992, the members of this small but effective group had succeeded in

- Making a detailed study of the bioregion, which, with the participation of the local people, is being used to define the possible uses of different areas

Q: What percentage of the world's population is under age 15?

- Creating a system of credits to help native inhabitants build better houses, establish small family-run restaurants, and manufacture hammocks for rent or sale to visitors
- Beginning the construction of systems for drainage, water collection, and latrines using low-impact technology and local materials and workers, as well as nurseries for local seeds and facilities for reforestation, and wildlife preservation projects
- Working with the community to promote Mazunte as a center for ecotourism—a place where visitors can experience unique ecosystems containing alligators, turtles, and hundreds of species of birds and fishes

In only 2 years, the native inhabitants of Mazunte and other neighboring communities completely changed their opinion about and perspective on ecology and environmentalists. In May 1992, Mazunte hosted the second annual gathering of "Earth Keepers," involving nearly 150 representatives of 35 organizations from 20 different countries.

For 1 week, these specialists shared their practical knowledge with the local people. They used their health skills to help the community set up an alternative clinic for healing arts, their skills in permaculture (a form of sustainable agriculture) and organic agriculture to improve local hatcheries and home-based plant nurseries, and their skills in ecotechnology to build both biodigestors for producing natural fertilizers from biomass and a village recycling center located at the school. In addition, artistic and cultural activities took place every night in the Center for Biological Investigations, which the people of Mazunte want to turn into Mexico's first Marine Turtles Museum. Run by local people, it will attract and educate visitors from around the world.

However, despite the legal protection of sea turtles in Mexico, illegal killing of turtles and removal of their eggs continues. A few days after the event concluded, the village of Mazunte called a general meeting attended by 150 heads of families to discuss ways to get community members to protect turtle nesting areas instead of illegally killing the turtles for food. Out of that meeting came a "Declaration of Mazunte" requesting that competent higher authorities and the president of Mexico put an immediate end to such destruction, which violates the earlier presidential decree forbidding the annihilation of turtles in Mexico.

The community went on to declare their village and neighboring environments to be Mexico's first *Farming and Fishing Reserve.* Its goals would be to protect the area's forests, water sources, wetlands, wildlife, shores, beaches, and scenic places, and to "establish new forms of relationship between humans and nature, for the well-being of today's and tomorrow's generations." This declaration has been presented to the government of Mexico and to many national and international organizations.

Mazunte is taking the lead in demonstrating that cooperation between local people and environmental experts can lead to ecologically sustainable communities that benefit local people and wildlife alike. This model can show farmers and indigenous communities everywhere how they can live sustainably on Earth and turn things around in a short time. It is a message of hope and empowerment for people seeking a better world for themselves and others.

Critical Thinking

1. What lessons that could be applied to your own life have you learned from this essay?

2. Could the rapid change toward sustainability brought about by environmentalists and local people in Mazunte be accomplished in your own community? If so, how? If not, why not?

A: 32% (35% in developing countries, 22% in the United States, and 19% in developed countries) in 1998

13 WATER

Water Wars in the Middle East

Because of differences in climate, some parts of the earth have an abundance of water, whereas other parts have a shortage. As population, agriculture, and industrialization grow, there is increasing competition for water, especially in dry regions such as the Middle East.

The next war in the Middle East may well be fought over water, not oil. Most water in this dry region comes from three shared river basins: the Jordan, the Tigris–Euphrates, and the Nile (Figure 13-1). Water in much of this arid region is already in short supply, and the human population in this region is projected to double within only 25 years.

Disputes among water-short Ethiopia, Sudan, and Egypt over access to the water from the Nile River basin are escalating rapidly. Ethiopia, which controls the headwaters that feed 86% of the Nile's flow, plans to divert more of this water; so does Sudan. This could reduce the amount of water available to desperately water-short Egypt, whose terrain is desert except for a thin green strip of irrigated cropland running down its middle along the Nile and its delta. About 95% of Egypt's 66 million people are jammed together in this thin stripe that makes up only 5% of the country's land area.

Figure 13-1 The Middle East, whose countries have some of the highest population growth rates in the world. Because of the dry climate, food production depends heavily on irrigation. Existing conflicts among countries in this region over access to water may soon overshadow both long-standing religious and ethnic clashes and attempts to take over valuable oil supplies.

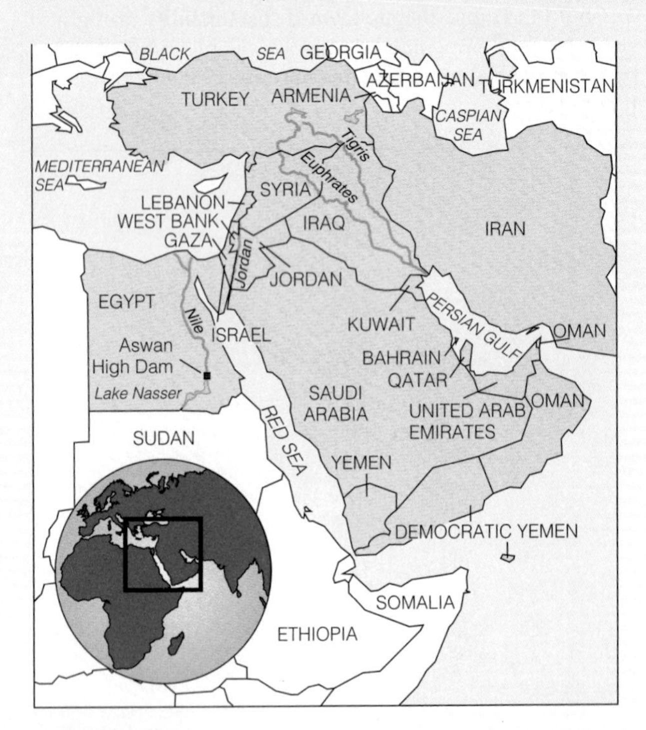

Between 1998 and 2025, Egypt's population (which is growing by 1 million every 9 months) is expected to increase from 66 million to 96 million, greatly increasing the demand for already scarce water. Egypt's options are to **(1)** go to war with Sudan and Ethiopia to obtain more water, **(2)** slash population growth, **(3)** save water by improving irrigation efficiency, repairing leaking pipes and lining irrigation ditches, reducing the growth of water-intensive crops such as rice, cotton, and sugar, and reducing water subsidies that discourage water conservation, **(4)** spend $2 billion to build the world's longest concrete canal and a massive pumping station to pump water out of Lake Nasser (the reservoir created from the Nile by the Aswan High Dam) and create a lush new valley of irrigated farmland in the middle of the desert, **(5)** import more grain to reduce the need for irrigation water, **(6)** work out water-sharing agreements with other countries, or **(7)** suffer the harsh human and economic consequences.

There is also fierce competition for water among Jordan, Syria, and Israel, which get most of their water from the Jordan River basin (Figure 13-1). Israel irrigates two-thirds of its croplands and uses water more efficiently than any other country in the world. However, within the next few years its supply is projected to fall up to 30% short of demand because of increased immigration.

Jordan, which gets about 75% of its water from the Jordan River system, must double its supply over the next 20 years just to keep up with projected population growth. In 1990, King Hussein declared that water was the only issue that could cause him to go to war against Israel.

Syria, which expects water shortages by the year 2000, plans to build a series of dams and to withdraw more water from the Jordan River to supplement what it gets from the Euphrates River. This will decrease the downstream water supply for Jordan and Israel. Israel warns that if the largest proposed dam is built, it will consider destroying it.

Some 90 million people currently live in the water-short basins of the Tigris and Euphrates rivers (Figure 13-1), and by the year 2020 the population there is projected to almost double to 170 million. Turkey, located at the headwaters of these two rivers, has abundant water, and it plans to build 22 dams along the upper Tigris and Euphrates to generate huge quantities of electricity and irrigate a large area of land. These dams will reduce the flow of water to downstream Syria and Iraq.

Syria, which gets 90% of its water from the Euphrates, plans to build a huge dam that could eventually divert half of the water arriving from Turkey, thereby leaving little water for Iraq. Such a cutoff of its water supply by Turkey and Syria is a severe threat to Iraq and could cause it to go to war.

Turkey hopes to become the region's water superpower. It has proposed building two pipelines to transport and sell water to parched Saudi Arabia and Kuwait, and perhaps to Syria, Israel, and Jordan as well. However, the downstream Arab nations (and Israel) are reluctant to place their water security in Turkey's hands or rely on pipelines that could easily be sabotaged.

Clearly, distribution of water will be a key issue in any future peace talks in this region. Resolving these problems will require a combination of regional cooperation in allocating water supplies, slowed population growth, improved efficiency in the use of water, and eliminating water subsidies and raising the price of water to encourage conservation and improve irrigation efficiency. Mideast nations could reduce their need for water to irrigate crops by importing more food and banding together to use their influence in the marketplace to reduce prices.

By the middle of the next century, almost twice as many people will be trying to share the same amount of fresh water the earth has today. Already, 1.2 billion people lack access to clean drinking water, 2.2 billion live without sewage systems, and two-thirds of the world's households don't have running water. As fresh water becomes scarcer, access to water resources will be a major factor in determining the economic, environmental, and military security of a growing number of countries around the globe (Guest Essay, p. 530).

Water, the lifeblood of the ecosphere, is truly a wondrous substance; it connects us to one another, to other forms of life, and to the entire planet. Despite its importance and unique properties (Spotlight, p. 113), water is one of our most poorly managed resources. We waste it and pollute it, and we also charge too little for making it available, thus encouraging greater waste and pollution of this potentially renewable resource for which there is no substitute.

Our liquid planet glows like a soft blue sapphire in the hard-edged darkness of space. There is nothing else like it in the solar system. It is because of water.

JOHN TODD

In this chapter we answer the following questions:

- How much fresh water is available to us, and how much of it are we using?
- What areas face water shortages, and what can be done about this problem?
- What are the pros and cons of using dams and reservoirs to supply more water?
- What are the pros and cons of transferring large amounts of water from one place to another?
- How can we waste less water?
- How can a large water basin be managed to provide water and protect aquatic life?
- What areas suffer from flooding, and what can be done to reduce the risk of flooding and flood damage?
- How can we use the earth's water more sustainably?

13-1 SUPPLY, RENEWAL, AND USE OF WATER RESOURCES

How Much Fresh Water Is Available? Only a tiny fraction of the planet's abundant water is available to us as fresh water. About 97% by volume is found in the oceans and is too salty for drinking, irrigation, or industry (except as a coolant).

The remaining 3% is fresh water. About 2.997% of it is locked up in ice caps or glaciers or is buried so deep that it costs too much to extract. Only about 0.003% of the earth's total volume of water is easily available to us as soil moisture, usable groundwater, water vapor, and lakes and streams. If the world's water supply were only 100 liters (26 gallons), our usable supply of fresh water would be only about 0.003 liter (one-half teaspoon) (Figure 13-2).

Fortunately, the available fresh water amounts to a generous supply that is continuously collected, purified, recycled, and distributed in the solar-powered *hydrologic cycle* (Figure 5-4) as long as we don't overload it with slowly degradable and nondegradable

Figure 13-2 The planet's water budget. Only a tiny fraction by volume of the world's water supply is fresh water that is available for human use.

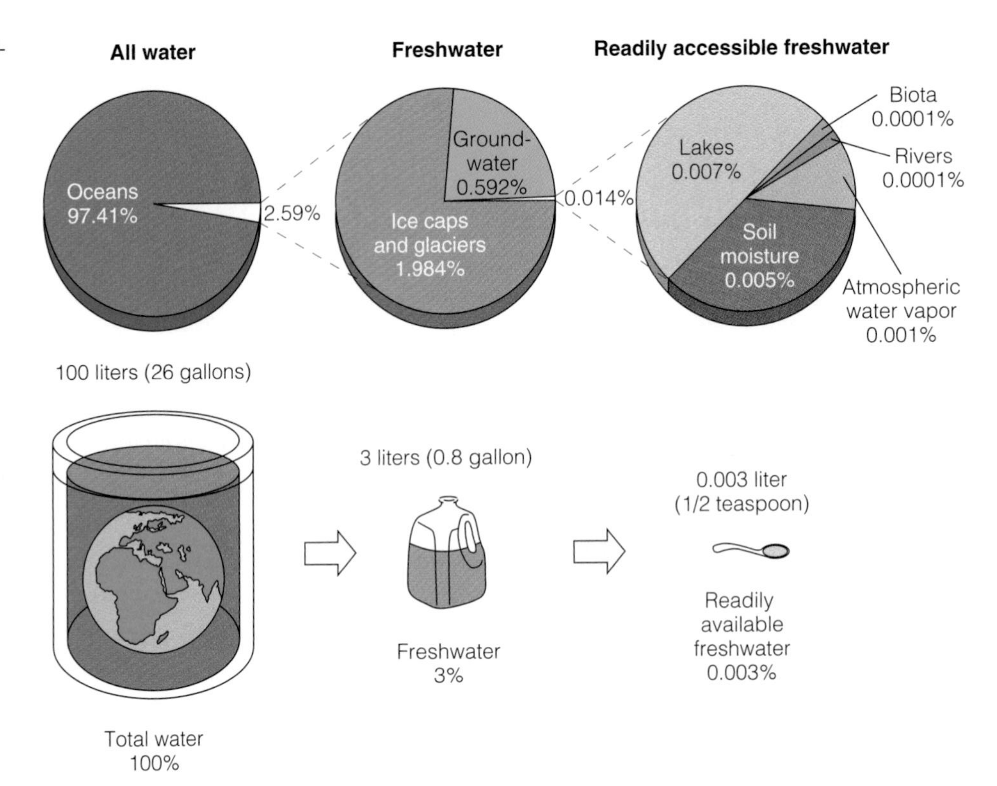

wastes or withdraw it from underground supplies faster than it is replenished. Unfortunately, we are doing both.

Differences in average annual precipitation divide the world's countries and people into water haves and have-nots. For example, Canada, with only 0.5% of the world's population, has 20% of the world's fresh water supply. By contrast, China, with 21% of the world's people, has only 7% of the world's fresh water supply. Already more than 300 Chinese cities are short of water, and 100 of them are very short. According to the administrator of China's Environmental Protection Agency, the country's water supplies are capable of sustainably supporting only about half of the country's current population of 1.2 billion.

As population, irrigation, and industrialization increase, water shortages in already water-short regions will intensify and wars over water may erupt (p. 310). Projected global warming also might cause changes in rainfall patterns and disrupt water supplies in unpredictable ways (Section 19-3).

What Is Surface Water? The fresh water we use first arrives as the result of precipitation (Figure 13-3). Precipitation that does not infiltrate the ground or return to the atmosphere by evaporation (including transpiration) is called **surface runoff** that flows into streams, lakes, wetlands, and reservoirs.

A **watershed**, also called a **drainage basin**, is a region from which water drains into a stream, stream system, lake, reservoir, or other water body.

What Is Groundwater? Some precipitation infiltrates the ground and percolates downward through voids (pores, fractures, crevices, and other spaces) in soil and rock (Figure 13-4). The water in these voids is called **groundwater**.

Close to the surface, the voids have little moisture in them. However, below some depth, in what is called the **zone of saturation**, they are filled with water except for an occasional air bubble. The surface of the zone of saturation, at the boundary with the unsaturated zone above, is the **water table**. The water table falls in dry weather and rises in wet weather.

Porous, water-saturated layers of sand, gravel, or bedrock through which groundwater flows are called **aquifers** (Figure 13-4). Any area of land through which water passes downward or laterally into an aquifer is called a **recharge area**. Aquifers are replenished naturally by precipitation that percolates downward through soil and rock in what is called **natural recharge**, but some are recharged from the side by *lateral recharge*.

Groundwater moves from the recharge area through an aquifer and out to a discharge area (well, spring, lake, geyser, stream, or ocean) as part of the hydrologic cycle. Groundwater normally moves from points of high elevation and pressure to points of lower elevation and pressure. This movement is quite slow, typically only a meter or so (about 3 feet) per year and rarely more than 0.3 meter (1 foot) per day.

Some aquifers get very little (if any) recharge. Often found fairly deep underground and formed tens

Q: What is the baby-boom generation in the United States?

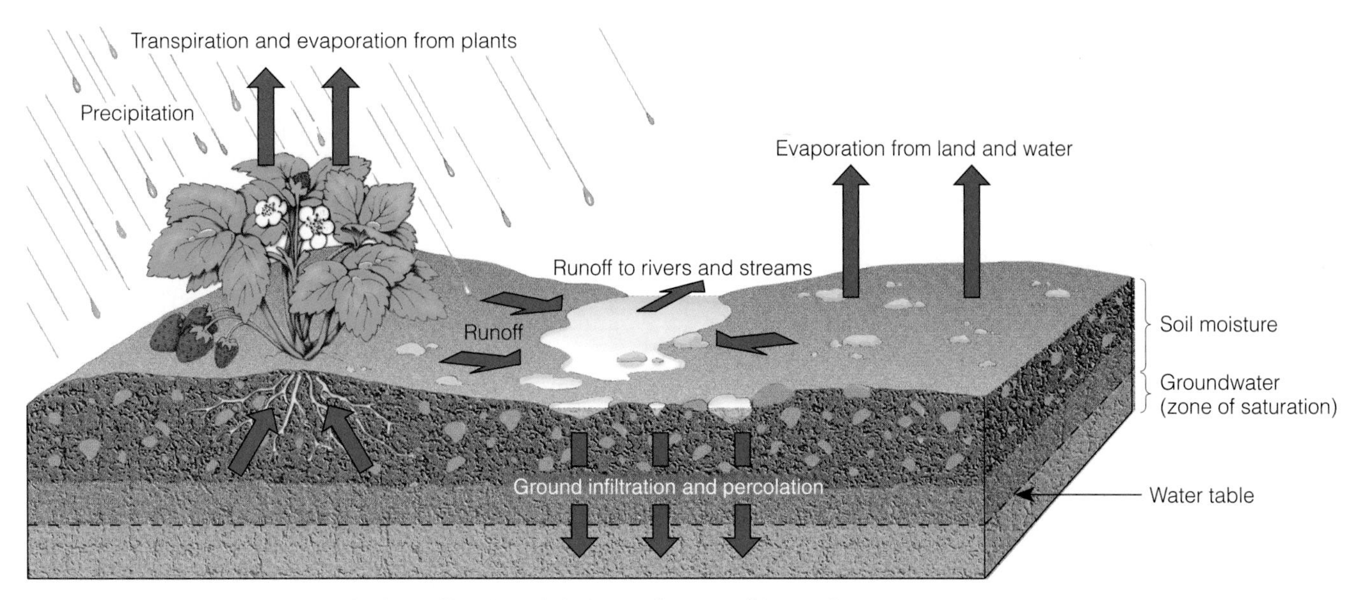

Figure 13-3 Main routes and destinations of local precipitation: surface runoff into surface waters, ground infiltration, and evaporation and transpiration into the atmosphere.

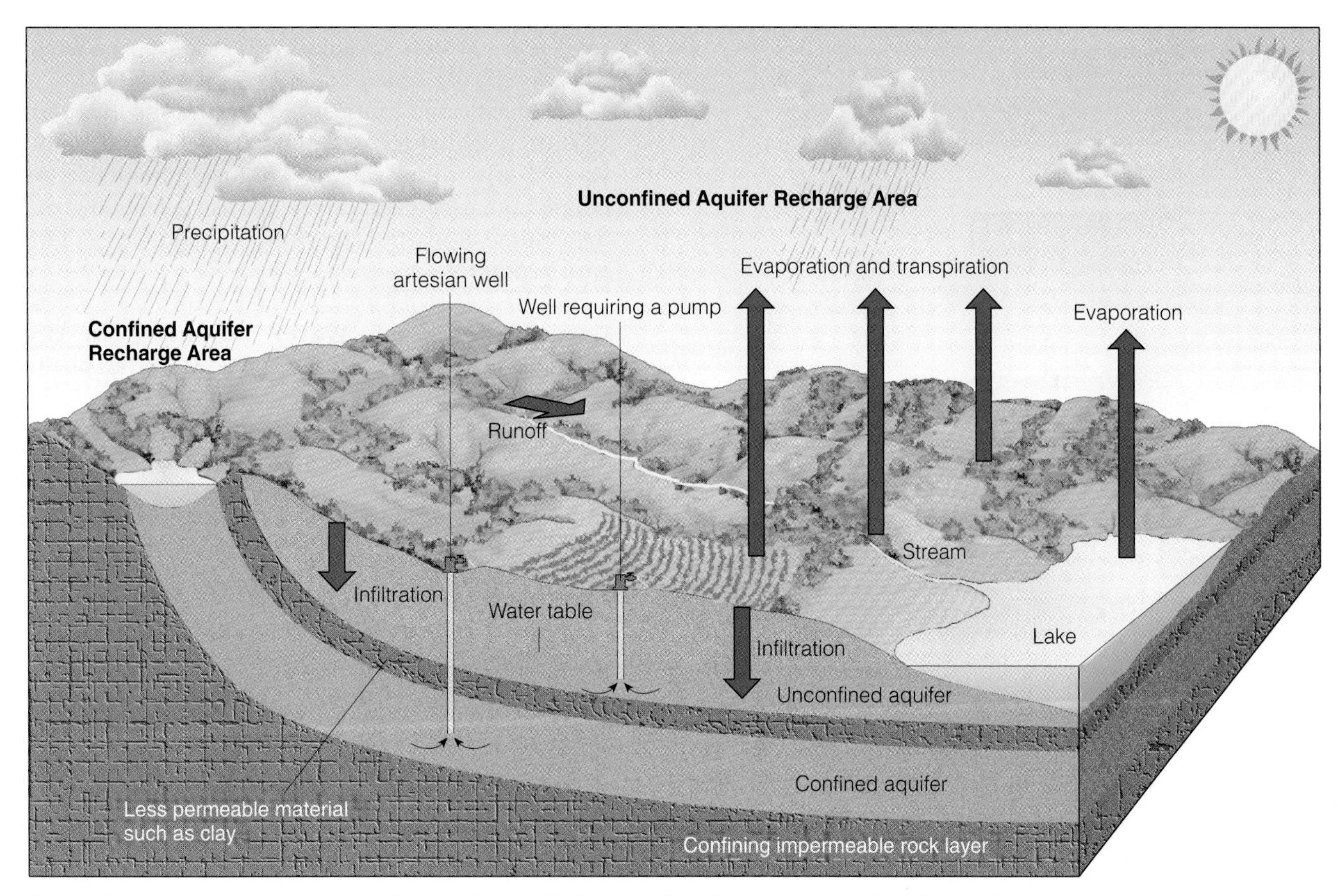

Figure 13-4 The groundwater system. An *unconfined aquifer* is an aquifer with a water table. A *confined aquifer* is bounded above and below by less permeable beds of rock. Groundwater in this type of aquifer is confined under pressure.

A: The 78 million people born between 1946 and 1964, when total fertility rates were high of thousands of years ago, they are (on a human time scale) nonrenewable resources. Withdrawals from such aquifers amount to *water mining* that, if kept up, will deplete these ancient deposits of liquid earth capital.

How Do We Use the World's Fresh Water? Since 1950, the global rate of water withdrawal from surface and groundwater sources has increased almost fivefold and per capita use has tripled. According to a 1996 study, humans currently use about 54% of the global surface runoff that is realistically available from the hydrologic cycle. Because of increased population growth and economic development, global withdrawal rates of surface water are projected to at least double in the next two decades and exceed the available surface runoff in a growing number of areas.

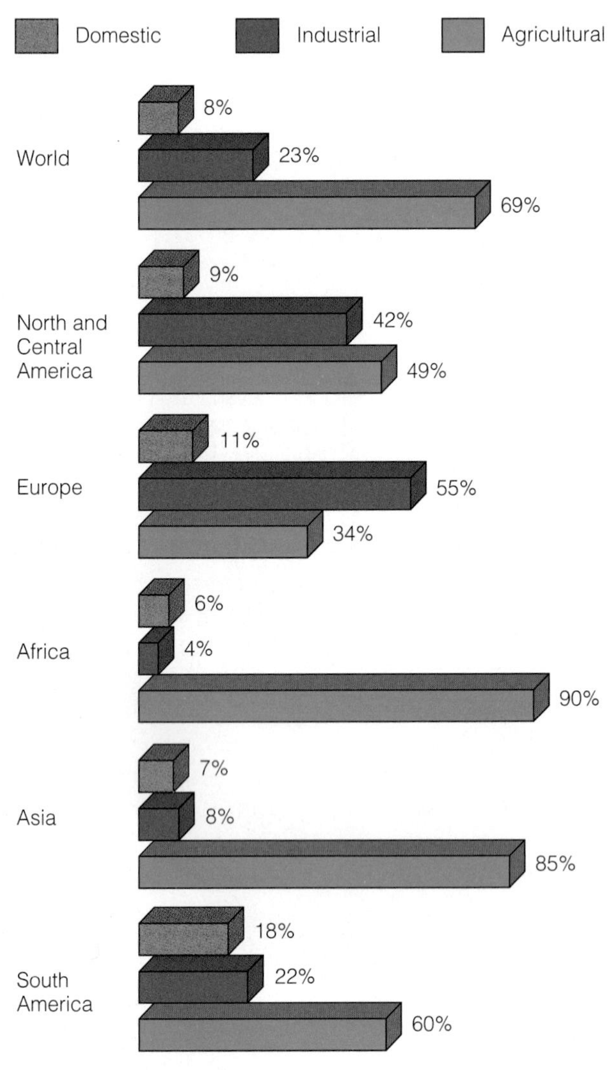

Figure 13-5 Water withdrawal by use and region. (Data from World Resources Institute)

Uses of withdrawn water vary from one region to another (Figure 13-5) and from one country to another (Figure 13-6). Averaged globally, about 65% of all water withdrawn each year from rivers, lakes, and aquifers is used to irrigate 16% of the world's cropland. Some 60–80% of this water either evaporates or seeps into the ground before reaching crops.

Worldwide, about 25% of the water withdrawn is used for energy production (oil and gas production and power-plant cooling) and industrial processing, cleaning, and waste removal. Water withdrawal for energy production and industrial use is highest in Europe and North America (Figure 13-5), especially in the United States (Figure 13-6).

Agricultural and manufactured products require large amounts of water, much of which could be used more efficiently and reused. It takes about 380,000 liters (100,000 gallons) to make an automobile, 3,800 liters (1,000 gallons) to produce 454 grams (1 pound) of aluminum, 3,000 liters (800 gallons) to produce 454 grams (1 pound) of grain-fed beef in a feedlot (where large numbers of cattle are confined to a fairly small area), and 100 liters (26 gallons) to produce 1 kilogram (2.2 pounds) of paper.

Domestic and municipal use accounts for about 10% of worldwide water withdrawals and about 13–16% of withdrawals in developed countries. As population, urbanization, and industrialization grow, the volume of wastewater needing treatment will increase enormously.

Case Study: Fresh Water Resources in the United States Although the United States has plenty of fresh water, much of it is in the wrong place at the wrong time or is contaminated by agricultural and industrial practices. The eastern states usually have

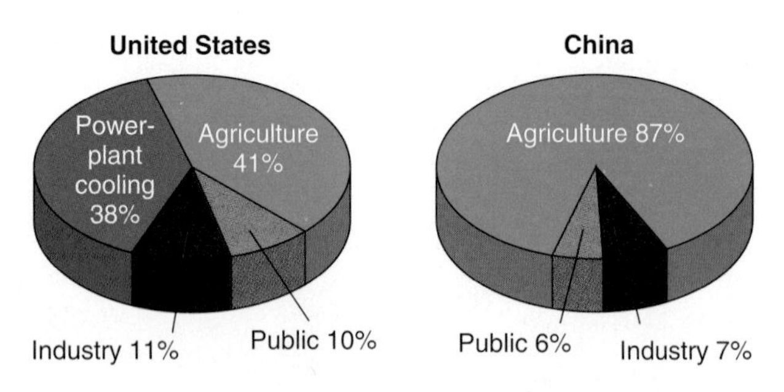

Figure 13-6 Use of water in the United States and China. The United States has the world's highest per capita use of water, amounting to an average of 6,000 liters (1,600 gallons) per person every day. About half of the water used in the United States is unnecessarily wasted. (Data from Worldwatch Institute and World Resources Institute)

 Johnson, Dan. 1998. "Averting a Water Crisis: The Era of Taking Water for Granted Is Ending." *Futurist,* vol. 32, no. 2, 7(1).

ample precipitation, whereas many of the western states have too little. In the East, the largest uses for water are for energy production, cooling, and manufacturing; in the West, the largest use by far is for irrigation (which accounts for about 85% of all water use).

In many parts of the eastern United States the most serious water problems are flooding, occasional urban shortages, and pollution. For example, the 3 million residents of Long Island, New York, get most of their water from an aquifer that is becoming severely contaminated. The major water problem in the arid and semiarid areas of the western half of the country is a shortage of runoff caused by low precipitation, high evaporation, and recurring prolonged drought. Water tables in many areas are dropping rapidly as farmers and cities deplete groundwater aquifers faster than they are recharged.

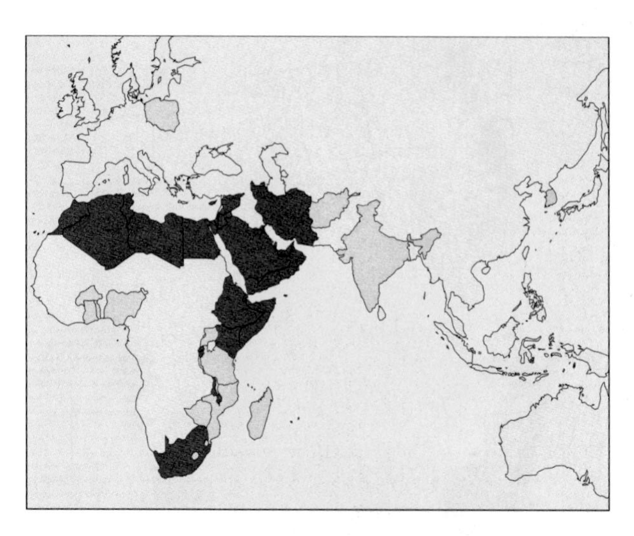

Figure 13-7 Water-short countries. Countries suffering extreme water stress (red) have fewer than 500 cubic meters (650 cubic yards) of fresh water per person; those suffering from severe water stress (yellow) have fewer than 1,000 cubic meters (1,300 cubic yards) of fresh water per person. (Data from Population Action International, Malin Falkenmark, and Peter Gleick)

13-2 TOO LITTLE WATER

What Causes Freshwater Shortages? According to water expert Malin Falkenmark, there are four causes of water scarcity: **(1)** a *dry climate* (Figure 7-2), **(2)** *drought* (a period in which precipitation is much lower and evaporation is higher than normal), **(3)** *desiccation* (drying of the soil because of such activities as deforestation and overgrazing by livestock), and **(4)** *water stress* (low per capita availability of water caused by increasing numbers of people relying on limited levels of runoff).

Since the 1970s water scarcity intensified by prolonged drought (mostly in areas with dry climates, where 40% of the world's people live) has killed more than 24,000 people per year and created many environmental refugees. In water-short areas, many women and children must walk long distances each day, carrying heavy jars or cans, to get a meager supply of sometimes contaminated water. Millions of poor people in developing countries have no choice but to try to survive on drought-prone land. If global warming occurs as projected, severe droughts may become more common in some areas of the world.

According to a 1995 World Bank study, 30 countries containing 40% of the world's population (2.4 billion people) now experience chronic water shortages that threaten their agriculture and industry and the health of their people (Figure 13-7). By 2025 at least 3 billion people in 90 countries are expected to face severe water stress. In most of these countries the problem is not a shortage of water but the wasteful and unsustainable use of normally available supplies.

In the United States, many major urban centers (especially those in the West and Midwest) are located in areas that don't have enough water or are projected to have water shortages by 2000 (Figure 13-8). Experts project that current shortages and conflicts over water supplies will get much worse as more industries and people migrate west and compete with farmers for scarce water. These shortages could worsen even more if the climate warms as a result of an enhanced greenhouse effect.

A number of analysts believe that *access to water resources, a key foreign policy and environmental security issue for water-short countries in the 1990s, will become even more important early in the next century* (Guest Essay, p. 530). Almost 150 of the world's 214 major river systems (57 of them in Africa) are shared by 2 countries; another 50 are shared by 3 to 10 nations. Some 40% of the world's population already clashes over water, especially in the Middle East (p. 310).

Some areas have lots of water, but the largest rivers (which carry most of the runoff) are far from agricultural and population centers where the water is needed. For example, South America has the largest annual water runoff of any continent, but 60% of the runoff flows through the Amazon River in remote areas where few people live. Africa has a total runoff much larger than that of Europe, but much of this runoff drains into the Atlantic and is not

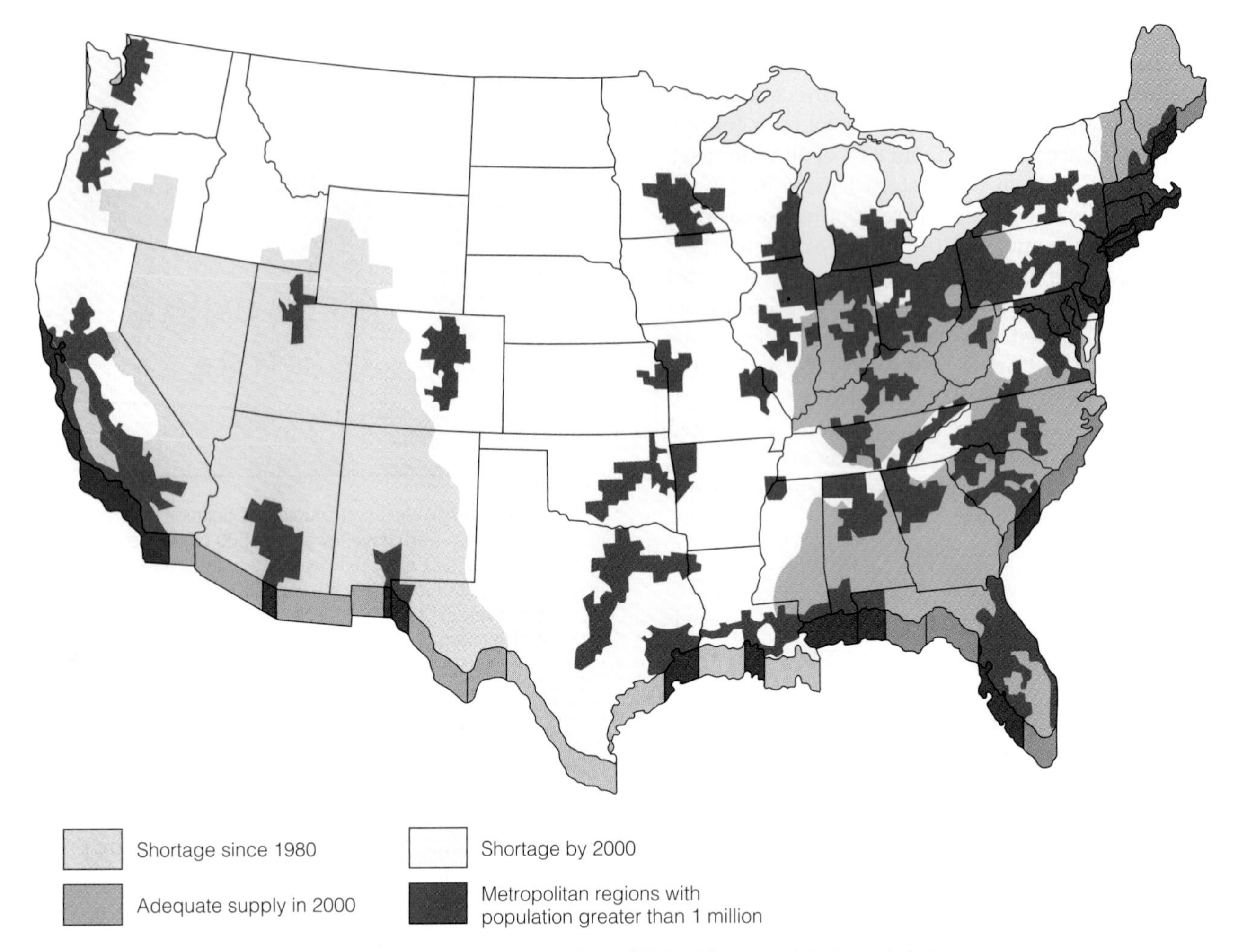

Figure 13-8 Current and projected water-deficit regions in the continental United States and their proximity to metropolitan areas having populations greater than 1 million. (Data from U.S. Water Resources Council and U.S. Geological Survey)

available to the water-poor regions in North, South, and East Africa.

How Can Water Supplies Be Increased? There are five ways to increase supply of fresh water in a particular area: **(1)** build dams and reservoirs to store runoff, **(2)** bring in surface water from another area, **(3)** withdraw groundwater, **(4)** convert salt water to fresh water (desalination), and **(5)** improve the efficiency of water use. Figure 13-9 shows the major components of a human-developed and -managed water supply system

In developed countries, people tend to live where the climate is favorable and then bring in water from another watershed. In developing countries, most people (especially the rural poor) must settle where the water is and try to capture and use as much precipitation as they can.

13-3 USING DAMS AND RESERVOIRS TO SUPPLY MORE WATER

What Are the Pros and Cons of Large Dams and Reservoirs? Large dams and reservoirs have benefits and drawbacks (Figure 13-10 and Case Study, p. 318). Large reservoirs created by damming streams can capture and store water from rain and melting snow. This water can then be released as desired to produce hydroelectric power at the dam site, irrigate land below the dam, control flooding of land below the reservoir, and provide water carried to towns and cities by aqueducts. Reservoirs also provide recreational activities such as swimming, fishing, and boating.

To maximize a river's benefits to humans, a growing number of the world's rivers are essentially elaborate plumbing systems consisting of a sequence of dams that time and control water flow much like

McManus, Reed. 1998. "Down Come the Dams." *Sierra*, vol. 83, n.3, 16(2).

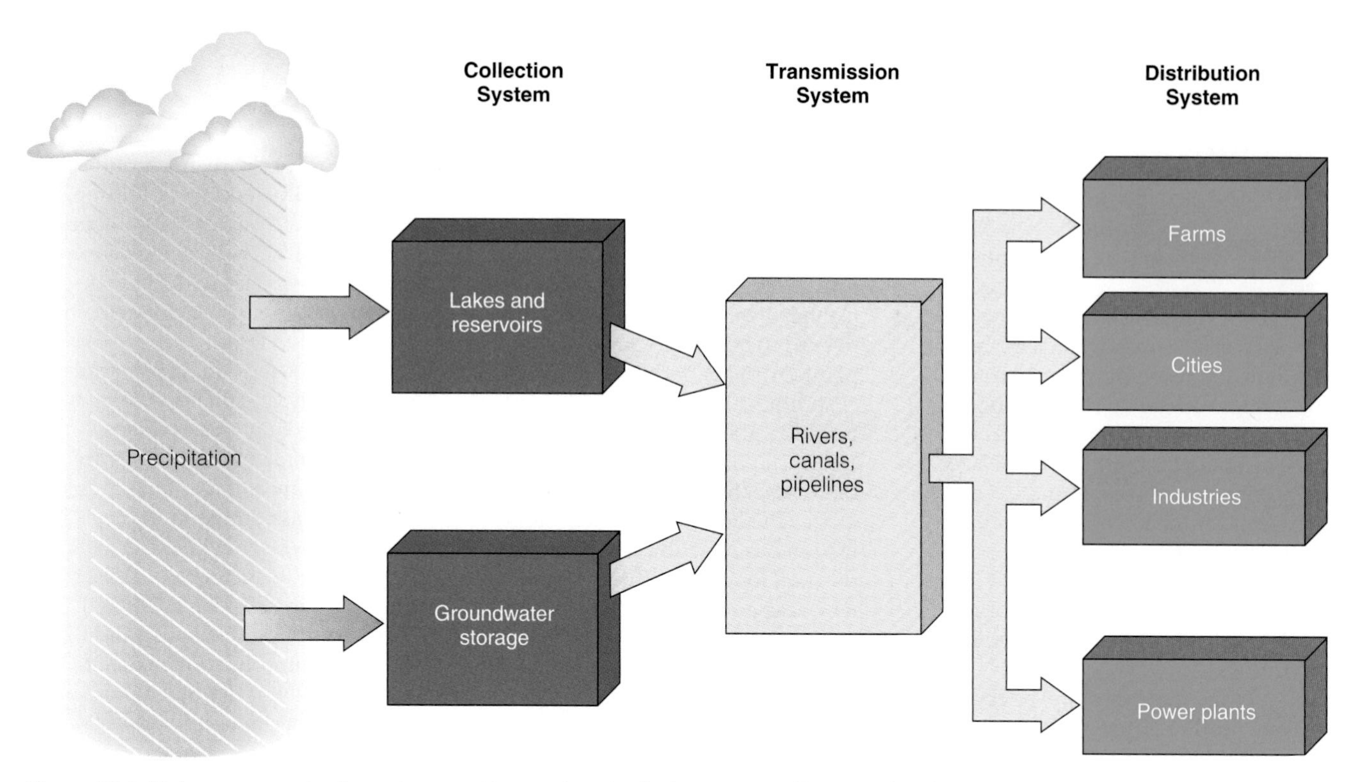

Figure 13-9 Major components of a system used to supply water for human uses. The capacity of the entire system is limited by the component with the lowest capacity.

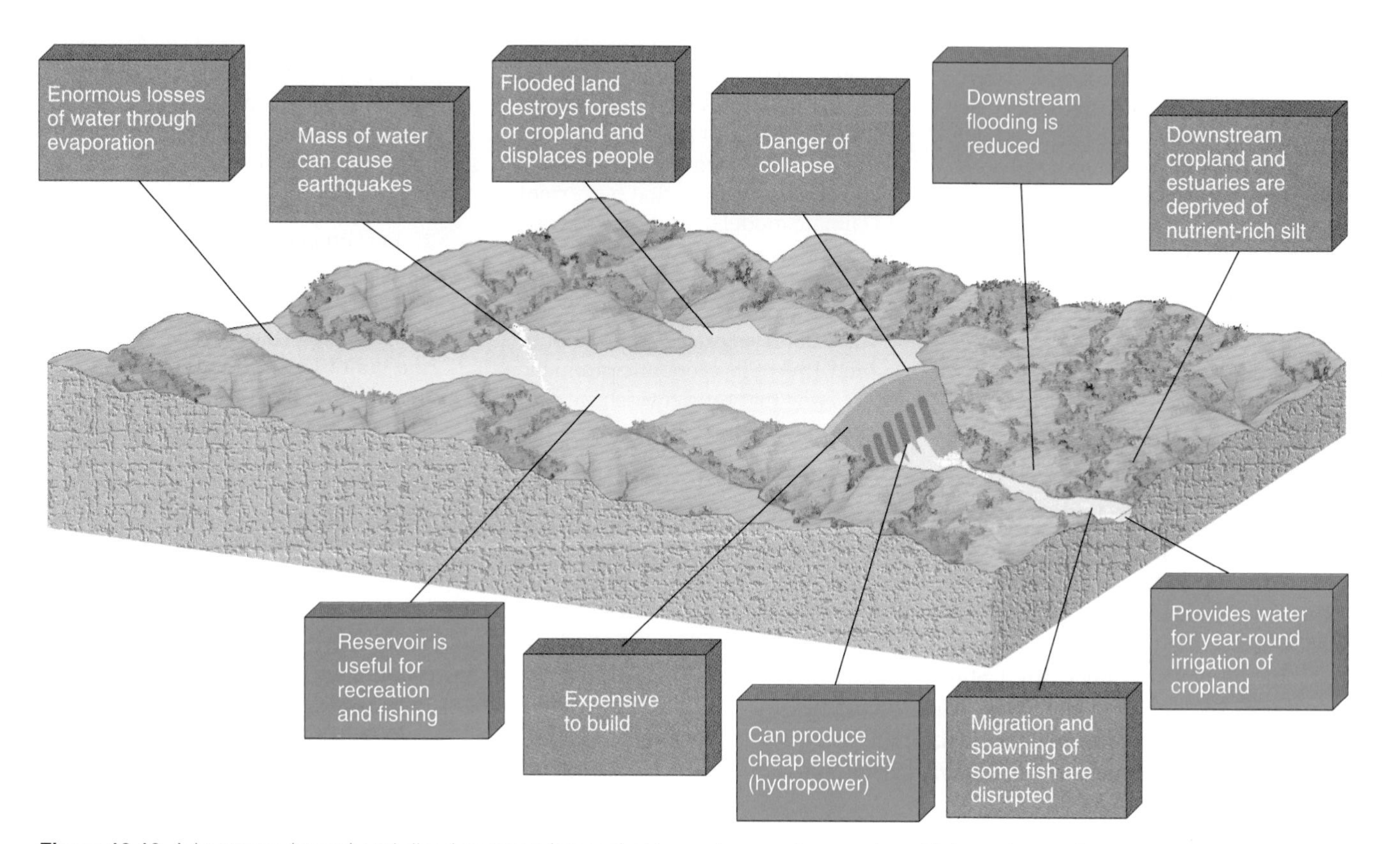

Figure 13-10 Advantages (green) and disadvantages (orange) of large dams and reservoirs, which can be used to produce electricity.

Hint: Enter the search terms *dams, economic aspects* using the Subject Guide.

Egypt's Aswan High Dam: Blessing or Disaster?

CASE STUDY

The billion-dollar Aswan High Dam on the Nile River in Egypt and its reservoir, Lake Nasser (Figure 13-1), demonstrate what can happen when a major dam and reservoir project is built without adequate consideration of long-term environmental effects. The dam was built in the 1960s to provide flood control and irrigation water for the lower Nile basin and electricity for Cairo and other parts of Egypt. All cropland in arid Egypt must be irrigated, and the country is totally dependent on the Nile for this water.

Today the dam supplies about one-third of Egypt's electrical power. Lake Nasser can store at least 2 years of the Nile's average annual flow; it saved Egypt's rice and cotton crops during the droughts of 1972 and 1973. Year-round irrigation in the lower Nile basin has increased food production: Farmers harvest cash crops such as water-intensive sugar cane and rice three times per year on land that once produced only one. Irrigation has also brought some 405,000 hectares (1 million acres) of desert land under cultivation.

Since the dam opened in 1964, however, it has also produced a number of harmful ecological effects. It ended the yearly flooding that had fertilized the Nile Delta with silt. Now the river's silt accumulates behind the dam, filling up Lake Nasser.

Cropland in the Nile Delta basin must now be treated with commercial fertilizer at an annual cost of over $100 million to make up for plant nutrients once available at no cost. Ironically, the country's new fertilizer plants use up much of the electrical power produced by the dam. Also, because salts are no longer flushed from the soil, salinization has offset three-fourths of the gain in food production from new, less productive land irrigated by water from the reservoir. Today the country loses about 10% of its crop production annually to declining soil fertility, mostly because of salinization.

Without the Nile's annual discharge of sediment, the sea is eroding the delta and advancing inland, reducing productivity on large areas of agricultural land. Lowering of land elevation by loss of the Nile's vital river delta could be one of the most threatening long-term consequences of the dam.

If global warming causes sea level rises (as projected by current climate models), roughly 60% of Egypt's habitable land could be flooded within about 60 years. This could displace as much as 16% of the country's projected population and 15% of its economic production.

Now that nutrient-rich silt no longer reaches the river's mouth, Egypt's sardine, mackerel, shrimp, and lobster industries have all but disappeared. Only 17 of the 47 commercial fish species in the Nile before the dam's construction remain. This has led to losses of approximately 30,000 jobs, millions of dollars annually, and an important source of protein for Egyptians. However, some of these losses are being offset by a new fishing industry taking bass, catfish, and carp from Lake Nasser behind the dam.

Flooding to create Lake Nasser uprooted 125,000 people. Today the reservoir is only about half full, and most authorities believe that the level will not rise much more in the next 100 years. Another problem is that about 80% of the water flowing into Lake Nasser comes from Ethiopia; this vital source of water may be reduced by dams being built by Ethiopia.

Some analysts believe that in the long run the benefits of the Aswan High Dam will outweigh its costs; others already consider it an economic and ecological disaster. Time will tell who is right.

Critical Thinking

1. Do you believe that the benefits of the Aswan High Dam outweigh its drawbacks? Explain.

2. Give two principles for designing and building large dams based on lessons from the Aswan High Dam.

water from a faucet. Between 25 and 50% of the total runoff on every continent is now captured and controlled by 40,000 large dams (over 15 meters high) and about 800,000 smaller ones, and many more large projects are planned.

In the continental United States, Norway, New Zealand, Mexico, and the Swiss Alps, most sites for dams and reservoirs have been developed. But many possible sites for large and small dams exist in parts of Latin America and Asia.

However, a series of dams on a river, especially in arid areas, can reduce downstream flow to a trickle and prevent it from reaching the sea as a part of the hydrologic cycle (Figure 5-4), as has happened with the Colorado River in the United States (Case Study, right). Recently several older small dams along various rivers in the United States have been dismantled to help restore fish habitats and more may be torn down as 550 dams come up for government relicensing over the next 15 years.

Dams can devastate fish life in downstream areas. A 1997 study in Sweden revealed that dams also reduce biodiversity. Plant communities along the banks of Swedish rivers dammed for hydroelectric power contain 15–50% fewer species than those alongside neighboring free-flowing rivers.

Q: What is the baby-bust generation in the United States?

A recent study indicates that the Iron Gates dam built in 1972 across the Danube River between Romania and Yugoslavia has devastated marine life in the Black Sea. Research suggests that the dam has blocked the flow of silicates—natural components of sand that enter the Danube as it flows to the sea—and disrupted Black Sea food webs. Levels of silicates in the river are one-sixtieth of what they were in 1969 and nitrates—a component of fertilizer and sewage—have increased by a factor of 600. As a result, Black Sea single-celled diatoms don't get enough silicates to build their bodies and are being outcompeted by toxic nitrate-using algae that poison fish. Food webs are disrupted because the zooplankton that feed on the diatoms—and the fish that in turn feed on the zooplankton—are deprived of their food sources.

India's irrigation systems, with much of the water provided by controlled release of water from dams, have enabled the country to be self-sufficient in food since 1974. However, India has had a history of environmental problems with many of its 1,500 large dam and irrigation projects (Case Study, left).

Despite this experience and protests by more than 100,000 villagers, the Indian government is going ahead with plans to build 30 large dams and thousands of smaller ones along the Namada River and 41 of its tributaries in the Indian state of Gujarat. This project will flood an estimated 4,000 square kilometers (1,500 square miles) of forests and farms and displace 1 million people.

This project is small compared to China's Three Gorges project on the mountainous upper reaches of the Yangtze River. When completed, it will be the world's largest hydroelectric dam and reservoir. This superdam, with the electric output of 18 large nuclear power plants, will supply power to industries and to 150 million Chinese. The project will also help China reduce its dependence on coal (Connections, p. 378) and help hold back the Yangtze River's floodwaters (which have claimed at least 500,000 lives in this century). Chinese officials expect the project to be completed by 2009 and cost at least $17 billion (with some estimates as high as $75 billion). However, estimates by independent experts put the completion date at around 2019 and the cost at $70 billion or more.

In addition, the project will flood large areas of some of the best farmland in the region and 800 existing factories; it will also displace 1.9 million people, including 100 towns and 2 cities (each with about 100,000 people), and flood some of China's most scenic land to create a reservoir 596 kilometers (370 miles) long. This will add to the estimated 10.2 million people previously displaced by reservoirs in China between 1950 and 1989. Government officials say they will relocate the people and find new land for the 300,000 displaced farmers. Critics charge that the dam will transform the Yangtze into a sewer for industrial wastes and the immense pressure from the huge volume of water in the reservoir could trigger landslides and earthquakes.

At best, the reservoir could store only a fraction of the floodwaters entering the Yangtze during a peak-flow year. Furthermore, as the reservoir fills up with sediment, the dam could overflow and expose half a million people to severe flooding (especially if the reservoir is kept filled at a high level, as planned, to provide maximum hydroelectric power). Opponents of the dam project that the river and reservoir will be poisoned with large amounts of toxic industrial chemicals—arsenic, cyanide, and methylmercury, among others—leached out of factories drowned by the reservoir.

Many analysts see a series of small dams as a better, cheaper, and less destructive alternative to large dams. Chinese officials claim that both big and small dams are needed. In 1997, Malaysia became one of the first developing countries to indefinitely postpone development of several large dams. This included the Bakun dam, which if built would be the highest dam in southeast Asia and flood an area as big as Singapore.

Japan has installed more than 1,000 small rubber dams that are inflated with air or water and anchored to the riverbed. These dams, which can span distances up to 135 meters (440 feet), can provide irrigation water, flood control, and groundwater recharge—and generate electricity. A major advantage of these dams is that they can be deflated to allow accumulated silt to flow downstream. Japan is now exporting this technology to Europe and Southeast Asia.

Case Study: The Colorado River Basin In 1869, after his boat trip down the Colorado River, John Wesley Powell proclaimed, "The West is an arid land, hostile to farming, and will never be settled. . . . unless the government dams the rivers and saves up the winter and spring runoff in artificial lakes and reservoirs." For the past 100 years this concept has been implemented by harnessing and using water from the Colorado River, which flows 2,300 kilometers (1,400 miles) from its glacial headwaters in north central Colorado to its mouth in the Gulf of California and ends in a rich delta in Mexico.

Along the way, the river passes through some of the world's most spectacular scenery, including the Grand Canyon, gouged out by the river during its 5- to 6-million-year previously uncontrolled rush to the sea. With hundreds of tributaries, the Colorado River basin drains a vast area in the southwestern United States and northern Mexico (Figure 13-11).

A: The people born since 1965, when total fertility rates have remained below 2.1

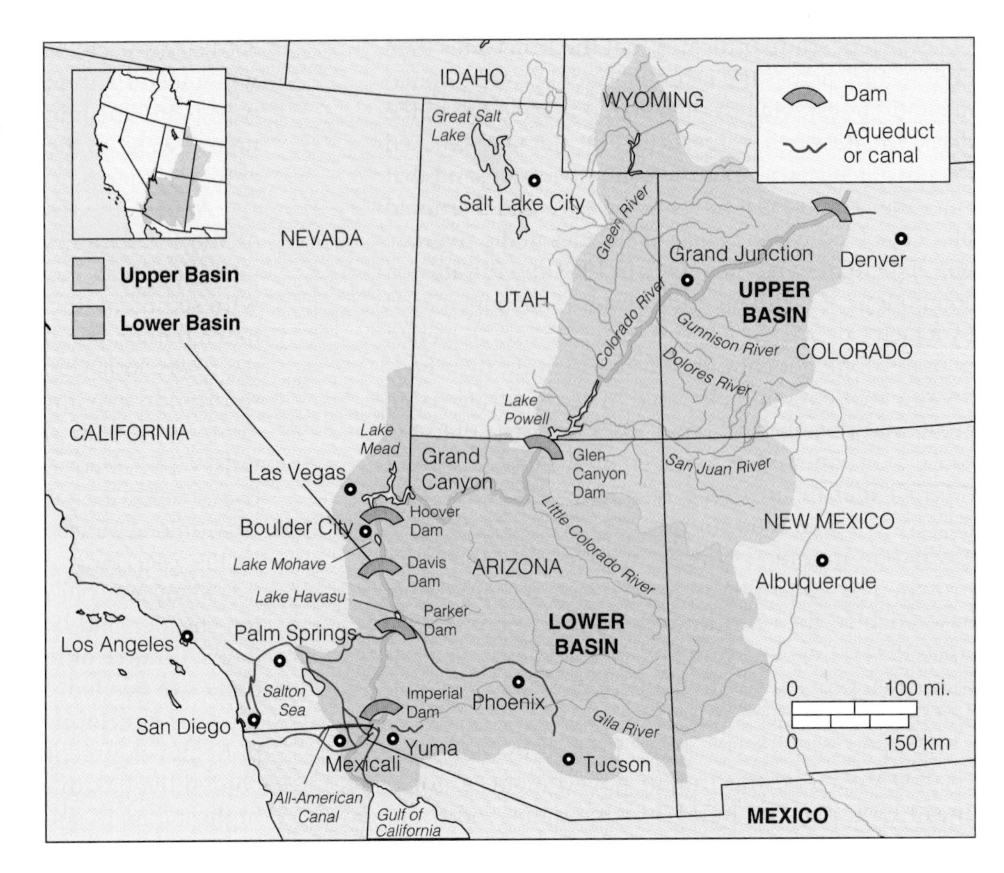

Figure 13-11 The Colorado River basin. The area drained by this basin is equal to more than one-twelfth of the land area of the lower 48 states.

During the past 50 years this free-flowing, gorge-cutting river has been tamed by a gigantic plumbing system consisting of several major dams and reservoirs (Figure 13-11), hundreds of smaller dams, and a network of aqueducts and canals that supply water to farmers, ranchers, and cities. Colorado River water is subsidized by federal taxpayers, who have provided the money for building and running the dams and water-distribution systems. Because of these government subsidies, the river's water costs only 25¢ per acre-foot—equal to about 1.2 million liters (326,000 gallons). Such low prices help farmers, industries, and consumers but encourage water waste.

Today the domesticated river dispenses water for more than 20 million people in seven states and for over 7% of the nation's cropland (producing about 15% of the nation's produce and livestock); its hydroelectric plants generate about 12 million kilowatts of electricity a year. Take away this tamed river and Las Vegas, Nevada, would be a mostly uninhabited desert area; San Diego (which gets 70% of its water from the Colorado) couldn't come close to supporting its present population; and California's Imperial Valley (which grows a major portion of the nation's vegetables) would consist mostly of cactus and mesquite plants.

Six national parks and recreation areas along the river's shores and in its huge reservoirs support a multi–billion-dollar recreation industry of whitewater rafting, boating, fishing, camping, and hiking enjoyed by more than 15 million people a year.

The basic problem is that the Colorado River basin includes some of the driest lands in the United States and Mexico (Figure 13-8). For 100 years there have been intense controversy and legal battles over how much of the river's limited and precious water could be used by towns, ranchers, farmers, Native Americans, and Mexicans and how much should be left for wildlife. Environmentalists have also fought developers in mostly unsuccessful attempts to keep more of the river wild by not building so many large dams.

For purposes of legal water distribution and control, the Colorado River basin has been divided into an upper basin (Wyoming, Utah, Colorado, and New Mexico) and lower basin (Arizona, Nevada, and California) (Figure 13-11). In 1922, the use of water from the river was portioned out among the states in both basins by the Colorado River Compact. In 1944, the U.S. government also guaranteed Mexico a certain amount of water per year from the river.

Despite these agreements, the once mighty Colorado rarely makes it to the Gulf of California. Instead it fizzles into a trickle that disappears into the Mexican desert or (in drought years) the Arizona desert. However, in drawing up the 1922 and 1944

Q: How many legal immigrants and refugees are admitted to the United States each year?

agreements, the average output of the river legally available for withdrawal was overestimated by 33% in normal years and by as much as 50% in drought years. So far the system has worked (except in drought years) only because Colorado (in the upper basin) and the states in the lower basin have not withdrawn their full allocations.

However, this is changing because of population growth and economic development in the mushrooming cities of the lower basin states. By 2000 all states are expected to be using their allocations, which together exceed the river's usable supply even in years without a drought.

In 1997, Arizona began taking its full share of water from the Colorado River, pumping it to more than a dozen reservoirs, where it will seep through the soil into natural underground aquifers. This has diverted cheap water that southern California once used mostly to irrigate crops. To replace their previous supply of government-subsidized water, California may have to pay $80 million a year to replace water that had cost only $100,000.

As each state takes all or more of its legally allocated share, there is increasing controversy among urban dwellers, farmers, ranchers, environmentalists, recreationists and the recreational industry, and floodplain residents over how to divide this "pie" of water that is smaller than the sum of "pieces" allocated to each state and Mexico. All of these groups are threatened by Native Americans living along the banks of the Colorado who want more development on their reservations. As senior owners of water rights dating back to the mid-1880s, they have the law on their side and have been making their case vigorously (and in many cases successfully) in courtrooms.

Traditionally, about 80% of the water withdrawn from the Colorado has been used to irrigate crops and raise cattle because ranchers and farmers (after Native Americans) got there first and established legal rights to use a certain amount of water each year. This large-scale use of water for agriculture was made possible because the government paid for the dams and reservoirs and under long-term contracts has supplied many of the farmers and ranchers with water at a very low price. This has led to wasteful irrigation practices and the growing of water-thirsty crops such as rice, cotton, and alfalfa (for cattle feed) in arid and semiarid areas along the Colorado.

Some cities (such as Tucson, Arizona, and Colorado Springs, Colorado) have been buying up the legal water rights of farmers and ranchers. Others are paying farmers to install less wasteful irrigation systems so that more water will be available to support urban areas. It is estimated that improving overall irrigation efficiency by about 10–15% would provide enough water to support projected urban growth in the areas served by the river to 2020.

There is also controversy over water quality, especially in the lower basin. As more water is withdrawn, the remaining flow gets saltier, largely because of evaporation and because water withdrawn to irrigate fields trickles back to the river laden with salts from the soil. Water reaching Mexico is often so salty that it can't be used for irrigation by Mexican farmers. To defuse this controversy, in 1974 the U.S. government agreed to build the world's largest desalination plant near Yuma, Arizona. At great cost, this plant (which went into operation in 1991) removes much of the salt from the water left in the Colorado before it flows into Mexico.

Engineers, economists, and developers see the taming of the Colorado River as a shining example of human ingenuity and mastery over nature. Environmentalists point out that this massive, mostly government-subsidized plumbing system reduces biodiversity and disrupts aquatic systems; it encourages often unplanned economic growth and development in a semiarid biome that (because of lack of water and frequent prolonged drought) is not well suited for irrigated agriculture and large urban areas.

These controversies illustrate the major problems that governments and people in other semiarid regions with river systems will face as population and economic growth place increasing demands on limited supplies of surface water.

13-4 TRANSFERRING WATER FROM ONE PLACE TO ANOTHER

What Are the Pros and Cons of Watershed Transfers? The California Experience Tunnels, aqueducts, and underground pipes can transfer stream runoff collected by dams and reservoirs from water-rich watersheds to water-poor areas. Although such transfers have benefits, they also create environmental problems. Indeed, most of the world's large-scale water transfers illustrate the important ecological principle that *you can't do just one thing.*

One of the world's largest watershed transfer projects is the California Water Project. In California, the basic water problem is that 75% of the population lives south of Sacramento but 75% of the rain falls north of it. The California Water Project uses a maze of giant dams, pumps, and aqueducts to transport water from water-rich northern California to heavily populated areas and to arid and semiarid agricultural regions, mostly in southern California (Figure 13-12).

For decades, northern and southern Californians have been feuding over how the state's water should

A: About 935,000 in 1998

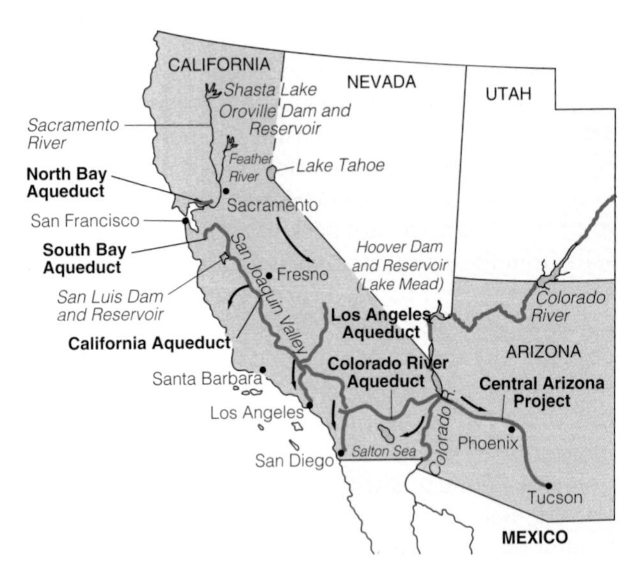

Figure 13-12 The California Water Project and the Central Arizona Project involve large-scale transfers of water from one watershed to another. Arrows show the general direction of water flow.

be allocated under this project. Southern Californians say they need more water from the north to support Los Angeles, San Diego, and other growing urban areas and to grow more crops. Although agriculture uses 82% of the water withdrawn in California, it accounts for only 2.5% of the state's economic output. Irrigation for just two crops, alfalfa (to feed cattle) and cotton, uses as much water as the homes of all 33 million Californians.

Opponents in the north say that sending more water south would degrade the Sacramento River, threaten fisheries, and reduce the flushing action that helps clean San Francisco Bay of pollutants. They also argue that much of the water already sent south is unnecessarily wasted and that making irrigation just 10% more efficient would provide enough water for domestic and industrial uses in southern California.

If water supplies in California were to drop sharply because of projected global warming, the amount of water delivered by the huge distribution system would plummet. According to EPA studies, global warming from a doubling of current atmospheric levels of CO_2 would reduce the water supplies in California's Central Valley basin by 7–16% annually. Most irrigated agriculture in California would have to be abandoned and much of the population of southern California might have to move to areas with more water. The 6-year drought that California experienced between 1986 and 1992 was perhaps just a small taste of the future.

Pumping out more groundwater is not the answer. Throughout much of California, groundwater is already being withdrawn faster than it is replenished. Most analysts see improving irrigation efficiency and allowing farmers to sell their legal rights to withdraw certain amounts of water from the river as much quicker and cheaper solutions.

Case Study: The James Bay Watershed Transfer Project Another major watershed transfer project is the James Bay project, a $60-billion, 50-year scheme to harness the wild rivers that flow into Quebec's James and Hudson Bays to produce electric power for Canadian and U.S. consumers (Figure 13-13). If completed, this megaproject would **(1)** construct 600 dams and dikes that will reverse or alter the flow of 19 giant rivers covering a watershed three times the size of New York State, **(2)** flood an area of boreal forest and tundra equal in area to Washington State or Germany, and **(3)** displace thousands of indigenous Cree and Inuit, who for 5,000 years have lived off James Bay by subsistence hunting, fishing, and trapping.

After 20 years and $16 billion, Phase I has been completed. The second and much larger phase was postponed indefinitely in 1994 because of an excess of power generated, opposition by the Cree (whose

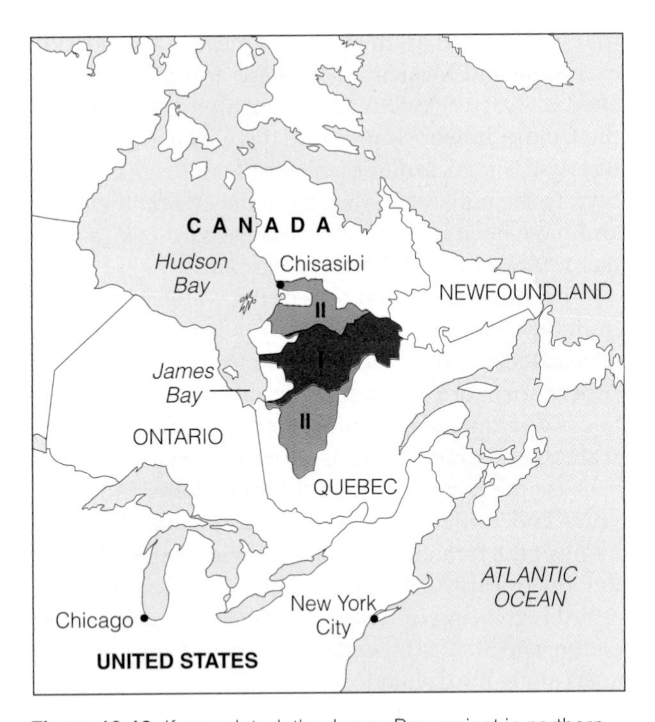

Figure 13-13 If completed, the James Bay project in northern Quebec will alter or reverse the flow of 19 major rivers and flood an area the size of Washington State to produce hydropower for consumers in Quebec and the United States, especially in New York State. Phase I of this 50-year project is completed; phase II was postponed indefinitely in 1994 because of a surplus of electricity and opposition by environmentalists and the indigenous Cree, whose ancestral hunting grounds would have been flooded.

Q: How many illegal immigrants enter the United States each year?

ancestral hunting grounds would have been flooded) and Canadian and U.S. environmentalists, and New York State's cancellation of two contracts to buy electricity produced by Phase II.

Case Study: The Aral Sea Watershed Transfer Disaster The shrinking of the Aral Sea (Figure 13-14) is a result of a large-scale water-transfer project in an area of the former Soviet Union with the driest climate in Central Asia. Since 1960, enormous amounts of irrigation water have been diverted from the inland Aral Sea and its two feeder rivers to irrigate cotton, vegetable, fruit, and rice crops to supply much of the region's needs. The irrigation canal, the world's longest, stretches over 1,300 kilometers (800 miles).

This water diversion (coupled with droughts) has caused a regional ecological, economic, and health disaster, described by one former Soviet official as "ten times worse than the 1986 Chernobyl nuclear power-plant accident." The sea's salinity has tripled, its surface area has shrunk by 54%, and its volume has decreased by almost 75%; the two supply rivers are now mere trickles. About 30,000 square kilometers (11,600 square miles) of former lake bottom have become a human-made salt desert. The process continues, and within a few decades the once enormous Aral Sea may be reduced to a few small brine lakes.

Some 20 of the 24 native fish species have become extinct, devastating the area's fishing industry, which once provided work for more than 60,000 people. Two major fishing towns are surrounded by a desert containing stranded fishing boats and rusting commercial ships that now lie more than 70 kilometers (44 miles) from the lake's receded shore. Roughly half of the area's bird and mammal species have also disappeared.

Winds pick up the salty dust that encrusts the lake's now-exposed bed and blow it onto fields as far away as 300 kilometers (190 miles) away. As the salt spreads, it kills crops, trees, and wildlife and destroys pastureland. This phenomenon has added a new term to our list of environmental ills: *salt rain*.

These changes have also affected the area's already semiarid climate. The once-huge Aral Sea acted as a thermal buffer, moderating the heat of summer and the extreme cold of winter. Now there is less rain, summers are hotter, winters are colder, and the growing season is shorter. Cotton and crop yields have dropped dramatically.

To raise yields, farmers have increased inputs of herbicides, insecticides, and fertilizers on some crops. Many of these chemicals have percolated downward and accumulated to dangerous levels in the groundwater, from which most of the region's drinking water comes. The low river flows have also concentrated salts, pesticides, and other toxic chemicals, making surface water supplies hazardous to drink.

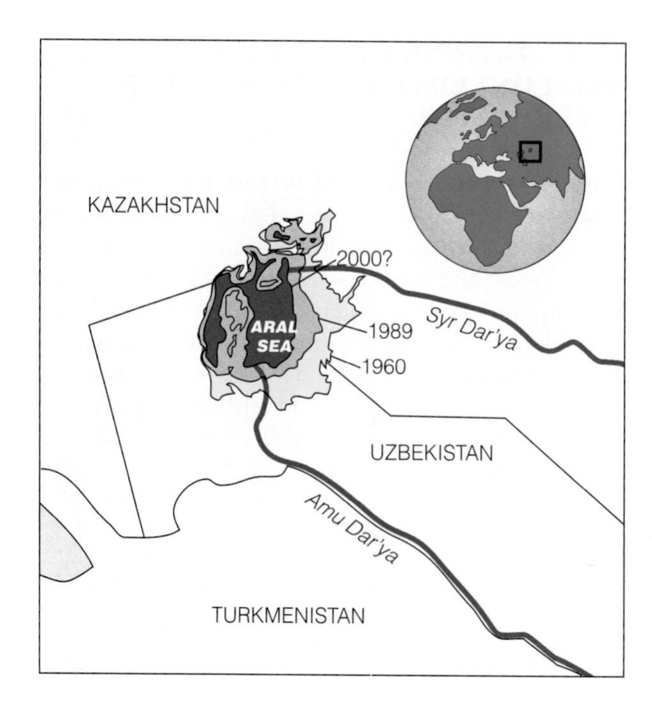

Figure 13-14 Once the world's fourth-largest freshwater lake, the Aral Sea has been shrinking and getting saltier since 1960 because most of the water from the rivers that replenish it has been diverted to grow cotton and food crops. As the lake shrinks, it leaves behind a salty desert, economic ruin, increasing health problems, and severe ecological disruption.

During the mid-1980s, the area near the shrunken Aral Sea had the highest levels of infant mortality and maternal mortality in the former Soviet Union. Between 1980 and 1993, kidney and liver diseases (especially cancers) increased 30- to 40-fold, arthritic diseases 60-fold, chronic bronchitis 30-fold, typhoid fever 30-fold, and hepatitis 7-fold.

Ways to deal with the ecological, economic, and health problems caused by this disaster include **(1)** charging farmers more for irrigation water to reduce waste and encourage a shift to less water-intensive crops, **(2)** decreasing irrigation water quotas, **(3)** introducing water-saving technologies, at a cost of at least $50 million, **(4)** developing a regional integrated water management plan, **(5)** planting protective forest belts, **(6)** using underground water to supplement irrigation water and to lower the water table to reduce waterlogging and salinization, **(7)** improving health services, and **(8)** slowing the area's rapid population growth.

Even with help from foreign countries, the United Nations, and agencies such as the World Bank, the money needed to save the Aral Sea and restore the ecological and economic services may not be available to this extremely poor area. What has happened to the Aral Sea basin is a stark reminder that everything in nature is connected and that preventing an ecological problem is much cheaper than trying to deal with its harmful consequences.

A: About 400,000

13-5 TAPPING GROUNDWATER AND CONVERTING SALT WATER TO FRESH WATER

What Are the Pros and Cons of Withdrawing Groundwater? In the United States, about half of the drinking water (96% in rural areas and 20% in urban areas) and 40% of irrigation water is pumped from aquifers. In Florida, Hawaii, Idaho, Mississippi, Nebraska, and New Mexico, more than 90% of the population depends on groundwater for drinking water.

Overuse of groundwater can cause or intensify several problems: *aquifer depletion*, *aquifer subsidence* (sinking of land when groundwater is withdrawn), and *intrusion of salt water into aquifers* (Figure 13-15). Groundwater can also become contaminated from industrial and agricultural activities, septic tanks, and other sources, as discussed in Section 20-5. Because groundwater is the source of about 40% of the stream flow in the United States, groundwater depletion also robs streams of water.

Currently, *groundwater in the United States is being withdrawn at four times its replacement rate*. The most serious overdrafts are occurring in parts of the huge Ogallala Aquifer, extending from southern South Dakota to central Texas (Figure 13-17 and Case Study, p. 326) and in parts of the arid southwestern United States (Figure 13-8), especially California's Central Valley, which is the country's vegetable basket. So much

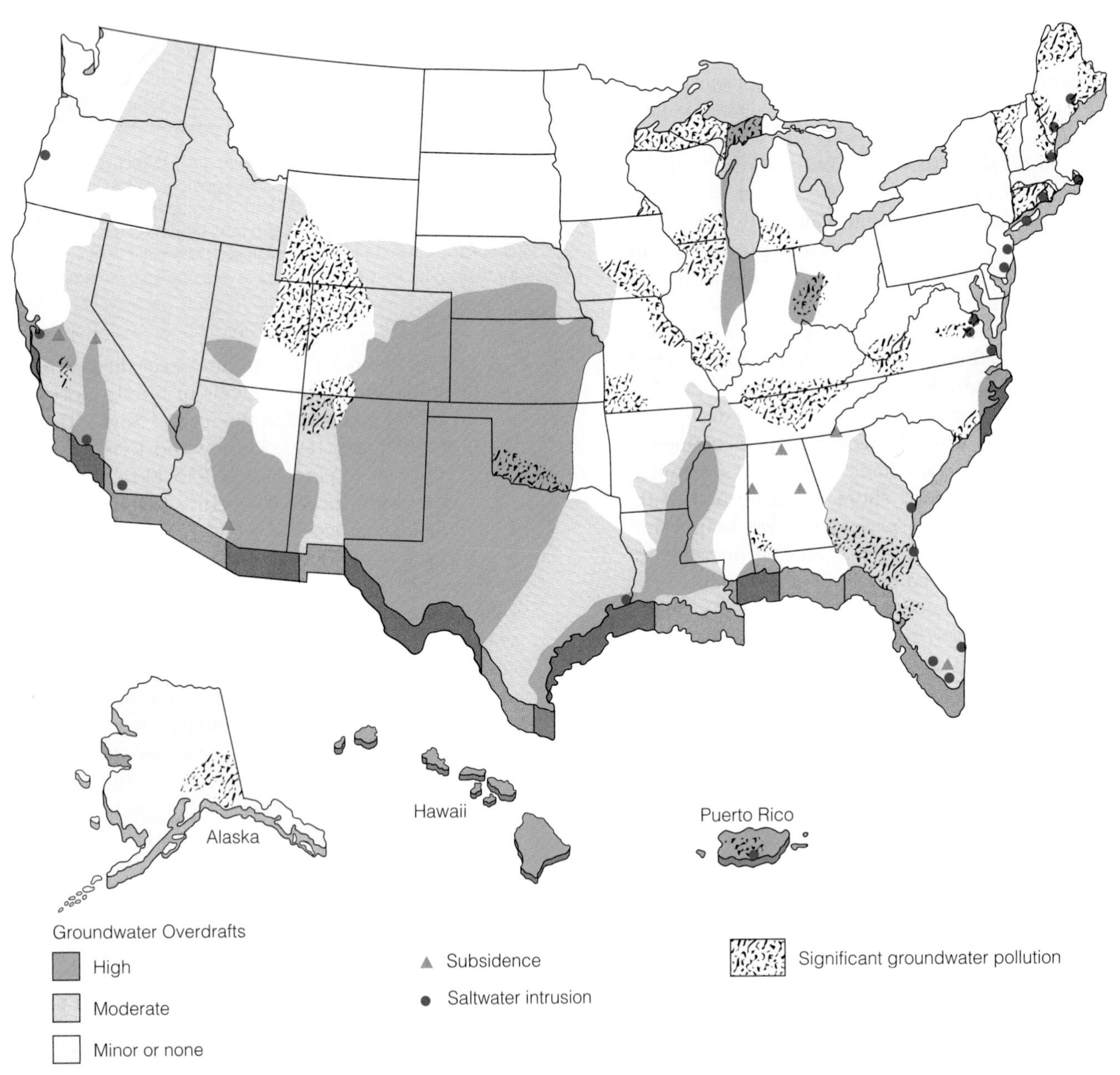

Figure 13-15 Areas of greatest aquifer depletion, subsidence, saltwater intrusion, and groundwater contamination in the United States. (Data from U.S. Water Resources Council and U.S. Geological Survey)

Q: What percentage of current annual U.S. population growth is due to legal and illegal immigration?

groundwater has been pumped from below Tucson, Arizona—the largest U.S. city to rely exclusively on groundwater—that in some places land in the city has sunk (subsided) by more than 2 meters (7 feet).

Aquifer depletion is also a problem in Saudi Arabia, northern China, northern Africa (especially Libya and Tunisia, where the rate of withdrawal is estimated to be 10 times the rate of discharge), southern Europe, the Middle East, and parts of Mexico, Thailand, and India. Saudi Arabia's remarkable increase in agricultural productivity is based on withdrawing water mostly from aquifers (with essentially negligible recharge rates) to irrigate crops in the desert. Currently, groundwater withdrawal is nearly three times greater than recharge. At this rate, the country's groundwater—more vital than the country's oil—could be depleted within 50 years.

When fresh water from an aquifer near a coast is withdrawn faster than it is recharged, salt water intrudes into the aquifer (Figure 13-16). Such intrusion threatens to irreversibly contaminate the drinking water of many towns and cities along the Atlantic and Gulf coasts (Figure 13-15) and in the coastal areas of Israel, Syria, and the Arabian Gulf states. Inland movement of salt water into aquifers also occurs when rivers that normally empty into the ocean have so much water diverted and managed by dams and irrigation projects that the rivers no longer empty into the sea.

Ways to slow groundwater depletion include controlling population growth, not planting water-thirsty crops in dry areas, developing crop strains that require less water and are more resistant to heat stress, and wasting less irrigation water.

How Useful Is Desalination? The removal of dissolved salts from ocean water or from brackish (slightly salty) groundwater, called **desalination**, is another way to increase fresh water supplies. Distillation and reverse osmosis are the two most widely used methods. *Distillation* involves heating salt water until it evaporates and condenses as fresh water, leaving salts behind in solid form. In *reverse osmosis*, salt water is pumped at high pressure through a thin membrane whose pores allow water molecules, but not dissolved salts, to pass through.

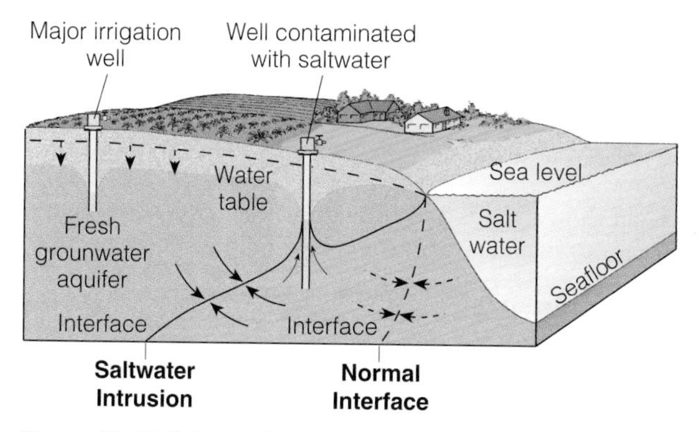

Figure 13-16 Saltwater intrusion along a coastal region. When the water table is lowered, the normal interface (dotted line) between fresh and saline groundwater moves inland (solid line).

About 7,500 desalination plants in 120 countries provide about 0.1% of the fresh water used by humans. Desalination plants in the arid Middle East (especially in Saudi Arabia and Kuwait) and North Africa produce about two-thirds of the world's desalinated water. Desalination is also used in parts of Florida. Soon Marin County (northern California) and Santa Barbara, San Diego, and Los Angeles may turn to desalination to supplement their water supplies.

However, desalination has a downside. Because it uses vast amounts of electricity, water produced in this way costs three to five times more than water from conventional sources. Distributing the water from coastal desalination plants costs even more in terms of the energy needed to pump the water uphill and inland.

Desalination also produces large quantities of brine containing high levels of salt and other minerals. Dumping the concentrated brine into the ocean near the plants might seem to be the logical solution, but this would increase the local salt concentration and threaten food resources in estuary waters. If these wastes were dumped on the land, they could contaminate groundwater and surface water.

Desalination can provide fresh water for coastal cities in arid countries (such as sparsely populated Saudi Arabia), where the cost of getting fresh water by any method is high. However, desalinated water will probably never be cheap enough to irrigate conventional crops or to meet much of the world's demand for fresh water unless affordable, efficient solar-powered distillation plants can be developed and someone can figure out what to do with the resulting mountains of salt. Instead of spending less money to import more wheat, Saudi Arabia uses desalinated water to cultivate wheat in the desert at an estimated cost of seven times the world price per bushel.

Can Cloud Seeding and Towing Icebergs Improve Water Supplies? For years several countries, particularly the United States, have been experimenting with seeding clouds with tiny particles of chemicals (such as silver iodide) to form water condensation nuclei and thus produce more rain over dry regions and more snow over mountains. However, cloud seeding is not useful in very dry areas, where it is most needed, because rain clouds are rarely available there. Furthermore, widespread cloud seeding would introduce large amounts of the cloud-seeding chemicals into soil and water systems, possibly harming people, wildlife, and agricultural productivity.

Another obstacle to cloud seeding is legal disputes over the ownership of water in clouds. During the 1977 drought in the western United States, the

A: About 44%

CASE STUDY

Mining Groundwater: The Shrinking Ogallala Aquifer

Vast amounts of water pumped from the Ogallala, the world's largest known aquifer (Figure 13-17), have helped transform vast areas of arid high plains prairie land into one of the largest and most productive agricultural regions in the United States. More than 90% of the water withdrawn from this aquifer irrigates crops, amounting for about 30% of all U.S. groundwater used for irrigation.

With yields two to three times those of dryland farming, irrigated farming is a profitable venture. As a result, this region now produces 20% of U.S. agricultural output (including 40% of its feedlot beef), valued at $32 billion. This production has brought prosperity to many farmers and merchants in this region, but the hidden environmental (and economic) cost has been increasing depletion of the aquifer in some areas.

Although this aquifer is gigantic, it is essentially a nonrenewable aquifer (stored during the retreat of the last ice age about 15,000–30,000 years ago) with an extremely slow recharge rate. In some areas, water is being pumped out of the aquifer 8–10 times faster than the natural recharge rate. The northernmost states (Wyoming, North Dakota, South Dakota, and parts of Colorado) still have ample supplies. However, supplies in parts of the southern states, where the aquifer is thinner (Figure 13-7) are being depleted rapidly.

Water experts project that at the current rate of withdrawal, one-fourth of the aquifer's original supply will be depleted by 2020, and much sooner in areas where it is shallow. It will take thousands of years to replenish the aquifer.

Government farm policies encourage depletion of the aquifer by subsidizing water-thirsty crops that require irrigation and by providing crop-disaster payments. Federal groundwater depletion allowances for High Plains farmers also encourage groundwater mining. The greater the use of groundwater, the greater this tax break—an example of positive feedback in action.

Depletion of this essentially nonrenewable water resource can be delayed. Farmers can use more efficient forms of irrigation (Section 13-6); in some areas they may have to switch to crops that require less water, or to dryland farming.

Total irrigated area is already declining in five of the seven states using this aquifer because of rising drilling and pumping costs as the water table drops. For example, between 1979 and 1994, the area of irrigated land in the Texas High Plains dropped by 26% because of depletion of this shallow end of the Ogallala aquifer. Many farmers in northwest Texas have reduced their water use by 20–25% by switching to more efficient irrigation technologies.

Cities using this groundwater can improve their water management by wasting less water and using less for lawns and golf courses. Individuals enjoying the benefits of this aquifer can pitch in by installing water-saving toilets and showerheads and converting their lawns to plants that can survive in an arid climate with little watering.

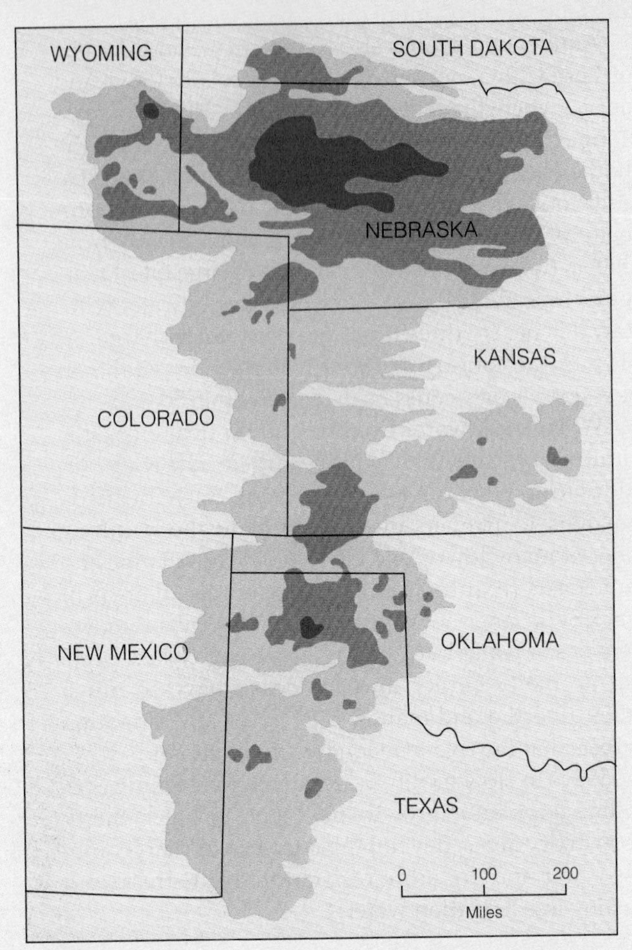

Figure 13-17 The Ogallala, the world's largest known aquifer. If the water in this aquifer were above ground, it could cover all 50 states with 0.5 meter (1.5 feet) of water. This fossil aquifer, which is renewed very slowly, is being depleted (especially at its thin southern end in parts of Texas, New Mexico, Oklahoma, and Kansas) to grow crops, raise cattle, and provide urban dwellers and industries with water. (Data from U.S. Geological Survey)

Critical Thinking

Do you believe that farmers and cattle ranchers using water withdrawn from the Ogallala should receive federal subsidies to grow crops and raise livestock that require large amounts of irrigation water? Explain. What are the alternatives?

Q: What is the estimated value of women's unpaid work at home?

attorney general of Idaho accused officials in neighboring Washington of "cloud rustling" and threatened to file suit in federal court.

There also have been proposals to tow massive icebergs to arid coastal areas (such as Saudi Arabia and southern California) and then to pump the fresh water from the melting bergs ashore. However, the technology for doing this is not available and the costs may be too high, especially for water-short developing countries.

13-6 USING WATER MORE EFFICIENTLY

Why Is Reducing Water Waste So Important? Increasing the water supply in some areas is important, but soaring population, food needs, and industrialization (and unpredictable shifts in water supplies) will eventually outstrip this approach. It makes much more sense, economically and environmentally, to use water more efficiently.

Mohamed El-Ashry of the World Resources Institute estimates that *65–70% of the water people use throughout the world is wasted through evaporation, leaks, and other losses*. The United States, the world's largest user of water, does slightly better but still loses about 50% of the water it withdraws. El-Ashry believes that it is economically and technically feasible to reduce water waste to 15%, thereby meeting most of the world's water needs for the foreseeable future.

Conserving water would have many other benefits, including reducing the burden on wastewater plants and septic systems, decreasing pollution of surface water and groundwater, reducing the number of expensive dams and water-transfer projects that destroy wildlife habitats and displace people, slowing depletion of groundwater aquifers, and saving energy and money needed to treat and distribute water.

Why Is So Much Water Wasted? A prime cause of water waste in the United States (and in most countries) is artificially low water prices resulting from government subsidies. Cheap water is the only reason farmers in Arizona and southern California can grow water-thirsty crops such as alfalfa in the middle of the desert. It also enables affluent people in desert areas to keep their lawns and golf courses green.

The federal Bureau of Reclamation supplies one-fourth of the water used to irrigate land in the western United States under long-term contracts (typically 40 years) at greatly subsidized prices. Such water subsidies are paid for by all taxpayers through higher taxes. Because the harmful environmental and groundwater-depletion costs caused by these perverse subsidies don't show up on monthly water bills, consumers have little incentive to use less water or to adopt water-conserving devices and processes.

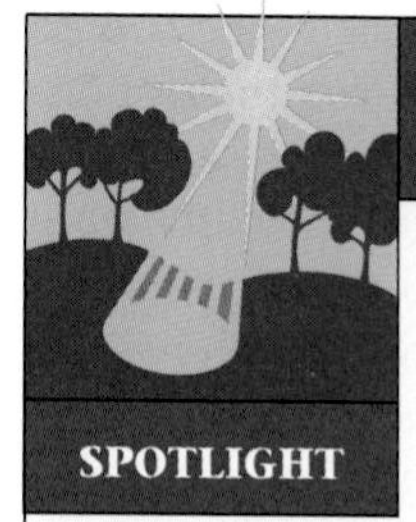

SPOTLIGHT

Water Rights in the United States

Laws regulating surface-water access and use differ in the eastern and western parts of the United States. In most of the East, water use is based on the doctrine of riparian rights. Basically, this system of water law gives anyone whose land adjoins a flowing stream the right to use water from the stream as long as some is left for downstream landowners. However, as population and water-intensive land uses grow, there is often too little water to meet the needs of all the people along a stream.

In the arid and semiarid West, the riparian system does not work because large amounts of water are needed in areas far from major surface-water sources. In most of this region the principle of prior appropriation regulates water use. In this first-come, first-served approach, the first user of water from a stream establishes a legal right for continued use of the amount originally withdrawn, a use-it-or-lose-it approach that discourages farmers from adopting water-conserving irrigation methods. If there is a shortage, subsequent users are cut off in order, one by one, until there is enough water to satisfy the demands of the earlier users. Some states have a combination of riparian and prior appropriation water rights.

Most groundwater use is based on common law, which holds that subsurface water belongs to whoever owns the land above such water. This means that landowners can withdraw as much water as they want to use on their land. When many users tap the same aquifer, that aquifer becomes a common-property resource; the largest users have little incentive to conserve and can deplete the aquifer for everyone, creating another tragedy of the commons.

Critical Thinking

Do you agree with the principles of riparian rights and prior appropriation? Explain. If you don't agree, how would you divide up water rights?

Some analysts believe that sharply raising the price of federally subsidized water would encourage investments in improving water efficiency, and many of the West's water supply problems could be eased; so far Western members of Congress have prevented such increases. Outdated laws governing access to and use of water resources also encourage unnecessary water waste (Spotlight, above). However, reforms of such laws are opposed by economically and politically powerful interests in the West.

A: About $11 trillion annually

Another reason for the water waste in the United States (and many other countries) is that the responsibility for water resource management in a particular watershed may be divided among many state and local governments rather than being handled by one authority. The Chicago metropolitan area, for example, has 349 water-supply systems divided among some 2,000 local units of government over a six-county area.

In sharp contrast is the regional approach to water management used in England and Wales. The British Water Act of 1973 replaced more than 1,600 agencies with 10 regional water authorities based on natural watershed boundaries. Each water authority owns, finances, and manages all water supply and waste treatment facilities in its region. The responsibilities of each authority include water pollution control, water-based recreation, land drainage and flood control, inland navigation, and inland fisheries. Each water authority is managed by a group of elected local officials and a smaller number of officials appointed by the national government.

Solutions: How Can We Waste Less Water in Irrigation? Worldwide, about 65% of the water that is diverted from rivers or pumped out of aquifers is used for irrigating about 16% of the world's cropland to produce almost 40% of the world's food. Because only about 40% of this water reaches crops, more efficient use of even a small amount of irrigation water would free large amounts of water for other uses. Some analysts estimate that irrigated cropland may have to produce 50–75% of our additional food in the future—something that won't happen without much more efficient use of irrigation water.

Most irrigation systems distribute water from a groundwater well or a surface canal by downslope or gravity flow through unlined ditches in cropfields so that the water can be absorbed by crops (Figure 13-18, left). However, this flood irrigation method delivers far more water than needed for crop growth, and because of evaporation, deep percolation (seepage), and runoff, only 50–60% of the water reaches crops.

Seepage can be reduced by placing plastic, concrete, or tile liners in the large irrigation canals that distribute water to unlined irrigation ditches. Lasers can be used to make sure that fields are level so that water flowing into unlined irrigation ditches is distributed more evenly. Small check dams of earth and stone can capture runoff from hillsides and channel it to fields. Holding ponds can store rainfall or capture irrigation water for recycling to crops.

Many U.S. farmers served by the dwindling Ogallala Aquifer now use center-pivot sprinkler systems (Figure 13-18, right), with which 70–80% of the water reaches crops. Some of these farmers are switching to low-energy precision-application (LEPA) sprinklers. These systems bring 75–85% (some claim 85–95%) of the water to crops by spraying it closer to the ground and in larger droplets than the center-pivot system; they also reduce energy use and costs by 20–30%. However, because of the high initial costs, such sprinklers are used on only about 1% of the world's irrigated cropland.

In the 1960s, highly efficient trickle or drip irrigation systems were developed in arid Israel. A network of perforated piping, installed at or below the ground surface, releases a trickle of water close to plant roots (Figure 13-18, center), minimizing evaporation and seepage and bringing 80–90% of the water to crops. Drip systems tend to be too expensive for most poor farmers and for use on low-value row crops. However, they are economically feasible for high-profit fruit, vegetable, and orchard crops and for home gardens. Researchers are testing drip irrigation systems that cost one-third to one-tenth as much as current drip systems.

Irrigation efficiency can also be improved by computer-controlled systems that monitor soil moisture and provide water only when necessary. The Phoenix-based U.S. Water Conservation Laboratory is developing an infrared monitoring device that can scan cropfields from a satellite and pinpoint when crops will most efficiently absorb water. Farmers can also switch to more water-efficient, drought-resistant, and salt-tolerant crop varieties and use nighttime irrigation to reduce loss to evaporation. In addition, farmers can use organic farming techniques, which produce higher crop yields per hectare and require only one-fourth of the water and commercial fertilizer of conventional farming.

Since 1950 Israel has used many of these techniques to slash irrigation water waste by about 84% while irrigating 44% more land. The government also gradually removed most government subsidies to raise the price of irrigation water to one of the highest rates in the world. Israel also imports most of its wheat and meat, concentrating on fruits, vegetables, and flowers that require less water. However, in other countries where government-funded water-supply schemes and other subsidies provide artificially cheap irrigation water, farmers have little incentive to invest in water-saving techniques.

As fresh water becomes scarce and cities consume water once used for irrigation, carefully treated urban wastewater (which is rich in nitrate and phosphate plant nutrients) could be used for irrigation. Israel now treats and reuses 55% of its municipal sewage water, mostly for irrigating nonedible crops such as cotton and flax, and plans to reuse 80% of this flow within the next few years.

Q: How many species of the world's plants are edible?

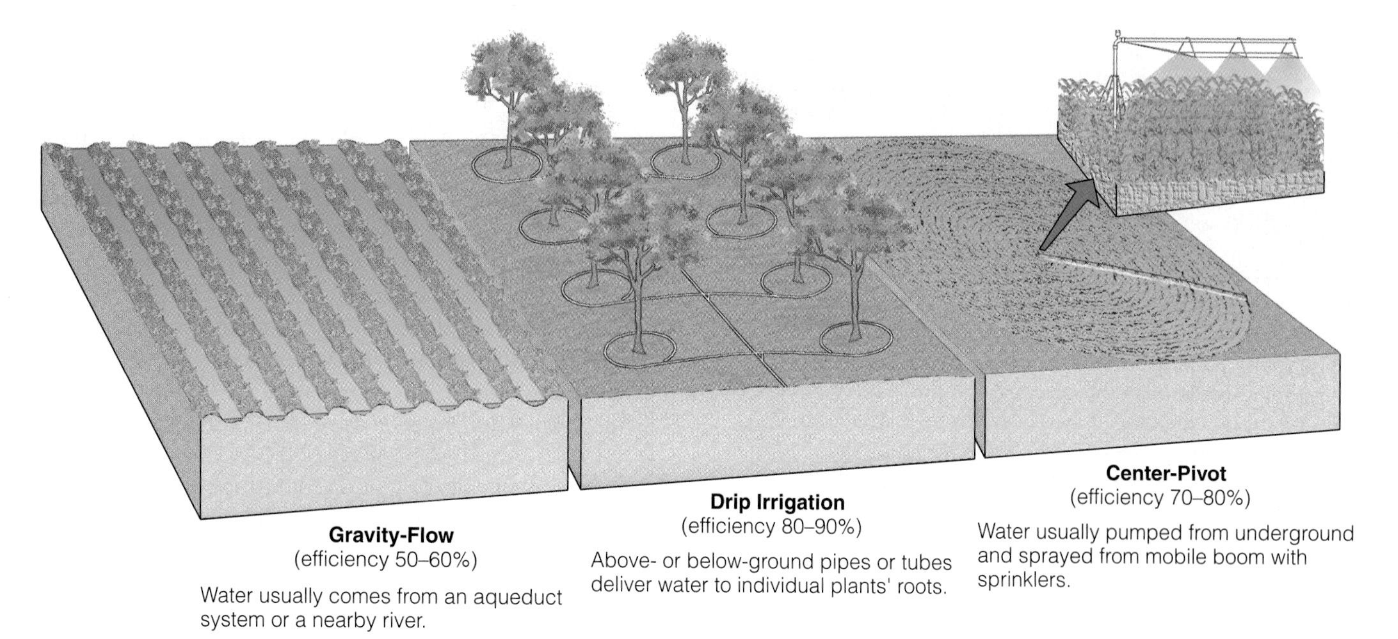

Figure 13-18 Major irrigation systems.

Solutions: How Can We Waste Less Water in Industry, Homes, and Businesses? Manufacturing processes can use recycled water or be redesigned to save water. Japan and Israel lead the world in conserving and recycling water in industry. One paper mill in Hadera, Israel, uses one-tenth as much water as most other paper mills. Manufacturing aluminum from recycled scrap rather than from virgin ores can reduce water needs by 97%.

Nearly half of the water supplied by municipal water systems in the United States is used to flush toilets and water lawns, and another 15% is lost through leaky pipes. In a typical U.S. home, flushing toilets, washing hands, and bathing account for about 78% of the water used. In the arid western United States and in dry Australia, lawn and garden watering can take 80% of a household's daily usage, and much of this water is unnecessarily wasted.

Green lawns in an arid or semiarid area can be replaced with vegetation adapted to a dry climate—a form of landscaping called *xeriscaping*, from the Greek word *xeros*, meaning "dry." A xeriscaped yard typically uses 30–80% less water than a conventional one. Drip irrigation can water gardens and other vegetation around homes.

More than half the water supply in Cairo, Lima, Mexico City, and Jakarta disappears before it can be used, mostly from leaks. In Cairo people often have to wade ankle deep across streets because of leaky water pipes. Leaky pipes, water mains, toilets, bathtubs, and faucets waste 20–35% of water withdrawn from public supplies in the United States and the United Kingdom.

Many cities offer no incentive to reduce leaks and waste. About one-fifth of all U.S. public water systems don't have water meters and charge a single low rate (often less than $100 a year for an average family) for virtually unlimited use of high-quality water. Many apartment dwellers have little incentive to conserve water because their water use is included in their rent.

In Boulder, Colorado, the introduction of water meters reduced water use by more than one-third. Tucson, Arizona, a desert city with ordinances that require conserving and reusing water, now consumes half as much water per capita as Las Vegas, a desert city where water conservation is still voluntary (Spotlight, p. 330).

A California water utility gives rebates for water-saving toilets; it also distributed some 35,000 water-saving showerheads, cutting per capita water use 40% in only 1 year. Converting to water-saving toilets can also provide jobs in low-income areas (Individuals Matter, p. 331). On January 1, 1994, the Comprehensive Energy Act of 1992 required that all new toilets sold in the United States use no more than 6 liters (1.6 gallons) per flush. Similar laws have been passed in Mexico and in Ontario, Canada.

A low-flow showerhead costing about $20 saves about $34–56 per year in water heating costs—a one-time investment that can give you a 50% return on your money with no financial risk. Audits conducted in Brown University's environmental studies program showed that the school could save $44,000 a year by using low-flow showerheads in dormitories.

Two decades of experience with droughts in northern California has shown that water demand can

A: About 30,000

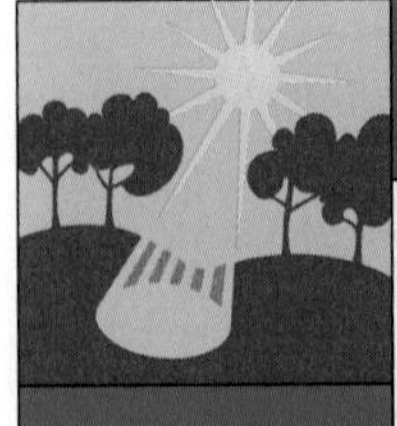

Running Out of Water in Las Vegas

SPOTLIGHT

Las Vegas, Nevada, located in the Mojave Desert, is an artificial aquatic wonderland of large trees, green lawns and golf courses, waterfalls, and swimming pools. Homeowners use about 64% of the metro area's water (two-thirds for lawns), followed by casinos (8%) and golf courses (8%). Residents in this arid area (Figure 13-8) have to apply 3 meters (10 feet) of water on a lawn each year to keep it green.

Las Vegas is one of the fastest-growing cities in the United States, with its population doubling from 550,000 in 1985 to 1.1 million in 1995. It is estimated that the city uses more water per person than any city in the world.

Whereas Tucson, in the Sonora Desert, gets only 30 centimeters (12 inches) of rainfall a year, Las Vegas averages only 10 centimeters (4 inches). Tucson, now a model of water conservation, began a strict water conservation program in 1976, including raising water rates 500% for some residents.

In contrast, Las Vegas only recently, and reluctantly, started to encourage water conservation. It has done that by raising water rates sharply (but they are still less than half those in Tucson) and encouraging desert landscaping with native plants that survive on little water rather than lawns. Studies show that such landscaping can reduce water use by as much as 80%.

Water experts project that even if these recent water conservation efforts are successful, the city will run short of water by 2007. It remains to be seen whether this city, whose residents have grown accustomed to using water as if they lived in a water-rich forest instead of a desert, can change their ways within the next decade.

Critical Thinking

If you were an elected official in charge of Las Vegas, what three things would you do to require or improve water conservation? What might be the political implications of instituting such a program? If water shortages by 2007 limit the growth of the area's population, would you consider this good or bad? Explain.

be cut by more than 50% for homes, 60% for parks, and 20% for businesses without economic hardships. Savings in water use—called *negaliters* or *negagallons*—can also save money. For instance, a 1994 New York City study showed that providing a $240 rebate to customers who convert to water-saving toilets and showerheads would cost only one-third of the $8- to $12-billion price tag for developing new water supplies for the city's residents.

In 1989, officials in Ashland, Oregon, needed more water and looked into building a small dam across a nearby stream. However, the price tag of $12 million was too high for a town of fewer that 20,000 people. On the advice of consultants, they were able to institute a water conservation program (involving free home water audits and rebates for efficient toilets and showerheads); this project will save as much water as the dam would have provided, at a cost of only $825,000. This *negadam* approach will also save Ashland's residents more than 500,000 kilowatt-hours a year on water heating and will reduce the annual volume of waste water to be treated by about 162 million liters (43 million gallons).

Used water (gray water) from bathtubs, showers, bathroom sinks, and clothes washers can be collected, stored, treated, and reused for irrigation and other purposes. California has become the first state to legalize reuse of gray water to irrigate landscapes. An estimated 50–75% of the water used by a typical house could be reused as gray water.

In some parts of the United States, people can lease systems that purify and completely recycle wastewater from houses, apartments, or office buildings. Such a system can be installed in a small outside shed and serviced for a monthly fee roughly equal to that charged by most city water-and-sewer systems. In Tokyo, all the water used in Mitsubishi's 60-story office building is purified for reuse by an automated recycling system. Some actions you can take to waste less water are listed in Appendix 5.

13-7 CASE STUDY: WATER MANAGEMENT IN THE COLUMBIA RIVER BASIN

How Has the Columbia River Basin Been Used and Abused? The Columbia River, flowing 1,900 kilometers (1,200 miles) through the Pacific Northwest, receives water from a huge basin extending across parts of seven U.S. states and two Canadian provinces (Figure 13-19). The Columbia River basin has the world's largest hydroelectric power system, consisting of more than 100 dams, 19 of them large hydroelectric dams. Most of these dams were built fairly cheaply by the U.S. government in the 1930s to furnish jobs, produce cheap electricity (40% below the national average), provide flood control, and help stimulate industrial and agricultural development. The river and its reservoirs are also used for recreation (including swimming,

Q: How many plants feed most of the world's people?

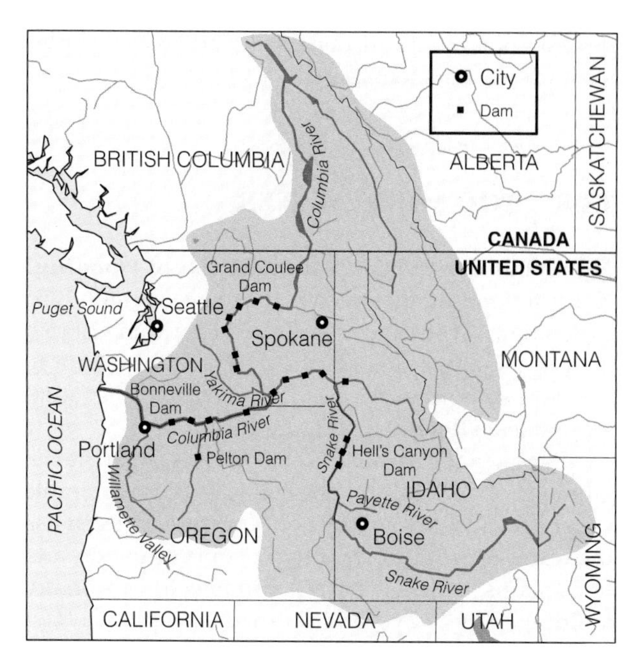

Figure 13-19 The Columbia River basin. (Data from Northwest Power Planning Council)

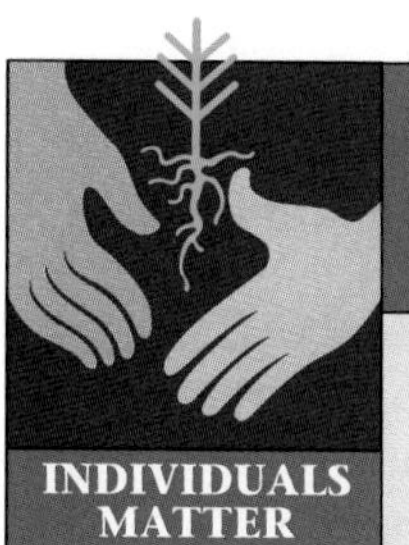

Saving Water and Providing Jobs in East Los Angeles

INDIVIDUALS MATTER

Some eco-pioneers are combating environmental problems with low-tech strategies that also provide social benefits. In a low-income neighborhood of East Los Angeles, resident Juana Beatriz Gutierrez developed a water conservation program that provides jobs in her community, where unemployment is high.

Working with city officials, she designed a program that distributes water-saving toilets free to local residents in exchange for their old, water-guzzling porcelain toilets. The old toilets are smashed to pieces and recycled as an underlayment for streets in Los Angeles.

The money she earns from administering the program is used to hire young people to go door-to-door urging families to have their children immunized against various infectious diseases and to get them tested for lead poisoning.

boating, and wind surfing) and for sport and commercial fishing for salmon, steelhead trout, and other fish.

What Are the Life Cycles of Wild and Hatchery-Raised Salmon? Since the dams were built, the Columbia River's wild salmon population has dropped by 94%, causing severe losses in the $1.2-billion-per-year Pacific Coast commercial salmon fishing industry. There were many causes of this decline: dams and reservoirs that hindered or prevented salmon migration, overfishing, destruction of spawning grounds by irresponsible logging and mining practices, and withdrawals of water for irrigation and other consumption uses.

Salmon are among the 1% of the world's fish species that are *anadromous*, living part of their lives in fresh water and part in salt water. Wild salmon hatch in clear, well-aerated headwater streams, develop for 1–2 years (depending on the species), and then migrate to the ocean to feed. When they reach sexual maturity after 1–3 years at sea, they return to the same gravel-bottomed streams in which they hatched to reproduce and die (Figure 12-19). Apparently each stream has unique chemical properties the fish can detect.

Without the installation of fish ladders and bypasses, mature salmon cannot pass over high dams and return to the streams where they were hatched in order to reproduce and die. Before the dams were built along the Columbia River, spring floods carried young salmon to the Pacific Ocean in less than 2 weeks; now water in the reservoirs is so deep and slow moving that the trip may take 2 months, exposing the fish to predators for a longer time.

Without screens and turbine bypass systems, many of the young fish die when they pass through the turbine blades of hydroelectric plants. Pollutants from riverside cities and towns can also weaken and kill young salmon migrating toward the sea. The American Fisheries Society says that 106 of the Columbia River's 400 salmon runs (each a genetically distinct wild stock) are now extinct and most of the rest are at moderate to high risk.

To get around this problem and to increase the efficiency of salmon harvesting, commercial fishing operations have modified the salmon's natural cycle by using *salmon ranching*: Salmon returning to the mouths of rivers are captured; their eggs are removed and hatched in hatcheries, and the mature fish are canned for sale. The young salmon are kept in tanks until they reach a certain size before being released to make their way to the ocean. By providing a protected environment, hatcheries allow a much larger proportion of young salmon to survive until they are ready to migrate to the ocean than is the case for wild salmon in their altered habitats. When, after several years, the surviving adult salmon return to breed and die, they can be harvested by placing nets across the streams (or by diverting them through chutes directly

A: 15 (especially wheat, rice, corn, and potato)

into canneries) to repeat this fairly efficient artificial cycle (Figure 12-19). Roughly two-thirds of the salmon now returning to the mouth of the Columbia River began their lives in hatcheries.

However, there are problems with hatchery-raised fish. Their close quarters permit rapid spread of diseases, and because of their genetic uniformity they are more susceptible to diseases and environmental stress after release. Ranch salmon that interbreed with wild ones reduce the genetic variability of the wild fish and their ability to survive.

What Is Being Done to Restore the Salmon? In 1980 Congress passed the Northwest Power Act, which set up a mechanism for developing and implementing long-range plans to meet the region's electricity needs and to rebuild wild and hatchery-raised salmon and other fish populations. This act is administered by the Northwest Power Planning Council, a government and management body that combines both federal and state authority.

With the approval of the council, publicly owned utilities formed a consortium called the Washington Public Power Supply System (WPPSS), which developed plans to build five nuclear power plants. But huge cost overruns, high interest rates, and a rapidly falling demand for electricity in the 1980s did the entire project in. In July 1983, WPPSS and the 100 utilities whose credit was used to borrow money defaulted on their bonds—the largest municipal bond default in the nation's history.

The Northwest Power Planning Council also began experimenting with ways to build up the Columbia River's salmon runs. This is being done by **(1)** building hatcheries upstream of the dams and releasing juveniles from these hatcheries to underpopulated streams (so they will return to them as adults to reproduce); **(2)** using trucks and other devices to transport young salmon around dams during periods of heavy downstream migration—and in some cases turning off turbines to allow the juveniles to swim safely over dams; **(3)** putting more than 64,000 kilometers (40,000 miles) of stream off limits for hydropower development; and **(4)** obliterating old logging roads and reducing the runoff of silt from existing dirt logging roads above salmon spawning streams.

The entire project is expected to cost at least $2 billion and take two decades of hard work. Will it work? No one knows because it will take decades to see whether the salmon populations can be rebuilt. In 1994, a federal appeals court ruled that the Northwest Power Planning Council must get more serious about saving wild salmon species. Despite problems, this program demonstrates that people with diverse and often conflicting economic, political, and environmental interests can work together to try new ideas and develop potentially sustainable solutions to complex resource management issues.

13-8 TOO MUCH WATER

What Are the Causes and Effects of Flooding? Natural flooding by streams, the most common type of flooding, is caused primarily by heavy rain or rapid melting of snow; this causes water in the stream to overflow its normal channel and to cover the adjacent area, called a **floodplain** (Figure 13-20).

People have settled on floodplains since the beginnings of agriculture. The soil is fertile, and ample water is available for irrigation. Communities can use the water for transportation of people or goods, and floodplains (being flat) are suitable for cropland, buildings, highways, and railroads. People may decide that all of these benefits outweigh the risk of flooding (if they are even aware of the risk).

On marine coasts, flooding results most often from wind-driven storm surges and rain-swollen streams associated with tropical cyclones (typhoons and hurricanes). Flooding can also occur on the shorelines of large inland lakes.

Floods are a natural phenomenon and have several benefits. They provide the world's most productive farmland because they are regularly covered with nutrient-rich silt left after flood waters recede. Floods also recharge groundwater under plains and refill wetlands, which help keep rivers flowing during droughts and provide important breeding and feeding grounds for fish and waterfowl.

Even though floods are usually considered to be natural disasters, human activities have contributed to the sharp rise in flood deaths and damages since the 1960s. The main way humans increase the probability and severity of flooding is by removing vegetation—through logging, overgrazing by livestock, forest fires, certain mining activities, and urbanization. Vegetation retards surface runoff and increases infiltration. When the vegetation is removed by human activities (or by natural occurrences), precipitation runs off into streams more rapidly, often with a large load of sediment, which increases the chance of flooding by making the stream shallower.

Each year, flooding (Figure 13-20) kills thousands of people and causes tens of billions of dollars in property damage, including massive flooding in the United States in 1993 along the flood plains of the Mississippi, Missouri, and Illinois rivers (Figure 13-21).

Floods, like droughts, are usually considered natural disasters, but since the 1960s human activities have contributed to the sharp rise in flood deaths and damages. Indeed, floods account for about 39% of all

Q: What percentage of the earth's land area is suitable for cultivation?

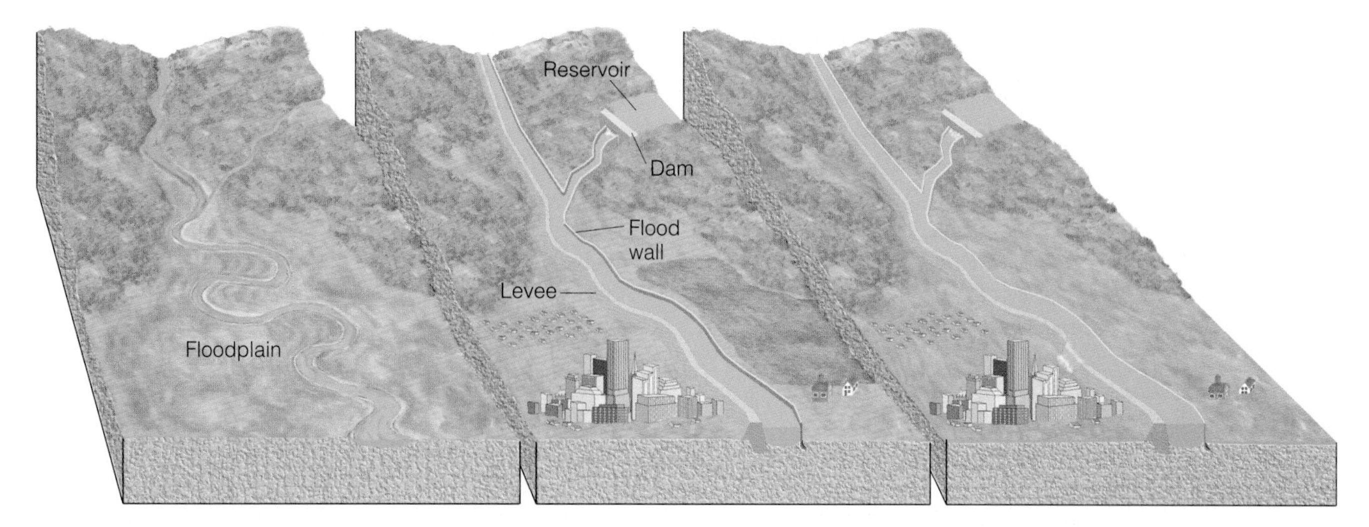

Figure 13-20 Land in a natural floodplain (left) is often flooded after prolonged rains. When the floodwaters recede, alluvial deposits of silt are left behind, creating a nutrient-rich soil. To reduce the threat of flooding (and thus allow people to live in floodplains), rivers have been dammed to create reservoirs, to store and release water as needed. They have also been narrowed, straightened, and equipped with protective levees and walls (middle). However, these alterations can give a false sense of security to floodplain dwellers, who actually live in high-risk areas. In the long run, such measures can greatly increase flood damage. Although dams, levees, and walls prevent flooding in most years, they can be overwhelmed by prolonged rains (right), as happened in the Midwestern United States during the summer of 1993 (Figure 13-21).

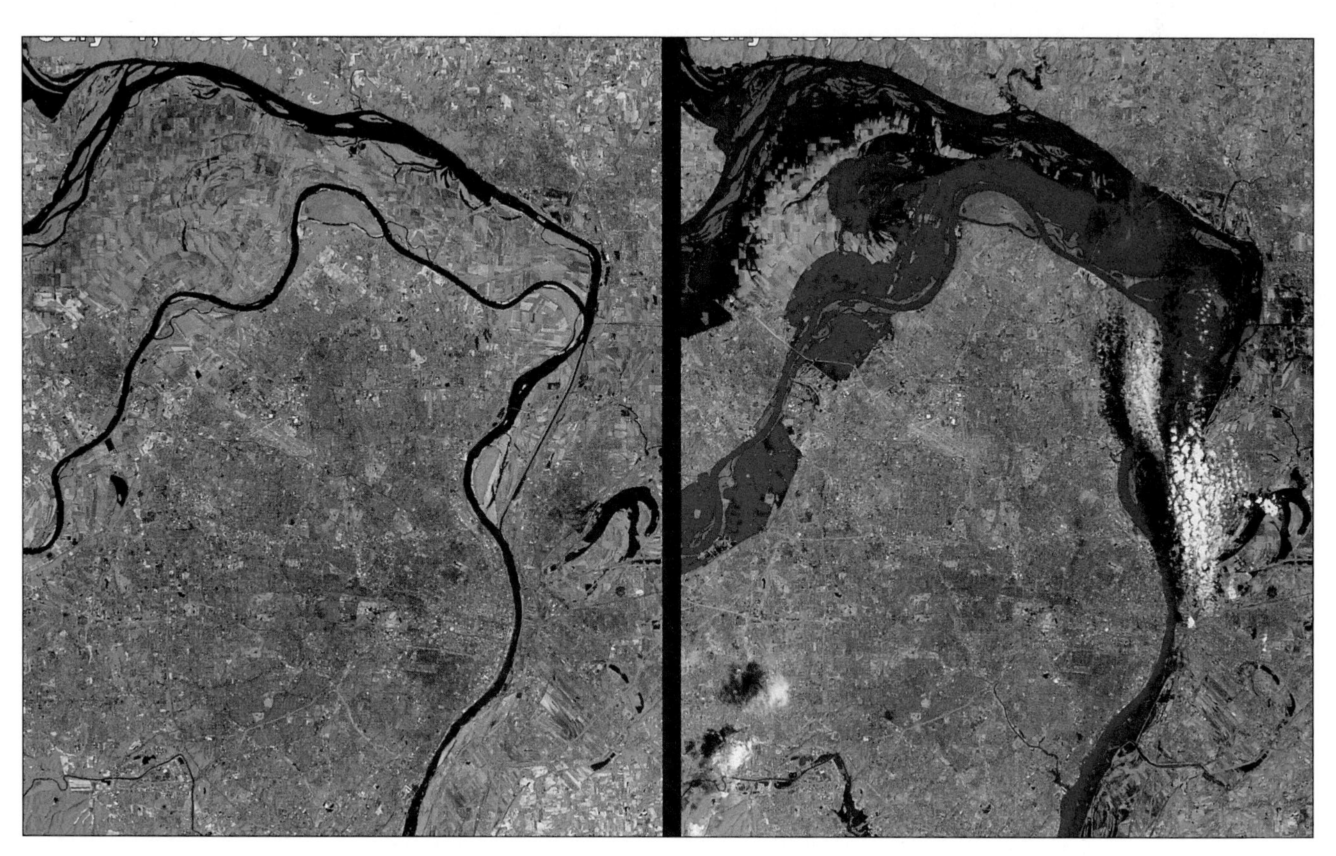

Figure 13-21 Satellite images of the area around St. Louis before flooding on July 4, 1988 (left), and on July 18, 1993, after severe flooding from prolonged rains. (Earth Satellite Corporation)

A: About 11%

deaths from natural disasters—more than any other type of natural disaster.

One way humans increase the likelihood of flooding and the resulting damage is by removing water-absorbing vegetation, especially on hillsides (Figure 13-22); another is by living on floodplains. In many developing countries, the poor have little choice but to try to survive in flood-prone areas. In developed countries, however, people deliberately settle on floodplains and then expect dams, levees, and other devices to protect them from flood waters—solutions that don't work when heavier-than-normal rains come (Figure 13-21).

Urbanization increases flooding (even with only moderate rainfall) by replacing vegetation and soil with highways, parking lots, and buildings, which leads to rapid runoff of rainwater. If sea levels rise during the next century, as projected, many low-lying coastal cities, wetlands, and croplands will be under water.

Case Study: Living Dangerously in Bangladesh
Bangladesh (Figure 11-16) is one of the world's most densely populated countries, with 123 million people—nearly one-half of the population of the United states—packed into an area roughly the size of Wisconsin. Its population is projected to reach 180 million by 2025. Bangladesh is also one of the world's poorest countries, with an average per capita GNP of about $260, or 71¢ per day.

More than 80% of the country consists of floodplains and shifting islands of silt formed by a delta at the mouths of three major rivers. Runoff from annual monsoon rains in the Himalaya Mountains of India, Nepal, Bhutan, and China flows down the rivers through Bangladesh into the Bay of Bengal.

The people of Bangladesh are used to moderate annual flooding during the summer monsoon season, and they depend on the floodwaters to grow rice, their primary source of food. The annual deposit of eroded Himalayan soil in the delta basin also helps maintain soil fertility.

In the past, great floods occurred every 50 years or so, but during the 1970s and 1980s they came about every 4 years. Bangladesh's flood problems begin in the Himalayan watershed, where a combination of rapid population growth, deforestation, overgrazing, and unsustainable farming on steep, easily erodible mountain slopes has greatly diminished the ability of the soil to absorb water. Instead of being absorbed and released slowly, water from the monsoon rains runs off the denuded Himalayan foothills, carrying vital topsoil with it (Figure 13-22). This runoff, combined with heavier-than-normal monsoon rains, has increased the severity of flooding along Himalayan rivers and in Bangladesh—another example of connections.

In 1988 a disastrous flood covered two-thirds of Bangladesh's land area for 3 weeks and leveled 2 million homes after the heaviest monsoon rains in

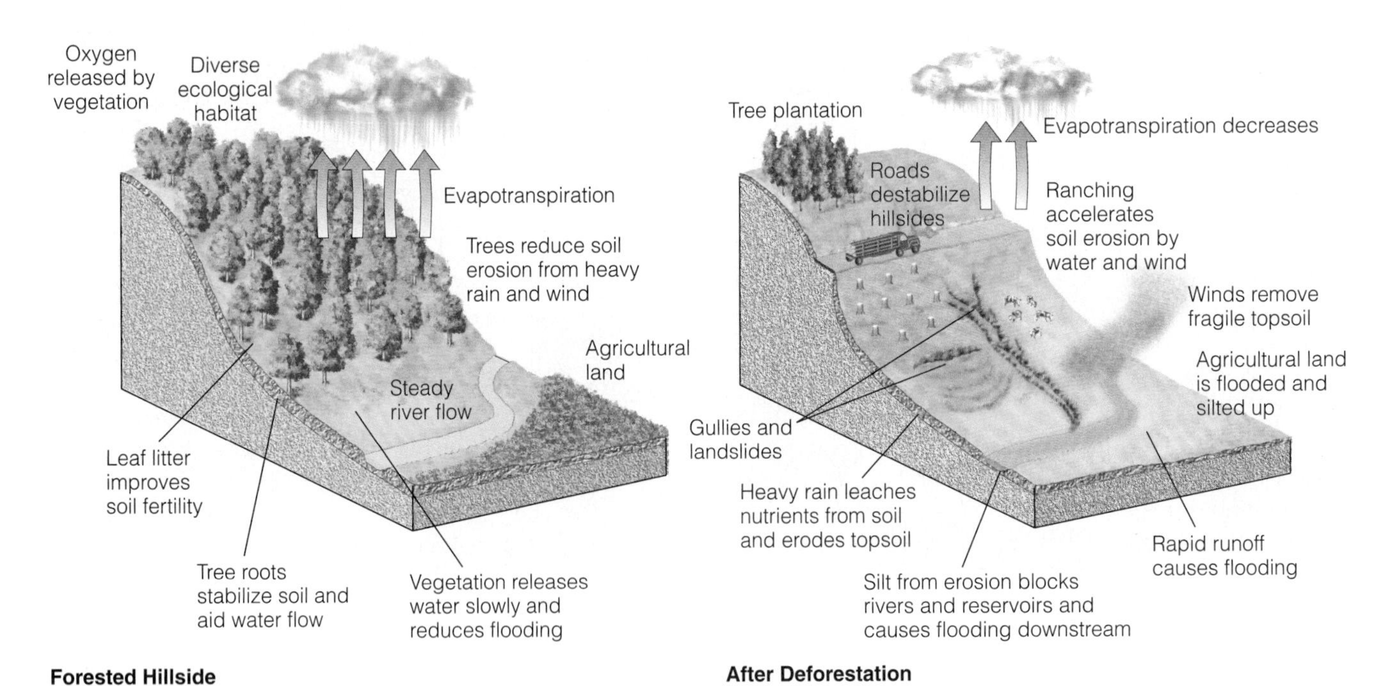

Figure 13-22 A hillside before and after deforestation. Once a hillside has been deforested—for timber and fuelwood, grazing livestock, or unsustainable farming—water from precipitation rushes down the denuded slopes, eroding precious topsoil and flooding downstream areas. A 3,000-year-old Chinese proverb says, "To protect your rivers, protect your mountains."

Q: What percentage of the world's food is produced on irrigated cropland?

70 years. At least 2,000 people drowned, and 30 million people—one of every four—were left homeless. More than a quarter of the country's crops were destroyed, costing at least $1.5 billion and causing thousands of people to die of starvation.

In their struggle to survive, the poor in Bangladesh have cleared many of the country's coastal mangrove forests for fuelwood, cultivation of crops, and aquaculture ponds for raising shrimp. This has led to more severe flooding because these coastal wetlands shelter the low-lying coastal areas from storm surges and cyclones.

Sixteen devastating cyclones have slammed into Bangladesh since 1961; in 1970, as many as 1 million people drowned in one storm. Another surge killed an estimated 130,000 people in 1991. Most of those who died had been forced by overcrowding and lack of land to try to farm sandbars and coastal islands. Damages and deaths from cyclones in areas still protected by mangrove forests are much lower than in areas where the forests have been cleared.

The severity of this problem can be reduced if Bangladesh, Bhutan, China, India, and Nepal cooperate in reforestation and flood control measures and reduce their population growth.

Solutions: How Can We Reduce Flooding Risks? Ways humans can reduce the risk include **(1)** straightening and deepening streams (channelization), **(2)** building levees and dams, **(3)** restoring wetlands to take advantage of the natural flood control provided by floodplains, and **(4)** instituting floodplain management to get people out of flood-prone areas.

One controversial way of reducing flooding is *channelization*, in which a section of a stream is deepened, widened, or straightened to allow more rapid runoff (Figure 13-20, middle). Although channelization can reduce upstream flooding, the increased flow of water can also increase upstream bank erosion and downstream flooding and deposition of sediment.

Artificial levees and embankments along stream banks can reduce the chances of water overflowing into nearby floodplains. They may be permanent or temporary (such as sandbags placed when a flood is imminent). Levees, like channelization, contain and speed up stream flow and increase the water's capacity for doing damage downstream. If a levee breaks or the water spills over it, floodwater may persist long after the stream discharge has decreased (Figure 13-20, right).

Many people in the Midwestern United States learned this harsh lesson during massive flooding from heavy rains during the summer of 1993 (Figure 13-21). After the floodwaters receded, a toxic sludge laced with pesticides, industrial wastes, and raw sewage was left on large areas of land along the floodplains of the Mississippi, Missouri, and Illinois rivers. Severe erosion from the 1993 flooding reduced crop yields on already heavily eroded fields.

A levee can contribute to increased destruction when major floods occur, but the destruction happens *downriver* from each levee. Thus, downriver landowners need levees to protect against upriver levees, which are needed to protect against flood crests caused by levees further upriver. The result is a spiraling *levee race* that can eventually cause destruction for almost everyone when the incredible force of a hugely flood-swollen river inevitably spills over or takes out levees.

A *flood control dam* built across a stream can hold back, store, and release water more gradually (Figure 13-20, middle). The dam and its reservoir may also provide such secondary benefits as hydroelectric power, water for irrigation, and recreational facilities. A reservoir can reduce floods only if its water level is kept low. However, it is much more profitable for dam operators to keep water levels high (for producing electricity and supplying irrigation water). As a result, after prolonged rains the reservoir can overflow, or operators may release large volumes of water to prevent overflow, thereby worsening the severity of flooding downstream. The reservoir gradually fills with sediment (that once fertilized downstream floodplains) until it is useless, and it also has other drawbacks (Figure 13-10).

Some flood control dams have failed for one reason or another, causing sudden, catastrophic flooding and threatening lives, property, and wildlife. According to the Federal Emergency Management Agency, the United States has about 1,300 unsafe dams in populated areas, and the dam safety programs of most states are inadequate because of weak laws and budget cuts.

Over the past 65 years the U.S. Army Corps of Engineers has spent $25 billion on channelization, dams, and levees. Even so, the financial losses from flood damage have increased steadily during this period. The massive floods in the Midwestern United States in 1993 caused an estimated $12 billion in damage to property and crops; billions more will have to be spent to repair the more than 800 levees and embankments that were topped or breached.

This disaster also demonstrated that dams, levees, and channelization can give a false sense of security and encourage people to settle on floodplains, and thus worsen the severity of flood damage. Some countries are realizing that the risks of managing natural water flow can outweigh the benefits; Germany, for example, plans to bulldoze through dikes and allow parts of floodplains to flood regularly.

From an environmental viewpoint, *floodplain management* is the best approach. The first step is to construct a *flood-frequency curve* (based on historical records and an examination of vegetation) to determine how often *on average* a flood of a certain size occurs in a

A: Almost 40%

particular area. This doesn't tell us exactly when floods will occur, but it provides a general idea of how often they might occur, based on the area's history.

Using these data, a plan is developed **(1)** to prohibit certain types of buildings or activities in high-risk zones, **(2)** to elevate or otherwise floodproof buildings that are allowed on the legally defined floodplain, and **(3)** to construct a floodway that allows floodwater to flow through the community with minimal damage. Floodplain management based on thousands of years of experience can be summed up in one idea: *Sooner or later the river (or the ocean) always wins.*

In the United States, the Federal Flood Disaster Protection Act of 1973 requires local governments to adopt floodplain development regulations in order to be eligible for federal flood insurance. It also denies federal funding to proposed construction projects in designated flood hazard areas. The federal flood insurance program underwrites $185 billion in policies because private insurance companies are unwilling to fully insure people who live in flood-prone areas. This federal program thus can encourage people to build on floodplains and low-lying coastal areas.

Some economists argue that it would make more economic sense if the government would buy the 2% of the country's land that repeatedly floods, instead of continuing to make disaster payments. Others believe that people should be free to live in flood-prone areas, but that those choosing to do so should accept the risk without the assistance of federal flood insurance. Attempts to reduce or eliminate federal flood insurance coverage (or to restrict development on floodplains) are usually defeated politically by intense protests from property owners who already live in these risky areas.

13-9 A SUSTAINABLE WATER FUTURE

An effective strategy for the sustainable use of water (or any resource) involves preserving the ecological integrity of water supply systems, wasting less water, allowing fair access to water supplies (equity), and giving people a say in how water resources are developed and used (participatory decision making). Sustainable water use is based on the commonsense principle stated in an old Inca proverb: "The frog does not drink up the pond in which it lives."

According to water resource expert Sandra Postel, none of the technological solutions for supplying more water—dams, watershed transfers, tapping groundwater, and using water more efficiently—deal with the three underlying forces that can lead humans to use such a potentially renewable resource in unsustainable ways: **(1)** depletion or degradation of a shared resource, which shrinks the resource pie to be shared (the tragedy of the commons); **(2)** population growth, which forces the resource pie to be divided into smaller slices; and **(3)** unequal distribution or access (primarily because of poverty), which means that some countries and people get larger slices than others.

Sustainable use of potentially renewable groundwater means that the rate of extraction should not exceed the rate of recharge. Determining what constitutes sustainable use of shared rivers is much more complex because available water from a river varies with the time of year and with conditions such as drought and higher-than-normal precipitation.

Sustainable water use also requires an integrated plan governing water use, sewage treatment, and water pollution among all users of a water basin (as is done in Great Britain). Such a plan must specify an absolute quantity and quality of water that must be preserved to protect a river's ecological functions, even in dry years. Instead of trying to allocate a fixed amount of water from a river to each country, Sandra Postel suggests that agreements specify what percentage of a river's runoff each country, state, or area gets, with the absolute amount tied to how much is available in a particular year.

Working together to share a limited resource is a difficult endeavor that requires delicate negotiations, especially with countries or areas that control the headwaters of a river and thus have the power to control the amount and quality of water available to downstream users. Despite the difficulties, negotiated agreements are necessary to avoid armed conflicts over water supplies and escalating depletion and degradation of available water resulting from increasing population and development.

In countries where there are clear property rights to surface and groundwater supplies (and laws to protect such rights), *water marketing* can ease shortages and save money. Instead of building an expensive new dam or water diversion system, cities and farmers can buy rights to supplies from others. However, unregulated or monopolistic water markets can lead to exploitive prices, overexploitation of water resources, and inequalities in water distribution.

Many analysts believe that a key to reducing water waste and providing more equitable access to water resources is for governments to phase out subsidies that reduce the price of water for those benefiting from such subsidies (often large farmers and industries). According to Sandra Postel, "By heavily subsidizing water, governments give out the false message that it is abundant and can afford to be wasted—even as rivers are drying up, fisheries are collapsing, and species are going extinct."

Developing and implementing sustainable water strategies for the world's major water basins and groundwater supplies are difficult and controversial

Q: What percentage of the U.S. population lives on the country's farms?

undertakings. However, water experts argue that not developing such strategies will eventually lead to economic and health problems, displace large numbers of environmental refugees, and threaten military and environmental security. Policies focusing mostly on short-term economic gain will lead to long-term economic and environmental pain.

It is not until the well runs dry that we know the worth of water.
BENJAMIN FRANKLIN

CRITICAL THINKING

1. How do human activities increase the harmful effects of prolonged drought? How can these effects be reduced?

2. Explain how dams and reservoirs can cause more flood damage than they prevent. Should all proposed large dam and reservoir projects be scrapped? Explain.

3. Should the World Bank and other international lending agencies (supported and dominated by the United States and other developed countries) continue to make loans for development of large dam and reservoir projects in China, India, Brazil, and other developing countries similar to those carried out on the Colorado and Columbia rivers in the United States? Explain your answer. What are the alternatives?

4. Do you agree or disagree with Dan Beard (former Commissioner of the U.S. Bureau of Reclamation, the federal agency responsible for much of the country's dams and other water development systems), who said, "It is a serious mistake for any region in the world to use what we did on the Colorado and Columbia Rivers as examples to be duplicated." Explain.

5. In the 1920s, U.S. President Herbert Hoover claimed that "every drop of water that runs to the sea without rendering a commercial return is a public waste." Do you agree or disagree with this statement? Explain. What are the ecological implications of putting such a concept into practice, as has been done in the United States in the Colorado River basin (Figure 13-11) and the Columbia River basin (Figure 13-19)?

6. Should federal subsidies of irrigation projects in the western United States (or in the country where you live) be gradually phased out? Explain.

7. Should the price of water for all uses be raised sharply to include more of its environmental costs and to encourage water conservation? Explain. What harmful and beneficial effects might this have on business and jobs, your lifestyle and the lifestyles of any children or grandchildren you might have, the poor, and the environment?

8. List five major ways to conserve water on a personal level (see Appendix 5). Which, if any, of these practices do you now use or intend to use?

9. How do human activities contribute to flooding and flood damage? How can these effects be reduced?

10. Do you agree or disagree with the statement that flooding is a natural part of a river's complex ecological cycle and is not a pathological state that needs to be fixed by building dams and reservoirs? Explain. How would adopting such an approach affect the large numbers of people living in floodplains along the world's river systems?

11. Some analysts argue that the U.S. Federal Flood Disaster Protection Act should be repealed because it encourages people to build on areas with a high risk of flooding by providing them with government-backed insurance (paid for mostly by people who do not live in high flood-risk areas) to rebuild after a flood. Instead, they would replace it with an act that provides emergency relief to victims of a disastrous flood but does not help flood victims rebuild in such risky areas. Do you agree or disagree with this proposal? Explain. Would your position be the same if you lived in an area prone to flooding?

12. Try to come up with analogies to complete the following statements: **(a)** An aquifer is like _______. **(b)** Building a large hydropower dam to prevent flooding is like _______. **(c)** Transferring large amounts of water from northern California to the desert area of southern California is like _______. **(d)** The Aral Sea watershed transfer project is like _______. **(e)** Keeping water prices artificially low is like _______.

PROJECTS

1. In your community,
- **a.** What are the major sources of the water supply?
- **b.** What percentage of the people rely on a community system for water, and what percentage rely on private wells?
- **c.** How is water use divided among agricultural, industrial, power-plant cooling, and public uses? Who are the biggest consumers of water?
- **d.** Where is most groundwater recharged?
- **e.** What has happened to water prices during the past 20 years? Are they too low to encourage water conservation and reuse?
- **f.** What water supply problems are projected?
- **g.** How is water being wasted?

2. Consult with local officials to identify any floodplain areas in your community. Develop a map showing these areas and the types of activities (such as housing, manufacturing, roads, recreational land) found on these lands.

3. Develop a water conservation plan for your school and submit it to school officials.

4. Make a concept map of this chapter's major ideas, using the section heads and subheads and the key terms (in boldface). Look at the inside back cover and in Appendix 4 for information about concept maps.

A: About 1.8% (only about 650,000 Americans are full-time farmers)

14 MINERALS AND SOIL RESOURCES

THE GREAT TERRAIN ROBBERY

Want to get rich at the taxpayers' expense? You can if you know how to make use of a little-known mining law passed in 1872 to encourage mining of gold, silver, lead, copper, uranium, and other hard-rock minerals on U.S. public lands.

Under this 1872 law, a person or corporation can assume legal ownership of any public land not classified as wilderness or park simply by *patenting* it. This involves declaring their belief that the land contains valuable hard-rock minerals, spending $500 to improve the land for mineral development, filing a claim, and then paying the federal government $6–12 per hectare ($2.50–5.00 an acre) for the land. So far, lands containing at least $385 billion of the public's mineral resources have been transferred to private interests at these bargain-basement prices.

In 1993 the Manville Corporation paid $10,000 for federal land in Montana that contains an estimated $32 billion of platinum and palladium. Secretary of Interior Bruce Babbitt was forced by the courts in 1994 to sell to a Canadian corporation for only $10,000 federal land in Nevada containing an estimated $10–15 billion worth of gold. In 1995 he had to sign over federal land in Idaho containing $1 billion worth of minerals to a Danish company that paid just $275 for the bonanza.

Mining companies operating under this law—*almost half of them controlled by foreign corporations*—annually remove mineral resources worth about $3.6 billion on land they have bought at absurdly low prices. They pay no royalties to the U.S. Treasury. If mining companies had to pay a 12.5% royalty on the minerals they have removed, as petroleum companies do on oil they remove, the government would have received at least $48 billion in revenues (in 1988 dollars).

The 1872 law does not even require that patented property be mined. Land speculators have often purchased such property at 1872 prices and then sold it or leased it for thousands of times what they paid. In 1986 a mining company paid the government $42,500 for oil shale land in Colorado and a month later sold it to Shell Oil for $37 million. Other owners have developed sites as casinos, ski resorts, golf courses, and vacation-home developments. According to Senator

Figure 14-1 Bear Trap Creek in Montana is one example of how gold mining can contaminate water with highly toxic cyanide or mercury used to extract gold from its ore. Air and water also convert the sulfur in gold ore to sulfuric acid, which releases toxic metals such as cadmium and copper into streams and groundwater. (Bryan Peterson)

Dale Bumpers, "The 1872 mining law is a license to steal and the biggest scam in America."

Once a company has obtained a patent right in valuable wilderness areas, national forests, or other public lands, all it has to do is threaten to begin mining. The only way the government can stop the mining is to buy out the claim at taxpayers' expense. Thus, a company purchasing a claim from taxpayers for only a few dollars can then blackmail the government (taxpayers) into buying the land back for millions of dollars. By holding such public lands hostage a mining company can make millions without even having to take the economic risks of actual mining.

There is also no provision in the 1872 law requiring reclamation of damaged land. Acids and toxic metal residues continue to leach from thousands of the almost 558,000 abandoned hard-rock mines and open pits, mostly in the West. One result is some 19,000 kilometers (12,000 miles) of polluted streams (Figure 14-1); another result is that 56 abandoned mine sites are now listed on the EPA's Superfund list of the nation's worst hazardous waste dump sites.

It's common for a company to mine a site, abandon it, file for bankruptcy, and leave the public with the cleanup bill (see Case Study, p. 353). It is estimated that the taxpayers' cleanup costs for all land damaged by hard-rock mining on existing or sold public lands will be $33–72 billion, depending on whether groundwater and toxic waste cleanup are included. Environmentalists have been trying, without success, to have this law revised to protect taxpayers and the environment.

We live on a dynamic planet. Energy from the sun and from the earth's interior, and the action of water, have created continents, mountains, valleys, plains, and ocean basins—an ongoing process that continues to change the landscape. Nonrenewable resources in the earth's crust (Figure 1-11) are among the foundations of modern civilization; no less important is potentially renewable *soil* (Figures 5-13 and 5-16). In addition to food, soil indirectly provides us with wood, paper, fiber, and medicines, and it helps to purify the water we drink and to decompose and recycle biodegradable wastes. There is no substitute for fertile soil. Unfortunately, human activities that lead to rapid soil erosion can convert this potentially renewable resource into a nonrenewable resource. Preventing such loss of vital earth capital should be one of our highest priorities.

Below that thin layer comprising the delicate organism known as the soil is a planet as lifeless as the moon.

G. Y. Jacks and R. O. Whyte

In this chapter we seek answers to the following questions:

- What are the major geologic processes occurring on and in the earth?
- How are nonrenewable mineral resources found and removed?
- How fast are nonfuel mineral supplies being used up?
- How can we increase supplies of key minerals?
- What are the environmental effects of extracting and using mineral resources?
- Why should we worry about soil erosion and soil degradation?
- How can we control erosion and reduce the loss of nutrients from topsoil?

14-1 GEOLOGIC PROCESSES

What Is the Earth's Structure? As the primitive earth cooled over eons, its interior separated into three major concentric zones, which geologists identify as the core, the mantle, and the crust (Figure 4-5). Various indirect measurements indicate that the earth's innermost zone, the **core**, is made mostly of iron (with perhaps some nickel). The core has a solid inner part, surrounded by a liquid core of molten material.

The earth's core is surrounded by a thick, solid zone called the **mantle**. This largest zone of the earth's interior is rich in iron (its major constituent), silicon, oxygen, and magnesium. Most of the mantle is solid rock, but under its rigid outermost part there is a zone of very hot, partly melted rock that flows like soft plastic. This plastic region of the mantle is called the *asthenosphere*.

The outermost and thinnest zone of the earth is called the **crust**. It consists of the *continental crust*, which underlies the continents (including the continental shelves extending into the oceans), and the *oceanic crust*, which underlies the ocean basins and covers 71% of the earth's surface (Figure 14-2). It is

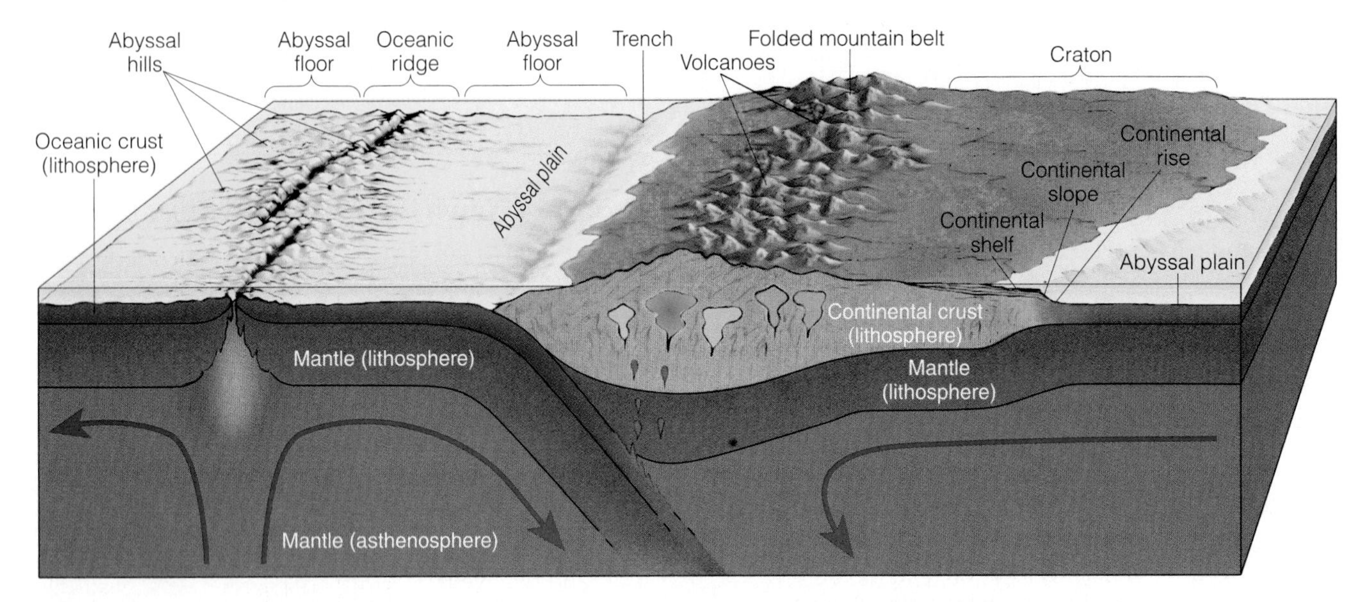

Figure 14-2 Major features of the earth's crust and upper mantle. The lithosphere, composed of the crust and outermost mantle, is rigid and brittle. The asthenosphere, a zone in the mantle, can be deformed by heat and pressure (that is, it is plastic).

from the earth's crust that mineral resources and soil come, as do the elements that make up living organisms. The three types of rocks—igneous, sedimentary, and metamorphic—found in the earth's mantle and core are changed from one type to another in the *rock cycle* (Figure 5-10).

The earth's crust, which is still forming in various places, is composed of minerals and rocks. It is the source of virtually all the nonrenewable resources we use: fossil fuels, metallic minerals, and nonmetallic minerals (Figure 1-11). It is also the source of soil (Figures 5-13 and 5-16) and the elements (Figure 3-4) that make up our bodies and those of other living organisms. Most of the over 2,000 identified minerals occur as inorganic compounds formed by various combinations of the eight elements that make up 98.5% by weight of the earth's crust (Figure 3-11).

What Is Plate Tectonics? Internal Geologic Processes We tend to think of the earth's crust, mantle, and core as fairly static and unchanging. However, these parts of the earth are constantly changing because of geologic processes taking place within the earth and on the earth's surface, most over thousands to millions of years.

A map of the earth's earthquakes and volcanoes shows that most of these phenomena occur along certain lines or belts on the earth's surface (Figure 14-3a). The areas of the earth outlined by these major belts are called **plates** (Figure 14-3b). They are about 100 kilometers (60 miles) thick and are composed of the crust and the rigid, outermost part of the mantle (above the asthenosphere)—a combination called the **lithosphere**. These plates move constantly, supported by the slowly flowing asthenosphere like large pieces of ice floating on the surface of a lake during the spring breakup. Some plates move faster than others, but a typical speed is about the rate at which fingernails grow.

The theory explaining the movements of the plates and the processes that occur at their boundaries is called **plate tectonics**. The concept, which became widely accepted by geologists in the 1960s, was developed from an earlier idea called *continental drift*. Throughout the earth's history, continents have split and joined as plates have drifted thousands of kilometers back and forth across the planet's surface (Figure 6-12).

Plate motion produces mountains (including volcanoes), the oceanic ridge system, trenches, and other features of the earth's surface (Figure 14-2); certain natural hazards are likely to be found at plate boundaries (Figure 14-3a); and plate movements and interactions also concentrate many of the minerals we extract and use.

The theory of plate tectonics also helps explain how certain patterns of biological evolution occurred (Section 6-3). By reconstructing the course of continental drift over millions of years (Figure 6-12), we can trace how life-forms migrated from one area to another when continents that are now far apart were still joined together. As the continents separated, speciation occurred as populations became geographically and reproductively isolated (Figure 6-11).

Q: In total annual sales, what is the biggest industry in the United States?

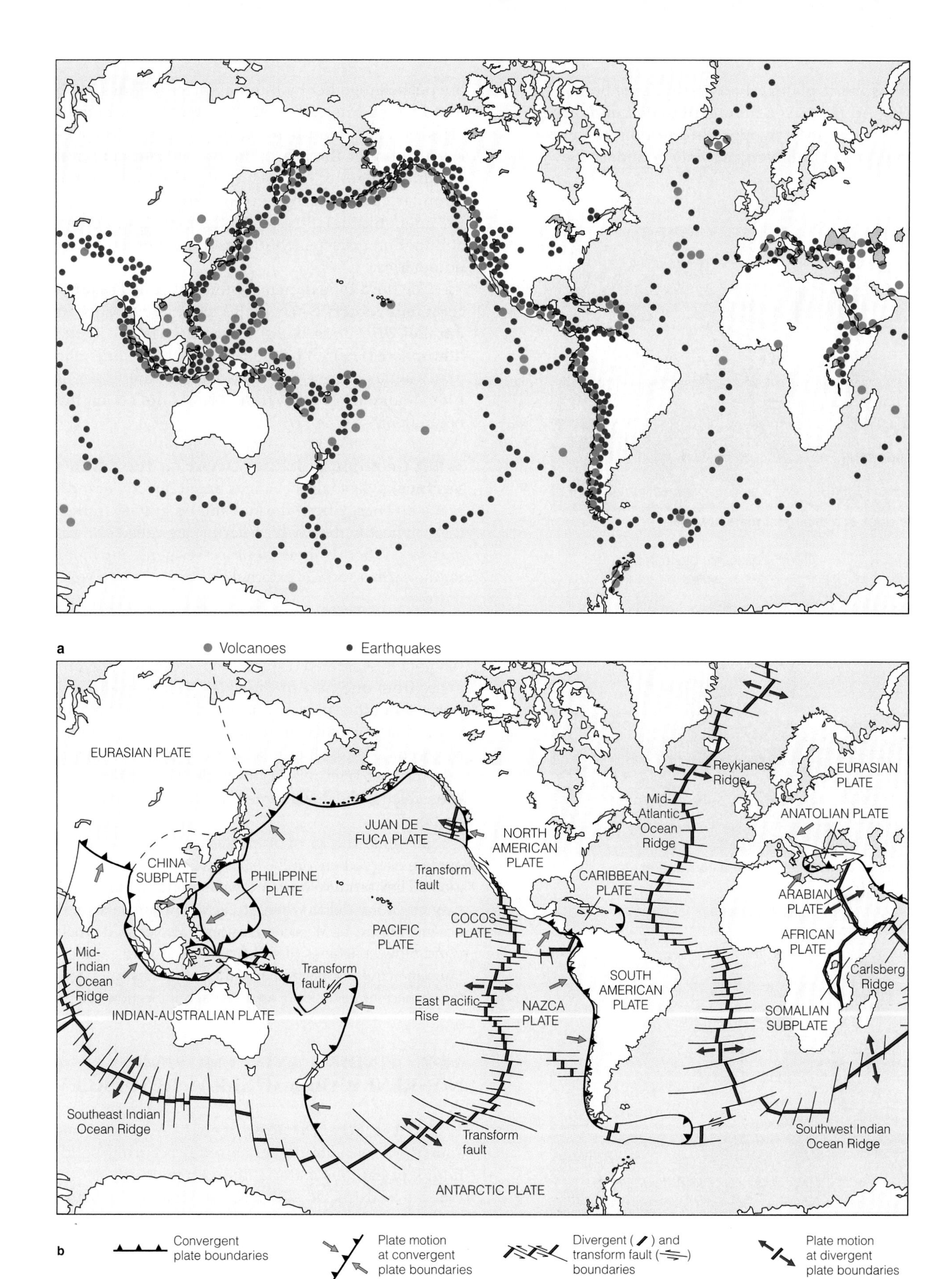

Figure 14-3 **(a)** Earthquake and volcano sites are distributed mostly in bands along the planet's surface. **(b)** These bands correspond to the patterns for the types of lithospheric plate boundaries shown in Figure 14-4.

Lithospheric plates have three types of boundaries (Figure 14-4). At a **divergent plate boundary** the plates move apart in opposite directions (Figure 14-4, top), and at a **convergent plate boundary** they are pushed together by internal forces (Figures 14-3b and 14-4, middle). At most convergent plate boundaries, oceanic lithosphere is carried downward (subducted) under the island arc or the continent at a **subduction zone**. A trench ordinarily forms at the boundary between the two converging plates (Figure 14-4, middle). Stresses in the plate undergoing subduction cause earthquakes at convergent plate boundaries.

The third type of plate boundary, called a **transform fault**, occurs where plates move in opposite but parallel directions along a fracture (fault) in the lithosphere (Figure 14-4, bottom). In other words, the plates slide past one another. Like the other types of plate boundaries, most transform faults are on the ocean floor.

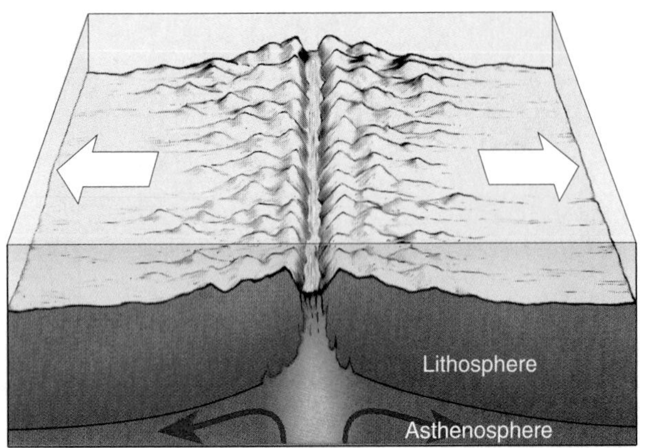

Oceanic ridge at a divergent plate boundary

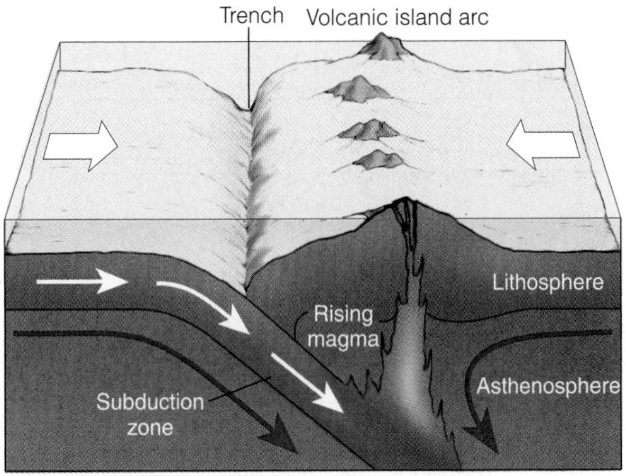

Trench and volcanic island arc at a convergent plate boundary

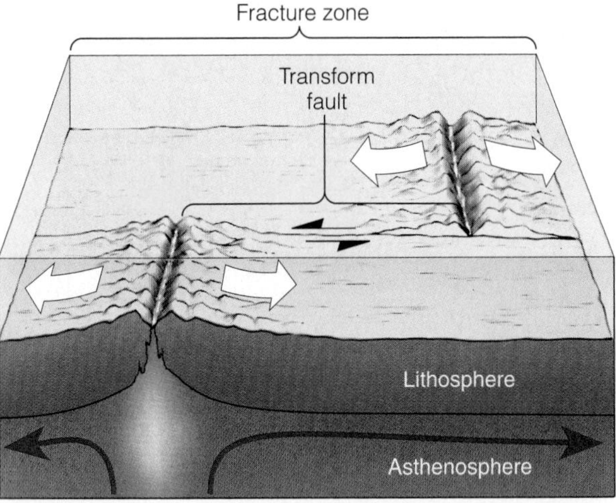

Transform fault connecting two divergent plate boundaries

Figure 14-4 Types of boundaries between the earth's lithospheric plates. All three boundaries occur both in oceans and on continents.

What Geologic Processes Occur on the Earth's Surface? Geological changes based directly or indirectly on energy from the sun and on gravity (rather than on heat in the earth's interior) are called *external processes*. Whereas internal processes generally build up the earth's surface, external processes tend to wear it down.

A major external process is **erosion**. It is the process by which loosened material (as well as material not yet separated) is dissolved, loosened, or worn away from one part of the earth's surface and deposited in other places. Loosened material that can be eroded is usually produced by *weathering*, which can occur as a result of mechanical processes, chemical processes (Figure 5-11), or both. Weathering is responsible for the development of soil, as discussed in Section 5-8.

Streams, the most important agent of erosion, operate everywhere on the earth except in the polar regions. They produce ordinary valleys and canyons, and they may form deltas where streams flow into lakes and oceans (Figure 14-5). Some erosion is also caused when wind blows particles of soil from one area to another. Human activities, particularly those that destroy vegetation, accelerate erosion, as discussed in Section 14-6.

14-2 FINDING AND REMOVING NONRENEWABLE MINERAL RESOURCES

What Are Mineral Resources? A **mineral resource** is a concentration of naturally occurring solid, liquid, or gaseous material in or on the earth's crust that can be extracted and processed into useful materials at an affordable cost. The earth's internal and external processes have produced numerous mineral resources, which on a human time scale are essentially nonrenewable because of the slowness of the rock cycle. Mineral resources (Figure 1-11) include *energy*

Q: How many people on average does one U.S. farmer feed?

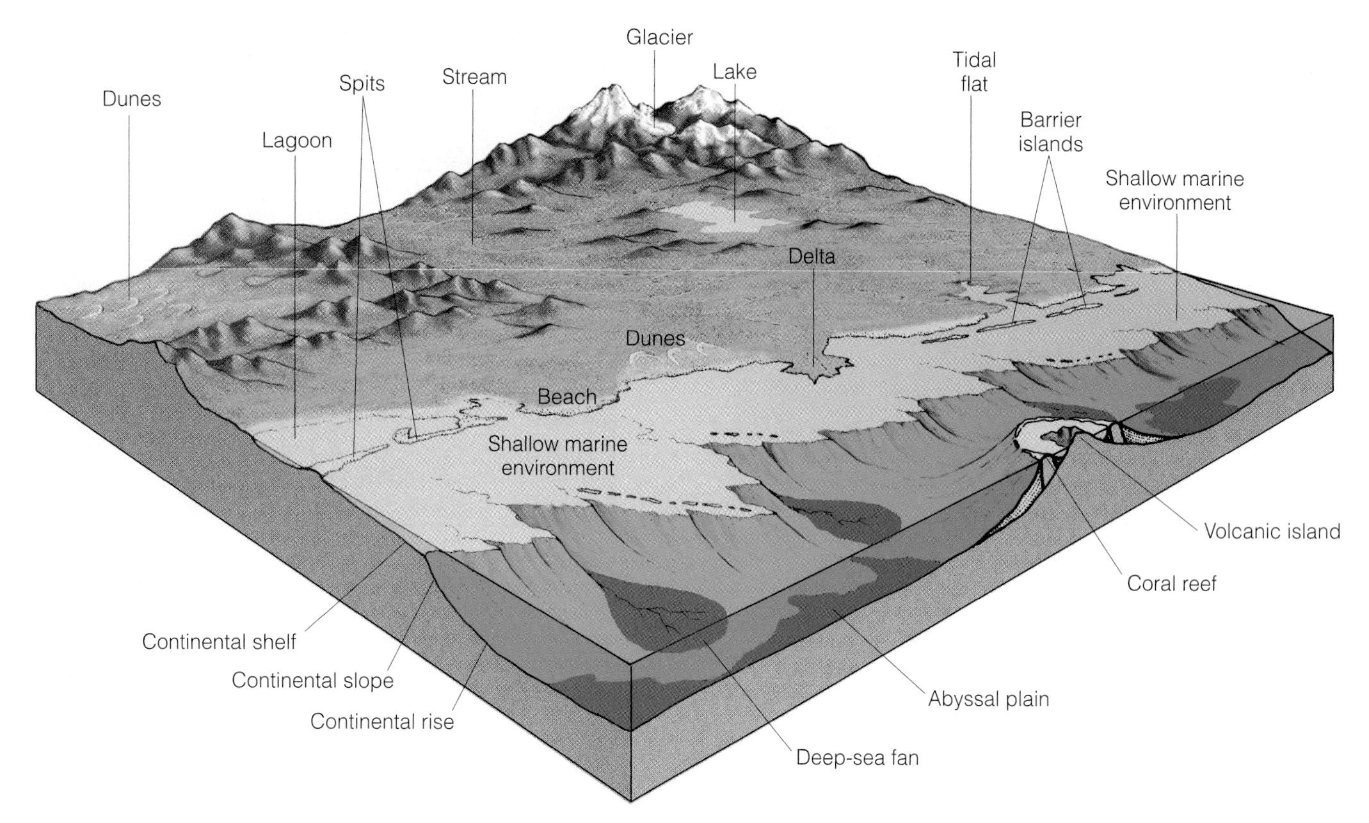

Figure 14-5 The variety of landforms and sedimentary environments depicted here is mainly the result of *external processes*, powered primarily by solar energy (as it drives the hydrologic cycle and wind) and gravity, with some assistance from organisms such as reef-building corals.

resources (coal, oil, natural gas, uranium, geothermal energy), *metallic mineral resources* (iron, copper, aluminum), and *nonmetallic mineral resources* (salt, gypsum, clay, sand, phosphates, water, and soil).

An **ore** is a metal-yielding material that can be economically extracted at a given time. To be profitable, copper in copper ore must be concentrated to 86 times its crustal average, gold 1,000 times, and mercury an astonishing 100,000 times.

We know how to find and extract more than 100 nonrenewable minerals from the earth's crust. We convert these minerals into many everyday items that we either use and discard (Figure 3-21) or learn to reuse, recycle, or use less wastefully (Figure 3-22).

The U.S. Geological Survey (USGS) divides mineral resources into two broad categories, *identified* and *undiscovered* (Figure 14-6). **Identified resources** are deposits of a particular mineral resource that have a *known* location, quantity, and quality, or deposits for which these parameters are estimated from direct geological evidence and measurements. **Undiscovered resources** are potential supplies of a particular mineral resource that are *assumed* to exist on the basis of geologic knowledge and theory (although specific locations, quality, and amounts are unknown).

Reserves are identified resources that can be extracted economically at current prices using current mining technology. **Other resources** are identified and undiscovered resources not classified as reserves.

Most published estimates of particular mineral resources refer only to reserves. Reserves can increase when exploration finds previously undiscovered economic-grade mineral resources, or when identified subeconomic-grade mineral resources become economically viable because of new technology or higher prices. Figure 14-6 shows a region labeled *potential reserves*, indicating how resources can be expanded. Theoretically, all of the *other resources* could eventually be converted to reserves, but this is highly unlikely.

How Do We Find and Remove Mineral Deposits? Mining companies use several methods to find promising mineral deposits. Geological information about plate tectonics and mineral formation suggests areas worthy of closer study. Aerial photos and satellite images sometimes reveal rock formations associated with certain minerals. Other instruments on aircraft and satellites can detect mineral deposits by their effects on the earth's magnetic or gravitational fields.

A: 140 (105 at home and 35 abroad)

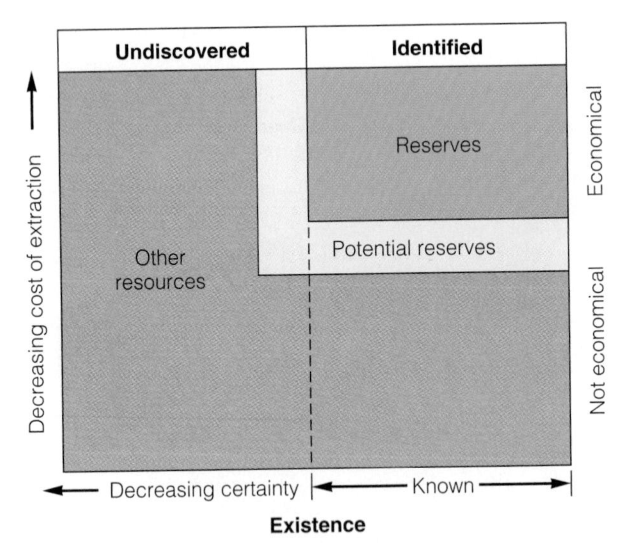

Figure 14-6 General classification of mineral resources. (The area shown for each class does not represent its relative abundance.) In theory, all mineral resources classified as *other resources* could become reserves because of rising mineral prices, improved mineral location and extraction technology, or both. In practice, geologists expect only a fraction of other resources to become reserves. The area labeled *potential reserves* shows the way reserves normally increase.

Figure 14-7 This open-pit copper mine in Bingham, Utah, is the largest human-made hole in the world—4.0 kilometers (2.5 miles) in diameter and 0.8 kilometer (0.5 mile) deep. The amount of material removed from this mine is seven times the amount moved to build the Panama Canal. (Don Green/Kennecott Copper Corporation, now owned by British Petroleum)

After profitable deposits of minerals are located, deep deposits are removed by **subsurface mining** and shallow deposits by **surface mining**. In surface mining, mechanized equipment strips away the **overburden** of soil and rock and usually discards it; such waste material is called **spoil**. Surface mining extracts about 90% of the mineral and rock resources and more than 60% of the coal by weight in the United States.

The type of surface mining used depends on the resource being sought and on local topography. In **open-pit mining** (Figure 14-7), machines dig holes and remove ores such as iron and copper. This method is also used for sand and gravel and for building stone such as limestone, sandstone, slate, granite, and marble. Another form of surface mining is **dredging**, in which chain buckets and draglines scrape up underwater mineral deposits.

Strip mining is surface mining in which bulldozers, power shovels, or stripping wheels remove the overburden in strips. It is used mostly for extracting coal and some phosphate rock. Most surface-mined coal is removed by area strip mining or contour strip mining, depending on the terrain.

Area strip mining is used where the terrain is fairly flat. An earthmover strips away the overburden, and then a power shovel digs a cut to remove the mineral deposit, such as coal. After the mineral is removed, the trench is filled with overburden and a new cut is made parallel to the previous one. The process is repeated over the entire site. If the land is not restored, area strip mining leaves a wavy series of highly erodible hills of rubble called *spoil banks* (Figure 14-8).

Contour strip mining is used in hilly or mountainous terrain (Figure 14-9). A power shovel cuts a series of terraces into the side of a hill. An earthmover removes the overburden and a power shovel extracts the coal; the overburden from each new terrace is dumped onto the one below. Unless the land is restored, a wall of dirt is left in front of a highly erodible bank of soil and rock called a *highwall*.

Sometimes giant augers are used to drill horizontally into a hillside to extract underground coal. There is also growing use of extracting coal in parts of West Virginia, Kentucky, Tennessee, Virginia, and Pennsylvania by a form of surface mining called *mountaintop removal*. A powerful $100-million machine called a dragline that can be as high as 20 stories is used to completely remove the top of a mountain to get at the coal below. The debris is dumped into the valleys. Aerial surveys reveal that 15–25% of the mountaintops in the south-central part of West Virginia are being removed by this technique. Indeed, if this type of mining continues unabated, environmentalists project that in two decades half the mountain peaks of southern West Virginia's blue-green skyline will be gone.

Q: How much of the world's food and fiber are produced by U.S. farmers?

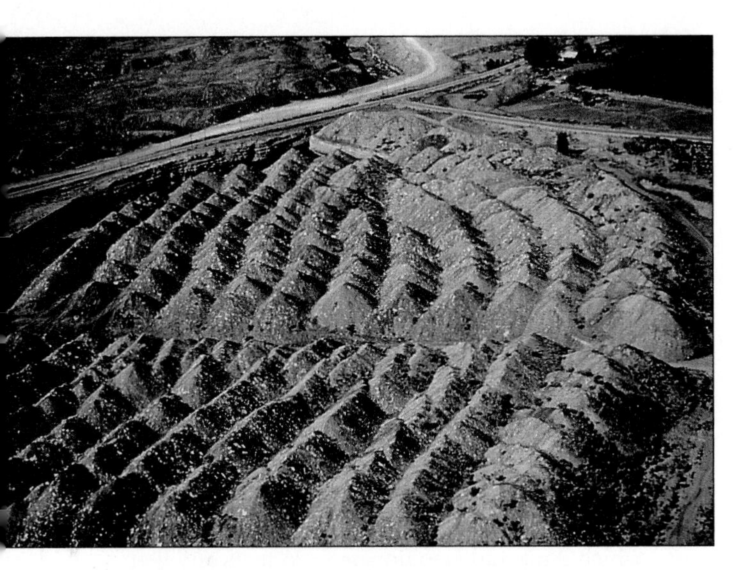

Figure 14-8 Spoil banks, the results of unrestored area strip mining of coal near Mulla, Colorado. Restoration of newly strip-mined areas is now required in the United States, but many previously mined areas have not been restored. (National Archives/EPA Documerica)

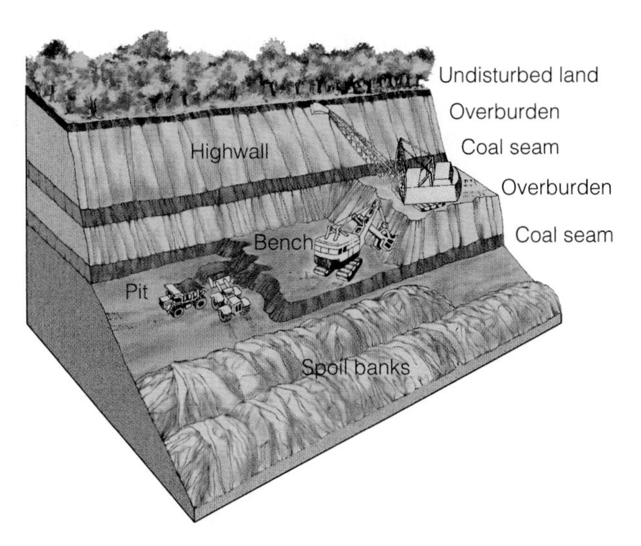

Figure 14-9 Contour strip mining of coal.

In the United States, contour strip mining and mountaintop removal is used mostly to remove coal in the mountainous Appalachian region. Restoring land on such surface mining sites can reverse mining's devastating effects. However, much of the land previously strip-mined for coal in the Appalachian region has not been restored. In addition, about three-fourths of the coal that can be surface mined in the United States is in arid and semiarid regions in the West, where the soil and climate usually prevent full restoration.

Subsurface mining is used to remove coal and various metal ores that are too deep to be extracted by surface mining. Miners dig a deep vertical shaft, blast subsurface tunnels and chambers to get to the deposit, and haul the coal or ore to the surface. In the *room-and-pillar method*, as much as half of the coal is left in place as pillars to prevent the mine from collapsing. In the *longwall method*, a narrow tunnel is dug and then supported by movable metal pillars. After a cutting machine has removed the coal or ore from part of the mineral seam, the roof supports are moved, allowing the earth behind the supports to collapse. No tunnels are left behind after the mining operation has been completed.

Subsurface mining disturbs less than one-tenth as much land as surface mining and usually produces less waste material. However, it leaves much of the resource in the ground and is more dangerous and expensive than surface mining. Roofs and walls of underground mines collapse, trapping and killing miners, explosions of dust and natural gas injure or kill them, and prolonged inhalation of mining dust causes lung diseases.

14-3 ESTIMATING SUPPLIES OF NONRENEWABLE MINERAL RESOURCES

Will There Be Enough Mineral Resources? The future supply of nonrenewable minerals depends on two factors: the actual or potential supply and the rate at which that supply is used. We never completely run out of any mineral. However, a mineral becomes *economically depleted* when the costs of finding, extracting, transporting, and processing the remaining deposits exceed the returns. At that point we have five choices: recycle or reuse existing supplies, waste less, use less, find a substitute, or do without.

As mentioned earlier, most published estimates of the supplies of a given resource refer to *reserves*: known deposits from which a usable mineral can be extracted profitably at current prices (Figure 14-6). **Depletion time** is the time it takes to use up a certain proportion (usually 80%) of the reserves of a mineral at a given rate of use. When experts disagree about depletion times, they are often using different assumptions about supply and rate of use (Figure 14-10). A traditional measure of the projected availability of nonrenewable resources is the **reserve-to-production ratio**: the number of years that proven reserves of a particular nonrenewable mineral will last at current annual production rates.

Estimates of reserves are continually changing because new deposits are often discovered, and new mining and processing can allow some of the minerals classified as other resources (Figure 14-6) to be converted to reserves. Under these circumstances, the

A: About 25%

reserve-to-production ratio is the best available projection of the current estimated supply and its estimated depletion time.

The shortest depletion time assumes no recycling or reuse and no increase in reserves (curve A, Figure 14-10). A longer depletion time assumes that recycling will stretch existing reserves and that better mining technology, higher prices, and new discoveries will increase reserves (curve B, Figure 14-10). An even longer depletion time assumes that new discoveries will further expand reserves and that recycling, reuse, and reduced consumption will extend supplies (curve C, Figure 14-10). Finding a substitute for a resource leads to a new set of depletion curves for the new resource.

While world population doubled between 1950 and 1993, global production of six key metals (aluminum, copper, lead, nickel, tin, and zinc) increased more than eightfold. During this same period, world reserves of copper increased almost fivefold, lead almost threefold, zinc fourfold, and aluminum almost ninefold. Furthermore, the prices of most metals today have changed little in constant dollars over the last 150 years, mostly because of government subsidies and failure to include the harmful environmental effects of metal mining and processing in their market prices.

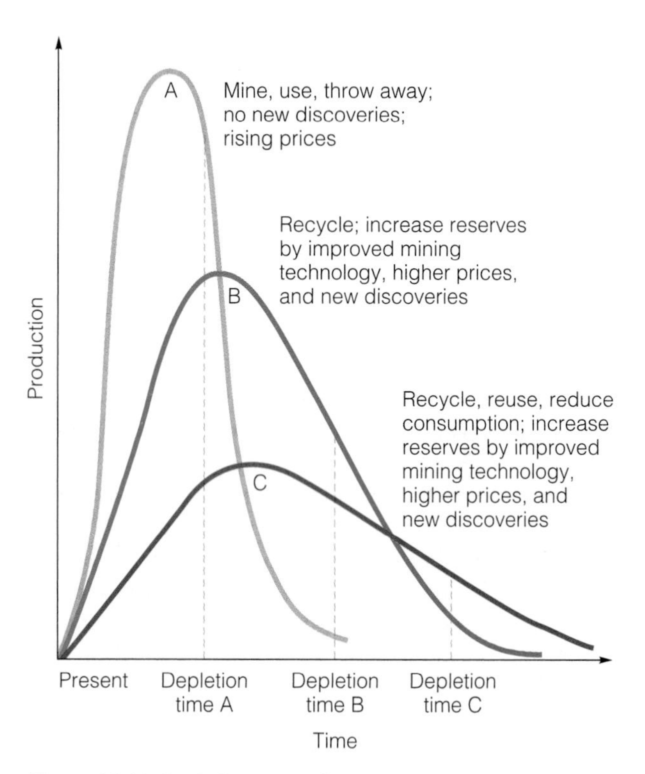

Figure 14-10 Depletion curves for a nonrenewable resource (such as aluminum or copper) using three sets of assumptions. Dashed vertical lines represent times when 80% depletion occurs.

Who Has the World's Nonfuel Mineral Resources? Global nonfuel mineral resources are unevenly distributed. The United States, Canada, Australia, South Africa, and republics making up the former Soviet Union supply most of the world's 20 most important nonfuel minerals. Current world reserve-to-production ratios for key nonrenewable metal resources range from about 20 years for zinc, lead, and mercury to well over 100 years for iron and aluminum.

No industrialized country is self-sufficient in mineral resources, although the former Soviet Union came close and was a major exporter of important minerals. Since its breakup, some of its republics have ample minerals, but others do not. By contrast, Japan, in addition to lacking coal, oil, and timber resources, has virtually no metals. Japan depends on resource imports, which it upgrades to finished products and then sells abroad to buy the resources it needs to sustain its economy. Most western European countries depend heavily on minerals from Africa.

Case Study: Mineral Resources in the United States The U.S. Bureau of Mines estimates that the total consumption of virgin materials in the United States increased 14-fold between 1900 and 1991; during this same period the country's population grew by just a little over 3-fold. This sharp rise in both total resource use and per capita resource use had a price: the rapid depletion of many of the country's reserves of nonrenewable energy resources (especially oil) and nonfuel mineral resources (such as lead, aluminum ore, and iron ore).

According to resource experts, the United States will never again be self-sufficient in oil or many key metal resources. Currently, the United States imports 50% or more of 24 of its 42 most important nonfuel minerals. Some are imported because they are used faster than they can be produced from domestic supplies; others are imported because foreign ore deposits are of a higher grade and cheaper to extract than remaining U.S. reserves.

Figure 14-11 shows both U.S. reserves projected to the year 2000 and major foreign sources for 20 important nonfuel minerals. Most U.S. mineral imports come from reliable and politically stable countries. However, experts are concerned about the availability of four minerals—manganese, cobalt, platinum, and chromium—of which the United States has little or no reserves and which it imports from potentially unstable countries in the former Soviet Union and Africa (South Africa, Zambia, Zaire). As the American Geological Institute notes, "Without manganese, chromium, platinum, and cobalt, there can be no automobiles, no airplanes, no jet engines, no satellites, and no sophisticated weapons—not even home appliances."

Q: What percentage of the commercial energy used in the United States is consumed by agriculture?

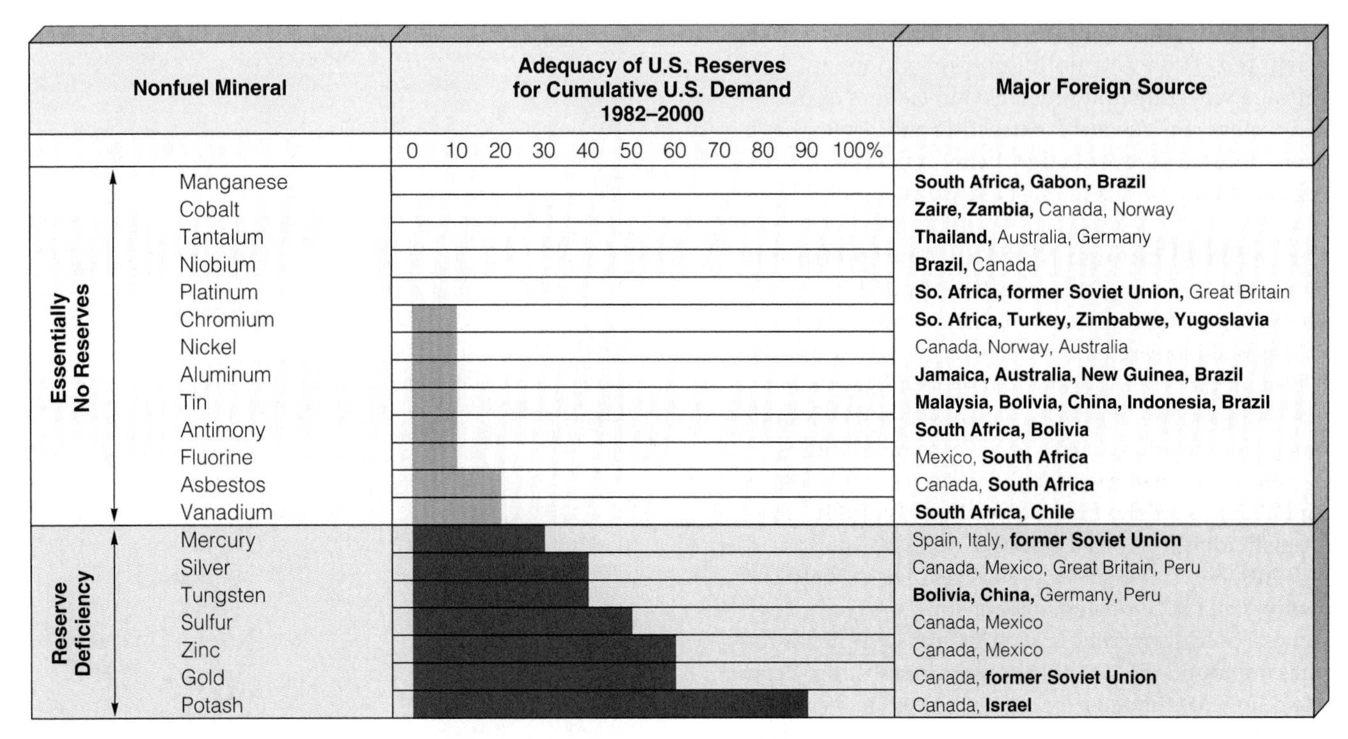

Figure 14-11 Evaluation of the supply of selected nonfuel minerals in the United States, 1982–2000, and major foreign sources of these minerals. Foreign sources subject to potential interruption of supply by political, economic, or military disruption are shown in boldface. Virtually all U.S. supplies of four important minerals—manganese, cobalt, platinum, and chromium—are imported from potentially unstable countries. (Data from U.S. Geological Survey and U.S. Bureau of Mines)

14-4 INCREASING MINERAL RESOURCE SUPPLIES

How Does Economics Affect Resource Supplies? Geologic processes determine the quantity and location of a mineral resource in the earth's crust; economics determines what part of the known supply is used.

According to standard economic theory, in a competitive free market a plentiful resource is cheap because supply exceeds demand; when a resource becomes scarce its price rises, stimulating exploration and development of better mining technology. Rising prices also make it profitable to mine ores of lower grades and encourage the search for substitutes. However, this theory may no longer apply to developed countries because in the economic systems of such countries, industry and government control supply, demand, and prices of minerals to such a large extent that a truly competitive free market does not exist.

Most mineral prices are low because countries subsidize development of their domestic mineral resources to help promote economic growth and national security. In the United States, for instance, mining companies get depletion allowances amounting to 5–22% of their gross income, depending on the mineral. In addition, the companies can deduct much of their costs for finding and developing mineral deposits. Moreover, hard-rock mining companies operating in the United States get public land and the minerals they extract essentially for free (p. 338). Between 1982 and 1995, these mining subsidies cost U.S. taxpayers about $5.5 billion.

Another problem is that the cost of nonfuel mineral resources is only a small part of the final cost of goods. Thus, because scarcity of minerals does not raise the market prices of products very much, industries and consumers have no incentive to reduce demand for products in time to avoid economic depletion of the minerals.

Most mineral prices are low because most of the harmful environmental costs of mining and processing are not included in their prices. As a result, mining companies and manufacturers have little incentive to reduce resource waste and pollution as long as they can pass many of the harmful environmental costs of their production on to society. Environmentalists and some economists argue that taxing rather than subsidizing the extraction of nonfuel mineral resources would provide governments with revenue, create

A: About 17%

incentives for more efficient resource use, promote waste reduction and pollution prevention, and encourage recycling and reuse. So far, leaders of politically powerful resource extraction industries have been able to prevent significant taxation of the resources they extract in most countries.

Can We Find Enough New Land-Based Mineral Deposits? Geologic exploration guided by better knowledge, satellite surveys, and other new techniques will increase current reserves of most minerals. Although most of the easily accessible, high-grade deposits are already known, new deposits will be found, mostly in unexplored areas of developing countries.

However, exploring for new resources takes lots of capital and is a risky financial venture. Typically, if geologists identify 10,000 possible deposits of a given resource, only 1,000 sites are worth exploring; only 100 justify drilling, trenching, or tunneling; and only 1 becomes a producing mine or well. Even if large new supplies are found, no nonrenewable mineral supply can stand up to continued exponential growth in its use.

One factor limiting production of nonfuel minerals is lack of investment capital. With today's fluctuating mineral markets and rising costs, investors are wary of tying up large sums for long periods with no assurance of a reasonable return.

Case Study: Should Mining Be Allowed in Antarctica? Antarctica (Figure 14-12), often called the earth's last great wilderness area, is the highest, coldest, driest, windiest, iciest, and most remote of the earth's seven continents. The vast majority of this beautiful and fragile frozen expanse remains unspoiled.

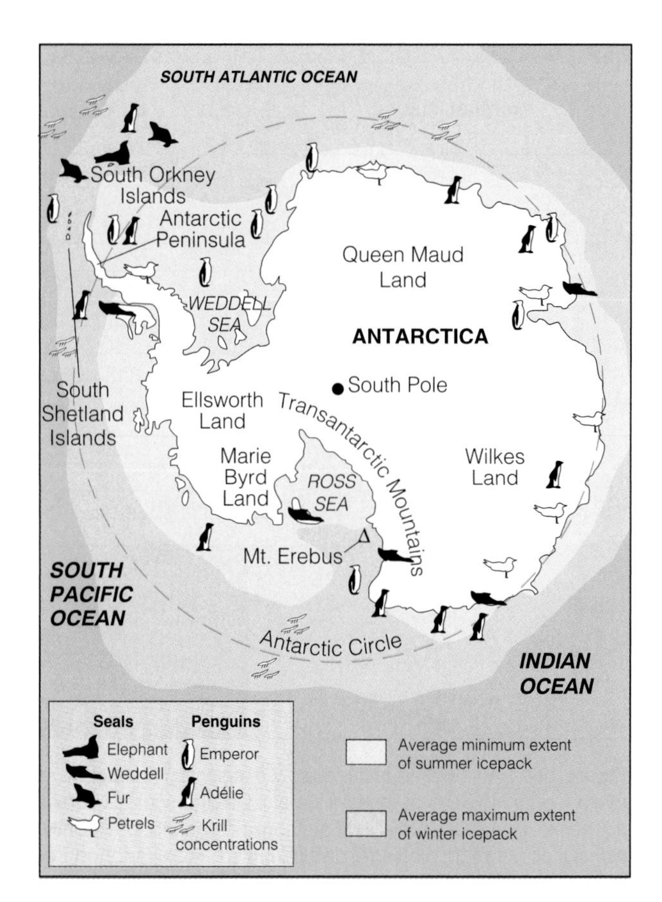

Figure 14-12 The ice-covered continent of Antarctica makes up 10% of the earth's land mass and is the world's last great wilderness. The ice—up to 3 kilometers (1.9 miles) thick—that covers 98% of the continent contains 90% of the earth's ice and 70% of its fresh water. Antarctica helps regulate the global climate and sea level. In winter parts of the sea freeze, and the continent doubles in size. In summer it shrinks, when some of the ice pack melts and huge icebergs break off the edges of the ice shelf.

Many geologists believe that the continent and its offshore waters may contain significant amounts of oil, natural gas, and coal, and metals such as lead, uranium, chromium, and manganese. However, no one knows whether these minerals are there or whether it would be physically possible and economically profitable to exploit them.

Moreover, environmentalists believe that we should protect this remaining last large wilderness area on the planet from development. Seven countries have unresolved ownership claims to parts of Antarctica and its offshore waters; 26 countries operate 69 scientific research stations there, which entitles them to vote on treaties concerning activities on the continent. Most of Antarctica's life and its research bases are crowded onto the less than 2% of the continent that is ice-free year-round.

In 1991, the 26 nations involved signed a new treaty that designates Antarctica as a natural reserve devoted to peace and science. The treaty bans mineral and oil exploration and mining in Antarctica for at least 50 years. At the end of this period a two-thirds majority of these nations would be needed to lift the ban. This historic agreement also includes new regulations for wildlife protection, marine pollution, and continued environmental monitoring. However, Greenpeace expeditions in 1993 and 1995 revealed an array of blatant breaches of the treaty, including siting fuel dumps in tern breeding areas, using lakes as landfills, and dumping waste into the sea. Run-down and abandoned bases litter the continent. Some countries have made efforts to clean up their bases but many have not.

The treaty is an important step, but environmentalists and the governments of France, New Zealand, and Australia favor declaring Antarctica a permanent World Park—an international wilderness area in which only scientific research and carefully controlled tourism would be allowed.

Q: How many units of energy are required to put one unit of food energy on the table in the United States?

Can We Get Enough Minerals by Mining Lower-Grade Ores? Some analysts contend that all we need to do to increase supplies of any mineral is to extract lower grades of ore. They point to new earth-moving equipment, improved techniques for removing impurities, and other technological advances in mineral extraction and processing during the past few decades.

In 1900 the average copper ore mined in the United States was about 5% copper by weight; today it is 0.5%, and copper costs less (adjusted for inflation). New methods of mineral extraction may allow even lower-grade ores of some metals to be used (Solutions, right).

Several factors limit the mining of lower-grade ores, however. As poorer-grade ores are mined, a point is reached at which it costs more to mine and process most such resources than they are worth (unless we have a virtually inexhaustible source of cheap energy). Availability of fresh water also may limit the supply of some mineral resources. To extract and process most minerals by conventional means requires large amounts of water and many mineral-rich areas lack fresh water. Finally, exploitation of lower-grade ores may be limited by the environmental impact of waste material produced during mining and processing. At some point, the costs of land restoration and pollution control exceed the current value of the minerals, unless we continue to pass these harmful costs on to society and to future generations.

Can We Get Enough Minerals by Mining the Oceans? Ocean mineral resources are found in three areas: seawater, sediments and deposits on the shallow continental shelf (Figure 14-5), and sediments and nodules on the deep-ocean floor. Most of the chemical elements found in seawater occur in such low concentrations that recovering them takes more energy and money than they are worth. Only magnesium, bromine, and sodium chloride are abundant enough to be extracted profitably at current prices with existing technology.

Deposits of minerals (mostly sediments) along the continental shelf and near shorelines are already significant sources of sand, gravel, phosphates, sulfur, tin, copper, iron, tungsten, silver, titanium, platinum, and even diamonds.

The deep-ocean floor at various sites may be a future source of manganese and other metals. Manganese-rich nodules may cover 20% of the world's ocean floors and have been found in large quantities at a few sites. These cherry- to potato-sized rocks are 30–40% manganese by weight; they also contain small amounts of other important metals such as nickel, copper, and cobalt. They might be sucked up from the ocean floor by giant vacuum pipes or scooped up by buckets on a continuous cable operated by a mining ship.

Mining with Microbes

SOLUTIONS

One way to improve mining technology is the use of microorganisms for in-place *(in situ)* mining, which would remove desired metals from ores while leaving the surrounding environment undisturbed. This biological approach to mining would also reduce both the air pollution associated with the smelting of metal ores and the water pollution associated with using hazardous chemicals such as cyanides to extract gold.

Once an ore deposit has been identified and deemed economically viable, wells are drilled into it and the ore is fractured. Then the ore is inoculated with either natural or genetically engineered bacteria to extract the desired metal. Next the ore is flooded with water, which is then pumped to the surface, where the desired metal is removed.

This technique permits economical extraction from low-grade ores, which are increasingly being used as high-grade ores are depleted. Since 1958, the copper industry has been using natural strains of the bacterium *Thiobacillus ferroxidans* to remove copper from low-grade copper ore. Currently, at least 25% of all copper produced worldwide, worth more than $1 billion a year, comes from such biomining.

Microbiological processing of ores is slow, however: It can take decades to remove the same amount of material that conventional methods can remove within months or years. So far, biological methods are economically feasible only with low-grade ore (such as gold and copper) for which conventional techniques are too expensive.

Critical Thinking

If you had a large sum of money to invest, would you invest it in the microbiological processing of aluminum ore? Explain.

Nodule beds in international waters are not being developed today because of squabbles over who owns them, how any profits should be distributed among the world's nations, and because land supplies of these minerals are more plentiful and accessible. An international Law of the Sea Treaty (signed by the United States in 1994) may resolve some of these issues.

Rich deposits of gold, silver, zinc, and copper are also produced as sulfide deposits around very hot volcanic springs and vents found at various areas in the deep ocean. Some of these crystallized deposits, called

A: About 10 (a loss of 9 units of energy)

rock chimneys, are as high as 15 stories. In 1998, Australian miners made the first claim to such rich deposits on the bottom of about 5,200 square kilometers (2,000 square miles) of the territorial waters of Papua New Guinea.

Some analysts doubt that the potential earnings can cover the high costs of removing such metal deposits so deep in the ocean. However, the Australian mining company says that making a profit is likely through use of advanced technologies such as sonar, robots, and giant claws lowered from ships. According to the company's CEO, A. Geoff Loudon, such deep-see mining is inevitable and will eventually turn conventional on-shore mining into a dinosaur.

Some environmentalists believe that seabed mining would probably cause less harm than mining on land. However, they are concerned that removing seabed mineral deposits and dumping back unwanted material will stir up ocean sediments, which could destroy seafloor organisms and have unknown effects on poorly understood ocean food webs. Surface waters might also be polluted by the discharge of sediments from mining ships and rigs.

Some biologists warn that deep-sea mining could have a devastating effect on mostly unknown marine biodiversity. Scientists now believe there may be 10 million—and perhaps as many as 100 million—previously unknown species of microbes and small animals residing in the waters and sediments of the deep. Instead of sunlight, these microbes rely on heat and dissolved chemicals vented from the earth's turbulent interior as their energy sources. These microbes themselves become the base of food webs involving anemones, sponges, crabs, fish, mussels, tube worms, and other animals not yet recognized that colonize these deep-ocean chimneys.

Even the lower estimates of deep-ocean species would equal the total estimated number of land-based species, giving the deep ocean as much genetic diversity as a tropical rain forest. Scientists contend that we should understand more about the nature and functions of such diversity before disturbing and killing off such marine life. As with extinction in rain forests, extinction of deep-sea organisms could disrupt ocean ecosystems and prevent future opportunities to develop valuable medicines and technologies. Preliminary research indicates that enzymes released by some of these sulfur-vent microbes can process chemicals toxic to land-based organisms. Some of these new enzymes might eventually be used to break down hazardous wastes.

Can We Find Substitutes for Scarce Nonrenewable Mineral Resources? The Materials Revolution Some analysts believe that even if supplies of key minerals become very expensive or scarce, human ingenuity will find substitutes. They point to the current materials revolution in which silicon and new materials, particularly ceramics and plastics, are being developed and used as replacements for metals.

Ceramics have many advantages over conventional metals. They are harder, stronger, lighter, and longer-lasting than many metals. They withstand intense heat and do not corrode. Within a few decades we may have high-temperature ceramic superconductors in which electricity flows without resistance. Such a development may lead to faster computers, more efficient power transmission, and affordable electromagnets for propelling magnetic levitation trains.

Plastics also have advantages over many metals. High-strength plastics and composite materials strengthened by lightweight carbon and glass fibers are likely to transform the automobile and aerospace industries. They cost less to produce than metals because they require less energy, don't need painting, and can be molded into any shape. New plastics and gels are also being developed to provide superinsulation without taking up much space. One new plastic can withstand extremely high temperatures and is not even affected by exposure to the most intense laser beams.

Substitutes can undoubtedly be found for many scarce mineral resources, but the search is costly, and phasing a substitute into a complex manufacturing process takes time. While a vanishing mineral is being replaced, people and businesses dependent on it may suffer economic hardships. Moreover, finding substitutes for some key materials may be difficult or impossible. Examples are helium, phosphorus for phosphate fertilizers, manganese for making steel, and copper for wiring motors and generators. Finally, some substitutes are inferior to the minerals they replace. For example, even though aluminum could replace copper in electrical wiring, producing aluminum takes much more energy than producing copper, and aluminum wiring presents a greater fire hazard than copper wiring.

14-5 ENVIRONMENTAL EFFECTS OF EXTRACTING AND USING MINERAL RESOURCES

What Are the Environmental Impacts of Using Mineral Resources? The mining, processing, and use of crustal resources require enormous amounts of energy and often cause land disturbance, erosion, and air and water pollution (Figure 14-13).

Mining can affect the environment in several ways. Most noticeable are scarring and disruption of the land surface (Figures 14-7–14-9). Underground

Bumpers, D. 1998. "Capitol Hill's Longest-Running Outrage." *Washington Monthly,* vol. 30, no. 1, 14(5).

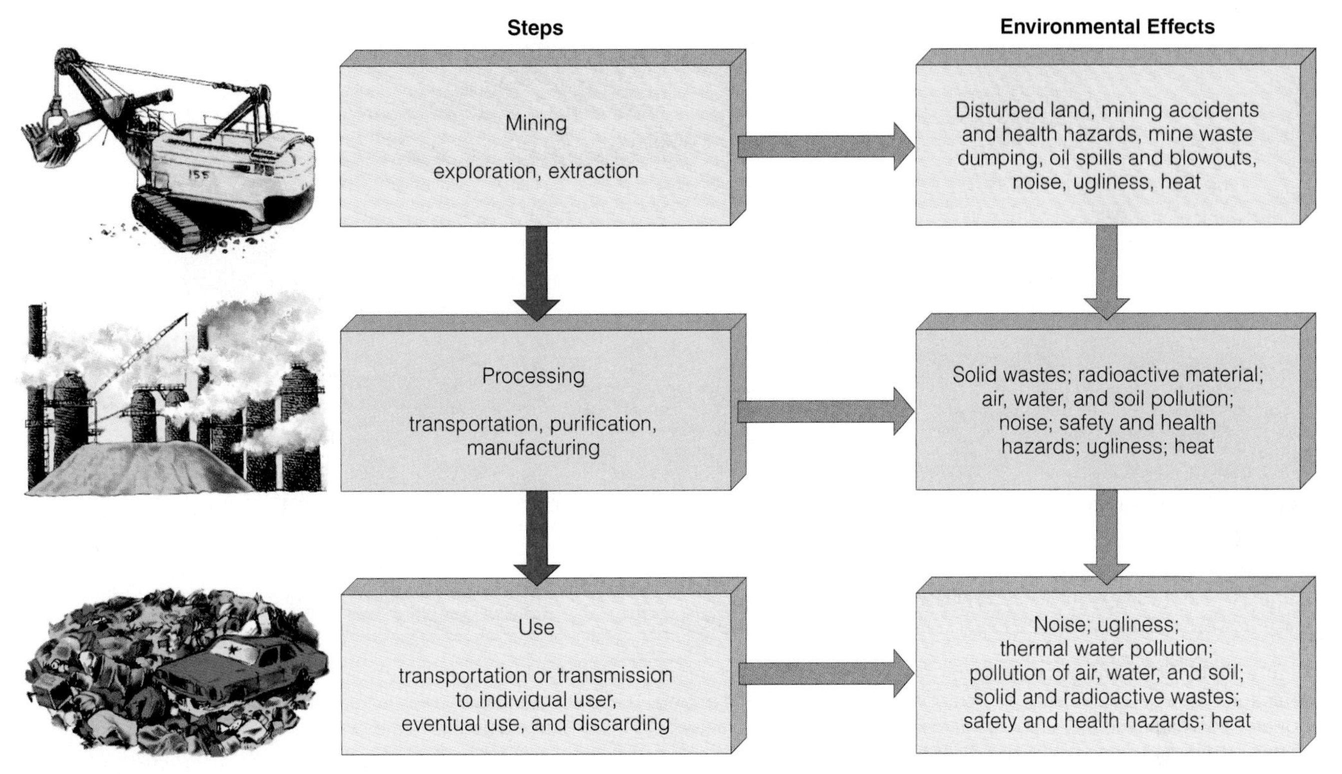

Figure 14-13 Some harmful environmental effects of resource extraction, processing, and use. The energy used to carry out each step causes additional pollution and environmental degradation. Harm could be minimized by requiring mining, processing, and manufacturing companies to include the full costs of the pollution and environmental degradation in the prices of their products. Many of these *external* costs are now passed on to society in the form of poorer health, increased health and insurance costs, and increased taxes to deal with pollution and environmental degradation.

fires in coal mines cannot always be put out. Land above underground mines collapses or subsides, causing houses to tilt, sewer lines to crack, gas mains to break, and groundwater systems to be disrupted. In addition, spoil heaps and tailings can be eroded by wind and water. The air can be contaminated with dust and toxic substances, and water pollution is a serious concern.

Past and present mining operations for metallic and nonmetallic minerals occupy only a small percentage of the total land area in any country (0.25% of the United States). However, the scars from mining are long lasting (Figure 14-7), and the resulting air and water pollution can extend beyond the limits of ground disturbance.

Rainwater seeping through a mine or mine wastes can carry sulfuric acid (H_2SO_4, produced when aerobic bacteria act on iron sulfide minerals in spoil) to nearby streams and groundwater (Figure 14-14). Such *acid mine drainage* can destroy aquatic life and contaminate water supplies. Other harmful materials that either run off or are dissolved from underground mines or aboveground mining wastes include radioactive uranium compounds and compounds of toxic metals such as lead, mercury, arsenic, and cadmium. To date, effluents from mines in the United States have contaminated 19,300 kilometers (12,000 miles) of streams and rivers, 73,000 hectares (180,000 acres) of reservoirs and lakes, and untold amounts of groundwater.

After extraction from the ground, many resources must be separated from other matter, a process that can pollute the air and water. Ore, for example, typically contains two parts: the ore mineral, which contains the desired metal, and the **gangue**, which is the waste mineral material. **Beneficiation**, or separation in a mill of the ore mineral from the gangue, produces waste called **tailings**. Piles of tailings are ugly, and toxic metals blown or leached from them by rainfall can contaminate surface and groundwater.

Most ore minerals do not consist of pure metal, so **smelting** is done to separate the metal from the other elements in the ore mineral. Without effective pollution control equipment, smelters emit enormous quantities of air pollutants, which damage vegetation and soils in the surrounding area. Pollutants include sulfur dioxide, soot, and tiny particles of arsenic,

Hint: Enter the search term *mines and mineral resources;* then select the subtopic *laws, regulations, etc.*

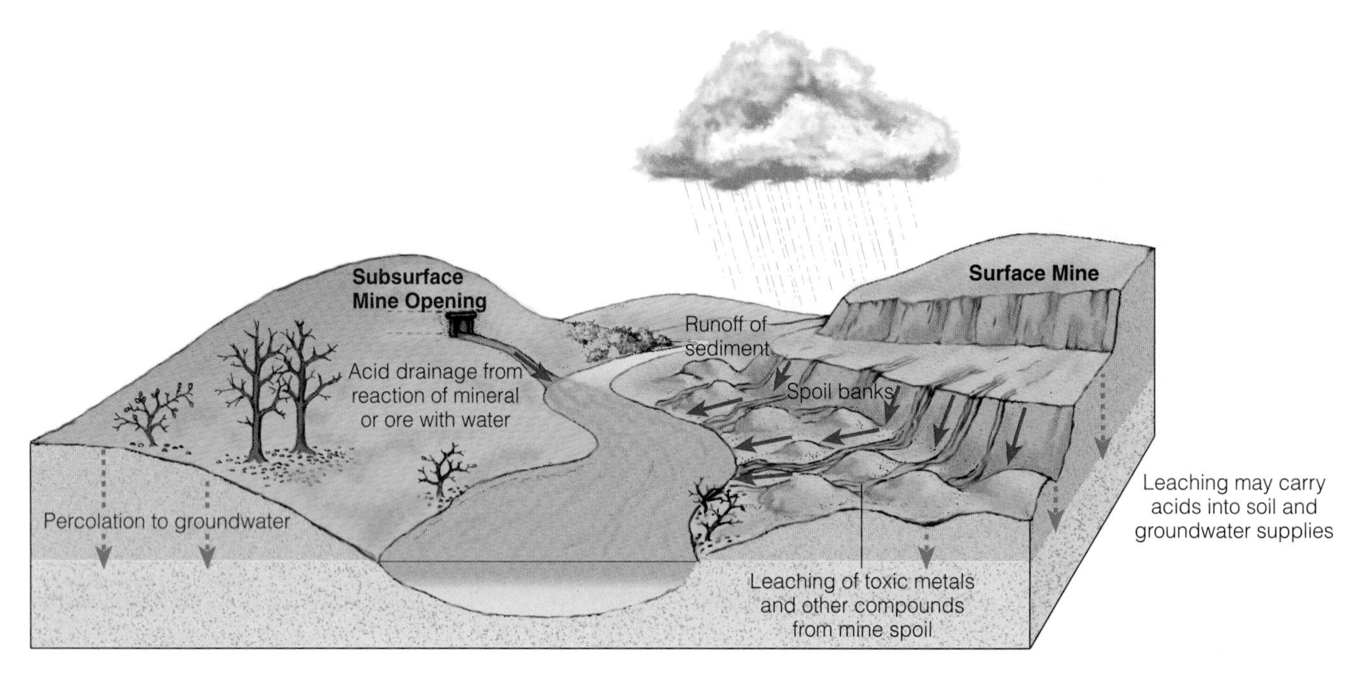

Figure 14-14 Pollution and degradation of a stream and groundwater by runoff of acids—called acid mine drainage—and by toxic chemicals from surface and subsurface mining. These substances can kill fish and other aquatic life. Acid mine drainage has damaged over 26,000 kilometers (16,000 miles) of streams in the United States, mostly in Appalachia and in the West.

cadmium, lead, and other toxic elements and compounds found in many ores.

Decades of uncontrolled sulfur dioxide emissions from copper-smelting operations near Copperhill and Ducktown, Tennessee, killed all the vegetation over a wide area around the smelter; another dead vegetation zone was created around the Sudbury, Ontario, nickel smelter (Figure 14-15). New dead vegetation zones have formed in parts of eastern Europe, the former Soviet Union, and Chile. Smelters also cause water pollution and produce liquid and solid hazardous wastes that must be disposed of safely.

Some companies are using improved technology to reduce pollution from smelting, reduce production costs, and save costly cleanup bills and liability for damages. For example, the new $880 million Kennecott smelter refinery at Bingham, Utah, is expected to be among the cleanest in the world. It also is projected to reduce production costs by 53%, making Kennecott one of the world's cheapest and cleanest copper producers.

Figure 14-15 The dead zone around a nickel smelter in Sudbury, Ontario, Canada. Sulfur dioxide and other fumes released from the smelter over several decades killed the forest once found on this land. (A. J. Copley/Visuals Unlimited)

Are There Environmental Limits to Resource Extraction and Use? Some environmentalists and resource experts believe that the greatest danger from high levels of resource consumption may not be the exhaustion of resources but the damage that their extraction, processing, and conversion to products do to the environment (Figure 14-13).

The minerals industry accounts for 5–10% of world energy use, making it a major contributor to air and water pollution and to emissions of greenhouse gases. As more remote, deeper deposits are mined, even more energy will be needed to dig bigger holes and to transport the metal ores over greater distances.

The *grade* of an ore—its percentage of metal content—largely determines the environmental impact of mining it (Case Study, right); more accessible and

Q: What percentage of the world's cropland is used to grow livestock feed?

CASE STUDY: The Environment and the New Gold Rush

Miners must extract and process massive quantities of soil and rock to end up with small quantities of gold. Gold miners typically remove ore equal to the weight of 50 automobiles to extract an amount of gold that would fit inside your clenched fist.

The mountains of solid waste remaining after gold is extracted from its ore are left piled near the mine sites and can pollute the air, surface water (Figure 14-1), and groundwater. Gold-bearing rock tends to contain large quantities of sulfur, which forms sulfuric acid when exposed to air and water (Figure 14-14). In addition to killing aquatic life, the acid puts highly toxic metals such as cadmium and copper into solution.

In Australia and North America, a new mining technology, called *cyanide heap leaching*, has been cheap enough to allow mining companies to level entire mountains containing very low-grade gold ore. To extract the gold, miners spray a cyanide solution (which reacts with gold) onto huge open-air piles of crushed ore. They then collect the solution in leach beds and overflow ponds, recirculate it a number of times, and extract gold from it. Unfortunately, cyanide is extremely toxic and can be harmful or lethal to people, plants, and wildlife, especially to birds and mammals drawn to cyanide collection ponds as a source of water.

Cyanide leach pads and collection ponds can also leak or overflow, posing threats to underground drinking water supplies and to wildlife (especially fish) in lakes and streams. Special liners beneath the ore heaps and in the collection ponds are supposed to prevent leaks, but some have failed; according to the EPA all such liners will eventually leak.

In the United States, companies have used the 1872 mining law to buy public land for practically nothing, mine a site, abandon it, file for bankruptcy, and leave the public with the cleanup bill (p. 338). A glaring example is the Summitville gold mine site in the San Juan Mountains of southern Colorado. A Canadian company bought the land from the federal government for a pittance, spent $1 million developing the site, and then removed $98 million worth of gold. Shoddy construction allowed acids and toxic metals to leak from the site and poison a 27-kilometer (17-mile) stretch of the Alamosa River, the source of irrigation water for farms and ranches in the San Luis Valley.

The company then declared bankruptcy and abandoned the property, but only after being allowed to retrieve $2.5 million of the $7.5 million reclamation bond it had posted with the state. Summitville is now a Superfund site; the EPA spends $40,000 a day just to contain the site's toxic wastes. Ultimately, the EPA expects to spend about $120 million to finish the cleanup.

The gold rush of the 1980s and 1990s has also caused millions of miners—many of them landless poor—in various Latin American, Asian, and African developing countries to stream into tropical forests and other areas in search of gold. These small-scale miners use destructive mining techniques such as digging large pits by hand, river dredging, and hydraulic mining (a technique, outlawed in the United States, in which water jets wash entire hillsides into sluice boxes). Highly toxic mercury is usually used to extract the gold from the other materials. In the process, much of the mercury ends up contaminating water supplies and fish consumed by people.

Critical Thinking

Do you believe that the harmful environmental impacts of gold mining should be much more strictly regulated in **(a)** the United States and **(b)** other countries? If so, what major regulations would you like to see? How would you see that such controls are implemented and enforced?

higher-grade ores are generally exploited first. As they are depleted, it takes more money, energy, water, and other materials to exploit lower-grade ores, and environmental effects increase accordingly.

Should the U.S. 1872 Mining Law Be Reformed to Reduce Environmental Harm and Save Taxpayers Money? Environmentalists point out that mining is the only natural resource industry in the United States that by law can buy public lands; it is also the only resource industry that pays no rents or royalties for resource extraction. Environmentalists and a growing number of citizens support a drastic reform of the 1872 mining law (p. 338), including the following changes:

- *Prohibiting buying of public land for mining but allowing such land to be leased for mining for up to 20 years.*
- *Requiring a full environmental impact assessment of the proposed mining activities before a mining lease is approved.*
- *Setting strict environmental standards for preventing and controlling pollution and environmental degradation resulting from mining activities during the period of the lease.*
- *Requiring companies leasing public land to post an environmental performance bond to cover estimated*

A: More than 50% (about 19% in the United States)

environmental damage and ecological restoration costs based on the environmental impact study of the project. After such costs are deducted as needed and restoration is complete, remaining funds plus interest would be returned to the mining company.

- *Requiring mining companies to pay rent to cover all government (taxpayer) costs in evaluating and monitoring any leased mining site.*
- *Requiring mining companies to pay a 12.5% royalty on the gross (not net) values of all minerals removed.*
- *Making mining companies legally and financially responsible for environmental cleanup and restoration of each site.*

Mining companies claim that charging royalties for minerals taken from public lands and requiring them to pay for cleanup will force them to do their mining in other countries, which would cost American jobs and reduce tax revenues. They also argue that their average cost for patenting public land under the 1872 law is about $42,000 per hectare ($17,000 per acre) when their mining development costs are included.

Environmentalists counter that mining companies would still make a reasonable profit on the high-value minerals such as gold and platinum they get from public lands and that threats to move operations elsewhere are a rarely implemented scare tactic (greenmail). For example, gold costs miners about $30 per ounce to extract, but in recent years it has been sold for $320–395 per ounce. Even with a 12.5% royalty and responsibility for cleanup costs—as required for oil, gas, and coal companies—hard-rock mining companies can turn a hefty profit on high-price minerals such as gold and platinum.

Environmentalists also point out that Canada, Australia, South Africa, and other countries that are major extractors of hard-rock minerals don't sell public lands to mining companies, and they require the companies to pay rent on any public land they lease and royalties on the minerals they extract.

Mostly because of the political influence of mining companies and their congressional allies, this law stands little chance of serious reform in the near future without intense pressure from citizens.

14-6 SOIL EROSION AND DEGRADATION

What Causes Soil Erosion? **Soil erosion** is the movement of soil components, especially surface litter and topsoil, from one place to another. The two main agents of erosion are flowing water and wind. Some soil erosion is natural—the long-term wearing down of mountains and building up of plains and deltas by the combined action of physical, chemical, and biological forces (Figure 14-5). In undisturbed vegetated ecosystems, the roots of plants help anchor the soil, and usually soil is not lost faster than it forms.

However, farming, logging, construction, overgrazing by livestock, off-road vehicles, deliberate burning of vegetation, and other activities that destroy plant cover leave soil vulnerable to erosion. Such human activities can speed up erosion and destroy in a few decades what nature took hundreds to thousands of years to produce. In 1937, U.S. President Franklin D. Roosevelt sent a letter to the governors of the states in which he said, *"The nation that destroys its soil destroys itself."*

Most soil erosion is caused by moving water. Soil scientists distinguish among three types of water erosion. *Sheet erosion* occurs when surface water moves down a slope or across a field in a wide flow and peels off fairly uniform sheets or layers of soil. Because the topsoil disappears evenly, sheet erosion may not be noticeable until much damage has been done. In *rill erosion* the surface water forms fast-flowing little rivulets that cut small channels in the soil (Figure 14-16). In *gully erosion*, rivulets of fast-flowing water join together and with each succeeding rain cut the channels wider and deeper until they become ditches or gullies (Figure 14-16). Gully erosion usually happens on steep slopes where all or most vegetation has been removed.

Losing topsoil makes a soil less fertile and less able to hold water. The resulting sediment—the largest source of water pollution—clogs irrigation ditches, boat channels, reservoirs, and lakes. The sediment-laden water is cloudy and tastes bad, fish die, and flood

Figure 14-16 Rill and gully erosion of vital topsoil from irrigated cropland in Arizona. (Soil Conservation Service)

Willis, K. J., et al. 1998. "Prehistoric Land Degradation in Hungary: Who, How and Why?" *Antiquity*, vol. 72, no. 275, 101(13).

risk increases. Rivers running brown with silt from human-accelerated soil erosion contain earth capital that is hemorrhaging from the land (Figure 8-5).

Soil, especially topsoil, is classified as a potentially renewable resource because it is continuously regenerated by natural processes. However, in tropical and temperate areas it takes 200–1,000 years (depending on climate and soil type) for 2.54 centimeters (1 inch) of new topsoil to form. If topsoil erodes faster than it forms on a piece of land, the soil becomes a nonrenewable resource. Annual erosion rates for farmlands throughout the world are 7–100 times the natural renewal rate (Guest Essay, p. 364). Soil erosion is milder on forestland and rangeland than on cropland, but forest soil takes two to three times longer to restore itself than does cropland. Construction sites usually have the highest erosion rates by far.

How Serious Is Global Soil Erosion? According to a 1990 UN Environment Programme survey, topsoil is eroding faster than it forms on about one-third of the world's cropland, causing an estimated 85% of the world's land degradation from human activities. Scientists at the U.S. Department of Agriculture estimate that 60% of water-eroded soil is removed from the land and ends up in streams.

A 1992 study by the World Resources Institute and the UN Environment Programme found that soil on an area equal to the size of China and India combined had been seriously eroded since 1945 (Figure 14-17). The study also found that about 15% of land scattered across the globe was too eroded to grow crops anymore because of a combination of overgrazing (35%), deforestation (30%), and unsustain-able farming (28%). Two-thirds of these seriously degraded lands are in Asia and Africa. In Africa, for example, soil erosion has increased about 14-fold in the past three decades.

Each year we must feed about 84 million more people with an estimated 24 billion metric tons (26 billion tons) less topsoil, eroded as a result of human activities. A 1995 study by soil expert David Pimentel (Guest Essay, p. 364) and his associates estimated that global soil erosion is 75 billion metric tons (82.5 billion tons) per year—three times the previous estimate. The topsoil that washes and blows into the world's streams, lakes, and oceans each year would fill a train of freight cars long enough to encircle the planet 150 times—450 times if the larger estimate is correct. At

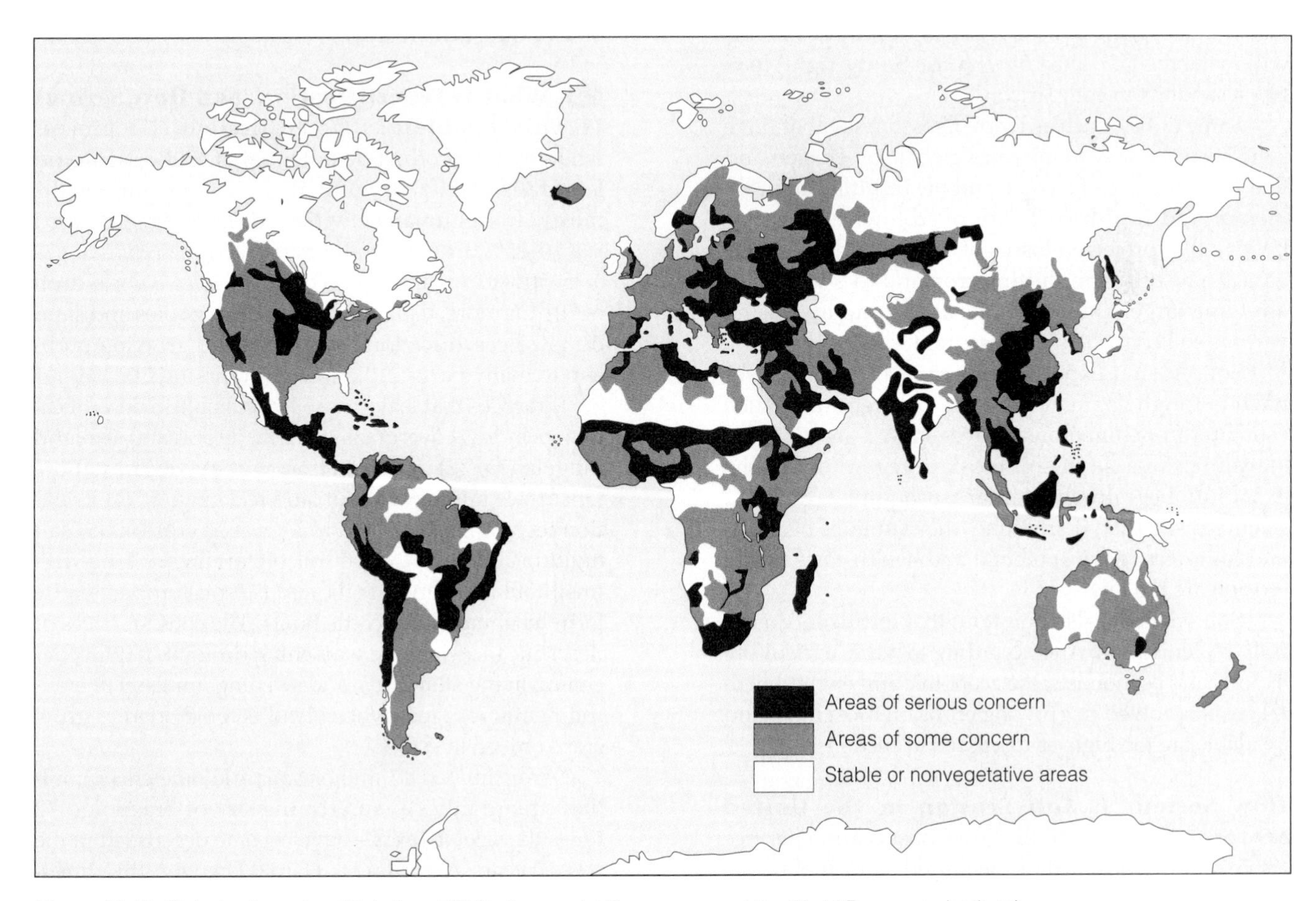

Figure 14-17 Global soil erosion. (Data from UN Environmental Programme and the World Resources Institute)

Hint: Enter the search term *soil degradation* using the Subject Guide.

that rate, the world is losing about 7–21% of its topsoil from actual or potential cropland each decade. According to studies by several soil scientists, if global soil erosion is not severely reduced, it could cause a 19–29% drop in food production from rain-fed cropland during the 25 years between 1985 and 2010.

In developing countries, poverty and erosion interact in a destructive positive feedback cycle. Many poor farmers in developing countries plow up marginal (easily erodible) lands to survive, and the resulting erosion and land degradation then increases poverty—which reinforces the cycle of destruction. According to a 1990 UN study, agricultural mismanagement, overgrazing, deforestation, and overharvesting of fuelwood mostly by the poor account of about 70% of the damage done to the world's soil. In 1995, David Pimentel estimated that soil erosion causes nearly $400 billion per year worldwide in direct damage to agricultural lands and indirect damage to waterways, infrastructure, and human health, an average of $46 million in damages per hour.

According to recent studies by Lester Brown (Guest Essay, p. 34) and David Pimentel, without better soil conservation practices between 1.4 million (equal to the land area of Alaska) and 2.0 million square kilometers (540,000–770,000 square miles) will lose much of their good-quality topsoil by 2015. This will make the U.S. Dust Bowl (Case Study, right) look like a sandbox in comparison.

Some critics, such as Pierre Crosson, say that there is no accurate way to measure global soil erosion and that the estimates of erosion and the resulting environmental and health costs are overblown. They also believe that projected losses of topsoil and soil nutrients can be offset by shifting cropland to other areas, applying larger amounts of fertilizer, and increasing various soil conservation measures (Section 14-7).

Soil erosion experts agree that all we can ever have is rough estimates but they contend that such estimates from numerous sources show a growing and alarming increase in soil erosion, regardless of the exact numbers involved. Many economists and soil scientists say that, if anything, the estimates of short- and long-term environmental and health costs of soil erosion are too low.

Soil scientists also point out that fertilizers do not fully replenish soil. And according to many agriculture experts and economists, the economic and environmental costs involved in growing crops on most new land available are too high, as discussed in Section 12-3.

How Serious Is Soil Erosion in the United States? According to the Soil Conservation Service (SCS), about one-third of the nation's original prime topsoil has been washed or blown into streams, lakes, and oceans, mostly as a result of overcultivation, overgrazing, and deforestation (Case Study, right).

Today, soil on cultivated land in the United States is eroding about 16 times faster than it can form. Erosion rates are even higher in heavily farmed regions, including the Great Plains, which has lost one-third or more of its topsoil in the 150 years since it was first plowed. Some of the country's most productive agricultural lands, such as those in Iowa, have lost about half their topsoil. California's soil is eroding about 80 times faster than it can be formed.

The estimated amount of topsoil that erodes away each day in the United States would fill a line of dump trucks 5,600 kilometers (3,500 miles) long. In 1995, soil expert David Pimentel (Guest Essay, p. 364) estimated that the direct and indirect costs of soil erosion and runoff in the United States exceed $44 billion per year—an average loss of $5 million per hour.

Critics say that estimates of soil erosion and damages from such erosion are overblown and that soil erosion does not pose a serious threat to the capacity of the global agricultural system to increase yields (Section 12-3). They point to studies by several soil scientists concluding that if current rates of cropland erosion in the United States continue for 100 years, crop yields will be only 3–10% less than they would be without such erosion.

What Is Desertification, and How Serious Is This Problem? **Desertification** is a process whereby the productive potential of arid or semiarid land falls by 10% or more; this phenomenon results mostly from human activities. *Moderate desertification* is a 10–25% drop in productivity, *severe desertification* is a 25–50% drop, and *very severe desertification* is a drop of 50% or more, usually creating huge gullies and sand dunes. Desertification is a serious and growing problem in many parts of the world (Figure 14-19).

Practices that leave topsoil vulnerable to desertification include **(1)** overgrazing on fragile arid and semiarid rangelands, **(2)** deforestation without reforestation, **(3)** surface mining without land reclamation, **(4)** irrigation techniques that lead to increased erosion, **(5)** salt buildup and waterlogged soil, **(6)** farming on land with unsuitable terrain or soils, and **(7)** soil compaction by farm machinery and cattle hoofs. The consequences of desertification include worsening drought, famine, declining living standards, and swelling numbers of environmental refugees whose land is too eroded to grow crops or feed livestock.

An estimated 8.1 million square kilometers (3.1 million square miles)—an area the size of Brazil and 12 times the size of Texas—have become desertified in the past 50 years. According to Harold Drengue, this threatens the livelihoods of at least 900 million people in 100

Q: What percentage of U.S. cropland is used to produce fruits and vegetables?

CASE STUDY: The Dust Bowl

In the 1930s, Americans learned a harsh environmental lesson when much of the topsoil in several Midwestern states was lost through a combination of poor cultivation practices and prolonged drought.

Windy and dry, the vast grasslands of the Great Plains stretch across 10 states, from Texas through Montana and the Dakotas. Before settlers began grazing livestock and planting crops there in the 1870s, the deep and tangled root systems of native prairie grasses anchored the fertile topsoil firmly in place (Figure 5-16). Plowing the prairie tore up these roots, and the agricultural crops the settlers planted annually in their place had less extensive root systems.

After each harvest, the land was plowed and left bare for several months, exposing it to the plains winds. Overgrazing also destroyed large expanses of grass, denuding the ground. The stage was set for severe wind erosion and crop failures; all that was needed was a long drought.

Such a drought occurred between 1926 and 1934. In the 1930s, dust clouds created by hot, dry windstorms darkened the sky at midday in some areas; rabbits and birds choked to death on the dust. During May 1934, the entire eastern United States was blanketed by a cloud of topsoil blown off the Great Plains, some 2,400 kilometers (1,500 miles) away. Journalists gave the Great Plains a new name: the *Dust Bowl* (Figure 14-18).

During the so-called Dirty Thirties, cropland equal in area to Connecticut and Maryland combined was stripped of topsoil and an area the size of New Mexico was severely eroded. Thousands of displaced farm families from Oklahoma, Texas, Kansas, and Colorado migrated to California or to the industrial cities of the Midwest and East. Most found no jobs because the country was in the midst of the Great Depression.

In May 1934 Hugh Bennett of the U.S. Department of Agriculture (USDA) went before a congressional hearing in Washington to plead for new programs to protect the country's topsoil. Lawmakers took action when Great Plains dust began seeping into the hearing room.

In 1935, the United States passed the Soil Erosion Act, which established the Soil Conservation Service (SCS) as part of the USDA. With Bennett as its first head, the SCS began promoting sound conservation practices, first in the Great Plains states and later elsewhere. Soil conservation districts were formed throughout the country, and farmers and ranchers were given technical assistance in setting up soil conservation programs.

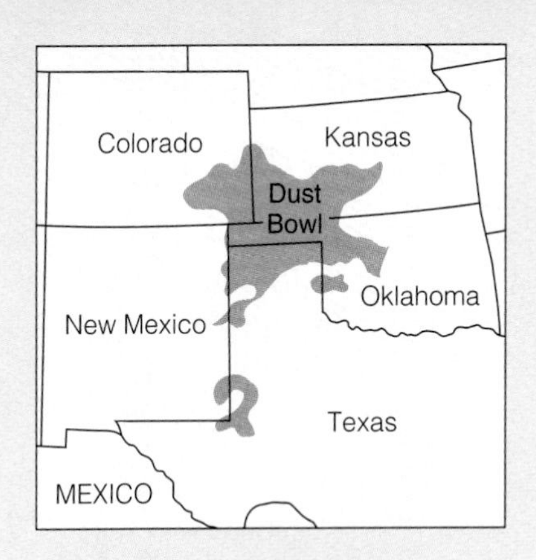

Figure 14-18 The Dust Bowl of the Great Plains, where a combination of extreme drought and poor soil conservation practices led to severe wind erosion of topsoil in the 1930s. Note the connection of this area with the Ogallala aquifer in Figure 13-7.

Unfortunately, these heroic efforts have not yet stopped human-accelerated erosion in the Great Plains. The basic problem is that much of the region is better suited for moderate grazing than for farming. If the earth warms as projected, the region could become even drier and farming might have to be abandoned.

Critical Thinking

Do you think Americans learned a lesson about protecting soil as a result of the Dust Bowl in the 1930s? Explain.

countries, with at least 135 million people suffering from the effects of very severe desertification. If current trends continue, within a few years desertification could threaten the livelihoods of 1.2 billion people.

Every year, low to moderate new desertification occurs on an estimated 60,000 square kilometers (23,000 square miles, an area the size of West Virginia) of agricultural land; another 200,000 square kilometers (77,000 square miles, an area the size of Kansas) undergo severe desertification and lose so much soil and fertility that they are no longer economically valuable for farming or grazing.

Solutions: How Can We Slow Desertification? The most effective way to slow desertification is to drastically reduce overgrazing, deforestation, and the destructive forms of planting, irrigation, and mining that are to blame. In addition, planting trees and grasses will anchor soil and hold water while slowing desertification and reducing the threat of global warming (Section 19-4).

Such prevention and rehabilitation would cost \$10–22 billion annually for the next 20 years. This expenditure is considerably less than the estimated \$42 billion annual loss in agricultural productivity from

A: About 2%

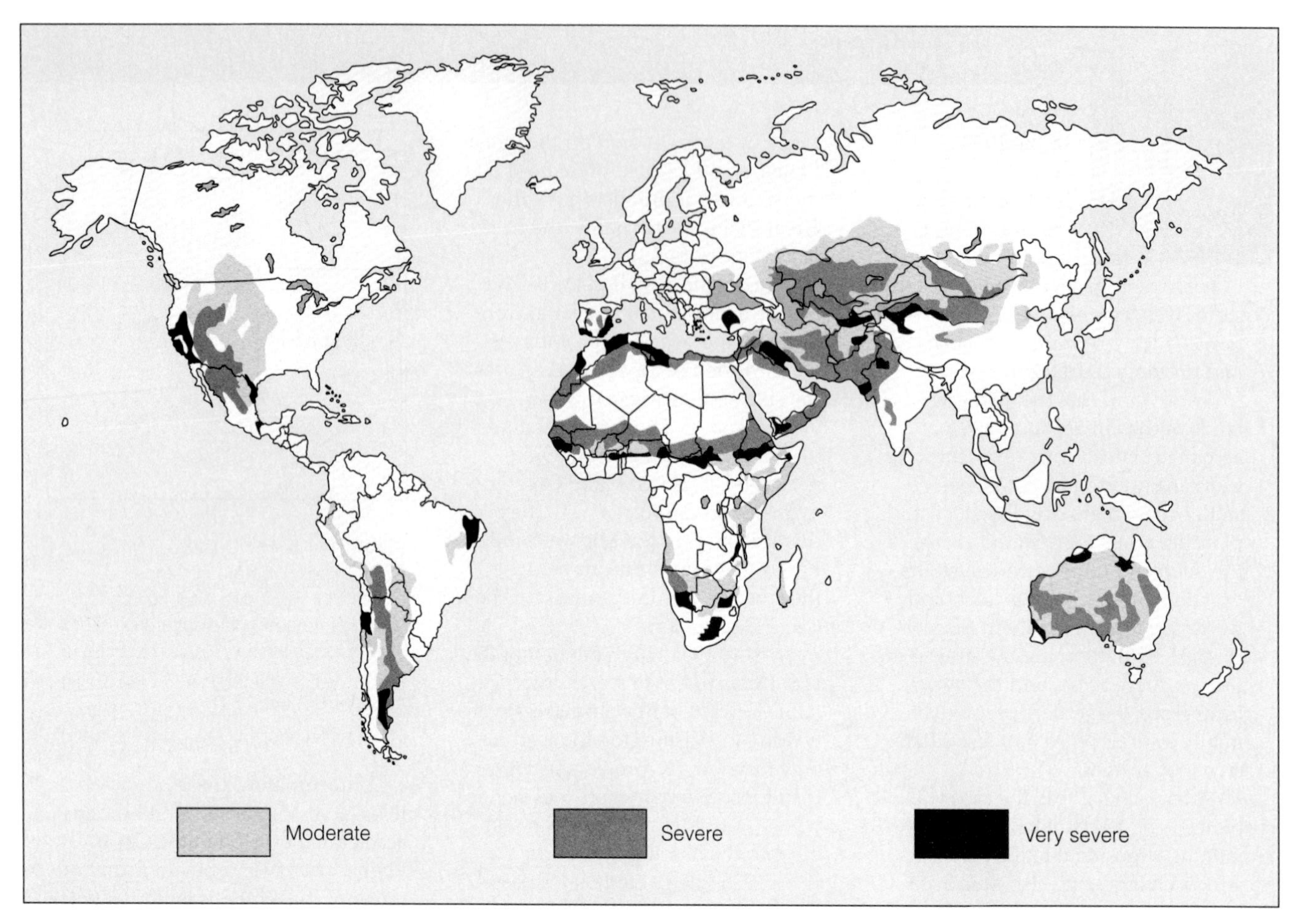

Figure 14-19 Desertification of arid and semiarid lands. (Data from UN Environmental Programme and Harold E. Drengue)

desertified land; once this potential productivity is restored, the cost of the program could be quickly recouped. So far, less than $1 billion per year is spent globally to halt this form of land degradation.

How Do Excess Salts and Water Degrade Soils? The approximately 16% of the world's cropland (almost equal to the area of India) that is now irrigated by various methods (Figure 13-18) produces almost 40% of the world's food and may have to be expanded to produce 50–75% of our food within the next 30 years.

Irrigated land can produce crop yields that are two to three times greater than those from rain watering, but irrigation has its downside. Most irrigation water is a dilute solution of various salts, picked up as the water flows over or through soil and rocks. Small quantities of these salts are essential nutrients for plants, but they are also toxic in large amounts.

Irrigation water not absorbed into the soil evaporates, leaving behind a thin crust of dissolved salts (such as sodium chloride) in the topsoil. The accumulation of these salts, called **salinization** (Figure 14-20), stunts crop growth, lowers yields, and eventually kills plants and ruins the land. According to a 1995 study, severe salinization has reduced yields on 20% of the world's irrigated cropland, and another 30% has been moderately salinized. The most severe salinization occurs in Asia, especially in China, India, and Pakistan.

Precipitation can desalinate soil, but this takes thousands of years in arid and semiarid areas where irrigation is used. Salts can be flushed out of soil by applying much more irrigation water than is needed for crop growth, but this practice increases pumping and crop-production costs, wastes enormous amounts of water, and waterlogs plants if the water table rises close to the surface.

Heavily salinized soil can also be renewed by taking the land out of production for 2 to 5 years, installing an underground network of perforated drainage pipes, and flushing the soil with large quantities of low-salt water. However, this costly scheme only slows the salt buildup; it does not stop the process. Flushing salts from the soil also makes downstream irrigation water saltier unless the saline water can be drained into evaporation ponds rather than returned to the stream or canal.

Q: What percentage of the world's grain production is consumed by livestock?

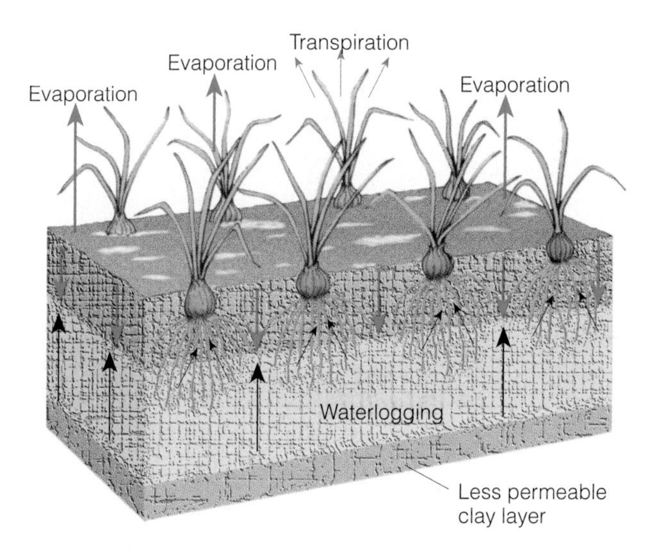

Salinization

1. Irrigation water contains small amounts of dissolved salts.
2. Evaporation and transpiration leave salts behind.
3. Salt builds up in soil.

Waterlogging

1. Precipitation and irrigation water percolate downward.
2. Water table rises.

Figure 14-20 Salinization and waterlogging of soil on irrigated land without adequate drainage lead to decreased crop yields.

Another problem with irrigation is **waterlogging** (Figure 14-20). Farmers often apply large amounts of irrigation water to leach salts deeper into the soil. Without adequate drainage, however, water accumulates underground, gradually raising the water table. Saline water then envelops the deep roots of plants, lowering their productivity and killing them after prolonged exposure. At least one-tenth of all irrigated land worldwide suffers from waterlogging, and the problem is getting worse.

14-7 SOLUTIONS: SOIL CONSERVATION

How Can Conservation Tillage Reduce Soil Erosion? **Soil conservation** involves reducing soil erosion and restoring soil fertility. For hundreds of years, farmers have used various methods to reduce soil erosion, most of which involve keeping the soil covered with vegetation.

In **conventional-tillage farming** the land is plowed and then the soil is broken up and smoothed to make a planting surface. In areas such as the Midwestern United States, harsh winters prevent plowing just before the spring growing season. Thus, cropfields are often plowed in the fall, baring the soil during the winter and early spring and leaving it vulnerable to erosion.

To reduce erosion, many U.S. farmers are using **conservation-tillage farming** (either *minimum-tillage* or *no-till farming*). The idea is to disturb the soil as little as possible while planting crops. With minimum-tillage farming, special tillers break up and loosen the subsurface soil without turning over the topsoil, previous crop residues, and any cover vegetation. In no-till farming, special planting machines inject seeds, fertilizers, and weed-killers (herbicides) into slits made in the unplowed soil.

Besides reducing soil erosion, conservation tillage saves fuel, cuts costs, holds more water in the soil, keeps the soil from getting packed down, and allows more crops to be grown during a season (multiple cropping). Yields are at least as high as those from conventional tillage. It also cuts release of carbon dioxide from the soil to the air, helping ease the threat of global warming (Sections 19-2 and 19-3). In addition, increased levels of CO_2 in the soil can help farmers grow more food to feed the expanding world population without using more fertilizers and pesticides.

At first, conservation tillage was thought to require more herbicides, but a 1990 USDA study of corn production in the United States found no real difference in levels of herbicide use between conventional and conservation tillage systems. However, no-till cultivation of corn does leave stalks, which can serve as habitats for the corn borer; this can potentially increase the use of pesticides.

By 1997 conservation tillage was used on about 40% of U.S. croplands and is projected to be used on over half of it by 2005. The USDA estimates that using conservation tillage on 80% of U.S. cropland would reduce soil erosion by at least half. So far, the practice is not widely used in other parts of the world.

How Can Terracing, Contour Farming, Strip Cropping, and Alley Cropping Reduce Soil Erosion? **Terracing** can reduce soil erosion on steep slopes, each of which is converted into a series of broad, nearly level terraces that run across the land contour (Figure 14-21a). Terracing retains water for crops at each level and reduces soil erosion by controlling runoff.

In mountainous areas such as the Himalayas and the Andes, farmers traditionally built elaborate systems of terraces to grow crops. Today, however, some of these slopes are being farmed without terraces, leaving the land too nutrient poor to grow crops or generate new forest after only 10–40 years. Although most poor farmers know the risk of not terracing, many have

A: About 37% (70% in the United States)

(a)

(b)

(c)

(d)

Figure 14-21 Soil conservation methods: **(a)** terracing in Bali, **(b)** contour planting and strip cropping in Illinois, **(c)** alley cropping in Peru, and **(d)** windbreaks in South Dakota. (Clockwise from top left, Prato/Bruce Coleman Ltd.; U.S. Soil Conservation Service; P. A. Sanchez/North Carolina State University; U.S. Soil Conservation Service)

Q: What percentage of the water withdrawn each year in the United States is consumed by livestock?

too little time and too few workers to build terraces; they must plant crops or starve. The resultant loss of protective vegetation and topsoil (Figure 13-22) also greatly intensifies flooding below these watersheds.

Soil erosion can be reduced by 30–50% on gently sloping land by means of **contour farming**: plowing and planting crops in rows across, rather than up and down, the sloped contour of the land (Figure 14-21b). Each row planted along the contour of the land acts as a small dam to help hold soil and slow water runoff.

In **strip cropping**, a row crop such as corn alternates in strips with another crop (such as a grass or a grass–legume mixture) that completely covers the soil and thus reduces erosion (Figure 14-21b). The strips of the cover crop trap soil that erodes from the row crop. They also catch and reduce water runoff and help prevent the spread of pests and plant diseases. Nitrogen-fixing legumes such as soybeans or alfalfa planted in some of the strips help restore soil fertility.

Erosion can also be reduced by **alley cropping**, or **agroforestry**, a form of *intercropping* in which several crops are planted together in strips or alleys between trees and shrubs that can provide fruit or fuelwood (Figure 14-21c). The trees provide shade (which reduces water loss by evaporation) and help to retain and slowly release soil moisture. The tree and shrub trimmings can be used as mulch (green manure) for the crops and as fodder for livestock.

How Can Gully Reclamation, Windbreaks, Land Classification, and PAM Reduce Soil Erosion? **Gully reclamation** can restore sloping bare land on which water runoff quickly creates gullies. Small gullies can be seeded with quick-growing plants such as oats, barley, and wheat for the first season, whereas deeper gullies can be dammed to collect silt and gradually fill in the channels. Fast-growing shrubs, vines, and trees can also be planted to stabilize the soil, and channels can be built to divert water from the gully and prevent further erosion.

Windbreaks, or **shelterbelts**, can reduce wind erosion. Long rows of trees are planted to partially block the wind (Figure 14-21d). Windbreaks, which are especially effective if uncultivated land is kept covered with vegetation, also help retain soil moisture, supply some wood for fuel, and provide habitats for birds, pest-eating and pollinating insects, and other animals. Unfortunately, many of the windbreaks planted in the upper Great Plains after the 1930s Dust Bowl disaster have been cut down to make way for large irrigation systems and modern farm machinery.

Land can be evaluated with the goal of identifying easily erodible (marginal) land that should be neither planted in crops nor cleared of vegetation. In the United States, the Soil Conservation Service (SCS) has set up a classification system to identify types of land that are suitable or unsuitable for cultivation. The SCS relies on voluntary compliance with its guidelines in the almost 3,000 local and state soil and water conservation districts it has established, and it provides technical and economic assistance through local district offices.

A chemical called polyacrylamide (PAM) has been used recently to sharply reduce erosion of some irrigated fields, at moderate cost in most soils (except desert soils). Tests show that adding just 10 parts per million (ppm) of this white crystal to water during the first hour of irrigation can reduce soil erosion by 70–99%. After that, irrigation can continue for 19–24 hours without further treatment. The negatively charged PAM particles may work by binding to the positively charged clay particles in soils, thereby reducing erosion by increasing the cohesiveness of surface soil particles.

Case Study: Slowing Soil Erosion in the United States Of the world's major food-producing countries, only the United States is reducing some of its soil losses through conservation tillage and government-sponsored soil conservation programs.

The 1985 Farm Act established a strategy for reducing soil erosion in the United States. In the first phase of this program, farmers are given a subsidy for highly erodible land they take out of production and replant with soil-saving grass or trees for 10 years. The land in such a *conservation reserve* cannot be farmed, grazed, or cut for hay. Farmers who violate their contracts must pay back all subsidies plus interest.

By 1994, the authorized limit of 15 million hectares (37 million acres) of highly erodible land had been placed in conservation reserves, cutting soil erosion on U.S. cropland by almost one-third. Between 1985 and 1995, this program has cut soil losses on cropland in the United States by about 60%—a shining example of good news—and could eventually cut such losses as much as 80%. In 1996 Congress reauthorized the Conservation Reserve Program until 2002.

The second phase of the program required all farmers with highly erodible land to develop SCS-approved 5-year soil conservation plans for their entire farms by the end of 1990. A third provision of the Farm Act authorizes the government to forgive all or part of farmers' debts to the Farmers Home Administration if they agree not to farm highly erodible cropland or wetlands for 50 years. The farmers must plant trees or grass on this land or to restore it to wetland.

In 1987, however, the SCS eased the standards that farmers' soil conservation plans must meet to keep them eligible for other subsidies. Environmentalists have also accused the SCS of laxity in enforcing the Farm Act's "swampbuster" provisions, which deny

A: More than 50%

federal funds to farmers who drain or destroy wetlands on their property. In 1996 Congress weakened the act by allowing farmers to end their contracts without USDA approval. Despite some weaknesses, the 1985 Farm Act makes the United States the first major food-producing country to make soil conservation a national priority.

Even though these efforts to slow soil erosion are an important step, effective soil conservation is practiced on only about half of all U.S. agricultural land and on less than half of the country's most erodible cropland.

How Can We Maintain and Restore Soil Fertility? Fertilizers partially restore plant nutrients lost by erosion, crop harvesting, and leaching. Farmers can use either **organic fertilizer** from plant and animal materials or **commercial inorganic fertilizer** produced from various minerals.

Three basic types of *organic fertilizer* are animal manure, green manure, and compost. **Animal manure** includes the dung and urine of cattle, horses, poultry, and other farm animals. It improves soil structure, adds organic nitrogen, and stimulates beneficial soil bacteria and fungi. Despite its effectiveness, the use of animal manure in the United States has decreased. One reason is that separate farms for growing crops and raising animals have replaced most mixed animal-raising and crop-farming operations. Animal manure is available at feedlots near urban areas, but transporting it to distant rural crop-growing areas usually costs too much. Thus, much of this valuable resource is wasted and can end up polluting nearby bodies of water. In addition, tractors and other motorized farm machinery have replaced horses and other draft animals that naturally added manure to the soil.

Green manure is fresh or growing green vegetation plowed into the soil to increase the organic matter and humus available to the next crop. It may consist of weeds in an uncultivated field, grasses and clover in a field previously used for pasture, or legumes such as alfalfa or soybeans grown to restore soil nitrogen.

Compost is a rich natural fertilizer and soil conditioner that aerates soil, improves its ability to retain water and nutrients, helps prevent erosion, and prevents nutrients from being wasted in landfills. Farmers, homeowners, and communities produce compost by piling up alternating layers of nitrogen-rich wastes (such as grass clippings, weeds, animal manure, and vegetable kitchen scraps), carbon-rich plant wastes (dead leaves, hay, straw, sawdust), and topsoil (Individuals Matter, right). This mixture provides a home for microorganisms that aid the decomposition of the plant and manure layers. Composting also reduces the amount of waste taken to landfills and incinerators.

Another method for conserving soil nutrients is **crop rotation**. Corn, tobacco, and cotton can deplete the topsoil of nutrients (especially nitrogen) if planted on the same land several years in a row. Farmers using crop rotation may plant areas or strips with such nutrient-depleting crops one year; the next year, however, they plant the same areas with legumes—whose root nodules (Figure 5-7) add nitrogen to the soil—or with crops such as soybeans, oats, barley, rye, or sorghum. This method helps restore soil nutrients and reduces erosion by keeping the soil covered with vegetation. It also helps reduce crop losses to insects by presenting them with a changing target.

Will Inorganic Fertilizers Save the Soil? Today, many farmers (especially in developed countries) rely on *commercial inorganic fertilizers* containing nitrogen (as ammonium ions, nitrate ions, or urea), phosphorus (as phosphate ions), and potassium (as potassium ions). Other plant nutrients may also be present in low or trace amounts.

Inorganic commercial fertilizers are easily transported, stored, and applied. Worldwide, their use increased about 10-fold between 1950 and 1989, but declined by 12% between 1990 and 1996. Today, the additional food they help produce feeds one of every three people in the world; without them, world food output would plummet an estimated 40%.

Commercial inorganic fertilizers have some disadvantages, however. They do not add humus to the soil. Unless animal manure and green manure are also added, the soil's content of organic matter, and thus its ability to hold water, will decrease, and the soil will become compacted and less suitable for crop growth. By decreasing the soil's porosity, inorganic fertilizers also lower its oxygen content and keep added fertilizer from being taken up as efficiently. In addition, most commercial fertilizers supply only 2 or 3 of the 20 or so nutrients needed by plants. Moreover, producing, transporting, and applying inorganic fertilizers require large amounts of energy and release nitrous oxide (N_2O), a greenhouse gas.

The widespread use of commercial inorganic fertilizers, especially on sloped land near streams and lakes, also causes water pollution as some fertilizer nutrients are washed into nearby bodies of water; the resulting plant-nutrient enrichment causes algae blooms that use up oxygen dissolved in the water, thereby killing fish. Rainwater seeping through the soil can also leach nitrates in commercial fertilizers

Q: What percentage of U.S. topsoil loss is directly associated with livestock grazing?

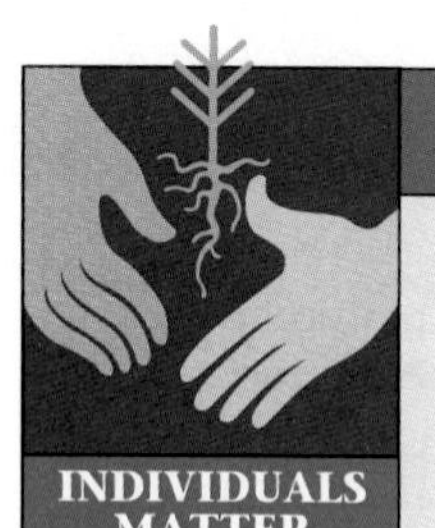

Fast-Track Composting

INDIVIDUALS MATTER

Compost piles must be turned over every few days for aeration to speed up decomposition. Ruth Beckner has invented a special drill bit that attaches to a cordless electric drill to inject air into a compost pile with very little physical exertion. Using her COMPOST AIR™* device for about 3 minutes per day, you can create good compost in about 3 weeks instead of waiting 12–18 months for the pile to decompose naturally.

Worms, the planet's champion recyclers, can also help create compost efficiently and without any odors. Recycling coordinators in Seattle, Washington, and Toronto, Canada, have given out thousands of worm bins to homeowners.

*For information contact Beckner & Beckner, 15 Portola Avenue, San Rafael, CA 94903, or call 1-800-58COMPOST.

into groundwater. Drinking water drawn from wells containing high levels of nitrate ions can be toxic, especially for infants.

Environmental historian Donald Worster reminds us that fertilizers are not a substitute for fertile soil:

> *We can no more manufacture a soil with a tank of chemicals than we can invent a rain forest or produce a single bird. We may enhance the soil by helping its processes along, but we can never recreate what we destroy. The soil is a resource for which there is no substitute.*

According to soil scientists, responsibility for reducing soil erosion should not be limited to farmers. Timber cutting, overgrazing, mining, and urban development carried out without proper regard for soil conservation cause at least 40% of soil erosion in the United States. Each of us has a role in seeing that these vital soil resources are used sustainably. Some things you can do to reduce soil erosion are listed in Appendix 5.

At some point, either the loss of topsoil from the world's croplands will have to be checked by effective soil conservation practices or the growth in the world's population will be checked by hunger and malnutrition.

LESTER R. BROWN

CRITICAL THINKING

1. **(a)** Explain what would happen if plate tectonics stopped. **(b)** Explain what would happen if erosion stopped.

2. Use the second law of energy (Section 3-5) to analyze the feasibility of each of the following processes. Which, if any, could be profitable without subsidies?

- **a.** Extracting most minerals dissolved in seawater
- **b.** Recycling minerals that are widely dispersed
- **c.** Mining increasingly lower-grade deposits of minerals
- **d.** Using inexhaustible solar energy to mine minerals
- **e.** Continuing to mine, use, and recycle minerals at increasing rates

3. Explain why you support or oppose each of the following proposals:

- **a.** Eliminating all tax breaks and depletion allowances for extraction of virgin resources by mining industries
- **b.** Stopping the grant of title to public lands for actual or claimed hard-rock mining (p. 338)
- **c.** Requiring mining companies to pay a royalty of 12.5% on their gross (not net) revenues from any hard-rock minerals they extract from public lands
- **d.** Making hard-rock mining companies responsible for restoring the land and cleaning up environmental damage caused by their activities

4. Should Antarctica be set aside as a permanent World Park—an international wilderness area in which only scientific research and carefully controlled tourism would be allowed? Explain your answer.

5. Why should everyone, not just farmers, be concerned with soil conservation?

6. Do you believe that soil erosion in the country where you live is a serious problem? Explain. What about in the area where you live?

7. What are the main advantages and disadvantages of commercial inorganic fertilizers? Why should both inorganic and organic fertilizers be used?

8. Try to come up with analogies to complete the following statements: **(a)** Allowing private companies to remove hard-rock minerals from public lands in the United States without paying royalties to the government and not being responsible for environmental damage such removal creates is like _______. **(b)** Not including the environmental costs of extracting and processing minerals in their market prices is like _______. **(c)** Soil erosion is like _______. **(d)** Desertification is like _______. **(e)** Composting is like _______. **(f)** Salinization of soil is like _______. **(g)** A tectonic plate is like _______. **(h)** A transform fault is like _______. **(i)** The reserves of a resource are like _______. **(j)** Open-pit mining is like _______.

A: About 14%

GUEST ESSAY

Land Degradation and Environmental Resources

David Pimentel

David Pimentel is professor of insect ecology and agricultural sciences in the College of Agriculture and Life Sciences at Cornell University. He has published over 440 scientific papers and 20 books on environmental topics, including land degradation, agricultural pollution and energy use, biomass energy, and pesticides. He was one of the first ecologists to use an interdisciplinary, holistic approach in investigating complex environmental problems.

At a time when the world's human population is rapidly expanding and its need for more land to produce food, fiber, and fuelwood is also escalating, valuable land is being degraded through erosion and other means at an alarming rate. Soil degradation is of great concern because soil reformation is extremely slow. Globally, it takes an average of 500 years (with a range of 220 to 1,000 years) to renew 2.5 centimeters (1 inch) of soil in tropical and temperate areas—a renewal rate of about 1 metric ton of topsoil per hectare of land per year. Worldwide annual erosion rates for agricultural land are about 20–100 times this average.

Erosion rates vary in different regions because of topography, rainfall, wind intensity, and the type of agricultural practices used. In China, for example, the average annual soil loss is reported to be about 40 metric tons per hectare (18 tons per acre), whereas the U.S. average is 18 metric tons per hectare (8 tons per acre). In states such as Iowa and Missouri, however, annual soil erosion averages more than 35 metric tons per hectare (16 tons per acre).

Worldwide, about 10 million hectares (25 million acres) of land—an area about the size of Virginia—are abandoned for crop production each year because of high erosion rates plus waterlogging, salinization, and other forms of soil degradation. In addition, according to the UN Environment Programme, crop production becomes uneconomical on about 20 million hectares (49 million acres) each year because soil quality has been severely degraded.

Soil erosion also occurs in forestlands, but it is not as severe as that in the more exposed soil of agricultural land. Soil erosion in managed forests is a primary concern because the soil reformation rate in forests is about two to three times slower than that in agricultural land. To compound this erosion problem, at least 24 million hectares (59 million acres) of forest are being cleared each year throughout the world; most of this land is used to grow food and graze cattle.

The effects of agriculture and forestry are interrelated in other ways. Deforestation reduces fuelwood supplies and forces the poor in developing countries to substitute crop residue and manure for fuelwood. When these plant and animal wastes are burned instead of being returned to the land as ground cover and organic fertilizer, erosion is intensified and productivity of the land is decreased. These factors, in turn, increase pressure to convert more forestland into agricultural land, further intensifying soil erosion.

One reason that soil erosion is not a high priority among many governments and farmers is that it usually occurs so slowly that its cumulative effects may take decades to become apparent. For example, the removal of 1 millimeter (1/25 inch) of soil is so small that it goes undetected. But over a 25-year period the loss would be 25 millimeters (1 inch), which would take about 500 years to replace by natural processes.

Besides reduced soil depth, soil erosion leads to reduced crop productivity because of losses of water, organic matter, and soil nutrients. A 50% reduction of soil organic matter on a plot of land has been found to reduce corn yields as much as 25%.

PROJECTS

1. Use the library or the Internet to find out what key mineral resources are found in the country where you live. How long could the estimated reserves of each of these meet the current needs of your country? How long will the estimated reserves last if the use of each of these resources increases by 2% a year?

2. As a class project, evaluate soil erosion on your school grounds. Use this information to develop a soil conservation plan for your school and then present it to school officials.

3. What mineral resources are extracted in your local area? What mining methods are used? Do local, state, or federal laws require restoration of the landscape after mining is completed? If so, how stringently are those laws enforced?

4. If feasible, visit a variety of farms. If you can't find actual farms to answer these questions, use the library or Internet to find and evaluate data from such farms.

a. What kind of tillage do they use?

b. See whether you can find a farm using conventional tillage and a nearby one using conservation tillage. Talk to the farmers and compare the two

Q: How many people are undernourished or malnourished?

When soil erodes, vital plant nutrients such as nitrogen, phosphorus, potassium, and calcium are also lost. With U.S. annual cropland erosion rates of about 18 metric tons per hectare (8 tons per acre), an estimated $18 billion of plant nutrients is lost annually. Using fertilizers to replace these nutrients substantially adds to the cost of crop production.

Some analysts who are unaware of the numerous and complex effects of soil erosion have falsely concluded that the damages are minor. For example, they report that soil loss causes an annual reduction in crop productivity of only 0.1–0.5% in the United States. However, we must consider all the ecological effects caused by erosion, including reductions in soil depth, availability of water for crops, and soil organic matter and nutrients. When this is done, agronomists and ecologists report a 15–30% reduction in crop productivity, leading to increased use of costly fertilizer. Because fertilizers are not a substitute for fertile soil, they can be applied only up to certain levels before crop yields begin to decline.

Reduced agricultural productivity is only one of the effects of soil erosion. In the United States, water runoff is responsible for transporting about 3 billion metric tons (3.3 billion tons) of sediment (about 60% from agricultural land) each year to waterways in the lower 48 states. Off-site damages to U.S. water storage capacity, wildlife, and navigable waterways from these sediments cost an estimated $6 billion each year. About 25% of new water storage capacity in U.S. reservoirs is built solely to compensate for sediment buildup.

When soil sediments that include pesticides and other agricultural chemicals are carried into streams, lakes, and reservoirs, fish production is adversely affected. These contaminated sediments interfere with fish spawning, increase predation on fish, and destroy fisheries in estuarine and coastal areas.

Increased erosion and water runoff on mountain slopes flood agricultural land in the valleys below, further decreasing agricultural productivity. Eroded land also does not hold water very well, again decreasing crop productivity. This effect is magnified in the 80 countries (with nearly 40% of the world's population) that experience frequent droughts.

Thus, soil erosion is one of the world's critical problems, and if not slowed, it will seriously reduce agricultural and forestry production and degrade the quality of aquatic ecosystems. Solutions that are not particularly difficult are often not implemented because erosion occurs so gradually that we fail to acknowledge its cumulative effects until damage is irreversible. Many farmers have also been conditioned to believe that losses in soil fertility can be remedied by applying more fertilizer or by using more fossil-fuel energy.

The principal way to control soil erosion and its accompanying runoff of sediment is to maintain adequate vegetative coverage on soils [by using various methods discussed in Section 14-7]. These methods are also cost-effective, especially when off-site costs of erosion are included. Scientists, policy makers, and agriculturists need to work together to implement soil and water conservation practices before world soils lose most of their productivity.

Critical Thinking

1. Some analysts contend that average soil erosion rates around the world are low and that the soil erosion problem can easily be solved with improved agricultural technology such as no-till cultivation and increased use of commercial inorganic fertilizers. Do you agree or disagree with this position? Explain.

2. What three major things do you believe elected officials should do to decrease soil erosion and the resulting water pollution by sediment in the United States or in the country where you live?

cultivation techniques in terms of soil erosion, soil fertility, crop yields, pest damage, use of herbicides and insecticides, and net profits per hectare on the same crops.

c. See whether you can find a conventional farm using inorganic fertilizers and pesticides and an organic farm using all or mostly organic fertilizer and little or no pesticides. Talk to the farmers and compare the two crop-growing techniques in terms of soil erosion, soil fertility, pest damage, crop yields, and net profits per hectare on the same crops.

5. Make a concept map of this chapter's major ideas, using the section heads and subheads and the key terms (in boldface). Look at the inside back cover and in Appendix 4 for information about concept maps.

A: About 840 million in 1995 (down from 940 million in 1970)

15 NONRENEWABLE ENERGY RESOURCES

Bitter Lessons from Chernobyl

Chernobyl is a chilling word recognized around the globe as the site of a horrendous nuclear disaster (Figure 15-1). On April 26, 1986, a series of explosions in one of the reactors in a nuclear power plant in Ukraine (then part of the Soviet Union) blew the massive roof off the reactor building and flung radioactive debris and dust high into the atmosphere to encircle the planet. Here are some consequences of this disaster caused by poor reactor design and human error:

- In 1998 the Ukrainian Health Ministry put the official death toll from the accident at 3,576. However, Greenpeace Ukraine estimates that by 1995 the total death toll from the accident was about 32,000.
- According to the United Nations, almost 400,000 people have been forced to leave their homes, probably never to return. Most were not evacuated until 10 days or more after the accident.
- According to a recent UN report, some 160,000 square kilometers (62,000 square miles) of the former Soviet Union—almost the size of the state of Florida or the combined areas of Denmark and Greece—remain contaminated with radioactivity.
- The 9 million people (about one-third of them children) still living in this area (and especially the almost 400,000 evacuated environmental refugees) live under constant stress.
- In 1994 analysis of samples taken from the reactor core revealed that the accident released the highly radioactive contents of 80% of the core, not the 3% announced to the world.
- Over half a million people were exposed to dangerous radioactivity, and some may suffer from cancers, thyroid tumors, and eye cataracts. Between 1986 and 1995 the rate of thyroid cancer (mostly among children) in Ukraine and nearby Belarus was more than 10 times the 1986 level and 30 times higher among those evacuated from the Chernobyl area. A 1996 study found that genetic mutations occur twice as often in children exposed to radioactive fallout from Chernobyl as in other families.
- Government officials say that the total cost of the accident will reach at least $358 billion—many times greater than the value of all the nuclear electricity that has ever been generated in the former Soviet Union.

The environmental refugees evacuated from the Chernobyl region had to leave their possessions behind and say good-bye to lush, green wheat fields and blossoming apple trees, to land their families had farmed for generations, to cows and goats that would

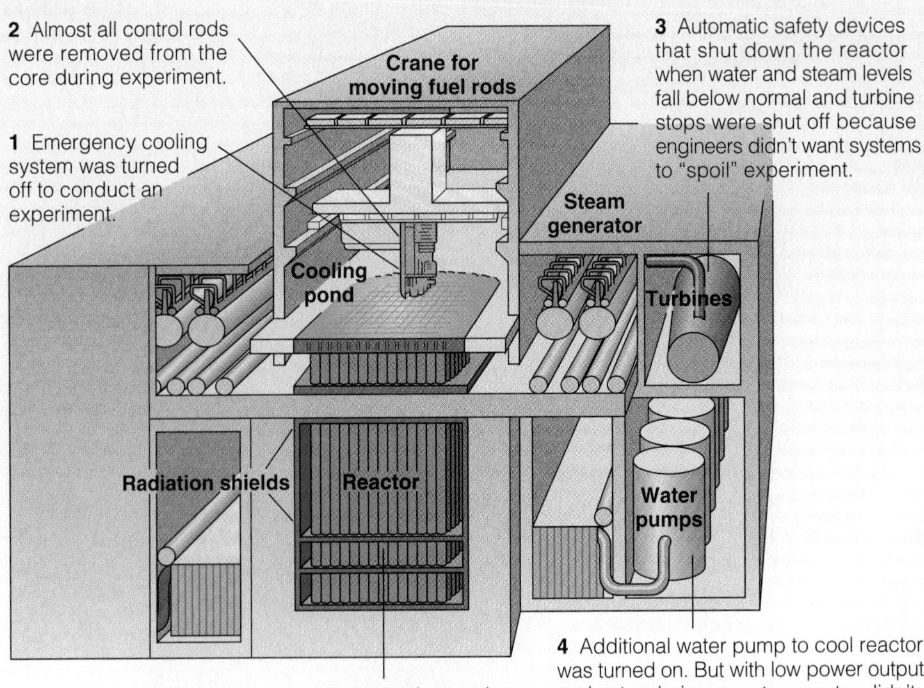

Figure 15-1 Major events leading to the Chernobyl nuclear power-plant accident on April 26, 1986, in the former Soviet Union. The accident happened because engineers turned off most of the reactor's automatic safety and warning systems (to keep them from interfering with an unauthorized safety experiment) and because of inadequate safety design of the reactor (no secondary containment shell, as in Western-style reactors, no emergency core-cooling system to prevent reactor core meltdown, and a design flaw that leads to unstable operation at low power). After the reactor exploded, crews exposed themselves to lethal levels of radiation to put out fires and encase the shattered reactor in a hastily constructed concrete tomb. This tomb is now sagging and full of holes that allow water to seep in and radioactive dust to drift out. Building a new tomb for the reactor will cost at least $1.5 billion—money the Ukrainian government doesn't have.

be shot because the grass they ate was radioactive, and to their radioactivity-poisoned cats and dogs. They will not be able to return.

World-famous gymnast Olga Korbut gave this account in 1991:

> *I was . . . in Minsk when Chernobyl happened, and they didn't tell us for three or four days. . . . We were all outdoors, because it was close to the May 1 celebration, and we were planting gardens and enjoying the spring. . . . It has been five years . . ., but people are still very frightened . . . and very angry. Our food and water supply is contaminated from radiation.*
>
> *When I went into the schools in Byelorussia, I learned that the first-graders have never been in the forest . . . because the trees were so contaminated. . . . When children want to see what nature used to be like, they go into a little courtyard inside the building, and the teacher says, "This is a bird and this is a tree," and they are plastic. Isn't that sad?*

Today the Chernobyl power plant remains one of the most dangerous places on the earth. Some areas in the burned out building are so radioactive that within a few seconds a human could absorb enough radiation to become extremely sick and possibly die.

Engineers fear that hastily built and crumbling, 20-story concrete shell (called a sarcophagus) built around the building to contain its radioactive core might collapse. This could allow potentially deadly radiation to escape into the atmosphere and increase cancer risks for millions of people worldwide. In 1997, the United States and six other developed countries agreed to provide $380 million to help stabilize the old shell by 2005 and build a new structure. It is a crucial race against time.

Chernobyl taught us that *a major nuclear accident anywhere is a nuclear accident everywhere.*

What role should nonrenewable nuclear energy play in our future? What are the pros and cons of continuing to depend mostly on nonrenewable oil, natural gas, and coal? For example, Chernobyl may have caused the premature deaths of 32,000 people between 1986 and 1995, but *each year* air pollution from coal burning causes the premature deaths of 65,000 to 200,000 people in the United States alone. Globally the *annual* death toll from coal burning is in the millions. This chapter is devoted to evaluating these nonrenewable energy resources.

We are an interdependent world and if we ever needed a lesson in that, we got it in the oil crisis of the 1970s.

ROBERT S. MCNAMARA

In this chapter we will seek answers to the following questions:

- How should we evaluate energy alternatives?
- What are the benefits and drawbacks of oil?
- What are the benefits and drawbacks of natural gas?
- What are the benefits and drawbacks of coal?
- What are the benefits and drawbacks of conventional nuclear fission, breeder nuclear fission, and nuclear fusion?

15-1 EVALUATING ENERGY RESOURCES

What Types of Energy Do We Use? *Some 99% of the energy used to heat the earth and all of our buildings comes directly from the sun.* Without this input of essentially inexhaustible solar energy, the earth's average temperature would be –240°C (–400°F) and life as we know it would not exist. Solar energy also helps recycle the carbon, oxygen, water, and other chemicals we and other organisms need to stay alive and healthy. This direct input of solar energy also produces several forms of renewable energy: wind, falling and flowing water (hydropower), and biomass (solar energy converted to chemical energy stored in chemical bonds of organic compounds in trees and other plants).

The remaining 1%, the portion we generate to supplement the solar input, is *commercial energy* sold in the marketplace. Most commercial energy comes from extracting and burning mineral resources obtained from the earth's crust, primarily nonrenewable fossil fuels (Figure 15-2).

Developed countries and developing countries differ greatly in their sources of energy (Figure 15-3) and average per capita energy use. The most important supplemental source of energy for developing countries is potentially renewable biomass, especially fuelwood and charcoal made from fuelwood, the main sources of energy for heating and cooking for roughly half the world's population. Within a few decades

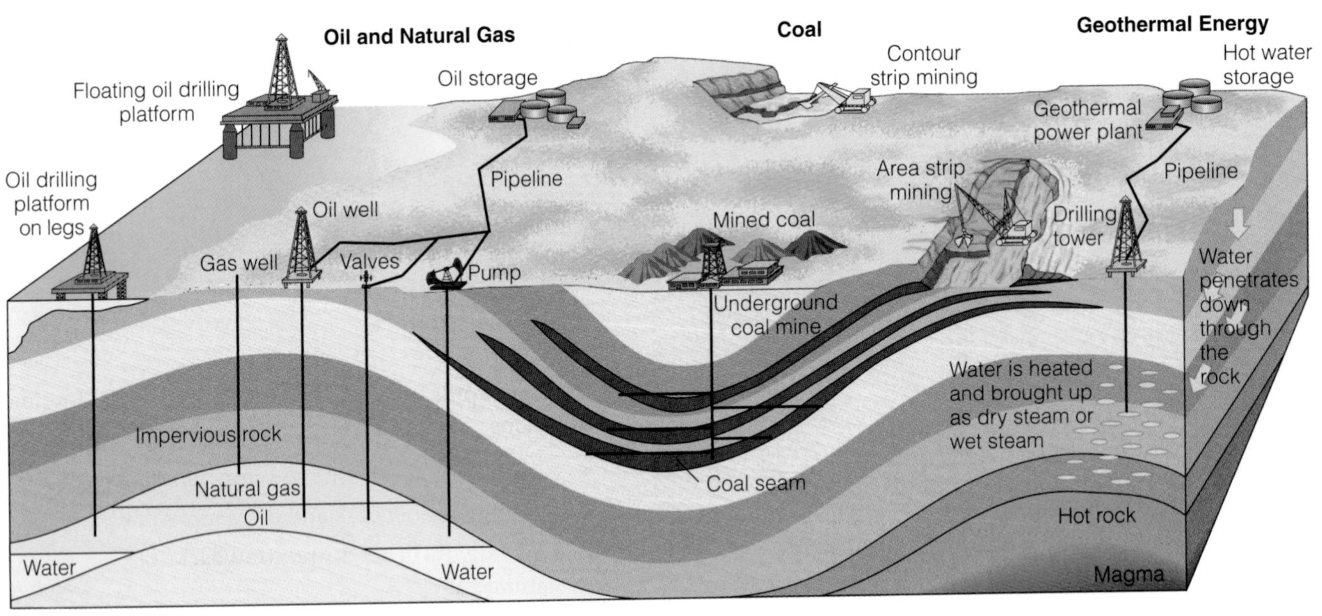

Figure 15-2 Important commercial energy resources from the earth's crust are geothermal energy, coal, oil, and natural gas. Uranium ore is also extracted from the crust and then processed to increase its concentration of uranium-235, which can be used as a fuel in nuclear reactors to produce electricity.

one-fourth of the world's population in developed countries may face an oil shortage, but half the world's population in developing countries already faces a fuelwood shortage.

The United States is the world's largest user (and waster) of energy. With only 4.6% of the population, it uses 24% of the world's commercial energy, 93% from *nonrenewable* fossil fuels (85%) and nuclear energy (8%). In contrast, India, with 17% of the world's people, uses only about 3% of the world's commercial energy.

How Should We Evaluate Energy Resources? The types of energy we use and how we use them are major factors determining our quality of life and our harmful environmental effects. Our current dependence on nonrenewable fossil fuels is the primary cause of air and water pollution, land disruption, and projected global warming. Moreover, affordable oil, the most widely used energy resource in developed countries, will probably be depleted within 40–80 years and must be replaced by other energy resources.

What is our best immediate energy option? The general consensus is to cut out unnecessary energy waste by improving energy efficiency, as discussed in Chapter 16. What is our next best energy option? There is disagreement about that. Some say we should get much more of the energy we need from the sun, wind, flowing water, biomass, heat stored in the earth's interior, and hydrogen gas by making the transition to a *renewable energy* or *solar age*. These renewable energy options are evaluated in Chapter 16.

Others say we should burn more coal and synthetic liquid and gaseous fuels made from coal. Some believe natural gas is the answer, at least as a transition fuel to a new solar age built around improved energy efficiency and renewable energy. Others think nuclear power is the answer. These nonrenewable energy options are evaluated in this chapter.

Experience shows that it usually takes at least 50 years and huge investments to phase in new energy alternatives (Figure 15-4), with the exception of nuclear power, which after almost 50 years still provides only a small proportion of the world's commercial energy. Thus, we must plan for and begin the shift to a new mix of energy resources now. To do so involves answering the following questions for *each* energy alternative:

- How much of the energy source will be available in the near future (the next 15 years), the intermediate future (the next 30 years), and the long term (the next 50 years)?
- What is this source's net energy yield?
- How much will it cost to develop, phase in, and use this energy resource?

Q: How many people die each year from hunger-related causes?

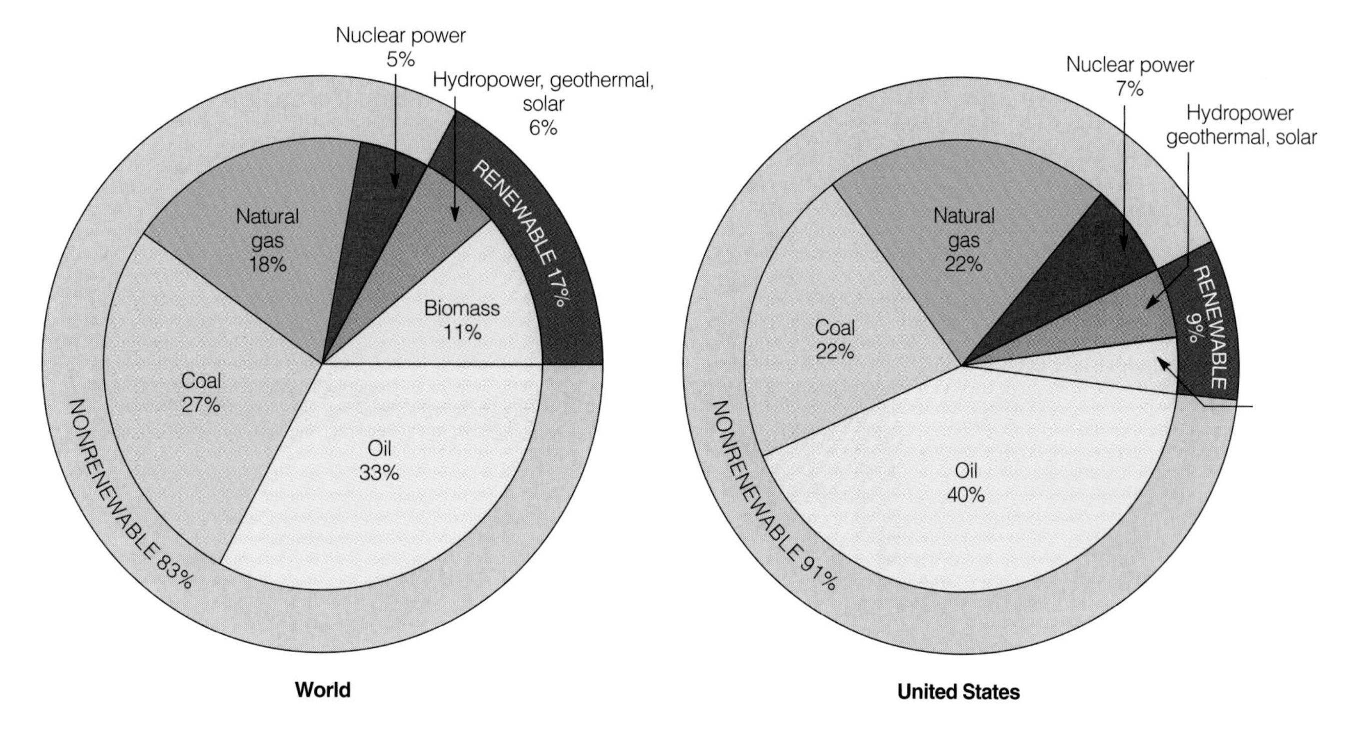

Figure 15-3 Commercial energy use by source for developed countries and developing countries. Commercial energy amounts to only 1% of the energy used in the world; the other 99% comes from the sun and is not sold in the marketplace. (Data from U.S. Department of Energy, British Petroleum, and Worldwatch Institute)

- How will extracting, transporting, and using the energy resource affect the environment?
- What will using this energy source do to help sustain the earth for us, for future generations, and for the other species living on this planet?

What Is Net Energy? The Only Energy That Really Counts It takes energy to get energy. For example, oil must be found, pumped up from beneath the ground, transported to a refinery and converted to useful fuels (such as gasoline, diesel fuel, and heating oil), then transported to users, and finally burned in furnaces and cars before it is useful to us. Each of these steps uses energy, and the second law of energy (Section 3-5) tells us that each time we use energy to perform a task, some of it is always wasted and is degraded to low-quality energy.

The usable amount of high-quality energy available from a given quantity of an energy resource is its **net energy**: the total useful energy available from the resource over its lifetime minus the amount of energy used (the first law of energy), automatically wasted (the second law of energy), and unnecessarily wasted in finding, processing, concentrating, and transporting it to users. Net energy is like your net spendable income (your wages minus taxes and other deductions). For example, suppose that for every

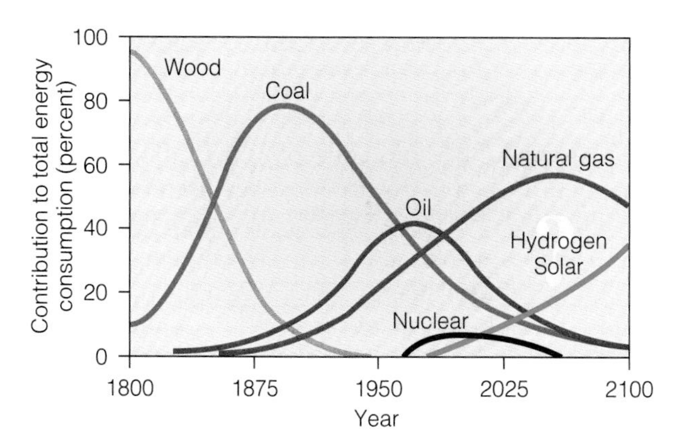

Figure 15-4 Shifts in the use of commercial energy resources in the United States since 1850, with projected changes to 2100. Shifts from wood to coal and then from coal to oil and natural gas have each taken about 50 years. Affordable oil is expected to be running out within 40–50 years, and burning fossil fuels is the primary cause of air pollution and projected warming of the atmosphere. For these reasons, most analysts believe we must make a new shift in energy resources over the next 50 years. Some believe that this shift should involve improved energy efficiency and greatly increased use of solar energy and hydrogen. (Data from U.S. Department of Energy)

10 units of energy in oil in the ground we have to use and waste 8 units of energy to find, extract, process, and transport the oil to users. Then we have only 2 units of useful energy available from every 10 units of energy in the oil.

A: At least 10 million (some say 18 million)

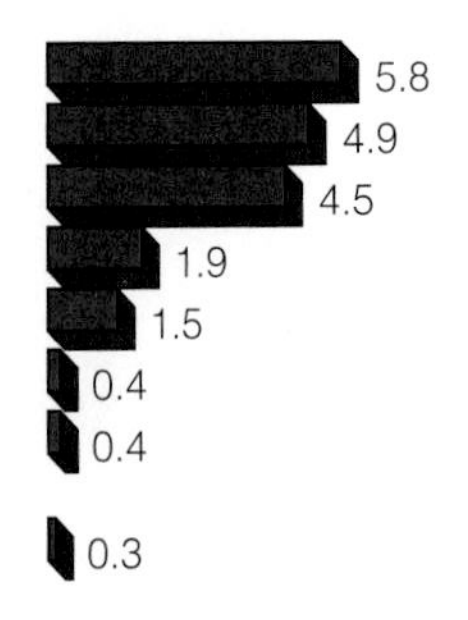

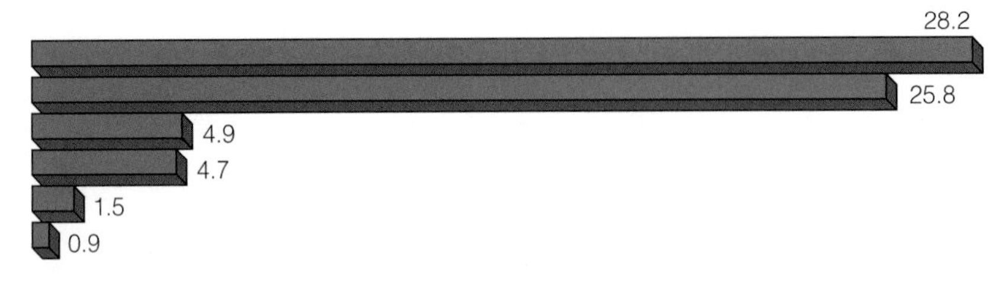

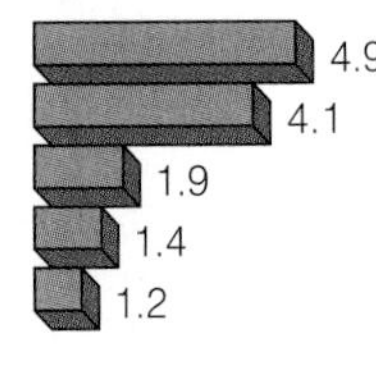

Figure 15-5 Net energy ratios for various energy systems over their estimated lifetimes. (Data from Colorado Energy Research Institute, *Net Energy Analysis*, 1976; and Howard T. Odum and Elisabeth C. Odum, *Energy Basis for Man and Nature*, 3d ed., New York: McGraw-Hill, 1981)

We can look at this concept as the ratio of useful energy produced to the useful energy used to produce it. In the example just given, the *net energy ratio* would be 10/8, or approximately 1.25. The higher the ratio, the greater the net energy yield. When the ratio is less than 1, there is a net energy loss.

Figure 15-5 shows estimated net energy ratios for various types of space heating, high-temperature heat for industrial processes, and transportation. Currently, oil has a high net energy ratio because much of it comes from large, accessible deposits such as those in the Middle East. When those sources are depleted, the net energy ratio of oil will decline and prices will rise. Then more money and high-quality fossil-fuel energy will be needed to find, process, and deliver new oil from widely dispersed small deposits, deposits buried deep in the earth's crust, or deposits located in remote areas.

Conventional nuclear energy has a low net energy ratio because large amounts of energy are required to extract and process uranium ore, convert it into a usable nuclear fuel, and build and operate power plants. In addition, more energy is needed to dismantle the plants after their 15–40 years of useful life and to store the resulting highly radioactive wastes for thousands of years.

15-2 OIL

What Is Crude Oil, and How Is It Extracted and Processed? **Petroleum**, or **crude oil** (oil as it comes out of the ground) is a fossil fuel produced by the decomposition of deeply buried dead organic matter from plants and animals under high temperatures and pressures over millions of years. Typically this gooey, smelly liquid consists mostly of hydrocarbons, with small amounts of sulfur, oxygen, and nitrogen impurities.

Crude oil and natural gas are often trapped together under a dome deep within the earth's crust (Figure 15-2). The crude oil is dispersed in pores and cracks in underground rock formations, like water in a sponge. If there is enough pressure from natural gas and water under the dome of oil-containing rock, some of the crude oil is pushed to the surface when a well is drilled (known as a *gusher*). Such wells are rare, however, and oil in most wells must be pumped to the surface.

Primary oil recovery involves drilling a well and pumping out the oil that flows by gravity into the bottom of the well. After the flowing oil has been removed, water can be injected into nearby wells to force some of the remaining heavy oil to the surface, a process known as **secondary oil recovery**.

Campbell, C. J. 1998. "Running Out of Gas: This Time the Wolf Is Coming." *National Interest*, no. 51, 47(9).

On average, producers get only about 35% of the oil out of a reservoir by primary and secondary recovery before they abandon it because the *heavy oil* that remains is too difficult or expensive to recover.

As oil prices rise, it may become economical to remove about 10–25% of this heavy oil by **enhanced**, or **tertiary, oil recovery**. Steam or CO_2 gas can be used to force some of the heavy oil into the well cavity for pumping to the surface. However, enhanced oil recovery is expensive: It takes the energy in one-third of a barrel of oil to retrieve each barrel of heavy oil. Researchers hope to use bacteria to increase tertiary recovery by 10–25% at a lower cost.

Crude oil is a complex mixture of various hydrocarbons with different boiling points. Most crude oil travels by pipeline to a *refinery* (Figure 15-6), where it is heated and distilled in gigantic columns to separate it into liquid components with different boiling points, such as naphtha, diesel oil, heating oil, aviation fuel, and gasoline and solids such as grease, wax, and asphalt.

Oil in the form of such processed products is the most widely used source of energy in developed countries (Figure 15-3, left) and the second most widely used source of energy (after biomass) in developing countries (Figure 15-3, right).

Some of the products of oil distillation, called **petrochemicals**, are used as raw materials in industrial organic chemicals, pesticides, plastics, synthetic fibers, paints, medicines, and many other products.

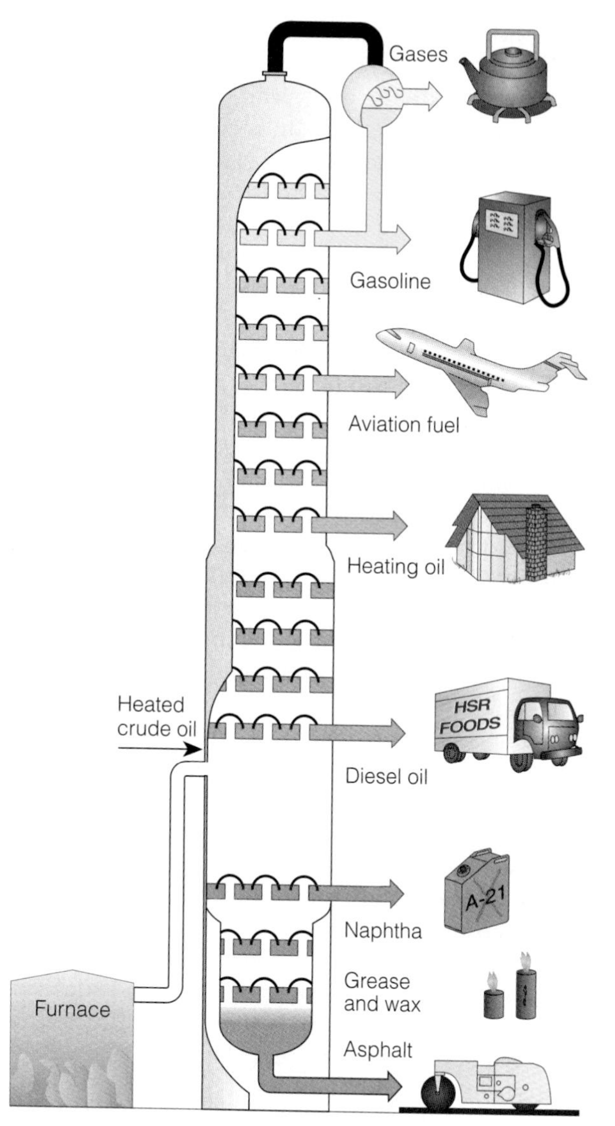

Figure 15-6 Refining crude oil. Components are removed at various levels, depending on their boiling points, in a giant distillation column. The most volatile components with the lowest boiling points are removed at the top of the column.

Who Has the World's Oil Supplies? Oil *reserves* are identified deposits from which oil can be extracted profitably at current prices with current technology. The 13 countries that make up the Organization of Petroleum Exporting Countries (OPEC)* have 67% of the world's reserves, which explains why OPEC is expected to have long-term control over world oil supplies and prices. Saudi Arabia, with 26%, has the world's largest known crude oil reserves, followed by Iraq with 10%.

In 1997, OPEC produced 40% of the world's oil (down from 65% in 1973) but by 2010 is expected to supply almost half of the world's oil. Oil production from large oil fields developed over the past two decades in Alaska, Siberia, and Mexico is still large but has stopped expanding, and in some cases production is declining. In the former Soviet Union oil reserves declined by 10% between 1984 and 1994.

Between 1973 and 1993, global oil reserves rose 60%. Geologists believe that the politically volatile Middle East—with more than 65% of the world's current reserves—contains the majority of the world's undiscovered oil. However, some additional oil is expected to be found from increased exploration in China, Mexico, and other developing countries.

Better technology such as three-dimensional seismic surveys, floating platforms, and horizontal drilling techniques has improved exploration and cut operating costs, allowing oil companies to profit even when oil prices are as low as $14 per barrel. Between 1981 and 1994, offshore drilling costs fell 38%.

*OPEC was formed in 1960 so that developing countries with much of the world's known and projected oil supplies could get a higher price for this resource. Today its members are Algeria, Ecuador, Gabon, Indonesia, Iran, Iraq, Kuwait, Libya, Nigeria, Qatar, Saudi Arabia, United Arab Emirates, and Venezuela.

Hint: Enter the search term *energy crisis* using the Subject Guide.

Currently, the United States has only 2.3% of the world's oil reserves but uses nearly 30% of the oil extracted worldwide each year, 65% of it for transportation, mostly because oil is cheap (Figure 15-7). Figure 15-8 shows the locations of the largest crude oil fields in the United States and Canada. Despite an upsurge in exploration and test drilling, U.S. oil extraction has declined since 1985, and the net energy yield for small new domestic oil supplies is low and falling.

Mostly because of declining oil reserves and increased oil use, the United States imported 55% of the oil it used in 1997 (up from 36% in 1973); by 2010 it could be importing 70% or more of the oil it uses. The cost of this imported oil contributes to the huge trade deficit that has made the United States the world's largest international debtor.

This dependence on imported oil (about half of it from OPEC countries) and the likelihood of much higher oil prices within 10–20 years could drain the United States and other major oil-importing nations of vast amounts of money. This could lead to severe inflation and widespread economic recession, perhaps even a major depression. Moreover, sabotage of the vulnerable Trans-Alaska oil pipeline, which the Department of Defense says is impossible to protect, could disrupt the entire U.S. economy.

When Iraq invaded Kuwait in the summer of 1990, the United States and other developed countries went to war, mostly to protect their access to oil in the Middle East, especially from Saudi Arabia. As Worldwatch Institute researchers Christopher Flavin and Nicholas Lenssen put it, "Not only is the world addicted to cheap oil, but the largest liquor store is in a very dangerous neighborhood."

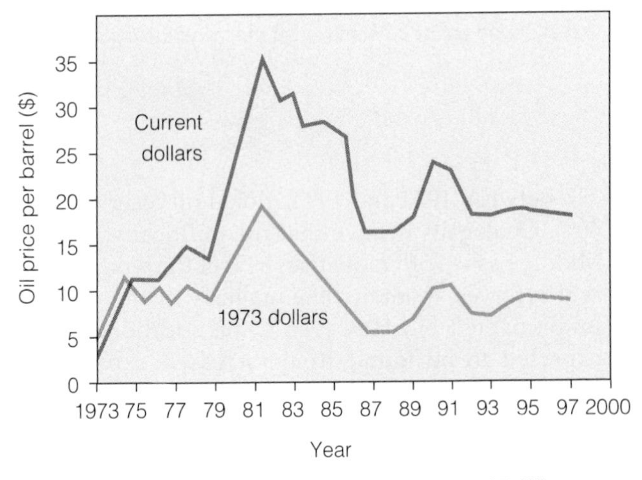

Figure 15-7 Average world crude oil prices, 1973–97. When the price is adjusted for inflation, oil has remained cheap since 1975. (Data from Department of Energy and Department of Commerce)

How Long Will Oil Supplies Last? The reserves—known and affordable supplies of a nonrenewable resource (Figure 14-6) such as oil—are considered economically depleted when 80% of the supply has been used; the remaining 20% is considered too expensive to extract (Figure 1-15).

Oil's fatal flaw is that its reserves may be 80% depleted within 35–84 years, depending on how rapidly it is used. Currently, each day the world consumes oil equal to what it took the earth's natural processes 10,000 days to produce. At the current rate of consumption, global oil reserves will last at least 44 years. Undiscovered oil that is thought to exist might last another 20–40 years. At today's consumption rate, U.S. oil reserves will be depleted in about 24 years (less if consumption increases as projected); potential reserves might yield an additional 24 years of production.

Instead of remaining at the current level, however, global oil consumption is projected to increase by about 25% by 2010. This will hasten depletion of global oil reserves.

Some analysts argue that rising oil prices (when oil consumption exceeds oil production) will stimulate exploration and lead to enough new reserves to meet future demand through the next century. Other analysts argue that such optimistic projections about future oil supplies ignore the consequences of exponentially increasing consumption of oil.

Even assuming that we continue to use crude oil at the current rate, Saudi Arabia, with the largest

Figure 15-8 Locations of the largest crude oil and natural gas fields in the United States and Canada. Little *new* oil and natural gas is expected to be found in the United States. (Data from Council on Environmental Quality)

Q: What is the chief cause of hunger, malnutrition, and premature death from hunger-related diseases?

known crude oil reserves, could supply all the world's oil needs for only 10 years; the estimated reserves under Alaska's North Slope (the largest ever found in North America) would meet world demand for only 6 months or U.S. demand for 3 years. In short, just to keep on using oil at the *current* rate and not run out, we must discover and add to global oil reserves the equivalent of a new Saudi Arabian supply *every 10 years*. According to the U.S. Geological Survey, global discovery of large oil fields peaked in 1962 and has been declining since.

In 1998 oil-industry experts Colin Campbell and Jean H. Laherrère projected that (barring a global recession) global oil production should peak during the first decade of the 21st century and then begin falling. They contend that projections of ample and affordable supplies of conventional oil for 35–84 years are flawed because they **(1)** rely on distorted estimates of reserves, **(2)** assume that production will remain constant, and **(3)** assume falsely that new technologies will allow tertiary recovery of 40–60% of the oil in existing wells instead of the usual 30%. They also project that sources of unconventional oil from oil shale and tar sands will not significantly increase global oil supplies within the next 10–20 years because they are too costly, can cause unacceptable environmental problems, and cannot be developed rapidly enough to meet the growing demand for oil over the next two decades. If they are correct and production does peak within the next 10–20 years, oil prices are expected to rise sharply and probably permanently.

What Are the Pros and Cons of Conventional Oil? Oil is still cheap; when adjusted for inflation it costs about the same as it did in 1975 (Figure 15-7). It is easily transported within and between countries and, when extracted from easily accessible deposits, it has a high net energy yield (Figure 15-5). Oil's low price has encouraged developed countries and developing countries alike to become heavily dependent on—indeed, addicted to—this important resource. Low prices have also encouraged waste of oil and discouraged improvements in energy efficiency and the switch to other sources of energy.

Oil also has some drawbacks. As mentioned, estimated reserves of oil may be 80% depleted within 44–84 years, depending on how rapidly it is used, and demand may exceed production within 10–20 years. Oil, like any other fossil fuel, can cause pollution and environmental degradation throughout its life cycle of extraction, processing, transportation, and use (Figure 14-13).

The oil-drilling process causes land disturbance, which can accelerate erosion. It also produces waste materials (soil, rock, and drilling muds removed from the hole), and can pollute soil and water if some of the oil is spilled. Leakage of drilling muds (usually stored in unlined pits until it is reinjected into the well) can contaminate nearby surface and groundwater. Burning fuel oil and gasoline releases heat-trapping carbon dioxide, which could alter global climate (Sections 19-2 and 19-3), and other air pollutants such as sulfur dioxide (SO_2) and nitrogen oxides (NO and NO_2) that harm people, crops, trees, fish, and other species (Section 18-5).

Figure 15-9 Oil shale and the shale oil extracted from it. Big U.S. oil shale projects have been canceled because of excessive cost. (U.S. Department of Energy)

Suppose that all of the harmful environmental effects of using oil were included in its market price (full-cost pricing) and that current government subsidies were phased out. Then analysts project that oil would become so expensive that it would probably be replaced by improved energy efficiency and by a variety of less harmful and cheaper renewable energy resources (also evaluated using full-cost pricing).

What Are the Pros and Cons of Using Heavy Oil Produced from Oil Shale? **Oil shale** is a fine-grained rock (Figure 15-9) that contains a solid, waxy mixture of hydrocarbon compounds called **kerogen**. After being removed by surface or subsurface mining, the shale is crushed and heated above ground in a retort to vaporize the kerogen (Figure 15-10). The kerogen vapor is condensed, forming heavy, slow-flowing, dark-brown **shale oil**. Before shale oil can be sent by pipeline to a refinery, it must be heated to increase its flow rate and processed to remove sulfur, nitrogen, and other impurities.

The shale oil potentially recoverable from U.S. deposits, mostly on federal lands in Colorado, Utah, and Wyoming, could probably meet the country's crude oil demand for 41 years at current use levels. Canada, China, and several republics in the former Soviet Union also have large oil-shale deposits. Indeed, according to energy expert John Harte, estimated

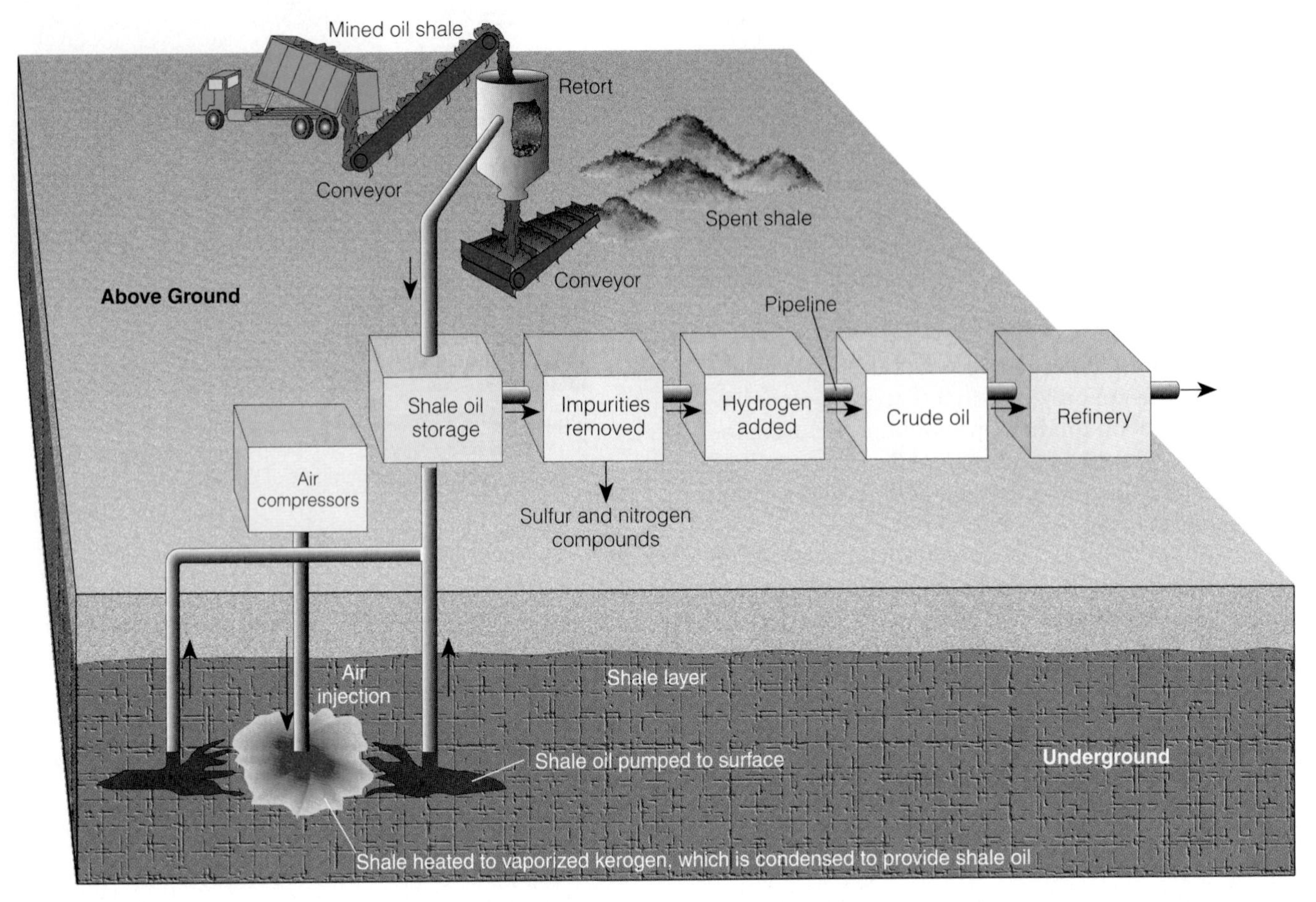

Figure 15-10 Aboveground and underground (*in situ*) methods for producing synthetic crude oil from oil shale.

potential global supplies of shale oil are 200 times larger than estimated global supplies of conventional oil.

However, there are problems with shale oil. It has a lower net energy yield than does conventional oil because it takes the energy from almost half a barrel of conventional oil to extract, process, and upgrade one barrel of shale oil. Processing the oil requires large amounts of water, which is scarce in the semiarid locales where the richest deposits are located. Surface mining of shale oil tears up the land, leaving mountains of shale rock (which expands somewhat like popcorn when heated). In addition, salts, cancer-causing substances, and toxic metal compounds can be leached from the processed shale rock into nearby water supplies. Some of these problems can be reduced by extracting shale oil underground (Figure 15-10), but this method is too expensive and produces more sulfur dioxide air pollution than does surface processing.

What Are the Pros and Cons of Using Heavy Oil Produced from Tar Sand? **Tar sand** (or oil sand) is a mixture of clay, sand, water, and **bitumen** (a gooey, black, high-sulfur heavy oil). It is usually removed by surface mining. It is then heated with pressurized steam until the bitumen fluid softens and floats to the top. The bitumen is then purified and chemically upgraded into a synthetic crude oil suitable for refining (Figure 15-11).

The world's largest known deposits of tar sands, the Athabasca Tar Sands, lie in northern Alberta, Canada. Currently, these deposits supply about 21% of Canada's oil needs. These deposits could supply all of Canada's projected oil needs for about 33 years at its current consumption rate, but they would last the world only about 2 years. Other large deposits of tar sands are in Venezuela, Colombia, and parts of the former Soviet Union.

Producing synthetic crude oil from tar sands has several disadvantages. The net energy yield is low because it takes the energy in almost one-half a barrel of conventional oil to extract and process one barrel of bitumen and upgrade it to synthetic crude oil. Processing requires large quantities of water, and upgrading bitumen to synthetic crude oil releases large quantities of air pollutants. The plants also create huge waste disposal ponds.

Q: What human activity has the most harmful overall environmental impact?

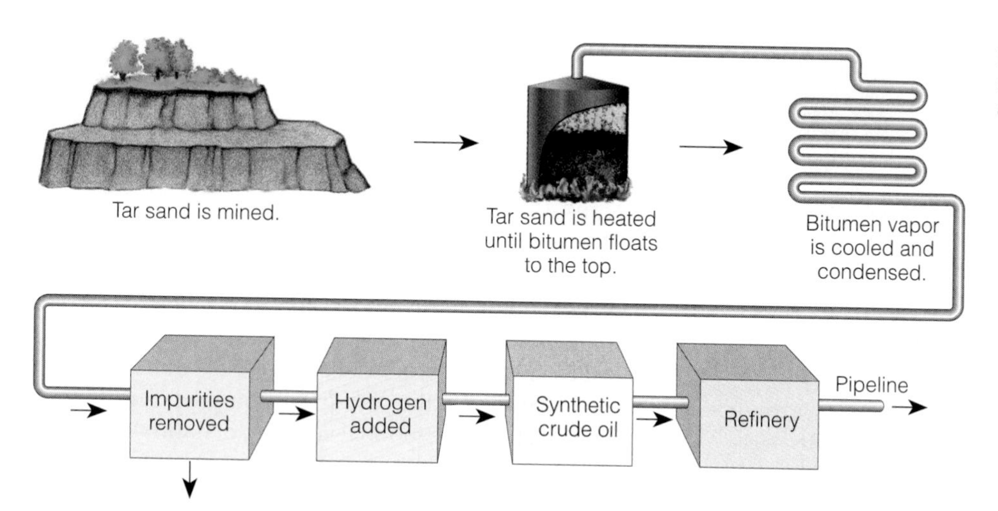

Figure 15-11 Generalized summary of how synthetic crude oil is produced from tar sand.

15-3 NATURAL GAS

What Is Natural Gas? In its underground gaseous state, **natural gas** is a mixture of 50–90% by volume of methane (CH_4), the simplest hydrocarbon; smaller amounts of heavier gaseous hydrocarbons such as ethane (C_2H_6), propane (C_3H_8), and butane (C_4H_{10}); and highly toxic hydrogen sulfide (H_2S), a byproduct of naturally occurring sulfur in the earth. Most natural gas deposits occur at or near hot spots, where high temperatures and pressures or catalytically active metals in the ground break down long-chain hydrocarbon molecules in petroleum to short hydrocarbons (methane, ethane, and propane) in natural gas.

Conventional natural gas lies above most reservoirs of crude oil (Figure 15-2). Natural gas is also found in *unconventional* deposits, including coalbeds, Devonian shale rock, deep underground deposits of tight sands, and deep zones that contain natural gas dissolved in hot water. Another source is gas hydrates, an icelike material that occurs in underground deposits all over the world and is composed largely of water and methane. Global deposits of gas hydrates are estimated to contain twice as much carbon as all other fossil fuels on earth. It is not yet economically feasible to get natural gas from such unconventional sources, but the extraction technology is being developed rapidly.

When a natural gas field is tapped, propane and butane gases are liquefied and removed as **liquefied petroleum gas (LPG)**. LPG is stored in pressurized tanks for use mostly in rural areas not served by natural gas pipelines. The rest of the gas (mostly methane) is dried to remove water vapor, cleansed of poisonous hydrogen sulfide and other impurities, and pumped into pressurized pipelines for distribution. Because natural gas and LPG are highly flammable and odorless, a compound is added to give these fuels a smell that warns of leakage.

Although natural gas is a vast and valuable natural resource, it is often cheaper to burn in the atmosphere rather than to use. Unless pipelines are already in place, removing natural gas from remote sites such as Alaska's north slope or Siberia costs more than the gas is worth. As a result, the gas released when oil wells are tapped (Figure 15-2) is often burned in the atmosphere (wasting this resource and adding CO_2 to the atmosphere that can enhance possible global warming) or pumped back into the ground.

In 1998, researchers developed a new catalyst that may sharply reduce this form of resource waste and air pollution. When methane is burned, the catalyst speeds up the conversion of about 70% of gas to a chemical that can easily be converted into methanol, a liquid fuel that can be transported in trucks and tankers, much like petroleum. If this process turns out to be financially feasible, it can open the door to some of the world's vast untapped reserves of natural gas and help slow projected global warming.

At a very low temperature of –184°C (–300°F), natural gas can be converted to **liquefied natural gas (LNG)**. This highly flammable liquid can then be shipped to other countries in refrigerated tanker ships.

Who Has the World's Natural Gas Supplies? Russia and Kazakhstan have almost 40% of the world's natural gas reserves. Other countries with large known natural gas reserves are Iran (15%), Qatar (5%), Saudi Arabia (4%), Algeria (4%), the United States (3%), Nigeria (3%), and Venezuela (3%). Geologists expect to find more natural gas, especially in unexplored developing countries.

Most U.S. natural gas reserves are located in the same places as crude oil (Figure 15-8). About 90–95% of the natural gas used in the United States is domestic and is distributed by about 411,000 kilometers (255,000 miles) of pipeline; the other 5–10% is imported by pipeline from Canada. Because of the current low price of natural gas, lack of potential drilling sites for large deposits, and stricter environmental regulations, there is little incentive for significantly

A: Agriculture

Combined-Cycle Natural Gas Systems

SOLUTIONS

Since 1980 there has been rapidly increasing use of a new technology—the *combined-cycle natural gas system*—to generate electricity. First, natural gas is burned in a gas turbine (which works like a jet engine, with pressurized natural gas injected and then combusted to spin the turbine) to produce electricity; then the excess heat from the turbine powers a second steam turbine (similar to those in conventional power plants) to produce more electricity. The energy efficiency of such systems is typically 40–60%, compared to 33% efficiency for a nuclear power plant and 36% for a coal-burning plant.

Such systems take half the time (about two and a half years) and cost about half as much per kilowatt-hour to build as a conventional coal-burning plant. These very reliable systems consist of small, factory-produced modular units, which allows capacity to be expanded as needed.

Without having to add expensive air pollution control equipment, combined-cycle gas plants emit virtually no sulfur dioxide and negligible amounts of particulates. Compared to coal-burning power plants, they also cut emissions of nitrogen oxides by up to 90% and carbon dioxide by up to 60%.

Because of the low cost and low pollutant emissions, it is expected that many utilities will convert hundreds of their aging coal plants into combined-cycle natural gas plants. Studies have also shown that it would be much cheaper to convert unfinished nuclear power plants throughout the world to combined-cycle gas plants than to finish them.

Smaller, even more versatile combined-cycle gas units are on the way. They could be used as cogenerators to supply all of the heat and electricity needs of an apartment building or office building at an overall efficiency of at least 85%.

Critical Thinking

What might be some disadvantages of the widespread use of combined-cycle natural gas turbine systems as power plants and as sources of electricity for buildings?

increased exploration and development of new natural gas supplies in the United States.

Algeria and some of the countries of the former Soviet Union use pipelines to supply Europe with natural gas. In 1994 Japan proposed that a 41,000-kilometer (25,500-mile) international pipeline network be built to link natural gas fields in Central and Southeast Asia, Siberia, and Alaska with the main markets of Asia, primarily Japan and China.

What Are the Pros and Cons of Natural Gas? Natural gas has a number of advantages over other nonrenewable energy sources. To date, it is cheaper than oil. At the current consumption rate, known reserves and undiscovered, potential reserves of conventional natural gas in the United States are projected to last 65–80 years; world reserves are expected to last at least 125 years at the current consumption rate. It is estimated that conventional supplies of natural gas, plus unconventional supplies available at higher prices, will last at least 200 years at the current consumption rate and 80 years if usage rates rise 2% per year.

Natural gas can be transported easily over land by pipeline; it has a high net energy yield (Figure 15-5), burns hotter, and produces less air pollution than any other fossil fuel. Burning natural gas produces 43% less heat-trapping carbon dioxide per unit of energy than coal and 30% less than oil (Figure 15-12). Extracting natural gas also damages the environment much less than extracting either coal or uranium ore.

Natural gas is easier to process than oil and, because it can be transported by pipeline, it is less expensive to transport by pipeline than coal (which is usually moved by rail). It can also be used to power vehicles, with over 100,000 vehicles now running on clean-burning natural gas in the United States.

In 1997, Honda designed an engine that runs on natural gas and is the cleanest internal combustion engine ever made. Emissions of hydrocarbons, nitrous oxide, and carbon monoxide of Honda's natural-gas Civic are at least 60 times below the 1997 federal limits for new cars in the United States. The car has a range of 350 kilometers (220 miles) and is now being

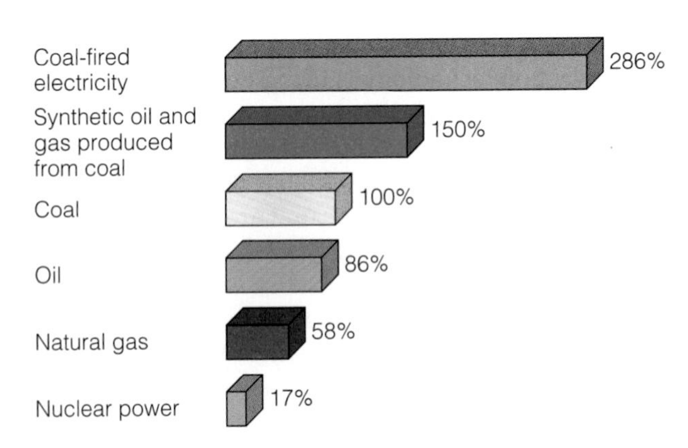

Figure 15-12 Carbon dioxide emissions per unit of energy produced by various fuels, expressed as percentages of emissions produced by coal.

Loeb, Penny. 1997. "Shear Madness." *U.S. News & World Report,* vol. 23, no. 6, 26(9).

produced in the United States. Two drawbacks are the car's $4,500 higher price and the fact that it is feasible only for companies with car fleets that have natural-gas refueling depots.

New *combined-cycle natural gas systems* (Solutions, left) can produce electricity much more efficiently and cheaply (and with less pollution and no risk of emissions of radioactive wastes) than burning coal or oil or using nuclear power. In addition, natural gas can be burned cleanly and efficiently by *cogeneration* to produce both high-temperature heat and electricity, further improving the efficiency of this fuel.

Natural gas can also be used in highly efficient fuel cells. For example, Southern California Gas and Tokyo Electric Power are using fuel cells to provide electricity and heat in several hotels, hospitals, and other commercial buildings. Such fuel cells are expensive, but their cost is expected to drop as they are mass produced in the late 1990s and beyond.

Natural gas has a few drawbacks, but they are minor compared to those of other fossil fuels and nuclear power. When it is processed, some highly toxic H_2S can be released into the atmosphere, along with some sulfur dioxide (SO_2). Natural gas must be converted to liquid form (LNG) before it can be shipped by tanker from one country to another overseas. This is expensive and dangerous; huge explosions in urban areas near LNG loading and unloading facilities could kill many people and cause much damage. Conversion of natural gas into LNG also reduces the net useful energy yield by one-fourth. However, the proposed Asian pipeline, if built, could reduce the need for much shipping of natural gas as LNG. Recently chemists have developed a new process for converting natural gas into methanol, an easily transported liquid fuel.

Another problem is leaks of natural gas into the atmosphere from natural gas pipelines, storage tanks, and distribution facilities. Methane, the major component of natural gas, is a greenhouse gas that is much more potent than CO_2 in causing global warming. For instance, Ukraine's natural gas grid currently leaks about 40% of its supply. Improved construction and maintenance of all pipelines and other gas-handling facilities could greatly reduce such leaks.

Because of its advantages over oil, coal, and nuclear energy, some analysts see natural gas as the best fuel to help us make the transition to improved energy efficiency and more renewable energy over the next 50 years. Additionally, hydrogen gas produced from water by solar-generated electricity could be mixed with natural gas to help smooth the shift to a solar-hydrogen economy (Section 16-7).

15-4 COAL

What Is Coal and How Is It Extracted and Processed? **Coal** is a solid, rocklike fossil fuel; it formed in several stages as the buried remains of ancient swamp plants that died during the Carboniferous period (a geologic era that ended 286 million years ago) were subjected to intense pressure and heat over many millions of years. Coal is mostly carbon (40–98% depending on the type), with much smaller amounts of water (0.2–1.2%), sulfur (0.2–2.5%), and trace amounts of radioactive materials found in the earth. During the fossilization process, sulfur from the corpses of decaying plants and from hydrogen sulfide (H_2S) produced by anaerobic decomposer bacteria became incorporated in the coal deposits.

Three types of increasingly harder coal formed over the eons: lignite (brown coal), bituminous coal (soft coal), and anthracite (hard coal) (Figure 15-13). As coal ages its carbon content increases and its water

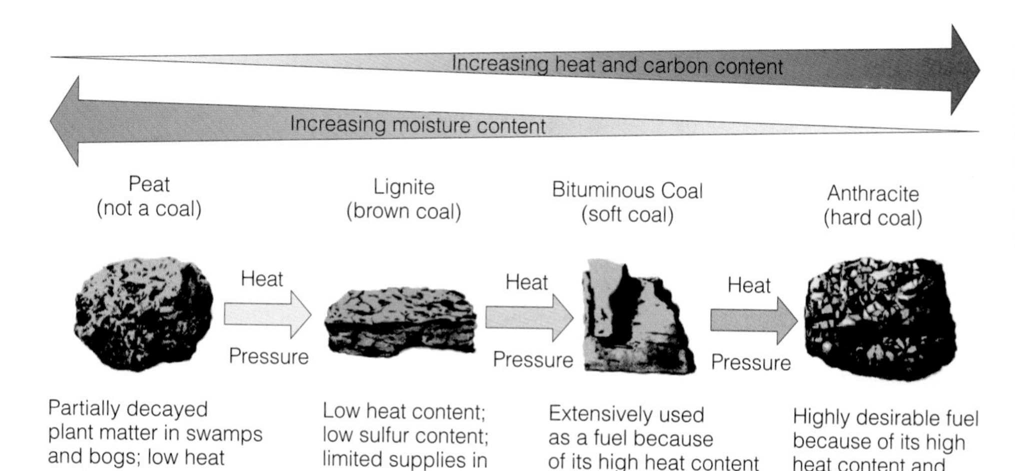

Figure 15-13 Stages in the formation of coal over millions of years. Peat is a soil material made of moist, partially decomposed organic matter. Lignite and bituminous coal are sedimentary rocks, whereas anthracite is a metamorphic rock (Figure 5-10).

Hint: Enter the search terms *coal industry, environmental aspects* using the Subject Guide.

Coal Burning and Destructive Synergies in China

China has huge quantities of coal (11% of the world's reserves) and is the world's largest producer of coal, accounting for 25% of global output. Currently, the country burns coal to provide 76% of its commercial energy.

Mostly because of coal burning, China now produces 10% of the world's emissions of CO_2 each year. By 2025, China is projected to emit more CO_2 than the current combined total of the United States, Canada, and Japan by relying on coal to provide 68% of its energy and on oil for 25% of its energy.

Several factors interact synergistically to cause China to burn more coal than it needs to. One is energy inefficiency. About 66% of the energy used in China is unnecessarily wasted, compared to about 43% in the United States (which is in turn only half as energy-efficient as Japan and most western European countries).

Another factor is that drinking water must usually be boiled because it is too contaminated to drink. This uses more energy, much of it produced by the inefficient burning of coal in homes. Destruction of forests for more cropland, fuelwood, and construction material for the country's growing population leads to more soil erosion and increased use of fertilizers, further polluting the water and requiring more coal burning to sterilize water.

For China's economic future, and perhaps for the world's climate, many analysts believe that it is crucial to defuse these destructive synergistic interactions, which lead to a cycle of harmful positive feedback. Part of the solution is for China to undertake an energy efficiency revolution and to shift from coal to natural gas and renewable energy such as wind, biomass, solar, and small-scale hydroelectric power.

At the same time, China could reduce coal use, improve public health, and increase economic productivity by instituting a nationwide program to provide safe drinking water.

Critical Thinking

1. How might your life and lifestyle be affected if China continues to burn coal as its primary fuel?

2. What role, if any, should the United States and other developed countries play in helping China reduce its dependence on coal?

content decreases. Peat (formed in the first stage), although not a coal, is burned in some places but has a low heat content. Lignite has a low heat value and a high moisture content. Low-sulfur coal (lignite and anthracite) produces less sulfur dioxide when burned than does high-sulfur bituminous coal. Anthracite, which is about 98% carbon, is the most desirable type of coal because of its high heat content and low sulfur content. However, because it takes much longer to form it is less common and therefore more expensive.

Some coal is extracted underground (Figure 15-2) by miners working in tunnels and shafts. Such *subsurface mining* is very labor intensive, requiring about five times as many workers per metric ton as surface mining. It is also one of the world's most dangerous occupations because of accidents and black lung disease (caused by prolonged inhalation of coal dust particles).

When coal lies close to the earth's surface it is extracted by *surface mining*. Bulldozers and huge earth-moving machines remove soil and rock, known as *overburden*, to recover underlying coal deposits. *Area strip mining* is used on fairly flat terrain (Figures 14-8 and 15-2), and *contour strip mining* on hilly or mountainous terrain (Figures 14-9 and 15-2). Thick beds of coal fairly near the surface are removed by digging a deep pit to remove the coal—a form of surface mining called *open-pit mining*.

After the coal is removed it is transported (usually by train) to a processing plant, where it is broken up, crushed, and then washed to remove impurities. The coal is then dried and shipped (again usually by train) to users, mostly power plants and industrial plants.

How Is Coal Used? Currently coal provides about 25% of the world's commercial energy (22% in the United States). It is used to generate almost 64% of the world's electricity and to make 75% of its steel. The remainder of the world's electricity is supplied by hydroelectric dams (18%, Figure 13-10), nuclear energy (17%), and other alternatives such as wind and solar energy (1%).

China, which gets 76% of its energy from coal, is the world's largest user of coal, followed closely by the United States. Other large coal users are the countries of the former Soviet Union, Germany, India, Australia, Poland, and Great Britain.

Coal supplies 57% of the electricity generated in the United States; the rest is produced by nuclear energy (19%), natural gas (11%), hydropower (9%), oil (3%), and renewable energy resources such as hydropower, geothermal, and wind energy (1%). Utilities consume 87% of U.S. coal production and industry 13%.

How Does a Coal-Fired Electric Power Plant Work? In a coal-fired power plant, large amounts of

Q: Between 1950 and 1996 by how much did the annual world fish catch increase?

coal are pulverized to a fine dust and then burned at very high temperatures in a huge boiler with up to 48 kilometers (30 miles) of stainless steel tubing running through it. The intense heat converts purified water continuously running through the tubes into high-pressure steam, which spins the shaft of a turbine. The rotating shaft turns the rotor of a generator (a large electromagnet) to produce electricity, which goes through a transmission and distribution system to customers.

Air pollutants are removed from the gases flowing through the smokestack by devices such as *electrostatic precipitators* (to remove much of the particulate matter) and *scrubbers* (to remove most of the sulfur dioxide, formed when oxygen gas combines with sulfur impurities in the burning coal) (Figure 18-15). Ash collected from the coal burning and the electrostatic precipitators and slurry from the scrubbers are usually disposed of in landfills. Highly damaging sulfur dioxide emissions can be reduced by using low-sulfur coal (lignite and anthracite) or by removing most of the sulfur from the coal before it is burned.

Who Has the World's Coal Supplies? About 66% of the world's proven coal reserves and 85% of the estimated undiscovered coal deposits are located in the United States, the former Soviet Union, and China (Connections, left). Asia (excluding China), western Europe, and Australia each have about 9% of the world's coal reserves, Africa 6%, and Latin America 1%.

Figure 15-14 shows the locations of major coal fields in the United States and Canada. Anthracite makes up only 2% of U.S. coal reserves (and accounts for only about 0.5% of the country's coal production). About 45% of U.S. reserves is high-sulfur bituminous coal with a high fuel value. More than half of these reserves are found west of the Mississippi River, far from the heavily industrialized and populated East, where most coal is consumed.

What Are the Pros and Cons of Solid Coal? Coal is both the world's most abundant and its dirtiest fossil fuel. *Identified* world reserves of coal should last at least 220 years at current usage rates, but only 65 years if usage rises 2% per year. The world's *unidentified* coal reserves are projected to last about 900 years at the current consumption rate, but only 149 years if the usage rate increases 2% per year. Identified U.S. coal reserves should last about 300 years at the current consumption rate; unidentified U.S. coal resources could extend those supplies for perhaps 100 years, at a much higher average cost. Coal also has a high net energy yield (Figure 15-5).

However, coal has a number of drawbacks, especially the harmful environmental effects associated with its extraction, processing, and use (Figure 14-13).

Coal mining is dangerous because of accidents and black lung disease, a form of emphysema caused by prolonged breathing of coal dust and other particulate matter. Since 1900, underground mining in the United States has killed more than 100,000 miners and permanently disabled at least 1 million, and mining safety laws in most countries are much weaker than those in the United States. Another 250,000 or more retired U.S. miners suffer from black lung disease.

Coal mining also harms land and causes water pollution. Underground mining causes land to sink when a mine shaft collapses during or after mining. Over 800,000 hectares (2 million acres) of land (much of it in the central Appalachian region of the United States) has subsided as a result of underground coal mining. Surface mining of coal causes severe land disturbance (Figures 14-8, 14-9, and 15-2) and soil erosion, which can pollute nearby streams.

Surface-mined land can be restored, which involves burying any toxic materials, returning the land to its approximate original contour, and planting vegetation to reduce erosion. However, this is expensive and is not done in many countries; in arid and semiarid areas the land cannot be fully restored. In the United States, there are nearly 405,000 hectares (1 million acres) of unrestored land that were surface mined before regulations requiring restoration went into effect.

Surface and subsurface coal mining can severely pollute nearby streams and groundwater. Acids and toxic metal compounds drain from piles of waste material (spoil), from both surface mines and abandoned underground mines (Figure 14-14), and from large piles of coal stored on the ground. Once coal is mined, it is expensive to move from one place to another. It cannot be used in solid form as a fuel for cars and trucks; it must be converted to liquid or gaseous fuels, with a significant drop in its net energy yield.

Coal is by far the dirtiest fossil fuel to burn, releasing carbon monoxide (CO), carbon dioxide (CO_2), sulfur dioxide (SO_2), nitrogen oxides (NO and NO_2), particulates (known as fly ash), toxic metals (lead, arsenic, nickel, and cadmium), and very small amounts of radioactive elements (such as uranium, radium, and thorium). Burning coal releases thousands of times more radioactive particles into the atmosphere per unit of energy produced than does a normally operating nuclear power plant. Without expensive air pollution control devices, burning coal produces more air pollution per unit of energy than any other fossil fuel. Because coal produces more carbon dioxide per unit of energy than do other fossil fuels (Figure 15-12), burning more coal accelerates projected global warming (Section 19-2).

Burning coal is also one of the greatest threats to human health. Each year in the United States alone,

A: 4.9-fold

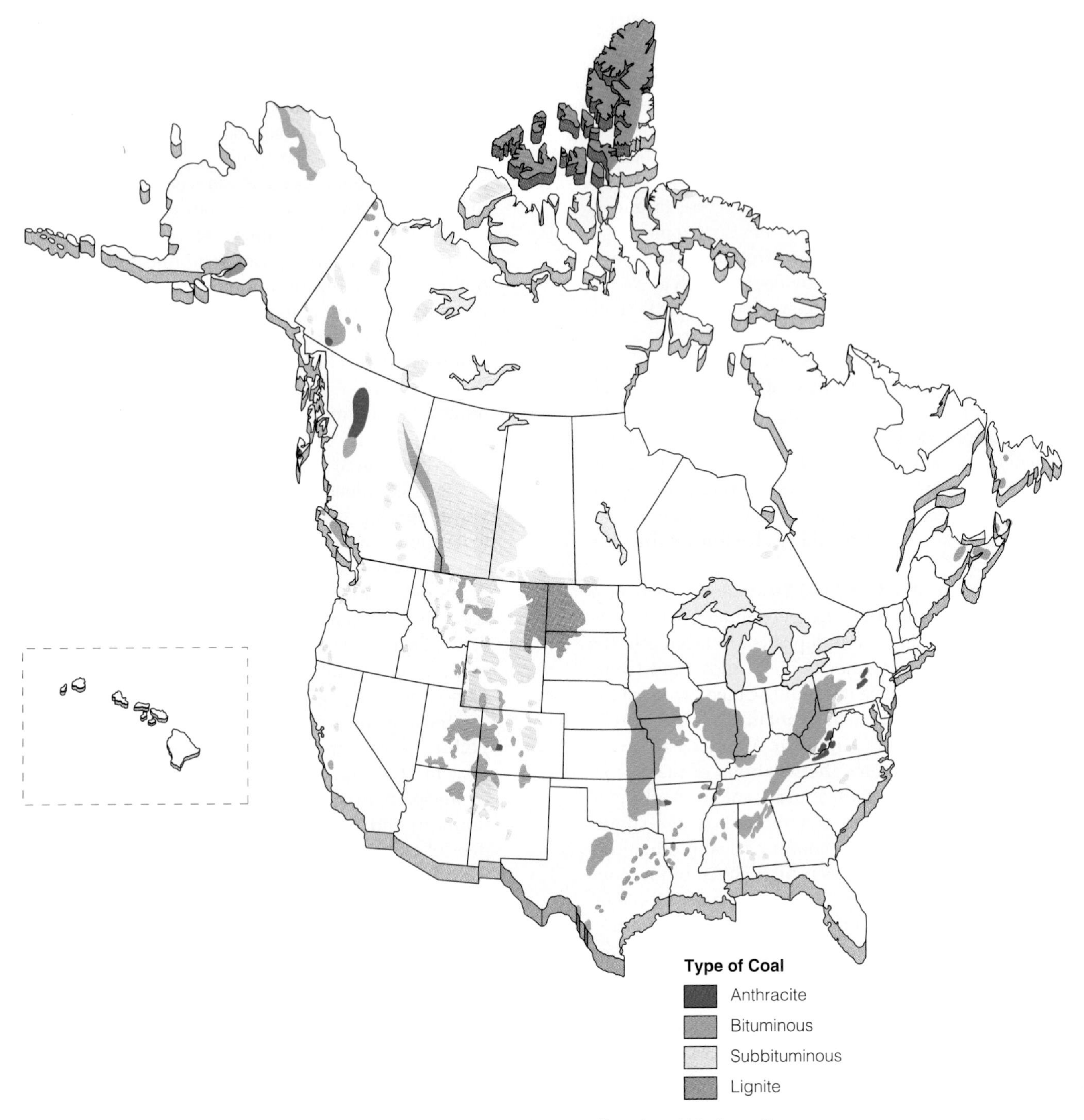

Figure 15-14 Locations of major coal fields in the United States and Canada. (Data from U.S. Council on Environmental Quality)

air pollutants from coal burning kill thousands of people (with estimates ranging from 65,000 to 200,000), cause at least 50,000 cases of respiratory disease, and result in several billion dollars of property damage.

However, new ways, such as *fluidized-bed combustion* (Figure 15-15), have been developed to burn coal more cleanly and efficiently. Successful small-scale fluidized-bed combustion plants have been built in Great Britain, Sweden, Finland, some republics of the former Soviet Union, Germany, and China. In the United States, commercial fluidized-bed combustion boilers are expected to slowly replace most conventional coal boilers.

A number of utilities are reducing emissions of sulfur dioxide by switching to low-sulfur forms of coal. Although such coal is usually more expensive, overall cost can be lower because utilities don't need expensive scrubbers and other air-pollution control devices to reduce sulfur dioxide emissions (Figure 18-14).

Q: How many of the world's 15 major marine fishing zones have been overfished?

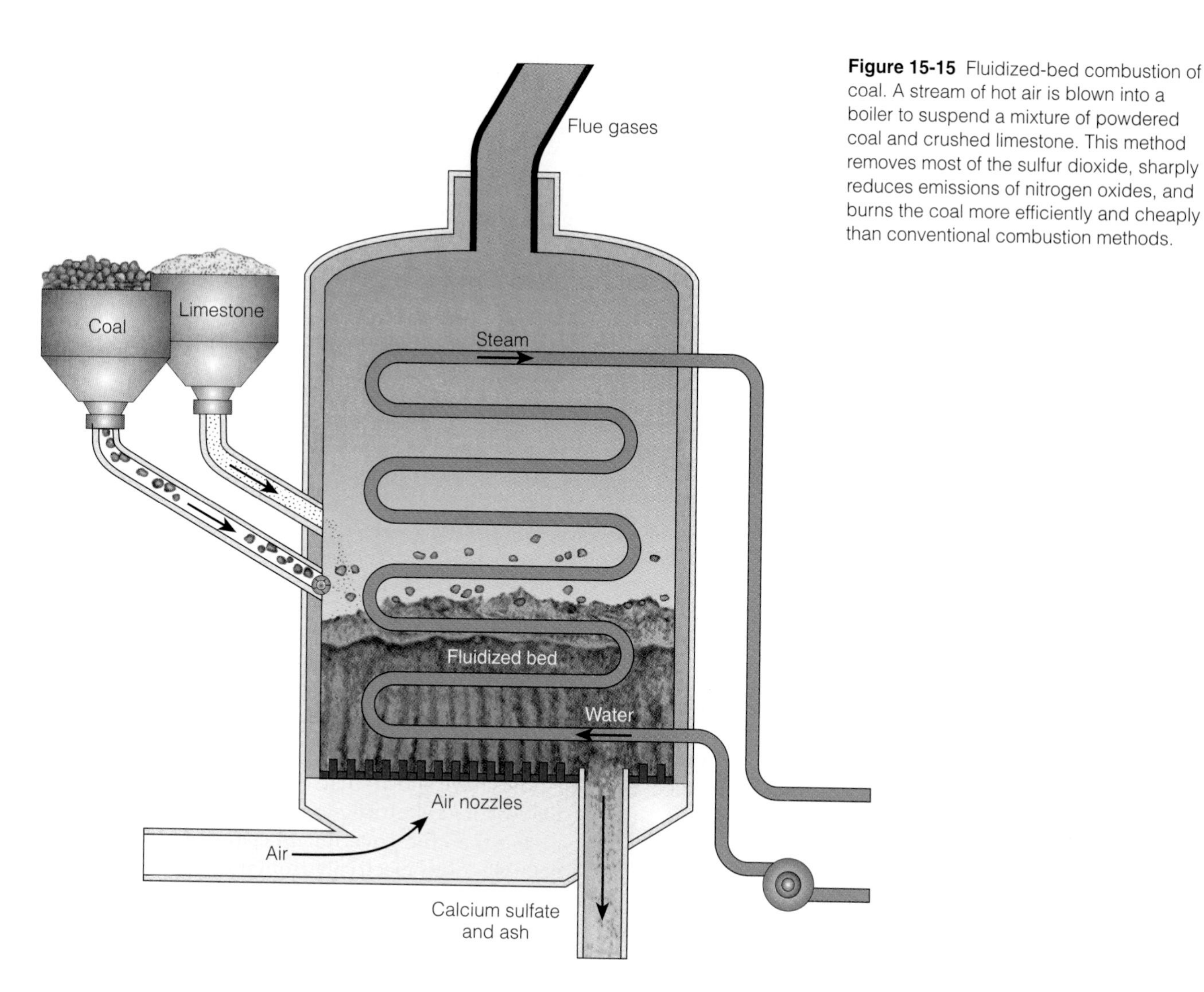

Figure 15-15 Fluidized-bed combustion of coal. A stream of hot air is blown into a boiler to suspend a mixture of powdered coal and crushed limestone. This method removes most of the sulfur dioxide, sharply reduces emissions of nitrogen oxides, and burns the coal more efficiently and cheaply than conventional combustion methods.

Recent research done for the U.S. Department of Energy concluded that the *environmental costs* of producing electricity were 5.7¢ per kilowatt-hour for coal, 5.0¢ for nuclear, 2.7¢ for oil, 1¢ for natural gas, under 0.7¢ for biomass, less than 0.4¢ for solar cells, and under 0.1¢ for wind and geothermal. Suppose that taxes (per unit of energy produced) added these costs to the market price of coal and that government subsidies that make coal artificially cheap were phased out. According to some analysts, coal would become so expensive that it would probably be replaced for most uses by a combination of improved energy efficiency and cheaper and less environmentally harmful renewable energy resources.

What Are the Pros and Cons of Converting Solid Coal into Gaseous and Liquid Fuels? Solid coal can be converted into **synthetic natural gas (SNG)** by *coal gasification* (Figure 15-16), into *hydrogen gas* (Section 16-7), or into a liquid fuel such as methanol or synthetic gasoline by *coal liquefaction*. These *synfuels* can be transported by pipeline, and they produce much less air pollution than solid coal. They can be burned to produce high-temperature heat and electricity, to heat houses and water, and to propel vehicles.

However, coal gasification has a low net energy yield (Figure 15-5), as does coal liquefaction. A synfuel plant costs much more to build and run than an equivalent coal-fired power plant fully equipped with air-pollution control devices. In addition, the widespread use of synfuels would accelerate the depletion of world coal supplies because 30–40% of the energy content of coal is lost in the conversion process. It would also lead to greater land disruption from surface mining because producing a unit of energy from synfuels uses more coal than burning solid coal does.

Producing synfuels also requires huge amounts of water. Additionally, synfuels release more carbon dioxide per unit of energy than coal does (Figure 15-12). For these reasons, most analysts expect synfuels to play only a minor role as an energy resource in the next 30–50 years.

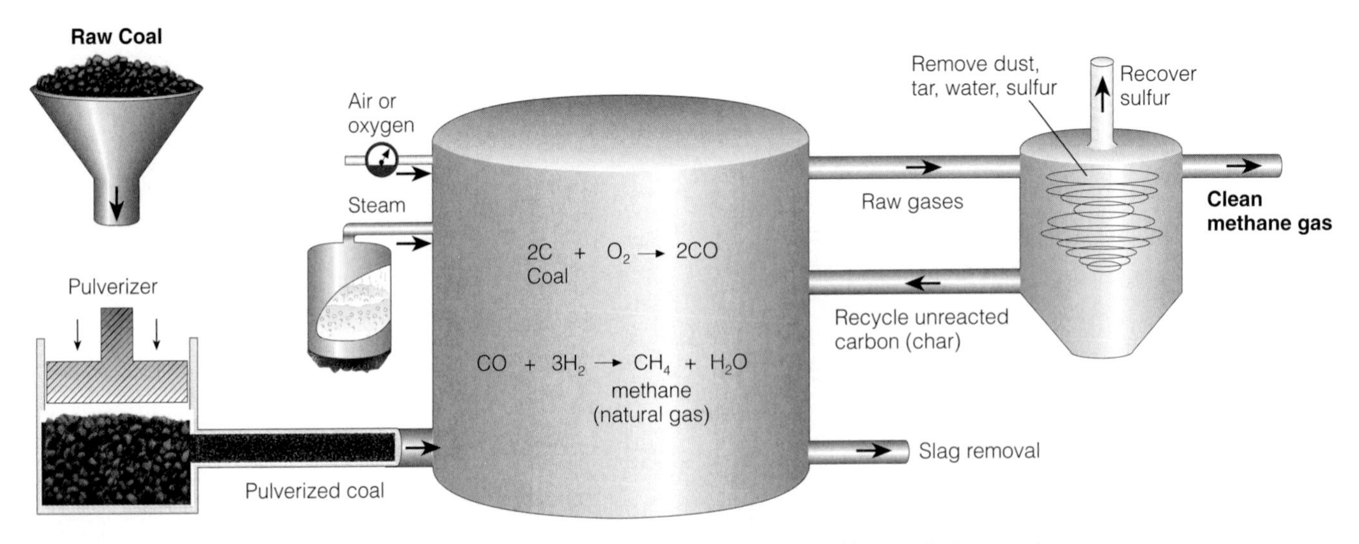

Figure 15-16 Coal gasification. Generalized view of one method for converting solid coal into synthetic natural gas (methane).

15-5 NUCLEAR ENERGY

What Happened to Nuclear Power? U.S. utility companies began developing nuclear power plants in the late 1950s for three reasons: **(1)** The Atomic Energy Commission (which had the conflicting roles of promoting and regulating nuclear power) promised them that nuclear power would produce electricity at a much lower cost than coal and other alternatives (with one early proponent touting nuclear power as "too cheap to meter"), **(2)** the government paid about a quarter of the cost of building the first group of commercial reactors, and **(3)** after U.S. insurance companies refused to cover more than a small part of the possible damages from a nuclear power-plant accident Congress passed the Price–Anderson act, which protects the U.S. nuclear industry and utilities from significant liability to the general public in case of accidents.

In the 1950s researchers predicted that by the end of the century 1,800 nuclear power plants would supply 21% of the world's commercial energy (25% in the United States) and most of the world's electricity. By 1996, after over 40 years of development, enormous government subsidies, and an investment of $2 trillion, 437 commercial nuclear reactors in 32 countries were producing only 6% of the world's commercial energy and 17% of its electricity. Little or no further growth in nuclear power is projected, and its capacity is expected to decline between 2000 and 2020 as existing plants wear out and are retired (decommissioned).

In western Europe, plans to build more new nuclear power plants have come to a halt—except in France, which gets about 78% of its electricity from nuclear power. However, France has more electricity than it needs (Guest Essay, p. 433), and the heavily subsidized government agency that builds and operates France's nuclear plants has lost money for 22 years, accumulating a huge $30 billion debt—a serious burden to the French economy. Since 1994 France's commitment to nuclear power has come under severe attack as studies have shown that, once subsidies are included, the cost of its electricity from nuclear power are 30–90% higher than the government claims. In Japan, which gets 34% of its electricity from nuclear power, public opposition has been so intense that only two sites for new nuclear plants have been approved since 1979.

In the United States, no new nuclear power plants have been ordered since 1978, and all 120 plants ordered since 1973 have been canceled—more than the total nuclear capacity existing in the country today. In 1997, the 105 licensed commercial nuclear power plants in the United States generated about 20% of the country's electricity. This percentage is expected to decline over the next two decades as many of the current reactors reach the ends of their useful lives. According to the U.S. Department of Energy, 40% of the current U.S. nuclear power generation capacity will be retired by 2015 and the rest by 2030.

What happened to nuclear power? The answer is multibillion-dollar construction cost overruns, high operating costs, frequent malfunctions, false assurances and cover-ups by government and industry officials, inflated estimates of electricity use, poor management, the Chernobyl (p. 366) and Three Mile Island accidents, and public concerns about safety, costs, and radioactive waste disposal.

How Does a Nuclear Fission Reactor Work? To evaluate the pros and cons of nuclear power, we first must know how a nuclear power plant and its

Q: How much does it cost the global fishing industry to catch $70 billion worth of fish per year?

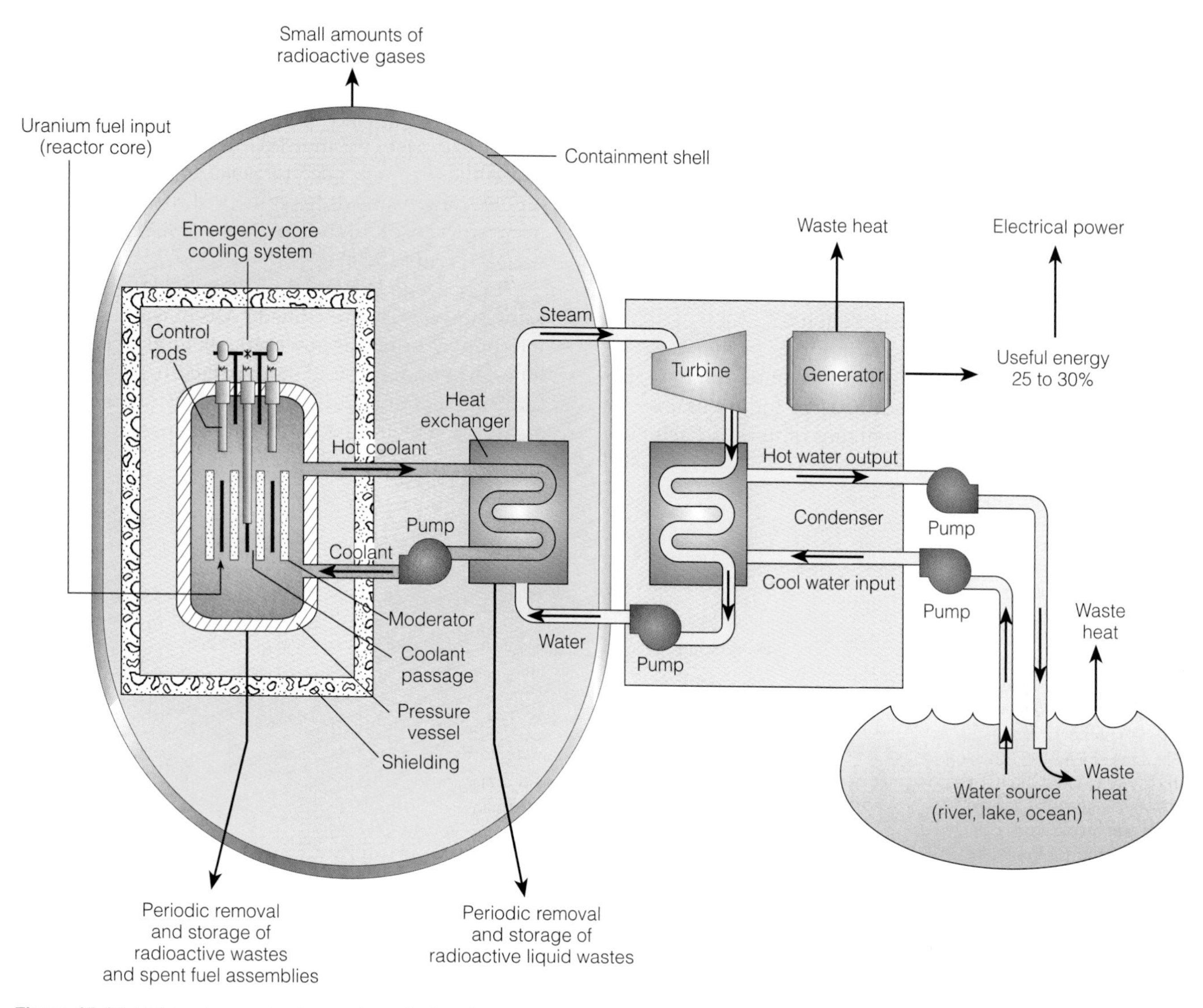

Figure 15-17 Light-water–moderated and –cooled nuclear power plant with a pressurized water reactor. The ill-fated Chernobyl nuclear reactor (Figure 15-1) in Ukraine was a graphite-moderated reactor with less extensive safety features than most reactors in the rest of the world.

accompanying nuclear fuel cycle work. When the nuclei of atoms such as uranium-235 and plutonium-239 are split by neutrons in a nuclear fission chain reaction, energy is released and converted mostly into high-temperature heat (Figure 3-20). The rate at which this happens can be controlled in the nuclear fission reactor of a nuclear power plant; the heat generated can produce high-pressure steam, which spins turbines that generate electricity.

Light-water reactors (LWRs) like the one diagrammed in Figure 15-17 produce about 85% of the world's nuclear-generated electricity (100% in the United States). An LWR has several key parts. One is its *core*, which contains 35,000–40,000 long, thin fuel rods, each of which is packed with pellets of uranium oxide fuel. Each pellet is about one-third the size of a cigarette. About 97% of the uranium in each fuel pellet is uranium-238, a nonfissionable isotope; the other 3% (versus 0.7% in nature) is uranium-235, which is fissionable. The concentration of uranium-235 in the ore is increased (enriched) from 0.7% to 3% by removing some of the uranium-238 to create a suitable fuel. Another important component is *control rods*, which are moved in and out of the reactor core to absorb neutrons and thus regulate the rate of fission and amount of power the reactor produces.

The *moderator* slows down the neutrons emitted by the fission process so that the chain reaction can be kept going. This is a material such as liquid water (75% of the world's reactors, called *pressurized water reactors*), solid graphite (20% of reactors), or heavy water (D_2O, 5% of reactors). Graphite-moderated reactors,

A: About $125 billion; most of the deficit is made up by government subsidies including the ill-fated one at Chernobyl (Figure 15-1), can also produce fissionable plutonium-239 for nuclear weapons. A *coolant*, usually water, circulates through the reactor's core to remove heat (to keep fuel rods and other materials from melting) and to produce steam for generating electricity.

Nuclear power plants, each with one or more reactors, are only one part of the nuclear fuel cycle (Figure 15-18). In evaluating the safety and economic feasibility of nuclear power, we need to look at this entire cycle, not just the nuclear plant itself.

After 3–4 years, the concentration of fissionable uranium-235 in a reactor's fuel rod becomes too low to keep the chain reaction going or the rod becomes damaged from exposure to ionizing radiation. Each year about one-third of the spent fuel assemblies in a reactor are removed and placed in large, concrete-lined pools of water at the plant site. The water serves as a radiation shield and a coolant. After losing some of their radioactivity and cooling down, the spent fuel assemblies are supposed to be shipped to spent fuel-reprocessing plants or to permanent sites for the long-term storage of high-level, long-lived radioactive wastes (Figure 15-18).

In the United States all spent fuel rods currently are being stored in concrete-lined pools of water at each of the country's nuclear power plants until a permanent long-term underground storage facility is developed (probably not until 2010 or later). Many of these plants are reaching their capacity for storing spent fuel.

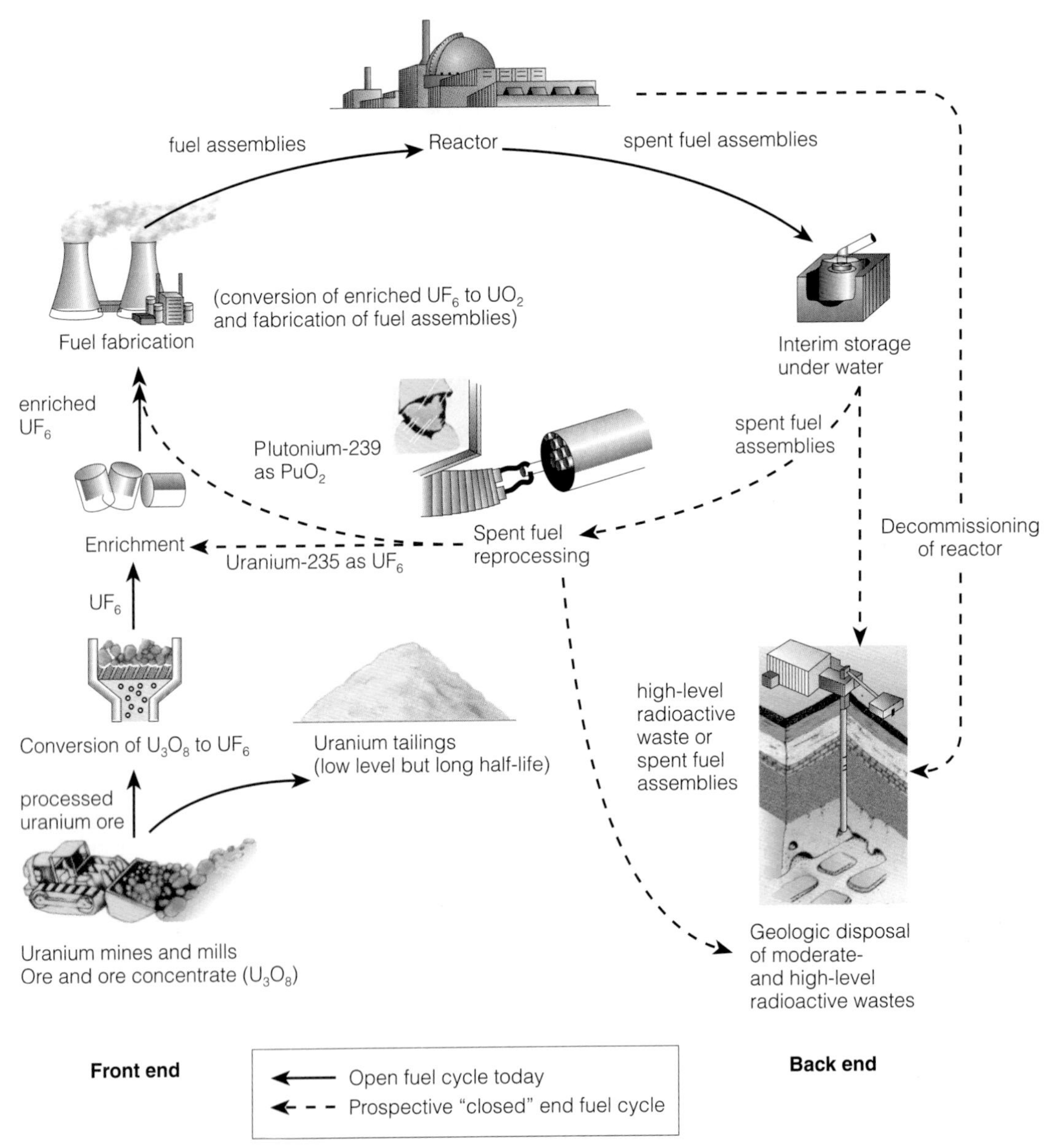

Figure 15-18 The nuclear fuel cycle. As long as it is operating properly, little pollution or radioactivity is released from a nuclear power plant (although the potential health, environmental, and economic consequences from an extremely unlikely major accident are significant). Each of the other steps in the nuclear fuel cycle can release significant amounts of toxic chemicals and radioactive materials into the environment.

Q: What percentage of the food produced in the United States is wasted?

If the fuel is reprocessed to remove plutonium and other radioactive isotopes with long half-lives, the remaining radioactive waste must be safely stored for at least 10,000 years; otherwise, the rods must be stored safely for at least 240,000 years (10 times the 24,000-year half-life of plutonium-239)—about four times as long as the latest species of humans has been on earth.

Commercial spent fuel-reprocessing plants have been built in the United States, France, Great Britain, Japan, and Germany, but all have had severe operating and economic problems. The United States has stopped development of commercial fuel-reprocessing plants because of policy concerns (including diversion of bomb-grade materials by employees or terrorists), technical difficulties, high construction and operating costs, and ample domestic or imported supplies of uranium. During fuel reprocessing some highly radioactive materials can be released into the air, water, and soil.

After approximately 15–40 years of operation, a nuclear reactor becomes dangerously contaminated with radioactive materials and many of its parts are worn out. Then the plant must be *decommissioned* or retired by **(1)** dismantling it and storing its large volume of highly radioactive materials in high-level nuclear waste storage facilities (which still don't exist), **(2)** putting up a physical barrier and setting by full-time security for 30–100 years before the plant is dismantled, or **(3)** enclosing the entire plant in a tomb that must last for several thousand years.

What Are the Advantages of Nuclear Power? Using nuclear power to produce electricity has some important advantages, especially when compared to mining, processing, and burning coal.

Nuclear plants don't emit air pollutants (as coal-fired plants do) as long as they are operating properly. The entire nuclear fuel cycle (Figure 15-18) adds about one-sixth as much heat-trapping carbon dioxide per unit of electricity as using coal does, thus making it more attractive than coal and other fossil fuels for reducing the threat of global warming.

Water pollution and disruption of land are low to moderate if the entire nuclear fuel cycle operates normally. Thick and strong reactor vessel walls, a steel-reinforced containment building, an emergency core cooling system, and other safety systems with automatic backups greatly decrease the likelihood of a catastrophic accident releasing deadly radioactive material into the environment.

Chernobyl may have caused the premature deaths of 32,000 people between 1986 and 1995, but as mentioned at the beginning of this chapter, *each year* air pollution from coal burning causes the premature deaths of 65,000 to 200,000 people in the United States alone. Globally the *annual* death toll from coal burning is in the millions.

How Safe Are Nuclear Power Plants and Other Nuclear Facilities? Because of the built-in safety features, the risk of exposure to radioactivity from nuclear power plants in the United States (and presumably in most other developed countries) is extremely low compared to other risks (Figures 17-13 and 17-14). However, a partial or complete meltdown or explosion is possible, as the Chernobyl (Figure 15-1) and Three Mile Island nuclear power-plant accidents have taught us.

On March 29, 1979, the number 2 reactor at the Three Mile Island (TMI) nuclear plant near Harrisburg, Pennsylvania, lost its coolant water because of a series of mechanical failures and human operator errors not anticipated in safety studies. The reactor's core became partially uncovered and about 50% of it melted and fell to the bottom of the reactor. Unknown amounts of radioactive materials escaped into the atmosphere, 50,000 people were evacuated, and another 50,000 fled the area on their own. Partial cleanup of the damaged TMI reactor, lawsuits, and payment of damage claims has cost $1.2 billion so far, almost twice the reactor's $700-million construction cost.

Most analysts say that the estimated radiation released was not enough to cause deaths or cancers. However, there is controversy over how much radiation was released and there has been an increase in certain cancers among residents who live near the plant. A 1991 study concluded that the excess cancers may have been caused by stress related to the accident. Stress is known to damage the immune system so that it may fail to prevent cancers. However, a 1997 study concluded that the increased cancer rates were caused by radiation released from the plant.

The Nuclear Regulatory Commission (NRC) estimates that there is a 15–45% chance of a complete core meltdown at a U.S. reactor during the next 20 years. The NRC also found that 39 U.S. reactors have an 80% chance of either containment failure from a meltdown or a tremendous explosion of gases inside the containment structures.

Nuclear scientists and government officials throughout the world are deeply concerned about 26 especially risky nuclear reactors in some republics of the former Soviet Union that have a Chernobyl-type design (15 reactors) or another flawed design (11 reactors). There is growing international consensus (including the Russian Academy of Sciences) that these 26 reactors, along with 10 other poorly designed and poorly operated nuclear plants in eastern Europe,

A: About 25%

should be shut down. However, without considerable government and private economic aid from the world's developed countries, it is extremely unlikely that these dangerous plants will be closed and replaced with safer nonnuclear alternatives.

A 1982 study by the Sandia National Laboratory estimated that a *worst-case accident* in a reactor near a large U.S. city might cause 50,000–100,000 immediate deaths, 10,000–40,000 subsequent deaths from cancer, and $100–150 billion in damages. Most citizens and businesses suffering injuries or property damage from a major nuclear accident would get little if any financial reimbursement because combined government and nuclear industry insurance covers only 7% of the estimated damage from such an accident. Critics of nuclear power contend that this 17-year-old study probably underestimates the deaths and damages. They also contend that plans for dealing with major nuclear accidents in the United States are inadequate.

In 1998, officials in Canada's province of Ontario decided to indefinitely shut down 7 of its 19 operating nuclear reactors—all within 160 kilometers (100 miles) of the U.S. border. This was done after an internal report concluded that the province's utility company, Ontario Hydro, was so badly managed that it compromised the safety of its nuclear power system. The utility will make up for the loss of electrical output by shifting to existing coal and oil plants at a cost of about $2.2 billion. The company will also spend $1.5 billion by 2001 to overhaul the 12 reactors that continue to operate.

All French pressurized water nuclear reactors are built using a cost-saving standardized design. However, plants built with this generic design have suffered from a growing number of faulty steam generator problems and cracked reactor vessel heads, which are having to be replaced at great cost to minimize the chances of a serious accident. Other countries using various versions of pressurized water reactors are also finding cracks in the reactor vessel heads. If the cracks rupture a reactor, nuclear engineers fear that it could not be shut down safely.

In the United States, there is widespread public distrust of the ability of the NRC and the Department of Energy (DOE) to enforce nuclear safety in commercial (NRC) and military (DOE) nuclear facilities. Congressional hearings in 1987 uncovered evidence that high-level NRC staff members destroyed documents and obstructed investigations of criminal wrongdoing by utilities, suggested ways utilities could evade commission regulations, and provided utilities and their contractors with advance notice of "surprise" inspections. The nuclear power industry vetoes NRC nominees it deems too hostile and many top NRC officials enjoy a revolving door to good jobs at nuclear power-plant utility companies.

In 1995, George Galatis and George Betancourt, two respected senior nuclear engineers working for Northeast Utilities (which operates five nuclear power plants in New England) revealed that for years the NRC has been allowing nuclear power plant operators to overload spent fuel rod storage pools. They warned that if the cooling capacity of the pools is exceeded, water in the pool could boil, filling the plant with radioactive steam. Additionally, if human error, mechanical failure, or an earthquake drained such pools, there could be a catastrophic meltdown of fuel rods *outside* the primary reactor containment structure. This could release massive amounts of radiation and render hundreds of square kilometers around the plant uninhabitable. In 1996 George Galatis said, "I believe in nuclear power but after seeing the NRC in action I'm convinced a serious accident is not just likely, but inevitable. . . . They're asleep at the wheel."

Since 1986 government studies and once-secret documents have revealed that for decades most of the nuclear weapon production facilities supervised by the DOE have been operated with gross disregard for the safety of their workers and the people nearby. Since 1957 these facilities have released huge quantities of radioactive particles into the air and dumped tons of radioactive waste and toxic substances into flowing creeks and leaking pits without informing the public. As former Senator John Glenn of Ohio summed up the situation, "We are poisoning our own people in the name of national security."

According to the nuclear power industry, nuclear power plants in the United States have not killed anyone. However, according to U.S. National Academy of Sciences estimates, U.S. nuclear plants cause 6,000 premature deaths and 3,700 serious genetic defects each year. If correct, this annual death toll is much smaller than the 65,000 to 200,000 deaths per year caused by coal-burning plants in the United States. However, critics point out the estimated annual deaths from both of these types of plants are unacceptable, given the much less harmful alternatives.

What Do We Do with Low-Level Radioactive Waste? Each part of the nuclear fuel cycle (Figure 15-18) produces low-level and high-level solid, liquid, and gaseous radioactive wastes with various half-lives (Table 3-1 and Figure 3-16). Wastes classified as *low-level radioactive wastes* give off small amounts of ionizing radiation and must be stored safely for 100–500 years before decaying to levels that, according to the NRC, don't pose an unacceptable risk to public health and safety.

Q: What percentage of the USDA's annual research budget is spent on development of sustainable agriculture?

From the 1940s to 1970, most low-level radioactive waste produced in the United States (and most other countries) was put into steel drums and dumped into the ocean; the United Kingdom and Pakistan still dispose of their low-level wastes in this way.

Since 1970, low-level radioactive wastes from U.S. military activities have been buried in commercial, government-run landfills. Today, low-level waste materials from commercial nuclear power plants, hospitals, universities, industries, and other producers are put in steel drums and shipped to the two remaining regional landfills run by federal and state governments (both scheduled for closure).

Attempts to build new regional dumps for low-level radioactive waste using improved technology (Figure 15-19) have met with fierce public opposition. According to the EPA, all landfills eventually leak. The best-designed landfills may not leak for several decades, which approximates the time when the companies running them would no longer be liable for leaks and problems; then any problems would be passed on to taxpayers and future generations.

Some environmentalists believe that low-level radioactive waste should be stored in carefully designed aboveground buildings that could capture and contain any leaks. They urge that all such buildings (or low-level waste dumps) be located at nuclear power-plant sites, which produce more than half of these wastes (as much as 80% in some states) and 80% of the radioactivity.

They point out that nuclear power-plant sites have already been carefully evaluated to meet strict geological standards. They have the personnel with the expertise and equipment to manage the low-level wastes and are strictly regulated by the federal government. Locating low-level waste at plant sites would also reduce the need to transport most such wastes long distances. Such proposals are opposed by utility companies and the nuclear power industry because this would force the main producers of such wastes to be financially and legally responsible for the wastes they produce.

In June 1990, the NRC proposed that most of the country's low-level radioactive waste be removed from federal regulation and handled like ordinary trash (dumped in landfills, incinerated, reused, or recycled into consumer products). According to the NRC, exposure to radiation from these unregulated waste sites would kill 2,500 Americans per year—1 out of every 100,000 citizens. Other estimates project as many as 12,412 more cancer deaths per year. Critics have called this proposal the *bureaucratic semantic solution*: Solve the problem of low-level radioactive waste by officially redefining it as essentially nonradioactive. Several environmental groups filed a suit to block this NRC policy; by early 1999 the policy had not been implemented.

In 1995 Westinghouse Electric began discussions about shipping low-level radioactive wastes in the United States for permanent burial in southern Russia in old mine shafts. This idea is facing strong opposition from antinuclear activists and many members of the Russian parliament, who charge

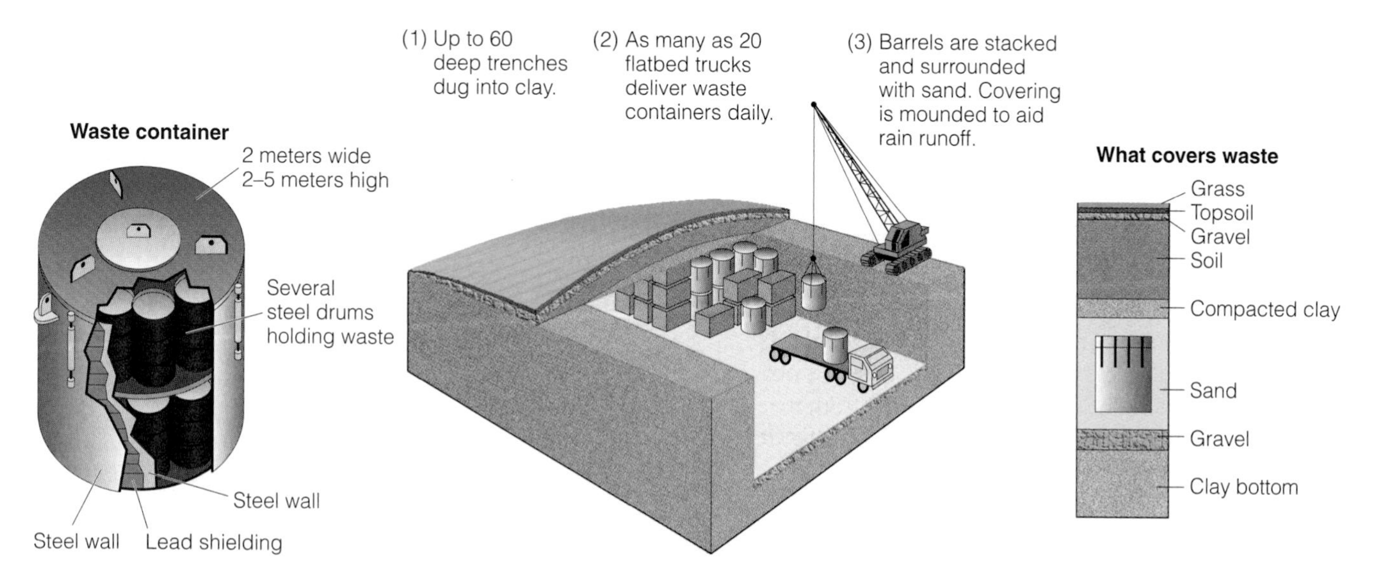

Figure 15-19 Proposed design of a state-of-the-art low-level radioactive waste landfill. (Data from U.S. Atomic Industrial Forum)

A: About 1%

the West with trying to turn Russia into a radioactive dumping ground.

What Should We Do with High-Level Radioactive Waste? *High-level radioactive wastes* give off large amounts of ionizing radiation for a short time and small amounts for a long time. Such wastes must be stored safely for thousands of years—about 240,000 years if plutonium-239 is not removed by reprocessing. Most high-level radioactive wastes are spent fuel rods from commercial nuclear power plants and an assortment of wastes from plants that produce plutonium and tritium for nuclear weapons.

After 50 years of research, scientists still don't agree on whether there is any safe method of storing these wastes. Some scientists believe that the long-term safe storage or disposal of high-level radioactive wastes is technically possible. Others disagree, pointing out that it is impossible to demonstrate that any method will work for the 10,000–240,000 years of fail-safe storage needed for such wastes. Here are some of the proposed methods and their possible drawbacks:

- *Bury it deep underground.* This favored strategy is under study by all countries producing nuclear waste. To reduce storage time from hundreds of thousands of years to about 10,000 years, very long-lived radioactive isotopes such as plutonium-239 must be extracted from spent fuel rods. The remainder would be fused with glass or a ceramic material and sealed in metal canisters for burial in a deep underground salt, granite, or other stable geological formation that is earthquake resistant and waterproof. However, according to a 1990 report by the U.S. National Academy of Sciences, "Use of geological information—to pretend to be able to make very accurate predictions of long-term site behavior—is scientifically unsound."
- *Shoot it into space or into the sun.* Costs would be very high, and a launch accident, such as the explosion of the space shuttle *Challenger*, could disperse high-level radioactive wastes over large areas of the earth's surface. This strategy has been abandoned for now.
- *Bury it under the Antarctic ice sheet or the Greenland ice cap.* The long-term stability of the ice sheets is not known. They could be destabilized by heat from the wastes, and retrieval of the wastes would be difficult or impossible if the method failed. This strategy is prohibited by international law and has been abandoned for now.
- *Dump it into descending subduction zones in the deep ocean* (Figure 14-2). Again, our geological knowledge is incomplete; wastes might eventually be spewed out somewhere else by volcanic activity. Waste containers might also leak and contaminate the ocean before being carried downward; retrieval would be impossible if the method did not work. This method is under active study by a consortium of 10 countries, but current international agreements and U.S. laws forbid dumping radioactive waste beneath the seas.
- *Bury it in thick deposits of mud on the deep ocean floor in areas that tests show have been geologically stable for 65 million years.* The waste containers would eventually corrode and release their radioactive contents. Some geologists contend that gravity and the powerful adhesive qualities of the mud would prevent the wastes from migrating, but this hypothesis has not been verified. This approach is banned under current international agreements and U.S. law.
- *Change it into harmless, or less harmful, isotopes.* Currently there is no way to do this, although some scientists are studying the use of a particle-beam laser (called a Baser) to convert radioactive wastes into nonradioactive materials. Even if a method were developed over the next few decades, costs would probably be extremely high. Moreover, depending on the process, any resulting toxic materials and low-level (but very long-lived) radioactive wastes would still need to be disposed of safely.

So far no country has come up with a scientifically and politically acceptable solution for the long-term storage of nuclear wastes.

High-level nuclear wastes from nuclear weapon production facilities are to be stored in the DOE's $2 billion Waste Isolation Pilot Plant (WIPP) on government-owned land near Carlsbad, New Mexico. This underground geologic repository, authorized by the U.S. Congress in 1980, has been built and is supposed to be put into operation in 1999.

In 1985 the DOE announced plans to build the first repository for underground storage of high-level radioactive wastes from commercial nuclear reactors (Figure 15-20) on federal land in the Yucca Mountain desert region, 160 kilometers (100 miles) northwest of Las Vegas, Nevada. The facility, which is expected to cost at least $26 billion, was scheduled to open by 2003, but in 1990 its opening was postponed to at least 2010. The site may never open, mostly because of rock faults, a nearby active volcano, 36 active earthquake faults on the site itself, and leakage of rainwater through the site faster than previous studies had predicted.

Fed up with the delay, some utilities are taking matters into their own hands and are trying to develop a temporary, aboveground, monitored retrievable

Q: What percentage of your body weight is due to water?

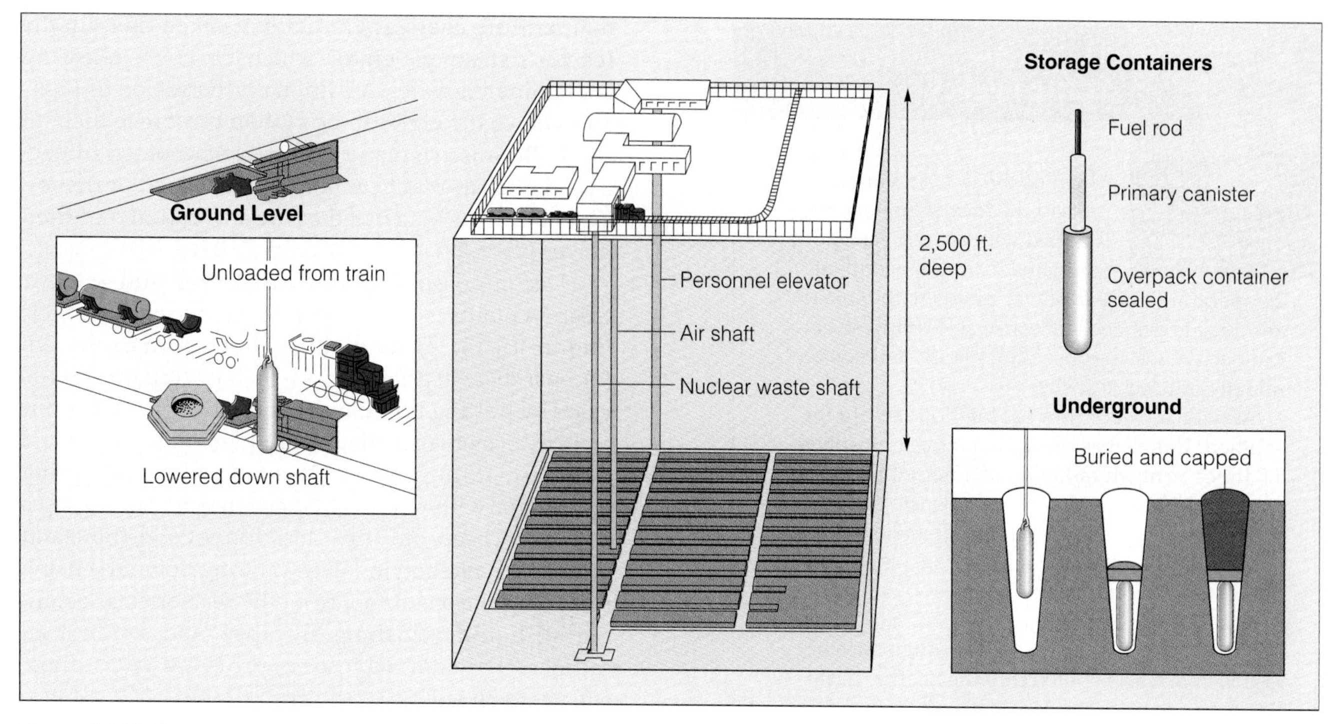

Figure 15-20 Proposed general design for deep-underground permanent storage of high-level radioactive wastes from commercial nuclear power plants in the United States. (Source: U.S. Department of Energy)

storage (MRS) facility for commercial high-level radioactive waste on the Mescalero Apache reservation in New Mexico. Because of the sovereignty of Native American nations, it is easier to site a radioactive storage facility on tribal lands than anywhere else in the United States. New Mexico officials have threatened to block the MRS facility, claiming it would be too risky to transport spent fuel from 71 reactors in 31 states across the country to the facility. Even if a permit is granted, the site would not be open until 2003 at the earliest.

Despite the lack of any accepted long-term disposal site, in 1996 the United States agreed to accept and manage highly radioactive spent fuel from 41 countries, with the $1 billion estimated costs to be paid by these countries. The DOE argued that if the United States did not accept the fuel, highly enriched uranium could be extracted from it and used to produce nuclear weapons.

How Widespread Are Contaminated Radioactive Sites? In 1992 the EPA estimated that as many as 45,000 sites in the United States may be contaminated with radioactive materials, 20,000 of them belonging to the DOE and the Department of Defense. According to the DOE it will cost taxpayers at least $230 billion over the next 75 years to clean up these facilities (with some estimating that the price will range from $400 to $900 billion).

Some critics doubt that the necessary cleanup funds will be provided by Congress as the cleanup process drags on and fades from serious public scrutiny. Others question spending so much money on a problem that is ranked by scientific advisers to the EPA as a low-risk ecological problem and not among the top high-risk health problems (Figure 17-13). However, the radioactive contamination situation in the United States pales in comparison to the post–Cold War legacy of nuclear waste and contamination in the republics of the former Soviet Union (Spotlight, p. 390).

What Can We Do with Worn-Out Nuclear Plants? The useful operating life of today's nuclear power plants is supposed to be 40 years, but many plants are wearing out and becoming dangerous faster than anticipated. Worldwide, 81 reactors have been shut down after being in operation for an average of only 17 years.

After many years of bombardment by neutrons released by nuclear fission, the walls of a reactor's pressure vessel become brittle and thus are more likely to crack, thereby exposing the highly radioactive reactor core. In addition, decades of pressure and

A: About 55% (60% women, 50% men)

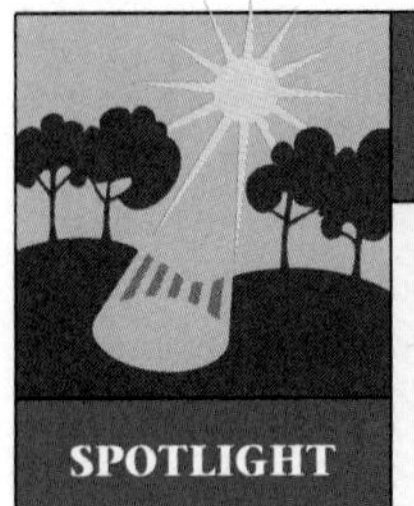

The Soviet Nuclear-Waste Legacy

SPOTLIGHT

Much of the land in various parts of the former Soviet Union is dotted with areas severely contaminated by nuclear accidents, 26 operating nuclear power plants with flawed and unsafe designs, nuclear-waste dump sites, radioactive waste-processing plants, and contaminated nuclear test sites.

Mayak, a plutonium production facility in southern Russia, has spewed into the atmosphere 2.5 times as much radiation as Chernobyl, mostly from the explosion of a nuclear-waste storage tank in 1957. About 2,600 square kilometers (1,000 square miles) of the surrounding area is still contaminated with radioactive wastes so dangerous that no one can live there.

Between 1949 and 1967, huge quantities of radioactive waste spilled into the nearby Techa River. For decades the 64,000 people living along this regional waterway shared by 30 villages drank the river's water, bathed in it, and washed their clothes in it. Release of once-secret studies of 28,000 Techa River villagers revealed that they had a statistically significant increase in the incidence of leukemia, an overall increase in cancer mortality, and more lymphatic genetic mutations than a control group not living in a contaminated area.

Large quantities of radioactive materials from the Mayak facility were also dumped into nearby Lake Karachay between 1949 and 1967. Today the lake is so radioactive that standing on its shores for about an hour would be fatal. Special machinery is now being used to fill and cap the lake.

At the Tomsk-7 plutonium production site in central Siberia, wastes with radioactivity 20 times that released at Chernobyl were dumped into lakes and underground formations in the region.

Around the shores of Novaya Zemyla, an island off the coast of Russia, 18 defunct nuclear-powered submarines were dumped (at least three of them loaded with nuclear fuel), along with as many as 17,000 containers of radioactive waste. If released by corrosion, radioactivity from these submarines and containers threatens Russian, Norwegian, Finnish, and perhaps Swedish citizens in the area. In 1993 Greenpeace caught Russians still dumping massive amounts of radioactive waste from nuclear submarines into the Sea of Japan.

Critical Thinking

Because of serious economic problems there is little money for cleanup of radioactive sites in parts of the former Soviet Union. Should the United States and other developed nations provide technical and financial aid to help deal with this problem? Explain.

temperature changes gradually weaken tubes in the reactor's steam generator, which can crack, releasing contaminated water. Additionally, corrosion of pipes and valves throughout the system can cause them to crack. Because so many of its parts become radioactive, a nuclear plant cannot be abandoned or demolished by a wrecking ball the way a worn-out coal-fired power plant can.

Decommissioning nuclear power and nuclear weapon plants is the last step in the nuclear fuel cycle (Figure 15-18). Three methods have been proposed: **(1)** *immediate dismantling,* **(2)** *mothballing* for 30–100 years by putting up a barrier and setting up a 24-hour security system (costing about $15 million a year) and then dismantling it, and **(3)** *entombment* by covering the reactor with reinforced concrete and putting up a barrier to keep out intruders for several thousand years. Each method involves shutting down the plant, removing the spent fuel from the reactor core, draining all liquids, flushing all pipes, and sending all radioactive materials to an approved waste storage site yet to be built.

By 1995 more than 30 commercial reactors worldwide (12 in the United States) had been retired and awaited decommissioning. Another 228 large commercial reactors (20 in the United States) are scheduled for retirement between 2000 and 2012. By 2030 all U.S. reactors will have to be retired, based on the life of their current operating licenses, and many may be retired early for safety or financial reasons.

Experience suggests that decommissioning old plants by dismantling and safely taking care of the resulting radioactive wastes sometimes exceeds the costs of building them in the first place. U.S. utilities have been setting aside funds for decommissioning more than 100 reactors, but to date these funds fall far short of the estimated costs. If sufficient funds are not available for decommissioning, the balance of the costs could be passed along to ratepayers and taxpayers.

What Is the Connection Between Nuclear Reactors and the Spread of Nuclear Weapons?

Since 1958 the United States has been giving away and selling to other countries various forms of nuclear technology, mostly in the form of nuclear power plants and research reactors. Today the United States and at least 14 other countries sell nuclear power technology in the international marketplace. Information, components, and materials used to build and operate such reactors can be used to produce fissionable isotopes such as uranium-235 and plutonium-239 (the explosive material in nuclear weapons).

We already live in a world with enough nuclear weapons to kill everyone on the earth 40 times over—20 times if current nuclear arms reduction agreements are

Q: What percentage of earth's enormous volume of water is available to us as usable fresh water?

carried out. By the end of this century, 60 countries—1 of every 3 in the world—are expected to have either nuclear weapons or the knowledge and ability to build them, with the fuel and knowledge coming mostly from research and commercial nuclear reactors. Dismantling thousands of Russian and American nuclear warheads can increase the threat from the resulting huge amounts of bomb-grade plutonium that must be safeguarded.

Once fissionable material is in hand, through stealing it or buying it in the growing international black market, no sophisticated technology is needed to make a nuclear weapon. Depending on the technology used, only 3–6 kilograms (7–14 pounds) of plutonium-239 (somewhere between the size of a hockey puck and a soda can) or 5–16 kilograms (11–35 pounds) of weapons-grade enriched uranium-235 are needed to make an atomic bomb the size of the one dropped on Nagasaki.

The amount of plutonium-239 needed to make a smaller nuclear bomb capable of leveling a city block would fit in only half of a typical soft-drink can. Any terrorists who steal plutonium-239 or buy it on the black market need not bother to make bombs; they could simply use a conventional explosive charge to disperse the plutonium into the atmosphere from atop any tall building. Dispersed in that way, 1 kilogram (2.2 pounds) of plutonium oxide powder theoretically would contaminate 8 square kilometers (3 square miles) with dangerous levels of radioactivity for several hundred thousand years. Terrorists could also put plutonium in a city's water supply.

So far no one has been able to come up with truly effective solutions to the serious problems of nuclear weapon proliferation and what to do with retired weapon-grade plutonium. All we can do is try to slow the spread of such weapons and hope we can find a way to keep plutonium removed from weapons out of the hands of those who want nuclear arms.

Can We Afford Nuclear Power? Experience has shown that nuclear power is an extremely expensive way to boil water to produce electricity, even when it is shielded partially from free-market competition with other energy sources by huge government subsidies. Thus, the major reason utility officials, investors, and most governments are shying away from nuclear power is not safety but the *extremely high cost* of making it a safe technology.

Some of these costs can be reduced by using standardized designs for all plants (which also cuts construction time in half), as France has done. Even so, the French government has run up a $30-billion debt subsidizing its government-run nuclear power industry.

Despite massive federal subsidies and tax breaks, the most modern nuclear power plants built in the United States produce electricity at an average of about 13.5¢ per kilowatt-hour, the equivalent of burning oil costing about $216 per barrel (compared to its recent prices of $12–20 per barrel) to produce electricity. All methods of producing electricity in the United States (except solar photovoltaic and solar thermal plants) have average costs below those of new nuclear power plants (Figure 15-21). By 2000–2005, even these methods (with few subsidies) are expected to be cheaper than nuclear power.

In the 1980s Great Britain tried to sell its government-run nuclear power plants to private interests, but no one would buy them or even accept them for free. Business interests knew that the plants were losers because of the high costs of maintenance, decommissioning, and radioactive waste storage.

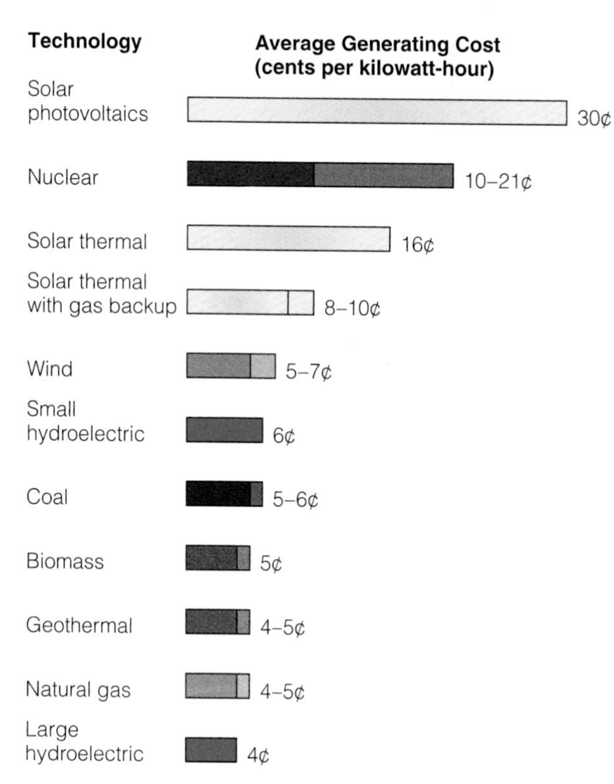

Figure 15-21 Generating costs of electricity per kilowatt-hour by various technologies in 1993. By 2000, costs per kilowatt-hour for wind are expected to decrease to 4–5¢, solar thermal with gas assistance 5–6¢, solar photovoltaic 6–12¢, and natural gas 3–4¢; coal 4-5¢; nuclear power is expected to remain at 10–21¢. Costs for the other technologies shown in this figure are expected to remain about the same. (Data from U.S. Department of Energy, Council for Renewable Energy Education, Investor Responsibility Research Center, California Energy Commission, and Worldwatch Institute)

A: About 0.003%

Banks and other lending institutions have become leery of financing new U.S. nuclear power plants. Abandoned reactor projects have cost U.S. utility investors over $100 billion since the mid-1970s, bankrupting U.S. utilities such as the Washington Public Power Supply System and the Public Service Company of New Hampshire. The Three Mile Island accident showed that utility companies could lose $1 billion worth of equipment in an hour and at least $1 billion more in cleanup costs, even without any established harmful effects on public health. A poll of U.S. utility executives found that only 2% would even consider ordering a new nuclear power plant. At the global level the World Bank said in 1995 that nuclear power is too costly and risky.

Forbes business magazine has called the failure of the U.S. nuclear power program "the largest managerial disaster in U.S. business history." The U.S. Department of Energy estimates that nearly one-fourth of the current 109 existing U.S. nuclear reactors may be closed prematurely for financial reasons by 2003, as utility companies write off their losses and move on to other, more profitable alternatives.

A growing number of the large companies that once built nuclear power plants are getting out of the business, and some of them are switching to developing energy conservation and solar energy technologies. The increasingly rapid global transition to smaller, easily expandable decentralized sources of electricity and heat may mean that large, centralized coal and nuclear power plants may soon be bypassed by more flexible and cheaper energy-efficiency and solar energy technologies (Chapter 16).

Since the Three Mile Island accident, the U.S. nuclear industry and utility companies have financed a $21-million-a-year public relations campaign by the U.S. Council for Energy Awareness to improve the industry's image and resell nuclear power to the American public.

Most of the council's ads advance the argument that the United States needs more nuclear power to reduce dependence on imported oil and to improve national security. But since 1979, only about 5% (3% in 1995) of the electricity in the United States has been produced by burning oil, and 95% of that is residual oil that can't be used for other purposes. The nuclear industry also does not point out that 73% of the uranium used for nuclear fuel in the United States in 1993 was imported (most from Canada).

Nuclear industry officials also claim that nuclear power, unlike coal burning, adds no greenhouse-enhancing CO_2 to the atmosphere. Nuclear power plants themselves don't emit CO_2, but processing of uranium fuel (Figure 15-12) produces about one-sixth the CO_2 per unit of electricity as that from a coal-burning plant. To offset just 5% of current global CO_2 emissions would require nearly doubling the worldwide nuclear capacity at a cost of more than $1 trillion. According to Amory Lovins (Guest Essay, p. 433), if we want to reduce CO_2 emissions using the least expensive methods, then investing in energy efficiency and renewable energy resources are at the top of the list and nuclear power is at the bottom.

The U.S. nuclear industry hopes to persuade the federal government and utility companies to build hundreds of new second-generation smaller plants using standardized designs with supposedly fail-safe features, which they claim are safer and can be built more quickly (in 3–5 years). However, according to *Nucleonics Week*, an important nuclear industry publication, "Experts are flatly unconvinced that safety has been achieved—or even substantially increased—by the new designs."

Furthermore, none of the new designs solves the problems of what to do with nuclear waste and worn-out nuclear plants and how to prevent the use of nuclear technology to build nuclear weapons. Indeed, these problems would become more serious if the number of nuclear plants increased from a few hundred to several thousand. None of these new designs changes the fact that nuclear power, even with huge government subsidies, is an incredibly expensive way to produce electricity compared to other methods. Most analysts agree that simple economics has proved to be the Achilles heel of nuclear power.

If nuclear power is essentially a dead industry, why does this book (and most other environmental science textbooks) spend so much space discussing it? The answer is that despite its increasingly obvious financial and other shortcomings, the nuclear industry has enormous political support in many governments, which still gives it massive research and development subsidies and tax breaks. This reduces funding for cheaper and less risky energy alternatives.

For example, in 1993 some 95% of Japan's energy research and development budget was devoted to nuclear power. And the International Energy Agency (supported by funds from the governments of many developed countries) spent nearly 55% of its 1994 energy research and development budget on nuclear power. Nuclear energy is also a major component of the energy research and development budget of the U.S. Department of Energy (Connections, right). One observer remarked that the nuclear power industry is like a giant dinosaur whose head has been cut off but with enough power to keep its tail thrashing around for decades.

Q: What percentage of the world's cropland is irrigated?

Should Conventional Nuclear Power Have a Future? Some analysts argue that we should continue some low-level government funding of research and development and pilot plant testing of new reactor designs to keep this option available for use in the future. There may be an urgent need to sharply reduce fossil-fuel greenhouse gas (because of serious global warming) or because improved energy efficiency and renewable energy options may fail to keep up with demands for electricity. What do you think?

Is Breeder Nuclear Fission a Feasible Alternative? Some nuclear power proponents urge the development and widespread use of **breeder nuclear fission reactors**, which generate more nuclear fuel than they consume by converting nonfissionable uranium-238 into fissionable plutonium-239. Because breeders would use over 99% of the uranium in ore deposits, the world's known uranium reserves would last at least 1,000 years, and perhaps several thousand years.

However, if the safety system of a breeder reactor fails, the reactor could lose some of its liquid sodium coolant, which ignites when exposed to air and reacts explosively if it comes into contact with water. This could cause a runaway fission chain reaction and perhaps a nuclear explosion powerful enough to blast open the containment building and release a cloud of highly radioactive gases and particulate matter. Leaks of flammable liquid sodium can also cause fires, as has happened with all experimental breeder reactors built so far.

In December 1986, France opened a commercial-size breeder reactor. Not only did it cost three times the original estimate to build, but the little electricity it produced was twice as expensive as that generated by France's conventional fission reactors. So far France has invested almost $13 billion in this breeder reactor, which was shut down for expensive repairs between 1989 and 1995.

Tentative plans to build full-size commercial breeders in Germany, some republics of the former Soviet Union, the United Kingdom, and Japan have been abandoned because of the French experience and an excess of electric-generating capacity and conventional uranium fuel. An experimental breeder reactor in Japan has been shut down since December 1995 because of a leak in its liquid sodium coolant system.

In addition, existing breeders produce plutonium fuel much too slowly. If this problem is not solved, it would take 100–200 years for breeders to begin producing enough plutonium to fuel a significant number of other breeder reactors. In 1994 U.S. Secretary of Energy Hazel O'Leary ended government-supported research for breeder technology after some $9 billion had been spent on it.

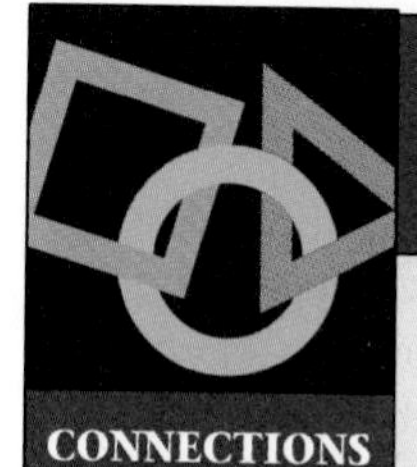

CONNECTIONS

U.S. Government Funding for Energy Research and Development

Since 1973 about 54% of the research and development (R&D) budget of the U. S. Department of Energy (DOE) has gone to nuclear power, 25% to fossil fuels, and 21% to energy conservation and renewable energy options.

In 1998, the DOE's R&D budget for development of energy supply options amounted to $1.6 billion (only 9.5% of the department's overall budget). Of this amount, 30% went for nuclear energy (divided about equally between conventional nuclear fission and nuclear fusion), 25% for energy conservation, 23% for fossil fuels, and 22% for solar and other forms of renewable energy. Most environmentalists believe that these R&D research priorities create an uneven economic playing field for energy efficiency and renewable energy, which should receive greatly increased funding.

Most Americans are unaware that typically 65–67% of the DOE's annual budget is used to produce and dismantle nuclear weapons and clean up the country's badly contaminated nuclear weapons production and test facilities. In other words, two-thirds of the DOE budget is an appendage to the Department of Defense's annual budget. Some analysts and members of Congress urge that this portion of the DOE budget be transferred to the Department of Defense.

Critical Thinking

1. Should more government R&D funding go to energy efficiency and renewable energy and less to fossil fuels and nuclear energy? Explain.
2. Should the roughly two-thirds of the DOE annual budget used to produce and dismantle nuclear weapons and clean up contaminated nuclear weapons and test facilities be transferred to the Department of Defense or some other separate agency? Would this tend to decrease or increase the amount of money Congress allocates for research and development of energy technologies?

Is Nuclear Fusion a Feasible Alternative? Scientists hope that someday controlled nuclear fusion will provide an almost limitless source of high-temperature heat and electricity. Research has focused

A: About 17%

on the D–T nuclear fusion reaction, in which two isotopes of hydrogen—deuterium (D) and tritium (T)—fuse at about 100 million degrees (Figure 3-19).

Despite 50 years of research and huge expenditures of mostly government funds, controlled nuclear fusion is still in the laboratory stage. Deuterium and tritium atoms have been forced together using electromagnetic reactors (the size of 12 locomotives), 120-trillion-watt laser beams, and bombardment with high-speed particles. So far, none of these approaches has produced more energy than it uses.

In 1989 two chemists claimed to have achieved deuterium–deuterium (D–D) nuclear fusion at room temperature using a simple apparatus. However, subsequent experiments have not substantiated their claims.

If researchers can eventually get more energy out of nuclear fusion than they put in, the next step would be to build a small fusion reactor and then scale it up to commercial size—one of the most difficult engineering problems ever undertaken. The estimated cost of a commercial fusion reactor is several times that of a comparable conventional fission reactor.

Proponents contend that with greatly increased federal funding, a commercial nuclear fusion power plant might be built by 2030, but many energy experts don't expect nuclear fusion to be a significant energy source until 2100, if then. Meanwhile, experience shows that we can produce more electricity than we need using several quicker, cheaper, and safer methods.

In the long run, humanity has no choice but to rely on renewable energy. No matter how abundant they seem today, eventually coal and uranium will run out.

DANIEL DEUDNEY AND CHRISTOPHER FLAVIN

CRITICAL THINKING

1. What is your understanding of the statement that "net energy is the only energy that really counts"? Explain the concept of net energy in terms of the second law of energy (thermodynamics) and our failure to use good ecological design, as discussed by David Orr in his Guest Essay on p. 746.

2. Just to continue using oil at the current rate (not the projected higher exponential increase in its annual use) and not run out, we must discover and add to global oil reserves the equivalent of a new Saudi Arabia supply (the world's largest) *every 10 years*. Do you believe this is possible? If not, what effects might this have on your life? On a child or grandchild you might have? List five things you can do to help reduce your dependence on oil and resources derived from oil (such as gasoline and most plastics) (see Appendix 5). Which of these things do you actually plan to do?

3. The United States now imports more than half of the oil it uses and could be importing 70% of its oil by 2010. How do you feel about this situation? List what you believe are the five best ways to reduce this dependence.

4. **(a)** Should air pollution emission standards for *all* new and existing coal-burning plants be tightened significantly? Explain. **(b)** Do you favor an energy strategy based on greatly increased use of coal-burning plants to produce electricity? Explain. What are the alternatives?

5. Explain your understanding of the statement that "a major nuclear power-plant accident anywhere is a nuclear accident everywhere." Is this a valid claim? Explain. If you believe that it is valid, what can you do as an individual to help ensure that a serious nuclear power-plant accident doesn't occur anywhere in the world? For example, would you support using taxpayer funds to help countries in the former Soviet Union and eastern Europe improve energy conservation and develop new energy alternatives so that the 26 potentially dangerous Chernobyl-type and other unsafe nuclear reactors in such countries could be shut down as soon as possible? Explain.

6. Explain why you agree or disagree with each of the following proposals made by the nuclear power industry:

a. A large number of new, better-designed nuclear fission power plants should be built in the United States to reduce dependence on imported oil and slow down projected global warming.

b. Federal subsidies to the commercial nuclear power industry (already totaling $1 trillion) should be continued so that it does not have to compete in the open marketplace with other energy alternatives receiving no, or smaller, federal subsidies.

c. A comprehensive program for developing the breeder nuclear fission reactor should be developed and funded largely by the federal government.

d. Current federal subsidies for developing nuclear fusion should be increased.

7. Try to come up with analogies to complete the following statements: **(a)** Using oil to supply most of our energy is like ______. **(b)** Using natural gas to supply most of our energy is like ______. **(c)** Using nuclear power to provide electricity is like ______. **(d)** Disposing of high-level radioactive waste is like ______. **(e)** Retiring a nuclear power plant is like ______. **(f)** Keeping the prices of fossil fuels and nuclear energy artificially low is like ______.

Q: Most of the water withdrawn from surface or groundwater sources worldwide is used for what purpose?

PROJECTS

1. Where does the electricity to light your room come from? Trace it back to its source as coal or uranium ore extracted from the ground and list the various steps involved in getting this energy to you as light. How many energy-conversion steps are needed to produce the light? Can you identify any ways to cut the number of steps?

2. Write a two-page scenario of what your life might be like without oil. Compare and discuss the scenarios developed by members of your class.

3. Throughout the United States, 42 nuclear reactors are operating on college campuses. Does your campus have a nuclear reactor? If so, find out how large it is, what it is used for, and who is in charge of operating it. Has it had any safety problems? Explain.

4. How is the electricity in your community produced? How has the cost of that electricity changed since 1970 compared to general inflation? Do your community and your school have a plan for improving energy efficiency? If so, what is this plan, and how much money has it saved during the past 10 years? If there is no plan, develop one and present it to the appropriate officials.

5. Make a concept map of this chapter's major ideas, using the section heads and subheads and the key terms (in boldface). Look at the inside back cover and in Appendix 4 for information about concept maps.

A: Irrigation (65% of global withdrawal)

16 ENERGY EFFICIENCY AND RENEWABLE ENERGY RESOURCES

The Energy-Efficiency and Renewable-Energy Revolution

Energy experts Hunter and Amory Lovins (Guest Essay, p. 433) have built a large, passively heated, superinsulated, partially earth-sheltered home in Snowmass, Colorado (Figure 16-1), where winter temperatures can drop to −40°C (−40°F).

This structure, which also houses the research center for the Rocky Mountain Institute, an office used by 40 people, gets 99% of its space and water heating and 95% of its daytime lighting from the sun. It uses one-tenth the usual amount of electricity for a structure of its size. Total energy savings repaid the cost of its energy-saving features after 10 months and are projected to pay for the entire facility in 40 years.

Superinsulating windows, already here, mean that a house can have large numbers of windows without much heat loss in cold weather or heat gain in hot weather. They pay for themselves in lower fuel bills within 2–4 years, and they save money every year thereafter.

Thinner insulation material now being developed will allow roofs and walls to be insulated far

Figure 16-1 The Rocky Mountain Institute in Colorado. This facility is a home and a center for the study of energy efficiency and sustainable use of energy and other resources. It is also an example of energy-efficient passive solar design. (Robert Millman/Rocky Mountain Institute)

better than in today's best superinsulated houses. Researchers are also experimenting with *smart walls*. These consist of wallboard containing materials that absorb or release heat to a room at certain temperatures.

In energy-efficient houses of the near future, sensors will monitor indoor temperatures, sunlight angles, and the occupants' locations. Then small microprocessors will send heat or cooled air where it is needed and open or close window shutters to let solar energy in or reduce heat loss from windows at night and on cloudy days.

A small but growing number of people in developed and developing countries are getting their electricity from *solar cells* that convert sunlight directly into electricity. They can be attached like shingles to a roof or applied to window glass as a coating. Currently, solar-cell prices are high but are expected to fall rapidly within a decade (Figure 15-21).

As the price of producing electricity from solar cells drops, this could usher in a *solar–hydrogen revolution*. Electricity produced by large banks of solar cells could be passed through water to make hydrogen gas (H_2), which could be used to fuel vehicles, industries, and buildings. Hydrogen gas produced by solar-cell power plants in deserts could be distributed throughout most of the world by pipeline. Another solution is to burn hydrogen in energy-efficient *fuel cells* that produce electricity to run cars and heat buildings and water.

If we can make the transition to an energy-efficient solar–hydrogen age, we could say goodbye to smog, oil spills, acid rain, most coal-burning and nuclear power plants, and perhaps the threat of global warming. The reason is simple: When hydrogen burns in air, it reacts with oxygen gas to produce water vapor—not a bad thing to have coming out of our tailpipes, chimneys, and smokestacks. If clean solar–hydrogen technology is implemented soon enough, it could also help heavily populated developing countries raise their living standards without severely disrupting the earth's life-support systems.

These are only a few of the components of the exciting *energy-efficiency and renewable-energy revolution* that many analysts believe will help us make the transition to more sustainable societies over the next 40–50 years.

If the United States wants to save a lot of oil and money and increase national security, there are two simple ways to do it: stop driving Petropigs and stop living in energy sieves.

AMORY B. LOVINS

These are the questions answered in this chapter:

- What are the benefits and drawbacks of improving energy efficiency; using solar energy to heat buildings and water and to produce electricity; using flowing water and solar energy stored as heat in water to produce electricity; using wind to produce electricity; burning plants and organic waste (biomass) for heating buildings and water, for producing electricity, and for transportation (biofuels); producing hydrogen gas and using it to make electricity, heat buildings and water, and propel vehicles; and extracting heat from the earth's interior (geothermal energy)?
- What are the best energy options?

16-1 THE IMPORTANCE OF IMPROVING ENERGY EFFICIENCY

What Is Energy Efficiency? Doing More with Less You may be surprised to learn that *84% of all commercial energy used in the United States is wasted* (Figure 16-2). About 41% of this energy is wasted automatically because of the degradation of energy quality imposed by the second law of energy. More important, about 43% is wasted unnecessarily, mostly by using fuel-wasting motor vehicles, furnaces, and other devices, and by living and working in leaky, poorly insulated, poorly designed buildings.

People in the United States unnecessarily waste as much energy as two-thirds of the world's population consumes. Indeed, according to energy expert Amory Lovins (Guest Essay, p. 433), unnecessary energy waste in the United States amounts to about $300 billion per year—an average of $570,000 per minute. In 1998 this waste was much more than the roughly $200-billion annual federal budget deficit and the $268-billion military budget. U.S. electric power plants waste more energy than Japan consumes. To Lovins the single most important way for the United States (and other countries) to improve its military, economic, and environmental security is to reduce energy waste.

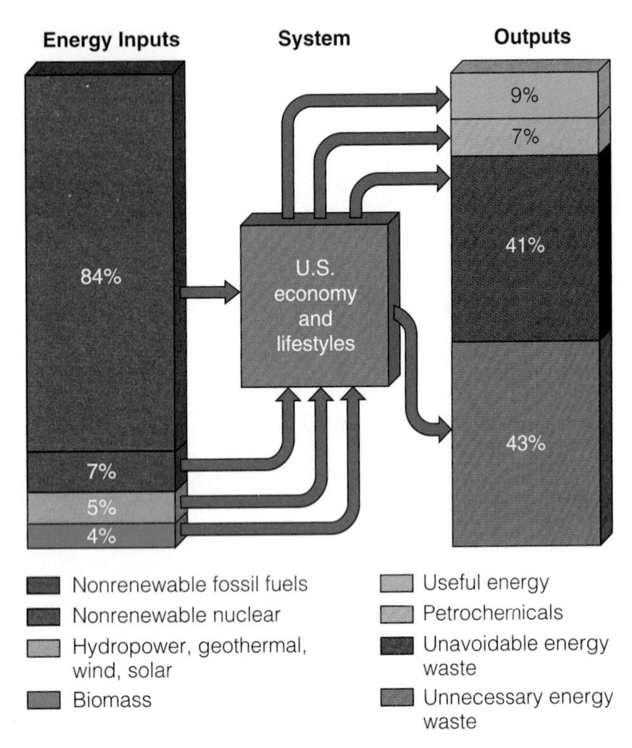

Figure 16-2 Flow of commercial energy through the U.S. economy. Note that only 16% of all commercial energy used in the United States ends up performing useful tasks or being converted to petrochemicals; the rest is either automatically and unavoidably wasted because of the second law of energy (41%) or wasted unnecessarily (43%).

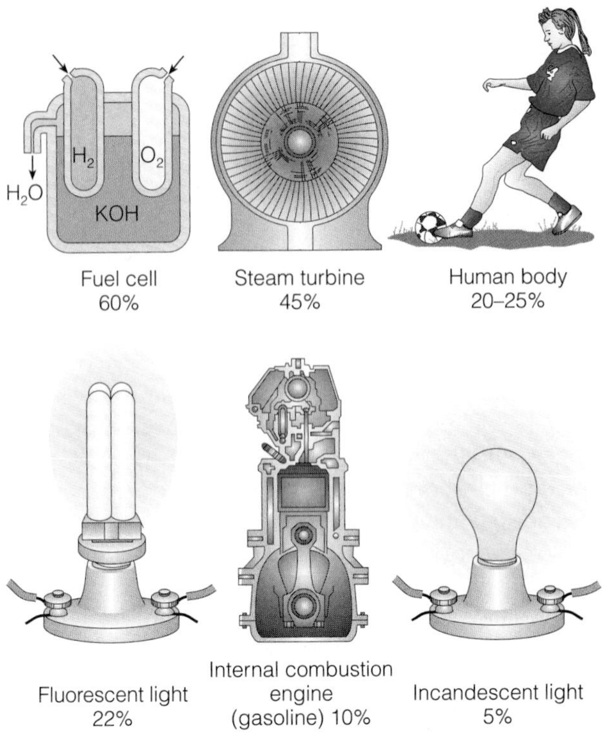

H_2 O_2 H_2O KOH
Fuel cell 60%
Steam turbine 45%
Human body 20–25%
Fluorescent light 22%
Internal combustion engine (gasoline) 10%
Incandescent light 5%

Figure 16-3 Energy efficiency of some common energy conversion devices.

According to Lovins, the easiest, fastest, and cheapest way to get more energy with the least environmental impact is to eliminate much of this energy waste by making lifestyle changes that reduce energy consumption: walking or biking for short trips, using mass transit, putting on a sweater instead of turning up the thermostat, and turning off unneeded lights.

Another equally important way is to increase the efficiency of the energy conversion devices we use. **Energy efficiency** is the percentage of total energy input that does useful work (is not converted to low-quality, essentially useless heat) in an energy conversion system. The energy conversion devices we use vary in their energy efficiencies (Figure 16-3). We can save energy and money by buying the most energy-efficient home heating systems, water heaters, cars, air conditioners, refrigerators, computers, and other appliances that are available, and by supporting research to invent even more energy-efficient devices. Some energy-efficient models may cost more initially, but in the long run they usually save money by having a lower **life cycle cost**: initial cost plus lifetime operating costs. Although the United States has a long way to go, improvements in energy efficiency since the oil embargo in 1973 have cut the country's energy bills by a whopping $275 billion a year.

The net efficiency of the entire energy delivery process for a space heater, water heater, or car is determined by the efficiency of each step in the energy conversion process. For example, the sequence of energy-using (and energy-wasting) steps involved in using electricity produced from fossil or nuclear fuels is Extraction → Transportation → Processing → Transportation to power plant → Electric generation → Transmission → End use.

Figure 16-4 shows the net energy efficiency for heating two well-insulated homes: one with electricity produced at a nuclear power plant, transported by wire to the home, and converted to heat (electric resistance heating), the other heated passively, with an input of direct solar energy through high-efficiency windows facing the sun, with heat stored in rocks or water for slow release. This analysis shows that the process of converting the high-quality energy in nuclear fuel to high-quality heat at several thousand degrees in the power plant, converting this heat to high-quality electricity, and then using the electricity to provide low-quality heat for warming a house to only

Q: What percentage of the water withdrawn in the United States is used for irrigation?

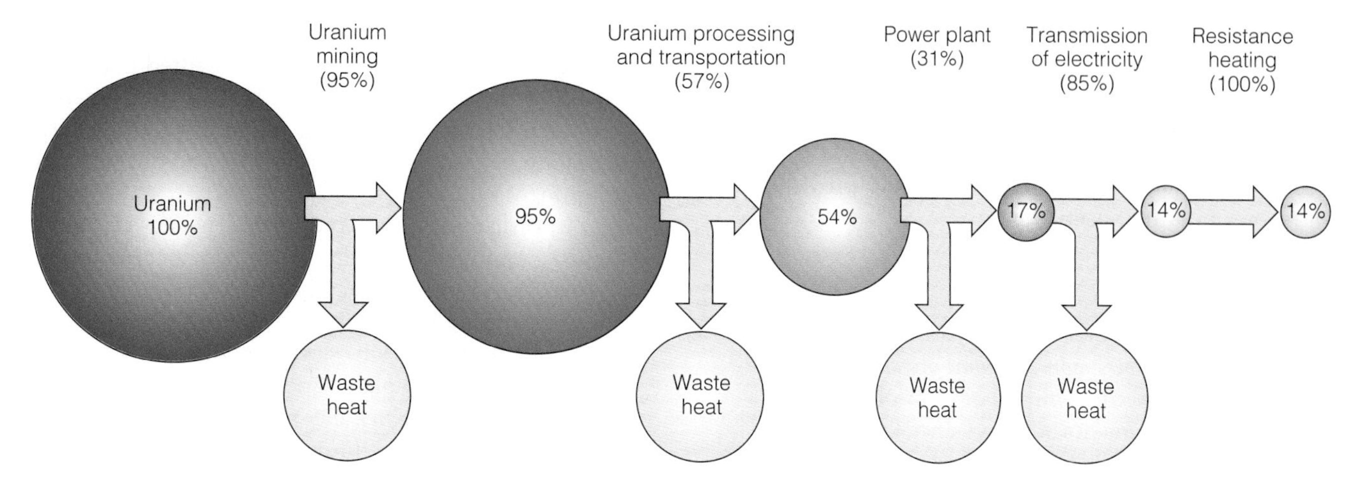

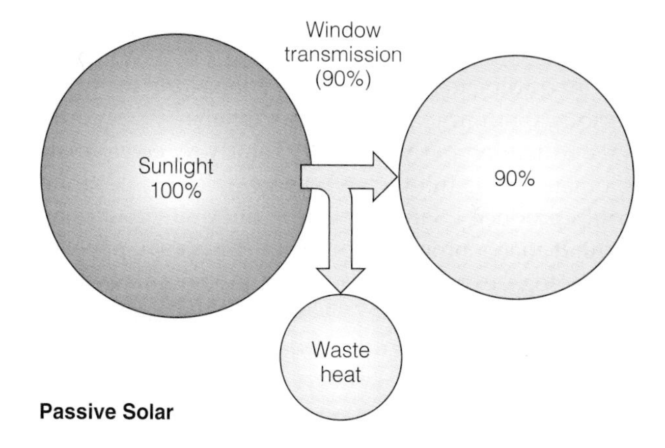

Figure 16-4 Comparison of net energy efficiency for two types of space heating. The cumulative net efficiency is obtained by multiplying the percentage shown inside the circle for each step by the energy efficiency for that step (shown in parentheses). Because of the second law of thermodynamics, in most cases the greater the number of steps in an energy conversion process, the lower its net energy efficiency. About 86% of the energy used to provide space heating by electricity produced at a nuclear power plant is wasted. If the additional energy needed to deal with nuclear wastes and to retire highly radioactive nuclear plants after their useful life is included, then the net energy yield for a nuclear plant is only about 8% (or 92% waste). By contrast, with passive solar heating, only about 10% of incoming solar energy is wasted.

about 20°C (68°F) is extremely wasteful of high-quality energy. Burning coal or any fossil fuel at a power plant to supply electricity for space heating is also inefficient. It is much less wasteful to collect solar energy from the environment, store the resulting heat in stone or water, and, if necessary, raise its temperature slightly to provide space heating or household hot water.

Physicist and energy expert Amory Lovins (Guest Essay, p. 433) points out that using high-quality electrical energy to provide low-quality heating for living space or household water is like using a chain saw to cut butter or a sledgehammer to kill a fly. As a general rule, he suggests that we not use high-quality energy to do a job that can be done with lower-quality energy (Figure 3-14).

The logic of this point is illustrated by looking at the costs of providing heat using various fuels. In 1995, the average price of obtaining 250,000 kilocalories (1 million Btu) for heating either space or water in the United States was $6.05 using natural gas, $7.56 using kerosene, $9.30 using oil, $9.74 using propane, and $24.15 using electricity. As these numbers suggest, if you don't mind throwing away hard-earned dollars, then use electricity to heat your house and bathwater. Yet almost one of every four homes in the United States is heated by electricity.

Perhaps the three least efficient energy-using devices in widespread use today are incandescent light bulbs (which waste 95% of the energy input), vehicles with internal combustion engines (which waste 86–90% of the energy in their fuel), and nuclear power plants producing electricity for space heating or water heating (which waste 86% of the energy in their nuclear fuel—and probably 92% when the energy needed to deal with radioactive wastes and retired nuclear plants is included). Energy experts call for us to replace these devices or greatly improve their energy efficiency over the next few decades.

Why Is Reducing Energy Waste So Important? According to Amory Lovins and other energy analysts, reducing energy waste is one of the planet's best and most important economic and environmental bargains. It

- *Makes nonrenewable fossil fuels last longer.*
- *Gives us more time to phase in renewable energy resources.*

A: 41% (85% in the West)

- *Decreases dependence on oil imports* (55% in 1997).
- *Lessens the need for military intervention in the oil-rich but politically unstable Middle East.*
- *Reduces local and global environmental damage* because less of each energy resource would provide the same amount of useful energy.
- *Is the cheapest and quickest way to slow projected global warming.*
- *Saves more money, provides more jobs, improves productivity, and promotes more economic growth per unit of energy than other alternatives.* For example, energy efficiency improvements adopted since the 1970s have already saved the United States over $1 trillion in energy costs.
- *Improves competitiveness in the international marketplace.*

According to Amory Lovins, if the world really got serious about improving energy efficiency, we could save $1 trillion per year. A 1993 study by economists estimated that a full-fledged energy-efficiency program could produce 1.3 million jobs in the United States by 2010. If such a program were implemented, a panel of technology experts projects that global energy efficiency could increase by 50% between 1997 and 2016.

Why isn't there more emphasis on improving energy efficiency? The primary reason is a glut of low-cost, underpriced fossil fuels. As long as such energy is artificially cheap because its harmful environmental costs are not included in its market prices, people are more likely to waste it and not make investments in improving energy efficiency.

Another cause is the use of huge government subsidies and tax breaks to promote the use of energy-wasting power plants, motor vehicles, and buildings and few economic incentives for consumers to invest in improving energy efficiency even if it saves them money. A third problem is that most consumers don't have adequate information about the availability of energy-saving devices and the amount of money such items can save them using *life cycle cost analysis*.

16-2 WAYS TO IMPROVE ENERGY EFFICIENCY

How Can We Use Waste Heat? Could we save energy by recycling energy? No. The second law of energy tells us that we cannot recycle energy. However, we can slow the rate at which waste heat flows into the environment when high-quality energy is degraded. For a house, the best way to do this is to insulate it thoroughly, eliminate air leaks, and equip it with an air-to-air heat exchanger to prevent buildup of indoor air pollutants. Many homes in the United States are so full of leaks that their heat loss in cold weather and heat gain in hot weather are equivalent to having a large window-size hole in the wall of the house (Figure 3-1).

In office buildings and stores, waste heat from lights, computers, and other machines can be collected and distributed to reduce heating bills during cold weather; during hot weather, this heat can be collected and vented outdoors to reduce cooling bills. Waste heat from industrial plants and electrical power plants can be used to produce electricity (cogeneration); it can also be distributed through insulated pipes to heat nearby buildings, greenhouses, and fish ponds, as is done in some parts of Europe.

How Can We Save Energy in Industry? There are a number of ways to save energy and money in industry. One is **cogeneration**, the production of two useful forms of energy (such as steam and electricity) from the same fuel source. Waste heat from coal-fired and other industrial boilers can produce steam that spins turbines and generates electricity at roughly half the cost of buying it from a utility company. By using the electricity or selling it to the local power company for general use, a plant can save energy and money. Cogeneration allows up to 90% of the energy in fuel to be used productively—far more than the 33% efficiency for central power plants owned by U.S. utilities.

Cogeneration has been widely used in western Europe for years, and its use in the United States is growing. Within 8 years cogeneration could produce more electricity than all U.S. nuclear power plants, and much more cheaply. In China, cogeneration provides about 12% of the country's electricity. Recently the Chinese government required that all large industrial boilers be converted to cogeneration, a policy that the Worldwatch Institute says could serve as a model for the rest of the world.

In Germany, small cogeneration units that run on natural gas or liquefied petroleum gas (LPG) supply restaurants, apartment buildings, and houses with all their energy. In 4–5 years they pay for themselves in saved fuel and electricity. In 1997, the U.S. company Allied Signal announced plans to market a small cogenerator (micro-powerplant).

Replacing energy-wasting electric motors is another strategy. About 60–70% of the electricity used in U.S. industry drives electric motors, most of which are run at full speed with their output throttled to match their task—somewhat like driving with the gas pedal to the floor and the brake engaged. Each year a heavily used electric motor consumes 10 times its purchase cost in electricity—equivalent to using $200,000 worth

Q: Worldwide, how much of the water withdrawn for irrigation is wasted?

of gasoline each year to fuel a $20,000 car. According to Amory Lovins, it would be cost-effective to scrap virtually all such motors and replace them with adjustable-speed drives. The costs would be paid back in 1–3 years, depending on how the motors were used.

Energy can also be saved by switching to high-efficiency lighting. Additionally, computer-controlled energy management systems can turn off lighting and equipment in nonproduction areas and make adjustments during periods of low production. Recycling and reuse and making products that last longer and that are easy to repair and recycle also saves energy compared to using virgin resources.

Industrial energy conservation in the United States still remains at an embryonic stage compared to efforts in countries such as Japan and Germany. The primary reason is that because of low energy costs, energy expenses for most U.S. industries are less than 2% of the value of their manufacturing shipments. Thus, they have little incentive to save energy.

How Can We Save Energy in Producing Electricity? The Negawatt Revolution The policies of U.S. utility companies have a major impact on the economy and on efforts to save energy. With assets exceeding $600 billion and annual sales greater than $210 billion, electric utilities are by far the nation's largest industry—roughly twice the size of the telecommunications industry and almost 30% larger than U.S.-based manufacturers of motor vehicles.

As regulated monopolies, U.S. utilities have had little incentive to save energy. These utilities make more money by increasing the demand for electricity. This process encourages the building of often unnecessary power plants to send electricity to inefficient appliances, heating and cooling systems, and industrial plants. To make more money, many utilities encourage their customers to use (and thus waste) even more electricity—a classic example of harmful positive feedback in action.

A small but growing number of utility companies in the United States and elsewhere are trying to reverse this wasteful process by reducing the demand for electricity. By helping customers use electricity more efficiently, these 20 utility companies can avoid financing and building expensive new power plants.

This approach is known as *demand-side management* or the *negawatt revolution*. To reduce demand, utilities give customers cash rebates for buying efficient lights and appliances, free home-energy audits, low-interest loans for home weatherization or industrial retrofits, and lower rates to households or industries meeting certain energy-efficiency standards. To make demand-side management feasible, state and other utility regulators must allow utility investors to make a reasonable return on their money, based on the amount of energy the utilities save. Such a policy allows utility companies to shift their emphasis from megawatts to energy-saving negawatts.

At their peak in 1993, U.S. utility companies spent nearly $3 billion on demand-side management programs. These efforts save electricity equal to the output of 44 large (1 megawatt) nuclear or coal-fired power plants at a low cost of 2.1¢ per kilowatt-hour, much cheaper than any of the various methods for producing electricity (Figure 15-21). However, since 1993 such programs have been cut back sharply because the U.S. electric power industry is in the process of being restructured and deregulated to help increase competition.

How Can We Save Energy in Transportation? The most important way to save energy (especially oil) and money in transportation is to *increase the fuel efficiency of motor vehicles*. Between 1973 and 1985, the average fuel efficiency doubled for new American cars and rose 37% for all passenger cars on the road because of government-mandated standards. However, since 1985 automakers have succeeded in opposing any increase in the government standards. Actually, the average fuel efficiency of new vehicles has fallen by about 5% in recent years because of the popularity of sport utility vehicles (SUVs), minivans, and light trucks (subject to much lower mileage standards than cars) and to larger, less-efficient autos.

According to the former U.S. Office of Technology Assessment, existing technology could raise current fuel efficiency of the entire U.S. automotive fleet to 15 kilometers per liter (35 miles per gallon) by 2010. Doing this would save U.S. consumers about $65 billion a year (about $576 per household), cause a sharp drop in emissions of CO_2 and other air pollutants, cut oil imports in half (which would significantly lower the annual U.S. trade deficit), and create about 244,000 new jobs throughout the U.S. economy.

Currently, the U.S. market includes more than 15 car models with fuel efficiencies of at least 17 kilometers per liter (40 miles per gallon), but they made up only 5% of U.S. car sales. Since 1985 at least 10 automobile companies have made nimble and peppy prototype cars that meet or exceed current safety and pollution standards, with fuel efficiencies of 29–59 kilometers per liter (67–138 miles per gallon) (Figure 16-5). If such cars were mass-produced, their slightly higher costs would be more than offset by their fuel savings.

The problem is that there is little consumer interest in fuel-efficient cars when the inflation-adjusted price of gasoline today is the lowest it has been since 1920 (Figure 16-6). This underpricing of gasoline encourages energy waste and pollution by failing to include in its market price the many harmful social

A: 70–80% evaporates or seeps into the ground before reaching crops

Figure 16-5 The "Ultralite," prototype car built by General Motors. It is fast and much safer than existing cars. When driven at 81 kph (50 mph), it gets 43 kpl (100 mpg). Replacing its engine with a 50% more efficient diesel engine (already available) would raise overall fuel efficiency to 81 kpl (190 mpg). It has a 218-kph (135-mph) top speed, is equipped with four air bags, and easily carries four adults. Its body is made from lightweight carbon-fiber composite plastics that won't dent, scratch, or corrode. With 200 fewer parts than a conventional engine, it costs about $400 less to make and costs consumers less in maintenance expenses. (General Motors)

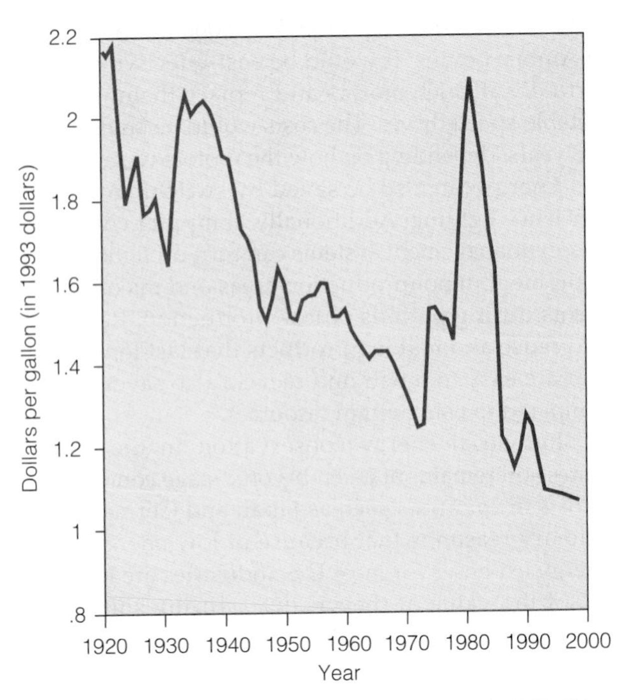

Figure 16-6 Real price of gasoline (in 1993 dollars), 1920–97. When the price is adjusted for inflation, U.S. consumers are paying less for gasoline than at any time since 1920. This underpricing of gasoline encourages energy waste and pollution by failing to include in its market price the many harmful social and environmental costs of gasoline, which consumers ultimately end up paying anyway.

and environmental costs of gasoline, which consumers ultimately end up paying anyway.

To encourage consumers to buy energy-efficient vehicles, Amory Lovins has suggested that a system of revenue-neutral rebates and "freebates" be established for motor vehicles. Buyers of a new vehicle would pay a fee or receive a rebate depending on its fuel efficiency. The fees on inefficient vehicles would be used to pay for the rebates on efficient ones.

Conventional battery-powered *electric cars* might help reduce dependence on oil, especially for urban commuting and short trips. Electric vehicles are extremely quiet, need little maintenance, and can accelerate rapidly. The cars themselves, called zero-emissions vehicles, produce no air pollution. However, using coal and nuclear power plants to produce the electricity needed to recharge their batteries daily does produce air pollution and nuclear wastes—something called *elsewhere pollution*.

Greatly increased manufacture, disposal, and recycling of lead batteries for widespread use of electric vehicles could also greatly increase the input of toxic lead into the environment (Section 22-7). If solar cells or wind turbines could be used for recharging the car's batteries, CO_2 and other air pollution emissions would be virtually eliminated.

On the negative side, today's electric cars are not very efficient; they are equivalent to gasoline-powered cars that get about 7–11 kilometers per liter (16–25 miles per gallon). Current electric cars can travel only 81–161 kilometers (50–100 miles) vs. 480–640 kilometers (300–400 miles) for gasoline cars. Their batteries must be recharged for 3–8 hours (although a new device may reduce recharge time to 10–20 minutes at a cost of $8–15 per recharge) and must be replaced about every 48,000 kilometers (30,000 miles) at a cost of at least $2,000. This requirement, plus the electricity costs for daily recharging and the cost of a recharger ($700–3,500), means that today's electric cars have twice the operating cost of gasoline-powered cars.

Drivers could get around the charging problem by exchanging a used battery pack for a charged one at a service station. Israeli engineers have developed an exchangeable zinc–air battery with a range comparable to that of conventional gasoline-powered cars. However, such batteries, which are being tested in Europe, are available only for buses and large vans. A growing number of analysts believe that the future lies in the development of *ecocars* (Solutions, right).

Another way to save energy is to *shift to more energy-efficient ways to move people* (Figure 16-7) *and*

Q: What percentage of the world's population lives in areas that have prolonged droughts?

Ecocars

SOLUTIONS

There is growing interest in developing hybrid electric–internal combustion *ecocars*, or *supercars*, getting 64–128+ kilometers per liter (150–300+ miles per gallon). They get most of their power from a small hybrid electric/fuel engine that makes its own electricity and uses a small battery or flywheel to provide the extra energy needed for acceleration and hill climbing.

The small gasoline engine keeps the batteries charged, thus reducing the number of batteries needed and greatly increasing the range. If needed, the electric engine could also operate emission-free in urban areas. Canada's HydroQuebec has developed a hybrid car system that places small electric motors in all four of a car's wheels.

In 1997 Toyota developed a hybrid electric/gasoline vehicle that is twice as fuel efficient as an equivalent gasoline-powered car, while significantly reducing air pollution emissions. This car, called the Prius, is on the market in Japan and gets 28 miles per liter (66 miles per gallon). In 1999, Honda began selling a fuel-efficient, hybrid-engine car.

A major limitation of gaseous fuels such as natural gas and hydrogen is that they require fairly heavy and bulky fuel tanks, which reduce efficiency and range. However, a hybrid car might be able to overcome this problem by going 700 kilometers (450 miles) on just 5.6–11 liters (1.5–3 gallons) of gas.

Such supercars will be lighter and safer than conventional cars because their bodies and many other parts will be made of lightweight composite materials (Figure 16-5) that absorb much more crash energy per unit of weight than the steel in today's cars. Today most race car bodies are made of carbon-fiber plastics, which is why race car drivers can usually walk away after hitting a wall at 320 kilometers (200 mph).

Car bodies and other parts made from composite materials don't rust and can be recycled. They also don't need to be painted (currently the most costly, slowest, and polluting step in car manufacture) because the desired color can be added to the molds that shape the composite materials.

Manufactured mostly by a fundamentally different process (molding), an ecocar's composite body would take twice as many worker-hours to make as its steel-bodied ancestors, yet cost a little less to make because of much lower tooling and equipment costs. Thus, switching to such ecocars would create more jobs. Switching to hybrid ecocars with composite bodies and far fewer parts would also sharply reduce this use of minerals and the resulting pollution and environmental degradation. In 1998 Chrysler was close to perfecting a plastic car body using a cheap, common beverage-bottle plastic called PET (for polyethylene terephthalate) that could halve the price of a car and be on the market between 2005 and 2010.

Amory Lovins (Guest Essay, p. 433) and other researchers project that with proper financial incentives, such low-polluting, ultralight, ultrasafe, ultraefficient hybrid electric/internal combustion engine cars could be on the market within a decade. With incentives such as tax credits or energy rebates, such cars could replace much of the existing car fleet within 10–12 years (the average road life of existing cars).

Lovins argues that such cars would sell "not just because they are more than 5 times as efficient and more than 100 times cleaner than current cars, but also because they will be markedly superior in comfort, quietness, performance, durability, beauty, and safety."

Another environmentally appealing type of ecocar is an electric vehicle that uses fuel cells powered by solar-produced hydrogen (Section 16-7). Fuel cells consist of two electrodes immersed in a solution (electrolyte) that conducts electricity. They produce electricity by combining hydrogen and oxygen ions, typically from hydrogen and oxygen gas supplied as fuel for the cell (Figure 16-3) or from gasoline.

In 1997, Toyota, which is spending $800 million a year on alternative-fuel research, built a prototype sport utility vehicle retrofitted with a hydrogen-based fuel-cell system. Ford plans to build such a car by 2000. In 1997 researchers announced the development of a new type of fuel-cell system that burns hydrogen produced from gasoline (or from other fuels such as methane or ethanol). Such cars, which would get about twice the gas mileage of a comparable gasoline-powered car and produce 90% less pollution, could be on the road by 2010.

Fuel cells are compact and safe, make no noise, and require little maintenance. They are also very efficient, converting 50% or more of the fuel energy to power, compared to 10–14% efficiency for gasoline-powered vehicles. More important, fuel-cell cars running on hydrogen come close to being true zero-emission vehicles because they emit only water vapor and trace amounts of nitrogen oxides, easily controlled with existing technology.

Critical Thinking

Do you believe that it is economically and politically important to build and phase in ecocars over the next 20 years? Explain. Can you think of any drawbacks from doing this?

A: Almost 40%

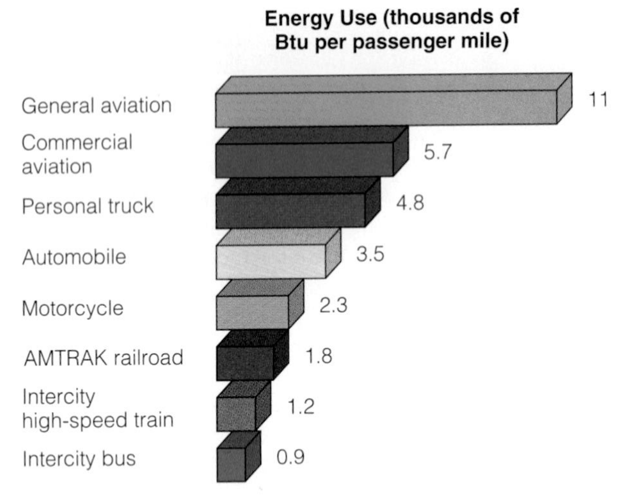

Figure 16-7 Energy use of various types of domestic passenger transportation: the lower the energy use, the greater the efficiency.

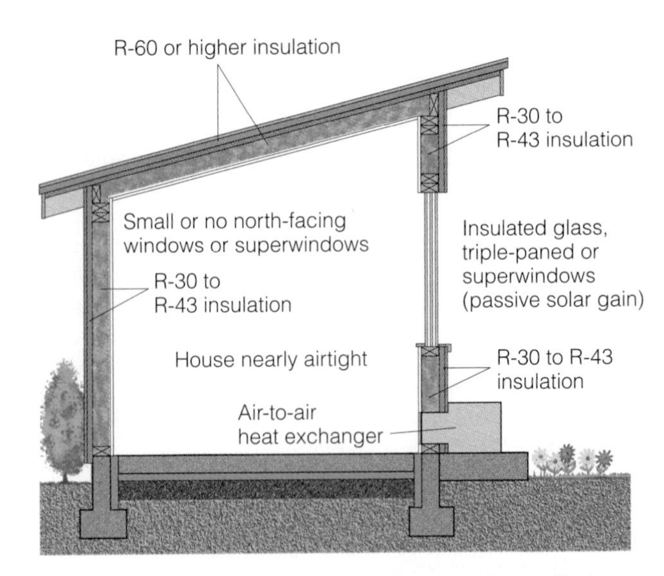

Figure 16-8 Major features of a superinsulated house. Such a house is heavily insulated and nearly airtight. Heat from direct solar gain, appliances, and human bodies warms the house, which requires little or no auxiliary heating. An air-to-air heat exchanger prevents buildup of indoor air pollution.

freight. In the United States trucking accounts for two-thirds of the energy used to ship freight in the United States. However, trucks are a far more polluting and less fuel-efficient way to move freight than pipelines, barges, trains, and planes. More freight could be shifted from trucks and planes to more energy-efficient trains and ships.

The fuel efficiency of other means of transport can also be increased. With improved aerodynamic design, turbocharged diesel engines, and radial tires, new transport trucks can be 50% more fuel efficient than today's conventional trucks. Boeing's new 777 jet uses about half the fuel per passenger seat as a 727. Existing advanced diesel engine technology could improve the fuel efficiency of ships by 30–40% over the next few decades.

How Can We Save Energy in Buildings?

Heating, cooling, and lighting buildings consume about one-third of the energy used by modern societies, with much of this energy unnecessarily wasted. The 110-story, twin-towered World Trade Center in Manhattan is a monument to energy waste: It uses as much electricity as a city of 100,000 people for only about 53,000 employees.

In contrast, Atlanta's 13-story Georgia Power Company building uses 60% less energy than conventional office buildings of the same size. The largest surface of the building faces south to capture solar energy. Each floor extends out over the one below it, blocking out the higher summer sun to reduce air conditioning costs but allowing warming by the lower winter sun. Energy-efficient lights focus on desks rather than illuminating entire rooms. The Georgia Power model and other existing cost-effective commercial building technologies could reduce energy use by 75% in U.S. buildings, cut carbon dioxide emissions in half, and save more than $130 billion per year in energy bills by 2010.

In 1987, the Internationale Nederlanden (ING) Bank in Amsterdam built a new headquarters that uses 90% less energy than its predecessor. This new building saves the bank $2.4 million per year in energy costs.

There are a number of ways to improve the energy efficiency of buildings, many of them discussed in the opening of this chapter (p. 396). One is to build more *superinsulated houses* (Figure 16-8). Such houses are so heavily insulated and so airtight that heat from direct sunlight, appliances, and human bodies warms them, with little or no need for a backup heating system. An air-to-air heat exchanger prevents buildup of indoor air pollution without wasting much heat. Although such houses typically cost 5% more to build than conventional houses of the same size, this extra cost is paid back by energy savings within 5 years and can save a homeowner $50,000–100,000 over a 40-year period. More than 100,000 superinsulated houses have been built in Scandinavia, Canada, and the northern United States.

Since the mid-1980s there has been growing interest in building long-lasting, affordable, and easily constructed superinsulated houses called *strawbale houses* with walls consisting of compacted bales of

Q: What percentage of the world's population clashes over rights to water?

Figure 16-9 An energy-efficient, environmentally healthy, and affordable Victorian-style strawbale house designed and built by Alison Gannett (Guest Essay, p. 407) in Crested Butte, Colorado. The left photo shows construction and the right photo is the completed house. Depending on the thickness of the bales, plastered strawbale walls have an insulating value of R-35 to R-60, compared to R-12 to R-19 in a conventional house. (The R value is a measure of resistance to heat flow.) (Alison Gannett)

certain types of straw (available at a low cost almost everywhere) covered with plaster or adobe (Figure 16-9).* By 1997 there were more than 1,000 such homes built or under construction in the United States (Guest Essay, p. 407). Using straw, an *annually* renewable agricultural residue often burned as a waste product, for the walls reduces the need for wood and thus slows deforestation. There is little ecological impact on land because the straw comes from land already converted to pasture or cropland. The main problem is getting banks and other money lenders to recognize the potential of this type of housing and provide homeowners with construction loans.

The Real Goods of Ukiah, California, a company that sells sustainable technology products, has built a demonstration Solar Living Center† that is one of the world's largest strawbale buildings. A combination of superinsulation from the strawbales, roof overhangs, a central fountain with a drip ring for evaporative cooling, and solar-powered evaporative coolers provide cooling during the area's 100°+ summer days. A south-facing glass wall provides all of the building's daytime lighting and most of its heat during cold weather. The building is hooked up to the power grid for backup purposes but the company sells more energy to the utility company than it buys.

Another way to save energy is to *use the most energy-efficient ways to heat houses* (Figure 16-10). The most energy-efficient ways to heat space are to build a superinsulated house, use passive solar heating, and use high-efficiency (85–98% efficient) natural gas furnaces. The most wasteful and expensive way is to use electric resistance heating with the electricity produced by a coal-fired or nuclear power plant.

Heat pumps can save energy and money for space heating in warm climates (where they aren't needed much), but not in cold climates; at low temperatures they automatically switch to wasteful, costly electric resistance heating (which is why utilities wanting to sell more electricity advocate their use). Some heat pumps in their air conditioning mode are also much less efficient than many individual air conditioning units. Most heat pumps also require expensive repair every few years. However, manufacturers have developed some improved models that produce warmer air and are more efficient than older models.

The energy efficiency of existing houses can be improved significantly by adding insulation, plugging leaks, and installing energy-saving windows. About one-third of heated air in U.S. homes and buildings escapes through closed windows and holes and cracks (Figure 3-1)—equal to the energy in all the oil flowing through the Alaska pipeline every year. During hot weather these windows also let heat in, increasing the

*For information on strawbale houses see Steen et al., *The Straw Bale House* (White River Junction, Vt.: Chelsea Green, 1994); and GreenFire Institute, 1509 Queen Anne Ave., North #606, Seattle, WA 98109, (206) 284-7470.

†For more information contact The Solar Living Center, 13771 South Highway 101, Hopland, CA 95449, (707) 744-2100.

A: About 40%

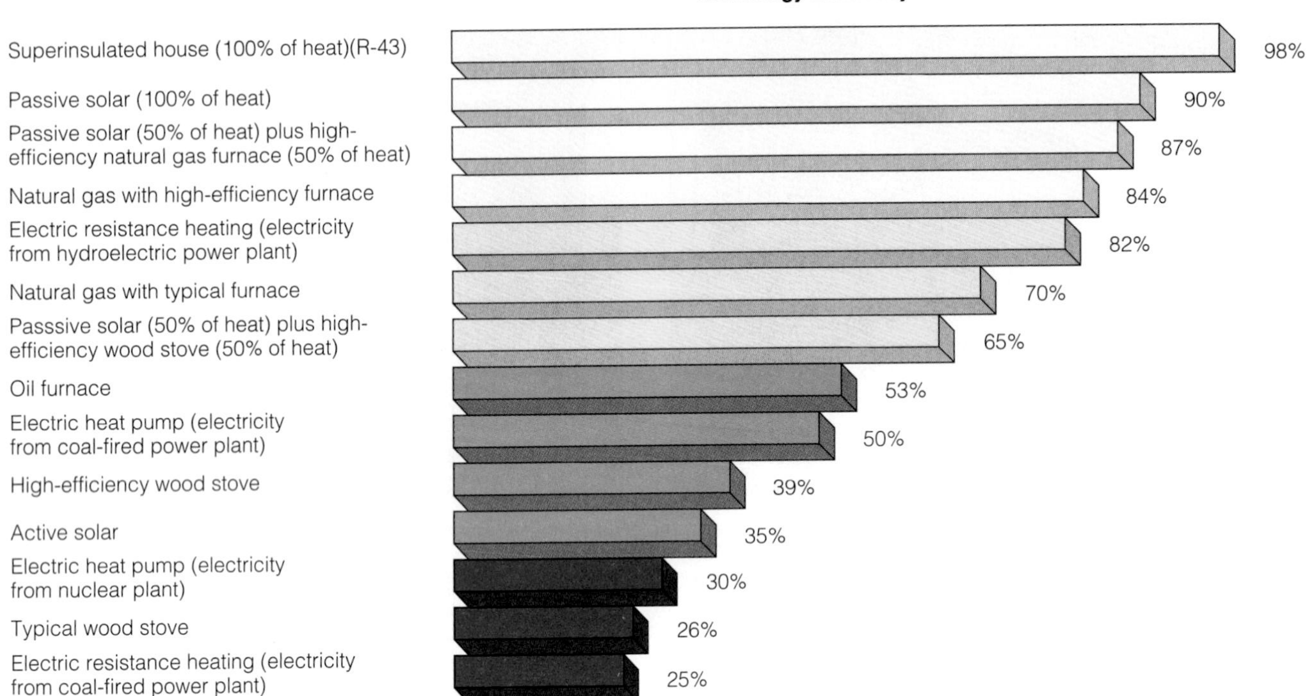

Figure 16-10 Net energy efficiencies for various ways to heat an enclosed space such as a house.

use of air conditioning. Caulking and weatherstripping leaky doors and windows costs about $40–50 for the average house, but the energy savings can amount to 10% a year.

We can also *use the most energy-efficient ways to heat household water*. An efficient method is to use tankless instant water heaters (about the size of bookcase loudspeakers) fired by natural gas or liquefied petroleum gas (LPG). These devices, widely used in many parts of Europe, heat the water instantly as it flows through a small burner chamber and provide hot water only when (and as long as) it is needed. A well-insulated, conventional natural gas or LPG water heater is also fairly efficient (although all conventional natural gas and electric resistance heaters waste energy by keeping a large tank of water hot all day and night and can run out after a long shower or two).

The most inefficient and expensive way to heat water for washing and bathing is to use electricity produced by any type of power plant. A $425 electric water heater can cost $5,900 in energy over its 15-year life, compared to about $2,900 for a comparable natural gas water heater over the same period.

Setting higher energy-efficiency standards for new buildings would also save energy. Building codes could require that all new houses use 60–80% less energy than conventional houses of the same size, as has been done in Davis, California (p. 720). Because of tough energy-efficiency standards, the average Swedish home consumes about one-third as much energy as the average American home of the same size.

Another way to save energy is to *buy the most energy-efficient appliances and lights* (Figure 16-11). Federal energy efficiency standards set for more than 20 appliances—including a tripling of refrigerator efficiency—since 1978 have saved over $15 billion in energy costs in the United States and are expected to save U.S. consumers some $56 billion between 1990 and 2015. Developing these standards has cost the government only $50 million, yielding a benefit-to-cost ratio for the government's investment of 2,600 to 1. For example, the new standard for refrigerator/freezers to take effect in 2001 will make new refrigerators 30% more efficient than today's models and save consumers $40 billion.

The good news is that replacing a standard incandescent bulb with an energy-efficient compact fluorescent bulb that uses only one-fourth as much energy as the standard bulb to achieve the same illumination saves about $48–70 per bulb over its 10-year life and

Q: What percentage of the drinking water in the United States is withdrawn from groundwater?

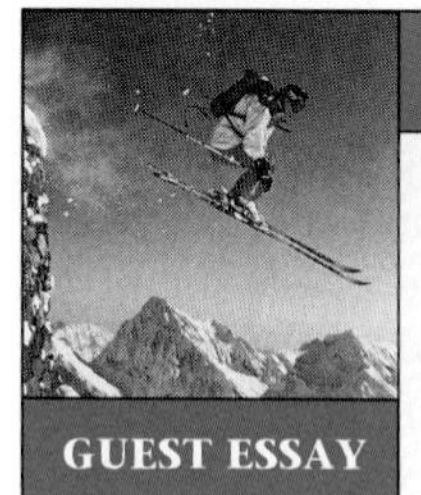

GUEST ESSAY

Building a Strawbale Home

Alison Gannett

Alison Gannett is a professional extreme skier residing in Crested Butte, Colorado. She has won many titles, including the 1998 Canadian World Extreme Championships and the South American Extreme Championships. When not traveling around the world, she runs Sunseekers Design, where she consults and designs solar, strawbale, and green building techniques. She is one of the pioneers in non-load-bearing strawbale construction. In the future, she hopes to help people in foreign countries build strawbale homes.

What is it about a strawbale home? I knew I was going to build one the first time I saw a picture of one. I was a student at the Solar Energy International in Carbondale, Colorado. Little did I know that years later I would have my own business, designing and consulting for solar, green, and strawbale homes.

I now have designed and general-contracted my own strawbale home in Crested Butte, Colorado (Figure 16-9). Our town is a National Historic District, which means that homes must conform to a strict rulebook that outlines the styles of the homes at the turn of the century.

My goal was to have the first historic/Victorian-style strawbale home built in Crested Butte. One of my other goals was to show that a strawbale home could be attractive to the average consumer or builder. Many strawbales are either very simple one-room cabins or are outlandishly weird. Another goal was to use high-quality appliances and green building materials while keeping the cost of the home low.

Major goals for my home and business include:

- Afford to build and live and work in town, not commute (or buy a new car).
- Show that building green is not necessarily more expensive and can be truly affordable. My home uses almost no formaldehyde (used in most building products, especially plywood, carpeting, flooring). I used nontoxic paints and stains and nontoxic glues, cotton, and straw insulation, etc.
- Demonstrate that an affordable home does not have to be "cheaply built" or use nondurable, low-quality materials. I used hydronic in-floor heat, R-6 top-quality superinsulating windows (three times as efficient as normal double-pane windows), and super-efficient appliances (Danish refrigerator, Swedish washer, British dishwasher, compact warm lighting fixtures and bulbs, custom cabinets, and exposed post-and-beam support system).
- Use local materials whenever possible, including local standing deadwood, straw, and clay adobe plaster inside and out.
- Use local labor to build custom cabinets, post and beam materials, countertops, and organically dyed concrete floors and thus support the local economy.
- Provide a food-growing area. The Victorian entry way doubles as an area to grow herbs, spices, lettuce, and tomatoes for summer and winter because the summer growing season is short at an elevation of 2,700 meters (9,000 feet).
- Demonstrate that passive solar design can be inconspicuous. Most people equate solar with ugly. Most people do not realize that my Victorian-style home is a solar home.

Here are some of the hard facts about my home:

- Cost including everything except land: Approximately \$6 per square meter (\$65 per square foot)—about half the typical cost in my area. I saved about 93¢ per square meter (approximately 10¢ per square foot) by acting as my own general contractor.
- Size: Approximately 112 square meters (1,200 square feet inside) and 130 square meters (1,400 square feet) outside. The house has two bedrooms and a loft, an open kitchen/living/dining great room on the second floor (because of snowfall covering downstairs windows, heat rising, and great views).
- Construction began on June 12, 1997, and was completed October 31, 1997 (4.5 months). Rapid construction time was due to several design features requiring fewer steps and the fact that I trained most of the subcontractors.

Critical Thinking

1. Why aren't more strawbale houses being built?
2. Would you build and live in a strawbale house? Explain.

saves enough electricity to avoid burning 180 kilograms (400 pounds) of coal. Thus, replacing 25 incandescent bulbs in a house with energy-efficient fluorescent bulbs saves \$1,250–1,750. These new bulbs cost much more than conventional bulbs but last up to 10 times longer, and pay their cost back in 2–4 years. The bad news is that by 1997 only 9% of U.S. homes used these bulbs.

Under the EPA's voluntary Green Lights Program, corporations signing an agreement with the agency to adopt more energy-efficient lighting retrofits receive free technical assistance. By 1996, the more than 2,500

A: About 50% (96% in rural areas and 20% in urban areas)

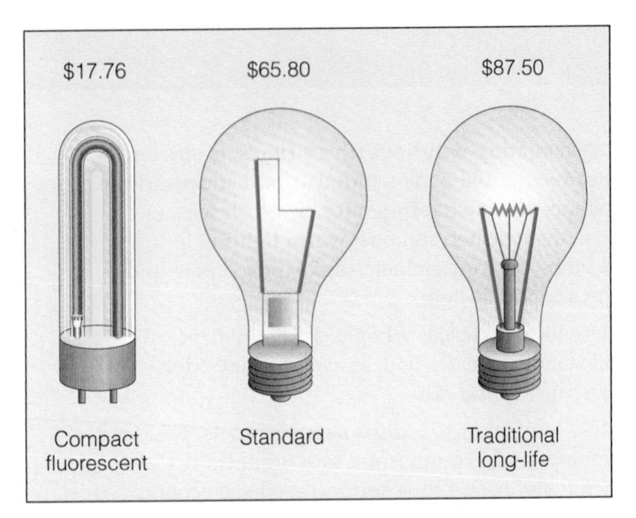

Figure 16-11 Cost of electricity for comparable light bulbs used for 10,000 hours. Because conventional incandescent bulbs are only 5% efficient and last only 750–1,500 hours, they waste enormous amounts of energy and money and they add to the heat load of houses during hot weather. Socket-type fluorescent lights use one-fourth as much electricity as conventional bulbs. Although these new bulbs cost about $10–20 per bulb, they last 10–20 times longer than conventional incandescent bulbs used the same number of hours, and they save considerable money (compared with incandescent bulbs) over their long life. In 1998 compact fluorescent bulbs that can be dimmed went on sale. (Data from Electric Power Research Institute)

participating companies had saved $250 million in electricity costs.

Energy-efficient lighting could save U.S. businesses $15–20 billion per year in electricity bills. Students in Brown University's environmental studies program showed that the school could save more than $40,000 per year just by replacing the incandescent light bulbs in exit signs with fluorescents.

If the most energy-efficient appliances and lights now available were installed in all U.S. homes over the next 20 years, the savings in energy would equal the estimated energy content of Alaska's entire North Slope oil fields. If all U.S. households used the most efficient frost-free refrigerator now available, 18 large (1,000 megawatt) power plants could close. Microwave ovens can cut electricity use for cooking by 25–50% (but not if used for defrosting food). Clothes dryers with moisture sensors cut energy use by 15%, and front-loading washers use 50% less energy than top-loading models yet cost no more. New microwave clothes dryers use 15% less energy than conventional electric dryers and 28% less energy than gas units while virtually eliminating wrinkles.

Improvements in energy efficiency could be encouraged by giving rebates or tax credits for building energy-efficient buildings, for improving the energy efficiency of existing buildings, and for buying high-efficiency appliances and equipment. Lack of information is a major reason why consumers have not taken full advantage of existing energy-saving products. Government, utilities, and industries acting together can lower the costs of getting energy-savings information to consumers.

It is encouraging that average American energy use per household fell 19% between 1973 and 1991. According to energy experts, however, residential and commercial energy use in the United States could be cut 25–50% using existing technology and save consumers hundreds of billions of dollars—a win–win situation for consumers and the environment.

For example, recently Pacific Gas & Electric built an ordinary-looking tract house in Davis, California (p. 720), that has no furnace and no air conditioning (despite summer temperatures of up to 113°F). It uses 67% less energy than comparable houses in the area, saving $490 annually. Its designers estimate that if the house were built in the same quantity as other comparable tract houses, it would cost $1,800 *less* than they do. In 1997, the U.S. Department of Energy began sponsoring part of a Building America program with the goal of reducing the energy consumption of new homes by 50% without adding *anything* to the purchase price for a comparable home.

Greener, more energy-efficient computers, copiers, and other forms of office equipment can also reduce energy use and pollution and save money. In the United States high-tech office equipment—most of it energy-inefficient—costs the business community $2 billion annually. In the process of providing electricity for the country's huge fleet of office machines, power plants emit as much carbon dioxide into the atmosphere as nearly 14 million automobiles.

16-3 DIRECT USE OF SOLAR ENERGY FOR HEAT AND ELECTRICITY

What Are the Advantages of Solar Energy? The Coming Renewable-Energy Age About 92% of the known reserves and potentially available energy resources in the United States are renewable energy from the sun, wind, flowing water, biomass, and earth's internal heat. The other 8% of potentially available domestic energy resources are coal (5%), oil (2.5%), and uranium (0.5%).

Developing these mostly untapped renewable energy resources could meet 50–80% of projected U.S. energy needs by 2040 or sooner, and it could meet virtually all energy needs if coupled with improvements in energy efficiency. In 1994 Shell International Petro-

Flavin, Christopher, and Molly O' Meara. 1998. "Solar Power Markets Boom." *World Watch*, vol. 11, no. 5, p23(5).

leum in London, which forecast the steep rise in oil prices in the 1970s (Figure 15-7), projected that renewable energy, especially solar, will account for 50% of world energy production by 2050.

In 1995, two major U.S. corporations, Bectel Enterprises (once a leading builder of nuclear power plants) and PacifiCorp (a giant utility that operates several large coal-burning power plants in the northwestern United States) announced that they were teaming up to invest in solar energy systems for the 21st century. Their new joint venture, called Energy Works, will develop projects around the world based on wind turbines, biomass generators, improved industrial energy efficiency, and other solar technologies.

Developing renewable energy resources would save money, create two to five times more jobs per unit of electricity produced than coal and nuclear power plants, eliminate the need for oil imports, cause much less pollution and environmental damage per unit of energy used, and increase military, economic, and environmental security. In the United States, wind farms and power plants using geothermal, wood (biomass), hydropower, and high-temperature solar energy (with natural gas backup) can already produce electricity more cheaply than can new nuclear power plants, and with far fewer federal subsidies (Figure 15-21).

How Can Solar Energy Be Used to Heat Houses and Water? Buildings and water can be heated by solar energy using two methods: passive and active (Figure 16-12). A **passive solar heating system** captures sunlight directly within a structure (Figures 16-1 and 16-13) and converts it into low-temperature heat for space heating. Energy-efficient windows, greenhouses, and sunspaces face the sun to collect solar energy by direct gain.

Thermal mass (heat-storing capacity), in the forms of walls and floors of concrete, adobe, brick, stone, salt-treated timber, walls made of used tires packed with dirt, and water in 55-gallon drums, stores much of the collected solar energy as heat and releases it slowly throughout the day and night. Buildup of moisture and indoor air pollutants is minimized by an air-to-air heat exchanger, which supplies fresh air without much heat loss or gain. A small backup heating system may be used but is not necessary in many climates. It's important to get the ratio of sun-capturing glass to thermal mass right because a house with too much sun-facing glass will overheat.

Engineer and builder Michael Sykes has designed a solar envelope house that is heated and cooled passively by solar energy and the slow storage and release of energy by massive timbers and the earth beneath the house (Figure 16-14). The front and back of this house are double walls of heavy timber impregnated with salt to increase the ability of the wood to store heat. The space between these two walls, plus the basement, forms a convection loop or envelope around the inner shell of the house. In summer, roof vents release heated air from the convection loop throughout the day; at night, these roof vents, with the aid of a fan, draw air into the loop, passively cooling the house. The interior temperature of the house typically stays within 2° of 21°C (70°F) year-round

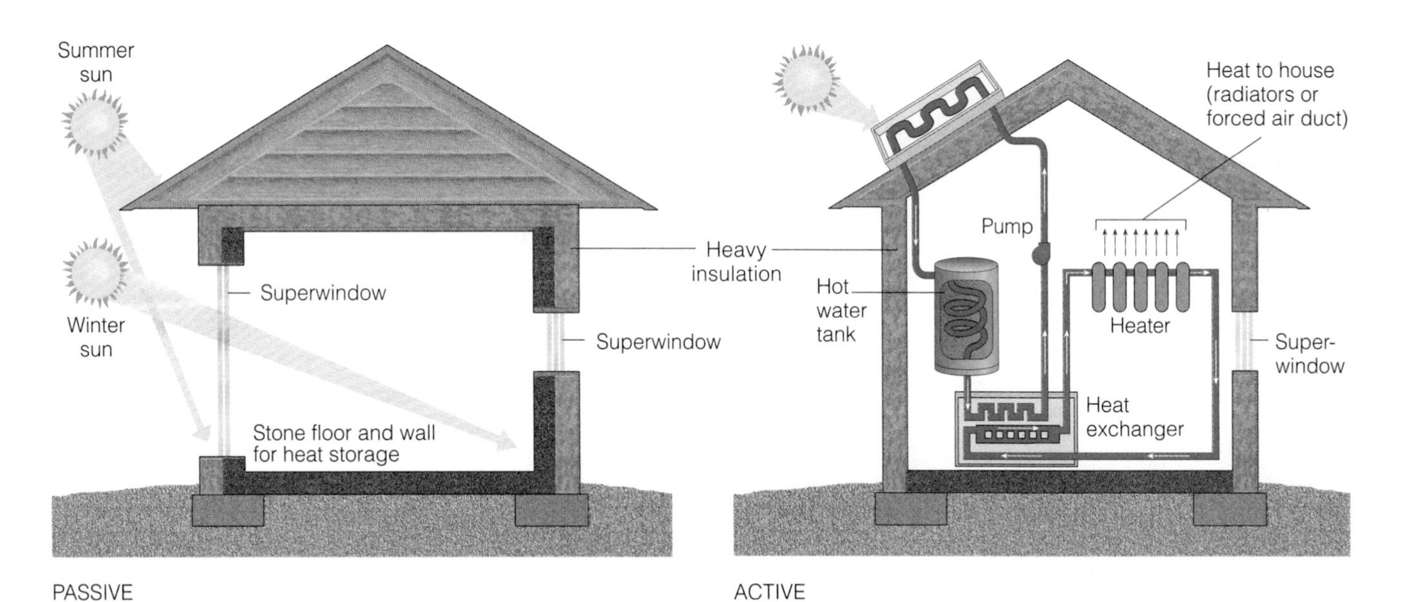

Figure 16-12 Passive and active solar heating for a home.

Hint: Enter the search term *solar energy* using the Subject Guide.

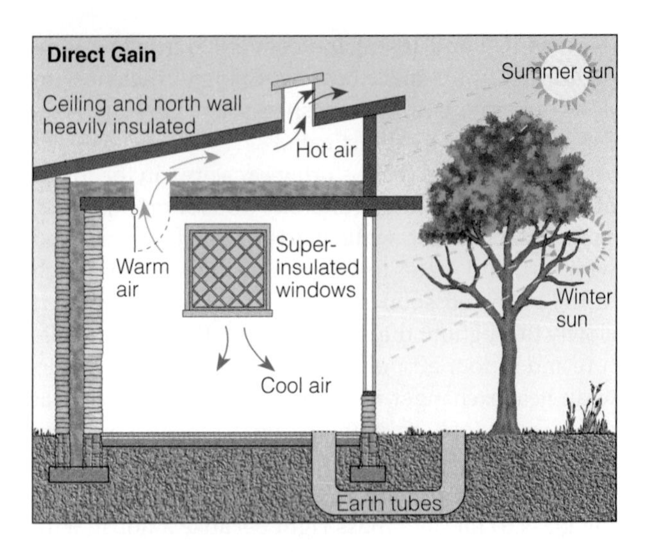

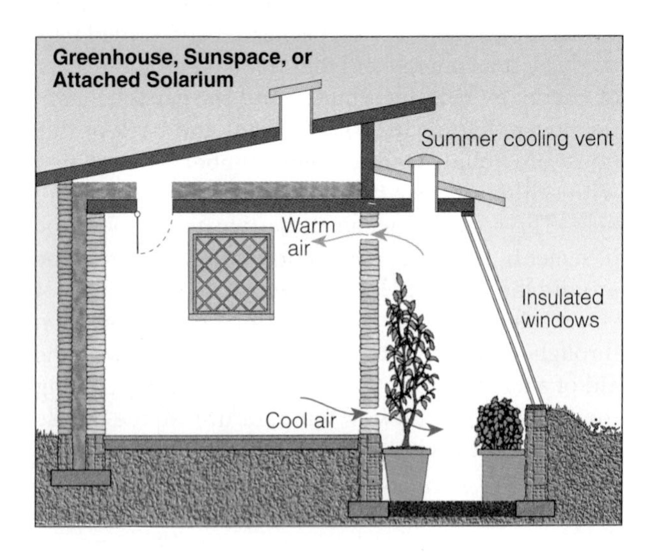

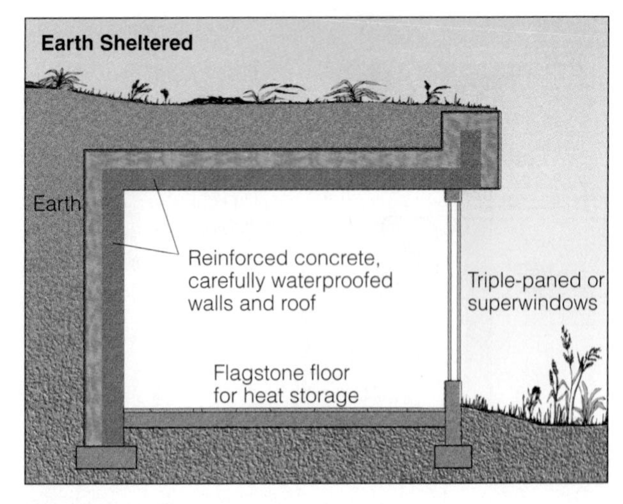

Figure 16-13 Three examples of passive solar design for houses.

without any conventional cooling or heating system. In cold or cloudy climates, a small wood stove or vented natural gas heater in the basement can be used as a backup to heat the air in the convection loop.

With available and developing technologies, passive solar designs can provide at least 70% of a *residential* building's heating needs and at least 60% of its cooling needs, and up to 60% of a *commercial* building's energy needs. In North America there are an estimated 250,000 fully passive solar homes and more than 1 million buildings that include some aspect of passive solar design. Roof-mounted passive solar water heaters, such as the thermosiphoning Copper Cricket, can supply all or most of the hot water for a typical house.

On a life cycle cost basis, good passive solar and superinsulated design is the cheapest way to heat a home or a small building in regions where ample sunlight is available more than 60% of daylight hours (Figure 16-15). Such a system usually adds 5–10% to the construction cost, but the life cycle cost of operating such a house is 30–40% lower. The typical payback time for passive solar features is 3–7 years. In 1994 the Esperanza del Sol subdivision of low-income energy-efficient and solar-oriented houses was built in Texas. The houses add only $13 per year to mortgage payments but save residents $480 per year in heating and cooling costs.

In an **active solar heating system**, specially designed collectors absorb solar energy and a fan or a pump supplies part of a building's space-heating or water-heating needs (Figure 16-12, right). Several connected collectors are usually mounted on a portion of the roof with an unobstructed exposure to the sun. Some of the heat can be used directly, and the rest can be stored in insulated tanks containing rocks, water, or a heat-absorbing chemical for release later as needed.

Active solar collectors can also supply hot water. In Cyprus and Jordan, active solar water heaters supply 25–65% of the hot water for homes. About 12% of houses in Japan, 37% in Australia, and 83% in Israel also use such systems. Currently there are about 1.3 million active solar hot-water systems in U.S. homes. At an average cost of $2,500, active solar water heating systems have been too expensive for most homeowners. However, in 1992 Solar Development in Riviera Beach, Florida, developed a system that costs less than $700. In Colombia, South America, an ecopioneer has spent 30 years building a solar village in an inhospitable wet savanna infested with mosquitoes and with soil so toxic that it could support only tough grass (Individuals Matter, p. 412).

Solar energy for low-temperature heating of buildings, whether collected actively or passively, is

Q: What percentage of the groundwater withdrawn in the United States is not replenished?

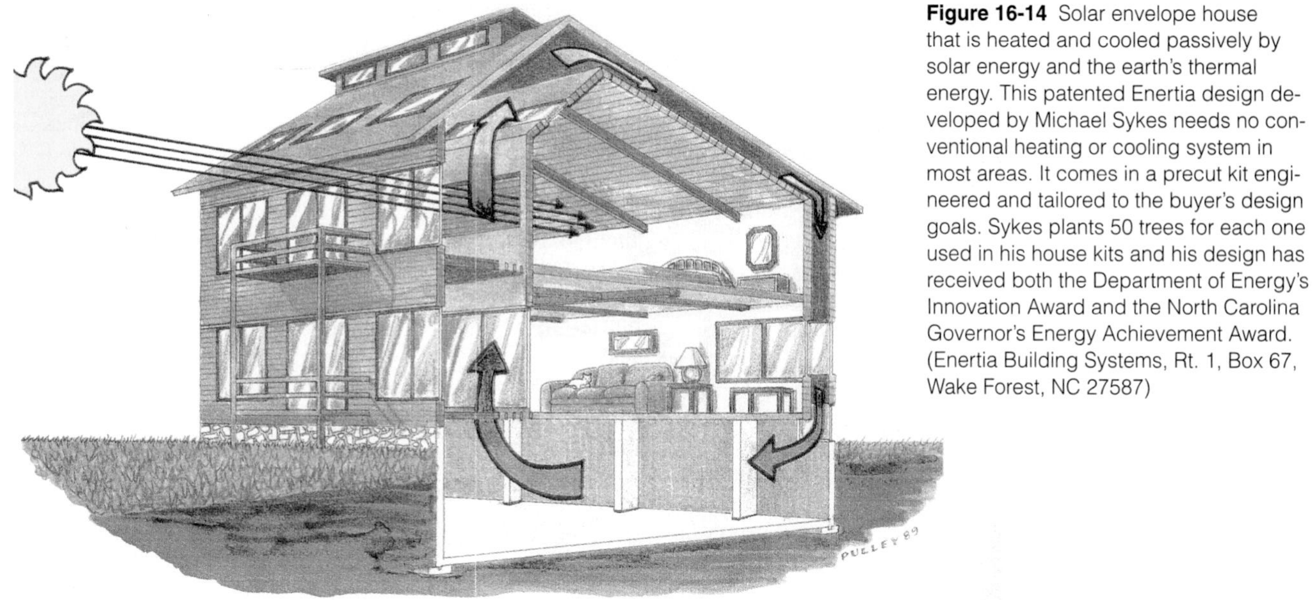

Figure 16-14 Solar envelope house that is heated and cooled passively by solar energy and the earth's thermal energy. This patented Enertia design developed by Michael Sykes needs no conventional heating or cooling system in most areas. It comes in a precut kit engineered and tailored to the buyer's design goals. Sykes plants 50 trees for each one used in his house kits and his design has received both the Department of Energy's Innovation Award and the North Carolina Governor's Energy Achievement Award. (Enertia Building Systems, Rt. 1, Box 67, Wake Forest, NC 27587)

free, and the net energy yield is moderate (active) to high (passive). Both active and passive technology are well developed and can be installed quickly. No heat-trapping carbon dioxide is added to the atmosphere and environmental effects from air and water pollution are low. Land disturbance is also minimal because passive systems are built into structures and active solar collectors are usually placed on rooftops. However, owners of passive and active solar systems need *solar legal rights* laws to prevent others from building structures that block their access to sunlight.

In 1996, modular home builder Avis/America of Avis, Pennsylvania, teamed with Fully Independent Residential Solar Technology in Hopewell, New Jersey, to build manufactured (factory-built) homes that use improved energy efficiency and various combinations of passive and active solar energy (depending on location) to provide most or all of the homes' heating and electricity.

With current technology, active solar systems usually cost too much for heating most homes and small buildings because they use more materials to build, need more maintenance, and eventually deteriorate and must be replaced. In addition, some people consider active solar collectors sitting on rooftops or in yards to be ugly. However, retrofitting existing buildings is often easier with an active solar system.

How Can We Cool Houses Naturally? Superinsulation and superinsulated windows help make buildings cooler. Passive cooling can also be provided by blocking the high summer sun with deciduous trees, window overhangs, or awnings (Figure 16-13,

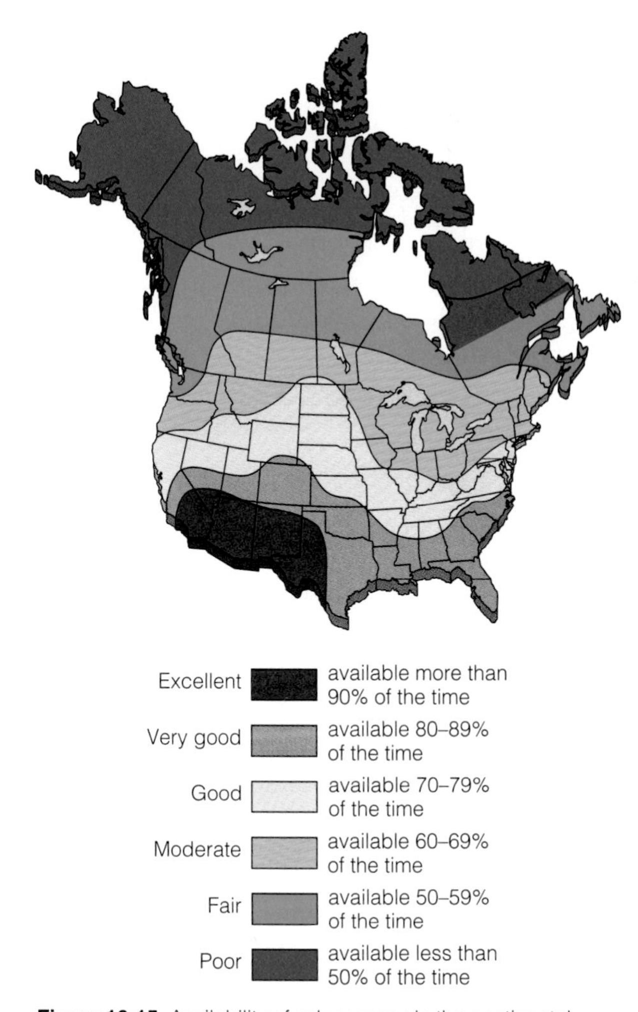

Figure 16-15 Availability of solar energy in the continental United States and Canada. (Data from U.S. Department of Energy and National Wildlife Federation)

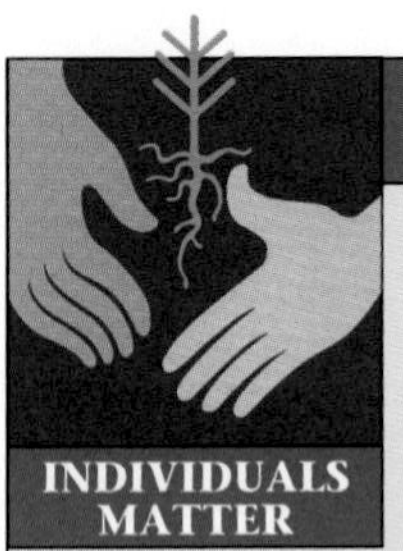

Gaviotas: A Solar Village in Colombia*

Thirty-three years ago ecopioneer and visionary Paolo Lugari began developing an ecologically sustainable solar village called Gaviotas in Colombia east of the Andes.

When he first saw this area in 1965 he thought that if people can live in this inhospitable area, they can live anywhere. With the help of professors and students from the universities in Bogota—16 hours away by jeep—he set out to build a sustainable ecovillage in this place.

They learned how to make a soil-cement for use in building dams and buildings by mixing 14 parts of the area's toxic soil with 1 part cement. They made water pipes for irrigation and buildings by lining ditches with the soil-cement. They attached water pumps to see-saws powered by kids having a good time.

They invented cheap solar water heaters that could work even on cloudy days and helped finance development of the village by building and selling them to thousands of people in Bogota. They also produced electricity by designing ultralight windmills to catch the area's mild but steady winds.

They built a village hospital heated by the sun and cooled by the wind for the area's several hundred inhabitants. People in rural areas for hundreds of miles around came there for medical care and sent their children to school there.

The villagers found enough nontoxic soil along the riverbanks to grow mangoes, cassava, and cashews. They got protein from fish in the river and cattle that could eat the area's tough grass.

They planted hundreds of pines, which have grown into forests. The villagers tap gum oozing from the pines and use solar energy to distill it into turpentine and a valuable resin used in paints, cosmetics, and medicines. The huge market for these products provided a new industry for Gaviotas.

Dropping needles from the pines built up the soil, which then spouted with hundreds of new kinds of plants. Now a rain forest that once grew in the area is returning from seeds carried by birds and roots emerging from the river edges.

Members of this inspiring ecovillage live in peace without guns and pesticides. They are willing to teach all comers about what they have learned about living more sustainably by tapping into the earth's sun, water, and wind. Their simple and affordable technologies, purposely unpatented, are spreading throughout the world.

These earth-citizens believe that technical solutions to learning how to work with the earth can come from people anywhere who are willing to use their ingenuity and work hard to convert their dreams into reality.

Critical Thinking

Why is it apparently more difficult to develop such sustainable communities in most developed countries? Give three important ways to overcome this problem.

*For more details see Alan Weisman, *Gaviotas: A Village to Reinvent the World* (White River Junction, Vt.: Chelsea Green Publishing, 1988).

top). Landscape architects estimate that a single 21-meter (70-foot) shade tree provides cooling equivalent to that of a 3-ton air conditioner. Studies show that planting or saving trees around some buildings can provide enough shade to cut cooling costs by 30%.

Windows and fans can be used to take advantage of breezes and keep air moving. A reflective insulating foil sheet suspended in the attic can block heat from radiating down into the house. Solar-powered evaporative air conditioners have been developed but so far are too expensive for residential use.

Earth tubes can also be used for indoor cooling (Figure 16-13, top). At a depth of 3–6 meters (10–20 feet), the soil temperature stays at about 5–13°C (41–55°F) all year long in cold northern climates and about 19°C (67°F) in warm southern climates. Several earth tubes, simple plastic (PVC) plumbing pipes with a diameter of 10–15 centimeters (4–6 inches) buried about 0.6 meter (2 feet) apart and 3–6 meters deep, can pipe cool and partially dehumidified air into an energy-efficient house at a cost of a few dollars per summer.* People allergic to pollen and molds should add an air purification system, but this is also necessary with a conventional cooling system.

How Can Solar Energy Be Used to Generate High-Temperature Heat and Electricity? Several so-called *solar thermal* systems collect and transform radiant energy from the sun into high-temperature thermal energy (heat), which can be used directly or converted to electricity (Figure 16-16). In one such *central receiver system*, called a *power tower*, huge arrays of computer-controlled mirrors called *heliostats* track the sun and focus sunlight on a central heat-collection tower (Figure 16-16a). Several govern-

*They work. I used them in a passively heated office for 15 years.

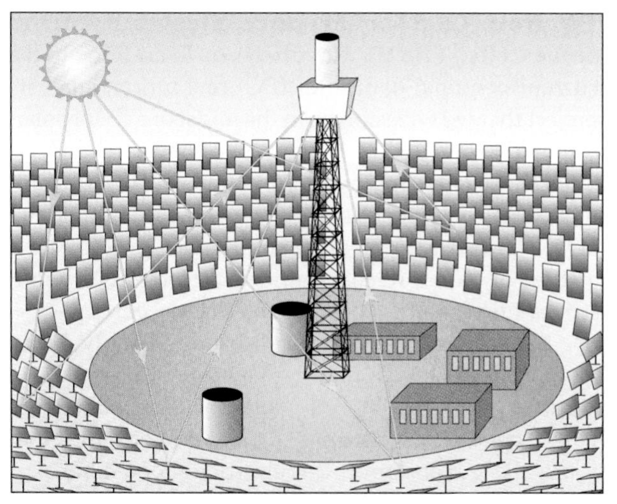

a. Solar Tower Power

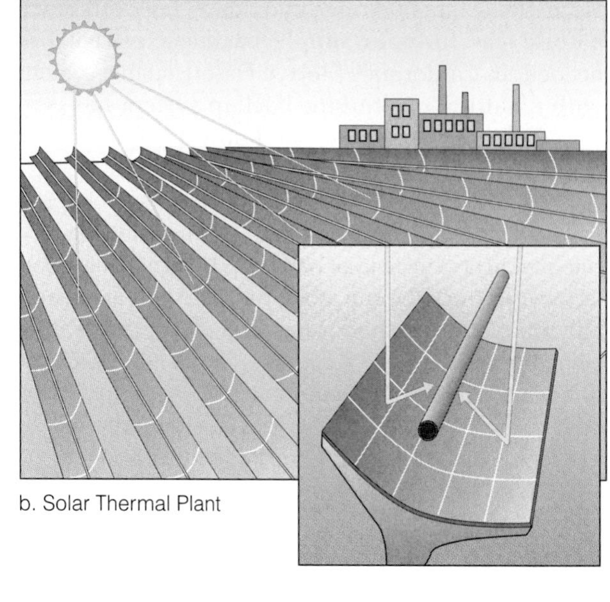

b. Solar Thermal Plant

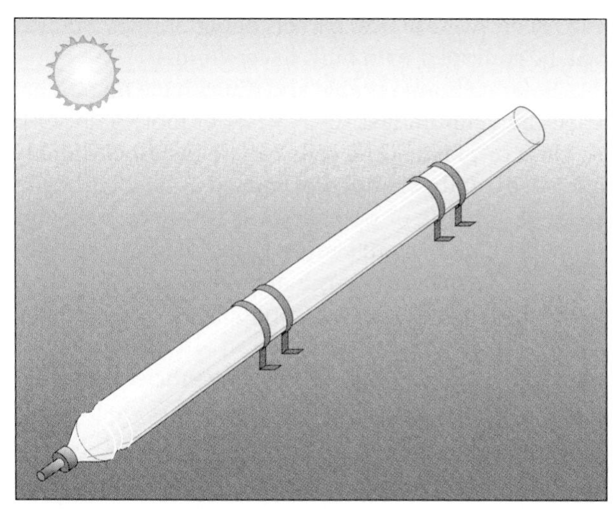

c. Nonimaging Optical Solar Concentrator

d. Solar Cooker

Figure 16-16 Several ways to collect and concentrate solar energy to produce high-temperature heat and electricity are in use. Today, solar plants are used mainly to supply reserve power for daytime peak electricity loads, especially in sunny areas with a high demand for air conditioning.

ment-subsidized experimental power towers are being tested in the United States (at the 10-megawatt Solar Two pilot plant in the desert near Barstow, California), Japan, Italy, France, and Spain.

At California's Solar Two plant, which opened in 1996, 1,900 mirrored heliostats reflect sunlight to the top of a tower. There the intense heat is stored in molten salt (sodium and potassium nitrate). The hot liquid salt then flows into an insulated tank, where it can be used right away or stored (for producing electricity at night or on cloudy days) to heat water in a steam generator to power turbines to produce electricity. The salt then goes to a cooling tank before returning to the tower in a continuous cycle. This experimental plant cost about eight times more to build than a coal-fired plant and produces electricity at about 10¢ per kilowatt-hour, about double that of a coal-fired plant. Supporters say the construction and operating costs will drop as the technology spreads. Moreover, if the DOE-estimated 5.7¢ per kilowatt-hour environmental cost of coal-fired electricity is included, electricity from Solar Two is comparable to that from coal.

In a *solar thermal plant* or *distributed receiver system*, sunlight is collected and focused on oil-filled pipes running through the middle of curved solar collectors (Figure 16-16b). This concentrated sunlight can generate temperatures high enough for industrial processes or for producing steam to run turbines and generate

A: About 0.1%

electricity. At night or on cloudy days, high-efficiency natural gas turbines supply backup electricity as needed. In California's Mojave Desert, such a system with a natural gas turbine backup system has produced power for 8–10¢ per kilowatt-hour—much cheaper than nuclear power plants.

Another type of distributed receiver system uses *parabolic dish collectors* (that look somewhat like TV satellite dishes) instead of parabolic troughs. These collectors can track the sun along two axes and are generally more efficient than troughs. A pilot plant is now being built in northern Australia. The U.S. Department of Energy projects that early in the 21st century parabolic dishes with gas turbine backup should be able to produce electrical power for 6¢ per kilowatt-hour.

Distributed receiver systems entered the commercial market before central receivers because central receivers cost more to operate. However, central receivers have the potential for greater efficiency in producing electricity because they can achieve higher temperatures.

Another promising approach to intensifying solar energy is a *nonimaging optical solar concentrator*. With this technology, the sun's rays are allowed to scramble instead of being focused on a particular point (Figure 16-16c). Experiments show that a nonimaging parabolic concentrator can intensify sunlight striking the earth 80,000 times. Because of its high efficiency and ability to generate extremely high temperatures, nonimaging concentrators may make solar energy practical for widespread industrial and commercial use within a decade.

Inexpensive solar cookers can focus and concentrate sunlight and cook food, especially in rural villages in sunny developing countries. They can be made by fitting an insulated box big enough to hold three or four pots inside with a transparent, removable top (Figure 16-16d). Solar cookers reduce deforestation for fuelwood, the time and labor needed to collect firewood, and indoor air pollution from smoky fires.

The impact of solar thermal power plants on air and water is low. They can be built as large or as small as needed in 1–2 years, compared to 5–7 years for a coal-fired plant and 12–15 years for a nuclear power plant. This saves investors millions in interest on construction loans. Distributed system solar thermal power plants with small turbines burning natural gas as backup can produce electricity at about half the cost of a nuclear power plant (Figure 15-21). According to the Worldwatch Institute, solar power plants installed on a total land area equal to that of Panama or South Carolina could provide as much electricity as the entire world now uses.

How Can We Produce Electricity from Solar Cells? The PV Revolution In addition to the current personal computer (PC) revolution, analysts project that we will soon be in the midst of a PV or solar-cell revolution. Solar energy can be converted directly into electrical energy by **photovoltaic (PV) cells**, commonly called **solar cells** (Figure 16-17). Sunlight falling on a solar cell, a transparent silicon wafer thinner than a sheet of paper, releases a flow of electrons when it strikes silicon atoms, creating an electrical current.

Because a single solar cell produces only a tiny amount of electricity, many cells are wired together in a panel providing 30–100 watts. Several panels are wired together and mounted on a roof or on a rack that tracks the sun and produces electricity for a home or a building. The direct current (DC) electricity produced can be stored in batteries and used directly or converted to conventional alternating current (AC) electricity. Currently, a separate inverter is used to convert the DC to AC, but new panels being built by Solarex will be equipped with built-in converters.

Solar cells are reliable and quiet, have no moving parts, and should last 20–30 years or more if encased in glass or plastic. The cells can be installed quickly and easily and expanded or moved as needed; maintenance consists of occasional washing to keep dirt from blocking the sun's rays. Arrays of cells can be located in deserts and marginal lands, along interstate highways, in yards, and on rooftops.

Expandable banks of cells can also produce electricity at a small power plant, using efficient turbines burning natural gas (Connections, p. 376) to provide backup power when the sun isn't shining. If one or several PV panels fail, this does not threaten production as much as the failure of an entire coal-burning or nuclear power plant. Small solar-cell power plants are also providing power when electricity needs peak. Theoretically, all U.S. electricity needs could be met by placing solar cells on only 0.6% of U.S. land (and on only half this area if solar-cell efficiencies increase as projected from about 12% to 21%).

Solar cells produce no heat-trapping CO_2. Air and water pollution during operation is extremely low, air pollution from manufacture is low, land disturbance is very low for roof-mounted systems, and producing the materials doesn't require strip mining. The net energy yield is fairly high and is increasing with new designs. Whereas solar thermal systems require direct sunlight, PV cells work in cloudy weather.

Traditional-looking solar-cell roof shingles and photovoltaic panels that resemble metal roofs are now available, reducing the cost of solar-cell installations by saving on roof costs. A California company is working with utilities to install prototype solar-cell panels

Q: Worldwide, what percentage of the water withdrawn is unnecessarily wasted?

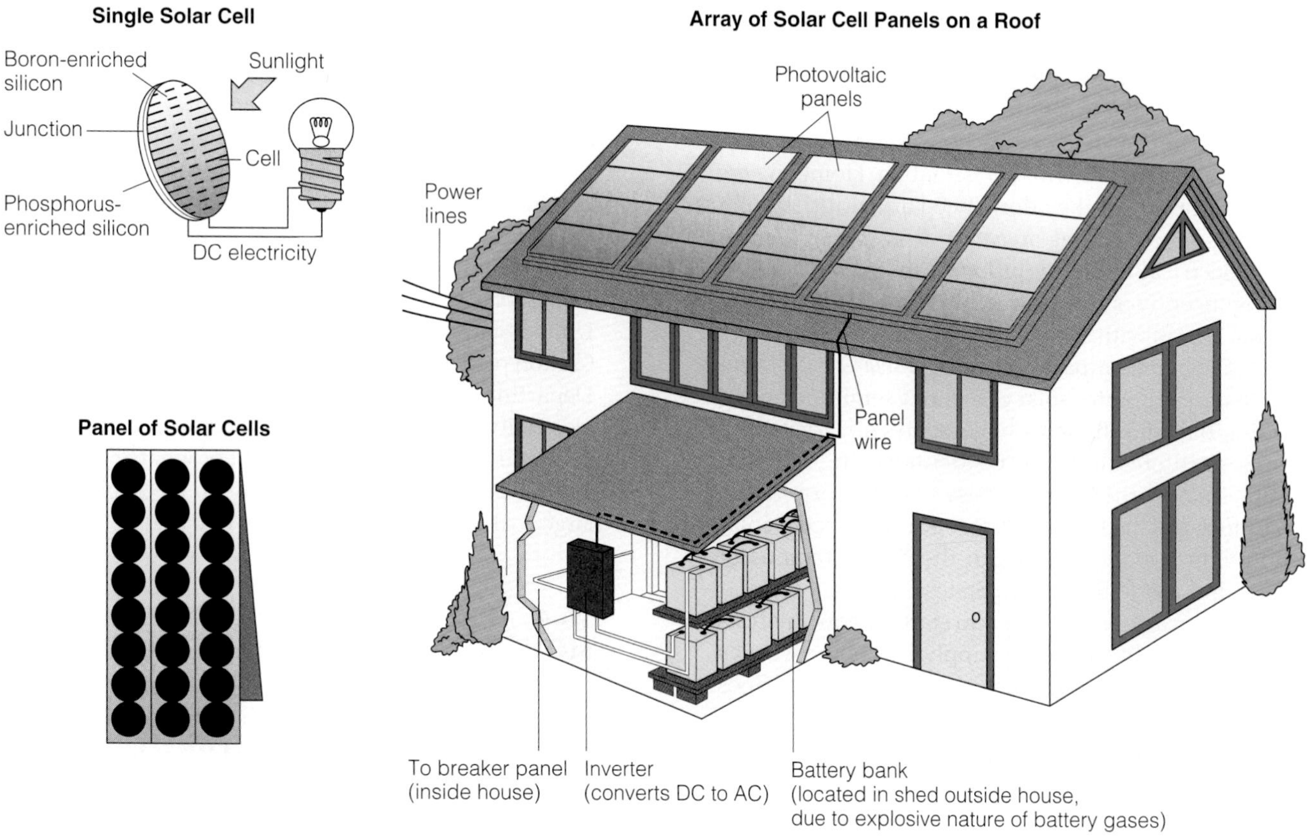

Figure 16-17 Photovoltaic (solar) cells can provide electricity for a house. Small and easily expandable arrays of such cells can provide electricity for urban villages throughout the world without large power plants or power lines. Massive banks of such cells can also produce electricity at a small power plant. Today, at least two dozen U.S. utility companies are using photovoltaic cells in their operations; as the price of such electricity drops, usage will increase dramatically. In 1990 a Florida builder began selling tract houses that get all of their electricity from roof-mounted solar cells. Although the solar-cell systems account for about one-third of the cost of each house, the savings in electric bills will pay this off over the term of a 30-year mortgage. Soon the racks of solar cell panels shown in this diagram will be incorporated in the roof, in the form of solar-cell roof shingles or PV panel roof systems that will look like metal roofs.

directly over the rubber or plastic roofing membranes of flat-roofed apartment and office buildings.

By 2000 to 2005, electricity produced by solar cells could drop from the current average of 30¢ per kilowatt-hour to 6–12¢ per kilowatt-hour, making it cost-competitive with most other technologies (Figure 15-21). In 1994, United Solar Systems announced development of a PV module that can generate power at 10–12¢ per kilowatt hour. Australian scientists have come up with a 21.5% efficient design for solar cells that could lower the cost of solar cells to 6–7¢. They hope to have this technology available between 2001 and 2004. If these and other new solar-cell technologies can be brought to the marketplace, the production, sale, and installation of photovoltaics will be one of the world's largest and fastest-growing businesses. Mass production and technological advances could bring costs down as low as 4¢ per kilowatt hour by 2020, causing a positive feedback loop of even more sales.

Solar cells are an ideal technology for providing electricity to 2 billion people in mostly sunny rural areas in most developing countries, which have not committed to expensive large-scale centralized nuclear and fossil fuel power plants and electric-line transmission systems. The biggest markets are expected to be in India, China, Indonesia, Philippines, Brazil, Mexico, and South Africa. With financing from the World Bank, India is installing solar-cell systems in 38,000 villages and Zimbabwe is bringing solar electricity to 2,500 villages.

By 1998, some 400,000 families worldwide had bought PV systems and this is only the tip of the iceberg as PV prices come down. In the United States there are over 36,000 PV-powered homes, and many more are being built. The municipal utility company in Sacramento, California, is putting shiny blue solar-cell

A: 65–70% (about 50% in the United States)

panels on 100 homes each year. These rooftop systems are connected to the utility's electric grid so that any excess power produced by the homes during the day can be sold to the utility for use by other customers. When the sun goes down, the houses draw electricity off the grid. This is a win–win situation. Homeowners get paid for power they don't need and utilities don't have to spend millions on new power plants to meet peak electrical demand. In Japan the government is giving incentives for contractors to build 70,000 residential homes with solar-cell roof tiles.

A German company is testing a *solar-electric window* that incorporates solar cells into a semitransparent glazing that simultaneously generates electricity and provides filtered light for residents during daylight hours. In 1994, the Houston-based Enron Corporation announced plans to build large, grid-connected solar photovoltaic power plants in the desert regions of China, India, and the United States.

With an aggressive program starting now, analysts project that solar cells could supply 17% of the world's electricity by 2010—as much as nuclear power does today—at a lower cost and much lower risk. With a strong push from governments and private investors, by 2050 solar cells could provide as much as 25% of the world's electricity (at least 35% in the United States).

There are some drawbacks, however. The current costs of solar-cell systems are high (Figure 15-21), but they should become competitive in 5–15 years and are already cost-competitive in some situations. The manufacture of solar cells produces moderate levels of water pollution, which can be eliminated by effective pollution controls. Some people find racks of solar cells on rooftops or in yards to be unsightly, but new thin and flexible rolls of cells (and solar shingles and solar-cell metal roofs) should eliminate this problem.

Another problem is storage of the electricity produced by PV systems during the day for use at night. Batteries can be used, but this is expensive. Flywheels, which should be commercially available in 10–15 years, are a promising way to store this electricity. Electricity from solar cells is fed to a small electric motor to set a disk (the flywheel) spinning at a high speed inside an airless container, almost eliminating resistance. The stored kinetic energy in the spinning wheel can then be converted back to electricity by a generator as needed, with an efficiency of 90%.

According to some analysts, unless federal and private research efforts on photovoltaics are increased sharply, the United States may lose out on a huge global market ($1.4 billion in 1997 and at least $5 billion per year by 2010) and may have to import solar cells from Japan, Germany, Italy, and other countries that have invested heavily in this promising technology since 1980. In 1997, U.S. manufacturers led the world in sales of solar cells but most U.S. solar-cell panels are exported. U.S. solar industry scientists and executives warn that the United States must develop a large domestic market for solar cells or watch other countries take over making and selling a technology originally developed in the United States.

In 1998, the U.S. Department of Energy launched the Million Solar Roofs Initiative to encourage utilities, the solar industry, and governments to work together to install solar-cell energy systems on 1 million roofs in the United States by 2010. To get the ball rolling, President Clinton pledged to solarize 20,000 federal buildings. The Department of Energy estimates that a million rooftop electricity generators could cut total U.S. emissions of carbon dioxide by the equivalent of what 850,000 cars emit annually, and furnish as much electricity as 3–5 large coal-burning or nuclear power plants.

16-4 PRODUCING ELECTRICITY FROM MOVING WATER AND FROM HEAT STORED IN WATER

What Is Hydroelectric Power? In a *large-scale hydropower project*, a high dam is built across a large river to create a reservoir. Some of the water stored in the reservoir is allowed to flow through huge pipes at controlled rates, spinning turbines and producing electricity. Such projects have advantages and disadvantages (Figure 13-10).

In a *small-scale hydropower project*, a low dam with no reservoir (or only a small one) is built across a small stream. Output in small systems can vary with seasonal changes in stream flow. A new small-scale hydropower technology invented by former Soviet hydropower expert Alexander Gorlov may be able to produce electricity from slow-moving rivers without damming them up.

Falling water can also produce electricity in *pumped-storage hydropower systems*, which supply extra power mainly during times of peak electrical demand. When demand is low (usually at night), pumps using surplus electricity from a conventional power plant pump water from a lake or a reservoir to another reservoir at a higher elevation. When a power company temporarily needs more electricity than its other plants can produce, water in the upper reservoir is released, flows through turbines, and generates electricity on its return to the lower reservoir. Another possibility may be to use solar-powered pumps to raise water to the upper reservoir.

Hydroelectric power supplies about 20% of the world's electricity (99% in Norway, 75% in New Zealand, 50% in developing countries, 25% in China, and 13% in the United States) and 6% of its total com-

Q: What do many analysts believe is the key to reducing water waste?

mercial energy. In 1998 the world's three largest producers of hydroelectric power were Canada, the United States, and Brazil. Any new large supplies of hydroelectric power in the United States will be imported from Canada, which gets more than 70% of its electricity from hydropower. Within a few years, China, with 10% of the world's hydropower potential, may become the largest producer of hydroelectricity.

Hydropower has a moderate to high net useful energy yield and fairly low operating and maintenance costs. Hydroelectric plants rarely need to be shut down, and emit very little heat-trapping carbon dioxide or other air pollutants during operation. They have life spans 2–10 times those of coal and nuclear plants. Large dams also help control flooding and supply a regulated flow of irrigation water to areas below the dam.

However, hydropower—especially huge dams and reservoirs—floods vast areas, destroys wildlife habitats, uproots people, decreases natural fertilization of prime agricultural land in river valleys below the dam, and decreases fish harvests below the dam. Dams can also keep fish such as salmon from migrating upstream to spawn, and they kill most juvenile salmon migrating downstream to the ocean (when they pass through dam turbines). Pumped storage systems can also kill large numbers of fish. Although the dam's generating plants emit only small amounts of CO_2, a 1997 study by Robert Goodland, environmental advisor to the World Bank, warned that a hydropower plant's large, shallow reservoirs, especially in the tropics, can generate large amounts of greenhouse gases from the decay of biomass. According to Goodland, "The worst hydropower projects may produce more greenhouse gases than a coal-fired equivalent."

Because of increasing concern about the environmental and social consequences of large dams, there has been growing pressure on the World Bank and other development agencies to stop funding new large-scale hydropower projects. Small-scale projects eliminate most of the harmful environmental effects of large-scale projects, but they can threaten recreational activities and aquatic life, disrupt the flow of wild and scenic rivers, and destroy wetlands.

Is Producing Electricity from Tides and Waves a Useful Option? Twice a day in high and low tides, water that flows into and out of coastal bays and estuaries can spin turbines to produce electricity (Figure 16-18). Two large tidal energy facilities are currently operating, one at La Rance in France and the other in Canada's Bay of Fundy. However, most analysts expect tidal power to make only a tiny contribution to world electricity supplies. There are few suitable sites, and construction costs are high.

The kinetic energy in ocean waves, created primarily by wind, is another potential source of electricity (Figure 16-18). Most analysts expect wave power to make little contribution to world electricity production, except in a few coastal areas with the right conditions (such as western England). Construction costs are moderate to high and the net energy yield is moderate, but equipment could be damaged or destroyed by saltwater corrosion and severe storms.

How Can We Produce Electricity from Heat Stored in Water? Japan and the United States have been evaluating the use of the large temperature differences (between the cold, deep waters and the sun-warmed surface waters) of tropical oceans for producing electricity. If economically feasible, this would be done in *ocean thermal energy conversion* (OTEC) plants anchored to the bottom of tropical oceans in suitable sites (Figure 16-18). However, most energy analysts believe that the large-scale extraction of energy from ocean thermal gradients may never compete economically with other energy alternatives. Despite 50 years of work, the technology is still in the research-and-development stage.

Saline solar ponds, usually located near inland saline seas or lakes in areas with ample sunlight, can produce electricity from heat stored in layers of increasing concentrations of salt (Figure 16-18). Heat accumulated during the day in the denser bottom layer can be used to produce steam that spins turbines, generating electricity. A small experimental saline solar pond power plant on the shore of the Israeli side of the Dead Sea operated for several years but was closed in 1989 because of high operating costs; smaller experimental systems have been built in California and Australia.

Freshwater solar ponds can be used to heat water and space (Figure 16-18). A shallow hole is dug and lined with concrete. A number of large, black plastic bags, each filled with several centimeters of water, are placed in the hole and then covered with fiberglass insulation panels. The panels let sunlight in but keep most of the heat stored in the water during the daytime from being lost to the atmosphere. When the water in the bags has reached its peak temperature in the afternoon, a computer turns on pumps to transfer hot water from the bags to large, insulated tanks for distribution.

Saline and freshwater solar ponds require no energy storage and backup systems. They emit no air pollution and have a moderate net energy yield. Freshwater solar ponds can be built in almost any sunny area and have moderate construction and operating costs. With adequate research and development support, proponents believe that freshwater solar ponds could supply 3–4% of U.S. electricity needs within 10 years.

A: Raising the price of water, the planet's most underpriced vital resource

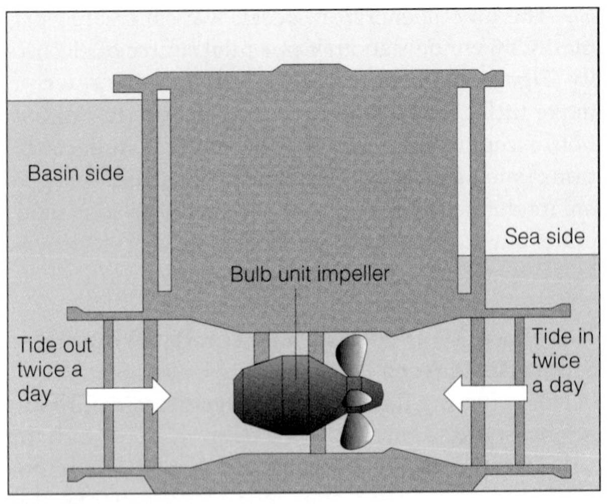

Tidal Power Plant

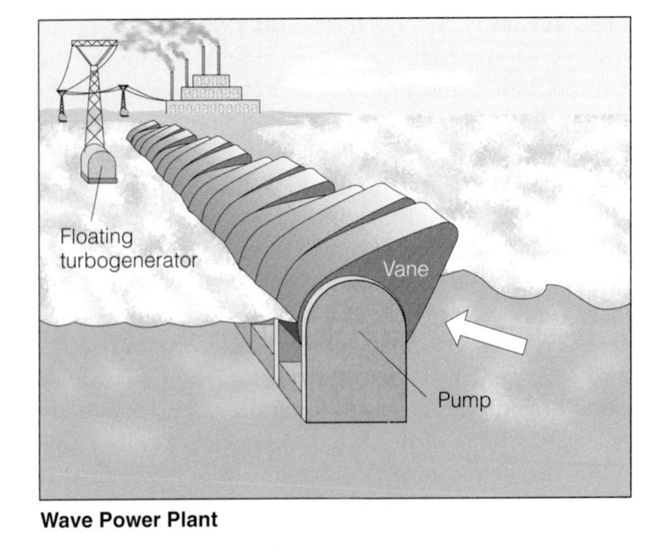

Wave Power Plant

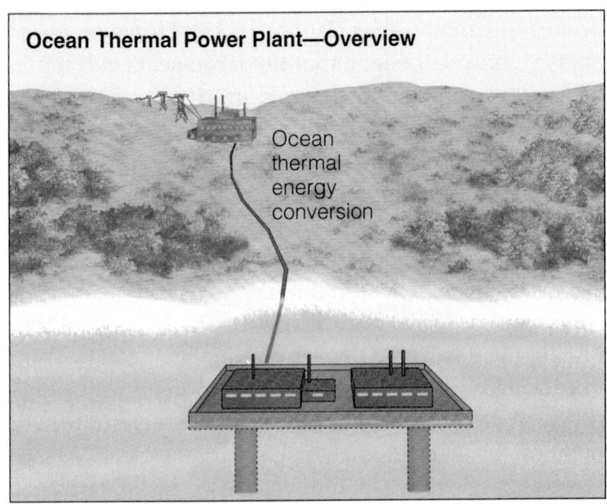

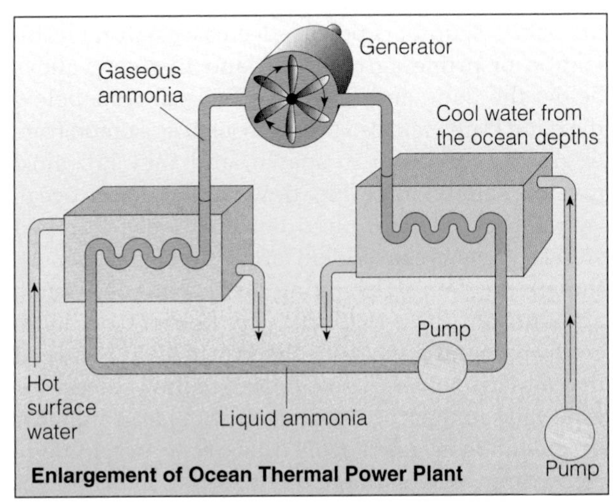

Ocean Thermal Electric Plant

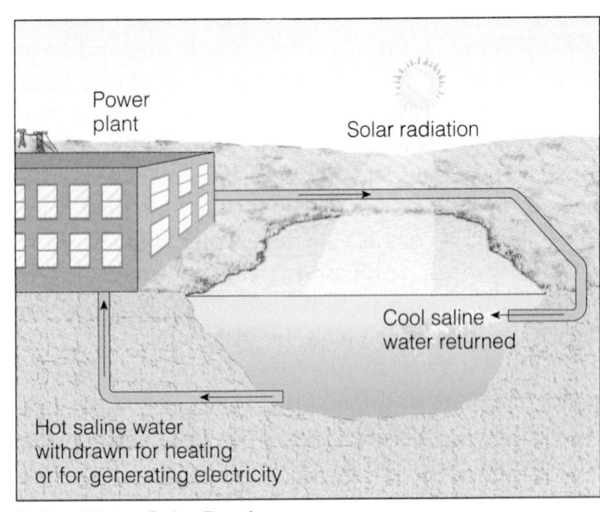

Saline Water Solar Pond

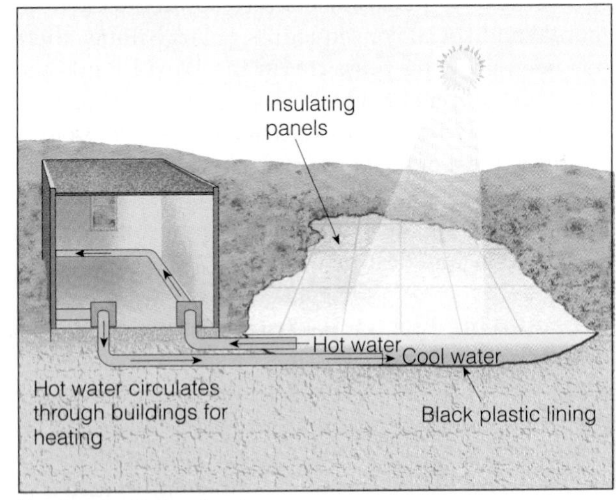

Freshwater Solar Pond

Figure 16-18 Ways to produce electricity from moving water and to tap into solar energy stored in water as heat. None of these systems are expected to be significant sources of energy in the near future.

Q: What percentage of U.S. toilets leak?

16-5 PRODUCING ELECTRICITY FROM WIND

Since 1980 the use of wind to produce electricity has grown rapidly and in 1996 was the world's fastest-growing energy resource. In 1997 there were more than 27,000 wind turbines worldwide, collectively producing 7,000 megawatts of electricity (about 1% of the world's electricity).

Figure 16-19 shows the potential areas for use of wind power in the United States. Wind farms (clusters of 20–100 wind turbines), most of which are highly automated, now provide about 1% of California's electricity—enough to power 280,000 homes. Sizable wind-farm projects are being planned in 12 other states. In principle, all the power needs of the United States could be provided by exploiting the wind potential of just three states: North Dakota, South Dakota, and Texas.

Wind power is a virtually unlimited source of energy at favorable sites; the global potential of wind power is about five times current world electricity use. Wind farms can be built in 6 months to a year and then easily expanded as needed. With a moderate to fairly high net energy yield, these systems emit no heat-trapping carbon dioxide or other air pollutants and need no water for cooling; manufacturing them produces little water pollution. The land under wind turbines can be used for grazing cattle or farming.

Wind power (with much lower government subsidies) has a significant cost advantage over nuclear power (Figure 15-21) and has become competitive with coal-fired power plants in many places. With new technological advances and mass production, projected cost declines should make wind power one of the world's cheapest ways to produce electricity within the next decade. In the long run, electricity from large wind farms in remote areas might be used to make hydrogen gas from water during off-peak periods. The hydrogen could then be fed into a pipeline and storage system.

One drawback of wind power is that it's economical primarily in areas with steady winds. When the wind dies down, backup electricity from a utility company or from an energy storage system becomes necessary. Backup power could also come from linking wind farms with a solar-cell or hydropower plant, from efficient natural-gas–burning turbines, or from flywheels.

Other drawbacks to wind farms include visual pollution and noise, although these can be overcome by improving their design and locating them in isolated areas. Large wind farms might also interfere with the flight patterns of migratory birds in certain areas, and they have killed large birds of prey (especially

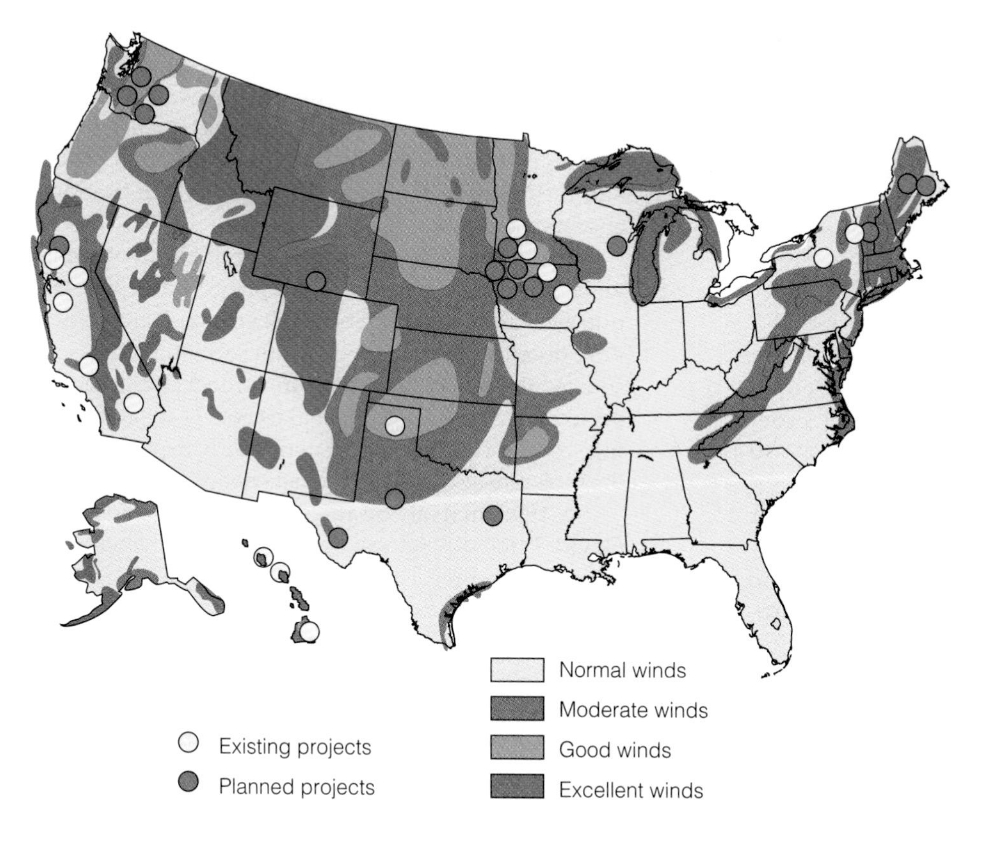

Figure 16-19 Potential for use of wind power in the United States. (Data from U.S. Department of Energy)

A: 25%; a leaky toilet can waste 83,000 liters (22,000 gallons) of water a year

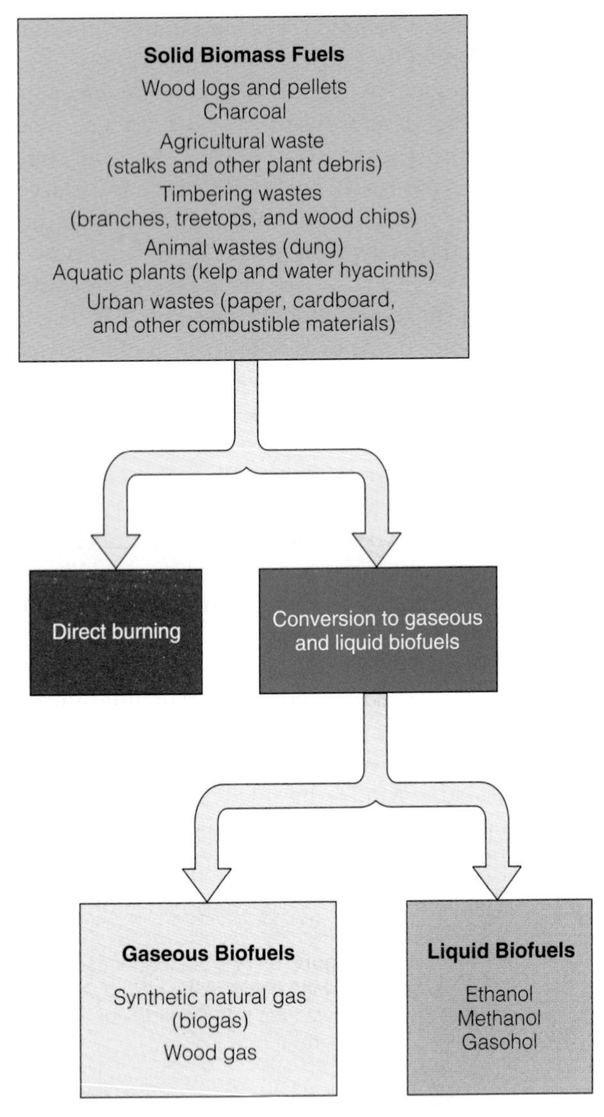

Figure 16-20 Principal types of biomass fuel.

hawks, falcons, and eagles) that prefer to hunt along the same ridge lines that are ideal for wind turbines. Researchers are evaluating how serious and widespread this problem is and are trying to find ways to deal with it.

Wind power experts project that by the middle of the 21st century wind power could supply more than 10–20% of the world's electricity and 10–25% of the electricity used in the United States. European governments are currently spending 10 times more for wind energy research and development than the U.S. government. Analysts warn that if this trend continues, U.S. consumers will be buying wind turbines from European countries and eventually from China and India, and American companies will lose out on a rapidly growing global business ($2 billion in 1997).

16-6 PRODUCING ENERGY FROM BIOMASS

How Can We Produce Energy from Plants? Plant Power Many forms of *biomass*—organic matter in plants produced through photosynthesis—can be burned directly as a solid fuel or converted into gaseous or liquid *biofuels* (Figure 16-20). The burning of wood and manure to heat buildings and cook food supplies about 14% of the world's energy (4–5% in Canada and the United States) and about 35% of the energy used in developing countries.

Biomass is a potentially renewable energy resource as long as trees and plants are not harvested faster than they grow back, a requirement that is not being met in many places. No net increase in atmospheric levels of heat-trapping carbon dioxide occurs as long as the rate of removal and burning of trees and plants, and the rate of loss of below-ground organic matter, do not exceed the rate of replenishment. Burning biomass fuels adds much less air pollution to the atmosphere per unit of energy produced than does the uncontrolled burning of coal.

According to a 1992 UN study, by 2050 biomass could supply as much as 55% of the total energy used globally today. Whether such projections become reality depends on the availability of large areas of productive land and adequate water and fertilizer—resources that may be in short supply in coming decades—and the ability to minimize the harmful environmental effects of such large-scale production of biomass. Currently, potentially renewable biomass is being exploited in nonrenewable and unsustainable ways, primarily because of deforestation, soil erosion, and the inefficient burning of wood in open fires and energy-wasting stoves.

Without effective land-use controls and replanting, widespread removal of trees and plants can deplete soil nutrients and cause excessive soil erosion, water pollution, flooding, and loss of wildlife habitat. Biomass resources also have a high moisture content (15–95%). The added weight of moisture makes collecting and hauling wood and other plant material fairly expensive, and the moisture also reduces the net energy yield.

A 1998 analysis by three researchers, including David Pimentel (Guest Essay, p. 364), concluded, "Large-scale biofuel production is not an alternative to the current use of oil and is not even an advisable option to cover a significant fraction of it."

Is Producing Gaseous and Liquid Fuels from Solid Biomass a Useful Option? Various forms of solid biomass can be converted by bacteria and various chemical processes into gaseous and liquid biofuels (Figure 16-20). Examples include *biogas*, a mixture of

Q: What is the monetary value of hard-rock minerals removed from U.S. public lands each year?

60% methane (the principal component of natural gas) and 40% carbon dioxide; *liquid ethanol* (ethyl, or grain, alcohol); and *liquid methanol* (methyl, or wood alcohol).

In China, anaerobic bacteria in more than 6 million *biogas digesters* convert organic plant and animal wastes into methane fuel for heating and cooking. These simple devices can be built for about $50 including labor. After the biogas has been separated, the solid residue is used as fertilizer on food crops or, if contaminated, on trees. When they work, biogas digesters are very efficient. However, they are also slow and unpredictable, a problem that could be corrected with development of more reliable models.

Some analysts believe that liquid ethanol and methanol produced from biomass could replace gasoline and diesel fuel when oil becomes too scarce and expensive. *Ethanol* can be made from sugar and grain crops (sugarcane, sugar beets, sorghum, and corn) by fermentation and distillation. Gasoline mixed with 10–23% pure ethanol makes *gasohol,* which can be burned in conventional gasoline engines and is sold as super unleaded or ethanol-enriched gasoline.

Another alcohol, *methanol,* is made mostly from natural gas but can also be produced at a higher cost from wood, wood wastes, agricultural wastes (such as corncobs), sewage sludge, garbage, and coal. The advantages and disadvantages of using ethanol, methanol, and several other fuels as alternatives to gasoline are summarized in Table 16-1. Recently a new water-based gasoline fuel has entered the picture and is currently undergoing extensive testing by state and federal agencies (Individuals Matter, p. 423). By 1997, an estimated 386,000 alternative-fuel vehicles were in use in the United States.

Is Producing Energy from Biomass Plantations a Useful Option? One way to produce biomass fuel is to plant large numbers of fast-growing trees (especially cottonwoods, poplars, sycamores, and leucaenas), shrubs, and water hyacinths in *biomass plantations*. After harvest, these "Btu bushes" can be burned directly, converted into burnable gas, or fermented into a liquid alcohol fuel. Such plantations can be located on semiarid land not needed for crops (although lack of water can limit productivity) and can be planted to reduce soil erosion and help restore degraded lands.

However, this industrialized approach to biomass production requires large areas of land (about 81 times as much land as solar cells to provide the same amount of electricity) as well as heavy use of pesticides and fertilizers, which can pollute drinking water supplies and harm wildlife. In some areas the plantations might compete with food crops for prime farmland. Conversion of large forested areas or natural grasslands into single-species biomass plantations also reduces biodiversity and ecological integrity.

Is Burning Wood a Useful Option? Almost 70% of the people living in developing countries heat their homes and cook their food by burning wood or charcoal. However, about 2 billion people in developing countries cannot find (or are too poor to buy) enough fuelwood to meet their needs.

Sweden leads the world in using wood as an energy source, mostly for district heating plants. In the United States, small wood-burning power plants located near sources of their fuel provide 23% of the electricity used in heavily forested Maine and about 4% of the electricity used in New England. However, nearly 70% of the heat energy produced from burning biomass is lost when it is converted to electricity.

Wood provides 8% of all energy used by U.S. industry, with the forest products industry (mostly paper companies and lumber mills) consuming almost two-thirds of the fuelwood used in the United States. About 20% of U.S. homes get some heat from burning wood, and about 4% burn wood as their main heating fuel (down from 7.5% in 1984).

Wood has a moderate to high net useful energy yield when collected and burned directly and efficiently near its source. However, in urban areas, to which wood must be hauled over long distances, wood can cost homeowners more per unit of energy produced than oil or electricity. Harvesting wood can cause accidents (mostly from chain saws) and burning wood in poorly maintained or operated wood stoves can cause house fires.

Wood smoke contains pollutants (especially particulates and polycyclic aromatic hydrocarbons or PAHs) known to cause bronchitis, emphysema, cancers, and other illnesses. According to the EPA, air pollution from wood burning causes as many as 820 cancer deaths a year in the United States. Since 1990 the EPA has required all new wood stoves sold in the United States to emit at least 70% less particulate matter than earlier models.

Fireplaces can also be used for heating, but they usually result in a net loss of energy from a house. The draft of heat and gases rising up the fireplace chimney pushes out warm inside air and pulls in cold outside air through cracks and crevices throughout a house. Fireplace inserts with glass doors and blowers help, but they still waste energy compared with efficient wood-burning stoves. Energy loss can be reduced by closing the room off and cracking a window so that the fireplace won't draw much heated air from other rooms. A better solution is to run a small pipe from outside into the front of the fireplace so that it gets the air it needs during combustion.

A: At least $1.2 billion (some say $4–6 billion)

Table 16-1 Evaluation of Alternatives to Gasoline

Advantages	Disadvantages
Compressed Natural Gas	
Fairly abundant, inexpensive domestic and global supplies Low emissions of hydrocarbons, CO, and CO_2 Vehicle development advanced; well suited for fleet vehicles Reduced engine maintenance	Large fuel tank required; one-fourth the range Expensive engine modification required ($2,000) New filling stations required Nonrenewable resource
Electricity	
Renewable if not generated from fossil fuels or nuclear power Zero vehicle emissions Electric grid in place Efficient and quiet	Limited range and power Batteries expensive Slow refueling (6–8 hours) Power-plant emissions if generated from coal or oil
Reformulated Gasoline (Oxygenated Fuel)	
No new filling stations required Low to moderate emission reduction of CO No engine modification required	Nonrenewable resource Dependence on imported oil perpetuated No emission reduction of CO_2 Higher cost Water resources contaminated by leakage and spills
Methanol	
High octane Emission reduction of CO_2 (total amount depends on method of production) Reduced total air pollution (30–40%)	Large fuel tank required; one-half the range Corrosive to metal, rubber, plastic Increased emissions of potentially carcinogenic formaldehyde High CO_2 emissions if generated by coal High capital cost to produce Hard to start in cold weather
Ethanol	
High octane Emission reduction of CO_2 (total amount depends on distillation process and efficiency of crop growing) Emission reduction of CO Potentially renewable	Large fuel tank required; lower range Much higher cost Corn supply limited Competition with food growing for cropland Smog formation possible Corrosive Hard to start in cold weather
Solar–Hydrogen	
Renewable if produced using solar energy Lower flammability Virtually emission-free Zero emissions of CO_2 Nontoxic	Nonrenewable if generated by fossil fuels or nuclear power Large fuel tank required No distribution system in place Engine redesign required Currently expensive

Is Burning Agricultural and Urban Wastes a Useful Option? In agricultural areas, crop residues (such as sugarcane residues, rice husks, cotton stalks, and coconut shells) and animal manure can be collected and burned or converted into biofuels. Since 1985, Hawaii has been burning *bagasse*—the residue left after sugarcane harvesting and processing—to supply almost 10% of its electricity. Brazil gets 10% of its electricity from burning bagasse. Denmark is using thousands of small-scale burners fueled by straw—its largest agricultural waste product—to provide heat for on-farm use and district heating for homes in 60 areas. Rapid heating of biomass in a low-oxygen container converts it to a gas that can be burned efficiently and cleanly in a gas turbine. The ash from biomass power plants and burners can sometimes be used as fertilizer.

This approach makes sense when crop residues are burned in small power plants or burners located

Q: What royalties do companies removing hard-rock minerals from U.S. federal public lands pay the federal government?

A New Car Fuel That Is Mostly Water

INDIVIDUALS MATTER

It seems too good to be true, but a breakthrough fuel that is more than half water could power cars, trucks, train locomotives, generators, and aircraft running on gasoline and diesel fuel within a few years.

This new, milky-looking fuel, called A-55, was developed by Reno, Nevada, inventor Rudolf Gunnerman. It consists of 30–55% water and 45% naphtha (a cheap-to-produce by-product of petroleum distillation). In addition, it contains lubricants, antirust and antifreeze agents, and small amounts of a blending agent developed by Gunnerman (which attracts the normally repellent water and oily naphtha to one another). The water in A-55 allows the fuel to burn cooler and more efficiently, producing more power and less pollution. In 1994 Gunnerman and diesel equipment giant Caterpillar formed Advanced Fuels, a joint venture to test and market the new fuel.

So far in every test the new fuel tops conventional gasoline and diesel fuel as a clean, safe, and cheap fuel that can be used in almost any combustion engine. In November 1995, Nevada certified A-55 as a clean alternative fuel. That means it can meet federal and Nevada laws requiring clean fuels in fleets and other vehicles. A-55 is especially attractive to the trucking industry, facing federal crackdowns on dirty diesel engine emissions.

Since 1996 A-55 has been tested in California and Illinois. If it passes these hurdles and even more stringent testing by the Department of Energy, A-55 could be used in all states. By early 1996 the new fuel was successfully being used in several city buses and a power generator in Reno, in generators at Caterpillar's Illinois plants, and in public and private vehicles in Sacramento, California.

Tests so far show that using A-55 leads to a 60% drop in EPA-monitored emissions of air pollutants such as carbon monoxide, nitrous oxide, and hydrocarbons. The vapor pressure of the fuel is about one-fifth that of gasoline. That means that vapor recovery systems won't be needed at the pump. Because gasoline vapor is what catches fire, A-55 is virtually immune to fire and explosions. Because it doesn't burn outside engine combustion chambers, the fuel can be stored in aboveground tanks without risking explosions.

Converting vehicles to use A-55 costs about $300 and requires only fairly simple adjustments to fuel injectors and timing. Once that's done, the vehicle could use either the water-based fuel or conventional gasoline or diesel fuels.

Car and diesel truck drivers could see their fuel costs cut in half. That's because the refining process for producing naphtha (currently used mostly as a hardener in road tars) is faster and cheaper than producing gasoline and diesel fuel. In addition to cutting costs, using naphtha instead of gasoline and diesel fuel would eliminate up to 90% of the air pollutants produced by oil refineries. Stay tuned to see whether this fuel passes its final tests.

Critical Thinking

Can you think of some economic and political reasons why this potentially promising fuel may not be used widely? How could such hurdles be overcome?

near areas where the residues are produced. Otherwise, it takes too much energy to collect, dry, and transport the residues to power plants. Some ecologists argue that it makes more sense to use animal manure as a fertilizer and to use crop residues to feed livestock, retard soil erosion, and fertilize the soil.

An increasing number of cities in Japan, western Europe, and the United States have built incinerators that burn trash and use the energy released to produce electricity or to heat nearby buildings (Figure 22-9). However, this approach has been limited by opposition from citizens concerned about emissions of toxic gases and disposal of the resulting toxic ash. Some analysts argue that more energy is saved by composting or recycling paper and other organic wastes than by burning them. For example, recycling paper saves at least twice as much energy as is produced by burning the paper.

16-7 THE SOLAR–HYDROGEN REVOLUTION

What Can We Use to Replace Oil? Good-bye Oil and Smog, Hello Hydrogen When oil is gone (or when what's left costs too much to use) what will we use to fuel vehicles, industry, and buildings? Some scientists say the fuel of the future is hydrogen gas (H_2) (Table 16-1).

There is very little hydrogen gas around, but we can get it from something we have plenty of: water. Water can be split by electricity into gaseous hydrogen and oxygen (Figure 16-21).

Hydrogen, the simplest chemical fuel, contains no carbon. Thus, when burned it doesn't produce carbon dioxide like all carbon-containing fossil fuels and biomass fuels (such as wood). Instead, when hydrogen

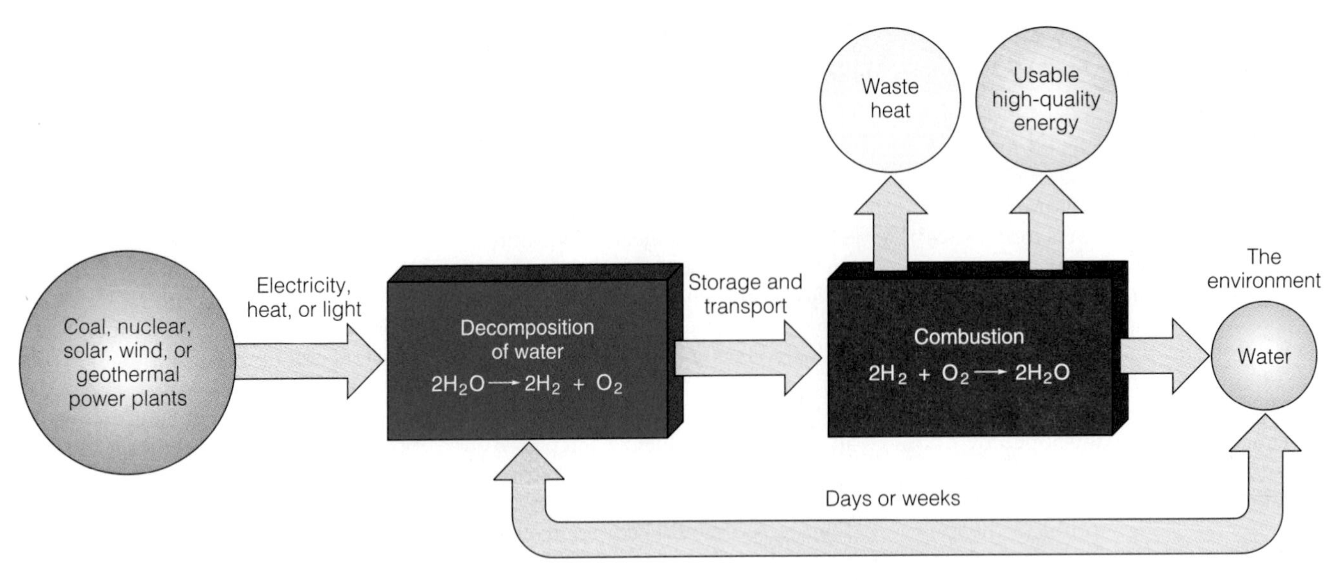

Figure 16-21 The hydrogen energy cycle. The production of hydrogen gas requires electricity, heat, or solar energy to decompose water, thus leading to a negative net energy yield. However, hydrogen is a clean-burning fuel that can replace oil, other fossil fuels, and nuclear energy. Using solar energy to produce hydrogen from water could also eliminate most air pollution and greatly reduce the threat of global warming.

burns it combines with oxygen gas in the air and produces nonpolluting water vapor and some nitrogen oxides, thus eliminating most of the air pollution problems we face today.

What's the Catch? If you think using hydrogen as an energy source sounds too good to be true, you're right. Several problems must be solved to make hydrogen one of our primary energy resources, but scientists are making rapid progress in finding solutions.

One problem is that it takes energy to produce this marvelous fuel. Most hydrogen used today is produced by reacting natural gas and steam to form carbon dioxide and hydrogen, at a cost equivalent to producing gasoline at 65–92¢ per gallon (compared to 50¢ per gallon to produce gasoline itself). Hydrogen can also be produced by gasifying coal at a cost equal to producing gasoline at $1.05 per gallon. We could use electricity from coal-burning and nuclear power plants to split water and produce hydrogen, but this subjects us to the harmful environmental effects associated with using these fuels, and it costs more than the hydrogen fuel is worth.

Most proponents of hydrogen gas believe that if we are to get its very low pollution benefits, the energy to produce the gas from water must come from the sun, in the form of electricity generated by sources such as hydropower, solar thermal and solar-cell power plants, and wind farms. If scientists and engineers can learn how to use sunlight to decompose water cheaply enough, they will set in motion a *solar–hydrogen revolution* over the next 50 years and change the world as much as the agricultural and industrial revolutions did.

Such a revolution would eliminate the air and water pollution caused by extracting, transporting, and burning fossil fuels. It would reduce the threat of global warming by sharply reducing CO_2 emissions and would also decrease the threat of wars over dwindling oil supplies. It would allow nuclear and coal-fired power plants to be phased out and let individuals produce most or all of their own energy. An important step in this direction occurred in 1998 when researchers at the U.S. National Renewable Energy Laboratory created a combined *photovoltaic–photoelectrochemical cell* that uses sunlight to split water into hydrogen and oxygen at an efficiency of 12.4%—a spectacular advance according to prominent researchers in this field.

Currently, using solar energy to produce hydrogen gas is too costly, but the costs of using solar energy to produce electricity are coming down. Moreover, if the health and environmental costs of using gasoline were included in its market prices (through gasoline taxes), it would cost about $1 per liter ($4 per gallon)—roughly the price in Japan and many European countries. At this price, hydrogen produced by any form of solar energy would be competitive.

Hydrogen for vehicles could initially be produced from natural gas. Blends of natural gas and hydrogen produced from solar sources could then be phased in gradually as reserves of natural gas fuel become depleted. Hydrogen produced by solar-cell power plants (in sunny, mostly desert areas) could be carried by pipeline wherever it was needed. However, such facili-

Q: Mining activities on federal public lands in the United States have caused how much monetary damage?

ties would have to be designed and managed to reduce the disruption of easily harmed desert ecosystems.

Hydrogen gas is much easier to store than electricity. It can be stored in a pressurized tank, metal powders, and activated carbon that absorb gaseous hydrogen and release it when heated for use as a fuel. If properly handled, hydrogen is a safer fuel than gasoline and natural gas. Unlike gasoline, compounds and powders containing absorbed hydrogen will not explode or burn if a vehicle's tank is ruptured in an accident. However, it's difficult to store enough hydrogen gas in a car as a compressed gas or a metal powder for it to run very far—a problem similar to the one current electric cars face. Scientists and engineers are seeking solutions to this problem.

Another possibility is to power a car with a *fuel cell* (Figure 16-3) in which hydrogen and oxygen gas combine to produce electrical current. Fuel cells produce no air pollution, have no moving parts, and have high energy efficiencies of up to 65%—several times the efficiency of conventional gasoline-powered engines and electric cars. When fuel cells are used to cogenerate both electricity and heat, their total system energy efficiency will approach 90%. Fuel cells could be resupplied with hydrogen in a matter of minutes, compared to the several hours needed to recharge the batteries in electric cars.

A major U.S. automobile manufacturer and the Department of Energy estimate that with mass production, fuel-cell cars would cost little more to build than gasoline cars. A number of prototype fuel-cell systems for cars, buses, homes, and buildings are being tested and evaluated. In 1996, three experimental buses powered by fuel cells built by Vancouver-based Ballard Power Systems began operating in Chicago. Currently, fuel cells are expensive and heavy, and they produce energy at a cost equivalent to gasoline at $2.50 per gallon, but this could change with more research and mass production.

In 1998, the Electric Power Institute and Analytic Power in Boston built a residential fuel cell that uses methane to produce hydrogen fuel for the cell. The $3,000 unit, now being tested, is about the size of a gas furnace. It could supply all the heating, cooling, cooking, refrigeration, and electrical needs of a home and provide hydrogen fuel for one or more cars at an affordable price. In 1997, a panel of technology experts projected that fuel cells could supply as much as 30% of the world's electricity by 2017.

What's Holding Up the Solar–Hydrogen Revolution? Politics and economics, not a lack of promising technology, are the main factors holding up a more rapid transition to a solar–hydrogen age. Phasing out dependence on fossil fuels and phasing in new solar–hydrogen technologies over the next 40–50 years involves convincing investors and energy companies with strong vested interests in fossil fuels to risk a lot of capital on hydrogen. It also involves convincing governments to put up some of the money for developing hydrogen energy (as they have done for decades for fossil fuels and nuclear energy).

In the United States, large-scale government funding of hydrogen research is generally opposed by powerful U.S. oil companies, electric utilities, and automobile manufacturers, who understandably see it as a serious threat to their profits. In contrast, the Japanese and German governments have been spending seven to eight times more on hydrogen research and development than the United States. Some analysts warn that without greatly increased government and private research and development, Americans may be buying solar–hydrogen equipment and fuel cells from Germany and Japan and may lose out on a huge global market and source of domestic jobs.

The solar–hydrogen revolution has already started. Soon a German firm plans to market solar–hydrogen systems that would meet all the heating, cooling, cooking, refrigeration, and electrical needs of a home, as well as providing hydrogen fuel for one or more cars. Germany and Saudi Arabia have each built a large solar–hydrogen plant, and Germany and Russia are jointly developing a prototype commercial airliner fueled by hydrogen. (Hydrogen has about 2.5 times the energy by weight of gasoline, making it an especially attractive aviation fuel.)

We don't need to invent hydrogen-powered vehicles. Mercedes, BMW, and Mazda already have prototypes being tested on the roads. In the United States, a *hydrogen corridor* of research and production facilities is beginning to develop in the desert east of Los Angeles. It is anchored by a $2.5-million facility built by Xerox for using solar energy to convert water into hydrogen.

16-8 GEOTHERMAL ENERGY

How Can We Tap the Earth's Internal Heat? Going Underground Heat contained in underground rocks and fluids is an important source of energy. Over millions of years, this **geothermal energy** from earth's mantle (Figure 15-2) has been transferred to underground concentrations of *dry steam* (steam with no water droplets), *wet steam* (a mixture of steam and water droplets), and *hot water* trapped in fractured or porous rock at various places in earth's crust. All three types are currently being used to heat space and water—and in some cases to produce electricity (Figure 15-2).

If such geothermal sites are close to the surface, wells can be drilled to extract the dry steam, wet steam

A: $33–72 billion, with most cleanup costs to be paid by taxpayers

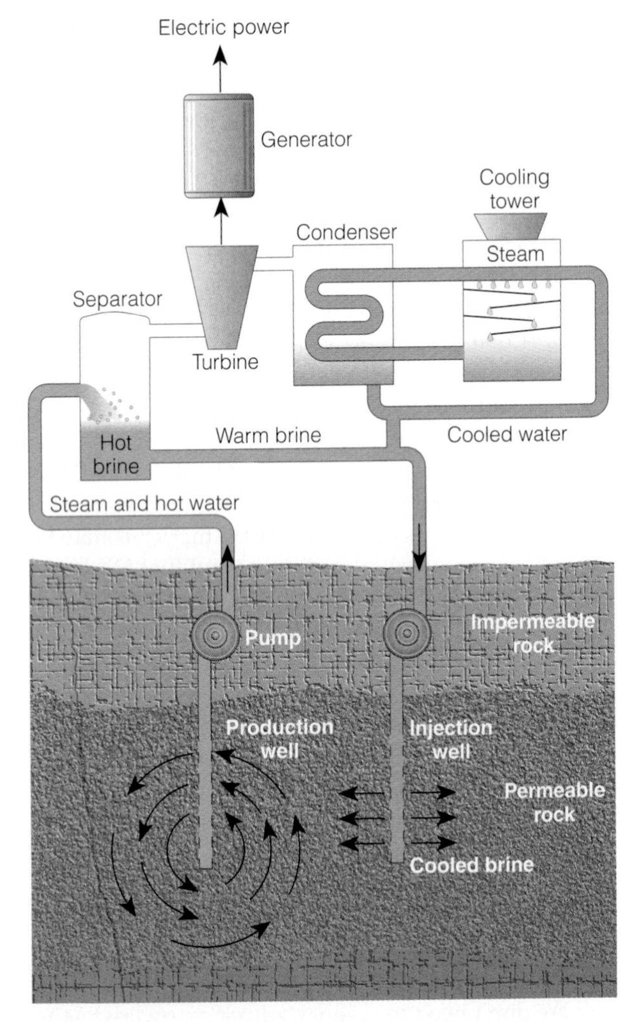

Figure 16-22 Tapping the earth's heat or geothermal energy in the form of wet steam to produce electricity.

(Figure 16-22), or hot water. This thermal energy can be used for space heating and to produce electricity or high-temperature heat for industrial processes.

Geothermal reservoirs containing dry steam, wet steam, or hot water can be depleted if heat is removed faster than it is renewed by natural processes. Thus, geothermal resources are nonrenewable on a human time scale, but the potential supply is so vast that it is usually classified as a potentially renewable energy resource. However, easily accessible concentrations of geothermal energy are fairly scarce.

Currently, about 22 countries (most of them in the developing world) are extracting energy from geothermal sites to produce less than 1% of the world's electricity. The United States accounts for 44% of the geothermal electricity generated worldwide, with most of the favorable sites in California, Hawaii, Nevada, and Utah.

Three other virtually nondepletable or potentially renewable sources of geothermal energy are *molten rock* (magma); *hot dry-rock zones*, where molten rock that has penetrated the earth's crust heats subsurface rock to high temperatures; and low- to moderate-temperature *warm-rock reservoir deposits*, which could be used to preheat water and run heat pumps for space heating and air conditioning.

Research is being carried out to see whether hot dry-rock zones, which can be found almost anywhere if one can drill deep enough, can provide affordable geothermal energy. According to the National Academy of Sciences, the energy potentially recoverable from such reservoirs would meet U.S. energy needs at current consumption levels for 600–700 years, but the projected cost is high.

What Are the Pros and Cons of Geothermal Energy? The biggest advantages of geothermal energy include a vast, reliable, and sometimes renewable supply of energy for areas near reservoir sites, moderate net energy yields for large and easily accessible reservoir sites, about 96% fewer CO_2 emissions per unit of energy than fossil fuels, and a competitive cost of producing electricity (Figure 15-21).

A serious limitation of geothermal energy is the scarcity of easily accessible reservoir sites. Dry steam, wet steam, and hot water geothermal reservoirs must be carefully managed or they can be depleted within a few decades. Furthermore, geothermal development in some areas can destroy or degrade forests or other ecosystems. In Hawaii, for example, environmentalists have successfully fought the construction of a large geothermal project in the only lowland tropical rain forest left in the United States. Additionally, in some areas the extraction of hot water from hot-water reservoirs causes land to sink.

Noise, odor, and local climate changes can also be problems. With proper controls, however, most experts consider the environmental effects of geothermal energy to be less (or no greater) than those of fossil fuel and nuclear power plants.

Economics is the main barrier to greater use of geothermal energy. Currently, the cost of tapping geothermal energy is too high for all but the most concentrated and accessible sources.

16-9 SOLUTIONS: A SUSTAINABLE ENERGY STRATEGY

What Are the Best Energy Alternatives for the United States? Table 16-2 summarizes the major advantages and disadvantages of the energy alternatives discussed in this and the previous chapter, with

Dunne, Seth. 1998. "Green Power Spreads to California." *World Watch*, vol. 11, no. 4, 7(1).

emphasis on their potential in the United States (and presumably many other countries). Energy experts argue over these and other projections, and new data and innovations may affect the status of certain alternatives. However, the data in the table do provide a useful framework for making decisions based on currently available information.

Many scientists and energy experts who have evaluated these energy alternatives have come to the following general conclusions:

- *The best short-term, intermediate, and long-term alternatives are a combination of improved energy efficiency and greatly increased use of locally available renewable energy resources.*
- *Future energy alternatives will probably have low to moderate net energy yields and moderate to high development costs.*
- *Because there is not enough financial capital to develop all energy alternatives, projects must be chosen carefully.*
- *We cannot and should not depend mostly on a single nonrenewable energy resource such as oil, coal, natural gas, or nuclear power.*

What Role Should Economics Play in Energy Resource Use? Cost is the biggest factor determining which commercial energy resources are widely used by consumers. Governments throughout the world use three basic economic and political strategies to stimulate or dampen the short-term and long-term use of a particular energy resource.

The first approach is *not attempting to control the price of an energy resource*, allowing all energy resources to compete in open, free-market competition. However, leaving energy pricing to the marketplace without any government interference is rarely politically feasible because of well-entrenched government intervention into the marketplace in the form of subsidies, taxes, and regulations. Furthermore, the free-market approach, with its emphasis on short-term gain, inhibits long-term development of new energy resources, which can rarely compete economically in their development stages without government support.

The second approach is *keeping energy prices artificially low*, which encourages use and development of a resource. In the United States (and most other countries), the energy marketplace is greatly distorted by huge government subsidies and tax breaks (such as depletion write-offs for fossil fuels) that make the prices of fossil fuels and nuclear power artificially low and help perpetuate the use of these energy resources, even when better and less costly alternatives are available (Figure 15-21). At the same time, programs for improving energy efficiency and solar alternatives receive much lower subsidies and tax breaks. This creates an uneven economic playing field that encourages waste and rapid depletion of nonrenewable energy resources and discourages the development of energy alternatives that are not getting at least the same level of subsidies and price control. For example, the U.S. fossil fuel industry gets almost $20 billion a year in taxpayer-supported government subsidies, compared to only about $200 million for renewable energy. Worldwide government subsidies for fossil fuel and nuclear energy amount to an estimated $200 billion per year, more than half the value of all the crude oil produced each year.

The third approach is *keeping energy prices artificially high*. This can discourage development, use, and waste of a resource. Governments can raise the price of an energy resource by withdrawing existing tax breaks and other subsidies or by adding taxes on its use. This provides increased government revenues, encourages improvements in energy efficiency, reduces dependence on imported energy, and decreases use of an energy resource that has a limited future supply. However, increasing taxes on energy use can dampen economic growth and put a heavy economic burden on the poor and lower middle class unless some of the energy tax revenues are used to help offset their increased energy costs.

In the United Kingdom the government eliminated fossil fuel subsidies amounting to $7 billion in 1989. Coal use fell by 31%, replaced mostly by North Sea natural gas. By contrast, in Germany government subsidies for coal rose by more than 50% between 1980 and 1995, reaching nearly $7 billion in 1995.

One popular myth is that higher energy prices wipe out jobs. Actually, low energy prices increase unemployment because farmers and industries find it cheaper to substitute machines run on artificially cheap energy for human labor. Raising energy prices stimulates employment because building solar collectors, adding insulation, and carrying out most forms of improving energy efficiency require a high input of human labor.

Solutions: How Can the United States Develop a More Sustainable Energy Future? Communities such as Osage, Iowa (p. 56), Davis, California (p. 720), and Gaviotas, Columbia (p. 412), and individuals are taking energy matters into their own hands. At the same time, most environmentalists urge citizens to exert intense pressure on elected officials to develop a national energy policy based on much greater improvements in energy efficiency and a more rapid transition to a mix of renewable energy resources. A variety of analysts have suggested strategies, some of them highly controversial, to achieve such goals (Figure 16-23).

Hint: Enter the search terms *alternative energy, environmental aspects* using the Subject Guide.

Table 16-2 Evaluation of Energy Alternatives for the United States (shading indicates favorable conditions)						
	Estimated Availability					
Resources	**Short Term (2000–2010)**	**Intermediate Term (2010–2020)**	**Long Term (2020–2050)**	**Estimated Net Useful Energy of Entire System**	**Projected Cost of Entire System**	**Actual or Potential Overall Environmental Impact of Entire System**
NONRENEWABLE RESOURCES						
Fossil fuels						
Petroleum	High (with imports)	Moderate (with imports)	Low	High but decreasing	High for new domestic supplies	Moderate
Natural gas	High	Moderate to high	Moderate (with imports)	High but decreasing	High for new domestic supplies	Low
Coal	High	High	High	High but decreasing	Moderate but increasing	Very high
Oil shale	Low	Low to moderate	Low to moderate	Low to moderate	Very high	High
Tar sands	Low	Fair? (imports only)	Poor to fair (imports only)	Low	Very high	Moderate to high
Synthetic naturall gas (SNG) from coal	Low	Low to moderate	Low to moderate	Low to moderate	High	High (increases use of coal)
Synthetic oil, H_2, and alochols from coal	Low	Moderate	High	Low to moderate	High	High (increases use of coal)
Nuclear energy						
Conventional fission (uranium)	Low to moderate	Low to moderate	Low to moderate	Low to moderate	Very high	Very high
Breeder fission (uranium and thorium)	None	None to low (if developed)	Low to moderate (if developed)	Unknown, but probably moderate	Very high	Very high
Fusion (deuterium and tritium)	None	None	None to low (if developed)	Unknown, but may be high	Very high	Unknown (probably moderate to high
Geothermal energy (Some are renewable)	Low	Low	Moderate	Moderate	Moderate to high	Moderate to high
RENEWABLE RESOURCES						
Improving energy efficiency	High	High	High	Very high	Low	Decreases impact of other sources
Hydroelectric						
New large-scale dams and plants	Low	Low	Very low	Moderate to high	Moderate to very high	Low to moderate

Q: In tropical and temperate areas, how long does it take to renew 2.54 centimeters (1 inch) of topsoil?

Table 16-2 Evaluation of Energy Alternatives for the United States (shading indicates favorable conditions)

Resources	Estimated Availability: Short Term (2000–2010)	Estimated Availability: Intermediate Term (2010–2020)	Estimated Availability: Long Term (2020–2050)	Estimated Net Useful Energy of Entire System	Projected Cost of Entire System	Actual or Potential Overall Environmental Impact of Entire System
RENEWABLE RESOURCES (continued)						
Hydroelectric (continued)						
Reopening abandoned small-scale plants	Moderate	Moderate	Low	Moderate	Moderate	Low to moderate
Tidal energy	Very low	Very low	Very low	Moderate	High	Low to moderate
Ocean thermal gradients	None	Low	Low to moderate (if developed)	Unknown (probably low to moderate)	High	Unknown (prob- moderate to high)
Solar energy						
Low-temperature heating (for homes and water)	Moderate to high	High	High	Moderate to high	Moderate	Low
High-temperature heating	Low	Moderate	Moderate to high	Moderate	High initially but probably delining fairly rapidly	Low to moderate
Photovoltaic production of electricity	Low to moderate	Moderate	High	Fairly high	High initially but delining fairly rapidly	Low
Wind energy	Low to moderate	Moderate	Moderate to high	Fairly high	Moderate	Low
Geothermal energy (low heat flow)	Very low	Very low	Low to moderate	Low to moderate	Moderate to high	Moderate to high
Biomass (burning of wood and agricultural wastes)	Moderate	Moderate	Moderate to high	Moderate	Moderate	Moderate to high
Biomass (urban wastes for incineration)	Low	Moderate	Moderate	Low to farily high	High	Moderate to high
Biofuels (alcohols and biogas from organic wastes)	Low to moderate	Moderate	Moderate to high	Low to farily high	Moderate to high	Moderate to high
Hydrogen gas (from coal or water)	Very low	Low to moderate	Moderate to high	Variable but probably low	Variable	Variable but low if produced with solar energy

A: 200–1,000 years depending on climate and soil type

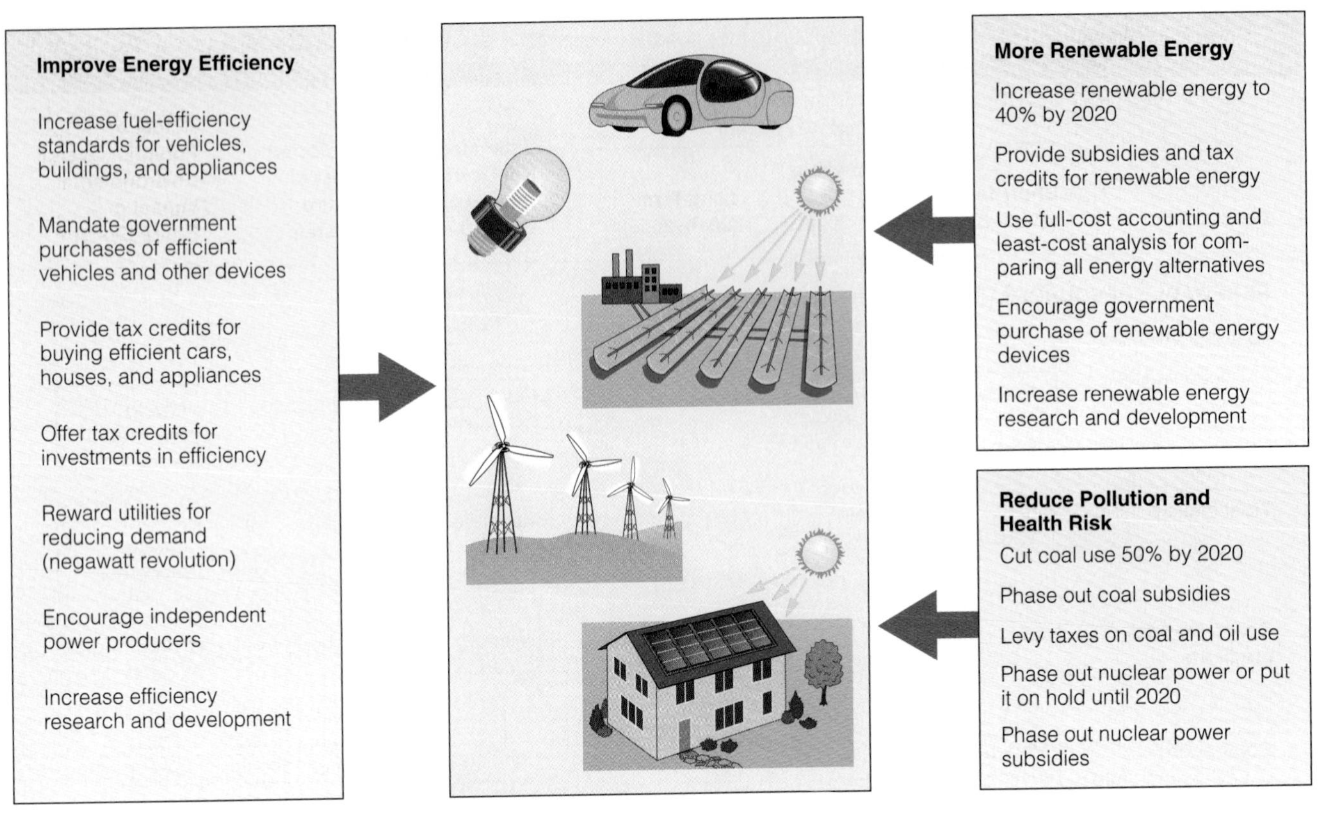

Figure 16-23 Solutions. Some ways various analysts have suggested to attain a sustainable energy future.

Government-based ways to improve energy efficiency include **(1)** *increasing fuel-efficiency standards for vehicles,* **(2)** *establishing energy-efficiency standards for buildings and appliances,* **(3)** *greatly increasing government-sponsored research and development to improve energy efficiency,* and **(4)** *giving tax credits and exemptions or government rebates for purchases of energy-efficient vehicles, houses, buildings, and appliances*. This last strategy could be accomplished with *freebates,* a self-financing mechanism that charges fees to purchasers of inefficient products and uses those funds to provide rebates to buyers of energy-efficient products. This approach relies on market forces and requires no new government taxes or subsidies.

Another suggested government-based approach involves *taxing energy*. This would help include some of the harmful costs of using energy (which consumers are now paying indirectly) in the market prices of energy; the tax revenues would be used to improve energy efficiency, encourage use of renewable energy resources, and provide energy assistance to poor and lower-middle-class Americans. Some economists believe that the public might accept these higher taxes if income or other taxes were lowered as gasoline or other fossil fuel taxes were raised.

Governments can also improve energy efficiency by *modifying electric utility regulations*. These changes, which a few state utility commissions are now putting into effect, would require utilities to produce electricity on a least-cost basis, permit them to earn money for their shareholders by reducing electricity demand, and allow rate increases based primarily on improvements in energy efficiency.

If utilities are rewarded for reducing electricity demand and cutting their customers' bills, the goal of the companies would be to maximize production of what Amory Lovins calls energy- and money-saving *negawatts* instead of megawatts. Lovins says we should "think of a 16-watt compact-fluorescent replacement for a 75-watt incandescent bulb, for example, as a 59-negawatt power plant."

Another suggested major goal would be to *rely more on renewable energy*. One way to encourage such a shift over the next 20–50 years would be to phase in *full-cost pricing*, in which environmental and health costs are included in the market price of energy. In other words, the prices of energy (and ideally all goods and services sold in the marketplace) should tell the ecological truth. This can provide consumers and power producers with more accurate information

Q: Worldwide, how much topsoil is eroded each year?

about the benefits and consequences of using various types of energy. Currently, Massachusetts, New York, and Wyoming are experimenting with environmental costing for electricity production.

Two other ways to encourage more reliance on renewable energy are *to provide subsidies or tax credits for homeowners and businesses that switch to various forms of renewable energy* and *to greatly increase government funding of renewable energy research and development* (Connections, p. 393).

Another innovation in the mostly monopolistic public utility industry is to *allow and encourage more competition from independent power producers.* A growing number of independent power companies will now build almost any kind of power plant anywhere in the world. Small decentralized devices for generating electricity and storing excess energy could allow developing countries to leapfrog into 21st-century power systems and bypass expensive central power plants and distribution systems.

Another suggested major goal is to *reduce the pollution and harmful health and ecological effects of relying on nonrenewable energy from coal and nuclear power.* This could be accomplished by phasing out most or all government subsidies and tax breaks for coal and nuclear power over a decade.

Within a decade or two many residential customers may be able to generate their own electricity using new technologies such as small cogenerators burning natural gas, hydrogen- or natural-gas–powered fuel cells, and roofs made of solar-cell tiles (already developed by several U.S. and Japanese companies). Any excess power they produce could be sold to other users on their electrical grid. Utilities might find it cheaper to buy much of their electricity from decentralized home-producers of electricity than to produce and distribute their own.

Energy experts estimate that implementing a carefully evaluated mix of such policies over the next two decades could save money, create a net gain in jobs, improve competitiveness in the global marketplace, slow projected global warming, and sharply reduce air and water pollution.

Making the Transition to a Sustainable Energy Future Industry, governments (national, state, and local), and individuals all have important roles to play in the development of a sustainable energy future. Brazil and Norway get more than half of their energy from hydropower, wood, and alcohol fuel. Israel, Japan, the Philippines, and Sweden plan to rely on renewable sources for most of their energy. India is greatly expanding its use of wind energy and offers full income tax deductions for renewable energy investments. California has become the world's showcase for solar and wind power. Some actions you can take to waste less energy (and thus reduce pollution and environmental degradation) and make more use of renewable energy are listed in Appendix 5.

The chief feature of the sustainable energy revolution over the next 40–50 years is likely to be significant decentralization, analogous to the computer industry's shift from large, centralized mainframes to small, widely dispersed PCs. As the rapidly growing forces for technological change in the energy industry gain momentum in the early 21st century, such a shift is likely to take place very rapidly.

Recent advances in two-way communications using fiber-optic cables and microprocessors will allow utility companies to monitor and control individual fuel cells, solar-cell rooftop systems, air conditioners, and water heaters of their customers. These "smart" control systems can turn such devices on or off, withdraw and buy excess power from customers, and sell them power as needed.

Such *energy management systems* can be integrated into high-speed information networks now being hooked up to millions of businesses and homes by the rapidly expanding telecommunication industry. Globally, such energy control systems should eliminate the need to build hundreds of large power plants over the next few decades. Alternatively, electric companies could play a major role in the information superhighway by installing their own fiber-optic lines and then leasing the excess capacity to companies that provide information services to homes and businesses.

By the middle of the next century the decentralized global energy system is likely to be based mostly on improved energy efficiency and much greater use of renewable forms of energy such as wind power, solar cells, and solar-produced hydrogen. Companies and investors trying to preserve large, centralized coal-burning and nuclear power plants may find themselves saddled with expensive technological dinosaurs. The nation that becomes the leader in the development of energy efficiency and solar technologies is likely to capture the huge future global market for this industry.

Between 1800 and 1970, the world shifted from dependence on solid fuels (wood and then coal) to a liquid fuel (oil) (Figure 15-4). Over the next 50 years we might see a shift from liquid oil to much cleaner gaseous fuels (methane, and solar-produced hydrogen).

In the long run, humanity has no choice but to rely on renewable energy. No matter how abundant they seem today, eventually coal and uranium will run out.

DANIEL DEUDNEY AND CHRISTOPHER FLAVIN

A: About 24 billion metric tons (26 billion tons)

CRITICAL THINKING

1. A homebuilder installs electric baseboard heat and claims that "it's the cheapest and cleanest way to go." Apply your understanding of the second law of energy (thermodynamics) to evaluate his claim.

2. Someone tells you that we can save energy by recycling it. How would you respond?

3. What are the five most important things an individual can do to save energy at home and in transportation (see Appendix 5)? Which, if any, of these do you currently do? Which, if any, do you plan to do?

4. What are the major reasons why most of the goods in the United States are transported by using energy-inefficient trucks instead of more energy-efficient trains? Relate this to the fact that most gasoline tax revenue is used to build roads.

5. Congratulations. You have just won $100,000 to build a house of your choice anywhere in the country. What type of house would you build? Where would you locate it? What types of materials would you use? What types of materials would you be sure *not* to use? How would you heat and cool your house? How would you heat your water? Considering fuel and energy efficiency, what sort of lighting, stove, refrigerator, washer, and dryer would you use? Which of these appliances could you do without? Would you use a dishwasher? A garbage disposal? A trash compactor? Suppose you learn that it will cost $140,000 to build the type of energy-efficient, passive solar house you want. Considering lifetime energy costs, would it make sense for you to borrow $40,000 to complete your house or to save the $40,000 by cutting back on energy-saving and solar design features?

6. Should the United States (or the country where you live) institute a crash program to develop solar photovoltaic cells and solar-produced hydrogen fuel? Explain.

7. Explain why you agree or disagree with the following proposals by various energy analysts:

- **a.** Federal subsidies for all energy alternatives should be eliminated so that all energy choices can compete in a true free-market system.
- **b.** All government tax breaks and other subsidies for conventional fuels (oil, natural gas, coal), synthetic natural gas and oil, and nuclear power (fission and fusion) should be removed and replaced with subsidies and tax breaks for improving energy efficiency and developing solar, wind, geothermal, and biomass energy alternatives.
- **c.** Development of solar and wind energy should be left to private enterprise and receive little or no help from the federal government, but nuclear energy and fossil fuels should continue to receive large federal subsidies.
- **d.** To solve present and future U.S. energy problems, all we need to do is find and develop more domestic supplies of oil, natural gas, and coal, and increase dependence on nuclear power.
- **e.** A heavy federal tax should be placed on gasoline and imported oil used in the United States.
- **f.** Between 2000 and 2020, the United States should phase out all nuclear power plants.

8. Explain why you agree or disagree with the proposals suggested in Figure 16-23 by various analysts as ways to promote a sustainable energy future for the United States.

9. Try to come up with analogies to complete the following statements: **(a)** A negawatt is like ______. **(b)** Living in a passively heated solar home is like ______. **(c)** Using solar cells to provide electricity is like ______.

10. Congratulations. You have just been put in charge of the world. List the five most important features of your energy policy.

PROJECTS

1. Make an energy-use study of your school or dorm, and use the findings to develop an energy-efficiency improvement program. Present your plan to school officials.

2. Do research to compare the direct and indirect federal, state, and local subsidies received by truck owners compared to those owning freight trains and railroad lines. Compare the use of trains to haul passengers and freight in most of Europe to their use for the same purposes in the United States and in Canada and Australia.

3. Make a concept map of this chapter's major ideas, using the section heads and subheads and the key terms (in boldface). Look at the inside back cover and in Appendix 5 for information about concept maps.

Q: Worldwide, how fast is soil eroding on farmland?

GUEST ESSAY

Technology Is the Answer (But What Was the Question?)

Amory B. Lovins

Physicist and energy consultant Amory B. Lovins is one of the world's most recognized and articulate experts on energy strategy. In 1989, he received the Delphi Prize for environmental work; in 1990 the Wall Street Journal *named him one of the 39 people most likely to change the course of business in the 1990s. He is research director at Rocky Mountain Institute, a nonprofit resource policy center that he and his wife, Hunter, founded in Snowmass, Colorado, in 1982. He has served as a consultant to over 200 utilities, private industries, and international organizations, and to many national, state, and local governments. He is active in energy affairs in more than 35 countries and has published several hundred papers and a dozen books on energy strategies and policies.*

The answers you get depend on the questions you ask. It is fashionable to suppose that we're running out of energy and that the solution is obviously to get lots more of it. But asking how to get more energy begs the question of how much we need and what are the cheapest and least environmentally harmful ways to meet these needs.

How much energy it takes to make steel, run a sewing machine, or keep ourselves comfortable in our houses depends on how cleverly we use energy, and the more it costs, the smarter we seem to get. It is now cheaper, for example, to double the efficiency of most industrial electric motor drive systems than to fuel existing power plants to make electricity. *(Just this one saving can more than replace the entire U.S. nuclear power program.)* We know how to make lights five times as efficient as those currently in use and how to make household appliances that give us the same work as now, but use one-fifth as much energy (saving money in the process).

Ten automakers have made good-sized, peppy, safe prototype cars averaging 29–59 kilometers per liter (67–138 miles per gallon) and within a decade automakers could have cars getting 64–128 kpl (150–300 mpg) on the road, if consumers demanded such cars. We know today how to make new buildings (and many old ones) so heat-tight (but still well ventilated) that they need essentially no outside energy to maintain comfort year-round, even in severe climates. In fact, I live and work in one [Figure 16-1].

These energy-saving measures are uniformly cheaper than going out and getting more energy. However, the old view of the energy problem included a worse mistake than forgetting to ask how much energy we needed: It sought more energy, in any form, from any source, at any price—as if all kinds of energy were alike.

Just as there are different kinds of food, so there are many different forms of energy, whose different prices and qualities suit them to different uses [Figure 3-14]. There is, after all, no demand for energy as such; nobody wants raw kilowatt-hours or barrels of sticky black goo. People instead want energy services: comfort, light, mobility, hot showers, cold beverages, and the ability to bake bread and make cement. We ought therefore to start at that part of the energy problem and ask, "What tasks do we want energy for, and what amount, type, and source of energy will do each task most cheaply?"

Electricity is a particularly high-quality, expensive form of energy. An average kilowatt-hour delivered in the United States in 1994 was priced at about 9.3¢, equivalent to buying the heat content of oil costing $154 per barrel—over eight times oil's average world price in 1994. The average cost of electricity from nuclear plants (including construction, fuel, and operating expenses) that began operation in 1988 was 13.5¢ per kilowatt-hour in 1994, equivalent on a heat basis to buying oil at about $216 per barrel.

Such costly energy might be worthwhile if it were used only for the premium tasks that require it, such as lights, motors, electronics, and smelters. But those special uses—only 8% of all delivered U.S. energy needs—are already met twice over by today's power stations. Two-fifths of our electricity is already spilling over into uneconomic, low-grade uses such as water heating, space heating, and air conditioning; yet no matter how efficiently we use electricity (even with heat pumps), we can never get our money's worth on these applications.

Thus, *supplying more electricity is irrelevant to the energy problem we have.* Even though electricity accounts for almost all the federal energy research-and-development budget and for at least half the national energy investment, it is the wrong kind of energy to meet our needs economically. Arguing about what kind of new power station to build—coal, nuclear, solar—is like shopping for the best buy in antique Chippendale chairs to burn in your stove, or for brandy to put in your car's gas tank. *It is the wrong approach.*

Indeed, *any kind of new power station is so uneconomical that if you have just built one, you will save the country money by writing it off and never operating it.* Why? Because its additional electricity can be used only for low-temperature heating and cooling (the premium "electricity-specific" uses being already filled up) and is the most expensive way of supplying those services. Saving electricity (creating negawatts) is much cheaper than making it.

The real question is, "What is the cheapest way to do low-temperature heating and cooling?" The answer is weather-stripping, insulation, heat exchangers, greenhouses, superwindows (which have as much insulating value as the outside wall of a typical house), window shades and overhangs, trees, and so on. These measures generally cost about half a penny to 2¢ per kilowatt-hour; the running costs alone for a new nuclear plant

(continued on the next page)

(continued from previous page)
will be nearly 4¢ per kilowatt-hour, so it's cheaper not to run it. In fact, under the crazy U.S. tax laws, the extra saving from not having to pay the plant's future subsidies is probably so big that by shutting the plant down society can also recover the capital cost of having built it!

If we want more electricity, we should get it from the cheapest sources first. In approximate order of increasing price, these include:

- Converting to efficient lighting equipment. This would save the United States electricity equal to the output of 120 large power plants, plus $30 billion a year in fuel and maintenance costs.
- Using more efficient electric motors to save up to half the energy used by motor systems. This would save electricity equal to the output of another 150 large power plants and repay the cost in about a year.
- Eliminating pure waste of electricity, such as lighting empty offices.
- Displacing with good architecture, weatherization, insulation, and passive and some active solar techniques the electricity now used for water heating and for space heating and cooling.
- Making appliances, smelters, and the like cost-effectively efficient.

Just these five measures can quadruple U.S. electrical efficiency, making it possible to run today's economy with no changes in lifestyles and using no power plants, whether old or new or fueled with oil, gas, coal, or uranium. We would need only the present hydroelectric capacity, readily available small-scale hydroelectric projects, and a modest amount of wind power.

If we still wanted more electricity, the next cheapest sources would include industrial cogeneration, combined heat-and-power plants, low-temperature heat engines run by industrial waste heat or by solar ponds, filling empty turbine bays and upgrading equipment in existing big dams, modern wind machines or small-scale hydroelectric turbines in good sites, steam-injected natural-gas turbines, and perhaps recent developments in solar cells with waste heat recovery.

It is only after we had clearly exhausted all these cheaper opportunities that we would even consider building a new central power station of any kind—the slowest and costliest known way to get more electricity (or to save oil).

To emphasize the importance of starting with energy end uses rather than energy sources, consider a sad little story from France involving a "spaghetti chart" (or energy flowchart)—a device energy planners often use to show how energy flows from primary sources via conversion processes to final forms and uses. In the mid-1970s, energy conservation planners in the French government started, wisely, on the right-hand side of the spaghetti chart. They found that their biggest need for energy was to heat buildings, and that even with good heat pumps, electricity would be the costliest way to do this. So they had a fight with their nationalized utility; they won, and electric heating was supposed to be discouraged or even phased out because it was so wasteful of money and fuel.

Meanwhile, down the street, the energy supply planners (who were far more numerous and influential in the French government) were starting on the left-hand side of the spaghetti chart. They said: "Look at all that nasty imported oil coming into our country! We must replace that oil. Oil is energy. . . . We need some other source of energy. Voilá! Reactors can give us energy; we'll build nuclear reactors all over the country." But they paid little attention to who would use that extra energy and no attention to relative prices.

Thus these two groups of the French energy establishment went on with their respective solutions to two different, indeed contradictory, French energy problems: *more energy of any kind* versus *the right kind to do each task in the most inexpensive way.* It was only in 1979 that these conflicting perceptions collided. The supply-side planners suddenly realized that the only thing they would be able to *sell* all that nuclear electricity for would be electric heating, which they had just agreed not to do.

Every industrial country is in this embarrassing position (especially if we include as "heating" air conditioning, which just means heating the outdoors instead of the indoors). Which end of the spaghetti chart we start on, or *what we think the energy problem is,* is not an academic abstraction; it *determines what we buy.* It is the most fundamental source of disagreement about energy policy.

People starting on the left side of the spaghetti chart think the problem boils down to whether to build coal or nuclear power stations (or both). People starting on the right realize that no kind of new power station can be an economic way to meet the needs for using electricity to provide low- and high-temperature heat and for the vehicular liquid fuels that are 92% of our energy problem.

So if we want to provide our energy services at a price we can afford, let's get straight what question our technologies are supposed to answer. Before we argue about the meatballs, let's untangle the strands of spaghetti, see where they're supposed to lead, and find out what we really need the energy *for*!

Critical Thinking

1. The author argues that building more nuclear, coal, or other electrical power plants to supply electricity for the United States is unnecessary and wasteful. Summarize the reasons for this conclusion, and give your reasons for agreeing or disagreeing with this viewpoint.

2. Do you agree or disagree that increasing the supply of energy, instead of improving energy efficiency, is the wrong answer to U.S. energy problems? Explain.

P A R T   I V

ENVIRONMENTAL QUALITY AND POLLUTION

In our every deliberation, we must consider the impact of our decisions on the next seven generations.

IROQUOIS CONFEDERATION, 18TH CENTURY

17 RISK, TOXICOLOGY, AND HUMAN HEALTH

The Big Killer

What is roughly the diameter of a 30-caliber bullet, can be bought almost anywhere, is highly addictive, and kills about 8,200 people every day? It's a cigarette (Figure 17-1). Currently, there are about 1.1 billion smokers in the world—about one-third of the global population above age 15. Most of these smokers (800 million) live in developing countries.

Cigarette smoking is the single most preventable major cause of death and suffering among adults. The World Health Organization (WHO) estimates that each year tobacco contributes to the premature deaths of at least 3 million people from heart disease, lung cancer, other cancers, bronchitis, emphysema, and stroke. The annual death toll from smoking-related diseases is projected by WHO to reach 10 million by 2020 and 12 million (primarily in developing countries) by 2050—an average of almost 33,000 preventable deaths per day.

In 1993 smoking killed about 419,000 Americans (up from 390,000 in 1985), an average of 1,150 deaths per day (Figure 17-1). This death toll is equivalent to three fully loaded jumbo jets crashing every day with no survivors. Smoking causes more deaths each year in the United States than do all illegal drugs, alcohol (the second most harmful drug after nicotine), automobile accidents, suicide, and homicide combined (Figure 17-1). According to a 1998 study, second-hand smoke (inhaled by nonsmokers) annually causes 30,000 to 60,000 premature deaths in the United States. A 1998 report estimated that more than 500,000 people in Europe and 750,000 people in China die from smoking-related disease every year.

The overwhelming consensus in the scientific community is that the nicotine (and probably the acetaldehyde) inhaled in tobacco smoke is highly addictive. Only 1 in 10 people who try to quit smoking succeed—about the same relapse rate as for recovering alcoholics and those addicted to heroin or crack cocaine. A British government study showed that adolescents who smoke more than one cigarette have an 85% chance of becoming smokers. About 80% of

Cause of Death	Deaths
Tobacco use	419,000
Alcohol use	150,000
Automobile accidents	42,000
AIDS	33,600
Suicides	30,000
Hard drug use	30,000
Homicides	22,500

Figure 17-1 Deaths in the United States from tobacco use and other causes in 1993. Smoking is by far the nation's leading cause of preventable death, causing more premature deaths each year than all the other categories in this figure combined. Cardiovascular disease causes about 180,000 of the 419,000 smoking-related deaths per year, followed by 120,000 from lung cancer and 9,000 from inhaling secondhand smoke. (Data from National Center for Health Statistics)

all smokers say they wish they had never started smoking and have tried to quit.

Government agencies and independent economists estimate that the country's 48 million smokers cost the United States up to $100 billion a year in medical bills, increased insurance costs, disability, lost earnings and productivity because of illness, and property damage from smoking-caused fires. This is an average of $4.20 per pack of cigarettes sold. Some put this figure much lower because when smokers die prematurely they don't draw benefits from social security and private pensions and cut down nursing home bills—certainly not a good argument for the benefits of smoking.

Many health experts urge that a $2–4 federal tax be added to the price of a pack of cigarettes. In England a pack of cigarettes costs about $5, versus $2 in the United States. Such a tax would mean that the users of cigarettes (and other tobacco products), not the rest of society, would pay a much greater share of the health, economic, and social costs associated with their smoking: a *user-pays* approach.

Other suggestions for reducing the death toll and health effects of smoking in the United States include **(1)** banning all cigarette advertising (as has been done in France), **(2)** forbidding the sale of cigarettes and other tobacco products to anyone under 21 (with strict penalties for violators), **(3)** banning all cigarette vending machines, **(4)** classifying nicotine as an addictive and dangerous drug (and placing its use in tobacco or other products under the jurisdiction of the Food and Drug Administration), **(5)** eliminating all federal subsidies and tax breaks to U.S. tobacco farmers and tobacco companies, and **(6)** using cigarette tax income to finance a massive antitobacco advertising and education program.

All of us face hazards to our health and well-being. Some of them, such as smoking, are avoidable, but others are not. When we evaluate the hazards we face, we must decide whether the risks of damage from each hazard outweigh the short- and long-term benefits, and how we can reduce the hazards and minimize the risks. Becoming better at evaluating the most serious risks we face and have control over (voluntary risks) can lead to a longer and healthier life. Such awareness can also keep us from being unnecessarily scared by minor risks.

For the first time in the history of the world, every human being is now subjected to dangerous chemicals from the moment of conception until death.

RACHEL CARSON

The discussion in this chapter answers several general questions:

- What types of hazards do people face?
- What is toxicology, and how do scientists measure toxicity?
- What chemical hazards do people face, and how can they be measured?
- What physical hazards do people face from earthquakes, volcanic eruptions, and ionizing radiation?
- What types of disease (biological hazards) threaten people in developing countries and developed countries?
- How can risks be estimated, managed, and reduced?

17-1 RISK AND HAZARDS

What Is Risk? **Risk** is the possibility of suffering harm from a *hazard* that can cause injury, disease, economic loss, or environmental damage. Risk is expressed in terms of **probability**: a mathematical statement about how likely it is that some event or effect will occur.

The probability of a risk is expressed as a fraction ranging from 0 (absolute certainty that there is no risk, which can never be shown) to 1.0 (absolute certainty that there is a risk). For example, the lifetime cancer risk from exposure to a particular chemical may have a probability of 0.000001, or 1 in 1 million. This means that one of every 1 million people exposed to the chemical at a specified average daily dose will develop cancer over a typical lifetime (usually considered to be 70 years).

Risk assessment involves using data, hypotheses, and models to estimate the probability of harm to human health, to society, or to the environment that may result from exposure to specific hazards.

What Are the Major Types of Hazards? The various kinds of hazards we face can be categorized as follows:

- *Cultural hazards* such as unsafe working conditions, smoking (p. 436), poor diet, drugs, drinking, driving, criminal assault, unsafe sex, and poverty
- *Chemical hazards* from harmful chemicals in the air (Chapter 18), water (Chapter 20), soil (Section 14-6), and food (Chapter 21)
- *Physical hazards* such as noise, fire, tornadoes, hurricanes, earthquakes, volcanic eruptions, floods (Section 13-8), and ionizing radiation
- *Biological hazards* from pathogens (bacteria, viruses, and parasites), pollen and other allergens, and animals such as bees and poisonous snakes

17-2 TOXICOLOGY

What Determines Whether a Chemical Is Harmful? Dose and Response The study of the adverse effects of chemicals on health is called **toxicology**. **Toxicity** is a measure of how harmful a substance is.

The amount of a potentially harmful substance that a person has ingested, inhaled, or absorbed through the skin is called the **dose** and the amount of the resulting type and amount of damage to health are called the **response**. Whether a chemical is harmful depends on the size of the dose over a certain time, how often an exposure occurs, who is exposed (adult or child, for example), and how well the body's detoxification systems (liver, lungs, and kidneys) work. *Acute exposure* involves a single dose, *chronic exposure* occurs over most or all of an entire lifetime, and *subchronic exposure* involves repeated exposures for some fraction of a lifetime.

There are two major types of responses to a harmful dose of a substance. An *acute effect* is an immediate or rapid harmful reaction to an exposure; it can range from dizziness or a rash to death. A *chronic effect* is a permanent or long-lasting consequence (kidney or liver damage, for example) of exposure to a harmful substance. Some individuals are more susceptible to specific toxins than others because of genetic differences, allergic responses, weakened immune systems, acute sensitivity, or age (children are usually more susceptible than adults).

Two factors affecting dose and response are *bioaccumulation* and *biomagnification*. **Bioaccumulation** is an increase in the concentration of a chemical in specific organs or tissues at a level higher than would normally be expected. Water-soluble toxic chemicals can be transported rapidly to most of the body's cells, but they rarely stay in the body very long because they are usually excreted in urine. However, oil- or fat-soluble toxins (usually organic molecules) don't move throughout the body; they can accumulate and be stored for years in lipid (fat) deposits. The time a chemical resides in the body (its *residence time*) is commonly given by its *biological half-life*—the time required for the quantity of the chemical in the body to be reduced by half.

The levels of some toxins in the environment can also be magnified as they pass through food chains and webs by a process called **biomagnification** (Figure 17-2). Examples of chemicals that can be biomagnified include long-lived, fat-soluble organic compounds such as the pesticide DDT and PCBs (oily chemicals used in electrical transformers) and some radioactive isotopes (such as strontium-90, Table 3-1). Stored in body fat, such chemicals can be passed along to offspring during gestation or egg laying and as mothers nurse their young.

Dose and especially response can also depend on *whether and how a chemical interacts with other chemicals*. Some chemicals can undergo *antagonistic interactions* that reduce their harmful effects or those of other chemicals. Vitamins A and E, for example, can interact to reduce the effects of some cancer-causing chemicals (carcinogens). Other chemicals can undergo *synergistic interactions* that increase the harmful effects of one or both chemicals. For example, nitrogen dioxide (NO_2), ozone (O_3), and several other common air pollutants apparently interact synergistically with allergens to increase the frequency and severity of asthma attacks.

The detection of trace amounts of a chemical in air, water, or food does not necessarily mean that it is there at a level harmful to most people. In some cases, all we may be doing is finding trace levels that could not be detected before. Indeed, practically any synthetic or natural chemical, even water, can be harmful if ingested in a large enough quantity. The critical question is how much exposure to a particular toxic chemical causes a harmful response.

Some people have the mistaken idea that all natural chemicals are safe and all synthetic chemicals are harmful. In fact, many synthetic chemicals are quite safe if used as intended, and many natural chemicals are deadly.

What Is a Poison? Legally, a **poison** is a chemical that has an LD_{50} of 50 milligrams or less per kilogram of body weight. An $\mathbf{LD_{50}}$ is the **median lethal dose**, or the amount of a chemical received in one dose that kills exactly 50% of the animals (usually rats and mice) in a test population (usually 60–200 animals) within a 14-day period. LD stands for lethal dose, and the

1998. "The Five Worst Environmental Threats to Children's Health." *Journal of Environmental Health*, vol. 60, no. 9, 46(2).

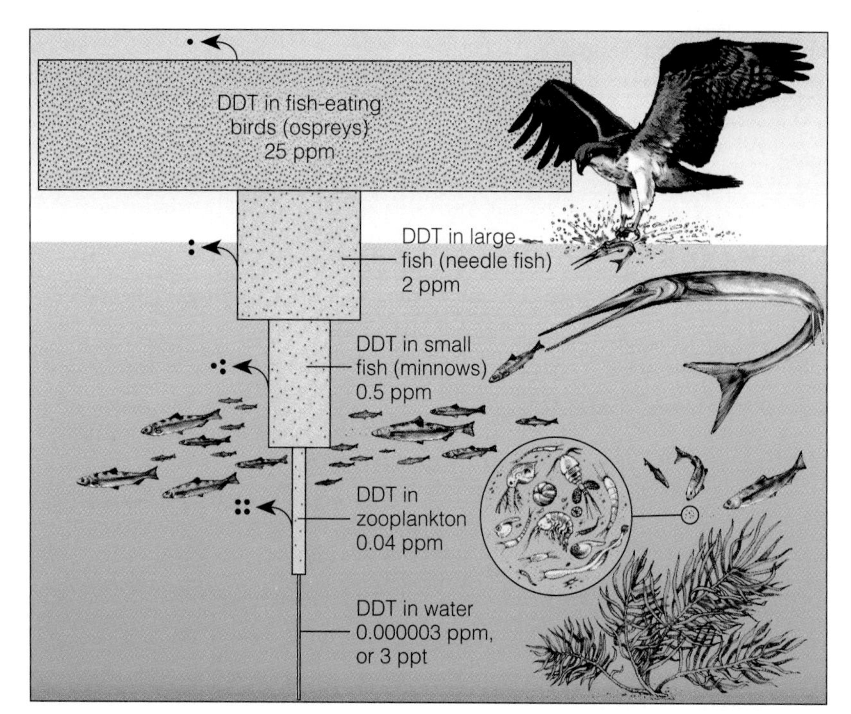

Figure 17-2 Bioaccumulation and biomagnification. DDT is a fat-soluble chemical that can bioaccumulate in the fatty tissues of animals. In a food chain or food web the accumulated concentrations of DDT can be biologically magnified in the bodies of animals at each higher trophic level. This diagram shows that the concentration of DDT in the fatty tissues of organisms was biomagnified about 10 million times in this food chain in an estuary near Long Island Sound. If each phytoplankton organism in such a food chain takes up from the water and retains one unit of DDT, a small fish eating thousands of zooplankton (which feed on the phytoplankton) will store thousands of units of DDT in its fatty tissue. Then each large fish that eats 10 of the smaller fish will ingest and store tens of thousands of units, and each bird (or human) that eats several large fish will ingest hundreds of thousands of units. Dots represent DDT, and arrows show small losses of DDT through respiration and excretion.

subscript 50 refers to the percentage of test organisms for which the dose was lethal.

Chemicals vary widely in their toxicity (Table 17-1); some poisons can cause serious harm or death after a single acute exposure at extremely low doses, whereas others cause such harm only at such huge doses that it is nearly impossible to get enough into the body. Most chemicals fall between these two extremes.

How Do Scientists Determine Toxicity? Three methods are used to determine the level at which a substance poses a health threat; each has certain limitations. One is *case reports* (usually made by physicians) about people suffering some adverse health effect or death after exposure to a chemical. Such information often involves accidental poisonings, drug overdoses, homicides, or suicide attempts.

A second method relies on *laboratory investigations* (usually on test animals) to determine toxicity, residence time, what parts of the body are affected, and (sometimes) how the harm takes place. Ideally, such a study also involves tracking and analyzing not only the original chemical but also its *metabolites*—potentially harmful chemicals to which the original chemical may be converted within the body. The third method is *epidemiology*, which involves studies of populations of humans exposed to certain chemicals or diseases.

Case reports are usually the least valuable source for determining toxicity because the actual dose and the exposed person's health status are often not known. However, such reports can provide clues about environmental hazards and suggest the need for laboratory investigations.

Acute toxicity and chronic toxicity are usually determined by tests on live laboratory animals (especially mice and rats, which are small and prolific and can be housed inexpensively in large numbers) and on bacteria. Such tests are also made on cell and tissue cultures and chicken egg membranes.

Acute toxicity tests are run to develop a **dose–response curve**, which shows the effects of various doses of a toxic agent on a group of test organisms (Figure 17-3). Such tests are *controlled experiments* in which the effects of the chemical on a *test group* are compared with the responses of a *control group* of organisms not exposed to the chemical. Care is taken to ensure that organisms in each group are as identical as possible in age, health status, and genetic makeup and that they are exposed to the same environmental conditions.

Fairly high dose levels are used in order to reduce the number of test animals needed, obtain results quickly, and lower costs. Otherwise, tests would have to be run on millions of laboratory animals for many years, and manufacturers couldn't afford to test most chemicals. For the same reasons, the results of high-dose exposures are usually extrapolated to low-dose levels using mathematical models. Then the extrapolated low-dose results on the test organisms are extrapolated to humans to estimate LD_{50} values for acute toxicity (Table 17-1).

Hint: Enter the search terms *pollution, health* using Key Words.

Table 17-1 Toxicity Ratings and Average Lethal Doses for Humans

Toxicity Rating	LD_{50} (milligrams per kilogram of body weight)*	Average Lethal Dose†	Examples
Supertoxic	Less than 0.01	Less than 1 drop	Nerve gases, botulism toxin, mushroom toxins, dioxin (TCDD)
Extremely toxic	Less than 5	Less than 7 drops	Potassium cyanide, heroin, atropine, parathion, nicotine
Very toxic	5–50	7 drops to 1 teaspoon	Mercury salts, morphine, codeine
Toxic	50–500	1 teaspoon to 1 ounce	Lead salts, DDT, sodium hydroxide fluoride, sulfuric acid, caffeine, carbon tetrachloride
Moderately toxic	500–5,000	1 ounce to 1 pint	Methyl (wood) alcohol, ether, phenobarbital, amphetamines (speed), kerosene, aspirin
Slightly toxic	5,000–15,000	1 pint to 1 quart	Ethyl alcohol, Lysol, soaps
Essentially nontoxic	15,000 or greater	More than 1 quart	Water, glycerin, table sugar

*Dosage that kills 50% of individuals exposed.
†Amounts of substances that are liquids at room temperature when given to a 70.4-kilogram (155-pound) human.

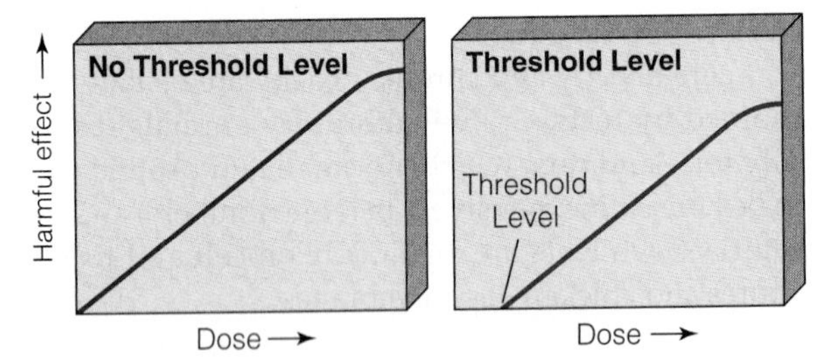

Figure 17-3 Two hypothetical dose–response curves. The curve on the left represents harmful effects that occur with increasing doses of a chemical or ionizing radiation; no dose is considered to be safe. The curve on the right shows the response of exposure to a chemical or ionizing radiation in which harmful effects appear only when the dose exceeds a certain threshold level. There is considerable uncertainty and controversy over which of these models applies to various harmful agents because of the difficulty in estimating the response to very low doses. Nonlinear threshold and nonthreshold dose–response curves, in which the graphs are curved instead of being straight lines, have also been observed.

According to the *linear dose–response model*, any dose of a toxic chemical or ionizing radiation has a certain risk of causing harm (Figure 17-3, left). With the *threshold dose–response model* (Figure 17-3, right) there is a threshold dose below which no detectable harmful effects occur, presumably because the body can repair the damage caused by low doses of some substances. It is extremely difficult to establish which of these models applies at low doses. To err on the side of safety, the linear or nonthreshold dose–response model is often assumed.

Some scientists challenge the validity of extrapolating data from test animals to humans because human physiology and metabolism are often different from those of the test animals. Others counter that such tests and models work fairly well (especially for revealing cancer risks) when the correct experimental animal is chosen or when a chemical is toxic to several different species of test animals.

Animal tests take 2–5 years and cost $200,000 to $2 million per substance tested. They are opposed by animal rights groups that want to limit or ban all use of test animals (or to ensure that experimental animals are treated in the most humane manner possible). Scientists are looking for substitute methods, but they point out that some animal testing is needed because the known alternatives cannot adequately mimic the complex biochemical interactions of a live animal.

Another approach to testing for toxicity and identifying the agents causing diseases is **epidemiology**: the study of the patterns of disease or toxicity to find out why some people get sick and others do not. Typically, the health of people exposed to a particular toxic agent or disease organism (the experimental group) is compared with the health of another group of statistically similar people not exposed to these conditions (the control group).

Q: How much money is lost each year because of the direct and indirect damage from soil erosion?

Epidemiology has limitations. For many toxic agents, too few people have been exposed to sufficiently high levels to allow detection of statistically significant differences. Because people are exposed to many different toxic agents and disease-causing factors throughout their lives, it is usually impossible to conclusively link an observed epidemiological effect with exposure to a particular toxic agent. All an epidemiological study can do is to establish strong, moderate, weak, or no statistical associations between a hazard and a health problem. Because epidemiology can be used to evaluate only the hazards to which people have already been exposed, it is rarely useful for predicting the effects of new technologies, substances, or diseases.

Thus, all methods for estimating toxicity levels and risks have serious limitations, but they are all we have. To take this uncertainty into account and minimize harm, standards for allowed exposure to toxic substances and radiation are typically set at levels 100 or even 1,000 times lower than the estimated harmful levels.

Despite their many limitations, carefully conducted and evaluated toxicity studies are important sources of information used to help us understand dose and response effects and to estimate and set exposure standards. But citizens, lawmakers, and regulatory officials must recognize the huge uncertainties and guesswork involved in all such studies.

17-3 CHEMICAL HAZARDS

What Are Toxic and Hazardous Chemicals? **Toxic chemicals** are generally defined as substances that are fatal to over 50% of test animals (LD_{50}) at given concentrations. **Hazardous chemicals** cause harm by **(1)** being flammable or explosive, **(2)** irritating or damaging the skin or lungs (strong acidic or alkaline substances such as oven cleaners), **(3)** interfering with or preventing oxygen uptake and distribution (asphyxiants such as carbon monoxide and hydrogen sulfide), or **(4)** inducing allergic reactions of the immune system (allergens).

What Are Mutagens, Teratogens, and Carcinogens? **Mutagens** are agents, such as chemicals and radiation, that cause random *mutations*, or changes in the DNA molecules found in cells. Mutations in a sperm or egg cell may be passed on to future generations and cause diseases such as manic depression, cystic fibrosis, hemophilia, sickle-cell anemia, Down's syndrome, and some types of cancer. Those in other cells are not inherited but may cause harmful effects such as tumors. Although some mutations are harmful, most are of no consequence, probably because all organisms have biochemical repair mechanisms that can find and correct mistakes or changes in the DNA code.

Teratogens are chemicals, radiation, or viruses that cause birth defects while the human embryo is growing and developing during pregnancy, especially during the first 3 months. Chemicals known to cause birth defects in laboratory animals include PCBs, thalidomide, steroid hormones, and heavy metals such as arsenic, cadmium, lead (Section 22-7), and mercury.

In the 1960s thalidomide was widely used as a nonprescription sleeping pill in Europe. However, as little as one pill of this drug during the first few weeks of pregnancy caused birth defects in which children had seal-like limbs with a hand or foot but no arm or leg. Before the drug was withdrawn from the market about 10,000 children were born with this defect. American women and children were protected because the U.S. Food and Drug Administration felt that laboratory tests of its safety were inadequate.

Carcinogens are chemicals, radiation, or viruses that cause or promote the growth of a malignant (cancerous) tumor, in which certain cells multiply uncontrollably. Many cancerous tumors spread by **metastasis** when malignant cells break off from tumors and travel in body fluids to other parts of the body. There, they start new tumors, making treatment much more difficult.

Because there are more than 100 types of cancer (depending on the types of cells involved), there are many different causes. These include genetic predisposition, viral infections, and exposure to various mutagens and carcinogens.

In the U.S. the incidence of all new cancers increased by 54% between 1950 and 1992, and the death rate for all cancers increased by 9.6%. There is good news and bad news about these rates. Both rates went *down* for four types of cancers (cervix, stomach, rectum, and uterus) but went *up* for 11 types of cancer (cancers of the ovaries, lung, skin, female breast, prostate, kidney, liver, brain, and pancreas, non-Hodgkin's lymphoma, and multiple myeloma). For eight other types of cancer (colon, larynx, testicular, bladder, and thyroid, Hodgkin's disease, leukemias, and all childhood cancers) the news is mixed, with the incidence increasing and the death rate declining (presumably because of earlier detection and improved treatment).

According to the World Health Organization, environmental and lifestyle factors play a key role in causing or promoting up to 80% of all cancers. Major sources of carcinogens are cigarette smoke (30–40% of cancers), diet (20–30%), occupational exposure (5–15%), and environmental pollutants (1–10%). About 10–20% of cancers are believed to be caused by inherited genetic factors (5%) or by certain viruses.

A: About $400 million (or $46 million per hour) worldwide and $44 billion in the United States

Typically, 10–40 years may elapse between the initial exposure to a carcinogen and the appearance of detectable symptoms. Partly because of this time lag, many healthy teenagers and young adults have trouble believing that their smoking (p. 436), drinking, eating, and other lifestyle habits today could lead to some form of cancer before they reach age 50.

How Can Chemicals Harm the Immune, Nervous, and Endocrine Systems? Since the 1970s a growing body of research on wildlife and laboratory animals and epidemiological studies of humans indicates that long-term (often low-level) exposure to various toxic chemicals in the environment can cause damage by disrupting the body's immune, nervous, and endocrine systems.

The *immune system* consists of numerous specialized cells and tissues that protect the body against disease and harmful substances. You can think of your immune system as having two levels of defense: **(1)** *antibodies*, which identify alien invaders in your bloodstream and mark them for other immune cells to attack, and **(2)** *cellular defenses*, such as killer T cells, which seek out and kill cells that have been invaded.

When it functions properly, the immune system fights off diseases caused by bacteria, viruses, fungi, parasites, and cancer cells. However, some synthetic chemicals, viruses such as HIV, ionizing radiation, and malnutrition can weaken the human immune system and leave the body wide open to attacks by allergens and infectious bacteria, viruses, and protozoans. Recent studies of laboratory animals and wildlife as well as epidemiological studies of humans (especially in developing countries) have linked suppression of the immune system to several widely used pesticides.

The human *nervous system* (brain, spinal cord, and peripheral nerves) is also being threatened by synthetic chemicals in the environment. Many poisons are *neurotoxins*, which attack nerve cells (neurons). Examples are **(1)** chlorinated hydrocarbons (DDT, PCBs, dioxins); **(2)** organophosphate pesticides (Section 21-2); **(3)** formaldehyde (Section 18-4); **(4)** various compounds of arsenic, mercury, lead (Section 22-7), and cadmium; and **(5)** widely used industrial solvents such as trichloroethylene (TCE), toluene, and xylene. In 1995, researchers in the Netherlands found a correlation between exposure of infants to low concentrations of PCBs and dioxin to weakened immunity because of suppressed levels of disease-fighting white blood cells.

The *endocrine system* is a complex set of organs and tissues whose actions are coordinated by chemical messengers called *hormones*. These hormones (which are secreted in extremely low levels into the bloodstream) control sexual reproduction, growth, development, and behavior in humans and other animals. The mix of estrogens (female hormones) and androgens (male hormones such as testosterone) is what creates one gender or another in developing embryos and fetuses.

Each type of hormone has a specific molecular shape that allows it to attach only to certain cell receptors; they fit together somewhat like a molecular key fitting into a specially shaped keyhole on a receptor molecule (Figure 17-4, left). Once bonded together, the hormone and its receptor molecule move onto the cell's nucleus to execute the chemical message carried by the hormone.

Case Study: Do Hormone Disrupters Threaten the Health of Wildlife and Humans? Over the last few years experts from a number of disciplines have been piecing together field studies on wildlife, studies on laboratory animals, and epidemiological studies of human populations indicating that a variety of human-made chemicals can act as *hormone disrupters*.

Some, called *hormone mimics*, are estrogen-like chemicals that disrupt the endocrine system by being able to attach to estrogen receptor molecules (Figure 17-4, center). Others, called *hormone blockers*, disrupt the endocrine system by preventing natural hormones such as androgens (male sex hormones) from attaching to their receptors (Figure 17-4, right). Recent research suggests that some chemicals can increase estrogen activity in one cell and block such activity in another, making it difficult to unravel their effects and establish safe levels of exposure. There is also growing concern about pollutants that can act as *thyroid disrupters* and cause growth, weight, brain, and behavioral disorders. Numerous studies indicate that extremely low levels of hormone disrupters, like trace levels of natural hormones, can affect developing embryos, human fetuses, and infants less than 6 months old.

So far 51 chemicals, many of them widely used, have been shown to act at extremely low levels as hormone disrupters in wildlife, laboratory animals, and some populations of humans. Examples include **(1)** dioxins (chlorinated hydrocarbons that are unwanted byproducts when chlorine-containing compounds are incinerated, Section 22-7), **(2)** certain PCBs (a group of chlorinated hydrocarbons widely distributed in the environment and capable of being biologically magnified in food chains), **(3)** various chemicals in certain plastics, **(4)** some pesticides, and **(5)** lead (Section 22-7) and mercury.

Numerous wildlife and laboratory studies reveal various possible effects of estrogen mimics and hormone blockers. Here are a few of many examples: Ranch minks fed Lake Michigan fish contaminated with endocrine disrupters such as DDT and PCBs failed to reproduce. Alligators in a Florida lake

Q: On what percentage of the world's cropland is topsoil eroding faster than it forms?

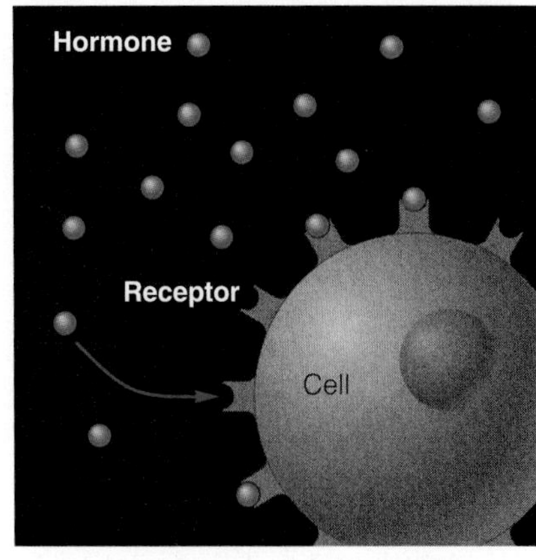

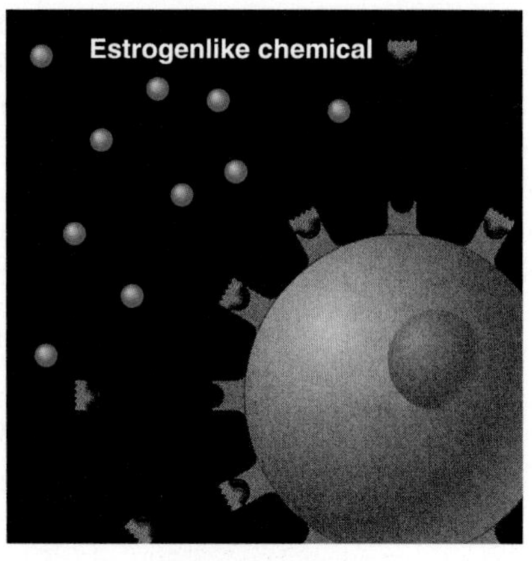

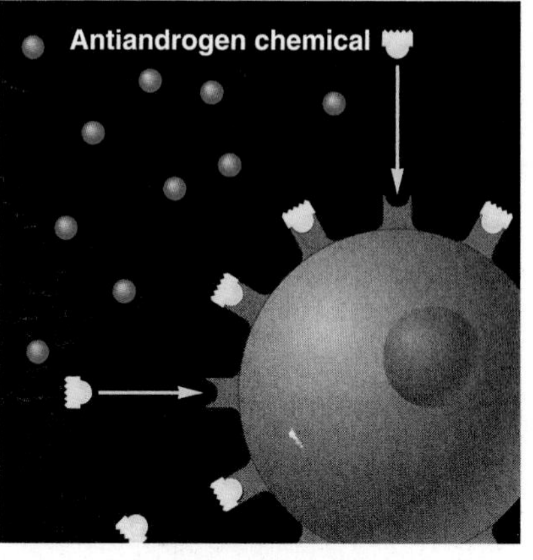

Normal Hormone Process **Hormone Mimic** **Hormone Blocker**

Figure 17-4 Hormones are molecules that act as messengers in the endocrine system to regulate various bodily processes, including reproduction, growth, and development. Each type of hormone has a unique molecular shape that allows it to attach to specially shaped receptors on the surface of or inside cells and transmit its chemical message (left). Molecules of certain pesticides and other molecules have shapes similar to those of natural hormones. Some of these hormone impostors, called *hormone mimics*, disrupt the endocrine system by attaching to estrogen receptor molecules (center). Others, called *hormone blockers*, prevent natural hormones such as androgens (male sex hormones) from attaching to their receptors (right). Some pollutants called *thyroid disrupters* may disrupt hormones released by thyroid glands and cause growth and weight disorders and brain and behavioral disorders.

polluted by a nearby hazardous-waste site had such abnormally small penises and low testosterone levels that they couldn't reproduce. Rats exposed to PCBs in the womb and infancy tend to be hyperactive. Estrogenlike p-nonylphenol (added to PVC plastic) can inhibit the growth of testicles in laboratory animals.

Most natural hormones are broken down or excreted. However, many of the synthetic hormone impostors are stable, fat-soluble compounds whose concentrations can be biomagnified as they move through food chains and webs (Figure 17-2). Thus, they can pose a special threat to humans and other carnivores dining at the top of food webs.

Here are the results of a few of the many frontier science studies suggesting that hormone disrupters can affect the human endocrine system:*

- Between the 1940s and early 1970s, several million pregnant women were given a synthetic estrogen called diethylstilbestrol (DES) to help prevent miscarriage. Later it was found that daughters of women taking DES had a much higher risk of vaginal cancer and their sons had a higher incidence of undescended testicles and testicular cancer.
- In 1973, estrogen mimics called PBBs accidentally got into cattle feed in Michigan, and from there into beef. Pregnant women who ate the beef (and whose breast milk had high levels of PBBs) had sons with undersized penises and malformed testicles.
- A higher than normal number of the children of women in Taiwan who ate rice oil contaminated with PCBs in 1979 have suffered growth retardation and slightly lower IQs, and many of the boys have abnormally small penises. In 1996, researchers found persistent decreases in intellect in children exposed before birth to much lower doses of PCBs than the children in Taiwan.
- During the past 50 years there have been dramatic increases in non-Hodgkin's lymphoma and cancers of the prostate, testicles, ovaries, and female breast in the United States and most industrialized countries. All of these cancers occur at hormonally sensitive receptor sites.
- There has been a sharp rise in endometriosis in increasingly younger women in the United States (from a handful of cases 50 years ago to 5 million cases today). This painful inflammation of the uterine lining often causes infertility. A recent study by German scientists found that women with endometriosis were more likely than others to have high levels of PCBs in their blood.
- In 1997, a reanalysis of sperm count data in the United States, Europe, and other parts of the world concluded that average sperm counts among men in the United States and Europe have declined by 50%

*For a readable summary of this research see Theo Coburn, *Our Stolen Future* (New York: Dutton, 1996).

A: About 33%

during the past 60 years. No drop was found in less industrialized countries in Asia and Latin America. In the early 1980s, researchers with the U.S. Environmental Protection Agency identified 16 industrial chemicals that reduce sperm counts. In 1995, European researchers found that exposure of mice to low levels of common industrial chemicals present in many U.S. foods cause smaller testicles and reduced sperm counts.

- Both childhood and adult-onset types of diabetes—serious diseases of the endocrine system—are on the rise worldwide and are expected to be one of the world's major killers by 2010. Much of this (especially adult-onset diabetes) may be caused by excess body fat and lack of exercise. However, a 1997 study found that veterans of the Vietnam War have an increased likelihood of getting diabetes if they have elevated levels of dioxin (Section 22-7) in their blood. Dioxin is a powerful hormone-disrupting chemical found in Agent Orange, an herbicide sprayed by American forces to defoliate forests during the Vietnam War.

Much more research is needed to verify such frontier science findings and to determine whether low levels of most hormone-disrupting chemicals in the environment pose a threat to the human population. In 1996, an international group of scientists and physicians issued a consensus statement expressing great concern about the effects of hormone-disrupting chemicals on the human brain and central nervous system. That same year the U.S. National Academy of Sciences began an evaluation of endocrine disrupters and the EPA made evaluation of this potentially serious problem a top research priority.

Why Do We Know So Little About the Harmful Effects of Chemicals? According to risk assessment expert Joseph V. Rodricks, "Toxicologists know a great deal about a few chemicals, a little about many, and next to nothing about most." The U.S. National Academy of Sciences estimates that only about 10% of the nearly 100,000 chemicals in commercial use have been thoroughly screened for toxicity and only 2% have been adequately tested to determine whether they are carcinogens, teratogens, or mutagens. Hardly any of these chemicals have been screened for damage to the nervous, endocrine, and immune systems.

Each year we introduce into the marketplace about 1,000 new chemicals about whose potentially harmful effects we have little knowledge. Currently, about 99.5% of the commercially used chemicals in the United States are not regulated by federal and state governments.

There are three major reasons for this lack of information. *First*, under existing laws most chemicals are considered innocent until proven guilty. No one is required to investigate whether they are harmful. *Second*, there are not enough funds, personnel, facilities, and test animals to provide such information for more than a small fraction of the many chemicals we encounter in our daily lives.

Third, even if we could make a reasonable estimate of the biggest risks associated with particular technologies or chemicals (a very difficult and expensive thing to do), we know little about their possible interactions with other technologies and chemicals or about the effects of such interactions on human health and ecosystems. For example, just to study the possible different three-chemical interactions among the top 500 most widely used industrial chemicals would require 20.7 million experiments—a physical and financial impossibility.

The difficulty and expense of getting information about the harmful effects of chemicals is one reason an increasing number of environmentalists and health officials are pushing for much greater emphasis on *pollution prevention*. This strategy greatly reduces the need for statistically uncertain and controversial toxicity studies and exposure standards. It also reduces the risk posed by potentially hazardous chemicals and products and their possible but poorly understood multiple interactions.

17-4 PHYSICAL HAZARDS: EARTHQUAKES, VOLCANIC ERUPTIONS, AND IONIZING RADIATION

What Are Earthquakes? Stress in the earth's crust can cause solid rock to deform until it suddenly fractures and shifts along the fracture, producing a *fault*. The faulting or a later abrupt movement on an existing fault causes an **earthquake**. An earthquake has certain features and effects (Figure 17-5). When the stressed parts of the earth suddenly fracture or shift, energy is released as shock waves, which move outward from the earthquake's focus like ripples in a pool of water.

One way of measuring the severity of an earthquake is by its *magnitude* on a modified version of the Richter scale. The magnitude is a measure of the amount of energy released in the earthquake, as indicated by the amplitude (size) of the vibrations when they reach a recording instrument (seismograph). Using this approach, seismologists rate earthquakes as *insignificant* (less than 4.0 on the Richter scale), *minor* (4.0–4.9), *damaging* (5.0–5.9), *destructive* (6.0–6.9), *major* (7.0–7.9), and *great* (over 8.0). Each unit on the Richter scale represents an amplitude that is 10 times greater than the next smaller unit, so a magnitude 5.0 earth-

Q: How much land has become desertified during the past 50 years?

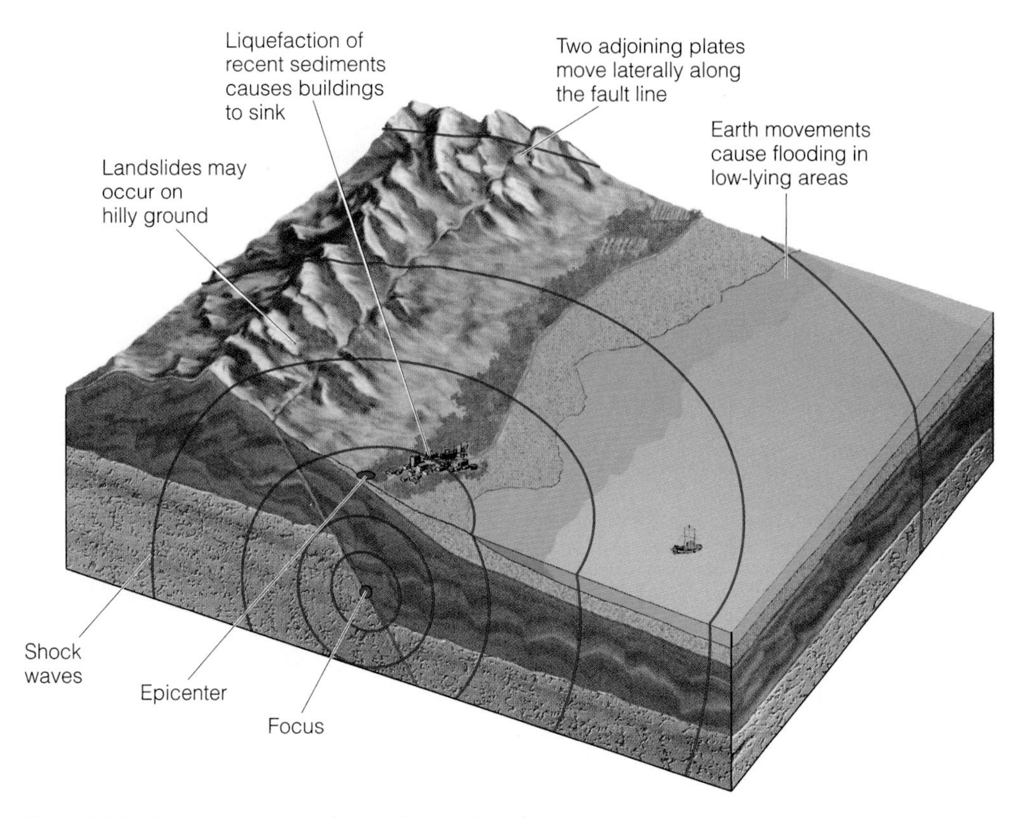

Figure 17-5 Major features and effects of an earthquake.

quake is 10 times greater than a magnitude 4.0, and a magnitude 6.0 quake is 100 times greater than a magnitude 4.0 quake.

The northern California earthquake of 1989 had a Richter magnitude of 7.1 and caused damage within a radius of 97 kilometers (60 miles) from its epicenter (Figure 17-6). Earthquakes often have *aftershocks* that gradually decrease in frequency over a period of up to several months, and some have *foreshocks* from seconds to weeks before the main shock.

The primary effects of earthquakes include shaking and sometimes a permanent vertical or horizontal displacement of the ground. These effects may have serious consequences for people and for buildings, bridges, freeway overpasses, dams, and pipelines. Secondary effects of earthquakes include rockslides, urban fires, and flooding caused by subsidence (sinking) of land. Coastal areas can also be severely damaged by large, earthquake-generated water waves, called *tsunamis* (misnamed "tidal waves," even though they have nothing to do with tides) that travel as fast as 950 kilometers (590 miles) per hour.

Solutions: How Can We Reduce Earthquake Hazards? Loss of life and property from earthquakes can be reduced. To do this we **(1)** examine historical records and make geologic measurements to locate active fault zones, **(2)** make maps showing areas in which ground conditions are more subject to shaking and thus are high-risk areas (Figure 17-7), **(3)** establish building codes that regulate the placement and design of buildings in areas of high risk, and **(4)** ideally, learn to predict when and where earthquakes will occur.

What Are Volcanoes? An active **volcano** occurs where magma (molten rock) reaches the earth's surface through a central vent or a long crack (fissure). Volcanic activity can release *ejecta* (debris ranging from large chunks of lava rock to ash that may be glowing hot), liquid lava, and gases (water vapor, carbon dioxide, sulfur dioxide, nitrogen oxides, and others) into the environment.

Some volcanoes, such as those at Mount St. Helens in Washington (which erupted in 1980) and Mount Pinatubo in the Philippines (which erupted in 1991), have a steep, flaring cone shape and usually erupt explosively. Others, such as those in the islands of Iceland and Hawaii (Figure 27-1), typically erupt more quietly. Most produce lava flows, which can cover roads and villages and ignite brush, trees, and homes.

Land-use planning, better prediction of volcanic eruptions, and development of effective evacuation plans can reduce the loss of human life (and sometimes property) from volcanic eruptions. Prediction of

A: About 8.1 million square kilometers (3.1 million square miles), an area about the size of Brazil

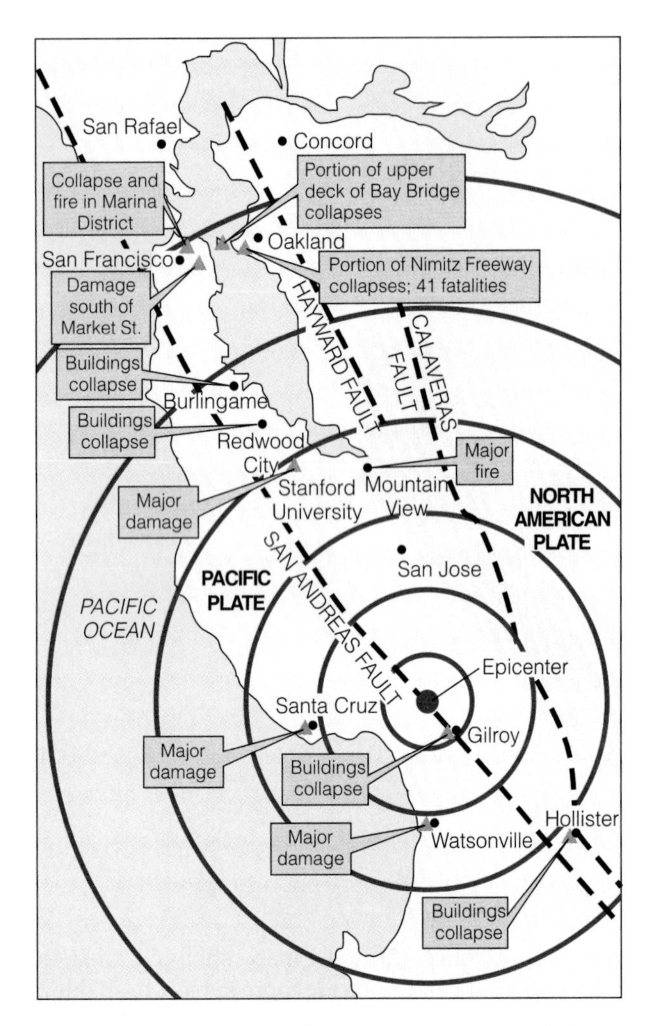

Figure 17-6 At 5:04 P.M. on October 17, 1989, a magnitude 7.1 earthquake occurred in northern California along the San Andreas Fault. It was the largest earthquake in northern California since 1906, when the great San Francisco quake and the resulting fires destroyed much of the city. In the 1989 quake, the most extensive damage was within a radius of 32 kilometers (20 miles) from the epicenter, but seismic waves caused major damage as far away as San Francisco and Oakland. Sixty-seven people were killed and official damage estimates were as high as $10 billion. This was North America's costliest natural disaster until Hurricane Andrew devastated Florida in 1992.

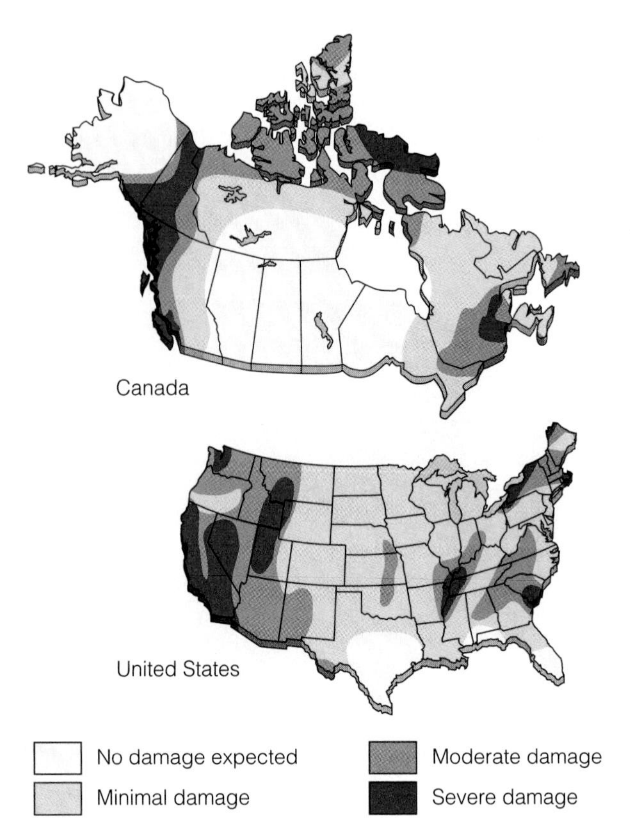

Figure 17-7 Expected damage from earthquakes in Canada and the contiguous United States. Except for a few regions along the Atlantic and the Gulf coasts, virtually every part of the continental United States is subject to some risk from earthquakes. Several areas have a risk of moderate to major damage. This map is based on earthquake records. (Data from U.S. Geological Survey)

volcanic activity has improved considerably during this century, but every volcano has its own personality. Like earthquake maps, the eruptive history of a volcano or volcanic center can provide some indication of where the risks are. Volcanologists are now studying precursor phenomena such as tilting or swelling of the cone, changes in magnetic and thermal properties of the volcano, changes in gas composition, and increased seismic activity.

We tend to think negatively of volcanic activity, but it also provides some benefits. One is outstanding scenery in the form of majestic mountains, some lakes (such as Crater Lake in Oregon; see photo in Table of Contents), and other landforms. Perhaps the most important benefit of volcanism is the highly fertile soils produced by the weathering of lava.

On May 18, 1980, Mount St. Helens, in the Cascade Range near the Washington–Oregon border, erupted in what has been described as the worst volcanic disaster in U.S. history. Fifty-seven people died in the eruption, and several hundred cabins and homes were destroyed or severely damaged. Tens of thousand of hectares of forest were obliterated, along with campgrounds and bridges. An estimated 7,000 big-game animals (bear, deer, elk, mountain lions) died, as did millions of smaller animals and birds and some 11 million fish. Salmon hatcheries were damaged and crops (including alfalfa, apples, potatoes, and wheat) were lost. Many people living in the area also lost their jobs.

On the plus side, trace elements from ash that were added to the soil may benefit agriculture in the long run.

Q: How much of the planet's land is threatened by desertification?

Furthermore, increased tourism to the area brought new jobs and income. By 1990 many biologists were surprised at how fast various forms of life had begun colonizing many of the most devastated areas. This rapid recovery has taught biologists important and often surprising lessons about nature's ability to recover from what seems to be almost total devastation.

How Much Ionizing Radiation Are We Exposed To? Ionizing radiation, a form of electromagnetic radiation (Figure 3-13), has enough energy to damage body tissues. Examples are X rays, ultraviolet radiation from the sun and sun lamps (Connections, p. 526), neutrons emitted by nuclear fission (Figure 3-18) and nuclear fusion (Figure 3-19), and alpha, beta, and gamma radiation emitted by radioactive isotopes (Figure 3-15).

Each year people are exposed to some ionizing radiation from natural or background sources and from human activities (Figure 17-8). Sources of natural ionizing radiation include cosmic rays from outer space, soil, rocks, air, water, and food. Nuclear power plants provide very low exposure if they are operating properly. However, a serious nuclear accident, such as the one that occurred in 1986 at the Chernobyl nuclear power plant in Ukraine (p. 366), can release large quantities of radioactive materials, killing and harming people and rendering large areas uninhabitable.

Most human-caused exposure to ionizing radiation comes from medical X rays and from diagnostic tests and treatments using radioactive isotopes. The federal government estimates that one-third of the X rays taken each year in the United States are unnecessary.

What Are the Effects of Ionizing Radiation? Ionizing radiation can cause harm by penetrating a human cell, knocking loose (ionizing) one or more electrons from a cellular chemical, and thus altering molecules needed for normal cellular functioning. Exposure to ionizing radiation can damage cells in two ways. One is *genetic damage* from mutations in DNA molecules (Figure 3-9) that alter genes and chromosomes (Figure 3-10). If the mutation is harmful, genetic defects can become apparent in the next generation of offspring or several generations later. The other way is *somatic damage* to tissues, which causes harm during the victim's lifetime. Examples include burns, miscarriages, eye cataracts, and certain cancers.

Most scientists assume that the dose–response curve for radioactivity is linear (Figure 17-3, left), but others believe that there may be a threshold level (Figure 17-3, right) because of the body's ability to repair some of the damage. However, a 1990 study by the U.S. National Academy of Sciences concluded that the likelihood of getting cancer from exposure to

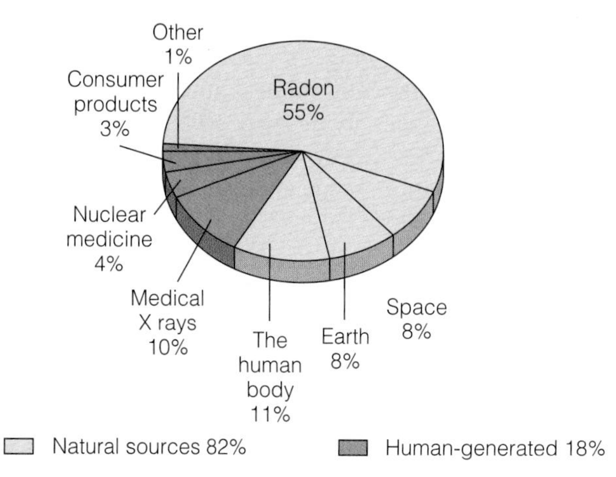

Figure 17-8 Natural and human sources of the average annual dose of ionizing radiation received by the U.S. population. Most studies indicate that there is no safe dose of ionizing radiation. (Data from National Council on Radiation Protection and Measurements)

a low dose of radiation is two to four times higher than previously thought, and that there is no threshold level for damage from ionizing radiation. A few scientists believe that there is evidence that low doses of radiation produce more cancer per unit of radiation than high doses—the *superlinear hypothesis*.

As more information on the effects of exposure to low levels of ionizing radiation has become available, laws have lowered the amount of occupational exposure to radiation allowed in the United States to almost one-eighth of its original level between 1949 and 1996. Some scientists believe that the current limits on occupational exposure to low-level radiation should be even lower. The legal limit of radiation exposure allowed for the public (which is one-twentieth that allowed for radiation workers) decreased to one-fifth of its original level between 1956 and 1996.

The effects of ionizing radiation vary with the type, penetrating power (Figure 3-15), source (outside or inside the body), and half-life of the radioisotope (Figure 3-16 and Table 3-1). Alpha particles lack the penetrating power of beta particles but have more energy; thus, alpha-emitting isotopes are particularly dangerous when breathed in or ingested with food or water. Alpha particles outside the body can cause skin cancer but cannot penetrate the skin and reach vital organs.

Because beta particles can penetrate the skin, a beta emitter outside the body can damage internal organs. In general, radioisotopes with intermediate half-lives pose the greatest threat to human health; they stay around long enough to reach and enter the human body but decay fast enough to cause considerable damage while they are around.

A: An area roughly the size of North and South America combined

According to the National Academy of Sciences, exposure over an average lifetime to average levels of ionizing radiation from natural and human sources (Figure 17-8) causes about 1% of all fatal cancers and 5–6% of all normally encountered genetic defects in the U.S. population. Some analysts believe that these estimates are too low.

Is Nonionizing Electromagnetic Radiation Harmful? The simple answer is that we don't know. *Electromagnetic fields (EMFs)* are low-energy, nonionizing forms of electromagnetic radiation (Figure 3-13) given off when an electric current passes through a wire or a motor. Sources of these weak electrical and magnetic fields include overhead power lines and household electrical appliances (such as microwave ovens, hair dryers, electric blankets, waterbed heaters, electric razors, and video display terminals of computer monitors and TV sets).

Since the late 1960s there has been growing public concern and controversy over the possibility that EMFs (especially the magnetic fields) could have harmful health effects on humans. Numerous epidemiological studies have suggested that prolonged exposure to EMFs could lead to increased risk from some cancers (including childhood leukemia, brain tumors, and breast cancer), miscarriages, birth defects, and Alzheimer's disease.

However, a number of the earlier studies have been criticized because of poor design, faulty statistical analysis, and very weak statistical correlation between EMFs and various health effects. Better-designed, more recent studies show some statistical correlation, but the connections are still weak.

Many respected scientists say that a statistical link between EMFs and various health effects has been established. At least an equal number of other respected scientists contend that no link has been established or that the links are statistically insignificant. It may be decades before these contradictory results and evaluations are resolved. In the meantime, scientists urge us not to panic but to take prudent actions, based on the *precautionary principle*, to reduce our exposure to EMFs. Such actions are summarized in Appendix 5.

17-5 BIOLOGICAL HAZARDS

What Are Nontransmissible Diseases? Diseases not caused by living organisms and that do not spread from one person to another are called **nontransmissible diseases**. Examples are cardiovascular (heart and blood vessel) disorders, most cancers, diabetes, bronchitis, emphysema, and malnutrition. Such diseases typically have multiple (and often unknown) causes and tend to develop slowly and progressively over time.

What Are Transmissible Diseases? A **transmissible disease** is caused by a living organism (such as a bacterium, virus, protozoa, or parasite) and can be spread from one person to another. The infectious agents, called *pathogens*, are spread by air, water, food, body fluids, and some insects (such as mosquitoes, flies, and ticks), animals (such as rodents and monkeys), and other nonhuman carriers called *vectors*. Table 17-2 gives information on the world's eight deadliest infectious diseases.

Typically, a *bacterium* is a one-celled microorganism capable of replicating itself by simple division. A *virus* is a microscopic, noncellular infectious agent. It is basically a small strand of nucleic acid, either DNA or RNA, wrapped in a protein coat. Its DNA contains instructions for making more viruses, but it has no apparatus to do so. To replicate itself, a virus must invade a host cell and take over the cell's DNA to create a factory for producing more viruses.

In developing countries, infectious diseases accounted for about 44% of the deaths, compared to only 5% in developed countries in 1997. In 1997, infectious diseases were the world's leading cause of death, resulting in 52.2 million deaths. According to the World Health Organization and UNICEF, every year in developing countries at least 11 million children under the age of 5 die of mostly preventable infectious diseases—an average of at least 27,000 premature deaths per day. About 80% of all illnesses in developing countries are caused by waterborne infectious diseases (such as diarrhea, hepatitis, typhoid fever, and cholera), mainly from unsafe drinking water and inadequate sanitation systems.

Table 17-2 shows that the biggest killers by far are infectious diseases—such as respiratory infections, diarrheal diseases, and tuberculosis (Case Study, p. 450)—that have been around for a long time. By comparison, the various forms of highly publicized and sensationalized *ebola viruses* killed about 650 people (mostly in the African countries of Zaire, Sudan, and Gabon) between 1976 and June 1996. In their present nonairborne forms, these viruses are not very contagious. Transmission requires intimate contact with the blood, feces, or other body fluids of infected victims. Most victims also don't live long enough to infect many people.

However, ebola, HIV, and other new *emerging diseases*, which have been rising for at least two decades and are likely to increase in the near future, should not be ignored. Since 1973 at least 28 such diseases have been identified. Many of these diseases have no known cure or treatment and thus could spread rapidly. In most cases, these "new" diseases do not

Q: What major food-producing country is doing the most to reduce soil erosion?

Table 17-2 The World's Eight Most Deadly Infectious Diseases

Disease	Cause	Estimated New Cases per Year	Estimated Deaths per Year
Acute respiratory infections*	Bacteria, viruses	1 billion	4.7 million
Diarrheal diseases†	Bacteria, viruses parasites	1.8 billion	3.1 million
Tuberculosis	Bacteria	9 million	3.1 million
Malaria	Parasitic protozoa	110 million	2.5–2.7 million
AIDS	Virus (HIV)	600,000 (AIDS) 4 million (HIV)	2.3 million
Measles	Viruses	200 million	1 million
Hepatitis B	Virus	200 million	1 million
Tetanus	Bacteria	1 million	500,000

Source: World Health Organization (WHO). *The World Health Report 1997* (Geneva, Switzerland: WHO, 1998).

*Includes pneumonia, influenza, and whooping cough.

†Includes amoebic dysentery, cryptosporisdosis, and gastroenteritis.

emerge through mutation, but because disease-causing bacteria, viruses, or parasites find a new pathway to infect previously unexposed populations.

Connections: What Factors Can Affect the Spread of Transmissible Diseases? Outbreaks of infectious diseases often occur because of a change in the physical, social, or biological environment of disease reservoirs, carrier vectors, or hosts. For example, travel, migration, deforestation, loss of biodiversity, agriculture, urbanization, climate change, and natural disasters all contribute to the introduction and spread of infectious agents in new populations.

In 1985, for instance, the Asian tiger mosquito that is a vector for dengue fever (called "breakbone fever" by those who experience the excruciating pain it causes in joints), yellow fever, and other viruses was brought accidentally to the United States inside used tires shipped from Asia. Today, this species has become established in Hawaii, the southeastern United States, and has begun extending its range northward toward Chicago and Washington, D.C. Some disease-carrying species of mosquitoes, flies, and beetles can survive extremely cold temperatures inside the wheel bays of aircraft for 6–9 hours and thus can rapidly spread infectious organisms between countries and across continents.

Migration to urban areas increases the probability of infection from diseases such as TB (Case Study, p. 450), cholera, and sexually transmitted diseases (Connections, p. 454). Conversely, migration to uninhabited rural areas and deforestation can expose people to new diseases and disease vectors.

In 1998, scientists warned that increased outbreaks of many tropical infectious diseases in developing countries (such as malaria, sleeping sickness, and river blindness) are related to reducing biodiversity by destroying forests (Chapter 24) and wiping out other species that help control disease vectors (Chapter 25). For example, as more people clear and move into tropical forests, they come into more contact with monkeys that harbor many mostly unknown viruses that humans are susceptible to. People catch the viruses when they eat monkeys or chimpanzees, capture them as pets, or use them in scientific laboratories. As species of primates are wiped out, such viruses will seek out new hosts, perhaps humans.

Another example is increases in tick-transmitted Lyme disease in the United States. Deer mice carry the spirochete bacterium that causes Lyme disease, which causes fever, lethargy, and (sometimes) long-lasting arthritis in humans. The mice are bitten by certain species of ticks that can then transfer the infection to people. As more people have moved into wooded suburbs (especially in the eastern United States), they have come into greater contact with ticks infested with Lyme disease bacteria.

Recent research shows that there are lower incidences of Lyme disease in states with high numbers of animals such as birds, rabbits, and raccoons, even though the tick carriers are common in all states. A suggested reason for this is that although most of these

A: United States

The Global Tuberculosis Epidemic

CASE STUDY

One of the world's most underreported stories in the 1990s has been the rapid spread of tuberculosis (TB), a highly infectious bacterial disease that currently kills about 3.1 million people per year—about one-fourth of all preventable deaths in developing countries (Figure 17-9). The current TB epidemic is so severe that in 1993 the World Health Organization (WHO) declared a global state of emergency.

Tuberculosis in humans is caused by the bacterium *Mycobacterium tuberculosis.* Infection moves from person to person, mainly in airborne droplets produced by coughing, sneezing, singing, or even talking. After entering the lungs, the bacteria multiply. The body's immune system checks the infections in about 90–95% of the people carrying the bacterium, leaving them with what is called *inert TB.* Currently, about one of every three people on earth has inert TB, which can sometimes become *active TB.* About half of the infected people who spread this highly contagious disease don't know they are infected. Because many of the infected people look and act normal, this global health problem has been called a *silent epidemic.*

In about 5–10% of the cases, the infection is not stopped; it multiplies in the lungs and spreads into the lymph nodes. Young children, the elderly, alcoholics, people with weakened immune systems (from diabetes, AIDS, or chemotherapy, for example), and those in close contact with people suffering from active TB are especially susceptible.

Until recently the incidence of TB had fallen sharply (except among the poor), mostly because of prevention programs (such as X ray screening and detection of people with active TB) and treatment with antibiotics, which began in the 1940s. Until antibiotics were developed, the spread of the disease was controlled by isolating patients with active TB in sanitariums until they either died or recovered.

Major reasons for the recent increase in TB are **(1)** poor TB screening and control programs (especially in developing countries, where about 95% of the new cases occur), **(2)** development of strains of the tuberculosis bacterium that are genetically resistant to virtually all effective antibiotics (typically leading to mortality rates of over 50%), **(3)** population growth and increased urbanization and crowding (which increase contacts among people), **(4)** poverty, and **(5)** the spread of AIDS, which greatly weakens the immune system and allows TB bacteria to multiply.

Slowing the spread of the disease involves early identification and treatment of people with active TB, usually those with a chronic cough. Treatment with a combination of four inexpensive drugs can cure 90% of those with active TB. To be successful, however, the drugs must be taken every day for 6 to 8 months. Because the symptoms disappear after a few weeks, many patients think they are cured and stop taking the drugs. This allows the disease to recur in a hard-to-treat form; it then spreads to other people and drug-resistant strains of TB bacteria develop.

In most developing countries it costs less than $30 to supply the drugs to cure one person and stop the transmission of the disease. Yet developing country governments spend on TB only 2¢ out of every $10 they assign to health. Researchers are closing in on a TB vaccine that has been successful in preventing TB infections in guinea pigs, but such a vaccine may not be available for years or decades.

According to the World Health Organization, a worldwide campaign to help control TB would cost only about $360 million to help save 30 million lives during the next decade.

Critical Thinking

Before you read this chapter were you aware of the serious global TB epidemic? Why do you think that this important story has gotten so little media attention compared to other diseases that cause many fewer deaths per year?

other animals carry the Lyme bacterium, only a few transmit it to ticks. Thus, preserving this animal biodiversity in forests spreads the Lyme bacterium over a large number of different species and can reduce its spread to humans.

The increased cultivation of rice in many parts of the world also increases populations of mosquitoes and other insects that transmit diseases to humans. When they plant rice, farmers often flood their fields or paddies, creating an ideal breeding ground for such insects.

Climate change can also affect the spread of infectious diseases. For example, scientists have established a link between the *El Niño–Southern Oscillation (ENSO,* Figure 7-8) that occurs every few years and spread of the bacterium that causes cholera. These bacteria can live inside tiny crustaceans called copepods that in turn live in microscopic aquatic animals (zooplankton) found in the coastal waters and the open sea. The warming of surface waters warm during an ENSO can produce plankton blooms and increase the populations of copepods containing cholera-causing bacteria. The bacteria can then infect populations of humans living along the Pacific coast of South America, as occurred during the 1991 ENSO. In 1991, a cholera

Q: What percentage of the world's irrigated cropland suffers reduced yields because of soil salinization?

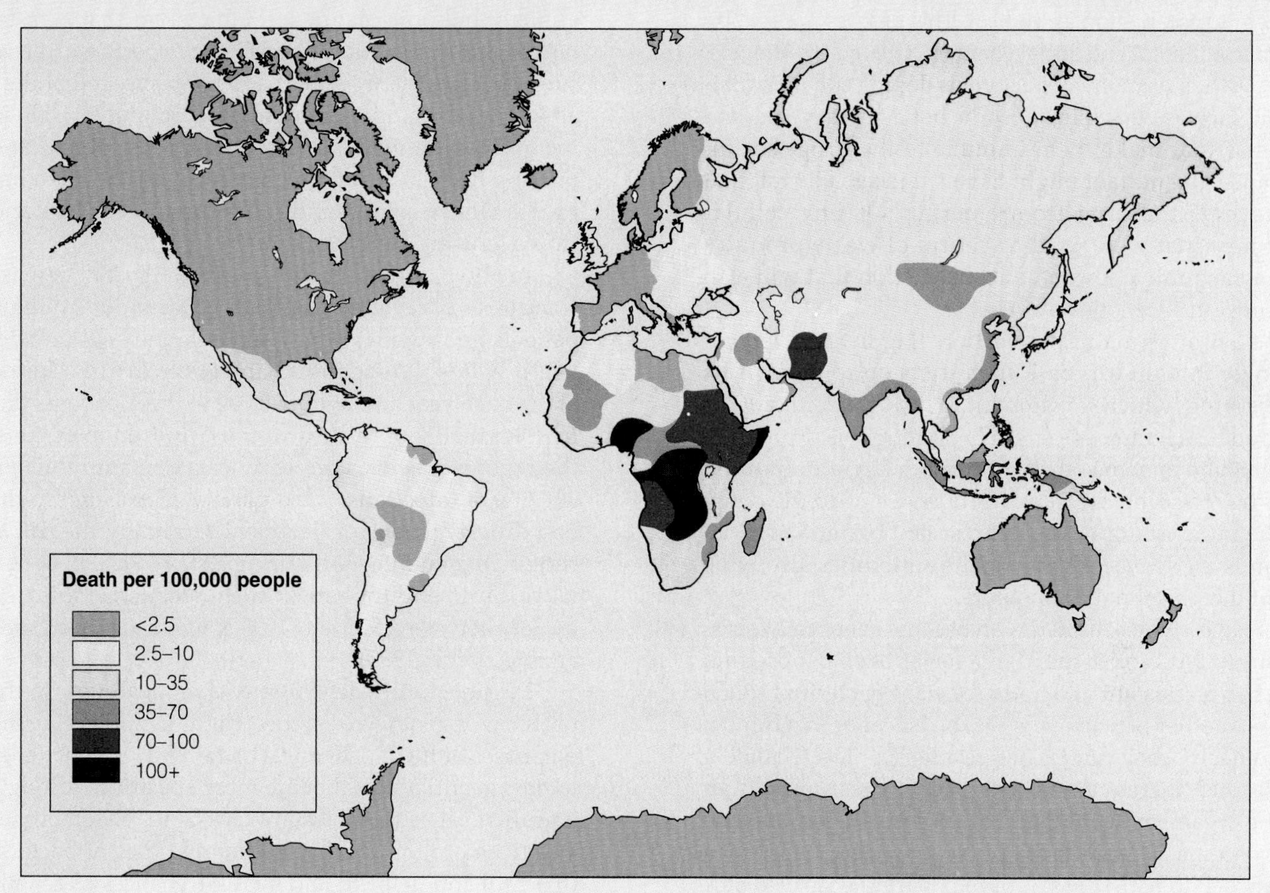

Figure 17-9
The current global tuberculosis (TB) epidemic. This easily transmitted disease is spreading rapidly and now kills about 3.1 million people a year. Without global effort to identify and fully treat people with active TB, the death toll is projected to rise to about 4 million a year by 2005. (Data from World Health Organization)

epidemic struck a 2,000-kilometer (1,200-mile) stretch of Peruvian coast, making 500,000 people ill and killing 5,000 people. Scientists also warn that projected global warming can lead to the spread of tropical infectious diseases such as malaria, dengue fever, and yellow fever to temperate areas.

Natural disasters such as floods (Section 13-8), earthquakes (Section 17-4), landslides, and hurricanes can also spread disease-causing organisms. Flooding is an especially dangerous threat because it often contaminates water supplies with raw sewage and creates areas of standing water and moist soil that are ideal breeding grounds for mosquitoes and other insects that spread infectious diseases.

Earthquakes and landslides can release fungi and other soil-residing infectious organisms into the air, where they can be spread by winds. The high winds of hurricanes can transfer infectious organisms and carriers of disease (such as insects) from tropical to temperate areas.

Are We Losing the War Against Infectious Bacteria? There is growing and alarming evidence that we may be losing our war against infectious bacterial

A: About 25% (expected to increase to at least 50% by 2000)

diseases because bacteria are among the earth's ultimate survivors. Microbes that cause disease are constantly and rapidly mutating and evolving in ways that allow them to escape human control. This process is aided by environmental and social changes initiated by humans.

When a colony of bacteria is dosed with an antibiotic such as penicillin, most die but a few have mutant genes that make them immune to the drug. For instance, the mutant might have a thicker cell wall that prevents the drug from entering. Or tiny cellular pumps that were used to get rid of waste products may acquire a new genetic instruction that will also dump an antibiotic out.

Through natural selection (Figure 6-9, left), a single mutant can pass such traits on to most of its offspring, which can amount to 16,777,216 in only 24 hours! Each time this strain of bacterium is exposed to penicillin or some other antibiotic, a larger proportion of its offspring are genetically resistant to the drug. The rapid multiplication of resistant bacteria in a victim is made easier because the antibiotics also wipe out their bacterial competitors.

Even worse, bacteria can become genetically resistant to antibiotics they have never been exposed to. When a resistant and a nonresistant bacterium touch one another (say in a hospital bedsheet or in a human stomach), they can exchange a loop of DNA called a plasmid, thereby transferring genetic resistance from one organism to another. This process allows bacteria to become resistant to antibiotics they have never been exposed to (Guest Essay, p. 466). Bacteria can also pick up genetic resistance to various antibiotics from viruses (that have acquired it while infecting other bacteria). Indeed, the exchange of genes among different bacterial cells is so pervasive that the entire bacterial world can be thought of as a single huge multicellular organism.

The incredible genetic adaptability of bacteria is one reason the world now faces a potentially serious rise in the incidence of some infectious bacterial diseases once controlled by antibiotics (Guest Essay, p. 466). Other factors also play a key role, including **(1)** spread of bacteria (some beneficial and some harmful) around the globe by human travel and the trade of goods; **(2)** overuse of antibiotics by doctors, often at the insistence of their patients; **(3)** failure of many patients to take all of their prescribed antibiotics, which promotes bacterial resistance; **(4)** availability of antibiotics in many countries without prescriptions; **(5)** overuse of pesticides (Section 21-3), which increases populations of pesticide-resistant insects and other disease carriers; and **(6)** widespread use of antibiotics in the livestock and dairy industries (Spotlight, right).

The result of these factors acting together is that every major disease-causing bacterium now has strains that resist at least one of the roughly 160 antibiotics we use to treat bacterial infections. In 1998, health officials were alarmed to learn of the existence of a strain of bubonic plague in Madagascar that is resistant to multiple antibiotics. Health experts warn that such drug-resistant "superbugs" or bacteria that defy all known antibiotics are virtually certain to pop up in increasing numbers unless doctors, hospitals, and nursing homes mount an emergency program to improve hand-washing and sanitation procedures and stop overusing antibiotics.

In 1998, officials at the Centers for Disease Control and Prevention estimated that about 2 million patients (most with a weakened immunity system) develop a hospital-acquired infection in the United States each year and about 90,000 of these patients die. In at least 70% of these hospital-acquired infections, the organism is resistant to at least one antibiotic. In 30–40% of infections, the organism is resistant to the best drug available for treatment. Currently, the risk of contracting an infection during a stay in a U.S. hospital is 1 in 15 and the rate of such infections increased by 36% between 1975 and 1995 (I was a victim of such an infection in 1997).

Possible future developments for dealing with the problem of genetic resistance might include **(1)** using databases of the biochemical codes of microbial genes to design antibiotics that are more specific in killing a harmful bacterium instead of having to rely on broad-spectrum antibiotics that kill many types of bacteria (harmful and helpful) and increase chances of genetic resistance, **(2)** designing drugs that disable antibiotic-resistance mechanisms (such as antibiotic dumping pumps) in bacteria, **(3)** using bioengineering to alter the genetic action inside bacteria that makes them resistant to various antibiotics, **(4)** finding the protein that controls all the toxins that make a bacterium dangerous and discovering ways to stop the protein's action (a method that would greatly reduce the chances of drug-resistance because unlike antibiotics it doesn't kill the target bacteria and useful bacteria), and **(5)** returning to remedies used before the introduction of antibodies such as treating infections with phage, a virus that grows in bacteria until it pops out and kills its host. Most of these possibilities (except the last one) are years or decades away from being available for human use.

How Rapidly Are Viral Diseases Spreading? Viral diseases include *influenza* or *flu* (transmitted by the bodily fluids or airborne emissions of an infected person), *ebola* (transmitted by the blood or other body fluids of an infected person), *rabies* (transmitted by dogs, coyotes, raccoons, skunks, and bats), and *AIDS* (Connections, p. 454). Viruses, like bacteria, can genetically adapt rapidly to different conditions.

Q: What percentage of the world's irrigated cropland suffers from waterlogging?

Drug-Resistant Bacteria in Meat

SPOTLIGHT

Traditionally, the United States has had the world's safest food supply. In recent years, however, outbreaks of foodborne illness have become more common. Each year up to 81 million Americans—almost one in three people—suffer a foodborne illness, and about 9,100 die.

According to the U.S. Department of Agriculture (USDA), about 70% of outbreaks of foodborne illness are traced back to meat and poultry products. About 80% of the illnesses and 75% of the deaths from meat and poultry products are caused by two bacteria: salmonella and cyanobacter. In March 1998, *Consumer Reports* found *Campylobacter* in 63% of randomly selected chickens, *salmonella* in 13%, and *E. coli* in almost all samples.

The Centers for Disease Control and Prevention estimate that there are 2–4 million salmonella cases in the United States each year, which result in 1,000–2,000 deaths. With poultry the main problem is not salmonella but *Campylobacter jejuni.* Unrecognized until 1977, this bacterium is now the most common cause of acute bacterial diarrhea in the United States, striking 2–8 million people and causing up to 800 deaths per year. Another culprit is *E. coli* found in some samples of beef (especially hamburger) that invades nearly every organ in the body and weakens the blood's ability to clot, killing about 10% of its victims.

Here are some of the causes of this decrease in food safety in the United States:

- *Decreased monitoring by federal enforcement agencies.* The Food and Drug Administration has 700 inspectors and lab personnel to monitor 53,000 food-processing plants and all imported produce. Plant inspections have gone from once every 3–5 years in 1992 to once every 10 years today. USDA inspectors rely only on visual inspections in U.S. slaughterhouses, where no microbiological testing is required. Nearly two-thirds of all winter produce eaten in the United States is imported from Mexico and enters the country through Nogales, Arizona. The FDA does a cursory random examination of about 30% of the trucks (whose numbers doubled between 1989 and 1997) and takes samples to be tested for about 300 pesticides from about 3%. There are no on-the-spot tests for infectious bacteria.

- *Crowding of cattle (see photo in Table of Contents) into feedlots and poultry into factory farms, multistep food processing, and broad distribution of perishable food.*

- *Emergence of new disease organisms resistant to antibiotics used in preventing disease in animals raised for food.* In 1987, cattle in Great Britain began dying from infection by a strain of salmonella bacteria called DT104. Two years after farmers treated the infected cattle with several types of antibiotics the bacterium became resistant to the antibiotics. Since then the resistant bacterium has spread to the United States and other countries and has infected people in contact with sick animals and who drank infected milk or meat. About one-third of the antibiotics manufactured in the United States are given to nonhuman animals, with most of these compounds mixed into feed to promote growth.

Food industry representatives say that food can never be risk free but that an American consumer's chances of contracting either *E. coli* or *salmonella* are less than being hit by a meteorite. Critics agree that most food in the United States is safe but warn of the increasing risk of bacterial infections because of changes in meat production and processing, lax inspection, and overuse of antibiotics by doctors and meat producers. By 1996, the Centers for Disease Control and Prevention reported that 34% of the meat samples tested in the United States contained drug-resistant salmonella bacteria (up from 0.6% in 1979).

Critical Thinking

Should growers of cattle, chickens, turkeys, and other forms of meat and meat products be allowed to use the increasingly fewer remaining effective antibiotics for human infectious diseases to treat livestock infected with diseases and thereby create multiresistant bacterial strains against which humans will have no defense? Explain. What might this do to the safety of meat and meat products? If you were a dairy farmer or raised livestock for a living, would you be in favor of such a regulation? Explain.

Although health officials worry about the emergence of new viral diseases (such as those caused by ebola viruses), they recognize that the greatest virus health threat to humans is the emergence of new, very virulent strains of influenza. Flu viruses move through the air and are highly contagious. In 1918–19 a flu epidemic infected more than half the world's population and killed 20–30 million people. These occasional massive epidemics, called *pandemics* (many originating from China), are believed to be the result of major viral genetic shifts in which genes from birds and mammalian viruses mix to create a new strain that can infect humans.

A: About 10%

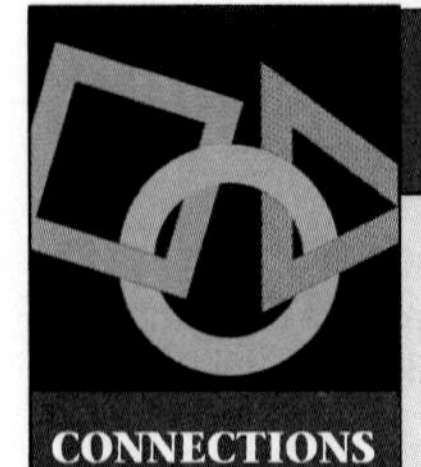

Sex Can Be Hazardous to Your Health

CONNECTIONS

Sexually transmitted diseases (STDs) are typically passed on during sexual activity; many STDs can also be transmitted from mother to infant during birth, from one intravenous (IV) drug user to another through shared needles, and by exposure to infected blood.

Worldwide there are about 330 million new infections and 750,000 deaths from STDs each year. By the year 2000 the number of deaths from STDs is expected to reach 1.5 million—an average of 4,100 per day. In the United States, STDs strike at least 8 million people (3 million of them teenagers) per year.

Major STDs caused by bacteria are *chlamydia, gonorrhea,* and *syphilis.* These bacterial diseases can be treated with antibiotics if caught in time. However, some of the bacteria causing these diseases are becoming genetically resistant to an increasing number of the antibiotics used to control them.

Major STDs caused by viruses include *genital warts, genital herpes, hepatitis B,* and *acquired immune deficiency syndrome or (AIDS),* which is caused by the human immunodeficiency virus (HIV). These viral diseases, with the possible exception of genital warts, are currently incurable. The AIDS virus itself isn't deadly, but it kills immune cells and leaves the body defenseless against all sorts of other infectious bacteria and viruses.

According to the World Health Organization, in 1997 at least 31 million people (21 million of them in sub-Saharan Africa) were infected with HIV. An average of 16,000 people are newly infected with the virus each day, 80% of them in Asia and Africa. Heterosexual transmission accounts for about 90% of the new infections worldwide and for about 13% in the United States (up from 2% in 1985).

Within about 7–10 years, 95% of those with HIV develop AIDS. This long incubation period means that infected people often spread the virus for several years before they learn that they are infected. There is as yet no cure for AIDS, although drugs may help some infected people live longer. By the end of 1997, 12.5 million of the 31 million people infected with HIV had acquired full-blown AIDS, with a record number (2.3 million) dying from AIDS in 1996.

Critical Thinking

What are the three most important things you would do to reduce the incidence of sexually transmitted diseases in the country where you live? Throughout the world?

Once a viral infection starts it is much harder to fight than infections by bacteria and protozoans. Only a few antiviral drugs exist because most drugs that will kill a virus also harm the cells of its host. Treating viral infections with antibiotics is useless and merely increases genetic resistance in disease-causing bacteria. Medicine's only effective weapon against viruses are preventive vaccines that stimulate the body's immune system to produce antibodies to ward off viral infections. The most effective vaccines contain weakened or dead versions of the invader agent to stimulate the immune system. Immunization with vaccines has helped control viral diseases such as smallpox, polio, rabies, influenza, and measles.

Case Study: Malaria, a Protozoal Disease According to a 1993 World Bank report, among infectious diseases malaria is a serious global health problem second only to tuberculosis. About 45% (or 2.6 billion people) of the world's population live in tropical and subtropical regions in which malaria is present (Figure 17-10). Currently, an estimated 300–500 million people are infected with malaria parasites worldwide, and at least 110 million new cases occur each year (97 million of them in Africa).

Malaria's symptoms come and go and include fever and chills, anemia, an enlarged spleen, severe abdominal pain and headaches, extreme weakness, and greater susceptibility to other diseases. The disease kills 2.5–2.7 million people each year, more than half of them children under age 5. Malaria can also be transmitted by blood transfusions or by sharing needles.

Malaria is caused by four species of protozoa of the genus *Plasmodium*. Most cases of the disease are transmitted when an uninfected female of any one of 60 species of *Anopheles* mosquito bites an infected person, ingests blood that contains the parasite, and later bites an uninfected person (Figure 17-11). When this happens, *Plasmodium* parasites move out of the mosquito and into the human's bloodstream, multiply in the liver, and then enter blood cells to continue multiplying.

The symptoms of the disease appear only when the parasite invades red blood cells (Stage 4 in Figure 17-11). There they clog small blood vessels and decrease the flow of oxygen to tissues, causing up to a 70% decrease in the body's ability to transport oxygen. The resulting chronic anemia reduces resistance to almost any infection and a malaria victim may die of the common cold or measles (which infects 43 million people and kills 1 million children annually).

The malaria cycle (Figure 17-11) repeats itself until immunity develops (usually within 3–5 years, unless the person is reinfected), treatment is given, or the victim dies. The disease is difficult to treat because the parasite takes on different forms and invades different

Q: How much of the commercial energy used in the world comes from nonrenewable resources?

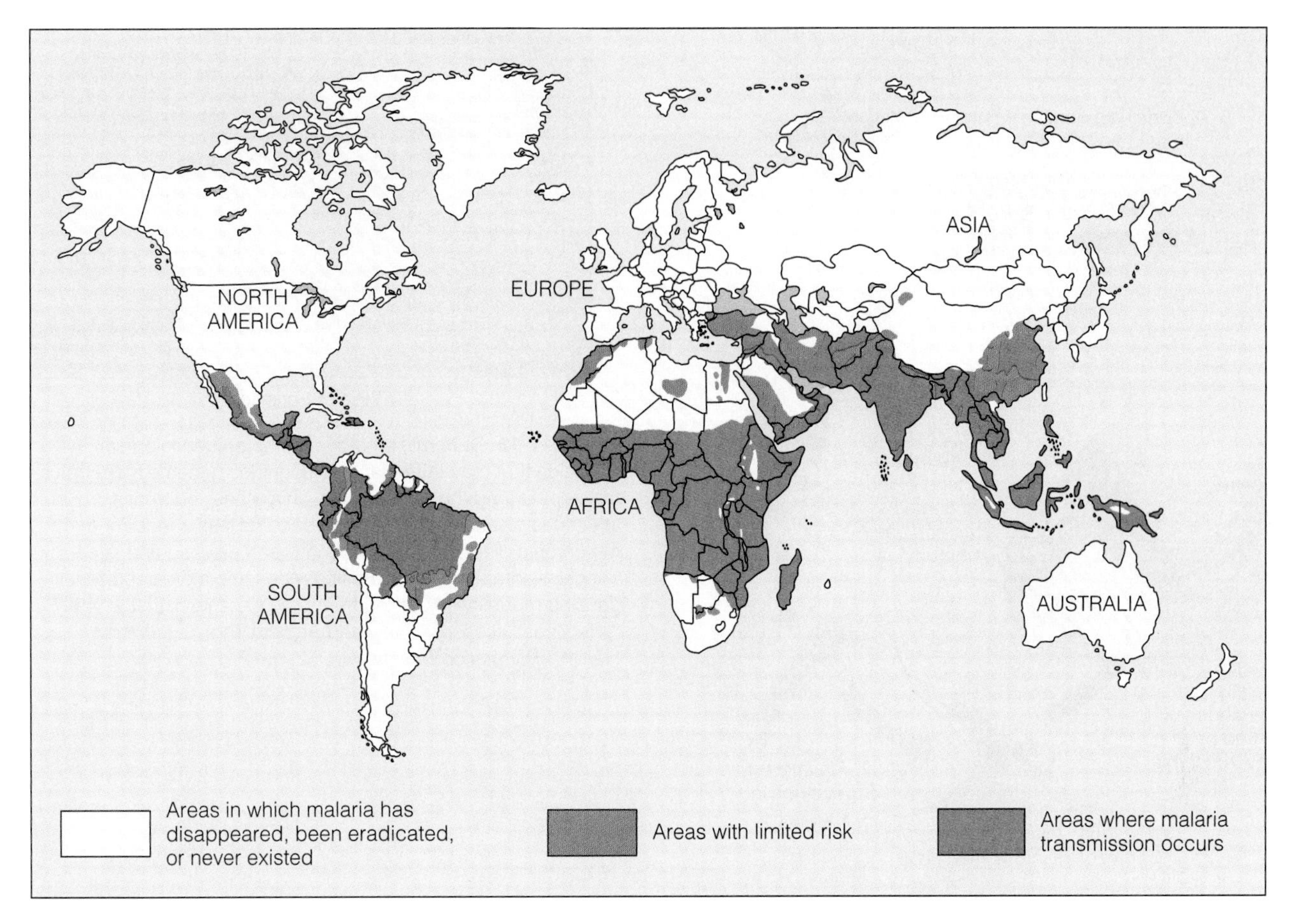

Figure 17-10 Worldwide distribution of malaria today (red) and its projected distribution in 2046 (orange). About 45% of the world's current population lives in areas in which malaria is present, with the disease killing 2.5–2.7 million people a year. If the world becomes warmer, as projected by current climate models, by 2046 malaria could affect 60% of the world's population. (Data from the World Health Organization)

areas during its life cycle. Over the course of human history, rapidly reproducing and highly adaptable malarial protozoa have probably killed more people than all the wars ever fought and may have killed one of every two human beings who have ever lived on this planet.

During the 1950s and 1960s, the spread of malaria was sharply curtailed by draining swamplands and marshes, spraying breeding areas with insecticides, and using drugs to kill the parasites in the bloodstream. Since 1970, however, malaria has come roaring back. Most species of the malaria-carrying *Anopheles* mosquitoes have become genetically resistant to most of the insecticides used.

Worse, the *Plasmodium* parasites have become genetically resistant to the common antimalarial drugs (among the most commonly prescribed drugs in the world) mostly because the disease is so widespread and because these drugs are often overused as a result of improper diagnoses. Today such resistance is spreading faster than new drugs are being developed, partly because most major pharmaceutical companies have discontinued or downsized their costly and unprofitable development programs for antimalarials.

The spread of irrigated agriculture and hydroelectric power has also produced vast areas of irrigation ditches and reservoirs behind dams that are breeding places for mosquitoes. If the earth's temperature rises (global warming) as projected by climate models (Section 19-1), by 2046 the range of malaria is projected to include 60% of the world's population (Figure 17-10).

Researchers are working to develop new antimalarial drugs, vaccines, and biological controls for *Anopheles* mosquitoes. Such approaches are underfunded and have proved more difficult than originally thought. Scientists estimate that an effective vaccine is years, if not decades, away. It is also likely to have a limited impact on preventing the disease because of the difficulty and expense of vaccinating almost half of the world's population (half of them infants under

A: 83% (78% from fossil fuels and 5% from nuclear power)

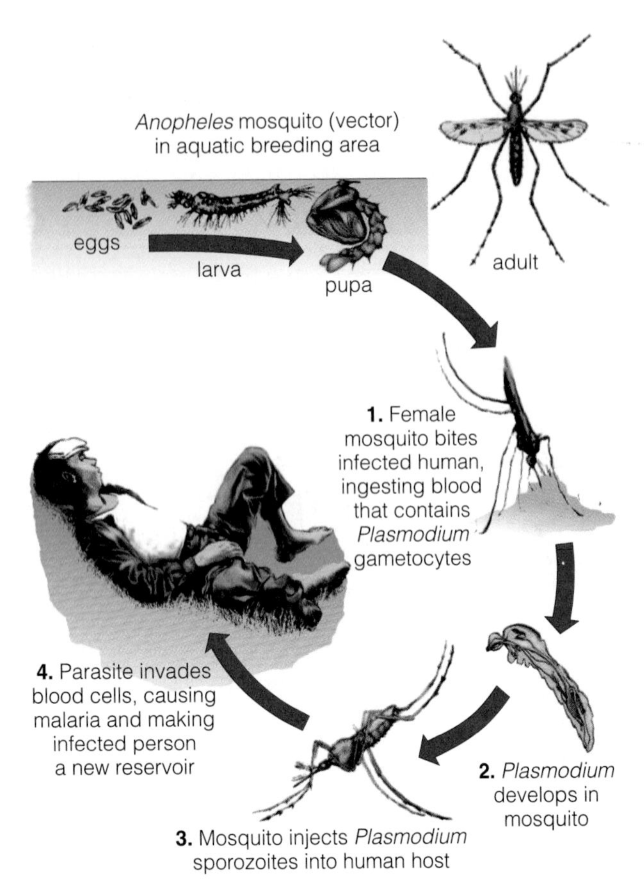

Figure 17-11 The life cycle of malaria. This life cycle of *Plasmodium* circles from mosquito to human and back to mosquito.

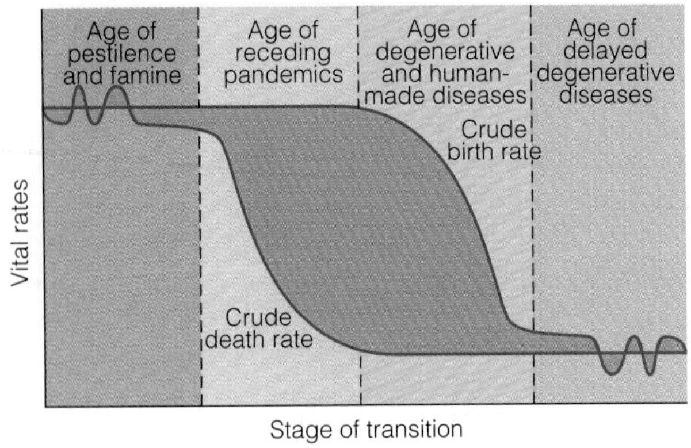

Age of pestilence and famine
Age of receding pandemics
Age of degenerative and human-made diseases
Age of delayed degenerative diseases
Crude birth rate
Crude death rate
Vital rates
Stage of transition

Figure 17-12 Generalized model of the epidemiological transition that may take place as countries become more industrialized. During the phases of this hypothesized transition, the infectious diseases of childhood become less important and the chronic diseases of adulthood (heart disease and stroke, cancer, and respiratory conditions) become more important in causing mortality. (Adapted from S. Jay Olshansky et al. 1997. "Infectious Diseases—New and Ancient Threats to World Health," *Population Bulletin*, vol. 52, no. 2, 1–52)

6 months of age) in poor rural areas. Researchers are also studying the feasibility of altering the mosquitoes' genetic makeup so that they cannot carry and transmit the parasite to humans.

According to health experts, prevention is the best approach to slowing the spread of malaria. Methods include **(1)** increasing water flow in irrigation systems to prevent mosquito larvae from developing (an expensive and wasteful use of water), **(2)** using mosquito nets dipped in a nontoxic, biodegradable pyrethroid insecticide (Table 21-1) in windows and doors of homes, **(3)** cultivating fish that feed on mosquito larvae (biological control), **(4)** clearing vegetation around houses, and **(5)** planting trees that soak up water in low-lying marsh areas where mosquitoes thrive (a method that can degrade or destroy ecologically important wetlands). Experts warn that without a massive education program and greatly increased funding from developed and developing countries, these preventive measures will not be widely used and deaths from malaria will increase.

What Are the Major Diseases in Developed Countries? As a country industrializes, it usually makes an *epidemiological transition* (Figure 17-12), somewhat analogous to the *demographic transition* (Figure 11-20). The *first phase* is characterized by extremely high death rates that vacillate between peaks and troughs in response to epidemics, famines, and wars. During the *second phase* epidemic peaks become less frequent and the crude death rate from infectious diseases drops because of a combination of factors including better sanitation, nutrition, vaccines, and medical advances.

During the past two centuries, humans have raised living standards, improved the health status, reduced infant and child mortality, lengthened the life expectancy of much of the world's population (from an average of 48 years in 1955 to 66 years in 1998), and increased the earth's carrying capacity for humans (Figure 1-22). As such changes decrease the threat of infectious diseases, a country enters a *third phase* in which the death rate levels off and the leading causes

Q: How much of the commercial energy used in the United States comes from nonrenewable resources?

of death are from nontransmissible diseases associated with aging (such as heart disease, strokes, and cancers). During this phase the age structure of the population shifts upward as people live longer and birth rates fall (Figure 11-13). In a proposed *fourth phase*, degenerative diseases associated with aging continue to cause most deaths, but medical advances allow an increasing number of people with such diseases to live longer.

Some analysts hypothesize that the emergence of new infectious diseases and reemergence of once controlled infectious diseases might signal the development of a *fifth phase* in which death rates might rise. Factors promoting such a possibility include overuse of antibiotics and pesticides, growing population and urbanization, increased migration and travel, and emergence of new infectious diseases as people clear forests and move into previously uninhabited areas.

Other analysts contend that the current rise in deaths from infectious diseases has not been going on long enough to be treated as a prolonged trend and that advances in medical technology and genetic engineering (Pro/Con, p. 462) may overcome this effect.

Industrialized countries such as the United States are considered to be in the third phase of the epidemiologic transition and industrialized countries that have stabilized their population sizes (Figure 11-13) are in the fourth phase. In 1994 about 2.3 million people died of all causes in the United States: 39% from heart attacks and strokes, 24% from cancer, 5% from infectious diseases (mostly pneumonia, influenza, and AIDS), and 4% from accidents (half from automobile accidents). Almost two-thirds of these deaths result from chronic diseases (such as heart attack, stroke, cancer, and diabetes) that typically take a long time to develop and have multiple causes. They are largely related to location (urban or rural), work environment, diet, smoking (p. 436), amount of exercise, sexual habits (Connections, p. 454), and the use of alcohol or other harmful drugs.

According to the Department of Health and Human Services, about 95% of the money spent on health care in the United States each year is used to *treat* rather than to *prevent* disease—one reason why health care costs are so high and continue to climb. Health experts estimate that changing harmful lifestyle factors could prevent 40–70% of all premature deaths, one-third of all cases of acute disability, and two-thirds of all cases of chronic disability in the United States.

The U.S. death rate from infectious diseases increased 58% between 1980 and 1992 (36% from AIDS and 22% from other infectious diseases). In 1980, infectious diseases ranked as America's fifth leading killer. By 1992 they were the third leading cause of death, representing a partial reversal of the epidemiological transition (Figure 17-12) in the United States.

Solutions: How Can We Reduce Infectious and Other Diseases? According to health scientists and public health officials, the majority of existing infectious diseases that affect humanity could be prevented or their incidence greatly reduced by

- Greatly increasing research on tropical diseases. The World Health Organization estimates that only 3% of the money spent worldwide each year for such research is devoted to malaria and other tropical diseases, even though more people suffer and die worldwide from these diseases than from all others combined.
- Mounting a global campaign to reduce overcrowding, unsafe drinking water, poor sanitation, inadequate health care systems, malnutrition, and poverty.
- Increasing funding for monitoring, diagnosing, and responding to disease outbreaks to reduce the chances that a small outbreak can turn into an epidemic.
- Using extreme caution in instituting strategies to battle disease-causing organisms that can reproduce and evolve at a rapid pace to avoid our attempts to control their populations.
- Not using powerful, broad-spectrum antibiotics to treat minor infections or undiagnosed symptoms.
- Educating the public to understand the dangers of overuse of antibiotics and the need to take all of the antibiotics in any prescription.
- Not using antibiotics that have caused widespread genetic resistance, coupled with rotating from one antibiotic to another.
- Not selling antibiotics without a prescription (allowed in many countries).
- Sharply reducing the use of antibiotics in livestock to promote growth.
- Establishing and enforcing much more rigorous anti-infection programs in hospitals, nursing homes, and doctor's offices, with special emphasis on having all health personnel wash their hands properly before any contact with a patient.
- Sharply reducing the use of pesticides to slow the increase in numbers of pesticide-resistant insects and other carriers of disease (Sections 21-3 and 21-5).

A: 93% (85% from fossil fuels and 8% from nuclear power)

Improving Health Care in Developing Countries

SOLUTIONS

With adequate funding, the health of people in developing countries (and the poor in developed countries) can be improved dramatically, quickly, and cheaply by providing the following forms of mostly preventive health care:

- Family-planning counseling.
- Better nutrition, prenatal care, and birth assistance for pregnant women. At least 600,000 women in developing countries die each year of mostly preventable pregnancy-related causes, compared with about 6,000 in developed countries.
- Better nutrition for children.
- Greatly improved postnatal care (including promotion of breast-feeding) to reduce infant mortality. Breast-fed babies get natural immunity to many diseases from antibodies in their mothers' milk.
- Immunization against the world's five largest preventable infectious diseases: tetanus, measles, diphtheria, typhoid fever, and polio. The average cost of immunizing a child in a developing country is $15, which avoids future medical costs of at least $150. Between 1971 and 1992, the percentage of children in developing countries immunized against these diseases rose from 10% to 80%, saving about 9 million lives a year
- Oral rehydration therapy for victims of diarrheal diseases, which cause about one-fourth of all deaths of children under the age of 5. A simple solution of boiled water, salt, and sugar or rice, at a cost of only a few cents per person, can prevent death from dehydration. According to the British medical journal *Lancet*, this simple treatment is "the most important medical advance of the century."
- Careful and selective use of antibiotics for infections.
- Clean drinking water and sanitation facilities for the one-third of the world's population that lacks them.

According to the World Health Organization, extending such primary health care to all the world's people would cost an additional $10 billion per year, a mere 4% of what the world spends every year on cigarettes or devotes every 4 days to military spending. The cost of this program is about $1 per child. Experience shows that the resulting drop in infant mortality would also help reduce population growth.

Critical Thinking

1. Do you believe that developed countries should foot at least half of the bill implementing such proposals? What major economic and environmental advantages would this have for developed countries?

2. How many dollars per year of your taxes would you be willing to spend for such a preventive health program in developing countries?

- Putting much more money into the development of vaccines to prevent infections by bacteria and viruses responsible for most disease and death.
- Educating consumers and food handlers to practice safe hygiene in handling and preparing meats and produce, to not undercook meat, and to wash raw fruit and vegetables thoroughly to remove resistant bacteria, antibiotic residues, and pesticide residues.
- Protecting biodiversity (Chapters 23–26) as a mechanism for reducing the spread of disease.
- Slow possible global warming (Section 19-4) to reduce the threat of the spread of tropical diseases such as malaria (Figure 17-10) to temperate areas.
- Emphasizing preventive health care in developing countries (Solutions, above) and developed countries.

17-6 RISK ANALYSIS

 **How Can We Estimate Risks?** Risk analysis involves identifying hazards and evaluating their associated risks (*risk assessment*), ranking risks (*comparative risk analysis*), determining options and making decisions about reducing or eliminating risks (*risk management*), and informing decision makers and the public about risks (*risk communication*).

Risk assessment involves determining the types of hazards involved, estimating the probability that each hazard will occur, and estimating how many people are likely to be exposed to it and how many may suffer serious harm. Statistical probabilities based on past experience, animal testing and other tests, and epidemiological studies are used to estimate risks from older technologies and products (Section 17-1). For new technologies and

Slovik, Paul. 1997. "Public Perception of Risk." *Journal of Environmental Health*, vol. 59, no. 9, 22(3).

Scientists (Not in rank order in each category)	Citizens (In rank order)
High-Risk Health Problems • Indoor air pollution • Outdoor air pollution • Worker exposure to industrial or farm chemicals • Pollutants in drinking water • Pesticide residues on food • Toxic chemicals in consumer products **High-Risk Ecological Problems** • Global climate change • Stratospheric ozone depletion • Wildlife habitat alteration and destruction • Species extinction and loss of biodiversity	**High-Risk Problems** • Hazardous waste sites • Industrial water pollution • Occupational exposure to chemicals • Oil spills • Stratospheric ozone depletion • Nuclear power-plant accidents • Industrial accidents releasing pollutants • Radioactive wastes • Air pollution from factories • Leaking underground tanks
Medium-Risk Ecological Problems • Acid deposition • Pesticides • Airborne toxic chemicals • Toxic chemicals, nutrients, and sediment in surface waters	**Medium-Risk Problems** • Coastal water contamination • Solid waste and litter • Pesticide risks to farm workers • Water pollution from sewage plants
Low-Risk Ecological Problems • Oil spills • Groundwater pollution • Radioactive isotopes • Acid runoff to surface waters • Thermal pollution	**Low-Risk Problems** • Air pollution from vehicles • Pesticide residues in foods • Global climate change • Drinking water contamination

Figure 17-13 Comparative risk analysis of the most serious ecological and health problems according to scientists acting as advisers to the U.S. Environmental Protection Agency (left column). Risks in each of these categories are not listed in rank order. The right side of this figure represents polls showing how U.S. citizens rank the ecological and health risks they perceive as being the most serious. Why do you think there is such a great difference between the ranking by risk experts and by the public? (Data from Science Advisory Board, *Reducing Risks*, Washington, D.C.: Environmental Protection Agency, 1990)

products, much more uncertain statistical probabilities, based on models rather than actual experience, must be calculated.

The left side of Figure 17-13 is an example of *comparative risk analysis,* summarizing the greatest ecological and health risks identified by a panel of scientists acting as advisers to the U.S. Environmental Protection Agency. Note the considerable difference between the comparison of relative risk by scientists (Figure 17-13, left) and the public (Figure 17-13, right). These differences result largely from failure of professional risk evaluators to communicate the nature of risks and their relative importance to the public, teachers, and members of the media. Much of our risk education is based on often misleading media reports on the latest risk scare of the week (based mainly on frontier science) that do not put such risks in perspective.

The key question is whether the estimated short- and long-term risks of using a particular technology or product outweigh the estimated short- and long-term benefits of other alternatives. One method for making such evaluations is **risk–benefit analysis**, which involves estimating such benefits and the risks involved.

What Are the Greatest Risks People Face? The greatest risks most people face today are rarely dramatic enough to make the daily news. In terms of reduced life span from malnutrition, exposure to disease-causing organisms and dangerous chemicals, and lack of basic health care, *the greatest risk by far is poverty* (Figure 17-14). These factors can interact synergistically. Poverty can lead to malnutrition, which makes one more susceptible to disease (Figure 12-10). Conversely, people debilitated by diseases such as malaria

Hint: Enter the search terms *risk assessment, public perception* using Key Words.

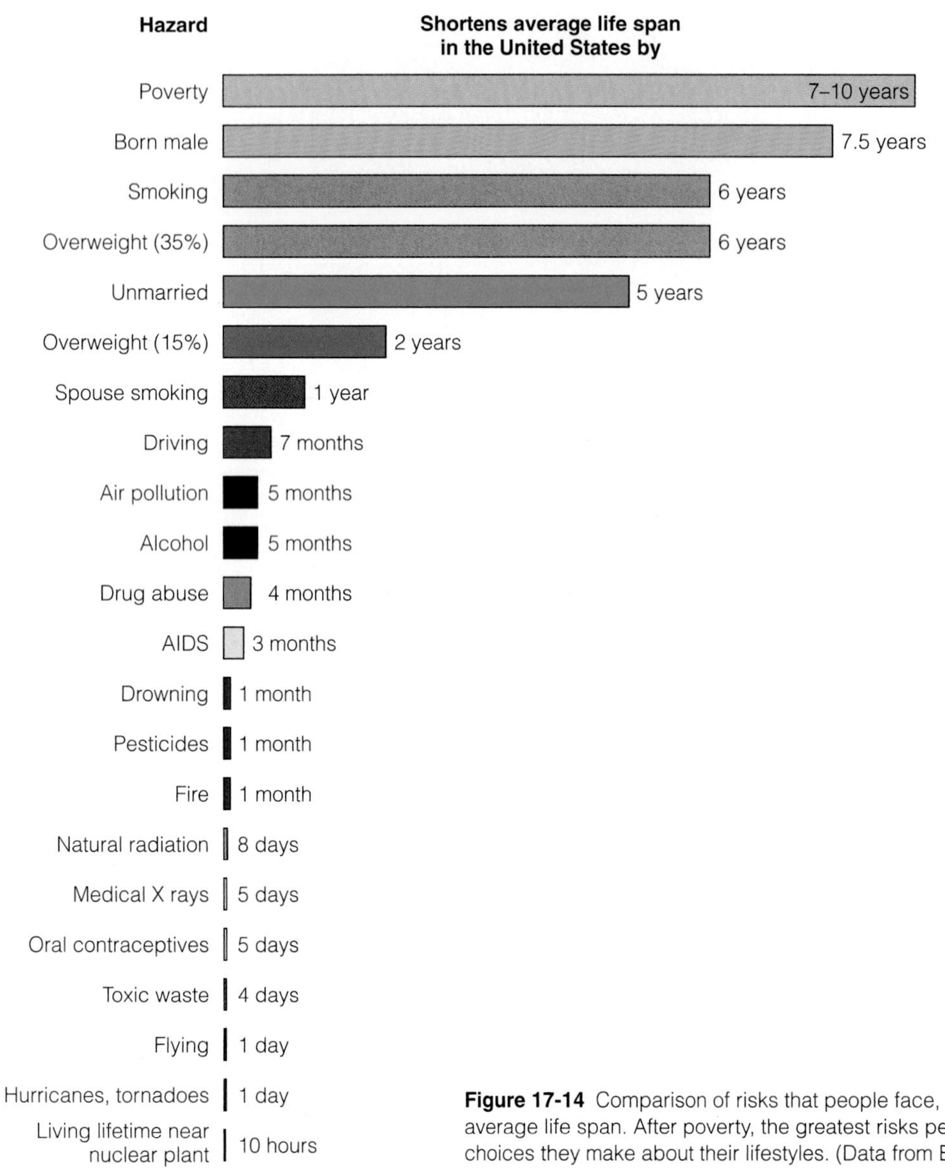

Figure 17-14 Comparison of risks that people face, expressed in terms of shorter average life span. After poverty, the greatest risks people face are mostly from voluntary choices they make about their lifestyles. (Data from Bernard L. Cohen)

and AIDS are more susceptible to malnutrition. In turn, people affected by poverty, malnutrition, and disease are likely to be more susceptible to the effects of pollution, drought, global warming, and other environmental stresses. If these four factors operate synergistically, the net effect could be 40 times greater than each factor operating alone.

After the health risks associated with poverty, the greatest risks of premature death are mostly the result of voluntary—and thus correctable—choices people make about their lifestyles (Figures 17-1 and 17-14). By far the best ways to reduce one's risk of premature death and serious health risks are to not smoke, avoid excess sunlight (which ages skin and causes skin cancer), not drink alcohol or drink only in moderation (no more than two drinks in a single day), reduce consumption of foods containing cholesterol and saturated fats, eat a variety of fruits and vegetables, exercise regularly, lose excess weight, and (for those who can afford a car) drive as safely as possible in a vehicle with the best available safety equipment.

How Can We Estimate Risks for Technological Systems? The more complex a technological system and the more people needed to design and run it, the more difficult it is to estimate the risks. The overall reliability of any technological system (expressed as a

Q: What percentage of the world's commercial energy is used by the United States?

percentage) is the product of two factors (multiplied by 100):

$$\text{System reliability \%} = \text{Technology reliability} \times \text{Human reliability} \times 100$$

With careful design, quality control, maintenance, and monitoring, a highly complex system such as a nuclear power plant or a space shuttle can achieve a high degree of technology reliability. However, human reliability is almost always much lower than technology reliability and is virtually impossible to predict; to err is human. Suppose, for example, that the technology reliability of a nuclear power plant is 95% (0.95) and that human reliability is 75% (0.75). Then the overall system reliability is only 71% ($0.95 \times 0.75 \times 100 = 71\%$). Even if we could make the technology 100% reliable (1.0), the overall system reliability would still be only 75% ($1.0 \times 0.75 \times 100 = 75\%$). The crucial dependence of even the most carefully designed systems on unpredictable human reliability helps explain essentially "impossible" tragedies such as the Chernobyl (p. 366) nuclear power-plant accident and the explosion of the space shuttle *Challenger*.

One way to make a system more foolproof or fail-safe is to move more of the potentially fallible elements from the human side to the technical side. However, chance events such as a lightning bolt can knock out an automatic control system. No machine or computer program can completely replace human judgment. Of course, the parts in any automated control system are manufactured, assembled, tested, certified, and maintained by fallible human beings. Computer software programs used to monitor and control complex systems can also contain human errors, or they can be deliberately modified by computer viruses to malfunction. The pros and cons of genetic engineering reveal the difficulty in evaluating a new technology (Pro/Con, p. 462).

What Are the Limitations of Risk Assessment and Risk–Benefit Analysis? Risk assessment is a young science that has many built-in uncertainties and limitations. It depends on toxicology assessments that have scientific and economic limitations (Section 17-2). Each additional step in risk assessment (and related risk–benefit analysis) also has uncertainties and economic limitations.

Here are some of the key questions involved in risk assessment:

- How reliable are risk assessment data and models?
- Who profits from allowing certain levels of harmful chemicals into the environment, and who suffers? Who decides this?
- Should estimates emphasize short-term risks, or should more weight be put on long-term risks? Who should make this decision?
- Should the primary goal of risk analysis be to determine how much risk is acceptable (the current approach) or to figure out how to do the least damage (a prevention approach)?
- Who should do a particular risk–benefit analysis or risk assessment, and who should review the results? A government agency? Independent scientists? The public?
- Should cumulative effects of various risks be considered, or should risks be considered separately, as is usually done? Suppose a pesticide is found to have an annual risk of killing one person in a million from cancer, the current EPA limit. Cumulatively, however, effects from 40 such pesticides might kill 40, or even 400, of every 1 million people. Is this acceptable?
- Should risk levels be higher for workers (as is almost always the case) than for the general public? What say should workers and their families have in this decision? According to government estimates, the exposure of workers to toxic chemicals in the United States causes 50,000–70,000 deaths, at least half from cancer, and 350,000 new cases of illness per year. In the United States, on-the-job accidents kill about 6,500 workers per year and injure about 13 million others each year. The situation is much worse in developing countries. Is this a necessary cost of doing business?

Some see risk analysis as a useful and much-needed tool. Others see it as a way to justify premeditated murder in the name of profit. According to the National Academy of Sciences, exposure to toxic chemicals is responsible for 2–4% of the 521,000 cancer deaths in the United States; this amounts to 10,400–20,800 premature cancer deaths per year. According to hazardous waste expert Peter Montague (Guest Essay, p. 70), "The explicit aim of risk assessment is to convince people that some number of citizens *must* be killed each year to maintain a national lifestyle based on necessities like Saran Wrap, throwaway cameras, and lawns without dandelions."

Such critics understand that all dangerous chemicals can't be banned, but they argue that the emphasis should shift from determining acceptable risk levels to trying to reduce the risks as much as possible by pollution prevention. Less toxic or nontoxic chemicals can often be substituted in manufacturing processes or in products, toxic chemicals can be recycled or reused instead of being buried or burned, and industrial

A: About 24% (compared to 3% used by India, with almost four times more people)

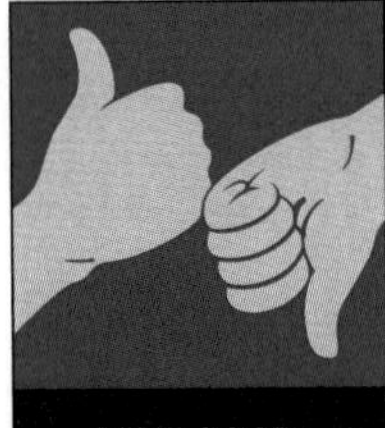

Genetic Engineering: Savior or a Potential Monster?

For thousands of years geneticists have used *crossbreeding* to transfer genes from closely related plants and animals that can sexually interbreed. Today *genetic engineers* have learned how to splice genes and recombine sequences of existing DNA molecules in organisms to produce DNA with new genetic characteristics (recombinant DNA). These advances allow the transfer of genes with certain traits from one species to another (Figure 17-15) without waiting for new genetic combinations to evolve through the process of natural selection and without the limited number of genetic combinations possible with conventional crossbreeding.

Global life-science companies are expected to introduce thousands of new genetically engineered organisms into the environment in the coming century. When the U.S. Supreme Court ruled in 1980 that genetically engineered organisms could be patented, investors began pouring billions of dollars into the fledgling biotech industry.

This rapidly developing technology excites many scientists and investors, who see it as a way to produce, patent, and sell high-yield crops and livestock with more protein and greater resistance to diseases, pests, herbicides, frost, and drought. Genetic manipulation could also result in tree strains with disease resistance and improved productivity. Plantations of such genetically improved forests could help in restoration of many denuded areas.

Genetic engineers also hope to create bacteria that can destroy oil spills, degrade toxic wastes, concentrate metals found in low-grade ores (Solutions, p. 349), and serve as biological factories for new vaccines, drugs, and therapeutic hormones. Gene therapy, its proponents argue, could also eliminate certain genetic diseases by removing, turning off, or blocking harmful genes.

Genetic engineering has produced a drug that reduces heart attack damage and agents that fight diabetes, hemophilia, and some forms of cancer; it has also been used to diagnose AIDS and cancer. Genetically altered bacteria have been used to manufacture more effective vaccines and hormones. In 1998, researchers bioengineered a potato to contain a protein that causes diarrhea (a major killer, Table 17-2). When subjects eat the potato, they develop antibodies to the diarrhea bug.

In agriculture, gene transfer has been used to develop strawberries and tomatoes that resist frost because they contain "antifreeze" genes from fish, tomatoes that stay fresh and tasty longer, potatoes that resist disease because they contain chicken genes, and smaller cows that produce more milk.

Some people worry that biotechnology may run amok. They argue that it should be kept under strict control because we don't understand well enough how nature works to allow unregulated genetic alteration of humans and other species. They also point to serious and widespread ecological problems that have resulted from the accidental or deliberate introduction of nonnative organisms into biological communities (Section 25-3).

To such critics the resulting *genetic pollution* may eventually prove to be more harmful than chemical pollution. Because they are alive, genetically engineered organisms may interact with other living things in the environment in more unpredictable ways than do chemicals, whose effects and interactions we still have little understanding of after decades of research. Unlike chemicals, genetically engineered organisms reproduce, grow, mutate, change their form and behavior, migrate to other areas, and alter the genetic traits of existing wild species; unlike defective cars and other products, they couldn't be recalled (especially rapidly multiplying microscopic organisms).

Global life-science companies project that within 10–15 years, all of the major crops grown in the world will be genetically engineered to include pest-, virus-, bacteria-, fungus-, herbicide-, and stress-resistant genes. Ecologists warn that a serious danger might lie in the flow of such genes from the altered crops to weedy relatives through cross-pollination. This could create superweeds that are resistant to herbicides, pests, viruses, and drought and reduce our ability to grow food crops. In 1996 Danish scientists observed this type of gene transfer, which for years biotech companies have dismissed as a remote or nonexistent possibility.

Critics are also concerned that one of the most serious effects of widespread use of biotechnology is reduction of the world's vital genetic diversity and thus biodiversity (Guest Essay, p. 674). Already the world's 20 major food crops have become 70% less genetically diverse because a wide range of wild strains have been replaced with only a few varieties developed by crossbreeding. Widespread use of bioengineered crop strains could hasten this loss of vital biodiversity.

Development of all new varieties by crossbreeding or genetic engineering is based on mixing certain traits found in wild varieties. Scientists can use gene splicing, cloning, and tissue culturing to rearrange genetic material but they cannot create it. Thus, the biotech industry is utterly dependent on nature's biologically diverse seed stock—germplasm—for its raw materials. Reducing the natural genetic diversity of wild plants and animals by

Q: What percentage of earth's oil reserves are in OPEC countries?

Figure 17-15
An example of genetic engineering. The 6-month-old mouse on the left is normal; the other mouse of the same age contains a human growth hormone gene in the chromosomes of all its cells. Mice with the human growth hormone gene normally grow two to three times faster and reach a size twice that of mice without the gene. (R. L. Brinster and R. E. Hammer/School of Veterinary Medicine, University of Pennsylvania)

substituting a much smaller number of bioengineered plants and animals undermines the genetic engineers' ability to produce new combinations in the laboratory and undermines the future success of the biotech industry.

Genetically engineered plants can also affect human health. Use of the genetically engineered bovine growth hormone (recently approved by the FDA to increase milk production in cows) raises the incidence of udder infections. This requires giving cows higher doses of antibiotics, which can accelerate the problem of genetic resistance to widely used antibiotics.

A 1996 study showed that genetically engineered soybeans containing a gene from a Brazil nut could create an allergic reaction in people allergic to the nuts. Consumers allergic to ingredients in various commonly eaten foods would be unprotected because in 1992 the U.S. Food and Drug Administration (FDA) announced that special labeling would not be required for genetically engineered foods.

In 1989, a committee of prominent ecologists appointed by the Ecological Society of America warned that the ecological effects of new combinations of genetic traits from different species would be difficult to predict. Their report called for a case-by-case review of any proposed environmental releases, as well as carefully regulated, small-scale field tests before any bioengineered organism is put into commercial use—practices that are not being adequately followed.

A growing number of ecologists warn that the environmental problems of the next century may include accelerated buildup of genetically resistant bacteria and superinsects, the uncontrollable spread of superweeds, the creation of novel viruses, harmful genetic contamination of food, gene pool depletion, and ecosystem destabilization.

Proponents contend that such effects could be minimized by developing microorganisms that would self-destruct by expressing a "suicide" gene after their task is accomplished. They also argue that more emphasis could be placed on developing environmentally benign bioengineered organisms that could help save endangered species and promote ecological sustainability.

Proponents argue that the risk of such biotech-caused ecological catastrophes is small. Critics agree that there is only a small chance of a genetically engineered organism running amok. However, it can happen and when it does the harmful consequences could be significant and irreversible. Insurance companies have refused to insure the release of genetically engineered organisms into the environment against the possibility of catastrophic environmental damage because they have no way to evaluate the risks of any given introduction.

This controversy illustrates the difficulty of balancing the actual and potential benefits of a technology with its actual and potential risks of harm. Proponents of this new technology may be right in their belief that the benefits far exceed the potential and mostly unknown risks. But what if they are wrong?

Critical Thinking

1. What government controls, if any, do you believe should be applied to the development and use of genetic engineering? How would you enforce any restrictions?

2. Use the library or the Internet to find out what controls now exist on biotechnology in the United States (or the country where you live) and how well such controls are enforced.

A: About 67% (26% in Saudi Arabia and 10% in Iraq)

How Useful Are Risk Assessment and Risk–Benefit Analysis?

SPOTLIGHT

When we add new chemicals or technologies we are mostly *flying blind* about their possible harmful effects. For example, a recent study has documented the significant uncertainties involved in even simple risk assessment. Eleven European governments established 11 different teams of their best scientists and engineers (including those from private companies) to assess the hazards and risks from a small plant storing only one hazardous chemical (ammonia). The 11 teams, consisting of world-class experts analyzing this very simple system, disagreed with one another on fundamental points and varied in their assessments of the hazards by a factor of 25,000!

Thus, the current built-in uncertainty in risk assessment and risk–benefit analysis (and cost–benefit analysis, Section 27-3) is analogous to a radar device that can detect a car speeding at 160 kilometers (100 miles) per hour, but can tell us only that the car is traveling somewhere between 0.16 kilometer (0.1 mile) per hour and 160,000 kilometers (100,000 miles) per hour. Such inherent uncertainty explains why regulators setting human exposure levels for toxic substances usually divide the best results by 100 to 1,000 to provide the public with a margin of safety.

Despite the inevitable uncertainties involved, risk assessment and risk–benefit analysis are useful ways to **(1)** organize available information, **(2)** identify significant hazards, **(3)** focus on areas that need more research, **(4)** help regulators decide how money for reducing risks should be allocated, and **(5)** stimulate people to make more informed decisions about health and environmental goals and priorities.

However, at best risk assessments and risk–benefit analyses yield only a range of probabilities and uncertainties based on different assumptions—not the precise numbers that decision makers and the public wish they had. Also, politics, economics, and value judgments are involved at every step of the risk analysis process. Environmentalists and health officials believe that all such analyses should contain the wide range of assumptions, probabilities, and uncertainties involved and be open to full public review.

Critics of risk assessment and risk–benefit analysis argue that the main decision-making tools we should rely on are *looking at the available alternatives to a particular chemical or technology and having a full public discussion of the major advantages (benefits) and disadvantages (costs) of those alternatives.*

The goal of this approach is not to find out how much risk is acceptable but to find out the least damaging reasonable alternatives by asking, "Which alternative will bring sufficient benefits and minimize damage to humans and to the earth?" They argue that if the alternatives are fairly examined, emphasis will shift from trying to determine "acceptable" risk levels to trying to reduce the risks as much as possible by *pollution prevention.*

Critical Thinking

Do you believe that **(a)** risk assessment, risk–benefit analysis, and cost–benefit analysis (Section 28-3) should be used as the primary tools for establishing any federal health, safety, or environmental regulation? or **(b)** that the emphasis should be placed on fair evaluation of alternatives with the goal of finding the least harmful (and most affordable) alternative? Explain.

processes can be redesigned to eliminate or sharply reduce the use and release of toxic chemicals into the environment.

These critics also accuse industries of favoring risk analysis because so little is known about health risks from pollutants and because the data that do exist are controversial. The result is that risk assessment and risk–benefit analysis can be crafted to support almost any conclusion and then called scientific decision making. The huge uncertainties in risk assessment and risk–benefit analysis (Spotlight, above) also allow industries to delay regulatory decisions for decades by challenging data in the courts.

How Should Risks Be Managed? Once an assessment of risk is made, decisions must be made about what to do about the risk. **Risk management** includes the administrative, political, and economic actions taken to decide whether and how to reduce a particular societal risk to a certain level and at what cost.

Risk management involves deciding **(1)** which of the vast number of risks facing society should be evaluated and managed and in what order or priority with the limited funds available, **(2)** how reliable the risk–benefit analysis or risk assessment performed for each risk is, **(3)** how much risk is acceptable, **(4)** how much money it will take to reduce each risk

Q: What percentage of the world's current oil reserves does the United States have?

to an acceptable level, **(5)** how much each risk will be reduced if available funds are limited (as is almost always the case), **(6)** and how the risk management plan will be communicated to the public, monitored, and enforced. Each step in this process involves making value judgments and weighing trade-offs to find some reasonable compromise among conflicting political, economic, health, and environmental interests.

How Well Do We Perceive Risks? Most of us do poorly in assessing the relative risks from the hazards that surround us (Figure 17-13 and 17-14). Many people deny or shrug off the high-risk chances of dying (or injury) from voluntary activities they enjoy, such as motorcycling (1 in 50 participating), smoking (1 in 300 participants by age 65 for a pack-a-day smoker), hang-gliding (1 in 1,250), and driving (1 in 2,500 without a seatbelt and 1 death in 5,000 with a seatbelt).

Yet some of these same people may be terrified about the possibility of dying from a commercial airplane crash (1 in 4.6 million), train crash (1 in 20 million), snakebite (1 in 36 million), shark attack (1 in 300 million), or exposure to trichloroethylene (TCE) in drinking water at the trace levels allowed by the EPA (1 in 2 billion).

Being bombarded with news about people killed or harmed by various hazards distorts our sense of risk. However, *the most important good news each year is that about 99.1% of the people on the earth didn't die*. But that's not what we see on TV, hear, or read about every day. Despite the greatly increased use of synthetic chemicals in food production and processing, the general health and average life expectancy of people in the United States (and most developed countries) have increased during the past 50 years.

Our perceptions of risk and our responses to perceived risks often have little to do with how risky the experts say something is (Figures 17-13 and 17-14). The public generally sees a technology or a product as being riskier than experts do when

- *It is new or complex rather than familiar.* Examples include genetic engineering or nuclear power, as opposed to large dams or coal-fired power plants.
- *It is perceived as being mostly involuntary.* Examples include nuclear power plants or food additives, as opposed to driving or smoking.
- *It is viewed as unnecessary rather than as beneficial or necessary.* Examples might include using chlorofluorocarbon (CFC) propellants in aerosol spray cans or using food additives that increase sales appeal, as opposed to cars or aspirin.
- *Its use involves a large, well-publicized death toll from a single catastrophic accident rather than the same or an even larger death toll spread out over a longer time.* Examples might include a severe nuclear power plant accident, an industrial explosion, or a plane crash, as opposed to coal-burning power plants, automobiles, or smoking.
- *Its use involves unfair distribution of the risks.* Citizens are outraged when government officials decide to put a hazardous-waste landfill or incinerator in or near their neighborhood, even when the decision is based on risk–benefit analysis. This is usually seen as politics, not science. Residents will not be satisfied by estimates that the lifetime risks of cancer death from the facility are no greater than, say, 1 in 100,000. Living near the facility means that they, not the 99,999 people living farther away, have a much higher risk of dying from cancer by having this risk involuntarily imposed on them.
- *The people affected are not involved in the decision-making process from start to finish.*
- *Its use does not involve a sincere search for and evaluation of alternatives.* People who believe that their lives and the lives of their children are being threatened want to know what the alternatives are and which alternative provides the least harm to them and the earth.

Better education and communication about the nature of risks will help bring the public's perceptions of various risks closer to those of professional risk evaluators. However, such education will not eliminate the emotional, cultural, and ethical factors that decision makers must take into account in determining the acceptability of a particular risk and in evaluating the possible alternatives.

> *Not all waste and pollution can be eliminated. . . . What is absolutely crucial, however, is to recognize that pollution prevention should be the first choice and the option against which all other options are judged. The burden of proof imposed on individuals, companies, and institutions should be to show that pollution prevention options have been thoroughly examined, evaluated, and used before lesser options are chosen.*
>
> JOEL HIRSCHORN

A: About 2.3% (but the United States uses about 30% of all oil extracted worldwide each year)

The Current Peril of Antibiotic Resistance

Paul F. Kamitsuka, M.D.

Paul Kamitsuka is an infectious disease expert and physician in Wilmington, North Carolina. He is also clinical associate professor at the University of North Carolina School of Medicine and a guest lecturer on infectious diseases at Harvard Medical School. He received his medical degree at Harvard Medical School, where he later served as a research fellow in infectious diseases. Recently I picked up an infection while in the hospital, and I credit Dr. Kamitsuka's expertise with saving my life.

When most of us seek medical attention for an infection, we naturally assume that there is an antibiotic that will speed up our recovery. After all, it has been seven decades since the discovery of penicillin. Whereas "new" infections such as Legionnaire's disease or AIDS occasionally arise for which effective therapies must be discovered, medical science has had decades of experience treating more common ailments such as strep throat, bacterial pneumonia, or staph infection of the skin. Surely with all that experience, and with all of the antibiotics that have been developed since penicillin, the treatment of such infections must by now be a routine, almost trivial matter. Or so it would seem.

In reality, microorganisms are constantly threatening the utility of our antibiotics, and in many cases have evolved ways to successfully disarm them. Consider pneumococcus, for example, which is a kind of strep bacteria that is the most common cause of pneumonia and meningitis in the United States. Just a decade ago, penicillin resistance among pneumococcus was almost unheard of in the United States. Now over 25% of all pneumococcus isolates are at least partially resistant, rendering penicillin useless in many instances. Just when we felt confident that we had conquered the pneumococcus, suddenly we were faced with the prospect of therapeutic failure.

Surely, one might think, there are new antibiotics that may take the place of penicillin. In the case of pneumococcus, there are other options but resistance to even these antibiotics is quickly emerging. The fact is that we are steadily losing the race to find new antibiotics to treat resistant organisms. Strains of bacteria already exist for which no effective treatments are left. When treating patients infected with such strains, we are like space travelers caught in a time warp, thrown back to the era before antibiotics with nothing to offer but supportive care.

How do we find ourselves in this situation? How are bacteria and other microorganisms winning? How does antibiotic resistance arise? How does it become widespread? How do humans contribute to this problem? What can be done about it?

The emergence of antibiotic resistance amounts to an example of biological evolution occurring at a highly accelerated pace. Biological evolution occurs when members of a species acquire a genetic trait that confers a survival advantage. As a consequence of this advantage, the trait is more likely to be passed on to the next generation and eventually predominates within the population as long as its presence remains advantageous. In the case of antibiotic resistance, the genetic trait enables the bacterium to survive exposure to the antibiotic. Only a tiny proportion of the bacterial population may harbor the resistant trait at first. Because these are the bacteria most likely to survive, however, over the next few generations these resistant bacteria become predominant.

Several factors have contributed to the speed at which antibiotic resistance has evolved. First is the fact that bacteria replicate with generation times as short as 20 minutes. Thus, resistance may evolve within the time course of even a single infection—that is, within only a *few days* as opposed to the prolonged evolutionary time frame we usually think of.

The second factor that accelerates the pace at which antibiotic resistance evolves is that bacteria exchange genetic information with extraordinary frequency, and not just when actual replication is occurring. Moreover, such exchange may occur not only within members of the same species, but also among members of unrelated species. In contrast, genetic exchange among mammals, for example, only occurs within members of the same species and only at the time of mating.

Bacteria harbor their genes not only on chromosomes, but also on mobile genetic elements, such as plasmids or transposons. Bacteria may transfer plasmids to each other through a process called conjugation. If the plasmid carries a gene encoding antibiotic resistance, then the recipient of the plasmid becomes antibiotic-resistant and passes this trait on to its progeny and to other bacteria. Transposons, or "jumping genes," are segments of genetic material that may jump from chromosome to plasmid, or vice versa. They are one mechanism by which resistance genes residing on chromosomes may be transferred to plasmids, and then to other bacteria via conjugation. It is a process akin to a computer software program being copied from the hard drive of one computer onto a floppy disk, then passed on to other computers.

There are numerous examples of bacterial plasmids that harbor antibiotic resistance genes. Gram-negative

Q: What percentage of the oil used in the United States is imported?

bacteria, which are common causes of infections of the urinary tract and lung, routinely carry resistance plasmids. These plasmids enable bacteria to inactivate antibiotics through a variety of different means. Some carry genes that encode enzymes that chemically destroy antibiotics such as penicillin. Others may harbor mutant genes that render antibiotics such as tetracycline ineffective by changing the target site to which the antibiotic binds and exerts its antibacterial activity.

One example of a transposon of particular medical significance is one that enables a bacteria called Enterococcus to be resistant to the antibiotic vancomycin, which acts by binding to the cell wall of the bacteria and interfering with replication. This transposon includes a set of genes that encodes the production of a slightly altered bacterial cell wall so that vancomycin can no longer bind and exert its antibacterial activity.

Another example involves highly virulent Staphylococcus bacteria, which is the number-one cause of infection in the United States. Some forms of Staphylococcus are resistant to all antibiotics except vancomycin. What would happen if the transposon encoding vancomycin resistance in Enterococcus were to jump onto a plasmid and be transferred to Staphylococcus? Such an event, as it turns out, has already occurred under laboratory conditions and outside the lab. In 1998, doctors were shocked to learn of three cases in the United States of this bacterium developing resistance to the last antibiotic (vancomycin) capable of readily killing it.

The third factor that favors the rapid emergence of antibiotic resistance among bacteria is our own antibiotic use. It has been shown that bacteria actually prefer not to be antibiotic resistant. Energy must be expended by the bacteria to maintain resistance plasmids, for example, and once the selective pressure of continued antibiotic pressure is removed, they tend to lose these plasmids.

The fact that so much antibiotic resistance exists today is testimony to the extraordinary amount of antibiotics we use. Unfortunately, much of this use is unnecessary. The sad legacy of antibiotic use has been one of overuse followed by resistance, followed by the introduction of new antibiotics and overuse of these agents, followed by the emergence of resistance to these, and so on. Unfortunately, emerging resistance is outpacing the ability of pharmaceutical companies to keep up.

Overuse of antibiotics occurs in several different settings. Antibiotics are vastly overused in clinics and hospitals. Most cases of the common cold, sore throat, or bronchitis, for example, are caused by viruses that do not respond to antibiotics. Nonetheless, patients expect to be treated with an antibiotic. Health care providers are leery of disappointing them, and reluctant to spend the time convincing the patient otherwise.

In the hospital setting, antibiotics are also used improperly. Rather than making the effort to obtain suitable specimens for culture and sensitivity determination to enable the selection of the narrowest-spectrum older agent that will effectively treat the infection, health care providers often choose the newest broad-spectrum antibiotic. Resistance to the newer agents then occurs and future antibiotic options are curtailed.

Another setting in which antibiotics are overused is in agriculture. Antibiotics are often added to animal feed. In the short term, this may reduce the incidence of infection among livestock, allowing animals to grow larger and faster. However, increasing data suggest that such practices may be increasing resistance levels of bacteria to which humans are ultimately exposed.

A final factor that favors the rapid emergence of resistance is the inadvertent spread of resistant strains from person to person or from place to place. Resistant bacteria may survive in the hospital environment. Furthermore, health care providers who fail to wash their hands between patient contacts may inadvertently spread resistant bacteria from one patient to the next. From a geographic standpoint, the advent of jet travel has meant that resistant organisms may spread rapidly from one locale to another.

So what can be done about the problem of antibiotic resistance? Evolutionary principles dictate that first and foremost we must reduce the selective pressure favoring resistance by curtailing the unnecessary use of antibiotics. In controlled settings where this has been done, such as in individual hospitals with regard to specific antibiotics, resistance levels have fallen. No longer can we assume that new and better agents will appear in time to rescue us from therapeutic peril. The answer will lie in the disciplined judicious use of existing agents and in measures that reduce the spread of resistant isolates in hospital settings.

Critical Thinking

1. What do you believe are the three most important things we should do to reduce the development of genetically resistant infectious bacteria?

2. Use the library or the Internet to study what progress is being made in dealing with this problem.

A: About 55% in 1997 (up from 36% in 1973)

CRITICAL THINKING

1. Explain why you agree or disagree with the proposals made by health officials for reducing the death toll and other harmful effects from smoking given on p. 436.

2. In 1997 U.S. tobacco company officials and attorney generals of states suing tobacco companies for state-related Medicaid costs met to negotiate a solution. One proposal was for the tobacco industry to pay $368 billion to states over 25 years to cover the states' Medicaid expenses related to harm from smoking. Tobacco companies could fund this payment by raising the price of a pack of cigarettes by about 50–75¢, an increase critics say would do little to hinder sales. In return for this payment tobacco companies want the U.S. Congress to guarantee that any individual who sues them for any health-related illness could collect no more than $250,000 in damages. The companies would be able to deduct the deal's costs from their taxable income. This would mean that the companies would pay only about two-thirds of the bill, with U.S. taxpayers paying the rest. Values of most tobacco stocks went up in response to this possible settlement. Are you for or against such an agreement? Explain. What has happened with this issue?

3. Do you think chemicals should be regulated based not only on their carcinogenic properties, but also their effects on the nervous, immune, and endocrine systems? Do the potential costs of doing this outweigh the potential benefits?

4. Do you believe that health and safety standards in the workplace should be strengthened and enforced more vigorously, even if this causes a loss of jobs when companies transfer operations to countries with weaker standards? Explain.

5. Evaluate the following statements:
   a. Because almost any chemical can cause some harm in a large enough dose, we shouldn't get so worked up about exposure to toxic chemicals.
   b. We shouldn't worry so much about exposure to toxic chemicals because through genetic adaptation we can develop immunity to such chemicals.
   c. We shouldn't worry so much about exposure to toxic chemicals because we can use genetic engineering to reduce or eliminate such problems.

6. Which type of natural disaster (flood, earthquake, or volcanic eruption) are you most likely to experience? Why? What can you do to escape or reduce the harm if such a disaster strikes? What actions can you take when it occurs?

7. How can changes in the age structure of a human population increase the spread of infectious diseases? How can the spread of infectious diseases affect the age structure of human populations?

8. What are the major social, economic, and demographic characteristics of human populations most susceptible to infectious diseases?

9. Give three examples of how technological innovations have contributed to the reemergence of infectious diseases. What three things might be done to reduce the harmful effects of new technologies on human health?

10. What are the five major risks that you face from your lifestyle, where you live, and what you do for a living? Which of these risks are voluntary and which are involuntary? List the five most important things you can do to reduce these risks. Which of these things do you actually plan to do?

11. Some analysts argue that the chemical industry is conducting a large-scale experiment on humanity without our informed consent by exposing us to a large number of chemicals whose potential harmful effects have not been established before they are widely used. What, if anything, do you believe should be done about this? Explain. Do the benefits outweigh the risks? Does this involuntary exposure violate the Universal Declaration of Human Rights, which was signed by the United States and most industrialized countries in 1948? Explain.

12. Do you believe that we should shift the emphasis from risk assessment and risk–benefit analysis to pollution prevention and risk reduction? Explain. What beneficial and harmful effects might such a shift have on your life? On the life of any child you might have?

13. How would you answer each of the questions raised about risk assessment and risk–benefit analysis on p. 461? Explain each of your answers.

14. Try to come up with analogies to complete the following statements: **(a)** Smoking is like _______. **(b)** Assessing risk is like _______. **(c)** A carcinogen is like _______. **(d)** Hormones are like _______. **(e)** Overusing antibiotics is like _______. **(f)** Using genetic engineering is like _______. **(g)** Setting risk levels for exposure to chemicals is like _______.

PROJECTS

1. Assume that members of your class (or small manageable groups in your class) have been appointed to a technology risk–benefit assessment board. As a group, decide why you would approve or disapprove of widespread use of each of the following: **(a)** drugs to retard the aging process; **(b)** electrical or chemical devices that would stimulate the brain to eliminate anxiety, fear, unhappiness, and aggression; **(c)** genetic engineering to produce people with superior intelligence, strength, and other traits.

2. Conduct a survey to determine the major physical, chemical, and biological hazards to human health at your school. Develop a plan for preventing or reducing these hazards. Present your findings to school officials.

3. Conduct a survey of officials, citizens, hospitals, and medical personnel to determine the effectiveness of

Q: What percentage of the oil used in the United States is for transportation?

efforts to control the transmission of infectious diseases in your community. List major improvements that could be made and submit them to the appropriate officials.

4. Use the library or the Internet to find recent articles describing the rise of genetic resistance of disease-causing bacteria to commonly used antibiotics. Evaluate the evidence and claims in these articles. What are the three most important strategies for reducing this threat to human health?

5. Pick a specific deadly viral disease and use the library or Internet to find out about how it spreads, its effects, strategies for controlling its spread, and possible treatments. Share evaluations of different viral diseases made by various members of your class.

6. Use examples from the health history of your family or others to determine what stages they have experienced in the epidemiologic transition (Figure 17-12). For example, at what ages did members of your family, beginning with your great grandparents, die and what did they die from?

7. Make a concept map of this chapter's major ideas, using the section heads and subheads and the key terms (in boldface). Look at the inside of the back cover and in Appendix 5 for information about concept maps.

A: About 66% (up from 53% in 1973)

18 AIR POLLUTION

When Is a Lichen Like a Canary?

Nineteenth-century coal miners took canaries with them into the mines—not for their songs, but for the moment when they stopped singing. Then the miners knew it was time to get out of the mine because the air contained methane, which could ignite and explode.

Today we use sophisticated equipment to monitor air quality, but living things such as lichens (Figure 18-1) still have a role in warning us of bad air. A lichen consists of a fungus and an alga living together, usually in a mutually beneficial (mutualistic) partnership. Typically, the fungus absorbs moisture that the alga needs, secretes acids that help the lichen stay attached to its substrate (say, a rock) and dissolves minerals from the rock that both the fungus and alga need. The alga, in turn, carries out photosynthesis and supplies itself and the fungus with carbohydrates.

With more than 20,000 known species, lichens can live almost anywhere—on rocks, trees, bare soil, buildings, gravestones, and even sun-bleached bones. Some lichens survive for more than 4,000 years.

These hearty pioneer species are good air-pollution detectors because they are always absorbing air as a source of nourishment. Certain lichen species are sensitive to specific air-polluting chemicals. Old man's beard (*Usnea trichodeas*) (Figure 18-1, right) and yellow *Evernia* lichens, for example, sicken or die in the presence of too much sulfur dioxide.

Because lichens also vary in their vulnerability to specific pollutants, they can help determine levels of individual pollutants and monitor those levels over time. For sulfur dioxide, for example, *Usnea trichodeas* is the most sensitive.

Because lichens are widespread, long-lived, and reliably anchored in place, they can also help track pollution to its source. The scientist who discovered sulfur dioxide pollution on Isle Royale in Lake Superior, where no car or smokestack had ever intruded,

Figure 18-1 Red and yellow crustose lichens growing on slate rock in the foothills of the Sierra Nevada near Merced, California (left), and *Usnea trichodea* lichen growing on a branch of a larch tree in Gifford Pinchot National Park, Washington (right). The vulnerability of various species of lichens to specific pollutants can help detect levels of specific air pollutants and track down their sources. (Left, Kenneth W. Fink/Ardea London; right, Milton Rand/Tom Stack & Associates)

used *Evernia* lichens to point the finger northward to coal-burning facilities at Thunder Bay, Canada.

Conversely, healthy lichens on damaged trees in Germany's Black Forest got French coal-burning power plants off the hook in the 1970s and allowed investigators to identify the true culprit: nitrogen oxides from car exhausts. The result was Germany's first auto emissions standards, which went into effect in 1992.

Radioactive particles spewed into the atmosphere by the Chernobyl nuclear power-plant disaster (p. 366) fell to the ground over much of northern Scandinavia, spread, and were absorbed by lichens that carpet much of Lapland. The area's Saami people depend on reindeer meat for food, and the reindeer feed on lichens. After Chernobyl more than 70,000 reindeer had to be killed and the meat discarded because it was too radioactive to eat. However, scientists helped the Saami identify which of the remaining reindeer to move by analyzing lichens (which absorbed some of the radioactive fallout) to pinpoint the most contaminated areas.

Last but not least, lichens can replace electronic monitoring stations that would cost more than $100,000 each. This is not so much a triumph of nature over technology as a partnership between the two, for technicians use highly sophisticated methods to analyze lichens for pollution and measure their rates of photosynthesis. The happy result, in any case, is a bad-air warning system that is definitely better than canaries, for it can warn us in advance, even before any visible damage to the lichen occurs.

Lichens are now sending us an urgent warning, with the toxicity and levels of air pollutants decimating lichen populations in some areas. According to a 1996 study, at least a dozen sites in North America have become *lichen deserts*, with up to 80% of native species having disappeared. Lichens can alert us to the danger, but as with all forms of pollution, the only real solution is prevention.

Because we all must breathe air from a global atmospheric commons, air pollution anywhere is a potential threat elsewhere, and in some cases, everywhere. Some air-pollution threats can be global, such as those leading to global warming and ozone depletion discussed in Chapter 19. Other forms of air pollution can cause damage at regional, urban, and household levels (indoor air pollution), as discussed in this chapter.

I thought I saw a blue jay this morning. But the smog was so bad that it turned out to be a cardinal holding its breath.

MICHAEL J. COHEN

This chapter is devoted to answering the following questions:

- What layers are found in the atmosphere?
- What are air pollutants, and where do they come from?
- What is smog?
- What is acid deposition?
- What are the harmful effects of air pollutants?
- How can we prevent and control air pollution?

18-1 THE ATMOSPHERE

What Is the Troposphere? Weather Breeder We live at the bottom of a "sea" of air called the **atmosphere**. This thin envelope of life-sustaining gases surrounding the earth is divided into several spherical layers characterized by abrupt changes in temperature, the result of differences in the absorption of incoming solar energy (Figure 18-2).

About 75% of the mass of the earth's air is found in the atmosphere's innermost layer, the **troposphere**, which extends only about 17 kilometers (11 miles) above sea level at the equator and about 8 kilometers (5 miles) over the poles. If the earth were an apple, this lower layer containing the air we breathe would be no thicker than the apple's skin. This thin and turbulent layer of rising and falling air currents and winds is the planet's weather breeder.

Throughout the earth's long history the composition of the troposphere has varied considerably (Section 6-2). Today, about 99% of the volume of clean, dry air in the troposphere consists of two gases: nitrogen (78%) and oxygen (21%). The remainder has slightly less than 1% argon (Ar), 0.036% carbon dioxide (CO_2), and trace amounts of several other gases. Air in the troposphere also holds water vapor in amounts varying from 0.01% by volume at the frigid poles to 5% in the humid tropics.

The average pressure exerted by the gases in the atmosphere decreases with altitude because the

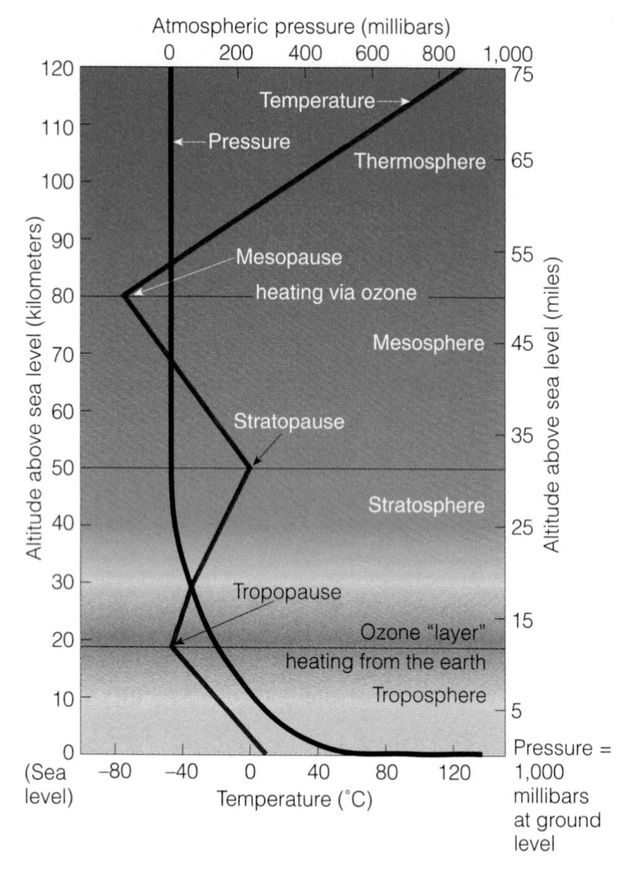

Figure 18-2 The earth's current atmosphere consists of several layers.

average density (mass of gases per unit volume) decreases with altitude. Temperature also declines with altitude in the troposphere, but abruptly begins to rise at the top of this zone, called the *tropopause*. This temperature change limits mixing between the troposphere and upper layers of the atmosphere.

What Is the Stratosphere? Earth's Global Sunscreen The tropopause marks the end of the troposphere and the beginning of the **stratosphere**, the atmosphere's second layer, which extends from about 17–48 kilometers (11–30 miles) above the earth's surface (Figure 18-2). Although the stratosphere contains less matter than the troposphere, its composition is similar, with two notable exceptions: Its volume of water vapor is about 1,000 times less, and its volume of ozone (O_3) is about 1,000 times greater.

The "global sunscreen" of ozone in the stratosphere keeps about 99% of the sun's harmful ultraviolet radiation from reaching the earth's surface (Section 19-5). This UV filter **(1)** allows humans and other forms of life to exist on land; **(2)** helps protect humans from sunburn, skin and eye cancer, cataracts, and damage to the immune system; and **(3)** prevents much of the oxygen in the troposphere from being converted to ozone, a harmful air pollutant. The trace amounts of ozone that do form in the troposphere are a component of urban smog and they damage plants, the respiratory systems of humans and other animals, and materials such as rubber.

Thus, our good health, and that of many other species, depends on having enough ozone in the stratosphere and as little ozone as possible in the troposphere. There is considerable evidence that some human activities are both increasing the amount of harmful ozone in the tropospheric air we must breathe and decreasing the amount of beneficial ozone in the stratosphere (Table 18-1 and Section 19-5).

Unlike air in the troposphere, that in the stratosphere is calm, with little vertical mixing. Pilots like to fly in this layer because it has so little turbulence and such excellent visibility (due to the almost complete absence of clouds). Flying in the stratosphere also improves fuel efficiency because the thin air offers little resistance to the forward thrust of the plane. Temperature rises with altitude in the stratosphere until there is another temperature reversal at the *stratopause*. This change marks the end of the stratosphere and the beginning of the atmosphere's next layer: the *mesosphere* (Figure 18-2).

What Are Two Important Global Processes Taking Place in the Atmosphere? Two natural processes that play crucial roles in the earth's climate and life forms (biodiversity) are the *greenhouse effect* (Figure 7-9), which traps heat in the troposphere, and the *ozone shield* in the stratosphere (Figure 18-2), which filters out most of the sun's ultraviolet radiation (Table 18-1). Both of these natural processes are necessary for life as we know it. However, there is considerable evidence that heat-trapping chemicals we add to the atmosphere can enhance the natural greenhouse effect and lead to **global warming** and reduce the concentration of ozone in the stratosphere, an effect called **ozone depletion** (Table 18-1).

How Are We Disrupting the Earth's Gaseous Nutrient Cycles? The earth's biogeochemical cycles (Chapter 5) work fine as long as we don't disrupt them by overloading them at certain points or removing too many vital chemicals at other points. However, our activities are disrupting the gaseous parts of some of these cycles.

We add about one-fourth as much CO_2 to the troposphere as the rest of nature does by burning up nature's one-time-only deposit of fossil fuels and clearing forests. This massive human intrusion into the

Q: Will the United States ever again be self-sufficient in oil?

Table 18-1 Characteristics of Two Important Global Processes		
	Greenhouse Effect	**Ozone Shield**
Where in the atmosphere does this occur?	Troposphere	Stratosphere
What process occurs?	Traps heat near the earth's surface (short-wave radiation in, long-wave) reradiated)	Filters ultraviolet (UV) radiation from the sun
What natural gases are involved?	Water (H_2O), carbon dioxide (CO_2), methane (CH_4)	Oxygen (O_2), ozone (O_3)
What are important human inputs?	Carbon dioxide (CO_2), methane (CH_4) chlorofluorocarbons (CFCs), nitrous oxide (N_2O)	Chlorofluorocarbons (CFCs), halons, carbon tetrachlorine, methyl choloroform (stable), halogen-containing gases)
What problem results?	Global warming	Ozone depletion.

carbon cycle (Figure 5-5) has the potential to warm the earth (Table 18-1) and alter global climate and food-producing regions (Sections 19-2 and 19-3). We also disrupt natural energy flows (Figure 4-7) by producing huge heated air masses (heat islands) and dust domes over urban areas.

By burning fossil fuels and using nitrogen fertilizers, we are releasing three times more nitrogen oxides (NO, NO_2, and N_2O) and gaseous ammonia (NH_3) into the troposphere than do natural processes in the nitrogen cycle (Figure 5-6). In the troposphere, most of these nitrogen oxides are converted to nitric acid vapor (HNO_3) and acid-forming nitrate salts that dissolve in water and return to the earth, where they can increase the acidity of soils, streams, and lakes and harm plant and animal life.

Our inputs of sulfur dioxide (SO_2) into the troposphere (mostly from petroleum refining and burning of coal and oil; see photo in table of contents) are about twice as high as natural inputs from the sulfur cycle (Figure 5-9). Much of this sulfur dioxide is converted to sulfuric acid (H_2SO_4) and sulfate salts that return to the earth's surface as components of acid deposition.

These are only a few of the chemicals we are spewing into the troposphere. Small amounts of toxic metals such as arsenic, cadmium, and lead also circulate in the ecosphere in chemical cycles. Scientists estimate that we now inject into the troposphere about twice as much arsenic as nature does, 7 times as much cadmium, and 17 times as much lead.

18-2 URBAN OUTDOOR AIR POLLUTION FROM SMOG

What Are the Major Types and Sources of Air Pollution? **Air pollution** is the presence of one or more chemicals in the atmosphere in quantities and duration that cause harm to humans, other forms of life, and materials. As clean air in the troposphere moves across the earth's surface, it collects the products of natural events (dust storms and volcanic eruptions) and human activities (emissions from cars and smokestacks). These potential pollutants, called **primary pollutants**, are mixed vertically and horizontally and are dispersed and diluted by the churning air in the troposphere. While in the troposphere, some of these primary pollutants may react with one another or with the basic components of air to form new pollutants, called **secondary pollutants** (Figure 18-3).

Long-lived primary and secondary pollutants can travel great distances before they return to the earth's surface as solid particles, droplets, or chemicals dissolved in precipitation. Pollutants are also found indoors from infiltration of polluted outside air and from various chemicals used or produced inside buildings, as discussed in Section 18-4. Risk analysis experts rate indoor and outdoor air pollution as high-risk human health problems (Figure 17-13, left). Air pollution is not new, but it has increased significantly during the industrial revolution (Spotlight, p. 475).

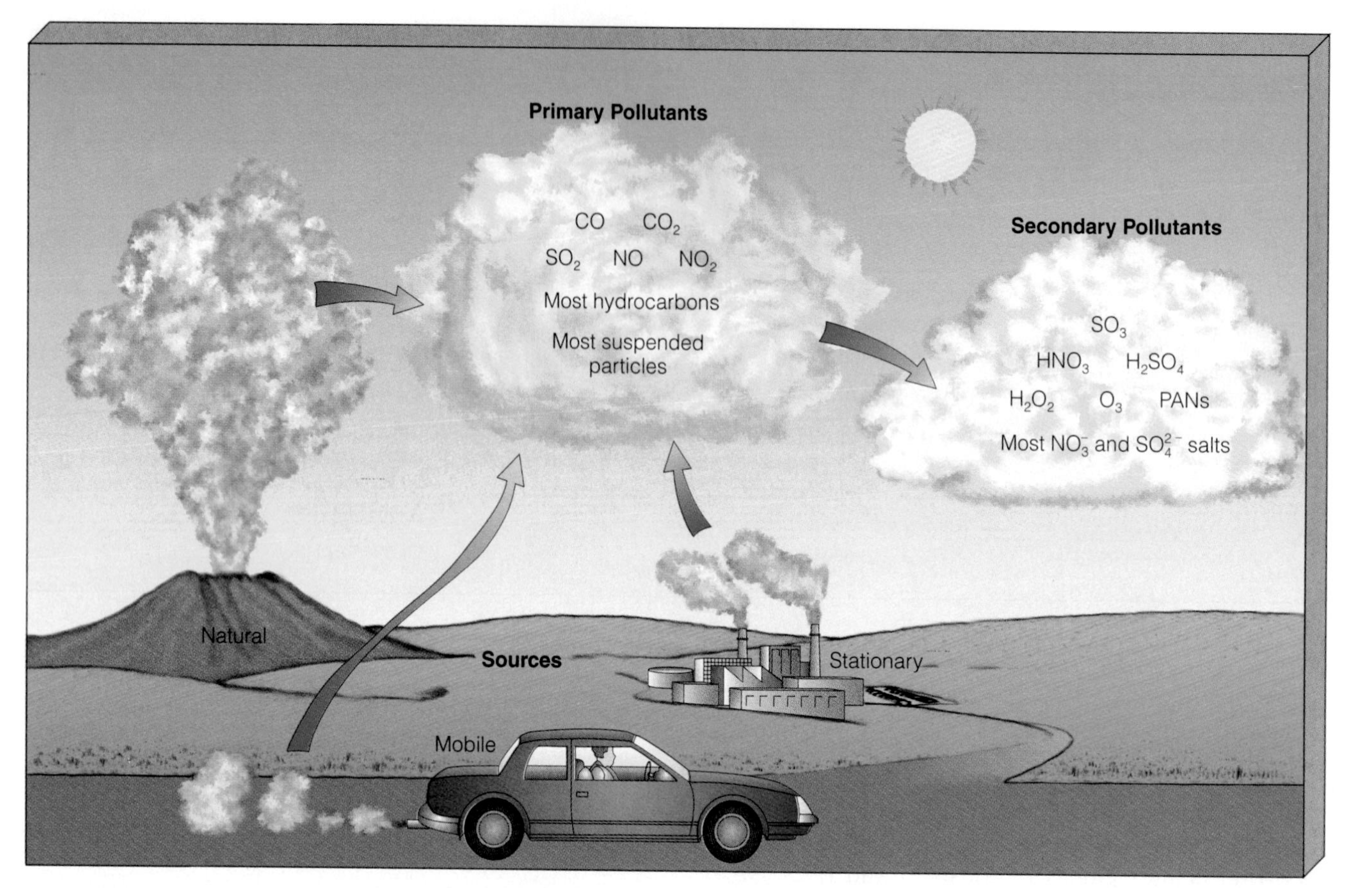

Figure 18-3 Sources and types of air pollutants. Human inputs of air pollutants may come from mobile sources (cars) and stationary sources (industrial and power plants). Some primary air pollutants may react with one another or with other chemicals in the air to form secondary air pollutants.

Table 18-2 lists the major classes of pollutants commonly found in outdoor air. Figure 18-4 shows the sizes and types of various forms of suspended particulate matter—solid particles and droplets of liquid small enough to remain suspended in the atmosphere.

Most pollutants in urban areas enter the atmosphere from the burning of fossil fuels in power plants and factories (*stationary sources*) and motor vehicles (*mobile sources*). In car-clogged cities such as Los Angeles, California; São Paulo, Brazil; Bangkok, Thailand (Figure 11-1); Rome, Italy; and Mexico City, Mexico, motor vehicles are responsible for 80–88% of the air pollution. According to the World Health Organization, more than 1.1 billion people—one of every five—live in urban areas where the air is unhealthy to breathe. Because they contain large concentrations of cars and factories, cities normally have higher air-pollution levels than rural areas. However, prevailing winds can spread long-lived primary and secondary air pollutants from emissions in urban and industrial areas to the countryside and to other downwind urban areas.

What Is Photochemical Smog? Brown-Air Smog Any chemical reaction activated by light is called a *photochemical reaction*. Air pollution known as **photochemical smog** is a mixture of primary and secondary pollutants formed under the influence of sunlight (Figure 18-5). This mixture of pollutants forms when some primary pollutants (mostly nitrogen oxides from motor vehicles and volatile hydrocarbons from various human and natural sources) interact under the influence of sunlight to produce a mixture of more than 100 secondary air pollutants, including ozone (O_3), aldehydes (such as formaldehyde, CH_2O), tear-producing chemicals called peroxyacyl nitrates or PANs, and nitric acid (HNO_3) (Figure 18-5).

Here is a simplified version of the complex chemistry involved in the formation of photochemical smog. It begins when nitrogen and oxygen in air react at the high temperatures found inside automobile engines and the boilers in coal-burning power and industrial plants to produce colorless nitric oxide

Q: How long will the world's oil reserves last at the current consumption rate?

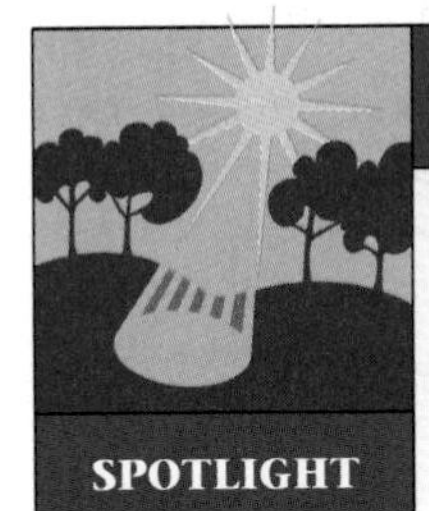

SPOTLIGHT

Air Pollution in the Past: The Bad Old Days

Modern civilization didn't invent air pollution. Recent analyses of Greenland ice cores reveal that ice samples from ancient Greek and Roman times (500 B.C. to A.D. 300) contain lead in concentrations four times those present before these civilizations began smelting metals and releasing lead into the atmosphere. Lead pollution was also observed in ice samples from medieval and Renaissance times.

Burning wood also caused air pollution. In A.D. 61, the Roman author Seneca wrote of the "stink, soot, and heavy air" from the burning of wood. In 1257 the queen of England moved away from the city of Nottingham because the heavy smoke from wood burning was unendurable.

The industrial revolution brought even worse air pollution as coal was burned to power factories and heat homes. In 1273, King Edward I of England banned the burning of coal in London to reduce air pollution (but this ban did not become permanent). More than 500 years later, the English poet Shelley observed, "Hell must be much like London, a smoky and populous city."

In 1911, more than 1,100 Londoners died from the effects of coal smoke. The authors of a report on this disaster coined the word *smog* for the deadly mixture of smoke and fog that blanketed the city. An even worse yellow fog killed 4,000 Londoners in 1952, and disasters in 1956, 1957, and 1962 killed 2,500 more. As a result, London again heeded the advice of Edward I and passed strong air-pollution control measures.

The first U.S. air pollution disaster occurred in 1948, when fog laden with sulfur dioxide and suspended particulate matter stagnated for 5 days over the town of Donora, in Pennsylvania's Monongahela Valley south of Pittsburgh. About 6,000 of the town's 14,000 inhabitants fell ill, and 20 of them died. This killer fog resulted from a combination of mountainous terrain surrounding the valley and weather conditions that trapped and concentrated deadly pollutants emitted by the community's steel mill, zinc smelter, and sulfuric acid plant (see Figure 18-6).

In 1963, high concentrations of air pollutants accumulated in the air over New York City, killing about 300 people and injuring thousands. Other episodes in New York, Los Angeles, and other large cities in the 1960s led to much stronger air-pollution control programs in the 1970s.

Critical Thinking

Why were the earlier air pollution disasters fairly easy to identify and react to? Today air pollution in the United States still causes the premature deaths of 150,000 to 350,000 people per year. Why is it more difficult to deal with the problem of reducing these deaths?

Table 18-2 Major Classes of Air Pollutants

Class	Examples
Carbon oxides	Carbon monoxide (CO), carbon dioxide (CO_2)
Sulfur oxides	Sulfur dioxide (SO_2), sulfur trioxide (SO_3)
Nitrogen oxides	Nitric oxide (NO), nitrogen dioxide (NO_2) (NO and NO_2 are often lumped together and labeled NO_x)
Volatile organic compounds	Methane (CH_4), propane (C_3H_8), benzene (C_6H_6), chlorofluorocarbons (CFCs)
Suspended organic particles	Solid particles (dust, soot, asbestos, lead, nitrate and sulfate salts), liquid droplets (sulfuric acid, PCBs, dioxins, pesticides)
Photochemical oxidants	Ozone (O_3), peroxyacyl nitrates (PANs), hydrogen peroxide (H_2O_2), aldehydes
Radioactive substances	Radon-222, iodine-131, strontium-90, plutonium-239 (Table 3-1)
Toxic compounds	Trace amounts of at least 600 toxic substances (many of them volatile organic compounds), 60 of them known to cause cancer in test animals

A: At least 44 years; undiscovered oil might add another 20–40 years

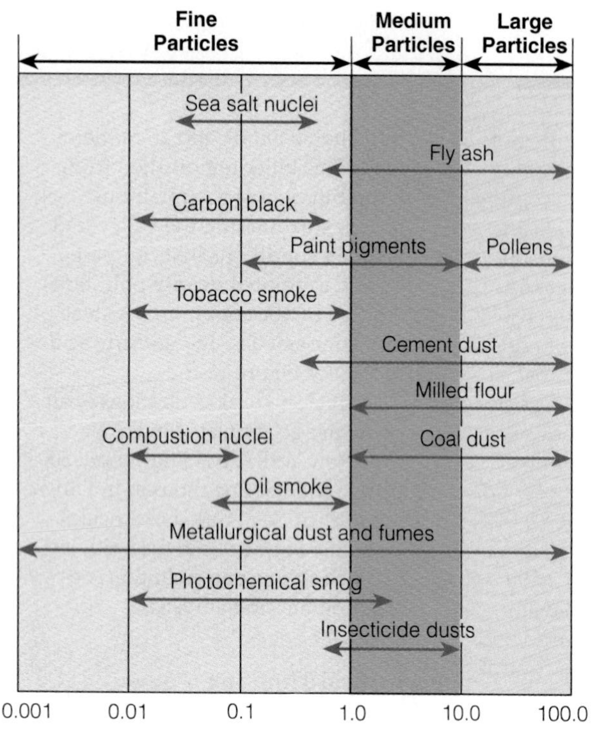

Figure 18-4 Suspended particulate matter consists of particles of solid matter and droplets of liquid that are small and light enough to remain suspended in the atmosphere for short periods (large particles) to long periods (small particles). Suspended particles are found in a wide variety of types and sizes, ranging from 0.001 micrometer to 100 micrometers (a micrometer or micron is one millionth of a meter, or about 0.00004 inches). Since 1987, the EPA has focused on *fine particles* smaller than 10 microns (known as PM-10) and in 1997 on *ultrafine particles* with diameters less than 2.5 microns (known as PM-2.5) because these particles are small enough to reach the lower part of human lungs and contribute to lung and respiratory diseases.

($N_2 + O_2 \rightarrow 2NO$). Once in the troposphere, the nitric oxide slowly reacts with oxygen to form nitrogen dioxide, a yellowish-brown gas with a choking odor ($2NO + O_2 \rightarrow 2NO_2$). The NO_2 is responsible for the brownish haze that hangs over many cities during the afternoons of sunny days, explaining why photochemical smog is sometimes called *brown-air smog*.

Some of the NO_2 reacts with water vapor in the atmosphere to form nitric acid vapor and nitric oxide ($3NO_2 + H_2O \rightarrow 2HNO_3 + NO$). When the remaining NO_2 is exposed to ultraviolet radiation from the sun, some of it is converted to nitric oxide and oxygen atoms (NO_2 + UV radiation $\rightarrow NO + O$). The highly reactive oxygen atoms then react with O_2 to produce ozone ($O_2 + O \rightarrow O_3$). Both the oxygen atoms and ozone then react with volatile organic compounds (mostly hydrocarbons released by vegetation, vehicles, gas stations, oil refineries, and dry cleaners) to produce aldehydes. In addition, hydrocarbons, oxygen, and nitrogen dioxide react to produce peroxyacyl nitrates (hydrocarbons + $O_2 + NO_2 \rightarrow$ PANs).

Collectively, NO_2, O_3, and PANs are called *photochemical oxidants* because they can react with (oxidize) certain compounds in the atmosphere (or inside your lungs) that normally are not oxidized by reaction with oxygen. Mere traces of these photochemical oxidants and aldehydes in photochemical smog can irritate the respiratory tract and damage crops and trees.

The hotter the day, the higher the levels of ozone and other components of photochemical smog. As traffic increases in the morning, levels of NO and NO_2 and unburned hydrocarbons rise and begin reacting in the presence of sunlight to produce photochemical smog. On a sunny day the photochemical smog builds up to peak levels by early afternoon, irritating people's eyes and respiratory tracts.

Virtually all modern cities have photochemical smog. However, it is much more common in cities with sunny, warm, dry climates and lots of motor vehicles such as Los Angeles, California; Denver, Colorado; and Salt Lake City, Utah in the United States, as well as Sydney, Australia; Mexico City, Mexico; and São Paulo and Buenos Aires in Brazil.

What Is Industrial Smog? Gray-Air Smog Thirty years ago cities such as London, England, and Chicago and Pittsburgh in the United States burned large amounts of coal and heavy oil (which contain sulfur impurities) in power plants and factories and for space heating. During winter, people in such cities were exposed to **industrial smog** consisting mostly of sulfur dioxide, suspended droplets of sulfuric acid (formed from some of the sulfur dioxide, Figure 18-3), and a variety of suspended solid particles and droplets (Figure 18-4).

The chemistry of industrial smog is fairly simple. When burned, the carbon in coal and oil is converted to carbon dioxide ($C + O_2 \rightarrow CO_2$) and carbon monoxide ($2C + O_2 \rightarrow 2CO$). Some of the unburned carbon also ends up in the atmosphere as suspended particulate matter (soot).

The sulfur compounds in coal and oil also react with oxygen to produce sulfur dioxide, a colorless, suffocating gas ($S + O_2 \rightarrow SO_2$). Sulfur dioxide also is emitted into the troposphere when metal sulfide ores (such as lead sulfide, PbS) are roasted or smelted to convert the metal ore to the free metal.

In the troposphere, some of the sulfur dioxide reacts with oxygen to form sulfur trioxide ($2SO_2 + O_2 \rightarrow 2SO_3$), which then reacts with water vapor in the air to produce tiny suspended droplets of sulfuric acid ($SO_3 + H_2O \rightarrow H_2SO_4$). Some of these droplets

Q: How long will U.S. oil reserves last at the current consumption rate?

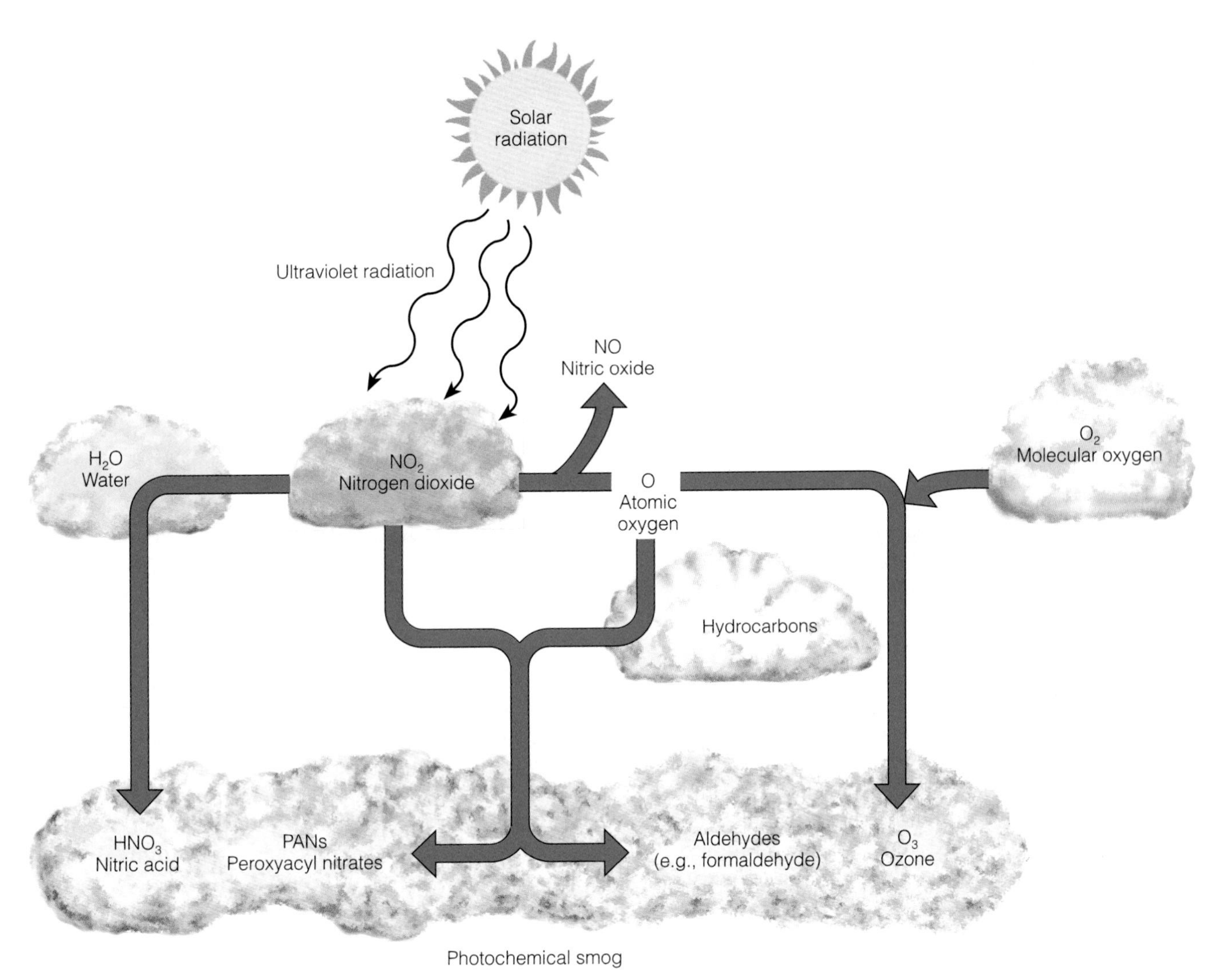

Figure 18-5 Simplified scheme of the formation of photochemical smog. The severity of smog is generally associated with atmospheric concentrations of ozone at ground level.

react with ammonia in the atmosphere to form solid particles of ammonium sulfate ($2NH_3 + H_2SO_4 \rightarrow (NH_4)_2SO_4$). The tiny suspended particles of such salts and carbon (soot) give the resulting industrial smog a gray color, explaining why it is sometimes called *gray-air smog*.

Urban industrial smog is rarely a problem today in most developed countries because coal and heavy oil are burned only in large boilers with reasonably good pollution control or with tall smokestacks. However, industrial smog is a problem in industrialized urban areas of China, India, Ukraine, and some eastern European countries, where large quantities of coal are burned with inadequate pollution controls.

What Factors Influence the Formation of Photochemical and Industrial Smog? The frequency and severity of smog in an area depend on several things: the local climate and topography, the population density, the amount of industry, and the fuels used in industry, heating, and transportation. In areas with high average annual precipitation, rain and snow help cleanse the air of pollutants. Winds help sweep pollutants away and bring in fresh air, but they may also transfer some pollutants to downwind areas.

Hills and mountains tend to reduce the flow of air in valleys below them and allow pollutant levels to build up at ground level. Buildings in cities generally slow wind speed, thereby reducing dilution and removal of pollutants.

During the day the sun warms the air near the earth's surface. Normally this heated air expands and rises, carrying low-lying pollutants higher into the troposphere (Figure 18-6, left). Colder, denser air from surrounding high-pressure areas then sinks into the low-pressure area created when the hot air rises. This

A: 24 years (undiscovered oil might add another 24 years)

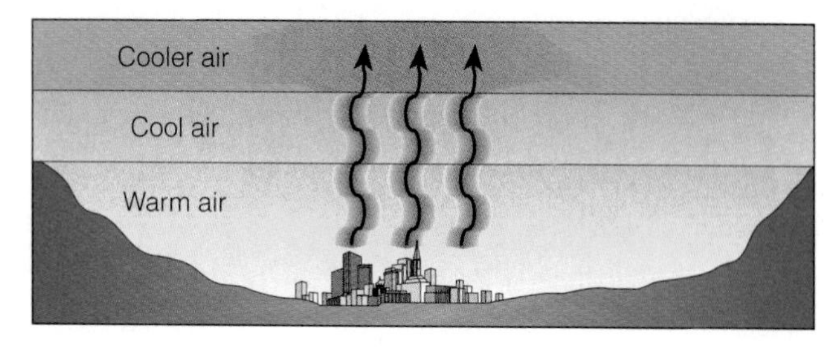

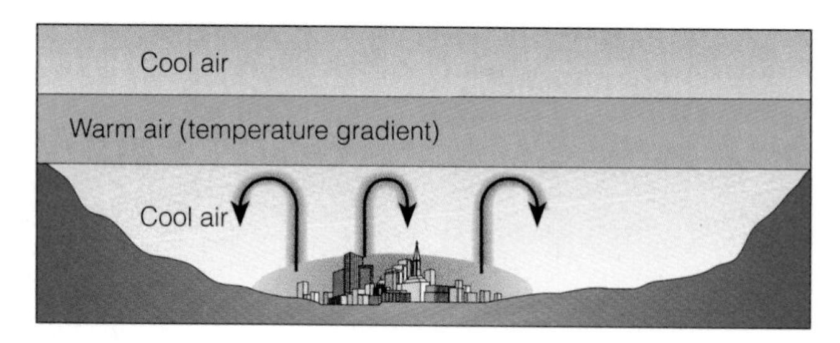

Figure 18-6 Thermal inversion. The change in temperature (temperature gradient) in the warm air (right) prevents ascending air currents from rising and dispersing and diluting pollutants in the cool pool near the ground. Because of their topography, Los Angeles in the United States and Mexico City in Mexico have frequent thermal inversions, many of them prolonged during the summer.

continual mixing of the air helps keep pollutants from reaching dangerous concentrations near the ground.

Sometimes, however, a layer of dense, cool air beneath can be trapped beneath a layer of less dense, warm air in an urban basin or valley, causing a phenomenon known as a **temperature inversion** or **thermal inversion** (Figure 18-6, right). The changing temperature (temperature gradient) in the warm air above the pool of cool air prevents ascending air currents (that would disperse and dilute pollutants) from developing. These inversions usually last for only a few hours but when a high-pressure air mass stalls over an area, they can last for several days, allowing air pollutants at ground level to build up to harmful and even lethal concentrations.

A city with several million people and motor vehicles in an area with a sunny climate, light winds, mountains on three sides, and the ocean on the other has ideal conditions for photochemical smog worsened by frequent thermal inversions. This describes California's Los Angeles basin, which has 14 million people, 23 million motor vehicles, thousands of factories, and thermal inversions at least half of the year. Despite having the world's toughest air-pollution control program, Los Angeles is the air-pollution capital of the United States. Other cities with frequent thermal inversions are Denver, Colorado, in the United States; Mexico City, Mexico; Rio de Janeiro and São Paulo in Brazil; and Beijing and Shenyang in China.

18-3 REGIONAL OUTDOOR AIR POLLUTION FROM ACID DEPOSITION

What Is Acid Deposition? To reduce local air pollution (and meet government standards without having to add expensive air-pollution control devices), most coal-burning power plants, ore smelters, and other industrial plants in developed countries use tall smokestacks to emit sulfur dioxide, suspended particles, and nitrogen oxides above the inversion layer. As this practice spread in the 1960s and 1970s, pollution in downwind areas began to increase. In addition to smokestack emissions, large quantities of nitrogen oxides are also released by motor vehicles.

This "dilution solution" reduces local air pollution. However, it increases pollution downwind because what goes up must come down—another example of connections or unintended consequences. As the primary pollutants, sulfur dioxide and nitrogen oxides are transported as much as 1,000 kilometers (600 miles) by prevailing winds, they form secondary pollutants such as nitric acid vapor, droplets of sulfuric acid, and particles of acid-forming sulfate and nitrate salts (Figure 18-3). In water, acidic substances produce hydrogen ions (H^+), and basic substances produce hydroxide ions (OH^-). An *acidic solution* has more hydrogen ions than hydroxide ions ($H^+ > OH^-$), and an *alkaline* or *basic* solution has more hydroxide ions than hydrogen ions ($OH^- > H^+$).

These chemicals descend to the earth's surface in two forms: *wet* (as acidic rain, snow, fog, and cloud vapor) and *dry* (as acidic particles). The resulting mixture is called **acid deposition** (Figure 18-7). Although this form of pollution is commonly called *acid rain, acid deposition* is a better term because the acidity can reach the earth's surface not only in rain but also as gases and as solid particles.

Acidity of substances in water is commonly expressed in terms of **pH** (a numerical measure of the concentration of hydrogen ions in a solution). Solutions with pH values less than 7 are acidic, and those with pH values greater than 7 are alkaline or basic (Figure 5-18). Natural precipitation is slightly acidic, with a pH of 5.0–5.6. However, primarily because of acid deposition, typical rain in the eastern United States is now about 10 times more acidic, with a pH of 4.3. In some areas it is 100 times more acidic, with a

Q: How much new oil must be discovered to allow global oil consumption to continue at the current rate?

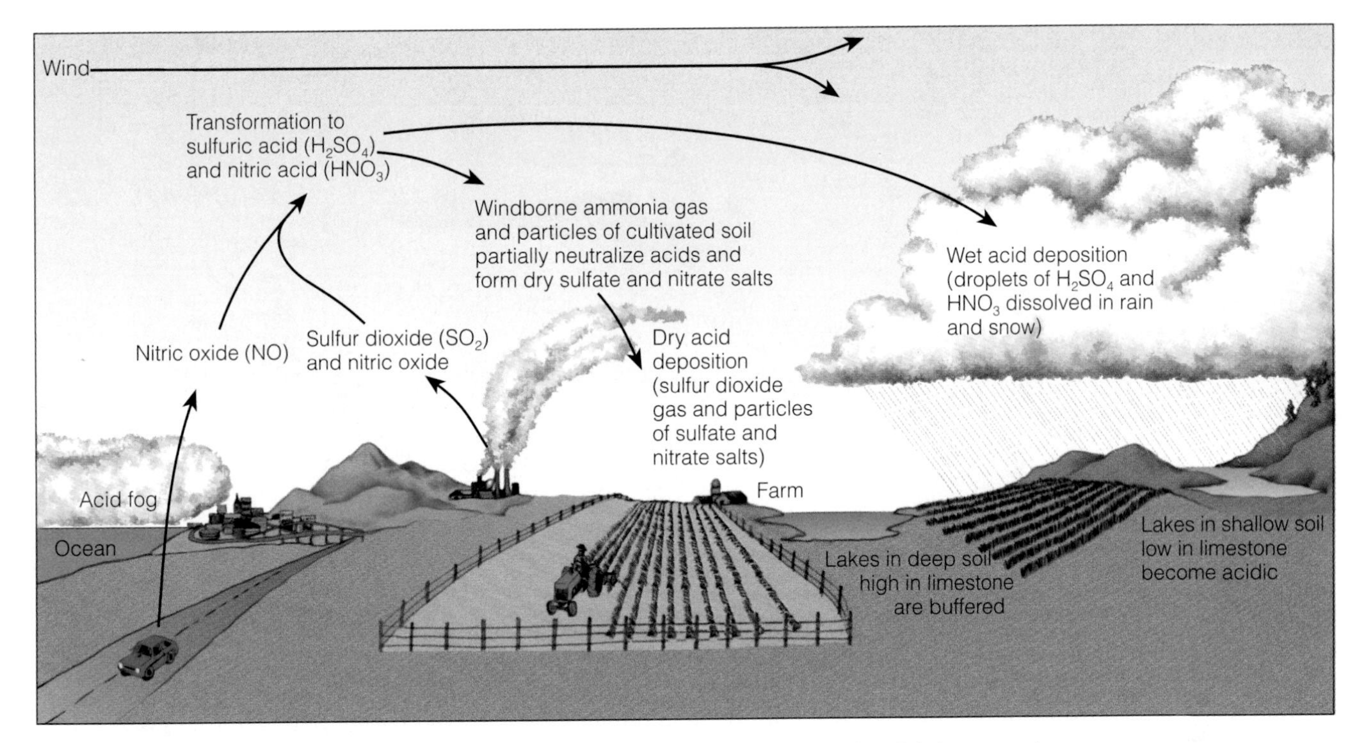

Figure 18-7 Acid deposition, which consists of rain, snow, dust, or gas with a pH lower than 5.6, is commonly called acid rain. Soils and lakes vary in their ability to buffer or remove excess acidity.

pH of 3—as acidic as vinegar. Some cities and mountaintops are bathed in a fog as acidic as lemon juice, with a pH of 2.3—about 1,000 times the acidity of normal precipitation.

What Areas Are Most Affected by Acid Deposition? Acid deposition occurs on a regional rather than a global basis because the acidic components remain in the atmosphere only for a few days. However, acid deposition is a serious regional problem (Figure 18-8) in many areas downwind from coal-burning power plants, smelters, factories, and large urban areas. How seriously vegetation and aquatic life in nearby lakes are affected by an area receiving acid deposition depends mostly on whether its soils are acidic or basic.

In some areas, soils contain enough calcium ions (Ca^{2+}) and magnesium ions (Mg^{2+})—attached to negatively charged soil minerals—to neutralize or buffer some inputs of acids. The ecosystems most harmed by acid deposition are those containing thin, acidic soils without such natural buffering (Figure 18-8, green areas) and those where the buffering capacity of soils has been depleted because of decades of exposure to acid deposition.

Many of the acid-producing chemicals generated by power plants, factories, smelters, and cars in one country may be exported to others by prevailing winds. For example, more than three-fourths of the acid deposition in Norway, Switzerland, Austria, Sweden, the Netherlands, and Finland is blown to those countries from industrialized areas of western Europe (especially the United Kingdom and Germany) and eastern Europe.

Acid deposition is also a growing problem in China (40% of its land), parts of the former Soviet Union, India, Nigeria, Brazil, Venezuela, and Colombia. In the Chinese city of Chongquing the rain typically has a pH of 3, about as acidic as vinegar (Figure 5-18).

Chemical detective work indicates that more than half the acid deposition in southeastern Canada and the eastern United States originates from coal- and oil-burning power plants and factories in the states of Ohio, Indiana, Pennsylvania, Illinois, Missouri, West Virginia, and Tennessee. In areas near and downwind from large urban areas, emissions of NO and NO_2 (mostly from cars) leading to the formation of nitric acid may be the main culprit.

Within a few decades, NO_x and SO_2 emissions from developing countries are expected to outstrip those from developed countries, leading to greatly increased damage from acid deposition over a much wider area, especially where soils are sensitive to acidification.

What Are the Effects of Acid Deposition? Risk analysis experts rate acid deposition as a medium-risk ecological problem and a high risk to human

A: A new supply equal to the reserves in Saudi Arabia (the world's largest) every 10 years

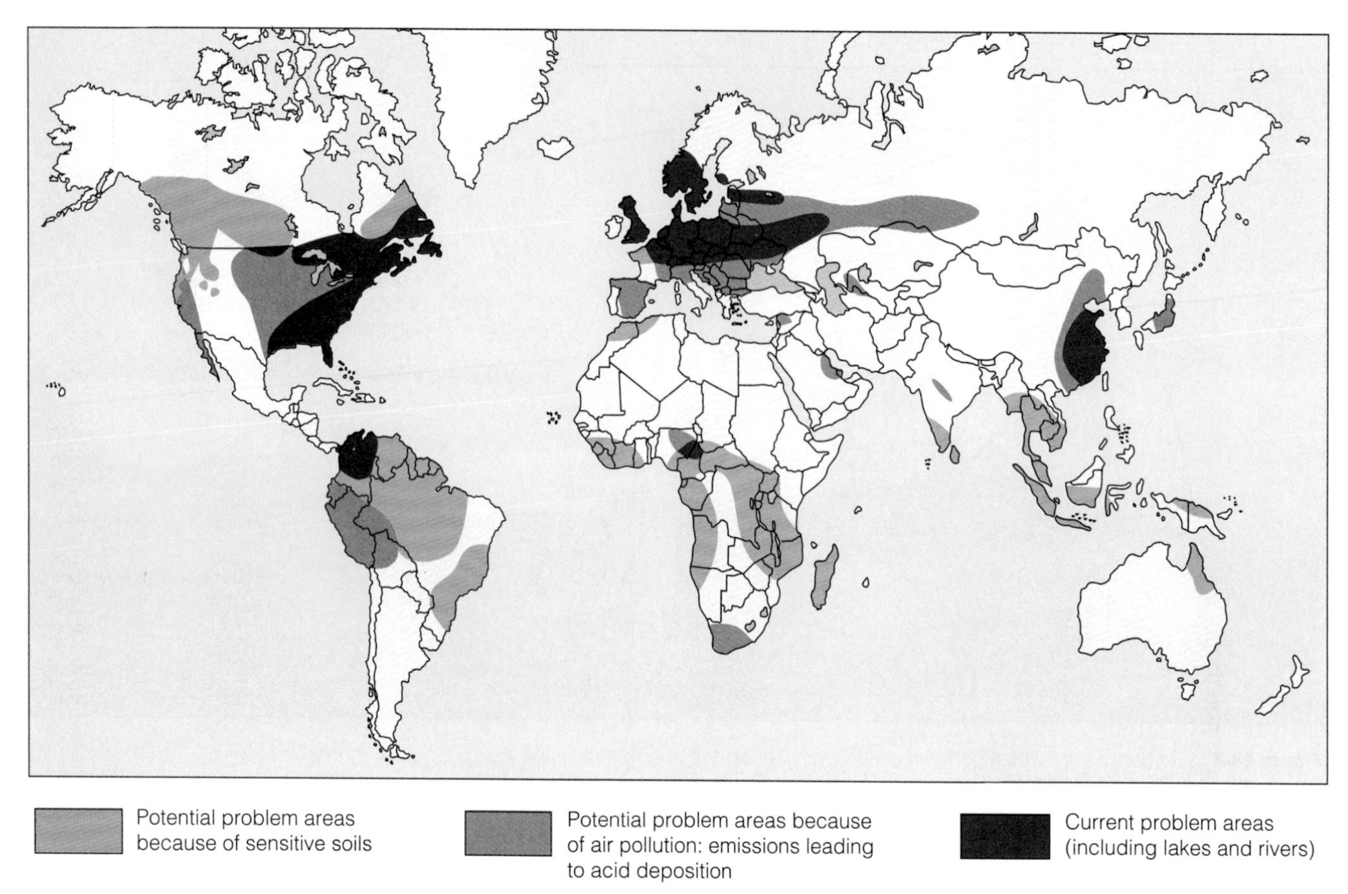

Figure 18-8 Regions where acid deposition is now a problem and regions with the potential to develop this problem, either because of increased air pollution (mostly from power plants, industrial plants, and ore smelters) or because of soils that cannot neutralize inputs of acidic compounds (green areas and most red areas). (Data from World Resource Institute and U.S. Environmental Protection Agency)

health (Figure 17-13, left). Acid deposition has many harmful ecological effects, especially when the pH falls below 5.1 for terrestrial systems and below 5.5 for aquatic systems. It also contributes to human respiratory diseases such as bronchitis and asthma (which can cause premature death) and it damages statues, buildings, metals, and car finishes.

Acid deposition and other air pollutants such as ozone (O_3) can damage tree foliage directly, but the most serious effect is weakening trees so they become more susceptible to other types of damage (Figure 18-9). The areas hardest hit by acid deposition are mountaintop forests, which tend to have thin soils without much buffering capacity. Trees on mountaintops, especially conifers such as red spruce that keep their leaves year-round, are bathed almost continuously in very acidic fog and clouds. The high humidity of this cloudy environment can also promote the growth of acid-loving mosses that reduce nutrient uptake by the trees by killing mycorrhizae fungi on the tree roots (Figure 9-15).

A combination of acid deposition and other air pollutants (especially ozone) can make trees more susceptible to stresses such as cold temperatures, diseases, insects, drought, and fungi (which thrive under acidic conditions) and to a drop in net primary productivity from loss of soil plant nutrients (Connections, p. 482). Although the final cause of tree damage or death may be mosses, insect attacks, diseases, and lack of plant nutrients, the underlying cause is often years of exposure to an atmospheric cocktail of air pollutants and soil overloaded with acids.

Acid deposition in acidic soils and in alkaline soil depleted of calcium and magnesium ions can also release aluminum ions (Al^{3+}) attached to soil minerals. Once released from soil particles, these water-soluble ions can damage tree roots. When washed into lakes, aluminum ions can also kill many kinds of fish by stimulating excessive mucus formation, which asphyxiates the fish by clogging their gills.

Excess acidity can contaminate fish in some lakes with highly toxic methylmercury (CH_3Hg). Increased acidity of lakes apparently converts moderately toxic inorganic mercury compounds in lake-bottom sediments into highly toxic methylmercury, which is more

Q: Adjusting for inflation, how did the price of gasoline in the United States in 1998 compare with its price in 1950? in 1973?

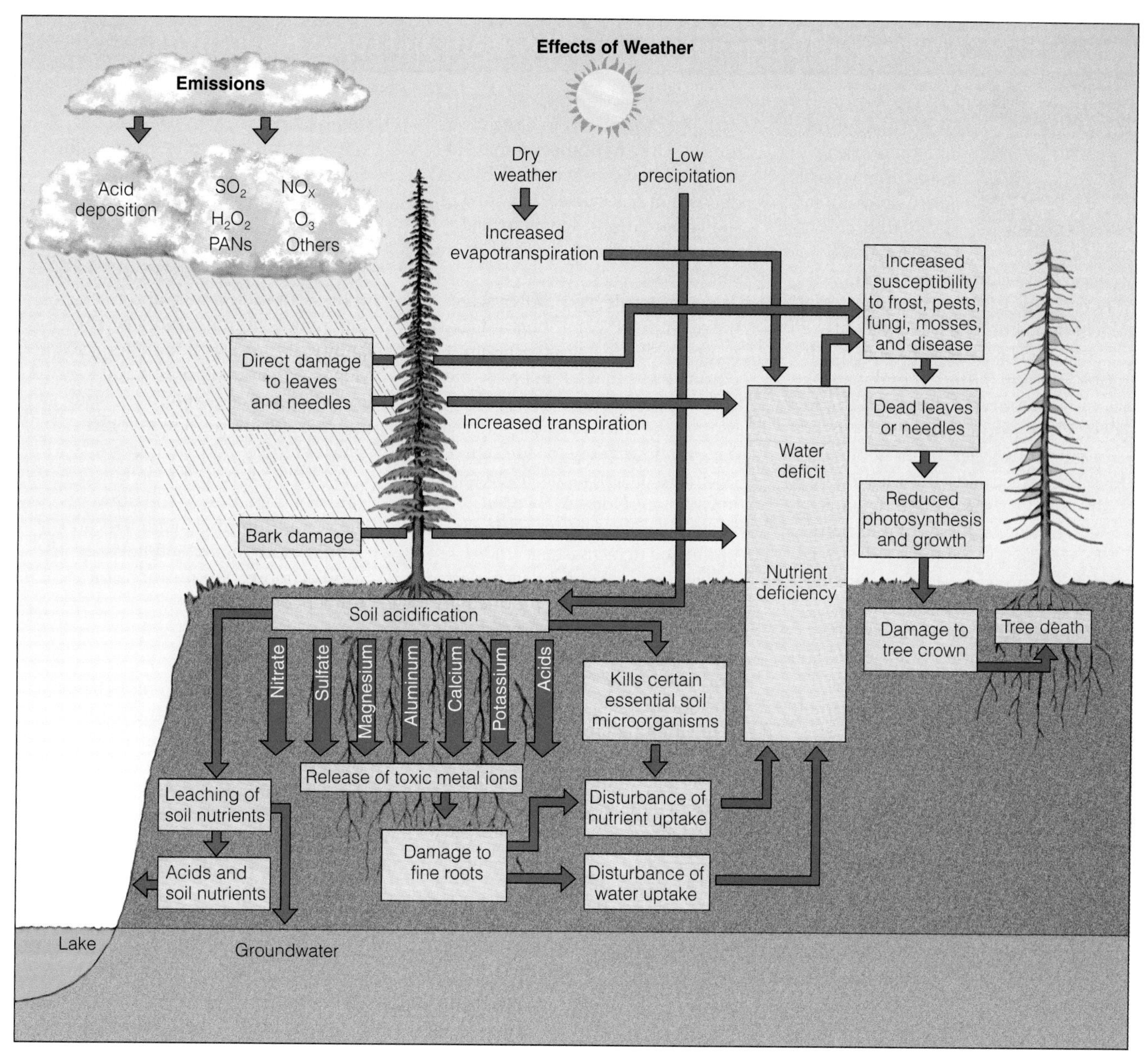

Figure 18-9 Possible or suspected harmful effects of prolonged exposure to an atmospheric cocktail of air pollutants on trees.

soluble in the fatty tissue of animals and can be biomagnified to higher concentrations in aquatic food chains and webs. However, acid runoff into lakes and streams is rated by risk analysis experts as a low-risk ecological problem (Figure 17-13, left).

How Serious Is Acid Deposition in the United States? A large-scale, government-sponsored research study on the ecological effects of acid deposition in the United States in the 1980s concluded that the problem was serious but not yet at a crisis stage. However, numerous health studies have shown that the effects from exposure to the chemical components of acid deposition are a serious health problem (Figure 17-13 and Section 18-4) and also damage materials (Table 18-3).

Representatives of coal companies and industries that burn coal and oil claim that adding expensive air-pollution control equipment or burning low-sulfur coal or oil costs more than the resulting health and environmental benefits are worth. According to the EPA, however, the actual cleanup costs of SO_2 in 1994 were about one-tenth of the estimate given by industry when they opposed the new standards set by the Clean Air Act of 1990. A comprehensive 1997 study by Resources for the Future found that the environmental

A: About the same in both cases

CONNECTIONS

Effects of Acid Deposition on Soil Nutrients and Forest Productivity

Recent analysis of data on nutrient losses from soils at Yale University's Hubbard Brook Experimental Forest in New Hampshire (Figure 5-2) over the past 45 years indicates that acid deposition has leached out more than 50% of the pool of available calcium in the generally alkaline soils found in this research forest. Plant productivity has dropped because of such nutrient depletion.

Because it takes decades to hundreds of years for soil to replenish such nutrients, losses in plant productivity could continue for decades even if emissions of sulfur dioxide and nitrogen oxides are reduced by air-pollution control programs. Another study found that since 1950 acidification has caused the loss of half of the calcium and magnesium stored in measured forest soils in Sweden.

Recently Dutch scientists have implicated acid deposition in the decline of titmouses, nuthatches, and other songbirds in Dutch forests. They hypothesize that calcium leached out of the soil by acid deposition has caused a decline in snail populations, which provide most of the calcium for snail-eating songbirds.

As the snails have grown scarce, the calcium-deficient songbirds have begun producing eggs with thin, porous shells that either break or dry out within days after they are laid. Most of the chicks that do hatch have bone malformations, a sure sign of calcium deficiency. Forest songbirds in Germany, Norway, Sweden, and Switzerland are also increasingly producing defective eggs.

Until recently scientists expected some of the lost forest productivity to be offset by increased productivity from the larger input of nitric acid and nitrate salts from acid deposition in areas where this nutrient is the limiting factor. However, recent research revealed that much of the nitrate raining down on forest areas in parts of Germany and Norway damaged by air pollution is not being taken up by the trees or by nitrogen-using microbes in the soil.

One explanation is that the excess nitrates reduce forest productivity by removing calcium and magnesium ions as the nitrate ions flow through forest soils. As nitrate saturation occurs and calcium and magnesium ions are lost, the ability of forest soils to buffer acid deposition and support plant growth progressively drops, leading to a further decline in forest productivity.

Drops in forest productivity because of depletion of soil nutrients and acid-buffering chemicals can reduce biodiversity and have significant economic implications for timber companies.

Critical Thinking

What, if anything, do you believe should be done about the depletion of soil nutrients and acid-buffering chemicals as a result of acid deposition?

and public health benefits of reductions in SO_2 from 1995 to 2030 will generate more than $12 in benefits for every $1 in compliance costs.

Many scientists support greatly reducing emissions from coal- and oil-burning facilities to reduce their harmful effects on human health and materials and to *prevent* acidic compounds in soil and aquatic systems from exceeding the tolerance levels of various species and eventually serious and costly ecological and economic damage.

Progress is being made. A 1993 study by the U.S. Geological Survey found that the concentration of sulfate ions, a key component of acid deposition, declined at 26 out of 33 U.S. rainwater collection sites between 1980 and 1991. In addition, U.S. sulfur dioxide emissions dropped 30% between 1970 and 1993 and are expected to fall further by 2000 because of the requirements of the Clean Air Act of 1990.

Solutions: What Can Be Done to Reduce Acid Deposition? According to most scientists studying acid deposition, the best solutions are *prevention approaches*. They include **(1)** reducing energy use and thus air pollution by improving energy efficiency (Section 16-2); **(2)** switching from coal to cleaner-burning natural gas (Section 15-3) and renewable energy resources (Chapter 16); **(3)** removing sulfur from coal before it is burned; **(4)** burning low-sulfur coal; **(5)** removing SO_2, particulates, and nitrogen oxides from smokestack gases; and **(6)** removing nitrogen oxides from motor vehicle exhaust.

Reducing coal use, the major culprit, will be economically and politically difficult. For example, China (the world's largest user of coal, Connections, p. 378) and India (the fourth largest user) are using their own coal reserves to fuel rapid industrial growth and have put little money into pollution control.

Some *cleanup approaches* can be used, but they are expensive and merely mask some of the symptoms temporarily without treating underlying causes. Acidified lakes can be neutralized by treating them or the surrounding soil with large amounts of limestone or lime ($CaCO_3$). However, such liming is an expensive and only temporary remedy that usually must be

Oliver, L. C., and B. W. Shackleton. 1998. "The Indoor Air We Breathe." *Public Health Reports,* vol. 113, no. 5, 398(12).

Table 18-3 Harmful Effects of Air Pollution on Materials

Material	Effects	Principal Air Pollutants
Stone and concrete	Surface erosion, discoloration, soiling	Sulfur dioxide, sulfuric acid, nitric acid, particulate matter
Metals	Corrosion, tarnishing, loss of strength	Sulfur dioxide, sulfuric acid, nitric acid, particulate matter, hydrogen sulfide
Ceramics and glass	Surface erosion	Hydrogen fluoride, particulate matter
Paints	Surface erosion, discoloration, soiling	Sulfur dioxide, hydrogen sulfide, ozone, particulate matter
Paper	Embrittlement, discoloration	Sulfur dioxide
Rubber	Cracking, loss of strength	Ozone
Leather	Surface deterioration, loss of strength	Sulfur dioxide
Textiles	Deterioration, fading, soiling	Sulfur dioxide, nitrogen dioxide, ozone, particulate matter

repeated annually. Using lime to reduce excess acidity in U.S. lakes would cost at least $8 billion per year.

Liming can also kill some types of plankton and aquatic plants and can harm wetland plants that need acidic water. It is also difficult to know how much of the lime to put where (in the water or at selected places on the ground). In addition, some recent research suggests that liming can increase populations of microbes that deplete carbon stored in the slowly decaying soil matter (humus) and thus reduce forest productivity. Recently, however, researchers in England found that adding a small amount of phosphate fertilizer can neutralize excess acidity in a lake.

18-4 INDOOR AIR POLLUTION

What Are the Types and Sources of Indoor Air Pollution? If you are reading this book indoors, you may be inhaling more air pollutants with each breath than if you were outside (Figure 18-10). According to EPA studies, in the United States levels of 11 common pollutants are generally 2–5 times higher inside homes and commercial buildings than outdoors, and as much as 70 times higher in some cases. A 1993 study found that pollution levels inside cars in traffic-clogged U.S. urban areas can be up to 18 times higher than those outside the vehicles.

The health risks from exposure to such chemicals are magnified because people spend 70–98% of their time indoors. In 1990 the EPA placed indoor air pollution at the top of the list of 18 sources of cancer risk, and it is rated by risk analysis scientists as a high-risk health problem for humans (Figure 17-13, left). At greatest risk are smokers, infants and children under age 5, the old, the sick, pregnant women, people with respiratory or heart problems, and factory workers.

Danish and U.S. EPA studies have linked pollutants found in buildings to dizziness, headaches, coughing, sneezing, nausea, burning eyes, chronic fatigue, and flulike symptoms, known as the *sick building syndrome*. A building is considered "sick" when at least 20% of its occupants suffer persistent symptoms that disappear when they go outside.

New buildings are more commonly "sick" than old ones because of reduced air exchange (to save energy) and chemicals released from new carpeting and furniture. According to the EPA, at least 17% of the 4 million commercial buildings in the United States are considered "sick" (including EPA's headquarters). Indoor air pollution in the United States costs an estimated $100 billion per year in absenteeism, reduced productivity, and health costs.

A 1994 study by Cornell University researchers suggested that the primary culprits in causing "sick building" symptoms may be mineral fibers falling from ceiling tiles and blowing in from the lining of air conditioning ducts. The study also suggests why workers who spend a lot of time in front of computer terminals may suffer more than coworkers who don't: Electrostatic fields generated by computer monitors attract the fibers, thus exposing people sitting in front of the screens to more of them.

According to the EPA and public health officials, cigarette smoke (p. 436), formaldehyde, asbestos, and radioactive radon-222 gas are the four most dangerous indoor air pollutants. A number of research studies on laboratory animals have also identified tiny fibers of *fiberglass* as a widespread and potentially potent carcinogen in indoor air.

The chemical that causes most people difficulty is *formaldehyde*, an extremely irritating gas. As many as

Hint: Enter the search term *indoor air quality* using the Subject Guide.

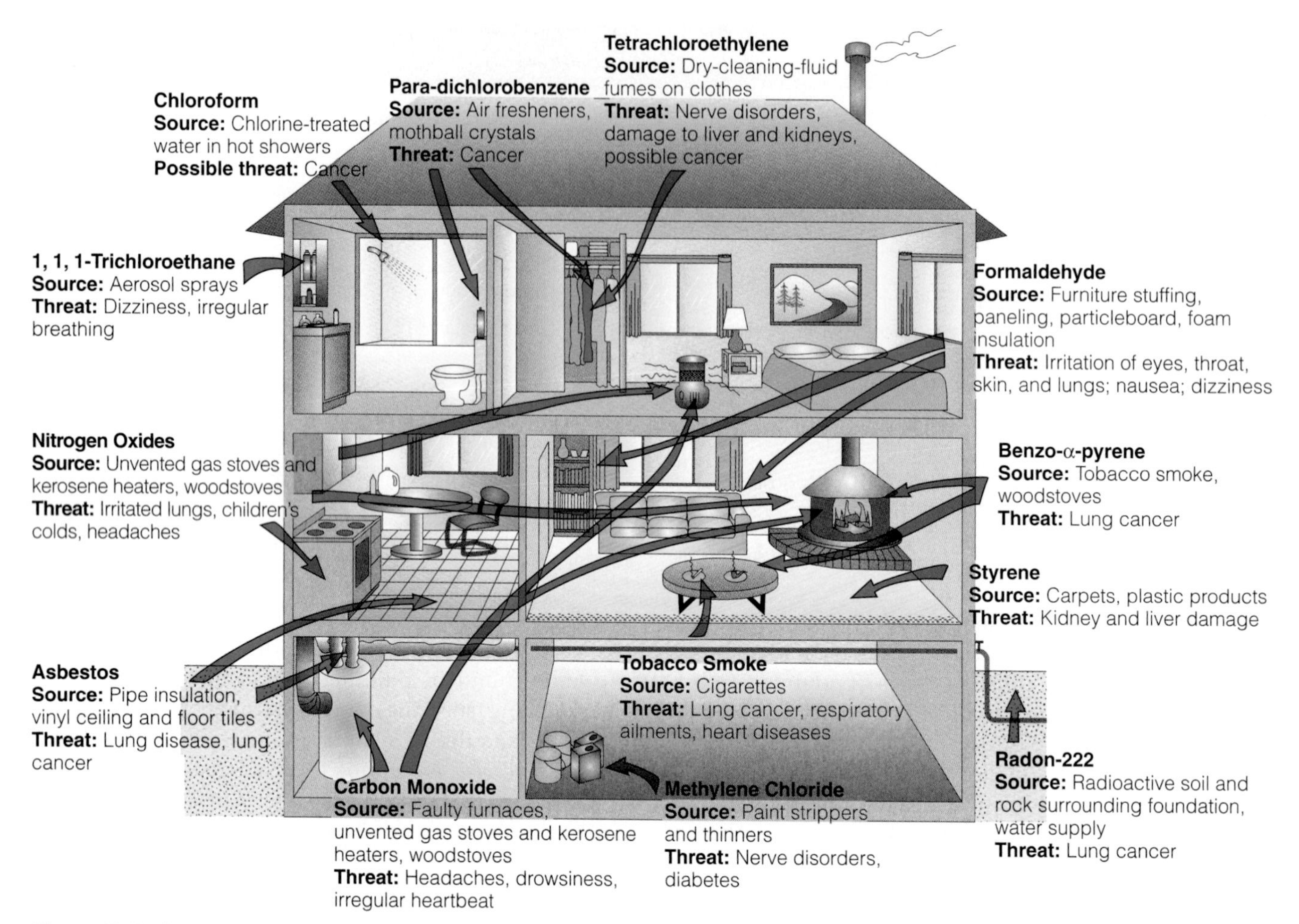

Figure 18-10 Some important indoor air pollutants. (Data from U.S. Environmental Protection Agency)

20 million Americans suffer from chronic breathing problems, dizziness, rash, headaches, sore throat, sinus and eye irritation, and nausea caused by daily exposure to low levels of formaldehyde emitted (outgassed) from common building materials (such as plywood, particleboard, and paneling), furniture, drapes, upholstery, and adhesives in carpeting and wallpaper (Figure 18-10). The EPA estimates that as many as 1 out of every 5,000 people who live in manufactured homes for more than 10 years will develop cancer from formaldehyde exposure.

In developing countries, the burning of wood, dung, and crop residues in open fires or in unvented or poorly vented stoves for cooking and heating exposes inhabitants, especially women and young children, to very high levels of particulate air pollution. Partly as a result, respiratory illnesses are a major cause of death and illness among the poor in most developing countries.

Case Study: What Should Be Done About Asbestos? There is intense controversy over what to do about possible exposure to tiny fibers of asbestos, a name given to several different fibrous forms of silicate minerals widely used since the 1940s for fireproofing and thermal insulation. Unless completely sealed within a product, asbestos easily crumbles into a dust of fibers tiny enough to become suspended in the air and to be inhaled deep into the lungs, where they remain for many years.

Prolonged exposure to asbestos fibers can cause *asbestosis* (a chronic, sometimes fatal disease that eventually makes breathing nearly impossible and was recognized as a hazard among asbestos workers as early as 1924), *lung cancer*, and *mesothelioma* (an inoperable cancer of the chest cavity lining). Epidemiological studies have shown that lung cancer death rates for nonsmoking asbestos workers were 5 times higher than for nonsmokers in a control group, and 53 times higher for asbestos workers who smoked.

Most of these diseases occur in people exposed for years to high levels of asbestos fibers. This group includes asbestos miners, insulators, pipefitters, shipyard employees, and workers in asbestos-producing factories. By the year 2000 it is estimated that 300,000 American workers will have died prematurely be-

Q: What would happen if oil's harmful effects were included in its price and government subsidies were removed?

cause of exposure to asbestos fibers. After being swamped with health claims from workers, most U.S. asbestos manufacturing companies have either declared bankruptcy or have moved their operations to other countries (such as Mexico and Brazil) with weaker environmental laws and lax enforcement.

In recent years the focus has shifted from asbestos workers to concern over possible health effects of inhalation of low levels of asbestos fibers by the public in buildings. In the United States between 1900 and 1984, asbestos was sprayed on ceilings and walls of schools and other public and private buildings for fireproofing, soundproofing, insulation of heaters and pipes, and wall and ceiling decoration. The EPA banned those uses in 1984.

In 1989 the EPA ordered a ban on almost all remaining uses of asbestos (such as brake linings, roofing shingles, and water pipes) in the United States by 1997. Representatives of the asbestos industry in the United States and Canada (which now produces most of the asbestos used in the United States) challenged the ban in court, contending that with proper precautions asbestos products can be used safely and that the costs of the ban outweigh the benefits. Industry officials also pointed to some controversial evidence that most of the harm from asbestos comes from inhalation of needle-shaped amphibole fibers, which are rarely found inside buildings. In 1991 a federal appeals court overturned the 1989 EPA ban.

In 1988 the EPA estimated that more than 760,000 buildings—one of every seven commercial and public buildings in the United States (including 30,000 schools)—contained asbestos that had crumbled or could crumble and release fibers. Removal of asbestos from such buildings could cost $50–200 billion, with about $10 billion being spent by 1993.

Critics contend that the health benefits of asbestos removal from many schools, homes, and other buildings are not worth the costs unless measurements (not just visual inspection) indicate that the buildings have high levels of airborne asbestos fibers, especially amphibole fibers. They call for sealing, wrapping, and other forms of containment instead of removal of most asbestos, and they point out that improper or unnecessary removal can release more asbestos fibers than sealing off asbestos that is not crumbling.

After much controversy and huge expenditures on asbestos removal, there is now general agreement that the degree of risk from low-level exposure to asbestos fibers is unclear and asbestos should not be removed from buildings where it has not been damaged or disturbed. Instead it should be sealed or wrapped, with removal only as a last, carefully conducted resort.

In 1998 chemists developed a foam that lets building owners treat asbestos-containing fireproofing material without removing it. The foam initiates chemical reactions that bind the minerals in the asbestos together to form a hard material that is nontoxic and still acts as a fireproofing material.

Critics of environmentalists charge that much of the government-required removal of asbestos from schools and other public buildings was unnecessary and wasted billions of dollars—an example of regulatory overkill.

Case Study: Is Your Home Contaminated with Radon Gas? Radon-222 is a colorless, odorless, tasteless, naturally occurring radioactive gas produced by the radioactive decay of uranium-238. Small amounts of uranium-238 are found in most soil and rock, but this isotope is much more concentrated in underground deposits of minerals such as uranium, phosphate, granite, and shale.

When radon gas from such deposits seeps upward through the soil and is released outdoors, it disperses quickly in the atmosphere and decays to harmless levels. However, when the gas is drawn into buildings through cracks, drains, and hollow concrete blocks (Figure 18-11), or seeps into groundwater in underground wells over such deposits, it can build up to high levels.

Radon-222 gas quickly decays into solid particles of other radioactive elements that, if inhaled, expose lung tissue to a large amount of ionizing radiation from

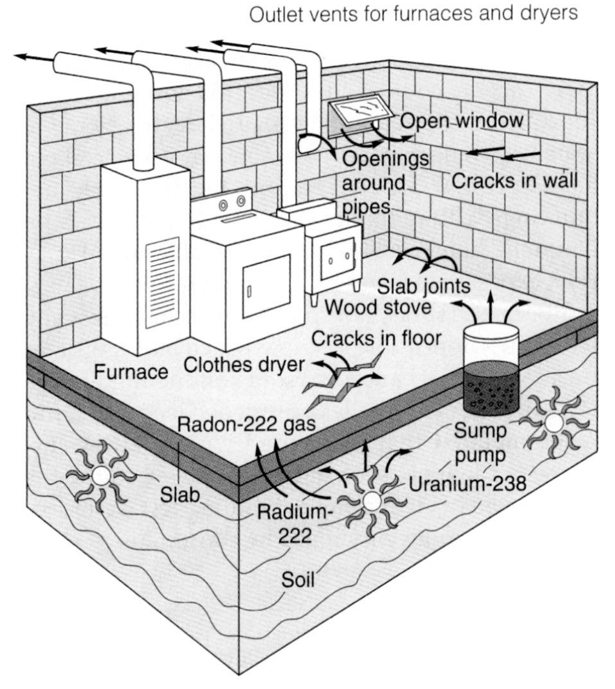

Figure 18-11 Sources and paths of entry for indoor radon-222 gas. (Data from U.S. Environmental Protection Agency)

A: It would be too expensive to use and most of its use would be phased out alpha particles (Figure 3-15). Radon and its airborne decay products account for an estimated 55% of the current radiation dose of the U.S. population (Figure 17-8).

Assuming that there is no threshold for harm caused by inhaled asbestos fibers (Figure 17-3, left), scientists have extrapolated the harmful effects of high-level exposure to radon on uranium miners to low levels of exposure in homes. Using this approach, they estimate that prolonged exposure (defined as 75% of one's time spent in the same home for a lifetime of 70 years) to low levels of radon or radon acting together with smoking is responsible for 6,000–36,000 of the 130,000 lung cancer deaths each year in the United States (13,600 deaths is the best estimate).

Scientists who believe there is a threshold dose before radon is harmful say that these estimates are too high. The results of several epidemiological studies in various countries are mixed and provide no clear evidence to support or refute a connection between lung cancer deaths and inhaled radon in homes.

According to the EPA, prolonged exposure to average radon levels above 4 picocuries* per liter of air in a closed house is considered unsafe. Other researchers cite evidence suggesting that radon becomes dangerous only if indoor levels exceed 20 picocuries per liter, the level accepted in Canada, Sweden, and Norway. Such controversy over acceptable radon levels demonstrates the uncertainties and problems inherent in risk assessment and risk management (Section 17-6).

EPA indoor radon surveys suggest that 4–5 million U.S. homes may have annual radon levels above 4 picocuries per liter of air and that 50,000–100,000 homes may have levels above 20 picocuries per liter. If the 4 picocuries per liter standard is adopted (as proposed by the EPA), the cost of testing and correcting the problem could run about $50 billion, with a 15–20% reduction in radon-related deaths. Some researchers argue that it makes more sense to spend perhaps only $500 million to find and fix homes and buildings with radon levels above 20 picocuries per liter until more reliable data are available on the threat from exposure to lower levels of radon.

Because radon "hot spots" can occur almost anywhere, it's impossible to know which buildings have unsafe levels of radon without conducting tests. In 1988 the EPA and the U.S. Surgeon General's Office recommended that everyone living in a detached house, a town house, a mobile home, or on the first three floors of an apartment building test for radon.† Ideally, radon levels should be continuously monitored in the main living areas (not basements or crawl spaces) for 2 months to a year. By 1997 only about 6% of U.S. households had conducted radon tests (most lasting only 2 to 7 days and costing $20–100 per home).

If testing reveals an unacceptable level, homeowners can consult the free EPA publication *Radon Reduction Methods* for ways to reduce radon levels and health risks. According to the EPA, radon control could add $350–500 to the cost of a new home and correcting a radon problem in an existing house could run $800–2,500.

*A picocurie is a trillionth of a curie, which is the amount of radioactivity emitted by a gram of radium.

†For information, see "Radon Detectors: How to Find Out if Your House Has a Radon Problem," *Consumer Reports*, July 1987.

18-5 EFFECTS OF AIR POLLUTION ON LIVING ORGANISMS AND MATERIALS

How Is Human Health Harmed by Air Pollutants? Your respiratory system (Figure 18-12) has a number of mechanisms that help protect you from air pollution. Hairs in your nose filter out large particles. Sticky mucus in the lining of your upper respiratory tract captures smaller (but not the smallest) particles and dissolves some gaseous pollutants. Sneezing and coughing expel contaminated air and mucus when your respiratory system is irritated by pollutants. The cells of your upper respiratory tract are also lined with hundreds of thousands of tiny, mucus-coated hairlike structures called *cilia* that continually wave back and forth, transporting mucus and the pollutants they trap to your throat (where they are then either swallowed or expelled).

Years of smoking and exposure to air pollutants can overload or break down these natural defenses, causing or contributing to respiratory diseases. Examples are **(1)** *lung cancer*, **(2)** *asthma* (typically an allergic reaction causing sudden episodes of muscle spasms in the bronchial walls, resulting in acute shortness of breath), **(3)** *chronic bronchitis* (persistent inflammation and damage to the cells lining the bronchi and bronchioles, causing mucus buildup, painful coughing, and shortness of breath), and **(4)** *emphysema* (irreversible damage to air sacs or alveoli leading to abnormal dilation of air spaces, loss of lung elasticity, and acute shortness of breath, Figure 18-13). Elderly people, infants, pregnant women, and people with heart disease, asthma, or other respiratory diseases are especially vulnerable to air pollution.

About 90% of the *carbon monoxide* (CO)—a colorless, odorless, poisonous gas—in the troposphere comes from natural sources. Most of this is produced by reaction in the upper troposphere between methane (emitted mostly by the anaerobic decay of organic matter in swamps, bogs, and marshes) and

 Tennesen, M. 1997. "Air Pollution Damages Waterways, Destroys Trees, and Clouds Views." *National Parks*, vol. 71, no. 11–12, 26(4).

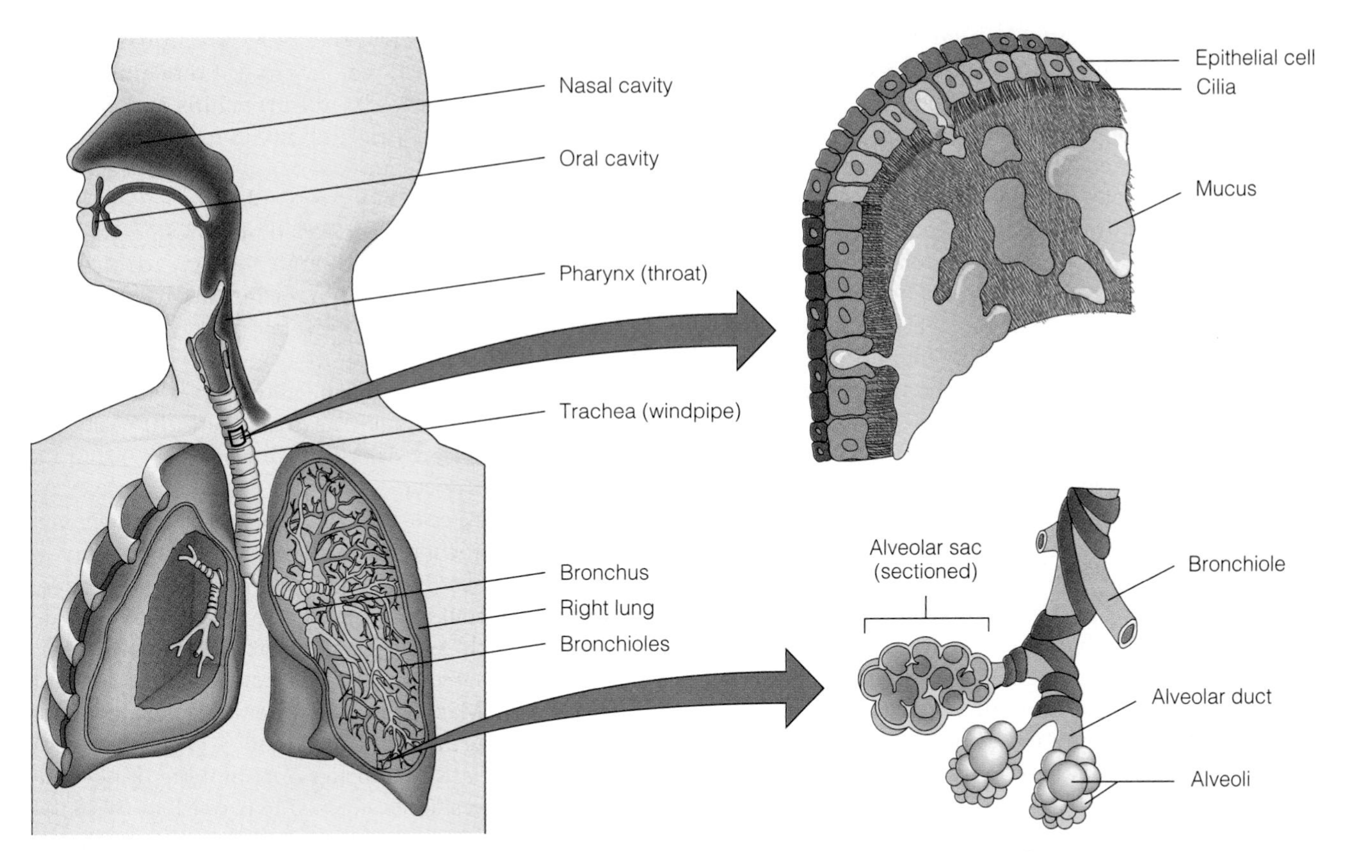

Figure 18-12 Major components of the human respiratory system. The figure shows the detailed structure of alveoli, the main sites where oxygen diffuses into the blood and carbon dioxide diffuses out into the lungs.

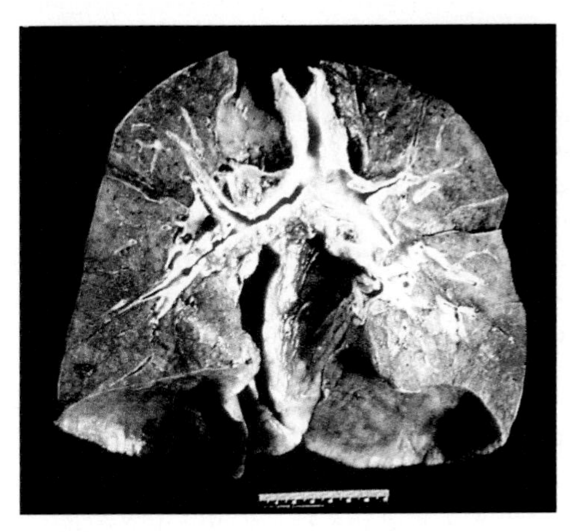

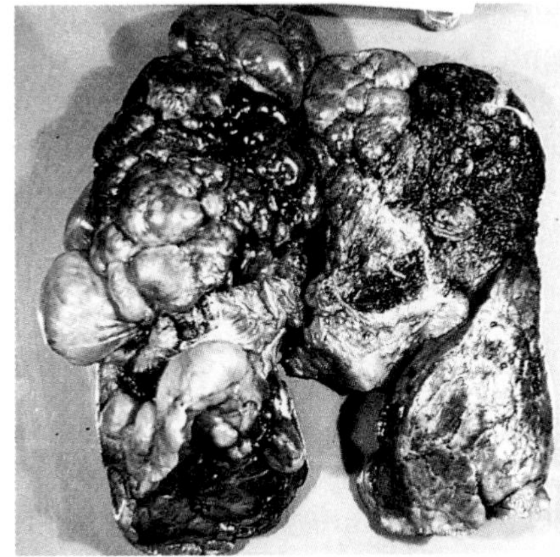

Figure 18-13 Normal human lungs (left) and the lungs of a person who died from emphysema (right). Prolonged smoking and exposure to air pollutants can cause emphysema in anyone, but about 2% of emphysema cases result from a defective gene that reduces the elasticity of the air sacs in the lungs. Anyone with this hereditary condition, for which testing is available, should not smoke and should not live or work in a highly polluted area. (O. Auerbach/Visuals Unlimited)

oxygen. Because this CO is diluted by the turbulent air flows in the troposphere, it does not build up to harmful levels.

However, the remaining 10% of the CO added to the atmosphere comes from the incomplete burning of carbon-containing chemicals (primarily fossil fuels). Cigarette smoking (p. 436) is responsible for the largest human exposure to CO, but this gas is also released by motor vehicles, kerosene heaters, woodstoves, fireplaces, and faulty heating systems.

CO reacts with hemoglobin in red blood cells and thus reduces the ability of blood to carry oxygen. This impairs perception and thinking, slows reflexes, and causes headaches, drowsiness, dizziness, and nausea. CO can also trigger heart attacks and angina attacks in people with heart disease. It can damage the development of fetuses and young children and aggravate chronic bronchitis, emphysema, and anemia. Exposure to high levels of CO causes collapse, coma, irreversible damage to brain cells, and even death.

Hint: Enter the search term *air pollution* using the Subject Guide.

Carbon monoxide detectors (similar in size to smoke detectors) are now available for about $50 and are a good safety investment for homeowners.

Inhaling *suspended particulate matter* aggravates bronchitis and asthma, and long-term exposure can contribute to development of chronic respiratory disease and cancer. Invisible particles (Figure 18-4)—especially *fine particles* with diameters less than 10 microns and *ultrafine particles* with diameters less than 2.5 microns—are viewed as especially hazardous. Such particles are emitted by incinerators, motor vehicles, radial tires, wind erosion, wood-burning fireplaces, and power and industrial plants.

Such tiny particles, much thinner than a human hair, are not effectively captured by modern air-pollution control equipment, and they are small enough to penetrate the respiratory system's natural defenses against air pollution. They can also bring with them droplets or other particles of toxic or cancer-causing pollutants that become attached to their surfaces. Once they are lodged deep within the lungs, evidence suggests that these fine and ultrafine particles can cause chronic irritation that can trigger asthma attacks, aggravate other lung diseases, cause lung cancer, and interfere with the blood's ability to take in oxygen and release CO_2. This strains the heart, increasing the risk of death from heart disease.

Several recent studies of air pollution in U.S. cities have indicated that fine and ultrafine particles prematurely kill 65,000–150,000 Americans each year. Factoring out smoking and other causes of heart and respiratory disease, researchers recently found that U.S. cities reporting the highest levels of ultrafine particulates (those with diameters less than 2.5 microns) had the highest death rates from lung disease and heart disease. To date there is no known threshold level below which the harmful effects of fine particles disappear.

Exposure to particulate air pollution is much worse in most developing countries, where urban air quality has generally deteriorated. The World Bank estimates that if particulate levels were reduced globally to WHO guidelines, 300,000–700,000 premature deaths per year could be prevented.

Sulfur dioxide causes some constriction of the airways in healthy people and severe restriction in people with asthma. Chronic exposure causes a condition similar to bronchitis. Sulfur dioxide and suspended particles react to form far more hazardous acid sulfate particles, which are inhaled more deeply into the lungs than SO_2 and remain there for long periods. According to the World Health Organization, at least 625 million people are exposed to unhealthy levels of sulfur dioxide from fossil-fuel burning.

Nitrogen oxides (especially NO_2) can irritate the lungs, aggravate asthma or chronic bronchitis, cause conditions similar to chronic bronchitis and emphysema, and increase susceptibility to respiratory infections such as the flu and common colds (especially in young children and elderly people). Recent evidence from test animals indicates that nitrogen dioxide exposure may also encourage the spread of some cancers—especially malignant melanoma (Connections, p. 526)—throughout the body.

Research indicates that any *volatile organic compounds* (such as benzene and formaldehyde) and *toxic particulates* (such as lead, cadmium, PCBs, and dioxins) can cause mutations, reproductive problems, or cancer.

Evidence also shows that inhaling *ozone*, a component of photochemical smog (Figure 18-5), causes coughing, chest pain, shortness of breath, and eye, nose, and throat irritation. It also aggravates chronic diseases such as asthma, bronchitis, emphysema, and heart disease and reduces resistance to colds and pneumonia. Many U.S. cities often exceed safe levels, especially during warm weather. A 1987 study also showed that after factoring out smoking, long-time Los Angeles residents exposed to ozone had twice the risk of cancer compared to residents of cleaner cities.

How Many People Die Prematurely from Air Pollution? No one knows how many people die prematurely from respiratory or cardiac problems caused or aggravated by air pollution. Such figures are difficult to estimate by risk analysis because people are exposed to so many different pollutants over their lifetimes.

In the United States, estimates of annual deaths related to outdoor air pollution range from 65,000 to 200,000 (most from exposure to fine and ultrafine particles). If indoor air pollution is included, estimated annual deaths from air pollution in the United States range from 150,000 to 350,000 people—equivalent to 1–2 fully loaded 400-passenger jumbo jets crashing *each day* with no survivors.

Millions more become ill and lose work time. According to the EPA and the American Lung Association, air pollution in the United States costs at least $150 billion annually in health care and lost work productivity, with $100 billion of that caused by indoor air pollution.

According to a 1997 study by the World Health Organization and the World Bank, worldwide at least 2.7 million people (most of them in Asia) die prematurely each year from the effects of indoor (2.2 million deaths) and outdoor (0.5 million deaths) air pollution.

Q: Who has the world's largest reserves of natural gas?

How Are Plants Damaged by Air Pollutants? Some gaseous pollutants (especially ozone) damage leaves of crop plants and trees directly when they enter leaf pores. Chronic exposure of leaves and needles to air pollutants can break down the waxy coating that helps prevent excessive water loss and damage from diseases, pests, drought, and frost. Such exposure also interferes with photosynthesis and plant growth, reduces nutrient uptake, and causes leaves or needles to turn yellow or brown and drop off (Figure 18-9). Spruce, fir, and other conifers, especially at high elevations, are most vulnerable to air pollution because of their long life spans and the year-round exposure of their needles to a mixture of air pollutants.

Prolonged exposure to high levels of several air pollutants from smelters (plants that extract metal from ores) can kill trees and most other vegetation in an area (Figure 14-15). Lengthy exposure to a mixture of air pollutants from coal-burning power and industrial plants and from cars can also damage trees and many other plants. However, the effects may not become visible for several decades, when large numbers of trees suddenly begin dying off because of depletion of soil nutrients and increased susceptibility to pests, diseases, fungi, and drought. This phenomenon, known as *Waldsterben* (forest death), has turned whole forests of spruce, fir, and beech into stump-studded meadows and mountainsides.

Trees at high elevations in mountain biomes (Section 7-6) (especially in Europe), with their thin and easily erodible soils and almost year-round exposure to air pollutants, have suffered the most damage. The diebacks of trees in such areas lead to extensive soil erosion, which in turn leads to increased flooding and avalanches (particularly in the heavily developed Alps).

Surveys in 1993 and 1994 revealed that almost one out of four trees in Europe had been damaged by air pollution, resulting in a loss of more than 25% of their leaves. Such damage occurs in virtually all European countries but is highest in the Czech Republic, where 57% of all trees suffer moderate or severe defoliation or have died (photo on p. 435). It is estimated that air pollution has been a key factor in reducing the overall productivity of European forests by about 16% and causing damage valued at roughly $30 billion per year.

Forest diebacks have also occurred in the United States, mostly on high-elevation slopes that face moving air masses and are dominated by red spruce. The most seriously affected areas are in the Appalachian Mountains, where research reported in 1994 found a rapid die-off of more than 70 tree species and standing tree skeletons of more than 20 species. Ozone and other air pollutants are believed to be responsible for this forest degradation and loss of biodiversity. In addition, ozone is suspected to be the main cause of the reduced growth rate observed in commercial yellow pine forests in the southern United States.

Air pollution, mostly by ozone, also threatens some crops—especially corn, wheat, and soybeans, the three most important U.S. crops—and is reducing U.S. food production by 5–10%. In the United States, estimates of economic losses to agriculture as a result of air pollution range from $1.9 to $5.4 billion per year.

Acid deposition related to SO_2 emissions from the burning of coal in China (Connections, p. 378) causes an estimated $5 billion in damage annually to Chinese crops and forests, not counting damage downwind to Japan and South Korea. At this level of economic damage to crops, China could save money by curbing emissions of SO_2.

How Can Air Pollutants Damage Aquatic Life? High acidity (low pH) can severely harm the aquatic life in freshwater lakes, both where the surrounding soils have little acid-neutralizing capacity and in the northern hemisphere, where there is significant winter snowfall. Much of the damage to aquatic life in such areas is a result of *acid shock* caused by the sudden runoff of large amounts of highly acidic water and aluminum ions into lakes and streams, when snow melts in the spring or after unusually heavy rains. The aluminum ions leached from the soil and lake sediment by this sudden input of acid can kill fish and inhibit their reproduction.

As the acidity of a lake increases and its food chain is disrupted, there is a decline in net primary productivity. This can turn a moderately eutrophic lake (Figure 8-13, top) into a clear blue oligotrophic lake (Figure 8-13, bottom).

Because of excess acidity, at least 16,000 lakes in Norway and Sweden contain no fish, and 52,000 more lakes have lost most of their acid-neutralizing capacity. In Canada some 14,000 acidified lakes are almost fish graveyards, and 150,000 more are in peril.

In the United States, about 9,000 lakes are threatened with excess acidity, one-third of them seriously. Most of them are concentrated in the Northeast and the upper Midwest—especially Minnesota, Wisconsin, and the upper Great Lakes—where 80% of the lakes and streams are threatened by excess acidity. Over 10% of some 200 lakes in New York's Adirondack Mountains are too acidic to support fish.

A: Russia and Kazakhstan (40%), Iran (15%), United States (3%)

What Are the Harmful Effects of Air Pollutants on Materials? Each year, air pollutants cause billions of dollars in damage to various materials we use (Table 18-3). The fallout of soot and grit on buildings, cars, and clothing requires costly cleaning. Air pollutants break down exterior paint on cars and houses, and they deteriorate roofing materials. Irreplaceable marble statues, historic buildings, and stained glass windows around the world have been pitted, gouged, and discolored by air pollutants. For example, the famous Greek ruins on the Acropolis in Athens have deteriorated more during the past 50 years than during the previous 2,000 years. Damage to buildings in the United States from acid deposition alone is estimated at $5 billion per year.

18-6 SOLUTIONS: PREVENTING AND REDUCING AIR POLLUTION

How Have Laws Been Used to Reduce Air Pollution in the United States? The U.S. Congress passed Clean Air Acts in 1970, 1977, and 1990, providing federal air-pollution regulations that are enforced by each state. These laws required the EPA to establish *national ambient air quality standards (NAAQS)* for seven outdoor pollutants: suspended particulate matter, sulfur oxides, carbon monoxide, nitrogen oxides, ozone, volatile organic compounds, and lead. Each standard specifies the maximum allowable level, averaged over a specific period, for a certain pollutant in outdoor (ambient) air.

Another strategy required by the Clean Air Act is *prevention of significant deterioration*. According to this policy, air quality in regions in which the air is cleaner than required by the NAAQS for suspended particulate matter and sulfur dioxide should not be allowed to deteriorate. Without such a policy, industries could move into those areas and gradually degrade air quality down to the national standards for these two pollutants.

The legislation also required the EPA to set *national emission standards for toxic air pollutants*, including 302 individual compounds and 20 categories of chemical compounds that toxicological evidence indicates are harmful to human health (including an EPA estimate of 1,500–3,000 premature cancer deaths per year). So far such standards have been set for only a few of these compounds. The agency cites lack of money and the difficulty in estimating toxicity levels with enough scientific certainty to withstand court challenges as the major causes of its delay in developing and implementing these standards.

By the year 2000 the Clean Air Act of 1990 also requires coal-burning power plants (responsible for 70% of U.S. SO_2 emissions) to cut their 1991 annual SO_2 emissions roughly in half and their 1991 NO_x emissions by 33%. However, the 1990 act still allows U.S. factories to produce about four times the amount of air pollution as German factories.

Congress also set a timetable for achieving reductions in emissions of carbon monoxide, hydrocarbons, and nitrogen oxides from motor vehicles. These standards forced automakers to build cars that emit six to eight times fewer pollutants than did the cars of the late 1960s. The 1990 act required a further reduction in hydrocarbon and nitrogen oxide emissions in new cars by 1994. Stricter emission standards for new cars will go into effect by 2003, and auto emission controls will also be required to last for 160,000 kilometers (100,000 miles) instead of the current 80,000 kilometers (50,000 miles). The 1990 act also required oil companies to sell cleaner-burning gasoline or other fuels in the nine cities with severe ozone problems (Baltimore, Chicago, Hartford, Houston, Los Angeles, Milwaukee, New York, Philadelphia, and San Diego) by 1995.

The Clean Air Act has worked. Between 1970 and 1997, levels of six major air pollutants decreased nationally by 31%, despite significant increases in population size and a doubling of economic growth and the average number of kilometers driven. Further decreases are projected by 2015. According to EPA data, between 1988 and 1997 lead levels in U.S. air decreased by 66% (98% since 1970), carbon monoxide 38%, sulfur dioxide 39%, suspended particulate matter 10 micrometers or less in diameter 17%, nitrogen dioxide 14%, and ground-level ozone 6%.

Between 1990 and 1995 ozone levels in U.S. urban areas fell by 50%. Result: 50 million people breathe cleaner air. Nitrogen dioxide levels have risen slightly since 1980 because of a combination of inadequate automobile emission standards and more vehicles traveling longer average distances.

Without the 1970 standards for emissions of pollutants, air-pollution levels would be much higher today. Even so, in 1997 the EPA estimates that 107 million Americans still live in areas that exceed at least one outdoor air-pollution standard.

A 1996 study by the EPA found that the benefits of the Clean Air Act greatly exceed costs. Between 1970 and 1990, the U.S. spent about $436 billion (in 1990 dollars) to comply with clean air regulations. Total human health and ecological benefits during the same 20-year period were estimated to range from

Koontz, Michael. 1998. "Clean Air and Transportation: The Facts May Surprise You" *Public Roads*, vol. 62, no. 1, 42(5).

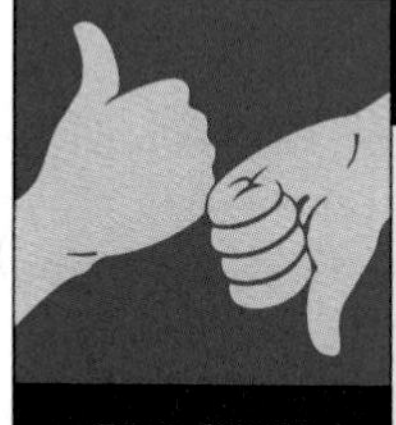

Controversy over Stricter Particle Emission Standards in the United States

In 1997, EPA chief Carol Browner proposed stricter national standards for emissions of ultrafine microscopic particles with diameters less than 2.5 microns (PM-2.5). This new standard was triggered largely by a lawsuit brought by the American Lung Association accusing the EPA of ignoring new scientific evidence showing that fine and ultrafine particles in the air are especially harmful to human health.

A federal judge ordered the agency to evaluate the evidence and, if warranted, come up with new standards. After reviewing 86 separate studies EPA scientists found 60 of the health effects such as deaths, hospitalization, and respiratory disease linked to fluctuating levels of particulate matter levels in a number of different cities. This strong circumstantial evidence, reviewed by two scientific advisory panels, led the agency to conclude that a new standard for ultrafine particles was in order, a decision that set off a storm of controversy.

According to Browner, the cost of implementing new standards could be as high as $9.7 billion a year for measures such as installing new equipment on power plants and diesel trucks. However, according to the EPA, the new standards will prevent as many as 15,000 premature deaths, 60,000 chronic bronchitis cases, 350,000 cases of aggravated asthma, and 1 million cases of significantly decreased lung function in children per year in the United States. Other benefits would be reduced haze in national parks and a $1-billion reduction in crop losses. The EPA estimates that the new regulation will save $19–104 billion per year in reduced health and other costs—2–12 times the estimated cost of compliance.

Industry groups mounted a major lobbying and advertising campaign to have the standards withdrawn or to influence Congress to delay or overturn them. Big-city mayors joined forces with the business lobby, fearing that the new regulations would cause factories to move from urban areas to places with less pollution.

Business leaders accused the EPA of basing their decision on flimsy scientific evidence. They also contended that it will cost industry billions of dollars and that the resulting costs (estimated to be as high as $200 billion per year) will not outweigh the benefits.

These opponents agree that many studies show that statistically cities with lots of fine and ultrafine particles in the air have more illnesses and premature deaths than cities and rural areas with lower levels of such particles. However, they say there are no smoking guns linking ultrafine particles to lung and heart disease and no definitive mechanisms explaining how such particles might cause such effects. Industry representatives claim that the harmful health effects could be from lifestyle factors or other pollutants.

EPA officials say that their review of the scientific evidence—one of the most exhaustive ones ever undertaken by the agency—supports the need for a standard for ultrafine particles. They and other scientists also point out that a number of the studies finding relationships between ultrafine particle levels and higher death rates from lung disease and heart disease factored out smoking and other causes of such diseases.

Browner and EPA scientists are aware of the uncertainties and gaps in the scientific evidence, as is the case in all of their regulatory decisions because science provides only probabilities, not absolute proof, and almost always involves evaluation of circumstantial evidence. She points out that if public officials had waited to uncover precisely how lead, certain pesticides, and tobacco smoke cause illness, millions of people would have died unnecessarily.

Most air pollution scientists agree there is enough circumstantial evidence for the EPA to set a standard for PM-2.5. However, some contend that more research is needed to set a specific allowed level. Browner agrees that more scientific studies are needed and points out that compliance for the new particulate standard will not be required until the EPA gathers additional scientific data on particulate concentrations and health effects, and would be enforced no earlier than 2005.

Proponents of the regulations point out that industry forecasts of the costs of compliance with virtually every air pollution emission standard proposed during the last 25 years have been shown to be greatly exaggerated. For example, opponents of reducing SO_2 emissions as required by the 1990 Clean Air Act said the new standards would cost up to $1,500 per ton. The actual market cost was $78 per ton.

Critical Thinking

Do you believe that a stricter standard for emissions of ultrafine particles should be put into effect? Explain. Use the library or the Internet to determine what has happened to this proposed new standard.

Hint: Enter search term *Clean Air Act* using the Subject Guide and select *Clean Air Act Amendments of 1990.*

$2.7 to $14.6 trillion (in 1990 dollars)—6 to 33 times higher than the costs. In 1997, the EPA proposed stricter national ambient air quality standards for emissions of nitrogen oxides and ultrafine particles (Pro/Con, p. 491).

How Could U.S. Air-Pollution Laws Be Improved? The Clean Air Act of 1990 was an important step in the right direction, but most environmentalists point to the following major deficiencies in this law:

- *Continuing to rely almost entirely on pollution cleanup rather than pollution prevention.* In the United States, the air pollutant with the largest drop (98% between 1970 and 1995) in its atmospheric level was lead, which was virtually banned in gasoline. This shows the effectiveness of the pollution prevention approach.
- *Failing to sharply increase the fuel efficiency standards for cars and light trucks.* According to environmental scientists, this would reduce oil imports and air pollution more quickly and effectively than any other method and would save consumers enormous amounts of money (Section 16-2).
- *Not requiring stricter emission standards for fine particulates.* (Pro/Con, p. 491).
- *Giving municipal trash incinerators 30-year permits,* which locks the nation into hazardous air-pollution emissions and toxic waste from incinerators well into the 21st century. This also undermines pollution prevention, recycling, and reuse (Chapter 22).
- *Setting weak standards for air-pollution emissions from incinerators,* thus allowing unnecessary emissions of mercury, lead, dioxins, and other toxic pollutants (Section 22-7).
- *Doing too little to reduce emissions of carbon dioxide and other greenhouse gases* (Section 19-4).

Should We Use the Marketplace to Reduce Pollution? To help reduce SO_2 emissions, the Clean Air Act of 1990 allows an *emission trading policy,* which enables the 110 most polluting power plants in 21 states (primarily in the Midwest and East) to buy and sell SO_2 pollution rights.

Each year a power plant is given a certain number of pollution credits or rights that allow it to emit a certain amount of SO_2. A utility that emits less SO_2 than its limit receives more pollution credits. It can use these credits to avoid reductions in SO_2 emissions from some of its other facilities, bank them for future plant expansions, or sell them to other utilities, private citizens, or environmental groups.

Proponents of this system argue that it allows the marketplace to determine the cheapest, most efficient way to get the job done instead of having the government dictate how to control pollution. If this market-based approach works for reducing SO_2 emissions, it could be applied to other air and water pollutants.

Some environmentalists see this market approach as an improvement over the current regulatory approach, as long as it achieves net reduction in SO_2 pollution. This would be done by limiting the total number of credits and gradually lowering the annual number of credits (to encourage pollution prevention and the development of better pollution control technologies), something that is not required by the 1990 Clean Air Act. Without such reductions, critics contend that the system of tradable pollution rights is essentially an economic shell game, with no continuing progress in reducing overall SO_2 emissions.

Some environmentalists also contend that marketing pollution rights allows utilities with older, dirtier power plants to buy their way out and keep on emitting unacceptable levels of SO_2. They also warn that this approach creates incentives to cheat. Because air quality regulation is based largely on self-reporting of emissions and because pollution monitoring is incomplete and imprecise, sellers of permits will benefit by understating their reductions (to get more permits), and permit buyers will benefit by underreporting emissions (to reduce their permit purchases).

Between 1994 and 1997, the emission trading system helped reduce SO_2 emissions in the United States by 30% at less than one-tenth the cost projected by industry because this market-based system has enabled companies to reduce emissions in efficient ways.

In 1997, the EPA proposed a voluntary emission trading program involving smog-forming nitrogen oxides (NO_x) for 22 eastern states and the District of Columbia. Each state would establish a cap on NO_x emissions and then give power plants and industries more flexibility in finding ways to cut NO_x emissions by allowing sources that reduce emissions by more than required to sell credits to facilities that cannot reduce emissions as quickly or cost-effectively. This program is designed to help states meet stricter NO_x emission standards in attempts to cut ozone pollution. Emission trading programs may also be implemented for particulates emissions.

Q: How long will the reserves of natural gas for the world and the United States last at current consumption rates?

How Can We Reduce Outdoor Air Pollution? Figure 18-14 summarizes ways to reduce emissions of sulfur oxides, nitrogen oxides, and particulate matter from stationary sources (such as electric power plants and industrial plants that burn coal). Until recently, emphasis has been primarily on dispersing and diluting the pollutants by using tall smokestacks or adding equipment that removes some of the particulate pollutants after they are produced (Figure 18-15). Under the sulfur-reduction requirements of the Clean Air Act of 1990, more utilities are switching to low-sulfur coal to reduce SO_2 emissions. Environmentalists call for taxes on air pollutant emissions and greater emphasis on prevention methods.

Figure 18-16 lists ways to reduce emissions from motor vehicles, the primary culprits in producing photochemical smog. Ford Motor Company and Englehard Corporation (which develops and makes catalysts) are working together to develop a catalyst-coated car radiator capable of destroying up to 90% of

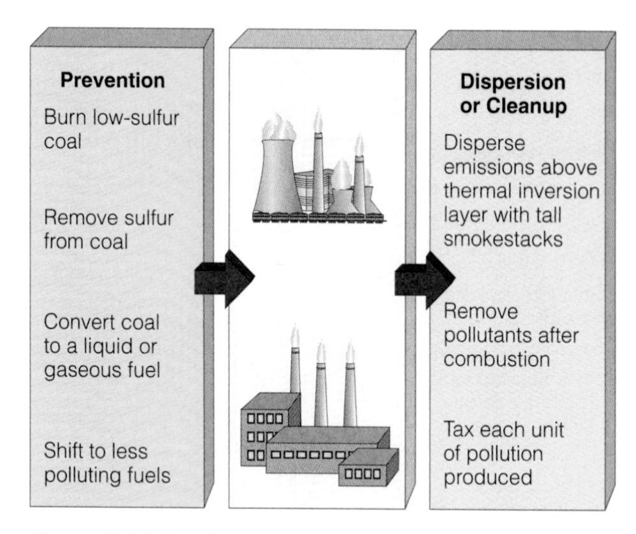

Figure 18-14 Solutions: methods for reducing emissions of sulfur oxides, nitrogen oxides, and particulate matter from stationary sources such as coal-burning electric power plants and industrial plants.

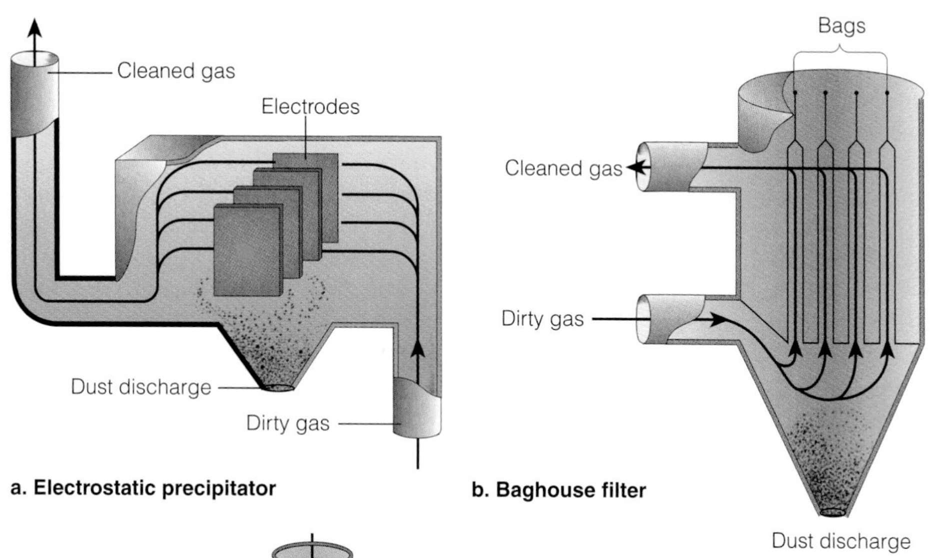

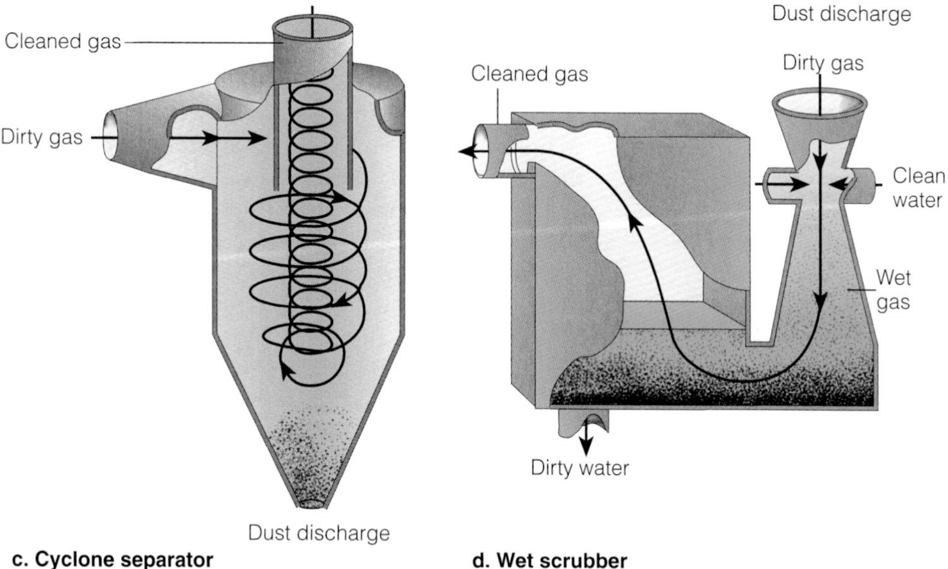

Figure 18-15 Solutions: four commonly used methods for removing particulates from the exhaust gases of electric power and industrial plants. Of these, only baghouse filters remove many of the more hazardous fine particles. All these methods produce hazardous materials that must be disposed of safely, and except for cyclone separators, all of them are expensive. The wet scrubber can also reduce sulfur dioxide emissions.

A: At least 125 years for the world, 65–80 years for the United States

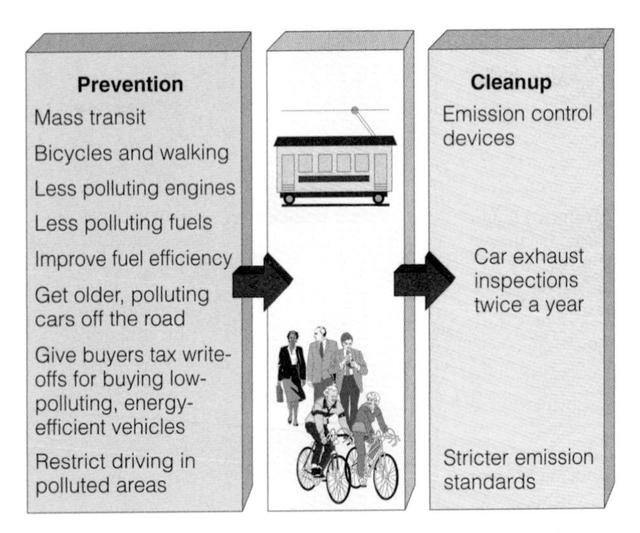

Figure 18-16 Solutions: methods for reducing emissions from motor vehicles.

the ozone and carbon monoxide in the atmosphere. Use of alternative vehicle fuels to reduce air pollution was evaluated in Table 16-1.

One way to reduce CO emissions from an automobile is to make the air-to-fuel ratio in the engine richer, to bring about more complete combustion of the hydrocarbons in gasoline to CO_2. Unfortunately, because of the nature of the internal combustion engine, this increases NO emissions. Catalytic converters have been developed that use catalysts (such as particles of platinum) to speed up the reaction of NO and CO produced during combustion to N_2 and CO_2 ($2NO + 2CO \rightarrow N_2 + 2CO_2$). However, catalytic converters are not as effective as many people think: When the car's accelerator is held down (as for passing), the sensors that maintain the chemical balance needed for the catalyst to work are bypassed and emission rates increase sharply.

It's currently estimated that the emission control systems on at least 50% of the U.S. car and light-truck fleet have been disconnected or are not working properly. Requiring car emission testing twice a year would detect faulty pollution control systems, discourage tampering with them, and encourage drivers to keep them in good working order.

Recently, a University of Colorado professor developed a 1-second highway test for auto emissions. A beam of infrared light is sent across a highway to an air-pollution detector that measures levels of carbon monoxide, hydrocarbons, and nitrogen oxides in a car's exhaust as it passes by. The information is fed into a computer and a video camera captures the car's license plate number and stores it in the computer. Drivers can be sent notices directing them to correct faulty emissions and have the vehicle rechecked at an official testing station or face large fines.

An important way to make significant reductions in air pollution is to get older, high-polluting vehicles off the road. A badly maintained older vehicle can emit 100 times more pollutants than a properly maintained modern vehicle. According to EPA estimates, 10% of the vehicles on the road in the United States emit 50–60% of the pollutants. A problem is that many old cars are owned by people who can't afford to buy a newer car. One suggestion would be to pay people to take their old cars off the road, which would result in huge savings in health and air-pollution control costs. Companies could buy up old cars (as oil companies in the Los Angeles area are now doing) to get emissions credits that enable them to temporarily extend the lives of their polluting facilities while new technology is installed.

Environmentalists also call for stricter emission standards for new outboard motors, chain saws, leaf blowers, weed trimmers, lawnmowers, and other devices with gasoline engines, which account for at least 10% of carbon monoxide and ozone pollution and about 5% of all U.S. air pollution. In 1996, the EPA began phasing in emission standards for new lawnmowers.

There is some good news. From 1982 to 1993, overall U.S. smog dropped by 8% even as population, traffic, and industry rose. In 1993, the number of Americans living in areas where ozone concentrations exceeded Clean Air Act standards was 54 million, the lowest total in 20 years and down from the 100 million people exposed to excessive ozone in 1992.

In 1995 Los Angeles had 103 unsafe air days compared to 239 in 1988. Despite such success, smog levels in the Los Angeles basin (which is subject to frequent thermal inversions) are too high much of the year (30% of the days in 1996) and could rise as population and consumption increase. The 1990 Clean Air Act gives Los Angeles until 2010 to meet federal air-pollution standards. The EPA estimates that meeting the standards will cost California $4–6 billion annually from 1995 through 2010. However, a 1989 study estimated that implementing the program will save California an estimated $9.4 billion a year in health costs, reduced crop yields, and lowered productivity, leading to a net economic gain of $4–5 billion per year.

Q: What fossil fuel produces the least amount of CO_2 per unit of energy?

California's South Coast Air Quality Management District Council developed a drastic program to produce an 80% reduction in ozone, photochemical smog, and other major air pollutants in the Los Angeles area by 2009. This plan would **(1)** sharply reduce use of gasoline-burning engines over two decades by converting cars, trucks, buses, chain saws, outboard motors, and lawnmowers to run on electricity or on alternative fuels; **(2)** outlaw drive-through facilities to keep vehicles from idling in lines; **(3)** substantially raise parking fees and assess high fees for families owning more than one car; **(4)** require gas stations to use a hydrocarbon-vapor recovery system on gas pumps and to sell alternative fuels (Table 16-1); **(5)** strictly control or relocate industrial plants and businesses that release large quantities of hydrocarbons and other pollutants; and **(6)** find substitutes for or ban consumer products that release hydrocarbons, including aerosol propellants, paints, household cleaners, and barbecue starter fluids. Such measures are a glimpse of what many other cities may have to do if people, cars, and industries continue to proliferate.

To ease the financial burden, the council provides free technical assistance and offers loans for the purchase of pollution-control equipment. An emission trading policy also allows businesses to reduce emissions themselves or buy emission credits from companies that have reduced their emissions below the standards.

Here is some more good news. Since the 1960s, Tokyo, Japan (with a current population of about 26 million) has implemented a strict air-pollution control program that has sharply reduced levels of sulfur dioxide, carbon monoxide, and ozone. During the past 25 years outdoor air quality in most western European cities has also improved. However, outdoor air quality has remained about the same or has gotten worse in most rapidly growing urban areas in developing countries.

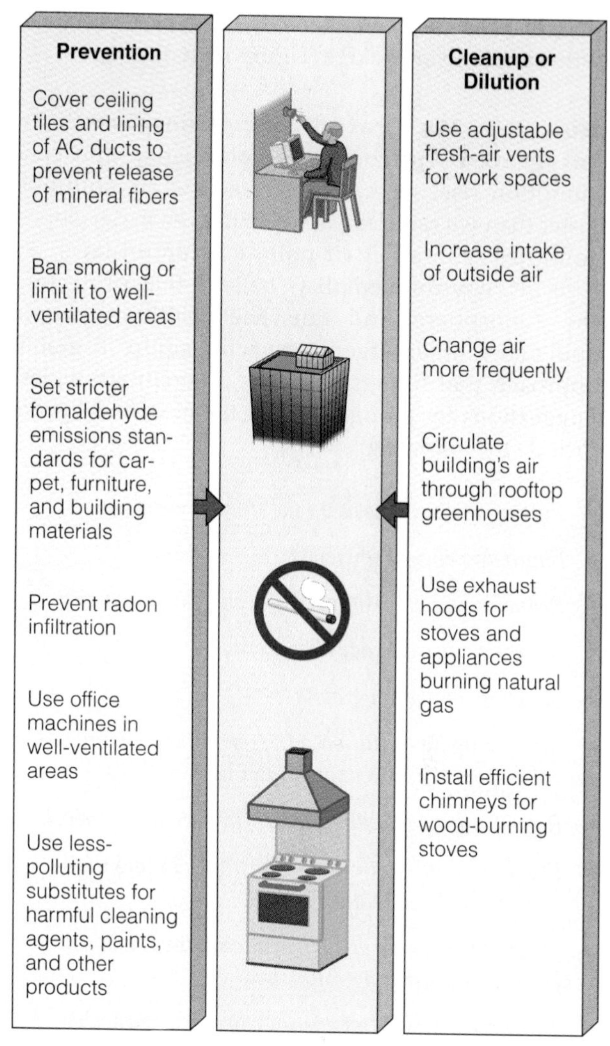

Figure 18-17 Solutions: ways to prevent and reduce indoor air pollution.

How Can We Reduce Indoor Air Pollution? In the United States indoor air pollution poses a greater threat to health for many people than outdoor air pollution. Yet the EPA spends about $500 million per year fighting outdoor air pollution and only about $13 million a year on indoor air pollution.

To reduce indoor air pollution, it's not necessary to impose indoor air quality standards and monitor the more than 100 million homes and buildings in the United States. Instead, air-pollution experts suggest that indoor air-pollution reduction can be achieved by several means (Figure 18-17). Another possibility for cleaner indoor air in high-rise buildings is rooftop greenhouses through which building air can be circulated. Canadian researchers are also developing an air-filtering system called a *breathing wall* that in effect absorbs indoor dirty air and exhales clean air. It is an indoor ecosystem composed of rocks, plants, fish, and microorganisms. Some actions you can take to reduce your exposure to indoor air pollutants are listed in Appendix 5.

In developing countries, indoor air pollution from open fires and leaky and inefficient wood- or charcoal-burning stoves (and the resulting high levels of respiratory illnesses) could be reduced if governments **(1)** gave people simple stoves that burn biofuels more efficiently (which would also reduce deforesta-

A: Natural gas (43% less than coal and 30% less than oil)

tion) and that are vented outside, or **(2)** provided them with simple solar cookers (Figure 16-16d).

How Can We Protect the Atmosphere? An Integrated Approach As population and consumption rise, we can generate new air pollution faster than we can clean up the old, even in developed countries with strict air-pollution control laws. As a result, environmentalists believe that protecting the atmosphere, and thus the health of people and many other organisms, will require a global approach that integrates many different strategies. Suggestions for doing this over the next 40–50 years include the following:

- *Putting more emphasis on pollution prevention*
- *Improving energy efficiency*
- *Reducing use of fossil fuels (especially coal and oil)*
- *Increasing use of renewable energy*
- *Slowing population growth*
- *Integrating air-pollution, water-pollution, energy, land-use, population, economic, and trade policies*
- *Regulating air quality for an entire region or airshed*
- *Phasing in full-cost pricing, mostly by taxing the production of air pollutants*
- *Distributing cheap and efficient cookstoves and solar cookstoves in developing countries*
- *Transferring the latest energy-efficiency, renewable-energy, pollution prevention, and pollution control technologies to developing countries*

Making such changes will be controversial and expensive. However, proponents argue that not implementing such an integrated approach will cost far more in money, poor human health and premature death, and ecological damage.

Turning the corner on air pollution requires moving beyond patchwork, end-of-pipe approaches to confront pollution at its sources. This will mean reorienting energy, transportation, and industrial structures toward prevention.

HILARY F. FRENCH

CRITICAL THINKING

1. Evaluate the pros and cons of the following statement: "Because we have not proven absolutely that anyone has died or suffered serious disease from nitrogen oxides, current federal emission standards for this pollutant should be relaxed."

2. Should all remaining uses of asbestos be banned in the United States? Explain.

3. Should annual government-held auctions of marketable trading permits be used as a primary way of controlling and reducing air pollution? Explain. What conditions, if any, would you put on this approach?

4. Evaluate your exposure to some or all of the indoor air pollutants in Figure 18-10 where you work and live. Come up with a plan for reducing your exposure to these pollutants.

5. Try to come up with analogies to complete the following statements: **(a)** A secondary pollutant is like _______. **(b)** Using tall smokestacks as the solution to air pollution is like _______. **(c)** Photochemical smog is like _______. **(d)** Acid deposition is like _______. **(e)** Relying mostly on pollution control instead of pollution prevention to reduce air pollution is like _______.

6. Do you agree or disagree with the possible weaknesses of the U.S. Clean Air Act listed on p. 492? Defend each of your choices. Can you identify other weaknesses?

7. Do you agree or disagree with the suggested components of an integrated approach to air pollution listed on the left? Defend your answers. How likely is it that such a plan can be implemented over the next 40–50 years? What can you do to spur the implementation of such an approach?

PROJECTS

1. What climate and topographical factors in your local community intensify air pollution, and which of these factors help reduce air pollution?

2. Everyone is downwind from somebody. Use a regional map and prevailing wind patterns to determine the main towns and cities, factories, and coal-burning power plants that produce long-lived air pollution that reaches where you live. Draw these flows on your map. Then draw flows showing where long-lived air pollution produced in your community goes.

Q: What nonrenewable energy resource may be best for achieving a transition to an energy-efficiency/renewable-energy era?

3. Do buildings in your school contain asbestos? If so, what are the indoor levels? Should this asbestos be removed? If indoor asbestos testing has not been done, talk with school officials about having it done.

4. Have dormitories and other buildings on your campus been tested for radon? If so, what were the results? What has been done about areas with unacceptable levels? If this testing has not been done, talk with school officials about having it done.

5. Consult local officials and records to determine whether outdoor air pollution in your community has gotten better or worse since 1980. Get information on specific major air pollutants and on emissions of toxic chemicals by industries. What is being done about indoor pollution in your community?

6. Use the library or the Internet to evaluate the effectiveness of the Clean Air Act of 1990. List its major accomplishments, disappointments, and weaknesses.

7. Make a concept map of this chapter's major ideas, using the section heads and subheads and the key terms (in boldface). Look at the inside back cover and in Appendix 4 for information about concept maps.

A: Natural gas

19 GLOBAL WARMING AND OZONE LOSS

A.D. *2060: Hard Times on Planet Earth**

Mary Wilkins sat in the living room of her underground home in Illinois (Figure 19-1), which she shared with her daughter Jane and her family. It was July 4, 2060: Independence Day. There would be no parade or barbecue or fireworks today; people didn't stay outside for long because of the searing heat from global warming. With the food riots and martial law in place since 2040, people stayed home.

Heating bills in the northern United States had gone down but air conditioning bills had soared and deaths from heatstroke in the Midwest had risen sharply. Much of Miami had been abandoned because of the searing heat year-round, lack of drinking water because of saltwater intrusion into the Biscayne aquifer, and heavy outbreaks of malaria and yellow fever.

Many of her friends and millions of other Americans had long ago migrated to Canada, to find a cooler climate and more plentiful food supply after America's Midwestern breadbasket and central and southern California had been mostly abandoned because of a lack of water and food. Her friend June had recently written, wondering where to go now that Canadian farmland was drying up.

By 2040 dengue fever had spread as far as Winnipeg, Canada, and much of Louisiana's coastline and wetlands was under water. The glaciers in Glacier National Park had melted by 2030 and Alaskan towns were spending millions to shore up sinking roads and houses as the permafrost in tundra melted and released methane gas that accelerated atmospheric warming. By 2030 no more oil was coming from Alaska because the pipeline had sunken and ruptured as the tundra thawed out.

Just then a door opened. Her daughter Jane came out, ready to go to work. Behind her, shouts and squeals erupted from her grandchildren, Lynn and Jeffrey, who were playing Refugees and Border Patrol. It had been a popular game since 2020, when the United States had built a "Great Wall" with armed guards along its border with Mexico in a mostly vain attempt to keep out millions of Latin Americans trying to find food and work in the north.

"They're playing that awful game again, Mother," said Jane. "Could you tell them a story or something?" Mary sighed as she went to corral her pale and undernourished grandchildren. She felt sorry for them; their father has been killed fighting last summer's almost endless forest fires, their mother was struggling to support them all on her job at the Refugee Center, and they had no place to play in the daytime except their underground bedroom.

With her two sweaty grandchildren gathered

*Compare this fictional worst-case scenario with the hopeful scenario that opens Chapter 29.

Figure 19-1 An earth-sheltered house in Will County, Illinois, in the United States. About 13,000 families across the United States have built such houses. Mary Wilkins's fictional house in 2060 could be similar to this one. (Pat Armstrong/Visuals Unlimited)

around her chair, Mary began her favorite story. It was about the old days, before the Warming. There were green parks to play in and green trees to climb, swimming pools full of water, and lakes and rivers everywhere. In winter, cold, white stuff called snow fell from the sky and could be gathered up in balls to throw at each other.

She also told them that almost everyone had a car. "What's a car?" asked Lynn, the oldest child. "Is it like the bus that Mommy rides to work?"

"Yes, only much smaller—just for one person or one family," Mary answered. "It could go fast and ran on a fuel called gasoline—much too rare to be used anymore. You could go anywhere you wanted in your car, even to the drive-in for a hamburger."

"What's a hamburger, Grandma?" asked Jeffrey, the younger child. "What's a drive-in?" asked Lynn.

Mary patiently explained to them about drive-ins and how good hamburgers tasted. "Why did people let things get so bad?" Lynn asked.

Mary's eyes filled with tears as she took the child in her arms. "Why didn't we listen to the warnings of scientists in the 1980s and 1990s?" she asked herself. Then she looked at Lynn and admitted, "Because we didn't want to believe anything bad could happen."

Although our species has existed for only an eye-blink of the earth's history, evidence indicates that we may be altering its atmosphere 10–100 times faster than the natural rate of change over the past 10,000 years.

Many scientists believe that global warming from our binge of fossil fuel burning and deforestation that enhances the earth's natural greenhouse effect (Figure 7-9) and depletion of stratospheric ozone from our use of chlorofluorocarbons and other chemicals (Table 18-1) will threaten life as we know it in the next century if we don't take action now. Other analysts believe that these problems are exaggerated or that we need more research before acting.

Slowing projected global warming involves wasting less energy (Section 16-2), decreasing use of polluting fossil fuels (Chapter 15), greatly increasing our use of renewable forms of energy (such as solar power and wind power, Chapter 16), and reducing deforestation (Chapter 24) as ways to support and strengthen global and national economies during the next 50 years. Most environmental scientists argue that even if projected global warming doesn't happen, these are things we should be doing anyway to save money, reduce pollution, and preserve biodiversity.

We are embarked on the most colossal ecological experiments of all time—doubling the concentration in the atmosphere of an entire planet of one of its most important gases—and we really have little idea of what might happen.

PAUL A. COLINVAUX

This chapter is devoted to answering the following questions:

- What is the difference between the greenhouse effect and global warming?
- What do we know about the earth's temperatures in the past and about possible temperatures in the future?
- Can we really make the earth warmer, and if so, will a few degrees matter?
- What are some possible effects of global warming?
- What can we do about possible global warming?
- Are we depleting ozone in the stratosphere, and why should we care?
- What can we do to slow ozone depletion?

19-1 THE GREENHOUSE EFFECT AND GLOBAL WARMING

What Do We Know About the Greenhouse Effect? In the *greenhouse effect* (Figure 7-9 and Table 18-1), certain gases in the atmosphere trap heat in the lower atmosphere (troposphere). This natural effect, first proposed by Swedish chemist Svante Arrhenius in 1896, has been confirmed by numerous laboratory experiments and atmospheric measurements. It is one of the most widely accepted theories in the atmospheric sciences. Indeed, without the thermal insulation provided by the natural greenhouse effect, the entire earth would be covered with ice and life as we know it would not exist.

In 1990 and 1995 the Intergovernmental Panel on Climate Change (IPCC, a network of about 2,500 of the world's leading climate experts from 70 nations) published several reports evaluating the best available evidence concerning the greenhouse effect, past changes in global temperatures, and climate models projecting future changes in global temperatures and climate. According to this scientific consensus, the amount of heat trapped in the troposphere depends mainly on the concentrations of heat-trapping or *greenhouse gases* and the length of time they stay in the atmosphere.

Figure 19-2 Increases in average concentrations of major greenhouse gases in the troposphere, mostly as a result of human activities. (Data from Electric Power Research Institute. Adapted and updated by permission from Cecie Starr and Ralph Taggart, *Biology: The Unity and Diversity of Life*, 6th ed., Belmont, Calif.: Wadsworth, 1992)

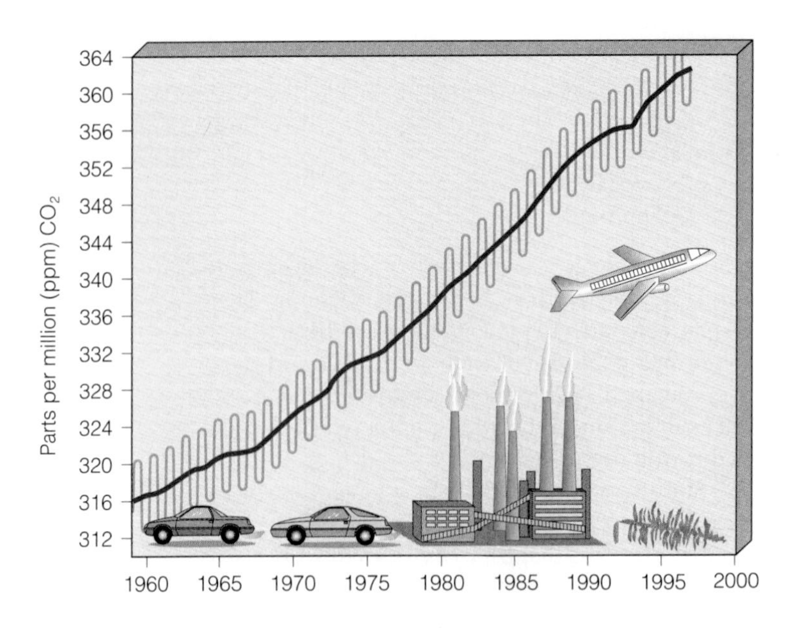

(a) Carbon dioxide (CO_2) is responsible for 50–60% of the global warming from greenhouse gases produced by human activities since preindustrial times. The main sources are fossil-fuel burning (70–75%) and land clearing and burning (20–25%). Most of the CO_2 comes from burning coal, but an increasing fraction is released from motor vehicle exhaust. CO_2 remains in the atmosphere for 50–200 years. The annual rise and fall of CO_2 levels shown in the graph result from less photosynthesis during winter and more during summer.

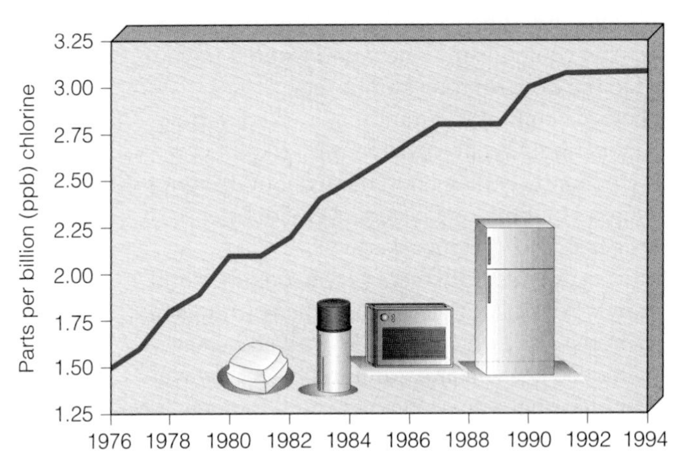

(b) Chlorofluorocarbons (CFCs) contribute to global warming in the troposphere and also deplete ozone in the stratosphere. The main sources are leaking air conditioners and refrigerators, evaporation of industrial solvents, production of plastic foams, and aerosol propellants. CFCs take 10–20 years to reach the stratosphere and generally trap 1,500–7,000 times as much heat per molecule as CO_2 while they are in the troposphere. This heating effect in the troposphere may be partially offset by the cooling caused when CFCs deplete ozone during their 65- to 135-year stay in the stratosphere. Their use is being phased out.

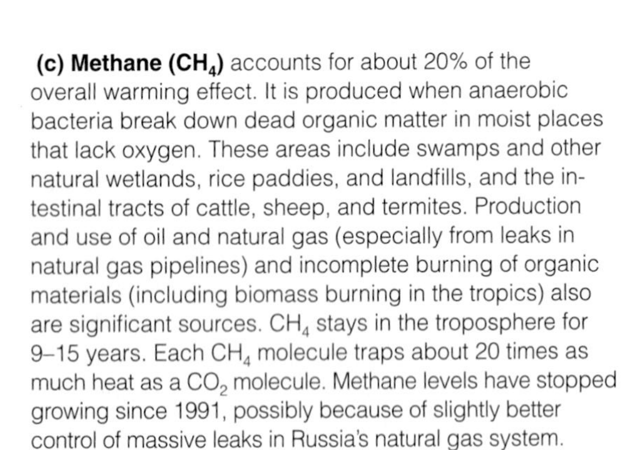

(c) Methane (CH_4) accounts for about 20% of the overall warming effect. It is produced when anaerobic bacteria break down dead organic matter in moist places that lack oxygen. These areas include swamps and other natural wetlands, rice paddies, and landfills, and the intestinal tracts of cattle, sheep, and termites. Production and use of oil and natural gas (especially from leaks in natural gas pipelines) and incomplete burning of organic materials (including biomass burning in the tropics) also are significant sources. CH_4 stays in the troposphere for 9–15 years. Each CH_4 molecule traps about 20 times as much heat as a CO_2 molecule. Methane levels have stopped growing since 1991, possibly because of slightly better control of massive leaks in Russia's natural gas system.

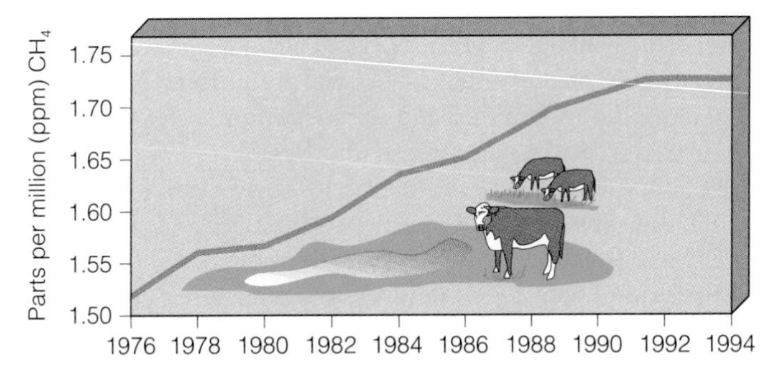

(d) Nitrous oxide (N_2O) can trap heat in the troposphere and can also deplete ozone in the stratosphere. It is released from nylon production, burning of biomass and nitrogen-rich fuels (especially coal), smog-fighting catalytic converters on motor vehicles, and the breakdown of nitrogen fertilizers in soil, livestock wastes, and nitrate-contaminated groundwater. Its life span in the troposphere is about 120 years, and it traps about 200 times as much heat per molecule as CO_2.

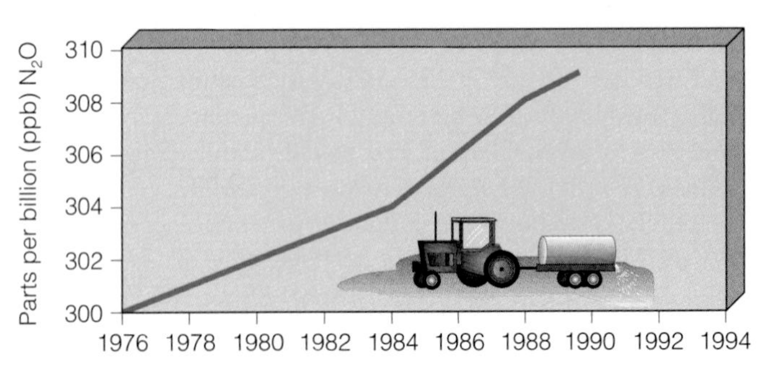

Kaiser, Jocelyn. 1998. New Network Aims to Take the World's CO_2 Pulse. *Science,* vol. 281, no. 5376, 506(2).

The major greenhouse gases are water vapor (H_2O), carbon dioxide (CO_2), ozone (O_3), methane (CH_4), nitrous oxide (N_2O), and chlorofluorocarbons (CFCs). Recently scientists have identified another greenhouse gas—perfluorocarbons (PFCs, such as CF_4)—emitted mostly from aluminum production. They remain in the atmosphere for 2,000–50,000 years.

The two predominant greenhouse gases in the troposphere are water vapor, controlled by the hydrologic cycle (Figure 5-4), and carbon dioxide, controlled by the global carbon cycle (Figure 5-5). The primary heat-trapping gas is water vapor; because its concentration in the atmosphere is fairly high (1–5%), inputs of water vapor from human activities have little effect on this chemical's greenhouse effects. By contrast, the concentration of carbon dioxide in the atmosphere is so small (0.036%) that a fairly large input of CO_2 from human activities can significantly affect the amount of heat trapped in the atmosphere.

What Is Global Warming? Measured atmospheric levels of certain greenhouse gases—CO_2, CFCs, methane, and nitrous oxide—have risen substantially in recent decades (Figure 19-2) and are projected to enhance the earth's natural greenhouse effect, a phenomenon called *global warming* (Table 18-1).

Most of the increased levels of these greenhouse gases since 1958 have been caused by human activities: burning fossil fuels, agriculture, deforestation, and use of CFCs. Although molecules of CFCs, methane, and nitrous oxide trap much more heat per molecule than CO_2, they make up only about 36% of greenhouse gases. The much larger input of CO_2 makes it the most important greenhouse gas produced by human activities.

Developed countries account for about 60% of current CO_2 emissions and developing countries for the remaining 40%. The United States alone accounts for about 23% of global CO_2 emissions from human activities, followed by China (14%), Russia (7%), and Japan (5%). However, emissions of CO_2 are increasing rapidly in developing countries such as China and India that are industrializing rapidly.

According to the EPA, emissions of greenhouse gases by the United States rose by 20% between 1990 and 1996. Energy-related activities accounted for about 86% of these emissions in 1996, mostly through burning fossil fuels.

Ice core analysis reveals that at the beginning of the industrial revolution the atmospheric concentration of CO_2 was about 280 parts per million (referred to as the *preindustrial level*). Between 1860 and 1997 the concentration of CO_2 in the atmosphere grew exponentially to 364 parts per million (Figure 19-2), higher than any time in the past 160,000 years. The atmospheric concentrations of CO_2 and other greenhouse gases are projected to double from preindustrial (1860) levels sometime during the next century—probably by 2050—and then continue to rise. There is widespread agreement that altered amounts of greenhouse gases in the atmosphere can affect the climate for many centuries.

What Is the Scientific Consensus About the Earth's Past Temperatures? Layers of ancient ice in Antarctic glaciers provide a time capsule whose contents can be analyzed to provide information about the temperature and contents of the atmosphere in the ancient past. Such analyses—much like counting tree rings—and other data show that the earth's average surface temperature has fluctuated considerably over geologic time. These data show that during the past 800,000 years several ice ages have covered much of the planet with thick ice. Each glacial period lasted about 100,000 years and was followed by a warmer interglacial period of 10,000–12,500 years (Figure 19-3).

For the past 10,000 years we have enjoyed the warmth of the latest interglacial period (called the Holocene). This climatic stability has prevented drastic changes in the nature of soils and vegetation patterns throughout the world, allowing large increases in food production and thus in population (Figure 1-22).

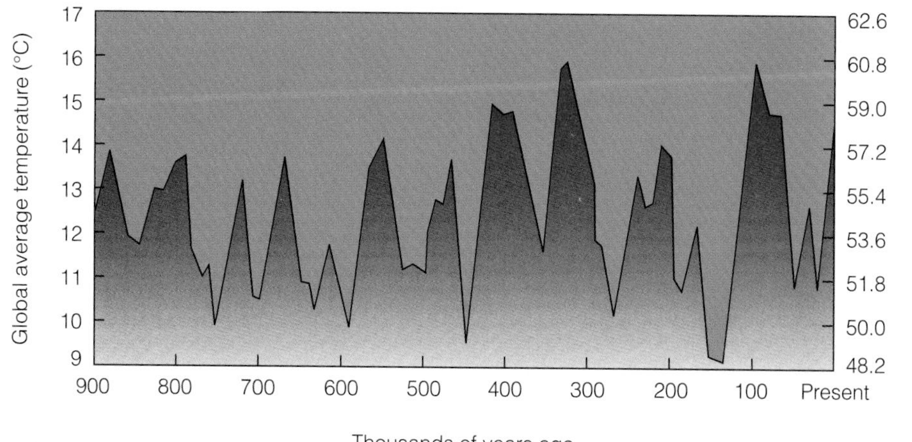

Figure 19-3 During the past 900,000 years the earth has experienced cycles of ice ages, each lasting about 100,000 years and then followed by warmer interglacial periods lasting 10,000–12,500 years. The warm interglacial period during the past 10,000 years has been a major factor in the development of agriculture, human civilizations, and population growth.

Hint: Enter the search term *carbon dioxide monitoring* using the Subject Guide.

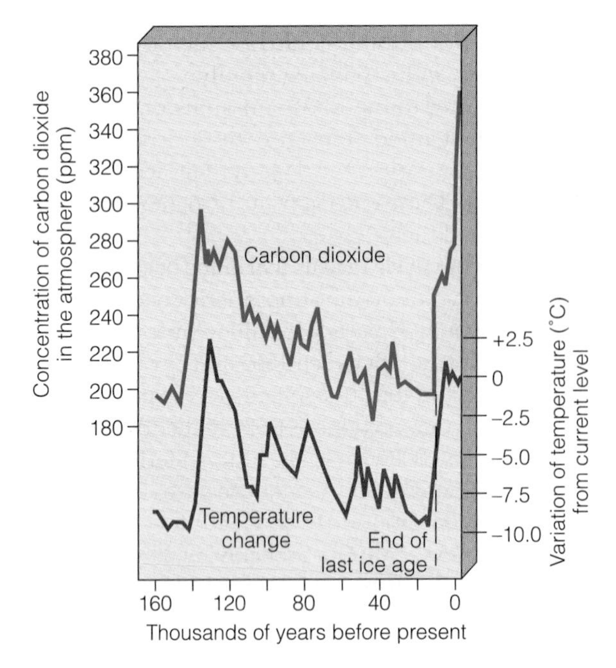

Figure 19-4 Estimated long-term variations in mean global surface temperature and average tropospheric carbon dioxide levels over the past 160,000 years. These CO_2 levels were obtained by inserting metal tubes deep into Antarctic glaciers, removing the ice, and analyzing bubbles of ancient air trapped in ice at various depths throughout the past. Such analyses reveal that since the last great ice age ended about 10,000 years ago, we have enjoyed a warm interglacial period. The rough correlation between tropospheric CO_2 levels and temperature shown in these estimates based on ice core data suggests a connection.

However, even small temperature changes during this period have led to hardship and premature deaths and to large migrations of peoples in response to changed agricultural and grazing conditions.

Analysis of gases in bubbles trapped in ancient ice show that over the past 160,000 years tropospheric water vapor levels (the dominant greenhouse gas) have remained fairly constant. During most of this period levels of CO_2 have fluctuated between 190 and 290 parts per million. Estimated changes in the levels of tropospheric CO_2 correlate fairly closely with estimated variations in the earth's mean surface temperature during the past 160,000 years (Figure 19-4).

Since 1860 (when measurements began), mean global temperature after correcting for excess heating over urban areas (urban heat island effects) has risen 0.3–0.6°C (0.5–1.1°F). The temperature rose about 0.3°C between 1946 and 1997 (Figure 19-5). Since 1860, the thirteen warmest years occurred between 1979 and 1998, with 1990, 1995, 1997, and 1998 being the four hottest years.

Many uncertainties remain. Some or even all of the roughly 0.5°C rise since 1860 could result from normal fluctuations in the mean global temperature. On balance, however, the Intergovernmental Panel on Climate Change concluded in its 1995 report that "the observed increase over the last century is unlikely to be entirely due to natural causes" and "the balance of evidence suggests that there is a discernible human influence on global climate."

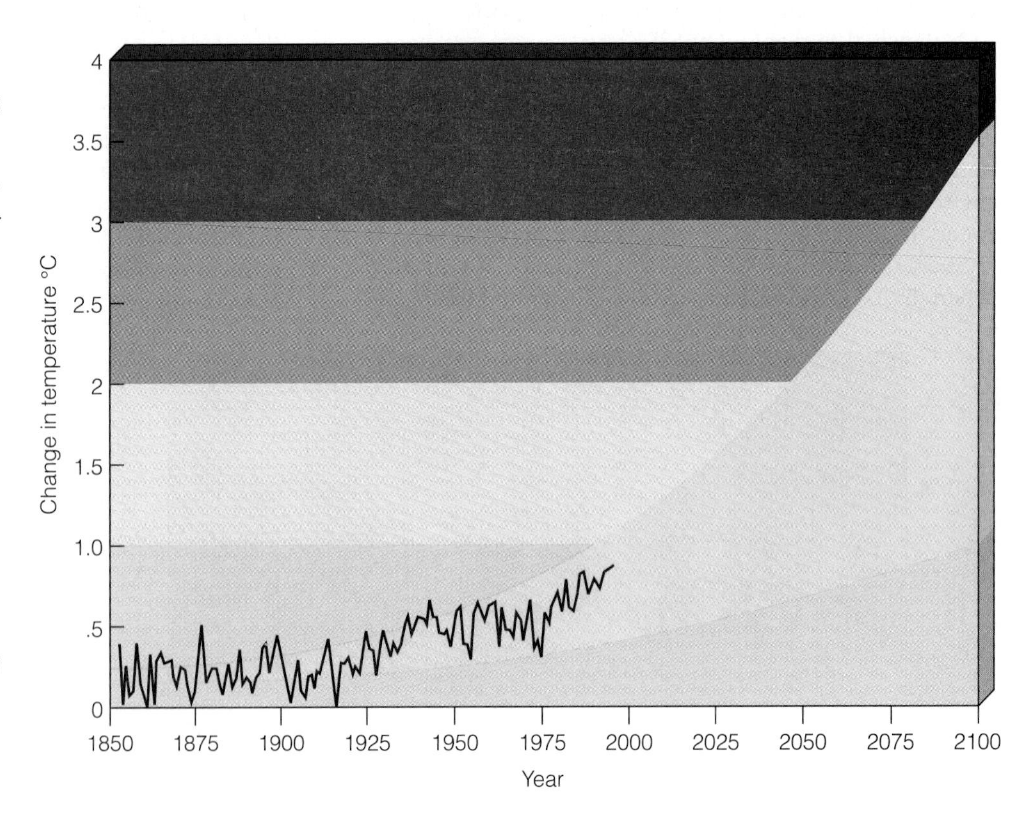

Figure 19-5 Recorded changes in the earth's mean surface temperature between 1860 and 1996 (dark line). The curved yellow region shows global warming projected by various computer models of the earth's climate systems. Note that the climate model projections roughly match the 0.3–0.6°C (0.5–1.1°F) recorded temperature increase between 1860 and 1996. Current models indicate that the average global temperature will rise by 1–3.5°C (1.8–6.3°F) sometime during the next century. However, this projection assumes that air pollution from sulfate aerosols will continue to increase and exert a slight cooling effect, despite the fact that sulfur emissions have leveled off since 1960. The projected warming shown here could be *overestimated* or *underestimated* by a factor of two. (Data from U.S. National Academy of Sciences and National Center for Atmospheric Research)

Q: What are the three major types of coal?

How Do Scientists Model Greenhouse Warming? Computer Models as Crystal Balls To project the effects of increases in greenhouse gases on average global temperature and changes in the earth's climate, scientists develop mathematical models of such systems and run them on supercomputers. The most sophisticated of these climate models are called *general circulation models* (GCMs). Recently they have been coupled with models of ocean circulation and inputs of aerosols (tiny particles and droplets) into the atmosphere by volcanoes and pollution from human activities (Spotlight, p. 504).

Figure 19-6 shows the series of models involved in making predictions about the effects of human activities and natural processes on levels of greenhouse gases (model 1), average global temperature (model 2), changes in regional climate in various parts of the world (model 3), and possible effects of such changes in these different areas (models 4 through 7). Current versions of models 1 and 2 project that the earth's average global temperature should increase within a certain range as atmospheric levels of greenhouse gases rise. However, most models disagree on how projected rises in average global temperature (global warming) might affect the climate in different areas (model 3) and the effects of such changes in these areas (models 4 through 7).

What Is the Scientific Consensus About Future Global Warming and Its Effects? According to the latest climate models, the IPCC projects that the earth's mean surface temperature should rise 1–3.5°C (1.8–6.3°F) between 1990 and 2100 (Figure 19-5); the most likely rise in temperature before 2100 would be about 2°C (3.6°F) if the atmospheric concentration of CO_2 doubles from its preindustrial level of 280 ppm by volume to 560 ppm by volume. This may not seem like much, but even at the lowest projected increase of 1.0°C, the earth would be warmer than it has been for 10,000 years (see cartoon). Current models project that climate change, once begun, will continue for hundreds of years.

According to the models, the northern hemisphere should warm more and faster than the southern hemisphere because the latter has more heat-absorbing ocean than land (Figure 8-3) and because water cools more slowly than land. Current climate models project a more pronounced warming at the earth's poles. Measurements reveal that the surface temperatures at nine stations north of the Arctic circle have risen by about 5.5°C (9.9°F) since 1968. Since 1947 the average summer temperature at Antarctica has risen by almost 2°C (3.6°F) and five of the nine massive, floating ice shelves surrounding the continent have broken up since 1950 (three between 1994 and 1998).

© Matt Wuerker. (Color added)

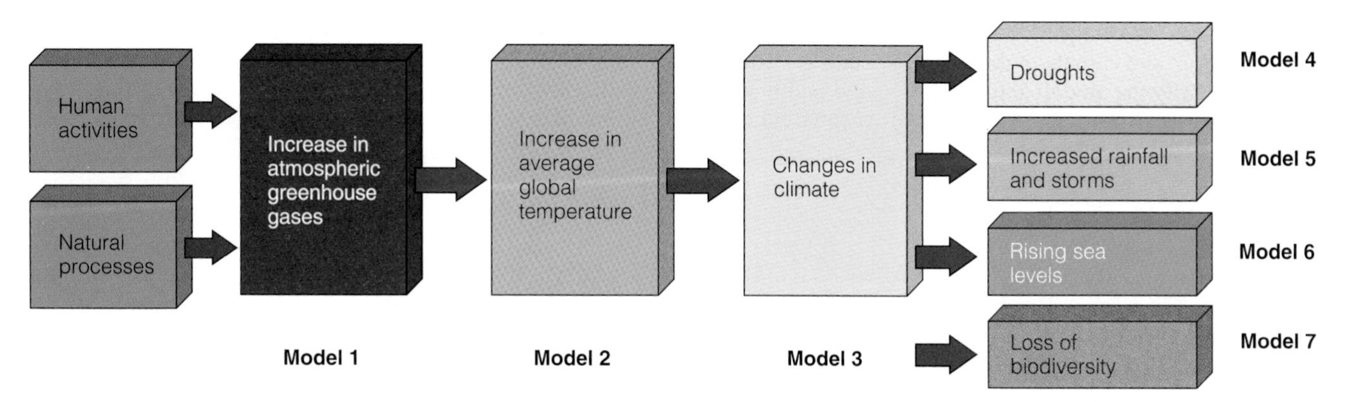

Figure 19-6 Generalized schematic of the greenhouse effect and models of its possible consequences. The greenhouse effect itself (model 1) is well established. The increase in average global temperature resulting from higher concentrations of greenhouse gases (model 2) is projected by crude climate models. Current models cannot consistently project either the specific climate changes (model 3) or their consequences (models 4–7) in various parts of the world.

A: Lignite (brown coal), bituminous (soft coal), and anthracite (hard coal)

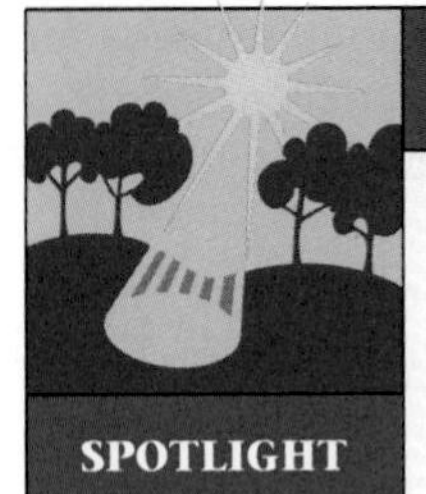

How Do Climate Models Work? Cloudy Crystal Balls

Here is a rough idea of how current general circulation models of global climate work. Imagine the earth's entire surface covered with gigantic cells or boxes, each several hundred kilometers on a side and about 3 kilometers (2 miles) high. Then 6–20 layers of such boxes are stacked on top of one another.

Initial conditions of a number of variables are assigned to each box. These variables include temperature, air pressure, wind speed and direction, and concentrations of key chemicals such as H_2O and CO_2.

Next comes a set of equations describing the expected flow of various types of energy and matter in and out of the box-filled atmosphere. These inputs, called *boundary conditions,* include variables such as solar radiation, precipitation, heat radiated by the earth, cloudiness, interactions between the atmosphere and ocean, greenhouse gases, and air pollutants (especially aerosols).

An elaborate set of equations in each model connects the cells so that when the value of any of these input variables changes in one cell, the values of variables in surrounding cells change in a realistic way. Then the model is run on a supercomputer to simulate changes in average global temperature and climate in the past (to verify the model) and to project future changes.

These models can provide us with scenarios of what *could* happen based on various assumptions and data fed into each model. How well the results correspond to the real world depends on **(1)** the design and assumptions of each model, **(2)** the accuracy of the data used, **(3)** magnification of tiny errors over time, **(4)** factors in the earth's climate system that amplify (positive feedback) or dampen (negative feedback) changes in average global temperatures, and **(5)** the effects of totally unexpected or unpredictable events (chaos).

Current models simulate the large-scale features of the current climate fairly well and have reproduced some climate features of the distant past fairly accurately. When atmospheric and ocean models are coupled and take into account atmospheric cooling from sulfate aerosols, they also reproduce fairly accurately the global warming that has taken place over the past century. The greatest uncertainties in current models involve incorporating the effects of clouds and the ecosphere on climate.

The models can also help us evaluate the possible effects of various ways of slowing or even halting possible global warming. For example, how much would we have to decrease the current worldwide use of fossil fuels to prevent the global warming projected during the next century? The answers are startling and highly unlikely to be implemented for economic and political reasons: a 20% cut by 2000 (35% in the United States), 50% by 2010, and 65–80% by 2030.

Despite their limitations such mathematical models are the most useful tools we have for understanding and projecting the behavior of complex systems such as global climate. In the global climate system there are 14 levels of scale, from the planetary scale to liquid cloud particles. Current models are able to include only the two largest levels: the planetary scale and the scale of weather disturbances. To add the next lowest level, which is that of a thunderstorm, to current models requires a computer a thousand times faster than those available now. Such computers may be available by 2000.

Climate models are cloudy crystal balls, but without them we are groping around in total darkness trying to understand and deal with serious issues. Most government and personal decision making is based on mental, mathematical, and other models (Section 2-2) that are far less sophisticated and reliable than current climate models.

Critical Thinking

Climate models have been criticized as being too crude. Are there better alternatives to help us make and evaluate the possibility of global warming and how various policies might slow projected global warming? Do you support greatly increased funding for development of better climate models? Explain.

Although climate warming has destroyed several of Antarctica's smaller ice shelves, a recent study indicates that global warming could thicken the region's two largest ice shelves.

Climate models also project that as the earth's atmosphere warms, the rate of water evaporation will increase and global average precipitation will rise. At the mid- to high latitudes (where most of North America, Europe, and Japan are located) average precipitation is projected to increase (especially during the cold season). Much of it will come in heavy showers or thunderstorms (thus increasing flooding) rather than in gentler, longer-lasting rainfalls. Such changes are already being observed in many parts of the northern hemisphere.

Other possible signs of global warming include **(1)** increased retreat of some glaciers on the tops of mountains in the Alps, Andes, Himalayas, and North-

Q: What is the most desirable type of coal?

ern Cascades of Washington during the last 30 years, **(2)** northward migrations of some warm-climate fish and trees, **(3)** spread of some tropical diseases away from the equator, and **(4)** bleaching of coral reefs in tropical areas with warmer water (p. 186).

With a warmer climate global sea levels will rise, mainly because water expands slightly when heated. Between 1900 and 1990, global sea levels have risen by 9–18 centimeters (3.5–7 inches), more than in the previous 1,000 years. Climate models estimate that two-thirds of this rise is the result of global warming (mostly from expansion of water when it becomes warmer). Recent studies indicate that the remaining third may result from deforestation (which transfers water in trees and soil to the atmosphere and eventually to the ocean), draining of wetlands, and large-scale pumping of groundwater out of deep aquifers (which transfers water to the earth's surface, from which some of it reaches the ocean). Current climate models of global warming project that global sea levels will rise by 15–95 centimeters (6–37 inches) between 1990 and 2100, with a best estimate of 48 centimeters (19 inches).

The models also indicate that if atmospheric warming at the poles caused ice sheets and glaciers to melt even partially, global sea level would rise much more. However, this is expected to take hundreds of years to occur.

IPCC scientists warn that warming or cooling by more than 1°C (1.8°F) over a few decades (instead of over many centuries, as has been the pattern during the last 10,000 years) causes serious disruptions of the current structure and functioning of earth's ecosystems and of human economic and social systems.

19-2 GLOBAL WARMING OR A LOT OF HOT AIR? HOW SERIOUS IS THE THREAT?

Will the Earth Really Get Warmer? There is much controversy over whether we are already experiencing global warming, how warm temperatures might be in the future, and the effects of such temperature increases. One problem is that many of the past measurements and estimates of the earth's average temperature are imprecise, and we have only about 100 years of fairly accurate data. With such limited data, it is difficult to separate out the normal short-term ups and downs of global temperatures (called *climate noise*) from an overall rise in average global temperature.

Scientists have also identified a number of factors that might amplify (positive feedback) or dampen (negative feedback) a rise in average atmospheric temperature. These factors influence both how fast temperatures might climb and what the effects might be on various areas. Let's look more closely at the possible effects of such factors.

How Do Changes in Solar Output Affect Earth's Temperatures? Solar output varies by about 0.1% over the 11-year and 22-year sunspot cycles—and over 80-year and other much longer cycles. Sunspots occur when strong solar magnetic fields periodically protrude through the sun's surface and slightly increase the sun's energy output. These up-and-down changes in solar output can temporarily warm or cool the earth and thus affect the projections of climate models.

In 1995, solar physicist Judith Lean at the Naval Research Laboratory estimated that solar warming accounted for roughly half of the global warming taking place between 1860 and 1970 and about one-third of the warming since 1970 (Figure 19-5). In 1998, two Danish scientists published an untested hypothesis suggesting that virtually all of rises and falls of the earth's temperatures are due to the effect of cosmic rays produced by solar cycles on cloud cover.

Other studies suggest that sunspot cycles may account for only 10–30% of the warming during the past century. Two 1992 studies concluded that the projected warming power of greenhouse gases should outweigh the climatic influence of the sun over at least the next 50 years.

Suppose that increases in atmospheric temperatures during this century were caused mostly by a slightly brighter sun. If this is a valid hypothesis, then analysts point out that a brighter sun can cause even more greenhouse warming as human activities continue adding greenhouse gases to the troposphere.

How Do the Oceans Affect Climate? The world's oceans might amplify global warming by releasing more CO_2 into the atmosphere (positive feedback) or might dampen it by absorbing more heat (negative feedback). We know that the oceans currently help moderate tropospheric temperature by removing about 29% of the excess CO_2 we pump into the atmosphere (negative feedback), but we don't know whether they can absorb more. If the oceans warm up enough, some of the dissolved CO_2 will bubble out into the atmosphere (just as in a glass of carbonated ginger ale left out in the sun), amplifying and accelerating global warming (positive feedback).

Global warming could be dampened if the oceans absorbed more heat, but this depends on how long the heat takes to reach deeper layers. Recent measurements indicate that deep vertical mixing in the ocean occurs extremely slowly (taking hundreds of years) in

A: Anthracite because of its high heat content and low sulfur content

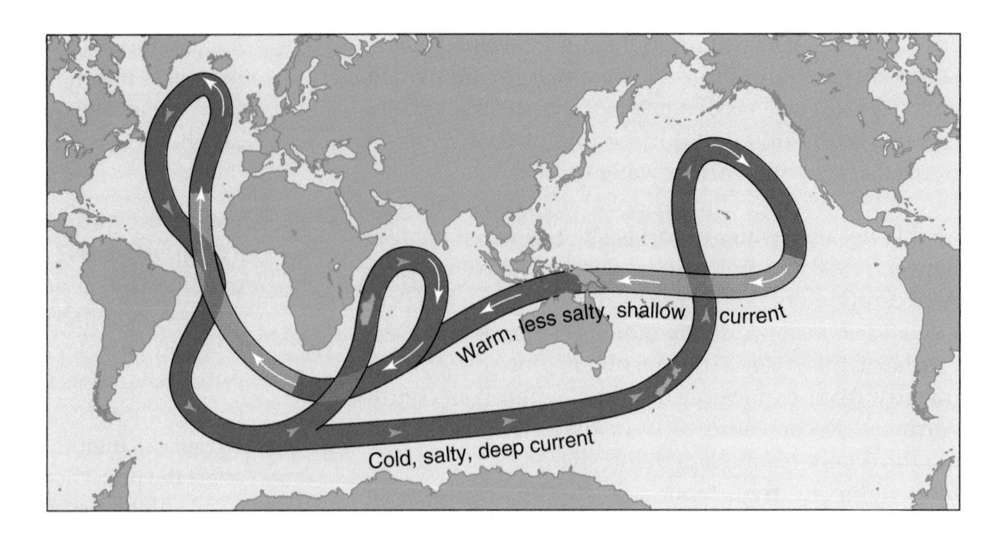

Figure 19-7 This loop of ocean water stores carbon dioxide in the deep sea and brings warmth to Europe. It occurs when ocean water in the North Atlantic is dense enough (because of its salt content and cold temperature) to sink to the ocean bottom and well up in the warmer Pacific (helping cool that part of the world). Then a shallower return current aided by winds brings warmer and less salty—and thus less dense—water to the Atlantic, which can then cool and sink to begin the cycle again. If this heat conveyor belt or loop stalls out because of a drop in density of ocean water in the North Atlantic (possibly from changes caused by global warming), massive climate changes over much of the earth's surface could occur within only a few decades. Models indicate that this oceanic heat conveyor belt would return, but only after hundreds or thousands of years.

most places because water density increases with depth, inhibiting mixing of different layers.

There is also concern that deep ocean currents could be disrupted. At present, these currents (driven largely by differences in water density and winds) act like a gigantic conveyor belt, transferring heat from one place to another and storing carbon dioxide in the deep sea (Figure 19-7). In the Atlantic Ocean, the resulting enormous amount of heat transported northward from tropical waters accounts for Europe's unusually warm climate relative to its latitude. There is concern that global warming could halt this thermal conveyor belt by reducing the density and salinity of water in the North Atlantic so that the water would not sink. If this loop stalls out, evidence from past climate changes indicates that this could trigger atmosphere temperature changes of more than 5°C (9°F) over periods as short as 40 years (positive feedback).

How Do Water Vapor Content and Clouds Affect Climate? Changes in the atmosphere's water vapor content and the amount and types of cloud cover also affect climate. Warmer temperatures would increase evaporation and the water-holding capacity of the air and create more clouds. Significant increases in water vapor, a potent greenhouse gas, could enhance warming (positive feedback).

However, it is difficult to predict the net effect of additional clouds on climate. They could have a warming effect (positive feedback) by trapping heat or a cooling (negative feedback) effect by reflecting sunlight back into space. The net result of these two opposing effects depends on whether it is day or night and on the type (thin or thick) and altitude of clouds. During the day clouds are reflective and have a cooling effect; at night they have an insulating effect and lead to warmer temperatures. In general, high and thin (cirrus) clouds have a warming effect; low and thick clouds have a cooling effect. Scientists don't know which of these factors might predominate or how cloud types and heights might vary in different parts of the world as a result of global warming. However, within a few years scientists hope to better understand these effects.

How Might Changes in Polar Ice Affect Climate? The ability of the earth's surface (land, water, or ice) to reflect light is called its **albedo**. Dark-colored surfaces (land and water) absorb heat from sunlight and warm up. Light-colored or shiny surfaces (polar ice caps and high, thin clouds) remain cool because they reflect back into space much of the sunlight that hits them.

Because of their high albedo, the light-colored Greenland and Antarctic ice sheets act like enormous mirrors, reflecting sunlight back into space. If warmer temperatures melted some of this ice and exposed darker ground or ocean, more sunlight would be absorbed and warming would be accelerated. Then more ice would melt, further accelerating the rise in atmospheric temperature (positive feedback).

On the other hand, the early stages of global warming might increase the amount of the earth's water stored as ice. Warmer air would carry more water vapor, which could drop more snow on some polar glaciers,

Q: Who has the world's largest reserves of coal?

especially the gigantic Antarctic ice sheet. If snow accumulated faster than ice was lost, the ice sheet would grow, reflect more sunlight (higher albedo), and help cool the atmosphere (negative feedback), perhaps leading to a new ice age within a thousand years.

How Might Air Pollution Affect Climate? Climate can also be affected by air pollution. Projected global warming might be partially offset by *aerosols* (tiny droplets and solid particles, Figure 18-4) of various air pollutants released or formed in the atmosphere by volcanic eruptions and human activities. It is hypothesized that SO_2 and tiny particles in the troposphere attract enough water molecules to form condensation nuclei, which leads to increased cloud formation. The resulting clouds have a high albedo and thus reflect more incoming sunlight back into space during daytime.

During daytime this helps counteract the heating effects of increased greenhouse gases. However, nights would be warmer because the clouds would still be there and prevent some of the heat stored in the earth's surface (land and water) during the day from being radiated back into space.

Because about 90% of SO_2 emissions (which can be transformed in the atmosphere to aerosol droplets of sulfuric acid and particles of sulfate salts) occur in the heavily industrialized northern hemisphere, this portion of the globe may undergo cooling that could offset or delay global warming. These pollutants may explain why most recent warming in the northern hemisphere occurs at night. Climate expert Thomas Karl at the National Climatic Data Center estimates that currently the United States is about 1°C (1.8°F) cooler because of aerosols formed from SO_2 emissions.

A similar cooling effect may be occurring in the southern hemisphere from particles in smoke emitted by large-scale burning of rain forests, grasslands, agricultural waste, and wood (used for heating and cooking). These aerosol emissions in both hemispheres may help explain why global warming observed to date is only about one-half of that projected by earlier climate models that did not include this effect.

However, these interactions are complex. Pollutants in the lower troposphere can either warm or cool the air, depending on the reflectivity of the underlying surface. We also know little about the effects of aerosols on the properties of clouds. These contradictory and patchy effects and uncertainties, plus improved air pollution control, make it unlikely that air pollutants will counteract projected global warming very much in the next half century. Aerosols also fall back to the earth or are washed out of the atmosphere within weeks or months, whereas CO_2 and other greenhouse gases remain in the atmosphere for decades to several hundred years.

These aerosols are major components of acid deposition (commonly known as *acid rain*, Section 18-3), which slows forest growth and weakens or kills many trees (Figure 18-9). This reduces the ability of trees to absorb some of the CO_2 we are putting into the atmosphere and could accelerate global warming (positive feedback).

We could maintain or increase levels of aerosol air pollutants to offset possible global warming. However, because these pollutants already kill hundreds of thousands of people a year and damage food crops, trees, and other forms of vegetation (Section 18-5), they must be reduced (Figure 18-6).

How Might Increased CO_2 Levels Affect Photosynthesis and Methane Emissions? Another uncertainty is the effect of increased CO_2 on photosynthesis. Some studies suggest that more CO_2 in the atmosphere is likely to increase the rate of photosynthesis in areas with adequate amounts of water and other soil nutrients. This would remove more CO_2 from the atmosphere and help slow global warming (negative feedback).

Other studies suggest that this effect varies with different types of plants and in different climate zones. For example, yields of wheat, rice, and corn might increase in areas with good soils and enough water but corn and sugarcane yields don't respond much to increased levels of CO_2.

Studies also indicate that under high CO_2 concentrations the additional growth may be concentrated in plant parts not consumed by people. Also, much of any increased plant growth could be offset by plant-eating insects and weeds that breed more rapidly and year-round in warmer temperatures.

Stomata are tiny pores in leaf surfaces that allow CO_2 to enter for photosynthesis and that let water out through evapotranspiration that cools the plant. A 1996 study by scientists at NASA's Goddard Flight Center found that increased CO_2 levels are likely to worsen global warming by causing stomata to remain closed for longer periods of time. Because the hot water inside plants can't escape, the plant gets hotter and in turn warms its surroundings. This adds to global warming and could cut rainfall over vegetated areas during the growing season.

Another complication is the overloading of terrestrial ecosystems with nitrogen, mostly from acid deposition of nitric acid and nitrate salts (Figure 18-7) and increased use of nitrate fertilizers. A 12-year field study of grasslands has shown that the double stress of too much carbon and too much nitrogen temporarily spurred plant growth. However, in the long run the excess nitrogen favored plant species that are less adept at taking up carbon.

A: United States (66%), republics of the former Soviet Union (9%), and China (9%)

Another factor that can affect CO_2 levels is *forest turnover*: how fast trees grow and die in a forest. A 1994 study indicated that the turnover in tropical forests worldwide is accelerating. Besides reducing the biodiversity of tree species in tropical forests, this change could enhance global warming (positive feedback) by reducing removal of CO_2 from the atmosphere because less dense, faster-growing trees require less CO_2 for growth.

Global warming could be accelerated by increased release of methane, a potent greenhouse gas, from wetlands. A 1994 study indicated that increased uptake of CO_2 by wetland plants could boost emissions of methane by providing more organic matter for methane-producing anaerobic bacteria to decompose (positive feedback). Some scientists also speculate that in a warmer world huge amounts of methane now tied up in arctic tundra soils and in muds on the bottom of the Arctic Ocean might be released if the blanket of permafrost in tundra soils melts and the oceans warm considerably. Conversely, some scientists believe that bacteria in tundra soils would rapidly oxidize the escaping methane to CO_2, a less potent but still important greenhouse gas.

How Rapidly Could Climate Shift? If moderate change takes place gradually over several hundred years, people in areas with unfavorable climate changes may be able to adapt to the new conditions. However, if the projected global temperature change takes place over several decades or all within the next century, we may not be able to switch food-growing regions and relocate the large portion of the world's population living near coastal areas fast enough. The result would be large numbers of premature deaths from lack of food, as well as social and economic chaos. Such rapid changes would also reduce the earth's biodiversity because many species couldn't move or adapt.

Recent data from analysis of ice cores and deep-sea sediments suggest that the earth's climate has shifted often and more drastically and quickly than previously thought. This new evidence indicates that average temperatures during the warm interglacial period that began about 125,000 years ago (Figure 19-4) varied as much as 10°C (18°F) in only a decade or two and that such warming and cooling periods each lasted 1,000 years or more.

If these findings are correct and also apply to the current interglacial period, fairly small rises in greenhouse gas concentrations could trigger rapid up-and-down shifts in average global temperatures. Such rapid shifts would be disastrous for humans and many other forms of life on the earth.

How Could Human Responses Accelerate Global Warming? If the earth warms, many people in urban areas (which tend to be warmer than nearby rural areas), especially in developed countries, would use more air conditioning. Burning more fossil fuels (especially coal) to supply the additional electricity would release more CO_2 and lead to additional warming. Then people would use more air conditioning and worsen the situation (positive feedback).

Polluting aerosols produced by burning more coal might temporarily offset some of this warming. As previously discussed, however, such pollution has serious harmful impacts on the health of humans, trees, food crops, and animal wildlife.

As a result of the factors discussed in this section, climate scientists estimate that their projections about *global warming and rises in average sea levels during the next 50–100 years could be half the current projections (the best-case scenario) or double them (the worst-case scenario).* In any event, possible climate change and its effects are likely to be erratic and mostly unpredictable (see the Guest Essay by John Harte on p. 510).

19-3 SOME POSSIBLE EFFECTS OF A WARMER WORLD

Why Should We Worry If the Earth's Temperature Rises a Few Degrees? So what's the big deal? Why should we worry about a possible rise of only a few degrees in the earth's average temperature? We often have that much change between May and July, or even between yesterday and today.

This is a common critical thinking trap that many people fall into. The key point is that we are not talking about normal swings in *local weather*, but about a projected *global* change in *climate*—weather averaged over decades, centuries, and millennia. A warmer global climate could have a number of possible effects, some harmful and some beneficial depending on where one lives.

How Might Food Production Be Affected? A warmer global climate could increase food production in some areas and lower it in others. Archeological evidence and computer models indicate that climate belts would shift northward by 100–150 kilometers (60–90 miles) or upward 150 meters (500 feet) in altitude for each 1°C (1.8°F) rise in global temperature.

Whether such poleward shifts lead to increased crop productivity in new crop-growing areas depends mainly on two factors: the fertility of the soil in such

Q: What percentage of the energy used in China comes from coal?

regions and the availability of enormous amounts of money to build a new agricultural infrastructure (for irrigation and food storage and distribution).

In parts of Asia, food production in more northern areas could increase because of favorable soils. In North America, however, the northward expansion of crop-growing regions from the Midwestern United States into Canada would be limited by the thinner and less fertile soils there. Siberia also lacks rich soils and couldn't make up for lower crop production in Ukraine.

Current climate models project 10–70% declines in the global yield of key food crops and a loss in current cropland area of 10–50%, especially in most poor countries. Other studies project a 1–8% drop in the global production of wheat, rice, and other grains by 2060 because of projected warming. With this scenario, developed countries could experience only a slight drop in grain production or perhaps an increase by as much as 11%, whereas grain production in developing countries could drop by 9–11%.

Currently, we can't predict where changes in crop-growing capacity might occur or how long such changes might last. However, we do know that drops in global crop yields of only 10% would cause large increases in hunger and starvation (especially in poor countries) and cause economic and social chaos.

Because winter cold kills insects and their eggs, temperature determines how far from the equator crop-destroying pests can live. As temperatures rise in food-producing areas in the United States and other currently temperate areas, food production could be lowered by exploding populations of insects, crop diseases (such as rusts and molds), and weeds.

Rising sea levels could flood low-lying coastal areas where crops are grown; they could decrease seafood supplies by flooding half of the world's productive coastal wetlands, marshes, and mangrove swamps sometime during the next century. Warmer water can also affect fish populations in some areas. In 1995, Scripps biologist John McGovern reported that Pacific zooplankton, a key food source for fish, declined by 70% between 1974 and 1994 because of a 1–1.7°C (2–3°F) increase in surface water temperatures.

Increases in the average temperature and depth of tropical ocean waters would reduce biodiversity by destroying highly productive coral reefs. These reefs, dubbed the "tropical rainforests of the sea" because of their rich biological diversity, are already suffering from stresses such as enhanced sedimentation from runoff, pollution, and overfishing (p. 186). Such stresses could increase the vulnerability of the coral animals in such reefs to only slight rises in ocean water temperatures. An increase of 1–2°C (1.8–3.6°F) can cause corals to become bleached (Figure 8-1, right) as they expel the algae that provide them with food and give them their vibrant colors. Prolonged increases in water temperature of 3–4°C (5–7°F) can cause significant death of coral animals. Research suggests that warmer than normal sea water has contributed to extensive coral bleaching near the Galápagos Islands off the coast of Ecuador and at several other areas in the eastern Pacific, Caribbean, and Atlantic oceans.

How Might Water Supplies Be Affected? Global warming would also *reduce water supplies* in some areas. Lakes, streams, and aquifers in some areas that have provided water to ecosystems, croplands, and cities for centuries could shrink or dry up altogether. This would force entire populations to migrate to areas with adequate water supplies—if they could. So far we can't say with much certainty where this might happen.

For example, environmental scientist Mauri Pelto has found that during the past several decades the number of alpine glaciers melting and retreating on mountaintops in the Alps of Switzerland and Italy and in the Northern Cascades of Washington has increased dramatically. Summer runoff from snowmelt from such alpine glaciers provides 20–50% of the total streamflow in the surrounding drainage basins. Retreat of such glaciers would reduce the area exposed to summer melting and thus reduce the availability of water to lowland areas. Research by Pelto has shown that in two basins in the North Cascades of Washington where glaciers have recently disappeared, summer streamflow dropped by 73%.

What Effects Might Global Warming Have on Forests and Biodiversity? According to scientists, global warming will change the makeup and location of many of the world's forests. Over time, populations of some trees and other plants can move as seeds, dispersed by animals and winds, are able to grow in a more favorable habitat. With a warmer climate, forests in temperate and subarctic regions would move toward the poles or to higher altitudes, leaving more grassland and shrubland in their wake.

However, tree species move slowly through the growth of new trees along forest edges—typically about 0.9 kilometer (0.5 mile) per year or 9 kilometers (5 miles) per decade. According to the 1995 report of the IPCC, midlatitude climate zones are projected to shift northward by 550 kilometers (340 miles) over the next century. At that rate some tree species such as

A: 76% (compared to 22% in the United States)

GUEST ESSAY

The Scientific Consensus About Global Warming

John Harte

John Harte is professor of energy and resources at the University of California, Berkeley. He is a Phi Beta Kappa distinguished visiting scholar, a fellow of the American Physical Society, a member of the California Academy of Sciences, a Guggenheim awardee, and a recipient of a Pew Scholars Prize in Conservation and the Environment. He is the author of five books, including The Green Fuse *and* Consider a Spherical Cow, *and over 100 journal articles in fields ranging from biodiversity and conservation to biogeochemistry, global warming, and sustainability. His current research focuses on three areas: ecological responses to global climate change, the relationship between ecological diversity and function, and the social role of ecological complexity.*

Global warming is a controversial political topic today, but a global network of the roughly 2,500 of the world's leading climate experts have reached a remarkable consensus about the underlying technical issues.

What Do Most Climate Scientists Agree On?

First, we know that over the past 200 years humanity has altered the composition of the atmosphere. In particular, we have loaded the air with gases such as carbon dioxide (up 30% since 1800, largely from combustion of coal, petroleum products, and natural gas) that trap heat and thus alter earth's climate. The basic mechanism is simple. Light from the sun largely penetrates the atmosphere and is absorbed at the planetary surface. There, it is converted from energy in the form of light to energy in the form of heat. As the surface temperature rises because of this heat, the earth radiates more and more heat back out to space, thereby maintaining an energy balance. But if the amount of heat-absorbing gas in the atmosphere rises, then some of the escaping heat is trapped and reflected back to the planetary surface, where it causes a further rise in temperature.

This is called the greenhouse effect and gases such as carbon dioxide are called greenhouse gases; other greenhouse gases whose concentrations have increased because of human activity are methane, nitrous oxide, and chlorofluorocarbons (which also are responsible for the thinning of the ozone layer).

Second, we know from temperature records maintained at many hundreds of sites around the world that the average global surface temperature has risen by about 0.6°C (1°F) during the past 100 years. It hasn't been a steady warming but the overall trend is upward; during the period from 1940 to 1970, the temperature actually fell a little. A warming of 0.6°C (1°F) is actually a substantial rise; the odds that this could be caused by natural variation in the strength of sunlight or by other natural causes are small.

This historic warming has not been uniform; it is greater at the poles than at the equator, it is greater at night than during the day, and it is much greater low in the atmosphere than it is higher up. All three of these patterns match quite well the predictions from mathematical models used to project the future strength of the greenhouse effect [Spotlight, p. 504], and this gives us confidence that the models are fairly reliable. They are by no means perfect, however, and much remains to be learned about the climate system. Most importantly, our understanding of clouds—their formation and their influence on climate—is less than adequate. That is why scientists give a range of predictions about the future climate change.

Third, if current trends in fossil fuel use and other sources of greenhouse gases continue, then around the middle of the 21st century the rise in the average surface temperature is predicted to be 1–3.5°C (1.8–6.3°F). About 20,000 years ago, when the earth was in the grip of the last ice age, the average surface temperature was roughly 5.6°C (10°F) colder than it is today. Do you see why this suggests that the projected warming is likely to have larger consequences?

What Are the Projected, Although Somewhat Speculative, Consequences of Global Warming?

Climatic extremes (drought, deluge, summer heat waves) are expected to be more common and possibly more intense. This has obvious implications for human health and welfare, prompting some insurance companies to rethink their policies in this area of risk.

Although total precipitation may increase in parts of the world, increased temperatures are likely to lead to drier soil conditions, which will stress our capacity to produce food and perhaps alter many of the earth's natural ecosystems. Moreover, winter snowfall in many parts of the world is likely to be replaced by rain, resulting in a reduced snowpack and earlier snowmelt; because agriculture in much of the western United States depends on the snowpack for irrigation water in summer, this could add to the stress on food production.

A rise in sea level is also projected, although the rate of rise is still conjectural. Most estimates suggest that the rise will be slight during the next 50 years but that within 100 or 200 years it could be as much as several meters. The effects of this will be felt particularly on small island nations in the South Pacific, on large areas in places such as Bangladesh and South Florida that lie near sea level, and on near-shore ecosystems and shoreline construction around the world.

Yet another possible consequence of global warming is the spread of tropical diseases [Section 17-5] to temperate-zone nations such as the United States. The reason is that insects that spread certain tropical diseases will be able to survive in the warmer climate of the temperate zones, where they are now virtually nonexistent.

Q: How long will the world's proven reserves of coal last at current consumption rates?

What Remedial Actions Might Society Take to Reduce the Risk?

The following remedial actions could be taken:

- Promote the development of affordable and clean solar energy [Chapter 16] through increased support of research and development and economic incentives. Economists sometimes refer to this as leveling the playing field because many nonsolar energy industries already receive large subsidies and incentives from governments. In contrast to the burning of coal, oil, and gas, solar energy results in very low greenhouse gas emissions (producing the cement for solar collector bases would release some carbon dioxide). The cost of solar energy is dropping fast and with some additional nurturing it could become competitive with conventional energy in the coming decades [Figure 15-21].
- Promote greater efficiency in the use of energy. It was once widely believed that a nation's economy could grow only if energy consumption grew. But during the late 1970s and early 1980s, energy use in the United States remained fairly constant, yet the economy grew. The reason was that people began driving more fuel-efficient cars, insulating their homes better, improving the efficiency of factories, and using more efficient electric appliances. The opportunities to further reduce energy consumption without sacrificing lifestyle are considerable [Sections 16-1 and 16-2].
- Reduce deforestation worldwide [Chapter 24]. Deforestation is the second most important source (after fossil fuel use) of carbon dioxide to the atmosphere. The reason is that forests store a large amount of carbon in the living trees; when the trees are cut and either burned or allowed to rot, the carbon is converted to carbon dioxide.

There is a growing consensus among economists and scientists that taking the first two steps would in the long run not only reduce the threat of global warming, but also strengthen our economy and save consumers money, reduce dependence on unstable fuel supplies, reduce urban smog, and reduce other hazards associated with fossil fuel use, such as coal mining accidents and oil spills. For that reason, taking the first two steps is sometimes called a no-regrets policy; even if global warming predictions are wrong, we wouldn't regret developing clean energy and learning to use less energy to achieve the same economic goals.

What Are Some Global Warming Myths?

As of this writing, the political battle over whether to take remedial actions to reduce the threat of global warming is heating up. An intense and well-publicized attack on the current consensus of climate scientists is under way. Among the most pervasive myths and distortions accompanying these efforts are the following.

- The historic global warming trend over the past 100 years is the result of an increasingly brighter sun. In fact, no mechanism is known today that could convert the observed very slight changes in solar output into a warming trend consistent with observations.
- Satellite data disprove the hypothesis that the historic warming is caused by greenhouse gases. In fact, the satellite data cover too short a time period to disprove current climate models, but the available satellite data (mostly on upper atmosphere temperatures) are reasonably consistent with model projections.
- Most of the warming occurred early in the century but the dramatic change in the atmosphere occurred more recently. In fact, the warming trend since 1970 has been even more dramatic than that early in the century, consistent with the fact that the rate of buildup of greenhouse gases is increasing.
- We will welcome global warming because an ice age is coming. Yes, another ice age will undoubtedly come, but not in the next few hundred years; perhaps over the next several thousand years, the earth will cool, but on millennial time scales the warming episode that human activity is now triggering will long have dissipated as the greenhouse gases largely disappear from the atmosphere.
- A cooling trend during the period 1940–1970 contradicts our climate models. In fact, the cooling trend resulted from increasing levels of light-obscuring fine particles (from fuel burning) and dust in the atmosphere during that period; although climate modelers initially failed to include that in the analyses, when the particles are included, the model predictions are consistent with observations.

The very nature of science is distorted by this debate. Scientists can never know with absolute certainty that a theory is true; all they can do is disprove a theory by making observations that contradict the theory.

By arguing that gaps in our current knowledge of global warming (of which there are many) obviate action on the things we are fairly confident about, those who oppose taking action undermine the public's understanding of the scientific process. This will not help future generations, who increasingly will need to base public policy on the best available scientific information. The way we deal with the threat of climate change will determine society's skill at blending science and policy in the decades ahead.

Critical Thinking

1. Can you explain why the greenhouse effect is actually a different warming mechanism than that which actually occurs in a glass greenhouse?

2. Do you believe that enhanced global warming from human activities is a serious problem? Explain.

A: About 220 years (65 years if use increases by 2% a year)

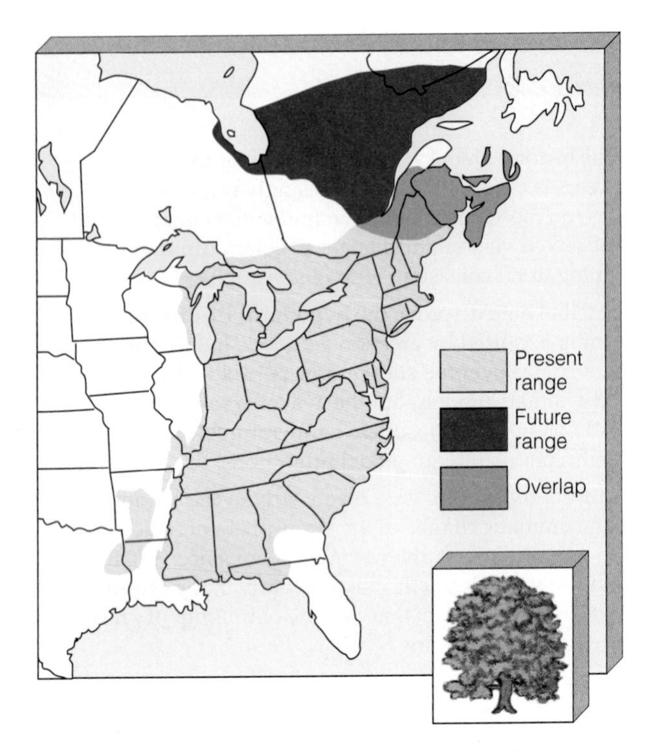

Figure 19-8 Possible effects of global warming on the geographic range of beech trees. According to one projection, if CO_2 emissions doubled between 1990 and 2050, beech trees (now common throughout the eastern United States) would survive only in a greatly reduced range in northern Maine and southeastern Canada. This is only one of a number of tree species whose geographic ranges would be drastically changed by global warming. (Data from Margaret B. Davis and Catherine Zabinski, University of Minnesota)

beech (Figure 19-8) might not be able to migrate fast enough and would die out.

Also, species that already live at high latitudes and on mountaintops would have nowhere to go, and many would become extinct. Ecologists at the University of Vienna have found that over the past 90 years some species of plants have moved up the Alps at roughly 3.7 meters (12 feet) per decade.

According to the IPCC, over the next century "entire forest types might disappear, including half of the world's dry tropical forests." Such forest diebacks would release carbon stored in their biomass and in surrounding soils and accelerate global warming (positive feedback).

Oregon State University scientists project that drying from global warming could cause massive *wildfires* in up to 90% of North American forests. If widespread fires occurred, large numbers of homes and large areas of wildlife habitats would be destroyed. Huge amounts of CO_2 would be injected into the atmosphere, which would accelerate global warming.

Climate change would lead to *reductions in biodiversity* in many areas (Connections, right). Large-scale forest diebacks would cause mass extinction of plant and animal species that couldn't migrate to new areas. Fish would die as temperatures soared in streams and lakes and as lowered water levels concentrated pesticides. Any shifts in regional climate would threaten many parks, wildlife reserves, wilderness areas, wetlands, and coral reefs, wiping out many current efforts to stem the loss of biodiversity (Chapter 25).

What Could Happen to Sea Levels? In a warmer world, sea levels are expected to rise, primarily because ocean water expands when heated—and to a lesser degree because some land-based glacial ice will melt. Melting of floating ice won't raise sea levels because the ice has already displaced ocean water, just as ice cubes melting in a glass of water don't raise the water level. However, if you add more ice cubes (analogous to glacial ice on land sliding into the water or melting), the water level rises.

According to current climate models, global sea levels are projected to rise by 15–95 centimeters (6–37 inches) between 1990 and 2100, with a best estimate of 48 centimeters (19 inches). This might seem like a trivial gain, but many of the world's major cities are located close to sea level; some are built below sea level and are protected from flooding by expensive levees, dikes, and other devices.

Even the best estimate of a 48-centimeter (19-inch) rise projected to occur by 2100 would overwhelm these defenses. Coastal regions where about one-third of the world's people and economic infrastructure are concentrated would be flooded. Most or all of low-lying islands such as the Cook and Marshall Islands in the Pacific, the Maldives, and some Caribbean island nations would be covered with water and disappear. Agricultural lowlands and deltas in parts of Bangladesh (p. 334), India, and China (where much of the world's rice is grown) would also be flooded. Much of the beaches on the U.S. East Coast might disappear within 25–50 years.

By spending trillions of dollars in building levees, dikes, and other flood control devices, wealthy countries could prevent much of this flooding. It is estimated that the Netherlands alone would have to spend $3.5 trillion (10 times its current annual GNP) to build its dikes high enough to prevent flooding by a rising North Sea. Most poorer countries won't have enough money to erect such defenses.

The projected modest increase in global sea level would also place the world's already dwindling fisheries at risk because of flooding of coastal estuaries, wetlands, and coral reefs. Rising sea levels would also move barrier islands farther inland, accelerate coastal erosion, contaminate coastal aquifers with salt water, and flood tanks storing oil and other hazardous chemicals in coastal areas.

Q: How long will proven reserves of coal in the United States last at current consumption rates?

If warming at the poles caused ice sheets and glaciers to melt even partially, the global sea level would rise far more. One comedian jokes that he plans to buy land in Kansas because it will probably become valuable beach-front property; another boasts that she isn't worried because she lives in a houseboat—the "Noah strategy."

How Might Weather Extremes Change? As more heat is retained in the earth's climate system, more air will move across the earth's surface. This will produce higher wind speeds, more clashing warm and cold fronts, and generally more violent weather conditions in parts of the world.

Some analysts project that this will increase the *intensity* of damaging hurricanes, typhoons, tornadoes, and violent storms in various areas. It is possible that the *number* of hurricanes, typhoons, tornadoes, and violent storms might also increase because of global warming, but currently there appears to be no evidence to support this hypothesis.

Any major increase in weather extremes could bankrupt the insurance and banking industries. Along the eastern U.S. coastline alone, some *$2 trillion* worth of insured real estate is at risk of flooding or other damages by storms and rising sea levels. Between 1990 and 1998, the global insurance industry paid out $159 billion in weather-related claims, compared to $83 billion during the 1980s. In 1998, weather-related disasters killed an estimated 32,000 people and displaced 300 million—more than the population of the United States.

A growing number of the world's insurance companies are sharply raising their premiums for coverage of damages from such events. Some companies are dropping all coverage of such risks, especially in coastal areas, wildfire-prone regions, and valleys with high risks of flooding.

Executives of some insurance companies are also pressuring government leaders to get more serious about slowing possible global warming. In November 1995, forty insurance companies (none of them American) signed a declaration, developed by the UN Environment Programme, committing themselves to help reduce the threat of global climate change. They are exerting political pressure on governments and spurring improvements in energy efficiency and increased use of renewable energy through their procurement and investment policies.

How Might Human Health Be Affected by Global Warming? According to the 1995 IPCC report, global warming would bring more heat waves. This would double or triple heat-related deaths among the elderly and people with heart disease; it would also increase suffering from respiratory ailments such as asthma and bronchitis.

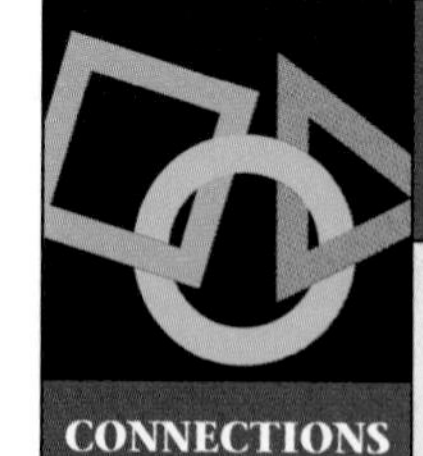

CONNECTIONS

Global Warming, Kirkland's Warblers, and Adélie Penguins

Projected global warming may sharply reduce populations of some species, especially those with specialized niches. For example, Kirkland's warbler, an endangered bird that nests exclusively under young jack pines in northern Michigan, would probably become extinct. The reason is that if northern Michigan ends up with less rain and more heat, jack pines there will probably die off in the next 30–90 years.

Another example involves Adélie penguins (Figure 4-17) that live in the western Antarctic (Figure 14-12). The western Antarctic has warmed up 4–5°C (8–9°F) in the last 50 years. During the past 22 years, University of Montana ecologist William Fraserhas observed a 40% drop in the population of this penguin species.

He suggests that sea ice melted by warmer temperatures has reduced populations of shrimplike krill (Figure 4-17) that are the Adélie's favorite food. Apparently populations of marine plankton that the krill eat have been dropping, perhaps because of melting of the sea ice and increased ultraviolet radiation because of drastic drops in concentrations of protective ozone in the stratosphere above Antarctica for several months each year (Figure 19-13). In addition, the penguins' ability to produce young may be decreased by increased spring snowfall (caused by warmer, moisture-laden air) that buries the Adélie's eggs under snowbanks.

Fraser and other scientists believe that the penguins' home is one of the first places on the earth to experience the effects of global warming that will eventually spread to temperate areas. If this hypothesis is correct, the plight of Adélie penguins serves as an early warning of changes we may experience on a much larger scale.

In 1997, 2,400 scientists signed an Ecologists' Statement on the Consequences of Rapid Climate Change that was delivered to U.S. President Bill Clinton. The statement included the following:

We are performing a global experiment on our natural ecosystems for which we have little information to guide us.... The rate of projected change is enough to threaten seriously the survival of many species.... While plant and animal communities may be able to eventually adapt to a stable climate system that is warmer than the existing one, many species may not be able to survive a transition to that new climate.

Critical Thinking

Give another species with a specialized niche whose population size could be sharply reduced because of projected global warming.

A: About 300 years

A warmer world would also disrupt supplies of food and fresh water, displacing millions of people and altering disease patterns in unpredictable ways. The spread of warmer and wetter tropical climates from the equator would bring malaria (Figure 17-10), encephalitis, yellow fever, dengue fever, and other insectborne diseases to formerly temperate zones. Scientists also project that higher ocean temperatures could trigger algal blooms, leading to cholera epidemics.

Higher humidity levels would also cause a rise in fungal skin diseases, yeast diseases, and prickly heat and heat rash (both of which impair the skin's ability to lose heat and increase the thermal stress on an overheated person). Atmospheric warming also affects the respiratory tract by increasing air pollution in winter and by increasing exposure to dusts, pollens, and smog in summer.

Warmer and more humid conditions would also speed the growth of bacteria and molds on stored foods, increasing spoilage and some types of food poisoning. Sea-level rise could spread infectious disease by flooding coastal sewage and sanitation systems.

According to Norman Myers (Guest Essay, p. 530), climate change is a major threat to global environmental security and would lead to a large number of *environmental refugees*. By 2050, he warns that global warming could produce 50–150 million environmental refugees (compared to 7 million war refugees in Europe after World War II). Most of these refugees would illegally migrate to other countries, causing much social disorder and political instability. Thus, global warming has serious implications for the foreign, military, and economic security policies of nations.

19-4 SOLUTIONS: DEALING WITH THE THREAT OF GLOBAL WARMING

Should We Do More Research or Act Now? There are three schools of thought concerning global warming. A *very small* group of scientists (many of them not experts in climate research or heavily funded by the oil and coal industries) contend that global warming is not a threat; some popular press commentators and writers even claim that it is a hoax. Widespread reporting of this *no-problem* minority view in the media has clouded the issue, cooled public support for action, and slowed international negotiations to deal with this threat.

A second group of scientists and economists believe we should wait until we have more information about the global climate system, possible global warming, and its effects before we take any action. Proponents of this *waiting strategy* question whether we should spend hundreds of billions of dollars phasing out fossil fuels and replacing deforestation with reforestation (and in the process risk disrupting national and global economies) to help ward off something that might not happen. They call for more research before making such far-reaching decisions.

A third group of scientists and economists point out that greatly increased spending on research about the possibility and effects of global warming will not provide the certainty decision makers want because the global climate system is so complex. These scientists urge us to adopt a *precautionary strategy*. They believe that when dealing with risky and far-reaching environmental problems such as possible global warming, the safest course is to take informed preventive action *before* there is overwhelming scientific knowledge to justify acting.

Economists at a 1997 meeting of the American Economics Association, led by Nobel laureates Kenneth Arrow and Robert Solow, declared, "As economists, we believe that global climate change carries with it significant environmental, economic, social, and geopolitical risks and that preventive steps are justified." In 1997, CEOs of several major oil companies, including British Petroleum, Royal Dutch Shell, and Sun Oil, joined most insurance company executives and expressed the view that there was enough evidence that human activities were contributing to global warming to begin taking precautionary action.

Those who favor doing nothing or waiting before acting point out that there is a 50% chance that we are *overestimating* the impact of rising greenhouse gases. However, those urging action point out that there is also a 50% chance that we are *underestimating* such effects.

If global warming does occur as projected, it will take place gradually over many decades until it crosses thresholds and triggers obvious and serious effects. By then it will be too late to take corrective action. In the early stages of projected global warming it will be easy for people to deny that anything serious is happening. Psychologist Robert Ornstein calls this denial the *boiled frog syndrome*. He describes it as like trying to alert a frog to danger as it sits in a pan of water very slowly being heated on the stove. If the frog could talk it would say, "I'm a little warmer, but I'm doing fine." As the water gets hotter, we would warn the frog that it will die, but it might reply, "The temperature has been increasing for a long time, and I'm still alive."

Eventually the frog dies because it has no evolutionary experience of the lethal effects of boiling water and thus cannot perceive its situation as dangerous. Like the frog, humans face a possible future without precedent, and our senses are unable to pick up warnings of impending danger.

Q: What percentage of the world's electricity is produced by burning coal?

Suppose that the threat of global warming does not materialize. Should we take actions that will cost enormous amounts of money and create political turmoil based on a cloudy crystal ball? Some say that we should take the actions needed to slow global warming even if there were no threat because of their important environmental, health, and economic benefits (Solutions, right). For example, a reduction in the combustion of fossil fuels, especially coal, will lead to sharp reductions in air pollution that prematurely kills and harms large numbers of people, lowers food and timber productivity, and decreases biodiversity (Section 18-5). This *no-regrets strategy* is an important part of the precautionary strategy.

How Can We Slow Possible Global Warming? According to the 1995 IPCC report, stabilizing CO_2 levels at the current level would require reducing current global CO_2 emissions by 66–83%. This is a highly unlikely and politically charged change. The International Energy Agency projects that CO_2 emissions will increase by nearly 50% between 1990 and 2010, with most of the increase coming from developing countries. According to climate models, even stabilizing CO_2 concentrations at 450 ppm requires cutting CO_2 emission by more than half.

Figure 19-9 presents a variety of solutions analysts have suggested to slow possible global warming; none of these solutions is being vigorously pursued. The quickest, cheapest, and most effective way to reduce emissions of CO_2 and other air pollutants over the next two to three decades is to use energy more efficiently (Section 16-2 and Solutions, right).

Some analysts call for increased use of nuclear power (Section 15-5) because it produces only about one-sixth as much CO_2 per unit of electricity as coal. Other analysts argue that the danger of large-scale releases of highly radioactive materials from nuclear power-plant accidents and the very high cost of nuclear power make it a much less desirable option than improving energy efficiency and relying more on renewable energy resources (Chapter 16).

Using natural gas (Section 15-3) could help make the 40- to 50-year transition to an age of energy efficiency and renewable energy. When burned, natural gas emits only half as much CO_2 per unit of energy as coal (Figure 15-12), and it emits far smaller amounts of most other air pollutants. Shifting from high-carbon fuels such as coal to low-carbon fuels such as natural gas could reduce CO_2 emissions by as much as 40%. However, without effective maintenance, more reliance on natural gas can increase inputs of methane (a potent greenhouse gas, Figure 19-2c) from leaking tanks and pipelines and thus increase global warming.

Energy Efficiency to the Rescue

SOLUTIONS

According to energy expert Amory Lovins (Guest Essay, p. 433), *the major remedies for slowing possible global warming are things we should be doing already even if there were no threat of global warming*. He argues that if we waste less energy, reduce air pollution by cutting down on our use of fossil fuels and by switching to renewable forms of energy, and harvest trees sustainably, we and other forms of life—in this and in future generations—would be better off even if these actions had nothing to do with global climate.

According to Lovins, improving energy efficiency (Sections 16-1 and 16-2) would be the fastest, cheapest, and surest way to slash emissions of CO_2 and most other air pollutants within two decades, using existing technology. He estimates that increased energy efficiency would also save the world up to $1 trillion per year in reduced energy costs—as much as the annual global military budget.

Using energy more efficiently would also reduce pollution, help protect biodiversity, and deter arguments among governments about how CO_2 reductions should be divided up and enforced. This approach would also make the world's supplies of fossil fuel last longer, reduce international tensions over who gets the dwindling oil supplies, and allow more time to phase in renewable energy.

According to a 1990 government report, controlling emissions of greenhouse gases in the United States will cost about $10 billion per year over the next century, for a total of $1 trillion. Cutting annual U.S. oil imports by 20% by wasting less energy would cover the annual $10 billion projected costs of reducing greenhouse gas emissions.

To Lovins and most environmentalists, greatly improving worldwide energy efficiency *now* is a money-saving, life-saving, biodiversity-saving, win–win, no-regrets proposition that we should not refuse, even if climate change were not an issue. Atmospheric scientist Fred Singer, a strong critic of global warming projections (as well as of ozone depletion), agrees.

Critical Thinking

1. Do you agree that improving energy efficiency **(a)** is an important way to reduce the input of CO_2 into the atmosphere and **(b)** should be done regardless of its impact on the threat of global warming?

2. Why do you think there has been little emphasis on improving energy efficiency? Explain. (See Sections 16-1 and 16-2.)

A: 64% (57% in the United States)

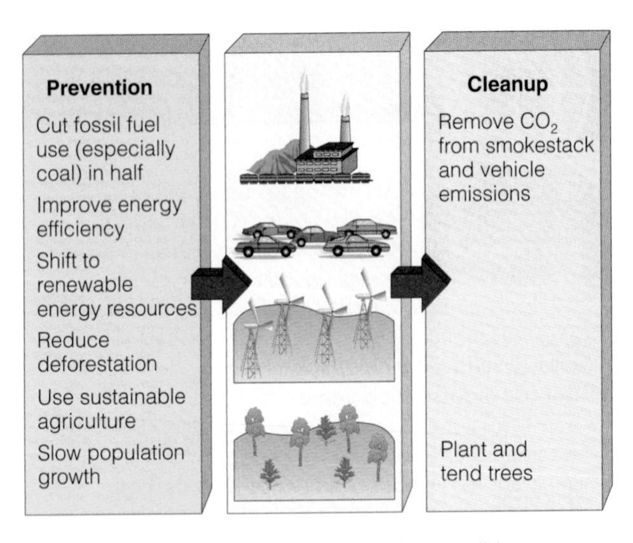

Figure 19-9 Solutions: methods for slowing possible global warming.

One method for reducing CO_2 emissions would be to phase out government subsidies for fossil fuels over a decade and gradually phase in *carbon taxes* on fossil fuels (especially coal and gasoline) based on their emissions of CO_2 and other air pollutants. To be politically feasible, analysts warn that these consumption tax increases should be matched by declines in taxes on income, labor, or capital. In 1997 more than 2,500 economists, including eight Nobel laureates, signed a statement **(1)** stating that sound economic analysis shows that greenhouse emissions can be cut without harming American living standards and **(2)** calling for carbon taxes as part of an international system of tradeable permits for greenhouse gas emissions.

Ideally, such tax revenues would be used to improve the energy efficiency of dwellings and heating systems for the poor in developed countries and developing countries, provide them with enough energy to offset higher fuel prices, and subsidize the transition to improved energy efficiency and renewable energy resources. Economists argue that such a carbon tax is based on the *polluter-pays* principle, which requires industries and consumers to pay directly for the full environmental costs of the fuels they use.

Another approach is to agree to global and national limits on greenhouse gas emissions and allow industries and countries to sell and trade emission permits in the marketplace. Companies and nations exceeding their limits could reduce their emissions or buy emissions permits from other companies or countries that have reduced their emissions below their allowed limits. Some believe that this approach is more politically feasible than carbon taxes.

Reducing deforestation (Sections 24-4 and 25-3) and switching to more sustainable agriculture (Section 12-6) would reduce CO_2 emissions and help preserve biodiversity. According to most analysts, slowing population growth (Section 11-3) is also crucial. If we cut per capita greenhouse gas emissions in half but world population doubles, we're back where we started. Some analysts argue that it is vital to global environmental security (Guest Essay, p. 530) that developed countries transfer energy efficiency, renewable energy, pollution prevention, and waste reduction technologies to developing countries as soon as possible.

It has also been suggested that we remove CO_2 from the exhaust gases of fossil fuel–burning vehicles, furnaces, and industrial boilers. However, available methods can remove only about 30% of the CO_2, and using them would double the cost of electricity.

Some call for a massive global reforestation program as a strategy for slowing global warming. However, some studies suggest that such a program (requiring each person in the world to plant and tend an average of 1,000 trees every year) would offset only about 3 years of our current CO_2 emissions from burning fossil fuels. Moreover, once the planted trees grow to maturity their net uptake of CO_2 is greatly reduced, and when they die and decompose, the CO_2 they removed from the atmosphere is returned by the respiration of decomposers.

However, a global program for planting and tending trees would help restore deforested and degraded land and reduce soil erosion, water pollution, and loss of biodiversity. Furthermore, a recent study suggests that forests store much more carbon than previously thought, primarily in peat and other organic matter in soils.

According to the IPCC, a global effort devoted to protecting existing forests, increasing the rate of reforestation, improving management of commercial forests, and raising the efficiency of wood-product use could offset as much as 15% of the excess CO_2 projected to be emitted between 1995 and 2050. The best potential for increasing uptake of CO_2 from the atmosphere is in tropical forests (80%), followed by temperate (17%) and boreal (3%) forests. Studies show that this would be a fairly inexpensive way to temporarily slow the buildup of CO_2 in the atmosphere.

Can Technofixes Save Us? Some scientists have suggested various *technofixes*: technological solutions for dealing with possible global warming. They include the following:

- Adding iron to the oceans to stimulate the growth of marine algae, which would remove more CO_2 through photosynthesis. Experiments done on small

Q: What percentage of global emissions of carbon dioxide come from burning coal?

patches of ocean suggest that if this were done on a continuous basis by several hundred ships over an area of ocean the size of Asia, it might offset only about one-third of global CO_2 emissions, at an estimated cost of $10–110 billion per year.

- Unfurling gigantic foil-surfaced sun mirrors in space to reduce solar input. The mirrors would have to be replaced frequently at great cost because of collisions with space debris. They would also cast undesirable shadows (like eclipses) on parts of the earth's surface.

- Injecting sunlight-reflecting sulfate particulates into the stratosphere to mimic the cooling effects of giant volcanic eruptions. However, this would also increase harmful air pollution in the troposphere and speed up destruction of ozone in the stratosphere.

Many of these costly schemes might not work, and most would probably produce unpredictable short- and long-term harmful environmental effects. Moreover, once started, those that work could never be stopped without a renewed rise in CO_2 levels. Instead of spending huge sums of money on such schemes, many scientists believe it would be much more effective and cheaper to improve energy efficiency (Solutions, p. 515) and shift to renewable forms of energy that don't produce carbon dioxide (Chapter 16).

What Has Been Done to Reduce Greenhouse Gas Emissions? At the 1992 Earth Summit in Rio de Janeiro, 106 nations approved a Convention on Climate Change in which developed countries committed themselves to reducing their emissions of CO_2 and other greenhouse gases to 1990 levels by the year 2000. However, the convention did not *require* countries to reach this goal and most will not achieve this goal.

In December 1997 more than 2,200 delegates from 161 nations met in Kyoto, Japan, to negotiate a new treaty to help slow global warming. The resulting treaty would

- Require 38 developed countries to cut greenhouse emissions to an average of 5.2% below 1990 levels between 2008 and 2012.

- Require European industrialized nations to lower greenhouse emission to 8% below 1990 levels. The United States would have to cut greenhouse emissions to 7% below 1990 levels and Japan would have to reduce its emissions to 6% below 1990 levels between 2008 and 2012. In 1996, U.S. greenhouse gas emissions were 9% above 1990 levels and by 2012 are projected to be 30% higher than they were in 1990.

- Not require developing countries to make any cuts in their greenhouse gas emissions because of an earlier treaty unless they choose to do so.

- Allow emissions trading in which a country that beats its goal for reducing greenhouse gas emissions can sell its excess reductions to countries that failed to meet their reduction goals.

- Allow forested countries to get a break in their quotas because trees absorb carbon dioxide.

- Allow penalties for countries that violate the treaty to be determined later.

Some analysts praise the agreement as a small but important step in dealing with the problem of potential global warming. However, there is strong opposition in the U.S. Senate for ratifying the treaty because developing countries are not required to meet any emission-reduction goals (Spotlight, p. 519). However, U.S. net emissions of greenhouse gases rose by 20% between 1990 and 1996 and global CO_2 emissions are projected to grow by 70% between 1995 and 2020.

In 1998, Costa Rica launched the world's first greenhouse gas trading project, following the guidelines set up by the Kyoto treaty. The government will raise funds to buy and protect more than 5,000 square kilometers (2,000 square miles) of rain forest (most of it privately owned) by selling companies in other countries allowances to emit CO_2 into the atmosphere. However, this scheme is viewed with suspicion by many in the developing world because they fear that rich industrialized countries will use their financial power to buy their way out of meeting their own emission goals.

In addition, climate scientists estimate that it would take at least a 60% reduction in global emissions of greenhouse gases below 1990 levels to slow projected global warming to an acceptable level—compared to the only 5.2% reduction goal set by the Kyoto treaty. Many analysts doubt that countries signing the agreement will even be able to meet their goals for this modest reduction.

Many energy specialists (Solutions, p. 515) and more than 2,000 economists say that we can cut CO_2 emissions to 30% below 1990 levels within 10–15 years by implementing existing methods for improving energy efficiency (Section 16-2) and at the same time save industries and consumers billions of dollars. Within another 10–15 years, energy experts project an additional 30% reduction in 1990 CO_2 levels by subsidizing and phasing in greatly increased use of renewable energy resources (Chapter 16).

A: About 42%

How Can We Prepare for Possible Global Warming? It seems clear that many (perhaps most) of the things climate experts have recommended (Figure 19-9) either will not be done or will be done too slowly. As a result, some analysts suggest that we should also begin to prepare for the effects of long-term global warming. Figure 19-10 shows some ways to do this.

Implementing the key measures for slowing or responding to climate change listed in Figures 19-9 and 19-10 will cost a lot of money. However, studies indicate that in the long run the savings would greatly exceed the costs. Some actions you can take to reduce the threat of global warming are given in Appendix 5.

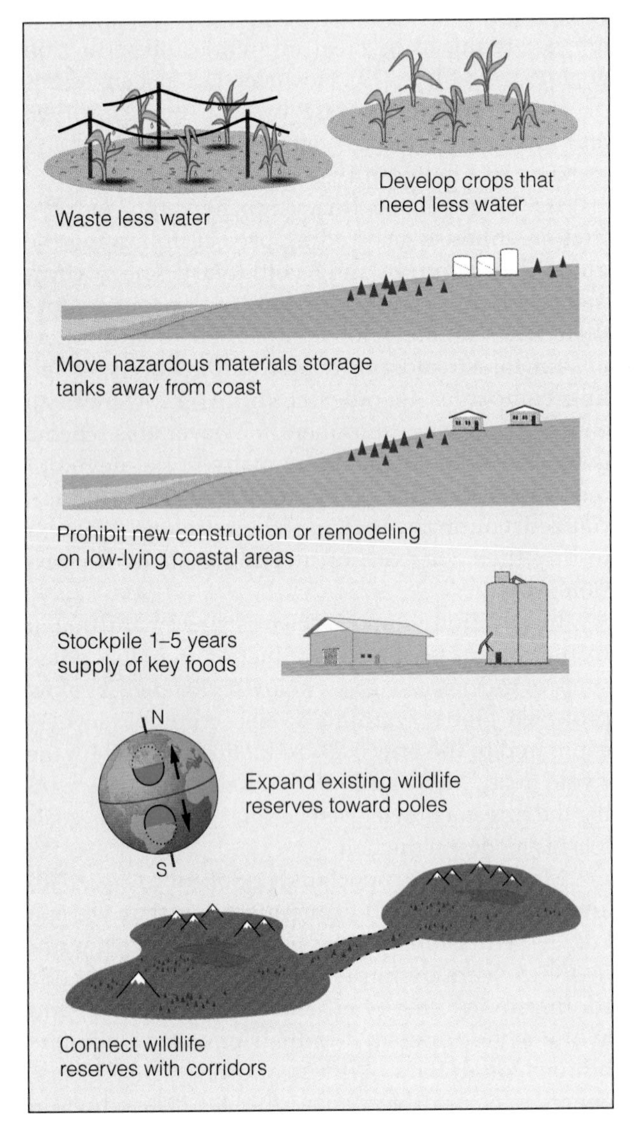

Figure 19-10 Solutions: ways to prepare for the possible long-term effects of global warming.

19-5 OZONE DEPLETION: IS IT A SERIOUS THREAT?

What Is the Threat from Ozone Depletion? Evolution of photosynthetic, oxygen-producing bacteria has produced a stratospheric global sunscreen: the *ozone layer*. Its presence for the past 450 million years has allowed life to develop and expand on land and in the surface layers of aquatic systems.

In a band of the stratosphere 17–26 kilometers (11–16 miles) above the earth's surface, oxygen is continuously converted to ozone and back to oxygen by a sequence of reactions initiated by ultraviolet radiation from the sun. The overall net equation for this reversible sequence of reactions is $3O_2 + UV \rightarrow 2O_3$. The result is a thin veil of renewable ozone at very low concentrations (up to 10 parts per million) that absorbs about 99% of the harmful incoming ultraviolet radiation from the sun and prevents it from reaching the earth's surface. Normally, the average levels of ozone in this layer don't change much because the rate of ozone destruction is equal to its rate of formation.

UV radiation reaching the stratosphere consists of three bands: A, B, and C. The ozone layer blocks out nearly all of the highest-energy, shortest-wavelength UV-C radiation, approximately half of the next highest-energy and biologically damaging UV-B band, but only a small part of the lowest-energy UV-A radiation (which can also damage living cells).

A handful of scientists dismiss the threat of ozone depletion from human-produced chemicals. Based on measurements and models (Figure 19-11), however, the overwhelming consensus of researchers in this field is that ozone depletion by certain chlorine- and bromine-containing chemicals emitted into the atmosphere by human activities is a serious long-term threat to human health, animal life, and the sunlight-driven primary producers (mostly plants) that support the earth's food chains and webs.

Ozone concentrations in the stratosphere have been measured since the mid-1960s at more than 30 locations around the world and also by satellites since 1970. These measurements show that during the 1980s normal ozone levels dropped 5–15% in winter above the temperate and tropical zones of both hemispheres—three times the losses measured in the 1970s. Globally, the earth lost an average of about 4% of its stratospheric ozone between 1979 and 1994. According to a 1995 report by prominent atmospheric scientists, average global ozone levels are projected to drop 7–13% during the 1990s.

Q: How many people in the United States die prematurely each year from coal burning?

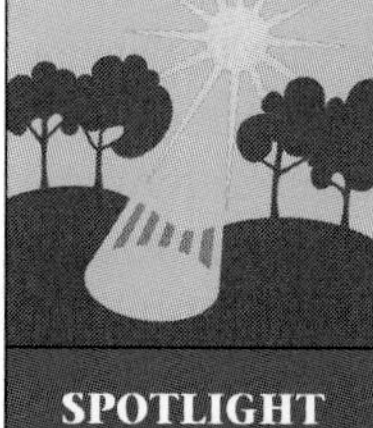

SPOTLIGHT

What Role Should Developing Countries Play in Reducing CO_2 Emissions?

There is controversy over what role developing countries should be required to make in reducing their CO_2 emissions. Currently, developing countries contribute about 40% of global CO_2 emissions. However, their total emissions are increasing exponentially by about 5% per year—a doubling time of only 14 years.

For example, if coal burning continues as expected, by 2025 China will emit three times more CO_2 than the current total of the United States, Japan, and Canada combined (even though China's per capita CO_2 emissions would still be less than those of Americans).

President Clinton and the majority of members of the U.S. Senate (which must ratify the climate treaty developed in Kyoto) argue that warming of the atmosphere by greenhouse gases is a global concern. Thus, they insist that all countries should share in the solution, especially because emissions by developing countries are expected to surpass those by developed countries in 20–30 years.

However, developing countries with a scarcity of money and technology don't like being told they must reduce their emissions so industrialized countries can continue to enjoy the benefits of their wasteful and polluting lifestyle, which has caused the threat of global warming.

Some leaders in developing countries are willing to commit to lower CO_2 emissions provided that any global treaty requires rich industrialized nations to help them leapfrog the fossil-fuel stage of economic development by transfer of more efficient energy and less-polluting technologies (Section 16-2) and technologies for relying on renewable energy resources (Chapter 16). Although some joint projects are being tried on a pilot basis, the Kyoto treaty does not require developed countries to make such a commitment.

At Kyoto and in most international environmental negotiations, leaders of developed countries refuse to talk about their overconsumption and leaders of developing countries don't want to talk about overpopulation. Thus, two major causes of environmental degradation and pollution (including greenhouse gas emissions) are not seriously considered.

In 1998, economist Herman Daly (Guest Essay, p. 774) suggested that this impasse might be overcome by all countries recognizing that total CO_2 emissions equals per capita emissions (typically high in developed countries) times population (typically high in many developing countries). Then all countries could agree to meet individual quotas for reducing CO_2 emissions by any combination of two ways: **(1)** reducing per capita emissions relative to some base or **(2)** keeping per capita emissions constant by reducing population growth relative to some base.

Critical Thinking

1. Do you believe that developing nations should agree to reduce their CO_2 emissions? Explain. If developing nations agree to this, should industrialized nations agree to transfer energy efficiency, pollution control, and renewable energy technologies to developing countries? Explain.

2. Do you agree or disagree with economist Herman Daly's proposal? Explain. Why do you think such an approach has been resisted so far? How might such resistance be overcome?

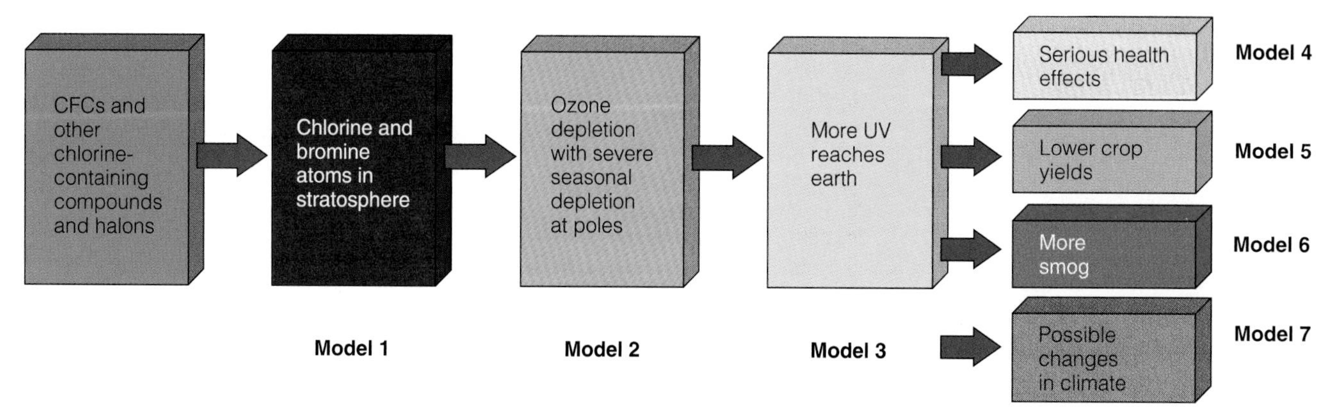

Figure 19-11 Generalized schematic of ozone depletion and models of its possible effects. Models 1–3 are widely accepted. The nature and extent of the effects in models 4–7 are still being studied.

A: 65,000 to 200,000

What Causes Ozone Depletion? From Dream Chemicals to Nightmare Chemicals This situation started when Thomas Midgley, Jr., a General Motors chemist, discovered the first chlorofluorocarbon (CFC) in 1930, and chemists then made similar compounds to create a family of highly useful CFCs. The two most widely used are CFC-11 (trichlorofluoromethane, CCl_3F) and CFC-12 (dichlorofluoromethane, CCl_2F_2), known by their trade name as **Freons**.

These chemically stable, odorless, nonflammable, nontoxic, and noncorrosive compounds seemed to be dream chemicals. Cheap to make, they became popular as coolants in air conditioners and refrigerators (replacing toxic sulfur dioxide and ammonia), propellants in aerosol spray cans, cleaners for electronic parts such as computer chips, sterilants for hospital instruments, fumigants for granaries and ship cargo holds, and bubbles in plastic foam used for insulation and packaging.

But CFCs were too good to be true. In 1974 calculations by chemists Sherwood Rowland and Mario Molina (building on earlier work by Paul Crutzen) indicated that CFCs were creating a global chemical time bomb by lowering the average concentration of ozone in the stratosphere. They shocked both the scientific community and the $28-billion-per-year CFC industry by calling for an immediate ban of CFCs in spray cans (for which substitutes were readily available).

Here's what Rowland and Molina found: Spray cans, discarded or leaky refrigeration and air conditioning equipment, and the production and burning of plastic foam products release CFCs into the atmosphere. Because these molecules are insoluble in water and are chemically unreactive, they are not removed from the troposphere. As a result—mostly through convection, random drift, and the turbulent mixing of air in the troposphere—they rise slowly into the stratosphere, taking 10–20 years to make the journey.

In the stratosphere, under the influence of high-energy UV radiation, these molecules break down and release highly reactive chlorine atoms, which speed up the breakdown of highly reactive ozone (O_3) into O_2 and O in a cyclic chain of chemical reactions (Figure 19-12). This causes ozone in the stratosphere to be destroyed faster than it is formed.

Each CFC molecule can last in the stratosphere for 65–110 years (depending on its type). During that time each chlorine atom released from these molecules can convert up to 100,000 molecules of O_3 to O_2 before it is removed from the stratosphere by forming HCl (which diffuses downward to the troposphere and is removed by rain). If Rowland and Molina's calculations and later models and atmospheric measurements of CFCs in the stratosphere are correct (as almost all scientists in this field believe), these dream molecules have turned into a nightmare of global ozone destroyers.

Although Rowland and Molina warned us of this problem in 1974, it took 15 years of interaction between the scientific and political communities before

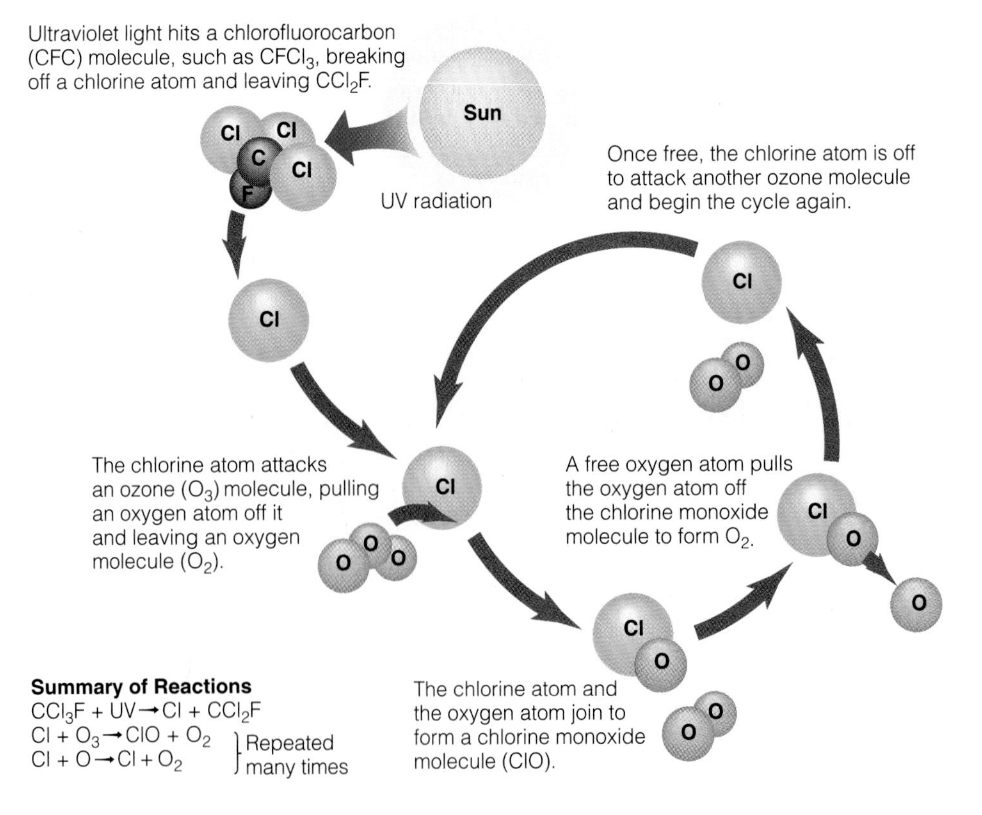

Figure 19-12 A simplified summary of how chlorofluorocarbons (CFCs) and other chlorine-containing compounds destroy ozone in the stratosphere. Note that chlorine atoms are continuously regenerated as they react with ozone. Thus, they act as *catalysts*, chemicals that speed up chemical reactions without themselves being used up by the reaction. Bromine atoms released from bromine-containing compounds that reach the stratosphere also destroy ozone by a similar mechanism.

Q: What would happen if coal's harmful effects were included in its market price and government subsidies were removed?

countries agreed to begin phasing out CFCs. The CFC industry (led by the Du Pont Company), a powerful, well-funded adversary with a lot of profits and jobs at stake, attacked Rowland and Molina. However, they held their ground, expanded their research, and explained the meaning of their calculations to other scientists, elected officials, and the media. It was not until 1988—14 years after Rowland and Molina's study—that Du Pont officials acknowledged that CFCs were depleting the ozone layer and agreed to stop producing them once they found substitutes.* In 1995, Rowland and Molina (along with Paul Crutzen) received the Nobel prize in chemistry for their work.

What Other Chemicals Deplete Stratospheric Ozone? CFCs are not the only ozone-eaters; other chemicals can release highly reactive chlorine and bromine atoms if they reach the stratosphere and are exposed to intense UV radiation. Collectively, all ozone-depleting compounds are called *ODCs*.

One group consists of long-lived bromine-containing compounds such as *halons* and *HBFCs*, both used in fire extinguishers. Another is *methyl bromide* (CH_3Br), a widely used fumigant. Another group consists of chlorine-containing compounds such as *carbon tetrachloride* (CCl_4), a cheap, highly toxic solvent, and toxic *methyl chloroform*, or 1,1,1-trichloroethane ($C_2H_3Cl_3$), used as a cleaning solvent for clothes and metals and as a propellant in more than 160 consumer products, such as correction fluid, dry-cleaning sprays, spray adhesives, and other aerosols. Another source of ozone depletion is the emission of hydrogen chloride (HCl) into the stratosphere by the U.S. space shuttle vehicles.

*For a fascinating account of how corporate stalling, politics, economics, and science can interact, see Sharon Roan's *Ozone Crisis: The 15-Year Evolution of a Sudden Global Emergency* (New York: John Wiley, 1989).

Why Is There Seasonal Thinning of Ozone over the Poles? At times the news about ozone loss has taken scientists by surprise. The first surprise came in 1984, when researchers analyzing satellite data discovered that 40–50% of the ozone in the upper stratosphere over Antarctica was being destroyed during the Antarctic spring and early summer (September–December), when sunlight returned after the dark Antarctic winter. Since then, this seasonal Antarctic *ozone thinning* (incorrectly called an ozone hole) has expanded in most years (Figure 19-13), with a typical seasonal loss of about 50% (but as high as 100% in some spots). In 1998, seasonal ozone thinning above Antarctica was the largest ever, surpassing the record set in 1993.

This pronounced seasonal loss of stratospheric ozone had not been predicted by computer models or detected by earlier satellite measurements. However, when NASA scientists looked at earlier satellite measurements they found that their computers had been programmed to regard very low ozone readings as errors. When they went back and looked at the earlier rejected data, they found that average seasonal ozone levels over Antarctica had been dropping since 1974 and that the problem had been getting worse (Figure 19-13).

The "smoking gun" considered by most atmospheric scientists to establish the link between ozone depletion and CFCs was found in 1987, when a

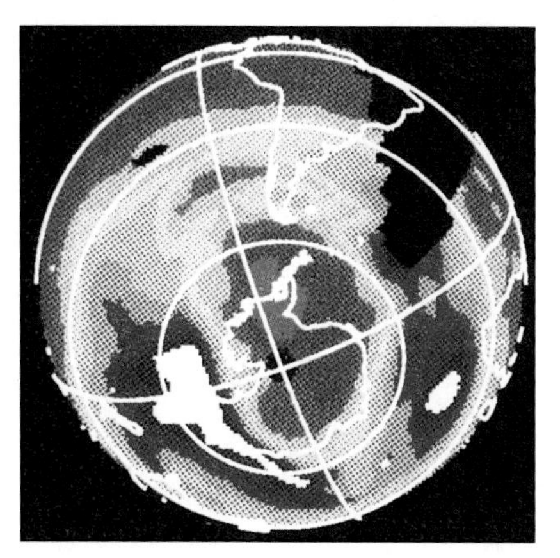

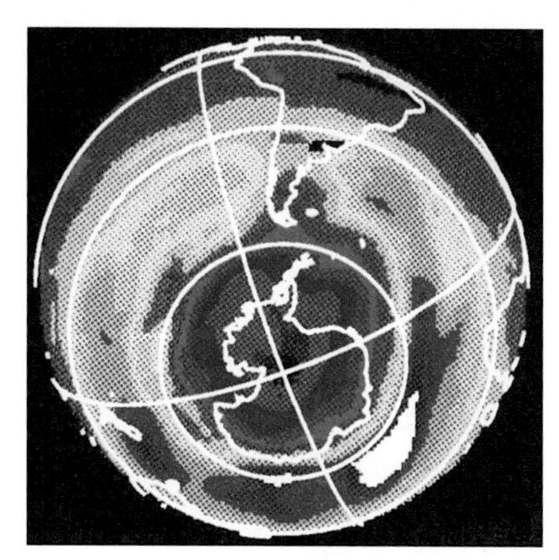

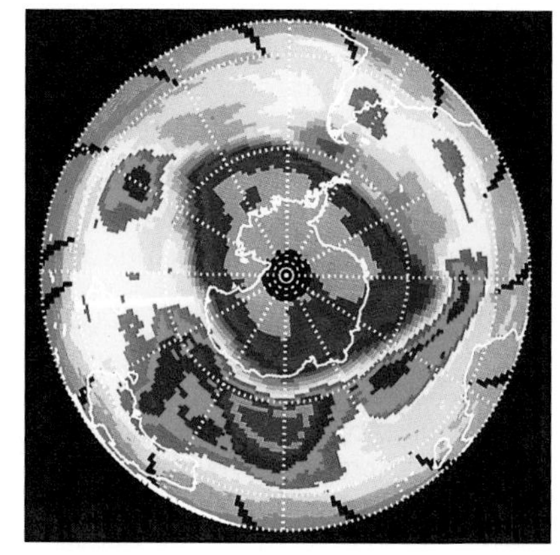

Figure 19-13 Seasonal thinning or loss of ozone (shown by shades of pink) in the upper stratosphere over Antarctica as measured by satellite on September 23 of 1979, 1991, and 1996. Since 1987 this area of seasonal thinning (in which the concentration of ozone is cut at least in half) has spanned an area over Antarctica larger than the continental United States or Europe. In 1998, the area above Antarctica affected by seasonal O_3 thinning was the largest ever, covering an area 2–5 times the size of Europe. Some good news is that a recent computer model suggests that this huge area of seasonal ozone thinning should not get any larger. It should also be decreasing in size over the next 50 years as ozone-depleting chemicals are phased out and those already in the atmosphere are eventually destroyed (NASA/GSFC and NOAA)

A: It would be too expensive to use and would be phased out for most uses research plane flew into the stratosphere above Antarctica and measured chlorine (as ClO derived primarily from CFCs) and ozone concentrations. As ClO concentrations rose, ozone concentrations dropped.

Measurements have indicated that CFCs are the primary culprits. Each sunless winter, steady winds blow in a circular pattern over the earth's poles. This creates a *polar vortex*: a huge swirling mass of very cold air that is isolated from the rest of the atmosphere until the sun returns a few months later. When water droplets in clouds enter this circling stream of extremely frigid air, they form tiny ice crystals. The surfaces of these ice crystals collect CFCs and other ozone-depleting chemicals in the stratosphere and speed up (catalyze) the chemical reactions that release Cl atoms and ClO. Instead of entering a chain reaction of ozone destruction (Figure 19-12), the ClO atoms combine with one another to form Cl_2O_2 molecules. In the dark of winter the Cl_2O_2 molecules cannot react with ozone, so they accumulate in the polar vortex.

When sunlight and the Antarctic spring return 2–3 months later (in October), the light breaks up the stored Cl_2O_2 molecules, releasing large numbers of Cl atoms and initiating the catalyzed chlorine cycle (Figure 19-12). Within weeks, this destroys 40–50% of the ozone above Antarctica (and 100% in some places). The returning sunlight gradually melts the ice crystals, breaks up the vortex of trapped south polar air, and allows it to begin mixing again with the rest of the atmosphere. Then new ozone forms over Antarctica until the next dark winter.

When the vortex breaks up, huge masses of ozone-depleted air above Antarctica flow northward and linger for a few weeks over parts of Australia, New Zealand, South America, and South Africa. This raises biologically damaging UV-B levels in these areas by 3–10%, and in some years by as much as 20%.

In 1988 scientists discovered that similar but less severe ozone thinning occurs over the north pole during the arctic spring and early summer (February–June), with a seasonal ozone loss of 10–38% (compared to a typical 50% loss above Antarctica). When this mass of air above the Arctic breaks up each spring, large masses of ozone-depleted air flow south to linger over parts of Europe, North America, and Asia.

So far, seasonal ozone loss over the north pole is much lower than that over the south pole. One reason is that air masses flowing toward and away from the Arctic pass alternately over both land and water instead of mostly water, as in the Antarctic (Figure 8-3). This produces intense atmospheric disturbances that cause the arctic vortex to wander, lurch, and be warmer and less stable than the Antarctic vortex.

However, the situation is changing. In 1992 atmospheric scientists warned that if rising levels of greenhouse gases warm the troposphere, the resulting insulating blanket could cool the stratosphere. Global warming changes wind patterns in the troposphere so that the Arctic becomes more isolated and stays colder in the winter, closer to what happens in the Antarctic. This could increase the size and duration of seasonal ozone-depleting vortexes over both poles, perhaps leading to large and regular arctic ozone thinning. As a result, ozone levels over parts of the northern hemisphere (including the United States) would decline sharply.

According to measurements by the U.S. National Oceanic and Atmospheric Administration (NOAA), atmospheric ozone over the northern hemisphere was depleted by 10–25% between 1979 and 1996. During the winters of 1996 and 1997, severe ozone loss was observed over the Arctic. According to a 1998 model developed by NASA scientists at the Goddard Institute for Space Studies, ozone destruction over the Arctic will be at its worst between 2010 and 2019 (Figure 19-14).

Is Ozone Depletion Really a Serious Problem? Political talk show commentator Rush Limbaugh, zoologist and former head of the Atomic Energy Commission Dixy Lee Ray (now deceased), physicist and climate scientist S. Fred Singer, and articles and books in the popular press have claimed that ozone depletion by CFCs is a hoax, or at least a vastly exaggerated problem. The Pro/Con box on the right summarizes some of their charges and the answers by prominent research scientists studying ozone depletion.

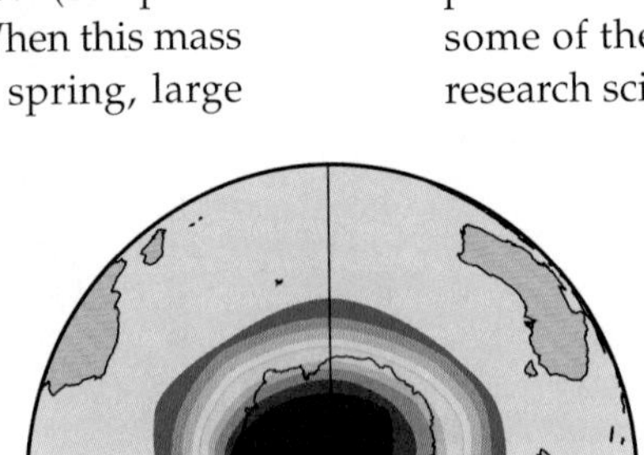
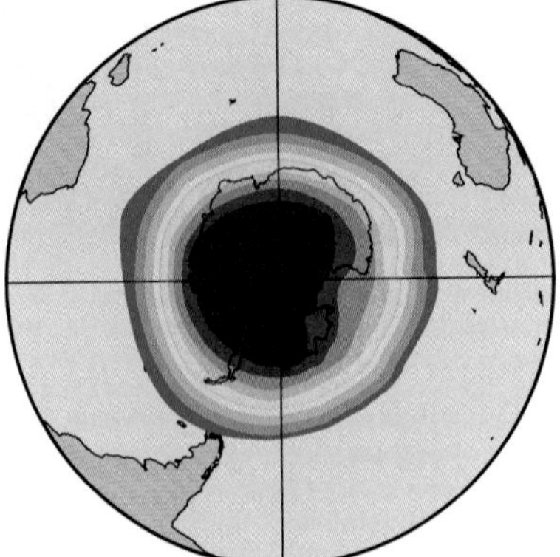
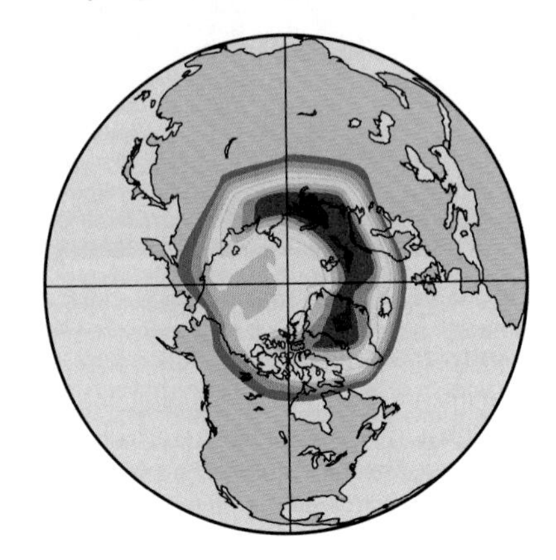

Figure 19-14 Projected total ozone loss, averaged over 2010–2019, during September for the Antarctic (left) and during March for the Arctic. According to the model used to make these projections, during this period the severity of ozone loss over the Arctic may approach that over the Antarctic. Dark red represents ozone depletion of 54% or more; light blue, 18–30%; and dark blue, 6–12%. (Data from NASA Goddard Institute for Space Studies)

Q: What percentage of the world's electricity is supplied by nuclear power?

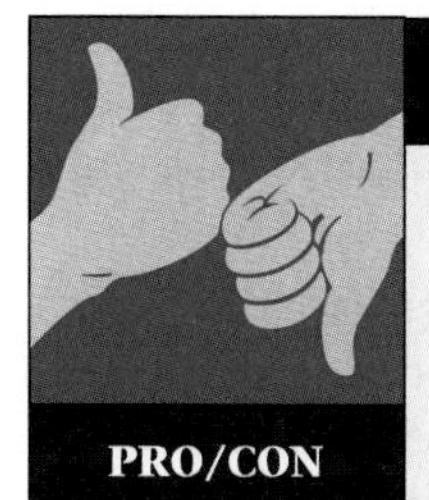

Is Ozone Depletion a Hoax?

Charge: There is no ozone hole, and the whole idea is merely a scientific theory.

Response: Technically, it is not correct to call this phenomenon an ozone *hole.* Instead, it is a thinning or partial depletion of normal ozone levels in the stratosphere. A scientific theory is a widely accepted idea or principle that has a very high degree of certainty because it has been supported by a great deal of evidence. Nothing in science can be proven absolutely, but there is overwhelming scientific evidence of ozone thinning in the stratosphere.

Charge: There is no scientific proof that an ozone hole exists.

Response: Ground-based and satellite measurements clearly show seasonal ozone thinning above Antarctica and the Arctic, and other measurements reveal a lower overall thinning everywhere except over the tropics.

Charge: CFC molecules can't reach the stratosphere because they are heavier than air, and no measurements have detected them in the stratosphere.

Response: The atmosphere is like a turbulent fluid in which gases, both lighter and heavier than air, are mixed thoroughly by the churning movements of large air masses. Since 1975 measuring instruments on balloons and satellites have clearly shown that CFCs are transported high into the stratosphere.

Charge: Sodium chloride (NaCl) from the evaporation of sea spray contributes more chlorine to the stratosphere than do CFCs.

Response: Particles of NaCl from sea spray (unlike CFCs) are soluble in water and are washed out of the lower atmosphere. Measurements have detected no sodium in the lower stratosphere.

Charge: Volcanic eruptions and biomass burning have added much more chlorine (Cl) to the stratosphere than have human-caused CFC emissions.

Response: Most of the water-soluble HCl injected into the troposphere from occasional large-scale volcanic eruptions is washed out by rain before it reaches the stratosphere. A serious calculation error by one of the skeptics (Dixy Lee Ray) greatly overestimated the amount of HCl injected into the stratosphere by recent volcanic eruptions. Measurements indicate that no more than 20% of the chlorine from biomass burning (in the form of methyl chloride, CH_3Cl) reaches the stratosphere; this amount of chlorine is about five times less than the current contribution from CFCs. Even Fred Singer (who is highly skeptical about some aspects of ozone depletion models) stated in 1993 that "CFCs make the major contribution to stratospheric chlorine." However, recent research indicates that aerosol particles of sulfate salts produced from sulfur dioxide emitted by volcanic eruptions can destroy stratospheric ozone for several years after an eruption. These volcanic aerosols, plus the human-related inputs of ozone-destroying chemicals, can seriously deplete ozone in the stratosphere.

Charge: Seasonal ozone thinning over Antarctica is a natural phenomenon because it appeared long before CFCs were in wide use.

Response: In the late 1950s, British scientist Gordon Dobson discovered that the Antarctic polar vortex leads to some seasonal ozone loss through natural causes. Measurements indicate that between 1956 and 1976 the natural pattern of slight ozone loss above Antarctica did not change significantly. Since 1976, however, seasonal losses of ozone above Antarctica have increased dramatically and have been linked by measurements and models to rising levels of CFCs in the stratosphere. A 1994 review of Dobson's 1958 data concluded that in 1958 there was no credible evidence for a significant ozone thinning (such as that observed since 1976) over Antarctica.

Charge: The best models of ozone production and destruction estimate that global ozone levels should be 10% less than we actually observe.

Response: Scientists recognize that their models must be improved. However, they point out that their models of ozone depletion by chlorine- and bromine-containing chemicals produced by human activities still account for about 90% of the observed ozone depletion.

Charge: The expected increase in UV-B radiation caused by stratospheric ozone loss has not been detected in urban areas in the United States and most other developed countries.

Response: Increased UV levels have been detected in urban areas in countries in the southern hemisphere such as Australia, New Zealand, South Africa, Argentina, and Chile (as well as in Toronto, Canada) that are exposed to ozone-depleted air drifting away from Antarctica after the seasonal polar vortex breaks up each year. Some ozone depletion experts hypothesize that significant ground-level increases in UV-B have not been observed in many urban areas (especially in industrialized countries in the northern hemisphere) because ozone-laden smog over most cities may filter out some of the UV-B. They argue that polluting our way out of increased UV-B levels in such cities is unacceptable; it will continue to cause premature death and widespread health problems for humans and damage trees and crops (Section 18-5).

Critical Thinking

Do you believe that ozone depletion in the stratosphere is a serious problem or one that has been greatly exaggerated? Explain.

A: 17% (6% of the world's total commercial energy use) in 1996

Why Should We Be Worried About Ozone Depletion? Life in the Ultraviolet Zone Why should we care about ozone loss? From a human standpoint the answer is that with less ozone in the stratosphere, more biologically damaging UV-B radiation will reach the earth's surface and give humans worse sunburns, more cataracts (a clouding of the lens that reduces vision and can cause blindness if not corrected) and more skin cancers (Connections, p. 526).

According to UN Environment Programme estimates, the additional UV-B radiation reaching the earth's surface resulting from an annual 10% loss of global ozone (already a likely possibility within a few years) could lead to 300,000 additional cases of squamous cell cancer (Figure 19-15, left) and basal cell cancer (Figure 19-15, center) worldwide each year, 4,500–9,000 additional cases of potentially fatal malignant melanoma (Figure 19-15, right) each year, and 1.5 million new cases of cataracts (which account for over half of the world's 25–35 million cases of blindness) each year.

Cases of skin cancer and cataracts are increasing in Australia, New Zealand, South Africa, Argentina, and Chile, where the ozone layer is very thin for several months each year. Australian television stations now broadcast daily UV levels, warning Australians (with the world's highest rate of skin cancer) to stay inside during bad spells and to protect themselves from the sun's rays with hats, clothing, and sunscreens when they go out during the daytime. These precautions are required by law for schoolchildren. Some newspapers and TV weather forecasts in the United States report the EPA and National Weather Service daily ultraviolet index.

Assuming that we phase out *all* ozone-destroying chemicals over the next three decades, the EPA estimates that projected ozone thinning during the 1980s and 1990s will lead to 12 million new cases of skin cancer and 200,000 additional skin cancer deaths in the United States alone over the next 50 years.

Other effects from increased UV exposure are

- Suppression of the immune system. This makes the body more susceptible to infectious diseases and some forms of cancer.
- An increase in eye-burning, highly damaging acid deposition (Section 18-3) and ozone in smog in the troposphere (Section 18-2).
- Lower yields of key crops such as corn, rice, cotton, soybeans, beans, peas, sorghum, and wheat, with estimated losses totaling $2.5 billion per year in the United States before the middle of the 21st century.
- A serious decline in forest productivity of the many tree species sensitive to UV-B radiation. This could reduce CO_2 uptake and enhance global warming.
- Increased breakdown and degradation of materials such as various types of paints, plastics, and outdoor materials. Such damage could cost billions of dollars per year.
- Reduction in the productivity of surface-dwelling phytoplankton, which could upset aquatic food webs, decrease yields of seafood eaten by humans, and possibly accelerate global warming by decreasing the oceanic uptake of CO_2 by phytoplankton.
- Damage to the ecological structure and function of lakes because of deeper penetration of UV light caused by a synergistic interaction between ozone depletion, global warming, and acid deposition (Section 18-3).

In a worst-case ozone depletion scenario, most people would have to avoid the sun altogether (see cartoon below). Even cattle could graze only at dusk, and farmers and other outdoor workers might need to limit their exposure to the sun to minutes. Fortunately, we are acting to prevent such a scenario.

Some critics who believe that the threat of ozone depletion has been exaggerated argue that we should not worry about a 10%, 20%, or even 50% increase in UV-B radiation reaching the earth's surface. According to Fred Singer, moving just 97 kilometers (60 miles) closer to the equator involves potential exposure to 10% more UV-B radiation.

Many scientists contend that this argument ignores two crucial points. First, the southward movement of the U.S. population (and people in many other countries) is one reason skin cancer rates have risen. Any increased exposure to UV-B radiation because of ozone depletion would add to this already serious health threat. Second, we are talking about the likely possibility of rapid (within decades), widespread, long-lasting, and essentially unpredictable ecological disruptions in species adapted to existing levels of background UV-B radiation.

©1988, *Los Angeles Times* Syndicate. Reprinted with permission. (Color added)

Kellner, Tomas. 1998. Cool Operators. *Sciences*, vol. 38, no. 5, 19(5).

Humans can quickly make cultural adaptations to increased UV-B radiation by staying out of the sun, protecting their skin with clothing, and applying sunscreens. However, plants and other animals that help support us and other forms of life can't make such changes except through the long process of biological evolution.

19-6 SOLUTIONS: PROTECTING THE OZONE LAYER

How Can We Protect the Ozone Layer? The scientific consensus of researchers in this field is that we should immediately stop producing all ozone-depleting chemicals. Even with immediate action, the models indicate that it will take 50–60 years for the ozone layer to return to 1975 levels and another 100–200 years for full recovery to pre-1950 levels.

Substitutes are already available for most uses of CFCs (Table 19-1), and others are being developed (Individuals Matter, p. 528). One substitute for CFCs is *hydrochlorofluorocarbons* or *HCFCs* (such as $CHClF_2$, containing fewer chlorine atoms per molecule than CFCs). Because of their shorter lifetimes in the stratosphere, these compounds should have only about 2.5% of the ozone-depleting potential of CFCs. If used in massive quantities, however, HCFCs would still cause ozone depletion and act as potent greenhouse gases. In addition, a 1996 study indicated that one widely

Table 19-1 CFC Substitutes

Types	Pros	Cons
HCFCs (hydrochlorofluorocarbons)	Break down faster (2–20 years). Pose about 90% less danger to ozone layer. Can be used in aerosol sprays, refrigeration, air conditioning, foam, and cleaning agents.	Are greenhouse gases. Will still deplete ozone, especially is used in large quantities. Health effects largely unknown. Expensive. May lower energy efficiency of appliances. Can be degraded to trifluoroacetate (TFA), which can inhibit plant growth in wetlands.
HFCs (hydrofluorocarbons)	Break down faster (2–20 years). Do not contain ozone-destroying chlorine. Can be used in aerosol sprays, refrigeration, air conditioning, and insulating foam.	Are potent greenhouse gases. Expensive. Safety questions about flammability and toxicity still unresolved. Can be degraded to trifluoroacetate (TFA), which can inhibit plant growth in wetlands. May lower energy efficiency of appliances. Production of HFC-134a, a refrigerant substitute, yields an equal amount of methyl chloroform, a serious ozone depleter.
Hydrocarbons (HCs) (such as propane and butane)	Cheap and readily available. Can be used in aerosol sprays, refrigeration, foam, and cleaning agents. Not patentable. Can be made locally in developing countries.	Can be flammable and poisonous if released. Some increase in ground-level air pollution.
Ammonia	Simple alternative for refrigerators; widely used before CFCs.	Toxic if inhaled. Must be handled carefully.
Water and steam	Effective for some cleaning operations and for sterilizing medical instruments	Creates polluted water that must be treated. Wastes water unless the used water is cleaned up and reused.
Terpenes (from the rinds of lemons and other citrus fruits)	Effective for cleaning electronic parts.	None.
Helium	Effective coolant for refrigerators, freezers, and air conditioners.	This rare gas may become scarce if use is widespread, but very little coolant is needed per appliance.

Hint: Enter the search term *ozone layer depletion* using the Subject Guide.

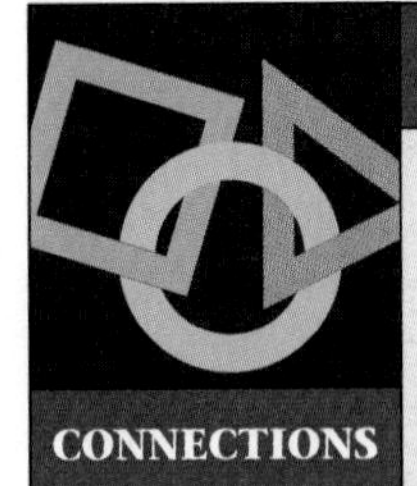

CONNECTIONS

The Cancer You Are Most Likely to Get

Considerable research indicates that years of exposure to UV-B ionizing radiation in sunlight is the primary cause of *squamous-cell* (Figure 19-15, left) and *basal-cell* (Figure 19-15, center) *skin cancers*, which together make up 95% of all skin cancers. Typically there is a 15- to 40-year lag between excessive exposure to UV-B and development of these cancers.

Caucasian children and adolescents who get only a single severe sunburn double their chances of getting these two types of cancers. Some 90–95% of these types of skin cancer can be cured if detected early enough, although their removal may leave disfiguring scars. In 1997, about 900,000 Americans developed such skin cancers. These cancers kill only 1–2% of their victims, but this still amounts to about 2,300 deaths in the United States each year.*

A third type of skin cancer, *malignant melanoma* (Figure 19-15, right), occurs in pigmented areas such as moles anywhere on the body's surface. This type of cancer can spread rapidly (within a few months) to other organs. It kills about one-fourth of its victims (most under age 40) within 5 years, despite surgery, chemotherapy, and radiation treatments. Each year it kills about 100,000 people (including 7,000 Americans in 1997), mostly Caucasians. It can often be cured if detected early enough, but recent studies show that some melanoma survivors have a recurrence more than 15 years later.

Recent evidence suggests that about 90% of sunlight's melanoma-causing effect may come from exposure to UV-A and 10% from UV-B. Most sunscreens don't protect from UV-A and tanning booth lights emit mostly UV-A.

Worldwide, the number of diagnosed cases of melanoma is growing by 4% a year. In 1998, researchers used human skin–grafted to rats to show that UV-B radiation, on its own, could cause a malignant tumor to grow on human skin that had never been exposed to sunlight before. The lag time between first substantial exposure to UV radiation (apparently mostly UV-A) and the occurrence of melanoma is 15–25 years.

Evidence indicates that people (especially Caucasians) who get three or more blistering sunburns before age 20 are five times more likely to develop malignant melanoma than those who have never had severe sunburns. About 10% of those who get malignant melanoma have an inherited gene that makes them especially susceptible to the disease.

To protect yourself, the safest course is to stay out of the sun (especially between 10 A.M. and 3 P.M., when UV levels are highest) and avoid tanning parlors. When you are in the sun, wear tightly woven protective clothing, a wide-brimmed hat, and sunglasses that protect against UV-A and UV-B radiation (ordinary sunglasses may actually harm your eyes by dilating your pupils so that more UV radiation strikes the retina). Because UV rays can penetrate clouds, overcast skies do not protect you; neither does shade, because UV rays can reflect off sand, snow, water, or patio floors. People who take antibiotics and women who take birth control pills are more susceptible to UV damage.

Use a sunscreen that offers protection against both UV-A and UV-B and has a protection factor of 15 or more (25 if you have light skin). Apply to all exposed skin and reapply it after swimming or excessive perspiration. Most people don't realize that the protection factors for sunscreens are based on using one full ounce of the product—far more than most people apply. Children who use a sunscreen with a protection factor of 15 every time they are in the sun from birth to age 18 decrease their chance of getting skin cancer by 80%; babies under a year old should not be exposed to the sun at all.

Become familiar with your moles and examine your skin at least once a month. The warning signs of skin cancer are a change in the size, shape, or color of a mole or wart (the major sign of malignant melanoma, which must be treated quickly); sudden appearance of dark spots on the skin; or a sore that keeps oozing, bleeding, and crusting over but does not heal. Be alert for precancerous growths (reddish-brown spots with a scaly crust). If you observe any of these signs, consult a doctor immediately.

Critical Thinking

What precautions, if any, do you take to reduce your chances of getting skin cancer from exposure to sunlight? Explain why you do or don't take such precautions.

*I have had six basal-cell cancers on my face and neck because of "catching too many rays" in my younger years. I wish I had known then what I know now.

used HCFC (HCFC-123) may be causing acute hepatitis and other liver abnormalities.

Another substitute for CFCs is *hydrofluorocarbons* or *HFCs* (such as CF_3, containing fluorine but no chlorine or bromine). Recent research indicates that HFCs have a negligible effect on ozone depletion. However, they may need to be restricted or phased out because they are powerful greenhouse gases that remain in the atmosphere much longer than CFCs or CO_2.

To a growing number of scientists, hydrocarbons (HCs) such as propane and butane are a better way to

Q: How many new nuclear power plants have been ordered in the United States since 1978?

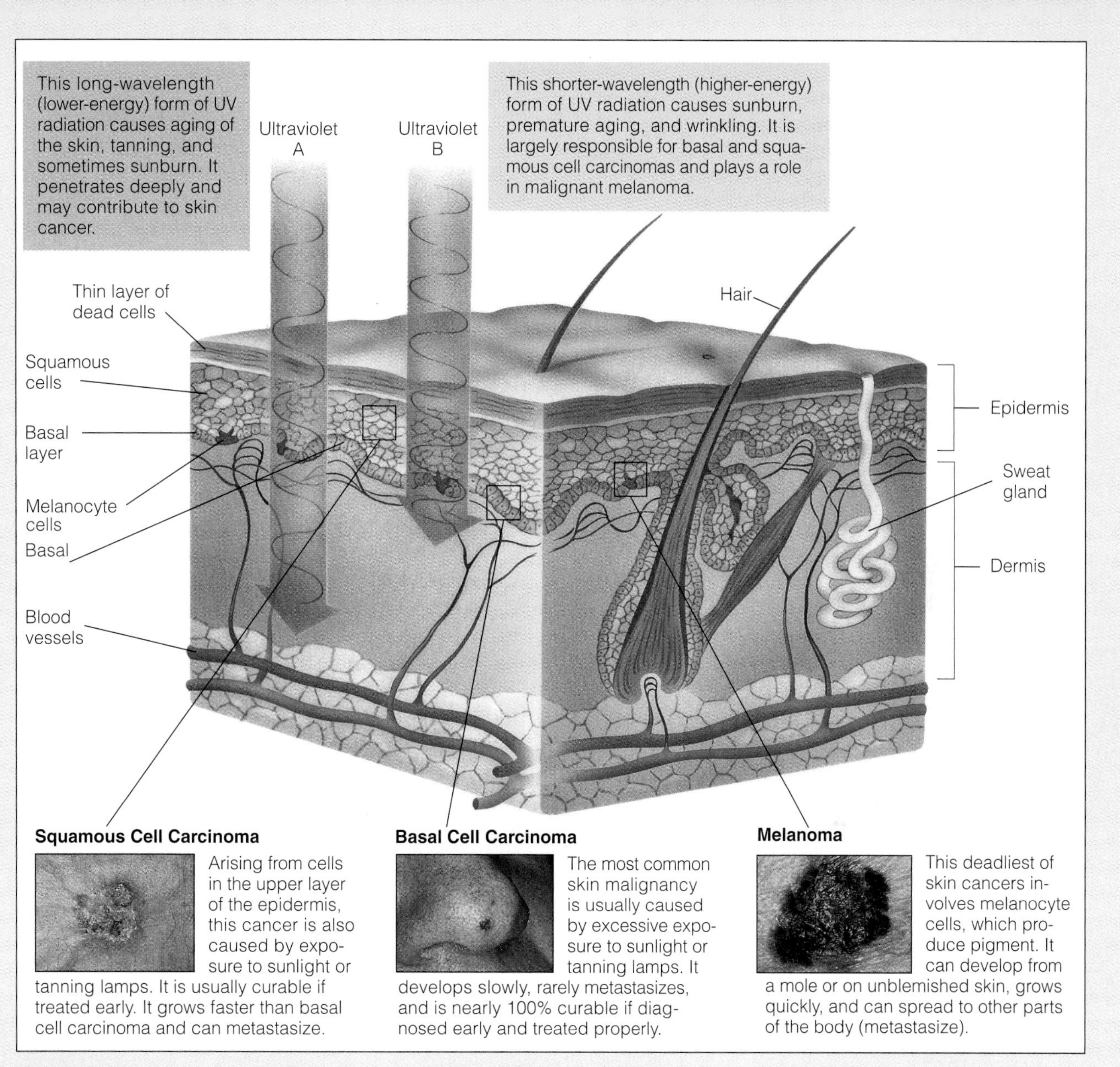

Figure 19-15
Structure of the human skin and the relationships between ultraviolet (UV-A and UV-B) radiation and the three types of skin cancer. The incidence of these types of cancer is rising, mostly because more fair-skinned people have increased their exposure to sunlight by moving to areas with sunnier climates and by spending more of their leisure time exposed to sunlight. If ozone-destroying chemicals continue to reduce stratospheric ozone levels, incidence of these types of cancers is expected to rise. (Source: The Skin Cancer Foundation)

reduce ozone depletion while doing little to enhance global warming. HCs are especially useful as coolants and insulating foam in refrigerators recently developed by German scientists.

Developing countries can use HC refrigerator technology to leapfrog ahead of industrialized countries without having to invest in costly HFC and HCFC technologies that will have to be phased out within a few decades. This approach is also less costly because HCs cannot be patented and can be manufactured locally, reducing the need to import expensive HFCs and HCFCs.

A: None, and all 120 plants ordered since 1973 have been canceled

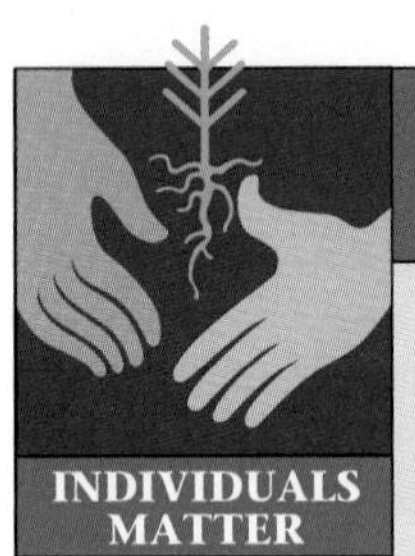

Ray Turner and His Refrigerator

Ray Turner, an aerospace manager at Hughes Aircraft in California, made an important low-tech, ozone-saving discovery by using his head—and his refrigerator. His concern for the environment led him to look for a cheap and simple substitute for the CFCs used as cleaning agents for removing films of oxidation from the electronic circuit boards manufactured at his plant.

He started his search by looking in his refrigerator for a better circuit board cleaner. He decided to put drops of various substances on a corroded penny to see whether any of them removed the film of oxidation. Then he used his soldering gun to see whether solder would stick to the surface of the penny, indicating that the film had been cleaned off.

First, he tried vinegar. No luck. Then he tried some ground-up lemon peel, also a failure. Next he tried a drop of lemon juice and watched as the solder took hold. The rest, as they say, is history.

Today, Hughes Aircraft uses inexpensive citrus-based solvents that are CFC-free to clean circuit boards. This new cleaning technique has reduced circuit board defects by about 75% at Hughes. And Turner got a hefty bonus. Now other companies, such as AT&T, clean computer boards and chips using acidic chemicals extracted from cantaloupes, peaches, and plums. Maybe you can find a solution to an environmental problem in your refrigerator, grocery or drugstore, or backyard.

Can Technofixes Save Us? What about a quick fix from technology, so that we can keep using CFCs? Physicist Alfred Wong has proposed that each year we launch a fleet of 20–30 football-field-long, radio-controlled blimps into the stratosphere above Antarctica. Hanging from each blimp would be a huge curtain of electrical wires that would inject negatively charged electrons into the stratosphere when exposed to high voltages (produced by electricity from huge panels of solar cells). Based on 4 years of laboratory experiments, Wong believes that ozone-destroying chlorine atoms (Cl) in the stratosphere would each pick up an electron and be converted to chloride ions ($Cl + e \rightarrow Cl^-$) that would not react with ozone. A second suspended sheet of positively charged wires could be used to attract the negatively charged ions and remove them from the stratosphere. Wong estimates that it would cost about $400 million a year to remove 10–30% of the chlorine atoms formed in the stratosphere each year.

However, atmospheric chemist Ralph Ciecerone believes that this plan won't work because other chemical species in the stratosphere snatch electrons more readily than does chlorine. This scheme could also have unpredictable side effects on atmospheric chemistry.

Others have suggested using tens of thousands of lasers to blast CFCs out of the atmosphere before they can reach the stratosphere. However, the energy required to do this would be enormous and expensive, and decades of research would be needed to perfect the types of lasers needed. Moreover, we can't predict the possible effects of such powerful laser blasts on climate, birds, or planes.

What Is Being Done to Reduce Ozone Depletion? Some Hopeful Progress In 1987, 36 nations meeting in Montreal developed a treaty, commonly known as the *Montreal Protocol*, to cut emissions of CFCs (but not other ozone depleters) into the atmosphere by about 35% between 1989 and 2000. After hearing more bad news about ozone depletion, representatives of 93 countries met in London in 1990, in Copenhagen in 1992, and again in Montreal in 1997 and adopted a protocol accelerating the phaseout of key ozone-depleting chemicals, with some phaseout schedules accelerated in 1995 and 1997 (Table 19-2).

The agreements reached so far are important examples of global cooperation in response to serious threats to global environmental security (Guest Essay, p. 530). Because of these agreements, CFC production fell by 85% between 1988 (its peak production year) and 1998. Global production of halons, carbon tetrachloride, and methyl chloroform has also dropped sharply, but that of methyl bromide and HCFCs continues to rise.

Developed countries have set up a $250-million fund to help developing countries make the switch away from CFCs, but much more money will be needed. It is encouraging that 58 developing countries have committed themselves to phasing out CFCs early. However, the world's two most populous countries, China and India, have refused to sign the ozone agreements. Even if the 1992 agreements are upheld, scientists estimate that the ozone layer will continue to be depleted until around 2080 and will cause ozone losses of 10–30% over the northern hemisphere (where most of the world's people live). However, without the 1992 international agreement, ozone depletion would be a much more serious threat.

Will the International Treaty to Slow Ozone Depletion Work? There is growing concern that the requirements of the Copenhagen agreement

Q: What percentage of current U.S. nuclear power generation will be retired by 2015?

Table 19-2 Phaseout of Ozone-Depleting Chemicals in Developed Countries Adopted by International Agreements in 1992, 1995, and 1997

Substances	Phaseout Deadline for Developed Countries*
Halons	1/1/94 (except for essential uses).
CFCs	1/1/96 (except for small amounts of imports to developing countries and for a few essential scientific and medical uses).
	All production to cease by 2006.
Methyl chloroform	1/1/96.
Methyl bromide	2005 (with possible exemptions for essential uses) in developed countries and 2015 in developing countries.
HCFCs (CFC substitutes)	2020 (freeze production at 1991 levels by 1/1/96).

*Developing countries signing the treaty are given a 10-year extension in meeting these deadlines, but this may be shortened.

(Table 19-2) might not be met. By 1995 there was a political and economic backlash in the United States against this treaty. This was caused by widely publicized (but scientifically refuted; see Pro/Con on p. 523) attacks on the overwhelming scientific consensus that ozone depletion is a very serious problem.

The effectiveness of the treaty is also being undermined by a rapidly growing black market in CFCs (apparently being smuggled into the United States and other developed countries from Russia, China, and Mexico). This black market is being stimulated by dwindling supplies of CFCs (phased out by 1996), the high prices of some CFC substitutes, and the costly conversion of some older air conditioners and refrigerators to use replacement chemicals. There have also been signs that some countries are cheating and not living up to the requirements of the ozone treaty. A cheating rate of only 10% can keep stratospheric levels of CFCs from declining as projected.

Even if the ozone treaty is only partially implemented, it has set an important precedent for global cooperation and action when faced with potential global disaster. However, international cooperation in dealing with projected global warming is much more difficult because the evidence for global warming is less clear-cut. Moreover, lowering our inputs of greenhouse gases by greatly reducing use of fossil fuels and greatly slowing deforestation is economically and politically difficult to do.

Some people have wondered whether there is intelligent life in other parts of the universe; others wonder whether there is intelligent life on the earth. To them, if we can seriously deal with the interconnected global problems of loss of biodiversity (Chapters 24, 25, and 26), possible climate change, and depletion of stratospheric ozone from human activities, then the answer is a hopeful yes. This means recognizing that *prevention* is the best (and in the long run, the least costly) way to deal with global environmental problems. Otherwise, they believe the answer is a tragic no.

The atmosphere is the key symbol of global interdependence. If we can't solve some of our problems in the face of threats to this global commons, then I can't be very optimistic about the future of the world.

MARGARET MEAD

CRITICAL THINKING

1. Do you believe that possible global warming from an enhanced natural greenhouse effect caused at least partly by human activities is a serious problem or one that has been greatly exaggerated? Explain.

2. Explain why you agree or disagree with each of the proposals listed in Figure 19-9 (for slowing down emissions of greenhouse gases into the atmosphere) and those given in Figure 19-10 (for preparing for the effects of global warming). What effects would carrying out these proposals have on your own lifestyle and on those of any grandchildren you might have? What might be the effects of *not* carrying out these actions?

3. What consumption patterns and other features of your lifestyle directly add greenhouse gases to the atmosphere? Which, if any, of these things would you be willing to give up to slow projected global warming and reduce other forms of air pollution?

4. Do you believe that ozone depletion in the stratosphere is a serious problem or one that has been greatly exaggerated? Explain. Are the actions now being taken to deal with this problem unnecessary and too costly? Are they inadequate to deal with the problem rapidly enough? Explain your answers.

5. Use the information in Table 19-2 to determine which chemicals should be given priority as substitutes for CFCs.

6. In 1995 and 1996, some members of the U.S. House of Representatives introduced legislation that would cancel U.S. participation in the international ozone treaty on the basis that there is no scientific proof of ozone thinning. Criticize this position on the basis of the nature of science (Section 2-1) and the information in the Pro/Con box on p. 523.

7. In preparation for the 1992 UN Conference on the Human Environment in Rio de Janeiro, President Bush's top White House economic adviser gave an address in Williamsburg, Virginia, to representatives of governments

A: At least 40% and the rest by 2030

GUEST ESSAY

Environmental Security

Norman Myers

Norman Myers is an international consultant in environment and development, with emphasis on conservation of wildlife species and tropical forests, and is one of the world's leading environmental experts. His research and consulting have taken him to 80 countries. He has served as a consultant for many development agencies and research organizations, including the U.S. National Academy of Sciences, the World Bank, the Organization for Economic Cooperation and Development, various UN agencies, and the World Resources Institute. Among his many publications (see Further Readings) are Conversion of Tropical Moist Forests *(1980),* A Wealth of Wild Species *(1983),* The Primary Source: Tropical Forests and Our Future *(1984),* The Gaia Atlas of Planet Management *(1992),* The Gaia Atlas of Future Worlds *(1990),* Ultimate Security: The Environmental Basis of Political Security *(1993), and* Scarcity or Abundance: A Debate on Environment *(1994).*

> *There is a new and different threat to our national security emerging—the destruction of our environments. I believe that one of our key national security objectives must be to reverse the accelerating pace of environmental destruction around the globe.*
>
> SENATOR SAM NUNN, former chairman of the U.S. Senate Armed Service Committee

We are already engaged in World War III. It is a war against nature, and it is simply no contest. As a result, the threat from the skies is no longer missiles but ozone-layer depletion and global warming, and the threat on land is from soil erosion.

These and other environmental threats, including overpopulation and widespread poverty in developing countries, have been described by a growing number of political leaders and military planners as the greatest threat we face short of nuclear war. So a new concept is emerging in councils of foreign policy makers and military strategists: the environmental dimension to security issues.

According to this idea, we should move beyond traditional thinking about security concepts and consider a series of environmental factors underpinning our material welfare. These factors include such natural resources as a nation's soil, water, forests, grasslands, and fisheries, and the climatic patterns and biogeochemical cycles that maintain the life-support systems of all nations [Figure 1-2].

If a nation's environmental foundations are degraded or depleted, its economy may well decline, its social fabric deteriorate, and its political structure become destabilized as growing numbers of people seek to sustain themselves from declining resource stocks. The likely outcome is tension and conflict within a nation and possibly with other nations.

Thus, national security is no longer about fighting forces and weaponry alone. It relates increasingly to watersheds, croplands, forests, genetic resources, climate, and other factors that, taken together, are as crucial to a nation's security as are military factors.

Consider Ethiopia and its agricultural decline as a source of conflict. The country's traditional farming area in the highlands was losing an estimated 1 billion tons of soil a year by the early 1970s (for comparison, the United States loses an estimated 2.8 billion tons a year, from a cropland area 20 times as large). The resulting drop in agricultural production led to food shortages in Ethiopian cities. Ensuing disorders precipitated the overthrow of Emperor Haile Selassie in 1974.

The new regime did not move fast enough, despite some efforts to restore agriculture, so throngs of impoverished peasants streamed into the country's lowlands, including an area that straddles the border with Somalia and is of long-standing dispute between the two countries. This resulted in an outbreak of hostilities in 1977, which threatened oil tankers heading from the Persian Gulf to the industrialized nations of the Western world.

Between 1976 and 1980, Ethiopia spent an average of $225 million a year on military activities. If this money had been used to safeguard topsoil, tree cover, and associated factors of the natural-resource base in traditional farmlands of Ethiopia, the migration to the lowlands would have been far less likely. Ironically, the amount that the United Nations has budgeted (but not spent) for Ethiopia for anti-erosion, reforestation, and related measures under its Anti-Desertification Plan suggests that no more than $50 million a year would have been needed to counter much of the problem, had this investment in environmental security been undertaken in due time.

from a number of countries. He told his audience not to worry about global warming because the average temperature increases scientists are predicting were much less than the temperature increase he experienced in coming from Washington, D.C., to Williamsburg. What is the fundamental flaw in this reasoning?

8. Try to come up with analogies to complete the following statements: **(a)** A greenhouse gas is like _______. **(b)** Adding CO_2 gas to the atmosphere by burning fossil fuels and cutting and burning forests is like _______. **(c)** Worrying about ozone depletion is like _______. **(d)** Chlorofluorocarbons are like _______. **(e)** Spending a lot of time in sunlight without protecting your skin from UV radiation is like _______.

PROJECTS

1. As a class, conduct a poll of students at your school to determine whether they believe that possible global warming from an enhanced greenhouse effect is a very

Q: How many people in the United States die prematurely each year from U.S. nuclear power plants?

By comparison, the amount required to counter Ethiopia's famine during 1985 amounted to $500 million for relief measures alone.

In many countries, decline of the natural-resource base that underpins agriculture has led to increased imports, rising prices, and ultimately outright shortages of food. In turn, these shortages have helped trigger civil disorders, military outbreaks, and downfalls of governments.

Food-supply disputes have also occurred among developed countries. For example, in the North Atlantic, Great Britain and Iceland have come to the edge of hostilities over declining cod stocks. At least 16 similar major clashes over declining fishery stocks have occurred in other parts of the world. Such disputes are likely to increase in the future in light of the failure of fisheries around the world to maintain sustainable yields.

Another source of conflict is water supplies. In the Middle East, competition for scarce water is a major factor in political confrontations, and shortages are projected to grow even more acute [p. 310 and Figure 13-1]. Israel went to war in 1967 in part because the Arabs were trying to divert the headwaters of the River Jordan. And currently, Israel—the world's most water-efficient country—is beginning to suffer from critical water deficits.

Of 200 major river systems, almost 150 are shared by two nations, and more than 50 by three to ten nations; all in all, these rivers support almost 40% of the world's population. As many as 80 countries, with 40% of the world's population, already experience water shortages.

Other conflicts occur because of deforestation. The Ganges River system is dependent on tree cover in its catchment zone in the Himalayas. Because of deforestation, flooding during the annual monsoon has become so widespread that it regularly imposes damages to crops, livestock, and property worth $1 billion a year among downstream communities in India, even though most of the deforestation occurs in Nepal.

In every category of environmental decline there are major implications for international relations. They all act, to a varying degree, as sources of economic disruption and political tension. Although they may not trigger outright confrontation, they help to destabilize societies in an already increasingly unstable world.

In the instances cited here, the environmental issues are readily apparent. In other cases, the impact is more deferred and diffuse, as in the case of species extinctions and gene depletion, with all this ultimately implies for future genetic contributions to agriculture, medicine, industry, and energy. Probably the most deferred and diffuse impact of all—although altogether the most significant—will prove to be climatic change. Buildup of carbon dioxide and other greenhouse gases in the global atmosphere will, if it persists as projected, cause far-reaching disruptions for temperature and rainfall patterns.

We cannot dispatch battalions to turn back the deserts, we cannot launch a flotilla to resist the rising seas, we cannot send fighter planes to counter the greenhouse effect. Instead, we can achieve more enduring, widespread, and true security by safeguarding our environments than we can by engaging in military buildups. Increasingly, it is becoming a case of trees now or tanks later.

Major environmental problems recognize no geographic boundaries; the winds carry no passports. The new security will be security for us all, or for none. This means working toward our new security *together*—another big break from the past. Peace with the earth means peace with each other, and vice versa. Hopefully, the prospect of catastrophic breakdown can motivate us to achieve ultimate peace on the earth, by learning how to cooperate with one another and with the rest of nature.

Critical Thinking

1. Do you agree or disagree with the position that we need to place much more emphasis on environmental security? Should we treat it with the same degree of seriousness, analysis, and funding as we do economic security and military security, or should environmental security have higher or lower priority? Defend your answers.

2. Some 300 million couples in developing countries want to reduce their family size but lack the family planning services to do so. If we were to meet this need, we would reduce the ultimate global population by at least 2 billion. The cost would be $6 billion per year, equivalent to two and one-half days of military spending worldwide. Why do you think it is not being done? What might be the environmental and economic security benefits?

serious problem, a moderately serious problem, or of little concern. Tally the results to see whether there are differences related to year in school, political leaning (liberal, conservative, independent), or sex of poll participants.

2. As a class, conduct a poll of students at your school to determine whether they believe that stratospheric ozone depletion is a very serious problem, a moderately serious problem, or of little concern. Tally the results to see whether there are differences related to year in school, political leaning (liberal, conservative, independent), or sex of poll participants.

3. Make a concept map of this chapter's major ideas, using the section heads and subheads and the key terms (in boldface). Look at the inside back cover and in Appendix 4 for information about concept maps.

A: About 6,000 according to the National Academy of Sciences

20 WATER POLLUTION

Learning Nature's Ways to Purify Sewage

Some communities and individuals are seeking better ways to purify contaminated water by working with nature. A low-tech, low-cost alternative to expensive waste treatment plants is to create an artificial wetland. This is what the residents of Arcata, California, did, led by Robert Gearheart and George Allen—teacher at Humboldt State University.

In this coastal town of 17,000, some 63 hectares (155 acres) of wetlands has been created between the town and the adjacent Humboldt Bay. The marshes, developed on land that was once a dump, act as an inexpensive, natural waste treatment plant (Figure 20-1). The project was completed in 1974 for less half than the cost of a conventional treatment plant.

Here's how it works: First, sewage is held in sedimentation tanks, where the solids settle out as sludge that is removed and processed for use as fertilizer. The liquid is pumped into oxidation ponds, where remaining wastes are broken down by bacteria. After a month or so, the water is released into the artificial marshes, where it is further filtered and cleansed by plants and bacteria. Although the water is clean enough to be discharged directly into the bay, state law requires that it first be chlorinated. So the town chlorinates the water and then dechlorinates it before sending it into the bay, where oyster beds thrive.

Some water from the marshes is piped into the city's salmon hatchery. Arcata hopes to establish a salmon-ranching operation and turn its marsh treatment plant into a moneymaker. The town plans to greatly expand its marsh system. Arcata also has a sustainable harvest plan for its town forest and a cooperative farm and is home for a number of nonprofit and environmentally oriented businesses.

The marshes and lagoons are an Audubon Society bird sanctuary and provide habitats for thousands of otters, seabirds, and marine animals; the treatment center is a city park and attracts more than 150,000 visitors a year. Residents, with the lowest sewage bills in the United States, celebrate their natural sewage treatment system with an annual "Flush with Pride" festival. Over 150 cities and towns in the United States now use natural and artificial wetlands for treating sewage.

How can natural processes be used for treating municipal sewage and industrial wastewater if there isn't a wetland available or enough land on which to create one? According to marine biologist John Todd, you set up a greenhouse lagoon and use natural food chains and sunshine the way nature does (Figure 20-2).

The purification process in Todd's *living machines* begins when sewage flows into a greenhouse containing rows of large tanks full of aquatic plants such as water hyacinths, cattails, and bulrushes. In these tanks, algae and microorganisms decompose wastes into nutrients absorbed by the plants. The decomposition is speeded up by sunlight streaming into the greenhouse. Then the water passes through an artificial

Figure 20-1 Marsh sewage treatment area in Arcata, California. This low-cost, natural sewage-treatment system is feasible in areas with lots of space to build artificial wetlands and the abscence of toxic heavy metals produced by industry. (George W. Allen)

marsh of sand, gravel, and bulrush plants, which filters out algae and organic waste. Next the water flows into aquarium tanks, where snails and zooplankton consume microorganisms and are in turn consumed by crayfish, tilapia, and other fish that can be eaten or sold as bait. After 10 days, the now-clear water flows into a second artificial marsh for final filtering and cleansing.

When working properly, such solar–aquatic treatment systems produce water fit for drinking. The chief by-products of such living machines are plants, trees, snails, and fish that can be sold as compost, ornamental plants, or baitfish. Todd's living machine purification systems now operate in 13 states around the United States and in seven other countries.

Water pollution is related to air pollution, land-use practices, and the number of people, farms, and industries producing sewage. These connections explain why solving water pollution problems should be integrated with air pollution, energy, land-use, and population policies that emphasize pollution prevention. Otherwise, we will continue shifting potential pollutants from one part of the ecosphere to another until threshold levels of damage are exceeded as more people and more industries produce more wastes.

Figure 20-2 At the Providence, Rhode Island, Solar Sewage Plant, biologist John Todd demonstrates how ecological waste engineering in a greenhouse can be used to purify wastewater. Todd and others are conducting research to perfect solar–aquatic systems based on working with nature. (Ocean Arks International)

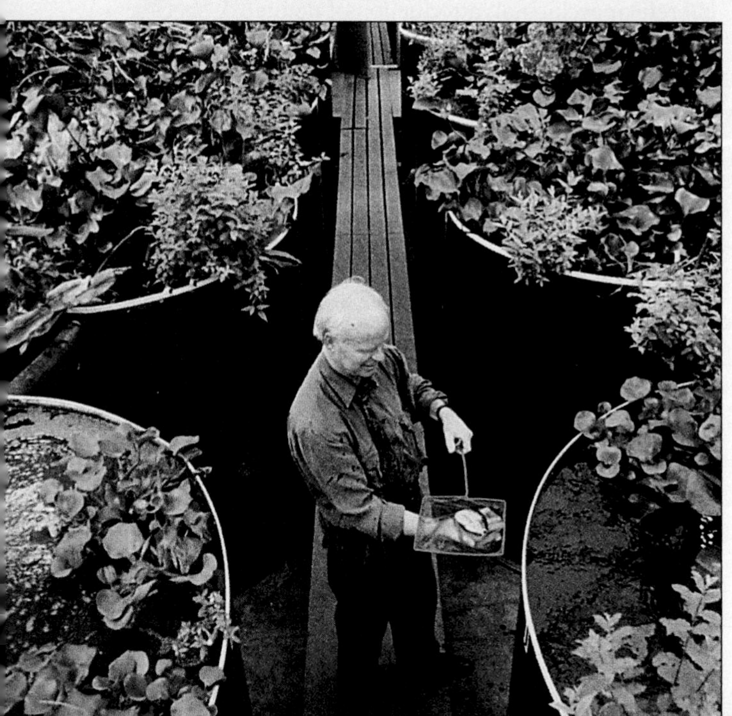

Today everybody is downwind or downstream from somebody else.

WILLIAM RUCKELSHAUS

In this chapter we will seek answers to the following questions:

- What pollutes water, where do the pollutants come from, and what effects do they have?
- What are the major water pollution problems of streams and lakes?
- What are the major water pollution problems of oceans?
- How can we prevent and reduce surface-water pollution?
- How is groundwater polluted and what can be done to prevent such pollution?
- How safe is drinking water?

20-1 TYPES AND SOURCES OF WATER POLLUTION

What Are the Major Water Pollutants? **Water pollution** is any chemical, biological, or physical change in water quality that has a harmful effect on living organisms or makes water unsuitable for desired uses. There are several classes of water pollutants. One is *disease-causing agents* (*pathogens*), which include bacteria, viruses, protozoa, and parasitic worms that enter water from domestic sewage and untreated human and animal wastes (Table 20-1). According to a 1995 World Bank study, contaminated water (lack of clean drinking water and lack of sanitation) causes about 80% of the diseases in developing countries and kills about 10 million people annually—an average of 27,000 premature deaths per day, more than half of them children under age 5.

A good indicator of the quality of water for drinking or swimming is the number of colonies of *coliform bacteria* present in a 100-milliliter (0.1-quart) sample of water. The World Health Organization recommends a coliform bacteria count of 0 colonies per 100 milliliters for drinking water, and the EPA recommends a maximum level for swimming water of 200 colonies per 100 milliliters. Because the average human excretes about 2 billion such organisms a day, we can see how easily untreated sewage can contaminate water.

Table 20-1 Common Diseases Transmitted to Humans Through Contaminated Drinking Water

Type of Organism	Disease	Effects
Bacteria	Typhoid fever	Diarrhea, severe vomiting, enlarged spleen, inflamed intestine; often fatal if untreated
	Cholera	Diarrhea, sever vomiting, dehydration; often fatal if untreated
	Bacterial dysentery	Diarrhea; rarely fatal except in infants without proper treatment
	Enteritis	Severe stomach pain, nausea, vomiting; rarely fatal
Viruses	Infectious hepatitis	Fever, severe headache, loss of appetite, abdominal pain, jaundice, enlarged liver; rarely fatal but may cause permanent liver damage
Parasitic protozoa	Amoebic dysentery	Severe diarrhea, headache, abdominal pain, chills, fever; if not treated can cause liver abscess, bowel perforation, and death
	Giardiasis	Diarrhea, abdominal cramps, flatulence, belching, fatigue
Parasitic worms	Schistosomiasis	Abdominal pain, skin rash, anemia, chronic fatigue, and chronic general ill health

A second category of water pollutants is *oxygen-demanding wastes*, organic wastes that can be decomposed by aerobic (oxygen-requiring) bacteria. Large populations of bacteria decomposing these wastes can degrade water quality by depleting water of dissolved oxygen (Figures 20-3 and 8-2), causing fish and other forms of oxygen-consuming aquatic life to die. The quantity of oxygen-demanding wastes in water can be determined by measuring the **biological oxygen demand (BOD)**: the amount of dissolved oxygen needed by aerobic decomposers to break down the organic materials in a certain volume of water over a 5-day incubation period at 20°C (68°F).

A third class of water pollutants is *water-soluble inorganic chemicals*, which include acids, salts, and compounds of toxic metals such as mercury and lead. High levels of these chemicals can make water unfit to drink, harm fish and other aquatic life, lower crop yields, and accelerate corrosion of metals exposed to such water.

Inorganic plant nutrients are another class of water pollutants. They are water-soluble nitrates and phosphates that can cause excessive growth of algae and other aquatic plants, which then die and decay, depleting water of dissolved oxygen and killing fish. Drinking water with excessive levels of nitrates lowers the oxygen-carrying capacity of the blood; this can kill unborn children and infants, especially those under 1 year old.

Water can also be polluted by a variety of *organic chemicals*, which include oil, gasoline, plastics, pesticides, cleaning solvents, detergents, and many other chemicals. They threaten human health and harm fish and other aquatic life.

By far the biggest class of water pollutants by weight is *sediment*, or *suspended matter*—insoluble particles of soil and other solids that become suspended in water, mostly when soil is eroded from the land. Sediment clouds water and reduces photosynthesis; it also disrupts aquatic food webs and carries pesticides, bacteria, and other harmful substances. Sediment that settles out destroys feeding and spawning grounds of fish. It also clogs and fills lakes, artificial reservoirs, stream channels, and harbors.

Water can also be polluted by *water-soluble radioactive isotopes*, some of which are concentrated or biologically magnified in various tissues and organs as they pass through food chains and webs (p. 438). Ionizing radiation emitted by such isotopes can cause birth defects, cancer, and genetic damage (Section 17-4).

Heat absorbed by water used to cool industrial and power plants can lower water quality. The resulting rise in water temperature, called *thermal pollution*, lowers dissolved oxygen levels and makes aquatic organisms more vulnerable to disease, parasites, and toxic chemicals.

Another form of water pollution, *genetic pollution*, occurs when aquatic systems are disrupted by the deliberate or accidental introduction of nonnative species. Some of these species can crowd out native species, reduce biodiversity, and cause economic losses.

The principal method by which nonnative species are introduced into marine systems is through the intake and discharge of ballast from ships. When ships take on ballast water at their point of departure, they also take on board thousands of microscopic organisms, as well as the early life stages of larger plants and animals. When this ballast water is emptied at the port of call, these stowaway organisms are discharged as well.

How Do We Detect Water Pollutants? *Chemical analysis* helps determine the presence and concentrations of most water pollutants. Living organisms can also be used as indicator species to monitor water pollution. For example, the tissues of filter-feeding mussels, harvested

Q: Is there a scientifically and politically acceptable method for the long-term disposal of high-level radioactive waste?

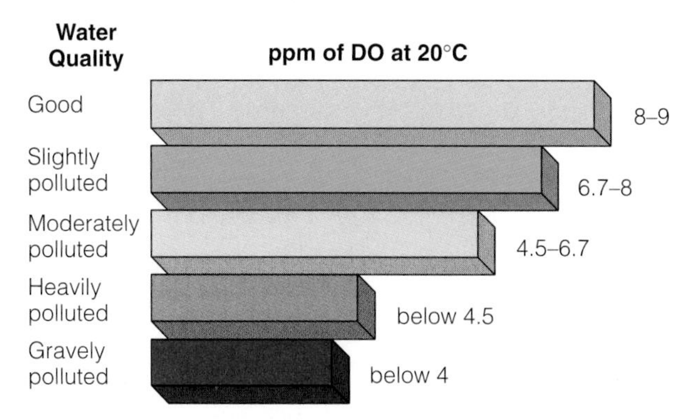

Figure 20-3 Water quality and dissolved oxygen (DO) content in parts per million (ppm) at 20°C (68°F). The solubility of oxygen decreases as the water temperature increases. Only a few species of fish can survive in water with fewer than 4 ppm of dissolved oxygen.

from the sediments of coastal waters, can be analyzed for the presence of various industrial chemicals, toxic metals (such as mercury and lead), and pesticides.

In recent years, computer models of large aquatic ecosystems have revealed complex inputs and interactions that couldn't be determined by conventional chemical and biological methods. Complex models show water flows, air flows, and the deposition of air pollutants into such systems. Such models have revealed the surprising finding that *air* pollutants (from nitrogen oxides, nitric acid, and nitrate salts) account for about 35% of the input of nitrogen plant nutrients into the Chesapeake Bay. This clearly illustrates the need to integrate both air and water pollution prevention and control programs.

Computer modeling can often save a lot of money. For example, running chemical analysis of the air and water in the Chesapeake Bay could cost at least $500 million, whereas the computer model of this system costs around $500,000.

What Are Point and Nonpoint Sources of Water Pollution? **Point sources** discharge pollutants at specific locations through pipes, ditches, or sewers into bodies of surface water. Examples include factories, sewage treatment plants (which remove some but not all pollutants), active and abandoned underground mines, offshore oil wells, and oil tankers. Because point sources are at specific places, they are fairly easy to identify, monitor, and regulate. In developed countries many industrial discharges are strictly controlled, whereas in most developing countries such discharges are largely uncontrolled.

Nonpoint sources are sources that cannot be traced to any single site of discharge. They are usually large land areas or airsheds that pollute water by runoff, subsurface flow, or deposition from the atmosphere. Examples include acid deposition (Figure 18-7), runoff of chemicals into surface water (including stormwater), and seepage into the ground from croplands, livestock feedlots, logged forests, streets, lawns, and parking lots.

In the United States, nonpoint pollution from agriculture—mostly in the form of sediment, inorganic fertilizers, manure, salts dissolved in irrigation water, and pesticides—is responsible for an estimated 64% of the total mass of pollutants entering streams and 57% of those entering lakes. According to the EPA, nonpoint runoff of stormwater causes 33% of all contamination in U.S. lakes and estuaries and 10% of all stream contamination. Little progress has been made in controlling nonpoint water pollution because of the difficulty and expense of identifying and controlling discharges from so many diffuse sources.

20-2 POLLUTION OF STREAMS AND LAKES

What Are the Pollution Problems of Streams? Flowing streams, including large ones called rivers, can recover rapidly from degradable, oxygen-demanding wastes and excess heat by a combination of dilution and bacterial decay. This natural recovery process works as long as streams are not overloaded with these pollutants and as long as their flow is not reduced by drought, damming, or diversion for agriculture and industry. However, these natural dilution and biodegradation processes do not eliminate slowly degradable and nondegradable pollutants.

The breakdown of degradable wastes by bacteria depletes dissolved oxygen, which reduces or eliminates populations of organisms with high oxygen requirements until the stream is cleansed of wastes. The depth and width of the resulting *oxygen sag curve* (Figure 20-4) (and thus the time and distance required for a stream to recover) depend on the stream's volume, flow rate, temperature, pH level, and the volume of incoming degradable wastes. Similar oxygen sag curves can be plotted when heated water from industrial and power plants is discharged into streams.

What Progress Has Been Made in Reducing Stream Pollution? Requiring cities to withdraw their drinking water downstream rather than upstream (as is done now) would dramatically improve water quality because each city would be forced to clean up its own waste outputs rather than passing them downstream. However, upstream users, who already have the use of fairly clean water without high cleanup costs, fight this pollution prevention approach.

Here is some good news. Water pollution control laws enacted in the 1970s have greatly increased the number and quality of wastewater treatment plants in

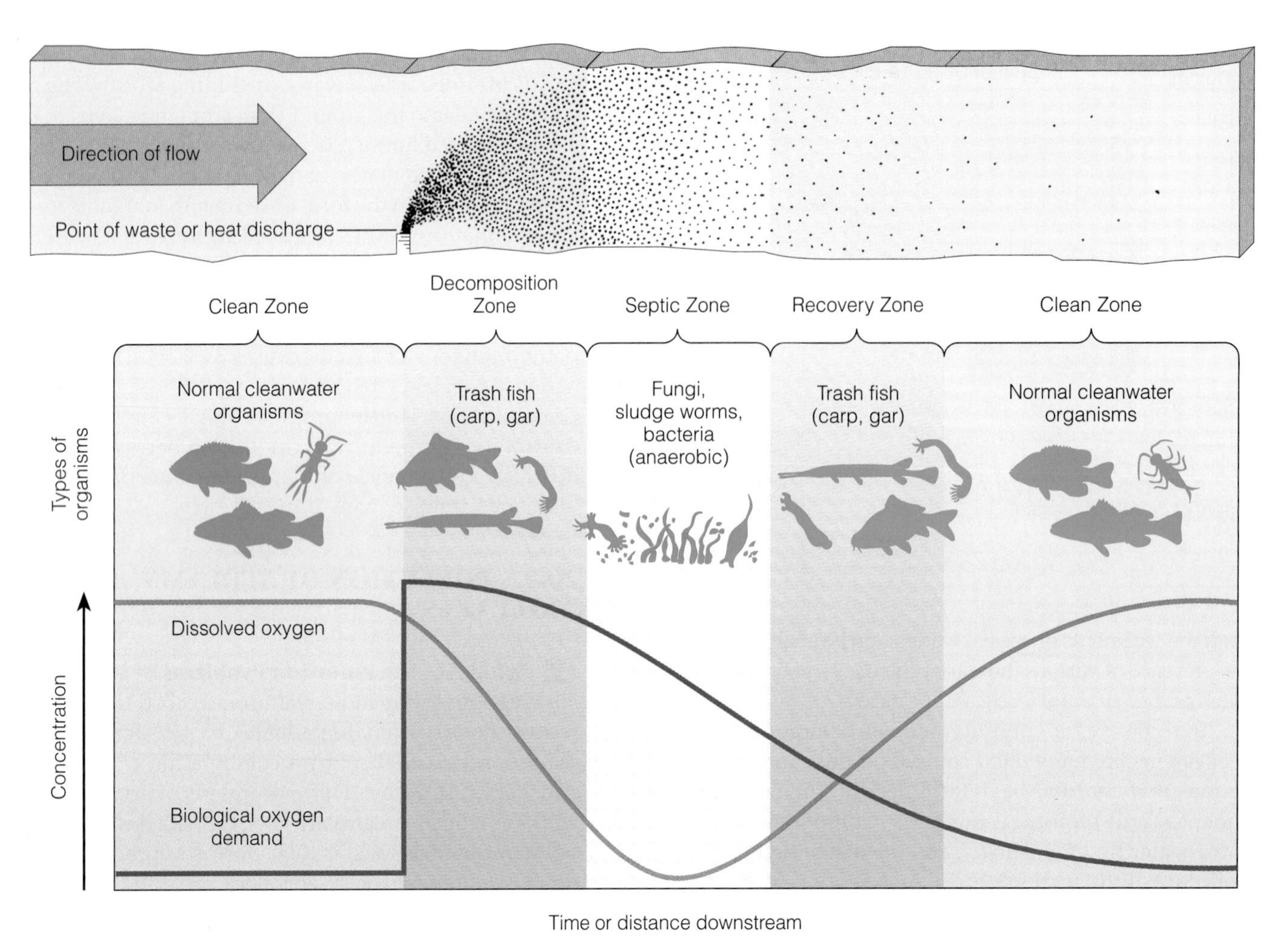

Figure 20-4 Dilution and decay of degradable, oxygen-demanding wastes and heat, showing the oxygen sag curve (orange) and the curve of oxygen demand (blue). Depending on flow rates and the amount of pollutants, streams recover from oxygen-demanding wastes and heat if they are given enough time and are not overloaded.

the United States and many other developed countries; laws have also required industries to reduce or eliminate point-source discharges into surface waters. These efforts have enabled the United States to hold the line against increased pollution of most of its streams by disease-causing agents and oxygen-demanding wastes. This is an impressive accomplishment considering the rise in economic activity and population since the laws were passed.

One success story is the cleanup of Ohio's Cuyahoga River, which was so polluted that in 1959 and again in 1969 it caught fire and burned for several days as it flowed through Cleveland, Ohio. The highly publicized image of this burning river prompted city, state, and federal officials to enact laws limiting the discharge of industrial wastes into the river and sewage systems and to appropriate funds to upgrade sewage treatment facilities. Today the river has made a comeback and is widely used by boaters and anglers.

Pollution control laws passed since 1970 have also led to improvements in dissolved oxygen content in many streams in Canada, Japan, and most western European countries. A spectacular cleanup has occurred in Great Britain. In the 1950s the Thames River was little more than a flowing anaerobic sewer, but after more than 30 years of effort, $250 million of British taxpayers' money, and millions more spent by industry, the Thames has made a remarkable recovery. Commercial fishing is thriving, and many species of waterfowl and wading birds have returned to their former feeding grounds.

What Is the Bad News About Stream Pollution? Despite progress in improving stream quality in most developed countries, large fish kills and contamination of drinking water still occur. Most of these disasters are caused by accidental or deliberate releases of toxic inorganic and organic chemicals by industries, malfunctioning sewage treatment plants, and nonpoint runoff of pesticides and nutrients (eroded soil, fertilizer, and animal waste) from cropland or animal feedlots (Individuals Matter, p. 538).

Q: How many sites in the United States are contaminated with radioactive materials?

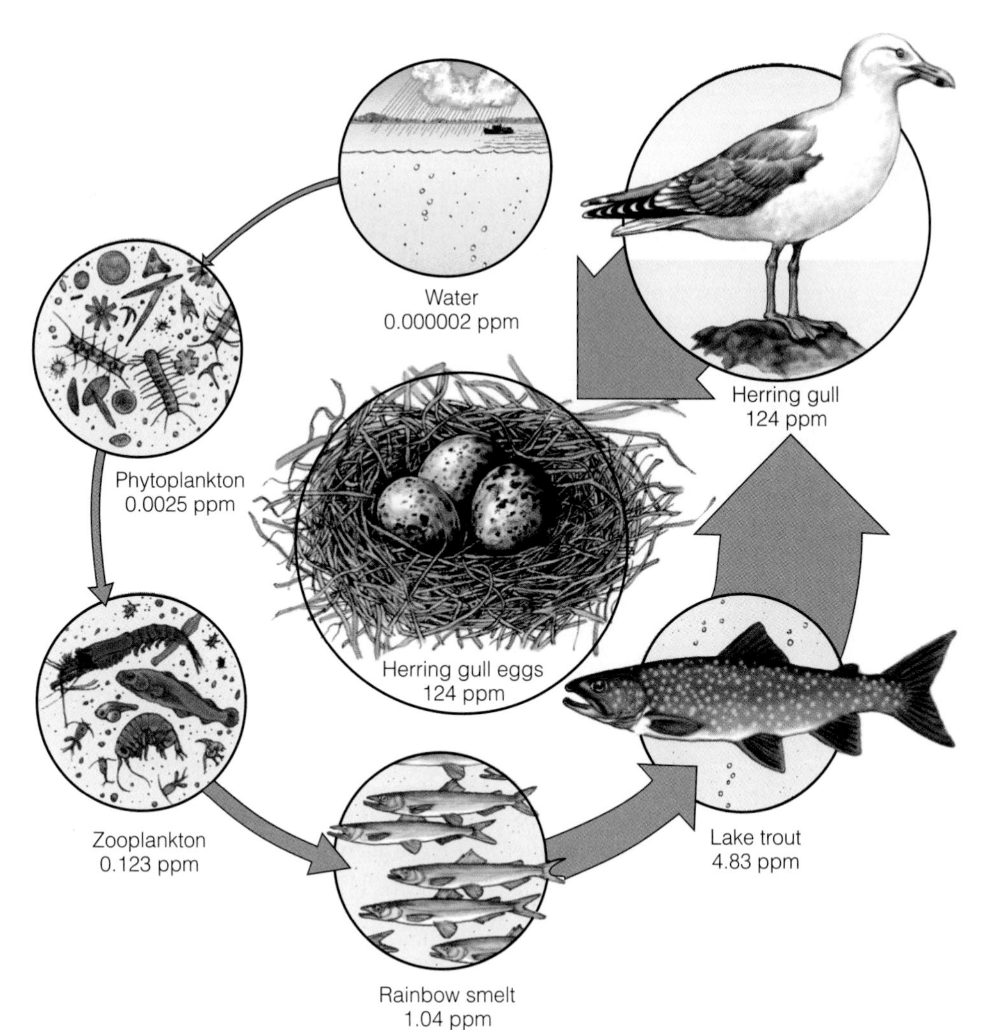

Figure 20-5 Biological magnification of PCBs (polychlorinated biphenyls) in an aquatic food chain in the Great Lakes. Most of the 209 different PCBs are insoluble in water, soluble in fats, and resistant to biological and chemical degradation—properties that result in their accumulation in the tissues of organisms and their biological amplification in food chains and webs. Although the long-term health effects on people exposed to low levels of PCBs are unknown, high doses of PCBs in laboratory animals produce liver and kidney damage, gastric disorders, birth defects, skin lesions, hormonal changes, smaller penis size, and tumors. Boys in Taiwan exposed to PCBs while in their mothers' wombs developed smaller penises than other Taiwanese boys. In the United States, manufacture and use of PCBs have been banned since 1976; before then, millions of metric tons of these long-lived chemicals were released into the environment, many of them ending up in bottom sediments of lakes, streams, and oceans.

Available data indicate that stream pollution from huge discharges of sewage and industrial wastes is a serious and growing problem in most developing countries, where waste treatment is practically nonexistent. Numerous streams in the former Soviet Union and in eastern European countries are severely polluted. Currently, more than two-thirds of India's water resources are polluted with industrial wastes and sewage. Of the 78 streams monitored in China, 54 are seriously polluted with untreated sewage and industrial wastes, and 20% of China's rivers are too polluted to use for irrigation. In Latin America and Africa, most streams passing through urban or industrial areas are severely polluted.

What Are the Pollution Problems of Lakes? In lakes, reservoirs, and ponds, dilution is often less effective than in streams. Lakes and reservoirs often contain stratified layers (Figure 8-12) that undergo little vertical mixing, and ponds contain small volumes of water. Stratification also reduces levels of dissolved oxygen, especially in the bottom layer (Figure 8-14). In addition, lakes, reservoirs, and ponds have little flow, further reducing dilution and replenishment of dissolved oxygen. The flushing and changing of water in lakes and large artificial reservoirs can take from 1 to 100 years, compared with several days to several weeks for streams.

Thus, lakes, reservoirs, and ponds are more vulnerable than streams to contamination by plant nutrients, oil, pesticides, and toxic substances such as lead, mercury, and selenium. These contaminants can destroy both bottom life and fish and birds that feed on contaminated aquatic organisms. For example, selenium-contaminated water flowing from irrigated croplands into ponds and lakes in and near the Kesterson National Wildlife Refuge in California's San Joaquin Valley has killed thousands of waterfowl and fish and has affected populations of other forms of wildlife that feed on such species. Atmospheric fallout and runoff of acids is a serious problem in lakes subject to acid deposition (Figure 18-8).

Concentrations of some chemicals, such as DDT, PCBs (Figure 20-5), some radioactive isotopes, and

A: About 45,000 (20,000 belong to the Department of Energy and Department of Defense)

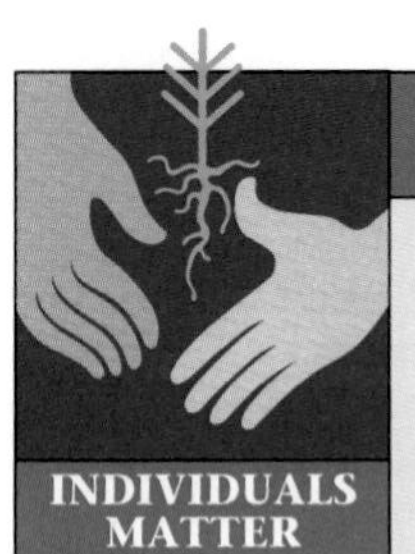

Tracking Down the Pfiesteria Cell from Hell

INDIVIDUALS MATTER

Dr. JoAnn M. Burkholder is an aquatic botanist at North Carolina State University. She knows what it is to be sickened by a fish-killing newly identified microbe and to experience the political heat when you go public with your research to alert people about a potentially serious health threat.

In 1986 she investigated why a colleague's laboratory research fish were dying mysteriously and discovered that the culprit was a new microbe so tiny that dozens could fit on the head of a pin. She named it Pfiesteria (pronounced fee-STEER-e-ah) piscida (pis-kuh-SEED-uh)—Latin for "fish killer" and dubbed by some biologists the "cell from hell."

She has discovered that this complex microscopic organism can behave as both a plant and an animal and assume at least 24 guises in its lifetime. Most of their life Pfiesteria masquerade as a photosynthesizing plant or lie dormant in a nontoxic cyst stage in the bottom sediment of rivers and bays. During most of its life stages the organism feeds on bacteria and bits of algae.

However, when conditions are just right—calm, warm, shallow waters with plentiful supplies of nitrogen and phosphorus nutrients—these benign microbes can turn vicious when they detect chemicals given off by schools of live fish or their excrement or by organisms ranging from blue crabs to eels. Then the microbes change from algae eaters into fish-killing dinoflagellates that sprout tails and move at speeds up to 480 kilometers per hour (300 miles per hour). These rapidly moving organisms then release a water-soluble neurotoxin that stuns the fish or other prey and usually kills them within 10 minutes to several hours.

Next the Pfiesteria excrete a lipid-soluble toxin that causes oozing skin lesions on their prey. After feeding on blood and skin sloughed off from the sores, the swollen and sated organisms reproduce furiously, change shape, and return to dormancy in bottom sediment, usually within 48 hours.

The neurotoxin can also form an aerosol above the water. In 1993, Dr. Burkholder and her chief research aid experienced nausea, burning eyes and cramps, weakness, slow-healing sores, difficulty breathing, and severe loss of memory and mental powers from breathing toxic fumes released in tanks of fish dying from Pfiesteria attacks. They eventually recovered, but still cannot exercise strenuously without severe shortness of breath and the onset of respiratory illness. Since then more than 100 researchers, fishermen, and waterskiers in North Carolina, Virginia, and Maryland have experienced one or more of these symptoms when exposed to water or air contaminated by Pfiesteria toxins.

She developed evidence through lab and field research that connected outbreaks or blooms of Pfiesteria with excessive levels of nitrogen (as nitrates) and phosphorus (as phosphates) nutrients in rivers and estuaries. High levels of such nutrients are found in runoff from fertilized croplands, industrial development, and feedlots (especially

some mercury compounds can be biologically magnified as they pass through food webs in lakes. Many toxic chemicals also enter lakes and reservoirs from the atmosphere.

Lakes receive inputs of nutrients and silt from the surrounding land basin as a result of natural erosion and runoff. This natural nutrient enrichment of lakes is called **eutrophication**. Over time, some of these lakes become more eutrophic (Figure 8-13, top), but others don't because of differences in the surrounding water basin. Near urban or agricultural areas, human activities can greatly accelerate the input of nutrients to a lake, which results in a process known as **cultural eutrophication**. Such a change is caused mostly by nitrate- and phosphate-containing effluents from sewage treatment plants, runoff of fertilizers and animal wastes, and accelerated erosion of nutrient-rich topsoil (Figure 20-6).

For example, a 1997 U.S. Senate study found that the amount of animal manure produced in the United States is 130 times greater than the amount of human waste, and there are no federal standards for dealing with this waste. A recently completed 202-square-kilometer (78-square-mile) hog farm in Utah will produce more than 2.5 million pigs a year and potentially put out more animal waste than the city of Los Angeles.

Some of the animal waste produced in the United States is used to fertilize cropland. But much of it is stored in pits or lagoons that can pollute the air, surface water, and groundwater (Individuals Matter, above). In 1996, more than 40 animal waste spills from feedlot lagoons killed 670,000 fish in Iowa, Minnesota, and Missouri.

During hot weather or drought, this nutrient overload produces dense growths of organisms such as algae, cyanobacteria, water hyacinths, and duckweed. Dissolved oxygen (in both the surface layer of water near the shore and in the bottom layer) is depleted when large masses of algae die, fall to the bottom, and are decomposed by aerobic bacteria. This oxygen depletion can kill fish and other aquatic ani-

Q: How much will it cost U.S. taxpayers to clean up contaminated nuclear weapon production facilities?

those used to raise hogs and chickens) into rivers flowing into coastal estuaries.

Eastern North Carolina is home to more than 3,500 industrial-scale hog farms and more than 16 million pigs—second only to Iowa in pork production. Untreated wastes from these hog farms and stored in lagoons have often spilled into nearby rivers. In 1995, five major lagoon spills dumped millions of gallons of hog wastes into the state's rivers. One spill caused by a heavy rain spilled a volume of hog waste three times the volume of oil spilled in the Exxon Valdez disaster (p. 547) into a nearby river. That same year more than 10 million fish died, many of their bodies covered with bleeding sores associated with attacks by Pfiesteria.

These outbreaks are of deep concern because North Carolina's Albemarle–Pamlico Sound provides half of the nursery waters for fish spawned on the East Coast between Maine and Florida. In 1997, *Pfiesteria* outbreaks cost $60 million in losses to U.S. fisheries and tourism.

In 1991, Dr. Burkholder went public with her findings and urged state legislators to put curbs on hog farming and enact much tougher laws to reduce the flow of nutrients and other pollutants into the state's rivers. Hog farmers, developers, farming interests, fishing industry officials (worried about whether it is safe to eat fish and shellfish from affected rivers and estuaries), tourist-industry officials (alarmed about a negative image on the state's huge coastal recreational industry), and some state officials reacted negatively to her political activism. Some challenged her character and competence and accused her of using the results of preliminary research to push for questionable policies. She also received some anonymous death threats.

But she did not back down and continued criticizing state health officials and legislators for not taking her concerns about public health seriously enough. Eventually, under the glare of state and national publicity, state officials softened their public criticism. The state now supports research on the problem and in 1997 the North Carolina legislature passed a bill putting a 2-year moratorium on new hog farms, reducing nutrient levels allowed in wastewater discharges and requiring better management of land draining into rivers.

Her initial concerns have been vindicated as Pfiesteria have been implicated in fish kills in rivers flowing into the Chesapeake Bay in Maryland, the Rappaha-nock River in Virginia, and other rivers in Alabama, Delaware, and Florida. With the proper nutrient environment and other conditions, the organism can live in waters as far south as the Gulf of Mexico and thrive throughout much of the world.

Research by other scientists has confirmed the link between Pfiesteria outbreaks and nutrient overloading of rivers. By 1997, a number of federal and state environmental, health, and agricultural agencies had set up a coordinated research effort to learn more about what triggers outbreaks of the organism and how they affect human health and other organisms.

mals. If excess nutrients continue to flow into a lake, anaerobic bacteria take over and produce gaseous decomposition products such as smelly, highly toxic hydrogen sulfide and flammable methane.

About one-third of the 100,000 medium to large lakes and about 85% of the large lakes near major population centers in the United States suffer from some degree of cultural eutrophication. One-fourth of China's lakes are classified as eutrophic.

Ways to *prevent* or reduce cultural eutrophication include advanced waste treatment (Section 20-5), bans or limits on phosphates in household detergents and other cleaning agents, and soil conservation and land-use control to reduce nutrient runoff.

Major *cleanup methods* are dredging bottom sediments to remove excess nutrient buildup, removing excess weeds, controlling undesirable plant growth with herbicides and algicides, and pumping air through lakes and reservoirs to avoid oxygen depletion (an expensive and energy-intensive method).

As usual, pollution prevention is more effective and usually cheaper in the long run than pollution control. If excessive inputs of limiting plant nutrients stop, a lake can usually return to its previous state.

Seattle's Lake Washington is a success story of recovery from severe eutrophication after decades of use as a sewage repository. The recovery took place after the sewage was diverted into Puget Sound. This worked for three reasons. First, a large body of water (Puget Sound) was available to receive the sewage wastes. Second, the lake had not yet filled with weeds and sediment because of its large size and depth. Third, preventive corrective action was taken before the lake had become a shallow, highly eutrophic lake (Figure 8-13, top).

Case Study: Chemical and Genetic Pollution in the Great Lakes The five interconnected Great Lakes (Figure 20-7) contain at least 95% of the

A: At least $400–900 billion over the next 75 years

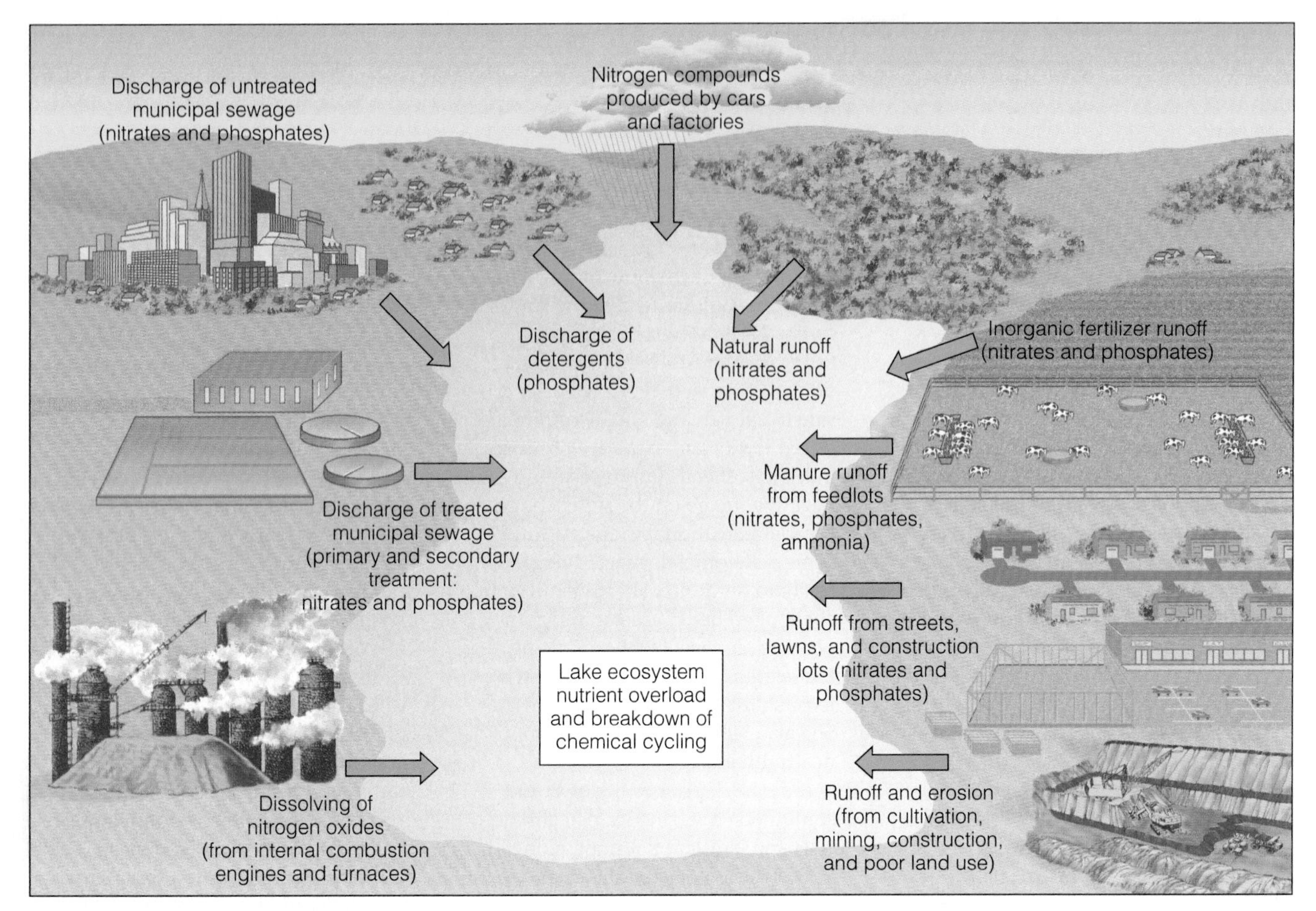

Figure 20-6 Principal sources of nutrient overload causing cultural eutrophication in lakes. The amount of nutrients from each source varies according to the types and amounts of human activities occurring in each airshed and watershed. Levels of dissolved oxygen (Figure 20-3) drop when enlarged populations of algae and plants (stimulated by increased nutrient input) die and are decomposed by aerobic bacteria. Lowered oxygen levels can kill fish and other aquatic life and reduce the aesthetic and recreational values of the lake.

fresh surface water in the United States and 20% of the world's fresh surface water. The Great Lakes basin is home for about 38 million people, about 30% of the Canadian population and 14% of the U.S. population.

Despite their enormous size, these lakes are vulnerable to pollution from point and nonpoint sources because less than 1% of the water entering the Great Lakes flows out to the St. Lawrence River each year. In addition to land runoff, these lakes receive large quantities of acids, pesticides, and other toxic chemicals by deposition from the atmosphere (often blown in from hundreds or thousands of kilometers away).

By the 1960s many areas of the Great Lakes were suffering from severe cultural eutrophication, huge fish kills, and contamination from bacteria and other wastes. The impact on Lake Erie was particularly intense because it is the shallowest of the Great Lakes. Many bathing beaches had to be closed, and by 1970 the lake had lost nearly all its native fish.

Since 1972, a $20-billion pollution-control program, carried out jointly by Canada and the United States, has significantly decreased levels of phosphates, coliform bacteria, and many toxic industrial chemicals in the Great Lakes. Algae blooms have also decreased, dissolved oxygen levels and sport and commercial fishing have increased, and most swimming beaches have reopened. These improvements were brought about mainly by new or upgraded sewage treatment plants, better treatment of industrial wastes, and banning of phosphate detergents, household cleaners, and water conditioners. Even so, less than 3% of the lakes' shoreline is clean enough for swimming or for supplying drinking water.

Levels of several toxic chlorinated hydrocarbon pollutants such as DDT and PCBs in Great Lakes water have dropped to their lowest levels in two decades. Despite this progress, contamination from toxic wastes flowing into the lakes (especially Lakes Erie and On-

Q: What countries have the most sites contaminated with radioactive materials?

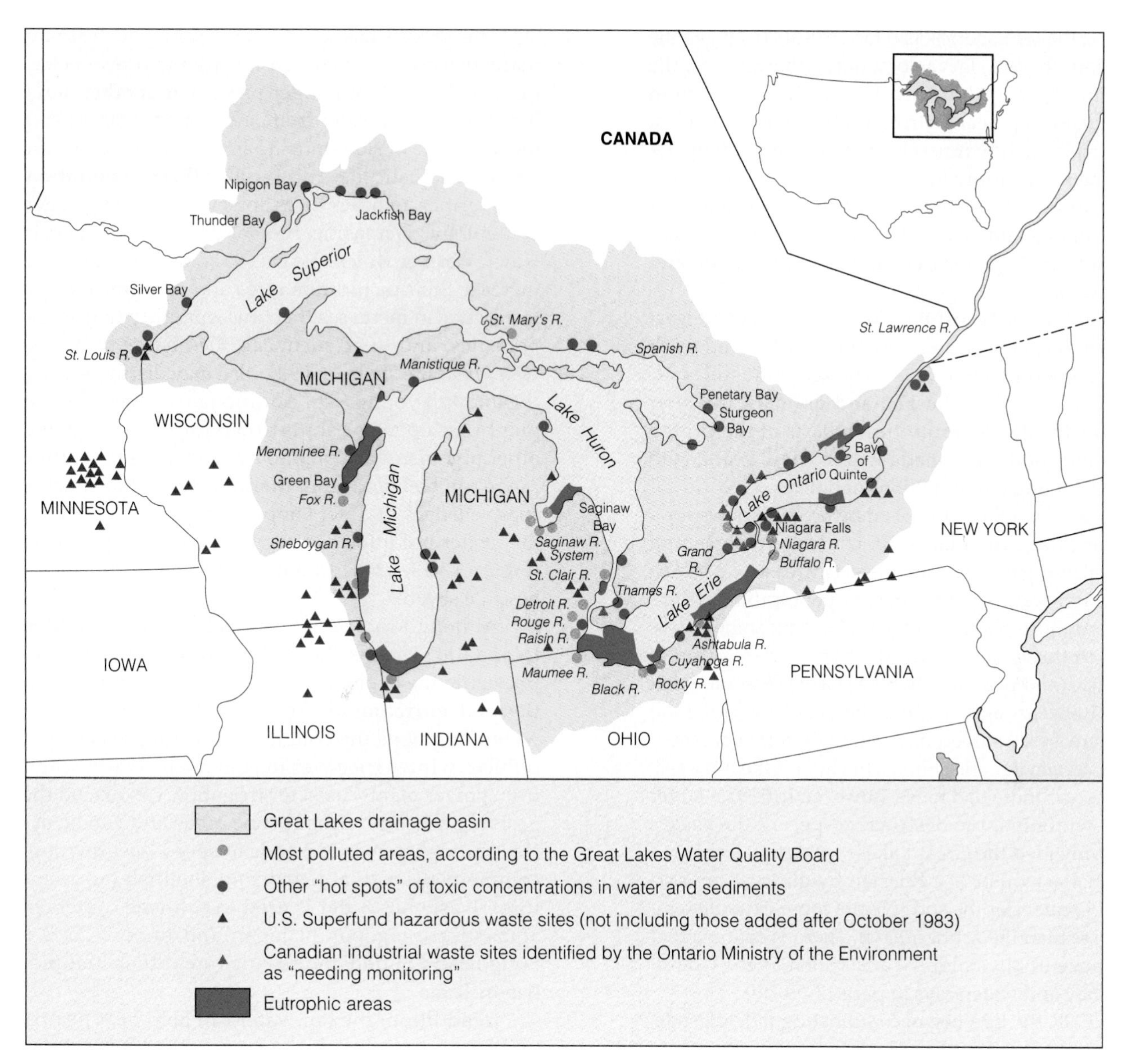

Figure 20-7 The Great Lakes basin and the locations of some of its water quality problems. (Data from Environmental Protection Agency)

tario) from land runoff, streams, and atmospheric deposition (which accounts for an estimated 50% of the input of toxic compounds) is still a serious problem.

Toxic chemicals such as PCBs have built up in food chains and webs (Figure 20-5), contaminated many types of sport fish, and depleted populations of birds, river otters, and other animals feeding on contaminated fish. There is growing evidence and concern about possible effects of exposure to small amounts of many of these substances on the hormone systems of wildlife and humans (Section 17-3). A survey by Wisconsin biologists revealed that one fish in four taken from the Great Lakes is unsafe for human consumption.

In 1991 the U.S. government passed a law requiring accelerated cleanup of the lakes, especially 42 toxic hot spots, and an immediate reduction in air pollutant emissions in the region. However, meeting these goals may be delayed by a lack of federal and state funds.

Some environmentalists call for a ban on the use of chlorine as a bleach in the pulp and paper industry around the Great Lakes, a ban on all new incinerators in the area, and an immediate ban on discharge into the lakes of 70 toxic chemicals that threaten human health and wildlife. Understandably, officials of these industries strongly oppose such bans.

A: The republics of the former Soviet Union

Great lakes fisheries also face threats from genetic pollution. In 1986, larvae of a nonnative species, the *zebra mussel*, arrived in ballast water discharged from a European ship near Detroit. With no known natural enemies, these tiny mussels have run amok; they deplete the food supply for other lake species, clog irrigation pipes, shut down water intake systems for power plants and city water supplies, foul beaches, and grow in huge masses on boat hulls, piers, and other surfaces.

Zebra mussels cost the Great Lakes basin at least $500 million per year, and the annual costs could reach $5 billion within a few years. The zebra mussel is expected to spread unchecked and dramatically alter most freshwater communities in parts of the United States and southern Canada within a few years, with damage costing tens of billions of dollars.

However, zebra mussels may be good news for a number of aquatic plants. By consuming algae and other microorganisms, the mussels increase water clarity. Clearer waters permit deeper penetration of sunlight and more photosynthesis, allowing some native plants to thrive and return the plant composition of Lake Erie (and presumably other lakes) closer to what it was 100 years ago. Because the plants provide food and increase dissolved oxygen, their comeback may benefit certain aquatic animals (including the mussels).

There is more bad news, however. In 1991 a larger and potentially more destructive species, the *quagga mussel*, invaded the Great Lakes, probably discharged in the ballast water of a Russian freighter. It can survive at greater depths and tolerate more extreme temperatures than the zebra mussel. There is concern that it may eventually colonize areas such as the Chesapeake Bay and waterways in parts of Florida.

In 1995, the EPA began coordinating federal, state, and nongovernment agencies in a 5-year strategy to improve the water quality and biological life of the Great Lakes. This plan is an integrated ecosystem management approach for problem solving and decision making in the Great Lakes region. Time will reveal how well this important new conceptual approach will work. Another lake under stress from pollution is Siberia's Lake Baikal (Case Study, right).

How Can Streams and Lakes Be Polluted by Heat? About two-thirds of the energy in the fuel used by coal-burning and nuclear power plants is converted to heat that must be dissipated into the environment. The cheapest and easiest method is to withdraw nearby surface water, pass it through the plant, and return the heated water to the same body of water (Figure 15-17). Almost half of all water withdrawn in the United States each year is for cooling electric power plants.

This process has several drawbacks. For one thing, many fish die on intake screens used to prevent clogging of the heat exchanger pipes. For another, large inputs of heated water from one or more plants using the same lake or slow-moving stream can have harmful effects on aquatic life. This is called **thermal pollution**.

Warmer temperatures lower dissolved oxygen content by decreasing the solubility of oxygen in water. Warmer water also causes aquatic organisms to increase their respiration rates and consume oxygen faster; it also increases their susceptibility to disease, parasites, and toxic chemicals. Discharge of heated water near the shore of a lake also may disrupt spawning and kill young fish. Additionally, when a power plant first opens or shuts down for repair, fish and other organisms adapted to a particular temperature range can be killed from **thermal shock**: the effect of sharp changes in water temperature. Experts rate thermal water pollution as a low-risk ecological problem (Figure 17-13, left), although its localized impact can be quite severe.

Although some scientists call excess heat added to aquatic systems thermal pollution, others—emphasizing the beneficial uses of heated water—call it **thermal enrichment**. They point out that heated water lengthens the commercial fishing season and reduces winter ice cover in cold areas. Warm water from power plants, used for irrigation, can extend the growing season in frost-prone areas and can be cycled through pens and ponds to speed the growth of commercially valuable fish and shellfish (aquaculture). Waste hot water is used to cultivate oysters in aquaculture lagoons in Japan and in New York's Long Island Sound and to cultivate catfish and redfish in Texas.

In addition, the hot water can help heat nearby buildings and greenhouses and desalinate ocean water, and it can be run under sidewalks to melt snow. However, because of dangers related to air pollution and release of radioactivity, most coal-burning and nuclear power plants are located too far from aquaculture operations, buildings, and industries to make some applications of thermal enrichment economically feasible.

Major ways to reduce or control thermal water pollution are **(1)** using and wasting less energy (Section 16-2), **(2)** limiting the amount of heated water discharged into a body of water, **(3)** returning the heated water some distance away from the ecologically vulnerable shore zone, **(4)** transferring the heat from the water to the atmosphere by means of huge cooling towers, and **(5)** discharging the heated water into shallow ponds or canals, allowing it to cool, and reusing it as cooling water (if enough land is available).

Q: Does using nuclear power add carbon dioxide to the atmosphere?

Protecting Lake Baikal

CASE STUDY

Lake Baikal, located in Siberia, contains the world's largest volume of fresh water. After a 20-million-year unbroken history of evolution, it is also one of the most biologically rich lakes on the earth. It contains about 1,700 species of plants and animals, 1,200 of which are not found anywhere else on the earth. The lake's huge watershed consists almost entirely of boreal forest or taiga.

The installation of two paper mills on the lake in the 1960s started an environmental battle that continues today. There was concern over whether the proposed wastewater treatment plants at the two paper mills would be adequate to preserve the quality of Lake Baikal's waters. This concern led to an upgrading of the waste treatment plants, appointment of a commission to monitor the lake's water quality, and the establishment of protected areas around parts of the lake to reduce the input of erosion from logging and other land development. Officials subsequently decided to send the wood pulp produced at the plants elsewhere for processing into paper, to reduce pollution of the lake.

Because of continuing pressure from scientists and environmentalists, one of the mills was converted in 1987 to a furniture factory (which is much less polluting), a closed-cycle water system was ordered for the other plant, and timber cutting was reduced. Since the breakup of the Soviet Union, the lake has faced serious new threats; unregulated Russian business interests and German, Japanese, American, and South Korean investors are seeking access to Siberia's vast timber resources (amounting to 26% of the world's remaining forests) and its oil and mineral deposits.

Because of its biological uniqueness, protecting the lake is ultimately a global concern. In 1992, a team of scientists, two-thirds Russian and one-third American, was assembled to evaluate the lake and its watershed and to develop a land-use zoning plan for ecologically sustainable development of the entire watershed.

In 1993, Russian President Boris Yeltsin created a Baikal Commission to help implement the team's recommendations, but so far it has not received enough money to carry out this mission. Although success is not assured, the Lake Baikal plan serves as a model for sustainable development of watersheds elsewhere.

Critical Thinking

Do you believe that the biodiversity of Lake Baikal should be protected in light of Siberia's need for jobs and capital to develop its timber, oil, and mineral resources? Explain. Use the literature and the Internet to find out the latest developments in the struggle to protect Lake Baikal.

20-3 OCEAN POLLUTION

How Much Pollution Can the Oceans Tolerate? The oceans are the ultimate sink for much of the waste matter we produce, as summarized in the African proverb, "Water may flow in a thousand channels, but it all returns to the sea."

Oceans can dilute, disperse, and degrade large amounts of raw sewage, sewage sludge, oil, and some types of industrial waste, especially in deep-water areas. Marine life has also proved to be much more resilient than some scientists had expected, leading some to suggest that it is generally safer to dump sewage sludge and most other hazardous wastes into the deep ocean than to bury them on land or burn them in incinerators.

Other scientists dispute this idea, pointing out that we know less about the deep ocean than we do about outer space. They add that dumping waste in the ocean would delay urgently needed pollution prevention and promote further degradation of this vital part of the earth's life-support system.

How Do Pollutants Affect Coastal Areas? Coastal areas—especially wetlands and estuaries, coral reefs, and mangrove swamps—bear the brunt of our enormous inputs of wastes into the ocean. This is not surprising because half the world's population lives on or within 100 kilometers (160 miles) of the coast, and coastal populations are growing at a more rapid rate than global population. Fourteen of the world's 15 largest metropolitan areas, each with 10 million people or more, are near coastal waters.

In most coastal developing countries (and in some coastal developed countries), municipal sewage and industrial wastes are often dumped into the sea without treatment. The most polluted seas lie off the densely populated coasts of Bangladesh, India, Pakistan, Indonesia, Malaysia, Thailand, and the Philippines. About 85% of the sewage from large cities along the Mediterranean Sea, which has a coastal population of 200 million people during tourist season, is discharged into the sea untreated, causing widespread beach pollution and shellfish contamination.

A: Yes, but it produces only one-sixth as much CO_2 per unit of electricity as a coal-burning plant

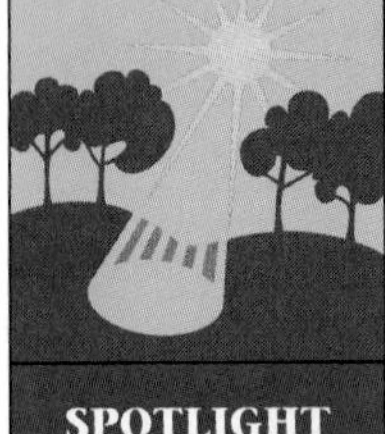

SPOTLIGHT

The Rise in Ocean Blooms of Toxic Tides

In recent years there has been a sharp increase in the frequency and magnitudes of harmful blooms of algae and other organisms in coastal waters of the United States and in other areas.

These blooms of toxic organisms are called red tides (caused by dinoflagellates), green tides (caused by cyanobacteria), or brown tides (caused by marine plankton called chrysophytes, also known as golden brown algae) depending on their color. The organisms responsible for red and green tides can release waterborne and airborne toxins that damage fisheries, kill some fishing-eating birds, reduce tourism, and poison seafood.

When the organisms die and decompose, coastal waters are depleted of oxygen and a variety of marine species die. A growing body of evidence links these harmful blooms to cultural eutrophication, mostly from the runoff of fertilizers and animal wastes into inland rivers and from there to coastal waters. For example, runoff of poultry waste is suspected in the bloom of a toxic microbe that killed about 480,000 fish in the coastal waterways of Mid-Atlantic states in 1997.

Large quantities of nutrients reaching coastal waters can create a hypoxic zone (sometimes inaccurately called a dead zone) so low in dissolved oxygen near the bottom (Figure 8-2) that some fish and most bottom dwellers (such as shrimp, crabs, and starfish) are killed or driven away. In the Gulf of Mexico, such a zone lasts about 8 months a year and covers an area of up to 18,000 square kilometers (7,000 square miles). This zone is overfertilized by plant nutrients in the Mississippi River, with a gigantic drainage basin receiving inputs from 40% of the land area of the contiguous states.

Studies show that since 1960 the amount of dissolved nitrogen in the outflow of the Mississippi River (and the adjacent Atchafalaya) has tripled and phosphorus levels have doubled. According to the U.S. Geological Survey, about 56% of the nitrogen is from fertilizer runoff, 25% from animal manure, and 6% from domestic wastes. Toxic algal blooms, fed by excessive nutrients, are becoming more common in the Gulf and increased more than fourfold in U.S. coastal waters between 1976 and 1996.

In the Baltic Sea such excessive cultural eutrophication has killed almost all bottom-dwelling animal life over an area of about 109,000 square kilometers (42,000 square miles)—about the size of Guatemala or Tennessee. In 1998 a deadly red tide appeared on the China coast, where none have ever been recorded before. Within a few hours it wiped out fish farms, leaving thousands of tons of rotting fish.

Critical Thinking

Why is so difficult to reduce the nonpoint pollution that causes cultural eutrophication? What three things would you do to sharply reduce cultural eutrophication and the accompanying toxic tides?

In the United States about 35% of all municipal sewage ends up virtually untreated in marine waters. Most U.S. harbors and bays are badly polluted from municipal sewage, industrial wastes, and oil. Scuba divers talk of swimming through clouds of half-dissolved feces, and of bay and harbor bottoms covered with foul and toxic sediment known as black mayonnaise. They see lobsters and crabs covered with mysterious burn holes, and fish with cancerous sores and rotting fins.

California's Santa Monica Bay is the filming site for the widely watched television show *Baywatch*, which gives an illusion of a clean California beach lifestyle. What viewers don't know is that the bay is so polluted that the actors get extra pay each time they enter the water and are chemically cleaned afterward.

Runoff of sewage and agricultural wastes into coastal waters and acid deposition from the atmosphere (Figure 18-7) introduce large quantities of nitrogen and phosphorus, which can cause explosive growth or bloom of algae or other organisms (Spotlight, above).

Case Study: The Chesapeake Bay The Chesapeake Bay, the largest estuary in the United States, is in trouble because of human activities. Between 1940 and 1995, the number of people living in the Chesapeake Bay area grew from 3.7 million to 15.5 million, and within a few years its population may reach 18 million.

The estuary receives wastes from point and nonpoint sources scattered throughout a huge drainage basin that includes 9 large rivers and 141 smaller streams and creeks in parts of six states (Figure 20-8). The bay has become a huge pollution sink because it is quite shallow and because only 1% of the waste entering it is flushed into the Atlantic Ocean.

Levels of phosphates and nitrates have risen sharply in many parts of the bay, causing algae blooms and oxygen depletion (Figure 20-8). Studies have shown that point sources, primarily sewage treatment plants, contribute about 60% by weight of the phosphates. Nonpoint sources—mostly runoff from urban, suburban, and agricultural land and deposition from

Q: Nuclear weapons existing today could kill everyone in the world how many times over?

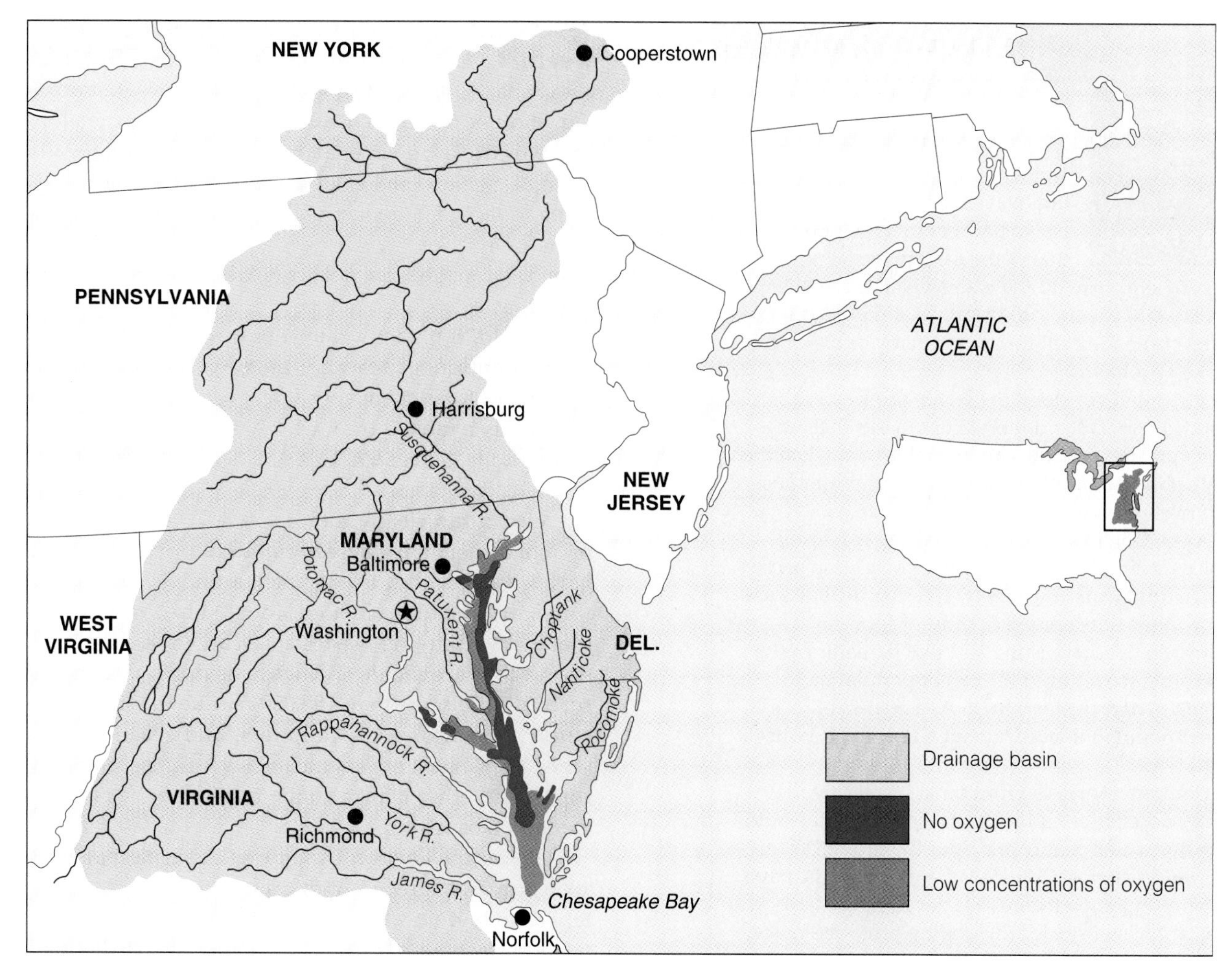

Figure 20-8 Chesapeake Bay, the largest estuary in the United States, is severely degraded as a result of water pollution from point and nonpoint sources in six states and from deposition of air pollutants.

the atmosphere—are the origins of about 60% by weight of the nitrates.

Air pollutants account for nearly 35% of the nitrogen entering the estuary. In addition, large quantities of pesticides run off cropland and urban lawns and industries discharge large amounts of toxic wastes, often in violation of their discharge permits. Commercial harvests of oysters (Solutions, p. 546), crabs, and several important fish have fallen sharply since 1960 because of a combination of overfishing, pollution, and disease.

In the 1980s the Chesapeake Bay Program, the country's most ambitious attempt at integrated coastal management, was implemented. Results have been impressive. Between 1987 and 1993, phosphorus levels declined 16% through a combination of bans on phosphorus-containing detergents, upgrades of municipal sewage treatment plants, and soil erosion controls and nutrient management on agricultural land. During this same period nitrogen levels dropped 7%—a significant achievement given the increasing population in the watershed and the fact that more than a third of the nitrogen inputs come from the atmosphere. These drops in nutrients led to a 75% increase in the abundance of submerged vegetation between 1978 and 1993 and a rebound of the striped bass population (also helped by strict limits on fishing).

In 1997, federal and state officials agreed to offer Maryland participating farmers more than $250 million in subsidies to leave potential cropland near the bay unplanted and to create natural buffers that would reduce contaminating runoff into the Chesapeake Bay.

Reaching the goals of a 40% reduction in nutrient levels and a significant improvement in habitat water quality throughout the bay will be especially difficult because the area's population is expected to grow by

A: About 40 (20 if current arms reduction proposals are carried out)

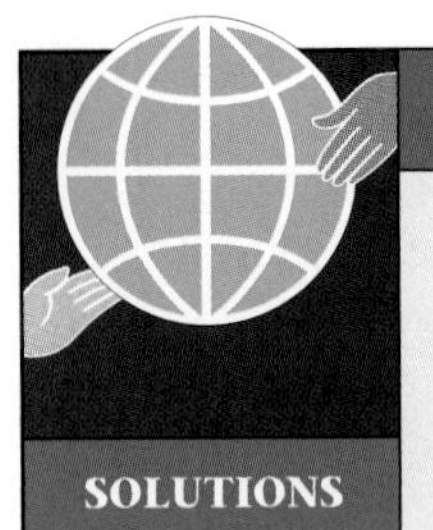

Bring Back the Oysters

SOLUTIONS

A growing number of scientists and environmentalists want to find ways to rebuild the Chesapeake Bay's once huge population of the eastern oyster as a way to help clean up the water. Oysters are filter feeders that vacuum up the algae and nutrient-laden suspended silt that cause many of the Chesapeake's problems.

Algae and silt cloud the water and prevent sunlight from reaching underwater grasses that are vital nurseries for crabs and fish. Further, when the algae populations bloom and die their decomposition robs the water of dissolved oxygen.

According to aquatic biologist Roger Newell, the bay's once prodigious oyster population served as a natural water purifier by filtering the bay's entire volume of water every three or four days. However, overharvesting and two parasitic oyster diseases have reduced the oyster population to about 1% of its historic high. As a result, today's oyster population needs about a year to filter the bay's water.

Computer models project that increasing the oyster population to 10% of its historic high would improve water quality and spur the growth of underwater sea grass. Methods for restoring the bay's oyster population include **(1)** developing disease-resistant oyster stocks, **(2)** seeding protected oyster beds with large, older oysters presumed to have some disease resistance, **(3)** dumping hundreds of millions of oyster shells on dozens of historic reefs areas with the goal of resurrecting some of the old oyster breeding reefs, **(4)** trying to find a reef-building substitute to hasten reef reconstruction, **(5)** setting aside 20–25% of the bay's oyster beds as sanctuaries to protect stocks from overfishing, and **(6)** greatly increasing funds for research and implementation of such a program.

Critical Thinking

Should more of the scarce funds for reducing pollution and degradation of the Chesapeake Bay be diverted from other efforts to increasing the oyster population? Explain.

25% between 1995 and 2020. Moreover, the bay will soon be invaded by zebra and quagga mussels. So far, however, the Chesapeake Bay Program shows what can be done when a diversity of interested parties work together to achieve goals that benefit both wildlife and people.

What Pollutants Are Dumped into the Ocean? Dumping of industrial waste off U.S. coasts has stopped, although it still occurs in a number of other developed countries and some developing countries. However, barges and ships still legally dump large quantities of **dredge spoils** (materials, often laden with toxic metals, scraped from the bottoms of harbors and rivers to maintain shipping channels) at 110 sites off the Atlantic, Pacific, and Gulf coasts.

In addition, many countries dump into the ocean large quantities of **sewage sludge**: a gooey, mudlike, mixture of toxic chemicals, infectious agents, and settled solids removed from wastewater at sewage treatment plants. Since 1992 this practice has been banned in the United States.

Fifty countries with at least 80% of the world's merchant fleet have agreed not to dump sewage and garbage at sea, but this agreement is difficult to enforce and is often violated. Most ship owners save money by dumping wastes at sea and risk only small fines if they are caught. Each year as many as 2 million seabirds and more than 100,000 marine mammals (including whales, seals, dolphins, and sea lions) die when they ingest or become entangled in fishing nets, ropes, and other debris dumped into the sea and discarded on beaches.

Under the London Dumping Convention of 1972, 100 countries agreed not to dump highly toxic pollutants and high-level radioactive wastes in the open sea beyond the boundaries of their national jurisdictions. Since 1983 these same nations have observed a moratorium on the dumping of low-level radioactive wastes at sea, which in 1994 became a permanent ban. However, France, Great Britain, Russia, China, and Belgium may legally exempt themselves from this ban. In 1992 it was learned that for decades the former Soviet Union had been dumping large quantities of high- and low-level radioactive wastes into the Arctic Ocean and its tributaries.

What Are the Effects of Oil on Ocean Ecosystems? *Crude petroleum* (oil as it comes out of the ground) and *refined petroleum* (fuel oil, gasoline, and other processed petroleum products; Figure 15-6) are accidentally or deliberately released into the environment from a number of sources.

Although tanker accidents and blowouts at offshore drilling rigs (when oil escapes under high pressure from a borehole in the ocean floor) get most of the publicity, more oil is released during normal operation of offshore wells, from washing tankers and releasing the oily water, and from pipeline and storage tank leaks. A 1993 Friends of the Earth study estimated that each year U.S. oil companies

Q: How much does it cost to decommission a worn out and highly radioactive nuclear power plant?

unnecessarily spill, leak, or waste an amount of oil equal to that shipped by 1,000 huge *Exxon Valdez* tankers (Case Study, below)—more oil than Australia uses. Oil pollution from shipping in the Mediterranean Sea is equivalent to 17 huge *Exxon Valdez* tankers emptying their tanks per year.

Natural oil seeps also release large amounts of oil into the ocean at some sites, but most ocean oil pollution comes from activities on land. Almost half (some experts estimate 90%) of the oil reaching the oceans is waste oil dumped, spilled, or leaked onto the land or into sewers by cities, individuals, and industries. Each year, a volume of oil equal to 20 times the amount spilled by the *Exxon Valdez* is improperly disposed of by about 50 million U.S. motorists changing their own motor oil. Worldwide, about 10% of the oil that reaches the ocean comes from the atmosphere, mostly from smoke emitted by oil fires.

The effects of oil on ocean ecosystems depend on a number of factors: type of oil (crude or refined), amount released, distance of release from shore, time of year, weather conditions, average water temperature, and ocean currents. Volatile organic hydrocarbons in oil immediately kill a number of aquatic organisms, especially in their vulnerable larval forms. Most of these toxic chemicals evaporate within a day or two in warm waters, but in cold waters this may take up to a week.

Some other chemicals form tarlike globs that float on the surface. This floating oil coats the feathers of birds (especially diving birds) and the fur of marine mammals, destroying the animals' natural insulation and buoyancy; many drown or die of exposure from loss of body heat. The globs of oil are broken down by bacteria over several weeks or months, although they persist much longer in cold polar waters because the chemical reactions involved in decomposition are slowed down. Heavy oil components that sink to the ocean floor or wash into estuaries can smother bottom-dwelling organisms such as crabs, oysters, mussels, and clams or make them unfit for human consumption. Some oil spills have killed reef corals. A recent study also showed that diesel oil spilled at sea becomes more toxic to marine life with the passage of time.

Research shows that most (but not all) forms of marine life recover from exposure to large amounts of crude oil within 3 years. However, recovery from exposure to refined oil, especially in estuaries, may take 10 years or longer. The effects of spills in cold waters and in shallow enclosed gulfs and bays generally last longer.

Oil slicks that wash onto beaches can have a serious economic impact on coastal residents, who lose income from fishing and tourist activities. Oil-polluted beaches washed by strong waves or currents become clean after about a year, but beaches in sheltered areas remain contaminated for several years. Estuaries and salt marshes suffer the most and longest-lasting damage. Despite their localized harmful effects, oil spills are rated by experts as a low-risk ecological problem (Figure 17-13, left).

How Can Oil Spills Be Cleaned Up? If they aren't too large, oils spills can be cleaned up by mechanical, chemical, fire, and natural methods. *Mechanical methods* include using **(1)** floating booms to contain the oil spill or keep it from reaching sensitive areas, **(2)** skimmer boats to vacuum up some of the oil into collection barges, and **(3)** absorbent pads or large feather-filled pillows to soak up oil on beaches or in waters too shallow for skimmer boats.

Chemical methods include **(1)** coagulating agents to cause floating oil to clump together for easier pickup or sink to the bottom, where it usually does less harm, and **(2)** dispersing agents to break up oil slicks. However, these agents can also damage some types of organisms. Fire can burn off floating oil, but crude oil is hard to ignite, and this approach produces air pollution. In time, the natural action of wind and waves mixes or emulsifies oil with water (like emulsified salad dressing) and bacteria biodegrade some of the oil.

These methods remove only part of the oil and none work well on a large spill. This explains why preventing oil pollution is the most effective and in the long run the least costly approach, as revealed by the large spill from the *Exxon Valdez* oil tanker in 1989. However, because oil is also responsible for most of the world's air pollution and the resulting health problems and premature deaths, to most environmentalists the key issue isn't just how to prevent water and air pollution from oil but to get away from using oil by reducing oil waste (Section 16-2) and shifting to various renewable energy resources (Chapter 16).

Case Study: The *Exxon Valdez* Oil Spill Crude oil from Alaska's North Slope fields near Prudhoe Bay is carried by pipeline to the port of Valdez and then shipped by tanker to the West Coast. On March 24, 1989, the *Exxon Valdez*, a tanker more than three football fields long, went off course in a 16-kilometer-wide (11-mile-wide) channel in Prince William Sound near Valdez, Alaska. It hit submerged rocks, creating the worst oil spill ever in U.S. waters (Figure 20-9).

The rapidly spreading oil slick coated more than 1,600 kilometers (1,000 miles) of shoreline, almost the length of the shoreline between New Jersey and South

A: At least as much as it cost to build it

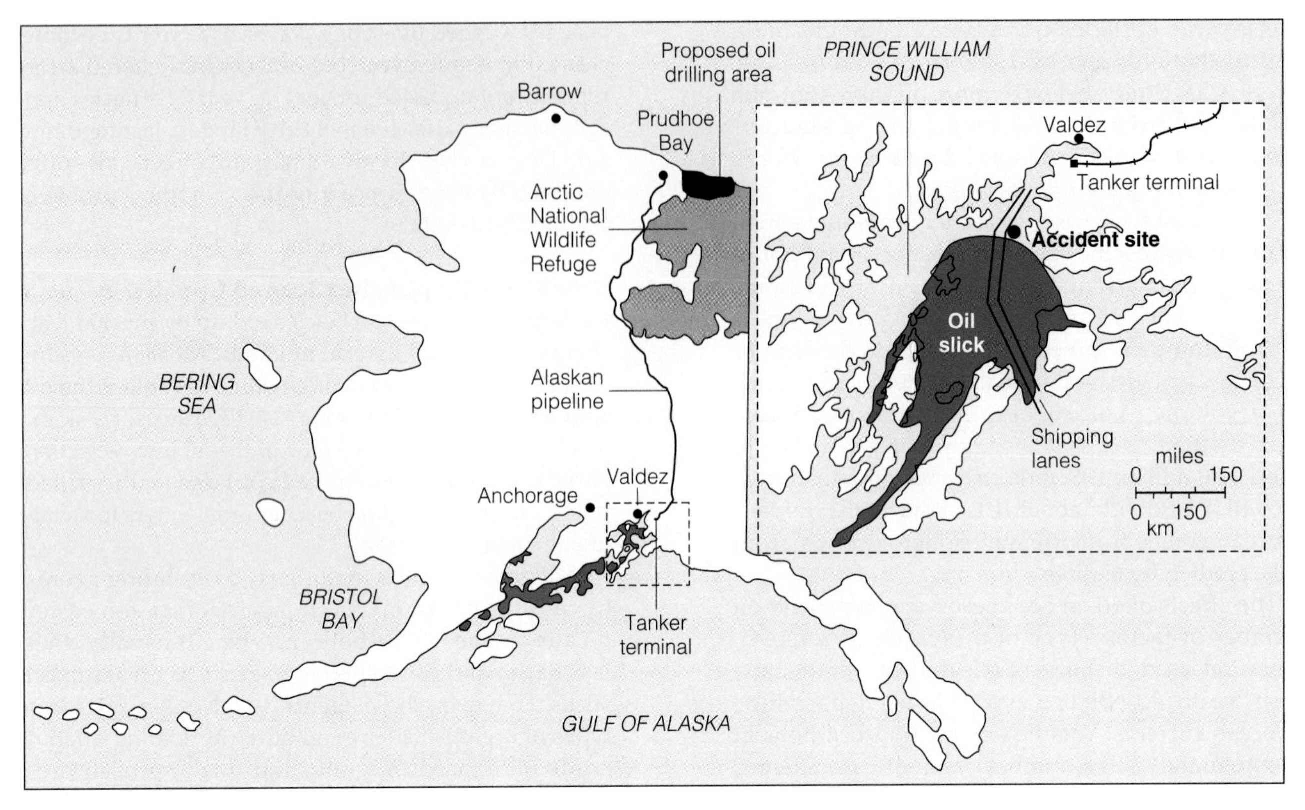

Figure 20-9 Alaska's Prince William Sound, the site of the oil spill from the tanker *Exxon Valdez* on March 24, 1989.

Carolina. The full loss of wildlife will never be known because most of the dead animals sank and decomposed without being counted.

In the early 1970s, environmentalists had predicted that a large, damaging spill might occur in these waters, made treacherous by icebergs, submerged reefs, and violent storms. But officials of Alyeska, a company formed by the seven oil companies extracting oil from Alaska's North Slope, assured Congress that they would be at the scene of any accident within five hours and have enough equipment and trained people to clean up any spill.

However, when the Valdez spill occurred, Alyeska and Exxon officials did not have enough equipment and personnel and responded with too little too late. Even though it became apparent that no amount of equipment and personnel could clean up such a large spill, a prompt response could have helped contain some of the leaking oil and reduce its environmental impact.

Exxon spent $2.2 billion directly on the cleanup, but some aspects of the cleanup effort did more harm than good. For example, the use of high-pressure jets of hot water to clean beaches killed coastal plants and animals that had survived the spill. As a result, a year after the spill the oiled sites had recovered more rapidly than the washed sites.

In 1990, the National Transportation Safety Board ruled that the accident was the result of drinking by the captain, a fatigued and overworked crew, and inadequate traffic control by the Coast Guard. In 1991 Exxon pleaded guilty to federal felony and misdemeanor charges and agreed to pay the federal government and the state of Alaska $1 billion in fines and civil damages.

In 1994, a federal jury found that Exxon had been reckless in allowing Captain Joseph Hazelwood, who had a history of alcohol abuse, to command the *Exxon Valdez* (now repaired, renamed the *Sea River Mediterranean*, and operating in the Mediterranean). The jury also found that Captain Hazlewood had been negligent and reckless in light of testimony that he had downed 14 shots of vodka on the afternoon before the ship left port.

In 1994 a jury awarded members of the fishing industry, landowners, and other Alaskan residents $5 billion in damages and penalties. However, Exxon has appealed this decision and it may be tied up in the courts for decades.

Schneider, Paul. 1997. Clear Progress: 25 Years of the Clean Water Act. *Audubon*, vol. 99, no. 5, 36(14).

This roughly $8.5-billion accident might have been prevented if Exxon had spent only $22.5 million to fit the tanker with a double hull (which it still has not done). In the early 1970s, then–Interior Secretary Rogers Morton told Congress that all oil tankers using Alaskan waters would have double hulls, but under pressure from oil companies the requirement was dropped.

Today, virtually all merchant ships have double hulls, but only 15% of oil tankers have such hulls. Legislation passed since the spill requires all new tankers to have double hulls and all existing large single-hulled oil tankers to be phased out between 1995 and 2015. However, the oil industry is working to weaken these and other stricter requirements enacted since the spill as the public memory of the accident fades. Oil company officials also hope to return the repaired but still single-hulled *Exxon Valdez* tanker (under a new name) to operation in Prince William Sound.

The Oil Protection Act of 1990, passed in the wake of the Valdez spill, was supposed to regulate supertankers and reduced the chances of supertanker oil spills. However, to get around the law many oil carriers have shifted their oil transport operations to lightly regulated oil barges pulled by tugboats. This reduction in oil spill safety has led to several barge oil spills, including the one in January 1996 in Rhode Island's Moonstone Bay and another in March 1996 in Texas's Galveston Bay.

Studies show that others must share the blame for this tragedy. State officials had been lax in monitoring Alyeska, and the Coast Guard did not effectively monitor tanker traffic because of inadequate radar equipment and personnel. Americans who drive fuel inefficient cars and don't insist that we use and waste less oil by requiring greatly increased fuel efficiency standards also share part of the blame.

This spill highlighted the importance of pollution prevention and the advantages of shifting to improving energy efficiency and renewable energy (Chapter 16) to reduce dependence on oil. Even with the best technology and a fast response by well-trained people, scientists estimate that no more than 11–15% of the oil from a major spill can be recovered.

Solutions: How Can We Protect Coastal Waters? The key to protecting oceans is to reduce the flow of pollution from the land and from streams emptying into the ocean. Such efforts must also be integrated with efforts to prevent and control air pollution because an estimated 33% of all pollutants entering the ocean worldwide comes from air emissions from land-based sources.

Some ways various analysts have suggested to prevent and reduce excessive pollution of coastal waters are as follows:

Prevention

- *Encourage or require separate sewage and storm runoff lines in coastal urban areas.*
- *Discourage ocean dumping of sludge and hazardous dredged materials.*
- *Protect sensitive and ecologically valuable coastal areas from development, oil drilling, and oil shipping.*
- *Use ecological land-use planning to control and regulate coastal development.*
- *Require double hulls for all oil tankers by 2002.*
- *Recycle used oil.*
- *Reduce genetic pollution from nonnative aquatic species by using heat to kill organisms in ballast water or developing filters to trap the organisms when ballast water is taken on or discharged from a ship.*

Cleanup

- *Improve oil spill cleanup capabilities.*
- *Require at least secondary treatment of coastal sewage, or use wetlands, solar aquatic, or other environmentally acceptable methods* (Figures 20-1 and 20-2)

20-4 SOLUTIONS: PREVENTING AND REDUCING SURFACE-WATER POLLUTION

What Can We Do About Water Pollution from Nonpoint Sources? The leading nonpoint source of water pollution is agriculture. Farmers can sharply reduce fertilizer runoff into surface waters and leaching into aquifers by using only moderate amounts of fertilizer and by using none at all on steeply sloped land. They can use slow-release fertilizers and alternate their plantings between row crops and soybeans or other nitrogen-fixing plants to reduce the need for fertilizer. Farmers can also plant buffer zones of permanent vegetation between cultivated fields and nearby surface water.

Applying pesticides only when needed can reduce pesticide runoff and leaching. Farmers can also reduce the need for pesticides by using biological control or integrated pest management (Section 21-5). Nonfarm uses of inorganic fertilizers and pesticides—on golf courses, lawns, and public lands, for

Hint: Enter the search term *Clean Water Act of 1977* using the Subject Guide.

example—can also be sharply reduced and replaced with organic methods.

Livestock growers can control runoff and infiltration of manure from feedlots and barnyards by managing animal density, planting buffers, and not locating feedlots on land near surface water when that land slopes steeply toward the water. Diverting the runoff into well-designed detention basins would allow this nutrient-rich water to be pumped out and applied as fertilizer to cropland or forestland.

Another way to reduce nonpoint water pollution, especially from eroded soil, is to reforest critical watersheds. Besides reducing water pollution from sediments, reforestation would reduce soil erosion and the severity of flooding; it would also help slow projected global warming (Section 19-2) and loss of wildlife habitat.

What Can We Do About Water Pollution from Point Sources? The Legal Approach In many developing countries and in some developed countries, sewage and waterborne industrial wastes are discharged without treatment into the nearest waterway or into wastewater lagoons. In Latin America, less than 2% of urban sewage is treated. Only 15% of the urban wastewater in China receives treatment, and in India treatment facilities protect water quality for less than a third of the urban population.

In developed countries, most wastes from point sources are purified to varying degrees. The Federal Water Pollution Control Act of 1972 (renamed the Clean Water Act when it was amended in 1977) and the 1987 Water Quality Act form the basis of U.S. efforts to control pollution of the country's surface waters. The main goals of the Clean Water Act were to make all U.S. surface waters safe for fishing and swimming by 1983 and to restore and maintain the chemical, physical, and biological integrity of the nation's waters. Progress has been made, but these goals have not been met.

The 1972 act required states to develop and execute plans to control nonpoint pollution, something that largely has not been done. It also established a federal wetlands protection program that has been partially successful but faces increasing attempts to weaken its provisions.

In 1995 the EPA developed a *discharge trading policy* designed to use market forces to reduce water pollution (as has been done with sulfur dioxide for air pollution control). The policy would allow a water pollution source, such as an industrial plant or a sewage treatment plant, to sell credits for its excess reductions to another facility that can't reduce its discharges as cheaply.

Here is some good news. The Clean Water Act of 1972 has led to significant improvements in U.S. water quality between 1972 and 1992. The percentage of U.S. rivers and lakes tested that have become fishable and swimmable increased from 36% to 62%.

Here is some bad news. Despite this significant progress, a 1994 report by the EPA revealed a number of problems. Antiquated sewage systems in 1,100 cities still dump poorly treated sewage into streams, lakes, and coastal waters. Between 1995 and 2015, aging U.S. municipal water and sewer systems will need $400–500 billion to comply with existing federal clean water regulations. About 44% of lakes, 37% of rivers, and 32% of estuaries in the United States are still unsafe for fishing, swimming, and other recreational uses. Fish caught in more than 1,400 different waterways are unsafe to eat because of high levels of pesticides and other toxic substances.

Some environmentalists call for the Clean Water Act to be strengthened by **(1)** increasing funding and the authority to control nonpoint sources of pollution; **(2)** strengthening programs to prevent and control toxic water pollution, including phasing out certain toxic discharges (such as many organic chemicals containing chlorine); **(3)** providing more funding and authority for integrated watershed and airshed planning to protect groundwater and surface water from contamination; **(4)** permitting states with good records of environmental stewardship to take over parts of the clean water program, under looser federal control; and **(5)** expanding the ability of citizens to bring lawsuits to ensure that water pollution laws are enforced.

Many industries, state and local officials, and farmers and developers oppose these proposals, contending that the Clean Water Act's regulations are already too restrictive and costly. Farmers and developers also see the law as a curb on their rights as property owners to fill in wetlands; they believe that they should be compensated for property value losses because of federal wetland protection regulations. State and local officials want more discretion in testing for and meeting water quality standards. They argue that in many communities it is too expensive and unnecessary to test for all the water pollutants required by federal law.

What Can We Do About Water Pollution from Point Sources? The Technological Approach In rural and suburban areas with suitable soils, sewage from each house is usually discharged into a **septic tank** (Figure 20-10). About 25% of all homes in the United States are served by septic tanks, which should be cleaned out every 3–5 years by a reputable contractor so that they won't contribute to groundwater pollution.

Q: Can switching to increased use of nuclear power in the United States save much oil?

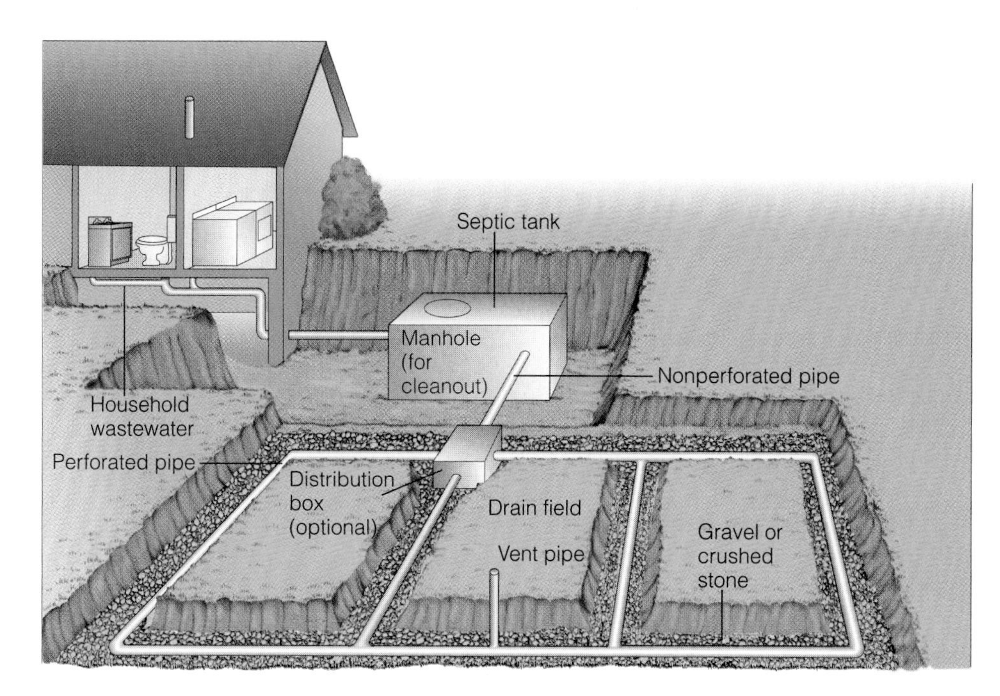

Figure 20-10 Septic tank system used for disposal of domestic sewage and wastewater in rural and suburban areas. This system traps greases and large solids and discharges the remaining wastes over a large drainage field. As these wastes percolate downward, the soil filters out some potential pollutants, and soil bacteria decompose biodegradable materials. To be effective, septic tank systems must be properly installed in soils with adequate drainage, not placed too close together or too near well sites, and pumped out when the settling tank becomes full.

In urban areas, most waterborne wastes from homes, businesses, factories, and storm runoff flow through a network of sewer pipes to wastewater treatment plants. Some cities have separate lines for stormwater runoff, but in 1,200 U.S. cities the lines for these two systems are combined because it is cheaper. When rains cause combined sewer systems to overflow, they discharge untreated sewage directly into surface waters.

When sewage reaches a treatment plant, it can undergo up to three levels of purification, depending on the type of plant and the degree of purity desired. **Primary sewage treatment** is a mechanical process that uses screens to filter out debris such as sticks, stones, and rags; suspended solids settle out as sludge in a settling tank (Figure 20-11). Improved primary treatment uses chemically treated polymers to remove suspended solids more thoroughly.

Secondary sewage treatment is a biological process in which aerobic bacteria are used to remove up to 90% of biodegradable, oxygen-demanding organic wastes (Figure 20-11). Some plants use *trickling filters*, in which aerobic bacteria degrade sewage as it seeps through a bed of crushed stones covered with bacteria and protozoa. Others use an *activated sludge process*, in which the sewage is pumped into a large tank and mixed for several hours with bacteria-rich sludge and air bubbles to facilitate degradation by microorganisms. The water then goes to a sedimentation tank, where most of the suspended solids and microorganisms settle out as sludge. The sludge produced from primary or secondary treatment is broken down in an anaerobic digester and either incinerated, dumped into the ocean or a landfill, or applied to land as fertilizer.

Even after secondary treatment, wastewater still contains about 3–5% by weight of the original oxygen-demanding wastes, 3% of the suspended solids, 50% of the nitrogen (mostly as nitrates), 70% of the phosphorus (mostly as phosphates), and 30% of most toxic metal compounds and synthetic organic chemicals. Virtually no long-lived radioactive isotopes or persistent organic substances such as pesticides are removed.

As a result of the Clean Water Act, most U.S. cities have secondary sewage treatment plants. In 1989, however, the EPA found that more than 66% of sewage treatment plants have either water quality or public health problems, and studies by the General Accounting Office have shown that most industries have violated regulations. Moreover, 500 cities have failed to meet federal standards for sewage treatment plants, and 34 East Coast cities simply screen out large floating objects from their sewage before discharging it into coastal waters.

Advanced sewage treatment is a series of specialized chemical and physical processes that remove specific pollutants left in the water after primary and secondary treatment (Figure 20-12). Types of advanced treatment vary according to the

A: No; in 1995, burning oil supplied only 3% of the country's electricity

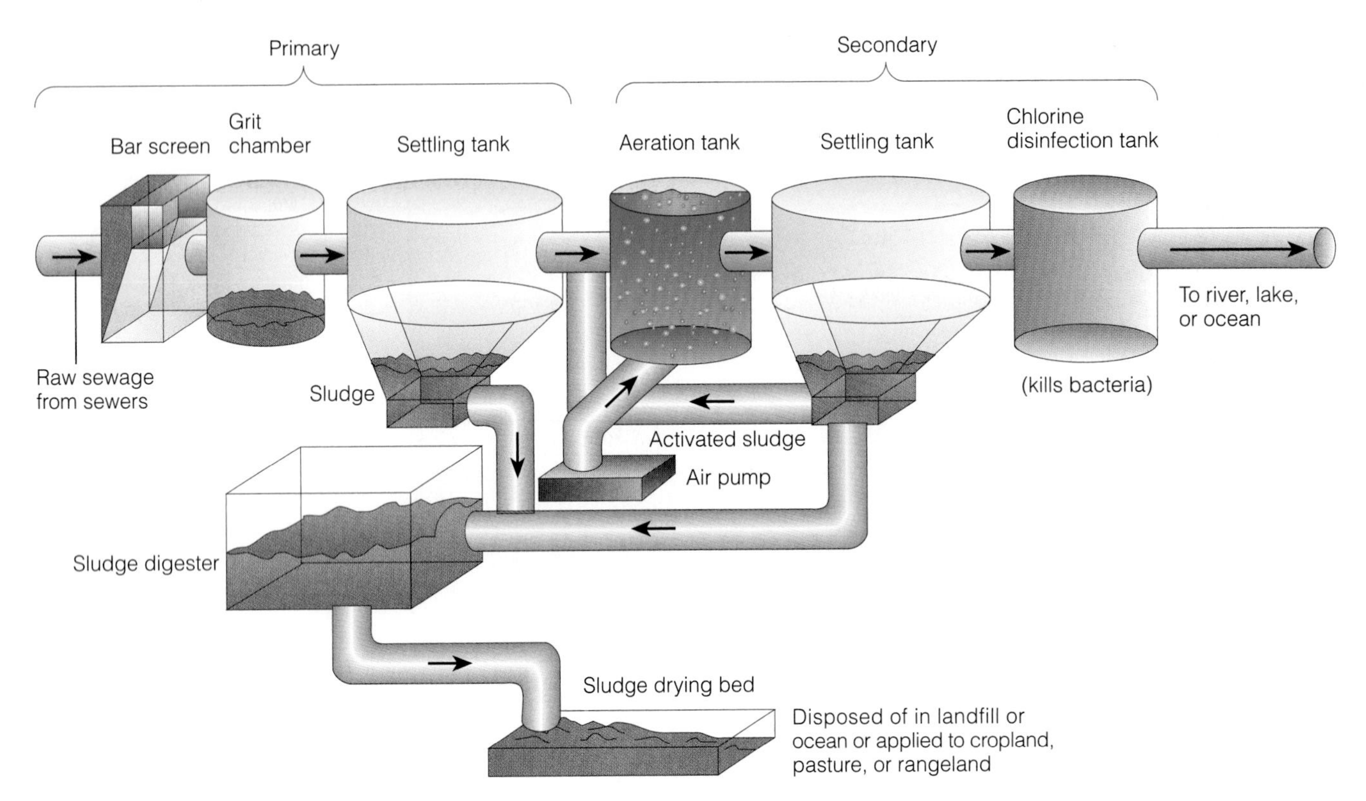

Figure 20-11 Primary and secondary sewage treatment.

specific contaminants to be removed. Without advanced treatment, sewage treatment plant effluents contain enough nitrates and phosphates to contribute to accelerated eutrophication of lakes, slow-moving streams, and coastal waters. Advanced treatment is rarely used because such plants typically cost twice as much to build and four times as much to operate as secondary plants.

Before water is discharged after primary, secondary, or advanced treatment, it is bleached (to remove water coloration) and disinfected (to kill disease-carrying bacteria and some but not all viruses). The usual method for doing this is *chlorination*. However, chlorine can react with organic materials in water to form small amounts of chlorinated hydrocarbons, some of which cause cancers in test animals. According to some preliminary research in 1992, chlorinated drinking water may cause 7–10% of all cancers in the United States. There is also growing evidence that some chlorinated hydrocarbons may damage the human nervous, immune, and endocrine systems (Section 17-3). Other disinfectants such as ozone and ultraviolet light are used in some places, but they cost more than chlorination and are not as long-lasting.

Sewage treatment produces a toxic, gooey sludge that must be disposed of or recycled as fertilizer. About 36% by weight of all municipal sludge produced in the United States is applied to farmland, forests, highway medians, and degraded land as fertilizer, and 9% is composted. Much of this contaminated sludge is used as fertilizer for crops intended for animal feed and for human food. About 38% is dumped in conventional landfills (where it can contaminate groundwater) and 16% is incinerated (which can pollute the air with traces of toxic chemicals, and the resulting toxic ash is usually buried in landfills that EPA experts say will eventually leak).

Before it is applied to land, sewage sludge can be heated to kill harmful bacteria, as is done in Switzerland and parts of Germany; it can also be treated to remove toxic metals and organic chemicals before application, but such treatment can be expensive. (Scientists say that the best and cheapest solution is to prevent these toxins from reaching sewage treatment plants.) Untreated sludge can be applied to land not used for crops or livestock, such as forests, surface-mined land, golf courses, lawns, cemeteries, and highway medians. However, this is a controversial issue (Spotlight, p. 554).

Q: What if nuclear power's harmful effects were included in its market price and government subsidies were removed?

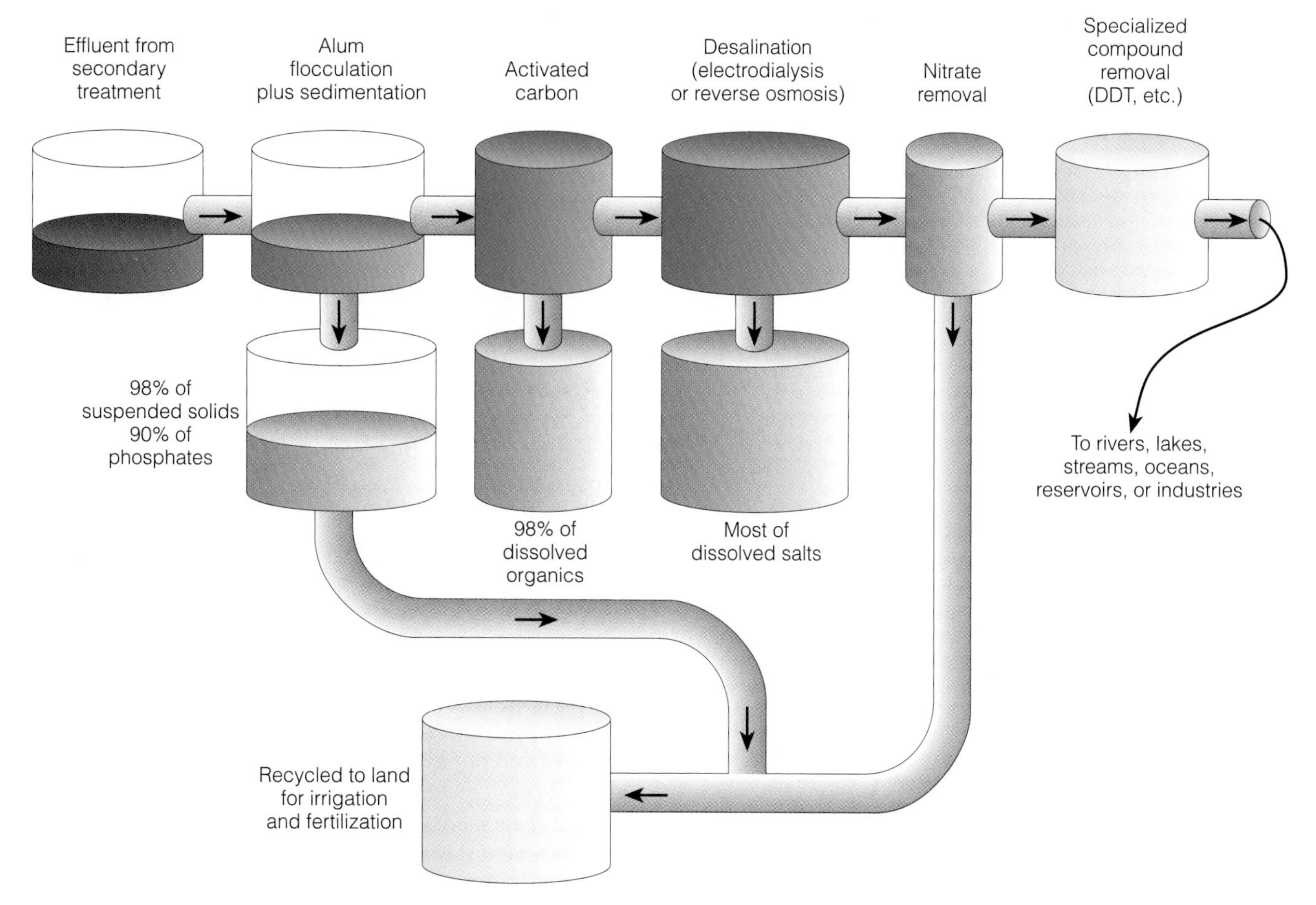

Figure 20-12 Advanced sewage treatment. Often only one (or two) of these processes is used to remove specific pollutants in a particular area. This expensive method is not widely used.

How Can We Treat Sewage by Working with Nature? Some communities and individuals are seeking better ways to purify contaminated water by working with nature (Figures 20-1 and 20-2). Scientists at Living Technologies in Vermont have built and installed 20 solar aquatic water purification systems similar to those developed by John Todd and William Jewell.

Scientists at Living Technologies are also developing neighborhood-level *sewage walls* that would run along the length of a residential block. Sewage would be channeled through a series of four terraced planters that progressively filter and purify the waste. Each planter would be capped with glass to allow use of sunlight and contain the bacteria and plants best suited for the various stages of treatment. The resulting effluent could be used on local gardens and the plants could be periodically harvested and converted to compost for use on neighborhood gardens.

In developing countries, a promising new technology for processing domestic waste is the SIRDO system. This double-vault waste treatment system uses one chamber in which human wastes and organic kitchen wastes are deposited and a second chamber is filled with such wastes and allowed to compost for several months. Solar heating and bacteria convert the waste in the second chamber into a safe and odorless soil conditioner that is sold to nearby farms. Such "dry" composting units that use no water are small enough to serve one or two families and save water.

Other SIRDO units are neighborhood miniplants that biologically process the "wet" or water-flushed wastes of up to 1,000 people. These systems separate gray water from solids. This water is then percolated through a bed of sand and gravel until it is pure enough to be used on gardens or to irrigate flowers, grass, or trees. Such systems typically cost one-seventh

A: It would be too expensive to use for producing electricity and would be phased out

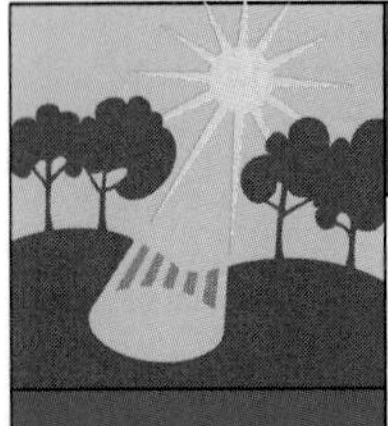

Should Sewage Sludge Be Recycled to the Land?

SPOTLIGHT

Since 1992, when Congress banned ocean dumping of sludge, its use as a fertilizer in the United States has increased sharply. However, most sewage sludge used as fertilizer in the United States is not adequately treated to kill harmful bacteria and remove toxic metals and organic chemicals, which may end up in food products or leach into groundwater.

In 1997 the EPA approved a plan to help recycle a greater share of U.S. sludge for "beneficial uses." This plan allows wastewater from the EPA cleanup of a Superfund toxic waste landfill near Denver to be dumped into sewers and treated at the local sewage treatment plant. The resulting sludge—containing numerous pesticides and other toxic organic chemicals, toxic metals such as lead and mercury, and nuclear waste—would be used to fertilize nearby farms.

Most environmentalists favor the recycling of carefully treated sewage sludge to the land. However, they say that the recycling of sewage sludge from cleaning up a toxic waste site makes a mockery out of the concept of recycling organic wastes.

A growing number of health problems and lawsuits have resulted from use of sludge to fertilize crops in the United States. To protect consumers and to avoid lawsuits, some food packers such as DelMonte and Heinz have banned produce grown on farms using sludge as a fertilizer. And Farm Credit Bureaus are refusing to finance slugged farms because of the financial risks from contaminated soils.

According to environmentalists, the current EPA policy that promotes beneficial use of sewage sludge as a fertilizer ignores the disastrous results of land application of contaminated sludge by Sweden in the 1980s.

Critical Thinking

Do you believe that sewage sludge should be recycled to the land for fertilizing the growth of food and nonfood plants? Explain. If you are opposed to this idea, under what controls and circumstances would you support this use of currently wasted plant nutrients?

as much as a conventional sewer and waste treatment plant system. Moreover, a cost–benefit analysis found that the start-up costs can be recovered through the sale of the resulting fertilizer (which can boost the income of families) within 2–20 years depending on the model used.

20-5 GROUNDWATER POLLUTION AND ITS PREVENTION

Why Is Groundwater Pollution Such a Serious Problem? Highly visible oil spills get a lot of media attention, but a much greater threat to human health is the out-of-sight pollution of groundwater (Figure 20-13), a prime source of water for drinking and irrigation. This vital form of earth capital is easy to deplete and pollute because much of it is renewed so slowly. Although experts rate groundwater pollution as a low-risk ecological problem, they consider pollutants in drinking water (much of it from groundwater) a high-risk health problem (Figure 17-13, left). Laws protecting groundwater are weak in the United States and nonexistent in most countries.

When groundwater becomes contaminated, it cannot cleanse itself of degradable wastes, as surface water can if it is not overloaded. Because groundwater flows are slow and not turbulent, contaminants are not effectively diluted and dispersed. Groundwater also has much smaller populations of decomposing bacteria than surface-water systems, and its cold temperature slows down decomposition reactions. Thus, it can take hundreds to thousands of years for contaminated groundwater to cleanse itself of degradable wastes; nondegradable wastes are there permanently on a human time scale.

Crude estimates indicate that up to 25% of usable groundwater in the United States is contaminated (and in some areas as much as 75%). In New Jersey, for example, every major aquifer is contaminated. In California, pesticides contaminate the drinking water of more than 1 million people. In Florida, where 92% of the residents rely on groundwater for drinking, over 1,000 wells have been closed. The EPA has documented groundwater contamination by 74 pesticides in 38 states.

Groundwater can be contaminated from a number of sources, including underground storage tanks, landfills, abandoned hazardous waste dumps, deep wells used to dispose of liquid hazardous wastes, and industrial waste and livestock waste storage lagoons located above or near aquifers (Figure 20-13). An EPA survey found that one-third of 26,000 industrial waste

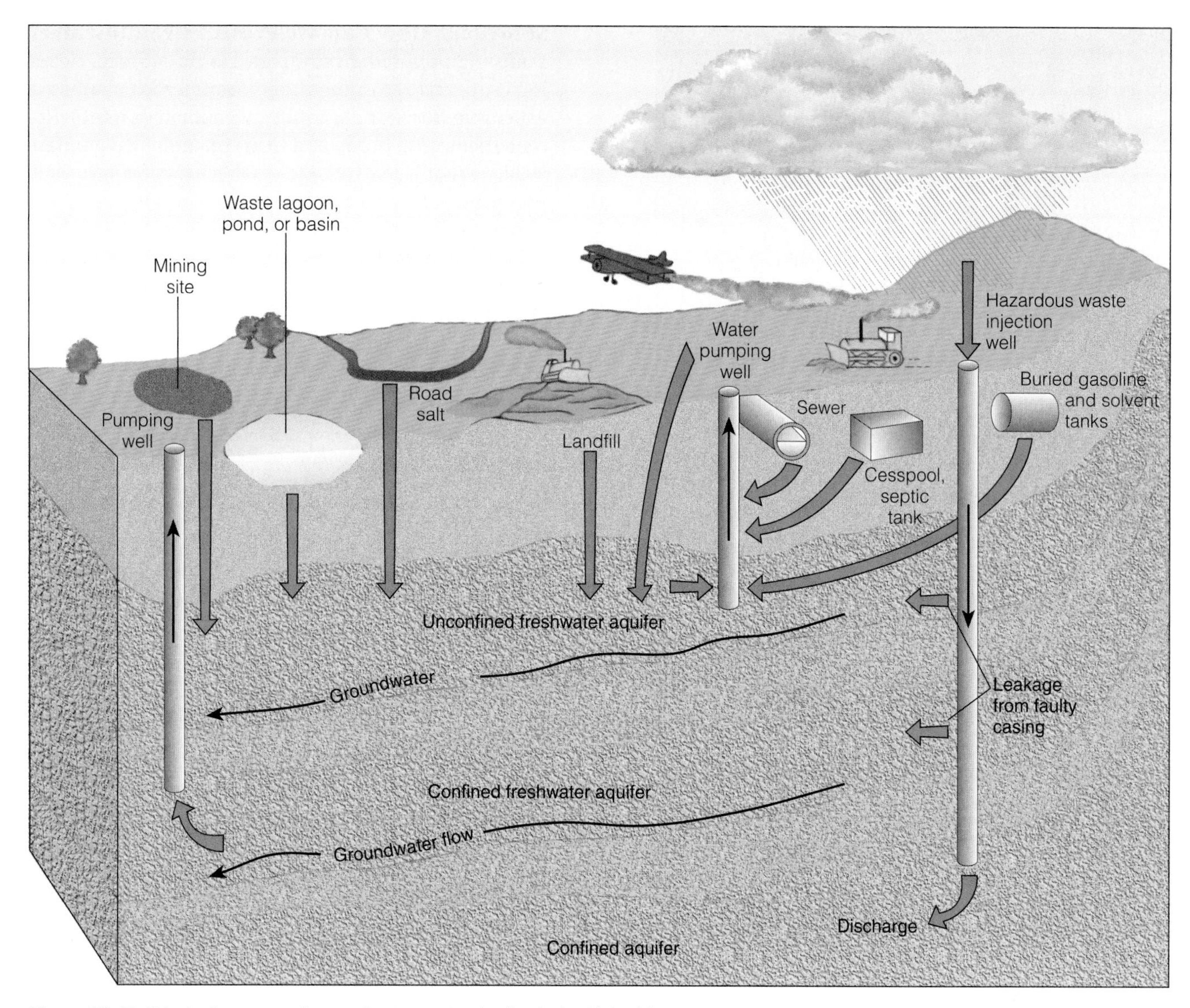

Figure 20-13 Principal sources of groundwater contamination in the United States.

ponds and lagoons have no liners to prevent toxic liquid wastes from seeping into aquifers, and one-third of those sites are within 1.6 kilometers (1 mile) of a drinking water well.

The 1,600 dairies in California's San Joaquin Valley that are the nation's largest milk supplier are regulated by only a single inspector. Studies show that nitrates from animal wastes produced by the area's dairy cows are seeping through the soil into precious groundwater.

The EPA estimates that at least 1 million underground tanks storing gasoline, diesel fuel, and toxic solvents are leaking their contents into groundwater. A slow gasoline leak of just 4 liters (1 gallon) per day can seriously contaminate the water supply for 50,000 people. Such slow leaks usually remain undetected until someone discovers that a well is contaminated.

Determining the extent of a leak can cost $25,000–250,000. Cleanup costs range from $10,000 for a small spill to $250,000 or more if the chemical reaches an aquifer, and complete cleanup is rarely possible. Replacing a leaking tank adds an additional $10,000–60,000. Legal fees and damages to injured parties can run into the millions.

Since 1993, all new tanks are required by the EPA to have a leak detection system and discovered leaks must be fixed right away. New tanks must also be made of noncorrosive material such as fiberglass and those holding petroleum products or any of 701

A: No; scientists estimate it could not become a significant producer of electricity until 2100, if then

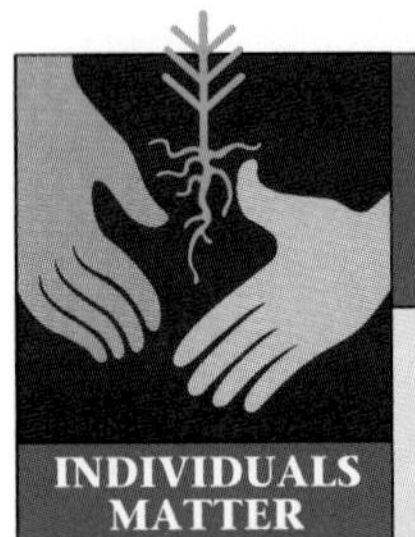

Using UV Light, Horseradish, and Slimes to Purify Water

Ashok J. Gadgil, a physicist at California's Lawrence Berkeley National Laboratory, and his colleagues have recently developed a simple device that uses ultraviolet light to kill disease-causing organisms in drinking water. When water from a well or hand pump (in a village or household in a developing country) is passed through this tabletop system, UV radiation from a mercury vapor lamp zaps germs in the water.

This $300 device weighs only 7 kilograms (15 pounds) and can disinfect 57 liters (15 gallons) of water per minute at a very low cost. It draws only 40 watts of power, supplied by solar cells, and can run unsupervised in remote areas of developing countries.

Recently, Pennsylvania State soil biochemists have discovered that chopped horseradish mixed with hydrogen peroxide (H_2O_2) helps rid contaminated water of organic pollutants called phenols. The horseradish contains an enzyme that speeds up the breakdown of the phenols.

Judith Bender and Peter Phillips at Clark Atlanta University in Georgia have found a way to use slime produced by cyanobacteria to decompose chlorinated hydrocarbons that contaminate drinking water. Within 3 weeks, a slimy, floating bacterial mat can surround and decompose a glob of toxic chlordane (a pesticide banned in the United States as a suspected carcinogen). Such slime mats can also remove lead, copper, chromium, cadmium, selenium, and other toxic metals from water.

hazardous chemicals must have overfill and spill prevention devices and double walls or concrete vaults to help prevent leaks into groundwater. Each owner of a commercial underground tank must also carry at least $1 million in liability insurance—a requirement that has driven many independent gasoline stations out of business.

In 1986, Congress passed legislation placing a tax on motor fuel to create a $500-million trust fund for cleaning up leaking underground tanks. However, this is a drop in the bucket compared to estimated costs of only partial cleanup of such spills, which run as high as $32 billion. Current regulations should reduce leakage from new tanks but would do little about the millions of older tanks that are toxic time bombs.

Solutions: How Can We Protect Groundwater? Pumping polluted groundwater to the surface, cleaning it up, and returning it to the aquifer is usually too expensive (for a single aquifer, $5 million or more). Recent attempts to pump and treat contaminated aquifers indicate that it may take 50–1,000 years of continuous pumping before all the contamination is forced to the surface and drinking water quality is achieved. Thus, *preventing contamination by various means is considered the only effective way to protect groundwater resources.* Ways to do this include

- *Monitoring aquifers near landfills and underground tanks*
- *Requiring leak detection systems for existing and new underground tanks used to store hazardous liquids*
- *Requiring liability insurance for old and new underground tanks used to store hazardous liquids*
- *Banning or more strictly regulating disposal of hazardous wastes in deep injection wells and landfills*
- *Storing hazardous liquids above ground in tanks with systems for detecting and collecting any leaks*

20-6 DRINKING WATER QUALITY

Is the Water Safe to Drink? Many rivers in eastern Europe, Latin America, and Asia are used as sources of drinking water but are severely polluted, as are some rivers in developed countries. Aquifers used as sources of drinking water in many developed countries and developing countries are becoming contaminated with pesticides, fertilizers, and hazardous organic chemicals. In China, 41 large cities get their drinking water from polluted groundwater. In Russia, half of all tap water is unfit to drink, and a third of the aquifers are too contaminated for drinking purposes.

In 1980, the UN recommended spending $300 billion to supply all the world's people with clean drinking water and adequate sanitation by 1990. The $30-billion annual cost of this program is about what the world spends every 10 days for military purposes. Only about $1.5 billion per year was actually spent. Researchers are trying to find cheap and simple ways to purify drinking water in developing nations (Individuals Matter, left).

How Is Drinking Water Purified? Treatment of water for drinking by city dwellers is much like wastewater treatment. Areas that depend on sur-

Baker, Beth. 1998. Keeping Aquaculture Environmentally Friendly. *BioScience*, vol. 48, no. 8, 592(1).

face water usually store it in a reservoir for several days to improve clarity and taste by allowing the dissolved oxygen content to increase and suspended matter to settle out. The water is then pumped to a purification plant, where it is treated to meet government drinking water standards. Usually the water is run through sand filters and activated charcoal before it is disinfected. In areas with very pure sources of groundwater, little treatment is necessary.

How Is Drinking Water Quality Protected? About 54 countries, most of them in North America and Europe, have safe drinking water standards. The U.S. Safe Drinking Water Act of 1974 requires the EPA to establish national drinking water standards, called maximum contaminant levels, for any pollutants that may have adverse effects on human health. This act has helped improve drinking water in much of the United States, but attempts to weaken this law continue. In addition, maximum contaminant levels have not been set for many potentially dangerous water pollutants such as certain synthetic organic compounds, radioactive materials, toxic metals, and pathogens.

Currently, operators of public drinking water systems must test the water for 64 contaminants. In 1997, the EPA proposed reducing the expensive testing burden on localities by basing monitoring requirements on the risks from specific contaminants in each area. For example, areas with a low risk of contamination from certain pesticides, volatile organic compounds, or metals would be required to test for such chemicals only once every 5 years.

Privately owned wells are not required to meet federal drinking water standards, primarily because of the costs of testing each well regularly (at least $1,000) and ideological opposition to mandatory testing and compliance by some homeowners.

According to a 1994 study by the Natural Resources Defense Council (NRDC), in 1992 the drinking water of 50 million people—one American in five—violated one or more EPA pollutant standards. In most cases people were not notified (as required by the 1974 law) when their drinking water was contaminated. This study also estimated that contaminated drinking water is responsible for 7 million illnesses and 1,200 deaths per year in the United States.

Such dangers were revealed when residents of Milwaukee, Wisconsin, were told on April 7, 1993, that water from their taps was unsafe to drink. Before this health crisis was over, 104 people had died and 400,000 people had become sick as a result of exposure to *Cryptosporidium*, a parasitic organism. Milwaukee spent an estimated $54 million to deal with this outbreak. In May 1994, another outbreak of this parasite killed 19 and sickened more than 100 people in Las Vegas, Nevada.

A survey in 1992 showed that nearly 40% of treated drinking water supplies in the United States contained either *Cryptosporidium* or *Giardia* (a protozoan that infects the small intestine, causing nausea and diarrhea). Mostly as a result of the Milwaukee crisis and pressure from environmentalists and health officials, the EPA began requiring municipal drinking water suppliers to begin testing for cryptosporidium in 1996.

According to the NRDC, U.S. drinking water supplies could be made safer at a cost of only about $30 a year per household. One might expect that such information would lead to public pressure to upgrade drinking water in the United States. Instead, Congress is being pressured by water-polluting industries to weaken the Safe Drinking Water Act by **(1)** eliminating national tests of drinking water, **(2)** eliminating the requirement that the media be advised of emergency water health violations and that water system officials notify their customers of such violations (a "don't ask–don't tell" policy), **(3)** allowing states to give drinking water systems a permanent right to violate the standard for a given contaminant if the provider claims it can't afford to comply, and **(4)** eliminating the requirement that water systems use affordable, feasible technology to remove cancer-causing contaminants.

Environmentalists call for the U.S. Safe Drinking Water Act to be strengthened by **(1)** improving treatment by combining at least half of the 50,000 water systems that serve fewer than 3,300 people each with larger ones nearby, **(2)** strengthening and enforcing public notification requirements about violations of drinking water standards, and **(3)** banning all lead in new plumbing pipes, faucets, and fixtures (current law allows fixtures with up to 10% lead to be sold as lead-free).

Is Bottled Water the Answer? The United States has one of the world's best drinking water supply systems. Yet about half of all Americans worry about getting sick from tap water contaminants and many either drink bottled water or install expensive water purification systems. Studies indicate that many of these consumers are being ripped off and in some cases may end up drinking water that is dirtier than they can get from their taps.

To be safe, consumers purchasing bottled water should determine whether the bottler belongs to the

Hint: Enter the search terms *Agriculture* using the Subject Guide.

International Bottled Water Association (IBWA) and adheres to its testing requirements.* The IBWA requires its members to test for 181 contaminants, and annually it sends an inspector to bottling plants to check all pertinent records and ensure that the plant is run cleanly. Some companies pay $2,500 annually to obtain more stringent certification by the National Sanitation Foundation, an independent agency that tests for 200 chemical and biological contaminants.

Before drinking only expensive bottled water and buying costly home water purifiers, health officials suggest that consumers have their water tested by local health authorities or private labs† to identify what contaminants, if any, must be removed; then they should buy a unit that does the required job. Independent experts contend that unless tests show otherwise, for most urban and suburban Americans served by large municipal drinking water systems, home water treatment systems aren't worth the expense and maintenance hassles.

Buyers should be suspicious of door-to-door salespeople, telephone appeals, scare tactics, and companies offering free water tests (which often are neither accurate nor carried out by certified labs). They should carefully check out companies selling such equipment and demand a copy of purifying claims by EPA-certified laboratories. Buyers should also be wary of claims that a treatment device has been *approved* by the EPA. Although the EPA does *register* such devices, it neither tests nor approves them.

Solutions: How Can We Use Water Resources More Sustainably? Many analysts believe that using the earth's water resources sustainably involves developing an integrated approach to managing water resources and water pollution throughout each watershed and airshed. They believe that doing this requires that we shift our emphasis from pollution cleanup to pollution prevention. This involves **(1)** *source reduction* to reduce the toxicity or volume of pollutants (for example, replacing organic solvent-based inks and paints with water-based materials), **(2)** *reuse* of wastewater instead of discharging it (for example, reusing treated wastewater for irrigation), and **(3)** *recycling* pollutants (for example, cleaning up and recycling contaminated solvents for reuse) instead of discharging them.

To make such a shift, we need to accept that the environment—air, water, soil, life—is an interconnected whole. Without an integrated approach to all forms of pollution, environmentalists argue that we will continue to shift environmental problems from one part of the environment to another. Some actions you can take to help reduce water pollution are listed in Appendix 5.

The care of rivers is not a question of rivers but of the human heart.

TANAKA SCHOZO

CRITICAL THINKING

1. Why is dilution not always the solution to water pollution? Give examples and conditions for which this solution is, and is not, applicable.

2. How can a stream cleanse itself of oxygen-demanding wastes? Under what conditions will this natural cleansing system fail?

3. Should all dumping of wastes in the ocean be banned? Explain. If so, where would you put the wastes instead? What exceptions would you permit, and why?

4. Why will banning ocean dumping alone not stop ocean pollution?

5. Your town (Town B) is located on a river between towns A and C. What are the rights and responsibilities of upstream communities to downstream communities? Should sewage and industrial wastes be dumped at the upstream end of a community that generates them?

6. Try to come up with analogies to complete the following statements: **(a)** Nonpoint sources of water pollution are like _______. **(b)** Lake Baikal is like _______. **(c)** Contaminating groundwater is like _______. **(d)** Trying to prevent nonpoint water pollution is like _______. **(e)** A septic tank is like _______. **(f)** A secondary sewage treatment plant is like _______. **(g)** Relying on bottled drinking water is like _______.

PROJECTS

1. In your community,
   a. What are the principal nonpoint sources of contamination of surface water and groundwater?
   b. What is the source of drinking water?
   c. How is drinking water treated?
   d. What contaminants are tested for?
   e. Has drinking water been analyzed recently for the presence of synthetic organic chemicals, especially chlorinated hydrocarbons? If so, were any found, and are they being removed?

*Check for the IBWA seal of approval on the bottle or contact the IBWA (113 North Henry Street, Alexandria, VA 22314; phone 703-683-5213) for a member list.

†Contact the state health or other appropriate department for help in finding laboratories that are certified to do tests. You can also call the EPA's Safe Drinking Water Hotline from 8:30 A.M. to 4:30 P.M. EST at 800-426-4791 or 202-382-5533.

Q: What percentage of the commercial energy used in the United States is wasted?

f. How many times during each of the past 5 years have levels of tested contaminants violated federal standards? Was the public notified about the violations?

g. Is fishing prohibited in any lakes or rivers in your region because of pollution? Are people warned about this?

h. Is groundwater contamination a problem? If so, where, and what has been done about the problem?

i. Is there a vulnerable aquifer or critical recharge zone that should be protected to ensure the quality of groundwater? Is your local government aware of this? What action (if any) has it taken?

2. Are storm drains and sanitary sewers combined or separate in your area? Are there plans to reduce pollution from stormwater runoff? If not, make an economic evaluation of the costs and benefits from developing separate storm drains and sanitary sewers, and then present your findings to local officials.

3. Use library research, the Internet, and user interviews to evaluate the relative effectiveness and costs of home water purification devices (Section 20-6). Determine the type or types of water pollutants each device removes and the effectiveness of this process.

4. Make a concept map of this chapter's major ideas, using the section heads and subheads and the key terms (in boldface). Look at the inside back cover and in Appendix 4 for information about concept maps.

A: About 84% (43% of this energy is wasted unnecessarily)

21 PROTECTING FOOD RESOURCES: PESTICIDES AND PEST CONTROL

Along Came a Spider

The longest war in human history is our ongoing war against insect pests. This war was declared about 10,000 years ago, when humans first got serious about agriculture. Today we are not much closer to winning this war than we were then.

Chinese farmers recently decided it's time to change strategies. Instead of spraying their rice and cotton fields with poisons, they build little straw huts around the fields in the fall.

If this sounds crazy, it's crazy like a fox. These farmers are giving aid and comfort to insects' worst enemy, one that has hunted them for millions of years: spiders (Figure 21-1). The little huts are for hibernating spiders. Protected from the worst of the cold by the huts, far more of the hibernating spiders become active in the spring. Ravenous after their winter fast, they scuttle off into the fields to stalk their insect prey.

Even without human help, the world's 30,000 known species of spiders kill far more insects every year than insecticides do. A typical acre of meadow or woods contains an estimated 50,000 to 2 million spiders, each devouring hundreds of insects per year.

No more than 2% of all the insecticide sprayed on a field finds its target, but every spider gets insect meals, or it does not live to reproduce. Spraying a field with some types of pesticides is like using saturation bombing: It kills almost everything, including the spiders and other species that help control those pests. Recruiting spiders for our war against insects is like sending in "smart bombs" to hit specific targets.

Figure 21-1 Spiders are insects' worst enemies. Most spiders, like the crab spider (left) and the wolf spider (right), found in many parts of the world, are harmless to humans. (Left, James C. Cokendolpher; right, Dan Kline/Visuals Unlimited)

Apparently, spiders are not only dedicated insect killers but astonishingly patient creatures. Unlike ladybugs or wasps, they do not abandon a field if the hunting turns poor; instead they wait, even for weeks, until the next wave of insects arrives.

Entomologist Willard H. Whitcomb found that leaving strips of weeds around cotton and soybean fields provides the kind of undergrowth favored by insect-eating wolf spiders (Figure 21-1, right). He also sings the praises of one type of banana spider, which lives in warm climates and can keep a house clear of cockroaches (Spotlight, p. 209). In Maine, Daniel Jennings of the U.S. Forest Service uses spiders to help control the spruce budworm, which devastates the Northeast's spruce and fir forests. Spiders also attack the much-feared gypsy moth, which devastates tree foliage.

The idea of encouraging populations of spiders in fields, forests, and even houses scares some people because spiders have bad reputations. Furthermore, perhaps because of their eight legs and numerous eyes, they look like dangerous space aliens instead of the helpful and mostly harmless creatures they are.

A few species of spiders, such as the black widow, the brown recluse, and eastern Australia's Sydney funnel web, are dangerous to people. However, the vast majority of spider species, including the ferocious-looking wolf spider, are harmless to humans. Even the giant tarantula rarely bites people, and its venom is too weak to harm us or other large mammals.

As biologist Thomas Eisner puts it, "Bugs are not going to inherit the earth. They own it now. So we might as well make peace with the landlord." As we seek new ways to coexist with the insect rulers of the planet, we would do well to be sure that spiders are in our corner.

This chapter looks first at the conventional chemical approach to pest control and then discusses alternatives based on working with nature. Some alternatives, such as using spiders and other natural predators to kill pest insects, are not new; their virtues merely wait to be rediscovered. Others, such as food crops genetically engineered for resistance to specific pests, did not exist a mere decade ago.

A weed is a plant whose virtues have not yet been discovered.

Ralph Waldo Emerson

This chapter answers the following questions:

- What types of pesticides are used?
- What are the pros and cons of using chemicals to kill insects and weeds?
- How well is pesticide use regulated in the United States?
- What alternatives are there to using pesticides?

21-1 PESTICIDES: TYPES AND USES

How Does Nature Keep Pest Populations Under Control? A **pest** is any species that competes with us for food, invades lawns and gardens, destroys wood in houses, spreads disease, or is simply a nuisance. In natural ecosystems and many polyculture agroecosystems, natural enemies (predators, parasites, and disease organisms) control the populations of 50–90% of pest species, thus constituting a crucial type of earth capital.

When we simplify natural ecosystems, we upset these natural checks and balances, which keep any one species from taking over for very long. Then we must devise ways to protect our monoculture crops, tree farms, and lawns from insects and other pests that nature once controlled at no charge.

We have done this primarily by developing a variety of **pesticides** (or *biocides*): chemicals to kill organisms we consider undesirable. Common types of pesticides include **insecticides** (insect-killers), *herbicides* (weed-killers), *fungicides* (fungus-killers), *nematocides* (roundworm-killers), and *rodenticides* (rat- and mouse-killers).

Humans didn't invent the use of chemicals to repel or kill other species; plants have been producing chemicals to ward off or poison herbivores that feed on them for about 225 million years. This is a never-ending, ever-changing process: Herbivores overcome various plant defenses through natural selection; then the plants use natural selection to develop new defenses. The result of these dynamic interactions between predator and prey species is what biologists call *coevolution* (Section 6-3).

What Was the First Generation of Pesticides and Repellents? As the human population grew and agriculture spread, people began looking for ways to protect their crops, mostly by using chemicals to kill or repel insect pests. Sulfur was used as an insecticide well before 500 B.C.; by the 1400s, toxic compounds of arsenic, lead, and mercury were being applied to crops as insecticides. Farmers abandoned this approach in the late 1920s, when the increasing number of human poisonings and fatalities encouraged a search for less toxic substitutes. However, traces of these nondegradable toxic metal compounds are still being taken up by tobacco, vegetables, and other crops grown on soil dosed with them long ago.

In the 1600s nicotine sulfate, extracted from tobacco leaves, came into use as an insecticide. In the mid-1800s two more natural pesticides were introduced: pyrethrum, obtained from the heads of chrysanthemum flowers, and rotenone, from the root of the derris plant and other tropical forest legumes. These *first-generation pesticides* were mainly natural substances, weapons borrowed from plants that had been at war with insects for eons.

In addition to protecting crops, people have used natural chemicals (produced mostly by plants) to repel or kill insects in their households, yards, and gardens (Solutions, right); they also save money and reduce health hazards associated with using commercial insecticides.

What Was the Second Generation of Pesticides? A major pest control revolution began in 1939, when entomologist Paul Mueller discovered that DDT (dichlorodiphenyltrichloroethane), a chemical known since 1874, was a potent insecticide. DDT, the first of the so-called *second-generation pesticides*, soon became the world's most-used pesticide and Mueller received the Nobel prize in 1948. Since 1945 chemists have developed hundreds of synthetic organic chemicals for use as pesticides.

Worldwide, about 2.3 million metric tons (2.5 million tons) of such pesticides are used yearly—0.45 kilogram (1 pound) for each person on the earth. About 75% of these chemicals are used in developed countries, but use in developing countries in Latin America, Asia, and Africa is soaring. In 1996 worldwide sales of pesticides were $30 billion ($11 billion in the United States alone).

In the United States, about 630 different biologically active (pest-killing) ingredients and about 1,820 inert (inactive) ingredients are mixed to make some 25,000 different pesticide products. Cultivation of two crops—cotton (55%) and corn (35%)—used about 90% of the insecticides and 80% of the herbicides applied to crops in the United States in 1995.

Manufacturers add pesticides to products as diverse as paints, shampoos, carpets, mattresses, wax on produce, and contact lenses. About 25% of pesticide use in the United States is for ridding houses, gardens, lawns, parks, playing fields, swimming pools, and golf courses of unwanted pests. According to the EPA, the average lawn in the United States is doused with more than 10 times more synthetic pesticides per hectare than U.S. cropland. Each year, more than 250,000 U.S. residents become ill because of household use of pesticides, and such pesticides are a major source of accidental poisonings and deaths for children under age 5.

Some pesticides, called *broad-spectrum* agents, are toxic to many species; others, called *selective* or *narrow-spectrum* agents, are effective against a narrowly defined group of organisms. Pesticides vary in their *persistence*, the length of time they remain deadly in the environment (Table 21-1). Most organophosphates (except malathion) are highly toxic to humans and other animals, and they account for most human pesticide poisonings and deaths. In 1962 biologist Rachel Carson warned against relying on synthetic chemicals to kill insects and other species we deem pests (Individuals Matter, p. 565).

21-2 THE CASE FOR PESTICIDES

Proponents of pesticides contend that their benefits outweigh their harmful effects.

Pesticides save human lives. Since 1945 DDT and other chlorinated hydrocarbon and organophosphate insecticides have probably prevented the premature deaths of at least 7 million people from insect-transmitted diseases such as malaria (carried by the *Anopheles* mosquito; Figure 17-11), bubonic plague (rat fleas), typhus (body lice and fleas), and sleeping sickness (tsetse fly).

Pesticides increase food supplies and lower food costs. About 55% of the world's potential human food supply is lost to pests before (35%) or after (20%) harvest. An estimated 37% of the potential U.S. food supply is destroyed by pests before and after harvest; insects cause 13% of these losses, plant pathogens 12%, and weeds 12%. Without pesticides, these losses would be worse, and food prices would rise (by 30–50% in the United States, according to pesticide company officials).

Pesticides increase profits for farmers. Pesticide companies estimate that every $1 spent on pesticides leads to an increase in U.S. crop yields worth approximately $4 (but studies have shown that this benefit drops to about $2 if the harmful effects of pesticides are included).

Pesticides work faster and better than alternatives. Pesticides can control most pests quickly and at a reasonable cost. They have a long shelf life, are easily

Q: What percentage of the energy input of an incandescent light bulb is converted to light?

Using Natural Chemicals to Control Common Household Pests

SOLUTIONS

If *ants* come indoors, you can usually persuade them to leave within about 4 days by sprinkling repellents such as red or cayenne pepper, crushed mint leaves, or boric acid (with an anticaking agent) along their trails inside a house, and by wiping off countertops with vinegar. (However, boric acid is poisonous and should not be placed in areas accessible to small children and pets.)

Repel *mosquitoes* by planting basil outside windows and doors and rubbing a bit of vinegar, basil oil, lime juice, or mugwort oil on exposed skin. You can also reduce mosquito attacks by not using scented soaps or wearing perfumes, colognes, and other scented products outdoors during mosquito season. Researchers have found that the $30 million U.S. consumers spend each year on electric bug zappers to kill mosquitoes is mostly wasted because only about 3% of the insects they kill on an average night are female mosquitoes—the kind that bite.

You can kill *cockroaches* (Spotlight, p. 209) by sprinkling boric acid under sinks and ranges, behind refrigerators, and in cabinets, closets, and other dark, warm places (but not in areas accessible to children and pets), or by establishing populations of banana spiders. You can also trap roaches by greasing the inner neck of a bottle or large jar with petroleum jelly, filling much of it with raw potato, stale beer, banana skins, or other food scraps (especially fruits), and placing a small ramp leading to it. Placing fresh or dried bay leaves in and around cupboards repels cockroaches.

Repel *flies* by planting sweet basil and tansy (a common herb) near doorways and patios and by hanging a series of polyethylene strips in front of entry doors (like the ones on some grocery-store coolers). You can make nontoxic flypaper by applying honey to strips of yellow paper (their favorite color) and hanging it from the ceiling in the center of rooms. You can also hang clusters of cloves.

Keep *fleas* off pets by using green dye or flea-repellent soaps; feeding them brewer's yeast or vitamin B; using flea powders made from eucalyptus, sage, tobacco, wormwood, or vetiver; or dipping or shampooing pets in a mixture of water and essential oils such as citronella, cedarwood, eucalyptus, fleabane, sassafras, geranium, clove, or mint. Researchers recently invented a trap that uses green-yellow light to attract fleas to an adhesive-coated surface. Another solution is to put a light over a shallow pan of water before going to bed at night and turn out all other lights, empty the water every morning, and continue for a month. (Fleas are attracted to heat and light but they can't swim.) Desiccant powders, such as Dri-Die, Perma-Guard, or SG-67, can also rid a house of flea infestations by drying-up the insects. Diatomaceous earth (or diatom powder) can also be sprinkled on carpets and pets to kill fleas.* This powder consists of the skeletal remains of microscopic algae and kills fleas and many other harmful insects by cutting their outer shell so they lose their body fluids.

You can control many *garden insect pests* by introducing their natural predators—such as various other insects, insect-eating birds, and toads and frogs—and by preserving or providing habitats for such predators. Garden plants can also be sprayed with one of the following solutions. Steep one handful of tobacco in warm water for 24 hours (handle with care because nicotine is a poison); or mix 2–3 very hot peppers, $^1/_2$ onion, and 1 clove of garlic in water, boil, let sit for 2 days, and strain; or mix 2 tablespoons liquid soap in 1 quart water or 2 ounces of dry soap in 1 quart water. Another way to kill most common garden insect pests is to sprinkle crops with diatomaceous earth powder. It does not harm animals, people, earthworms, spiders, and other helpful creatures.

Control *lawn weeds* by raising the cutting level of your lawn mower so the grass can grow 8–10 centimeters (3–4 inches) high. This gives it a strong root system that can hinder weed growth; the higher grass also provides habitats for spiders and other insects that eat insect pests. Pull weeds and douse the hole with soap solution or human urine (which is high enough in nitrogen to burn the weed).

Critical Thinking

Have you or someone you know tried any of the natural approaches to pest control listed in this box? If so, how well did they work? Which, if any, of the methods suggested in this box do you plan to use?

*Diatom powder can be purchased in bulk at stores that sell it for use in swimming pool filters. It is also found in some gardening stores under the name Permatex. Because it contains fine particles of silicate, you should wear a dust mask when applying this powder to avoid inhaling tiny particles of silicate.

shipped and applied, and are safe when handled properly. When genetic resistance occurs, farmers can use stronger doses or switch to other pesticides.

The health risks of pesticides are insignificant compared with their benefits. According to Elizabeth Whelan, director of the American Council on Science and Health (ACSH), which presents the position of the pesticide industry, "The reality is that pesticides, when used in the approved regulatory manner, pose no risk to either farm workers or consumers." Pesticide

A: 5% (the rest is heat; these light bulbs are really heat bulbs)

Table 21-1 Major Types of Pesticides

Type	Examples	Persistence	Biologically Magnified?
Insecticides			
Chlorinated hydrocarbons	DDT, aldrin, dieldrin, toxaphene, lindane, chlordane, methoxychlor, mirex	High (2–15 years)	Yes (Figure 17-2)
Organophosphates	Malathion, parathion, diazinon, TEPP, DDVP, mevingphos	Low to moderate (1–2 weeks), but some can last several years	No
Carbamates	Aldicarb, carbaryl (Sevin), propoxur, maneb, zineb	Low (days to weeks)	No
Botanicals	Rotenone, pyrethrum, and camphor extracted from plants, synthetic pyrethroids (variations of pyrethrum) and rotenoids (variations of rotenone)	Low (days to weeks)	No
Microbotanicals	Various bacteria, fungi, protozoa	Low (days to weeks)	No
Herbicides			
Contact chemicals	Atrazine, simazine, paraquat	Low (days to weeks)	No
Systemic chemicals	2,4-D, 2,4,5-T, Silvex, diruon, daminozide (Alar), alachlor (Lasso), glyphosate (Roundup)	Mostly low (days to weeks)	No
Soil sterilants	Tribualin, diphenamid, dalapon, butylate	Low (days)	No
Fungicides			
Various chemicals	Captan, pentachlorphenol, zeneb, methyl bromide, carbon bisulfide	Most low (days)	No
Fumigants			
Various chemicals	Carbon tetrachloride, ethylene dibromide, methyl bromide	Mostly high	Yes (for most)

proponents consider media reports describing pesticide health scares to be distorted science and irresponsible reporting. They also point out that about 99.99% of the pesticides we consume in our food are natural chemicals produced by plants.

Proponents point out some beneficial changes in pesticides and their use. For example, *safer and more effective pesticides are being developed.* Industry scientists are developing pesticides, such as botanicals and microbotanicals (Table 21-1), that are safer to users and less damaging to the environment. Genetic engineering also holds promise in developing pest-resistant crop strains. However, total research and development and government approval costs for a new pesticide have risen from $6 million in 1976 to $80–120 million today. As a result, new chemicals are being developed only for crops such as wheat, corn, and soybeans with large markets.

Many new pesticides are used at very low rates per unit area compared to older products. For example, application amounts per hectare for many new herbicides are one hundredth those of older ones.

Scientists continue to search for the ideal pest-killing chemical, which would

- *Kill only the target pest*
- *Harm no other species*
- *Disappear or break down into something harmless after doing its job*
- *Not cause genetic resistance in target organisms*
- *Be cheaper than doing nothing*

Unfortunately, no known pesticide meets all these criteria, and most don't even come close.

21-3 THE CASE AGAINST PESTICIDES

How Do Pesticides Cause Genetic Resistance and Kill Natural Pest Enemies? Opponents of widespread use of pesticides believe that their harmful

 Graffy, E.A. 1998. "Low-Level Detection of Pesticides . . . So What?" *Journal of Soil and Water Conservation*, vol. 53, no. 1, 11(2).

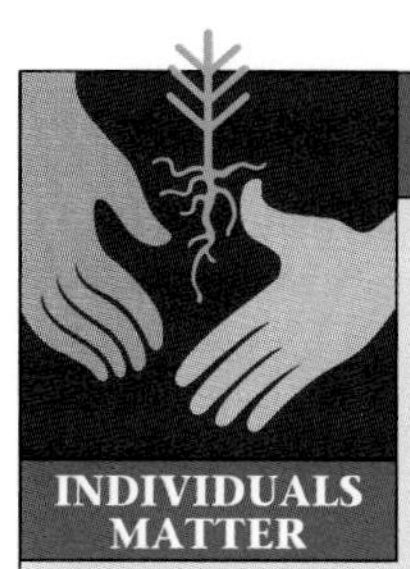

INDIVIDUALS MATTER

Rachel Carson

Rachel Carson (Figure 21-2) began her professional career as a biologist for the Bureau of U.S. Fisheries (later to become the U.S. Fish and Wildlife Service). In that capacity, she carried out research on oceanography and marine biology, wrote articles about the oceans and topics related to the environment, and became editor-in-chief of the bureau's publications in 1949.

In 1951, she wrote *The Sea Around Us,* which described in easily understandable terms the natural history of oceans and the harm that humans were doing them. The book was on the best-seller list for 86 weeks, sold more than 2 million copies, was translated into 32 languages, and won a National Book Award.

During the late 1940s and throughout the 1950s, the use of DDT and related compounds—to kill insects that ate food crops, attacked trees, bothered people, and transmitted diseases such as malaria—expanded rapidly.

In 1958, DDT was sprayed to control mosquitoes near the home and private bird sanctuary of Olga Huckins, a good friend of Carson. After the spraying, Huckins witnessed the agonizing deaths of several of her birds, and in distress she asked Carson whether she could find someone to investigate the effects of pesticides on birds and other wildlife.

Carson decided to look into the issue herself and quickly found that almost no independent critical research on the environmental effects of pesticides existed. As a well-trained scientist, Carson surveyed the scientific literature and methodically built her case against the widespread use of pesticides.

In 1962, she published her findings in popular form in *Silent Spring,* an allusion to the silencing of "robins, catbirds, doves, jays, wrens, and scores of other bird voices" because of their exposure to pesticides. She pointed out that "for the first time in the history of the world, every human being is now subjected to dangerous chemicals, from the moment of conception until death."

Carson's book was read by many scientists, politicians, and policy makers and was embraced by the public. However, the chemical industry viewed the book as a serious threat to booming pesticide sales and mounted a $250,000 campaign to discredit Carson. A parade of critical reviewers and industry scientists claimed that her book was full of inaccuracies, made selective use of research findings, and failed to give a balanced account of the benefits of pesticides.

Some critics even claimed that, as a woman, she was incapable of understanding the highly scientific and technical subject of pesticides. Others charged that she was a hysterical woman and a radical nature lover trying to scare the American public in order to sell books.

During this period of intense controversy Carson was suffering from terminal cancer, but she was able to defend her research and strongly counter her critics. She died in 1964—18 months after the publication of *Silent Spring*—without knowing that her efforts were a driving force in the birth of what is now known as the environmental movement in the United States. To environmentalists and most citizens, Rachel Carson is an outstanding example of a dedicated scientist and effective communicator of complex scientific information.

Figure 21-2 Biologist Rachel Carson (1907–64) was a pioneer in increasing public awareness of the importance of nature and the threat of pollution. She died without knowing that her efforts were a key in starting today's environmental movement. (Eric Hartmann/Magnum)

effects outweigh their benefits. The biggest problem is the development of *genetic resistance* to pesticides by pest organisms. Insects breed rapidly (Figure 21-3), and within 5–10 years (much sooner in tropical areas) they can develop immunity to pesticides through natural selection (Section 6-3) and come back stronger than before. Weeds and plant-disease organisms also become resistant, but more slowly.

Since 1950 at least 520 insect and mite species, 273 weed species, 150 plant diseases, and 10 rodent species (mostly rats) have developed genetic resistance to one or more pesticides. At least 17 insect pest species are

Figure 21-3 A boll weevil, just one example of an insect capable of rapid breeding. In the cotton fields of the southern United States, these insects lay thousands of eggs, producing a new generation every 21 days and as many as six generations in a single growing season. Attempts to control the cotton boll weevil account for at least 25% of insecticide use in the United States. However, farmers are now increasing their use of natural predators and other biological methods to control this major pest. (U.S. Department of Agriculture)

resistant to all major classes of insecticides, and several fungal plant diseases are now immune to most widely used fungicides. Because of genetic resistance, most widely used insecticides no longer protect people from insect-transmitted diseases in many parts of the world, leading to resurgences of diseases such as malaria (Figure 17-10).

Another problem is that *broad-spectrum insecticides kill natural predators and parasites that may have been maintaining the population of a pest species at a reasonable level.* With wolf spiders (Figure 21-1, right), wasps, predatory beetles, and other natural enemies out of the way, a rapidly reproducing insect pest species can make a strong comeback only days or weeks after initially being controlled. Although natural predators can also develop genetic resistance to pesticides, most predators can't reproduce as quickly as their insect prey.

Wiping out natural predators can also unleash new pests whose populations the predators had previously held in check, causing other unexpected effects (Connections, p. 247). Of the 300 most destructive insect pests in the United States, 100 were secondary pests that became major pests through the effects of insecticides.

What Is the Pesticide Treadmill? When genetic resistance develops, pesticide sales representatives usually recommend more frequent applications, larger doses, or a switch to new (usually more expensive) chemicals to keep the resistant species under control. This can put farmers on a **pesticide treadmill** whereby they pay more and more for a pest control program that often becomes less and less effective. Between the mid-1950s and late 1970s, for example, cotton growers in Central America increased the frequency of insecticide applications from 10 to 40 times per growing season; still, declining yields and falling profits forced many of the farmers into bankruptcy.

In 1989 David Pimentel (Guest Essay, p. 364), an expert in insect ecology, evaluated data from more than 300 agricultural scientists and economists and came to the following conclusions:

- Although the use of synthetic pesticides has increased 33-fold since 1942, it is estimated that more of the U.S. food supply is lost to pests today (37%) than in the 1940s (31%). Losses attributed to insects almost doubled (from 7% to 13%) despite a 10-fold increase in the use of synthetic insecticides.
- The estimated environmental, health, and social costs of pesticide use in the United States range from $4 billion to $10 billion per year. The International Food Policy Research Institute puts the estimate much higher, at $100–200 billion per year, or $5–10 in damages for every dollar spent on pesticides.
- Alternative pest control practices (Section 21-5) could halve the use of chemical pesticides on 40 major U.S. crops without reducing crop yields.
- A 50% cut in U.S. pesticide use would cause retail food prices to rise by only about 0.2% but would raise average income for farmers about 9%.

Although pesticides can't be completely eliminated, numerous studies and experience show that they can be sharply reduced without decreasing yields, and in some cases yields would increase. Over the past few years, Sweden has cut pesticide use in half with virtually no decrease in the harvest. Campbell Soup uses no pesticides on tomatoes it grows in Mexico, and yields have not dropped. After a 65% cut in pesticide use on rice in Indonesia, yields increased by 15%.

Where Do Pesticides Go, and How Can They Harm Wildlife? *Pesticides don't stay put.* According to the U.S. Department of Agriculture, no more than 2% (and often less than 0.1%) of the insecticides applied to crops by aerial spraying (Figure 21-4) or by ground spraying actually reaches the target pests; less than 5% of herbicides applied to crops reaches the target weeds.

Q: What percentage of the energy input of a screw-in fluorescent light bulb is converted to light?

Figure 21-4 A crop duster spraying an insecticide on grapevines south of Fresno, California. Aircraft apply about 25% of the pesticides used on U.S. cropland, but only 0.1–2% of these insecticides actually reach the target pests. To compensate for the drift of pesticides from target to nontarget areas, aircraft apply up to 30% more pesticide than ground-based application does. (National Archives/EPA Documerica)

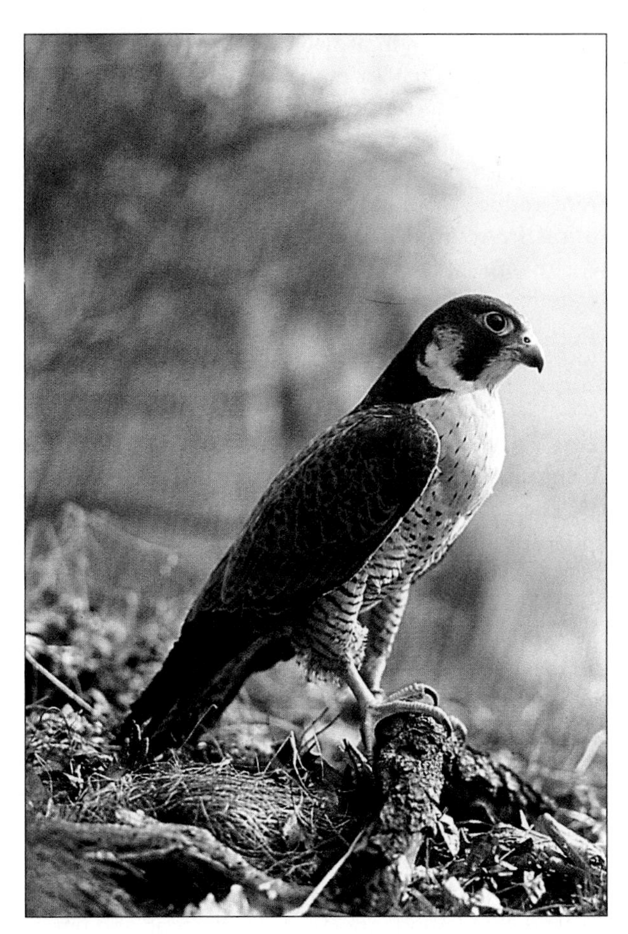

Figure 21-5 Until 1994, the peregrine falcon was listed as an endangered species in the United States, mostly because DDT caused their young to die before hatching because the eggshells were too thin to protect them. Only about 120 peregrine falcons were left in the lower 48 states in 1975; today there are about 1,990, most of them bred in captivity and then released into the wild in a $2.7-million-per-year recovery program. Because of the success of this program, the status of the peregrine falcon in the United States has been changed from endangered to threatened and it may soon be removed from the threatened list. (Hans Reinhard/Bruce Coleman Ltd.)

Pesticides that miss their target pests end up in the air, surface water, groundwater, bottom sediments, food, and nontarget organisms, including humans and wildlife. Pesticide waste and mobility can be reduced by using recirculating sprayers, covering spray booms to reduce drift, and using rope-wick applicators (which deliver herbicides directly to weeds and reduce herbicide use by 90%).

Some pesticides can harm wildlife. During the 1950s and 1960s, populations of fish-eating birds such as the osprey, cormorant, brown pelican, and bald eagle plummeted. Research indicated that a chemical derived from DDT, when biologically magnified in food webs (Figure 17-2), made the birds' eggshells so fragile that they could not successfully reproduce. Also hard-hit were such predatory birds as the prairie falcon, sparrow hawk, and peregrine falcon (Figure 21-5), which help control rabbits, ground squirrels, and other crop-eaters.

Since the U.S. ban on DDT in 1972, most of these species have made a comeback. In 1980, however, DDT levels were again rising in the peregrine falcon and the osprey. Scientists believe that the birds may be picking up DDT and other banned pesticides in Latin America, where they winter and where use of such chemicals is still legal. Illegal use of banned pesticides in the United States, as well as airborne drift of long-lived pesticides from developing countries onto U.S. land

A: 22% (over four times as much as an incandescent bulb)

and surface water, may also play a role. Recently, the EPA found DDT in 99% of the freshwater fish it tested.

Each year 20% of U.S. honeybee colonies are wiped out by pesticides, and another 15% are damaged, costing farmers at least $200 million per year from reduced pollination of vital crops. Pesticide runoff from cropland, a leading cause of fish kills worldwide, kills 6–14 million fish each year in the United States. According to the U.S. Fish and Wildlife Service, pesticides menace about 20% of the endangered and threatened species in the United States.

How Can Pesticides Threaten Human Health? According to the World Health Organization and the UN Environment Programme, an estimated 25 million agricultural workers in developing countries are seriously poisoned by pesticides each year, resulting in an estimated 220,000 deaths. Health officials believe that the actual number of pesticide-related illnesses and deaths among the world's farm workers is probably greatly underestimated because of poor records, lack of doctors and disease reporting in rural areas, and faulty diagnoses.

Farm workers in developing countries are especially vulnerable to pesticide poisoning because three-quarters of all pesticides are applied by hand. In addition, educational levels are low, warnings are few, pesticide regulations are lax or nonexistent, and use of protective equipment is rare (especially in the hot and humid tropics). Workers' clothing also spreads pesticide contamination to family members, and it is common in developing countries for children to work or play in fields treated with pesticides. To make matters worse, some families reuse pesticide containers to store food and drinking water.

Some of these same conditions exist for farm workers in parts of the United States and in other developed countries. In the United States, for example, it is estimated that at least 300,000 farm workers suffer from pesticide-related illnesses each year. At least 30,000 of these pesticide poisonings are acute, and typically 25 U.S. farm workers die every year from such poisonings. Every year about 250,000 Americans, mostly children, get sick from home misuse or unsafe storage of pesticides.

DDT and related persistent chlorinated hydrocarbon pesticides have been banned in the United States and most developed countries. However, they were largely replaced by organophosphate pesticides (Table 21-1) that, although considerably less persistent, are much more toxic to humans and have led to increased poisonings and deaths among farm workers.

In 1995 the Environmental Working Group estimated that about 13% of vegetables and fruits consumed in the United States may contain illegal pesticides and levels of approved pesticides above their legally allowed limits. These estimates may be too low because U.S. Food and Drug Administration (FDA) inspectors check less than 1% (about 12,000 samples) of domestic and imported food for pesticide contamination and measuring techniques by FDA laboratories can detect only about half of the legal and illegal pesticides in use. Moreover, the FDA's turnaround time for food analysis is so long that the food has often been sold and eaten before contamination is detected. Even when contaminated food is found, the growers and importers are rarely penalized.

In 1987 the EPA ranked pesticide residues in foods as the third most serious environmental health threat in the United States in terms of cancer risk. At least 75 of the active ingredients approved for use in U.S. pesticide products cause cancer in test animals. A 1993 study of Missouri children revealed a statistically significant correlation between childhood brain cancer and use of various pesticides in the home, including flea and tick collars, no-pest strips, and chemicals used to control pests such as roaches, ants, spiders, mosquitoes, and termites.

A 1995 study also found that children whose yards were treated with pesticides (mostly herbicides) were four times more likely to suffer from cancers of muscle and connective tissues than children whose yards were not treated. The study also found that children whose homes contained pest strips (with dichlorovos) faced 2.5–3 times the risk of leukemia as children whose homes didn't contain such strips.

Some scientists are becoming increasingly concerned about possible genetic mutations, birth defects, nervous system disorders (especially behavioral disorders), and effects on the immune and endocrine systems from long-term exposure to low levels of various pesticides (Section 17-3). Very little research has been conducted on these potentially serious threats to human health.

Accidents and unsafe practices in pesticide manufacturing plants can expose workers, their families, and sometimes the general public to harmful levels of pesticides or toxic chemicals used in their manufacture (Case Study, right).

21-4 PESTICIDE REGULATION IN THE UNITED STATES

Is the Public Adequately Protected? The Federal Insecticide, Fungicide, and Rodenticide Act (FIFRA) requires that all commercial pesticides be approved by the EPA for general or restricted use. Pesticide companies first evaluate the biologically active ingredients in their products. EPA officials then review these data. In the late 1970s the EPA discovered that

Q: What percentage of the energy in gasoline is used to move a motor vehicle powered by an internal combustion engine?

CASE STUDY

A Black Day in Bhopal

December 2, 1984, will long be a black day on the Indian calendar because on that date *the world's worst industrial accident* occurred at a Union Carbide pesticide plant in Bhopal, India.

Some 36 metric tons (40 tons) of highly toxic methyl isocyanate (MIC) gas, used in the manufacture of carbamate pesticides, leaked from an underground storage tank. When water accidentally entered the tank, its cooling system failed, which caused the reaction mixture to overheat and explode. Once in the atmosphere, some of the toxic MIC was converted to even more deadly hydrogen cyanide gas.

The toxic cloud of gas settled over about 78 square kilometers (30 square miles), exposing up to 600,000 people, many of them illegal squatters living near the plant because they had no other place to go.

According to Indian officials, at least 5,100 people (some say 7,000–15,000, based on the sales of shrouds and cremation wood) were killed; according to a 1996 report by the International Medical Commission on Bhopal, another 50,000 to 60,000 sustained permanent injuries such as blindness or lung damage.

Indian officials claim that the accident was caused by negligence, whereas Union Carbide officials claim that it was caused by sabotage (but has presented no evidence in court to back up this charge).

The Indian Supreme Court ordered Union Carbide to pay a $470-million settlement. However, the Indian government challenged the ruling, arguing that the settlement was inadequate. In 1991, the court upheld the settlement amount.

Leaving aside fair compensation to the victims, the economic damage from the accident was estimated at $4.1 billion, so Union Carbide got off extremely lightly, explaining why the company's stock price rose when the settlement amount was announced.

After the accident, Union Carbide reduced the corporation's liability risks for compensating victims by selling off a portion of its assets and giving much of the profits to its shareholders in the form of special dividends. In 1994, Union Carbide sold its holdings in India.

Union Carbide could probably have prevented this tragedy, which cost billions of dollars (not including the tragic loss of life and serious health effects) by spending no more than $1 million to improve plant safety.

Critical Thinking

Did Union Carbide behave responsibly or irresponsibly in this matter with respect to **(a)** those killed and injured and **(b)** its stockholders, who saw their investment rise because of the accident? Explain. How could future disasters like this be prevented?

data used to support registrations for more than 200 pesticide active ingredients had been falsified by a now-defunct laboratory.

When a pesticide is legally approved for use on fruits or vegetables, the EPA sets a *tolerance level* specifying the amount of toxic pesticide residue that can legally remain on the crop when the consumer eats it. According to the National Academy of Sciences, "Tolerance levels are not based primarily on health considerations." In 1993 the National Academy of Sciences recommended that the legally allowed tolerance levels for all active ingredients in pesticides be reduced to one-tenth of previous levels to help protect infants and children, who are more vulnerable to such chemicals than adults. By 1999 this had not been done.

Traditionally, pesticide makers have also been required to evaluate the effect of active pesticide ingredients on wildlife. Recently, however, the EPA has stopped requiring them to conduct field tests on birds and fish for newly developed pesticides.

Between 1972 and 1996 the EPA canceled or severely restricted the use of 55 active pesticide ingredients. The banned chemicals include most chlorinated hydrocarbon insecticides, several carbamates and organophosphates, and the systemic herbicides 2,4,5-T and Silvex (Table 21-1). However, banned or unregistered pesticides may be manufactured in the United States and exported to other countries (Connections, p. 570).

FIFRA required the EPA to reevaluate the more than 600 active ingredients approved for use in pre-1972 pesticide products to determine whether any of them caused cancer, birth defects, or other health risks. However, by 1998—27 years after Congress ordered the EPA to evaluate these chemicals—less than 10% of these active ingredients had been fully evaluated. The EPA contends that it hasn't been able to complete this evaluation because of the difficulty and expense of evaluating the health effects of chemicals (Section 17-2).

According to a National Academy of Sciences study, federal laws regulating the use of pesticides in the United States are inadequate and poorly enforced by the EPA, FDA, and USDA. The study also concluded that up to 98% of the potential risk of developing cancer from pesticide residues on food grown in the United States would be eliminated if EPA standards were as strict for pre-1972 pesticides as they are for later ones.

A: About 10% (the other 90% is given off to the environment as waste heat)

The Circle of Poison

CONNECTIONS

U.S. pesticide companies can make and export to other countries pesticides that have been banned or severely restricted—or never even approved—in the United States. But what goes around comes around.

In what environmentalists call a *circle of poison*, residues of some of these banned or unapproved chemicals can return to the United States in or on imported items such as coffee, cocoa, pineapples, and out-of-season melons, tomatoes, and grapes. More than one-fourth of the produce (fruits and vegetables) consumed in the United States is grown overseas. Persistent pesticides such as DDT can also be carried by winds from other countries to the United States.

Environmentalists have urged Congress—without success—to break this deadly circle. Supporters of pesticide exports argue that such sales increase economic growth and provide jobs. They also contend that if the United States didn't export pesticides, other countries would.

In 1998, more than 100 countries began talks to finalize an international treaty curbing trade in potential lethal chemicals such as mercury and its compounds and pesticides such as DDT and organophosphates.

Critical Thinking

Should U.S. companies be allowed to export pesticides to other countries that have been banned, severely restricted, or not registered for use in the United States? Explain.

According to a 1987 study by the National Academy of Sciences, exposure to pesticide residues in food causes 4,000–20,000 cases of cancer per year in the United States. Because roughly 50% of those getting cancer die prematurely, this amounts to about 2,000–10,000 premature deaths per year in the United States from exposure to legally allowed pesticide residues in foods. Representatives from pesticide companies dispute these findings. Indeed, the food industry denies that anyone in the United States has ever been harmed by eating food that has been grown using pesticides for the past 50 years.

The 1,820 so-called *inert* ingredients (such as chlorinated hydrocarbon solvents) in pesticide products normally don't kill pest organisms. However, many of them may be active chemically or biologically in or on other organisms, including humans and various wildlife species. In fact, because inert ingredients make up 80–99% by weight of a pesticide product, they can pose a higher health risk than some active ingredients.

Inert ingredients are not listed on pesticide product labels because they are legally classified as business trade secrets. Indeed, government officials are forbidden by law—under penalty of prison—to reveal the names or any information about the inert ingredients used in pesticide products.

Since 1987 the EPA has been evaluating these inert ingredients; thus far it has labeled 100 of them "of known or potential toxicological concern" but has not yet banned their use. By 1997 the EPA still had no toxicity information on more than 75% of the inert ingredients used in pesticides.

FIFRA allows the EPA to leave inadequately tested pesticides on the market and to license new chemicals without full health and safety data. It also gives the EPA unlimited time to remove a chemical, even when its health and environmental risks are shown to outweigh its economic benefits. The built-in appeals and other procedures often keep a dangerous chemical on the market for up to 10 years.

The EPA can ban a chemical immediately in an emergency. Until 1990, however, the law required the EPA to use its already severely limited funds to compensate pesticide manufacturers for their remaining inventory and for all storage and disposal costs. Because compensation costs for a single chemical could exceed the agency's annual pesticide budget, the only economically feasible solution was for the EPA to allow existing stocks of a dangerous chemical to be sold. One of the 1988 amendments to FIFRA shifted some of the costs of storage and disposal of banned pesticides from the EPA to the manufacturers. This law is the only major environmental statute that does not allow citizens to sue the EPA for not enforcing the law, an essential tool to ensure government compliance.

According to a 1996 study by Charles M. Benbrook, former executive director of the U.S. National Academy of Sciences' Board of Agriculture, despite the expenditure of more than $1 billion per year of taxpayers' money to regulate pesticides, the public health hazards and ecological damage created by pesticides have not diminished in the past 30 years.

A 1993 study of pesticide safety by the U.S. National Academy of Sciences urged the government to do the following things:

- Make human health the primary consideration for setting limits on pesticide levels allowed in food.
- Collect more and better data on exposure to pesticides for different groups, including farm workers, adults, and children.

Freeman, Joe. 1997. The ABCs of IPM. *Flower & Garden Magazine*, vol. 41, no. 4, 14(2).

- Develop new and better test procedures for evaluating the toxicity of pesticides, especially for children.

- Consider cumulative exposures of all pesticides in food and water, especially for children, instead of basing regulations on exposure to a single pesticide. Currently, for example, tiny quantities of nearly 100 pesticides are legally allowed in milk, which typically makes up one-fifth of a toddler's diet. A child may also be exposed to trace amounts of 20 different pesticides on grapes, 20 on oranges, and 13 on apples. Are the cumulative levels of these pesticides safe for children?

Some progress was made with the passage of the 1996 Food Quality Protection Act:

- It requires new standards for pesticide tolerance levels in foods, based on a reasonable certainty of no harm to human health (defined for cancer as producing no more than 1 additional cancer per million people exposed to a certain pesticide over a lifetime).

- It requires manufacturers to demonstrate that the active ingredients in their pesticide products are safe for infants and children.

- It allows the EPA to apply an additional 10-fold safety factor to pesticide tolerance levels to protect infants and children.

- It requires the EPA to consider exposure to more than one pesticide when setting pesticide tolerance levels.

- It requires the EPA to develop rules for a program to screen all active and inactive ingredients for their estrogenic and endocrine effects by 1999.

21-5 SOLUTIONS: OTHER WAYS TO CONTROL PESTS

How Can Cultivation Practices Control Pests? A number of cultivation practices can help reduce pest damage. The type of crop planted in a field each year can be changed (crop rotation). Rows of hedges or trees can be planted around fields to hinder insect invasions and provide habitats for their natural enemies (with the added benefit of reduced soil erosion). Planting times can be adjusted so that major insect pests either starve or get eaten by their natural predators. Recent research by USDA scientists suggests that plowing at night reduces the growth of certain types of weeds by 50–60%, by preventing their buried seed from being sprouted by exposure to a short flash of light. Crops can also be grown in areas where their major pests do not exist.

Trap crops can be planted to lure pests away from the main crop. For example, in Nicaraguan cotton fields several rows of cotton are planted several months ahead of the regular crop to attract boll weevils, which can then be destroyed by hand or with small doses of pesticides.

Growers can switch from vulnerable monocultures to intercropping, agroforestry, and polyculture, which use plant diversity to reduce losses to pests (Section 12-6). Diseased or infected plants and stalks and other crop residues that harbor pests can be removed from cropfields. Photodegradable plastic can keep weeds from sprouting between crop rows. Vacuum machines can be used to gently remove harmful bugs from plants.

With the rise of industrial agriculture and the greatly increased use of synthetic pesticides over the last 30 years, many farmers in developed countries no longer use many of these traditional cultivation methods for controlling pest populations. However, as the problems and expense of using pesticides have risen since 1980, more farmers are returning to these methods.

How Can Genetically Resistant Plants and Crops Help Lower Pest Losses? Plants and animals that are genetically resistant to certain pest insects, fungi, and diseases can be developed. However, resistant varieties usually take a long time (10–20 years) and lots of money to develop by conventional crossbreeding methods. Moreover, insects and plant diseases can develop new strains that attack the once-resistant varieties, forcing scientists to continually develop new resistant strains. Genetic engineering is now helping to speed the process (Figure 21-6 and Pro/Con, p. 462).

Figure 21-6 The results of one example of genetic engineering against pest damage. Both tomato plants were exposed to destructive caterpillars. The normal plant's leaves are almost gone (left), whereas the genetically altered plant (right) shows little damage. (Monsanto)

Hint: Enter the search term *integrated pest management* using the Subject Guide.

There has been much talk about using genetic engineering to build pest resistance into crops and thus reduce the need for pesticides. However, some biotech companies have been focusing much of their genetic engineering research on developing crop strains that are resistant to herbicides—a strategy designed to increase the use and sale of their patented seeds and herbicides. For example, Monsanto's new "Roundup Ready" patented seeds are resistant to Roundup—its best-selling chemical herbicide.

Critics warn that farmers using new herbicide-tolerant crop strains are likely to use even greater quantities of herbicides to control weeds. This in turn increases the possibility of weeds developing resistance to the herbicides

In addition, some resistance factors bred or engineered into crops can be toxic to beneficial insects, humans, and other animals. For example, certain alkaloids bred into potatoes to defend against the Colorado potato beetle are hazardous to humans.

How Can Natural Enemies Help Control Pests? Biological control using predators (Figures 21-1 and 21-7), parasites, and pathogens (disease-causing bacteria and viruses) can be encouraged or imported to regulate pest populations. More than 300 biological pest control projects worldwide have been successful, especially in China (p. 560) and Cuba. In Nigeria, crop-duster planes release parasitic wasps instead of pesticides to fight the cassava mealybug; farmers get a $178 return for every $1 they spend on the wasps. In the United States, natural enemies have been used to control more than 70 insect pests.

Biological control has several advantages. It focuses on selected target species and is nontoxic to other species, including people. Once a population of natural predators or parasites is established, biological pest control can often be self-perpetuating. Development of genetic resistance is minimized because pest and predator species interact and change together (coevolution). In the United States, biological control has saved farmers an average of $25 for every $1 invested.

Figure 21-7 Biological control of pests: An adult convergent ladybug (right) is consuming an aphid (left). (Peter J. Bryant/Biological Photo Service)

Biological control is likely to play an increasingly important role as genetic engineering alters pest organisms to render them beneficial rather than harmful. Genetic engineering may also help develop pesticide-resistant natural enemies of common insect pests, incorporate pesticide activity into plants, or allow food crop varieties to tolerate higher applications of pesticides, but such approaches are controversial (Pro/Con, p. 462).

One drawback is that years of research may be needed first, both to understand how a particular pest interacts with its various enemies and to choose the best biological control agent. Biological agents can't always be mass-produced, and farmers find them slower acting and more difficult to apply than pesticides. Biological agents must also be protected from pesticides sprayed in nearby fields.

Once released into the environment, biological control agents can multiply, cause unpredictable harmful ecological effects, and be impossible to recapture. Some may even become pests themselves; others (such as praying mantises) devour beneficial and pest insects alike. Indeed, species introduced for biological pest control have been strongly implicated in the extinction of nearly 100 insect species worldwide.

How Can Biopesticides Help Control Pests? Botanicals such as synthetic pyrethroids (Table 21-1) are an increasingly popular method of pest control, and scientists are looking for new plant toxins to synthesize for mass production. One promising new synthetic botanical is a modified seed extract from the neem tree.

Microbes are also being drafted for insect wars, especially by organic farmers. For example, *Bacillus thuringensis (Bt)* toxin is a registered pesticide sold commercially as a dry powder. Each of the thousands of strains of this common soil bacterium kills a specific pest. Various strains of *Bt* are used by almost all organic farmers as a nonchemical pesticide.

Recently Monsanto transferred a *Bacillus thuringensis* gene to cotton and other plants. The *Bt* gene produces a protein that disrupts the digestive system of pests; insects that bite the plant die within hours. However, preliminary trials indicate that it is not very effective because of the low doses of *Bt* produced by the cotton. More bad news is

Q: If the world really got serious about improving energy efficiency, how much money could be saved?

that genetic resistance is already developing to some *Bt* toxins.

In addition, environmentalists have charged Monsanto with deliberately sabotaging the growing use of *Bt*, the main nonchemical pesticide used by organic farmers, to eliminate this competition. As more crops containing *Bt* genes are used, insect pests will become genetically resistant to the various forms of *Bt* and they will no longer be useful to farmers. Insect experts say this will probably happen in 3–10 years.

How Can Insect Birth Control, Sex Attractants, and Hormones Help Control Pests? Males of some insect pest species can be raised in the laboratory, sterilized by radiation or chemicals, and then released into an infested area to mate unsuccessfully with fertile wild females. This technique works best if the females mate only once, the infested area is isolated (so that it can't be repopulated with nonsterilized males), and the insect pest population has already been reduced to a fairly low level by weather, pesticides, or other factors. Problems with this approach include high costs, the difficulties in knowing the mating times and behaviors of each target insect, the large number of sterile males needed, and the few species for which this strategy works.

The U.S. Department of Agriculture (USDA) used the sterile-male approach to essentially eliminate the screwworm fly, a major livestock pest (Figure 21-8), from the southeastern states between 1962 and 1971. To prevent resurgence of this pest, new strains of sterile male flies must be released every few years. Currently, the USDA has a cooperative male release program with the Mexican government in southern Mexico.

In many insect species, a female that is ready to mate releases a minute amount (typically about one-millionth of a gram) of a chemical sex attractant called a *pheromone*. Whether extracted from insects or synthesized in the laboratory, pheromones can lure pests into traps or attract their natural predators into cropfields (usually the more effective approach). More than 50 companies worldwide sell about 250 pheromones to control pests (Figure 21-9). These chemicals attract only one species, work in trace amounts, have little chance of causing genetic resistance, and are not harmful to nontarget species. However, it is costly and time-consuming to identify, isolate, and produce the specific sex attractant for each pest or predator.

Each step in the insect life cycle is regulated by the timely natural release of juvenile hormones (JH)

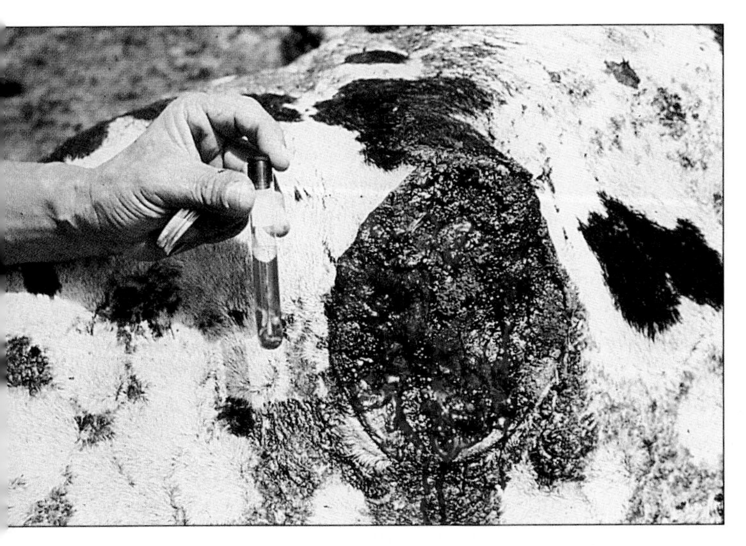

Figure 21-8 Infestation of a steer by screwworm fly larvae in Texas (U.S.A.). An adult steer can be killed in 10 days by thousands of maggots feeding on a single wound. (U.S. Department of Agriculture)

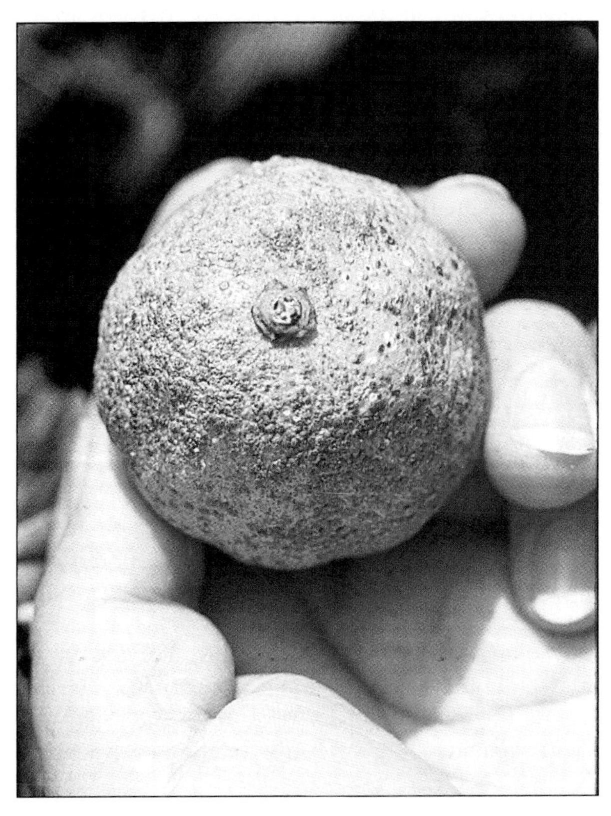

Figure 21-9 Pheromones can help control populations of pests, such as the red scale mites that have infested this lemon grown in Florida (U.S.A.). (Agricultural Research Service/USDA)

A: About $1 trillion per year (at least $400 billion per year in the United States)

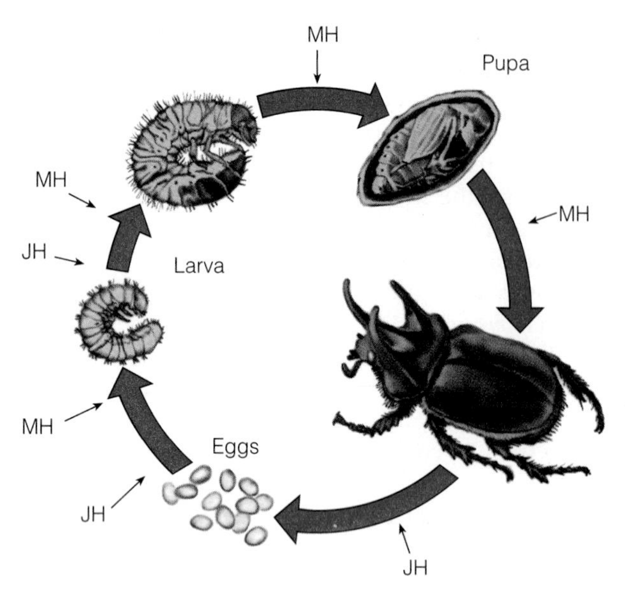

Figure 21-10 For normal insect growth, development, and reproduction to occur, certain juvenile hormones (JH) and molting hormones (MH) must be present at genetically determined stages in the insect's life cycle. If applied at the proper time, synthetic hormones disrupt the life cycles of insect pests and control their populations.

Figure 21-11 A use of hormones to prevent insects from maturing completely, making it impossible for them to reproduce. The stunted tobacco hornworm (left) was fed a compound that prevents production of molting hormones; a normal hornworm is shown on the right. (Agricultural Research Service/USDA)

and molting hormones (MH) (Figure 21-10). These chemicals, which can either be extracted from insects or synthesized in the laboratory, can disrupt an insect's normal life cycle, causing the insect to fail to reach maturity and reproduce (Figure 21-11). Insect hormones have the same advantages as sex attractants, but they take weeks to kill an insect, are often ineffective with large infestations of insects, and sometimes break down before they can act. They must also be applied at exactly the right time in the target insect's life cycle. Moreover, they sometimes affect the target's predators and other nonpest species, and they can kill crustaceans if they get into aquatic ecosystems. Finally, like sex attractants, they are difficult and costly to produce.

How Can Hot Water Be Used to Zap Pests? Some farmers have begun using the *Aqua Heat* machine, which sprays boiling water on crops to kill weeds and insects. Water is boiled and drawn from a large stainless steel tank mounted on a tractor and sprayed on crops using a long boom.

So far, the system has worked well on cotton, alfalfa, and potato fields and in citrus groves in Florida, where the machine was invented. The costs are roughly equal to those of using chemical pesticides.

How Can Zapping Foods with Radiation Help Control Pests? Certain foods can be exposed after harvest to gamma rays emitted by radioactive isotopes such as cobalt-60 (Table 3-1). Such exposure extends food shelf life and kills harmful insects, parasitic worms (such as trichinae in pork), and bacteria (such as salmonella, campylobacter, and *E. coli*, Spotlight, p. 453). A food does not become radioactive when it is irradiated, just as being exposed to X rays does not make the body radioactive.

According to the U.S. Food and Drug Administration and the World Health Organization, over 2,000 studies show that foods exposed to low doses of ionizing radiation are safe for human consumption. Currently, 37 countries (8 of them in western Europe) allow irradiation of one or more food items. Proponents of this technology contend that irradiation of food is likely to lower health hazards to people and reduce pesticide use and that its potential benefits greatly exceed the risks.

Meat and other foods are zapped after they are packaged, but before they are transported to a supermarket or restaurant. This gives food processors a strong defense in any potential liability cases.

Critics contend that irradiating food destroys some of its vitamins and other nutrients and that the irradiation process may form trace amounts of toxic, possibly carcinogenic compounds in the food. However, proponents of food irradiation point out that fresh foods lose similar amounts of vitamins when they are cooked, canned, or frozen.

Critics also point to some studies indicating that research animals fed irradiated foods have shown increases in kidney damage, tumors, and reproductive failures. However, other studies with rats, mice, guinea pigs, and monkeys have shown no greater incidence of cancers than with animals fed nonirradiated

Q: What is the most inefficient way to produce electricity for heating an interior space or water?

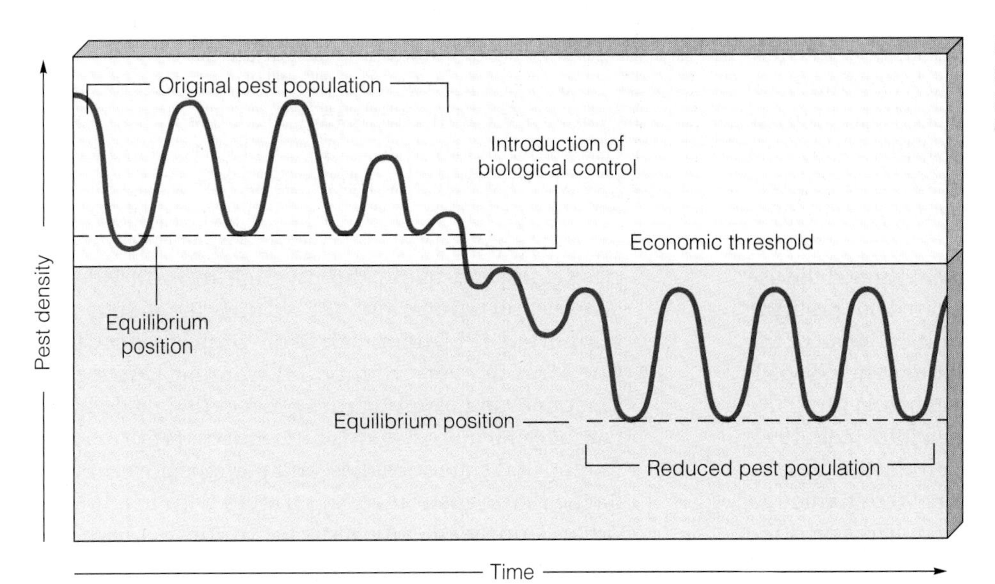

Figure 21-12 The goal of integrated pest management is to keep each pest population just below the size at which it causes economic loss.

foods. These opponents contend that more studies are needed to ensure the safety and wholesomeness of irradiated food and that it is too soon to see possible long-term effects, which might not show up for 30–40 years.

Critics also argue that Americans want fresh, wholesome food, not old, possibly less nutritious food made to appear fresh and healthy by irradiation. And people prefer food that is truly clean to food contaminated by fecal matter even if it is germ-free.

Irradiating food will also raise the price of some foods such as hamburger by 3–7¢ per pound. It will also endanger the health and safety of food-irradiation workers. Critics say that safer and cheaper methods are available for killing harmful bacteria in meat. Examples are **(1)** *steam pasteurization* (already in use in some meat processing operations), in which meat carcasses are suspended and passed through a cabinet, where they are blasted with steam to kill surface microbes; **(2)** development of vaccines to protect poultry from salmonella; and **(3)** use of ozone (long used to disinfect bottled water and water in some municipal water systems) to kill harmful bacteria.

Irradiation could also increase deaths from botulism poisoning. Current levels of irradiation do not destroy botulinum spores, but they do destroy the bacteria that give off the rotten odor warning us of their presence. Furthermore, some microorganisms could mutate when exposed to radiation, possibly creating new and more dangerous forms of botulism.

Irradiated foods sold in the United States are supposed to bear a characteristic logo and a label stating that the product has been *picowaved*. But consumer advocates say that the logo and uninformative label mean little to most consumers. The label is not required for prepared or packaged food that contains irradiated ingredients or for irradiated food sold in restaurants and cafeterias.

The FDA has approved the use of irradiation on poultry, pork, beef, spices, and all domestic fruits and vegetables to delay sprouting, kill insect pests, and slow ripening. However, New York, New Jersey, and Maine have prohibited the sale and distribution of irradiated food, as have Germany, Austria, Denmark, Sweden, Switzerland, Sudan, Singapore, Australia, and New Zealand.

Of course, the best food processing system in the world can't protect us from ourselves. More than 75% of reported food-poisoning cases can be traced to improper food handling by consumers or food preparers in restaurants.

How Can Integrated Pest Management Help Control Pests? An increasing number of pest control experts and farmers believe that the best way to control crop pests is a carefully designed **integrated pest management (IPM)** program. In this approach, each crop and its pests are evaluated as parts of an ecological system. Then a control program is developed that includes a mix of cultivation and biological and chemical methods applied in proper sequence and with the proper timing.

The overall aim of IPM is not eradication of pest populations, but reduction of crop damage to an economically tolerable level (Figure 21-12). Fields are carefully monitored; when a damaging level of pests is reached, farmers first use biological methods

A: Nuclear power (only about 14% efficient)

(natural predators, parasites, and disease organisms) and cultivation controls, including vacuuming up harmful bugs. Small amounts of insecticides (mostly botanicals or microbotanicals) are applied only as a last resort, and different chemicals are used to slow development of genetic resistance and to avoid killing predators of pest species.

In 1986 the Indonesian government banned the use of 57 of 66 pesticides used on rice, phased out pesticide subsidies over a 2-year period, and used some of the money to help launch a nationwide program to switch to IPM, including a major farmer education program. The results were dramatic: Between 1987 and 1992, pesticide use dropped by 65%, rice production rose by 15%, and more than 250,000 farmers were trained in IPM techniques. By 1993 the program had saved the Indonesian government over $1.2 billion (most from elimination of $120 million per year pesticide subsidies)—more than enough to fund its IPM program.

In Australia use of IPM on Queensland citrus crops saves growers about $1 million per year for every 1,000 hectares (2,500 acres). Overall pesticide use fell 90% between 1980 and 1990. In 1994 farmers using IPM in Bangladesh used 80–90% less pesticide and still got yields 12–13% higher than those of non-IPM fields. In the Philippines nearly 4,000 farmers trained to use IPM used 60–98% less pesticide in 1993 and 1994 and had yield increases of 5–15%. In 1997 Mexico unveiled a program to phase out all uses of the pesticides DDT and chlordane (Table 21-1) within 10 years.

The experiences of countries such as China, Brazil, Indonesia, Australia, and the United States show that a well-designed IPM program can reduce pesticide use and pest control costs by 50–90%. IPM can also reduce preharvest pest-induced crop losses by 50%. It can improve crop yields, reduce inputs of fertilizer and irrigation water, and slow the development of genetic resistance because pests are assaulted less often and with lower doses of pesticides. Thus, IPM is an important form of pollution prevention that reduces risks to wildlife and human health.

However, IPM requires expert knowledge about each pest situation, and it is slower acting than conventional pesticides. Methods developed for a crop in one area might not apply to areas with even slightly different growing conditions. Although long-term costs are typically lower than the costs of using conventional pesticides, initial costs may be higher.

Despite its promise and growth, widespread use of IPM is hindered by government subsidies of conventional chemical pesticides and by opposition from agricultural chemical companies, whose sales would drop sharply. In addition, farmers get most of their information about pest control from pesticide salespeople (and in the United States from USDA county farm agents), few of whom have adequate training in IPM.

In 1996 a study by the National Academy of Sciences recommended that the United States shift from chemically based approaches to ecologically based approaches to pest management. A growing number of scientists urge the USDA to promote IPM in the United States by **(1)** adding a 2% sales tax on pesticides and using the revenue to fund IPM research and education; **(2)** setting up a federally supported IPM demonstration project on at least one farm in every county; **(3)** training USDA field personnel and county farm agents in IPM so that they can help farmers use this alternative; **(4)** providing federal and state subsidies, and perhaps government-backed crop insurance, to farmers who use IPM or other approved alternatives to pesticides; and **(5)** gradually phasing out subsidies to farmers who depend almost entirely on pesticides, once effective IPM methods have been developed for major pest species. Some actions you can take to reduce your use of and exposure to pesticides are listed in Appendix 5.

We need to recognize that pest control is basically an ecological, not a chemical problem.

ROBERT L. RUDD

CRITICAL THINKING

1. Should DDT and other pesticides be banned from use in malaria control efforts throughout the world? Explain. What are the alternatives?

2. Do you believe that because essentially all pesticides eventually fail, their use should be phased out or sharply reduced and that farmers should be given economic incentives for switching to integrated pest management? Explain your position.

3. Do you agree or disagree that because DDT and the other banned chlorinated hydrocarbon pesticides pose no demonstrable threat to human health and have saved millions of lives, they should again be approved for use in the United States? Explain.

4. Should certain types of foods used in the United States be irradiated? Explain. If so, should such foods be required to carry a clear label stating that they have been irradiated? Explain.

5. What changes, if any, do you believe should be made in the Federal Insecticide, Fungicide, and Rodenticide Act regulating pesticide use in the United States?

6. Try to come up with analogies to complete the following statements: **(a)** Using pesticides to control pests is like _______. **(b)** Excessive fear of bugs is like _______. **(c)** Being on a pesticide treadmill is like _______.

Q: What are the two most energy-efficient ways to heat interior space?

(d) Using integrated pest management to help control pests is like _______.

PROJECTS

1. How are bugs and weeds controlled in your yard and garden? On the grounds of your school? On public school grounds, parks, and playgrounds near your home? Consider organizing an effort to have integrated pest management and organic fertilizers used on school and public grounds. Do the same thing for your yard and garden.

2. Make a survey of all pesticides used in or around your home. Compare the results for your entire class.

3. Some research shows that although many people agree that we need to make greater use of alternatives to conventional pesticides for controlling pests, when they are faced with an actual infestation from insects or rodents the first thing they do is spray with pesticides. Survey members of your class and other groups to help determine the validity of this hypothesis.

4. Make a concept map of this chapter's major ideas, using the section heads and subheads and the key terms (in boldface). Look at the inside back cover and in Appendix 4 for information about concept maps.

A: Passive solar and natural gas

22 SOLID AND HAZARDOUS WASTE

There Is No "Away": The Love Canal Tragedy

Between 1942 and 1953, Hooker Chemicals and Plastics (owned by OxyChem since 1968) sealed chemical wastes containing at least 200 different chemicals into steel drums and dumped them into an old canal excavation (called Love Canal after its builder, William Love) near Niagara Falls, New York.

In 1953 Hooker Chemicals filled the canal, covered it with clay and topsoil, and sold it to the Niagara Falls school board for $1. The company inserted in the deed a disclaimer denying legal liability for any injury caused by the wastes. In 1957 Hooker warned the school board not to disturb the clay cap because of the possible danger from toxic wastes.

By 1959 an elementary school, playing fields, and 949 homes had been built in the 10-square-block Love Canal area (Figure 22-1, left). Roads and sewer lines crisscrossed the dump site, some of them disrupting the clay cap covering the wastes. An expressway built at one end of the dump in the 1960s blocked groundwater from migrating to the Niagara River. This created a "bathtub effect" that allowed contaminated groundwater and rainwater to build up and overflow the disrupted cap.

Residents began complaining to city officials in 1976 about chemical smells and chemical burns their children received playing in the canal area, but these complaints were ignored. In 1977 chemicals began leaking from the badly corroded steel drums into storm sewers, gardens, basements of homes next to the canal, and the school playground.

Alarmed residents conducted an informal health survey. They found an unusually high incidence of birth defects, miscarriages, assorted cancers, and nerve, respiratory, and kidney disorders among people who lived near the canal.

Considerable media publicity and pressure from residents led by Lois Gibbs (a mother galvanized into action as she watched her children come down with one illness after another; Guest Essay, p. 586) eventually led state officials to conduct a preliminary health survey and medical tests. They found that pregnant women living near the canal had a miscarriage rate four times higher than average. They also found that the air, water, and soil of the canal area and the base-

Figure 22-1 The Love Canal housing development near Niagara Falls, New York, was built near a hazardous-waste dump site. The photo on the left shows the area when it was abandoned in 1980; the photo on the right shows the area in 1990, when people were allowed to buy some of the remaining houses and move back into the area despite protests from environmentalists. (Left, NY State Department of Environmental Conservation; right, J. Goerg/NY State Department of Environmental Conservation)

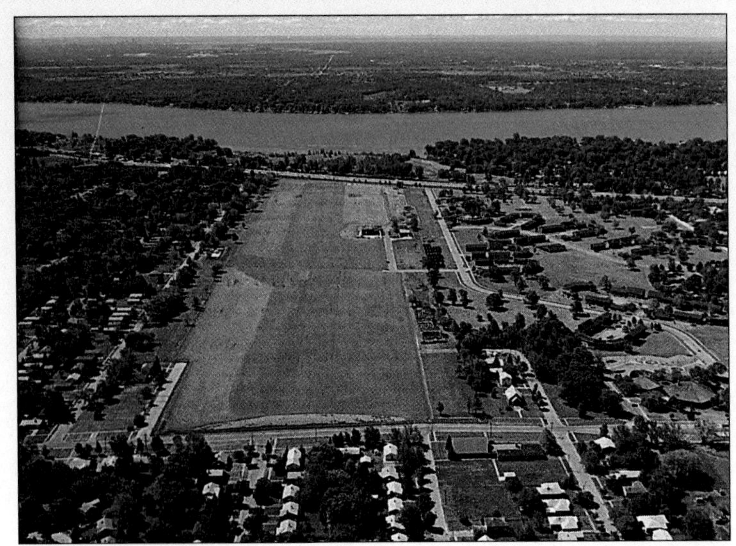

ments of nearby houses were contaminated with several toxic and carcinogenic chemicals.

In 1978, after more pressure from residents and unfavorable publicity, state officials closed the school and arranged for the 239 homes closest to the dump to be evacuated, purchased, and destroyed. Two years later, after protests from families still living fairly close to the landfill, President Jimmy Carter declared Love Canal a federal disaster area, had the remaining families relocated, and offered federal funds to buy 564 more homes. Residents of all but 72 of the homes moved out.

The dump site was covered with a new clay cap and surrounded by a drainage system that pumps leaking wastes to a new treatment plant. Some of the residents who remained claim that the entire problem was overblown by other residents, environmentalists, and the media.

After more than 15 years of court cases, OxyChem agreed in 1994 to a $98-million settlement with New York State and agreed to be responsible for all future treatment of wastes and wastewater at the Love Canal site. In 1996 the company agreed to reimburse the federal government $129 million for the Love Canal cleanup.

Because of the difficulty in linking exposure to a variety of chemicals to specific health effects (Section 17-2), the long-term health effects of exposure to hazardous chemicals on Love Canal residents remain unknown and controversial. However, the psychological damage to the evacuated families is enormous: For the rest of their lives they will worry about the possible effects of the chemicals on themselves and their children and grandchildren.

In June 1990 the EPA declared the area (renamed Black Creek Village) safe and allowed state officials to begin selling 234 of the remaining houses at 10–20% below market value. (Figure 22-1, right). Still, the dump has not yet been cleaned up but only fitted with a drainage system. The EPA acknowledges that the dump site will leak again, sooner or later. Buyers must sign an agreement stating that New York State and the federal government make no guarantees or representations about the safety of living in these homes.

The Love Canal incident is a vivid reminder that we can never really throw anything away, that wastes don't stay put, and that preventing pollution is much safer and cheaper than trying to clean it up.

Solid wastes are only raw materials we're too stupid to use.

Arthur C. Clarke

This chapter is devoted to answering the following questions:

- What are solid waste and hazardous waste, and how much of each type is produced?
- What can we do to reduce, reuse, and recycle solid waste and hazardous waste?
- What is being done to recycle aluminum, paper, and plastics?
- What are the advantages and disadvantages of burning or burying wastes?
- What can we do to reduce exposure to lead, dioxins, and hazardous chlorine compounds?
- How can we make the transition to a low-waste society?

22-1 WASTING RESOURCES: THE HIGH-WASTE APPROACH

What Is Solid Waste, and How Much Is Produced? The United States, with only 4.6% of the world's population, produces about 33% of the world's **solid waste**: any unwanted or discarded material that is not a liquid or a gas. The United States generates about 10 billion metric tons (11 billion tons) of solid waste per year—an average of 40 metric tons (44 tons) per person. Although garbage produced directly by households and businesses is a significant problem, *about 98.5% of the solid waste in the United States comes from mining, oil and natural gas production, agriculture, and industrial activities used to produce goods and services for consumers* (Figure 22-2).

Although mining waste is the single largest category of U.S. solid waste, the EPA has done little to regulate its disposal, mostly because the U.S. Congress has exempted it from regulation as a hazardous waste. Most mining waste is left piled near mine sites and can pollute the air, surface water, and groundwater. In developing countries there is even less regulation of mining procedures and wastes.

Industrial solid waste includes scrap metal, plastics, paper, fly ash (removed by air-pollution control equipment in industrial and electrical power plants,

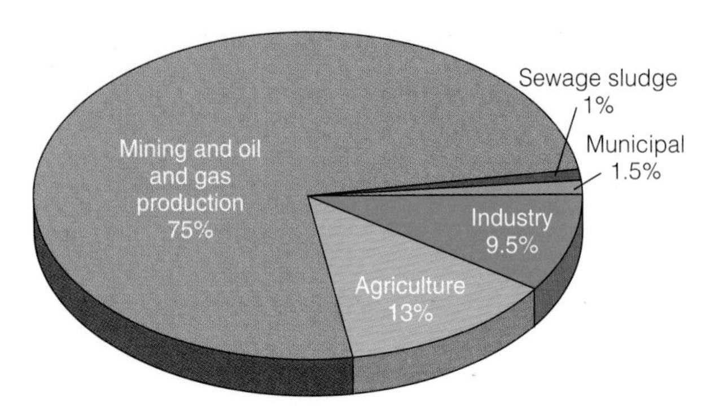

Figure 22-2 Sources of the estimated 10 billion metric tons (11 billion tons) of solid waste produced each year in the United States. Mining and industrial activities produce 65 times as much solid waste as household activities. (Data from U.S. Environmental Protection Agency and U.S. Bureau of Mines)

Figure 18-15), and sludge from industrial waste treatment plants (Figure 20-11). Most of it is buried or incinerated at the plant sites where it is produced.

The remaining 1.5% of solid waste produced in the United States is **municipal solid waste** (MSW) from homes and businesses in or near urban areas. The amount of municipal solid waste, often called *garbage*, produced in the United States in 1996 was enough to fill a bumper-to-bumper convoy of garbage trucks encircling the globe almost eight times. This amounted to an average of 680 kilograms (1,500 pounds) per person in the United States—two to three times that in most other developed countries and many times that in developing countries.

About 27% of the resources in MSW produced in the United States in 1996 was recycled or composted (up from 7% in 1960); the other 73% was hauled away and either dumped in landfills (58%) or burned in incinerators and waste-to-energy plants (15%) at a cost of about $40 billion (projected to rise to $75 billion by 2005).

What Does It Mean to Live in a High-Waste Society? U.S. consumers throw away astounding amounts of solid waste, including the following quantities:

- Enough aluminum to rebuild the country's entire commercial airline fleet every 3 months
- Enough tires each year to encircle the planet almost three times
- About 18 billion disposable diapers per year, which if linked end-to-end would reach to the moon and back seven times
- About 2 billion disposable razors, 10 million computers, and 8 million television sets each year
- About 2.5 million nonreturnable plastic bottles each hour
- Some 14 billion catalogs (an average of 54 per American) and 38 billion pieces of junk mail each year

This is only part of the 1.5% of all solid waste labeled "municipal" in Figure 22-2.

What Is Hazardous Waste, and How Much Is Produced? In the United States, **hazardous waste** is legally defined as any discarded solid or liquid material that **(1)** contains one or more of 39 toxic, carcinogenic, mutagenic, or teratogenic compounds at levels that exceed established limits (including many solvents, pesticides, and paint strippers); **(2)** catches fire easily (gasoline, paints, and solvents); **(3)** is reactive or unstable enough to explode or release toxic fumes (acids, bases, ammonia, chlorine bleach); or **(4)** is capable of corroding metal containers such as tanks, drums, and barrels (industrial cleaning agents and oven and drain cleaners).

This narrow official definition of hazardous wastes (mandated by Congress) does *not* include the following materials: **(1)** radioactive wastes (Sections 15-5 and 17-4), **(2)** hazardous and toxic materials discarded by households (Table 22-1), **(3)** mining wastes, **(4)** oil- and gas-drilling wastes (routinely discharged into surface waters or dumped into unlined pits and landfills), **(5)** liquid waste containing organic hydrocarbon compounds (80% of all liquid hazardous waste), **(6)** cement kiln dust produced when liquid hazardous wastes are burned in a cement kiln (a practice classified as recycling by the EPA but considered dangerous *sham recycling* by environmentalists), and **(7)** wastes from the thousands of small businesses and factories that generate less than 100 kilograms (220 pounds) of hazardous waste per month.

Environmentalists call these omissions from the list of hazardous wastes *linguistic detoxification* designed to save industries and the government money and to mislead the public. Designating these excluded categories as hazardous waste would shift efforts from waste management and pollution control to waste reduction and pollution prevention. Industry representatives say that having to manage these wastes would bankrupt them or force them to move their operations to other countries with less strict waste regulations.

The EPA estimates that at least 5.5 billion metric tons (6 billion tons) of hazardous waste are produced each year in the United States—an average of 21 metric tons (23 tons) per person. However, because only about 6% of the total is *legally* defined as hazardous waste, *94% of the country's hazardous waste is not regulated by hazardous-waste laws.* In most other countries, especially developing countries, even less of the hazardous waste is regulated.

Q: What is the most energy-efficient fuel for powering a motor vehicle?

22-2 PRODUCING LESS WASTE AND POLLUTION: REDUCING THROUGHPUT

What Are Our Options? There are two ways to deal with the solid and hazardous waste we create: *waste management* and *pollution (waste) prevention*. Waste management is a *high-waste approach* that views waste production as an unavoidable product of economic growth. It attempts to manage the resulting wastes in ways that reduce environmental harm, mostly by burying them, burning them, or shipping them off to another state or country. The goal is to move increasing amounts of matter and energy resources through the economy to enhance economic growth (Figure 3-21).

Preventing pollution and waste is a *low-waste approach* that views most solid and hazardous waste either as potential resources (that we should be recycling, composting, or reusing) or as harmful substances that we should not be using in the first place (Figures 3-22, 22-3, and 22-4). With this approach, taxes and subsidies are used to discourage waste production and encourage waste prevention (Guest Essays, pp. 70 and 586).

According to the U.S. National Academy of Sciences (Figures 22-3 and 22-4), the low-waste approach should have the following hierarchy of goals: **(1)** *reduce* waste and pollution, **(2)** *reuse* as many things as possible, **(3)** *recycle and compost* as much waste as possible, **(4)** *chemically or biologically treat or incinerate* waste that can't be reduced, reused, recycled, or composted, and **(5)** *bury* what is left in state-of-the-art landfills or aboveground vaults after the first four goals have been met.

Scientists estimate that in a low-waste society 60–80% of the solid and hazardous waste produced could be eliminated through reduction, reuse, and recycling (including composting). The remaining 20–40% of such wastes would then be treated to reduce their toxicity, and what's left would be burned or buried under carefully regulated conditions. Currently, the order of priorities shown in Figures 22-3 and 22-4 for dealing with solid and hazardous wastes is reversed in the United States (and in most other countries), mostly because the costs of producing and dealing with these wastes are not included in the market prices of products.

Table 22-1 Common Household Toxic and Hazardous Materials

Cleaning Products
Disinfectants
Drain, toilet, and window cleaners
Oven cleaners
Bleach and ammonia
Cleaning solvents and spot removers
Septic tank cleaners
Paint and Building Products
Latex and oil-based paints
Paint thinners, solvents, and strippers
Stains, varnishes, and lacquers
Wood preservatives
Acids for etching and rust removal
Asphalt and roof tar
Gardening and Pest Control Products
Pesticide sprays and dusts
Weed killers
Ant and rodent killers
Flea powder
Automotive Products
Gasoline
Used motor oil
Antifreeze
Battery acid
Solvents
Brake and transmission fluid
Rust inhibitor and rust remover
General Products
Dry-cell batteries (mercury and cadmium)
Artists' paints and inks
Glues and cements

Why Is Producing Less Waste and Pollution the Best Choice? A small but increasing number of companies are learning that reducing waste and pollution can be good for corporate profits, worker health and safety, the local community, consumers, and the environment as a whole. Such methods **(1)** save energy and virgin resources by keeping the quality of matter resources high (Figure 3-12) with a lower input of high-quality energy; **(2)** reduce the environmental effects of extracting, processing, and using resources (Figure 14-13); **(3)** improve worker health and safety by reducing exposure to toxic and hazardous materials; **(4)** decrease pollution control and waste management costs (which are now rising faster than the rate of industrial production) and future liability for toxic and hazardous materials; and **(5)** are usually less

A: Natural gas (unless affordable fuel cells are developed)

Figure 22-3 Solutions: priorities suggested by prominent scientists for dealing with material use and solid waste. To date, these priorities have not been followed in the United States (and in most countries); most efforts are devoted to waste management (bury it or burn it). (U.S. Environmental Protection Agency and U.S. National Academy of Sciences)

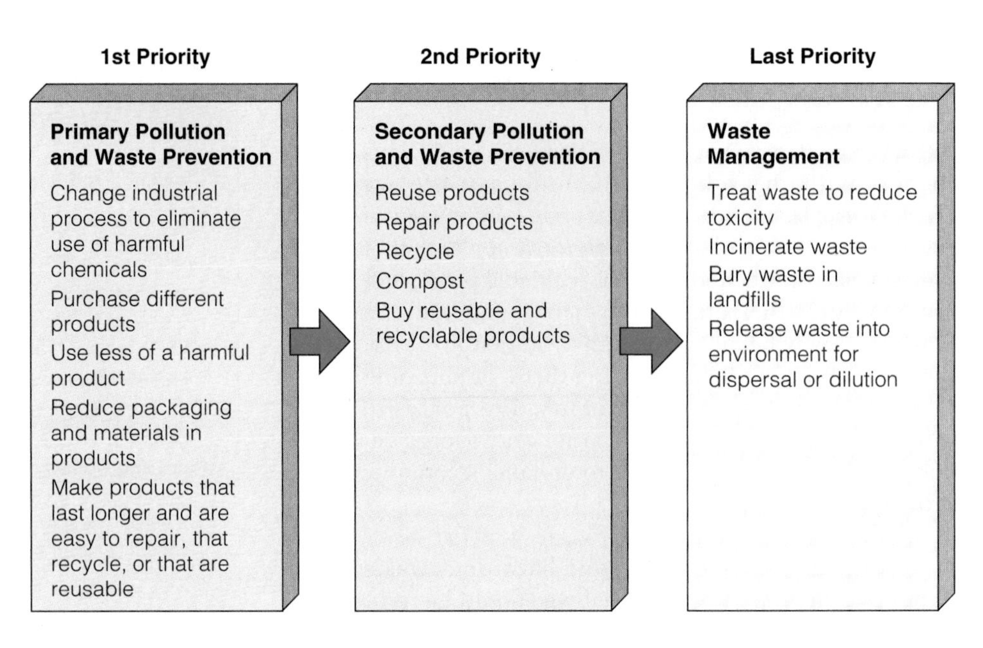

Figure 22-4 Solutions: priorities suggested by prominent scientists for dealing with hazardous waste. To date, these priorities have not been followed in the United States (and in most countries). (U.S. National Academy of Sciences)

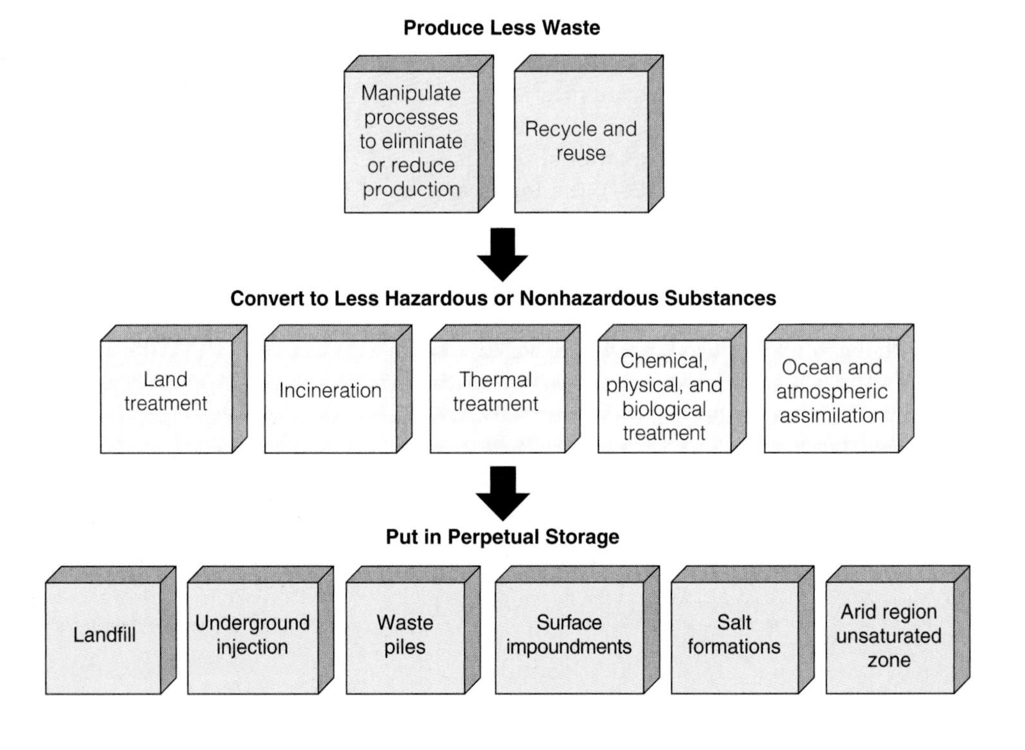

costly on a life cycle basis than trying to clean up pollutants and manage wastes once they are produced.

A 1992 study of 181 waste reduction initiatives in 27 U.S. firms found that two-thirds of the initiatives took 6 months or less to implement. One-fourth required no capital investment, two-thirds paid back their capital investments in 6 months or less, and 93% got them back within 3 years.

In 1975 the Minnesota Mining and Manufacturing Company (3M), which makes 60,000 different products in 100 manufacturing plants, began a Pollution Prevention Pays (3P) program. It redesigned equipment and processes, used fewer hazardous raw materials, identified hazardous chemical outputs (and recycled or sold them as raw materials to other companies), and began making more nonpolluting products. By 1995 3M's overall waste production was down by one-third, emissions of air pollutants were reduced by 70%, and the company had saved over $750 million in waste disposal costs.

Q: What is the best way to save oil, slow ozone depletion and global warming, and reduce air pollution?

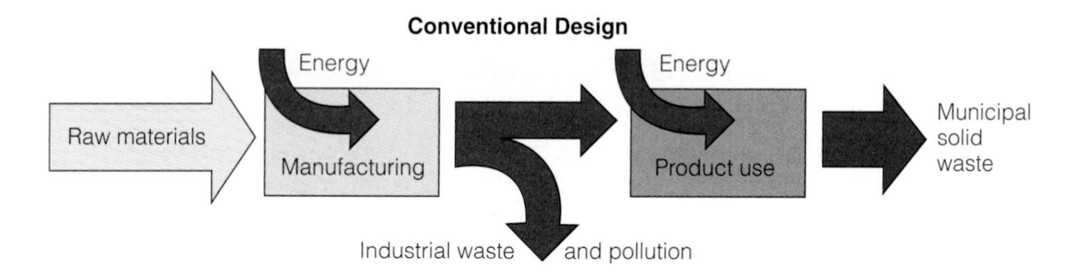

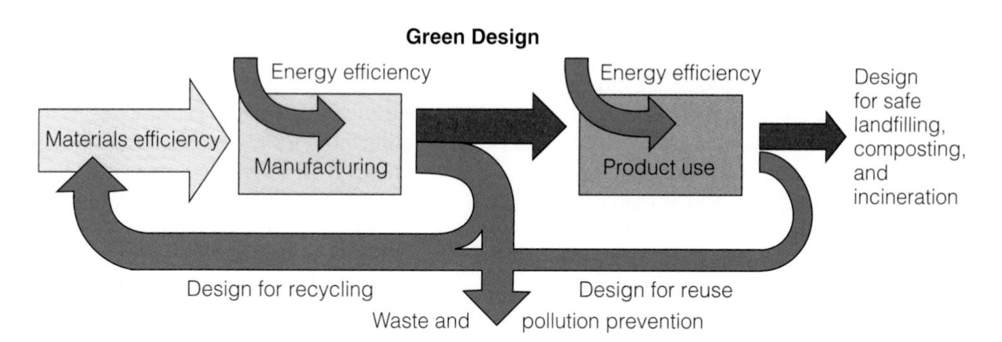

Figure 22-5 How product design affects throughputs of matter and energy resources and out-puts of pollution and solid waste. Most conventional manufacturing processes involve a one-way flow of materials from raw materials to waste and pollution. Green design reduces the overall impact by emphasizing efficient use of matter and energy resources, reduction of outputs of solid and hazardous waste, and reuse and recycling of materials. (Adapted from U.S. Congress Office of Technology Assessment, *Green Products by Design: Choices for a Cleaner U.S. Environment*, Washington, D.C.: Government Printing Office, 1992)

Solutions: How Can We Reduce Waste and Pollution? There are several ways to reduce waste and pollution or resource throughput (Figure 3-22). One is to *decrease consumption*, which begins by asking whether we really need a particular product (sometimes called *precycling*).

Another way to reduce throughput is to *redesign manufacturing processes and products to use less material* (Figure 22-5). For example, cars made today are lighter because an increasing proportion of their steel parts are being replaced with aluminum (80% from recycled metal) and plastic parts. Another strategy is to *design products that produce less pollution and waste fewer resources when used* (Figure 22-5). Examples include paints (which are responsible for about 8% of the volatile organic compounds emitted into the atmosphere) that use fewer volatile solvents and more energy-efficient cars, lights, and appliances (Section 16-2).

Manufacturing processes can also be redesigned to produce less waste and pollution (Figure 22-5). Most toxic organic solvents can be recycled within plants or replaced with water-based or citrus-based solvents (Individuals Matter, p. 528). Some processes can be redesigned to eliminate washing altogether. Hydrogen peroxide can be used instead of toxic chlorine to bleach paper and other materials. For clothes, wet cleaning (with water and steam or with liquid carbon dioxide) and microwave drying can replace dry cleaning with toxic organic solvents. In effect, over the next two decades we need to change the ways we make things (Solutions, p. 584).

Individuals can use less hazardous (and usually cheaper) cleaning products (Table 22-2).Three inexpensive chemicals—baking soda (which can also be used as a deodorant and a toothpaste), vinegar, and borax—can be used for most cleaning and clothes bleaching. People can also use pesticides and other hazardous chemicals only when absolutely necessary and in the smallest amount possible.

Green design and life cycle assessment can help develop products that last longer and that are easy to repair, reuse, remanufacture, compost, or recycle. Several European auto manufacturers now design their cars for easy disassembly and for reuse and recycling of up to 80% of their parts (75% in the United States), and they are trying to minimize the use of nonrecyclable or hazardous materials.

Products can be designed to last longer. For example, although tires are now being produced with an average life of 97,000 kilometers (60,000 miles), researchers believe this could be extended to at least 160,000 kilometers (100,000 miles).

Eliminating or reducing unnecessary packaging is another important strategy. Here are some key questions that environmentalists believe manufacturers and consumers should ask about packaging: Is it necessary? Can it use fewer materials? Can it be reused? Are the resources that went into it nonrenewable or potentially renewable? Does it contain the highest feasible amount of consumer-discarded (postconsumer) recycled material? Is it designed to be easily recycled? Can it be incinerated without producing harmful air pollutants or a toxic ash? Can it be buried and decomposed

An Ecoindustrial Revolution: Cleaner Production

SOLUTIONS

Some analysts urge us to bring about an *ecoindustrial revolution* over the next 50 years as a way to help achieve industrial, economic, and environmental sustainability. All industrial products and processes would be redesigned and integrated into an essentially closed system of cyclical material flows.

The goals of this emerging concept of *cleaner production* (Guest Essay, p. 70), or *industrial ecology*, are to reduce resource throughput, waste, and pollution (Figure 3-22) by **(1)** minimizing the input of energy and matter resources and the output of wastes, **(2)** using less material and energy per unit of output, **(3)** substituting less toxic or nontoxic chemicals in manufacturing processes or products, **(4)** recycling or reusing toxic chemicals used or produced in manufacturing processes, and **(5)** developing materials exchange systems in which one company's wastes become another company's resources and companies take back packaging and used products from consumers for reuse, recycling, repair, or remanufacturing.

In effect, companies would mimic natural chemical cycles (Chapter 5) and interact in complex *resource exchange webs* similar to food webs in natural ecosystems. With such an exchange and chemical cycling network, producing a large amount of easily reusable waste might sometimes be preferable to designing a more efficient process that produces a small amount of unusable waste.

A prototype of this concept exists in Kalunborg, Denmark, where a coal-fired power plant, an oil refinery, a municipal heating authority, a producer of sulfuric acid, a sheet rock plant, a biotechnology company, local farms, some greenhouses, a fish farm, and several other businesses are working together to exchange and convert their wastes into resources. Ecoindustrial parks are in the planning stages in areas such as Baltimore, Rochester (New York), Chattanooga, Halifax (Nova Scotia), and the Brownsville/Matamoros region along the Texas–Mexico border.

Such an ecoindustrial revolution, carried out over the next 50 years, will challenge designers, engineers, and scientists to revolutionize the way we design and make things (Guest Essay, p. 746). This will reduce pollution, improve human health, and help preserve biodiversity and ecological integrity.

This important form of learning how to work with nature will also provide economic benefits to businesses by **(1)** reducing the costs of controlling pollution and complying with pollution regulations, **(2)** improving the health of workers (thus reducing company health-care insurance expenses), **(3)** reducing legal liability for toxic and hazardous wastes, **(4)** stimulating companies to come up with new, environmentally friendly processes and products that can be sold worldwide, and **(5)** giving companies a better image among consumers, based on results rather than on public relations campaigns. Such a revolution would be a win–win situation for everyone and for the earth.

Critical Thinking

What short- and long-term disadvantages (if any) might there be with an ecoindustrial revolution? Do you believe that it will be possible to phase in such a revolution in the country where you live over the next 2–3 decades? Explain. What are the three most important strategies for doing this?

in a landfill without producing chemicals that can contaminate groundwater?

Trash taxes can also be used to reduce waste. For example, in 1992 the city of Victoria, British Columbia, instituted a trash tax of $1.20–2.10 per bag, along with a strong recycling program. Within a year, household waste generation fell by 18%. In 1986 Denmark imposed a tax on many types of solid waste to promote reuse and recycling and reduce solid waste. By 1996 the country had reduced the amount of waste brought to its municipal landfills and incinerators by 26% and achieved an overall recycling rate of 61%. By 1998, six other countries in Europe (Austria, Belgium, Finland, France, the Netherlands, and the United Kingdom) had adopted taxes on solid waste and two other countries (Norway and Sweden) were considering them. A related *pay-as-you-throw* system that bases garbage collection charges on the amount of waste a household generates for disposal is being used in an increasing number of communities in the United States for reducing solid waste and encouraging recycling.

22-3 REUSE

What Are the Advantages of Refillable Containers? *Reuse*, a form of waste reduction, extends

Q: What is a negawatt?

Table 22-2 Alternatives to Some Common Household Chemicals

Chemical	Alternative	Chemical	Alternative
Deodorant	Sprinkle baking soda on a damp wash cloth and wipe skin; dab on vanilla extract.	General surface cleaner	Mixture of vinegar, salt, and water, or mixture of borax, soap, lemon juice, and water.
Oven cleaner	Baking soda and water paste, scouring pad.	Bleach	Baking soda or borax.
Toothpaste	Baking soda.	Mildew remover	Mix ½ cup vinegar, ½ cup borax, and warm water.
Drain cleaner	Pour ½ cup salt down drain, followed by boiling water; or pour 1 handful baking soda and ½ cup white vinegar and cover lightly for one minute.	Disinfectant and general cleaner	Mix ½ cup borax in 1 gallon hot water.
Window cleaner	Add 2 teaspoons white vinegar to 1 quart warm water.	Furniture or floor polish	Mix ½ cup lemon juice and 1 cup vegetable or olive oil.
Toilet bowl, tub, and tile cleaner	Mix a paste of borax and water; rub on and let set one hour before scrubbing. Can also scrub with baking soda and a brush.	Carpet and rug shampoos	Sprinkle on cornstarch, baking soda, or borax and vacuum.
Floor cleaner	Add ½ cup vinegar to a bucket of hot water; sprinkle a sponge with borax for tough spots.	Detergents and detergent boosters	Washing soda or borax and soap powder.
Shoe polish	Polish with inside of a banana peel, then buff.	Spray starch	In a spray bottle, mix 1 tablespoon cornstarch in a pint of water.
Silver polish	Clean with baking soda and warm water.	Fabric softener	Add 1 cup white vinegar or ¼ cup baking soda to final rinse.
Air freshener	Set vinegar out in an open dish. Use an opened box of baking soda in closed areas such as refrigerators and closets. To scent the air, use pine boughs or make sachets of herbs and flowers.	Dishwasher soap	1 part borax and 1 part washing soda.
		Pesticides (indoor and outdoor)	Use natural biological controls (Section 21-5).

resource supplies, keeps high-quality matter resources from being reduced to low-quality matter waste (Figure 3-12), and reduces energy use and pollution even more than recycling. Two examples of reuse are refillable glass beverage bottles and refillable soft drink bottles made of polyethylene terephthelate (PET) plastic. Unlike throwaway and recyclable cans and bottles, refillable beverage bottles create local jobs related to their collection and refilling. Moreover, studies by Coca-Cola and PepsiCo of Canada show that 0.5-liter (16-ounce) bottles of their soft drinks cost one-third less in refillable bottles.

In 1964, 89% of all soft drinks and 50% of all beer in the United States were sold in refillable glass bottles. By 1995, such bottles made up only about 7% of the beer and soft drink market, and only 10 states even had refillable glass bottles. The disappearance of most local bottling companies has led to a loss of local jobs, income, and tax revenues. Some call for reinstatement of this bottling reuse system in the United States; others say it isn't practical because the system of collections and returns has been dismantled.

Denmark has led the way by banning all beverage containers that can't be reused. To encourage use of refillable glass bottles, Ecuador has a refundable beverage container deposit fee that is 50% of the cost of the drink. In Finland, 95% of the soft drink, beer, wine, and spirits containers are refillable, and in Germany, 73% are refillable.

Another reusable container is the metal or plastic lunchbox that most workers and schoolchildren once used. Sandwiches and refrigerator leftovers can be put in small reusable plastic containers instead of in plastic wrap and aluminum foil (most of which is not recycled). This practice and most forms of reuse save money and reduce resource waste.

What Kind of Bags Should You Use for Groceries? When you're offered a choice between plastic or paper bags for your groceries, which should you choose? The answer is *neither*. Both are environmentally harmful, and the question of which is the more damaging has no clear-cut answer.

A: A watt of electricity that is not needed (saved) because of better energy efficiency

GUEST ESSAY

We Have Been Asking the Wrong Questions About Wastes

Lois Marie Gibbs

In 1977, Lois Marie Gibbs was a young housewife with two children living near the Love Canal toxic dump site. She had never engaged in any sort of political action until her children began experiencing unexplained illnesses and she learned that toxic chemicals were oozing from the dump site into many of the area's yards and basements. Then she organized her neighborhood and became the president and major strategist for the Love Canal Homeowners Association. This dedicated grassroots political action by "amateurs" brought hazardous-waste issues to national prominence and spurred passage of the federal Superfund legislation to help clean up abandoned hazardous-waste sites. Lois Gibbs then moved to Washington, D.C., and formed Citizens' Clearinghouse for Hazardous Wastes (renamed the Center for Health, Environment, and Justice), an organization that has helped over 7,000 community organizations protect themselves from hazardous wastes. Her story is told in her autobiography, Love Canal: My Story *(State University of New York Press, 1982); she was also the subject of a CBS movie, "Lois Gibbs: The Love Canal," which aired in 1982. Her latest book is* Dying from Dioxin *(Boston: South End Press, 1995). She is an inspiring example of what a dedicated citizen can do to change the world.*

Just about everyone knows our environment is in danger. One of the most serious threats is the massive amount of waste we put into the air, water, and ground every year. All across the United States and around the world are thousands of places that have been, and continue to be, polluted by toxic chemicals, radioactive waste, and just plain garbage.

For generations, the main question people have asked is, "Where do we put all this waste? It's got to go somewhere." That is the wrong question, as has been shown by a series of experiments in waste disposal and by the simple fact that there is no "away" in "throwaway."

We tried dumping our waste in the oceans. That was wrong. We tried injecting it into deep, underground wells. That was wrong. We've been trying to build landfills that don't leak, but that doesn't work. We've been trying to get rid of waste by burning it in high-tech incinerators; that only produces different types of pollution, such as air pollution and toxic ash. We've tried a broad range of "pollution" controls, but all that does is allow legalized, high-tech pollution. Even recycling, which is a very good thing to do, suffers from the same problem as all the other methods: It addresses waste *after* it has been produced.

For many years, people have been assuming that "it's got to go somewhere," but now many people, especially young people, are starting to ask why. Why do we produce so much waste? Why do we need products and services that have so many toxic by-products? Why can't industry change the way it makes things so that it stops producing so much waste?

These are the *right* questions. When you start asking them, you start getting answers that lead to *pollution prevention* and *waste reduction* instead of *pollution control* and *waste management.* People, young and old, who care about pollution prevention are challenging companies to stop making products with gases that reduce ozone in the ozone layer [Section 19-6] and contribute to the threatening possibility of global warming [Section 19-3]. They are asking why so many goods are wrapped in excessive, throwaway packaging. They are challenging companies that sell pesticides, cleaning fluids, batteries, and other hazardous products to either remove the toxins from those products or take them back for recovery or recycling rather than disposing of them in the environment. They are demanding alternatives to throwaway materials in general.

Since 1988, hundreds of student groups have contacted my organization to get help and advice in taking these effective types of actions. Many of these groups

On one hand, plastic bags degrade slowly in landfills and can harm wildlife if swallowed, and producing them pollutes the environment. On the other hand, producing the brown paper bags used in most supermarkets uses trees and pollutes the air and water. Overall, white or clear polyethylene plastic bags require less energy for manufacture and cause less damage to the environment than paper bags not made mostly from recycled paper.

Instead of having to choose between paper and plastic bags, you can bring your own *reusable* canvas or string containers to the store—and save and reuse any paper or plastic bags you get. Using a renewable bag just five times displaces the pollution caused by the manufacture of the bag. To encourage people to bring their own reusable bags, stores in the Netherlands charge for paper or plastic bags.

What Should We Do About Diapers? About 95% of the 18 billion disposable diapers discarded each year in the United States end up entombed in landfills for centuries. Even the new biodegradable diapers take 100 years to break down in a landfill. Production of disposable diapers uses trees, consumes plastic resources (about a third of each diaper is plastic), and creates air and water pollution.

Q: How much of the heat in U.S. homes and other buildings escapes through closed windows?

begin by working to get polystyrene food packaging out of their school cafeterias and local fast-food restaurants.

Oregon students even took legal action to get rid of cups and plates made from bleached paper because the paper contains the deadly poison dioxin. They were asking the right questions, and they got the right answer when the school systems switched to reusable cups, plates, and utensils.

Dozens of student groups have joined with local grass-roots organizations to get toxic-waste sites cleaned up or to stop construction of new toxic-waste sites, radioactive-waste sites, or waste incinerators.

Waste issues are not simply environmental issues; they are also tied up with our economy, which is geared to producing and then disposing of waste. *Somebody* is making money from every scrap of waste and has a vested interest in keeping things the way they are. Environmentalists and industry officials constantly argue about "cost–benefit analysis," which, simply stated, poses the question of whether the benefit of controlling pollution or waste will be greater than the cost. But this is another example of asking the wrong question. The right questions are, "Who will benefit, and who will pay the cost?"

Waste issues are also issues of *justice and fairness.* Again, there's a lot of debate between industry officials and environmentalists, especially those in federal and state environmental agencies, about so-called "acceptable risk" [Section 17-6]. These industry officials and environmentalists decide the degree of people's exposure to toxic chemicals, but don't ask the people who will actually be exposed how they feel about it. Again, asking what is an "acceptable risk" of exposure is the wrong question. It's simply not just to expose people to chemical poisons without their consent.

Risk analysts often say, "But there's only a one in a million chance of increased death from this toxic chemical." That may be true. But suppose I took a pistol and went to the edge of your neighborhood and began shooting into it. There's probably only a one in a million chance that I'd hit somebody, but would you issue me a license to do that? As long as we don't stand up for our rights and demand that "bullets" in the form of hazardous chemicals not be "fired" in our neighborhoods, we are giving environmental regulators and waste producers a license to kill a certain number of us without our even being consulted.

From my personal experience, I know that decisions made to dump wastes at Love Canal and in thousands of other places were not made purely on the basis of the best available scientific knowledge. The same holds true for decisions about how to manage the wastes we produce today and how to produce less waste.

Instead, the world we live in is shaped by decisions based on money and power. If you really want to understand what's behind any given environmental issue, the first question you should ask is, "Who stands to profit from this?" Then ask, "Who is going to pay the price?" You can then identify both sides of the issue and decide whether you want to be part of the problem or part of the solution.

Critical Thinking

1. Do you believe that we should put primary emphasis on pollution prevention and waste reduction? Explain. What changes would you be willing to make in your own lifestyle to prevent pollution and reduce waste?

2. What political and economic changes, if any, do you believe must be made so that we shift from primary emphasis on waste production and waste management to primary emphasis on pollution prevention and waste reduction? What actions are you taking to bring about such changes?

Cloth diapers can be washed and reused 80–200 times and then retired into lint-free rags. In addition to not ending up in landfills, cloth diapers save trees and money. For example, the disposable diapers needed for a typical baby cost about $1,716. A cloth diaper service costs about $1,326, and washing cloth diapers at home costs about $318.

Does the choice seem simple? Consider this: Laundering cloth diapers produces 9 times as much air pollution and 10 times as much water pollution as disposable diapers do in their lifetimes. Over their lifetimes cloth diapers also consume six times more water and three times more energy than disposables. And because of their ability to absorb large amounts of urine, parents may end up using a smaller number of disposable diapers than less-absorbent cloth ones.

So the choice between disposable diapers and reusable cloth diapers is not clear-cut. Many people opt for cheaper cloth diapers and use disposable ones for trips or at child-care centers that don't allow use of cloth diapers.

What Can We Do with Used Tires? Another big solid-waste problem is used tires. Some 2.5–4 billion used tires are now heaped in landfills, old mines, abandoned houses, and other dump sites throughout

A: About 33%—an energy loss equal to all the oil flowing through the Alaska pipeline every year the United States; the pile grows by about 250 million tires per year. Tire dumps are fire hazards and breeding grounds for mosquitoes. A fire in a tire dump is extremely difficult to put out and produces air pollution and toxic runoff that can pollute nearby surface water and groundwater.

About 47 million of the tires thrown away each year in the United States are suitable for retreading, a form of reuse. Discarded tires can also be packed with dirt and reused for the foundations and walls of low-cost passive solar homes (although without effective sealants the tires can release potentially dangerous indoor air pollutants). Some worn-out tires have been reused in constructing artificial reefs to attract fish. In Vienna, Austria, discarded tires are being used to build vine-covered walls along highways to reduce noise.

Used tires can be burned to produce electricity or to fuel cement kilns (as is done in Japan), although without strict controls this can pollute the air. Tires usually have a higher heating value per unit of weight than coal and can be less polluting. Tires can also be recycled to make resins for products ranging from car bumpers, garbage cans, and doormats to road-building materials, but this is not always cost-effective. Preliminary research indicates that a bacterium that dines on sulfur in the hot springs of Yellowstone National Park may be used in fermentation reactors to break down the rubber in used tires for recycling.

22-4 RECYCLING

How Can We Recycle Organic Solid Wastes? Community Composting **Compost** is a sweet-smelling, dark-brown, humuslike material that is rich in organic matter and soil nutrients. It is produced when microorganisms (mostly fungi and aerobic bacteria) in soil break down organic matter such as leaves, food wastes, paper, and wood in the presence of oxygen.

Biodegradable wastes make up about 35% by weight of the municipal solid waste output in the United States. Such wastes can be composted by consumers in backyard bins or indoor containers or collected and composted in centralized community facilities (as is done in many western European countries).

To compost such wastes, we mix them with soil, put the mixture into a pile or container, stir it occasionally, and let it rot for several months. Heat generated by microbial decomposition rises inside the pile. The pile is periodically turned and mixed to ensure that the temperature is high enough throughout to kill pathogens and weed seeds (but not hot enough to kill the decomposing microbes). There are also ways to speed up the composting process (Individuals Matter, p. 363).

The resulting compost can be used as an organic soil fertilizer or conditioner, as topsoil, or as a landfill cover. Compost can also be used to help restore eroded soil on hillsides and along highways, strip-mined land, overgrazed areas, and eroded cropland. Between 1988 and 1996 the number of facilities for composting yard trimmings in the United States increased from about 700 to 3,260.

To be successful, a large-scale composting program must overcome siting problems (few people want to live near a giant compost pile or plant), control odors, and exclude toxic materials that can contaminate the compost and make it unsuitable for use as a fertilizer on crops and lawns. Three ways to control or reduce odors for large-scale composting operations are **(1)** enclosing the facilities and filtering the air inside (but residents near large composting plants still complain of unacceptable odors), **(2)** creating municipal compost operations near existing landfills or at other isolated sites, and **(3)** decomposing biodegradable wastes in a closed metal container in which air is recirculated to give precise control of available oxygen and temperature (a technique that has been used successfully in the Netherlands for 20 years).

What Are the Two Types of Recycling? There are two types of recycling for materials such as glass, metals, paper, and plastics: primary and secondary. The most desirable type is *primary* or *closed-loop recycling*, in which wastes discarded by consumers (*postconsumer wastes*) are recycled to produce new products of the same type (such as newspaper into newspaper and aluminum cans into aluminum cans).

A still useful but less desirable type is *secondary*, or *open-loop, recycling*, in which waste materials are converted into different products. Primary recycling reduces the amount of virgin materials in a product by 20–90%, whereas secondary recycling reduces virgin material by only 25% at most.

Environmentalists urge us not to be misled by labels claiming that paper and plastic bags or other items are recyclable. Just about anything is in theory recyclable. What counts is whether an item is actually recycled, and whether we complete the recycling loop by buying products using the maximum feasible content of postconsumer recycled materials.

Case Study: Recycling Municipal Solid Waste in the United States In 1996, about 27% of U.S. municipal solid waste was recycled or composted (up from 10% in 1980); 16 states have adopted goals of recycling or composting at least half of their municipal solid waste by 2000. By 1996 the United States had more than 8,800 municipal curbside recycling programs serving 51% of the population. Recent pilot

Q: What is life-cycle cost?

Fighting City Hall and Winning

INDIVIDUALS MATTER

Frank Schiavo, an environmental studies professor at San Jose State University, and his wife, Linda Munn, a sixth-grade teacher, recycle so efficiently that they haven't set out a curbside garbage can in 18 years. By recycling or composting almost everything, they produce only about two handfuls of trash a week, which Frank deposits at a private recycling center.

He and his wife run their home like an ecological laboratory. In their yard they have drought-resistant plants and a highly efficient compost pile that helps fertilize their organic garden. They use reusable containers and cloth shopping bags, buy recycled paper and other recycled products, and use substitutes for common toxic household chemicals (Table 22-2).

A passive solar porch heats water stored in metal cans that radiate heat into the house during cold weather. Thermal curtains, double-pane windows, and extra wall insulation keep the house comfortable year-round. Energy-efficient compact fluorescent lights are used throughout the house. Frank rides the bus to and from work, only occasionally using his 34-year-old Chevy Nova, which runs on propane.

You would think that San Jose officials would give Frank and Linda medals for being such outstanding environmental citizens. Instead, in a city that prides itself on recycling, Frank has been in a garbage-fee fight with city officials. It began in July 1993, when the city began its new Recycle Plus program, which Frank strongly supported and gave speeches to help pass. In this pay-as-you-throw system based on garbage-can size, the more you recycle, the smaller the garbage can you need, and the less you pay; the smallest can costs $13.95 a month.

Under the new program Frank and Linda were charged for the smallest can, even though they don't use a can at all. At first they decided to go along. Then they decided to petition city officials to be exempt from the minimum user fee, arguing that they were not users and had been exempted from all garbage collection fees for 15 years before the new program was established. He also pointed out that the purpose of the program was to encourage composting and recycling and that they were being penalized for setting an outstanding example of achieving this goal. To Frank and Linda, it wasn't the money but the principle of the matter.

City officials turned them down, arguing that the minimum garbage user fees, like school taxes, aren't based on a resident's use of a service but are a way to distribute the cost of a service more fairly among the people who do use it. Frank and Linda disagreed, saying that the garbage collection fees are user fees and are not classified by the city as taxes.

They began depositing their monthly garbage charge into an escrow account and continued lobbying city officials to financially reward their house as a model garbage-free zone. After much unfavorable nationwide publicity on TV and in newspapers, city officials have given in. They no longer send the Schiavos garbage bills. It's not easy, but you can fight city hall and win.

studies in several U.S. communities show that a 60–80% recycling and composting rate is possible.

One way to spur recycling is a *pay-as-you-throw* program that bases garbage collection charges on the amount of waste a household generates for disposal; materials sorted out for recycling are hauled away free. Currently, more than 2,800 communities in the United States have curbside pay-as-you-throw systems. Studies have shown that this is the single best way to boost recycling.

Recycling also benefits communities by creating more jobs than burying or burning wastes. Recycling 1 million tons of solid waste in the United States requires about 1,800 workers, compared to 600 workers for landfilling these wastes and only 80 jobs for incinerating them. Unfortunately, the effective pay-as-you-throw system in San Jose, California, has penalized the city's best recycling family (Individuals Matter, above).

Is Centralized Recycling of Mixed Solid Waste the Answer? Large-scale recycling can be accomplished by collecting mixed urban waste and transporting it to centralized *materials-recovery facilities (MRFs)*. There, machines shred and automatically separate the mixed waste to recover glass (which can be melted and converted to new bottles or fiberglass insulation), iron, aluminum, and other valuable materials (Figure 22-6); these are then sold to manufacturers as raw materials. The remaining paper, plastics, and other combustible wastes are recycled or burned. The resulting heat produces steam or electricity to run the recovery plant or to sell to nearby industries or homes.

By 1996, the United States had more than 220 materials-recovery facilities, and at least 60 more were in the planning stages. However, such plants are expensive to build and maintain, and once trash is mixed it takes a lot of money and energy to separate it, which

A: The initial cost plus lifetime operating costs of a house, vehicle, appliance, or other item

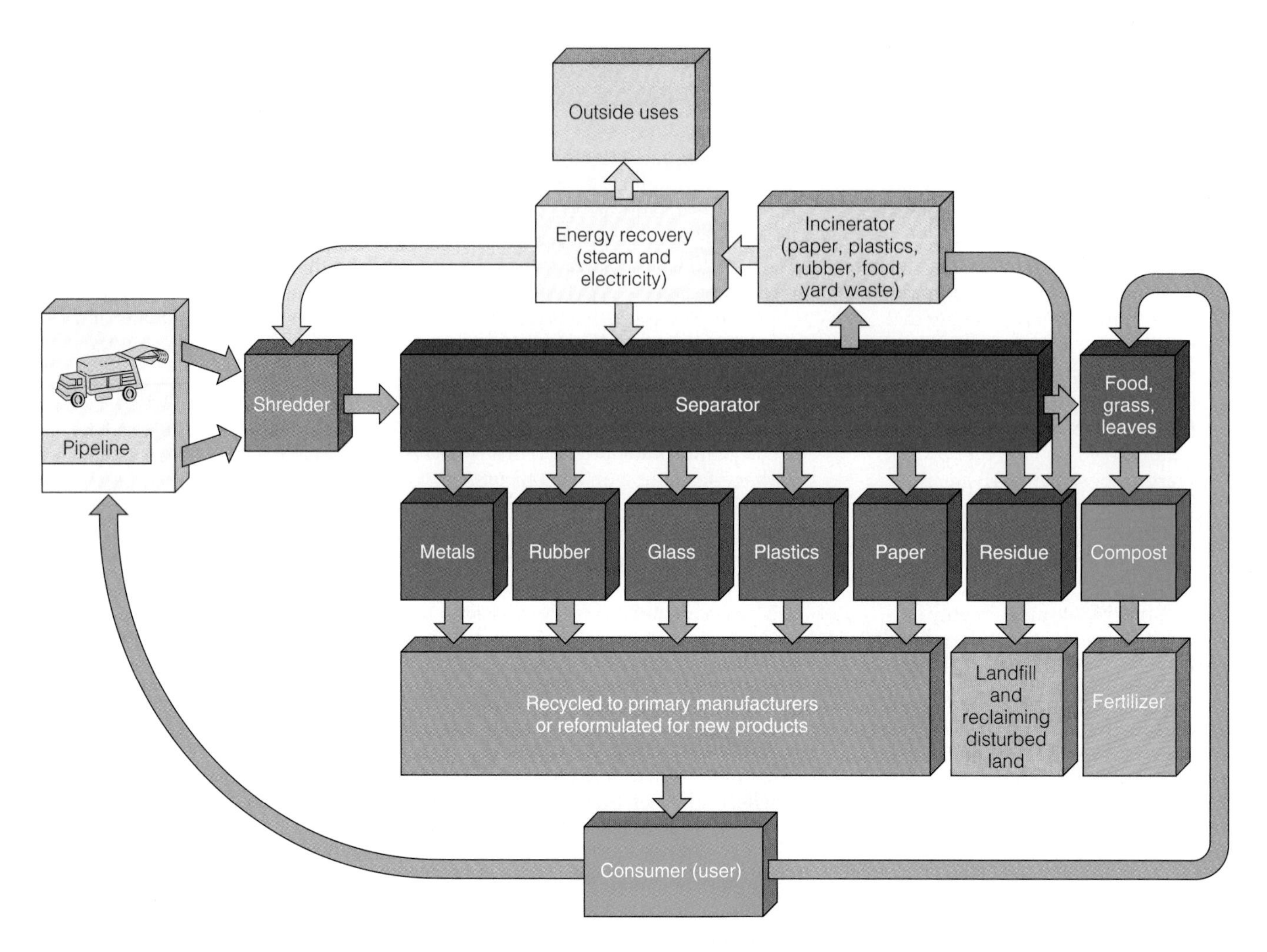

Figure 22-6 Schematic of a generalized materials-recovery facility used to sort mixed wastes for recycling and burning to produce energy. Because such plants require high volumes of trash to be economical, they discourage reuse and waste reduction.

is why some MRFs have shut down. MRFs must have a large input of garbage to make them financially successful. Thus, their owners have a vested interest in increased *throughput* of matter and energy resources to produce more trash—the reverse of what prominent scientists believe we should be doing (Figure 22-3).

These facilities also can emit toxic air pollutants and produce a toxic ash that must be disposed of safely. Other problems include health and accident threats to workers in poorly designed and managed plants and increased truck traffic, odor, and noise. Most communities collecting mixed solid waste can't afford to build or operate high-tech resource recovery facilities. Instead, to meet government-required recycling goals, many communities hire workers to sort the trash by hand. This is a hazardous, low-pay job.

Is Separating Solid Wastes for Recycling the Answer? Many solid-waste experts argue that it makes more sense economically and environmentally for households and businesses to separate trash into recyclable and reusable categories (such as glass, paper, metals, certain types of plastics, and compostable materials) before it is picked up. Compartmentalized city collection trucks, private haulers, or volunteer recycling organizations then pick up the segregated wastes and sell them to scrap dealers, compost plants, and manufacturers. Another alternative (especially in less populated areas) is the establishment of a network of drop-off centers, buyback centers, and deposit-refund programs in which people deliver and either sell or donate their separated recyclable materials.

The source-separation approach produces little air and water pollution, reduces litter, and has low start-up costs and moderate operating costs. It also saves more energy and provides more jobs for unskilled workers than centralized MRFs, and it creates three to six times more jobs per unit of material than landfilling or incineration. In addition, separated recyclables

Q: How much money does switching from an incandescent bulb to a fluorescent bulb save?

are cleaner and can usually be sold for a higher price. Source separation also educates people about the need for waste reduction, reuse, and recycling (Individuals Matter, right).

In the United States, many small- and medium-scale source-separation operations—pioneers in the recycling business—are being squeezed out by large waste management companies operating MRFs, which need large inputs of trash to be profitable. In some communities, elected officials have signed long-term (20- to 30-year) contracts giving companies exclusive ownership of all garbage. Now owners of some small-scale recycling businesses are being sued for recycling materials that large waste management companies claim they own. The fight used to be over what to do with garbage; now, as garbage is becoming more valuable, the fight is over who owns it.

Aluminum and paper separated out for recycling are worth a lot of money. As a result, in a growing number of cities people steal these materials—from curbside containers set out by residents and from unprotected recycling drop-off centers—and sell them. This undermines municipal recycling programs by lowering the income available from selling these high-value materials.

Does Recycling Make Economic Sense? The answer is yes and no, depending on different ways of looking at the economic and environmental costs and benefits of recycling. Critics contend that recycling **(1)** has become almost a religion that is above criticism regardless of how much it costs communities, **(2)** does not make sense if it costs more to recycle materials than to send them to a landfill or incinerator (as is the case in some areas), **(3)** is often not needed to save landfill space because most areas in the United States are not running out of landfill space, and **(4)** may make economic sense for valuable and easy-to-recycle materials (such as aluminum, mercury, paper, and steel) but not for cheap or plentiful resources (such as glass from silica) and most plastics (which are expensive to recycle).

Many communities established recycling programs with the idea that they should pay for themselves. But recycling proponents argue that *recycling programs should not be judged on whether they pay for themselves any more than are conventional garbage disposal systems based on land burial or incineration.* Management of solid wastes, regardless of the method used, is simply a cost society must bear.

Moreover, recycling proponents contend that with full-cost accounting, the net economic, health, and environmental benefits of recycling far outweigh the costs. A study by MIT economist Robert F. Stone revealed that recycling in Massachusetts yields a net social benefit of about of $254 per metric ton ($231 per ton) if both the roughly $146 per metric ton ($133 per ton) environmental benefits of recycling and the avoided solid-waste disposal costs are taken into account.

An analysis of various recycling programs revealed that cities and private waste collection firms

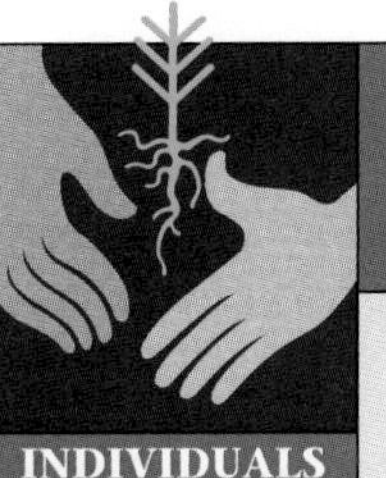

INDIVIDUALS MATTER

Source-Separation Recycling in Some Georgia Schools

In Rome, Georgia, environmental educator Steve Cordle has set up an exciting project that combines the important concept of source-separation recycling with environmental education.

He has designed and implemented a source-separation recycling program for 17 Floyd County Schools in Georgia. Before his program, which began in 1992, the schools were a major contributor to the county landfill and hardly any of their waste output was recycled.

During a 7-month period in 1996, approximately 114 metric tons (250,000 pounds) of paper, corrugated cardboard, and aluminum and steel cans were collected in separate containers. The school system stored and sold the items for recycling. During this period the project added $10,139.69 to school funds, although school officials consider any money made incidental compared to the educational value of the program.

This important experiment in *environmental education* teaches students (and teachers) from kindergarten through high school about the need for recycling and engages them in hands-on participation in source separation. Mixing and throwing waste materials in garbage cans (which should be called *resource containers*) and hauling them off to a landfill, incinerator, or mixed-resource recycling plant is an out-of-sight, out-of-mind approach that does little to further environmental education.

Cordle is working hard to expand this idea to a larger area and, he hopes, to Georgia's entire school system. With proper backing and support, this program could become a model for use throughout the United States and other parts of the world. In 1998, Cordle's project received a $250,000 grant from federal and state sources.

Critical Thinking

Use the second law of energy (thermodynamics) to explain why a *properly designed* source-separation recycling program takes less energy and produces less pollution than a centralized program that collects mixed waste over a large area and hauls it to a centralized facility where workers or machinery separate the wastes for recycling.

A: $20–25 over the life of the fluorescent bulb

Recycling, Reuse, and Waste Reduction in Germany

CASE STUDY

In 1991, Germany enacted the world's toughest packaging law, which was designed to **(1)** reduce the amount of waste being landfilled or incinerated, **(2)** reduce waste production, and **(3)** recycle or reuse 65% of the nation's packaging by 1995. Product distributors must take back their boxes and other containers for reuse or recycling, and incineration of packaging is not allowed. Later the government added a sliding fee (based on the types rather than just weight of packaging) that charges manufacturers more for plastic and composite packaging than for glass and cardboard packaging.

To implement the system, over 600 German manufacturers and distributors pay fees to a nonprofit company they formed to collect, sort, and reprocess the packaging discards (coded with a green dot) of member firms.

The system has been so successful that in 1996, 80% of all packaging in Germany was recycled. Plastic packaging has lost one-third of its market share to glass and cardboard, and four out of five German manufacturers have reduced their use of packaging.

Critical Thinking

What might be some long-term disadvantages of Germany's program? Do you believe that a similar program should be put into place in the country where you live? Explain.

tend to make money if they have high recycling rates and a single pickup system (for both materials to be recycled and garbage that can't be recycled). Cities cited by critics as losing money on recycling often have expensive dual collection systems and have not designed programs to encourage more cost-effective, high rates of recycling.

Finally, environmentalists point out that the primary benefit from recycling is *not* a reduction in the use of landfills and incinerators. Instead, *the major reasons for recycling are reduced use of virgin resources, reduced throughput of matter and energy resources, and reduced pollution and environmental degradation* (Figure 3-22).

Why Don't We Have More Reuse and Recycling? Three factors that hinder recycling (and reuse) are **(1)** failure to include the environmental and health costs of raw materials in the market prices of consumer items, **(2)** more tax breaks and subsidies for resource-extracting industries than for recycling industries, and **(3)** lack of large, steady markets for recycled materials.

Analysts have suggested various ways to overcome the obstacles to recycling, including the following:

- Taxing virgin resources and phasing out subsidies for extraction of virgin resources
- Lowering or eliminating taxes on recycled materials based on postconsumer waste content
- Providing subsidies for reuse and postconsumer waste recycling
- Requiring households and businesses to pay directly for garbage collection based on how much they throw away, with lower charges or no charges for materials separated for recycling and composting (a *pay-as-you-throw* system)
- Encouraging or requiring government purchases of recycled products to help increase demand and lower prices
- Viewing landfilling and incineration of solid wastes as last resorts to be used only for wastes that can't be reused, composted, or recycled (Figure 22-3)
- Requiring ecolabels on all products, evaluating life cycle environmental costs and listing preconsumer and postconsumer recycled content

Countries such as Germany are leading the way in recycling, reuse, and waste reduction (Case Study, left).

22-5 CASE STUDIES: RECYCLING ALUMINUM, WASTEPAPER, AND PLASTICS

How Much Aluminum Is Being Recycled? Worldwide, the recycling rate for aluminum in 1996 was about 35% (34% in the United States). Recycling aluminum produces 95% less air pollution and 97% less water pollution, and it requires 95% less energy than mining and processing aluminum ore.

In 1994, 62% (compared to 15% in 1973) of aluminum beverage cans in the United States were recycled. Despite this progress, about 38% of the 95 billion aluminum cans produced in 1994 in the United States was still thrown away. Laid end to end, these cans would wrap around the planet more than 120 times.

Recycling aluminum cans is great, but many environmentalists believe that these cans are an example of an unnecessary item that could be replaced by more energy-efficient and less polluting refillable glass or

Larane, Andre. 1998. "Recycling is Standard in French Auto Industry." *World Wastes*, vol. 41, no. 1, 8(3).

PET plastic bottles—a switch from recycling to reuse that also creates local jobs. One way to encourage this change would be to place a heavy tax on nonrefillable containers and no tax on reusable beverage containers, as is done in at least nine countries. What do you think?

How Much Wastepaper Is Being Recycled? Paper (especially newspaper and cardboard) is one of the easiest materials to recycle. To make paper, a chemical such as caustic soda (sodium hydroxide) is used to convert wood chips into a soft mush (pulp) that is pressed into a thin sheet and then dried. Recycling paper involves removing its ink, glue, and coating and reconverting it to pulp that is pressed again into new paper. This process breaks down some of the paper fibers, requiring addition of some new pulp to maintain paper strength.

In 1996 the United States recycled about 40% of its wastepaper (up from 25% in 1989) and is projected to recycle 50% by the year 2000. At least 10 other countries recycle 50–98% of their wastepaper.

Recycling the Sunday newspapers in the United States alone would save 500,000 trees per week. In addition to saving trees, recycling paper **(1)** saves energy because it takes 30–64% less energy to produce the same weight of recycled paper as to make the paper from trees, **(2)** reduces air pollution from pulp mills by 74–95%, **(3)** lowers water pollution by 35%, **(4)** helps prevent groundwater contamination by toxic ink left after paper rots in landfills over a 30- to 60-year period, **(5)** conserves large quantities of water, **(6)** can save landfill space, **(7)** creates five times more jobs than harvesting trees for pulp, and **(8)** can save money.

Buying recycled paper products does not necessarily reduce solid waste. Only products made from *postconsumer waste*—waste intercepted on its way from consumer to the landfill or incinerator—do that.

Most recycled paper is actually made from *preconsumer waste:* scraps and cuttings recovered from paper and printing plants. Because paper manufacturers have always recycled this waste, it has never contributed to landfill problems. Now this paper is labeled "recycled" as a marketing ploy, giving the false impression that people who buy such products (often at higher prices) are helping the solid-waste problem. Most "recycled" paper has no more than 50% recycled fibers, with only 10% from postconsumer waste. Environmentalists propose that the government require companies to report the amount of postconsumer recycled materials in paper and other products and reserve the term *recycled* only for items using postconsumer recycled materials.

Is It Feasible to Recycle Plastics? Plastics are various types of polymer molecules made by chemically linking monomer molecules (called *petrochemicals*) produced mostly from oil and natural gas (Figure 22-7). The plastics industry is one of the leading producers of hazardous waste.

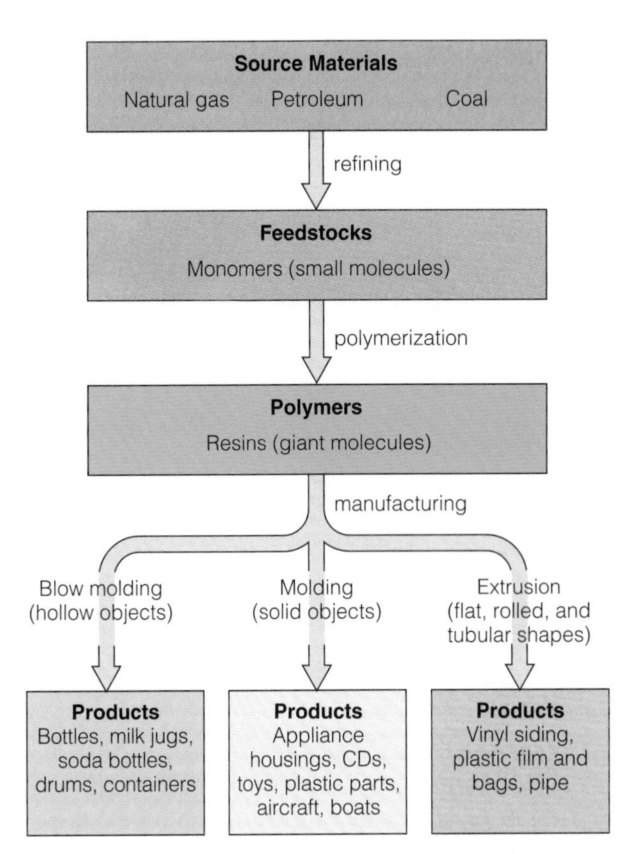

Figure 22-7 How plastics are made. (Adapted from the Society of the Plastics Industry)

Plastics now account for about 9% by weight and 22% by volume of municipal solid wastes in the United States and about 60% of the debris found on U.S. beaches. In landfills, toxic cadmium and lead compounds used as binders, colorants, and heat stabilizers can leach out of plastics and ooze into groundwater and surface water. Most plastics used today are nondegradable or take 200–400 years to degrade. Even biodegradable plastics take decades to partially decompose in landfills because they lack sufficient oxygen and moisture. When plastic products are thrown away as litter, they can harm animals that swallow or get entangled in them (Figure 22-8).

Currently, only about 5% by weight of all plastic wastes and 6% of plastic packaging used in the United States are recycled because there are so many different types of plastic resin. Before they can be recycled, plastics in trash must be separated into different types of resins by consumers or separated from mixed trash (a costly, labor-intensive procedure unless automated separating technologies can be developed). Another

Figure 22-8 Before this discarded piece of plastic was removed, this Hawaiian monk seal was slowly starving to death. Each year plastic items dumped from ships and left as litter on beaches threaten the lives of millions of marine animals and seabirds that ingest, become entangled in, or choke on such debris. (Doris Alcorn/National Marine Fisheries)

problem is that the current price of oil is so low that the price of virgin plastic resins (except for PET, used mostly in plastic drink bottles) is about 40% lower than that of recycled resins. However, in 1998 Chrysler Corporation built an experimental automobile with a body made from plastic derived entirely from recycled PET bottles and expects to sell such vehicles within a decade.

Environmentalists recognize the beneficial qualities of plastics: durability (in products such as car and machine parts, carpeting, toys, furniture, reusable tubs and containers, and refillable bottles), light weight, unbreakability (compared to glass), and in some cases reusability as containers. But many environmentalists believe that some widespread uses of plastics—especially excessive and often unnecessary single-use packaging and throwaway beverage and food containers—should be sharply reduced and replaced with less harmful and wasteful alternatives. What do you think?

22-6 DETOXIFYING, BURNING, BURYING, AND EXPORTING WASTES

How Can Hazardous Waste Be Detoxified? According to the U.S. National Academy of Sciences, the next priority in hazardous-waste management (after waste reduction, reuse, and recycling) is to convert any remaining waste into less hazardous or nonhazardous materials (Figure 22-4). Denmark has the most comprehensive and effective hazardous-waste detoxification program. Each Danish municipality has at least one facility that accepts paints, solvents, and other hazardous wastes from households. Hazardous and toxic waste from industries is delivered to 21 transfer stations throughout the country. All waste is then transferred to a large treatment facility, where about 75% of it is detoxified; the rest is buried in a carefully designed and monitored landfill.

Some consider biological treatment of hazardous waste, or *bioremediation*, to be the wave of the future for cleaning up some types of toxic and hazardous waste. In this process, microorganisms, usually bacteria, are used to destroy toxic or hazardous substances or convert them to harmless forms. This approach mimics nature by using decomposers to recycle matter.

If toxin-degrading bacteria or fungi can be found or engineered for specific hazardous chemicals, they could clean up contaminated sites at less than half the cost of disposal in landfills, and at only about one-third the cost of on-site incineration. Bioremediation might also clean up contaminated groundwater at an affordable cost by pumping it to the surface, treating it with microorganisms, and then returning it to its aquifer.

Preliminary testing reveals that bioremediation is effective for a number of specific organic wastes, but it does not appear to work very well for toxic metals, highly concentrated chemical wastes, or complete digestion of some complex mixtures of toxic chemicals.

Another biological way to treat hazardous wastes is *phytoremediation*, which involves using natural or genetically engineered plants to remove contaminants—again mimicking nature. This can be done in artificial marshes or solar greenhouses. So far emphasis is on plants that remove toxic metals such as lead and mercury from wastewater and contaminated soil because less is known about how plants handle organic contaminants such as oils and solvents.

Is Burning Solid and Hazardous Waste the Answer? In 1996 about 15% of the municipal solid waste and 7% of the officially regulated hazardous waste in the United States was burned in incinerators. Most of this mixed municipal solid waste was burned in 150 *mass-burn incinerators* (Figure 22-9), which burn mixed trash without separating out hazardous materials (such as batteries and polyvinylchloride or PVC plastic materials) or noncombustible materials that can interfere with combustion and pollute the air.

Incinerators are costly to build, operate, and maintain and they create very few long-term jobs. Without continual maintenance and good operator training and supervision, the air pollution control equipment on incinerators often fails, and emission standards are exceeded. According to a 1996 epidemiological study of 14 million people living near 72 incinerators in Great Britain, those living within 7.5 kilometers (4.6

Murray, Frank J. 1997. "Load of Rubbish Clogs Interstates." *Insight on the News*, vol. 13, no. 17, 39(1).

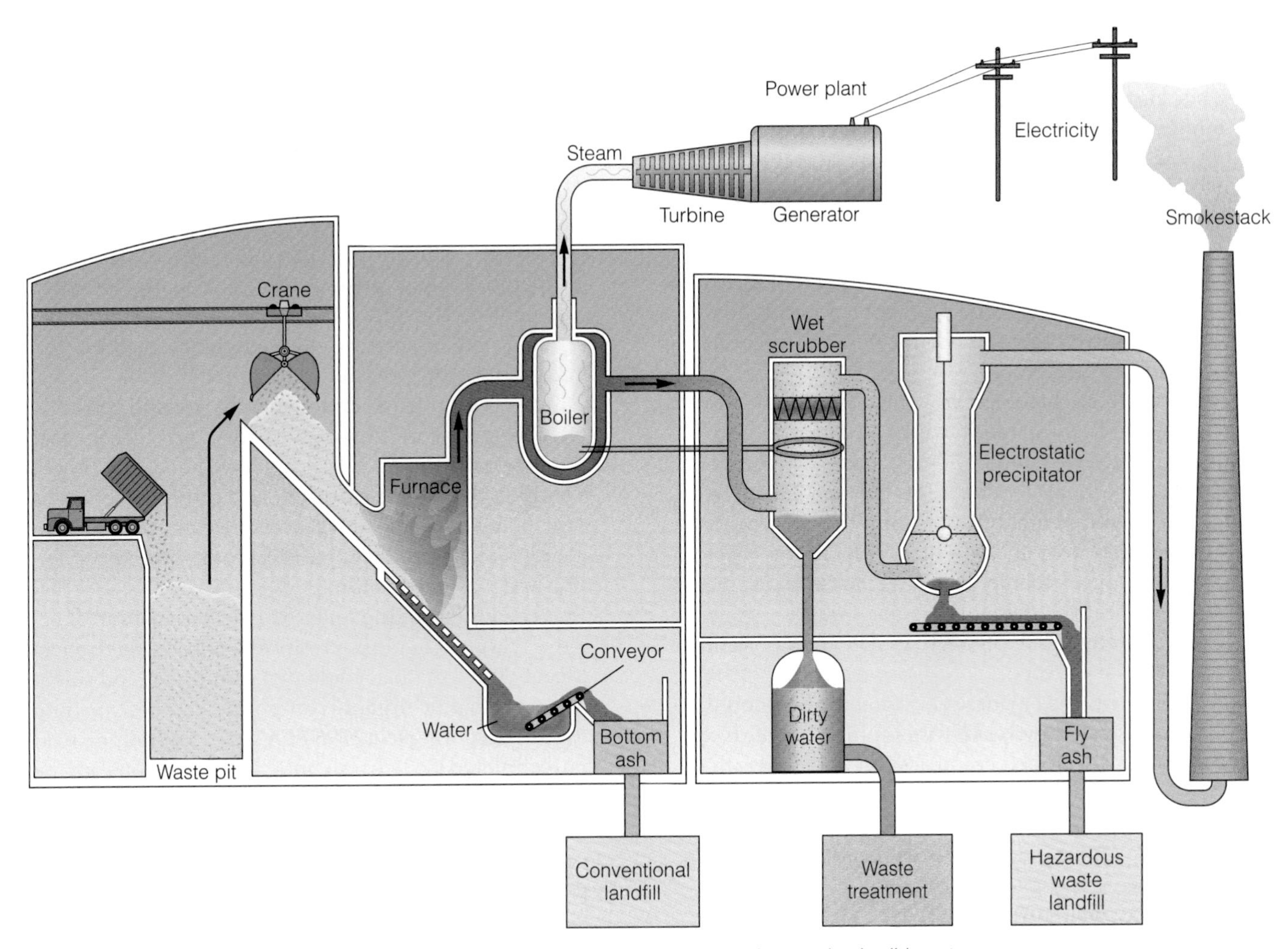

Figure 22-9 Schematic of a waste-to-energy incinerator with pollution controls that burns mixed solid waste and recovers some of the energy to produce steam used for heating or producing electricity. (Adapted from EPA, *Let's Reduce and Recycle*)

miles) of a municipal solid waste incinerator have an increased likelihood of getting several cancers

Hazardous-waste experts Peter Montague (Guest Essay, p. 70), Ellen and Paul Connett, and some EPA scientists point out that even with advanced air pollution control devices (mostly a scrubber followed by a baghouse filter or electrostatic precipitators; Figures 18-15 and 22-9), *all incinerators burning hazardous or solid waste release toxic air pollutants.* Examples are very harmful fine particles of toxic lead and mercury (which cannot be removed by scrubbers) and dioxins (Section 22-7). They also leave a highly toxic ash to be disposed of in landfills that even the EPA concedes will eventually leak.

According to EPA hazardous-waste expert William Sanjour, EPA incinerator regulations don't work because

> *the regulations require no monitoring of the outside air in the vicinity of the incinerator. Because operators maintain the records, they can easily cheat. . . . Government inspectors are poorly trained and have low morale and high turnover. . . . Government inspectors typically work from nine to five Monday through Friday. So if there is anything particularly nasty to burn, it will be done at night or on weekends. When complaints come in, . . . the inspector may visit the plant but rarely finds anything. The enforcement officials tend to view the incinerator operator as their client and the public as a nuisance. . . . There is no reward to inspectors for finding serious violations.*

In 1993 the *Wall Street Journal* warned that using incinerators to burn municipal trash spells financial disaster for local governments in the United States. Since 1992 Rhode Island and West Virginia have banned solid-waste incineration because of its health threats and high cost. Sweden banned the construction of new incinerators in 1985.

Between 1985 and 1997, several solid-waste incinerators in the United States were shut down because of excessive costs and pollution, and over 280 new incinerator projects were blocked, delayed, or canceled because of intense public opposition and high costs. Most of the plants still in the planning stage may not be built as communities discover that recycling, reuse, composting, and waste reduction are cheaper, safer, and less environmentally harmful.

Japan depends on incinerators more than any country, burning about 76% of its municipal solid waste in more than 1,850 government-operated incinerators and more than 3,300 privately owned industrial incinerators. Many of these incinerators are built near populous areas.

Recently there has been growing concern in Japan about reported rising infant deaths and health problems in areas downwind of many Japanese incinerators from emissions of toxic dioxins (Section 22-7). The head of a government advisory committee on dioxins found that the concentration of dioxins in the air in Japan is three times that in the United States and some European countries. Dioxins are also found at high levels in the fatty tissue of fish that play an important role in the Japanese diet.

According to the Japanese government, the amount of dioxins ingested by residents is somewhat higher than in other developed countries but still generally within levels considered to be safe. Critics contend that Japan is far behind other industrialized countries in regulating dioxin emissions from incinerators.

Is Land Disposal of Solid Waste the Answer? Currently, about 57% by weight of the municipal solid waste in the United States is buried in sanitary landfills. A **sanitary landfill** is a garbage graveyard in which solid wastes are spread out in thin layers, compacted, and covered daily with a fresh layer of clay or plastic foam.

Modern state-of-the-art landfills on geologically suitable sites are lined with clay and plastic before being filled with garbage (Figure 22-10). The bottom is covered with a second impermeable liner, usually made of several layers of clay, thick plastic, and sand. This liner collects *leachate* (rainwater contaminated as it percolates through the solid waste) and is intended to prevent its leakage into groundwater. Collected leachate ("garbage juice") is pumped from the bottom of the landfill, stored in tanks, and sent either to a regular sewage treatment plant or to an on-site treatment plant. When full, the landfill is covered with clay, sand, gravel, and topsoil to prevent water from seeping in. Several wells are drilled around the landfill to monitor any leakage of leachate into nearby groundwater.

Sanitary landfills for solid wastes offer certain benefits. Air-polluting open burning is avoided. Odor is seldom a problem, and rodents and insects cannot thrive. If located properly they can reduce water pollution from leaching, but proper siting is not always achieved. A sanitary landfill can be put into operation fairly quickly, has low operating costs, and can handle a huge amount of solid waste. After a landfill has been filled, the land can be graded, planted, and used as a park, golf course, ski hill, athletic field, or wildlife area, or for some other recreational purpose.

Solid-waste landfills also have drawbacks. While in operation, they cause traffic, noise, and dust; most also emit toxic gases. Paper and other biodegradable wastes break down very slowly in today's compacted and water- and oxygen-deficient landfills. Newspapers dug up from some landfills are still readable after 30 or 40 years; hot dogs, carrots, and chickens that have been dug up after 10 years have not decomposed. Landfills also deprive present and future generations of valuable reusable and recyclable resources and encourage waste production instead of pollution prevention and waste reduction.

The underground anaerobic decomposition of organic wastes at landfills produces explosive methane (a greenhouse gas), toxic hydrogen sulfide gas, and smog-forming volatile organic compounds that escape into the air. According to a 1996 EPA study, landfills give off an estimated 36% of all methane emissions in the United States. Large landfills can be equipped with vent pipes to collect these gases, and the collected methane can be burned in small power plants or in fuel cells to produce steam or electricity (Figure 22-10). Methane must be collected in all new landfills in the United States, but thousands of older and abandoned landfills don't have such systems and will emit methane for decades. By 1995 only 17% of trash was disposed of in landfills equipped with gas-collection systems. In 1996, 140 U.S. landfills collected gas for use by industry, power plants, or buildings (Individuals Matter, p. 598).

Contamination of groundwater and nearby surface water by leachate from both unlined and lined landfills is another serious problem. Some 86% of the U.S. landfills studied have contaminated groundwater, and 250 of the 1,360 Superfund hazardous-waste sites are former municipal landfills that will cost billions of dollars to clean up. Once groundwater is contaminated, it is extremely expensive and difficult—often impossible—to clean up.

Modern double-lined landfills (Figure 22-10), required in the United States since 1996, delay the release of toxic leachate into groundwater below landfills, but they do not prevent it. These landfills are designed to accept waste for 10–40 years, and current EPA regulations require owners to maintain and monitor landfills for at least 30 years after they are closed. However, they could begin to leak after this period, passing the health risks and costs of contamination to future generations.

Q: What percentage of the world's energy is provided by renewable resources?

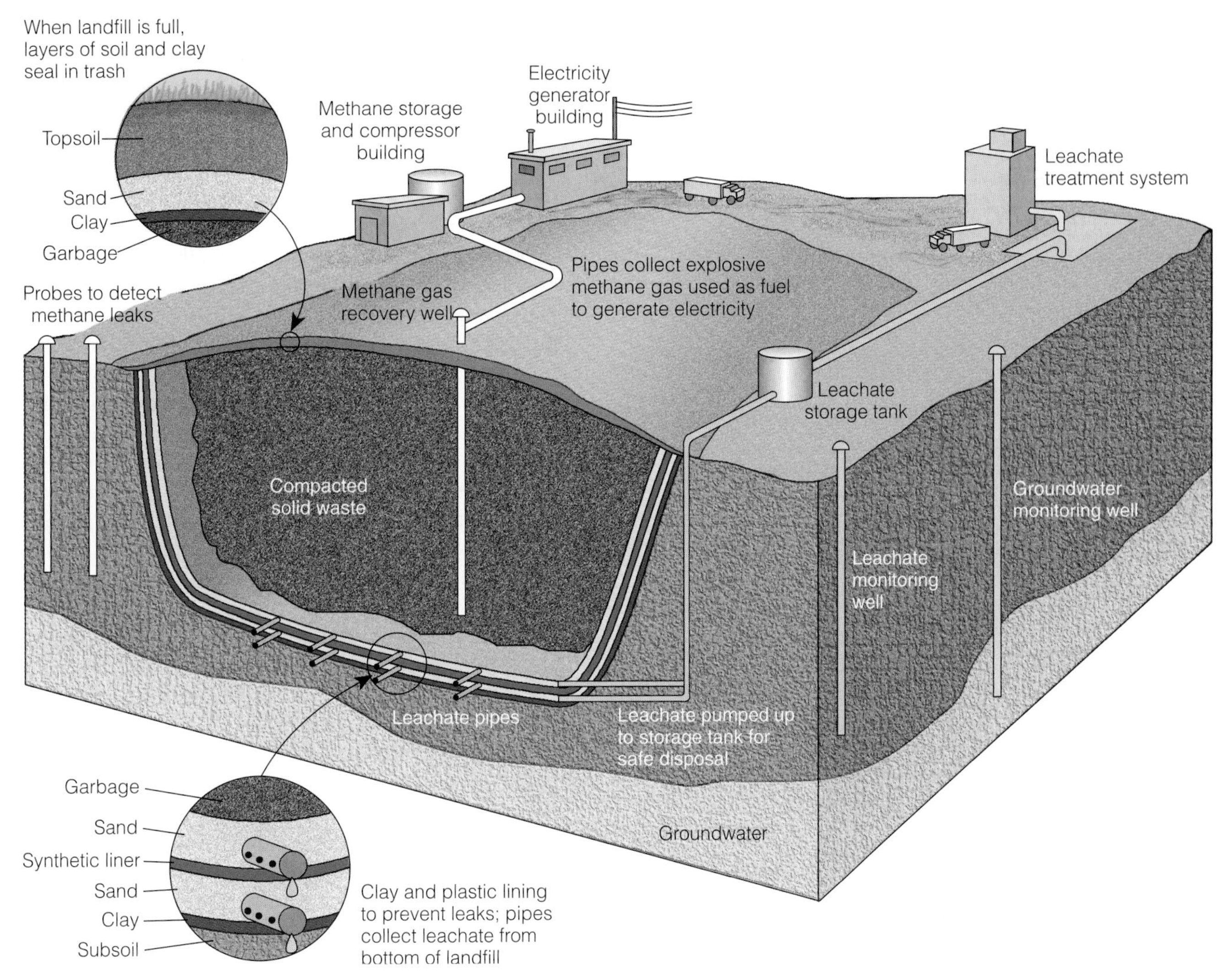

Figure 22-10 State-of-the-art sanitary landfills are designed to eliminate or minimize environmental problems that plague older landfills. Only a few of the 7,500 municipal and industrial landfills in the United States have such a state-of-the-art design, and 85% of U.S. landfills are unlined. Even state-of-the-art landfills will eventually leak, passing both the effects of contamination and cleanup costs on to future generations.

According to G. Fred Lee, an experienced landfill consultant, the best solution to the leachate problem is to apply clean water to landfills continuously and then collect and treat the resulting leachate in carefully designed and monitored facilities. He contends that after 10–20 years of such washing, little potential for groundwater pollution should remain. This wetting would also hasten the breakdown of wastes and thus allow old landfills to be dug out and used again.

Since 1979 there has been a sharp drop in the number of solid-waste landfills in the United States as existing landfills reached their capacity or closed; since 1997 only more expensive, state-of-the art landfills (Figure 22-10) are allowed to operate. Although a few U.S. cities, notably Philadelphia and New York, are having trouble finding nearby landfills, there is no national shortage of landfill volume because most of the smaller local landfills are being replaced by larger local and regional landfills.

Is Land Disposal of Hazardous Waste the Answer? Most hazardous waste in the United States is disposed of by deep-well injections, surface impoundments, and state-of-the-art landfills. In *deep-well disposal,* liquid hazardous wastes are pumped under pressure through a pipe into dry, porous geologic formations or into fracture zones of rock far beneath aquifers tapped for drinking and irrigation water (Figure 20-13). In theory, these liquids soak into the porous rock material and are isolated from overlying groundwater by essentially impermeable layers of rock.

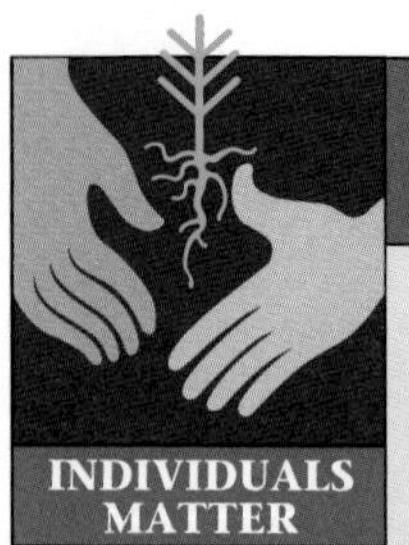

INDIVIDUALS MATTER

Using Landfill Methane to Heat a School

In January 1997, Pattonville High School in Maryland Heights, Missouri (near St. Louis), became the first public school to use methane gas produced by a nearby landfill as a source of heat for the school's 117 classrooms and two gymnasiums.

Most students at the high school looked at the nearby landfill and saw garbage. In 1993, however, members of the school's Ecology Club saw it as an opportunity to reduce resource waste and pollution.

After doing their homework, they convinced school officials to build a pipeline to transfer waste methane gas from the landfill to the school and to rebuild two basement heating system boilers to burn the methane to heat the school. The school is expected to recoup its investment in about 5 years and save money thereafter.

This project reduces resource waste and pollution because before the landfill was burning off the methane gas, which emitted carbon dioxide, a greenhouse gas (Figure 19-2a), into the atmosphere. In addition, students at the school are getting a lesson in applied environmental science by studying the operation and design of the system.

Critical Thinking

This project may not have wide application because most schools are not near landfills. Are there any renewable energy resources (Chapter 16) available at your school site that could be used to provide all or part of the school's heat or electricity? If so, consider evaluating such resources and presenting a plan to school officials.

Because this method is simple and cheap, its use is increasing rapidly as other methods are legally restricted or become too expensive. If sites are chosen according to the best geological and seismic data, a number of scientists believe deep wells may be a reasonably safe way to dispose of fairly dilute solutions of organic and inorganic waste; with proper site selection and care, they may be safer than incineration. Furthermore, if some use eventually is found for the waste, it could be pumped back to the surface.

However, many scientists believe that current regulations for geologic evaluation, long-term monitoring, and long-term liability if wells contaminate groundwater are inadequate. The regulations could allow injected wastes to spill or leak at the surface and leach into groundwater, to escape into groundwater from corroded pipe casing or leaking seals in the well (Figure 20-13), or to migrate from the porous layer of rock to aquifers (either through existing fractures or through new ones caused by earthquakes or stresses caused by introduction of the wastes). Until deep-well disposal is more carefully evaluated and regulated, most environmentalists believe that it should not be allowed to proliferate. What do you think?

Surface impoundments such as ponds, pits, or lagoons (Figure 20-13) used to store hazardous waste are supposed to be sealed with a plastic liner on the bottom. Solid wastes settle to the bottom and accumulate, whereas water and other volatile compounds evaporate into the atmosphere. According to the EPA, however, 70% of these storage basins have no liners, and as many as 90% may threaten groundwater. Eventually all liners will leak, and waste will percolate into groundwater. Moreover, volatile compounds (such as hazardous organic solvents) can evaporate into the atmosphere, promote smog formation, and return to the ground to contaminate surface water and groundwater in other locations.

About 5% by weight of the legally regulated hazardous waste produced in the United States is concentrated, put into drums, and buried, either in one of 21 specially designed and monitored commercial hazardous-waste landfills or in one of 35 such landfills run by companies to handle their own waste. Sweden goes further and buries its concentrated hazardous wastes in underground vaults (Figure 22-11). In the United Kingdom, most hazardous wastes are mixed with household garbage and stored in hundreds of conventional landfills all over the country.

When current and future commercial hazardous-waste landfills in the United States eventually leak and threaten water supplies, many of their operators will declare bankruptcy. Then the EPA will put the landfills on the Superfund list and taxpayers will pick up the tab for cleaning them up. As EPA hazardous-waste expert William Sanjour points out,

> *The real cost of dumping is not borne by the producer of the waste or the disposer, but by the people whose health and property values are destroyed when the wastes migrate onto their property and by the taxpayers who pay to clean it up. . . . It is better for liners to leak sooner rather than later, because then there will be responsible parties that they can get to clean it up. Liners don't protect communities. They protect the people who put the waste there and the politicians who let them put the waste there, because they are long since gone when the problem comes up.*

Some engineers and environmentalists have proposed storing hazardous wastes above ground in

Q: What is the largest untapped energy source in the United States?

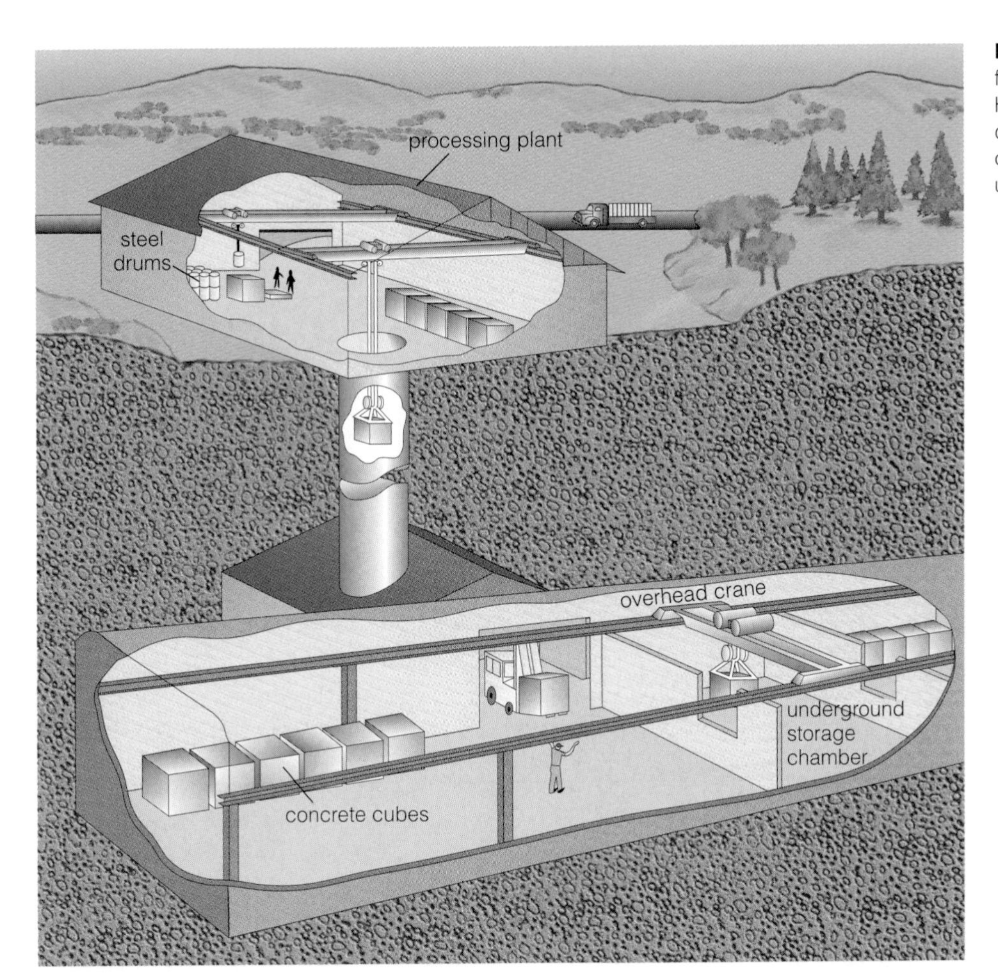

Figure 22-11 Swedish method for handling hazardous waste. Hazardous materials are placed in drums, which are embedded in concrete cubes and stored in an underground vault.

large, two-story, reinforced-concrete buildings until better technologies are developed (Figure 22-12). The first floor would contain no wastes but would have inspection walkways so people could easily check for leaks from the upper story. Any leakage would be collected, treated, solidified, and returned to the storage building. The buildings would have negative air pressure—caused by fans blowing in air and then filtering it—to prevent the release of toxic gases. Proponents believe that such an *in-sight* approach would be cheaper and safer than *out-of-sight* landfills, deep wells, or incinerators for many hazardous wastes. What do you think?

There is also growing concern about accidents during some of the more than 500,000 shipments of hazardous wastes (mostly to landfills and incinerators) in the United States each year. Between 1988 and 1992, approximately 34,500 toxic chemical accidents caused 100 deaths, over 11,000 injuries, and evacuation of over 500,000 people. Few communities have the equipment and trained personnel to deal with hazardous-waste spills.

Is Exporting Waste the Answer? Some U.S. cities that lack sufficient landfill space are shipping some of their solid waste to other states or other countries, especially developing countries. However, a few of these states and some developing countries are refusing to accept delivery.

To save money and to escape regulations and local opposition, some U.S. cities and waste disposal companies legally ship large quantities of hazardous waste to other countries. U.S. exports of hazardous wastes can take place without EPA approval because U.S. laws allow exports for recycling; some exported wastes designated for recycling are instead dumped after reaching their destination.

Waste disposal firms can charge high prices for picking up hazardous wastes. If they can then dispose of them (legally or illegally) at low costs, they pocket huge profits. Often officials of developing countries find it difficult to resist the income (often bribes) derived from receiving these wastes.

In 1994, 64 nations met in Basel, Switzerland, and developed a convention banning on all exports of

Figure 22-12 An aboveground hazardous-waste storage facility would help keep wastes from contaminating groundwater.

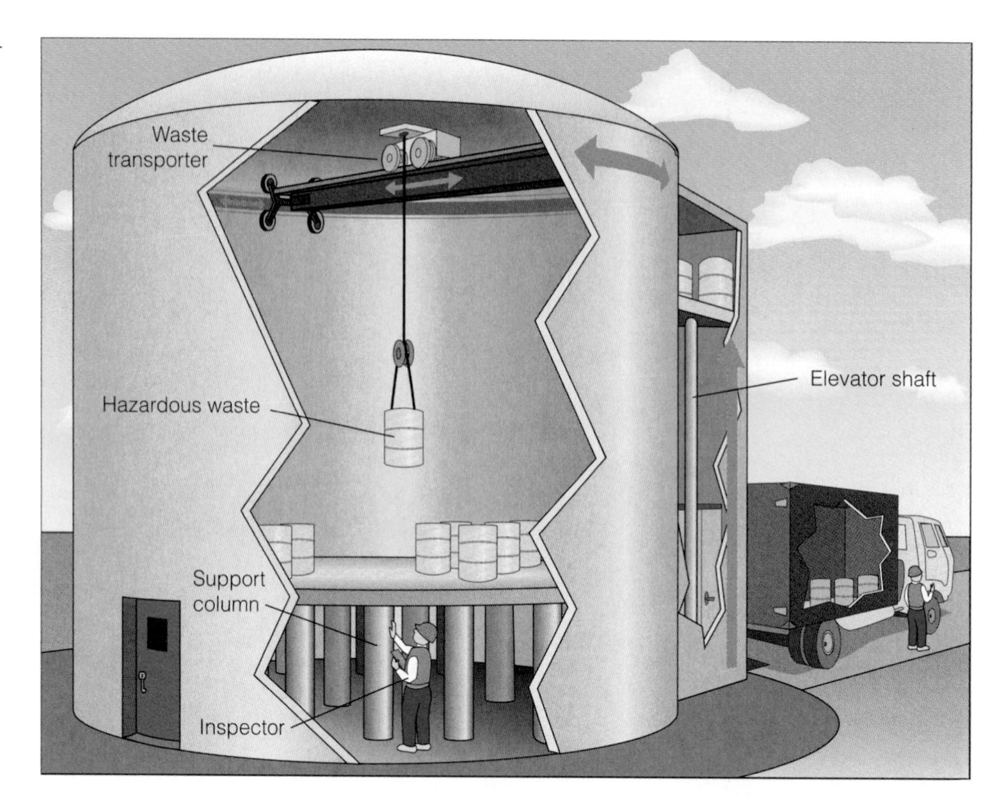

hazardous wastes for disposal from developed countries to developing countries (including eastern Europe and the former Soviet Union). By 1998 the convention had been ratified by 118 countries, but not the United States.

In 1994 the Clinton administration recommended that Congress ban exports of all hazardous wastes outside North America (with some exceptions, including materials containing toxic metals such as lead exported for recycling). This ban would do little to change U.S. policy because it would still allow export of hazardous waste to Canada and Mexico, which now receive almost 98% of U.S. hazardous waste exports.

An effective worldwide ban on all hazardous-waste exports, including those slated to be recycled (often a sham), would help reduce this transfer of risk and encourage developed countries to reduce their production of such wastes. But this would not end illegal trade in these wastes because the potential profits are much too great. To most environmentalists, the only real solution to the hazardous-waste problem is not to produce it in the first place (pollution prevention). It is encouraging that 34 industrialized nations adopted agreements in 1998 to sharply reduce the use of 16 persistent organic pollutants and toxic metals (such as lead, mercury, and cadmium).

22-7 CASE STUDIES: LEAD, DIOXINS, AND CHLORINE

How Can We Reduce Exposure to Lead?

Each year 12,000–16,000 American children under age 9 are treated for acute lead poisoning, and about 200 die. About 30% of the survivors suffer from palsy, partial paralysis, blindness, and mental retardation.

Lead can also cause damage at levels far below those that cause acute lead poisoning, especially in children and unborn fetuses. In the United States, the current maximum legal level in the blood is 10 micrograms (μg) of lead per deciliter (dL, or about half a cup) of blood, or 10 μg/dL. Research indicates that children under age 6 and unborn fetuses with blood levels greater than this standard are especially vulnerable to nervous system impairment, a lowered IQ (by 4–6 points), a shortened attention span, hyperactivity, hearing damage, and various behavior disorders.

It is great news that between 1976 and 1992 the percentage of U.S. children ages 1–5 with lead levels above the 10 μg/dL standard dropped from 85% to 6% for white children and from 98% to 21% for black children, preventing at least 9 million childhood lead poisonings. The primary reason is government

Q: What percentage of U.S. and world energy needs could be provided by renewable energy resources by 2040 or sooner?

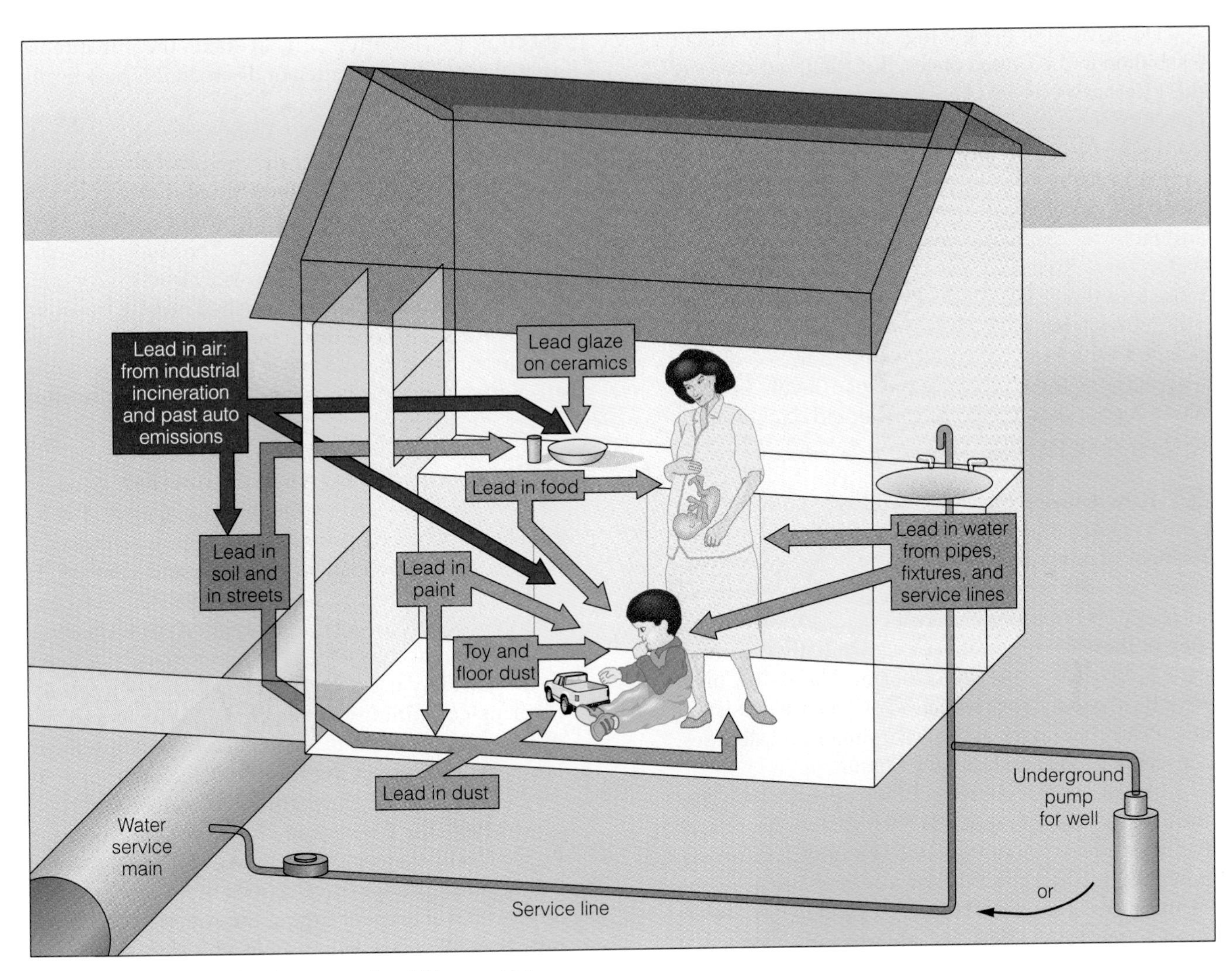

Figure 22-13 Sources of lead exposure for children and fetuses.

regulations that phased out leaded gasoline and lead-based solder, a pollution prevention approach instituted primarily because of the pioneering research on lead toxicity in children published in 1979 by Dr. Herbert Needleman.

Even with the encouraging drop in average blood levels of lead, a large number of U.S. children and unborn fetuses still have unsafe levels of lead from exposure to a number of sources (Figure 22-13). Health scientists have proposed a number of ways to help protect children from lead poisoning:

- *Testing all children for lead by age 1.*
- *Banning incineration of solid and hazardous waste or greatly increasing current pollution control standards for old and new incinerators.*
- *Phasing out leaded gasoline worldwide over the next decade.*
- *Testing older housing and buildings for leaded paint and lead dust and removing this hazard.* According to government estimates, 74% of U.S. homes contain lead-based paint.
- *Banning all lead solder in plumbing pipes and fixtures and in food cans.*
- *Removing lead from municipal drinking water systems within 10 years.*
- *Washing fresh fruits and vegetables and hands thoroughly.*
- *Testing ceramicware used to serve food for lead glazing.*
- *Reevaluating the proposed increase in electric cars propelled by lead batteries.* A 1995 study estimated that the manufacture, handling, disposal, and recycling of lead batteries for widespread use in electric vehicles could create up to 60 times the lead pollution caused by vehicles burning leaded gasoline.

A: 50–80% with an aggressive program (20–40% with a moderate program)

Doing most of these things will cost an estimated $50 billion in the United States. But health officials say the alternative is to keep poisoning and mentally handicapping large numbers of children.

In the cities of many developing countries where car use is rising, lead levels in air, water, and human blood are rising because leaded gasoline is still used and factories discharge large quantities of lead into waterways. Rivers in Asia typically have 20 times more lead than rivers in developed countries. In Chinese cities such as Shanghai, where the roads are clogged with new cars, studies have found that at least 65% of the children have lead levels in their blood higher than the point considered dangerous to mental development.

How Dangerous Are Dioxins? **Dioxins** are a family of 75 chlorinated hydrocarbon compounds formed as unwanted by-products in chemical reactions (usually at high temperatures, especially in incinerators) involving chlorine and hydrocarbons. One of these compounds, TCDD—sometimes simply called dioxin—is the most harmful and the most widely studied. Dioxins such as TCDD are persistent chemicals that linger in the environment for decades, especially in soil and human fat tissue.

In 1990 representatives of the paper and chlorine industries claimed that the inconsistent results of health studies exonerated TCDD and other dioxins, and they pushed the EPA for a reassessment of the health risks of dioxins. This strategy may have backfired when EPA's 1994 reevaluation indicated that dioxins are an even greater health threat than previously thought. This new review concluded that dioxin is "likely to present a cancer threat to humans." Unlike most carcinogens, TCDD does not damage DNA; it apparently promotes cancer by activating DNA already damaged by other carcinogens.

This finding explains why researchers found a variety of cancers rather than a single type, as is usually the case with most carcinogens. According to the EPA review, TCDD and several other dioxins in food and air may already be a major cancer hazard; existing levels of dioxins in the U.S. population are now believed to cause an estimated 2.5–25% of all new cancers, or 26,000–260,000 new cancers each year.

This comprehensive, 3-year, $4-million review by more than 100 scientists around the world also revealed that the most powerful effects of exposure to various dioxins at or near levels already occurring in the U.S. population are seen in the reproductive system, endocrine (hormone) system, and immune system. These effects may pose an even greater threat to human health (especially for fetuses and newborns) than the cancer-promoting ability of various dioxins (Section 17-3). With profits at stake, there is intense industry pressure to soften or discredit this new health reassessment of dioxins.

In 1996, biologist Barry Commoner and other researchers released a study showing that about 86% of the dioxins produced in the United States could be eliminated without economic sacrifice and with important environmental gains. The study also concluded that doing this could possibly lead to economic gains. This would be accomplished mostly by eliminating the sources of dioxin from medical waste incinerators, municipal solid waste generators, paper mills, iron ore sintering plants, and cement kilns burning hazardous wastes.

What Should We Do About Chlorine? Modern society depends heavily on chlorine (Cl_2) and chlorine-containing compounds. This highly reactive gas is used to purify water, bleach paper and wood pulp, and produce household bleaching agents; it also combines with various organic compounds to form about 11,000 different chlorinated organic compounds, many of them widely used. The problem is that some of the chlorine-containing chemicals we produce and use in fairly large amounts are persistent, accumulate in body fat, and (according to animal and other toxicity studies; Section 17-2) are harmful to human health.

In 1993, the Governing Council of the American Public Health Association (APHA), one of the premier U.S. scientific and medical associations, unanimously approved a statement urging the American chemical industry to phase out use of chlorine except for producing some pharmaceuticals and for disinfection of public water supplies. The goal would be to replace most chlorine compounds with environmentally and economically acceptable substitutes over the next 10–20 years.

The three largest single uses, which together make up 60% of all chlorine use, are plastics (mostly to make PVC), solvents, and paper and pulp bleaching. Most of these uses have substitutes (although some are more expensive as long as full-cost pricing is not used) and could be phased out, as a small but growing number of companies are doing. Nonchlorine plastics and other materials could replace most uses of PVC, and incineration of PVCs could be forbidden (which would also reduce production of dioxins).

Studies estimate that 60% of the chlorinated organic solvents currently used could be phased out over a decade and replaced with less harmful and affordable substitutes, such as soap and water, steam cleaning, citrus-based solvents (Individuals Matter, p. 528), and physical cleaning (including elbow grease and blasting with plastic beads and a pressurized solution of water and baking soda). Such a phaseout

Q: What percentage of the world's electricity could solar cells supply by 2050?

has already begun, not just for environmental reasons but also to reduce legal liability for possible effects of such chemicals on workers and consumers.

Using chlorine to bleach wood pulp and paper could be replaced with less harmful processes that rely on oxygen, ozone, chlorine dioxide, and hydrogen peroxide, as several paper companies are doing. By the early 1990s, the German paper industry had achieved totally chlorine-free paper production (which is 30% cheaper than processes using chlorine); the rest of Europe is following Germany's lead.

About 5% of the chlorine produced in the United States is used to purify drinking water (1%) and wastewater from sewage treatment plants (4%). Much of this could be gradually replaced with ozone and other nonchlorine purification processes. Drinking water may still need to be purified by chlorination because it continues to kill harmful microorganisms in water lines.

The remaining 35% of chlorine is used to manufacture a variety of organic and inorganic chemicals, many of them for small and specialized uses. Scrutiny of each of these chemicals is required to determine which are essential (many pharmaceuticals) and which might be replaced by affordable and environmentally acceptable substitutes.

Understandably, producers of chlorine chemicals strongly oppose such threats to their profits. However, if given proper financial incentives and 10–20 years to develop and phase in substitutes, chemical producers could profit from new markets resulting from this important part of the ecoindustrial revolution based on pollution prevention (Solutions, p. 584); they would also protect themselves from adverse publicity, possible future lawsuits, and cleanup costs.

22-8 HAZARDOUS-WASTE REGULATION IN THE UNITED STATES

What Is the Resource Conservation and Recovery Act? In 1976, the U.S. Congress passed the Resource Conservation and Recovery Act (RCRA, pronounced "RICK-ra"), which was amended in 1984. This law requires the EPA to identify hazardous wastes and set standards for their management, and it provides guidelines and financial aid for states to establish waste management programs. The law also requires all firms that store, treat, or dispose of more than 100 kilograms (220 pounds) of hazardous wastes per month to have a permit stating how such wastes are to be managed.

To reduce illegal dumping, hazardous-waste producers that are granted disposal permits by the EPA must use a cradle-to-grave system to keep track of waste transferred from point of origin to approved off-site disposal facilities. However, the EPA and state regulatory agencies don't have sufficient personnel or money to review the documentation of more than 750,000 hazardous-waste generators and 15,000 haulers each year, let alone detect and prosecute offenders. Environmentalists argue that fines are too small and that culprits don't get jail time for serious violations that can cause severe health problems and even death, sending violators the clear message that pollution pays.

What Is the Superfund Act? In 1980, the U.S. Congress passed the Comprehensive Environmental Response, Compensation, and Liability Act, commonly known as the Superfund program. This law (plus amendments in 1986 and 1990) established a $16.3-billion Superfund financed jointly by federal and state governments and by special taxes on chemical and petrochemical industries (which provide 86% of the funding). The purpose of the Superfund is to identify and clean up abandoned hazardous-waste dump sites such as Love Canal (p. 578) and leaking underground tanks that threaten human health and the environment.

To keep taxpayers from footing most of the bill, cleanups are based on the *polluter-pays principle*. The EPA is charged with locating dangerous dump sites, finding the potentially liable culprits, ordering them to pay for the entire cleanup, and suing them if they don't. When the EPA can find no responsible party, it draws money out of the Superfund for cleanup.

To implement the polluter-pays principle, the Superfund legislation considers all polluters of a site to be subject to strict joint and several liability. This controversial strategy means that each individual polluter (no matter what its contribution) can be held liable for the entire cost of cleaning up a site if the other parties can't be found or have gone bankrupt. Once the EPA sues any parties they consider liable, these parties typically try to reduce their cleanup costs by suing any other contributors to the site that they can find. Once stuck with the cleanup bill, responsible parties file claims with their casualty and property insurers, who usually fight such claims in court. Although this process holds identified polluters (or their insurance companies) responsible for the cleanup, the resulting lengthy legal battles slow cleanup.

Currently about 1,360 sites are on a National Priority List for cleanup because they pose a real or potential threat to nearby populations; the number may eventually grow to 2,000–3,000. By June 1997 emergency cleanup had been carried out at virtually all sites and more than 400 sites had been cleaned up by containing the wastes to prevent leakage. The EPA has plans to stabilize an additional 500 sites by 2000.

A: 25% (35% in the United States) with an aggressive program

How Can the Superfund Law Be Improved?

SOLUTIONS

Even though the Superfund was created with noble goals, there is widespread agreement that the program can and should be improved. Since the program began, polluters and their insurance companies have been working hard to do away with the polluter-pays principle at the heart of the program and make it mostly a *public-pays* approach.

This strategy, which is working, has three components: **(1)** Deny responsibility (stonewall) in order to tie up the EPA in expensive legal suits for years, **(2)** sue local governments and small businesses to make them responsible for cleanup, both as a delay tactic and to turn local governments and small businesses into opponents of Superfund's strict liability requirements, and **(3)** mount a public relations campaign declaring that toxic dumps pose little threat, that cleanup is too expensive compared to the risks involved, and that Superfund is unfair to polluters and is wasteful and ineffective.

The strict joint and several liability of the Superfund law may seem unfair to polluters, but the authors of the legislation argued that any other liability scheme wouldn't work. If the EPA had to identify and bring an enforcement action against every party liable for a dump site, the agency would be overwhelmed with lawsuits, and the pace of cleanup would be still slower. Administrative costs would also rise sharply, causing the program to become more of a *public-pays program* favored by the polluters.

The EPA also points out that the strict polluter-pays principle in the Superfund Act has been effective in making illegal dump sites virtually a thing of the past—an important form of pollution prevention for the future. It has also forced waste producers who are fearful of future liability claims to reduce their production of such waste and to recycle or reuse much more of what they do generate.

A generally accepted proposal for reducing lawsuits and speeding cleanup is to set up an $8-billion Environmental Insurance Resolution Fund funded by insurers for a 10-year period. Companies found liable for cleanup of wastes they disposed of before 1986 would be able to use money from this fund rather than going to their individual insurance companies.

Although most experts agree that the Superfund's ultimate goal—permanent cleanup—should be maintained, many argue that sites should be ranked in three general categories: **(1)** sites requiring immediate full cleanup, **(2)** sites considered to pose serious hazard but located sufficiently far from concentrations of people or endangered ecosystems (these sites would receive emergency cleanup and then be isolated by barriers and signs, with more complete cleanup to come later), and **(3)** lower-risk sites requiring only stabilization (capping and containment) and monitoring.

There is also considerable pressure to involve people and local governments in communities with contaminated sites in cleanup decisions. Too often the people affected by risks from Superfund sites are not consulted until the EPA and the polluters have worked out a plan, which may or may not be in the best interests of local citizens.

Critical Thinking

1. Do you support the *polluter-pays principle* used in the original Superfund legislation? Defend your position. Use the library or Internet to find out whether there have been any recent changes in the Superfund law. If so, evaluate these changes and decide which of them you support. Defend your position.

2. Explain why you agree or disagree with the other suggestions listed in this box for improving the Superfund law. Use the library or the Internet to determine what changes (if any) have been made in the Superfund law and explain why you agree or disagree with these changes.

Once a site is stabilized, final cleanup (which often involves pumping up and cleaning contaminated groundwater) can take decades. Attempts are being made to find ways to improve the Superfund Act without seriously weakening it (Solutions, above).

The U.S. Office of Technology Assessment and the Waste Management Research Institute estimate that the Superfund list could eventually include at least 10,000 priority sites, with cleanup costs of up to $1 trillion, not counting legal fees. Other studies project only 2,000 sites with estimated cleanup costs of $100–165 billion.

Cleaning up toxic military dumps will cost another $100–200 billion and take at least 30 years. Cleaning up contaminated Department of Energy sites used to make nuclear weapons will cost an additional $200–400 billion and take 30–50 years. In addition, the Department of Interior will need to spend at least $100 billion to clean up 300,000 active and abandoned mine sites, 200,000 oil and gas leases, 3,000 landfills, and

Q: What percentage of the world's electricity is supplied by hydropower?

4,200 leaking underground storage tanks on public lands under its jurisdiction.

These estimated costs are only for cleanup to prevent future damage; they don't include the health and ecological costs already associated with such wastes. To environmental scientists and some economists, it is difficult to imagine a more convincing reason for emphasizing waste reduction and pollution prevention (Figures 22-3 and 22-4).

22-9 SOLUTIONS: ACHIEVING A LOW-WASTE SOCIETY

What Is the Role of Grassroots Action? Bottom-Up Change In the United States, local citizens have been stirred into action by proposals to bring incinerators, landfills, or treatment plants for hazardous and radioactive wastes into their communities. Outrage has grown as numerous studies have shown that such facilities have traditionally been located in communities populated by African Americans, Asian Americans, Hispanics, and poor whites (Guest Essay, p. 606).

This loose coalition of grassroots organizations offers the following guidelines for helping achieve environmental justice for all:

- *Don't compromise our children's futures by cutting deals with polluters and regulators.*
- *Don't be bulldozed by scientific and risk analysis experts.* Risks from incinerators and landfills, when averaged over the entire country, are quite low, but the risks for the people near these facilities are much higher. These people, not the rest of the population, are the ones whose health, lives, and property values are being threatened.
- *Hold polluters—and elected officials who go along with them—personally accountable because what they are doing is wrong.*
- *Don't fall for the argument that protesters against hazardous-waste landfills, incinerators, and injection wells are holding up progress in dealing with hazardous wastes.* Instead, recognize that the best way to deal with waste and pollution is to produce much less of it. For example, after a charge for each unit of hazardous waste produced in Germany was imposed in 1991, hazardous-waste production fell more than 15% in 3 years.
- *Oppose all hazardous-waste landfills, deep-disposal wells, and incinerators.* This will sharply raise the cost of dealing with hazardous materials, discourage location of such facilities in poor neighborhoods often populated by minorities, and encourage waste producers and elected officials to get serious about pollution prevention.
- *Recognize that there is no such thing as safe disposal of a toxic or hazardous waste.* For such materials, the goal should be "Not in Anyone's Backyard" (NIABY) or "Not on Planet Earth" (NOPE).
- *Pressure elected officials to pass legislation requiring that unwanted industries and waste facilities be distributed more widely instead of being concentrated in poor and working-class neighborhoods, many populated mostly by minorities.*
- *Ban all hazardous-waste exports from one country to another.*

It is not surprising that representatives of the solid- and hazardous-waste management industries oppose these tactics and have been able to persuade U.S. federal and state legislators not to put such ideas into practice.

How Can We Make the Transition to a Low-Waste Society? According to physicist Albert Einstein, "A clever person solves a problem, a wise person avoids it." To prevent pollution and reduce waste, many environmental scientists urge us to understand and live by four key principles: **(1)** Everything is connected, **(2)** there is no "away" for the wastes we produce, **(3)** dilution is not the solution to most pollution, and **(4)** the best and cheapest way to deal with waste and pollution is to produce less of it and then reuse and recycle most of the materials we use (Figures 22-3 and 22-4). Some actions you can take to reduce your production of solid waste and hazardous waste are listed in Appendix 5.

Unless we learn to depend on fewer virgin resources—by development of secondary materials industries . . . and by redesigning goods, services, and communities—we will continue propelling ourselves toward economic and ecological disaster.

JOHN E. YOUNG AND AARON SACHS

CRITICAL THINKING

1. Explain why you support or oppose the following:
   a. Requiring that all beverage containers be reusable
   b. Requiring all households and businesses to sort recyclable materials into separate containers for curbside pickup
   c. Requiring consumers to pay for plastic or paper bags at grocery and other stores to encourage the use of reusable shopping bags

A: 20% (99% in Norway, 70% in Canada, 50% in developing countries, 13% in the United States)

GUEST ESSAY

Environmental Justice for All

Robert D. Bullard

Robert D. Bullard is professor of sociology and director of the Environmental Justice Center at Clark Atlanta University. For more than a decade, he has conducted research in the areas of urban land use, housing, community development, industrial facility siting, and environmental justice. His scholarship and activism have made him one of the leading experts on environmental racism: the systematic selection of communities of color as sites for waste facilities and polluting industries. He is the author of four books and more than three dozen articles, monographs, and scholarly papers that address equity concerns. His book, Dumping in Dixie: Race, Class, and Environmental Quality, *2d ed. (Westview Press, 1994), has become a standard text in the field. Other books are* Confronting Environmental Racism *(South End Press, 1993) and* Unequal Protection: Environmental Justice and Communities of Color *(Sierra Club Books, 1994).*

Despite widespread media coverage and volumes written on the U.S. environmental movement, environmentalism and social justice have seldom been linked. Nevertheless, an environmental revolution is now taking shape in the United States that combines the environmental and social justice movements into one framework.

People of color (African Americans, Latinos, Asians, Pacific Islanders, and Native Americans), working-class people, and poor people in the United States suffer disproportionately from industrial toxins, dirty air and drinking water, unsafe work conditions, and the location of noxious facilities such as municipal landfills, incinerators, and toxic-waste dumps. Despite the government's attempts to level the playing field, all communities are not created equal.

The environmental justice movement attempts to dismantle exclusionary zoning ordinances, discriminatory land-use practices, differential enforcement of environmental regulations, disparate siting of risky technologies, and the dumping of toxic waste on the poor and people of color in the United States and in developing countries.

All communities are not treated as equals when it comes to resolving environmental and public health concerns, either. Over 300,000 farm workers (over 90% of whom are people of color) and their children are poisoned by pesticides sprayed on crops in the United States. Some 3–4 million children (many of them African Americans or Latinos living in the inner city) are poisoned by lead-based paint in old buildings, lead-soldered pipes and water mains, lead-tainted soil contaminated by industry, and air pollutants from smelters.

All communities do not bear the same burden or reap the same benefits from industrial expansion. This is true in the case of the mostly African-American Emelle, Alabama (home of the nation's largest hazardous-waste landfill); Navajo lands in Arizona where uranium is mined; and the 2,000 factories known as *maquiladores*, located just across the U.S. border in Mexico.

Communities, states, and regions that contain hazardous-waste disposal facilities (importers) receive far fewer economic benefits (jobs) than the geographic locations that generate the wastes (exporters). Nationally, 60% of African Americans and 50% of Latinos live in communities with at least one uncontrolled toxic-waste site. Three of the five largest hazardous-waste landfills are located in communities that are predominantly African American or Latino.

2. Explain why you are not recycling if you're not buying products made from the maximum feasible content of postconsumer waste.

3. Would you oppose having a hazardous-waste landfill, waste treatment plant, deep-injection well, or incinerator in your community? Explain. If you oppose these disposal facilities, how do you believe that hazardous waste generated in your community and your state should be managed?

4. Give your reasons for agreeing or disagreeing with each of the following proposals for dealing with hazardous waste:

- **a.** Reducing the production of hazardous waste and encouraging recycling and reuse of hazardous materials by levying on producers a tax or fee for each unit of waste generated
- **b.** Banning all land disposal and incineration of hazardous waste to encourage recycling, reuse, and treatment and to protect air, water, and soil from contamination
- **c.** Providing low-interest loans, tax breaks, and other financial incentives to encourage industries producing hazardous waste to recycle, reuse, treat, destroy, and reduce generation of such waste
- **d.** Banning the shipment of hazardous waste from the United States to any other country
- **e.** Banning the shipment of hazardous waste from one state to another

5. What ethical deliberations do we face with creating landfills that may leak in the future? Whose quality of life is being affected? For what purpose? Relate this to the concept of environmental justice.

Q: What percentage of the world's electricity could wind power supply by 2050?

The marginal status of many people of color in the United States makes them prime actors in the movement for environmental and social justice. For example, the organizing theme of the 1991 First National People of Color Environmental Summit, held in Washington, D.C., was justice, fairness, and equity. More than 650 delegates from all 50 states, as well as from Puerto Rico, Mexico, Chile, Colombia, and the Marshall Islands, participated in this historic 4-day gathering.

Environmental justice does not stop at the U.S. border. Environmental injustices exist from the *favelas* of Rio de Janeiro, Brazil, to the shantytowns of Johannesburg, South Africa. Members of the environmental justice movement are also questioning the wasteful and nonsustainable development models being exported to the developing world.

It is no mystery why grassroots environmental justice groups in Louisiana's "Cancer Alley," Chicago's south side, and Los Angeles's East and South Central neighborhoods are attacking the institutions they blame for their underdevelopment, disenfranchisement, and poisoning. Some people see these threats to their communities as a form of genocide.

Grass-roots leaders are demanding justice. Residents of communities such as West Dallas and Texarkana (Texas), West Harlem (New York), Rosebud (South Dakota), Kettleman City (California), and Sunrise, Lions, and Wallace (Louisiana) see their struggle for environmental justice as a life-and-death matter. Unfortunately, their stories of environmental racism are not broadcast into the nation's living rooms during the nightly news, nor are they splashed across the front pages of national newspapers and magazines. To a large extent, the communities that are the victims of environmental injustice remain invisible to the larger society.

The environmental justice movement is led, planned, and to a large extent funded by people who are not part of the established environmental community or the "Big 10" environmental organizations. Most environmental justice groups are small and operate with resources generated from the local community.

For too long these groups and their leaders have been invisible and their stories muted. This is changing as these grassroots groups are forcing their issues onto the nation's environmental agenda.

The United States has a long way to go in achieving environmental justice for all its citizens. The membership of decision-making boards and commissions still does not reflect the racial, ethnic, and cultural diversity of the country, and token inclusion of people of color on boards and commissions does not necessarily mean that their voices will be heard or their cultures respected. The ultimate goal of any inclusion strategy should be to democratize the decision-making process and empower disenfranchised people to speak and do for themselves.

Critical Thinking

1. Does your lifestyle and political involvement help promote or reduce environmental racism in your community and in society as a whole?

2. How would you go about helping prevent polluting factories and hazardous-waste facilities from being located in or near communities made up largely of people of color, working-class people, and poor people?

6. What hazardous and solid wastes do you create? What specific things could you do to reduce, reuse, or recycle as much of these wastes as possible? Which of these things do you actually plan to do? Do you use any of the alternative chemicals shown in Table 22-2?

7. Explain why you agree or disagree with each of the suggestions on p. 605 made by a coalition of grassroots organizations in the United States seeking environmental justice for all.

8. Try to come up with analogies to complete the following statements: **(a)** A high-waste society is like _______. **(b)** Reducing waste by not producing so much in the first place is like _______. **(c)** Burning and burying hazardous or toxic waste is like _______. **(d)** Recycling waste materials is like _______. **(e)** An ecoindustrial revolution is like _______. **(f)** A sanitary landfill is like a _______. **(g)** Exporting hazardous waste from one country to another is like _______. **(h)** The polluter-pays principle is like _______.

9. What good things about your quality of life are built on past and current environmental injustice to others (especially the poor and lower middle class)? Have you ever experienced environmental injustice? Explain. If so, how did it make you feel? If not, try to imagine how it might feel and describe these feelings.

PROJECTS

1. For one week, keep a list of the solid waste you throw away. What percentage is materials that could be recycled, reused, or burned for energy? What percentage of the items could you have done without in the first place? Tally and compare the results for your entire class.

A: At least 10% (10–25% in the United States)

2. Tour a supermarket, a drug store, a hardware store, and a large store such as K-Mart or Wal-Mart and determine what items contain unnecessary packaging. For example, does a plastic or glass bottle of something really need to be encased in a paper container, and which items really need to be shrink-wrapped with plastic? Make a list of the overpackaged products, then send the list (with brand names) to local store managers, the chief executive officers of chain stores, and the companies that made each item.

3. Determine whether **(a)** your school and your city have recycling programs, **(b)** your school sells soft drinks in throwaway cans or bottles, and **(c)** your school bans release of throwaway helium-filled balloons at sporting events and other activities.

4. What happens to solid waste in your community? How much is landfilled? Incinerated? Composted? Recycled? What technology is used in local landfills and incinerators? What leakage and pollution problems have local landfills or incinerators had? Does your community have a recycling program? Is it voluntary or mandatory? Does it have curbside collection? Drop-off centers? Buyback centers?

5. What hazardous wastes are produced at your school? In your community? What happens to these wastes?

6. Are there any active or abandoned hazardous-waste dumps in your community? Where are they located? What has been dumped there? Do they have one or more liners? Has there been any testing to determine whether wastes have leaked from the sites? If so, what were the results of the tests? Who owns the sites?

7. Make a concept map of this chapter's major ideas, using the section heads and subheads and the key terms (in boldface). Look at the inside back cover and in Appendix 4 for information about concept maps.

PART V

LAND USE, BIODIVERSITY, AND CONSERVATION

In our every deliberation, we must consider the impact of our decisions on the next seven generations.

IROQUOIS CONFEDERATION (18TH CENTURY)

23 SUSTAINING ECOSYSTEMS: LAND USE, CONSERVATION, AND MANAGEMENT*

Frontier Expansion, Native Americans, and Bison

In 1500, before Europeans settled North America, 60–125 million North American bison grazed the plains, prairies, and woodlands over much of the continent. A single herd on the move might thunder past for hours. Several Native American tribes depended heavily on bison (Figure 23-1), and typically they killed only the animals they needed for food, clothing, and shelter. By 1906, however, the once-vast range of the bison had shrunk to a tiny area and the species had been driven nearly to extinction (Figure 23-2).

How did this happen? First, as settlers moved west after the Civil War, the sustainable balance between Native Americans and bison was upset. The Sioux, Cheyenne, Comanches, and other plains tribes traded bison skins to settlers for steel knives and firearms, so they began killing more bison.

*Paul M. Rich, associate professor of Ecology and Evolutionary Biology and Environmental Studies at the University of Kansas, is coauthor of this chapter.

The most relentless slaughter, however, was caused by the new settlers. As railroads spread westward in the late 1860s, railroad companies hired professional bison hunters, including Buffalo Bill Cody, to supply construction crews with meat. Passengers also gunned down bison from train windows for sport, leaving the carcasses to rot. Commercial hunters shot millions of bison for their hides and tongues (considered a delicacy), leaving most of the meat to rot. "Bone pickers" collected the bleached bones that whitened the prairies and shipped them east to be ground up as fertilizer.

Farmers shot bison because they damaged crops, fences, telegraph poles, and sod houses. Ranchers killed them because they competed with cattle and sheep for grass. The U.S. Army killed at least 2.5 million bison each year between 1870 and 1875 as part of their campaign to subdue the plains tribes by killing off their primary source of food.

By 1892 only 85 bison were left. They were given refuge in Yellowstone National Park and protected by an 1893 law against the killing of wild animals in national parks. In 1905, 16 people formed the American Bison Society to protect and rebuild the captive

Figure 23-1 Plains Indians hunting bison. (*Buffalo Chase with Bows and Arrows.* 1832–33. George Catlin Museum of American Art)

population. Soon thereafter, the federal government established the National Bison Range near Missoula, Montana. Today there are an estimated 250,000 bison, about 97% of them on privately owned ranches, where many are being raised for a growing $500-million-a-year market in bison meat and hides.

Some wildlife conservationists have suggested restoring large herds of bison on public lands in the North American plains. Not surprisingly, this idea has been strongly opposed by ranchers with permits to graze cattle and sheep on federally managed lands.

Every country and region has three forms of wealth: material, cultural, and biological. So far we have considerable knowledge of the first two types of wealth but little understanding of the biological wealth or biodiversity that is so vital to all life and economies. Forests, rangelands, parks, and wilderness are crucial sources of biological wealth that are coming under increasing pressure from population growth and economic development. Conservation biologists believe that learning how to use these potentially renewable forms of earth capital sustainably—by protecting some from exploitation and helping to heal those we have degraded—is an urgent priority.

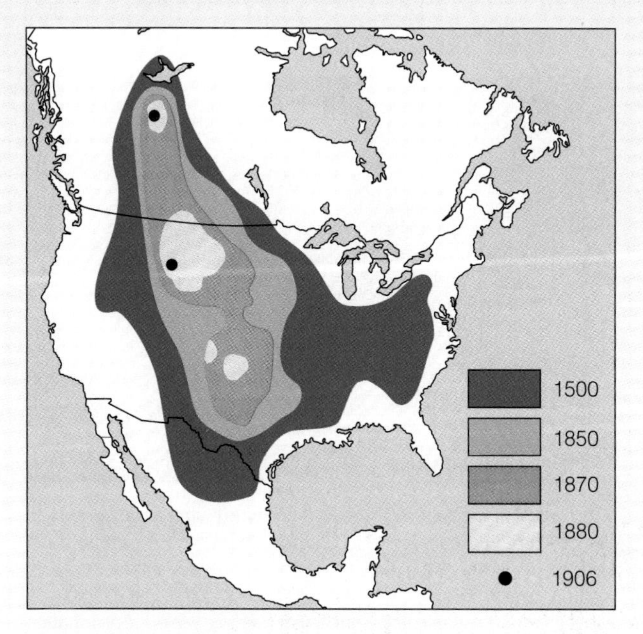

Figure 23-2 The historical range of the bison shrank severely between 1500 and 1906.

We abuse land because we regard it as a commodity belonging to us. When we see land as a community to which we belong, we may begin to use it with love and respect.

ALDO LEOPOLD

This chapter is devoted to answering the following questions:

- What are the major phases in the history of land use, resource conservation, and environmental protection in the United States?
- What is conservation biology, and why is it important?
- What are the types of public lands in the United States, and how are they used?
- Why are rangelands important, and how should they be managed?
- What problems do parks face, and how should they be managed?
- Why is wilderness important, and how much wilderness and other biodiversity sanctuaries should be preserved?

23-1 LAND USE, CONSERVATION, AND PUBLIC HEALTH IN THE UNITED STATES: 1400–1960

How Were Resources Used Between 1400 and 1900? Europeans first came to North America in the 15th and 16th centuries. They found it populated with diverse groups of indigenous people, called Indians by the Europeans and now often called Native Americans. For at least 10,000 years most of these peoples had practiced mostly sustainable forms of hunting (Figure 23-1), gathering, and use of wild resources. Most Native American cultures had a deep respect for the land and its plants and animals.

When European colonists began settling North America in 1607, they found a vast continent with seemingly unlimited resources of forests, wildlife, and some of the world's richest soils. Not surprisingly, these settlers had a **frontier worldview**. They saw a hostile wilderness to be conquered (cleared, planted) and exploited for its resources as quickly as possible. This frontier attitude, and the vast size of the continent, led to enormous resource waste because the pioneers believed there would always be more.

In 1850, about 80% of the total land area of the territorial United States was government owned, most of it taken from Native Americans. As a Sioux elder said in 1891, "They made us many promises, more than I can remember, but they never kept but one; they promised to take our land and they took it." By 1900, more than half of the country's public land had been given away or sold cheaply. These low prices encouraged widespread abuse of the country's resources.

Between 1832 and 1870, some people became alarmed at the scope of this resource depletion and degradation. These early conservationists included George Catlin, Horace Greeley, Ralph Waldo Emerson, Frederick Law Olmsted, Charles W. Eliot, Henry David Thoreau, and George Perkins Marsh. They urged that part of the unspoiled wilderness owned jointly by all people (but managed by the government) be protected as a legacy to future generations. Their warnings were either ignored or vigorously opposed. Many citizens (and their elected officials) believed that the country's forests and wildlife would last forever and that people had the right to use public and private land as they pleased.

Between 1870 and 1900, concerns about environmental and public health hazards began increasing in the country's rapidly growing and industrializing cities. These environmental problems included air pollution from coal burning, contaminated water supplies, mounds of horse manure in the streets, inadequate garbage collection, overcrowded neighborhoods and tenements, unsafe and hazardous working conditions in factories and sweatshops, and epidemics of waterborne infectious diseases, such as yellow fever, typhoid fever, and cholera, which prematurely killed large numbers of people.

1872	Yellowstone National Park established
1891	Forest Reserve Act enacted
1892	Sierra Club founded by John Muir
1891–97	Timber cutting on large tracts of public land banned
1900	Lacey Act enacted; transport of live or dead animals across state borders and imports of wildlife banned
1902	Reclamation Act enacted to promote irrigation and water development projects in arid West
1903	First national wildlife refuge established
1905	U.S. Forest Service created; Audubon Society founded
1906	Pure Food and Drug Act enacted
1912–16	National Park System created
1920–27	Public health boards established in most cities

Figure 23-3 Examples of the increased role of the federal government in resource conservation and public health and establishment of two key private environmental groups, 1872–1927.

What Role Did the Government Play in Resource Conservation Between 1900 and 1960? Between 1872 and 1927, a number of important actions increased the role of the federal government and citizens in resource conservation and public health (Figure 23-3). However, effective protection of forests and wildlife didn't begin until Theodore Roosevelt, an ardent conservationist, became president. His term of office, 1901–9, has been called the country's Golden Age of Conservation. Roosevelt persuaded Congress to give him the power to designate public land as federal wildlife refuges. In 1903, he established the first federal refuge at Pelican Island off the east coast of Florida for preservation of the endangered brown pelican (Figure 23-4). Roosevelt also tripled the size of the forest reserves and transferred their administration from the Department of the Interior, known for lax enforcement, to the Department of Agriculture.

In 1905, Congress created the U.S. Forest Service to manage and protect the forest reserves. Roosevelt appointed Gifford Pinchot as its first chief. Pinchot pioneered scientific management of forest resources on public lands, using the principles of *sustainable yield* (cutting trees no faster than they could regenerate) and *multiple use* (using the lands for a variety of purposes, including resource extraction, recreation, and wildlife protection). That same year the Audubon Society was founded to preserve the country's bird species.

In 1907, Congress, upset because Roosevelt had added vast tracts to the forest reserves, banned further executive withdrawals of public forests. On the day before the bill became law, Roosevelt defiantly reserved another 6.5 million hectares (16 million acres).

Early in the 20th century the U.S. conservation movement split over how the beautiful Hetch Hetchy

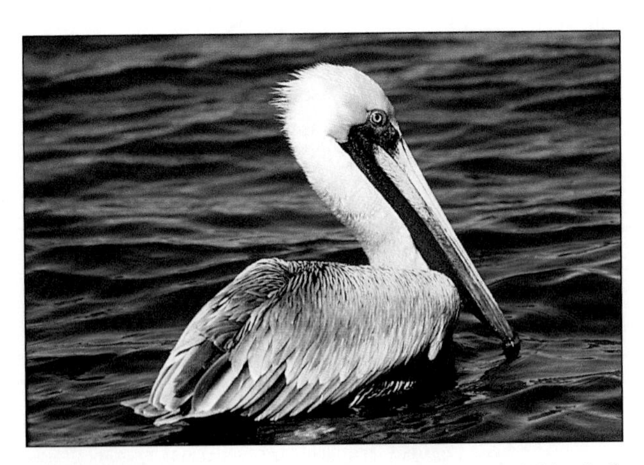

Figure 23-4 The first national wildlife refuge was established off the coast of Florida in 1903 to protect the brown pelican from overhunting and loss of habitat. In the 1960s this species was again threatened with extinction when exposure to DDT and other persistent pesticides in the fish it eats caused reproductive failures. Now it is making a comeback. (E. R. Degginger)

Smith, M. B. 1998. "The Value of a Tree: Public Debates of John Muir and Gifford Pinchot." *Historian*, vol. 60, no. 4, 757(22).

Preservation or Wise Use of Public Resources?

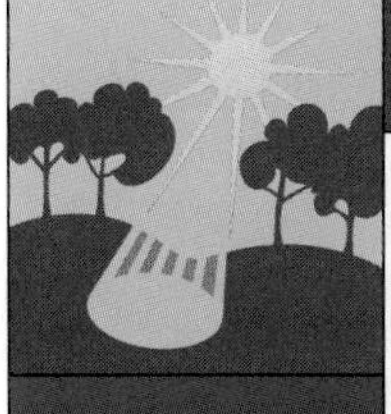

SPOTLIGHT

In 1901, resource managers, led by Gifford Pinchot and San Francisco mayor James D. Phelan, proposed to dam the Tuolumne River running through Hetch Hetchy Valley to supply drinking water for San Francisco. Preservationists, led by John Muir, were opposed. After a bitter 12-year battle, the dam was built and the valley was flooded. Today the controversy continues, with preservationists pressing to have the dam removed.

Preservationists would keep large areas of public lands untouched so they can be enjoyed today and passed on unspoiled to future generations. After Muir's death in 1914, preservationists were led by forester Aldo Leopold, who said that the role of the human species should be to protect nature, not conquer it.

Another effective supporter of wilderness preservation was Robert Marshall of the U.S. Forest Service. In 1935, he and Leopold founded the Wilderness Society. More recent preservationist leaders include David Brower (former head of the Sierra Club and founder of both Friends of the Earth and Earth Island Institute), Ernest Swift, Stewart L. Udall (a former Secretary of the Interior), and Howard Zahniser (who as head of the Wilderness Society beginning in 1945 helped draft the Wilderness Act of 1964 and lobby Congress for its passage).

In contrast, *wise-use* or *utilitarian conservation* managers see wilderness and other public lands as resources to be used to enhance the nation's economic growth and to provide the greatest benefit to the greatest number of people. In their view the government should protect these lands from harm by managing them efficiently and scientifically using the principles of sustainable yield and multiple use. Early wise-use advocates included Theodore Roosevelt, Gifford Pinchot, John Wesley Powell, and Charles Van Hise.

Roosevelt and Pinchot thought conservation experts should form an elite corps of resource managers within the federal bureaucracy. Shielded there from political pressure, they could develop scientific management strategies. Pinchot angered Muir and other preservationists when he stated his "wise use" principle:

> *The first great fact about conservation is that it stands for development. There has been a fundamental misconception that conservation means nothing but the husbanding of resources for future generations. There could be no more serious mistake. . . . The first principle of conservation is the use of the natural resources now existing on this continent for the benefit of the people who live here now.*

Despite their basic differences, both schools oppose delivering public resources into the hands of a few for profit. Both groups have been disappointed. Since 1910 rights to extract minerals, graze livestock, and harvest trees in public lands have routinely been given away or sold at below-market prices to large corporate farms, ranches, mining companies, and timber companies. Taxpayers have subsidized this use of resources by absorbing the loss of potential revenue and paying for most of the resulting environmental damage.

Critical Thinking

1. Why do the philosophies of *utilitarian conservation* and *preservation* lead to different land management practices?

2. What do you think should be the balance between *utilitarian conservation* and *preservation* in management of public lands?

Valley (in what is now Yosemite National Park) was to be used (Spotlight, above). The *wise-use* or *utilitarian conservation* school believed that all public lands should be managed wisely and scientifically to provide needed resources. The *preservationist* school believed that remaining wilderness areas on public lands should be left untouched.

In 1912, Congress created the U.S. National Park System, and in 1916 it passed the National Park System Organic Act. The act declared that the parks were to be maintained in a manner that leaves them unimpaired for future generations and established the National Park Service (within the Department of the Interior) to manage the system. Its first head, Stephen Mather, began establishing grand hotels and other tourist facilities in parks with spectacular scenery, mostly for use by the wealthy. The dominant park policy was to encourage tourism by allowing private concessionaires to operate facilities within the parks.

During the early 1900s there were also improvements in public health. Between 1900 and 1927, public health boards were established in most cities, and governments began paying more attention to public health issues. Most efforts to pressure government officials to improve public health were led by women such as Jane Adams, Mary McDowell, and Alice Hamilton (Individuals Matter, p. 614).

After World War I (1914–18) the country entered a new era of economic growth and expansion. During the Republican administrations headed by presidents Warren Harding, Calvin Coolidge, and Herbert Hoover (1921–33), the federal government promoted

Hint: Enter the search terms *environmental movement, history* using the Subject Guide.

Alice Hamilton

Alice Hamilton (1869–1970) was the country's first influential expert in industrial medicine (Figure 23-5). After graduating from medical school she became professor of pathology at the Woman's Medical School of Northwestern University. She became interested in the neglected and poorly understood field of industrial medicine after hearing numerous stories about health hazards in stockyards and factories.

Hamilton began investigating various hazardous industries. Despite little information, company resistance, and workers' failure to report health problems from fear of losing their jobs, Hamilton's persistence and resourcefulness as a researcher paid off. During the next several decades she became the country's leading investigator of occupational hazards.

In 1919, Hamilton was appointed assistant professor of industrial medicine at Harvard University, the first teaching appointment of a woman at this institution. In the 1920s she published her classic text *Industrial Poisons in the United States* and became the country's most effective advocate for investigating and dealing with the environmental consequences of industrial activity.

Four decades before the widespread concern about pesticides and other industrial chemicals in the 1960s and 1970s, Hamilton was warning workers that they were being exposed to a variety of new chemicals whose effects on human health were unknown.

She unsuccessfully opposed the use of tetraethyl lead in gasoline in the 1920s, arguing that there is no safe exposure to lead. Her position was vindicated in the late 1980s, when tetraethyl lead was phased out of gasoline in the United States after decades of denial and delay by the industries involved. Her efforts were an important factor in the introduction of workers' compensation laws.

Alice Hamilton was a strong advocate of pollution prevention, an orientation that has been rediscovered in the 1990s. In a 1925 article she expressed the hope "that the day is not far off when we shall take the next step and investigate a new danger in industry before it is put to use, before any fatal harm has been done."

Figure 23-5 Alice Hamilton (1869–1970) was the first and foremost expert on industrial disease in the United States. (Schliesinger Library, Radcliffe College)

increased resource removal from public lands at low prices to stimulate economic growth. This represented a shift from a policy of scientific wise-use management of public lands to one of exploitation, with most of the benefits enjoyed by the wealthy.

President Hoover (1929–33) went even further and proposed that the federal government either return all remaining federal lands to the states or sell them to private interests for economic development. However, the Great Depression (1929–41) made owning such lands unattractive to state governments and private investors.

The second wave of national resource conservation and improvements in public health began in the early 1930s, as President Franklin D. Roosevelt (1933–45) strove to prod the country out of the Great Depression. Cash-poor landowners eagerly sold large tracts of land cheaply to the government. Several government programs initiated during the Roosevelt administration had lasting effects on U.S. resource and environmental policy.

The Civilian Conservation Corps (CCC) was created in 1933 to provide jobs for 2 million unemployed people. The CCC planted trees, developed parks and recreation areas, restored silted waterways, provided flood control, controlled soil erosion, and protected wildlife. The federal government built and operated many large dams in the arid western states, including Hoover Dam on the Colorado River. These projects provided jobs, flood control, cheap irrigation water, and cheap electricity.

The Taylor Grazing Act of 1934 required permits and fees for the use of federal grazing lands and placed limits on the number of livestock that could be grazed. However, western members of Congress had enough power to keep the Grazing Service and its successor, the Bureau of Land Management (BLM), poorly funded, understaffed, and without enforcement

Q: What do many scientists believe is the best fuel to replace oil and other fossil fuels within the next 40 years?

authority for another 40 years. As a result, many ranchers and mining and timber companies continued to abuse western public lands and to extract publicly owned resources at very low prices (p. 338).

Passage of the Migratory Bird Hunting Stamp Act of 1934 required waterfowl hunters to buy a federal duck hunting license. Since then, the sale of these permits has raised over $300 million for use in waterfowl research and the purchase of land for waterfowl refuges.

The Soil Conservation Service (now called the Natural Resource Conservation Service) was created in 1935 as part of the Department of Agriculture to correct the enormous erosion problems that had ruined many farms in the Great Plains states (Case Study, p. 357). In 1937, the Federal Aid in Wildlife Restoration Act (also known as the Pittman–Robertson Act) levied a federal tax on all sales of guns and ammunition. States have received more than $2.2 billion to buy land for wildlife conservation (mostly for game species), to support wildlife research, and to reintroduce wildlife in depleted areas. A similar law, the Federal Aid in Fish Restoration Act of 1950, helps state agencies restock and conserve game fish through a tax on fishing equipment.

In 1940, two bureaus in the Departments of Commerce and Agriculture were merged to form the U.S. Fish and Wildlife Service. This new agency was placed in the Department of Interior and given the roles of managing the National Wildlife Refuge System and protecting wild species in danger of extinction.

There were few new developments in federal resource conservation and public health policy during the 1940s and 1950s, mostly because of preoccupation with World War II (1941–45) and economic recovery after the war. However, during these years some highly visible episodes of air pollution (Spotlight, p. 475) and water pollution occurred, providing early warnings of the environmental problems the 1960s and 1970s would bring. By the 1950s most U.S. cities had sewer systems, but the sewage was simply dumped untreated into holding ponds or nearby waters.

During the late 1940s and early 1950s, several writers, including William Voight, Fairfield Osborn, Aldo Leopold, Jane Jacobs, and Vance Packard, warned of the environmental problems that would result from the rapid economic and population following World War II. Few people took these warnings seriously; most considered pollution and population growth as minor irritants that could be overcome by technological innovations and wealth resulting from ever-increasing economic growth.

The years between 1933 and 1960 saw a gradual improvement in public health. Public health boards and agencies were established at the municipal, state, and federal levels. Public education about health issues was emphasized in schools and communities. Vaccination programs were introduced, and improved sanitation and garbage collection led to a sharp reduction in waterborne diseases. In 1938, Congress enacted a modern version of the Food, Drug, and Cosmetic Act after 100 people died from acute kidney failure by ingesting certain tainted lots of the sulfa drug sulfanilamide.

At the same time, rapid industrial growth after World War II began exposing workers and the general public to an increasing array of pesticides and solvents whose health effects were unknown. Industries also created thousands of mostly unregulated dumpsites for toxic and hazardous chemicals (p. 578).

23-2 THE ENVIRONMENTAL MOVEMENT IN THE UNITED STATES: 1960–1998

What Major Environmental Developments Took Place in the 1960s? In 1962, biologist Rachel Carson (1907–64) published *Silent Spring*, which documented the pollution of air, water, and wildlife from pesticides such as DDT (Individuals Matter, p. 565). This influential book helped broaden the concept of resource conservation to include preservation of the *quality* of air, water, soil, and wildlife.

In 1964, Congress passed the Wilderness Act. The act authorized the government to protect undeveloped tracts of public land as part of the National Wilderness System unless Congress later decides they are needed for the national good. Land in this system is to be used only for nondestructive forms of recreation such as hiking and camping.

Between 1965 and 1970, the emerging science of ecology received widespread media attention. At the same time, the popular writings of biologists such as Paul Ehrlich, Barry Commoner, and Garrett Hardin (Guest Essay, p. 266) awakened people to the interlocking relationships among population growth, resource use, and pollution (Figure 1-18).

During that same period, a number of events increased public awareness of pollution. In 1963, high concentrations of pollutants accumulated in the air above New York City, killing about 300 people and injuring thousands. Foam from nonbiodegradable additives in laundry detergents and cleansers began appearing on creeks and rivers in the mid-1960s (Figure 23-6). In 1969, the oil-polluted Cuyahoga River, in Cleveland, Ohio, caught fire and burned for 8 days. Also in 1969, oil leaking from an offshore well near Santa Barbara, California, coated beaches and wildlife. By the late 1960s, Lake Erie had become so severely polluted that millions of fish died and many beaches had to be closed. And during the late 1960s and early 1970s, the North American bald eagle, the grizzly bear,

A: Hydrogen gas produced from water by using solar energy

Figure 23-6 Foam on a creek caused by nonbiodegradable additives in synthetic laundry detergents in 1966. (Soil Conservation Service)

the whooping crane, the peregrine falcon, and other wildlife species were threatened with extinction from pollution and loss of habitat.

What Major Environmental Developments Took Place in the 1970s? Media attention, public concern about environmental problems, scientific research, and action to address these concerns grew rapidly during the 1970s—sometimes called the *first decade of the environment.* Figure 23-7 summarizes some important events during this decade. The first annual Earth Day was held on April 20, 1970. During this event, proposed by Senator Gaylord Nelson, some 20 million people in more than 2,000 communities took to the streets to heighten awareness and demand improvements in environmental quality.

An eye-opening event occurred in 1973 when the Arab members of the Organization of Petroleum Exporting Countries (OPEC) reduced oil exports to the West and banned oil shipments to the United States because of its support for Israel in the 18-day Yom Kippur War with Egypt and Syria. This embargo, lasting until March 1974, sharply raised the price of crude oil (Figure 15-7). The result was double-digit inflation in the United States and many other countries, high interest rates, soaring international debt, and a global economic recession. In 1979, a second reduction in oil supplies and a sharp price increase occurred when Iran's Islamic Revolution shut down most of Iran's oil production.

The Federal Land Policy and Management Act of 1976 gave the Bureau of Land Management (BLM) its first real authority to manage the public land under its control, 85% of which is in 12 western states. This law angered a number of western interests whose use of these lands was being restricted for the first time. Thus, in the late 1970s a coalition of ranchers, miners, loggers, developers, farmers, politicians, and others launched a political campaign known as the *sagebrush rebellion.* Its primary goal was to remove most western lands from federal ownership and turn them over to the states. The plan was to persuade state legislatures to sell or lease the resource-rich lands at low prices to ranching, mining, timber, land development, and other private interests.

When Jimmy Carter was president between 1977 and 1981, he persuaded Congress to create the Department of Energy to develop a long-range energy strategy to reduce the country's heavy dependence on imported oil and to face up to the projected end of the oil era over the next 40–80 years. However, most efforts to have the United States reduce energy waste and rely more on renewable energy sources have been undermined by the temporary oil glut since 1980 and the resultant low price of oil. (When adjusted for inflation, oil and gasoline cost no more in the United States today than they did in 1950, Figures 15-7 and 16-6.)

Carter also appointed a number of competent and experienced administrators, drawn heavily from environmental and conservation organizations, to key posts in the EPA, the Department of the Interior, and the Department of Energy. He consulted with environmental leaders on environmental and resource policy matters. He also helped create a Superfund to clean up abandoned hazardous waste sites, including the Love Canal near Niagara Falls, New York (p. 578). Just before he left office President Carter (somewhat like Teddy Roosevelt) used the Antiquities Act of 1906 to triple the amount of land in the National Wilderness System and double the area in the National Park System (primarily by adding vast tracts in Alaska).

Q: What do most scientists believe are our best energy options?

1970–80	Many environmental laws passed (see Appendix 3)
1970	First Earth Day; EPA established by President Richard Nixon
1972	Oregon passes first beverage-bottle recycling law; *The Limits to Growth* published, leading to global debate on pros and cons of economic growth as usual; U.N. Conference on the Human Environment held in Stockholm, Sweden
1973	OPEC oil embargo
1974	Chemists Sherwood Roland and Mario Molina suggest CFCs may be depleting ozone in stratosphere
1978	Love Canal, N.Y., housing development evacuated because of leaks of toxic wastes from old dumpsite
1979	Oil supplies decreased because of Iran's Islamic Revolution; Three Mile Island nuclear power plant accident

Figure 23-7 Some important environmental events during the 1970s, sometimes called the environmental decade.

What Major Environmental Developments Took Place in the 1980s? In 1981, Ronald Reagan, a self-declared sagebrush rebel and advocate of less federal control, became president. During his 8 years in office, he appointed to key federal positions people who opposed existing environmental, resource conservation, and land use legislation and policies—an action environmentalists likened to putting foxes in charge of the chicken house.

Reagan lived up to his sagebrush rebel philosophy by greatly increasing private energy and mineral development and timber cutting on public lands, allowing these resources to be sold at near giveaway prices. He cut federal funding for energy conservation research and development by 70% and funding for renewable resources by 85%, and he eliminated tax incentives for residential solar energy and for energy conservation enacted during the Carter administration. He also lowered automobile gas mileage standards and relaxed federal air- and water-quality pollution standards.

Although Reagan was an immensely popular president, many people strongly opposed his environmental and resource policies. These policies led to strong opposition in Congress, public outrage, and legal challenges by environmental and conservation organizations, whose memberships soared during this period.

Upon his election in 1989, George Bush promised to be "the environmental president." However, he received criticism from environmentalists for failing to provide leadership on such key environmental issues as population growth, global warming, and loss of biodiversity. President Bush also continued support of exploitation of valuable resources on public lands at giveaway prices, and allowing some environmental laws to be undercut by the powerful influence of industry, mining, ranching, and real estate development officials. Figure 23-8 lists some other significant environmental events in the 1980s and 1990s.

1986	Chernobyl nuclear power plant in Ukraine explodes; Times Beach, Missouri, evacuated and bought up by EPA because of dioxin contamination
1987	Montreal Protocol to halve emissions of ozone-depleting CFCs signed by 24 countries
1988	Industry-backed Wise Use movement established to weaken and destroy U.S. environmental movement
1989	*Exxon Valdez* oil tanker accident in Alaska's Prince William Sound
1990	Twentieth annual Earth Day observed by 200 million people in 141 nations
1991	Persian Gulf War to protect oil supplies in Middle East; 39 nations agree to 50-year moratorium on mining in Antarctica; National People of Color Environmental Summit to promote environmental justice
1992	U.N. Earth Summit in Rio de Janeiro, Brazil
1994	U.N. Conference on Population and Development held in Cairo, Egypt
1995–98	Efforts in U.S. Congress to seriously weaken or do away with major U.S. environmental laws and sharply reduce spending on environmental protection—most of these bills were vetoed by President Clinton
1997	Meeting of 161 nations in Kyoto, Japan, to negotiate a new treaty to help slow global warming
1997	Evaluation of results of 1992 Earth Summit meeting showed little progress in meeting goals

Figure 23-8 Some important environmental events during the 1980s and 1990s.

What Major Environmental Developments Took Place in the 1990s? In 1993, Bill Clinton became president and promised to provide national and global environmental leadership. In 1998, after over 6 years in office, the new administration's environmental report card was mixed. Overall most environmentalists gave Clinton a B for **(1)** selecting Al Gore (considered to have a comprehensive understanding of environmental problems, but not as active as expected) for vice president, **(2)** appointing respected environmentalists to key positions in environmental and resource agencies, **(3)** consulting with environmentalists about environmental policy, and **(4)** vetoing many of the bills passed during 1995 through 1998 by a Republican-dominated Congress that were designed to weaken or overturn many of the key environmental laws passed in the 1970s and 1980s (see Appendix 3).

A: Energy-efficient renewable energy (especially solar-produced hydrogen gas)

Environmentalists gave Clinton a B for proposing some new environmental initiatives, but they gave him a C– for failing to put his political weight behind many of his environmental proposals and being too willing to compromise or even retreat when his proposed environmental policies come under attack.

The June 1992 *Rio Earth Summit*, held in Rio de Janeiro, Brazil (Section 28-7), opened the decade with international attention on major environmental problems concerning pollution, deforestation, biodiversity loss, and global change. However, by 1997 little had been done to fulfill the promises made at the Rio meeting.

The December 1997 *Kyoto Climate Change Summit*, held in Kyoto, Japan, again brought international focus to environmental issues, bringing together representatives of more than 160 nations to sign a historic protocol aimed at decreasing global emissions of greenhouse gases. Most environmentalists view the Kyoto protocol as a step in the right direction, but see a long road ahead before reductions in emissions of greenhouse gases are actually realized (p. 517).

For the most part the 1990s have been disappointing to environmentalists. They have had to spend much of their time and funds fighting efforts to discredit the environmental movement (Section 28-4 and Case Study, p. 787) and to weaken or eliminate most environmental laws passed during the 1970s (Appendix 3) and countering claims that major environmental problems such as global warming (Section 19-2) and ozone depletion (Pro/Con, p. 523) are either hoaxes or not very serious.

23-3 BIODIVERSITY, CONSERVATION BIOLOGY, AND ECOLOGICAL INTEGRITY

Why Is Biodiversity Loss Considered the Key Environmental Problem? Every country and region has three forms of wealth: material, cultural, and biological (biodiversity). The forests, rangelands, parks, wilderness, and aquatic systems, where the genes, species, and ecosystems making up biodiversity are found, are crucial sources of biological wealth that are coming under increasing pressure from population growth and economic development.

Since 1980 there has been considerable growth in **(1)** the scientific understanding of biological wealth or biodiversity that is so vital to all life and economies, **(2)** the ecological processes such as matter cycling, energy flow, and species interactions that sustain biodiversity, and **(3)** the biological consequences of biodiversity loss through environmental degradation.

As a result, most ecologists and other environmental scientists consider the loss of biodiversity the key environmental problem. This revolution in awareness has led to a much deeper understanding and appreciation of *interrelationships or connections in nature*—a major theme of this book.

What Are Conservation Biology and Ecological Integrity? The science-based study of biodiversity has led to the development of **conservation biology**. It is a multidisciplinary science created in the late 1970s to deal with the crisis of maintaining the genes, species, communities, and ecosystems that make up the earth's *biological diversity*. Its goals are to investigate the human impacts on biodiversity and to develop practical approaches to preserving biodiversity.

Conservation biology uses scientific data and concepts to find practical ways to protect critical ecosystems and biodiversity-rich areas and prevent the premature extinctions of species. Conservation biology differs from *wildlife management*, which is devoted primarily to the manipulation of the population sizes of various animal species, especially game species prized by hunters and fishers (Section 25-5).

Since the mid-1980s conservation biologists and ecologists have increasingly emphasized the preservation of **ecological integrity**, the conditions and natural processes (such as the energy flow and matter cycling in ecosystems and species interactions) that generate and maintain biodiversity and allow evolutionary change (Section 6-3) as a key mechanism for adapting to changes in environmental conditions. The **ecological health** of an area can be described in terms of the degree to which its biodiversity and ecological integrity remain intact.

Conservation biology rests on the following principles: **(1)** Biodiversity and ecological integrity are necessary to all life on earth and should not be reduced by human actions, **(2)** humans should not cause or hasten the premature extinction of populations and species by disrupting evolutionary processes and critical ecological processes, **(3)** the best way to preserve biodiversity and ecological integrity is to preserve habitats, niches, and ecological interactions, and **(4)** goals and strategies for preserving biodiversity and ecological integrity of an area must be based on a deep understanding of the ecological properties and processes of that system.

This *scientific approach* recognizes that saving wildlife means saving their habitats and not disrupting the complex interactions among species in an ecosystem (ecological integrity). It is also based on Aldo Leopold's ethical principle that something is right when it tends to maintain the earth's life-support systems for us and other species, and wrong when it doesn't.

Q: What natural hazards causes the most deaths per year?

23-4 PUBLIC LANDS IN THE UNITED STATES

What Are the Major Types of U.S. Public Lands? No nation has set aside as much of its land—about 42%—for public use, enjoyment, and wildlife as has the United States. Almost one-third of the country's land belongs to every American and is managed for them by the federal government; 73% of this public land is in Alaska and another 22% is in the western states (where 60% of all land is public land). These public lands are classified as multiple-use lands, moderately restricted-use lands, and restricted-use lands.

Multiple-Use Lands

- The 156 forests (Figure 23-9) and 20 grasslands of the *National Forest System* are managed by the U.S. Forest Service, an agency of the U.S. Department of Agriculture. Except for wilderness areas (15%), this land is supposed to be managed using two principles: **(1)** the **principle of sustainable yield**, which states that potentially renewable resources (such as trees) should not be harvested or used faster than they are replenished; and **(2)** the **principle of multiple use**, which states that the same land should be managed simultaneously for a variety of uses, such as sustainable timber harvesting, grazing recreation, and wildlife conservation.

  Today, national forests are used for logging (the dominant use in most cases), mining, livestock grazing, farming, oil and gas extraction, recreation, sport hunting, sport and commercial fishing, and conservation of watershed, soil, and wildlife resources. Off-road vehicles are usually restricted to designated routes.
- *National resource lands* in the western states and Alaska are managed by the Bureau of Land Management under the principle of multiple use. Emphasis is on providing a secure domestic supply of energy and strategic minerals and on preserving rangelands for livestock grazing under a permit system.

Moderately Restricted-Use Lands

- The 508 *national wildlife refuges* (Figure 23-9) are managed by the U.S. Fish and Wildlife Service. About 24% of this land is designated as wilderness. Most refuges protect habitats and breeding areas for waterfowl and big game to provide a harvestable supply for hunters; a few protect endangered species from extinction (Figure 23-4). Sport hunting, trapping, sport and commercial fishing, oil and gas development, mining, logging, grazing, some military activities, and farming are permitted as long as the Department of the Interior finds such uses compatible with the purposes of each unit.

Restricted-Use Lands

- The 375 units of the *National Park System* include 54 major parks (mostly in the West) and 321 national recreation areas, monuments, memorials, battlefields, historic sites, parkways, trails, rivers, seashores, and lakeshores (Figure 23-9) managed by the National Park Service. Its goals are to preserve scenic and unique natural landscapes, preserve and interpret the country's historic and cultural heritage, protect wildlife habitats and wilderness areas, and provide certain types of recreation.

  National parks may be used only for camping, hiking, sport fishing, and boating. Motor vehicles are permitted only on roads. In national recreation areas, these same activities, plus sport hunting, mining, and oil and gas drilling, are allowed. About 49% of the National Park System land is designated as wilderness.
- The 630 roadless areas of the *National Wilderness Preservation System*, which lie within the national parks, national wildlife refuges, and national forests, are managed by the National Park Service (42%), Forest Service (33%), Fish and Wildlife Service (20%), and Bureau of Land Management (5%). These areas are open only for recreational activities such as hiking, sport fishing, camping, nonmotorized boating, and, in some areas, sport hunting and horseback riding. Roads, logging, livestock grazing, mining, commercial activities, and buildings are banned, except when they predate the wilderness designation.

Between 1970 and 1994 the area within all public land systems (excluding national forests) increased significantly (2.8-fold in the National Park System, 3-fold in the National Wildlife Refuge System, and 9-fold in the National Wilderness Preservation System). Most of the additions, made by President Carter just before he left office in 1980, are in Alaska. Since then little land has been added. An exception was President Clinton's executive action in 1996 to add a new national monument area in Utah that is almost as large as Yellowstone National Park. Because of the resources they contain, there has been intense controversy over how public lands should be used and managed (Spotlight, p. 621).

Case Study: The Takings/Property Rights Controversy Since 1991 the Wise-Use movement (formed in 1988, Case Study, p. 787) has helped spearhead a legal tactic called the *takings/property rights movement* to paralyze government regulation of public lands. One approach, called the County Movement, has involved lobbying state legislatures to pass laws that allow county zoning, land-use plans, and environmental ordinances to take precedence over any

A: Floods (40% of such deaths), typhoons and hurricanes (36%), and earthquakes (13%)

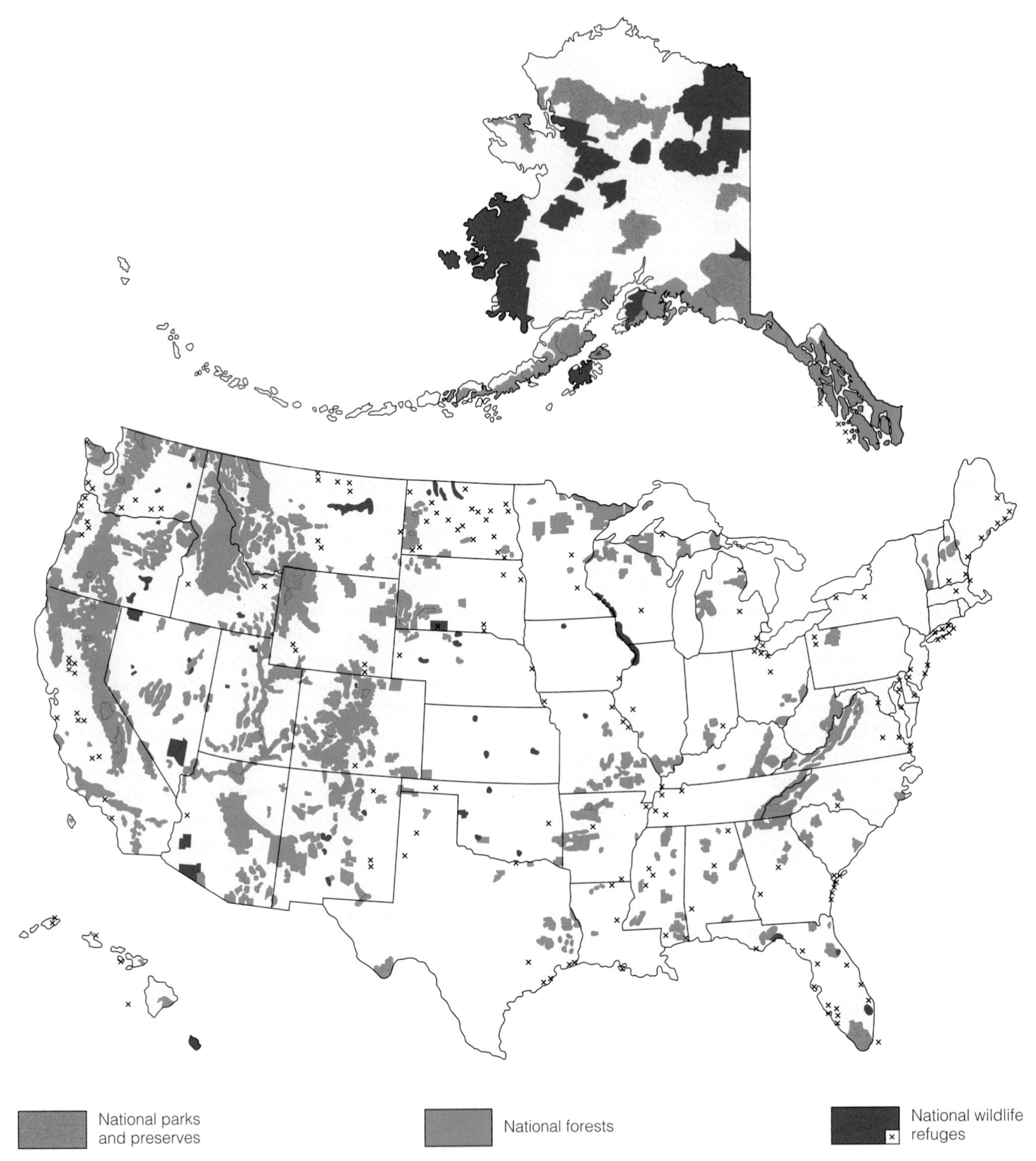

Figure 23-9 National forests, national parks, and wildlife refuges managed by the U.S. federal government. (Data from U.S. Geological Survey)

federal laws that affect private property rights on public land within county borders.

If successful, such legislation at the state level would severely limit the ability of the federal government to enforce grazing, logging, mining, wildlife protection, zoning, environmental, and health laws on public lands. Government regulatory agencies would be tied up in lengthy court cases, required to make complex evaluations of possible financial losses from restrictions on the use of public land, and forced to

SPOTLIGHT

How Should U. S. Public Lands Be Managed?

Economists, developers, and resource extractors tend to view areas of the earth's surface in terms of their usefulness in providing mineral and other resources, their potential for development, and their ability to increase short-term economic growth.

Ecologists and conservation biologists take a scientific approach and view the earth's ecosystems in terms of their usefulness in maintaining the planet's biodiversity and life-sustaining processes. To conservation biologists and ecologists, destroying or degrading the earth's remaining intact ecosystems for short-term economic gain will make them less useful to humans and other forms of life.

In recent years, the total cost of subsidies (in public funds spent and taxes and user fees not collected) given to mining (p. 338), logging, and grazing interests using public lands has exceeded $1 billion per year. In addition to losing potential revenue for resource extraction on lands they own, taxpayers must also pay for dealing with most of the resulting environmental damage.

Most environmentalists and many free-market economists believe that the following four principles should govern use of public land:

- Protection of biodiversity and the ecological integrity of vital public lands should be the primary goal.
- No one should be given subsidies or tax breaks for using or extracting resources on public lands.
- The American people deserve fair compensation for the use of their property.
- All users or extractors of resources on public lands should be fully responsible for any environmental damage they cause.

According to U.S. Representative George Miller, "It is time for the resource industries to grow up and get off the federal bottle."

Since 1995, however, there have been increasing attempts by some members of the U.S. Congress (influenced by ranching, timber, mining, and development interests) to pass laws that would **(1)** sell public lands or their resources to corporations or individuals, usually at less than market value; **(2)** give public lands to the states (where these interests often have more influence); **(3)** slash federal funding for administration of public lands so that unlawful exploiters have less chance of getting caught; **(4)** weaken, waive, or eliminate federal laws that regulate use of public lands; **(5)** require the government to reimburse private landowners for regulatory takings; **(6)** allow highways to be built through national parks and wilderness areas; **(7)** redefine protected wetlands so that about half of them would no longer be protected; and **(8)** repeal or seriously weaken the Endangered Species Act (p. 705).

Between 1995 and 1998, most of these laws were not passed or were vetoed by President Clinton. However, each year there are new attempts in Congress to pass such laws.

Critical Thinking

Explain why you agree or disagree with the four principles listed.

compensate parties for any losses incurred. This would leave the agencies little money for enforcing environmental and public land-use laws passed by Congress.

Opponents argue that the County Movement is a thinly disguised effort by mining, logging, energy, ranching, and development interests to sharply reduce or eliminate regulations governing their use of resources on public land. If regulations are imposed, then these resource extractors demand that the federal government pay them for not carrying out potentially harmful activities on land owned by all the American people. Meanwhile, the industry-supported takings/property rights movement has carried its crusade to Congress, attempting to attach takings/property rights amendments to every important piece of environmental and other legislation.

A different controversy in the takings/property rights issue focuses on government action that affects *private* lands. It involves the issue of whether federal and state governments must compensate property owners whenever environmental, health, safety, or zoning laws limit how the owners can use their property or decrease its financial value—something called *regulatory taking*.

The Fifth Amendment of the U.S. Constitution gives the government the power, known as *eminent domain*, to force a citizen to sell property needed for a public good. If your land is needed for a road, for example, the government can take your land, but must pay its fair market value in exchange for your loss of its use.

The unanswered legal question is whether the Constitution requires the government to compensate you if instead of taking your property (a *physical taking*) it reduces its value by not allowing you to do certain things with it (a *regulatory taking*). For example, you might not be allowed to build on some or all of your property because it is a wetland protected by law, or you may not be able to harvest trees on all or part of your land because it is a habitat for an endangered species.

The problem of regulatory takings is a complicated and highly controversial private property issue. On one hand, most people justifiably feel that they should be compensated for losses in property values as a result of government regulation. The problem is that requiring compensation for regulatory takings could cripple the financial ability of state and federal government to protect the public good in enforcing environmental, land use, health, and safety laws by imposing massive costs.

This would undermine the long-standing rule of law in the United States that landowners must not use their land in any way that creates a public or private nuisance (in other words, harms neighbors or the public). Most government land-use, environmental, health, zoning, and safety regulations are designed to protect the community from harmful actions by individual property owners.

Environmentalists warn that requiring compensation for regulatory takings will do little to protect the private property rights and values of ordinary landowners and citizens. Indeed, such takings laws could lower their property values because of increased pollution and environmental degradation. Instead, they contend that government compensation for regulatory takings will mostly benefit the 5% of the nation's largest private landowners (mostly timber and mining companies, agribusinesses, energy companies, and developers) who own almost three-quarters of all privately owned land in the United States. Instead of paying, federal regulatory agencies could take the cheaper and more likely route of not enforcing the regulations and not enacting new regulations.

Environmentalists argue that this law requires taxpayers to pay people not to do something they shouldn't (or don't really plan to) be doing in the first place: *Pay me and I won't pollute, develop this land, build an incinerator or landfill here, or fill in this wetland.* Indeed, some people may go into the no-pollution, no-development business by buying up land at a low price and then make a profit by getting taxpayers to pay them not to develop it.

Environmentalists recognize the need to respect private property rights. Indeed, the desire to protect health and property values is the driving force behind the opposition of grassroots environmental groups to incinerators, landfills, toxic waste dumps, and some forms of development in their communities. However, environmentalists don't believe that the solution is to replace the *polluter- or degrader-pays* principle found in most environmental laws with a *taxpayer-pays* approach.

Some government officials and environmentalists are looking for solutions to this difficult and highly controversial problem without having to spend large sums of increasingly scarce public funds or being forced not to enforce most environmental, safety, and health regulations. One possibility is to give tax deductions to landowners who agree to allow certain types of regulatory takings on their land. Another possibility is to eliminate or modify environmental regulations that go too far or that are unjustly enforced.

How Have Utilitarian Conservation and Preservationist Philosophies Led to Different Land Management Practices? Management of multiple-use public lands, such as National Forests and National Resource Lands, has its roots in the philosophy of *utilitarian conservation* or *wise use* (Spotlight, p. 613). Although sustainability is specifically a part of the utilitarian concept, a major emphasis is put on the economic benefits of logging, grazing, and mining. Typically problems of land management result because of economic and political pressures to extract resources in manners that are not sustainable and that degrade public lands.

By contrast, management of restricted-use lands, such as national parks and wilderness areas, has its roots in the *preservationist* philosophy. Even with policies that don't permit major resource extraction, these lands are being degraded because of heavy use, disruption of ecological processes, and threats originating from degradation of surrounding lands and air pollution.

The next section examines the management of rangelands, most of which are in public lands managed by the Bureau of Land Management and the U.S. Forest Service, and which have been degraded by overgrazing. Sections 23-6 and 23-7 examine management issues in National Parks and wilderness areas, respectively. Chapter 24 gives an in-depth look at forests and forest management.

23-5 MANAGING AND SUSTAINING RANGELANDS

What Is Rangeland, and Why Is It Important? Almost half of the earth's ice-free land is **rangeland**: land that supplies forage or vegetation for grazing (grass-eating) and browsing (shrub-eating) animals and that is not intensively managed. Most rangelands are grasslands in arid and semiarid areas too dry for nonirrigated crops. Such lands can support sparse populations of domesticated grazing animals on open ranges (unfenced, natural grazing lands) or through nomadic herding, in which animals are moved from one area to another. Livestock also graze in *pastures*, consisting of domes-

Q: Worldwide, how many people die prematurely each year of causes related to smoking?

tic meadows or managed grazing lands, and in some open woodlands.

About 42% of the world's rangeland is used for grazing livestock; much of the rest is too dry, cold, or remote from population centers to be grazed by large numbers of livestock. About 34% of the total U.S. land area is rangeland, most of it short-grass prairie in the arid and semiarid western half of the country (Figure 7-11). About 2% of the cattle and 10% of the sheep in the United States graze on public rangelands.

Most rangeland grasses have deep and complex root systems (Figure 5-16) that help anchor the plants. Leaves of grass grow from the base, not the tip. Thus, as long as only its upper portion is eaten and its lower portion remains, rangeland grass is a renewable resource that can be grazed again and again. Range plants vary in their ability to recover, however, and grazing pressure changes the balance of plant species in grassland communities (Figure 23-10).

Rangeland has a number of important ecological functions. It provides forage for large numbers of wild herbivores and essential habitats for a variety of wild plant and animal species. Indeed, an estimated 84% of the wild mammal species and 74% of the wild bird species in the United States are supported by rangeland ecosystems. Rangelands also act as crucial watersheds that help replenish surface and groundwater resources by absorbing and slowly releasing rainfall.

In addition to vast numbers of wild herbivores, the world has about 10 billion domesticated animals. About 3 billion of these are *ruminants* that can digest the cellulose in grasses and convert it to meat and milk. Most domesticated ruminants are cattle (1.3 billion worldwide and 100 million in the United States), sheep, and goats. Three-fourths of these animals forage on rangeland vegetation before being slaughtered for meat; the rest are fattened on grain in feedlots before slaughter. Seven million *nonruminants* (mostly pigs, chickens, and other livestock) cannot feed on rangeland vegetation and eat mostly cereal grains grown on cropland.

Rangeland is also a valuable resource for recreation. Since 1975, use of U.S. rangeland for activities such as hiking, camping, and hunting has risen dramatically. Some have suggested that some rangelands could be used to raise wild grazing animals for meat instead of conventional livestock (Solutions, below).

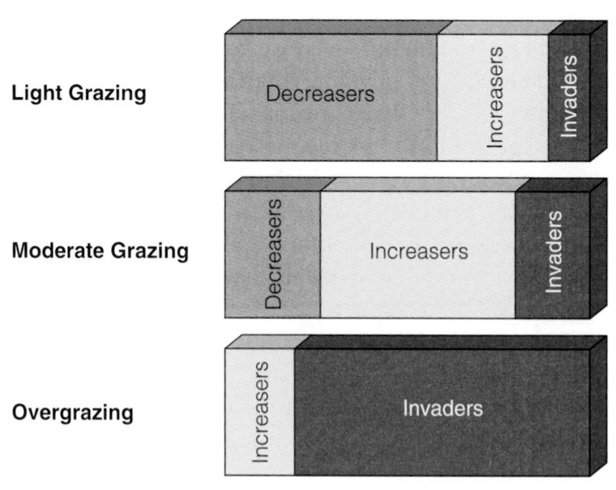

Figure 23-10 Effects of various degrees of grazing on the relative amounts of three major types of grassland plants. *Decreasers* are grass species that decline in abundance with moderate grazing; *increasers* are those that increase with moderate to heavy grazing pressure. *Invaders* are plants that colonize an area because of overgrazing or other changes in rangeland conditions.

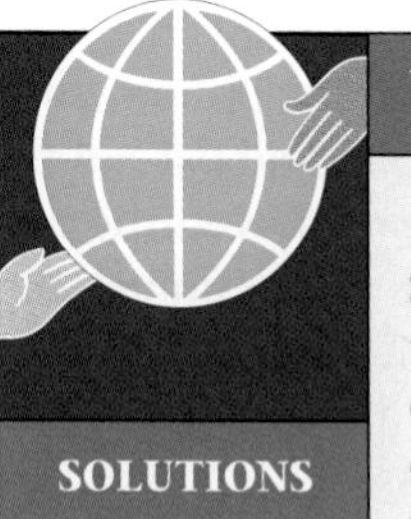

Wild Game Ranching

SOLUTIONS

Some ecologists have suggested that wild herbivores such as eland, oryx, and Grant's gazelle could be raised on ranches on tropical savanna (Figure 7-18). Because many wild herbivores have a more diversified diet than cattle, they can make more efficient use of available vegetation and thus reduce the potential for overgrazing.

In addition, they need less water than cattle and are more resistant than cattle to animal diseases native to savanna grasslands. These animals also achieve much of their own predator control because of their long evolutionary experience with lions, leopards, cheetahs, and wild dogs.

Since 1978 David Hopcraft has carried out a successful game-ranching experiment on the Athi Plains near Nairobi, Kenya. The ranch is stocked with various native grazers and browsers, including antelope, zebras, giraffes, and ostriches.

Cattle once raised as a comparison group are being phased out and may be replaced with Cape buffalo. The yield of meat from the native herbivores has been rising steadily, the condition of the range has improved, and costs are much lower than those for raising cattle in the same region.

Critical Thinking

Do you believe that wild game ranching on public and private rangeland would be useful in the western United States? Explain. If so, what animals might be raised successfully? What policies might support such wild game ranching?

A: 3 million per year (8,200 per day), including 419,000 Americans per year (1,150 per day)

Figure 23-11 Rangeland: overgrazed (left) and lightly grazed (right). (Soil Conservation Service)

What Is Overgrazing? Overgrazing occurs when too many animals graze for too long and exceed the carrying capacity of a grassland area. It lowers the productivity of vegetation and changes the number and types of plants in an area (Figure 23-10). Usually the first symptom of overgrazing is a sharp decline in the most palatable herbs and grasses. Large populations of wild ruminants can overgraze rangeland during prolonged dry periods, but most overgrazing is caused by excessive numbers of domestic livestock feeding for too long in a particular area.

Heavy overgrazing compacts the soil, which diminishes its capacity to hold water and to regenerate itself. Overgrazing also converts continuous grass cover into patches of grass and thus exposes the soil to erosion, especially by wind; then woody shrubs such as mesquite and prickly pear cactus take over. Overgrazing is the major cause of desertification in arid and semiarid lands (Figure 14-19).

Figure 23-11 shows lightly grazed and overgrazed grassland side by side. A UN assessment of earth's dryland regions showed that livestock production worth $23.2 billion ($15 billion of this in Africa and Asia) was lost in 1990 as a result of rangeland degradation.

What Is the Condition of the World's Rangelands? *Range condition* is usually classified as excellent (containing more than 75% of its potential forage production), good (51–75%), fair (26–50%), and poor (0–25%). Limited data from surveys in various countries indicate that most of the world's rangelands have been degraded to some degree, mostly by desertification (Figure 14-19).

In 1990, 68% of nonarctic U.S. public rangeland was rated by the Bureau of Land Management and the General Accounting Office as being in unsatisfactory (fair or poor) condition, compared with 84% in 1936. Since 1936 U.S. public range area in excellent or good condition has doubled while the area rated as poor has shrunk by half. This is a considerable improvement, but there's still a long way to go.

There is controversy, however, over estimates of rangeland condition. According to the National Academy of Sciences, past measurements of range quality in the United States used too many different methods and included too little data. Such uncertainty has allowed ranchers to claim that public rangelands are in the best condition they have been during this century and that their health continues to improve.

On the other hand, environmentalists claim that although the condition of public rangelands has improved, they are in much worse shape from overgrazing than they should be. They contend that overgrazing on many federal lands by livestock and by herds of feral burros and horses (a *feral animal* is a domestic animal that has adopted a wild existence) has allowed populations of unpalatable or inedible plants (such as mesquite, sage, cheatgrass, and cactus) to invade large areas of private and public rangeland in the West. Livestock have also caused severe damage to vital strips of vegetation along streams (called *riparian areas*). Neither critics nor advocates of grazing on public lands can produce unambiguous data to support their positions.

What Are Riparian Zones, and Why Are They Important? According to some conservation biologists and rangeland experts, overall estimates of rangeland condition obscure severe damage to certain heavily grazed areas, especially vital **riparian zones**—thin strips of lush vegetation along streams (Figure 23-12, right).

These zones help prevent floods (and help keep streams from drying out during droughts) by storing and releasing water slowly from spring runoff and summer storms. These centers of biodiversity also provide habitats, food, water, and shade for wildlife in the arid and semiarid western lands. Studies indicate that 65–75% of the wildlife in the western United States is totally dependent on riparian habitats.

Because cattle need lots of water, they congregate near riparian zones and feed there until the grass and shrubs are gone (Figure 23-12, left). As a result, riparian vegetation is destroyed by trampling and overgrazing. It takes only a few head of cattle to degrade a riparian zone.

The denuded banks erode, making the streams wider and shallower and muddying the water with

Q: What is the projected annual global death toll from smoking by 2020?

Figure 23-12 Cattle on a riparian zone of a public rangeland along Arizona's San Pedro River (left) in the mid-1980s just before this section of waterway was protected by banning domestic livestock grazing for 15 years, eliminating sand and shovel operations and water pumping rights in nearby areas, and limiting access by off-highway vehicles. The photo on the right shows the recovery of this riparian area at the same time of year after 10 years of protection. (Bureau of Land Management)

sediment. The depletion of dissolved oxygen when the sun warms the shallower water threatens cold-water fish such as wild trout. Runoff rushes over the compacted and denuded riparian land, adding to floods and speeding the drying up of streams during drought.

A 1988 General Accounting Office report concluded that "poorly managed livestock grazing is the major cause of degraded riparian habitat on federal rangelands." Arizona and New Mexico have already lost 90% of their riparian areas, primarily to grazing.

Riparian areas can be restored by using fencing to restrict access to degraded areas and by developing off-stream watering sites for livestock. Sometimes protected areas can recover in a few years (Figure 23-12, right). Despite the threat to soil and wildlife, little has been done to protect or repair riparian zones on U.S. public land, mostly because of political opposition by ranchers and lack of awareness of the problem by the general public.

How Can Rangelands Be Managed? The primary goal of rangeland management is to maximize livestock productivity without overgrazing rangeland vegetation. The most widely used method to prevent overgrazing is controlling the *stocking rate*—the number of each kind of animal placed in a given area—so that it doesn't exceed an area's carrying capacity. But determining the carrying capacity of a range site is difficult and costly, and carrying capacity varies with factors such as climatic conditions (especially drought), past grazing use, soil type, invasions by new species, kinds of grazing animals, and intensity of grazing.

Both the numbers and distribution of livestock on a rangeland must be controlled to prevent overgrazing. Ranchers can control livestock distribution by fencing off damaged rangeland and riparian zones, rotating livestock from one grazing area to another, providing supplemental feed at selected sites, and situating water holes and salt blocks in strategic places.

Continuous grazing occurs throughout the year (or appropriate season). This method, used in many range areas with favorable climatic conditions (especially in the southwestern United States) is popular because it's easy to manage and it reduces costs by requiring little livestock handling and fencing. But it has several disadvantages. Left to their own devices, livestock tend to overgraze flat areas and areas near water supplies, especially vulnerable riparian areas (Figure 23-12, left). They also can overgraze desirable species of forage grass, facilitating invasion by less desirable vegetation.

Another common grazing system is *deferred-rotation grazing*, which involves moving livestock between two or more range areas, to allow perennial grasses to recover from the effects of grazing. This method requires more intense management than continuous grazing, but it is usually a better way to protect and improve range quality.

A: 10 million (an average of almost 33,000 preventable deaths per day)

Figure 23-13 For decades, ranchers in the western United States have waged war on the coyote because they kill some livestock. Each year, almost 100,000 coyotes are poisoned, trapped, or shot in the western United States in a government-run predator control program paid for by U.S. taxpayers. (Ken W. Davis/Tom Stack & Associates)

A more expensive and less widely used method of rangeland management involves suppressing the growth of unwanted plants (mostly invaders, Figure 23-10) by herbicide spraying, mechanical removal, or controlled burning. A cheaper and less harmful way to discourage unwanted vegetation is controlled, short-term trampling by large numbers of livestock.

Seeding and applying fertilizer can increase growth of desirable vegetation. Even though this method usually costs too much, reseeding is an excellent way to restore severely degraded rangeland.

Case Study: Livestock and Coyotes Another aspect of rangeland management for livestock is *predator control*. For decades, hundreds of thousands of predators have been shot, trapped, and poisoned by U.S. ranchers, farmers, hunters, and federal officials in a little-known agency called Animal Damage Control (ADC).

Federally subsidized predator control programs have so reduced gray wolf (Case Study, p. 630) and grizzly bear populations that they are now endangered species. Today the U.S. Fish and Wildlife Service spends additional funds to protect and help revive them.

Coyotes (*Canis latrans*), the main target of livestock predator control programs, are smaller relatives of wolves. The effect of coyotes on livestock, especially sheep, is a controversial issue in the western United States. Sheep ranchers claim that coyotes kill large numbers of sheep on the open range and should be exterminated (Figure 23-13). However, many wildlife experts maintain that although coyotes kill some sheep, their net effect is to increase rangeland vegetation for livestock and wild herbivores by reducing grazing pressure from rodents.

Because coyotes are so prolific, adaptable, and efficient, environmentalists contend that any attempt to control them is doomed to failure. They also argue that it would cost taxpayers much less to pay ranchers for each head of livestock killed by a coyote than to spend millions of dollars each year for predator control.

Environmentalists suggest using a combination of fences, repellents, and trained guard dogs to keep predators away. Such methods have given sheep producers in Kansas one of the country's lowest rates of livestock losses to coyotes.

In 1986 USDA researchers reported that predation can be sharply reduced by penning young lambs and cattle together for 30 days and then allowing them to graze together on the same range. When predators attack, cattle butt and kick them, protecting both themselves and the sheep. Llamas and donkeys are also tough fighters against predators and act in the same way to protect sheep. Livestock raisers have little incentive to use such methods, however, when a local ADC or state predator control official is only a phone call away and takes care of any predator problems at little or no charge.

Case Study: Livestock and U.S. Public Rangeland About 29,000 U.S. ranchers hold what are essentially lifetime permits to graze about 4 million livestock (3 million of them cattle) on Bureau of Land Management (BLM) and Forest Service rangelands in 16 western states. About 10% of these permits are held by small livestock operators; the other 90% belong to ranchers with large livestock operations, including four U.S. billionaires, and to corporations such as Metropolitan Life Insurance, Union Oil, Getty Oil, and the Vail Ski Corporation.

Q: How many Americans die because of exposure to other people's smoke (passive smoke)?

Permit holders pay the federal government a grazing fee for this privilege. Since 1981, grazing fees on public rangeland have been set by Congress at only one-fourth to one-eighth the going rate for comparable private land. This means that taxpayers give the roughly 1% of U.S. ranchers with federal grazing permits subsidies amounting to $60–200 million a year, depending on how the costs of federal rangeland management are calculated.

The economic value of a permit is a part of the overall worth of a ranch. It can serve as collateral for a loan and is usually renewed automatically every 10 years, allowing permit holders to treat federal land as part of their ranches. It's not surprising that politically influential permit holders have fought so hard to block any change in this system.

The public subsidy does not end with low grazing fees, however. When other costs such as water pipelines, stock ponds, weed control, livestock predator control, clearing of undesirable vegetation, planting of grass for livestock, erosion, and loss of biodiversity are added, it's estimated that taxpayers are providing 29,000 ranchers with an annual subsidy of about $2 billion (an average of $69,000 per rancher) to produce only 2% of the country's beef. In 1991 one rancher alone received subsidies worth almost $950,000, which explains why many critics charge that the public lands grazing programs are little more than rancher welfare, mostly for the well-to-do ranchers who hold 90% of the permits.

However, ranchers with permits say that grazing fees on public lands should be low because most private rangeland is more productive than public rangeland. Environmentalists contend that this varies with the land involved and that even when this is true, current grazing fees on public rangelands are much too low.

Economic studies show that federal expenditures on grazing programs exceed not only grazing fee receipts but also the profits ranchers make from grazing their livestock on public lands. Thus, some analysts suggest that the government could save money by paying ranchers with permits not to graze their livestock on public lands.

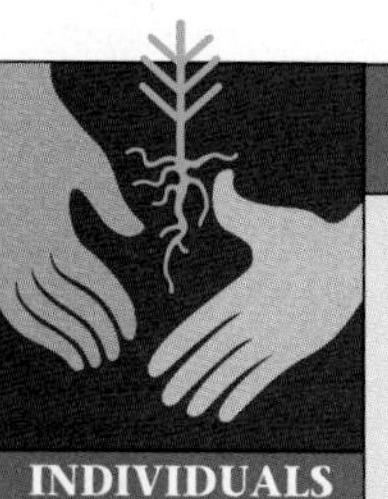

The Eco-Rancher

Wyoming rancher Jack Turnell is one of a new breed of cowpuncher who gets along with environmentalists. He talks about riparian ecology and biodiversity as fluently as he talks about cattle: "I guess I have learned how to bridge the gap between the environmentalists, the bureaucracies, and the ranching industry."

Turnell grazes cattle on his 32,000-hectare (80,000-acre) ranch south of Cody, Wyoming, and on 16,000 hectares (40,000 acres) of Forest Service land on which he has grazing rights. For the first decade after he took over the ranch, he raised cows the conventional way. Since then, he's made some changes.

Turnell disagrees with the proposals by environmentalists to raise grazing fees and remove sheep and cattle from public rangeland. He believes that if ranchers are kicked off the public range, ranches like his will be sold to developers and chopped up into vacation sites, irreversibly destroying the range for wildlife and livestock alike.

At the same time, he believes that ranches can be operated in more ecologically sustainable ways. To demonstrate this, Turnell began systematically rotating his cows away from the riparian areas, gave up most uses of fertilizers and pesticides, and crossed his Hereford and Angus cows with a French breed that tends to congregate less around water. Most of his ranching decisions are made in consultation with range and wildlife scientists, and changes in range condition are carefully monitored with photographs.

The results have been impressive. Riparian areas on the ranch and Forest Service land are lined with willows and other plant life, providing lush habitat for an expanding population of wildlife, including pronghorn antelope, deer, moose, elk, bear, and mountain lions. And this eco-rancher now makes more money because the higher-quality grass puts more meat on his cattle.

Solutions: How Can U.S. Public Rangeland Be Managed More Sustainably? Some environmentalists believe that all commercial grazing of livestock on western public lands should be phased out over the next 10–15 years. They contend that the water-poor western range (Figure 13-8) is not a very good place to raise cattle and sheep, which require a lot of water.

However, some ranchers have demonstrated that western rangeland can be grazed sustainably; they make the case that such practices should not be prohibited on public rangeland (Individuals Matter, above). Some environmentalists agree that with proper management, ranching on western rangeland is a potentially sustainable operation and that encouraging sustainable ranching practices keeps the

A: At least 3,000 and as many as 60,000 per year

Figure 23-14 Majestic scenery (in this case, in Yosemite National Park, California) is a prime attraction of the parks in the U.S. National Park System. These parks are also important repositories of biodiversity. (M. P. L. Fogden/Bruce Coleman Ltd.)

land from being developed and destroyed. They believe that the following measures can promote sustainable use of public rangeland:

- *Allow no or only limited grazing on riparian areas.*
- *Ban grazing on rangeland in poor condition.*
- *Use competitive bidding for all grazing permits.* The existing noncompetitive system gives ranchers with essentially lifetime permits an unfair economic advantage over ranchers who can't get such permits, which amount to government subsidies. Competitive bidding would let free enterprise work and might allow conservation and wildlife groups to obtain grazing permits they could then use to protect such lands from overgrazing.
- *Allow individuals and environmental groups to purchase grazing permits and not use the land for grazing.*
- *Until a competitive bidding system is implemented, raise grazing fees to fair market value.*
- *Give ranchers with small holdings grazing fee discounts to reduce the financial pressure for them to sell their ranch lands to developers.*
- *Abolish rancher-dominated grazing advisory boards.*

The problem is that despite their small numbers, western ranchers with grazing permits wield enough political power to see that measures designed to promote sustainability are not enacted by elected officials. There are also efforts in the U.S. Congress to turn management of public rangelands over almost completely to ranchers and exempt them from most environmental laws and any legal challenges.

23-6 MANAGING AND SUSTAINING NATIONAL PARKS

How Popular Is the Idea of National Parks? Today, over 1,100 national parks larger than 1,000 hectares (2,500 acres) each are located in more than 120 countries, together covering an area equal to that of Alaska, Texas, and California combined. This important achievement in global conservation was spurred by the creation of the first national park system in the United States in 1912.

The U.S. National Park System is dominated by 54 national parks, most of them in the West (Figure 23-9). These repositories of majestic beauty and biodiversity (Figure 23-14), sometimes called *America's crown jewels*, are supplemented by state, county, and city parks. Most state parks are located near urban areas and thus are more heavily used than national parks. Nature walks, guided tours, and other educational services offered by U.S. Park Service employees have given many visitors a better appreciation for and understanding of nature, but recently these services have diminished because of budget cutbacks.

Q: Is the nicotine in tobacco smoke a highly addictive chemical?

How Are Parks Being Threatened? Parks everywhere are under siege. In developing countries parks are often invaded by local people who desperately need wood, cropland, and other resources. Poachers kill animals to obtain and sell rhino horns, elephant tusks, and furs. Park services in developing countries typically have too little money and too few personnel to fight these invasions, either by force or by education. In addition, most of the world's national parks are too small to sustain many of the larger animal species.

Popularity is one of the biggest problems of national and state parks in the United States (and other developed countries). Because of increased numbers of roads, cars, and affluent people, annual recreational visits to major national parks increased fourfold (and visits to state parks sevenfold) between 1950 and 1997; visits are projected to increase from 271 million in 1997 to 500 million by 2010.

During the summer, the most popular U.S. national and state parks are often choked with cars and trailers and hour-long entrance backups. They are also plagued by noise, traffic jams, litter, vandalism, poaching, deteriorating trails, polluted water, rundown visitor centers, garbage piles, and crime. Many visitors to heavily used parks leave the city to commune with nature, only to find the parks as noisy, congested, and stressful as where they came from.

In winter, the quiet solitude one might expect at Yellowstone National Park has been replaced with the noise and pollution of snowmobiles. As many as 1,000 snowmobiles per day roar into the park, spewing air pollutants equivalent to the tailpipe emissions of 1.7 million cars.

U.S. Park Service rangers now spend an increasing amount of their time on law enforcement instead of conservation, management, and education. Since 1976 the number of federal park rangers has not changed while the number of visitors to park units has risen by 85 million; the number of visitors is expected to rise by another 162 million in the next 20 years. Currently, there is one ranger for every 84,200 visitors to the major national parks. Because overworked rangers earn an average salary of less than $28,000, many leave for better-paying jobs.

Wolves (Case Study, p. 630), bears, and other large predators in and near various parks have all but vanished because of excessive hunting, poisoning by ranchers and federal officials, and the limited size of most parks. As a result, populations of species these predators once helped control have exploded, destroying vegetation and crowding out other species.

Nonnative species have moved into or been introduced into many parks. Wild boars (imported to North Carolina in 1912 for hunting) are threatening vegetation in part of the Great Smoky Mountains National Park. The Brazilian pepper tree has invaded Florida's Everglades National Park. Mountain goats in Washington's Olympic National Park trample native vegetation and accelerate soil erosion. At the same time that some species have moved into parks, other native species of animals and plants (including many threatened or endangered species) are being killed or removed illegally in almost half of U.S. national parks (p. 693).

Some say that the greatest threat to many U.S. parks today is posed by nearby human activities. Wildlife and recreational values are threatened by mining, logging, livestock grazing, coal-burning power plants, water diversion, and urban development. Polluted air drifts hundreds of kilometers, killing ancient trees in California's Sequoia National Park and blurring the awesome vistas at Arizona's Grand Canyon. According to the National Park Service, air pollution affects scenic views in national parks more than 90% of the time.

That's not all. Mountains of trash wash ashore daily at Padre Island National Seashore in Texas. Water use in Las Vegas threatens to shut down geysers in the Death Valley National Monument. Visitors to Sequoia National Park complain of raw sewage flowing through a parking lot; similar contamination has occurred for years at Kentucky's Mammoth Cave. Unless a massive ecological restoration project is successful, Florida's Everglades National Park may dry up, making it the country's most endangered national park (Case Study, p. 632).

Case Study: Why is Yellowstone's Ecosystem Unraveling? To a growing number of ecologists, river experts, and foresters, the gravest danger to many U.S. national parks is coming from *within*. For example, they point to Yellowstone National Park as an ecosystem that is unraveling as stands of aspens and willows die, woody shrubs disappear, populations of once abundant animal species dwindle or disappear, and stream banks erode.

These scientists point to the explosion of the elk population as the major culprit. They have collected data indicating that between 1968 and 1996, the National Park Service's hands-off management policy of *natural regulation* has been the key factor in allowing the park's elk population to increase more than sixfold from 3,100 to about 20,000.

According to Charles Kay, a wildlife biologist at Utah State University, the rising elk population

CASE STUDY

Who's Afraid of the Big Gray Wolf?

At one time, the gray wolf, *Canis lupus* (Figure 23-15), ranged over most of North America, but between 1850 and 1900 2 million wolves were systematically shot, trapped, and poisoned by ranchers, hunters, and government employees. The idea was to make the West and the Great Plains safe for livestock and for big-game animals prized by hunters.

In the Yellowstone region most wolves had been purged by the late 1920s, and the remaining few survivors were gone by the 1940s. By the 1960s, the gray wolf was found mostly in Alaska (which has about 6,000 wolves) and Canada; a few hundred remained in Minnesota and nearby Isle Royale in Lake Superior as well. The species is now listed as endangered in all 48 lower states except Minnesota (which has an estimated 2,500 wolves).

Ecologists now recognize the important role these predators once played in parts of the West and the Great Plains by culling herds of bison, elk, caribou, and mule deer of weaker animals, thereby strengthening the genetic pool of the survivors. In recent years, these herds have proliferated, devastating some of the area's vegetation, increasing erosion, and threatening the niches of other wildlife species.

In 1987, the U.S. Fish and Wildlife Service proposed that gray wolves be reintroduced into the Yellowstone ecosystem. This proposal to reestablish ecological connections that humans had eliminated brought outraged protests.

Some objections came from ranchers who feared the wolves would attack their cattle and sheep; one enraged rancher said that the idea was "like reintroducing smallpox." Other protests came from hunters who feared that the wolves would kill too many big game animals and from miners and loggers who worried that the government would force them to cease operations on wolf-populated federal lands.

National Park Service officials promised to trap or shoot any wolves that killed livestock outside designated areas and to reimburse ranchers from a private fund established by Defenders of Wildlife for any livestock verified to have been killed by wolves. However, these promises fell on deaf ears, and many ranchers and hunters worked hard to unsuccessfully delay or defeat the plan in Congress.

Since 1995 federal wildlife officials have been catching about 30–40 gray wolves per year in Canada and relocating them in Yellowstone National Park and northern Idaho. The goal is to allow the wolf populations to increase to sustainable levels by 2002. By 1998 there were about 90 gray wolves in the Yellowstone ecosystem.

Under this relocation plan, the wolves, even though officially listed as endangered, are treated as an "experimental" population with reduced protection; if they venture outside designated areas, they can be killed by ranchers and hunters. Environmental groups have strongly supported reintroducing the wolves. However, they believe that all reintroduced wolves should be given full legal protection as an endangered species. They argue that without full protection, the reintroduced wolves and those that have already naturally migrated to Idaho don't stand a chance.

Some ranchers and hunters say that either way they'll take care of the wolves quietly—what they call the "shoot, shovel, and shut up" solution. Meanwhile, there are continuing efforts in Congress to eliminate the program and its funding.

In 1997 the recovery program received a serious setback when a

is consuming tree shoots and other forms of vegetation and changing the once diverse park into a simplified lawn by disrupting a series of important ecological connections. He has compared the vegetation in an area, called the Mammoth Enclosure that has been fenced off for 40 years, with vegetation outside the enclosure.

Inside the enclosed area there is a dense woody understory that provides habitats and food for deer, mice, and many other small mammals; outside the enclosure there is little left of the woody understory. Inside the fenced area there are dense stands of aspen, willows, and woody shrubs; outside such species are almost nonexistent. Outside the enclosed area, centuries-old stands of aspen, which reproduce by sending up shoots or suckers, have died out as their shoots have been eaten by elk.

This has resulted in an interconnected cascade of other ecological changes. Loss of these tree species has eliminated food and dam-building materials for beavers, whose numbers have plummeted in the last 60 years. As the beaver populations have dropped, so have their dams and ponds, which trapped eroded silt and slowly built up the park's streambeds. With a stronger flow of water and without aspens, cottonwoods, and willows to hold streambed soil in place, stream banks have eroded and collapsed.

Q: What three ways are used to determine the toxicity of substances?

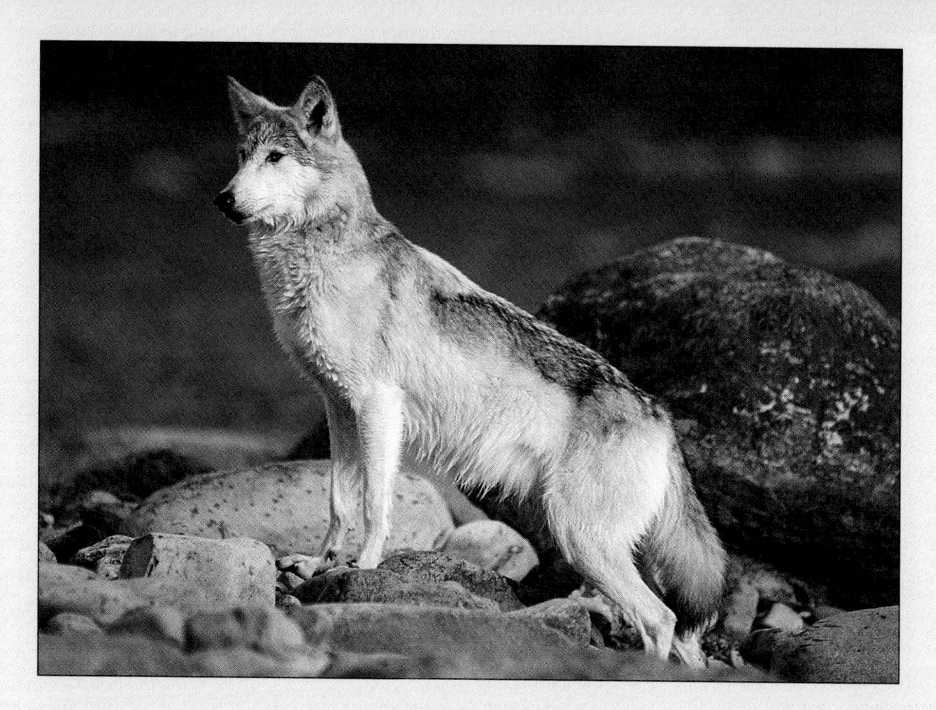

Figure 23-15 The gray wolf is an endangered species in the lower 48 states (except Minnesota). Efforts to return this species to its former habitat in the Yellowstone National Park area have been vigorously opposed by ranchers, hunters, miners, and loggers, but wolves were reintroduced beginning in 1995. (Tom J. Ulrich/Visuals Unlimited)

Wyoming federal appeals court judge found the wolf recovery program illegal. He said that the U.S. Fish and Wildlife Service had violated the law by declaring the wolves an "experimental population." That designation denies them the protection of the Endangered Species Act. The judge ordered that the Canadian wolves reintroduced into Yellowstone National Park and central Idaho be removed. The case has been appealed by several wildlife conservation organizations.

Critical Thinking

Do you agree or disagree with the program to reestablish populations of the gray wolf in the Yellowstone ecosystem? Explain. What are the major drawbacks of this program? Could the money be better spent on other wildlife programs? If so, what programs would you suggest?

Since adopting their policy of natural regulation in the early 1970s, park service officials have claimed that the large increase in elk represents the park ecosystem returning to its natural equilibrium state. Many ecologists say this is an outdated idea because recent ecological research has revealed that ecosystems are in a constant state of change and rarely attain an equilibrium point (Section 9-6). Park officials also attribute the decline in woody streamside vegetation to climate change or the long-standing policy of suppressing fires within the park.

Kay challenges these conclusions with his data on changes inside the park and the status of vegetation in many areas outside the park. For example, Eagle Creek on national forest land just north of the park near the town of Gardiner, Montana, has the same climate, no cattle grazing, and the same fire history as Yellowstone.

In this area about 3,000 elk are shot by hunters each year. Stands of aspen have regenerated because their shoots have not been devoured by elk. There are also thick stands of shrubby undergrowth, rarely found inside the park. In private and national forest land north of Gardiner along the Yellowstone River, the river is lined with a thick band of aspens, willows, and cottonwoods, in sharp contrast to the almost barren river banks in most parts of the park.

A: Case reports, laboratory investigations, and epidemiology

Partially Restoring the Florida Everglades

CASE STUDY

Most people think of the Florida Everglades as a swamp; it actually is a slow-moving, shallow river that flows south through Everglades National Park to an estuary at Florida Bay. Without the Everglades' rain distribution and aquifer recharge systems, Miami and the rest of south Florida would be uninhabitable.

Much of the Everglades' water needed to sustain the park's wildlife has been diverted for crops and cities by a network of canals, levees, spillways, and pumping stations to the north. Because of a lack of water and the resulting ecological changes, nesting birds in the park declined by 90%, from about 90,000 in 1931 to fewer than 9,000 in 1994. All other vertebrates, from deer to turtles, are down 75–95%.

By the 1970s state and federal officials recognized that this massive plumbing project had been a serious ecological blunder. After over 20 years of political haggling, in 1990 state and federal governments agreed on a massive restoration project to undo some of the damage.

Whether this 20-year project will work and be funded adequately remains to be seen. The high cost of undoing some of the environmental harm humans have done is another example of a fundamental lesson from nature: Prevention is the cheapest and best way to go.

Critical Thinking

Are you in favor of spending public funds to partially restore the Florida Everglades ecosystem and help protect the endangered Everglades National Park? Explain.

Under such mounting criticism, the National Park Service has hinted that it may reexamine its long-standing natural regulation policy. However, this may not be politically feasible because polls indicate that the public (mostly unaware of the ecological decline of the park) favors natural regulation of the park over the park's old policy of allowing hunters to kill thousands of elk each winter. What do you think should be done?

Solutions: How Can Park Management in Developing Countries Be Improved? Some park managers, especially in developing countries, are developing integrated management plans that combine conservation practices with sustainable development of resources in the park and surrounding areas. In such plans, the inner core and vulnerable areas of each park are protected from development and treated as wilderness. Restricted numbers of people are allowed to use these areas for hiking, nature study, ecological research, and other nondestructive recreational and educational activities.

In buffer areas surrounding the core, controlled commercial logging, sustainable grazing by livestock, and sustainable hunting and fishing by local people are allowed. Money spent by park visitors adds to local income. By involving nearby residents in developing park management and restoration plans, managers and conservationists seek to help them see the park as a vital resource to protect and sustain rather than ruin. Costa Rica has led the way in protecting much of its parklands and other biodiversity sanctuaries (p. 673).

Integrated park management plans look good on paper, but they cannot be carried out without adequate funding and the support of nearby landowners and users. Moreover, in some cases the protected inner core may be too small to sustain the park's larger animal species.

Solutions: How Can Management of U.S. National Parks Be Improved? The 54 major U.S. national parks are managed under the principle of *natural regulation*, as if they were wilderness ecosystems that would sustain themselves if left alone. Many ecologists consider this a misguided policy. Most parks are far too small to even come close to sustaining themselves. Even the biggest ones, such as Yellowstone, cannot be isolated from the harmful effects caused by activities in nearby areas and from destruction from within by exploding populations of some plant-eating species (Case Study, p. 629).

The U.S. National Park Service has two goals that increasingly conflict: to preserve nature in parks and to make nature more available to the public. The Park Service must accomplish these goals with a $1.5-billion annual budget and a $6-billion backlog of maintenance, repairs, and high-priority construction projects at a time when park usage and external threats to the parks are increasing. Private developers continually pressure Congress and the Park Service to turn national parks into high-tech, heavily developed theme or amusement parks instead of the repositories of natural ecosystems and centers for learning about nature they were intended to be.

Currently the Park Service spends about 92% of its budget on visitor services; only 7% is spent on

Q: How many U.S. workers die prematurely from exposure to toxic substances?

protecting natural resources, and a mere 1% funds environmental research that could help park officials implement better ecological management of parklands. Doubling the Park Service's current budget would cost as much as one B-2 bomber and cost each American only $5. This would be a good investment because according to a Park Service model, national park units pump about $10 billion annually into local areas.

In 1988 the Wilderness Society and the National Parks and Conservation Association made a number of suggestions for sustaining and expanding the national park system, including the following:

- *Have all user and entrance fees for national park units be used for the management, upkeep, and repair of the national parks.* Currently, most of these fees go into the general treasury.
- *Require integrated management plans for parks and other nearby federal lands.*
- *Increase the budget for adding new parkland near the most threatened parks.*
- *Increase the budget for buying private lands inside parks.*
- *Locate all new and some existing commercial facilities and visitor parking areas outside parks.*
- *Require private concessionaires who provide campgrounds, restaurants, hotels, and other services for park visitors to compete for contracts, raise their fees to 22% of their gross (not net) receipts, and pour receipts back into parks.* In 1996, for instance, private concessionaires in national parks reported gross receipts of $700 million but paid only 6.7% of their gross receipts in franchise fees to the government. Many large concessionaires have long-term contracts by which they pay the government as little as 0.75% of their gross receipts.
- *Allow concessionaires to lease but not own facilities inside parks.*
- *Provide more funds for backlog of park system maintenance and repairs.*
- *Raise fees for park visitors and pour receipts back into parks.*
- *Restrict the number of visitors to crowded park areas.*
- *Increase the number and pay of park rangers.*
- *Survey condition and types of wildlife species in parks.*
- *Encourage volunteers to give tours and lectures to visitors.*
- *Encourage individuals and corporations to donate money for park maintenance and repair.*

23-7 PROTECTING AND MANAGING WILDERNESS AND OTHER BIODIVERSITY SANCTUARIES

How Much of the Ecosphere Should We Protect from Exploitation? Most wildlife and conservation biologists believe that the best way to preserve biodiversity and ecological integrity—or wildness—is through a worldwide network of reserves, parks, wildlife sanctuaries, wilderness, and other protected areas. Currently, about 6% of the world's land area is either strictly or partially protected in more than 20,000 nature reserves, parks, wildlife refuges, and other areas around the globe (Figure 23-16). North and Central America have set aside the highest percentage of land in parks and reserves (almost 12%), followed by Oceania (10%). The countries of the former Soviet Union have set aside the least, only 1.1%.

This is an important beginning, but conservation biologists say that to keep biodiversity and ecological integrity from being depleted, a minimum of 10% of the globe's land area should be protected. Moreover, many existing reserves are too small to provide any real protection for the wild species that live on them. Many existing preserves receive little protection. Currently, less than 2% of the world's protected areas are at least 10,000 square kilometers (3,900 square miles) in size, a conservative estimate of the minimum habitat needed to maintain viable populations of the largest mammals.

In 1981 the UN Educational, Scientific, and Cultural Organization (UNESCO) proposed that at least one (and ideally five or more) *biosphere reserves* be set up in each of the earth's 193 biogeographical zones. Each reserve should be large enough to prevent gradual species loss and should combine both conservation and sustainable use of natural resources. To date, more than 300 biosphere reserves have been established in 76 countries.

A well-designed biosphere reserve has three zones (Figure 23-17): **(1)** a *core area* containing an important ecosystem that has had little or no disturbance from human activities, **(2)** a *buffer zone* where activities and uses are managed in ways that help protect the core, and **(3)** a second buffer or *transition zone*, which combines conservation and sustainable forestry, grazing, agriculture, and recreation. Buffer zones can also be used for education and research.

Proponents of large reserves say they are the only way to maintain viable populations of large, wide-ranging species (such as panthers, elephants, and grizzly bears). Also, large reserves can sustain more species, minimize edge effects, and provide

A: 50,000–70,000 per year

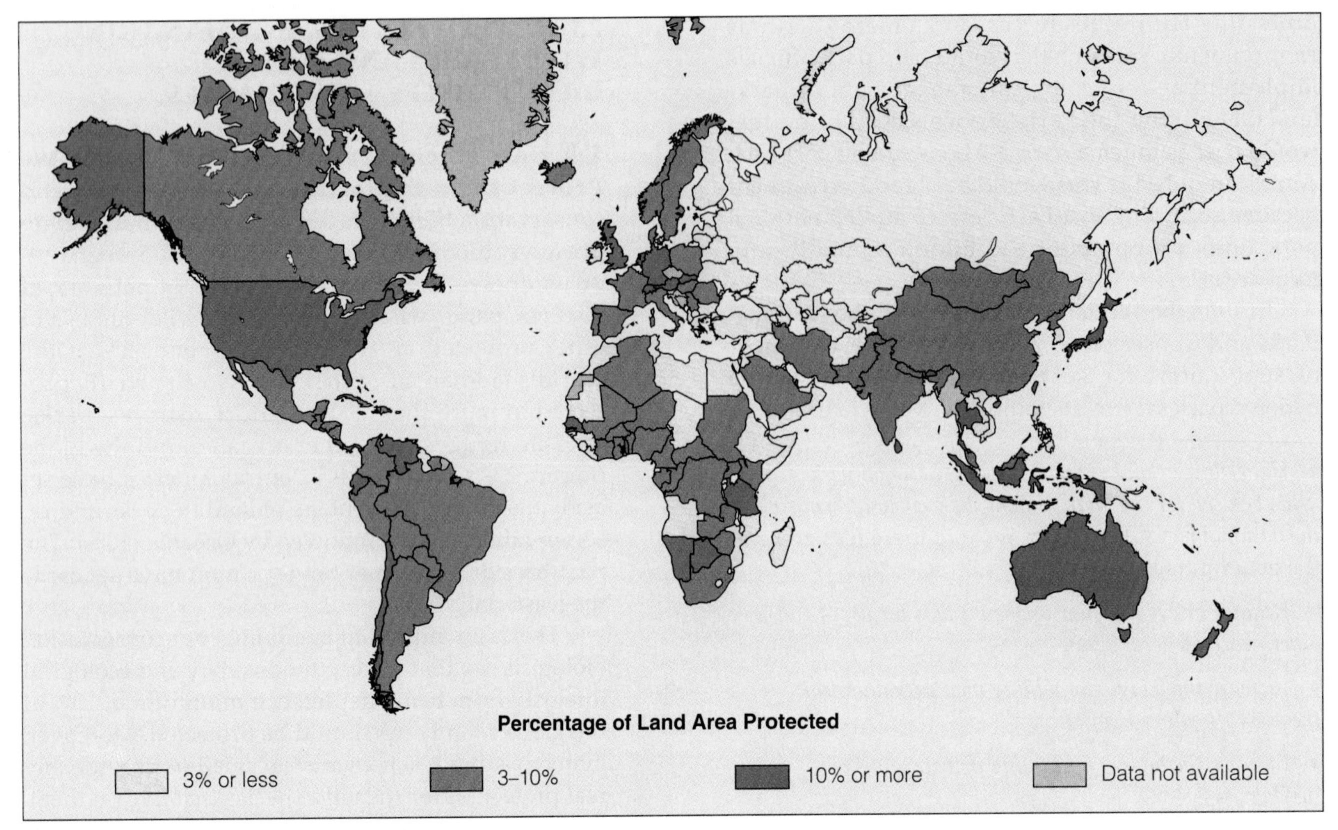

Figure 23-16 Percentage of land area set aside as protected parks, wildlife refuges, and nature preserves in various regions of the world. A large percentage of these parks and preserves are established in mountainous areas covered by rock and ice. The International Union for the Conservation of Nature (IUCN) has identified an additional 3,000 sites worthy of being designated as national parks or wildlife refuges. (Data from World Conservation Monitoring Center, 1993)

greater habitat diversity than small preserves. These advantages of large parks or reserves are based on the theory of island biogeography developed by Robert MacArthur and Edward O. Wilson (Connections, p. 245, and Figure 9-24), in which the reserves are assumed to be habitat islands surrounded by expanses of often inhospitable development or other terrain.

In reality, few countries are physically or politically able to set aside and protect large biosphere reserves or other protected areas. Moreover, some research indicates that in some locales several well-placed medium-sized, isolated reserves may better protect a greater variety of habitat types and more populations of rare species than a single large one. If the population of a species is wiped

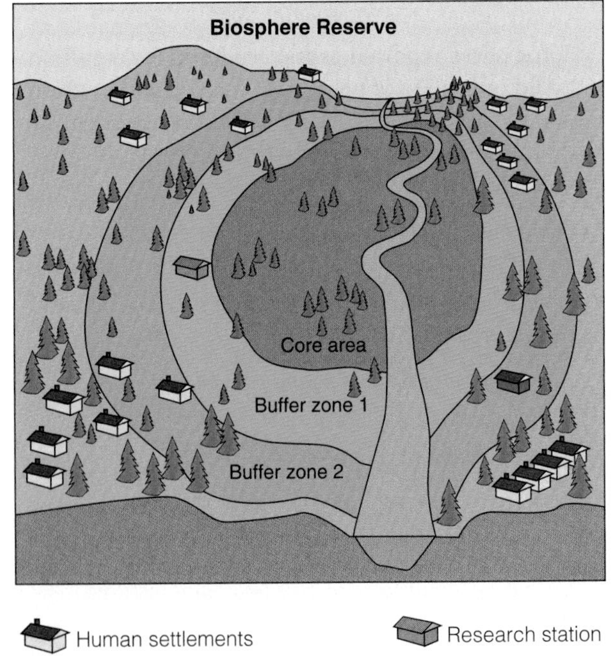

Figure 23-17 Design of a model biosphere reserve. In traditional parks and wildlife reserves, well-defined boundaries keep people out and wildlife in. By contrast, biosphere reserves recognize people's needs for access to sustainable use of various resources in parts of the reserve.

Q: What percentage of the 72,000 chemicals in commercial use have been thoroughly screened for toxicity?

out in one reserve because of fire, epidemic, or some other disaster, the species might still survive in the other reserves.

Conservation biologists also suggest establishing protected habitat corridors running between reserves to help support more species, allow migrations when environmental conditions in a reserve deteriorate, and reduce loss of genetic diversity from inbreeding. Such corridors might also help preserve animals that must make seasonal migrations to obtain food.

However, there are some drawbacks to corridors. They can allow the movement of pest species, disease, and exotic species between reserves. Also, animals moving along corridors may be exposed to greater risks of predation by both natural predators and human hunters. Although most conservation biologists support the development of wildlife corridors between reserves, they recognize the need to evaluate them on a park-by-park basis.

So far most biosphere reserves fall short of the bioregional ideal (Figure 23-17) and too little funding has been provided for their protection and management. An international fund to help developing countries protect and manage bioregions and biosphere reserves would cost $100 million per year—about what the world's nations spend on arms every 90 minutes.

Because there won't be enough money to protect most of the world's biodiversity, environmentalists believe that efforts should be focused on the most biodiverse countries (Figure 23-18) and threatened species-rich areas within such countries. In the United States, conservation biologists have urged Congress to pass an Endangered Ecosystems or Biodiversity Protection Act as an important step toward surveying and preserving the country's biodiversity.

In 1996 the World Wildlife Fund conducted a survey identifying 217 terrestrial, marine, and freshwater ecoregions throughout the world that were in the greatest need of protection. These ecoregions were identified on the basis of rareness of their habitat types, species found only in them (endemic species), total number of species (species biodiversity), and unusual ecological phenomena (such as the mass migration of caribou in the Canadian tundra). In addition to government action, private groups have helped establish biodiversity and wildlife sanctuaries (Solutions, p. 636).

Most economists, developers, and resource extractors disagree with protecting even the 6% of the earth's remaining undisturbed ecosystems. They contend that such areas contain valuable resources that would add to economic growth. They view protected areas as "useless" and "empty" because they are not being

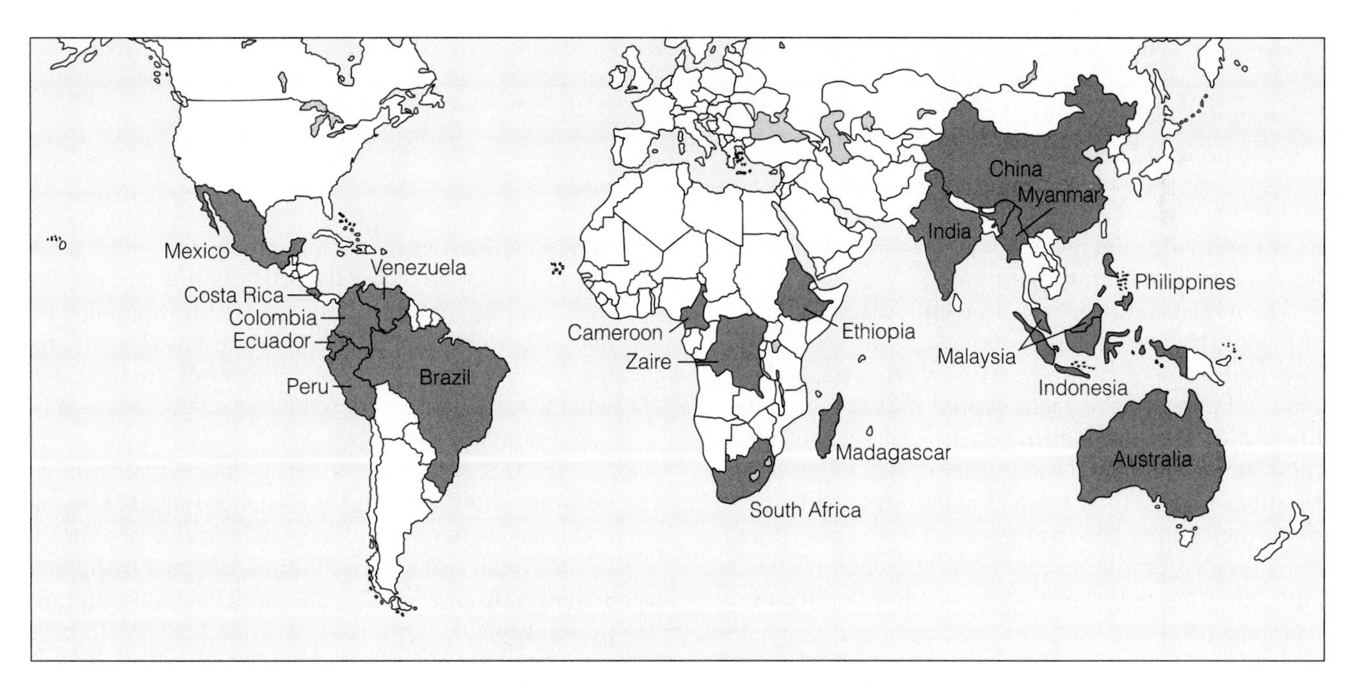

Figure 23-18 The earth's most biodiverse countries. Conservation biologists believe that efforts to preserve repositories of biodiversity as protected wilderness areas should be concentrated in these species-rich countries. The 80% of the world's people who live in developing countries have 80% of the world's biodiversity but only 15% of the world's income and financial wealth. Protecting the world's biodiversity will require considerable financial and scientific help from the developed countries. (Data from Conservation International and World Wildlife Fund)

A: About 10% (only 2% have been adequately screened as carcinogens, teratogens, or mutagens)

The Nature Conservancy

SOLUTIONS

Private groups play an important role in establishing wildlife refuges and other protected areas as a major strategy to protect biological diversity and ecological integrity. For example, since 1951 the Nature Conservancy has established more than 1,500 preserves in the United States, providing protection to over 38,000 square kilometers (14,700 square miles) of areas with unique ecological or aesthetic significance. Outside the United States, it has helped protect 170,000 square kilometers (65,600 square miles) of wildlife habitat. The Nature Conservancy now has over 820,000 members and programs in all 50 states, Latin America, and the Caribbean.

For several years, the Conservancy has been rated among the top conservation organizations, with one of the lowest overhead rates of any nonprofit organization: 85% of all contributions go directly to its conservation programs.

The Nature Conservancy began in 1951 when an association of professional ecologists wanted to use their scientific knowledge of nature to conserve natural areas. Since its beginning this science-based organization has used the most sophisticated scientific knowledge available to identify and rank sites that are unique and ecologically significant and whose biodiversity or existence is threatened by development or other human activities.

Once sites are identified, a variety of techniques are used to see that they receive legal protection. Then science-based management plans are drawn up to maintain or restore the ecological health of each site and to provide long-term stewardship.

This conservation organization uses private and corporate donations to maintain a fund of over $150 million, which can be used to buy ecologically important pieces of land or wetlands threatened by development when no other option is available. Sometimes this land is then sold or donated to federal or state agencies.

If it cannot buy land for the protection of habitat, the Conservancy helps landowners obtain tax benefits in exchange for accepting legal restrictions or conservation easements preventing development. Other techniques include long-term management agreements and debt-for-nature swaps. Landowners have also received sizable tax deductions by donating their land to the Nature Conservancy in exchange for lifetime occupancy rights.

Through such efforts, this organization has created the world's largest system of private natural areas and wildlife sanctuaries, using the guiding principle of land conservation through private action.

Critical Thinking

Why do you think the anti-environmental Wise Use movement (Case Study, p. 787) has singled out the Nature Conservancy as the most important environmental group to weaken or destroy? Do you agree or disagree with this strategy? Explain.

used to provide mineral and other resources and land for development for humans.

Ecologists and conservation biologists disagree and view protected areas as islands of biodiversity and ecological integrity—or wildness—that are vital parts of the earth capital that sustains all life and economies (Figure 1-2). In ecological terms, they view these areas as perhaps the most useful places on earth as centers of evolution. Instead of being empty or nearly empty of life (like most greatly simplified cropfields, tree plantations, and urban areas), these oases of biodiversity are teeming with life.

What Is Wilderness and How Much Is Left? Protecting biodiversity and ecological integrity means protecting *wildness*. One way to do this is to protect undeveloped lands from exploitation by setting them aside as **wilderness**. According to the U.S. Wilderness Act of 1964, wilderness consists of areas "where the earth and its community of life are untrammeled by man, where man himself is a visitor who does not remain." President Theodore Roosevelt summarized what we should do with wilderness: "Leave it as it is. You cannot improve it."

The U.S. Wilderness Society estimates that a wilderness area should contain at least 4,000 square kilometers (1,500 square miles); otherwise it can be affected by air, water, and noise pollution from nearby human activities. To provide the minimum habitat to support populations of the largest mammals, a wilderness area should contain at least 10,000 square kilometers (3,900 square miles).

A 1987 survey sponsored by the Sierra Club revealed that only about 34% of the earth's land area is undeveloped wilderness in blocks of at least 4,000

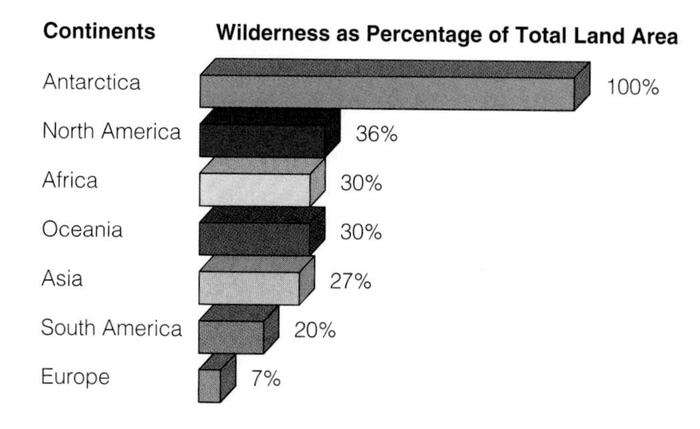

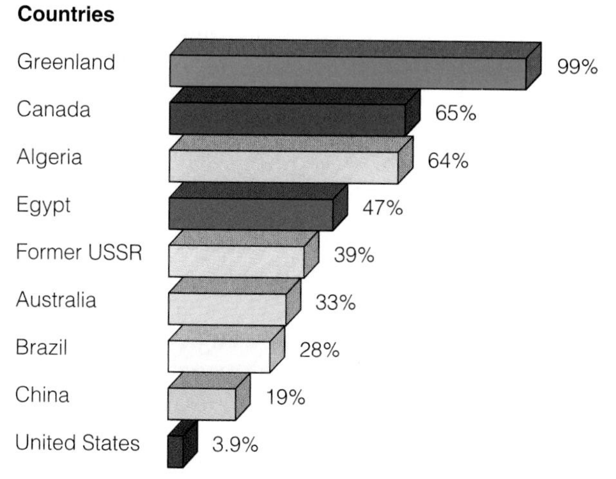

Figure 23-19 Wilderness areas by major geographical regions. (Data from J. Michael McCloskey and Heather Spalding, "A Reconnaissance-Level Inventory of the Wilderness Remaining in the World," Sierra Club, 1987)

square kilometers (1,500 square miles) (Figure 23-19). About 30% of these remaining wildlands are forests, many of them threatened tropical forest; tundra, ice-covered land, and desert make up most of the rest. Only about 20% of the wildlands identified in this survey are protected.

Conservation biologists urge that remaining wilderness be protected by law everywhere, focusing first on the most endangered spots in wilderness- and species-rich countries (Figure 23-18). Conservation biologists urge that more wilderness areas be protected in the United States (Case Study, right).

Why Preserve Wilderness? According to wilderness supporters, we need wild places where we can experience the beauty of nature and observe natural biological diversity, where we can enhance our mental and physical health by getting away from

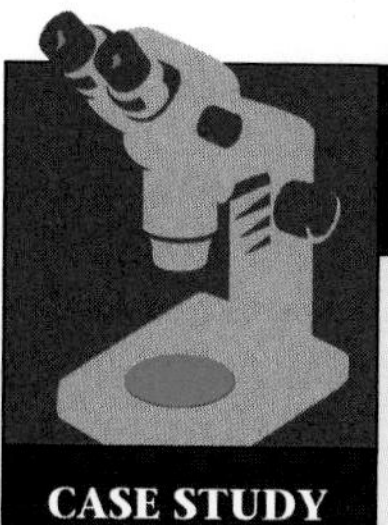

CASE STUDY

The U.S. National Wilderness Preservation System

In the United States, preservationists have been trying to save wild areas from development since 1900. On the whole, they have fought a losing battle. Not until 1964 did Congress pass the Wilderness Act, which allowed the government to protect undeveloped tracts of public land from development as part of the National Wilderness Preservation System.

Only about 4% of U.S. land area is protected as wilderness, with almost three-fourths of it in Alaska. Only 1.8% of the land area of the lower 48 states is protected, most of it in the West.

Of the 413 wilderness areas in the lower 48 states, only four are larger than 4,000 square kilometers. Furthermore, the present wilderness preservation system includes only 81 of the country's 233 distinct ecosystems. Like the national parks, most wilderness areas in the lower 48 states are habitat islands in a sea of development.

There remain almost 400,000 square kilometers (150,000 square miles) of scattered blocks of public lands that could qualify for designation as wilderness. Conservation biologists would like to see all this land protected as wilderness. Such efforts are strongly opposed by timber, mining, ranching, energy, and other interests who want to extract resources from these and most other public lands or convert them to private ownership (Case Study, p. 787).

Some conservation biologists also urge that *wilderness recovery areas* be created by closing and obliterating nonessential roads in large areas of public lands, restoring habitats, allowing natural fires to burn, and reintroducing species that have been driven from such areas. However, resource developers lobby elected officials and government agencies to build roads into national forests (and other areas being evaluated for inclusion in the wilderness system) so that they can't be designated as wilderness; they also strongly oppose the idea of wilderness recovery areas.

Critical Thinking

1. Do you believe that all or most of the remaining potential wilderness area in the United States (or in the country where you live) should be protected? Explain.

2. Should a large portion of U.S. Forest Service and Bureau of Land Management land be reclassified as wilderness recovery areas, closed off for many uses, and allowed to undergo natural restoration as wildlife habitat? Explain. If so, how would you go about doing this?

A: Essentially none

noise, stress, and large numbers of people. Wilderness preservationist John Muir advised us to

> *Climb the mountains and get their good tidings. Nature's peace will flow into you as the sunshine into the trees. The winds will blow their freshness into you, and the storms their energy, while cares will drop off like autumn leaves.*

Even those who never use the wilderness may want to know it is there, a feeling expressed by novelist Wallace Stegner:

> *Save a piece of country . . . and it does not matter in the slightest that only a few people every year will go into it. This is precisely its value. . . . We simply need that wild country available to us, even if we never do more than drive to its edge and look in. For it can be a means of reassuring ourselves of our sanity as creatures, a part of the geography of hope.*

Recently some critics have argued that protecting wilderness for its scenic and recreational value for a small number of people is an outmoded concept that keeps some areas of the planet from being economically useful to humans. They also contend that wilderness protection diverts money, energy, and talent from learning how to better manage urban and other landscapes disturbed by human activities.

Scientists argue that this idea is outmoded because it fails to take into account the research that has revealed the ecological importance of wilderness areas for all people and all life. To most biologists, *the most important reason for protecting wilderness and other areas from exploitation and degradation is to preserve the biodiversity and ecological integrity they contribute as a vital part of earth capital* (Figure 1-2).

Wilderness areas provide mostly undisturbed habitats for wild plants and animals, protect diverse biomes from damage, and provide a laboratory in which we can discover more about how nature works. In other words, wilderness is a biodiversity and wildness bank and an ecoinsurance policy.

Instead of decreasing the ecologically inadequate 5–6% of the world's wilderness areas currently under some degree of protection, most ecologists and conservation biologists believe there is an urgent need to protect at least 10–12% of the earth's surface as wilderness and to allow some degraded areas to return to a more wild state. They agree that we need to learn much more about how to manage and restore land disturbed by human activities. However, they don't believe that this should be done by decreasing or unraveling wilderness protection, which will lead to even more areas disturbed by human activities.

To protect more of the earth's biodiversity and ecological integrity, conservation biologists and other environmental scientists believe the designation of future wilderness (and other protected) areas should focus on prairies, lowland forests, and wetlands that are largely absent from the current system. This will bring wilderness advocates into increasing conflict with developers, farmers, ranchers, and resource extractors who view remaining lands in these categories as prime sites for human use.

Some analysts believe that wilderness should be preserved because the wild species it contains have a right to exist (or struggle to exist) and to play their roles in the earth's ongoing saga of biological evolution and ecological processes, without human interference.

How Can Wilderness Be Protected? To protect the most popular areas from damage, wilderness managers must designate sites where camping is allowed and limit the number of people using these sites at any one time. Managers have also increased the number of wilderness rangers to patrol vulnerable areas, and they have enlisted volunteers to pick up trash discarded by thoughtless users.

Environmental historian and wilderness expert Roderick Nash suggests that wilderness areas be divided into three categories. The easily accessible, popular areas would be intensively managed and have trails, bridges, hiker's huts, outhouses, assigned campsites, and extensive ranger patrols. Large, remote wilderness areas would be used only by people who get a permit by demonstrating their wilderness skills. The third category, biologically unique areas, would be left undisturbed as gene pools of plant and animal species, with no human entry allowed.

What Is the U.S. National Wild and Scenic Rivers System? In 1968 Congress passed the National Wild and Scenic Rivers Act, which allows rivers and river segments with outstanding scenic, recreational, geological, wildlife, historical, or cultural values to be protected in the National Wild and Scenic Rivers System. These waterways are to be kept free of development; they may not be widened, straightened, dredged, filled, or dammed along the designated lengths. The only activities allowed are camping, swimming, nonmotorized boating, sport hunting, and sport and commercial fishing. New mining claims are permitted in some areas, however.

Lackey, R. T. 1998. "Ecosystem Management." *Journal of Soil and Water Conservation,* vol. 53, no. 2, 92(3).

Currently, only 0.2% of the country's 6 million kilometers (3.5 million miles) of waterways along some 150 stretches are protected by the Wild and Scenic River System. In contrast, 17% of the country's wild river length has been tamed and land flooded by dams and reservoirs. Environmentalists have urged Congress to add 1,500 additional river segments to the system, a goal that is vigorously opposed by some local communities and anti-environmental groups. If that goal is achieved, about 2% of the country's river systems would be protected. Environmentalists also urge that a permanent federal administrative body be established to manage the Wild and Scenic Rivers System and that states develop their own wild and scenic river programs.

What Is the U.S. National Trails System? In 1968, Congress passed the National Trails Act, which protects scenic and historic hiking trails in the National Trails System. However, no designated scenic or historic trail is complete because the Trails System has a low priority and receives little funding.

To fulfill the promise of the Trails Act, environmentalists propose that the Trails System be managed by a single agency. Armed with a comprehensive study of current and future trail needs, the agency would plan the development of a countrywide network of trails in the next decade.

23-8 LAND MANAGEMENT, BIODIVERSITY, AND SUSTAINABLE ECOSYSTEMS

Should Conservation Efforts Focus on Sustaining Ecosystems or Species? The consensus among most conservation biologists is that protecting biodiversity requires both whole ecosystem and species-by-species approaches for conservation efforts to be successful. The whole ecosystem approach has the advantage that it focuses on ensuring that sufficient land is protected to ensure ecological integrity and provide sufficient habitat for the majority of wild species.

By contrast, a species-by-species approach (Chapter 25) has the advantage that it involves identifying which species are at greatest risk, gaining a detailed understanding of those species, and focusing efforts on protecting them. Much of the world's biodiversity is not known, especially in the species-rich tropics. Even in the temperate zone our knowledge of populations of wild organisms is often limited. Thus, even though a species-by-species approach may be desirable, an ecosystem approach is often more practical and likely to succeed.

What is Gap Analysis? In recent years a scientific method called **gap analysis** has been developed to determine how adequately native plant and animal species and natural communities are protected by the existing network of conservation lands. Species and communities not adequately represented in existing conservation lands constitute conservation gaps. The idea is to identify these gaps and then eliminate them in a proactive manner through the establishment of new reserves or changing land management practices (Figure 23-20).

Gap analysis originated from the concept that a species-by-species approach to biodiversity loss is neither effective nor efficient. In 1987, Dr. J. Michael Scott observed that most of the native birds of Hawaii live in habitat that is outside existing reserves. He recognized that such gaps are widespread.

The Biological Resources Division of the U.S. Geological Survey (formerly called the National Biological Survey) has initiated a Gap Analysis Program to provide broad geographic information on the status of biodiversity protection in the United States. Additional support has been provided by the Department of Defense, the Environmental Protection Agency, the National Mapping Division of the U.S. Geological Survey, and the Nature Conservancy. The goal is to provide land managers, planners, scientists, and policy makers with the information they need to make better-informed decisions. The Gap Analysis Program involves state-by-state and national efforts to complete the following:

- Map existing natural vegetation
- Map distribution of native vertebrate species
- Map public land ownership and private conservation lands
- Show the current network of conservation lands
- Compare distributions of any native vertebrate species, group of species, or vegetation communities of interest with the network of conservation lands
- Provide an objective base of information for local, state, and national management of land and biological resources

Gap analysis works by overlaying maps of vegetation and species occurrence onto maps of protected areas using geographic information system (GIS) technology (Figure 4-26). The Gap Analysis Program has been acclaimed as "a truly visionary program that brings modern computer technology to bear on questions of biodiversity and related changes in land management."

Hint: Enter the search terms *ecosystem management* using the Subject Guide.

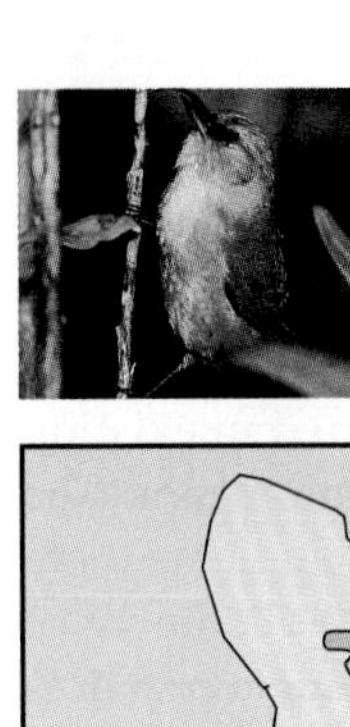

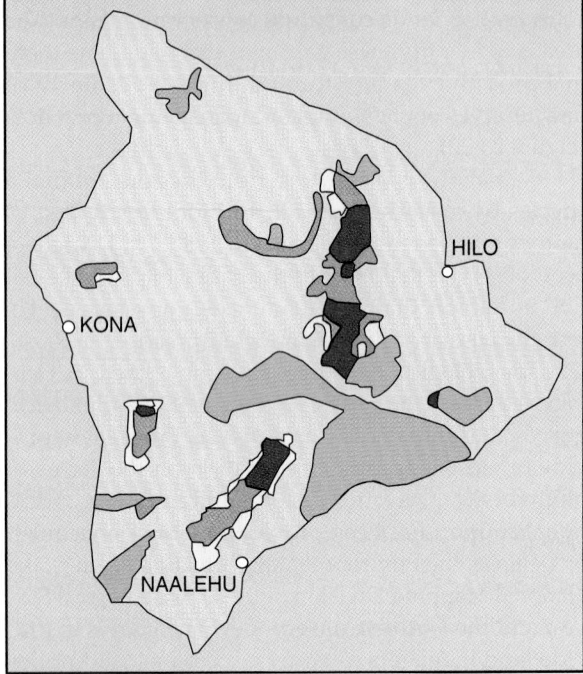

One Species
Three Species Overlap
Two Species Overlap
Existing Nature Reserves

Figure 23-20 Gap analysis involves identifying how well species and communities are represented in the existing network of conservation reserves. In 1986, gap analysis was used to determine that less than 10% of the ranges of endangered Hawaiian birds on the island of Kauai were in protected reserves. Since then, several areas of high endangered bird richness have been protected by the Nature Conservancy and state and federal agencies. Pictured here are examples of endangered and potentially extinct species. Top, left to right: Akiapolaau (Hemignathus munroi), Hawaii Creeper (Oreomystis mana), Palila (Loxioides bailleui); bottom, left to right: Akepa (Loxops coccineus) and Ou (Psittirostra psittacea). The Ou is believed to be extinct since the late 1980s. (Adapted from J. M. Scott, S. Mountainspring, F. L. Ramsey, and C. B. Kepler, "Forest Bird Communities of the Hawaiian Islands: Their Dynamics, Ecology, and Conservation," *Studies in Avian Biology*, vol. 9, 1986) (All images courtesy of Jack Jeffrey except Ou courtesy of Michael Furuay.)

In this chapter we have seen that the task of protecting the earth's biodiversity and ecological integrity and restoring lands we have damaged is enormous. However, environmentalists believe that it can be done if enough of us become involved in this vital earth-sustaining activity.

Thank God, they cannot cut down the clouds!

Henry David Thoreau

CRITICAL THINKING

1. Why are the majority of U.S. public lands in the west or Alaska? Which do you think are more important: historical reasons or differences in the types of land in the east and west? Explain.

2. Should owners of private property be able to do anything they want with their land? Explain. If not, what specific restrictions on use should be imposed?

3. Is the principle of sustainable yield sufficient to ensure ecological integrity? Explain. Should the principle of multiple use be applied to all public lands? Explain.

4. Explain why you agree or disagree with the proposals given on p. 628 for providing more sustainable use of public rangeland in the United States.

5. Should cattle be banned from grazing on western public rangeland? Explain. If not, what new restrictions (if any) would you place on ranchers allowed to graze livestock on U.S. public lands? Explain.

6. Should trail bikes, dune buggies, and other off-road vehicles be banned from public rangeland to reduce damage to vegetation and soil? Explain. Should such a ban also include national forests, national wildlife refuges, and national parks? Explain.

7. **(a)** Should individuals and corporations with grazing or mineral rights on U.S. public lands be compensated financially if these rights are reduced or taken away by the federal government? Explain. **(b)** Should individuals and corporations be compensated financially if they are prevented from using land they own or from extracting timber or other resources from land they own because the land is classified by state or federal government as protected wetlands or habitats for endangered or threatened wildlife species? Explain.

8. Explain why you agree or disagree with each of the proposals given on p. 633 concerning the U.S. national park system.

9. Should more wilderness areas and wild and scenic rivers be preserved in the United States, especially in the lower 48 states? Explain.

Q: What percentage of commercially used chemicals in the United States are regulated by federal and state governments?

10. Should a large portion of U.S. Forest Service and Bureau of Land Management land be reclassified as wilderness recovery areas, closed off for many uses, and allowed to undergo natural restoration as wildlife habitat? Explain.

11. Is a protected area "useless" because it does not add significantly to economic growth? Explain. Describe the usefulness of an area in ecological terms.

12. How might the database of species, communities, and land protection status be used to establish new reserves and change land management practices? Do you believe that having the information available will actually lead to successful conservation efforts? Explain.

13. Try to come up with analogies to complete the following statements: **(a)** Returning wolves to the Yellowstone National Park area is like _______. **(b)** Exterminating coyotes to keep them from killing sheep is like _______. **(c)** Subsidizing ranchers, miners, and timber cutters who remove resources from public lands is like _______. **(d)** Overgrazing rangeland is like _______. **(e)** The Gap Analysis Program is like _______.

14. Do you have a favorite natural place that no longer exists? What happened? How did this make you feel?

PROJECTS

1. Obtain a topographic map of the region where you live and use it to identify local, state, and federally owned lands in the form of parks, rangeland, forests, and wilderness areas. Identify the government agency responsible for management of each of these areas and evaluate how well these agencies are preserving the natural resources on this land on your behalf.

2. Evaluate cattle grazing on private and public rangeland and pastures in your local area. Try to document any harmful environmental impacts. Have the economic benefits to the community outweighed any harmful environmental effects?

3. Make a survey of the national, state, and local parks within a 97-kilometer (60-mile) radius of your local community. How widely are they used by local residents and by outside visitors? What is their condition? Develop a plan for their improved management.

4. Make a concept map of this chapter's major ideas, using the section heads and subheads and the key terms (in boldface). Look at the inside back cover and in Appendix 4 for information about concept maps.

A: About 0.5%

24 SUSTAINING ECOSYSTEMS: DEFORESTATION, BIODIVERSITY, AND FOREST MANAGEMENT*

How Farmers and Loud Monkeys Saved a Forest

It's early morning in a tropical forest in the Central American country of Belize. Suddenly, loud roars that trail off into wheezing moans—territorial calls of black howler monkeys (Figure 24-1)—wake up everyone in or near the wildlife sanctuary by the Belize River. These long-tailed primates, who live in small troops headed by a dominant male, travel slowly among the treetops, feeding on leaves, flowers, and fruits.

This species is the centerpiece of an experiment recruiting peasant farmers to preserve tropical forests and wildlife. The project is the brainchild of an American biologist, Robert Horwich, who in 1985 met with villagers and suggested that they establish a sanctuary that would benefit the local black howlers and themselves.

He proposed that the farmers leave strips of forest along the edges of their fields to provide the howlers with food and a travel route through the sanctuary's patchwork of active garden plots and young and mature forest. Leaving strips of forest along the river, he noted, would also reduce soil

*Paul M. Rich, associate professor of Ecology and Evolutionary Biology and Environmental Studies at the University of Kansas, is coauthor of this chapter.

Figure 24-1 A black howler monkey, shown in a tropical forest in Belize. The populations of six howler monkey species have declined as tropical forests have been cleared. The black howler, like more than two-thirds of the world's 150 known species of primates, is threatened with extinction. (Carol Farnetti/Planet Earth Pictures)

erosion and river silting, yielding more fish for the villagers.

To date, more than 100 farmers have participated, and the 47-square-kilometer (18-square-mile) sanctuary is home for an estimated 1,100 black howlers. The idea has spread to seven other villages.

Now, as many as 6,000 ecotourists visit the sanctuary each year to see its loud monkeys and other wildlife. Villagers serve as tour guides, cook meals for the visitors, and lodge tourists overnight in their spare rooms.

As long as this ecotourism doesn't disturb the sanctuary ecosystem, this experiment can demonstrate the benefits of integrating ecology and economics, allowing villagers to make money by helping sustain the forest and its wildlife.

A diverse array of forests (Figure 7-11 and Section 7-5) is a vital part of the earth capital (Figure 1-2) that helps sustain the ecosystems and economies. The trees and soils in forests also absorb emissions of carbon dioxide gas (mostly from burning fossil fuels) and thus help slow the projected increase in the earth's average temperature through an enhanced greenhouse effect (Figure 7-9 and Sections 19-1 and 19-2). Forests are home to 50–90% of the earth's species and are also potentially renewable resources, if used sustainably.

However, forests—especially tropical forests (Figures 7-23 and 7-24) and old-growth forests in the northwestern United States and southwestern Canada—are disappearing faster than any other biome. They are being cleared for timber, paper pulp, or fuelwood and are being replaced by tree farms (see photo in the table of contents), cropland, grazing land, mines, reservoirs, and cities. They are also threatened by air pollution (see photo on p. 435 and Figure 18-9) and by possible climate change (Section 19-3) that their own destruction is helping to cause.

If incredibly diverse and important tropical forest (Figure 7-23) and temperate old-growth forest (Figure 7-25) biomes continue to vanish at the current rate, most will be gone within 30–50 years. Preventing further losses of these vital forms of earth capital and havens of biodiversity, learning how to use them more sustainably, and speeding the healing of forests we have degraded (Section 10-6) are urgent priorities. Each of us can play a role in this important challenge.

Forests precede civilizations, deserts follow them.

FRANCOIS-AUGUSTE-RENÉ DE CHATEAUBRIAND

In this chapter we answer the following questions:

- What are the major types of forests, and why are they such important ecosystems?
- How has deforestation occurred in North America, why are the remaining old-growth forests important, and what can be done to prevent their destruction?
- How fast is tropical deforestation proceeding, and why should we care about this problem?
- How can forest resources be managed and conserved?
- How can we solve problems of tropical deforestation and fuelwood shortages?
- What can individuals do to help protect and sustain forests?

24-1 FORESTS: TYPES AND IMPORTANCE

What Are the Major Types of Forests? There are three general types of forests, depending primarily on climate: tropical, temperate, and polar (Figure 7-12). Since agriculture began about 10,000 years ago, human activities have reduced the earth's forest cover by about one-quarter—from about 34% to 26% of the world's land area (excluding Antarctica and Greenland)—and only 12% consists of intact forest ecosystems.

Slightly more than half of the world's forests are in the tropics; the rest are in temperate and boreal zones (Figure 7-11). More than 60% of these remaining forests are in seven countries: Brazil, Russia, Canada, the United States, China, Indonesia, and Congo (formerly Zaire).

If the rate of cutting and degradation does not exceed the rate of regrowth, and if protecting biodiversity is emphasized, forests are renewable resources. However, forests are disappearing or are being fragmented and degraded almost everywhere, especially in tropical countries (Figure 24-2). At least 2 million square kilometers (770,000 square miles) of forest were lost between 1990 and 1995—an area three times larger

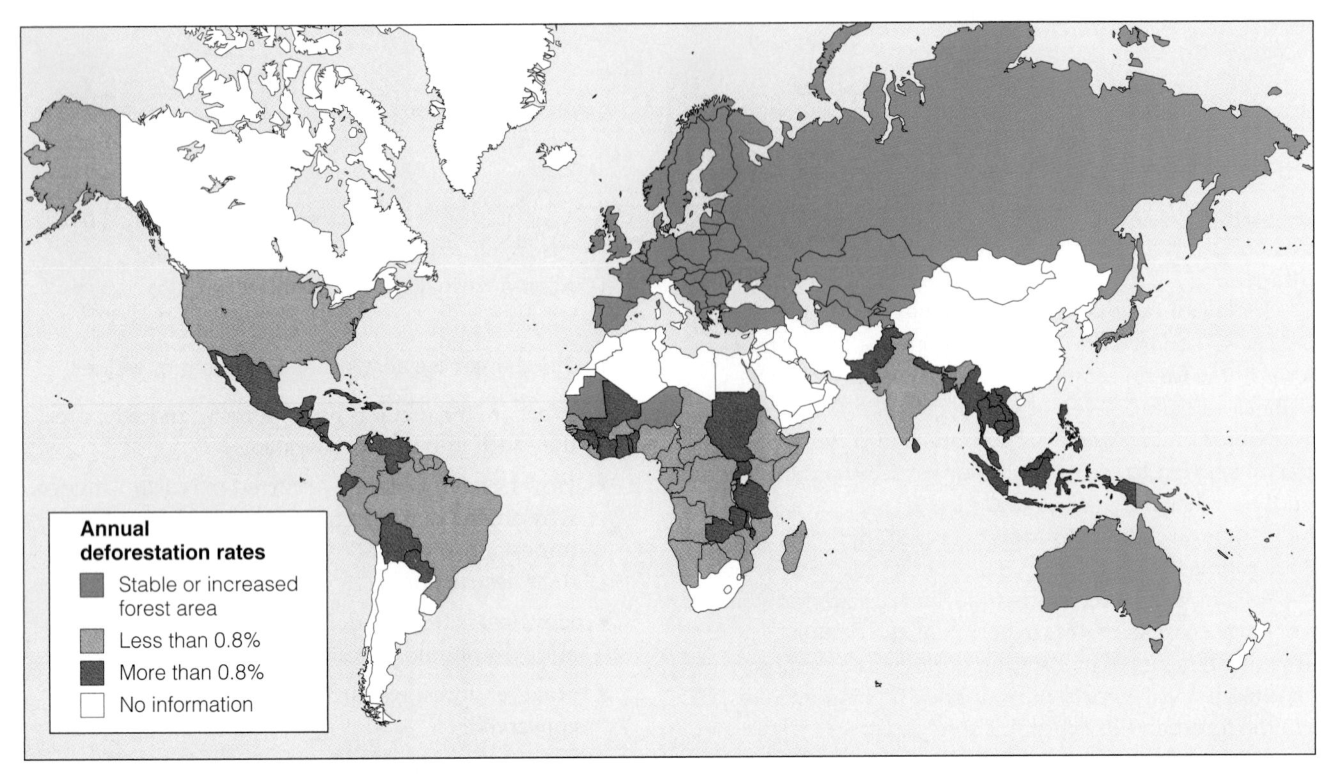

Figure 24-2 Estimated annual rates of deforestation, 1981–90. During this period the area of the world's tropical forest decreased by 8% (an average of 0.8% per year). Similar rates of deforestation have occurred during the 1990s. (Data from UN Food and Agriculture Organization)

than the state of Texas. Each year there is a net loss of another 160,000 square kilometers (62,000 square miles) of forest. At least 107 countries had a net loss of forest between 1990 and 1995.

Old-growth forests are uncut forests and regenerated forests that have not been seriously disturbed for several hundred or thousands of years. Examples include forests of Douglas fir, western hemlock, giant sequoia, and coastal redwoods in the western United States; loblolly pine in the Southeast; boreal forests in Russia, western Canada, and Alaska; and about 60% of the world's tropical forests. Two-thirds of Europe's old-growth forests are gone and in the United States 95–98% are gone.

Old-growth forests provide ecological niches for a multitude of wildlife species (Figure 7-24). These forests also have large numbers of standing dead trees (snags) and fallen logs, which are habitats for a variety of species. Decay of this dead vegetation returns plant nutrients to the soil and helps build fertile soil (Figure 4-13).

Second-growth forests are stands of trees resulting from secondary ecological succession after cutting (Figure 9-21). Most forests in the United States and other temperate areas are second-growth forests that grew back after old-growth forests were cleared for timber or to create farms that were later abandoned. About 40% of tropical forests are second-growth forests.

Some second-growth stands have remained undisturbed long enough to become old-growth forests, but many are not diverse forests at all, but rather are *tree farms* or *plantations* (see photo in table of contents). They are managed tracts with uniformly aged trees of one species that are harvested by clear-cutting as soon as they become commercially valuable. Then they are replanted and clear-cut again on regular cycles. Between 1980 and 1995 the world's area of tree farms doubled and by 1995 occupied a total area slightly smaller than the land area of Mexico. The area devoted to tree plantations is projected to double between 1995 and 2010.

What Is the Economic Importance of Forests? Forests provide lumber for housing, biomass for fuelwood, pulp for paper, medicines, and many other products (Figure 24-3) worth more than $300 billion a year. Many forestlands are also used for mining, grazing livestock, and recreation. Over half of the world's timber harvested for industrial use is consumed by the 20% of the world's people who live in the United States, Western Europe, and Japan. Each year Americans consume enough wood and wood products to fill a train of 2 million boxcars encircling the earth at the equator.

Worldwide, about 55% of the timber cut each year is used as fuelwood and charcoal for heating and cooking, mostly in developing countries. Some of this fuel

Q: What percentage of fatal cancers in the U.S. are caused by ionizing radiation from natural and human sources?

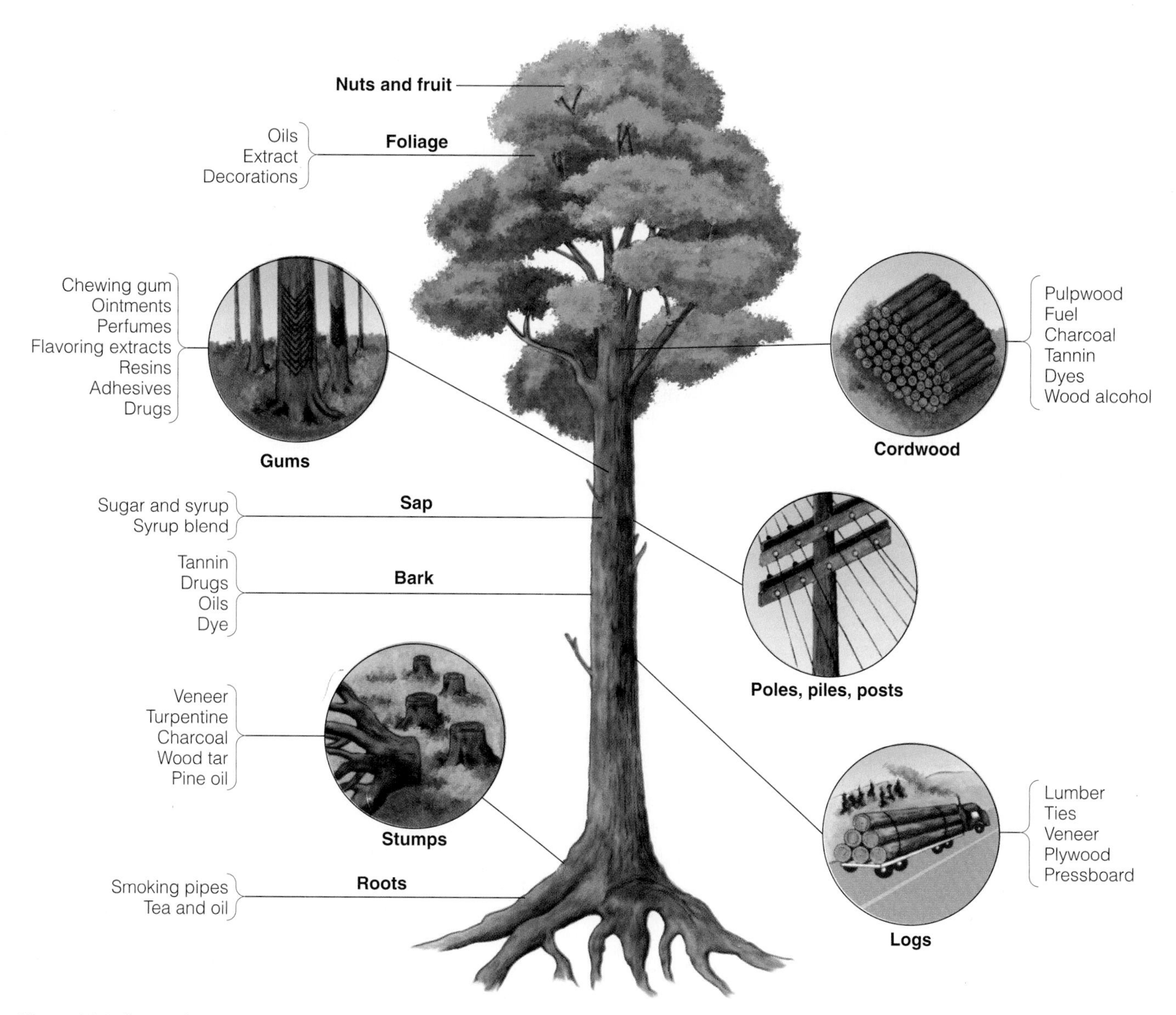

Figure 24-3 Some of the many useful products obtained from trees.

is burned directly as firewood. Some is converted into charcoal, which is widely used by urban dwellers in developing countries and by some industries. One-third of the world's annual harvest is sawlogs that are converted into building materials: lumber, plywood, hardboard, particleboard, and chipboard.

One-sixth is converted into pulp used in a variety of paper and other products. About 66% of the paper produced worldwide is made from harvested logs, 30% comes from recycled wastepaper, and 4% from nonwood sources such as cotton or rice straw.

Since 1950, the global demand for wood has doubled and paper use has increased more than fivefold. Between 1995 and 2010, global demand for paper is expected to double again. The United States has the world's highest per capita use of paper—about seven times the average global per capita use. Other countries with very high per capita uses of paper are Japan and Germany.

The United States is the world's largest importer of wood products. Nearly 20% of its domestic consumption is imported, most of it softwood (conifer) lumber from Canada. The United States also imports hardwood from tropical forests, mostly in Indonesia, Malaysia, and Brazil. The United States is also one of the world's largest exporters of wood, second only to Canada. Most U.S. wood exports go to Japan, Canada, Germany, South Korea, and Mexico.

What Is the Ecological Importance of Forests? Forested watersheds act as giant sponges, slowing down runoff and holding water that recharges springs, streams, and groundwater. Thus, they regulate the

A: About 1%

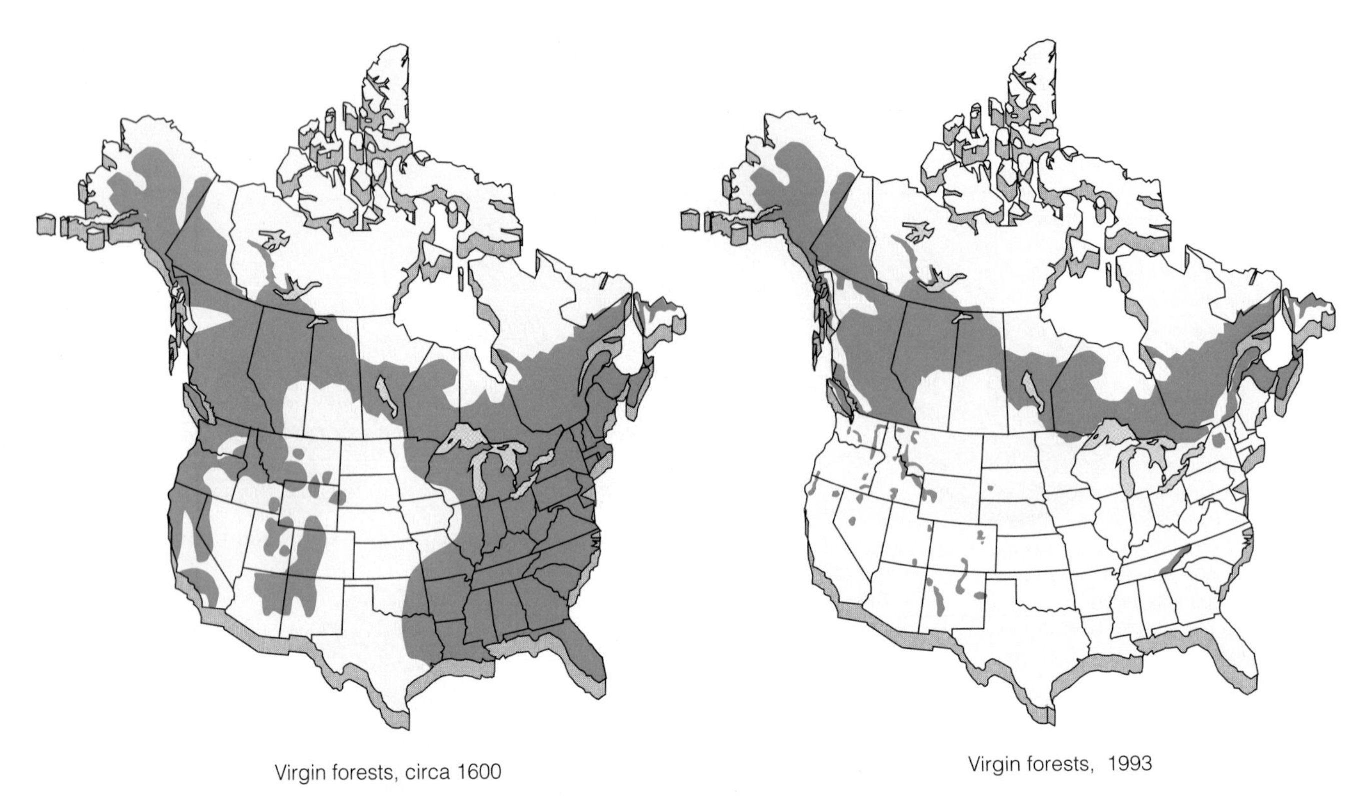

Figure 24-4 Vanishing old-growth forests in the United States and Canada. Since about 1600 most of the old-growth forests that once covered much of the lower 48 states have been cleared away; most of the remaining old-growth forests in those states are on public lands (especially in the Pacific Northwest). About 60% of old-growth forests in western Canada have been cleared, and much of what remains is slated for cutting. (Data from the Wilderness Society, the U.S. Forest Service, and *Atlas Historique du Canada*, vol. 1)

flow of water from mountain highlands to croplands and urban areas, and they reduce the amount of sediment washing into streams, lakes, and reservoirs by reducing soil erosion.

Forests also influence climate. For example, 50–80% of the moisture in the air above tropical forests comes from trees via transpiration and evaporation. If large areas of these lush forests are cleared, average annual precipitation drops and the region's climate gets hotter and drier; then soils become depleted of already-scarce nutrients (Figure 5-16), and they are baked and washed away. This process can eventually convert a diverse tropical forest into a sparse grassland or even a desert.

Forests are vital to the global carbon cycle (Figure 5-5). They take up around 90% of the carbon removed from the atmosphere as CO_2 by terrestrial ecosystems and provide habitats for more wildlife species than any other biome. They also buffer us against noise, absorb air pollutants, and nourish the human spirit.

According to one calculation, a typical tree provides $196,250 worth of ecological benefits during its lifetime, in the form of oxygen, air purification, soil fertility and erosion control, water recycling and humidity control, and wildlife habitats. Sold as timber, the same tree is worth only about $590. Even if such estimates are off by a factor of 100, the long-term ecological benefits of a tree still clearly exceed its short-term economic benefits.

The ecological benefits of forests are undervalued in the marketplace. Until this situation is changed, most analysts believe that we will continue to sacrifice these forests and their long-term ecological services for short-term economic gain (Spotlight, right).

24-2 OLD-GROWTH DEFORESTATION IN THE UNITED STATES AND CANADA

How Rapidly Are Old-Growth Forests Being Cleared in the United States? Today, American forests are generally bigger and often healthier than they were in 1900, when the country's population was below 100 million. Indeed, since 1950 the country's total volume of timber has increased by 30%. Reasons for this include reversion of marginal farmlands to forests, planting of tree farms, more efficient use of wood products, paper and wood recycling, and substitution of other materials for construction lumber.

This is an important accomplishment, but it can be misleading. An estimated 85–95% of the untouched, temperate-zone old-growth forests that once covered

 Q: What percentage of genetic defects in the U.S. population are caused by ionizing radiation from natural and human sources?

SPOTLIGHT

How Much Is a Forest Worth?

In 1997 a group of 13 ecologists, economists, and geographers from 12 prestigious laboratories and universities attempted to estimate how much nature's ecological services (Figure 1-2) are worth. According to this ambitious and crude appraisal led by ecological economist Robert Costanza of the University of Maryland, the economic value of Mother Nature's bounty ranges from $16 trillion to $54 trillion. In other words, each year nature provides free goods and services worth somewhere between $2,800 and $9,000 per person.

Their medium estimate for nature's life-support services for humans is $33 trillion a year. This is nearly twice the $18 trillion per year economic output of the earth's 194 nations and almost five times the $6.9 trillion U.S. GDP in 1996.

To make these estimates the researchers divided the earth's surface into 16 biomes (Figure 7-11) and aquatic life zones such as various types of forests, grasslands, cropland, open oceans, estuaries, wetlands, and lakes and rivers (they omitted deserts and tundra because of a lack of data). Then they agreed on a list of 17 goods and services provided by nature (Figure 1-2) and sifted through more than 100 studies that attempted to put a dollar value on such services in the 16 different types of ecosystems.

Some analysts believe that such estimates are misleading and dangerous because they amount to putting a dollar value on ecosystem services with an infinite value because they are irreplaceable. In the 1970s economist E. F. Schumacher warned that "to undertake to measure the immeasurable is absurd" and is a "pretense that everything has a price." Biologist and conservationist David Ehrenfield's response to this evaluation was, "I am afraid that I don't see much hope for a civilization so stupid that it demands a quantitative estimate of the value of its own umbilical cord." Those who cringe at the thought of putting a dollar value on nature's ecosystem services believe that they should be protected on moral grounds instead of being reduced to easily manipulated cost–benefit analyses (Spotlight, p. 762).

The researchers admit that their estimates rely on many assumptions and omissions and could easily be too low by a factor of 10 to 1 million or more. For example, their calculations include only estimates of the ecosystem services themselves, not the natural or earth capital that generates them, and omitted the value of nonrenewable minerals and fuels. They also recognize that as the supply of ecosystem services declines their value will rise sharply and that such services can be viewed as having an infinite value.

However, they contend that their estimates are much more accurate than the zero value the market assigns to these ecosystem services. They hope such estimates will call people's attention to the fact that the earth's ecosystem services are absolutely essential for all humans and their economies and that their economic value is huge. They believe that failure to estimate economic values for ecosystem services helps ensure that such vital life-support services will continue to be degraded for short-term economic gain.

Critical Thinking

Do you agree or disagree with the idea of trying to estimate the economic value of the earth's ecosystem services? Explain. What are the alternatives?

much of what is now the lower 48 states when European settlers arrived has been cleared away (Figure 24-4). Most of the old-growth forests that were cleared or partially cleared between 1800 and 1960 grew back as fairly diverse second-growth (and in some cases third-growth) forests. However, since the mid-1960s a rapidly increasing area of the nation's remaining old-growth and fairly diverse second-growth forests has been clear-cut and replaced with tree plantations (especially in the South). As a result, forest cover in the Pacific Northwest declined by 20% between 1952 and 1992. Although this doesn't reduce the nation's overall tree cover, it does reduce overall forest biodiversity and ecological integrity.

Most of the remaining old-growth forests are in fragmented sections on U.S. public lands in Washington, Oregon (Figure 24-5), and northern California. The huge conifers that dominate much of this region often live for 500 years and can survive for as long as 3,500 years—15 times longer than the United States has been in existence.

Large areas of these ancient forests have been destroyed and fragmented by roads and clear-cuts for short-term economic gain. At the current lower rate of cutting, most unprotected ancient forests in western Washington and Oregon will be gone within 2–3 decades.

Connections: Ecology of Old-Growth Forests in the Northwest In the Pacific Northwest it typically takes 350 years for an old-growth forest to reach its prime in terms of growth and diversity. These forests have a number of important ecological functions. They have the

A: About 5–6%

Figure 24-5 Fragmentation of old-growth forest is evident in this computer-enhanced satellite image of a 1,300-square-kilometer (500-square-mile) area of Mount Hood National Forest in Oregon in 1991. Old-growth forest patches are dark red; clear-cut old-growth forests regenerating as tree farms are lighter red; and recently clear-cut areas are blue-green and white. Most of the same area of Amazonian rain forest shown in Figure 24-14 is still undisturbed, whereas this old-growth forest has been extensively cleared and fragmented into vulnerable patches. (NASA/Goddard Space Flight Center)

world's largest accumulation of dead standing trees (snags) and fallen dead trees, which decompose slowly over 200–400 years and recycle nutrients in the forest ecosystem (Figure 4-13). For their latitude, these forests are unusually rich in wildlife species, and they act as giant sponges that hold and slowly release enough moisture to help protect against fires and floods and recharge nearby streams and aquifers.

Fragmenting these forests into small patches, or clearing and replacing them with tree plantations and second-growth forests, disrupts the vital network of symbiotic relationships that help sustain the complex, old-growth forest communities. Without snags and fallen trees, about 29% of the wildlife in old-growth forests would disappear.

The survival of old-growth forests in the Pacific Northwest is also connected to the survival of the region's salmon stocks (Figure 12-19); annually, the $1.25-billion sport and commercial salmon fishing industry provides jobs for 62,000 people. Several factors—including damming of rivers, pollution, and overfishing—are involved in the decline of salmon stocks, but a key factor has been excessive logging on public lands throughout the region. Sediment deposited in streams from the logging of old-growth forests and associated road building smothers spawning beds and disrupts the feeding patterns of young salmon.

Case Study: Should Remaining Old-Growth Forests on U.S. Public Lands Be Cut or Preserved? To officials of timber companies, the giant living trees and rotting dead trees in the Pacific Northwest's old-growth forests are valuable resources that could provide jobs and should be harvested for profit, not left alone to please environmentalists. These officials also point out that the timber industry annually pumps millions of dollars into the region's economy and provides jobs for about 100,000 loggers and millworkers. Timber officials claim that protecting large areas of remaining old-growth forests on public lands will cost as many as 53,000 jobs and hurt the economies of logging and milling towns in such areas.

To conservation biologists, the remaining ancient forests on the nation's public lands are a treasure whose ecological, scientific, aesthetic, and recreational values far exceed the economic value of cutting them down for short-term economic gain. They argue that the fate of these forests is a national issue because these forests are owned by all U.S. citizens. They also view this as a global issue because these forests are important reservoirs of the earth's biodiversity and ecological integrity.

Conservation biologists and economists point out that the major causes of past and projected job losses in the timber industry in this region are automation, export of logs overseas (depriving U.S. millworkers of jobs while providing jobs for millworkers in other countries), and timber imports from Canada. Loggers, millworkers, and store owners living in logging communities are caught in the middle of a highly controversial situation.

Environmentalists charge the U.S. Forest Service with playing a key role in destroying old-growth forests on public lands. One reason is that Congress allows the Forest Service to use most of the income from timber sales in national forests to supplement its budget; any financial losses it incurs are passed on to taxpayers. This gives the Forest Service a powerful incentive to make timber sales on public lands its primary objective. Many elected officials in local communities near national forests also push for increased timber harvesting because their communities get 25% of the gross receipts from such cutting. Some dynamics of the U.S. economic system also reward timber companies for liquidating old-growth timber on their own lands instead of harvesting it sustainably by selective cutting (Connections, right).

Q: What percentage of the annual deaths worldwide are caused by infectious diseases?

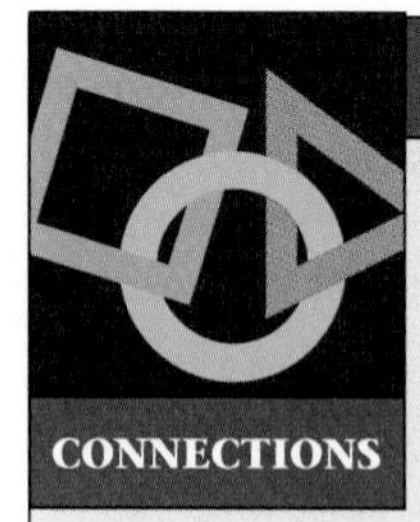

Junk Bonds Versus Redwood Trees

About 97% of original redwood forests found along the coasts of California to Oregon have been cleared. The philosophy of such overexploitation was summed up by Ronald Reagan when he was governor of California: "When you've seen one redwood tree, you've seen them all." Of the 3% of these forests left, about three-quarters are protected in parks.

Almost half of the remaining unprotected ancient redwood trees are owned by one company, Pacific Lumber. Until 1986 it was a model logging company: highly profitable and debt-free, with a $55-million surplus in its employee pension fund. To make its redwood forests—located some 400 kilometers (250 miles) north of San Francisco—last indefinitely, it had for decades selectively harvested individual trees no faster than they grew back.

In 1986, the company was taken over by Maxxam Corporation, headed by Texas corporate raider Charles Hurwitz, in a leveraged buyout arranged by Ivan Boesky (later jailed for illegal financial dealings). In such a buyout, investors borrow huge sums of money, usually through the sale of high-risk "junk" bonds, to buy a company they think is undervalued in the stock market. After using the company's assets as collateral for the borrowed funds, they obtain the cash needed to pay off the interest on the debt and the junk bonds by selling off the company's undervalued assets.

Timber companies are often attractive takeover targets because their forest assets can be cut and quickly converted into cash that can be invested to take over other companies. After raising $900 million in junk bonds to buy up Pacific Lumber's stock, Hurwitz drained the surplus from the pension fund and sold off the company's welding division and office building for $417 million.

Then, to help pay off the interest on the junk bonds, he began liquidating the company's timberlands, estimated to be worth more than $1.4 billion. He did this mainly by increasing the rate of logging and by switching from sustainable selective cutting to unsustainable clear-cutting of the company's towering 800-year-old redwoods—the earth's largest stand of privately owned virgin redwood trees. The timber from a single large redwood tree can be worth as much as $200,000.

Some of what's left of these forests may get a reprieve. Hurwitz owes the federal government $250 million for the bailout of his failed United Savings Association of Texas, which cost taxpayers $1.6 billion. If an agreement by California lawmakers and Congress goes through, Hurwitz will be paid $460 million in cash or in state and federal property (with $250 million coming from the federal government and $210 from the California government). In return Hurwitz would give 1,400 hectares (3,500 acres) of ancient redwoods and 1,600 hectares (4,000 acres) of additional land (most of it heavily logged) to the state of California and the federal government. Hurwitz's company would still own about 81,000 hectares (200,000 acres) of timberland (some of it clear-cut and scarred).

Although this is a step in the right direction, environmentalists want protection for all of the company's irreplaceable old-growth redwood forests, or at least a binding agreement to harvest what's left using a sustainable rate of selective cutting instead of clear-cutting. To support complete protection, environmental activist Julia Hill has been illegally camping out on a platform near the top of a giant 1,000-year-old redwood tree for over a year.

She has survived high winds and freezing rain. Timber company officials tried to get her to quit by buzzing the tree with a helicopter, trying to starve her out with a 10-day security blockade, and harassing her with air horns and flood lights and cutting down trees around her. She and many environmentalists say Maxxam is being grossly overpaid by taxpayers for the land they are giving up through a form of environmental blackmail and believe that the entire tract should be preserved.

Other timber companies that had been sustaining their forests are being forced to liquidate (clear-cut) their timber assets to avoid being taken over. Environmentalists call for changes in tax laws and lending regulations to halt such rapid depletion of natural resources, but such measures are opposed by most lending agencies. What do you think should be done?

Critical Thinking

1. Do you believe that tax laws and lending regulations should be changed to help halt rapid depletion of forests and other natural resources? Explain. What disadvantages might such laws have? If you favor such laws, how would you see that they are passed and implemented?

2. Do you believe that all of the redwood forests owned by Maxxam should be protected from harvesting? Explain. If you agree, how would you accomplish this goal?

A: About 32% (44% in developing countries and 5% in developed countries)

Figure 24-6 The threatened northern spotted owl, which lives almost exclusively in old-growth forests of the Pacific Northwest. Environmentalists have used the owl's threatened status both to save it and to halt or decrease logging in the national forests in which it lives, thereby advancing their wider goal of preserving biodiversity. (Pat & Tom Leeson/ Photo Researchers, Inc.)

The threatened northern spotted owl (Figure 24-6) lives almost exclusively in 200-year-old Douglas fir forests in western Oregon, Washington, and northern California (mostly in 17 national forests and 5 Bureau of Land Management parcels). The species is vulnerable to extinction because of its low reproductive rate and the low survival rate of juveniles through their first 5 years. Only 2,000–3,600 pairs remain. However, logging company officials contend that the owls do not require old-growth forest and can instead adapt to younger second-growth forests.

In July 1990 the U.S. Fish and Wildlife Service added the northern spotted owl to the federal list of threatened species. This requires that its habitat be protected from logging or other practices that would decrease its chances of survival. The timber industry hopes to persuade Congress to revise the Endangered Species Act to allow for economic considerations that will reverse this protection.

The owl and other threatened species are symbols of the broader clash between timber companies who want to clear-cut most remaining old-growth stands in the national forests and conservation biologists who want to protect their priceless biodiversity and ecological integrity, or at least allow only sustainable harvesting in some areas. The question is, "Should the biodiversity and ecological integrity of the last 5–8% of remaining ancient forests on public lands in the Northwest be preserved to protect their biodiversity and ecological integrity or cut to save a final decade or so of logging jobs?" What do you think?

It is encouraging to conservationists that the amount of timber harvested from public and private lands in western Oregon and Washington dropped by 38% between the late 1980s and 1996 (and 87% on federal lands). Some of this drop has been made up by increased softwood harvests elsewhere, especially industrial pine plantations in the Southeast. The rest has been made up by increased imports, mostly from Canada.

Case Study: Destruction of Old-Growth Forests in Western Canada With nearly 50% of its land forested (much of it in second-growth forest), Canada has 10% of the world's forests. One of every 10 jobs in Canada is directly related to forestry. The country is by far the world's largest exporter of timber products valued at more than $30 billion per year.

Even though provincial governments own and manage 94% of Canada's forests, the country has lost an estimated 60% of its old-growth forests (most of them publicly owned) to logging (Figure 24-4); less than 20% of what remains is in protected areas. Ninety percent of logging in Canada involves clear-cutting, a practice expected to double over the next few years.

Western Canada's Texas-sized province of British Columbia produces more than half of the country's timber and pulp. More than half of this province's ancient forests have already been cut, with most of the logs exported to make lumber, newspapers, and phone books. Each year clear-cutting of old-growth boreal and diverse temperate rain forest (some containing trees up to 2,000 years old) in this province removes more than twice as many trees as the harvest from all U.S. national forests combined. One-half of the world's remaining temperate rain forests—most of them slated to be clear-cut—are also located in British Columbia.

Critics charge that the Canadian government is giving massive subsidies to logging companies (many of them Japanese owned) without providing many jobs for Canadians. Most of the profit is not even made in Canada. For example, Canada's western province of Alberta collects only $0.90 for the 16 aspen trees that will make $590 worth of pulp shipped to Japan, where a Japanese mill and workers will turn the pulp into about $1,300 worth of paper.

Manser-Fonds, Bruno. 1998. "Why Are Forests Burning?" *Ecologist*, vol. 28, no. 1, 8(1).

Environmentalists and a growing number of Canadians were outraged when the provincial government spent $50 million to buy a 4% share in MacMillan Bloedel, British Columbia's largest international forest products company. Moreover, it did this only a few weeks before it awarded the company a huge contract to clear-cut the magnificent ancient forest of Clayoquot (pronounced "clack-what") Sound on the spectacular west coast of Vancouver Island. Under intense pressure from environmental groups and subject to the glare of unfavorable international publicity, the British Columbia government announced in 1995 that it would end clear-cutting in Clayoquot Sound.

In response to widespread national and international protests about its rapid degradation of forest resources, British Columbia also enacted a Forest Practices Code in 1995. However, surveys and audits by environmental organizations revealed that the law is largely being ignored. In June 1997, the government even eased the code.

To environmentalists, the situation is grim, but there is some hope. In the 1990s citizens' groups throughout Canada formed a powerful grassroots network of environmentalists, labor union members, and citizens called Canada's Future Forest Alliance. Their goal is to direct international attention to Canada's rapid deforestation and to pressure the provincial and federal governments to restructure forest policies to emphasize truly sustainable forestry and social justice for native peoples.

They also encourage people to boycott companies and governments buying timber from Canada. Several companies in Europe and in the United States (including the *New York Times*) have announced that they will no longer buy wood products and paper from MacMillan Bloedel. Such critics point out that it is hypocritical for Americans to tell Canadians how to manage their forestlands when the United States increases Canadian imports of timber to make up for timber harvest losses from protected U.S. national forests.

As Canadian forestry activist Colleen McCrory (awarded the Goldman Environmental Prize for her efforts to protect Canadian forests and founder of Canada's Future Forest Alliance) puts it:

> *We've got the Brazil of the North happening right here. . . . In the Amazon region, one acre of forest is lost every nine seconds to logging and forest clearing. In Canada, one acre every twelve seconds is lost to logging or burning. . . . In Brazil, the desperation of poverty and rapid population growth are the central driving factors in deforestation, whereas in Canada, the central cause of deforestation is the greed of multinational corporations, and their ability to dominate debate about forest policy because of their wealth and political influence.*

24-3 TROPICAL DEFORESTATION AND LOSS OF BIODIVERSITY

How Fast Are Tropical Forests Being Cleared and Degraded? Tropical forests cover about 6% of the earth's land area (roughly the area of the lower 48 states of the United States) and grow in equatorial Latin America, Africa, and Asia (Figure 7-11). Four countries—Brazil, Indonesia, Zaire, and Peru—contain more than half the world's total. Tropical forests come in several varieties (Figure 7-12): rain forests (which receive rainfall almost daily), tropical deciduous forests (with one or two dry seasons each year), dry and very dry deciduous forests, and forests on hills and mountains.

Climatic and biological data suggest that mature tropical forests once covered at least twice as much area as they do today, with most of the destruction occurring since 1950. Between 1960 and 1990, about one-fifth of all tropical forest cover was lost (with Asia losing one-third). Between 1975 and 1997, the area deforested in the Amazon increased 20-fold and is roughly equal to the area of Germany. Between 1995 and 1997 satellite pictures revealed that an area of tropical forest twice the size of Belgium was deforested

Satellite scans and ground-level surveys used to estimate forest destruction indicate that large areas of tropical forests are being cut (Figure 24-2). It's estimated that an equivalent area of these forests is seriously degraded and fragmented without being destroyed outright each year.

There is considerable debate over the current rates of tropical deforestation and degradation because of difficulties in interpreting satellite images, different ways of defining deforestation and forest degradation, and political and economic factors that cause countries to hide or exaggerate deforestation. The lowest estimated rate of loss and degradation of remaining tropical forests is 62,000 square kilometers (24,000 square miles) per year. This is equivalent to a loss and degradation of *14 city blocks of tropical forest per minute.* The highest estimated rate of destruction and degradation is 308,000 square kilometers (118,000 square miles) per year—equivalent to a loss and degradation of *68 city blocks of tropical forest per minute.*

Whether the best estimate of tropical deforestation and degradation is 14 or 68 city blocks per minute (or somewhere in between), it is clear that these reservoirs of biodiversity and ecological integrity are being lost and degraded at a high rate. Scientists estimate that this annual rate of destruction and degradation could well double within another decade. Indeed, recent data indicate that the rate of clearing and burning tropical forests in the Amazon Basin increased by about 34% between 1991 and 1994.

Hint: Enter the search terms *rain forests, environmental aspects* using the Subject Guide.

About 40% of current tropical deforestation is taking place in South America (especially in the vast Amazon Basin). However, the *rates* of such deforestation in Southeast Asia and Central America are about 2.7 times higher than those in South America. Haiti has lost 99% of its original forest cover, the Philippines 97%, and Madagascar 84%. Most attention has focused on the Amazon Basin, but 92% of the once-vast rain forests on Brazil's Atlantic coast and 98% of the coniferous Araucaria forests of the south have been devastated by logging and urban expansion.

Case Study: Madagascar: A Threatened Jewel of Biodiversity Madagascar, the world's fourth largest island, lies in the Indian Ocean off the east African coast (Figure 24-7). Most of its species have evolved in near isolation for at least 40 million years, after the continental drift of the earth's lithospheric plates (Figure 6-12) moved the island far enough from Africa's mainland to prevent migrations to and from Africa.

The result is an estimated 160,000 species that are unique to this island, mostly in its vanishing eastern rain forests. Unique species include 80% of its 10,000 species of flowering plants (including 1,000 orchids), 60% of the world's species of chameleons (Figure 24-8), 800 butterfly species, half of the island's birds, and all its reptiles and mammals, including the ring-tailed lemur (Figure 24-9). Because of its astounding biological diversity, this Texas-size island is considered a crown jewel among the earth's ecosystems (Figure 23-18).

Madagascar's plant and animal species are among the world's most endangered, mostly because of loss of habitat from slash-and-burn agriculture (Figure 1-23) on poor soils, fueled by rapid population growth. Since humans arrived about 1,500 years ago, 84% of the island's tropical seasonal forests and over 66% of its rain forests have been cut for cropland, fuel, and lumber. The resulting eroded fields and hillsides make Madagascar the world's most eroded country, with huge quantities of eroded sediment flowing in its rivers and emptying into its coastal areas (Figure 8-5).

Since 1984 the government, conservation organizations, and scientists worldwide have united to slow

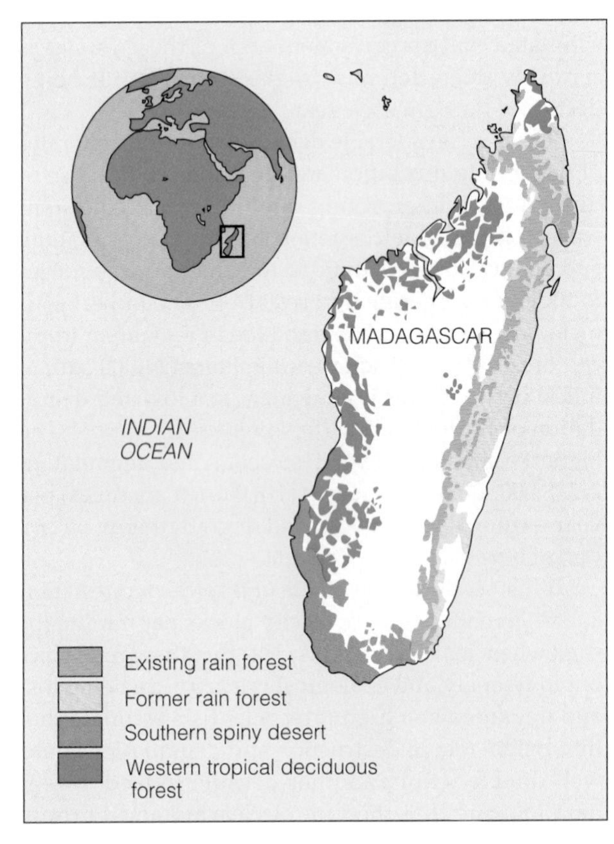

Figure 24-7 Some biomes in Madagascar, and its location off the coast of Africa.

Figure 24-8 Parson's giant chameleon, in a tropical forest in Madagascar. The largest of the chameleons, an adult Parson's can reach 70 centimeters (28 inches) in length. Among its prey are small birds. Madagascar is the only home for 60% of the world's known species of chameleons, some as small as a thumbnail. (Gérand Locz/NHPA)

Figure 24-9 Endangered ring-tailed lemur in Madagascar. The island's 30 species of lemurs, the oldest distant relative of the human species, are found nowhere else in the wild. Half of them are endangered. (Robert and Linda Mitchell)

Sekhran, Nik. 1997. "Green or Greed." *Geographical Magazine*, vol. 69, no. 10, 75(7).

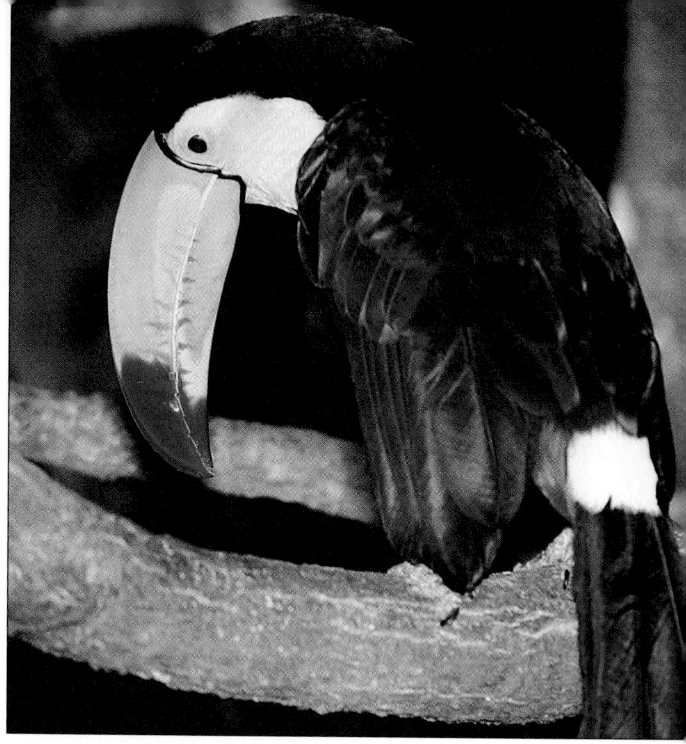

Figure 24-10 Tropical forests are the planet's largest storehouse of biological diversity. Two of the species found there are the red uakari monkey (left) from the Peruvian portion of the Amazon basin, and the keel-billed toucan from Belize in Central America (right). Because most tropical forest species have specialized niches, they are highly vulnerable to extinction when their forest habitats are removed or damaged. (Left, Roy P. Fontaine/ Photo Researchers, Inc.; Right, William Grenfell/Visuals Unlimited)

the island's plunge into wasteland. For such efforts to succeed, population growth, projected to double between 1998 and 2025, will have to slow drastically. Local people must participate in the planning and management of conservation areas and of land they occupy and the trees they plant, and learn to make a living from reforestation, ecotourism, and sustainable use of forest, wildlife, and soil resources.

The hour is late, but if fully supported and implemented, this internationally funded effort can serve as a model for other areas. Even then, Madagascar will probably lose half of its remaining plant and animal species. But doing nothing would be much worse.

Why Should We Care About Tropical Forests? Biologists consider the plight of tropical forests to be one of the world's most serious environmental problems, primarily because these forests are home to 50–90% of the earth's terrestrial species (Figure 24-10), most of them still unknown and unnamed. Biologist Edward O. Wilson estimates that by 2022 at least 20% of tropical forest species could be gone, and as many as 50% by 2042, if current rates of tropical deforestation and degradation continue. If these estimates are correct, no extinction of this size has occurred for 65 million years.

Tropical forests touch the daily lives of everyone on the earth through the products and ecological services they provide. These forests supply half of the world's annual harvest of hardwood, hundreds of food products (including coffee, tea, cocoa, spices, nuts, chocolate, and tropical fruits), and materials such as natural latex rubber (Figure 24-11), resins, dyes, and essential oils that can be harvested sustainably. A 1988 study showed that sustainable harvesting of such nonwood products as nuts, fruits, herbs, spices, oils, medicines, and latex rubber in Amazon rain forests over 50 years would generate twice as much revenue per hectare as timber production and three times as much as cutting them and converting them to grassland for cattle ranching.

The active ingredients for 25% of the world's prescription drugs are derived from plants, most of which grow in tropical rain forests. Commercial sales of drugs with active ingredients derived from tropical forests total at least $100 billion per year worldwide and $15.5 billion per year in the United States. Some analysts put the economic value of plant-based drugs from all types of forests at $200 million to $1.2 billion annually.

Of the 3,000 plants identified by the National Cancer Institute (NCI) as sources of cancer-fighting chemicals, 70% come from tropical forests (Figure 24-12). Some tropical tree species can be used for many purposes (Solutions, p. 655).

Tropical plants are not the only source of important drugs. Recent research has discovered a number of useful drugs in the skins of the world's declining populations of frogs (Connections, p. 211), many of them in

Hint: Enter the search terms *rain forests, environmental aspects* using the Subject Guide.

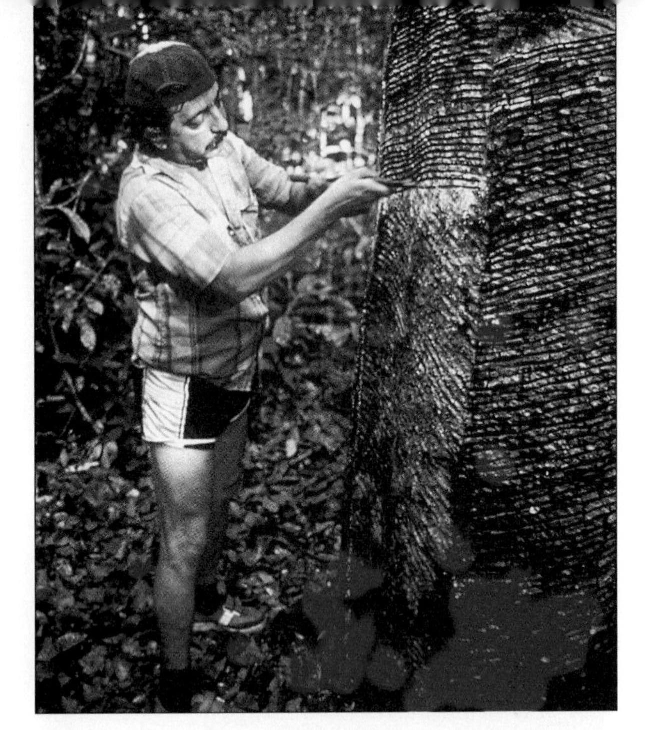

Figure 24-11 Rubber tapping is a potentially sustainable use of tropical forests. This 1987 photo shows the activist Chico Mendez (murdered in 1988 for his efforts to protect these forests) making a cut in a rubber tree. The cut allows milky liquid latex to trickle into a collecting cup. The latex is processed to make rubber, and the scars heal without killing the tree. During the seasons when latex is not flowing, rubber tappers make a living by gathering and selling nuts. (Randall Hyman)

Figure 24-12 Rosy periwinkle in the threatened tropical forests of Madagascar. Two compounds extracted from this plant have been used to cure most victims of two deadly cancers: lymphocytic leukemia (which was once almost always fatal for children) and Hodgkin's disease (mostly affecting young adults). Worldwide sales of these two drugs are about $180 million per year, none of which is returned to Madagascar. Only a tiny fraction of the world's tropical plants has been studied for such potential uses, and many will become extinct before we can study them. (Heather Angel)

tropical forests. Examples include a painkiller hundreds of times more potent than morphine, a whole new class of powerful and versatile antibiotics, and a growth hormone that helps detect cancer in humans. Less than 5% of the world's frog species has been evaluated as a source of chemicals that are important to humans.

Most of the original strains of rice, wheat, and corn that supply more than half of the world's food were developed from wild tropical plants. Botanists believe that tens of thousands of plant strains with potential food value await discovery in tropical forests.

Despite this immense potential, fewer than 1% of the estimated 125,000 flowering plant species in tropical forests (and less than 3% of all the world's 240,000 such species) have been examined closely for possible use as human resources. Biologist Edward Wilson warns that destroying these forests and the species they contain for short-term economic gain is like burning down an ancient library before reading the books.

Case Study: Cultural Extinction in Tropical Forests An estimated 250 million people—1 person out of every 20 on the earth—belong to indigenous cultures found in about 70 countries. Many of these peoples, such as the Kuana of Panama, the Kenyah in Indonesia, the Yánesha in Peru, and the Yanomami in Brazil, have been living in and using forests sustainably for centuries. They get most of their food from hunting and gathering, trapping, and sustainable slash-and-burn and shifting cultivation (Figure 1-23).

Many of the earth's remaining tribal peoples, representing 5,000 cultures, are vanishing as their lands are taken over for economic development. They are seeing their homelands bulldozed, cut, burned, mined, flooded, and contaminated. They are forced to adopt new ways while reeling from shock and hunger, and they are often killed by introduced diseases to which they have no immunity. Those who resist development are often killed by ranchers, miners, and settlers.

About 500 years ago some 6 million tribal people lived in the Amazon Basin. Today there are only about 200,000, and many of them are threatened. In remote Amazon Basin rain forests of Brazil and Venezuela, the Yanomami, a Stone Age tribe now reduced to fewer than 20,000 people (9,000 in Brazil), have been struggling for 20 years to preserve their land and way of life. Since 1985, some 45,000–100,000 gold miners have invaded Yanomami territory in Brazil, bringing with them diseases that killed some of the tribespeople. The miners carved more than 100 airstrips out of the rain forests. They cleared large tracts of land, turned streams into sewers, and poisoned nearby soil, water, and food webs with mercury used to extract gold from its ore. Since 1974 more than 1,500 Yanomami have been massacred, mostly by invading miners.

In 1991, Brazil and Venezuela recognized an area of Amazon Basin forest about the size of South Dakota as the permanent homeland of the beleaguered Yanomami people. However, these governments have not adequately protected the Yanomami and the land from thousands of new intruders. Indeed, in 1996 Brazilian

 Nelson, G. C., and D. Hellerstein. 1997. "Do Roads Cause Deforestation?" *American Journal of Agricultural Economics*, v. 79, no. 1, 80(9).

The Incredible Neem Tree

SOLUTIONS

Wouldn't it be nice if there were a single plant that could quickly reforest bare land, provide fuelwood and lumber in dry areas, provide alternatives to toxic pesticides, be used to treat numerous diseases, and help control population growth? There is: the neem tree, a broadleaf evergreen member of the mahogany family.

This remarkable tropical species, native to India and Burma, is ideal for reforestation because it can grow 9 meters (30 feet) tall in only 5–7 years! It grows well in poor soil in semiarid lands such as those in Africa, providing an abundance of fuelwood, lumber, and lamp oil in the process.

It also contains various natural pesticides. Chemicals from its leaves and seeds can repel or kill over 200 insect species, including termites, gypsy moths, locusts, boll weevils, and cockroaches.

Extracts from neem seeds and leaves can be used to fight bacterial, viral, and fungal infections. Indeed, the tree's chemicals have allegedly relieved so many different afflictions that the tree has been called a "village pharmacy." Its twigs are used as an antiseptic toothbrush, and oil from its seeds is used to make toothpaste and soap.

But that's not all. Neem-seed oil evidently acts as a strong spermicide that might be used in producing a much-needed male birth control pill. According to a study by the U.S. National Academy of Sciences, the neem tree "may eventually benefit every person on the planet."

Since 1985 numerous U.S. patents have been awarded to U.S. and Japanese firms for neem-based products. This has raised a legal controversy. Scientists in India have been researching the neem tree since the 1920s, and farmers in India have freely shared processes they developed over decades for extracting neem-seed oil.

Indian officials and farmers claim that no company should be able to receive exclusive patents on such widespread and general knowledge based on the genetic resources of a particular country. And if such patents are granted, they contend that the courts should award the country where the genetic resources were discovered and first developed a certain percentage of all profits. Vandana Shiva (Guest Essay, p. 674) calls the granting of patents for neem tree products economic, intellectual, and biological "piracy of the third world."

Patent holders say that they have taken neem material and improved it through innovative extraction and formulation processes. They believe that their investment of capital and the resulting unique products entitle them to a patent and to any profits made from such patents.

Critical Thinking

Do you believe that exclusive patents should be awarded to companies that have developed neem-based products for sale? Explain. Should Indian farmers and scientists, who originally developed and shared their knowledge of the neem tree's many benefits, receive a percentage of the profits from such products? Explain.

president Fermando Cardoso issued a decree allowing commercial interests to challenge the assignment of lands as protected reserves for indigenous communities.

Many analysts believe that eliminating indigenous peoples, besides being wrong or even genocidal, causes a tragic loss of earth wisdom and cultural diversity. People in these cultures know more about living sustainably in tropical forests (and in other biomes) and about what plants can be used as foods and medicines than anyone.

Others believe that economic development of these mostly unused and almost empty areas should not be held back by small numbers of primitive tribal peoples who don't hold legal titles to the land they live on. Environmentalists and others, led by organizations such as Survival International and Cultural Survival, disagree. Instead of being "unused" and "empty," they see such areas as important centers of cultural diversity, biodiversity, and ecological integrity that are used sustainably by indigenous people as a vital part of the planet's earth capital (Figure 1-2).

They call on governments to protect the rights and the earth wisdom of remaining indigenous peoples by **(1)** establishing a UN Declaration on the Rights of Indigenous Peoples enforceable by international law, **(2)** mapping their homelands and giving them full ownership of their land and all mineral rights, **(3)** protecting their lands from intrusion and illegal resource extraction, **(4)** giving them legal control over and just compensation for marketable drugs and other products derived from their lands, and **(5)** creating an international organization to fight for their legal rights. What do you think?

What Causes Tropical Deforestation? Tropical deforestation results from a number of interconnected causes, all of which are related to population growth, poverty, and certain government policies that encourage deforestation (Figure 24-13). Population growth and poverty combine to drive subsistence farmers and the landless poor to tropical forests, where they try to grow enough food to survive.

Hint: Enter the search terms *deforestation, causes of* using the Subject Guide.

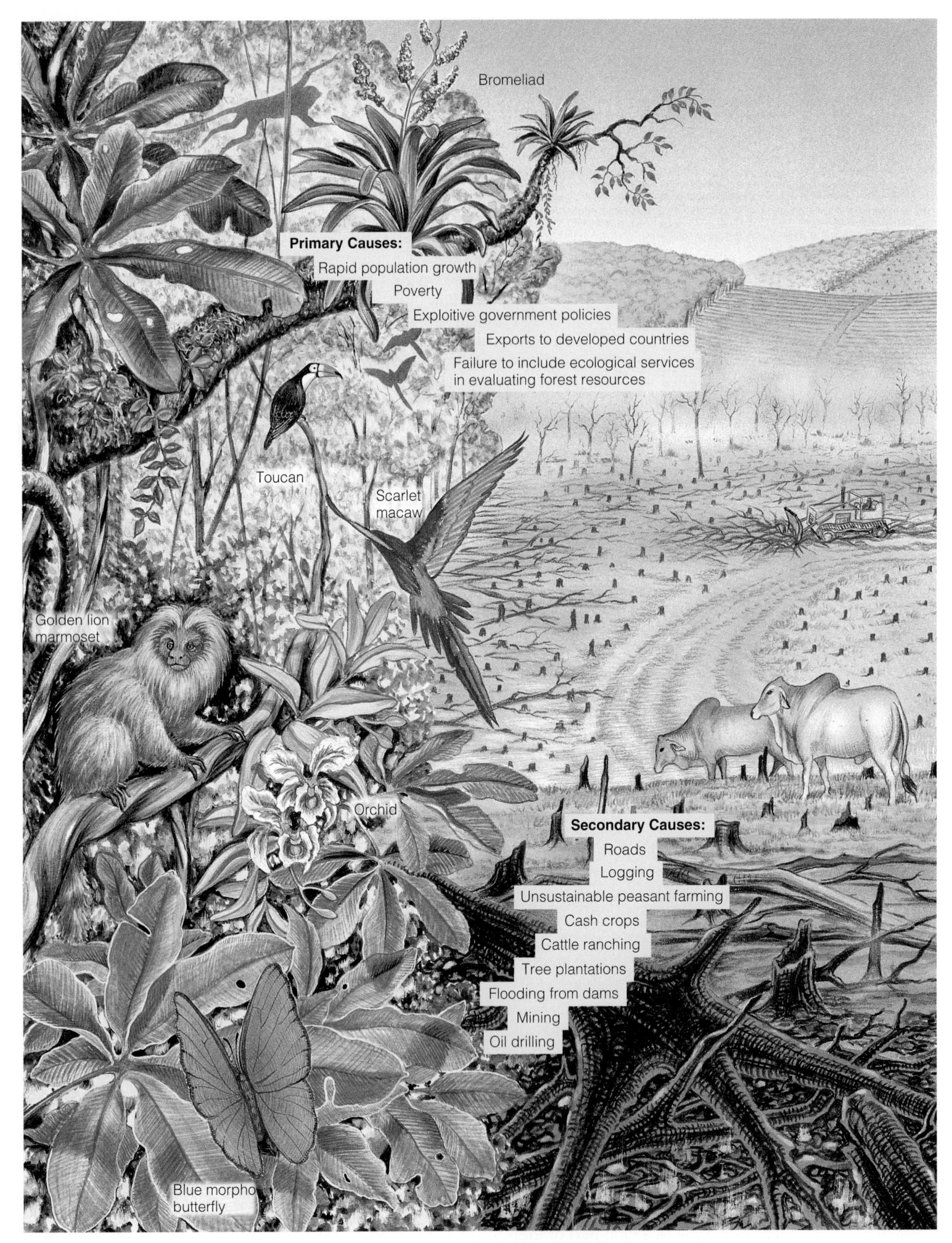

Figure 24-13 Major interconnected causes of the destruction and degradation of tropical forests, all of which are ultimately related to population growth, poverty, and government policies that encourage deforestation.

Q: How many children under age 5 die each year in developing countries from mostly preventable infectious diseases?

Government subsidies can accelerate deforestation by making timber or other resources cheap relative to their full ecological value and by encouraging the poor to colonize tropical forests, by giving them title to land they clear (as is done in Brazil, Indonesia, and Mexico). International lending agencies encourage developing countries to borrow huge sums of money from developed countries to finance projects such as roads, mines, logging operations, oil drilling, and dams in tropical forests. To stimulate economic development (and in some cases to pay the interest on loans from developed countries), these countries often sell off some of their timber and other natural resources, mostly to developed countries.

The process of degrading a tropical forest begins with a road (Figures 24-14 and 24-15), usually cut by logging companies. Once the forest becomes accessible, it can be cleared and increasingly degraded by a number of factors. One is unsustainable forms of small-scale farming. Hordes of poor people follow logging roads into the forest to build homes and plant crops on small cleared plots. Instead of practicing various methods of traditional and potentially sustainable shifting cultivation, these inexperienced *shifted cultivators* often practice unsustainable farming that depletes soils and destroys large tracts of forests.

According to environmental scientist Norman Myers (Guest Essay, p. 530), these forest newcomers are responsible for well over half of all tropical forest destruction. Driven by rapidly growing population and poverty, their numbers and impact are rising rapidly.

Cattle ranching also degrades tropical forests. Cattle ranches, sometimes aided by government subsidies, are often established on cropland that has been exhausted by small-scale farmers. Some farmers simply abandon their plots to ranchers; others seed their plots with grass and then sell them to ranchers. When torrential rains and overgrazing turn the usually thin and nutrient-poor tropical forest soils (Figure 5-16) into eroded wastelands, ranchers move to another area and repeat the destructive process known as *shifting ranching*.

In Brazil's Amazon Basin, the grazing of cattle allows ranchers (supported by government subsidies averaging $5.6 million per ranch) to claim large tracts

Figure 24-14 The effects of road building on deforestation are evident in this 1991 computer-enhanced satellite image of a 1,300-square-kilometer (500-square-mile) area of tropical rain forest north of Manus, in Brazil's Amazon Basin. Remaining old-growth forest is dark red, regenerating forest is orange or light red, and recently deforested areas are blue-green or white. (NASA/Goddard Space Flight Center)

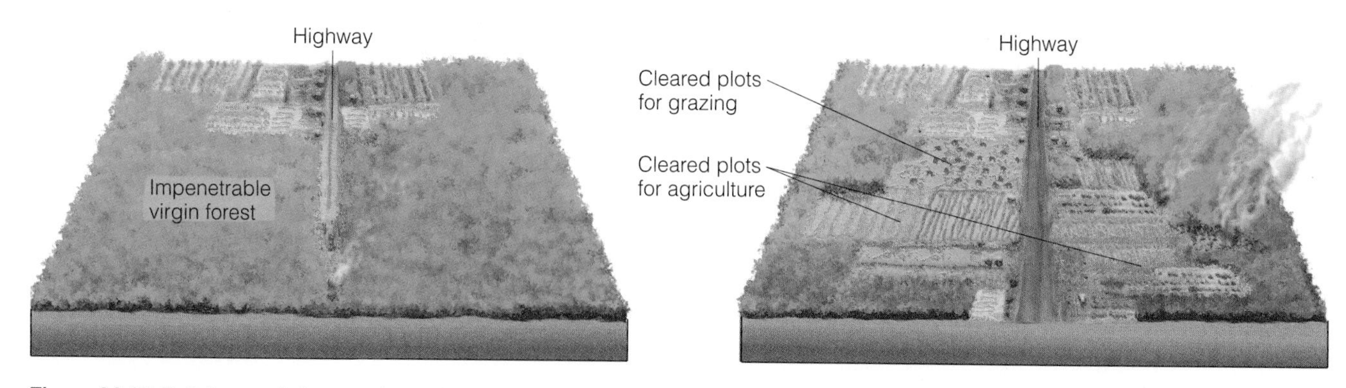

Figure 24-15 Building roads into previously inaccessible tropical (and other) forests paves the way to fragmentation, destruction, and degradation.

A: About 10 million (an average of 27,000 per day)

of land and the mineral rights below it. Some then sell both land and mineral rights for quick profits, a process called *ghost ranching*. At least half of the 600 large ranches in the Brazilian Amazon Basin have never sent a single head of cattle to market, and about 30% are now abandoned. Although this program has been a financial success for wealthy ranchers, it's been an economic and ecological disaster for the national economy, costing the government at least $2.5 billion in lost revenue.

Between 1965 and 1983, Central America lost two-thirds of its tropical forestland, much of it cleared to raise beef for export to the United States, Canada, and western Europe. Many large fast-food chains claim they no longer buy beef from tropical countries, but proof is difficult: Once the U.S. government inspects imported beef, the meat enters the domestic market with no origin labels.

Indonesia is clearing and burning millions of hectares of tropical forest for low-quality cattle pastures and to grow low-yielding corn and soybeans on highly erodible soils to feed chickens. Clearing large areas of tropical forest for raising cash crops such as cotton, tea, coffee, and bananas—mostly for export to developing countries—severely degrades these reservoirs of biodiversity.

As forests become opened up by roads, logging, agricultural plots, and cattle ranches they become drier and more susceptible to fires. This phenomenon became apparent when huge areas of forests (dried out more than normal by a very strong El Niño-Southern Oscillation, Figure 7-8) burned in Indonesia, Brazil, Guatemala, Nicaragua, and Mexico in 1997 and 1998. Most of these fires were caused by farmers using fire to prepare fields for planting or cattle-grazing. In Indonesia evidence shows that settlers were being paid by large corporations to convert tropical forest land into corporate-owned palm or rice plantations.

Tens of millions of people were sickened by the highly polluted air and hundreds died and the resulting damage and loss of timber from the fires amounted to billions of dollars. These mostly human-caused fires also released large amounts of heat-trapping CO_2 into the atmosphere. In coming years more fires are expected in Latin America, Africa, and Asia, destroying large areas of forest, reducing biodiversity and ecological integrity, and adding to greenhouse gases. Some conservationists warn that mostly because of human actions coupled with prolonged droughts (also made more likely by large-scale land clearing) we may be entering an age of fire.

Mining and oil drilling also degrade tropical forests. Widespread gold mining has been particularly damaging because it erodes soils and pollutes streams and soils with toxic mercury (Case Study, p. 349). Dams built on rivers also flood large areas of tropical forests.

Commercial logging also degrades tropical forests. Since 1950 the consumption of tropical timber has risen 15-fold. Japan alone accounts for 53% of the world's tropical timber imports, followed by Europe (32%) and the United States (15%). Currently, almost three-fourths of exported tropical timber comes from Southeast Asia. As tropical timber in Asia is depleted in the 1990s, cutting is shifting to Latin America and Africa. In 1996 alone, the area of Amazonian forest under concession to Asian timber companies quadrupled. Although timber exports to developed countries contribute significantly to tropical forest depletion and degradation, over 80% of the trees cut in developing countries are used at home, mostly for firewood and construction.

Mature tropical forest produce little wood suitable for harvest and export because most of the trees are too small or poorly formed (not straight enough), or are undesirable species. Loggers typically take only the best large and medium trees by bulldozing a road into the forest and then clearing a tract to each tree they take. As a result, they topple up to 17 other trees for each 1 they remove. The falling trees damage many others, with up to 70% of them eventually dying from their injuries.

Logging is directly responsible for only a small portion of tropical deforestation and degradation when compared to small-scale agriculture. However, the construction of logging and other resource access roads initiates and helps accelerate the destruction by opening up these forests to small-scale farmers, ranchers, and miners.

Case Study: Japan: Ecovillain or Ecosavior? Japan illustrates both the harmful effects a developed country can have on tropical deforestation and environmental degradation and the beneficial effects in reducing pollution and conserving resources. Japan (Figure 11-16), with 126 million people squeezed into an area slightly larger than California, has few minerals or fossil fuels and a limited timber supply; it must import vast quantities of raw materials. As Japan has developed into a leading economic power, its impact on the global environment has grown. It has a mixed environmental record.

On the negative side, Japan

- Has logged much of its forests, drained wetlands, destroyed offshore reefs, and built nuclear power plants in earthquake zones.
- Accounts for 53% of the world's tropical timber imports, resulting in the deforestation of vast tracts in Asia and displacement of indigenous

Q: Worldwide, how many people die each year from sexually transmitted diseases?

peoples. Much of this wood is used to make throwaway items such as concrete molds, packing crates, and chopsticks.

- Has bought logging rights to clear-cut 63% of remaining old-growth coniferous forests in Alberta, Canada, and plans to buy rights to cut the vast and largely untapped coniferous forests of Siberia.
- Is a major illegal importer of endangered and threatened species and of products made from them.
- Lobbies to have the international ban on commercial whaling lifted (at least for species whose populations are not in danger) and illegally kills several hundred whales a year.
- Finances large-scale environmentally harmful projects such as roads, dams, power plants, and mines in developing countries.
- Is the only developed country without a strong environmental movement.

On the positive side, Japan

- Is a leader in control of industrial and urban air pollution.
- Makes and sells the world's finest and most cost-effective incinerators and air pollution control scrubbers.
- Is the world's most energy-efficient country and has led the world in developing and selling fuel-efficient cars and energy-efficient industrial processes.
- Has a high recycling rate using a sophisticated resource recovery system.
- Slashed its birth rate during the 1960s and now has an annual population growth rate one-third that of the United States.
- Devotes more than $1 billion of its annual foreign aid budget to environmental projects in developing countries.
- Assumed a major leadership role at the 1992 Earth Summit in Rio de Janeiro and committed at least $7 billion for environmental assistance to developing countries (compared to $1.2 billion pledged by the United States).
- Plans to stabilize its carbon dioxide emissions at 1990 levels by 2000, even though it contributes only 5% of global CO_2 emissions (compared to 23% by the United States).
- Unveiled a 100-year plan in 1990 to protect and restore the earth's natural functions, dominate global markets for new ecotechnologies, and lead the way in what it calls an era of global citizenry.

Although Japan has become a world leader in pollution control, waste management, and energy efficiency, it is a major destroyer of biodiversity. If Japan becomes a major force in helping sustain the earth through both improved technology and biodiversity protection, it can turn its critics into admirers and stimulate other countries to follow its lead.

How Serious Is the Fuelwood Crisis In Developing Countries? In 1998 about 2.2 billion people in 63 developing countries either could not get enough fuelwood to meet their basic needs or were forced to meet their needs by using wood faster than it was being replenished. The UN Food and Agriculture Organization projects that by the end of this century this already serious shortage of fuelwood will affect 2.7 billion people in 77 developing countries (Figure 24-16). As burning wood (or charcoal derived from wood) to boil water becomes an unaffordable luxury, waterborne infectious diseases and death will spread.

Besides deforestation and accelerated soil erosion, fuelwood scarcity has other harmful effects. It places an additional burden on the rural poor, especially women and children, who often must walk long distances searching for firewood. Buying fuelwood or charcoal can take 40% of a poor family's meager income.

An estimated 800 million poor people who can't get enough fuelwood burn dried animal dung and crop residues for cooking and heating (Figure 24-17). These natural fertilizers thus never reach the soil, cropland productivity is reduced, the land is degraded still further, and hunger and malnutrition increase—another example of harmful positive feedback in action.

Fuelwood shortages, like most environmental problems in developing countries, are driven by a combination of rapid population growth and poverty. Urbanization is another factor. City dwellers in many

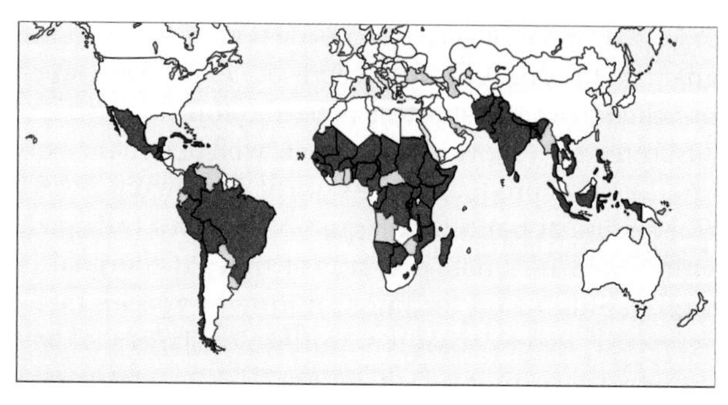

Figure 24-16 Scarcity of fuelwood, 1985 and 2000 (projected). (Data from UN Food and Agriculture Organization)

A: About 750,000—an average of 2,100 deaths per day

Figure 24-17 Making fuel briquettes from cow dung in India. As fuelwood becomes scarce, more dung is collected and burned, depriving the soil of an important source of plant nutrients. (Pierre A. Pittet/UN Food and Agriculture Organization)

developing countries burn charcoal because it is much lighter than fuelwood and is thus much cheaper to transport. But burning wood in traditional earthen pits to produce charcoal consumes more than half of the wood's energy. Thus, each city dweller who burns charcoal uses twice as much wood for a given amount of energy as a rural dweller who burns firewood. This helps explain the expanding rings of deforested land that surround many cities that use charcoal as the major source of fuel.

24-4 MANAGING AND SUSTAINING FORESTS

What Are the Major Types of Forest Management? About 25% of the world's forests are managed for wood production. Whereas agricultural crops can be harvested annually or even more often, trees take ten to hundreds of years to mature, depending on the species.

The total volume of wood produced by a particular stand of forest varies as it goes through different stages of growth and ecological succession (Figure 24-18). If the goal is to produce fuelwood or fiber for paper production in the shortest time, the forest is usually harvested on a short rotation cycle, well before the volume of wood produced peaks (point A in Figure 24-18). Harvesting at point B in Figure 24-18 gives the maximum yield of wood per unit of time. If the goal is high-quality wood for fine furniture or veneer, managers use longer rotations to develop larger, older-growth trees (point C in Figure 24-18), whose rate of growth has leveled off and is much lower than that of young trees.

There are two basic forest management systems: even-aged and uneven-aged. With **even-aged management**, trees in a given stand are maintained at about the same age and size. A major goal of even-aged management, sometimes called *industrial forestry*, is to grow and harvest trees using monoculture techniques. This is achieved by replacing a biologically diverse natural forest with a simplified tree farm of one or two fast-growing and economically desirable species that can be harvested every 10–100 years, depending on the species. Crossbreeding and genetic engineering can improve the quality and the quantity of tree-farm wood.

Even-aged management begins with one or two cuttings of all or most trees from an area; then the site is usually replanted with seedlings of one or more species of the same age. In a natural forest, dead and fallen trees are seen as vital wildlife habitats and integral parts of a natural cycle of decay and forest renewal. In most industrial forests, they are viewed as debris to be removed and burned to make way for the growth of planted tree seedlings.

Even-aged management of an area leads to more even-aged management of that area. Once a diverse forest has been cleared and replaced with even-aged stands, the only economical thing for timber companies to do is to keep repeating the process of cutting down a stand of trees before it turns into a true forest. In even-aged management, forests are viewed primarily as lumber and fiber factories, and old-growth forests are seen as timber going to waste instead of as essential centers of the earth's biodiversity and ecological integrity.

In **uneven-aged management**, a variety of tree species in a given stand are maintained at many ages and sizes to foster natural regeneration. Here the goals are biological diversity, long-term production of high-quality timber, a reasonable economic return, and multiple use. Mature trees are selectively cut; the removal of all trees is used only on small patches of species that benefit from such a practice.

Noble, Ian R. 1997. "Forests as Human-Dominated Ecosystems." *Science*, vol. 275, no. 5325, 522(4).

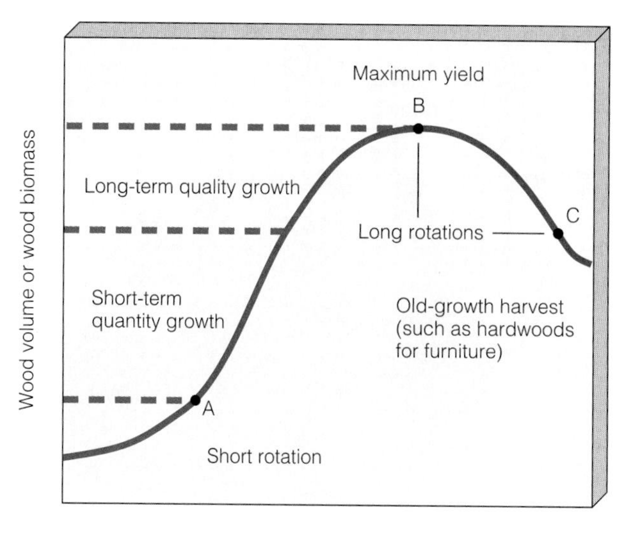

Figure 24-18 Changes in wood volume over various growth–harvest cycles in forest management. Forest management occurs over a cycle of decisions and events called a *rotation.* The most important steps in a rotation include taking an inventory of the site, developing a forest management plan, building roads into the site, preparing the site for harvest, harvesting timber, and regenerating and managing the site until the next harvest.

The rate of return owners expect on a forest asset is the key factor determining whether most will use short-term, even-aged management or longer-term, uneven-aged management. Before the 1960s forest owners figured a rate of return on their investment of 2–3%. Because the trees grew about as fast as the money invested in them, owners could afford to wait fairly long times before they harvested them, normally using uneven-aged management.

In the 1960s and 1970s, though, forest owners began basing their management decisions on returns of almost 10%, reflecting rising interest rates on money they could put into other investments. Thus, it was more profitable to turn trees into money as fast as possible. This meant clear-cutting a forest, investing the profits in something else, growing a new even-aged stand of trees as quickly as possible, cutting them down, and reinvesting the money—and repeating this process until the soil was exhausted. Then some of the profits could be used to buy another area of diverse forest and repeat this ecologically destructive process, which some call *grow-and-slash deforestation* (Connections, p. 649). As long as no monetary value is assigned to the ecological and recreational services trees and forests provide, this short-term economically profitable—but long-term ecologically and economically unsustainable—approach to forest management will continue.

How Are Trees Harvested? Before larger amounts of timber can be harvested, roads must be built for access and removal of timber. Even with careful design, logging roads have a number of harmful impacts (Figure 24-15). They cause erosion and increase sedimentation of waterways and cause severe habitat fragmentation and loss of biodiversity. They can expose forests to invasion by exotic pests, diseases, and introduced wildlife species. They can also open up once-inaccessible forests to farmers, miners, ranchers, hunters, and off-road vehicle users. In addition, logging roads on public lands in the United States disqualify the land for protection as wilderness.

Once loggers can reach a forest, they use various methods for harvesting the trees (Figure 24-19). With **selective cutting**, intermediate-aged or mature trees in an uneven-aged forest are cut singly or in small groups (group selection), creating gaps no larger than the height of the standing trees (Figure 24-19a). Selective cutting reduces crowding, encourages the growth of younger trees, and maintains an uneven-aged stand of trees of different species. It also allows natural regeneration from the surrounding trees, thereby avoiding the financial and environmental costs of site preparation (usually by bulldozer or herbicides) and planting of trees after clear-cutting. If done properly, selective cutting also helps protect the site from soil erosion and wind damage.

Many industry and government foresters contend that selective cutting is not profitable and that many of the trees that bring the best prices don't get enough sunlight for regeneration with selective cutting. But numerous private foresters and timber companies have been making satisfactory profits by using selective cutting of small and large areas of forests containing such species.

An unsound type of selective cutting is *high grading*, or *creaming*, which removes the most valuable trees. This practice, common in many tropical forests, ends up injuring one-third to two-thirds of the remaining trees when they are knocked over by logging equipment or by the large target trees when they are felled and removed.

Some tree species grow best in full or moderate sunlight in moderate to large clearings. Such sun-loving species are usually harvested by shelterwood cutting, seed-tree cutting, or clear-cutting. **Shelterwood cutting** removes all mature trees in two or typically three cuttings over a period of about 10 years (Figure 24-19b). The first cut removes most mature canopy trees, unwanted tree species, and diseased, defective, and dying trees. This opens the forest floor to light but leaves enough mature trees to cast seed and to shelter growing seedlings. Some years later, after enough seedlings take hold, a second cut removes

Hint: Enter the search terms *forest ecology* using the Subject Guide.

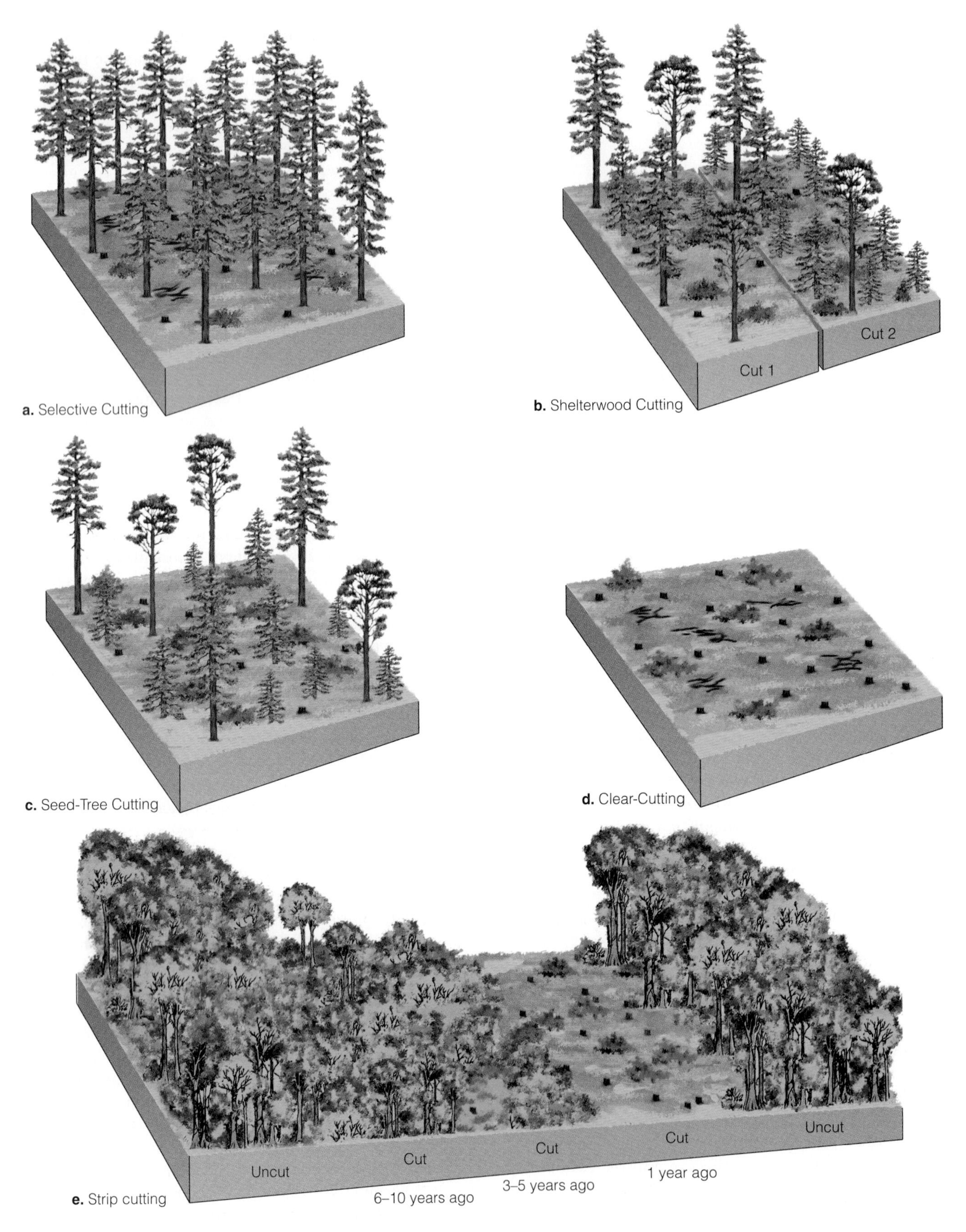

Figure 24-19 Tree harvesting methods.

Q: Worldwide, how many people die each year from diarrheal diseases?

more canopy trees, although some of the best mature trees are left to shelter the young trees. After these are well established, a third cut removes the remaining mature trees, and the even-aged stand of young trees then grows to maturity.

This method allows natural seeding from the best seed trees and keeps seedlings from being crowded out. It leaves a fairly natural-looking forest that can serve a variety of purposes; it also helps reduce soil erosion and provides a good habitat for wildlife. The danger is that loggers may take too many trees in the first cut.

Seed-tree cutting harvests nearly all of a stand's trees in one cutting, leaving a few uniformly distributed seed-producing trees to regenerate the stand (Figure 24-19c). After the new trees become established, the seed trees may be harvested. By allowing several species to grow at once, seed-tree cutting leaves an aesthetically pleasing forest that is useful for recreation, deer hunting, erosion control, and wildlife conservation. Leaving the best trees for seed can also lead to genetic improvement in the new stand.

Clear-cutting is the removal of all trees from an area in a single cutting. The clear-cut area may be a whole stand (Figure 24-19d), a strip, or a series of patches. After all trees are cut, the site is usually reforested; seed is released naturally, by the harvest, or artificially as foresters either broadcast seed over the site or plant seedlings raised in a nursery. If clear-cut areas are small enough, seeding may occur from nearby uncut trees.

On the positive side, clear-cutting increases timber yield per hectare, permits reforesting with genetically improved stocks of fast-growing trees, and shortens the time needed to establish a new stand of trees. It also requires much less skill and planning than other harvesting methods, and it usually provides timber companies the maximum economic return in the shortest time.

However, clear-cutting leaves ugly, unnatural forest openings and eliminates any potential recreational value for several decades. It also destroys and fragments some wildlife habitats and thus reduces biodiversity and disrupts ecological integrity. Environmental degradation from clear-cutting above ground is obvious, but equally serious damage occurs underground from the loss of fungi, worms, bacteria, and other microbes that condition soil and help protect plants from disease (Figures 5-14 and 5-16). Furthermore, trees in stands bordering clear-cut areas are more vulnerable to being blown down by windstorms. Large-scale clear-cutting on steep slopes leads to severe soil erosion, sediment water pollution, and flooding (Figure 13-22).

If done carefully and responsibly, clear-cutting is often the best way to harvest tree farms and stands of some tree species that require full or moderate sunlight for growth. However, for economic reasons it is often done irresponsibly and used on species that don't require this method.

A variation of clear-cutting that can allow a sustainable timber yield without widespread destruction is **strip cutting**. A strip of trees is clear-cut along the contour of the land, with the corridor narrow enough to allow natural regeneration within a few years (Figure 24-19e). After regeneration, another strip is cut above the first, and so on. This allows a forest to be clear-cut in narrow strips over several decades with minimal damage.

Another variation of clear-cutting is **whole-tree harvesting**, in which a machine cuts trees at ground level or uproots entire trees. These trees are usually transported to a chipping machine in which huge blades reduce the wood to small chips. This approach is used primarily to harvest stands for use as pulpwood or fuelwood chips.

Many foresters and ecologists oppose whole-tree harvesting because it removes all tree materials, including standing dead timber and fallen logs, which deprives the soil of plant nutrients and removes numerous wildlife habitats. Research is under way to determine how whole-tree harvesting methods might be modified to reduce such harmful environmental effects.

How Can Forests Be Protected from Pathogens and Insects? In a healthy and diverse forest, tree diseases and insect pest populations are usually controlled by interactions with other species. Thus, they rarely get out of control and seldom destroy many trees. By contrast, a less diverse tree farm is vulnerable to attack by pathogens and insects and sometimes must be protected with pesticides.

The most destructive tree diseases are caused by parasitic fungi. Three deadly tree diseases were accidentally introduced into the United States from other countries: chestnut blight (from China), Dutch elm disease (from Asia via Europe), and white-pine blister rust (from Europe). Chestnut blight has almost eliminated the once-abundant and valuable chestnut tree from eastern hardwood forests.

The fungus leading to Dutch elm disease has killed more than two-thirds of the elm trees in the United States. In 1997, however, the U.S. National Arboretum developed an American elm tree that has an extraordinary genetic ability to withstand the fungus. The tree, named the Valley Forge elm, is the result of 60 years of government research and will be available to the public by 2000. By 2010, the elm may again be the most popular tree in the United States.

Since the 1980s a fungus called *dogwood blight* has been damaging various species of flowering dogwood trees in 19 states, mostly in the eastern United States. When damaged trees don't flower—and thus don't

A: About 3.1 million (most of them young children)

produce dogwood fruit in the fall—the result is a severe reduction in an important food supply for populations of animals, especially migratory and overwintering bird species.

A few insect species can cause havoc, especially in tree farms or simplified forests, in which their natural predators may not exist. For example, bark beetles that bore channels through the layer beneath the bark of spruce, fir, and pine trees have killed large expanses of forest in the western and southern United States. The larvae of insects, such as spruce budworm and gypsy moth larvae, eat the needles or leaves of trees. Repeated attacks over several years can kill trees by eliminating the foliage they need to carry out photosynthesis and produce food. In diverse forests, populations of such leaf-eating insects are usually kept under control by predator populations.

To conservation biologists, preserving biodiversity is the best and cheapest defense against tree diseases and insects. Other methods include banning imported timber that might introduce harmful new pathogens, removing infected trees or clear-cutting infected areas and burning all debris, treating diseased trees with antibiotics, developing disease-resistant tree species, applying pesticides (Section 21-1), and using integrated pest management (Section 21-5).

How Do Fires Affect Forest Ecosystems? Intermittent natural fires set by lightning are an important part of the ecological cycle of some types of forests (but not tropical rain forests). Some species actually need occasional fires. The seeds of the giant sequoia, the lodgepole pine, and the jack pine, for instance, are released from cones and germinate only after being exposed to intense heat.

Forest ecosystems can be affected by different types of fires. Some, called *surface fires*, usually burn only undergrowth and leaf litter on the forest floor. These fires kill seedlings and small trees but spare most mature trees, and most wild animals can escape.

In forests in which ground litter accumulates rapidly, a surface fire every 5 or so years burns away flammable material and helps prevent more destructive fires. Surface fires also release valuable mineral nutrients tied up in slowly decomposing litter and undergrowth. They increase the activity of underground nitrogen-fixing bacteria, stimulate the germination of certain tree seeds, and help control pathogens and insects. Some wildlife species—deer, moose, elk, muskrat, woodcock, and quail, for example—depend on occasional surface fires to maintain their habitats and to provide food in the form of vegetation that sprouts after fires.

Some extremely hot fires, called *crown fires*, may start on the ground but eventually burn whole trees and leap from treetop to treetop. They usually occur in forests in which no surface fires have occurred for several decades, allowing dead wood, leaves, and other flammable ground litter to build up. These rapidly burning fires can destroy most vegetation, kill wildlife, lead to accelerated soil erosion, and set back the clock of ecological recovery by up to hundreds of years.

Sometimes surface fires go underground and burn partially decayed leaves or peat (Figure 15-13). Such *ground fires* are most common in northern peat bogs. They may smolder for days or weeks before being detected and are difficult to extinguish.

Protecting forest resources from fire can involve four approaches: *prevention*, *prescribed burning* (setting controlled ground fires to prevent buildup of flammable material), *presuppression* (early detection and control of fires), and *suppression* (fighting fires once they have started). Means of forest-fire prevention include requiring burning permits, closing all or parts of a forest to travel and camping during periods of drought and high fire danger, and public education. The Smokey the Bear educational campaign of the U.S. Forest Service and the National Advertising Council, for example, has prevented countless forest fires in the United States, saving many lives and preventing billions of dollars in losses.

However, ecologists contend that because it allows litter to accumulate in some forests, prevention increases the likelihood of highly destructive crown fires. Since 1972 U.S. Park Service policy has been to allow most lightning-caused fires to burn themselves out, as long as they don't threaten human lives, park facilities, private property, or endangered wildlife.

During the hot, dry summer of 1988 more than 200 fires triggered by lightning and people ravaged parts of Yellowstone National Park, burning about 36% of the park. After the Yellowstone fires and many other fires that broke out in a number of national forests in the West during the summer of 1994, some people called for a reversal of the federal let-it-burn policy.

However, biologists contend that these fires caused more damage than they should have because the previous policy of fighting all fires in the park had allowed the buildup of flammable ground litter and small plants. They believe that many fires in national parks, national forests, and wilderness areas should be allowed to burn as part of the natural ecological cycle of succession and regeneration (Figure 9-21). Exceptions would include fires that posed a serious threat to human-built structures.

Since 1988 research has shown that the ecological effects of the massive 1988 fires in Yellowstone National Park were minor, with no land made unfit for regrowth. Critics of putting out all fires in national parks also point out that in 1988 firefighters caused

Q: Worldwide, how many people die each year from tuberculosis?

more damage in Yellowstone National Park by bulldozing fire lines and felling trees than if they had let the fires burn. Moreover, it would have been much cheaper for the government to rebuild burned buildings in Yellowstone than to spend $300 million to protect them. Timber companies argue that they can help prevent fires and tree disease by being allowed to make extensive salvage cuts in national forests and national parks.

Many foresters managing public lands would like to make more use of carefully controlled prescribed burning to reduce the threat of highly damaging crown fires in some types of forests. But they can't carry out many prescribed burns because of complaints from people who have been taught to see all fire as a harmful and destructive force.

How Are Forests Threatened by Air Pollution and Climate Change? Forests at high elevations and those downwind from urban and industrial centers are exposed to a variety of air pollutants that can harm trees, especially conifers. Besides direct harm, prolonged exposure to multiple air pollutants makes trees much more vulnerable to drought, diseases, and insects (Figure 18-9). The solution is to reduce emissions of the offending pollutants from coal-burning power plants, industrial plants, and motor vehicles (Section 18-6).

In coming decades an even greater threat to forests (especially temperate and boreal forests) may come from regional climate changes brought about by projected global warming. Ways to deal with projected global warming are discussed in Section 19-4.

Is Industrial Forestry Sustainable? To timber companies, sustainable forestry means producing a maximum sustainable yield of commercial timber in as short a time as possible. This usually means clearing diverse, uneven-aged forests by using clear-cutting, seed-tree cutting, or shelterwood cutting (Figure 24-19b, c, and d) and replacing them with intensively managed, even-aged tree plantations that can be clear-cut again as quickly as possible.

To conservation biologists and some foresters such *maximum sustained yield forestry* does not result in long-term *sustainable forestry*. Research reveals that industrial or monoculture forestry has led to a decrease in the biodiversity of forests, elimination of competing species, suppression of ecologically important natural fires, draining of wetlands, and increased soil erosion and loss of soil nutrients.

For example, decades of experience with intensive even-aged management of German forests has shown that in a forest in which virtually all of the trees are repeatedly cut and removed, the soil is depleted of nutrients. Trees then have little resilience for coping with stresses such as drought, diseases, and pests (Figure 18-9).

Ecologists also point out that in nature's economy nothing goes to waste. What timber company officials call "wasted trees" are the ecological raw materials for regenerating and maintaining a forest; the decomposition of dead and rotting trees returns nutrients to the soil slowly enough so that growing trees can take them up again to renew life. While they decay dead trees and moss-covered logs help sustain biodiversity by providing habitats and food for a variety of species.

To conservation biologists the most devastating ecological damage occurs when large stands of trees are clear-cut and all debris is removed on steep slopes in rainy areas. The rain rapidly erodes the exposed soil, slowing regeneration by secondary ecological succession for centuries or longer. Landslides cover lower-lying land areas, clog streams with sediment, and destroy fisheries.

Recently, many government and industry foresters have been advocating *New Forestry*, a variation of clear-cutting on public lands based on the principle of *ecosystem management*, with the basic goal of protecting ecosystem integrity. A few dead trees (snags) are left, and sometimes a few live seed trees are left as well. Instead of piling up and burning all the unmarketable logs and debris, some of this so-called *slash* is left scattered around to create wildlife habitats and encourage biodiversity. Sometimes the sizes and shapes of the clear-cut patches are varied. Many foresters see this approach as an important step in the right direction, provided it is widely implemented and not used primarily as a smoke screen for continuing unsustainable, even-aged industrial forestry.

However, other foresters contend that this practice does not constitute sustainable forestry. Instead of relying primarily on selective cutting, it replaces one-step clear-cutting with two-step shelterwood and seed-tree cutting, which still leads to a drastic loss of habitat, biodiversity, and ecological integrity compared to a natural forest or a selectively cut forest. Most of the woody material is still removed from the ecosystem instead of decomposing to supply nutrients for new growth. Most of the soil is exposed to the sun and storms and is compacted by heavy machinery. Herbicides are often used to kill plants that compete with the new crop of plantation trees, further reducing biodiversity and disrupting ecological integrity. According to John Dennington, head of the Sportsman's Association, "Ecosystem Management is about how to grow the most pine and fool the most people."

Solutions: What Is Sustainable Forestry? Instead of conventional industrial forestry and New Forestry, conservation biologists and a growing

A: About 3.1 million

number of foresters call for sustainable forest management that emphasizes the following:

- *Recycling more paper to reduce the harvest of pulpwood trees* (Section 22-5), *using fibers from fast-growing plants such as kenaf to make tree-free paper, and reducing wood waste* (p. 667)
- *Growing more timber on long rotations,* generally about 100–200 years (point C, Figure 24-18), depending on the species and soil quality
- *Practicing selective cutting of individual trees or small groups of most tree species* (Figure 24-19a)
- *Minimizing fragmentation of remaining larger blocks of forest*
- *Using road building and logging methods that minimize soil erosion and compaction*
- *Practicing strip cutting* (Figure 24-19e) *instead of conventional clear-cutting, and banning clear-cutting (including seed-tree and shelterwood cutting) on land that slopes more than 15-20°.*
- *Leaving most standing dead trees (snags) and fallen timber to maintain diverse wildlife habitats and to be recycled as nutrients* (Figure 4-13)
- *Including the ecological and recreational services provided by trees and a forest in determining their economic value*

Some foresters have suggested that the United States meet most of its demand for wood and wood products by growing genetically improved trees on carefully managed tree plantations. If done with minimum use of pesticides and commercial fertilizer and carried out mostly on already cleared marginal farmland and pastureland that already has access roads, this practice could be sustainable. This would reduce the need for the United States to import wood that causes environmental damage in other countries. It would also reduce pressures to cut timber in ancient forests and help protect wildlife habitats. In 1903 New Zealand adopted such an approach. Today, the country meets all of its domestic wood needs from plantations and has protected about 30% of its original native forest.

Case Study: How Are U.S. National Forests Managed? About 22% of the commercial forest acreage in the United States is located within the 156 national forests (Figure 23-9) managed by the U.S. Forest Service and occupying about 10% of the U.S. land area. These forests serve as grazing lands for more than 3 million cattle and sheep each year. They support multibillion-dollar mining operations, supply about 4% of the nation's timber (down from 15% in the 1980s), contain a network of roads equal in area to the entire U.S. interstate highway system, and receive more recreational visits than any other federal public lands. The resulting networks of roads and clear-cuts severely fragment wildlife habitats (Figure 24-5), isolating many wildlife populations and making them vulnerable to genetic deterioration and predation.

The Forest Service is required by law to manage national forests according to the principles of *sustained yield* (which states that potentially renewable tree resources should not be harvested or used faster than they are replenished) and *multiple use* (which says that each of these forests should be managed simultaneously for a variety of uses such as sustainable timber harvesting, recreation, and wildlife conservation).

Environmentalists charge that especially since 1980 the Forest Service has allowed timber harvesting to become the dominant use in most national forests. To back up this charge, they point out that until 1998 about 70% of the Forest Service budget was devoted directly or indirectly to the sale of timber.

A major reason for this emphasis on timber cutting is that Congress allows the agency to keep most of the money it makes on timber sales, whereas any losses are passed on to taxpayers. Because logging increases its budget, the Forest Service has a powerful built-in incentive to encourage timber sales, another example of positive feedback in action. Local county officials also lobby Congress and encourage Forest Service officials to keep timber harvests high because counties get 25% of the gross receipts from national forests within their boundaries.

A drive through the 122 national forests open to logging, or a boat ride along streams in such forests, does not reveal the emphasis on timber cutting because the Forest Service leaves thin buffers of uncut trees, called *beauty strips,* along roads and streams. However, a flight over these forests reveals that many of the trees have been clear-cut and most of the forests have been highly fragmented (Figure 24-5).

Environmentalists and the U.S. General Accounting Office have accused the Forest Service of poor financial management of public forests. By law, the Forest Service must sell timber for no less than the cost of reforesting the land from which it was harvested. However, the cost of access roads is not included in this price but is provided as a subsidy to logging companies. Usually the companies also get the timber itself for less than they would pay a private landowner for an equivalent amount of timber.

Studies have shown that between 1978 and 1994, national forests lost at least $6.6 billion from below-cost timber sales. Each year typically only about 5–25 of the 122 national forests open to logging make money. In effect, publicly owned national forests provide about 4% of the nation's timber at a *net loss* of about $75–500 million a year. Timber sales from U.S.

Q: Worldwide, what three infectious diseases each kill about 1 million people a year?

federal lands have turned a profit for taxpayers in only 3 of the last 100 years. In the early 1990s it was also discovered that timber companies were illegally cutting trees worth hundreds of millions of dollars per year on federal lands.

For decades the Forest Service juggled its books to disguise timber sales losses by defraying its annual up-front costs of road building over 99 years. In 1998 the Forest Service began using more realistic bookkeeping and announced that it lost $88.6 million in timber sales in 1997 and $230 million in 1995.

In 1998 Forest Service chief Mike Dombeck also announced a policy designed to put much greater emphasis on overall forest health and recreational use. He pointed out that annual timber harvest levels in the national forests have dropped from 12 billion board feet in the 1980s to about 4 billion board feet in 1997. If this new policy remains in effect, it would recognize that in 1997 three times more people made recreational visits to national forests than they did to national parks. Also, in 1995 recreational use of national forests generated 2.6 million jobs and added $97.8 billion to the national economy. By contrast, logging in these forests added only 76,000 jobs and $3.5 billion.

Timber company officials argue that being able to get timber from federal lands fairly cheaply benefits taxpayers by keeping lumber and paper prices down. Environmentalists respond that below-cost timber prices discourage investments to produce additional timber on private lands and put more pressure for timber cutting on national and state forests.

Congressional supporters of the timber industry threatened to cut the Forest Service budget if it moves away from logging as its top priority. In addition, Senator Larry Craig (R-Idaho) wants to revise the 1976 National Forest Management Act to **(1)** make timber harvesting the primary priority and make such harvests mandatory, **(2)** allow the Forest Service to impose fines up to $10,000 on individuals or groups who file appeals for an undefined improper purpose, **(3)** prevent forest projects and timber sales from being stopped in court, regardless of how much damage they do, **(4)** allow timber, mining, and grazing interests to meet in secret with officials of the Forest Service or the Bureau of Land Management to cut deals about extraction of forests and other resources, and **(5)** eliminate the requirement that the Forest Service and Bureau of Land Management consult with wildlife officials on the impacts of proposed logging on fish and wildlife.

Instead of providing economic stability for local industries and communities, logging in national forests tends to have the opposite effect. Communities that rely on proceeds from national forest timber sales experience boom-and-bust cycles as the timber is depleted and timber companies move to other areas. Analysts have proposed a number of reforms in Forest Service policy that would lead to more sustainable use and management of the national forests (Solutions, p. 668).

Solutions: How Can We Reduce Wood Waste and Use Other Fibers to Make Paper? According to the Worldwatch Institute and forestry analysts, up to 60% of the wood consumed in the United States is unnecessarily wasted. Because only 4% of the total U.S. timber production comes from the national forests, reducing the waste of wood and paper products by only 4% could eliminate the need to remove any timber from the national forests. Then these public lands could be used primarily for recreation and protection of biodiversity and ecological integrity.

Ways to reduce such waste include **(1)** reducing construction waste, **(2)** using laminated boards instead of solid wood, **(3)** using less packaging, **(4)** reducing junk mail, **(5)** recycling more paper to make products with the highest possible content of postconsumer waste, and **(6)** using paper made from fibers obtained from rapidly growing plants other than trees.

Some have suggested that wood can be saved by switching to nonwood home building and remodeling materials such as steel studs and beams, aluminum and vinyl siding, concrete slabs instead of floor joists and plywood, and plastics instead of paper products. However, this involves shifting from a potentially renewable resource to nonrenewable resources such as limestone, iron ore and other mineral deposits, and crude oil used to make these products and plastics.

Analyses show that switching to these nonrenewable resources usually requires more energy for transportation and manufacture and produces more pollution than use of wood and wood products. For example, steel framing requires 13 times more energy than wood framing, aluminum framing for walls takes nearly 20 times as much energy as wood framing, steel floor joists require 50 times more than those made of wood, and wall-to-wall carpeting and padding is four times more energy intensive than wood floors.

Because of deforestation and the rising costs of wood pulp, an increasing number of countries use other sources of fiber for making paper. Paper can be made from rags, flax, straw (a by-product of grain cultivation), hemp, bagasse (a by-product of sugar production), kenaf, and a number of other materials. In China, for example, 60% of the paper is made with nonwood pulp such as rice straw and other agricultural wastes left after harvest. Globally about 4% of the world's paper is produced from nonwood sources.

In Canada, Al Wong is running a pilot mill to convert wheat straw or rice straw left over after harvesting into paper. Currently, California rice farmers are frantically looking for alternatives to burning their rice-straw wastes. It can be tilled back into the soil, but

A: Measles, hepatitis B, and AIDS

How Can U.S. Federal Forest Management Be Improved?

SOLUTIONS

Foresters and conservation biologists have proposed a number of management reforms that would allow national forests in the United States to be used more sustainably without seriously impairing their important ecological functions. These suggestions include the following:

- *Make sustaining biodiversity and ecological integrity the first priority of national forest management policy.*
- *Use full-cost accounting to include the ecological services provided by old-growth forests in the market price of timber harvested from national forests.*
- *Prohibit timber harvesting on at least half of remaining old-growth forests on public lands.*
- *Ban all timber cutting in national forests and fund the Forest Service completely from recreational user fees.* The Forest Service estimates that recreational user fees (as low as $3 per day) would generate three times what it earns from timber sales.
- *Sharply reduce all new road building in national forests.*
- *Allow individuals or groups to buy conservation easements that prevent timber harvesting on designated areas of public old-growth forests.* In such conservation-for-tax-relief swaps, purchasers would be allowed tax breaks for the funds they put up.
- *Require that timber from national forests be sold at a price that includes the costs of road building, site preparation, and site regeneration, and that all timber sales in national forests yield a profit for taxpayers.*
- *Do not use money from timber sales in national forests to supplement the Forest Service budget* because it encourages overexploitation of timber resources.
- *Eliminate the provision that returns 25% of gross receipts from national forests to counties containing the forests, or base such returns on recreational user-fee receipts only.*
- *Strictly enforce and close loopholes in the current federal ban on exporting raw logs harvested from public lands.*
- *Tax exports of raw logs but not finished wood products* (which provide and create domestic milling and other jobs).
- *Establish federally funded reforestation and ecological restoration projects on degraded public lands.* In addition to the ecological benefits, this will provide jobs for displaced forest harvesting and wood products workers.
- *Provide increased aid and job retraining for displaced workers in the forest harvesting and products industry.*
- *Require use of sustainable forestry methods* (p. 666).

Understandably, timber company officials, most county officials, and many Forest Service officials oppose such policies and have prevented such reforms.

Critical Thinking

Explain why you agree or disagree with each of these proposals for reforming federal forest management in the United States.

this has led to an increase in rice stem rot. According to Wong, North America produces enough rice and wheat straw and other agricultural wastes to more than meet all of its paper needs without having to ever cut another tree. The problem is convincing the $130-billion-a-year paper industry to invest in the technology needed to produce such ecopaper.

Currently, most of the small amount of treeless paper produced in the United States is made from the fibers of a woody annual plant called *kenaf* (pronounced kuh-NAHF; Figure 24-20). The tall, bamboo-like stalks of this annual woody plant grow to heights of 4 meters (12 feet) in only 4–5 months, compared to 7–60 years for pine trees. Large-scale commercial production of Kenaf has occurred in India and Asia since 1900. Currently, it is the major fiber used to make paper in Thailand, which has depleted many of its forests.

Kenaf thrives in tropical and temperate climates and herbicides are rarely needed because it grows faster than most weeds; few insecticides are needed because its outer fibrous covering is nearly insect-proof. Because kenaf is a nitrogen fixer, cultivating it does not deplete soil nitrogen. It also takes fewer chemicals and less energy to break kenaf down into fibers used for making paper, so less toxic wastewater is produced and water treatment costs are lower than when pulpwood is used. The U.S. Department of Agriculture considers kenaf the best renewable fiber to replace trees in paper making and predicts that the market for kenaf should expand rapidly in coming years.

There are some drawbacks, but they can be overcome. Because the moisture content of kenaf is much higher than that of wood, the plants must be dried to prevent spoilage. Kenaf paper currently costs three to five times more than either old-growth or recycled paper stock, which is why this book is not printed on tree-free kenaf paper. However, as demand increases and producers cut production costs, the price of kenaf

Q: How many people are expected to be infected with HIV by 2000?

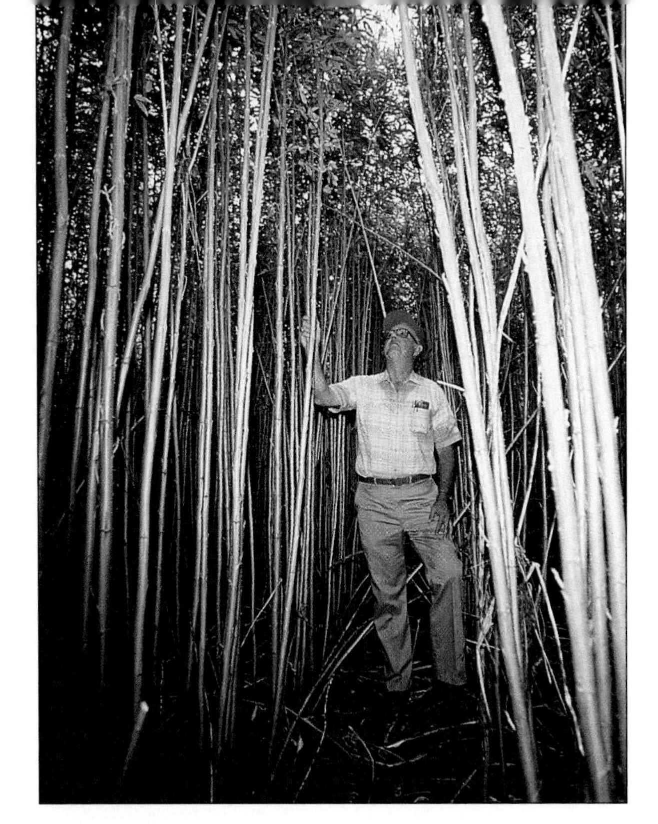

Figure 24-20 Kenaf growing in a field in south Texas. This fast-growing plant can be harvested annually and its fibers used to make paper. Its widespread cultivation would decrease the need to grow and harvest trees. (U.S. Department of Agriculture)

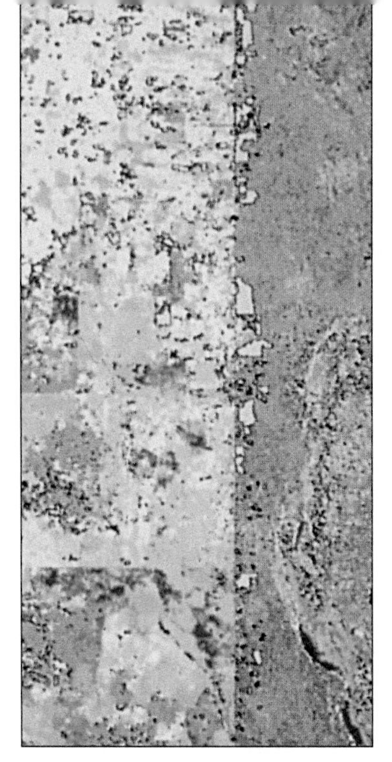

Figure 24-21 The results of two governments' different forest policies are evident in this 1986 computer-enhanced satellite view of a portion of the border between Mexico (left) and Guatemala (right). Undisturbed forests are shown in brown and green, lakes and rivers in dark blue, and cleared areas in white, light blue, and yellow. In the 1970s the Mexican government began giving settlers forestland they had cleared and planted with corn. When the soil gave out, the settlers turned to grazing cattle, resulting in almost complete loss of forest in this part of Mexico. Most forests in Guatemala, which doesn't have a settlement program, are still intact, except for areas illegally cleared by Mexican farmers and cattle grazers (yellow areas). (Landsat 5, John C. Stennis Space Center, NASA)

should come down. The basic problem is getting investors to gamble money on converting or constructing mills to produce rice, wheat, hemp, or kenaf paper without an established market.

24-5 SOLUTIONS: REDUCING TROPICAL DEFORESTATION AND FUELWOOD SHORTAGES

How Can We Reduce Tropical Deforestation and Degradation? A number of analysts have suggested ways to protect tropical forests, use them more sustainably, and restore degraded areas of such forests. An important first step is to make a detailed survey to determine how much of the world is covered with tropical (and other) forests, how much has been deforested or degraded, and where. This could be done by a combination of remote-sensing satellites and ground-level evaluations, at a cost roughly equivalent to what the world spends for military purposes *every 3 minutes*.

Conservation biologists urge us to move rapidly to protect areas of tropical forests (and other biomes) that are rich in unique species and in imminent danger—so called *hot spots* or *critical ecosystems*. Many governments oppose such protection because of political opposition by local people who are in a desperate struggle to survive.

Some biologists believe that hot spots should be identified for all of the earth's major biomes and areas. Otherwise, they fear that most money and efforts will go into protecting endangered areas of tropical forests while equally important endangered and unique ecosystems elsewhere are neglected.

Environmentalists urge governments to reduce the flow of the landless poor to tropical forests by slowing population growth and discouraging the poor from migrating to undisturbed tropical forests. Figure 24-21 shows the effects of such government policies on tropical deforestation in Mexico and in Guatemala. Environmentalists also call for a global effort to sharply reduce the poverty that leads the poor to use forests (and other resources) unsustainably for

A: About 40 million (80% of them in Africa and Asia)

Sustainable Agriculture and Forestry in Tropical Forests

SOLUTIONS

A combination of the knowledge of indigenous peoples, ecological research, and modern technology can show people how to grow crops and harvest timber in tropical forests in more sustainable ways.

One approach would be to show newcomers how to grow crops using more sustainable agroforestry techniques developed by various indigenous peoples throughout the world. The Lacandon Maya Indians of Chipas, Mexico, for example, have developed a multilayered system of agroforestry that allows them to cultivate up to 75 crop species on 1-hectare (2.5-acre) plots for up to 7 consecutive years. After that another plot must be planted to allow regeneration of the soil in the original plot.

Another approach being used by Yánesha Indians in the lush rain forests of Peru's Palcazú Valley is strip-shelterbelt harvesting of tropical trees for lumber. The idea was based on research by tropical ecologist Gary Hartschorn. He discovered that many tropical trees grow only in natural gaps in the forest. He suggested that people can mimic such gaps by harvesting trees in narrow strips and leaving wide areas of forest intact. Natural reseeding from surrounding trees and seed-carrying animals can regenerate each strip in about 30 years—a fraction of the time it takes for a large area of clear-cut tropical forest to replenish itself.

Today more than 200 Indians earn a living as members of the Yánesha Forestry Cooperative. They use chain saws to cut strips 27 meters (30 yards) wide and 270 meters (300 yards) long through the forest. The small strips are barely visible from an airplane. Oxen haul the cut logs to a processing plant, where other tribe members convert them to charcoal fuel, boards, and fence posts. Tribe members market their ecologically harvested wood products in the global marketplace and act as consultants to help other forest dwellers set up similar systems.

Critical Thinking

What applications (if any) might these practices have for growing food and harvesting timber in the United States and other developed counties? How would you encourage such approaches for more sustainable use of forests?

short-term survival. They also call for establishing programs to help new settlers in tropical forests learn how to practice small-scale sustainable agriculture and forestry (Solutions, left).

Another suggestion is to use economic policies to protect and sustain tropical forests. One approach is to phase in *full-cost pricing* over the next two decades. For example, a study of the mangrove forest in Indonesia's Bintuni Bay found that when fish, locally produced forest products, and erosion control were calculated, the most profitable option was to keep the forest standing. However, as long as tropical forests and other forms of the earth capital are valued only for the resources that can be removed from them, and not for the ecological services they provide (Spotlight, p. 647), we will continue to destroy and degrade these forests.

Governments can also phase out subsidies that encourage unsustainable deforestation and phase in tariffs, use taxes, user fees, and subsidies that favor sustainable forestry and protection of biodiversity. Another approach is to pressure banks and international lending agencies (controlled by developed countries) not to lend money for environmentally destructive projects, especially road building and dams, involving old-growth tropical forests.

In 1993, the Forest Stewardship Council developed a list of "Principles and Criteria for Forest Stewardship" that can be used to establish an international system for evaluating and labeling timber produced by valid sustainable methods. A recent proposal to raise the area of forest under certified sustainable management from 45,000 square kilometers (17,000 square miles) in 1997 to 2 million square kilometers (770,000 square miles) by 2005 has been supported by the World Bank and various environmental and business groups.

Debt-for-nature swaps and conservation easements can be used to encourage countries to protect tropical forests. In a *debt-for-nature swap*, participating countries act as custodians for protected forest reserves in return for foreign aid or debt relief (Case Study, right); in *conservation easements*, a private organization, country, or group of countries compensates other countries for protecting selected forest areas. Currently less than 5% of the world's tropical forests are part of parks and preserves, and many of these are protected in name only.

In 1988 a new form of tropical forest protection and reforestation was initiated to help slow global warming by reducing CO_2 emissions. With this tool polluters (mostly in developed countries) can continue to emit CO_2 by entering into an agreement with

Q: How many people die from malaria each year?

CASE STUDY

Debt-for-Nature Swap in Bolivia

In 1984, biologist Thomas Lovejoy suggested that debtor nations should be rewarded for protecting their natural resources. In such debt-for-nature swaps, Lovejoy proposed that a certain amount of foreign debt be canceled in exchange for spending a certain sum on better natural resource management.

Typically, a conservation organization buys a specified amount of a country's debt from a bank at a discounted rate and negotiates the swap. A government or private agency must agree to enact and supervise the conservation program.

In 1987, Conservation International, a private U.S. banking consortium, purchased $650,000 of Bolivia's $5.7-billion national debt from a Citibank affiliate in Bolivia for $100,000. In exchange for not having to pay back this part of its debt, the Bolivian government agreed to expand and protect from harmful forms of development 1.5 million hectares (3.7 million acres) of tropical forest around its existing Beni Biosphere Reserve in the Amazon Basin, which contains some of the world's largest reserves of mahogany and cedar. The government was to establish maximum legal protection for the reserve and create a $250,000 fund, with the interest to be used to manage the reserve.

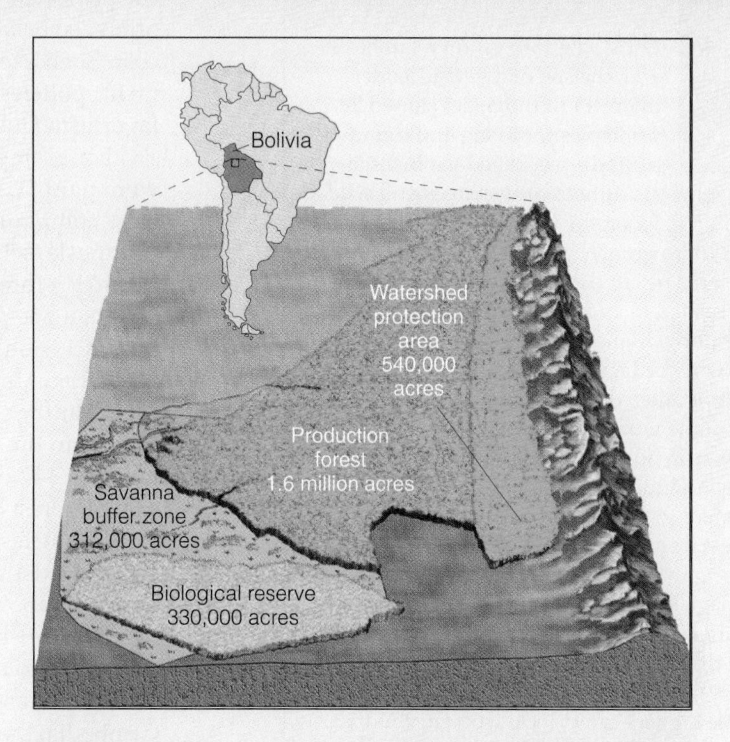

Figure 24-22 Blending economic development and conservation in a 1.5-million-hectare (3.7-million-acre) tract in Bolivia. A U.S. conservation organization arranged a debt-for-nature swap to help protect this land from destructive development.

The plan was intended to be a model of how conservation of forest and wildlife resources could be compatible with sustainable economic development. Central to the plan was a virgin tropical forest to be set aside as a biological reserve, surrounded by a protective buffer of savanna used for sustainable grazing of livestock (Figure 24-22). Controlled commercial logging, as well as hunting and fishing by local inhabitants, would be permitted in some parts of the forest but not in the area above the tract, which was to be set aside to protect the area's watershed and prevent erosion.

By 1999, 12 years after the agreement was signed, the Bolivian government still had not provided legal protection for the reserve. Meanwhile, timber companies (with government approval) have cut thousands of mahogany trees from the area; most of this lumber has been exported to the United States. The area's 5,000 native inhabitants were not consulted about the swap plan, even though they are involved in a land-ownership dispute with logging companies.

One lesson from this first debt-for-nature swap is that *legislative and budget requirements must be met before the swap is executed.* Another is that such swaps must be carefully monitored by environmental organizations to ensure that proposals for sustainable development are not disguises for unsustainable development.

Critical Thinking

Do you agree or disagree with the charge that many of the debt-for-nature swaps support international (mostly corporate) control over debtor countries' development while giving banks in developed countries a way to make good on part of essentially bad loans? Explain. Overall, do you believe that debt-for-nature swaps are a good idea? Explain.

A: About 2.1 million (some say 5 million), more than half of them children under age 5

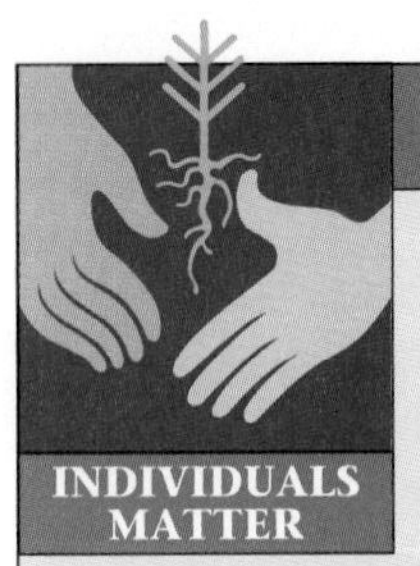

The Chipko Movement

In the late 15th century, Jambeshwar, the son of a village leader in northern India, developed 29 principles for living and founded a Hindu sect called the Bishnois, based on a religious duty to protect trees and wild animals. In 1730, when the Maharajah of Jodhpur in northern India ordered that the few remaining trees in the area be cut down, the Bishnois forbade it. Women rushed in and hugged the trees to protect them, but the Maharajah's minister ordered the work to proceed anyway. According to legend, 363 Bishnois women died that day.

In 1973, some women in the Himalayan village of Gopeshwar in northern India started the modern Chipko (an Indian word for "hug" or "cling to") movement to protect the remaining trees in a nearby forest from being cut down to make tennis rackets for export. It began when Chandi Prasad Bhatt, a village leader, urged villagers to run into the forest ahead of loggers and to hold on to and protect the trees, thus reviving the tradition started hundreds of years earlier by the Bishnois.

When the loggers came, local women, children, and men rushed into the forests, flung their arms around trees, and dared the loggers to let the axes fall on their backs. They were successful, and their action inspired women in other Himalayan villages to protect their forests. As a result, the commercial cutting of timber in the hills of the Indian state of Uttar Pradesh has been banned.

Since 1973, the Chipko movement has widened its efforts. Now the women—who still protect trees from loggers—also plant trees, prepare village forestry plans, and build walls to stop soil erosion.

another country or company guaranteeing that an equal amount is absorbed from the atmosphere through photosynthesis by young growing trees. By April 1998, 25 such offset projects had generated about $45 million for conservation and reforestation around the world.

Some conservationists and developing countries oppose this offset program. They contend that it is a way for developed nations to continue producing large amounts of CO_2 and reduces the pressure for such countries to get their own houses in order. What do you think about this policy?

Tropical timber-cutting regulations and practices can be reformed. New logging contracts could charge more for timber-cutting concessions and require companies to post adequate bonds for reforestation and restoration. However, such restrictions are hard to enact and enforce because of the economic and political power of timber companies, whose executives often see bribes and fines as a very minor business cost. Such executives often shape or ignore government policies by legal or illegal means through favoritism and corruption.

For example, in the Solomon Islands landowners were paid $2.70 per cubic meter for timber that foreign companies sold for $350 per cubic meter. In Indonesia, where the government owns 74% of the country's forests, the government typically collects less than one-fifth of the potential value of timber cut by concessions, amounting to a loss of $2.5 billion in 1990 alone. In the Philippines the cutting of 90% of the country's primary forest during the Marcos regime in the 1960s and 1970s made a few hundred families $42 billion richer while impoverishing 18 million forest dwellers.

Countries also lose money because of failure to enforce existing timber cutting laws. In Ghana, for example, about one-third of the timber is harvested illegally, causing a loss of about $65 million per year. According to the Brazilian government, about 80% of timber harvesting in the Amazon is done illegally. In Cambodia, the estimated amount of money lost to the national treasury because of illegal logging is equal to the entire national budget.

Gentler methods could also be used for harvesting trees. Cutting canopy vines (lianas) before felling a tree can reduce damage to neighboring trees by 20–40%, and using the least obstructed paths to remove the logs can halve the damage to other trees.

Pressure for clearing old-growth tropical forests can be reduced by concentrating peasant farming, tree and crop plantations, and ranching activities on already-cleared tropical forest areas that are in various stages of secondary ecological succession. In addition, global efforts can be mounted to reforest and rehabilitate degraded tropical forests and watersheds.

Another solution is transfer control of forests from governments to communities. For example, the condition of forests declined in India when the government took over their control more than a century ago. Because of protests (Individuals Matter, left) the policy was modified in the late 1980s to give thousands of communities control over state forestlands.This has helped improve the prospects for more sustainable use of the forests and the quality of life for people living in or near such forests.

Finally, lowering the waste and overconsumption of forest products by consumers would play a major role in reducing the pressures on the world's forests. For example, 20% of the world's people living in the

Q: How many offspring can a single bacterium have in 24 hours?

United States, Japan, and Western Europe consume more than half of the world's industrial timber and 70% of the paper. Reducing consumption and waste in these countries by even a small fraction would greatly reduce pressures on the world's forests.

Case Study: Tropical Forest Protection and Ecological Restoration in Costa Rica Costa Rica, smaller in area than West Virginia, was once almost completely covered with tropical forests. Between 1963 and 1983, politically powerful ranching families cleared much of the country's forests to graze cattle, and they exported most of the beef to the United States and western Europe. By 1983, only 17% of the country's original tropical forest remained, and soil erosion was rampant.

Despite widespread forest loss, tiny Costa Rica (like Madagascar) is a superpower of biodiversity (Figure 23-18), with an estimated 500,000 species of plants and animals. A single park in Costa Rica is home for more bird species than all of North America.

In the mid-1970s, Costa Rica established a system of national parks and conservation reserves that by 1997 included 12% of its land (6% of it in reserves for indigenous peoples), compared to only 1.8% set aside as reserves in the lower 48 United States. As a result Costa Rica now has a larger proportion of land devoted to biodiversity conservation than any other country. The government has also established a 15% national fuel tax, with part of the income used for reforestation.

Carefully selected parts of the country's protected areas are to be used for regulated forms of sustainable farming, hunting, and logging. They may also become profitable centers for regulated ecotourism, education, and research.

One reason for this outstanding accomplishment in biodiversity protection was the establishment in 1963 of the Organization of Tropical Studies, a consortium of more than 50 U.S. and Costa Rican universities with the goal of promoting research and education in tropical ecology. The resulting infusion of several thousand scientists has helped Costa Ricans appreciate their country's great biodiversity. It also led to the establishment in 1989 of the National Biodiversity Institute (INBio), a private nonprofit organization set up to survey and catalog the country's biodiversity.

In 1991, Merck & Co. paid INBio $1.35 million to locate and collect specimens of tropical organisms as possible sources of drugs. If Merck develops a marketable product from this *chemical prospecting*, the company will retain all patent rights but will pay INBio an undisclosed royalty (believed to be 1–3%). Since the Merck contract, INBio has signed seven more bioprospecting agreements with other companies looking for products such as antibiotics, flavors, fragrances, and natural alternatives to pesticides.

Critics of such arrangements argue that some (if not most) of the money should go to local or indigenous people without whose land or knowledge the new medicine or other products would not have come to pass. Some analysts go further and question whether patents should be given for development of genetically engineered varieties (Guest Essay, p. 674).

Costa Rica's biodiversity conservation strategy has paid off. Today, revenue from tourism (almost two-thirds of it from ecotourists) is the country's largest source of outside income. However, legal and illegal deforestation threatens this plan because of the country's population growth and poverty, which still affects 10% of its people. Currently, this small country still loses roughly 400 square kilometers (150 square miles) of primary forest per year—four times the rate of loss in Brazil. If Costa Rica's attempts to protect and use biodiversity in sustainable ways are successful, it could be a model for the entire world.

Costa Rica is also the site of one of the world's largest and most innovative ecological restoration projects. In the lowlands of the country's Guanacaste National Park, a small tropical seasonal forest is being restored and relinked to the rain forest on adjacent mountain slopes.

Daniel Janzen, professor of biology at the University of Pennsylvania and a leader in the growing field of rehabilitation and restoration of degraded ecosystems, has helped galvanize international support and has raised more than $10 million for this restoration project. Janzen's vision is to make the nearly 40,000 people who live near the park an essential part of the restoration of 740 square kilometers (285 square miles) of degraded forest—a concept he calls *biocultural restoration*. By actively participating in the project, local residents will reap enormous educational, economic, and environmental benefits. Local farmers have been hired to sow large areas with tree seeds and to plant seedlings started in Janzen's lab.

Students in grade schools, high schools, and universities study the ecology of the park in the classroom and then go on field trips to the park itself. Educational programs for civic groups and tourists from Costa Rica and elsewhere will stimulate the local economy.

The project will also serve as a training ground in tropical forest restoration for scientists from all over the world. Research scientists working on the project give guest classroom lectures and lead some of the field trips.

Janzen recognizes that in 20–40 years today's children will be running the park and the local political

The War Against Species

GUEST ESSAY

Vandana Shiva

Vandana Shiva has a background in physics and the philosophy of science, and she is director of the Research Foundation for Science, Technology, and Natural Resource Policy in Dehra Dun, India. She is also a consultant to the UN Food and Agriculture Organization and in recent years has served as a visiting professor at Oslo University in Norway, Schumacher College in England, and Mount Holyoke College in the United States. She has published 11 books and 74 articles concerned with environmental research and analysis. In 1993 she received several international environmental awards: the Global 500 Roll of Honor, the Earth Day International Award, the VIDA SANA International Award, and the Alternative Nobel Peace Prize (Right Livelihood Award).

Biodiversity, the richness of the living diversity, is severely threatened because of a worldview that perceives the value of all species only in terms of human exploitation. Because we are ignorant of the role that most species play in maintaining the intricate web of life, and because most species are "useless" from the perspective of the dominant economy, such a worldview leads to extinction of a growing number of species.

The main threat to biodiversity comes from the habit of thinking in terms of monocultures, from what I have called "monocultures of the mind." The resulting disappearance of diversity is also a disappearance of alternatives and gives rise to the TINA (there is no alternative) syndrome. Alternatives do exist, but are often excluded. Their inclusion requires an emphasis on preserving all kinds of diversity, biological and cultural, to allow the possibility of multiple choices.

This mindset underlies industrial forestry, in which tree species are useful only if they can provide raw material for construction and to make pulp and paper. Diverse natural forests are thus clear-cut and converted to monoculture plantations.

This worldview also permeates industrial agriculture. Thousands of crop varieties have disappeared because they are declared "uneconomic" and "unproductive," even though they enrich ecosystems and are nutritionally productive. Millets and legumes that require very low resource inputs and produce high nutrient value are declared "inferior" and "marginal" and pushed to extinction. In their place are introduced large-scale monocultures of globally traded crops such as wheat, rice, soybeans, and maize.

In our drive to dominate nature for our purposes, tree crop and animal varieties are engineered as if they were machines, not living organisms. Reducing living organisms to machines removes their ecological resilience. Thus, crop varieties engineered to be dwarfs in order to take up high doses of chemical fertilizers to produce high yields are especially vulnerable to pests and disease. Animals engineered to be either "milk machines" or "meat factories" are also highly vulnerable

system. If they understand the importance of their environment, they are more likely to protect and sustain its biological resources. He believes that education, awareness, and involvement—not guards and fences—are the best ways to protect ecosystems from unsustainable use.

Can Establishing Extractive Reserves Reduce Tropical Deforestation? One profitable way to use tropical forests sustainably is to harvest potentially renewable nonwood products such as nuts, fruits, herbs, spices, oils, medicines, and latex rubber. According to a 1988 study by economists, sustainable harvesting of such products in Amazon rain forests over the next 50 years would generate twice as much revenue per hectare as timber production and three times as much as cutting the forests and converting them to grassland for cattle ranching.

Chico Mendes, the rubber tapper and labor union activist shown in Figure 24-11, proposed that one-fourth of the Brazilian Amazon rain forest be set aside as *extractive reserves*—areas reserved for sustainable harvesting of potentially renewable commercial forest products by traditional resident populations. Making such proposals can be dangerous. Mendes was one of more than 1,500 Brazilian rubber tappers, priests, lawyers, union officials, and opponents of rain forest destruction who have been killed over the past 25 years, allegedly by gunmen working for wealthy ranchers and cash-crop farmers trying to amass large land holdings.

In 1988, before his death Chico Mendes said, "First I thought I was fighting for the rubber tappers, then I thought I was fighting for the Amazon, then I realized I was fighting for humanity." His actions forged links between environmental issues and issues of human

Q: In terms of deaths, what is the world's most dangerous viral disease?

to disease and need heavy doses of antibiotics to survive in factory farming environments.

It is in this context of the production of uniformity that conservation of biodiversity must be understood. Conservation of biodiversity is, above all, the preservation of alternatives, of keeping alive alternative forms of production for the biotechnology industry, and more important for future biological evolution on the earth.

The rapidly growing threat to biodiversity emerges from the myths that monocultures are essential for solving problems of scarcity and that there is no option but to destroy biodiversity to increase productivity. It is not true that without monoculture tree plantations there will be shortages of fuel wood, and without monocultures in agriculture there will be famines of food. Monocultures are in fact a source of scarcity and poverty, both because they destroy diversity and alternatives and because they destroy decentralized control of production and consumption systems.

Humans' escalating war against and control of other species for maximization of profits and production also influence concepts of ownership and control. People have always owned their farm animals and pets. However, the new trend toward "patenting of life forms" has changed the notion of ownership. A patent is a property right derived from an innovation; it is also called an "intellectual property right." However, when applied to the domain of life forms, this concept has far-reaching implications. A patent term is usually 20 years. If a company has genetically modified a sheep or pig or cow and gotten a patent for it, by implication the patent holder has "ownership" of future generations of the modified animal. The machine view of a living organism thus leads to the ethically outrageous and biologically absurd idea that future generations of genetically engineered plants and animals are solely the products of human innovation and are somehow divorced from the billions of years of evolution that produced all of their genes.

If diverse species on this planet are to survive, we must shift our worldview from a monoculture paradigm that views life as a machine to one based on biodiversity and on allowing the earth's life forms to play their roles in evolution under ever-changing conditions.

It is up to citizens and activists around the world, who see reverence for life as an essential part of the ecology movement, to ensure that technology and law serve the objective of conserving the rich biodiversity of our planet, and not merely serve the objective of protecting profits through unrestrained manipulation and monopoly control of life.

Critical Thinking

1. Do you believe that monocultures of crops and trees are essential for our survival? Explain.

2. Do you believe that a company or individual should be awarded a patent for a form of life that has been genetically engineered from existing forms of life? Explain.

rights and social justice and helped spearhead the rapidly growing international movement for *environmental justice*.

Because of the international outcry over Mendes's murder in 1988, Brazilian officials established 14 extractive reserves covering about 0.8% of the Brazilian Amazon and three new national parks since 1988. More areas have been added and settlers are now required to keep 80% of their land forested.

Environmentalists applaud these actions but fear that they are mostly window dressing calculated to weaken local and international pressures to protect much larger areas of the Amazon Basin from development. For example, between 1991 and 1997 Brazil's government reinstated tax breaks for cattle ranches and other forms of unsustainable development in the Amazon Basin and cut funds for identifying and setting aside indigenous and extractive reserves by 80%.

In 1997 the Brazilian government ended its subsidy program and announced plans to protect an area of the Amazon rain forest equal in size to the state of Oregon by the year 2000. But ranchers and farmers keep burning forests, the laws are not enforced, and in 1998 the government cut its rain forest protection budget by 90%. Brazil's environmental protection agency has only about 80 enforcement officers in the whole of the Amazon. Worse, Brazil's courts have ruled that the agency does not have the authority to enforce the law.

Extractive reserves can help sustain some areas of tropical forests, but this approach has limitations. Brazil's roughly 300,000 rubber tappers are greatly outnumbered by some 6 million farmers, ranchers, and loggers who make a living largely from destructive, unsustainable use of the land. Furthermore, local extractors of renewable forest resources often

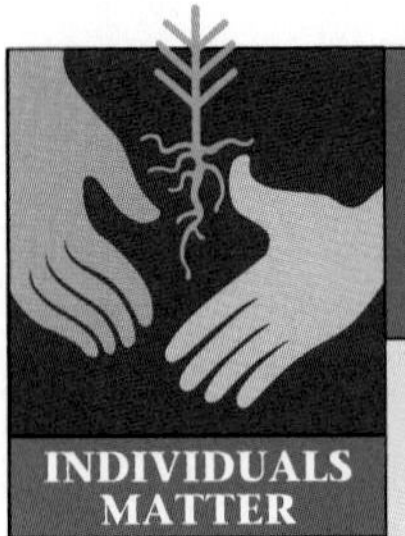

New Stoves Help Save India's Forests and Improve Women's Health

Since 1984 Lalita Balakrishnan, supported by the All-India Women's Conference (formed in 1926 with its early mission to educate Indian women), has put together an army of 300,000 women dedicated to spreading efficient and smokeless stoves throughout much of India. These women have helped design, build, and install 300,000 smokeless wood stoves, called chulhas, across the country.

These stoves cost about $5 and are made of cow-dung, mud, and hay slapped together by hand, with a metal pipe leading to the roof. The Indian government pays one-third of the stoves' cost, and all materials (except the pipe) are provided by the stove buyer.

The program now has 60 different models, with designs modified for local cultures and diets. Local women are hired to sell, install, and maintain the stoves.

About half of India's annual loss of 11,800 square kilometers (4,600 square miles) of forest is caused by use of fuelwood. During the past decade these cheap, fuel-efficient stoves have helped reduce deforestation and saved India more than 182,000 metric tons (200,000 tons) of fuelwood.

Replacement of smoky indoor fires with such stoves has also led to improved health for an estimated 1 million Indian women who previously suffered from chronic bronchitis, asthma, and almost certain premature death from inhalation of smoke from indoor open fires used for cooking with wood. This program has also raised the self-esteem of the women participating.

In recognition of her efforts, the UN Environment Programme added the name of Lalita Balakrishnan to its 1996 Global 500 Roll of Honor.

In 1996 the Indian government developed a plan to distribute 2.7 million chulhas, mostly through state governments.

receive little of the profits from their efforts. About $20 million worth of Brazil nuts are exported to the United States each year, but Brazil nut gatherers get less than 3% of what the nuts sell for wholesale in New York City.

Solutions: What Can Be Done About the Fuelwood Crisis? Developing countries can reduce the severity of the fuelwood crisis by planting more fast-growing fuelwood trees or shrubs, burning wood more efficiently, and switching to other fuels. One fast-growing fuelwood shrub is *Prosopis*, a genus of shrubs (including mesquite in North America) native to dry-land regions where trees are scarce. However, fast-growing tree and shrub species used to establish fuelwood plantations must be selected carefully to prevent harm to local ecosystems.

Eucalyptus trees, for example, are being used to reforest areas threatened by desertification and to establish fuelwood plantations in some parts of the world. Because these species grow fast even on poor soils, this might seem like a good idea, but environmentalists and local villagers see it as an ecological disaster. In their native Australia these trees thrive in areas with good rainfall; when planted in arid areas, however, the trees suck up so much of the scarce soil water that most other plants can't grow. Then farmers don't have fodder to feed their livestock, and groundwater is not replenished. The eucalyptus trees also deplete the soil of nutrients and produce toxic compounds that accumulate in the soil because of low rainfall. In Karnata, India, villagers became so enraged over a government-sponsored project to plant these trees that they uprooted the saplings.

Experience shows that planting projects are most successful when local people participate in their planning and implementation. Programs work best that give village farmers incentives, such as ownership of the land or ownership of any trees they grow on village land. Emphasis should be placed on community woodlots, which are easy to tend and harvest, rather than on creating large fuelwood plantations located far from where the wood is needed.

The governments of China, Nepal, Senegal, and South Korea have established successful tree-planting programs in selected areas, typically joint ventures by the government and locally elected village forestry associations. Government foresters supply villagers with seeds or seedlings of fast-growing fuelwood trees and shrubs, and they advise them on the planting and care of the trees; the villagers do the planting. They are encouraged to plant these species in fields along with crops (agroforestry), on unused patches of land around homes and farmland, and along roads and waterways.

Another promising method is to encourage villagers to use the sun-dried roots of various gourds and squashes as cooking fuel. These rootfuel plants, which regenerate themselves each year, produce large quantities of burnable biomass per unit area on dry deforested lands. They also help reduce soil erosion and produce an edible seed with a high protein content.

Q: What percentage of the money spent worldwide each year for biomedical research is devoted to tropical diseases?

The traditional three-stone fire is a very inefficient way to burn wood, typically wasting about 94% of the wood's energy content. New, cheap, more efficient, less polluting stoves can provide both heat and light while reducing indoor air pollution, a major health threat. However, such stoves won't be widely used if they don't take into account local needs, resources, and cultural practices. The stoves must be easy and cheap to build, and the materials used to make them must be easily accessible locally (Individuals Matter, left).

Cheap and easily made solar ovens (Figure 16-16e) that capture sunlight can also be used to reduce wood use and air pollution in sunny and warm areas; however, they are often not accepted by cultures that also use fires for light and heat at night and as centers for social interaction.

Despite encouraging success in some countries (such as China, Nepal, Senegal, and South Korea), most developing countries suffering from fuelwood shortages have inadequate forestry policies and budgets, and they lack trained foresters. Such countries are cutting trees for fuelwood and forest products 10–20 times faster than new trees are being planted.

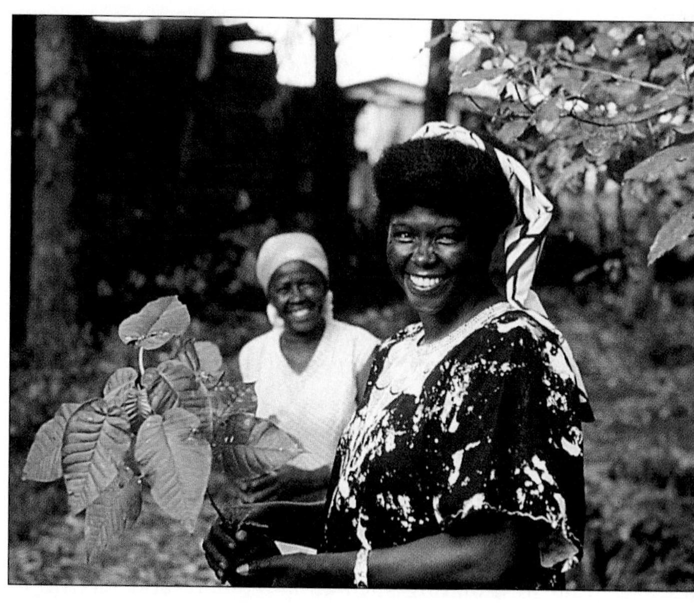

Figure 24-23 Wangari Maathai, the first Kenyan woman to earn a Ph.D. (in anatomy) and to head an academic department (veterinary medicine) at the University of Nairobi, organized the internationally acclaimed Green Belt Movement in 1977. The goals of this highly regarded women's self-help community are to establish tree nurseries, raise seedlings, and plant a tree for each of Kenya's 27 million people. Each of the 50,000 members of this grassroots group receives a small fee for every tree that survives. By 1997 more than 15 million trees had been planted, producing income for over 80,000 people. The success of this project has sparked the creation of similar programs in more than 30 other African countries. (William Campbell/*TIME* magazine)

24-6 SOLUTIONS: INDIVIDUAL ACTION

How Can Change Be Brought About from the Bottom Up? People the world over are working to protect forests. Penan tribespeople in Sarawak, Malaysia, have joined forces with environmentalists to protest destructive logging, the effects of which have reduced their population from 10,000 to less than 500. In Brazil, 500 conservation organizations have formed a coalition to preserve the country's remaining tropical forests. In India, members of the Chipko movement (Individuals Matter, p. 672) have helped protect forests.

The Awá Indians living in tropical forests along the border of Ecuador and Colombia have fought hard to protect a quarter-million acres of some the world's most biologically diverse land from invading ranchers, plantation farmers, loggers, and hunters. In the United States, environmental activists have perched in the tops of giant Douglas firs and redwood trees and lain in front of logging trucks and bulldozers to prevent the felling of ancient trees in national forests and in privately owned ancient forests scheduled for clear-cutting (Connections, p. 649).

In 1985 the Rainforest Action Network was virtually alone in the United States in opposing tropical deforestation. Today there are nearly 200 Rainforest Action groups around the United States, most on college campuses, and a new one is formed about every 10 days. In 1991 Find Fensen, a first-year student at Princeton University, became concerned about the world's tropical forests and decided to buy and protect some rain forest. His dorm room operation grew into the Rainforest Conservancy. By 1993 such efforts to preserve tropical forests had spread to 45 campuses, attracted contributions from individuals and corporations, and resulted in the purchase and protection of hundreds of acres of rain forest in Belize.

In Kenya, Wangari Maathai (Figure 24-23) started the Green Belt Movement, a national self-help community action effort by 50,000 women and half a million schoolchildren to plant trees for firewood and to help hold the soil in place. This inspiring leader has said,

> *I don't really know why I care so much. I just have something inside me that tells me that there is a problem and I have got to do something about it.*

A: About 3%

And I'm sure it's the same voice that is speaking to everyone on this planet, at least everybody who seems to be concerned about the fate of the world, the fate of this planet.

In response to her efforts to sustain the earth and fight for human rights, the Kenyan government closed down her Green Belt Movement offices (she moved the headquarters into her home) and jailed her twice. In 1992 she was severely beaten by police while leading a peaceful protest against government imprisonment of several environmental and political activists.

Despite the urgency, world leaders have failed to develop an international agreement to curb deforestation, protect the biodiversity and other ecological services provided by the planet's forests, and help reforest degraded areas. Environmentalists believe that formulating such an agreement and mounting and funding a global program to reduce the threats to much of the world's remaining tropical (and other old-growth) forests (Figure 24-14) should be one of our most urgent priorities.

The inspiring actions of such *earth citizens* show us that working with the earth to preserve biodiversity and ecological integrity will come mostly from the bottom up, through the actions of ordinary people, not from the top down. Some actions you can take to help sustain the earth's biodiversity and ecological integrity are listed in Appendix 5.

We can still save much of these forests and the species they contain. After all, we are trying to save the greatest celebration of nature in the biosphere. So should we not feel privileged that we, alone of all past and future human generations, should face this challenge? . . . If we don't convert this profound problem into a glorious opportunity, there will be nothing left for our successors to do but pick up all the pieces. It is in our hands. . . . Those tropical forests are waiting to hear from you.

NORMAN MYERS

CRITICAL THINKING

1. What difference could the loss of essentially all the remaining old-growth tropical forests and old-growth forests in North America have on your life and on the lives of your descendants?

2. Explain how eating a hamburger from a fast-food chain might indirectly contribute to the destruction of old-growth tropical forests. What, if anything, do you believe should be done about this?

3. Should developed countries provide most of the money to preserve remaining tropical forests in developing countries? Explain.

4. What five actions can you take to help preserve some of the world's tropical forests? Which, if any, of these actions do you plan to carry out?

5. Explain why you agree or disagree with each of the proposals listed in Section 24-5 concerning protection of the world's tropical forests.

6. Should all cutting of remaining old-growth forests in U.S. National Forests be banned? Explain.

7. Explain why you agree or disagree with each of the ways of improving the management of forests on public lands in the United States listed in the box on p. 668.

8. Do you believe that Americans have a right to criticize the forest management policies of Canada and tropical countries when the United States has cut most of its ancient forests, has the world's highest per capita wood and paper consumption, and makes up for increased demand for wood and protection of some of its remaining ancient forests by importing more wood (thus transferring the resulting environmental problems to such countries)? Explain. What three things would you do about this situation?

9. Although they support prescribed burning in some national parks and national forests, many conservationists oppose plans to use such burns in federal wilderness areas. They argue that humans know too little about fire to mimic nature and believe that lightning should be the only source of fire in federal wilderness areas. What do you think about this issue? Explain.

10. Should the United States meet most of its demand for wood and wood products by growing genetically improved trees on carefully managed tree plantations? Explain. If you agree, what regulations and safeguards would you impose to help ensure that such an approach would be sustainable and have a low environmental impact?

11. Evaluate Japan's environmental record. Is Japan more of an ecohero or an ecovillain? Can you identify any consistent guiding principles that result in this environmental record? Use the library or Internet to determine whether Japan's record has recently improved or worsened.

12. Try to come up with analogies to complete the following statements: **(a)** An old-growth forest is like _______. **(b)** Clear-cutting an old-growth forest is like _______. **(c)** Not caring about the destruction and degradation of tropical forests is like _______. **(d)** Displacing an indigenous culture is like _______. **(e)** Shifting cultivation in tropical forests is like _______. **(f)** Providing government subsidies to encourage the clear-cutting of trees in national forests is like _______.

Q: How much of the money spent on health care in the United States is used to prevent disease?

13. Make a list of the ecosystem services that a forest provides. Make another list of the forest products you use in your daily life. Based on your lists, what do you see as the balance between forest protection and extraction of forest products?

14. Congratulations. You have just been put in charge of the world. List the five most important features of your forest policy.

PROJECTS

1. If possible, try to visit **(a)** a diverse old-growth forest, **(b)** an area that has been recently clear-cut, and **(c)** one that was clear-cut 5–10 years ago. Compare the biodiversity, soil erosion, and signs of rapid water runoff in each of the three areas.

2. Go to a large building supply chain and try to determine which of its wood products come from tropical forests or from old-growth forests in the United States and Canada. Make the same evaluation for a large department store chain such as Wal-Mart. Share your evaluation with the managers of these stores.

3. Determine what paper and wood products derived from tropical forests or from old-growth forests in the United States and Canada your school purchases. Present the results to top administrative officials.

4. Since 1903 New Zealand has had a policy of meeting all of its demand for wood and wood products by growing timber on intensively managed tree plantations. Use the library or Internet to evaluate the effectiveness of this approach and its major advantages and disadvantages.

5. Make a concept map of this chapter's major ideas, using the section heads and subheads and the key terms (in boldface). Look at the inside back cover and in Appendix 4 for information about concept maps.

A: About 5%

25 SUSTAINING WILD SPECIES*

The Passenger Pigeon: Gone Forever

In the early 1800s, bird expert Alexander Wilson watched a single migrating flock of passenger pigeons darken the sky for over 4 hours. He estimated that this flock was more than 2 billion birds strong, some 386 kilometers (240 miles) long, and 1.6 kilometers (1 mile) wide.

By 1914 the passenger pigeon (Figure 25-1) had disappeared forever. How could a species that was once the most common bird in North America become extinct in only a few decades?

The answer is humans. The main reasons for the extinction of this species were uncontrolled commercial hunting and loss of the bird's habitat and food supply as forests were cleared to make room for farms and cities.

*__Paul M. Rich__, associate professor of Ecology and Evolutionary Biology and Environmental Studies at the University of Kansas, is coauthor of this chapter.

Figure 25-1 Passenger pigeons, extinct in the wild since 1900. The last known passenger pigeon died in the Cincinnati Zoo in 1914. (John James Audubon/The New York Historical Society)

Passenger pigeons were good to eat, their feathers made good pillows, and their bones were widely used for fertilizer. They were easy to kill because they flew in gigantic flocks and nested in long, narrow colonies.

Commercial hunters would capture one pigeon alive, sew its eyes shut, and tie it to a perch called a stool. Soon a curious flock would land beside this "stool pigeon," and the birds would then be shot or ensnared by nets that might trap more than 1,000 birds at once.

Beginning in 1858, passenger pigeon hunting became a big business. Shotguns, traps, artillery, and even dynamite were used. Birds were sometimes suffocated by burning grass or sulfur below their roosts. Live birds were even used as targets in shooting galleries. In 1878 one professional pigeon trapper made $60,000 by killing 3 million birds at their nesting grounds near Petoskey, Michigan.

By the early 1880s commercial hunting had ceased because only a few thousand birds were left. At that point, recovery of the species was doomed because the females laid only one egg per nest. On March 24, 1900, a young boy in Ohio shot the last known wild passenger pigeon. The last passenger pigeon on earth, a hen named Martha after Martha Washington, died in the Cincinnati Zoo in 1914. Her stuffed body is now on view at the National Museum of Natural History in Washington, D.C.

Eventually, all species become extinct or evolve into new species, but humans have become the primary factor in the premature extinction of more and more species. Conservation biologists estimate that every day at least 10 (and perhaps as many as 200) species become extinct because of human activities, and studies indicate that the rate of this loss of biodiversity is increasing rapidly.

Many biologists consider this epidemic of extinction an even more serious problem than depletion of stratospheric ozone and global warming (Chapter 19) because it is happening ever more quickly and is irreversible. To these biologists, stemming this hemorrhage of biodiversity, protecting wildlife habitats throughout the world, and restoring species that we have decimated are planetary emergencies to which we should respond now.

The last word in ignorance is the person who says of an animal or plant: "What good is it?" . . . If the land mechanism as a whole is good, then every part of it is good, whether we understand it or not. . . . Harmony with land is like harmony with a friend; you cannot cherish his right hand and chop off his left.

ALDO LEOPOLD

This chapter is devoted to answering the following questions:

- Why should we care about wildlife?
- Are human activities causing a new mass extinction?
- What activities and traits of humans endanger wildlife?
- How can we prevent premature extinction of species?
- Can game animals be managed sustainably?
- Can freshwater and marine fish be managed sustainably?

25-1 WHY PRESERVE WILD SPECIES?

Why Not Let Wild Species Die Out? If all species eventually become extinct, why should we worry about losing a few more because of our activities? Does it matter that the passenger pigeon (Figure 25-1), the black rhinoceros (Figure 9-14), the northern spotted owl (Figure 24-6), the green sea turtle (p. 79), the 70 remaining Florida panthers (p. 609), or some unknown plant or insect in a tropical forest becomes prematurely extinct because of human activities?

Biologists contend that the answer is *yes* because of the economic, medical, scientific, ecological, aesthetic, and recreational value of all species. Some environmentalists go further and contend that each species has an inherent right to play its role in the ongoing evolution of life on earth until it becomes extinct without interference by humans.

What Is the Economic and Medical Importance of Wild Species? Some 90% of today's food crops were domesticated from wild tropical plants (Figure 12-5). Moreover, agricultural scientists and genetic engineers need existing wild plant species to derive today's crop strains and to develop the new crop strains of tomorrow.

Figure 25-2 Many species of wildlife, such as this bird of paradise in a New Guinea tropical forest, are a source of beauty and pleasure. This species is vulnerable to extinction because its mating ritual requires a throng of males to assemble and "strut their stuff" in front of females. When population numbers decline too much, mating cannot occur. (A. J. Deane/Bruce Coleman Ltd.)

Wild plants and plants domesticated from wild species supply rubber, oils, dyes, fiber, paper, lumber, and other useful products (Figure 24-3). Nitrogen-fixing microbes in the soil and in plants' root nodules (Figure 5-7) supply nitrogen to grow food crops worth almost $50 billion per year worldwide ($7 billion in the United States alone). Pollination by birds and insects is essential to many food crops, including 40 U.S. crops valued at approximately $30 billion per year.

About 80% of the world's population relies on plants or plant extracts for medicines. At least 40% of all pharmaceuticals, worth at least $100 billion per year, owe their existence to the genetic resources of wild plants (Figure 24-12), mostly from tropical developing countries. Plant-derived anticancer drugs save an estimated 30,000 lives per year in the United States. Over 3,000 antibiotics, including penicillin and tetracycline, are derived from microorganisms. Only about 5,000 of the 250,000 known plant species have been studied thoroughly for their possible medical uses.

What Is the Scientific and Ecological Importance of Wild Species? Every species can help scientists understand how life has evolved and functions, and how it will continue to evolve on this planet. Wild species also provide many of the ecological services that make up earth capital (Figure 1-2) and thus are key factors in sustaining the earth's biodiversity and ecological integrity.

They supply us (and other species) with food, recycle nutrients essential to agriculture, and help generate and maintain soils. They also produce oxygen and other gases in the atmosphere, absorb pollution, moderate the earth's climate, help regulate local climates and water supplies, reduce erosion and flooding, and store solar energy. Moreover, they detoxify poisonous substances, break down organic wastes, control potential crop pests and disease carriers, and make up a vast gene pool for future evolutionary processes.

What Is the Aesthetic and Recreational Importance of Wildlife? Wild plants and animals are a source of beauty, wonder, joy, and recreational pleasure for many people (Figure 25-2). Americans spend about $18.2 billion a year to watch wildlife—over three times the $5.8 billion they spend each year on movie tickets and the $5.9 billion they spend annually on professional sporting events.

Wildlife tourism, sometimes called *ecotourism*, is the fastest growing segment of the global travel industry and generates an estimated $30 billion in revenues each year. Conservation biologist Michael Soulé estimates that one male lion living to age 7 generates $515,000 in tourist dollars in Kenya; by contrast, if killed for its skin the lion would bring only about $1,000. Similarly, over a lifetime of 60 years, a Kenyan elephant is worth close to $1 million in ecotourist revenue. Florida's coral reefs are worth an estimated $1.6 billion a year in tourism revenue.

However, according to Emily Young, professor of geography at the University of Arizona, "ecotourism is fast becoming tourism without the eco." New hotels are devouring wildlife habitat. Beachfront hotels on Mexico's Pacific coast lure visitors to watch endangered sea turtles lay their eggs, but their bright lights disorient the turtles as they lumber ashore and many fail to lay eggs.

Q: What percentage of cancers are caused or promoted by environmental and lifestyle factors?

Before embarking on an ecotour, wildlife conservationists urge us to seek answers *in writing* to several questions:

- What precautions are taken to reduce the tour's impact on local ecosystems?
- How much time is spent in the field, versus in the city or traveling in a vehicle?
- How will the tour's garbage be disposed of?
- What percentage of the people involved in planning, organizing, and guiding the tour are local?
- Are any of the guides trained naturalists?
- Will you stay in locally owned hotels or other facilities, or will you be staying in internationally or nationally owned accommodations? Do these facilities practice environmentally sound sewage treatment and waste disposal?
- Does the tour operation respect local customs and cultures? If so, how?
- What percentage of the tour's gross income goes into the salaries and businesses of local residents and what percentage of this gross does the tour company donate to local environmental and social projects?

Kenya has developed national guidelines for ecotourism. All revenue raised by Kenyan national parks and reserves remains under the jurisdiction of the Kenyan Wildlife Service and is used primarily to provide better conservation of the protected areas. The government requires that tourist lodges preferentially hire Kenyans for all but the most senior positions and use Kenyan food products as much as possible. Furthermore, every game lodge visitor pays a $5 tax that goes into a trust fund for local use. Parks without lodges deposit a percentage of gate receipts into local trusts used to fund schools and hospitals and to compensate local landowners.

Why Is It Ethically Important to Preserve Wild Species? Some people believe that each wild species has an inherent right to exist, or to struggle to exist. This ethical stance is based on the view that each species has *intrinsic value* unrelated to its usefulness to humans.

According to this view, we have an ethical responsibility to protect species from becoming prematurely extinct as a result of human activities. Biologist Edward O. Wilson believes that deep within, most people feel obligated to protect other species and the earth's biodiversity because of the natural affinity for nature built into our genes (Spotlight, right).

SPOTLIGHT

Biophilia

Biologist Edward O. Wilson contends that because of the billions of years of biological connections leading to the evolution of the human species (Figure 6-6), we have an inherent affinity for the natural world—a phenomenon he calls *biophilia* (love of life). He points out that we cannot erase this evolutionary imprint in our genes through a few generations of urban living.

Evidence of this natural affinity for life is seen in the preference people have for almost any natural scene over one from an urban environment. Given a choice, people prefer to live in an area where they can see water, grassland, or a forest. More people visit zoos and aquariums than attend all professional sporting events combined.

In the 1970s I was touring the space center at Cape Canaveral in Florida. During our bus ride the tour guide pointed out each of the abandoned multimillion-dollar launch sites and gave a brief history of each launch. Most of us were utterly bored. All of a sudden people started rushing to the front of the bus and staring out the window with great excitement. What they were looking at was a baby alligator—a dramatic example of how *biophilia* can triumph over *technophilia.*

Critical Thinking

Do you have a built-in affinity for wildlife and wild ecosystems (biophilia)? If so, how do you display this inherent love of wildlife in your daily actions? What patterns of your consumption help destroy and degrade wildlife?

Some people distinguish between the survival rights of plants and those of animals, mostly for what they consider practical reasons. Poet Alan Watts once said that he was a vegetarian "because cows scream louder than carrots."

Other people distinguish among various types of species. They might think little about killing a fly, mosquito, cockroach (Spotlight, p. 209), or rat or ridding the world of disease-causing bacteria. Unless they are strict vegetarians, they might also see no harm in having others kill domesticated animals in slaughterhouses to provide them with meat, leather, and other products. However, these same people might deplore the killing of wild animals such as deer, squirrels, or rabbits.

A: Up to 80%, most caused by smoking and poor diet

Some proponents go further and assert that each individual organism, not just each species, has a right to survive without human interference, just as each human being has the right to survive. Others emphasize the importance of preserving the whole spectrum of biodiversity by protecting entire ecosystems rather than individual species or organisms, as discussed in Chapters 23 and 24. This view is based on the principle that humans have an ethical obligation to prevent premature extinction of wildlife by saving their habitats and not disrupting the complex ecological interactions that sustain all life.

25-2 THE RISE AND FALL OF SPECIES

How Does Background Extinction Differ from Mass Extinction? Extinction is a natural process and eventually all species become extinct. David Raup and several other evolutionary biologists estimate that more than 99.9% of all the species that have ever existed are now extinct because of a combination of background and mass extinctions. Each year, a small number of species become extinct naturally at a low rate, a phenomenon called the **background** or **natural rate of extinction**. Based mostly on the fossil record, evolutionary biologists estimate that the current average background rate of extinction is 3 species per year if there are about 10 million species and 30 species per year if there are 100 million species.

In contrast, **mass extinction** is an abrupt rise in extinction rates above the background level. It is a catastrophic, widespread (often global) event in which large groups of existing species (perhaps 25–70%) are wiped out. Most mass extinctions are believed to result from one or a combination of global climate changes that kill many species and leave behind those able to adapt to the new conditions.

Fossils and geological evidence indicate that the earth's species have experienced five great mass extinctions (20–60 million years apart) during the past 500 million years in which large numbers of species became extinct each year for tens of thousands to millions of years (Figure 6-13). Evidence also shows that these mass extinctions were followed by other periods, called *adaptive radiations*, when the diversity of life increased and spread for 10 million years or more (Figures 6-13 and 6-14). The last mass extinction took place about 65 million years ago, when the dinosaurs became extinct, for reasons that are hotly debated, after thriving for 140 million years.

Why Do Conservation Biologists Believe There Is a New Mass Extinction Crisis? Imagine that you have built a two-story house using wood as the basic structural material and that termites are slowly destroying various parts of the structure. How long will it be before they destroy enough of the structure for parts of the house to collapse?

Conservation biologists believe that this urgent question, when applied to the earth, is one that we as a species should be asking ourselves. As we tinker with the only home for us and other species, we are rapidly removing parts of the earth's natural biodiversity and ecological integrity on which we and other species depend in ways we know little about. We are not heeding Aldo Leopold's warning: "To keep every cog and wheel is the first precaution of intelligent tinkering."

So far we have little understanding of the ecological roles of the world's identified 1.75 million species, much less of the 100 million more species that may exist. Until we have such information, most biologists believe that we should use the *precautionary principle*, based on Aldo Leopold's warning, to prevent the premature extinction of species as a result of our activities.

It's difficult to document extinctions, and most go unrecorded. However, fossil and other evidence indicate that since agriculture began about 10,000 years ago, human activities have caused a growing number of species to become extinct (Figure 25-3). According to a 1995 comprehensive assessment of global biodiversity by the UN Environment Programme, at least 484 animal (mainly birds and mammals) and 654 plant species have become extinct mostly as a result of human activities since 1600. The actual total is believed to be much higher.

Some conservation biologists estimate that currently 18,000–73,000 species become extinct each year (on average, 50–200 species per day)—thousands of times the estimated natural background extinction rate of 3–30 species per year. Even if this estimate of the erosion of biodiversity is 100 times too high, the current extinction rate is many times the estimated rate of background extinction.

Scientists expect this rate of extinction to accelerate as the human population grows and takes over more of the planet's wildlife habitats and net primary productivity (Figure 1-5). They warn that within the next few decades we could easily lose at least 1 million of earth's species, most of them in tropical regions. According to biodiversity expert Edward O. Wilson, "Clearly, we are in the midst of one of the great extinction spasms of geological history. . . . As a biologist, I sometimes feel like an art curator watching the Louvre burn down."

Mass extinctions occurred long before humans evolved (Figure 6-13), but there are three important differences between the current mass extinction most biologists believe we are bringing about and those of the past. First, *the current extinction crisis is the first to be caused by a single species: our own.* By using or

Q: In terms of reduced life span, what is the world's greatest risk?

Figure 25-3 Some species that have become extinct largely because of human activities, mostly habitat destruction and overhunting.

wasting approximately 40% of the earth's terrestrial net primary productivity (25% of the world's total net primary productivity), we are crowding out other terrestrial species. What will happen to wildlife and the services they provide for humans and other species if our population doubles in the next 40 years?

Second, *the current mass wildlife extinction is taking place in only a few decades, rather than over thousands to millions of years.* Such rapid extinction cannot be balanced by speciation because it typically takes 2,000–100,000 generations for new species to evolve. Fossil and other evidence related to past mass extinctions indicates that it takes millions of years to recover biodiversity loss through adaptive radiations (Figures 6-13 and 6-14). Thus, repercussions for humans and other species from the current human-caused mass extinction will affect the future course of evolution for 5–10 million years and could lead to the premature extinction of the human species.

Third, *besides killing off species, we are eliminating many biologically diverse environments such as tropical forests, tropical coral reefs, wetlands, and estuaries that in the past have served as evolutionary centers for the recovery of biodiversity after a mass extinction.* In the words of Norman Myers (Guest Essay, p. 530), "Within just a few human generations, we shall—in the absence of greatly expanded conservation efforts—impoverish the biosphere to an extent that will persist for at least 200,000 human generations or twenty times longer than the period since humans emerged as a species."

Mathematical models recently developed by ecologists indicate a time lag of several generations between habitat loss and extinction, primarily because habitat loss also removes potential colonization sites. If this model is correct, biologists may be grossly underestimating the magnitude of the current biodiversity meltdown.

One of the models also projected that the top competitors best adapted to make use of a habitat's resources are the most vulnerable to habitat loss. If this unexpected result is correct, many keystone species, which help keep ecosystems productive and sustainable, are especially at risk. Their loss would lead to the loss of many other species—a cascading effect similar to removing one card from a house of cards.

Even without extinctions, ecosystems may lose their ability to support many forms of life because of the disappearance of local populations of key organisms. According to a 1997 study by biologist Jennifer Hughes and her colleagues at Stanford University, every year an estimated 16 million local plant and animal populations—an average of 1,800 populations every hour—disappear from the world's tropical forests.

Is There Really an Extinction Crisis? Some social scientists and a few biologists question the existence of an extinction crisis caused by our activities. They point to several problems in estimating species loss. First, we don't know how many species there are; estimates range between 5 million and 100 million. Second, it is difficult to observe species extinction, especially for species we know little or nothing about. A species is generally considered extinct when it has not been seen for at least 50 years or when the last of a few closely monitored individuals dies. This means that species extinctions recorded today really pertain to the 1940s rather than the 1990s.

As a result, biologists have to use models and field data to make estimates of current and future extinction rates. Using such approaches, the annual loss of tropical forest habitat is estimated at about 1.8% per year. Edward O. Wilson and several tropical biologists who have counted species in patches of tropical forest before and after destruction or degradation estimate that this 1.8% loss in habitat results in roughly a 0.5% loss of species. However, biomes vary in the number of species they contain (species diversity), and the ratio of habitat loss to species loss varies in different biomes.

Do such estimates add up to an extinction crisis? Let us assume, as Wilson and many other biologists do, that a loss of 1 million species over several decades represents an extinction crisis. If we assume the global decline in species to be 0.5% per year, then we will lose 25,000 species per year if there are 5 million species, 100,000 per year if there are 20 million species, and 500,000 per year if there are 100 million species. If these assumptions are correct, we will lose 1 million species in 40 years if there are 5 million species, in 10 years with 20 million species, and in only 2 years with 100 million species.

Let's assume, however, that the estimate of 0.5% species loss per year is too high for the earth as a whole. If it is 0.25% per year, then we will lose 1 million species in 80 years with 5 million species, in 20 years with 20 million species, and in 4 years with 100 million species. Even if we halve the estimated species loss again, to 0.125% per year, we can still lose 1 million species within 8–160 years, enough to easily qualify the situation as an extinction crisis.

According to a 1998 survey, 70% of the biologists polled believe that we are in the midst of a mass extinction and that this loss of species will pose a major threat to human existence in the next century. Biologists don't contend that their estimates of extinction

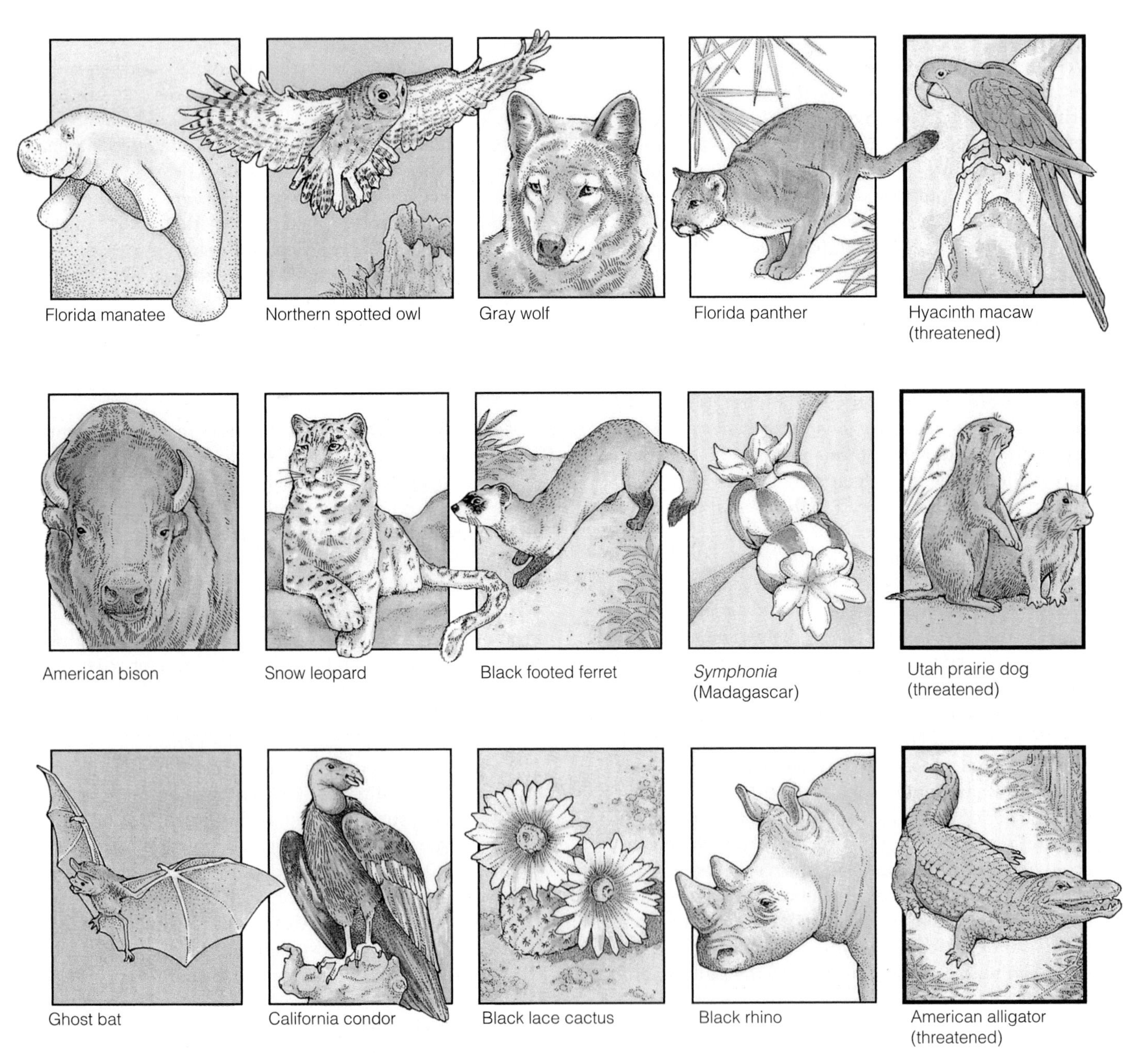

Figure 25-4 Species that are endangered or threatened largely because of human activities. Populations of some of the species are recovering, such that some endangered species may be relisted as threatened and some threatened species (such as the bald eagle and peregrine falcon) may be delisted.

Q: What does the EPA consider to be the three most dangerous indoor air pollutants in the United States?

rates are precise enough to make a firm prediction. Instead, they argue that there is ample evidence that we are destroying and degrading wildlife habitats at an increasing rate and that our actions certainly lead to a significant loss of species, even though the number and rate will vary in different parts of the world.

What Are Endangered and Threatened Species? Biologists distinguish among three levels of extinction: **(1)** *Local extinction* occurs when a species is no longer found in an area it once inhabited but is still found elsewhere in the world, **(2)** *ecological extinction* occurs when there are so few members of a species left that it can no longer play its ecological roles in the biological communities where it is found, and **(3)** *biological extinction* occurs when a species is no longer found anywhere on the earth. Biological extinction is forever.

Species heading toward biological extinction are classified as either *endangered* or *threatened* (Figure 25-4). An **endangered species** has so few individual survivors that the species could soon become extinct over all or most of its natural range. Examples are the California condor in the United States (40 in the wild), the

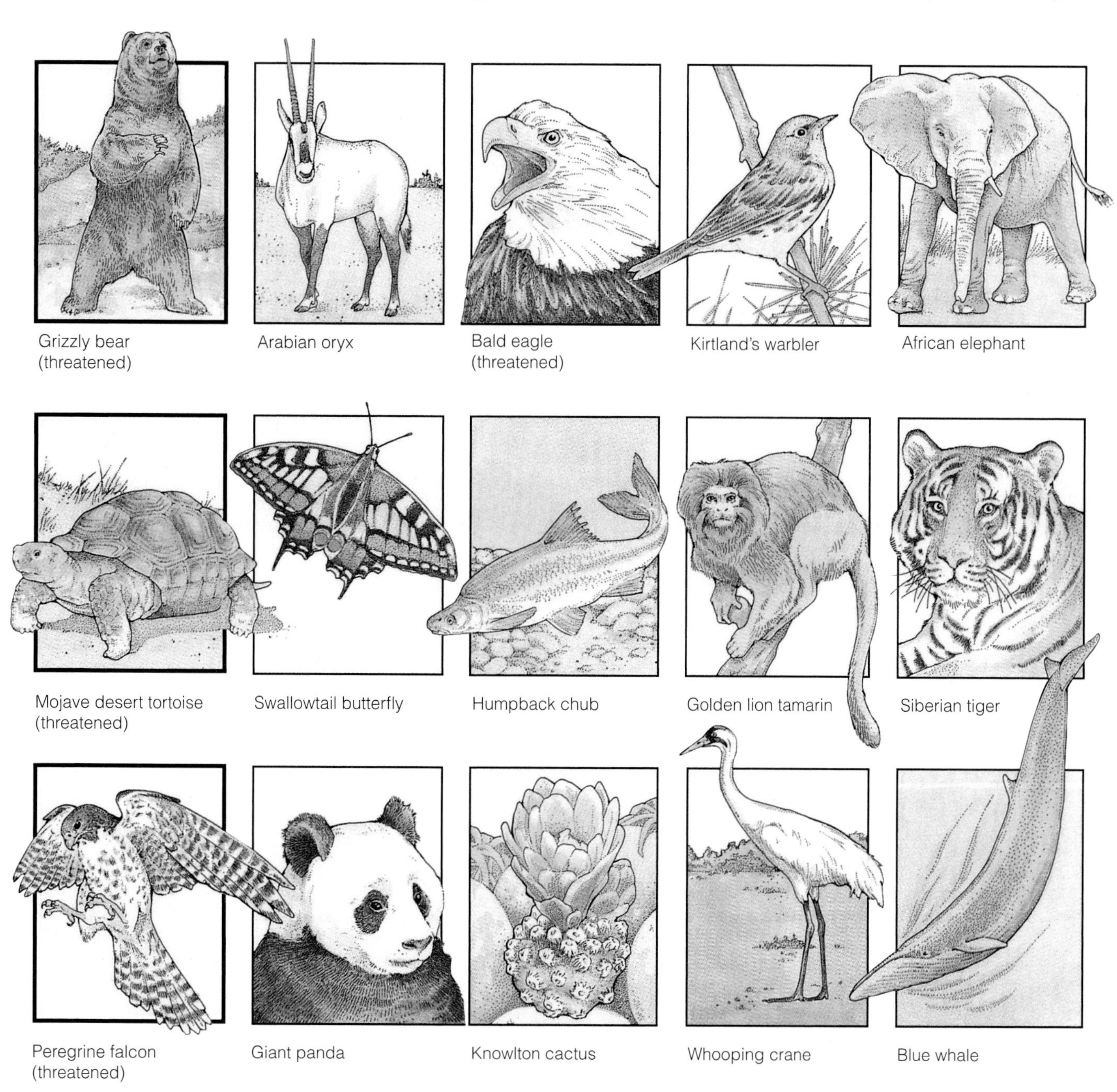

whooping crane in North America (about 288 left), the giant panda in central China (about 1,000 left, Figure 9-2), the snow leopard in central Asia (about 2,500 left), and the black rhinoceros in Africa (about 2,400 left).

A **threatened species** is still abundant in its natural range but is declining in numbers and is likely to become endangered. Examples are the grizzly bear, the southern sea otter (Figure 10-1), and the American alligator (Figure 6-1). Endangered and threatened species are ecological smoke alarms.

Some species have characteristics that make them more vulnerable than others to premature extinction (Table 25-1). In general, species more vulnerable to extinction are found in a limited area and have a small population size, a low population density, large body size, a specialized niche (Section 9-1), and a low reproductive rate (K-strategists; Figure 10-8). Such species also tend to undergo fairly fixed migrations, feed atop long food chains or webs, and have high economic values to people.

According to a 1996 joint study by the International Union for the Conservation and Conservation International, more than 5,200 known animal species are at risk of extinction. They include 34% of the world's fish, 25% of amphibians (such as salamanders and frogs; Connections, p. 211), 25% of mammals (Figure 25-5), 20% of reptiles, and 11% of the bird species. The study also estimated that at least one of every eight plant species in the world—and nearly one of three in the United States—are under threat of extinction. According to biologists, such figures are probably underestimated because of lack of data from many areas, including many species-rich tropical countries (Figure 23-18) where areas are being rapidly cleared (Figure 24-2).

Each animal species has a *critical population density* and a *minimum viable population size*, below which survival may be jeopardized because males and females have a difficult time finding each other. Once a population falls below its critical size, it continues to decline, even if the species is protected, because its death rate exceeds its birth rate. The remaining small population can easily be wiped out by fire, flood, landslide, disease, or some other single catastrophe. As population size drops, genetic diversity also decreases because the resulting smaller gene pool and inbreeding reduce a population's ability to respond to environmental changes through natural selection.

According to a 1995 study of the genetics of small populations by population geneticist Russell Lande, an endangered species must number at least 10,000, and often more, to maintain its evolutionary potential for survival. On average, only about 1,000 individuals of an animal species and 120 of a plant species remain

Table 25-1 Characteristics of Extinction-Prone Species

Characteristic	Examples
Low reproductive rate	Blue whale, polar bear, California condor, Andean condor, passenger pigeon, giant panda, whooping crane
Specialized feeding habits	Everglades kite (eats apple snail of southern Florida), blue whale (krill in polar upwellings), black-footed ferret (prairie dogs and pocket gophers), giant panda (bamboo), koala (certain eucalyptus leaves)
Feed at high trophic levels	Bengal tiger, bald eagle, Andean condor, timber wolf
Large size	Bengal tiger, lion, elephant, Javan rhinoceros, American bison, giant panda, grizzly bear
Limited or specialized nesting or breeding areas	Kirtland's warbler (6- to 15-year-old jack pine trees), whooping crane (marshes), orangutan (only on Sumatra and Borneo), green sea turtle (lays eggs on only a few beaches), bald eagle (prefers forested shorelines), nightingale wren (only on Barro Colorado Island, Panama)
Found in only one place or region	Woodland caribou, elephant seal, Cooke's kokio, many unique island species
Fixed migratory patterns	Blue whale, Kirtland's warbler, Bachman's warbler, whooping crane
Preys on livestock or people	Timber wolf, some crocodiles
Behavioral patterns	Passenger pigeon and white-crowned pigeon (nest in large colonies), redheaded woodpecker (flies in front of cars), Carolina parakeet (when one bird is shot, rest of flock hovers over body), Key deer (forages for cigarette butts along highways—it's a "nicotine addict")

Q: What is the most dangerous indoor air pollutant in developing countries?

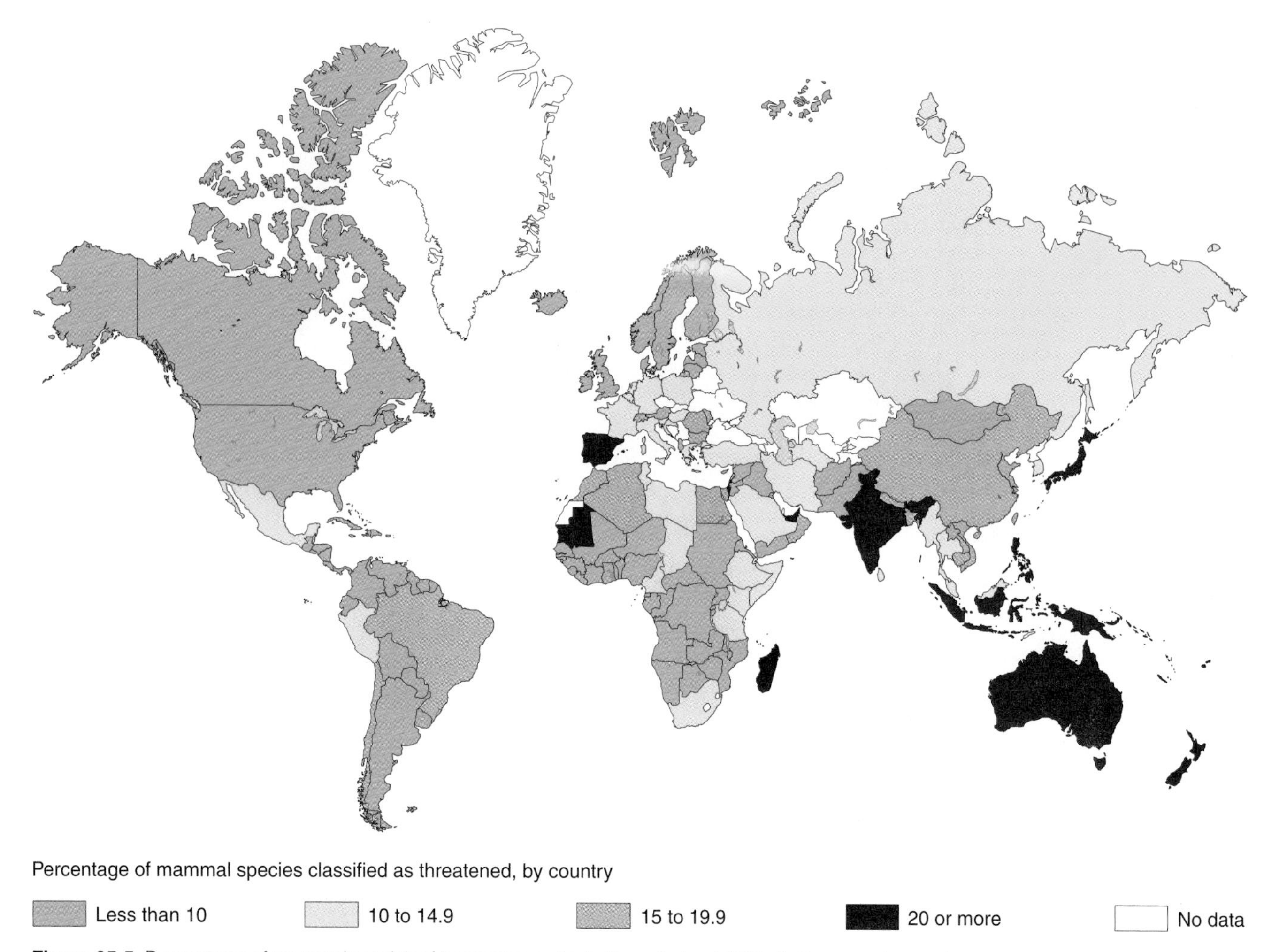

Figure 25-5 Percentage of mammals at risk of becoming extinct. Overall about 25% of all mammals are at risk. Of the 4,327 mammal species, 1,096 are at risk and 169 are critically endangered, with an extremely high risk of becoming extinct in the wild in the very near future. The highest percentage of mammal species at risk are apes and monkeys (46%); moles and shrews (36%); antelopes and cattle (33%); bats, wild dogs, bears, and cats (26%); and rodents (17%). Countries with the highest numbers of mammal species threatened with extinction are Indonesia (128), China (75), India (75), Brazil (71), Mexico (64), and Australia (58). (Data from International Union for the Conservation of Nature and Conservation International)

when they are listed as endangered, according to a 1993 study by the Environmental Defense Fund.

The populations of many wild species that are not yet in danger of extinction have diminished locally or regionally to the point where they are locally or ecologically extinct. Such species may be a better indicator of the condition of entire ecosystems than endangered and threatened species. These *indicator species* can serve as early warnings so that we can prevent species extinction rather than respond to emergencies, often with little chance of success.

Case Study: What Is the Status of Wild Species and Ecosystems in the United States? According to a 1995 study by the Nature Conservancy, about 32% of the 20,500 U.S. animal and plant species studied by scientists are vulnerable to premature extinction, mostly because of human activities. The animals that are most at risk are those that depend on aquatic ecosystems, such as freshwater fish and amphibians; a large number of flowering plants are also in trouble. An estimated 29% of all higher plants in the United States are threatened—a higher percentage than any other country in the world. Approximately one-third of all plants and animals that are endangered or threatened in North America make their homes in wetlands, which are under growing pressure for draining and development and stress from pollution (Sections 8-2 and 8-3). Figure 25-6 shows the estimated number of species extinctions by state in the lower 48 states.

A: Smoke from unvented or poorly vented stoves for cooking and heating

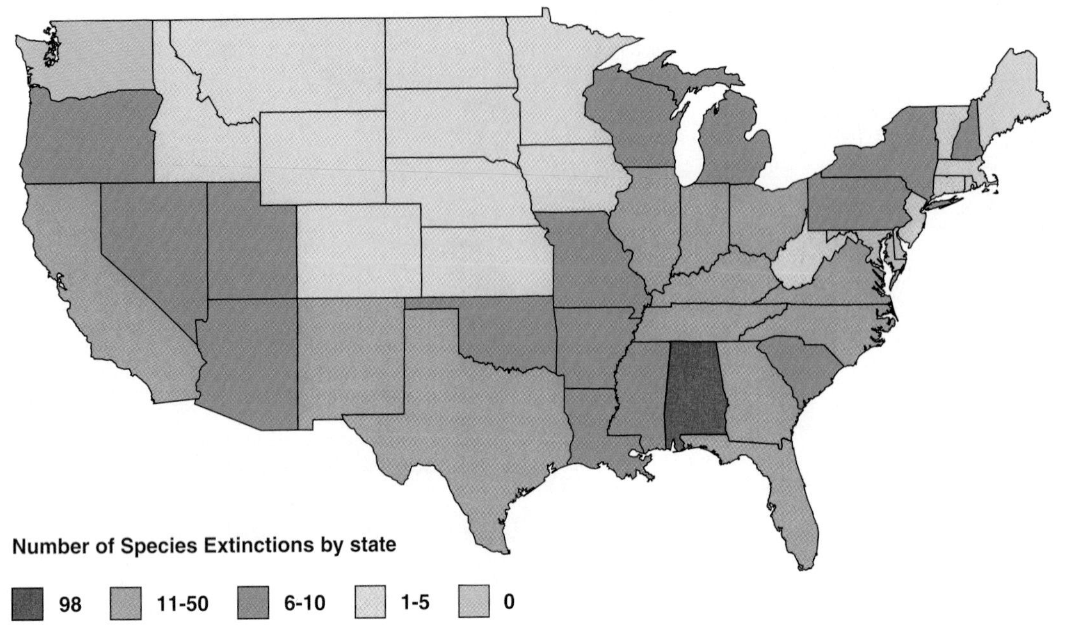

Figure 25-6 Estimated number of species extinctions of plants and animals by state in the lower 48 states. Alabama leads in extinct species. (Data from the Nature Conservancy)

Another survey published in 1995 by Defenders of the Wildlife focused on the status of ecosystems in the United States. According to the report, ecosystem health is especially threatened in 10 states: California, Hawaii, Texas, and the Southeastern states, particularly Florida. However, the report noted that "natural areas are at a biological breaking point all across the country."

Case Study: Bats Are Getting a Bad Rap Despite their variety (950 known species) and worldwide distribution, bats—the only mammals that can fly—have certain traits that place them at risk because of human activities. Bats reproduce slowly, and many bat species that live in huge colonies in caves and abandoned mines become vulnerable to destruction when people block the passageways or disturb their hibernation.

Because of unwarranted fears of bats and lack of knowledge about their vital ecological roles, several species have been driven to extinction. Currently, 26% of the world's bat species, including the ghost bat (Figure 25-7), are listed as endangered or threatened.

Conservation biologists urge us not to kill bats and also to protect them because of their important ecological roles. About 70% of all bat species feed on crop-damaging nocturnal insects and other pest species such as mosquitoes, making them the primary control agents for such insects.

In some tropical forests and on many tropical islands, pollen-eating bats pollinate flowers and fruit-eating bats distribute plants throughout tropical forests by excreting undigested seeds (Figure 9-1). As keystone species, they are vital for maintaining plant biodiversity and for regenerating large areas of tropical forest cleared by human activities. If you enjoy bananas, cashews, dates, figs, avocados, or mangos, you can thank bats.

Many people mistakenly view bats as fearsome, filthy, aggressive, rabies-carrying bloodsuckers. But most bat species are harmless to people, livestock, and crops. In the United States, only 10 people have died of bat-transmitted disease in four decades of recordkeeping; more Americans die each year from falling coconuts.

One way to protect bats from human disturbance and killing is to prevent human access to caves and mines that are major bat roosts and hibernation sites. *Bat gates*—vandalproof metal grids that allow bats to enter and leave but keep out people—are one option. Another is to educate people about the importance of bats to humans and other species. We need to see bats as valuable allies, not as enemies.

25-3 CAUSES OF DEPLETION AND PREMATURE EXTINCTION OF WILD SPECIES

What Are the Root Causes of Wildlife Depletion and Extinction? Three underlying causes of population reduction and extinction of wildlife are **(1)** human population growth (Figures 1-1 and

Q: How does the human input of SO_2 into the atmosphere compare to that from the natural sulfur cycle?

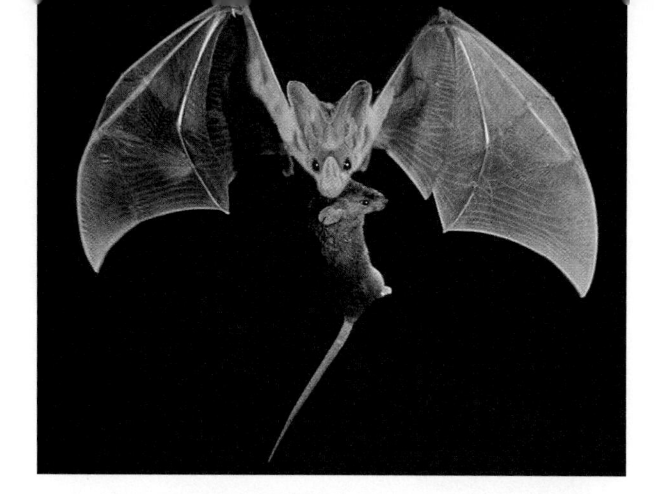

Figure 25-7 An endangered ghost bat carrying a mouse in tropical northern Australia. This nocturnal carnivore is harmless to people. Bats are considered keystone species in many ecosystems because of their roles in pollinating plants, dispersing seeds, and controlling insect and rodent populations. (G. B. Barker/A.N.T. Photo Library)

1-7); **(2)** economic systems and policies that fail to value the environment and its ecological services (Figure 1-2), thereby promoting unsustainable exploitation; and **(3)** greater per capita resource use as a result of increasing affluence and economic growth, which is a prime factor in the exploitation and degradation of wildlife habitats for human uses (Figure 1-5). In developing countries, the combination of rapid population growth (Figure 11-3) and poverty push the poor to cut forests, grow crops on marginal land, overgraze grasslands, deplete fish species, and kill endangered animals for their valuable furs, tusks, or other parts in order to survive.

These underlying causes lead to other more direct causes of the endangerment and extinction of wild species, such as **(1)** habitat loss and degradation, **(2)** habitat fragmentation, **(3)** commercial hunting and poaching, **(4)** overfishing, **(5)** predator and pest control, **(6)** sale of exotic pets and decorative plants, **(7)** climate change and pollution, and **(8)** deliberate or accidental introduction of nonnative (exotic or alien) species into ecosystems.

What Is the Role of Habitat Loss and Fragmentation? The greatest threat to all types of wild species is reduction of habitats as we increasingly occupy or degrade more of the planet (Figures 1-5 and 25-8). According to conservation biologists, tropical deforestation (Figure 24-2) is the greatest eliminator of species, followed by destruction of coral reefs and wetlands (Section 8-2), plowing of grasslands (Figure 7-21), and pollution of freshwater and marine habitats (Chapter 20).

In the lower 48 states of the United States, 98% of the tall-grass prairies have been plowed, half of the wetlands drained, and 85–95% of old-growth forests cut (Figure 24-4). Overall forest cover has been reduced by 33%. At least 500 native species have been driven to extinction, and others to near-extinction (Figures 25-4 and 25-6), mostly because of habitat loss and fragmentation.

Island species, many of them *endemic species* found nowhere else on earth, are especially vulnerable to extinction. Most have no other place to go and few have evolved defenses against predators or diseases accidentally or deliberately introduced onto islands. Roughly half of the plants and animals known to have become extinct since 1600 were island species, even though islands make up only a small fraction of the earth's surface. Almost 900 bird species—10% of the world's known bird species—exist on only one island. The endangered *Symphonia* (Figure 25-4) clings to life on the large island of Madagascar, a megadiversity country where 90% of the original vegetation has been destroyed; hundreds of this island's endemic species are also threatened with extinction (Case Study, p. 652).

The theory of island biogeography (Figure 9-24) developed by Robert MacArthur and Edward O. Wilson has been used to predict the number and percentage of species that would become extinct when habitats are destroyed or seriously degraded. This model has been extended from islands to national parks, tropical rain forests, lakes, and nature reserves (Connections, p. 245), which can be viewed as *habitat islands* in an inhospitable sea of unsuitable habitat. Such models predict that when half of the habitat on an island or a habitat island is destroyed, approximately 10% of the native species will become extinct. When 90% of the habitat is destroyed, about 50% of the species will be lost. Such calculations are one tool conservation biologists use to estimate the number of current and future extinctions in various areas.

Most national parks and other protected areas are habitat islands, many of them surrounded by potentially damaging logging, mining, energy extraction, and industrial activities. Freshwater lakes are also habitat islands that are especially vulnerable when nonnative species are introduced, as has happened in the Great Lakes (Case Study, p. 539).

Migrating species face a double habitat problem. Nearly half of the 700 U.S. bird species spend two-thirds of the year in the tropical forests of Central or South America or on Caribbean islands, returning to North America during the summer to breed. A U.S. Fish and Wildlife study showed that between 1978 and 1987, populations of 44 of the 62 surveyed species of insect-eating, migratory songbirds in North America declined; 20 species experienced drops of 25–45%.

A: Our input is about twice as much

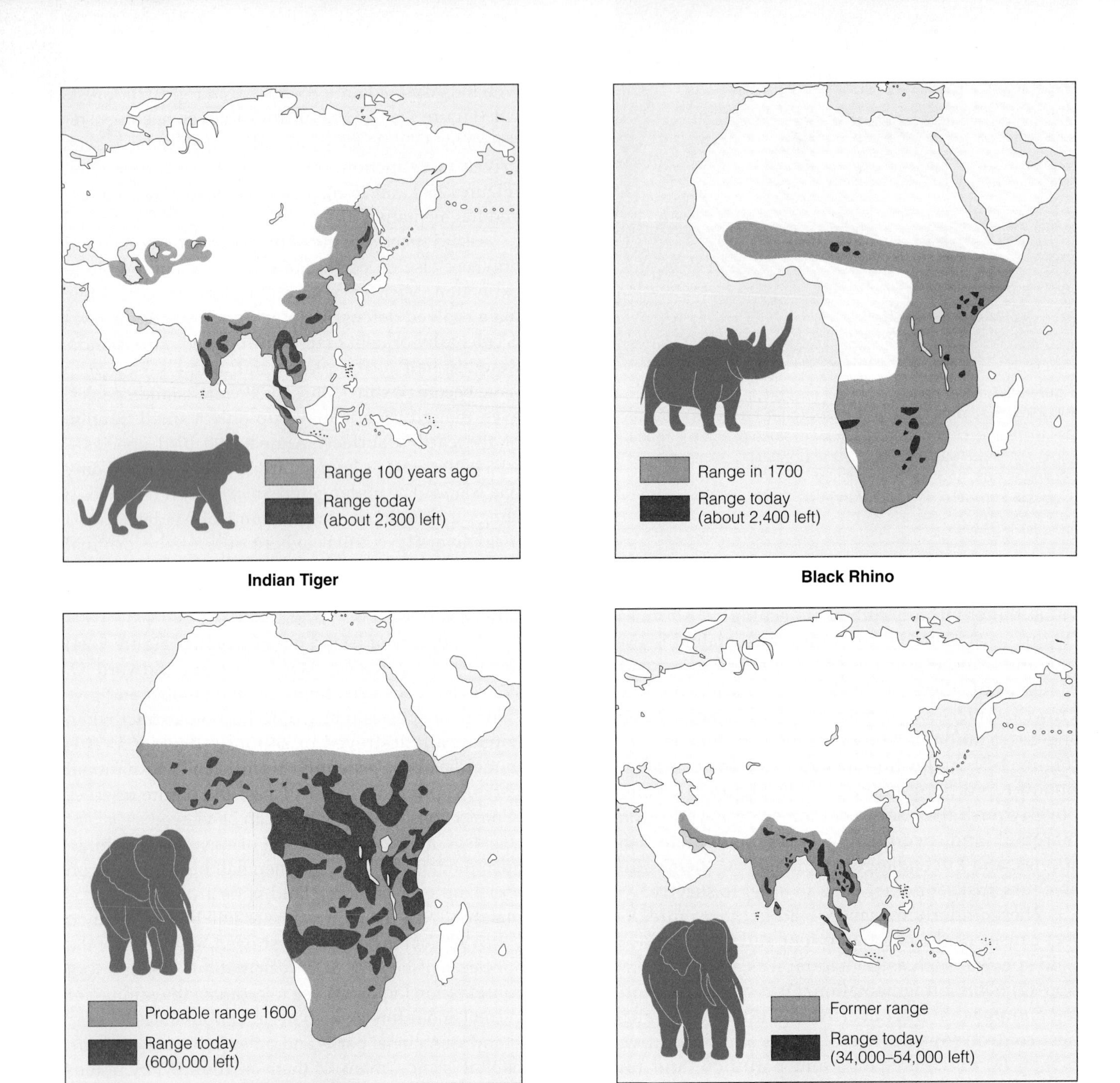

Figure 25-8 Reductions in the ranges of four wildlife species, mostly the result of a combination of habitat loss and hunting. What will happen to these and millions of other species when the global human population doubles in the next few decades? (Data from International Union for the Conservation of Nature and World Wildlife Fund)

The main culprits are logging of tropical forests in the birds' winter habitats and fragmentation of their summer forest habitats in North America. The intrusion of farms, freeways, and suburbs break forests into patches. This makes it easier for opossums, skunks, squirrels, raccoons, and blue jays to feast on the eggs and the young of migrant songbirds. Another serious threat to songbirds from such habitat fragmentation comes from parasitic cowbirds that lay their eggs in the nests of various songbird species and have those birds raise their young for them.

Approximately 69% of the world's 9,600 known bird species are declining in numbers (58%) or are threatened with extinction (11%), mostly because of

Q: How does the human input of toxic lead into the atmosphere compare to that from natural sources?

habitat loss and fragmentation. Conservation biologists view this loss and decline of bird species as an early warning of the widespread and greater loss of biodiversity and ecological integrity to come. Birds are excellent *environmental indicators* because they live in every climate and biome, respond quickly to environmental changes in their habitats, and are easy to track and count.

In addition to serving as indicator species, birds play important ecological roles: They help control populations of insects (including the spruce budworm, gypsy moth, and tent caterpillar, which decimate many tree species) and rodents, pollinate a wide variety of flowering plants, and spread plants throughout their habitats by consuming plant seeds and excreting them in their droppings.

According to the World Conservation Union, populations of about 25% of all known mammal species are in decline and are in danger of becoming extinct, with habitat loss and fragmentation accounting for at least 75% of these declines. Nearly half of all primate species are in decline, followed by hoofed animals (37%).

Habitat fragmentation (Figure 24-5) reduces biodiversity in several ways. *First*, it increases the exposed surface area (edge), which makes many species vulnerable to predators, fires, changes in microclimate, winds, and outbreaks of disease. *Second*, many of the patches are too small to support the minimum breeding populations of some species. As a rule, the number of species is halved with each 10-fold decrease in habitat area. *Third*, fragmentation can create barriers that limit the ability of some species to disperse and colonize new areas, find enough to eat, and find mates.

Figure 25-9 Rhinoceros horns are carved into ornate dagger handles that sell for $500–12,000 in Yemen and other parts of the Middle East. In China and other parts of Asia, powdered rhino horn is used for medicinal purposes, particularly as a proven fever reducer, and occasionally as an alleged aphrodisiac. All five species of rhinoceros are threatened with extinction because of poachers (who kill them for their horns) and loss of habitat. Between 1973 and 1990 the population of African black rhinos (Figure 25-8) dropped from approximately 63,000 to about 2,400. (R. F. Porter/Ardea London)

What Is the Role of Commercial Hunting and Poaching? The international trade in wild plants and animals is big business, bringing in up to $12 billion a year worldwide, with almost half of this trade involving the illegal sale of endangered and threatened species or their parts. Organized crime has moved into wildlife smuggling because of the huge profits involved (second only to drug smuggling). An estimated 60–80% of all live animals smuggled around the world die in transit.

Worldwide, some 622 species of animals and plants face extinction, mostly because of illegal trade (Case Study, p. 694). Populations of at least one in five species of mammals are declining because of excessive hunting for meat, hides, tusks, and medicinal products.

A live mountain gorilla is worth $150,000, a live chimpanzee $50,000, and a live Imperial Amazon macaw $30,000. Bengal tigers are at risk because a tiger fur sells for $100,000 in Tokyo. The skin of a snow leopard (Figure 25-4) can bring $14,000. Rhinoceros horn (Figure 25-9) sells for as much as $28,600 per kilogram. The pulverized bones of a rare Siberian tiger (only about 450 left; Case Study, p. 694) used as a medicinal powder is worth more than $500,000.

As more species become endangered, the demand for them on the black market soars, hastening their chances of extinction—another example of runaway positive feedback. Poaching of endangered or threatened species (many for markets in Asia) is increasing in the United States, especially in western national parks and wilderness areas covered by only 200 federal wildlife protection officers.

According to the U.S. Fish and Wildlife Service, a poached gyrfalcon sells for $120,000, a bighorn sheep head for $10,000–60,000, an 87-gram (3-ounce) bear gallbladder (used in Asia for medicinal purposes) for $22,000, a large saguaro cactus for $5,000–15,000, a

A: It is about 17 times higher

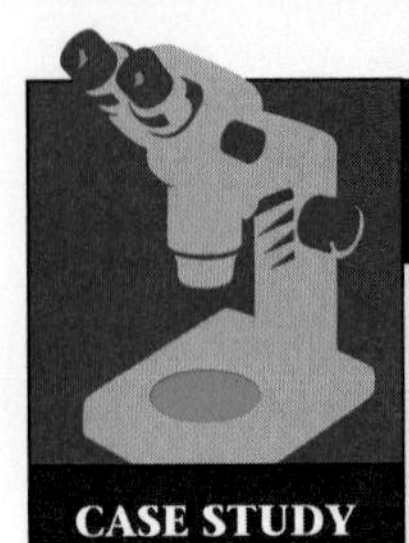

CASE STUDY

Can Tigers Survive Much Longer?

There are five species of tigers (India–Chinese, South China, Bengal, Sumatra, and Siberian), all protected as endangered species. Despite international protection, the world's total tiger population has dwindled to 4,600–8,000. Without emergency action, there may be few or no tigers left in the wild within 20 years.

The main reasons for the rapid decline in the world's tigers are habitat loss (Figure 25-8) and poaching for their furs and bones. Illegally poached tiger skins bring lots of money in the black market, but tiger bones and penises bring even more money and are much easier to smuggle and sell. The bones are ground into a powder that is widely used in China, Taiwan, and South Korea as a topical treatment for arthritis and rheumatism; tiger penises are ground up and cooked to make "tiger wine" and a soup (costing as much as $350 per bowl in China and Taiwan), which are consumed as an alleged aphrodisiac.

With the body parts of a single tiger potentially worth as much as $5 million, it is not surprising that illegal hunting has skyrocketed, especially in India. As the number of tigers drops, the already high black market prices of tiger skins and parts rise further, creating even greater incentives for poachers to kill the remaining animals—another example of runaway positive feedback.

Critical Thinking

What difference would it make if all tigers prematurely disappeared from the earth as a result of human actions? How would you go about protecting the world's remaining tigers?

peregrine falcon (Figure 21-5) for $10,000, a polar bear for $6,000, a grizzly bear for $5,000, and a bald eagle for $2,500. Most poachers are not caught, and the money to be made far outweighs the risk of fines and the much smaller risk of imprisonment.

Case Study: How Should We Protect Elephants from Extinction? For decades poachers have slaughtered elephants for their valuable ivory tusks (Figure 25-10). Habitat loss (Figure 25-8) and legal and illegal trade in elephant ivory have reduced African elephant numbers from 2.5 million in 1970 to about 580,000 today.

Figure 25-10 Vultures are feeding on this elephant carcass poached in Tanzania for its ivory tusks, which are used to make jewelry, piano keys, ornamental carvings, and art objects. (E. R. Degginger)

Since 1989 this decline has been slowed by an international ban on the sale of ivory from African elephants, but things are not quite that simple. Increases in elephant populations in areas where their habitat has shrunk has resulted in widespread destruction of vegetation by these animals. This in turn reduces the niches available for other wild species. Some analysts also argue that sustainable legal harvesting of elephants for their ivory, meat, and hides, with the profits going mostly to local people, will be more successful in the long run than banning all ivory sales.

In several southern African nations in which elephant populations were high, the governments allowed a certain number of elephants to be killed each year. Income from this sustainable harvesting of elephants encouraged local people to protect elephants from unsustainable and illegal poaching, and it also provided governments with funds from the sale of ivory tusks to help pay for conservation of elephants and other wildlife species. With the ban on ivory trade, such income dried up.

To restore these important sources of income, some wildlife conservationists and leaders of several southern African nations have called for a partial lifting of the elephant ivory ban in areas in which the elephant populations are not endangered. Before being sold in the international marketplace, ivory from the sustainable culling of elephants in these areas would be marketed in a way that certifies that it was obtained legally; most of the profits would go to local people. There would also be strict measures to help prevent poaching, enforced largely by local people, who would have an economic stake in preventing illegal killing of elephants.

Q: How much air pollution is emitted into the atmosphere each year by a typical motor vehicle in the United States?

In Africa, governments have a stockpile of at least 450 metric tons (500 tons) of ivory, most of which was confiscated from poachers. Government officials in these countries ask why they should be forced by an international ban to sit on a stockpile of ivory worth hundreds of millions of dollars while their people beg and starve. Selling this ivory could provide much needed revenue for rural development, wildlife habitat acquisition, and wildlife protection.

However, some analysts worry that most of the funds from the sale of ivory would end up in the hands of corrupt bureaucrats. One proposal being considered by the World Wildlife Fund is a *debt-for-ivory swap* in which developed countries would write off part of an African country's debt in return for destruction of its stockpile of ivory tusks.

In most Western countries elephants are viewed as benign. However, according to David Western, head of the Kenya Wildlife Service, in some countries (such as Kenya) elephants are highly destructive of ecosystems and some wildlife preserves (such as Kenya's Amboseli National Park) where elephant populations are too high relative to available vegetation and space. Elephants also destroy fences, crops, and houses and sometimes kill people.

The ban on elephant ivory sales has increased the killing of other species with ivory parts such as bull walruses (for their ivory tusks, worth $500–1,500 a pair) and hippos (for their ivory teeth, worth $70 per kilogram)—another example of the principle that we can never do only one thing. Eastern Zaire's hippo population, once one of the world's largest, has dropped from about 23,000 to 11,000 since the 1989 elephant ban, mostly because of poaching for hippo teeth.

Others argue that opening up the legal elephant ivory market again, even partially, would lead to a renewal of massive poaching of elephants to supply a new ivory black market driven by increased consumption of ivory. They see continuing the ban as a way to help prevent African elephants from becoming threatened or endangered by keeping the price of ivory down.

In 1997 the Convention on International Trade in Endangered Species voted to let Zimbabwe, Botswana, and Namibia sell a total of almost 55 metric tons (60 tons) of stockpiled ivory under strict monitoring. Profits from the sale will be used for wildlife conservation and rural development programs. What do you think about this policy?

What Is the Role of Overfishing in Reducing Aquatic Biodiversity? Concentrations of particular aquatic species suitable for commercial harvesting in a given ocean area or inland body of water are called **fisheries**. Some commercially important marine species of fish and shellfish are shown in Figure 25-11.

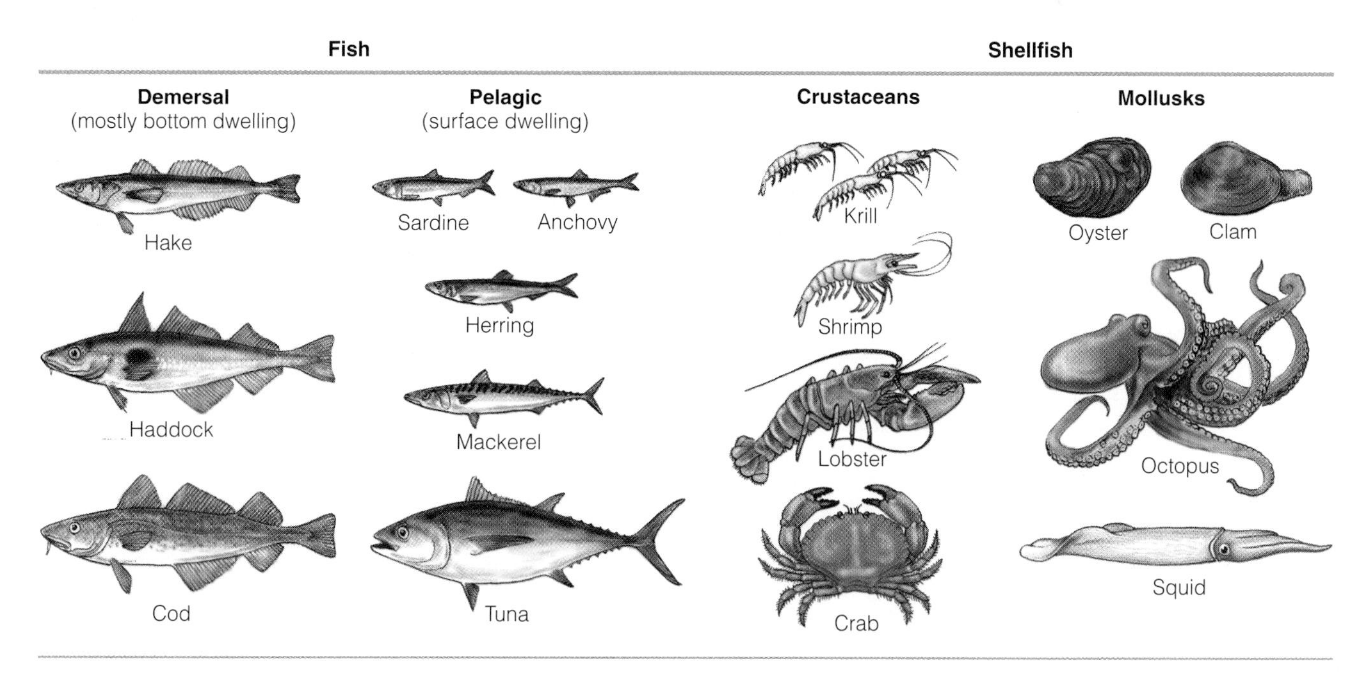

Figure 25-11 Some major types of commercially harvested marine fish and shellfish.

Various methods for harvesting fish are shown in Figure 25-12. Some fishing boats, called *trawlers*, catch demersal fish by dragging a funnel-shaped net held open at the neck along the ocean bottom. Newer trawling nets are large enough to swallow 12 jumbo jets in a single gulp, and even larger ones are on the way. The large mesh of the net allows most small fish to escape but can capture and kill other species such as seals and endangered and threatened sea turtles. Only the large fish are kept, with most of the fish and other aquatic species thrown back into the ocean either dead or dying.

Pelagic species such as tuna, which feed in schools near the surface or in shallow areas, are often caught by *purse-seine fishing* (Figure 25-12). After a school is found, it's surrounded by a large purse-seine net. The net is then closed like a drawstring purse to trap the fish, and the catch is hauled aboard. Nets used to capture yellowfin tuna in the eastern tropical Pacific Ocean have also killed large numbers of dolphins, which swim on the surface above schools of the tuna.

Another increasingly used method for catching open-ocean fish species such as swordfish, tuna, and sharks (Case Study, p. 217) is *longlining*, in which fishing vessels put out lines up to 130 kilometers (80 miles) long, hung with thousands of baited hooks. This practice has caused such a severe drop in the population of Atlantic swordfish that many restaurant owners and chefs have agreed to stop serving this fish and many consumers are refusing to order it. Long lines also hook pilot whales, dolphins, endangered sea turtles (Figure 25-13), and sea-feeding albatross birds.

One of the greatest threats to marine biodiversity is *drift-net fishing* (Figure 25-12). These monster nets drift in the water and catch fish when their gills become entangled in the nylon mesh. Each net descends as much as 15 meters (50 feet) below the surface and is up to 65 kilometers (40 miles) long. Almost anything

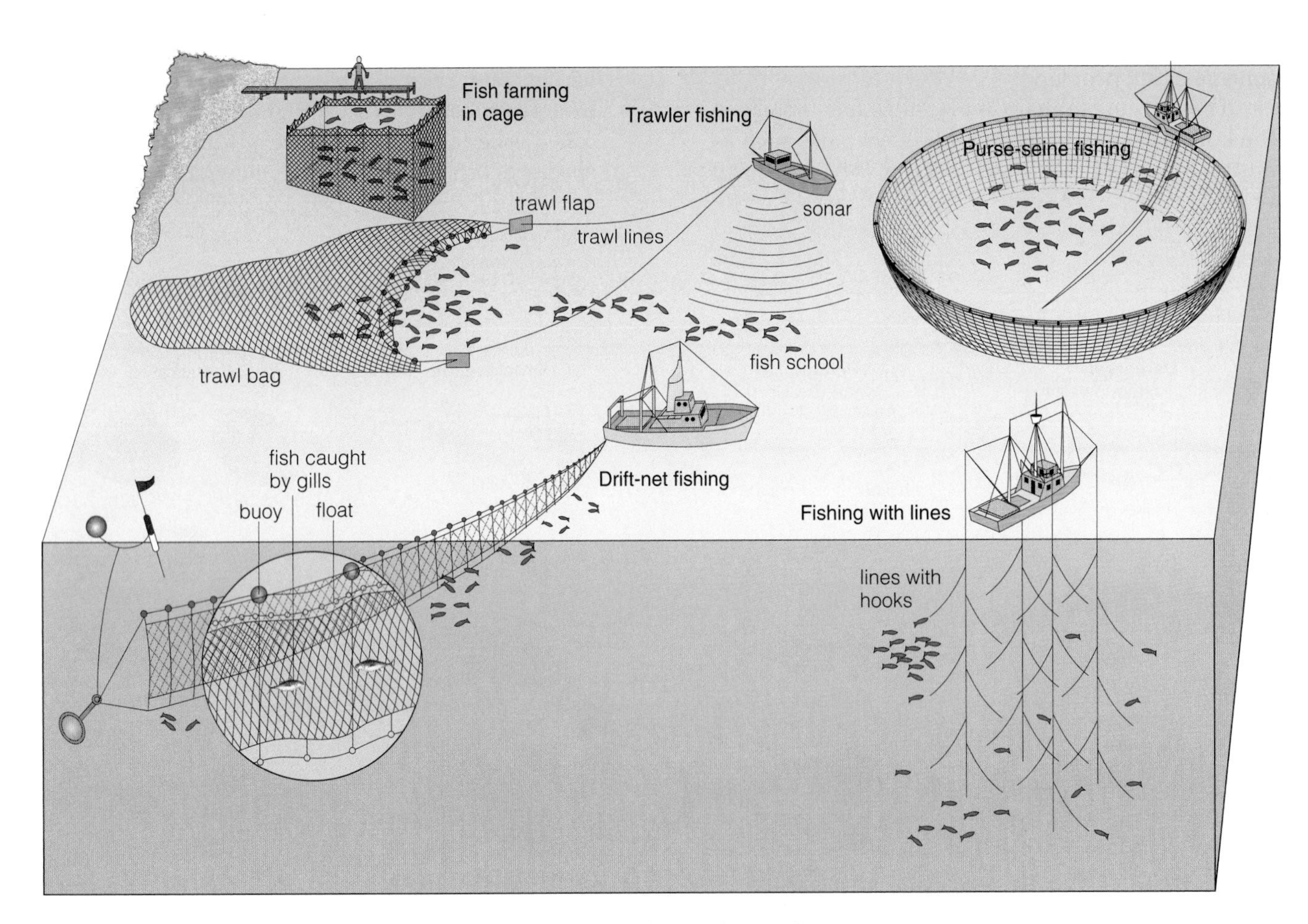

Figure 25-12 Major commercial fishing methods used to harvest various marine species.

Q: How many people in the United States die prematurely each year because of outdoor and indoor air pollution?

that comes in contact with these nearly invisible "curtains of death" becomes entangled.

In 1990 the UN General Assembly declared a moratorium on the use of drift nets longer than 2.5 kilometers (1.6 miles) in international waters after December 31, 1992, to help reduce overfishing and reduction of marine biodiversity. However, compliance is voluntary and there is no effective mechanism for monitoring fleets' activities over vast ocean areas and no structure for enforcement or punishment of violators. The financial rewards of illegal harvesting using drift nets far outweigh the very slim chances of being caught and having to pay small fines.

Even if illegal drift-netting could somehow be effectively banned, thousands of kilometers of *ghost nets*—stretches of netting that have been lost or abandoned and are adrift at sea—will continue to kill marine animals until the nets sink with the weight of decomposing bodies.

How Serious Are the Threats to the Biodiversity of Freshwater and Marine Systems? According to a number of aquatic scientists, the economic pressures to catch more and more fish pose a serious threat to freshwater and marine diversity. Most people are unaware how vulnerable freshwater environments are to environmental degradation and depletion. Although only 1% of the earth's surface is fresh water, these aquatic systems are home for 12% of all known animal species (including 41% of the known fish species).

Currently, about 34% of the known fish species are at risk of becoming extinct. According to a recent study by the World Resources Institute, coastal developments threaten marine biodiversity along roughly half of the world's coasts. The greatest threats occur in Europe (86% of coastal waters at high or moderate risk) and in Asia (69% threatened).

The major causes of declines in fish populations are habitat loss and degradation (mostly from pollution), invasions by nonnative species, and overharvesting. Although loss and degradation of forests get most of the attention, freshwater habitats are even more seriously altered by human activities. More than 40,000 large dams (Figures 13-10, 13-11, 13-12, and 13-19) and hundreds of thousands of smaller barriers alter the seasonal flows, temperatures, and sediment loads, and other river properties to which native fish are adapted. For example, in the heavily altered Columbia River basin (Figure 13-19), 29 of 50 native fish species are either extinct or endangered mostly because of habitat alteration. Flood control levees (Figure 13-20, middle) disconnect rivers from their floodplains and thus eliminate wetlands and backwaters that are important spawning grounds for fish. Pollution from industries, farms, and urban areas further reduces and degrades fish habitat (Chapter 20).

Populations of many marine species are also in decline. The most obvious signs of this degradation of marine life include **(1)** dramatic declines in many of the world's fish stocks, mostly because of overfishing (Section 12-4), **(2)** destruction and bleaching of coral reefs (Figure 8-1, right), **(3)** massive die-offs of dolphins and seals, **(4)** loss of mangrove forests, **(5)** increasing incidences of huge blooms of toxic red tides (Connections, p. 544), and **(6)** pollution and sedimentation.

Overfishing leads in most cases to *commercial extinction*, which is usually only a temporary depletion of fish stocks, as long as depleted areas are allowed to recover. However, large-scale fish harvesting can lead to serious depletion and extinction of species such as sea turtles (Figure 25-13) and other marine mammals that are unintentionally caught.

Case Study: Where Have All the Anchovies Gone? In 1953, Peru began commercial fishing for anchovies off its coast. The size of the fishing fleet increased rapidly, from 100 boats in 1953 to 1,700 by 1969. The tiny fish were processed into fish meal and sold to developed countries as livestock feed. Between 1965 and 1971, Peruvian anchovies made up about 20% of the world's commercial fish catch. The enormous anchovy populations there depend on an upwelling of cold, nutrient-rich bottom water near the Peruvian coast (Figure 7-8, left). At unpredictable intervals—when the productivity of these upwellings declines sharply because of a natural weather change called the El-Niño–Southern Oscillation or ENSO (Figure 7-8, right)—the anchovy populations decline.

The Peruvian anchovy bonanza was short-lived. Biologists with the UN Food and Agriculture Organization warned that during 7 of the 8 years between 1964 and 1971, the anchovy harvest exceeded the estimated sustainable yield. Peruvian fishery officials ignored those warnings. When a strong ENSO arrived in 1972, the anchovy population, already decimated by overfishing, could not recover, and the annual yield plummeted (Figure 25-14, p. 700). By putting short-term profits above a long-term anchovy fishery, Peru lost a major source of income and jobs and suffered an increase in its foreign debt.

Since 1983, the Peruvian anchovy fishery has made only a slight recovery. The species may never recover to former levels because the severe

A: 150,000–350,000, an average of 410–960 per day

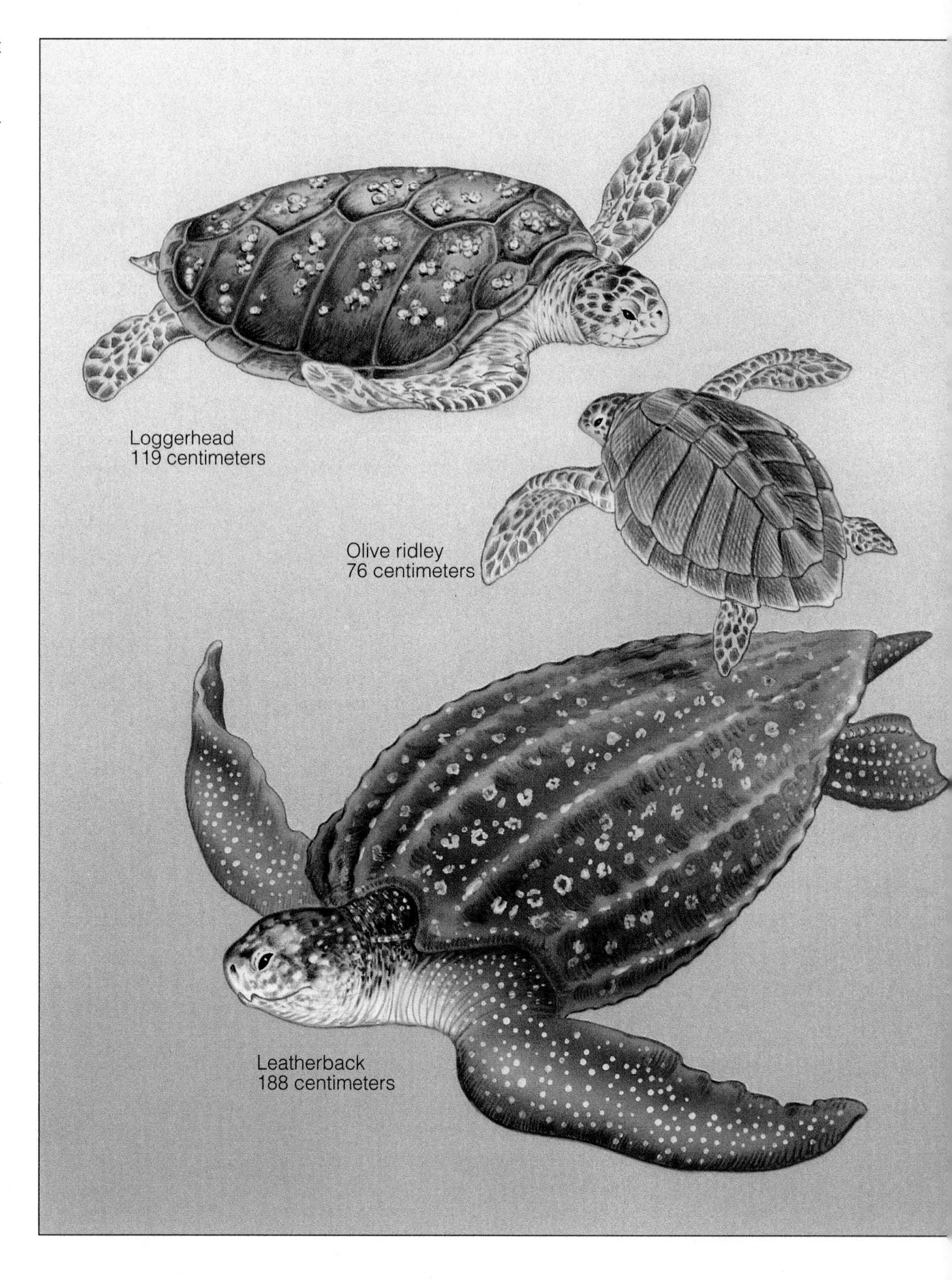

Figure 25-13 Major species of sea turtles that have roamed the seas for 150 million years, showing their maximum adult lengths. These species are becoming endangered mostly because of loss or degradation of beach habitat (where they come ashore to lay their eggs), legal and illegal taking of their eggs (Guest Essay, p. 308), and unintentional capture by commercial fishing boats.

overfishing allowed much of the Peruvian anchovy's niche to be taken over by sardines, which now prey on the anchovy.

What Is the Role of Predator and Pest Control? People try to exterminate species that compete with them for food and game animals. For example, U.S. fruit farmers exterminated the Carolina parakeet around 1914 because it fed on fruit crops. The species was easy prey because when one member of a flock was shot, the rest of the birds hovered over its body, making themselves easy targets (Table 25-1).

African farmers kill large numbers of elephants to keep them from trampling and eating food crops. Ranchers, farmers, and hunters in the United States support the killing of coyotes (p. 626), wolves, and other species that can prey on livestock and on species prized by game hunters. Since 1929, U.S. ranchers and government agencies have poisoned 99% of North America's prairie dogs because horses

Q: Worldwide, how many people live in cities where outdoor air is unhealthy to breathe?

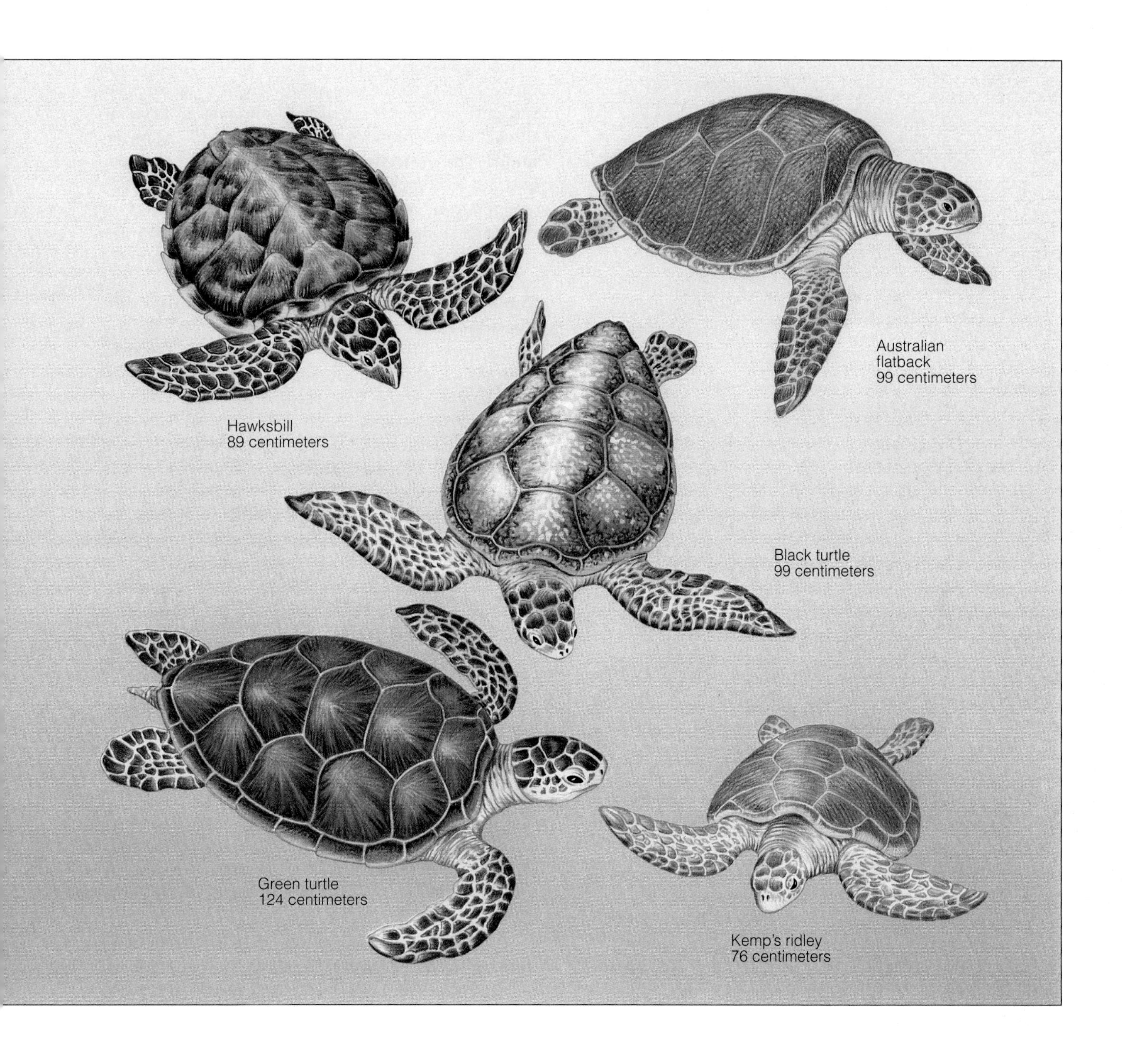

and cattle sometimes step into the burrows and break their legs. This has also nearly wiped out the endangered black-footed ferret (Figure 25-4), which preyed on the prairie dog. However, as a result of captive breeding started in 1981, 316 of these ferrets have been released into the wild in Montana, Wyoming, and South Dakota.

What Is the Role of the Market for Exotic Pets and Decorative Plants? The global legal and illegal trade in wild species for use as pets is a huge and very lucrative business. However, for every live animal captured and sold in the pet market an estimated 50 other animals are killed.

Worldwide, over 5 million live wild birds are captured and sold legally each year and an estimated 2.5 million more are captured and sold illegally, primarily in Europe, Japan, and the United States. Over 40 species, mostly parrots, are endangered or threatened because of this wild bird trade. Collectors of exotic

A: About 1.3 billion, or 1 in 5

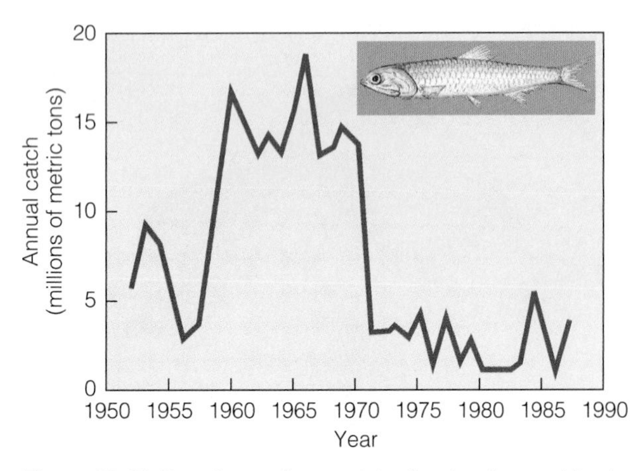

Figure 25-14 Peruvian anchovy catch, showing the combined effects of overfishing and a natural climate change, the El Niño–Southern Oscillation, that occurs every few years. (Data from UN Food and Agriculture Organization and Peter Muck)

birds may pay up to $12,000 for a pair of golden-shouldered parakeets from Australia or $10,000 for a threatened hyacinth macaw (Figure 25-4) smuggled out of Brazil; however, during its lifetime, a single macaw left in the wild might yield as much as $165,000 in tourist income.

About 25 million U.S. households have exotic birds as pets, 85% of them imported. For every wild bird that reaches a pet shop legally or illegally, as many as 10 others die during capture or transport. Keeping pet birds may also be hazardous to human health. A 1992 study suggested that keeping indoor pet birds for more than 10 years doubles a person's chances of getting lung cancer.

Other wild species whose populations are depleted because of the pet trade include amphibians, reptiles, mammals, and tropical fish. A rare albino python might bring $20,000 in Germany. About three-quarters of all saltwater tropical aquarium fish sold come from the coral reefs of Indonesia and the Philippines. Divers catch the fish by using plastic squeeze bottles of cyanide to stun them. For each fish caught alive, many more die. In addition, the cyanide solution kills the coral animals that create the reef, which is a center for marine biodiversity (Figures 8-1 and 8-10). This loss of biodiversity could be reduced if people stopped buying tropical fish or bought only tropical fish that were certified to have been caught with nets.

Some exotic plants, especially orchids and cacti such as the black lace cactus (Figure 25-4), are endangered because they are gathered (often illegally) and sold to collectors to decorate houses, offices, and landscapes. A collector may pay $5,000 for a single rare orchid. A single rare mature crested saguaro cactus can earn cactus rustlers as much as $15,000. To thwart cactus rustlers, Arizona has put 222 species under state protection, with penalties of up to $1,000 and jail sentences of up to 1 year. However, only seven people are assigned to enforce this law over the entire state, and the fines are too small to discourage poaching.

What Are the Roles of Climate Change and Pollution? A potential problem for many species is the possibility of fairly rapid (50–100 years) changes in climate, accelerated by deforestation and emissions of heat-trapping gases into the atmosphere (Sections 19-2 and 19-3). Wildlife in even the best-protected and best-managed reserves could be depleted in a few decades if such changes in climate take place.

Toxic chemicals degrade wildlife habitats, including wildlife refuges, and kill some terrestrial plants and animals and aquatic species. Slowly degradable pesticides, such as DDT, can be biologically magnified to very high concentrations in food chains and webs (Figure 17-2). The resulting high concentrations of DDT or other slowly biodegraded, fat-soluble organic chemicals can kill some organisms, reduce their ability to reproduce, or make them more vulnerable to diseases, parasites, and predators.

What Is the Role of Deliberately Introduced Species? Deliberate or accidental introduction of nonnative or alien species into habitats where they are not native is the second biggest cause of animal and plant extinctions (after habitat alteration and degradation).

Travelers sometimes collect plants and animals intentionally and introduce them into new geographical regions. Many of these introduced species ultimately provide food, game, and aesthetic beauty and may even help control pests in their new environments.

However, some introduced species have no natural predators, competitors, parasites, or pathogens to control their numbers in their new habitats. This can allow them to dominate their new ecosystem and reduce or wipe out the populations of many native species, making them a form of *biological pollution* (Table 25-2).

The United States is home to at least 4,000 nonnative plant species and 2,300 alien animal species. Damages and pest control costs for these unwanted species are estimated at $122 billion per year and future losses are expected to be much higher. About

Q: What percentage of the trees in Europe have been damaged by air pollution?

Table 25-2 Damage Caused by Plants Imported into the United States

Name	Origin	Mode of Transport	Type of Damage
Mammals			
European wild boar	Russia	Intentionally imported (1912), escaped captivity	Destroys habitat by rooting; damages crops
Nutria (cat-sized rodent)	Argentina	Intentionally imported (1940), escaped captivity	Alters marsh ecology; damages levees and earth dams; destroys crops
Birds			
European starling	Europe	Intentionally released (1890)	Competes with native songbirds; damages crops; transmits swine diseases; causes airport nuisance
House sparrow	England	Intentionally released by Brooklyn Institute (1853)	Damages crops; displaces native songbirds
Fish			
Carp	Germany	Intentionally released (1877)	Displaces native fish; uproots water plants; lowers waterfowl populations
Sea lamprey	North Atlantic Ocean	Entered Great Lakes via Welland Canal (1829)	Wiped out lake trout, lake whitefish, and sturgeon in Great Lakes
Walking catfish	Thailand	Imported into Florida	Destroys bass, bluegill, and other fish
Insects			
Argentine fire ant	Argentina	Probably entered via coffee shipments from Brazil (1918)	Damages crops; destroys native ant species
Camphor scale insect	Japan	Accidentally imported on nursery stock (1920s)	Damaged nearly 200 plant species in Louisiana, Texas, and Alabama
Japanese beetle	Japan	Accidentally imported on irises or azaleas (1911)	Defoliates more than 250 species of trees and other plants, including many of commercial importance
Plants			
Water hyacinth	Central America	Accidentally introduced (1882)	Clogs waterways; shades out other aquatic vegetation
Chestnut blight (fungus)	Asia	Accidentally imported on nursery plants (1900)	Killed nearly all eastern U.S. chestnut trees; disturbed forest ecology
Dutch elm disease (fungus)	Europe	Accidentally imported on infected elm timber used for veneers (1930)	Killed millions of elms; disturbed forest ecology

*Source: *Biological Conservation* by David W. Ehrenfeld. Copyright ©1970 by Holt, Rinehart & Winston, Inc. Modified and reprinted by permission.

49% of the species in the United States on the official list of endangered and threatened species are there in part because of population declines caused by nonnative species. In Hawaii, more than 95% of the 282 imperiled plant and bird species are threatened by alien species.

In the 1930s, the *kudzu vine* was imported from Japan and planted in the southeastern United States to help control soil erosion. In the 1940s, the government paid incentives to farmers to grow it. It does control erosion, but it is so prolific and difficult to kill that it engulfs hillsides, trees, abandoned houses and cars, stream banks, patches of forest, and anything else in its path (Figure 25-15). Loss of forest productivity and eradication efforts cost the United States at least $500 million a year. Kudzu could spread as far north as the Great Lakes by 2040 if projected global warming occurs.

Although kudzu is considered a menace in the United States, Asians use a powdered kudzu starch in

A: About 25%

Figure 25-15 Kudzu taking over a field and trees near Lyman, South Carolina. This vine can grow 0.3 meter (1 foot) per day and is now found from East Texas to Florida and as far north as southeastern Pennsylvania. Kudzu was deliberately introduced into the United States for erosion control, but can't be stopped by being dug up or burned. Grazing by goats and repeated doses of herbicides can destroy it, but goats and herbicides also destroy other plants, and herbicides can contaminate water supplies. (Angelina Lax/Photo Researchers, Inc.)

beverages, gourmet confections, and herbal remedies for a range of diseases. A Japanese firm is building a large kudzu farm and processing plant in Alabama and will ship the extracted starch to Japan, where demand exceeds supply.

U.S. researchers are currently testing a kudzu-based drug made for reducing an alcoholic's craving for alcohol and may seek FDA approval for use in a few years. In an ironic twist, kudzu—which can engulf and kill trees—could eventually help save trees from loggers. Research at Georgia Institute of Technology indicates that kudzu may join kenaf (Figure 24-20) as a source of tree-free paper.

If you thought kudzu was bad, look out for the *tropical soda apple*, a plant that rivals kudzu as a pest. This thorny plant with a sweet-smelling fruit has moved in on a zoo and at least 5 farms in the lowcountry of South Carolina. This rapidly growing plant was first discovered in southwest Florida in 1988, where it brought viruses to Florida vegetable farms, cost tomato farmers millions of dollars, and choked melon fields.

An estimated 1 million *wild hogs* are roaming the state of Florida. They arrived with Spanish conquistadors more than 450 years ago. The state's panthers (see photo p. 609), bears, and coyotes used to help control the populations of these pigs, but the hogs now outbreed them. They are hogging food from endangered animals, rooting up farm fields, and causing traffic accidents. Game and wildlife officials have had little success in controlling their numbers with hunting and trapping and say there is no way of stopping them.

An ecological and economic disaster at Lake Victoria in East Africa has become a warning of the dangers of meddling with nature by deliberately introducing alien species. This lake, which is surrounded by Kenya, Uganda, and Tanzania, is one of the world's largest freshwater lakes, roughly equal in area to Switzerland.

In 1957, British colonials deliberately introduced the Nile perch into the lake to improve the fishing. Instead, within 20 years this predator fish helped eliminate nearly half of the lake's 400 native fish species, and the remaining species are endangered. To make matters worse, the Nile perch's oily flesh is edible only if it is smoked. Providing wood and charcoal for doing this led to logging of nearby forests. Now the population of the perch appears to be declining because of overfishing and a lack of prey. If the lake continues to deteriorate and the perch population crashes, poor people living near the lake will lose a source of protein, and the lake's commercial fishing industry will collapse.

According to Chris Bright of the Worldwatch Institute, "To date, the study of exotics can be summed up in three negative statements: It's impossible to predict where an exotic will establish itself, or what it will do afterwards, or when it will do it." This is another lesson from nature indicating the need to use the *precautionary principle*, which recognizes that deliberately introducing any nonnative or exotic species into an ecosystem is risky. Once an introduced species has become established in an ecosystem its wholesale removal is almost impossible—somewhat like trying to get smoke back into a chimney or unscramble an egg.

What Is the Role of Accidentally Introduced Species? Two examples of the harmful effects of accidentally introduced exotic species discussed earlier in this book are the fire ant (Case Study, p. 213) and the invasion of the Great Lakes by zebra and quagga mussels (Case Study, p. 539). Another example of the effects of the accidental introduction of a nonnative species is

Gayton, Don. 1997. "Terms of Endangerment." *Canadian Geographic,* vol. 117, no. 3, 30(12).

the fast-growing water hyacinth, which is native to Central and South America. In 1884, a woman took one from a New Orleans exhibition and planted it in her back yard, from which it spread to Florida's waterways.

With no natural enemies and the ability to double their population every two weeks, the plants took advantage of nutrient-rich waters and spread rapidly (see photo in table of contents). Within 10 years water hyacinths had smothered and displaced many native plants and fish and clogged many ponds, streams, and canals, first in Florida and later elsewhere in the southeastern United States.

Mechanical harvesters and herbicides have failed to keep the plant in check, although grazing Florida manatees (Figure 25-16) found in the coastal rivers of Florida and southern Georgia can control water hyacinths better than mechanical or chemical methods. However, these gentle and playful herbivores are threatened with extinction because of a combination of pollution, habitat loss, and slashing by powerboat propellers. To help control its spread, scientists have introduced water hyacinth–eaters, including a weevil from Argentina, a water snail from Puerto Rico, and the grass carp, a fish from the former Soviet Union. These species can help, but water snails and grass carp also feed on other, desirable aquatic plants.

The good news is that water hyacinths can provide several benefits. They absorb toxic chemicals in sewage treatment lagoons; they can be fermented into a biogas fuel similar to natural gas, added as a mineral and protein supplement to cattle feed, and applied to the soil as fertilizer; and they can be used to clean up polluted ponds and lakes if their growth can be kept under control.

About 50 years ago the *brown tree snake* arrived accidentally on the island of Guam in the cargo of a military plane. Since then this highly aggressive, nocturnal snake with no natural enemies has eliminated or decimated nine of Guam's native forest bird species. It has also snatched chickens and pets from yards and has even attacked babies asleep in cribs. These climbing snakes often short out power lines, causing an average of one electric outage every 3 days. According to Mike Pitzler, a biologist with the U.S. Department of Agriculture in Guam, "Never in history has a snake done as much ecological damage as this snake."

This light-sensitive snake can select an airplane wheel or outgoing cargo as a hiding place during daylight hours and thus spread to other areas. Since 1981 seven brown tree snakes have been found in Hawaii. Those found alive were captured when discovered. As home for 41% of all endangered birds in the United States, Hawaii has a lot to lose from an invasion by this snake.

Figure 25-16 The Florida manatee, or sea cow, could help control water hyacinth abundance, but it is one of America's most endangered species. Only an estimated 3,000 of these gentle mammals are left in the sluggish and increasingly polluted bays and streams of Florida and southern Georgia. Most manatee deaths result from habitat loss, cold weather, or entanglement in fishing gear, but about 25% are killed when they are hit by boat hulls or propellers. To help protect manatees, Florida's Department of Natural Resources has set low speed limits for boats in parts of waterways containing manatee habitats. (Florida Marine Research Institute/Florida Department of Natural Resources)

25-4 SOLUTIONS: PROTECTING WILD SPECIES FROM DEPLETION AND EXTINCTION

How Can We Protect Wildlife and Biodiversity? There are three basic approaches to managing wildlife and protecting biodiversity. The *ecosystem approach*, discussed in Chapter 23, aims to preserve balanced populations of species in their native habitats, establish legally protected wilderness areas and wildlife reserves, and eliminate or reduce the populations of nonnative species.

Biologists consider protecting ecosystems to be the best way to preserve biological diversity and ecological integrity. The basic problem is that fully or partially protected wildlife sanctuaries make up only 6% of the world's land area (Figure 23-16), and the human population is expected to double in 40 years.

The *species approach* is based on protecting endangered species by identifying them, giving them legal protection, preserving and managing their crucial habitats, propagating them in captivity, and reintroducing them into suitable habitats.

Hint: Enter the search terms *extinction, prevention* using the Subject Guide.

Biointormatics: Reinventing Museums and Systematics

SPOTLIGHT

A new buzz word is *bioinformatics.* By providing new computer technologies for organizing, managing, accessing, and analyzing information about biodiversity, bioinformatics is transforming the way biologists and ecologists work.

Traditionally, natural history museums, arboreta, botanical gardens, herbaria, and zoos have been much more than places where the public could view exhibits and learn about nature. They have housed research institutions dedicated to exploration of the natural world. They have served as repositories of *biological collections.*

They have also been centers for the study of *systematics,* or taxonomy, the branch of biology that deals with description, naming, classification, and evolutionary relationships of the earth's organisms. Biological collections provide systematists with the formal documentation and basis for naming and classifying the earth's known biodiversity.

In recent years, the role of these institutions and the importance of systematics have been underappreciated. Some have even viewed systematics and biological collecting as old-fashioned, and a shortage of qualified systematists has resulted. But this is changing because of a growing appreciation of the need for fundamental biodiversity information.

Bioinformatics is playing an important role in this reinvention of institutions that house biological collections and the role of systematists who work with the collections. Most institutions that house biological collections are busily entering information about their collections into computerized databases. Most systematists are building databases concerning characteristics of the organisms they study and using these databases to reconstruct phylogenies, the evolutionary lineages of organisms, used to understand evolution and develop modern classification.

Species 2000 is an international project that has the lofty goal of providing information about all known species of plants, animals, fungi, and microbes on the earth as the baseline dataset for studies of global biodiversity. Users worldwide will be able to verify the scientific name, status, and classification of any known species via the Species Locator, which will provide online access to authoritative species data drawn from an array of participating databases.

Species 2000 is one of many projects initiated to take advantage of advances in bioinformatics. Other projects include online biodiversity databases concerning species distributions, gap analysis (p. 639), and introduced species.

Critical Thinking

Do you think bioinformatics is an important new applied science that will help with efforts to protect biodiversity, or do you think it is just a fad that is wasting the efforts of many talented people? Explain.

The *wildlife management approach* manages game species for sustained yield by using laws to regulate hunting, establishing harvest quotas, developing population management plans, and using international treaties to protect migrating game species such as waterfowl. These two approaches are discussed in the remainder of this chapter.

How Can Bioinformatics Help Protect Biodiversity? It has been said that knowledge is power. Fundamental information about the biology and ecology of wild species is required to give the power to be able to protect biodiversity. In particular, information is needed concerning species names and descriptions, distributions, status of populations, habitat requirements, and interactions with other species.

Bioinformatics is the applied science of managing, analyzing, and communicating biological information. It involves **(1)** *building computer databases* to organize and store useful biological information, **(2)** *providing computer tools to find, visualize, and analyze* the information, and **(3)** *providing means for communicating* the information, especially using the Internet. Bioinformatics is being applied to many aspects of biology, ranging from storing DNA sequences to storing names, descriptions, and locations of collections of biological organisms in museums (Spotlight, above).

Various government and private organizations have initiated major efforts to organize and communicate biodiversity information. In particular, the Biological Research Division (BRD) of the United States Geological Survey has initiated a program known as the National Biological Information Infrastructure (NBII). NBII is intended to be an electronic gateway to biological information maintained by local, state, and federal government agencies, private-sector organizations, and other national and international partners.

How Can International Treaties Help Protect Endangered Species? Several international treaties and conventions help protect endangered or threatened wild species. One of the most far-reaching is the 1975 Convention on International Trade in Endangered Species (CITES). This treaty, now signed by 136 countries, lists almost 700 species that cannot be commercially traded as live specimens or wildlife products because they are endangered or threatened.

However, enforcement of this treaty is spotty, convicted violators often pay only small fines, and member countries can exempt themselves from protecting any listed species. Furthermore, much of the estimated $6-billion annual illegal trade in wildlife and wildlife products goes on in countries that have not signed the treaty. Centers of illegal animal trade are Singapore, Argentina, Indonesia, Spain, Taiwan, and Thailand.

Terrestrial and aquatic species are also given some protection under the Convention on Biological Diversity, a comprehensive binding agreement that went into effect in 1993. In addition to CITES and the Convention on Biological Diversity, several other international agreements help protect marine biodiversity.

How Can Laws Help Protect Endangered Species? The United States controls imports and exports of endangered wildlife and wildlife products through two important laws. The Lacey Act of 1900 prohibits transporting live or dead wild animals or their parts across state borders without a federal permit. The U.S. Endangered Species Act of 1973 (amended in 1982 and 1988) makes it illegal for Americans to import or trade in any product made from an endangered or threatened species unless it is used for an approved scientific purpose or to enhance the survival of the species.

The Endangered Species Act of 1973 is one of the world's toughest environmental laws. It authorizes the National Marine Fisheries Service (NMFS) to identify and list endangered and threatened ocean species; the U.S. Fish and Wildlife Service (USFWS) identifies and lists all other endangered and threatened species. These species cannot be hunted, killed, collected, or injured in the United States.

Any decision by either agency to add or remove a species from the list must be based on biology only, not on economic or political considerations. However, economic factors can be used in deciding whether and how to protect endangered habitat and in developing recovery plans for listed species. The act also forbids federal agencies to carry out, fund, or authorize projects that would either jeopardize an endangered or threatened species or destroy or modify the critical habitat it needs to survive. On private lands, fines and even jail sentences can be imposed to ensure protection of the habitats of endangered species.

Between 1973 and 1998, the number of U.S. species included in the official endangered and threatened list increased from 92 to over 1,100 species (of which approximately 60% are plants and 40% animals). Hawaii has the largest number of species on the list (298). More than 200 other species are being evaluated as candidates for listing and each year about 85 species are added to the list.

Getting listed is only half the battle. Next, the USFWS or the NMFS is supposed to prepare a plan to help the species recover. However, because of a lack of funds, final recovery plans have been developed and approved for only about two-thirds of the endangered or threatened U.S. species, and half of those plans exist only on paper.

The act requires that all commercial shipments of wildlife and wildlife products enter or leave the country through one of nine designated ports. Few illegal shipments are confiscated (Figure 25-17) because the 60 USFWS inspectors can examine only about one-fourth of the 90,000 shipments that enter and leave the United States each year. Even if caught, many violators are not prosecuted, and convicted violators often pay only a small fine.

In 1998, Secretary of Interior Bruce Babbitt announced that the U.S. Fish and Wildlife Service

Figure 25-17 Confiscated products made from endangered species. Because of a scarcity of funds and inspectors, probably no more than one-tenth of the illegal wildlife trade in the United States is discovered. The situation is even worse in most other countries. (Steve Hillebrand/U.S. Fish and Wildlife Service)

A: 5–10% (with economic losses of $1.5–5.4 billion per year)

would remove (delist) 17 animals and 12 plants from the endangered and threatened species list over 2 years, subject to final approval by federal biologists. Some would be downgraded from endangered to threatened and others removed from the law's protection. Candidates for this delisting include well-known species such as the bald eagle, peregrine falcon, brown pelican, and gray wolf, as well as lesser-known species such as Pahrump poolfish and the Missouri bladder-pod plant.

Some environmentalists believe this is a major victory for the U.S. Fish and Wildlife Service and the Endangered Species Act because populations of these species are making a steady recovery. Others believe that, even though the populations are recovering, the delisting is premature and it would be better to wait for a more complete recovery.

Should the Endangered Species Act Be Weakened? Since 1995 there have been intense efforts to seriously weaken the Endangered Species Act by **(1)** making the protection of endangered species on private land voluntary, **(2)** having the government pay landowners if it forces them to stop using part of their land to protect endangered species (the issue of regulatory takings, p. 619), **(3)** making it harder and more expensive to list newly endangered species by requiring government wildlife officials to navigate through a series of hearings and peer-review panels, **(4)** giving the secretary of interior the power to permit a listed species to become extinct without trying to save it, **(5)** allowing the secretary of interior to give any state, county, or landowner permanent exemption from the law, with no requirement for public notification or comment, **(6)** allowing landowners to lock in long-term endangered species management plans—known as *habitat conservation plans* (HCPs) that exempt the owners from further obligations for 100 years or more, and **(7)** prohibiting the public from commenting on or bringing lawsuits on any changes in habitat conservation plans for endangered species.

Those who favor weakening or eliminating the Endangered Species Act argue that it has been a failure because only 7 species have been removed from the list and only 20 species have recovered enough to be reclassified from endangered to threatened. Some of these critics have spread horror stories about how environmentalists have used endangered species (many of them small and unfamiliar) to block development and resource extraction, violate public property rights, and waste tax dollars trying to save useless creatures that are on the verge of extinction anyway. Here are several examples of misleading or false anecdotal stories widely circulated by opponents of the Endangered Species Act:

- *Fiction*: Hundreds of homes in Southern California were destroyed in 1993 wildfires because residents were prohibited from clearing brush around their homes because it was a habitat for the protected kangaroo rat (Figure 7-16). *Fact*: The General Accounting Office (GAO), the watchdog arm of Congress, investigated and found no basis for this claim. The GAO also said clearing brush would have been useless against a fire so strong that it jumped canals and highways.
- *Fiction*: Bart Dye, an Indiana farmer, may lose the 391-hectare (965-acre) farm that has been in his family since 1865 to protect two mussel species in a river adjacent to his land. *Fact*: The Farmer's Home Administration had foreclosed on Dye's farm in 1984, before the two endangered mussel species were discovered during the resale procedure. Subsequently, only 4.5 hectares (11 acres) or 1% of the farm was taken out of production to protect the mussels.
- *Fiction*: The cattle industry submitted testimony to Congress that a Texas widow with endangered songbirds on her property had been threatened with criminal prosecution if she tried to clear a hedgerow. *Fact*: A U.S. Fish and Wildlife Service biologist visited the property and determined that the woman could clear a 9-meter-wide (30-foot-wide) 1.6 kilometer (1 mile) hedgerow without harming the birds.

Before his death California Representative Sonny Bono joked that the best way to deal with endangered species is to "give them all a designated area and then blow it up." Washington Senator Slade Gordon suggests that all endangered species be removed from the wild and bred in zoos as "a way to preserve animals without blocking economic development."

Should the Endangered Species Act Be Strengthened? Most conservation biologists and wildlife scientists contend that the Endangered Species Act has not been a failure (Spotlight, right). They also refute the charge that the act has caused severe economic losses.

The truth is that the Endangered Species Act has had virtually no impact in the nation's overall economic development. For example, between 1989 and 1992 only 55 of more than 118,000 projects evaluated by the USFWS were blocked or withdrawn as a result of the Endangered Species Act.

Moreover, *the act does allow for economic concerns*. By law, a decision to list a species must be based solely on science. But once a species is listed, economic

Has the Endangered Species Act Been a Failure?

SPOTLIGHT

Critics of the Endangered Species Act call it an expensive failure because only a few species have been removed from the endangered list. Most of these critics are ranchers, developers, and officials of timber and mining companies who want more access to resources on public lands (Spotlight, p. 621).

Most biologists strongly disagree that the act has been a failure, for several reasons. *First,* species are listed only when they are already in serious danger of extinction. This is like setting up a poorly funded hospital emergency room that takes only the most desperate cases, often with little hope for recovery, and then saying it should be shut down because it has not saved enough patients.

Second, it takes decades for most species to become endangered or threatened. Thus, it should not be surprising that it usually takes decades to bring a species in critical condition back to the point where it can be removed from the list. Expecting the Endangered Species Act (which has been in existence only since 1973) to quickly repair the biological depletion of centuries is unrealistic. The most important measure of the law's success is that the conditions of almost 40% of the listed species are stable or improving. A hospital emergency room taking only the most desperate cases yet stabilizing or improving the condition of 40% of its patients would be considered an astounding success.

Third, the federal endangered species budget was only $93 million in 1999 (up from $23 million in 1993). This funding is equal to about what it takes to build 3.2 kilometers (1.5 miles) of urban interstate highway, or about 36¢ a year per U.S. citizen—a pittance to help save some of the country's irreplaceable biodiversity. One C-17 transport airplane costs $300 million—almost three times what was spent on protecting endangered species in the United States during 1998.

To most biologists it's amazing that so much has been accomplished in stabilizing or improving the condition of almost 40% of nearly terminal species on a shoestring budget. However, critics call the act an expensive failure.

Critical Thinking

1. Do you believe that the Endangered Species act should be weakened or strengthened? Explain.

2. Do you agree or disagree with each of the proposals made to **(a)** weaken the act given on p. 706 and **(b)** strengthen the act listed below? Defend each of your answers.

considerations can be weighed against species protection in protecting critical habitat and in designing and implementing recovery plans. The act also allows a special Cabinet-level panel, called the "God Squad," to exempt any federal project from having to comply with the act if the economic costs are too high. In addition, the act allows the government to issue permits and exemptions to landowners with listed species living on their property.

Most biologists and wildlife conservationists believe that we should develop a new system to protect and sustain biological diversity and ecological integrity based on three principles: **(1)** Find out what species and ecosystems we have, **(2)** locate and protect the most endangered ecosystems and species, and **(3)** give private landowners financial carrots (tax breaks and write-offs) for helping protect endangered species and ecosystems.

A step toward implementing the first principle was made in 1993 when Interior Secretary Bruce Babbitt launched a *national biological survey* to undertake a complete census of U.S. wild species, along with an inventory of the biological communities and ecosystems in which these species live. This is designed to help identify centers of biodiversity and the species and ecosystems most in need of protection. However, its budget has been severely limited by Congress, mostly because of pressure by economic interests favoring weakening or elimination of the Endangered Species Act.

Conservation biologists suggest that the second principle be implemented by greatly increasing currently minuscule funding for implementing the Endangered Species Act. They also urge Congress to pass a *National Biodiversity Protection Act* to identify, protect, and in some cases restore the country's most threatened wildlife habitats and ecosystems. The act would emphasize emergency protection of the most threatened hot spots of biodiversity, as is being done in Florida (Connections, p. 708). This prevention or ecosystem management approach would protect large numbers of species in functioning ecosystems and thus help reduce the number of species needing emergency care under the Endangered Species Act.

A: Warming of earth's lower atmosphere (troposphere) because of the presence of heat-trapping gases

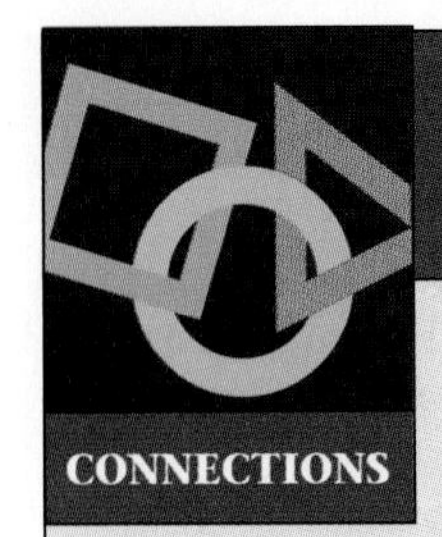

Protecting Hot-Spot Wildlife Reserves in Florida

CONNECTIONS

Florida is the fastest-growing state in the United States, and its wildlife (a vital component of its multibillion tourism industry) is under a severe threat. In 1985, conservation biologists Reed Noss and Larry Harris at the University of Florida designed a nature reserve system for Florida, consisting of core reserves surrounded by buffer zones (Figure 23-17) and linked by habitat corridors.

When this proposal, calling for 60% of Florida to be protected in such a reserve system, was first published, critics dismissed it as an impractical, pie-in-the sky plan of naive and politically out-of-touch college professors. However, since 1985 this visionary practical application of conservation biology has been refined and is now helping protect Florida's wildlife.

State agencies and the Nature Conservancy are now using the refined version (based on computer mapping technology) to identify biodiversity hot spots called Strategic Habitat Conservation Areas, occupying about 1.9 million hectares (4.8 million acres). Currently, Florida and Nature Conservancy officials are working with private landowners to protect (and where necessary to purchase) these areas. Where it is feasible and desirable, there are plans to use habitat corridors to connect many of these reserves.

Critical Thinking

Do you favor a program for protecting hot-spot wildlife reserves in every state? Explain.

Should We Try to Protect All Endangered and Threatened Species? Because of limited funds and trained personnel, only a few endangered and threatened species can be saved. Many wildlife experts suggest that the limited funds available for preserving threatened and endangered wildlife be concentrated on species that **(1)** have the best chance for survival, **(2)** have the most ecological value to an ecosystem, and **(3)** are potentially useful for agriculture, medicine, or industry.

Others oppose such ideas on ethical grounds or contend that currently we don't have enough biological information to make such evaluations. Proponents argue that, in effect, we are already deciding by default which species to save and that, despite limited knowledge, the selective approach is more effective and a better use of limited funds than the current one. What do you think?

How Can Wildlife Refuges and Other Protected Areas Help Protect Endangered Species? In 1903, President Theodore Roosevelt established the first U.S. federal wildlife refuge at Pelican Island off Florida's Atlantic coast to protect the brown pelican (Figure 23-4) from overhunting and loss of habitat. Before the end of Roosevelt's first term he had created 49 more refuges. Since then the U.S. National Wildlife Refuge System has grown to 508 refuges (Figure 23-9). About 85% of the area included in these refuges is in Alaska.

Over three-fourths of the refuges are wetlands for protection of migratory waterfowl. About 20% of the species on the U.S. endangered and threatened list have habitats in the refuge system, and some refuges have been set aside for specific endangered species. These have helped Florida's key deer, the brown pelican, and the trumpeter swan to recover. Conservation biologists urge the establishment of more refuges for endangered plants. They are also urging Congress and state legislatures to allow abandoned military lands that contain significant wildlife habitat to become national or state wildlife refuges.

So far Congress has not established guidelines (such as multiple use or sustained yield) for management within the National Wildlife Refuge System, as it has for other public lands. A 1990 report by the General Accounting Office found that activities considered harmful to wildlife occur in nearly 60% of the nation's wildlife refuges. In addition, a 1986 USFWS study estimated that one federal refuge in five is contaminated with chemicals from old toxic-waste dump sites (including military bases) and runoff from nearby agricultural land.

Since 1986, the World Conservation Union has helped several countries identify and set up marine protected areas (MPAs). The eventual goal is to have a global network of such sites. Currently, about 1,300 MPAs exist (almost half of them located in waters near Asia, Australia, and New Zealand). However, at least half of them are too small to sustain the species found in them. There is also limited information about the effectiveness of MPAs in protecting marine biodiversity.

South Africa has a success story in its efforts to bring back the southern white rhino from the brink of extinction by loss of habitat and overhunting. In the early 1900s the 20 remaining southern white rhinos were confined to a single protected area. By 1961 the rhino numbers had risen to the point where individuals could be relocated to protected areas in other parts

Q: Is there doubt about the validity of the greenhouse effect?

of the country. Now there are more than 7,000 white rhinos, and some are being relocated to start new populations in more than 30 other countries.

Case Study: Oil and Gas Development in the Arctic National Wildlife Refuge? The Arctic National Wildlife Refuge on Alaska's North Slope (Figure 25-18) contains more than one-fifth of all the land in the National Wildlife Refuge System and has been called the crown jewel of the system. The refuge's coastal plain, its most biologically productive part, is the only stretch of Alaska's arctic coastline not open to oil and gas development.

For years, U.S. oil companies have been working to change this because they believe that this coastal area *might* contain oil and natural gas deposits. They argue that such exploration is needed to increase U.S. oil and natural gas supplies and to reduce dependence on oil imports. However, oil companies plan to export most of any oil they find in the reserve to Japan.

Conservation biologists oppose this proposal and urge Congress to designate the entire coastal plain as wilderness. They cite Department of Interior estimates that there is only a 19% chance of finding as much oil there as the United States consumes every 6 months. Even if the oil does exist, environmentalists do not believe the potential degradation of any portion of this irreplaceable wilderness area would be worth it, especially when improvements in energy efficiency would save far more oil at a much lower cost (Sections 16-1 and 16-2).

Oil company officials claim they have developed Alaska's Prudhoe Bay oil fields without significant harm to wildlife; they also contend that the area they seek to open to oil and gas development is less than 1.5% of the entire coastal plain region. However, the huge 1989 oil spill from the tanker *Exxon Valdez* in Alaska's Prince William Sound (Figure 20-9) casts serious doubt on such claims.

Moreover, a study leaked from the U.S. Fish and Wildlife Service in 1988 revealed that oil drilling at Prudhoe Bay has caused much more air and water pollution than was anticipated before drilling began in 1972. According to this study, oil development in the coastal plain could cause the loss of 20–40% of the area's 180,000-head caribou herd, 25–50% of the remaining musk oxen, and 50% or more of the wolverines that live there part of the year. A 1988 EPA study also found that "violations of state and federal environmental regulations and laws are occurring at an unacceptable rate" in portions of the Prudhoe Bay area in which oil fields and facilities have been developed.

Do you believe that possible oil and gas reserves should be developed in this wildlife refuge? Why?

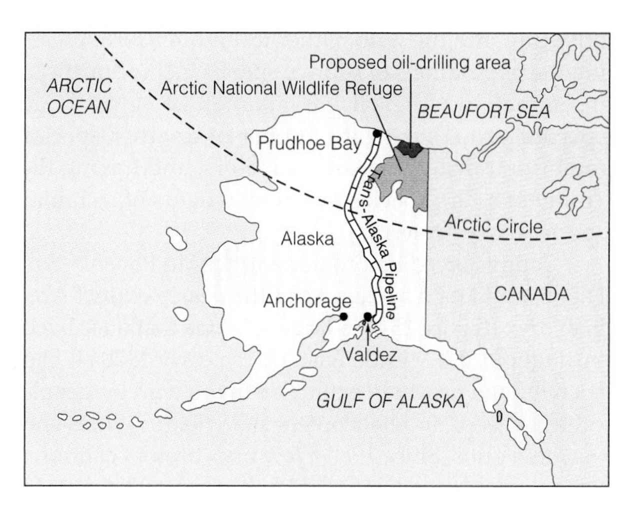

Figure 25-18 Proposed oil-drilling area in Alaska's Arctic National Wildlife Refuge. (Courtesy U.S. Fish and Wildlife Service)

Can Gene Banks and Botanical Gardens Help Save Most Endangered Species? Botanists preserve genetic information and endangered plant species by storing their seeds in gene or seed banks: refrigerated, low-humidity environments (p. 295). Scientists urge that many more such banks be established, especially in developing countries. But some species can't be preserved in gene banks, and maintaining the banks is very expensive.

The world's 1,500 botanical gardens and arboreta maintain at least 40,000 plant species. However, these sanctuaries have too little storage capacity and too little funding to preserve most of the world's rare and threatened plants.

Can Zoos Help Protect Most Endangered Species? Worldwide, 1,000 zoos house about 540,000 terrestrial vertebrate animals from 3,000 species of mammals, amphibians, birds, and reptiles. Many of them are neither threatened nor endangered.

Traditionally, zoos have focused on large, charismatic invertebrates because they are of the greatest interest to the public and thus provide the most income. However, zoos and animal research centers are increasingly being used to preserve some individuals of critically endangered animal species, with the long-term goal of reintroducing the species into protected wild habitats.

Two techniques for preserving such species are egg pulling and captive breeding. *Egg pulling* involves collecting wild eggs laid by critically endangered bird species and then hatching them in zoos or research centers. In *captive breeding*, some or all wild individuals of a critically endangered species are captured for breeding in captivity, with the aim of reintroducing the

A: No; without it, earth would be too cold for life as we know it to exist offspring into the wild. Other techniques for increasing the populations of captive species include artificial insemination, surgical implantation of eggs of one species into a surrogate mother of another species (embryo transfer), use of incubators, and having the young of a rare species raised by parents of a similar species (cross-fostering).

Captive breeding programs at zoos in Phoenix, San Diego, and Los Angeles saved the nearly extinct Arabian oryx (Figure 25-4), a large antelope that once lived throughout the Middle East. By the early 1970s it had been hunted nearly to extinction in the wild by people riding in jeeps and helicopters and wielding rifles and machine guns. Since 1980 a few oryx bred in captivity have been returned to the wild in protected habitats in the Middle East; the wild population is now about 120. Endangered U.S. species now being bred in captivity and returned to the wild include the peregrine falcon (Figure 21-5), changed from endangered to threatened in 1995, and the black-footed ferret (Figure 25-4). Endangered golden lion tamarins (Figure 25-4) bred at the National Zoo in Washington, D.C., have been released in Brazilian rain forests.

For every successful reintroduction, however, there are many more species that can't go home again because their habitats no longer exist or have been too degraded to support the minimum population size needed for survival. Thus, many zoos are becoming arks that cannot set free their ecologically valuable and endangered passengers.

Efforts to maintain populations of endangered species in zoos and research centers are limited by lack of space and money. The captive population of each species must number 100 to 500 individuals to avoid extinction through accident, disease, or loss of genetic diversity through inbreeding. Recent genetic research indicates that 10,000 or more individuals are needed for an endangered species to maintain its capacity for biological evolution.

The world's zoos now house only 27 endangered animal species that have populations of 100 or more individuals. It is estimated that today's zoos and research centers have space to preserve healthy and sustainable populations of only 925 of the 2,000 large vertebrate species that could vanish from the planet. It is doubtful that the more than $6 billion needed to care for these animals for 20 years will become available. According to one estimate, if all of the space in U.S. zoos were used for captive breeding, only about 100 species of large animals could be sustained on a long-term basis.

Some critics see zoos as prisons for once wild animals. They also contend that zoos foster the notion that we don't need to preserve large numbers of wild species in their natural habitats.

Whether one agrees or disagrees with this position, it is clear that zoos and botanical gardens are not a biologically or economically feasible solution for most of the world's current endangered species and the much larger number expected to become endangered over the next few decades.

25-5 WILDLIFE MANAGEMENT

How Can Wildlife Populations Be Managed? **Wildlife management** entails manipulating wildlife populations (especially game species) and their habitats for their welfare and for human benefit. It includes preserving endangered and threatened wild species and enforcing wildlife laws.

In the United States, funds for state game management programs come from the sale of hunting and fishing licenses and from federal taxes on hunting and fishing equipment. Two-thirds of the states also have provisions on state income tax returns that allow individuals to contribute money to state wildlife programs. Only 10% of all government wildlife dollars are spent to study or benefit nongame species, which make up nearly 90% of the country's wildlife species. Since the passage of the Wildlife Restoration Act in 1937 there has been a spectacular increase in the number of game animals (such as white-tailed deer, wild turkeys, Rocky Mountain elk, and pronghorn antelope) sought by many sport hunters.

The first step in wildlife management is to decide which species are to be managed in a particular area, a decision that is a source of much controversy. Ecologists and conservation biologists emphasize preserving biodiversity and ecological integrity, wildlife conservationists are concerned about endangered species, bird-watchers want the greatest diversity of bird species, and hunters want sufficiently large populations of game species for harvest during hunting season. In the United States and other developed countries, most wildlife management is devoted to producing surpluses of game animals and game birds for hunters.

After goals have been set, the wildlife manager must develop a management plan. Ideally, this is based on the principles of ecological succession (Figures 9-19 and 9-21), wildlife population dynamics (Figures 10-3, 10-4, and 10-6), and an understanding of the cover, food, water, space, and other habitat requirements of each species to be managed. The manager must also consider the number of potential hunters, their likely success rates, and the regulations for preventing excessive hunting.

This information is difficult, expensive, and time-consuming to obtain. It involves much educated

Q: What are the principal greenhouse gases?

guesswork and trial and error, which is why wildlife management is as much an art as a science. Management plans must also be sensitive to political pressures from conflicting groups and to budget constraints.

How Can Vegetation and Water Supplies Be Manipulated to Manage Wildlife? Wildlife managers can encourage the growth of plant species that are the preferred food and cover for a particular animal species in a given area by controlling the ecological succession of the vegetation in that area.

Wildlife species can be classified into four types, according to the stage of ecological succession at which they are most likely to be found: early successional, midsuccessional, late successional, and wilderness (Figure 9-22). *Early successional species* find food and cover in weedy pioneer plants that invade an area that has been disturbed, whether by human activities or natural phenomena (fires, volcanic activity, or glaciation).

Midsuccessional species are found around abandoned croplands and partially open areas created by logging of small stands of timber, controlled burning, and clearing of vegetation for roads, firebreaks, oil and gas pipelines, and electrical transmission lines. Such openings of the forest canopy promote the growth of vegetation favored by midsuccessional mammal and bird species; they also increase the amount of edge habitat, where two communities such as a forest and a field come together. This transition zone or ecotone (Figure 4-9) allows animals such as deer to feed on vegetation in clearings and then quickly escape to cover in the nearby forest.

Late successional species rely on old-growth and mature forests to produce the food and cover on which they depend. These animals require the protection of moderate-sized, old-growth forest refuges.

Wilderness species flourish only in undisturbed, mature vegetation communities, such as large areas of old-growth forests, tundra, grasslands, and deserts. They can survive only in large wilderness areas and wildlife refuges.

Various types of habitat management can be used to attract a desired species and encourage its population growth. Examples are planting seeds, transplanting certain types of vegetation, building artificial nests, and deliberately setting controlled, low-level ground fires to help control vegetation. Wildlife managers often create or improve ponds and lakes in wildlife refuges to provide water, food, and habitat for waterfowl and other wild animals.

How Useful Is Sport Hunting in Managing Wildlife Populations? Most developed countries use sport hunting laws to manage populations of game animals. Licensed hunters are allowed to hunt only during certain portions of the year so as to protect animals during their mating season. Limits are set on the size, number, and sex of animals that can be killed, as well as on the number of hunters allowed in a given area.

In the United States and in some other countries, a growing controversy surrounds sport hunting. Sport hunters, hunting groups, and game officials believe that people should be free to hunt as long as they obey state and local game regulations and don't damage wildlife resources or endanger humans. They argue that without carefully regulated sport hunting, deer and other large game animals will exceed the carrying capacity of their habitats and destroy vegetation they and other species need.

Between 1900 and 1997, for example, the estimated population of white-tailed deer in the United States increased from 500,000 to 30 million and is still growing rapidly. These deer are now invading subdivisions and eating shrubs and raiding farmers' fields and orchards. In some areas, hordes of the deer are threatening rare plants and animals, and even entire forests. As the deer population grows, so does the spread of Lyme disease carried by deer ticks. In 1996, there were 16,461 reported cases of Lyme disease. Wildlife biologist William Porter estimates that without hunting, the U.S. deer population would be five times its current level and cause extensive ecological damage. In the United States, each year more than 500,000 deer–vehicle collisions kill more Americans than any other wild animal, injure thousands, and cost at least $1.1 billion.

Raccoons, geese, beavers, and even coyotes and bears are increasingly turning up in rapidly spreading U.S. suburban areas. In Massachusetts, beaver dams threaten hundreds of septic tanks and wells each year. In many areas droppings from Canada geese pollute yards, ball fields, and freshwater supplies. Cougars and bears rarely attack humans but they often prey on household pets. As a result, a growing number of suburban dwellers support killing of animals invading their homes and gardens, even though in most cases it is humans that have invaded the territories of such animals.

Sport hunting also provides recreational pleasure for millions of people (16 million in the United States) and stimulates local economies. Its defenders point out that sales of hunting licenses and taxes on firearms and ammunition have provided more than $1.7 billion since 1937; this money constitutes the vast majority of all funds used to buy, restore, and maintain wildlife habitats and to support wildlife research in the United States.

Most dedicated hunters argue that they respect the land and wildlife and eat any animals they kill. They don't like being lumped together with irresponsible

A: Water vapor (H_2O), carbon dioxide (CO_2), CFCs, methane (CH_4), and nitrous oxide (N_2O)

killers who carelessly go into the woods to derive pleasure by shooting at almost anything that moves.

Environmental groups such as the Sierra Club and Defenders of Wildlife support carefully controlled sport hunting as a way of preserving biological diversity by helping to prevent depletion of other native species of plants and animals. But some individuals and groups, including the Humane Society, oppose sport hunting, claiming it inflicts unnecessary pain and suffering on wild animals, few of which are killed to supply food humans need for survival.

Opponents of hunting also argue that game managers create a surplus of game animals by deliberately eliminating their natural predators (such as wolves) and that game managers then claim that the surplus must be harvested by hunters to prevent habitat degradation or starvation of the game. Instead of eliminating natural predators, say opponents, wildlife managers should reintroduce them to reduce the need for sport hunting (Case Study, p. 630).

Supporters of hunting point out that populations of many game species (for example, deer) are so large that predators such as wolves cannot possibly control them, and that because most wildlife habitats are fragmented, introduction of predators can lead to the loss of nearby livestock. However, critics of hunting point out that deer make up only 2% of the 200 million animals hunters kill each year in the United States. They contend that deer are being used as a smoke screen argument to allow killing of many other game species that don't threaten vegetation. What do you think?

How Can Populations of Migratory Waterfowl Be Managed? Migratory birds—including ducks, geese, swans, and many songbirds—make north-south journeys from one habitat to another each year, usually to find food, suitable climate, and other conditions necessary for reproduction. Such bird species use many different north-south routes called **flyways**, but only about 15 are considered major routes (Figure 25-19). Some countries along such flyways have entered into agreements and treaties to protect crucial habitats needed by such species, both along their migration routes and at each end of their journeys.

In North and South America, migratory waterfowl (ducks, geese, and swans) nest in Canada during the summer. Then during the fall they migrate to the United States and Central America along generally fixed flyways.

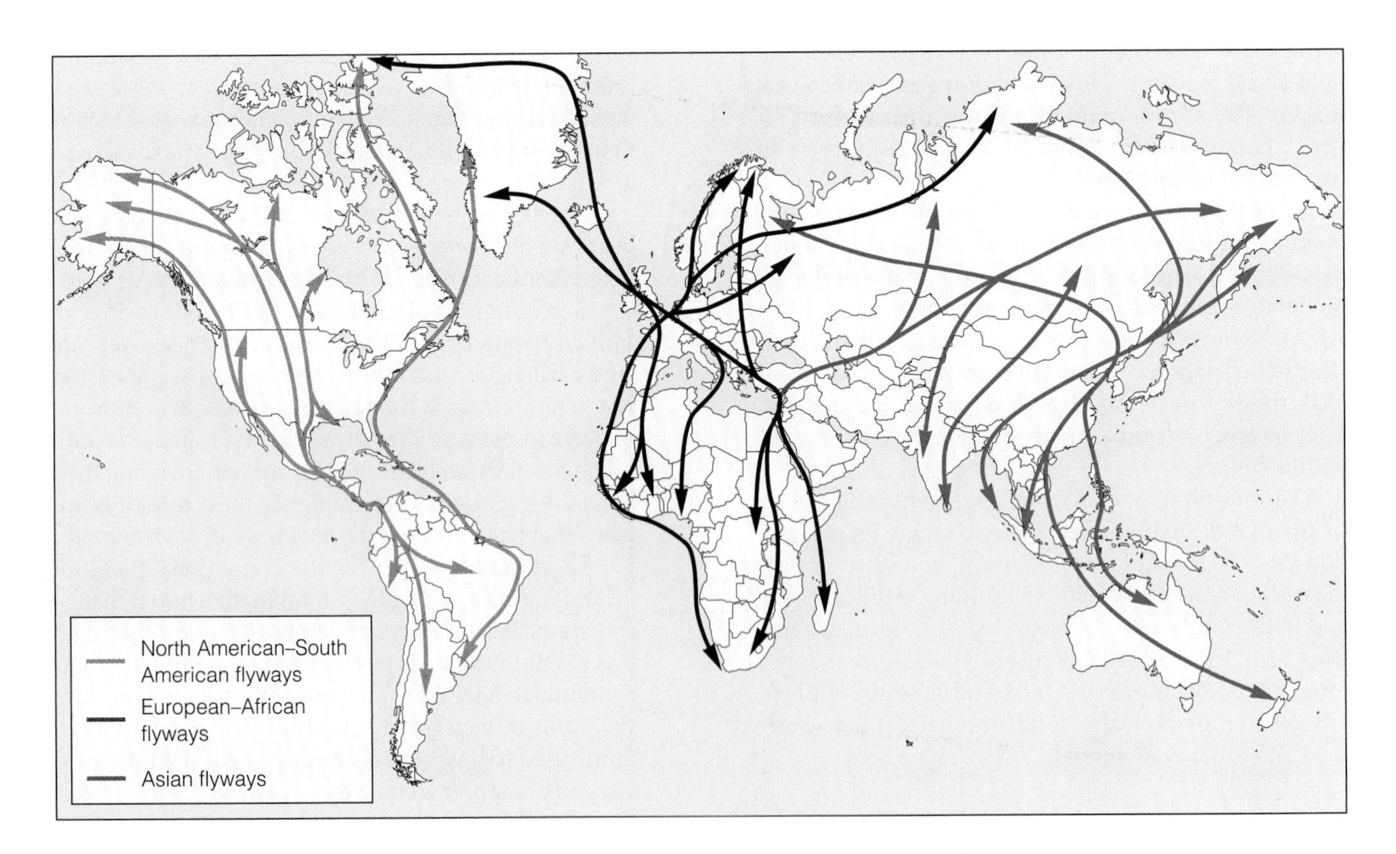

Figure 25-19 Major flyways used by migratory birds, mostly waterfowl. Each route has a number of subroutes.

Botsford, Louis W. 1997. "The Management of Fisheries and Marine Ecosystems." *Science,* vol. 275, no. 5325, 509(7).

Canada, the United States, and Mexico have signed agreements to prevent habitat destruction and overhunting of migratory waterfowl. However, between 1972 and 1992 the estimated breeding population of ducks in North America dropped 38%, mostly because of prolonged drought in key breeding areas and degradation and destruction of wetland and grassland nesting sites by farmers. What wetlands remain are used by flocks of ducks and geese (Connections, right), whose high densities in shrinking wetlands make them more vulnerable to diseases and to predators such as skunks, foxes, coyotes, raccoons, and hunters. Waterfowl in wetlands near croplands are also exposed to pollution from pesticides and other toxic chemicals such as selenium in irrigation runoff.

Wildlife officials manage waterfowl by regulating hunting, protecting existing habitats, and developing new habitats, including artificial nesting sites, ponds, and nesting islands. More than 75% of the federal wildlife refuges in the United States are wetlands used by migratory birds. Local and state agencies and private conservation groups such as Ducks Unlimited, the Audubon Society, and the Nature Conservancy (Solutions, p. 636) have also established waterfowl refuges.

In 1986 the United States and Canada agreed to spend $1.5 billion over a 24-year period, with the goal of almost doubling the continental duck-breeding population, mostly through purchase, improvement, and protection of additional waterfowl habitats in five priority areas. According to Ducks Unlimited, the number of migratory waterfowl in North America increased by 20% between 1986 and 1994. Conservationists fear that this gain may be undermined by congressional efforts to eliminate funding for this program.

Since 1934 the Migratory Bird Hunting and Conservation Stamp Act has required waterfowl hunters to buy a duck stamp each season they hunt. Revenue from these sales goes into a fund to buy land and easements for the benefit of waterfowl.

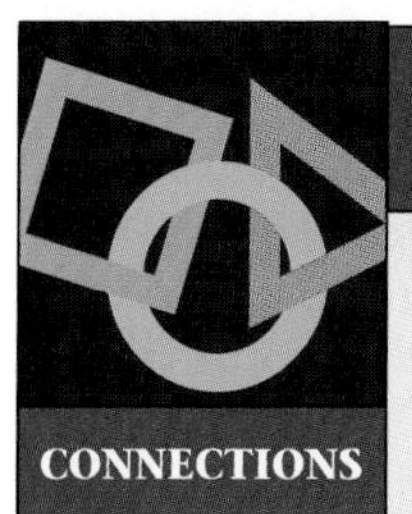

What Should Be Done About Snow Geese?

The population of the lesser snow goose (Figure 10-11) has grown so large that it threatens its arctic tundra and wetlands habitats. Over the past three decades, the population has risen from about 800,000 to at least 3 million. Hunters kill about 500,000 of the birds annually, but their population is still growing at a rate of about 250,000 birds per year.

Their massive numbers have converted vast tracts of arctic tundra in Canada and Alaska to highly saline, bare soil where few plants can grow, creating what has been called a "botanical desert." Snow geese have also destroyed about one-third of the salt marshes along the Hudson and James bays and have heavily damaged another third.

The snow geese population has thrived because each year they can gorge themselves off of the abundant grains in the U.S. agricultural heartland from the Great Plains to the Gulf Coast. As a result, they arrive at their winter breeding grounds in better condition, leading to higher rates of reproduction.

There are several options for dealing with this problem. Eventually the geese will destroy their winter breeding ground and their population will crash. However, this may take decades and lead to widespread ecosystem degradation.

Another strategy is to encourage hunters to kill up to 3 million of these birds each year—about 6 times the current hunting rate. A third suggestion is to have U.S. and Canadian troops gather and destroy about 2 million snow geese eggs a year in the arctic tundra.

This problem that connects high agricultural productivity with overpopulation of a species and ecosystem degradation is another reminder that we can never do one thing in nature.

Critical Thinking

What do you think should be done about the snow geese problem? Explain.

25-6 FISHERY MANAGEMENT AND PROTECTING MARINE BIODIVERSITY

How Can Freshwater Fisheries Be Managed and Sustained? Managing freshwater fish involves encouraging populations of commercial and sport fish species and reducing or eliminating populations of less desirable species. Several techniques are used. It is usually done by regulating time and length of fishing seasons and the number and size of fish that can be taken.

Other techniques include building reservoirs and farm ponds and stocking them with fish, fertilizing nutrient-poor lakes and ponds, and protecting and creating spawning sites. Habitats can also be protected from buildup of sediment and other forms of pollution, and debris can be removed. Excessive growth of aquatic plants from cultural eutrophication (Figure 20-6) can be prevented, and small dams can be built to control water flow.

Hint: Enter the search term *fisheries management* using the Subject Guide.

Predators, parasites, and diseases can be controlled by improving habitats, breeding genetically resistant fish varieties, and using antibiotics and disinfectants judiciously. Hatcheries can be used to restock ponds, lakes, and streams with prized species such as trout and salmon, and entire river basins can be managed to protect such valued species as salmon (Figure 12-19).

How Can Marine Fisheries Be Managed? By international law, a country's offshore fishing zone extends to 370 kilometers (200 nautical miles, or 230 statute miles) from its shores. Foreign fishing vessels can take certain quotas of fish within such zones, called *exclusive economic zones*, but only with a government's permission. Ocean areas beyond the legal jurisdiction of any country are known as the *high seas*; any limits on the use of the living and mineral common-property resources in these areas are set by international maritime law and international treaties.

Managers of marine fisheries use several techniques to help prevent overfishing and commercial extinction and allow depleted stocks to recover. Fishery commissions, councils, and advisory bodies (with representatives from countries or states using a fishery) can set annual quotas and establish rules for dividing the allowable catch among the participating countries or states, limiting fishing seasons, and regulating the type of fishing gear that can be used to harvest a particular species.

As voluntary associations, however, fishery commissions don't have any legal authority to compel their members to follow their rules. Furthermore, it is very difficult to estimate the sustainable yields of various marine species. Experience has shown that many fishery commissions, most of them dominated by fishing industry representatives, have not prevented overfishing.

In 1976, the U.S. Congress passed the Magnuson Fisheries Management and Conservation Act. This law gives the federal government authority to manage fisheries in the zone between 5 and 320 kilometers (3 and 200 miles) off U.S. shores. Fish in this zone are classified as a public resource, owned in common by all U.S. citizens. Most fisheries in this zone are managed under an open-access system: There are no limits on the number of U.S. fishing vessels, but quotas can be imposed on the quantity of fish taken.

The act achieved its goal of keeping foreign fishing fleets from overfishing species in this zone, but it did not prevent the U.S. fishing fleet from expanding and overfishing these waters. One reason is that other laws enacted around the same time provided federal subsidies for upgrading fishing fleets and equipment, which increased the number and efficiency of U.S. fishing boats. Another reason is that the eight regional fishing councils established by the Magnuson Act to oversee U.S. fisheries have usually been dominated by fishing industry representatives who have paid more attention to the short-term economic needs of commercial fishers than to the long-term sustainability of fish stocks.

Various suggestions have been made to reduce overfishing in U.S. waters. *First*, gradually phase out government subsidies of the fishing industry, such as low-interest loans and direct subsidies for boats and fishing operations.

Second, impose fees for harvesting fish and shellfish from publicly owned and managed offshore waters, similar to fees now imposed for grazing and logging on U.S. public lands. Currently, the U.S. fishing industry, like the hard-rock mining industry (p. 338), pays no fees or royalties on the catch it harvests from this publicly owned resource. Fishing harvest fees could be adjusted to changes in the estimated size of fish stocks, with fees rising as stocks become more depleted. Australia imposes fish harvesting fees ranging from 11% to 60% of the value of the gross catch, with an average of about 30%. Understandably, the U.S. fishing industry strongly opposes such changes. *Third*, give each fishing vessel owner a marketable quota for removing a certain amount of fish (Spotlight, right).

Why Is It Difficult to Protect Marine Biodiversity? One reason why it is difficult to protect marine biodiversity is that shore-hugging species are adversely affected by coastal development and the accompanying massive inputs of sediment and other wastes from land (Figure 8-5). This poses a severe threat to biologically diverse and highly productive coastal ecosystems such as coral reefs (Figures 8-1 and 8-10), marshes (Figure 8-6), and mangrove swamps (see photo in table of contents).

Protecting marine biodiversity is also difficult because much of the damage is not visible to most people. In addition, the seas are viewed by many as an inexhaustible resource, capable of absorbing an almost infinite amount of waste and pollution. Finally, most of the world's ocean area lies outside the legal jurisdiction of any country and is thus an open-access resource, subject to overexploitation because of the tragedy of the commons.

Protecting marine biodiversity requires countries to enact and enforce tough regulations to protect coral reefs, mangrove swamps, and other coastal ecosystems from unsustainable use and abuse. Furthermore, much more effective international agreements are needed to protect biodiversity in the open seas.

Q: Is the possibility of global warming from human inputs of greenhouse gases a serious threat?

25-7 CASE STUDY: THE WHALING INDUSTRY

Why Is Whaling an Example of the Tragedy of the Commons? *Cetaceans* are an order of mostly marine mammals ranging in size from the 0.9-meter (3-foot) porpoise to the giant 15- to 30-meter (50- to 100-foot) blue whale. They are divided into two major groups: toothed whales and baleen whales (Figure 25-20).

Toothed whales, such as the porpoise, sperm whale, and killer whale (orca), bite and chew their food; they feed mostly on squid, octopus, and other marine animals. *Baleen whales*, such as the blue, gray, humpback, and finback, are filter feeders. Instead of teeth, they have several hundred horny plates made of baleen, or whalebone, that hang down from their upper jaw. These plates filter plankton from the seawater, especially shrimplike krill, which are smaller than your thumb (Figure 4-17). Baleen whales are the most abundant group of cetaceans.

Whales are fairly easy to kill because of their large size and their need to come to the surface to breathe. Mass slaughter has become especially efficient since the advent of fast propeller-driven ships, harpoon guns, and inflation lances (which pump dead whales full of air and make them float).

Whale harvesting—mostly in international waters, a commons open to all resource extractors—has followed the classic pattern of a tragedy of the commons. Between 1925 and 1975, whalers killed an estimated 1.5 million whales; this overharvesting drove the populations of 8 of the 11 major species to commercial extinction, the point where it no longer paid to hunt and kill them.

Case Study: Near Extinction of the Blue Whale In addition to commercial extinction, the populations of some prized species were reduced to the brink of biological extinction. A prime example is the endangered blue whale (Figure 25-20), the world's largest animal. Fully grown, it's more than 30 meters (100 feet) long—longer than three train boxcars—and weighs more than 25 elephants. The adult has a heart the size of a Volkswagen Beetle car, and some of its arteries are so big that a child could swim through them.

Blue whales spend about 8 months of the year in antarctic waters. There they find an abundant supply of krill, which they filter daily by the trillions from seawater. During the winter they migrate to warmer waters, where their young are born.

Before commercial whaling began, an estimated 200,000 blue whales roamed the Antarctic Ocean. Today, the species has been hunted to near biological extinction for its oil, meat, and bone. Its decline was caused by a combination of prolonged overharvesting and certain natural characteristics of blue whales. Their huge size made them easy to spot. They were caught in large numbers because they grouped together in their Antarctic feeding grounds. They also take 25 years to mature sexually and have only one offspring every 2–5 years, a reproductive rate that makes it difficult for the species to recover once its population falls beneath a certain threshold.

Blue whales haven't been hunted commercially since 1964 and have been classified as an endangered species since 1975. Despite this protection, some

SPOTLIGHT

Is Using Transferable Quotas to Privatize Fishing Rights a Good Idea?

In 1994, the National Marine Fisheries Service (NMFS) claimed that it could no longer manage and conserve U.S. public marine fisheries under an open-access policy. In 1996 it initiated a controversial system of individual transferable quotas (ITQs), in which the government gives each fishing vessel owner a specified percentage of the total allowable catch for each fishery. The largest quotas go to those who have historically taken the most fish, and ITQ owners are permitted to buy, sell, or lease their quotas like private property.

On the surface, this example of free-market environmentalism sounds like a good idea, but most environmentalists generally oppose this approach for several reasons. *First*, it would in effect transfer ownership of publicly owned fisheries to the private commercial fishing industry but still use taxpayer dollars to pay for NMFS oversight of this newly privatized resource.

Second, the scheme would shift fishing quotas from small-scale operators to large-scale operators (whose unsustainable harvesting caused most of the problem in the first place), as has happened with the ITQ system now used in New Zealand. Small fishers would be squeezed out because they don't have the capital to buy the ITQs from others.

Third, the New Zealand experience also shows that the ITQ system increases poaching and sales of illegally caught fish on the black market by small-scale fishers who received no quota or too small a quota to make a living.

Critical Thinking

Do you believe that privatizing fishing in publicly owned waters by using transferable quotas is a good or a bad idea? Explain.

A: Yes, according to the overwhelming consensus of scientists in this field

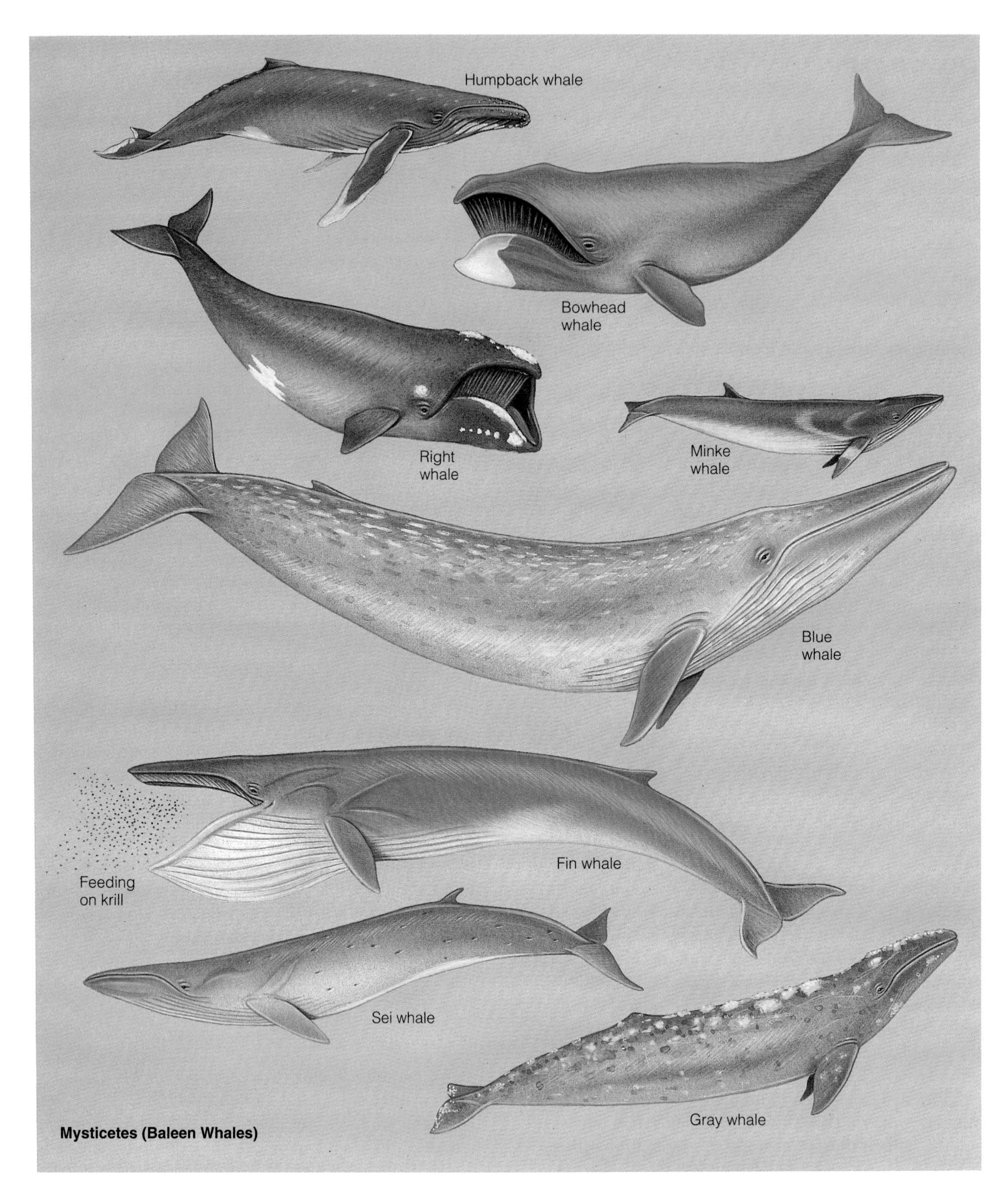

Figure 25-20 Examples of cetaceans, which can be classified as baleen whales and toothed whales.

Q: What are the four principal human sources of greenhouse gas emissions?

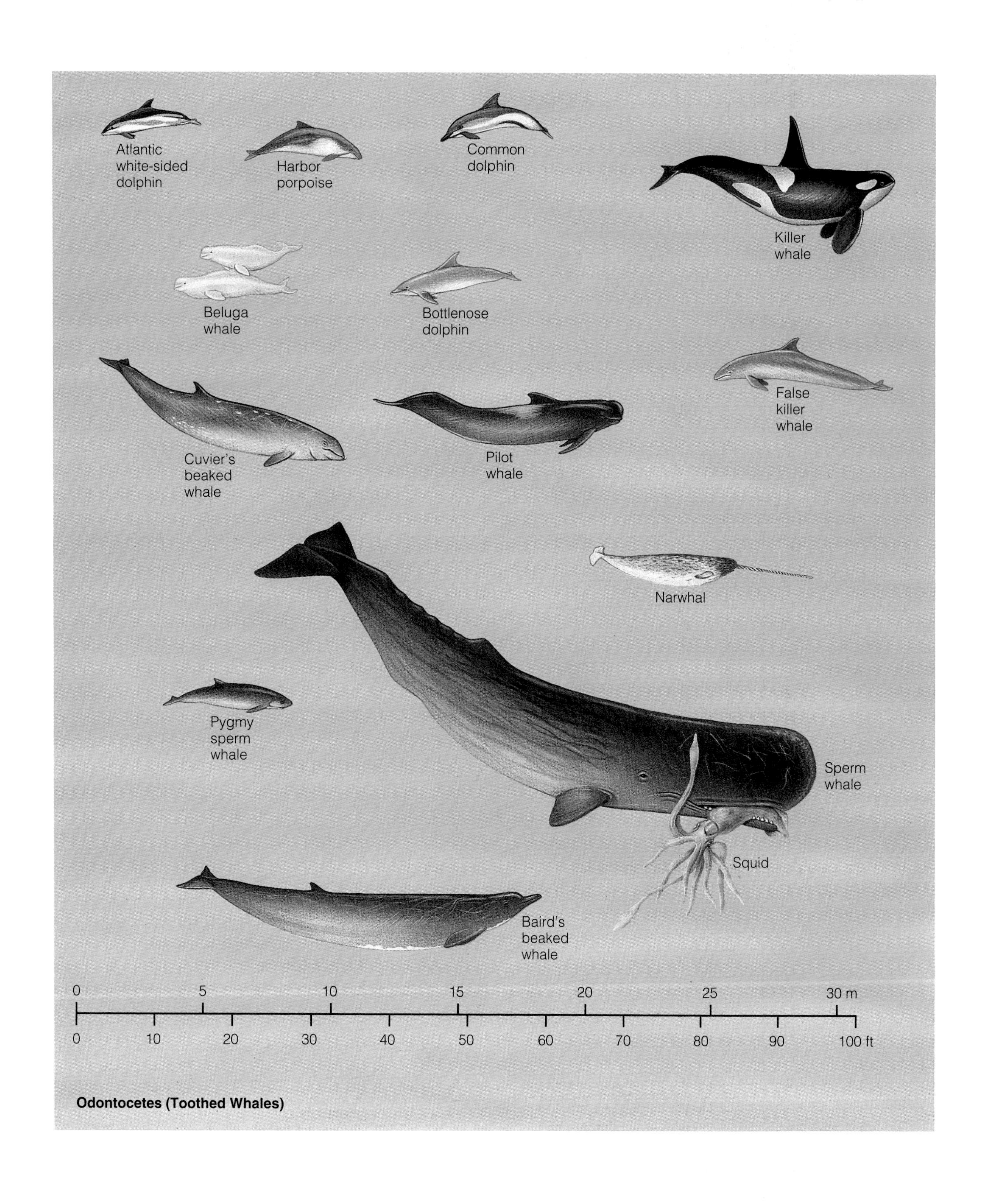
Atlantic white-sided dolphin
Harbor porpoise
Common dolphin
Killer whale
Beluga whale
Bottlenose dolphin
False killer whale
Cuvier's beaked whale
Pilot whale
Narwhal
Pygmy sperm whale
Sperm whale
Squid
Baird's beaked whale
0
5
10
15
20
25
30 m
0
10
20
30
40
50
60
70
80
90
100 ft
Odontocetes (Toothed Whales)

marine biologists believe that too few blue whales—an estimated 3,000-10,000—remain for the species to recover and avoid extinction.

How Have Whale Populations Been Managed? In 1946, the International Whaling Commission (IWC) was established to regulate the whaling industry by setting annual quotas to prevent overharvesting and commercial extinction. However, these quotas often were based on inadequate data or were ignored by whaling countries. Without any powers of enforcement, the IWC has been unable to stop the decline of most whale species.

In 1970, the United States stopped all commercial whaling and banned all imports of whale products. Under intense pressure from environmentalists and governments of many countries, led by the United States, the IWC has imposed a moratorium on commercial whaling since 1986. As a result, the estimated number of whales killed commercially worldwide dropped from 42,480 in 1970 to several hundred per year.

Some whaling has continued because Japan, Norway, Peru, and the former Soviet Union have exempted themselves from the moratorium. Since 1985, Norway, Iceland, and Japan have each killed several hundred whales a year by exploiting a loophole in the IWC treaty that allows whales to be killed for scientific research. Opponents of whaling call this a sham. With whale meat fetching more than $600 per kilogram ($273 per pound) in Japan, there is a great incentive to cheat.

Should the International Ban on Whaling Be Lifted? Whaling is a traditional part of the economies and cultures of countries such as Japan, Iceland, and Norway. To many people in these cultures hunting whales is no more immoral than hunting deer or elk. These and other whaling nations believe that the international ban on commercial whaling should be lifted, arguing that the moratorium is based on emotion, not science.

Most environmentalists disagree. Some argue that whales are peaceful, intelligent, sensitive, and highly social mammals that pose no threat to humans and should not be killed; others fear that opening the door to any commercial whaling may eventually lead to widespread harvests of whales by weakening current international disapproval and economic sanctions against commercial whaling. They cite the earlier failure of the IWC to enforce quotas that prevent commercial extinction of most whale species. They also contend that comparing whale populations to deer and elk populations is misleading because deer and elk are abundant in the United States and in no danger of becoming endangered or extinct.

In 1994 the IWC voted 23 to 1 to establish a permanent whale sanctuary in the Antarctic Ocean. However, it will be almost impossible to monitor and prevent illegal whaling in this huge sanctuary. Moreover, Japan and Russia filed objections to the IWC decision and Japan announced its intentions to continue commercial whaling in the sanctuary.

Opponents of commercial whaling have urged the U.S. Congress to impose sanctions against countries defying the IWC ban on commercial whaling. They also urge consumers to support economic boycotts of all imports from offending nations, to avoid using air and sea travel vessels owned by such countries, and to cease traveling to offending countries. By 1994 a boycott of Norwegian fish products in the United Kingdom and Germany had cost Norway more than $30 million in canceled contracts. What is your view about such boycotts?

During our short time on this planet we have gained immense power over which species, including our own, live or die. We named ourselves the wise (*sapiens*) species. Most conservation biologists believe that in the next few decades, we will learn whether we are indeed a wise species—whether we have the wisdom to learn from and work with nature to protect ourselves and other species. Some actions you can take to help protect wildlife and preserve biodiversity are listed in Appendix 5.

A greening of the human mind must precede the greening of the earth. A green mind is one that cares, saves, and shares. These are the qualities essential for conserving biological diversity now and forever.

M. S. SWAMINATHAN

CRITICAL THINKING

1. Discuss your gut-level reaction to the following statement: "It doesn't really matter that the passenger pigeon is extinct and that the blue whale, the whooping crane, the California condor, and the world's remaining species of rhinoceros and tigers are endangered mostly because of human activities, because eventually all species become extinct anyway." Be honest about your reaction and give arguments for your position.

2. **(a)** Do you accept the ethical position that each *species* has the inherent right to survive without human interference, regardless of whether it serves any useful purpose for humans? Explain. **(b)** Do you believe that each *individual* of an animal species has an inherent right to survive? Explain. Would you extend such rights to individual plants and microorganisms? Explain.

3. Should U.S. energy companies be allowed to drill for oil and natural gas in the Arctic National Wildlife Refuge? Explain your position.

Q: What greenhouse gas is being emitted into the atmosphere in the largest quantity by human activities?

4. Are you for or against sport hunting? Explain your position.

5. Should the international ban on whaling be lifted? Explain. How would you monitor and enforce annual commercial whaling quotas?

6. Should the U.S. government impose an economic boycott on products imported from any nation that illegally resumes whaling? Explain.

7. Cowbird parasites that lay eggs in the nests of other birds are a serious threat to the populations of many songbird species, especially where fragmentation of forests by human activities has allowed them to invade once protected nesting areas. Do you favor killing cowbirds in areas where they are parasitizing the nests of endangered bird species, as well as in the cowbird's winter roosting areas? Defend your position. If you don't believe that members of these (or other species) should be killed, what is your solution to this growing wildlife problem?

8. Since 1972, when California outlawed the hunting of mountain lions, the lion population has more than doubled from an estimated 2,200 to 5,000. Increased expansion of suburbs and increased competition for food have driven some lions to seek food (such as dogs) in suburban areas. In 1994, mountain lions killed two women in a remote state park, the first such deaths in California since 1909. Do you believe that California should legalize hunting of mountain lions? Explain.

9. Which of the following statements best describes your feelings toward wildlife: **(a)** As long as it stays in its space, wildlife is OK; **(b)** as long as I don't need its space, wildlife is OK; **(c)** I have the right to use wildlife habitat to meet my own needs; **(d)** when you've seen one redwood tree, fox, elephant, or some other form of wildlife you've seen them all, so lock up a few of each species in a zoo or wildlife park and don't worry about protecting the rest; **(e)** wildlife should be protected. Explain your choice.

10. List your three favorite species. Examine why they are your favorites. Are they cute and cuddly-looking, like the giant panda and the koala? Do they have humanlike qualities, like apes or penguins that walk upright? Are they large, like elephants or blue whales? Are they beautiful, like tigers and monarch butterflies? Are any of them plants? Are any of them species such as bats, sharks, snakes, or spiders that most people are afraid of? Are any of them microorganisms that help keep you alive? Reflect on what your choice of favorite species tells you about your attitudes toward most wildlife.

11. Try to come up with analogies to complete the following statements: **(a)** Killing an elephant for its ivory tusks is like _______. **(b)** Causing the premature extinction of a species is like _______. **(c)** Destroying or degrading the habitat (home) of a wild species is like _______. **(d)** Deliberately introducing a new species into an ecosystem without thinking about the possible consequences is like _______. **(e)** Using zoos to help protect endangered species is like _______. **(f)** Wearing or collecting clothes or other items made from endangered species is like _______. **(g)** Dressing up monkeys, dogs, cats, or other animals in clothes and teaching them to do tricks that make them act like humans is like _______. **(h)** Opening the coastal plain of Alaska's Arctic National Wildlife Refuge to oil exploration and drilling is like _______.

PROJECTS

1. Make a log of your own consumption of all products for a single day. Relate your level and types of consumption to the decline of wildlife species and the increased destruction and degradation of wildlife habitats in the United States, in tropical forests, and in aquatic ecosystems.

2. Identify examples of habitat destruction or degradation in your community that have had harmful effects on the populations of various wild plant and animal species. Develop a management plan for the rehabilitation of these habitats and wildlife.

3. Use the Internet to compile a list of bioinformatics projects that supply biodiversity information. Evaluate what kind of information is available, how complete it appears to be, and whether it is likely to be useful to managers and decision makers.

4. Make a concept map of this chapter's major ideas, using the section heads and subheads and the key terms (in boldface). Look at the inside back cover and in Appendix 4 for information about concept maps.

A: Carbon dioxide (CO_2)

26 SUSTAINABLE CITIES: URBAN LAND USE AND MANAGEMENT*

Is Your City Green? The Ecocity Concept in Davis, California

Few of today's cities are sustainable. They have become dependent on distant sources for their food, water, energy, and materials. Their massive use of resources damages nearby and distant air, water, soil, and wildlife. Virtually every city can become more environmentally and economically self-sufficient by relying more on locally available resources and using them more sustainably.

In a sustainable and ecologically healthy city—called an *ecocity* or *green city*—matter and energy resources are used efficiently. Far less pollution and waste are produced than in conventional cities. Emphasis is placed on pollution prevention, reuse, recycling, and efficient use of energy and matter resources. Per capita solid waste production is greatly reduced, and at least 60% of what is produced is reused, recycled, or composted. An ecocity takes advantage of locally available energy sources and requires that all buildings, vehicles, and appliances meet high energy-efficiency standards.

Trees and plants adapted to the local climate and soils are planted throughout an ecocity to provide shade and beauty, to reduce pollution and noise, and to supply habitats for wildlife. Abandoned lots and polluted creeks are cleaned up and restored. Nearby forests, grasslands, wetlands, and farms are preserved instead of being devoured by urban sprawl. Much of an ecocity's food comes from nearby organic farms, solar greenhouses, community gardens, and small gardens on rooftops, in yards, and in window boxes.

An ecocity is a people-oriented, not car-oriented city. Its residents are able to walk or bike to most places, including work, and to take low-polluting mass transit. It is designed, retrofitted, and managed to provide a sense of community.

The ecocity is not a futuristic dream. The citizens and elected officials of Davis, California—a city of about 54,000 people about 130 kilometers (80 miles) northeast of San Francisco—committed themselves in the early 1970s to making their city ecologically sustainable.

City building codes encourage the use of solar energy for water and space heating. All new homes

*__Paul M. Rich__, associate professor of Ecology and Evolutionary Biology and Environmental Studies at the University of Kansas, is coauthor of this chapter.

Figure 26-1 Solar home in Village Homes in Davis, California, a neighborhood of 200 houses. This community, designed by Peter Calthorpe and Michael Corbett and completed in 1982, is an experiment in urban sustainability. Direct solar gain provides 50–75% of the heating needs in winter. In the hot summers the focus shifts from harvesting solar energy to blocking it; most residents in the community rarely need air conditioners because carefully sited trees shade the houses and paved areas. (Virginia Thigpen)

must meet high standards of energy efficiency, and when an existing home changes hands the buyer must bring it up to the energy conservation standards for new homes. In Davis's Village Homes, America's first solar neighborhood (Figure 26-1), houses are heated by solar energy. They face into a common open space reserved for people and bicycles; cars are restricted to streets, which are located only on the periphery of the development. The neighborhood also has commonly shared orchards, vineyards, gardens, playgrounds, playing fields, and a solar-heated community center used for day care, meetings, and social gatherings.

Since 1975 Davis has cut its use of energy for heating and cooling in half. It has a solar power plant, and some of the electricity it produces is sold to the regional utility company. Eventually the city plans to generate all of its own electricity.

The city discourages the use of automobiles and encourages the use of bicycles by closing some streets to automobiles, building bike lanes on major streets, and building bicycle paths. Any new housing tract must have a separate bike lane, and some city employees are given bikes. As a result, more than 28,000 bicycles account for 40% of all in-city transportation, and less land is needed for parking spaces. This heavy dependence on the bicycle is aided by the city's warm climate and flat terrain.

Davis limits the type and rate of its growth, and it maintains a mix of homes for people with low, medium, and high incomes. Development of the fertile farmland surrounding the city for residential or commercial use is restricted. Davis also limits the size of shopping centers to encourage smaller neighborhood shopping centers, each easily reached by foot or bicycle.

Davis and other cities—San Jose, California; Osage, Iowa (p. 56); Horsen, Denmark; Curitiba, Brazil (Case Study, p. 736); Gavioras, Columbia (p. 412); Tuggelite, Sweden; and Tapiola, Finland (Case Study, p. 743)—are blazing a trail by trying to make cities more sustainable and enjoyable places to live.

The quality of urban life affects virtually every person and species on the planet. By the year 2000, about half of the world's population will be living in urban areas and by 2025 about two-thirds of humanity will be urban dwellers. As the planet becomes more urbanized, reshaping existing cities and designing new ones to work with nature must become top priorities.

The test of the quality of life in an advanced economic society is now largely in the quality of urban life. Romance may still belong to the countryside—but the present reality of life abides in the city.

John Kenneth Galbraith

This chapter focuses on answering the following questions:

- How is the world's population distributed between rural and urban areas?
- What factors determine how urban areas develop?
- What are the pros and cons of living in an urban area?
- How do transportation systems shape urban areas and growth?
- How can we manage land use in urban areas?
- How can cities be made more livable and sustainable?

26-1 URBANIZATION AND URBAN GROWTH

How Fast Are Urban Areas Growing? For more than 6,000 years cities—often called the cradles of civilization—have been centers of commerce, communication, technological developments, education, religion, social change, political power, and progress. They have also been centers of crowding, pollution, and disease.

For almost 300 years, since the beginning of the industrial revolution, cities have been growing rapidly in size. They are now called **urban areas**—towns or cities plus their adjacent suburban fringes with populations of more than 2,500 people (although some countries set the minimum at 10,000–50,000 residents). A **rural area** is usually defined as an area with a population of less than 2,500 people.

A country's **degree of urbanization** is the percentage of its population living in an urban area. **Urban growth** is the rate of increase of urban populations.

Between 1950 and 1998, the number of people living in the world's urban areas increased 12-fold, from 200 million to 2.6 billion. By 2025 it is projected to reach 5.5 billion, almost equal to the world's current population. About 90% of this urban growth will occur in developing countries. At current rates the world's population will double in 47 years, the urban population in 22 years, and the urban population of developing countries in 20 years.

Several trends are important in understanding the problems and challenges of urban growth on this rapidly urbanizing planet. *First*, the proportion of the global population living in urban areas increased between 1850 and 1998 from 2% to 44%. This degree of urbanization varies in major areas of the world (Figure 26-2).

During the 1990s, more than 70% of the world's population increase is expected to occur in urban areas, adding about 63 million people a year to these already overburdened areas. This is equivalent to adding about four cities the size of New York each year. If UN projections are correct, by 2025 about 63% of the world's people will be living in urban areas.

Second, the number of large cities is mushrooming. In 1960 there were 111 cities with populations of more than 1 million; today there are 293, and some analysts expect this number to increase to at least 400 by 2025. Currently, 1 person out of every 10 lives in a city with a million or more inhabitants, and many live in the world's 15 *megacities* (Figure 26-3) with 10 million or more people. The United Nations projects that by 2000 the world will have 21 megacities, 17 of them in developing countries.

However, the move to urban areas is not limited to the world's 100 or so largest urban areas. The fastest urban growth is occurring in the 30,000 or so medium-size cities in developing countries.

Third, developing countries, with 36% urbanization, contain 1.7 billion urban dwellers—more than the total populations of Europe, North America, Latin America, and Japan combined. The urban population in developing countries is growing at 3.5% per year and they are projected to reach at least 57% urbanization by 2025. Most of this growth will occur in large cities, which already have trouble supplying their residents with water, food, housing, jobs, sanitation, and basic services.

Fourth, in developed countries (with 73% urbanization), urban growth is less than 1% per year, much slower than in developing countries. Still, developed countries should reach 84% urbanization by 2025.

Finally, poverty is becoming increasingly urbanized as more poor people migrate from rural to urban areas. The United Nations estimates that at least 1 billion people, 17% of the world's current population, live in the crowded *slums* of inner cities or in vast, mostly illegal *squatter settlements* and *shantytowns*, where people move onto undeveloped land (usually without the owner's permission) and build shacks made of packing crates, plastic sheets, corrugated metal pipes, or whatever they can find.

Shantytowns and squatter settlements, called *barrios*, *favelas*, or *turgios* in Latin America, *bustees* in India, and *bidonvilles* in Africa, ring the outskirts of most large cities in developing countries. They usually lack clean water supplies, sewers, electricity, and roads. Often the land on which such settlements are built is not suitable for human habitation because of air and

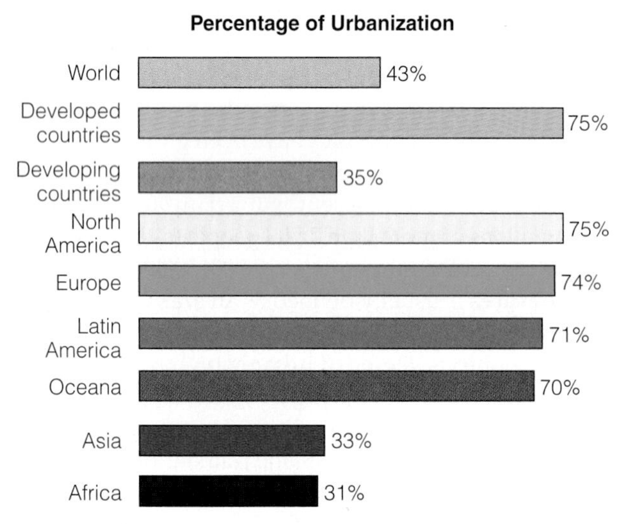

Figure 26-2 Degree (percentage) of urbanization for various groupings of countries in 1996. (Data from Population Reference Bureau)

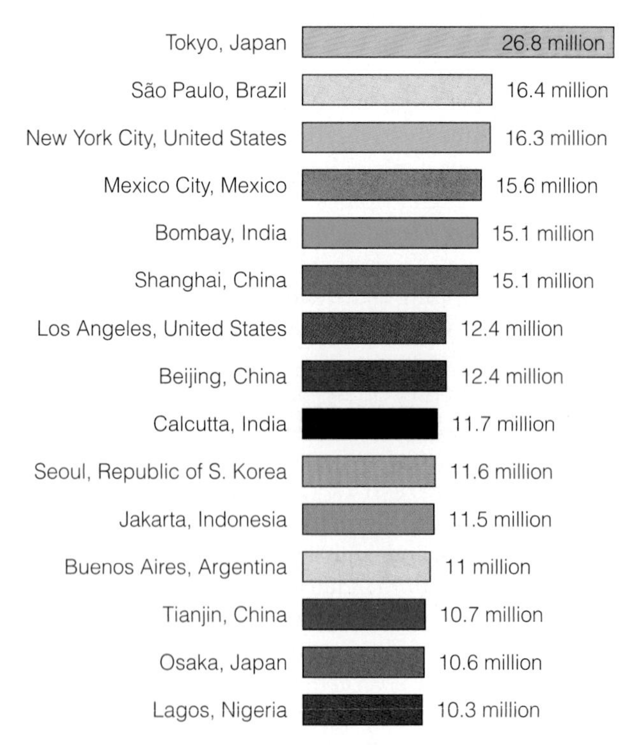

Figure 26-3 The world's 15 megacities—defined as 10 million or more inhabitants—as of 1995. (Data from United Nations)

Macie, E. 1997. "The Environmental Consequences of Urbanization." *American Forests*, vol. 103, no. 3, 28(2).

water pollution and hazardous wastes from nearby factories, or because the land is especially subject to landslides, flooding, earthquakes, or volcanic eruptions.

In Manila, Philippines, for example, some 20,000 people live in city dumps in shacks built on huge mounds of garbage and burning industrial waste. In 1984 the world's worst industrial accident occurred at the Union Carbide factory in Bhopal, India. The release of toxic gas killed at least 5,100 people and caused at least 200,000 serious injuries (including blindness). Most of these victims lived in shantytowns near the chemical factory.

An estimated 100 million people are homeless and sleep on the streets (Figure 1-10) or wherever they can. Half of all urban children under age 15 in developing countries live in conditions of extreme poverty, and about one-fifth of them are street children with little or no family support. In Cairo, Egypt, children of kindergarten age can be found digging through clods of ox dung, looking for undigested kernels of corn to eat.

Many cities do not provide squatter settlements and shantytowns with adequate drinking water, sanitation facilities, electricity, food, health care, housing, schools, or jobs. Not only do these cities lack the needed money, but their officials fear that improving services will attract even more of the rural poor. Many city governments regularly bulldoze squatter shacks and send police to drive the illegal settlers out. The people either move back in or develop another shantytown somewhere else.

Shantytowns and squatter settlements are also found in some developed countries. For example, in the *colonias* shantytowns in Texas along the southern Rio Grande river, living conditions are as bad as in similar settlements in the cities of developing countries. In the United States, most inner cities have concentrations of poor people.

Despite joblessness, squalor, overcrowding, environmental hazards, and rampant disease, most squatter and slum residents are better off than the rural poor. With better access to family-planning programs, they tend to have fewer children, who have better access to schools. Most residents are adaptable and resilient and have hope for a better future. Many squatter settlements provide a sense of community and a vital safety net of neighbors, friends, and relatives for the poor. A few squatter communities have organized to improve their living conditions.

In Villa El Salvador outside Lima, Peru, for example, a network of women's groups and neighborhood associations planted half a million trees, trained hundreds of door-to-door health workers, and built 300 community kitchens, 150 day-care centers, and 26 schools. Through their own efforts the 300,000 residents now have homes with electricity, over half of which have running water and sewers. In addition, illiteracy has fallen to 3%—one of the lowest rates in Latin America—and infant mortality is 40% below the national average. Such success stories are rare, but they show what can be done by individuals organizing from the bottom up.

What Causes Urban Growth? Urban populations grow in two ways: by *natural increase* (more births than deaths) and by *immigration* (mostly from rural areas). Improved food supplies and better sanitation and health care in urban areas lower the death rate and cause urban populations to grow. Today, in developing countries the natural increase of the urban population is at least as important as migration.

Because cities are the main centers for new jobs, higher income, education, innovation, culture, better health care, and trade, people are *pulled* to urban areas in search of jobs, a better life, and freedom from the constraints of village cultural life. They may also be *pushed* from rural areas into urban areas by factors such as poverty, lack of land, declining agricultural work, famine, and war.

Modern mechanized agriculture, for example, uses fewer farm laborers and allows large landowners to buy out subsistence farmers who cannot afford to modernize. Without jobs or land, these people are forced to move to cities. The urban poor fortunate enough to find employment usually must work long hours for low wages at jobs that may expose them to dust, hazardous chemicals, excessive noise, and dangerous machinery.

Urban growth in developing countries is also fueled by government policies that distribute most income and social services to urban dwellers (especially in capital cities, where a country's leaders live) at the expense of rural dwellers. Various studies suggest that the vast majority of rural migrants feel that their move to the city improved their situation, but often not as much as they had hoped.

Case Study: Mexico City About 18 million people—about one of every six Mexicans—live in Mexico City, the world's fourth most populous city (Figure 26-4). Immigration, not reproduction, is the main reason for Mexico City's high rate of population growth. Every day an additional 2,000 poverty-stricken rural peasants pour into the city, hoping to find a better life. This adds about 750,000 new people per year, equivalent to having to provide food, water, sanitation, jobs, and other services for a new city the size of Baltimore or San Francisco every year.

Mexico City suffers from severe air pollution, high unemployment (close to 50%), deafening noise, congestion, and a soaring crime rate. More than one-third

Hint: Enter the search terms *urbanization, environmental aspects* using the Subject Guide.

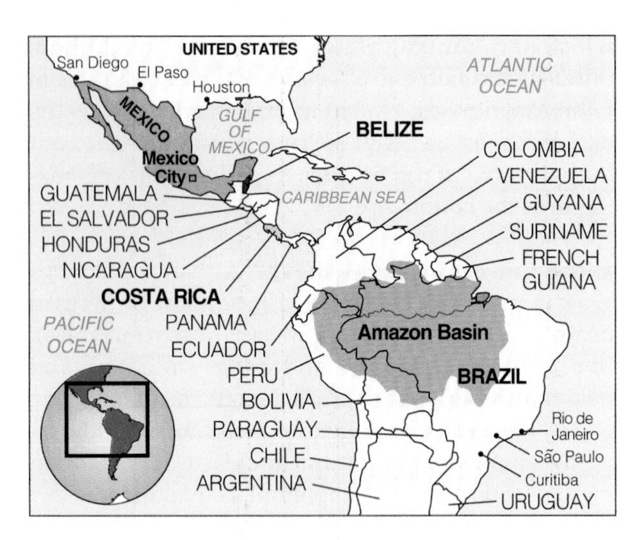

Figure 26-4 The locations of Mexico, Brazil, Belize, and Costa Rica. Many other countries highlighted here are discussed in other chapters.

of its residents live in crowded slums (called *barrios*) or squatter settlements, without running water or electricity. At least 8 million people have no sewer facilities. This means that huge amounts of human waste are deposited in gutters and vacant lots every day, attracting armies of rats and swarms of flies. When the winds pick up dried excrement, a *fecal snow* often falls on parts of the city, leading to widespread salmonella and hepatitis infections, especially among children.

People living in the slums of the city must buy water by the bucket from vendors at a high cost; many who have access to city water won't drink it. About 1.5 million of the city's poor live in the drained bed of Lake Texcoco, which floods or becomes a bog when it rains. Excessive withdrawal of groundwater has caused parts of the city to sink more than 9 meters (30 feet) since 1900.

Some 4 million motor vehicles, 30,000 factories, and leaking unburned liquefied petroleum gas (LPG) from stoves and heaters spew pollutants into the atmosphere. Air pollution is intensified because the city lies in a basin surrounded by mountains, and frequent thermal inversions trap pollutants at ground level (Figure 18-6). The city's high altitude causes even greater air pollution because automobile engines burn fuel less efficiently at high altitudes. Since 1982 the amount of contamination in the city's air (Figure 18-5) has more than tripled, making breathing the air equivalent to smoking about two packs of cigarettes a day. The city's health costs from air pollution are estimated at $1.5 billion per year.

Pediatricians estimate that 85% of childhood illnesses in the city are related to air pollution and believe that the only way parents can improve the health of such children is to get them out of the city. Writer Carlos Fuentes has nicknamed this megacity "Makesicko City." Because of the air pollution, many foreign companies and governments give imported workers additional hazard pay for working in Mexico City.

Few urban areas in the world are as far from being self-sufficient as the Basin of Mexico, which contains Mexico City. Raw materials and energy generated in the basin cannot meet the needs of even a small fraction of the city's 18 million residents. Most of the nearby forest has been cut, much of the nearby cropland has been converted to urban development, practically all of the lakes have dried up, and almost one-third of the city's water comes from other increasingly dry basins.

The Mexican government is industrializing other parts of the country in an attempt to slow migration to Mexico City. In 1991, the government closed the city's huge state-run oil refinery and ordered many of the industrial plants in the basin to go elsewhere by 1994. Cars have been banned from a 50-block central zone. Taxis built before 1985 have been taken off the streets, and trucks can run only on liquefied petroleum gas (LPG). The government began phasing in unleaded gasoline in 1991, but it will be years before millions of older vehicles, which burn leaded gas, are eliminated. Since 1993, all new cars in Mexico must have catalytic converters (devices for reducing pollution emissions), but new cars represent only about 5% of the cars on the road each year. The government has also planted 25 million trees to help clean the air and has purchased some land to provide green space for the city.

The city's air and water pollution cause an estimated 100,000 premature deaths a year. These problems, already at crisis levels, will become even worse if the city grows as projected to 25.8 million within the next decade.

If you were in charge of Mexico City, what would you do?

How Urbanized Is the United States? In 1800 only 5% of Americans lived in cities. Since then, four major internal population shifts have taken place in the United States. As a result of the first shift, *migration to large central cities*, about 75% of Americans live in 350 *metropolitan areas* (cities and towns with at least 50,000 people). Nearly half of the country's population lives in consolidated metropolitan areas containing 1 million or more residents (Figure 26-5).

In the second shift, more people began *migrating from large central cities to suburbs and smaller cities*. Since 1970 this type of migration has followed new jobs to such areas. Today about 41% of the country's urban dwellers live in central cities and 59% live in suburbs.

Q: What four countries account for 70% of the world's CO_2 emissions?

The third shift, which has been taking place for several decades, is *migration from the North and East to the South and West* (Figure 26-6). Since 1980, about 80% of the U.S. population increase has occurred in the South and West, particularly near the coasts. This shift is expected to continue.

In a recent fourth shift, people have been *migrating from urban areas back to rural areas*. This movement has reversed a decline of 1.4 million in the rural population in the 1980s. According to the Census Bureau, between 1900 and 1997 rural counties had a net influx of about 2 million people, virtually all of it from domestic migration.

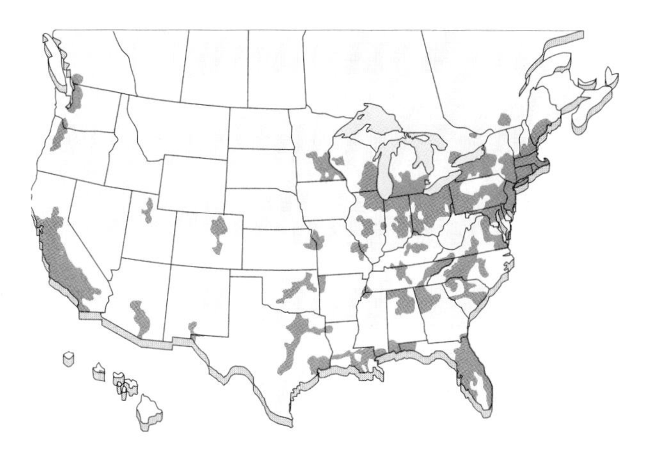

Figure 26-5 Major urban regions in the United States by the year 2000. Nearly half (48%) of Americans live in *consolidated metropolitan areas* with 1 million or more people. (Data from U.S. Census Bureau)

What Are the Major Urban Problems in the United States? Here is some great news. Since 1920, many of the worst urban environmental problems in the United States (and other developed countries) have been significantly reduced. Most people have better working and housing conditions and air and water quality have improved. Better sanitation, public water supplies, and medical care have slashed death rates and the prevalence of sickness from malnutrition and transmittable diseases such as measles, diphtheria, typhoid fever, pneumonia, and tuberculosis. Furthermore, concentrating most of the population in urban areas has helped protect the country's biodiversity by reducing the destruction and degradation of wildlife habitat.

The biggest problems facing numerous cities in the United States are deteriorating services, aging infrastructures (streets, schools, bridges, housing, sewers), budget crunches from lost tax revenues and rising costs as businesses and more affluent people move out, and rising poverty in many central city areas. As a result, violence, drug traffic and abuse, crime, decay, and blight have increased in parts of central cities. Unemployment rates in some inner-city areas are typically 50% or higher.

What Are the Major Spatial Patterns of Urban Development? Three generalized models of urban structure are shown in Figure 26-7. A *concentric-circle city*, such as New York City, develops outward from its central business district (CBD) in a series of rings as the area grows in population and size. Typically industries and businesses in the CBD

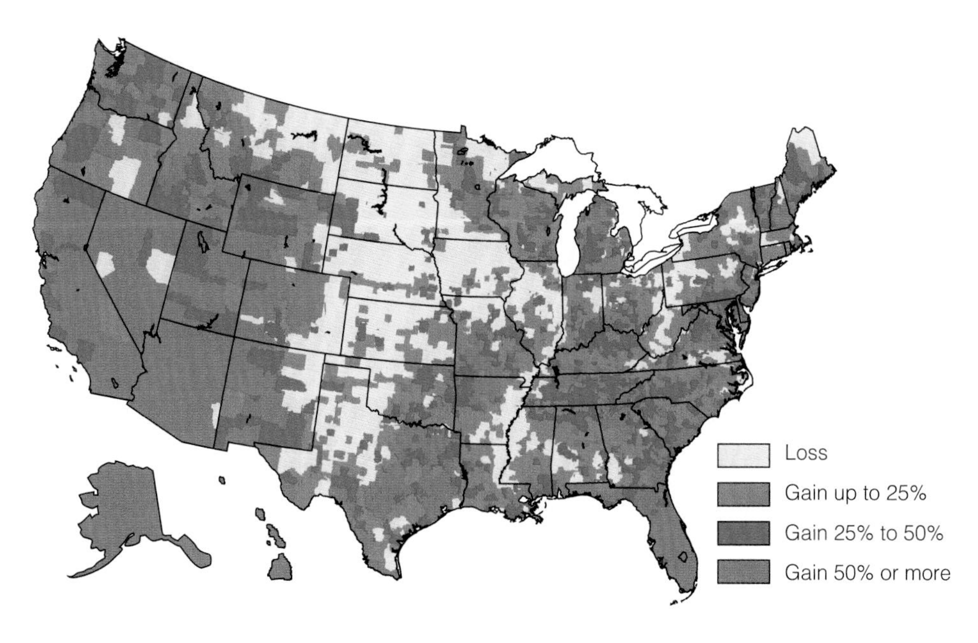

Figure 26-6 Changes in population in the United States by county, 1970–95. Notice the general trend toward migration from the North and East to the West and South. (Data from U.S. Bureau of Census)

A: United States (22%), China (12%), Russia (9.4%), and Japan (5%)

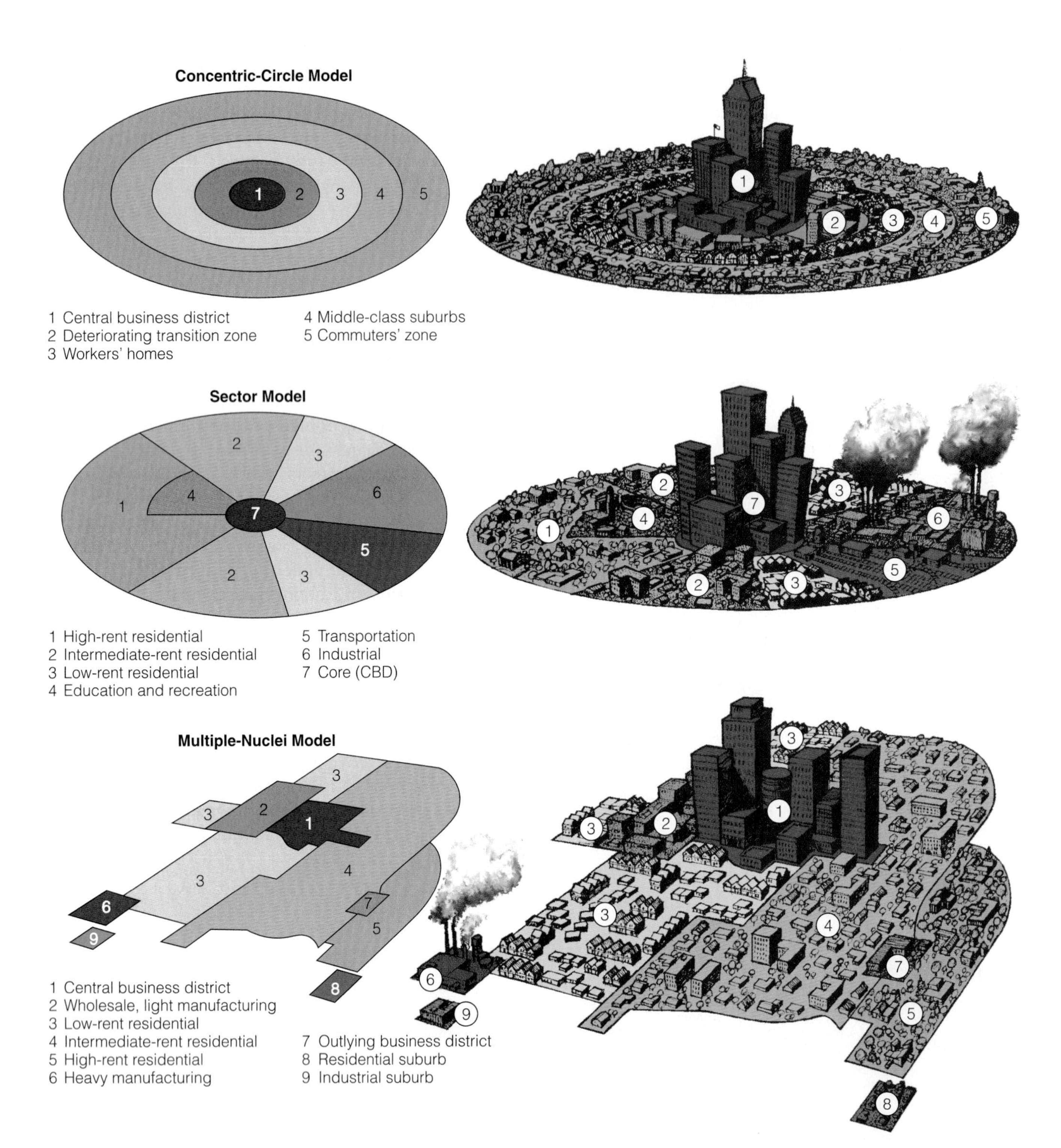

Figure 26-7 Three models of urban spatial structure. Although no city perfectly matches any of them, these simplified models can be used to identify general patterns of urban development. (Modified with permission from Harm J. deBlij, *Human Geography*, New York: Wiley, 1977)

and poverty-stricken inner-city housing areas are ringed by housing zones that become more affluent toward the suburbs. In many developing countries, affluent residents cluster in the central city, and many of the poor live in squatter settlements that spring up on the outskirts.

A *sector city* grows in pie-shaped wedges or strips. Growth sectors develop when commercial, industrial, and housing districts push outward from the CBD along major transportation routes. An example is the large urban area extending from San Francisco to San Jose in California.

Parlange, M. 1998. "The City as Ecosystem." *BioScience*, vol. 48, no. 8, 581(5).

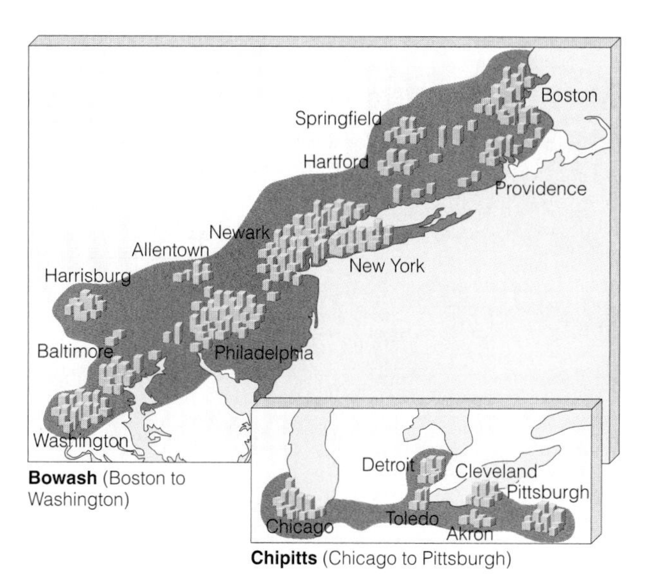

Figure 26-8 Two megalopolises: Bowash, consisting of urban sprawl and coalescence between Boston and Washington, D.C., and Chipitts, extending from Chicago to Pittsburgh.

A *multiple-nuclei city* develops around a number of independent centers, or satellite cities, rather than a single center. Metropolitan Los Angeles comes fairly close to this pattern. Of course, some cities develop in various combinations of these three patterns.

As they grow outward, separate urban areas may merge to form a *megalopolis*. For example, the remaining open space between Boston and Washington, D.C., is rapidly urbanizing and coalescing. This 800-kilometer-long (500-mile-long) urban area, sometimes called *Bowash* (Figure 26-8), contains almost 60 million people, more than twice Canada's entire population.

Megalopolises have developed all over the world—in Europe between Amsterdam and Paris; in Japan the Tokyo–Yokohama–Osaka–Kobe corridor (with nearly 50 million people); and in the Brazilian Industrial Triangle made up of São Paulo, Rio de Janeiro, and Belo Horizonte.

26-2 URBAN RESOURCE AND ENVIRONMENTAL PROBLEMS

What Are the Environmental Pros and Cons of Urban Areas? Most of today's urban areas don't even come close to being self-sustaining; they survive only by importing food, water, energy, minerals, and other resources from farms, forests, mines, and watersheds. They also produce enormous quantities of wastes that can pollute air, water, and land within and outside their boundaries (Figure 26-9).

The 44% of the world's people currently living in urban areas occupy only about 5% of the planet's land area but consume 75% of the world's resources. Supplying these urban dwellers with resources is a major reason that humans use or have disturbed about 73% of the earth's habitable land area (Figure 1-5).

However, urbanization has some environmental benefits. Recycling is more economically feasible because of the large concentration of recyclable materials. The environmental pressures from population growth are reduced because birth rates in urban areas usually are three to four times lower than in rural areas. Cities provide better opportunities to educate people about environmental issues and to mobilize residents to deal with environmental problems. In addition, per capita expenditures on environmental protection are higher in urban areas.

Concentrating people in urban areas also helps preserve biodiversity by reducing the stress on wildlife habitats. This effect is important but is not as great as it first appears because large areas of the earth's land area must be disturbed and degraded to provide urban dwellers with food, water, energy, minerals, and other resources (Figure 26-9). Furthermore, rural or wildlife areas downwind or downstream from urban areas are receptacles for much of the pollution produced in urban areas.

Some analysts call for seeking more sustainable relationship between cities and the living world. To do this will require converting high-waste, unsustainable cities with a *linear metabolism* (based on an ever-increasing throughput of resources and output of wastes; Figure 3-21) to low-waste, sustainable cities with a *circular metabolism* (based on efficient use of resources, reuse, recycling, pollution prevention, and waste reduction; Figure 3-22).

Why Are Trees and Food Production Important in Cities? *Most cities have few trees, shrubs, or other plants* that absorb air pollutants, give off oxygen, help cool the air (as water transpires from their leaves), muffle noise, provide wildlife habitats, and give aesthetic pleasure. As one observer remarked, "Most cities are places where they cut down the trees and then name the streets after them." This is a tragic loss. According to the American Forestry Association, one city tree provides over $57,000 worth of air conditioning, erosion and stormwater control, wildlife shelter, and air pollution control over a 50-year lifetime.

Most cities produce little of their own food. However, people can grow their own food by planting community gardens in unused lots, using window-box and balcony planters, and creating gardens or greenhouses on the roofs of apartment buildings. Urban gardens currently provide about 15% of the world's food and

Hint: Enter the search term *ecosystem, city* using Key Words.

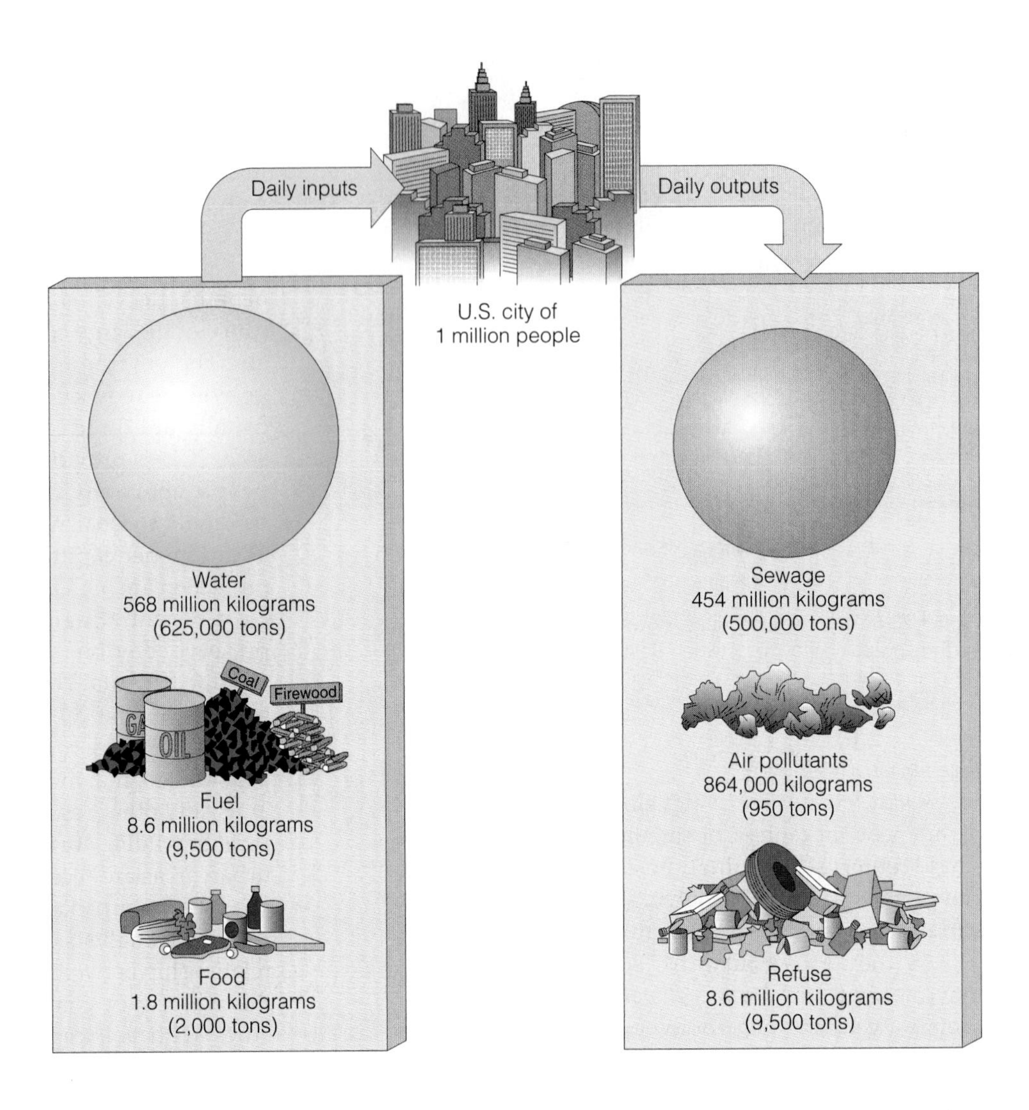

Figure 26-9 Typical daily inputs and outputs of matter and energy for a U.S. city of I million people.

this proportion could be increased. Cities can also encourage farmers' markets, which lower food prices by allowing farmers to sell directly to customers. This also helps prevent nearby farmland from being swallowed up by urban sprawl.

What Are the Water Supply Problems of Cities?

Many cities have water supply and flooding problems. As cities grow and their water demands increase, expensive reservoirs and canals must be built and deeper wells drilled. The transfer of water to urban areas deprives rural and wild areas of surface water and sometimes depletes groundwater faster than it is replenished.

Covering land with buildings, asphalt, and concrete causes precipitation to run off quickly; it can overload sewers and storm drains, contributing to water pollution and flooding in cities and downstream areas. In car-dominated cities, stormwater running off roads and parking lots is contaminated with oil, road salt, and toxic liquids; in suburbs large amounts of fertilizers and pesticides run off lawns and golf courses. Large unbroken expanses of concrete or asphalt can also prevent precipitation from entering the soil to renew groundwater.

Many cities are built on floodplain areas subject to natural flooding (Figures 13-20 and 13-21). Floodplains are considered prime land for urbanization because they are flat, accessible, and near rivers. The poor often have little choice but to live on areas experiencing frequent flooding and landslides.

Many of the world's largest cities are in coastal areas. If an enhanced greenhouse effect increases the average atmospheric temperature as projected, a rise in average sea level of even a meter or so could flood many of these cities, perhaps sometime during the 21st century.

What Are the Pollution Problems of Cities?

Urban residents are generally subjected to much higher concentrations of pollutants than are rural residents. Litter and garbage accumulate in slums and squatter villages, where solid-waste pickup services often don't exist. These conditions promote the spread of disease.

Q: What has been the warmest decade since 1860?

According to the World Health Organization, more than 1.1 billion people—about one-fifth of humanity—live in urban areas where air pollution levels exceed healthful levels. Smog is now a virtually unavoidable aspect of urban life in most of the world. Other major sources of air pollution, especially in developing countries, are smoky factories and the burning of wood, charcoal, and coal for cooking and heating.

Air pollution control in most cities in developing countries is lax because of lenient pollution laws, lack of enforcement, corrupt officials, inadequate testing equipment, and a shortage of funds. In many urban areas of developed countries, air pollution control since 1970 has helped keep levels of some air pollutants from rising. However, without improved controls a combination of more people, more cars, more factories, and other activities that burn fossil fuels could lead to rising air pollution in many of these urban areas.

Water purification and wastewater treatment plants (Figure 20-11) and fairly strict pollution control laws (Section 20-4) have reduced water pollution in most developed countries. In developing countries, however, few cities can afford to build and maintain such systems for their rapidly growing urban populations. The World Bank estimates that almost two-thirds of urban residents in developing countries don't have adequate sanitation facilities, with more than 420 million urban dwellers not having access to even the simplest outdoor latrines.

In the developing world, it is estimated that 90% of all sewage is discharged directly into rivers, lakes, and coastal waters without treatment of any kind. In Latin America, 98% of the urban sewage receives no treatment. The 50-year-old sewer system in Cairo, Egypt, built to serve 2 million people, is completely inadequate for Cairo's current population of 10 million.

According to the World Bank, at least 220 million people in the urban areas of developing countries don't have safe drinking water. Those without access to safe drinking water must buy water from vendors at 4–100 times the cost of water from a piped city supply. Many rivers and streams in such countries are essentially open sewers. Most poor people have little choice but to use them for bathing and washing clothes—and in some cases as a source of drinking water.

Air pollution is discussed in more detail in Chapter 18, water pollution in Chapter 20, and solid and hazardous wastes in Chapter 22.

How Do Urban and Rural Climates Differ?

Urbanization alters the local (and sometimes the regional) climate. Cities are generally warmer, rainier, foggier, and cloudier than suburbs and nearby rural areas. The enormous amounts of heat generated by cars, factories, furnaces, lights, air conditioners, and people in cities create an **urban heat island** (Figure 26-10) surrounded by cooler suburban and rural areas. The dome of heat also traps pollutants, especially tiny solid particles (suspended particulate matter), creating a **dust dome** above urban areas. If wind speeds increase, the dust dome elongates downwind to form a **dust plume**, which can spread the city's pollutants for hundreds of kilometers.

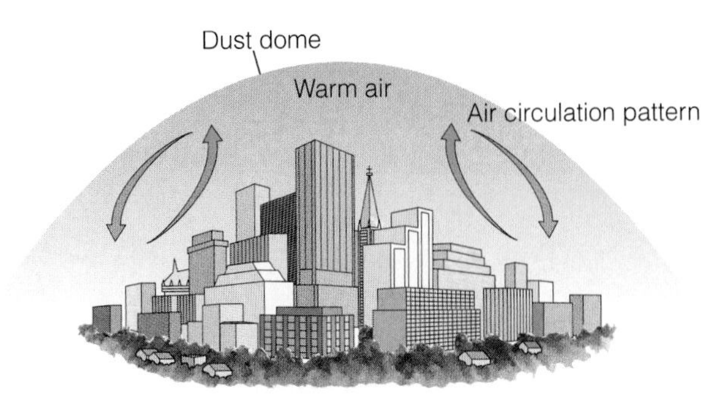

Figure 26-10 An urban heat island causes patterns of air circulation that create a dust dome over the city. Winds can elongate the dome toward downwind areas. A strong cold front can blow the dome away; this lowers local pollution levels but increases pollution in downwind areas.

As urban areas grow and merge, individual heat islands also merge, which can affect the climate of a large area and keep polluted air from being diluted and cleansed. Cities can save money and counteract the heat-island effect by instituting tree-planting programs, requiring lighter and more reflective paints and building materials, adding light-colored sand to asphalt to increase reflectivity, and establishing high energy-efficiency standards for vehicles, buildings, and appliances.

How Serious Is Urban Noise Pollution?

Most urban dwellers are subjected to excessive noise. According to the U.S. Environmental Protection Agency, nearly half of all Americans, mostly urban residents, are regularly exposed to **noise pollution**—any unwanted, disturbing, or harmful sound that impairs or interferes with hearing, causes stress, hampers concentration and work efficiency, or causes accidents. Noise is the country's most widespread occupational hazard, with about 9 million workers in the United States exposed to potentially hazardous levels of noise. Millions of people who listen to loud music using home and car stereos ("boom cars"), portable stereos ("boom boxes") held close to the ear, and earphones are also damaging their hearing.

Harmful effects from prolonged exposure to excessive noise include permanent hearing loss, high blood pressure (hypertension), muscle tension,

A: 1988 through 1998

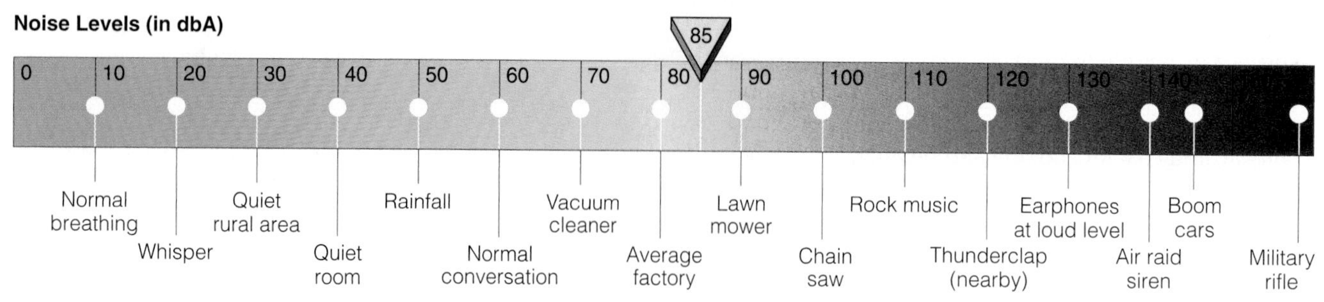

Figure 26-11 Noise levels (in decibel-A or dbA sound pressure units) of some common sounds.

migraine headaches, higher cholesterol levels, gastric ulcers, irritability, insomnia, and psychological disorders, including increased aggression.

Sound pressure is measured in decibel-A (dbA) units (Figure 26-11). Sound pressure becomes damaging at about 75 dbA and painful around 120 dbA; it can kill at 180 dbA. Because the db and dbA scales are logarithmic, sound pressure is multiplied 10-fold with each 10-decibel rise. Thus, a rise from 30 dbA (quiet rural area) to 60 dbA (normal restaurant conversation) represents a 1,000-fold increase in sound pressure on the ear.

You are being exposed to a sound level high enough to cause permanent hearing damage if you need to raise your voice to be heard above the racket, if a noise causes your ears to ring, or if nearby speech seems muffled. Prolonged exposure to lower noise levels and occasional loud sounds may not damage your hearing but can greatly increase internal stress.

There are five major ways to control noise: **(1)** Modify noisy activities and devices to produce less noise, **(2)** shield noisy devices or processes, **(3)** shield workers or other receivers from the noise, **(4)** move noisy operations or things away from people, and **(5)** use antinoise, a new technology that cancels out one noise with another.

How Does Urban Life Affect Human Health? *Urban areas have beneficial and harmful effects on human health.* Many aspects of urban life benefit human health, including better access to education, social services, and medical care. In many parts of the world urban populations live longer and have lower infant mortality rates than do rural populations.

On the other hand, high-density city life increases the spread of infectious diseases (especially if adequate drinking water and sewage systems are not available, Section 17-5), physical injuries (mostly from industrial and traffic accidents), and health problems caused by increased exposure to pollution and noise.

About 220 million people in cities in developing countries don't have safe drinking water. Where the poor are forced to use contaminated water, diarrhea, dysentery, typhoid, and cholera are widespread, and infant mortality is high. The World Health Organization estimates that 600 million urban dwellers in developing countries live or work in life- and health-threatening environments.

How Does Urban Growth Affect Nearby Rural Land and Small Towns? Another problem is the *loss of rural land, fertile soil, and wildlife habitats as cities expand.* As urban areas expand they swallow up rural land, especially flat or gently rolling land with well-drained, fertile soil. Each year in the United States about 526,000 hectares (1.3 million acres) of rural land—mostly prime cropland and forestland—is converted to urban development, rights-of-way, highways, and airports. This is equivalent in area to building a 1-kilometer-wide highway between New York City and Los Angeles each year. In 1952 Los Angeles was the top-producing county in the United States. Today, 70% of its land is devoted to cars. According to a 1997 study by the American Farmland Trust, the United States may lose 13% of its prime farmland by 2050.

Once prime agricultural land or forestland is paved over or built upon, it is lost for food production and habitat for most of its former wildlife. As a city expands, more energy is needed to transport food to its people; this in turn causes more pollution. In coastal areas, urban growth destroys or pollutes ecologically valuable wetlands.

As land values near urban areas rise, taxes on nearby farmland increase so much that many farmers are forced to sell their land. They can make much more money selling to developers than raising corn or cows.

Gardner, G. 1998. "When Cities Take Bicycles Seriously." *World Watch*, vol 11, no. 5, 16(7).

The outward expansion of cities creates numerous problems for the nearby rural areas. Narrow country roads and small-town streets become congested with traffic. Town and county health, school, police, fire, water, sanitation, and other services are overwhelmed. Air and water pollution, crime, noise, and congestion increase.

Suburbanized towns and counties must raise taxes to meet the demand for new public services. Some townspeople profit from the increased economic growth, and new arrivals from urban areas escape some of the inner-city problems they once faced. However, other long-time small-town residents are forced out because of rising prices, higher property taxes, decreased environmental quality, and disruption of their way of life. Unless growth is carefully controlled (which is rare), old and new residents eventually experience the inner-city problems they sought to avoid—another example of the harmful results of positive feedback.

26-3 TRANSPORTATION AND URBAN DEVELOPMENT

How Do Transportation Systems Affect Urban Development? If a city cannot spread outward, it must grow vertically—upward and downward (below ground)—so that it occupies a small land area with a high population density. Most people living in such compact cities walk, ride bicycles, or use energy-efficient mass transit. Residents often live in multistory apartment buildings; with few outside walls in many apartments, heating and cooling costs are reduced. Many European cities and urban areas such as Hong Kong and Tokyo are compact and tend to be more energy-efficient than the dispersed cities in the United States, Canada, and Australia, where ample land is often available for outward expansion.

A combination of cheap gasoline (Figure 16-6), plentiful land, and a network of highways produces dispersed, automobile-oriented cities with low population density, often called *urban sprawl*. Most people living in such urban areas live in single-family houses with unshared walls that lose and gain heat rapidly unless they are well insulated and airtight. Urban sprawl also gobbles up unspoiled natural habitats, paves over fertile farmland, and promotes heavy dependence on the automobile.

In the United States urban sprawl from the spread of suburbs is encouraged by both deliberate and unintended government subsidies. One involves using federal funds to build most of the country's highways, which is the major factor in opening up new land for development. In addition, sprawl causes the cost of most government services to rise. Sewer lines must be longer, new schools must be built, school buses must travel further, and more fire stations and roads are needed. Residents of low-density developments pay taxes, of course, but they rarely are charged the full cost of the government services they use.

Who Has Most of the World's Motor Vehicles? There are two main types of ground transportation: *individual* (such as cars, motor scooters, bicycles, and walking) and *mass* (mostly buses and rail systems). About 89% of the world's 501 million cars and trucks are in developed countries. The new cars produced each day worldwide could form a traffic jam 604 kilometers (375 miles) long.

Despite such production, only about 8% of the world's population own cars, and only 10% can afford to. In developing countries as few as 1% of the people can afford a car; they travel mostly by foot, bicycle, or motor scooter. However, because of projected increases in economic growth, the world is projected to have almost 1 billion motor vehicles by 2020 (especially in Asia, Latin America, and Eastern Europe), greatly increasing congestion, pollution, land disruption, and use of energy and matter resources. There are already 4.5 million cars in São Paulo, Brazil, more than twice the 2.1 million in New York City (which has about the same number of people). The number of cars in São Paulo is increasing by about 1,000 a day.

Despite having only 4.6% of the world's people, the United States has 35% of the world's cars and trucks. In the United States the car is used for 98% of all urban transportation and for 86% of travel to work (with 73% of Americans driving to work alone). Currently, the number of Americans commuting from suburb to suburb is more than double the number commuting from suburb to downtown, with suburb-to-suburb trips now accounting for 44% of all commuting. This trend makes it difficult for workers in a suburban neighborhood to use carpools or mass transit because they are typically headed in a number of directions instead of downtown.

Americans drive 3 billion kilometers (2 billion miles) each year—as far as the rest of the world combined. No wonder British author J. B. Priestley remarked, "In America, the cars have become the people." Such "automania" is spreading to developing countries. Despite their many advantages, there are serious drawbacks to relying on motor vehicles as the major form of transportation (Pro/Con, p. 732).

Hint: Enter the search term *transportation planning* using the Subject Guide.

Good and Bad News About Motor Vehicles

The automobile provides convenience and unprecedented mobility. To many people, cars are also symbols of power, sex, excitement, social status, and success. Moreover, much of the world's economy is built on producing motor vehicles and supplying roads, services, and repairs for them. In the United States, $1 of every $4 spent and one of every six nonfarm jobs is connected to the automobile.

Despite their benefits, motor vehicles have many destructive effects on people and the environment. Since 1885, when Karl Benz built the first automobile, almost 18 million people have been killed in motor vehicle accidents. According to the World Health Organization, this global death toll increases by an estimated 885,000 people per year (an average of 2,400 deaths per day—equal to 8 fatal jumbo jet crashes each day), and annually about 15 million people are injured or permanently disabled in motor vehicle accidents.

In the United States alone, 16 million motor vehicle accidents (up from 7 million in 1970) kill about 42,000 people each year and injure about 4 million people, at least 300,000 of them severely. *More Americans have been killed by cars than by all the country's wars.* In 1994 the costs from motor vehicle accidents in the United States exceeded $150.5 billion—about $574 per citizen.

Motor vehicles are also the largest source of air pollution (including 15% of global CO_2 emissions), laying a haze of smog over the world's cities. In the United States, motor vehicles produce at least 50% of the air pollution, even though emission standards are as strict as any in the world. Gains in fuel efficiency and emission reductions have been largely off-set by the increase in cars and a more than doubling of the distance Americans traveled by car between 1970 and 1997. Two-thirds of the oil used in the United States and one-third of the world's total oil consumption are devoted to transportation.

By making long commutes and shopping trips possible, automobiles and highways have helped create urban sprawl and have reduced use of more efficient forms of transportation. Worldwide, at least a third of urban land is devoted to roads, parking lots, gasoline stations, and other automobile-related uses. In the United States, more land is now devoted to cars than to housing. Half the land in an average U.S. city is used for cars, prompting urban expert Lewis Mumford to suggest that the U.S. national flower should be the concrete cloverleaf.

Car-culture cities have not delivered the promised convenience and speed of travel. In 1907 the average speed of horse-drawn vehicles through the borough of Manhattan was 18.5 kilometers (11.5 miles) per hour; today cars and trucks creep along Manhattan streets at an average speed of 5 kilometers (3 miles) per hour. If current trends continue, U.S. motorists will spend an average of 2 years of their lifetimes in traffic jams, imprisoned in metal boxes that were supposed to provide speed, freedom, and mobility. The U.S. economy loses at least $100 billion a year because of time lost in traffic delays. In London, the average speed of a car today is about the same as that of a horse-drawn carriage in 1900.

Bangkok, Thailand, has become known for its air pollution (Figure 11-1) and infamous traffic jams, where at peak hours speeds average a little over 1 kilometer (0.6 miles) per hour—about half as fast as a leisurely walk. It can take 3–6 hours to get to the airport in Bangkok. The average car in Bangkok spends the equivalent of 44 days per year stuck in traffic.

Even if the money is available, building more roads is not the answer because, as economist Robert Samuelson put it, "cars expand to fill available concrete." According to the U.S. General Accounting Office, at the current rate of growth traffic congestion on U.S. roads will triple in 15 years, even if road capacity is increased by 20%.

Critical Thinking

If you own a car (or hope to own one), what conditions (if any) would encourage you to rely less on the automobile and instead encourage you to travel to school or work by bicycle or motor scooter, on foot, by mass transit, or by a car or van pool?

Are Motor Scooters the Answer? A growing number of people in developing countries who cannot afford cars are using motor scooters, which produce more air pollution than cars. Most burn a mixture of oil and kerosene in small, inefficient, and noisy engines that emit clouds of air pollutants. Because they are cheap, their numbers are increasing three times faster than cars and trucks in developing countries.

One solution is to replace these smog machines with zero pollution (except at power plants supplying the electricity for recharging the batteries) and quiet electric scooters. Recently Taiwan and Indonesia have introduced air pollution control legislation that may spur the use of electric scooters. The major weakness of electric cars is their limited range. However, this is less of a problem with much lighter electric scooters.

Are Riding Bicycles and Walking Alternatives to the Car? Globally, bicycles outsell cars by almost 3 to 1 because most people can afford a bicycle whereas fewer than 10% can afford a car. Besides being inexpensive to buy and maintain, bicycles produce no pollution, are rarely a serious danger to pedestrians or cyclists, take few resources to make, and are the most energy-efficient form of transportation (including walking).

In 1996 a California firm developed a simple $15 bicycle that eliminates the chain by attaching the pedals to the front wheel. This inexpensive design could greatly increase bicycle use in developing countries.

In urban traffic, cars and bicycles move at about the same average speed. Using separate bike paths or lanes running along roads, cyclists can make most trips shorter than 8 kilometers (5 miles) faster than drivers.

In China, at least 50% of urban trips are made by bicycle and the government gives subsidies to those who bicycle to work. Many cities in western Europe and Japan have taken back the streets for pedestrians, cyclists, and children by banning cars or slowing motor traffic in residential and shopping areas. In Copenhagen, Denmark, bike lanes enable residents to make 25% of all urban trips by cycling, a figure that increases during summer. People there can rent bikes from special token-fed racks; when a bike is returned to any rack, the renter gets a full refund. Studies in the United States, the United Kingdom, and Germany have shown that creating automobile-free zones in city centers increases local business sales by 25% or more. An increasing number of European cities are closing off large downtown areas to all motor vehicles.

For longer trips, secure bike parking spaces can be provided at mass transit stations, and buses and trains can be equipped to carry bicycles. Such *bike-and-ride* systems are widely used in Japan, Germany, the Netherlands, and Denmark. In Seattle, Washington, all 1,250 city buses are equipped with bicycle racks. In the Netherlands bicycle travel makes up 30% of all urban trips, and in Japan 15% of all commuters ride bicycles to work or to commuter-rail stations. Many train stations in Japan have high-rise parking towers with lifts for bicycles. In Canada, the province of Quebec is constructing a huge network of bikeways within and between cities. In Paris bikeways now line most of the major boulevards.

In the car-dependent, dispersed urban areas of the United States and Australia only about 1% of all trips are made by bicycle. However, bike-and-ride commuting is beginning to catch on in Atlanta, Boston, Milwaukee, San Francisco, Seattle, and Washington, D.C. So far, however, only a handful of U.S. cities—such as Davis, California (p. 720); Seattle; Eugene, Oregon; Boulder, Colorado; and Palo Alto, California—actively encourage bicycling.

Only about 2% of commuters in the United States bicycle to work (the "no-pollute commute"), even though half of all U.S. commutes are under 8 kilometers (5 miles). However, according to recent polls, 20% of Americans say they would bicycle to work if safe bike lanes were available and if their employers provided secure bike storage and showers at work.

As developed countries embrace the bicycle as an ecologically sustainable form of transportation, the reverse may happen in developing countries. As such countries experience economic growth, some governments are discouraging bicycle use, viewing it as a sign of backwardness. In 1993, for example, Chinese officials banned bicycles from Shanghai's main street and prestigious shopping area so they wouldn't be seen by tourists and wealthy Chinese shoppers. Similarly, officials in Jakarta, Indonesia, have confiscated thousands of cycle rickshaws to project a more modern image for visitors.

In addition, unless efficient and affordable forms of mass transportation are already in place, many urban residents in developing countries abandon bikes and walking as soon as they can afford to buy motorscooters or cars, as is happening in China (currently, the world's most bicycle-centered country), India, and Indonesia. Even in the Netherlands, which has a long tradition of bicycle use and ample bicycle paths, more and more people are using cars instead of bicycles.

One development that may increase bike use is the explosive growth in the use of electric bicycles. Battery-powered bikes allow cyclists to go as fast as 24 kilometers per hour (15 miles per hour) without pedaling and much faster if both pedaling and electric power are used. Electric bikes can extend the range of bike trips, allow cyclists to travel over hilly terrain, and may increase the number of people using bicycles for commuting. Electric bikes can also replace mopeds and motor scooters, which are noisy, highly polluting, and more costly.

Case Study: Mass Transit in the United States Another alternative to the car is *mass transit*. In the United States mass transit accounts for only 3% of all passenger travel, compared with 15% in Germany and 47% in Japan.

In 1917, all major U.S. cities had efficient electric trolley or streetcar systems. Many people think of Los Angeles as the original car-dominated city, but early in this century Los Angeles had the largest electric rail mass transit system in the United States.

By 1950, a holding company called National City Lines (formed by General Motors, Firestone Tire, Standard Oil of California, Phillips Petroleum, and Mack Truck, which also made buses) had purchased

A: Shifts in food-growing regions and water supplies, rising sea levels, and decreased biodiversity privately owned streetcar systems in 83 major cities. It then dismantled these systems in order to increase sales of buses and cars. The courts found the companies guilty of conspiracy to eliminate the country's light-rail system, but the damage had already been done. The executives responsible were fined $1 each, and each company paid a fine of $5,000, less than the profit returned by replacing a single streetcar with a bus. General Motors alone had made $25 million in additional bus and car sales by the time the case was tried. Rebuilding these dismantled streetcar (light-rail) systems today would cost at least $300 billion.

During this same period National City Lines worked to convert electric-powered commuter locomotives to much more expensive and less reliable diesel-powered locomotives. The resulting increased costs contributed significantly to the sharp decline of the nation's railroad system.

In the United States, only 20% of the federal gasoline tax goes to mass transit; the remaining 80% goes to highways. This encourages states and cities to invest in highways instead of mass transit.

The federal tax code also penalizes mass transit users and those who cycle or walk to work. In the United States only 10% of commuting employees pay for parking, mainly because employers can deduct from their taxes the expense of providing parking for workers. This gives such auto commuters a tax-free fringe benefit worth $200–400 a month in major cities. On the other hand, employers can write off only about $15 a month for employees who use public transit, and nothing for those who walk or bike.

In downtown Los Angeles, employer-paid parking increases drive-alone commuting by an estimated 44%. One remedy for this situation would be for the government to provide a general transportation tax deduction to employers for their employees. Instead of getting free parking, all employees would receive a certain amount of money per month that could be used for any type of transportation to work.

What Are the Pros and Cons of Rail Systems?

Rail systems, usually operated by electric engines, fall into three categories: *rapid rail* (also called the underground, tube, metro, or subway), which operates on exclusive rights-of-way in tunnels or on elevated tracks; *suburban or regional trains*, which connect the central city with surrounding areas or provide transportation between major cities in a region; and *light rail (such as trolleys) or trams*, more modern versions of streetcars, which can run either with other traffic or on exclusive rights-of-way.

Rail systems have a number of advantages over highway and air transport. They are much more energy-efficient, produce less air pollution, cause fewer injuries and deaths, and take up less land. They are also available to people who are too young, old, or disabled to drive or can't afford a car. However, such train systems are efficient only where many people live along a narrow corridor and can easily reach properly spaced stations.

In recent years some U.S. cities have built successful rapid-rail systems. Since Atlanta's system opened in 1979, it has steadily added riders and opened new stations; it has also attracted $70 billion worth of new apartments, office buildings, and other facilities to the vicinity of the rapid-rail lines in an otherwise sprawling city. Pittsburgh now has cleaner air and renewed business vitality, partly because of its new rapid-rail system, which opened in 1985.

One of the world's most successful rapid-rail systems is in Hong Kong. Several factors contribute to its success. Hong Kong is densely populated, making it ideal for a rapid-rail system running through its corridor, and half of the population can walk to a subway station in 5 minutes. Also, most people depend on public transportation because only one person in 30 owns a car, which is an economic liability in this crowded city even for those who can afford one.

Over the past two decades, 21 large cities in developing countries, including Cairo, Shanghai, and Mexico City, have built rapid-rail systems that have improved transport service in dense city centers, but often at great cost. Critics believe that this money could have been better spent expanding and modernizing bus systems that could carry many more people at a much lower cost.

In the United States 21 cities—including San Diego, Sacramento, San Jose, Los Angeles, Seattle, Buffalo, and Portland (Oregon)—have built light-rail systems and seven more cities are planning to build them. In Portland, 43% of all people commuting downtown ride buses and a light-rail system, a shift to mass transit that has dramatically improved the city's air quality.

Three Canadian cities—Toronto, Edmonton, and Calgary—have recently built light-rail systems. In Toronto more than three-fourths of downtown commuters use public transportation to get to work. Zurich, Switzerland, has a highly efficient light-rail system and new lines are being planned in a number of European cities, including London, Dublin, Bordeaux, and Stockholm.

A light-rail line costs about one-tenth as much to build per kilometer as a highway or a heavy-rail system. Although the start-up cost of a light-rail system is higher than for a comparable bus system, its operating costs are much lower. By linking trolley cars together, a light-rail system can carry up to 400 people for each driver, compared with 40–50 passengers on a typical bus. Trolleys are also cleaner and quieter than buses.

Q: What are the major ways to reduce the threat of global warming?

And unlike rapid rail, light rail does not require an exclusive right-of-way; it can run down the side or middle of roads, utility corridors, or even alleys.

What Are the Pros and Cons of High-Speed Regional Trains? In western Europe and Japan, a new generation of streamlined, comfortable, and low-polluting high-speed rail (HSR) lines is being used for medium-distance travel between cities. These *bullet* or supertrains travel on new or upgraded tracks at speeds up to 330 kilometers (200 miles) per hour. They are ideal for trips of approximately 200–1,000 kilometers (120–620 miles). For every kilometer of travel, such trains consume only one-third as much energy per rider as a commercial airplane and one-sixth as much as a car carrying only one driver. High-speed train systems are expensive to run and maintain, however, and they must operate along heavily used transportation routes to be profitable.

In Japan, bullet trains began operating in 1964. They have carried over 3 billion passengers between Tokyo and Osaka without a single injury or fatality, and they depart and arrive on schedule 99% of the time. Recently, 12 European countries agreed to spend $76 billion to link major cities with nearly 30,000 kilometers (18,600 miles) of high-speed rail lines. Within the next decade passengers in Australia, Taiwan, China, Pakistan, and Brazil will also be able to ride new high-speed trains between cities in less time than it takes by plane. The United States has lagged behind in the development of a high-speed train network. Such a system could be developed at a reasonable cost by upgrading existing intercity tracks and train systems on key routes (Figure 26-12), rather than building new and expensive rail rights-of-way. Criticism that such a system would cost too much in government subsidies ignores the fact that motor vehicle transportation receives subsidies of $300–600 billion per year in the United States. Some analysts believe that phasing out some of these subsidies and applying them to developing a national rail system might be a far more efficient use of increasingly limited government funds. What do you think?

A second type of high-speed rail for travel between cities, the magnetic levitation (MAGLEV) train, is still under development. It uses powerful superconducting electromagnets built into the track to suspend the train on a cushion of air a few centimeters above a guiding rail. Such trains zoom along without rail friction at speeds up to 400 kilometers (250 miles) per hour. Because these trains never touch the track, they make little noise, require little maintenance, and can be elevated over existing median strips or other rights-of-way along highways, thereby avoiding disruptive and costly land acquisitions.

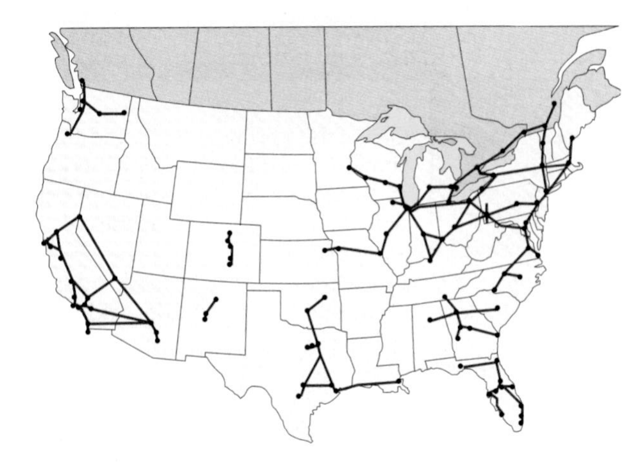

Figure 26-12 Potential routes for high-speed bullet trains in the United States and parts of Canada. Such a system would allow rapid, comfortable, safe, and affordable travel between major cities in a region. It would greatly reduce dependence on cars, buses, and airplanes for trips among these urban areas. (Data from High-Speed Rail Association)

Germany and Japan have been developing prototype MAGLEV trains for 20 years, but so far no passenger service is operating. Concerns over economic feasibility because of high construction and maintenance costs, safety (there was a fire on a prototype model in Japan), and the possibility of harm to passengers from exposure to high electromagnetic fields have held back development of MAGLEV. Critics say the system is too costly and that investment in high-speed bullet trains makes more sense. Over a typical 500-kilometer (310-mile) run a typical MAGLEV train might save only 30–60 minutes compared to a typical bullet train.

What Are the Pros and Cons of Buses? Bus systems are more flexible than rail systems; they can run throughout sprawling cities and be rerouted overnight if transportation patterns change. Bus systems require less capital and have lower operating costs than heavy-rail systems. In Canada, the city of Ottawa, Ontario, is developing an extensive bus system instead of a subway system because of its much lower cost and greater flexibility in serving medium- and low-density urban areas. Curitiba, Brazil, has developed one of the world's best bus systems (Solutions, p. 736).

However, because they must offer low fares to attract riders, bus systems often cost more to operate than they bring in. To reduce losses, bus companies tend to cut service, skimp on maintenance, and seek government subsidies. Furthermore, unless they operate in separate express lanes, buses often get caught in traffic.

Because buses are cost-effective only when full, they are sometimes supplemented by more flexible carpools, van pools, jitneys (small vans or minibuses

A: Improve energy efficiency, shift from fossil fuels to renewable energy, and slow deforestation

Curitiba, Brazil

SOLUTIONS

One of the world's showcase ecocities is Curitiba, Brazil, with a population of 2.2 million. It is a prosperous, booming, clean, and mostly economically self-sufficient city.

Trees are everywhere in Curitiba because city officials have given neighborhoods over 1.5 million trees to plant and care for. No tree in the city can be cut down without a permit; for every tree cut, two must be planted.

After persistent flooding in the 1950s and 1960s, city officials put certain flood-prone areas off-limits for building, passed laws protecting natural drainage systems, built artificial lakes to contain floodwaters, and converted many riverbanks and floodplains into parks. This design-with-nature strategy has made costly flooding a thing of the past and has greatly increased the amount of open or green space per capita (which increased 100-fold during the city's period of rapid population growth between 1950 and 1998).

The air is clean because the city is not built around the car. There are 145 kilometers (90 miles) of bike paths, and more are being built. With the support of shopkeepers, many streets in the downtown shopping district have been converted to pedestrian zones in which no cars are allowed. Abandoned factories and buildings have been recycled into sports and other recreational facilities.

A key feature of Curitiba's success is its integration of transportation and land use planning. City officials decided to develop a sophisticated bus system instead of a more expensive and less flexible subway or light-rail system.

The key concept was to channel the city's development along five major transportation and high-density corridors that extend outward from the central city like spokes on a bicycle wheel. Each corridor has exclusive lanes for express buses.

Curitiba probably has the world's best bus system. Each day a network of clean and efficient buses carries over 1.5 million passengers—75% of the city's commuters and shoppers—at a low cost (20–40¢ per ride with unlimited transfers) on express bus lanes. Only high-rise apartment buildings are allowed near major bus routes, and each building must devote the bottom two floors to stores, which reduces the need for residents to travel. Since 1974 traffic has declined by 30% while the city's population doubled.

At peak times the city also runs double- and triple-length buses. This self-financing bus system cost the city $200,000 per kilometer to build, compared to $60–70 million per kilometer for a subway system. Instead of being discarded, old buses are used as mobile schools and to provide transportation to parks. As a result, Curitiba's gasoline use per person is 25% than that of eight comparable Brazilian cities, and the city has one of Brazil's lowest outdoor air pollution rates.

The city recycles roughly 70% of its paper (equivalent to saving nearly 1,200 trees per day) and 60% of its metal, glass, and plastic, which is sorted by households for collection. Recovered materials are sold mostly to the city's more than 340 major industries. The city's Free University for the Environment offers practical short courses at no cost to sensitize people to environmental concerns by teaching the environmental effects of even the most commonplace jobs.

The city bought a plot of land downwind of downtown as an industrial park. The city put in streets, services, housing and schools, ran a special worker's bus line to the area, and enacted stiff air and water pollution control laws. This has attracted clean national and foreign corporations.

To equip the poor with basic technical training skills needed for jobs, the city set up old buses as roving technical training schools; courses cost the equivalent of two bus tokens. Each bus gives courses for 3 months in a particular area and then moves to another area.

The poor can swap sorted trash for food or bus tokens and receive free medical, dental, and child care; there are also 40 feeding centers for street children. As a result, the infant mortality rate has fallen by more than 60% since 1977.

All of these things have been accomplished despite Curitiba's enormous population growth from 300,000 in 1950 to 2.2 million in 1998, as rural poor people have flocked to the city.

Curitiba has slums, shantytowns, and most of the problems of other cities. But most of its citizens have a sense of vision, solidarity, pride, and hope, and they are committed to making their city even better.

One secret of Curitiba's success is the willingness of citizens to work together to create a better future. Another is city officials who genuinely care about providing a high quality of life for *all* of the city's inhabitants. The entire program is the brainchild of architect and former college teacher Jaime Lerner, an energetic and charismatic leader, who has served as the city's mayor three times since the 1970s.

City leaders and citizens have worked together to make Curitiba a living laboratory for sustainable living by favoring public transportation over the private automobile, planning carefully, and working with the environment instead of against it.

Critical Thinking

1. Why do you think Curitiba has been so successful in its efforts to become an ecocity, compared to the generally unsustainable cities found in most developed and developing countries?

2. What is the city or area where you live doing to make itself more ecologically and economically sustainable?

Q: By what percentage must global CO_2 emissions be cut by 2030 to stabilize CO_2 at current levels?

that run along regular and stop-on-demand routes), and dial-a-ride systems. In dial-a-ride systems, passengers (most without a car or access to mass transportation) call for a van, minibus, or tax-subsidized taxi, which picks them up at their doorstep, usually in 20–50 minutes. Two-way radios and computerized routing can increase the efficiency of these systems, but the rather expensive dial-a-ride systems are the least efficient mass transportation system in terms of energy costs per passenger-kilometer.

In Mexico City, Caracas, and Cairo, large fleets of jitneys carry millions of passengers each day. Laws banning jitney service in the United States were repealed in 1979, despite objections by taxi and transit companies. Since then privately owned jitney services have flourished in San Diego, San Francisco, and Los Angeles and may spread to other cities.

Is It Feasible to Reduce Automobile Use? The major hidden costs of driving include deaths and injuries from accidents, the value of time wasted in traffic jams, air pollution, increased threats from global warming, the drop in property values near roads because of noise and congestion, and the cost of maintaining a formidable military presence in the Middle East to ensure access to oil. Two recent estimates by economists put these hidden costs of driving in the United States at roughly $300–350 billion per year—about 5% of the country's GDP.

Environmentalists and a number of economists suggest that one way to break this increasingly destructive cycle of positive feedback is to make drivers pay directly for most of the true costs of automobile use. This could be done by including the current harmful hidden costs of driving as a tax on gasoline and by phasing out government subsides for motor vehicle owners, a *user-pays* approach.

Currently, U.S. drivers pay low gasoline taxes (compared to most other developed countries) that are used to build roads and other transportation infrastructures, but these taxes cover only 60–69% of the total costs. The remainder is subsidized by federal, state, and local governments. Governments also subsidize drivers by allowing tax write-offs for business-related car mileage or other car-related costs. U.S. employers who provide parking facilities for their workers can deduct these expenses from their taxes. This gives U.S. auto commuters a tax-free benefit worth $2,400–4,800 per year.

Making heavy trucks pay for the road damage they cause would shift more freight to energy-efficient rail systems. It's estimated that heavy trucks cause 95% of all damage to U.S. highways, with one heavy truck causing as much highway wear and tear as 9,600 cars. Current U.S. government subsidies for trucks not only drain public funds but also give trucking an unfair economic advantage over more efficient and less damaging rail freight.

According to a study by the World Resources Institute, federal, state, and local government automobile subsidies in the United States amount to $300–600 billion a year (depending on the costs included), an average subsidy of $1,600–3,200 per vehicle. Taxpayers (drivers and nondrivers alike) foot this bill mostly unknowingly. If drivers had to pay these hidden costs directly in the form of a gasoline tax, the tax on each gallon would be about $5–7. Pollution emission fees, toll charges on roads (especially during peak traffic times), and higher parking fees are also ways to have motorists pay for the environmental and social costs of driving—a user-pays approach.

Deciding to include the hidden costs in the market prices of cars, trucks, and gasoline up front makes economic and environmental sense. However, this approach faces massive political opposition from the public (mostly because they are unaware of the huge hidden costs they are already paying) and from the powerful transportation-related industries. In addition, such tools for encouraging more use of mass transportation will not work unless fast, efficient, reliable, and affordable transportation alternatives are available. Furthermore, most people who can afford cars are virtually addicted to them and most people who can't afford them hope to buy one someday.

Developing countries now have a unique economic and environmental opportunity to avoid the "car trap" by not subsidizing motor vehicle transport. Instead, they can invest most of their money in developing modern, efficient, low-polluting public transit systems and providing city-wide paths for bicycling and walking (Solutions, left).

26-4 URBAN LAND-USE PLANNING AND CONTROL

What Is Conventional Land-Use Planning? Most urban areas and some rural areas use some form of **land-use planning** to determine the best present and future use of each parcel of land in the area. Zoning regulations or other means are then used to control how the land is used.

Because land is such a valuable economic resource, land-use planning is a complex and controversial process involving competing values and intense power struggles. Much land-use planning is based on the assumption that substantial future population growth and economic development should be encouraged, regardless of the environmental and other consequences. Typically this leads to uncontrolled or poorly controlled urban growth and sprawl.

A major reason for this often destructive process is that in the United States 90% of the revenue that local

A: 66–83%

SPOTLIGHT

GIS: A New Tool for Urban Planning

"The problems we are solving with GIS could not have been solved using non-computerized mapping techniques. Unfortunately many urban planners don't realize the potential," says Santa Cruz County, California, principal planner Gale Conley.

Since 1991 Santa Cruz County has been committed to building a geographic information system (GIS) (Figure 4-26). It was one of the first counties in the United States to develop GIS capabilities and its experience illustrates the benefits and problems of investing in this expensive and complex technology.

The Santa Cruz County GIS did not emerge overnight. Feasibility planning started in 1985. In 1991 custom GIS software was installed on the existing mainframe computer, and the first maps were produced by a professional firm contracted to scan existing maps and place them in a common coordinate system using a technique called rubbersheeting.

The first application was to generate mailing lists for notification of public hearings concerning permit applications and requests for land divisions. In the past generating mailing lists was an arduous and time-consuming process. With GIS such mailing lists can be generated in less than a minute. Soon the county began using GIS for other routine mapping tasks involving queries about parcel identification, ownership, zoning, land use, and location of associated planning records.

"This was a simple beginning. But it represented a huge savings by our county government," says Gale Conley.

The current Santa Cruz County database has maps of approximately 92,000 land parcels, along with approximately 100 layers of other geographic information, including land use (residential, business, agriculture, parks, greenbelt), political and tax jurisdictions, population census data, rare and endangered species locations, and watershed characteristics.

In recent years the GIS database has been expanded to include many environmental map layers, including maps of critical fire hazards (steep slope and combustible vegetation), water supply and watershed boundaries, streams, groundwater recharge, biological resources, and archaeological sites. Applications are continually expanding and include analyzing soils and geology to assess development proposals, mapping election districts, mapping police beats, siting health facilities, compiling census data, locating facilities along bus routes, and protecting greenbelts.

Most of the maps are made available to the public. Some, such as maps of rare and endangered species, are no longer released, to prevent deliberate damage to plant and animal populations. The GIS has been used primarily to produce informational and analytical reports for the county administrators, the planning commission, and the planning board. The county also provides map information based

governments use to provide schools, police and fire protection, public water and sewer systems, and other public services comes from *property taxes* levied on all buildings and property based on their economic value. When an area is economically developed, property values—and thus local tax revenues—go up.

However, the costs of providing more basic services to accommodate economic and population growth often exceed the tax revenues accompanying increased property values. Because local governments can rarely raise property taxes enough to meet expanding needs, they often try to raise more money by promoting economic growth. Typically the long-term result is a destructive positive feedback loop of economic growth leading to environmental degradation. If taxes get too high, businesses and residents move away, decreasing the tax base and reducing tax revenues. This causes further environmental and social decay as governments are forced to cut the quantity and quality of services or to raise the tax rate again (which drives more people away and further worsens the situation).

What Is Ecological Land-Use Planning? Environmentalists challenge the all-growth-is-good dogma that is the basis of most land-use planning. They urge communities to use comprehensive, regional **ecological land-use planning**, in which additional variables are integrated into a model designed to anticipate a region's present and future needs and problems. It is a complex process that takes into account geological, ecological, economic, health, and social factors. Six steps are involved:

1. *Make an environmental and social inventory.* Experts survey geological factors (soil type, floodplains, water availability), ecological factors (wildlife habitats, stream quality, pollution), economic factors (housing, transportation, industrial development), and health and social factors (disease, crime rates, poverty). A top priority is to identify and protect areas that are critical for preserving water quality, supplying drinking water, and reducing erosion, and that are most likely to suffer from toxic wastes and natural hazards such as flooding.

Q: Globally, how many trees per person would have to be planted each year to absorb the CO_2 we put into the atmosphere?

Figure 26-13 In 1997 Santa Cruz County used its GIS to help a citizen group, the Arana Gulch Watershed Alliance, to develop maps to provide baseline information in developing a resource management plan. This effort involved calculation of runoff into Monterey Bay based on analysis of land use in the watershed.

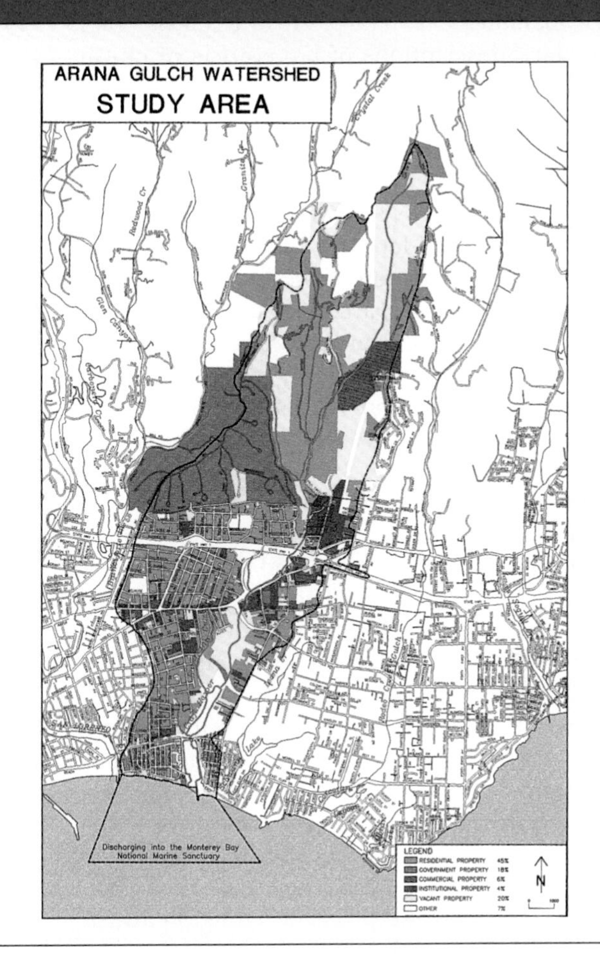

on requests by individuals and citizen groups (Figure 26-13).

On the negative side, GIS has proven to be expensive to develop and maintain, data quality is only as good as the source maps, the technology changes so rapidly that current investment is quickly outmoded, and precautions must be taken to ensure the security of data from accidental or intentional corruption. GIS is not the solution to every problem. However, GIS has proven to be a powerful and useful urban planning tool.

Critical Thinking

Do you believe that using GIS for urban planning is worth the effort and expense? Explain. Is such a system being used where you live? If so, how effective has it been?

2. *Identify and prioritize goals.* What are the primary goals? To encourage or discourage further economic development (at least some types) and population growth? To preserve prime cropland, forests, and wetlands from development? To reduce soil erosion?

3. *Develop individual and composite maps.* Data for each factor surveyed in the environmental and social inventory are plotted on separate transparent plastic maps. The transparencies are then superimposed or combined by computer into three composite maps, one each for geological, ecological, and socioeconomic factors (Spotlight, above).

4. *Develop a master composite.* The three composite maps are combined to form a master composite, which shows how the variables interact and indicates the suitability of various areas for different types of land use.

5. *Develop a master plan.* The master composite (or a series of alternative master composites) is evaluated by experts, public officials, and the public, and a final master plan is drawn up and approved.

6. *Implement the master plan.* The plan is set in motion, monitored, updated, and revised as needed by the appropriate government, legal, environmental, and social agencies.

Ecological land-use planning sounds good on paper, but it is not widely used for several reasons. First, local officials seeking reelection every few years usually focus on short-term rather than long-term problems and can often be influenced by economically powerful developers. Second, officials (and often the majority of citizens) are often unwilling to pay for costly ecological land-use planning and implementation, even though a well-designed plan can prevent or ease many urban problems and save money in the long run. Finally, it's difficult to get municipalities within a region to cooperate in planning efforts. As a result, an ecologically sound development plan in one area may be undermined by unsound development in nearby areas.

Most cities and counties in the United States have initiated efforts to convert all of the planning maps into digital form using geographical information system (GIS) technology (Figure 4-26, Spotlight, above).

A: About 1,000 (4,500 per American because of their larger input)

In developing countries, most cities don't have the information or funding to carry out ecological land use planning. Urban maps are often 20–30 years old and lack descriptions of large areas of cities, especially those growing rapidly because of squatter settlements.

How Can Land Use Be Controlled? Once a plan is developed, governments control the uses of various parcels of land by legal and economic methods. The most widely used approach is **zoning**, in which various parcels of land are designated for certain uses. Principal zoning categories include commercial, residential, industrial, utilities, transport, recreation (parks and forest preserves), bodies of water, wetlands, floodplains, and wildlife preserves.

Zoning can be used to control growth and protect areas from certain types of development. Some cities, such as Portland, Oregon, and Curitiba, Brazil (Solutions, p. 736), have used zoning to encourage high-density development along major mass transit corridors to reduce automobile use and air pollution. Portland has used zoning to create large areas of green space (Figure 26-14). To reduce auto use and the costs of providing services for cars, the city has **(1)** developed an efficient mass transportation system, **(2)** used zoning to encourage high-density development along major transit lines, **(3)** allowed mixed development of offices, shops, and residences in the same area, **(4)** and placed a ceiling on downtown parking spaces. By making downtown more inviting, these policies have helped create 30,000 new inner-city jobs and attract $900 million in private investment.

Zoning is useful, but it can be influenced or modified by developers because local governments depend on property taxes for revenue. Thus, zoning often favors high-priced housing and factories, hotels, and other businesses, a feedback loop that works against preserving cropland, wetlands, forests, and other wildlife habitats. Overly strict zoning can also discourage innovative approaches to solving urban problems.

Local governments can also control the rate of development by limiting the number of building permits, sewer hookups, roads, and other services. Local, state, and federal governments can also take other measures to protect cropland, forestland, and wetlands near expanding urban areas from degradation and ecologically unsound development. They can require an environmental impact analysis for proposed roads and development projects. Land can be taxed on the basis of its *actual* use as agricultural land or forestland, rather than on the basis of its most profitable *potential* use. This would keep farmers and other landowners from being forced to sell land to pay their tax bills, but would decrease tax revenues for cities. Tax breaks can also be given to landowners who agree in legally binding conservation easements to use their land only for specified purposes, such as agriculture, wildlife habitat, or nondestructive forms of recreation.

Development rights that restrict land use can be purchased and land trusts can be used to buy and protect ecologically valuable land. Such purchases can be made by private groups such as the Nature Conservancy (Spotlight, p. 636) or the Audubon Society, by local nonprofit, tax-exempt, charitable organizations, and by government agencies.

Another problem is that most zoning sets up separate areas for residential, commercial, and industrial activities (which made sense in the smokestack days when factories polluted cities and few suburbs existed). The resulting separation of homes, jobs, and shops by long distances requires increased car and energy use and promotes urban sprawl. Changing zoning laws to encourage such reintegration of homes, workplaces, and shopping areas would reduce urban sprawl, energy waste, and loss of community—an example of instituting a corrective or negative feedback process.

Controlling where highways and streets are built may be far more influential in determining where development will take place. For example, concentrating development along bus lines or other forms of mass transportation can reduce car use and urban sprawl, as has happened in Portland, Oregon; Toronto, Canada; Vienna, Austria; Stockholm, Sweden; Copenhagen, Denmark; and Curitiba, Brazil (Solutions, p. 736).

Japan and western Europe have the world's most comprehensive land-use controls, and North America and Australia the weakest. In the United States, only Oregon has developed a comprehensive land-use plan (Solutions, right). Recently China began developing a comprehensive national zoning plan with the prime goal of achieving zero net loss of farmland to residential and industrial development.

26-5 SOLUTIONS: MAKING URBAN AREAS MORE LIVABLE AND SUSTAINABLE

What Urban Maintenance and Repair Problems Does the United States Face? As philosopher–longshoreman Eric Hoffer observed, "History

Q: What is the average amount of CO_2 each person in the United States adds to the atmosphere each year?

Land-Use Planning in Oregon

SOLUTIONS

Since the mid-1970s Oregon has had a comprehensive statewide land-use planning process based on three principles.

- *All rural land in Oregon has been permanently zoned as forest, agricultural, or urban land.*
- *An urban growth line has been drawn around each community in the state, with no urban development allowed outside the boundary.*
- *Control over the process has been placed in state hands through the Land Conservation and Development Commission.*

Not surprisingly, this last principle has been the most controversial. It is based on the idea that public good takes precedence over private property rights—a well-established principle in most European countries that is generally opposed in the United States. However, the success of Oregon's plan has helped people recognize that the land is not for owners alone to use in any way they see fit. Oregon's plan has worked because it is not designed to "just say no" to development; instead, it encourages certain kinds of development, such as dense urban development.

Because of the plan, most of the state's rural areas remain undeveloped, and the state and many of its cities are consistently rated among the best places in the United States to live. Portland, for example, has plenty of greenways (Figure 26-14), open spaces, and affordable housing; a healthy inner city; and one of the country's best public transit systems. It has been voted the most livable U.S. city (1988), one of the top 10 cities in which to do business (1989), one of the best places to raise children (1990), and best in the Green Index, which rates U.S. cities in terms of pollution, public health, and environmental health (1990).

Critical Thinking

Why don't more cities adopt the model for land management in Portland, Oregon? How could you improve on Portland's model?

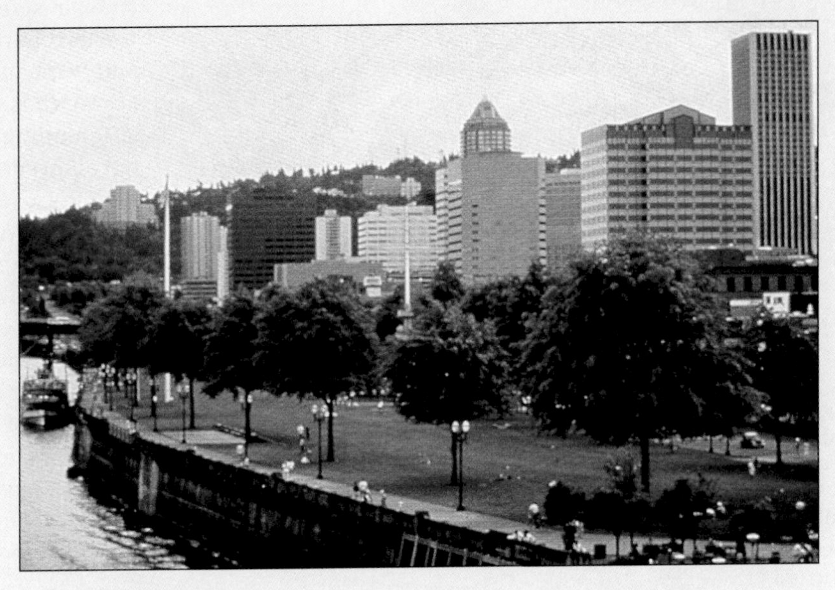

Figure 26-14 In Portland, Oregon, citizens decided to develop a parkland or greenway along the banks of the Willamette River. (Tri-County Metropolitan Transportation District of Oregon)

shows that the level achieved by a civilization can be measured by the degree to which it performs maintenance." America's older cities have enormous maintenance and repair problems, most of them aggravated by decades of neglect.

When it rains in Chicago, sewage backs up into basements of about 25% of the homes. An estimated 46% of Boston's water supply and 25% of Pittsburgh's are lost through leaky pipes. Some 39% percent of America's bridges are unsafe. About 56% of the paved highways in the United States are in poor or fair condition and need expensive repairs. Highways in western Europe are designed to last 40 years—twice as long as American roads.

Maintenance, repair, and replacement of existing U.S. bridges, roads, mass transit systems, water supply systems, sewers, and sewage treatment plants during the next decade could cost a staggering $2 trillion or more—an average expenditure of $2.1 million per minute during the next 10 years. These huge bills, born of neglect, are coming due in a time of record budget deficits, cutbacks in federal funds for building and maintaining public works, and strong citizen opposition to increases in federal, state, and local taxes.

A: 16.7 metric tons (18.4 tons)—six times more than the average citizen of a developing country

How Can Urban Open Space Be Preserved? Some cities have had the foresight to preserve significant blocks of open space in the form of municipal parks. Central Park in New York City, Golden Gate Park in San Francisco, Fairmont Park in Philadelphia, and Grant Park in Chicago are examples of large urban parks in the United States.

In 1883, Minneapolis officials vowed to create "the finest and most beautiful system of Public Parks and Boulevards of any city in America." This goal has been achieved. Today the city has 170 parks spaced by design so that every home in Minneapolis is within six blocks of a green space.

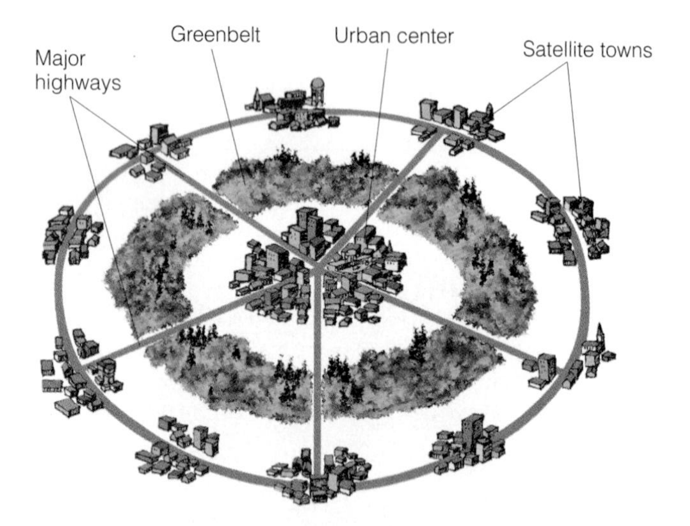

Figure 26-15 A greenbelt around a large city. This arrangement can control urban growth and provide open space for recreation and other nondestructive uses. Satellite towns are sometimes built outside the belt. Highways or rail systems can be used to transport people around the periphery or into the central city.

Unfortunately, older cities that did not plan for such parks early in their development have little or no chance of getting them now. However, as newer cities expand they can develop large or medium-size parks. Even older cities can create small or medium-size parks by removing buildings and streets, planting trees and grass, and establishing ponds and lakes in areas where buildings have been abandoned.

Another way to provide open space and control urban growth is to surround a large city with a *greenbelt* (Figure 26-15): an open area used for recreation, sustainable forestry, or other nondestructive uses. Satellite towns can be built outside the belt. Ideally, the outlying towns and the central city are linked by an extensive public transport system. Many cities in western Europe and Canadian cities such as Toronto and Vancouver have used this approach. Another approach is to set up an *urban growth boundary*: a line surrounding a city beyond which new development is not allowed. Portland, Oregon, has used this approach with great success.

Since World War II the typical approach to suburban housing development in the United States has been to bulldoze a tract of woods or farmland and build rows of standard houses on standard-size lots. Many of these developments and their streets are named after the trees and wildlife they displaced (Oak Lane, Cedar Drive, Pheasant Run, Fox Fields). In recent years builders have increasingly used a new pattern, known as *cluster development*, which provides areas of open or commonly shared space within housing developments (Figure 26-16).

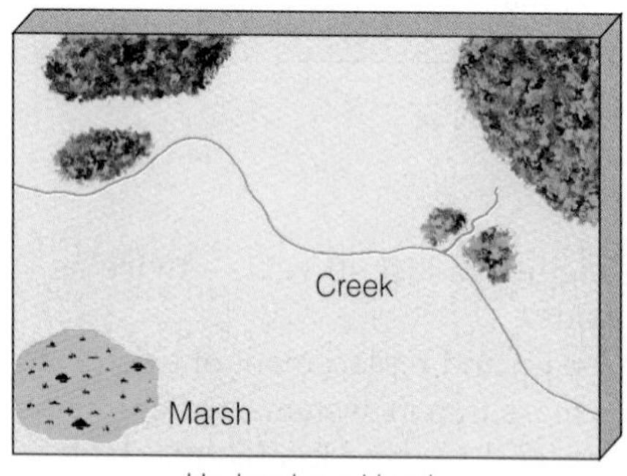

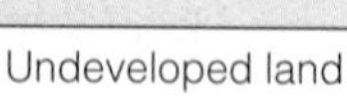

Undeveloped land

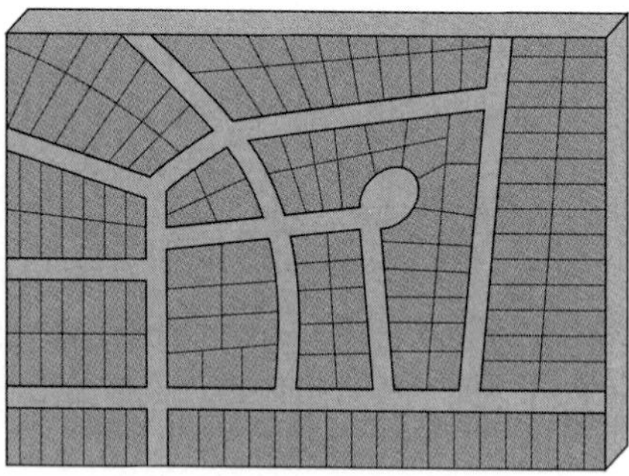

Typical housing development

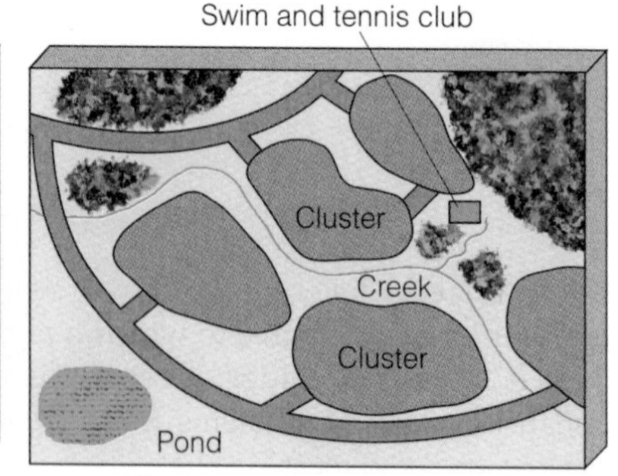

Cluster housing development

Figure 26-16 Conventional and cluster housing developments as they might appear if constructed on the same land area. With cluster development, houses, townhouses, condominiums, and two- to six-story apartments are built on part of the tract. The rest, typically 30% of the area, is left as open space, parks, and cycling and walking paths. Parking spaces and garages can also be clustered so that cars are not used within residential areas; access is achieved by walking, bicycling, or driving small electric or methane-powered golf carts.

Q: What is the role of ozone (O_3) gas in the stratosphere?

Some cities have converted abandoned railroad rights-of-way and dry creek beds into bicycle, hiking, and jogging paths, often called *greenways*. More than 500 new greenway projects, developed largely by citizens' groups, are under way in the United States. Many German, Dutch, and Danish cities are connected by extensive networks of footpaths.

What Are the Pros and Cons of Building New Cities and Towns? Although urban problems must be solved in existing cities, building new cities and towns could take some of the pressure off overpopulated and economically depressed urban areas.

Great Britain has built 16 new towns and is building 15 more. New towns have also been built in Singapore, Hong Kong, Japan, Finland, Sweden, France, the Netherlands, Venezuela, Brazil, and the United States. They are of three types: **(1)** *satellite towns*, located fairly close to an existing large city (Figure 26-15); **(2)** *freestanding new towns*, located far from any major city; and **(3)** *in-town new towns*, located within existing urban areas. Typically, new towns are designed for populations of 20,000 to 100,000 people (Case Study, right). Japan is building 18 new high-technology cities that are supposed to be centers for scientific and economic growth during the next century.

New towns rarely succeed without government financial support. Some don't succeed even then, primarily because of poor planning and management. In 1971 the Department of Housing and Urban Development (HUD) provided more than $300 million in federally guaranteed loans for developers to build 13 new towns in the United States. By 1980 HUD had to take title to nine of those projects, which had gone bankrupt. HUD no longer funds new towns.

Private developers of new towns must put up large amounts of money to buy the land and install facilities; they must pay heavy taxes and interest charges for decades before they see profit. In the United States two privately developed new towns—Reston, Virginia, and Columbia, Maryland—have been in financial difficulty since they were established about two decades ago. However, their situations are improving.

A successful new town called Las Colinas has been built 8 kilometers (5 miles) from the Dallas–Fort Worth Airport on land that was once a ranch. The community is built around an urban center, with a lake and water taxis to transport people across it. The town is laced with greenbelts and open spaces to separate high-rise office buildings, warehouses, and residential buildings. People working in high-rise buildings park their cars outside the core and take a computer-controlled personal transit system to their offices. New or rebuilt smaller-scale village communities have been created in Seaside, Florida; Gainesville, Florida; Harbor Town, Tennessee (developed on an island in the Mississippi River, 5 minutes from downtown Memphis); Laguna West, California (a middle-class development on the far fringes of suburban Sacramento); Valencia, California; and

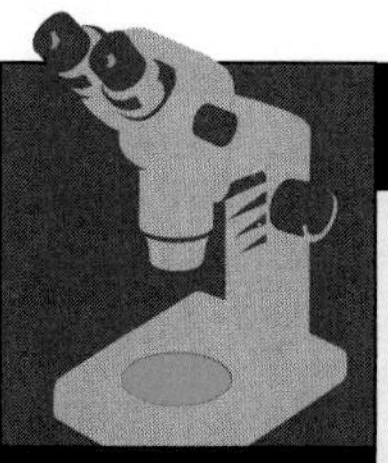

CASE STUDY

Tapiola, Finland

Tapiola, Finland, a new satellite town not far from Helsinki, is internationally acclaimed for its ecological design, beauty, and high quality of life for its residents. Designed in 1951, it is located in a beautiful setting along the shores of the Gulf of Finland. It is being built gradually in seven sections, with an ultimate projected population of 80,000. Today many of its more than 50,000 residents work in Helsinki, but its long-range goal is industrial and commercial independence.

Tapiola is divided into several villages separated by greenbelts. Each village consists of several neighborhoods clustered around a shopping and cultural center that can be reached on foot. Playgrounds and parks radiate from this center, and walkways lead to the various residential neighborhoods.

Each neighborhood has a social center and contains a mix of about 20% high-rise apartments and 80% single-family houses nestled among lush evergreen forests and rocky hills. Housing is not segregated by income. Because housing is clustered, more land is available for open space, recreation areas, and parks.

Industrial buildings and factories, with strict pollution controls, are located away from residential areas and screened by vegetation to reduce noise and visual pollution, but they are close enough so people can walk or bicycle to work. Finland plans to build six more new satellite towns around Helsinki.

Critical Thinking

What might be some drawbacks to living in a new town such as Tapiola? Would you like to live in such a town? Explain.

A: It filters out high-energy ultraviolet (UV) radiation from the sun

Eco-Village, New York (a potentially sustainable village being developed near Ithaca, New York).

How Can We Make Cities More Sustainable? An important goal in coming decades should be to make urban areas more self-reliant, sustainable, and enjoyable places to live. In a sustainable and ecologically healthy city, called an *ecocity* or *green city*, people walk or cycle for most short trips; they walk or bike to bus, metro, or trolley stops for longer urban trips. Rapid-rail transport between cities would replace many long drives and medium-distance airplane flights.

In such cities emphasis is placed on pollution prevention, recycling and reuse (at least 60% of all municipal wastes), use of renewable energy resources, encouraging rather than assaulting biodiversity, and use composting to help create rather than destroy soil. They are based on good ecological design (Guest Essay, p. 746). According to farmer and philosopher Wendell Berry,

> *The only sustainable city is a city in balance with its countryside: a city, that is, that would live off the net ecological income of its supporting region, paying as it goes all its ecological and human debts. Some cities can never be sustainable, because they do not have a countryside around them, or near them, from which they can be sustained.*

Ways to make existing and new suburbs more sustainable and livable include **(1)** giving up big lawns, **(2)** building houses and apartments in small, dense clusters so that more community open space is available, **(3)** developing a town center (a plaza, square, or green) that is a focus of civic life and community cohesiveness, **(4)** planting lots of new trees and not cutting down existing ones, and **(5)** discouraging excessive dependence on the automobile and encouraging walking and bicycling. This could be done by closing off some streets to automobiles and making others narrower, building a connected network of sidewalks and bike paths, planning areas to take advantage of mass transit, and bringing back local neighborhood stores so that most people can walk or bicycle to get food, medicines, and other essential items.

In a few decades people may be wondering why they allowed cars to dominate their lives and degrade the environment for so long. They may take seriously the advice Lewis Mumford gave Americans over three decades ago: "Forget the damned motor car and build cities for lovers and friends."

Examples of cities that have attempted to become more ecologically sustainable include Davis, California (p.720), Curitiba, Brazil (Solutions, p. 736), and Chattanooga, Tennessee.

Case Study: Chattanooga, Tennessee In the 1950s Chattanooga was known as one of the dirtiest cities in the United States. Its air was so polluted that people sometimes had to turn on their headlights in the middle of the day, and the EPA warned that when women walked outside their nylon stockings were likely to disintegrate. The Tennessee River, flowing through the city's industrial wasteland, bubbled with toxic waste.

Since the mid-1980s the combined efforts of thousands of Chattanooga citizens have helped clean up the city's air, revitalize its river front, and diversify its economy. The overall goal has been to show how environmental improvement and economic development can coexist.

Efforts began in 1984 when civic leaders used town meetings as part of a Vision 2000 process, a 20-week series of community meetings that brought together thousands of citizens from all walks of life to build a consensus about what the city could be like at the turn of the century. Groups of citizens and civic leaders identified the city's main problems, set goals, and brainstormed thousands of ideas for dealing with these problems. These meetings produced a consensus for a greener, cleaner, safer city, with rehabilitated low-income housing and nonpolluting jobs.

To help achieve the overall goal of sustainable development, city officials and 1,700 citizens spent over 5 months reducing the results of the Vision 2000 process into 34 specific goals and 223 projects. The city launched a number of combined public–private partnerships for green development, generating a total investment of $739 million (two-thirds of it from private investors), 1,300 new permanent jobs, and 7,800 temporary construction jobs.

The city has used a variety of approaches and projects. They include **(1)** enticing zero-emission industries to relocate in Chattanooga; **(2)** renovating existing low-income housing and building new low-income rental units; **(3)** building the nation's largest freshwater aquarium, which became the centerpiece for the city's downtown renewal (and in its first year of operation in 1992 injected an estimated $133 million into the local economy); **(4)** replacing all of its diesel buses with a fleet of quiet, nonpolluting electric buses, most made by a new local firm; **(5)** developing an

Q: Worldwide, how many people die each year from skin cancer?

8-kilometer-long (5-mile-long) riverfront park, which will be extended 35 kilometers (22 miles) along both sides of the Tennessee River; and **(6)** launching an innovative recycling program (after citizen activists and environmentalists blocked construction of a new garbage incinerator).

By 1992, 29 of the 34 goals had been met. In 1993 the community began the process again in Revision 2000. One ambitious goal calls for transforming a blighted, mostly abandoned industrial area in South Chattanooga into an environmentally advanced, mixed community of neighborhoods and zero-emission industries so that employees can live near their workplaces. The new ecoindustrial area will contain an ecology center that will treat sewage, wastewater, and contaminated soils with a series of interconnected tanks heated by the sun and filled with wetland plants (algae, snails, trees, fish and other organisms that can naturally filter and break down wastes), a process invented by ecologist John Todd (Figure 20-2).

These accomplishments show what citizens, environmentalists, and business leaders can do when they work together to develop and achieve common goals. However, the city does not even come close to being an ecocity and still faces serious environmental problems. One is the controversy over how to handle at least 11 Superfund toxic waste sites on the once heavily industrialized Chattanooga Creek, where many African-American communities are located. It took minority residents decades of action just to attract attention to this problem.

Peter Montague (Guest Essay, p. 70) contends that Chattanooga should not be held up as a model environmental city until it provides environmental justice for all by cleaning up South Chattanooga and Chattanooga Creek, which threaten the health of many of the city's minority population. Cleaning up South Chattanooga may prove to be the most important test of the city's commitment to improving environmental quality for all of its citizens.

How Can We Improve Urban Living? Because most people around the world now live, or will live, in urban areas, improving the quality of urban life is an urgent priority. Increased urbanization and urban density are better than spreading people out over the countryside, which would destroy more of the planet's biodiversity. The primary problem is not urbanization, but our failure to make most cities more sustainable and livable and to provide economic support for rural areas.

Urban areas that fail to become more ecologically sustainable over the next few decades are inviting economic depression and increased unemployment, pollution, and social tension. We have no time to waste in making urban areas—where most of the world's people will live by 2000—better places to live and work, with less stress on natural systems and people. Each of us has a vital role to play in converting this vision into reality.

The city is not an ecological monstrosity. It is rather the place where both the problems and the opportunities of modern technological civilization are most potent and visible.

PETER SELF

CRITICAL THINKING

1. What conditions, if any, would encourage you to rely less on the automobile, and instead encourage you to travel to school or work by bicycle or motor scooter, on foot, by mass transit, or by a car or van pool?

2. Do you believe that cities should continue expanding into nearby agricultural land, forests, and wetlands? If not, how would you control this expansion?

3. Should squatters around cities of developing countries be given title to land they don't own? Explain. What are the alternatives?

4. Do you prefer living in a rural, suburban, small-town, or urban environment? Describe the ideal environment in which you would like to live.

5. Should most, half, or a small percentage (the current situation in the United States) of gasoline tax revenues be used to support mass transit and bike paths in urban areas, where 75% of all people in the United States live? Explain your position.

6. Do you believe that the United States (or the country where you live) should develop a comprehensive and integrated mass transit system over the next 20 years, including building an efficient rail network for travel within and among its major cities? How would you pay for such a system?

7. Do you believe that the approach to land-use planning used in Oregon (Solutions, p. 741) should be used in the state or area where you live? Explain your position.

8. Why do you think that Curitiba, Brazil (Solutions, p. 736), has been so successful in its efforts to become an ecocity, compared to the generally unsustainable cities found in more developed countries?

9. In June 1996, representatives from the world's countries met in Istanbul, Turkey, at the Second UN

A: About 100,000 (7,000 in the United States)

The Ecological Design Arts

GUEST ESSAY

David W. Orr

David W. Orr is professor of environmental studies at Oberlin College and one of the nation's most respected environmental educators. He is author of numerous environmental articles and three books, including Ecological Literacy *and* Earth in Mind. *He is education editor for* Conservation Biology *and a member of the editorial advisory board of* Orion Nature Quarterly.

If *Homo sapiens sapiens* entered its industrial civilization in an intergalactic design competition, it would be tossed out in the qualifying round. It doesn't fit. It won't last. The scale is wrong. And even its defenders admit that it's not very pretty. The most glaring design failures of industrial/technologically driven societies are the loss of diversity of all kinds, impending climate change, pollution, and soil erosion.

Industrial civilization, of course, wasn't designed at all; it was mostly imposed by single-minded individuals, armed with one doctrine of human progress or another, each requiring a homogenization of nature and society. These individuals for the most part had no knowledge of "ecological design arts"—the set of perceptual and analytic abilities, ecological wisdom, and practical wherewithal needed to make things that fit into a world of microbes, plants, animals, and energy laws.

Good ecological design incorporates understanding about how nature works into the ways we design, build, and live. It is required in our designs of farms, houses, neighborhoods, cities, transportation systems, technologies, economies, energy policies, and just about anything that directly or indirectly requires energy or materials or governs their use.

When human artifacts and systems are well designed, they are in harmony with the ecological patterns in which they are embedded. When poorly designed, they undermine those larger patterns, creating pollution, higher costs, and social stress. Bad design is not simply an engineering problem, although better engineering would often help. Its roots go deeper.

Good ecological design has certain common characteristics, including correct scale, simplicity, efficient use of resources, a close fit between means and ends, durability, redundancy, and resilience. These characteristics are often place-specific, or, in John Todd's words, "elegant solutions predicated on the uniqueness of place." Good design also solves more than one problem at a time and promotes human competence, efficient and frugal use of resources, and sound regional economies. Where good design becomes part of the social fabric at all levels, unanticipated positive side effects multiply. When people fail to design with ecological competence, unwanted side effects and disasters multiply.

The pollution, violence, social decay, and waste all around us indicate that we have designed things badly, for, I think, three primary reasons. First, as long as land and energy were cheap and the world was relatively empty, we did not need to master the discipline of good design. The result was sprawling cities, wasteful economies, waste dumped into the environment, bigger and less efficient automobiles and buildings, and conversion of entire forests into junk mail and Kleenex—all in the name of economic growth and convenience.

Second, design intelligence fails when greed, narrow self-interest, and individualism take over. Good design is a cooperative community process requiring people who share common values and goals that bring them together and hold them together. American cities, with

Conference on Human Settlements (nicknamed the City Summit). One of the areas of controversy was over whether housing is a universal *right* (a position supported by most developing countries) or just a *need* (supported by the United States and several other developed countries). What is your position on this issue? Defend your choice.

10. Try to come up with analogies to complete the following statements: **(a)** Living in a city is like _______. **(b)** Living in a suburban area is like _______. **(c)** Cutting down most of the trees in an urban area is like _______. **(d)** Owning and using a car is like _______. **(e)** Living in a car-culture society is like _______. **(f)** Riding a bus is like _______. **(g)** Living in a sustainable city such as Davis, California, or Curitiba, Brazil, would be like _______.

11. Congratulations—you have just been put in charge of the world. List the five most important features of your urban policy.

PROJECTS

1. Consult local officials to determine how land use is decided in your community. What roles do citizens play in this process?

Q: Is ozone depletion from CFCs and other human-made chemicals reaching the stratosphere a serious threat?

their extremes of poverty and opulence, are products of people who believe they have little in common with one another. Greed, suspicion, and fear undermine good community and good design alike.

Third, poor design results from poorly equipped minds. Good design can be done only by people who understand harmony, patterns, and systems. Industrial cleverness, on the contrary, is mostly evident in the minutiae of things, not in their totality or in their overall harmony. Good design requires a breadth of view that causes people to ask how human artifacts and purposes fit within a particular culture and place. It also requires ecological intelligence, by which I mean an intimate familiarity with how nature works.

An example of good ecological design is found in John Todd's "living machines," which are carefully orchestrated ensembles of plants, aquatic animals, technology, solar energy, and high-tech materials to purify wastewater, but without the expense, energy use, and chemical hazards of conventional sewage treatment technology. Todd's "living machines" resemble greenhouses filled with plants and aquatic animals [Figure 20-2]. Wastewater enters at one end and purified water leaves at the other. In between, an ensemble of organisms driven by sunlight use and remove nutrients, break down toxins, and incorporate heavy metals in plant tissues.

Ecological design standards also apply to the making of public policy. For example, the Clean Air Act of 1970 required car manufacturers to install catalytic converters to remove air pollutants. Two decades later, emissions per vehicle are down substantially, but because more cars are on the road, air quality is about the same—an example of inadequate ecological design. A sounder design approach to transportation would create better access among housing, schools, jobs, stores, and recreation areas; build better public transit systems; restore and improve railroads; and create bike trails and walkways.

An education in the ecological design arts would foster the ability to see things in their ecological context, integrating firsthand experience and practical competence with theoretical knowledge about how nature works. It would aim to equip people to build households, institutions, farms, communities, corporations, and economies that **(1)** do not emit carbon dioxide or other heat-trapping gases, **(2)** operate on renewable energy, **(3)** preserve biological diversity, **(4)** recycle material and organic wastes, and **(5)** promote sustainable local and regional economies.

The outline of a curriculum in ecological design arts can be found in recent work in ecological restoration, ecological engineering, solar design, landscape architecture, sustainable agriculture, sustainable forestry, energy efficiency, ecological economics, and least-cost, end-use analysis. A program in ecological design would weave these and similar elements together around actual design objectives that aim to make students smarter about systems and about how specific things and processes fit in their ecological context. With such an education we can develop the habits of mind, analytical skills, and practical competence needed to help sustain the earth for us and other species.

Critical Thinking

1. Does your school offer courses or a curriculum in ecological design? If not, suggest some reasons why it does not.

2. Use the principles of good ecological design to evaluate how well your campus is designed. Suggest ways to improve its design.

---

2. Consult local officials to identify any floodplain areas in your community. How are these areas used? Are people allowed to build in such areas? If so, assess their awareness of the risks of property loss.

3. For a class or group project, try to borrow one or more decibel meters from your school's physics or engineering department (or from a local electronics repair shop). Make a survey of sound pressure levels at various times of day and at several locations. Plot the results on a map. Also, measure sound levels in a room with a stereo and from earphones at several different volume settings. If possible, measure sound levels at an indoor concert, a nightclub, and inside and outside a boom car various distances from the sound system speakers. Correlate your findings with those in Figure 26-11.

4. As a class project, evaluate land use and land-use planning by your school, draw up an improved plan based on ecological principles, and submit the plan to school officials.

5. As a class project, use the following criteria to rate your city on a green index from 0 to 100. Are existing trees protected and new ones planted throughout the city? Do you have parks to enjoy? Can you swim in any nearby lakes and rivers? What is the quality of your water and air? Is there an effective noise pollution reduction program? Does your city have a recycling program,

A: Yes, according to the overwhelming consensus of scientists in this field a composting program, and a hazardous waste collection program, with the goal of reducing the current solid waste output by at least 60%? Is there an effective mass transit system? Are there bicycle paths? Are all buildings required to meet high energy-efficiency standards? Is there an effective program for the sustainable use of locally available matter and energy resources and a buy-local program? How much of the energy is obtained from locally available renewable resources? Are environmental regulations for existing industry tough enough and enforced well enough to protect citizens? Do local officials look carefully at an industry's environmental record and plans before encouraging it to locate in your city or county? Are land-use decisions made by using ecological planning with active participation by citizens? Are city officials actively planning to improve the quality of life for all of its citizens now and in the future? If so, what is their plan?

6. Make a concept map of this chapter's major ideas, using the section heads and subheads and the key terms (in boldface). Look at the inside back cover and in Appendix 4 for information about concept maps.

PART VI

ENVIRONMENT AND SOCIETY

When it is asked how much it will cost to protect the environment, one more question should be asked: How much will it cost our civilization if we do not?

GAYLORD NELSON

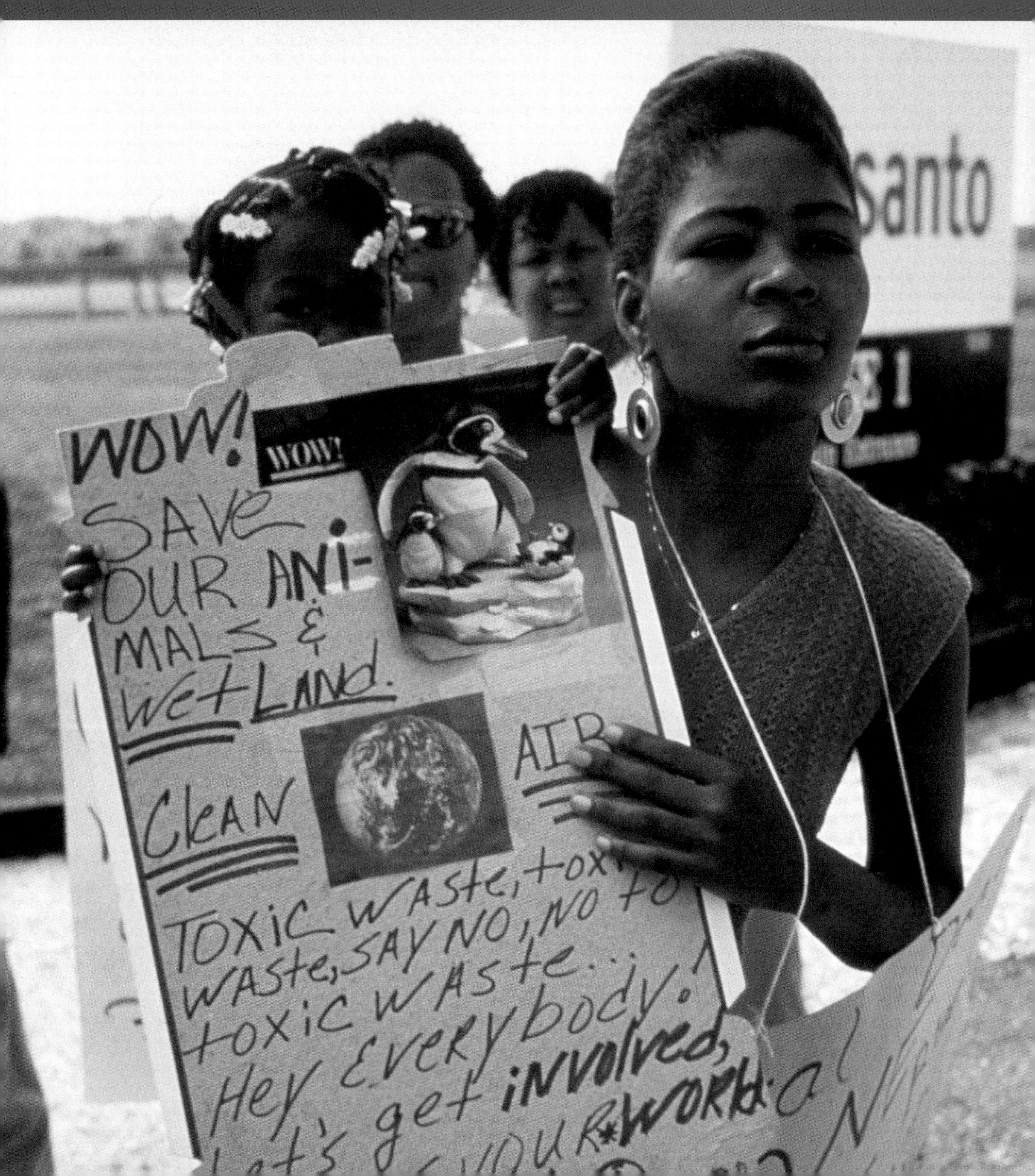

27 ECONOMICS AND ENVIRONMENT

To Grow or Not to Grow: Is That the Question?

Most economists, investors, and business leaders argue that we must have unlimited economic growth to create jobs, satisfy people's economic needs and desires, clean up the environment, and help reduce poverty.

These analysts see the earth as an essentially unlimited source of raw materials and the environment as an infinite sink for wastes; they believe technological innovation can overcome any resource or environmental limits. To people with this view, environmentalists put endangered species above endangered people; they threaten jobs and oppose the economic growth needed for human survival and gains in the quality of life.

On the other hand, environmentalists and a small but growing number of economists and business leaders argue that economic systems depend on resources and services provided by the sun and by the earth's basic components and processes (Figures 27-1 and 1-2). They maintain that a healthy economy ultimately depends on a healthy ecosphere. They believe that if we continue to support economic growth by consuming earth capital instead of living off sustainable earth income, such forms of growth will threaten

Figure 27-1 Earth, air, fire, water, and life at a volcanic lava-flow site in Hawaii. Most environmentalists and a growing number of economists believe that these basic components of earth capital, which sustain us and other species and all economies, are undervalued in the economic marketplace. (Greg Vaughn/Tom Stack & Associates)

business and impair the planet's life-support systems for humans and many other species.

If these beliefs are correct, then over the next few decades we must replace the economics of unlimited growth with the economics of sustainability. These modified economic systems would unite ecology and commerce by giving rewards (subsidies) to earth-sustaining businesses and activities and penalizing (taxing and regulating) earth-degrading activities, with the overall goal of providing both economic and environmental security.

The question is not so much, "To grow or not to grow?" but rather, "How can we grow without plundering the planet?" or "How can we grow as if the earth matters?"

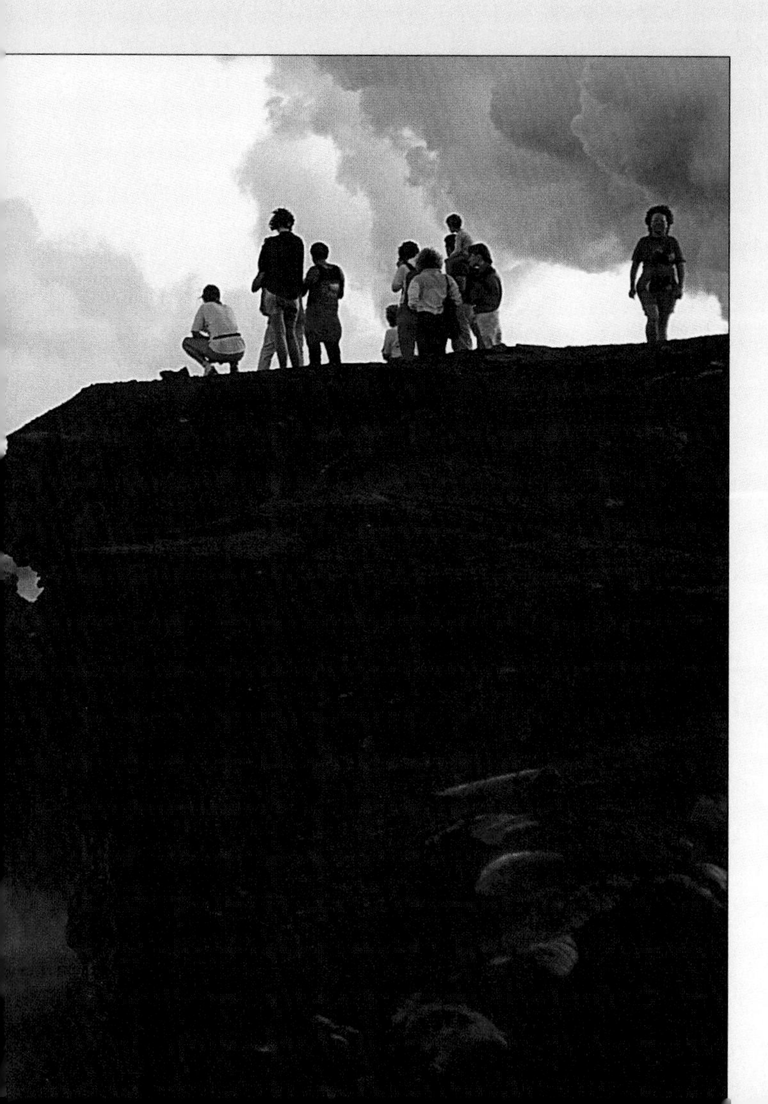

The century to come will be the environmental century. . . . The issue is not growth versus no growth, but what kind of growth and where. . . . Converting the economy of the twentieth century into one that is environmentally sustainable represents the greatest investment opportunity in history. . . . No challenge is greater, or more satisfying.

LESTER R. BROWN AND CHRISTOPHER FLAVIN

In this chapter we seek answers to the following questions:

- What are economic goods and resources and how are they provided?
- How should we measure economic growth?
- How can economics help improve environmental quality?
- How can we sharply reduce poverty?
- Should we gradually shift to an earth-sustaining economy, and if so how might this be done?

27-1 ECONOMIC GOODS, RESOURCES, AND SYSTEMS

What Supports and Drives Economies? An **economy** is a system of production, distribution, and consumption of *economic goods*: any material items or services that satisfy people's wants or needs. In an economy, individuals, businesses, and societies make **economic decisions** about what goods and services to produce, how to produce them, how much to produce, how to distribute them, and what to buy and sell.

The kinds of capital that produce material goods and services in an economy are called **economic resources**. They fall into three groups:

- **Earth capital** or **natural resources**: goods and services produced by the earth's natural processes (Figure 1-2). These include the planet's air, water, and land; nutrients and minerals in the soil and deeper in the earth's crust; wild and domesticated plants and animals (biodiversity); and nature's dilution, waste disposal, pest control, and recycling services. There are no other sources for these materials, which support all economies and lifestyles.
- **Manufactured capital**: items made from earth capital with the help of human capital. This type of capital includes tools, machinery, equipment,

factory buildings, and transportation and distribution facilities.

- **Human capital**: people's physical and mental talents. Workers sell their time and talents for wages. Managers take responsibility for combining earth capital, manufactured capital, and human capital to produce economic goods. In market-based systems, entrepreneurs and investors put up the money needed to produce an economic good with the intention of making a profit on their investment.

What Are the Major Types of Economic Systems? There are two major types of economic systems: centrally planned and market based. In a **pure command economic system**, or **centrally planned economy**, all economic decisions are made by the government. This command-and-control system assumes that government control and ownership of the means of production are the most efficient and equitable way to produce, use, and distribute goods and services.

In a **pure market economic system**, also known as **pure capitalism**, all economic decisions are made in *markets*, in which buyers (demanders) and sellers (suppliers) of economic goods freely interact without government or other interference. All economic resources are owned by private individuals and institutions rather than by the government. All buying and selling is based on *pure competition*, in which no seller or buyer is powerful enough to control the supply, demand, or price of a good. All sellers and buyers have full access to the market and enough information about the beneficial and harmful aspects of economic goods to make informed decisions.

Economists often depict pure capitalism as a circular flow of economic goods and money between households and businesses operating essentially independently of the ecosphere (Figure 27-2). By contrast, environmentalists and a small but growing number of economists emphasize the dependence of this or any economic system on the ecosphere (Figure 27-3).

In a pure capitalist system, a business has no legal allegiance to a particular nation, no obligation to supply any particular good or service, and no obligation to provide jobs, safe workplaces, or environmental protection. A company's only obligation is to produce

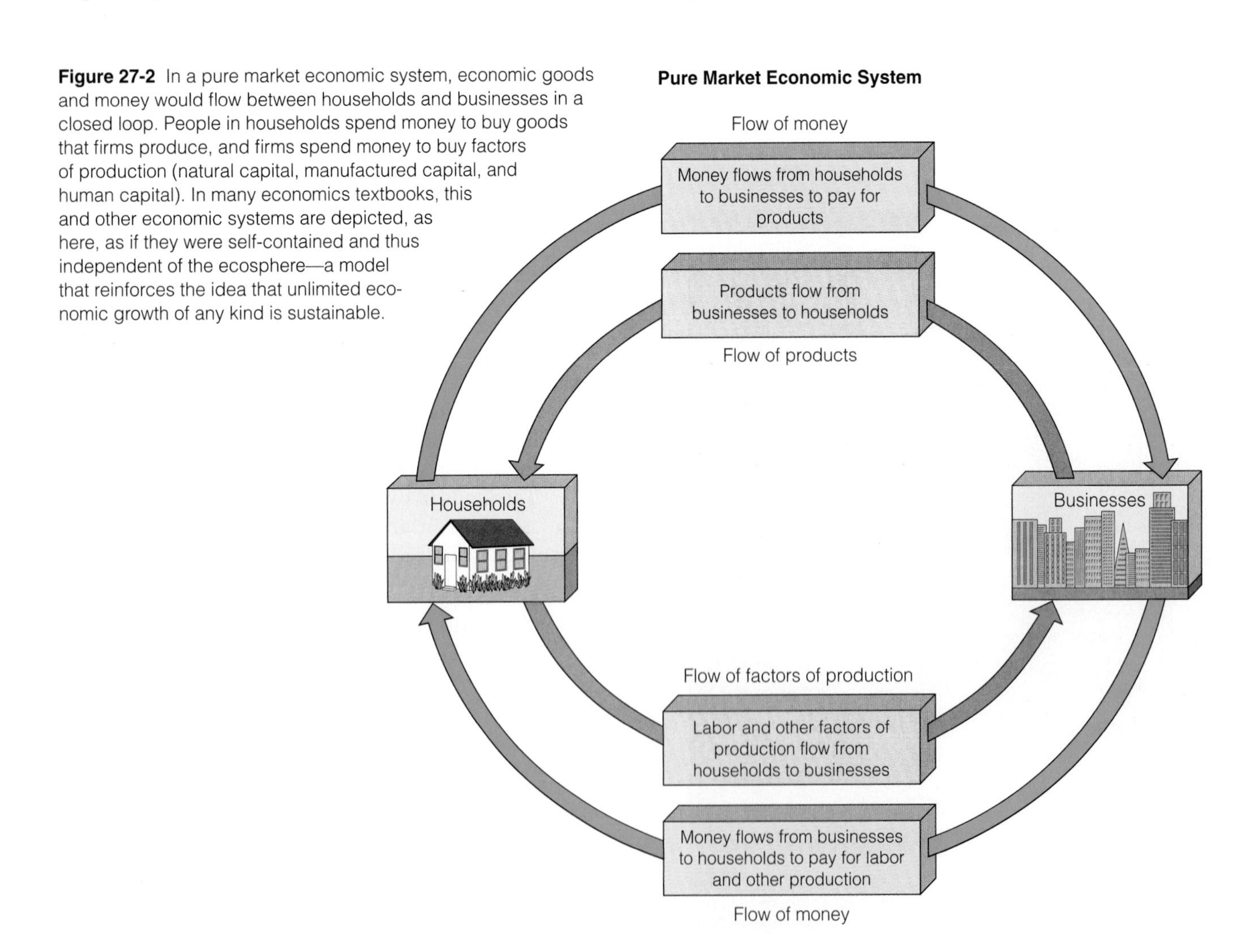

Figure 27-2 In a pure market economic system, economic goods and money would flow between households and businesses in a closed loop. People in households spend money to buy goods that firms produce, and firms spend money to buy factors of production (natural capital, manufactured capital, and human capital). In many economics textbooks, this and other economic systems are depicted, as here, as if they were self-contained and thus independent of the ecosphere—a model that reinforces the idea that unlimited economic growth of any kind is sustainable.

Q: How long do CFCs stay in the atmosphere?

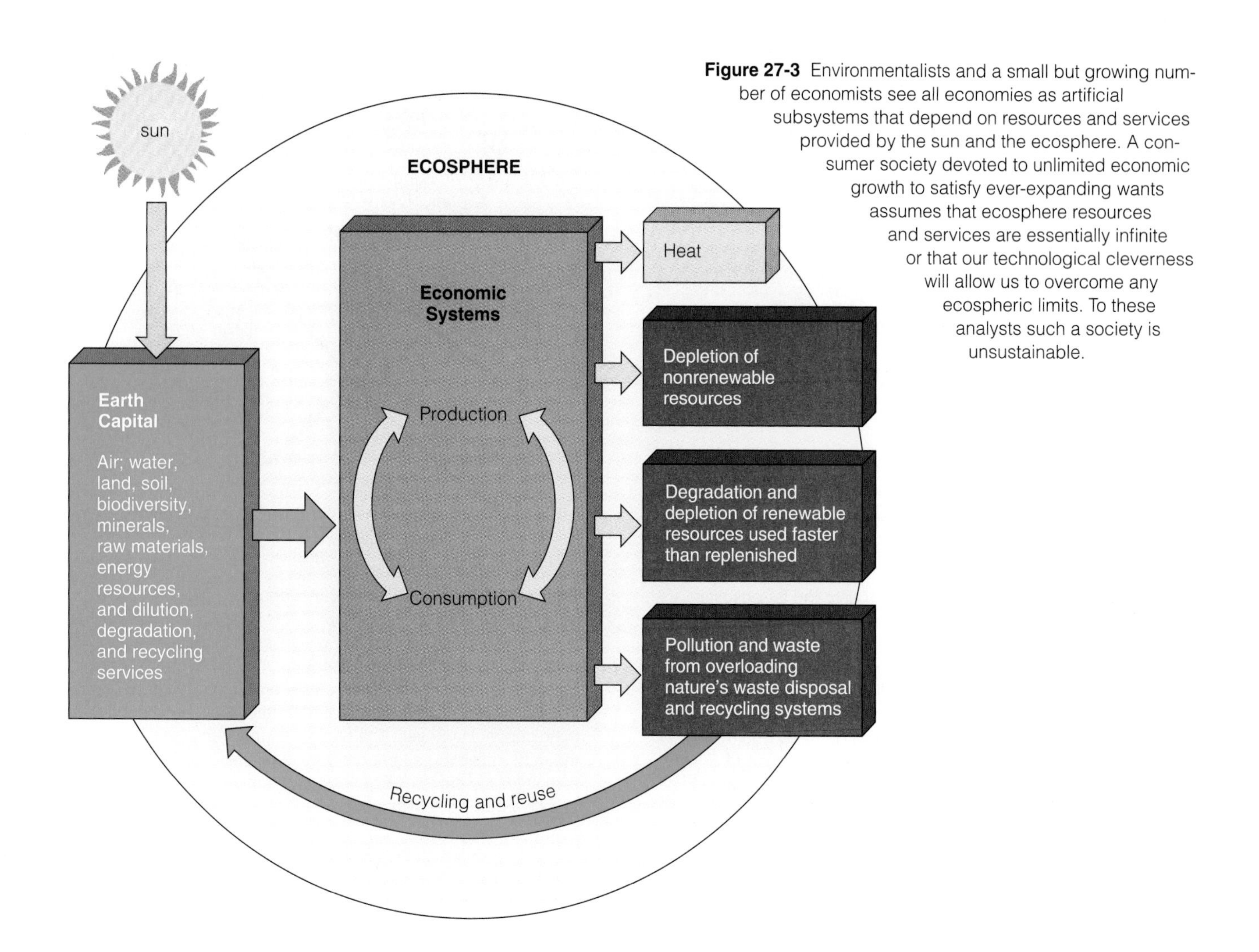

Figure 27-3 Environmentalists and a small but growing number of economists see all economies as artificial subsystems that depend on resources and services provided by the sun and the ecosphere. A consumer society devoted to unlimited economic growth to satisfy ever-expanding wants assumes that ecosphere resources and services are essentially infinite or that our technological cleverness will allow us to overcome any ecospheric limits. To these analysts such a society is unsustainable.

the highest possible short-term economic return (profit) for the owners or stockholders whose financial capital the company is using to do business.

Economic decisions in a pure market system are governed by interactions of *demand*, *supply*, and *price*. Buyers want to pay as little as possible for an economic good, and sellers want to set the highest price possible. **Market equilibrium** occurs when the quantity supplied equals the quantity demanded, and the price is no higher than buyers are willing to pay and no lower than sellers are willing to accept. If price, supply, and demand are the only factors involved, the demand and supply curves for an economic good intersect at the *market equilibrium point* (Figure 27-4). However, factors other than price can shift the original demand curves (Figure 27-5) and supply curves (Figure 27-6), upsetting the market equilibrium and establishing new equilibrium points.

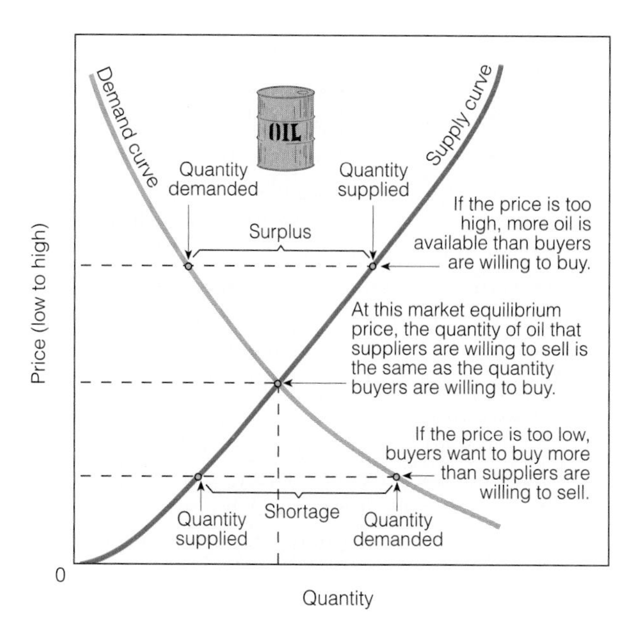

Figure 27-4 Supply, demand, and market equilibrium for gasoline in a pure market system. If price, supply, and demand are the only factors involved, market equilibrium occurs at the point at which the demand and supply curves intersect.

Why Do We Find Mixed Economic Systems in the Real World? In reality, all countries have **mixed economic systems** that fall somewhere between the pure market and pure command systems. The economic systems of countries such as China and North

A: 65–135 years depending on the type

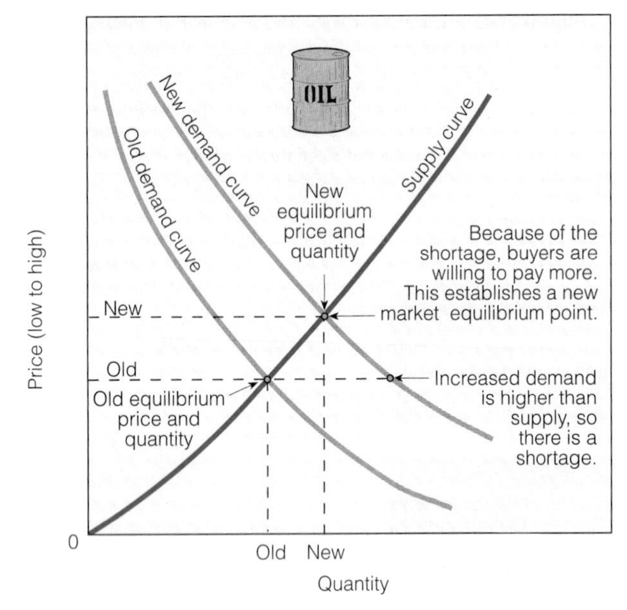

Figure 27-5 Short-term effects of a rising demand for gasoline because of more drivers, a switch to less fuel-efficient cars, more disposable income for travel, or decreased use of mass transit. If the demand fell because of fewer drivers, a switch to more fuel-efficient cars, less disposable income for travel, or increased use of mass transit, the original demand curve would shift to the left. The decreased demand would create a temporary surplus, stimulating sellers competing for consumers' money to charge less until the price reached a new market equilibrium point.

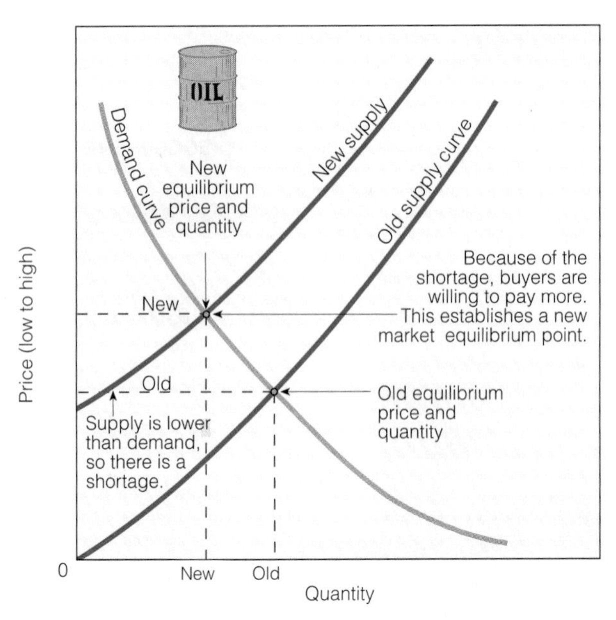

Figure 27-6 Short-term effects of a declining gasoline supply, which might occur if the costs of finding, extracting, and refining crude oil increase or if existing oil deposits are economically depleted. If oil producers intend to charge higher prices in the future, they may reduce production, with the hope of making larger profits later. A parallel situation occurs if the supply increases: The original supply curve shifts to the right, reflecting a temporary surplus; then competition stimulates sellers to charge less, and the price moves down to a new market equilibrium point.

Korea fall toward the command-and-control end of the economic spectrum, whereas those of countries such as the United States and Canada fall toward the market-based end of the spectrum. Most other countries fall somewhere in between.

Pure free-market economies don't exist because they have flaws that require government intervention in the marketplace. Such intervention can prevent a single seller or buyer (*monopoly*) or a single group of sellers or buyers (*oligopoly* or *cartel*) from dominating the market and thus controlling supply or demand and price. Indeed, virtually all business lobbyists work to subvert free-market competition by trying to get legislators and leaders to give the businesses they represent a special subsidy, tax break, or exemption from certain government regulations.

Governments also intervene in economies to provide national security, education, and other public goods; to help redistribute some income and wealth (especially to people unable to attain their basic needs); to protect people from fraud, trespass, theft, and bodily harm; to protect the health and safety of workers and consumers; and to help ensure economic stability by trying to control boom-and-bust cycles that afflict a pure market system.

Government intervention also helps compensate owners for large-scale destruction of their assets by floods, earthquakes, hurricanes, and other natural disasters; to prevent or reduce pollution and depletion of natural resources; and to manage public land resources (Chapter 23).

Pure command economies don't exist, either. Experience has shown that governments cannot efficiently control all economic activity from the top down. In recent years, countries in eastern Europe, the former Soviet Union, and China have moved away from command economic systems and toward market-based approaches.

27-2 ECONOMIC GROWTH AND EXTERNAL COSTS

How Is Economic Growth Measured? Virtually all economies seek **economic growth**: an increase in the capacity of the economy to provide goods and services for people's final use. Such growth is usually accomplished by maximizing the flow of matter and energy resources (throughout) by means of population

 Q: What percentage of the stratospheric ozone over Antarctica is destroyed from September to December each year?

growth (more consumers), more consumption per person, or both.

Economic growth is usually measured by the increase in a country's **gross domestic product (GDP)**, the market value (in current dollars) of all goods and services produced by an economy within its borders for final use during a year, and by its **gross national product (GNP)**, the GDP plus net income from abroad. To get a clearer picture, economists use the **real GNP** or **GDP**: the GNP or GDP adjusted for *inflation* (any increase in the average price level of final goods and services).

To show how the average person's slice of the economic pie is changing, economists use the **real per capita GNP** or **GDP**: the real GNP or GDP divided by the total population. If the population expands faster than economic growth, then real per capita GNP or GDP falls. This statistic can hide the fact that the wealthy few might have enormous slices of the economic pie, whereas the many poor have only a few crumbs.

Is Economic Growth Sustainable? Most economists believe that the capacity for economic growth is virtually unlimited because of the earth's vast amount of resources and the ability of human ingenuity to overcome resource shortages and environmental problems through science and technology. In other words, they believe that continued economic growth based on ever-increasing throughputs of matter and energy (Figure 3-21) is sustainable. Other economists accept the idea of limits to economic growth but believe that we will not reach such limits in the foreseeable future.

To environmentalists and a small but growing number of economists and business leaders, the notion of *sustainable economic growth* is nonsense because nothing that is based on the consumption of the earth capital that sustains all economies (Figure 27-3) can grow indefinitely. According to business leader Paul Hawken (Guest Essay, p. 756),

> *We have reached a point where the value we do add to our economy is now being outweighed by the value we are removing, not only from future generations in terms of diminished resources, but from ourselves in terms of unlivable cities, deadening jobs, deteriorating health, and rising crime. In biological terms, we have become a parasite and are devouring our host.*

However, many economists believe that we can use technology to extend or overcome such limits, or that we will not reach such limits in the foreseeable future.

Instead of unlimited economic growth, such critics call for **ecologically sustainable development**. This occurs when the total human population size and resource use in the world (or in a region) are limited to a level that does not exceed the carrying capacity of the existing natural capital, and are therefore sustainable (Guest Essay, p. 756). This is accomplished by reducing the throughput of matter and energy resources through economies by *doing the same with less* (Figure 3-22). It also includes a more equitable distribution of the world's wealth and resources with the goal of improving the quality of life for all (Guest Essay, p. 774).

Lately there has been much talk about so-called *sustainable development*. There is no problem when this term is used as another way to describe *ecologically* sustainable development. However, some environmentalists contend that governments and businesses have jumped on the sustainability bandwagon and used this term as a thinly disguised way to justify or promote unsustainable forms of economic growth and development. To these critics, what we need is the development of ecological sustainability, not sustainable development.

According to systems analysis expert and environmentalist Donella Meadows,

> *What we need is smart development, not dumb growth. . . . When something grows it gets quantitatively bigger. When something develops it gets qualitatively better. . . . Smart development invests in insulation, efficient cars, and ever-renewed sources of energy. It ensures that forests and fields continue to produce wood, paper, and food, recharge wells, harbor wildlife, and attract tourists. Dumb growth crashes around looking for more oil. It clearcuts forests to keep loggers and sawmills going just a few more years until the trees run out. . . . It covers the landscape with the same kind of honkytonk ugliness tourists leave home to escape. . . . We need to meet dumb growth with smart questions. What really needs to grow? Who will benefit? Who will pay? What will last?*

Suppose this view is correct and our collective consumption approaches or runs into some of the earth's natural limits. Then the difficult political and ethical question becomes how to share a roughly fixed economic pie instead of how to increase the size of the pie. According to Lester Brown (Guest Essay, p. 34) if this happens, the growth-driven economics and politics of the last century will be replaced with a new economics and politics of scarcity and sustainability.

Are GNP and GDP Useful Measures of Quality of Life and Environmental Degradation? We are urged to buy and consume more and more so that the GNP and GDP will rise, making the country where one lives and the world a better place for everyone. The truth is that GNP and GDP indicators are poor

A: About 50%, creating an area of ozone depletion larger in area than the continental United States

Natural Capital

GUEST ESSAY

Paul G. Hawken

Paul G. Hawken is a practical visionary who understands both business and ecology and can communicate what is needed to make the transition to an earth-sustaining, or restorative, economy. In addition to founding Smith & Hawken, a retail company known for its environmental initiatives, he has written seven widely acclaimed books, including Growing a Business *(1987),* The Ecology of Commerce *(1993), and* Factor 10, The Next Industrial Revolution *(1998, with Amory and Hunter Lovins). He produced and hosted* Growing a Business, *a series for public television shown nationwide on 210 stations and now shown in 115 countries. His book* The Ecology of Commerce *was hailed as the best business book of 1993 and one of the most important books of this century. In 1987, Inc. magazine named him one of the 12 best entrepreneurs of the 1980s, and in 1995 he was named by the* Utne Reader *as one of the 100 visionaries who could change our lives. He has also been described as the poet laureate of American capitalism.*

Great ideas, in hindsight, seem obvious. The concept of natural capital [or earth capital, Figure 1-2] is such an idea. Natural capital refers to the myriad necessary and valuable resources and ecological processes that we rely upon to produce our food, products, and services. These include the obvious contributions of clean air and water, and the lesser noted functions of the environment such as processor and locus of our industrial waste.

The concept of natural capital is not a new one. Economists have long noted that natural capital is a factor in industrial production, albeit a marginal factor. A new view is emerging, the proposition that our economic systems cannot long endure without taking the flow of renewable and nonrenewable resources through economies into account. The "value" of natural capital is becoming paramount to the success of all business. While economists may still insist that its value is less than that of labor, wealth, or technology, it is doubtful that this view can be profitably supported by business in the long term.

This revision of neo-classical economics, yet to be accepted by mainstream academicians, stabilizes both the theory and practice of free-market capitalism and provides business and public policy with a powerful new tool for growth, profitability, and the public good. It is as if we had been sitting on a two-legged stool for the past century, wondering why our economies have become increasingly unbalanced and unsteady.

The concept of natural capital, when intelligently linked to human and manufactured capital, provides the critical connection between the satisfaction of human needs, the continued prosperity of business, and the preservation (some might even suggest the restoration) of the earth's living natural systems.

Most Americans are filled with cornucopian fantasies of technological prowess, where human ingenuity bypasses natural limits and creates unimagined abundance. Optimism easily intertwines with the belief that nanotechnology, biotechnology, computers, and technologies yet to be developed will eliminate hunger, disease, and want.

Dreams of alleviating human suffering are worthy, but what they usually overlook is the absolute necessity for fertile soil, ocean fisheries, a stable climate, biological diversity, and pure water, all of which we are losing, and none of which can be substantively created by any human-made technology known or imagined.

In our pursuit of dominance over the natural world, we have not taken into account the basic principle that industrialism, for all its sophistication, is enormously inefficient with respect to resources, energy, and waste. It is difficult for neo-classical economists, whose hypotheses and theories originated in a time of resource abundance, to understand that the very success of linear industrial systems [Figure 3-21] has laid the groundwork for the next stage in economic evolution. This next stage, whatever it may be called, is being brought about by powerful and much-delayed feedback from high levels of inefficiency and waste. Information from extractive and destructive activities going back a hundred years or more is being incorporated into the market and political economy. As that happens, the foundation of industrialism is giving way while the basis for the next industrial revolution is being established.

This shift is profoundly biological. It's not about the celebration of nature, although that is certainly part of it; it's about the incorporation of natural systems into our industrial life, into our way of making things, our way of processing things and deprocessing things. The reason this shift is going to happen is because cyclical

measures of human welfare, environmental health, or even economic health, for the following reasons.

GNP and GDP hide the negative effects (on humans and on the rest of the ecosphere) of producing many goods and services. Pollution, crime, sickness, death, and depletion of natural resources are all counted as positive gains in the GDP or GNP. Every time an irreplaceable old-growth forest is felled (Chapter 23) or a wetland is drained, the GDP and GNP go up. Every time a chemical or radiation causes cancer and the victim is treated, the GNP and GDP go up. They also rise because of the funeral expenses for the 150,000–350,000 people killed prematurely each year in the United States from air pollution. The $2.2 billion that Exxon

Q: If all ozone-depleting substances were banned now, when might the ozone return to 1985 levels?

industrial systems work better than linear ones. They close the loop and reincorporate wastes as part of the production cycle [Figures 3-22 and 27-3]. There are no landfills in a cyclical society.

If there is so much inefficiency in our system, why isn't it more apparent? The inefficiencies are masked by a financial system in which money, prices, and markets give us improper information. Markets are not giving us proper information about how much our suburbs, spandex, and plastic drinking water bottles truly cost. Instead, we are getting such proper information from our beleaguered air- and watersheds, the overworked and eroded soils, the life-degrading inner cities and rural counties, the breakdown of stability worldwide, and the conflicts based on resource shortages; all these are providing the information that our prices should be giving us but don't.

Prices don't give us good information for a simple reason: improper accounting. Natural capital has never been placed on the balance sheets of companies or the countries of the world. To paraphrase G. K. Chesterton, it could fairly be said that capitalism might be a good idea except that we have never tried it yet. And try we must and will, for capitalism cannot be fully attained or practiced until, as any accounting student will tell us, we have an accurate balance sheet.

As it stands, our economic system is based on accounting principles that would bankrupt a company. Not surprisingly, it is posing problems for the world as a whole. When natural capital is placed on the balance sheet, not as a free amenity of infinite supply, but as an integral and valuable part of the production process, everything changes. The near obsessive pursuit of improvement in human productivity becomes balanced by the need for improved resource productivity. Using more and more resources to make fewer people more productive flies in the face of what we now need and require to improve our society and the environment. After all, it is people we have more of, not natural resources, so it is people we must use in order to reduce the flow (throughput) of matter and energy resources through economies.

And that is what can happen when we move from linear extractive systems [Figure 3-21] to cyclical ones [Figure 3-22]. Instead of the huge capital investments required to find, extract, and process nonrenewable oil, when we shift to renewable wind energy we spend our money on windmill maintenance instead of hydrocarbons. We get more people gainfully employed, reduce capital requirements, and produce almost no pollution.

Many people sincerely believe that an economic system based on the integrity of natural systems is unworkable. To answer that concern, we may well want to reverse the question and ask, "How have we created an economic system that tells us it is cheaper to destroy the earth than to maintain it?" We know this is not the way to take care of our cars, houses, and bridges, but somehow we have managed to overlook a pricing system that discounts the future and sells off the past. Or to put it another way, "How did we create an economic system that confuses capital with income?"

Can we come up with and implement a more rational economic system? I think so. It is right before us. It requires no new theories, only common sense. It is based on the simple but powerful proposition that *all capital must be valued.*

While there may be no *right* way to value a forest or a river, there is a *wrong* way, which is to give it no value at all. If we have doubts about how to value a 500-year-old tree, we need only ask how much it would cost to make a new one from scratch. Or a new river. Or even a new atmosphere.

The work of the future is the absorption and integration of the worth of living systems (Spotlight, p. 647) into every aspect of our culture and commerce, so that human systems mimic natural systems. Only by doing this can our cultures reflect growth and harmony rather than damage and discord.

Critical Thinking

1. If you were placed totally in charge of the world's economy, what are the three most important things you would do? Compare your answers with those of other members of your class.

2. Do you agree with the author of this essay that we must absorb and integrate the worth of living systems into every aspect of our culture and commerce? How would you go about doing this?

spent partially cleaning up the oil spill from the *Exxon Valdez* tanker (p. 547) also raised the GDP and GNP, as did the $1 billion spent because of the accident at the Three Mile Island nuclear power plant in Pennsylvania.

Pollution is counted as a *triple positive gain* even though it decreases the quality of life for hundreds of millions of people and should be subtracted from the GNP. It is counted as a gain in GDP when it is first produced, counted again when society pays to clean it up partially, and counted as a third gain when people become sick or die from exposure to the pollution.

GNP and GDP don't include the depletion and degradation of natural resources or earth capital on which all

A: In about 50 years

economies depend. A country can be headed toward ecological bankruptcy, exhausting its mineral resources, eroding its soils, cutting down its forests, destroying its wetlands and estuaries, and depleting its wildlife and fisheries. At the same time it can have a rapidly rising GNP and GDP, at least for a while, until its environmental debts come due. Any business (or nation) that counted depletion of its capital (assets) as current income would always have a rosy but false picture of its true financial condition.

GNP and GDP hide or underestimate some of the positive effects of responsible behavior on society. More energy-efficient light bulbs, appliances, and cars reduce electric and gasoline bills and pollution, but these beneficial effects register as a decline in GNP and GDP. GDP and GNP indicators also exclude the labor we put into volunteer work, the health care we give loved ones, the food we grow for ourselves, and the cooking, cleaning, and repairs we do for ourselves.

GNP and GDP tell us nothing about economic justice (Guest Essay, p. 774). They don't reveal how resources, income, or the harmful effects of economic growth (pollution, waste dumps, land degradation) are distributed among the people in a country. UNICEF suggests that countries should be ranked not by average per capita GNP or GDP but by average or median income of the poorest 40% of their people.

Solutions: How Can Environmental Accounting Help? Economists have never claimed that GNP and GDP indicators are good measures of environmental health and human welfare, but most governments and business leaders use them that way. Environmentalists and a growing number of economists believe that GNP and GDP indicators should be replaced or supplemented with widely publicized *environmental* and *social indicators* that give a more realistic picture by subtracting from the GDP and GNP things that lead to a lower quality of life and depletion of earth capital.

In 1972, economists William Nordhaus and James Tobin developed an indicator called *net economic welfare (NEW)* to estimate the annual change in a country's quality of life. They calculate a price tag for pollution and other "negative" goods and services included in the GNP and GDP and then subtract them to give the NEW. Economist David Pearce estimates that pollution and natural resource degradation subtract 1–5% from the GDP of developed countries (and 5–15% for developing countries), and that in general these percentages are rising. Other studies tally harmful environmental costs (excluding those from global warming and fossil-fuel depletion) at 2% of the GDP in Japan and Australia, 12–15% in China, 23% in Germany, 40% in Sweden, and 45% in the United States.

Dividing a country's NEW by its population gives the *per capita NEW.* Since 1940, the real per capita NEW in the United States has risen at about half the rate of the real per capita GNP, and since 1968 the gap between these two indicators has been widening.

Economist Robert Repetto and other researchers at the World Resources Institute have developed a *net national product (NNP)* that includes the depletion and destruction of natural resources as a factor in GNP. They have applied this indicator to Indonesia and Costa Rica.

Herman E. Daly, John B. Cobb, Jr., and Clifford W. Cobb have developed an *index of sustainable economic welfare (ISEW)* and applied it to the United States. This comprehensive indicator of human welfare measures per capita GNP adjusted for inequalities in income distribution, depletion of nonrenewable resources, loss of wetlands, loss of farmland from soil erosion and urbanization, the cost of air and water pollution, and estimates of long-term environmental damage from ozone depletion and possible global warming. After rising by 42% between 1950 and 1976, this indicator fell 14% between 1977 and 1990.

A more recent similar indicator is called the *genuine progress indicator (GPI).* When this indicator is applied to the United States, the GPI per person has steadily declined since 1973. Indeed, between 1973 and 1994 the GDP rose from $12,500 to $17,000 per person while the GPI fell from $6,500 to $4,000 per person (Figure 27-7).

Such indicators are far from perfect and require many crude estimates. However, they are more accurate than GNP and GDP as measures of life quality and environmental quality and even economic well-being per person. Without such indicators, we don't know much about what is happening to people, the environment, and the planet's natural resource base. We fail to see what needs to be done, and we don't have a way to measure what types of policies work. In effect, we are trying to guide national and global economies through treacherous economic and environmental waters at ever-increasing speeds using faulty radar.

The *good news* is that such indicators exist. The *bad news* is that so far they are not widely used.

Case Study: Kerala: Improving Life Quality Without Conventional Economic Growth The state of Kerala in southwest India (Figure 11-16) has shown how quality of life can be improved without emphasizing economic growth. Instead, Kerala has sought to better the lot of its people by economic redistribution.

By conventional measures Kerala is one of the world's poorest areas. In 1994, its per capita GDP was $340, compared to $1,090 for all developing countries

Q: How many people don't have a safe supply of drinking water?

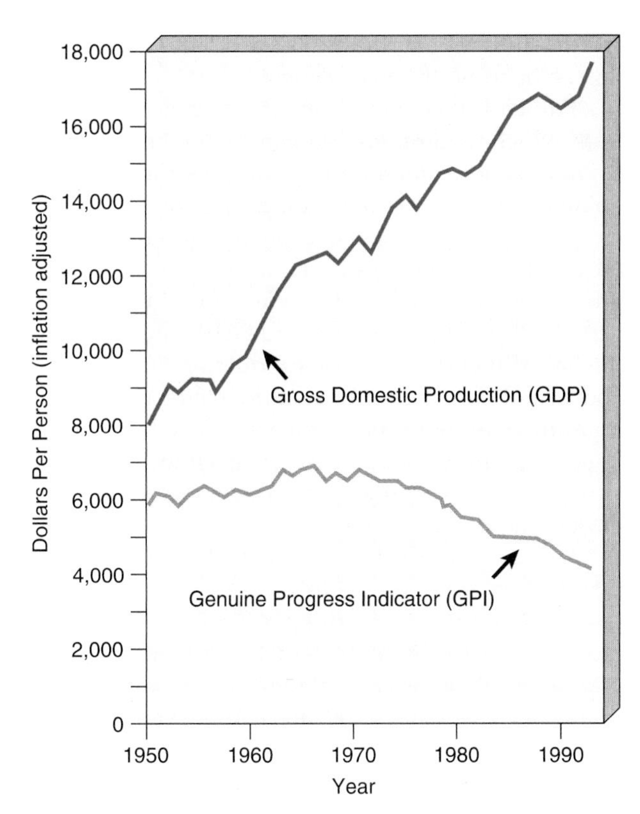

Figure 27-7 Comparison of per capita gross domestic product (GDP) and per capita genuine progress indicator (GPI) in the United States, 1950–94. Units of per capita GDP and GPI are inflation adjusted using 1982 dollars. (Data from Clifford Cobb, Ted Halstead, and Jonathan Rowe)

and $18,130 for all developed countries ($25,220 for the United States). However, in terms of quality of life it has some of the highest scores among developing countries.

Life expectancy in Kerala is 70 years, compared with 64 years in developing countries and 59 years in India. The infant mortality rate is 17 per 1,000 births, compared with 68 for developing countries and 79 for India. The total fertility rate is 2.3 in Kerala, compared with 3.4 both for developing countries and India, and Kerala's population is growing at a slower rate than that of the United States.

Essentially 100% of Kerala's citizens can read and write, compared with fewer than 50% for the rest of India. Indeed, the literacy rate in Kerala is higher than in the United States, where per capita GNP is 75 times higher. Kerala's literate and hard-working labor force is in demand throughout the world. At least one-quarter of its citizens work in the Persian Gulf, sending most of their wages home.

Kerala also leads India in the quality of its roads, schools, hospitals, public housing, drinking water, sanitation, immunization programs, women's rights, and nutrition programs for infants and for pregnant and lactating women. All of its people have access to free or inexpensive medical care, and all households receive ration cards allowing them to buy rice and certain basic commodities at subsidized prices. Since 1960 a land reform program has distributed small plots of land to more than 3 million tenants and landless poor.

Kerala manages to do all of this while being extremely poor and being one of the most crowded places on earth. It is twice as crowded as India as a whole and squeezes its 30 million people (roughly the same as the population of Canada) into an area the size of Vancouver Island.

Many analysts see Kerala as a shining example of people working together to help themselves by choosing self-reliance and economic redistribution of local wealth over conventional economic growth and dependence on outside loans. Kerala demonstrates that a very low-level economy can provide its citizens with education, health services, and a sense of community and hope.

It is tempting to romanticize Kerala, but its extremely poor people need much more than they have. However, they have figured out ways to do better with what they do have for more of their people than almost any other place on the earth.

What Are Internal and External Costs? All economic goods and services have both internal and external costs. For example, the price a consumer pays for a car reflects the costs of the factory, raw materials, labor, marketing, and shipping, as well as a markup to allow the car company and its dealers some profits. After a car is purchased, the buyer must pay for gasoline, maintenance, and repair. All these direct costs, which are paid for by the seller and the buyer of an economic good, are called **internal costs**.

Making, distributing, and using any economic good or service also involve **externalities**: social costs or benefits not included in the market price. For example, if a car dealer builds an aesthetically pleasing showroom and grounds, that is an **external benefit** to people who enjoy the sight at no cost.

On the other hand, extracting and processing raw materials to make and propel cars depletes nonrenewable energy and mineral resources, produces solid and hazardous wastes, disturbs land, pollutes the air and water, contributes to depletion of stratospheric ozone and possible global climate change, and reduces biodiversity and ecological integrity. These harmful effects are **external costs** passed on to workers, the public, and in some cases future generations. Car owners add to these external costs when they throw trash out of a car, drive a gas-guzzler (which produces more air pollution than a more efficient car), disconnect or don't maintain a car's air-pollution control devices, or don't keep the engine tuned.

Because these harmful costs aren't included in the market price, people don't connect them with car ownership. Still, everyone pays these hidden costs sooner or later, in the form of higher costs for health care and health insurance and higher taxes for pollution control.

To conventional economists, external costs are minor defects in the flow of production and consumption in a self-contained economy not significantly dependent on earth capital (Figure 27-2). They assume that these defects can be rectified using the profits made from additional economic growth in a free-market economy.

To environmentalists and an increasing number of economists and business leaders, harmful externalities are a warning sign that our economic systems are stressing the ecosphere and depleting earth capital (Figures 1-2 and 27-3). They believe that these harmful external costs should be included in the market prices of goods and services, a process economists call *internalizing the external costs* (Guest Essay, p. 756). According to economist Harold M. Hubbard, "The progress of civilization can be charted in terms of the internalization of costs formerly viewed as external."

27-3 SOLUTIONS: USING ECONOMICS TO IMPROVE ENVIRONMENTAL QUALITY

Should We Shift to Full-Cost Pricing? As long as businesses receive subsidies and tax breaks for extracting and using virgin resources and are not taxed for the pollutants they produce, few will volunteer to reduce short-term profits by becoming more environmentally responsible. Assume you own a company and believe it's wrong to subject your workers to hazardous conditions and pollute the environment beyond what natural processes can handle. Suppose you voluntarily improve safety conditions for your workers and install pollution controls, but your competitors don't. Then your product will cost more than theirs, and you will be at a competitive disadvantage. Your profits will decline; you may eventually go bankrupt and have to lay off your employees.

One way of dealing with the problem of harmful external costs is for the government to levy taxes, pass laws, provide subsidies, or use other strategies that encourage or force producers to include all or most of these costs in the market prices of economic goods and services. Then that price would be the **full cost** of these goods and services: internal costs plus short- and long-term external costs. The two main goals are to **(1)** close the gap between real and false prices by having prices that tell the environmental truth and **(2)** have people and businesses pay the full costs of the harm they do to others and the environment.

Full-cost pricing involves *internalizing the external costs*, which requires government action because few companies will intentionally increase their cost of doing business unless their competitors must do so as well. If the market prices of economic goods reflected all or most of their full estimated cost, economic growth would be redirected. We would increase the beneficial parts of the GNP and GDP (and decrease the harmful parts), increase production of beneficial goods, and raise the net economic welfare. Preventing pollution would become more profitable than cleaning it up, and waste reduction, recycling, and reuse would be more profitable than burying or burning most of the waste we produce.

Because external costs would be internalized, the market prices for most goods and services would rise. But the total price we would pay would be about the same because the hidden external costs related to each product would already be included in its market price. Using full-cost pricing to internalize external costs provides consumers with information needed to make informed economic decisions about the effects of their lifestyles on the planet's life-support systems.

However, as external costs are internalized, economists and environmentalists warn that governments must reduce income, payroll, and other taxes and must withdraw subsidies formerly used to hide and pay for these external costs. Otherwise, consumers will face higher market prices without tax relief—an unjust and politically unacceptable policy guaranteed to fail.

Some goods and services would cost less because internalizing external costs encourages producers to find ways to cut costs (by inventing more resource-efficient methods of production); it also encourages producers to offer more earth-sustaining (or *green*) products. Jobs would be lost in earth-degrading businesses, but at least as many (some analysts say more) jobs could be created in earth-sustaining businesses. If this change in the way market prices are established took place over several decades, most current earth-degrading businesses would have time to transform themselves into profitable earth-sustaining businesses.

Full-cost pricing seems to make a lot of sense. Why isn't it more widely done? One reason is that many producers of harmful and wasteful goods would have to charge so much that they couldn't stay in business, or they would have to give up government subsidies and tax breaks that have helped hide the external costs of their goods and services. Another problem is that it is hard to put a price tag on many of the harmful environmental and health costs.

Studies estimate that governments around the globe spend about $600 billion per year to subsidize

Portney, P. 1998. "Counting the Cost." *Environment*, vol. 40, no. 2, 14(8).

deforestation, overfishing, overgrazing, unsustainable agriculture, nonrenewable fossil fuels and nuclear energy, groundwater depletion, and other environmentally destructive activities—what environmental expert Norman Myers calls *perverse subsidies*. It is estimated that eliminating these earth-degrading subsidies that distort the global economy subsidies would allow about a 7% cut in the global tax burden of $7.5 trillion per year and encourage job creation and investment.

Despite the difficulties, proponents believe that full-cost pricing for harmful environmental and health effects deserves a serious try. They argue that doing the best we can to estimate and internalize current external costs is far better than continuing the current pricing system, which gives too little or misleading information about the environmental and health effects of goods and services. The key question is whether the problems with the *tell-the-truth full-cost* pricing system are worse than those with the current *hide-the-true-cost* pricing system. What do you think?

How Useful Is Cost–Benefit Analysis? One of the chief tools corporations and governments use in making economic decisions is **cost–benefit analysis**. This approach involves comparing the estimated short-term and long-term costs (losses) with the estimated benefits (gains) for various courses of action. Cost–benefit analyses can be useful guides and can indicate the cheapest way to go, but they can also be misused (Spotlight, p. 762).They can even do great harm if they aren't carefully conducted and scrutinized.

Furthermore, as Worldwatch Institute researcher David Rodman reminds us, *Environmental problems, like most important policy issues, involve more than costs and benefits; they also involve rights and wrongs, values and visions. (If crime paid, cost–benefit analysis would endorse it.)*

To minimize possible abuses, environmentalists and economists advocate the following guidelines for all cost–benefit analyses: **(1)** Use uniform standards; **(2)** clearly state all assumptions; **(3)** evaluate the reliability of all data inputs as high, medium, or low; **(4)** make projections using low, medium, and high discount rates; **(5)** show the estimated range of costs and benefits based on various sets of assumptions; **(6)** estimate the short- and long-term benefits and costs to all affected population groups; **(7)** estimate the effectiveness of the project or form of regulation instead of assuming (as is often done) that all projects and regulations will be executed with 100% efficiency and effectiveness; and **(8)** open the evaluations to public review and discussion.

Should We Rely Mostly on Regulations or Market Forces? Most economists agree that controlling or preventing pollution and reducing resource waste require government intervention in the marketplace. Such government action can take the form of regulation, the use of market forces, or some combination of these approaches.

Regulation is a *command-and-control* approach. It involves enacting and enforcing laws, for example, that set pollution standards, regulate harmful activities, ban the release of toxic chemicals into the environment, and require that certain irreplaceable or slowly replenished resources be protected from unsustainable use (or from any use at all).

Most studies of the effects of environmental regulations in the United States have found that they do businesses very little harm. Indeed, in many cases they have led to improvements in resource use efficiency, which reduce costs, and to innovative products and industrial processes, which increase profits. A recent study by economist Robert Repetto at the World Resources Institute showed that between 1970 and 1990, the U.S. industries that spent the most on pollution control fared significantly better than average in the global marketplace.

However, business leaders and many environmentalists in the United States agree that some pollution control regulations discourage innovation by being too prescriptive and costly, and need to be modified. They propose using regulations to set goals but then freeing industries to meet such goals in any way that works.

Market forces can help improve environmental quality and reduce resource waste, mostly by encouraging the internalization of external costs. This is based on a fundamental principle of the marketplace in today's mixed economic systems: *What we reward (mostly by subsidies and tax breaks) we tend to get more of, and what we discourage (mostly by regulations and taxes) we tend to get less of.*

One way to put this principle into practice would be *to phase in government subsidies that encourage earth-sustaining behavior and phase out current perverse subsidies that encourage earth-degrading behavior*. The difficulty with this ecologically and economically appealing *carrot approach* is that removing or adding subsidies involves political decisions; these are easily swayed by powerful economic interests that want to preserve ecologically unsound subsidies to increase their short-term profits.

Another market approach is for the government to *grant tradable pollution and resource-use rights*. For example, a total limit on emissions of a pollutant or use of a resource could be set, and the total would then be allocated among manufacturers or users by permit. Permit holders not using their entire allocation could use it as a credit against future expansion, use it in another part of their operation, or sell it to other

Hint: Enter the search terms *environmental economics, evaluation* using the Subject Guide.

Some Problems with Cost–Benefit Analysis

SPOTLIGHT

There are several controversies about cost–benefit analysis. One involves the **discount rate,** an estimate of a resource's future economic value compared to its present value. *The size of the discount rate (usually given as a percentage) is a primary factor affecting the outcome of any cost–benefit analysis.*

At a zero discount rate, a stand of redwood trees worth $1 million today will still be worth $1 million 50 years from now; thus there is no need to cut them down for short-term economic gain. However, at a 10% annual discount rate (normally used by most businesses and by the U.S. Office of Management and Budget), the same stand will be worth only $10,000 in 50 years. As a result, it makes short-term economic sense to cut them down now and invest the profits in something else (Connections, p. 649).

Proponents of high (5–10%) discount rates argue that inflation will reduce the value of their future earnings. They also fear that innovation or changes in consumer preferences will make a product or service obsolete.

Environmentalists question this belief, pointing out that current economic systems are based on depleting the natural capital that supports them. High discount rates worsen this situation by encouraging such rapid exploitation of resources for immediate payoffs that sustainable use of most potentially renewable natural resources is virtually impossible. They believe that unique and scarce resources should be protected by having a 0% or even a negative discount rate and that discount rates of 1–3% would make it profitable to use other resources sustainably or slowly. At its core, the choice of a discount rate is really an *ethical decision* about our responsibility to future generations.

Another problem with cost–benefit analysis is determining *who benefits and who is harmed.* In the United States, an estimated 100,000 employees die each year because of exposure to hazardous chemicals and other safety hazards at work, and an additional 400,000 are seriously injured by such exposure. In many other countries (especially developing countries), the situation is much worse. Is this a necessary or an unnecessary cost of doing business?

Another limitation of cost–benefit analysis is that *many things we value cannot easily be reduced to dollars and cents.* We can put estimated price tags on human life, good health, clean air and water, pollution and accidents that are prevented, wilderness, the northern spotted owl, and various forms of earth capital (Figure 1-2). However, the dollar values different people assign to such things vary widely because of different assumptions, discount rates, and value judgments, leading to a wide range of projected costs and benefits.

Because these and other estimates of costs and benefits are so variable, *figures can easily be weighted to achieve the outcome desired by proponents or opponents of a proposed project or action.* For example, one industry-sponsored cost–benefit study estimated that compliance with a standard to protect U.S. workers from vinyl chloride would cost $65–90 billion; in fact, less than $1 billion was actually needed to comply with the standard.

In 1996, 11 prominent U.S. economists published a joint statement concluding, "We suggest that benefit–cost analysis has a potentially important role to play in helping inform regulatory decision-making, although it should not be the sole basis for such decision-making." Analysis of various cost–benefit analyses indicates that using this method as the primary way to make decisions is somewhat like trying to detect a car speeding at 160 kilometers per hour (100 miles per hour) with a radar device so unreliable that at best it can tell us only that the car's speed is somewhere between 80 kph (50 mph) and 8,000 kph (5,000 mph).

Critical Thinking

Do you believe that cost–benefit analysis should be used as the primary way to evaluate whether any new environmental law or regulation should be put into effect and whether any existing environmental law or regulation should be weakened or strengthened? Explain. What should be the role of cost–benefit analysis? Why?

companies. Tradable rights could also be established among countries to preserve biodiversity and to reduce emissions of greenhouse gases, ozone-destroying chemicals, lead, and air and water pollutants chemicals with harmful regional (or global) effects.

Some environmentalists support tradable pollution rights as an improvement over the current regulatory approach. Other environmentalists believe that allowing companies to buy and trade rights to pollute is wrong because it allows the wealthiest companies to continue polluting, thereby excluding smaller companies from the market. Others point out that although pollution rights charge for the right to pollute up to a certain limit, the command-and-control approach gives away the right to pollute up to a certain level.

Many environmentalists contend that pollution rights trading does not reduce overall pollution, but

Swift, B. 1998. "A Low-Cost Way to Control Climate Change." *Issues in Science and Technology*, vol. 14, no. 3, 75(7).

allows polluters to shift harm from one place to another. Critics also argue that this approach creates an incentive for fraud because most pollution control regulations are based on self-reporting of pollution outputs (and government monitoring of such outputs is inadequate).

Another market-based method is to *enact green taxes or effluent fees* that would help internalize many of the harmful external costs of production and consumption. This method could include taxes on each unit of pollution discharged into the air or water, each unit of hazardous or nuclear waste produced, each unit of virgin resources used, and each unit of fossil fuel used. Currently, 90% of the world's $7.5 trillion yearly tax burden is levied on income and investment, while less than 5% comes from taxes on environmentally harmful activities. Fully taxing pollution would raise about $1 trillion per year worldwide, enough to allow a 15% cut in taxes on wages and profits.

Experience in the Netherlands and Germany shows that phasing in such taxes can encourage creativity in solving environmental problems and reducing costs by preventing pollution, using fewer resources, and developing more earth-sustaining technologies and products. For example, in the Netherlands fees for emissions of toxic metals (such as cadmium, lead, and mercury) into waterways have been gradually increased since 1970. According to studies, these taxes were the major factor in reducing emissions of toxic metals 86–97% between 1976 and 1994.

There are two major problems with this tax punishment, or stick, approach. *First*, because the taxes or effluent fees are set politically rather than by markets, elected officials find it easier to aim for popular approval rather than for economic and ecological efficiency. As a result, the taxes are too low to be effective and thus undermine the entire concept.

Second, elected officials are likely to see such taxes as ways of raising revenue instead of improving economic and ecological efficiency. However, economists point out that *green taxes* on pollution output, resource depletion, and environmental degradation use would work if they reduced or replaced income, payroll, or other taxes and if the poor were given a basic safety net to reduce the regressive nature of consumption taxes on essentials such as food, fuel, and housing. In other words, environmental taxes must be seen as a *tax-shifting* instead of a *tax-burden* approach.

In 1991, Sweden made such a shift by instituting effluent taxes on each metric ton of SO_2 and CO_2 produced and compensating for this tax increase by reducing income tax and other tax revenues by 1.9%. In 1994 and 1996, Denmark instituted various green taxes and reduced income tax and other taxes by 3%. Spain, the Netherlands, and the United Kingdom have also made modest tax shifts. In 1998, Germany announced plans to tax energy sales and use the revenue to reduce taxes on wages by 2.6%.

Charging user fees is another market-based method. For example, users would pay fees to cover all or most costs for grazing livestock, extracting lumber and minerals from public lands, using water provided by government-financed projects, and using public lands for recreation. In principle, this *user-pays* approach is favored by the public. However, it is opposed by ranchers, timber harvesters, miners, and tourists who benefit from having all taxpayers subsidize their low-cost use of public lands and resources.

Another market approach would *require businesses to post a pollution prevention or assurance bond* when they plan to develop a new mine, plant, incinerator, landfill, or development and before they introduce a new chemical or new technology. The size of the performance bond would be based on estimates of the worst-case consequences of each project, chemical, or technology as determined by the producer and reviewed by an independent panel of risk experts. Each deposit would be kept in an interest-producing escrow account. After a set length of time, the deposit (with interest) would be returned *minus* environmental costs. If harm occurred, all or part of the bond would be used for cleanup and environmental restoration and to pay damages to those who were harmed. This approach is similar to the performance bonds contractors are now required to post for major construction projects. Understandably, this approach is opposed by businesses that would have to post such bonds.

Each of these approaches has advantages and disadvantages (Table 27-1). Currently, in the United States private industry and local, state, and federal government spend about $130 billion a year to comply with federal environmental regulations. Studies estimate that greater reliance on a variety of market-based policies could cut these expenditures by one-third to one-half.

Most analysts see a combination of command-and-control and market-based approaches as the best solutions to most environmental problems. Regulatory abuses and excessively expensive regulations should be eliminated. However, regulations play an important environmental role and sometimes benefit polluters by stimulating companies to innovate in ways that make them more competitive. Much more research must be done to determine the most effective mix of regulatory and market-based strategies for each type of environmental problem.

Should We Emphasize Pollution Control or Pollution Prevention? Shouldn't our goal be zero pollution? Ideally, yes; in the real world, not

Hint: Enter the search term *emissions credit trading* using the Subject Guide.

Table 27-1 Economic Solutions to Pollution and Resource Waste

Solution	Internalizes External Costs	Innovation	International Competitiveness	Administrative Costs	Increases Government Revenue
Regulation	Partially	Can encourage	Decreased*	High	No
Subsidies	No	Can encourage	Increased	Low	No
Withdrawing harmful subsidies	Yes	Can encourage	Decreased*	Low	Yes
Tradable rights	Yes	Encourages	Decreased*	Low	Yes
Green taxes	Yes	Encourages	Decreased*	Low	Yes
User fees	Yes	Can encourage	Decreased*	Low	Yes
Pollution-prevention bonds	Yes	Encourages	Decreased*	Low	No

*Unless more cost-effective and productive technologies are developed.

necessarily. First, natural processes can handle some of our wastes, as long as we don't destroy, degrade, or overload these processes. However, environmentalists argue that harmful chemicals that either cannot be degraded by natural processes or that break down very slowly should not be released into the environment, or should be released only in small amounts and regulated by special permit.

Second, as long as we continue to rely on pollution control, we can't afford zero pollution. After we've removed a certain proportion of the pollutants in air, water, or soil, the cleanup cost per additional unit of pollutant rises sharply (Figure 27-8). Beyond a certain point, the cleanup costs exceed the harmful costs of pollution. Some businesses could then go bankrupt, and some people could lose jobs, homes, and savings. On the other hand, if we don't go far enough, dealing with the harmful external effects of pollution can cost more than pollution reduction.

To find the breakeven point, economists plot two curves: a curve of the estimated economic costs of cleaning up pollution and a curve of the estimated social (external) costs of pollution. Adding the two curves together, we get a third curve showing the total costs. The lowest point on this third curve is the point of the *optimal level of pollution* (Figure 27-9).

On a graph, this looks neat and simple, but environmentalists and business leaders often disagree in their estimates of the harmful costs of pollution. This approach assumes that we know which substances are harmful and how much each part of the environment can handle without serious environmental harm—things we probably will never know precisely.

Most environmentalists and a growing number of economists and business leaders believe that environmental laws should emphasize pollution prevention, which avoids most of the regulatory problems and excessive costs of end-of-pipe pollution control. Some environmentalists also call for reversing the current assumption that a chemical or new technology is safe until it is shown to be harmful. They argue that if a chemical or new technology were considered potentially harmful until shown to be safe—and if this approach were coupled with full-cost pricing—emphasis would shift from costly pollution cleanup to less costly and more effective pollution prevention (Guest Essays, pp. 70 and 586). However, most economists and business leaders argue that the *harmful-until-shown-safe* approach would wreck the economy and put large numbers of people out of work. What do you think?

Is Encouraging Global Free Trade Environmentally Helpful or Harmful? On April 15, 1994, representatives of 120 nations signed the Uruguay Round of the General Agreement on Tariffs and Trade (GATT). This is a revised version of the 1948 GATT convention, which attempted to lower tariff barriers to world trade among member nations. The new GATT establishes a World Trade Organization (WTO), giving it the status of a major international organization (similar to the United Nations and the World Bank) and the power to oversee and enforce the agreement.

The 1989 Free Trade Agreement (FTA) between Canada and the United States and the 1993 North American Free Trade Agreement (NAFTA) among Canada, the United States, and Mexico are also designed to remove trade barriers among participating nations.

Proponents argue that agreements to reduce global trade barriers have a number of important benefits. First, *such agreements will benefit developing*

Q: How many people die prematurely every year from drinking contaminated water?

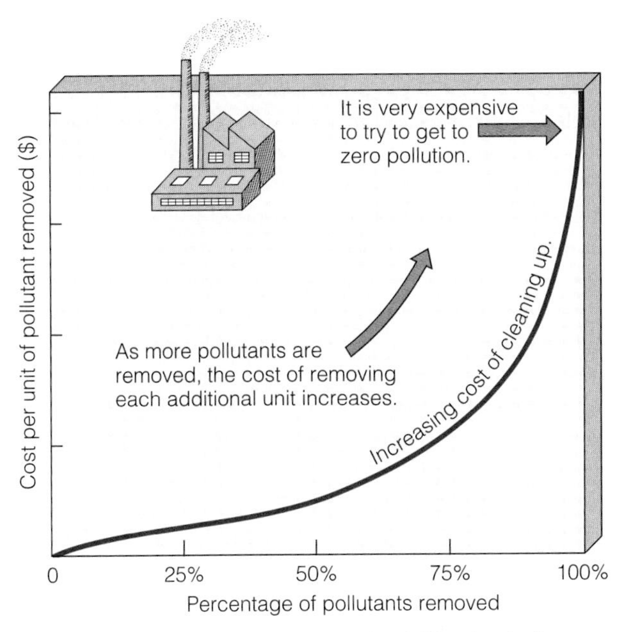

Figure 27-8 The cost of removing each additional unit of pollution rises exponentially, which explains why it is usually cheaper to prevent pollution than to clean it up.

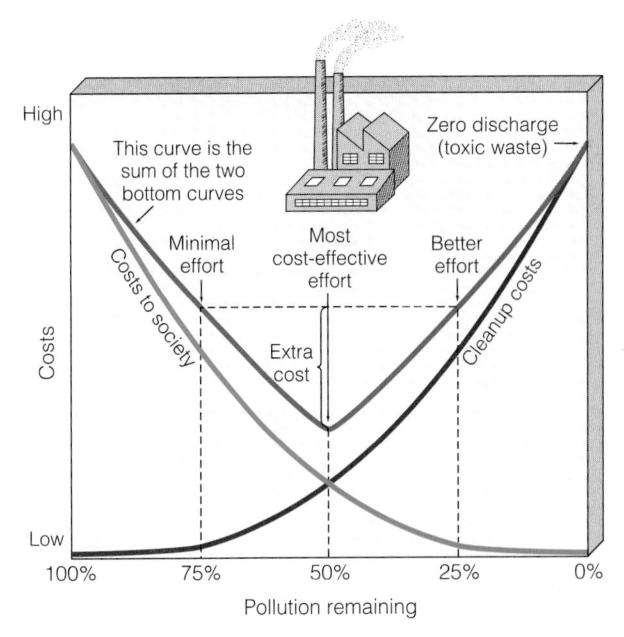

Figure 27-9 Finding the optimal level of pollution. This graph shows the optimal level at 50%, but the actual level varies depending on the pollutant.

countries, whose products are often at a competitive disadvantage in the global marketplace because of trade barriers erected by developed countries. Second, *they can allow consumers to buy more things at cheaper prices, thus stimulating economic growth in all countries.* Using this idea, proponents of GATT contend that it will save U.S. consumers $35 billion a year ($300 per household), boost the U.S. economy by as much as $219 billion a year after 10 years, and add millions of jobs.

Finally, such agreements *can raise the overall global levels of environmental protection and worker health and safety.* For example, agricultural trade barriers that protect crop prices from free-market competition can cause farmers to produce large surpluses because of guaranteed prices. This can cause farmers to reduce biodiversity and ecological integrity by cutting more forests, draining more wetlands, and bringing more easily eroded land into production. Ranching subsidies in Brazil have stimulated destruction of tropical rain forests (Section 24-3). Current subsidies for industrialized agriculture also reduce the incentive to shift to more sustainable agriculture (Section 12-6). Similarly, timber subsidies in the United States and other countries lead governments to sell timber from public forests for much less than the private market price (Section 24-5). This accelerates timber harvesting and reduces the incentive to use more sustainable methods of timber harvesting, such as selection cutting and strip cutting (Figure 24-19).

Because of such potential benefits, some environmental groups generally supported FTA and NAFTA. However, most environmental groups, and those concerned with consumer protection and worker health and safety, oppose the new version of GATT for several reasons.

First, *they believe that GATT will not provide ample economic benefits for everyone.* This hypothesis is valid if only products, but not factories or workers, cross borders. Currently, 84% of the benefits of world trade flow to the richest one-fifth of the world's population, but only 0.9% to the poorest one-fifth. At best, the new version of GATT would not change this situation, and it could worsen it.

Second, they contend that *GATT will increase the economic and political power of multinational corporations and decrease the power of small businesses, citizens, and democratically elected governments.* The revised GATT treaty was developed mostly in secret by government and business officials dominated by the interests of multinational corporations based in developed countries. Under the new agreement, any nation not abiding by the ruling of a GATT panel (whose deliberations are not open to public scrutiny) could be charged heavy fines by the WTO or by a country whose complaint is upheld.

In his 1995 book, *When Corporations Rule the World,* global business expert David Korten pointed out that during the 1980s, when the U.S. Fortune 500 companies downsized by eliminating millions of jobs, their sales increased 1.4 times, their assets 2.3 times, and the average annual compensation of their chief executive officers 6.1 times, averaging $3.8 million.

A: About 10 million (more than half of them children under age 5)

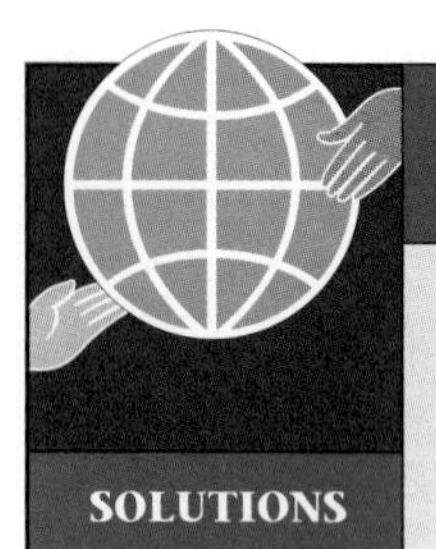

Improving Trade Agreements

SOLUTIONS

Critics of current trade agreements have suggested a number of ways to encourage free trade without sacrificing environmental protection or health and worker safety. They would rewrite and correct the serious weaknesses in GATT (and in FTA and NAFTA) and turn it into GAST: the *General Agreement for Sustainable Trade.* They offer the following suggestions for doing this:

- Having low or no trade tariffs on imports from nations that promote ecologically sustainable development by using environmental accounting and full-cost pricing
- Judging GATT or any trade agreement primarily on how it benefits the environment, workers, and the poorest 40% of humanity
- Setting minimum environmental, consumer protection, and worker health and safety standards for all participating countries
- Allowing any country to set more stringent versions of such standards without being penalized by fines or other means
- Requiring all panels or bodies setting and enforcing GATT standards to have environmental, labor, consumer, and health representatives from developed countries and developing countries alike
- Opening all discussions and findings of any GATT panel or other WTO body to global public scrutiny
- Allowing international environmental agreements to prevail when they conflict with GATT or any other trade agreement

Unless citizens exert intense pressure on legislators, critics warn that such safeguards will not be incorporated into international trade agreements.

Critical Thinking

1. Do you believe that the current GATT treaty will help or hinder environmental protection **(a)** globally, **(b)** in the country where you live, and **(c)** in your community? Explain.

2. Explain why you agree or disagree with each of the suggestions in this box. If you agree with all or most of these proposals, how could they be implemented? What role could you play in making such changes?

Third, *GATT will probably weaken environmental and health and safety standards in developed countries.* Under the new GATT, any country could be fined (or be subjected to tariff fees) if any portion of it adopted stricter environmental, health, worker safety, resource use, or other standards than the uniform global standards established by the WTO (dominated by multinational companies based in developed countries).

Faced with cheaper foreign products, domestic businesses operating in the international marketplace will have three choices: **(1)** go out of business, **(2)** move some or all of their operations abroad to take advantage of cheaper labor and less restrictive environmental and worker safety regulations, or **(3)** lobby to weaken domestic environmental, health, and worker and consumer safety laws. Under such pressures resulting from GATT, elected officials in the United States and other developed countries are likely to find it politically and economically necessary to weaken such laws to stem a flow of businesses, capital, tax revenue, and jobs out of their countries.

Environmentalists point out that the existing FTA, NAFTA, and GATT trade agreements have lowered some environmental standards. British Columbia discontinued a government-funded tree-planting program because the United States (under pressure from U.S. lumber companies) argued that as an unfair subsidy to Canada's timber industry, it violated the 1989 FTA agreement. Canada had to relax its regulations concerning pesticides and food irradiation to bring them in line with the weaker ones in the United States. The Canadian timber industry is urging its government to challenge a U.S. law requiring the use of recycled fiber in newsprint on the grounds that it is a trade barrier.

Under the new GATT, environmentalists fear that government bans on the export or import of raw logs, intended to slow the destruction of rain forests or ancient forests, could be overturned. Germany could be forced to repeal its law requiring recycling of all beverage containers. Countries might not be able to restrict imports of hazardous wastes, banned medicines, and dangerous pesticides. International moratoriums and quotas on whale harvests, as well as bans on trading ivory to protect elephants, could be overturned. A developing country banning dirty industries could be accused of violating free trade.

In 1998, a WTO panel ruled against a U.S. environmental law protecting sea turtles from being drowned in shrimp nets. The panel sided with Thailand, Malaysia, Pakistan, and India, who argued that the U.S. law barring imports of shrimp from countries that do not have sea turtle protection policies discriminated against them.

The new GATT wouldn't prohibit all such activities, but countries could be fined heavily for engaging

Q: What is the largest source of water pollution in the United States?

Figure 27-10 Extreme poverty forces hundreds of millions of people to live in slums such as this one in Rio de Janeiro, Brazil, where adequate water supplies, sewage disposal, and other services don't exist. (United Nations)

in them. Because this would reduce profits, the net effect would be to discourage the implementation of these and many other national environmental laws.

Under the old GATT a country receiving an unfavorable ruling by a GATT panel could block the decision unless the ruling was unanimous. Under the new GATT, panel rulings cannot be blocked. Critics of the latest version of GATT call for it to be improved (Solutions, left).

27-4 SOLUTIONS: REDUCING POVERTY

Does the Trickle-Down Approach to Reducing Poverty Work? **Poverty** is usually defined as the inability to meet one's basic economic needs. Currently, an estimated 1.3 billion people (70% of them women) in developing countries—one of every five people on the planet—have an annual income of less than $370 per year, with many of them living in urban slums (Figure 27-10). This income of roughly $1 per day is the World Bank's definition of poverty.

Poverty causes premature deaths and preventable health problems. It also tends to increase birth rates and often pushes people to use potentially renewable resources unsustainably in order to survive.

Most economists believe that a growing economy is the best way to help the poor. This is the so-called *trickle-down hypothesis*: Economic growth creates more jobs, enables more of the increased wealth to reach workers, and provides greater tax revenues that can be used to help the poor help themselves.

However, the facts suggest that either the hypothesis is wrong or it has not been applied. Instead of *trickling down*, most of the benefits of economic growth as measured by income have *flowed up* since 1960, making the top one-fifth of the world's people much richer and the bottom one-fifth poorer; most of those in between have lost, or gained only slightly, in real per capita GNP (Figure 1-8). This trend has accelerated in the 1980s and 1990s (Figure 27-11). In 1997, the combined $1 trillion wealth of the world's 225 richest people equalled the combined annual income of the world's poorest 2.5 billion people.

A major component of the flow-up system is government subsidies that encourage resource depletion, pollution, and environmental degradation—what environmentalist Norman Myers (Guest Essay, p. 530) calls *perverse subsidies*. Examples in the United States are giveaways or below-market prices for timber ($200–350

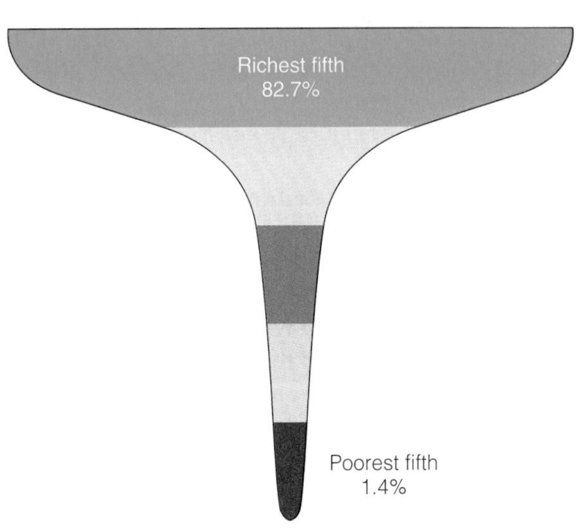

Figure 27-11 Data on the global distribution of income showing that, instead of trickling down, most of the world's income has flowed up, with the richest 20% of the world's population receiving more of the world's income than all of the remaining 80% in 1991. Each horizontal band in this diagram represents one-fifth of the world's population. This upward flow of global income has increased since 1960 and has accelerated in the 1980s and 1990s. Do you favor or oppose this trend? Explain. If you oppose this trend, what do you believe are the most important ways to correct it? (Data from UN Development Programme)

A: Agriculture (responsible for almost two-thirds)

million per year and $40 billion worldwide), grazing rights ($150–200 million per year), and extraction of minerals on public lands ($2–6 billion per year); predator control ($45 million per year); corporate tax breaks ($53 billion per year); and farm subsidies ($29 billion per year and $300 billion worldwide). According to Norman Myers, global subsidies for agriculture, fossil fuels, nuclear energy, water, fisheries, forestry, and other smaller categories amount to at least $500 billion per year—equal to 9% of all government revenue. Such massive subsidies create an uneven economic playing field that works against a free-market economy and rewards earth-degrading activities. Eliminating these environmentally harmful subsidies could allow an 8% cut in the global tax burden on income and investment. In the United States, this could lead to a net annual tax cut of $500 per person or $2,000 for a family of four.

In 1995, almost every proposal to cut corporate welfare in the United States—supported by a broad coalition of environmental and public health leaders as well as conservative think tanks—met a quiet death in congressional committees. However, under the banner of saving money and cutting the deficit, in 1995 Congress made drastic cuts in heating fuel and legal subsidies for the poor, housing subsidies for the elderly and disabled, and tuition grants for national service volunteers. As environmentalist and system analysis expert Donella Meadows puts it,

> *The goal can't be cutting since the fattest opportunities for cutting are bypassed. . . . The only real goal I can see in the new budget agenda is a strong shift from the long-term good of the nation to the short-term good of the rich and powerful. . . . It shows not a shred of compassion, community, or rationality.*

How Can Poverty Be Reduced? Analysts point out that reducing poverty requires the governments of most developing countries to make policy changes, including shifting more of the national budget to help the rural and urban poor work their way out of poverty and giving villages, villagers, and the urban poor title to common lands and to crops and trees they plant on them.

Analysts also urge developed countries and the wealthy in developing countries to help reduce poverty. Several controversial ways to do this have been suggested. First, *forgive at least 60% of the almost $2 trillion that developing countries owe to developed countries and international lending agencies.* Some of this debt can be forgiven in exchange for carefully monitored agreements by the governments of developing countries to increase expenditures for family planning, health care, education, land redistribution, protection and restoration of biodiversity, and more sustainable use of renewable resources.

Another proposal is to *increase the nonmilitary aid to developing countries from developed countries.* Proponents say that this aid should go directly to the poor to help them become more self-reliant. Analysts also recommend that we *shift most international aid from large-scale to small-scale projects intended to benefit local communities of the poor.* They would also *encourage banks and other organizations to make small loans to poor people wanting to increase their income* (Solutions, right).

Analysts also suggest that *international lending agencies should be required to use a standard environmental and social impact analysis to evaluate any proposed development project.* Proponents believe that no project should be supported unless its net environmental impact using full-cost accounting is favorable, most of its benefits go to the poorest 40% of the people affected, and the local people it affects are involved in planning and executing the project. They call for careful monitoring of all projects and an immediate halt in funding whenever environmental safeguards are not followed. A final proposal is to *establish policies that encourage both developed countries and developing countries to slow population growth and stabilize their populations* (Section 11-3).

27-5 SOLUTIONS: CONVERTING TO EARTH-SUSTAINING ECONOMIES

How Can We Make Working with the Earth Profitable? Greening Business In 1994, entrepreneur and business leader Paul Hawken (Guest Essay, p. 756) wrote *Ecology and Commerce*, a widely acclaimed book describing what's wrong with business and how it can be transformed to work with the earth. Hawken and several other business leaders and economists have laid out the following principles for transforming the planet's current earth-degrading economic systems into earth-sustaining, or restorative, economies over the next several decades:

- *Reward (subsidize) earth-sustaining behavior.*
- *Discourage (tax and don't subsidize) earth-degrading behavior—the polluter or degrader pays principle*
- *Use full-cost accounting to include the ecological value of natural resources in their market prices.*
- *Use environmental and social indicators to measure progress toward environmental and economic sustainability and human well-being.* (Figure 27-7).
- *Use full-cost pricing to include the external costs of goods and services in their market prices.*
- *Replace taxes on income and profits with taxes on throughput of matter and energy.*
- *Use low discount rates for evaluating future worth of irreplaceable or vulnerable resources* (Spotlight, p. 762).
- *Establish public utilities to manage and protect public lands and fisheries.*

Hartman, S. et al. 1997. "The Cost of Air Pollution Abatement." *Applied Economics*, vol. 29, no. 6, 759(16).

Microloans to the Poor

SOLUTIONS

Most of the world's poor desperately want to earn more, become more self-reliant, and have a better life. However, they have no credit record and few if any assets to use for collateral to secure a loan. Thus, they can't go to traditional banks or other money-lending institutions to buy seeds and fertilizer or to buy tools and materials for a small business.

During the last 20 years an innovative tool called *microlending* has increasingly helped deal with this problem. Several dozen organizations now make tiny loans to the poor as an important tool for grassroots economic development.

For example, since it was first started by economist Muhammad Yunus in 1976, the Grammeen (Village) Bank in Bangladesh has provided more than $16 million in microloans to mostly poor, rural, and landless women in 35,000 villages. The loans vary from $50 to $500 but average around $100. About 90% of the loans are to women who start their own small businesses as sewers, weavers, bookbinders, peanut fryers, or vendors.

To stimulate repayment, the Grammeen Bank organizes microborrowers into five-member peer groups. If one member of the group misses a weekly payment or defaults on the loan, the other members of the group must make the payments.

Since 1990 over 1,000 community banks (two-thirds of them in rural areas) have been established in Nigeria to make microloans to the poor. In effect, residents pool their money and then lend it to one another. Communities raise funds to establish each bank and are responsible for appointing its board of directors and hiring its staff. In parts of Nigeria, where community ties are strong, one's honor and standing in the community can take the place of collateral in securing a loan.

Since 1991, Massachusetts-based ACCION International has provided more than $200 million a year in microloans, mostly in Latin America. In the United States, more than 100 organizations assist microenterprises by providing loans, grants, or training. Since 1992, the U.S. Small Business Administration has earmarked $15 million a year for microloans.

Such experience has shown that microlending is both successful and profitable. Microborrowers have a much higher repayment rate than conventional borrowers, and microborrowers who repay their loans become eligible for further and larger loans. For example, in Bangladesh the repayment rate on microloans made by the Grammeen Bank is an astounding 96%, compared to a national repayment rate of conventional loans of only 30%.

Critical Thinking

Why do you think there has been little use of microloans by international development and lending agencies such as the World Bank and the International Monetary Fund? How might this situation be changed?

- *Revoke the government-granted charters of environmentally and socially irresponsible businesses.*
- *Make environmental concerns a key part of all trade agreements and of all loans made by international lending agencies* (Solutions, p. 766).
- *Reduce waste of energy, water, and mineral resources.*
- *Preserve biodiversity.*
- *Reduce future ecological damage and repair past ecological damage.*
- *Reduce poverty.*
- *Slow population growth.*

Paul Hawken's simple golden rule for such an economy is, *"Leave the world better than you found it, take no more than you need, try not to harm life or the environment, and make amends if you do."*

Most of these principles underlying an earth-sustaining economy have already been discussed, but a few are worth discussing more fully. Hawken and Amory Lovins (Guest Essay, p. 433) point out that one of the results of using full-cost pricing is that many businesses would make a substantial portion of their profits selling less of their current product or service and instead sell information or services about how to reduce energy and matter throughputs.

For example, utility companies would sell energy audits and improvements in energy efficiency (negawatts) instead of additional power. Energy companies would sell improvements in the energy efficiency of cars (negabarrels) and power plants (negacoal). Water companies would make more money by reducing water waste (negaliters) than by selling more water. Agricultural companies would make much of their money by selling information and providing services for farmers concerning soil conservation, sustainable agriculture, and integrated pest management (negafertilizers and negapesticides). Forest management companies would make money by selling sustainable tree-harvesting practices or alternatives to using trees for lumber or paper (negatrees).

Hint: Enter the search terms *cost benefit analysis, pollution planning* using Key Words.

Hawken and business experts such as David Korten argue that states should *revoke the charters of irresponsible corporations.* Although corporations can be fined for engaging in illegal activities, they can write off the fines (and the legal expenses incurred in fighting them) on their taxes, effectively passing these costs on to shareholders and customers. Thus, corporations are largely protected from economic sanctions intended to make them behave more responsibly. Hawken and Korten believe that this serious problem could be overcome if state legislatures began revoking the charters of corporations that continue to do major environmental and social harm, using a "three strikes and you're out" policy. Upon its third felony conviction, a corporation would automatically lose its charter and could not reopen.

Hawken also proposes that we *transform the entire tax system by taxing the throughput of energy and materials more, and labor and income less.* The present system taxes what we want more of (income, payrolls, profits, and creativity) instead of what we want less of (resource depletion and waste, pollution, and environmental degradation). A sustainable economy would reverse this situation over a 20-year period by replacing taxes on income, payrolls, and profits with green taxes on environmentally harmful activities.

Hawken argues that in making this shift in what is taxed, *there must be no net increase in taxes for consumers* and *governments should not view green taxes as ways to raise additional revenue;* otherwise, this shift in what we tax (discourage) and don't tax (reward) would be politically untenable and counterproductive. The government would also need to provide tax breaks or other types of aid to less affluent citizens to help offset the disproportionate bite out of their buying power resulting from the higher prices of basic consumer goods and services.

Sweden, Denmark, and the Netherlands (Section 27-6) are the leaders in the use of tax policy to build an environmentally sustainable economy. In most cases, income taxes in these countries have been reduced as taxes on environmentally destructive activities have been increased.

Hawken has also proposed a means of managing, protecting, and restoring public lands and fisheries that he believes will prevent the twin problems of overexploitation by private ownership and ineffective management by government agencies (Sections 23-3 and 25-6). He suggests that we *turn public land and fishery resource systems into public resource utilities that are publicly regulated, privately managed, and market-based.* By law, the primary purpose of such utilities would be **(1)** to ensure that public lands (except wilderness, which would remain under full government protection) and aquatic resources are used sustainably and **(2)** to provide funds for ecological restoration of degraded lands and fisheries and wildlife populations. These public resource utilities would receive income by charging fees for sustainable harvesting or other uses of resources from public lands and fisheries. They would issue stock and bonds (just as electric utilities do) to raise funds for investments in long-term restoration projects and additional purchases of public lands.

Public utility stockholders would be allowed a guaranteed profit of 10–12% depending on a utility's performance in achieving sustainable resource use and ecological restoration. The utility would be required by law to use 88–90% of its annual revenue *strictly* for ecological restoration, wildlife habitat protection, reduced resource waste and environmental degradation, pollution prevention, ecological research, and land and aquatic habitat acquisition.

Another important element in shifting to an earth-sustaining economy is to *improve environmental management in businesses.* To do this, colleges and universities training economists, business leaders, and lawyers must educate their students and those already in these professions about how the earth works, what we are doing to it, how things are connected, how we need to change the way we think and do business, and what the principles of good environmental management are (Spotlight, right).

How Can We Make the Transition to an Earth-Sustaining Economy? Even if people believe that an earth-sustaining economy is desirable, is it possible to make such a drastic change in the way we think and act? Some environmentalists, economists, and business leaders say that it's not only possible but imperative and that it can be done over the next 40–50 years. They point out that *the environmental revolution is also an economic revolution that uses a mix of regulations and market-based approaches to reward earth-sustaining behavior by businesses and consumers and discourage earth-degrading behavior.*

According to Paul Hawken, this new approach to economic thinking and actions recognizes that most business leaders are not evil, earth-degrading ogres. Instead, they are trapped in a system that by design rewards them (with the highest profits and salaries, and best chances for promotion) for maximizing short-term profits for owners and investors, regardless of the harmful short- and long-term environmental and social impacts.

Hawken argues that an earth-sustaining economy would free business leaders, workers, and investors from this ethical dilemma and allow them to be financially compensated and respected for doing socially and ecologically responsible work, improving

 Q: What percentage of all municipal sewage in the United States ends up essentially untreated in coastal waters?

environmental quality, and still making hefty profits for owners and stockholders. Making this shift should also create jobs (Connections, p. 772).

In 1996, Dow Chemical, the largest U.S. chemical producer, announced that it would invest $1 billion over the next 10 years to achieve an ambitious set of environmental, health, and safety goals at its plants worldwide. By 2005, it seeks **(1)** a 90% reduction in workplace-related illnesses and injuries, **(2)** a 50% cut in overall chemical emissions, and **(3)** a 50% drop in generation of waste and wastewater per pound of production.

In 1994, Ray Anderson, CEO of a $1 billion carpet company (Atlanta-based Interface, Inc.) announced plans to develop the nation's first totally sustainable green corporation, as a part of the emerging eco-industrial revolution (Solutions, p. 584). Since 1994 he has cut carpet scrap by 60%, saving $67 million. He has plans to stop selling carpet and lease it a way to control recycling and produce near zero factory waste. One factory in California is being converted to solar energy. He is one of a growing number of business leaders committed to improving environmental quality while still making a profit for stockholders.

Hawken and others offer the following outline of how the shift from current unsustainable economies to sustainable ones could be made over the next 40–50 years. Over the first 20 years, *all* government subsidies encouraging resource depletion, waste, pollution, and environmental degradation would be phased out and replaced with taxes on such activities. During that same period, taxes on incomes and profits would be phased out and new government subsidies would be phased in for businesses built around recycling and reuse, reducing waste, preventing pollution, improving energy efficiency, using renewable energy, protecting biodiversity, and facilitating ecological restoration. The gradual and mostly predictable change would encourage a shift to more sustainable economies while allowing businesses to plan ahead.

This system for change represents a shift in determining which economic actions are rewarded (profitable) and which ones are discouraged. These plans would be well publicized and would take effect over decades, giving businesses time to adjust. Because businesses go where the profits are, many of today's earth-degrading businesses might well be tomorrow's earth-sustaining businesses—a win–win solution for business, future generations, and the earth. Economic models indicate that after this first phase is completed the entire economy would be transformed within another 30–40 years.

The problem in making this shift is not economics, but politics. It involves the difficult task of convincing business leaders and elected officials to begin changing the current system of rewards and penalties, the profits from which have given them economic and political power. Without the active participation of forward-thinking business and political leaders and strong political pressure from citizens, there is little hope of making the transition to an earth-sustaining economy.

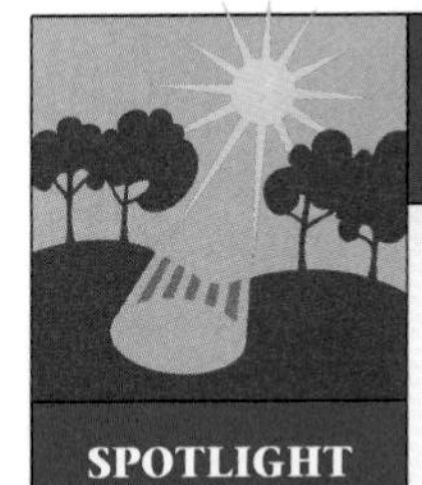

SPOTLIGHT

What Is Good Environmental Management?

According to business leaders and environmentalists, the key practices in good corporate environmental management are to:

- Provide leadership, beginning with the board of directors.
- Give a clear statement of the company's environmental principles and objectives, which have full backing of the board.
- Involve employees, environmental groups, customers, and members of local communities in developing and evaluating the business's environmental policies, strategies for improvement, and progress.
- Make improving environmental quality and worker safety and health a major priority for every employee.
- Conduct an annual cradle-to-grave environmental audit of all operations and products, including a detailed strategy for making improvements. Disseminate the results to employees, stockholders, and the public. Currently, some U.S. companies shun environmental audits because of fear of lawsuits or prosecution by state and federal environmental agencies. However, several states have passed laws protecting the results of a company's voluntary environmental audit as privileged information.
- Help customers safely distribute, store, use, and dispose of or recycle company products.
- Recognize that carrying out these policies is the best way to encourage innovation, expand markets, improve profit margins, develop happy and loyal customers, attract and keep the best-qualified employees, and help sustain the earth—a win–win strategy.

Critical Thinking

Why haven't more companies instituted the environmental management principles listed here? What could be done to change this situation?

A: About 35%

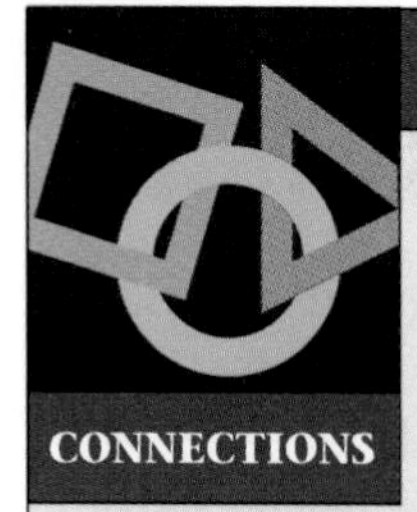

Jobs and the Environment

Critics of environmental laws and regulations contend that they cause the loss of large numbers of jobs in the United States and other developed countries. In fact, studies show that the opposite is true. U.S. Department of Labor statistics reveal that only 0.1% of the jobs lost in the United States between 1987 to 1990 were the result of environmental regulations.

In 1989, the Business Roundtable, an association of chief executive officers of U.S. companies, projected that a minimum of 200,000 and possibly as many as 1–2 million jobs would be lost if the Clean Air Act of 1990 were passed. However, a 1995 study by economist Eban Goodstein showed that by June 1994 only 2,363 jobs had been lost because of this act.

Instead of causing a net loss of jobs, environmental protection is a major growth industry that creates new jobs. Some 1 million jobs are expected to be added to the U.S. environmental protection industry alone between 1995 and 2005; increasing the aluminum recycling rate in the United States to 75% would create 350,000 more jobs. Collecting and refilling reusable containers creates many more jobs per dollar of investment than using throwaway containers, and most of the new jobs are created in local communities. Reforestation, ecological restoration, sustainable agriculture, and integrated pest management are all labor-intensive activities requiring low to moderate skill levels.

A congressional study concluded that investing $115 billion per year in solar energy and improving energy efficiency in the United States would eliminate about 1 million jobs in oil, gas, coal, and electricity production but would create 2 million other new jobs; investment of the money saved by reducing energy waste could create another 2 million jobs.

Although a host of new jobs would be created in an earth-sustaining economy, jobs would be lost in some industries, regions, and communities. Ways suggested by various analysts to ease the transition include providing tax breaks to make it more profitable for companies to keep or hire more workers instead of replacing them with machines; using incentives to encourage location of new, emerging industries in hard-hit communities, helping such areas diversify their economic base; and providing income and retraining assistance for workers displaced from environmentally destructive businesses (a *Superfund for Workers*).

Critical Thinking

1. What major things (if any) has the federal government in the United States (or the country where you live) done to stimulate the growth of environmental jobs? What major things (if any) has the government done to discourage the growth of environmental jobs?

2. Do you believe that the government should have a significant role in stimulating environmental jobs? Explain.

27-6 CASE STUDIES: ECOLOGICAL AND ECONOMIC POLICIES IN GERMANY AND THE NETHERLANDS

How Is Germany Investing in the Future and the Earth? German political and business leaders see sales of environmental protection goods and services—already a roughly $460 billion-per-year business in 1997 (and expected to rise to $572 billion by 2001)—as a major source of new markets and income in the next century because environmental standards and concerns are expected to rise everywhere.

Stricter environmental standards and regulations in Germany have paid off in a cleaner environment and the development of innovative green technologies that can be sold at home and abroad in a rapidly growing market. Mostly because of stricter air pollution regulations, German companies have developed some of the world's cleanest and most efficient gas turbines, and they have invented the world's first steel mill that uses no coal to make steel. Germany sells these and other improved environmental technologies globally.

Since 1992 Germany has been the world's leading exporter of environmental technologies, with export income from such technologies of $23.3 billion in 1992. According to a 1995 study by the German Institute for Economic Research, between 1990 and 2000 the number of jobs related to environmental protection and improvement in Germany is projected to increase from 600,000 to 1 million.

On the other hand, Germany continues to heavily subsidize its coal-mining industry, which increases environmental degradation and pollution and adds large amounts of carbon dioxide to the atmosphere. According to environmental expert Norman Myers (Guest Essay, p. 530), these perverse subsidies are now so high that the German government would save money by closing all of its coal mines and sending the miners home on full pay for the rest of their lives.

In 1977 the German government started the Blue Angel product-labeling program to inform consumers about products that cause the least environmental harm. Most international companies now use the German market to test and evaluate *green* products.

Q: What is the largest source of water pollution from oil?

Germany has also revolutionized the recycling business. German car companies are required to pick up and recycle all domestic cars they make. Bar-coded parts enable disassembly plants to dismantle an auto for recycling in 20 minutes. Such *take-back* requirements are being extended to almost all products to reduce use of energy and virgin raw materials. Germany plans to sell its newly developed recycling technologies to other countries.

The German government has also supported research and development aimed at making Germany the world's leader in solar-cell technology (Section 16-3) and hydrogen fuel (Section 16-7), which it expects will provide a rapidly increasing share of the world's energy. Finally, Germany provides about $1 billion per year in green foreign aid to developing countries. Much of the aid is designed to stimulate demand for German technologies such as solar-powered lights, solar cells, and wind-powered water pumps.

What Is the Netherlands' Green Plan to Achieve an Ecologically Sustainable Economy? In 1989, the Netherlands began implementing their National Environmental Policy Plan (or Green Plan) for creating an economy that doesn't destroy the environment. By 2010, the Dutch have the goals of slashing their production of many types of pollution by 70–90% and achieving the world's first ecologically sustainable economy.

The government identified eight major areas for improvement: climate change, acid deposition, eutrophication, toxic chemicals, waste disposal, groundwater depletion, unsustainable use of renewable and nonrenewable resources, and local nuisances (mostly noise and odor pollution). Target groups, consisting of major industrial, government, and citizens' groups, were identified and formed for each of the eight areas, and the government set up timetables and targets for each group. Each group could pursue whatever policies or technologies it wanted, but there were stiff penalties for failure.

The government identified four general themes for each group to focus on: **(1)** integrated life cycle management, which makes producers responsible for the remains of their products after users are through with them (a goal that increased the design of products that can be reused or recycled); **(2)** improving energy efficiency (Sections 16-1 and 16-2), with the government committing $385 million per year to energy conservation programs; **(3)** invention of new or improved more sustainable technologies, supported by a government program to help develop such technologies; and **(4)** improving public awareness through a massive government-sponsored public education program.

Many of the country's leading industrialists like the Green Plan for several reasons. *First*, they can make investments in pollution prevention and pollution control with less financial risk because they have a high degree of certainty about long-term environmental policy. *Second*, they have freedom to deal with the problems in the way that makes the most sense for their businesses. This has helped the Dutch avoid the route of sometimes too costly and excessive environmental regulations that have plagued many industries in the United States and have led to confrontation and political gridlock rather than cooperation between environmentalists and industrial leaders.

Finally, developing and implementing the plan has helped industrial leaders in the Netherlands (like those in Germany) learn that creating more efficient and environmentally sound products and processes enable them to reduce costs and increase profits by selling such innovations at home and abroad.

Is the plan working? The news is mixed but encouraging, given the goal of bringing about such radical changes in only a short time. Many of the target groups are meeting their goals on schedule, and some have even exceeded them. A huge amount of environmental research by the government and private sector has taken place. This has led to an increase in organic agriculture, greater reliance on bicycles in some cities, and new, more ecologically sound housing developments.

However, officials concede privately that some of the more ambitious goals such as decreasing CO_2 levels may have to be revised downward or even abandoned. Some environmentalists who strongly support the plan are not happy with the compromises they have had to make and with the mild backlash in some industry circles against some of the more ambitious goals, especially an energy tax to reduce CO_2 emissions.

Despite its shortcomings, the Netherlands plan is the first attempt by any country to deliberately foster a national debate on the vital issue of ecological sustainability and environmental security and to encourage innovative solutions to environmental problems. Many analysts see it as a daring and inspiring venture that other nations would do well to emulate.

Can We Change Economic Gears in the Next Few Decades? Critics claim that a shift toward an earth-sustaining economy won't happen because it would be opposed by people whose subsidies were being eliminated and whose activities were being taxed. However, investors and businesspeople have just as much at stake in helping sustain the earth (and thus the human species and economies) as anyone else. Forward-looking investors, corporate executives, and the governments of countries such as Japan, Germany, the Netherlands, and the United States recognize that earth-sustaining businesses with good environmental management will prosper as the environmental revolution proceeds. Companies and countries that fail to invest in a green future may find that they do not have a future. *The environmental revolution is also an economic revolution.*

A: Waste oil dumped onto land by cities, individuals, and industries (50–90% of the input)

The Steady State Economy in Outline

GUEST ESSAY

Herman E. Daly

Herman E. Daly is senior research scholar at the School of Public Affairs, University of Maryland. Between 1989 and 1993 he was senior economist at the Environmental Department of the World Bank. Before joining the World Bank he was Alumni Professor of economics at Louisiana State University. He has been a member of the Committee on Mineral Resources and Environment of the National Academy of Sciences and has served on the boards of advisers of numerous environmental organizations. He is cofounder and associate editor of the journal Ecological Economics. *He has written many articles and several books, including* Steady-State Economics *(2d ed., 1991),* Beyond Growth: The Economics of Sustainable Development *(1996), and, with coauthor John Cobb Jr.,* For the Common Good: Redirecting the Economy Towards Community, the Environment, and a Sustainable Future *(1994). He is one of a small but growing number of economists seriously thinking about earth-sustaining economics.*

The steady state economy is basically a physical concept with important social and moral implications. It is defined as a constant stock of physical wealth and people. This wealth and population size are maintained at some desirable, chosen level by a low rate of throughput of matter and energy resources so as a result the longevity of people and goods is high.

Throughput is roughly equivalent to GNP, the annual flow of new production. It is the cost of maintaining the stocks of final goods and services by continually importing high-quality matter and energy resources from the environment and exporting waste matter and low-quality heat energy back to the environment [Figure 3-21].

Currently we try to maximize the growth of the GNP, but the reasoning just given suggests that we should relabel it "gross national cost," or GNC. We should minimize it, subject to maintaining stocks of essential items. For example, if we can maintain a desired, sufficient stock of cars with a lower throughput of iron, coal, petroleum, and other resources, we are better off, not worse.

To maximize GNP throughput for its own sake is absurd. Physical and ecological limits to the volume of throughput suggest that a steady state economy will be necessary. Less recognizable but probably more stringent social and moral limits suggest that a steady state economy will become desirable long before it becomes a physical necessity.

Consider the development and use of nuclear reactors to produce electricity, technology that is heavily subsidized by the government. Since the mid-1970s the growth of this technology has declined sharply—not because of a shortage of uranium fuel, but because of social and economic pressures. Poor management, excessive costs, and serious accidents such as the one at the Chernobyl nuclear plant [p. 366] have seriously undermined public support of this technology.

Nuclear plants exist only because of huge government subsidies. If the nuclear power industry were forced to operate in an open market without government subsidies, it probably would never have been developed because of its low return on investment. Some have advocated reviving the nuclear industry by providing even more subsidies, arguing that nuclear power is needed to reduce the rate of projected global warming. However, this course of action wastes enormous amounts of limited economic and human resources on an uneconomical way of producing electricity [Figure 15-21]. Moreover, it

In a sustainable economy, the fish catch does not exceed the sustainable yield of fisheries, the amount of water from underground aquifers does not exceed acquifer recharge, soil erosion does not exceed the natural rate of new soil formation, tree cutting does not exceed tree planting, carbon emissions do not exceed the capacity of nature to fix atmospheric CO_2, and plant and animal species are not destroyed faster than new ones evolve.

LESTER R. BROWN AND CHRISTOPHER FLAVIN

CRITICAL THINKING

1. The primary goal of all current economic systems is to maximize economic growth by producing and consuming more and more economic goods. Do you agree with that goal? Explain. What are the alternatives?

2. Do you believe that cost–benefit analysis should be the *primary method* for making all decisions about how limited federal, state, and local government funds are used? Explain. If not, what are the alternatives?

3. Do you favor internalizing the external costs of pollution and unnecessary resource waste? Explain. How might it affect your lifestyle? Wildlife? Any children you might have?

4. (a) Do you believe that we should establish optimal levels or zero-discharge levels for most of the chemicals we release into the environment? Explain. What effects would adopting zero-discharge levels have on your lifestyle? **(b)** Should we assume that all chemicals we release or propose to release into the environment are potentially harmful until proven otherwise? Explain. What effects would adopting this principle have on your life?

5. Do you agree or disagree with the proposals various analysts have made for sharply reducing poverty as discussed on p. 768? Explain.

Q: How much water can be contaminated by 0.9 liter (1 quart) of oil?

will do little to slow global warming compared with other, more cost-effective alternatives.

Once we have attained a steady state economy at some level of stocks, we are not forever frozen at that level. Moral and technological changes may make it both possible and desirable to grow (or decline) to a different level. However, growth will be seen as a temporary process necessary to move from one steady state level to another, not as an economic norm. This requires a substantial shift in present economic thought; most current economic ideas and models must be either replaced or drastically modified. Most economists strongly resist this radical change in the way they think and act.

The greatest challenges facing us today are

- For physical and biological scientists to define more clearly the limits and interactions within ecosystems and the ecosphere (which determine the feasible levels of the steady state) and to develop technologies more in conformity with such limits
- For social scientists to design institutions that will bring about the transition to a steady state and permit its continuance
- For philosophers, theologians, and educators to stress the neglected traditions of stewardship and distributive justice that exist in our cultural and religious heritage.

The last item is of paramount importance because the problem of sharing a fixed amount of resources and goods is much greater than that of sharing a growing amount. Indeed, this has been the primary reason for giving top priority to growth. If the pie is always growing, it is said there will always be crumbs—and the hope of a slice—for the poor. This avoids the moral question of a more equitable distribution of the world's resources and wealth.

The kinds of economic institutions needed to make this transition follow directly from the definition of a steady state economy. We need an institution for maintaining a constant population size within the limits of available resources. For example, economic incentives can be used to encourage each woman or couple to have no more than a certain number of children, or each woman or couple could be given a marketable license to have a certain number of children, as economist Kenneth Boulding has suggested.

We also need an institution for maintaining a constant stock of physical wealth and for limiting resource throughput. For example, the government could set and auction off transferable annual depletion quotas for key resources. Finally, there must be an institution for limiting inequalities in the distribution of the constant physical wealth among the constant population in a steady state economy [Figure 27-11]. For example, there might be minimum and maximum limits on personal income and maximum limits on personal wealth.

Many such institutions could be imagined. The problem is to achieve the necessary global and societal (macro) control with the least sacrifice of freedom at the individual (micro) level.

Critical Thinking

1. Do you favor establishing institutions for maintaining population size and resource throughput to avoid depleting and degrading earth capital? Explain.

2. Should minimum and maximum limits on personal income and maximum limits on personal wealth be established? Explain.

6. Do you agree or disagree with the guidelines for an earth-sustaining economy discussed on pp. 768–769? Explain.

7. Do you favor making a shift to an earth-sustaining economy over the next 40–50 years? Explain. How might this affect your lifestyle? The lifestyle of any children you might have? Any grandchildren you might have?

8. Try to come up with analogies to complete the following statements: **(a)** Thinking of economic growth as not being dependent on earth capital is like _______. **(b)** Continuing to allow business and consumers not to include environmental and social costs in the prices of goods and services is like _______. **(c)** Using market forces as the primary method for improving environmental quality is like _______. **(d)** Green taxes are like _______. **(e)** Using government subsidies that encourage harmful environmental behavior is like _______. **(f)** Using the trickle-down approach to deal with poverty is like _______.

PROJECTS

1. List all the economic goods you use, then identify those that meet your basic needs and those that satisfy your wants. Identify any economic wants you would be willing to give up, those you believe you should give up but are unwilling to give up, and those you hope to give up in the future. List what is likely to make you happy and improve the quality of your life. Relate the results of this analysis to your personal impact on the environment. Compare your results with those of your classmates.

2. Make a concept map of this chapter's major ideas, using the section heads and subheads and the key terms (in boldface). Look at the inside back cover and in Appendix 4 for information about concept maps.

A: About 940,00 liters (250,000 gallons)

28 POLITICS AND ENVIRONMENT

Rescuing a River

When Marion Stoddart first moved to Groton, Massachusetts, on the Nashua River in the early 1960s, the riverside location was nothing to brag about. In fact, the Nashua was considered one of the nation's filthiest rivers. Industries and towns along the river's 92-kilometer (57-mile) length had used it as a dump for decades. Sludge piled up at every dam. Dead fish bobbed on its waves, and at times the water was red, green, or blue from pigments discharged by paper mills.

Marion Stoddart was appalled. Instead of thinking nothing could be done, she committed herself to restoring the Nashua and establishing public parklands along its banks. She didn't start by filing lawsuits or organizing demonstrations; instead she created a careful cleanup plan and approached state officials with it in 1962. They laughed, but she was not deterred and began practicing the most time-honored skill of politics: one-on-one persuasion. She identified the power brokers in the riverside communities and began to educate them, to win them over, and she got them to cooperate in cleaning up the river.

One of her converts was William Flynn, former mayor of Fitchburg, a town whose paper mills were a major source of the Nashua's pollution. When Stoddart got the state to ban open

Figure 28-1 Earth citizen Marion Stoddart canoeing down the Nashua River near Groton, Massachusetts. She spent over two decades spearheading successful efforts to have this river cleaned up. (Seth Resnick)

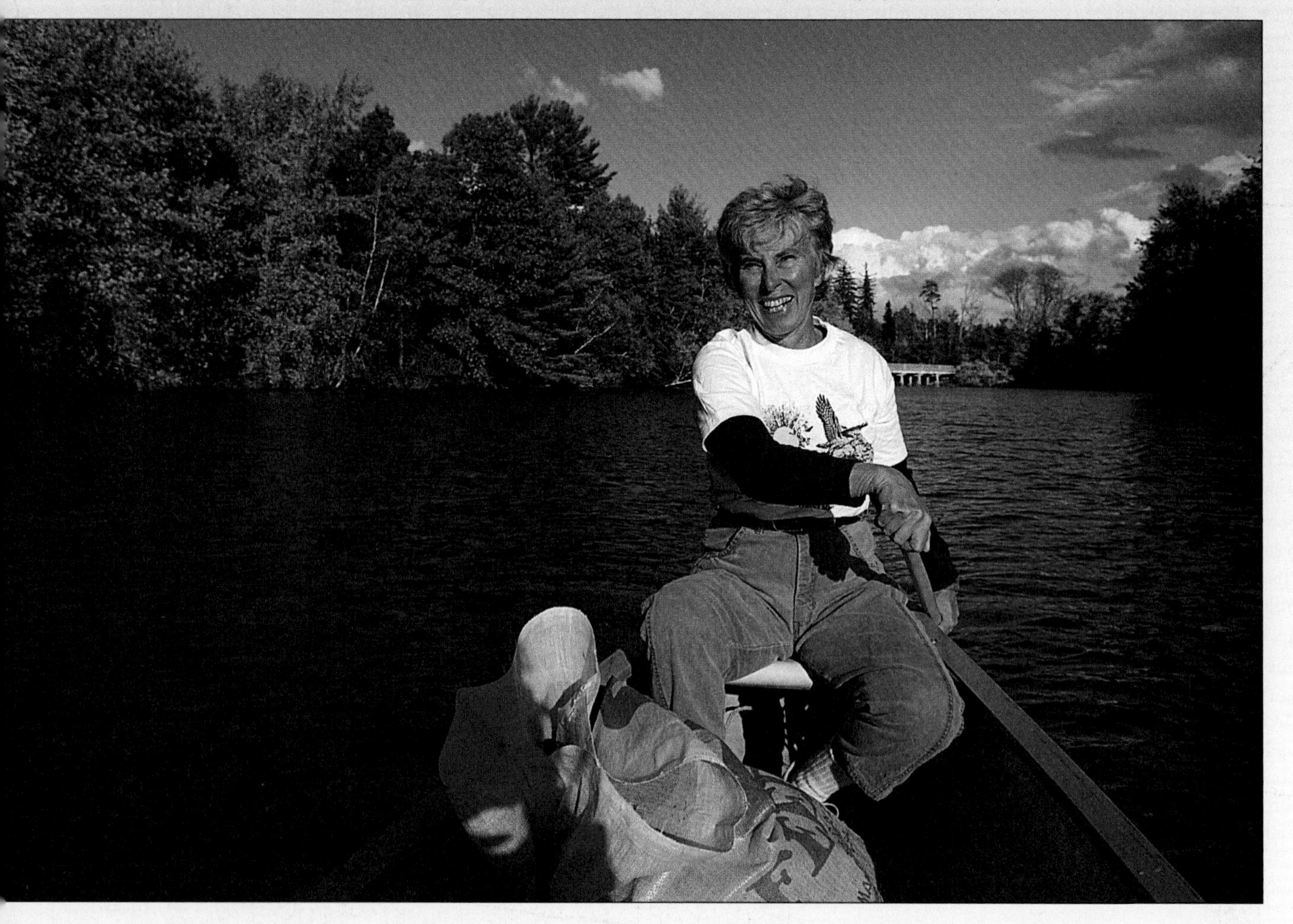

dumping in the river, one factory threatened to close down rather than help pay for cleanup facilities. Flynn helped Stoddart persuade Fitchburg paper mills to cooperate with authorities in building a wastewater treatment plant.

When promised federal matching funds for building the treatment plant failed to materialize, Stoddart gathered 13,000 signatures on a petition sent to President Nixon. The funds arrived in a hurry.

Stoddart's next success was getting a federal grant to beautify the river. She hired high school dropouts to clear away mounds of debris. When the river cleanup was completed, she persuaded communities along the river to create some 2,400 hectares (6,000 acres) of riverside park and woodlands along both banks.

Now, three decades later, the Nashua is still clean. Several new water treatment plants have been built, and a citizens' group founded by Stoddart keeps watch on water quality. The river supports many kinds of fish and other wildlife, and its waters are used for canoeing (Figure 28-1) and other kinds of recreation. The project is considered a model for other states and is testimony to what a committed individual can do to bring about change from the bottom up by getting people to work together.

For her efforts, Stoddart has been named by the UN Environment Programme as an outstanding worldwide worker for the environment. However, she might say that the blue and canoeable Nashua itself is her best reward.

Politics is the process by which individuals and groups try to influence or control the policies and actions of governments at the local, state, national, or international levels. Politics is concerned with who has power over the distribution of resources and benefits—who gets what, when, and how. Thus, it plays a significant role in regulating and influencing economic decisions (Chapter 27) and persuading people to work together toward a common goal, as Marion Stoddart did.

Some say we can keep on doing business and politics as usual. Others argue that the old ways of doing business and politics are so threatening to our environmental, economic, and military security that we must have the wisdom and courage to reshape them into earth-sustaining economic policies (discussed in Chapter 27) and earth-sustaining political policies discussed in this chapter.

National security is a meaningless concept if it does not include the preservation of livable conditions within a country—or on the planet as a whole. Environmental degradation imperils nations' most fundamental aspect of security by undermining the natural support systems on which all of human activity depends.

MICHAEL G. RENNER

This chapter is devoted to seeking answers to the following questions:

- How is environmental policy made in the United States?
- How can people affect environmental policy?
- What are the major types and roles of environmental groups?
- What are the goals and tactics of anti-environmental groups?
- How might the U.S. political system be improved?
- How might global environmental policy be improved?

28-1 POLITICS AND ENVIRONMENTAL POLICY

How Does Social Change Occur in Representative Democracies? **Democracy** is government "by the people" through elected officials and representatives. In a *constitutional democracy,* a constitution provides the basis of government authority, limits government power by mandating free elections, and guarantees free expression of public opinion.

Political institutions in constitutional democracies are designed to allow gradual change in order to ensure economic and political stability. Rapid change in the United States, for example, is curbed by the system of checks and balances that distributes power among the three branches of government—executive, legislative, and judicial—and among federal, state, and local governments.

In passing laws, developing budgets, and formulating regulations, government decision makers must deal with pressure from many competing *special-interest groups*. Each group advocates passing laws favorable to its cause and weakening or repealing laws unfavorable to its position. This includes cheap

SOLUTIONS

Types of Environmental Laws in the United States

Environmentalists and their supporters have persuaded the U.S. Congress to enact a number of important federal environmental and resource protection laws, as discussed throughout this text and listed in Appendix 3. These laws seek to protect the environment using the following approaches:

- Setting standards for pollution levels or limiting emissions or effluents for various classes of pollutants (Federal Water Pollution Control Act and Clean Air Acts)
- Screening new substances for safety before they are widely used (Toxic Substances Control Act)
- Requiring comprehensive evaluation of the environmental impact of an activity before it is undertaken by a federal agency (National Environmental Policy Act)
- Setting aside or protecting various ecosystems, resources, and species from harm (Wilderness Act and Endangered Species Act)
- Encouraging resource conservation (Resource Conservation and Recovery Act and National Energy Act)

Some environmental laws contain glowing rhetoric about goals but little guidance about how to meet them, leaving this task to regulatory agencies and the courts. In other cases, the laws or presidential executive orders specify one or more of the following general principles for setting regulations:

- *No unreasonable risk:* food regulations in the Food, Drug, and Cosmetic Act
- *No risk:* the zero-discharge goals of the Safe Drinking Water and Clean Water Acts
- *Standards based on best available technology:* the Clean Air, Clean Water, and Safe Drinking Water Acts
- *Risk–benefit balancing* (Section 17-6 and Spotlight, p. 464): pesticide regulations and an executive order signed by President Clinton in 1993 that requires all government regulatory agencies to include risk assessment in all of their decisions (something similar to what anti-environmental Vice President Dan Quayle tried but failed to accomplish during the Bush administration)
- *Cost–benefit balancing* (Spotlight, p. 762): the Toxic Substances Control Act and a presidential executive order (initiated by President Reagan) that gives the Office of Management and Budget the power to delay indefinitely, or even veto, any federal regulation not proven to have the least cost to society

Critical Thinking

Pick one of the U.S. environmental laws listed in Appendix 3 (or a law in the country where you live). Use the library or the Internet to evaluate the law's major strengths and weakness. Decide whether the law should be weakened, strengthened, or abolished and explain why. List the three most important ways you believe the law should be strengthened or weakened.

and better access to resources, subsidies, relief from taxes, and the shaping of regulations to foster the group's goals.

Some special-interest groups are *profit-making organizations* such as corporations, and others are *nonprofit, nongovernment organizations (NGOs)*. Examples of NGOs are educational institutions, labor unions, and mainstream and grassroots environmental organizations.

Most political decisions made in democracies result from bargaining, accommodation, and compromise among leaders of competing *elites*, or power brokers. The overarching goal of government by competing elites is to maintain the overall economic and political stability of the system (status quo) by making only gradual change; this goal does *not* involve questioning or changing the rules of the game (the fundamental societal beliefs) that gave the elites their political or economic power. Ralph Waldo Emerson once said, "Democracy is a raft which will never sink, but then your feet are always in the water."

One disadvantage of this deliberate design for stability is that democratic governments tend to *react* to crises instead of acting to prevent them. In other words, the emphasis is on a narrow, short-sighted approach instead of the broad (holistic), long-sighted approach needed to deal with environmental problems.

This means that there is a built-in bias against policies for protecting the environment because they often call for prevention of crises instead of reaction to them. Such policies require integrated planning now and into the future, and they sometimes call for fundamental changes in societal beliefs that can threaten the power of government and business elites.

Case Study: How Is Environmental Policy Made in the United States? The major function of the federal government in the United States is to develop and implement *policy* for dealing with various issues. This policy is typically composed of various *laws* passed by the legislative branch, *regulations* instituted

Q: How much might it have cost Exxon to prevent the $8.5-billion *Exxon Valdez* spill?

by the executive branch to put laws into effect, and enough *funding* to implement and enforce the laws and regulations (Figure 28-2).

The first step in establishing federal environmental policy (or any other policy) is to persuade lawmakers that a problem exists and that the government has a responsibility to find solutions to it. Once over that hurdle, lawmakers try to pass laws to deal with the problem. Most environmental bills are evaluated by as many as 10 committees in both the House of Representatives and the Senate. Effective proposals are often weakened by this fragmentation and by lobbying from groups opposing the law. Nonetheless, since the 1970s a number of environmental laws have been passed in the United States (Appendix 3 and Solutions, p. 778).

Even if an environmental (or other) law is passed, Congress must appropriate enough funds to implement and enforce it. Indeed, developing and adopting a budget is the most important and controversial thing the executive and legislative branches do. Developing a budget involves answering two key questions: What resource use and distribution problems will be addressed, and how much of the limited tax revenues will be used to address each problem?

Next, regulations for implementing the law are drawn up by the appropriate government department or agency. Groups try to influence how the regulations are written and enforced; some of the affected parties may even challenge the final regulations in court. Finally, the department or agency implements and enforces the approved regulations. Proponents or affected groups may take the agency to court for failing to implement and enforce the regulations or for enforcing them too rigidly.

Businesses affected by regulations try to influence regulatory agencies by setting up a *revolving-door* relationship between government and business officials. They try to have people sympathetic to their cause appointed to administrative positions in regulatory agencies. They also offer regulatory officials lucrative jobs and use their inside knowledge to find ways to weaken or get around the regulations. The hope of getting such a job motivates some agency officials to develop regulations favorable to the businesses they are regulating, or not to enforce existing regulations.

The net result of this fragmentation, lack of funding, and influence of the regulated on the regulators is *incremental decision making*. As a result, only small changes are usually made in existing policies and programs.

How Can the Courts Be Used to Implement or Weaken Environmental Regulations? Almost every major environmental regulation is challenged in court by industries, environmental organizations, or both. In any court case the **plaintiff** is the individual, group of individuals, corporation, or government agency bringing the charges; the **defendant** is the

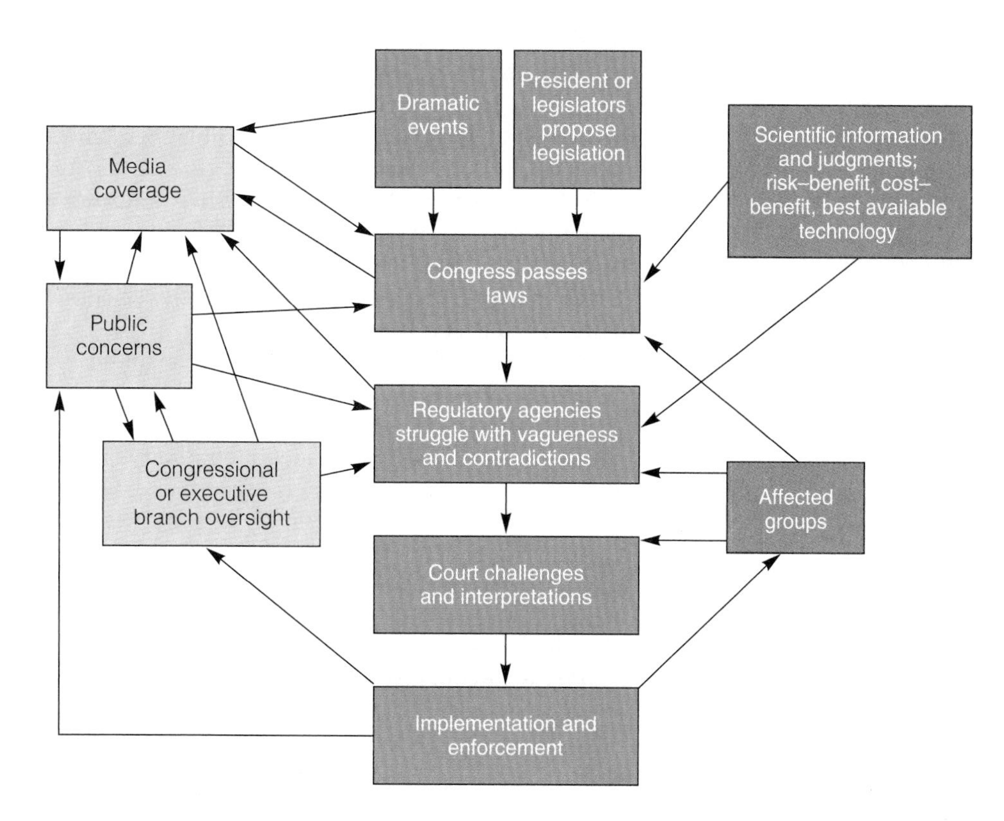

Figure 28-2 Primary forces involved in making environmental policy at the federal level in the United States.

A: About $22.5 million to equip the vessel with a double hull

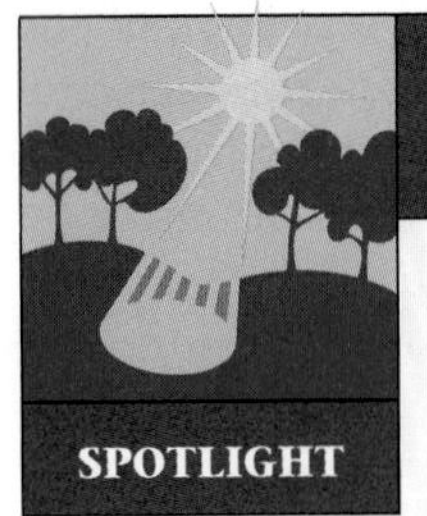

SLAPPs: Intimidating Citizen Activists

SPOTLIGHT

In recent years, some corporations and developers have begun filing strategic lawsuits against public participation (SLAPPs) against individuals and activist environmental groups.* Such suits have two goals: (1) to use up the time and financial resources of private citizens and environmental groups and (2) to deter citizens from becoming involved.

About 90% of SLAPPs (which range from $100,000 to $100 million, but average $9 million per suit) that go to court are thrown out by judges who recognize them for what they are. But those hit with SLAPPs must spend considerable money on lawyers and typically spend 1–3 years defending themselves.

Most SLAPPs aren't meant to be won; they are intended to intimidate individual citizens and activist groups so that they won't exercise their democratic rights. Once the victim of a SLAPP is shaken by fear and rising defense costs, the plaintiff often offers a voluntary dismissal, provided that the citizen or group agrees never to discuss the case or oppose the plaintiff again.

Some citizen activists have fought back with countersuits and have been awarded damages. For example, a Missouri woman who was sued for criticizing a medical-waste incinerator won an $86.5 million judgment against the incinerator's owner.

Even after paying such awards, corporations and developers generally save money by filing such suits. Unlike the people they are suing, these businesses can count such legal and insurance liability costs as a business expense and write them off on their taxes. In other words, they get all taxpayers to pay much of the cost of such lawsuits against a few taxpayers who are merely exercising their rights as citizens.

Critical Thinking

Are you in favor of SLAPPs? Explain. If not, what do you think should be done to discourage such tactics?

*Free information on SLAPPs can be obtained from the SLAPP Resource Center, University of Denver College of Law, 1900 Olive St., Denver, CO 80220; 303-871-6266. The Citizens' Clearinghouse for Hazardous Waste (Appendix 1) has a SLAPP Back Fact Pack available for $3.29.

individual, group of individuals, corporation, or government agency being charged.

In a **civil suit** the plaintiff seeks to collect damages for injuries to health or for economic loss, to have the court issue a permanent injunction against any further wrongful action, or both. Such suits may be brought by an individual or a clearly identified group. A **class action suit** is a civil suit filed by a group, often a public interest or environmental group, on behalf of a larger number of citizens who allege similar damages but who need not be listed and represented individually.

The effectiveness of environmental lawsuits is limited by several factors. First, permission to file a damage suit is granted only if the harm to an individual plaintiff is clearly unique or different enough to be distinguished from that to the general public. For example, you could not sue the Department of the Interior for actions leading to the commercialization of a wilderness area on the grounds that you do not want your taxes used to harm the environment; in this case the harm to you could not be separated from that to the general public. However, if the government damaged property you own, you would have grounds to sue.

Second, bringing any suit is expensive. According to EPA employee and whistle-blower Hugh Kaufman, "The citizens are armed with rubber bands and chewing gum against major polluters armed with the best-paid engineers and lawyers in the country. It ain't a fair fight." Third, public interest law firms cannot recover attorneys' fees unless Congress has specifically authorized such recovery in the laws the firms seek to have enforced. By contrast, corporations can reduce their taxes by deducting their legal expenses.

Fourth, it is often difficult for a plaintiff to prove that a defendant is liable and responsible for a harmful action. Suppose that one company (the defendant) is charged with harming individuals by polluting a river. If hundreds of other industries and cities dump waste into that river, establishing that the defendant is the culprit is very difficult, requiring costly scientific research and expert testimony.

Fifth, the court (or series of courts if the case is appealed) may take years to reach a decision. During that time a defendant may continue the allegedly damaging action unless the court issues a temporary injunction against the action until the case is decided. Finally, plaintiffs sometimes abuse the system by bringing frivolous suits that delay and run up the costs of projects. Recently some corporations and developers have begun filing lawsuits for damages against citizen activists (Spotlight, p. 780).

Despite many handicaps, proponents of environmental law have accomplished a great deal since the 1960s. More than 20,000 attorneys in 100 public inter-

Helvarg, D. et al. 1998. "Has Environmentalism Become Too Extreme?" *Sierra*, vol. 83, no. 1, 124(1).

est law firms and groups now specialize partly or entirely in environmental and consumer law; many other lawyers and scientific experts participate in environmental and consumer lawsuits as needed. Indeed, environmental law is the fastest-growing sector of the American legal profession.

28-2 INFLUENCING ENVIRONMENTAL POLICY

Solutions: How Can Individuals Affect Environmental Policy? A major theme of this book is that individuals matter. History shows that significant change comes from the *bottom up*, not the top down. Without grassroots political action by millions of individual citizens and organized groups, the air you breathe and the water you drink today would be much more polluted and much more of the earth's biodiversity would have disappeared. As grassroots populist leader Jim Hightower puts it,

> *To move America from greed to greatness we must once again tap into the genius and gumption of the . . . workaday majority of Americans who sweat, plow, invent, repair, teach, construct, nurse, and do the myriad of other productive tasks that sustain society from the bottom up. You can't keep a tree alive by fertilizing it at the top. . . . We want a government that quits doing things to us. But neither do we want a government that does things for us. We want a government that is us, that involves us and empowers us so we can do for ourselves and do for the country.*

Individuals can influence and change government policies in constitutional democracies in several ways. They can **(1)** vote for candidates and ballot measures; **(2)** contribute money and time to candidates seeking office; **(3)** lobby, write, fax, e-mail, or call elected representatives, asking them to pass or oppose certain laws, establish certain policies, and fund various programs (Appendix 5); **(4)** use education and persuasion; **(5)** expose fraud, waste, and illegal activities in government (whistleblowing); **(6)** file lawsuits; and **(7)** participate in grassroots activities to bring about change or enforce existing laws and regulations.

Solutions: What Are the Three Types of Environmental Leadership? There are three types of environmental leadership. One is *leading by example*, in which people use their lifestyles to show others that change is possible and beneficial.

Another involves *working within existing economic and political systems to bring about environmental improvement, often in new, creative ways*. People can influence political elites by campaigning and voting for candidates and by communicating with elected officials (Appendix 5). They can also work within the system by choosing an environmental career (Individuals Matter, p. 782).

A third type of leadership involves *challenging the system and basic societal values, as well as proposing and working for better solutions to environmental problems*. Leadership is more than being against something; it also involves coming up with better ways to accomplish various goals. All three types of leadership are needed. For an inspiring example of what can be done in a short time, see the Guest Essay on p. 308.

28-3 ENVIRONMENTAL GROUPS

What Are the Roles of Mainstream Environmental Groups? Many types of environmental groups work at the local, state, national, and international levels (Appendix 1). Environmental organizations range from multimillion-dollar *mainstream* groups, led by chief executive officers and staffed by experts, to *grassroots* neighborhood groups formed to do battle on local environmental issues. More than 8 million U.S. citizens belong to environmental organizations, which together received $894 million in contributions and income in 1994 (up from $630 million in 1990).

Mainstream environmental groups are active primarily at the national level and to a lesser extent at the state level; often they form coalitions to work together on issues. Some mainstream organizations (such as Greenpeace, Figure 28-3) funnel substantial funds to local activists and projects. The mainstream Sierra Club, for example, prefers grassroots action but still works to influence national environmental legislation. The Environmental Defense Fund and the Natural Resources Defense Council prefer legal action against corporations that degrade the environment or against government agencies that fail to enforce environmental laws and regulations.

Some groups focus much of their efforts on specific issues, such as population (Zero Population Growth), protecting habitats (Wilderness Society and Nature Conservancy; Solutions, p. 636), and wildlife conservation (National Audubon Society, National Wildlife Federation, and the World Wildlife Fund). Other organizations concentrate on education and research (Worldwatch Institute, Rocky Mountain Institute, Population Reference Bureau, and World Resources Institute). Still other groups provide information, training,

Environmental Careers

INDIVIDUALS MATTER

Besides committed earth citizens, the environmental movement needs dedicated professionals working to help sustain the earth. In the United States (and in other developed countries), the green job market is one of the fastest-growing segments of the economy. Already some 3 million people in the United States are working in the environmental field, and more than 125,000 new positions are being created each year.

Many employers are actively seeking environmentally educated graduates. They are especially interested in people with scientific and engineering backgrounds and in people with double majors (business and ecology, for example) or double minors.*

Environmental career opportunities exist in a large number of fields: environmental engineering (currently the fastest growing job market), sustainable forestry and range management, parks and recreation, air and water quality control, solid-waste and hazardous-waste management, recycling, urban and rural land-use planning, computer modeling, ecological restoration, and soil, water, fishery, and wildlife conservation and management.

Environmental careers can also be found in education, environmental planning, environmental management, environmental health, toxicology, geology, ecology, conservation biology, chemistry, climatology, population dynamics and regulation (demography), law, risk analysis, risk management, accounting, environmental journalism (Guest Essay, p. 792), design and architecture, energy conservation and analysis, renewable-energy technologies, hydrology, consulting, public relations, activism and lobbying, economics, diplomacy, development and marketing, publishing (environmental magazines and books), and law enforcement (pollution detection and enforcement teams).

Critical Thinking

1. Is the green job market really one of the fastest-growing segments of the economy? Use the library or the Internet to help answer this question.

2. Have you considered an environmental career? Why or why not?

*For details, consult the Environmental Careers Organization (see Appendix 1), *The New Complete Guide to Environmental Careers* (Covelo, Calif.: Island Press, 1993); Nicholas Basta, *The Environmental Career Guide* (New York: Wiley, 1991); Joan Moody and Richard Wizansky, eds., *Earth Works: Nationwide Guide to Green Jobs* (New York: Harper-Collins-West, 1994); and the *Environmental Career Directory* (Detroit, Mich.: Visible Ink Press, 1993). Environmental jobs are listed in publications such as Earth Work, Environmental Career Opportunities, Environmental Opportunities, and EcoNet, all listed in Appendix 1.

and assistance to localities and grassroots organizations (Citizens' Clearinghouse for Hazardous Waste and the Institute for Local Self-Reliance).

Recently, one major environmental group, Greenpeace USA, drastically cut its staff and budget because of declining membership and donations. At its height in 1991, it had 1.2 million members, a budget of $60 million, and 390 staff members. In 1997, after its membership fell to 420,000 and contributions sank to about $26 million, the organization cut its budget to $21 million, closed all of its field offices, and reduced its staff to 65 people. However, membership in the World Wildlife Fund has held steady at about 1.2 million since 1995 and the Sierra Club, with 600,000 members in 1996, has gained members every year since 1993.

Mainstream groups work within the political system; many have been major forces in persuading Congress to pass and strengthen environmental laws, as well as fighting off attempts to weaken or repeal such laws. However, these groups must continually guard both against having their efforts subverted by the political system they work to change and against losing touch with ordinary people and nature in the insulated atmosphere of national and state capitals.

All of the 10 largest U.S. mainstream environmental organizations—the "Group of 10"—rely heavily on corporate donations, and many of them have corporate executives as board members, trustees, or council members. Proponents of this corporate involvement argue that it is a way to raise much-needed funds and to influence industry; opponents believe that it is a way for corporations to unduly influence environmental organizations. Regardless of the relative merits of these arguments, the net effect has been to cause some internal conflict within these environmental organizations and to drive a wedge between them and many grassroots environmental activists.

What Are the Roles of Grassroots Environmental Groups? The base of the environmental movement in the United States consists of at least 6,000 grassroots citizens' groups organized to protect them-

Q: What percentage of the sewage in Latin America is treated?

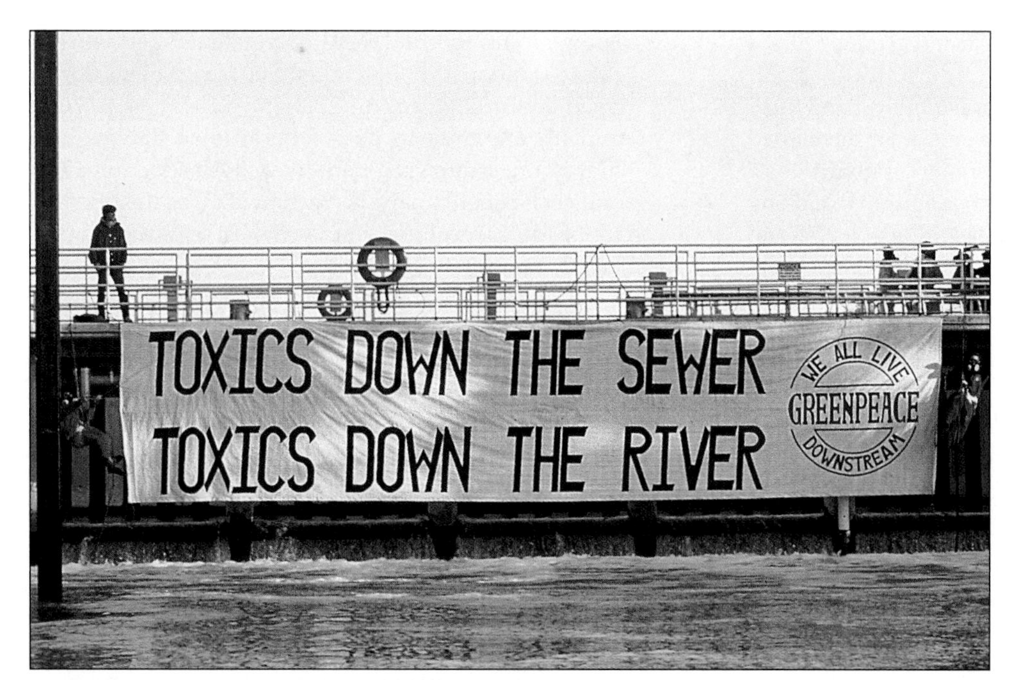

Figure 28-3 Greenpeace activists protesting discharge of toxic waste from the Pig's Eye sewage treatment plant at St. Paul, Minnesota. This plant is the largest source of toxic chemicals dumped in the Mississippi River north of St. Louis. (Sam Kittner/Greenpeace)

selves against local pollution and environmental damage. According to political analyst Konrad von Moltke, "There isn't a government in the world that would have done anything for the environment if it weren't for the citizen groups."

John W. Gardner, former cabinet official and founder of Common Cause, summarized the basic rules for effective political action by grassroots organizations:

- *Have a full-time continuing organization.*
- *Limit the number of targets and hit them hard.* Groups dilute their effectiveness by taking on too many issues.
- *Organize for action, not just for study, discussion, or education.*
- *Form alliances with other organizations on a particular issue.*
- *Communicate your positions in an accurate, concise, and moving way.*
- *Persuade and use positive reinforcement.*
- *Concentrate efforts mostly at the state and local levels.*

Individuals and grassroots groups have worked to restore wetlands (Individuals Matter, p. 249), save forests, and restore degraded rivers (p. 776). In the United States people of color, working-class people, and poor people, who often bear the brunt of pollution and environmental degradation, have formed a growing coalition known as the *grassroots movement for environmental justice* (Guest Essay, p. 606) to protect their human and environmental rights (Figure 28-4 and photo on p. 749). Some unions and nonunionized workers are also forming coalitions with environmental groups to improve worker safety and health. Environmental groups are also active on college campuses and in public schools across the United States (Individuals Matter, p. 785).

Many grassroots environmental organizations are unwilling to compromise or negotiate (p. 605). Instead of dealing with environmental goals and abstractions, they are fighting perceived threats to their lives, the lives of their children and grandchildren, and the value of their property. They want pollution and environmental degradation, environmental injustices, and violations of their human rights stopped and prevented—not merely controlled.

28-4 THE ANTI-ENVIRONMENTAL MOVEMENT

What Are the Goals of the Anti-Environmental Movement? Increasingly the *East–West* Cold War polarization is rapidly being replaced by two other interconnected clashes. One is the *North–South* or *rich–poor* clash between developed countries and developing countries over how the earth's limited resources and wealth should be shared; the other is the related *green–brown* clash between proponents of various versions of the earth wisdom and planetary-management worldviews (Section 1-8 and Chapter 29). These two clashes, which are fundamentally clashes over human values, will dominate our political and economic lives in coming decades.

A small but growing number of political and business leaders in the United States and other developed countries see improving environmental quality as a way to stimulate innovation, increase profits, and create jobs while working with the earth (Section 27-5 and Guest Essay, p. 756). However, leaders of some corporations and many people in positions of economic and political power see environmental laws and regulations as threats to their wealth and power, and they vigorously oppose such efforts.

Since 1980, anti-environmentalists in the United States have mounted a massive campaign to weaken or repeal existing environmental laws and to destroy the credibility and effectiveness of the environmental movement (Figure 28-5). This well-funded attack includes **(1)** stepped-up lobbying efforts against environmental laws and regulations in both Washington and state capitals; **(2)** similar efforts by mayors and government officials fed up with having to implement federal environmental laws without federal funding (unfunded mandates); **(3)** a coalition of grassroots front groups—calling themselves the Wise Use Movement—organized and funded mostly by anti-environmental interest groups such as ranchers, logging and mining companies, oil and gas companies, and real estate developers; **(4)** a global trade agreement (GATT) that many environmentalists warn could weaken existing environmental, health, consumer, and worker safety standards in the United States and other developed countries (Section 27-3 and Solutions, p. 766); and **(5)** attempts to pass federal laws that require highly uncertain risk–benefit analysis (Section 17-6) and cost–benefit analysis (Section 27-3) as the primary tools for formulating government environmental, health, and safety regulations.

What Are the Tactics of the Anti-Environmental Movement? Whatever your own position, it is useful to understand the following general legal and political tactics being used against the environmental movement. Start by *establishing an enemy to create fear and to divert people's attention and energy away from the real issues*. Prey on the fears of ordinary people by labeling environmentalists as the *green menace*—antibusiness, antipeople, antireligious, scaremongering radical extremists who are crippling the economy, hurting small business owners, costing taxpayers billions of dollars in unnecessary regulations, robbing small landowners of the right to do what they want with their land, trying to lock up the natural resources on public lands, and threatening jobs, national security, and traditional values.

When accused of using wild exaggeration and smear tactics against environmentalists to raise money and spread distrust, hate, and fear, Ron Arnold, one of the leaders of the anti-environmental Wise-Use

Figure 28-4 The Mississippi River between Baton Rouge and New Orleans, Louisiana, is lined with oil refineries and petrochemical plants. Along this corridor, known as "Cancer Alley" because of its abnormally high cancer rates, tons of carcinogenic and mutagenic chemicals leak into groundwater or are discharged into the river. In 1988 an environmental alliance of residents protested chemical dumping and groundwater pollution in their communities by marching the 137 kilometers (85 miles) from Baton Rouge to New Orleans. (Sam Kittner/Greenpeace)

Q: What percentage of the sewage in China is treated?

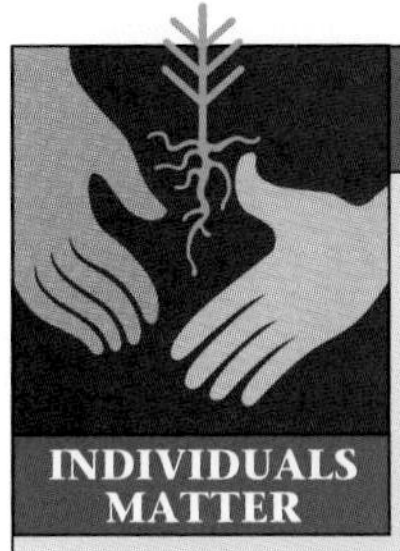

Environmental Action on Campuses

Since 1988 there has been a boom in environmental awareness on college campuses and some public schools across the United States. Much of this momentum began in 1989, when the Student Environmental Action Coalition (SEAC) at the University of North Carolina at Chapel Hill held the first national student environmental conference on the UNC campus.

SEAC groups are active on 700 campuses, and the National Wildlife Federation's Campus Ecology Program (launched in 1989) has groups on about 600 campuses.* Most student environmental groups work with members of the faculty and administration to bring about environmental improvements on their own campuses and in their local communities.

Many of these groups focus on making an environmental audit of their own campuses or schools†; they then use the data gathered to propose changes that will make their campuses or schools more ecologically sustainable, usually saving them money in the process. In addition, students who have learned to do careful research on and develop solutions for environmental problems will be able to use these skills the rest of their lives.

Such audits have resulted in numerous improvements. For example, an energy management plan developed by Morris A. Pierce, a graduate student at the University of Rochester, is projected to save the university $60 million over 20 years. Student-spurred energy conservation programs at the State University of New York at Buffalo had saved the university more than $60 million by 1997.

Students have induced almost 80% of universities and colleges in the United States to develop recycling programs. At Dartmouth College, students have set up a composting program and an organic farm to supply the food service with vegetables and herbs.

In the 1980s, David Wedlin, an undergraduate at St. Olaf College in Minnesota, submitted research findings showing how the college could improve its management of agricultural lands and natural habitat. By 1988, the college had adopted his ideas and by 1997 had restored native woodland, prairie grasses, and wetlands on college property (with more restoration to come), with students helping plant and tend more than 20,000 tree seedlings and nursery stock trees. College-owned farmland has also been converted from a conventional to more sustainable form of agriculture, with students helping to design and evaluate more sustainable farming practices.

At Bowdin College in Maine, chemistry Professor Dana Mayo and student Caroline Foote developed the concept of microscale experiments, in which smaller amounts of chemicals are used. This has reduced toxic wastes and saved the chemistry department more than $34,000. Today more than 50% of all undergraduates in chemistry in the United States use such microscale techniques, as do universities in a growing number of other countries.

Students at Oberlin College in Ohio (led by David Orr, Guest Essay, p. 746) helped design a new sustainable environmental studies building. The building, to be completed in 1999, will (1) be energy-efficient and use passive solar design, (2) be powered by solar cells that generate more electricity than they use, (3) house living machines (Figure 20-2) to discharge wastewater at least as pure as the water taken in, (4) incorporate materials grown or manufactured sustainably, (5) include no materials known to cause human health problems, and (6) be landscaped in a way that promotes biodiversity.

A 1997 report by the National Wildlife Foundation's Campus Ecology Program found that 23 student-initiated projects had together saved the participating universities and colleges $16.3 million. According to this study, implementing similar programs in the nation's 3,700 universities and colleges could not only help improve environmental quality and environmental education but also lead to a savings of more than $2.6 billion.

Such student-spurred environmental activities and research studies are spreading to universities in other countries such as Brazil and Columbia, with support from an environmental declaration signed in 1990 by some 250 university heads from 42 countries.

---

*These efforts are described in the book *Ecodemia: Campus Environmental Stewardship at the Turn of the 21st Century* (Washington, D.C.: National Wildlife Federation, 1995) and in the Campus Environmental Yearbook published annually by the National Wildlife Federation.

†Details for conducting campus environmental audits can be found in April Smith and the Student Environmental Action Coalition, *Campus Ecology: A Guide to Assessing Environmental Quality and Creating Strategies for Change.* (Los Angeles, Calif.: Living Planet Press, 1993) and Jean Heinze-Fry, *Green Lives, Green Campuses* (Belmont, Calif.: Wadsworth, 1995), a supplement designed to be used with this textbook.

A: About 15%

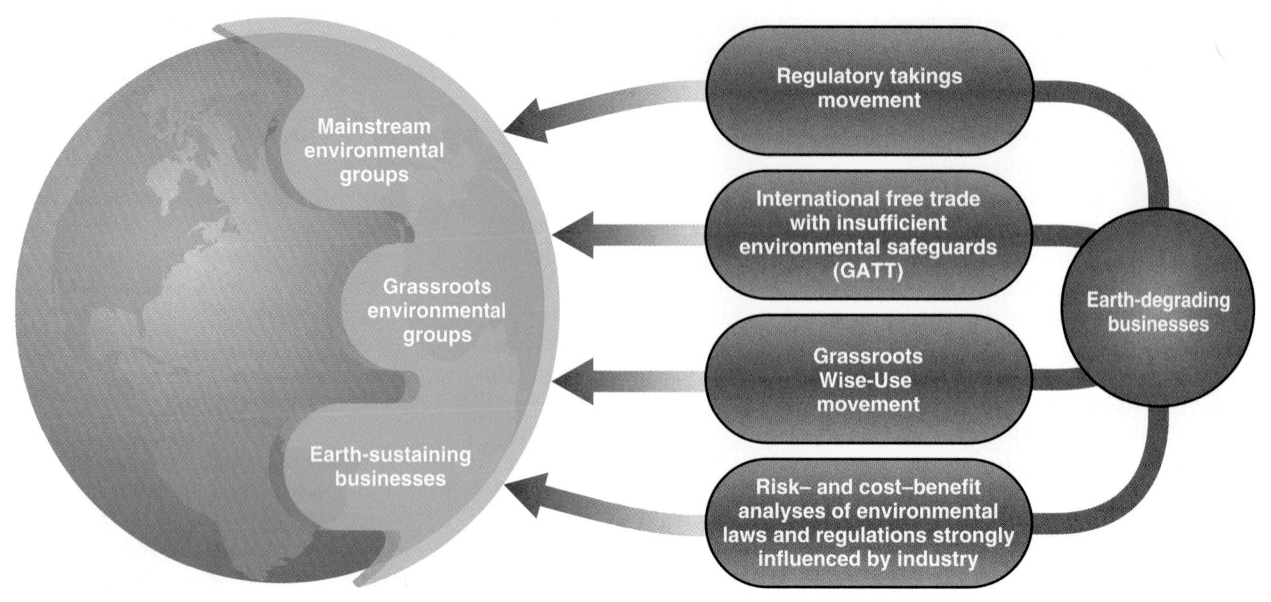

Figure 28-5 Greens vs. browns. Types of groups involved in strengthening (shown in green) and weakening (shown in brown) environmental protection in the United States.

movement (Case Study, p. 787), said, "Facts don't really matter. In politics, perception is reality."

Here are some of the basic tactics used. *Weaken and intimidate.* Infiltrate environmental groups to learn their plans and to create dissension and mistrust. One important public relations group proudly boasts to its potential business clients that one of its major strengths is spying on activists in universities, labor unions, churches, and environmental groups. Slap individual activists and small grassroots groups with nuisance lawsuits (Spotlight, p. 780). Another effective tactic is to fire whistle-blowers who expose harmful or illegal environmental practices, or shift them to less influential jobs.

Threaten or use violence. If environmental activists become too effective, harass them with phone calls threatening their lives or the lives of their family members. Try to have them fired, kill their pets, trash their homes and offices, cut phone lines, slash tires, sabotage their cars, plant drugs in their cars and notify the police, burn their houses or businesses, assault them, fire warning shots into their houses or cars—all things that have happened to environmentalists in the United States in recent years. Former Interior Secretary James Watt publicly declared in 1990, "If the troubles from the environmentalists cannot be solved in the jury box or the ballot box, perhaps the cartridge box should be used."

Influence public opinion. Commission books and articles, establish slick magazines, set up think tanks, and hire PR firms to deny that environmental problems exist or are serious. Use fax and letter-writing campaigns against corporations that provide financial support for TV or radio programs critical of environmentally harmful business practices or anti-environmental concerns. Flood schools with free pamphlets, videos, and other thinly disguised teaching materials that propagandize your position.

Either don't collect data, or keep them secret. Then claim that nothing can be done about a problem because of a lack of information. Exert political pressure to weaken government studies that are unfavorable to your position, limit their availability to the public, and decrease funding for their agencies.

*Falsify data, use misleading PR to manipulate research results, support and co-opt researchers to control the research agenda and get the desired research results, and attack independent scientists whose research threatens your products or processes.**

Exploit the built-in limitations of science and public ignorance about the nature of science to sow seeds of doubt. Mislead the public by saying that claims against your position have not been scientifically proven. If that doesn't work, imply that scientific findings that are contrary to your position should not be taken seriously because science can't really prove anything and can only put forth theories (Section 2-1). Ignore consensus science and instead focus on frontier science when it supports your position. When frontier science goes against your position, point out that such results are only preliminary and should not be taken

*For details on these strategies, see Dan Fagin and Marianne Lavelle, *Toxic Deception* (Secaucus, N.J.: Carol Publishing Group, 1996).

Q: What percentage of usable groundwater in the United States is contaminated?

The Wise-Use Movement

CASE STUDY

Since 1988, several hundred local and regional grassroots groups in the United States have formed a national anti-environmental coalition called the Wise-Use movement. Much of their money comes from real estate developers and from timber, mining, oil, coal, and ranching interests and traditional supporters of right-wing causes.

According to Ron Arnold, who played a key role in setting up this movement, the specific goals of the Wise-Use movement are as follows:

- *Cut all old-growth forests in the national forests and replace them with tree plantations.*
- *Modify the Endangered Species Act so that economic factors override preservation of endangered and threatened species.*
- *Eliminate government restrictions on wetlands development.*
- *Open all national parks, national wildlife refuges, and wilderness areas to oil drilling, mining, off-road vehicles (ORVs), and commercial development.*
- *Do away with the National Park Service and launch a 20-year construction program of new concessions run by private firms in the national parks.*
- *Continue mining on public lands under the provisions of the 1872 Mining Law (p. 000), which allows mining interests not only to pay no royalties to taxpayers for hard-rock minerals they remove, but also to buy public lands for a pittance.*
- *Recognize private property rights to mining claims, water, grazing permits, and timber contracts on public lands; do not raise fees for these activities. Ideally, sell resource-rich public lands to private enterprise.*
- *Provide civil penalties against anyone who legally challenges economic action or development on federal lands.*
- *Allow pro-industry (Wise-Use) groups or individuals to sue as "harmed parties" on behalf of industries threatened by environmentalists.*

Most anti-environmental groups either affiliated with or generally supportive of the Wise-Use movement and its offshoot, the *Alliance for America* (formed in 1991), use environmentally friendly-sounding names.* Here are a few examples: the *U.S. Council for Energy Awareness* (nuclear power industry), *America the Beautiful* (packaging industry), *Partnership for Plastics Progress* (plastics industry), *American Council on Science and Health* (food and pesticide industries), *National Wetlands Coalition* (real estate developers and oil and gas companies), *Global Climate Coalition* (50 corporations and trade associations opposed to reducing fossil-fuel use to slow the rate of global warming), *Blue Ribbon Coalition* (off-road vehicle users and manufacturers who want all public lands opened up to ORVs), *American Forest Resource Alliance* (timber and logging companies), *American Farm Bureau Federation* (agricultural chemicals industries), and *People for the West* (mining industry).

The Wise-Use movement also includes the *Sahara Club*, which advocates violence environmentalists in defense of its members' "right" to dirt-bike in wilderness areas. One of its leaders tells audiences to "Throw environmentalists off the bridge. Water optional," and that "You can't reason with eco-freaks, but you sure can scare them." Another group is *Citizens for the Environment* (an industry-backed education group that counters environmental ideas with such slogans as "Recycling doesn't save forests," "Packaging prevents waste," and "Global warming and ozone depletion are hoaxes").

Recently the Wise-Use movement has targeted the Nature Conservancy (Solutions, p. 636) as its number-one enemy. In addition to being the richest environmental group, it focuses its efforts on acquiring critical land to protect threatened and endangered species from development. It also helps private landowners set up tax-saving conservation easements to protect their land from ecologically damaging development.

According to Jay D. Hair, former president of the National Wildlife Federation, the Wise-Use movement is *"merely a wise disguise for a well-financed, industry-backed campaign that preys on the economic woes and fears of U.S. citizens.... These organizations with benign-sounding names are not grassroots—they are astro-turf laid down with big corporate money. And they are out to ... ensure that certain special interests will be allowed to continue to pollute and exploit public resources for private profit."*

According to environmentalist B. J. Bergman, the basic principles of the Wise Use movement are, "If it grows, cut it; if it flows, dam it; if it's underground, extract it; if it's swampy, fill it; if it moves, kill it."

Critical Thinking

Do you agree or disagree with each of the major goals of the Wise-Use movement? Explain. What positive effects has the Wise Use movement had on the environmental movement?

*For lists and information about these organizations, see *The Greenpeace Guide to Anti-Environmental Organizations* (1993, Odonian Press, Box 7776, Berkeley, CA 94707); *Masks of Deception: Corporate Front Groups in America* by Mark Megalli and Andy Friedman (1993, Essential Information, P.O. Box 19367, Washington, DC 20036); *Let the People Judge: A Reader on the Wise-Use Movement* (Covelo, Calif.: Island Press, 1993); and *The War Against the Greens* (San Francisco: Sierra Club Books, 1994).

A: About 25% (mostly contaminated by pesticides or nitrates from fertilizers) seriously. These tactics should drive scientists and environmentalists crazy and undermine public support of environmental causes.

Build up your public environmental image. Set up one or a few showcase projects and then run print and TV ads portraying what you are doing to protect the environment. Form or donate money to well-publicized partnerships with major environmental groups to build up your image and increase profits.

Delay and wear out reformers. Set up time-consuming and costly legal and bureaucratic roadblocks to forestall change and tire reformers out. Wear out, financially drain, and tie up your adversaries in court. Use paralysis by analysis to delay action by appointing blue ribbon panels to study problems and make recommendations. Pay little attention to their advice, and then call for more research (which is usually needed but can also be an excuse for inaction).

Urge the president and Congress to require that all government environmental decisions and regulations be evaluated primarily by cost–benefit and risk–benefit analysis. Then try to load the evaluation panels with experts who support your position. If this doesn't work, challenge the analyses in courts. The built-in limitations (Spotlight, pp. 484 and 762), information gaps, and vulnerable assumptions of many such analyses could take years to unravel. The end result of this version of the *paralysis-by-analysis* strategy will be fewer and weaker environmental, health, and worker safety regulations. All of this would be achieved without trying to repeal existing environmental laws (a politically unpopular cause).

Support unenforceable legislation and regulations. This undermines confidence in government, splits up opposition groups, and helps convince the public that real change is impossible.

When unfavorable environmental laws are passed, urge legislators not to fund the laws. Passing the cost of federally unfunded environmental mandates on to financially strapped states and cities will generate intense political pressure against such laws by citizens, mayors, and governors. Because Congress is unlikely to come up with funding for these laws, they will probably be weakened, unenforced, or repealed.

Require federal and state governments to compensate property owners whenever environmental, health safety, or zoning laws either limit how the owners can use their property or decrease its financial value—something called regulatory taking (p. 619). This will make such laws and regulations too expensive to enforce, and will keep new ones from being established.

A final tactic is *divide and conquer.* Keep people and interest groups fighting with one another so they can't get together on vital issues that threaten the status quo (such as election campaign financing; Solutions, p. 791). Act as a "good cop" by giving corporate money to mainstream environmental organizations to influence them and to create dissension between mainstream and grassroots organizations. At the same time, act as a "bad cop" by donating funds to elected officials to influence them to weaken or eliminate environment, health, and worker safety laws and by funding mainstream and grassroots anti-environmental groups (Case Study, p. 787).

Environmentalists share some of the blame for the increasing power of the anti-environmental movement. Some analysts point out that environmental organizations have neglected the grassroots citizens that have always been their main source of political and economic support. Environmentalists fought hard to have environmental laws passed and enforced, but a growing number of environmentalists believe that they haven't fought hard enough to ensure that these laws are effective without being excessive or oppressive.

Polls show that three-fourths of the general public are strong supporters of environmental laws and regulations and don't want them weakened. But polls also show that fewer than 5% of the U.S. public views the environment as one of the nation's most pressing problems. As a result, environmental concerns increasingly don't get transferred to the ballot box. As one political scientist put it, "Environmental concerns are like the Everglades, a mile wide but only a few inches deep."

28-5 EVALUATING CLAIMS OF ENVIRONMENTALISTS AND ANTI-ENVIRONMENTALISTS

Are Environmental Threats Exaggerated? In some cases environmental dangers have been exaggerated and overregulated, and just a few examples of environmental overkill and overregulation can play into the hands of anti-environmentalists. The anti-environmental movement has skillfully used a few such examples to fuel a public backlash against virtually *all* environmental, health, and worker safety laws and regulations.

One example of overkill cited by anti-environmentalists is the statement by some environmentalists that acid rain is causing widespread destruction of forests and aquatic life in lakes and streams. Even though tree diebacks and acidic lakes are serious problems (especially in parts of Europe), they involve a mixture of poorly understood natural and human-related factors and interactions, only one of which is deposition of acidic compounds (Figure 18-9). Even so, decreasing the emissions of air pollutants into the atmosphere is justifiable on the basis of improving

Q: In the United States, how many underground tanks storing gasoline and other hazardous chemicals are leaking?

human health and reducing costly corrosion and degradation of materials, in addition to their possible roles in degrading forests and streams.

Another controversy that some regarded as regulatory overkill was the government policy requiring the removal of all forms of asbestos from buildings (Section 18-4). Experience has shown that removing asbestos that is not crumbling or releasing fibers can both increase levels of asbestos fibers inside buildings and cost enormous amounts of money. In 1992 the EPA reversed its policy of requiring asbestos removal in most buildings (largely because of court rulings against this approach), but only after billions of dollars had been spent.

Many business leaders and state and local officials also complain that many government environmental, health, and worker safety regulations are excessive and too costly. Environmentalists agree that some government laws and regulations go too far and that bureaucrats sometimes develop and impose ridiculous and excessively costly regulations. But they argue that the solution is to stop regulatory abuse, not throw out or seriously weaken the body of laws and regulations that help protect the public good.

According to environmentalist William Ashworth,

> *If we wish to make progress, it will do us no good to replace one failed system with another that failed just as badly. Government regulation, after all, didn't fall out of the sky; it was erected, piece by piece, as an attempt to deal with the damage caused by unrestrained property rights and the unregulated free-market system. . . . We do not need to deconstruct regulation, but to reconstruct it.*

To accomplish this, a growing number of analysts urge environmentalists to take a hard look at existing environmental laws and regulations. Which laws (or parts of laws) have worked and why? Which ones have failed, and why? Which government bureaucracies concerned with developing environmental and resource regulations have either abused their power or have not been responsive enough to the needs of ordinary people? How can such abuses be corrected? What existing environmental laws (or parts of laws) and regulations should be repealed or modified? What environmental problems lend themselves to market-based approaches (free-market environmentalism), and which ones do not? How can the principles of pollution prevention and waste reduction become the guiding principles for environmental legislation and regulation? What is the minimum amount of environmental legislation and regulation that is needed? How can a balanced program of both regulation and market-based approaches (Table 27-1) be used to achieve environmental goals? How can the government oversee integrated environmental laws and regulations and set environmental goals but encourage innovation by giving regulated parties more flexibility in meeting such goals? How can government and private capital flows to developing countries be used to promote more sustainable forms of development? How can the principles of systems analysis, industrial ecology (Solutions, p. 584), and ecosystem management be used to develop, integrate, implement, and evaluate local, national, regional, and global environmental policies that recognize the inherent interdependence of all life systems?

We are just beginning to realize that environmental issues such as energy, food production, water resources, soil conservation, global warming, ozone depletion, solid and hazardous wastes, pollution, population growth, poverty, urbanization, transportation, world trade, land use, economic growth, and preservation of biodiversity are all interconnected and must be dealt with in an integrated and holistic way using a systems approach. This will require new thinking and a new set of integrated environmental laws and policies to replace the current fragmented approach that treats each environmental problem in isolation. Such fragmentation typically focuses **(1)** on a single chemical instead of how the chemicals we use interact and which ones can be eliminated by better design and **(2)** on protecting a single species instead of protecting and rehabilitating entire ecosystems.

The confrontational environmental politics of the last 28 years pitted the economy against the environment. Now we are learning that environmental protection cannot be boiled down simply to a struggle for power between environmental activists (the "good guys") and corporate leaders (the "bad guys"). A small but growing number of companies have enlightened environmental policies but others pollute with abandon and support attempts to eliminate or weaken environmental laws and regulations (Case Study, p. 787).

The next generation of environmental policies must create a flexible mix of positive incentives for earth-sustaining leaders and practices and penalties for earth-degrading activities. Such policies must go beyond focusing mostly on large point-source polluters and reward smaller firms, farms, households, and car owners for working with the earth.

These are important issues that environmentalists, business leaders, elected officials, government regulators, and citizens need to address with a cooperative, problem-solving spirit. Too much is at stake in terms of economic and environmental health to fall into the old confrontational patterns that inhibit change and a creative search for win–win solutions.

It is encouraging that a small but growing number of scientists, environmentalists, economists,

A: At least 1 million

business leaders, farmers, and citizens are beginning to work together to pioneer more sustainable **(1)** forms of agriculture (Sections 12-6 and 21-5), **(2)** use of water resources (Section 13-9), **(3)** renewable energy resources (Sections 16-9 and 16-10), **(4)** forms of pollution and waste prevention and control (Sections 18-6, 19-4, 19-6, 20-4, 20-5, 22-2, and 22-9), **(5)** protection of biodiversity (Chapters 23, 24, and 25), **(6)** urban areas (Section 26-5), and **(7)** economic systems and policies (Section 27-5). They are beginning to integrate such policies with those for controlling population growth (Section 11-3) and reducing poverty (Section 27-4). The efforts of such *eco-pioneers* must be publicized and encouraged so that such hopeful seedlings of change can multiply and spread.

Whom Should We Believe? We need to ask tough questions of people on both sides of environmental issues and gather and carefully evaluate the evidence for each position by using the techniques of critical thinking (Guest Essay, p. 460). In addition, we need to distinguish between frontier and consensus science when claims are made (Section 2-1).

To evaluate statements by environmentalists and anti-environmentalists alike, citizens, elected officials, and environmental reporters are urged to identify the consensus of most of the scientists in the particular field involved and not give equal weight to a small minority of scientists in such fields who disagree with the current consensus view. Such reporting under the guise of balance may make a story more interesting, but it represents an unbalanced approach that can mislead the public.

One problem is that as the *easy work* has been done, the focus of environmental issues has shifted to more complex and controversial environmental problems that are harder to understand and solve. Examples are global climate change, ozone depletion, biodiversity protection, nonpoint water pollution (such as runoff from farms and lawns), and protection of unseen groundwater. Explaining such complex issues to the public and mobilizing support for the often controversial solutions are quite difficult (Guest Essay, p. 792).

Another problem is the complexity of evaluating the relative risks associated with the constant barrage of environmental and health problems reported by the media. All of us need to become better informed about the relative degree of potential harm posed by various risk factors, and we must try to rank risks as best we can (Figures 17-13 and 17-14). Risk analysis is a difficult, uncertain, and controversial undertaking (Section 17-6), but we can become better at understanding and evaluating the results of risk analyses and cost–benefit analyses (Section 27-3 and Solutions, p. 762).

One reason why an increasing number of environmentalists emphasize pollution and waste prevention (Guest Essays, pp. 70 and 586) using the *precautionary principle* is that this approach can save individuals and businesses money. It can also eliminate or reduce certain risks and the need for certain types of command-and-control regulations that, if too prescriptive, can hinder innovation and market-based solutions (Section 27-3).

28-6 IMPROVING THE U.S. POLITICAL SYSTEM

How Can We Make Government More Responsive to Ordinary Citizens? According to most political analysts, *the biggest problem that keeps elected officials from being more responsive to the environmental and other needs and problems of ordinary citizens is that the enormous amounts of money needed to run for office or to get reelected come mostly from wealthy and powerful individuals and corporate interests.* Most analysts and about 80% of citizens polled agree that the U.S. political system is based more on "money talks" than on citizens' vote—one reason so many Americans don't bother to vote. Even those who do vote often feel they're simply choosing between the lesser of two evils.

About 90% of the time elections are won by the candidate raising the most money. In 1992, the average U.S. senator spent $3.9 million campaigning for reelection—an amount that requires raising an average of $12,600 per week for 312 weeks (6 years). Unless they are immensely wealthy and willing to spend their own money, candidates can get this kind of money only from wealthy individuals and corporations. Many analysts see drastic reform in the way elections to public office are financed as the key to making the government more responsive to ordinary citizens (Solutions, p. 791).

How Can Bureaucracies Be Improved? In its early stages, a small, vigorous bureaucratic agency with dynamic leadership can do some good. However, over the years an agency can become rigid and more concerned with its own survival and getting an increasing share of the federal or state budget than with its original mission. It may also come under the influence of the businesses it is supposed to regulate—the *foxes-guarding-the-henhouse dilemma.*

Another problem is that responsibility for managing the nation's environmental and resource policy is divided among many federal and state agencies. This situation often leads to contradictory policies, duplicated efforts, and wasted funds. It prevents an effective, integrated approach to interrelated environmental problems.

Q: What can be done when groundwater becomes contaminated?

Election Finance Reform*

SOLUTIONS

A growing number of analysts of all political persuasions see election finance reform as the single most important way to reduce the influence of so-called money votes. They urge citizens not to get diverted and fragmented over other issues but to focus their efforts on this crucial issue as the key to significant political reform designed to make government more responsive to ordinary citizens.

According to former senator, presidential candidate, and arch-conservative leader Barry Goldwater, "The sheer cost of running for office is having a corrosive effect not only on American politics but on the quality of American government. . . . Something must be done to liberate candidates from their dependence on special-interest money."

One suggestion for reducing undue influence by powerful special interests would be to let the people (taxpayers) alone finance all federal, state, and local election campaigns, with low spending limits. Candidates and parties could not accept direct or indirect donations from any other individuals, groups, or parties, with *absolutely no exceptions.*

Once elected, officials could not use free mailings, staff employees, or other privileges to aid their election campaigns, nor could they accept donations or any kind of direct or indirect financial aid from any individual, corporation, political parties, or interest group for their future election campaigns—or for any other reason that could even remotely influence their votes on legislation. Violators of this new Public Funding Elections Act would be barred from the campaign or removed from office. Anyone making illegal donations would be subject to large fines and possible jail sentences.

Having all elections financed entirely by public funds would get the corrupting influence of private money out of elections. It would cost each U.S. taxpayer only about $5–10 per year (as part of their income taxes) for all federal elections (and a much smaller amount for state and local elections)—a small investment for making democracy more responsive to ordinary people.

Environmentalist and newspaper columnist Donella Meadows gives the following summary of the goals of such election campaign reform:

> *Money shall not dominate our democracy. The Congress shall not be a millionaires' club, nor a club of people obligated to millionaires. The ability to spend money shall not swing elections. Decisions shall not favor special contributors over the public good.*

With such a reform, elected officials could spend their time governing instead of raising money and catering to powerful special interests. Office seekers would not need to be wealthy, and public service could become a way to help the country and the earth, not a lifelong profession or route to wealth and power. Special-interest groups would be heard because of the validity of their ideas, not the size of their pocketbooks.

Proponents of this reform contend that this is an issue that ordinary people of all political persuasions could work together on. From this fundamental political reform, other political, economic, and environmental reforms could then flow.

The problem is getting members of Congress (and state legislators) to pass a virtually foolproof and constitutionally acceptable plan that would put them on equal financial footing with their challengers—and that might possibly decrease their chances of getting reelected. Supporters of public financing of all elections argue that the way out of this dilemma is to band together to find and to elect candidates who pledge to bring about this fundamental political reform, and then vote them out of office if they don't.

Critical Thinking

1. Do you agree or disagree with this approach to election reform? Explain. What role would you take in bringing about or opposing such a reform?
2. Some of these election reform proposals—especially preventing candidates from using their own money to support their election campaigns—might be unconstitutional. How could the reform process be modified to accommodate such an objection?

*For an insightful and in depth analysis of this issue see Moti Nissani, "Brass-Tacks Ecology," *The Trumpeter* 14, no. 3 (Summer 1997), pp. 143–148 and Chapter 10 in Moti Nissani, *Lives in the Balance: The Cold War and American Politics, 1945–1991* (Carson City, Nev.: Dowser Publishing Group, 1992).

Several suggestions have been made for improving the responsiveness of elected officials and government agencies to the people. One is to *pass a sunset law that automatically terminates any government agency or program after, say, 6 years.* After 6 years, the General Accounting Office and an outside commission would evaluate each agency and program and recommend to Congress whether it should be reinstated, reorganized, assigned a new mission, or terminated. If renewal were recommended, the evaluating bodies would list suggested improvements and deadlines for implementation.

Others urge us to *reward whistle-blowers.* This would involve providing better job protection,

A: Very little that is affordable

energy-efficient living arrangements. Every brick in an abandoned urban building represents an energy waste equivalent to burning a 100-watt light bulb for 12 hours. Each new suburb means replacing farmland or reservoirs of natural biodiversity with dispersed, energy- and resource-wasting roads, houses, and shopping centers.

Crime also makes people less willing to use walking, bicycles, and energy-efficient public transit systems. Additionally, it forces people to use more energy by leaving lights, TVs, and radios on to deter burglars and to clear away trees and bushes near houses that can reduce solar heat gain in the summer and provide a windbreak in the winter. Because of crime, many items are overpackaged to deter shoplifting or adding poisons to food or drug items. Thus, supporters of environmental causes have a vested interest in supporting efforts to reduce crime.

28-7 GLOBAL ENVIRONMENTAL POLICY

Should We Expand the Concept of Security? Countries have legitimate *national security* interests, but without adequate soil, water, clean air, and biodiversity, no nation can be secure. Thus, to a growing number of policy analysts, national security ultimately depends on *global environmental security* based on sustainable use of the ecosphere.

Countries are also legitimately concerned with *economic security*. However, because all economies are supported by the ecosphere (Figure 27-3 and Guest Essay, p. 756), economic security also depends on environmental security.

Proponents of emphasizing environmental security propose that national governments have a council of advisers made up of highly qualified experts in environmental, economic, and military security. Any major decision would require integrating all three security concerns. These analysts also call for all countries to make environmental security a major focus of diplomacy and government policy at all levels. What do you think?

What Progress Has Been Made in Developing International Environmental Cooperation and Policy? Since the 1972 UN Conference on the Human Environment was held in Stockholm, Sweden, some progress has been made in addressing environmental issues at the global level. Today, 115 nations have environmental protection agencies and more than 215 international environmental treaties have been signed concerning issues such as endangered species, ozone depletion, ocean pollution, global warming, biodiversity, and export of hazardous waste. The 1972 conference also created the UN Environment Programme (UNEP) to negotiate environmental treaties and to help implement them.

In June 1992 the second UN Conference on the Human Environment, known as the *Rio Earth Summit*, was held in Rio de Janeiro, Brazil. More than 100 heads of state, thousands of officials, and more than 1,400 accredited nongovernment organizations (NGOs) from 178 nations met to develop plans for addressing environmental issues.

The major official results included **(1)** an *Earth Charter*, a nonbinding statement of broad principles for guiding environmental policy that commits countries that sign it to pursue sustainable development and to work toward eradicating poverty; **(2)** *Agenda 21*, a nonbinding detailed action plan to guide countries toward sustainable development and protection of the global environment during the 21st century; **(3)** a *forestry agreement* that is a broad, nonbinding statement of principles of forest management and protection; **(4)** a *convention on climate change* that requires countries to use their best efforts to reduce their emissions of greenhouse gases; **(5)** a *convention on protecting biodiversity* that calls for countries to develop strategies for the conservation and sustainable use of biological diversity; and **(6)** establishment of the *UN Commission on Sustainable Development*, composed of high-level government representatives charged with carrying out and overseeing the implementation of these agreements.

Most environmentalists were disappointed because these accomplishments consist of nonbinding agreements without sufficient incentives for their implementation. The convention on climate change, for example, lacks the targets and timetables for stabilizing CO_2 emissions favored by all major industrial countries except the United States (which signed the treaty only after such items were eliminated). The forest protection statement was so watered down that most environmentalists consider it virtually useless. In addition, countries did not commit even the minimum amount of money that conference organizers said was needed to begin implementing Agenda 21.

By 1997, 5 years after the conference, leaders meeting to review progress found little improvement in the major environmental problems discussed at the Rio summit. According to a UN Environmental Programme report prepared for the meeting and other independent evaluations, **(1)** emissions of CO_2 rose substantially in all but three countries, **(2)** air pollution in most of the world's cities has worsened, **(3)** freshwater supplies are much more imperiled, **(4)** an area of forest about the size of Iowa is burned or cut each year, **(5)** the loss of biodiversity has not

Bolin, B. 1998. "The Kyoto Negotiations on Climate Change: A Science Perspective." *Science*, vol. 279, no. 5349, 330(2).

slowed, **(6)** little has been done to reduce poverty, a major cause of environmental degradation, **(7)** the gaps between the rich and the poor have widened (Figures 1-8 and 27-11), **(8)** no national government has developed a policy for sustainable consumption and production, and **(9)** since the Earth Summit the World Bank has provided $9.4 billion for fossil-fuel projects that increase the likelihood of global warming.

There are a few bright spots. Leaders have agreed on a worldwide phaseout of lead additives in gasoline and to set up an intergovernmental forum to work out what can be done to slow the cutting and burning of the world's forests. A few countries have made firm commitments to increase their government assistance for sustainable development in developing countries.

However, developing countries feel betrayed by the developed world's failure to uphold the promises made at the Rio summit. Their leaders argue that faced with massive poverty and limited budgets, they cannot reduce pollution and protect biodiversity without outside assistance and transfer of new more earth-sustaining technologies.

It is discouraging that so little has been done but there is hope for progress in the slowly moving arena of international cooperation.

First, the conference gave the world a forum for discussing and seeking solutions to environmental problems. This led to general agreement on some key principles, which with enough political pressure from the bottom up could be implemented or improved upon.

Second, paralleling the official meeting was a Global Forum that brought together 18,000 people from more than 1,400 NGOs in 178 countries. These NGOs worked behind the scenes to influence official policy, formulated their own agendas and treaties for helping sustain the earth, learned from one another, and developed a series of new global networks, alliances, and projects. In the long run, these newly formed networks and alliances may play the greatest role in helping monitor, support, and implement the commitments and plans developed by the formal conference.

Bringing about environmental improvement at local and national levels is difficult enough, but doing this at the international level is like trying to change the direction of a huge floating iceberg. However, history shows that in the long run persistent grassroots pressure can bring about important social and environmental changes. The situation is difficult and challenging, but not hopeless.

How Can We Help Ensure Environmental Justice for All? In recent years there have been growing efforts to forge an alliance between the environmental and human rights movements to form a global *environmental justice* or *eco-justice* movement. Traditionally, these two movements have not worked together and often have been suspicious of one another's motives. Some human-rights activists have difficulty understanding why environmental activists are willing to spend so much energy on protecting endangered species and habitats while human beings are being tortured and killed because they oppose vested political and financial interests. Some environmental groups have also been accused of fighting to set aside protected wildlife reserves without consulting or working with the native peoples whose homes and livelihoods would be affected.

Similarly, environmental activists have criticized the narrow focus of human-rights groups on small numbers of cases of abuse and murder when millions of people are dying prematurely each year because of pollution, environmental degradation, and lack of food.

However, this situation began changing with the widely publicized 1988 assassination of Chico Mendes (Figure 24-11) because of his opposition to cutting and developing the Amazon rain forests. This has been followed by the increasing assassination, execution, brutalization, and intimidation of grassroots environmental leaders throughout the world and more recently in the United States. In 1995, for example, the Nigerian government executed Ken Saro-Wiwa—a Nigerian writer, a former commissioner of education, a Nobel Peace Prize nominee, an environmentalist, and a champion of minority rights—for his ideas and his effectiveness as a leader in opposing government human rights and environmental policies. In 1995, there were two unsuccessful bombing attempts to kill Guy Pence, a U.S. Forest Service Ranger who has been uncompromising in enforcing existing federal grazing regulations.

Since 1990, these and hundreds of other incidents have made environmental and human rights activists recognize that they have much common ground. Traditionally, both groups have focused on protecting fundamental civil and political freedoms: free speech, fair elections, freedom to organize in groups and criticize government and business leaders, a free press to expand access to information, and the rights of communities to participate in economic and political decisions that are likely to affect their well-being.

There is an increasingly powerful environmental justice movement in the United States (Guest Essay, p. 606). In 1992, Human Rights Watch and the Natural Resources Defense Council released a joint report documenting some of the most vicious attacks against environmental activists throughout the world. In 1995 the Sierra Club and Amnesty International issued their first joint letter dealing with the link between environmental degradation and human rights abuses

Hint: Enter the search term *Kyoto protocol* using Key Words.

in Nigeria. Although still small in number, such collaborations are growing.

Can We Develop Earth-Sustaining Political and Economic Systems Over the Next Few Decades? Environmentalists say we can if we care enough to make the necessary commitment. Ecosphere politics, earth-sustaining economics, and earth-sustaining living are based on the *green vision* that learning how to work with the earth is not a special interest, but rather the ultimate interest of all people.

Despite the claims of anti-environmentalists, environmentalists are not people-hating, antibusiness people (although any diverse movement has a very small number of people who take such radical positions). Instead, environmentalists have an exciting and positive vision of a new earth-sustaining, people-sustaining, and people-empowering society. It would be built on a new ecoindustrial revolution (Solutions, p. 584) and a shift to earth-sustaining economies (Sections 27-5 and 27-6 and Guest Essays, pp. 756 and 774).

Environmentalists believe that making this new cultural transition over the next 40–50 years should be our most urgent goal. Making this new cultural change will be controversial, and like all significant change it will not be predictable, orderly, or painless.

According to business leader Paul Hawken (Guest Essay, p. 756), making this change means

> *Thinking big and long into the future, It also means doing something now. It means electing people who really want to make things work [Case Study, p. 736], and who can imagine a better world. It means writing to companies and telling them what you think. It means never forgetting that the cash register is the daily voting booth in democratic capitalism.*

Most ordinary citizens may feel powerless in making government more responsive to their needs and problems. However, by acting together, consumers have enormous power to change harmful forms of corporate behavior by boycotting (or threatening to boycott) certain products—what some environmentalists and advocates for environmental justice (Guest Essay, p. 606) call the key environmental tactic for the 1990s and beyond.

In making this new cultural change, many environmentalists urge us not to fall into the trap of *thinking that we can save the earth.* We are absolutely dependent on the earth, but the earth's existence and functioning does not depend on us. We can impair some of the planet's natural processes at local, regional, and perhaps at global levels, and we can cause the premature extinction of species. However, fossils and other evidence show that on a time scale of hundreds to millions of years the earth is resilient and adaptable and will continue functioning and changing with or without us. What is at stake is not the earth, but the future quality of life for most human beings and—if we go too far—the existence of our species and many other species.

In working with the earth we should be guided by historian Arnold Toynbee's observation, "If you make the world ever so little better, you will have done splendidly, and your life will have been worthwhile," and by George Bernard Shaw's reminder that "indifference is the essence of inhumanity."

As the wagon driver said when they came to a long, hard hill, "Them that's going on with us, get out and push. Them that ain't, get out of the way."

ROBERT FULGHUM

CRITICAL THINKING

1. What are the greatest strengths and weaknesses of the system of government in your country with respect to protecting the environment, working with the earth, and ensuring environmental justice for all? What three major changes, if any, would you make in this system?

2. Do you agree or disagree with proposals various analysts have made to improve the U.S. political system through reforming the elective process (Solutions, p 791), reforming government agencies, and leveling the legal system for citizens? Explain.

3. Suppose that a presidential candidate ran on a platform calling for the federal government to phase in a tax on gasoline so that, over 5–10 years, the price of gasoline would rise to $3–5 a gallon (as is the case in Japan and most western European nations). The candidate argues that this tax increase is necessary to encourage conservation of oil and gasoline, reduce air pollution, and enhance future economic, environmental, and military security. Some of the tax revenue would be used to provide tax relief or other aid to people with incomes below a certain level (the poor and lower middle class), who would be hardest hit by such a consumption tax. Income taxes on the poor and middle class would be reduced by an amount roughly equal to the increase in gasoline taxes. Would you vote for this candidate who wants to triple the price of gasoline? Explain.

4. Do you agree or disagree with each of the major goals of the Wise-Use movement (Case Study, p. 787)? Explain.

5. Which of the tactics, if any, used by the anti-environmental movement (pp. 784–788) do you think are appropriate? Explain. Which, if any, of these same tactics have been used by the environmental movement?

6. Try to come up with analogies to complete the following statements: **(a)** Trying to influence environmental policy is like _______. **(b)** Mainstream environmental

Q: What percentage of the world's potential food supply is lost to pests?

groups are like _______. **(c)** Grassroots environmental groups are like _______. **(d)** The anti-environmental movement is like _______. **(e)** Trying to convert money-based politics to people-based politics is like _______. **(f)** Trying to improve a bureaucracy is like _______.

PROJECTS

1. Evaluate the 1992 Rio Earth Summit. Use the library or Internet to find out what has happened with respect to the official proclamations and treaties signed at the conference and to the People's Treaties. Summarize your major findings.

2. A 1990 national survey by the Roper Organization found that even though 78% of Americans believe that a major national effort is required for environmental improvement (ranking it fourth among national priorities), only 22% were making significant efforts to improve the environment. The poll identified five categories of citizens: **(1)** *true-blue greens* (11%) are involved in a wide range of environmental activities, **(2)** *greenback greens* (11%) don't have time to be involved but will pay more for a cleaner environment, **(3)** *grousers* (24%) aren't involved in environmental action, mainly because they don't see why they should be if everybody else isn't, **(4)** *sprouts* (26%) are concerned but don't believe individual action will make much difference, and **(5)** *basic browns* (28%) are the most apathetic and least involved—or actually oppose the environmental movement. To which category do you belong? As a class, conduct a similar poll on your campus.

3. Use a combination of the major societal trends, possible new trends, and your imagination to construct three different scenarios of what the world might be like in 2020. Identify the scenario you favor and outline a program for achieving this alternative future. Compare your scenarios with those of your classmates.

4. Conduct a survey of environmental education programs at your school and in your community. Develop a plan for making environmental education a priority at your school and in your community and develop a set of principles for presenting such education. Present your plan to school officials and the local school board. Call for public meetings to discuss this issue.

5. What student environmental groups (if any) are active at your school? How many people actively participate in these groups? What environmentally beneficial things have they done? What actions (if any) taken by such groups do you disagree with? Why?

6. Make a concept map of this chapter's major ideas, using the section heads and subheads and the key terms (in boldface). Look at the inside back cover and in Appendix 4 for information about concept maps.

A: About 55% (35% before harvest and 20% after harvest)

29 ENVIRONMENTAL WORLDVIEWS, ETHICS, AND SUSTAINABILITY

*A.D. 2060: Green Times on Planet Earth**

Mary Wilkins sat in the living room of the passive solar earth-sheltered house (Figure 19-1) she shared with her daughter Jane and her family. It was July 4, 2060: Independence Day.

She walked into the greenhouse and gazed out at the large blue and green earth flag (Figure 29-1) they flew with the American flag. She recalled the day in 2040 when President Elizabeth Jordan had declared that July 4 would also celebrate global independence from fossil and nuclear fuels.

Her eyes shifted to her grandchildren, Lynn and Jeffrey, running and hollering as they played tag on the neighborhood commons. She heard the hum of solar-powered pumps trickling water to rows of organically grown vegetables and glanced at the fish in the aquaculture and waste treatment tanks. She was glad the tanks provided fish year-round, but it was much more fun to take her grandchildren fishing at the pond that she, with help from their Bioregional Advisory Council, had helped restore 20 years ago as part of a neighborhood wetlands restoration project.

Things began changing rapidly in 2004 when environmentalists, workers, and ordinary citizens began working together to sustain the earth's life-support systems (and thus the human species) by learning how to work with the earth. They decided to live by one of nature's most fundamental lessons: Sustainability depends on cooperating and pursuing diverse strategies adapted to local conditions, problems, and cultural beliefs.

Mary returned to the coolness of her earth-sheltered house and began putting the finishing touches on the children's costumes for this afternoon's pageant in Rachel Carson Park. It would honor the earth heroes who began the Age of Ecology in the 20th century, as well as those who continued this tradition in the 21st century. She smiled when she thought of Jeffrey's delight at being chosen to play Aldo Leopold, who in the late 1940s began urging people to work with the earth. Her pride swelled when Lynn was chosen to play Rachel Carson (Spotlight, p. 565), who in the 1960s alerted us to threats from increasing exposure to pesticides and other harmful chemicals.

*Compare this hopeful scenario with the worst-case scenario that opens Chapter 19.

Figure 29-1 Some people display the earth flag as a symbol of their commitment to working with the earth at the individual, local, national, and international levels. (Courtesy of Earth Flag Co., 33 Roberts Road, Cambridge, MA 02138)

Mary finished a seam and leaned back with a sigh of satisfaction. She closed her eyes and remembered when, as a concerned 20-year-old college student in 1999, she had been confronted by environmentalists saying we needed to drastically change our ways and by anti-environmentalists saying that more economic growth and technology would protect the earth. She had decided to join the Earth Conservation Corps after graduation, to do what she could to work with the earth.

Even in her most idealistic dreams, she had never guessed she would see the hemorrhaging loss of global biodiversity slowed to a trickle. There were no more wars over oil, and most air pollution gradually disappeared when energy from the sun, wind, and hydrogen (produced by using solar energy to decompose water) replaced fossil and nuclear fuels. Now there was greatly increased emphasis on pollution prevention and waste reduction. Walking and bicycling had increased in cities and towns designed as vibrant communities for people instead of cars. Low-polluting and safe ecocars got 128 kilometers per liter (300 miles per gallon), and there was efficient mass transportation.

World population had stabilized at 8 billion in 2030 and then had begun a slow decline. The rate of global warming had slowed significantly, and international treaties enacted in the 1990s effectively banned the chemicals that had begun depleting ozone in the stratosphere.

Two hours later, she, her daughter Jane, and her son-in-law Gene watched with pride as 40 beautiful children honored the leaders of the Age of Ecology. At the end, Lynn stepped forward and said, "Today we have honored many earth heroes, but the real heroes are the ordinary people in this audience and around the world who worked to help sustain the earth. Thank you, Grandma, Mom, Dad, and everyone here for giving us such a wonderful gift. We promise to leave the earth even better for our children and grandchildren and all living creatures."

Environmentalists and many of the world's prominent scientists believe that we are at a critical turning point: a time to make crucial choices and to act on them. One option is to start on a path that could lead to a world like the one just described; another is to stay on our present path, which could result in a world like the one described on p. 498. The choice is ours; to refuse to decide is to decide.

The main ingredients of an environmental ethic are caring about the planet and all of its inhabitants, allowing unselfishness to control the immediate self-interest that harms others, and living each day so as to leave the lightest possible footprints on the planet.

ROBERT CAHN

This chapter answers the following questions:

- What major human-centered environmental worldviews guide most industrial societies?
- What are some life-centered and earth-centered worldviews?
- What ethical guidelines might be used to help us work with the earth?
- How can we live more sustainably?
- How can we bring about an earth-wisdom revolution?

29-1 ENVIRONMENTAL WORLDVIEWS IN INDUSTRIAL SOCIETIES

How Shall We Live? A Clash of Cultures and Values As a powerful species, what role should we play on the earth? What obligations do we have to the human species? To other species? To future generations? How serious are the environmental problems we face?

There are conflicting answers to these crucial questions. These conflicts arise mostly out of differing **environmental worldviews**: ways people think the world works, what they think their role in the world should be, and what they believe is right and wrong environmental behavior (**environmental ethics**).

People with widely differing environmental worldviews can take the same data, be logically consistent, and arrive at quite different conclusions (Table 1-2) because they start with different assumptions and are often seeking answers to different questions.

There are many different types of environmental worldviews, as summarized in Figure 29-2. Most can be divided into two groups according to whether they are *individual centered* (atomistic) or *earth centered* (holistic). Atomistic environmental worldviews tend to be *human centered* (anthropocentric) or *life centered* (biocentric, with the primary focus on either

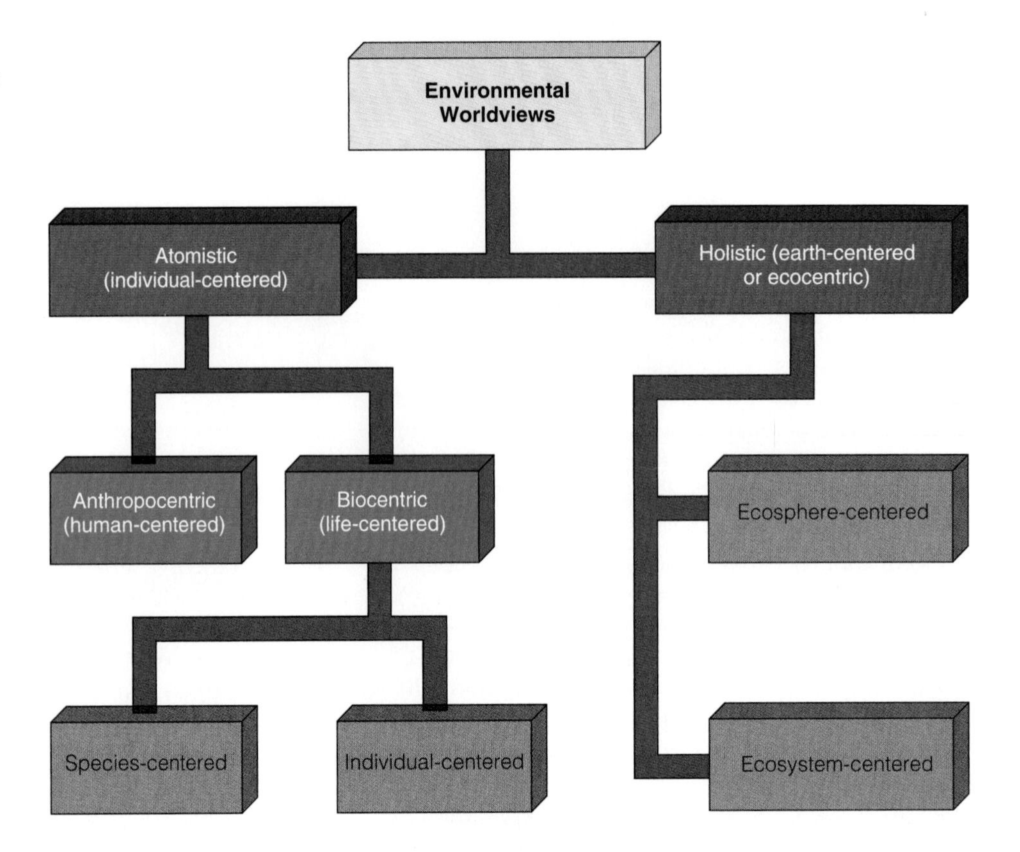

Figure 29-2 General types of environmental worldviews. (Diagram developed by Jane Heinze-Fry)

individual species or individual organisms). Holistic or ecocentric environmental worldviews are either *ecosystem centered* or *ecosphere* (life-support system) *centered*.

What Are the Major Human-Centered Environmental Worldviews? Most people in today's industrial–consumer societies have a **planetary management worldview**, which has become increasingly accepted during the past 50 years. According to this human-centered environmental worldview, human beings, as the planet's most important and dominant species, can and should manage the planet mostly for their own benefit. Other species are seen as having only *instrumental value*; that is, their value depends on whether they are useful to us. The basic environmental beliefs of this worldview include the following:

- *We are the planet's most important species, and we are in charge of the rest of nature.* This idea crops up when people talk about "our" planet or "our" earth and when people talk about "saving the earth."
- *There is always more.* The earth has an essentially unlimited supply of resources, to which we gain access via science and technology. If we deplete a resource, we will find substitutes. To deal with pollutants, we can invent technology to clean them up, dump them into space, or move into space ourselves. If we extinguish other species, we can use genetic engineering to create new and better ones.
- *All economic growth is good, more economic growth is better, and the potential for economic growth is essentially limitless.*
- *Our success depends on how well we can understand, control, and manage the earth's life-support systems for our benefit.*

People with this (and related) environmental worldviews seek answers to questions such as, "How can we keep economic growth or throughput of resources growing exponentially Figure 3-21)? How can we become better managers of the entire planet? How can we control and manage the pollutants and wastes we produce and the environmental degradation we cause?" This worldview is widely supported because it is said to be the primary driving force behind the major improvements in the human condition since the beginning of the industrial revolution (Table 1-2, left).

There are several variations of this environmental worldview. Some people belong to what might be called the *no-problem school*: There are no environmental, population, or resource problems that cannot be solved by more economic growth, better management, and better technology.

Another group, *the free-market school*, believes that the best way to manage the planet for human benefit is through a free-market global economy with minimal government interference and regulations. Free-market advocates would convert all public property

Angus, I. 1997. "Free Nature." (Interview with philosopher Arne Naess.) *Alternatives Journal*, 23 n3 p18(4).

resources to private property resources and let the global marketplace, governed by free-market competition (pure capitalism), decide essentially everything.

Another variation, held by many people, is that we do have serious environmental, resource, and population problems but that we can deal with them by engaging in *responsible planetary management*. People holding this view follow the pragmatic principle of *enlightened self-interest*: Better earth-care is better self-care. They believe that we can sustain our species with a mixture of market-based competition, better technology, and some government intervention to promote ecologically sustainable forms of economic development, protect environmental quality and private property rights, and protect and manage public and common property resources.

Still another variation of the responsible planetary management environmental worldview is the *spaceship-earth worldview*. Earth is seen as a spaceship: a complex machine that we can understand, dominate, change, and manage in order to prevent environmental overload and provide a good life for everyone.

Other people advocate that our management of the earth be guided by the principle of *stewardship*: We have an ethical responsibility to be caring and responsible managers or stewards, who tend the earth as if it were a garden. According to this view, we can and should make the world a better place for ourselves and other species through love, care, knowledge, and technology.

29-2 LIFE-CENTERED AND EARTH-CENTERED ENVIRONMENTAL WORLDVIEWS

Can We Manage the Planet? Some people believe that any human-centered worldview will eventually fail because it wrongly assumes that we now have (or can gain) enough knowledge to become effective managers or stewards of the earth. They compare our pursuit of unlimited economic growth on a finite planet to being on a treadmill that moves faster and faster. Sooner or later, they believe, we will either fall off the treadmill or damage it because our knowledge and managerial skills are so limited compared to the incredible complexity and adaptability of earth's life-support systems.

These people argue that the unregulated free-market approach won't work because it is based on mushrooming losses of earth capital (Figures 1-2, 1-5, and 1-14) and because it focuses on short-term economic benefits regardless of the harmful long-term consequences. They also contend that the spaceship-earth and stewardship versions of planetary management will not work because such human constructs are oversimplified and misleading ways to view an incredibly complex and ever-changing planet.

For example, these critics point out that we do not even know how many species live on the earth, much less what their roles are and how they interact with one another and their nonliving environment. We have only an inkling of what goes on in a handful of soil, a meadow, a patch of forest, a pond, or any other part of the earth. Such analysts liken us to technicians who think we can build and repair automobile engines after a couple of minutes of training with a vacuum cleaner motor.

As biologist David Ehrenfeld puts it, "In no important instance have we been able to demonstrate comprehensive successful management of the world, nor do we understand it well enough to manage it even in theory." Environmental educator David Orr (Guest Essay, p. 746) says we are losing rather than gaining the knowledge and wisdom needed to adapt creatively to continually changing environmental conditions:

> *On balance, I think, we are becoming more ignorant because we are losing cultural knowledge about how to inhabit our places on the planet sustainably, while impoverishing the genetic knowledge accumulated through millions of years of evolution. . . . Most research is aimed to further domination of the planet. Considerably less of it is directed at understanding the effects of domination. Less still is aimed to develop ecologically sound alternatives that enable us to live within natural limits.*

To critics of the planetary management environmental worldview, our task is not to learn how to pilot spaceship earth, but instead to give up our fantasies of omnipotence and base our actions on earth wisdom—on learning to work with the earth by becoming more responsible earth citizens. Even if we had enough knowledge and wisdom to manage spaceship earth, some critics see this approach as requiring us to give up individual freedom in order to survive. Life on spaceship earth under a comprehensive system of planetary management or world government might be very much like the regimented life of astronauts in their capsule. The astronauts have virtually no individual freedom; essentially all of their actions are dictated by a central command (ground control) without which they cannot survive. To these analysts, managing earth as a stripped-down spaceship to prevent disaster leaves little room for human freedom, novelty, adaptability, or long-term sustainability.

Theologian Thomas Berry calls the industrial consumer society built on the human-centered, planetary management environmental worldview the "supreme pathology of all history":

> *We can break the mountains apart; we can drain the rivers and flood the valleys. We can turn the most lux-*

Hint: Enter the search term *deep ecology* using Key Words.

uriant forests into throwaway paper products. We can tear apart the great grass cover of the western plains, and pour toxic chemicals into the soil and pesticides onto the fields, until the soil is dead and blows away in the wind. We can pollute the air with acids, the rivers with sewage, the seas with oil—all this in a kind of intoxication with our power for devastation. . . . We can invent computers capable of processing ten million calculations per second. And why? To increase the volume and speed with which we move natural resources through the consumer economy to the junk pile or the waste heap. . . . If, in these activities, the topography of the planet is damaged, if the environment is made inhospitable for a multitude of living species, then so be it. We are, supposedly, creating a technological wonderworld. . . . But our supposed progress . . . is bringing us to a wasteworld instead of a wonderworld.

What Are Some Major Biocentric and Ecocentric Worldviews? Critics of human-centered environmental worldviews believe that such worldviews should be expanded to recognize the *inherent value* of all forms of life (that is, value that exists regardless of these life-forms' potential or actual use to us). According to the broadest form of this *life-centered (biocentric)* worldview, all forms of life have an inherent right to struggle to exist.

Ecologist Stan Rowe offers the following criticism of human-centered (anthropocentric) worldviews:

From the precept that only humans matter, a disastrous corollary follows: The world is for exploiting. Parks are for people, animals are for shooting, forests are for logging, soils are for mining. The sole basis for ethical action is the greatest good for "the greatest number of people." . . . This becomes "People First." Five billion people going for ten, all believing in People First, increasing their wants without limit, are a sure recipe for species suicide.

Most people with a life-centered (biocentric) worldview argue that our actions should not lead to the *premature* extinction of species. Some analysts give species a hierarchy of values. Some, for example, believe we have greater responsibility to protect animal species than plant species. But critics point out that plants are what keep most animals alive. Others determine the survival rights of various species depending on the harm they do to humans. For example, they see nothing inherently wrong in trying to wipe out pest species such as rats, cockroaches (Spotlight, p. 209), mosquitoes, and indeed most insects (p. 80); species of disease-carrying bacteria; and species that we fear such as ecologically important alligators (p. 133), sharks (Case Study, p. 217), spiders (p. 561), and bats (Case Study, p. 690).

Some proponents go further and believe that each individual organism, not just each species, has an inherent right to survive. As in the species-centered approach, some people (the animal rights movement) place more value on the individuals of animal species than on plant species and microorganisms. Even then, the emphasis is usually on large mammals such as dogs, cats, elephants, and bears. Animal rights advocates have not mounted major campaigns to protect individual bats, spiders, sharks, or snakes from being killed, nor have they taken a strong stand that doing so is wrong.

Trying to decide what types of species or individuals should be protected from premature extinction or death resulting from human activities is an ethical dilemma. It is hard to know where to draw the line and be ethically consistent.

Others believe that we must go beyond this biocentric worldview, which focuses on species and individual organisms. They see our primary role as limiting our actions to those that do not degrade or destroy the earth's life-support systems (and thus threaten the existence of our own species). In other words, they have an *earth-centered*, or *ecocentric*, environmental worldview, devoted to preserving earth's biodiversity and ecological integrity (Figure 29-2).

Their view is that we are part of, not apart from, the community of life and the ecological processes that sustain all life. Aldo Leopold summed up this idea in 1948: "All ethics rest upon a single premise: that the individual is a member of a community of interdependent parts."

Recently there has been a trend among conservationists to move from protection of individual species (Chapter 25) to protection of ecosystems (Chapter 24). Proponents of this *ecosystem-centered worldview* argue that it is more effective and less expensive than waiting for individual species to be threatened with extinction and then trying to give them often ineffective and very costly emergency treatment. The idea of this prevention approach is not to depend primarily on 911 emergency treatment for the protection of life.

Other analysts move further up on the holistic ladder (Figure 4-2) and emphasize our ethical obligation not to impair or degrade the ecosphere itself—the life-support systems for all life (Figure 29-2). They see this larger umbrella of concern as an even broader prevention approach that is necessary for two reasons: **(1)** to preserve the global biodiversity and ecological integrity that supports all forms of life and **(2)** to deal with global environmental problems such as climate change (Sections 19-3 and 19-4) and ozone depletion

Q: What percentage of U.S. crops are lost to pests?

Cartoon © William K. Day, Tribune Media Service. Used by permission (Color added)

(Sections 19-5 and 19-6) that result in loss of biodiversity at the ecosystem level (Chapter 24) and the species level (Chapter 25). Most of us have a mixture of parts of various environmental worldviews that we pull out of our conceptual toolbox and apply depending on the level of life we are focusing on (Figure 4-2).

There are many life-centered and earth-centered environmental worldviews, and several of them overlap in some of their beliefs. One ecocentric environmental worldview is the **earth-wisdom worldview** It is based on the following major beliefs, which are the opposite of those making up the planetary management worldview:

- *Nature exists for all of the earth's species, not just for us.* We need the earth, but the earth does not need us.
- *There is not always more.* The earth's limited resources should not be wasted, but instead used efficiently and sustainably for us and all species.
- *Some forms of economic growth are environmentally beneficial and should be encouraged, but some are environmentally harmful and should be discouraged.*
- *Our success depends on learning to cooperate with one another and with the rest of nature by learning how to work with the earth.* Management of resources is essential to human survival. However, such management should involve learning as much as we can about how the earth works, sustains itself, and adapts to changing conditions and then using these lessons from nature to guide our actions.

People with such environmental worldviews seek answers to the following questions: How can we design and use economic and political systems to encourage earth-sustaining forms of development and discourage earth-degrading ones? What is sustainability, and what do we need to do to live sustainably on the planet? How can we produce fewer pollutants and wastes and not cause so much environmental degradation? Some analysts call for us to shift our image and worldview of nature as a foe to be conquered and exploited to nature as a nurturing mother that we should work with rather than against (see cartoon). A related ecocentric environmental worldview is the *deep ecology worldview* (Spotlight, p. 804).

These and other life-centered and earth-centered environmental worldviews have their roots in the ways of life of many primal peoples. Earth-wisdom principles have also been articulated by Saint Francis of Assisi, Benedict deSpinoza, Henry David Thoreau, Ralph Waldo Emerson, John Muir, Aldo Leopold, Rachel Carson, Alan Watts, Gary Snyder, Charles Reich, Theodore Roszak, Arne Naess (developer of what is called the *deep ecology worldview*; Spotlight, p. 804), Bill Devall, George Sessions, and many others. Many of the basic ideas of these worldviews are also found in teachings of Hinduism, Taoism, Zen Buddhism, and

A: About 37% (7% higher than in the 1940s despite a 33-fold increase in pesticide use since then)

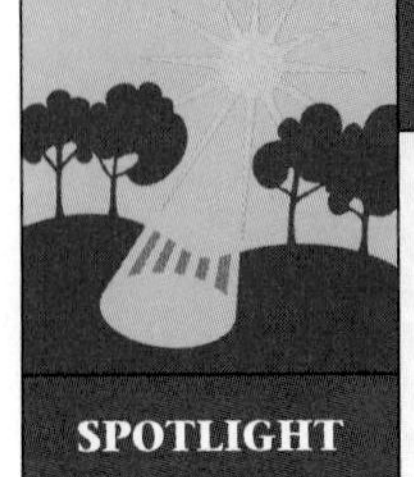

Deep Ecology

SPOTLIGHT

Deep ecology is an ecocentric environmental set of beliefs developed in 1972 by Norwegian philosopher Arnie Naess. In 1984 Naess, philosopher George Sessions, and sociologist Bill Devall drew up a list of eight beliefs of deep ecology. In 1993 these beliefs were modified somewhat by Naess.* Here is the modified list:

1. The well-being and flourishing of human and nonhuman life on earth have inherent value in themselves. These values are independent of the usefulness of the nonhuman world for human purposes.

2. The fundamental interdependence, richness, and diversity of life-forms contribute to the flourishing of human and nonhuman life on earth.

3. Humans have no right to reduce this interdependence, richness, and diversity except to satisfy vital needs.

4. Present human interference with the nonhuman world is excessive, and the situation is worsening rapidly.

5. Because of point 4, it would be better for humans, and much better for nonhumans, if there were a substantial decrease in the human population.

6. Policies must therefore be changed. These policies affect basic economic, technological, and ideological structures. The resulting state of affairs will be deeply different from the past.

7. The ideological change is mainly that of appreciating *life quality* (involving situations of inherent value) rather than adhering to an ever-higher material standard of living.

8. Those who subscribe to these points have an obligation directly or indirectly to try to implement the necessary changes.

Naess has also described some lifestyle guidelines compatible with the basic beliefs of deep ecology. They include **(1)** appreciating all forms of life, **(2)** protecting or restoring local ecosystems, **(3)** using simple means, **(4)** consuming less, **(5)** emphasizing satisfying vital needs rather than desires, **(6)** attempting to live in nature and promote community, **(7)** appreciating ethnic and cultural differences, **(8)** working to improve the standard of living for the world's poor, **(9)** working to eliminate injustice toward fellow humans or other species, and **(10)** acting nonviolently.

Since 1984, deep ecology has become a major philosophical and ethical movement, used by a growing number of people from different philosophical and religious backgrounds to bring about social and environmental change. Hundreds of articles and books have been written interpreting, analyzing, praising, and criticizing this new ecocentric environmental worldview.

Deep ecology is not an ecoreligion, nor is it antireligious or antihuman, as some of its critics have claimed. Instead, it is a set of beliefs designed to have us think more deeply about the inherent value of all life on the earth and about our obligations toward both human and nonhuman life.

Critical Thinking

1. Which, if any, of the eight basic beliefs of the deep ecology environmental worldview do you agree with? Explain.

2. List five major effects on your life and lifestyle if virtually everyone lived by the beliefs of deep ecology.

*For a set of readings about the nature of deep ecology see George Sessions, ed. *Deep Ecology for the 21st Century* (Boston: Shambhala, 1995).

the Judeo-Christian tradition. Various biocentric and ecocentric environmental worldviews also emphasize beliefs found in the ecofeminism, social ecology, and environmental justice movements (Guest Essay, p. 606).

Are Biocentrists and Ecocentrists Antihuman and Antireligious? Many anti-environmentalists accuse people who promote various biocentric and ecocentric worldviews of being antihuman or against celebrating humanity's special qualities and achievements. However, those with life-centered and earth-centered worldviews consider their environmental worldviews to be profoundly prohuman. To them, recognizing the inherent value of all life and not degrading the earth's life-support systems are ways to benefit all people in this and future generations (Connections, p. 805).

Others say we do not need to be biocentrists or ecocentrists to value life or the earth because human-centered stewardship and planetary-management environmental worldviews also call for us to value individuals, species, and the earth's life-support systems as part of our responsibility as earth's caretakers.

What Is the Ecofeminist Worldview? The term *ecofeminism*, coined in 1974 by French writer Françoise d'Eaubonne, includes a spectrum of views on the relationships of women to the earth and to male-dominated societies (patriarchies). Although most ecofeminists agree that we need a life-centered or

Lerner, S. 1998. "The New Environmentalists." *Futurist* v32 n4 p35(5).

earth-centered environmental worldview, they believe that a main cause of our environmental problems isn't just human-centeredness, but specifically male-centeredness (*androcentrism*).

Many ecofeminists argue that the rise of male-dominated societies and environmental worldviews since the advent of agriculture is primarily responsible for our violence against nature (and for the oppression of women and minorities as well). To such ecofeminists, this led to a shift from an image of nature as a nurturing mother (see cartoon, p. 803) to one of nature as a foe to be conquered.

As evidence of male domination, ecofeminists note that women earn less than 10% of all wages, own less than 1% of all property, and in most societies have far fewer rights than men. These analysts argue, moreover, that to become primary players in the male power-and-domination game, most women are forced to emphasize the characteristics deemed masculine and become "honorary men."

Some ecofeminists suggest that oppression by men has driven women closer to nature and made them more compassionate and nurturing. As oppressed members of society, they argue, women have considerable experience in dealing with interpersonal conflicts, bringing people together, acting as caregivers, and identifying emotionally with injustice, pain, and suffering.

Ecofeminists argue that women should be given the same rights as men, allowed to have their views heard and respected, and treated as equal partners. They do not want just a fair share of the patriarchal pie; they want to work with men to bake an entirely new pie that helps heal the rift between humans and nature and ends oppression based on sex, race, class, and cultural and religious beliefs. In doing this, they do not want to be given token roles or co-opted into the male power game.

Ecofeminists are not alone in calling for us to encourage the rise of *life-centered people* who emphasize the best human characteristics: gentleness, caring, compassion, nonviolence, cooperation, and love.

What Is the Social Ecology Worldview? According to anarchist philosopher Murray Bookchin, the ecological crisis we currently face results from the power of our hierarchical and authoritarian social, economic, and political structures, and from the various technologies used to dominate people and nature. In other words, our current environmental situation has been created by industrialized societies driven by the conventional planetary management environmental worldview (p.800).

To alleviate the ecological crisis, Bookchin believes we must adopt a *social ecology environmental worldview*, which would involve decentralizing political and economic systems and corporations and altering the types of technology we use. Bookchin urges us to create better versions of democratic communities, new forms of earth-sustaining production, and new types of ecotechnology that are smaller in scale, consume fewer resources and less earth capital, and are geared to the carrying capacities of local ecological regions.

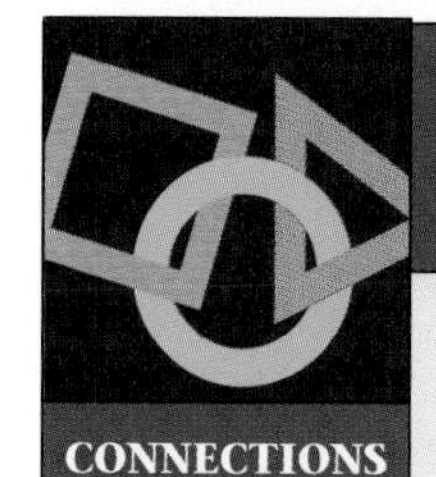

CONNECTIONS

Why Should We Care About Future Generations?

According to biologist David W. Ehrenfeld, caring about future generations enough not to degrade the earth's life-support systems is important because it gives future generations options for dealing with the problems they will face. He points out that if our ancestors had left for us the ecological devastation we are leaving our descendants, our options for enjoyment—perhaps even for survival—would be quite limited.

And in response to the question, "What can future generations do for us?" Ehrenfeld gives the following answer: "They give us a reason for treating our ecological home respectfully, so that our lives as well as theirs will be enriched."

In thinking about our responsibility toward future generations, some analysts believe that we should consider the wisdom given to us in the 18th century by the Iroquois Confederation of Native Americans: *In our every deliberation, we must consider the impact of our decisions on the next seven generations.*

Critical Thinking

What obligations, if any, concerning the environment do you have to future generations? Be honest about your feelings. To how many future generations do you have responsibilities? List the most important environmental benefits and harmful conditions passed on to you by the previous two generations.

29-3 SOLUTIONS: LIVING SUSTAINABLY

How Should We Evaluate Sustainability Proposals? *Sustainability* has become a buzzword. Hundreds of programs have been proposed or implemented in its name, some of them useful and some questionable or

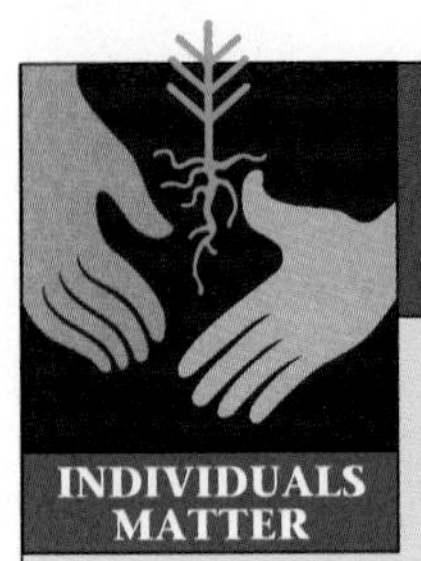

Mindquake: Evaluating One's Environmental Worldview

INDIVIDUALS MATTER

Questioning and perhaps even changing one's environmental worldview can be difficult and threatening and can set off a cultural *mindquake* that involves examining many of one's most basic beliefs. However, once individuals change their worldview, it no longer makes sense for them to do things in the old ways. If enough people do this, then tremendous cultural change, once considered impossible, can take place rapidly.

Most environmentalists urge us to think about what our basic environmental beliefs are and why we have them. They believe that evaluating our beliefs, and being open to the possibility of changing them, should be one of our most important lifelong activities.

As this book emphasizes, most environmental issues are filled with controversy and uncertainty. A clearly right or wrong path is not easy to discover and is usually strongly influenced by our environmental worldview. As philosopher Hegel pointed out nearly two centuries ago, tragedy is not the conflict between right and wrong, but the conflict between right and right.

Critical Thinking

What are the basic beliefs of your current environmental worldview? At the end of this course evaluate your answer to learn which (if any) of your basic environmental beliefs have changed.

harmful. Evaluating such proposals requires using our critical thinking skills (Guest Essay, p. 460).

Lester Brown (Guest Essay, p. 343) has this simple test for any sustainability proposal: "Does this policy or action lower carbon emissions? Does it reduce the generation of toxic wastes? Does it slow population growth? Does it increase the earth's tree cover? Does it cut ozone-depleting CFC emissions? Does it reduce air pollution? Does it reduce radioactive waste generation? Does it lead to less soil erosion? Does it protect the planet's biodiversity?"

Additional questions also help us evaluate sustainability proposals. Does it deplete earth capital? Does it diminish cultural diversity? Does it reduce poverty, hunger, and disease? Does it promote individual and community self-reliance? Does it prevent pollution? Does it reduce resource waste? Does it save energy? Does it transfer the most resource-efficient and environmentally benign technologies to developing countries?

The search for sustainability presumes that for any given proposed course of action, we know (or can discover) what is sustainable—that is, the levels and thresholds of environmental carrying capacity for a region or for the entire world. Unfortunately, we cannot do this very well with current knowledge. Until we can, a growing number of environmentalists believe that we should follow what is called the *precautionary principle*: playing it safe by using various prevention guidelines and strategies for developing sustainable societies.

What Are Some Ethical Guidelines for Working with the Earth? Various ethicists and philosophers have developed a variety of ethical guidelines for living more sustainably on the earth. Such guidelines can be used by anyone, whether they have a human-centered stewardship environmental worldview or a biocentric or ecocentric environmental worldview.

Ecosphere and Ecosystems

- We should try to understand and work with the rest of nature to help sustain the ecological integrity, biodiversity, and adaptability of the earth's life-support systems.
- When we must alter nature to meet our needs or wants, we should carefully evaluate our proposed actions and choose methods that do the least possible short- and long-term environmental harm.

Species and Cultures

- We should work to preserve as much of the earth's genetic variety as possible because it is the raw material for all future evolution and genetic engineering.
- We have the right to defend ourselves against individuals of species that do us harm and to use individuals of species to meet our vital needs, but we should strive not to cause the premature extinction of any wild species.
- The best ways to protect species and individuals of species are to protect the ecosystems in which they live and to help restore those we have degraded.
- No human culture should become extinct because of our actions.

Individual Responsibility

- We should not inflict unnecessary suffering or pain on any animal we raise or hunt for food or use for scientific or other purposes.
- We should leave the earth as good as or better than we found it.

Buttimar, A. 1998. "Close to Home: Making Sustainability Work at the Local Level." *Environment*, vol. 40, no. 3, 12(13).

- We should use no more of the earth's resources than we need.
- We should work with the earth to help heal ecological wounds we have inflicted.

What Is Earth Education? Most environmentalists believe that learning how to live sustainably requires a foundation of earth education that relies heavily on an interdisciplinary and holistic approach to learning and a lifelong commitment to such education (Guest Essay, p.812). Among the most important goals of such an education are the following:

- *Developing respect or reverence for all life.*
- *Understanding as much as we can about how the earth works and sustains itself, and using such earth wisdom to guide our lives, communities, and societies.*
- *Understanding as much as we can about connections and interactions*—those within nature, between people and the rest of nature, among people with different cultures and beliefs, among generations, among the problems we face, and among the solutions to these problems. Ecological literacy is primarily the ability to connect and synthesize knowledge from a spectrum of disciplines.
- *Becoming wisdom seekers instead of information vessels.* We need to learn how to sift through mountains of facts and ideas to find the nuggets of knowledge and wisdom that are worth knowing.
- *Understanding and evaluating one's worldview and seeing this as a lifelong process* (Individuals Matter, p. 806).
- *Learning how to evaluate the beneficial and harmful consequences of one's lifestyle and profession on the earth.*
- *Using critical thinking skills* (Guest Essay, p. 46) *to evaluate and resist advertising.* As humorist Will Rogers put it, "Too many people spend money they haven't earned to buy things they don't want, to impress people they don't like."
- *Fostering a desire to make the world a better place and to act on this desire.* As David Orr puts it, education should help students "make the leap from 'I know' to 'I care' to 'I'll do something.'"

According to environmental educator Mitchell Thomashow, four basic questions should be at the heart of environmental education:

- Where do the things I consume come from?
- What do I know about the place where I live?
- How am I connected to the earth and other living things?
- What are my purpose and responsibility as a human being?

How we answer these questions determines our *ecological identity.*

How Can Direct Experiences Help Us Learn How to Work with the Earth? In addition to top-down formal education, some analysts believe that we need to learn by experience how to walk more lightly on the earth (Connections, p. 808). One source of wisdom is bottom-up education obtained by *listening to children.* Listen to young children and you will find that many of them believe that much of what we are doing to the earth (and thus to them) is stupid and wrong, and they do not accept the excuses we give for not changing the harmful ways we act toward the earth.

Another goal should be *learning how to live sustainably in a place*—a piece of the earth or a bioregion (Connections, p. 809) to which we are rooted or emotionally attached and whose sustainability and adaptability we feel driven to nurture and protect. As poet–philosopher Gary Snyder puts it, "Find our place on the planet, dig in, and take responsibility from there."

Earth care also means *not using guilt and fear to motivate other people to work with the earth and other people.* We need to nurture, reassure, understand, and love, rather than threaten, one another. Finally, we need to *have fun and take time to enjoy life.* Every day we should laugh and enjoy wild nature, beauty, friendship, and love.

How Can We Live More Simply? Gandhi's Philosophy of Enoughness Living sustainably does not mean unplugging everything, moving to the country, growing our own food, or becoming a modern hunter–gatherer. The world is far too populated and developed to do that, even if we wanted to. And doing so would accelerate loss of biodiversity as more land would be paved over and built on, and more wildlife habitats destroyed or degraded. Instead, it involves each of us learning how to live more simply in the places we temporarily occupy during our short stay on the earth.

Although seeking happiness through the single-minded pursuit of material things is considered folly by virtually every major religion and philosophy, it is preached incessantly by modern advertising. Some affluent people in developed countries, however, are adopting a lifestyle of *voluntary simplicity,* doing and enjoying more with less.

Such people have learned that buying more products and luxuries to satisfy artificially created desires doesn't provide security, freedom, or joy; that is what Paul Wachtel calls "the poverty of affluence." They agree with Edward Goldsmith that "the more we look at it, the more it is apparent that economic growth is a device for providing us with the superfluous at the cost of the indispensable."

Hint: Enter the search terms *sustainable development, environmental aspects* using the Subject Index.

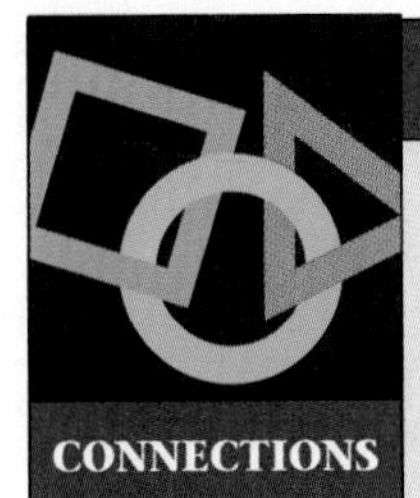

Learning from the Earth

Formal earth education is important, but Aldo Leopold, Henry David Thoreau, Gary Snyder, David Brower, Stephen Jay Gould, David Orr, Mitchell Thomashow, and many others believe it's not enough.

Such earth thinkers urge us to take the time to escape the cultural and technological body armor we use to insulate ourselves from nature and to experience nature directly. They suggest that we reenchant our senses and kindle a sense of awe, wonder, and humility by standing under the stars, sitting in a forest, taking in the majesty and power of an ocean, or experiencing a stream, lake, or other part of untamed nature.

We might pick up a handful of soil and try to sense the teeming microscopic life in it that keep us alive. We might look at a tree, mountain, rock, or bee and try to sense how they are a part of us, and we a part of them as interdependent participants in the earth's life-sustaining recycling processes.

Earth thinker Michael J. Cohen suggests that each of us recognize who we really are by saying,

> *I am a desire for water, air, food, love, warmth, beauty, freedom, sensations, life, community, place, and spirit in the natural world. . . . I have two mothers: my human mother and my planet mother, Earth. The planet is my womb of life.*

Many psychologists believe that consciously or unconsciously we spend much of our lives in a search for roots—something to anchor us in a bewildering and frightening sea of change. As philosopher Simone Weil observed, "To be rooted is perhaps the most important and least recognized need of the human soul."

Earth philosophers say that to be rooted, each of us needs to find a *sense of place*—a stream, a mountain, a yard, a neighborhood lot, or any piece of the earth we feel at one with as a place we know, experience emotionally, and love. It can be a place where we live or a place we occasionally visit and experience in our inner being. When we become part of a place, it becomes a part of us. Then we are driven to defend it from harm and to help heal its wounds.

To many earth thinkers, emotionally experiencing our connectedness with the earth leads us to recognize that the healing of the earth and the healing of the human spirit are one and the same. They call for us to discover and tap into what Aldo Leopold calls "the green fire that burns in our hearts" and use this as a force for respecting and working with the earth and with one another.

Critical Thinking

Some analysts believe that learning earth wisdom by experiencing the earth and forming an emotional bond with its life forms and processes is unscientific, mystical poppycock based on a romanticized view of nature. They believe that better scientific understanding of how the earth works and improved technology are the only ways to achieve sustainability. Do you agree or disagree? Explain.

As environmental expert Donella Meadows puts it,

> *People need respect, identity, community, challenge, acknowledgment, beauty, love, joy. To try to fill these needs with material things is to set up an unquenchable appetite for false solutions to real and never-satisfied problems. The resulting psychological emptiness is one of the major forces behind the desire for material growth.*

Substituting earth-friendly (green) technologies and goods for earth-degrading ones is highly desirable. However, this is only a short-term strategy because green consuming (especially to meet frivolous wants) is still consuming. Furthermore, if population and per capita consumption (even if it's all green) continue rising, they will eventually overwhelm the environmental benefits of such consumption.

Voluntary simplicity is based on Mahatma Gandhi's *principle of enoughness*: "The earth provides enough to satisfy every person's need but not every person's greed. . . . When we take more than we need, we are simply taking from each other, borrowing from the future, or destroying the environment and other species." It means asking oneself, "How much is enough?" This is not an easy thing to do because people in affluent societies are conditioned to want more and more, and they often think of such wants as vital needs.

However, voluntary simplicity by those who have more than they need should not be confused with the *forced simplicity* of the poor, who do not have enough to meet their basic needs for food, clothing, shelter, clean water and air, and good health.

After a lifetime of studying the growth and decline of the world's human civilizations, historian Arnold Toynbee summarized the true measure of a civilization's growth in what he called the *law of progressive simplification*: "True growth occurs as civilizations transfer an increasing proportion of energy and attention from the material side of life to the nonmaterial side and thereby develop their culture, capacity for compassion, sense of community, and strength of democracy."

Q: What percentage of insecticides applied to crops in the United States reaches the target pests?

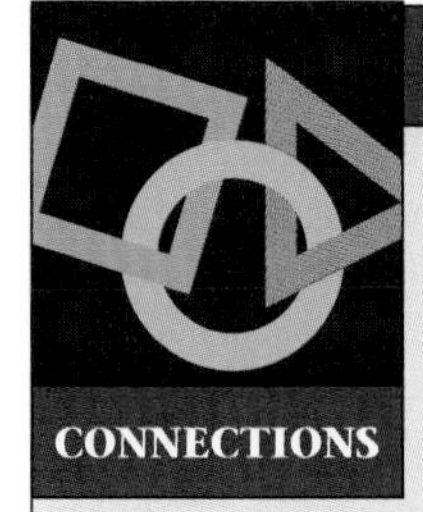

Bioregional Living

Some environmentalists believe that another aid to working with the earth is to view the geographic area where we live as part of a natural region or **bioregion,** a unique life territory with its own soils, landforms, watersheds, microclimates, native plants and animals, and other distinctive natural characteristics.

A bioregion is an area of land not defined by political boundaries—cities, states, countries—but by the natural, biological, and geological features that make up the real identity of a place. A bioregion is identified by its landforms, watersheds, climate and weather patterns, soil types, wildlife habitats, and the forms of life in these habitats.

The major bioregions of North America include areas such as the Sonora Desert, Cascadia (the Pacific Northwest), the Ozarks, the Maritimes (northern Maine and eastern Canada), and Appalachia. Bioregionalists further subdivide such larger regions into individual mountain ranges and watersheds.

On a bioregional scale it becomes easier for people to determine the most ecologically sound and sustainable ways for people to live in a certain place. *Bioregional living* is an attempt to understand and live sustainably within the natural cycles, flows, and rhythms of a particular place. It means becoming *dwellers in the land* who establish an ecologically and socially sustainable pattern of existence within it.

Most attempts at bioregional living involve *reinhabitation:* learning to live in an area that has been injured and disrupted by human exploitation. This means asking, What was this region like before we came? What might nature put here if we were not present? How can we begin healing the wounds inflicted on our bioregion by cooperating with the natural processes that shape and sustain it?

At the individual level, bioregional living means being able to answer fundamental ecological questions about the corner of the world we inhabit. Where does our water come from? Where does our energy come from? What kinds of soils are under our feet? How long is the growing season? What types of wildlife are our neighbors? Where does our food come from? Where does our waste go? This is real knowledge. Can you answer these questions about where you live?

How Can We Move Beyond Blame, Guilt, and Denial to Responsibility? According to many psychologists, when we first encounter an environmental problem, our initial response is often to find someone or something to blame: greedy industrialists, uncaring politicians, misguided worldviews. It is the fault of such villains, and we are the victims.

This can lead to despair, denial, and inaction because we feel powerless to stop or influence these forces. There are also so many complex and interconnected environmental problems and conflicting views about their seriousness and possible solutions that we feel overwhelmed and wonder whether there is any way out—another emotion leading to denial and inaction.

Upon closer examination we may realize that we all make some direct or indirect contributions to the environmental problems we face. We don't want to feel guilty or bad about all of the things we are not doing, so we avoid thinking about them—another path leading to denial and inaction.

How do we move beyond immobilizing blame, fear, and guilt to engaging in more responsible environmental actions in our daily lives? Analysts have suggested several ways to do this.

First, we need to recognize and avoid common mental traps that lead to denial, indifference, and inaction. These traps include **(1)** *gloom-and-doom pessimism* (it's hopeless), **(2)** *blind technological optimism* (science and technofixes will always save us), **(3)** *fatalism* (we have no control over our actions and the future), **(4)** *extrapolation to infinity* (if I can't change the entire world quickly, I won't try to change any of it), **(5)** *paralysis by analysis* (searching for the perfect worldview, philosophy, solutions, and scientific information before doing anything), and **(6)** *faith in simple, easy answers.*

Second, we should recognize that no one can even come close to doing all of the things people suggest (or that we know we should be doing) to work with the earth. Focus your energy on the few things that you feel most strongly about and that you can do something about. Instead of focusing on and feeling guilty about the things we haven't done, rejoice in the good things we have done, then jump in and do more to make the earth a better place.

Third, we should base our actions on a sense of *hope,* which history has shown to be the major energizing force for bringing about change. The secret is to keep our empowering feelings of hope and joy slightly ahead of our immobilizing feelings of despair. We need to acknowledge our *grief* about the harm the human species has done to the earth, ourselves, and future generations. Then we can move through the various stages of grief and become actively engaged in working with the earth in our daily lives.

A: No more than 2% and often less than 0.1%

Fourth, it is important to recognize that there is no single correct or best solution to the environmental problems we face. Indeed, one of nature's most important lessons (from evolution) is that preserving diversity or a rainbow of possibilities is the best way to adapt to earth's largely unpredictable, ever-changing conditions. Each human culture and environmental worldview provides different outlooks, wisdom, and insights for helping us learn how to work with the earth and make cultural changes in response to changes in environmental conditions. We are all in this together and need to work together to find a spectrum of flexible and adaptable solutions to the problems we face. To be effective in bringing about change, environmental worldviews must be broadly inclusive, not exclusive.

What Are the Major Components of an Earth-Wisdom Revolution? Many environmentalists call for us to make a new cultural change in the way we think about and use the earth's endowment of resources. Such an *earth-wisdom revolution* would have several phases. One is an *efficiency revolution* that involves not wasting matter and energy resources, using a combination of technological advances, lifestyle changes, recycling, and reuse.

A second phase is a *pollution prevention or ecoindustrial revolution* (Guest Essays, pp. 70 and 586 and Solutions, p. 589) built on the efficiency revolution. It reduces pollution and environmental degradation by reducing the waste of matter and energy resources; it does so by mimicking the earth's chemical cycling processes (Chapter 5), in which each organism's wastes serve as resource inputs for other organisms. Pollution prevention also involves keeping highly toxic substances from being released into the environment by recycling or reusing them within industrial processes, trying to find less harmful or easily biodegradable substitutes, or not producing such substances at all.

A third phase is a *sufficiency revolution*, which means being sure that the basic needs of all people on the planet are met and asking how many material things we really need to have a decent and meaningful life.

A fourth phase of this new cultural change is a *demographic revolution* based on bringing the size and growth rate of the human population into balance with the earth's carrying capacity for humans and other life forms.

Opponents of such an environmental cultural change like to paint environmentalists as messengers of gloom, doom, and hopelessness. However, *the real message of environmentalism is not gloom and doom, fear, and catastrophe but hope and a positive vision of the future*. This is an exciting message of challenge and adventure as we struggle to find better and more responsible ways to live on this planet.

Rejoice in our connections with all human beings and with all other forms of life—past, present, and future—in the magnificent tapestry of evolution that produces a diversity of solutions to ever-changing environmental conditions. Delight in being part of this unfolding drama and in living at a hinge of cultural history—a time when we have a unique opportunity and responsibility to become participants in bringing about a new, earth-sustaining cultural change.

Envision the world as a system of all kinds of matter cycles and energy flows. See these life-sustaining processes as a beautiful and diverse web of interrelationships—a kaleidoscope of patterns, rhythms, and connections whose very complexity and multitude of possibilities remind us that cooperation, sharing, honesty, humility, and love should be the guidelines for our behavior toward one another and the earth.

What has gone wrong, probably, is that we have failed to see ourselves as part of a large and indivisible whole. . . . . We have failed to understand that the earth does not belong to us, but we to the earth.

ROLF EDBERG

CRITICAL THINKING

1. What obligations, if any, concerning the environment do you have to future generations? To how many future generations do you have responsibilities? List the most important environmental benefits and harmful conditions passed on to you by the previous two generations.

2. What are the basic beliefs of your environmental worldview? Has taking this course changed your environmental worldview? In what ways?

3. Do you agree with the cartoon character Pogo that "we have met the enemy and he is us"? Explain. Criticize this statement from the viewpoint of the poor. Criticize it from the viewpoint that large corporations and government are the really big polluters and resource depleters and degraders.

4. **(a)** Would you accept employment on a project that you knew would kill or harm people? Degrade or destroy a wild habitat? Explain. **(b)** If you were granted three wishes, what would they be? **(c)** If you didn't have to work for a living, what would you do with your time? **(d)** Could you live without an automobile? Explain. **(e)** Could you live without TV? Explain. **(f)** If you won $10 million in a lottery, how would you spend the money? **(g)** Do you believe that wolves have as much right to eat sheep as people do? Explain.

5. How do you feel about **(a)** carving huge faces of people in mountains; **(b)** carving your initials into a tree; **(c)** driving an off-road motorized vehicle in a desert, grassland, or forest; **(d)** using throwaway paper towels, tissues, napkins, and plates; **(e)** wearing furs; **(f)** hunting

Q: Since 1945, how many premature deaths from insect-transmitted diseases have been prevented by using insecticides?

for recreation; **(g)** fishing for recreation; **(h)** having tropical fish, birds, snakes, or other wild animals as pets?

6. Do you believe that everyone has the right to have as many children as they want? Explain. If not, how would you limit such a right, including your own?

7. Do you believe that each member of the human species has a right to pollute and to use as many resources as they want? Explain. If not, how would you limit such a right, including your own?

8. Do you believe that the poor have a right to a transfer of enough wealth from the rich to meet their basic needs? Explain. If so, how would you implement this right?

9. Do you believe that individuals should have the right to do anything they want with land they own? Explain. If not, what specific limitations would you put on such private property ownership rights?

10. Which (if any) of the ethical guidelines on p. 806 do you disagree with? Explain. Can you suggest any additional ethical guidelines for working with the earth?

11. Review your experience with the mental traps described on p. 809. Which of these traps have you fallen into, if any? Were you aware that you had been ensnared by these mental traps? Do you plan to free yourself from these traps? How?

PROJECTS

1. Make a list of the goods and services you use each day. Trace each one back to see how it is ultimately obtained from the environment and thus originates from some part of the earth. Try to show the types and general levels (high, medium, low) of environmental impacts resulting from each step in this process: obtaining and transporting the resources from the earth, purifying or processing them to provide raw materials, transporting and converting the raw materials to finished products, transporting the products to you for use, your use of the good or service, and what happens to the things you use when you no longer need or want them.

2. Which of the goods and services you just listed meet primary needs, and which fulfill wants? Which ones could you do without? Which ones are you willing to do without? Which goods that you now throw away could be reused or recycled? What percentage of those do you actually reuse or recycle?

3. Make an environmental audit of your school.* What proportion of each of the major types of matter resources it uses are recycled, reused, or composted? What priority does your school give to the purchase of recycled materials? How much emphasis does it put on energy efficiency, use of renewable forms of solar energy, and environmental design in developing new buildings and renovating existing ones? Does it use ecologically sound planning in deciding how its grounds and buildings are managed and used? Does your school limit the use of toxic chemicals in its buildings and on its grounds? What proportion of its food purchases comes from nearby farmers? What proportion of the food it purchases is grown by sustainable agriculture?

4. Does your school's curriculum provide *all* graduates with the basic elements of ecological literacy? To what extent are the funds in any endowment invested in enterprises that are working to develop or encourage environmental sustainability? Over the past 20 years, what important roles have its graduates played in making the world a better and more sustainable place to live? Using such information, rate your school on a 1–10 scale in terms of its contributions to environmental awareness and sustainability. Develop a detailed plan illustrating how your school could become better at achieving such goals; present this information to school officials, alumni, parents, and financial backers.

5. If you knew you were going to die and had an opportunity to address everyone in the world for 5 minutes, what would you say? Write out your 5-minute speech and compare it with those of other members of your class.

6. Write an essay in which you try to identify key environmental experiences that have influenced your life and thus helped form your current ecological identity. Examples may include fond childhood memories of special places where you connected with the earth through emotional experiences; places that you knew and cherished that have been polluted, developed, or destroyed; key events that forced you to think about environmental values; individuals or educational experiences that influenced your understanding and concern about environmental problems and challenges; and direct experience and contemplation of wild places. Have you arrived at your current ecological identity mostly through formal education, directly experiencing nature, or both? Share your experiences with other members of your class.

7. Make a detailed list of *everything* you own. Then write a short essay about how these possessions affect your sense of self-esteem, comfort, security, and happiness. Do you feel burdened by the need to take care of so many possessions and worrying that they may be lost, stolen, or destroyed? Examine whether you feel guilty about owning too many things, especially those that mostly fill wants, not needs. Consider the general environmental impact of your possessions by evaluating the resources used in making them and how much pollution and environmental degradation are involved. Share these feelings about your possessions with other members of your class.

8. Make a concept map of this chapter's major ideas, using the section heads and subheads and the key terms (in boldface). Look at the inside back cover and in Appendix 4 for information about concept maps.

**Green Lives, Green Campuses*, written by Jane Heinze-Fry, is a supplement available for use with this book that provides guidelines for carrying out such an audit and for evaluating the environmental impact of your lifestyle.

A: At least 7 million

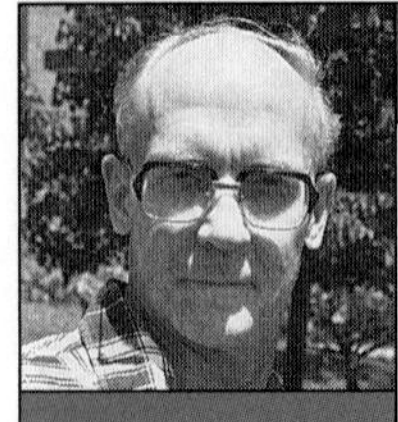

GUEST ESSAY

Envisioning a Sustainable Society

Lester W. Milbrath

Lester W. Milbrath is director of the Research Program in Environment and Society and professor emeritus of political science and sociology at the State University of New York at Buffalo. During his distinguished career he has served as director of the Environmental Studies Center at SUNY/Buffalo (1976–87) and taught at Northwestern University, Duke University, the University of Tennessee, National Taiwan University in Taipei, and Aarhus University in Denmark. He has also been a visiting research scholar at the Australian National University and at Mannheim University in Germany. His research has focused on the relationships among science, society, and citizen participation in environmental policy decisions, with emphasis on environmental perceptions, beliefs, attitudes, and values. He has written numerous articles and books. His book Envisioning a Sustainable Society: Learning Our Way Out *(1989) summarizes a lifetime of studying our environmental predicament. It is considered one of the best analyses of what we can do to learn how to work with the earth. His most recent book is* Learning to Think Environmentally While There Is Still Time *(1995).*

The 1992 Earth Summit at Rio popularized the goal of sustainable development. Most of the heads of state meeting there believed that goal could be achieved by developing better technology and by writing better laws, agreements, and treaties, and by enforcing them. Unfortunately, their approach was flawed and will not achieve sustainability because they do not understand the nature of the crisis in our earthly home.

Try this thought experiment: Imagine that, suddenly, all the humans disappeared, but all the buildings, roads, shopping malls, factories, automobiles, and other artifacts of modern civilization were left behind. What then? After three or four centuries, buildings would have crumbled, vehicles would have rusted and fallen apart, and plants would have recolonized fields, roads, parking lots, even buildings. Water, air, and soil would gradually clear up; some endangered species would flourish. Nature would thrive splendidly without us.

That mental experiment makes it clear that we do not have an environmental crisis; we have a crisis of civilization. Heads of state meeting at the Earth Summit neither understood nor dealt with civilization's most crucial problems: Humans are reproducing at such epidemic rates that world population is expected to almost double to 10–11 billion in 44 years; resource depletion and waste generation could easily triple or even quadruple over that period; waste discharges are already beginning to change the way the biosphere works; climate change and ozone loss will reduce the productivity of ecosystems just when hordes of new humans will be looking for sustenance, and will destroy the confidence people need in order to invest in the future.

Without intending to, we have created a civilization that is headed for destruction. Either we learn to control our growth in population and in economic activity, or nature will use death to control it for us.

Present-day society is not capable of producing a solution because it is disabled by the values our leaders constantly trumpet: economic growth, jobs, consumption, competitiveness, power, and domination. Societies pursuing these goals cannot avoid depleting their resources, degrading nature, poisoning life with wastes, and upsetting biospheric systems. *We have no choice but to change; resisting change will make us victims of change.*

But how do we transform to a sustainable society? My answer, which I believe is the only answer, is that *we must learn our way.* Nature, and the imperatives of its laws, will be our most powerful teacher as we learn our way to a new society. Most crucially, we must learn how to think about values.

Life in a viable ecosystem must become the core value of a sustainable society; that means all life, not just human life. Ecosystems function splendidly without humans (or any animals for that matter), but human society would die without viable ecosystems. Individuals seeking life quality require a well-functioning society living in well-functioning ecosystems. We must give top priority to the ecosystems that support us, and second priority to our societies.

A sustainable society would affirm love as a primary value and extend it not only to those near and dear, but to people in other lands, to future generations, and to other species. A sustainable society emphasizes partnership rather than domination, cooperation over competition, love over power. A sustainable society affirms justice and security as primary values.

A sustainable society would encourage self-realization—helping people to become all they are capable of being, rather than spending and consuming—as the key to a fulfilling life. A sustainable society would make

Q: What is the most serious drawback to using chemicals to control pests (especially insects)?

long-lasting products to be cherished and conserved. People would learn a love of beauty and simplicity.

A sustainable society would use both planning and markets as basic and supplementary information systems. Markets fail us because they can neither anticipate the future nor make moral choices between objects and between policies. Markets also cannot provide public goods such as schools, parks, and environmental protection, which are just as important for life quality as private goods.

A sustainable society would continue further development of science and technology because we need practical creative solutions that are both environmentally sound and economically feasible. However, we should recognize that those who control science and technology can use them to dominate all other creatures; we must learn to develop social controls of science and technology to make our society more sustainable. We should not allow the deployment of powerful new technologies that can induce sweeping changes in economic patterns, lifestyles, governance, and social values without careful forethought regarding their long-term impacts.

Conscious social learning would become the dynamic of social change in a sustainable society—not only to deal with pressing problems, but also to realize a vision of a good society. Meaningful and lasting social change occurs when nearly everyone learns the necessity of change and the value of working toward it.

Ecological thinking is different from most thinking that guides modern society. For example, the following key maxims derived from the law of conservation and matter, the laws of thermodynamics, and the workings of ecosystems are routinely violated in contemporary thinking and discourse: **(1)** Everything must go somewhere (there is no away); **(2)** energy should not be wasted because all use of energy produces disorder in the environment; **(3)** we can never do just one thing (everything is connected); and **(4)** we must constantly keep asking, "and then what?" Every schoolchild and every adult should learn these simple truths; we need to reaffirm the tradition that knowledge of nature's workings and a respect for all life are basic to a true education. We should require such environmental education of all students, just as we now require every student to study history.

Ecological thinking recognizes that a proper understanding of the world requires people to learn how to think holistically, systematically, and futuristically. Because everything is connected to everything else, we must learn to anticipate second-, third-, and higher-order consequences for any contemplated major societal action. A society learning to be sustainable would redesign government to maximize its ability to learn. It would use the government learning process to promote social learning. It would require that people who govern listen to citizens, not only to keep the process open for public participation, but also to cultivate mutual learning between officials and citizens.

In the recognition that our health and welfare are vitally affected by how people, businesses, and governments in other lands behave, a sustainable society would strive for an effective system of planetary politics. It would encourage transnational social movements and political parties. It would nurture global social learning.

Learning our way to a new society cannot occur, however, until enough people become aware of the need for major societal change. So long as contemporary society is working reasonably well, and leaders keep telling us that society is on the right track, the mass of people will not listen to a message urging significant change. For that reason, urgently needed change will probably be delayed, and conditions on our planet are likely to get worse before they can get better. Nature will be our most powerful teacher, especially when biospheric systems no longer work the way they used to. In times of great system turbulence, social learning can be extraordinarily swift.

Our species has a special gift: the ability to recall the past and foresee the future. Once we have a vision of the future, every decision becomes a moral decision. Even the decision not to act becomes a moral judgment. Those who understand what is happening to the only home for us and other species are not free to shrink from the responsibility to help make the transition to a sustainable society.

Critical Thinking

1. Do you agree or disagree that we can only learn our way to a sustainable society? Explain.

2. Do you think we will learn our way to a sustainable society? Explain. What role, if any, do you intend to play in this process?

A: Development of genetic resistance to the chemicals by target pests

EPILOGUE

PRINCIPLES FOR UNDERSTANDING AND WORKING WITH THE EARTH

Nature of Science

- Science is an attempt to discover order in nature and then to use that knowledge to describe, explain, and predict what happens in nature.
- Scientific theories and laws are well tested and widely accepted principles with a high degree of certainty.

Matter

- Matter cannot be created or destroyed; it can only be changed from one form to another. Everything we think we have thrown away is still with us in one form or another; there is no "away" (*law of conservation of matter*).
- Organized and concentrated matter is high-quality matter that can usually be converted into useful resources at an affordable cost; disorganized and dispersed matter is low-quality matter that often costs too much to convert to a useful resource (*principle of matter quality*).

Energy

- Energy cannot be created or destroyed; it can only be changed from one form to another. We can't get energy for nothing; in terms of energy quantity, it takes energy to get energy (*first law of energy or thermodynamics* or *law of conservation of energy*).
- Organized or concentrated energy is high-quality energy that can be used to do things; disorganized or diluted energy is low-quality energy that is not very useful (*principle of energy quality*).
- In any conversion of energy from one form to another, high-quality, useful energy is always degraded to lower-quality, less useful energy that can't be recycled to give high-quality energy; we can't break even in terms of energy quality (*second law of energy or thermodynamics*).
- High-quality energy should not be used to do something that can be done with lower-quality energy; we don't need to use a chain saw to cut butter (*principle of energy efficiency*).

Life

- Life on earth depends on the one-way flow of high-quality energy from the sun, through earth's life-support systems, and eventually back into space as low-quality heat; gravity; and the recycling of vital chemicals by a combination of biological, geological, and chemical processes (*principle of energy flow, gravity, and matter recycling*).
- Each species and each individual organism can tolerate only a certain range of environmental conditions (*range-of-tolerance principle*).
- Too much or too little of a physical or chemical factor can limit or prevent the growth of a population in a particular place (*limiting factor principle*).
- Every species has a specific role to play in nature (*ecological niche principle*).
- Species interact through competition for resources, predation, parasitism, mutualism (mutually beneficial interactions), and commensalism (interactions beneficial to one species with no harm to the other) (*principle of species interactions*).
- When possible, species reduce or avoid competition with one another by dividing up scarce resources so that species with similar requirements use them at different times, in different ways, or in different places (*principle of resource partitioning*).
- As environmental conditions change, the number and types of species present in a particular area change and, if not disturbed, can often form more complex communities (*principle of ecological succession*).
- Average precipitation and temperature are the major factors determining whether a particular land area supports a desert, grassland, or forest (*climate-biome principle*).
- All living systems contain complex networks of interconnected negative and positive feedback loops that interact to provide some degree of stability or sustainability over each system's expected life span (*principle of stability*).
- The size, growth rate, age structure, density, and distribution of a species's population are controlled by it's interactions with other species and with it's nonliving environment (*principle of population dynamics*).
- No population can keep growing indefinitely (*carrying capacity principle*).
- Individuals of a population of a species that possess genetically controlled characteristics enhancing their ability to survive under existing environmental conditions have a greater chance of surviving and producing more offspring than do those

lacking such traits (*principle of adaptation and natural selection*).

- All species eventually become extinct by disappearing or by evolving into one or more new species in response to environmental changes brought about by natural processes or by human action (*principle of evolution*).
- Over billions of years, changes in environmental conditions have led to development of a variety of species (species diversity), genetic variety within species (genetic diversity), and a variety of natural systems (ecosystem diversity) through a mixture of extinction and formation of new species (*biodiversity principle*).
- The earth's crust and upper mantle are made up of gigantic floating plates; their movement over millions of years reshapes the earth's crust, causes continents to move, and concentrates some of the minerals we extract and use. This continental drift is an important factor in the distribution and evolution of species (*theory of plate tectonics*).
- The earth's atmosphere, hydrosphere, lithosphere (upper crust and mantle), and forms of life are continually changing in response to changes in solar input, heat flows from the earth's interior, movements of the earth's crust, other natural changes, and changes brought about by humans and other living organisms (*principle of adaptability*).
- The earth's life-support systems can withstand much stress and abuse, but there are limits to how much can be tolerated (*principle of limits*).

Humans and Environment

- Our survival, life quality, and economies are totally dependent on the sun and the earth; the earth can get along without us, but we can't get along without the earth (*principle of earth capital*).
- The earth does not belong to us; we belong to the earth (*humility principle*).
- We should try to understand and work with the rest of nature to sustain the ecological integrity, biodiversity, and adaptability of earth's life-support systems for us and other species (*sustainability principle*).
- When we alter nature to meet our needs or wants, we should choose the method that does the least possible harm to us and other living things now and in the future (*least-harm principle*).
- The best way to protect species and individual organisms is to protect the ecosystems in which they live and to help restore those we have degraded (*principle of ecosystem protection and restoration*).
- We should not inflict unnecessary suffering or pain on any animal we raise or hunt for food or use for scientific or other purposes (*principle of humane treatment of animals*).
- We can learn a lot about how nature works, but nature is so incredibly complex and dynamic that such knowledge will always be limited (*principle of complexity*).
- In nature, we can never do just one thing; everything we do creates effects that are often unpredictable (*first law of human ecology*).
- Everything is connected to and intermingled with everything else; we are all in this together; we need to understand these connections and discover which connections are most important for sustaining life on earth (*principle of interdependence or connectedness*).
- Most resources are limited and should not be wasted; there is not always more (*principle of resource conservation*).
- Renewable resources should be used no faster than they are replenished by natural processes (*principle of sustainable use*).
- Living off renewable solar energy and renewable matter resources is a sustainable human lifestyle. Using renewable matter resources faster than they are replenished and living off nonrenewable matter and energy resources degrade and deplete earth capital and ultimately constitute an unsustainable lifestyle (*principle of sustainable living*).
- Increases in population, resource use, or both can eventually overwhelm attempts to control pollution and manage wastes (*environmental impact principle*).
- The best and cheapest way to reduce pollution and waste is to not produce so much (*principle of pollution prevention and waste reduction*).
- Pollution and wastes should not be put into the environment faster than the environment can degrade and recycle them or render them harmless (*principle of optimum pollution*).
- We should change earth-degrading and earth-depleting manufacturing processes, products, and businesses into earth-sustaining ones by using economic incentives and penalties (*principle of economic–ecological sustainability*).
- The market price of a product should include all estimated present and future costs of any pollution, environmental degradation, or other harmful effects connected with it that are passed on to society, the environment, and future generations (*principle of full-cost pricing*).
- Anticipating and preventing problems is cheaper and more effective than reacting to and trying to rectify them; an ounce of prevention is worth a pound of cure (*precautionary principle*).
- History shows that the most important changes brought about by human actions come from the bottom up, not from the top down (*individuals matter principle*).
- We should think globally and act locally (*principle of change*).

PUBLICATIONS, ENVIRONMENTAL ORGANIZATIONS, AND FEDERAL AND INTERNATIONAL AGENCIES

PUBLICATIONS

The following publications can help you keep well informed and up-to-date on environmental and resource problems. Subscription prices, which tend to change, are not given.

Alliance to Save Energy 1200 18th St. NW, Suite 900, Washington, DC 20036; Tel: 202-857-0666

Alternate Sources of Energy Alternate Sources of Energy, Inc., 107 South Central Ave., Milaca, MN 56353

Ambio: A Journal of the Human Environment Pergamon Press, Fairview Park, Elmsford, NY 10523

American Biology Teacher Journal of the National Association of Biology Teachers, 11250 Roger Bacon Dr., Rm. 319, Reston, VA 22090; Tel: 703-471-1134

American Council for an Energy Efficient Economy 1001 Connecticut Ave. NW, Suite 801, Washington, DC 20036; Tel: 202-429-8873

American Forests American Forestry Association, 1516 P St. NW, Washington, DC 20005; Tel: 202-667-3300

American Journal of Alternative Agriculture 9200 Edmonston Rd., Suite 117, Greenbelt, MD 20770

American Rivers Newsletter 801 Pennsylvania Ave. SE, Suite 400, Washington, DC 20003-2167; Tel: 202-547-6900, Fax: 202-543-6142

Amicus Journal Natural Resources Defense Council, 40 W. 20th St., New York, NY 10011; Tel: 212-727-2700

Annual Review of Energy Department of Energy, Forrestal Building, 1000 Independence Ave. SW, Washington, DC 20585

Audubon National Audubon Society, 950 Third Ave., New York, NY 10022; Tel: 212-832-3200, Web site: http://magazine.audubon.org/

BioCycle Journal of Waste Recycling J.G. Press, Inc., 419 State Ave., Emmaus, PA 19049; Tel: 215-967-4135

BioScience American Institute of Biological Sciences, Central Station, P.O. Box 27417, Washington, DC 20077-0038; Tel: 202-628-1500 or 800-992-2427, Fax: 202-628-1509; E-mail: bioscienc@aibs.org

Clean Water Action News Clean Water Action, 1320 18th St. NW, Washington, DC 20036; Tel: 202-457-1286

Climate Alert The Climate Institute, 324 4th St. NW, Washington, DC 20002; Tel: 202-547-0104, Fax: 202-547-0111

Conservation Biology Blackwell Scientific Publications, Commerce Place, 350 Maine St., Malden, MA 02148-5018; Tel: 888-661-5800, Fax: 617-388-8255, Web site: http://www.conbio.rice.edu/scb/journal

Demographic Yearbook Department of International Economic and Social Affairs, Statistical Office, United Nations Publishing Service, United Nations, New York, NY 10017; Tel: 617-253-2889, Fax: 617-258-6779

Discover 114 Fifth Avenue, New York, NY 10011-5690; Tel: 800-829-9132; E-mail: letters@discover.com

Earth Ethics Center for Respect of Life and Environment (CRLE), 2100 L St. NW, Washington, DC 20037

Earth Island Journal Earth Island Institute, 300 Broadway, Suite 28, San Francisco, CA 94133; Tel: 415-788-2666

Earthwatch Magazine P.O. Box 403N, Mt. Auburn St., Watertown, MA 02272; Tel: 800-776-0188 (in MA only: 617-926-8200)

Earth Work Student Conservation Association, P.O. Box 550, Charlestown, NH 03603; Tel: 603-543-1700, Fax: 603-543-1828

Ecodecision Environment and Policy Society Magazine, 176 Rue Saint Jacques Quest, Bureau 924, Montreal, Quebec, Canada H2Y 1N3; Tel: 514-284-3043, Fax: 514-284-3045

Ecological Economics Elsevier Science Publishing Co., 655 Avenue of the Americas, New York, NY 10010

The Ecologist MIT Press Journals, 55 Hayward St., Cambridge, MA 02142

Ecology Ecological Society of America, Dr. Duncan T. Patten, Center for Environmental Studies, Arizona State University, Tempe, AZ 85281; Tel: 602-965-3979

EcoNet Institute for Global Communication, 18 De Boom St., San Francisco, CA 94107; Tel: 415-422-0220

EDF Letter Environmental Defense Fund, Inc., 257 Park Ave. South, New York, NY 10010; Tel: 212-505-2100

Electric Power Research Institute Journal Electric Power Research Institute, Environmental Control Systems Dept., 3412 Hillview Ave., P.O. Box 10412, Palo Alto, CA 94303

E Magazine 28 Knight St., Westport, CT 06881; Tel: 203-854-5559, Fax 203-866-0602, E-mail: axgm65a@prodigy.com, Web site: http://www.emagazine.com

Endangered Species Technical Bulletin Division of Endangered Species and Habitat Conservation, U.S. Fish and Wildlife Service, Washington, DC 20240

Endangered Species Update School of Natural Resources, University of Michigan, Ann Arbor, MI 48109; Tel: 313-763-3243, Fax: 313-936-2195, E-mail: jfwatson@umich.edu

Energy Efficiency and Renewable Energy Clearinghouse P.O. Box 3048, Merrifield, VA 22116; Tel: 800-363-3732

Environment Heldref Publications, 1319 Eighteenth St. NW, Washington, DC 20036-1802; Tel: 800-365-9753

Environment Abstracts Congressional Information Services, Inc., 4520 East–West Hwy., Bethesda, MD 20814-3389; Tel: 800-638-8380 or 301-654-1550, Fax: 301-654-4033 (in most libraries)

Environmental Abstracts Annual Bowker A & I Publishing, 121 Chanlon Rd., New Providence, NJ 07974

Environmental Career Opportunities Brubach Publishing Co., Box 15629, Chevy Chase, MD 20825; Tel: 301-986-5545

Environmental Communicator North American Association for Environmental Education, 1255 23rd St. NW, Suite 300, Washington, DC 20037; Tel: 202-467-8754

Environmental Engineering News School of Civil Engineering, Purdue University, West Lafayette, IN 47907

Environmental Ethics Department of Philosophy, University of North Texas, Denton, TX 76203

Environmental Health Perspectives: Journal of the National Institute of Environmental Health Sciences Government Printing Office, Washington, DC 20402

Environmental History Review American Society for Environmental History, Center for Technology Studies, New Jersey Institute of Technology, Newark, NJ 07102; Tel: 201-596-3270

Environmental Law Northwestern School of Law, Lewis and Clark College, 10015 SW Terwilliger Blvd., Portland, OR 97219; Tel: 503-768-6700, Fax: 503-768-6671

Environmental Opportunities (Jobs) P.O. Box 78, Walpole, NH 03608

Environmental Outlook Institute for Environmental Studies, University of Washington, FM-12, Seattle, WA 98165; Tel: 216-543-1812

The Environmental Professional Editorial Office, Department of Geography, University of Iowa, Iowa City, IA 52242

Environmental Science and Technology American Chemical Society, 1155 16th St. NW, Washington, DC 20036; Tel: 202-872-4582, Fax: 202-872-6060

Environment Reporter Bureau of National Affairs, Inc., 1231 25th St. NW, Washington, DC 20037

Everyone's Backyard Center for Health, Environment, and Justice, P.O. Box 926, Arlington, VA 22216; Tel: 703-237-2249

Family Planning Perspectives Planned Parenthood–World Population, 666 Fifth Ave., New York, NY 10019; Tel: 212-541-7800

The Futurist World Future Society, P.O. Box 19285, Twentieth Street Station, Washington, DC 20036

Great Lakes Reporter Center for the Great Lakes, 35 E. Wacker Dr., Chicago, IL 60601; Tel: 312-263-0785; and 77 Harbor Square, Suite 2408, Toronto, Ontario, Canada M5J 2H2; Tel: 416-868-0550

Greenpeace Magazine Greenpeace USA, 1436 U St. NW, Washington, DC 20009

Green Teacher Box 1431, Lewiston, NY 14092

Hydrogen Letter 4104 Jefferson St., Hyattsville, MD 20781; Tel: 301-779-1561, Fax: 301-927-6345

In Business: The Magazine for Environmental Entrepreneuring Subscription Dept., 419 State Ave., Emmaus, PA 18049

Inner Voice Association of Forest Service Employees for Environmental Ethics, P.O. Box 11615, Eugene, OR 97440-9958; Tel: 503-484-2692, Fax: 503-484-3004

International Alliance for Sustainable Agriculture University of Minnesota, 1701 University Ave. SE, Minneapolis, MN 55414

International Environmental Affairs University Press of New England, 171/2 Lebanon St., Hanover, NH 03755

International Journal of Ecoforestry United States Ecoforestry Institute, P.O. Box 12543, Portland, OR 97212

International Rivers Network 1847 Berkeley Way, Berkeley, CA 94703; Tel: 510-848-1155, Fax: 510-848-1008, E-mail: irn@igc.apc.org

International Wildlife National Wildlife Federation, 8925 Leesburg Pike, Vienna, VA 22184; Tel: 703-790-4524

IPM Practitioner Bio-Integral Resource Center, P.O. Box 7414, Berkeley, CA 94707; Tel: 510-524-2567

Issues in Science and Technology National Academy of Sciences, 2101 Constitution Ave. NW, Washington, DC 20077-5576; Tel: 213-883-6325, Web site: http://utdallas.edu/research/issues

IUCN Bulletin International Union for Conservation of Nature and Natural Resources (IUCN), World Conservation Union, 1400 16th St. NW, Washington, DC 20036; Tel: 202-797-5454, Fax: 202-797-5461

Japan Environment Monitor 1941 Ogden Rd, Wilmington, OH 45177

Journal of Agricultural and Environmental Ethics Room 039, MacKinnon Building, University of Guelph, Guelph, Ontario, Canada N1G 2W1

Journal of the American Public Health Association 1015 18th St. NW, Washington, DC 20036

Journal of Energy and Development International Research Center for Energy and Economic Development, Campus Box 263, University of Colorado, Boulder, CO 80309

Journal of Environmental Health National Environmental Health Association, 720 S. Colorado Blvd., Suite 970, Denver, CO 80222; Tel: 303-756-9090

Journal of Environmental Science and Health Marcel Dekker Journals, P. O, Box 10018, Church Street Station, New York, NY 10249

Journal of Forestry Society of American Foresters, 5400 Grosvenor Lane, Bethesda, MD 20814

Journal of Geography National Council for Geographic Education, 16A Leonard Hall, Indiana University of Pennsylvania, Indiana, PA 15705; Tel: 703-471-1134

Journal of Pesticide Reform P.O. Box 1393, Eugene, OR 97440

Journal of Range Management Society for Range Management, 1839 York St., Denver, CO 80206; Tel: 303-355-7070

Journal of Soil and Water Conservation Soil and Water Conservation Society, 7515 NE Ankeny Rd., Ankeny, IA 50021; Tel: 515-289-2331, 800-843-7645

Journal of Sustainable Agriculture Food Products Press, 10 Alice Street, Binghampton, NY 13904

Journal of the Water Pollution Control Federation 601 Wythe St., Alexandria, VA 22314

Journal of Wildlife Management Wildlife Society, 5410 Grosvenor Lane, Bethesda, MD 20814; Tel: 301-897-9770

Kids for Saving Earth 620 Mendelssohn, Suite 145, Golden Valley, MN 55427; Tel: 612-525-0002, Fax: 612-525-0243

Marine Conservation News Center for Marine Conservation, 1725 DeSales St. NW, Suite 500, Washington, DC 20036; Tel: 202-429-5609, Fax: 202-872-0619

Mother Jones P.O. Box 50032, Boulder, CO 80322

National Geographic National Geographic Society, P.O. Box 2895, Washington, DC 20077-9960; Tel: 202-857-7000

National Library for the Environment National Institute for the Environment online library. Web site: http://www.cnie.org

National Parks National Parks and Conservation Association, 1015 31st St. NW, Washington, DC 20007; Tel: 202-223-6722, Fax: 202-659-0650

National Wetlands Newsletter Environmental Law Institute, 1616 P St. NW, Washington, DC 20036; Tel: 202-328-5150 or 328-5002

National Wildlife National Wildlife Federation, 1400 16th St. NW, Washington, DC 20036; Tel: 202-790-4524

Natural Resources Journal University of New Mexico School of Law, 1117 Stanford NE, Albuquerque, NM 87131

Nature 711 National Press Building, Washington, DC 20045

Nature Conservancy 1815 N. Lynn St., Arlington, VA 22209; Tel: 703-841-5300, Fax: 703-841-1283

New Economics New Economics Foundation, 1st Floor, Vine Court, 112-116 Whitechapel Road, London E1 1JE, England

The New Farm Rodale Press, 33 Minor St., Emmaus, PA 18049

New Scientist 128 Long Acre, London, WC2, England

Newsline Natural Resources Defense Council, 122 E. 42nd St., New York, NY 10168; Tel: 212-727-2700

Not Man Apart Friends of the Earth, 530 7th St. SE, Washington, DC 20003; Tel: 202-544-2600, Fax: 202-543-4710

Nucleus Union of Concerned Scientists, 26 Church St., Cambridge, MA 02238; Tel: 617-547-5552

Oceanus Woods Hole Oceanographic Institution, Woods Hole, MA 02543

Ocean Watch The Oceanic Society, 1536 16th St. NW, Washington, DC 20036

OIKO: The Alternative Environmental Digest P.O. Box 115, Greenwood, VA 22943

On the Wild Side Journal American Wildlands, 3609 S. Wadsworth Blvd., Suite 123, Lakewood, CO 80235; Tel: 303-988-2291, Fax: 303-988-6573

Organic Gardening & Farming Magazine Rodale Press, Inc., 33 E. Minor St., Emmaus, PA 18049

Orion Nature Quarterly P.O. Box 3000, Denville, NJ 07834-9797

Our Planet United Nations Environment Programme (U.S.), 2 United Nations Plaza, Room 803, New York, NY 10017

Park Science 4150 SW Fairhaven Dr., Corvallis, OR 97333

Pesticides and You National Coalition Against the Misuse of Pesticides, 701 E St. SE, Washington, DC 20003; Tel: 703-471-1134

Plain: The Magazine of Life, Land, and Spirit P.O. Box 200, Burton, OH 44021

Pollution Abstracts Cambridge Scientific Abstracts, 7200 Wisconsin Ave., Bethesda, MD 20814 (in many libraries)

Popline The Population Institute, 107 Second St. NE, Suite 207, Washington, DC 20002; Tel: 202-544-3300

Population and Vital Statistics Report United Nations Environment Programme, New York North American Office, Publication Sales Section, United Nations, New York, NY 10017

Population Bulletin Population Reference Bureau, 1875 Connecticut Ave. NW, Suite 520, Washington, DC 20009; Tel: 202-483-1100

Professional Geographer Association of American Geographers, 1710 16th St. NW, Washington, DC 20009; Tel: 202-234-1450

Rachel's Environment and Health Weekly Environmental Research Foundation, P.O. Box 4878, Annapolis, MD 21403; Tel: 410-263-1584

Rainforest News P.O. Box 140681, Coral Gables, FL 33115

Real World: The Voice of Ecopolitics 91 Nuns Moor Road, Newcastle upon Tyne, NE4 9BA, United Kingdom

Renewable Energy News Solar Vision, Inc., 7 Church Hill, Harrisville, NH 03450

Renewable Resources 5430 Grosvenor Lane, Bethesda, MD 20814; Tel: 301-493-9101

Resources Resources for the Future, 1616 P St. NW, Washington, DC 20036; Tel: 202-328-5000

Restoration and Management News University of Wisconsin Press, 144 North Murray St., Madison, WI 53715

River Voices River Network, P.O. Box 8787, Portland, OR 97207; Tel: 503-241-3506, 800-423-6747, Fax: 503-241-9256

Rocky Mountain Institute Newsletter 1739 Snowmass Creek Rd., Snowmass, CO 81654

Save America's Forests 4 Library Ct. SE, Washington, DC 20003; Tel: 202-544-9219

Science American Association for the Advancement of Science, 1333 H St. NW, Washington, DC 20005; Tel: 202-326-6400; Web site: http://www.aaas.org

Science News Science Service, Inc., 1719 N St. NW, Washington, DC 20036; Tel: 800-247-2160, Web site: http://www.sciencenews.org

Scientific American 415 Madison Ave., New York, NY 10017-1111; Web site: http://www.sciam.com

SE Journal Society of Environmental Journalists, 7904 Germantown Ave., Philadelphia, PA 19118; Tel: 215-247-9710

Sierra 730 Polk St., San Francisco, CA 94108

Simple Living 2319 N. 45th St., Box 149, Seattle, WA 98103; Tel: 206-464-4800

Solar Age Solar Vision, Inc., 7 Church Hill, Harrisville, NH 03450

Solar Industry Journal 777 N Capitol St. NE, Suite 805, Washington, DC 20002

Solar Today American Solar Energy Society, 2400 Central Ave., Unit B-1, Boulder, CO 80301

State of the Environment OECD Publications and Information Center, 1750 Pennsylvania Ave., Suite 1207, Washington, DC 20006 (published annually)

State of the World Worldwatch Institute, 1776 Massachusetts Ave. NW, Washington, DC 20036-1904; Tel: 202-452-1999, Fax: 202-296-7365 (published annually)

Statistical Yearbook Department of International Economic and Social Affairs, Statistical Office, United Nations Publishing Service, United Nations, New York, NY 10017

Technology Review P.O. Box 489, Mount Morris, IL 61054; Tel. 800-877-5320, Fax: 815-734-1127, E-mail: trsubscriprions@mit.edu, Web site: http://web.mit.edu/techreview/www/

Terra NOVA: Nature & Culture MIT Press Journals, 55 Hayward Street, Cambridge, MA; Tel: 617-253-2889, Fax: 617-258-6779, E-mail: journals-orders@mit.edu

Toxic Times National Toxics Campaign, 37 Temple Place, 4th Floor, Boston, MA 02111

Transition Laurence G. Wolf, ed., Department of Geography, University of Cincinnati, Cincinnati, OH 45221

Tropicus Conservation International, 1015 18th St. NW, Suite 1000, Washington, DC 20036; Tel: 202-429-5660

Vegetarian Journal P.O. Box 1463, Baltimore, MD 21203

Vegetarian Times 141 S. Oak Park St., P.O. Box 570, Oak Park, IL 60603

Vital Signs: The Trends That Are Shaping Our Future Worldwatch Institute, 1776 Massachusetts Ave. NW, Washington, DC 20036-1904; Tel: 202-452-1999, Fax: 202-296-7365 (published annually)

Waste Not Work on Waste USA, 82 Judson, Canton, NY 13617

Water Resources Bulletin American Water Resources Association, 5410 Grosvenor Lane, Suite 220, Bethesda, MD 20814; Tel: 301-493-8600

Wild Earth Cenozoic Society, Inc., P.O. Box 455, Richmond, VT 05477

Wilderness The Wilderness Society, 900 17th St. NW, Washington, DC 20006-2596; Tel: 202-833-2300

Wildlife Conservation New York Zoological Society, 185th St. & Southern Blvd., Bronx, NY 10460; Tel: 718-220-5100, Fax: 718-220-7114

Wildlife News African Wildlife Foundation, 1717 Massachusetts Ave. NW, Washington, DC 20036; Tel: 202-265-8393

World Rainforest Report Rainforest Action Network, 450 Sansome, Suite 700, San Francisco, CA 94111; Tel: 415-398-4404, Fax: 415-398-2732

World Resources World Resources Institute, 1709 New York Ave. NW, Washington, DC 20006; Tel: 202-638-6300 (published every 2 years)

Worldviews: Environment, Culture, and Religion White Horse Press, 10 High Street, Knapwell, Cambridge, CB3 8NR, England

World Watch Worldwatch Institute, 1776 Massachusetts Ave. NW, Washington, DC 20036-1904; Tel: 202-452-1999, Fax: 202-296-7365, E-mail: wwpub@worldwatch.org

Worldwatch Papers Worldwatch Institute, 1776 Massachusetts Ave. NW, Washington, DC 20036-1904; Tel: 202-452-1999, Fax: 202-296-7365

Yearbook of World Energy Statistics Department of International Economic and Social Affairs, Statistical Office, United Nations Publishing Service, United Nations, New York, NY 10017

ENVIRONMENTAL AND RESOURCE ORGANIZATIONS

For a more detailed list of national, state, and local organizations, see *Conservation Directory* (published annually by the National Wildlife Federation, 1400 16th St. NW, Washington, DC 20036), *Your Resource Guide to Environmental Organizations* (Irvine, CA: Smiling Dolphin Press, 1991), *National Environmental Organizations* (published annually by U.S. Environmental Directories, Inc., P.O. Box 65156, St. Paul, MN 55165), and *World Directory of Environmental Organizations* (published by the California Institute of Public Affairs, P.O. Box 10, Claremont, CA 91711). Also see Environmental Organizations Web site (http://www.econet.apc.org/econet/en.orgs.html/).

Adopt-a-Stream Foundation P.O. Box 5558, Everett, WA 98206; Tel: 206-388-3487

African Wildlife Foundation 1717 Massachusetts Ave. NW, Washington, DC 20036; Tel: 202-265-8393, E-mail: awnews@aol.com

Air Pollution Control Association P.O. Box 2861, Pittsburgh, PA 15230

Alan Guttmacher Institute 120 Wall Street, New York NY 10005; Tel: 212-254-5956

Alliance for Chesapeake Bay 6600 York Rd., Baltimore, MD 21212; Web site: http://www.gmu.edu/bios/Bay/

Alliance for Environmental Education P.O. Box 368, The Plains, VA 22171; Tel: 703-253-5812

Alliance to Save Energy 1725 K St. NW, Suite 914, Washington, DC 20006-1401

American Association for the Advancement of Science 1333 H St. NW, Washington, DC 20005; Tel: 202-326-6400, Web site: http://www2.nas.edu/cuselib/23ca.html

American Cetacean Society P.O. Box 1391, San Pedro, CA 9073-0391; Tel: 310-548-6279, Fax: 310-548-6950, Web site: http://www.acsonline.org/

American Conservation Association, Inc. 30 Rockefeller Plaza, Rm. 5402, New York, NY 10112; Tel: 212-649-5822

American Council for an Energy-Efficient Economy 1001 Connecticut Ave. NW #801, Washington, DC 20036; Tel: 202-429-8873, Web site: http://www.aceee.org/

American Farmland Trust 1920 N St. NW, Suite 400, Washington, DC 20036; Tel: 202-659-5170, Fax: 202-659-8334

American Fisheries Society 5410 Grosvenor Lane, Suite 110, Bethesda, MD 20814; Tel: 301-897-8616, Fax: 303-897-8096

American Forests (formerly American Forestry Association) 1516 P St. NW, Washington, DC 20005; Tel: 202-667-3300

American Geographical Society 156 Fifth Ave., Suite 600, New York, NY 10010; Tel: 212-242-0214

American Humane Society 63 Inverness Dr. E, Englewood, CO 80112; Tel: 303-792-9900

American Hydrogen Association P.O. Box 15075, Phoenix, AZ 85060; Tel: 602-921-0433, Fax: 602-967-6601, Web site: http://www.getnet.com/charity/aha/

American Institute of Biological Sciences, Inc. 730 11th St. NW, Washington, DC 20001-4521; Tel: 202-628-1500, Fax: 202-628-1509, Web site: http://www.yahoo.com/Science/Biology/Organizations/Professional/American_Institute_of_Biological_Sciences/

American Littoral Society Sandy Hook, Highlands, NJ 07732; Tel: 201-291-0055

American Lung Association 1740 Broadway, New York, NY 10019-4374; Tel: 1-800-586-4872

American Public Health Association 1015 15th St. NW, Washington, DC 20005; Tel: 202-789-5600, Web site: http://www.apha.org/5

American Rivers 801 Pennsylvania Ave. SE, Suite 400, Washington, DC 20003-2167; Tel: 202-547-6900, Fax: 202-543-6142, E-mail: amrivers@igc.apc.org, Web site: http://www.amrivers.org/

American Society for the Prevention of Cruelty to Animals (ASPCA) 441 E. 92nd Street, New York, NY 10128, Web site: http://www.aspca.org

American Solar Energy Society 2400 Central Ave., Suite G, Boulder, CO 80301; Tel: 303-443-3130, Web site: http://www.sni.net/solar/

American Water Resources Association 950 Herndon Parkway, Suite 300, Herndon, VA 22070-5528; Tel: 703-904-1255, Fax: 703-904-1228

American Wildlands 3609 S. Wadsworth Blvd., Suite 123, Lakewood, CO 80235; Tel: 303-988-2291, Fax: 303-988-6573

American Wind Energy Association 777 N. Capitol St. NE, Suite 805, Washington, DC 20002; Tel: 202-408-8988; Web site: http://www.igc.apc.org/awea/index.html

Animal Rights Information and Education Service P.O. Box 332, Rowayton, CT 06583

Association of American Geographers 1710 16th St. NW, Washington, DC 20009; Tel: 202-234-1450

Association of Forest Service Employees for Environmental Ethics P.O. Box 11615, Eugene, OR 97440-9958; Tel: 503-484-2692, Fax: 503-484-3004, Web site: http://www.afseee.org/

Bat Conservation International P.O. Box 162603, Austin, TX 78716; Tel: 512-327-9721, Web site: http://www.batcon.org/

Bio-Integral Resource Center P.O. Box 7414, Berkeley, CA 94707; Tel: 510-524-2567

Bioregional Institute 233 Miramar Street, Santa Cruz, CA 95060; Tel: 408-425-0264

Bioregional Project (North American Bioregional Congress) Turtle Island Office, 1333 Overhulse Rd. NE, Olympia, WA 98502

Business Council on Sustainable Development 1825 K St. NW, Washington, DC 20006; Tel: 202-833-9659

Buy Recycled Business Alliance c/o National Recycling Coalition, 30th St. NW, Washington, DC 20007; Tel: 202-625-6406

Campus Ecology National Wildlife Federation, 8925 Leesburg Pike, Vienna, Va. 22184; Tel: 703-790-4318, Web site: http://www.nwf.org/nwj/campus

Carrying Capacity Network, Inc. 2000 P St. NW, Suite 240, Washington, DC 20036-5915; Tel: 202-296-4548 or 800-466-4866, Fax: 202-296-4609

Center for Conservation Biology Alice Blandin, *Conservation Biology*, University of Washington, Box 351800, Seattle, WA 98195-1800; Web site: http://www.conbio.rice.edu/

Center for Energy and Environmental Studies The Engineering Quadrangle, Princeton University, Princeton, NJ 08544

Center for Environmental Education 1725 DeSales Street NW, Suite 500, Washington, DC 20036; Tel: 202-429-5609

Center for Global Change University of Maryland at College Park, The Executive Building, Suite 401, 7100 Baltimore Ave., College Park, MD 20740

Center for the Great Lakes 35 E. Wacker Dr., Chicago, IL 60601; Tel: 312-263-0785; and 77 Harbor Square, Suite 2408, Toronto, Ontario, Canada M5J 2H2; Tel: 416-868-0550

Center for Health, Environment, and Justice P.O. Box 6806, Falls Church, VA 22040; Tel: 703-237-2249, Web site: http://www.essential.org/orgs/cchw/cchw.html

Center for Holistic Resource Management 5820 4th St. NW, Albuquerque, NM 87107

Center for Marine Conservation 1725 DeSales St. NW, Suite 500, Washington, DC 20036; Tel: 202-429-5609, Fax: 202-872-0619, Web site: http://www.cmc-ocean.org/

Center for a New American Dream Jacqueline Hamilton, c/o NRDC Suite 300, 11350 New York Ave. NW, Washington, DC 20005; E-mail: jacqueline.hamilton@together.org

Center for Plant Conservation P.O. Box 299, St. Louis, MO 63166; Tel: 314-577-9450, Fax: 314-577-9465, Web site: http://www.mobot.org/CPC/welcome.html

Center for Science in the Public Interest 1875 Connecticut Ave. NW, Suite 300, Washington, DC 20009; Tel: 202-332-9110, Fax: 202-265-4594, Web site: http://www.cspinet.org/

Chesapeake Bay Foundation 162 Prince George St., Annapolis, MD 21401; Tel: 410-268-8816, Web site: http://www.baylink.org/baywatch.html

Chesapeake Bay Trust 60 West St., Suite 200-A, Annapolis, MD 21401

Chipko P.O. Silyara via Ghansale, Tehri-Garwhal, Uttar Pradesh, 249155 India

Clean Water Action Project 1320 18th St. NW, Suite 310, Washington, DC 20036; Tel: 202-457-0336, Fax: 202-457-0287, E-mail: schecter@essential.org

Climate Institute 324 4th St. NW, Washington, DC 20002; Tel: 202-547-0104, Fax: 202-547-0111

Coalition for Environmentally Responsible Economies 711 Atlantic Ave., 5th Fl., Boston, MA 02111; Tel: 617-451-0927

Coastal Resources Center P.O. Box 3084, San Rafael, CA 94912

Coastal Society P.O. Box 2081, Gloucester, MA 01930-2081; Tel: 508-281-9209

Common Cause 2030 M Street NW, Washington, DC 20036; Tel: 202-833-1200, E-mail: 75300.3120@compuserve.com, Web site: http://www.commoncause.org/

Concern, Inc. 1794 Columbia Road NW, Washington, DC 20009; Tel: 202-328-8160, E-mail: concern@igc.aoc.org

Conservation International 1015 18th St. NW, Suite 1000, Washington, DC 20036; Tel: 202-429-5660; Web site: http://www.conservation.org

Cool It National Wildlife Federation 1400 16th Street NW, Washington, DC 20036; Tel: 202-797-5435

Corporate Conservation Council c/o National Wildlife Federation, 1400 16th St. NW, Washington, DC; Tel: 202-797-6870

Council for Economic Priorities 30 Irving Place, New York, NY 10003; Tel: 212-420-1133. Maintains corporate environmental data clearinghouse.

Cousteau Society 870 Greenbrier Circle, Suite 402, Chesapeake, VA 23320; Tel: 804-523-9335

Critical Mass Energy Project 215 Pennsylvania Ave. SE, Washington, DC 20003; Tel: 202-546-4996, E-mail: cmep@citizen.org, Web site: http://www.essential.org

Cultural Survival 11 Divinity Ave., Cambridge, MA 02138; Tel: 617-495-2562

Defenders of Wildlife 1101 14th St. NW, Suite 1400, Washington, DC 20005; Tel: 202-682-9400, Fax: 202-682-1331, E-mail: information@defenders.org

Desert Protective Council, Inc. P.O. Box 4294, Palm Springs, CA 92263

Ducks Unlimited One Waterfowl Way, Memphis, TN 38120-2351; Tel: 901-758-3825, Fax: 901-758-3850

Earth First! 305 N. Sixth St., Madison, WI 53704

Earth Island Institute 300 Broadway, Suite 28, San Francisco, CA 94133: Tel: 415-788-3666, Fax: 415-788-7324, Web site: http://www.earthisland.org/ei

Earthjustice Legal Defense Fund (formerly the Sierra Club Legal Defense Fund) 180 Montgomery St., San Francisco, CA 94104-4209; Tel: 415-627-6700, Fax: 415-627-6740

EarthSave Foundation P.O. Box 949, Felton, CA 95018

Earthwatch 680 Mt. Auburn St., Box 403N, Watertown, MA 02272; Tel: 800-776-0188 (in MA only: 617-926-8200)

Elmwood Institute P.O. Box 5765, Berkeley, CA 94705; Tel: 510-845-4595

Energy Conservation Coalition 1525 New Hampshire Ave. NW, Washington, DC 20036

Energy Efficient Building Association 1000 W. Campus Dr., Wausau, WI 54401; Tel: 715-675-6331

Envirolink Network E-mail: admin@envirolink.org, Web site: http://www.envirolink.org

Environmental Action, Inc. 6930 Carroll Park, Suite 600, Takoma Park, MD 20912; Tel: 301-891-1100, Fax: 301-891-2218

Environmental Careers Organization, Inc. 286 Congress St., 3rd Fl., Boston, MA 02210; Tel: 617-426-4375

Environmental Defense Fund, Inc. 257 Park Ave. South, New York, NY 10010; Tel: 212-505-2100

Environmental and Energy Study Institute 122 C St. NW, Suite 700, Washington, DC 20001; Tel: 202-628-1400, Fax: 628-1825

Environmental Law Institute 1616 P St. NW, Suite 200, Washington, DC 20036; Tel: 202-328-5150, Fax: 202-328-5002

Farralones Institute 55C Gate Five Road, Sausalito, CA 94965; Tel: 415-332-3267

Fish and Wildlife Reference Service 5430 Grosvenor Ln., Bethesda, MD 20814; Tel: 301-492-6403 or 800-582-3421, Fax: 301-564-4059

Florida Solar Energy Center 300 State Road #401, Cape Canaveral, FL 32920

Friends of Animals 777 Post Rd., Suite 205, Darien, CT 06820; Tel: 203-656-1522, Fax: 656-0267, E-mail: soebc@igc.apc.org

Friends of the Earth The Global Building, 1025 Vermont Ave., NW, Suite 300, Washington, DC 20005; Tel: 202-783-7400, Fax: 202-783-0444, E-mail: foedc@igc.apc.org, Web site: http://www.foe.co.uk/

Friends of the River Fort Mason Center, Bldg. C, San Francisco, CA 94123; Tel: 415-771-0400

Fund for Animals, Inc. 200 W. 57th St., New York, NY 10019; Tel: 212-246-2096

Global Greenhouse Network 1130 17th St. NW, Suite 530, Washington, DC 20036

Global Tomorrow Coalition 1325 G St. NW, Suite 1010, Washington, DC 20005-3104; Tel: 202-628-4016, Fax: 202-628-4018

Green Corps 1724 Gilpin Street, Denver, CO 80218; Tel: 303-355-1881

Greenhouse Crisis Foundation 1130 17th St. NW, Suite 630, Washington, DC 20036; Tel: 202-466-2823

GreenNet 23 Bevenden St., London N1 6BH, England

Greenpeace, Canada 427 Bloor St., West Toronto, Ontario M5S 1X7

Greenpeace, USA, Inc. 1436 U St. NW, Washington, DC 20009; Tel: 202-462-1177, E-mail: your.name@green2.greenpeace.org and www.greenpeace.org, Web site: http://www.greenpeace.org/

Green Seal 1733 Rhode Island Ave. NW, Suite 1050, Washington, DC 20036-3101; Tel: 202-331-7337, Fax: 202-331-7533

Habitat for Humanity 121 Habitat Street, Americus, GA 31709-3498; Tel: 912-924-6935, E-mail: Frank_Purvs@habitat.org

Humane Society of the United States, Inc. 2100 L St. NW, Washington, DC 20037; Tel: 202-452-1100, Fax: 301-258-3077

INFORM 381 Park Ave. South, New York, NY 10016; Tel: 212-689-4040, Fax: 212-447-0689

Institute for Alternative Agriculture 9200 Edmonston Rd., Suite 117, Greenbelt, MD 20770

Institute for Earth Education Cedar Cove, Box 115, Greenville, WV 24945; Tel: 304-832-6404, Fax: 304-832-6077

Institute for Local Self-Reliance 2425 18th St. NW, Washington, DC 20009; Tel: 202-232-4108, E-mail: ilsr@itp.apc.org

Institute for Social Ecology P.O. Box 89, Plainsfield, VT 85667

International Alliance for Sustainable Agriculture 1701 University Ave. SE, Minneapolis, MN 55414

International Association of Fish and Wildlife Agencies 444 N. Capitol St. NW, Suite 544, Washington, DC 20001; Tel:. 202-624-7890

International Council for Outdoor Education P.O. Box 17255, Pittsburgh, PA 15235

International Institute for Energy Conservation 750 First St. NE, Washington, DC 20002; Tel: 202-842-3388, Fax: 202-842-1565, E-mail: iiec@igc.apc.org

International Planned Parenthood Federation 105 Madison Ave., 7th Floor, New York, NY 10016

International Rivers Network 1847 Berkeley Way, Berkeley, CA 94703; Tel: 510-848-1155, Fax: 510-848-1008, E mail: irn@igc.apc.org

International Society for Ecological Economics P.O. Box 1589, Solomons, MD 20688; Tel: 410-326-0794, E-mail: button@cbl.umd.edu

International Union for the Conservation of Nature and Natural Resources (IUCN) 1400 16th St. NW, Washington, DC 20036; Tel: 202-797-5454, Fax: 202-797-5461, E-mail: mail@hg.iucn.ch

Izaak Walton League of America 1401 Wilson Blvd., Level B, Arlington, VA 22209; Tel: 703-528-1818

Kids for Saving Earth 620 Mendelssohn, Suite 145, Golden Valley, MN 55427; Tel: 612-525-0002, Fax: 612-525-0243

Land Institute 2440 E. Well Water Road, Salina, KS 67401; Tel: 913-823-5376

Land Trust Alliance 1319 F St. NW, Suite 501, Washington, DC 20004; Tel: 202-638-4725, Fax 202-638-4730, Web site: http://www.lta.org/

League of Conservation Voters 1707 L Street, NW, Suite 750, Washington, DC 20036; Tel: 202-785-8683, Fax: 202-835-0491, Web site: http://www.lcv.org/

League of Women Voters of the U.S. 1730 M St. NW, Washington, DC 20036; Tel: 202-429-1965, Fax: 202-429-0854, Web site: http://acm.cs.umn.edu/~lisi/lwv/main.html

Mineral Policy Center 1325 Massachusetts Ave. NW, Suite 550, Washington, DC 20005; Tel: 202-737-1872

National Association of Biology Teachers 11250 Roger Bacon Drive, Rm. 19, Reston, VA 22090; Tel: 703-471-1134, Fax: 703-435-5582, E-mail: nabter@aol.com

National Association of Environmental Management 1440 New York Ave. NW, Suite 300, Washington, DC 20005; Tel: 202-737-3415

National Audubon Society 700 Broadway, New York, NY 10003-9501; Tel. 212-979-3000, E-mail: mis@audubon.org; Web site: http://www.audubon.org./audubon/

National Center for Urban Environmental Studies 516 North Charles St., Suite 501, Baltimore, MD 21201

National Coalition Against the Misuse of Pesticides 701 E St. SE, Suite 200, Washington, DC 20003; Tel: 202-543-5450, Web site: http://www.ncamp.org/

National Environmental Health Association 720 S. Colorado Blvd., South Tower 970, Denver, CO 80222; Tel: 303-756-9090

National Environmental Trust Tel: 202-887-8800, E-mail: netinfo@acpa.com, Web site: http://www.eic.org/

National Fish and Wildlife Foundation 1120 Connecticut Ave. NW, Suite 900, Washington, DC 20236; Tel: 202-857-0166, Fax: 202-857-0162

National Geographic Society 1145 17th St. NW, Washington, DC 20036; Tel: 202-857-7000, Web site: http://www.nationalgeographic.com/main.html

National Hydrogen Association 1800 M Street NW, Suite 300, Washington, DC 20036: Tel: 202-223-5547, Fax: 202-223-5537, Web site: http://www.ttcorp.com/nha/

National Park Foundation 1101 17th St. NW, Suite 1102, Washington, DC 20036; Tel: 202-785-4500, Web site: http://www.nationalparks.org/

National Parks and Conservation Association 1776 Massachusetts Ave. NW, Suite 200, Washington, DC 20036; Tel: 202-223-6722, Fax: 202-659-0650, E-mail: natparks@aol.com, Web site: http://www.npca.org/home/npca/

National Recreation and Park Association 22377 Belmont Ridge Road, Ashburn, VA 20148; Tel: 703-858-0784, Fax: 703-858-0794, E-mail: info@nrpa.org, Web site: http://www.nrpa.org/

National Recycling Coalition 1727 King Street, Suite 105, Alexandria VA 22314; Tel: 703-683-9025, Fax: 703-683-9026, Web site: http://www.recycle.net/recycle/index.html

National Science Teachers Association 1840 Wilson Blvd., Arlington, VA 22201; Tel: 703-7100, Fax: 703-243-7177, E-mail: maiser@nsta.org, Web site: http://www.nsta.org/

National Solid Waste Management Association 1730 Rhode Island Ave. NW, Suite 100, Washington, DC 20036

National Toxics Campaign 37 Temple Place, 4th Floor, Boston, MA 02111

National Wildlife Federation 8925 Leesburg Pike, Vienna, VA 22184; Tel 703-790-4000, E-mail: feedback@nws.org, Web site: http://www.nwf.org/nwf/home.html

Natural Resources Defense Council 40 W. 20th St., New York, NY 10011; Tel: 212-727-2700; and 1350 New York Ave. NW, Suite 300, Washington, DC 20005; Tel: 202-783-7800, E-mail: nrdcinfo@igc.apc.org, Web site: http://www.nrdc.org/nrdc/

Nature Conservancy 1814 N. Lynn St., Arlington, VA 22209; Tel: 703-841-5300, Fax: 703-841-1283, Web site: http://www.tnc.org/

New York Zoological Society/Wildlife Conservation Society (formerly the New York Zoological Society) 185th & Southern Blvd., Bronx, NY 10460-1099; Tel: 718-220-5100, Fax: 718-220-7114

North American Association for Environmental Education 1255 23rd St. NW, Suite 300, Washington, DC 20037; Tel: 202-884-8912, Fax 202-884-8701

Northwest Coalition for Alternatives to Pesticides P.O. Box 1393, Eugene, OR 97440

Nuclear Information and Resource Service 1424 16th St. NW, Suite 601, Washington, DC 20036

Ocean Arks International One Locust St., Falmouth, MA 02540

The Oceanic Society 218 D St. SE, Washington, DC 20003

Partners for Global Justice 4920 Piney Branch Road NW, Washington, DC 20011; Tel: 202-723-8273

People for the Ethical Treatment of Animals (PETA) Box 42516, Washington, DC 20015

Permaculture Association P.O. Box 202, Orange, MA 01364

Permaculture Institute of North America 4649 Sunnyside Ave. N, Seattle, WA 98103

Pesticide Action Network 965 Mission St., No. 514, San Francisco, CA 94103; Web site: http://www.panna.org/panna/

Pesticide Information Center P.O. Box 420870, San Francisco, CA 94142-0870; Tel: 415-391-8511

Physicians for Social Responsibility 1101 14th St. NW, 7th Floor, Washington, DC 20005; Tel: 202-898-0150

Planetary Citizens 325 Ninth St., San Francisco, CA 94103

Planet/Drum Foundation P.O. Box 31251, San Francisco, CA 94131; Web site: http://www.ic.org/fic/cdir/res/PlanetDrum.html

Planned Parenthood Federation of America 810 Seventh Ave., New York, NY 10019; Tel: 212-541-7800, Web site: http://www.plannedparenthood.org

Population Action International (formerly Population Crisis Committee) 1120 19th St. NW, Suite 550, Washington, DC 20036-3605; Tel: 202-659-1833, Web site: http://www.populationaction.org/

Population Council 1 Dag Hammarskjold Plaza, New York, NY 10017

Population–Environment Balance 2000 P St. NW, Suite 210, Washington, DC 20036-5915; Tel: 202-955-5700, Fax: 202-955-6161, E-mail: uspop@balance.org; Web site: http://www.balance.org

Population Institute 107 2nd St. NE, Washington, DC 20002; Tel: 202-544-3300

Population Reference Bureau 1875 Connecticut Ave. NW, Suite 520, Washington, DC 20009-5728; Tel: 202-483-1100, Fax: 202-328-3937, E-mail:

popref@prb.org, Web site: http://www.prb.org/prb/

Public Citizen 215 Pennsylvania Ave. SE, Washington, DC 20003; Web site: http://www.essential.org/orgs/public_citizen/

Public Interest Research Groups The Fund for Public Interest Research, 29 Temple Place, Boston, MA 02111; Tel: 617-292-4800

Rainforest Action Network 3450 Sansome St., Suite 700, San Francisco, CA 94111; Tel: 415-398-4404, Fax: 415-389-2732, Web site: http://www.ran.org

Rainforest Alliance 65 Bleecker St., New York, NY 10012; Tel: 212-677-1900, Fax: 212-677-2187, E-mail: canopy@cdp.apc.or, Web site: http://www.rainforest-alliance.org/

Recycled Products Information Clearinghouse 5528 Hempstead Way, Springfield, VA 22151; Tel: 703-941-4452

Renewable Natural Resources Foundation 5430 Grosvenor Lane, Bethesda, MD 20814; Tel: 301-493-9101, Fax: 301-493-6148, E-mail: mrf@aol.com

Resources for the Future 1616 P St. NW, Washington, DC 20036; Tel: 202-328-5000, Fax: 202-939-3460, E-mail: info@rrf.org, Web site: /wwwriff.org/

Rocky Mountain Institute 1739 Snowmass Creek Rd., Snowmass, CO 81654-9199; Tel: 970-927-3851, Fax: 970-927-3420, Web site: http://www.rmi.org/index.html

Rodale Institute 222 Main St., Emmaus, PA 18098; Web site: http://www.envirolink.org/seel/rodale/

Save America's Forests 4 Library Court SE, Washington, DC 20003; Tel: 202-544-9219

Save the Whales, Inc. P.O. Box 2397, 1426 Main St., Unit E, Venice, CA 90291

Scientists' Institute for Public Information 355 Lexington Ave., New York, NY 10017; Tel: 212-661-9110

Sea Shepherd Conservation Society P.O. Box 628, Venice, CA 90294; Tel: 310-301-7325, Fax: 310-574-3161, Web site: http://www.seashepherd.org/

Sierra Club 730 Polk St., San Francisco, CA 94109; Tel 415-776-2211, E-mail: information@sierraclub.org, Web site: http://www.sierraclub.org

Simple Living 355 E. 19th Ave., Columbus, OH

Smithsonian Institution 1000 Jefferson Dr. SW, Washington, DC 20560; Tel: 202-357-2700, Web site: http://www.yahoo.com/Government/Agencies/Independent/Smithsonian_Institution/

Social Investment Forum C.E.R.E.S. Project, 711 Atlantic Ave., Boston, MA 02111

Society for Ecological Restoration University of Wisconsin–Madison Arboretum, 1207 Seminole Hwy., Madison, WI 53711; Tel: 608-262-9547, Fax: 608-262-9547

Society for Range Management 1839 York St., Denver, CO 80206; Tel: 303-355-7070

Society of American Foresters 5400 Grosvenor Lane, Bethesda, MD 20814; Tel: 301-897-8720

Soil and Water Conservation Society 7515 NE Ankeny Rd., Ankeny, IA 50021-9764; Tel: 515-289-2331 or 800-863-7645, Fax: 515-289-1227

Solar Energy International P.O. Box 715, Carbondale, CO 81623; Tel: 303-963-8855

Solar Information Center The University of Oregon, Dept. of Agriculture, Eugene, OR 97403; Tel: 503-346-3696

Solid Waste Association of North America P.O. Box 7219, Silver Spring, MD 20907-7219, Tel: 301-585-2898, Fax: 301-589-7068, Web site: http://www.swana.org/ or http://bianca.com/lolla/politics/seac/seac.html; Conservation Association, Inc., P.O. Box 550, Charlestown, NH 03603; Tel: 603-543-1700, Fax: 603-543-1828

Student Environmental Action Coalition (SEAC) P.O. Box 1168, Chapel Hill, NC 27514; Tel: 919-967-4600 or 800-700-SEAC, Fax: 919-967-4648, Web site: http://www.seac.org/

Survival International 2121 Decatur Place NW, Washington, DC 20008; Web site: http://www.survival.org.uk/

Union of Concerned Scientists Two Brattle Square, Cambridge, MA 02238-9105; Tel: 617-547-5552, Web site: http://www.ucsusa.org/

United Nations Population Fund 220 East 42nd St., New York, NY 10017; Tel: 212 297-5020, Fax: 212-557-6416, Web site: http://www.unfpa.org/

The Urban Agriculture Network 1711 Lamont St. NW, Washington, DC 20010; Tel: 202-483-8130, Fax: 202-986-6732, E-mail: 72144.3466@compuserve.com

Urban Ecology P.O. Box 10144, Berkeley, CA 94709; Tel: 415-549-1724

U.S. Congress Web Server Web site: http://thomas.loc.gov

U.S. Public Interest Research Group 215 Pennsylvania Ave. SE, Washington, DC 20003; Tel: 202-546-9707

Voluntary Simplicity Study Circle Network Ruth Pickering, P.O. Box 1203, Fall City, WA 98024; Tel: 206-392-2354

Water Pollution Control Federation 601 Wythe St., Alexandria, VA 22314

White House Web Server http://www.whitehouse.gov

The Wilderness Society 900 17th St. NW, Washington, DC 20006-2596; Tel: 202-833-2300, Web site: http://www.wilderness.org/

Wildlife Conservation Society 185th St. and Southern Blvd., Bronx, NY 10460-1099; Tel 718-220-5100, Fax: 718-220-7114, Web site: http://www.wcs.org/

Wildlife Society 5410 Grosvenor Lane, Bethesda, MD 20814; Tel: 301-897-9770

Work on Waste 82 Judson St., Canton, NY 13617

World Future Society 4916 St. Elmo Ave., Bethesda, MD 20814

World Rainforest Movement International Secretariat, 87 Cantonment Rd., 10250, Penang, Malaysia

World Resources Institute 1709 New York Ave. NW, 7th Floor, Washington, DC 20006; Tel: 202-638-6300, Web site: http://www.wri.org/

Worldwatch Institute 1776 Massachusetts Ave. NW, Washington, DC 20036-1904; Tel: 202-452-1999, Fax: 202-296-7365, E-mail: worldwatch@igc.apc.org and wwpub@igc.apc.org, Web site: http://www.worldwatch.org/

World Wildlife Fund 1250 24th St. NW, Suite 500, Washington, DC 20037; Tel: 202-293-4800, Web site: http://www.worldwildlife.org/

Zero Population Growth 1400 16th St. NW, Suite 320, Washington, DC 20036; Tel: 202-332-2200, Fax: 202-332-2302, E-mail: zpg@apc.org, Web site: http://www.zpg.org

FEDERAL, SCIENTIFIC, AND INTERNATIONAL AGENCIES

Agency for International Development (USAID) State Building, 320 21st St. NW, Washington, DC 20523-0016; Tel: 202-647-1850, Fax: 202-647-8321, Web site: http://www.info.usaid.gov/

Bureau of Land Management U.S. Department of Interior, 1620 L St. NW, Rm. 5600, Washington, DC 20240; Tel: 202-208-3801, Web site: http://www.blm.gov

Bureau of Reclamation Washington, DC 20240; Web site: http://www.usbr.gov/main/

Congressional Research Service 101 Independence Ave. SW, Washington, DC 20540

Conservation and Renewable Energy Inquiry and Referral Service P.O. Box 8900, Silver Spring, MD 20907; Tel: 800-523-2929

Department of Agriculture 14th St. and Independence Ave. SW, Washington, DC 20250; Tel: 202-720-8732, Web site: http://www.usda.gov

Department of Commerce Herbert C. Hoover Bldg., Rm. 5610, 15th & Constitution Ave. NW, Washington, DC 20230; Tel: 202-219-3605, Web site: http://www.doc.gov

Department of Energy Forrestal Building, 1000 Independence Ave. SW, Washington, DC 20585; Web site: http://www.doe.gov and http://www.ener.doe.gov

Department of Health and Human Services 200 Independence Ave. SW, Washington, DC 20585; Web site: http://www.yahoo.com/Government/Executive_Branch/Departments_and_Agencies/Department_of_Health_and_Human_Services/

Department of Housing and Urban Development 451 7th St. SW, Washington, DC 20410; Tel: 202-755-5111, Web site: http://www.hud.gov

Department of the Interior 1849 C St. NW, Washington, DC 20240; Tel: 202-208-3100, Web site: http://www.usga.gov.doi

Department of Transportation 400 7th St. SW, Washington, DC 20590; Tel: 202-366-4000, Web site: http://www.dot.gov

Energy Information Administration Dept. of Energy, National Energy Information Center, Forrestal Bldg., Washington, DC 20585; Tel: 202-586-8800, Web site: http://www.yahoo.com/Government/Executive_Branch/Departments_and_Agencies/Department_of_Energy/Energy_Information_Administration/

Environmental Protection Agency 401 M St. SW, Washington, DC 20460; Tel: 202-260-2090, Web site: http://www.epa.gov and gopher.epa.gov

Federal Energy Regulatory Commission 888 First St. NE, Washington, DC 20426; Web site: http://www.ferc.fed.us/

Food and Agriculture Organization (FAO) of the United Nations 101 22nd St. NW, Suite 300,

Washington, DC 20437; Web site: http://www.fao.org/

Food and Drug Administration Department of Health and Human Services, 5600 Fishers Lane, Rockville, MD 20857; Tel: 410-433-1544, Web site: http://www.fda.gov

Forest Service P.O. Box 96090, Washington, DC 20090-6090; Tel: 202-205-0957, Web site: http://www.fs.fed.us/

Great Lakes Commission The Argus II Bldg., 400 Fourth St., Ann Arbor, MI 48103-4816; Tel: 313-665-9135, Fax: 313-665-4370, Web site: http://www.glc.org/

International Whaling Commission The Red House, 135 Station Rd., Histon, Cambridge CB4 4NP, England (0223 233971); Web site: http://av.yahoo.com/bin/query?p=International+Whaling+Commission&hc=0&hs=0

Marine Mammal Commission 1825 Connecticut Ave. NW, Rm. 512, Washington, DC 20009; Tel: 202-606-5504, Fax: 202-606-5510, Web site: http://www.citation.com/hpages/mmc.html

Mine Safety and Health Administration Ballston Tower 3, 4015 Wilson Blvd., Arlington, VA 22203; Tel: 703-235-1452, Web site: http://www.msha.gov/

National Academy of Sciences 2101 Constitution Ave. NW Washington, DC 20418; Web site: http://www.nas.edu/5

National Aeronautics and Space Administration 400 Maryland Ave. SW, Washington, DC 20546

National Cancer Institute 9000 Rockville Pike, Bethesda, MD 20892; Web site: http://www.nci.nih.gov/

National Center for Atmospheric Research P.O. Box 3000, Boulder, CO 80307; Web site: http://www.ucar.edu/

National Lead Information Center Tel: 1-800-LEAD-FYI, Web site: http://www.nsc.org/ehc/lead.htm

National Marine Fisheries Service U.S. Dept. of Commerce, Silver Spring Metro Center 1, 1335 East-West Hwy., Silver Spring, MD 20910; Tel: 301-713-2239, Web site: http://kingfish.ssp.nmfs.gov/

National Oceanic and Atmospheric Administration Herbert C. Hoover Bldg., Rm. 5128, 14th & Constitution Ave. NW, Washington, DC 20230; Tel: 202-482-3384, Web site: http://www.yahoo.com/Government/Executive_Branch/Departments_and_Agencies/Department_of_Commerce/National_Oceanic_and_Atmospheric_Administration/

National Park Service Interior Bldg., P.O. Box 37127, Washington, DC 20013-7127; Tel: 202-208-4747, Web site: http://www.nps.gov/

National Renewable Energy Laboratory 1617 Cole Blvd., Golden, CO 80401; Web site: http://www.nrel.gov/

National Science Foundation 4201 Wilson Blvd., Arlington, VA 22230; Tel: 703-306-1234, Web site: http://www.nsf.gov/

National Solar Heating and Cooling Information Center P.O. Box 1607, Rockville, MD 20850

National Technical Information Service U.S. Department of Commerce, 5285 Port Royal Rd., Springfield, VA 22161; Web site: http://www.ntis.gov/

Nuclear Regulatory Commission Washington, DC 20555; Tel: 301-415-7000, Web site: http://www.nrc.gov/

Occupational Safety and Health Administration Department of Labor, 200 Constitution Ave. NW, Washington, DC 20210; Web site: http://www.nrc.gov/

Office of Ocean and Coastal Resource Management 1825 Connecticut Ave. NW, Suite 700, Washington, DC 20235; Tel: 202-208-2553, Web site: http://wave.nos.noaa.gov/ocrm/

Office of Surface Mining Reclamation and Enforcement Interior South Bldg., 1951 Constitution Ave. NW, Washington, DC 20240; Web site: http://www.osmre.gov/astart3.htm

Organization for Economic Cooperation and Development (U.S. Office) 2001 L St. NW, Suite 700, Washington, DC 20036; Web site: http://www.oecd.org/

Soil Conservation Service (now called Natural Resource Conservation Service) USDA, 14th and Independence Ave. SW, P.O. Box 2890, Washington, DC 20013; Tel: 202-720-3210, Web site: http://www.nrcs.usda.gov/

United Nations 1 United Nations Plaza, New York, NY 10017; Web site: http://www.yahoo.com/Government/International_Organizations/United_Nations/

United Nations Environment Programme Regional North American Office, United Nations Rm. DC 20803, New York, NY 10017; and 1889 F St. NW, Washington, DC 20006; Web site: http://www.yahoo.com/Government/International_Organizations/United_Nations/Programs/United_Nations_Environment_Programme__UNEP_/

U.S. Fish and Wildlife Service Department of the Interior, Washington, DC 20240; Web site: http://www.fws.gov/

U.S. Geological Survey National Center, Reston, VA 22092; Tel: 703-648-4000

U.S. The Man and the Biosphere (U.S. MAB) Program U.S. MAB Secretariat, OES/EGC/MAB, Rm. 608, SA-37, Dept. of State, Washington, DC 20522-3706; Web site: http://www.mabnetamericas.org/home2.html

World Bank 1818 H St. NW, Washington, DC 20433; Web site: http://www.worldbank.org/

APPENDIX 2

UNITS OF MEASUREMENT

LENGTH

Metric

1 kilometer (km) = 1,000 meters (m)
1 meter (m) = 100 centimeters (cm)
1 meter (m) = 1,000 millimeters (mm)
1 centimeter (cm) = 0.01 meter (m)
1 millimeter (mm) = 0.001 meter (m)

English

1 foot (ft) = 12 inches (in)
1 yard (yd) = 3 feet (ft)
1 mile (mi) = 5,280 feet (ft)
1 nautical mile = 1.15 miles

Metric-English

1 kilometer (km) = 0.621 mile (mi)
1 meter (m) = 39.4 inches (in)
1 inch (in) = 2.54 centimeters (cm)
1 foot (ft) = 0.305 meter (m)
1 yard (yd) = 0.914 meter (m)
1 nautical mile = 1.85 kilometers (km)

AREA

Metric

1 square kilometer (km^2) = 1,000,000 square meters (m^2)
1 square meter (m^2) = 1,000,000 square millimeters (mm^2)
1 hectare (ha) = 10,000 square meters (m^2)
1 hectare (ha) = 0.01 square kilometer (km^2)

English

1 square foot (ft^2) = 144 square inches (in^2)
1 square yard (yd^2) = 9 square feet (ft^2)
1 square mile (mi^2) = 27,880,000 square feet (ft^2)
1 acre (ac) = 43,560 square feet (ft^2)

Metric-English

1 hectare (ha) = 2.471 acres (ac)
1 square kilometer (km^2) = 0.386 square mile (mi^2)
1 square meter (m^2) = 1.196 square yards (yd^2)
1 square meter (m^2) = 10.76 square feet (ft^2)
1 square centimeter (cm^2) = 0.155 square inch (in^2)

VOLUME

Metric

1 cubic kilometer (km^3) = 1,000,000,000 cubic meters (m^3)
1 cubic meter (m^3) = 1,000,000 cubic centimeters (cm^3)
1 liter (L) = 1,000 milliliters (mL) = 1,000 cubic centimeters (cm^3)
1 milliliter (mL) = 0.001 liter (L)
1 milliliter (mL) = 1 cubic centimeter (cm^3)

English

1 gallon (gal) = 4 quarts (qt)
1 quart (qt) = 2 pints (pt)

Metric-English

1 liter (L) = 0.265 gallon (gal)
1 liter (L) = 1.06 quarts (qt)
1 liter (L) = 0.0353 cubic foot (ft^3)
1 cubic meter (m^3) = 35.3 cubic feet (ft^3)
1 cubic meter (m^3) = 1.30 cubic yards (yd^3)
1 cubic kilometer (km^3) = 0.24 cubic mile (mi^3)
1 barrel (bbl) = 159 liters (L)
1 barrel (bbl) = 42 U.S. gallons (gal)

MASS

Metric

1 kilogram (kg) = 1,000 grams (g)
1 gram (g) = 1,000 milligrams (mg)
1 gram (g) = 1,000,000 micrograms (μg)
1 milligram (mg) = 0.001 gram (g)
1 microgram (μg) = 0.000001 gram (g)
1 metric ton (mt) = 1,000 kilograms (kg)

English

1 ton (t) = 2,000 pounds (lb)
1 pound (lb) = 16 ounces (oz)

Metric-English

1 metric ton (mt) = 2,200 pounds (lb) = 1.1 tons (t)
1 kilogram (kg) = 2.20 pounds (lb)
1 pound (lb) = 454 grams (g)
1 gram (g) = 0.035 ounce (oz)

ENERGY AND POWER

Metric

1 kilojoule (kJ) = 1,000 joules (J)
1 kilocalorie (kcal) = 1,000 calories (cal)
1 calorie (cal) = 4,184 joules (J)

Metric-English

1 kilojoule (kJ) = 0.949 British thermal unit (Btu)
1 kilojoule (kJ) = 0.000278 kilowatt-hour (kW-h)
1 kilocalorie (kcal) = 3.97 British thermal units (Btu)
1 kilocalorie (kcal) = 0.00116 kilowatt-hour (kW-h)
1 kilowatt-hour (kW-h) = 860 kilocalories (kcal)
1 kilowatt-hour (kW-h) = 3,400 British thermal units (Btu)
1 quad (Q) = 1,050,000,000,000,000 kilojoules (kJ)
1 quad (Q) = 2,930,000,000,000 kilowatt-hours (kW-h)

TEMPERATURE CONVERSIONS

Fahrenheit (°F) to Celsius (°C): °C = F((°F – 32.0) ÷ 1.80)
Celsius (°C) to Fahrenheit (°F): °F = F((°C × 1.80) + 32.0)

APPENDIX 3

MAJOR U.S. RESOURCE CONSERVATION AND ENVIRONMENTAL LEGISLATION

GENERAL

National Environmental Policy Act of 1969 (NEPA)
International Environmental Protection Act of 1983

ENERGY

Energy Policy and Conservation Act of 1975
National Energy Act of 1978, 1980
National Appliance Energy Conservation Act of 1987
Energy Policy Act of 1992

WATER QUALITY

Water Quality Act of 1965
Water Resources Planning Act of 1965
Federal Water Pollution Control Acts of 1965, 1972
Ocean Dumping Act of 1972
Safe Drinking Water Act of 1974, 1984, 1996
Water Resources Development Act of 1986
Clean Water Act of 1977, 1987
Ocean Dumping Ban Act of 1988

AIR QUALITY

Clean Air Act of 1963, 1965, 1970, 1977, 1990
Pollution Prevention Act of 1990

NOISE CONTROL

Noise Control Act of 1965
Quiet Communities Act of 1978

RESOURCES AND SOLID WASTE MANAGEMENT

Solid Waste Disposal Act of 1965
Resource Recovery Act of 1970
Resource Conservation and Recovery Act of 1976
Marine Plastic Pollution Research and Control Act of 1987

TOXIC SUBSTANCES

Hazardous Materials Transportation Act of 1975
Toxic Substances Control Act of 1976
Resource Conservation and Recovery Act of 1976
Comprehensive Environmental Response, Compensation, and Liability (Superfund) Act of 1980, 1986
Nuclear Waste Policy Act of 1982

PESTICIDES

Federal Insecticide, Fungicide, and Rodenticide Control Act of 1972, 1988

WILDLIFE CONSERVATION

Lacey Act of 1900
Migratory Bird Treaty Act of 1918
Migratory Bird Conservation Act of 1929
Migratory Bird Hunting Stamp Act of 1934
Pittman-Robertson Act of 1937
Anadromous Fish Conservation Act of 1965
Fur Seal Act of 1966
Species Conservation Act of 1966, 1969
National Wildlife Refuge System Act of 1966, 1976, 1978
Marine Mammal Protection Act of 1972
Marine Protection, Research, and Sanctuaries Act of 1972
Endangered Species Act of 1973, 1982, 1985, 1988
Whale Conservation and Protection Study Act of 1976
Fishery Conservation and Management Act of 1976, 1978, 1982
Fish and Wildlife Improvement Act of 1978
Fish and Wildlife Conservation Act of 1980 (Nongame Act)

LAND USE AND CONSERVATION

Taylor Grazing Act of 1934
Wilderness Act of 1964
Multiple Use Sustained Yield Act of 1968
Wild and Scenic Rivers Act of 1968
National Trails System Act of 1968
National Coastal Zone Management Act of 1972, 1980
Forest Reserves Management Act of 1974, 1976
Forest and Rangeland Renewable Resources Act of 1974, 1978
Federal Land Policy and Management Act of 1976
National Forest Management Act of 1976
Soil and Water Conservation Act of 1977
Surface Mining Control and Reclamation Act of 1977
Antarctic Conservation Act of 1978
Endangered American Wilderness Act of 1978
Alaskan National Interests Lands Conservation Act of 1980
Coastal Barrier Resources Act of 1982
Food Security Act of 1985

APPENDIX 4

MAPPING CONCEPTS AND CONNECTIONS*

This textbook emphasizes the connections between environmental principles, problems, and solutions. One way to organize the material in various chapters and to understand such connections is to map the concepts in a particular chapter or portion of a chapter. Inside the back cover you will find a general concept map showing how the parts of this entire textbook are related. At the end of each chapter you will find a question asking you to make a chapter concept map. This appendix provides you with information about how to create such maps. Here are the basic steps in making a concept map:

1. List the key general and specific concepts by indenting specific concepts under more general concepts. Basically, make an outline of the material using section heads, subsection heads, and key terms.
2. Arrange the clusters of concepts with the more general concepts at the top of a page and the more specific concepts at the bottom of the page.
3. Circle key concepts, draw lines connecting them, and write labels on the lines that describe linkages between the concepts.
4. Draw in crosslinking connections (lines that relate various parts of the map) and write labels on the line, describing these connections.
5. Study the map to see how you might improve it. You might add clustering concepts and connections of your own. You might eliminate detail to make the big picture stand out. You might add more details to clarify certain concepts. You might add or alter broad crosslinking connections on the map as you think and learn more about the material.

Let's go through these steps using the concept of living systems discussed in Chapters 7 and 8 as an example.

1. List the appropriate general and specific concepts. Note that Section 7-2 is about land systems (called biomes) and Section 8-1 is about aquatic systems. Use the list of section titles, subtitles, words in bold and italics, key drawings, and other terms that you think are particularly important. An outline of these two sections in which specific concepts are indented under more general ones would look something like Figure 1.
2. Arrange clusters of the concepts in Figure 1 with the more general ones at the top and more specific ones at the bottom. Concepts of the same general level should be at about the same height on the map. Figure 2 shows what this might look like. You might be able to do this on a single sheet of paper turned sideways, or you might find it easier to use large sheets of paper. Use a pencil to draw the map so that you can make changes. Note that the diagram has been simplified by eliminating some of the concepts in Figure 1 such as latitude and altitude, various types of temperate grasslands, and tropical scrub forests.
3. Circle key concepts, draw lines connecting them, and write labels on the lines that describe linkages between the concepts. You might decide to add or delete some of the concepts to make the map more logical, clear, and helpful to you. Figure 3 shows one way to do this. Note that in this figure the subcategories of deserts, grasslands, and forests are put together to improve clarity.
4. Draw in crosslinks that relate various parts of the map. Figure 4 illustrates one possibility. At this point you are trying to understand how the various concepts are connected to and interact with one another. Note that many of the human systems concepts interact positively or negatively with ecosystems and natural resource concepts (in terms of human value judgment). Note that Chapter 5 focuses on natural systems. As you learn more about how human systems interact with natural systems in this chapter and in other chapters, you should be able to add additional crosslinkages.
5. Study your map and try to improve it. You might choose to eliminate excessive detail or add additional concepts and linkages to improve clarity. You might also find it useful to rearrange the position of concepts vertically or horizontally to eliminate crowding or to reduce confusion over too many connecting links. In Figure 5 note that the general concepts of *fresh water* and *salt water* have been added, and aquatic systems has been divided into two subcategories based on salinity (dissolved salt content) instead of four. To improve clarity, the concepts of *moving* and *still* under *fresh water* have also been added because these characteristics greatly affect the types of organisms living in such systems. Finally, concepts have been connected with some of those in Chapter 4 by including details about photosynthetic and chemosynthetic productivity because these activities form the basis of the food web relationships in terrestrial and aquatic systems.

Remember the following things about concept maps:

- They are particularly useful for showing the big picture and for clarifying particularly difficult concepts. They can also be used in studying for tests to help you sort out what is important and how concepts are related.
- They take time to make. Sometimes they are easy and seem almost to write themselves. Other times, they can be difficult and frustrating to draw.
- There is no single "best" map. Rarely do concept maps developed by different individuals look alike.
- They are works in progress that should change and improve over time as you get better at making them and develop a deeper understanding of how things are connected.
- Comparing your maps with those developed by others is a good way to learn how to improve your mapmaking skills and to come up with more connections.
- Making them is challenging and fun.

---

*This appendix was developed by **Jane Heinze-Fry** with assistance from G. Tyler Miller, Jr.

LAND SYSTEMS (Biomes)
- climate
  - precipitation
  - temperature
  - latitude
  - altitude
- deserts
  - tropical
  - temperate
  - cold
  - semidesert
- grasslands
  - tropical
    - savanna
  - temperate
    - tall-grass
    - short-grass
    - pampas
    - veld
    - steppes
  - polar (arctic tundra)
    - permafrost
- forests
  - tropical
    - rain
    - deciduous
    - scrub
  - temperate deciduous
  - evergreen coniferous

WATER SYSTEMS
- ocean
  - coastal
    - coral reefs
    - estuaries
    - wetlands
    - barrier islands
    - beaches
      - rocky shore
      - barrier beach
- open sea
  - euphotic zone
  - bathyal zone
  - abyssal zone
- freshwater lakes
  - eutrophic
  - oligotrophic
  - mesotrophic
  - littoral zone
  - limnetic zone
  - profundal zone
  - benthic zone
  - thermal stratification
    - thermoclines
    - turnover
- freshwater streams
  - surface water
  - runoff
  - watershed
- inland wetlands
  - year-round
  - seasonal

Figure 1 List of concepts for Living Systems map.

LIVING SYSTEMS

Land Systems (Biomes)
- climate
  - precipitation
  - temperature
  - ~~latitude~~
  - ~~altitude~~

Water Systems

ocean

- deserts
  - tropical
  - temperate
  - cold
  - ~~semidesert~~
- grasslands
  - tropical
    - ~~savanna~~
  - temperate
    - ~~tall grass~~
    - ~~short grass~~
    - ~~pampas~~
    - ~~veld~~
    - ~~steppes~~
  - polar (arctic tundra)
    - ~~permafrost~~
- forests
  - tropical
    - rain
    - deciduous
    - ~~scrub~~
  - temperate deciduous
  - evergreen coniferous
- coastal
  - coral reefs
  - estuaries
  - wetlands
  - barrier islands
  - beaches
    - rocky shore
    - barrier beach
- open sea
  - euphotic zone
  - bathyal zone
  - abyssal zone
- freshwater lakes
  - eutrophic
  - oligotrophic
  - mesotrophic
  - littoral zone
  - limnetic zone
  - profundal zone
  - benthic zone
  - thermal stratification
    - thermoclines
    - turnover
- inland wetlands
  - year-round
  - seasonal
- freshwater streams
  - ~~surface water~~
  - ~~runoff~~
  - ~~watershed~~

Figure 2 Clusters of concepts are arranged in a hierarchy from more general at the top to more specific at the bottom.

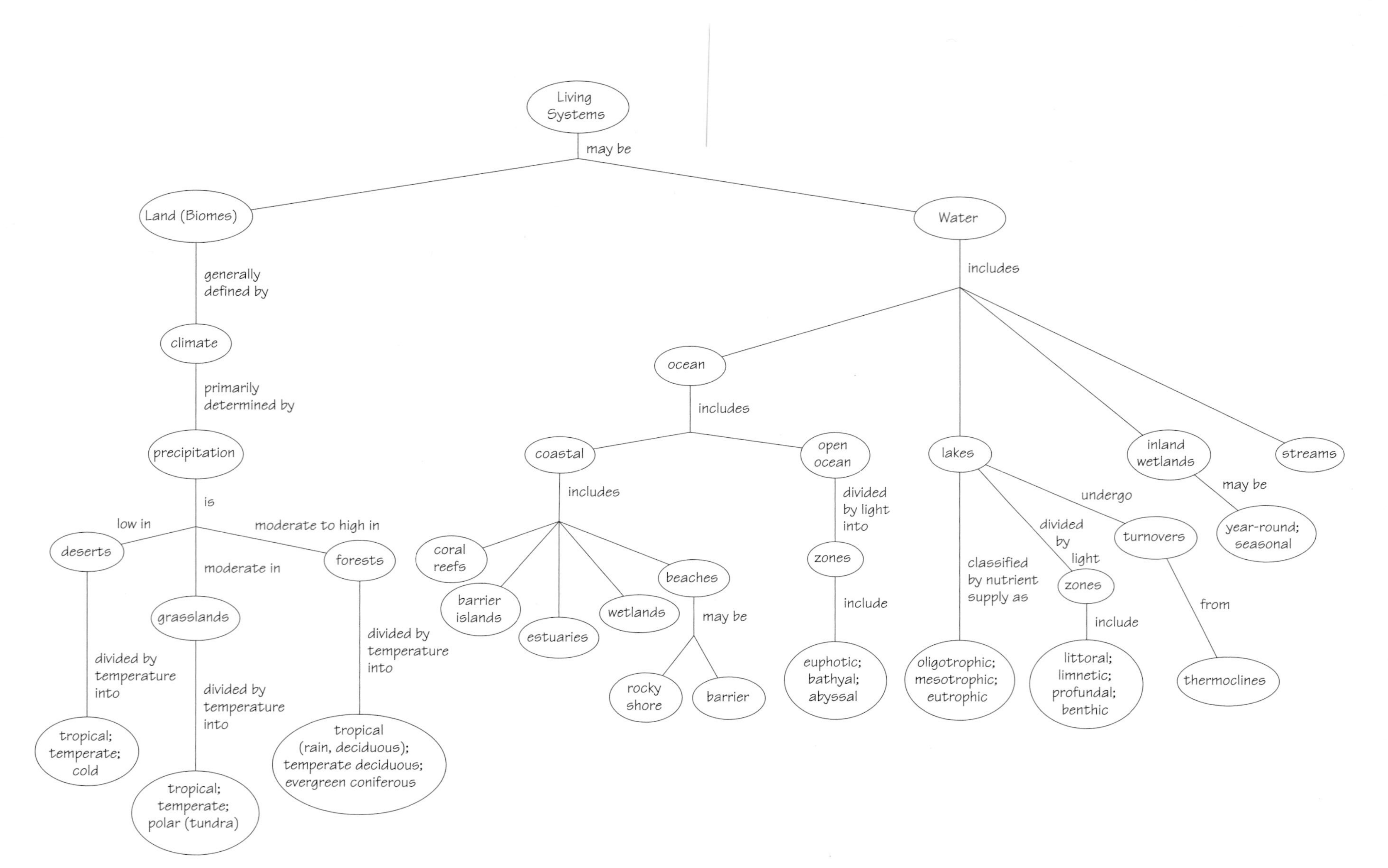

Figure 3 Concepts are circled and linkages are drawn.

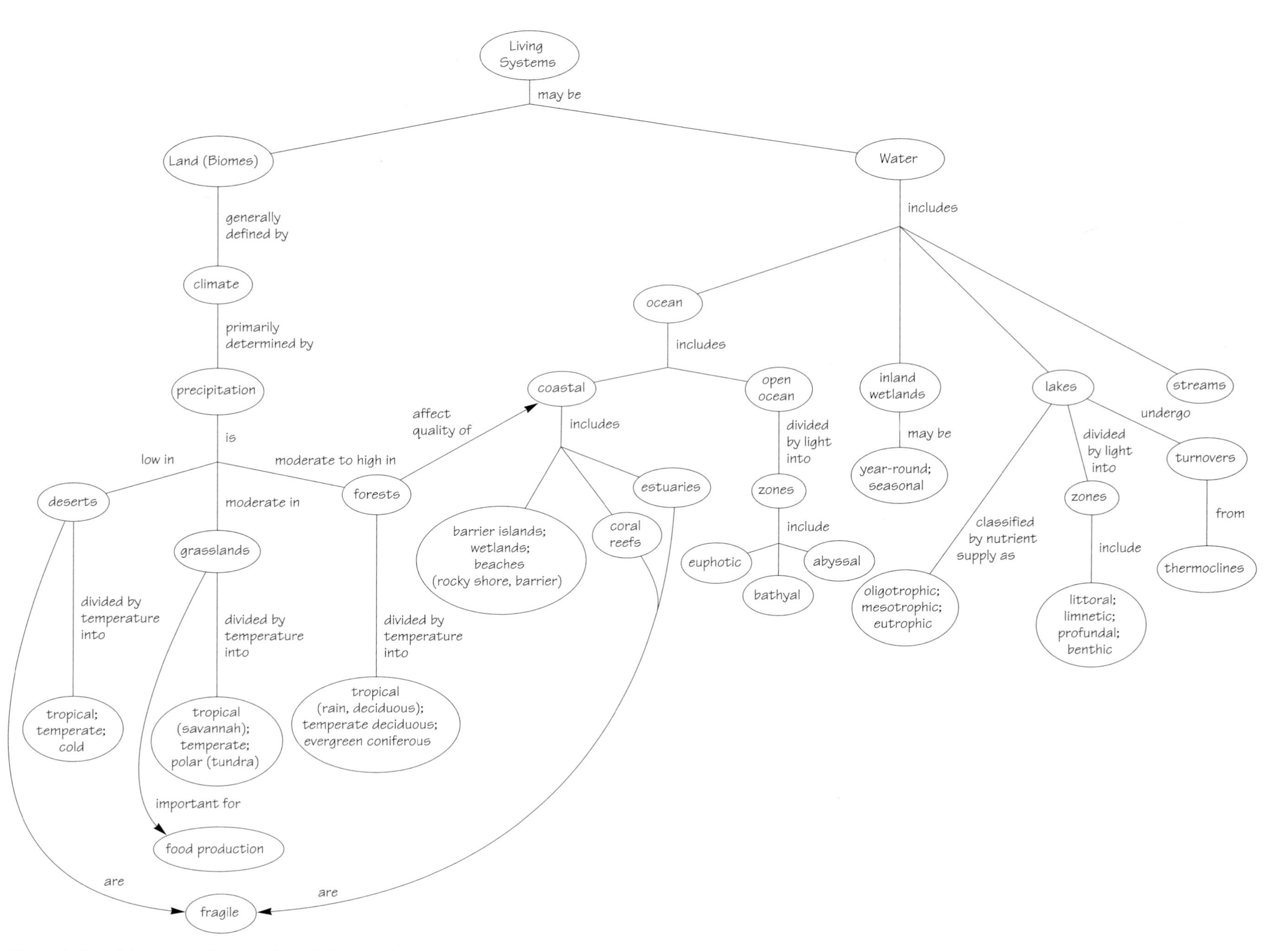

Figure 4 Crosslinkages are drawn and map is improved.

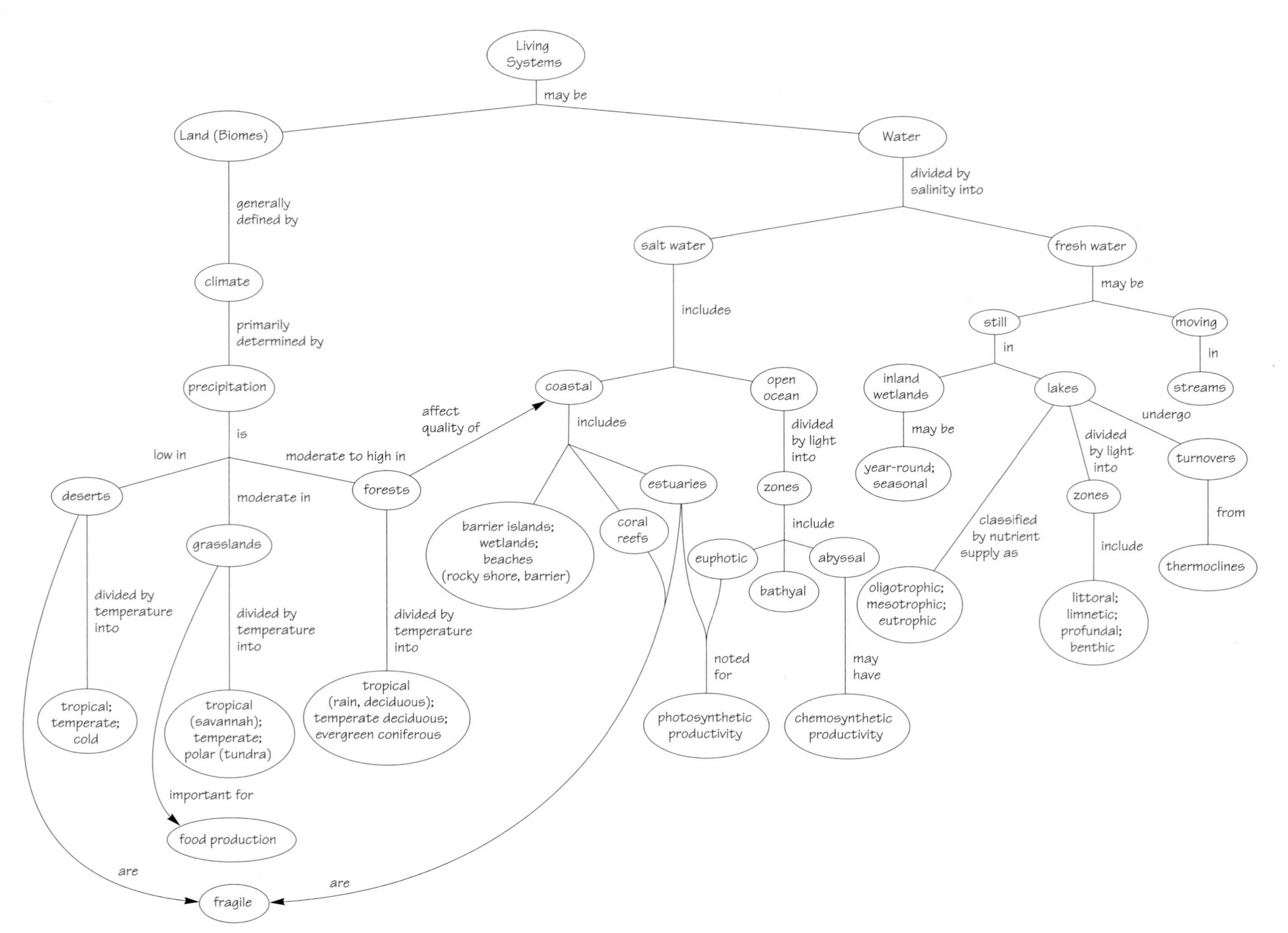

Figure 5 Map is further refined. New hierarchical clusters may emerge.

APPENDIX 5

INDIVIDUALS MATTER: WORKING WITH THE EARTH

This is a list of things individuals can do based on suggestions from a wide variety of environmentalists. It is not meant to be a list of things you must do but a list of actions you might consider. Don't feel guilty about the things you are not doing. Start off by picking the ones that you are willing to do and that you feel will have the most impact. Many of these suggestions are controversial. Carefully evaluate each action to see whether it fits in with your beliefs. Based on practicality and my own beliefs, I do some of these things. Each year I look over this list and try to add several new items.

PRESERVING BIODIVERSITY AND PROTECTING THE SOIL

- *Develop a plan for the sustainable use of any forested area you own.*
- *Plant trees on a regular basis and take care of them.*
- *Reduce the use of wood and paper products, recycle paper products, and buy recycled paper products.*
- *Don't buy furniture, doors, flooring, window frames, paneling, or other products made from tropical hardwoods such as teak or mahogany.* Look for the Good-Wood seal given by Friends of the Earth, or consult the *Wood User's Guide* by Pamela Wellner and Eugene Dickey (San Francisco: Rainforest Action Network, 1991).
- *Don't purchase wood and paper products produced by cutting remaining old-growth forests in the tropics and elsewhere.* Information on such products can be obtained from the Rainforest Action Network, Rainforest Alliance, and Friends of the Earth (Appendix 1).
- *Help rehabilitate or restore a degraded area of forest near your home.*
- *Don't buy furs, ivory products, items made of reptile skin, tortoiseshell jewelry, and materials of endangered or threatened animal species.*
- *When building a home, save all the trees possible.* Require that the contractor disturb as little soil as possible, set up barriers that catch any soil eroded during construction, and save and replace any topsoil removed instead of hauling it off and selling it.
- *Landscape areas not used for gardening with a mix of wildflowers, herbs (for cooking and for repelling insects), low-growing ground cover, small bushes, and other forms of vegetation natural to the area.*
- *Set up a compost bin and use it to produce soil conditioner for yard and garden plants.*

PROMOTING SUSTAINABLE AGRICULTURE AND REDUCING PESTICIDE USE

- *Waste less food.* An estimated 25% of all food produced in the United States is wasted.
- *Eat lower on the food chain* by reducing or eliminating meat consumption to reduce its environmental impact.
- *If you have a dog or a cat, don't feed it canned meat products.* Balanced-grain pet foods are available.
- *Reduce the use of pesticides on agricultural products by asking grocery stores to stock fresh produce and meat produced by organic methods.*
- *Grow some of your own food using organic farming techniques and drip irrigation to water your crops.*
- *Compost your food wastes.*
- *Think globally, eat locally.* Whenever possible, eat food that is locally grown and in season. This supports your local economy, gives you more influence over how the food is grown (by either organic or conventional methods), saves energy required to transport food over long distances, and reduces the fossil-fuel use and pollution. If you deal directly with local farmers, you can also save money.
- *Give up the idea that the only good bug is a dead bug.* Recognize that insect species keep most of the populations of pest insects in check and that full-scale chemical warfare on insect pests wipes out many beneficial insects.
- *Don't insist on perfect-looking fruits and vegetables.* These are more likely to contain high levels of pesticide residues.
- *Use pesticides in your home only when absolutely necessary, and use them in the smallest amount possible.*
- *Don't become obsessed with having the perfect lawn.* About 40% of U.S. lawns are treated with pesticides. These chemicals can cause headaches, dizziness, nausea, eye trouble, and even more acute effects in sensitive people (including children who play on treated lawns and in parks).
- *If you hire a lawn-care company, use one that relies only on organic methods, and get its claims in writing.*

SAVING ENERGY AND REDUCING OUTDOOR AIR POLLUTION

- *Reduce use of fossil fuels.* Drive a car that gets at least 15 kilometers per liter (35 miles per gallon), join a car pool, and use mass transit, walking, and bicycling as much as possible. This reduces emissions of CO_2 and other air pollutants, saves energy and money, and can improve your health.
- *Plant and care for trees to help absorb CO_2.* During its lifetime, the average tree absorbs enough CO_2 to offset the amount produced by driving a car 42,000 kilometers (26,000 miles).
- *Insulate new or existing houses heavily, caulk and weatherstrip to reduce air infiltration and heat loss, and use energy-efficient windows.* Add an air-to-air heat exchanger to minimize indoor air pollution.
- *Obtain as much heat and cooling as possible from natural sources,* especially sun, wind, geothermal energy, and trees.
- *Buy the most energy-efficient homes, lights, cars, and appliances available. Evaluate them only in terms of lifetime cost.*
- *Turn down the thermostat on water heaters to 43–49°C (110–120°F) and insulate hot water pipes.*
- *Lower the cooling load on an air conditioner by increasing the thermostat setting, installing energy-efficient lighting, using floor and ceiling fans, and using whole-house window or attic fans to bring in outside air (especially at night, when temperatures are cooler).*

REDUCING EXPOSURE TO INDOOR AIR POLLUTANTS

- *Test for radon and take corrective measures as needed.*
- *Install air-to-air heat exchangers or regularly ventilate your house by opening windows.*
- *At the beginning of the winter heating season, test your indoor air for formaldehyde when the house is closed up.* To locate a testing laboratory in your area, write to Consumer Product Safety Commission, Washington, DC 20207, or call 301-492-6800.
- *Don't buy furniture and other products containing formaldehyde. Use low-emitting formaldehyde or nonformaldehyde building materials.*
- *Reduce indoor levels of formaldehyde and other toxic gases by growing certain house plants.* Examples are the spider or airplane plant (removes 96% of carbon monoxide), aloe vera (90% of formaldehyde), banana (89% of formaldehyde), elephant ear philodendron (86% of formaldehyde), ficus (weeping fig, 47% of formaldehyde), golden porthos (67% of formaldehyde and benzene and 75% of carbon monoxide), Chinese evergreen (92% of toluene and 81% of benzene), English ivy (90% of benzene), peace lily (80% of benzene and 50% of trichloroethylene), and Janet Craig (corn plant, 79% of benzene). (Toxic removal figures indicate percentage of toxin removed by one plant in a 24-hour period in a 3.4-cubic-meter (12-cubic-foot) space). Plants should be potted with a mixture of soil and granular charcoal (which absorbs organic air pollutants).
- *Consider not using carpeting and using wood or linoleum floors instead.* New synthetic carpeting releases vapors from more than 100 volatile organic compounds. New and old carpeting is a haven for microbes (many of them highly allergenic), dust, and traces of lead and pesticides brought in by shoes.
- *Remove your shoes before entering your house.* This reduces inputs of dust, lead, and pesticides.
- *Test your house or workplace for asbestos fiber levels and for any crumbling asbestos materials if it was built before 1980.* Don't buy a pre-1980 house without having its indoor air tested for asbestos and lead. To get a free list of certified asbestos laboratories that charge $25–50 to test a sample, call the EPA's Toxic Substances Control Hotline at 202-554-1404.
- *Don't store gasoline, solvents, or other volatile hazardous chemicals inside a home or attached garage.*
- *Don't use aerosol spray products and commercial room deodorizers or air fresheners.*
- *If you smoke, do it outside or in a closed room vented to the outside.*
- *Make sure that wood-burning stoves, fireplaces, and kerosene- and gas-burning heaters are properly installed, vented, and maintained.* Install carbon monoxide detectors in all sleeping areas.

SAVING WATER

- *For existing toilets, reduce the amount of water used per flush* by putting into each tank a tall plastic container weighted with a few stones or by buying and inserting a toilet dam.
- *Install water-saving toilets that use no more than 6 liters (1.6 gallons) per flush.*
- *Flush toilets only when necessary.* Consider using the advice found on a bathroom wall in a drought-stricken area: "If it's yellow, let it mellow; if it's brown, flush it down."
- *Install water-saving showerheads and flow restrictors on all faucets.* If a 3.8-liter (1-gallon) jug can be filled by your showerhead in less than 15 seconds, you need a more efficient fixture.
- *Check frequently for water leaks in toilets and pipes, and repair them promptly.* A toilet must be leaking more than 940 liters (250 gallons) per day before you can hear the leak. To test for toilet leaks, add a water-soluble vegetable dye to the water in the tank, but don't flush. If you have a leak, some color will show up in the bowl's water within a few minutes.
- *Turn off sink faucets while brushing teeth, shaving, or washing.*
- *Wash only full loads of clothes;* if smaller loads must be used, use the lowest possible water-level setting.
- *When buying a new washer, choose one that uses the least amount of water and that fills up to different levels for loads of different sizes.* Front-loading clothes models use less water and energy than comparable top-loading models.
- *Use automatic dishwashers for full loads only.* Also, use the short cycle and let dishes air dry to save energy and money.
- *When washing many dishes by hand, don't let the faucet run.* Instead, use one filled dishpan or sink for washing and another for rinsing.
- *Keep one or more large bottles of water in the refrigerator rather than running water from the tap until it gets cold enough for drinking.*
- *Don't use a garbage disposal system—a large user of water.* Instead, consider composting your food wastes.
- *Wash a car from a bucket of soapy water, and use the hose for rinsing only. Use a commercial car wash that recycles its water.*
- *Sweep walks and driveways instead of hosing them off.*
- *Reduce evaporation losses by watering lawns and gardens in the early morning or evening, rather than in the heat of midday or when it's windy.*
- *Use drip irrigation and mulch for gardens and flower beds.* Better yet, landscape with native plants adapted to local average annual precipitation so that watering is unnecessary.

REDUCING WATER POLLUTION

- *Use manure or compost instead of commercial inorganic fertilizers to fertilize garden and yard plants.*
- *Use biological methods or integrated pest management to control garden, yard, and household pests.*
- *Use low-phosphate, phosphate-free, or biodegradable dishwashing liquid, laundry detergent, and shampoo.*
- *Don't use water fresheners in toilets.*
- *Use less harmful substances instead of commercial chemicals for most household cleaners* (Table 22-2).
- *Don't pour pesticides, paints, solvents, oil, antifreeze, or other products containing harmful chemicals down the drain or onto the ground.* Contact your local health department about disposal.
- *If you get water from a private well or suspect that municipal water is contaminated, have it tested by an EPA-certified laboratory for lead, nitrates, trihalomethanes, radon, volatile organic compounds, and pesticides.*
- *If you have a septic tank, have it cleaned out every 3–5 years by a reputable contractor so that it won't contribute to groundwater pollution.*

REDUCING SOLID WASTE AND HAZARDOUS WASTE

- *Buy less by asking yourself whether you really need a particular item.*
- *Buy things that are reusable, recyclable, or compostable, and be sure to reuse, recycle, and compost them.*
- *Buy beverages in refillable glass containers instead of cans or throwaway bottles. Urge companies and legislators to make refillable plastic (PET) bottles available in the United States.*
- *Use reusable plastic or metal lunch boxes and metal or plastic garbage containers*

without throwaway plastic liners (unless such liners are required for garbage collection).

- *Carry sandwiches and store food in the refrigerator in reusable containers instead of wrapping them in aluminum foil or plastic wrap.*
- *Use rechargeable batteries and recycle them when their useful life is over.* In 1993 Rayovac began selling mercury-free, rechargeable alkaline batteries (Renewal batteries) that outperform conventional nickel-cadmium rechargeable batteries.
- *Carry groceries and other items in a reusable basket, a canvas or string bag, or a small cart.*
- *Use sponges and washable cloth napkins, dish towels, and handkerchiefs instead of paper ones.*
- *Stop using throwaway paper and plastic plates, cups, and eating utensils, and other disposable items when reusable or refillable versions are available.*
- *Buy recycled goods, especially those made by primary recycling, and then make an effort to recycle them.* If you're not buying recycled materials, you're not recycling.
- *Reduce the amount of junk mail you get.* Do this (as several million Americans have done) at no charge by contacting the Mail Preference Service, Direct Marketing Association, 11 West 42nd St., P.O. Box 3681, New York, NY 10163-3861 (212-768-7277) and asking that your name not be sold to large mailing-list companies. Of the junk mail you do receive, recycle as much of the paper as possible.
- *Buy products in concentrated form whenever possible.*
- *Choose items that have the least packaging—or better yet, no packaging* ("nude products").
- *Don't buy helium-filled balloons that end up as litter. Urge elected officials and school administrators to ban balloon releases except for atmospheric research and monitoring.*
- *Lobby local officials to set up a community composting program.*
- *Use pesticides and other hazardous chemicals* (Table 21-1) *only when absolutely necessary and in the smallest amount possible.*
- *Use less hazardous (and usually cheaper) cleaning products* (Table 22-2).
- *Don't dispose of hazardous chemicals by flushing them down the toilet, pouring them down the drain, burying them, throwing them into the garbage, or dumping them down storm drains.* Consult your local health department or environmental agency for safe disposal methods.

COMMUNICATING WITH ELECTED OFFICIALS

- Find out their names and addresses. Then write, call, fax, or e-mail them. Contact a senator by writing The Honorable, U.S. Senate, Washington, DC 20510; Tel: 202-224-3121, Web site: http://thomas.loc.gov. Contact a representative by writing The Honorable, U.S. House of Representatives, Washington, DC 20510; Tel: 202-225-3121, Web site: http://thomas.loc.gov; Contact the president by writing President, The White House, 1600 Pennsylvania Ave. NW, Washington, DC 20500; Tel: 202-456-1414, Comment line: 202-456-1111, Fax: 202-456-2461, E-mail: president@whitehouse.gov, Web site: http://www.whitehouse.gov.
- When you write a letter or e-mail, use your own words, be brief and courteous, address only one issue, and ask the elected official to do something specific (such as cosponsoring, supporting, or opposing certain bills). Give reasons for your position, explain its effects on you and your district, try to offer alternatives, share any expert knowledge you have, and ask for a response. Be sure to include your name and return address.
- After your representatives have cast votes* supporting your position, send them a short note of thanks.
- Call and ask to speak to a staff member who works on the issue you are concerned about: for the White House, 202-456-1414 (Web site: http://www.whitehouse.gov); for the U.S. Senate, 202-224-3121; for the House of Representatives, 202-225-3121. The Web site address for the U.S. Congress is http://thomas.loc.gov.
- Once a desirable bill is passed, call or write to urge the president not to veto it. Urge the members of the appropriations committee to appropriate enough money to implement the law—a crucial decision.
- Monitor and influence action at the state and local levels, where all federal and state laws are either ignored or enforced. As Thomas Jefferson said, "The execution of laws is more important than the making of them."
- Get others who agree with your position to contact their elected officials.

*Each year the League of Conservation Voters (P.O. Box 500, Washington, DC 20077; 202-785-VOTE) publishes an *Environmental Scorecard* that rates all members of Congress on how they voted on environmental issues.

FURTHER READINGS

INTERNET

For a list of useful Internet sites, hyperlinks to material in this book, and other helpful and interesting information, please visit our Web site at http://www.brookscole.com/biology.

GENERAL SOURCES OF ENVIRONMENTAL INFORMATION

Readings

Ashworth, William. 1991. *The Encyclopedia of Environmental Studies.* New York: Facts on File.

Beacham, W., ed. 1993. *Beacham's Guide to Environmental Issues and Sources.* 5 vols. Washington, D.C.: Beacham.

British Petroleum. Annual. *BP Statistical Review of World Energy.* New York: BP America.

Brown, Lester R., et al. Annual. *State of the World.* New York: Norton.

Brown, Lester R., et al. Annual. *Vital Signs.* New York: Norton.

Harms, Valerie, et al., 1993. *Almanac of the Environment: The Ecology of Everyday Life.* New York: National Audubon Society.

Katz, Michael, and Dorothy Thornton. 1997. *Environmental Management Tools on the Internet: Assessing the World of Environmental Information.* Delray Beach, Calif.: St. Lucie Press.

Kirdon, Michael, and Ronald Segal. 1991. *The New State of the World Atlas.* 4th ed. New York: Simon & Schuster.

Kurland, Daniel J., and Jane Heinze-Fry. 1996. *Introduction to the Internet.* Belmont, Calif.: Wadsworth. Available as a supplement for use with this book.

Lean, Geoffrey, and Don Hinrichsen, eds. 1994. *Atlas of the Environment.* 2d ed. New York: HarperCollins.

Marien, Michael, ed. 1996. *Environmental Issues and Sustainable Futures: A Critical Guide to Recent Books, Reports, and Periodicals.* Bethesda, Md.: World Future Society.

Paehilke, Robert, ed. 1995. *Conservation and Environmentalism: An Encyclopedia.* New York: Garland.

Population Reference Bureau. Annual. *World Population Data Sheet.* Washington, D.C.: Population Reference Bureau.

Rittner, Don. 1992. *Ecolinking: Everyone's Guide to Online Environmental Information.* Berkeley, Calif.: Peachpit Press.

Schupp, Jonathan F. 1995. *Environmental Guide to the Internet.* Rockville, Md.: Government Institutes, Inc.

Seager, Joni, et al. 1995. *The New State-of-the-Earth Atlas.* New York: Simon & Schuster.

United Nations. Annual. *Demographic Yearbook.* New York: United Nations.

United Nations Children's Fund (UNICEF). Annual. *The State of the World's Children.* New York: UNICEF.

United Nations Environment Programme (UNEP). Annual. *State of the Environment.* New York: UNEP.

United Nations Population Fund. Annual. *The State of World Population.* New York: United Nations Population Fund.

U.S. Bureau of the Census. Annual. *Statistical Abstract of the United States.* Washington, D.C.: U.S. Bureau of the Census.

World Health Organization (WHO). Annual. *World Health Statistics.* Geneva, Switzerland: WHO.

World Resources Institute and International Institute for Environment and Development. Biannual. *World Resources.* New York: Basic Books.

1 / ENVIRONMENTAL PROBLEMS, THEIR CAUSES, AND SUSTAINABILITY

Readings

Bailey, Ronald, ed. 1995. *The True State of the Planet.* New York: The Free Press.

Barney, Gerald G., et al. 1993. *Global 2000 Revisited: What Shall We Do?* Arlington, Va.: Millennium Institute.

Bast, Joseph L., et al. 1994. *Eco Sanity: A Common-Sense Guide to Environmentalism.* Lantham, Md.: Madison Books.

Bowler, Peter J. 1993. *The Environmental Sciences.* New York: HarperCollins.

Brown, Halina S., et al. 1998. "Environmental Reforms in Poland: A Case for Cautious Optimism." *Environment,* vol. 40, no. 10, 10–13, 33–37.

Brown, Lester R., and Hal Kane. 1994. *Full House: Reassessing the Earth's Population Carrying Capacity.* New York: Norton.

Brundtland, G. H., et al. 1987. *Our Common Future: World Commission on Environment and Development.* New York: Oxford University Press.

Chivian, Eric, et al. 1993. *Critical Condition: Human Health and the Environment.* Cambridge, Mass.: MIT Press.

Cole, Daniel H. 1997. *Institutional Environmental Protection in Poland.* New York: St. Martin's Press.

Commoner, Barry. 1994. *Making Peace with the Planet.* Rev. ed. New York: New Press.

Daily, Gretchen, ed. 1997. *Nature's Services: Society's Dependence on Natural Ecosystems.* Covelo, Calif.: Island Press.

Dunn, Terry L. 1997. *Guide to Global Environmental Issues.* Golden, Colo.: Fulcrum

Easterbrook, Gregg. 1995. *A Moment on the Earth: The Coming Age of Environmental Optimism.* New York: Viking Press.

Edberg, R., and A. Yablokov. 1991. *Tomorrow Will Be Too Late: East Meets West on Global Ecology.* Tucson: University of Arizona Press.

Ehrlich, Paul R., and Anne H. Ehrlich. 1990. *The Population Explosion.* New York: Doubleday.

Ehrlich, Paul R., and Anne H. Ehrlich. 1991. *Healing the Planet.* Reading, Mass.: Addison-Wesley.

Ehrlich, Paul R., and Anne H. Ehrlich. 1996. *Betrayal of Science and Reason: How Anti-Environmental Rhetoric Threatens Our Future.* Covelo, Calif.: Island Press.

Feshbach, Murray, and Alfred Friendly Jr. 1992. *Ecocide in the USSR: Health and Nature Under Siege.* New York: Basic Books.

Flattau, Edward. 1998. *Tracking the Charlatans: Countering the Eco-Bashers.* New York: Global Horizons Press.

Goldsmith, Edward, et al. 1990. *Imperiled Planet: Restoring Our Endangered Ecosystems.* Cambridge, Mass.: MIT Press.

Gordon, Anita, and David Suzuki. 1991. *It's a Matter of Survival.* Cambridge, Mass.: Harvard University Press.

Gore, Al. 1992. *Earth in the Balance: Ecology and the Human Spirit.* Boston: Houghton Mifflin.

Hardin, Garrett. 1968. "The Tragedy of the Commons." *Science,* vol. 162, 1243–48.

Hardin, Garrett. 1993. *Living Within Limits: Ecology, Economics, and Population Taboos.* New York: Oxford University Press.

Harrison, Paul. 1992. *The Third Revolution: Environment, Population, and a Sustainable World.* New York: St. Martin's Press.

Hinrichsen, Don. 1998. "On a Slow Trip Back from Hell." *International Wildlife,* January/February, 36–43.

Kanieniecki, Sheldon, et al. 1993. *Controversies in Environmental Policy.* Ithaca: State University of New York Press.

Jones, Gareth, and Graham Hollier. 1997. *Resources, Society, and Environmental Management.* London: Paul Chapman.

Keating, Michael. 1997. *Canada and the State of the Planet: The Social, Economic, and Environmental Trends That Are Shaping Our Lives.* New York: Oxford University Press.

Lehr, Jay, ed. 1994. *Rational Readings on Environmental Concerns.* New York: Van Nostrand Reinhold.

Lubchenco, Jane. 1998. "Entering the Century of the Environment: A New Social Contract for Science." *Science,* vol. 279, 491–97.

Markley, Oliver W., and Walter R. McCuan, eds. 1996. *21st Century Earth: Opposing Viewpoints.* Boston: Greenhaven Press.

Mazur, Laurie Ann, ed. 1994. *Beyond the Numbers: A Reader on Population, Consumption, and the Environment.* Covelo, Calif.: Island Press.

McMichael, A. J. 1993. *Planetary Overload: Global Environmental Change and the Health of the Human Species*. New York: Cambridge University Press.

Meadows, Donella. 1991. *The Global Citizen*. Covelo, Calif.: Island Press.

Meadows, Donella H., et al. 1992. *Beyond the Limits: Confronting Global Collapse, Envisioning a Sustainable Future*. White River Junction, Vt.: Chelsea Green.

Meyers, Norman. 1997. "Consumption in Relation to Population, Environment, and Development." *The Environmentalist*, vol. 17, 33–44.

Mitchell, Bruce, ed. 1995. *Resource and Environmental Management in Canada: Addressing Conflict and Uncertainty*. New York: Oxford University Press.

Myers, Norman, ed. 1993. *Gaia: An Atlas of Planet Management*. Garden City, N.Y.: Anchor Press/Doubleday.

Myers, Norman. 1996. *Ultimate Security: The Environmental Basis of Political Stability*. New York: Norton.

Myers, Norman, and Julian Simon. 1994. *Scarcity or Abundance? A Debate on the Environment*. New York: Norton.

Parkinson, Claire L. 1997. *Earth from Above: Using Color-Coded Satellite Images to Examine the Global Environment*. New York: University Science Press.

Peterson, D. J. 1993. *Troubled Lands: The Legacy of Soviet Environmental Destruction*. Boulder, Colo.: Westview Press.

Porritt, Jonathan. 1991. *Save the Earth*. Atlanta, Ga.: Turner.

Pryde, Philip R., ed. 1995. *Environmental Resources and Constraints in the Former Soviet Republics*. Boulder, Colo.: Westview Press.

Quinn, Daniel. 1992. *Ishmael*. New York: Bantam/Turner.

Sadler, A. E. 1996. *The Environment: Opposing Viewpoints*. San Diego, Calif.: Greenhaven Press.

Saign, Geoffrey. 1994. *Green Essentials*. San Francisco: Mercury House.

Simon, Julian L., ed. 1995. *The True State of the Planet*. Cambridge, Mass.: Basil Blackwell.

Simon, Julian L. 1996. *The Ultimate Resource 2*. Princeton, N.J.: Princeton University Press.

Smith, Courtland L. 1995. "Assessing the Limits to Growth." *BioScience*, vol. 45, no. 7, 478–83.

Suzuki, David. 1994. *Time to Change*. Toronto: Stoddard.

Tobias, Michael. 1994. *World War III: Population and the Biosphere at the End of the Millennium*. Santa Fe, N.M.: Bear.

United Nations. 1997. *Global Environmental Outlook*. New York: Oxford University Press.

Wagner, Travis. 1993. *In Our Backyard: A Guide to Understanding Pollution and Its Effects*. New York: Van Nostrand Reinhold.

Ward, Barbara, and René Dubos. 1972. *Only One Earth: The Care and Maintenance of a Small Planet*. New York: Norton.

World Conservation Union (IUCN). 1993. *National Environmental Status Reports*. Vol. 1, *Czechoslovakia, Hungary, Poland*. Covelo, Calif.: Island Press.

2 / CRITICAL THINKING: SCIENCE, MODELS, AND SYSTEMS

Readings

Bahn, Paul, and John Flenly. 1992. *Easter Island, Earth Island*. New York: Thames & Hudson.

Bauer, Henry H. 1992. *Scientific Literacy and the Myth of the Scientific Method*. Urbana: University of Illinois Press.

Bono, de Edward. 1992. *Serious Creativity*. New York: HarperCollins.

Bronowski, J. 1964. *Science and Human Values*. New York: Harper & Row.

Carey, Stephen S. 1998. *A Beginner's Guide to Scientific Method*. 2d ed. Belmont, Calif.: Wadsworth. Available as a supplement for use with this book.

Diamond, Jared. 1995. "Easter's End." *Discovery*, August, 63–69.

Feynman, Richard. 1965. *The Character of Physical Law*. Cambridge, Mass.: M.I.T. Press.

Heinze-Fry, Jane. 1996. *An Introduction to Critical Thinking*. Belmont, Calif.: Wadsworth. Available as a supplement for use with this book.

Kuhn, Thomas S. 1970. *The Structure of Scientific Revolutions*. 2d ed. Chicago: University of Chicago Press.

Lewin, Roger. 1992. *Complexity: Life at the Edge of Chaos*. New York: Macmillan.

Lubchenco, Jane. 1998. "Entering the Century of the Environment: A New Social Contract for Science." *Science*, vol. 279, 491–97.

McCain, Garvin, and Erwin M. Segal. 1998. *The Game of Science*. 5th ed. Pacific Grove, Calif.: Brooks/Cole. Available as a supplement for use with this book.

Meadows, Donella H., et al. 1992. *Beyond the Limits: Confronting Global Collapse, Envisioning a Sustainable Future*. White River Junction, Vt.: Chelsea Green.

Rothman, Milton A. 1993. *The Science Gap: Dispelling the Myths and Understanding the Reality of Science*. Buffalo, N.Y.: Prometheus Books.

Strahler, Arthur N. 1993. *Understanding Science: An Introduction to Concepts and Issues*. Buffalo, N.Y.: Prometheus Books.

Vance, Mike, and Diane Deacon. 1996. *Break Out of the Box*. New York: Career Press.

Wynne, B., and S. Mayer. 1993. "How Science Fails the Environment." *New Scientist*, vol. 138, June 5, 1876–80.

Zimmerman, Michael. 1995. *Science, Nonscience, and Nonsense: Approaching Environmental Literacy*. Baltimore, Md.: Johns Hopkins University Press.

3 / MATTER AND ENERGY PRINCIPLES

Readings

Fowler, John M. 1984. *Energy and the Environment*. 2d ed. New York: McGraw-Hill.

Miller, G. Tyler, Jr., and David G. Lygre. 1991. *Chemistry: A Contemporary Approach*. 3d ed. Belmont, Calif.: Wadsworth.

Rifkin, Jeremy. 1989. *Entropy: Into the Greenhouse World: A New World View*. New York: Bantam.

4 / ECOLOGY, ECOSYSTEMS, AND FOOD WEBS

Readings

Baskin, Yvonne. 1997. *The Work of Nature: How the Diversity of Life Sustains Us*. Covelo, Calif.: Island Press.

Begon, Michael, John L. Harper, and Colin R Townsend. 1996. *Ecology: Individuals, Populations, and Communities*. 3d ed. Boston: Blackwell Scientific.

Berenbaum, May R. 1995. *Bugs in the System: Insects and Their Impact on Human Affairs*. Reading, Mass.: Addison-Wesley.

Colinvaux, Paul A. 1975. *Why Big Fierce Animals Are Rare*. New York: Wiley.

Colinvaux, Paul A. 1992. *Ecology*. 2d ed. New York: Wiley.

Daily, Gretchen, ed. 1997. *Nature's Services: Society's Dependence on Natural Ecosystems*. Covelo, Calif.: Island Press.

Dodson, Stanley I., et al. 1998. *Ecology*. New York: Oxford University Press.

Friedrich, Robert L., and Robert V. Blystone. 1998. "Internet Teaching Resources for Remote Sensing and GIS." *BioScience*, vol. 48, no. 3, 187–93.

Golley, Frank Benjamin. 1993. *A History of the Ecosystem Concept in Ecology: More Than the Sum of the Parts*. New Haven, Conn.: Yale University Press.

Grant, William E., Ellen K. Pedersen, and Sandra L. Marin. 1997. *Ecology and Natural Resource Management: Systems Analysis and Simulation*. New York: Wiley.

Heathcote, Isabel W. 1997. *Environmental Problem Solving: A Case Study Approach*. New York: McGraw-Hill.

Krebs, Charles J. 1994. *Ecology: The Experimental Analysis of Distribution & Abundance*. 4th ed. New York: HarperCollins.

Lubchenco, Jane. 1998. "Entering the Century of the Environment: A New Social Contract for Science." *Science*, vol. 279, 491–97.

McIntosh, Robert P. 1985. *The Background of Ecology: Concept and Theory*. New York: Cambridge University Press.

Odum, Eugene P. 1997. *Ecology: A Bridge Between Science and Society*. Sunderland, Mass.: Sinauer.

Ricklefs, Robert E. 1993. *Ecology*. 4th ed. New York: W.H. Freeman.

Ricklefs, Robert E. 1997. *The Economy of Nature: A Textbook in Basic Ecology*. 4th ed. New York: W.H. Freeman.

Smith, Robert L. 1998. *Elements of Ecology*. 4th ed. New York: Harper & Row.

Starr, Cecie, and Ralph Taggart. 1998. *Biology: The Unity and Diversity of Life*. 8th ed. Belmont, Calif.: Wadsworth.

Tudge, Colin. 1991. *Global Ecology*. New York: Oxford University Press.

Vitousek, Peter M., Paul R. Ehrlich, Anne H. Ehrlich, and Pamela Matson. 1986. "Human Appropriation of the Products of Photosynthesis." *BioScience*, June, 368–73.

Whittaker, Robert H. 1975. *Communities and Ecosystems*. 2d ed. New York: Macmillan.

Worster, Donald. 1985. *Nature's Economy: A History of Ecological Ideas*. New York: Cambridge University Press.

5 / NUTRIENT CYCLES AND SOILS

Readings

Berner, Elizabeth K., and Robert A. Berner. 1996. *Global Environment: Water, Air, and Geochemical Cycles*. Upper Saddle River, NJ: Prentice Hall.

Bormann, F. H., and Gene E. Likens. 1994. *Pattern and Process in a Forested Ecosystem: Disturbance, Development and the Steady State Based on the Hubbard Brook Ecosystem Study*. New York: Springer-Verlag.

Bradbury, Ian K. 1991. *The Biosphere*. New York: Belhaven Press.

Chapin, F. S. "Terry," and Pamela A. Matson. 1998. *Principles of Ecosystem Ecology*. New York: Springer-Verlag.

Ehrlich, Anne H., and Paul R. Ehrlich. 1987. *Earth*. New York: Franklin Watts.

Hemond, Harold F., and Elizabeth J. Fechner. 1994. *Chemical Fate and Transport in the Environment*. San Diego: Academic Press.

Hillel, Daniel. 1992. *Out of the Earth: Civilization and the Life of the Soil*. New York: Free Press.

Lovelock, James. 1988. *The Ages of Gaia*. New York: Norton.

Mackenzie, Fred T. 1998. *Our Changing Planet*. 2d ed. Englewood Cliffs, N.J.: Prentice-Hall.

Post, Wilfred M., et al. 1990. "The Global Carbon Cycle." *American Scientist*, vol. 78, 310–26.

Rowell, David I. 1994. *Soil Science: Methods and Applications*. New York: Wiley.

Scientific American. 1970. *The Biosphere*. San Francisco. W.H. Freeman.

Smil, Vaclav. 1993. *Global Ecology*. London: Routledge.

Smil, Vaclav. 1997. *Cycles of Life: Civilization and the Biosphere*. New York. Scientific American Library.

Smil, Vaclav. 1997. "Global Population and the Nitrogen Cycle." *Scientific American*, July, 76–81.

Vernadsky, Vladimir Ivanovich. 1998. *The Biosphere*. (Translated by Dave B. Langmuir; revised and annotated by Mark A. S. McMenamin). New York: Copernicus.

Vitousek, Peter M., et al. 1997. "Human Alteration of the Global Nitrogen Cycle: Sources and Consequences." *Ecological Applications*, vol. 7, no. 3, 737–50.

Volk, Tyler. 1998. *Gaia's Body: Toward a Physiology of Earth*. New York: Copernicus.

Winegardner, Duane L. 1996. *An Introduction to Soils for Environmental Professionals*. Boca Raton, Fla.: Lewis.

6 / EVOLUTION AND BIODIVERSITY: ORIGINS, NICHES, AND ADAPTATION

Readings

Allègre, Claude J., and Stephen H. Schneider. 1994. "The Evolution of the Earth." *Scientific American*, October, 66–75.

Attenborough, David. 1984. *The Living Planet: A Portrait of the Earth*. Boston: Little, Brown.

Baskin, Yvonne. 1997. *The Work of Nature: How the Diversity of Life Sustains Us*. Covelo, Calif.: Island Press.

Benton, M. J. 1995. "Diversification and Extinction in the History of Life." *Science*, vol. 268, 52–58.

Brown, Bruce, and Lane Morgan. 1990. *The Miracle Planet*. Edison, N.J.: W.H. Smith.

Chapin, F. Stuart, et al. 1998. "Ecosystem Consequences of Changing Biodiversity." *BioScience*, vol. 48, no. 1, 45–51.

Darwin, Charles. 1859. *On the Origin of Species*. New York: D. Appleton.

Dawkins, Richard. 1989. *The Selfish Gene*. New York: Oxford University Press.

Elderidge, Niles. 1991. *The Miner's Canary: Unravelling the Mysteries of Extinction*. Englewood Cliffs, N.J.: Prentice Hall.

Elderidge, Niles. 1995. *Dominion*. New York: Holt.

Glen, W. 1990. "What Killed the Dinosaurs?" *American Scientist*, vol. 78, 354–70.

Gould, Stephen Jay. 1977. *Ever Since Darwin: Reflection in Natural History*. New York: Norton.

Leakey, Richard, and Roger Lewin. 1995. *The Sixth Extinction: Patterns of Life and the Future of Humankind*. New York: Doubleday.

Lewontin, Richard C. 1995. *Human Diversity*. New York: W.H. Freeman.

Lovelock, James E. 1988. *The Ages of Gaia: A Biography of Our Living Earth*. New York: Norton.

Margulis, Lynn, and Lorraine Olendzenski. 1992. *Environmental Evolution: Effects of the Origin and Evolution of Life on Planet Earth*. Cambridge, Mass.: MIT Press.

Myers, Norman. 1997. "Mass Extinction and Evolution." *Science*, vol. 278, 597–98.

Pimentel, David, et al. 1997. "Economic and Environmental Benefits of Biodiversity." *BioScience*, vol. 47, no. 11, 747–57.

Potts, Rick. 1997. *Humanity's Descent: The Consequences of Ecological Instability*. New York: Avon.

Reaka-Kudla, Marjorie I., Don E. Wilson, and Edward O. Wilson, eds. 1997. *Biodiversity II: Understanding and Protecting Our Biological Resources*. Washington, D.C.: Joseph Henry Press.

Ridley, Mark. 1993. *Evolution*. Boston: Blackwell Scientific.

Savage, Jay M. 1995. "Systematics and the Biodiversity Crisis." *BioScience*, November, 673–79.

Scientific American. 1974. *Ecology, Evolution, and Population Biology*. San Francisco: W.H. Freeman.

Scientific American. 1978. *Evolution*. San Francisco: W.H. Freeman.

Wilson, Edward O., ed. 1988. *Biodiversity*. Washington, D.C.: National Academy Press.

Wilson, Edward O. 1992. *The Diversity of Life*. Cambridge, Mass.: Belknap Press of Harvard University Press.

7 / GEOGRAPHICAL ECOLOGY, CLIMATE, AND BIOMES

Readings

See also the readings for Chapters 5 and 6.

Aber, John, and Jerry Melilo. 1991. *Terrestrial Ecosystems*. Philadelphia: Saunders.

Ahrens, C. Donald. 1994. *Meteorology Today: An Introduction to Weather, Climate, and the Environment*. 5th ed. Belmont, Calif.: Wadsworth.

Akin, Wallace E. 1991. *Global Patterns: Climate, Vegetation, and Soils*. Norman: University of Oklahoma Press.

Allan, Nigel J. R., Gregory W. Knapp, and Christoph Stadel, eds., 1988. *Human Impact on Mountains*. Totowa, N.J.: Rowman & Littlefield.

Allan, Tony, and Andrew Warren, eds., 1993. *Deserts: The Encroaching Wilderness*. New York: Oxford University Press.

Attenborough, David, et al. 1989. *The Atlas of the Living World*. Boston: Houghton Mifflin.

Bailey, Robert G. 1998. *The Ecosystem Geography of the Oceans and Continents*. New York: Springer-Verlag.

Brown, David E., et al. 1998. *A Classification of North American Biotic Communities*. Salt Lake City: University of Utah Press.

Brown, James. H., and Arthur C. Gibson. 1983. *Biogeography*. St. Louis: Mosby.

Brown, Lauren. 1985. *Grasslands*. New York: Random House.

Denniston, Derek. 1995. *High Priorities: Conserving Mountain Ecosystems and Cultures*. Washington, D.C.: Worldwatch Institute.

Gates, David M. 1980. *Biophysical Ecology*. New York: Springer-Verlag.

Goldsmith, Edward, et al. 1990. *Imperiled Planet: Restoring Our Endangered Ecosystems*. Cambridge, Mass.: MIT Press.

Golley, Frank B., ed. 1983. *Tropical Rain Forest Ecosystems: Structure and Function*. New York: Elsevier.

Haines-Young, Roy, David R. Green, and Steven Cousins, eds. 1993. *Landscape Ecology and Geographic Information Systems*. New York: Taylor & Francis.

Hare, Tony, ed. 1994. *Habitats*. New York: Macmillan.

Hinrichsen, Don. 1997. "Coral Reefs in Crisis." *BioScience*, vol. 47, no. 9, 554–58.

Hinrichsen, Don. 1997. *The World's Coastal Seas: Trends, Threats, and Strategies*. Covelo, Calif.: Island Press.

Ives, Jack D., ed. 1994. *Mountains: The Illustrated Library of the Earth*. Emmaus, Pa.: Rodale Press.

Lincare, Edward, and Bart Geerts. 1997. *Climate and Weather Explained*. New York: Routledge.

MacArthur, Robert H. 1972. *Geographical Ecology*. San Francisco: Harper & Row.

Myers, Norman. 1993. *The Primary Source: Tropical Forests and Our Future*. New York: Norton.

Rosenberg, Norman J., Blaine L. Blad, and Shashi B. Verma. 1983. *Microclimate: The Biological Environment*. 2d ed. New York: Wiley.

Schimper, A. F. W. 1960. *Plant-Geography Upon a Physiological Basis*. New York: Hafner.

Whitmore, T. C. 1990. *An Introduction to Tropical Rain Forests*. New York: Oxford University Press.

Williams, Jack. 1997. *The Weather Book*. 2d ed. New York: Vintage Books.

8 / AQUATIC ECOLOGY

Readings

See also the readings for Chapters 5 and 6.

Abramovitz, Janet N. 1996. "Sustaining Freshwater Ecosystems." In Lester R. Brown et al., *State of the World 1996*. Washington, D.C.: Worldwatch Institute, 60–77.

Allan, J. David. 1995. *Stream Ecology: Structure and Function of Running Waters*. New York: Chapman & Hall.

Beatley, Timothy, et al. 1994. *An Introduction to Coastal Zone Management*. Covelo, Calif.: Island Press.

Beer, Tom. 1997. *Environmental Oceanography*. 2d ed. Boca Raton, Fla.: CRC Press.

Brown, Barbara E., and John C. Ogden. 1993. "Coral Bleaching," *Scientific American*, January, 64–70.

Chapin, F. Stuart, et al. 1997. "Biotic Control Over the Functioning of Ecosystems." *Science*, vol. 277, 500–504.

Council for Agricultural Science and Technology. 1994. *Wetland Policy Issues*. Ames, Iowa: Council for Agricultural Science and Technology.

Couper, Alastair, ed. 1990. *The Times Atlas and Encyclopedia of the Sea*. Hagerstown, Md.: Lippincott.

Daiber, Franklin C. 1986. *Conservation of Tidal Marshes*. New York: Van Nostrand Reinhold.

Dugan, Patrick, ed. 1993. *Wetlands in Danger: A World Conservation Atlas*. New York: Oxford University Press.

Elder, Danny, and John Pernetta, eds. 1991. *Oceans*. London: Michael Beazley.

Garrison, Tom. 1995. *Essentials of Oceanography*. Belmont, Calif.: Wadsworth.

Goldman, C., and A. Horne. 1994. *Limnology*. New York: McGraw-Hill.

Goldman-Carter, Jan. 1989. *A Citizen's Guide to Protecting Wetlands.* Washington, D.C.: National Wildlife Federation.

Goldsmith, Edward, et al. 1990. *Imperiled Planet: Restoring Our Endangered Ecosystems.* Cambridge, Mass.: MIT Press.

Hinrichsen, Don. 1997. *Living on the Edge: Managing Coasts in Crisis.* Covelo, Calif.: Island Press.

Hutchinson, G. Evelyn. 1993. *A Treatise on Limnology.* New York: Wiley.

Kusler, Jon A., et al. 1994. "Wetlands." *Scientific American,* January, 64–70.

Lampert, Winfried, and Ulrich Sommer. 1997. *Limnoecology.* New York: Oxford University Press.

Mitsch, William J., and James G. Gosselink. 1993. *Wetlands.* 2d ed. New York: Van Nostrand Reinhold.

National Academy of Sciences. 1994. *Understanding Marine Biodiversity.* Washington, D.C.: National Academy Press.

National Academy of Sciences. 1995. *Wetlands: Characteristics and Boundaries.* Washington, D.C.: National Academy Press.

Niering, William A. 1991. *Wetlands of North America.* Charlottesville, Va.: Thomasson-Grant.

Pilkey, Orin H., Jr., et al. 1984. *Coastal Design: A Guide for Builders, Planners, & Homeowners.* New York: Van Nostrand Reinhold.

Stevenson, Robert F., and Frank H. Talbot. 1993. *Oceans.* Emmaus, Pa.: Rodale Press.

Teal, John, and Mildred Teal. 1969. *Life and Death of a Salt Marsh.* New York: Ballantine.

Thorne-Miller, Boyce, and John Catena. 1990. *The Living Ocean: Understanding and Protecting Marine Diversity.* Covelo, Calif.: Island Press.

Vileisis, Ann. 1997. *Discovering the Unknown Landscape: A History of America's Wetlands.* Covelo, Calif.: Island Press.

Weber, Michael L., and Judith A. Gradwohl. 1995. *The Wealth of Oceans: Environment and Development on Our Ocean Planet.* New York: Norton.

Weber, Peter. 1993. *Abandoned Seas: Reversing the Decline of the Oceans.* Washington, D.C.: Worldwatch Institute.

Wells, Sue, and Nick Hanna. 1992. *The Greenpeace Book of Coral Reefs.* New York: Sterling.

9 / COMMUNITY PROCESSES: SPECIES INTERACTIONS AND SUCCESSION

Readings

Agosta, William. 1995. *Bombardier Beetles and Fever Trees: A Close-Up Look at Chemical Warfare and Signals in Animals and Plants.* New York: Addison-Wesley.

Andrewartha, H. G., and L. C. Birch. 1986. *The Ecological Web: More on the Distribution and Abundance of Animals.* Chicago: University of Chicago Press.

Blaustein, Andrew R., and David W. Wake. 1995. "The Puzzle of Declining Amphibians." *Scientific American,* April, 52–57.

Botkin, Daniel. 1990. *Discordant Harmonies: A New Ecology for the Twenty-First Century.* New York: Oxford University Press.

Burton, Robert, ed. 1991. *Animal Life.* New York: Oxford University Press.

Colinvaux, Paul A. 1975. *Why Big Fierce Animals Are Rare.* New York: Wiley.

Connell, Joseph H., and R. O. Slatyer. 1977. "Mechanisms of Succession in Natural Communities and Their Role in Community Stability and Organization." *American Naturalist,* vol. 11, 1119–44.

Foster, David R., et al. 1997. "Forest Response to Disturbance and Anthropogenic Stress." *BioScience,* vol. 47, no. 7, 437–45.

Fujita, Marty S. 1991. "Flying Foxes (Chiroptera: Pteropodidae): Threatened Animals of Key Ecological and Economic Importance." *Conservation Biology,* vol. 5, 455–63.

Gause, G. F. 1934. *The Struggle for Existence.* Baltimore, Md.: Williams & Wilkins.

Gleason, H. A. 1926. "The Individualistic Concept of Plant Association." *Bulletin of the Torrey Botanical Club,* vol. 53, 7–26.

Gordon, David G. 1996. *The Compleat Cockroach: A Comprehensive Guide to the Most Despised (and Least Understood) Creature on Earth.* Berkeley, Calif.: Ten Speed Press.

Holt, Robert D., George R. Robinson, and Michael S. Gaines. 1995. "Vegetation Dynamics in an Experimentally Fragmented Landscape." *Ecology,* vol. 76, 1610–24.

Howe, Henry, and Lynn Westley. 1988. *Ecological Relationships of Plants and Animals.* New York: Oxford University Press.

Korpimäki, Erkhi, and Charles J. Kreps. 1996. "Predation and Population Cycles of Small Mammals." *BioScience,* vol. 46, no. 10, 754–64.

MacArthur, Robert H., and E. O. Wilson. 1967. *The Theory of Island Biogeography.* Princeton, N.J.: Princeton University Press.

Orians, Gordon H. 1990. "Ecological Concepts of Sustainability." *Environment,* November, 10–39.

Pimm, Stuart L. 1992. *The Balance of Nature?* Chicago: University of Chicago Press.

Power, Mary E., et al. 1996. "Challenges in the Quest for Keystones." *BioScience,* vol. 46, 606–19.

Rennie, J. 1992. "Living Together." *Scientific American,* September, 56–64.

Rickleffs, Robert E. 1996. *The Economy of Nature: A Textbook of Ecology.* New York: W.H. Freeman.

Rosenthal, G. A. 1986. "The Chemical Defenses of Higher Plants." *Scientific American,* vol. 254, 94–99.

Stebbins, Robert. 1994. *The Theory of Island Biogeography.* Princeton, N.J.: Princeton University Press.

Turner, Monica G., et al. 1997. "Fires, Hurricanes, and Volcanoes: Comparing Large Disturbances." *BioScience,* vol. 47, no. 11, 758–67.

Wickler, W. 1968. *Mimicry in Plants and Animals.* London: World University Library.

Whittaker, Robert H., and Paul P. Feeny. 1971. "Allelochemics: Chemical Interactions Between Species." *Science,* vol. 171, 757–71.

Windsor, Donald A. 1996. "Endangered Interrelationships: The Ecological Cost of Parasites Lost." *Wild Earth,* Winter, 78–83.

10 / POPULATION DYNAMICS, CARRYING CAPACITY, AND CONSERVATION BIOLOGY

Readings

Baldwin, A. Dwight, Jr., Judith de Luce, and Carl Pletsch, eds. 1994. *Beyond Preservation: Restoring and Inventing Landscapes.* Minneapolis: University of Minnesota Press.

Berger, John J. 1986. *Restoring the Earth.* New York: Knopf.

Berger, John J., ed. 1990. *Environmental Restoration.* Covelo, Calif.: Island Press.

Cairns, John Jr., ed. 1994. *Rehabilitating Damaged Ecosystems.* 2d ed. Toledo, Ohio: CRC Press.

Cappuccino, Naomi, and Peter W. Price, eds. 1995. *Population Dynamics: New Approaches and Syntheses.* New York: Academic Press.

Cook, L. M. 1991. *Genetic and Ecological Diversity.* New York: Chapman & Hall.

Daily, Gretchen C. 1995. "Restoring Value to Degraded Lands." *Science,* vol. 269, 350–54.

Dobson, Andy P., et al. 1997. "Hopes for the Future: Restoration Ecology and Conservation Biology." *Science,* vol. 277, 515–22.

Fielder, Peggy L., and Subodh K. Jain, eds. 1992. *Conservation Biology: The Theory and Practice of Nature Conservation, Preservation and Management.* New York: Chapman & Hall.

Freedman, Bill. 1989. *Environmental Ecology: The Impacts of Pollution and Other Stresses on Ecosystem Structure and Function.* San Diego: Academic Press.

Hardin, Garrett. 1985. "Human Ecology: The Subversive, Conservative Science." *American Zoologist,* vol. 25, 469–76.

Harte, John. 1993. *The Green Fuse: An Ecological Odyssey.* Los Angeles: University of California Press.

Hunter, Malcolm L. Jr. 1996. *Fundamentals of Conservation Biology.* Cambridge, Mass.: Blackwell Scientific.

Matson, P. A., et al. 1997. "Agricultural Intensification and Ecosystem Properties." *Science,* vol. 277, 502–9.

Meffe, Gary K., and C. Ronald Carroll. 1994. *Principles of Conservation Biology.* Sunderland, Mass.: Sinauer.

Mlot, Christine. 1990. "Restoring the Prairie." *BioScience,* vol. 40, no. 11, 804–9.

Monschke, Jack. 1990. "How to Heal the Land." *Whole Earth Review,* Spring, 72–79.

Myers, Norman. 1993. "The Question of Linkages in Environment and Development." *BioScience,* vol. 43, no. 5, 302–9.

National Academy of Sciences. 1986. *Ecological Knowledge and Environmental Problem-Solving.* Washington, D.C.: National Academy Press.

National Academy of Sciences. 1992. *Restoration of Aquatic Ecosystems.* Washington, D.C.: National Academy Press.

Nilsen, Richard, ed. 1991. *Helping Nature Heal: An Introduction to Environmental Restoration.* Berkeley, Calif.: Ten Speed Press.

Primack, Richard B. 1995. *A Primer of Conservation Biology.* Sunderland, Mass.: Sinauer.

Reaka-Kudla, Marjorie, Don E. Wilson, and Edward O. Wilson, eds. 1996. *Biodiversity II: Understanding and Protecting Our Biological Resources.* New York: John Henry.

Rosenberg, Daniel K., et al. 1997. "Biological Corridors: Form, Function, and Efficacy." *BioScience,* vol. 47, no. 10, 672–87.

Slobodkin, Laurence B. 1980. *Growth and Regulation of Animal Populations.* New York: Dover.

Solbrig, Otto T. 1991. "The Origin and Function of Biodiversity." *Environment,* vol. 33, no. 5, 17–20, 34–38.

Vitousek, Peter M., et al. 1997. "Human Domination of Earth's Ecosystems." *Science,* vol. 277, 494–99.

Wilcove, David S., and Michael J. Bean, eds. 1994. *The Big Kill: Declining Biodiversity in America's Lakes and Rivers.* Washington, D.C.: Environmental Defense Fund.

Wilson, E. O. 1984. *Biophilia.* Cambridge, Mass.: Harvard University Press.

Wilson, E. O. 1992. *The Diversity of Life.* Cambridge, Mass.: Harvard University Press.

Wilson, E. O., and W. H. Bossert. 1971. *Primer of Population Biology.* Sunderland, Mass.: Sinauer.

Wilson, E. O., ed. 1988. *Biodiversity.* Washington, D.C.: National Academy Press.

11 / HUMAN POPULATION: GROWTH, DEMOGRAPHY, AND CARRYING CAPACITY

Readings

Abernathy, Virginia D. 1993. *Population Politics: The Choices That Shape Our Future.* New York: Plenum.

Arizpe, Lourdes, et al., eds. 1994. *Population and Environment: Rethinking the Debate.* Boulder, Colo.: Westview Press.

Ashford, Lori S. 1995. "New Perspectives on Population: Lessons from Cairo." *Population Bulletin,* vol. 50, no. 1, 1–44.

Barna, George. 1992. *The Invisible Generation: Baby Busters.* New York: Barna Research Group.

Beck, Roy. 1996. *The Case Against Immigration: The Moral, Economic, Social, and Environmental Reasons for Reducing U.S. Immigration.* New York: Norton.

Bongarts, John. 1993. *The Fertility Impact of Family Planning Programs.* New York: Population Council.

Bongarts, John. 1994. "Population Policy Options in the Developing World." *Science,* vol. 263, 771–76.

Bouvier, Leon F., and Carol J. De Vita. 1991. "The Baby Boom—Entering Midlife." *Population Bulletin,* vol. 6, no. 3, 1–35.

Bouvier, Leon F., and Lindsey Grant. 1994. *How Many Americans? Population, Immigration, and the Environment.* San Francisco: Sierra Club Books.

Brower, Michael. 1994. *Population Complications: Understanding the Population Debate.* Cambridge, Mass.: Union of Concerned Scientists.

Brown, Lester R. 1994. *Who Will Feed China?* New York: Norton.

Brown, Lester R., Gary Gardner, and Brian Halweil. 1998. *Beyond Malthus: Sixteen Dimensions of the Population Problem.* Washington, D.C.: Worldwatch Institute.

Brown, Lester R., and Hal Kane. 1994. *Full House: Reassessing the Earth's Population Carrying Capacity.* New York: Norton.

Coates, Joseph et al. 1997. "The Promise of Genetics." *The Futurist,* September–October, 18–22.

Cohen, Joel E. 1995. *How Many People Can the Earth Support?* New York: Norton.

Cornelius, Wayne A., et al. 1994. *Controlling Immigration: A Global Perspective.* Stanford, Calif.: Stanford University Press.

Day, Lincoln H. 1992. *The Future of Low-Birthrate Populations.* New York: Routledge.

DeVita, Carol J. 1996. "The United States at Mid-Decade." *Population Bulletin,* vol. 50, no. 4, 1–48.

Donaldson, Peter J., and Amy Ong Tsui. 1990. "The International Family Planning Movement." *Population Bulletin,* vol. 43, no. 3, 1–42.

During, Alan Thein, and Christopher D. Crowther. 1997. *Misplaced Blame, The Real Roots of Population Growth.* Seattle: Northwest Environment Watch.

Ehrlich, Paul R., and Anne H. Ehrlich. 1990. *The Population Explosion.* New York: Doubleday.

Ehrlich, Paul R., Anne H. Ehrlich, and Gretchen C. Daily. 1995. *The Stork and the Plow: The Equity Answer to the Human Dilemma.* New York: Putnam.

Erickson, Jon. 1995. *The Human Volcano: Population Growth as a Geologic Force.* New York: Facts on File.

Grant, Lindsey. 1992. *Elephants and Volkswagens: Facing the Tough Questions About Our Overcrowded Country.* New York: W.H. Freeman.

Grant, Lindsey. 1996. *Juggernaut: Growth on a Finite Planet.* Santa Ana, Calif.: Seven Locks Press

Hall, Charles A. H., et al. 1995. "The Environmental Consequences of Having a Baby in the United States." *Wild Earth,* Summer, 78–87.

Hardin, Garrett. 1993. *Living Within Limits: Ecology, Economics, and Population Taboos.* New York: Oxford University Press.

Hardin, Garrett. 1995. *Immigration Reform: Avoiding the Tragedy of the Commons.* Washington, D.C.: Federation for American Immigration Reform.

Harrison, Paul. 1992. *The Third Revolution: Environment, Population, and a Sustainable World.* New York: I.B. Tauris.

Hartmann, Betsy. 1995. *Reproductive Rights and Wrongs: The Global Politics of Population Control and Contraceptive Choice.* Boston: South End Press.

Haub, Carl. 1987. "Understanding Population Projections." *Population Bulletin,* vol. 42, no. 4, 1–41.

Haupt, Arthur, and Thomas T. Kane. 1997. *The Population Handbook.* 4th ed. Washington, D.C.: Population Reference Bureau.

Hollingsworth, William J. 1996. *Ending the Explosion: Population Policies and Ethics.* Santa Ana, Calif.: Seven Locks Press.

Jacobson, Jodi. 1991. *Women's Reproductive Health: The Silent Emergency.* Washington, D.C.: Worldwatch Institute.

Jiggins, Janice. 1994. *Changing the Boundaries: Women-Centered Perspectives on Population and the Environment.* Covelo, Calif.: Island Press.

Keyfitz, Nathan. 1992. "Completing the Worldwide Demographic Transition: The Relevance of Past Experience." *Ambio,* no. 21, 26–30.

Leisinger, Klaus M., and Karin Schmitt. 1994. *All Our People: Population Policy with a Human Face.* Covelo, Calif.: Island Press.

Lutz, Wolfgang. 1994. "The Future of World Population." *Population Bulletin,* vol. 46, no. 2, 1–43.

Martin, Philip, and Elizabeth Midgley. 1994. "Immigration to the United States: Journey to an Uncertain Destination." *Population Bulletin,* vol. 49, no. 2, 1–47.

Mazur, Laurie Ann, ed. 1994. *Beyond the Numbers: A Reader on Population, Consumption, and the Environment.* Covelo, Calif.: Island Press.

McFalls, Joseph A., Jr. 1998. "Population: A Lively Introduction." *Population Bulletin,* vol. 53, no. 3, 1–48.

McKibben, Bill. 1998. *Maybe One: A Personal and Environmental Argument for Single-Child Families.* New York: Simon & Schuster.

Moffett, George D. 1994. *Critical Masses: The Global Population Challenge.* New York: Viking Press.

Myers, Norman. 1993. "Population, Environment, and Development." *Environmental Conservation,* vol. 20, no. 3, 1–12.

National Academy of Sciences. 1997. *The New Americans: Economic, Demographic, and Fiscal Effects of Immigration.* Washington, D.C.: National Academy Press.

Olshansky, S. Jay, et al. 1993. "The Aging of the Human Species." *Scientific American,* April, 46–52.

Population Reference Bureau. 1990. *World Population: Fundamentals of Growth.* Washington, D.C.: Population Reference Bureau.

Population Reference Bureau. Annual. *World Population Data Sheet.* Washington, D.C.: Population Reference Bureau.

Repetto, Robert. 1994. *"Second India" Revisited: Population, Poverty, and Environmental Stress Over Two Decades.* Washington, D.C.: World Resources Institute.

Robey, Bryant, et al. 1993. "The Fertility Decline in Developing Countries." *Scientific American,* December, 60–67.

Ross, John A., et al. 1993. *Family Planning and Population: A Compendium of International Statistics.* New York: Population Council.

Russell, Cheryl. 1993. *The Master Trend: How the Baby Boom Generation Is Remaking America.* New York: Plenum.

Simon, Julian L. 1989. *Population Matters: People, Resources, Environment, and Immigration.* New Brunswick, N.J.: Transaction.

Simon, Julian L. 1996. *The Ultimate Resource 2.* Princeton, N.J.: Princeton University Press.

Smil, Vaclav. 1993. *China's Environmental Crisis: An Inquiry into the Limits of National Development.* Armonk, N.Y.: Sharpe.

Tien, H. Yuan, et al. 1992. "China's Demographic Dilemmas." *Population Bulletin,* vol. 47, no. 1, 1–44.

United Nations. 1991. *Consequences of Rapid Population Growth in Developing Countries.* New York: United Nations.

United Nations. 1992. *Long-Range World Population Projections: Two Centuries of Population Growth, 1950–2150.* New York: United Nations.

United Nations. Annual. *Demographic Yearbook.* New York: United Nations.

United Nations Children's Fund (UNICEF). Annual. *The State of the World's Children.* New York: UNICEF.

Visaria, Leela, and Pravin Visaria. 1995. "India's Population in Transition." *Population Bulletin,* vol. 50, no. 3, 1–51.

Weeks, John R. 1996. *Population: An Introduction to Concepts and Issues.* 6th ed. Belmont, Calif.: Wadsworth.

World Bank. 1997. *China 2020: Development Challenges in the New Century.* Washington, D.C.: World Bank.

Zero Population Growth. 1990. *Planning the Ideal Family: The Small Family Option.* Washington, D.C.: Zero Population Growth.

12 / FOOD RESOURCES

Readings

Aberley, Doug. 1993. *Greening the Garden: A Guide to Sustainable Growing.* Philadelphia: New Society.

Aliteri, Miguel A. 1995. *Agroecology: The Scientific Basis of Alternative Agriculture.* 2d ed. Boulder, Colo.: Westview Press.

Ausubel, Kenny. 1994. *Seeds of Change.* New York: HarperCollins.

Avery, Dennis T. 1997. "Saving Nature's Legacy Through Better Farming." *Issues in Science and Technology,* Fall, 59–64.

Berrill, Michael. 1997. *The Plundered Seas: Can the World's Fish Be Saved?* San Francisco, Calif.: Sierra Club Books.

Berry, Wendell. 1990. *Nature as Measure.* Berkeley, Calif.: North Point Press.

Berstein, Henry, et al., eds. 1990. *The Food Question: Profits Versus People.* East Haven, Conn.: Earthscan.

Bongarts, John. 1994. "Can the Growing Human Population Feed Itself?" *Scientific American,* March, 36–42.

Bray, Francesca. 1994. "Agriculture for Developing Nations." *Scientific American,* July, 30–37.

Brookfield, Harold, and Christine Padoch. 1994. "Appreciating Biodiversity: A Look at the Dynamism and Diversity of Indigenous Farming Practices." *Environment,* June, 6–42.

Brown, Lester R. 1994. *Who Will Feed China?* New York: Norton.

Brown, Lester R. 1996. *Tough Choices: Facing the Challenge of Food Scarcity*. New York: Norton.

Brown, Lester R., and Hal Kane. 1994. *Full House: Reassessing the Earth's Population Carrying Capacity*. New York: Norton.

Carroll, C. Ronald, et al. 1990. *Agroecology*. New York: McGraw-Hill.

Clemings, Russell. 1996. *The False Promise of Desert Agriculture*. San Francisco, Calif.: Sierra Club Books.

Coleman, Elliot. 1992. *The New Organic Grower's Four-Season Harvest*. White River Junction, Vt.: Chelsea Green.

Conaway, Gordon R., and Jules N. Petty. 1991. *Unwelcome Harvest: Agriculture and Pollution*. East Haven, Conn.: Earthscan.

Curtis, Jennifer, et al. 1991. *Harvest of Hope: The Potential of Alternative Agriculture to Reduce Pesticide Use*. New York: Natural Resources Defense Council.

Denckla, Tanya. 1994. *The Organic Gardener's Home Reference*. Pownal, Vt.: Garden Way Publishing.

Dower, Roger, et al. 1997. *Frontiers of Sustainability: Environmentally Sound Agriculture, Forestry, Transportation, and Power Production*. Covelo, Calif.: Island Press.

Dunning, Alan B., and Holly W. Brough. 1991. *Taking Stock: Animal Farming and the Environment*. Washington, D.C.: Worldwatch Institute.

***The Ecologist*. 1995,** Vol. 25, no. 2/3. Entire issue devoted to overfishing.

Edwards, Clive A., et al., eds. 1990. *Sustainable Agricultural Systems*. Ankeny, Iowa: Soil and Water Conservation Society.

Ehrlich, Paul R., et al. 1995. *The Stork and the Plow: The Equity Answer to the Human Dilemma*. New York: Putnam.

Faeth, Paul. 1995. *Growing Green: Enhancing the Economic and Environmental Performance of U.S. Agriculture*. Washington, D.C.: World Resources Institute.

Faeth, Paul, et al. 1993. *Agricultural Policy and Sustainability: Case Studies from India, Chile, the Philippines, and the United States*. Washington, D.C.: World Resources Institute.

Foster, Phillip. 1992. *The World Food Problem*. Boulder, Colo.: Rienner.

Fowler, Cary, and Pat Mooney. 1990. *Shattering: Food, Politics, and the Loss of Genetic Diversity*. Tucson: University of Arizona Press.

Francis, Charles A., et al., eds. 1990. *Sustainable Agriculture in Temperate Zones*. New York: Wiley.

Fukuoka, Masanobu. 1985. *The Natural Way of Farming: The Theory and Practice of Green Philosophy*. New York: Japan Publications.

Gardner, Gary. 1996. *Shrinking Fields: Cropland Loss in a World of Eight Billion*. Washington, D.C.: Worldwatch Institute.

Goering, Peter, et al. 1993. *From the Ground Up: Rethinking Industrial Agriculture*. Atlantic Highlands, N.J.: Zed Books.

Goldsmith, James. 1994. *The Trap*. New York: Carroll & Graf.

Gordon, R. Conway, and Edward R. Barbier. 1990. *After the Green Revolution: Sustainable Agriculture for Development*. East Haven, Conn.: Earthscan.

International Food Policy Research Institute. 1997. *China's Food Economy for the Twenty-First Century: Supply, Demand, and Trade*. Washington, D.C.: International Food Policy Research Institute.

Jackson, Wes, et al., eds. 1985. *Meeting the Expectations of Land: Essays in Sustainable Agriculture and Stewardship*. Berkeley, Calif.: North Point Press.

Jacobson, Michael, et al. 1991. *Safe Food: Eating Wisely in a Risky World*. Washington, D.C.: Planet Earth Press.

Jeavons, John, and Carol Cox. 1993. *Lazy-Bed Gardening: The Quick and Dirty Guide*. Willits, Calif.: Ecology Action.

Kane, Hal. 1993. "Growing Fish in Fields." *World Watch*,. September/October, 20–27.

Kiley-Worthington, Marthe. 1993. *Eco-agriculture: Food First Farming*. London: Souvenir Press.

Klopenburg, Jack, Jr., and Beth Burrows. 1996. "Biotechnology to the Rescue: Twelve Reasons Why Biotechnology Is Incompatible with Sustainable Agriculture." *The Ecologist*, vol. 26, no. 2, 61–67.

Krimsky, Sheldon, and Roger Wrubel. 1996. *Agricultural Biotechnology and the Environment: Science, Policy, and Social Issues*. Urbana: University of Illinois Press.

Landau, Matthew. 1992. *Introduction to Aquaculture*. New York: Wiley.

League of Women Voters. 1991. *U.S. Farm Policy: Who Benefits? Who Pays? Who Decides?* Washington, D.C.: League of Women Voters.

Logsdon, Gene. 1994. *At Nature's Pace: Farming and the American Dream*. New York: Pantheon.

Lowrance, Richard, ed. 1984. *Agricultural Ecosystems: Unifying Concepts*. New York: Wiley.

Mann, Charles. 1997. "Reseeding the Green Revolution." *Science*, vol. 277, 1038–43.

Matson, P. A., et al. 1997. "Agricultural Intensification and Ecosystem Properties." *Science*, vol. 277, 502–9.

McGinn, Anne Platt. 1998. *Rocking the Boat: Conserving Fisheries and Protecting Jobs*. Washington, D.C.: Worldwatch Institute.

McKinney, Tom. 1987. *The Sustainable Farm of the Future*. Snowmass, Colo.: Rocky Mountain Institute.

Mollison, Bill. 1990. *Permaculture: A Practical Guide for a Sustainable Future*. Covelo, Calif.: Island Press.

Motavalli, Jim. 1998. "The Trouble with Meat: Dangerous Diseases Emerge in a Planet Eating High on the Food Chain." *E Magazine*, May/June, 29–35.

Myers, Norman. 1997. "Population and Food Security." In S. R. Johnson, ed. *Food Security: New Solutions for the 21st Century*. Ames: Iowa State University Press.

National Academy of Sciences. 1990. *Sustainable Agriculture Research and Education in the Field*. Washington, D.C.: National Academy Press.

National Academy of Sciences. 1992. *Marine Aquaculture: Opportunities for Growth*. Washington, D.C.: National Academy Press.

National Academy of Sciences. 1993. *Soil and Water Quality: An Agenda for Agriculture*. Washington, D.C.: National Academy Press.

National Academy of Sciences. 1993. *Sustainable Agriculture and the Environment in the Humid Tropics*. Washington, D.C.: National Academy Press.

Pauly, Daniel, et al. 1998. "Fishing Down Marine Food Webs." *Science*, vol. 279, 860–63.

Pimentel, David, and Carl W. Hall. 1989. *Food and Natural Resources*. San Diego: Academic Press.

Pimentel, David, and Marcia Pimentel, eds. 1996. *Food, Energy, and Society*. Niwot: University Press of Colorado.

Poisson, Leandre, and Gretchen Vogel Poisson. 1994. *Solar Gardening: Growing Vegetables Year-Round the American Intensive Way*. White River Junction, Vt.: Chelsea Green.

Postel, Sandra L. 1998. "Water for Food Production: Will There Be Enough in 2025?" *BioScience*, vol. 48, no. 8, 629–37.

Pretty, Jules N. 1995. *Regenerating Agriculture: Policies and Prospects for Sustainability and Self-Reliance*. London: Earthscan.

Prosterman, Roy L., et al. 1996. "Can China Feed Itself?" *Scientific American*, November, 90–96.

Raeburn, Paul. 1995. *The Last Harvest: The Genetic Gamble That Threatens to Destroy American Agriculture*. New York: Simon & Schuster.

Rau, Bill. 1991. *From Feast to Famine: Official Cures and Grassroots Remedies to Africa's Food Crisis*. Atlantic Highlands, N.J.: Zed Books.

Rifkin, Jeremy. 1992. *Beyond Beef: The Rise and Fall of the Cattle Culture*. New York: Dutton.

Rissler, Jane, and Margaret Mellon. 1996. *The Ecological Risks of Engineered Crops*. Cambridge, Mass.: MIT Press.

Ritchie, Mark. 1990. "GATT, Agriculture, and the Environment." *The Ecologist*, vol. 20, no. 6, 214–20.

Robbins, John. 1992. *May All Be Fed: Diet for a New World*. New York: William Morrow.

Rosegrant, Mark W., and Robert Livernash. 1996. "Growing More Food, Doing Less Damage." *Environment*, vol. 38, no. 7, 6–11, 28–30.

Ruttan, Vernon W., ed. 1993. *Agriculture, Environment and Health*. Minneapolis: University of Minnesota Press.

Safina, Carl. 1994. "Where Have All the Fishes Gone?" *Issues in Science and Technology*, Spring, 37–43.

Safina, Carl. 1998. *Song for the Blue Ocean: Encounters Along the World's Coast and Beneath the Seas*. New York: Holt.

Shiva, Vandana. 1991. *The Violence of the Green Revolution*. Atlantic Highlands, N.J.: Zed Books.

Smith, Katherine R. 1995. "Time to 'Green' U.S. Farm Policy." *Issues in Science and Technology*, Spring, 71–78.

Soil and Water Conservation Society. 1990. *Sustainable Agricultural Systems*. Ankeny, Iowa: Soil and Water Conservation Society.

Solbrig, Otto T., and Dorothy Solbirg. 1994. *So Shall You Reap: Farming and Crops in Human Affairs*. Covelo, Calif.: Island Press.

Soule, Judith D., and Jon Piper. 1992. *Farming in Nature's Image: An Ecological Approach to Agriculture*. Covelo, Calif.: Island Press.

Swanson, Louis E., and Frank B. Clearfield, eds. 1994. *Agricultural Policy and the Environment: Iron Fist or Open Hand?* Ankeny, Iowa: Soil and Water Conservation Society.

Tamsey, Geoff, and Tony Worsley. 1995. *The Food System*. East Haven, Conn.: Earthscan.

Tivy, Joy. 1991. *Agricultural Ecology*. New York: Wiley.

Todd, Nancy J., and John Todd. 1984. *Bioshelters, Ocean Arks, City Farming: Ecology as a Basis for Design*. San Francisco: Sierra Club Books.

Vietmeyer, Noel D. 1986. "Lesser-Known Plants of Potential Use in Agriculture." *Science*, vol. 232, 1379–84.

Weber, Peter. 1994. *Net Loss: Fish, Jobs, and the Marine Environment*. Washington, D.C.: Worldwatch Institute.

Welch, Ross M. 1997. "Toward a 'Greener' Revolution." *Issues in Science and Technology*, Fall, 50–58.

Yang, Linda. 1990. *The City Gardener's Handbook: From Balcony to Backyard*. New York: Random House.

13 / WATER

Readings

Abramovitz, Janet N. 1995. "Freshwater Failures: The Crises on Five Countries." *World Watch*, September/October, 27–35.

Abramovitz, Janet N. 1996. "Sustaining Freshwater Ecosystems." In Lester R. Brown et al., *State of the World 1996*. New York: Norton, 60–77.

Allaby, Michael. 1992. *Water: Its Global Nature*. New York: Facts on File.

Bates, Sarah F., et al. 1993. *Searching out the Headwaters: Change and Rediscovery in Western Water Policy*. Covelo, Calif.: Island Press.

Biswas, Asit K., et al. 1997. *Core and Periphery: A Comprehensive Approach to Middle Eastern Water*. New York: Oxford University Press.

Bolling, David M. 1994. *How to Save a River: A Handbook for Citizen Action*. Covelo, Calif.: Island Press.

Christopher, Thomas. 1994. *Water-Wise Gardening: America's Backyard Revolution*. New York: Simon & Schuster.

Clarke, Robin. 1993. *Water: The International Crisis*. Cambridge, Mass.: MIT Press.

Cone, David. 1995. *A Common Fate: Endangered Salmon and the Fate of the Pacific Northwest*. New York: Holt.

Davis, S., and J. Ogden, eds. 1993. *Everglades: The Ecosystem and Its Restoration*. Delray Beach, Fla.: St. Lucie Press.

Denver Water and American Water Works Administration. 1996. *Xeriscape Plant Guide*. Golden, Colo.: Fulcrum.

Doppelt, Bob, et al. 1993. *Entering the Watershed: A New Approach to Save America's River Ecosystems*. Covelo, Calif.: Island Press.

Dzurik, Andrew A. 1993. *Water Resources Planning*. Lantham, Md.: University Press of America.

Echeverria, John D., et al. 1990. *Rivers at Risk: The Concerned Citizen's Guide to Hydropower*. Covelo, Calif.: Island Press.

El-Ashry, Mohamed, and Diana C. Gibbons, eds. 1988. *Water and the Arid Lands of the Western United States*. New York: Cambridge University Press.

Engleman, Robert, and Pamela LeRoy. 1993. *Sustaining Water: Population and the Future of Renewable Water Supplies*. Washington, D.C.: Population Action International.

Environmental and Energy Study Institute. 1993. *New Policy Directions to Sustain the Nation's Water Resources*. Washington, D.C.: Environmental and Energy Study Institute.

Euphrat, F. D., and B. P. Warkentin. 1994. *A Watershed Assessment Primer*. Springfield, Va.: National Technical Information Service.

Falkenmark, Malin, and Carl Widstand. 1992. "Population and Water Resources: A Delicate Balance." *Population Bulletin*, vol. 47, no. 3, 1–36.

Feldman, David Lewis. 1991. *Water Resources Management: In Search of an Environmental Ethic*. Baltimore: Johns Hopkins University Press.

Gardner, Gary. 1995. "From Oasis to Mirage: The Aquifers That Won't Replenish." *World Watch*, May/June, 30–36, 40–41.

Gleick, Peter H. 1993. *Water in Crisis: A Guide to the World's Fresh Water Resources*. New York: Oxford University Press.

Gleick, Peter H. 1994. "Water, War and Peace in the Middle East." *Environment*, April, 6–41.

Gottlieb, Robert. 1989. *A Life of Its Own: The Politics and Power of Water*. New York: Harcourt Brace Jovanovich.

Graves, William K., ed. 1993. *Water: The Power, Promise, and Turmoil of North America's Fresh Water*. Washington, D.C.: National Geographic Society.

Hillel, Daniel. 1995. *Rivers of Eden: The Struggle for Water and the Quest for Peace in the Middle East*. New York: Oxford University Press.

Hundley, Norris, Jr. 1992. *The Great Thirst: Californians and Water, 1770s–1990s*. Berkeley: University of California Press.

Ives, J. D., and B. Messeric. 1989. *The Himalayan Dilemma: Reconciling Development and Conservation*. London: Routledge.

Joyce, Stephanie. 1997. "Is It Worth a Dam?" *Environmental Health Perspectives*, vol. 105, no. 10, 1050–55.

Kotlyakov, V. M. 1991. "The Aral Sea Basin: A Critical Environmental Zone." *Environment*, vol. 33, no. 1, 4–9, 36–39.

Kourik, Robert. 1992. *Drip Irrigation*. Santa Rosa, Calif.: Edible Publications.

Lee, Kai N. 1993. *Compass and Gyroscope: Integrating Science and Politics for the Environment*. Covelo, Calif.: Island Press.

Matthiessen, Peter. 1992. *Lake Baikal: Siberia's Sacred Sea*. San Francisco: Sierra Club Books.

McCully, Patrick. 1996. *Silenced Rivers: The Ecology and Politics of Large Dams*. London: Zed.

McCutcheon, Sean. 1992. *Electric Rivers: The Story of the James River Project*. New York: Paul.

Micklin, Philip P., and William D. Williams. 1996. *The Aral Sea Basin*. New York: Springer-Verlag.

Munasinghe, Mohan. 1992. *Water Supply and Environmental Management*. Boulder, Colo.: Westview Press.

Myers, Mary Fran, and Gilbert F. White. 1993. "The Challenge of the Mississippi Flood." *Environment*, vol. 35, no. 10, 6–36.

Naiman, Robert J., ed. 1992. *Watershed Management: Balancing Sustainability and Environmental Change*. New York: Springer-Verlag.

National Academy of Sciences. 1992. *Water Transfers in the West: Efficiency, Equity, and the Environment*. Washington, D.C.: National Academy Press.

National Academy of Sciences. 1997. *Valuing Ground Water*. Washington, D.C.: National Academy Press.

Opie, John. 1993. *Ogallala: Water for a Dry Land*. Lincoln: University of Nebraska Press.

Palmer, Tim. 1994. *Lifelines: The Case for River Conservation*. Covelo, Calif.: Island Press.

Pearce, Fred. 1992. *The Dammed: Rivers, Dams, and the Coming World Water Crisis*. London: Bodley Head.

Philippi, Nancy. 1995. "Plugging the Gaps in Flood-Control Policy." *Issues in Science and Technology*, Winter, 71–78.

Poff, N. LeRoy et al. 1997. "The Natural Flow Regime: A Paradigm for River Conservation and Restoration." *BioScience*, vol. 47, no. 11, 769–84.

Postel, Sandra. 1996. *Dividing the Waters: Food Security, Ecosystem Health, and the New Politics of Scarcity*. Washington, D.C.: Worldwatch Institute.

Postel, Sandra. 1996. "Forging a Sustainable Water Strategy." In Lester R. Brown et al., *State of the World 1996*. New York: Norton, 40–59.

Postel, Sandra, et al. 1996. "Human Appropriation of Renewable Fresh Water." *Science*, vol. 271, 785–87.

Prager, Herman. 1993. *Global Marine Environment: Does the Water Planet Have a Future?* Lanham, Md.: University Press of America.

Prillwitz, Marsha, and Larry Farwell. 1994. *Graywater Guide*. Sacramento: California Department of Water Resources.

Qing, Dai. 1994. *Yangtse! Yangtze! Debate Over the Three Gorges Project*. East Haven, Conn.: Earthscan.

Reisner, Marc, and Sara Bates. 1990. *Overtapped Oasis: Reform or Revolution for Western Water*. Covelo, Calif.: Island Press.

Rocky Mountain Institute. 1990. *Catalog of Water-Efficient Technologies for the Urban/Residential Sector*. Snowmass, Colo.: Rocky Mountain Institute.

Rogers, Peter and Peter Lydon, eds. 1994. *Water in the Arab World: Perspectives and Prognoses*. Cambridge, Mass.: Harvard University Press.

Shady, Aly M., et al., eds. 1996. *Management and Development of Major Rivers*. New York: Oxford University Press.

Starr, Joyce R. 1991. "Water Wars." *Foreign Policy*, Spring, 12–45.

Waggoner, Paul E., ed. 1990. *Climate Change and U.S. Water Resources*. New York: Wiley.

Weber, Michael L., and Judith A. Gradwohl. 1995. *The Wealth of Oceans: Environment and Development on Our Ocean Planet*. New York: Norton.

Weber, Peter. 1993. *Abandoned Seas: Reversing the Decline of the Oceans*. Washington, D.C.: Worldwatch Institute.

Wilkinson, Charles F. 1993. *Crossing the Next Meridian: Land, Water, and the Future of the West*. Covelo, Calif.: Island Press.

Worster, Donald. 1985. *Rivers of Empire: Water, Aridity, and the Growth of the American West*. New York: Pantheon.

14 / MINERAL AND SOIL RESOURCES

Readings

Brady, Nyle C. 1989. *The Nature and Properties of Soils*. 10th ed. New York: Macmillan.

Brown, Bruce, and Lane Morgan. 1990. *The Miracle Planet*. Edison, N.J.: W.H. Smith.

Campbell, Stu. 1990. *Let It Rot: The Gardener's Guide to Composting*. Pownal, Vt.: Storey Communications.

Crosson, Pierre. 1997. "Will Erosion Threaten Agricultural Productivity?" *Environment*, vol. 39, no. 8, 4–9, 29–31.

Daily, Gretchen C. 1995. "Restoring Value to the World's Degraded Lands." *Science*, vol. 269, 350–54.

Debus, Keith H. 1990. "Mining with Microbes." *Technology Review*, August/September, 50–57.

Donahue, Roy, et al. 1990. *Soils and Their Management*. 5th ed. Petaluma, Calif.: Inter Print.

Eggert, R. G., ed. 1994. *Mining and the Environment*. Washington, D.C.: Resources for the Future.

Grainger, Alan. 1990. *The Threatening Desert: Controlling Desertification*. East Haven, Conn.: Earthscan.

Greer, Jed. 1993. "The Price of Gold: Environmental Costs of the New Gold Rush." *The Ecologist*, vol. 23, 91–96.

Hillel, Daniel. 1992. *Out of the Earth: Civilization and the Life of the Soil*. New York: Free Press.

Hodges, Carroll A. 1995. "Mineral Resources, Environmental Issues, and Land Use." *Science*, vol. 268, 1305–11.

Huston, Michael. 1993. "Biological Diversity, Soils, and Economics." *Science*, vol. 262, 1676–80.

Little, Charles E. 1987. *Green Fields Forever: The Conservation Tillage Revolution in America.* Covelo, Calif.: Island Press.

Mainguet, Monique. 1991. *Desertification: Natural Background and Human Mismanagement.* New York: Springer-Verlag.

May, John. 1989. *The Greenpeace Book of Antarctica.* New York: Doubleday.

Mollison, Bill. 1990. *Permaculture.* Covelo, Calif.: Island Press.

National Academy of Sciences. 1986. *Soil Conservation.* 2 vols. Washington, D.C.: National Academy Press.

National Academy of Sciences. 1990. *Our Seabed Frontier: Challenges and Choices.* Washington, D.C.: National Academy Press.

National Academy of Sciences. 1993. *Soil and Water Quality: An Agenda for Agriculture.* Washington, D.C.: National Academy Press.

Nelson, Ridley. 1988. *Dryland Management: The Desertification Problem.* Washington, D.C.: World Bank.

Paddock, Joe, et al. 1987. *Soil and Survival: Land Stewardship and the Future of American Agriculture.* San Francisco: Sierra Club Books.

Pimentel, David, ed. 1993. *World Soil Erosion and Conservation.* New York: Cambridge University Press.

Pimentel, David, et al. 1995. "Environmental and Economic Costs of Soil Erosion and Conservation Benefits." *Science,* vol. 267, 1117–23.

Ripley, Earle A., et al. 1996. *Environmental Effects of Mining.* Delray Beach, Fla.: St. Lucie Press.

Suter, Keith. 1991. *Antarctica: Private Property or Public Heritage?* Atlantic Highlands, N.J.: Zed Books.

Tapp, B. A., and J. R. Watkins. 1989. *Energy and Mineral Resource Systems: An Introduction.* New York: Cambridge University Press.

Tato, Kebede, and Hans Hurni, eds. 1992. *Soil Conservation for Survival.* Ankeny, Iowa: Soil and Water Conservation Society.

Wild, Alan. 1993. *Soils and the Environment: An Introduction.* New York: Cambridge University Press.

Wilson, G. F., et al. 1986. *The Soul of the Soil: A Guide to Ecological Soil Management.* 2d ed. Montreal: Gaia Services.

Young, John E. 1992. *Mining the Earth..* Washington, D.C.: Worldwatch Institute.

Young, John E., and Aaron Sachs. 1994. *The Next Efficiency Revolution: Creating a Sustainable Materials Revolution.* Washington, D.C.: Worldwatch Institute.

Youngquist, Walter. 1990. *Mineral Resources and the Destinies of Nations.* Portland, Oreg.: National Book.

15 / NONRENEWABLE ENERGY RESOURCES

Readings

Ahearne, John F. 1993. "The Future of Nuclear Power." *American Scientist,* vol. 81, 24–35.

Ballonoff, Paul. 1997. *Energy: The Never-Ending Crisis.* Washington, D.C.: Cato Institute.

Beck, Peter. 1994. *Prospects and Strategies for Nuclear Power: Global Boon or Dangerous Diversion?* East Haven, Conn.: Earthscan.

Campbell, Colin J., and Jean H. Laherrère. 1998. "The End of Cheap Oil." *Scientific American,* March, 78–83.

Cannon, James S. 1993. *Paving the Way to Natural Gas Vehicles.* New York: INFORM.

Chernousenko, Vladimir M. 1991. *Chernobyl: Insight from the Inside.* New York: Springer-Verlag.

Cohen, Bernard L. 1990. *The Nuclear Energy Option: An Alternative for the 90s.* New York: Plenum.

Eisenbud, Merril, and Thomas Gesell. 1997. *Environmental Radioactivity: From Natural, Industrial, and Military Sources.* San Diego, Calif.: Academic Press.

Flynn, James, et al. 1992. "Time to Rethink Nuclear Waste Storage." *Issues in Science and Technology,* Summer, 42–48.

Ford, Daniel F. 1986. *Meltdown.* New York: Simon & Schuster.

George, Richard L. 1998. "Mining for Oil." *Scientific American,* March, 84–88.

Gershey, Edward L., et al. 1992. *Low-Level Radioactive Waste: From Cradle to Grave.* New York: Van Nostrand Reinhold.

Golay, Michael W., and Neil E. Todreas. 1990. "Advanced Light-Water Reactors," *Scientific American,* April, 82–89.

Gould, Jay M., and Benjamin A. Goldman. 1991. *Deadly Deceit: Low-Level Radiation, High-Level Cover-Up.* New York: Four Walls Eight Windows.

Herman, Robin. 1990. *Fusion: The Search for Endless Energy.* New York: Cambridge University Press.

Hollister, Charles D., and Steven Nadis. 1998. "Burial of Radioactive Waste Under the Seabed." *Scientific American,* January, 60–65.

Hughes, Barry B., et al. 1985. *Energy in the Global Arena: Actors, Values, Policies, and Futures.* Durham, N.C.: Duke University Press.

Huizenga, John R. 1992. *Cold Fusion: The Scientific Fiasco of the Century.* Rochester, N.Y.: University of Rochester Press.

Jacop, Gerald. 1990. *Sight Unseen: The Politics of Siting a Nuclear Waste Repository.* Pittsburgh: University of Pittsburgh Press.

Lenssen, Nicholas. 1991. *Nuclear Waste: The Problem That Won't Go Away.* Washington, D.C.: Worldwatch Institute.

Lenssen, Nicholas. 1993. "All the Coal in China." *World Watch,* March/April, 22–28.

Lenssen, Nicholas, and Christopher Flavin. 1996. "Meltdown." *World Watch,* May/June, 23–31.

Lovins, Amory B. 1986. "The Origins of the Nuclear Power Fiasco." *Energy Policy Studies,* vol. 3, 7–34.

Lu, Yingzhong. 1993. *Fueling One Billion: An Insider's Story of Chinese Energy Policy Development.* Washington, D.C.: In Depth Books.

MacKenzie, James J. 1996. "Heading Off the Permanent Oil Crisis." *Issues in Science and Technology,* Summer, 48–54.

Makhijani, Arjun, et al., eds. 1995. *Nuclear Wastelands: A Global Guidebook to Nuclear Weapons Production and Its Health and Environmental Effects.* Cambridge, Mass.: MIT Press.

May, John. 1990. *The Greenpeace Book of the Nuclear Age.* New York: Pantheon.

Murray, Raymond L., and Judith Powell, eds. 1994. *Understanding Radioactive Waste.* 4th ed. Columbus, Ohio: Battelle Press.

National Academy of Sciences. 1991. *Nuclear Power: Technical and Institutional Options for the Future.* Washington, D.C.: National Academy Press.

National Academy of Sciences. 1991. *Undiscovered Oil and Gas Resources.* Washington, D.C.: National Academy Press.

National Academy of Sciences. 1995. *Coal: Energy for the Future.* Washington, D.C.: National Academy Press.

Office of Technology Assessment. 1991. *Complex Cleanup: The Environmental Legacy of Nuclear Weapons Production.* Washington, D.C.: U.S. Government Printing Office.

Oppenheimer, Ernest J. 1990. *Natural Gas, the Best Energy Choice.* New York: Pen & Podium.

President's Commission on the Accident at Three Mile Island. 1979. *Report of the President's Commission on the Accident at Three Mile Island.* Washington, D.C.: U.S. Government Printing Office.

Public Citizen. 1987. *The Price–Anderson Act: Multi–Billion-Dollar Nuclear Subsidy.* Washington, D.C.: Public Citizen.

Read, Piers Paul. 1993. *Ablaze: The Story of Chernobyl.* New York: Random House.

Rhodes, Richard. 1993. *Nuclear Renewal: Common Sense About Energy.* New York: Viking/Penguin.

Savchenko, V. K. 1995. *The Ecology of the Chernobyl Catastrophe.* Pearl River, N.Y.: Parthenon.

Shea, Cynthia Pollock. 1989. "Decommissioning Nuclear Plants: Breaking Up Is Hard to Do." *World Watch,* July/August, 10–16.

Solar Energy Research Institute. 1991. *Compressed and Liquefied Natural Gas: Just the Facts.* Golden, Colo.: Solar Energy Research Institute.

Squillace, Mark. 1990. *Strip Mining Handbook.* Washington, D.C.: Friends of the Earth.

Sthcherbak, Yuri M. 1996. "Ten Years of the Chernobyl Era." *Scientific American,* April, 44–49.

Strohmeyer, John K. 1993. *Extreme Conditions: Big Oil and the Transformation of Alaska.* New York: Simon & Schuster.

Toke, David. 1995. *Energy and Environment: The Political and Economic Debate.* Boulder, Colo.: Westview Press.

Union of Concerned Scientists. 1990. *Safety Second: The NRC and America's Nuclear Power Plants.* Bloomington: Indiana University Press.

Whipple, Chris G. 1996. "Can Nuclear Waste Be Stored Safely at Yucca Mountain?" *Scientific American,* June, 72–79.

Wolfe, Bertram. 1996. "Why Environmentalists Should Promote Nuclear Energy." *Issues in Science and Technology,* Summer, 55–60.

Yergin, Daniel. 1990. *The Prize: The Epic Quest for Oil, Money, and Power.* New York: Simon & Schuster.

Zorpette, Glenn. 1996. "Hanford's Nuclear Wasteland." *Scientific American,* May, 88–97.

16 / ENERGY EFFICIENCY AND RENEWABLE ENERGY RESOURCES

Readings

Alliance to Save Energy et al. 1991. *America's Energy Choices: Investing in a Strong Economy and a Clean Environment. (Executive Summary).* Cambridge, Mass: Union of Concerned Scientists.

American Council for an Energy Efficient Economy. 1991. *Energy Efficiency and Environment: Forging the Link.* Washington, D.C.: American Council for an Energy Efficient Economy.

American Council for an Energy Efficient Economy. Annual. *The Most Energy-Efficient Appliances.* Washington, D.C.: American Council for an Energy Efficient Economy.

American Solar Energy Society. 1993. *Economics of Solar Energy Technologies.* Boulder, Colo.: American Solar Energy Society.

Anderson, Bruce. 1990. *Solar Building Architecture.* Cambridge, Mass.: MIT Press.

Anderson, Victor. 1993. *Energy Efficiency Policies.* New York: Routledge.

Berger, John. 1997. *Charging Ahead: The Business of Renewable Energy and What It Means to America.* New York: Holt.

Berman, Daniel M., and John T. O'Connor. 1996. *Who Owns the Sun? People, Politics, and the Struggle for a Solar Economy.* River Junction, Vt.: Chelsea Green.

Blackburn, John O. 1987. *The Renewable Energy Alternative: How the United States and the World Can Prosper Without Nuclear Energy or Coal.* Durham, N.C.: Duke University Press.

Brower, Michael. 1992. *Cool Energy: Renewable Solution to Environmental Problems.* 2d ed. Cambridge, Mass.: MIT Press.

Butti, Ken, and John Perlin. 1980. *A Golden Thread: 2500 Years of Solar Architecture.* Palo Alto, Calif.: Cheshire Press.

Carless, Jennifer. 1993. *Renewable Energy: A Concise Guide to Green Alternatives.* New York: Walker.

Cole, Nancy, and P. S. Skerrett. 1995. *Renewables Are Ready: People Creating Energy Solutions.* White River Junction, Vt.: Chelsea Green.

Colorado Energy Research Institute. 1976. *Net Energy Analysis: An Energy Balance Study of Fossil Fuel Resources.* Golden, Colo.: Colorado Energy Research Institute.

Dower, Roger et al. 1997. *Frontiers of Sustainability: Environmentally Sound Agriculture, Forestry, Transportation, and Power Production.* Covelo, Calif.: Island Press.

Dunn, Seth. 1997. "Power of Choice." *World Watch*, September/October, 30–38.

Echeverria, John, et al. 1989. *Rivers at Risk: The Concerned Citizen's Guide to Hydropower.* Covelo, Calif.: Island Press.

Evans, Poppy. 1997. *The Complete Guide to Eco-Friendly Design.* Cincinnati, Ohio: North Light Books.

Feldman, David L., ed. 1996. *The Energy Crisis: Unresolved Issues and Enduring Legacies.* Baltimore, Md.: Johns Hopkins University Press.

Field, Frank R. III, and Joel P. Clark. 1997. "A Practical Road to Lightweight Cars." *Technology Review*, January, 28–36.

Flavin, Christopher. 1996. "Power Shock: The Next Energy Revolution." *World Watch*, January/February, 10–19.

Flavin, Christopher, and Seth Dunn. 1997. *Rising Sun, Gathering Winds: Policies to Stabilize the Climate and Strengthen Economies.* Washington, D.C.: Worldwatch Institute.

Flavin, Christopher, and Nicholas Lenssen. 1994. *Power Surge: Guide to the Coming Energy Revolution.* New York: Norton.

Gever, John, et al. 1991. *Beyond Oil: The Threat to Food and Fuel in Coming Decades.* Boulder: University of Colorado Press.

Giampietro, Mario et al. 1998. "Feasibility of Large-Scale Biofuel Production." *BioScience*, vol. 47, no. 9, 587–600.

Gipe, Paul. 1995. *Wind Energy Comes of Age.* New York: Wiley.

Grady, Wayne. 1993. *Greenhome: Planning and Building the Environmentally Advanced House.* Charlotte, Vt.: Camden House Books.

Harland, Edward. 1994. *Eco-Renovation: The Ecological Home Improvement Guide.* White River Junction, Vt.: Chelsea Green.

Hart, David. 1997. *Hydrogen Power.* London: FT Energy Publishing.

Hubbard, Harold M. 1991. "The Real Cost of Energy." *Scientific American*, April, 36–42.

Kachadorian, James. 1997. *The Passive Solar House: Using solar Design to Heat and Cool Your Home.* White River Junction, Vt.: Chelsea Green.

Kozloff, Keith, and Roger C. Dower. 1993. *A New Power Base: Renewable Energy Policies for the Nineties and Beyond.* Washington, D.C.: World Resources Institute.

Krupnick, Alan. 1990. *The Environmental Costs of Energy: A Framework for Estimation.* Washington, D.C.: Resources for the Future.

Lenssen, Nicholas. 1992. *Empowering Development: The New Energy Equation.* Washington, D.C.: Worldwatch Institute.

Lovins, Amory B. 1977. *Soft Energy Paths.* Cambridge, Mass.: Ballinger.

Lovins, Amory B. 1990. *The Negawatt Revolution.* Snowmass, Colo.: Rocky Mountain Institute.

Lovins, Amory B., and L. Hunter Lovins. 1995. "Reinventing the Wheels." *Atlantic Monthly*, January, 75–94.

MacKenzie, James J. 1994. *The Keys to the Car: Electric and Hydrogen Vehicles for the 21st Century.* Washington, D.C.: World Resources Institute.

McKeown, Walter. 1991. *Death of the Oil Age and the Birth of Hydrogen America.* San Francisco: Wild Bamboo Press.

Munson, Richard, and Tina Kaarsberg. 1998. "Unleashing Innovation in Electricity Generation." *Issues in Science and Technology*, Spring, 51–58.

Nansen, Ralph. 1995. *Sun Power: The Global Solution for the Coming Energy Crisis.* Ocean Shores, Wash.: Ocean Press.

National Academy of Sciences. 1990. *Energy: Production, Consumption, and Consequences.* Washington, D.C.: National Academy Press.

National Academy of Sciences. 1992. *Automotive Fuel Efficiency: How Far Can We Go?* Washington, D.C.: National Academy Press.

Odum, Howard T. 1995. *Environmental Accounting: Energy and Environmental Decision Making.* New York: Wiley.

Office of Technology Assessment. 1991. *Improving Automobile Fuel Economy: New Standards, New Approaches.* Washington, D.C.: U.S. Government Printing Office.

Office of Technology Assessment. 1992. *Building Energy Efficiency.* Washington, D.C.: U.S. Government Printing Office.

Office of Technology Assessment. 1992. *Fueling Development: Energy Technologies for Developing Countries.* Washington, D.C.: U.S. Government Printing Office.

Ogden, Joan M., and Robert H. Williams. 1989. *Solar Hydrogen: Moving Beyond Fossil Fuels.* Washington, D.C.: World Resources Institute.

Oppenheimer, Michael, and Robert H. Boyle. 1990. *Dead Heat: The Race Against the Greenhouse Effect.* New York: Basic Books.

Perrin, Noel. 1994. *Life with an Electric Car.* San Francisco: Sierra Club Books.

Pimentel, David, et al. 1994. "Environmental and Social Costs of Biomass Energy," *BioScience*, February, 89–93.

Potts, Michael. 1993. *The Independent Home: Living Well with Power from the Sun, Wind, and Water.* White River Junction, Vt.: Chelsea Green.

Rifkin, Jeremy. 1989. *Entropy: Into the Greenhouse World: A New World View.* New York: Bantam.

Robbins, Elaine, and Stephen Beers. 1998. "Lights Out: The Case for Energy Conservation: It Works, so Why Aren't We Using It?" *E Magazine*, 36–41.

Rocky Mountain Institute. 1995. *Community Energy Workbook.* Snowmass, Colo.: Rocky Mountain Institute.

Rocky Mountain Institute. 1995. *Homemade Energy: How to Save Energy and Dollars in Your Home.* Snowmass, Colo.: Rocky Mountain Institute.

Rocky Mountain Institute. 1997. *Primer on Sustainable Building.* Snowmass, Colo.: Rocky Mountain Institute.

Roodman, David M., and Nicholas Lenssen. 1995. *A Building Revolution: How Ecology and Health Concerns Are Transforming Construction.* Washington, D.C.: Worldwatch Institute.

Rosen, Harold A., and Deborah R. Castleman. 1997. "Flywheels in Hybrid Vehicles." *Scientific American*, October, 75–77.

Schipper, Lee, et al. 1992. *Energy Efficiency and Human Activity: Past Trends, Future Prospects.* New York: Cambridge University Press.

Sierra Club. 1991. *Kick the Oil Habit: Choosing a Safe Energy Future for America.* San Francisco: Sierra Club Books.

Solar Energy Research and Education Foundation. 1991. *Renewable Energy: Facts and Figures.* Washington, D.C.: Solar Energy Research and Education Foundation.

Sperling, Daniel. 1996. "The Case for Electric Vehicles." *Scientific American*, November, 54–59.

Stalahlkopf, Karl. 1995. *Powering the Future.* Palo Alto, Calif.: Electric Power Research Institute.

Steen, Athena Swentzell, et al. 1994. *The Straw Bale House.* White River Junction, Vt.: Chelsea Green.

St. John, Andrew, ed. 1992. *The Sourcebook for Sustainable Design.* Boston: Boston Society of Architects.

Strong, Steven, and William G. Scheller. 1993. *The Solar Electric House.* Washington, D.C.: Sustainability Press.

Swift, A., et al. 1990. "Commercialization of Solar Ponds." *Solar Energy*, July/August, 17–18.

Tenenbaum, David. 1995. "Tapping the Fire Down Below." *Technology Review*, January, 38–47.

Underground Space Center, University of Minnesota. 1979. *Earth-Sheltered Housing Design.* New York: Van Nostrand Reinhold.

U.S. Department of Energy. 1995. *Tips for Energy Savers.* Washington, D.C.: U.S. Department of Energy. Also see web site: www.ener.doe.gov

Vale, Brenda, and Robert Vale. 1991. *Green Architecture.* Boston: Little, Brown.

Weisman, Alan. 1998. *Gaviotas: A Village to Reinvent the World.* White River Junction, Vt.: Chelsea Green.

Wells, Malcolm. 1991. *How to Build an Underground House.* Brewster, Mass.: Malcolm Wells.

Wilkinson, Stephan. 1993. "The Automobile and the Environment: Our Next Car?" *Audubon*, May/June, 56–67.

Williams, Robert H. 1994. "The Clean Machine." *Technology Review*, April, 21–30.

Wouk, Victor. 1997. "Hybrid Electric Vehicles." *Scientific American*, October, 70–74.

Yeang, Ken. 1995. *Designing with Nature: The Ecological Basis for Architectural Design.* New York: McGraw-Hill.

Zweibel, Ken. 1990. *Harnessing Solar Power: The Challenge of Photovoltaics.* New York: Plenum.

17 / RISK, TOXICOLOGY, AND HUMAN HEALTH

Readings

American Council on Science and Health. 1997. *Cigarettes: What the Warning Label Doesn't Tell You.* Amherst, N.Y.: Prometheus.

Andur, M., et al., eds. 1991. *Casarett & Doull's Toxicology: The Basic Science of Poisons.* New York: Pergamon.

Baggs, Sydney, and Joan C. Baggs. 1997. *The Healthy House: Creating a Safe, Healthy, and Environmentally Friendly Home.* New York: HarperCollins.

Bartecchi, Carl E., et al. 1995. "The Global Tobacco Epidemic." *Scientific American,* May, 44–47.

Bates, David V. 1995. *Environmental Health Risks and Public Policy: Decision Making in Free Societies.* Seattle: University of Washington Press.

Bates, Roger. 1997. *What Risk?* Newton, Mass.: Butterworth-Heinemann.

Bernarde, Melvin A. 1989. *Our Precarious Habitat: Fifteen Years Later.* New York: Wiley.

Blumenthal, Daniel S., and James Ruttenberg, eds. 1995. *Introduction to Environmental Health.* 2d ed. New York: Springer-Verlag.

Brown, Phil. 1993. "When the Public Knows Better: Popular Epidemiology." *Environment,* October, 16–40.

Chivian, Eric, et al. 1993. *Critical Condition: Human Health and the Environment.* Cambridge, Mass.: MIT Press.

Clarke, Lee. 1989. *Acceptable Risk? Making Decisions in a Toxic Environment.* Berkeley: University of California Press.

Colborn, Theo, et al. 1996. *Our Stolen Future.* New York: Dutton.

Coogan, Patricia, and Terry Greene. 1992. *Environment and Health: How to Investigate Community Health Problems.* Boston: JSI Center for Environmental Health Studies.

Dadd-Redalla, Debra. 1994. *Sustaining the Earth: Choosing Consumer Products That Are Environmentally Safe for You.* New York: Hearst Books.

Davies, Clarence, ed. 1996. *Comparing Environmental Risks: Tools for Setting Government Guidelines.* Washington, D.C.: Resources for the Future.

Desowitz, Robert S. 1997. *Who Gave Pinta to the Santa Maria: Torrid Diseases in a Temperate World.* New York: Norton.

Dixon, Bernard. 1994. *Power Unseen: How Microbes Rule the World.* San Francisco: W.H. Freeman.

Drlica, Karl L. 1996. *Double-Edged Sword: The Promises and Risks of the Genetic Revolution.* New York: Addison-Wesley.

Eisenbud, Merril, and Thomas Gesell. 1997. *Environmental Radioactivity: From Natural, Industrial, and Military Sources.* San Diego, Calif.: Academic Press.

Emsley, John. 1994. *The Consumers' Good Chemical Guide: A Jargon-Free Guide to the Chemicals of Everyday Life.* New York: W.H. Freeman.

Environmental Protection Agency. 1987. *Unfinished Business: A Comparative Assessment of Environmental Problems.* Washington, D.C.: Environmental Protection Agency.

Environmental Protection Agency. 1990. *Reducing Risk: Setting Priorities and Strategies for Environmental Protection.* Washington, D.C.: Environmental Protection Agency.

Environmental Protection Agency. 1992. *Respiratory Health Effects of Passive Smoking: Lung Cancer and Other Disorders.* Washington, D.C.: Environmental Protection Agency.

Ewald, Paul W. 1993. "The Evolution of Virulence." *Scientific American,* vol. 268, no. 4, 56–62.

Finkel, Adam M. 1996. "Who's Exaggerating?" *Discover,* May, 48–55.

Fisher, Jeffrey A. 1994. *The Plague Makers: How We Are Creating Catastrophic New Epidemics and What We Must Do to Avert Them.* New York: Simon & Schuster.

Foster, Kenneth R., et al. 1993. *Phantom Risk: Scientific Inference and the Law.* Cambridge, Mass.: MIT Press.

Fox, Michael W. 1992. *Superpigs and Wondercorn: The Brave New World of Biotechnology and Where It All May Lead.* New York: Lyons & Burford.

Fox, Nicols. 1997. *Spoiled: The Dangerous Truth About a Food Chain Gone Haywire.* New York: Basic Books.

Francis, B. M. 1994. *Toxic Substances in the Environment.* New York: Wiley.

Freedman, B. 1994. *Environmental Ecology.* New York: Academic Press.

Freudenburg, William R. 1988. "Perceived Risk, Real Risk: Social Science and the Art of Probabilistic Risk Assessment." *Science,* vol. 242, 44–49.

Freudenthal, Ralph I., and Susan Loy Freudenthal. 1989. *What You Need to Know to Live with Chemicals.* Green Falls, Conn.: Hill & Garnett.

Garrett, Laurie. 1994. *The Coming Plague: Newly Emerging Diseases in a World Out of Balance.* New York: Farrar, Straus & Giroux.

Gillis, Anna Maria. 1993. "Toxicity Tests Minus Animals?" *BioScience,* vol. 43, no. 3, 137–40.

Gofman, John W. 1990. *Radiation-Induced Cancer from Low-Dose Exposure.* San Francisco: Committee for Nuclear Responsibility.

Gould, Jay, and Benjamin Goldman. 1990. *Deadly Deceit: Low-Level Radiation—High-Level Cover-Up.* New York: Four Walls Eight Windows.

Graham, John D., and Jonathan Baert Weiner, eds. 1995. *Risk vs. Risk: Tradeoffs in Protecting Health and the Environment.* Cambridge, Mass.: Harvard University Press.

Harris, John. 1992. *Wonderwoman and Superman: The Ethics of Human Biotechnology.* New York: Oxford University Press.

Harte, John, et al. 1992. *Toxics A to Z: A Guide to Everyday Pollution Hazards.* Berkeley: University of California Press.

Hileman, Bette. 1995. "Views Differ Sharply Over Benefits, Risks, of Agricultural Biotechnology." *Chemical and Engineering News,* August 21, 8–17.

Imperato, P. J., and Greg Mitchell. 1985. *Acceptable Risks.* New York: Viking Press.

Kitcher, Philip. 1996. *The Lives to Come: The Genetic Revolution and Human Possibilities.* New York: Simon & Schuster.

Knudson, Mary. 1998. "The Hunt Is On for New Ways to Overcome Bacterial Resistance." *Technology Review,* January/February, 22–29.

Krantz, Les. 1993. *What the Odds Are.* New York: HarperPerennial.

Lappé, Marc. 1991. *Chemical Deception: The Toxic Threat to Health and Environment.* San Francisco: Sierra Club Books.

Lappé, Marc. 1995. *The Evolving Threat of Drug-Resistant Disease.* San Francisco: Sierra Club Books.

Laudan, Larry. 1997. *Danger Ahead: The Risks You Really Face on Life's Highway.* New York: Wiley.

Levy, Stuart B. 1998. "The Challenge of Antibiotic Resistance." *Scientific American,* March, 46–53.

Lewis, H. W. 1990. *Technological Risk.* New York: Norton.

Louvar, Joseph. 1997. *Health and Environmental Risk Analysis.* Englewood Cliffs, N.J.: Prentice Hall.

Mayo, Deborah G., and Rachelle D. Hollander, eds. 1992. *Acceptable Evidence: Science and Values in Risk Management.* New York: Oxford University Press.

Merrell, Paul, and Carol Van Strum. 1990. "Negligible Risk or Premeditated Murder?" *Journal of Pesticide Reform,* vol. 10, Spring, 20–22.

Meyers, Norman. 1997. "Development, Environment, and Health." *Environment and Development Economics,* vol. 2, no. 1, 367–70.

Miller, Robert V. 1998. "Bacterial Gene Swapping in Nature." *Scientific American,* January, 67–71.

Misch, Ann. 1994. "Assessing Environmental Health Risks." In Lester R. Brown et al,. *State of the World 1994.* New York: Norton, 117–36.

Moeller, Dade W. 1997. *Environmental Health.* Revised ed. Cambridge, Mass.: Harvard University Press.

Montague, Peter, ed. *Rachel's Environment and Health Weekly.* (P.O. Box 5036, Annapolis, MD 21403-7036). (A two-page weekly newsletter explaining health, risk, and environmental issues in an easily understandable manner.)

Morgan, M. Granger. 1993. "Risk Analysis and Management." *Scientific American,* July, 32–41.

Motavalli, Jim. 1998. "The Trouble with Meat: Dangerous Diseases Emerge in a Planet Eating High on the Food Chain." *E Magazine,* May/June, 29–35.

Nadakavukaren, Anne. 1995. *Our Global Environment: A Health Perspective.* 4th ed. New York: Waveland Press.

National Academy of Sciences. 1990. *Health Effects of Exposure to Low Levels of Ionizing Radiation.* Washington, D.C.: National Academy Press.

National Academy of Sciences. 1991. *Malaria: Obstacles and Opportunities.* Washington, D.C.: National Academy Press.

National Academy of Sciences. 1992. *Eat for Life: The Food and Nutrition Board's Guide to Reducing Your Risk of Chronic Disease.* Washington, D.C.: National Academy Press.

National Academy of Sciences. 1993. *Issues in Risk Assessment.* Washington, D.C.: National Academy Press.

National Academy of Sciences. 1993. *Science and Judgment in Risk Assessment.* Washington, D.C.: National Academy Press.

National Academy of Sciences. 1996. *Understanding Risk: Informing Decisions in a Democratic Society.* Washington, D.C.: National Academy Press.

Neese, Randolph M., and George C. Williams. 1994. *Why People Get Sick.* New York: Random House.

Ottoboni, M. Alice. 1991. *The Dose Makes the Poison: A Plain-Language Guide to Toxicology.* 2d ed. New York: Van Nostrand Reinhold.

Perrow, Charles. 1985. *Normal Accidents: Living with High-Risk Technologies.* New York: Basic Books.

Piller, Charles. 1991. *The Fail-Safe Society.* New York: Basic Books.

Pimentel, David, et al. 1989. "Benefits and Risks of Genetic Engineering in Agriculture." *BioScience,* vol. 39, no. 9, 606–14.

Platt, Anne. 1996. "Water-Borne Killers." *World Watch,* March/April, 28–35.

Platt, Anne. 1996. *Infecting Ourselves: How Environmental and Social Disruptions Trigger Disease.* Washington, D.C.: Worldwatch Institute.

Preston, Richard. 1994. *The Hot Zone.* New York: Random House.

Real, Leslie A. 1996. "Sustainability and the Challenge of Infectious Disease." *BioScience,* vol. 46, no. 2, 88–97.

Regenstein, Lewis G. 1993. *Cleaning Up America the Poisoned.* Washington, D.C.: Acropolis Books.

Rhodes, Richard. 1997. *Deadly Feasts: Tracking the Secrets of a Terrifying New Plague.* New York: Simon & Schuster.

Rifkin, Jeremy. 1998. *The Biotech Century: Harnessing the Gene and Remaking the World.* New York: Tarcher/Plenum.

Rissler, Jane, and Margaret Mellon. 1996. *The Ecological Risks of Engineered Crops.* Cambridge, Mass.: MIT Press.

Rockett, Ian R. H. 1994. "Population and Health: An Introduction to Epidemiology." *Population Bulletin*, vol. 49, no. 3, 1–47.

Rodricks, Joseph V. 1992. *Calculated Risks: Understanding the Toxicity and Human Health Risks of Chemicals in the Environment.* New York: Cambridge University Press.

Shrader-Frechette, K. S. 1991. *Risk and Rationality.* Berkeley: University of California Press.

Steingraber, Sandra. 1997. *Living Downstream: An Ecologist Looks at Cancer and the Environment.* Reading, Mass.: Addison-Wesley.

Suzuki, David, and Peter Knudtson. 1989. *Genethics: The Clash Between the New Genetics and Human Values.* Cambridge, Mass.: Harvard University Press.

Taube, Gary. 1994. "Fields of Fear." *The Atlantic Monthly*, November, 94–108.

Taube, Gary. 1998. "Malarial Dreams." *Discover*, March, 109–16.

Tenner, Ward. 1996. *Why Things Bite Back: Technology and the Revenge of Unintended Consequences.* New York: Knopf.

Tokar, Brian. 1996. "Biotechnology vs. Biodiversity." *Wild Earth*, Spring, 50–55.

Trichopoulos, Dimitrios, et al. 1996. "What Causes Cancer?" *Scientific American*, September, 80–96.

United Nations. 1990. *Radiation: Doses, Effects, and Risks.* New York: United Nations Publications.

U.S. Department of Health and Human Services. Annual. *The Health Consequences of Smoking.* Rockville, Md.: U.S. Department of Health and Human Services.

Whelan, E. M. 1993. *Toxic Terror: The Truth Behind the Cancer Scares.* Buffalo, N.Y.: Prometheus.

Wildavsky, Aaron. 1995. *But Is It True? A Citizen's Guide to Environmental Health and Safety Issues.* Cambridge, Mass.: Harvard University Press.

Wilson, Richard, and E. A. C. Crouch. 1987. "Risk Assessment and Comparisons: An Introduction." *Science*, vol. 236, 267–70.

Wirth, Dyann F., and Jacqueline Cattani. 1997. "Winning the War Against Malaria." *Technology Review*, August/September, 52–61.

World Health Organization (WHO). Annual. *The World Health Report.* Geneva, Switzerland: WHO.

World Resources Institute. 1994. *Human and Ecosystem Health.* Washington, D.C.: World Resources Institute.

18 / AIR POLLUTION

Readings

Alleman, James E., and Brooke T. Mossman. 1997. "Asbestos Revisited." *Scientific American*, July, 70–75.

Bennett, Michael J. 1991. *The Asbestos Racket.* Bellevue, Wash.: Free Enterprise Press.

Bridgman, Howard. 1991. *Global Air Pollution: Problems for the 1990s.* New York: Belhaven Press.

Brookins, Douglas G. 1990. *The Indoor Radon Problem.* New York: Columbia University Press.

Brouder, Paul. 1985. *Outrageous Misconduct: The Asbestos Industry on Trial.* New York: Pantheon.

Bryner, Gary. 1992. *Blue Skies, Green Politics: The Clean Air Act of 1990.* Washington, D.C.: Congressional Quarterly Press.

Coffel, Steve, and Karyn Feiden. 1991. *Indoor Pollution.* New York: Random House.

Cohen, Bernie. 1988. *Radon: A Homeowner's Guide to Detection and Control.* Mt. Vernon, N.Y.: Consumer Report Books.

Cole, Leonard D. 1994. *Element of Risk: The Politics of Radon.* New York: Oxford University Press.

Elson, Derek. 1987. *Atmospheric Pollution: Causes, Effects, and Control Policies.* Cambridge, Mass.: Basil Blackwell.

Environmental Protection Agency. 1992. *What You Can Do to Reduce Air Pollution.* Washington, D.C.: Environmental Protection Agency.

Harrington, Winston, et al. 1995. "Using Economic Incentives to Reduce Auto Pollution." *Issues in Science and Technology*, Winter, 26–32.

Hedin, Lars O., and Gene E. Likens. 1996. "Atmospheric Dust and Acid Rain." *Scientific American*, December, 88–92.

Lents, James M., and William J. Kelly. 1993. "Clearing the Air in Los Angeles." *Scientific American*, October, 32–39.

Leslie, G. B., and F. W. Linau. 1994. *Indoor Air Pollution.* New York: Cambridge University Press.

MacKenzie, James J., and Mohamed T. El-Ashry. 1990. *Air Pollution's Toll on Forests and Crops.* New Haven, Conn.: Yale University Press.

McCormick, John. 1997. *Acid Earth: The Politics of Acid Pollution.* London: Earthscan.

Mossman, B. T., et al. 1990. "Asbestos: Scientific Developments and Implications for Public Policy." *Science*, vol. 251, 247–300.

Munton, Don. 1998. "Dispelling the Myths of the Acid Rain Story." *Environment*, vol. 40, no. 6, 4–7, 29–34.

National Academy of Sciences. 1995. *Health Effects of Exposure to Radon: Time for Reassessment?* Washington, D.C.: National Academy Press.

Smith, William H. 1991. "Air Pollution and Forest Damage." *Chemical and Engineering News*, November 11, 30–43.

Soroos, Marvin S. 1997. *The Endangered Atmosphere: Preserving a Global Commons.* Columbia: University of South Carolina Press.

Steidlmeier, Paul. 1993. "The Morality of Pollution Permits." *Environmental Ethics*, vol. 15, 133–50.

Wagner, Travis. 1993. *In Our Backyard: A Guide to Understanding Pollution and Its Effects.* New York: Van Nostrand Reinhold.

Warde, John. 1997. *The Healthy Home Handbook: All You Need to Know to Rid Your Home of Health and Safety Hazards.* New York: Times Books.

19 / GLOBAL WARMING AND OZONE LOSS

Readings

American Forestry Association. 1993. *Forests and Global Warming.* Vol. 1. Washington, D.C.: American Forestry Association.

Ayres, Ed, and Hilary French. 1996. "The Refrigerator Revolution." *World Watch*, September/October, 15–21.

Balling, Robert C., Jr. 1992. *The Heated Debate: Greenhouse Predictions Versus Climate Reality.* San Francisco: Pacific Research Institute.

Bates, Albert K. 1990. *Climate in Crisis: The Greenhouse Effect and What We Can Do.* Summertown, Tenn.: Book Publishing Company.

Benedick, Richard Eliot. 1991. *Ozone Diplomacy: New Directions in Safeguarding the Planet.* Cambridge, Mass.: Harvard University Press.

Berks, John W., et al., eds. 1993. *The Chemistry of the Atmosphere: Its Impact on Global Change.* Washington, D.C.: American Chemical Society.

Bernard, Harold W., Jr. 1993. *Global Warming Unchecked: Signs to Watch For.* Bloomington: University of Indiana Press.

Broecker, Wallace S. 1995. "Chaotic Climate." *Scientific American*, November, 62–68.

Burroughs, William. 1997. *Does the Weather Really Matter? The Social Implications of Climate Change.* New York: Cambridge University Press.

California Energy Commission. 1989. *The Impacts of Global Warming on California.* Sacramento: California Energy Commission.

Charlson, Robert J., and Tom M. L. Wigley. 1994. "Sulfate Aerosol and Climate Change." *Scientific American*, February, 48–57.

Cogan, Douglas. 1992. *The Greenhouse Gambit: Business and Investment Responses to Climate Change.* Washington, D.C.: Investor Responsibility Research Center.

Council for Agricultural Science and Technology. 1992. *Preparing U.S. Agriculture for Global Climate Change.* Ames, Iowa: Council for Agricultural Science and Technology.

Edgerton, Lynne T. 1990. *The Rising Tide: Global Warming and World Sea Levels.* Covelo, Calif.: Island Press.

Ewan, C. E., et al., eds. 1997. *Health in the Greenhouse: The Medical and Environmental Health Effects of Global Environmental Change.* Portland, Ore.: International Specialized Book Services.

Fermann, Gunnar, ed. 1997. *International Politics of Climate Change: Key Issues and Critical Actors.* Oslo, Norway: Scandinavian University Press.

Flavin, Christopher. 1996. "Facing Up to the Risks of Climate Change." In Lester R. Brown et al. *State of the World 1996.* Washington, D.C.: Worldwatch Institute, 21–39.

Flavin, Christopher, and Seth Dunn. 1997. *Rising Sun, Gathering Winds: Policies to Stabilize Climate and Strengthen Economies.* Washington, D.C.: Worldwatch Institute.

Flavin, Christopher, and Odil Tunali. 1996. *Climate of Hope: New Strategies for Stabilizing the World's Atmosphere.* Washington, D.C.: Worldwatch Institute.

Gates, David M. 1993. *Climate Change and Its Biological Consequences.* Sunderland, Mass.: Sinauer.

Gelbspan, Ross. 1997. *The Heat is On: The High Stakes Battle over Earth's Threatened Climate.* Reading, Mass.: Addison Wesley.

Graedel, T. C., and Paul J. Cutzen. 1995. *Atmosphere, Climate, and Change.* New York: W.H. Freeman.

Hoffman, Andrew J., ed. 1997. *Global Climate Change.* San Francisco: New Lexington Press.

Houghton, J. T. 1997. *Global Warming: The Complete Briefing.* New York: Cambridge University Press.

Intergovernmental Panel on Climate Change (IPCC). 1990. *Climate Change: The IPCC Assessment.* New York: Cambridge University Press.

Intergovernmental Panel on Climate Change (IPCC). 1992. *The Supplementary Report to the IPCC Scientific Assessment.* New York: Cambridge University Press.

Intergovernmental Panel on Climate Change (IPCC). 1994. *The Supplementary Report to the IPCC Scientific Assessment.* New York: Cambridge University Press.

Intergovernmental Panel on Climate Change (IPCC). 1995. *The Supplementary Report to the IPCC Scientific Assessment.* New York: Cambridge University Press.

IUCN Global Change Programme. 1993. *Impact of Climate Change on Ecosystems and Species.* Covelo, Calif.: Island Press.

Karplus, Walter J. 1992. *The Heavens Are Falling: The Scientific Prediction of Catastrophes in Our Time.* New York: Plenum.

Kimball, Norman J., et al. 1990. *Impact of Carbon Dioxide, Trace Gases, and Climate Change on Global Agriculture.* Madison, Wis.: Soil Science Society of America.

Krause, Florentin, et al. 1992. *Energy Policy in the Greenhouse.* New York: Wiley.

Lee, Henry, ed. 1995. *Shaping Responses to Climate Change.* Covelo, Calif.: Island Press.

Leffell, David J., and Douglas E. Brash. 1996. "Sunlight and Skin Cancer." *Scientific American,* July, 52–59.

Leggett, Jeremy, ed. 1994. *The Climate Time Bomb: Signs of Climate Change from the Greenpeace Database.* Amsterdam: Stichting Greenpeace Council.

Lovins, Amory B., et al. 1989. *Least-Cost Energy: Solving the CO_2 Problem.* 2d ed. Andover, Mass.: Brick House.

Lyman, Francesca, et al. 1990. *The Greenhouse Trap: What We're Doing to the Atmosphere and How We Can Slow Global Warming.* Washington, D.C.: World Resources Institute.

Mabey, Nick et al., eds. 1997. *Argument in the Greenhouse: The International Economics of Controlling Global Warming.* New York: Routledge.

Mahlman, J. D. 1997. "Uncertainties in Projections of Human-Caused Climate Warming." *Science,* vol. 278, November 21, 1416–17.

Makhijani, Arjun, and Kevin Gurney. 1995. *Mending the Ozone Hole: Science, Technology, and Policy.* Cambridge, Mass.: MIT Press.

Manne, Alan S., and Richard G. Richels. 1992. *Buying Greenhouse Insurance: The Economic Costs of CO_2 Emission Limits.* Cambridge, Mass.: MIT Press.

McKibben, Bill. 1989. *The End of Nature.* New York: Random House.

Michaels, Patrick J. 1992. *Sound and Fury: The Science and Politics of Global Warming.* Washington, D.C.: Cato Institute.

Mintzer, Irving, et al. 1990. *Protecting the Ozone Shield: Strategies for Phasing Out CFCs During the 1990s.* Washington, D.C.: World Resources Institute.

Mintzer, Irving, and William R. Moomaw. 1991. *Escaping the Heat Trap: Probing the Prospects for a Stable Environment.* Washington, D.C.: World Resources Institute.

Moore, Thomas G. 1998. *Climate of Fear: Why We Shouldn't Worry About Global Warming.* Washington, D.C.: Cato Institute.

Mueller, Frank. 1996. "Mitigating Climate Change: The Case for Energy Taxes." *Environment,* March, 13–43.

National Academy of Sciences. 1990. *Confronting Climate Change.* Washington, D.C.: National Academy Press.

National Academy of Sciences. 1990. *Sea Level Change.* Washington, D.C.: National Academy Press.

National Academy of Sciences. 1991. *Policy Implications of Greenhouse Warming.* Washington, D.C.: National Academy Press.

National Academy of Sciences. 1992. *Global Environmental Change.* Washington, D.C.: National Academy Press.

National Audubon Society. 1990. *CO_2 Diet for a Greenhouse Planet: A Citizen's Guide to Slowing Global Warming.* New York: National Audubon Society.

Nissani, Moti. 1996. "The Greenhouse Effect: An Interdisciplinary Perspective." *Population and Environment: A Journal of Interdisciplinary Studies,* vol. 17, no. 6, 459–98.

Nordhaus, William D. 1994. *Managing the Commons: The Economics of Climate Change.* Cambridge, Mass.: MIT Press.

Office of Technology Assessment. 1991. *Changing by Degrees: Steps to Reduce Greenhouse Gases.* Washington, D.C.: U.S. Government Printing Office.

Office of Technology Assessment. 1993. *Preparing for an Uncertain Climate.* Washington, D.C.: U.S. Government Printing Office.

O'Meara, Molly. 1997. "The Risks of Disrupting Climate." *World Watch,* November/December, 16–24.

Oppenheimer, Michael, and Robert H. Boyle. 1990. *Dead Heat: The Race Against the Greenhouse Effect.* New York: Basic Books.

Parry, Martin. 1990. *Climate Change and World Agriculture.* London: Earthscan.

Peters, Robert L., and Thomas E. Lovejoy. 1992. *Global Warming and Biological Diversity.* New Haven, Conn.: Yale University Press.

Philander, S. George. 1998. *Is the Temperature Rising?: The Uncertain Science of Global Warming.* Princeton, N.J.: Princeton University Press.

Ray, Dixy Lee, and Lou Guzzo. 1993. *Environmental Overkill: Whatever Happened to Common Sense?* Washington, D.C.: Regnery Gateway.

Revkin, Andrew. 1992. *Global Warming: Understanding the Forecast.* New York: Abbeville Press.

Roan, Sharon L. 1989. *Ozone Crisis: The 15-Year Evolution of a Sudden Global Emergency.* New York: Wiley.

Rowland, F. Sherwood, and Mario J. Molina. 1994. "Ozone Depletion: 20 Years After the Alarm." *Chemical and Engineering News,* August 15, 8–13.

Rowlands, Ian H. 1996. *The Politics of Global Atmospheric Change.* London: Manchester University Press.

Ryan, John C. 1997. *Over Our Heads: A New Look at Global Climate.* Seattle: Northwest Environment Watch.

Sommerville, Richard C. J. 1996. *The Forgiving Air: Understanding Environmental Change.* Los Angeles: University of California Press.

Turekian, Karl K. 1996. *Global Environmental Change: Past, Present, and Future.* Upper Saddle River, N.J.: Prentice Hall.

UN Environment Programme, World Meteorological Organization, U.S. National Aeronautics and Space Administration, and National Oceanic and Atmospheric Administration. 1994. *International Assessment of Ozone Depletion.* Washington, D.C.: UN Environment Programme.

Waggoner, Paul E., ed. 1990. *Climate Change and U.S. Water Resources.* New York: Wiley.

Watson, R. B. et al., eds. 1996. *Climate Change 1995: Impacts, Adaptations, and Mitigation of Climate Change: Scientific Technical Analysis.* New York: Cambridge University Press.

Weiner, Jonathan. 1990. *The Next One Hundred Years: Shaping the Fate of Our Living Earth.* New York: Bantam.

Wittwer, Sylvan H., ed. 1995. *Food, Climate, and Carbon Dioxide: The Global Environment and World Food Production.* Boca Raton, Fla.: CRC Press.

Woodwell, George M., and Fred T. Mackenzie, eds. 1995. *Biotic Feedbacks in the Global Climatic System. Will the Warming Feed the Warming?* New York: Oxford University Press.

Young, Louise B. 1990. *Sowing the Wind: Reflections on Earth's Atmosphere.* Englewood Cliffs, N.J.: Prentice Hall.

Zurer, Pamela S. 1993. "Ozone Depletion's Recurring Surprises Challenge Atmospheric Scientists." *Chemical and Engineering News,* May 24, 8–18.

20 / WATER POLLUTION

Readings

Abel, P. D. 1997. *Water Pollution Biology.* 2d ed. London: Taylor & Francis.

Abramovitz, Janet N. 1996. "Sustaining Freshwater Ecosystems." In Lester R. Brown et al., *State of the World 1996.* New York: Norton, 60–77.

Adler, Robert W., et al. 1993. *The Clean Water Act Twenty Years Later.* Covelo, Calif.: Island Press.

Ashworth, William. 1986. *The Late, Great Lakes: An Environmental History.* New York: Knopf.

Burger, Joanna. 1997. *Oil Spills.* Rutgers, N.J.: Rutgers University Press.

Costner, Pat, and Glenna Booth. 1986. *We All Live Downstream: A Guide to Waste Treatment That Stops Water Pollution.* Berkeley, Calif.: Bookpeople.

Davidson, Art. 1990. *In the Wake of the Exxon Valdez.* San Francisco: Sierra Club Books.

Edmonson, W. T. 1991. *The Uses of Ecology: Lake Washington and Beyond.* Seattle: University of Washington Press.

Environmental Protection Agency. 1990. *Citizen's Guide to Ground-Water Protection.* Washington, D.C.: Environmental Protection Agency.

Gardner, Gary. 1998. "Recycling Human Waste: Fertile Ground or Toxic Legacy?" *World Watch,* January/February, 28–34.

Gray, N. F. 1992. *Biology of Wastewater Treatment.* New York: Oxford University Press.

Hansen, Nancy R., et al. 1988. *Controlling Nonpoint-Source Water Pollution.* New York: National Audubon Society and The Conservation Society.

Hileman, Bette. 1997. "Pfiesteria Health Concerns Realized." *Chemical and Engineering News,* October 13, 14–15.

Hinrichsen, Don. 1998. *Coastal Waters of the World: Trends, Threats, and Strategies.* Covelo, Calif.: Island Press.

Horne, A. J., and C. R. Goldman. 1994. *Limnology.* New York: McGraw-Hill.

Horton, Tom, and William Eichbaum. 1991. *Turning the Tide: Saving the Chesapeake Bay.* Covelo, Calif.: Island Press.

Jefferies, Michael, and Derek Mills. 1991. *Freshwater Ecology: Principles and Applications.* New York: Belhaven Press.

Jewell, William J. 1994. "Resource-Recovery Wastewater Treatment." *American Scientist,* vol. 82, 366–75.

Keeble, John. 1991. *Out of the Channel: The Exxon Valdez Oil Spill in Prince William Sound.* San Francisco: HarperCollins.

Knopman, Debra S., and Richard A. Smith. 1993. "20 Years of the Clean Water Act." *Environment,* vol. 35, no. 1, 17–20, 34–41.

Laws, Edward A. 1993. *Aquatic Pollution: An Introductory Text.* 2d ed. New York: Wiley.

Lewis, Scott A. 1995. *Guide to Safe Drinking Water.* San Francisco: Sierra Club Books.

Malle, Karl-Geert. 1996. "Cleaning Up the River Rhine." *Scientific American*, January, 70–75.

Marx, Wesley. 1991. *The Frail Ocean: A Blueprint for Change in the 1990s and Beyond.* San Francisco: Sierra Club Books.

Mason, C. F. 1991. *Biology of Freshwater Pollution*, 2d ed. New York: Wiley.

Montgomery, Ted. 1990. *On-Site Wastewater Treatment Systems.* East Falmouth, Mass.: New Alchemy Institute.

Outwater, Alice. 1996. *Water: A Natural History.* New York: Basic Books.

Patrick, Ruth, et al. 1992. *Surface Water Quality: Have the Laws Been Successful?* Princeton, N.J.: Princeton University Press.

Platt, Anne. 1995. "Dying Seas." *World Watch*, January/February, 10–19.

Platt, Anne. 1996. "Water-Borne Killers." *World Watch*, March/April, 28–35.

Sierra Club Defense Fund. 1989. *The Poisoned Well: New Strategies for Groundwater Protection.* Covelo, Calif.: Island Press.

Steingraber, Sandra. 1997. *Living Downstream: An Ecologist Looks at Cancer and the Environment.* Reading, Mass.: Addison-Wesley.

U.S. Geological Survey. 1988. *Groundwater and the Rural Homeowner.* Denver: U.S. Geological Survey.

Wagner, Travis. 1993. *In Our Backyard: A Guide to Understanding Pollution and Its Effects.* New York: Van Nostrand Reinhold.

21 / PROTECTING FOOD RESOURCES: PESTICIDES AND PEST CONTROL

Readings

Benbrook, Charles M., et al. 1996. *Pest Management at the Crossroads.* Yonkers, N.Y.: Consumer's Union.

Bormann, F. Herbert, et al. 1993. *Redesigning the American Lawn.* New Haven, Conn.: Yale University Press.

Briggs, Shirley, and the Rachel Carson Council. 1992. *Basic Guide to Pesticides: Their Characteristics and Hazards.* Washington, D.C.: Taylor & Francis.

Care, James R., and Maureen K. Hinkle. 1994. *Integrated Pest Management: The Path of a Paradigm.* Washington, D.C.: National Audubon Society.

Carson, Rachel. 1962. *Silent Spring.* Boston: Houghton Mifflin.

Cassels, Jamie. 1993. *The Uncertain Promise of Law: Lessons from Bhopal.* Toronto: University of Toronto Press.

Daar, Sheila, et al. 1991. *Common-Sense Pest Control.* Newtown, Conn.: Taunton Press.

Dinham, Barbara. 1993. *The Pesticide Hazard: A Global Health and Environmental Audit.* Atlantic Highlands, N.J.: Zed Books.

Environmental Working Group. 1995. *Forbidden Fruits: Illegal Pesticides in the U.S. Food Supply.* Washington, D.C.: Environmental Working Group.

Flint, Mary Louise. 1990. *Pests of the Garden and a Small Farm: Grower's Guide to Using Less Pesticide.* Oakland, Calif.: ANR.

Friends of the Earth. 1990. *How to Get Your Lawn and Garden Off Drugs.* Ottawa, Ontario: Friends of the Earth.

Gustafson, David. 1993. *Pesticides in Drinking Water.* New York: Van Nostrand Reinhold.

Heylin, Michael, ed. 1991. "Pesticides: Costs Versus Benefits," *Chemical and Engineering News*, January 7, 27–56.

Horn, D. J. 1988. *Ecological Approach to Pest Management.* New York: Guilford.

Hussey, N. W., and N. Scopes. 1986. *Biological Pest Control.* Ithaca, N.Y.: Cornell University Press.

Jenkins, Virginia Scott. 1994. *The Lawn: A History of an American Obsession.* Washington, D.C.: Smithsonian Institution Press.

League of Women Voters. 1989. *America's Growing Dilemma: Pesticides in Food and Water.* Washington, D.C.: League of Women Voters.

Leslie, Anne R., and Gerritt W. Cuperus. 1993. *Successful Implementation of Integrated Pest Management for Agricultural Crops.* Boca Raton, Fla.: Lewis.

Marquardt, Sandra. 1989. *Exporting Banned Pesticides: Fueling the Circle of Poison.* Washington, D.C.: Greenpeace.

Mollison, Bill. 1990. *Permaculture.* Covelo, Calif.: Island Press.

National Academy of Sciences. 1993. *Pesticides in the Diet of Infants and Children.* Washington, D.C.: National Academy Press.

National Academy of Sciences. 1996. *Ecologically Based Pest Management: New Solutions for a New Century.* Washington, D.C.: National Academy Press.

Natural Resources Defense Council. 1993. *After Silent Spring: The Unsolved Problems of Pesticide Use in the United States.* New York: Natural Resources Defense Council.

Olkowski, William, et al. 1991. *Common Sense Pest Control: Least Toxic Solutions for Your Home, Garden, Pets, and Community.* Newtown, Conn.: Taunton Press.

Pimentel, David, et al. 1992. "Environmental and Economic Cost of Pesticide Use." *BioScience*, vol. 42, no. 10, 750–60.

Pimentel, David, and Hugh Lehman, eds. 1993. *The Pesticide Question: Environment, Economics, and Ethics.* New York: Chapman & Hall.

Platt, Anne. 1996. "IPM and the War on Pests." *World Watch*, March/April, 21–27.

Preston-Mafham, Rod, and Ken Preston-Mafham. 1984. *Spiders of the World.* New York: Facts on File.

Schultz, Warren. 1994. *Natural Insect Control.* Handbook No. 139. Brooklyn, N.Y.: Brooklyn Botanic Garden.

Skerrett, P. J. 1997. "Food Irradiation: Will It Keep the Doctor Away?" *Technology Review*, November/December, 28–36.

Stein, Sara. 1993. *Noah's Garden: Restoring the Ecology of Our Own Back Yards.* New York: Houghton Mifflin.

United Nations Development Fund. 1996. *Urban Agriculture: Food, Jobs, and Sustainability.* New York: United Nations.

Vandeman, Ann, et al. 1994. *Adoption of Integrated Pest Management in U.S. Agriculture.* Washington, D.C.: USDA.

Wargo, J. 1996. *Our Toxic Legacy: How Science and Law Fail to Protect Us From Pesticides.* New Haven, Conn.: Yale University Press.

Wiles, Richard, and Christopher Campbell. 1993. *Pesticides in Children's Food.* Washington, D.C.: Environmental Working Group.

Winston, Mark L. 1997. *Nature Wars: People vs. Pests.* Cambridge, Mass.: Harvard University Press.

World Bank. 1996. *Integrated Pest Management.* Washington, D.C.: World Bank.

Wrubel, Roger. 1994. "The Promise and Problems of Herbicide-Resistant Crops." *Technology Review*, May/June, 56–61.

Yepsen, Roger B., Jr. 1987. *The Encyclopedia of Natural Insect and Pest Control.* Emmaus, Pa.: Rodale Press.

22 / SOLID AND HAZARDOUS WASTE

Readings

Ackerman, Frank. 1997. *Why Do We Recycle? Markets, Values, and Public Policy.* Covelo, Calif.: Island Press.

Alexander, Judd H. 1993. *In Defense of Garbage.* Westport, Conn.: Praeger.

Allen, Robert. 1992. *Waste Not, Want Not.* London: Earthscan.

Atlas, Ronald M. 1995. "Bioremediation." *Chemical and Engineering News*, April 3, 32–42.

Barnett, Harold C. 1994. *Toxic Debts and the Superfund Dilemma.* Chapel Hill: University of North Carolina Press.

Brown, Phil, and Edwin J. Mikkelsen. 1997. *No Safe Place: Toxic Waste, Leukemia, and Community Action.* Berkeley: University of California Press.

Bullard, Robert D., ed. 1993. *Confronting Environmental Racism: Voices from the Grassroots.* Boston: South End Press.

Bullard, Robert D. 1994. *Dumping in Dixie: Race, Class, and Environmental Quality.* 2d ed. Boulder, Colo.: Westview Press.

Bullard, Robert D., ed. 1994. *Unequal Protection: Environmental Justice and Communities of Color.* San Francisco: Sierra Club Books.

Calow, Peter, 1997. *Controlling Environmental Risks from Chemicals: Principles and Practice.* New York: Wiley.

Carless, Jennifer. 1992. *Taking Out the Trash: A No-Nonsense Guide to Recycling.* Covelo, Calif.: Island Press.

Centers for Disease Control and Prevention. 1991. *Preventing Lead Poisoning in Young Children.* Atlanta: Centers for Disease Control and Prevention.

Christopher, Tom, and Marty Asher. 1994. *The Art of Composting for Your Yard, Your Community, and the Planet.* San Francisco: Sierra Club Books.

Clarke, Marjorie J., et al. 1991. *Burning Garbage in the U.S.: Practice vs. State of the Art.* New York: INFORM.

Cohen, Gary, and John O'Connor. 1990. *Fighting Toxics: A Manual for Protecting Family, Community, and Workplace.* Covelo, Calif.: Island Press.

Commoner, Barry. 1994. *Making Peace with the Planet.* Rev. ed. New York: New Press.

Commoner, Barry, et al. 1996. *Zeroing Out Dioxin in the Great Lakes: Within Our Reach.* Flushing College, Queens, N.Y.: Center for the Biology of Natural Systems.

Connett, Paul H. 1992. "The Disposable Society." In F. H. Bormann, and Stephen R. Kellert, eds., *Ecology, Economics, Ethics.* New Haven, Conn.: Yale University Press, 99–122.

Connett, Paul, and Ellen Connett. 1994. "Municipal Waste Incineration: Wrong Question, Wrong Answer." *The Ecologist*, January/February, vol. 24, 14–20.

Costner, Pat, and Joe Thornton. 1989. *Sham Recyclers, Part 1: Hazardous Waste Incineration in Cement and Aggregate Kilns.* Washington, D.C.: Greenpeace.

Denison, Richard A., and John Ruston. 1997. "Recycling Is Not Garbage." *Technology Review*, October, 55–60.

Devito, Stephen C., and Roger L. Garrett, eds. 1996. *Designing Safe Chemicals: Green Chemistry for Pollution Prevention*. Washington, D.C.: American Chemical Society.

Dorfman, Mark H., et al. 1992. *Environmental Dividends: Cutting More Chemical Wastes*. New York: INFORM.

Durning, Alan Thein. 1992. *How Much Is Enough? The Consumer Society and the Future of the Earth*. New York: Norton.

Earth Works Group. 1990. *The Recycler's Handbook: Simple Things You Can Do*. Berkeley, Calif.: Earth Works Press.

Environmental Protection Agency. 1992. *The Consumer's Handbook for Reducing Solid Waste*. Washington, D.C.: Environmental Protection Agency.

Finkel, Adam M., and Dominic Golding. 1994. *Worst Things First? The Debate Over Risk-Based National Environmental Priorities*. Washington, D.C.: Resources for the Future.

Frosch, Robert A. 1995. "Industrial Ecology." *Environment*, December, 16–37.

Gardner, Gary. 1997. *Recycling Organic Waste: From Urban Pollutant to Farm Resource*. Washington, D.C.: Worldwatch Institute.

Gertler, Nicholas, and John R. Ehrenfeld. 1996. "A Down-to-Earth Approach to Clean Production." *Technology Review*, February/March, 48–54.

Gibbs, Lois. 1995. *Dying from Dioxin*. Boston: South End Press.

Gibbs, Lois. 1997. *Love Canal: The Story Continues*. Philadelphia: New Society Publishers.

Gordon, Ben, and Peter Montague. 1989. *Zero Discharge: A Citizen's Toxic Waste Manual*. Washington, D.C.: Greenpeace.

Gordon, Wendy, and Jane Bloom. 1985. *Deeper Problems: Limits to Underground Injection as a Hazardous Waste Disposal Method*. New York: Natural Resources Defense Council.

Gottlieb, Robert, ed. 1995. *Reducing Toxics: A New Approach to Policy and Industrial Decisionmaking*. Covelo, Calif.: Island Press.

Gourlay, K. A. 1992. *World of Waste: Dilemmas of Industrial Development*. Atlantic Highlands, N.J.: Zed Books.

Graedel, T. E., and B. R. Allenby. 1995. *Industrial Ecology*. Englewood Cliffs, N.J.: Prentice Hall.

Harte, John, et al. 1991. *Toxics A to Z: A Guide to Everyday Pollution Hazards*. Berkeley: University of California Press.

Hellberg, Tom. 1995. "Incineration by the Back Door: Cement Kilns as Waste Sinks." *The Ecologist*, vol. 25, no. 6, 4–8, 232–37.

Hilz, Christoph. 1993. *The International Toxic Waste Trade*. New York: Van Nostrand Reinhold.

Hird, John A. 1994. *Superfund: The Political Economy of Environmental Risk*. Baltimore: Johns Hopkins University Press.

Hofrichter, Richard, ed. 1993. *Toxic Struggles: The Theory and Practice of Environmental Justice*. Philadelphia: New Society.

Kane, Hal. 1996. "Shifting to Sustainable Industries." In Lester R. Brown et al., *State of the World 1996*. Washington, D.C.: Worldwatch Institute, 152–67.

Kenworthy, Lauren, and Eric Schaeffer. 1990. *A Citizen's Guide to Promoting Toxic Waste Reduction*. New York: INFORM.

Kessel, Irene, and John T. O'Connor. 1997. *Getting the Lead Out: The Complete Resource on How to Prevent and Cope with Lead Poisoning*. New York: Plenum Press.

Lappé, Marc. 1991. *Chemical Deception: The Toxic Threat to Health and Environment*. San Francisco: Sierra Club Books.

League of Women Voters. 1993. *The Garbage Primer: A Handbook for Citizens*. New York: Lyons & Burford.

League of Women Voters. 1993. *A Plastic Waste Primer*. New York: Lyons & Burford.

Lipsett, B., and D. Farrell. 1990. *Solid Waste Incineration Status Report*. Arlington, Va.: Citizen's Clearinghouse for Hazardous Waste.

Love Canal Homeowners Association. 1984. *Love Canal: A Chronology of Events That Shaped a Movement*. Arlington, Va.: Citizen's Clearinghouse for Hazardous Waste.

Mazmanian, Daniel, and David Morrell. 1992. *Beyond Superfailure: America's Toxics Policy for the 1990s*. Boulder, Colo.: Westview Press.

McLenighan, Valjean. 1991. *Sustainable Manufacturing: Saving Jobs, Saving the Environment*. Chicago: Center for Neighborhood Technology.

Minnesota Mining and Manufacturing. 1988. *Low- or Non-Pollution Technology Through Pollution Prevention*. St. Paul, Minn.: 3M Company.

Mowrey, Marc, and Tim Redmond. 1993. *Not in Our Backyard*. New York: William Morrow.

Moyers, Bill. 1990. *Global Dumping Ground: The International Traffic in Hazardous Waste*. Cabin John, Md.: Seven Locks Press.

National Academy of Sciences. 1991. *Environmental Epidemiology*. Vol. 1, *Public Health and Hazardous Wastes*. Washington, D.C.: National Academy Press.

National Academy of Sciences. 1994. *The Greening of Industrial Ecosystems*. Washington, D.C.: National Academy Press.

National Academy of Sciences. 1994. *Industrial Ecology: U.S.–Japan Perspectives*. Washington, D.C.: National Academy Press.

Needleman, Herbert L., and Philip J. Landrigan. 1995. *Raising Children Toxic Free*. New York: Avon Books.

Nemerow, Nelson L. 1995. *Zero Pollution for Industry: Waste Minimization Through Industrial Complexes*. New York: Wiley.

Office of Technology Assessment. 1992. *Green Products By Design*. Washington, D.C.: Government Printing Office.

Ortbal, John. 1991. *Buy Recycled! Your Practical Guide to the Environmentally Responsible Office*. Chicago: Services Marketing Group.

Pellerano, Maria B. 1995. *How to Research Chemicals: A Resource Guide*. Annapolis, Md.: Environmental Research Foundation.

Piasechi, Bruce, et al. 1998. "Is Combustion of Plastics Desirable?" *American Scientist*, vol. 86, July/August, 364–73.

Platt, Brenda, et al. 1991. *Beyond 40 Percent: Record-Setting Recycling and Composting Programs*. Covelo, Calif.: Island Press.

Portney, Kent E. 1992. *Siting Hazardous Waste Treatment Facilities: The NIMBY Syndrome*. New York: Auburn House.

Puckett, Jim. 1994. "Disposing of the Waste Trade: Closing the Recycling Loophole." *The Ecologist*, vol. 24, no. 2, 53–58.

Rathje, William, and Cullen Murphy. 1992. *Rubbish! The Archaeology of Garbage: What Our Garbage Tells Us About Ourselves*. San Francisco: HarperCollins.

Regenstein, Lewis G. 1993. *Cleaning Up America the Poisoned*. Washington, D.C.: Acropolis Books.

Revesz, Richard L., and Richard B. Stewart. eds. 1995. *Analyzing Superfund: Economics, Science, and Law*. Washington, D.C.: Resources for the Future.

Reynolds, Michael. 1990-91. *Earthship*. vols. 1, 2, and 3 Taos, N.M.: Solar Survival Architecture.

Rocky Mountain Institute. 1997. *A Primer on Sustainable Building*. Snowmass, Colo.: Rocky Mountain Institute.

Sachs, Aaron. 1995. *Eco-Justice: Linking Human Rights and the Environment*. Washington, D.C.: Worldwatch Institute.

Schecter, Arnold, ed. 1994. *Dioxins and Health*. New York: Plenum.

Schmidt, Karen. 1994. "Can Superfund Get on Track?" *National Wildlife*, April/May, 10–17.

Schwab, Jim. 1994. *Deeper Shades of Green: The Rise of Blue Collar and Minority Environmentalism in America*. San Francisco: Sierra Club Books.

Setterberg, Fred, and Lonny Shavelson. 1993. *Toxic Nation: The Fight to Save Our Communities from Chemical Contamination*. New York: Wiley.

Shulman, Seth. 1992. *The Threat at Home: Confronting the Toxic Legacy of the U.S. Military*. Boston: Beacon Press.

Stapleton, Richard M. 1995. *Lead Is a Silent Hazard*. New York: Walker.

Stein, Kathy. 1998. *Beyond Recycling: A Reuser's Guide*. Santa Fe, N.M.: Clear Light Publishers.

Szasz, Andrew. 1993. *EcoPopulism: Toxic Waste and the Movement for Environmental Justice*. Minneapolis: University of Minnesota Press.

Theodore, Louis, and Young C. McGuinn. 1992. *Pollution Prevention*. New York: Van Nostrand Reinhold.

Thomas, William. 1995. *Scorched Earth: The Military's Assault on the Environment*. Philadelphia: New Society.

Thornton, Joe. 1997. *The PVC Lifecycle: Dioxin from Cradle to Grave*. Washington, D.C.: Greenpeace USA.

Tibbs, Hardin B. C. 1992. "Industrial Ecology: An Environmental Agenda for Industry." *Whole Earth Review*, Winter, 4–19.

Wagner, Travis. 1993. *In Our Backyard: A Guide to Understanding Pollution and Its Effects*. New York: Van Nostrand Reinhold.

Walsh, Edward J., et al. 1997. *Don't Burn It Here: Grassroots Challenges to Trash Incineration*. University Park: Pennsylvania State University Press.

Whelan, E. M. 1993. *Toxic Terror: The Truth Behind the Cancer Scares*. Buffalo, N.Y.: Prometheus.

Whitaker, Jennifer Seymour. 1994. *Salvaging the Land of Plenty: Garbage and the American Dream*. New York: William Morrow.

Williams, Joy, et al. 1998. *Toxic Turnaround*. San Diego, Calif.: Environmental Health Coalition.

Young, John E. 1991. *Discarding the Throwaway Society*. Washington, D.C.: Worldwatch Institute.

Young, John E. 1995. "The Sudden New Strength of Recycling." *World Watch*, July/August, 20–25.

23 / SUSTAINING ECOSYSTEMS: LAND USE, CONSERVATION, AND MANAGEMENT

Readings

Baskin, Yvonne. 1997. *The Work of Nature: How the Diversity of Life Sustains Us*. Covelo, Calif.: Island Press.

Beatley, Timothy. 1994. *Ethical Land Use: Principles of Policy and Planning*. Baltimore: Johns Hopkins University Press.

Boucher, Norman. 1995. "Back to the Everglades." *Technology Review*, August/September, 25–35.

Brick, Philip D., and R. McGreggor Cawley. 1996. *A Wolf in the Garden: The Land Rights Movement and Renewing American Environmentalism*. Lanham, Md.: Rowman & Littlefield.

Cairns, John, Jr., and B. R. Niederlehner. 1995. "Ecosystem Health Concepts as a Management Tool." *Journal of Aquatic Ecosystem Health*, vol. 4, 91–95.

Caldwell, Lynton Keith, and Kristin Shrader-Frechette. 1993. *Policy for Land: Law and Ethics*. Washington, D.C.: World Resources Institute.

Callenbach, Ernest. 1995. *Bring Back the Buffalo: A Sustainable Future for America's Great Plains*. Covelo, Calif.: Island Press.

Chapin, F. Stuart, et al. 1998. "Ecosystem Consequences of Changing Biodiversity." *BioScience*, vol. 48, no. 1, 45–51.

Cohen, Michael P. 1984. *The Pathless Way: John Muir and American Wilderness*. Madison: University of Wisconsin Press.

Costanza, Robert, et al., eds. 1992. *Ecosystem Health: New Goals for Environmental Management*. Covelo, Calif.: Island Press.

Cronnon, W., ed. 1995. *Uncommon Ground: Toward Reinventing Nature*. New York: Norton.

Daily, Gretchen, ed. 1997. *Nature's Services: Society's Dependence on Natural Ecosystems*. Covelo, Calif.: Island Press.

Defenders of Wildlife. 1996. *A Status Report on America's Vanishing Habitat and Wildlife*. Washington, D.C.: Defenders of Wildlife.

DiSilvestro, Roger L. 1993. *Reclaiming the Last Wild Places: A New Agenda for Biodiversity*. New York: Wiley.

Dobson, Andrew. 1996. *Conservation and Biodiversity*. New York: Scientific American Library.

Dunlap, Riley E., and Angela G. Mertig, eds. 1992. *American Environmentalism: The U.S. Environmental Movement, 1970–1990*. Washington, D.C.: Taylor & Francis.

Dyson, Freeman. 1992. *From Eros to Gaia*. New York: Pantheon.

Eisler, Riane. 1987. *The Chalice and the Blade: Our History, Our Future*. New York: Harper & Row.

Elderige, Niles. 1995. *Dominion*. New York: Holt.

Fielder, Peggy L., and Subodh K. Jain, eds. 1992. *Conservation Biology: The Theory and Practice of Nature Conservation, Preservation and Management*. New York: Chapman & Hall.

Fleischner, Thomas L. 1994. "Ecological Costs of Livestock Grazing in Western North America." *Conservation Biology*, vol. 8, no. 3, 629–44.

Fox, Stephen. 1981. *John Muir and His Legacy: The American Conservation Movement*. Boston: Little, Brown.

Freemuth, John C. 1991. *Islands Under Siege: National Parks and the Politics of External Threats*. Lawrence: University of Kansas Press.

Goklany, Indur M. 1998. "Saving Habitat and Conserving Biodiversity on a Crowded Planet." *BioScience*, vol. 48, no. 11, 941–52.

Graf, William L. 1990. *Wilderness Preservation and the Sagebrush Rebellion*. Savage, Md.: Rowman & Littlefield.

Graham, Frank. 1971. *Man's Dominion: The Story of Conservation in America*. New York: M. Evans.

Grifo, F., and J. Rosenthal, eds. 1998. *Biodiversity and Human Health*. Covelo, Calif.: Island Press.

Grumbine, R. Edward. 1996/97. "Using Biodiversity as a Justification for Nature Protection in the U.S." *Wild Earth*, Winter 1996/97, 71–80.

Halvorson, William L., and Gary E. Davis, eds. 1996. *Science and Ecosystem Management in the National Parks*. Tucson: University of Arizona Press.

Hamilton, Alice. 1943. *Exploring the Dangerous Trades*. Boston: Little, Brown.

Harwell, Mark A. 1997. "Ecosystem Management of South Florida." *BioScience*, vol. 47, no. 8, 500–511.

Hays, Samuel. 1987. *Beauty, Health, and Permanence: Environmental Politics in the United States: 1955–1985*. New York: Cambridge University Press.

Heady, Harold F., and R. Dennis Child. 1994. *Rangeland Ecology and Management*. Boulder, Colo.: Westview Press.

Hendee, John, et al. 1991. *Principles of Wilderness Management*. Golden, Colo.: Fulcrum.

Hess, Karl, Jr. 1992. *Visions Upon the Land: Man and Nature on the Western Range*. Covelo, Calif.: Island Press.

Heywood, V. H., ed. 1995. *Global Biodiversity Assessment*. New York: Cambridge University Press.

Hughes, J. Donald. 1975. *Ecology in Ancient Civilizations*. Albuquerque: University of New Mexico Press.

Hughes, J. Donald. 1983. *American Indian Ecology*. El Paso: Texas Western Press.

Hyams, Edward. 1976. *Soils and Civilization*. New York: Harper & Row.

Jacobs, Lynn. 1992. *Waste of the West: Public Lands Ranching*. Tucson, Ariz.: Lynn Jacobs.

Jacobsen, Judith, and John Firor, eds. 1992. *Human Impact on the Environment: Ancient Roots, Current Challenges*. Boulder, Colo.: Westview Press.

Kellert, Stephen R. 1996. *The Value of Life: Biological Diversity and Human Society*. Covelo, Calif.: Island Press.

Knight, Richard L., and Sarah F. Bates, eds. 1994. *A New Century for Natural Resources Management*. Covelo, Calif.: Island Press.

Kuletz, Valerie L. 1998. *The Tainted Desert: Environmental Ruin in the American West*. New York: Routledge.

Leopold, Aldo. 1949. *A Sand County Almanac*. New York: Oxford University Press.

Loomis, John D. 1993. *Integrated Public Lands Management: Principles and Applications to National Forest, Parks, Wildlife Refuges, and BLM Lands*. New York: Columbia University Press.

Lowry, William R. 1994. *The Capacity for Wonder: Preserving National Parks*. Washington, D.C.: The Brookings Institution.

Manning, Richard. 1997. *Grassland: The History, Biology, Politics, and Promise of the American Prairie*. New York: Penguin.

Marsh, George Perkins. 1964. *Man and Nature*. New York: Scribner.

Martin, Calvin Luther. 1992. *In the Spirit of the Earth: Rethinking History and Time*. Baltimore: Johns Hopkins University Press.

McCormick, John. 1989. *Reclaiming Paradise: The Global Environmental Movement*. Bloomington: Indiana University Press.

Meffe, Gary K., and C. Ronald Carroll. 1994. *Principles of Conservation Biology*. Sunderland, Mass.: Sinauer.

Meine, Curt. 1988. *Aldo Leopold: His Life and Work*. Madison: University of Wisconsin Press.

Miller, Kenton R. 1994. *Balancing the Scales: Managing Biodiversity at the Bioregional Level*. Washington, D.C.: World Resources Institute.

Mowery, Marc, and Tim Redmond. 1993. *Not in Our Backyard: The People and Events That Shaped America's Modern Environmental Movement*. New York: William Morrow.

Naar, Jon, and Alex J. Naar. 1993. *This Land Is Your Land: A Guide to North America's Endangered Ecosystems*. New York: Harper & Row.

Nabokov, Peter, ed. 1992. *Native American Testimony: A Chronicle of Indian–White Relations from Prophecy to the Present, 1492–1992*. New York: Viking Press.

Nash, Roderick. 1982. *Wilderness and the American Mind*. 3d ed. New Haven, Conn.: Yale University Press.

Nash, Roderick. 1990. *American Environmentalism: Readings in Conservation History*. 3d ed. New York: McGraw-Hill.

National Academy of Sciences. 1993. *Setting Priorities for Land Conservation*. Washington, D.C.: National Academy Press.

National Academy of Sciences. 1996. *Biodiversity II: Understanding and Protecting Our Biological Resources*. Washington, D.C.: National Academy Press.

National Parks and Conservation Association. 1988. *Blueprint for National Parks*. 9 vols. Washington, D.C.: National Parks and Conservation Association.

National Parks and Conservation Association. 1994. *Our Endangered Parks: What You Can Do to Protect Our National Heritage*. San Francisco: Foghorn Press.

National Park Service. 1992. *National Parks for the 21st Century*. Washington, D.C.: National Park Service.

Noss, Reed F., and Allen Y. Cooperrider. 1994. *Saving Nature's Legacy: Protecting and Restoring Biodiversity*. Covelo, Calif.: Island Press.

Oelschlager, Max. 1991. *The Idea of Wilderness from Prehistory to the Age of Ecology*. New Haven, Conn.: Yale University Press.

Osborn, Fairfield. 1948. *Our Plundered Planet*. Boston: Little, Brown.

Palmer, Tim. 1993. *The Wild and Scenic Rivers of America*. Covelo, Calif.: Island Press.

Patton, David R. 1992. *Wildlife Habitat Relationships in Forested Ecosystems*. Covelo, Calif.: Island Press.

Payne, Daniel G. 1996. *Voices in the Wilderness: American Nature Writing and Environmental Politics*. Hanover, N.H.: University of New England Press.

Pearce, David, and Dominic Moran. 1994. *The Economic Value of Biodiversity*. East Haven, Conn.: Earthscan.

Ponting, Clive. 1992. *A Green History of the World. The Environment and the Collapse of Great Civilizations*. New York: St. Martin's Press.

Primack, Richard B. 1993. *Essentials of Conservation Biology*. Sunderland, Mass.: Sinauer.

Rapport, David, et al. 1998. *Ecosystem Health*. Cambridge, Mass: Blackwell.

Reed, P., ed. 1990. *Preparing to Manage Wilderness in the 21st Century*. Washington, D.C.: U.S. Forest Service.

Riebsame, William E. 1996. "Ending the Range Wars?" *Environment*, vol. 38, no. 4, 4–29.

Rifkin, Jeremy. 1992. *Beyond Beef: The Rise and Fall of the Cattle Culture*. New York: Dutton.

Robbins, Jim. 1994. *Last Refuge: The Environmental Showdown in the American West*. San Francisco: HarperCollins.

Roe, Frank G. 1970. *The North American Buffalo*. Toronto: University of Toronto Press.

Rosenberg, Daniel K., et al. 1997. "Biological Corridors: Form, Function, and Efficacy." *BioScience*, vol. 47, no. 10, 672–87.

Rudzitis, Gundars. 1996. *Wilderness and the Changing American West.* New York: Wiley.

Runte, Alfred. 1990. *Yosemite: The Embattled Wilderness.* Lincoln: University of Nebraska Press.

Runte, Alfred. 1991. *Public Lands, Public Heritage: The National Forest Idea.* Niwot, Colo.: Roberts Rinehart.

Russell, Sharman Apt. 1993. *Kill the Cowboy: A Battle of Mythology in the New West.* New York: Addison-Wesley.

Sale, Kirkpatrick. 1993. *The Green Revolution: The American Environmental Movement 1962–1992.* New York: Hill & Wang.

Scheffer, Vincent B. 1991. *The Shaping of Environmentalism in America.* Seattle: University of Washington Press.

Schmitz, Don C., and Daniel Simberloff. 1997. "Biological Invasions: A Growing Threat." *Issues in Science and Technology*, vol. 13, no. 4, 3–41.

Scott, J. M., et al. 1993. "Gap Analysis: A Geographic Approach to Protection of Biological Diversity." *Journal of Wildlife Management*, vol. 57, no. 1 supplement, Wildlife Monographs No. 123.

Shabecoff, Philip. 1993. *A Fierce Green Fire: The American Environmental Movement.* New York: Hill & Wang.

Shephard, Paul. 1991. *Man in the Landscape: A Historic View of the Esthetics of Nature.* 2d. ed. College Station: Texas A&M University Press.

Shiva, Vandana. 1993. *Monocultures of the Mind: Perspectives on Biodiversity and Biotechnology.* Atlantic Highlands, N.J.: Zed Books.

Shiva, Vandana. 1997. *Biopiracy: The Plunder of Nature and Knowledge.* Boston: South End Press.

Short, C. Brant. 1989. *Ronald Reagan and the Public Lands: America's Conservation Debate: 1979–1984.* College Station: Texas A&M University Press.

Simon, Noel. 1997. *Nature in Danger: Threatened Habitats and Species.* New York: Oxford University Press.

Soulé, Michael E. , and Gary Lease, eds. 1995. *Reinventing Nature? Responses to Postmodern Deconstruction.* Covelo, Calif.: Island Press.

Sutherland, William, ed. 1998. *Conservation Science and Action.* Cambridge, Mass.: Blackwell.

Toynbee, Arnold. 1972. *A Study of History.* New York: Oxford University Press.

Turner, B. L., et al., eds. 1990. *The Earth as Transformed by Human Action: Global and Regional Changes in the Biosphere Over the Last 300 Years.* New York: Cambridge University Press.

Udall, Stewart L. 1963. *The Quiet Crisis.* New York: Holt, Rinehart & Winston. (Reprint, with updating. Salt Lake City: Gibbs Smith, 1991.)

United Nations Development Programme. 1992. *Global Biodiversity Strategy.* Washington, D.C.: World Resources Institute.

U.S. Department of Interior. 1995. *Our Living Resources: A Report to the Nation on the Distribution, Abundance, and Health of U. S. Plants, Animals, and Ecosystems.* Covelo, Calif.: Island Press.

Valentine, John E., ed. 1990. *Grazing Management.* San Diego, Calif.: Academic Press.

Vasey, Daniel E. 1992. *An Ecological History of Agriculture, 10,000 B.C.–A.D. 10,000.* Ames: Iowa State University Press.

Vig, Norman J., and Michael J. Craft. 1984. *Environmental Policy in the 1980s.* Washington, D.C.: Congressional Quarterly Press.

Vogt, William. 1948. *The Road to Survival.* New York: Sloane.

Wagner, Frederick H., et al. 1995. *Wildlife Policies in the U.S. National Parks.* Covelo, Calif.: Island Press.

Wald, Joanna, et al. 1991. *How Not to Be Cowed: Livestock Grazing on Public Lands: An Owner's Manual.* Salt Lake City: Southern Utah Wilderness Alliance.

Wall, Derek., ed. 1994. *Green History: An Anthology of Environmental Literature, Philosophy and Politics.* New York: Routledge.

Wallach, Bret. 1991. *At Odds with Progress: Americans and Conservation.* Tucson: University of Arizona Press.

Waller, Donald M. 1996/97. "Wilderness Redux: Can Biodiversity Play a Role?" *Wild Earth*, Winter 1996/97, 36–45.

Wilcove, David S., and Michael J. Bean, eds. 1994. *The Big Kill: Declining Biodiversity in America's Lakes and Rivers.* Washington, D.C.: Environmental Defense Fund.

Wilderness Society. 1991. *Keeping It Wild: A Citizen Guide to Wilderness Management.* Washington, D.C.: Wilderness Society.

Woodley, Stephen, et al., eds. 1993. *Ecological Integrity and the Management of Ecosystems.* Covelo, Calif.: Island Press.

Worster, Donald, ed. 1988. *The Ends of the Earth: Perspectives on Modern Environmental History.* New York: Cambridge University Press.

Worster, Donald. 1992. *Under Western Skies: Nature and History in the American West.* New York: Oxford University Press.

Worster, Donald. 1993. *The Wealth of Nature: Environmental History and the Ecological Imagination.* New York: Oxford University Press.

Worster, Donald. 1994. *An Unsettled Country: Changing Landscapes of the American West.* Albuquerque: University of New Mexico Press.

Wright, R. Gerald. 1992. *Wildlife Research and Management in the National Parks.* Covelo, Calif.: Island Press.

Zaslowsky, Dyan, and T. H. Watkins. 1994. *These American Lands: Parks, Wilderness, and the Public Lands.* Rev. ed. Covelo, Calif.: Island Press.

24 / SUSTAINING ECOSYSTEMS: DEFORESTATION, BIODIVERSITY, AND FOREST MANAGEMENT

Readings

Abramovitz, Janet N. 1998. *Taking a Stand: Cultivating a New Relationship With the World's Forests.* Washington, D.C.: Worldwatch Institute.

Altman, Nathaniel. 1993. *Sacred Trees.* San Francisco: Sierra Club Books.

Alverson, William S., et al. 1994. *Wild Forests: Conservation Biology and Public Policy.* Covelo, Calif.: Island Press.

Amakrishna, Kilaparti, and George M. Woodwell, eds. 1993. *World Forests for the Future: Their Use and Conservation.* New Haven, Conn.: Yale University Press.

Anderson, Anthony B., et al., eds. 1990. *Alternatives to Deforestation: Steps Toward Sustainable Use of the Amazon Rain Forest.* New York: Columbia University Press.

Anderson, Patrick. 1989. "The Myth of Sustainable Logging: The Case for a Ban on Tropical Timber Imports." *The Ecologist*, vol. 19, no. 5, 166–68.

Aplet, Greg, et al., eds. 1993. *Defining Sustainable Forestry.* Covelo, Calif.: Island Press.

Barber, Charles V., et al. 1993. *Breaking the Deadlock: Obstacles to Forest Reform in Indonesia and the United States.* Washington, D.C.: World Resources Institute.

Beattie, Mollie, et al. 1983. *Working with Your Woodland.* Hanover, N.H.: University Press of New England.

Berkmuller, Klaus. 1992. *Environmental Education About the Rain Forest.* Covelo, Calif.: Island Press.

Booth, Douglas E. 1993. *Valuing Nature: The Decline and Preservation of Old Growth Forests.* Lanham, Md.: University Press of America.

Bormann, F. Herbert, et al. 1993. *World Forests for the Future: Their Use and Conservation.* New Haven, Conn.: Yale University Press.

Browder, John O. 1992. "The Limits of Extractivism." *BioScience*, vol. 42, no. 3, 174–82.

Brush, Stephen B., and Doreen Stabinsky. 1995. *Valuing Local Knowledge: Indigenous People and Intellectual Property Rights.* Covelo, Calif.: Island Press.

Burger, J. 1990. *The Gaia Atlas of First Peoples: A Future for the Indigenous World.* New York: Anchor.

Carroll, Mathew S. 1995. *Community and the Northwestern Logger.* Boulder, Colo.: Westview Press.

Chagnon, Napoleon A. 1992. *Yanomamo: The Last Days of Eden.* New York: Harcourt Brace Jovanovich.

Colomeda, Lori. 1997. *Keepers of the Central Fire: Issues of Health and Ecology for Indigenous People.* New York: National League for Nursing.

Cubbage, Frederick W., et al. 1993. *Forest Resource Policy.* New York: Wiley.

Davidson, Art. 1993. *Endangered Peoples.* San Francisco: Sierra Club Books.

Davis, Mary Byrd, ed. 1995. *Eastern Old-Growth Forests: Prospects for Rediscovery and Recovery.* Covelo, Calif.: Island Press.

Dekker-Robinson, Donna L., and William J. Libby. 1998. "American Forest Policy: Global Ethical Tradeoffs." *BioScience*, vol. 48, no. 6, 471–77.

de Onis, Juan. 1992. *The Green Cathedral: Sustainable Development of Amazonia.* New York: Oxford University Press.

Denniston, Derek. 1995. *High Priorities: Conserving Mountain Ecosystems and Cultures.* Washington, D.C.: Worldwatch Institute.

Devall, Bill, ed. 1994. *Clearcut: The Tragedy of Industrial Forestry.* San Francisco: Sierra Club Books/Earth Island Press.

Dietrich, William. 1993. *The Final Forest: The Last Great Trees of the Pacific Northwest.* New York: Penguin.

DiSilvestro, Roger L. 1993. *Reclaiming the Last Wild Places: A New Agenda for Biodiversity.* New York: Wiley.

Dower, Roger et al. 1997. *Frontiers of Sustainability: Environmentally Sound Agriculture, Forestry, Transportation, and Power Production.* Covelo, Calif.: Island Press.

Drengson, Alan, and Duncan Taylor, eds. 1997. *Ecoforestry: The Art & Science of Sustainable Forest Use.* Gabriola Island, Canada: New Society Publishers.

Durning, Alan Thein. 1992. *Guardians of the Land: Indigenous Peoples and the Health of the Earth.* Washington, D.C.: Worldwatch Institute.

Durning, Alan Thein. 1992. *Saving the Forests: What Will It Take?* Washington, D.C.: Worldwatch Institute.

Ellison, John, ed. 1993. *Beloved of the Sky: Essays and Photographs on Clearcutting.* Seattle: Broken Moon Press.

Foster, David R., et al. 1997. "Forest Response to Disturbance and Anthropogenic Stress." *BioScience*, vol. 47, no. 7, 437–45.

Friends of the Earth. 1992. *The Rainforest Harvest: Sustainable Strategies for Saving Tropical Forests.* Washington, D.C.: Friends of the Earth.

Goodland, Robert, ed. 1990. *Race to Save the Tropics: Ecology and Economics for a Sustainable Future.* Covelo, Calif.: Island Press.

Goodman, Steven M., and Bruce D. Patterson, eds. 1997. *Natural Change and Human Impact in Madagascar.* Washington, D.C.: Smithsonian Institution Press.

Greenpeace. 1990. *The Greenpeace Guide to Paper.* Washington, D.C.: Greenpeace.

Hammond, Herb. 1991. *Seeing the Forest Among the Trees: The Case for Wholistic Forest Use.* Vancouver, Canada: Raincoast Books.

Head, Suzanne, and Robert Heinzman. 1990. *Lessons of the Rainforest.* San Francisco: Sierra Club Books.

Hirt, Paul W. 1994. *A Conspiracy of Optimism: Management of the National Forests Since World War Two.* Lincoln: University of Nebraska Press.

Holloway, Marguerite. 1993. "Sustaining the Amazon." *Scientific American,* July, 90–99.

Hunter, Malcolm L., Jr. 1990. *Wildlife, Forests, and Forestry: Managing Forests for Biological Diversity.* Englewood Cliffs, N.J.: Prentice Hall.

International Union for Conservation of Nature and Natural Resources (IUCN). 1992. *Tropical Deforestation and the Extinction of Species.* Gland, Switzerland: IUCN.

Johnson, Nels, and Brice Cabarle. 1993. *Looking Ahead: Sustainable Natural Forest Management in the Humid Tropics.* Washington, D.C.: World Resources Institute.

Jolly, Alison, and Frans Lanting. 1990. *Madagascar: A World Out of Time.* New York: Aperture.

Kelly, David, and Gary Braasch. 1988. *Secrets of the Old Growth Forest.* Salt Lake City: Peregrine Smith.

Kemf, Elizabeth, ed. 1993. *The Law of the Mother: Protecting Indigenous Peoples in Protected Areas.* San Francisco: Sierra Club Books.

Kidd, Charles V., and David Pimentel, eds. 1992. *Integrated Resource Management: Agroforestry for Development.* New York: Academic Press.

Kimmins, James P. 1997. *Forest Ecology: A Foundation for Sustainable Management.* 2d ed. Upper Saddle River, N.J.: Prentice Hall.

Klyza, Christopher M., and Stephen C. Trombulak, eds. 1996. *The Future of the Northern Forest.* Hanover, N.H.: University Press of New England.

Kohn, Kathryn A., and Jerry M. Franklin. 1997. *Creating a Forestry for the 21st Century.* Covelo, Calif.: Island Press.

Kramer, Randall, et al. 1997. *Last Stand: Protected Areas and the Defense of Tropical Biodiversity.* New York: Oxford University Press.

Lansky, Mitch. 1992. *Beyond the Beauty Strip: Saving What's Left of Our Forests.* Gardiner, Maine: Tilbury House.

Lien, Carsten. 1991. *Olympic Battleground: The Power Politics of Timber Preservation.* San Francisco: Sierra Club Books.

Little, Charles E. 1995. *The Dying of the Trees: The Pandemic in America's Forests.* New York: Viking Press.

Living Earth Foundation. 1990. *The Rainforests: A Celebration.* San Francisco: Chronicle Books.

Lyle, John Tillman. 1994. *Regenerative Design for Sustainable Development.* New York: Wiley.

Mahony, Rhona. 1992. "Debt-for-Nature Swaps: Who Really Benefits?" *The Ecologist,* vol. 22, no. 3, 97–103.

Manning, Richard. 1993. *Last Stand.* New York: Penguin.

Margolis, Marc. 1992. *The Last New World: The Conquest of the Amazon Frontier.* New York: Norton.

Maser, Chris. 1989. *Forest Primeval.* San Francisco: Sierra Club Books.

Maser, Chris, et al. 1994. *Sustainable Forestry: Philosophy, Science and Economics.* San Francisco: Rainforest Action Network.

Maybury-Lewis, David. 1992. *Millennium: Tribal Wisdom and the Modern World.* New York: Viking Press.

McInnis, Doug. 1997. "The Burning Season." *Earth,* August, 36–41.

Miller, Kenton, and Laura Tangley. 1991. *Trees of Life: Protecting Tropical Forests and Their Biological Wealth.* Boston: Beacon Press.

Minckler, Leon S. 1980. *Woodland Ecology.* 2d ed. Syracuse, N.Y.: Syracuse University Press.

Myers, Norman. 1993. *The Primary Source: Tropical Forests and Our Future.* New York: Norton.

Nabhan, Gary. 1997. *Cultures of Habitat: On Nature, Culture, and Story.* Washington, D.C.: Counterpoint Press.

Newman, Arnold. 1990. *The Tropical Rainforest: A World Survey of Our Most Valuable Endangered Habitats.* New York: Facts on File.

Nichol, John. 1990. *The Mighty Rainforest.* London: David & Charles.

Norse, Elliot A. 1990. *Ancient Forests of the Pacific Northwest.* Covelo, Calif.: Island Press.

Office of Technology Assessment. 1992. *Combined Summaries: Technologies to Sustain Tropical Forest Resources and Biological Diversity.* Washington, D.C.: Office of Technology Assessment.

O'Toole, Randal. 1987. *Reforming the Forest Service.* Covelo, Calif.: Island Press.

Panayotou, Theodore, and Peter S. Ashton. 1992. *Not By Timber Alone: Economics and Ecology for Sustaining Tropical Forests.* Covelo, Calif.: Island Press.

Park, Chris C. 1993. *Tropical Rainforests.* New York: Routledge.

Patterson, Alan. 1990. "Debt for Nature Swaps and the Need for Alternatives." *Environment,* vol. 32, no. 10, 5–13, 31–32.

Perlin, John. 1989. *A Forest Journey: The Role of Wood in the Development of Civilization.* New York: Norton.

Plotkin, Mark J., and Lisa M. Famolare. 1992. *Sustainable Harvest and Marketing of Rain Forest Products.* Covelo, Calif.: Island Press.

Posey, Darrell L. 1996. "Protecting Indigenous Peoples' Rights to Biodiversity." *Environment,* vol. 38, no. 8, 6–9, 37-45.

Ramakrisna, K., and G. M. Woodward, eds. 1993. *World Forests for the Future: Their Use and Conservation.* New Haven, Conn.: Yale University Press.

Raphael, Ray. 1994. *More Tree Talk: The People, Politics, and Economics of Timber.* Covelo, Calif.: Island Press.

Reaka-Kudla, Marjorie, Don E. Wilson, and Edward O. Wilson, eds. 1996. *Biodiversity II: Understanding and Protecting Our Biological Resources.* New York: John Henry.

Reid, Walter V. C., and Kenton R. Miller. 1989. *Keeping Options Alive: The Scientific Basis for Conserving Biodiversity.* Washington, D.C.: World Resources Institute.

Reid, Walter, et al. 1993. *Biodiversity Prospecting: Using Genetic Resources for Sustainable Development.* Washington, D.C.: World Resources Institute.

Revkin, Andrew. 1990. *The Burning Season: The Murder of Chico Mendes and the Fight for the Amazon.* Boston: Houghton Mifflin.

Rice, R. E. 1990. "Old-Growth Logging Myths." *The Ecologist,* vol. 20, no. 4, 141–46.

Rietbergen, Simon, ed. 1994. *The Earthscan Reader in Tropical Forestry.* San Francisco: Rainforest Action Network.

Robinson, Gordon. 1987. *The Forest and the Trees: A Guide to Excellent Forestry.* Covelo, Calif.: Island Press.

Romme, William H., and Don G. Despain. 1989. "The Yellowstone Fires." *Scientific American,* vol. 261, no. 5, 37–46.

Rush, James. 1991. *The Last Tree: Reclaiming the Environment in Tropical Asia.* Boulder, Colo.: Westview Press.

Ryan, John C. 1992. *Life Support: Conserving Biological Diversity.* Washington, D.C.: Worldwatch Institute.

Sedjo, Roger A., and Daniel Botkin. 1997. "Using Forest Plantations to Spare National forests." *Environment,* vol. 39, no. 10, 12–20, 29–30.

Shiva, Vandana. 1997. *Biopiracy: The Plunder of Nature and Knowledge.* Boston: South End Press.

Shoumatff, Alex. 1990. *The World of Burning: The Tragedy of Chico Mendes.* Boston: Little, Brown.

Spurr, Stephen H., and Burton V. Barnes. 1992. *Forest Ecology.* 3d ed. Melbourne, Fla.: Kreiger.

Stoddard, Charles H., and Glenn M. Stoddard. 1987. *Essentials of Forestry Practice.* 4th ed. New York: Wiley.

Taylor, David A. 1997. "Saving the Forest for the Trees: Alternative Products from Woodlands." *Environment,* vol. 39, no. 1, 6–11, 33–36.

Teitel, Martin. 1992. *Rain Forest in Your Kitchen: The Hidden Connection Between Extinction and Your Supermarket.* Covelo, Calif.: Island Press.

Tennebaum, David. 1995. "The Greening of Costa Rica." *Technology Review,* October, 42–52.

Terborgh, John. 1992. *Diversity and the Tropical Rain Forest.* New York: Scientific American Library.

Vandermeer, John, and Ivette Perfecto. 1995. *Breakfast of Biodiversity: The Truth About Rain Forest Destruction.* Oakland, Calif.: Food First.

Weber, Thomas. 1990. *Hugging the Trees: The Story of the Chipko Movement.* New York: Penguin.

Whelan, Robert J. 1995. *The Ecology of Fire.* New York: Cambridge University Press.

Wilkinson, Todd. 1998. *Science Under Siege: The Politicians' War on Nature and Truth.* New York: Johnson Books.

Wilson, E. O., ed. 1988. *Biodiversity.* Washington, D.C.: National Academy Press.

Wilson, E. O. 1992. *The Diversity of Life.* Cambridge, Mass.: Harvard University Press.

World Resources Institute (WRI), World Conservation Union (IUCN), and United Nations Development Programme. 1992. *Global Biodiversity Strategy.* Washington, D.C.: World Resources Institute.

Yaffee, Steven Lewis. 1994. *The Wisdom of the Spotted Owl: Policy Lessons for a New Century.* Covelo, Calif.: Island Press.

25 / SUSTAINING WILD SPECIES

Readings

Ackerman, Diane. 1991. *The Moon by Whalelight.* New York: Random House.

Baker, Ron. 1985. *The American Hunting Myth.* New York: Vantage Press.

Barker, Rocky. 1993. *Saving All the Parts: Reconciling Economics and the Endangered Species Act*. Covelo, Calif.: Island Press.

Bolen, Eric G., and William J. Robinson. 1998. *Wildlife Ecology and Management*. 4th ed. Englewood Cliffs, N.J.: Prentice Hall.

Bonner, Raymond. 1993. *At the Hand of Man: Peril and Hope for African Wildlife*. New York: Knopf.

Boo, Elizabeth. 1990. *Ecotourism: The Potential and the Pitfalls*. Vols. 1 and 2. Washington, D.C.: World Wildlife Fund.

Bright, Chris. 1995. "Understanding the Threat of Bioinvasions." In Lester R. Brown et al., *State of the World 1996*. Washington, D.C.: Worldwatch Institute, 95–113.

Burton, Robert, ed. 1991. *Nature's Last Strongholds*. New York: Oxford University Press.

Caughley, Graeme, and A. R. E. Sinclair. 1994. *Wildlife Ecology and Management*. Cambridge, Mass.: Basil Blackwell.

Causey, Ann S. 1989. "On the Morality of Hunting." *Environmental Ethics*, Winter, 327–43.

Clark, Tim W., et al., eds. 1994. *Endangered Species Recovery: Finding the Lessons, Improving the Process*. Covelo, Calif.: Island Press.

Cox, George W. 1993. *Conservation Ecology*. Dubuque, Iowa: Wm. C. Brown.

Credlund, Arthur G. 1983. *Whales and Whaling*. New York: Seven Hills Books.

DeBlieu, Jan. 1991. *Meant to Be Wild: The Struggle to Save Endangered Species Through Captive Breeding*. New York: Fulcrum.

DiSilvestro, Roger L. 1992. *Rebirth of Nature*. New York: Wiley.

Dixon, John A., and Paul B. Sherman. 1990. *Economics of Protected Areas*. Covelo, Calif.: Island Press.

Durrell, Lee. 1986. *State of the Ark: An Atlas of Conservation in Action*. New York: Doubleday.

Elderidge, Niles. 1991. *The Miner's Canary*. Englewood Cliffs, N.J.: Prentice Hall.

Falk, Donald, et al. 1996. *Restoring Biodiversity: Strategies for Reintroduction of Endangered Plants*. Covelo, Calif.: Island Press.

Gibbons, Walt. 1993. *Keeping All the Pieces: Perspectives on Natural History and the Environment*. Washington, D.C.: Smithsonian Institution Press.

Gilbert, Frederick F., and Donald G. Dodds. 1992. *The Philosophy and Practice of Wildlife Management*. 2d ed. Melbourne, Fla.: Kreiger.

Gray, Gary G. 1993. *Wildlife and People: The Human Dimensions of Wildlife Ecology*. Champaign: University of Illinois Press.

Grumbine, R. Edward. 1992. *Ghost Bears: Exploring the Biodiversity Crisis*. Covelo, Calif.: Island Press.

Hargrove, Eugene C., ed. 1992. *The Animal Rights/Environmental Ethics Debate: The Environmental Perspective*. Albany: State University of New York Press.

Howes, Chris. 1997. *The Spice of Life: Biodiversity and the Extinction Crisis*. London: Blandford.

Kaufman, Les, and Kenneth Malay, eds. 1993. *The Last Extinction*. 2d ed. Cambridge, Mass.: MIT Press.

Kellert, Stephen, and E. O. Wilson, eds. 1995. *The Biophilia Hypothesis*. Covelo, Calif.: Island Press.

Klenig, John. 1992. *Valuing Life*. Princeton, N.J.: Princeton University Press.

Leopold, Aldo. 1933. *Game Management*. New York: Scribner.

Livingston, John A. 1981. *The Fallacy of Wildlife Conservation*. Toronto: McClelland and Stewart.

Luoma, Jon. 1987. *A Crowded Ark: The Role of Zoos in Wildlife Conservation*. Boston: Houghton Mifflin.

Mann, Charles C., and Mark L. Plummer. 1995. *Noah's Choice: The Future of Endangered Species*. New York: Knopf.

Mathiessen, Peter. 1992. *Shadows of Africa*. New York: Frank Abrams.

McGinn, Anne Platt. 1998. "Promoting Sustainable Fisheries." In Lester R. Brown et al. *State of the World 1998*. Washington, D.C.: Worldwatch Institute, 59–78.

Myers, Norman. 1994. *A Wealth of Wild Species: Storehouse for Human Welfare*. Boulder, Colo.: Westview Press.

Myers, Norman. 1997. "Mass Extinction and Evolution." *Science*, vol. 278, 597–98.

Nash, Roderick F. 1988. *The Rights of Nature: A History of Environmental Ethics*. Madison: University of Wisconsin Press.

National Academy of Sciences. 1994. *Understanding Marine Biodiversity*. Washington, D.C.: National Academy Press.

National Academy of Sciences. 1995. *Science and the Endangered Species Act*. Washington, D.C.: National Academy of Sciences.

National Academy of Sciences. 1997. *Striking a Balance: Improving Stewardship of Marine Areas*. Washington, D.C.: National Academy Press.

National Wildlife Federation. 1987. *The Arctic National Wildlife Refuge Coastal Plain: A Perspective for the Future*. Washington, D.C.: National Wildlife Federation.

Norse, Elliott A., ed. 1993. *Global Marine Biological Diversity*. Covelo, Calif.: Island Press.

Norton, B. G., ed. 1986. *Why Preserve Natural Variety?* Princeton, N.J.: Princeton University Press.

Noss, Reed F., et al. 1997. *The Science of Conservation Planning: Habitat Conservation Under the Endangered Species Act*. Covelo, Calif.: Island Press.

Office of Technology Assessment. 1989. *Oil Production in the Arctic National Wildlife Refuge*. Washington, D.C.: U.S. Government Printing Office.

Oldfield, Margery L., and Janis B. Alcorn, eds. 1992. *Biodiversity: Culture, Conservation, and Ecodevelopment*. Boulder, Colo.: Westview Press.

Passmore, John. 1974. *Man's Responsibility for Nature*. New York: Scribner.

Pimentel, David, et al. 1992. "Conserving Biological Diversity in Agricultural/Forestry Systems." *BioScience*, vol. 42, no. 5, 354–62.

Primack, Richard B. 1993. *Essentials of Conservation Biology*. Sunderland, Mass.: Sinauer.

Pringle, Laurence. 1989. *The Animal Rights Controversy*. New York: Harcourt Brace Jovanovich.

Pringle, Laurence. 1991. *Exploring the World of Bats*. New York: Scribner.

Reagan, Tom. 1983. *The Case for Animal Rights*. Berkeley: University of California Press.

Reid, Walter V. C., and Kenton R. Miller. 1989. *Keeping Options Alive: The Scientific Basis for Conserving Biodiversity*. Washington, D.C.: World Resources Institute.

Reisner, Marc. 1991. *Game Wars: The Undercover Pursuit of Wildlife Poachers*. New York: Viking Press.

Ryan, John C. 1992. *Life Support: Conserving Biological Diversity*. Washington, D.C.: Worldwatch Institute.

Schaller, George. 1993. *The Last Panda*. Chicago: University of Chicago Press.

Schmitz, Don C., and Daniel Simberloff. 1997. "Biological Invasions: A Growing Threat." *Issues in Science and Technology*, Summer, 33–40.

Shafer, Craig L. 1990. *Nature Reserves: Island Theory and Conservation Practice*. Washington, D.C.: Smithsonian Institution.

Shilling, Fraser. 1997. "Do Habitat Conservation Plans Protect Endangered Species?" *Science*, vol. 276, 1662–63.

Soulé, Michael E. 1991. "Conservation: Tactics for a Constant Crisis." *Science*, vol. 253, 744–50.

Tenenbaum, David. 1996. "Weeds from Hell." *Technology Review*, August/September, 31–40.

Terborgh, John. 1992. "Why American Songbirds Are Vanishing." *Scientific American*, May, 98–104.

Tobin, Richard J. 1990. *The Expendable Future: U.S. Politics and the Protection of Biodiversity*. Durham, N.C.: Duke University Press.

Tudge, Colin. 1992. *Last Animals at the Zoo: How Mass Extinction Can Be Stopped*. Covelo, Calif.: Island Press.

Tuttle, Merlin D. 1988. *America's Neighborhood Bats: Understanding and Learning to Live in Harmony with Them*. Austin: University of Texas Press.

Tuxill, John. 1998. *Losing Strands in the Web of Life: Vertebrate Declines and the Conservation of Biological Diversity*. Washington, D.C.: Worldwatch Institute.

Vitali, Theodore. 1990. "Sport Hunting: Moral or Immoral?" *Environmental Ethics*, Spring, 69–81.

Vitousek, Peter M., et al. 1996. "Biological Invasions as Global Environmental Change." *American Scientist*, Vol. 84, 468–78.

Ward, Peter, 1995. *The End of Evolution: Mass Extinctions and the Preservation of Biodiversity*. New York: Bantam.

Watkins, T. H. 1988. *Vanishing Arctic: Alaska's National Wildlife Refuge*. New York: Aperture.

Whelan, Tensie, ed. 1991. *Nature Tourism: Managing for the Environment*. Covelo, Calif.: Island Press.

Wilson, E. O., ed. 1988. *Biodiversity*. Washington, D.C.: National Academy Press.

Wilson, E. O. 1992. *The Diversity of Life*. Cambridge, Mass.: Harvard University Press.

World Resources Institute (WRI) et al. 1992. *Global Biodiversity Strategy*. Washington, D.C.: World Resources Institute.

26 / SUSTAINABLE CITIES: URBAN LAND USE AND MANAGEMENT

Readings

Aaberley, Doug, ed. 1994. *Futures by Design: The Practice of Ecological Planning*. Philadelphia: New Society.

Alexander, Christopher W. 1993. *A New Theory of Urban Design*. New York: Oxford University Press.

Badshah, Akhtar A. 1997. *Our Urban Future: Paradigms for Equity and Sustainability*. London: Zed Books.

Berg, Peter, et al. 1989. *A Green City Program for San Francisco Bay Area Cities and Towns*. San Francisco: Planet/Drum Foundation.

Bookchin, Murray. 1986. *The Limits of the City*. Montreal: Black Rose.

Cadman, D., and G. Payne, eds. 1990. *The Living City: Towards a Sustainable Future*. New York: Routledge.

Calthorpe, Peter. 1993. *The Next American Metropolis: Ecology, Community, and the American Dream*. Princeton, N.J.: Princeton Architectural Press.

Carlson, Daniel, et al. 1995. *At Road's End: Transportation and Land Use Choices for Communities*. Covelo, Calif.: Island Press.

Coates, Gary. 1981. *Resettling America: Energy, Ecology, and Community*. Andover, Mass.: Brick House.

Corbett, Michael. 1990. *A Better Place to Live.* Davis, Calif.: agAccess.

Dower, Roger et al. 1997. *Frontiers of Sustainability: Environmentally Sound Agriculture, Forestry, Transportation, and Power Production.* Covelo, Calif.: Island Press.

Drakakis-Smith, D. 1990. *The Third World City.* New York: Routledge.

Eisner, S., et al. 1993. *The Urban Pattern.* 6th ed. New York: Van Nostrand Reinhold.

Engwicht, David. 1993. *Reclaiming Our Cities and Towns: Better Living with Less Traffic.* Philadelphia: New Society.

Ezcurra, Exequiel, and Marisa Mazari-Hriart. 1996. "Are Megacities Viable? A Cautionary Tale from Mexico City." *Environment*, vol. 38, no. 1, 6–15, 25–34.

Feder, Eben. 1998. *Better Not Bigger: How to Take Care of Urban Growth and Improve Your Community.* Philadelphia: New Society.

Garreau, Joel. 1992. *Edge City: Life on the New Frontier.* New York: Doubleday.

Garreau, Joel. 1995. "Ten Commandments for Planners." *Whole Earth Review*, Winter, 64–66.

Gilbert, Alan, ed. 1996. *The Mega-City in Latin America.* New York: United Nations University Press.

Giradet, Herbert. 1993. *The Gaia Atlas of Cities: New Directions for Sustainable Urban Living.* New York: Anchor Books.

Gordon, D. 1990. *Green Cities: Ecologically Sound Approaches to Urban Space.* New York: Black Rose.

Gordon, Deborah. 1991. *Steering a New Course: Transportation, Energy, and the Environment.* Covelo, Calif.: Island Press.

Gratz, Roberta B. 1989. *The Living City.* New York: Simon & Schuster.

Hardoy, Jorge E., et al. 1993. *Environmental Problems in Third World Cities.* East Haven, Conn.: Earthscan.

Hart, John. 1993. *Saving Cities, Saving Money: Environmental Strategies that Work.* Sausalito, Calif.: Resource Renewal Institute.

Honachefsky, William B. 1992. *Land Planner's Environmental Handbook.* New York: Noyes.

Kay, Jane Holtz. 1997. *Asphalt Nation: How the Automobile Took Over America and How We Can Take It Back.* New York: Crown.

Lowe, Marcia D. 1990. *Alternatives to the Automobile: Transport for Living Cities.* Washington, D.C.: Worldwatch Institute.

Lowe, Marcia D. 1991. *Shaping Cities: The Environmental and Human Dimensions.* Washington, D.C.: Worldwatch Institute.

Lowe, Marcia D. 1994. *Back on Track: The Global Rail Revival.* Washington, D.C.: Worldwatch Institute.

Lyle, John T. 1993. *Regenerative Design for Sustainable Development.* New York: Wiley.

Lyman, Francesca. 1997. "Twelve Gates to the City." *Sierra*, May/June, 29–35.

MacKenzie, James J., et al. 1992. *The Going Rate: What It Really Costs to Drive.* Washington, D.C.: World Resources Institute.

Makower, Joel. 1992. *The Green Commuter.* Washington, D.C.: National Press.

Mantrell, Michael L., et al. 1989. *Creating Successful Communities: A Guidebook to Growth Management Strategies.* Covelo, Calif.: Island Press.

McHarg, Ian L. 1992. *Designing with Nature.* Reprint. New York: Wiley.

McKibben, Bill. 1995. *Hope, Human and Wild: True Stories of Living Lightly on the Earth.* Boston: Little, Brown.

Mumford, Lewis. 1968. *The Urban Prospect.* New York: Harcourt Brace Jovanovich.

Nadis, Steve, and James J. MacKenzie. 1993. *Car Trouble.* Boston: Beacon Press.

Papanek, Victor. 1985. *Design for the Real World: Human Ecology and Social Change.* Chicago: Academy Chicago.

Rabinovitch, Jonas, and Josef Leitman. 1996. "Urban Planning in Curitiba." *Scientific American*, March, 46–53.

Raoul, Jean-Claude. 1997. "How High-Speed Trains Make Tracks." *Scientific American*, October, 100–105.

Register, Richard. 1992. *Ecocities.* Berkeley, Calif.: North Atlantic Books.

Rocky Mountain Institute. 1995. *A Primer on Sustainable Building.* Snowmass, Colo.: Rocky Mountain Institute.

Rocky Mountain Institute. 1997. *Green Development: Integrating Ecology and Real Estate.* Snowmass, Colo.: Rocky Mountain Institute.

Roseland, Mark. 1992. *Toward Sustainable Communities.* Philadelphia: New Society.

Ryn, Sin van der, and Stuart Cowan. 1995. *Ecological Design.* Covelo, Calif.: Island Press.

Simon, Joel. 1997. *Endangered Mexico: An Environment on the Edge.* San Francisco, Calif.: Sierra Club Books.

Steele, James. 1997. *Sustainable Architecture: Principles, Paradigms, and Case Studies.* New York: McGraw-Hill.

Stokes, Samuel N., et al. 1989. *Saving America's Countryside: A Guide to Rural Conservation.* Baltimore: Johns Hopkins University Press.

Stren, Richard, et al., eds. 1991. *Sustainable Cities: Urbanization and the Environment in International Perspective.* Boulder, Colo.: Westview Press.

Tarr, Joel A. 1996. *The Search for the Ultimate Sink: Urban Pollution in Historical Perspective.* Akron, Ohio: University of Akron Press.

Thompson, George F., and Frederick R. Steiner, eds. 1997. *Ecological Design and Planning.* New York: Wiley.

Todd, John, and Nancy Jack Todd. 1993. *From Ecocities to Living Machines: Precepts for Sustainable Technologies.* Berkeley, Calif.: North Atlantic Books.

Todd, John, and George Tukel. 1990. *Reinhabiting Cities and Towns: Designing for Sustainability.* San Francisco: Planet/Drum Foundation.

Todd, Nancy Jack, and John Todd. 1984. *Bioshelters, Ocean Arks, City Farming: Ecology As the Basis of Design.* San Francisco: Sierra Club Books.

Tolley, Rodney, ed. 1991. *The Greening of Urban Transport: Planning for Walking and Cycling in Western Cities.* New York: Pinter.

Tunali, Odil. 1996. "A Billion Cars: The Road Ahead." *World Watch*, January/February, 24–33.

Urban Ecology. 1997. *Blueprint for a Sustainable Bay Area.* Oakland, Calif.: Urban Ecology.

Vranich, Joseph. 1991. *Supertrains: Solutions to America's Transportation Gridlock.* New York: St. Martin's Press.

Walter, Bob, et al., eds. 1992. *Sustainable Cities: Concepts and Strategies for Eco-City Development.* Los Angeles: Eco-Home Media.

Wann, David. 1994. *Biologic: Environmental Protection by Design.* 2d ed. Boulder, Colo.: Johnson Books.

Wann, David. 1995. *Deep Design: Pathways to a Livable Future.* Covelo, Calif.: Island Press.

Weisman, Alan. 1998. *Gaviotas: A Village to Reinvent the World.* White River Junction, Vt.: Chelsea Green.

Weston, Anthony. 1994. *Is There Life on Earth?* Philadelphia: Temple University Press.

White, Rodney R. 1994. *Urban Environmental Management. Environmental Change and Urban Design.* New York: Wiley.

Wieman, Clark. 1996. "Downsizing Infrastructure." *Technology Review*, May/June, 49–55.

World Resources Institute et al. 1996. *World Resources 1996–97.* New York: Oxford University Press.

Yang, Linda. 1990. *The City Gardener's Handbook.* New York: Random House.

Zuckerman, Wolfgang. 1991. *End of the Road: The World Car Crisis and How We Can Solve It.* White River Junction, Vt.: Chelsea Green.

27 / ECONOMICS AND ENVIRONMENT

Readings

Aaseng, Nathan. 1994. *Jobs vs. the Environment: Can We Save Both?* Hillside, N.J.: Enslow.

Anderson, Terry, and Donald Leal. 1990. *Free Market Environmentalism.* San Francisco: Pacific Research Institute.

Anderson, Victor. 1991. *Alternative Economic Indicators.* New York: Routledge.

Ashworth, William. 1995. *The Economy of Nature: Rethinking the Connections Between Ecology and Economics.* Boston: Houghton Mifflin.

Athansiou, Tony. 1996. *Divided Planet: The Ecology of Rich and Poor.* Boston: Little, Brown.

Ayres, Robert U. 1998. *Turning Point: The End of the Growth Paradigm.* New York: St. Martin's.

Barlett, Donald L, and James B. Steele. 1994. *America: Who Really Pays the Taxes?* New York: Simon & Schuster.

Barnet, Richard J., and John Cavanagh. 1994. *Global Dreams: Imperial Corporations and the New World Order.* New York: Simon & Schuster.

Bennett, Steven J., et al. 1994. *Corporate Realities and Environmental Truths.* New York: Wiley.

Berry, Wendell. 1987. *Home Economics.* Berkeley, Calif.: North Point Press.

Bormann, F. H., and Stephen R. Kellert, eds. 1992. *Ecology, Economics, Ethics.* New Haven, Conn.: Yale University Press.

Bowden, Elbert V. 1997. *Principles of Economics: Theory, Problems, Policies.* 7th ed. Cincinnati: South-Western.

Bromley, Daniel W. 1991. *Environment and Economy: Property Rights and Public Policy.* Cambridge, Mass.: Basil Blackwell.

Brown, Lester R., et al. 1991. *Saving the Planet: How to Shape an Environmentally Sustainable Global Economy.* New York: Norton.

Brundtland, G. H., et al. 1987. *Our Common Future: World Commission on Environment and Development.* New York: Oxford University Press.

Cairncross, Frances. 1992. *Costing the Earth.* Boston: Harvard Business School.

Cairncross, Frances. 1995. *Green, Inc.: A Guide to Business and the Environment.* Covelo, Calif.: Island Press.

Callenbach, Ernest, et al. 1993. *EcoManagement: The Elmwood Guide to Ecological Auditing and Sustainable Business.* San Francisco: Berrett-Koehler.

Carson, Patrick, and Julia Moulden. 1991. *Green is Gold: Talking to Business About the Environmental Revolution*. New York: HarperBusiness.

Certain, Marvin. 1991. *Crystal Globe: The Haves and the Have-Nots of the New World Order*. New York: St. Martin's Press.

Chomsky, Noam. 1993. *The Prosperous Few and the Restless Many*. Berkeley: University of California Press.

Cobb, Clifford, et al. 1995. *The Genuine Progress Indicator*. San Francisco, Calif.: Redefining Progress.

Cobb, John B., Jr., 1992. *Sustainability: Economics, Ecology and Justice*. Maryknoll, N.Y.: Orbis Books.

Costanza, Robert, et al. 1997. *An Introduction to Ecological Economics: The Science and Management of Sustainability*. Boca Raton, Fla.: Lewis.

Court, T. de la. 1990. *Beyond Brundtland: Green Development in the 1990s*. Atlantic Highlands, N.J.: Zed Books.

Daily, Gretchen C., ed. 1997. *Nature's Services: Societal Dependence on Natural Ecosystems*. Covelo, Calif.: Island Press.

Daly, Herman E. 1991. *Steady-State Economics*. 2d ed. Covelo, Calif.: Island Press.

Daly, Herman E. 1992. *Environmentally Sustainable Development: Building on Brundtland*. Covelo, Calif.: Island Press.

Daly, Herman E. 1996. *Beyond Growth: The Economics of Sustainable Development*. Boston: Beacon Press.

Daly, Herman E., and John B. Cobb, Jr. 1994. *For the Common Good: Redirecting the Economy Toward Community, the Environment, and a Sustainable Future*. 2d ed. Boston: Beacon Press.

Daly, Herman E., and Kenneth N. Townsend, eds. 1993. *Valuing the Earth: Economics, Ecology, Ethics*. Cambridge, Mass.: MIT Press.

Davis, John. 1991. *Greening Business: Managing for Sustainable Development*. Cambridge, Mass.: Basil Blackwell.

DeSimone, Livio, and Frank Popoff. 1997. *Eco-Efficiency: The Business Link to Sustainable Development*. Cambridge, Mass.: MIT Press.

Douthwaite, Richard. 1992. *The Growth Illusion*. Tulsa, Okla.: Council Oaks Books.

Durning, Alan T. 1989. *Poverty and the Environment: Reversing the Downward Spiral*. Washington, D.C.: Worldwatch Institute.

Durning, Alan T. 1992. *How Much Is Enough? The Consumer Society and the Earth*. New York: Norton.

Earth Island Books. 1993. *The Case Against Free Trade: GATT, NAFTA, and the Globalization of Corporate Power*. Berkeley, Calif.: North Atlantic Books.

Eden, Sally. 1996. *Environmental Issues and Business: Implications of a Changing Agenda*. New York: Wiley.

Elkington, John, et al. 1990. *The Green Consumer*. New York: Penguin Books.

Epping, Randy Charles. 1992. *A Beginner's Guide to the World Economy*. New York: Random House.

Etzioni, A. 1988. *The Moral Dimension: Toward a New Economics*. New York: Free Press.

Fischer, Kurt, and Johan Schot, eds. 1993. *Environmental Strategies for Industry*. Covelo, Calif.: Island Press.

Flavin, Christopher, and John E. Young. 1993. "Shaping the Next Industrial Revolution." In Lester R. Brown et al., *State of the World 1993*. New York: Norton, 180–99.

Foster, John Bellamy. 1995. *The Vulnerable Planet: A Short Economic History of the Planet*. New York: Monthly Review Press.

French, Hilary F. 1993. *Costly Trade-Offs: Reconciling Trade and the Environment*. Washington, D.C.: Worldwatch Institute.

French, Hilary F. 1994. "Rebuilding the World Bank." In Lester R. Brown et al., *State of the World 1994*. Washington, D.C.: Worldwatch Institute, 156–76.

George, Susan. 1992. *The Debt Boomerang: How Third World Debt Harms Us All*. Boulder, Colo.: Westview Press.

Goldsmith, Edward, et al. 1992. *The Future of Development: Reflections on Environment and Development*. Berkeley, Calif.: International Society for Ecology and Culture.

Goodland, Robert, et al. 1992. *Environmentally Sustainable Economic Development: Building on Brundtland*. Paris: UNESCO Press.

Goodstein, E. B. 1994. *Economics and the Environment*. Englewood Cliffs, N.J.: Prentice Hall.

Goudy, John, and Sabine O'Hara. 1995. *Economic Theory for Environmentalists*. New York: St. Lucie Press.

Grossman, Richard, and Frank T. Adams. 1992. *Taking Care of Business: Citizenship and the Charter of Incorporation*. Cambridge, Mass.: Charter, Inc.

Hackett, Stephen C. 1997. *Environmental and Natural Resource Economics: Theory, Policy, and Sustainability*. Armonk, N.Y.: M.E. Sharpe.

Hawken, Paul. 1993. *The Ecology of Commerce*. New York: HarperCollins.

Henderson, Hazel. 1991. *Paradigms in Progress: Life Beyond Economics*. Indianapolis: Knowledge Systems.

Henderson, Hazel. 1996. *Building a Win–Win World: Life Beyond Global Economic Warfare*. San Francisco: Berrett-Koehler.

Hirschorn, Joel S., and Kirsten U. Oldenberg. 1990. *Prosperity Without Pollution: The Prevention Strategy for Industry and Consumers*. New York: Van Nostrand Reinhold.

Institute for Local Self-Reliance. 1990. *Proven Profits from Pollution Prevention*. Washington, D.C.: Institute for Local Self-Reliance.

Jacobs, Jane. 1993. *Systems of Survival*. New York: Random House.

Jacobs, Michael. 1993. *The Green Economy: Environment, Sustainable Development, and Politics*. New York: Pluto Press.

Jacobson, Jodi L. 1992. *Gender Bias: Roadblock to Sustainable Development*. Washington, D.C.: Worldwatch Institute.

Jansson, Ann Marie, et al., eds. 1994. *Investing in Natural Capital: The Ecological Economics Approach to Sustainability*. Covelo, Calif.: Island Press.

Kane, Hal, and Linda Starke. 1992. *Time for Change: A New Approach to Environment and Development*. Covelo, Calif.: Island Press.

Kassiola, Joel Jay. 1990. *The Death of Industrial Civilization*. Albany: State University of New York Press.

Kazis, Richard, and Richard L. Grossman. 1991. *Fear at Work: Job Blackmail, Labor, and the Environment*. Philadelphia: New Society.

Kennedy, Paul. 1993. *Preparing for the Twenty-First Century*. New York: Random House.

Korten, David. 1995. *When Corporations Rule the World*. West Hartford, Conn.: Kumarian Press.

Krishman, Bajaram, et al., eds. 1995. *A Survey of Ecological Economics*. Covelo, Calif.: Island Press.

Lang, Tim, and Colin Hines. 1994. *The New Protectionism: Protecting the Future Against Free Trade*. New York: New Press.

MacDonald, Mary. 1998. *Agendas for Sustainability: Environment and Development into the 21st Century*. New York: Routledge.

Makower, Joel. 1993. *The E-Factor: The Bottom-Line Approach to Environmentally Responsible Business*. New York: Times Books.

Mander, Jerry, and Edward Goldsmith, eds. 1996. *The Case Against the Global Economy*. San Francisco: Sierra Club Books.

Marchak, Patricia. 1998. "Environment and Resource Protection: Does NAFTA Make a Difference?" *Organization & Environment*, vol. 11, no. 2, 133–54.

Maser, Chris. 1997. *Sustainable Community Development*. Boca Raton, Fla.: Lewis.

McConnell, Campbell R. 1997. *Economics: Principles, Problems, and Policies*. 13th ed. New York: McGraw-Hill.

Mehta, Shekhar, et al. 1997. *Controlling Pollution: Incentives and Regulation*. Thousand Oaks, Calif.: Sage.

Mikesell, Raymond F., and Lawrence F. Williams. 1992. *International Banks and the Environment*. San Francisco: Sierra Club Books.

Monks, Robert A. G., and Nell Minow. 1991. *Power and Accountability*. New York: HarperCollins.

Moore, Curtis, and Alan Miller. 1994. *Green Gold: Japan, Germany, and the United States, and the Race for Environmental Technology*. Boston: Beacon Press.

National Academy of Sciences. 1993. *Greening Industrial Ecosystems*. Washington, D.C.: National Academy Press.

National Academy of Sciences. 1995. *Assigning Economic Value to Natural Resources*. Washington, D.C.: National Academy Press.

Neber, Philip A. 1990. *Natural Resource Economics: Conservation and Exploitation*. New York: Cambridge University Press.

Office of Technology Assessment. 1992. *Green Products by Design*. Washington, D.C.: U.S. Government Printing Office.

Office of Technology Assessment. 1992. *Trade and Environment: Conflicts and Opportunities*. Washington, D.C.: Office of Technology Assessment.

O'Riordan, ed. 1997. *Ecotaxation*. New York: St. Martin's Press.

Paepke, C. Owen. 1993. *The Evolution of Progress: The End of Economic Growth and the Beginning of the Human Transformation*. New York: Random House.

Panayotou, Theodore. 1993. *Green Markets: The Economics of Sustainable Development*. San Francisco: Institute for Contemporary Studies Press.

Pearce, David. 1993. *Economic Values and the Natural World*. Cambridge, Mass.: MIT Press.

Pearce, David, et al. 1991. *Blueprint 2: Greening the World Economy*. East Haven, Conn.: Earthscan.

Peet, John. 1992. *Energy and the Ecological Economics of Sustainability*. Covelo, Calif.: Island Press.

Peng, Martin Khor Kok. 1992. *The Future of North–South Relations: Conflict or Co-operation*. London: WEC Press.

Peterson, Tarla Rai. 1997. *Sharing the Earth: The Rhetoric of Sustainable Development*. Columbia: University of South Carolina Press.

Piasecki, Bruce W. 1995. *Corporate Environmental Strategy: The Avalanche of Change Since Bhopal*. New York: Wiley.

Plant, Christopher, and Judith Plant, eds. 1991. *Green Business: Hope or Hoax?* Philadelphia: New Society.

Power, Thomas M. 1996. *Extraction and the Environment: The Economic Battle to Control Our Natural Landscapes*. Covelo, Calif.: Island Press.

President's Council on Sustainable Development. 1996. *Sustainable America: A New Consensus for the Future.* Washington, D.C.: U.S. Government Printing Office.

Prince, R., and P. Gordon. 1994. *Greening the National Accounts.* Washington, D.C.: U.S. Congressional Budget Office.

Prugh, Thomas, et al. 1995. *Natural Capital and Human Economic Survival.* Solomons, Md.: International Society for Ecological Economics.

Public Health Institute. 1994. *Jobs and the Environment Workbook.* New York: Public Health Institute.

Renner, Michael. 1991. *Jobs in a Sustainable Economy.* Washington, D.C.: Worldwatch Institute.

Repetto, Robert. 1992. "Accounting for Environmental Assets." *Scientific American,* June, 94–100.

Repetto, Robert. 1995. *Jobs, Competitiveness, and Environmental Regulation: What Are the Real Issues?* Washington, D.C.: World Resources Institute.

Repetto, Robert, et al. 1991. *Transforming Technology: An Agenda for Environmentally Sustainable Growth in the Twenty-First Century.* Washington, D.C.: World Resources Institute.

Repetto, Robert, et al. 1992. *Green Fees: How a Tax Shift Can Work for the Environment and the Economy.* Washington, D.C.: World Resources Institute.

Rich, Bruce. 1994. *Mortgaging the Earth.* Boston: Beacon Press.

Roddick, Anita. 1991. *Body and Soul: Profits with Principles.* New York: Crown.

Roodman, David M. 1996. *Paying the Piper: Subsidies, Politics, and the Environment.* Washington, D.C.: Worldwatch Institute.

Roodman, David M. 1997. *Getting the Signals Right: Tax Reform to Protect the Environment and the Economy.* Washington, D.C.: Worldwatch Institute.

Rotman, David M. 1998. *The Natural Wealth of Nations: Harnessing the Market for the Environment.* New York: Norton.

Rugman, Alan M., et al., eds. 1998. *Trade and the Environment.* Northhampton, Mass.: Edgar Elgar.

Runge, C. Ford. 1994. *Freer Trade, Protected Environment: Balancing Trade Liberalization and Environmental Interests.* New York: Council on Foreign Relations Press.

Saunders, Tedd, and Loretta McGovern. 1994. *The Bottom Line of Green Is Black.* New York: HarperCollins.

Schmidheiny, Stephan. 1992. *Changing Course: A Global Business Perspective on Development and the Environment.* Cambridge, Mass.: MIT Press.

Schumacher, E. F. 1973. *Small Is Beautiful: Economics as if the Earth Mattered.* New York: Harper & Row.

Seabrook, Jeremy. 1990. *The Myth of the Market.* Bideford, U.K.: Green Books.

Smart, Bruce, ed. 1992. *Beyond Compliance: A New Industry View of the Environment.* Washington, D.C.: World Resources Institute.

Smith, Denis, ed. 1993. *Business and the Environment: Restoring the Dream.* New York: St. Martin's Press.

Smith, Joseph Wayne, et al. 1997. *Healing a Wounded World: Economics, Ecology, and Health for a Sustainable Life.* Westport, Conn.: Praeger.

Smith, Joseph, et al. 1998. *The Bankruptcy of Economics: Ecology, Economics, and the Sustainability of the Earth.* New York: St. Martin's Press.

Stead, W. Edward, and John Garner Stead. 1992. *Management for a Small Planet.* New York: Sage.

Steffen, Alex, and Alan Atkisson. 1995. "The Netherlands' Radical Practical Plan." *Whole Earth Review,* Fall, 94–98.

Strong, Susan. 1995. *The GDP Myth: How It Harms Our Quality of Life and What Communities Are Doing About It.* Mountain View, Calif.: Center for Economic Conversion.

Tibbs, Hardin B. C. 1992. "Industrial Ecology: An Environmental Agenda for Industry." *Whole Earth Review,* Winter, 4–19.

Tietenberg, Tom. 1992. *Environmental and Resource Economics.* 3d ed. Glenview, Ill.: Scott, Foresman.

Toffler, Alvin, and Heidi Toffler. 1990. *Powershift.* New York: Bantam.

Tokar, Brian. 1997. *Earth for Sale: Reclaiming Ecology in the Age of Corporate Greenwash.* Boston: South End Press.

Turner, R. Kerry, et al. 1994. *Environmental Economics: An Elementary Introduction.* Baltimore: Johns Hopkins University Press.

Wachtel, Paul. 1988. *The Poverty of Affluence.* Santa Cruz, Calif.: New Society.

Ward, Barbara. 1979. *Progress for a Small Planet.* New York: Norton.

World Resources Institute. 1995. *Corporate Environmental Accounting.* Washington, D.C.: World Resources Institute.

Yu, Douglas. 1994. "Free Trade is Green, Protectionism Is Not." *Conservation Biology,* vol. 8, no. 4, 989–96.

28 / POLITICS AND ENVIRONMENT

Readings

Abbey, Edward. 1986. *The Monkey Wrench Gang.* New York: Avon.

Adler, Jonathan. 1996. *Environmentalism at the Crossroads: Green Activism in America.* Washington, D.C.: Capital Research Center.

Arnold, Ron, and Alan Gottlieb. 1993. *Trashing the Economy: How Runaway Environmentalism Is Wrecking America.* Bellevue, Wash.: Free Enterprise Press.

Arrow, Kenneth J., et al. 1996. "Is There a Role for Benefit–Cost Analysis in Environmental, Health, and Safety Regulation? *Science,* vol. 272, April 12, 221–22.

Atchia, Michael, and Shawn Tropp, eds. 1995. *Environmental Management: Issues and Solutions.* New York: Wiley.

Bast, Joseph L., et al. 1994. *Eco-Sanity: A Common-Sense Guide to Environmentalism.* Lanham, Md.: Madison Books.

Basta, Nicholas. 1991. *The Environmental Career Guide: Job Opportunities with the Earth in Mind.* New York: Wiley.

Beder, Sharon. 1997. *Global Spin: The Corporate Assault on Environmentalism.* White River Junction, Vt.: Chelsea Green.

Berry, Joyce K., and John C. Gordon, eds. 1993. *Environmental Leadership: Developing Effective Skills and Styles.* Covelo, Calif.: Island Press.

Bramwell, Anna. 1994. *The Fading of the Greens: The Decline of Environmental Politics in the West.* New Haven, Conn.: Yale University Press.

Brick, Phil. 1995. "Determined Opposition: The Wise Use Movement Challenges Environmentalism." *Environment,* vol. 27, no. 8, 17–41.

Brockmeier, Jons, et al. 1998. *Greenspeak: A Study of Environmental Discourse.* Thousand Oaks, Calif.: Sage.

Bryant, Bunyan, ed. 1995. *Environmental Justice: Issues, Policies, and Solutions.* Covelo, Calif.: Island Press.

Bullard, Robert D., ed. 1994. *Unequal Protection: Environmental Justice and Communities of Color.* San Francisco: Sierra Club Books.

Caldwell, Lynton K. 1997. *International Environmental Policy: From the Twentieth to the Twenty-First Century.* Durham, N.C.: Duke University Press.

Carty, Winthrop P., and Elizabeth Lee. 1992. *The Rhino Man and Other Uncommon Environmentalists.* Washington, D.C.: Seven Locks Press.

Chetlow, Marian R., and Daniel C. Esty. 1997. *Thinking Ecologically: The Next Generation of Environmental Policy.* New Haven, Conn.: Yale University Press.

Chomsky, Noam. 1994. *Secrets, Lies, and Democracy.* Tucson, Ariz.: Odonian Press.

Chomsky, Noam, and Edward Herman. 1988. *Manufacturing Consent: The Political Economy of the Mass Media.* New York: Pantheon.

Choucri, Nazli. 1991. "The Global Environment and Multinational Corporations." *Technology Review,* April, 52–59.

Choucri, Nazli, ed. 1993. *Global Accord: Environmental Challenges and International Responses.* Cambridge, Mass.: MIT Press.

Cohn, Susan. 1995. *Green at Work: Finding a Business Career That Works for the Environment.* Rev. ed. Covelo, Calif.: Island Press.

Coleman, Daniel A. 1994. *Ecopolitics: Building a Green Society.* New Brunswick, N.J.: Rutgers University Press.

Conca, Ken, and Geoffrey D. Dabelko, eds. 1998. *Green Planet Blues: Environmental Politics from Stockholm to Tokyo.* 2d ed. Boulder, Colo.: Westview Press.

Conley, Verena. 1997. *Ecopolitics: The Environment in Poststructuralist Thought.* New York: Routledge.

Council on Environmental Policy. 1997. *The National Environmental Policy Act: A Study of Its Effectiveness After Twenty-Five Years.* Washington, D.C.: Council on Environmental Quality, Executive Office of the President.

Cousins, Norman. 1987. *The Pathology of Power.* New York: Norton.

Day, Martyn. 1997. *Environmental Action: A Citizen's Guide.* Chicago: Pluto Press.

Dobson, Andrew. 1990. *Green Political Thought: An Introduction.* New York: Routledge.

Dowie, Mark. 1995. *Losing Ground: American Environmentalism at the Close of the Twentieth Century.* Cambridge, Mass.: MIT Press.

Dragun, Andrew, and Kristin M. Jakobsson, eds. 1997. *Sustainability and Global Environmental Policy: New Perspectives.* Lyme, N.H.: Edward Elgar.

Durnil, Gordon K. 1995. *The Making of an Environmental Conservative.* Bloomington: Indiana University Press.

Eagen, David J., and David W. Orr. 1994. *The Campus and Environmental Responsibility.* San Francisco: Jossey-Bass.

Echeverria, John, and Raymond Booth Eby, eds. 1995. *Let the People Judge: A Reader on the Wise Use Movement.* Covelo, Calif.: Island Press.

Ehrlich, Paul R., and Anne H. Ehrlich. 1996. *Betrayal of Science and Reason: How Anti-Environmental Rhetoric Threatens Our Future.* Covelo, Calif.: Island Press.

Elkins, Paul, and Jakob von Uexhull. 1992. *Grassroots Movements for Global Change.* New York: Routledge.

Ember, Lois. 1995. "EPA at 25." *Chemical and Engineering News,* October 30, 16–49.

Environmental Careers Organization. 1993. *The New Complete Guide to Environmental Careers.* Covelo, Calif.: Island Press.

Erickson, Brad, ed. 1990. *Call to Action: Handbook for Ecology, Peace, and Justice.* San Francisco: Sierra Club Books.

Fagin, Dan, and Marianne Lavelle. 1996. *Toxic Deception.* Secaucus, N.J.: Carol Publishing.

Fanning, Odum. 1995. *Opportunities for Environmental Careers.* Lincolnwood, Ill.: VCM Career Horizons.

Farber, Daniel, ed. 1998. *The Struggle for Ecological Democracy. Environmental Justice Movements in the United States.* New York: Guilford.

Finley, Roger W., and Daniel A. Farber. 1996. *Environmental Law in a Nutshell.* St. Paul, Minn.: West.

Flattau, Edward. 1998. *Tracking the Charlatans: Countering the Eco-Bashers.* Washington, DC: Global Horizins.

Fiorino, Daniel J. 1995. *Making Environmental Policy.* Berkeley: University of California Press.

Fisher, Julie. 1993. *The Road from Rio: Sustainable Development and the Nongovernmental Movement in the Third World.* Westport, Conn.: Praeger.

Flavin, Christopher, and John E. Young. 1993. "Shaping the Next Industrial Revolution." In Lester R. Brown et al., *State of the World 1993.* New York: Norton, 180–99.

Foreman, Dave. 1990. *Confessions of an Eco-Warrior.* New York: Crown.

French, Hilary E. 1992. *After the Earth Summit: The Future of Environmental Governance.* Washington, D.C.: Worldwatch Institute.

French, Hilary E. 1995. *Partnership for the Planet: An Environmental Agenda for the United Nations.* Washington, D.C.: Worldwatch Institute.

Gandhi, M. K. 1961. *Non-Violent Resistance.* New York: Schocken.

Gardner, Richard N. 1993. *Negotiating Survival: Four Priorities After Rio.* Washington, D.C.: Council on Foreign Relations.

Gartner, Bob. 1995. *Careers Inside the World of Environmental Science.* New York: Rosen.

Gay, Kathlyn. 1994. *Pollution and the Powerless: The Environmental Justice Movement.* New York: F. Watts.

Gedicks, Al. 1993. *The New Resource Wars: Native and Environmental Struggles Against Multinational Corporations.* Boston: South End Press.

Gorz, Andre. 1989. *Ecology as Politics.* Boston: South End Press.

Greenpeace. 1993. *The Greenpeace Guide to Anti-Environmental Organizations.* Berkeley, Calif.: Odonian Press.

Greider, William. 1992. *Who Will Tell the People?* New York: Simon & Schuster.

Grove, Richard. 1994. *Green Imperialism.* New York: Cambridge University Press.

Gupta, Avijit, and Mukul Asher. 1998. *Environment and the Developing World: Principles, Policies, and Management.* New York: Wiley.

Hall, Bob. 1990. *Environmental Politics: Lessons from the Grassroots.* Durham, N.C.: Institute for Southern Studies.

Helvarg, David. 1994. *The War Against the Greens.* San Francisco: Sierra Club Books.

Hempel, Lamont C. 1995. *Environmental Governance: The Global Challenge.* Covelo, Calif.: Island Press.

Hoffman, Andrew J. 1997. *From Heresy to Dogma: An Institutional History of Corporate Environmentalism.* San Francisco, Calif.: New Lexington Press.

Hunter, J. Robert. 1997. *Simple Things Won't Save the Earth.* Austin: University of Texas Press.

Isaac, Katherine. 1992. *Ralph Nader Presents Civics for Democracy: A Journey for Teachers and Students.* Washington, D.C.: Essential Books.

Johnston, Barbara R. 1994. *Who Pays the Price? The Sociocultural Context of Environmental Crisis.* Covelo, Calif.: Island Press.

Kahn, S. 1982. *A Guidebook for Grassroots Leaders.* New York: McGraw-Hill.

Keene, Ann T. 1993. *Observers and Protectors of Nature.* New York: Oxford University Press.

Kempton, Willett, et al. 1995. *Environmental Values in American Culture.* Cambridge, Mass.: MIT Press.

Keniry, Julian. 1995. *Ecodemia: Campus Environmental Stewardship at the Turn of the Century.* Washington, D.C.: National Wildlife Federation.

Killingsworth, M. J., and Jacqueline S. Palmer. 1992. *Ecospeak: Rhetoric and Environmental Politics in America.* Washington, D.C.: Campus Outreach Division, National Wildlife Federation.

LaMay, Craig L., and Everette E. Dennis, eds. 1992. *Media and the Environment.* Covelo, Calif.: Island Press.

Landy, Marc K., et al. 1994. *The Environmental Protection Agency: Asking the Wrong Questions, from Nixon to Clinton.* New York: Oxford University Press.

Learner, Steve. 1992. *Beyond the Earth Summit: Conversations with Advocates of Sustainable Development.* New York: Common Knowledge Press.

Learner, Steve. 1997. *Eco-Pioneers: Practical Visionaries Solving Environmental Problems.* Cambridge, Mass.: MIT Press.

Lee, Kai N. 1993. *Compass and Gyroscope: Integrating Science and Politics for the Environment.* Covelo, Calif.: Island Press.

Lewis, Martin W. 1992. *Green Delusions: An Environmentalist Critique of Radical Environmentalism.* Durham, N.C.: Duke University Press.

Lipschutz, Ronnie D., and Judith Mayer. 1996. *Global Civil Society and Global Environmental Governance: The Politics of Nature from Place to Planet.* Albany: State University of New York Press.

Loudiyi, Dounia, and Alison Meares. 1993. *Women in Conservation: Tools for Analysis and a Framework for Action.* Covelo, Calif.: Island Press.

Luke, Timothy W. 1997. *Ecocritique: Contesting the Politics of Nature, Economy, and Culture.* Minneapolis: University of Minnesota Press.

Manes, Christopher. 1990. *Green Rage: Radical Environmentalism and the Unmaking of Civilization.* Boston: Little, Brown.

Mansfield, William H. III. 1998. "Taking the University to Task." *World Watch,* May/June, 24–30.

Mathews, Christopher. 1988. *Hardball: How Politics Is Played—Told by One Who Knows the Game.* New York: Summit Books.

McCoy, Michael, and Patrick McCully. 1994. *The Road from Rio: An NGO Action Guide to Environment and Development.* Oxford, U.K.: Jon Carpenter.

Meadows, Donella H. 1991. *Global Citizen.* Covelo, Calif.: Island Press.

Megalli, Mark, and Andy Friedman. 1993. *Masks of Deception: Corporate Front Groups in America.* Washington, D.C.: Essential Books.

Merchant, Carolyn. 1992. *Radical Ecology: The Search for a Livable World.* New York: Routledge.

Middleton, Neil, et al. 1993. *Tears of the Crocodile: From Rio to Reality in the Developing World.* Boulder, Colo.: Westview Press.

Mintz, Joel A. 1996. *Enforcement at the EPA: High Stakes and Hard Choices.* Austin: University of Texas Press.

Moody, Joan, and Richard Wisansky, eds. 1994. *Earth Work: Resource Guide to Nationwide Green Jobs.* New York: HarperCollins.

Morgan, Bradley J., and Joseph M. Palmisano. 1993. *Environmental Career Directory.* Detroit: Visible Ink Press.

Myers, Norman. 1993. *Ultimate Security: The Environmental Basis of Political Stability.* New York: Norton.

National Roundtable on the Environment and Economy. 1996. *The Green Guide: A User's Guide to Sustainable Development for Canadian Colleges.* Ottawa, Ontario, Canada: National Roundtable on the Environment and Economy.

National Wildlife Federation. Annual. *Campus Environmental Yearbook.* Washington, D.C.: National Wildlife Federation.

Nissani, Moti. 1992. *Lives in the Balance: The Cold War and American Politics.* Wakefield, N.H.: Hollowbrook.

Ophuls, William, and A. Stephen Boyan, Jr. 1992. *Ecology and the Politics of Scarcity Revisited: The Unravelling of the American Dream.* San Francisco: W.H. Freeman.

Pearce, Fred. 1991. *Green Warriors: The People and Politics Behind the Environmental Revolution.* London: Bodley Head.

Peavey, Fran, et al. 1986. *Heart Politics.* Philadelphia: New Society.

Piasecki, Bruce, and Peter Asmus. 1990. *In Search of Environmental Excellence: Moving Beyond Blame.* New York: Simon & Schuster.

Pick, Maritza. 1993. *How to Save Your Neighborhood, City or Town: The Sierra Club Guide to Community Organizing.* San Francisco: Sierra Club Books.

Plant, Judith, and Christopher Plant, eds. 1992. *Putting Power in Its Place: Create Community Control.* Philadelphia: New Society.

Porter, Gareth, and Janet Welsh Brown. 1994. *Global Environmental Politics.* 2d ed. Boulder, Colo.: Westview Press.

Powell, Frona M. 1997. *Law and the Environment.* Cincinnati, Ohio: South-Western.

Prins, Gwaya, ed. 1993. *Threats Without Enemies: Facing Environmental Insecurity.* East Haven, Conn.: Earthscan.

Rai, Verso. 1995. *Chomsky's Politics.* London: Verso.

Raskin, Jamin B., and John Bonifaz. 1994. *The Wealth Primary.* Washington, D.C.: Center for Responsive Politics.

Rifkin, Jeremy. 1991. *Biosphere Politics: A New Consciousness for a New Century.* New York: Crown.

Rogers, Adam. 1993. *The Earth Summit: A Planetary Reckoning.* Los Angeles: Global View Press.

Romm, Joseph J. 1992. *The Once and Future Superpower: How to Restore America's Economic, Energy, and Environmental Security.* New York: William Morrow.

Romm, Joseph J. 1993. *Defining National Security: The Nonmilitary Aspects.* New York: Council on Foreign Relations Press.

Roseland, Mark. 1997. *Toward Sustainable Communities: Resources for Citizens and Their Governments.* Philadelphia: New Society Publishers.

Rosenbaum, Walter A. 1998. *Environmental, Politics and Solutions.* Washington, D.C.: Congressional Quarterly.

Rowell, Andrew. 1997. *Green Backlash: Subversion of the Environmental Movement.* New York: Routledge.

Ruckelshaus, William D., and Karl Hausker, eds. 1998. *The Environmental Protection System in Transition: Toward a More Desirable Future.* Washington, D.C.: Center for Strategic and International Studies.

Sachs, Aaron. 1995. *Eco-Justice: Linking Human Rights and the Environment.* Washington, D.C.: Worldwatch Institute.

Sale, Kirkpatrick. 1993. *The Green Revolution: The American Environmental Movement 1962–1992.* New York: Hill & Wang.

Sanjor, William. 1992. *Why the EPA Is Like It Is and What Can Be Done About It.* Washington, D.C.: Environmental Research Foundation.

Scarce, Rik, 1990. *Eco-Warriors: Understanding the Radical Environmental Movement.* Chicago: Noble Press.

Schnaiberg, Allan, and Kenneth Alan Gould. 1993. *Environment and Society: The Enduring Conflict.* New York: St. Martin's Press.

Schwab, Jim. 1994. *Deeper Shades of Green: The Rise of Blue Collar and Minority Environmentalism in America.* San Francisco: Sierra Club Books.

Selcraig, Bruce. 1998. "Reading, 'Riting, and Ravaging: The Three R's Brought to You by Corporate America and the Far Right." *Sierra,* May/June, 60–65, 86–88.

Shabecoff, Philip. 1993. *A Fierce Green Fire: The American Environmental Movement.* New York: Hill & Wang.

Shabecoff, Philip. 1996. *A New Name for Peace: International Environmentalism, Sustainable Development, and Democracy.* Hanover, N.H.: University of New England Press.

Sitarz, Daniel, ed. 1993. *Agenda 21, The Earth Summit Strategy to Save Our Planet.* Boulder, Colo.: Earth Press.

Sivard, Ruth Leger. Annual. *World Military and Social Expenditures.* Washington, D.C.: World Priorities.

Smil, Vaclav. 1994. *Global Ecology: Environmental Change and Social Flexibility.* New York: Routledge.

Smith, April, and the Student Environmental Action Coalition. 1993. *Campus Ecology: A Guide to Assessing Environmental Quality and Creating Strategies for Change.* Los Angeles, Calif.: Living Planet Press.

Smookler, Andrew Bard. 1984. *The Parable of the Tribes.* Berkeley: University of California Press.

Snow, Donald. 1992. *Inside the Environmental Movement: Meeting the Leadership Challenge.* Covelo, Calif.: Island Press.

Stauber, John C., and Sheldon Rampton. 1995. *Toxic Sludge Is Good for You: Lies, Damn Lies, and the Public Relations Industry.* Madison, Wis.: PR Watch.

Stone, Christopher D. 1993. *The Gnat Is Older Than Man: Global Environment and Human Agenda.* Princeton, N.J.: Princeton University Press.

Switzer, Jacqueline, and Gary Byner. 1998. *Environmental Politics: Domestic and Global Dimensions.* New York: St. Martin's Press.

Taylor, Ann. 1992. *A Practical Politics of the Environment.* New York: Routledge.

Taylor, Bob Pepperman. 1992. *Our Limits Transgressed: Environmental Political Thought in America.* Lawrence: University Press of Kansas.

Theobald, Robert. 1992. *Turning the Century: Personal and Organizational Strategies for Your Changed World.* Indianapolis: Knowledge Systems.

Theodore, Lewis, et al., eds. 1998. *Environmental Management: Problems and Solutions.* Berlin, Germany: Springer-Verlag

Tokar, Brian. 1995. "The 'Wise Use' Backlash: Responding to Militant Anti-Environmentalism." *The Ecologist,* vol. 25, no. 4, July/August, 150–56.

Vig, Norman, and Michael Kraft. 1990. *Environmental Policy in the 1990s.* Washington, D.C.: Congressional Quarterly.

Voight, Kristina, et al. 1997. *Ecosystems: Balancing Science with Management.* New York: Springer.

Warner, David J. 1992. *Environmental Careers: A Practical Guide to Opportunities in the 1990s.* Boca Raton, Fla.: Lewis.

Watson, Paul. 1994. *Ocean Warrior.* Marina del Ray, Calif.: Key Porter Books.

Weinstein, Mirriam. 1993. *Making a Difference College Guide: Education for a Better World.* San Anselmo, Calif.: Sage.

Werbach, Adam. 1997. *Act Now, Apologize Later.* New York: Cliff Street Books.

Westra, Laura, and Peter Wenz, eds. 1995. *Facing Environmental Racism: Confronting Issues of Global Justice.* Lantham, Md.: Rowman & Littlefield.

Wilkinson, Todd. 1998. *Science Under Siege: The Politicians' War on Nature and Truth.* New York: Johnson Books.

World Bank. 1997. *Five Years After Rio: Innovations in Environmental Policy.* Washington, D.C.: World Bank.

World Resources Institute. 1993. *A New Generation of Environmental Leadership: Action for the Environment and the Economy.* Washington, D.C.: World Resources Institute.

Zimmerman, Michael. 1995. *Science, Nonscience, and Nonsense: Approaching Environmental Literacy.* Baltimore, Md.: Johns Hopkins University Press.

Zimmerman, Richard. 1992. *What Can I Do to Make a Difference?* New York: Plume.

29 / ENVIRONMENTAL WORLDVIEWS, ETHICS, AND SUSTAINABILITY

Readings

Anderson, E. N. 1996. *Ecologies of the Heart: Emotion, Belief, and the Environment.* New York: Oxford University Press.

Andruss, Van, et al. 1990. *Home! A Bioregional Reader.* Philadelphia: New Society.

Armstrong, Susan J., and Richard G. Botzler, eds. 1993. *Environmental Ethics: Divergence and Convergence.* New York: McGraw-Hill.

Attfield, Robin. 1991. *The Ethics of Environmental Concern.* 2d ed. Athens: University of Georgia Press.

Ausbel, Kenny. 1997. *Restoring the Earth: Visionary Solutions from the Bioneers.* Tiburon, Calif.: H.J. Kramer.

Baer, Donald M., and Elsie M. Pinkston, eds. 1997. *Environment and Behavior.* Boulder, Colo.: Westview Press.

Barbour, Ian G. 1990. *Religion in an Age of Science.* New York: Harper & Row.

Bazerman, Max H., et al., eds. 1997. *Environment, Ethics, and Behavior: The Psychology of Environmental Valuation and Degradation.* San Francisco: Jossey-Bass.

Berman, Morris. 1989. *Coming to Our Senses: Body and Spirit in the Hidden History of the West.* New York: Bantam.

Berry, Thomas. 1988. *The Dream of the Earth.* San Francisco: Sierra Club Books.

Berry, Wendell. 1990. *What Are People For?* Berkeley, Calif.: North Point Press.

Bookchin, Murray. 1990. *Remaking Society: Pathways to a Green Future.* Boston: South End Press.

Bossel, Hartmut. 1998. *Paths to a Sustainable Future.* New York: Cambridge University Press.

Bowers, C. A. 1993. *Education, Cultural Myths, and the Ecological Crisis: Toward Deep Changes.* Albany: State University of New York Press.

Bowers, C. A. 1997. *The Culture of Denial: Why the Environmental Movement Needs a Strategy for Reforming Universities and Schools.* Albany: State University of New York Press.

Broder, Bill. 1992. *The Sacred Hoop: A Cycle of Earth Tales.* San Francisco: Sierra Club Books.

Brower, David, and Steve Chapple. 1995. *Let the Mountains Talk, Let the Rivers Run.* New York: HarperCollins.

Caldicott, Helen. 1992. *If You Love This Planet: A Plan to Heal the Earth.* New York: Norton.

Callicott, J. Baird. 1994. *Earth's Insights (A Multicultural Survey of Ecological Ethics from the Mediterranean Basin to the Australian Outback).* Berkeley: University of California Press.

Campbell, Joseph. 1988. *The Power of Myth.* New York: Doubleday.

Capra, Fritjof. 1983. *The Turning Point: Science, Society, and the Rising Culture.* New York: Bantam.

Carley, Michael, and Philippe Spapens. 1998. *Sharing the World: Sustainable Living and Global Equity in the 21st Century.* London: Earthscan.

Clark, John, ed. 1990. *Renewing the Earth: The Promise of Social Ecology.* London: Green Print.

Clark, Mary E. 1989. *Ariadne's Thread: The Search for New Models of Thinking.* New York: St. Martin's Press.

Clark, Mary E., and Sandra A. Wawrytko, eds. 1990. *Toward an Integrated Interdisciplinary College Education.* Westport, Conn.: Greenwood Press.

Coates, Peter. 1998. *Nature: Western Attitudes Since Ancient Times.* Berkeley: University of California Press.

Cobb, John B., Jr. 1996. *Is It Too Late? A Theology of Ecology.* Rev. ed. Denton, Tex.: Environmental Ethics Books.

Cohen, Michael J. 1989. *Connecting with Nature: Creating Moments That Let Earth Teach.* Eugene, Oreg.: World Peace University.

Cohen, Michael J. 1997. "The Secrets-of-Nature Trail and Game: A Hands-On Challenge for Multi-Sensory Learning and Relating." *Trumpeter,* vol. 14, no. 1, 10–13.

Collett, Jonathan, and Stephen Karakashian, eds. 1995. *Greening the College Curriculum: A Guide to Environmental Teaching in the Liberal Arts Colleges.* Covelo, Calif.: Island Press.

Cornell, Joseph. 1989. *Sharing the Joy of Nature.* Nevada City, Calif.: Dawn.

Crocker, David A., and Toby Linden, eds. 1997. *The Ethics of Consumption.* Savage, Md.: Rowman & Littlefield.

Desjardins, Joseph R. 1993. *Environmental Ethics.* Belmont, Calif.: Wadsworth.

Devall, Bill, and George Sessions. 1985. *Deep Ecology: Living as if Nature Mattered.* Salt Lake City: Gibbs M. Smith.

Diamond, Irene, and Gloria F. Orenstein, eds. 1990. *Reweaving the World: The Emergence of Ecofeminism.* San Francisco: Sierra Club Books.

Drengson, Alan. 1989. *Beyond the Environmental Crisis: From Technology to Planetary Person.* New York: Peter Lang.

Drengson, Alan, and Yuichi Inoue, eds. 1995. *The Deep Ecology Movement: An Introductory Anthology.* Berkeley: North Atlantic Books.

Durning, Alan. 1996. *This Place on Earth: Home and the Practice of Permanence.* Seattle: Sasquatch Books.

Earley, Jay. 1997. *Transforming Human Culture: Social Evolution and the Planetary Crisis.* Albany: State University of New York Press.

Ehrenfeld, David. 1978. *The Arrogance of Humanism.* New York: Oxford University Press.

Ehrenfeld, David. 1993. *Beginning Again: People and Nature in the New Millennium.* New York: Oxford University Press.

Ehrlich, Paul R., and Anne H. Ehrlich. 1991. *Healing the Planet*. Reading, Mass.: Addison-Wesley.

Eisenberg, Evan. 1998. *The Ecology of Eden*. New York: Knopf.

Eisler, Riane. 1987. *The Chalice and the Blade*. New York: HarperCollins.

Eldredge, Niles. 1995. *Dominion*. Berkeley: University of California Press.

Elgin, Duane. 1993. *Voluntary Simplicity: Toward a Way of Life That Is Outwardly Simple, Inwardly Rich*. Rev. ed. New York: William Morrow.

Ereira, Allan. 1992. *The Elder Brothers*. New York: Knopf.

Evernden, Neil. 1993. *The Social Creation of Nature*. Baltimore: Johns Hopkins University Press.

Ferkiss, Victor. 1993. *Nature, Technology, and Society: Cultural Roots of the Current Environmental Crisis*. New York: New York University Press.

Fox, Stephen. 1981. *John Muir and His Legacy: The American Conservation Movement*. Boston: Little, Brown.

Fox, Warrick. 1990. *Toward a Transpersonal Ecology: Developing New Foundations for Environmentalism*. New York: Shambhala.

Gilman, Robert. 1993. "What Time Is It?" *In Context*, no. 36, 11–17.

Glacken, Clarence. 1967. *Traces on the Rhodian Shore: Nature and Culture in Western Thought*. Berkeley: University of California Press.

Glendinning, Chellis. 1994. *My Name is Chellis, and I'm in Recovery from Western Civilization*. Boston: Shambhala.

Goldsmith, Edward. 1996. *The Way*. 2d ed. Cornwall, U.K.: WEC Books.

Golley, Frank B. 1989. "Deep Ecology: An Analysis from the Perspective of Ecological Science." *Trumpeter*, vol. 6, no. 1, 24–28.

Gore, Al. 1992. *Earth in the Balance: Ecology and the Human Spirit*. Boston: Houghton Mifflin.

Gray, Elizabeth. 1982. *Green Paradise Lost*. Wellesley, Mass.: Roundtable Press.

Griffin, Susan. 1978. *Woman and Nature. The Roaring Inside Her*. New York: Harper & Row.

Gruen, Lori, and Dale Jamieson, eds. 1994. *Reflections on Nature: Readings in Environmental Philosophy*. New York: Oxford University Press.

Hallman, David G. 1994. *Ecotheology: Voices from South and North*. New York: Orbis Books.

Hardin, Garrett. 1977. *The Limits of Altruism: An Ecologist's View of Survival*. Bloomington: Indiana University Press.

Hardin, Garrett. 1978. *Exploring New Ethics for Survival*. 2d ed. New York: Viking Press.

Hardin, Garrett. 1986. *Filters Against Folly*. New York: Penguin.

Hargrove, Eugene C. 1989. *Foundations of Environmental Ethics*. Englewood Cliffs, N.J.: Prentice Hall.

Hayden, Tom. 1996. *The Lost Gospel of the Earth: A Call for Renewing Nature, Spirit, and Politics*. San Francisco: Sierra Club Books.

Hunter, J. Robert. 1997. *Simple Things Won't Save the Earth*. Austin: University of Texas Press.

Irvine, Sandy. 1989. *Beyond Green Consumerism*. London: Friends of the Earth.

Jacob, Jeffrey. 1997. *New Pioneers: The Back-to-the-Land Movement and the Search for a Sustainable Future*. Penn State Press.

Jackson, Wes. 1994. *Becoming Native to This Place*. Lexington: University of Kentucky Press.

Johnson, Huey D. 1995. *Green Plans: Greenprint for Sustainability*. Lincoln: University of Nebraska Press.

Johnson, Warren. 1985. *The Future Is Not What It Used to Be: Returning to Traditional Values in an Age of Scarcity*. New York: Dodd, Mead.

Jordan, Carl F. 1995. *Conservation: Replacing Quantity with Quality as a Goal for Global Management*. New York: Wiley.

Kaza, Stephanie. 1993. *The Attentive Heart: Conversations with Trees*. New York: Ballantine-Fawcett Columbine.

Kellert, Stephen R. 1996. *The Value of Life: Biological Diversity and Human Society*. Covelo, Calif.: Island Press.

Kellert, Stephen R., and Edward O. Wilson, eds. 1993. *The Biophilia Hypothesis*. Covelo, Calif.: Island Press.

Klenig, John. 1991. *Valuing Life*. Princeton, N.J.: Princeton University Press.

LaChapelle, Dolores. 1992. *Sacred Land, Sacred Sex: Rapture of the Deep*. Durango, Colo.: Kivaki Press.

Lemons, John, et al., eds. 1998. *Ecological Sustainability and Integrity: Concepts and Approaches*. Boston: Kluwer.

Leopold, Aldo. 1949. *A Sand County Almanac*. New York: Oxford University Press.

Lewis, Martin W. 1992. *Green Delusions: An Environmentalist Critique of Radical Environmentalism*. Durham, N.C.: Duke University Press.

List, Peter C. 1993. *Radical Environmentalism: Philosophy and Tactics*. Belmont, Calif.: Wadsworth.

Little, Charles E. 1992. *Hope for the Land*. Newark, N.J.: Rutgers University Press.

Maguire, Daniel C., et al. 1998. *New Ethics for a Small Planet*. Albany: SUNY Press.

Mahilton, Laurence S., ed. 1994. *Ethics, Religion and Biodiversity*. Cambridge, U.K.: White Horse Press.

Marshall, Peter. 1992. *Nature's Web: An Exploration of Ecological Thinking*. New York: Simon & Schuster.

Matre, Steve Van. 1990. *Earth Education*. Greenville, W.V.: Institute for Earth Education.

McKibben, Bill. 1997. *Hope, Human, and Wild: Treading Lightly on the Earth*. St. Paul, Minn.: Hungry Mind Press.

McLaughlin, Andrew. 1993. *Regarding Nature: Industrialism and Deep Ecology*. Albany: State University of New York Press.

McPhee, John, 1989. *The Control of Nature*. New York: Putnam.

Meeker, Joseph W. 1988. *Minding the Earth: Thinly Disguised Essays on Human Ecology*. Berkeley, Calif.: Latham Foundation.

Meeker, Joseph W. 1996. *The Comedy of Survival: Literary Ecology and a Play Ethic*. 3d ed. Tucson: University of Arizona Press.

Meine, Curt. 1988. *Aldo Leopold: His Life and Work*. Madison: University of Wisconsin Press.

Menzel, Peter. 1994. *Material World: A Global Family Portrait, Thirty Statistically Average Families in Thirty Countries*. San Francisco: Sierra Club Books.

Merchant, Carolyn. 1992. *Radical Ecology: The Search for a Livable World*. New York: Routledge.

Merchant, Carolyn. 1995. *Earthcare*. New York: Routledge.

Meyer, Christine, and Faith Moosang, eds. 1992. *Living with the Land: Communities Restoring the Earth*. Philadelphia: New Society.

Meyers, Norman. 1997. "Consumption in Relation to Population, Environment, and Development." *The Environmentalist*, vol. 17, 33–44.

Meyers, Norman. 1997. "Consumption: Challenge to Sustainable Development." *Science*, vol. 17276, 53–57.

Mies, Maria, and Vandana Shiva. 1993. *Ecofeminism*. Atlantic Highlands, N.J.: Zed Books.

Milbrath, Lester W. 1989. *Envisioning a Sustainable Society*. Albany: State University of New York Press.

Milbrath, Lester W. 1995. *Learning to Think Environmentally While There Is Still Time*. Albany: State University of New York Press.

Naar, Jon. 1990. *Design for a Livable Planet*. New York: Harper & Row.

Nabhan, Gary Paul, and Stephen Trimble. 1994. *The Geography of Childhood: Why Children Need Wild Places*. Boston: Beacon Press.

Naess, Arne. 1989. *Ecology, Community, and Lifestyle*. New York: Cambridge University Press.

Nagpal, Tanvi, and Camilla Foltz, eds. 1995. *A World of Difference: Giving Voices to Visions of Sustainability*. Washington, D.C.: World Resources Institute.

Nash, Roderick. 1988. *The Rights of Nature: A History of Environmental Ethics*. Madison: University of Wisconsin Press.

National Commission on the Environment. 1993. *Choosing a Sustainable Future*. Covelo, Calif.: Island Press.

Newman, Peter, et al. 1990. *Case Studies in Environmental Hope*. Perth: Australia: E.P.A. Support Services.

Norton, Bryan G. 1991. *Toward Unity Among Environmentalists*. New York: Oxford University Press.

Norwood, Vera. 1993. *Made from this Earth: American Women and Nature*. Chapel Hill: University of North Carolina Press.

Olsen, Marvin E., et al. 1991. *Viewing the World Ecologically*. Boulder, Colo.: Westview Press.

O'Riordan, Timothy. 1995. "Frameworks for Choice: Core Beliefs and the Environment." *Environment*, vol. 37, no. 8, 4–29.

O'Riordan, Timothy, and James Cameron, eds. 1994. *Interpreting the Precautionary Principle*. East Haven, Conn.: Earthscan.

Orr, David. 1992. *Ecological Literacy*. Ithaca: State University of New York Press.

Orr, David. 1994. *Earth in Mind: On Education, Environment, and the Human Prospect*. Covelo, Calif.: Island Press.

Orr, David. 1994. "Love It or Lose It: The Coming Biophilia Revolution." *Orion*, Winter, 8–15.

Palmer, Joy. 1997. *Environmental Education in the 21st Century: Theory, Practice, Progress, and Promise*. New York: Routledge.

Passmore, John. 1980. *Man's Responsibility for Nature: Ecological Problems and Western Traditions*. New York: Scribner.

Pearce, David. 1995. *Blueprint 4: Sustaining the Earth*. East Haven, Conn.: Earthscan.

Pepper, David. 1993. *Eco-Socialism: From Deep Ecology to Social Justice*. New York: Routledge.

Peterson's Guides. 1994. *Education for the Earth: A Guide to the Top Environmental Studies Programs*. Princeton, N.J.: Peterson's Guides.

Plant, Judith, ed. 1989. *Healing the Wounds: The Promise of Ecofeminism*. Philadelphia: New Society.

Plant, Judith, and Christopher Plant, eds. 1990. *Turtle Talk: Fifteen Voices for a Sustainable Future*. Philadelphia: New Society.

Plumwood, Val. 1992. *Gender and Ecology: Feminism and the Mastery of Nature*. New York: Routledge.

Pojman, L. P., ed. 1994. *Environmental Ethics: Readings in Theory and Application*. Boston: Bartlett.

Quinn, Daniel. 1992. *Ishmael*. New York: Bantam/Turner.

Rampal, Shridath. 1992. *Our Country, The Planet: Forging a Partnership for Survival.* Covelo, Calif.: Island Press.

Rasmussen, Larry. 1998. *Earth Community, Earth Ethics.* Maryknoll, N.Y.: Orbis Books.

Reed, P., and D. Rothenberg, eds. 1993. *Wisdom in the Open Air: The Norwegian Roots of Deep Ecology.* Duluth: University of Minnesota Press.

Rees, William, and Mathis Wackernagel. 1995. *Our Ecological Footprint.* Philadelphia: New Society.

Regenstein, Lewis. 1991. *Replenish the Earth.* London: SCM Press.

Rifkin, Jeremy. 1989. *Entropy: Into the Greenhouse World: A New World View.* New York: Bantam.

Rodda, Annabel, ed. 1992. *Women and the Environment.* Atlantic Highlands, N.J.: Zed Books.

Rolston, Holmes, III. 1988. *Environmental Ethics: Duties to and Values in the Natural World.* Philadelphia: Temple University Press.

Rolston, Holmes, III. 1994. *Conserving Natural Values.* New York: Columbia University Press.

Roszak, Theodore. 1978. *Person/Planet.* New York: Doubleday.

Roszak, Theodore. 1992. *The Voice of the Earth.* New York: Simon & Schuster.

Rothburg, Paul, and Robert L. Olson, eds. 1990. *Mending the Earth: A World for Our Grandchildren.* Berkeley, Calif.: North Atlantic Books.

Rubin, Charles T. 1994. *The Green Crusade: Rethinking the Roots of Environmentalism.* New York: Free Press.

Russell, Colin A. 1994. *The Earth, Humanity, and God.* Bristol, Pa.: Taylor & Francis.

Ryan John C. and AlanThein. 1997. *Stuff: The Secret Lives of Everyday Things.* Seattle, Wash.: Northwest Environmental Press.

Sale, Kirkpatrick. 1985. *Dwellers in the Land: The Bioregional Vision.* San Francisco: Sierra Club Books.

Sale, Kirkpatrick. 1990. *Conquest of Paradise.* New York: Knopf.

Salleh, Ariel. 1992. "The Ecofeminism/Deep Ecology Debate: A Reply to Patriarchal Reason." *Environmental Ethics*, Fall, 195–216.

Scharper, Stephen B., and Hilary Cunningham. 1993. *The Green Bible.* Maryknoll, N.Y.: Orbis Books.

Seager, Joni. 1993. *Earth Follies: Coming to Feminist Terms with the Global Environmental Crisis.* New York: Routledge.

Seidel, Peter. 1998. *Invisible Walls: Why We Ignore the Damage We Inflict on the Planet and Ourselves.* New York: Prometheus.

Sessions, George, ed. 1994. *Deep Ecology for the Twenty-First Century.* Boston: Shambhala.

Shiva, Vandana. 1989. *Staying Alive: Women, Ecology, and Development.* Atlantic Highlands, N.J.: Zed Books.

Snyder, Gary. 1990. *The Practice of the Wild.* San Francisco: North Point Press.

Snyder, Gary. 1995. *A Place in Space: Ethics, Aesthetics, and Watersheds.* Washington, D.C.: Counterpoint.

Soulé, Michael L., and Gary Lease, eds. 1994. *Reinventing Nature?* Covelo, Calif.: Island Press.

Sowell, Thomas. 1987. *A Conflict of Visions.* New York: William Morrow.

Suzuki, David. 1994. *Time to Change.* Toronto: Stoddard.

Suzuki, David. 1998. *The Sacred Balance: Rediscovering Our Place in Nature.* New York: Prometheus.

Suzuki, David, and Peter Knudtson. 1992. *Wisdom of the Elders: Sacred Native Visions of Nature.* New York: Bantam.

Swimme, Brian, and Thomas Berry. 1992. *The Universe Story: From the Primordial Flaring Forth to the Ecozoic Era.* New York: HarperCollins.

Sylvan, Richard, and David Bennett. 1994. *The Greening of Ethics.* Cambridge, U.K.: White Horse Press.

Taylor, Paul W. 1986. *Respect for Nature: A Theory of Environmental Ethics.* Princeton, N.J.: Princeton University Press.

Temple, Lannis. 1993. *Dear World.* New York: Random House.

Thayer, Robert L., Jr. 1994. *Gray World, Green Heart: Technology, Nature and Sustainability in the Landscape.* New York: Wiley.

Thomas, Lewis. 1992. *The Fragile Species.* New York: Scribner.

Thomashow, Mitchell. 1995. *Ecological Identity: Becoming a Reflective Environmentalist.* Cambridge, Mass.: MIT Press.

Tokar, Brian. 1996. *Renewing the Environmental Revolution.* Boston: South End Press.

Tucker, Mary Evelyn, and John A. Grim, eds. 1994. *Worldviews and Ecology: Religion, Philosophy, and the Environment.* Maryknoll, N.Y.: Orbis Books.

Van DeVeer, Donald, and Christine Pierce. 1994. *The Environmental Ethics and Policy Book: Philosophy, Ecology, Economics.* Belmont, Calif.: Wadsworth.

Vonnegut, Kurt. 1990. *Hocus Pocus.* Berkeley, Calif.: Berkeley Books.

Wenz, Peter. 1988. *Environmental Justice.* Albany: State University of New York Press.

Weston, Anthony. 1994. *Back to Earth: Tomorrow's Environmentalism.* Philadelphia: Temple University Press.

Westra, Laura. 1997. *Living in Integrity: A Global Effort to Restore a Fragmented Earth.* Savage, Md.: Rowman & Littlefield.

Westra, Laura, and Patricia H. Werhane, eds. 1997. *The Business of Consumption: Environmental Ethics and the Global Economy.* Savage, Md.: Rowman & Littlefield.

White, Lynn, Jr. 1967. "The Historical Roots of Our Ecologic Crisis." *Science*, vol. 155, 1203–7.

Wilkinson, Loren, ed. 1996. *Earthkeeping in the '90s.* Grand Rapids, Mich.: Calvin Center for Christian Scholarship.

Willers, Bill, ed. 1991. *Learning to Listen to the Land.* Covelo, Calif.: Island Press.

Young, Richard A. 1994. *Healing the Earth: A Theocentric Perspective on Environmental Problems and Their Solutions.* Nashville, Tenn.: Broadman & Holman.

Zimmerman, Michael E., ed. 1993. *Environmental Philosophy: From Animal Rights to Radical Ecology.* Englewood Cliffs, N.J.: Prentice Hall.

GLOSSARY

abiotic Nonliving. Compare *biotic*.

absolute humidity Amount of water vapor found in a certain mass of air (usually expressed as grams of water per kilogram of air). Compare *relative humidity*.

acclimation Adjustment to slowly changing new conditions. Compare *threshold effect*.

accumulation Buildup of matter, energy, or information in a system.

accuracy Extent to which a measurement agrees with the accepted or correct value for that quantity, based on careful measurements by many people over a long time. Compare *precision*.

acid deposition The falling of acids and acid-forming compounds from the atmosphere to earth's surface. Acid deposition is commonly known as *acid rain*, a term that refers only to wet deposition of droplets of acids and acid-forming compounds.

acid rain See *acid deposition*.

acid solution Any water solution that has more hydrogen ions (H^+) than hydroxide ions (OH^-); any water solution with a pH less than 7. Compare *basic solution, neutral solution*.

active solar heating system System that uses solar collectors to capture energy from the sun and store it as heat for space heating and heating water. Liquid or air pumped through the collectors transfers the captured heat to a storage system such as an insulated water tank or rock bed. Pumps or fans then distribute the stored heat or hot water throughout a dwelling as needed. Compare *passive solar heating system*.

adaptation Any genetically controlled structural, physiological, or behavioral characteristic that helps an organism survive and reproduce under a given set of environmental conditions. It usually results from a beneficial mutation. See *biological evolution, differential reproduction, mutation, natural selection*.

adaptive radiation Period of time (usually millions of years) during which numerous new species evolve to fill vacant and new ecological niches in changed environments, usually after a mass extinction.

adaptive trait See *adaptation*.

advanced sewage treatment Specialized chemical and physical processes that reduce the amount of specific pollutants left in wastewater after primary and secondary sewage treatment. This type of treatment is usually expensive. See also *primary sewage treatment, secondary sewage treatment*.

aerobic respiration Complex process that occurs in the cells of most living organisms, in which nutrient organic molecules such as glucose ($C_6H_{12}O_6$) combine with oxygen (O_2) and produce carbon dioxide (CO_2), water (H_2O), and energy. Compare *photosynthesis*.

age structure Percentage of the population (or the number of people of each sex) at each age level in a population.

agricultural revolution Gradual shift from small, mobile hunting and gathering bands to settled agricultural communities, in which people survived by learning how to breed and raise wild animals and to cultivate wild plants near where they lived. It began 10,000–12,000 years ago. Compare *industrial revolution*.

agroforestry Planting trees and crops together.

air pollution One or more chemicals in high enough concentrations in the air to harm humans, other animals, vegetation, or materials. Excess heat and noise can also be considered forms of air pollution. Such chemicals or physical conditions are called air pollutants. See *primary pollutant, secondary pollutant*.

albedo Ability of the earth's surface (land, water, or ice) to reflect light.

alien species See *nonnative species*.

alleles Slightly different molecular forms found in a particular gene.

alley cropping Planting of crops in strips with rows of trees or shrubs on each side.

alpha particle Positively charged matter, consisting of two neutrons and two protons, that is emitted as a form of radioactivity from the nuclei of some radioisotopes. See also *beta particle, gamma rays*.

altitude Height above sea level. Compare *latitude*.

anaerobic respiration Form of cellular respiration in which some decomposers get the energy they need through the breakdown of glucose (or other nutrients) in the absence of oxygen. Compare *aerobic respiration*.

ancient forest See *old-growth forest*.

animal manure Dung and urine of animals that can be used as a form of organic fertilizer. Compare *green manure*.

animals Eukaryotic, multicelled organisms such as sponges, jellyfishes, arthropods (insects, shrimp, lobsters), mollusks (snails, clams, oysters, octopuses), fish, amphibians (frogs, toads, salamanders), reptiles (turtles, lizards, alligators, crocodiles, snakes), birds, and mammals (kangaroos, bats, cats, rabbits, elephants, whales, porpoises, monkeys, apes, humans). See *carnivores, herbivores, omnivores*.

annual Plant that grows, sets seed, and dies in one growing season. Compare *perennial*.

aquaculture Growing and harvesting of fish and shellfish for human use in freshwater ponds, irrigation ditches, and lakes, or in cages or fenced-in areas of coastal lagoons and estuaries. See *fish farming, fish ranching*.

aquatic Pertaining to water. Compare *terrestrial*.

aquatic life zone Marine and freshwater portions of the ecosphere. Examples include freshwater life zones (such as lakes and streams) and ocean or marine life zones (such as estuaries, coastlines, coral reefs, and the deep ocean).

aquifer Porous, water-saturated layers of sand, gravel, or bed rock that can yield an economically significant amount of water.

arable land Land that can be cultivated to grow crops.

arid Dry. A desert or other area with an arid climate has little precipitation.

asexual reproduction Reproduction in which a mother cell divides to produce two identical daughter cells that are clones of the mother cell. This type of reproduction is common in single-celled organisms. Compare *sexual reproduction*.

atmosphere The whole mass of air surrounding the earth. See *stratosphere, troposphere*.

atomic number Number of protons in the nucleus of an atom. Compare *mass number*.

atoms Minute units made of subatomic particles that are the basic building blocks of all chemical elements and thus all matter; the smallest unit of an element that can exist and still have the unique characteristics of that element. Compare *ion, molecule*.

autotroph See *producer*.

background extinction Normal extinction of various species as a result of changes in local environmental conditions. Compare *mass extinction*.

bacteria Prokaryotic, one-celled organisms. Some transmit diseases. Most act as decomposers and get the nutrients they need by breaking down complex organic compounds in the tissues of living or dead organisms into simpler inorganic nutrient compounds.

barrier islands Long, thin, low offshore islands of sediment that generally run parallel to the shore along some coasts.

basic solution Water solution with more hydroxide ions (OH^-) than hydrogen ions (H^+); water solution with a pH greater than 7. Compare *acid solution, neutral solution*.

beneficiation Separation of an ore mineral from the waste mineral material (gangue). See *tailings*.

benthos Bottom-dwelling organisms in aquatic systems. Compare *nekton, plankton*.

beta particle Swiftly moving electron emitted by the nucleus of a radioactive isotope. See also *alpha particle, gamma rays*.

bioaccumulation An increase in the concentration of a chemical in specific organs or tissues at a level higher than would normally be expected. Compare *biomagnification*.

biodegradable Capable of being broken down by decomposers.

biodegradable pollutant Material that can be broken down into simpler substances (elements and compounds) by bacteria or other decomposers. Paper and most organic wastes such as animal manure are biodegradable but can take decades to biodegrade in modern landfills. Compare *degradable pollutant, nondegradable pollutant, slowly degradable pollutant*.

biodiversity See *biological diversity*.

biofuel Gas or liquid fuel (such as ethyl alcohol) made from plant material (biomass).

biogeochemical cycle Natural processes that recycle nutrients in various chemical forms from the nonliving environment to living organisms, and then back to the nonliving environment. Examples are the carbon, oxygen, nitrogen, phosphorus, sulfur, and hydrologic cycles.

bioinformatics Applied science of managing, analyzing, and communicating biological information. It involves **(1)** *building computer databases* to organize and store useful biological information, **(2)** *providing computer tools to find, visualize, and analyze* the information, and **(3)** *providing means for communicating* the information, especially using the Internet.

biological community See *community*.

biological diversity Variety of different species (*species diversity*), genetic variability among individuals within each species (*genetic diversity*), and variety of ecosystems (*ecological diversity*).

biological evolution Change in the genetic makeup of a population of a species in successive generations. If continued long enough, it can lead to the formation of a new species. Note that populations—not individuals—evolve. See also *adaptation, differential reproduction, natural selection, theory of evolution*.

biological oxygen demand (BOD) Amount of dissolved oxygen needed by aerobic decomposers to break down the organic materials in a given volume of water at a certain temperature over a specified time period.

biological pest control Control of pest populations by natural predators, parasites, or disease-causing bacteria and viruses (pathogens).

biomagnification Increase in concentration of DDT, PCBs, and other slowly degradable, fat-soluble chemicals in organisms at successively higher trophic levels of a food chain or web. Compare *bioaccumulation*.

biomass Organic matter produced by plants and other photosynthetic producers; total dry weight of all living organisms that can be supported at each trophic level in a food chain or web; dry weight of all organic matter in plants and animals in an ecosystem; plant materials and animal wastes used as fuel.

biome Terrestrial regions inhabited by certain types of life, especially vegetation. Examples are various types of deserts, grasslands, and forests.

biosphere Zone of the earth where life is found. It consists of parts of the atmosphere (the troposphere), hydrosphere (mostly surface water and groundwater), and lithosphere (mostly soil and surface rocks and sediments on the bottoms of oceans and other bodies of water) where life is found. It is also called the *ecosphere*.

biotic Living. Compare *abiotic*.

biotic potential Maximum rate at which the population of a given species can increase when there are no limits on its rate of growth. See *environmental resistance*.

birth rate See *crude birth rate*.

bitumen Gooey, black, high-sulfur, heavy oil extracted from tar sand and then upgraded to synthetic fuel oil. See *tar sand*.

breeder nuclear fission reactor Nuclear fission reactor that produces more nuclear fuel than it consumes by converting nonfissionable uranium-238 into fissionable plutonium-239.

broadleaf deciduous plants Plants such as oak and maple trees that survive drought and cold by shedding their leaves and becoming dormant during such periods. Compare *broadleaf evergreen plants, coniferous evergreen plants*.

broadleaf evergreen plants Plants that keep most of their broad leaves year-round. Examples are the trees found in the canopies of tropical rain forests. Compare *broadleaf deciduous plants, coniferous evergreen plants*.

cancer Group of more than 120 different diseases, one for each type of cell in the human body. Each type of cancer produces a tumor in which cells multiply uncontrollably and invade surrounding tissue.

capitalism See *pure market economic system*.

carbon cycle Cyclic movement of carbon in different chemical forms from the environment to organisms and then back to the environment.

carcinogen Chemicals, ionizing radiation, and viruses that cause or promote the development of cancer. See *cancer, mutagen, teratogen*.

carnivore Animal that feeds on other animals. Compare *herbivore, omnivore*.

carrying capacity (K) Maximum population of a particular species that a given habitat can support over a given period of time.

cell Smallest living unit of an organism. Each cell is encased in an outer membrane or wall and contains genetic material (DNA) and other parts to perform its life function. Organisms such as bacteria consist of only one cell, but most of the organisms we are familiar with contain many cells. See *eukaryotic cell, prokaryotic cell*.

centrally planned economy See *pure command economic system*.

CFCs See *chlorofluorocarbons*.

chain reaction Multiple nuclear fissions, taking place within a certain mass of a fissionable isotope, that release an enormous amount of energy in a short time.

chaos Behavior that never repeats itself exactly. Examples of chaotic behavior are the waves of an ocean, the movement of leaves in the wind, and day-to-day variations in weather.

character displacement Development of different physical or behavioral characteristics or adaptations by similar species that allow them to use different resources and thus reduce competition through resource partitioning. In birds, for instance, bill sizes of similar species found in the same ecosystem often differ.

chemical One of the millions of different elements and compounds found naturally or synthesized by humans. See *compound, element*.

chemical change Interaction between chemicals in which there is a change in the chemical composition of the elements or compounds involved. Compare *nuclear change, physical change*.

chemical evolution Formation of the earth and its early crust and atmosphere, evolution of the biological molecules necessary for life, and evolution of systems of chemical reactions needed to produce the first living cells. These processes are believed to have occurred about 1 billion years before biological evolution. Compare *biological evolution*.

chemical formula Shorthand way to show the number of atoms (or ions) in the basic structural unit of a compound. Examples are H_2O, NaCl, and $C_6H_{12}O_6$.

chemical reaction See *chemical change*.

chemosynthesis Process in which certain organisms (mostly specialized bacteria) extract inorganic compounds from their environment and convert them into organic nutrient compounds without the presence of sunlight. Compare *photosynthesis*.

chlorofluorocarbons (CFCs) Organic compounds made up of atoms of carbon, chlorine, and fluorine. An example is Freon-12 (CCl_2F_2), used as a refrigerant in refrigerators and air conditioners and in making plastics such as Styrofoam. Gaseous CFCs can deplete the ozone layer when they slowly rise into the stratosphere and their chlorine atoms react with ozone molecules. Use of these molecules is being phased out.

chromosome A grouping of various genes and associated proteins in plant and animal cells that carry certain types of genetic information. See *genes*.

civil suit Lawsuit in which a plaintiff seeks to collect damages for injuries or for economic loss, to have the court issue a permanent injunction against any further wrongful action, or both. Compare *class action suit*.

class action suit Civil lawsuit in which a group files a suit on behalf of a larger number of citizens who allege similar damages but who need not be listed and represented individually. Compare *civil suit*.

clear-cutting Method of timber harvesting in which all trees in a forested area are removed in a single cutting. Compare *seed-tree cutting, selective cutting, shelterwood cutting, strip cutting*.

climate Physical properties of the troposphere of an area based on analysis of its weather records over a long period (at least 30 years). The two main factors determining an area's climate are *temperature*, with its seasonal variations, and the amount and distribution of *precipitation*. Compare *weather*.

closed system System in which energy—but not matter—is exchanged between the system and its environment. Compare *open system*.

coal Solid, combustible mixture of organic compounds with 30–98% carbon by weight, mixed with various amounts of water and small amounts of sulfur and nitrogen compounds. It is formed in several stages as the remains of plants are subjected to heat and pressure over millions of years.

coal gasification Conversion of solid coal to synthetic natural gas (SNG).

coal liquefaction Conversion of solid coal to a liquid hydrocarbon fuel such as synthetic gasoline or methanol.

coastal wetland Land along a coastline, extending inland from an estuary that is covered with salt water all or part of the year. Examples are marshes, bays, lagoons, tidal flats, and mangrove swamps. Compare *inland wetland*.

coastal zone Warm, nutrient-rich, shallow part of the ocean that extends from the high-tide mark on land to the edge of a shelflike extension of continental land masses known as the continental shelf. Compare *open sea*.

coevolution Evolution in which two or more species interact and exert selective pressures on each other that can lead each species to undergo various adaptations. See *evolution, natural selection*.

cogeneration Production of two useful forms of energy, such as high-temperature heat or steam and electricity, from the same fuel source.

cold front Leading edge of an advancing mass of cold air. Compare *warm front*.

commensalism An interaction between organisms of different species in which one type of organism benefits and the other type is neither helped nor harmed to any great degree. Compare *mutualism*.

commercial extinction Depletion of the population of a wild species used as a resource to a level at which it is no longer profitable to harvest the species.

commercial inorganic fertilizer Commercially prepared mixtures of plant nutrients such as nitrates, phosphates, and potassium applied to the soil to restore fertility and increase crop yields. Compare *organic fertilizer*.

common-property resource Resource that people are normally free to use; each user can deplete or degrade the available supply. Most are potentially renewable and are owned by no one. Examples are clean air, fish in parts of the ocean not under the control of a coastal country, migratory birds, gases of the lower atmosphere, and the ozone content of the upper stratosphere. See *tragedy of the commons*.

community Populations of all species living and interacting in an area at a particular time.

community development See *ecological succession.*

competition Two or more individual organisms of a single species (*intraspecific competition*) or two or more individuals of different species (*interspecific competition*) attempting to use the same scarce resources in the same ecosystem.

competitive exclusion No two species can occupy exactly the same fundamental niche indefinitely in a habitat where there is not enough of a particular resource to meet the needs of both species. See *ecological niche, fundamental niche, realized niche.*

compost Partially decomposed organic plant and animal matter that can be used as a soil conditioner or fertilizer.

compound Combination of atoms, or oppositely charged ions, of two or more different elements held together by attractive forces called chemical bonds. Compare *element.*

concentration Amount of a chemical in a particular volume or weight of air, water, soil, or other medium.

condensation Conversion of a gas to a liquid.

condensation nuclei Tiny particles on which droplets of water vapor can collect.

conditions Physical or chemical attributes of the environment that, while not being consumed, influence biological processes and population growth. Examples are temperature, salinity, and acidity. Compare *resources.*

coniferous evergreen plants Cone-bearing plants (such as spruces, pines, and firs) that keep some of their narrow, pointed leaves (needles) all year. Compare *broadleaf deciduous plants, broadleaf evergreen plants.*

consensus science Scientific data, models, theories, and laws that are widely accepted. This aspect of science is very reliable. Compare *frontier science.*

conservation biology Multidisciplinary science created to deal with the crisis of maintaining the genes, species, communities, and ecosystems that make up earth's biological diversity. Its goals are to investigate human impacts on biodiversity and to develop practical approaches to preserving biodiversity and ecological integrity.

conservation-tillage farming Crop cultivation in which the soil is disturbed little (minimum-tillage farming) or not at all (no-till farming) to reduce soil erosion, lower labor costs, and save energy. Compare *conventional-tillage farming.*

constancy Ability of a living system, such as a population, to maintain a certain size. Compare *inertia, resilience.* See *homeostasis.*

consumer Organism that cannot synthesize the organic nutrients it needs and gets its organic nutrients by feeding on the tissues of producers or of other consumers; generally divided into *primary consumers* (herbivores), *secondary consumers* (carnivores), *tertiary (higher-level) consumers, omnivores,* and *detritivores* (decomposers and detritus feeders). In economics, one who uses economic goods.

contour farming Plowing and planting across the changing slope of land, rather than in straight lines, to help retain water and reduce soil erosion.

controlled burning Deliberately set, carefully controlled surface fires to reduce flammable litter and decrease the chances of damaging crown fires.

conventional-tillage farming Crop cultivation that involves making a planting surface by plowing land, breaking up the exposed soil, and then smoothing the surface. Compare *conservation-tillage farming.*

convergence Resemblance among species belonging to different taxonomic groups as the result from adaptation to similar environments.

convergent plate boundary Area where earth's lithospheric plates are pushed together. See *subduction zone.* Compare *divergent plate boundary, transform fault.*

coral reef Formation produced by massive colonies containing billions of tiny coral animals, called *polyps*, that secrete a stony substance (calcium carbonate) around themselves for protection. When the corals die, their empty outer skeletons form layers that cause the reef to grow. They are found in the coastal zones of warm tropical and subtropical oceans.

core Inner zone of the earth. It consists of a solid inner core and a liquid outer core. Compare *crust, mantle.*

corridors Long areas of land that connect habitat that would otherwise become fragmented.

cost–benefit analysis Estimates and comparison of short-term and long-term costs (losses) and benefits (gains) from an economic decision. If the estimated benefits exceed the estimated costs, the decision to buy an economic good or provide a public good is considered worthwhile.

critical mass Amount of fissionable nuclei needed to sustain a nuclear fission chain reaction.

crop rotation Planting a field, or an area of a field, with different crops from year to year to reduce depletion of soil nutrients. A plant such as corn, tobacco, or cotton, which removes large amounts of nitrogen from the soil, is planted one year. The next year a legume such as soybeans, which adds nitrogen to the soil, is planted.

crude birth rate Annual number of live births per 1,000 people in the population of a geographic area at the midpoint of a given year. Compare *crude death rate.*

crude death rate Annual number of deaths per 1,000 people in the population of a geographic area at the midpoint of a given year. Compare *crude birth rate.*

crude oil Gooey liquid consisting mostly of hydrocarbon compounds and small amounts of compounds containing oxygen, sulfur, and nitrogen. Extracted from underground accumulations, it is sent to oil refineries, where it is converted to heating oil, diesel fuel, gasoline, tar, and other materials.

crust Solid outer zone of the earth. It consists of oceanic crust and continental crust. Compare *core, mantle.*

cultural eutrophication Overnourishment of aquatic ecosystems with plant nutrients (mostly nitrates and phosphates) because of human activities such as agriculture, urbanization, and discharges from industrial plants and sewage treatment plants. See *eutrophication.*

cyanobacteria Single-celled, prokaryotic, microscopic organisms. Before being reclassified as monera, they were called blue-green algae.

DDT Dichlorodiphenyltrichloroethane, a chlorinated hydrocarbon that has been widely used as a pesticide but is now banned in some countries.

death rate See *crude death rate.*

debt-for-nature swap Agreement in which a certain amount of foreign debt is canceled in exchange for local currency investments that will improve natural resource management or protect certain areas in the debtor country from harmful development.

deciduous plants Trees that survive during dry seasons or cold seasons by shedding their leaves. See *broadleaf deciduous plants.* Compare *broadleaf evergreen plants, coniferous evergreen plants.*

decomposer Organism that digests parts of dead organisms and cast-off fragments and wastes of living organisms. A decomposer breaks down the complex organic molecules in those materials into simpler inorganic compounds and then absorbs the soluble nutrients. Most of these chemicals are returned to the soil and water for reuse by producers. Decomposers consist of various bacteria and fungi. Compare *consumer, detritivore, producer.*

deductive reasoning Using logic to arrive at a specific conclusion based on a generalization or premise. It goes from the general to the specific. Compare *inductive reasoning.*

defendant The individual, group of individuals, corporation, or government agency being charged in a lawsuit. Compare *plaintiff.*

deforestation Removal of trees from a forested area without adequate replanting.

degradable pollutant Potentially polluting chemical that is broken down completely or reduced to acceptable levels by natural physical, chemical, and biological processes. Compare *biodegradable pollutant, nondegradable pollutant, slowly degradable pollutant.*

degree of urbanization Percentage of the population in the world, or a country, living in areas with a population of more than 2,500 people (higher in some countries). Compare *urban growth.*

democracy Government by the people through their elected officials and appointed representatives. In a *constitutional democracy*, a constitution provides the basis of government authority and puts restraints on government power through free elections and freely expressed public opinion.

demographic transition Hypothesis that countries, as they become industrialized, have declines in death rates followed by declines in birth rates.

depletion time How long it takes to use a certain fraction—usually 80%—of the known or estimated supply of a nonrenewable resource at an assumed rate of use. Finding and extracting the remaining 20% usually costs more than it is worth.

desalination Purification of salt water or brackish (slightly salty) water by removing dissolved salts.

desert Biome in which evaporation exceeds precipitation and the average amount of precipitation is less than 25 centimeters (10 inches) a year. Such areas have little vegetation or have widely spaced, mostly low vegetation. Compare *forest, grassland.*

desertification Conversion of rangeland, rainfed cropland, or irrigated cropland to desertlike land, with a drop in agricultural productivity of 10% or more. It is usually caused by a combination of overgrazing, soil erosion, prolonged drought, and climate change.

detrital food webs Food webs in which organic waste material or detritus is the major food source, and energy flows mainly from plants to decomposers and detritivores. Compare *grazing food webs.*

detritivore Consumer organism that feeds on detritus, parts of dead organisms and cast-off fragments and wastes of living organisms. The two principal types are *detritus feeders* and *decomposers.*

detritus Parts of dead organisms and cast-off fragments and wastes of living organisms.

detritus feeder Organism that extracts nutrients from fragments of dead organisms and their cast-off parts and organic wastes. Examples are earthworms, termites, and crabs. Compare *decomposer.*

deuterium (D; hydrogen-2) Isotope of the element hydrogen, with a nucleus containing one proton and one neutron, and a mass number of 2. Compare *tritium.*

developed country Country that is highly industrialized and has a high per capita GNP. Compare *developing country.*

developing country Country that has low to moderate industrialization and low to moderate per capita GNP. Most are located in Africa, Asia, and Latin America. Compare *developed country.*

development Change from a society that is largely rural, agricultural, illiterate, and poor, with a rapidly growing population, to one that is mostly urban, industrial, educated, and wealthy, with a slowly growing or stationary population. See *developed country, developing country, economic development, ecologically sustainable development.*

dew point Temperature at which condensation occurs for a given amount of water vapor.

differential reproduction Phenomenon in which individuals with adaptive genetic traits produce more living offspring than do individuals without such traits. See *natural selection.*

dioxins Family of 75 different toxic chlorinated hydrocarbon compounds formed as by-products in chemical reactions involving chlorine and hydrocarbons, usually at high temperatures.

discount rate The economic value a resource will have in the future compared with its present value.

dissolved oxygen (DO) content Amount of oxygen gas (O_2) dissolved in a given volume of water at a particular temperature and pressure, often expressed as a concentration in parts of oxygen per million parts of water.

disturbance A discrete event in time that disrupts an ecosystem or community. Examples of *natural disturbances* include fires, hurricanes, tornadoes, droughts, and floods. Examples of *human-caused disturbances* include deforestation, overgrazing, and plowing.

divergent plate boundary Area where earth's lithospheric plates move apart in opposite directions. Compare *convergent plate boundary, transform fault.*

DNA (deoxyribonucleic acid) Large molecules in the cells of organisms that carry genetic information in living organisms.

domesticated species Wild species tamed or genetically altered by crossbreeding for use by humans for food (cattle, sheep, and food crops), pets (dogs and cats), or enjoyment (animals in zoos and plants in gardens).

dose The amount of a potentially harmful substance an individual ingests, inhales, or absorbs through the skin. Compare *response.* See *dose–response curve, lethal dose, median lethal dose.*

dose–response curve Plot of data showing effects of various doses of a toxic agent on a group of test organisms. See *dose, lethal dose, median lethal dose, response.*

doubling time The time it takes (usually in years) for the quantity of something growing exponentially to double. It can be calculated by dividing the annual percentage growth rate into 70.

drainage basin See *watershed.*

dredge spoils Materials scraped from the bottoms of harbors and streams to maintain shipping channels. They are often contaminated with high levels of toxic substances that have settled out of the water. See *dredging.*

dredging Type of surface mining in which chain buckets and draglines scrape up sand, gravel, and other surface deposits covered with water. It is also used to remove sediment from streams and harbors to maintain shipping channels. See *dredge spoils.*

drift-net fishing Catching fish in huge nets that drift in the water.

drought Condition in which an area does not get enough water because of lower-than-normal precipitation, higher-than-normal temperatures that increase evaporation, or both.

dust dome Dome of heated air that surrounds an urban area and traps pollutants, especially suspended particulate matter. See also *urban heat island.*

dust plume Elongation of a dust dome by winds that can spread a city's pollutants hundreds of kilometers downwind.

early successional plant species Plants that grow close to the ground, can establish large populations quickly under harsh conditions, and have short lives. Compare *late successional plant species* and *midsuccessional species.*

earth capital The earth's natural resources and processes that sustain us and other species. Compare *human capital, manufactured capital, solar capital.*

earthquake Shaking of the ground resulting either from the fracturing and displacement of rock, producing a fault, or from subsequent movement along the fault.

earth-sustaining economy Economic system in which the number of people and the quantity of goods are maintained at a constant level. This level is ecologically sustainable over time and meets at least the basic needs of all members of the population.

earth-wisdom revolution Cultural change involving halting population growth and altering lifestyles, political and economic systems, and the way we treat the environment so that we can help sustain the earth for ourselves and other species. This involves working with the rest of nature by learning more about how nature sustains itself. See *earth-wisdom worldview.* Compare *matter-recycling society, planetary management worldview.*

earth-wisdom society Society based on working with nature by recycling and reusing discarded matter; preventing pollution; conserving matter and energy resources by reducing unnecessary waste and use; not degrading renewable resources; building things that are easy to recycle, reuse, and repair; not allowing population size to exceed the carrying capacity of the environment; and preserving biodiversity and ecological integrity. See *earth-wisdom worldview.* Compare *matter-recycling society, planetary management worldview.*

earth-wisdom worldview Beliefs that nature exists for all of the earth's species, not just for us; that we are not in charge of the rest of nature; that there is not always more, and it's not all for us; that some forms of economic growth are beneficial, and some are harmful; and that our goals should be to design economic and political systems that encourage earth-sustaining forms of growth and discourage or prohibit earth-degrading forms; our success depends on learning to cooperate with one another and with the rest of nature instead of trying to dominate and manage earth's life-support systems primarily for our own use. Compare *planetary management worldview.*

ecological diversity The variety of forests, deserts, grasslands, oceans, streams, lakes, and other biological communities interacting with one another and with their nonliving environment. See *biological diversity.* Compare *genetic diversity, species diversity.*

ecological health The degree to which an area's biodiversity and ecological integrity remain intact. See *biological diversity, ecological integrity.*

ecological integrity The conditions and natural processes (such as the flow of materials and energy through ecosystems) that generate and maintain biodiversity and allow evolutionary change as a key mechanism for adapting to changes in environmental conditions.

ecological land-use planning Method for deciding how land should be used; development of an integrated model that considers geological, ecological, health, and social variables.

ecologically sustainable development Development in which the total human population size and resource use in the world (or in a region) are limited to a level that does not exceed the carrying capacity of the existing natural capital and is therefore sustainable. Compare *economic development, economic growth, sustainable development.*

ecological niche Total way of life or role of a species in an ecosystem. It includes all physical, chemical, and biological conditions a species needs to live and reproduce in an ecosystem. See *fundamental niche, realized niche.*

ecological population density Number of individuals of a population per unit area of habitat. Compare *population density.*

ecological resource Anything required by an organism for normal maintenance, growth, and reproduction. Examples include habitat, food, water, and shelter. Compare *economic resource.*

ecological succession Process in which communities of plant and animal species in a particular area are replaced over time by a series of different and often more complex communities. See *primary succession, secondary succession.*

ecology Study of the interactions of living organisms with one another and with their nonliving environment of matter and energy; study of the structure and functions of nature.

economic decision Deciding what goods and services to produce, how to produce them, how much to produce, and how to distribute them to people.

economic depletion Exhaustion of 80% of the estimated supply of a nonrenewable resource. Finding, extracting, and processing the remaining 20% usually costs more than it is worth; may also apply to the depletion of a potentially renewable resource, such as a species of fish or tree.

economic development Using economic systems to improve the *quality* of people's lives and the environment. Also refers to improvements in the efficiency of resource use so that the same or greater output of goods and services is produced with smaller throughputs of natural, manufactured, and human capital. Compare *ecologically sustainable development, economic growth, sustainable development.*

economic growth Increase in the real value of all final goods and services produced by an economy; an increase in real GNP or GDP. Compare *ecologically sustainable development, economic development, sustainable development.*

economic resource Anything obtained from the environment (the earth's life-support systems) to meet human needs and wants. Examples include food, water, shelter, manufactured goods, transportation, communication, and recreation.

economic resources Natural resources, capital goods, and labor used in an economy to produce material goods and services. See *earth capital, human capital, manufactured capital.*

economic system Method that a group of people uses to choose what goods and services to produce, how to produce them, how much to produce, and how to distribute them to people. See *mixed economic system, pure command economic system, pure market economic system.*

economy System of production, distribution, and consumption of economic goods.

ecosphere The earth's collection of living organisms interacting with one another and their nonliving environment (energy and matter) throughout the world; all of earth's ecosystems. Also called the *biosphere.*

ecosystem Community of different species interacting with one another and with the chemical and physical factors making up its nonliving environment.

ecosystem services Natural services or earth capital that support life on the earth and are essential to the quality of human life and to the functioning of the world's economies. See *earth capital.*

ecotone Transitional zone in which one type of ecosystem tends to merge with another ecosystem. See *edge effect.*

edge effect The existence of a greater number of species and a higher population density in an ecotone than in either adjacent ecosystem. See *ecotone.*

electromagnetic radiation Forms of kinetic energy traveling as electromagnetic waves. Examples are radio waves, TV waves, microwaves, infrared radiation, visible light, ultraviolet radiation, X rays, and gamma rays. Compare *ionizing radiation, nonionizing radiation.*

electron (e) Tiny particle moving around outside the nucleus of an atom. Each electron has one unit of negative charge ([–]) and almost no mass. Compare *neutron, proton.*

element Chemical, such as hydrogen (H), iron (Fe), sodium (Na), carbon (C), nitrogen (N), or oxygen (O), whose distinctly different atoms serve as the basic building blocks of all matter. There are 92 naturally occurring elements. Another 18 have been made in laboratories. Two or more elements combine to form compounds, which make up most of the world's matter. Compare *compound.*

endangered species Wild species with so few individual survivors that the species could soon become extinct in all or most of its natural range. Compare *threatened species.*

energy Capacity to do work by performing mechanical, physical, chemical, or electrical tasks or to cause a heat transfer between two objects at different temperatures.

energy efficiency Percentage of the total energy input that does useful work and is not converted into low-quality, usually useless heat in an energy conversion system or process. See *energy quality, net energy.*

energy quality Ability of a form of energy to do useful work. High-temperature heat and the chemical energy in fossil fuels and nuclear fuels are concentrated high-quality energy. Low-quality energy such as low-temperature heat is dispersed or diluted and cannot do much useful work. See *high-quality energy, low-quality energy.*

enhanced oil recovery Removal of some of the heavy oil left in an oil well after primary and secondary recovery. Compare *primary oil recovery, secondary oil recovery.*

entropy A measure of the disorder or randomness of a system. The greater the disorder of a system, the higher its entropy; the greater its order, the lower its entropy.

environment All external conditions and factors, living and nonliving (chemicals and energy), that affect an organism or other specified system during its lifetime; the earth's life-support systems for us and for all other forms of life—another term for solar capital and earth capital.

environmental degradation Depletion or destruction of a potentially renewable resource such as soil, grassland, forest, or wildlife by using it faster than it is naturally replenished. If such use continues, the resource can become nonrenewable (on a human time scale) or nonexistent (extinct). See also *sustainable yield.*

environmental ethics Our beliefs about what is right or wrong environmental behavior.

environmental resistance All the limiting factors acting jointly to limit the growth of a population. See *biotic potential, limiting factor.*

environmental revolution See *earth-wisdom revolution.*

environmental science Study of how we and other species interact with one another and with the nonliving environment (matter and energy). It is a physical and social science that integrates knowledge from a wide range of disciplines including physics, chemistry, biology (especially ecology), geology, geography, resource technology and engineering, resource conservation and management, demography (the study of population dynamics), economics, politics, sociology, psychology, and ethics. In other words, it is a study of how the parts of nature and human societies operate and interact—a study of connections and interactions.

environmental worldview How individuals think the world works, what they think their role in the world should be, and what they believe is right and wrong environmental behavior (environmental ethics). See *earth-wisdom worldview, planetary management worldview, spaceship-earth worldview.*

EPA Environmental Protection Agency; responsible for managing federal efforts in the United States to control air and water pollution, radiation and pesticide hazards, environmental research, hazardous waste, and solid-waste disposal.

epidemiology Study of the patterns of disease or other harmful effects from toxic exposure within defined groups of people to find out why some people get sick and some do not.

epiphytes Plants that use their roots to attach themselves to branches high in trees, especially in tropical forests.

erosion Process or group of processes by which loose or consolidated earth materials are dissolved, loosened, or worn away and removed from one place and deposited in another. See *weathering.*

estuary Partially enclosed coastal area at the mouth of a river where its fresh water, carrying fertile silt and runoff from the land, mixes with salty seawater.

eukaryotic cell Cell containing a *nucleus,* a region of genetic material surrounded by a membrane. Membranes also enclose several of the other internal parts found in a eukaryotic cell. Compare *prokaryotic cell.*

euphotic zone Upper layer of a body of water through which sunlight can penetrate and support photosynthesis.

eutrophication Physical, chemical, and biological changes that take place after a lake, estuary, or slow-flowing stream receives inputs of plant nutrients—mostly nitrates and phosphates—from natural erosion and runoff from the surrounding land basin. See *cultural eutrophication.*

eutrophic lake Lake with a large or excessive supply of plant nutrients, mostly nitrates and phosphates. Compare *mesotrophic lake, oligotrophic lake.*

evaporation Conversion of a liquid into a gas.

even-aged management Method of forest management in which trees, sometimes of a single species in a given stand, are maintained at about the same age and size and are harvested all at once. Compare *uneven-aged management.*

evergreen plants Plants that keep some of their leaves or needles throughout the year. Examples are ferns and cone-bearing trees (conifers) such as firs, spruces, pines, redwoods, and sequoias. See *broadleaf evergreen plants, broadleaf deciduous plants, coniferous evergreen plants.* Compare *deciduous plants, succulent plants.*

evolution See *biological evolution.*

exhaustible resource See *nonrenewable resource.*

exotic species See *nonnative species.*

experiment Procedure a scientist uses to study some phenomenon under known conditions. Some experiments are conducted in the laboratory, but others are conducted in nature. The resulting scientific data or facts must be verified or confirmed by repeated observations and measurements, ideally by several different investigators.

exploitation competition Situation in which two competing species have equal access to a specific resource but differ in how quickly or efficiently they exploit it. See *interference competition, interspecific competition.* Compare *intraspecific competition.*

exponential growth Growth in which some quantity, such as population size or economic output, increases by a fixed percentage of the whole in a given time period; when the increase in quantity over time is plotted, this type of growth yields a curve shaped like the letter *J.* Compare *linear growth.*

external benefit Beneficial social effect of producing and using an economic good that is not included in the market price of the good. Compare *external cost, full cost, true cost.*

external cost Harmful social effect of producing and using an economic good that is not included in the market price of the good. Compare *external benefit, full cost, true cost.*

externalities Social benefits ("goods") and social costs ("bads") not included in the market price of an economic good. See *external benefit, external cost.* Compare *full cost, internal cost.*

extinction Complete disappearance of a species from the earth. This happens when a species cannot adapt and successfully reproduce under new environmental conditions or when it evolves into one or more new species. Compare *speciation.* See also *endangered species, threatened species.*

family planning Providing information, clinical services, and contraceptives to help people choose the number and spacing of children they want to have.

famine Widespread malnutrition and starvation in a particular area because of a shortage of food, usually caused by drought, war, flood, earthquake, or other catastrophic events that disrupts food production and distribution.

feedback loop Circuit of sensing, evaluating, and reacting to changes in environmental conditions as a result of information fed back into a system; it occurs when one change leads to some other change, which then eventually either reinforces or slows the original change. See *negative feedback loop, positive feedback loop.*

feedlot Confined outdoor or indoor space used to raise hundreds to thousands of domesticated livestock. Compare *rangeland.*

fermentation See *anaerobic respiration.*

fertilizer Substance that adds inorganic or organic plant nutrients to soil and improves its ability to grow crops, trees, or other vegetation. See *commercial inorganic fertilizer, organic fertilizer.*

first law of energy See *first law of thermodynamics.*

first law of human ecology We can never do merely one thing. Any intrusion into nature has numerous effects, many of which are unpredictable.

first law of thermodynamics In any physical or chemical change, no detectable amount of energy is created or destroyed, but in these processes energy can be changed from one form to another. You can't get more energy out of something than you put in; in terms of energy quantity, you can't get something for nothing (there is no free lunch). This law does not apply to nuclear changes, in which energy can be produced from small amounts of matter. See also *second law of thermodynamics.*

fishery Concentrations of particular aquatic species suitable for commercial harvesting in a given ocean area or inland body of water.

fish farming Form of aquaculture in which fish are cultivated in a controlled pond or other environment and harvested when they reach the desired size. See also *fish ranching.*

fish ranching Form of aquaculture in which members of a fish species such as salmon are held in captivity for the first few years of their lives, released, and then harvested as adults when they return from the ocean to their freshwater birthplace to spawn. See also *fish farming.*

fissionable isotope Isotope that can split apart when hit by a neutron at the right speed and thus undergo nuclear fission. Examples are uranium-235 and plutonium-239.

floodplain Flat valley floor next to a stream channel. For legal purposes, the term is often applied to any low area that has the potential for flooding, including certain coastal areas.

flow See *throughput*.

flyway Generally fixed route along which waterfowl migrate from one area to another at certain seasons of the year.

food chain Series of organisms in which each eats or decomposes the preceding one. Compare *food web*.

food web Complex network of many interconnected food chains and feeding relationships. Compare *food chain*. See *detrital food webs*, *grazing food webs*.

forest Biome with enough average annual precipitation (at least 76 centimeters, or 30 inches) to support growth of various species of trees and smaller forms of vegetation. Compare *desert*, *grassland*.

fossil fuel Products of partial or complete decomposition of plants and animals that occur as crude oil, coal, natural gas, or heavy oils as a result of exposure to heat and pressure in earth's crust over millions of years. See *coal*, *crude oil*, *natural gas*.

fossils Skeletons, bones, shells, body parts, leaves, seeds, or impressions of such items that provide recognizable evidence of organisms that lived long ago.

Freons See *chlorofluorocarbons*.

freshwater life zones Aquatic systems where water with a dissolved salt concentration of less than 1% by volume accumulates on or flows through the surfaces of terrestrial biomes. Examples are *standing* (lentic) bodies of fresh water such as lakes, ponds, and inland wetlands and *flowing* (lotic) systems such as streams and rivers. Compare *biomes*.

front The boundary between two air masses with different temperatures and densities. See *cold front*, *warm front*.

frontier science Preliminary scientific data, hypotheses, and models that have not been widely tested and accepted. Compare *consensus science*.

frontier worldview Viewing undeveloped land as a hostile wilderness to be conquered (cleared, planted) and exploited for its resources as quickly as possible. See *earth-wisdom worldview*, *planetary management worldview*.

full cost Cost of a good when its internal costs and its estimated short- and long-term external costs are included in its market price. Compare *external cost*, *internal cost*.

fundamental niche The full potential range of the physical, chemical, and biological factors a species can use if there is no competition from other species. See *ecological niche*. Compare *realized niche*.

fungi Eukaryotic, mostly multicelled organisms such as mushrooms, molds, and yeasts. As decomposers, they get the nutrients they need by secreting enzymes that speed up the breakdown the organic matter in the tissue of other living or dead organisms. Then they absorb the resulting nutrients.

fungicide Chemical that kills fungi.

Gaia hypothesis Proposal that earth is alive and can be considered a system that operates and changes by feedback of information between its living and nonliving components.

game species Type of wild animal that people hunt or fish for, for sport and recreation and sometimes for food.

gamma rays A form of ionizing electromagnetic radiation with a high energy content emitted by some radioisotopes. They readily penetrate body tissues.

gangue Waste or undesired material in an ore. See *ore*.

gap analysis Scientific method used to determine how adequately native plant and animal species and natural communities are protected by the existing network of conservation lands. Species and communities not adequately represented in existing conservation lands constitute conservation gaps. The idea is to identify these gaps and then eliminate them through the establishment of new reserves or changing land management practices.

GDP See *gross domestic product*.

gene flow Movement of genes between populations, which can lead to changes in the genetic composition of local populations.

gene mutation See *mutation*.

gene pool The sum total of all genes found in the individuals of the population of a particular species.

generalist species Species with a broad ecological niche. They can live in many different places, eat a variety of foods, and tolerate a wide range of environmental conditions. Examples are flies, cockroaches, mice, rats, and human beings. Compare *specialist species*.

genes Coded units of information about specific traits that are passed on from parents to offspring during reproduction. They consist of segments of DNA molecules found in chromosomes.

genetic adaptation Changes in the genetic makeup of organisms of a species that allow the species to reproduce and gain a competitive advantage under changed environmental conditions. See *differential reproduction*, *evolution*, *mutation*, *natural selection*.

genetic diversity Variability in the genetic makeup among individuals within a single species. See *biodiversity*. Compare *ecological diversity*, *species diversity*.

genetic drift Change in the genetic composition of a population by chance. It is especially important for small populations.

geographic isolation Separation of populations of a species for fairly long times into areas with different environmental conditions.

geothermal energy Heat transferred from the earth's underground concentrations of dry steam (steam with no water droplets), wet steam (a mixture of steam and water droplets), or hot water trapped in fractured or porous rock.

global warming Warming of the earth's atmosphere as a result of increases in the concentrations of one or more greenhouse gases. See *greenhouse effect*, *greenhouse gases*.

GNP See gross national product.

grassland Biome found in regions where moderate annual average precipitation (25 to 76 centimeters, or 10 to 30 inches) is enough to support the growth of grass and small plants, but not enough to support large stands of trees. Compare *desert*, *forest*.

grazing food webs Food webs in which the energy flows from plants to herbivores (grazers), then through an array of carnivores, and eventually to decomposers. Compare *detrital food webs*.

greenhouse effect A natural effect that traps heat in the atmosphere (troposphere) near earth's surface. Some of the heat flowing back toward space from earth's surface is absorbed by water vapor, carbon dioxide, ozone, and several other gases in the lower atmosphere (troposphere) and then radiated back toward the earth's surface. If the atmospheric concentrations of these greenhouse gases rise and are not removed by other natural processes, the average temperature of the lower atmosphere will increase gradually.

greenhouse gases Gases in the earth's lower atmosphere (troposphere) that cause the greenhouse effect. Examples are carbon dioxide, chlorofluorocarbons, ozone, methane, water vapor, and nitrous oxide.

green manure Freshly cut or still-growing green vegetation that is plowed into the soil to increase the organic matter and humus available to support crop growth. Compare *animal manure*.

green revolution Popular term for introduction of scientifically bred or selected varieties of grain (rice, wheat, maize) that, with high enough inputs of fertilizer and water, can greatly increase crop yields.

gross domestic product (GDP) Total market value in current dollars of all goods and services produced *within* a country for final use usually during a year. Compare *gross national product*.

gross national product (GNP) Total market value in current dollars of all goods and services produced by an economy for final use usually during a year. Compare *gross domestic product*.

gross primary productivity (GPP) The rate at which an ecosystem's producers capture and store a given amount of chemical energy as biomass in a given length of time. Compare *net primary productivity*.

groundwater Water that sinks into the soil and is stored in slowly flowing and slowly renewed underground reservoirs called aquifers; underground water in the zone of saturation, below the water table. Compare *runoff*, *surface water*.

gully reclamation Restoring land suffering from gully erosion by seeding gullies with quick-growing plants, building small dams to collect silt and gradually fill in the channels, and building channels to divert water away from the gully.

habitat Place or type of place where an organism or a population of organisms lives. Compare *ecological niche*.

habitat fragmentation Breakup of a habitat into smaller units, usually as a result of human activities.

half-life Time needed for one-half of the nuclei in a radioisotope to emit their radiation. Each radioisotope has a characteristic half-life, which may range from a few millionths of a second to several billion years.

hazardous chemical Chemical that can cause harm because it is flammable or explosive, or that can irritate or damage the skin or lungs (such as strong acidic or alkaline substances) or cause allergic reactions of the immune system (allergens). See *toxic chemical*.

hazardous waste Any solid, liquid, or containerized gas that can catch fire easily, is corrosive to skin tissue or metals, is unstable and can explode or release toxic fumes, or has harmful concentrations of one or more toxic materials that can leach out. See also *toxic waste*.

heat Total kinetic energy of all the randomly moving atoms, ions, or molecules within a given substance, excluding the overall motion of the whole object. This form of kinetic energy flows from one body to another when there is a temperature difference between the two bodies. Heat always flows spontaneously from a hot sample of matter to a colder sample of matter. This is one way to state the second law of thermodynamics. Compare *temperature*.

herbicide Chemical that kills a plant or inhibits its growth.

herbivore Plant-eating organism. Examples are deer, sheep, grasshoppers, and zooplankton. Compare *carnivore*, *omnivore*.

heterotroph See *consumer*.

high-input agriculture See *industrialized agriculture*.

high-quality energy Energy that is organized or concentrated and has great ability to perform useful work. Examples are high-temperature heat and the energy in electricity, coal, oil, gasoline, sunlight, and nuclei of uranium-235. Compare *low-quality energy*.

high-quality matter Matter that is organized and concentrated and contains a high concentration of a useful resource. Compare *low-quality matter*.

high-throughput society See *high-waste society*.

high-waste society The situation in most advanced industrialized countries, in which ever-increasing economic growth is sustained by maximizing the rate at which matter and energy resources are used, with little emphasis on pollution prevention, recycling, reuse, reduction of unnecessary waste, and other forms of resource conservation. Compare *earth-wisdom society, matter-recycling society*.

homeostasis Maintenance of favorable internal conditions in a system despite fluctuations in external conditions. See *constancy, inertia, resilience*.

host Plant or animal on which a parasite feeds.

human capital Physical and mental talents of people used to produce, distribute, and sell an economic good. Compare *earth capital, manufactured capital, solar capital*.

humus Slightly soluble residue of undigested or partially decomposed organic material in topsoil. This material helps retain water and water-soluble nutrients, which can be taken up by plant roots.

hunter–gatherers People who get their food by gathering edible wild plants and other materials and by hunting wild animals and fish.

hydrologic cycle Biogeochemical cycle that collects, purifies, and distributes the earth's fixed supply of water from the environment to living organisms, and then back to the environment.

hydrosphere The earth's liquid water (oceans, lakes and other bodies of surface water, and underground water) earth's frozen water (polar ice caps, floating ice caps, and ice in soil known as permafrost), and small amounts of water vapor in the atmosphere.

identified resources Deposits of a particular mineral-bearing material of which the location, quantity, and quality are known or have been estimated from direct geological evidence and measurements. Compare *undiscovered resources*.

igneous rock Rock formed when molten rock material (magma) wells up from earth's interior, cools, and solidifies into rock masses. Compare *metamorphic rock, sedimentary rock*. See *rock cycle*.

immature community Community at an early stage of ecological succession. It usually has a low number of species and ecological niches and cannot capture and use energy and cycle critical nutrients as efficiently as more complex, mature ecosystems. Compare *mature community*.

immigrant species Species that migrate into an ecosystem or are deliberately or accidentally introduced into an ecosystem by humans. Some of these species are beneficial, whereas others can take over and eliminate many native species. Compare *indicator species, keystone species, native species*.

immigration Migration of people into a country or area to take up permanent residence.

indicator species Species that serve as early warnings that a community or an ecosystem is being degraded. Compare *immigrant species, keystone species, native species*.

inductive reasoning Using observations and facts to arrive at generalizations or hypotheses. It goes from the specific to the general and is widely used in science. Compare *deductive reasoning*.

industrialized agriculture Using large inputs of energy from fossil fuels (especially oil and natural gas), water, fertilizer, and pesticides to produce large quantities of crops and livestock for domestic and foreign sale. Compare *subsistence farming*.

industrial revolution Use of new sources of energy from fossil fuels and later from nuclear fuels, and use of new technologies, to grow food and manufacture products. Compare *agricultural revolution*.

industrial smog Type of air pollution consisting mostly of a mixture of sulfur dioxide, suspended droplets of sulfuric acid formed from some of the sulfur dioxide, and a variety of suspended solid particles. Compare *photochemical smog*.

inertia Ability of a living system to resist being disturbed or altered. Compare *constancy, resilience*.

infant mortality rate Number of babies out of every 1,000 born each year that die before their first birthday.

infiltration Downward movement of water through soil.

information revolution Use of new technologies such as the telephone, radio, television, and computers to enable people to deal with more and more information more rapidly.

inland wetland Land away from the coast, such as a swamp, marsh, or bog, that is covered all or part of the time with fresh water. Compare *coastal wetland*.

inorganic fertilizer See *commercial inorganic fertilizer*.

input Matter, energy, or information entering a system. Compare *output, throughput*.

input pollution control See *pollution prevention*.

insecticide Chemical that kills insects.

integrated pest management (IPM) Combined use of biological, chemical, and cultivation methods in proper sequence and timing to keep the size of a pest population below the size that causes economically unacceptable loss of a crop or livestock animal.

intercropping Growing two or more different crops at the same time on a plot. For example, a carbohydrate-rich grain that depletes soil nitrogen and a protein-rich legume that adds nitrogen to the soil may be intercropped. Compare *monoculture, polyculture, polyvarietal cultivation*.

interference competition Situation in which one species limits access of another species to a resource, regardless of whether the resource is abundant or scarce. See *exploitation competition, interspecific competition*. Compare *intraspecific competition*.

intergenerational equity Ethical concept that future generations should receive undiminished earth capital and economic opportunity.

intergenerational fairness See *intergenerational equity*.

intermediate goods See *manufactured capital*.

internal cost Direct cost paid by the producer and the buyer of an economic good. Compare *external cost*.

interplanting Simultaneously growing a variety of crops on the same plot. See *agroforestry, intercropping, polyculture, polyvarietal cultivation*.

interspecific competition Members of two or more species trying to use the same limited resources in an ecosystem. See *competition, competitive exclusion, exploitation competition, intraspecific competition, interference competition*.

intertidal zone The area of shoreline between low and high tides.

intraspecific competition Two or more organisms of a single species trying to use the same limited resources in an ecosystem. See *competition, interspecific competition*.

intrinsic rate of increase (r) Rate at which a population could grow if it had unlimited resources. Compare *environmental resistance*.

inversion See *thermal inversion*.

invertebrates Animals that have no backbones. Compare *vertebrates*.

ion Atom or group of atoms with one or more positive (+) or negative ([−]) electrical charges. Compare *atom, molecule*.

ionizing radiation Fast-moving alpha or beta particles or high-energy radiation (gamma rays) emitted by radioisotopes. They have enough energy to dislodge one or more electrons from atoms they hit, forming charged ions (in tissue) that can react with and damage living tissue. Compare *nonionizing radiation*.

isotopes Two or more forms of a chemical element that have the same number of protons but different mass numbers because of different numbers of neutrons in their nuclei.

***J*-shaped curve** Curve with a shape similar to that of the letter *J*; can represent prolonged exponential growth. See *exponential growth*.

kerogen Solid, waxy mixture of hydrocarbons found in oil shale rock. When the rock is heated to high temperatures, the kerogen is vaporized. The vapor is condensed, purified, and then sent to a refinery to produce gasoline, heating oil, and other products. See also *oil shale, shale oil*.

keystone species Species that play roles affecting many other organisms in an ecosystem. Compare *immigrant species, indicator species, native species, nonnative species*.

kinetic energy Energy that matter has because of its mass and speed or velocity. Compare *potential energy*.

K-strategists Species that produce a few, often fairly large offspring but invest a great deal of time and energy to ensure that most of those offspring reach reproductive age. Compare *r-strategists*.

kwashiorkor Type of malnutrition that occurs in infants and very young children when they are weaned from mother's milk to a starchy diet low in protein. See *marasmus*.

labor See *human capital*.

lake Large natural body of standing fresh water formed when water from precipitation, land runoff, or groundwater flow fills a depression in the earth created by glaciation, earth movement, volcanic activity, or a giant meteorite. See *eutrophic lake, mesotrophic lake, oligotrophic lake*.

landfill See *sanitary landfill*.

land-use planning Process for deciding the best present and future use of each parcel of land in an area.

late successional plant species Mostly trees that can tolerate shade and that form a fairly stable complex forest community. Compare *early successional plant species* and *midsuccessional plant species*.

latitude Distance from the equator. Compare *altitude*.

law of conservation of energy See *first law of thermodynamics*.

law of conservation of matter In any physical or chemical change, matter is neither created nor destroyed, but merely changed from one form to another; in physical and chemical changes, existing atoms are rearranged into different spatial patterns (physical changes) or different combinations (chemical changes).

law of tolerance The existence, abundance, and distribution of a species in an ecosystem are determined by whether the levels of one or more physical or chemical factors fall within the range tolerated by the species. See *threshold effect*.

LD_{50} See *median lethal dose*.

LDC See *developing country*.

leaching Process in which various chemicals in upper layers of soil are dissolved and carried to lower layers and, in some cases, to groundwater.

less developed country (LDC) See *developing country*.

life cycle cost Initial cost plus lifetime operating costs of an economic good.

life expectancy Average number of years a newborn infant can be expected to live.

limiting factor Single factor that limits the growth, abundance, or distribution of the population of a species in an ecosystem. See *limiting factor principle.*

limiting factor principle Too much or too little of any abiotic factor can limit or prevent growth of a population of a species in an ecosystem, even if all other factors are at or near the optimum range of tolerance for the species.

linear growth Growth in which a quantity increases by some fixed amount during each unit of time. Compare *exponential growth.*

liquefied natural gas (LNG) Natural gas converted to liquid form by cooling to a very low temperature.

liquefied petroleum gas (LPG) Mixture of liquefied propane (C_3H_8) and butane (C_4H_{10}) gas removed from natural gas.

lithosphere Outer shell of the earth, composed of the crust and the rigid, outermost part of the mantle outside of the asthenosphere; material found in the earth's plates. See *crust, mantle.*

loams Soils containing a mixture of clay, sand, silt, and humus. Good for growing most crops.

logistic growth Exponential population growth when the population is small and results in a steady decrease in population growth with time as the population approaches the carrying capacity.

low-input agriculture See *sustainable agriculture.*

low-quality energy Energy that is disorganized or dispersed and has little ability to do useful work. An example is low-temperature heat. Compare *high-quality energy.*

low-quality matter Matter that is disorganized, dilute, or dispersed or that contains a low concentration of a useful resource. Compare *high-quality matter.*

low-throughput society See *earth-wisdom society.*

low-waste society See *earth-wisdom society.*

LPG See *liquefied petroleum gas.*

macroevolution Long-term, large-scale evolutionary changes among groups of species. Compare *microevolution.*

macronutrients Chemical elements organisms need in fairly large amounts to live, grow, or reproduce. Examples are carbon, oxygen, hydrogen, nitrogen, phosphorus, sulfur, potassium, calcium, magnesium, and iron. Compare *micronutrients.*

magma Molten rock below the earth's surface.

malnutrition Faulty nutrition. Caused by a diet that does not supply a person with enough protein, essential fats, vitamins, minerals, and other nutrients needed for good health. Compare *overnutrition, undernutrition.*

mangrove swamps Swamps found on the coastlines in warm tropical climates. They are dominated by mangrove trees, any of about 55 species of trees and shrubs that can live partly submerged in the salty environment of coastal swamps.

mantle Zone of the earth's interior between its core and its crust. Compare *core, crust.* See *lithosphere.*

manufactured capital Manufactured items made from earth capital and used to produce and distribute economic goods and services bought by consumers. These include tools, machinery, equipment, factory buildings, and transportation and distribution facilities. Compare *earth capital, human capital, solar capital.*

manure See *animal manure, green manure.*

marasmus Nutritional-deficiency disease caused by a diet that does not have enough calories and protein to maintain good health. See *kwashiorkor, malnutrition.*

market equilibrium State in which sellers and buyers of an economic good agree on the quantity to be produced and the price to be paid.

mass The amount of material in an object.

mass extinction A catastrophic, widespread, often global event in which major groups of species are wiped out over a short time compared to normal (background) extinctions. Compare *background extinction.* See *adaptive radiation.*

mass number Sum of the number of neutrons and the number of protons in the nucleus of an atom. It gives the approximate mass of that atom. Compare *atomic number.*

mass transit Buses, trains, trolleys, and other forms of transportation that carry large numbers of people.

matter Anything that has mass (the amount of material in an object) and takes up space. On earth, where gravity is present, we weigh an object to determine its mass.

matter quality Measure of how useful a matter resource is, based on its availability and concentration. See *high-quality matter, low-quality matter.*

matter-recycling society Society that emphasizes recycling the maximum amount of all resources that can be recycled. The goal is to allow economic growth to continue without depleting matter resources and without producing excessive pollution and environmental degradation. Compare *earth-wisdom society, high-waste society.*

mature community Fairly stable, self-sustaining community in an advanced stage of ecological succession; usually has a diverse array of species and ecological niches; captures and uses energy and cycles critical chemicals more efficiently than simpler, immature communities. Compare *immature community.*

maximum sustainable yield See *sustainable yield.*

MDC See *developed country.*

median lethal dose (LD_{50}) Amount of a toxic material per unit of body weight of test animals that kills half the test population in a certain time.

megacities Cities with 10 million or more people.

meltdown The melting of the core of a nuclear reactor.

mesosphere Third layer of the atmosphere; found above the stratosphere. Compare *stratosphere, thermosphere, troposphere.*

mesotrophic lake Lake with a moderate supply of plant nutrients. Compare *eutrophic lake, oligotrophic lake.*

metabolism Ability of a living cell or organism to capture and transform matter and energy from its environment to supply its needs for survival, growth, and reproduction.

metamorphic rock Rock produced when a preexisting rock is subjected to high temperatures (which may cause it to melt partially), high pressures, chemically active fluids, or a combination of these agents. Compare *igneous rock, sedimentary rock.* See *rock cycle.*

metastasis Spread of malignant (cancerous) cells from a cancer to other parts of the body.

microclimates Local climatic conditions that differ from the general climate of a region. They are typically caused by various topographic features of the earth's surface such as mountains and cities.

microevolution The small genetic changes that a population experiences. Compare *macroevolution.*

micronutrients Chemical elements organisms need in small or even trace amounts to live, grow, or reproduce. Examples are sodium, zinc, copper, chlorine, and iodine. Compare *macronutrients.*

microorganisms Organisms that are so small that they can be seen only by using a microscope.

midsuccessional plant species Grasses and low shrubs that are less hardy than early successional plant species. Compare *early successional plant species* and *late successional plant species.*

mineral Any naturally occurring inorganic substance found in the earth's crust as a crystalline solid. See *mineral resource.*

mineral resource Concentration of naturally occurring solid, liquid, or gaseous material in or on the earth's crust, in a form and amount such that extracting and converting it into useful materials or items is currently or potentially profitable. Mineral resources are classified as *metallic* (such as iron and tin ores) or *nonmetallic* (such as fossil fuels, sand, and salt).

minimum-tillage farming See *conservation-tillage farming.*

mixed economic system Economic system that falls somewhere between pure market and pure command economic systems. Virtually all of the world's economic systems fall into this category, with some closer to a pure market system and some closer to a pure command system. Compare *pure command economic system, pure market economic system.*

mixture Combination of two or more elements and compounds.

model An approximate representation or simulation of a system being studied.

molecule Combination of two or more atoms of the same chemical element (such as O_2) or different chemical elements (such as H_2O) held together by chemical bonds. Compare *atom, ion.*

monera See *bacteria, cyanobacteria.*

monoculture Cultivation of a single crop, usually on a large area of land. Compare *polyculture, polyvarietal cultivation.*

more developed country (MDC) See *developed country.*

municipal solid waste Solid materials discarded by homes and businesses in or near urban areas. See *solid waste.*

mutagen Chemical or form of ionizing radiation that causes inheritable changes in the DNA molecules in the genes found in chromosomes (mutations). See *carcinogen, mutation, teratogen.*

mutation A random change in DNA molecules making up genes that can yield changes in anatomy, physiology, or behavior in offspring. See *mutagen.*

mutualism Type of species interaction in which both participating species generally benefit. Compare *commensalism.*

native species Species that normally live and thrive in a particular ecosystem. Compare *immigrant species, indicator species, keystone species, nonnative species.*

natural gas Underground deposits of gases consisting of 50–90% by weight methane gas (CH_4) and small amounts of heavier gaseous hydrocarbon compounds such as propane (C_3H_8) and butane (C_4H_{10}).

natural greenhouse effect Trapping of heat in the troposphere by certain gases, called greenhouse gases. Without this atmospheric thermal blanket, the earth would be nearly as cold as Mars and life as we know it could not exist. There is considerable evidence that we are enhancing this natural effect by excess additions of greenhouse gases from human activities.

natural ionizing radiation Ionizing radiation in the environment from natural sources.

natural radioactive decay Nuclear change in which unstable nuclei of atoms spontaneously shoot out particles (usually alpha or beta particles), energy (gamma rays), or both at a fixed rate.

natural recharge Natural replenishment of an aquifer by precipitation that percolates downward through soil and rock. See *recharge area.*

natural resource capital See *earth capital.*

natural resources Nutrients and minerals in the soil and deeper layers of the earth's crust, water, wild and domesticated plants and animals, air, and other resources produced by the earth's natural processes. Compare *human capital, manufactured capital, solar capital.* See *earth capital.*

natural selection Process by which a particular beneficial gene (or set of genes) is reproduced more than other genes in succeeding generations. The result of natural selection is a population that contains a greater proportion of organisms better adapted to certain environmental conditions. See *adaptation, biological evolution, differential reproduction, mutation.*

negative feedback loop Situation in which a change in a certain direction provides information that causes a system to change less in that direction. Compare *positive feedback loop.*

negawatt A watt of electrical power saved by improving energy efficiency.

nekton Strongly swimming organisms found in aquatic systems. Compare *benthos, plankton.*

nematocide Chemical that kills nematodes (roundworms).

net energy Total amount of useful energy available from an energy resource or energy system over its lifetime minus the amount of energy used (the first energy law), automatically wasted (the second energy law), and unnecessarily wasted in finding, processing, concentrating, and transporting it to users.

net primary productivity (NPP) Rate at which all the plants in an ecosystem produce net useful chemical energy; equal to the difference between the rate at which the plants in an ecosystem produce useful chemical energy (primary productivity) and the rate at which they use some of that energy through cellular respiration. Compare *gross primary productivity.*

neutral solution Water solution containing an equal number of hydrogen ions (H^+) and hydroxide ions (OH^{-}); water solution with a pH of 7. Compare *acid solution, basic solution.*

neutron (n) Elementary particle in the nuclei of all atoms (except hydrogen-1). It has a relative mass of 1 and no electric charge. Compare *electron, proton.*

niche See *ecological niche.*

nitrogen cycle Cyclic movement of nitrogen in different chemical forms from the environment to organisms and then back to the environment.

nitrogen fixation Conversion of atmospheric nitrogen gas into forms useful to plants by lightning, bacteria, and cyanobacteria; it is part of the nitrogen cycle.

noise pollution Any unwanted, disturbing, or harmful sound that impairs or interferes with hearing, causes stress, hampers concentration and work efficiency, or causes accidents.

nondegradable pollutant Material that is not broken down by natural processes. Examples are the toxic elements lead and mercury. Compare *biodegradable pollutant, degradable pollutant, slowly degradable pollutant.*

nonionizing radiation Forms of radiant energy such as radio waves, microwaves, infrared light, and ordinary light that do not have enough energy to cause ionization of atoms in living tissue. Compare *ionizing radiation.*

nonnative species Species that migrate into an ecosystem or are deliberately or accidentally introduced into an ecosystem by humans.

nonpersistent pollutant See *degradable pollutant.*

nonpoint source Large or dispersed land areas such as cropfields, streets, and lawns that discharge pollutants into the environment over a large area. Compare *point source.*

nonrenewable resource Resource that exists in a fixed amount (stock) in various places in the earth's crust and has the potential for renewal only by geological, physical, and chemical processes taking place over hundreds of millions to billions of years. Examples are copper, aluminum, coal, and oil. We classify these resources as exhaustible because we are extracting and using them at a much faster rate than they were formed. Compare *potentially renewable resource.*

nontransmissible disease A disease that is not caused by living organisms and does not spread from one person to another. Examples are most cancers, diabetes, cardiovascular disease, and malnutrition. Compare *transmissible disease.*

no-till farming See *conservation-tillage farming.*

nuclear change Process in which nuclei of certain isotopes spontaneously change, or are forced to change, into one or more different isotopes. The three principal types of nuclear change are natural radioactivity, nuclear fission, and nuclear fusion. Compare *chemical change, physical change.*

nuclear energy Energy released when atomic nuclei undergo a nuclear reaction such as the spontaneous emission of radioactivity, nuclear fission, or nuclear fusion.

nuclear fission Nuclear change in which the nuclei of certain isotopes with large mass numbers (such as uranium-235 and plutonium-239) are split apart into lighter nuclei when struck by a neutron. This process releases more neutrons and a large amount of energy. Compare *nuclear fusion.*

nuclear fusion Nuclear change in which two nuclei of isotopes of elements with a low mass number (such as hydrogen-2 and hydrogen-3) are forced together at extremely high temperatures until they fuse to form a heavier nucleus (such as helium-4). This process releases a large amount of energy. Compare *nuclear fission.*

nucleus Extremely tiny center of an atom, making up most of the atom's mass. It contains one or more positively charged protons and one or more neutrons with no electrical charge (except for a hydrogen-1 atom, which has one proton and no neutrons in its nucleus).

nutrient Any atom, ion, or molecule an organism needs to live, grow, or reproduce.

nutrient cycle See *biogeochemical cycle.*

oil See *crude oil.*

oil shale Fine-grained rock containing various amounts of kerogen, a solid, waxy mixture of hydrocarbon compounds. Heating the rock to high temperatures converts the kerogen into a vapor that can be condensed to form a slow-flowing heavy oil called shale oil. See *kerogen, shale oil.*

old-growth forest Virgin and old, second-growth forests containing trees that are often hundreds, sometimes thousands of years old. Examples include forests of Douglas fir, western hemlock, giant sequoia, and coastal redwoods in the western United States. Compare *second-growth forest, tree farm.*

oligotrophic lake Lake with a low supply of plant nutrients. Compare *eutrophic lake, mesotrophic lake.*

omnivore Animal that can use both plants and other animals as food sources. Examples are pigs, rats, cockroaches, and people. Compare *carnivore, herbivore.*

open-pit mining Removal of minerals such as gravel, sand, and ores of metals (such as iron and copper) by digging them out of the earth's surface and leaving an open pit.

open sea The part of an ocean that is beyond the continental shelf. Compare *coastal zone.*

open system A system, such as a living organism, in which both matter and energy are exchanged between the system and the environment. Compare *closed system.*

ore Part of a metal-yielding material that can be economically extracted at a given time. An ore typically contains two parts: the ore mineral, which contains the desired metal, and waste mineral material (gangue).

organic farming Producing crops and livestock naturally by using organic fertilizer (manure, legumes, compost) and natural pest control (bugs that eat harmful bugs, plants that repel bugs, and environmental controls such as crop rotation) instead of using commercial inorganic fertilizers and synthetic pesticides and herbicides.

organic fertilizer Organic material such as animal manure, green manure, and compost, applied to cropland as a source of plant nutrients. Compare *commercial inorganic fertilizer.*

organism Any form of life.

other resources Identified and undiscovered resources not classified as reserves. See *identified resources, reserves, undiscovered resources.*

output Matter, energy, or information leaving a system. Compare *input, throughput.*

output pollution control See *pollution cleanup.*

overburden Layer of soil and rock overlying a mineral deposit; removed during surface mining.

overconsumption Situation in which some people consume much more than they need at the expense of those who cannot meet their basic needs and at the expense of earth's present and future life-support systems.

overfishing Harvesting so many fish of a species (especially immature fish) that there is not enough breeding stock left to replenish the species, such that it is not profitable to harvest them.

overgrazing Destruction of vegetation when too many grazing animals feed too long and exceed the carrying capacity of a rangeland area.

overnutrition Diet so high in calories, saturated (animal) fats, salt, sugar, and processed foods and so low in vegetables and fruits that the consumer runs high risks of diabetes, hypertension, heart disease, and other health hazards. Compare *malnutrition, undernutrition.*

oxygen-demanding wastes Organic materials that are usually biodegraded by aerobic (oxygen-consuming) bacteria if there is enough dissolved oxygen in the water. See also *biological oxygen demand.*

ozone depletion Decrease in concentration of ozone in the stratosphere. See *ozone layer.*

ozone layer Stratospheric layer of gaseous ozone (O_3) that protects life on the earth by filtering out most harmful ultraviolet radiation from the sun.

PANs Peroxyacyl nitrates. Group of chemicals found in photochemical smog.

parasite Consumer organism that lives on or in and feeds on a living plant or animal, known as the host, over an extended period of time. The parasite draws nourishment from and gradually weakens its host; it may or may not kill the host. See *parasitism.*

parasitism Interaction between species in which one organism, called the parasite, preys on another organism, called the host, by living on or in the host. See *host, parasite.*

parts per billion (ppb) Number of parts of a chemical found in one billion parts of a particular gas, liquid, or solid.

parts per million (ppm) Number of parts of a chemical found in one million parts of a particular gas, liquid, or solid.

parts per trillion (ppt) Number of parts of a chemical found in one trillion parts of a particular gas, liquid, or solid.

passive solar heating system System that captures sunlight directly within a structure and converts it into low-temperature heat for space heating or for heating water for domestic use, without the use of mechanical devices. Compare *active solar heating system.*

pathogen Organism that produces disease.

PCBs See *polychlorinated biphenyls.*

per capita GDP Annual gross domestic product (GDP) of a country divided by its total population. See *gross domestic product, real per capita GDP.*

per capita GNP Annual gross national product (GNP) of a country divided by its total population. See *gross national product, real per capita GNP.*

percolation Passage of a liquid through the spaces of a porous material such as soil.

perennial Plant that can live for more than 2 years. Compare *annual.*

permafrost Perennially frozen layer of the soil that forms when the water there freezes. It is found in arctic tundra.

permeability The degree to which underground rock and soil pores are interconnected with each other, and thus a measure of the degree to which water can flow freely from one pore to another. Compare *porosity.*

perpetual resource See *renewable resource.*

persistence See *inertia.*

persistent pollutant See *slowly degradable pollutant.*

pest Unwanted organism that directly or indirectly interferes with human activities.

pesticide Any chemical designed to kill or inhibit the growth of an organism that people consider to be undesirable. See *fungicide, herbicide, insecticide.*

pesticide treadmill Situation in which the cost of using pesticides increases while their effectiveness decreases, mostly because the pest species develop genetic resistance to the pesticides.

petrochemicals Chemicals obtained by refining (distilling) crude oil. They are used as raw materials in the manufacture of most industrial chemicals, fertilizers, pesticides, plastics, synthetic fibers, paints, medicines, and many other products.

petroleum See *crude oil.*

pH Numeric value that indicates the acidity or alkalinity of a substance on a scale of 0 to 14, with the neutral point at 7. Acid solutions have pH values lower than 7, and basic or alkaline solutions have pH values greater than 7.

phosphorus cycle Cyclic movement of phosphorus in different chemical forms, from the environment to organisms and then back to the environment.

photochemical smog Complex mixture of air pollutants produced in the lower atmosphere by the reaction of hydrocarbons and nitrogen oxides under the influence of sunlight. Especially harmful components include ozone, peroxyacyl nitrates (PANs), and various aldehydes. Compare *industrial smog.*

photosynthesis Complex process that takes place in cells of green plants. Radiant energy from the sun is used to combine carbon dioxide (CO_2) and water (H_2O) to produce oxygen (O_2) and carbohydrates (such as glucose, $C_6H_{12}O_6$) and other nutrient molecules. Compare *aerobic respiration, chemosynthesis.*

photovoltaic cell (solar cell) Device in which radiant (solar) energy is converted directly into electrical energy.

physical change Process that alters one or more physical properties of an element or compound without altering its chemical composition. Examples are changing the size and shape of a sample of matter (crushing ice and cutting aluminum foil) and changing a sample of matter from one physical state to another (boiling and freezing water). Compare *chemical change, nuclear change.*

phytoplankton Small, drifting plants, mostly algae and bacteria, found in aquatic ecosystems. Compare *plankton, zooplankton.*

pioneer community First integrated set of plants, animals, and decomposers found in an area undergoing primary ecological succession. See *immature community, mature community.*

pioneer species First hardy species, often microbes, mosses, and lichens, that begin colonizing a site as the first stage of ecological succession. See *ecological succession, pioneer community.*

plaintiff The individual, group of individuals, corporation, or government agency bringing the charges in a lawsuit. Compare *defendant.*

planetary management worldview Beliefs that we are the planet's most important species; we are in charge of the rest of nature; there is always more, and it's all for us; all economic growth is good, more economic growth is better, and the potential for economic growth is limitless; our success depends on how well we can understand, control, and manage earth's life-support systems for our own benefit. Compare *earth-wisdom worldview.*

plankton Small plant organisms (phytoplankton) and animal organisms (zooplankton) that float in aquatic ecosystems. Compare *benthos, nekton.*

plantation agriculture Growing specialized crops such as bananas, coffee, and cacao in tropical developing countries, primarily for sale to developed countries.

plants (plantae) Eukaryotic, mostly multicelled organisms such as algae (red, blue, and green), mosses, ferns, flowers, cacti, grasses, beans, wheat, rice, and trees. These organisms use photosynthesis to produce organic nutrients for themselves and for other organisms feeding on them. Water and other inorganic nutrients are obtained from the soil for terrestrial plants and from the water for aquatic plants.

plates Various-sized areas of earth's lithosphere that move slowly around with the mantle's flowing asthenosphere. Most earthquakes and volcanoes occur around the boundaries of these plates. See *asthenosphere, lithosphere, plate tectonics.*

plate tectonics Theory of geophysical processes that explains the movements of lithospheric plates and the processes that occur at their boundaries. See *lithosphere, plates.*

point source A single identifiable source that discharges pollutants into the environment. Examples are the smokestack of a power plant or an industrial plant, the drainpipe of a meat-packing plant, the chimney of a house, or the exhaust pipe of an automobile. Compare *nonpoint source.*

poison A chemical that in one dose kills exactly 50% of the animals (usually rats and mice) in a test population (usually 60 to 200 animals) within a 14-day period. See *median lethal dose.*

politics Process through which individuals and groups try to influence or control government policies and actions that affect the local, state, national, and international communities.

pollutant A particular chemical or form of energy that can adversely affect the health, survival, or activities of humans or other living organisms. See *pollution.*

pollution An undesirable change in the physical, chemical, or biological characteristics of air, water, soil, or food that can adversely affect the health, survival, or activities of humans or other living organisms.

pollution cleanup Device or process that removes or reduces the level of a pollutant after it has been produced or has entered the environment. Examples are automobile emission-control devices and sewage treatment plants. Compare *pollution prevention.*

pollution prevention Device or process that prevents a potential pollutant from forming or from entering the environment, or sharply reduces the amounts entering the environment. Compare *pollution cleanup.*

polyculture Complex form of intercropping in which a large number of different plants maturing at different times are planted together. See also *intercropping.* Compare *monoculture, polyvarietal cultivation.*

polyvarietal cultivation Planting a plot of land with several varieties of the same crop. Compare *intercropping, monoculture, polyculture.*

population Group of individual organisms of the same species living within a particular area.

population change An increase or decrease in the size of a population. It is equal to (Births + Immigration) [–] (Deaths + Emigration).

population density Number of organisms in a particular population found in a specified area. Compare *ecological population density.*

population dispersion General pattern in which the members of a population are arranged throughout its habitat.

population distribution Variation of population density over a particular geographic area. For example, a country has a high population density in its urban areas and a much lower population density in rural areas.

population dynamics Major abiotic and biotic factors that tend to increase or decrease the population size and the age and sex composition of a species.

population size Number of individuals making up a population's gene pool.

porosity Percentage of space in rock or soil occupied by voids, whether the voids are isolated or connected. Compare *permeability.*

positive feedback loop Situation in which a change in a certain direction provides information that causes a system to change further in the same direction. This can lead to a runaway or vicious cycle. Compare *negative feedback loop.*

potential energy Energy stored in an object because of its position or the position of its parts. Compare *kinetic energy.*

potentially renewable resource Resource that can be replenished fairly rapidly (hours to several decades) through natural processes. Examples are trees in forests, grasses in grasslands, wild animals, fresh surface water in lakes and streams, most groundwater, fresh air, and fertile soil. If such a resource is used faster than it is replenished, it can be depleted and converted into a nonrenewable resource. Compare *nonrenewable resource,* and *renewable resource.* See also *environmental degradation.*

poverty Inability to meet basic needs for food, clothing, and shelter.

ppb See *parts per billion.*

ppm See *parts per million.*

ppt See *parts per trillion.*

precipitation Water in the form of rain, sleet, hail, and snow that falls from the atmosphere onto the land and bodies of water.

precision A measure of reproducibility, or how closely a series of measurements of the same quantity agree with one another. Compare *accuracy.*

predation Situation in which an organism of one species (the predator) captures and feeds on parts or all of an organism of another species (the prey).

predator Organism that captures and feeds on parts or all of an organism of another species (the prey).

predator–prey relationship Interaction between two organisms of different species in which one organism, called the *predator*, captures and feeds on parts or all of another organism, called the *prey*.

prey Organism that is captured and serves as a source of food for an organism of another species (the predator).

primary consumer Organism that feeds directly on all or parts of plants (herbivore) or on other producers. Compare *detritivore, omnivore, secondary consumer*.

primary oil recovery Pumping out the crude oil that flows by gravity or under gas pressure into the bottom of an oil well. Compare *enhanced oil recovery, secondary oil recovery*.

primary pollutant Chemical that has been added directly to the air by natural events or human activities and occurs in a harmful concentration. Compare *secondary pollutant*.

primary sewage treatment Mechanical treatment of sewage in which large solids are filtered out by screens and suspended solids settle out as sludge in a sedimentation tank. Compare *advanced sewage treatment, secondary sewage treatment*.

primary succession Sequential development of communities in a bare area that has never been occupied by a community of organisms. Compare *secondary succession*.

prior appropriation Legal principle by which the first user of water from a stream establishes a legal right to continued use of the amount originally withdrawn. Compare *riparian rights*.

probability A mathematical statement about how likely it is that something will happen.

producer Organism that uses solar energy (green plant) or chemical energy (some bacteria) to manufacture the organic compounds it needs as nutrients from simple inorganic compounds obtained from its environment. Compare *consumer, decomposer*.

prokaryotic cell Cell that doesn't have a distinct nucleus. Other internal parts are also not enclosed by membranes. Compare *eukaryotic cell*.

protists Eukaryotic, mostly single-celled organisms such as diatoms, amoebas, some algae (golden brown and yellow-green), protozoans, and slime molds. Some protists produce their own organic nutrients through photosynthesis. Others are decomposers and some feed on bacteria, other protists, or cells of multicellular organisms.

proton (p) Positively charged particle in the nuclei of all atoms. Each proton has a relative mass of 1 and a single positive charge. Compare *electron, neutron*.

pure capitalism See *pure market economic system*.

pure command economic system System in which all economic decisions are made by the government or some other central authority. Compare *mixed economic system, pure market economic system*.

pure market economic system System in which all economic decisions are made in the market, where buyers and sellers of economic goods interact freely, with no government or other interference. Compare *mixed economic system, pure command economic system*.

pyramid of biomass Diagram representing the biomass (total dry weight of all living organisms) that can be supported at each trophic level in a food chain or food web. See *pyramid of energy flow, pyramid of numbers*.

pyramid of energy flow Diagram representing the flow of energy through each trophic level in a food chain or food web. With each energy transfer, only a small part (typically 10%) of the usable energy entering one trophic level is transferred to the organisms at the next trophic level. Compare *pyramid of biomass, pyramid of numbers*.

pyramid of numbers Diagram representing the number of organisms of a particular type that can be supported at each trophic level from a given input of solar energy at the producer trophic level in a food chain or food web. Compare *pyramid of biomass, pyramid of energy flow*.

radiation Fast-moving particles (particulate radiation) or waves of energy (electromagnetic radiation).

radioactive decay Change of a radioisotope to a different isotope by the emission of radioactivity.

radioactive isotope See *radioisotope*.

radioactivity Nuclear change in which unstable nuclei of atoms spontaneously shoot out "chunks" of mass, energy, or both, at a fixed rate. The three principal types of radioactivity are gamma rays and fast-moving alpha particles and beta particles.

radioisotope Isotope of an atom that spontaneously emits one or more types of radioactivity (alpha particles, beta particles, gamma rays).

rain shadow effect Low precipitation on the far side (leeward side) of a mountain when prevailing winds flow up and over a high mountain or range of high mountains. This creates semiarid and arid conditions on the leeward side of a high mountain range.

rangeland Land that supplies forage or vegetation (grasses, grasslike plants, and shrubs) for grazing and browsing animals and is not intensively managed. Compare *feedlot*.

range of tolerance Range of chemical and physical conditions that must be maintained for populations of a particular species to stay alive and grow, develop, and function normally. See *law of tolerance*.

real GDP Gross domestic product adjusted for inflation.

real GNP Gross national product adjusted for inflation.

realized niche Parts of the fundamental niche of a species that are actually used by that species. See *ecological niche, fundamental niche*.

real per capita GDP Per capita GDP adjusted for inflation.

real per capita GNP Per capita GNP adjusted for inflation.

recharge area Any area of land allowing water to pass through it and into an aquifer. See *aquifer, natural recharge*.

recycling Collecting and reprocessing a resource so it can be made into new products. An example is collecting aluminum cans, melting them down, and using the aluminum to make new cans or other aluminum products. Compare *reuse*.

reforestation Renewal of trees and other types of vegetation on land where trees have been removed; can be done naturally by seeds from nearby trees or artificially by planting seeds or seedlings.

relative humidity The amount of water vapor in a certain mass of air, expressed as a percentage of the maximum amount it could hold at that temperature. Compare *absolute humidity*.

renewable resource An essentially inexhaustible resource on a human time scale. Solar energy is an example. See *potentially renewable resource*.

replacement-level fertility Number of children a couple must have to replace themselves. The average for a country or the world is usually slightly higher than 2 children per couple (2.1 in the United States and 2.5 in some developing countries) because some children die before reaching their reproductive years. See also *total fertility rate*.

reproduction Production of offspring by one or more parents. See *asexual reproduction, sexual reproduction*.

reproductive isolation Long-term geographic separation of members of a particular sexually reproducing species.

reproductive potential See *biotic potential*.

reserves Resources that have been identified from which a usable mineral can be extracted profitably at present prices with current mining technology. See *identified resources, other resources, undiscovered resources*.

reserve-to-production ratio Number of years that reserves of a particular nonrenewable mineral will last at current annual production rates. See *reserves*.

resilience Ability of a living system to restore itself to original condition after being exposed to an outside disturbance that is not too drastic. See *constancy, inertia*.

resource Anything obtained from the living and nonliving environment to meet human needs and wants. It can also be applied to other species.

resource partitioning Process of dividing up resources in an ecosystem so that species with similar requirements (overlapping ecological niches) use the same scarce resources at different times, in different ways, or in different places. See *ecological niche, fundamental niche, realized niche*.

resources Substances that can be consumed by an organism and, as a result, become unavailable to other organisms. Examples include food, water, and nesting sites for animals, and water, nutrients, and solar radiation for plants. Compare *conditions*.

respiration See *aerobic respiration*.

response The amount of health damage caused by exposure to a certain dose of a harmful substance or form of radiation. See *dose, dose–response curve, median lethal dose*.

reuse To use a product over and over again in the same form. An example is collecting, washing, and refilling glass beverage bottles. Compare *recycling*.

riparian rights System of water law that gives anyone whose land adjoins a flowing stream the right to use water from the stream as long as some is left for downstream users. Compare *prior appropriation*.

riparian zones Thin strips and patches of vegetation that surround streams. They are very important habitats and resources for wildlife.

risk The probability that something undesirable will happen from deliberate or accidental exposure to a hazard. See *risk analysis, risk assessment, risk–benefit analysis, risk management*.

risk analysis Identifying hazards, evaluating the nature and severity of risks (*risk assessment*), using this and other information to determine options and make decisions about reducing or eliminating risks (*risk management*), and communicating information about risks to decision makers and the public (*risk communication*).

risk assessment Process of gathering data and making assumptions to estimate short- and long-term harmful effects on human health or the environment from exposure to hazards associated with the use of a particular product or technology. See *risk, risk–benefit analysis*.

risk–benefit analysis Estimate of the short- and long-term risks and benefits of using a particular product or technology. See *risk*.

risk communication Communicating information about risks to decision makers and the public. See *risk, risk analysis, risk–benefit analysis*.

risk management Using risk assessment and other information to determine options and make decisions about reducing or eliminating risks. See *risk, risk analysis, risk–benefit analysis, risk communication*.

rock Any material that makes up a large, natural, continuous part of earth's crust. See *mineral*.

rock cycle Largest and slowest of the earth's cycles, consisting of geologic, physical, and chemical processes that form and modify rocks and soil in the earth's crust over millions of years.

rodenticide Chemical that kills rodents.

r-strategists Species that reproduce early in their life span and produce large numbers of usually small and short-lived offspring in a short period of time. Compare *K-strategists.*

rule of 70 Doubling time (in years) = 70/percentage growth rate. See *doubling time, exponential growth.*

runoff Fresh water from precipitation and melting ice that flows on the earth's surface into nearby streams, lakes, wetlands, and reservoirs. See *surface runoff, surface water.* Compare *groundwater.*

rural area Geographic area in the United States with a population of less than 2,500. The number of people used in this definition may vary in different countries. Compare *urban area.*

salinity Amount of various salts dissolved in a given volume of water.

salinization Accumulation of salts in soil that can eventually make the soil unable to support plant growth.

saltwater intrusion Movement of salt water into freshwater aquifers in coastal and inland areas as groundwater is withdrawn faster than it is recharged by precipitation.

sanitary landfill Waste disposal site on land in which waste is spread in thin layers, compacted, and covered with a fresh layer of clay or plastic foam each day.

scavenger Organism that feeds on dead organisms that were killed by other organisms or died naturally. Examples are vultures, flies, and crows. Compare *detritivore.*

science Attempts to discover order in nature and use that knowledge to make predictions about what should happen in nature. See *consensus science, frontier science, scientific data, scientific hypothesis, scientific law, scientific methods, scientific model, scientific theory.*

scientific data Facts obtained by making observations and measurements. Compare *model, scientific hypothesis, scientific methods, scientific law, scientific theory.*

scientific hypothesis An educated guess that attempts to explain a scientific law or certain scientific observations. Compare *model, scientific data, scientific law, scientific methods, scientific theory.*

scientific law Description of what scientists find happening in nature over and over in the same way, without known exception. See *first law of thermodynamics, second law of thermodynamics, law of conservation of matter.* Compare *scientific data, scientific hypothesis, scientific methods, scientific model, scientific theory.*

scientific methods The ways scientists gather data and formulate and test scientific hypotheses, models, theories, and laws. See *model, scientific data, scientific hypothesis, scientific law, scientific theory.*

scientific model See *model.*

scientific theory A well-tested and widely accepted scientific hypothesis. Compare *model, scientific data, scientific hypothesis, scientific methods.*

secondary consumer Organism that feeds only on primary consumers. Most secondary consumers are animals, but some are plants. Compare *detritivore, omnivore, primary consumer.*

secondary oil recovery Injection of water into an oil well after primary oil recovery to force out some of the remaining, usually thicker, crude oil. Compare *enhanced oil recovery, primary oil recovery.*

secondary pollutant Harmful chemical formed in the atmosphere when a primary air pollutant reacts with normal air components or other air pollutants. Compare *primary pollutant.*

secondary sewage treatment Second step in most waste treatment systems, in which aerobic bacteria break down up to 90% of degradable, oxygen-demanding organic wastes in wastewater. This is usually done by bringing sewage and bacteria together in trickling filters or the activated sludge process. Compare *advanced sewage treatment, primary sewage treatment.*

secondary succession Sequential development of communities in an area in which natural vegetation has been removed or destroyed but the soil is not destroyed. Compare *primary succession.*

second-growth forest Stands of trees resulting from secondary ecological succession. Compare *ancient forest, old-growth forest, tree farm.*

second law of energy See *second law of thermodynamics.*

second law of thermodynamics In any conversion of heat energy to useful work, some of the initial energy input is always degraded to a lower-quality, more dispersed, less useful energy, usually low-temperature heat that flows into the environment; you can't break even in terms of energy quality. See *first law of thermodynamics.*

sedimentary rock Rock that forms from the accumulated products of erosion and in some cases from the compacted shells, skeletons, and other remains of dead organisms. Compare *igneous rock, metamorphic rock.* See *rock cycle.*

seed-tree cutting Removal of nearly all trees on a site in one cutting, with a few seed-producing trees left uniformly distributed to regenerate the forest. Compare *clear-cutting, selective cutting, shelterwood cutting, strip cutting.*

selective cutting Cutting of intermediate-aged, mature, or diseased trees in an uneven-aged forest stand, either singly or in small groups. This encourages the growth of younger trees and maintains an uneven-aged stand. Compare *clear-cutting, seed-tree cutting, shelterwood cutting, strip cutting.*

selective pressure A factor in a population's environment that causes natural selection to occur.

septic tank Underground tank for treatment of wastewater from a home in rural and suburban areas. Bacteria in the tank decompose organic wastes, and the sludge settles to the bottom of the tank. The effluent flows out of the tank into the ground through a field of drain pipes.

sewage sludge See *sludge.*

sexual reproduction Reproduction in organisms that produce offspring by combining sex cells or *gametes* (such as ovum and sperm) from both parents. This produces offspring that have combinations of traits from their parents. Compare *asexual reproduction.*

shale oil Slow-flowing, dark brown, heavy oil obtained when kerogen in oil shale is vaporized at high temperatures and then condensed. Shale oil can be refined to yield gasoline, heating oil, and other petroleum products. See *kerogen, oil shale.*

shelterbelt See *windbreak.*

shelterwood cutting Removal of mature, marketable trees in an area in a series of partial cuttings to allow regeneration of a new stand under the partial shade of older trees, which are later removed. Typically, this is done by making two or three cuts over a decade. Compare *clear-cutting, seed-tree cutting, selective cutting, strip cutting.*

shifting cultivation Clearing a plot of ground in a forest, especially in tropical areas, and planting crops on it for a few years (typically 2–5 years) until the soil is depleted of nutrients or the plot has been invaded by a dense growth of vegetation from the surrounding forest. Then a new plot is cleared and the process is repeated. The abandoned plot cannot grow crops successfully for 10–30 years. See also *slash-and-burn cultivation.*

slash-and-burn cultivation Cutting down trees and other vegetation in a patch of forest, leaving the cut vegetation on the ground to dry, and then burning it. The ashes that are left add nutrients to the nutrient-poor soils found in most tropical forest areas. Crops are planted between tree stumps. Plots must be abandoned after a few years (typically 2–5 years) because of loss of soil fertility or invasion of vegetation from the surrounding forest. See also *shifting cultivation.*

slowly degradable pollutant Material that is slowly broken down into simpler chemicals or reduced to acceptable levels by natural physical, chemical, and biological processes. Compare *biodegradable pollutant, degradable pollutant, nondegradable pollutant.*

sludge Gooey mixture of toxic chemicals, infectious agents, and settled solids, removed from wastewater at a sewage treatment plant.

smelting Process in which a desired metal is separated from the other elements in an ore mineral.

smog Originally a combination of smoke and fog, but now used to describe other mixtures of pollutants in the atmosphere. See *industrial smog, photochemical smog.*

soil Complex mixture of inorganic minerals (clay, silt, pebbles, and sand), decaying organic matter, water, air, and living organisms.

soil conservation Methods used to reduce soil erosion, prevent depletion of soil nutrients, and restore nutrients already lost by erosion, leaching, and excessive crop harvesting.

soil erosion Movement of soil components, especially topsoil, from one place to another, usually by exposure to wind, flowing water, or both. This natural process can be greatly accelerated by human activities that remove vegetation from soil.

soil horizons Horizontal zones that make up a particular mature soil. Each horizon has a distinct texture and composition that varies with different types of soils.

soil permeability Rate at which water and air move from upper to lower soil layers. Compare *porosity.*

soil porosity See *porosity.*

soil profile Cross-sectional view of the horizons in a soil.

soil structure Ways in which the particles that make up a soil are organized and clumped together. See also *soil permeability, soil texture.*

soil texture Relative amounts of the different types and sizes of mineral particles in a sample of soil.

solar capital Solar energy from the sun reaching the earth. Compare *earth capital.*

solar cell See *photovoltaic cell.*

solar collector Device for collecting radiant energy from the sun and converting it into heat. See *active solar heating system, passive solar heating system.*

solid waste Any unwanted or discarded material that is not a liquid or gas. See *municipal solid waste.*

specialist species Species with a narrow ecological niche. They may be able to live in only one type of habitat, tolerate only a narrow range of climatic and other environmental conditions, or use only one or a few types of food. Compare *generalist species.*

speciation Formation of two species from one species as a result of divergent natural selection in response to changes in environmental conditions; usually takes thousands of years. Compare *extinction.*

species Group of organisms that resemble one another in appearance, behavior, chemical makeup and processes, and genetic structure. Organisms that reproduce sexually are classified as members of the same species only if they can breed with one another and produce fertile offspring.

species diversity Number of different species and their relative abundances in a given area. See *biological diversity.* Compare *ecological diversity, genetic diversity.*

species equilibrium model See *theory of island biogeography*.

spoils Unwanted rock and other waste materials produced when a material is removed from the earth's surface or subsurface by mining, dredging, quarrying, or excavation.

***S*-shaped curve** Leveling off of an exponential, *J*-shaped curve when a rapidly growing population exceeds the carrying capacity of its environment and ceases to grow.

stability Ability of a living system to withstand or recover from externally imposed changes or stresses. See *constancy, inertia, resilience*.

stratosphere Second layer of the atmosphere, extending from about 17–48 kilometers (11–30 miles) above the earth's surface. It contains small amounts of gaseous ozone (O_3), which filters out about 99% of the incoming harmful ultraviolet (UV) radiation emitted by the sun. Compare *troposphere*.

strip cropping Planting regular crops and close-growing plants, such as hay or nitrogen-fixing legumes, in alternating rows or bands to help reduce depletion of soil nutrients.

strip cutting A variation of clear-cutting in which a strip of trees is clear-cut along the contour of the land, with the corridor narrow enough to allow natural regeneration within a few years. After regeneration, another strip is cut above the first, and so on. Compare *clear-cutting, seed-tree cutting, selective cutting, shelterwood cutting*.

strip mining Form of surface mining in which bulldozers, power shovels, or stripping wheels remove large chunks of the earth's surface in strips. See *surface mining*. Compare *subsurface mining*.

subduction zone Area in which oceanic lithosphere is carried downward (subducted) under the island arc or continent at a convergent plate boundary. A trench ordinarily forms at the boundary between the two converging plates. See *convergent plate boundary*.

subsidence Slow or rapid sinking down (not slope related) of part of earth's crust.

subsistence farming Supplementing solar energy with energy from human labor and draft animals to produce enough food to feed oneself and family members; in good years there may be enough food left over to sell or put aside for hard times. Compare *industrialized agriculture*.

subsurface mining Extraction of a metal ore or fuel resource such as coal from a deep underground deposit. Compare *surface mining*.

succession See *ecological succession*.

succulent plants Plants, such as desert cacti, that survive in dry climates by having no leaves, thus reducing the loss of scarce water. They store water and use sunlight to produce their food in the thick, fleshy tissue of their green stems and branches. Compare *deciduous plants, evergreen plants*.

sulfur cycle Cyclic movement of sulfur in different chemical forms from the environment to organisms and then back to the environment.

surface mining Removing soil, subsoil, and other strata, and then extracting a mineral deposit found fairly close to the earth's surface. Compare *subsurface mining*.

surface runoff Water flowing off the land into bodies of surface water.

surface water Precipitation that does not infiltrate the ground or return to the atmosphere by evaporation or transpiration. See *runoff*. Compare *groundwater*.

survivorship curve Graph showing the number of survivors in different age groups for a particular species.

sustainability Ability of a system to survive for some specified (finite) time. See *sustainable system*.

sustainable agriculture Method of growing crops and raising livestock based on organic fertilizers, soil conservation, water conservation, biological control of pests, and minimal use of nonrenewable fossil-fuel energy.

sustainable development Forms of economic development and activities that do not deplete or degrade natural resources on which present and future economic growth and life depend. Compare *economic growth, economic development, ecologically sustainable development*.

sustainable earth Situation in which the earth's supplies of resources and the processes that make up earth capital are used and maintained over a specified period in ways that don't deplete or degrade these resources and processes.

sustainable living Taking no more potentially renewable resources from the natural world than can be replenished naturally and not overloading the capacity of the environment to cleanse and renew itself by natural processes.

sustainable resource harvest Concept that a certain quantity of a potentially renewable resource such as fish or timber can be harvested each year (or other time interval) over a specified period without depleting the resource.

sustainable society A society that manages its economy and population size without doing irreparable environmental harm by overloading the planet's ability to absorb environmental insults, replenish its resources, and sustain human and other forms of life over a specified period, usually hundreds to thousands of years. During this period it satisfies the needs of its people without depleting earth capital and thereby jeopardizing the prospects of current and future generations of humans and other species.

sustainable system A system that survives and functions over some specified (finite) time; a system that attains its full expected lifetime.

sustainable yield (sustained yield) Highest rate at which a potentially renewable resource can be used without reducing its available supply throughout the world or in a particular area. See also *environmental degradation*.

symbiosis Any intimate relationship or association between members of two or more species. See *symbiotic relationship*.

symbiotic relationship Species interaction in which two kinds of organisms live together in an intimate association. Members of the participating species may be harmed by, benefit from, or be unaffected by the interaction. See *commensalism, interspecific competition, mutualism, parasitism, predation*.

synergistic interaction Interaction of two or more factors or processes so that the combined effect is greater than the sum of their separate effects.

synergy See *synergistic interaction*.

synfuels Synthetic gaseous and liquid fuels produced from solid coal or sources other than natural gas or crude oil.

synthetic natural gas (SNG) Gaseous fuel containing mostly methane produced from solid coal.

system A set of components that function and interact in some regular and theoretically predictable manner. See *closed system, open system*.

tailings Rock and other waste materials removed as impurities when waste mineral material is separated from the metal in an ore.

tar sand Deposit of a mixture of clay, sand, water, and varying amounts of a tarlike heavy oil known as bitumen. Bitumen can be extracted from tar sand by heating. It is then purified and upgraded to synthetic crude oil. See *bitumen*.

technology Creation of new products and processes intended to improve our efficiency, chances for survival, comfort level, and quality of life. Compare *science*.

temperature Measure of the average speed of motion of the atoms, ions, or molecules in a substance or combination of substances at a given moment. Compare *heat*.

temperature inversion See *thermal inversion*.

teratogen Chemical, ionizing agent, or virus that causes birth defects. See *carcinogen, mutagen*.

terracing Planting crops on a long, steep slope that has been converted into a series of broad, nearly level terraces (with short vertical drops from one to another) that run along the contour of the land to retain water and reduce soil erosion.

terrestrial Pertaining to land. Compare *aquatic*.

tertiary (higher-level) consumers Animals that feed on animal-eating animals. They feed at high trophic levels in food chains and webs. Examples are hawks, lions, bass, and sharks. Compare *detritivore, primary consumer, secondary consumer*.

tertiary oil recovery See *enhanced oil recovery*.

tertiary sewage treatment See *advanced sewage treatment*.

theory of evolution Widely accepted idea that all life-forms developed from earlier life-forms. Although this theory conflicts with the creation stories of most religions, it is the way biologists explain how life has changed over the past 3.6–3.8 billion years and why it is so diverse today.

theory of island biogeography The number of species found on an island is determined by a balance between two factors: the *immigration rate* (of species new to the island) from other inhabited areas and the *extinction rate* (of species established on the island). The model predicts that at some point the rates of immigration and extinction will reach an equilibrium point that determines the island's average number of different species (species diversity).

thermal enrichment Beneficial effects in an aquatic ecosystem from a rise in water temperature. Compare *thermal pollution*.

thermal inversion Layer of dense, cool air trapped under a layer of less dense, warm air. This prevents upward-flowing air currents from developing. In a prolonged inversion, air pollution in the trapped layer may build up to harmful levels.

thermal pollution Increase in water temperature that has harmful effects on aquatic life. See *thermal shock*. Compare *thermal enrichment*.

thermal shock A sharp change in water temperature that can kill or harm fish and other aquatic organisms. See *thermal pollution*. Compare *thermal enrichment*.

thermocline Zone of gradual temperature decrease between warm surface water and colder deep water in a lake, reservoir, or ocean.

threatened species Wild species that is still abundant in its natural range but is likely to become endangered because of a decline in numbers. Compare *endangered species*.

threshold effect The harmful or fatal effect of a small change in environmental conditions that exceeds the limit of tolerance of an organism or population of a species. See *law of tolerance*.

throughput Rate of flow of matter, energy, or information through a system. Compare *input, output*.

throwaway society See *high-waste society*.

time delay Time lag between the input of a stimulus into a system and the response to the stimulus.

tolerance limits Minimum and maximum limits for physical conditions (such as temperature) and concentrations of chemical substances beyond which no members of a particular species can survive. See *law of tolerance*.

total fertility rate (TFR) Estimate of the average number of children that will be born alive to a woman during her lifetime if she passes through all her childbearing years (ages 15–44) conforming to age-specific fertility rates of a given year. In simpler terms, it is an estimate of the average number of children a woman will have during her childbearing years.

totally planned economy See *pure command economic system.*

toxic chemical Chemical that is fatal to humans in low doses or fatal to over 50% of test animals at stated concentrations. Most are neurotoxins, which attack nerve cells. See *carcinogen, hazardous chemical, mutagen, teratogen.*

toxicity Measure of how harmful a substance is.

toxicology Study of the adverse effects of chemicals on health.

toxic waste Form of hazardous waste that causes death or serious injury (such as burns, respiratory diseases, cancers, or genetic mutations). See *hazardous waste.*

traditional intensive agriculture Producing enough food for a farm family's survival and perhaps a surplus that can be sold. This type of agriculture requires higher inputs of labor, fertilizer, and water than traditional subsistence agriculture. See *traditional subsistence agriculture.*

traditional subsistence agriculture Production of enough crops or livestock for a farm family's survival and, in good years, a surplus to sell or put aside for hard times. Compare *traditional intensive agriculture.*

tragedy of the commons Depletion or degradation of a resource to which people have free and unmanaged access. An example is the depletion of commercially desirable species of fish in the open ocean beyond areas controlled by coastal countries. See *common-property resource.*

transform fault Area where earth's lithospheric plates move in opposite but parallel directions along a fracture (fault) in the lithosphere. Compare *convergent plate boundary, divergent plate boundary.*

transmissible disease A disease that is caused by living organisms (such as bacteria, viruses, and parasitic worms) and can spread from one person to another by air, water, food, or body fluids (or in some cases by insects or other organisms). Compare *nontransmissible disease.*

transpiration Process in which water is absorbed by the root systems of plants, moves up through the plants, passes through pores (stomata) in their leaves or other parts, and then evaporates into the atmosphere as water vapor.

tree farm Site planted with one or only a few tree species in an even-aged stand. When the stand matures it is usually harvested by clear-cutting and then replanted. These farms are normally used to grow rapidly growing tree species for fuelwood, timber, or pulpwood. See *even-aged management.* Compare *old-growth forest, second-growth forest, uneven-aged management.*

trophic level All organisms that are the same number of energy transfers away from the original source of energy (for example, sunlight) that enters an ecosystem. For example, all producers belong to the first trophic level and all herbivores belong to the second trophic level in a food chain or web.

troposphere Innermost layer of the atmosphere. It contains about 75% of the mass of earth's air and extends about 17 kilometers (11 miles) above sea level. Compare *stratosphere.*

true cost See *full cost.*

undernutrition Consuming insufficient food to meet one's minimum daily energy requirement for a long enough time to cause harmful effects. Compare *malnutrition, overnutrition.*

undiscovered resources Potential supplies of a particular mineral resource, believed to exist because of geologic knowledge and theory, although specific locations, quality, and amounts are unknown. Compare *identified resources, other resources, reserves.*

uneven-aged management Method of forest management in which trees of different species in a given stand are maintained at many ages and sizes to permit continuous natural regeneration. Compare *even-aged management.*

upwelling Movement of nutrient-rich bottom water to the ocean's surface. This can occur far from shore but usually occurs along certain steep coastal areas, where the surface layer of ocean water is pushed away from shore and replaced by cold, nutrient-rich bottom water.

urban area Geographic area with a population of 2,500 or more. The number of people used in this definition may vary, with some countries setting the minimum number of people at 10,000–50,000.

urban growth Rate of growth of an urban population. Compare *degree of urbanization.*

urban heat island Buildup of heat in the atmosphere above an urban area. This heat is produced by the large concentration of cars, buildings, factories, and other heat-producing activities. See also *dust dome.*

urbanization See *degree of urbanization.*

vertebrates Animals with backbones. Compare *invertebrates.*

volcano Vent or fissure in the earth's surface through which magma, liquid lava, and gases are released into the environment.

warm front The boundary between an advancing warm air mass and the cooler one it is replacing. Because warm air is less dense than cool air, an advancing warm front rises up over a mass of cool air. Compare *cold front.*

water cycle See *hydrologic cycle.*

waterlogging Saturation of soil with irrigation water or excessive precipitation so that the water table rises close to the surface.

water pollution Any physical or chemical change in surface water or groundwater that can harm living organisms or make water unfit for certain uses.

watershed Land area that delivers water, sediment, and dissolved substances via small streams to a major stream (river).

water table Upper surface of the zone of saturation, in which all available pores in the soil and rock in the earth's crust are filled with water.

weather Short-term changes in the temperature, barometric pressure, humidity, precipitation, sunshine, cloud cover, wind direction and speed, and other conditions in the troposphere at a given place and time. Compare *climate.*

weathering Physical and chemical processes in which solid rock exposed at earth's surface is changed to separate solid particles and dissolved material, which can then be moved to another place as sediment. See *erosion.*

wetland Land that is covered all or part of the time with salt water or fresh water, excluding streams, lakes, and the open ocean. See *coastal wetland, inland wetland.*

whole-tree harvesting A variation of clear-cutting in which a machine cuts trees at ground level or uproots entire trees. These trees are usually transported to a chipping machine in which huge blades reduce the wood to small chips.

wilderness Area where the earth and its community of life have not been seriously disturbed by humans and where humans are only temporary visitors.

wildlife All free, undomesticated species. Sometimes the term is used to describe only free, undomesticated species of animals.

wildlife management Manipulation of populations of wild species (especially game species) and their habitats for human benefit, the welfare of other species, and the preservation of threatened and endangered wildlife species.

wildlife resources Species of wildlife that have actual or potential economic value to people.

wildness Existence of wild gene pools, species, and ecosystems that are completely or mostly undisturbed by human activities. Another term for biodiversity.

wild species Species found in the natural environment.

windbreak Row of trees or hedges planted to partially block wind flow and reduce soil erosion on cultivated land.

worldview How people think the world works and what they think their role in the world should be. See *earth-wisdom worldview, planetary management worldview.*

zero population growth (ZPG) State in which the birth rate (plus immigration) equals the death rate (plus emigration) so that the population of a geographic area is no longer increasing.

zone of saturation Area where all available pores in soil and rock in the earth's crust are filled by water. See *water table.*

zoning Regulating how various parcels of land can be used.

zooplankton Animal plankton. Small floating herbivores that feed on plant plankton (phytoplankton). Compare *phytoplankton.*

INDEX

Note: Italicized letters *f* and *t* following page numbers indicate figures and tables, respectively.

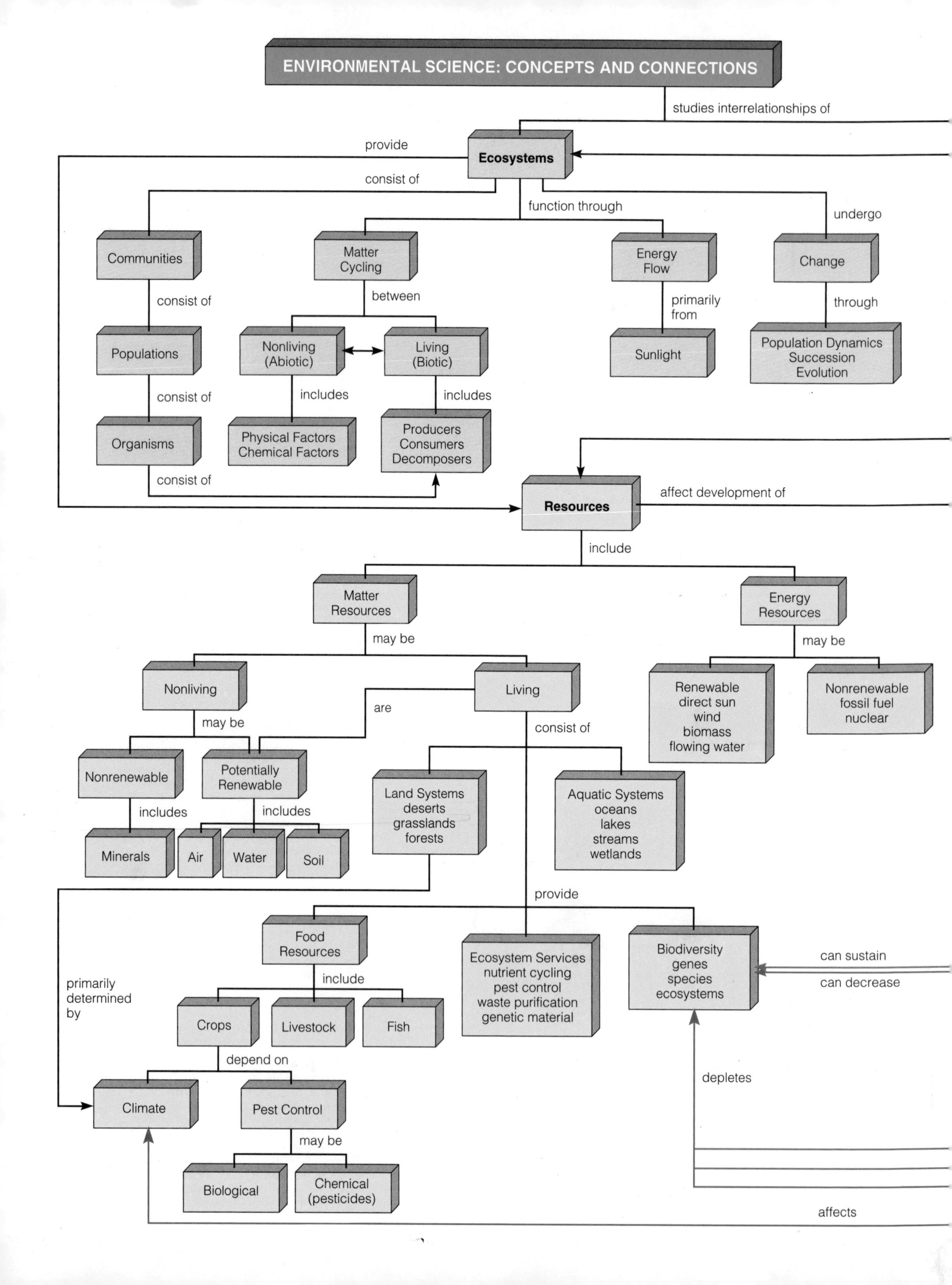

ENVIRONMENTAL SCIENCE: CONCEPTS AND CONNECTIONS
studies interrelationships of
Ecosystems
provide
consist of
function through
undergo
Communities
consist of
Populations
consist of
Organisms
consist of
Matter Cycling
between
Nonliving (Abiotic)
Living (Biotic)
includes
includes
Physical Factors Chemical Factors
Producers Consumers Decomposers
Energy Flow
primarily from
Sunlight
Change
through
Population Dynamics Succession Evolution
Resources
affect development of
include
Matter Resources
may be
Nonliving
may be
Nonrenewable
includes
Minerals
Potentially Renewable
includes
Air
Water
Soil
Living
are
consist of
Land Systems deserts grasslands forests
Aquatic Systems oceans lakes streams wetlands
Energy Resources
may be
Renewable direct sun wind biomass flowing water
Nonrenewable fossil fuel nuclear
provide
Food Resources
include
Crops
Livestock
Fish
depend on
Climate
Pest Control
may be
Biological
Chemical (pesticides)
primarily determined by
Ecosystem Services nutrient cycling pest control waste purification genetic material
Biodiversity genes species ecosystems
can sustain
can decrease
depletes
affects